The
Environmental
Resource
Handbook

2005/06
Third Edition

The
Environmental
Resource
Handbook

Associations	Environmental Statistics
Research Centers	Green City Rankings
Environmental Health	Grants
Publications	Government Agencies
Educational Programs	Consultants
Environmental Law	Green Product Catalogs
Trade Shows	Web Sites

A UNIVERSAL REFERENCE BOOK

Grey House
Publishing

PUBLISHER:	Leslie Mackenzie
EDITORIAL DIRECTOR:	Laura Mars-Proietti
STATISTICS EDITOR:	David Garoogian
PRODUCTION MANAGER:	Neil O'Connor
PRODUCTION ASSISTANTS:	Isabel Gottlieb, Karynn Ketiinq, Maria Knox, Marge Lutz, Sharon Moskiewicz, Jane Murphy, Jael Powell, Erica Schneider, Bobbie Jo Scutt, Jennie Taylor, Leah Wessel
MARKETING DIRECTOR:	Jessica Moody

A Universal Reference Book
Grey House Publishing, Inc.
185 Millerton Road
Millerton, NY 12546
518.789.8700
FAX 518.789.0545
www.greyhouse.com
e-mail: books @greyhouse.com

The environmental resource handbook. -- 1st
ed. (2001)-
    1037 p. 27.5 cm.
  Annual
  Spine title: ERH

1. Environmental protection--United States--Directories. 2. Environmental agencies--United States--Directories. 3. Conservation of natural resources--Societies, etc.--Directories. 4. Nature conservation--Societies, etc.--Directories. 5. Environmental protection--United States--Bibliography. 6. Conservation of natural resources--Bibliography. 7. Nature conservation--Bibliography. I. Grey House Publishing, Inc. II. Title: ERH.

GE20 .E586
363'7—dc21   2001238714
ISBN: 1-59237-090-X

Table of Contents

SECTION TWO: TABLES

Introduction

This is the third edition of *The Environmental Resource Handbook*. It offers immediate access to a unique combination of 6,756 environmental resources and 102 tables and charts of environmental statistics and rankings, all revised with the most current information available.

This revised edition includes the most important information available in a logical, carefully organized way. It is designed to provide, in one place, must-have resources for environmentalists, educators, researchers and students of the environment. Our research efforts for this edition included Internet, phone and fax campaigns. We updated more than 70% of the listings, verified the currency of the rest, and added hundreds of new resources. This edition includes 5,458 fax numbers, 3,538 e-mails, 4,921 web sites, 6,418 key contacts, and 3 indexes, so users can access this wealth of data in a variety of ways.

Section One: Resources

Here you will find 15 chapters and 70 subchapters. Many chapters are similarly organized by the following specific environmental issues:

Air Quality . . . Business . . . Disaster Preparedness . . . Energy . . . Health . . . Habitat Recycling . . . Pollution . . . Development . . . Travel . . . Water Resources

- **Associations & Organizations** disseminate information, host seminars, provide educational literature and promote studies. These listings are defined by the above categories, plus *Environmental Health* and *Environmental Law*.
- **Awards & Honors** lists organizations that recognize education and business professionals for excellence in environmental sciences.
- **Conferences & Trade Shows** includes large conventions, as well as small, specialized conferences. Listings include who, what, when, where and how many.
- **Consultants** offer information on environmental consulting services, including hazardous material screening, construction requirements, and habitat conservation.
- **Environmental Health** addresses health issues caused by the environment. It includes **Pediatric Health**, with resources that focus on how the environment affects children.
- **Environmental Law** offers resources involved in the legalities of the environment.
- **Financial Resources** includes grants, foundations and scholarships that give money for educational programs, research and environmental clean-up programs.
- **Government Agencies & Programs** offers Federal and State agencies, including a separate state section of **U.S. National Forests, Parks, and Refuges.**
- **Publications** offer books and periodicals that focus on the environment as a business, course of study, the subject of activism, and scientific research. Plus you'll find **Environmental Library Collections** and **Publishers.**
- **Research Centers** helps those looking to research environmental issues, and includes groups that operate within universities, as well as those that are commercially based.
- **Educational Programs** includes universities, private institutions and camps that offer environmental educational experiences for professionals and students of all ages.

- **Industry Web Sites, Online Databases**, and **Videos** offer ways to connect electronically to environmental resources.
- **Green Product Catalogs** lists catalogs of products that are environmentally friendly, from tree-free paper to cruelty-free aromatherapy products.

Section Two: Statistics & Rankings

The 102 tables and charts comprise 17 main topics – all updated with the most recent data available. Many tables include state rankings. Topics covered include:

*Air Quality . . . Brownfields . . . Children's Environmental Health . . . Drinking Waste
Endangered Species . . . Energy . . . Environmental Finances . . . Erosion
Global Warming . . . Habitats . . . Hazardous Waste . . . Land Use . . . Municipal Waste
Noise Pollution . . . Pesticides . . . Recycling . . . Toxic Exposures . . . Chemicals*

This section is designed to provide statistical snapshots of state and county environmental atmospheres, and show which communities are taking an active role in protecting our natural resources. Using the most current data available from more than 15 sources, this section helps to complete the picture of environmentalists doing research, conducting business, or providing education and consulting services. Worth noting in this section are **Green Metro Area Rankings** and a **UV Index.**

Section Three: Appendices

- **Glossary of Terms** from the Environmental Protection Agency has been recently updated and defines 1,800 commonly used environmental terms in non-technical language.
- **Abbreviations & Acronyms**, also updated by the EPA, lists 1,200 abbreviations that help identity the political and educational language of the industry.

Section Four: Indices

- **Entry**: Lists all entries alphabetically, identified by record number.
- **Geographic**: Lists entries alphabetically by state.
- **Subject**: Facilitates a fine-tuned search of resources and statistical tables by more than 250 specific environmental categories.

User Guide

Descriptive listings in *The Environmental Resource Handbook (ERH)* are organized into 15 chapters and 70 subchapters. You will find the following types of listings throughout the book:

- Associations & Organizations
- Conferences & Trade Shows
- Print & Electronic Media
- Foundations
- Government Agencies
- Research Centers
- Educational Programs
- Catalogs

Below is a sample listing illustrating the kind of information that is or might be included in an Association entry. Each numbered item of information is described in the paragraphs on the following page.

---

(1) **12345**

(2) **Water Environment Association of South Central US**

(3) 1762 South Major Drive
Suite 200
New Orleans, LA 98087

(4) 800-000-0000

(5) 058-884-0709

(6) 058-884-0568

(7) info@wenvi.com

(8) www.wenvi.com

(9) Barbara Pierce, Executive Director
Diane Watkins, Marketing Director
Robert Goldfarb, Administrative Assistant
Ann Klein, Wastewater Consultant

(10) The mission of the Association is to develop and disseminate information concerning waste quality management and the nature, collection, treatment, and disposal of wastewater. The Association publishes information, including a monthly magazine, manages a web site, and offers workshops and consultation on health and legal issues. A variety of educational programs are offered throughout the year on the history, ecology and culture of local rivers and streams, both on site, and in community schools.

(11) Founded 1964

(12) *18 pages*

(13) *Monthly*

User Key

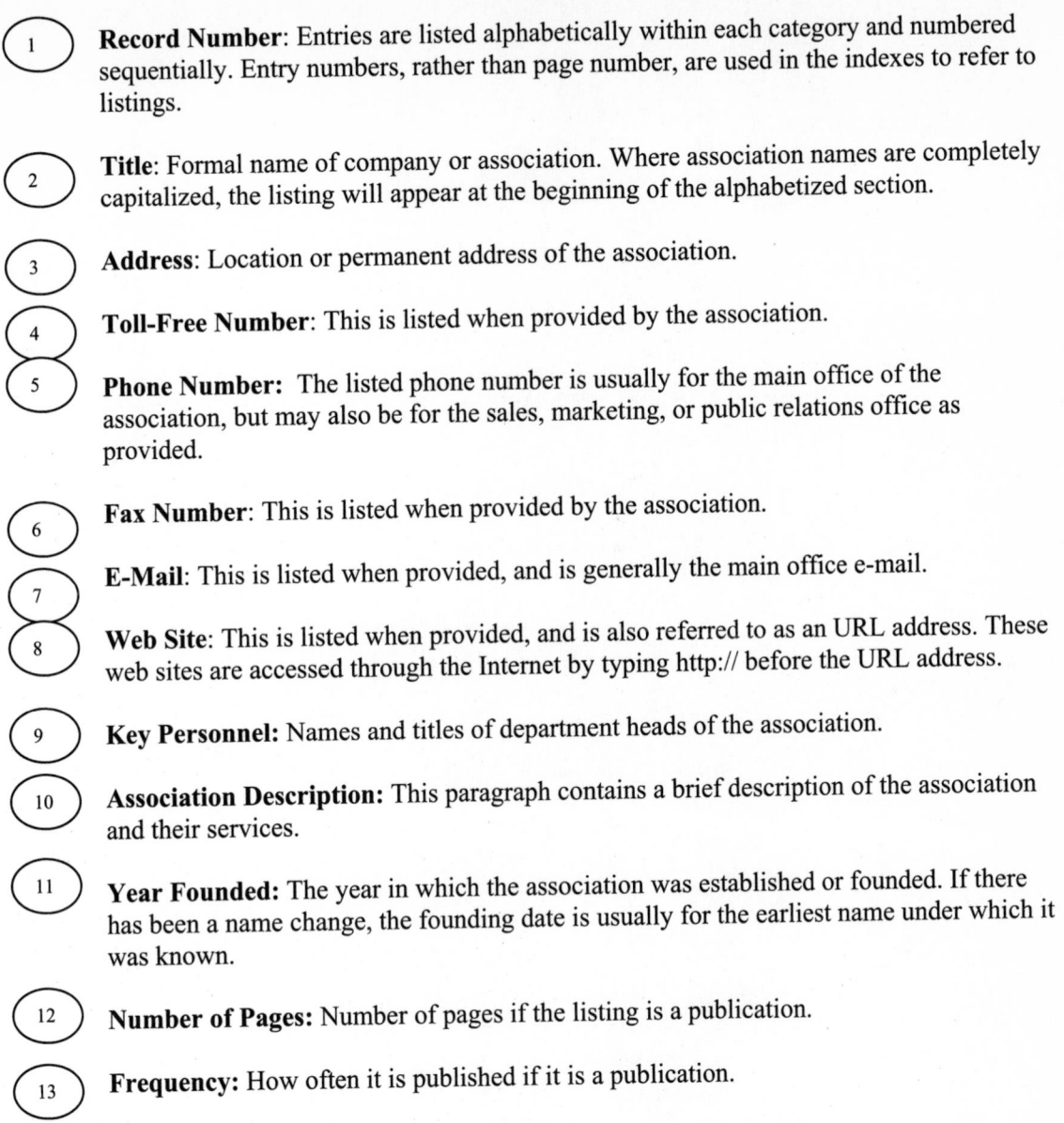

Record Number: Entries are listed alphabetically within each category and numbered sequentially. Entry numbers, rather than page number, are used in the indexes to refer to listings.

Title: Formal name of company or association. Where association names are completely capitalized, the listing will appear at the beginning of the alphabetized section.

Address: Location or permanent address of the association.

Toll-Free Number: This is listed when provided by the association.

Phone Number: The listed phone number is usually for the main office of the association, but may also be for the sales, marketing, or public relations office as provided.

Fax Number: This is listed when provided by the association.

E-Mail: This is listed when provided, and is generally the main office e-mail.

Web Site: This is listed when provided, and is also referred to as an URL address. These web sites are accessed through the Internet by typing http:// before the URL address.

Key Personnel: Names and titles of department heads of the association.

Association Description: This paragraph contains a brief description of the association and their services.

Year Founded: The year in which the association was established or founded. If there has been a name change, the founding date is usually for the earliest name under which it was known.

Number of Pages: Number of pages if the listing is a publication.

Frequency: How often it is published if it is a publication.

National: Air & Climate

1 Air and Waste Management Association
One Gateway Center, 3rd Floor   412-232-3444
420 Fort Duquesne Boulevard   Fax: 412-232-3450
Pittsburgh, PA 15222-1435   E-mail: info@awma.org
http://www.awma.org
The Air and Waste Management Association (A&WMA) is a nonprofit, nonpartisan professional organization that provides training, information, and networking opportunities to thousands of environmental professionals in 65 countries.
Edith M Ardiente, President
Susan S Wierman, Executive Director

2 American Meteorological Society
45 Beacon Street   617-227-2425
Boston, MA 02108-3693   Fax: 617-742-8718
E-mail: amsinfo@ametsoc.org
http://www.ametsoc.org
The society actively promotes the development and dissemination of information on the atmospheric and related oceanic and hydrologic sciences.
Walter A Lyons, President

3 Association for Arid Land Studies: Texas Tech University
601 Indiana Avenue   806-742-2218
PO Box 41036   Fax: 806-742-1954
Lubbock, TX 79409   E-mail: gay.riggan@ttu.edu
http://www.iaff.ttu.edu
The International Center for Arid and Semiarid Land Studies' (ICASALS') mission is to promote multidisciplinary research and education of topics relative to the quality of life in arid and semiarid regions of the world.
Ambassador Nagy, Director ICASALS

4 Association of Local Air Pollution Control Officials
STAPPA/ALAPCO
444 North Capitol Street Northwest   202-624-7864
Suite 307   Fax: 202-624-7863
Washington, DC 20001   E-mail: 4clnair@4cleanair.org
http://www.cleanairworld.org
Two national associations representing air pollution control agencies in the 54 states and territories and metropolitan areas across the United States.
Dennis J McLerran, President

5 Center for Clean Air Policy
750 1st Street Northeast   202-408-9260
Suite 940   Fax: 202-408-8896
Washington, DC 20002   E-mail: communications@ccap.org
http://www.ccap.org
To promote and implement innovative solutions to major environmental and energy problems which balance both environmental and economic interests.
Ned Helme, Executive Director
Tony Earl, Governor, Chair

6 Clean Air Council
135 S 19th Street   215-567-4004
Suite 300   Fax: 215-567-5791
Philadelphia, PA 19103   E-mail: rruiz@cleanair.org
http://www.cleanair.org
Dedicated to protecting everyone's right to breathe clean air. Works through public education, community advocacy, and government oversight to ensure enforcement of environmental laws. Five main issue areas are clean air, clean energy, sustainable transportation, waste reduction and recycling, and indoor air quality.
Joseph Otis Minott Esq., Executive Director

7 Climate Institute
Coping with Climate Change
1785 Massachusetts Avenue Northwest   202-547-0141
Washington, DC 20036   Fax: 202-547-0111
E-mail: info@climate.org
http://www.climate.org
The Institute works to protect the balance between climate and life on earth by facilitating dialogue among scientists, policy makers, business executives and citizens. In all its efforts, the institute strives to be a source of objective, reliable information.
William Becker, Executive Director

8 Global Climate Coalition
1275 K Street Northwest   202-628-9161
Washington, DC 20036   Fax: 202-628-8685
E-mail: gcc@globalclimate.org
http://www.globalclimate.org
The Coalition is an organization of trade associations established in 1989 to coordinate business participation in the international policy debate on the issue of global climate change and global warming.
Gail MacDonald, Executive Director

9 Greenhouse Crisis Foundation
1130 17th Street NW   202-429-9602
Suite 630   Fax: 202-775-0074
Washington, DC 20036   http://www.nndb.com
International organization of scientists, environmental activists, and policymakers working to establish international cooperation to mitigate global warming.
Jeremy Rifkin, President

10 Institute for Global Environmental Strategies
1600 Wilson Boulevard   703-312-0823
Suite 901   Fax: 703-312-8657
Arlington, VA 22209   E-mail: info@strategies.org
http://www.strategies.org
Conducts research on climate change. Informs the public about emerging issues.

11 Institute of Clean Air Companies
1660 L Street NW   202-457-0911
Suite 1100   Fax: 202-331-1388
Washington, DC 20036   E-mail: dfoerter@icac.com
http://www.icac.com
To promote the air pollution control industry and encourage improvement of engineering and technical systems. Members are leading manufacturers of equipment to monitor and control emissions of particulate, VOC, SO2, NOX, and airtoxics.
Jeffrey C Smith, Executive Director
Robert G Hilton, President

12 Institute of Global Environment and Society
4041 Powder Mill Road   301-595-7000
Suite 302   Fax: 301-595-9793
Claverton, MD 20705   E-mail: www@cola.iges.org
http://www.iges.org
Conducts research on climate varibility and service to society.

13 International Research Institute for Climate Prediction
PO Box 1000   914-680-4468
Palisades, NY 10964   Fax: 914-680-4666
http://http://iri.columbia.edu/
IRI's goal is to assess and develop climate forecasts, and to foster the application of such climate forecasts to the explicit benefit of society. At this site you will find training and visitor programs, a climate data library, climate monitoring data, forecasts, annual reports, a list of speakers for theIR lecture series, employment opportunities, and a link to the IRI Ocean Expo 98 site in Portugal.
Steve Zebiak, Director

14 State and Territorial Air Pollution Program Administrators
STAPPA/ALAPCO
444 N Capitol Street NW   202-624-7864
Suite 307   Fax: 202-624-7863
Washington, DC 20001   E-mail: 4clnair@4cleanair.org
http://www.4cleanair.org
To encourage the exchange of information among air pollution control officials, to enhance communication and cooperation among federal, state and local regulatory agencies, and to promote good management of our air resources.
Nancy L Seidman, President
Eddie Terrill, Vice President

15 Synthetic Organic Chemical Manufacturers Association
1850 M Street Northwest   202-721-4100
Suite 700   Fax: 202-296-8120
Washington, DC 20036-5810   E-mail: info@socma.com
                                                   http://www.socma.com
SOCMA member companies make the products and refine the raw materials that make our standard of living possible. From pharmaceuticals to cosmetics, soaps to plastics, and all manner of industrial and construction products, materialsmanufactured by SOCMA firms save lives, make our food supply safe and abundant, and enable the manufacture of literally thousands of other products.
Nick Shackley, Chairperson

16 The Clean Air Washington Act
2940-B Limited Lane, Northwest   360-586-1044
Olympia, WA 98502   Fax: 360-491-5308
                                                   E-mail: info@orcaa.org
                                                   http://www.orcaa.org
The Clean Air Washington Act states that it is public policy to preserve, protect, and enhance the air quality for current and future generations. The purpose of ORCAA is to carry out these public policies as specified by the StateLegislature. The agency regulates more than 600 air polution sources. The agency also administers laws and regulations regarding such programs as solid fuel burning devices, asbestos abatement, and open burning.
Richard Stedman, Executive Director

17 The International Center For Arid And SemiArid Land Studies
Office of Internal Affairs
601 Indiana Avenue   806-742-2218
PO Box 41036   Fax: 806-742-1954
Lubbock, TX 79409   E-mail: gay.riggan@ttu.edu
                                   http://www.iaff.ttu.edu/aals/default.htm
Ambassador Nagy, Director

National: Business & Education

18 American Association of Zoo Keepers
3601 29th Street Southwest   785-273-9149
Suite 133   Fax: 785-273-1980
Topeka, KS 66614-2054 E-mail: aazkoffice@zk.kscoxmail.com
                                                   http://www.aazk.org
AAZK is a nonprofit 501(c)(3) organization dedicated to the promotion of zoo keeping as a profession; support of worth conservation projects; provision of reference materials for those in the zoo field; and support of continuingeducation for zoo professionals.
Ed Hansen, Executive Director
Denise Wagner, President

19 American Chemistry Council
1300 Wilson Boulevard   703-741-5000
Arlington, VA 22209   Fax: 703-741-6000
                  E-mail: webmaster@americanchemistry.com
                          http://www.americanchemistry.com
Represents the chemical industry on public policy issues, coordinates the industry's research and testing programs and administers the industry's environmental, health and safety performance improvement initiative known as ResponsibleCare.
Jack Gerard, President

20 American Federation of Teachers
555 New Jersey Avenue Northwest   202-879-4463
Washington, DC 20001-2079   Fax: 202-879-4597
                                                   http://www.aft.org
The American Federation of Teachers is a union that represents K-12 teachers and other school employees, health care professionals and public employees. The union considers itself an advocacy organization for children and the public.
Edward J McElroy, President
Antonia Cortese, Executive Vice President

21 American Forest & Paper Association
1111 19th Street Northwest   202-463-2700
Suite 800   800-878-8878
Washington, DC 20036   Fax: 202-463-2471
                                                   E-mail: info@afandpa.org
                                                   http://www.afandpa.org
National trade association for the forest products industry. AF&PA represents more than 200 companies and related associations engaged in the manufacture of pulp, paper, paperboard and wood products.
Kris Seeger, VP Communications
W Henson Moore, President/CEO

22 American National Standards Institute
25 West 43rd Street   212-642-4900
4th Floor   Fax: 212-398-0023
New York, NY 10036   E-mail: ansionline@ansi.org
                                                   http://www.ansi.org
The Institute's mission is to enhance both the global competitivness of US business and the US quality of life by promoting and facilitating voluntary consensus standards and conformity assessment systems, and safeguarding theirintegrity.
Mark W Hurwitz, President

23 American Public Works Association
1401 K Street NW   202-408-9541
11th Floor   Fax: 202-408-9542
Washington, DC 20005   http://www.apwa.net
The Association serves its members by promoting professional excellence and public awareness through education, advocacy and the exchange of knowledge.
Peter King, Executive Director

24 American Society for Testing and Materials
100 Barr Harbor Drive   610-832-9585
West Conshohocken, PA 19428-2959   Fax: 610-832-9555
                                                   E-mail: service@astm.org
                                                   http://www.astm.org
To provide the optimum environment and support for technical committees to develop needed standards and related information. To ensure ASTM products and services are provided in a timely manner and meet current needs. To increase theawareness of the ASTM consensus process, the benefits of participation, and the value of ASTM standards and services in the global marketplace.
James Thomas, President

25 American Zoo and Aquarium Association
8403 Colesville Road   301-562-0777
Suite 710   Fax: 301-562-0888
Silver Springs, MD 20910-3314   http://www.aza.org
Syd Butler, Executive Director
Kris Vehrs, Deputy Director

26 Association of State Wetland Managers
2 Basin Road   208-892-3399
Windham, ME 04062   Fax: 207-892-3089
                                                   E-mail: aswm@aswm.org
                                                   http://www.aswm.org
Nonprofit organization dedicated to the protection and restoration of our nations wetlands. Our goal is to help public and private wetland decision-makers utilize the best possible scientific information and techniques in wetlanddelineation, assessment, mapping, planning, regulation, acquisition, restoration, and other management.
Jon Kusler, Executive Director

27 Bank Information Center
1100 H Street Northwest   202-737-7752
Suite 650   Fax: 202-737-1155
Washington, DC 20005   E-mail: info@bicusa.org
                                                   http://www.bicusa.org
Independent nonprofit, non-governmental organization that provides information and strategic support to NGOs and social movements throughout the world on the projects, policies and practices of the World Bank and other MultilateralDevelopment Banks.
Jonathan Fox, Chairperson

28 Biotechnology Industry Organization
1225 Eye Street Northwest   202-962-9200
Suite 400   Fax: 202-857-0237
Washington, DC 20005   E-mail: info@bio.org
                                                   http://www.bio.org
James C Greenwood, Executive Director

29 Business for Social Responsibility
111 Sutter Street
12th Floor
San Francisco, CA 94104
415-984-3200
Fax: 415-984-3201
http://www.bsr.org

BSR seeks to create a more just and sustainable world by working with companies to promote more responsible business practices, innovation and collaboration.

30 CERES (Coalition for Environmentally Responsible Economies)
99 Chauncy Street
6th Floor
Boston, MA 02111
617-247-0700
Fax: 617-267-5400
E-mail: jacobs@ceres.org (Lisa Jacobs)
http://www.ceres.org

Formed in 1989 out of an unprecedented partnership among some of America's largest socially responsible institutional investors and environmental groups, CERES (Coalition for Environmentally Responsible Economies) has pioneered aninnovative, practical approach to advancing corporate accountability through public reporting and stakeholder engagement.
Norman L Dean, Executive Director
Joan L Bavaria, President

31 CO-OP America
1612 K Street NW
Suite 600
Washington, DC 20006
202-872-5307
800-58G-REEN
Fax: 202-331-8166
E-mail: ccafaq.htm
http://www.coopamerica.org

Provides the economic strategies, organizing power and practical tools for businesses and individuals to address today's social and environmental problems.
Alisa Grovitz, Executive Director

32 Center for Policy Alternatives
1875 Connecticut Avenue NW
Suite 710
Washington, DC 20009
202-387-6030
Fax: 202-986-2539
E-mail: info@cfpa.org
http://www.cfpa.org

The nation's leading nonpartisan progressive public policy and leadership development center serving state legislators, state policy organizations, and state grassroots leaders. CPA is a 501 (c)(3) nonprofit corporation with a staff of35 and an annual budget of $4 million, supported by foundations, unions, corporations and individuals.
Founded: 1976
Juan Sepulveda, President
Nan Grogan-Orrock, Co Chairperson

33 Chemical Producers and Distributors Association
1430 Duke Street
Alexandria, VA 22314
703-548-7700
Fax: 703-548-3149
E-mail: cpda@cpda.com
http://www.cpda.com

To eliminate barriers to market entry for the generic pesticide manufacturer. To see that those concerns are fully addressed and considered both within the Congress and the US Environmental Protection Agency. Also seeks to promote thefact that generic pesticide products results in the economic benefits of healthy competition in the US business community and lower prices for the pesticide consumer.
Warren Stickle, President

34 Chlorine Institute
1300 Wilson Boulevard
Arlington, VA 22209
703-741-5760
Fax: 703-741-6068
E-mail: tkerns@c12.com
http://www.chlorineinstitute.org

Exists to support the chlor-alkali industry and serve the public by fostering continuous improvements to safety and the protection of human health and the environment connected with the production, distribution and use of chlorine,sodium and potassium hydroxides, and sodium hypochlorite; and the distribution and use of hydrogen chloride.
Kathleen Shaver, President

35 Consumer Specialty Products Association
900 17th Street, Northwest
Suite 300
Washington, DC 20006
202-872-8110
Fax: 202-872-8114
E-mail: info@cspa.org
http://www.cspa.org

To foster high standards for the industry; and concern for the health, safety and environmental impacts of its products; address legislative and regulatory challenges at the federal, state and local level; meet the needs of industryfor technical and legal guidance; provide a forum to share ideas for scientific and marketing excellence.
Christopher Cathcart, President/CEO

36 Council of State Governments
2760 Research Park Drive
PO Box 1190
Lexington, KY 40578-1910
859-244-8000
Fax: 859-244-8001
E-mail: info@csg.org
http://www.csg.org

CSG provides a network for identifying and sharing ideas with state leaders. To this end CSG: builds leadership skills to improve decision making; advocates multi-state problem solving and partnerships; interprets changing national andinternational conditions to prepare states for the future; and promotes the sovereignty of the states and their role in the American federal system.
Daniel Sprague, Executive Director

37 Earth First!
PO Box 3023
Tucson, AZ 85702
520-620-6900
Fax: 413-254-0057
E-mail: collective@earthfirstjournal.org
http://www.earthfirstjournal.org/

Earth First! was founded in 1979 in response to a lethargic, compromising and increasingly corporate environmental community. Earth First! takes a decidedly different tack toward environmental issues. We believe in using all the toolsin the toolbox, ranging from grassroots organizing and involvement in the legal process to civil disobedience and monkeywrenching.
Founded: 1979

38 Earth Share
7735 Old Georgetown Road
Suite 900
Bethesda, MD 20814
240-333-0300
Fax: 240-333-0301
E-mail: info@earthshare.org
http://www.earthshare.org

Earth Share, a federation of America's leading nonprofit environmental and conservation organizations, works to promote environmental education and charitable giving through workplace giving campaigns. Earth Share is an opportunity, asystem, and an answer for environmentally conscious employees and workplaces to support dozens of environmental groups at once through a charitable giving drive.
Kalman Stein, President

39 Ecological Society of America
1707 H Street Northwest
Suite 400
Washington, DC 20006
202-833-8773
Fax: 202-833-8775
E-mail: esahq@esa.org
http://www.esa.org

To promote ecological science by improving communication among ecologists; raise the public's level of awareness of the importance of ecological science; increase the resources available for the conduct of ecological science; andensure the appropriate use of ecological science in environmental decision making by enhancing communication between the ecological community and policy-makers.
Jerry Melillo, President

40 Environmental Careers Organization
30 Winter Street
Boston, MA 02108
617-426-4375
Fax: 617-423-0998
E-mail: info@eco.org
http://www.eco.org

ECO protects and enhances the environment through the development of diverse leaders, the promotion of careers, and the inspiration of the individuals action.
Jeff Cook, President

41 Environmental Council of the States
444 N Capitol Street
Suite 445
Washington, DC 20001
202-624-3660
Fax: 202-624-3666
E-mail: ecos@sso.org
http://www.sso.org/ecos/

To improve the environment of the United States, the Environmental Council of States will: champion the role of States in environmental management; provide for exchange of ideas, views and experiences among States; foster cooperationand coordination in environmental mangaement; and articulate state positions to Congress, federal agenices and the public on environmental issues.
Robert E Roberts, Executive Director

42 Environmental Industry Association
Environmental Industry Associates
4301 Connecticut Avenue NW   202-244-4700
Suite 300   800-424-2869
Washington, DC 20008   Fax: 202-966-4818
                                    http://www.envasns.org
The Environmental Industry Association represents about 2,000 companies that manage soild, hazardous, and medical wastes; manufacture and distribute waste equipment; and offer related pollution-prevention services.
Michael Frischkorn, President

43 Federal Wildlife Association
FWOA Secretary
PO Box 550060   703-358-1949
Waltham, MA 02455-1949   http://www.fwoa.org
This Association is an organization dedicated to the protection of wildlife, the enforcement of federal wildlife law, the fostering of cooperation and communication among federal wildlife officers, and the perpetuation, enhancement anddefense of the wildlife enforcement profession. Membership is open to all federal law enforcement officers charged with the protection of wildlife and to fill friends and supporters of this goal.
Mark Webb, President
Ellen Kiley, Secretary

44 Federation of Environmental Technologists
PO Box 624   262-644-0700
Slinger, WI 53086-0624   Fax: 262-644-7106
                                    E-mail: info@fetinc.org
                                    http://www.fetinc.org
A nonprofit organization formed to assist industry in interpretation of and compliance with environmental regulations. Membership is open to all industries, municipalities, organizations and individuals concerned about environmentalregulations. Currently there are approximately 1000 members and 125 patron companies.
        Founded: 1982
Triese Haase, Administrator

45 Forestry Conservation Communications Association
National Office
PO Box 3217   717-338-1505
Gettysburg, PA 17325   Fax: 717-334-5656
                                    E-mail: nfc@fcca-usa.org
                                    http://www.fcca-usa.org
Association for manufacturers or suppliers of forestry and conservation communications equipment, systems and procedures.
Ralph Haller, Executive Manager

46 Get America Working!
1700 North Moore Street   703-527-8300
Suite 200   Fax: 703-527-8383
Arlington, VA 22209   http://www.getamericaworking.org
Mission is to research and promulgate innovative solutions to joblessness in America, with special emphasis on developing solutions based on structural change in the nation's economic environment.
Robert Walker, President

47 Global Climate Coalition
1275 K Street Northwest   202-682-9161
Washington, DC 20005   Fax: 202-639-8685
                                    E-mail: gcc@globalclimate.org
                                    http://www.globalclimate.org
This Coalition is an organization of trade associations established in 1989 to coordinate business participation in the international policy debate on the issue of global climate change and global warming.
        Founded: 1989
Gail MacDonald, Executive Director
Frank Maisano, Media Contact

48 Global Environmental Management Initiative
1 Thomas Circle NW   202-296-7449
10th Floor   Fax: 202-296-7442
Washington, DC 20005   E-mail: info@gemi.org
                                    http://www.gemi.org
To help business achieve environmental, health and safety excellence and corporate citizenship.
Steven B Hellem, Executive Director
Amy Goldman, Director

49 Green Seal
1001 Connecticut Avenue   202-872-6400
Suite 827   Fax: 202-872-4324
Washington, DC 20036   E-mail: greenseal@greenseal.org
                                    http://www.greenseal.org
Green Seal is the independent nonprofit organization dedicated to protecting the environment by promoting the manufacture and sale of environmentally responsible consumer products. It sets environmental standards and awards a GreenSeal Approval to products that causes less harm to the environment than other similar products.
Arthur Weissman, President

50 Greenpeace USA
702 H Street Northwest   202-462-1177
Washington, DC 20001   800-326-0959
                                    Fax: 202-483-8683
                                    E-mail: info@wdc.greenpeace.org
                                    http://www.greenpeace.org
Greenpeace is the leading independent campaigning organization that uses non-violent direct action and creative communication to expose global environmental problems and to promote solutions that are essential to a green and peacefulfuture.
Elizabeth Feore, Media Associate

51 Halogenated Solvents Industry Alliance
1300 Wilson Boulevard   703-741-5780
12th Floor   Fax: 703-741-6077
Arlington, VA 22209   E-mail: contact@hsia.org
                                    http://www.hsia.org
The mission is to present the interests of users and producers of chlorinated solvents. To promote the continued safe use of these products and to promote the use of sound science in assessing their potential health effects.
Steven Risotto, Executive Director

52 Heinz Center
1001 Pennsylvania Avenue   202-737-6307
Suite 735 South   Fax: 202-737-6410
Washington, DC 20004   http://www.heinzctr.org
Facilitates communication between industrial, governmental and environmental agencies. Conducts research on environmental issues.
Thomas E Lovejoy, President

53 Institute for Resource Management
1104 Ashton Avenue   801-466-3600
Suite 210   Fax: 801-466-6800
Salt Lake City, UT 84106-2374   http://www.crm.org
The Center for Resource Management (CRM) is a national organization founded in 1981. The center was formed as a "safe harbor" where business executives, environmental leaders, citizens and government officials could work outenvironmental problems and conflicts in a collaborative setting.
        Founded: 1981
Terrell Minger, President

54 Institute of Clean Air Companies
1660 L Street Northwest   202-457-0911
Suite 1100   Fax: 202-331-1388
Washington, DC 20036   E-mail: jsmith@icac.com
                                    http://www.icac.com
To promote the air pollution control industry and encourage improvement of engineering and technical systems. Members are leading manufacturers of equipment to monitor and control emissions of particulate, VOC, SO2, NOx, and airtoxics.
David C Foerter, Executive Director
Robert G Hilton, President

55 **Institute of Hazardous Materials Management**
11900 Parklawn Drive   301-984-8969
Suite 450   Fax: 301-984-1516
Rockville, MD 20852   E-mail: ihmminfo@ihmm.org
                                        http://www.ihmm.org
Mission is to provide recognition for professionals engaged in the management and engineering control of hazardous materials who have attained the required level of education, experience and competence; foster continued professionaldevelopment of Certified Hazardous Materials Managers (CHMM).
John H Frick PhD, CHMM, Executive Director

56 **Interstate Mining Compact Commission**
445A Carlisle Drive   703-709-8654
Herndon, VA 20170   Fax: 703-709-8655
                                        http://www.imcc.isa.us
A multi-state governmental organization which represents its 20 member states on issues of mining and environmental regulations. It works closely with several federal agencies i.e. Office of Surface Mining, US EPA, and the Bureau ofland management.
Gregory E Conrad, Executive Director
Beth A Botsis, Diretor of Programs

57 **Land Improvement Contractors of America**
3080 Ogden Avenue   630-548-1984
Lisle, IL 60532   Fax: 630-548-9189
                                E-mail: nlica@aol.com
                                http://www.licanational.com
Organization composed primarily of small contractors whose activities related to the conservation, use and improvement of land and water resources ranging from terrace and pond building to septic tank installation to loggingoperations. Strives to improve the climate within which members conduct their businesses by working for better legislation and regulations.
Wayne March, Executive VP

58 **National Association for Environmental Management**
1612 K Street Northwest   202-986-6616
Suite 1102   800-391-6236
Washington, DC 20006-1903   Fax: 202-530-4408
                                E-mail: programs@naem.org
                                        http://www.naem.org
This Association is a nonprofit educational association composed of professional private and public sector environmental managers. NAEM provides professional development, recognition, and opportunities.
Carol Singer Neuvett, Executive Director

59 **National Association for Humane and Environmental Education**
67 Norwich Essex Turnpike   860-434-8666
East Haddam, CT 06423-1736   Fax: 860-434-6282
                                E-mail: nahee@nahee.org
                                        http://www.nahee.org
As the youth education division of the Humane Society of the US (HSUS), NAHEE works to establish humane education as a broadbased initiative that promotes environmental responsibility and good character as well as humane treatment ofanimals. Our mission is to help children embody the ideal of respect and kindness toward people, animals and the earth.
William DeRosa, Executive Director
Dorothy Weller, Director

60 **National Association of Biology Teachers**
12030 Sunrise Valley Drive   703-264-9696
Suite 110   800-406-0775
Reston, VA 20191   Fax: 703-264-7778
                                E-mail: office@nabt.org
                                        http://www.nabt.org
Includes 9,000 educators who share experience and expertise with colleagues from around the globe; keep up with trends and developments and grow professionally. The NABT empowers educators to provide the best biology and life scienceeducation for all students.
Rebecca Ross, President
Catherine Wilcoxson, Secretary/Treasurer

61 **National Association of Environmental Professionals**
PO Box 2086   301-860-1140
Bowie, MD 20718   800-406-0775
                                        Fax: 301-860-1141

E-mail: office@naep.org
http://www.naep.org
NAEP is the multidisciplinary association dedicated to the advancement of environmental professions in the US and abroad. A forum for the state-of-the-art information on environmental planning, research and management. A net work ofprofessional contacts and exchange on information among colleagues in industry, government, academic, and private sector.
Gary Kelman, President

62 **National Association of Environmental Risk Auditors**
6645 Colerain Avenue   513-674-1109
Cincinnati, OH 45253

63 **National Association of Service and Conservation Corps**
666 11th Street NW   202-737-6272
Suite 1000   Fax: 202-737-6277
Washington, DC 20001   http://www.nascc.org
The National Association of Service and Conservation Corps unites and supports youth corps as a preeminent strategy for achieving the nation's youth development, community service and environmental restoration goals.

64 **National Conference of State Legislatures**
7700 E First Place   303-364-7700
Denver, CO 80230   Fax: 303-364-7800
                                        http://www.ncsl.org
The National Conference of State Legislatures serves the legislators and staffs of the nation's 50 states and its commonwealths and territories. NCSL is a bipartisan organization with three objectives: to improve the quality andeffectiveness of state legislatures; to foster interstate communications and cooperation; and to ensure states a strong cohesive voice in the federal system.
Douglas Farquhar, Program Principal
Jeff Greenwalt, Staff Chair

65 **National Council for Science and the Environment**
1707 H Street Northwest   202-530-5810
Suite 212   Fax: 202-628-4311
Washington, DC 20006   E-mail: contact@NCSEonline.org
                                        http://www.ncseonline.org
National nonprofit organization working to improve the scientific basis for making decisions on environmental issues. The Council promotes a new crosscutting approach to environmental science that integrates interdisciplinary research,scientific assessment, environmental education and communication of science-based information to decision makers and the public.
Peter Saundry, Executive Director
Richard Benedict, President

66 **National Council of Forestry Association Executives**
146 State Street   207-622-9288
Augusta, ME 04330   http://www.agrinet.tamu.edu
Ted Johnston

67 **National Environmental Development Association**
1440 New York Avenue NW   202-638-1230
Suite 300   Fax: 202-639-8685
Washington, DC 20005
Companies and others concerned with balancing environmental and economic interests to obtain both a clean environment and a strong economy.
Andrew McElwaine, Director

68 **National Environmental Education**
1707 H Street Northwest   202-833-2933
Suite 900   Fax: 202-261-6464
Washington, DC 20006-3915   E-mail: netf@neetf.org
                                        http://www.neetf.org
Chartered by Congress in 1990, we are a private nonprofit organization that plays a unique role in the environmental learning programs to meet social goals, such as improved health, better education, and greener, more profitablebusiness. In particular, we address the needs of disadvantaged communities requiring cleaner local environments.
Karen Bates Kress, President
Dwight Minton, Chairman

69   National Environmental Trust

1200 18th Street NW   202-887-8800
5th Floor   Fax: 202-887-8877
Washington, DC 20036   http://www.eic.org
Environmental Information Center manages comprehensive
media and public policy campaigns around national environmental issues.
Phil Clapp, President

70   National FFA Organization

Future Farmers of America
6060 FFA Drive   317-802-6060
PO Box 68960   Fax: 317-802-6061
Indianapolis, IN 46268-0960   http://www.ffa.org
Future Farmers of America is dedicated to making a positive dif-
ference in the lives of young people by developing their poten-
tial for premier leadership, personal growth and career success
through agricultural education. The organization's motto is:
Learning to Do; Doing to Learn; Earning to Live; Living to
Serve.
Larry Case, CEO
C Coleman Harris, Executive Secretary

71   National Geographic Society

1145 17th Street Northwest   202-857-7000
Washington, DC 20036-4688   800-647-5463
                        Fax: 202-775-6141
            http://www.nationalgeographic.com
World's largest nonprofit scientific and educational organiza-
tion.

72   National Governors Association

44 N Capitol Street NW   202-624-5300
Suite 267   Fax: 202-624-5313
Washington, DC 20001-1512   http://www.nga.org
Founded in 1908, this Association works closely with the administration and Congress on state and federal policy issues, serves
as a vehicle for sharing knowledge of innovative programs
among states, and provides technical assistance and consultant
services to governors on a wide range of management and policy
issues. Part of the organization is the Center for Best Practices
which undertakes demonstration projects and provides anticipa-
tory research on important policy issues.
Founded: 1908
Raymond Scheppach, Director

73   National Institute for Global Environmental Change

University of California, Davis
1490 Drew Avenue   530-757-3350
Suite 140   Fax: 530-756-6499
Davis, CA 95616-8756   E-mail: nigec@ucdavis.edu
                    http://nigec.ucdavis.edu
Sponsored by the Office of Health and Environmental Research
and the Office of Environmental Analysis of the US Department
of Energy at the University of California, Davis. Offers congressional briefings and education materials for schools, universities, and public education programs.
W Lawrence Gates, Director

74   National Institute for the Environment

1725 K Street NW   202-530-5810
Suite 212   Fax: 202-628-4311
Washington, DC 20006-1401   E-mail: cnie@cnie.org
                    http://www.cnie.org
The committe for the National Institute for the Environment is a
national, nonprofit organization working to improve the scien-
tific basis for making decisions in environmental issues through
creation of a new, non-regulatory environmental science and ed-
ucation entity, The National Institute for the Environment
(NIA).
Peter Saundry, Executive Director
Richard Benedict, President

75   National Network of Forest Practitioners

305 Main Street   401-273-6507
Providence, RI 02903   Fax: 401-273-6508
                    E-mail: info@nnfp.org
                    http://www.nnfp.org

Grassroots alliance of rural people, organizations and businesses finding practical ways to integrate economic development, environmental protection, and social justice. At its heart,
the network is about connecting rural people with the goal of
sharing knowledge and ideas, building a national voice, and contributing to the well-being or rural communties and the forest on
which they depend.
Thomas Brendler, Executive Director
Susan Beech, Office Administrator

76   National Parent Teachers Association

541 North Fairbanks Court   312-670-6782
Suite 1300   800-307-4782
Chicago, IL 60611   Fax: 312-670-6783
                    E-mail: info@pta.org
                    http://www.pta.org
The mission of the National PTA is to support and speak on behalf of children and youth in the school, in the community and
before governmental bodies and other organizations; to assist
parents in developing the skills they need to raise and protect
their children and to encourage parent and public involvement in
the public schools.
Warlene Gary, Executive Officer

77   National Religious Partnership for the Environment

49 South Pleasant Street   413-253-1515
Suite 301   Fax: 413-253-1414
Amherst, MA 01002   E-mail: nrpc@nrpc.org
                    http://www.nrpe.org
The mission of the National Religious Partnership for the Environment is to permanently integrate issues of environmental
sustainability and justice across all aspects of organized religious life.

78   National Solid Waste Management Association

Environmental Industry Associates
4301 Connecticut Avenue Northwest   202-244-4700
Suite 300   800-424-2869
Washington, DC 20008   Fax: 202-364-3792
                    http://www.envasns.org
To advance the safe and environmentally protective manage-
ment of non-hazardous and hazardous wastes through the pri-
vate waste services industry.
Bruce Parker, CEO

**79   North American Association for Environmental Educa-
tion**

2000 P Street, Northwest   202-419-0412
Suite 540   Fax: 202-419-0415
Washington, DC 20036   E-mail: email@naaee.org
                    http://www.naaee.org
The NAAEE, originally the National Association for Environmental Education, is a professional association of educators,
communicators, and managers established to assist and support
the work of individuals and groups engaged in envronmental education, research, and service. In addition, they specifically promote environmental education at the post-secondary,
secondary, and elementary levels. The NAAEE has over 1,800
members.
Founded: 1971
William H Dent Jr, Executive Director
Joseph Baust, President

80   Orion Society

187 Main Street   413-528-4422
Great Barrington, MA 01230   888-909-6568
                    Fax: 413-528-0676
                    E-mail: orion@orionsociety.org
                    http://www.orionsociety.org
The Orion Society is an award-winning publisher, an environmental education organization, and a communications and support network for grassroots environmental and community
organization across North America. It is nonprofit
membership organization with 8000 members, individual and
organizational, representing all fifty states and thirty-one foreign countries.
Founded: 1982
Jason Houston, Managing Editor

81   Public Citizen

1600 20th Street NW
Washington, DC 20009   202-588-1000
                    Fax: 202-588-7796
                    http://www.citizen.org

Founded by Ralph Nader, Public Citizen is the consumer's eyes in Washington. With the support of more than 15,000 people like you, we fight for safer drugs and medical devices, cleaner and safer energy sources, a cleaner environment, fair trade and a more open and democratic government.
Joan Claybrook, President

82 Public Employees for Environmental Responsibility
PEEReview
2001 S Street NW   202-265-7337
Suite 570   Fax: 202-265-4192
Washington, DC 20009   E-mail: info@peer.org
                                    http://www.peer.org
A nonprofit organization protecting employees who are protecting the environment. PEER works to create fundamental change. Environmental agencies conduct the public's business. Above all, PEER provides environmental ethics and government accountability.
Howard Wilshire, Chairperson
Jeff Ruch, Executive Director

83 Renew America
1200 18th Street NW   202-721-1545
Suite 1100   Fax: 202-467-5780
Washington, DC 20036   http://www.solstice.crest.org
Renew America is a nonprofit organization founded in 1989. They coordinate a network of community and environment groups, businesses, government leaders and civic activists to exchange ideas and expertise for improving the enviroment. By finding and promoting programs that work, Renew America helps inspire communities and businesses to meet today's environmental challenges.
Founded: 1989
Kenneth Brown, Executive Director
L Hunter Lovins, President

84 Renewable Fuels Association
1 Massachusetts Avenue NW   202-289-3835
Suite 820   Fax: 202-289-7519
Washington, DC 20001-1401   E-mail: info@ethanolrfa.org
                                    http://www.ethanolrfa.org
As the national trade association for the US fuel ethanol industry, the Association has been working on behalf of the industry to secure a strong marketplace for ethanol. With a proven track record as a leader, the RFA is recognized nationwide as an organization dedicated to the continued vitality and growth of ethanol in the fuel marketplace.
Founded: 1981
Bob Dinneen, President/CEO

85 Rocky Mountain Institute
1739 Snowmass Creek Road   970-927-3851
Snowmass, CO 81654-9199   http://www.rmi.org
Rocky Mountain Institute is an entrepreneurial, nonprofit organization that fosters the efficient use of resources for sustainability.
Amory B Lovins, CEO
Marty Pickett, Executive Director

86 Silicone Health Council
1850 M Street, Northwest   202-721-4100
Suite 700   Fax: 202-296-8120
Washington, DC 20036   E-mail: info@socma.com
                                    http://www.socma.com
Coordinates health, environmental and safety programs. Conveys scientifically sound information about silicones.
Joseph Acker, President

87 Silicones Environmental Health and Safety Council of North America
2325 Dulles Center Boulevard   703-788-6570
Suite 500   Fax: 703-788-6345
Herndon, VA 20171   E-mail: sehsc@sehsc.com
                                    http://www.sehsc.com
An organization of organosilicones manufacturers. Formed to coordinate programs dealing with health, environmental and safety issues.
Reo Menning, Executive Director

88 Society of American Foresters
5400 Grosvenor Lane   301-897-8720
Bethesada, MD 20814   Fax: 301-897-3690
                                    E-mail: safweb@safnet.orgt
                                    http://www.safnet.org
To advance the science, education, technology, and practice of forestry; to enhance the competency of its members; to establish professional excellence; and to use the knowledge, skills, and conservation ethic of the profession to ensure the continued health and use of forest ecosystems present and future of availability of forest resources to benefit society.
Michael T Goergen, Jr., Executive VP, CEO

89 Student Pugwash USA
2029 P Street Northwest   202-429-8900
Suite 301   800-969-2784
Washington, DC 20036   Fax: 202-429-8905
                                    E-mail: spusa@spusa.org
                                    http://www.spusa.org
To promote the socially responsible application of science and technology in the 21st century. Encourages young people to examine the ethical, social, and global implications of science and technology, and to make these concerns a guiding focus of their academic and professional endeavors.
Christine Rovener, Executive Director

90 Together Foundation
130 S Willard Street   802-862-2030
Burlington, VT 05401   http://www.bestpractices.org
Promotes partnerships among governments, NGOs, and the private sector in efforts to deal with the world's social problems. Facilitates nonprofit management through its computer, information and networking services. Makes informational and communication functions and linkages more universally accessible to nonprofit organizations.
Ella Cisneros, President

91 U.S. Global Change Research Information Office
1717 Pennsylvania Avenue, Northwest 202-223-6262
Suite 250   Fax: 202-223-3065
Washington, DC 20006   E-mail: information@gcrio.org
                                     http://www.gcrio.org
The U.S. Global Change Research Information Office (GCRIO) provides access to data and information on climate change research, adaptation/mitigation strategies and technologies and global change-related educational resources on behalf of the various US federal agencies that are involved in the US global Change Research Program.
John S Perry, Director

92 US Dye Manufacturers Operating Committee
2001 Pennsylvania Avenue NW   202-833-2131
Suite 760   Fax: 202-659-1699
Washington, DC 20006   http://www.ainc.org
C Tucker Helmes PhD, Executive Director

93 United Nations Environment Programme New York Office
2 United Nations Plaza   212-963-8210
Room DC2-0803   Fax: 212-963-7341
New York, NY 10017   E-mail: info@nyo.unep.org
                                     http://www.nyo.unep.org
Was created to be the environmental conscience of the UN system-to make other agencies aware of the environmental impact of their activities. They leveraged a small budget into a program of major significance and lasting influence. It assesses the state of the world's environment; helps formulate international environmental law; strengthens the environmnetal management capacity of developing countries; and promotes the implementation of the dimension of sustainable development.
Adnan Amin, Director

National: Design & Architecture

94 American Bureau of Shipping
16800 Greenspoint Park Drive   281-673-2800
Suite 300 South   Fax: 281-673-2801
Houston, TX 77060-2393   http://www.abs-group.com

The mission of ABS is to serve the public interest as well as the needs of our clients by promoting the security of life, property, and the natural environment primarily through the development and verification of standards for thedesign, construction and operational maintenance of marine-related facilities.
T Peter Pappas, President

95 American Society of Landscape Architects
636 Eye Street, Northwest   202-898-2444
Washington, DC 20001-3736   Fax: 202-898-1185
http://www.asla.org
Founded in 1899, the American Society of Landscape Architects is the national professional association representing landscape architects. ASLA promotes the landscape architecture profession and advances the practice through advocacy,education, communication and fellowship. ASLA works to increase the public's awareness of and appreciation for the profession of landscape architecture.
Kristiana Farnsworth, President

96 American Society of Landscape Architects'
908 North 2nd Street   717-441-6041
Harrisburg, PA 17102 E-mail: info@landscapearchitects.org
http://www.landscapearchitects.org
Landscape architects design the built environment of neighborhoods, towns and cities while also protecting and managing the natural environment, from its forests and fields to rivers and coasts. Members of the profession have aspecial commitment to improving the quality of life through the best design of places for people and other living things.
Founded: 1899
Lisa Kunst Vavro, President

97 Center for Resourceful Building Technology NCAT
PO Box 100   406-549-7678
Missoula, MT 59806   Fax: 406-549-4100
E-mail: crbt@ncat.org
http://www.crbt.org
Dedicated to promoting environmentally responsible practices in construction. Works to serve as both catalyst and facilitator in encouraging building technologies which realize a sustainable and efficient use of resources.
Steve Loken, Founder

98 Environmental Action Foundation
333 John Carlyle Street   703-548-3118
Suite 200   Fax: 703-548-3119
Alexandria, VA 22314   E-mail: info@agc.org
http://www.agc.org
The mission is demonstrating the construction industry's commitment to improving the environment and quality of the life through: improving environmental education, supporting sensible application of environmental laws, promotingenvironmental awareness campaigns and assisting in environmental litigation.
Leah Wood, Senior Council

99 Environmental Design Research Association
PO Box 7146   405-330-4863
Edmond, OK 73083-7146   Fax: 405-330-4150
E-mail: edra@edra.org
http://www.edra.org
Supports sustainable development and environmental engineering.
Janet Singer, Executive Director

100 Land Improvement Contractors of America
3080 Ogden Avenue   630-548-1984
Lisle, IL 60532   Fax: 630-548-9189
E-mail: nlica@aol.com
Organization composed primarily of small contractors whose activities related to the conservation, use and improvement of land and water resources ranging from terrace and pond building to septic tank installation to loggingoperations. Strives to improve the climate within which members conduct their businesses by working for better legislation and regulations.
Wayne Maresch, Executive VP

National: Disaster Preparedness & Response

101 Chemical Emergency Preparedness Program
401 M Street SW
Washington, DC 20460   800-535-0202

102 Global Response
PO Box 7490   303-444-0306
Boulder, CO 80306-7490   Fax: 303-449-9794
E-mail: action@globalresponse.org
http://www.globalresponse.org
Global Response has created a worldwide network of people and organizations that refer environmental issues. In collaboration with the local people who are working on the problem, GR issues an Action to our members. The Action givesbackground information about the environmental threat and efforts of local organizations to resolve it. The Action asks GR members to use this information to write personal letters to specific officials who have the power to make positive changes.
Paula Palmer, Executive Director

103 International Association of Wildland Fires
PO Box 261   605-890-2348
Hot Springs, SD 57747-0261   Fax: 206-600-5113
http://www.iawfonline.org
To facilitate communication in the global wildland fire community. Has several programs to facilitate this mission.
Dick Mangan, President

104 Marine Mammal Stranding Center
3625 Brigantine Boulevard   609-266-0538
PO Box 773   Fax: 609-266-6300
Brigantine, NJ 08203   E-mail: mmsc@bellatlantic.net
http://marinemammalstrandingcenter.org
A private nonprofit organization fully licensed by the state and federal governments to rescue, rehabilitate and release stranded or otherwise distressed marine mammals and sea turtles that come ashore along the New Jersey coast.
Founded: 1978
Robert C Schoelkopf, Director
Sheila M Dean, Co-Director

105 National Association of Flood and Storm Water Management Agencies
1301 K Street Northwest   202-218-4122
Suite 800 East   Fax: 202-478-1734
Washington, DC 20004
They are an organziation of public agencies whose function is the protection of lives, property and economic activity from the adverse impacts of storm and flood waters. The mission of the association is to advocate public policy,encourage technologies and conduct education programs which facilitate and enhance the adchievement of the public service functions of its members.
Susan Gilson, Executive Director

106 National Council on Radiation Protection and Measurements
7910 Woodment Avenue   301-657-2652
Suite 400   800-229-2652
Bethesda, MD 20814   Fax: 301-907-8786
E-mail: ncrp@ncrp.com
http://www.ncrp.com
They represent the interests of professionals with responsibilities for measuring and providing protection from ionizing and non-ionizing radiation. It was started in 1964 and now has about 300 participants.
William Beckner, Executive Director
Thomas S Tenforde, President

107 National Fire Protection Association
1 Batterymarch Park   617-770-3000
Quincy, MA 02169   Fax: 617-770-0700
E-mail: public_affairs@nfpa.org
http://www.nfpa.org

The international NFPA, reduces the worldwide burden of fire and other hazards on the quality of life by developing and advocating scientifically-based consensus codes and standards, research, training and education. It is a private, independent and voluntary nonprofit organization with more than 75,000 members worldwide. Members come from fire department personnel, equipment manufacturers, healthcare fields, insurance companies, business and industry and state and local governments.
James M Shannon, President

108 National Oceanic and Atmospheric Administration
Office of Response and Restoration
National Ocean Service   301-713-2989
1305 East West Highway   Fax: 301-718-4389
Silver Spring, MD 20910 http://response.restoration.noaa.gov/
Responds to spills of oil and other hazardous materials. Helps emergency planners prepare for potential accidents. Creates software and other products to help people respond to hazardous materials accidents. Works to find remedies for the environmental damage caused by hazardous waste sites in coastal areas. Assesses injury to coastal resources from release of oil and other hazardous materials. Pursues restoration from those responsible for the harm.
William F Broglie, Chief Administrative Officer

109 Natural Hazards Research and Applications Information Center
University of Colorado
482-UCB   303-492-6818
Boulder, CO 80309   Fax: 303-492-2151
                                          E-mail: hazctr@colorado.edu
                                     http://www.colorado.edu/hazards
The Center serves as a national clearinghouse for social science research data on natural disasters, related technological events, and programs to reduce damage from natural disasters.
Kathleen Tierney, Director

110 Porpoise Rescue Foundation
2040 Harbor Island Drive   619-574-1573
Suite 201
San Diego, CA 92101
Jack Bowland, President

111 Safety Equipment Institute
1307 Dolley Madison Boulevard   703-442-5732
Suite 3A   Fax: 703-442-5756
McLean, VA 22101   E-mail: info@seinet.org
                                                http://www.seinet.org
The Institute is a private, nonprofit organization established in 1981 to administer the first nongovernmental, third-party certification programs to test and certify a broad range of safety equipment products. The purpose of SEI's certification programs is to assist government agencies along with users and manufacturers of safety equipment in meeting their mutual goals of protecting those who use safety equipment on or off the job and in keeping with recognized standards.
Founded: 1981
Patricia Gleason, President

National: Energy & Transportation

112 Alliance to Save Energy
1200 18th Street NW   202-857-0666
Suite 900   Fax: 202-331-9588
Washington, DC 20036   E-mail: info@ase.org
                                                    http://www.ase.org
Nonprofit coalition of prominent business, governmental, environmental, and consumer leaders who promote the efficient and clean use of energy worldwide to benefit consumers, the environment, economy, and natural security.
Kateri Callahan, President

113 Alternative Energy Resources Organization
432 N Last Chance Gulch   406-443-7272
Helena, MT 59601   Fax: 406-442-9120
                                              E-mail: aero@aeromt.org
                                                 http://www.aeromt.org

AERO is a statewide grassroots group whose members work together to strengthen communities through promoting sustainable agriculture, local food production and citizen-based Smart Growth community planning.
Jennifer Rasmuson, Office Manager

114 American Coal Ash Association
15200 East Girard Avenue   720-870-7897
Aurora, CO 80014-3050   E-mail: info@acaa-usa.org
                                               http://www.acca-usa.org
ACAA's mission is to advance the management and use of coal combustion products in ways that are technically sound, commercially competitive and environmentally safe.
Samuel Tyson, Executive Director

115 American Council for an Energy-Efficient Economy
1001 Connecticut Avenue NW   202-429-8873
Suite 801   Fax: 202-429-2248
Washington, DC 20036-5525   E-mail: info@aceee.org
                                                    http://aceee.org
A nonprofit organization dedicated to advancing energy efficiency as a means of promoting both economic prosperity and environmental protection. Fulfills its mission by: conducting technical and policy assessments; advising governments and utilities; publishing books, conference proceedings and research reports and informing consumers.
Carl Blumstein, Chairperson
Bill Valentino, President

116 American Gas Association
400 North Capitol Street Northwest   202-824-7000
Suite 450   Fax: 202-824-7115
Washington, DC 20001   http://www.aga.org
They advocate the interests of their natural gas utility members and their customers, provide information and services promoting demand and supply growth and operational excellence in the safe, reliable and cost - competitive delivery of natural gas.
David Parker, Director

117 American Petroleum Institute
1220 L Street NW   202-682-8000
Washington, DC 20005   http://www.api.org
API is the major national trade association representing the entire petroleum industry: exploration and production, transportation, refining, and marketing. With headquarters in Washington, DC, and petroleum councils in 33 states, it is a forum for all parts of the oil and natural gas industry to pursue priority public policy objectives and advance the interests of the industry in a legally appropriate manner.
Red Cavaney, Executive Director

118 American Public Power Association
2301 M Street NW   202-467-2900
Washington, DC 20037   Fax: 202-467-2910
                                                http://www.appanet.org
The American Public Power Association (APPA) is the service organization for the nation's more than 2,000 community-owned electric utilities that serve more than 43 million Americans. It was created in 1940 as a non-profit, non-partisan organization. Its purpose is to advance the public policy interests of its members and their consumers, and provide member services to ensure adequate, reliable electricity at a reasonable price with the proper protection of the environment.

119 American Solar Energy Society
2400 Central Avenue   303-443-3130
Suite A   Fax: 303-443-3212
Boulder, CO 80301   E-mail: ases@ases.org
                                                   http://www.ases.org
National organization dedicated to advancing the use of solar energy for the benefit of US citizens and the global environment. ASES promotes the widespread near-term and long-term use of solar energy. Sponsors the National Solar Energy Conference, publishes Solar Today magazine and Advances in Solar Energy. Publishes white papers, sponsors issue Roundtables in Washington, DC, distributes solar publications, organizes a Solar Action Network and has regional chapters throughout the country.
Thomas Starrs, Chairperson

120 Civil Engineering Research Foundation
1801 Alexander Bell Drive   703-295-6314
Suite 630   Fax: 202-833-6315
Reston, VA 20191-4400   E-mail: corporate@cerf.org
                                   http://www.cerf.org
A non-governmental organization aiming to promote the global adoption of energy efficient policies, technologies and practices in order to promote economically and ecologically sustainable development. Operates US based programs,regional offices (Europe, Asia, Latin America) and a publishing program.
Harvey M Bernstein, President/CEO

121 Clean Fuels Development Coalition
4641 Montgomery Avenue   703-276-2332
Suite 350   Fax: 703-276-8447
Bethesda, MD 20814   E-mail: cfdcinc@aol.com
                                   http://www.cleanfuelsdc.org
Innovative not-for-profit organization that actively supports the development and production of fuels that can reduce air pollution and lessen our dependence on imported oil.
Douglas A Durante, Executive Director

122 Compressed Gas Association
4221 Walney Road   703-788-2700
5th Floor   Fax: 703-461-1831
Chantilly, VA 20151-2923   E-mail: ega@cganet.com
                                   http://www.cganet.com
The mission of the CGA is to promote the safe manufacture, transportation, storage, transfilling, and disposal of industrial and medical gases and their containers.
Carl T Johnson, President

123 Conservation and Renewable Energy Inquiry and Referral Service
PO Box 8900
Silver Spring, MD 20907   800-523-2929
Provides general information on alternative energy resources and energy conservation, and offers referral services to other government and private organizations who can provide more specific and technical information.
Grace Gilden, Project Manager

124 Consumer Energy Council of America Research Foundation
2000 L Street NW   202-659-0404
Suite 802   Fax: 202-659-0407
Washington, DC 20036   E-mail: eberman@cecarf.org
                                   http://www.cecarf.org
CECA is the senior public interest organization in the US focusing on energy, telecommunications and other network industries that provide essential services to consumers.
        Founded: 1973
Ellen Berman, President

125 Energy Research Institute
6850 Rattlesnake Hammock Road   727-793-1922
Highway 951   Fax: 727-793-1260
Naples, FL 33962
JC Caruthers, President/Executive Director

126 Environmental Coalition on Nuclear Power
433 Orlando Avenue   814-237-3900
State College, PA 16803-3477   Fax: 814-237-3900
                                   E-mail: johnstrud@link.net
Groups and individuals concerned with the nuclear power and energy policies. Maintains speakers bureau and conducts research and educational programs.
Dr. Judith Johnsrud, Executive Officer

127 Fossil Fuels Policy Action Institute
1062 G Street, Suite K   707-826-7775
PO Box 4347   Fax: 707-822-7007
Arcata, CA 95518   E-mail: alliance@tidepool.com
                                   http://www.tidepool.com/alliance
The Institute aims to educate the general public about making an immediate, fundamental shift from oil dependence toward sustainable modes of transportation and production, such as sail transport and organic farming.

128 Institute for Transportation and the Environment
85 E Roanoke Street   206-322-5463
Seattle, WA 98102
Christopher K Leman, Executive Director

129 International Association for Energy Economics
28790 Chagrin Boulevard   212-464-5365
Suite 350   Fax: 216-464-2737
Cleveland, OH 44122   E-mail: iaee@iaee.org
                                   http://www.iaee.org
A nonprofit professional organization to provide an interdisciplinary forum for the exchange of ideas and experiences among energy professionals and others interested in shaping opinions and preparing for the events which effect energyindustries.
David Williams, Executive Director

130 Interstate Oil and Gas Compact Commission
PO Box 53127   405-525-3556
Oklahoma City, OK 73152-3127   800-822-4015
                                   Fax: 405-525-3592
                          E-mail: iogcc@iogcc.state.ok.us
                          http://www.iogcc.state.ok.us
An organization of states which promotes conservation and efficient recovery of domestic oil and natural gas resources while protecting health, safety and the environment.
Frank Murkowski, Chairperson

131 NE Sustainable Energy Association
50 Miles Street   413-774-6051
Greenfield, MA 01301   Fax: 413-774-6053
                                   E-mail: nesea@nesea.org
                                   http://www.nesea.org
NESEA is the nation's leading regional membership organization focused on promoting the understanding, development, and adoption of energy conservation and non-polluting renewable energy technologies. We work to bring cleanelectricity, green transportation, and healthy, efficient buildings into everyday use.
Nancy Hazard, Executive Director

132 NW Energy Coalition
219 1st Avenue S   206-621-0094
Suite 100   Fax: 206-621-0097
Seattle, WA 98104   E-mail: nwec@nwenergy.org
                                   http://www.nwenergy.org
The Coalition is an alliance of over 100 environmental, civic and human service organizations, progressive utilities and businesses from Oregon, Washington, Idaho, Montana, Alaska and British Columbia. We promote energy resources,consumer and low-income protection and fish and wildlife restoration on the Columbia and Snake Rivers.
Jay Formick, Chairman

133 NW Power Planning and Conservation Council
851 6th Avenue Southwest   503-222-5161
Suite 1100   800-452-5161
Portland, OR 97204-1348   Fax: 503-820-2370
                          E-mail: comments@nwppc.org
                                   http://www.nwcouncil.org
An interstate compact comprising Idaho, Montana, Oregon and Washington, created by the Pacific Northwest Electric Power Planning and Conservation Act of 1980. Responsibilities are to protect, mitigate and enhance fish and wildlife, andrelated spawning grounds and habitat of the Columbia River Basin that have been affected by hydropower, while assuring the region an adequate, efficient, economical and reliable power supply and informing and involving the public in decision-making.
Melinda Eden, Chairperson

134 National BioEnergy Industries Association
122 C Street NW   202-383-2540
4th Floor   Fax: 202-383-2670
Washington, DC 20001-2109   E-mail: www.newenergy.org
                                   http://www.ia-usa.org
Trade association for the US biomass industry. Members include landowners, foresters, harvesters, fuel transporters, processors and wood fuel users, as well as developers of biomass feedstock, equipment manufacturers, constructioncontractors, consultants, biomass energy project developers and owners/operators of biomass direct-combustion facilities.
Scott Sklar, Executive Director

135 National Energy Foundation
3676 California Avenue
Suite A117
Salt Lake City, UT 84104
801-908-5800
Fax: 801-908-5400
E-mail: info@nef1.org
http://www.nef1.org
A nonprofit educational organization that has become a national leader in teacher training and student programs, instructional materials development and distribution and has done so in cooperation with sponsors from business,government agencies and the education community.
Edward Dalton, President/CEO

136 North American Benthological Society
Savannah River Ecology Laboratory
Drawer E
Aiken, SC 29802
803-725-2472
Fax: 803-725-3309
E-mail: frazer@srel.edu
The Savannah River Ecology Laboratory is located at the Savannah River Site near Aiken, SC. Its mission is to conduct environmental studies in support of US Department of Energy SRS operations and defense programs.
Founded: 1951
Cheryl R Black, Executive Officer

137 Renewable Fuels Association
1 Massachusetts Avenue Northwest
Suite 820
Washington, DC 20001-1401
202-289-3835
Fax: 202-289-7519
E-mail: info@ethanolrfa.org
http://www.ethanolrfa.org
As the national trade association for the US fuel ethanol industry, the Association has been working on behalf of the industry to secure a strong marketplace for ethanol. With a proven track record as a leader, the RFA is recognizednationwide as an organization dedicated to the continued vitality and growth of ethanol in the fuel marketplace.
Founded: 1981
Bob Dinneen, President/CEO

138 Safe Energy Communication Council
1717 Massachusetts Avenue NW
Suite 805
Washington, DC 20036
202-483-8491
Fax: 202-234-9194
E-mail: safeenergy@erols.com
http://www.safeenergy.org
Christopher Sherry, Research Director

National: Environmental Engineering

139 American Academy of Environmental Engineers
130 Holiday Court
Suite 100
Annapolis, MD 21401
410-266-3311
Fax: 410-266-7653
E-mail: academy@aaee.net
http://www.aaee.net
They are dedicated to excellence in the practice of environmental engineering to ensure the public health, safety, and welfare to enable humankind to co-exist in harmony with nature.
David A Asselin, Executive Director

140 American Federation of Mineralogical Societies
AFMS Central Office
PO Box 302
Glyndon, MD 21071-0302
410-833-7926
E-mail: central_office@amfed.org
http://www.amfed.org
A nonprofit educational federation of seven similar regional organizations of gem, mineral and lapidary societies.
Founded: 1947
Bill Smith, President

141 American Insitute of Chemical Engineers
3 Park Avenue
New York, NY 10016-5991
212-591-8100
800-242-4363
Fax: 212-591-8888
E-mail: xpress@aiche.org
http://www.aiche.org
AICHE mission is to promote excellence in chemical engineering education and global practice to advance the development and exchange of relevant knowledge and to uphold and advance the profession's standards, ethics and diversity.
John Sofranko, Executive Director

142 American Institute of Chemists
315 Chestnut Street
Philadelphia, PA 19106-2702
215-873-8224
Fax: 215-925-1954
E-mail: info@theaic.org
http://www.theaic.org
To advance the chemical sciences by establishing high professional standards of practice and to emphasize the professional, ethical, economic, and social status of its members for the benefit of society as a whole.
Sharon Dobson, Executive Director

143 American Society of Agricultural Engineers
2950 Niles Road
Saint Joseph, MI 49085-9659
616-429-0300
Fax: 616-429-3852
http://www.asae.org
The Society is a professional and technical organization dedicated to the advancement of engineering applicable to agricultural, food, and biological systems. ASAE comprises of 9,000 members representing more than 90 countries. ASAEmembers serve in industry, academia, and public service and are uniquely qualified to determine and develop more efficient and environmentally sensitive methods.
Founded: 1907
M Melissa Moore, Executive VP

144 American Society of Civil Engineers
1801 Alexander Bell Road
Reston, VA 20191-4400
800-548-2723
Fax: 703-295-6444
http://www.asce.org
Our mission is to provide essential value to our members, our partners, and the public through developing leadership, advancing technology, advocating lifelong learning, and promoting the profession. This vision and mission are to berealized through the achievement of plan's four goals, which echo the mission's four main principles.
William P Henry, President

145 American Society of Safety Engineers
1800 E Oakton Street
Des Plaines, IL 60018-2187
847-699-2929
Fax: 847-768-3434
E-mail: customerservice@asse.org
http://www.asse.org
ASSE is a global association providing professional development and representation for those engaged in the practice of safety, health and environmental issues. Provides services to the private and public sectors to protect people,property and the environment.
Jack H Dobsonld, President
Fred J Fortman, Executive Director

146 American Society of Sanitary Engineering
901 Canterbury
Suite A
Westlake, OH 44145
440-835-3040
Fax: 440-835-3488
E-mail: info@asse_plumbing.org
http://www.asse_plumbing.org
They are comprised of members from all segments of the plumbing industry. The ASSE is a unique organization because its membership is a cross-section of the industry, including contractors, engineers, inspectors, journeymen,apprentices and others who are involved in various segments of the industry.
Shannon Corcoran, Executive Director
Randy Ackroyd, President

147 Association for Environmental Health of Soils
150 Fearing Street
Amherst, MA 01002
413-549-5170
Fax: 413-549-0579
http://www.aehs.com
The AEHS was created to facilitate communications and foster cooperation among professionals concerned with the challenge of soil protection and cleanup. Members represent the many disciplines involved in making decisions and solvingproblems affecting soils. AEHS recognizes that widely acceptable solutions to the problem can be found only through the integration of scientific and technological discovery, social and political judgement and hands on practice.
Paul Kostecki, Executive Director
Cindy Langlois, Managing Editor

148 Association of Conservation Engineers
PO Box 180
Jefferson City, MO 65102
573-522-2323
Fax: 573-522-2324

E-mail: mihalg@mail.conservation.stste.mo.us
http://www.conservation.state.mo.us

An organization of conservation engineers and technicians who are working to conserve and improve our nation's natural heritage. Brings together engineers and allied personnel employed by conservation and recreation agencies andconsultants who have a community of specialized interests in the areas of fish, wildlife, parks, forests and related conservation/recreation fields. Members pool experience and information pertaining to conservation engineering to make naturalresources more accessible.
Greg Mihalevich, President

149 Association of Energy Engineers
4025 Pleasantdale Road
Suite 420
Atlanta, GA 30340-4264
404-761-0509
Fax: 770-446-3969
E-mail: info@aeecenter.org
http://www.aeecenter.org
Association for those involved in air and water pollution contrasts, waste-to-energy services information, asbestos abatement and monitoring instruments and equipment.
Albert Thumann, Executive Director

150 Association of Environmental Engineering Professors
Michigan Technical University
1400 Townsend Drive
Houghton, MI 49931
906-487-2530
Fax: 906-487-2782
http://www.arcat.com/arcatcos/cos36
Peter Campbell, President

151 Association of Ground Water Scientists and Engineers
601 Dempsy Road
Westerville, OH 43087
800-557-7379
Fax: 614-808-7791
http://www.arcat.com
Jacqueline Mack

152 Environmental and Engineering Geophysical Society
1720 South Colorado Boulevard
Suite 110
Denver, CO 80222-4303
303-531-7517
Fax: 303-820-3844
E-mail: staff@eegs.org
http://www.eegs.org
The Environmental and Engineering Geophysical Society is a professional nonprofit society chartered in 1992. Its goal is to promote the science of geophysics especially as it is applied to environmental and engineering problems; tofoster common scientific interests of geophysicists and their colleagues in other related sciences and engineering; to maintain a high professional standing among its members; and to promote fellowship and coperation among persons interested in thescience.
Barbara Luke, President

153 Institute for Alternative Futures
100 North Pitt Street
Suite 235
Alexandria, VA 22314
703-684-5880
Fax: 703-684-0640
http://www.altfutures.com
Promotes environmental engineering through the use of environmentally safe technology.
Willis B Goldbeck, Chairman

154 Institute of Noise Control Engineering
210 Marston Hall
Ames, IA 50011-2153
515-294-6142
Fax: 515-294-3528
E-mail: ibo@inceusa.org
http://www.incusa.org
A primary purpose of the Institute is to promote engineering solutions to environmental noise problems.
Gerald C Lauche, President
Joseph M Cuschieri, Executive Director

155 Inter-American Association of Sanitary Engineering and Environmental Sciences
AIDIS-USA
PO Box 7737
McLean, VA 22106-7737
703-247-8730
Fax: 703-243-9004
E-mail: turnerje@cdm.com
http://www.aidis-usa.org
To further the goals of AIDIS Interamericana through programs and services that promote sound environmental practices, policies, management, and education to improve the quality of life throughout the Americas.

J Ellis Turner, President

National: Environmental Health

156 Acid Rain Foundation
1410 Varsity Drive
Raleigh, NC 27606-2010
919-828-9443
Fax: 919-515-3593
Dr. Harriett S Stubbs, Executive Director

157 Action on Smoking and Health
2013 H Street NW
Washington, DC 20016
202-659-4310
http://http://ash.org
Organized to use the power of the law to protect the rights of nonsmokers. Emphasis is placed on legal efforts to protect nonsmokers and to get courts to support the rights of nonsmokers. Also conducts educational and awarenesscampaigns regarding the problem of smoking and the rights of nonsmokers.
John Banzhaf, Executive Director

158 Air and Waste Management Association
420 Fort Duquesne Boulevard
Pittsburgh, PA 15222-1435
412-232-3444
Fax: 412-232-3450
http://www.awma.org
The Air and Waste Management Association is a nonprofit professional organization that enhances knowledge and expertise by providing a neutral forum for technology exchange, professional development, networking opportunities, publiceducation, and outreach to more than 9000 environmental professionals in 65 countries. A&WMA also promotes global environmental responsibility and increases the effectiveness of organizations to make critical decisions that benefit society.
Adrian Corolla, Director

159 All 4 Inc.
2393 Kimberton Road, Suite 100
PO Box 299
Kimberton, PA 19442-0299
610-933-5246
Fax: 610-933-5127
E-mail: jegan@all4inc.com
http://www.all4inc.com
All 4 is an environmental consulting company specializing in the air quality field.
John Egan, Principal Consultant

160 Alliance for Acid Rain Control and Energy Policy
444 N Capitol Street
Suite 602
Washington, DC 20001
202-624-5475
Fax: 202-508-3829

161 American Academy of Environmental Medicine
7701 E Kellogg
Suite 625
Wichita, KS 67207
316-684-5000
Fax: 316-684-5709
E-mail: Administrator@aaem.com
http://www.aaem.org
The Academy is interested in expanding the knowledge of interactions between human individuals and their environment, as these may be demonstrated to be reflected in their total health. The Academy is comprised primarily of medicalprofessionals who sponsor publications, seminars and courses. A newsletter and journal are among the organization's publications.
James W Willoughby, President
James F. Coy M.D., President Elect

162 American Association of Poison Control Centers
3201 New Mexico Avenue Northwest
Suite 310
Washington, DC 20016
202-362-6346
800-222-1222
Fax: 202-362-8377
E-mail: flanagan@aapcc.org
http://www.aapcc.org
Nationwide organization of poison centers and interested individuals. Provides a forum for poison centers and interested individuals to promote the reduction of morbidity and mortality from poisonings through public and professionaleducation and scientific research.
Anne Flanagan, Chairperson

163 American Board of Environmental Medicine
65 Wehrle Drive
Buffalo, NY 14225
716-833-2213
Fax: 716-833-2244
http://www.americanboardofenvironmentalmedicine.com

An independent non-profit organization built to certify the competency of physicians who practice the discipline of Environmental Medicine
Dr. K. Patel, President

164 American Board of Industrial Hygiene
6015 W St Joseph Highway   517-321-2638
Suite 102   Fax: 517-321-4624
Lansing, MI 48917   E-mail: abih@abih.org
                                                    http://www.abih.org
The American Board of Industrial Hygiene, a nonprofit corporation, was organized to improve the practice and educational standards of industrial hygiene through voluntary certification.
Lynn O'Donnell, CIH, Executive Director

165 American Cancer Society
1599 Clifton Road NW   404-329-7686
Atlanta, GA 30329   800-ACS-2345
                                                    Fax: 404-321-4669
                                                    http://www.cancer.org
The American Cancer Society is the nationwide, community-based, voluntary health organization dedicated to eliminating cancer as a major health problem by preventing cancer, saving lives from cancer, and diminishing suffering from cancer through research, education and service.
Clark W Heath Jr MD, VP
Virginia Krawiec, Program Director

166 American Conference of Governmental Industrial Hygienists
1330 Kemper Meadow Drive   513-742-2020
Cincinnati, OH 45240   Fax: 513-742-3355
                                                    E-mail: pubs@acgih.org
                                                    http://www.acgih.org
The American Conference of Governmental Industrial hygienists is a member-based organization and community of professionals that advances worker health and saftey through education and the development and dissemination of scientificand technical knowledge. Examples include annual editions of the TLVs and BELs and work practice guides in ACGIH's Signature Series of Publications.
            90 pages Magazine
Cindy Coe Laseter, Chair
Robert D Soule, Vice Chair

167 American Council on Science and Health
1995 Broadway   212-362-7044
2nd Floor   Fax: 212-362-4919
New York, NY 10023-5860   E-mail: acsh@acsh.org
                                                    http://www.acsh.org
A consumer education organization based in New York City. The council is an independent association that promotes scientifically balanced evaluations of food, chemicals and the environment, and their relationship to human health.
Eliabeth M Whelen, President

168 American Industrial Health Council
2001 Pennsylvania Avenue NW   202-833-2131
Suite 760   Fax: 202-833-2201
Washington, DC 20006-1850   http://www.ainc.org
Mission is to promote the sound use of scientific principles and procedures in public policy for the assessment and regulation of risks associated with human health effects and ecological effects. Members include major companies indiverse industries advocating reliance on the best available science for identifying, evaluating, and controlling health risks.
Gaylen Camera, Executive Director
Ronald A Lang, President

169 American Industrial Hygiene Association
2700 Prosperity Avenue   703-849-8888
Suite 250   Fax: 703-207-3561
Fairfax, VA 22031   E-mail: infonet@aiha.org
                                                    http://www.aiha.org
To promote the highest quality of occupational and environmental health and safety within the workplace and the community through advocacy and the provision of services and tools to enhance the professional practice of our members.
Donna M Doganiero, Chair

170 American Lung Association
61 Broadway   212-315-8100
6th Floor   800-LUN-GUSA
New York, NY 10006   http://www.lungusa.org
The American Lung Association has been fighting lung disease for nearly 100 years. The work continues as we strive to make breathing easier for everyone through education, community service, advocacy and research programs. Our missionis to prevent lung disease and promote lung health.
F Jack Sutter, Chair-Elect

171 American Public Health Association
800 I Street Northwest   202-777-APHA
Washington, DC 20001-3710   Fax: 202-777-2532
                                                    E-mail: comments@apha.org
                                                    http://www.apha.org
The APHA brings together researchers, health service providers, administrators, teachers and other health workers in a unique, multidisciplinary environment of professional exchange, study and action. APHA is concerned with a broad setof issues affecting personal and environmental health, including federal and state funding for health programs, pollution control, programs and policies related to chronic and infectious diseases, a smoke-free society and a professional education inpublic health.
Georges C Benjamin, Executive Director
Patricia D Mail, President Elect

172 American Society for Microbiology
1752 N Street NW   202-737-3600
Washington, DC 20036-2804   Fax: 202-942-9333
                                                    http://www.asm.org
The mission of the American Society for Microbiology is to promote the microbiological sciences and their applications for the common good. To meet the needs of our members, our activities include: publishing journals and books,convening meetings; conducting and supporting education, training and public information programs to facilitate the dissemination and application of new microbiological knowledge.
James Tiedje, President

173 Appalachian States Low-Level Radioactive Waste Commission
Department of Environment
1800 Washington Boulevard   410-537-3000
Baltimore, MD 21230-1729   800-633-6101
                                                    Fax: 410-537-4133
                                                    E-mail: bdye@mde.state.md.us
                                                    http://www.mde.state.md.us
The mission of the Maryland Department of the Environment (MDE) is to protect and restore the quality of Maryland's air, water, and land resources, while fostering smart growth, economic development, healthy and safe communities, andquality environmental education for the benefit of the environment, public health, and future generations.
Harold L Dye, Jr., MDE Contact
Kendl P Philbrick, Secretary

174 Asbestos Information Association of North America
1235 Jefferson Davis Highway   703-560-2980
PMB 114   Fax: 703-560-2981
Arlington, VA 22202   E-mail: aiabjpigg@aol.com
The Asbestos Information Association/North America was founded in 1970 to represent the interest of the asbestos industry and to collect and disseminate information about asbestos and asbestos products, with emphasis on safety, health,and environmental issues. The Association appears before Federal regulatory bodies and works with Government agencies to develop and implement standards for worker protection.
Bob Pigg, President

175 Association of American Pesticide Control Officials
PO Box 1249   802-472-6956
Hardwick, VT 05843   Fax: 802-472-6957
                                                    E-mail: aapco@vtlink.net
                                                    http://www.aapco.ceris.purdue.edu/
To encourage uniformity among the states in their pesticide regulatory programs.
Paul Liemandt, President

176 Association of State and Territorial Health Officials

1275 K Street Northwest 202-371-9090
Suite 800 Fax: 202-371-9797
Washington, DC 20005-4006 http://www.astho.org

Dedicated to formulating and influencing sound public health policy, and to assuring excellence in state-based public health practice.
George E Hardy Jr., Executive Director

177 Association of University Environmental Health/Sciences Centers

Health Science Institute 732-932-0200
PO Box 1179 Fax: 732-932-0131
Piscataway, NJ 08855 http://www.niehs.nih.gov

Provide a focused research effort on a particular environmental health problem, bring together many scientific disciplines to solve an environmental problem, attract and train young investigators, answer questions from the public on environmental problems, and help NIEHS identify emerging problems in the environmental health field.
Dr. Bernard Goldstein, President

178 Asthma and Allergy Foundation of America

1233 20th Street Northwest 202-466-7643
Suite 402 Fax: 202-466-8940
Washington, DC 20036 E-mail: info@aafa.org
http://www.aafa.org

Dedicated to improving the quality of life for people with asthma and allergies through education, advocacy and research.
Chris Ward, President

179 Beyond Pesticides/National Coalition Against the Misuse of Pesticides

701 E Street SE 202-543-5450
Suite 200 Fax: 202-543-4791
Washington, DC 20003 E-mail: info@beyondpesticides.org
http://www.beyondpesticides.org

Nonprofit organization formed in February 1981 to bring together health, environmental, labor, farm, consumer and church groups, as well as concerned individuals, to focus attention on the potential hazards associated with pesticide use.
Greg Small, President
Audrey Their, Vice President

180 Biomass Users Network

Bloomfield, NJ 07003 973-680-9100
Fax: 973-680-8066

Founded: 1985
David Mazambani, Chairman

181 Center for Science in the Public Interest

1875 Connecticut Avenue NW 202-332-9110
Suite 300 Fax: 202-265-4954
Washington, DC 20009-5728 E-mail: cspi@cspinet.org
http://www.cspinet.org

A nonprofit education and advocacy organization that focuses on improving the safety and nutritional quality of our food supply and on reducing the carnage caused by alcoholic beverages. CSPI seeks to promote health through educating the public about nutrition and alcohol; it represents citizen's interests before legislative, regulatory and judicial bodies; and it works to ensure advances in science are used for the public good.
Kathleen O'Reilly, President

182 Children of the Green Earth

1863 Ocean Drive 707-839-5013
McKinleyville, CA 866-98E-ARTH
http://www.childrenofthegreenearth.com

Children of the Green Earth is committed to offering quality, organic, earth friendly products for children of all ages. From grandmothers to babies, we've got something for everyone. All of our products are ethical, both environmentally and socially, meaning only organically grown products and all of the work involved in making the products is sweat shop and child labor free.

183 Chlorine Institute

2001 L Street NW 202-775-2790
Suite 506 Fax: 202-223-7225
Washington, DC 20036-4919 E-mail: tkerns@c12.com
http://www.cl2.com

Exists to support the chlor-alkali industry and serve the public by fostering continuous improvements to safety and the protection of human health and the environment connected with the production, distribution and use of chlorine, sodium and potassium hydroxides, and sodium hypochlorite; and the distribution and use of hydrogen chloride.
Rober Smerko, President

184 Citizens for Alternatives to Chemicals

8735 Maple Grove 517-544-3318
Lake, MI 48632-9511

185 Columbia Analytical Services

PO Box 479 360-501-3312
Kelso, WA 98626 Fax: 360-425-9096
http://www.caslab.com

Columbia Analytical Services (CAS), established in 1986, is an employee-owned, full service analytical laboratory network specializing in environmental testing.
Stephen W Vincent, President, Director and CEO

186 Commonweal

PO Box 316 415-868-0970
Bolinas, CA 94924 Fax: 415-868-2230
E-mail: commonweal@commonweal.org
http://www.commoanweal.org

A research institute primarily concerned with the environment and its relation to human health. Interests include organic pollutants.
Michael Lerner, President

187 Consumer Pesticide Project

425 Mississippi Street 415-826-6314
San Francisco, CA 94107

Craig Merrlees, Director

188 Corporate Accountability Intern

46 Plympton Street 617-695-2525
3rd Floor 800-688-8797
Boston, MA 02118 Fax: 617-695-2626
E-mail: info@infact.org
http://www.infact.org

Since 1977, Infact has been exposing life-threatening abuses by transnational corporations and organizing successful grassroots campaigns to hold corporations accountable to consumers and society at large. We are a nonprofit, national membership organization building an active, aware public and a core of well-trained organizers to lead the grassroots challenge to unwarranted corporate influence for years to come.
Founded: 1977
Kathryn Mulvey, Executive Director

189 Dangerous Goods Advisory Council

1100 H Street NW 202-289-4450
Suite 740 Fax: 202-289-4074
Washington, DC 20005 E-mail: info@dgac.org
http://www.hmac.org

HMAC promotes improvement in the safe transportation of hazardous materials/dangerous goods globally by: providing education, assistance, and information to the private and public sectors; through our unique status with regulatory bodies; and the adversity and technical strengths of our membership.
Alan I Roberts, President
Mike Morrissette, Vice President

190 ENSR International Headquarters

2 Technology Park Drive 978-589-3000
Westford, MA 01886-3140 800-722-9637
Fax: 978-589-3100
E-mail: rrozene@ensr.com
http://www.ensr.com

ENSR International is a global provider of environmental and energy development servies to industries and the government.
Dick Rozene, Director

191 Earth Regeneration Society

1442A Walnut Street 510-849-4155
Number 57 Fax: 510-849-0183
Berkeley, CA 94709 E-mail: csiri@igc.apc.org
http://www.imaja.com

The Earth Regeneration Society does research and education on climate change, ozone, and pollution, and calls for full employment and full social support based on surival programs and national and international networking.
Alden Bryant, President

192 EarthSave Foundation
Po Box 96   718-459-7503
New York, NY 10108   800-362-3648
                                                  Fax: 718-228-2491
               E-mail: information@earthsave.org
                         http://www.earthsave.org
Educators on the powerful effects our food choices have on the environment, our health, and all life on Earth, and supports people in moving toward a plant-based diet. Founded by John Robbins, author of Diet for a New America.
Patricia Carney, Executive Director

193 Environmental Hazards Management Institute
10 New Market Road   603-868-1496
PO Box 732   800-558-3464
Durham, NH 03824   Fax: 603-868-1547
                              E-mail: info@ehmi.org
                                    http://www.ehmi.org
An independent, nonprofit organization dedicated to understanding enhancement and preservation of our environment. A catalyst for informed environmental decision making by gathering, refining, and disseminating objective information toall stakeholders with emphasis on the role played by individuals and communities of individuals.
Alan J Borner, President

194 Environmental Health Network
PO Box 16267   757-546-0663
Chesapeake, VA 23328-6267   Fax: 757-482-9457
                              E-mail: ehn33@aol.com
                                    http://ehn.enviroweb.org

Linda King, Director

195 Environmental Information Association
6935 Wisconsin Avenue   301-961-4999
Suite 306   Fax: 301-961-3094
Chevy Chase, MD 20815-6112   E-mail: info@eia-usa.org
                                    http://www.eia-usa.org
A nonprofit organization dedicated to providing environmental information to individuals, members, and industry. We specialize in the dissemination of information about the abatement of asbestos and lead based paint, and about safetyand health issues, analytical issues and environmental site assessments.
Tom Broido, President

196 Environmental Resource Management (ERM)
3352 128th Avenue   616-399-3500
Holland, MI 49424-9236   Fax: 616-399-3777
                    E-mail: ronald_vriesman@erm.com
                                    http://www.erm.com
ERM works around the world with the private sector assessing how their business is likely to be impacted by environmental and social issues, new regulations, consumer concerns and supply chain issues and help companies developappropriate policies and management systems to manage these business risks.
Ronald Vriesman, Principal

197 Food and Water
389 Vermont Route 215   802-563-3300
Walden, VT 05873   Fax: 802-563-3310
                         http://www.foodandwater.org/
National political action organization that combines an eclectic mix of strategic and innovative grassroots campaigns with provocative publishing endeavors aimed at the underlying cultural, political, and economic roots of ourhuman-centered injustices. Food and Water has led tenacious and effective public campaigns against toxic food technologies such as food irradiation, pesticides, and genetically modified crops.
Michael Colby, Director

198 GBMC & Associates
219 Brown Lane   501-847-7077
Bryant, AK 72022   Fax: 501-847-7943
                    E-mail: kgreen@gbmcassoc.com
                         http://www.gbmcassoc.com

GBMC & Associates are a consulting firm providing strategic environmental services to industrial clients and air permitting support, water quality and toxicity studies, storm water management, environmental program development andreporting.
Ken Green, Principal

199 Halogenated Solvents Industry Alliance
1300 Wilson Boulevard   703-741-5780
12th Floor   Fax: 703-741-6077
Arlington, VA 22209   E-mail: contact@hsia.org
                                    http://www.hsia.org
The mission is to present the interests of users and producers of chlorinated solvents. To promote the continued safe use of these products and to promote the use of sound science in assessing their potential health effects.
Steven Risotto, Executive Director

200 Hazardous Waste Resource Center Environmental Technology Council
734 15th Street NW   202-783-0870
Suite 720   Fax: 202-737-2038
Washington, DC 20005   E-mail: mail@etc.org
                                    http://www.etc.org
The Environmental Technology Council (ETC) is a trade association of commercial environmental firms the recycle, treat and dispose of industrial and hazardous wastes; and firms involved in cleanup of contaminated sites.

201 Healthy Mothers, Healthy Babies
121 North Washington Street   703-836-6110
Suite 300   Fax: 703-836-3470
Alexandria, VA 22314   E-mail: info@hmhb.org
                                    http://www.hmhb.org
Healthy Mothers, Healthy Babies is a coalition of national, state and local providers, advocates and administrators concerned about health of pregnant women, infants and families. The coalition serves as a forum for informationexchange and as a catalyst to encourage collaborative partnerships amoung its members and colleagues.
George Guido, Chairperson
Leith Merrow Mullaly, Vice Chair

202 Home Chemical Awareness Coalition
Michigan State University   517-355-9578
Natural Resources Building   Fax: 517-353-8994
E Lansing, MI 48824-1222
Cynthia Frigden, Chairperson

203 Household Hazardous Waste Project
1031 E Battlefield   417-889-5000
Suite 214B   Fax: 417-889-5012
Springfield, MO 65807   http://outreach.missouri.edu
HHW provides assistance and education to promote consumer awareness, source reduction, safe handling, and sound management for hazardous wastes generated from households. This site contains guidesheets, technical bulletins, a link toour online training course (From Awareness to Action) and other resources.
Marie Steinwachs, Director

204 Household Products Disposal Council
1201 Connecticut Avenue NW   202-659-5535
Suite 300
Washington, DC 20036

205 Human Ecology Action League (HEAL)
PO Box 29629   404-248-1898
Atlanta, GA 30359   Fax: 404-248-0162
                              E-mail: HEALNatnl@aol.com
               http://members.aol.com/HEALNatnl/index/html
The HEAL is a nonprofit organization to serve those whose health has been adversely affected by environment exposures; to provide information to those who are concerned about the health effects of chemicals; and to alert the generalpublic about the potential dangers of chemicals. Referrals to local HEAL chapters and other support groups are available from the League.
Katherine P Collier, Contact

206 Human Environment Center
1930 18th Street NW 202-588-8036
Suite 24 Fax: 202-588-9422
Washington, DC 20009
Hector Eriksen-Mendoza, Executive Director

207 INFORM
120 Wall Street 212-361-2400
14th Floor Fax: 212-361-2412
New York, NY 10005-4001 http://www.informinc.org
INFORM is an independent research organization that examines the effects of business practices on the environment and on human health. Our goal is to identify ways of doing business that ensure environmentally sustainable economicgrowth. Our reports are used by government, industry, and environmental leaders around the world.
Joanna Underwood, President

208 Institute for the Human Environment
Po Box 395 707-935-9335
Sausalito, CA 94966 E-mail: info@iie.org
http://www.iie.org
Norman Gilroy, Executive Director
Shelley Arrowsmith, Vice President

209 Institute of Hazardous Materials Management
11900 Parklawn Drive 301-984-8969
Suite 450 Fax: 301-984-1516
Rockville, MD 20852 E-mail: ihmminfo@ihmm.org
http://www.ihmm.org
Mission is to provide recognition for professionals engaged in the management and engineering control of hazardous materials who have attained the required level of education, experience and competence; foster continued professionaldevelopment of Certified Hazardous Materials Managers (CHMM).
John H Frick PhD, CHMM, Executive Director

210 International Society for Environmental Toxicology and Cancer
PO Box 134 312-755-2080
Park Forest, IL 60466 Fax: 312-755-2096
Dr. George Scherr, Executive Officer

211 Kids for a Clean Environment
PO Box 158254 615-331-7381
Nashville, TN 37215 Fax: 615-333-9879
E-mail: kidsface@mindspring.com
http://www.kidsface.org
Established to help children who wanted to learn more about the world in which they live, provide a way for children to be involved in the protection of nature and connect children with other children who share their concerns aboutglobal environmental issues.
Founded: 1989
Trish Poe

212 Midwest Center for Environmental Science and Public Policy
1845 North Farewell Avenue 414-271-7280
Suite 100 Fax: 414-273-7293
Milwaukee, IL 53202 E-mail: mcespp@mcespp.org
http://www.mcespp.org
Protecting human health and the environment through research, advocacy, public education and citizen empowerment.
Jeffrey Foran, President

213 National Accreditation Council for Environmental Health Science and Protection
2632 Southeast 25th Avenue 503-235-6047
Suite D Fax: 503-235-7300
Portland, OR 97202 E-mail: ehacinfo@aehap.org
http://www.ehacoffice.org
Their purpose is to enhance the education and training of students who intend to become environmental health science and protection practitioners/ professionals. The council, originally known as the Health Curricula, is composed ofhighly qualified proessionals.
Founded: 1967
Sharon LaFollette, Chairperson
Richard Rowe, Director

214 National Alliance for Hispanic Health
1501 16th Street NW 202-387-5000
Washington, DC 20036 Fax: 202-265-8027
E-mail: alliance@hispanichealth.org
http://www.hispanichealth.org
The mission of the National Alliance for Hispanic Health is to improve the health and well-being of Hispanics. Issues covered include the full range of health and human services issues, including environmental health.
Jane L Delgado PhD, MS, President/CEO

215 National Association of City and County Health Officials
1100 17th Street NW 202-783-5550
2nd Floor Fax: 202-783-1583
Washington, DC 20036 http://www.naccho.org
The National Assoication of County and City Health Officials is a nonprofit, membership organization serving all 3,000 local health departments nationwide. NACCHO is dedicated to improving the health of people and communities byassuring an effective local public health system. As the Voice of local public health officials at the national level, NACCHO is able to promote the local perspective on national health programs and policies.
Michael C Caldwell, President
Poki S Namkung, Vice President

216 National Association of Noise Control Officials
53 Cubberly Road 609-586-2684
West Windsor, NJ 08550 Fax: 609-799-2616
http://www.arcat.com
The association consists of employees of the federal and state governments, consultants, scientists, and students concerned with acoustical control in the environment. It now has about 70 members.
Founded: 1978
Edward J DiPolzere, Executive Director

217 National Association of Physicians for the Environment
1643 Prince Street 703-684-1760
Alexandria, VA 22314-2818 Fax: 703-549-2772
E-mail: nape@napenet.org
http://www.greenlink.org
The National Association of Physicians for the Environment works to involve physicians and other health care professionals, particularly through their geographic and medical specialty organizations, to deal with the impact ofpollutants on organs and systems of the human body.
Jerome Goldsmith, Founder

218 National Cancer Institute
National Institutes of Health
9000 Rockville Pike
Health Building 10, Room 2A33 800-422-6237
Bethesda, MD 20892 E-mail: cancergovstaff@mail.nih.gov
http://www.cancer.gov
Leads the Nation's fight against cancer by supporting and conducting ground-breaking research in cancer biology, causation, prevention, detection, treatment and survivorship.
Andrew Eschenbach, Director

219 National Center for Environmental Health Strategies
1100 Rural Avenue 856-429-5358
Voorhees, NJ 08043 Fax: 856-429-5358
E-mail: mary@ncehs.org
http://www.ncehs.org/
Fosters the development of creative solutions to environmental health problems with a focus on indoor air quality, chemical sensitivites and environmental disabilities.
Mary Lamielle, Executive Director

220 National Center for Lead-Safe Housing
10227 Wincopin Circle 410-992-0712
Suite 205 Fax: 410-715-2310
Columbia, MD 21044 http://www.leadsafehousing.org
The National Center for Lead-Safe Housing was founded to bring the public health, housing and environmental communities together to combat our nation's epidemic of childhood lead poisoning.
Don Ryan, Executive Director

221 National Conference of Local Environmental Health Administrators
University of Washington
Department of Environmental Health   206-616-2097
Box 357234   Fax: 206-543-8123
Seattle, WA 98195-7234   E-mail: webmaster@ncleha.org
                                            http://www.ncleha.org
The NCLEHA's purpose is to provide a forum for local administrators to share common concerns and solutions to mutual problems, and to provide a professional organization for environmental health administrators, focused on the issuesand problems of local environmental health programs.
Mel Knight, Chairperson

222 National Education Association Health Information Network
1201 16th Street NW   202-822-7570
Suite 521   800-718-8387
Washington, DC 20036   Fax: 202-822-7775
                                       E-mail: info@neahin.org
                                            http://www.neahin.org
The National Education Assoication Health Information Network believes that sound public education must begin with school employees and students who are healthy and free of preventable diseases and supported with information, materialsand training opportunities that reaffirm these values.
Jerald Newberry, Executive Director

223 National Environmental Coalition of Native Americans
Po Box 988   405-567-4297
Claremore, OK 74018   Fax: 405-567-4297
                              E-mail: norteno_84@hotmail.com
                              http://http://oraibi.alphacdc.com
Nonprofit organization formed to educate Indians and Non-Indians about the health dangers of radioactivity and the transport of nuclear waste on America's rails and roads. Networks with environmentalists to develop grassrootscounter-movement to the efforts of the nuclear industry and develop Tribal nuclear free zones across the nation.
Grace Thorpe, President

224 National Institute for Occupational Safety and Health
Humphrey Boulevard Room 715 H   202-401-0721
200 Indiana Avenue   800-356-4674
Washington, DC 20201   http://www.cdc.gov
The National Institute for Occupational Safety and Health was created to conduct research and training and make recommendations for the prevention of work-related illnesses and injuries. NIOSH provides scientific information andrecommendations that may be used in setting workplace regulations.
Bill Halpherin, Director

225 National Institute of Child Health and Human Development
Po Box 3006
Rockville, MD 20847   800-370-2943
                                            Fax: 301-984-1473
     E-mail: NICHDinformationresourcecenter@mail.nih.gov
                                   http://www.nih.gov/nichd
National Institute of Child Health and Human Development supports and conducts basic, clinical and epidemiological research on the reproductive, neurobiological, developmental and behavioral processes that determine and maintain thehealth of children, adults, families and populations.
Duane Alexander, Director

226 National Institute of Environmental Health Sciences
Po Box 12233   301-402-3378
Research Triangle Park, NC 27709   Fax: 301-496-0563
                                       http://www.niehs.nih.gov
The mission of the National Institute of Environmental Health Sciences is to reduce the burden of environmentally associated diseases and dysfunctions.
Kenneth Olden, Director

227 National Pesticide Information Center
Oregon State University
333 Weniger Hall
Corvallis, OR 97331-6502   800-858-7378
                                            Fax: 541-737-0761

E-mail: npic@ace.orst.edu
http://http://npic.orst.edu
NPIC provides objective science based information about a wide variety of pesticide-related topics, including: pesticide product information, toxicology, environmental chemistry, referrals for lab analyses, investigation of pesticideincidents, emergency treatment information, safety information, health and environmental effects and clean-up/disposal procedures. NPIC is a cooperative effort between Oregon State University and the US EPA.
Terry Miller, Director

228 National Safety Council
Environmental Health Center
1025 Connecticut Avenue NW   202-293-2270
Suite 1200   Fax: 202-293-0032
Washington, DC 20036   E-mail: ehc@nsc.org
                                            http://www.nsc.org
The Mission of the Environmental Health Center is to foster improved public understanding of significant health risk and challenges facing modern society. This goal reinforces the National Safety Council's commitment to increased andmore effective citizen involvement in safety, health and environmental decision-making.
Jeffrey Shavelson, Policy Analyst

229 Natural Resources Defense Council
40 W 20th Street   212-727-2700
New York, NY 10011   Fax: 212-727-1773
                                       E-mail: nrdcinfo@nrdc.org
                                            http://www.nrdc.org
Dedicated to protecting endangered natural resources and improving the quality of the human environment. Areas of concentration are air and water pollution, nuclear safety, land use, urban environment, toxic substances control,resource management, wilderness and wildlife protection, Alaska, coastal zone management, energy conservation, soil erosion, and forestry. Founded in 1970, it has a membership of more than 500,000.
Francis Beinecke, Executive Director

230 Navy Environmental Health Center
620 John Paul Jones Circle   757-953-0700
Suite 1100   http://www.-nehc.med.navy.mil
Portsmouth, VA 23708-2103
Ensures Navy and Marine Corps readiness through leadership in prevention of disease and promotion of health.
Captain David Hiland, Commanding Officer

231 Novozymes North America Inc.
77 Perry Chapel Church Road   919-494-3039
Franklinton, NC   Fax: 919-494-3422
                                       E-mail: ggde@novozymes.com
                                            http://www.novozymes.com
Novozymes is the biotech bases world leader in enzymes and microorganisms. Using nature's own technologies, they continuously expand the frontiers of biological solutions to improve industrial performance everywhere.
Greg DeLozier, Scientist

232 Physicians for Social Responsibility
1875 Connecticut Avenue, Northwest   202-667-4260
Suite 1012   Fax: 202-667-4201
Washington, DC 20005   E-mail: psrnatl@psr.org
                                            http://www.psr.org/crh.htm
Physicians for Social Responsibility-the US affiliate of the International Physicians for the Prevention of Nuclear War is committed to the elimination of nuclear weapons of mass destruction, the achievement of a sustainableenvironment, and the reduction of violence and its causes. The active conscience of American medicine.
Robert Musil, Executive Director

233 RMT Inc.
PO Box 25000   864-281-0030
Greenville, SC 29616   Fax: 864-281-0288
                                       E-mail: info@rmtinc.com
                                            http://www.rmtinc.com
RMT delivers environmental engineering health and safety and construction solutions that help industrial companies solve complex problems while improving their bottom line.
Daniel Curry, Client Service Manager

234 Rachel Carson Council
PO Box 10779
Silver Spring, MD 20914   301-593-7507
                                            Fax: 301-593-6251
E-mail: rccouncil@aol.com
http://www.members.aol.com/rccouncil/ourpage
The Rachel Carson Council promotes a sense of wonder and respect for nature and helps society realize biologist Rachel Carson's vision of a healthy and diverse environment. The Council collects and disseminates information aboutchemical pesticides and their effects and about alternative pest control methods. The Council also provides information about Rachel Carson and her work.
Dr. Diana Post, Executive Director

235 Rene Dubos Center for Human Environments
81 Pondfield Road   914-337-1636
Suite 387   Fax: 914-771-5206
Bronxville, NY 10708   http://www.dubos.org
The Rene Dubos Center for Human Environments is a non-profit education and research organization focused on the social and humanistic aspects of environmental problems.
Ruth A Eblen, President

236 Rodale Institute
611 Siegfriedale Road   610-683-1400
Kutztown, PA 19530-9320   Fax: 215-967-8959
E-mail: info@rodaleinstitute.org
The Institute offers creative opportunities and solutions that contribute to regenerating environmental and human health worldwide. Our mission statement is clear: The Rodale Institute works worldwide to achieve a regenerative foodsystem that improves environmental and human health.
John Haberern, President

237 Silicone Health Council
1850 M Street,Northwest   202-721-4100
Suite 700   Fax: 202-296-8120
Washington, DC 20036   E-mail: info@socma.com
http://www.socma.com
Coordinates health, environmental and safety programs. Conveys scientifically sound information about silicones.
Jospeh Acker, President

238 Silicones Environmental Health and Safety Council of North America
2325 Dulles Center Boulevard   703-788-6570
Suite 500   Fax: 703-788-6345
Herndon, VA 20171   E-mail: sehsc@sehsc.com
http://www.sehsc.com
An organization of organosilicones manufacturers. Formed to coordinate programs dealing with health, environmental and safety issues.
Reo Menning, Executive Director

239 Society for Environmental Geochemistry and Health
4698 S. Forest Avenue   417-885-1166
Springfield, MO 65810   E-mail: drbgwixson@aol.com
http://ww.scgh.nct
John G Farmer, President
Andrew Hunt, Vice President

240 US Department of Energy
1000 Indiana Avenue Southwest
Washington, DC 20585   800-736-3282
                                            Fax: 202-586-4403
E-mail: eminfo@cemi.org
http://www.em.doe.gov/public/cemi
Provides information on the US Department of Energy's Environmental Management Program which is responsible for safely managing Federal nuclear waste from weapons production.

241 Western Occupational and Environmental Medical Association
575 Market Street   415-927-5736
Suite 2125   Fax: 415-927-5726
San Francisco, CA 94105   E-mail: woema@hp-assoc.com
http://www.woema.org

The mission is to represent and be a resource to members in the profession and practice of occupational and environmental medicine and to enhance their efforts to promote and improve health in the workplace.
Robert Orford, President

National: Gaming & Hunting

242 American Bass Association
402 N Prospect Avenue   310-376-1026
Redondo Beach, CA 90277   Fax: 310-376-5072
E-mail: feedback@americanbass.com
http://www.americanbass.com
The mission is to provide the highest value to sponsors, advertisers, and members through quality run events, promotions and media while developing and instituting programs that will protect, enhance and improve their environmental andnatural resources for future generations.
David Plotnik, President

243 American Birding Association
PO Box 6599   719-578-9703
Colorado Springs, CO 80934   Fax: 719-578-1480
E-mail: member@aba.org
http://www.americanbirding.org
The Association represents the interests of birdwatchers in various arenas, and helps birders increase their knowledge, skills, and enjoyment of birding. ABA also contributes to bird conservation by linking the skills of its members toon-the-ground projects. ABA promotes field-birding skills through meetings, workshops, equipment, and guided involvement in birding, promoting national and international birders networks and publications.
Paul Green, Executive Director
Lynn Yeager, Director Finance

244 American Fisheries Society
5410 Grosvenor Lane   301-897-8616
Bethesda, MD 20814   Fax: 301-897-8096
E-mail: main@fisheries.org
http://www.fisheries.org
The American Fisheries Society (AFS) is an international professional and scientific organization of nearly 9,000 fisheries managers and aquatic scientists. Founded in 1870, AFS is the worlds oldest and largest organization dedicatedto strengthening the fisheries profession, advancing fisheries science and conserving fisheries resource. AFS chapters exist throughout North America and members reside in 75 countries.
Gus Rassam, Executive Director

245 American Fisheries Society: International Fisheries Section
Port Washington Marina
1805 Thompson Drive   202-205-0878
Bremarton, WA 98337   Fax: 202-205-1054
E-mail: mmeyer@att.met
http://www.fisheries.org
Robert Meyer, President

246 American Friends of the Wildfowl and Wetlands Trust
69063 Wallowa Road   616-651-6417
White Pigeon, MI 49099-9745   Fax: 616-651-3679
David Gosling, Director

247 American Humane Association
68 Inverness Drive East   303-792-9900
Englewood, CO 80112-5117   Fax: 303-792-5333
http://www.americanhumane.org
The Associations mission, as a network of individuals and organizations, is to prevent cruelty, abuse, neglect and exploitation of children and animals and to assure that their interests and well-being are fully, effectively, andhumanely guaranteed by an aware and caring society.
John Nobil, President
David Gies, Executive Director

248 American Livestock Breeds Conservancy
PO Box 477   919-542-5704
Pittsboro, NC 27312   Fax: 919-545-0022

E-mail: albc@albc-usa.org
http://www.albc-usa.org

Protects genetic diversity in livestock and poultry species through the conservation and promotion of endangered breeds. These rare breeds are a part of our national heritage and represent a unique piece of the earth's bio-diversity.The loss of these breeds would impoverish agriculture and diminish the human spirit. We have inherited a rich variety of livestock breeds. For the sake of future generations we must work together to safeguard these treasures.
Charles Bassett, Executive Director
Don Bixby, Technical Program Director

249 American Medical Fly Fishing Association

PO Box 760 570-769-7375
Lock Haven, PA 17745 E-mail: amffa@cub.kcnet.org
http://www.amffa.org

The Association began with the idea of combining our professional medical interests with that of our interest in fly fishing and to promote conservation of our natural resources as pertains to our sport and support those causes andefforts oriented towards these latter goals. The organization has grown steadily over the years and presently is a strong and viable organization.
Founded: 1969
Veryl Frye MD, Secretary/Treasurer

250 American Pheasant and Waterfowl Society

W2270 US Highway 10 715-238-7291
Granton, WI 54436 E-mail: alpat@apexotics.com
http://www3.upatsix.com/apws

To promote the rights and interest of the members to keep pheasants, waterfowl, and other upland aquatic and ornamental birds. To collect and distribute pertinent and scientific data and information relating to these rights. Thesociety advocates and encourages public appreciation and understanding of wildlife conservation and publishes a monthly magazine.
Al Novosad, President

251 American Society of Ichthyologists and Herpetologists

College of Arts & Sciences
Florida International University 305-919-5651
N Miami, FL 33181-3000 Fax: 305-919-5964
E-mail: parenti.lunne@nmnh.si.edu
http://www.asih.org

The Society is dedicated to the scientific study of fishes, amphibians and reptiles. Its mission is to increase knowledge about these organisms, to disseminate that knowledge through publications, conferences, symposia, and othermeans, and to encourage and support young scientists who will make future advances in these fields.
Founded: 1913
Lynne R Parenti, President

252 American Society of Mammalogists

PO Box 7060 785-843-1235
Lawrence, KS 66044 800-627-0629
Fax: 785-843-1274
E-mail: asm@aibs.org
http://www.mammalsociety.org

Established for the purpose of promoting interest in the study of mammals.
Founded: 1919

253 Animal Protection Institute

PO Box 22505 916-447-3085
Sacramento, CA 95822 Fax: 916-447-3070
E-mail: info@api4animals.org
http://www.api4animals.org

A national animal advocacy nonprofit organization established to advocate for the protection of animals from cruelty and exploitation.
Founded: 1968
Gary Pike, Chairperson

254 Birds of Prey Rehabilitation Association

2290 S 104th Street 303-460-0674
Broomfield, CO 80020-9705
Sigrid Ueblacker, Executive Officer

255 Cetacean Society International

PO Box 953 203-770-8615
Georgetown, CT 06829 860-561-0187

Fax: 860-561-0187
E-mail: rossiter@csiwhalesalive.org
http://www.csiwhalesalive.org

All volunteer, nonprofit conservation, educational and research organization to benefit whales, dolphins, porpoises and the marine environment. Promotes education and conservation programs, including whale and dolphin watching, andnoninvasive, benign research. Advocates for laws and treaties to prevent commercial whaling, habitat destruction and other harmful or destructive human interactions. CSI's world goal is to minimize cetacean killing and captures and to enhance publicawareness.
William W Rossiter, President

256 Federal Wildlife Officers Association

PO Box 550060 703-358-1949
Waltham, MA 02455-0060 http://www.fwoa.org

This Association is an organization dedicated to the protection of wildlife, the enforcement of federal wildlife law, the fostering of cooperation and communication among federal wildlife officers, and the perpetuation, enhancement anddefense of the wildlife enforcement profession. Membership is open to all federal law enforcement officers charged with the protection of wildlife and to friends and supporters of this goal.
Mark Webb, President

257 Foundation for North American Wild Sheep Headquarters

720 Allen Avenue 307-527-6261
Cody, WY 82414-3402 Fax: 307-527-7117
E-mail: fnaws@fnaws.org
http://http://fnaws.org

The Foundation's mission is to promote and enhance populations of indigenous wild sheep on the North American continent, and to fund programs for the professional management of these populations, while keeping administrative costs to aminimum.
Raymond Lee, President

258 Friends of the Australian Koala Foundation

224 W 29th Street 212-967-8200
15th Floor Fax: 212-967-7292
New York, NY 10001
Debbie Tabart, Executive Director

259 Friends of the Earth

1717 Massachusettes Avenue 202-783-7400
Northwest 600 877-843-8687
Washington, DC 20036-2002 Fax: 202-783-0444
E-mail: foe@foe.org
http://www.foe.org

National nonprofit advocacy organization dedicated to protecting the planet from environmental degradation; preserving biological, cultural and ethnic diversity, and empowering citizens to have an influential voice in decisionsaffecting the quality of their environment and their lives.
Brent Blackwelder, President

260 Fund for Animals

200 W 57th Street 212-246-2096
New York, NY 10019 Fax: 212-246-2633
E-mail: hdquarters@fund.org
http://www.funforanimals.org

The Fund for Animals was founded in 1967 by author/humanitarian Cleveland Amory, and works in advocacy fields, using legislation, litigation and education as tools to advance the cause of animals.
Marian Probst, President

261 Game Conservancy USA

PO Box 922 203-655-5915
Darien, CT 06820 Fax: 203-655-9110
Christopher Van Munching, Executive Director

262 Hawk Migration Association of North America

164 1/2 Washington Street 973-335-0674
Carbondale, PA 18407-2483 E-mail: 6595@email.msn.com
http://www.hmana.org

The mission is to conserve raptor populations through the scientific study, enjoyment and appreciation of raptor migration.
William Weber, Chairperson

263 Hawkwatch International
1800 SW Temple   801-484-6808
Suite 226   800-726-4295
Salt Lake City, UT 84115   Fax: 801-484-6810
E-mail: info@hawkwatch.org
http://www.hawkwatch.org
Our mission is to monitor and protect hawks, eagles, other birds of prey and their environment through research, education and conservation.
Echo Thatcher, Membership Coordinator
Thom Benedict, Executive Director

264 Humane Society of the United States
2100 L Street NW   202-452-1100
Washington, DC 20037   Fax: 202-778-6132
http://www.hsus.org
HSUS is the nation's largest animal-protection organization with more than 5 million constituents. The HSUS was founded to promote the humane treatment of animals, to foster respect, understanding, and compassion for all creatures.Today our message of care and protection embraces not only the animal kingdom but also the Earth and its environment. To achieve our goals we work through legal, educational, legislative and investigative means.
Founded: 1954

265 International Association for Bear Research and Management
1300 College Road   907-459-7238
Fairbanks, AK 99071   Fax: 907-451-9723
http://www.bearbiology.com
A Volunteer organization open to professional biologists, wildlife managers and others dedicated to the conservation of all species of bears. Consists of several hundred members from over 20 countries. Supports the scientificmanagement of bears through research and distribution of information.
Harry Reynolds, President

266 International Society for Endangered Cats
3070 Riverside Drive   614-487-8760
Suite 160   Fax: 614-487-8769
Columbus, OH 43221   http://www.isec.org/isec-left-index.htm
A not-for-profit organization dedicated to the conservation of wild cats throughout the world. Most species of wild cats are threatened or endangered in all or parts of their native ranges and, unless action is taken in the nearfuture, many will be lost to extinction. The society's goals are to raise public understanding and knowledge of wild cats and support and facilitate research on ecology, captive breeding and reintroduction of cats to their native habitats.
Bill Simpson, President
Patricia Currie, Executive Director

267 International Wild Waterfowl Association
10114 54th Place NE   425-334-8223
Everett, WA 98205   Fax: 425-397-8136
E-mail: dye@greatnorthern.net
http://www.wildwaterfowl.org
IWWA is committed to protecting and enhancing wild waterflow habitats. Supports the captive breeding and restoration of endangered species, and supports the establishment and maintenance of genetically diverse and disease-free captivepopulations of endangered waterfowl.
Walter Sturgeon, President

268 International Wildlife Coalition
Whale Adoption Program
70 East Falmouth Highway   508-548-8328
PO Box 388   Fax: 508-548-8542
East Falmouth, MA 02536   http://www.iwc.org
Adoption fees are used to help purchase rescue and research vessels, hire crews, and rescue whales and other marine mammals from fishing nets, conduct crucial research, and help enforce wildlife protection laws and treaties.
Daniel J Morast, President

269 International Wildlife Conservation Society
New York Zoological Society   718-364-4275
Bronx, NY 10460   Fax: 212-220-7114
John G Robinson, Director

270 International Wolf Center
1396 Highway 169   218-365-4695
Ely, MN 55731-8129   Fax: 218-365-3318
E-mail: mplspac@wolf.org
http://www.wolf.org
The International Wolf Center advances the survival of the wolf populations by teaching about wolves, their relationship to wildlands and the human role in their future.
Nancy Ojo Tubbs, Chairperson

271 Kids for Conservation: Today and Tomorrow
Illinois Department of Conservation
One Natural Way   217-524-4126
Springfield, IL 62702-1271   http://www.dnr.state.il.us
Provides a variety of programs for young Illinois hunters.
Joel Brunsrold, Director

272 Last Chance Forever
506 Avenue A   210-499-4080
PO Box 460993   E-mail: raptor@ddc.net
San Antonio, TX 78246-099 http://www.lastchanceforever.org
Nonprofit organization dedicated to the rehabilitation and release of injured and orphaned raptors, birds of prey such as hawks, owls, eagles, falcons, and vultures. LCF has been responsible for treating thousands of birds of prey,including endangered species. Generally up to 80% of these cases received are released back to their natural home.
Founded: 1978
Melissa Hill, President

273 Muskies
PO Box 120870   701-239-9540
New Brighton, MN 58112   E-mail: info@muskiesinc.org
http://www.muskiesinc.org
Supports growth and interest in the sport of muskie fishing. We are reaching out to protect our existing fisheries and to develop new fisheries.
Greg Wells, President

274 National Bison Association
1400 West 122nd Avenue   303-292-2833
Westminster, CO 80234   Fax: 303-292-2564
E-mail: info@bisoncentral.com
http://www.bisoncentral.com
Created from a merger of the American Bison Association and the National Buffalo Association, the NBA has over 2400 members in all 50 states and 16 foreign countries. It is a nonprofit association which promotes the preservation,production and marketing of bison. NBA activities and services serve to better inform and educate our members and the general public about bison.
Founded: 1995
Dave Carter, Executive Director

275 National Endangered Species Act Reform Coalition
1050 Thomas Jefferson Street NW   202-333-7481
Suite 700   Fax: 202-338-2361
Washington, DC 20007   E-mail: nesarc@vnf.com
http://www.nesarc.org
A broad based coalition of roughly 150 member organizations, representing millions of individuals across the United States, that is dedicated to bringing balance back to Endangered Species Act. Our membership includes rural irrigators,municipalities, farmers, electric utilities and many other individuals and organizations that are directly affected by the ESA.
Nancy Macan McNally, Executive Director

276 National Foundation to Protect America's Eagles
PO Box 333   865-429-0157
Pigeon Forge, TN 37868   800-232-4537
Fax: 865-429-4743
E-mail: eaglemail@eagles.org
http://www.eagles.org
Not-for-profit organization of concerned citizens and professionals to develop and conduct bald eagle and environmental recovery programs in the United States and to assist private, state and federal projects that do the same.
Founded: 1985
Al Louis Cecere, President/CEO

277 National Hunters Association
PO Box 820   919-365-7157
Knightdale, NC 27545   Fax: 919-366-2142
                          E-mail: nhadvs@worldnet.att.net
                          http://www.nationalhunters.com
Our organization is dedicated to preserving the rights of hunters, promoting hunter safety among the youth and all hunters by demanding game controls and hunting laws, maintaining the rights to use and own firearms and protecting the environment and maintaining a healthy habitat.
D Smith, President
Faye Smith, Secretary/Treasurer

278 National Marine Fisheries Service Office of Protective Resources
1315 East-West Highway   301-713-2319
Silver Spring, MD 20910-3282   Fax: 301-713-0376
                          E-mail: don.knowles@noaa.gov
                          http://www.noaa.gov

James H Lucky, Director

279 National Rifle Association of America
11250 Waples Mill Road   703-267-1000
Fairfax, VA 22030   800-336-7402
                                    Fax: 703-267-3909
                                    http://www.nra.org

Chris Cox, Executive Director

280 National Shooting Sports Foundation
11 Mile Hill Road   203-426-1320
Newtown, CT 06470-2359   Fax: 203-426-1087
                          E-mail: info@nssf.org
                          http://www.nssf.org
Leading trade association of the firearms and recreational shooting sports industry. A nonprofit communications and marketing organization, the NSSF manages a variety of programs designed to promote a better understanding of and a more active participaqtion in the shooting sports.
Founded: 1961
Bill Brassard, Editorial Director

281 National Trappers Association
524 5th Street   812-277-9670
Bedford, IN 47421-2447   Fax: 812-277-9672
                  E-mail: ntaheadquarters@nationaltrappers.com
                          http://www.nationaltrapper.com
To promote conservation, legislation and administrative procedures; to save and faithfully defend from waste the natural resources of the United States; to promote environmental education programs; and to promote a continued annual furharvest using the best tools presently available for that purpose.
Steve Fitzwater, President

282 National Walking Horse Association
PO Box 100   903-564-3747
Whitesboro, TX 76237   Fax: 903-564-3902
                                    http://www.nwha.com
An alliance of people committed to preserving and fostering the natural abilities and welfare of the Walking Horse. Improves the lives of horses and people by encouraging responsibility and sportsmanship. Promotes educational and recreational activities, while preserving the unique qualities of the Walking Horse.
Charles Fulton, Vice President

283 National Wildlife Federation
1400 16th Street NW   202-797-6800
Suite 501   Fax: 202-797-6646
Washington, DC   http://www.nwf.org
The National Wildlife Federation is the largest member-supported conservation group, uniting individuals, organizations, businesses and government to protect wildlife, wild places and the environment.
Jim Lyon, Director

284 National Wildlife Federation Corporate Conservation Council
1400 16th Street NW
Washington, DC 20036   202-872-7772

Barbara Haas, Director

285 Native American Fish and Wildlife Society
From The Eagle's Nest
750 Burbank Street   303-466-1725
Broomfeild, CO 80020   Fax: 303-466-5414
National tribal organization to develop a national communications network for the exchange of information and management techniques related to self-determined tribal fish and wildlife management.
Founded: 1983
Matthew Vanderhoop, President

286 North American Bear Society
4061 E Hartford Avenue   602-971-2338
PO Box 555774   Fax: 602-971-2100
Phoenix, AZ 85078   E-mail: bearsociety@nonline.com
                          http://www.nonprofitnet.com/nabs
To support the conservation and management of bear populations throughout North America for the benefit of the general public and future generations. Dedicated to the conservation and management of the indigenous bears of North America.
William T Smaltz, Founder

287 North American Native Fishes Association
123 W Mount Airy Avenue
Philadelphia, PA 19119   E-mail: nanfa@att.net
                          http://www.nanfa.org
NANFA is dedicated to the enjoyment, study and conservation of the continent's native fishes. The objectives are: to increase and disseminate knowledge about fishes and their habitats among aquarium hobbists, biologists, fish and wildlife officals, anglers, educators, students and others, through publications, electronic media, regional and national meetings, and other means; to promote the conservation and the protection/restoration of natural habitats; to advance the captive husbandry.

288 Organization of Wildlife Planners
500 Lafayette Road
PO Box 25   http://www.owpweb.org
Saint Paul, MN 55155
A professional organization of creative, committed people concerned with the management and future of government agencies that manage fish and wildlife populations and habitat. The purpose of OWP is to improve fish and wildlife resource management capabilities through informed decision-making.
Founded: 1978
Larry Gigliotti, President

289 POWS Wildlife Rehabilitation Center
15305 44th Avenue West   425-743-1884
PO Box 1037   E-mail: kparker@paws.org
Lynnwood, WA 98046   http://www.paws.org/wildlife
Each wildlife patient requires individualized, specialized care. But for every animal the goal is identical: rehabilitation and release into the wild.
Kip Parker, Director

290 Pacific Whale Foundation
300 Maalaea Road   808-249-8811
Suite 211   800-942-5311
Wailuku, HI 96793   Fax: 808-243-9021
                          E-mail: info@pacificwhale.org
                          http://www.pacificwhale.org
Mission is to promote appreciation, understanding and protection of whales, dolphins, coral reefs, and our planet's oceans. We accomplish this by educating the public from a scientific perspective about the marine environment.
Gregory Kaufman, President

291 People for the Ethical Treatment of Animals
501 Front Street   757-622-7382
Norfolk, VA 23510   Fax: 757-622-0457
                          E-mail: info@peta.org
                          http://www.peta.org
With more than 850,000 members, PETA is the largest animal rights organization in the world. Also dedicated to establishing and protecting the rights of all animals. Operates under the simple principal that animals are not ours to eat, wear, experiment on, or use for entertainment.
Founded: 1980
Ingrid Newkirk, President

292 Pheasants Forever
1783 Buerkle Circle 651-773-2000
St. Paul, MN 55110 877-773-2070
Fax: 651-773-5500
E-mail: contact@pheasantsforever.org
http://www.pheasantsforever.org
A non profit organization dedicated to the protection and enhancement of pheasant and other wildlife populations in North America. This mission is carried out through habitat improvement, land management, public awareness, andeducation.
Founded: 1983
Howard K Vincent, Executive Officer
Joseph J Duggan, VP

293 Pope and Young Club
273 Mill Creek Road 507-867-4144
PO Box 548 Fax: 507-867-4144
Chatfield, MN 55923 E-mail: pyclub@isl.net
http://www.pope-young.org
The Pope and Young Club is one of North America's leading bowhunting and conservation organizations. Founded in 1961 as a nonprofit scientific organization, the Club is patterned after the prestigious Boone and Crockett Club. The Clubadvocates and encourages responsible bowhunting by promoting quality, fair chase hunting, and sound conservation practices. Today it fosters and nourishes bowhunting excellence and acts in the best interest of our bowhunting heritage everywhere.
Founded: 1961
Donald Ace Morgan, President

294 Purple Martin Conservation Association
Edinboro University of Pennsylvania 814-734-4420
219 Meadville Street Fax: 814-734-5803
Edinboro, PA 16444 E-mail: pmca@edinboro.edu
http://www.purplemartin.org
The only organization devoted exclusively to the scientific study of purple martins, their biology, and habitat requirements.
Founded: 1986
James Hill, Founder
John Tautin, Executive Director

295 Scientists Center for Animal Welfare
7833 Walker Drive 301-345-3500
Suite 410 Fax: 301-345-3503
Greenbelt, MD 20770 E-mail: INFO@SCAW.COM
http://WWW.SCAW.COM
A nonprofit educational association of individuals whose mission is to promote humane care, use, and management of animals involved in research, testing or education in laboratory, agricultural, wildlife or other settings.
Gregory R Reinhard, President
Joseph T Bielitzki, Vice President

296 Sea Shepherd Conservation Society
PO Box 2616 360-370-5650
Friday Harbor, WA 98250 Fax: 360-370-5651
E-mail: info@seasheperd.org
http://www.seashepherd.org
The mandate of this organization was mammal protection and conservation with an immediate goal of shutting down illegal whaling and sealing operations. We are committed to the eradication of private whaling, poaching, shark finning,unlawful habitat destruction, and violations of established laws in the world's oceans.
Founded: 1977
Paul Watson, Founder

297 Sea Turtle Restoration Project
PO Box 400 415-488-0370
Forest Knolls, CA 94933 Fax: 415-488-0372
E-mail: info@seaturtles.org
http://www.seaturtles.org
Fights to protect endangered sea turtle populations in ways that meet the ecological needs of the sea turtles and the oceans and the needs of the locla communities who share the beaches and waters with these gentle, beautifulcreatures.
Founded: 1989
Erica Heimberg, Managing Director

298 Society for Animal Protective Legislation
Georgetown Station 202-337-2334
PO Box 3719 Fax: 202-338-9478
Washington, DC 20007 E-mail: sapl@saplonline.org
http://www.saplonline.org
Prepares information for use by Members of Congress and their staffs to protect animals. Also sends action alerts to individuals and organizations interested in animal protective legislation, informing them of ways in which they mayhelp.
Founded: 1955
Madeleine Bemelmans, President

299 Society of Tympanuchus Cupido Pinnatus
College of Natural Resources 715-346-3859
University of Wisconsin-Stevens Point Fax: 715-346-3624
Stevens Point, WI 54481 E-mail: eanderson@uwsp.edu
http://www.uwsp.edu/wildlife/programs
Russell Schallert, President

300 Sportsmans Network
111 S Main Street 859-824-6526
PO Box 257 800-680-8058
Dry Ridge, KY 41035 Fax: 606-824-0556
E-mail: sportsman@sportsmansnetwork.org
http://www.sportsmansnetwork.org
To raise awareness of animal wildlife conservation issues through various forms of public media and to address the subject in all appropriate ways including hunting, fishing, trapping, and related activities.
William Krebs, Treasurer

301 Trout Unlimited
1300 North 17th Street 703-552-0200
Suite 500 800-834-2419
Arlington, VA 22209 Fax: 703-284-9400
E-mail: trout@tu.org
http://www.tu.org
To conserve, protect and restore North America's trout and salmon fisheries and their watersheds.
Founded: 1959
Charles Gauvin, President/CEO
Kenneth Mendez, Executive VP/COO

302 Trumpeter Swan Society
3800 County Road 24 763-694-7851
Maple Plain, MN 55359 Fax: 763-476-1514
E-mail: ttss@hennepinparks.org
http://www.trumpeterswansociety.org
Private, nonprofit organization dedicated to assuring the vitality and welfare of wild Trumpeter Swan populations, and to restoring the species to as much of its former range as possible.
Founded: 1968
Madeleine Linck, Editor
Ruth E Shea, Executive Director

303 US Sportsman's Alliance
801 Kingsmill Parkway 614-888-4868
Columbus, OH 43229 Fax: 614-888-0326
E-mail: info@USSPORTSMEN.org
http://www.ussportsmen.org
Protects and defends America's wildlife conservation programs and the pursuits of hunting, fishing, and trapping which generates the money to pay for them. Is also responsible for public education, legal defense and research.
Founded: 1978
Walter P Pidgeon Jr, President/CEO

304 US Sportsmen's Alliance
801 Kingsmill Parkway 614-888-4868
Columbus, OH 43229 Fax: 614-888-0326
E-mail: info@USSPORTSMEN.org
http://www.ussportsmen.org
Provides direct lobbying and grassroots coalition support to protect and advance the rights of hunters, fishermen, trappers and scientific wildlife though coalition, building, ballot issue campaigning and legislative and governmentrelations.
Founded: 1978
Walter J Pidgeon Jr, President/CEO

305 Waterfowl USA
Waterfowl Building   803-637-5767
Box 50   Fax: 803-637-6983
Edgefield, SC 29824   E-mail: president@waterfowlusa.org
                            http://www.waterfowlusa.org
The only national non profit, conservation organization, dedicated to using funds in the areas in which they were raised for local and state waterfowl projects. Founded by biologists for the purpose of preserving and improvingwintering and breeding habitat within the United States.
Founded: 1983
Roger L White, President/CEO

306 Whitetails Unlimited
2100 Michigan Street   920-743-6777
PO Box 720   800-274-5471
Strugeon Bay, WI 54235   Fax: 920-743-4658
                            E-mail: wtu@itol.com
                      http://www.whitetailsunlimited.org
A national conservation organization that has remained true to its mission and has made great strides in the field of conservation. We have gained the reputation of being the nation's premier organization dedicating our resources tothe betterment of the white-tail deer and its environment.
Founded: 1982
Jeffrey Schinkten, President
Peter Gerl, Executive Director

307 Wilderness Society
4077 Mission Inn Avenue   909-781-0938
Riverside, CA 92501   E-mail: twold@tws.org
Founded in 1935, The Wilderness Society works to protect America's wilderness and to develop a nationwide network of wild lands through public education, scientific analysis and advocacy. Our goal is to ensure that future generationsenjoy the clean air and water, beauty, wildlife and opportunities for recreation and spritual renewal provided by the nation's pristine forests, rivers, deserts and mountains.
Terry Wold, Regional Conservation Rep

308 Wildlife Trust
460 West 34th Street   212-380-4460
17th Floor   Fax: 212-380-4465
New York, NY 10001   E-mail: homeoffice@wildlifetrust.org
                            http://www.wildlifetrust.org
Works in the United States and worldwide to save threatened species from extinction, protect habitat, and link nature protection with health through collaborative projects with local scientists. We also communicate our results toeducators, health professionals, policy experts and civic leaders.
Founded: 1971
Mary C Pearl PhD, President
Fred W Koontz PhD, Executive VP

309 Xerces Society
4828 SE Hawthorne Boulevard   503-232-6639
Portland, OR 97215   Fax: 503-233-6794
                            E-mail: info@xerces.org
                            http://www.xerces.org
An international non profit organization dedicated to protecting biological diversity through inverterate conservation. The Society advocates for invertebrates and their habitats by working with scientists, land managers, educators,and citizens on conservation and education projects. Our core programs focus on endangered species, natice pollinators, and watershed health.
Scott Hoffman Black, Executive Director

National: Habitat Preservation & Land Use

310 Abundant Life Seed Foundation
PO Box 157   541-767-9606
Saginaw, OR 97472   Fax: 866-514-7333
                      E-mail: als@abundantlifeseeds.com
                      http://www.abundantlifeseed.org
A nonprofit organization dedicated to the preservation of rare, heirloom and native seeds. ALSF grows and distributes open-pollinated seeds and offers them for sale through an annual catalog. Seeds are also sent for free to people inneed through the World Seed Fund. ALSF teaches seed saving through workshops, apprenticeships and school programs.
Founded: 1965
Mark Tranel, Director
Elsa Golts, Board President

311 Agricultural Resources Center
115 W Main Street   919-967-1886
Carrboro, NC 27510   http://www.ibiblio.org
Allen Spalt, Director

312 Aldo Leopold Foundation
PO Box 77   608-355-0279
Baraboo, WI 53913-0077   Fax: 608-356-7309
                            E-mail: mail@aldoleopold.org
                            http://www.aldoleopold.org
A nonprofit organization founded in 1982, works to promote the philosophy of Aldo Leopold and the land ethic he so eloquently defined in his writing. The foundation actively integrates programs on land stewardship, environmentaleducation and scientific research to promote care of natural resources and have an ethical relationship between people and land.
Steve Swenson, Ecologist
Buddy Huffaker, Executive Director

313 America the Beautiful Fund
1511 K Street NW   202-638-1649
Suite 1002   Fax: 202-204-0028
Washington, DC 20006   http://www.freeseeds.org
America the Beautiful Fund has been preserving the historic and natural beauty of America since 1965. Operation provides free seeds and bulbs for environmental education and preservation, community gardens and hunger relief. Americanlandscapes protects delicate water sheds while Rediscover America provides seed grants to restore and protect historic buildings and sites.
Nanine Bilski, Executive Director
Penny Pagano, Secretary

314 American Association of Botanical Gardens and Arboreta
351 Longwood Road   610-925-2500
Kennett Square, PA 19348   Fax: 610-925-2700
                            http://www.aabga.org
AABGA's mission is to support North American botanical gardens and arboreta in fulfilling their missions to study, display, and conserve living plant collections for public benefit. Support for this mission comes in many forms:aquarterly magazine, a monthly newsletter, special publications, and programming at annual and regional meetings.
Eric Tschanz, President
Kathleen Socolofsky, Secretary

315 American Cave Conservation Association
119 E Main Street   270-786-1466
PO Box 409   Fax: 270-786-1467
Horse Cave, KY 42749   E-mail: debraheavers@caven.org
                            http://www.cavern.org
A national nonprofit association dedicated to the protection of caves, karstlands and groundwater. The ACCA operates the American Cave Museum and Karst Center, an educational center that includes the American Cave Museum and HiddenRiver Cave.
David G Foster, Executive Director

316 American Conservation Association
1200 New York Avenue NW   202-289-2431
Suite 400   Fax: 202-289-1396
Washington, DC 20005
Operative private organization dedicated to promote the knowledge and understanding of conservation, to preserve the live beauty of the landscape and the natural resources and organisms in areas of the United States and others, and toeducate to the public in the adapted use of these areas.
Charles Clusen, Executive Director
R Greathead, Secretary

317 American Council on the Environment
1301 20th Street NW
Suite 113   202-659-1900
Washington, DC 20036
John H Gullett, Executive Officer

318 American Institute of Fishery Research Biologists
1315 E West Highway
Silver Spring, MD 20910
Gary Sakagawa, President
Barbara Warkentine, Secretary

319 American Land Conservancy
1388 Sutter Street   415-749-3010
Suite 810   Fax: 415-749-3011
San Francisco, CA 94109   E-mail: mail@alcnet.org
                                     http://www.alcnet.org
ALC is dedicated to the preservation of land and water as endur-
ing public resources, to protect and enhance our nation's natural,
ecological historical, recreational, and scenic heritage.
Harriet Burgess, President

320 American Lands
726 7th Street SE   202-547-9400
Washington, DC 20003   Fax: 202-547-9213
                        E-mail: wafcdc@americanlands.org
                                http://www.americanlands.org
America's wildlife and wildlands continue to be threatened by
logging, roadbuilding, grazing, off road vechicles and mining. A
broad coalition of activists and organizations is dedicated to protecting our forest heritage and restoringecological integrity to
the landscape.
Jim Jontz, Executive Director
Randi Spivak, President

321 American Littoral Society
468 S Green Street   609-294-3111
PO Box 1306   Fax: 609-294-8044
Tuckerton, NJ 08087   E-mail: alssj@earthlink.net
                                http://www.littoralsociety.org
The American Littoral Society is an environmental organization
concerned about issues that affect the littoral zone: that area on
the beach between low and high tide. The American Littoral Society is a national, nonprofit,public-interest organization comprised of over 6,000 professional and amateur naturalists.
Michael Hubertt, President
Angela Cristini, Secretary

322 American Shore and Beach Preservation Association
1724 Indian Way   510-339-2818
Oakland, CA 94611   Fax: 510-339-6710
                            E-mail: business@asbpa.org
                                     http://www.asbpa.org
Federal, state and local government agencies and individuals interested in conservation, development and restoration of
beaches and shorefronts.
Gregory Woodell, President

323 Antarctica Project
1630 Connecticut Avenue NW   202-234-2480
3rd Floor   Fax: 202-387-4823
Washington, DC 20009   E-mail: antarctica@igc.org
                                     http://www.asoc.org
The only conservation organization in the world that works exclusively for Antarctica. Leads domestic and international campaigns to protect Antarctica's pristine wilderness and
environment, as the international secretariat for theAntarctic
and Southern Ocean Coalition (ASOC). ASOC is a global coalition with 214 member organizations working in 44 countries on
six continents.
Beth Clark, Director
Josh Stevens, Campaign Associate

324 Arctic Network
PO Box 102252   907-272-2452
Anchorage, AK 99510   Fax: 907-272-2453
Works to the conserve Arctic ecosystem. Focus includes indige-
nous cultures.

325 Association for Conservation Information
1000 Assembly Street   843-762-5032
PO Box 167   Fax: 843-734-3951
Columbia, SC 29202
ACI, the Association for Conservation Information, is a nonprofit association of information and education professionals
representing state, federal and Canadian agencies and private
conservation organizations.
Bob Campbell, President

326 Association for Environmental Health of Soils
150 Fearing Street   413-549-5170
Amherst, MA 01002   Fax: 413-549-0579
                                     http://www.aehs.com
The AEHS was created to facilitate communications and foster
cooperation among professionals concerned with the challenge
of soil protection and cleanup. Members represent the many disciplines involved in making decisions and solvingproblems affecting soils. AEHS recognizes that widely acceptable solutions
to the problem can be found only through the integration of scientific and technological discovery, social and political judgement and hands on practice.
Paul Kostecki, Executive Director
Marc A Nascarella, Managing Ed/Conference Coor

**327 Association for Natural Resources Enforcement
Training**
Missouri Department of Conservation   573-751-4115
Box 180   E-mail: yamnil@mail.conservation.state.mo.us
Jefferson City, MO 65102   http://www.dirdid.com\anret
Dave Windsor, VP

328 Association of Consulting Foresters of America, Inc
732 N Washington Street   703-548-0990
Suite 4-A   888-540-8733
Alexandria, VA 22314   Fax: 703-548-6395
                        E-mail: director@acf-foresters.com
                                http://www.acf-foresters.com
To protect the public welfare and property in the practice of forestry. To raise the professional standards and work of ACF Consultants and all other consulting foresters. To develop and
expand the services of ACF Consultants. To serveas a forum for
the exchange of information. To certify qualified foresters for
competency in specific areas of professional practice.
Lynn C Wilson, Executive Director

329 Association of Field Ornithologists
University of Maine at Machias
9 O'Brian Avenue
Machias, ME 04654   http://www.afonet.org
Society of professional and amateur ornithologists dedicated to
the scientific study and dissemination of information about
birds in their natural habitats. Especially active in bird-banding
and development of field techniques.Encourages participation
of amateurs in research, and emphasizes conservation biology
of birds.
    Founded: 1922
Charles Duncan, President

330 Association of Great Lakes Outdoor Writers
301 Cross Street   212-268-6232
Sullivan, IN 47882
Members are writers who cover environmental, outdoor and wilderness issues of the Great Lakes in the US and Canada. Membership is $50.00 per year for individuals and $110.00 per year
for companies.

331 Association of Partners for Public Lands
8375 Jumpers Hole Road   410-647-9001
Suite 104   Fax: 410-647-9003
Millersville, MD 21108   E-mail: appl@appl.org
                                     http://www.appl.org
To provide the highest levels of program and service to public
agencies entrusted with the care of America's natural and cul-
tural heritage.
Donna Asbury, Executive Director

332 Association of State Wetland Managers
1434 Helderberg Trail
PO Box 269   518-872-1804
Berne, NY 12023-9746   Fax: 518-872-2171
                                      E-mail: aswm@aswm.org
                                      http://www.aswm.org
Nonprofit organization dedicated to the protection and restoration of our nations wetlands. Our goal is to help public and private wetland decision-makers utilize the best possible scientific information and techniques in wetlanddelineation, assessment, mapping, planning, regulation, acquisition, restoration, and other management.
Jon Kusler, Executive Director

333 Bat Conservation International
PO Box 162603   512-327-9721
Austin, TX 78716   Fax: 512-327-9724
                                      http://www.batcon.org
Dedicated to preserving and restoring bat populations and habitats around the world. Uses a non-confrontational approach to educate the public about the ecological and economic value of bats, advance scientific knowledge about bats andthe ecosystems that rely on them and preserve critical bat habitats through win-win solutions that benefit both humans and bats. Subscription to magazine is included with membership. Senior, student or educator, $30; Basic $35; Friends of BCI $45;Supporting $60.
Dr Merlin D Tuttle, President/Founder
Robert J (Robb) Hankins, Executive Director

334 Beaver Defenders Unexpected Wildlife Refuge
PO Box 765   856-697-3541
Newfield, NJ 08344-0765   Fax: 856-697-5081
                                      E-mail: Sarah Summerville qdi@snip.net
                                      http://www.animalplace.org
Sarah Summerville, Contact

335 Big Thicket Association
Highway 770   936-274-5000
Box 198   E-mail: fhallen@swbell.net
Saratoga, TX 77585   http://www.in2000.net
Maintains interest in conservation of natural resources through state parks and nature sanctuaries in the Big Thicket area. To encourage tourism and promote research and publication.
Fred Allen, President

336 Birds of Prey Rehabilitation Association
2290 S 104th Street   303-460-0674
Broomfield, CO 80020-9705
Sigrid Ueblacker, Executive Officer

337 Boone and Crockett Club
250 Station Drive   406-542-1888
Missoula, MT 59801   Fax: 406-542-0784
                             E-mail: bccclub@boone-crockett.org
                                      http://www.boone-crockett.org
Oldest and most influential conservation policy group continues its second century of service in a new home in the heart of the Rockies.
Jack Reneau

338 Brotherhood of the Jungle Cock
PO Box 576   410-761-7727
Glen Burnie, MD 21061   Fax: 410-553-0575
                          E-mail: mbosleyco@mindspring.com
Bosley Wright, Executive VP

339 Camp Fire Conservation Fund
4607 Madison Avenue   914-941-0199
Kansas City, MO 64112-1278   Fax: 816-756-0258
                                      E-mail: info@campfire.org
                                      http://www.campfire.org
George R Lamb, President

340 Canvasback Society
PO Box 101   216-443-2340
Gates Mills, OH 44040
Keith C Russell, Chairman

341 Canyonlands Field Institute
PO Box 68   435-259-7750
Moab, UT 84532   800-860-5262
                                      Fax: 435-259-2335
                       E-mail: cfiinfo@canyonlandsfieldinst.org
                              http://www.canyonlandsfieldinst.org
To increase understanding of, connection to and care for the Colorado Plateau, expand perception of and appreciation for the beauty and integrity of the natural world, improve the quality of field oriented, experiential teaching andlearning for students and adults and to encourage individuals to be involved in the care of their own home places.
Karla Vander Zanden, Executive Director

342 Caribbean Conservation Corporation
4424 NW 13th Street   352-373-6441
Suite A1   800-678-7853
Gainesville, FL 32609   Fax: 352-375-2449
                                      E-mail: ccc@cccturtle.org
                                      http://cccturtle.org
Caribbean Conservation Corporation is a nonprofit membership organization based in Gainesville. CCC was the first marine turtle conservation organization in the world, and has more than 40 years of experience in national andinternational sea turtle conservation, research and educational endeavors.
David Godfrey, Executive Director

343 Carrying Capacity Network
2000 P Street NW   202-296-4548
Suite 240   Fax: 202-296-4609
Washington, DC 20036   http://www.carryingcapacity.org
An information exchange network. Interests include resource conservation, population stabilization and environmental protection.

344 Center for Plant Conservation
PO Box 299   314-577-9450
St. Louis, MO 63166   Fax: 314-577-9465
                                      http://www.mobot.org/CPC/
The CPC is a consortium of 28 American botanical gardens and arboreta whose mission is to conserve and restore the rare native plants of the US To meet this end, they are involved in plant conservation, research and education. Thissite includes information about the National Collection of Endangered Plants, which is maintained by the group.

345 Center for the Study of Tropical Birds
218 Conway   210-828-5306
San Antonio, TX 78209-1716   Fax: 210-828-9732
                                      E-mail: office@cstBbnc.org
                                      http://www.cstbinc.org
Devoted to the conservation of neotropical birdlife through collaborative programs of research and education.
Jack Eitniear, Director

346 Clean Sites
1199 N Fairfax Street   703-519-2135
Suite 400   Fax: 703-548-8733
Alexandria, VA 22314   E-mail: cses@cleansites.com
                                      http://www.cleansites.com
We apply sound project management principles, real-world experience, and cost control measures to find creative solutions to environmental remediation and land reuse problems.
Douglas Ammon, Contact

347 Committee for Conservation and Care of Chimpanzees
3819 48th Street NW   202-362-1993
Washington, DC 20016   Fax: 202-686-3402
Dr. Geza Teleki, Chairman

348 Committee for the Preservation of the Tule Elk
PO Box 3696
San Diego, CA 92103
Jolene W Steigerwalt, Secretary

349 Community Greens Shared Parks in Urban Blocks
1700 N Moore Street   703-527-8300
Arlington, VA 22209   Fax: 703-527-8383
                                      http://www.communitygreens.org

Community Greens aims to transform the interiors of urban blocks across the US into resplendent shared parks and gardens that are owned and managed by the residents who live around them. These community greens would provide a greenoasis for hassled city dwellers, offer a safe place for children to interact, increase home values and improve a sense of community. They would also help the environment by providing microhabitats for birds and small wildlife, clean the air andreduce urban sprawl.
Kate Herrod, Director

350 Conservation International

1919 M Street NW 202-912-1000
Suite 600 800-406-2306
Washington, DC 20036 Fax: 202-887-5188
http://www.conservation.org

Our mission is to conserve the Earth's living natural heritage, our global biodiversity, and to demonstrate that human societies are able to live harmoniously with nature.
Russell A Mittermeier, President

351 Conservation Management Institute

School of Natural Resources
203 W Roanoke Street 540-231-7348
Blackburg, VA 24061 Fax: 540-231-7019
E-mail: fwiexchg@vt.edu
http://fwie.fw.vt.edu/

To better address multi-disciplinary research questions that affect conservation management effectiveness in Virginia, North America and the world. Faculty from research institutions work collaboratively on projects ranging fromendangered species propagation to natural resource-based satellite imagery interpretation.
Jefferson Waldon, Assistant Director

352 Conservation Treaty Support Fund

3705 Cardiff Road 301-654-3150
Chevy Chase, MD 20815-5943 800-654-3150
Fax: 301-652-6390
E-mail: ctsf@conservationtreaty.org
http://www.conservationtreaty.org

CTSF's mission is to support the major inter-governmental treaties which conserve wild natural resources and habitat for their own sake and the benefit of people. These include the Endangered Species Convention and the InternationalWetlands Convention. CTSF raises support for treaty projects from individuals, corporations, foundations and government agencies. It also develops educational and informational materials including videos and the CITES Endangered Species Book.
George A Furness Jr, President

353 Counterpart International

1200 18th Street NW 888-296-9676
Suite 1100 Fax: 888-296-9676
Washington, DC 20036 http://www.counterpart.org

Promotes socioeconomics, health care, biodiversity, and natural resource management in over 60 countries.

354 Defenders of Wildlife

1101 14th Street NW 202-682-9400
Suite 1400 E-mail: info@defenders.org
Washington, DC 20005 http://www.dfenders.org

Defenders of Wildlife is dedicated to the protection of all native wild animals and plants in their natural communities. We focus our programs on what scientists consider two of the most serious environmental threats to the planet: theaccelerating rate of extinction of species and the associated loss of biological diversity, and habitat alteration and destruction. Long known for our leadership on endangered species issues.
Rodger Schlickeisen, President

355 Delta Wildlife

PO Box 276 601-686-3370
Stoneville, MS 38776 Fax: 601-686-4780
E-mail: teycoo@yahoo.com
http://www.nawf.org/dwf/

Recognizing the need for an agressive, but reasonable effort to develop wildlife, in 1990, one-hundred agri-business leaders, representing every country in Mississippi Delta, had the vision and dedication to form Delta Wildlife so theyand others could do more for conservation.
Bill Kennedy, Chairman

356 Desert Fishes Council

PO Box 337 760-872-8751
Bishop, CA 93515 Fax: 760-872-8751
E-mail: phidesfish@telis.org

The mission of the Desert Fisheries Council is to preserve the biological integrity of desert aquatic ecosystems and their associated life forms, to hold symposia to report related research and management endeavors, and to effect rapiddissemination of information concerning activities of the Council and its members.
E P Pister, Executive Secretary

357 Desert Protective Council

PO Box 3635 858-397-4264
San Diego, CA 92163-1635 E-mail: dpcinc@san.rr.com
http://www.dpcinc.org

The Desert Protective Council was founded in 1954. The DPC's purpose is to protect appreciate and enjoy our deserts.

358 Desert Tortoise Council

PO Box 3141 909-884-9700
Wrightwoodt, CA 92397 E-mail: info@deserttortoise.org
http://www.deserttortoise.org

The Council is a private, nonprofit organization made up of hundreds of professionals, and lay-persons from all walks of life, from across the United States, and several continents. We share a common fascination with wild deserttortoises and environment they depend upon.
Terrie Correll, Secretary

359 Dragonfly Society of the Americas

Bulletin of American Odonatology
2091 Partridge Lane 607-722-4939
Bingahamton, NY 13903 E-mail: tdonnel@binghamton.edu
http://www.afn.org/~iori/dsaintro.html

The Dragonfly Society of the Americas was organized in 1988. It is a nonprofit society whose purpose is to encourage scientific research, habitat preservation and aesthetic enjoyment of Odonata (dragon flies).
TW Donnelly, Editor

360 Ducks Unlimited

One Waterfowl Way 901-758-3825
Memphis, TN 38120 Fax: 847-438-9236
http://www.ducks.org/

The mission is to fulfill the annual life cycle needs of North American waterfowl by protecting, enhancing, restoring and managing important wetlands and associated uplands.
James Kennedy, President

361 Eagle Nature Foundation

300 E Hickory 815-594-2306
Apple River, IL 61001 Fax: 815-594-2305
E-mail: eaglenature.tni@juno.com
http://eaglenature.org

A nonprofit organization dedicated to the preservation of the bald eagle, our national symbol, and other endangered species from extinction and to increase public awareness of unique endangered plants and animals. We monitor bald eagleand other endangered species populations and strive to preserve habitat essential to their survival. We develop materials for schools to inform students about the needs of the bald eagle and how we can help preserve and protect their naturalenvironment.
Terrence N Ingram, Executive Director

362 Earth Ecology Foundation

612 N 2nd Street 559-442-3034
Fresno, CA 93702

Erik Wunstell, Director

363 Earth Island Institute

300 Broadway 415-788-3666
Suite 28 Fax: 415-788-7324
San Francisco, CA 94133 http://www.earthisland.org

Incubates and supports over 35 projects working on environmental issues worldwide. Publishes quarterly Earth Island Journal.
John A Knox, Executive Director

364 Elephant Interest Group
106 E Hickory Grove   313-540-3947
Bloomfield Hills, MI 48304   http://www.webdirectory.com
Hezy Shoshani, Contact

365 Elsa Wild Animal Appeal USA
PO Box 675   630-833-8896
Elmhurst, IL 60126   http://www.2ca.com/fullefx/2king4.htm
Donald A Rolla, President

366 Endangered Species Coalition
1101 14th Street NW   202-756-2804
Suite 1400   Fax: 202-429-3958
Washington, DC 20005   E-mail: esc@stopextinction.org
The Coalition is one of the most unique organizations in the
United States. An organization of ogranizations that supports
endangered species issues for over 430 environmental, religious, scientific, humane, and business groups aroundthe country. The vast majority of our member groups are small, local
grassroots organizations, who struggle to protect species and
habitats in their region.
Michael Bean

367 Federation of New York State Bird Clubs
PO Box 95
Durhanville, NY 13054   http://www.fnysbc.org
The objectives of the Federation are to document the ornithol-
ogy of New York State; to foster interest in and appreciation of
birds; and to protect birds and their habitats.
Sue Adadair, Treasurer

368 Ford Foundation
320 E 43rd Street   212-573-5000
New York, NY 10017   Fax: 212-599-4584
                                http://www.fordfound.org
Concerned with natural resource preservation. Provides funding
for environmental projects throughout the world.
Janet Maughan, Rural Poverty and Resources

369 Forest History Society
701 Vickers Avenue   919-682-9319
Durham, NC 27701   E-mail: stevena@duke.edu
                             http://www.lib.duke.edu/forest/
Brings lessons of forest and conservation history to bear on current issues in natural resource management. Identifies, collects,
preserves, interprets and disseminates information on forest and
conservation history with the goals ofimproving the public's understanding of the forest industry and forest products; providing
original resource material to researchers writing forest history,
to assure accuracy on facts and interpretation; and facilitating
development of naturalresource policy.
Steven Anderson, President

370 Forest Trust
PO Box 519   505-983-8992
Santa Fe, NM 87504-0519 E-mail: forest@theforesttrust.org
                             http://www.theforesttrust.org
The Trust's projects elevate the national dialogue over forests
by including people who were previously excluded from such
debates. At the same time, the Trust has helped to build a visible
and vocal rural constituency for protectingforests. By addressing issues of inclusive decision-making from these angles, the
Trust's programs ensure that national decision-making is accessible to the mainstream.
Henry H Carey, Director

371 Friends of Acadia
43 Cottage Street   207-288-3340
PO Box 45   Fax: 207-288-8938
Bar Harbor, ME 04609   E-mail: info@friendsofacadia.org
                             http://www.friendsofacadia.org
The mission of Friends of Acadia is to preserve and protect the
outstanding natural beauty, ecological vitality and cultural dis-
tinctiveness of Acadia National Park and the surrounding communities. Their methods are: raises and donatesprivate funds to
park and communities; advocates before legislatures and agencies; counters threats to park; and represents users in betterment
of its operations.
Ken Olson, President/CEO
Kelly S Dickson, Director of Development

372 Friends of the Sea Lion Marine Mammal Center
20612 Laguna Canyon Road   949-494-3050
Laguna Beach, CA 92651   Fax: 949-494-2802
                             E-mail: info@fslmmc.org
                                http://www.fslmmc.org
Friends of the Sea Lion Marine Mammal Center is a nonprofit
organization staffed by dedicated volunteers and funded by donations. Its mission is to rescue, medically treat and rehabilitate
seals and sea lions that are stranded alongOrange County, California beaches due to injury and illness; release healthly animals
back to their natural habitat; and increase public awareness of
the marine environment through education and research.
WH Ford, Executive Officer

373 Friends of the Sea Otter
2150 Garden Road   831-373-2747
Suite A-3   Fax: 831-373-2749
Monterey, CA 93922   E-mail: info@seaotters.org
                                http://www.seaotters.org
Friends of the Sea Otter is a nonprofit organization founded in
1968 dedicated to the protection of a threatened species, the
Southern Sea Otter, as well as Sea Otters throughout their north
pacific range, and all sea otter habitat.
Matt Rutishauser, Science Director
Mailee Flower, Education Director

374 Friends of the Trees Society
PO Box 4469   509-486-4726
Bellingham, WA 98229   Fax: 509-486-4726
                             E-mail: tern@geocities.com
                        http://www.geocities.com/RainForest/4663
Friends of the Trees Society is dedicated to doubling the world's
forest. This involves doubling the area covered by trees, dou-
bling the number of trees and most importantly, doubling the
weight of the world's forest biomass.
Michael Pilarski, Executive Officer

375 Garden Club of America
14 E 60th Street   212-753-8287
New York, NY 10022   Fax: 212-753-0134
                             E-mail: info@gcamerica.org
                                http://www.gcamerica.org
To stimulate the knowledge and love of gardening, to share the
advantages of association by means of educational meetings,
conferences, correspondence and publications, and to restore,
improve, and protect the quality of theenvironment through educational programs and action in the fields of conservation and
civic improvement.
Frederik Hansen, President

376 George Wright Society
PO Box 65   906-487-9722
Hancock, MI 49930   Fax: 906-487-9405
                                http://www.georgewright.org
Concerned with the preservation of natural and cultural parks in
the United States.

377 Global Coral Reef Alliance
324 Bedford Road   914-238-8788
Chappaqua, NY 10514   Fax: 914-238-8768
                             E-mail: goreau@bestweb.net
                        http://www.people.fas.harvard.edu/~goreau
A nonprofit organization for the protection and sustainable
management of coral reef.
Dr. Thomas J Goreau, President

378 Gopher Tortoise Council
Florida Museum of Natural History
PO Box 117800   904-362-1721
Gainesville, FL 32611   http://www.gohertortoisecouncil.org
The Council offers professional advice for management, conservation, and protection of gopher tortoises; encourages the
study of the life history, ecology, behavior, physiology and
management of gopher tortoises and other uplandspecies, conducts active public information and conservation education programs, and seeks effective protection of the gpphertortoise and
other upland species throughout the southeastern United States.
George Heinrich, Co-Chair

379 Grand Canyon Trust
2601 N Fort Valley   928-774-7488
Flagstaff, AZ 86001   888-428-5550
                                        Fax: 928-774-7570
                        E-mail: info@grandcanyontrust.org
                        http://www.grandcanyontrust.org
The mission of The Grand Canyon Trust is to protect and restore
the canyon country of the Colorado Plateau. Our vision for this
unique region 100 years from now is of a landscape still characterized by vast open spaces and dominated bywildness and of
healthy and restored natural ecosystems.
                        Founded: 1985
Bill Hedden, Executive Director
Darcy Allen, Director of Administration

380 Grassland Heritage Foundation
PO Box 394   913-262-3506
Shawnee Mission, KS 66201http://www.grasslandheritage.org
Grassland Heritage Foundation is a nonprofit membership orga-
nization to prairie preservation and education.
                        Founded: 1976
Sue Holcomb, Office Manager

381 Great Bear Foundation
802 E Front St   406-721-5420
PO Box 9383   Fax: 406-721-9917
Missoula, MT 59807   E-mail: awr@wildrockies.org
                              http://www.greatbear.org
The Foundation was established to promote conservation of wild
bears and their natural habitat worldwide.
Liz Sedler, President
Brian Horejsi, Vice President

382 Hawkwatch International
1800 SW Temple   801-484-6808
Suite 226   800-726-4295
Salt Lake City, UT 84115   Fax: 801-484-6810
                          E-mail: info@hawkwatch.org
                          http://www.hawkwatch.org
Our mission is to monitor and protect hawks, eagles, other birds
of prey and their environment through research, education and
conservation.
                        Founded: 1986
Yaeko Bryner, Chairman
Joan Deigiogio, Vice Chairman

383 Henry A Wallace Institute for Alternative Agriculture
9200 Edmonston Road   301-441-8777
Suite 117   Fax: 301-220-0164
Greenbelt, MD 20770-1551 E-mail: hawiaa@access.digex.net
                              http://www.hawiaa.org
The Institute is a nonprofit, tax-exempt, research and education
organization established in 1983 to encourage and facilitate the
adoption of low-cost, resource-conserving, and environmentally sound farming systems. Providesleadership, policy research and analysis to influence national agricultural
educational and research institutions, producer groups, farmers,
scientists, advocates, and other organizations that provide agricultural research, education, and informationservices.
                        Founded: 1983
I Garth Youngberg, Executive Director

384 Historic Trust for Historic Preservation
1785 Massachusetts Avenue NW   202-588-6000
Washington, DC 20036   800-944-6847
                                        Fax: 202-588-6038
                          E-mail: feedback@nthp.org
                          http://www.nthp.org
Provides leadership, education and advocacy to save America's
diverse historic places and revitalize our communities.
                        Founded: 1949
Richard Moe, Executive Director
David Brown, Executive Vice President

385 Holy Land Conservation Fund
969 Park Avenue   718-965-1057
New York, NY 10028
Bertel Bruun, President

386 Humane Society of the United States
2100 L Street NW   202-452-1100
Washington, DC 20037   Fax: 202-778-6132
                        E-mail: corprelations@hsus.org
                        http://www.hsus.org
HSUS is the nation's largest animal-protection organization
with more than 5 million constituents. The HSUS was founded
in 1954 to promote the humane treatment of animals, to foster respect, understanding, and compassion for allcreatures. Today
our message of care and protection embraces not only the animal
kingdom but also the Earth and its environment. To achieve our
goals we work through legal, educational, legislative and investigative means.
                        Founded: 1954
Wayne Pacelle, President/CEO
Andrew Rowan, Executive VP Operations

387 Hummingbird Society
249 E. Main Street   302-369-3699
Suite 9   800-529-3699
Newark, DE 19715   Fax: 302-369-1816
                        E-mail: info@hummingbird.org
                        http://www.hummingbird.org
Nonprofit corporation organized in 1996 for the purpose of encouraging international understanding and conservation of
hummingbirds by publishing and disseminating information,
promoting and supporting scientific study and
protectinghabitat. Publishers of The Hummingbird Connection.
                        Founded: 1996
Ross Hawkins, President

388 Inland Bird Banding Association
1409 Childs Road East   419-447-0005
Bellevue, NE 68005   E-mail: jingold@pilot.lsus.edu
                          http://www.aves.net/inlandbba/
Inland Bird Banding Association was organized in 1922, and
now supports the largest membership of any bird banding association in America. Inland Bird Banding Association is an organization for all individuals interested in the seriousstudy of
birds, their life-history, ecology, and conservation.
                        Founded: 1922
James Ingold, President
Vernon Kleen, Vice President

389 Institute for Conservation Leadership
2000 P Street NW   202-466-3330
Suite 413   Fax: 202-659-3897
Washington, DC 20036   http://www.icl.org
The mission of this Institute is to train and empower volunteer
leaders and to build volunteer institutions that protect and conserve the Earth's environment. We lend a helping hand to our
dedicated friends in pursuit of a better worldfor everyone.

**390 International Association for Bear Research and Man-
agement**
2841 Forest Avenue   510-549-3116
Berkley, CA 94705   E-mail: ucumari@aol.com
                          http://www.bearbiology.com
The International Association for Bear Research and Management (IBA)is a nonprofit tax-exempt volunteer organization
open to professional biologists, wildlife managers and others
dedicated to the conservation of all species of bears.The organization consists of several hundred members from over 20 countries. It supports the scientific management of the bears through
research and distrubition of information.
Bernie Peyton, Secretary

**391 International Association of Theoretical and Applied
Limnology**
University of Alabama
Department of Biology   205-348-1793
Tuscaloosa, AL 35487-0344   http://www.limnology.org
To further the study and understanding of all aspects of limnol-
ogy. Promotes and communicates new and emerging knowl-
edge among limnologists to advance the understanding of inland
aquatic ecosystems and their management.
Dr. Robert G Wetzel, Gen. Secretary/Treasurer

392 International Bird Rescue Research Center
4369 Cordelia Road   707-207-0380
Suisun City, CA 94585   818-222-9453

Fax: 707-207-0395
E-mail: tonya@ibrrc.org
http://www.ibrrc.org
Dedicated to mitigating the human impact on aquatic birds and other wildlife, worldwide. This is achieved through emergency response, education, research and planning.
Founded: 1972
Jay Holcomb, Executive Director

393 International Council for Bird Preservation, US Section
World Wildlife Fund
1250 24th Street NW   **202-467-8348**
Washington, DC 20037   **Fax: 202-293-9342**
E-mail: abc@mnsinc.com
http://www.worldwildlife.org
Founded: 1994
William K Reilly, Chairman
Edward P Bass, Vice Chairman

394 International Crane Foundation
E-11376 Shady Lane Rd.   **608-356-9462**
P.O. Box 447   **Fax: 608-356-9465**
Baraboo, WI 53913-0447   E-mail: cranes@savingcranes.org
http://www.savingcranes.org
Works worldwide to conserve cranes and the wetland and grasslands communities on which they depend. Dedicated to providing experience, knowledge, and inspiration to involve people in resolving threats to these ecosystems.
Founded: 1973
Jim Harris, President
Susan Finn, Office Administrator

395 International Ecology Society
R Kramer
1471 Barclay Street   **651-774-4971**
St Paul, MN 55106   **http://www.businessfinance.com**
Richard Kramer, President
Corey Pierce, Founder/COO

396 International Erosion Control Association
3001 S Lincoln Ave   **970-879-3010**
Suite A   **Fax: 970-879-8563**
Steamboat Springs, CO 80487   E-mail: ecinfo@ieca.org
http://www.ieca.org
An organization dedicated to minimizing accelerated soil erosion. IECA offers an annual conference, chapter events, training courses as well as a variety of topic specific publications and a quarterly news letter.
Founded: 1972
Doug Wimble, President
Becky Gauthier, Secretary

397 International Fund for Animal Welfare
411 Main Street   **508-744-2000**
PO Box 193   **800-932-4329**
Yarmouth Port, MA 02675   **Fax: 508-744-2009**
E-mail: info@ifaw.org
http://www.ifaw.org
IFAW's mission is to improve the welfare of the wild and domestic animals throughout the world by reducing commerical exploitation of animals, protecting wildlife habitats, and assisting animals in distress. We seek to motivate the public to prevent cruelty to animals and to promote animal welfare and conservation policies that advance the well-being of both animals and people.
Brian Davies, Founder

398 International Osprey Foundation
PO Box 250   **941-472-1862**
Sanibel Island, FL 33957   **http://www.sancap.com**
Nonprofit corporation dedicated to the continuing recovery and preservation of the osprey, others in the raptor family, wildlife and the environment as a whole. Conducts monitoring activities and accumulates data specific to the breeding activities of the osprey population. Publishes newsletter, issues grants for researchers whose studies involve environmental concerns. Directs and participates in all areas of wildlife and habitat maintenance and restoration.
Tim Gardner, President
Anne Mitchell, VP

399 International Snow Leopard Trust
4649 Sunnyside Avenue N   **206-632-2421**
Suite 325   **Fax: 206-632-3967**
Seattle, WA 98103   **E-mail: info@snowleopard.org**
http://www.snowleopard.org
The trust is dedicated to the conservation of the endangered Snow Leopard and its mountain ecosystem. Since its founding by Helen Freeman, ISLT has worked on more than 100 projects with local people throughout Central Asia. Focus is onsmall, creative and sustainable programs to make conservation happen now and in the future. Has hosted snow leopard symposia and developed the Snow Leopard Information Mangement System. This allows for range-wide comparison and sharing ofinformation.
Founded: 1981
Brad Rutherford, Executive Director
Tom McCarthy, Conservation Director

400 International Society for the Preservation of the Tropical Rainforest
3302 N Burton Avenue   **626-572-0233**
Rosemead, CA 91770   **800-932-4329**
Fax: 818-990-3333
E-mail: forest@nwc.net
http://www.isptr-pard.org/orderlist.html
We have developed a cooperative among 10 villages as a sustainable means of livelihood for the villagers. By providing an alternative and sustainable means of self-support, the project reduces the impact on the ecosystem; people cannow have chicken farms, and purchase or grow food rather than killing monkeys for food, destroying the forest by logging, or overfishing the rivers.
Founded: 1994
Arnold Newmann, Co-Director
Roxanne Kremer, Co-Director

401 International Society for the Protection of Mustangs and Burros
PO Box 14194   **602-502-7900**
Scottsdale, AZ 85267   **Fax: 602-502-2205**
E-mail: 103053.1112@compuserve.com
http://www.ispmb.com
Dedicated to preserving wild horses/burros and their habitats. We also run a rescue program and purchase slaughter bound wild horses and burros, placing them in permanent loving homes, or transporting entire herds of horses.
Karen Sussman, President

402 International Union for Conservation of Nature and Natural Resources
World Conservation Union
1630 Connecticut Avenue NW   **202-387-4826**
Washington, DC 20009   **Fax: 202-387-4823**
E-mail: postmaster@iucnus.org
http://www.iucn.org
Founded: 1948
Russell Mittermeier, Chairman
Gary Allport, Senior Conservation Policy A

403 International Union for the Conservation of Nature's Primate Specialist Group
1015 18th Street NW   **202-429-5660**
Suite 1000   **Fax: 202-887-5188**
Washington, DC 20036   **http://www.conservation.org**
The mission is to conserve the Earth's living natural heritage, our global biodiversity, and to demostrate that human societies are able to live harmoniously with nature.
Russell A Mittermeier

404 International Wildlife Rehabilitation Council
PO Box 8187   **408-271-2685**
San Jose, CA 95155   **Fax: 408-271-9285**
E-mail: office@iwrc-online.org
http://www.iwrc-online.org
A nonprofit international membership organization. Founded in 1972, the IWRC works to enhance the integrity of native wildlife systems and conserve biological diversity worldwide, through rehabilitation of wildlife, support ofrehabilitators, public education and advocacy.
Founded: 1974
Penny Elliston, President
Susan Heckly, Secretary

405 Island Conservation Effort

90 Edgewater Drive   305-666-5381
Suite 901   Fax: 305-663-9941
Coral Gables, FL 33133 E-mail: tropbird@unspoledqueen.com
Martha Walsh-McGehee, President
Rosemarie Gnam, Secretary/Tresurer

406 Izaak Walton League of America

707 Conservation Lane   301-548-0150
Gaithersburg, MD 20878   800-453-5463
                                      Fax: 301-548-0146
                                      E-mail: general@iwla.org
                                      http://www.iwla.org
The mission is to conserve, maintain, protect and restore the soil, forest water and other natural resources of the United States and other lands; to promote means and opportunities for the education of the public with respect to such resources and their enjoyment and wholesome utilization.
Founded: 1922
Paul Hansen, Executive Director
Georgia Townsend, President, Maryland Division

407 Jackson Hole Preserve

30 Rockefeller Plaza   212-649-5819
Room 5600   Fax: 212-649-5729
New York, NY 10112   http://www.undueinfluence.com
Founded: 1940
Laurance Rockefeller, Chairman of the Board
Clayton Frye, President

408 Life of the Land

76 N King Street   808-533-3454
Suite 203   Fax: 808-533-0993
Honolulu, HI 96817   E-mail: henry@lifeoftheland.net
                          http://www.lifeoflandhawaii.org
Hawaii's own environmental and community action group protecting our fragile natural and cultural resources through research, education, advocacy and litigation.
Founded: 1970
Kapua Sproat, President
Henry Curtis, Executive Director

409 Lighthawk

PO Box 29231   415-561-6250
The Presidio, Building 1007   Fax: 415-561-6251
San Francisco, CA 94129-0231   E-mail: sfo@lighthawk.org
                                http://www.lighthawk.org
Nonprofit organization founded in 1979, addresses critical environmental issues by providing an aerial perspective on areas of concern in the US, Canada and Central America. Using small aircraft, we fly partner organizations, elected officials, industry and media representatives, activists, and indigenous groups over protected and threatened regions. Our program flights give passengers both intellectual and visceral understanding of what is at stake.
Founded: 1979
Rick Durden, President

410 Marin Conservation League

55 Mitchell Boulevard   415-472-6170
Suite 21   Fax: 415-472-1404
San Rafael, CA 94903   E-mail: mcl@conservationleague.org
                         http://www.conservationleague.org
The Marin Conservation League is an a nonprofit organization founded in 1934 to preserve, protect and enhance the natural assets of Marin County for all people. MCL is Marin's oldest, locally based environmental organization, championing a sound balance between the needs of Marin's citizens and its beautiful and fragile environment.
Susan Stompe, President
Robert Berner, Secretary

411 Mineral Policy Center

1612 K Street NW   202-887-1872
Suite 808   Fax: 202-887-1875
Washington, DC 20006   E-mail: info@earthworksaction.org
                         http://www.mineralpolicy.org

Mining causes serious environmental problems for local communities across the United States and throughout the world. From the perpetual water pollution caused by mine drainage to cyanide spills and heavy metals contamination; from the desecration of scared sities to the creation of toxic waste rock- mining creates devastating environmental consequences. Mineral Policy Center carries out research and publishes comprehensive reports on the environmental impacts of mining.
Stephen D'Esposito, President/CEO
Cathy Carlson, Policy Advisor

412 Monitor Consortium of Conservation Groups

1506 19th Street NW   202-234-6576
Washington, DC 20036
Craig VanNote, Executive VP

413 National Association of Conservation Districts

509 Capitol Court NE   202-547-6223
Washington, DC 20002-4946   Fax: 202-547-6450
                          E-mail: Eugene-Lamb@NACDnet.org
                                 http://www.nacdnet.org
NACD, founded in 1946, is a nongovernmental, nonprofit organization representing nearly 3000 local soil and water conservation districts and their state associations in the 50 states, Puerto Rico, the Virgin Islands, Guam/Northern Mariana Islands, American Samoa, Federated States of Micronesia, Republic of Palau, and DC. It is financed and controlled by member districts and state associations. The primary function of NACD is to serve the member districts in the conservation of natural resources.
Wade Troutman, President
Paul Stoker, Vice President

414 National Association of University Fisheries and Wild-life Programs

University of Minnesota
1980 Folwell Avenue   612-624-3600
200 Hodson Hall   Fax: 612-625-5299
St Paul, MN 55108   E-mail: jperry@tc.umn.edu
                             http://www.fw.umn.edu
Develops effective communications and liaisons with other natural resource, conservation, environmental, scientific and educational organizations. These include the National Assoication of State universities and Land-Grant Colleges, American Fisheries Society, International Association of Fish and Wildlife Agencies. Establishing ongoing interaction with offices and agencies of the government that determine research, education and extension policies, support and fund university programs.
Founded: 1972
Jim Perry, Department Head

415 National Audubon Society

700 Broadway   212-979-3000
New York, NY 10003   800-274-4201
                                      Fax: 212-979-3188
                          E-mail: AudubonAtHome@Audubon.org
                                 http://www.audubon.org
The National Audubon Society is one of the oldest, largest, and most powerful nature appreciation and conservation organizations in the country. NAS works on a broad range of concerns related to the protection of the world's ecosystems; preserving wetlands, population planning, eliminating acid rain and reducing air pollution, promoting environmental justice, and protecting water quality.
Founded: 1905
John Flicker, President
Monique Quinn, Chief Finance Officer

416 National Audubon Society: Project Puffin

159 Sapsucker Woods Road   607-257-7308
Ithaca, NY 14850   Fax: 607-257-6231
                          E-mail: puffin@audubon.org
                          http://www.projectpuffin.org
Established in 1973 in an effort to learn how to restore puffins to historic nesting islands in the Gulf of Maine. Although puffins are abundant in Newfoundland, Iceland and Britain, they are rare in Maine. Project Puffin has a yearround staff of six which increases to include more than 50 biologists and researchers during the seabird breeding season in spring and summer. The project is based in Ithaca, New York and the Todd Wildlife Sanctuary in mid-coast Maine.
Founded: 1973
Stephen Kress, Director

417 National Conservation Foundation
509 Capitol Court NE   202-547-6223
Washington, DC 20002   Fax: 202-547-6450
E-mail: libbysimon@conservationfoundation.com
http://www.nacdnet.org
Founded: 1982
Donald Spickler, Chairman
Clear Spring, Managing Director

418 National Council for Environmental Balance
4169 Westport Road   502-896-8731
PO Box 7732   Fax: 502-339-1745
Louisville, KY 40257-0732 http://www.exxonsecrets.org
Irwin Tucker, President

419 National Fish and Wildlife Foundation
1120 Connecticut Avenue, NW   202-857-0166
Suite 900   Fax: 202-857-0162
Washington, DC 20036   E-mail: decarolis@nfwf.org
http://www.nfwf.org
The National Fish and Wildlife Foundation is a nonprofit organization dedicated to the conservation of fish, wildlife, and plants and the habitats on which they depend. Among its goals are species habitat protection, environmentaleducation, public policy development, natural resource management, habitat and ecosystem rehabilitation and restoration, and leadership training for conservation professionals.
Founded: 1984
John Berry, Executive Director
Peter Bower, CFO

420 National Forest Foundation
2715 M Street NW   202-298-6740
Suite 100   Fax: 202-298-6758
Washington, DC 20007   E-mail: info@natlforests.org
http://www.natlforests.org
The official nonprofit partner of the USDA Forest Service. To accept and administer private contributions, undertaking activities that further the purposes for which the National Forest System was established and conductingeducational, technical, and other activities that support the multiple use, research, and forestry programs administered by the Forest Service.
Bill Possiel, President
David Bell, Chief Operating Officer

421 National Garden Clubs
4401 Magnolia Avenue   314-776-7574
St Louis, MO 63110-3492   800-550-6007
Fax: 314-776-5108
E-mail: headquarters@gardenclub.org
http://www.gardenclub.org
National Garden Clubs is a nonprofit educational organization with its headquarters in St. Louis, Missouri, US. It is composed of 50 State Garden Clubs and the National Capital Area, 8,858 member garden clubs and 235,316 members. Inaddition, NGC proudly recognizes 200 Internationl Affiliates from Canada to Mexico and South America.
Founded: 1891
June P Wood, President
Renee Blaschke, Vise President

422 National Grange
1616 H Street NW   202-628-3507
Washington, DC 20006-4999   888-447-2643
Fax: 202-347-1091
E-mail: rfrederick@nationalgrange.org
http://www.nationalgrange.org
The Grange is a family based community organization with a special interest in agriculture and rural America, as well as in legislative efforts regarding these issues.
Founded: 1867
William Steel, President
Shirley Lawson, Secretary

423 National Institute for Urban Wildlife
10921 Trotting Ridge Way   301-596-3311
Columbia, MD 21044

424 National Military Fish and Wildlife Association
12428 Pinecrest Lane   540-663-4186
Newburg, MD 20664   Fax: 540-663-4016

E-mail: nmfwa@nmfwa.org
http://www.nmfwa.org
Thomas Wray II, President

425 National Park Foundation
1101 17th Street NW   202-785-4500
Suite 1102   E-mail: parkspass@nationalparks.org
Washington, DC 20036-4704 http://www.nationalparks.org
Honors, enriches and expands the legacy of private philanthropy that helped create and continues to sustain America's National Parks.

426 National Park Trust
51,Monroe Street   301-279-7275
Suite 110   866-281-5971
Rockville, MD 20850   Fax: 301-279-7211
E-mail: npt@parktrust.org
http://www.parktrust.org
Founded: 1983
Paul C Pritchard, President
William Brownell, Chairman

427 National Parks Conservation Association
1300 19th Street NW   800-628-7275
Suite 300   Fax: 202-659-0650
Washington, DC 20036   E-mail: npca@npca.org
http://www.npca.org
Since 1919, the National Parks Conservation Association has been the voice of the American people in the fight to safeguard the scenic beauty, wildlife, and historical and cultural treasures of the largest and most diverse park systemin the world.
Founded: 1916
Thomas Kiernan, President
Tom Martin, Executive Vice President

428 National Prairie Grouse Technical Council
Wildlife Research Center
317 W Prospect Road   970-484-2836
Fort Collins, CO 80526   Fax: 970-490-2621
Kenneth M Giesen, Executive Officer

429 National Recreation and Park Association
Parks and Recreation Magazine
22377 Belmont Ridge Road   703-858-0784
Ashburn, VA 20148-4150   Fax: 703-858-0794
E-mail: info@nrpa.org
http://www.nrpa.org
The NRPA, headquartered in Ashburn Virginia, is a national nonprofit organization devoted to advancing park, recreation and conservation efforts that enhance the quality of life for all Americans. The Association works to extendsocial, health, cultural and economic benefits of parks and recreation, through its network of 23,000 recreation and park professionals and civic leaders. NRPA encourages recreation initiatives for youth in high-risk environments.
Founded: 1965
John A Thorner, Executive Director
M Lauren Yost, Human Resources Manager

430 National Speleological Society
2813 Cave Avenue   256-852-1300
Huntsville, AL 35810-4431   Fax: 256-851-9241
E-mail: nss@caves.org
http://www.caves.org
Founded for the purpose of advancing the study, conservation, exploration, and knowledge of caves. More than 12,000 members in 200 grottos conduct regular meetings to bring cavers trogether within their general area to coordinateactivities which may include mapping, cleaning and investigating sensitive caves.
Paul Meyer, President
Stephanie Searles, Operations Manager

431 National Tree Society
PO Box 10808   805-589-6912
Bakersfield, CA 93389   Fax: 775-248-6035
E-mail: tnc@natural-connection.com
http://www.natural-connection.com

Organization to preserve the earth's biosphere by planting and caring for trees. Seeks to raise public understanding of the need for trees and the role they play in maintaining a healthy environment; works to acquire forest and otherlands to ensure the continued growth of trees on such lands; establishes nurseries to supply the trees.
Gregory W Davis, Contact

432 National Trust for Historic Preservation

1785 Massachusetts Avenue NW   202-588-6000
Washington, DC 20036-2117   800-944-6847
    Fax: 202-588-6038
    E-mail: pr@nthp.org
    http://www.nthp.org

The National Trust for Historic Preservation provides leadership, education and advocacy to save America's diverse historic places and revitalize our communities.
Founded: 1949
David J Brown, Executive VP
Richard Moe, President

433 National Waterfowl Council

Ohio Department of Natural Resources
2045 Morse Road   614-265-6565
Columbus, OH 43229   http://www.ohioiodnr.com

To ensure a balance between wise use and protection of our natural resources for the benefit of all.
Founded: 1949
Samuel Speck, Director

434 National Wildflower Research Center

Lady Bird Johnson Wildflower Center
4801 Lacross Avenue   512-471-1525
Austin, TX 78739-1702   Fax: 512-471-1551
    E-mail: facilityrentals@wildflower.org
    http://www.wildflower.org

Combines native plants with local culture, reflecting the specifics and peculiarities of Central Texas Hill Country ecosystems. Walking through the center, you'll find native plants in gardens and natural areas, an unparalleddrainwater collection and storage system, recycled building materials, American folk art, environmentally conscious construction and engaging educational faciltities— all designed to learn to live more gently on the land.
Founded: 1897
Robert.A. Wooster, President
Larry McNeill, Vice President

435 National Wildlife Federation

11100 Wildlife Center Drive   703-438-6000
Reston, VA 20190-5362   800-822-9919
    Fax: 703-438-6468
    E-mail: schweiger@nwf.org
    http://www.nwf.org

The nation's largest member-supported conservation education and advocacy group, the Nationa Wildlife Federation unites people from all walks of life to protect nature, wildlife and the world we all share. The Federation has educatedand inspired families to uphold America's conservation tradition since 1936.
Founded: 1936
Eileen Morgan Johnson, Secretary
Larry Schweiger, President

436 National Wildlife Productions

11100 Wildlife Center Drive   703-438-6077
Reston, VA 20190   1 8-0 8-2 99
    Fax: 703-438-6076
    http://www.nwf.org

Founded: 1934
Jerome.C. Ringo, Chairman
Stephen E Petron, Vice Chairman

437 National Wildlife Refuge Association

1010 Wisconsin Avenue NW   202-333-9075
Suite 200   877-396-NWRA
Washington, DC 20007   Fax: 202-333-9077
    E-mail: nwra@refugenet.org
    http://www.refugenet.org

Aims to protect, enhance and expand the National Wildlife Refuge System lands, set aside by the American public to protect our country's diverse wildlife heritage.

Founded: 1975
Charlie Estes, Executive Director
William H Meadows, President

438 National Wildlife Rehabilitators Association

2625 Clearwater Rd   320-230-9920
Suite 110   Fax: 320-230-3077
St. Cloud, MN 56301-4766   E-mail: nwra@nwrawildlife.org
    http://www.nwrawildlife.org

The National Wildlife Rehabilitators Association is a nonprofit international membership organization committed to promoting and improving the integrity and professionalism of wildlife rehabilitation and contributing to thepreservation of natural ecosystems. Please, call before you fax!
Curtiss Clumpner, President
Di Conger, Vice President

439 National Woodland Owners Association

374 Maple Avenue E   800-476-8733
Suite 310   888-GRN-TREE
Vienna, VA 22180-4751   Fax: 703-281-9200
    E-mail: info@woodlandowners.org
    http://www.woodlandowners.org/

NWOA is independent of the forest products industry and forestry agencies. We work with all organizations to promote non-industrial forestry and the best interested of woodland owners.
Founded: 1983
Keith A Argow, President
Eric Johnson, National Woodlands Editor

440 Natural Area Council

725 15th street NW   202-638-1649
Suite 605   Fax: 202-638-2175
Washington, DC 20005   http://www.america-the-beautiful.org
Founded: 1965
Nanine Bilski, President

441 Natural Areas Association

PO Box 1504   541-317-0199
Bend, OR 97709   Fax: 541-317-0140
    E-mail: mail@naturalarea.org
    http://www.naturalarea.org/

To advance the preservation of natural diversity. To inform, unite and support persons engaged in identifying, protecting, managing, and studying natural areas and biological diversity across landscapes and ecosystems.
Founded: 1980
Reid Schuller, Executive Director

442 Natural Resources Defense Council

40 W 20th Street   212-727-2700
New York, NY 10011   Fax: 212-727-1773
    E-mail: nrdcinfo@nrdc.org
    http://www.nrdc.org

This organization is dedicated to protecting endangered natural resources and improving the quality of the human environment. Areas of concentration are air and water pollution, nuclear safety, land use, urban environment, toxicsubstances control, resource management, wilderness and wildlife protection, Alaska, coastal zone management, energy conservation, soil erosion, and forestry. Founded in 1970, it has a membership of more than 500,000.
Founded: 1990
Francis G Beinecke, Executive Director
John H Adams, President

443 North American Bluebird Society (NABS)

P.O. Box 244   330-359-5511
Wilmot, OH 44689-0244   Fax: 330-359-5455
    E-mail: info@nabluebirdsociety.org
    http://www.nabluebirdsociety.org

Nonprofit conservation, education and research organization, promotes the recovery of the bluebirds and other native cavity-nesting bird species. NABS supports conservation through such continent wide programs as the TranscontinentalBluebird Trail and the NABS Nestbox Approval Process. NABS also produces award-winning educational materials.
Founded: 1977
Steve Garr, President
Julie Kutruff, Vice President

444 North American Coalition and Ecology
5 Thomas Circle NW
Washington, DC 20005-4104
202-462-2591
Fax: 202-234-0307
E-mail: webmaster@bostontheological.org
http://www.businessfinance.com

Donald Conroy, President

445 North American Crane Working Group
North American Bluebird Society
341 W Olympic Place
Seattle, WA 98119
NACWG is an organization of professional biologists, aviculturists, land managers and other interested individuals dedicated to the conservation of cranes and their habitats in North America. They sponsor a North American Crane Workshop every 3-4 years, promulgates technical information including a published Proceedings of a North American Workshop and a semi-annual newsletter, address conservation issues affecting cranes and their habitat, promote appropriate research on crane conservation.

Tom Hoffman, Contact

446 North American Falconers Association
Route 1
Box 82A
Ellinwood, KS 67526-9801
E-mail: sokolnik@lycos.com
http://www.n-a-f-a.org
Our purpose is to provide communication amoung and to disseminate relevant information to interested members; to provide scientific study of raptorial species, their care, welfare and training; to promote conservation of the birds of prey and an appreciation of their value in nature and in wildlife conservation programs; to urge recognition of falconry as a legal field sport; and to establish traditions which will aid, perpetuate and further the welfare of falconry and raptors it employs.

Sue Cecchini, Trasurer

447 North American Loon Fund
PO Box 329
Holderness, NH 03245
603-528-4711
800-462-5666
http://www.facstaff.uww.edu
The North American Loon Fund's mission is to promote the preservation of loons and their lake habitats through research, public education, and the involvement of people who share their lakes with loons.

Linda O'Bara, Director

448 North American Wildlife Park Foundation Wolf Park
4008 East 800 North
Battle Ground, IN 47920-9776
765-567-2265
Fax: 765-567-4299
E-mail: wolfpark@wolfpark.org
http://www.wolfpark.org/

Founded: 1972
Erich Klinghammer, President (Founder)

449 North American Wolf Society
PO Box 3243
Boulder, CO 80307

450 Open Space Institute
1350 Broadway
Room 201
New York, NY 10018
212-629-3981
Fax: 212-244-3441
E-mail: info@osiny.org
http://www.osiny.org
Protects land for public benefit and supports the efforts of citizen activists working to improve environmental regulations in their communities.

Christopher J Elliman, Chief Executive Officer
Joseph Martens, President

451 Openlands Project
25 E Washington Street
Suite 1650
Chicago, IL 60602-1708
312-427-4256
Fax: 312-427-6251
E-mail: info@openlands.org
http://www.openlands.org
Nonprofit project, founded in 1963, is an independent, nonprofit organization dedicated to preserving and enhancing public open space in northeastern Illinois. Openlands bridges political boundaries and build consensus on open space goals and regional growth strategies.

Founded: 1963
Gerald Adelmann, Executive Director
Joyce O'Kese, Deputy Director

452 Organization for Bat Conservation
39221 Woodward Ave
P.O. Box 801
Bloomfield Hills, MI 48303
248-645-3232
Fax: 517-339-5618
E-mail: obcbats@aol.com
http://www.batconservation.org
A nonprofit organization, our mission is to preserve bats and their habitats through education, collaboration, and research. We also work with local health departments and government agencies to aid in public health issues associated with bats, and we have trained field biologists to research endangered bats.

Kim Williams, Director
Rob Mies, Director

453 Ozark Society
PO Box 2914
Little Rock, AR 72203
501-847-3738
Fax: 501-372-5569
http://www.ozarksociety.net
The Ozark Society has remained a strong regional organization because is has not allowed itself to be diverted from its principal purpose: the preservation of wild and scenic rivers, wilderness and unique natural areas. It's primary focus is the Ozark-Ouachita region and its associated bottom land habitat.

Founded: 1962
Dana Steward, Senior Manager

454 Ozarks Resource Center
PO Box 3
Brixey, MO 65618
417-679-4773
Fax: 417-679-4773
E-mail: jlorrain@goin.missouri.org
http://www.ic.org/resources
The purpose of the organization is to provide research, education, technical assistance and dissemination of information on: renewable resource-based appropriate technology, environmentally responsible practices, sustainable agriculture, community economic development and self-reliance for the family, farm community, the Ozarks and their bioregions.

Janice Lorrain, Executive Director

455 Partners in Parks
Department of Parks and Recreation
City-County Building
Room 452
Pittsburgh, PA 15219
202-364-7244
Fax: 202-255-2364
E-mail: partpark@cqi.com
http://www.partnersinparks.org
Organize and direct volunteers for work in city parks and trails. Outdoor work can be as simple as litter pick-up to trail cleaning, tree planting, etc.

Founded: 1988
David Kekil, Secretary/Treasurer

456 Peregrine Fund
5668 W Flying Hawk Lane
Boise, ID 83709
208-362-3716
Fax: 208-362-2376
E-mail: tpf@peregrinefund.org
http://www.peregrinefund.org
Founded in 1970, The Peregrine Fund works nationally and internationally to conserve biological diversity and enhance environmental health by working with birds of prey through management and conservation of species and their habitat and through education and scientific investigation. Although best known nationally for species restoration, they have assisted on conservation projects in over 40 countries.

Founded: 1970
William Burnham, President
J Peter Jenny, Vice President

457 Public Lands Foundation
PO Box 7226
Arlington, VA 22207
703-790-1988
Fax: 703-893-1500
E-mail: leaplf@erols.com
http://www.publicland.org
Founded in 1987, the Public Lands Foundation is a private nonprofit organization dedicated to the proper use and protection of the public lands administrated by the Bureau of Land Management, implementation of the Federal Land Policy and Management Act, and professional land management by professional employees.

George Lea, President
Carl Enix, Secretary

458 Quail Unlimited

PO Box 610
Edgefield, SC 29824

803-637-5731
Fax: 803-637-0037
http://www.qu.org

Quail Unlimited was established in 1981 to battle the problem of dwindling quail and wildlife habitat. Quail Unlimited is the only national conservation organization dedicated to the wise management of America's wild quail as avaluable and renewable resource.

Rocky Evans, Executive VP
Roger Wells, National Habitat Coordinator

459 RARE Center for Tropical Conservation

1840 Wilson Boulevard
Suite 204
Arlington, VA 22201

703-522-5070
Fax: 703-522-5027
E-mail: rare@rareconservation.org
http://www.rarecenter.org

Our mission is to protect wildlands of globally significant biological diversity by empowering local people to benefit from their preservation. Publishes a triannual newsletter.

Founded: 1975
Wendy J Paulson, Chairman
Angus Parker, Vice Chairman

460 Rainforest Alliance

665 Broadway
Suite 500
New York, NY 10012-2420

212-677-1900
Fax: 212-677-2187
E-mail: canopy@ra.org
http://www.rainforest-alliance.org

Dedicated to tropical forest conservation for the benefit of the global community.

Founded: 1987
Daniel Katz, President
Tensie Whelan, Executive Director

461 Raptor Education Foundation

PO Box 200 400
Denver, CO 80220

303-680-8500
Fax: 303-680-8502
E-mail: raptor2@usaref.org
http://www.usaref.org

One of the most important challenges facing the world today is the preservation, protection and appropriate use of natural resources. The environmental decisions made today will have a monumental impact on wildlife on a global scaleand effect the overall quality of life for mankind. We must strive to restore and maintain a level of dynamic balance in nature and minimize the rate at which species of animals, plants and other natural resources are declining.

Founded: 1980
Peter Reshetniak, Chairman
Ann Price, Secretary

462 Ruffed Grouse Society

451 McCormick Road
Coraopolis, PA 15108

412-262-4044
888-564-6747
Fax: 412-262-9207
E-mail: RGS@ruffedgrousesociety.org
http://www.ruffedgrousesociety.org

The Ruffed Grouse Society's role in conservation of wildlife habitat is to enhance the environment for the Ruffed Grouse, American Woodcock, and other forest wildlife that require or utilize thick, young forests. Since forests aredynamic and constantly changing and man has virually eliminated the fires that shaped much of the forested land we know today, forests must be managed.

Founded: 1961
Robert L Patterson Jr, Executive Director
Tracy M Greene, Manager Chapter Programs

463 Save the Dunes Council

444 Barker Road
Michigan City, IN 46360

219-879-3937
Fax: 219-872-4875
E-mail: std@savedunes.org
http://www.savedunes.org

The Save the Dunes Council of northwest Indiana was founded in 1952, one of the oldest grassroots conservation organizations in the country. Its objectives are to maintain and restore the integrity and quality of the naturalenvironment of the Indiana Dunes region. The hard work of their members led to the establishment of the Indiana Dunes National Lakeshore in 1966; the group continues to work on a wide variety of issues concerning the Dunes and the environmentalquality of the area.

Founded: 1952
Thomas Serynek, President
Dorothy Potucek, Vice President

464 Save the Manatee Club

500 N Maitland Avenue
Maitland, FL 32751

407-539-0990
800-432-5646
Fax: 407-539-0871
E-mail: education@savethemanatee.org
http://www.savethemanatee.org

Save the Manatee Club (SMC) is a nonprofit organization, established in 1981 by US Senator Bob Graham and singer/songwriter Jimmy Buffett so the general public could participate in conservation efforts to save endangered manatees fromextinction. The purpose of SMC is to promote public awareness and education; fund manatee research, rescue, and rehabilitation efforts; lobby for the protection of manatees and their habitat, and take appropriate legal action.

Founded: 1981
Judith Vallee, Executive Director
Patti Thompson, Director of Science

465 Save the Whales

Animal Welfare Institute
Georgetown Station
PO Box 3650
Washington, DC 20027

703-836-4300
Fax: 703-836-0400
E-mail: awi@awionline.org
http://www.awionline.org

Save the Whales' purpose is to educate children and adults about marine mammals, their environment and their preservation. Founded in 1977, Save the Whale is a 801(c)(3) Educational Nonprofit Corporation.

Founded: 1951
Cathy Liss, President

466 Scenic America

1634 I Street NW
Suite 510
Washington, DC 20006

202-638-0550
Fax: 202-638-3171
E-mail: Scenic@scenic.org
http://www.scenic.org

The only national nonprofit organization dedicated solely to protecting natural beauty and distinctive community character. We provide technnical assistance across the nation and through our state affiliates on scenic byways, billboardand sign control, context sensitive highway design, wireless telecommunications tower location, transportation enhancements, and other scenic conservation. We advance our number one goal, to build a citizen movement for scenic conservation througheducation.

Kevin Fry, President
Peggy Lint, Office Manager

467 Sierra Club

1416 33rd Street NW
Washington, DC 20007

202-547-1141
Fax: 202-547-6009
E-mail: information@sierraclub.org
http://www.sierraclub.org

To advance the preservation and protection of the natural environment by empowering the citizenry, especially democratically-based grassroots organizations, with charitable resources to further the cause of environmental protection.The vehicle through which The Sierra Club Foundation generally fulfills its charitable mission.

Founded: 1912
Lisa Rentstrom, President
Barnard Zaleha, Vice President

468 Smithsonian Institution

Independence Avenue & 6th Street SW
Washington, DC 20560

202-357-2700
Fax: 202-357-2426
E-mail: info@si.edu
http://www.si.edu

Joel J Cohen, Chairman

469 Society for Marine Mammalogy

7600 San Point Way NE   PO Box
Seattle, WA 98115

206-526-4016
Fax: 206-526-6615
http://www.marinemammalogy.org/

To evaluate and promote the educational, scientific and managerial advancement of marine mammal science. Gather and disseminate to members of the Society, the public and private institutions, scientific, technical and managementinformation through publications and meetings. Provide scientific information, as required, on matters related to the conservation and management of marine mammal resources.
Founded: 1981
Daniel P Costa, Secretary
Kit M Kovacs, President

470 Society for the Conservation of Bighorn Sheep
PO Box 94182   323-256-0463
Pasadena, CA 91109-4182   E-mail: MRalles@msn.com
            http://http://desertbighorn.cjb.net/
Mission and ultimate goal is the full restoration of the California Desert Bighorn to its historic habitat and the establishment of self-sustaining populations throughout those ranges.
Founded: 1964
John Nelson, President

471 Society for the Preservation of Birds of Prey
P.O. Box 66070   310-285-5815
Los Angeles, CA 90066
J Richard Hilton, President

472 Sonoran/Rincon Institutes
7650 E Broadway   520-290-0828
Suite 203   Fax: 520-290-0969
Tucson, AZ 85710   E-mail: sonoran@sonoran.org
            http://www.sonoran.org
Nonprofit organization that works collaboratively with local people and interests to conserve and restore important natural landscapes in western North America, engaging partners such as landowners, public land managers, local leaders,community residents and nongovernmental organizations. Community Stewardship is an innovative approach to conservation.
Founded: 1990
Luther Propst, Executive Director
John Shepard, Associate Director

473 Tall Timbers Research Station
Tall Timbers
13093 Henry Beadel Drive   850-893-4153
Tallahassee, FL 32312-0918   Fax: 850-668-7781
            E-mail: rose@ttrs.org
            http://www.talltimbers.org
Dedicated to protecting wildlands and preserving natural habitats. Promotes public education on the importance of natural disturbances to the environment and the subsequent need for wildlife and land management. Conducts fire ecologyresearch and other biological research programs through the Tall Timbers Research Station. Operates museum.
Lane Green, Executive Director
Rose Rodriguez, Information Resources Mgr

474 Theodore Roosevelt Conservation Alliance
27 Fort Missola Road   406-549-0101
Suite 4K   877-770-8722
Missoula, MT 59804   Fax: 406-549-7402
            E-mail: info@trca.org
            http://www.trca.org
To inform and engage Americans to foster our conservation legacy while working to nurture, enhance and protect our fish, wildlife and habitat resources in our National Forest System.
Founded: 1991
Kristen Munson, Executive Assistant
Robert Munson, Executive Director

475 Threshold
Drawer CU   877-818-1881
Bisbee, AZ 85603
John P Milton, President

476 Tread Lightly!
298 24th Street   801-627-0077
Suite 325   800-966-9900
Ogden, UT 84401   Fax: 801-621-8633
            E-mail: tlinc@xmission.org
            http://www.treadlightly.org

Not-for-profit organizaiton dedicated to protecting the great outdoors. In an age where outdoor recreation is the sport of choice, it is our responsibility to exercise responsible outdoor practices.
Founded: 1990
Lori Davis, Executive Director
Lori Davis, Executive Director

477 TreePeople
12601 Mulholland Drive   818-753-4600
Beverly Hills, CA 90210   Fax: 818-753-4635
            E-mail: info@treepeople.org
            http://www.treepeople.org
To inspire the people of Los Angeles to take personal responsibility for their environment, training and supporting them as they plant and care for trees and improve the neighborhoods in which they live, work and play. Throngheducation, planting projects, policy development and research, the organization is helping lead the promotion of integrated urban watershed management.
Founded: 1950
Andi Lipkis, President

478 Trees for Life
3006 St. Louis   316-945-6929
Wichita, KS 67203-5129   Fax: 316-945-0909
            E-mail: info@treesforlife.org
            http://www.treesforlife.org
Empowers people by demonstrating that in helping each other, we can unleash extraordinary power that impacts our lives. By planting fruit trees in developing countries, we protect the environment and provide a low-cost, self-renewingsource of food for a large number of people. Activities include three elements: education, health and environment.
Founded: 1984
Balbir Mathur, President
David Kimble, Executive Director

479 Trust for Public Land
116 New Montgomery Street   415-495-4014
4th Floor   Fax: 415-495-4103
San Francisco, CA 94105   E-mail: info@tpl.org
            http://www.tpl.org
Founded in 1972, the Trust for Public Land is the only national nonprofit working exclusively to protect land for human enjoyment and well-being. TPL helps conserve land for recreation and spiritual nourishment and to improve thehealth and quality of life of American communities.
Founded: 1972
William Rogers, President
David Reed, Vice President, Finance

480 Trust for the Future
2704 12 Avenue South   615-297-2269
Nashville, TN 37204   Fax: 615-298-1611
            E-mail: info@tenngreen.org
            http://www.tenngreen.org
To promote conservation of natural resources. Encourage and perform research and analysis on conservation issues. Support public education for conservation and publish studies for the advancement of conservation and environmentalprotection.
Charles A Howell III, President

481 Western Hemisphere Shorebird Reserve Network
81 Stage Point Road   508-224-6521
PO Box 1770   Fax: 508-224-9220
Manomet, MA 02345   E-mail: whsrn@manomet.org
            http://www.manomet.org/whsrn
WHSRN is a voluntary, community-based coalation of over 185 organizations across the US and other counries in the Western Hemisphere that have joined together to protect, restore and manage critical wetland habitats for migratorybirds. Established in 1986, we now have 49 key sites in 7 countries as Network partners responsible for over 20 million acres. WHSRN links these sites into a collaborative network based on one ecological feature they all share-millons of migratorybirds.
Founded: 1843
Charles Duncan, Executive Director
Nan Harris, Secretary

482 Western Society of Naturalists
California State University
Department of Biology   818-677-3256
Northridge, CA 91330-8303   Fax: 818-677-2034
                    E-mail: jolene.koester@csun.edu
                              http://www.csun.edu
Members include researchers, educators, academics and others
with an interest in the area's biology, particularly its marine life.
Membership is $15.00 per year for individuals and $7.00 for students.
Jolene Koester, President
Claire Cavallaro, Chief of Staff

483 Whooping Crane Conservation Association
1393 Henderson Highway   337-234-6339
Breaux Bridge, LA 70517
The mission is to advance conservation, protection and propagation of the Whooping Crane population, to prevent its extinction, to establish and maintain a captive management program for the perpetuation of the species. We collect anddisseminate knowledge of this species; and advocate and encourage public appreciation and understanding of the Whooping Crane's educational, scientific and economic values.
Mary L Courville, Secretary/Treasurer

484 Wild Canid Survival and Research Center
PO Box 760   636-938-5900
Eureka, MO 63025   Fax: 636-938-6490
                    E-mail: wildcanidcenter@onemain.com
                              http://www.wolfsanctuary.org
A private nonprofit conservation organization dedicated to the preservation of the wolf and other endangered canids through education, research and captive breeding.
                    Founded: 1971
Susan Lindsey PhD, Executive Director
Kim Scott, Assistant Director

485 Wild Horse Organized Assistance
PO Box 555   702-851-4817
Reno, NV 89504 http://www.wildhorseorganizedassistance.org
The mission is to save diminishing herds of wild horses in the Western United States.
                    Founded: 1971
Dawn Y Lappin, Executive Director

486 Wild Horses of America Registry
6212 E Sweetwater   602-991-0273
Scottsdale, AZ 85254   Fax: 602-991-2920
                    E-mail: 103053.1112@campuserve.com
                              http://www.ispmb.com
Recognizes wild horses and burros of America that have been removed from public lands.
Karen Sussman, Registrar

487 Wilderness Society
1615 M Street NW   202-833-2300
Washington, DC 20036   800-THE-WILD
                                    Fax: 202-429-3958
                    E-mail: tws@wilderness.org
                              http://www.wilderness.org
The goal is to establish the land ethic as a basic element of the American culture and to educate people on the importance of wilderness preservation and land protection.
                    Founded: 1935
William H Meadows, President
Donald Barry, Executive Vice President

488 Wilderness Watch
PO Box 9175   406-542-2048
Missoula, MT 59807   Fax: 406-542-7714
                    E-mail: wild@wildernesswatch.org
                              http://www.wildernesswatch.org
Wilderness Watch is a national nonprofit conservation organization dedicated solely to the preservation and enhancement of America Wilderness and Wild & Scenic Rivers.
                    Founded: 1989
Joe Fontaine, President
Howie Wolke, Vice President

489 Wildlife Action
PO Box 866   843-464-8473
Mullins, SC 29574   800-753-2264
                                    Fax: 843-464-8859
                    E-mail: info@wildlifeaction.com
                              http://www.wildlifeaction.com
Raises public awareness about wildlife habitat, security, protection and management; protects the rivers and wetlands from unnecessary destruction and development and works to reduce poaching, trespassing and other illegal outdooractivities.
                    Founded: 1977
M Gault-Beeson, President
Ian Beeson, Secretary

490 Wildlife Conservation Society
2300 Southern Boulevard   718-220-5100
Bronx, NY 10460   Fax: 718-220-2685
                    E-mail: feedback@wcs.org
                              http://www.wcs.org
WCS is at work in 53 nations across Africa, Latin America and North America, protecting wild landscapes that are home to a variety of species from butterflies to tigers. We uniquely combine the resources of wildlife parks in New Yorkwith field projects around the globe to inspire care for nature, provide leadership in environmental eduation, and help sustain our planet's biological diversity.
                    Founded: 1895
Steven Sanderson, President
Richard Lattios, Vice President

491 Wildlife Disease Association
PO Box 1897   785-843-1235
Lawrence, KS 66044-8897   800-627-0629
                                    Fax: 785-843-1274
                    E-mail: wda@allenpress.com
                              http://www.wildlifedisease.org
Our mission is to acquire, disseminate and apply knowledge of the health and diseases of wild animals in relation to their biology, conservation and interactions with human and domestic animals.
                    Founded: 1951
Torsten Morner, President
Ed Addison, Executive Manager

492 Wildlife Forever
2700 Freeway Boulevard   952-833-1522
Minneapolis, MN 55430   Fax: 952-833-0804
                    E-mail: info@wildforever.com
                              http://www.wildforever.com
Wildlife Forever conserves America's wildlife heritage through preservation of habitat, conservation education and management of fish and wildlife.
                    Founded: 1987
Douglas Grann, Executive Director

493 Wildlife Habitat Enhancement Council
8737 Colesville Road   301-588-8994
Suite 800   Fax: 301-588-4629
Silver Spring, MD 20910   E-mail: Whc@wildlifehc.org
                              http://www.wildlifehc.org
The Wildlife Habitat Council is a nonprofit groups of corporations, conservations, and individuals dedicated to protecting and enhensing wildlife habitat.
                    Founded: 1998
Bill Howard, President
Emer O'Broin, Vice President

494 Wildlife Information Center
PO Box 198   610-760-8889
Slatington, PA 18080   Fax: 610-760-8889
                    E-mail: wiclgap@ptd.net
                              http://www.wildlifeinfo.org
                    Founded: 1986
Michal Kubik, President
Dan Kunkle, Executive Director

495 Wildlife Society
5410 Grosvenor Lane   301-897-9700
Suite 200   Fax: 301-530-2471
Bethesda, MD 20814   E-mail: tom@wildlife.org
                              http://www.wildlife.org

A nonprofit scientific and educational organization that serves professionals such as government agencies, academia, industry, and non-government organizations in all areas related to the conservation of wildlife and natural resourcesmanagement.
Founded: 1937
Thomas Franklin, Executive Director
Jane Pelkey, Financial Coordinator

496 Wilson Ornithological Society
816 Park Avenue
PO Box 8420212
Richmond, VA 23284-2012
804-828-1562
Fax: 804-828-0503
E-mail: cblem@cabell.vcv.edu
Charles Blem, VP

497 Windstar Foundation
2317 Snowmass Creek Road
Snowmass, CO 81654-9198
303-927-4777
Fax: 970-927-4779
E-mail: webhelp@wstar.org
http://www.wstar.org
Windstar is a nonprofit environmental education organization which promotes a holistic approach to addressing environmental concerns. Founded in 1976 for singer/songwriter and environmentalist John Denver along with Aikido Master TomCrum.
Founded: 1976
Ron Deutschendorf, President
Pam Peterson, Secretary

498 Wolf Education and Research Center
518 Joseph Ave.
PO Box 217
Winchester, ID 83555
208-924-6960
Fax: 208-924-6959
E-mail: werc@camasnet.com
http://www.wolfcenter.org
Dedicated to providing public education and scientific research concerning the gray wolf and its habitats in the Northern Rocky Mountains. It is our goal to be an exclusive organization that offers factual and balanced information. Weseek to enhance public awareness of threatened species in the region and to develop ways to coexist with these species.
Roy Farrar, President
Douglas M. Christensen, Executive Director

499 Wolf Fund
PO Box 471
Moose, WY 83012
307-733-0740
Fax: 307-733-0962
Renee Askins, Executive Officer

500 Wolf Haven International
3111 Offut Lake Road
Tenino, WA 98589
360-264-4695
800-448-9653
Fax: 360-264-4639
E-mail: wolfhaven@olywa.net
http://www.wolfhaven.org
Wolf Haven International's objectives include protection of the remaining wild wolves and their habitat, promotion of wolf re-establishment in historic ranges, provision of a sanctuary for captive wolves, and public education on thevalue of all wildlife.
Founded: 1982
Carole Russo, Executive Director
Kate Joki, Director of Development

501 Woodlands Mountain Institute
1707 L Street NW
Suite 1030
Washington, DC 26807
202-452-1636
Fax: 202-452-1635
E-mail: summit@mountain.org
http://www.mountain.org
Woodlands Mountain Institute is a nonprofit organization based in Franklin, West Virginia and with project offices in Kathmandu and Khandbari, Nepal, announces the search for qualified community development professional to work in theMakalu-Barum Project of Woodlands Mountain Institute's Himalaya Program.The Makalu-Barun Project works with local people, user groups and national park staff in eastern Nepal to establish an innovative model of community development and environmentalprotection.
Founded: 1972
Bob Davis, CEO
Elfy Walker, Managing Director

502 World Bird Sanctuary
PO Box 270270
St. Louis, MO 63127
636-938-6193
Fax: 636-938-9464
E-mail: info@worldbirdsanctuary.org
http://www.worldbirdsanctuary.org
The World Bird Sanctuary's mission is to preserve the earth's biological diversity and to secure the future of threatened bird species in their natural environment. We work to fulfill that mission through education, propagation, fieldstudies and rehabilitation.
Dennis Breite, Treasurer

503 World Forestry Center
4033 SW Canyon Road
Portland, OR 97221
503-228-1367
Fax: 503-228-4608
E-mail: mail@worlforestry.org
http://www.worldforestry.org
We educate and inform people about the world's forests and trees and their importance to all life, in order to promote a balanced and sustainable future.
Founded: 1966
Gary Hartshorn, President
Mark Reed, Executive Director

504 World Nature Association
PO Box 673
Silver Spring, MD 20901
301-593-2522
Fax: 301-593-2522
Donald H Messersmith, President

505 World Wildlife Fund
1250 24th Street NW
Washington, DC 20037
202-293-4800
800-225-5993
Fax: 202-293-9211
http://www.worldwildlife.org
Dedicated to protecting the world's wildlife and wildlands. The largest privately supported international conservation organization in the world. WWF has invested in over 13,100 projects in 157 countries.
Founded: 1960
Karter Roberts, President
Edward P Bass, Vice Chairman

506 Worldwide Network
1331 H Street NW
Suite 903
Washington, DC 20005
202-347-1514
Fax: 202-347-1524
Waafas Ofosu-Amaah, Manager/Director

National: Recycling & Pollution Prevention

507 Abandoned Mined Lands Reclamation Council
524 S 2nd Street
Springfield, IL 62701-1787
217-782-0588
Fax: 217-524-4819
Founded: 1974
Timothy J Hickmann, Executive Director

508 Acoustical Society of America
Suite 1NO1
2 Huntington Quadrangle
Melville, NY 11747-4502
516-576-2360
Fax: 516-576-2377
E-mail: asa@aip.org
http://http://asa.aip.org/
Focus includes noise pollution prevention.
Founded: 1929
Charles Schmid, Executive Director
Elaine Moran, Office Manager

509 Air and Waste Management Association
One Gateway Center 3rd Floor
420 Fort Duquesne Blvd.
Pittsburgh, PA 15222-1435
412-232-3444
800-270-3444
Fax: 412-232-3450
E-mail: info@awma.org
http://www.awma.org
Promotes pollution prevention.
Founded: 1907
Adrianne Carolla, Executive Director
Edith Ardiente, President

510 Alliance for Acid Rain Control and Energy Policy
444 N Capitol Street
Suite 602
Washington, DC 20001
202-624-5475
Fax: 202-508-3829

511 Aluminum Recycling Association
1000 16th Street NW
Suite 400
Washington, DC 20036
202-785-0951
Fax: 202-785-0210
Richard M Cooperman, Executive Director

512 Americans for the Environment
1400 16th Street NW
Box 24
Washington, DC 20036-2266
202-797-6665
Fax: 202-797-6563
E-mail: afedc@AforE.org
http://www.AforE.org
A community of environmental activists.
Founded: 1982
Roy Morgan, President

513 Aquatic Nuisance Species Task Force
4401 N Fairfax Drive
Suite 840
Arlington, VA 22203-1622
703-358-2308
Fax: 703-358-2044
E-mail: sharon_gross@fws.gov
http://www.anstaskforce.gov/25-7-2000pr.htm
Mamie Parker, Co-Chairman
Timothy Keeny, Co-Chairman

514 Association of Battery Recyclers
PO Box 290286
Tampa, FL 33687
813-626-6151
Fax: 813-622-8388
E-mail: joycemorales@aol.com
http://www.batteryrecyclers.com
To keep members abreast of environmental, health, and safety requirements that affect our industry.
Founded: 1984
Joyce Moralis, Sectrary Treasurer
Earl Cornette, Chairman

515 Association of Energy Engineers
4025 Pleasantdale Rd
Suite 420
Atlanta, GA 30340-4264
770-447-5083
Fax: 770-446-3969
E-mail: info@aeecenter.org
http://www.aeecenter.org
Association for those involved in air and water pollution contrasts, waste-to-energy services information, asbestos abatement and monitoring instruments and equipment.
Stephen A Roosa, President
Laurie Wiegand-Jackson, Secretary

516 Association of Responsible Recyclers
12429 Cedar Road
Suite 26
Cleveland, OH 44106
216-791-7316
Fax: 216-791-6047
http://www.liquidrecyclers.org
The mission of NORA is to promote the proper recycling of used oil, wastewater, oil filters, used antifreeze, and related petroleum streams in an environmentally responsible manner. The purpose of NORA is to represent agencies, andcourts, as well as to expand the success of its members by providing information, education, and networking services.
Martha Peckinpaugh, Executive Director, Admin.
Sandy Hunt, Executive Director

517 Association of State and Territorial Solid Waste Management Officials
444 N Capitol Street NW
Suite 315
Washington, DC 20001
202-624-5828
Fax: 202-624-7875
E-mail: swmtrina@sso.org
http://www.astswmo.org
To enhance and promote effective state and territorial waste management programs, and affect national waste management policies.
Founded: 1974
Thomas J Kennedy, Executive Director

518 C&D Debris Recycling
29 N Wacker Drive
Chicago, IL 60606-3203
312-726-2802
Fax: 312-726-2574

Strives to inform the recycling industry about innovations in the recycling of construction and demolition debris. Publishes a quarterly magazine that contains features on emerging equipment and processes for construction anddemolition waste recycling, as well as company profiles.
Robert Dimond, VP

519 Center for Hazardous Materials Research
320 William Pitt Way
Pittsburgh, PA 15238
412-826-5320
800-334-CHMR
Fax: 412-826-5552
Provides technical, business and public policy solutions to complex environmental issues.
Edgar Berkey, President

520 Coalition Against Pipeline Pollution
2335 Gatewood Street
Los Angeles, CA 90031
213-222-5951
Fax: 213-664-6115
Richard Adams, Executive Officer

521 Container Recycling Institute
1601 N kent Street
Suite 803
Arlington, VA 22209
703-276-9800
Fax: 703-276-9587
E-mail: cri@container-recycling.org
http://www.container-recycling.org
Founded in 1991, CRI is a nonprofit organization that studies and promotes policies and practices that shift the social and environmental costs associated with manufacturing, recycling, and disposal of container and packaging wastefrom government and taxpayers to producers and consumers. CRI plays a vital role in educating policy makers, government officials and the general public regarding the social and environmental impacts of the production and disposal of no-deposit,no-return.
Founded: 1991
Pat Franklin, Executive Director
Jenny Gitlitz, Research Director

522 Ducks Unlimited
One Waterfowl Way
Memphis, TN 38120
901-758-3825
800-453-8257
Fax: 847-438-9236
http://www.ducks.org/
The mission is to fulfill the annual life cycle needs of North American waterfowl by protecting, enhancing, restoring and managing important wetlands and associated uplands.
James C Kennedy, President
Don A Young, Chief Operating Officer

523 Eagle Nature Foundation
300 E Hickory
Apple River, IL 61001
815-594-2306
Fax: 815-594-2305
E-mail: eaglenature.tni@juno.com
http://www.eaglenature.com
A nonprofit organization dedicated to the preservation of the bald eagle, our national symbol, and other endangered species from extinction and to increase public awareness of unique endangered plants and animals. We monitor bald eagleand other endangered species populations and strive to preserve habitiat essential to their survival. We develop materials for schools to inform students about the needs of the bald eagle and how we can help preserve and protect their naturalenvironment.
Founded: 1995
Terrence N Ingram, President
Eugene Small, Vice President

524 Environmental Industry Association
Environmental Industry Associates
4301 Connecticut Avenue NW
Suite 300
Washington, DC 20008-2304
202-244-4700
800-424-2869
Fax: 202-966-4818
E-mail: membership@envasns.org
http://www.envasns.org
The Environmental Industry Association represents about 2,000 compaines that manage soild, hazardous, and medical wastes; manufacture and distribute waste equipment; and offer related pollution-prevention services.
Founded: 1981
Bruce Parker, President/CEO

525 Environmental Technology Council
374 15th Street NW
Suite 720
Washington, DC 20005

202-783-0870
Fax: 202-737-2038
E-mail: comments@etc.org
http://www.etc.org

This Council is a trade association of commerical environmental firms that recycle, treat and dispose of industrial and hazardous wastes; and firms involved in cleanup of contaminated sites.
Founded: 1982
David R Case, Executive Director
Scott Flsinger, Vice President

526 Get Oil Out
914 Anacapa Street
Santa Barbara, CA 93102

805-965-1519
E-mail: getoilout.goo@verizon.net
http://www.getoilout.org

A Santa Barbara public group dedicated to the protection of the Santa Barbara Channel and coastline from the environmental, economic and esthetic impact of oil development. The organization was formed in response to a 1969 accidentthat blackened the beaches, poisoned the ocean's water and killed many creatures that depend on clean water for survival.
Henry Feniger, President

527 Governmental Refuse Collection and Disposal Association
1100 Wayne Avenue
PO Box 7219
Silver Spring, MD 20910

800-467-9262
800-467-9262
Fax: 301-589-7068
E-mail: info@swana.org
http://www.swana.org

Founded: 1965
John H Skinner, Executive Director
Kathy Lane, Associate Director

528 Hazardous Waste Resource Center Environmental Technology Council
734 15th Street NW
Suite 720
Washington, DC 20005

202-783-0870
Fax: 202-737-2038
E-mail: mail@etc.org
http://www.etc.org

The Environmental Technology Council (ETC) is a trade association of commercial environmental firms the recycle, treat and dispose of industrial and hazardous wastes; and firms involved in cleanup of contaminated sites.

529 Institute of Clean Air Companies
1660 L Street NW
Suite 1100
Washington, DC 20036

202-457-0911
Fax: 202-331-1388
E-mail: jsmith@icac.com
http://www.icac.com

To promote the air pollution control industry and encourage improvement of engineering and technical systems. Members are leading manufacturers of equipment to monitor and control emissions of particulate, VOC, SO2, NOX, and airtoxics.
Founded: 1960
Robert G Hilton, President
Joseph Langone, Secretary

530 Institute of Scrap Recycling Industries
1325 G Street NW
Suite 1000
Washington, DC 20005

202-466-4050
Fax: 202-775-9109
http://www.isri.org

Trade association of the scrap processing and recycling industry.
Dr. Herschel Cutler, Executive Director

531 Kids Against Pollution
311 Main Street
3rd Floor
Utica, NY 13501

315-266-0185
Fax: 315-266-0186
E-mail: christine@kidsagainstpollution.org
http://www.kidsagainstpollution.org

Nonprofit organization of active youth dedicated to solving and preventing pollution problems through educational projects and events in order to protect children's health and the planet.
Founded: 1987
Christine Shahin, Director

532 Manufacturers of Emission Controls Association
1660 L Street NW
Suite 1100
Washington, DC 20036

202-296-4797
Fax: 202-331-1388
E-mail: info@meca.org
http://www.meca.org

The Manufacturers of Emission Controls Association offers current and relevant information on air pollution control technology.
Founded: 1976
Bruce I Bertelsen, Executive Director

533 Municipal Waste Management Association
The U.S. Conference of Mayors
1620 Eye Street NW
Washington, DC 20006

202-293-7330
Fax: 202-293-2352
E-mail: info@usmayors.org
http://www.usmayors.org/uscm/mwma

MWMA's purpose is to assist local communities in the development of comprehensive waste management systems that will maximize the reduction and recycling of materials and energy recovery, while meeting local government's principalresponsibility of protecting human health and environment through well regulated and operated waste disposal facilities.
Founded: 1982
Judy Sheahan, Assistant Executive
Ted Fischer, Director of Membership Servi

534 National Association for Plastic Container Recovery (NAPCOR)
PO Box 1327
Sonoma, CA 95476

707-996-4207
800-762-7267
Fax: 707-935-1998
E-mail: information@napcor.com
http://www.napcor.com

The National Association for Plastic Container Recovery, a nonprofit trade association formed in 1987, helps communities establish recycling programs and conducts promotional and educational activities to promote PET plastic containerrecycling. The 17 members of NAPCOR are manufacturers of polyester resins and bottles.
Founded: 1987
Dennis Sabourin, Executive Director
Mike Schedler, Technical Director

535 National Association of Chemical Recyclers
1900 Main Street NW
Suite 750
Washington, DC 20036-3508

202-296-1725
Fax: 202-296-2530
E-mail: info@nacr-r2.org

The National Association of Chemical Recyclers is comprised of companies that recycle solvents and other chemicals for reuse by industry. Its members include both large conglomerates and smaller companies. The association's responsiblerecycling program ensures that all recyclers adhere to the same standards and regulations members in the assocation pledge to meet the ten principles of responsible recycling.
Brenda Pulley, Executive Director
Christopher Goebel, Director

536 National Council of the Paper Industry for Air and Stream Improvements
PO BOX 13318
Durham, NC 27709-3318

919-558-1999
Fax: 919-558-1998
E-mail: ryeske@ncasi.org
http://www.ncasi.org

The Council is a technical organization devoted to finding solutions to environmental protection problems in the manufacture of pulp, paper, and wood products in industrial forestry. It was started in 1943 and now has about 100 membercompanies.
Founded: 1943
Ronald Yeske, President

537 National Office Paper Recycling Project
United States Conference of Mayors
1620 Eye Street NW
Washington, DC 20006

202-293-7330
Fax: 202-293-2352
E-mail: jwelfley@usmayors.org
http://www.usmayors.org

To maximize the recycling of office paper and minimize its disposal. This goal is to be achieved by implementing the now completed National Office Paper Recycling Strategy.

Founded: 1990
J Thomas Cochran, Executive Director

538 National Recycling Coalition
1101 30th Street NW   202-625-6406
Suite 305   Fax: 202-625-6409
Washington, DC 20007
David Loveland, Executive Director

539 Noise Pollution Clearinghouse
PO Box 1137   888-200-8332
Montpelier, VT 05601-1137   888-200-8332
                    E-mail: freenpc@nonoise.org
                        http://www.nonoise.org
The Noise Pollution Clearinghouse is a nonprofit organization
with extensive online noise related resources. The mission is to
create more civil cities and more natural rural and wilderness areas by reducing noise pollution and itssources.

540 One Person's Impact
PO Box 751   508-366-0146
Westborough, MA 01581   Fax: 508-336-0146
Maria Valenti, President

541 Paper Stock Institute
Institute of Scrap Recycling Industries
1325 G Street NW   202-737-1770
Suite 1000   Fax: 202-626-0900
Washington, DC 20005-3104   E-mail: ednabland@isri.org
                                http://www.isri.org
Founded: 1913
Joel Denbo, President
George Adams, Executive Director

542 Plastics Recycling Foundation
135 E State Street   215-444-0659
Kenette Square, PA 19348   Fax: 215-444-0923
        http://www.plastix.com/assn/rs000175.html
Wayne E Pearson, Executive Director

543 Public Citizen
1600 20th Street NW   202-588-1000
Washington, DC 20009   Fax: 202-588-7796
                    E-mail: pcmail@citizen.org
                        http://www.citizen.org
Founded by Ralph Nader in 1971, Public Citizen is the consumer's eyes in Washington. With the support of more than
15,000 people like you, we fight for safer drugs and medical devices, cleaner and safer energy sources, a cleanerenvironment,
fair trade and a more open and democratic government.
Founded: 1971
Joan Claybrook, President
Joseph Zllo, Office Manager

544 Secondary Materials and Recycled Textiles Association
7910 Woodmont Avenue   301-656-1077
Suite 1130   Fax: 301-656-1079
Bethesda, MD 20814   E-mail: smartasn@erols.com
                                http://www.smartasn.org
Since 1932, SMART has represented the interests of companies
dealing with pre-consumer and post-consumer recycable textile
materials. This material includes fibers, remnants, recycled
clothing and shoes, and other related materials.SMART members also manufacture and distribute industrial and commercial
wipers.
Bernard Brill, VP

545 Solid Waste Association of North America
1100 Wayne Ave   301-585-2898
Silver Spring, MD 20910   800-467-9262
                                     Fax: 301-589-7068
                        E-mail: info@swana.org
                            http://www.swana.org
SWANA is dedicated to education and training of its members
by advancing the practice of environmentally and economically
sound management of solid waste in North America.
Founded: 1965
Dr John Skinner, Executive Director
Lori Scozzafava, Deputy Executive Director

**546 State and Territorial Air Pollution Program Adminis-
trators**
444 N Capitol Street NW   202-624-7864
Suite 307   Fax: 202-624-7863
Washington, DC 20001   E-mail: 4cleanair@4cleanair.com
                            http://www.4cleanair.org
To encourage the exchange of information among air pollution
control officials, to enhance communication and cooperation
among federal, state and local regulatory agencies, and to pro-
mote good management of our air resources.
Founded: 1975
S William Becker, Executive Director
Nancy Seidman, President

547 Steel Recycling Institute
680 Andersen Drive   412-922-2772
Pittsburgh, PA 15220-2700   800-937-1226
                                     Fax: 412-922-3213
                            http://www.recycle-steel.org
The Steel Recycling Institute, a unit of the American Iron and
Steel Institute, is an industry association that promotes and sus-
tains the recycling of all steel products. The SRI educates the
solid waste industry, government, bussinesand ultimately the
consumer about the benefit of steel's infinite recycling cycle.
Founded: 1988
Bill Heenan, President
Gregory L Crawford, VP Operations

National: Sustainable Development

548 Alliance for Sustainability
1521 University Avenue SE   612-331-1099
Minneapolis, MN 55414   Fax: 612-379-1527
                        E-mail: iasa@mtn.org
                            http://www.mtn.org/iasa
The Mission of the Alliance is to bring about personal, organizational and planetary sustainability through support of projects
that are ecologically sound, economically viable, socially just
and humane. The Alliance for Sustainabilityis a Minne-
sota-based, tax-deductible nonprofit supporting model
sustanability projects on the local, national and international
levels.
Founded: 1983
Terry Gips, President

549 American Crop Protection Association
1156 15th Street NW   202-296-1585
Suite 400   Fax: 202-463-0474
Washington, DC 20005   http://www.acpa.org
ACPA promotes the environmentally sound use of crop protec-
tion products for the economical production of safe, high qual-
ity, abundant food, fiber and other crops.
Founded: 1933
Allan Noe, Director

550 American Forest Foundation
1111 19th Street NW   202-463-2462
Suite 780   888-889-4466
Washington, DC 20036   Fax: 202-463-2461
                    E-mail: info@forestfoundation.org
                        http://www.affoundation.org
Offers programs, professionals, volunteers and ideas. The vig-
orous growth of the American Forest Foundation is shaped by its
clear and certain mission: to encourage long term stewardship
and sustainable use of our natural resources.
Founded: 1982
Larry Wiseman, President
Caroline Alston, Director

551 American Forests
734 15th Street NW, Suite 800   202-955-4500
Washington, DC 20005   Fax: 202-955-4588
                        E-mail: info@amfor.org
                        http://www.americanforests.org
Oldest national citizens conservation organization in the US.
The organization has three program areas to address todays en-
vironmental challenges: the Global Releaf Center; the Urban
Forest Center and the Forest Policy Center.
Founded: 1875
Gerald Gray, VP Forest Policy Center
Deborah Gangloff, Executive Director

552 American Planning Association
122 S Michigan Avenue   312-431-9100
Suite 1600   Fax: 312-431-9985
Chicago, IL 60603   http://www.planning.org

The American Planning Association is a nonprofit public interest and research organization representing 30,000 practicing officials, and citizens involved with urban and rural planning issues. Sixty-five percent of APA's members areemployed by state and local government agencies. These members are involved, on a day-to-day basis, in formulating planning policies and preparing land use regulations.
Frank So, Executive Director

553 American Society for Environmental History
Forest History Society
701 Vickers Avenue   919-682-9319
Durham, NC 27701   Fax: 919-682-2349
                                        E-mail: recluce2@duke.edu
                                        http://www.foresthistory.org

Concerned with human ecology. Interests include history and the humanities.
Founded: 1946
Steven Anderson, President
Tom Marshall, Development Officer

554 American Society of Agronomy
677 S Segoe Road   608-273-8080
Madison, WI 53711   Fax: 608-273-2021
                                        E-mail: headquarters@agronomy.org
                                        http://www.agronomy.org

ASA is dedicated to development of agricultural interests in harmony with environmental and human values. The Society supports scientific, educational and professional activities that enhance communication and technology transfer amongagronomists and those in related disciplines on topics of local, regional, national and international significance.
David A Sleper, President
Ellen Bergfeld, Executive Vice President

555 Ancient Forest International
PO Box 1850   707-923-3015
Redway, CA 95560   Fax: 707-923-3015
                                        E-mail: afi@ancientforests.org
                                        http://www.ancientforests.org

Since 1989, Ancient Forest International has been instrumental in the protection of primary forests around the world. With the help of its international ancient forest network, AFL develops opportunities for wildlands philanthropistsand communities to work together to acquire and protect strategic and invaluable forestlands. AFL has helped coordinate the purchase of nearly a million acres of ecologically critical forested land, primarily along the Pacific coast of North andSouth America.
Founded: 1989
Rick Klein, President

556 Atlantic Center for the Environment Quebec-Labrador Foundation
QLF-Atlantic Center for the Environment
55 S Main Street   978-356-0038
Ipswich, MA 01938   Fax: 978-356-7322
                                        E-mail: atlantic@qlf.org
                                        http://www.qlf.org

QLF is a private, nonprofit organization with a mission to support the rural communities and environment of New England and eastern Canada (the Atlantic Region) and to provide models for stewardship that can be applied worldwide. QLFuses international and regional exchanges, internship, workshops, networks and technical assistance to promote sustainable communities, natural, cultural heritage, biodiversity, wildlife, and health.
Founded: 1963
Lawrence B Morris, President

557 CONCERN
1794 Columbia Road NW   202-328-8160
Washington, DC 20009   Fax: 202-387-3378
                                        E-mail: concern@concern.org
                                        http://www.sustainable.org

CONCERN, founded in 1970 is a national nonprofit environmental education organization with a focus on sustainable communities. CONCERN disseminates examples of successful initiatives, offers numerous resources and guidelines foraction, serves as a clearinghouse for information and collaborates with others to carry out its programs. Its Sustainable Communities Program CONCERN seeks to increase public understanding of and participation in initatives that are environmentallyand socially sound.
Founded: 1970
Susan Boyd, Executive Director

558 Conservation Fund
1800 North Kent Street   703-525-6300
Suite 1120   Fax: 703-525-4610
Arlington, VA 22209   http://www.conservationfund.org

Forges partnerships to protect America's legacy of land and water resources. Through land acquisition, community initiatives and leadership training, the Fund and its partners demonstrate sustainable conservation solutions emphasizingthe integration of economic and environmental goals.
Founded: 1985
Charles R Jordan, Chairman
Lawrence A Selzer, President

559 Earth Island Institute
300 Broadway   415-788-3666
Suite 28   Fax: 415-788-7324
San Francisco, CA 94133   http://www.earthisland.org

Incubates and supports over 35 projects working on environmental issues worldwide. Publishes quarterly Earth Island Journal.
Founded: 1982
John A Knox, Executive Director Operation
Bob Wilkinson, President

560 Environmental Concern
POW-The Planning of Wetlands
201 Boundary Lane   410-745-9620
PO Box P   Fax: 410-745-3517
St Michaels, MD 21663   E-mail: order@wetland.org
                                        http://www.wetland.org

Since its founding in 1972, EC has been specializing in consulting, planning design, education services, construction services and research related to all aspects of wetlands. As wetlands and contiguous upland forests and meadows areinteracting ecosystems EC specializes in consulting, planning, design, and project supervision services for such upland ecosystem constructions and restorations for the purpose of wetland buffers, reforestation, wildlife habitat and critical areas ofpreservation.
Founded: 1972
Suzanne Slear, President
Gene Slear, Vice President

561 Environmental Policy Center Global Cities Project
2962 Fillmore Street   415-775-0791
San Francisco, CA 94123   Fax: 415-775-4159
                                        E-mail: epc@globalcities.org
                                        http://www.globalcities.org

Provides assistance and information on sustainable development and environmental conservation to communities within North America.

562 Environmental and Energy Study Institute
122 C Street NW   202-628-1400
Suite 630   888-788-3378
Washington, DC 20001-2109   Fax: 202-628-1825
                                        E-mail: eesi@eesi.org
                                        http://www.eesi.org

They are a nonprofit organization dedicated to promoting environmentally sustainable societies. EESI believes meeting this goal requires transitions to social and economic patterns that sustain people, the environment and the naturalresources upon which present and future generations depend.
Founded: 1984
Richard Ottinger, Chairman
John Sheehan, Vice Chairman

563 Friends of the Earth
Global Building
1717 Massachusetts Avenue, NW, 600   202-783-7400
Suite 300   877-843-8687
Washington, DC 20036-2002   Fax: 202-783-0444

41

E-mail: foe@foe.org
http://www.foe.org

National nonprofit advocacy organization dedicated to protecting the planet from environmental degradation; preserving biological, cultural and ethnic diversity, and empowering citizens to have an influential voice in decisionsaffecting the quality of their environment and their lives.
Founded: 1969
Brent Blackwelder, President
Norman Dean, Executive Director

564   Global Action Network

50 California Street   415-477-2303
Suite 3325   Fax: 415-477-2334
San Francisco, CA 94111http://www.globalactionnetwork.org

A new model of relationship development and community building for the reproductive health field that employs an Internet-based model of networking proven successful in other sectors. Our goal is to provide resources and opportunitiesfor Network members to share knowledge, collaborate with other individuals working in the field, engage in mentoring relationships, and utilize online contacts and information to build networks with communities around the world.
Founded: 1998
Andrea Johnston, Co-Director
Jessica Klein, Program Associate

565   Global Committee of Parliamentarians on Population and Development

345 E 45th Street   212-953-7947
12th Floor   Fax: 212-557-2061
New York, NY 10017

566   Global Tomorrow Coalition

1325 G Street NW   202-628-4016
Suite 1010   Fax: 202-628-4018
Washington, DC 20005

Global Tomorrow Coalition - a national leadership alliance on sustainable development, with membership in business and industry, conservation and environment, education and community planning, and social issues and development.
Donald R Lesh, President

567   Institute for Agriculture and Trade Policy

2105 First Avenue South   612-870-0453
Minneapolis, MN 55404   Fax: 612-870-4846
                              E-mail: iatp@iatp.org
                              http://www.iatp.org

The mission of the Institute for Agriculture and Trade Policy is to foster socially, economically and environmentally sustainable rural communities and regions.
Founded: 1980
Mark Ritchie, President
Kristin Dawkins, Vice President

568   Institute for Sustainable Communities

535 Stone Cutters Way   802-229-2900
Montpelier, VT 05602   Fax: 802-229-2919
                              E-mail: isc@iscvt.org
                              http://www.iscvt.org

Promotes sustainable environmental practices in the US, Central and Eastern Europe and Russia.
Founded: 1991
George Hamilton, President
D Jill Arace, Vice President

569   International Association of Fish and Wildlife Agencies

444 N Capitol Street NW   202-624-7890
Suite 544   Fax: 202-624-7891
Washington, DC 20001   E-mail: iafwa@sso.org
                              http://www.iafwa.org

Mission is to promote the sustainable use of natural resources; encourage cooperation and coordination of fish and wildlife management at all levels of government; develop coalitions among conservation organizations or promote fish andwildlife interests; encourage the professional management of fish and wildlife; foster public understanding of the need for conservation.
Founded: 1902
Rachel Brittin, Public Affairs Director

570   International Mountain Society

PO Box 1978
Davis, CA 95617

An association registered in Berne, Switzerland, for the purpose of advancing knowledge and disseminating information about mountain research and development throughout the world. Aims to promote sustainable mountain developmentthrough improved communication among institutions and individuals, with a particular focus on mountain eco-regions in the developing world. Collaborates with like-minded institutions, and is a joint publisher of the journal Mountain Research andDevelopment.
Dr. Jack D Ives, President

571   International Society of Arboriculture

PO Box 3129   217-355-9411
Champaign, IL 61826-3129   888-472-8733
                              Fax: 217-355-9516
                              E-mail: isa@isa-arbor.com
                              http://www.isa-arbor.com

Through research, technology, and education promote the professional practice of arboriculture and foster a greater public awareness of the benefits of trees.
Founded: 1924
Jim Skiera, Executive Director

572   International Society of Tropical Foresters

5400 Grosvenor Lane   301-897-8720
Bethesda, MD 20814   Fax: 301-897-3690
                             E-mail: goergenm@safnet.org
                             http://www.safnet.org

The International Society of Tropical Foresters, Inc. (ISTF) is a nonprofit organization committed to the protection, wise management and rational use of the world's tropical forests. Established in 1950, ISTF has about 1500 members inmore than 100 countries. Financial support comes from membership dues, donations and grants. ISTF sponsors meetings, promotes chapters in other countries, maintains a web site and has chapters at universities.
Founded: 1900   24 pages
Michael Goergen, Chief Executive
Larry Burner, CFO

573   Interstate Mining Compact Commission

445A Carlisle Drive   703-709-8654
Herndon, VA 20170   Fax: 703-709-8655
                             http://www.imcc.isa.us

A multi-state governmental organization which represents its 20 member states on issues of mining and environmental regulations. It works closely with several federal agencies i.e. Office of Surface Mining, US EPA, and the Bureau ofland management.
Founded: 1970
Gregory E Conrad, Executive Director
Beth A Botsis, Diretor of Programs

574   Island Resources Foundation

1718 P Street NW   202-265-9712
Suite T-4   Fax: 202-232-0748
Washington, DC 20036   E-mail: irf@irf.org
                             http://www.irf.org

Island Resources Foundation is a private, nonprofit research and education organization based at Red Hook in St. Thomas, US Virgin Islands, dedicated to solving the environmental problems of developing in small tropical island.
Founded: 1972
Bruce Potter, President
Charles Consolvo, Secretary

575   Kids for Saving Earth Worldwide

PO Box 421118   763-559-1234
Minneapolis, MN 55442   Fax: 651-674-5005
                             E-mail: keseww@aol.com
                             http://www.kidsforsavingearth.org/

To help protect the Earth through kids and adults. To educate and inspire them to participate in Earth-saving actions.
Tessa Hill, President

576   Kids for Saving the Earth Worldwide

PO Box 421118   763-559-1234
Minneapolis, MN 55442   Fax: 651-674-5005
                             E-mail: keseww@aol.com
                             http://www.kidsforsavingearth.org/

To help protect the Earth through kids and adults. To educate and inspire them to participate in Earth-saving actions.
Tessa Hill, President/Director

577 Land Institute
2440 E Water Well Road   785-823-5376
Salina, KS 67401   Fax: 785-823-8728
                                      E-mail: theland@landinstitute.org
                                      http://www.landinstitute.org
This nonprofit research and education organization is engaged in a 25-year research program that marries ecology and agronomy to produce a Natural Systems Agriculture. The Land Institute is developing perennial grain plants to be grownin fields of mixed species patterned after native prairies. These domestic prairies will be plowed rarely, need few manufactured inputs because they will provide their own fertility and manage pests and diseases. Year-round roots will hold soils fromerosion.
Founded: 2000
Ken Warren, Executive Director
Strachan Donnelley, Speaker

578 Land Trust Alliance
1331 H Street NW   202-638-4725
Suite 400   Fax: 202-638-4730
Washington, DC 20005   E-mail: lta@lta.org
                                      http://www.lta.org
National leader of the private land conservation movement, promoting voluntary land conservation across the country and providing resources, leadership and training to the nation's 1200 plus nonprofit, grassroots land trusts, helpingthem to protect important open spaces. Provides an array of pograms, including direct grants to land trusts, training programs, answers to more than 3,000 inquiries for technical assistance each year, and one-on-one mentoring.
Founded: 1983   40 pages
ISBN: 1-046145-8 -
Will Shafroth, Chairman
Peter Hausmann, Vice Chairman

579 Manomet Center for Conservation Sciences
81 Stage Point Road   508-224-6521
PO Box 1770   Fax: 508-224-9220
Manomet, MA 02345   http://www.manomet.org
Manomet mission is to conserve natural resources for the benefit of wildlife and human populations. Through research and collaboration. Builds science-based, cooperative solutions to environmental problems.
Founded: 1970
Jeptha H Wade, Chairperson
D Reid Weedon, Jr., Vice Chairman

580 Manufacturers of Emission Controls Association
1660 L Street NW   202-296-4797
Suite 1100   Fax: 202-331-1388
Washington, DC 20036-5603   E-mail: info@meca.org
                                      http://www.meca.org
The Manufacturers of Emission Controls Association offers current and relevant information on air pollution control technology.
Founded: 1976
Bruce I Bertelsen, Executive Director

581 NE Sustainable Energy Association
50 Miles Street   413-774-6051
Greenfield, MA 01301   Fax: 413-774-6053
                                      E-mail: nesea@nesea.org
                                      http://www.nesea.org
NESEA is the nation's leading regional membership organization focused on promoting the understanding, development, and adoption of energy conservation and non-polluting renewable energy technologies. We work to bring cleanelectricity, green transportation, and healthy, efficient buildings into everyday use.
Founded: 1974
Nancy Hazard, Executive Director
Janet Nokes, Business Manager

582 National Association of State Departments of Agriculture
1156 15th Street NW   202-296-9680
Suite 1020   Fax: 202-296-9686
Washington, DC 20005   E-mail: nasda@nasda.org
                                      http://www.nasda.org
This organization's mission is to support and promote the American agriculture industry, while protecting consumers and the environment, through the development, implementation, and communication of sound public policy and programs.There are twenty national organizations affiliated with NASDA. They are made up of persons of similar responsibilities with the state departments of agriculture and other agencies of state government.
Founded: 1915
Richard Kirchhoff, Chief Executive
Steve Cox, Chief Operating Officer

583 National Association of State Land Reclamationists
Southern Illinois University   618-536-5521
Carbondale, IL 62901-4623   Fax: 618-453-7346
                                      E-mail: President@notes.siu.edu
                                      http://www.siu.edu
As a nationally recognized authority on the reclamation of mined lands, the National Association of State Land Reclamationists (NASLR) advocates the use of research, innovative technology and professional dicource to foster therestoration of lands and waters affected by mining related activities.
Founded: 1869
James Walker, President

584 National Audubon Society: Everglades Campaign
444 Brickell Avenue   305-371-6399
Suite 850   Fax: 305-371-6398
Miami, FL 33131   E-mail: danderson@audubon.org
                                      http://www.audubonofflorida.org
The mission of the NAS Everglades Conservation Office is to ensure the restoration and conservation of the Greater Everglades Ecosystem in order to achieve an ecologically and economically sustainable South Florida. Our Miami-basedoffice has a five-part program including science, education, advocacy, outreach and grassroots action.
Founded: 1886
David Anderson, Executive Director
John Flicker, President

585 National Environmental Development Association
1440 New York Avenue NW   202-638-1230
Suite 300   Fax: 202-639-8685
Washington, DC 20005
Companies and others concerned with balancing environmental and economic interests to obtain both a clean environment and a strong economy.
Andrew McElwaine, Director

586 National FFA Organization
National FFA Center
6060 FFA Drive   317-802-6060
PO Box 68960   Fax: 317-802-6061
Indianapolis, IN 46268-0960   E-mail: bstagg@ffa.org
                                      http://www.ffa.org
Future Farmers of America is dedicated to making a positive difference in the lives of young people by developing their potential for premier leadership, personal growth and career success through agricultural education. Theorganization's motto is: Learning to Do; Doing to Learn; Earning to Live; Living to Serve.
Founded: 1928
Larry Case, Chief Executive
C Coleman Harris, Executive Secretary

587 National Forestry Association
374 Maple Avenue East   703-255-2700
Suite 310   800-476-8733
Vienna, VA 22180   Fax: 703-281-9200
                                      E-mail: info@woodlandowners.org
                                      http://www.woodlandowners.org
Nation's largest referral program to link up private forest owners with professional foresters.To be supplemented with a new innovative Forest Practices Certification program for landowners. Landowners who complete the review processand follow designated practices will be certified by the National Forestry Association.
Founded: 1981
Keith Argow, President
Bob Playfair, Vise President

588 National Gardening Association

1100 Dorset Street

South Burlington, VT 05403

800-538-7476

800-538-7476

Fax: 802-864-6889

E-mail: barbara@garden.org

http://www.garden.org

The mission of the National Gardening Association is to sustain and renew the fundamental links between people, plants and the earth. NGA achieves its mission through youth and community gardening programs, industry research, freegardening information and memberships.

Founded: 1973

Valerie Kelsey, President

Larry Sommers, VP

589 National Mining Association

101 Constitution Avenue NW

Washington, DC 20001

202-463-2600

Fax: 202-463-2666

E-mail: craulston@nma.org

http://www.nma.org

The mission of the National Mining Association is to create and maintain a broad base of political support in Congress, the administration and the media for the mining industry of the US. In doing so, a secondary goal is to help ournation and the world realize the full promise and potential of the natural resources derived from America's mining industry.

Founded: 1917

Hal Quinn, Acting President

Nori Jones, Finance Head

590 Native Forest Council

PO Box 2190

1455 East Briarcliff Lane

Eugene, OR 97402

541-688-2600

Fax: 541-461-2156

E-mail: webmaster@forestcouncil.org

http://www.forestcouncil.org

To provide visionary leadership and to ensure the integrity of public land ecosystems without compromising people or forests.

Founded: 1987

Timothy Hermach, President

Deborah Ortuno, Administrative Director

591 Native Seeds/SEARCH

526 N 4th Avenue

Tucson, AZ 85705-8450

520-622-5561

866-622-5561

Fax: 520-622-5591

E-mail: info@nativeseeds.org

http://www.nativeseeds.org

Promote the use of ancient crops and their wild relatives by gathering, safeguarding, and distributing their seeds, while sharing benefits with tradtional communities. Work to preserve knowledge about their uses. Through research,training, and community education, works to protect biodiversity and to celebrate cultural diversity.

Founded: 1983

Todd Horst, Chairman

Kevin Dahl, Executive Director

592 Natural Land Institute

320 South 3rd Street

Rockford, IL 61104

815-964-6666

Fax: 815-964-6661

E-mail: nli@aol.com

http://www.naturalland.org

Dedicated to preserving land and natural diversity for future generations. Since 1958, NLI has protected, managed, and restored thousands of acres throughout Illinois and southern Wisconsin. These include prairies, forests, wetlands,and river corridors.

Founded: 1958

Jerry Paulson, Executive Director

Jill Kennay, Assisstant Director

593 Natural Resources Council of America

11100 Wiodlise Centre Drive

Reston, VA 20190-5362

703-438-6000

Fax: 703-438-3570

E-mail: nrca@naturalresourcescouncil.org

http://www.naturalresourcescouncil.org

The Council dedicated to strengthing the conservation movement as a whole. For more than 50 years the Council has been the Crossroads of Conservation, keeping conservationists connected, informed and prepared to face the challenges ofthe future. The Council provides their membership- more than 85 conservation groups and nearly 100 individual supporters- with unique networking opportunities, valuable leadership training and cost-saving services.

Founded: 1946

Andrea Yank, Executive Director

Carlton Gleed, Program Coordinator

594 Negative Population Growth

2861 Duke St

Suite 36

Alexandria, VA 22314

703-370-9510

Fax: 703-370-9514

E-mail: npg@npg.org

http://www.npg.org

Leader in the movement for a sound population policy and advocates a smaller and truly sustainable population through voluntary incentives for smaller families and reduced immigration levels.

Founded: 1972

Donald Mann, President

595 New England Coalition for Sustainable Population

PO Box 194

Sullivan, NH 03445

603-847-9798

E-mail: d9cat@cheshire.net

http://www.cheshire.net/~d9cat/necsp.html

Annie Faulkner, Coordinator

596 Pacific Institute for Studies in Development, Environment and Security

654 13th Street

Oakland, CA 94612

510-251-1600

Fax: 510-251-2203

E-mail: info@pacinist.org

http://www.pacinst.org

Nonprofit policy research group, bringing knowledge to power on issues of environmental, economical development, and international peace and security.

Founded: 1987

Catherine.E. Fox, Associate Director

Anne H Ehrlich, Chairperson

597 Panos Institute

1322 18th Street,NW

Suite 26

Washington, DC 20036

202-429-0730

Fax: 202-223-7947

E-mail: panoswashington@aol.com

http://www.panosinst.org

Founded in 1986, the Panos Institute is an international, nonprofit, nongovernmental organization with offices in Budapest, London, Paris, and Washington DC, working to raise public understanding of sustainable development issues.

Founded: 1986

Melanie Oliviero, Executive Director

598 Pinchot Institute for Conservation

1616 P Street NW

Suite 100

Washington, DC 20036-1434

202-797-6580

Fax: 202-797-6583

E-mail: pinchot@pinchot.org

http://www.pinchot.org

Recognized as a leader in forest conservation thought, policy and action. Its objectives are realized annually through its programs: Community Based Forest Stewardship, Conservation Policy and Organizational Change, ConservationLeadership and Executive Development, Conservation and the Arts, International Forest Policy and Planning and the Milford Experimental Forest.

Founded: 1963

V Alaric Sample, President

Frank Tugwell, Chairman

599 Population Communications International

777 United Nations Plaza

5th Floor

New York, NY 10017-3521

212-687-3366

Fax: 212-661-4188

E-mail: info@population.org

http://www.population.org

PCI's mission is to work creatively with the media and other organizations to motivate individuals and communities to make choices that influence population trends encouraging sustainable development and environmental protection.

Founded: 1985

Victoria A Staebler, Senior Financial Advisor

600 Population Connection
1400 16th Street NW   202-332-2200
Suite 320   1 8-0 7-7 19
Washington, DC 20036   Fax: 202-332-2302
E-mail: info@populationconnection.org
http://www.populationconnection.org/
A national nonprofit organization working to slow population growth and achieve a sustainable balance between the Earth's people and its resources. We seek to protect the environment and ensure a high quality of life for present andfuture generations.
Founded: 1965
John R Lazarus, Chairman
Kathleen L Bonk, Executive Director

601 Population Crisis Committee
1120 19th Street NW   202-659-1833
Suite 550   Fax: 202-293-1795
Washington, DC 20036
J Joseph Speidel, President

602 Population Institute
107 2nd Street NE   202-544-3300
Washington, DC 20002   888-787-0038
      Fax: 202-544-0068
E-mail: web@populationinstitute.org
http://www.populationinstitute.org
The Population Institute is the World's largest independent nonprofit, educational organization dedicated exclusively to achieving a more equitable balance between the worlds population, environment, and resources. Established in 1969,the Institute, with members in 172 countries, is headquartered on Capitol Hill in Washington DC. The Institute uses a variety of resources and programs to bring its concerns about the consequences of rapid poulation growth to the forefront of thenational agenda.
Founded: 1969
Werner Fornos, President
Victor Morgan, Executive Director

603 Population Reference Bureau
1875 Connecticut Avenue NW   202-483-1100
Suite 520   800-877-9881
Washington, DC 20009-5728   Fax: 202-328-3937
E-mail: popref@prb.org
http://www.prb.org
The Population Reference Bureau is the leader in providing timely and objective information on US and international population trends and their implications. PRB informs policymakers, educators, the media, and concerned citizensworking in the public interest around the world through a broad range of activities, including publications, information services, seminars and workshops, and technical support. We work with both public-sector and private-sector partners.
Founded: 1929
William Butz, President
Ellen Carnevale, Director of Communications

604 Population Resource Center
15 Roszel Road   609-452-2822
Princeton, NJ 08540   Fax: 609-452-0010
E-mail: prc@prcnj.org
http://www.prcdc.org
The mission of the Population Resource Center is to promote the use of accurate population data and sound, objective analysis of these data in the making of public policy.
Founded: 1985
Jane S De Lung, President
Linda Rosen, Director of Policy Analysis

605 Population: Environment Balance
2000 P Street Northwest   202-955-5700
Suite 600   800-866-8269
Washington, DC 20036   Fax: 202-955-6161
E-mail: uspop@us.net
http://www.balance.org
Population Environment Balance is a national, nonprofit membership organization dedicated to maintaing the quality of the United States population stabilization.
Founded: 1973
Aaron Beckwith, Vice President

606 Population: Environmental Council
2000 P Street NW 600   202-955-5700
Washington, DC 20036-5915   Fax: 202-955-6161
E-mail: uspop@balance.org
http://www.balance.org
Population-Environment Balance is dedicated to public education regarding the adverse effects of population growth on the environment. Founded in 1973, Population-Environment Balance has 8,800 members. It advocates measures that wouldencourage population stabilization.
Founded: 1973
Aaron Beckwith, Vise President

607 Rainforest Relief
PO Box 150566   718-398-3760
Brooklyn, NY 11215-0566   Fax: 718-398-3760
E-mail: relief@igc.org
http://www.rainforestrelief.org
Rainforest relief works to end the loss of the world's tropical and temperate rainforests by reducing the demand for materials for which rainforests are destroyed. These include rainforest woods such as mohogany, lauan and cedar,agricultural products such as bananas, chocolate, coffee and cut flowers, and mining products such as petroleum, gold, aluminum and copper. Rainforest relief works through research, education and non-violent direct action campaigns.
Founded: 1989
Tim Keating, Executive Director
Jeff Lockwood, West Coast Chapter Director

608 Rural Advancement Foundation International USA
PO Box 640   919-542-1396
Pittsboro, NC 27312   Fax: 919-542-0069
E-mail: info@raiusa.org
http://www.rafiusa.org
RAFI USA is a nonprofit organization promoting community, equity and sustainability for family farmers and rural communities. Our headquarters serves as a model for green building. In addition to daylighting, solar and energyconservation features, the building showcases the use of salvaged materials from the deconstruction of an 1830s farmhouse.
Founded: 1990
Helen Vinton, President
Betty Emarita, Vice President

609 Save America's Forests
4 Library Court SE   202-544-9219
Washington, DC 20003   Fax: 202-544-7462
E-mail: info@saveamericasforests.org
http://www.saveamericasforests.org
A nationwide campaign to end clearcutting and protect and restore our nation's wild and natural forests. A coalition of groups throught America working together to protect each other's local forests and to protect our nation's forestsand forests throughout the world. A network of individual citizens from the country, the cities, and the suburbs who love forests and want to save them.
Founded: 1990
Carl Ross, Executive Director

610 Society for Ecological Restoration
285 W. 18th Street   520-622-5485
Suite 1   Fax: 520-622-5491
Tucson, AZ 85701   E-mail: info@ser.org
http://www.ser.org
To serve the growing field of Ecological Restoration through facilitating dialogue among restorationists; encouraging research, promoting awareness and public support for restoration and restorative management; contributing to publicpolicy discussions; recognizing those who have made outstanding contributions to the field of restoration; and promoting ecological restoration around the globe.
Founded: 1987
M K LeSeeour, Executive Director

611 Southface Energy Institute
241 Pine Street NE   404-872-3549
Atlanta, GA 30308   Fax: 404-872-5009
E-mail: info@southface.org
http://www.southface.org
An environmental nonprofit working to promote sustainable homes, workplaces and communities through education, research, advocacy and technical assistance.
Founded: 1978
Dennis Creech, President

612 Urban Initiatives
530 W 25th Street
New York, NY 10001   212-620-9773
                    E-mail: placeprjct@aol.com
                    http://www.apa.org/pi/urban

Gianni Longo, President

613 World Environment Center
1300 Pennsylvania Avenue NW   212-683-4700
Suite 550   Fax: 202-682-1682
Washington, DC 20004   http://www.wec.org/
The World Environment Center contributes to sustainable development worldwide by strengthening industrial and urban environment, health, and safety practices.
            Founded: 1974
Dr. Bernard Tramier, Executive Director
Elizabeth Lowery, Vice President

614 World Population Society
1333 H Street   202-898-1303
Suite 106   Fax: 202-861-0621
Washington, DC 20005

Frank Oram, Executive Director

615 Worldwatch Institute
1776 Massachusetts Avenue NW   202-452-1999
Suite 800   Fax: 202-296-7365
Washington, DC 20036   http://www.worldwatch.org
The Worldwatch Institute is an independent, nonprofit environmental research organization in Washington DC. Its mission is to foster a sustainable society in which human needs are met in ways that do not threaten the health of the natural environment or future generations. To this end, this Institute conducts interdisciplinary research on emerging global issues, the results of which are published and disseminated to decision-makers and the media.
            Founded: 1974
Christopher Flavin, President
Barbara Fallin, Director Finance/Administrat

National: Travel & Tourism

616 American Association for Leisure and Recreation
1900 Association Drive   703-476-3400
Reston, VA 20191-1598   800-213-7193
                    Fax: 703-476-9527
                    E-mail: mail@aalr.org
                    http://www.aahperd.org/aaalf
The American Association for Leisure and Recreation serves recreation professionals, practitoners, educators, and students who advance the profession and enhance the quality of life of all Americans through creative and meaningfulleisure and recreation experiences.
            Founded: 1885
Connie Fox, President
Janet Seaman, Executive Director

617 American Hiking Society
1422 Fenwick Lane   301-565-6704
Silver Spring, MD 20910   Fax: 301-565-6714
                E-mail: info@americanhiking.org
                http://www.americanhiking.org
American Hiking Society is a recreation based conservation organization working to cultivate a nation of hikers dedicated to establishing, protecting, and maintaining foot trails in America. Our 10,000 individual members and 160 hikingclub members contribute to this national effort.
Bob Papp, Chairman
Gregory A Miller, President

618 American Recreation Coalition
1225 New York Avenue NW   202-682-9530
Suite 450   Fax: 202-682-9529
Washington, DC 20005-6405 E-mail: arc@funoutdoors.com
                http://www.funoutdoors.com
Dedicated to the protection and enhancement of the right to health and happiness through recreation.
            Founded: 1979
Derrick Crandall, President
Catherine Ahern, Vice President

619 American Whitewater
PO Box 1540   828-293-9791
Cullowhee, NC 28723   866-262-8429
                    Fax: 828-227-7422
                E-mail: info@amwhitewater.org
                http://www.americanwhitewater.org/
Pre-eminent conservation and access organization focused on whitewater recreation in the United States.
Mark Singleton, Executive Director
Jason Robertson, Managing Director

620 American Zoo and Aquarium Association
8403 Colesville Road   301-562-0777
Suite 710   Fax: 301-562-0888
Silver Spring, MD 20910 E-mail: generalinquiry@aza.org
                    http://www.aza.org

Sydney Buttler, Executive Director

621 Citizens for a Scenic Florida
4401 Emerson Street   904-396-0037
Suite 10   Fax: 904-398-4647
Jacksonville, FL 32207   E-mail: scenicfl@scenicflorida.org
                    http://www.scenicflorida.org
Scenic America is the only national nonprofit organization dedicated to protecting natural beauty and distinctive community character. We provide technical assistance across the nation and through affiliates on scenic byways,billboards and sign control, context sensitive highway design, wireless telecommunications tower location, transportation enhancements and other scenic conservation issues.
William C Jonson, President
William D Brinton, Secretary

622 Federation of Western Outdoor Clubs
PO Box 129
Selma, OR 97538

The Federation is composed of organizations that engage in hiking, camping, birding and other similar activities that rely on an outdoor environment where natural conditions predominate. Organizations in the West that have suchprograms, and that have an active interest in protecting the natural environment, are invited to affiliate.

623 Green Hotels Association
PO Box 420212   713-789-8889
Houston, TX 77242-0212   Fax: 713-789-9786
                E-mail: green@greenhotels.com
                http://www.greenhotels.com
Helping hotels save water, energy, solid and money world wide.
            Founded: 1993
Patricia Griffin, President

624 Lighthawk
PO Box 653   307-332-3242
Lander, WY 82520   Fax: 307-332-1641
                E-mail: info@lighthawk.org
                http://www.lighthawk.org
Nonprofit organization founded in 1979, addresses critical environmental issues by providing an aerial perspective on areas of concern in the US, Canada and Central America. Using small aircraft, we fly partner organizations, electedofficials, industry and media representatives, activists, and indigenous groups over protected and threatened regions. Our program flights give passengers both intellectual and visceral understanding of what is at stake.
            Founded: 1979
Maureen Smith, Executive Director
J Marlow Schmauder, Associate Director

625 National Association of Recreation Resource Planners
4200 Smith School Road   512-912-7109
Austin, SC 78744-3291   Fax: 512-707-2742
NARRP, is an organization comprised of outdoor recreation professionals and others interested in recreation resource planning. It is a nationwide organization with members in nearly every state representing federal and state agencies,land managers, consultants, and academic institutions. The mission of NARRP is to Advance the Art, the Science and the Profession of Recreation Resource Planning: and enhance the provision of recreation opportunities for all Americans.
Gordon Kimball, President
Robert Sammon, Vice President

626 National Association of State Outdoor Recreation Liason Officers
PO Box 7921
Madison, WI 53707

302-739-4401
Fax: 302-739-3817
E-mail: csalkin@state.de.us

Ney Landrum, Executive Director
Eleanor Cravan, Secretary

627 National Council of State Tourism Directors
TIA
1100 New York Avenue NW
Suite 450
Washington, DC 20005-3934

202-408-8422
Fax: 202-408-1255
E-mail: feedback@tia.org
http://www.tia.org

Founded: 1990
Roger Dow, President
Dexter Koehl, Vice President

628 Outdoors Unlimited
University of California Sanfrancisco
500 Parnassus Avenue
PO Box 0234-A
San Francisco, CA 94143-0234

415-476-2078
Fax: 502-620-7415
E-mail: outdoors@cls.ucsf.edu
http://www.outdoors.ucsf.edu

We offer people a connection to the outdoor experience, providing skills training and outings, equipment rental and an adventure community where people can network to create their own group experiences.
Founded: 1970
Steve Siskin, President

629 Rails-To-Trails Conservancy
1100 17th Street
10th Floor, NW
Washington, DC 20036

202-331-9696
Fax: 202-331-9680
E-mail: railtrails@railtrails.org
http://www.railtrails.org

Founded: 1986
Heath J Meriweather, Chairman
Joe Louis Barrow, Jr, Vice Chairman

630 Ranger Rick's Nature Club: National Wildlife Federation
8925 Leesburg Pike
Vienna, VA 22184-0001

800-822-9919
800-822-9919
http://www.nwf.org/kids/

Encounter lion families, leaping kangaroos, dueling giraffes, diving whales, acrobic eagles, and many more kinds of animals. To learn about animal armor, camouflage, migration, endangered habitats, incredible food chains. And toexplore the world hike across glaciers, track a rare spirit bear, face a raging hurricane, photograph creatures in the Galapagos Islands.
Founded: 1936
Jerome C Ringo, Chairman
Stephen E Petron, Vice Chairman

631 Safari Club International
4800 W Gates Pass Road
Tucson, AZ 85745-9490

520-620-1220
888-486-8724
Fax: 520-622-1205
http://www.safariclub.org/

Mike Simpson, President
Kevin Anderson, Vice President

632 Wilderness Education Association
900 E 7th Street
Bloomington, IN 47405

812-855-4095
Fax: 812-855-8697
E-mail: wea@indiana.edu
http://www.weainfo.org/

WEA provides professional instruction, leadership training and wilderness travel. Promoting safe, ethical and professional wilderness leaders.

National: Water Resources

633 Adopt-A-Stream Foundation
600-128th Street SE
Everett, WA 98208

425-316-8592
Fax: 425-338-1423

E-mail: aasf@streamkeeper.org
http://www.streamkeeper.org

The mission of the AASF is to train people how to become stewards of their watersheds, by providing stream and wetland restoration technical asistance to community groups. To increase its training capacity, AASF is devoloping theNorthwest Stream Center: a regional teaching facility with stream and wetland ecology and fish and wetland habitat restoration as its central them.
Founded: 1986
Tom Murdoch, Executive Director

634 Alliance for a Clean Rural Environment
1155 15th Street NW
Suite 900
Washington, DC 20005

800-545-5410
800-545-5410
Fax: 202-463-0474
http://www.aces.edu

ACRE is an organization supported by the makers of agricultural chemicals. It encourages environmental stewardship and protection of water quality through printed literature for individuals, community groups, and dealers.
John Thorne, Director

635 American Canal Society
117 Main Street
Freemansburg, PA 18018

http://www.americancanalsociety.org

Terry Woods, President
Charles W Derr, Secretary/Treasurer

636 American Clean Water Association
7308 Birch Avenue
Takoma Park, MD 20912

301-495-0746

Larry Silverman, Executive Officer

637 American Ground Water Trust
16 Centre Street
Concord, NH 03301

603-228-5444
Fax: 603-228-6557
E-mail: trustinfo@agwt.org
http://www.agwt.org

A not-for-profit education organization. Promotes opportunity, cooperation and action among individuals, groups and organizations in order to educate the public, and further its mission: to protect ground water, promote publicawareness of the environment and economic importance of groundwater and provide accurate information to assist public participation in water resources decisions.
Founded: 1986
Ron Peterson, Chairman
Gregory Zlotnick, Vice Chairman

638 American Rivers
1025 Vermont Avenue NW
Suite 720
Washington, DC 20005

202-347-7550
Fax: 202-347-9240
E-mail: amrivers@amrivers.org
http://www.americanrivers.org

We are a nonprofit conservation organization dedicated to protecting and restoring rivers nationwide. Founded over 25 years ago in dusty Denver office, today we have offices in Washington, DC and across the country.
Founded: 1973
Rebecca R Wodder, President
Leslie Beck, Associate Dir Finance/Admin

639 American Shore and Beach Preservation Association
American Shore and Beach Preservation Association
1100 Caswell Beach Road
Caswell Beach, NC 28465

910-200-7867
Fax: 310-821-6345
E-mail: president@asbpa.org
http://www.asbpa.org

This Association is formed in recognition of the fact that shores of our oceans, lakes and rivers constitute important assets for promoting the health and physical well-being of the people of this nation, and that their contiguity togo out great centers of population affords an opportunity for wholesome and necessary rest and recreation not equally available in any other form. The purpose of the Association is to bring together for cooperation and mutural helpfulness.
Founded: 1926
Harry Simmons, President
Kate & Ken Gooderham, Executive Directors

640 American Water Resources Association
4 W Federal Street
PO Box 1626
Middleburg, VA 20118

540-687-8390
Fax: 540-687-8395
E-mail: info@awra.org
http://www.awra.org

AWRA is a nonprofit, scientific educational association for individuals and organizations involved in all aspects of water resources. Its goal is to advance multidisciplinary water resources management and research through itsconferences, publications, technical commettees, state sections and student chapters.
Founded: 1964
Kenneth D Reid, Executive VP
Melinda M Lalor, President

641 American Water Works Association
6666 W Quincy Avenue
Denver, CO 80235

303-794-7711
800-926-7337
Fax: 303-347-0804
http://www.awwa.org

Dedicated to the promotion of public health and welfare in the provision of drinking water of unquestionable quality and sufficient quantity. AWWA must be proactive and effective in advancing the technology, science, management andgovernment policies relative to the stewardship of water.
Founded: 1881
Darcy Burke, Executive Director
JoAnn Taniguchi, Secretary

642 Ancient Forest International
PO Box 1850
Redway, CA 95560

707-923-3015
Fax: 707-923-3015
E-mail: afi@igc.org
http://www.ancientforests.org

Supports forest conservation in southern Chile, Ecuador and California.
Founded: 1989

643 Association of Ground Water Scientists and Engineers
National Ground Water Association
601 Dempsey Road
Westerville, OH 43081

614-898-7379
800-551-7379
Fax: 614-898-7786

The mission of the National Ground Water Association is to enhance the skills and credibility of all ground water professionals, develop and exchange industry knowledge, and promote the ground water industry and understanding of groundwater resources.
Kevin B. McCray, CAE, Executive Director

644 Association of Metropolitan Sewerage Agencies
1816 Jefferson Place NW
Washington, DC 20036-2505

202-833-2672
Fax: 202-833-4657
E-mail: info@nacwa.org
http://www.amsa-cleanwater.org

Represents the interests of the country's wastewater treatment agencies, true environmental practitioners that serve the majority of the sewered population in the US, and collectively treat and reclaim more than 17 billion gallons ofwastewater each day. Maintains a key role in the development of environmental legislation, and works closely with federal regulatory agencies in the implementation of environmental programs.
Ken Kirk, Executive Director

645 Association of Metropolitan Water Agencies
1620 I Street
Suite 500
Washington, DC 20006

202-331-2820
Fax: 202-785-1845
E-mail: info@amwa.net
http://www.amwa.net

Founded: 1981
Diane Van De Hei, Executive Director
David Rager, President

646 Association of State Floodplain Managers
2809 Fish Hatchery Road
Suite 204
Madison, WI 53713

608-274-0123
Fax: 608-274-0696
E-mail: asfpm@floods.org
http://www.floods.org

The Association is an organization of professionals involved in floodplain management, flood hazard mitigation, the National Flood Insurance Program, and flood preparedness, warning and recovery.

Founded: 1977
Larry Larson, Executive Director
Diane Brown, Administrative Manager

647 Association of State Wetland Managers
1434 Helderberg Trail
Berne, NY 12023-9746

518-872-1804
Fax: 518-872-2171
E-mail: aswm@aswm.org
http://www.aswm.org

Nonprofit organization dedicated to the protection and restoration of our nations wetlands. Our goal is to help public and private wetland decision-makers utilize the best possible scientific information and techniques in wetlanddelineation, assessment, mapping, planning, regulation, acquisition, restoration, and other management.
Jon Kusler, Associate Director
Jenne Christie, Executive Director

648 Association of State and Interstate Water Pollution Control Administrators
750 1st Street NE
Suite 1010
Washington, DC 20002

202-898-0905
Fax: 202-898-0929
E-mail: admin1@aswipca.org
http://www.asiwpca.org

Maintain and enhance the quality of the nation's water resources and protect the public health through improving the State's capability to develop and implement effective Federal and State water management programs.
Roberta Savage, Executive Director
Linda Eichmiller, Deputy Director

649 CEDAM International
1 Fox Road
Croton on Hudson, NY 10520

914-271-5365
Fax: 914-271-4723
E-mail: cedamint@aol.com
http://www.cedam.org

Conservation, Education, Diving, Awareness and Marine research International is a nonprofit organization dedicated to the understanding, protection and preservation of the world's marine resources. Through our expeditions, CEDAMInternational volunteer divers actively participate in scientific research and conservation-oriented education projects. The results of our findings and efforts are disseminated to both the scientific and lay communities. The CEDA also publishes anannual newsletter.
Susan Sammon, Director

650 Center for Coastal Studies
59 Commercial Street
Provincetown, MA 02657

508-487-3623
Fax: 508-487-4695
E-mail: ccs@coastalstudies.org
http://www.coastalstudies.org

Private nonprofit organization for research, conservation and education in the coastal and marine environments.
Roslyn Garfield, President

651 Center for Marine Conservation
2029 K Street, NW
Washington, DC 20006

202-429-5609
800-519-1541
Fax: 202-872-0619
E-mail: info@oceanconservancy.org
http://www.cmc-ocean.org

The mission of the CMC is to protect ocean ecosystems and conserve the global abundance and diversity of marine wildlife. Through sciencebased advocacy, research and public education, CMC informs, inspires and empowers people to speakand act for the oceans.
Roger Rufe Jr, President
Thomas J Tepper, Senior VP Operations

652 Center for Watershed Protection
8390 Main Street
Second Floor
Ellicott City, MD 21043-4605

410-461-8323
Fax: 410-461-8324
E-mail: center@cwp.org
http://www.cwp.org

The center is a nonprofit 501(c)3 organization dedicated to finding new ways to protect and restore our nation's streams, lakes, rivers and estuaries. The center publishes numerous technical publications on all aspects of watershedprotection, including stormwater management, watershed planning and better site design. Publications are available online.

Founded: 1992
Hye Yeong Kwon, Executive Director
Jessica Brooks, Publications Manager

653 Center for the Great Lakes
435 N Michigan Avenue   312-263-0785
Suite 1408   Fax: 312-201-0683
Chicago, IL 60611
Daniel K Ray, Director

654 Clean Harbors Cooperative
4601 Tremley Point Road   908-862-7500
Linden, NJ 07036   Fax: 908-862-7560
E-mail: chcllc@aol.com
http://www.apicom.org/members.html
Edward M Wirkowski, Manager
Dennis J McCarthy, Director

655 Clean Water Action
4455 Connecticut Avenue NW   202-895-0420
Suite A300   Fax: 202-895-0438
Washington, DC 20008   E-mail: cwa@cleanwater.org
http://www.cleanwateraction.org
National citizens' organization working for clean, safe and affordable water, prevention of health-threatening pollution, creation of environmentally-safe jobs and businesses, and empowerment of people to make democracy work.Organizes strong grassroots groups, coalitions and campaigns to protect our environment, health, economic well-being and community quality of life.
Founded: 1972
David Zwick, President
David Tykulsker, Chairman

656 Clean Water Fund
4455 Connecticut Avenue NW   202-895-0432
Suite A300-16   Fax: 202-895-0438
Washington, DC 20008-2328   E-mail: cwf@cleanwater.org
http://www.cleanwaterfund.org
Brings diverse communities together to work for changes that improve our lives, promoting sensible solutions for people and the environment.
Founded: 1972
David Vwick, President

657 Clean Water Network
Spills and Kills
1200 New York Avenue NW   202-298-2421
Suite 400   Fax: 202-289-1060
Washington, DC 20005   E-mail: info@cwn.org
http://www.cwn.org
A nonprofit network of over 1,000 organizations that deal with clean water issues covered by the Clean Water Act. Our member organizations consist of a variety of organizations representing environmentalists, family farmers, recreationanglers, commercial fishermen, surfers, boaters, faith communities, labor unions and civic associations. We publish a monthly newsletter and various reports.
Katherine Smitherman, Executive Director

658 Coast Alliance
600 Pennsylvania Avenue SE   202-546-9554
Suite 340   Fax: 202-546-9609
Washington, DC 20003   E-mail: coast@coastalliance.org
http://www.coastalliance.org
Nonprofit organization formed by a number of groups and individuals concerned about the effects of unprecedented development pressure and pollution on the coasts. Coast Alliance was founded to increase public awareness of the coast'simmense value; to bring new and important scientific facts about coastal ecology to public attention; to encourage groups across the country to work to protect valuable coastal resources and to urge federal, state and local governments to strengthenprotections.
Founded: 1979
Dery Bennet, Chair Person
Dawn Hamilton, Executive Director

659 Coastal Conservation Association
4801 Woodway W   713-626-4234
Suite 220   800-201-3474
Houston, TX 77056   Fax: 713-951-3801

E-mail: ccantl@joincca.org
http://www.joincca.org/
A national organization dedicated to the conservation and preservation of marine resources.
Founded: 1977
Dick Brane, Executive Director

660 Coastal Society
PO Box 25408   703-768-1599
Alexandria, VA 22313-5408   Fax: 703-933-1596
E-mail: coastalsoc@aol.com
http://www.thecoastalsociety.org
Organization of private sector, academic, government professionals and students dedicated to actively addressing emerging coastal issues by fostering dialogue, forging partnerships and promoting communication and education. NextConference: May 23-26, 2004.
Founded: 1975
Paul Ticco, President
Judy Tucker, Executive Director

661 Cook Inlet Keeper
State of the Inlet Report
PO Box 3269   907-235-4068
3734 Ben Walters Lane   Fax: 907-235-4069
Homer, AK 99603   E-mail: keeper@inletkeeper.org
http://www.inletkeeper.org
Dedicated to protecting the vast Cook Inlet watershed and the life it sustains.
Founded: 1995
Rob Ernest, President
Bob Shavelson, Executive Director

662 Coral Reef Alliance
417 Montgomery Street   415-834-0900
Suite 205   Fax: 415-834-0999
San Francisco, CA 94104   E-mail: coralmail@aol.com
http://www.coralreefalliance.org
CORAL is a nonprofit organization created to address the worldwide problems that coral reefs face. They provide information to divers and interested persons and support conservation efforts. This site contains information about thecurrent state of coral reefs around the world and The Coral Reef NGO Directory, a directory of groups worldwide involved in coral reef research and conservation efforts.
Jason De Salvo, Chairman
Elizabeth Ulmer, Vice President

663 Earth Island Institute
300 Broadway   415-788-3666
Suite 28   Fax: 415-788-7324
San Francisco, CA 94133   http://www.earthisland.org
Incubates and supports over 35 projects working on environmental issues worldwide. Publishes quarterly Earth Island Journal.
Founded: 1970
John A Knox, Executive Director
Dave Phillips, Executive Director

664 Ecological Society of America
1707 H Street NW   202-833-8773
Suite 400   Fax: 202-833-8775
Washington, DC 20006   E-mail: esahq@esa.org
http://www.esa.org
To promote ecological science by improving communication among ecologists; raise the public's level of awareness of the importance of ecological science; increase the resources available for the conduct of ecological science; andensure the appropriate use of ecological science in environmental decision making by enhancing communication between the ecological community and policy-makers.
Founded: 1915
Katherine McCarter, Executive Director
Elizabeth Biggs, Chief Financial Officer

665 Emergency Committee to Save America's Marine Resources
1552 Osprey Court   732-223-5729
Manasquan Park, NJ 08736   Fax: 732-528-1056
E-mail: cristori@aol.com
Founded: 1955
Allan J Ristori, Chairman

666 Foresta Institute for Ocean and Mountain Studies

2400 E Speedway 520-881-6174
Suite 118-293 Fax: 520-323-2751
Tucson, AZ 85716

Works to educate teachers and youth about environmental conservation.

667 Fresh-Water Foundation

2500 Shadywood Road 952-471-9773
Excelsior, MN 55331-9578 888-471-9773
Fax: 952-471-7685
E-mail: firstwater@firstwater.org
http://www.freshwater.org
Founded: 1975
Dan Brauer, Executive Director

668 Future Fisherman Foundation

225 Reinekers Lane 703-519-9691
Suite 420 Fax: 703-519-1872
Alexandria, VA 22314 E-mail: info@futurefisherman.org
http://www.asafishing.org
The Future Fisherman Foundation supports groups that offers training in fishing and aquatic resource stewardship by developing curriculum materials for environmental and angler education. Working through state fish and wildlifeagencies, the foundation strives to increase aquatic resource education by using available federal funds.
Anne Glick, Executive Director
John Bryan, Chief Philanthropy Officer

669 Gaia Institute

400 City Island Avenue 718-885-1906
Bronx, NY 10464 Fax: 718-885-0882
E-mail: Gaia@gaia-inst.org
http://www.gaia-inst.org
The purpose of the Gaia Institute is to test through demostration the means by which ecological components of backyards, communities, towns and cities, as well as watersheds and estuaries, can be enhanced through integratedwastes-into-resources technologies.
Founded: 1972
Paul S Mankiewicz, Director
Julie A Mankiewicz, PhD, Director

670 Great Lakes United

4525 Rue DeRouen 514-396-3333
Montreal Quebec, QC H1V 1 Fax: 514-396-0297
E-mail: glu@glu.org
http://www.glu.org
Great Lakes United is an international coalition dedicated to preserving and restoring the Great Lakes-St. Lawrence River ecosystem. Great Lakes United is made up of member organizations representing environmentalists,conservationists, hunters and anglers, labor unions, community groups, and programs, and promotes citizen action and grassroots leadership to assure.
Founded: 1982
Derek Stack, Executive Director
Patty O'Donnell, President

671 Groundwater Foundation

PO Box 22558 402-434-2740
Lincoln, NE 68542-2558 800-858-4844
Fax: 402-434-2742
E-mail: info@groundwater.org
http://www.groundwater.org
Nonprofit organization that is dedicated to informing the public about one of our greatest hidden resources, groundwater. Since 1985, programs and publications present the benefits everone receives from groundwater and the risks thatthreaten groundwater quality.
Founded: 1985
Susan Seacrest, President
Cindy Kreifels, Executive Director

672 International Association for Environmental Hydrology

PO Box 35324 210-344-5418
San Antonio, TX 78235 Fax: 210-344-9941
E-mail: hydroweb@mail.org
http://www.hydroweb.com

673 International Desalination Association

PO Box 387 978-887-0410
Topsfield, MA 01983 Fax: 978-877-0411
E-mail: info@idadesal.org
http://www.idadesal.org
Goals of the Association are the development and promotion of the appropriate use of desalination (the method of removing salts and other impurities from water) and desalination technology worldwide.
Founded: 1972
Patricia A Burke, Secretary General
Abdul hamid Al Mansour, Director General

674 International Oceanographic Foundation

4600 Rickenbacker Causeway 305-361-4000
PO Box 499900 Fax: 305-361-4711
Miami, FL 33149 http://www.rsmas.miami.edu
International Oceanographic Foundation (IOF) is a nonprofit supporting organization to the University of Miami's Rosenstiel School of Marine and Atmospheric Science. IOF was chartered in 1953 to encourage scientific investigation ofthe sea, to provide the public with current, accurate and unbiased information pertaining to marine environments and to promote awareness of the importance of Earth's oceans to humankind. IOF is a 501(c)(3) nonprofit organization, contributions aretax deductable.
Founded: 1953
Otis B Brown, Dean

675 International Rivers Network

1847 Berkeley Way 510-848-1155
Berkeley, CA 94703 Fax: 510-848-1008
E-mail: info@irn.org
http://www.irn.org
IRN supports local communities working to protect their rivers and watersheds. We work to halt destructive river development projects, and to encourage equitable and sustainable methods of meeting needs for water, energy and floodmanagement. Publishes a bimonthly newsletter.
Founded: 1985
Patrick McCully, Executive Director
Anne Carey, Office Manager

676 International Water Resources Association

Southern Illinois University Carbondale
4535 Faner Hall 618-453-5138
Carbondale, IL 62901-4516 Fax: 618-453-6465
E-mail: iwra@siu.edu
http://www.iwra.siu.edu
IWRA strives to improve water management worldwide through dialogue, education, and research. It seeks to improve water resources outcomes by improving our collective understanding of the physical, biological, chemical, institutonal,and socioeconomic aspects of water.
Founded: 1972 160 pages
ISSN: 0250-8060
Benedykt Dziegielewski, Executive Director
John W Nicklow, Treasurer

677 Interstate Council on Water Policy

955 L'Enfant Plaza SW 202-466-7287
6th Floor Fax: 202-646-6210
Washington, DC 20024 http://www.icwp.org
The Interstate Council on Water Policy is the national organization of state and regional water resource management agencies. It provides a means to exchange information, ideas, and experience and to work with federal agencies whichshare water management responsibilities. In paricular, ICWP focuses on water quality and water quantity issues, and on the dynamic interface state and federal roles.
Founded: 1959
Joe Hoffman, Executive Director

678 Marine Technology Society

5565 Sterrett Place 410-884-5330
Suite 108 Fax: 410-884-9060
Columbia, MD 21044 E-mail: mtsdir@aeros.com
http://www.mtsociety.org
Addresses coastal zone management, marine mineral and energy resources, marine environmental protection, and ocean engineering issues.

Founded: 1963
Jerry Streeter, President
Judith Krauthamer, Executive Director

679 National Association for State and Local River Conservation Programs

8630 Fenton Street 301-589-9455
Suite 910 Fax: 301-589-6121
Silver Spring, MD 20910
Barry Beasley, President

680 National Association of Flood and Storm Water Management Agencies

1301 K Street 202-218-4122
NW, Suite 800 East Fax: 202-478-1734
Washington, DC 20005 E-mail: info@nafsma.org
http://www.nafsma.org

The National Association of Flood and Stormwater Management Agencies is an organziation of public agencies whose function is the protection of lives, property and economic activity from the adverse impacts of storm and flood waters. The mission of the association is to advocate public policy, encourage technologies and conduct education programs which facilitate and enhance the adchievement of the public service functions of its members.
Founded: 1978
Susan Gilson, Executive Director
Kerry Wilson, President

681 National Audubon Society: Living Oceans Program

550 S Bay Avenue 516-859-3032
Islip, NY 11751 Fax: 516-581-5268
E-mail: mlee@audubon.org
http://www.audubon.org/campaign/10

Living Oceans is the marine conservation program of National Audubon Society. Audubon's Living Oceans uses science-based policy analysis, education and grassroots advocacy to put science to work on behalf of marine fish and oceanecosystems.
Carl Safina, Director
Mercedes Lee, Assistant Director

682 National Boating Federation

PO Box 4111 410-626-8566
Annapolis, MD 21403-4111 E-mail: rpdavid@capecod.net
http://www.n-b-f.org/faq.html

Founded in 1966, the NBF is the largest nationwide alliance of recreational boating organizations. The Federation can best be described as a volunteer legislative watchdog and educational organization, beholden to no otherconstituency and free to assess and comment on boating laws and regulations objectively.
Founded: 1966
Penelope Orth, President
Marlene Barrington, Vice President

683 National Coalition for Marine Conservation

5105 Paulsen Street 912-354-0441
Suite 243 Fax: 912-354-0234
Savannah, GA 31405 http://www.savethefish.org

National Coalition for Marine Conservation is dedicated exclusively to conserving ocean fish, preventing overfishing, reducing bycatch, and protecting habitat.
Founded: 1975
Ken Hinman, Executive Director

684 National Council of the Paper Industry for Air and Stream Improvements

PO Box 13318 919-558-1999
Research Triangle Park, NC 27709 Fax: 919-558-1998
E-mail: bob.carroll@amec.com
http://www.ncasi.org

The National Council of the Paper Industry for Air and Stream Improvements is a technical organization devoted to finding solutions to environmental protection problems in the manufacture of pulp, paper, and wood products in industrialforestry. It was started in 1943 and now has about 100 member companies.
Founded: 1943
Ronald Yeske, President

685 National Ground Water Association

601 Dempsey Road 614-898-7791
Westerville, OH 43081 800-551-7379
Fax: 614-898-7786
E-mail: ngwa@ngwa.org
http://www.ngwa.org

The mission of the National Ground Water Association is to provide professional and technical leadership for the advancement of the ground water industry and for the protection, promotion, and responsible development and use of groundwater resources.
Founded: 1948
John S Christ, Director

686 National Institutes for Water Resources

University of Massachusetts Water Resources Center
47 Harkness Road 413-253-5686
Pelham, MA 01002 Fax: 413-253-1309
E-mail: godfrey@tei.umass.edu
http://www.niwr.montana.edu/

The National Institutes for Water Resources is a network of Research Institutes in every state. They conduct basic and applied research to solve water problems unique to their area. The programmatic responsibilities stipulated by theWater resources Research Act provide a unified focus for the federal and non-federal components of the Institute Program.
Founded: 1970
Paul Godfrey, Executive Secretary
James Moncor, President

687 National Oceanic and Atmospheric Administration

14th Street/Constitution Avenue NW 202-482-6090
Room 6217 Fax: 202-482-3154
Washington, DC 20230 E-mail: answers@noaa.gov
http://www.noaa.gov

Describes and predicts changes in the Earth's environment and conserves and wisely manages the nation's coastal and marine resources. Goals and objectives include advance short-term warning and forecast services.
Comrad C Lautenbacher, Vice President

688 National Water Resources Association

3800 North Fairfax Drive 703-524-1544
Suite 4 Fax: 703-524-1548
Arlington, VA 22203 E-mail: nwra@nwra.org
http://www.nwra.org

The National Water Resources Association consists of individuals or groups, such as irrigation districts, canal companies and conservancy districts, municipalities and the public in general, who are interested in water resourcedevelopment projects. It was started in 1932 and now has about 5,000 members.
Founded: 1932
Norm M Semanko, President

689 National Water Supply Improvement Association

PO Box 102 301-855-1173
St. Leonard, MD 20685 Fax: 410-586-2844
Jack C Jorgensen, Executive Director

690 National Watershed Congress National Association of Conservation Districts

509 Capitol Court 202-547-6223
5th Floor, Fax: 202-547-6450
Washington, DC 20002-4937 http://www.nacdnet.org
Founded: 1937
Sarah Hickling, Executive Assistant
Krysta Harden, Chief Executive Officer

691 National Waterways Conference

1130 17th Street NW 202-296-4415
Suite 200 Fax: 202-835-3861
Washington, DC 20036-4676 E-mail: info@waterways.org
http://www.waterways.org

To encourage a better understanding of the public value of the American waterways system and to document the importance of far-sighted navigation and water resources policies to a sound economy, industrial and agriculturalproductivity, regional development, environmental quality, energy conservation, international trade, defense preparedness and the overall national interest.
Worth Hager, President
J Scott Robinson, Vice President

692 National Wetlands Technical Council Research Center
700 Cajundome Boulevard   337-266-8500
Lafayette, LA 70506   Fax: 337-266-8513
                        E-mail: nwrcweb@usgs.gov
                        http://www.nwrc.usgs.gov
To develop and disseminate scientific information needed for understanding the ecology and values of our nation's wetlands and for managing and restoring wetland habitats and associated plant and animal communities.
Founded: 1975
Gerald A Grau, Assistant Director
Dolores Richardson, Administrative Officer

693 National Xeriscape Council
PO Box 163172   512-392-6225
Austin, TX 78716   http://www.xeriscape.org
Systematic concept for saving water in landscaped areas.

694 North American Lake Management Society
1 Progress Boulevard   904-462-2554
Box 27   Fax: 904-462-2568
Alachua, FL 32615-9536
The North American Lake Management Society's mission is to forge partnerships among citizens, scientists and professionals to foster the management and protection of lakes and reservoirs for today and tomarrow.
Lorraine R Duncan, Administrative Assistant

695 North Plains Ground Water
603 E 1st Street   806-935-6401
Dumas, TX 79029-3221   Fax: 806-935-6633
                                        http://www.npwd.org
Founded: 1949
David Moore, President

696 Ocean Conservancy
2029 K Street   202-429-5609
Washington, DC 20006   800-519-1541
                                        Fax: 202-872-0619
                        E-mail: info@oceanconservancy.org
                        http://www.oceanconservancy.org
To conserve and protect the oceans. Advocate for the oceans, with an emphasis on conserving and protecting significant parts of our oceans.
Founded: 1972
Roger Rufe, President
Tom Tepper, Senior VP Operations

697 Oceana
2501 M Street NW   202-833-3900
Suite 300   877-762-3262
Washington, DC 20037   Fax: 202-833-2070
                        E-mail: info@oceana.org
                        http://www.oceana.org
Oceana is a nonprofit international advocacy organization dedicated to protecting and restoring the world's oceans through policy advocacy, science, law and public education. Founded in 2001, Oceana's constituency includes members andactivists from more than 190 countries and territories who are committed to saving the world's marine environment. In 2002, American Oceans Campaign became part of Oceana's international effort to protect ocean eco-systems and sustain the circle oflife.
Andy Sharpless, CEO
Jim Simon, Executive Director

698 Oceanic Society
Fort Mason Center   415-441-1106
Building E   800-326-7491
San Francisco, CA 94123   Fax: 415-474-3395
                        E-mail: office@oceanic-society.org
                        http://www.oceanic-society.org
This nonprofit society was founded in 1969 for the protection of the marine environment, environmental education and conservation-based field research.
Founded: 1972
Birgit Winning, President

699 River Network
520 SW 6th Avenue   503-241-3506
Suite 1130   800-423-6747
Portland, OR 97204   Fax: 503-241-9256

E-mail: info@rivernetwork.org
http://www.rivernetowrk.org
River Network is a national nonprofit organization for citizen groups working for river and watershed protection. Our mission is to help people understand, protect and restore rivers and their watersheds. We provide our conservationpartners with with information training, consultation, grants, referrals to other service organizations and networking opportunities including the annual national River Rally.
Founded: 1988
Clarence Alexander, Grand Cheif
Katherine Luscher, Dir Partnership Programs

700 Scientific Committee on Oceanic Research: Department of Earth and Planetary Science
Johns Hopkins University   410-516-4070
Baltimore, MD 21218   Fax: 410-516-4019
                        E-mail: scor@jhu.edu
                        http://www.jhu.edu
The Scientific Committee on Oceanic Research (SCOR), is an international nonprofit organization whose purpose is to encourage international cooperation in a branch of ocean research.
Founded: 1957
Edward Urban, Executive Director
Elizabeth Gross, Finance Officer

701 Seacoast Anti-Pollution League
163 Court Street   603-431-5089
PO Box 1136   E-mail: info@sapl.org
Portsmouth, NH 03802   http://www.sapl.org
Founded: 1969
Herb Moyer, President
Jean Lincoln, Field Director

702 Steamboaters
1665 Evergreen Drive   541-688-4980
Eugene, OR 97404   Fax: 541-607-3763
                        E-mail: steamboaters@rosenet.net
                        http://www.steamboaters.org
Nonprofit organization dedicated to work to restore the North Umpqua river system's wild fish stocks, particularly steelhead, to a sustainable level that is consistent with optimum natural population levels. Protect, preserve andrestore fish habitat, including adequate and consistent flows of high quality water in the North Umpqua and its tributaries.
Founded: 1975
Richard Bauer, Executive Director

703 United Citizens Coastal Protection League
PO Box 46   760-753-7477
Cardiff By The Sea, CA 92007
Founded: 1982
Robert Bonde, Executive Director

704 Water Center
Route 3   501-253-9431
Box 716
Eureka Springs, AR 72632
Barbara Harmony, President

705 Water Environment Federation
601 Wythe Street   703-684-2400
Alexandria, VA 22314-1994   800-666-0206
                                        Fax: 703-684-2492
                        E-mail: webfeedback@wef.org
                        http://www.wef.org
Founded in 1928, the Water Environment Federation (WEF) is a not-for-profit technical and educational organization with members from varied disciplines who work toward the WEF vision of preservation and enhancement of the global waterenvironment. The WEF network includes more than 100,000 water quality professionals from 77 member associations in 31 countries.
Founded: 1928
Phyllis Eastman, Managing Director
William Bertera, Executive Director

706 Water Quality Association
4151 Naperville Road   630-505-0160
Lisle, IL 60532-1088   Fax: 630-505-9637

E-mail: info@mail.wqa.org
http://www.wqa.org

Founded: 1974
Peter J Censky, Executive Director
Margit Fotre, Director Memebership/Marketi

707   Water Resources Congress
2300 Claredon Boulevard   703-525-4881
Suite 404   Fax: 703-527-1693
Arlington, VA 22201-3367
Kathleen A Phelps, Executive Director

708   Western Hemisphere Shorebird Reserve Network
Manomet Centre For Conservation Sciences
81 Stage Point Road   508-224-6521
PO Box 1770   Fax: 508-224-9220
Manomet, MA 02345   E-mail: whsrn@manomet.org
                                http://www.manomet.org
WHSRN is a voluntary, community-based coalation of over 185
organizations across the US and other counries in the Western
Hemisphere that have joined together to protect, restore and
manage critical wetland habitats for migratorybirds. Established in 1986, we now have 49 key sites in 7 countries as Net-
work partners responsible for over 20 million acres. WHSRN
links these sites into a collaborative network based on one ecological feature they all share-millons of migratorybirds.
Founded: 1985
Charles Duncan, Director
Linda Leddy, President

709   Wildlands Conservancy
3701 Orchid Place   610-965-4397
Emmaus, PA 18049-1637   Fax: 610-965-7223
                              E-mail: info@wildlandspa.org
                                http://www.wildlandspa.org
Wildlands Conservancy is a nonprofit organization dedicated to
preserving precious land, keeping our waterways healthy, teach-
ing the community about nature and caring for orphaned or in-
jured wildlife.
Founded: 1973
Nelson G Markley, Chairman
David N Shaffer, Vice Chairman

710   World Aquaculture Society
Louisiana State University
143 JM Parker Coliseum   225-388-3137
Louisiana State University   Fax: 225-388-3493
Baton Rouge, LA 70803   E-mail: wasmas@aol.com
                                http://www.was.org
The World Aquaculture Society is an international nonprofit so-
ciety, whose commitment to excellence in science, technology,
education and information exchange, will contribute to progres-
sive and sustainable development of aquaculturethroughout the
world.
Founded: 1970
Dr. Jeff Hinshaw, President
Jimmy Avery, Vice President

711   World Association of Soil and Water Conservation
945 SW Ankeny Road   515-289-2331
Ankeny, IA 50023-9723   800-843-7645
                                           Fax: 515-289-1227
                              E-mail: swes@swes.org
                                http://www.swcs.org
WASWC is an international non-governmemt organization of
professionals and informed laypersons dedicated to promoting
the sustained use of the earth's soil and water resources.
Founded: 1943
Craig Cox, President

Alabama

**712   Alabama Association of Soil & Water Conservation Dis-
tricts**
PO Box 304800   334-242-2622
Montgomery, AL 36130-4800   Fax: 334-242-0551
                              E-mail: vpayne@swcc.state.al.us
                                http://www.swcc.state.al.us

Founded: 1937
George Robertson, President
Stephen M Cauthen, Executive Director

**713   Alabama Chapter, National Safety Council (Birming-
ham)**
2168 Greensprings Highway   205-328-7233
Birmingham, AL 35205   800-457-7233
                                           Fax: 205-328-1467
                              E-mail: alabama@nsc.org
                                http://www.alabama.nsc.org
Donny Ward, Executive Director

714   Alabama Environmental Council
City Council Office   205-254-2294
710 20th Street   800-982-4364
Birmingham, AL 35203   Fax: 205-254-2603
                              E-mail: stateoffice@aeconline.ws
                                http://www.aeconline.ws
Statewide grassroots. The oldest environmental advocacy and
education organization dedicated to the preservation and pro-
tection of Alabama's natural heritage.
Founded: 1967
Ouida Fritschi, President
Larry Crenshaw, Secretary

715   Alabama Waterfowl Association
PO Box 67   205-259-2509
Guntersville, AL 35768   http://www.alabamawaterfowl.org
Jerry Davis, CEO
Gary Benefield, Executive Director

716   Alabama Wildlife Federation
3050 lanark road   334-285-4550
Millbrook, AL 36054   800-822-9453
                                           Fax: 334-285-4959
                              E-mail: awf@alabamawildlife.org
                                http://www.alabamawildlife.org
Founded: 1935
Tim Gothard, Executive Director
Rebecca Prichett, Representative

717   American Lung Association of Alabama
American Lung Association
3125 Independence Drive   205-933-8821
Suite 325   Fax: 205-930-1717
Birmingham, AL 35209   E-mail: aodom@alabamalung.org
                                http://www.lungusa2.org
Founded: 1904
John Kirkwood, President/CEO
Norman Edelman, Executive Vice President

718   BAMA Backpaddlers Association
307 Madison Place   205-951-0320
Trussville, AL 35173   E-mail: backpaddlers@aol.com
                   http://www.members.aol.com/backpaddlers
To promote recreation, conservation and education of Ala-
bama's rivers.
Founded: 1978
Pam Belrose, President

719   BASS Anglers Sportsman Society
P.O. Box 17900   334-272-9530
Montgomery, AL 36141   Fax: 334-270-7148
                              E-mail: bassmail@mindspring.com
                                http://www.bassmaster.com
Founded: 1967
Gary Jones, Federation Director

720   National Safety Council (Tennessee Valley Office)
2042 Beltline Road SW   256-308-1133
Building A, Suite 110   Fax: 256-308-1161
Decatur, AL 35601   E-mail: alabama@nsc.org
Donny Ward, Executive Director

721   Nature Conservancy: Colorado Field Office
2100 1st Ave. North   205-251-1155
Suite 500   800-628-6860
Birmingham, Al 35203   Fax: 205-251-4444

E-mail: comment@tnc.org
http://www.nature.org

The mission of the Nature Conservancy is to preserve plants, animals and natural communities that represent the diversity of life on earth by protecting the lands and waters they need to survive. To date, the conservation and its morethan one million members have been responsible for the protection of more than 12 million acres in 50 states.
Founded: 1951
Jeff Danter, Executive Director

722 Sierra Club
1330 21st Way S
Suite 110
Birmingham, AL 35205
205-933-9111
Fax: 202-939-1020
E-mail: se-al.field@sierraclub.org
http://http://alabama.sierraclub.org

To advance the preservation and protection of the natural environment by empowering the citizenry, especially democratically-based grassroots organizations, with charitable resources to further the cause of environmental protection.The vehicle through which The Sierra Club Foundation generally fulfills its charitable mission.
Founded: 1908
Carl Pope, Executive Director

Alaska

723 ASLA: Alaska Chapter
c/o Tryck Nyman Hayes
911 West 8th Avenue
Suite 300
Anchorage, AK 99501
907-343-0276
Fax: 907-276-7679
E-mail: LuanneU@tnh-inc.com
http://www.asla.org/states/chpr.htm
Founded: 1899
Luanne Urfer, President
Eric K Ouderkirk, President-Elect

724 Alaska Audubon Society
Nation Audubon Society
715 L Street
Suite 200
Anchorage, AK 99501
907-276-7034
Fax: 907-276-5069
E-mail: bdennerlein@audubon.org
http://www.audubonalaska.org
Founded: 1977
Matthew Kirchhoff, Chairman
Stanley E Senner, Executive Director

725 Alaska Conservation Alliance
810 N Street
Suite 203
Anchorage, AK 99501
907-258-6171
Fax: 907-258-6177
E-mail: unite@akvoice.org
http://www.akvoice.org
Founded: 1979
Tom Atkinson, Executive Director
Trish Rolfe, Associate Director

726 Alaska Conservation Voters
810 N Street
Suite 203
Anchorage, AK 99501
907-258-6171
Fax: 907-258-6177
E-mail: unite@akvoice.org
http://www.akvoice.org
Tom Atkinson, Executive Director
Trish Rolfe, Associate Director

727 Alaska Natural Resource & Outdoor Education
101 12th Avenue
Box 17, Room 222
Fairbanks, AK 99701
907-292-1772
Fax: 907-207-1795
E-mail: admin@anroe.org
http://www.anroe.org
Founded: 1982
Courtney Sullivan, President
Kristen Romanoff, Secretary

728 Alaska Wildlife Alliance
PO Box 202022
Anchorage, AK 99520
907-277-0897
Fax: 907-277-7423
E-mail: awa@alaska.net
http://www.akwildlife.org

Founded: 1978
John Toppenberg, Executive Director
Jenna White, Associate Director

729 American Lung Association of Alaska
500 West International Airport Road
Suite 1
Anchorage, AK 99518
907-276-5864
Fax: 907-565-5587
E-mail: marge@aklung.org
http://www.aklung.org/index2.htm
Founded: 1934
Marge Lawson, CEO

730 Northern Alaska Environmental Center
830 College Road
Fairbanks, AK 99701-2806
907-452-5021
Fax: 907-452-3100
E-mail: info@northern.org
http://www.northern.org
David van den Berg, Executive Director
Nikki Braem, Membership/Communications Di

731 Sierra Club
85 Second Street, 2nd Floor
San Francisco, CA 94105-0467
415-977-5500
Fax: 415-977-5799
E-mail: information@sierraclub.org
http://www.sierraclub.org/contact/

To advance the preservation and protection of the natural environment by empowering the citizenry, especially democratically-based grassroots organizations, with charitable resources to further the cause of environmental protection.The vehicle through which The Sierra Club Foundation generally fulfills its charitable mission.
Founded: 1892

732 Trustees for Alaska
1026 W 4th Avenue
Suite 201
Anchorage, AK 99501-2101
907-276-4244
Fax: 907-276-7110
E-mail: ecolaw@trustees.org
http://www.trustees.org

A public interest law firm whose mission is to provide legal counsel to sustain and protect Alaska's natural environment. We represent local and national environmental groups, Alaska Native villages, nonprofit organizations, communitygroups, hunters, fishers and others where the outcome of our advocacy could benefit Alaska's environment.
Founded: 1974
Danny Consenstein, Executive Director
Rebecca Bernard, Director

733 Wildlife Federation of Alaska
750 W 2nd Avenue
Suite 200
Anchorage, AK 99501-2168
907-258-4800
800-822-9919
Fax: 907-258-4811
E-mail: wfa@micronet.net
http://www.nwf.org
Founded: 1938
Jerome C Ringo, Chairman
Stephen E Petron, Vice Chairman

734 Wildlife Society
3298 Douglas Place
Homer, AK 99603
907-235-1492
Fax: 907-235-2448
E-mail: nicky_szarzi@fishgame.state.ak.us
http://www.mercury.bio.uaf.edu/ak-tws/

A nonprofit scientific and educational organization that serves professionals such as government agencies, academia, industry, and non-government organizations in all areas related to the conservation of wildlife and natural resourcesmanagement.
Founded: 1937
Tara Wertz, Secretary/Treasurer
Eric Taylor, President

Arizona

735 ASLA: Arizona Chapter
1745 East River Road
Tucson, AZ 85718
520-325-9977
Fax: 520-322-6956
E-mail: ljwoods@RECONAZ.com
http://www.azasla.org

Is the national professional association representing landscape architects.

Diana Smith, Executive Director
James D Coffman, President-Elect

736 American Environmental Health Foundation
8345 Walnut Hill Lane
Suite 225 800-428-2343
Dallas, TX 75231 Fax: 214-361-2534
E-mail: aehf@aehf.com
http://www.aehf.com
The Environmental Health Foundation's mission is two-fold: to fund scientific and/or medical research into the causes of environmentally linked disease; and to educate the public about environmentally linked illness and how to preventexposure through lifestyle changes.
Founded: 1975
William Rea, Founder

737 American Lung Association of Arizona/ New Mexico
The American Lung Association
102 West McDowell Road 602-258-7505
Phoenix, AZ 85003-1299 800-586-4872
http://www.lungusa2.org
Founded: 1912
Bill Pfeifer, President
Nancy Cohrs, Vice President

738 American Rivers, Southwest Regional Office
4120 N 20th Street 602-234-3946
Suite G Fax: 602-234-2217
Phoenix, AZ 85016 E-mail: amrivsw@aol.com
http://www.americanrivers.org
Founded: 1973
Mary Orton, Regional Director

739 Arizona Automotive Recyclers Association
PO Box 20662 480-391-2888
Phoenix, AZ 85036 Fax: 480-391-2390
E-mail: info@aara.com
http://www.aara.com
Select group of recyclers providing quality recycled parts for the benefit of our customers, communities and environment. There are 90 member companies. AARA is affiliated with the Automotive Dismantlers and Recyclers Association.
Robert Sibbio, President
Dan Rush, VIce President

740 Arizona BASS Chapter Federation
PO Box 84006 623-434-3520
Phoenix, AZ 85071-4006 http://www.azod.org/azbassfed
Founded: 1972
Mike Johnson, President
Kip Pollay, Executive Director

741 Arizona Chapter, National Safety Council
1606 W Indian School Road 602-264-2394
Phoenix, AZ 85015 Fax: 602-277-5485
E-mail: main@acnsc.org
http://www.acnsc.org

Jed Bullard, Chairman

742 Arizona Solar Energy Industries Association
Solar Energy Industries Association
2034 N 13th Street 602-253-8180
Phoenix, AZ 85006 Fax: 602-258-3422
E-mail: Solar-guy@msn.com
Michael Neary, Executive Director

743 Arizona Solar Update
Arizona Solar Energy Industries Association
2034 N 13th Street 602-253-8180
Phoenix, AZ 85006 Fax: 602-258-3422
Michael Neary, Executive Director

744 Arizona Water Well Association
4035 E Fanfol Drive 602-569-9169
Phoenix, AZ 85028-5103 Fax: 602-996-3966
Promotes protection and wise development of underground water resources.

745 Arizona-Sonora Desert Museum
2021 N Kinney Road 520-883-1380
Tucson, AZ 85743 Fax: 520-883-2500
E-mail: info@desertmuseum.org
http://www.desertmuseum.org
A museum, zoo and botanical garden center. The Sonora Desert of New Mexico and North America is the primary focus.
Founded: 1952
Kathryn Riser, Chief Financial Officer

746 Wildlife Society
130 W Calle Melendrez 435-688-3239
Green Valley, AZ 85614 Fax: 435-688-3258
E-mail: micheal_herder@blm.gov
A nonprofit scientific and educational organization that serves professionals such as government agencies, academia, industry, and non-government organizations in all areas related to the conservation of wildlife and natural resourcesmanagement.
Micheal Herder, Recording Secretary

Arkansas

747 ASLA: Arkansas Chapter
American Society of Landscape Architecture
636 Eye Street, NW 202-898-2444
Washington, DC 20001-3736 800-999-2752
Fax: 202-898-1185
E-mail: membership@asla.org
http://www.asla.org
Founded: 1899
Jerry Veaulieu, CFO
Nancy Somerville, Executive Director

748 American Lung Association of Arkansas
211 Natural Resources Drive 501-224-5864
Little Rock, AR 72205-1539 Fax: 501-224-5645
E-mail: fplunkett@lungark.org
http://www.lungusa2.org/arkansas/index.html

749 Arkansas Association of Conservation Districts
101 E Capitol 501-682-2915
Suite 350 Fax: 501-682-3991
Little Rock, AR 7220 E-mail: Randy.Young@mail.state.ar.us
http://www.aracd.org
Affiliated with the National Association of Conservation Districts. Membership is $25.00 per year for individuals and $1,200.00 per year for organizations and companies.
Randy Young, President

750 Arkansas Environmental Education Association
PO Box 488 301-346-5715
Hackett, AR 72937 Fax: 501-638-7151
E-mail: info@aeea.es
http://www.aeea.org

Rober McAfee, Executive Director

751 Arkansas Environmental Federation
1400 W Markham Street 501-374-0263
Suite 250 Fax: 501-374-8752
Little Rock, AR 72201 http://www.environmentark.org
Formerly the Arkansas Federation of Water and Air Users. Membership is $25.00 per year for individuals and $200.00 to $2,070.02 per year for organizations.
Founded: 1967
Randy Thurman, Executive Director

752 Northern Arkansas: Safety Council of the Ozarks
1111 South Glenstone 417-869-2121
Springfield, MO 800-334-1349
Fax: 417-869-2133
E-mail: dbiggs@nscoazarks.org
http://www.nscozarks.org
Dedicated to interpreting current safety & health issues and developing practical, cost effective, methodologies based on best practices for protecting human life, property and the environment.
Debora S Biggs, Executive Director

753 Southern Arkansas: National Safety Council, Ark-La-Tex Chapter
8101 Kingston Road   318-687-7550
#107   Fax: 318-687-7298
Shreveport, LA 71108 E-mail: altasafetycouncil@wnonline.net
http://www.nscarklatexsafetycouncil.com
Specialists in professional safety training.
Sally Head, Executive Director

California

754 ASLA: California-Sierra Chapter
PO Box 162776   916-447-7400
Sacramento, CA 95816   Fax: 916-447-8270
E-mail: sierra@sbcglobal.net
http://www.asla-sierra.org

Keith P Wilson, President
J Marq Truscott, President-Elect

755 ASLA: Northern California Chapter
American Society of Landscae Architects
5 Third Street   415-974-5430
Suite 724   Fax: 415-543-2112
San Francisco, CA 94103   E-mail: staff@asla-ncc.admn.org
http://www.host.asla.org/chapters/norca
Founded: 1899
Christopher Kent, President
Joe Owen, Executive Director

756 ASLA: San Diego Chapter
PO Box 81521   619-225-8155
San Diego, CA 92138-1521   Fax: 619-225-8151
E-mail: aslasd@att.net
http://www.asla-sandiego.org

Stephen Copley, Director
Tracy Morgan, Executive Director

757 ASLA: Southern California Chapter
1100 Irvine Boulevard   714-838-3615
Suite 371   Fax: 714-730-6296
Tustin, CA 92780   E-mail: sccasla@aol.com
http://host.asla.org/chapters/southca
Tal Jackson, President
Vicki Phillipy, Executive Director

758 American Fisheries Society-Fish Health Section
California State University
Department of Biological Sciences   510-885-3000
Hayward, CA 94542   Fax: 510-885-4747
E-mail: bdixon@csuhayward.edu
http://www.csuhayward.edu

Michael Kent, President

759 American Lung Association of California
American Lung Association
424 Pendleton Way   510-638-5864
Oakland, CA 94621-2189   Fax: 510-638-8984
E-mail: contact@californialung.org
http://www.californialung.org
Founded: 1904
Charles A Heinrich, Chairman
Robert A Green, Vice Chairman

760 American Lung Association of California- Redwood Empire Branch
115 Talbot Avenue   707-527-5864
Santa Rosa, CA 95404   Fax: 707-542-6111
E-mail: redwood@alac.org
http://www.californialung.org
Founded: 1953

761 American Lung Association of California- East Bay Branch
American Lung Association
1900 Powell Street   510-893-5474
Suite 800   Fax: 510-893-9008
Emeryville, CA 94608   E-mail: eastbaylung@alaebay.org
http://www.alaebay.org
Founded: 1904
karen Fulton, President
Dennis Thomatos, HR Director

762 American Lung Association of California— Superior Branch
10 Landing Circle   530-345-5864
Suite 1   800-586-4872
Chico, CA 95973   Fax: 530-345-6035
E-mail: contact@californialung.org
http://www.californialung.org
Founded: 1904
Stephen R O'Kane, Chairman
Don Brunson, Vice Chairman

763 American Lung Association of Central California
American Lung Association
4948 North Arthur   559-222-4800
Fresno, CA 93705   800-586-4872
Fax: 559-221-2081
E-mail: info@amerilungcencal.org
http://www.amerilungcencal.org
Founded: 1904
Michael Peterson MD, Chairman of the Board
Josette Merce Bello, CEO

764 American Lung Association of Los Angeles County
5858 Wilshire Boulevard   323-935-5864
Suite 300   Fax: 323-935-1873
Los Angeles, CA 90036-0926 E-mail: lalung@lalung.org
http://www.lalung.org
Giovanni M Smith, MD, Chair

765 American Lung Association of Orange County - Santa Ana
1570 East 17th Street   714-835-5864
Santa Ana, CA 92705   800-lun- usa
Fax: 714-835-0169
E-mail: alaoc1@aol.com
http://www.oclung.org
Founded: 1909
Glenn Maddalon, Executive Director
Dianna Sicialano, President

766 American Lung Association of Sacramento- Emigrant Trails
909 12th Street   916-444-5864
Sacramento, CA 95814-2997   800-LUN- USA
Fax: 916-444-6661
E-mail: staff@saclung.org
http://www.saclung.org
Founded: 1917
Earl Wisthycombe, President
Jane Hagerdorn, CFO

767 American Lung Association of San Diego & Imperial Counties
American Lung Association
2750 Fourth Avenue   619-297-3901
San Diego, CA 92103   800-LUN- USA
Fax: 619-297-8402
E-mail: info@lungsandiego.org
http://www.lungsandiego.org
Founded: 1904
Janie Davis, Chief Executive Officer

768 American Lung Association of San Francisco & San Mateo Counties
2171 Junipero Serra Boulevard   650-994-5864
Suite 720   Fax: 650-994-4601
Daly City, CA 94014-1980

E-mail: lung@alasfsm.org
http://www.californialung.org
Linda Civitello-Joy, Executive Director
S.Robert Politzer, Board President

769   American Lung Association of Santa Barbara and Ventura Counties
American Lung Association
1510 San Andres Street   805-963-1426
Santa Barbara, CA 93101-4104   800-586-4872
                                Fax: 805-962-2843
                         E-mail: Advocate@lungsbvc.org
                         http://www.californialung.org
Founded: 1904
John Kirkwood, President
Charles A Heinrich, Chairman

770   American Lung Association of Santa Clara - San Benito Counties
1469 Park Avenue   408-998-5864
San Jose, CA 95126   Fax: 408-998-0578
                         E-mail: info@lungsrus.org
                         http://www.lungsrus.org
Founded: 1904
Margo Leathers Sidener, Executive Director
Sheila Blash, Director Office Administrati

771   American Lung Association of the Central Coast
550 Camino El Estero   831-373-7306
Suite 100   800-586-4872
Monterey, CA 93940-3231   Fax: 831-373-5530
                         E-mail: admin@alaccoast.org
                         http://www.californialung.org
Founded: 1953
Karen Fulton, Chief Executive Officer
John Morrison, Vice President

772   American Lung Association of the Inland Counties
441 Mac Kay Drive   909-884-5864
San Bernadino, CA 92408-3230   800-586-4872
                                Fax: 909-884-6249
                         E-mail: pat@alaic.org
                         http://www.californialung.org
Founded: 1916
Patrick Kudell, Executive Director
Terry Robert, Program Associate

773   Asian Pacific Environmental Network
310 8th Street   510-834-8920
Suite 309   Fax: 510-834-8926
Oakland, CA 94607-4253   E-mail: apen@apen4ej.org
                           http://www.apen4ej.org
The Asian Pacific Environmental Network (APEN) empowers low-income Asian Pacific Islander (API) communities to take action on environmental and social justice issues. APEN builds organizations in dis-empowered API communities todevelop lasting capacity of the community to achieve solutions to problems affecting people's lives.
Founded: 1991
Vivian Chang, Executive Director
Lisa DeCastro, Development Associate

774   Bio Integral Resource Center
PO Box 7414   510-524-2567
Berkeley, CA 94707   Fax: 510-524-1758
                         E-mail: birc@igc.org
                         http://www.birc.org
The goal of the Bio Integral Resources Center is to reduce pesticide use by educating the public about effective, less toxic alternatives for pest problems.
Founded: 1979
Dr. William Quarles, Executive Director

775   California Academy of Sciences Library
875 Howard Street   415-321-8000
San Francisco, CA 94103-3009   Fax: 415-321-8633
                         E-mail: info@calacademy.org
                         http://www.calacademy.org

Founded: 1853
Patrick Kociolek, Executive Director
Allison Brown, CFO

776   California Air Resources Board
State of California
PO Box 2815   916-322-2990
Sacremento, CA 95812   800-242-4450
                                Fax: 916-322-3906
                         E-mail: helpline@arb.ca.gov
                         http://www.arb.ca.gov
The California ARB is the state agency charged with coordinating efforts to attain and maintain ambient air quality standards, conduct research into the causes of and solutions to air pollution and its adverse health impacts, andattack systematically the serious problem caused by motor vehicles, which are the major source of air pollution in many areas of the state. The California ARB's mission is to promote and protect public health, welfare and ecological resources.
Founded: 1967
Catherine Weatherspoon, CEO
Cindy K Tuck, Chairman

777   California Association of Environmental Health
3700 Chaney Court   916-944-7315
Carmichael, CA 95608   Fax: 916-944-2256
                         E-mail: justin@ccdeh.com
                         http://www.ccdeh.com

Justin Malan, Executive Director

778   California Association of Resource Conservation Districts
801 K Street   916-447-7237
Suite 1318   Fax: 916-447-2532
Sacramento, CA 95814-3500   http://www.carcd.org
Tom Wehri, Executive Director

779   California BASS Chapter Federation
21517 Appaloosa Court   909-244-6320
Canyon Lake, CA 92587   E-mail: fsh4bss@dellepro.com
                          http://www.californiabass.org
Founded: 1969
Gary Bradford, President

780   California Birth Defects Monitoring Program
1917 Fifth Street   510-549-4155
Berkeley, CA 94710   Fax: 510-549-4175
                         E-mail: info@cbdmp.org
                         http://www.cbdmp.org
The California Birth Defects Monitoring Program is a public health program devoted to finding the causes of birth defects so they can be prevented. The program is funded through the California Department of Health Services and isjointly operated with the March of Dimes Birth Defects Foundation.
Founded: 1982
John Harris, Program Chief

781   California Certified Organic Farmers
1115 Mission Street   831-423-2263
Santa Cruz, CA 95060   Fax: 831-423-4528
                         E-mail: ccof@ccof.org
                         http://www.ccof.org
Premier organic certification since 1973, for growers, processors, handlers, packers, retailers. Public education of organic, and advocacy for public policy to support organic. Newsletter subscriptions available through a supportingmembership program.
Founded: 1973
Vanessa Bogenholm, Chairman
Will Daniels, Vice Chaiman

782   California Chapter, National Safety Council
600 Wilshire Boulevard   213-385-6461
Suite 1263   800-421-9585
Los Angeles, CA   Fax: 213-385-8405
                         E-mail: socal@nsc.org
                         http://www.list.nsc.org/socal/

Jim Boyle, Director

783 California Council for Environmental and Economic Balance

100 Spear Street   415-512-7890
Suite 805   Fax: 415-512-7897
San Francisco, CA 94105   E-mail: cceeb@cceeb.org
     http://www.cceeb.org

Lobbying organization.
Founded: 1976
Victor Weisser, President
Bill Quinn, Executive Director

784 California Renewable Fuels Council

3310 Yorba Linda Boulevard   714-990-3333
Suite 249   Fax: 714-990-0418
Fullerton, CA 92831

785 California Solar Energy Industries Association

P.O. BOX 782   949-837-7430
Rio Vista, CA 94571   Fax: 949-709-8043
     E-mail: info@calseia.or
     http://www.calseia.org

Promotes the growth of California's solar energy industry. Membership is $325-650 per year.
Founded: 1977
Les Nelson, President
Pat Redgate, Vice President

786 California Trappers Association (CTA)

1705 Murchison Drive   650-697-1400
Burlingame, CA 94010   Fax: 650-552-5002
     E-mail: webmaster@cta.org
     http://www.cta.org

CTA's primary concern is Californias furbearing mammals, and the category of small predatory mammals. Approximately 20 of the associations 2,400 members do lengthy studies and assist State and Federal Agencies in this field with theirbiologists. Educational forums and talks are hosted at local universities. Farming agencies and timber companies commonly use their services. Members commonly work for numerous environmental public projects, both public and governmental.
Founded: 1863
Keith Carly, President
Donald Stehsel, Executive Secretary

787 California Waterfowl Association

4630 Northgate Boulevard   916-648-1406
Suite 150   Fax: 916-648-1665
Sacramento, CA 95834   E-mail: cwa_hq@calwaterfowl.org
     http://www.calwaterfowl.org

Founded: 1945
Sylvia Done, Secretary

788 California Wildlife Federation

PO Box 1527   916-441-7563
Sacramento, CA 95812-1527
Randy Walker, President

789 Californians for Population Stabilization (CAPS)

1129 State Street   805-564-6626
Suite 3-D   Fax: 805-564-6636
Santa Barbara, CA 93101   E-mail: info@capsweb.org
     http://www.capsweb.org

A nonprofit, public interest organization that works to protect California's environment and quality of life by turning the tide of population growth.
Founded: 1986
Diana Hull PhD, President
Ben Zuckerman, Vice President

790 Colorado River Board of California

770 Fairmont Avenue   818-543-4676
Suite 100   Fax: 818-543-4685
Glendale, CA 91203-1035   E-mail: crb@crb.ca.gov
     http://www.crb.ca.gov
Founded: 1937
Gerald Zimmerman, Executive Director

791 Communities for a Better Environment

1440 Broadway   510-302-0430
Suite 701   Fax: 510-302-0437
Oakland, CA 94612   E-mail: cbeca@mail.com
     http://www.cbecal.org

Nonprofit statewide, multiracial, urban environmental health and justice organization that works with urban communities and grassroots organizations, using science based research, legal tactics and organizing strategies to prevent airand water pollution, eliminate toxic hazards and improve public health. Long-term goals are to develop an environmentally sustainable manufacturing base, minimize the use of toxins, expand pollution prevention strategies and involve people most atrisk.
Jose T Bravo, President
Luke Cole, Vice President

792 Concerned Citizens of South Central Los Angeles

4111 S Central Avenue   213-846-2505
Suite 101   Fax: 213-846-2508
Los Angeles, CA 90011

The mission of the Concerned Citizens of South Central Los Angeles is to fight for social, economic and environmental justice and to encourage resident participation in the process.
Melodie Dove, Environmental Organizer

793 Council for Planning and Conservation

PO Box 228   310-276-2685
Beverly Hills, CA 90213   E-mail: esharris@earthlink.net
     http://www.beverlyhillscitizen.org
Ellen S Harris, President

794 Desert Tortoise Preserve Committee

Tortoise T-R-A-C-K-S
4067 Mission Inn Avenue   909-683-3872
Riverside, CA 92501   Fax: 909-683-6949
     E-mail: dtpc@pacbell.net
     http://www.tortise-tracks.org
Michael Connor, Executive Director

795 Ecology Center

2530 San Pablo Avenue   510-548-2220
Berkeley, CA 94702   Fax: 510-548-2240
     E-mail: info@ecologycenter.org
     http://www.ecologycenter.org

The Ecology Center promotes environmentally and socially responsible practices through programs that educate, demonstrate and provide direct services. The Ecology Center currently runs Ecology Center Bookstore, 3 Berkeley farmers'markets, a curbside recycling program and an Environmental Resource Center/Library.
Founded: 1969
Martin Bourque, Executive Director
Martin Bourque, Executive Director

796 Environmental Defense Center

906 Garden Street   805-963-1622
Santa Barbara, CA 93101   Fax: 805-962-3152
     E-mail: edc@rain.org
     http://www.rain.org

The Environmental Defense Center is a nonprofit, public interest organization that provides legal, educational and advocacy support to advance environmental quality. EDC primarily serves community groups on California's South CentralCoast. EDC selects cases and projects that preserve the environment for furture generations, protect human health, promote appropriate management and use of natural resources, and enhance the character of the community.
Linda Krop, Executive Director

797 Environmental Health Coalition

401 Mile of Cars Way   619-474-0220
Suite 310   Fax: 619-474-1210
National City, CA 91950http://www.environmentalhealth.org

One of the oldest and most effective grassroots organizations in the US, using social change strategies to achive environmental and social justice. We believe that justice is accomplished by empowered communities acting together tomake social change. We organize and advocate to protect public health and the environment threatened by toxic pollution. EHC supports broad efforts that create a just society which foster a healthy and sustainable quality of life.

Founded: 1985
Diane Takvorian, Executive Director
Lilia Escalante, Financial Manager

798   Environmental Health Network
PO Box 1155   415-541-5075
Larkspur, CA 94977-1155   E-mail: fdadockets@oc.fda.gov
                                          http://www.ehnca.org
Nonprofit, volunteer organization, whose main goal is to promote public awareness of evironmental sensitivities and causative factors. EHN's focus is on issues of access and
developments relating to the health and welfare of
theenvironmentally sensitive.
Barb Wilkie, President

799   Friends of the River
915 20th Street   916-442-3155
Sacramento, CA 95814   888-464-2477
                                          Fax: 916-442-3396
                        E-mail: info@friendsoftheriver.org
                            http://www.friendsoftheriver.org
Founded: 1973
Peter Ferenbach, Executive Director
Steve Evans, Conservation Director

800   Institute for the Human Environment
Institute of International Education, New Yourk
530 Bush Street, Suite 1000   415-362-6520
San Francisco, CA 94108   Fax: 415-392-4667
                                     E-mail: iiesf@iie.org
                                          http://www.iie.org
Founded: 1919
Allen Goodman, President
Karin Eisele, Executive Director

801   Laotian Organizing Project
220 25th Street   510-236-4616
Richmond, CA 94804-1808   Fax: 510-236-4572
                                     E-mail: apen@apen4ej.org
                                          http://www.apen4ej.org
The Laotian Organization Project (LOP), a project of the Asian
Pacific Environmental Network, is a membership-based organization of Laotian residents in Richmond and San Pablo, California. LOP works to bring people from the Laotiancommunity
together to identify problems, develop solutions, and take action
for a more healthy, safe, and just community.
Grace Kong

802   Lead Safe California
100 Pine Street   415-397-9401
26th Floor   Fax: 415-397-5159
San Francisco, CA 94111   E-mail: cehn@cehn.org
                                          http://www.cehn.org
Lead Safe California is a nonprofit, public interest organization
dedicated to the prevention of childhood lead poisoning and the
preservation of affordable, safe housing.
Ellen Widess, Executive Director

803   League to Save Lake Tahoe
955 Emerald Bay Road   530-541-5388
South Lake Tahoe, CA 96150   Fax: 530-541-5454
                                  E-mail: info@keeptahoeblue.org
                                     http://www.keeptahoeblue.org
Our mission is the preservation and restoration of the magnifi-
cent natural attributes of the Tahoe Basin's waters, forests, wild-
life and landscape for the enjoyment of present and future
generations.
Founded: 1957
Rochelle Nason, Executive Director
John Jorgenson, President

804   Mountain Lion Foundation
PO Box 1896   916-442-2666
Sacramento, CA 95812   800-319-7621
                                          Fax: 916-442-2871
                             E-mail: mlf@mountainlion.org
                                  http://www.mountainlion.org

A nonprofit conservation and education organization dedicated
to protecting the mountain lion, its wild habitat, and the wildlife
that shares that habitat- for present and future generations
throughout California and the west. TheFoundation is dedicated
to the proposition that much can be done to preserve the cougar
as a viable species and that the success of this effort can assure
the survival of other species.
Founded: 1986
Lynn Sadler, Executive Director
Tim Dunbar, Associate Director

**805   National Institute for Global Environmental Change:
Western Regional Center**
University of Califonia-Davis
1 Shields Avenue   530-752-7300
Davis, CA 95616   Fax: 530-752-7302
                                     E-mail: westgec@ucdavis.edu
                            http://http://nigec.ucdavis.edu/westgec/
Founded: 1990
Dr. Susin Ustin, Director

806   National Wildlife Federation
Western Natural Resource Center
3500 5th Avenue   206-285-8707
Suite 101   Fax: 619-296-8355
San Diego, CA 92103   E-mail: wnrc@nwf.org
                                     http://www.nwf.org/western
The Western Natural Resource Center of NWF connects the peo-
ple and issues of the West - educating, inspiring and empower-
ing individuals from all walks of life to protect and conserve the
regions diverse wildlife and natural resources.
Founded: 1934
Jerome C Ringo, Chairman
Stephen E Petron, Vice Chairman

807   Nature Conservancy: Western Division Office
201 Mission Street   415-777-0487
4th Floor   Fax: 415-777-0244
San Francisco, CA 94105   E-mail: comment@tnc.org
                                          http://www.tnc.org

Mark Burgett, President
Mick Sweeney, CFO

808   Northcoast Environmental Center
575 H Street   707-822-6918
Arcata, CA 95521   Fax: 707-822-0827
                                     E-mail: nec@yournec.org
                                          http://www.yournec.org
A nonprofit educational organization devoted to illuminating
people concerning the Biosphere. It has a library, information
and referral, radio show, nationally circulated newsletter and
national membership. It is focused on theredwood region and
northwest California and the Bioregoin along the California-Or-
egon border.
Founded: 1971
Tim McKay, Executive Director
Sid Dominitz, Editor

809   Pesticide Action Network North America
49 Powell Street   415-981-1771
Suite 500   Fax: 415-981-1991
San Francisco, CA 94102   E-mail: panna@panna.org
                                          http://www.panna.org
Pesticide Action Network North American advocates the adop-
tion of ecologically sound practices in place of pesticide use.
PANNA works with more than 100 affiliated oragnizations in
Canada, Mexico and US, as well as with PesticideAction Net-
work partners around the world to demand that development
agencies and governments redirect support from pesticides to
safe alternatives.
Monica Moore, Co-Director
Stephen Scholl Buckwald, Managing Director

810   Pesticide Education Center
PO Box 225279   415-665-4722
San Francisco, CA 94122-5279   Fax: 415-665-2693
                                     E-mail: pec@igc.org
                                          http://www.pesticides.org
The mission of the Pesticide Education Center is to educate the
public about the adverse health effects of exposure to pesticides
in the home, community and at work.

Founded: 1988
Marion Moses, President

811 Rainforest Action Network
1985
221 Pine Street   415-398-4404
Suite 500   Fax: 415-398-2732
San Francisco, CA 94104   E-mail: rainforest@ran.org
                                              http://www.ran.org
Rainforest Action Network works to protect the Earth's rainforest and support the rights of their inhabitants through education, grassroots organizing, and non-violent direct action.
Michael Brune, Executive Director

812 Save San Francisco Bay Association
350 Frank Ogawa Plaza   510-452-9261
Suite 900   Fax: 510-452-9266
Oakland, CA 94612   E-mail: savebay@savesfbay.org
                                              http://www.savesfbay.org
Save the Bay has worked for over 40 years to protect the San Francisco Bay-Delta from pollution, fill, shoreline destruction and fresh water diversion. We have launched a century of renewal to restore bay fish and wildlife, reclaimtidal wetlands and make the bay safe and accessible to all.
Founded: 1961
David Lewis, Executive Director
Robin Erickson, Director Finance/Administrat

813 Scenic California
2215 5th Street   510-883-0390
Berkeley, CA 94710   Fax: 510-883-0391
            E-mail: sceniccalifornia@lsa-assoc.com
                                    http://www.sceniccalifornia.org
Scenic America is the only national nonprofit organization dedicated to protecting natural beauty and distinctive community character. We provide technical assistance across the nation and through affiliates on scenic byways, billboardand sign control, context sensitive highway design, wireless telecommunications tower location, transportation enhancements, and other scenic conservation issues.
Sheila Brady, Manager

814 Sierra Club
PO Box 3357   805-822-4371
Bakersfield, CA 93385   http://www.sierraclub.org
To advance the preservation and protection of the natural environment by empowering the citizenry, especially democratically-based grassroots organizations, with charitable resources to further the cause of environmental protection,the vehicle through which The Sierra Club Foundation generally fulfills its charitable mission.
Founded: 1894
Lorraine Unger, Chairman
Harry Love, Vice Chairman

815 Sierra Club: San Diego Chapter
Sierra Club Nationals
3820 Ray Street   619-299-1743
San Diego, CA 92104-3623   Fax: 619-299-1742
            E-mail: san-diego.chapter@sierraclub.org
                          http://www.sierraclub.org/chapters/sandiego/
A nonprofit organization dedicated to preserving, protecting and enjoying the earth.
Founded: 1892
Cheryl Reiff, Office Administrator
Richard Miller, Chairman

816 Sierra Club: San Francisco Chapter
85 Second Street   415-977-5500
2nd Floor   Fax: 415-977-5799
San Francisco, CA 94105E-mail: information@sierraclub.org
                                              http://www.sierraclub.org
Nonprofit, member-supported, public interest organization that promotes enjoyment and preservation of the nation's forests, waters, wildlife and wilderness. In addition to a wide variety of environmental and conservation interests andactivities, active in the areas of pollution prevention and control and population policy.
Founded: 1892
Carl Pope, Executive Director
Lisa Remstrom, President

817 Sierra Club: San Gorgonio Chapter
4079 Mission Inn Avenue   909-684-6203
Riverside, CA 92501-3204   Fax: 909-684-6172
            E-mail: gorgonio@pe.net
            http://www.microbyte.net/sierra
To explore, enjoy and protect the wild places of the earth, to practice and promote the responsible use of the earth's ecosystems and resources; to educate and enlist humanity to protect and restore the quality of the natural and humanenvironment; and to use all lawful means to carry out these objectives.
Ralph Salisbury, Chapter Chair

818 Southwestern Herpetologists Society
PO Box 7469   818-503-2052
Van Nuys, CA 91409   E-mail: maven@post.com
                                              http://www.swhs.org
Founded: 1954
Bud James, President
Sabine Bradley Phillips, Vice President

819 Trout Unlimited
828 San Pablo Avenue   510-528-5390
Suite 244   Fax: 510-528-7880
Albany, CA 94706   E-mail: info@tucalifornia.org
                                              http://www.tucalifornia.org
Founded: 1959
Charles Gauvin, President/CEO
Kenneth Mendez, Executive Vice President

820 Urban Habitat Program
Presido Station
PO Box 29908   415-561-3336
San Francisco, CA 94129-9908   Fax: 415-561-3334
            E-mail: contact@urbanhabitatprogram.org
                          http://www.urbanhabitatprogram.org
Dedicated to building a multicultural majority that provides urban environmental leadership in order to create socially just, ecologically sustainable communities in the Bay Area.
Founded: 1989
Carl Anthony, Director

821 Western Occupational and Environmental Medical Association
575 Market Street   415-927-5736
Suite 2125   Fax: 415-927-5726
San Francisco, CA 94105   E-mail: woema@hp-assoc.com
                                              http://www.woema.org
The mission is to represent and be a resource to members in the profession and practice of occupational and environmental medicine and to enhance their efforts to promote and improve health in the workplace.
Founded: 1941
Robert Orford, President

822 Wildlife Society, San Joaquin Valley Chapter
1234 E Shaw Avenue   559-243-4005
Fresno, CA 93710   Fax: 559-243-4022
            E-mail: ekleinfe@dfg.ca.gov
A nonprofit scientific and educational organization that serves professionals such as government agencies, academia, industry and non-government organizations in all areas related to the conservation of wildlife and natural resourcesmanagement.
Eric Kleinfelter, President

Colorado

823 ASLA: Colorado Chapter
6456 South Niagra Court   303-830-6616
Centennial, CO 80111   Fax: 303-220-5833
            E-mail: mngmntplus@qwest.net
                                              http://www.ccasla.org
Founded: 1973
John Birkey, President
Donna Ralston, Association Manager

824 American Lung Association of Colorado
American Lung Association
5600 Greenwood Plaza Boulevard   303-388-4327
Suite 100   Fax: 303-377-1102
Greenwood Village, CO 80111   http://www.lungusa.org
Founded: 1904
Curt Hubbert, President

825 Arkansas River Compact Administration
122 E Elm Street   719-336-2732
PO Box 1600
Lamar, CO 81052-2702
Janet Anderson, Executive Director

826 Association of Midwest Fish and Game Law Enforcement Officers
Division of Wildlife
6060 Broadway   303-291-7223
Denver, CO 80216
Founded: 1944

827 Bison World
National Bison Association
1400 W. 122nd Ave.   303-292-2833
Suite 106   Fax: 303-292-2564
Westminster, CO 80234   E-mail: info@bisoncentral.com
                        http://www.bisoncentral.com
An organization of bison producers dedicated to awareness of the healthy properties of bison meat and bison production.
Founded: 1978
Dave Carter, Executive Director
Merle Maas, Chairman of Board

828 Colorado Association of Conservation Districts
3000 Youngfield Street   303-232-6242
Suite 163   Fax: 303-232-1624
Lakewood, CO 80215   http://www.cascd.com
Promotes soil and water conservation. Membership is $15.00 per year for individuals and $250.00 per year for organizations.
12 pages Quarterly
Jerry Schwien, Author
Callie Hendrickson, Executive Vice President

829 Colorado BASS Chapter Federation
4485 Enchanted Circle North   719-597-2304
Colorado Springs, CO 80917 E-mail: nozlnut36@comcast.net
                http://www.coloradobassfederation.org
John Bentz, President
Dave Lishman, Vice President

830 Colorado Forestry Association
Colorado Forestry
603 Atwood Court   970-223-3255
Ft. Collins, CO 80525   E-mail: pjhoefer@aol.com
                    http://www.coloradoforestry.org/
Founded: 1880
Philip Hoefer, President
Ken Ashley, Secretary

831 Colorado Safety Association
4730 Oakland Street   303-373-1937
Suite 500   800-727-0519
Denver, CO 80239   Fax: 303-373-1955
               E-mail: melodye@coloradosafety.org
                    http://www.coloradosafety.org
Not-for-profit, non-governmental educational organization specializing in occupational safety and health issues.
Melodye Turek, President

832 Colorado Solar Energy Industries Association
PO Box 18191   303-333-7342
Boulder, CO 80308-1191   866-633-9764
                                Fax: 970-491-7736
                             E-mail: info@coseia.org
                             http://www.coseia.org
Jon Klima, President
Thom Johnson, Secretary

833 Colorado Trappers Association
PO Box 397
Empire, CO 80438
Annual membership (individual or family) $25.00
Debra Watts, Secretary

834 Colorado Water Congress
1580 Logan Street   303-837-0812
Suite 400   Fax: 303-837-1607
Denver, CO 80203   E-mail: macravey@cowatercongress.org
                        http://www.cowatercongress.org
Protects and conserves Colorado's water resources by means of advocacy and education.
Founded: 1958
David Robbins, President
Richard D MacRavey, Executive Director

835 Colorado Wildlife Federation
Colorado Wildlife
4045 Wadsworth Blvd   303-987-0400
Suite 20   Fax: 303-987-0200
Wheat Ridge, CO 80033   E-mail: cwf@coloradowildlife.org
                        http://www.coloradowildlife.org
Founded: 1953
Wayne East, Executive Director
Nancy O'Brien, Manager

836 Keystone Center and Keystone Science School
Keystone Center & Science School
1628 Sts. John Road   970-513-5800
Keystone, CO 80435   Fax: 970-262-0152
                        E-mail: info@keystone.org
                        http://www.keystone.org
Nonprofit public policy and educational organization founded in 1975. Strives to develop creative problem-solving processes that assist diverse parties address issues of importance and to provide qualified science education through hands-on inquiry of the natural world. Keystone Center pursues this end through its two divisions, the Science and Public Policy Program and the Keystone Science School.
Founded: 1975
Robert W Craig, President
Dirk Forrister, Managing Director

837 Lighthawk, Southern Rocky Mountain Field Office
PO Box 653   307-332-3242
Lander, WY 82520   Fax: 307-332-1641
                        E-mail: info@lighthawk.org
                        http://www.lighthawk.org
Founded: 1979
Maureen Smith, Executive Director

838 National Wildlife Federation
11100 Wildlife Center Drive   303-786-8001
Reston, VA 20190-5362   800-822-9919
                                Fax: 303-786-8911
                             http://www.nwf.org
Founded: 1934
Eileen Morgan Johnson, Secretary
Larry Schweiger, President

839 Nature Conservancy
1881 9th Street   303-444-2950
Suite 200   800-628-6860
Boulder, CO 80302   Fax: 303-444-2986
                             E-mail: comment@tnc.org
                             http://www.nature.org
Founded: 1915
Steven J McCormick, Chief Executive

840 Sierra Club
2260 Baseline Road   303-449-5595
Suite 105   Fax: 303-449-6520
Boulder, CO 80302-7737 E-mail: information@sierraclub.org
                        http://www.sierraclub.org
To advance the preservation and protection of the natural environment by empowering the citizenry, especially democratically based grassroots organizations with charitable resources to further the cause of environmental protection, the vehicle through which The Sierra Club Foundation generally fulfills its charitable mission.

Adriana Raudvens, Regional Representative

841 Trout Unlimited
1320 Pearl Street
Suite 320   303-440-2937
Boulder, CO 80302   Fax: 303-440-7933
E-mail: dnickum@tu.org
http://www.tu.org
Founded: 1959
Dave Nickum, Executive Director
Charles Gauvin, President

842 Wildlife Society
317 W Prospect
Fort Collins, CO 80526   970-472-4461
E-mail: rsell@co.blm.gov
A nonprofit scientific and educational organization that serves professionals such as government agencies, academia, industry, and non-government organizations in all areas related to the conservation of wildlife and natural resourcesmanagement.
Robin Sell, Secretary

843 Yellowstone Grizzly Foundation
104 Hillside Court
Boulder, CO 80302-9452   303-939-8126

Connecticut

844 ASLA: Connecticut Chapter
87 Willow Street   800-878-1474
New Haven, CT 06511   800-878-1474
E-mail: roderickcameron@ccaengineering.com
http://www.ctasla.org
Roderick E Cameron, President
Brian Robinson, Vice President

845 American Association in Support of Ecological Initia-tives
150 Coleman Road   860-364-2967
Middletown, CT 06457   Fax: 860-347-8459
E-mail: wwasch@wesleyan.edu
http://www.wesleyan.edu
AASEI is a US 501 nonprofit organization which suports international environmental initiatives in Russian Nature Reserves. In cooperation with Russia and foreign scientists, students, and universities, AASEI organizes scientificresearch projects, academic internships, work camps, environmental exchanges, and eco-tourism. Our aim is to provide practical support to Russian Reserves, expand opportunities for international scientific research, and promote internationalunderstanding.
Founded: 1994
Stephanie Hitztaler, Acting Executive Director
Brendan Sweeney, President

846 American Lung Association of Connecticut
American Lung Association
45 Ash Street   860-289-5401
East Hartford, CT 06108-3272   800-586-4872
Fax: 860-289-5405
E-mail: bcase@alact.org
http://www.alact.org
Founded: 1904
John Zinn, Chief Executive Officer
Margaret LaCroix, Vice President

847 Connecticut Audubon Society
2325 Burr Street   203-259-6305
Fairfield, CT 06824   Fax: 203-254-7673
E-mail: casfairfield@ctaudubon.org
http://www.ctaudubon.org
A statewide, nonprofit, membership organization dedicated to protecting the Connecticut environment by providing citizens of all ages with top-quality education and outdoor experiences. Each year, through school programs, teachertraining workshops, youth activities, adult and family trips, community events and legislative initiatives, the Society reaches more than 175,000 people.
Robert Martinez, President
Barbara Strickland, Chairman

848 Connecticut Botanical Society
55 Harvest Lane   860-633-7557
Glastonbury, CT 06033   http://www.ct-botanical-society.org
Founded: 1903
Casper Ultee, President

849 Connecticut Forest and Park Association
16 Meriden Road   860-346-2372
Rockfall, CT 06481-2961   Fax: 860-347-7463
E-mail: conn.forest.assoc@snet.net
http://www.ctwoodlands.org
An organization for forest and wildlife conservation. Develops outdoor recreation and natural resources. Provides forest management, construction of hiking trails and consultation in the areas of forestry and environment.
Founded: 1895
Dick Whitehouse, President
Adam Moore, Executive Director

850 Connecticut Fund for the Environment
Fact Sheet
205 Whitney Ave   203-787-0646
1st floor   Fax: 203-787-0246
New Haven, CT 06511-3725   http://www.cfenv.org
Founded: 1978
Donald Strait, Executive Director
Curt Johnson, Senior Staff Attorney

851 Friends of Animals
777 Post Road   203-656-1522
Suite 205   Fax: 203-656-0267
Darien, CT 06820   E-mail: info@friendsofanimals.org
http://www.friendsofanimals.org
Founded: 1957
Priscilla Feral, President

852 Litchfield Environmental Council: Berkshire
Oriion Grassroots Network
PO Box 552   860-435-2004
Lakeville, CT 06039   E-mail: BLEC1@earthlink.net
http://www.oriononline.org/
M.G.H. Gilliam, Chairman

853 Save the Sound
185 Magee Avenue   203-327-9786
Stamford, CT 06902-5906   888-728-3547
Fax: 203-967-2677
E-mail: savethesound@snet.net
http://www.savethesound.org
Save the Sound Inc is a nonprofit membership organization dedicated to the restoration, protection and appreciation of Long Island Sound and its watershed through education, research and advocacy.
Founded: 1972
John Atkin, President

854 Trout Unlimited
Trout Unlimited
P.O. Box 151   860-464-8254
Ledyard, CT 06339   E-mail: rguarino@thamesvalleytu.org
http://www.thamesvalleytu.org/
Richard Guarino, President
Mike Goodwin, Vice President

District of Columbia

855 ASLA: Potomac Chapter
PO Box 18184
Washington, DC 20036-8184   E-mail: info@potomacasla.org
http://www.potomacasla.org
Kristiana Farnsworth, President
Barbara L Deutsch, President-Elect

856 American Lung Association of the District of Columbia
475 H Street Northwest   202-682-5864
Washington, DC 20001-2617   Fax: 202-682-5874

E-mail: info@aladc.org
http://www.aladc.org

Katrina Jones, Project Cordinator

857 Environmental Working Group
1436 U Street NW   202-667-6982
Suite 100   Fax: 202-232-2592
Washington, DC 20009   http://www.ewg.org
The Environmental Working Group is a small, computer powered research organization dedicated to improving environmental protection through the analysis of federal and state regulatory policies and performance and through technicalassistance and education.
Founded: 1993
Ken Cook, Presudent
Richard Wiles, Senior Vice President

858 Human Environment Center
1930 18th Street NW   202-588-8036
Suite 24   Fax: 202-588-9422
Washington, DC 20009
Hector Eriksen-Mendoza, Executive Director

859 Washington, DC: Chesapeake Region Safety Council
Rutherford Business Center   410-298-4770
17 Governor's Court   800-875-4770
Baltimore, MD 21244   Fax: 410-281-1350
E-mail: safety@chesapeakesc.org
http://www.chesapeakesc.org

Delaware

860 American Lung Association of Delaware
American Lung Association
1021 Gilpin Avenue   302-655-7258
Suite 202   800-586-4872
Wilmington, DE 19806-3280 Fax: 302-655-8546
E-mail: dbrown@alade.org
http://www.alade.org/main.htm
Founded: 1904
Peter Shanley, Chairman
John D'Angelo, Secretary

861 Atlantic Waterfowl Council
Division of Fish and Wildlife
PO Box 1401   302-739-5295
Dover, DE 19903   Fax: 302-739-6157
William C Wagner II, Chairman

862 Delaware Association of Conservation Districts
PO Box 242   302-789-4411
Dover, DE 19903-0242   Fax: 302-739-6724
Terry Pepper, President
Martha Pileggi, Staff Assistant

863 Delaware BASS Chapter Federation
2453 S State Street   302-698-9257
Camden, DE 19901   800-463-6062
                          Fax: 720-302-1230
http://www.easyy.net\'delbass
Founded: 1975
Jim Fields, Cheif Executive

864 Delaware Greenways
100 W. 10th St., Suite 1001   302-655-7275
P.O Box 2095   Fax: 302-655-7274
Wilmington, DE 19899   E-mail: greenways@dca.net
http://www.delawaregreenways.org/
Committed to the preservation and advencement of Delaware's natural, scenic, historic, cultural, and recreational resources. Works to accomplish this in preserving and connecting open space greenways, increasing opportunities forwalking and biking and creating more livable communities.
Founded: 1990
Robert J Valihura, Jr, President
Tim Plemmons, Executive Director

865 Delaware: Chesapeake Region Safety Council
Rutherford Business Center   410-298-4770
17 Governor's Court   800-875-4770
Baltimore, MD 21244   Fax: 410-281-1350
E-mail: safety@chesapeakesc.org
http://www.chesapeakesc.org

866 Nature Conservancy
260 Chapman Road   303-369-4144
Suite 210D   Fax: 302-369-4143
Newark, DE 19702   E-mail: comment@tnc.org
http://www.nature.org
Founded: 1984
Roger Jones, State Director
Terrence Scanlon, President

867 Save Wetlands and Bays
41 Beaver Circle   302-945-8578
Lewes, DE 19958
Henry Glowiak, President

868 Sierra Club
PO Box 160   302-664-0627
Nassau, DE 19969   Fax: 302-644-9712
E-mail: mike.damico@sierraclub.org
To advance the preservation and protection of the natural environment by empowering the citizenry, especially democratically-based grassroots organizations, with charitable resources to further the cause of environmental protection.The vehicle through which The Sierra Club Foundation generally fulfills its charitable mission.

Florida

869 ASLA: Florida Chapter
PO Box 770219
Naples, FL 34107-0219   E-mail: info@flasla.org
http://www.flasla.org
Jeff Caster, President
Kevin R Boyett, President-Elect

870 American Fisheries Society: Agriculture Economics Section
University of California
PO Box 240   352-392-4991
Gainsville, FL 32611   Fax: 352-392-3646
E-mail: adams@fred.ifas.ufl.edu
Charles Adams, President

871 American Lung Association of Florida
American Lung Association
5526 Arlington Road   904-743-2933
Jacksonville, FL 32211-5216   800-940-2933
                          Fax: 904-743-2916
E-mail: alaf@lungfla.org
http://www.lungfla.org
Founded: 1904
Pablo Mila, Director

872 American Lung Association: Central Area Office
American Lung Association
1333 West Colonial Drive 407-425-5864
Orlando, FL 32804-7133   800-586-4872
                          Fax: 407-425-2876
E-mail: info@alacf.org
http://www.lungfla.org
Founded: 1904
Martha Bogdan, Chief Executive

873 American Lung Association: Gulfcoast Area
American Lung Association
110 Carillon Parkway   727-347-6133
Saint Petersburg, FL 33716   Fax: 727-345-0287
E-mail: alagf@alagf.org
http://www.gulflung.org
Founded: 1904
Shirley Westrate, Area Executive Director

874 American Lung Association: Gulfcoast Area— Southwest Office
American Lung Association
12734 Kenwood Lane   239-275-7577
Suite 25   800-586-4872
Fort Meyers, FL 33907   Fax: 239-275-6739
                               E-mail: alagf@alagf.org
                               http://www.gulflung.org
Founded: 1904
Darius Joseph, President
Shirley M Westrate, Chief Operating Officer

875 American Lung Association: North Area- Northwest Office
American Lung Association
4300 Bayou Boulevard   850-478-5864
Suite 2   800-586-4872
Pensacola, FL 32503-2677   Fax: 850-474-6354
                          E-mail: alafnw@networktel.net
                               http://www.lungfla.org
Founded: 1904
John L Kirkwood, President

876 American Lung Association: North Area- Big Bend Office
539 Silver Slipper Lane   850-386-2065
Suite A   Fax: 850-422-1894
Tallahassee, FL 32303-4873   E-mail: alaf@earthlink.net
                               http://www.lungfla.org
Brenda Olsen, Director of Governmental Aff
Kelsey Ryan, President

877 American Lung Association: North Area- Daytona Office
American Lung Association
412 S. Palmetto Avenue   386-255-6447
Daytona Beach, FL 32114-4922   800-LUN- USA
                                   Fax: 386-253-2410
                        E-mail: alafspaceport@cfl.rr.com
                               http://www.lungusa.org/
Founded: 1904
Charles A Heinrich, Chairman
Robert A Green, Vice Chairman

878 American Lung Association: South Area Office
2020 South Andrews Avenue   954-524-4657
Fort Lauderdale, FL 33316-3430   800-524-8010
                                   Fax: 954-524-3162
                               E-mail: info@sflung.org
                               http://www.sflung.org
Founded: 1904
Denise Grimsley, President

879 American Lung Association: Southeast Area Office
American Lung Association : Florida
2090 Palm Beach Lakes Boulevard   561-659-7644
Suite 900   800-330-5864
West Palm Beach, FL 33409   Fax: 561-835-8967
                        E-mail: amlungsefl@inhaleexhale.org
                               http://www.inhaleexhale.com
Founded: 1937
James Sugarman, Executive Director

880 American Lung Association: Southeast Area- Belle Glade Office
American Lung Association of Florida
136 South Main Street   561-993-3632
Belle Glade, FL 33430   800-586-4872
                                   Fax: 561-993-3433
                       E-mail: amlungself@enhaleexhale.org
                               http://www.lungusa.org
Founded: 1904
Paul Polisena, President
James Sugarman, Executive Director

881 American Lung Association:Gulfcoast Area- Nature Coast Office
American Lung Association
PO Box 1445   352-860-0616
Inverness, FL 34451-1445   800-586-4872
                                   Fax: 352-860-0336
                               http://www.lungusa.org
Founded: 1904
Heinrich, Chairman

882 American Lung Association:Gulfcoast Area- East Bay Office
110 Carolon Parkway   813-962-4448
St. Petersburg, FL 33716   Fax: 727-345-0287

883 American Lung Association:Gulfcoast Area— South Bay Office
3333 Clark Road   941-377-5864
Suite 100   Fax: 941-342-6099
Sarasota, FL 34231

884 Association of Battery Recyclers
PO Box 290286   813-626-6151
Tampa, FL 33687   Fax: 813-622-8388
                        E-mail: info@batteryrecyclers.com
                               http://www.batteryrecyclers.com
To keep members abreast of environmental, health, and safety
requirements that affect our industry.
Founded: 1984
Earl Cornette, President
Joyce Morales Caramella, Secretary

885 Audubon of Florida
Travenier Science Center
115 Indian Mound Trail   305-852-5318
Tavernier, FL 33070   Fax: 305-852-8012
                            E-mail: wmones@audubon.org
                               http://www.audubon.org
The science center is the research arm of the Everglades Campaign. The mission of Audubon of Florida is to ensure the restoration and conservation of the Greater Everglades Ecosystem in order to achieve an ecologically and economicallysustainable South Florida. Our Everglades Conservation office has a five-part program including science, education, advocacy, outreach and grassroots action.
Founded: 1938
David Anderson, Executiv Director
Mark Krawse, Deputy Director

886 Central and North Florida Chapter, National Safety Council
7001 Lake Ellenor Drive   407-370-4098
Suite 120   800-427-2713
Orlando, FL 32809   Fax: 407-370-9902
                             E-mail: floridacn@nsc.org
                               http://www.floridacn.nsc.org
Bob Wilson, Executive Director

887 Florida Defenders of the Environment
4424 NW 13th Street   352-378-8465
Suite C-8   Fax: 352-377-0869
Gainesville, FL 32609-1885   E-mail: fde@fladefenders.org
                               http://www.fladefenders.org
One of the oldest and most accomplished conservation organizations in Florida with a network of scientists, economists and other professionals dedicated to preserving and protecting the state's natural resources. FDE's top priority iscurrently the restoration of a 16-mile stretch of the Ocklawaha River and its 9,000-acre floodplain forest by removal of Rodman Dam- the last vestige of the Cross-Florida Barge Canal.
Founded: 1969
Nick Williams, Executive Director

888 Florida Environmental Health Association
PO Box 271823   863-499-2550
Tampa, FL 33688-1823   Fax: 863-499-2654
                               E-mail: fehaweb@feha.org
                               http://www.feha.org

Promotes public health by means of advanced environmental control.
Founded: 1967
Timothy Mayer, President
Jennifer Williams, Executive Director

889 Florida Forestry Association
PO Box 1696   850-222-5646
Tallahassee, FL 32302   Fax: 850-222-6179
E-mail: info@forestfla.org
http://www.floridaforest.org
Founded: 1923
Jeff Doran, Executive Vice President
Debbie Bryant, Director of Member Services

890 Florida Keys Wild Bird Rehabilitation Center
93600 Overseas Highway   305-852-4486
Tavernier, FL 33070   Fax: 305-852-3186
E-mail: fkwbc@terrnova.net
http://www.florida-keys.fl.us

Horn Bruce, President
Laura Quinn, Executive Director

891 Florida Ornithological Society
143 Beacon Lane   850-942-2489
Jupiter, FL 33469   E-mail: necox@nettally.com
http://www.fosbirds.org
Founded: 1972
Jack Hailman, President
Susan B Whiting, Vice President

892 Florida Public Interest Research Group
926 E. Park Ave.   850-224-3321
Tallahassee, FL 32301   Fax: 850-224-1310
E-mail: info@floridapirg.org
http://www.floridapirg.org
State wide, nonprofit, public interest advocacy organization that focuses primarily on environmental and consumer protection.
Mark Ferrulo, Executive Director
Holly Binns, Field Director

893 Florida Solar Energy Industries Association
231 W Bay Avenue   407-339-2010
Longwood, FL 32750   800-426-5899
Fax: 407-260-1582
E-mail: bruce@flaseia.org
http://www.flaseia.org
A nonprofit professional association of companies involved in the solar energy industry.
Founded: 1977
R Bruce Kershner, Executive Director

894 Florida Trail Association
5415 SW 13th Street   352-378-8823
Gainesville, FL 32608   Fax: 352-378-4550
E-mail: fta@floridatrail.org
http://www.florida-trail.org
Builds, maintains and preserves the Florida Trail, a 1300 mile trail from Big Cypress Preserve to Gulf Islands National Seashore for hikers and backpackers. Sponsors hikes and canoe trips.
Founded: 1964
Deborah Stewart-Kent, Executive Director
Diane Wilkins, Office Manager

895 International Association for Hydrogen Energy
PO Box 248266   305-284-4666
Coral Gables, FL 33124   Fax: 305-284-4792
E-mail: ayfer@iahe.org
http://www.iahe.org
Advances the day when hydrogen energy will become the principal means by which the world will achieve its long-sought goal of abundant clean energy. Toward this end, the Association endeavors to inform scientists and the public of the important role of hydrogen energy in the planning of an inexhaustible and clean energy system through its publications (International Journal of Hydrogen Energy) and conferences.
Founded: 1974
T Nejat Veziroglu, President
Tokio Ohta, Vice President

896 International Game Fish Association
300 Gulf Stream Way   954-927-2628
Dania Beach, FL 33004   Fax: 954-924-4299
E-mail: hq@igfa.org
http://www.igfa.org
Founded as record-keeper and to maintain fishing rules. Today, emphasis is on conservation and education. Encourages youngsters to enter the sport and maintains a huge library on the subject of fishing. Has a network of well over 300 representatives around the world, many of whom are conservation leaders in their communities.
Founded: 1939
Rob Kramer, President
Phil Hott, Finance Manager

897 Keep Florida Beautiful
201 Park Avenue   850-385-1528
Tallahassee, FL 32301   Fax: 850-385-4020
http://www.keepflbeautiful.org

Jeff Koons, Chairman

898 Legal Environmental Assistance Foundation (LEAF)
1114 Thomasville Road   850-681-2591
Suite E   Fax: 850-224-1275
Tallahassee, FL 32303   E-mail: cvalencic@leaflaw.org
http://www.leaflaw.org
To protect human health and life-sustaining natural resources from pollution in Florida, Georgia and Alabama.
Founded: 1979
David Lupder, President
Cynthia Valencic, Vice President

899 National Wildlife Federation: Everglades Project
2590 Golden Gate Parkway   239-643-4111
Suite 109   Fax: 239-643-5130
Naples, FL 34105   E-mail: adamsk@nwf.org
http://www.nwf.org
To advocate and support restoration of the greater Everglades ecosystem and protection of the western Everglades through planning, education and management activities. We seek to re-create a more natural hydrologic flow through the greater Everglades that ensures the long-term viability of native habitats, threatened and endangered species and associated wildlife.
Founded: 1934
Larry Schweiger, President
Dulce Zormelo, Chief Finance Officer

900 Pelican Man's Bird Sanctuary
1708 Ken Thompson Parkway   941-388-4444
Sarasota, FL 34236   Fax: 941-388-3258
E-mail: mail@pelicanman.org
http://www.pelicanman.org
Southwest Florida's largest rescue and rehabilitation center for wildlife, emphasizing, but not limited to, birds.
Mona Schonbrunn PhD, President
Tonya Clauss, Director of Veterinary Medic

901 Reef Relief
Reef Relief Environmental Center   305-294-3100
201 Williams Street, Suite 5   Fax: 305-293-9515
Key West, FL 33040   E-mail: reef@bellsouth.net
http://www.reefrelief.org
Reef Relief is a nonprofit membership organization dedicated to preserve and protect living coral reef ecosystems through local, regional and global efforts.
Founded: 1986
Michael Blades, Project Director
Celia Stearns, Program Assistant

902 Sanibel-Captiva Conservation Foundation
3333 Sanibel-Captiva Road   239-472-2329
PO Box 839   Fax: 239-472-6421
Sanibel, FL 33957   E-mail: sccf@sccf.org
http://www.sccf.org
Land aquisition, native plant nursery, environmental education, habitat management, marine laboratory.
Founded: 1967
Erick Lindblad, Executive Director
Jean Laswell, Business Manager

903 Sierra Club

475 Central Avenue
Suite M-1
Saint Petersburg, FL 33701

727-824-8813
Fax: 727-824-0936
http://www.florida.sierraclub.org

To advance the preservation and protection of the natural environment by empowering the citizenry, especially democratically-based grassroots organizations, with charitable resources to further the cause of environmental protection.The vehicle through which The Sierra Club Foundation generally fulfills its charitable mission.
Founded: 1892
Lisa Renstrom, President
Bernard Zaleha, Vice President

904 Society of Environmental Toxicology and Chemistry

SETAC N America Office
1010 N 12th Avenue
Pensacola, FL 32501-3370

850-469-1500
Fax: 850-469-9778
E-mail: setac@setac.org
http://www.setac.org

Provides a forum for individuals and institutions engaged in study of environmental issues, management and conservation of natural resources, environmental education and environmental research and development.
Founded: 1979
Mimi Meredith, Sr Manager

905 South Florida Chapter, National Safety Council

4171 West Hillsborro Boulevard
Suite 5
Coconut Creek, FL 33073

954-422-5757
800-392-5101
Fax: 954-418-9290
E-mail: occupational@safetycouncil.com
http://www.safetycouncil.com

Not-for-profit, non-governmental, public service organization dedicated to the safety and health of the Broward, Dade, Palm Beach and surrounding communities.
Michael Walters, VP Operations

906 Southeastern Association of Fish and Wildlife Agencies

8005 Freshwater Farms Road
Tallahassee, FL 32308

850-893-1204
Fax: 850-893-6204
E-mail: seafwa@aol.com
http://www.seafwa.org

John Frampton, Director
Robert Cook, Executive Director

907 Suncoast Seabird Sanctuary

18328 Gulf Blvd
Indian Shores, FL 33785-2097

727-391-6211
800-406-3400
Fax: 727-399-2923
E-mail: seabird@seabirdsanctuary.org
http://www.seabirdsanctuary.org

The sanctuary is the largest wild bird hospital in the United States dedicated to the rescue, repair, rehabilitation and hopeful release of sick and injured native birds. Over 600 birds permanently reside at our beachfront sanctuary.We are open free of charge to the public 365 days a year. Tours and educational programs are available.
Founded: 1971
Ralph Heath Jr, Founder/Director

908 Tallahassee Museum of History and Natural Science

3945 Museaum Drive
Tallassee, FL 32310-6325

850-575-8684
Fax: 850-574-8243
E-mail: rdaws@tallasseemuseum.org
http://www.tallahasseemuseum.org

Founded: 1957
Russell Daws, Director/CEO
Jennifer Golden, Director Education

909 Wildlife Foundation of Florida

PO Box 11010
Tallahassee, FL 32302

850-922-1066
Fax: 850-921-5786
E-mail: foundation@myfwc.com
http://www.wildlifefoundationofflorida.com

Founded: 1994
William G Bostick, Chairman
C Martin Wood III, Vice Chairwoman

Georgia

910 ASLA: Georgia Chapter

PO Box 18622
Atlanta, GA 31126

404-250-0414
Fax: 404-250-0414
E-mail: sawhill@uga.edu
http://www.gaasla.org

Founded: 1899
Ron Sawhill, President
David Spooner, Secretary

911 Agency for Toxic Substances and Disease Registry

Centre for Disease Control
1600 Clifton Road NE
Mail Stop F-32
Atlanta, GA 30333

404-639-3311
800-311-3435
Fax: 770-488-4178
E-mail: atsdric@cdc.gov
http://www.cdc.gov

To prevent exposure and adverse human health effects and diminished quality of life associated with exposure to hazardous substances from waste sites, unplanned releases, and other sources of pollution present in the enviroment. ATSDRis an operating division of the US Department of Health and Human Services. It divides its activities between those related to a particular site and those related to a specific hazardous substance.
Founded: 1987
Julie Gerberding, Director
Henry Falk, Associate Director

912 American Academy of Sanitarians

3815 Stone Briar Ct
Duluth, GA 30097-2240

770-623-5691
http://www.sanitarians.org

Gary Noonan, Executive Secretary/Treas.

913 American Cancer Society

1599 Clifton Road NW
Atlanta, GA 30329

404-329-7686
Fax: 404-321-4669
http://www.cancer.org

The American Cancer Society is the nationwide, community-based, voluntary health organization dedicated to eliminating cancer as a major health problem by preventing cancer, saving lives from cancer, and diminishing suffering fromcancer through research, education and service.
Clark W Heath Jr MD, VP

914 American Lung Association of Georgia

American Lung Association
2452 Spring Road
Smyrna, GA 30080

770-434-5864
800-586-4872
Fax: 770-319-0349
E-mail: mail@alaga.org
http://www.lungusa2.org/georgia/index.html

Founded: 1904
Charles A Heinrich, Chairman
Robert A Green, Vice-Chairman

915 Center for a Sustainable Coast

221 Mallory Street
Suite B
St. Simons Island, GA 31522

912-638-3612
Fax: 912-638-3615
E-mail: susdev@gate.net
http://www.sustainablecoast.com

Founded: 1997
David Kyler, Executive Director
Helen Alexander, Administrative Assistant

916 Centers for Disease Control and Prevention

United States of health and human services
1600 Clifton Road NE
Atlanta, GA 30333

404-639-3311
800-311-3435
Fax: 404-639-7111
http://www.cdc.gov

Protects the public health of the nation by providing leadership and direction in the prevention and control of diseases and other preventable conditions and responding to public health emergencies.
Founded: 1946
Julie Louise Gerberding, Director
William Gimson, Chief Operating Officer

917 Coastal Conservation Association of Georgia
Coastal Conservation Association
515 Demark St   912-764-6222
Suite 300   800-266-0693
Statesboro, GA 30458   Fax: 912-764-6497
E-mail: info@ccaga.org
http://www.ccaga.org

Founded: 1986
Jonathan Cartwright, Executive Director

918 Coosa River Basin Initiative
408 Broad Street   706-232-2724
Rome, GA 30161   Fax: 706-235-9066
E-mail: info@coosa.org
http://www.coosa.org
We are a nonprofit environmental advocacy organization dedicated to creating a cleaner, healthier, more economically viable Coosa River Basin.
Carolyn Turner, Executive Director
Ben Harrison, President

919 Council of State and Territorial Epidemiologists
2872 Woodcock Boulevard   770-458-3811
Suite 303   Fax: 770-458-8516
Atlanta, GA 30341-4015   E-mail: fellowship@cste.org
http://www.cste.org
The Council of State and Territorial Epidemiologists is an organization of epidemiologists working together to establish effective relationships among state and other epidemiologists, to consult and advise with appropriate disciplinesin other health agencies, and to provide technical assistance to the Association of State and Territorial Health Officials.
Founded: 1951
Sewell Mack, President
Hahn Christine, Vice President

920 Earth Share of Georgia
1447 Peachtree Street   404-873-3173
Suite 214   Fax: 404-873-3135
Atlanta, GA 30309   E-mail: info@earthsharega.org
http://www.earthsharega.org
Nonprofit federation of local, national and global environmental groups addressing the critical environmental issues. ESGA raises funds for these groups through workplace giving campaigns, special events and individual contributions.
Founded: 1992
Richard Judy, Executive Director
Douglas Abramson, President

921 Environmental Justice Resource Center
223 James Brawley Drive SW   404-880-6911
Atlanta, GA 30314   Fax: 404-880-6909
E-mail: ejrc@cau.edu
http://www.ejrc.cau.edu
Since 1994, a research, policy and information clearinghouse on issues related to environmental justice, race and the environment, civil rights, facility siting, land use planning, brownfields, transportation equity, suburban sprawland Smart Growth. The overall goal of the center is to assist, support, train and educate people of color, students, professionals and grassroots community leaders with the goal of facilitating their inclusion into the mainstream of environmentaldecision-making.
Robert Bullard, Director
Michelle Dawkins, Program Manager

922 Georgia Association of Conservation District Supervisors
PO BOX 111   706-542-3065
Athens, GA 30603   E-mail: info@gacds.org
http://www.gacds.org
Founded: 1937
Jim Ham, President
David Bennett, Executive Director

923 Georgia Chapter, National Safety Council
3300 NE Expressway   770-457-5100
Suite 5A   800-441-5103
Atlanta, GA 30341-3941   Fax: 770-457-6189
E-mail: georgia@nsc.org
http://georgia.nsc.org
Bob Wilson, Executive Director

924 Georgia Conservancy
817 West Peachtree Street   404-876-2900
Suite 200   Fax: 404-872-9229
Atlanta, GA 30338   E-mail: mail@gaconservancy.org
http://www.gaconservancy.org
Founded: 1967
Lisa Patrick, Executive Assistant
Jim Strokes, President

925 Georgia Environmental Health Association
Golden Isles Parkway   478-892-8343
Route 2, Box 1140   E-mail: clerk@gehaorg.net
Hawkinsville, GA 31036   http://www.geha-online.org
Our mission is to provide excellence in services and regulatory functions, to protect and promote agriculture and consumer interests, and to ensure an abundance of safe food and fiber for Georgia, America and the world by usingstate-of-the-art technology and a professional workforce. Affiliated with NEHA.
Tonya Grey, President
Travis Sheppard, Vice President

926 Georgia Environmental Organization
3185 Center Street   404-605-0000
Smyma, GA 30080-7039   Fax: 404-350-9997
E-mail: geoco@geoco.org
http://www.geoco.org
Olin Ivey, Executive Director

927 Georgia Federation of Forest Owners
2402 Manchester Drive   912-283-0871
Waycross, GA 31501   E-mail: info@woodlandowners.org
http://www.woodlandowners.org
William Hubbard, South Regional Editor

928 Georgia Trappers Association
PO Box 95314   912-782-5417
Atlanta, GA 30347   http://www.geocities.com
Tom Ethridge, President

929 Georgia Water and Pollution Control Association
PO Box 6129   770-618-8690
Marietta, GA 30065-0129   Fax: 770-618-8695
E-mail: info@gwpca.org
http://www.gwpca.org
The GW+PCA is dedicated to education, dissemination of technical and scientific information, increased public understanding and promotion of sound public laws and programs in the water resources and related environmental fields.Founded in 1932.
Jack C Dozier, Executive Director
Terry Cole, President

930 Georgia Wildlife Federation
11600 Hazelbrand Road   770-787-7887
Covington, GA 30094-5046   Fax: 770-787-9229
E-mail: gwf@gwf.org
http://www.gwf.org
Founded: 1936
Jerry McCollum, President/CEO
Glenn Dowling, Executive Vice President

931 Human Ecology Action League (HEAL)
PO Box 29629   404-248-1898
Atlanta, GA 30359-0629   Fax: 404-248-0162
E-mail: healnatnl@aol.com
http://http://members.aol.com/healnatnl/index.html
The Human Ecology Action League Inc (HEAL) is a nonprofit organization founded in 1977 to serve those whose health has been adversely affected by environment exposures; to provide information to those who are concerned about the healtheffects of chemicals; and to alert the general public about the potential dangers of chemicals. Referrals to local HEAL chapters and other support groups are available from the League.
Founded: 1977
Katherine P Collier, Manager

932 Mountain Conservation Trust of Georgia
104 N Main Street   706-253-4077
Suite B3   Fax: 706-253-4078
Jasper, GA 30143   E-mail: mctgainfo@mctga.org
http://www.mctga.org

Dedicated to the permanent conservation of the scenic beauty and natural resources of the mountains and foothills of North Georgia through land protection, education and collaborative partnerships.
Founded: 1994
Dan Pool, Vice President
Clay Johnston, President

933 National Wildlife Federation
1330 W Peachtree Street
Suite 475
Atlanta, GA 30309
404-876-8733
Fax: 404-892-1744
http://www.nwf.org
Founded: 1936
Larry J Schweiger, President/CEO
Jerome C Ringo, Chairman

934 Nature Conservancy: Georgia Chapter
1330 W Peachtree Street
Suite 410
Atlanta, GA 30309-2904
404-873-6946
Fax: 404-873-6984
E-mail: comment@tnc.org
http://www.nature.org
Founded: 1951
Tavia McCuean, VP
Allen Harrison, Director of Operation

935 Sierra Club: Georgia Chapter
1401 Peachtree Street NE
Suite 345
Atlanta, GA 30309   E-mail: georgia.chapter@sierraclub.org
404-607-1262
Fax: 404-876-5260
http://http://georgia.sierraclub.org
A grassroots organization dedicated to the preservation, protection and enjoyment of our environment. We work towards those goals through public education, political advocacy, an active outings program and litigation when necessary.
Genie Strickland, Administrative Coordinator
Harry Truman, President

936 Trees Atlanta
96 Poplar Street NW
Atlanta, GA 30303
404-522-4097
Fax: 404-522-6855
E-mail: info@treesatlanta.org
http://www.treesatlanta.org
Founded: 1985
Marcia Bansley, Executive Director

937 Upper Chattahoochee Riverkeeper
3 Puritan Mill
916 Joseph Lowery Boulevard
Atlanta, GA 30318   E-mail: sbethea@ucriverkeeper.org
404-352-9828
Fax: 404-352-8676
http://www.chattahoochee.org
Founded: 1994
Sally Bethea, Executive Director
Birgit Bolton, Programs Coordinator

938 Wildlife Society
5410 Grosvenor Lane
Suite 200
Bethesda, MD 20814-2144
301-897-9770
Fax: 301-530-2471
E-mail: tws@Wildlife.org
http://www.wildlife.org
A nonprofit scientific and educational organization that serves professionals such as government agencies, academia, industry, and non-government organizations in all areas related to the conservation of wildlife and natural resourcesmanagement.
Founded: 1937
William R. Rooney, President
Jane Pelky, Financial Coordinator/Office

Hawaii

939 ASLA: Hawaii Chapter
1001 Bishop St.
Suite 650
Honolulu, HI 96813
808-521-5631
Fax: 808-523-1402
E-mail: info@hawaiiasla.org
http://www.hawaiiasla.org
Founded: 1969
Stan Duncan, President
Matt M Flach, President-Elect

940 American Lung Association of Hawaii
245 North Kukui Street
Suite 100
Honolulu, HI 96817
808-537-5966
Fax: 808-537-5971
E-mail: lung@ala-hawaii.org
http://www.ala-hawaii.org
Malcolm Koga, President
Sterling Yee, Vice President

941 American Lung Association of Hawaii- Maui Office
American Lung Association of Hawaii
95 Mahalani Street
Cameron Center
Wailuku, HI 96793
808-244-5110
Fax: 808-242-9041
E-mail: alahmaui@ala-hawaii.org
http://www.ala-hawaii.org
Founded: 1929
Mary Miller, President
Didier Decler, VP Finance

942 American Lung Association of Hawaii: East Hawaii Office
39 Ululani Street
Hilo, HI 96720
808-935-1206
Fax: 808-935-7474
E-mail: alahbi@ala-hawaii.org
http://www.ala-hawaii.org
Founded: 1929
Mary Miller, CEO
Malcolm Koga, President

943 American Lung Association of Hawaii: Kauai Office
29992 Umi Street
Lihue, HI 96766
808-245-4142
Fax: 808-245-8488
E-mail: alahkaui@pixi.com
http://www.ala-hawaii.org/
Malcolm Koga, President
Sterling Yee, Vice President

944 American Lung Association of Hawaii: West Hawaii Office
American Lung Association
74-5588 Pawai Place
Building P
Kailua-Kona, HI 96740
808-326-4755
800-LUN- USA
Fax: 808-326-9149
E-mail: alahkona@pixi.com
http://www.lungusa.org/site/
Founded: 1904
James M Anderson, Secretary

945 Big Island Rain Forest Action Group
PO Box 341
Kurtistown, HI 96760
808-966-7622
E-mail: ja@interpac.net
Founded: 1989
Jim Albertini, Co-ordinator

946 EarthTrust
Windward Environmental Center
1118 Maunawili Road
Kailua, HI 96734
808-261-5339
Fax: 206-202-3893
E-mail: sue@flipperfund.com
http://www.earthtrust.org
EarthTrust is the impossible missions"" team for wildlife and the environment. Its low-overhead high-tech campaigns are always positive and effective. Dedicated to saving marine mammals reforming unsustainable fisheries and ending thetrade in endangered species around the world. EarthTrust is a relatively small organization which may have directly saved more marine wildlife biomass than any other organization in history. Continually redefining the cutting edge of environmentalprotection.""
Founded: 1976
Don White, President

947 Flipper Foundation
Windward Environmental Center
1118 Maunawili Road
Kailua, HI 96734
808-261-5339
Fax: 815-333-1158
E-mail: sue@flipperfund.com
http://www.earthtrust.org

The Flipper Foundation exists to save dolphins and to revolutionize consumer control over environmental destruction by world fisheries. Its mission is to establish and maintain the highest world standard of dolphin safety"" andfisheries sustainability; to educate consumers worldwide while directly granting funds to save marine mammals and their habitats. Its primary way of accomplishing this is to engage fishery firms in voluntary partnerships to phase out destructivefishing technologies.""
Founded: 1992
Don White, President

948 Greenpeace Foundation
Windward Environmental Center
1118 Maunawili Road   808-263-4388
Kailua, HI 96734   Fax: 630-604-6129
                    E-mail: sw@gpfdn.com
              http://www.greenpeacefoundation.com
The oldest and original greenpeace"" organization in the US Greenpeace Foundation is dedicated to peaceful no-nonsense environmental advocacy. Greenpeace Foundation seeks to preserve biodiversity on a green and peaceful planet.Proudly unaffiliated with ""Greenpeace International"" we make no apologies for standing up for the earth; a human voice for the majority of earth's life which has none so that citizens may have a voice in what sort of planet we will leave to ourchildren and theirs."
Founded: 1976
Sharon White, President

949 Hawaii Association of Conservation Districts
PO Box 404   808-248-7725
Hana, HI 96713   Fax: 808-248-7725
          http://www.westerncoalition.org/members/hawaii.html
David Nobriga, President
Mike Tulang, Executive Director

950 Hawaii Nature Center
2131 Makiki Heights Drive   808-955-0100
Honolulu, HI 96822   Fax: 808-955-0116
          E-mail: hawaiinaturecenter@hawaii.rr.com
              http://www.hawaiinaturecenter.org
The Hawaii Nature Center is a private nonprofit organization specializing in environmental education field program for children, adults and families. Its mission is to foster awareness appreciation and understanding of Hawaii andencourage wise stewardship of the Islands. The Nature Center provides full day field trips for 25,000 students on two islands each year and features an interactive nature museum at its field site on Maui.
Founded: 1981
Jeffery Dinsmore, President
Meredith Ching, First Vice Chairman

951 Hawaiian Botanical Society
University of Hawaii
3190 Maile Way   808-956-8072
Honolulu, HI 96822   Fax: 808-956-3923
              http://www.botany.hawaii.edu
Founded: 1924
Eileen Helmstetter, President
Vickie Caraway, Vice-President:

952 Nature Conservancy: Hawaii Chapter
Nature Conservancy
923 Nuuanu Avenue   808-537-4508
Honolulu, HI 96817   E-mail: egoldstein@tnc.org
                    http://www.nature.org
Founded: 1903
Suzanne Case, Executive Director
Steven McCormick, President/CEO

953 Sierra Club: Hawaii Chapter
PO Box 2577   808-538-6616
Honolulu, HI 96803-2577   Fax: 808-537-9019
          E-mail: hawaii.chapter@sierraclub.org
              http://www.hi.sierraclub.org

To explore, enjoy and protect wild places and the environment. We use a multi-pronged approach to protecting and restoring Hawaii's environmental quality. Through the volunteer efforts of group leaders we conduct interpretive andeducational outings; lead fun and challenging hikes; conduct service projects involving fencing, cleaning streams, trail building and noxious plant control; and advocate and lobby for environmental protection.
Jeff Mikulina, Director
Jack Kelly, Chairman

Idaho

954 American Lung Association of Idaho/Nevada: Boise Office
1111 South Orchard   208-345-5864
Suite 245   Fax: 208-345-5896
Boise, ID 83705-1966   http://www.lungs.org
Founded: 1904

955 Idaho Association of Soil Conservation Districts
Box 697   208-338-5900
Lava Hot Springs, ID 83246   Fax: 208-338-9537
                    E-mail: kenc.fofter@idaho.us
                 http://www.iascd.state.id.us
Provides action at the local level for promoting wise and beneficial conservation of natural resources with emphasis on soil and water.
Founded: 1944
Kyle Hawley, President
Darwin Josephson, Secretary

956 Idaho Conservation League
PO Box 844   208-345-6933
Boise, ID 83701   877-345-6933
                           Fax: 208-344-0344
                    E-mail: icl@wildidaho.org
                 http://www.wildidaho.org
Founded: 1973
Rick Johnson, Executive Director
Suki Molina, Deputy Director

957 Idaho Forest Owners Association
233 E. Palouse River Road   208-883-4488
PO Box 9748   Fax: 208-883-1098
Moscow, ID 83843   http://www.consulting-foresters.com
Founded: 1983
Vincent Corrao, President
Tom Richards, Vice President

958 Rocky Mountain Elk Foundation
426 Rapid Creek Rd   208-775-3263
Incom, ID 83245   800-225-5355
                           Fax: 208-775-3263
                    E-mail: info@rmef.org
                 http://www.rmef.org
Founded: 1984
Buddy Smith, Chairman
Bob Wellman, Vice-Chairman

959 Trout Unlimited
Trout Unlimited
1300 N. 17th St.   703-522-0200
Suite 500   Fax: 703-284-9400
Arlington, VA 22209-2404   E-mail: trout@tu.org
                           http://www.tu.org/
Founded: 1959
Charles Gauvin, President/CEO
Kenneth Mendez, Executive Vice President

960 Wildlife Society
1215 Highway 93 N   208-756-2271
PO Box 1336   Fax: 208-756-6274
Salmon, ID 83467   E-mail: aowsiak@idfg.state.id.us
A nonprofit scientific and educational organization that serves professionals such as government agencies, academia, industry, and non-government organizations in all areas related to the conservation of wildlife and natural resourcesmanagement.

Anna Owsiak, Secretary

Illinois

961 ASLA: Illinois Chapter
PO Box 4566 630-833-4516
Oak Brook, IL 60522 Fax: 630-833-4030
E-mail: info@il-asla.org
http://www.il-asla.org

Founded: 1899
Brian Hopkins, President
Amy Olson, Secretary

962 American College of Occupational and Environmental Medicine
25 Northwest Point Blvd 847-818-1800
Suite 700 Fax: 847-818-9266
Elk Grove Village, IL 60007 E-mail: dcaplick@acoem.org
http://www.acoem.org
Made up of physicians in industry, government, academia, private practice and the military, who promote the health of workers through preventive medicine, clinical care, research and education.
Cheryl Barbanel, President
Barry Eisenberg, Executive Director

963 American Lung Association of Illinois-Iowa
American Lung Association
3000 Kelly Lane 217-787-5864
Springfield, IL 62707 800-586-4872
Fax: 217-787-5916
E-mail: info@lungil.org
http://www.lungusa.org/site

Founded: 1904
Harold Wimmer, CEO
Lori Younker, Manager

964 American Lung Association of Metropolitan Chicago
American Lung Association of New York
1440 West Washington Boulevard 312-243-2000
Chicago, IL 60607-1878 800-586-4872
Fax: 312-243-3954
E-mail: ccaplinger@alamc.org
http://www.lungchicago.org

Founded: 1906
Jole Africk, President
Marry Christofidis, Finanace Manager

965 American Lung Association: Chicagoland Collar Counties
American Lung Association
1749 South Naperville Road 630-260-9600
Suite 202 Fax: 630-260-1111
Wheaton, IL 60187 E-mail: info@lungil.org
http://www.lungil.org

Founded: 1904
Herald Wimmer, Chief Executive Officer
Ted Schlake, Senior Manager

966 American Lung Association: Northern Illinois
1330 East State Street 815-962-6412
Rockford, IL 61104 Fax: 815-962-6413
E-mail: info@lungil.org
http://www.lungusa.org

Founded: 1904
James M Anderson, Secretary

967 American Lung Association: Southwestern Illinois
American Lung Association
1600 Golfview Drive 618-344-8891
Suite 260 800-586-4872
Colinsville, IL 62234 Fax: 618-344-8933
E-mail: info@lungil.org
http://www.lungfla.org

Founded: 1917
Tina Barnard, President
Harold Wimmer, Executive Director

968 American Medical Association
515 N State Street 312-464-5000
Chicago, IL 60610 800-262-3211
Fax: 312-464-4184
http://www.ama-assn.org
Medical doctors concerned with environmentally related health issues.
Founded: 1847
John C Nelson, President
Michael D Maves, Managing Director

969 Audubon Council of Illinois
PO Box 813 815-389-4775
Bryon, IL 61010-0813 E-mail: stirli@hughestech.net
http://www.audubon.org/chapter/il/il/
Ed Stirling, President
Mary Ann Hahn, Vice President:

970 Chicago Chapter, National Safety Council
1121 Spring Lake Drive 630-775-2213
Suite 100 800-621-2855
Itasca, IL 60143-3201 Fax: 630-775-2136
E-mail: chicago@nsc.org
http://www.chicago.nsc.org

Alan McMillian, President

971 Chicago Zoological Society
Brookfield Zoo 708-485-0263
3300 Golf Road 800-201-0784
Brookfield, IL 60513 Fax: 708-485-3140
E-mail: bzadmin@brookfieldzoo.org
http://www.brookfieldzoo.org

Founded: 1934
Stuart Strahl, President

972 Eagle Nature Foundation
Eagle Nature Foundation, LTD
300 E. Hickory St. 815-594-2306
Apple River, IL 61001 Fax: 815-594-2305
E-mail: eaglenature.tni@juno.com
http://www.eaglenature.com
A nonprofit organization dedicated to the preservation of the bald eagle, our national symbol, and other endangered species from extinction and to increase public awareness of unique endangered plants and animals. We monitor bald eagleand other endangered species populations and strive to preserve habitiat essential to their survival. We develop materials for schools to inform students about the needs of the bald eagle and how we can help preserve and protect their naturalenvironment.
Founded: 1995
Terrence N Ingram, Chief Executive Officer
Eugene Small, Vice President

973 Environmental Education Association of Illinois
2112 Behan Road 815-479-5779
Crystal Lake, IL 60014 Fax: 815-479-5766
E-mail: dchap19265@aol.com
http://www.web.stclair.k12.il.us/eeai
Karen Zuckerman, President

974 Great Lakes Sport Fishing Council
PO Box 297 630-941-1351
Elmhurst, IL 60126 Fax: 630-941-1196
E-mail: hdqtrs@great-lakes.org
http://www.great-lakes.org
Our mission is to inform and educate the outdoor recreational community (sport fishing, boating and general public) through educational outreach programs about natural resource conservation and enhancement, wise conservation andboating policies, and the spread of unintentional introductions on nonindigenous aquatic nuisance species (exotics).
Founded: 1971
Dan Thomas, President
Robert Mitchell, Vice President

975 Illinois Association of Conservation Districts
9313 Bull Valley Road 815-338-7664
Woodstock, IL 60098 Fax: 815-338-2773
E-mail: conserveone@aol.com

Founded: 1972
Kathy Merner, President
Ken Fiske, Assistant Secretary

976 Illinois Association of Environmental Professionals
PO Box 81551   773-325-2771
Chicago, IL 60681-0551   E-mail: kkopija@cbbel.com
                                    http://www.iaepnetwork.org
Members include environmental planners, managers and impact assessors.
Michael Headrick, Vice President
Tom Guist, President

977 Illinois Audubon Society
425 B N Gilbert Street   217-446-5085
Danville, IL 61832   Fax: 217-446-6375
                                E-mail: director@pdnt.com
                              http://www.illinoisaudubon.org
A membership organization dedicated to the preservation of Illinois Wildlife and the habitats which support them. Has sanctuaries, conservation education and land acquisition programs and publishes quarterly magazines and newsletters.
Founded: 1897
Marilyn F Campbell, Executive Director

978 Illinois Environmental Council
107 W Cook   217-544-5954
Suite E   Fax: 217-544-5958
Springfield, IL 62704   E-mail: iec@ilenviro.org
                                    http://www.ilenviro.org
Coalition of over 70 environmental, conservation and health groups.
Founded: 1975
Jonathan Goldman, Executive Director
Rand , Sparling, President

979 Illinois Prairie Path
P.O. Box 1086   630-752-0120
Wheaton, IL 60189   http://www.ipp.org
Founded: 1905
Don Kirchenberg, President
Dick Wilson, VP

980 Illinois Recycling Association
PO Box 3717   708-358-0050
Oak Park, IL 60303-3717   Fax: 708-358-0051
        E-mail: executivedirector@illinoisrecycles.org
                            http://www.illinoisrecycles.org
The association's mission is to encourage the responsible use of resources and protecting the environment by promoting effective programs and practices regarding waste reduction, re-use of materials, and recycling.
Founded: 1980
Michael Mitchell, Executive Director

981 Illinois Solar Energy Association
PO Box 1592   630-260-0424
Wheaton, IL 60189-0634   Fax: 630-420-1517
                              E-mail: info@illinoissolar.org
                                  http://www.illinoissolar.org
Founded: 1975
Mark Burger, President
Howard Alan, Vice President

982 Kids for Conservation: Today and Tomorrow
Illinois Department of Conservation
524 S 2nd Street   217-782-6302
Room 510   http://www.dnr.state.il.us
Springfield, IL 62701-1787
Provides a variety of programs for young Illinois hunters.
Joel Brunsvold, Director

983 Lake Michigan Federation
220 S State Street   312-939-0838
Suite 1900   Fax: 313-939-2708
Chicago, IL 60609-2177   E-mail: chicago@greatlakes.org
                                    http://www.lakemichigan.org

Works to restore fish and wildlife habitat, conserve land and water and eliminate pollution in the watershed of America's largest lake. We achieve these through education, research, law, science, economics and strategic partnerships.
Founded: 1971
Cameron Davis, Executive Director
Susan Campbell, Manager, Communications

984 Midwest Center for Environmental Science and Public Policy
1845 North Farewell Avenue   414-271-7280
Suite 100   Fax: 414-273-7293
Milwaukee, IL 53202   E-mail: mcespp@mcespp.org
                                      http://www.mcespp.org
Protecting human health and the environment through research, advocacy, public education and citizen empowerment.
Jeffery Foran, President
Patrice Ann Morrow, Chairman

985 Nature Conservancy: Illinois Chapter
8 S Michigan Avenue   312-580-2100
Suite 900   800-628-6860
Chicago, IL 60603   Fax: 312-346-5606
                                E-mail: comment@tnc.com
                                      http://www.nature.org
Founded: 1915
Steven J McCormick, Chief Executive Officer

986 Outside Chicagoland: Iowa-Illinois Safety Council
8013 Douglas Avenue   515-276-4724
Urbandale, IA 50322-2453   Fax: 515-276-8038
                               E-mail: iiscadmin@iisc.org
                                         http://www.iisc.org
Our mission is to educate society to adopt safety, health, and environmental practices and to provide high quality, value added training and services.
Neil Longseth, Executive Director

987 Prairie Rivers Network
809 S 5th Street   217-344-2371
Champaign, IL 61820   Fax: 217-344-2381
                             E-mail: info@prairierivers.org
                                  http://www.prairierivers.org
The only statewide river conservation organization in Illinois. They strive to protect rivers and streams of Illinois and to promote the lasting health and beauty of watershed communities by providing information, sound science andhands-on assistance. They also help individuals and community groups become effective river conservation leaders.
Founded: 1967
Eric Frey, President
Michael Rosenthal, Secretary

988 Safer Pest Control Project
25 E. Washington   312-641-5575
Suite 2 1515   Fax: 312-641-5454
Chicago, IL 60602-1849   E-mail: info@spcpweb.org
                                      http://www.spcpweb.org
The mission of the Safer Pest control Project is to reduce pesticide use throughout rural, urban and suburban Illinios and to stimulate and help implement widespread adoption of safer alternatives to routine pesticide use in Illinois.
Founded: 1994
Rachel Lerner Rosenberg, Executive Director
Kim Stone, Associate Director

989 Sierra Club Illinois Chapter
200 N. Michigan Avenue   312-251-1680
Suite 505   Fax: 312-251-1780
Chicago, IL 60601   E-mail: illinois.chapter@sierraclub.org
                                http://www.illinois.sierraclub.org
A statewide part of the national sierra club, a nonprofit conservation organization. Members and staff of the Illinois Chapter seek to preserve and protect the state's natural areas and to secure cleaner land, air and water forIllinois through grassroots activism.
Founded: 1892
Jack Darin, Director

990 St. Louis Metropolitan Area: Safety Council of Greater St. Louis, MO
1015 Locust Street 314-621-9200
Suite 902 Fax: 314-621-9204
St. Louis, MO 63101 E-mail: director@stlsafety.org
http://ww.stlsafety.org
Promote the health and safety of the community by providing programs, resources, and educational services focused on the reduction and prevention of accidental injuries and adverse occupational environmental exposures.
Leslie Foran, Executive Director

991 Upper Mississippi River Conservation Committee
4469 48th Avenue Court 309-793-5800
Rock Island, IL 61201 Fax: 309-793-5804
E-mail: umrcc@mississippi-river.com
http://www.mississippi-river.com/umrcc
A 5 state organization of river biologists and managers dedicated to the wise use and preservation of Upper Mississippi River fish and wildlife natural resources. This is accomplished through workshops, publications and annualmeetings.
Founded: 1943
Mike McGhee, Chairman
Patrick Short, Secretary/Treasurer

Indiana

992 ASLA: Indiana Chapter
PO Box 441777 317-767-9375
Indianapolis, IN 46244 E-mail: info@inasla.org
http://www.inasla.org
Founded: 1972
Scott Siefker, President
Julie Zigler, President-Elect

993 Acres Land Trust
2000 N Wells Street 219-422-1004
Fort Wayne, IN 46808-2474 Fax: 219-422-1004
E-mail: acreslt@fwi.com
http://www.acres-land-trust.org
Forty-three nature Preserves in 12 counties in northeast Indiana. Offers canoe trips, hikes, festivals and concerts throughout the year for members and the public. More than 900 members-family, individual and business.
Founded: 1960
Carolyn McNagny, Executive Director
David Van Gilder, President

994 American Fisheries Society: Equal Opportunities
5410 Grosvenor Lane 301-897-8616
Bethesda, MD 20814 Fax: 301-897-8096
E-mail: afspubs@pbd.com
http://www.fisheries.org
The section of the American Fisheries Society promotes the representation and involvement of diverse ethnic, racial and cultural groups and women in the fisheries profession. The group fosters mentoring of under-represented groups,administers awards for travel and academic achievement and provides information on social and professional diversity in fisheries.
Gus Rassam, Executive Director

995 American Lung Association of Indiana, Northern Office
American Lung Association
802 West Wayne Street 260-426-1170
Fort Wayne, IN 46802-3996 Fax: 317-573-3909
E-mail: info@lungin.org
http://www.lungin.org
Nancy Turner, President
Kathy Such, Office Manager

996 American Lung Association of Indiana, State Office & Support Office
American Lung Association
9445 Delegates Row 317-573-3900
Indianapolis, IN 46240 800-586-4872
Fax: 317-573-3909
E-mail: info@lungin.org
http://www.lungin.org

Founded: 1904
Nancy Turner, President/CEO

997 Central/Southern Indiana: National Safety Council, Kentucky Office
3176 Richmond Road 859-294-4242
Suite 236 Fax: 859-297-7860
Lexington, KY 40509 E-mail: conleyg@nsc.org
Greg Conley, Manager

998 Conservation Technology Information Center
1220 Potter Drive 765-494-9555
West Lafayette, IN 47906 Fax: 765-494-5969
E-mail: ctic@ctic.purdue.edu
http://www.ctic.purdue.edu
A nonprofit organization dedicated to environmentally responsible and economically viable agricultural decision-making.
Founded: 1982
John Hassell, Executive Director
Karen Scanlon, Communications Director

999 Indiana Audubon Society
Indiana Audubon Society Inc.
3497 S Bird Sanctuary Road 765-827-0908
Connersville, IN 47331-8721 Fax: 765-825-9788
E-mail: indianaaudubon@yahoo.com
http://www.indianaudubon.org
Founded: 1898
Dan Leach, President
Tom Goldsmith, Director

1000 Indiana Forestry and Woodland Owners Association
5578 S 500 W 317-758-4735
Atlanta, IN 46031-9363 Fax: 317-758-4280
E-mail: steward@inwoodlands.org
http://www.inwoodlands.org
Founded: 1977
John Seifert, President

1001 Indiana State Trappers Association
6420 Street Road 47 N 812-939-3215
Crawfordsville, IN 47933 http://www.geocities.com
Ken Brosman, President

1002 Indiana Water Environment Association
1209 Polk Street 317-685-0009
Indianapolis, IN 46202-3677 Fax: 317-684-9457
http://www.iwpca.org
Ricky Dodd, Vice President
Herb Corn, President

1003 Northwestern Indiana: National Safety Council, Chicago Chapter
1121 Spring Lake Drive 630-775-2213
Suite 100 800-621-2855
Itasca, IL 60143-3201 Fax: 630-775-2136
E-mail: chicago@nsc.org
http://www.chicago.nsc.org
Alan McMillian, President

1004 Wildlife Society
1010 Yeardley Lane 219-258-0100
Mishawaka, IN 46544-6766 Fax: 219-258-0189
E-mail: phil@djcase.com
A nonprofit scientific and educational organization that serves professionals such as government agencies, academia, industry, and non-government organizations in all areas related to the conservation of wildlife and natural resourcesmanagement.
Phil Seng, President

Iowa

1005 ASLA: Iowa Chapter
American Society of Landscape Architecture
8345 University Boulevard   515-225-2323
Suite F-1   Fax: 515-225-6363
Clive, IA 50325   E-mail: ia-asla@assoc-serv.com
                                   http://www.iaasla.org
Founded: 1899
Christopher J Seeger, President
Kim Wagner, Association Manager

1006 American Lung Association of Iowa
5601 Douglas Avenue   515-278-5864
Des Moines, IA 50310   Fax: 515-334-9564
                                  E-mail: info@lungil.org
                                  http://www.lungil.org
Founded: 1904
Harold Wimmer, Chief Executive Officer
Shane Johnson, Executive Director

1007 Indian Creek Nature Center
6665 Otis Road SE   319-362-0664
Cedar Rapids, IA 52403   Fax: 319-362-2876
                                  E-mail: naturecenter@aol.com
                                  http://www.indiancreeknaturecenter.org
Founded: 1973
Rich Patterson, President
Dana Wood, Office Manager

1008 Iowa Academy of Science
Iowa Academy of Science
174 Baker Hall   319-273-2021
Cedar Falls, IA 50614-0508   Fax: 319-273-2807
                                  E-mail: iascience@uni.edu
                                  http://www.iacad.org
Iowa Academy of Science is a professional scientific organization.
Founded: 1875
Paul Bartelt, President
Craig Johnson, Executive Director

1009 Iowa Association of Soil and Water Conservation District Commissioners
38995 Honeysuckle Road   641-774-4461
Oakland, IA 52560-9686   Fax: 712-482-3386

Bernie Bolton, Secretary

1010 Iowa BASS Chapter Federation
3282 Midway   319-393-1481
Marion, IA 52302   E-mail: mail@iabass.com
                                  http://www.iabass.com

Tom Bowler, President

1011 Iowa Native Plant Society
Iowa State University
Botany Department, 341A Bessey Hall   515-294-9499
Iowa State University   Fax: 515-294-1337
Ames, IA 50011-1020   E-mail: mottl@grinnell.edu
                                      http://www.public.iastate.edu
An organization of amateur and professional botanists and native plant enthusiasts who are interested in the scientific, educational and cultural aspects, as well as the preservation and conservation of the native plants of Iowa. TheSociety was organized in 1995 to create a forum where plant enthusiasts, gardners and professional botanists could exchange ideas and coordinate activities such as field trips, work shops, and restoration of natural areas.
Larissa Mottl, President
Connie Mutel, Vice President

1012 Iowa Trappers Association
Gene Purdy   641-682-3937
122 2nd Street   Fax: 641-682-9092
Fontanelle, IA 50846   E-mail: cegrillo@fbcom.net
                                  http://www.iowatrappers.com
Spencer Hill, President
Chris Grillot, Secretary

1013 Iowa Wildlife Rehabilitators Association
1005 Harken Hill Drive   515-342-2783
PO Box 217   http://www.earthweshare.org
Osceola, IA 50213
Marlene Ehresman, President
Wendy DeWalle, Secretary

1014 Iowa-Illinois Safety Council
8013 Douglas Avenue   515-276-4724
Urbandale, IA 50322-2453   Fax: 515-276-8038
                                  E-mail: iiscadmin@iisc.org
                                  http://www.iisc.org
Our mission is to educate society to adopt safety, health and environmental practices and to provide high quality, value added training and services.
Neil Longseth, Executive Director

1015 MacBride Raptor Project
6301 Kirkwood Boulevard SW   319-398-5495
Cedar Rapids, IA 52406   Fax: 319-398-5611
                                  E-mail: iaraptor@avalon.net
                                  http://www.macbrideraptorproject.org
A nonprofit organization jointly sponsored by the University of Iowa and Kirkwood Community College. The project has two main facilities, the educational display facility and rehabilitation flight cage at the MacBride NatureRecreational Area and the medical clinic on the Kirkwood Campus.
Founded: 1985
Jodeane Cancilla, Director

1016 Nature Conservancy: Iowa Chapter
108 3rd Street   515-244-5044
Des Moines, IA 50309   800-628-6860
                                  Fax: 515-244-8890
                                  E-mail: comment@tnc.org
                                  http://http://nature.org
Founded: 1951
Margaret Collison, State Director
Rolf Koford, President

1017 Soil and Water Conservation Society
945 SW Ankeny Road   515-289-2331
Ankeny, Io 50023-9723   Fax: 515-289-1227
                                  E-mail: swcs@swcs.org
                                  http://www.swcs.org
Fosters the science and the art of soil, water and related natural resource management to achieve sustainability. Promote and practice an ethic recognizing the interdependence of people and the environment.
Founded: 1943
Craig Cox, Executive Director
Sue Ann Lynes, Executive Assistant

1018 State of Iowa Woodlands Associations
2404 S Duff   515-233-1161
Ames, IA 50010   Fax: 515-233-1131
Al Manning, President

Kansas

1019 American Lung Association of Kansas
4300 Southwest Drury Lane   785-272-9290
Topeka, KS 66604-2419   800-586-4872
                                  Fax: 785-272-9297
                                  E-mail: jkeller@kslung.org
                                  http://www.kslung.org
Judy Keller, Executive Director
Kris Scothorn, Office Manager

1020 Audubon of Kansas
210 Southwind Place   785-537-4385
Manhattan, KS 66503   Fax: 785-537-4395
                                  E-mail: aok@audubonofkansas.org
                                  http://www.audubonofkansas.org
Founded: 1999
Richard Seaton, Chairman
Ron Klataske, Executive Director

Associations & Organizations/Kentucky

1021 Conservation and Research Foundation
PO Box 909
Shelburne, VT 05482   913-268-0076
                             Fax: 913-268-0076
Founded: 1953
Mary Wetzel, President

1022 Kansas Academy of Science
1700 SW College Avenue   785-231-1010
Topeka, KS 66621   http://www.washburn.edu/kas
Founded: 1868
Mike Everhart, President
Pieter Berendsen, Secretary

1023 Kansas Association for Conservation and Environmental Education
2610 Claflin Road   785-532-3322
Manhatten, KS 66502-2743   Fax: 785-532-3305
                             E-mail: ldowney@oznet.ksu.edu
                             http://www.kacee.org
Statewide no-profit dedicated to promoting quality sound, non-brased environmental education in Kansas through professional development and technical assistance.
Kate Grove, President
Laura Downey, Executive Director

1024 Kansas Natural Resources Council
PO Box 2635   316-265-0767
Topeka, KS 66601 E-mail: robert.c.haughawout@boeing.com
                             http://www.knrc.ws
Protect the quality and supplies of Kansas' water. Support sustainable family farming practices that respect and restore the land and the community. Ensure a competitive energy market where renewable resources and conservation canflourish. Reduce the exposure to hazardous and nuclear wastes. Encourage environmentally sound industrial practices.
Founded: 1980
Bob Haughawout, President
Jay Barnes, Executive Director

1025 Kansas Wildflower Society
2045 Constant Avenue   785-864-3453
Lawrence, KS 66047-3729   Fax: 785-864-5093
                             http://www.naturalkansas.org
Dwight Platt, President
Cynthia Ford, Secretary

1026 Kansas Wildscape Foundation
Riverfront Plaza   785-843-9453
Suite 311   Fax: 785-843-6379
Lawrence, KS 66044   E-mail: wildscape@sunflower.com
                             http://www.kansaswildscape.org
Founded: 1991
Jim Huntington, President
Harland Priddle, Executive Director

1027 North Dakota Natural Science Society
Department of Biological Sciences
600 Park Street   785-628-4214
Hays, KS 67601-4099   Fax: 785-628-4156
                             E-mail: efinck@fhsu.edu
                             http://www.fhsu.edu/biology/pn/prarienat.htm
Regional organization with interests in the natural history of grasslands and the Great Plains.
Founded: 1967
Chris Deperno, President

1028 Safety & Health Council of Western Missouri & Kansas
5829 Troost Avenue   816-842-5223
Kansas City, MO 64110   Fax: 816-842-6226
                             E-mail: shc@safetycouncilmoks.com
                             http://www.safetycouncilmoks.com
Is a private not-for-profit community service organization which has been helping to make our community a safer place to live, work and play. We are dedicated to preventing unintentional injuries where ever they occur.
Kathy Zents, Executive Director

1029 Wildlife Society
Kansas State University
127 Call Hall   758-532-5734
Manhattan, KS 66505-1600   Fax: 785-532-5681
                             E-mail: clee@oz.oznet.ksu.edu
A nonprofit scientific and educational organization that serves professionals such as government agencies, academia, industry, and non-government organizations in all areas related to the conservation of wildlife and natural resourcesmanagement.
Charles Lee, Secretary

Kentucky

1030 American Lung Association of Kentucky
American Lung Association
4100 Churchman Avenue (40215)   502-363-2652
PO Box 9067   800-586-4872
Louisville, KY 40209-0067   Fax: 502-363-0222
                                 E-mail: info@kylung.org
                                 http://www.kylung.org
Founded: 1905
Jim Sugarman, Executive Director

1031 Kentucky Association for Environmental Education
PO Box 176055   606-578-0312
Covington, KY 41017   Fax: 606-341-9237
                             http://www.kall.org
Karen P Reagor, Executive Director
Joe Baust, President

1032 Kentucky Audubon Council
306 Hoover Hill Road   270-298-4237
Hartford, KY 42347   Fax: 502-425-4667
                             E-mail: kac@kentuckyaudubon.org
                             http://www.kentuckyaudubon.org
James Cain, President
Maggie Selvidge, Secretary

1033 Kentucky Resources Council
PO Box 1070   502-875-2428
Frankfort, KY 40602   Fax: 502-875-2845
                             E-mail: fitzKRC@aol.com
                             http://www.kyrc.org
Tom Fitzserald, Director

1034 Land Between the Lakes Association
100 Van Morgan Drive   502-924-5897
Golden Pond, KY 42211-9001
Charles Matheny, President
Gaye Luber, Director

1035 National Safety Council, Kentucky Office: Central/Southern Indiana & Cincinnati
3176 Richmond Road   859-294-4242
Suite 236   Fax: 859-294-7860
Lexington, KY 40509   E-mail: conleyg@nsc.org
Greg Conley, Manager

1036 Nature Conservancy: Kentucky Chapter
642 W Main Street   859-259-9655
Lexington, KY 40508   800-628-6860
                             Fax: 859-259-9678
                             E-mail: comment@tnc.org
                             http://nature.org
Founded: 1951
Steven J McCormick, Chief Executive Officer

1037 Scenic Kentucky
PO Box 2646   502-459-9497
Louisville, KY 40201   Fax: 502-459-5278
                             E-mail: keitheiken@msn.com
                             http://www.scenickentucky.org

Scenic America is the only national nonprofit organization dedicated to protecting natural beauty and distinctive community character. We provide technical assistance across the nation and through affiliates on scenic byways, billboardand sign control, context sensitive highway design, wireless telecommunications tower location, transportation enhancements, and other scenic conservation issues.
Keith P Eiken, Executive Director
Frederic H Davis, President

1038 Sierra Club
259 W Short Street
Lexington, KY 40507
859-255-7946
Fax: 859-233-4099
E-mail: chair@kentucky.sierraclub.org
http://www.sierraclub.org/chapters/ky/
Advance the preservation and protection of the natural environment by empowering the citizenry, especially democratically-based grassroots organizations, with charitable resources to further the cause of environmental protection. Thevehicle through which The Sierra Club Foundation generally fulfills its charitable mission.
Founded: 1968
Ray Barry, Chapter Chair
Sherry Otto, Chapter Coordinator

1039 Southeastern Association of Fish and Wildlife Agencies
1 Game Farm Road
Frankfort, KY 40601
405-521-4660
800-858-1549
E-mail: seafwa@aol.com
http://http://fw.ky.gov/

John Frampton, President
Robert Cook, Vice-President

Louisiana

1040 ASLA: Lousiana Chapter
636 Eye Street, NW
Washington, DC 20001-3736
202-898-2444
Fax: 202-898-1185
E-mail: rrichard@hntb.com
http://www.asla.org
Founded: 1899
Gerald Beaulieu, Managing Director
Alexis Borrero, Manager

1041 American Lung Association of Louisiana
2325 Severn Avenue
Suite 8
Metairie, LA 70001-6918
504-828-5864
800-586- 872
Fax: 504-828-5867
E-mail: aline@bellsouth.net
http://www.louisianalung.org
Founded: 1904
Janet Goforth, Director of Development

1042 Calcasieu Parish Animal Control and Protection Department
Department of Animal Services
5500A Swift Plant Road
Lake Charles, LA 70615
337-439-8879
Fax: 337-437-3343
E-mail: dmorales@cppj.net
http://cpac.cppj.net

David Marcantel, Operations Supervisor

1043 Louisiana Association of Conservation Districts
663 Holmes Road
Keatchie, LA 71046
318-933-5375
Fax: 318-872-3178

1044 Louisiana BASS Chapter Federation
603 Terri Drive
Luling, LA 70070
504-785-9069
E-mail: kevgobear@home.com
http://www.louisanabass.org
Kevin Gaubert, President

1045 Louisiana Wildlife Federation
PO Box 65239
Baton Rouge, LA 70896-5239
504-344-6707
Fax: 504-344-6707
E-mail: www.lawildfed@aol.com
Kathy Wascom, President/Alternative Rep
Edgar Veillon, Representative

1046 Nature Conservancy: Louisiana Chapter
PO Box 4125
Baton Rouge, LA 70821
225-338-1040
Fax: 225-338-0103
E-mail: lafo@tnc.org
http://http://nature.org
Founded: 1957
Keith Ouchley, Executive Director
Steve McCormack, President

1047 Tulane Environment Law Clinc
Tulane University
6329 Freret Street
New Orleans, LA 70118
504-865-5789
Fax: 504-862-8721
http://www.tulane.edu
Since 1989, the Tulane Law School, through its Environmental Law Clinic, has provided free legal assistance on wide varitey of environmental issues. In addition, the Clinic assists community groups with scientific and organizationalissues.
Founded: 1989
Adam Baeich, Executive Director

1048 Wildlife Society
LDWF PO Box 9800
Baton Rouge, LA 70898-9000
225-765-2346
Fax: 225-765-5476
E-mail: jlangilol@wlf.lousiana.gov
http://www.wlf.state.la.us
A nonprofit scientific and educational organization that serves professionals such as government agencies, academia, industry and non-government organizations in all areas related to the conservation of wildlife and natural resourcesmanagement.
Fred Kimmel, President

Maine

1049 American Lung Association of Maine
122 State Street
Augusta, ME 04330
207-622-6394
800-499-5864
Fax: 639-426-2919
E-mail: emiller@mainelung.org
http://www.mainelung.org
Founded: 1911
Edward Miller, CEO
Lee Scott, Deputy Director

1050 Atlantic Salmon Federation
PO Box 807
Calais, ME 04619-4581
506-529-4581
Fax: 506-529-4438
http://www.asf.ca
Promotes conservation of the Atlantic Salmon and related vital resources.
Bill Taylor, President

1051 Maine Association of Conservation Commissions
PO Box 702
Bath, ME 04530
207-443-2925
Fax: 207-443-6913
E-mail: macc@clinic.net
http://www.clinic.net/usa/macc
Founded: 1973
Robert C Cummings, Executive Director

1052 Maine Association of Conservation Districts
PO Box 152
Hallowell, ME 04347
207-622-4443
Fax: 207-623-3748
http://me.nacdnet.org

William A Bell, Executive Director

1053 Maine Audubon
20 Gilsland Farm Road
Falmouth, ME 04105
207-781-2330
Fax: 207-781-0974
E-mail: info@maineaudubon.org
http://www.maineaudubon.org
Maine Audubon works to conserve Maine's wildlife and wildlife habitat. It is an affiliate of Audubon's national organization, and has seven local chapters throughout the state.
Founded: 1843
Elyse Tipton, Communications Director
Kevin Karley, Executive Director

1054 Maine Coast Heritage Trust
1 Main Street
Suite 201
Topsham, ME 04086
207-729-7366
Fax: 207-729-6863
E-mail: info@mcht.org
http://www.mcht.org

Founded: 1970
Jay Espy, President
Karin Ponte, General Counsel

1055 National Association of School Nurses
PO Box 1300
Scarbough, ME 04070-1300
207-883-2117
877-627-6476
Fax: 207-883-2683
E-mail: gdurgin@nasn.org
http://www.nasn.org

The mission of The National Association of School Nurses is to advance the practice of school nursing and provide leadership in the delivery of quality health programs to school communities.
Founded: 1979
Wanda Miller, Executive Director
Gloria Durgin, Administrator

1056 Small Woodland Owners Association of Maine
PO Box 836
Augusta, ME 04332-0833
207-626-0005
877-467-9626
Fax: 207-626-7992
E-mail: info@swoam.com
http://www.swoam.com

Founded: 1975
Tom Doak, Executive Director
Everett Towle, President

Massachusetts

1057 ASLA: Boston Chapter
ASLA
19 Harrison Street
Framingham, MA 01702
508-620-5018
Fax: 508-879-4892
E-mail: info@bslaweb.org
http://www.bslaweb.org

Joseph Geller, President
Beth Foster, Secretary

1058 Alternatives for Community and Environment
2181 Washington St
Suite 301
Roxbury, MA 02119
617-442-3343
Fax: 617-442-2425
E-mail: info@ace-ej.org
http://www.ace-ej.org

A community-based, nonprofit, environmental justice, law and education center. Works in partnership with community groups from low income communities and communities of color to help them address their environmental and environmentalheath issues by providing free legal, educational and organizing services.
Founded: 1996
Pehn Loh, Executive Director
Warren Goldstein-Gelb, Program Director

1059 Association for Environmental Health of Soils
150 Fearing Street
suite number 21
Amherst, MA 01002
413-549-5170
Fax: 413-549-0579
E-mail: info@aehs.com
http://www.aehs.com

The AEHS was created to facilitate communications and foster cooperation among professionals concerned with the challenge of soil protection and cleanup. Members represent the many disciplines involved in making decisions and solvingproblems affecting soils. AEHS recognizes that widely acceptable solutions to the problem can be found only through the integration of scientific and technological discovery, social and political judgement and hands on practice.
Founded: 1985
Paul Kostecki, Executive Director/President
Marc A Nascarella, Managing Ed/Conference Coor

1060 Earthwatch Institute
3 Clock Tower Place
Suite 100, Box 75
Maynard, MA 01754
978-461-0081
800-776-0188
Fax: 978-461-2332
E-mail: info@earthwatch.org
http://www.earthwatch.org

Scientists, volunteers and amateurs conduct field research throughout the world. Expeditions relate to the natural environment.
Founded: 1971
Ed Wilson, Chief Operating Officer

1061 Environmental League of Massachusetts
14 Beacon
Suite 714
Boston, MA 02108
617-742-2553
Fax: 617-742-9656
E-mail: elm@environmentalleague.org
http://www.environmentalleague.org

Founded: 1898
James Gomes, President
Nancy Goodman, Vice President

1062 Ethnobotany Specialist Group
Oxford Street
Cambridge, MA 02138
617-495-2326
Fax: 617-495-5667

Prof. Richard Evans Schultes, President

1063 Genesis Fund/National Birth Defects Center: Pregnancy Environmental Hotline
40 2nd Avenue
Suite 520
Waltham, MA 02451
781-466-8474
800-322-5014
Fax: 781-487-2361
E-mail: peh@thegenesisfund.org
http://www.thegenesisfund.org/peh.htm

General information service that provides information regarding exposure to environmental factors during pregnancy and the effects on the developing fetus.
Founded: 1982
Jane E O'Brien, Managing Director
Murray Feingold, Founder

1064 Massachusetts Association of Conservation Districts
319 Littleton Road
Suite 205
Westford, MA 01886
978-692-9395
Fax: 978-392-1305
http://www.middlesexconservation.org

Founded: 1947
David Williams, Chairman

1065 Massachusetts Association of Conservation Commissions
10 Juniper Road
Belmont, MA 02478
617-489-3930
Fax: 617-489-3935
E-mail: staff@maccweb.org
http://www.maccweb.org

Protects open spaces, biodiversity, wetlands and other water resources through education, community support and advocacy. Provides a bi-monthly newsletter and telephone helpline for members. Publishes or distributes over 100 relevantpublications including the Environmental Handbook for Massachusetts Conservation Commissioners by Sally A Zielinski PhD and Alexandra Dawson JD.
Founded: 1961
Helen Bethell, President
Kenneth Pruitt, Executive Director

1066 Massachusetts Audubon Society
208 S Great Road
Lincoln, MA 01773
781-259-9500
800-AUD-UBON
Fax: 781-259-8899
E-mail: info@massaudubon.org
http://www.massaudubon.org

The largest conservation organization in New England, concentrating its efforts on protecting the nature of Massachusetts for people and wildlife. The society protects nearly 30,000 acres of conservation land, conducts nature educationprograms for over 250,000 school children and adults and works for sound environmental policies at the state, local and federal level.
Founded: 1896
Laura A Johnson, President
Bancroft R Poor, Chief Financial Officer

1067 Massachusetts Environmental Education Society
290 Turnpike Road
PO Box 105
Westboro, MA 01581
508-792-7270
Fax: 508-792-7275

Patti Steinman, President
Jennifer Wiest, Secretary

1068 Massachusetts Forestry Association
PO Box 1096   413-323-7326
Belchertown, MA 01007-1096   Fax: 413-323-9594
                                   http://www.businessfinance.com
Gregory Cox, Executive Director

1069 Massachusetts Trapper's Association
155 Williams Road   508-369-5065
Concord, MA 01742   http://www.eelink.net
William Andrade, President
Irene Hayes, Secretary

1070 Massachusetts Water Pollution Control Association
P.O. Box 221   978-374-0170
Groveland, MA 01834-0221   Fax: 978-521-4083
                           E-mail: mwpca1965@verizon.net
                                        http://www.mwpca.org
John Connor, Secretary/Treasurer

1071 Mount Grace Land Conservation Trust
1461 Old Keene Road   978-248-2043
Athol, MA 01331   Fax: 978-248-2053
                           E-mail: landtrust@mountgrace.org
                                   http://www.mountgrace.org
Protects significant natural, agricultural, and scenic areas and
encourages land stewardship in N Central Massachusetts for the
benefit of the environment, the economy, and future generations. Mount Grace has protected over 11,000acres of land. We
currently own 1,270 acres of land and we hold conservation restrictions on 2,164 acres.
            Founded: 1986
Richard French, President
Mary Williamson, Vice President

1072 New England Water Environment Association
100 Tower Office Park   781-939-0908
Suite K   Fax: 781-939-0907
Woburn, MA 01801   E-mail: mail@newea.org
                                        http://www.newea.org
We are a Regional Member Association of the Water Environ-
mental Federation. We provide technical and education for the
waste water industry.
            Founded: 1929
Elizabeth Haffner, Executive Director
Douglas Miller, President

1073 Save the Harbor/Save the Bay
286 Congress Street,7th Floor   617-451-2860
Suite 304   Fax: 617-451-0496
Boston, MA 02210   http://www.savetheharbor.org
Organization whose mission is to keep Boston Harbor and Massachusetts Bay clean forever and ensure that everyone has access to the cleaner Harbor. We believe the best way to keep the
Harbor and Bay clean forever is to capture theeconomic, educational and recreational benefits of the recovered Harbor and Bay
for everyone. We use education, outreach, advocacy and cele-
brations as well as publications, public forums, boat tours, youth
programs and special initiatives toaccomplish this.
            Founded: 1986
Patricia A Folly, President
Matt Wolfe, VIce President

1074 Walden Forever Wild
44 Baker Farm Rd   781-259-4700
Lincoln, MA 01773   800-554-3569
                                         Fax: 781-259-4710
                           E-mail: wwproject@walden.org
                                        http://www.walden.org
            Founded: 1990
Mary P Sherwood, Chairman

1075 Walden Pond Advisory Committee
Page Road   781-259-9544
Lincoln, MA 01773   http://www.concordnet.org
            Founded: 1975

Maryland

1076 ASLA: Maryland Chapter
407 East Cross Street
Baltimore, MD 21230   E-mail: stephdemchik@cs.com
                                   http://www.mdasla.org/contact.html
Stephanie Demchik, President
Valorie L Hennigan, President-Elect

1077 Alliance for the Chesapeake Bay: Baltimore Office
6600 York Road   410-377-6270
Suite 100   Fax: 410-377-7144
Baltimore, MD 21212   E-mail: mailat@acb-online.org
                                        http://www.acb-online.org
            Founded: 1607
David Bancroft, Executive Director
Darlin Hicks, CFO

1078 American Bass Association of Maryland
622 Powhattan Beach Road   410-255-0499
Pasadena, MD 21122
Clancy Thorn, Presidentt

1079 American Lung Association of Maryland
Executive Plaza 1, Suite 600   410-560-2120
11350 McCormick Road   800-642-1184
Hunt Valley, MD 21031   Fax: 410-560-0829
                           E-mail: info@marylandlung.org
                                        http://www.marylandlung.org
            Founded: 1919
Stephen Peregoy, President
Linda Marston, Vice President

1080 Audubon Naturalist Society of the Central Atlantic States
8940 Jones Mill Road   301-652-9188
Chevy Chase, MD 20815   Fax: 301-951-7179
                           E-mail: hq@audubonnaturalist.org
                                        http://www.audubonnaturalist.org
The Audubon Naturalist Society is an independent environmental education and conservation organization with over 10,000
members in the Washington DC area. The society offers a wide
variety of natural history classes and campaigns forthe protection and renewal of the Mid-Atlantic regions natural resources.
            Founded: 1897
Joseph Coleman, President:
Gilbert Gude, Honorary Vice President

1081 Center for Chesapeake Communities
229 Hanover Street   410-267-8595
Suite 101   Fax: 410-267-8597
Annapolis, MD 21401http://www.chesapeakecommunities.org
Technical assistance on environmental, land use, energy and
water quality issues for local government in Chesapeake Bay
Watershed.
Gary Allen, Executive Director

1082 Chesapeake Bay Foundation
Save The Bay Maryland Office
Philip Merrill Environmental Center   410-268-8833
6 Herndon Avenue   Fax: 410-280-3513
Annapolis, MD 21403   http://www.cbf.org
            Founded: 1964
Ronald Reagan, President

1083 Chesapeake Wildlife Heritage
Habitat Works
PO Box 1745   410-822-5100
Easton, MD 21601   Fax: 410-822-4016
                           E-mail: info@cheswildlife.org
                                        http://www.cheswildlife.org
Creates, restores and protects wildlife habitat on private and
public lands within the Chesapeake Bay watershed. Our mission
is to increase the amount and variety of wildlife habitat and to
educate the public about the need forwildlife habitat and sustainable farming oractices. CWH is dedicated to reversing the
trend of disappearing wildlife habitat and those wildlife species
that depend on it for survival.

Founded: 1980
John E Gerber, Executive Director
Chris Pupke, Development Director

1084 Colonial Waterbird Society: United States Fish & Wild-life Service

Patuxent Research Center	301-498-0380
Laurel, MD 20708	Fax: 301-498-0438
	http://www.biology.usgs.gov

Founded: 1976
B. A. Schreiber, President
Robert W Colburn, Secretary

1085 Conservation Federation of Maryland

Keep It Country

300 Lenora St., PMB-b156	206-441-3137
Seattle, WA 98121	Fax: 206-374-0858
	E-mail: f.a.r.m@erols.com
	http://www.darnet.com

Dolores Milmoe, President

1086 Eastern Shore Land Conservancy

PO Box 169	410-827-9756
Queenstown, MD 21658	Fax: 410-827-5765
	E-mail: info@elsc.org
	http://www.elsc.org

Aims to conserve those lands which will preserve the Eastern Shore's farms, forests, fisheries and rich rural heritage for the benefit of future generations. Mission statement: Preserving Land for Our Future.
Founded: 1990
Hon. Harry Hughes, Chairman
Leigh Sands, Secretary

1087 Environmental Health Education Center

University Of Maryland

28 East Ostend Street	410-706-1849
Baltimore, MD 21230	Fax: 410-706-0295
	E-mail: cehn@cehn.org
	http://www.cehn.org/cehn/resourceguide/ehec.html

The overall mission of the Center is to engage in research and provide training and education programs on topics related to occupational and environmental health and safety. Our focus is broad and the workplace, community and home areall included in our defination of environment. The audiences for our training and education programs include professionals, labor, industry and community members. Through our efforts we hope to prevent occupation and/or environment related injuriesand illnesses.
Lynn Goldman, Chairman
Dick J Batchelor, Vice Chairman

1088 Institute of Hazardous Materials Management

11900 Parklawn Drive	301-984-8969
Suite 450	Fax: 301-984-1516
Rockville, MD 20852	E-mail: ihmminfo@ihmm.org
	http://www.ihmm.org

Mission is to provide recognition for professionals engaged in the management and engineering control of hazardous materials who have attained the required level of education, experience and competence; foster continued professionaldevelopment of Certified Hazardous Materials Managers (CHMM).
Founded: 1984
John H Frick, Executive Director
Betty Fishman, Assistant Executive Director

1089 Izaak Walton League of America

707 Conservation Lane	301-548-0150
Gaithersburg, MD 20878	800-453-5463
	Fax: 301-548-0146
	E-mail: general@iwla.org
	http://www.iwla.org

Conducts research and education on river ecosystems and healthy fisheries.
Founded: 1931
William Henke, Executive Director
Charlotte Brooker, National Director

1090 Maryland Association of Soil Conservation Districts

53 Slama Road	410-956-5771
Edgewater, MD 21037-1423	Fax: 410-956-0161

E-mail: lynnehoot@aol.com
http://www.mascd.net
Founded: 1989
Lynne Hoot, Executive Director

1091 Maryland BASS Chapter Federation

PO Box 3620	301-842-3200
Baltimore, MD 21214	E-mail: mdbass@mdbass.com
	http://www.mdbass.com

Roger Trageser, President
Bill Sanders, 1st Vice President

1092 Maryland Native Plant Society

Native Plant Society

PO Box 4877	410-286-2928
Silver Spring, MD 20914	E-mail: mnps@toad.net
	http://www.mdflora.org/

Founded: 1992
Karyn Molinas, President
Samuel Jones, Secretary

1093 Maryland Recyclers Coalition

2105 Laurel Bush Road	443-640-1050
Suite 200	Fax: 443-640-1031
Bel Air, MD 21015-6185	E-mail: info@marylandrecyclers.org
	http://www.marylandrecyclers.org

Virginia Lipscomb, Vice President
Brian Ryerson, President

1094 Multiple Chemical Sensitivity Referral and Resources

508 Westgate Road	410-362-6400
Baltimore, MD 21229-2343	Fax: 410-448-3317
	E-mail: donnaya@rtk.net
	http://www.mcsrr.org

The mission of MCS Refferal and Resources is to further the diagnosis, treatment, accomodation and prevention of multiple chemical sensitivity (MCS) disorders.
Founded: 1994
Dr. Grace Ziem, President
Ann McCampbell, Director

1095 Nature Conservancy: Maryland/DC Chapter

5410 Grosvenor Lane	301-897-8570
Suite 100	800-628-6860
Bethesda, MD 20814	Fax: 301-897-0858
	E-mail: comment@tnc.org
	http://www.tnc.org

Founded: 1950
Steven J McCormick, President/CEO
Henry M Paulson, Jr, Chairman/CEO

1096 Rachel Carson Council

Po Box 10779	301-652-1877
Silver Spring Marlyand, MD 20914	Fax: 301-593-7508
	E-mail: rccouncil@aol.com
	http://www.rachelcarsoncouncil.com

Independent nonprofit scientific organization dedicated to protecting the environment against toxic and chemical threats, particularly those of pesticides.
Founded: 1965
Diana Post, Executive Director
Martha Collins, VP Marketing

1097 Sierra Club: Maryland Chapter

Sierra Club Nationals

7338 Baltimore Avenue	301-277-7111
Suite 101A	Fax: 301-277-6699
College Park, MD 20740	http://www.sierraclub.org/chapters/md/

A grassroots environmental organization which promotes appreciation of nature with hikes and outtings. We work to protect the environment in Maryland through legislative and grassroots organizing efforts.
Founded: 1865
Elizabeth Johnson, Chairman

1098 Trout Unlimited

3509 Pleasant Plains Drive	410-239-8468
Reisterstown, MD 21136-4417	Fax: 410-374-5719
	E-mail: tedgodfrey@erols.com
	http://www.mda.state.md.us

Founded: 1989
George Gaines, Chairman

1099 White Lung Association
PO Box 1483   410-243-5864
Baltimore, MD 21203-1483   Fax: 410-254-4602
                                          E-mail: jfite@whitelung.org
                                          http://www.whitelung.org
National nonprofit organization dedicated to the education of the public to the hazards of asbestos exposure. Has developed programs of public education and consults with victims of asbestos exposure, school boards, building owners, government agencies and others interested in identifying asbestos hazards and developing control programs.
James Fite, Executive Director

1100 Wildfowl Trust of North America
600 Discovery Lane   410-827-6694
PO Box 519   Fax: 410-827-6713
Grasonville, MD 21638   E-mail: cbec@cbec-wtna.org
                                          http://www.cbec-wtna.org
A nonprofit membership organization, the Wildfowl Trust of North America, Inc. is dedicated to promoting environmental education, research and conservation through an emphasis on sterwardship of the Chesapeake Bay Biodiversity at theChesapeake Bay Environmental Center.
Founded: 1979
Judy Wink, Executive Director
Torrey Brown, Secretary

1101 Wildlife Society
5410 Grosvenor Lane   301-897-9770
Suite 200   Fax: 301-530-2471
Bethesda, MD 20814   E-mail: tws@wildlife.org
                                          http://www.wildlife.org
Equipment, products and publications, as well as consultants, groups and associations related to wildlife conservation and management. Annual Conference of wildlife professionals; several hundred technical presentations and posters onall aspects of wildlife conservation and management.
Founded: 1937
Greg Moore, President
Deidre DeRoia, Secretary

Michigan

1102 ASLA: Michigan Chapter
Michigan Chapter of the American Society of Landsc
1000 W. St. Joseph Hwy.   517-485-4116
Suite 200   Fax: 517-485-9408
Lansing, MI 48915   E-mail: manager@michiganasla.org
                                          http://www.michiganasla.org
Founded: 1899
Timothy Britain, President
SuLin Ellerbrook, Secretary

1103 American Lung Association of Michigan
American Lung Association
25900 Greenfield   248-784-2000
Suite 401   800-543-5864
Oak Park, MI 48237   Fax: 248-784-2008
                                          http://www.alam.org
Founded: 1904
Rose Adams, Chief Executive Officer
Carol Christner, Chief Operating Operator

1104 American Lung Association of Michigan— Northern Region
American Lung Association
153-1/2 East Front Street   248-784-2000
Suite D   800-586-4872
Traverse City, MI 49684-2508   Fax: 231-946-0150
                                          E-mail: alam@alam.org
                                          http://www.alam.org
Founded: 1904
Paul Munzenberger, Secretary/Treasurer

1105 American Lung Association of Michigan— Upper Peninsula Region
227 West Washington Street   906-228-9833
Marquette, MI 49855-4321   800-586-4872
                                          Fax: 906-228-3430
                                          http://www.lungusa.org
Founded: 1904
Robert A Green, Chairman

1106 American Lung Association of Michigan—Capital Region Office
American Lung Association
403 Seymour Avenue   517-484-4541
Lansing, MI 48933-1179   800-678-5864
                                          Fax: 517-484-2118
                                          E-mail: alam@alam.org
                                          http://www.alam.org
Founded: 1904
Kevin M Chan, Managing Director
Stephen D Moore, President

1107 American Lung Association of Michigan—Genesee Valley Region
American Lung Association
519 South Saginaw   810-232-3177
Flint, MI 48502   800-lun- usa
                                          Fax: 810-232-6257
                                          http://www.alam.org
Founded: 1906
Rose Adams, President
Carol Christner, COO

1108 American Lung Association of Michigan—Grand Valley Region
2815 Michigan Street, Northeast   616-942-0513
Suite B   Fax: 616-942-0650
Grand Rapids, MI 49506

1109 Association of Midwest Fish and Wildlife Agencies
PO Box 30028   517-373-1263
Lansing, MI 48909   Fax: 517-373-6705
                                          http://www.mafwa.iafwa.org
Founded: 1934
Steve Gray, President
Becky Humphries, Director-at-Large

1110 Ecology Center of Ann Arbor
117 North Division Street   734-761-3186
Ann Arbor, MI 48104   Fax: 734-663-2414
                                          E-mail: info@ecocenter.org
                                          http://www.ecocenter.org/
To develop and conduct environmental education, advocacy, information and technical programs on a wide range of issues which encourage the development of sustainable communities.
Founded: 1970
Mike Wallad, President
Steven Wilcoxen, Vice President

1111 Great Lakes Commission
2805, South Industrial Hwy   734-971-9135
Suite 100   Fax: 734-971-9150
Ann Arbor, MI 48104-6791   E-mail: eschmidt@glc.org
                                          http://www.glc.org
Founded: 1955
Thomas R Crane, Interim Executive Director
Matthew Doss, Program Manager

1112 Home Chemical Awareness Coalition
Michigab State University
Natural Resource Building   517-355-9578
E Lansing, MI 48824-1222   Fax: 517-353-8994
Cynthia Frigden, Chairperson

1113 Michigan Association of Conservation Districts
PO Box 539   616-839-3360
Lake City, MI 49651-0539   Fax: 616-639-3361
                                          E-mail: mdistricts@aol.com
                                          http://www.macd.org

Associations & Organizations/Minnesota

Tom Middleton, President
Lester Langeland, Vice President

1114 Michigan BASS Chapter Federation
1010 S W Avenue   517-789-1008
Jackson, MI 49203   Fax: 517-789-5603
E-mail: jrice@co.jackson.mi.us
http://www.michiganbass.org

Dennis Beltz, President

1115 Michigan Forest Association
1558 Barrington   734-665-8279
Ann Arbor, MI 48103-5603   Fax: 734-913-9167
E-mail: mfa@i-star.com
http://www.michiganforests.com
An organization composed mainly of private owners of small woodlands. Our purpose is to promote good forest management and stewardship of all forest lands.
Founded: 1972
McClain B Smith Jr, Executive Director
Ken Serfass, President

1116 Michigan Natural Areas Council
1800 N Dixboro Road   734-975-7800
Ann Arbor, MI 48109-9741   Fax: 734-975-2424
E-mail: mnac@cyberspace.org
http://www.cyberspace.org/~mnac
Founded: 1946
Sylvia Taylor, Director
Rose Treppa, Secretary

1117 Michigan United Conservation Clubs
2101 Wood Street   517-371-1041
Lansing, MI 48912-3728   Fax: 517-371-1505
E-mail: mucc@mucc.org
http://www.mucc.org
Founded: 1937
Fran Yeager, President
Sam Washington, Executive Director

1118 National Wildlife Federation: Great Lakes Natural Resource Center
213 W Liberty   734-769-3351
Suite 200   Fax: 734-769-1449
Ann Arbor, MI 48104-1398   E-mail: greatlakes@nwf.org
http://www.nwf.org/greatlakes
For two decades the National Wildlife Federation's Great Lakes Natural Resource Center has been an on-the-ground header in protecting the region's natural legacy, focusing on issues critical to the region. The nation's largest member-supported conservation education and advocacy group, NWF unites people from all walks of life to protect nature, wildlife and the world we all share.
Founded: 1936
Andy Buchsbaum, Executive Director
Larry Schweiger, President

1119 Nature Conservancy: Michigan Chapter
101 E Grand River   517-316-0300
Lansing, MI 48906   Fax: 517-316-9886
E-mail: michigan@tnc.org
http://www.nature.org/michigan
The mission of The Nature Conservancy is to preserve the plants, animals and natural communities that represent the diversity of life on Earth by protecting the lands and waters they need to survive.
Founded: 1952
Helen Taylor, State Director
Melissa Soule, Communications Director

1120 Scenic Michigan
445 E Mitchell Street   231-347-1171
Petoskey, MI 49770   Fax: 231-347-1185
E-mail: info@scenicmichigan.org
http://www.scenicmichigan.org

The only national nonprofit organization dedicated to protecting natural beauty and distinctive community character. Provides technical assistance across the nation and through affiliates on scenic byways, billboard and sign control, context sensitive highway design, wireless telecommunications tower location, transportation enhancements, and other scenic conservation issues.
Deborah Rohe, President
Julie Metty Bennett, Vice President

1121 Sierra Club
109 E Grand River   517-484-2372
Lansing, MI 48906   Fax: 517-484-3108
E-mail: mackinac.chapter@sierraclub.org
http://http://michigan.sierraclub.org
To advance the preservation and protection of the natural environment by empowering the citizenry, especially democratically-based grassroots organizations, with charitable resources to further the cause of environmental protection. The vehicle through which The Sierra Club Foundation generally fulfills its charitable mission.
Founded: 1967
Anne Woiwode, Director

1122 Spill Control Association of America
615 Griswold   313-962-8255
7th Floor, Ford Building   Fax: 313-962-2937
Detroit, MI 48226   E-mail: marcs@scaa-spill.org
http://www.scaa-spill.org
SCAA actively promotes the interests of the spill response community.
Founded: 1973   Newsletter
David Usher, President
Ralph Bianchi, Vice President

1123 Trout Unlimited
7 Trowbridge NE   616-460-0477
Grand Rapids, MI 49503-1528
Richard Bowman, Executive Director

1124 Wildflower Association of Michigan
3853 Farrell Road   269-948-2496
Hastings, MI 49058   700-333-6459
Fax: 269-948-2957
E-mail: wam@iserv.net
http://www.wildflowersmich.org
Nonprofit organization to promote and preserve Michigan native plants and habitats.
Founded: 1986
Cheryl Smith Tolley, President
Jewel Richardson, Vice President

1125 Wildlife Society: Michigan Chapter
Michigan State University
Department of Natural Resources   517-353-2042
Room 13   Fax: 517-432-1699
East Lansing, MI 48824   E-mail: campa@msu.edu
http://www.wildlife.org/chapters/mi/index.cfm?tname=officers
A nonprofit scientific and educational organization that serves professionals such as government agencies, academia, industry, and non-government organizations in all areas related to the conservation of wildlife and natural resources management.
Founded: 1982
Brent Rudolph, President-Elect
Scott Winterstein, President

Minnesota

1126 ASLA: Minnesota Chapter
American Society of Landscape Architects
275 Market Street   612-339-0797
Suite 54   Fax: 612-338-7981
Minneapolis, MN 55405   http://www.masla.org/
John Slack, President
Bruce L Chamberlain, President-Elect

1127 American Lung Association of Minnesota
490 Concordia Avenue   651-227-8014
Saint Paul, MN 55103-2441   Fax: 651-227-5459

E-mail: info@alamn.org
http://www.alamn.org
Founded: 1903
Jerry Orr, CEO

1128 American Lung Association of Minnesota— Greater Minnesota Branch Office
American Lung Association
424 West Superior Street   218-726-4721
Suite 203   800-548-8252
Duluth, MN 55802-1532   Fax: 218-726-4722
E-mail: info@alamn.org
http://www.alamn.org/
Founded: 1903
Jerry Orr, Chief Executive Officer

1129 Institute for Agriculture and Trade Policy
2105 1st Avenue S   612-870-0453
Minneapolis, MN 55404   Fax: 612-870-4846
E-mail: iatp@iatp.org
http://www.sustain.org
The mission of the Institute for Agriculture and Trade Policy is to foster socially, economically and environmentally sustainable rural communities and regions.
Founded: 1987
Arie van den Brand, Board Chairman
Rod Leonard, Executive Director

1130 Minnesota Association of Soil and Water Conservation Districts
Soil and Water Conservation Districts
790 Cleveland Avenue S   651-690-9028
Suite 201   Fax: 651-690-9065
St. Paul, MN 55116   E-mail: lbuck@pioneerplanet.infi.net
http://www.maswcd.org
Founded: 1952
Carol Berg, President
Loyal Fisher, Vice President

1131 Minnesota BASS Chapter Federation
PO Box 225   612-339-5609
Howard Lake, MN 55349   E-mail: mmcdonou@isd.net
http://www.mnbf.org
Jay Green, President
Pat Corrigan, Vice President

1132 Minnesota Conservation Federation
551 S Snelling Avenue   651-690-3077
Suite B   Fax: 651-690-3077
Saint Paul, MN 55116-1525   E-mail: mncf@mtn.org
Gordy Meyer, President/Representative
Chris Vokaty, Alternate Representative

1133 Minnesota Ground Water Association
4779 126th Street North   651-296-7822
White Bear Lake, MN 55110-5910   http://www.mgwa.org
Founded: 2000
Laurel Reeves, President
Jon Pollock, Secretary

1134 Minnesota Native Plant Society
University of Minnesota
220 Biological Sciences Center
Saint Paul, MN 55108   E-mail: MNPS@HotPOP.com
http://www.stolaf.edu/depts/biology/mnps
A nonprofit organization dedicated to the conservation of the native plants of Minnesota through public education and advocacy. Offered are monthly meetings, field trips, symposia and a regular newsletter.

1135 Minnesota Wings Society
Bobwhite Quail Society of Minnesota
PO Box 11323   612-588-2966
Minneapolis, MN 55411   E-mail: wtcn.nature@att.net
http://www.nmu.edu/sbp/us_off.html
Founded: 1975
Thurman Tucker, President
Martin Hanson, Secretary

1136 National Flyway Council: Mississippi Office Section of Wildlife Natural Resources
North American Flyways
PO Box 30444   517-373-1263
Lansing, MI 48909-7944   Fax: 517-373-6705
E-mail: humphrir@state.mi.us
http://www.npwrc.usgs.gov/info/flyway/flychair.htm
Roger Holmes, Chairman
Joshua L Sandt, Deputy Director

1137 Nature Conservancy: Minnesota Chapter
Nature Conservancy
1101 W River Park Way   612-331-0750
Suite 200   800-628-6860
Minneapolis, MN 55415   Fax: 612-331-0770
E-mail: minnesota@tnc.org
http://www.nature.org
Aims to preserve plants, animals, and natural communities that represent the diversity of life on Earth by protecting the lands and waters they need to survive.
Founded: 1951
Steven McCormick, President/CEO

1138 Parks and Trails Council of Minnesota
275 E 4th Street   651-726-2457
Suite 642   Fax: 651-726-2458
Saint Paul, MN 55101   E-mail: info@parksandtrails.org
http://www.parksandtrails.org
Mission: to aquire, protect and enhance critical lands for the public's enjoyment now and in the future.
Founded: 1954
Grant Merritt, President
Dorian Grilley, Executive Director

1139 Raptor Center
The College of Veterinary Sciences
1920 Fitch Avenue   612-624-4745
St.Paul, MN 55108   Fax: 612-624-8740
E-mail: raptor@umn.org
http://www.raptorcenter.net
Our mission is to preserve biological diversity among raptors and other avian species through medical treatment, scientific investigation, education and management of wild populations.
Founded: 1974
Pat Redig, Director
Dr Julie Ponder, Interim Associate Director

1140 Sierra Club
2327 E Franklin Avenue #1   612-659-9124
Suite 1   Fax: 612-659-9129
Minneapolis, MN 55406 http://www.northstar.sierraclub.org
To advance the preservation and protection of the natural environment by empowering the citizenry, especially democratically based grassroots organizations, with charitable resources to further the cause of environmental protection,the vehicle through which The Sierra Club Foundation generally fulfills its charitable mission.
Founded: 1892
Russ Adams, Vice Chairman
Scott Elkins, State Director

Mississippi

1141 ASLA: Mississippi Chapter
Madison Planting & Design Group
PO Box 2171   601-898-0775
Madison, MS 39130   Fax: 601-898-9112
E-mail: mpdg@bellsouth.net
http://www.msasla.org
Founded: 1899
Robert Gammill, President
Jason B Walker, President-Elect

1142 American Lung Association of Mississippi
American Lung Association
PO Box 2178   601-206-5810
Ridgeland, MS 39158   800-586-4872
Fax: 601-206-5813

Founded: 1904
Charles A Heinrich, Chairman
Robert A Green, Vice Chairman

1143 Crosby Arboretum
Mississippi State University
PO Box 1639   601-799-2311
Picayune, MS 39466   Fax: 601-799-2372
E-mail: crosbyar@datastar.net
http://www.msstate.edu/dept/cree.camain.html
Regional arboretum dedicated to the natural and cultural history of the gulf coast. Located in Picayune Mississippi, the arboretum features a 104 acre public center and over 1,000 acres of natural lands.
Patricia Knight, Director
Melinda Lyman, Senior Curator

1144 Mississippi BASS Chapter Federation
295 County Road   601-455-6477
Meridian, MS 39301-9725   http://www.msbass.com
Harvey Cherry, President

1145 Mississippi Native Plant Society
2148 Riverside Drive   601-354-7303
Jackson, MS 39202   Fax: 601-354-7227
E-mail: manndl@millsap.edu
Debora Mann, Secretary/Treasurer

1146 Mississippi Wildlife Federation
855 South Pear Orchard Rd   601-206-5703
Suite 500   Fax: 601-206-5705
Ridgeland, MS 39157   E-mail: mstarnes@mswf.org
http://www.mswildlife.org
Founded: 1946
R Clarke Stewart, President
Cathy Shropshire, Executive Director

1147 Sierra Club
P.O. Box 4335   601-352-1026
Jackson, MS 39296-4335   Fax: 601-355-1506
E-mail: sierrams@bellsouth.net
http://www.mississippi.sierraclub.org
To advance the preservation and protection of the natural environment by empowering the citizenry, especially democratically-based grassroots organizations, with charitable resources to further the cause of environmental protection.The vehicle through which The Sierra Club Foundation generally fulfills its charitable mission.
Founded: 1892
Rose Johnson, Chairman

1148 Wildlife Society
PO Box 820161   601-631-7133
Vicksburg, MS 39182   Fax: 601-631-7133
E-mail: julie.b.marcy@usaace.army.mil
http://www.wildlife.org
A nonprofit scientific and educational organization that serves professionals such as government agencies, academia, industry, and non-government organizations in all areas related to the conservation of wildlife and natural resourcesmanagement.
Founded: 1936
Jane Pelkey, Finance Coordinator

Missouri

1149 ASLA: Saint Louis Chapter
The American Society of Landscape Architects
7722 Big Bend Boulevard   314-644-5700
Saint Louis, MO 63119   Fax: 314-644-6378
E-mail: hunterb@spaid-swt.com
http://www.stlouisasla.org
Founded: 1899
Hunter Beckham, President
Andrew Kilmer, Vice President

1150 American Fisheries Society: North Central Division
420 New Haven Road   573-875-5399
Columbia, MO 65201-9634   Fax: 573-876-1896
E-mail: pamela_haverland@usgs.gov
Pamela Haverland, President

1151 American Lung Association of Missouri
American Lung Association
1118 Hampton Avenue   314-645-5505
Saint Louis, MO 63139   800-586-4872
Fax: 314-645-7128
http://www.lungusa.org
Founded: 1904
Cynthia Erickson, CEO
Sarah Kitchen, Operating

1152 American Lung Association of Missouri- Southeast Missouri Office
American Lung Association
PO Box 482   573-651-3313
Cape Girardeau, MO 63702   Fax: 573-651-1883
http://www.lungusa2.org/missouri/
Founded: 1907
Lori Pickens, Chief Executive Officer
Charles A Heinrich, ALA Chair

1153 American Lung Association of Missouri— Southwest Missouri Office
2053-D South Waverly   417-883-7177
Springfield, MO 65804   Fax: 417-883-7026
http://www.lungusa2.org/missouri
Founded: 1907
Lori Pickens, Chief Executive
Barry Freedman, Vice President

1154 American Lung Association of Missouri—Kansas City Office
American Lung Association
2400 Troost Ave   816-842-5242
Suite 4300   Fax: 816-842-5470
Kansas City, MO 64108   E-mail: kcmo@alawmo.com
http://www.lungusa.
Founded: 1904
Joe Maxwell, President

1155 Kansas BASS Chapter Federation
PO Box 330   417-525-4940
Alba, MO 64830   E-mail: onemorefish@ckt.net
http://www.kbcf.com
Eric Strong, President
Dave Bond, Conservation Director

1156 Missouri Audubon Council
1001 Walnut   573-442-2583
Suite 200   Fax: 573-442-4378
Columbia, MO 65201   E-mail: criley@audubon.org
http://www.audubon.org/chapter/mo
Founded: 1990
Cheryl Riley, Executive Director
Charles Burwick, President

1157 Missouri Forest Products Association
611 E Capitol   573-634-3252
Jefferson City, MO 65101   Fax: 573-636-2591
E-mail: moforest@moforest.org
http://www.moforest.org
Steve Moore, President
Lynn Gastineau, Vice President

1158 Missouri Prairie Foundation
111A E Walnut   573-449-4805
Columbia, MO 65203   Fax: 573-442-0260
E-mail: gfreeman@coin.org
http://www.moprairie.org
Founded: 1966
George Nichols, President
Gary Freeeman, Membership Coordinator

1159 Missouri Public Interest Research Group
310A North Euclid
314-454-9560
Saint Louis, MO 63108
Fax: 314-454-0787
E-mail: info@mopirg.org
http://www.mopirg.org

Founded: 1972
Sarah Gaudette, Director

1160 Missouri Stream Team: Missouri Department of Conservation
PO Box 180
573-751-4115
Jefferson City, MO 65102

1161 Prarie Gateway Chapter
Water Works Building
201 Main
Suite 201
E-mail: jschuessler@bnim.com
Kansas City, MO 64105
http://www.pgasla.org
Jim Schuessler, President
John L Lutz, President-Elect

1162 Rocky Mountain Elk Foundation
311 Salt Lick Circle
816-240-2846
Napoleon, MO 64074
800-225-5355
E-mail: info@rmef.org
http://www.rmef.org

Buddy Smith, Chairman
Charlie Decker, Director

1163 Scenic Missouri
3610 Buttonwood Dr.
573-886-8954
Suite 200
Fax: 573-886-8901
Columbia, MO 65201
E-mail: info@scenicmissouri.org
http://www.scenicmissouri.org
Fights billboard blight and visual pollution. Works to protect scenic roadways. Help citizens and town planners develop tree preservation, landscaping and signage to protect communities and counties. Affiliate of Scenic America. Memberof Earth Share of Missouri.
Tom Nelson, Vice Chairman
Gene Bushmann, Chairman

1164 Society for Environmental Geochemistry and Health
1870 Miner Circle
573-341-4831
Rolla, MO 65409-0001
Fax: 303-556-4822
http://www.segh.net/contact.htm
Founded: 1971
Paula Lutz, Secretary/Treasurer

1165 Trout Unlimited
2010 Daisy Lane
573-634-3096
Jefferson, MO 65109-1810
Fax: 573-634-3096
E-mail: jdwenzlick@juno.com
http://www.midmotu.org
Founded: 1959
John Wenzlick, Chairman
Bill Lamberson, Vice Chairman

Montana

1166 American Lung Association of the Northern Rockies
825 Helena Avenue
406-442-6556
Helena, MT 59601-3459
Fax: 406-442-2346
E-mail: ala-nr@ala-nr.org
http://www.ala-nr.org

1167 American Rivers Montana Field Office
215 Woodland Estates
406-454-2076
Great Falls, MT 59404
Fax: 406-454-2530
E-mail: malbers@amrivers.org
http://www.americanrivers.org
Mark Albers, Office Director

1168 Chemical Injury Information Network
PO Box 301
406-547-2255
White Sulphur Springs, MT 59645
Fax: 406-547-2455

E-mail: chemicalinjury@ciin.org
http://www.ciin.org
Nonprofit tax-exempt support and advocacy organization run by the chemically injured for the benefit of the chemically injured. CIIN serves an international membership, and focuses primarily on eductaion, credible multiple sensitivityresearch and the empowerment of the chemically injured.

1169 Craighead Environmental Research Institute
201 S Wallace Avenue
405-585-8705
Suite B2D
Fax: 406-587-5951
Bozeman, MT 59715
E-mail: cer@avicom.net
http://www.avicom.net/ceri
The mission of the Institute is to increase humankind's understanding, appreciation, and protection of our natural environment; particulary wildlife populations and wild landscapes. Our goal is to enable human beings to live in harmonywith other species.
Lance Craighead, Director

1170 Craighead Wildlife: Wetlands Institute
5200 Upper Miller Creek Road
406-251-3867
Missoula, MT 59803
Fax: 406-251-5069
http://www.grizzlybear.org
John A Mitchell PhD, Director

1171 Greater Yellowstone Coalition
13 S Willson
406-586-1593
PO Box 1874
800-775-1834
Bozeman, MT 59771
Fax: 406-586-0851
E-mail: gyc@greateryellowstone.org
http://www.greateryellowstone.org
People working to protect America's first national park and the lands surrounding it.
Michael Scott, Executive Director
Al Jaeger, Communications Director

1172 Lighthawk: Northern Rocky Mountain Field Office
31845 Frontage Road
307-332-1642
Bozeman, MT 59715
Fax: 307-332-1641
E-mail: sarahd@lighthawk.org
http://www.lighthawk.org
Susan Benepe, Program Coordinator

1173 Montana Association of Conservation Districts
501 N Sanders
406-443-5711
Suite 2
Fax: 406-443-0174
Helena, MT 59601-4528
E-mail: mail@macd.org
http://www.macd.org

1174 Montana Audubon
PO Box 595
406-443-3949
Helena, MT 59624
Fax: 406-443-7144
E-mail: mtaudubon@mcn.net
http://www.mtaudubon.org
Janet Ellis, Executive Director
Chuck Carlson, Secretary

1175 Montana Environmental Information Center
PO Box 1184
406-443-2520
Helena, MT 59624
Fax: 406-443-2507
E-mail: meic@meic.org
http://www.meic.org
MEIC's purpose is to protect and restore Montana's natural environment. It works to do this by: monitoring and influencing the decisions and activities of the state, local and federal governments; educating individuals and by assistingindividuals and other nonprofit organizations.
Jim Jensen, Executive Director
Paul Edwards, President

1176 Montana Land Reliance
324 Fuller Avenue
406-443-7027
PO Box 355
Fax: 406-443-7061
Helena, MT 59624-0355
E-mail: info@mtlandreliance.org
http://www.mtlandreliance.org

Montana's only private, statewide land trust, an apolitical, non-profit corporation. Our mission is to provide permanent protection for private lands that are ecologically significant for agricultural production, fish and wildlifehabitat and scenic open space.
Bill Long, Managing Director
Rock Ringling, Managing Director

1177 Montana Water Environment Association
516 N Park Street
Suite A
Helena, MT 59601
406-449-7913
Fax: 406-449-6350

1178 Montana Wildlife Federation
PO Box 1175
Helena, MT 59624-1175
406-449-7604
800-517-7256
Fax: 406-449-8946
E-mail: mwf@mtwf.org
http://www.montanawildlife.com
Josh Turner, President/Alternate Rep
Kathy Hadley, Representative

1179 National Wildlife Federation
240 N Higgins
Suite 2
Missoula, MT 59802
406-721-6705
Fax: 406-721-6714
Thomas France, Director

1180 Rocky Mountain Elk Foundation
2291 W Broadway
PO Box 8249
Missoula, MT 59807
406-523-4500
800-225-5355
Fax: 406-523-4550
E-mail: info@elkfoundation.org
http://www.rmef.org
Tom Toman, Conservation VP

1181 Sierra Club
PO Box 1290
Bozeman, MT 59771
406-582-1281
406-582-9417
E-mail: WILDGRIZ@AOL.COM
http://WWW.SIERRACLUB.ORG
To advance the preservation and protection of the natural environment by empowering the citizenry, especially democratically-based grassroots organizations, with charitable resources to further the cause of environmental protection.The vehicle through which The Sierra Club Foundation generally fulfills its charitable mission.
Kathryn Hohmann, Sr. Regional Representative

1182 Trout Unlimited
PO Box 7186
Missoula, MT 59807
406-543-0054
Fax: 406-543-0054
E-mail: montrout@montanatu.org
http://www.montanatu.org
Bruce Farling, Executive Director

1183 Wildlife Society
3630 Columbus
Butte, MT 59701
406-533-3445
Fax: 406-533-3600
E-mail: montanatws@montanatws.org
http://www.montanatws.org
A nonprofit scientific and educational organization that serves professionals such as government agencies, academia, industry, and non-government organizations in all areas related to the conservation of wildlife and natural resourcesmanagement.
Founded: 1937
Tom Carlsen, President
Barb Pitman, Secretary

Nebraska

1184 American Lung Association of Nebraska
American Lung Association
7101 Newport Avenue
Suite 303
Omaha, NE 68152
402-572-3030
800-586-4872
Fax: 402-572-3028
E-mail: ala@lungnebraska.org
http://www.lungnebraska.org

Founded: 1904
Mary Peterson, CEO

1185 Iowa Prairie Network
6736 Laurel
Omaha, NE 68104
402-571-6230
Fax: 402-571-6230
E-mail: pollockg@top.net
http://www.iowaprairienetwork.org
Glenn Pollock, President
David Hansen, Director

1186 Nature Conservancy: Nebraska Chapter
1019 Leavenworth Street
Suite 100
Omaha, NE 68102
402-342-0282
Fax: 402-342-0474
E-mail: comment@tnc.org
http://www.nature.org
The mission of the Nature Conservancy is to preserve the plants, animals and natural communities that represent the diversity of life on Earth by protecting the lands and waters they need to survive.
Founded: 1915
Steven J McCormick, President
Henry M Paulson Jr, Chairman

1187 Nebraska Association of Resource Districts
601 S 12th Street
Suite 201
Lincoln, NE 68508
402-471-7670
Fax: 402-471-7677
E-mail: nard@nrcdec.nrc.state.ne.org
http://www.nrd.net.org
Dean E Edson, Executive Director
Jeanne Dryburgh, Office Manager

1188 Nebraska BASS Chapter Federation
National B.A.S. S. Chapter Federation
1518 Kozy Drive
Columbus, NE 68601
402-563-2297
E-mail: admin@nebraskabass.com
http://www.nebraskabass.com
Joe Citta, President
Dave Knuth, Vice President

1189 Nebraska Wildlife Federation
PO Box 81437
Lincoln, NE 68501-1437
402-477-1008
Fax: 402-994-2021
E-mail: nebraskawildlife@altel.net
http://www.nebraskawildlife.org
Founded: 1970
Duane Hovorka, Executive Director
Marian Maas, President

1190 Wildlife Society
The Wildlife Society
45090 Elm Island Road
Gibbon, NE 68840
308-865-5308
Fax: 308-865-5309
E-mail: mhumpert@lycosmail.com
http://www.wildlifeconsult.com/netws/
A nonprofit scientific and educational organization that serves professionals such as government agencies, academia, industry, and non-government organizations in all areas related to the conservation of wildlife and natural resourcesmanagement.
Founded: 1937
Chris Helzer, President
Renae Held, Secretary

Nevada

1191 American Lung Association of Idaho/Nevada
PO Box 7056
6275 Neil Road, Suite 300
Reno, NV 89510-7056
775-829-5864
Fax: 775-829-5850
E-mail: dszabo@lungs.org

1192 American Lung Association of Idaho/Nevada: Las Vegas Office
1800 East Sahara Avenue
Suite 106
Las Vegas, NV 89104
702-431-6333
Fax: 702-431-6630

1193 Nevada Wildlife Federation
PO Box 71238
Reno, NV 89570
775-885-0405
Fax: 775-885-0405
E-mail: dupree@pyramid.net
http://www.nvwf.org

Frank Maxwell, President

1194 Sierra Club
PO Box 8096
Reno, NV 89507
702-323-3162
Advances the preservation and protection of the natural environment by empowering the citizenry, especially democratically-based grassroots organizations, with charitable resources to further the cause of environmental protection. Thevehicle through which The Sierra Club Foundation generally fulfills its charitable mission.

1195 Tahoe Regional Planning Agency
308 Dorla Court, Suite 103
PO Box 1038
Zephyr Cove, NV 89448
702-588-4547
Fax: 702-588-4527
E-mail: mrhoades@trpa.org
http://www.trpa.org

Jim Baetge, Executive Director

New Hampshire

1196 American Bass Association of New Hampshire
235 Ridgeview Road
Weare, NH 03281
603-529-2642
John Cowan, President

1197 American Lung Association of New Hampshire
American Lung Association
9 Cedarwood Drive
Unit 12
Bedford, NH 03110
603-669-2411
800-835-8647
Fax: 603-645-6220
E-mail: info@nhlung.org
http://www.nhlung.org
Founded: 1904
Dan Fortin, President/CEO
Kent Taylor, VP Development

1198 Audubon Society of New Hampshire
3 Silk Farm Road
Concord, NH 03301-8200
603-224-9909
Fax: 603-226-0902
E-mail: asnh@nhaudubon.org
http://www.nhaudubon.org
Founded: 1923
Richard Moore, President

1199 Nature Conservancy: New Hampshire Chapter
Nature Conservancy
22 Bridge Street
4th Floor
Concord, NH 03301
603-224-5853
800-628-6860
Fax: 603-228-2459
E-mail: klabnville@tnc.org
http://www.nature.org
The mission of the Nature Conservancy is to preserve the plants, animals, and nature communities that represent the diversity of life on earth by protecting the lands and waters they need to survive.
Founded: 1987
Alice Chamberlin, Chairman

1200 New Hampshire Association of Conservation Districts
357 Prospect Hill Road
Rumney, NH 03266
603-786-9601
Fax: 603-747-3477
Calvin Perkins, President

1201 New Hampshire Association of Conservation Commissions
54 Portsmouth Street
Concord, NH 03301-5486
603-224-7867
Fax: 603-228-0423
E-mail: info@nhacc.org
http://www.nhacc.org

Founded: 1970
Mason Westfall, President
Deb Hinman, Vice-President

1202 New Hampshire Lakes Association (NHLA)
5 S State Street
Concord, NH 03301
603-226-0299
Fax: 603-226-0299
E-mail: info@nhlakes.org
http://www.nhlakes.org
A statewide nonprofit, member supported organization founded in 1992 to protect New Hampshire's lakes and ponds. We provide educational services and materials and engage in advocacy opportunities on lake issues such as water quality,balanced use, boater safety, education and invasive species prevention.
Founded: 1994
Nancy Christie, President
Thomas S Deans, Chairman

1203 New Hampshire Wildlife Federation
54 Portsmouth Street
Concord, NH 03301
603-224-5953
Fax: 603-228-0423
E-mail: info@nhwf.org
http://www.nhwf.org
The purpose of NHWF: to ensure the wise use and proper protection for the natural resources of New Hampshire and the United States. To inform and educate the public concerning the problems and solutions involved in the restoration,wise use, scientific management and conservation of wildlife and other natural resources.
Founded: 1933
Kenneth Kreis, Sr., President
Dennis Boyko, Director

1204 Northeast Resource Recovery Association
9 ,Bailey Road
Chichester, NH 03258
603-798-5777
Fax: 603-798-5744
E-mail: nrra@tds.net
http://www.recyclewithus.org
Founded: 1981
Elizabeth Bedard, Executive Director
Paula Dow, Finance Manager

1205 Trout Unlimited
9 Hilltop Drive
Hudson, NH 03051
603-886-5304
E-mail: trout@tu.org
http://www.tu.org
Founded: 1959
Charles Gauvin, President/CEO
Kenneth Mendez, Executive VP/COO

New Jersey

1206 ASLA: New Jersey Chapter
American Society of Landscape Architects
414 River View Plaza
Trenton, NJ 08611-3420
609-393-7500
Fax: 609-393-9891
E-mail: ntufaro@earthlink.net
http://www.njasla.org
Founded: 1901
Jeffrey A Tandul, President-Elect
Joseph Simonetta, Executive Director

1207 American Bass Association of Eastern Pennsylvania/New Jersey
7 Logan Drive
Somerville, NJ 08876
908-526-7721
Fax: 908-685-0970
E-mail: ehargraves@sdamechanical.com
http://www.aba-of-eastern-pa-nj.com
Founded: 1974
Ed Hargraves, President
Frank Pannick, Secretary

1208 American Lung Association of New Jersey—Main Office
1600 Route 22 East
Union, NJ 07083-3407
908-687-9340
Fax: 908-851-2625
E-mail: info@alanewjersey.org
http://www.alanewjersey.org
John A Rutkowski, President

1209 Association of New Jersey Environmental Commissions
PO Box 157 973-539-7547
Mendham, NJ 07945 Fax: 973-539-7713
E-mail: info@anjec.org
http://www.anjec.org
A statewide organization that offers training, information, guidance and policy analysis on practical approaches to natural resources preservation, reclamation and sustainable development through conferences, workshops, the quarterlyANJEC Report, handbooks and manuals.
Sandy Batly, Executive Director

1210 Biomass Users Network: Central America
Bloomfield, NJ 07003 973-680-9100
Fax: 973-680-8066

David Mazambani, Chairman

1211 Edison Facilities
2890 Woodbridge Avenue 732-321-6754
Ms 100 Fax: 732-321-4381
Edison, NJ 08837-3679

1212 Environmental and Occupational Health Science Institute
170 Frelinghuysen Road 732-445-0200
PO Box 1179 Fax: 732-445-0131
Piscataway, NJ 08855
Sponsors research, education and service programs in a setting that fosters interaction among experts in environmental health, toxicology, occupational health, exposure assessment, public policy and health education. The Institute alsoserves as an unbiased source of expertise about environmental problems for communities, employers and government in all areas of occupational and environmental health, toxicology and risk assessment.

1213 New Jersey Association of Conservation Districts
Trenton, NJ 08625 973-398-2511
Fax: 973-398-2511
E-mail: clifford-lundin@nj.nacdnet.org
Clifford R Lundin, President

1214 New Jersey BASS Chapter Federation
77 Kenvil Avenue 201-584-9387
Succasunna, NJ 07876 E-mail: mike@warwick.net
http://www.njbassfed.org
Founded: 1917
Tony Going, President

1215 New Jersey Department of Health and Senior Services
P. O. Box 360 609-984-5940
Trenton, NJ 08625-0360 888-865-8387
Fax: 609-292-3580
E-mail: aids@doh.state.nj.us
http://www.state.nj.us/health/
The mission of the Child and Adolescent Health Program is to promote optimum health and development of the children of New Jersey through the promotion of preventive services, linkages with primary medical care and healthy physcial andpsychosocial environments.
Fred M Jacobs, Managing Director
Richard J Codey, Acting Governor

1216 New Jersey Environmental Lobby
204 W State Street 609-396-3774
Trenton, NJ 08608 Fax: 609-396-4521
E-mail: njel@earthlink.net
http://www.njenvironment.org
Nonprofit organization devoted to lobbying for legislation and/or regulations that will preserve and protect New Jersey's natural resources and environment — both natural and built — and protect the public health.
Founded: 1969
Anne Poole, President
Marie A Curtis, Executive Director

1217 New Jersey Public Interest Research Group
11 N Willow Street 609-394-8155
Trenton, NJ 08608 Fax: 609-989-9013
E-mail: info@njpirg.org
http://www.njpirg.org
Dena Mottola, Executive Director
Doug O'Malley, Environmental Advocate

1218 New Jersey Society for Environmental Economic Development
30 W Lafayette Street 609-392-8899
Trenton, NJ 08608-0000 Fax: 609-396-0891
James C Morford, President

1219 Passaic River Coalition
246 Madisonville Road 908-766-7550
Basking Ridge, NJ 07920-1097 Fax: 908-766-7550
E-mail: prch2o@aol.com
http://www.passaicriver.org
Interested in preserving, maintaining and/or enhancing the water quality and quantity in the Passaic River Basin. Advocates on related issues, carries out projects that further its goals, and participates in land acquisition activitiesin order to provide open space.
Founded: 1972
Ella F Filippone, Executive Administrator
Warren Victor, Chairman

1220 Sierra Club: NJ Chapter
139 W Hanover Street 609-656-7612
Trenton, NJ 08618 Fax: 609-656-7618
http://www.sierraactivist.org
Our country's oldest and most effective grassroots environmental organization. Hikes and outings are scheduled throughout the year. We are dedicated to fighting sprawl and over-development.
Founded: 1892
Jeff Tittle, Director
Dennis Schvejda, Conservation Coordinator

New Mexico

1221 Center for Holistic Resource Management
PO Box 7128 505-344-3445
Albuquerque, NM 87194 Fax: 505-344-9079
Seeks to improve the human environment and quality of life through holistic management.
Shannon Horst, President

1222 National Parks Conservation Association
National Parks & Conservation
1300 19th Street, NW, Suite 300 505-247-1221
Washington, DC 20036 800-628-7275
Fax: 202-659-0650
E-mail: npca@npca.org
http://www.npca.org
Founded: 1919
Tom Kiernan, Director

1223 Nature Conservancy: New Mexico Chapter
Nature Conservancy
212 E Marcy Street 505-988-3867
Suite 200 800-628-6860
Santa Fe, NM 87501-2081 Fax: 505-988-4095
E-mail: nm@tnc.org
http://www.nature.org
Founded: 1951
Bill Waldman, State Director
Terry Sullivan, Assistant State Director

1224 New Mexico Association of Conservation Districts
163 Trail Canyon Road 505-981-2400
Carlsbad, NM 88220 Fax: 505-981-2400
http://www.nm.nacdnet.org
Founded: 1946
Leedrue Hyatt, VP
Cyle Sharp, President

1225 New Mexico Association of Soil and Water Conservation
163 Trail Canyon Road   505-981-2400
Carlsbad, NM 88220   Fax: 505-981-2422
Debbie Hughes, Executive Director

1226 New Mexico Center for Wildlife Law
University of New Mexico School of Law
1117 Stanford NE   505-277-5006
Albuquerque, NM 87131   Fax: 505-277-7064
E-mail: dfarris@unm.edu
http://http://ipl.unm.edu
The only national center dedicated to education, research and analysis of state, national and international wildlife laws. Established at the University of New Mexico Law School in 1990, the Center's mission is to raise the level ofunderstanding and discussion about wildlife issues through interdisciplinary education, training, analysis and information about wildlife law and policy issues. Not equipped to handle inquiries of day-to-day wildlife situations or personal legalmatters.
Founded: 1969
Ruth Musgrave, Director
Caroline Byers, President

1227 New Mexico Rural Water Association
3413 Carlisle Boulevard NE   505-884-1031
Albuquerque, NM 87110-1648   Fax: 505-884-1032
E-mail: nmrwa@nmrwa.org
http://www.nmrwa.org
Dionne Shirley, Program Manager

1228 Sierra Club
85 Second Street, 2nd Floor   415-977-5500
San Francisco, CA 94105-2519   Fax: 415-977-5799
E-mail: information@sierraclub.org
http://www.sierraclub.org/
The mission is to advance the preservation and protection of the natural environment by empowering the citizenry, especially democratically-based grassroots organizations, with charitable resources to further the cause of environmentalprotection. They are the vehicle through which The Sierra Club Foundation generally fulfills its charitable mission.
Founded: 1892
Bernard Zaleha, Vice President
Lisa Renstrom, President

1229 Wildlife Society
PO Box 35936   505-992-8651
Albuquerque, NM 87176-3593   800-299-0196
E-mail: triley@trcp.org
http://www.leopold.nmsu.edu
A nonprofit scientific and educational organization that serves professionals such as government agencies, academia, industry, and non-government organizations in all areas related to the conservation of wildlife and natural resourcesmanagement.
Founded: 1937
Terry Z Riley, President
Valerie A Williams, Secretary

New York

1230 ASLA: New York Chapter
Center for Architecture
536 LaGuardia Place   212-473-0620
New York, NY 10022   Fax: 212-473-1104
E-mail: edh@hollanderdesign.com
http://http://host.asla.org/chapters/nyasla/information.htm
Edmund D Hollander, President
Ellen LaCompte, Executive Director

1231 ASLA: New York Upstate Chapter
c/o Peter Auyer,Appel Osborne Landscape Architects
102 West Division Street   315-476-1022
Suite 400   Fax: 315-479-7573
Syracuse, NY 13204   E-mail: pauyer@appelosborne.com
http://www.asla.org
Peter V Auyer, President
Rick Rivers, Executive Director

1232 Adirondack Council
103 Hand Avenue   518-873-2240
PO Box D-2   877-873-2240
Elizabethtown, NY 12932   Fax: 518-873-6675
E-mail: info@adirondackcouncil.org
http://www.adirondackcouncil.org
Nonprofit environmental group with 18,000 members that has been working since 1975 to ensure the wild character and ecological integrity of New York State's six million acre Adirondack Park.
Founded: 1975
Brian L Houseal, Executive Director
Patricia Winterer, Chairman

1233 Adirondack Land Trust
PO Box 65   518-576-2082
Keene Valley, NY 12943-0065   Fax: 518-576-4203
Insures the preservation of land and natural resources.
Michael Carr, Executive Director

1234 American Council on Science and Health
1995 Broadway   212-362-7044
2nd Floor   Fax: 212-362-4919
New York, NY 10023-5860   E-mail: acsh@acsh.org
http://www.acsh.org
A consumer education organization based in New York City. The council is an independent association that promotes scientifically balanced evaluations of food, chemicals and the environment, and their relationship to human health.
Eliabeth M Whelan, President
Gilbert Ross, Executive Director

1235 American Lung Association of Mid-New York
587 Main Street   315-736-6099
Suite 109   Fax: 315-736-5976
New York Mills, NY 13417   E-mail: ALAMNY@aol.com
http://www.lungusa2.org
John Storey, Executive Director

1236 American Lung Association of New York State
American Lung Association
3 winners Circle   518-453-0172
Suite 300   800-LUN-GUSA
Albany, NY 12205-1187   Fax: 518-489-5864
E-mail: shudson@alanys.org
http://www.alanys.org/index.htm
Founded: 1904
Stanton H Hudson Jr, CEO
Deborah Carioto, Chief Operating Officer

1237 American Lung Association of New York State- Northeastern Region
3 Winners Circle   518-459-4197
Suite 300   Fax: 518-489-5864
Albany, NY 12205   E-mail: postmaster@alaneny.org

1238 Catskill Forest Association
PO Box 336   845-586-3054
Arkville, NY 12406   845-586-3054
Fax: 845-586-4071
E-mail: cfa@catskill.net
http://www.catskillforest.org
Founded: 1982
Joseph Kraus, President
Jim Waters, Executive Director

1239 Cornell Lab of Ornithology
Birdscope
159 Sapsucker Woods Road   607-254-2473
Ithaca, NY 14850   800-843-2473
Fax: 607-254-2415
E-mail: cornellbirds@cornell.edu
http://www.birds.cornell.edu
Founded: 1950
John Fitztatrizk, Director
Nancy Rice, Director Administration/Oper

1240 Environmental Action Coalition

625 Broadway   212-677-1601
9th Floor   Fax: 212-505-8613
New York, NY 10012-2611   E-mail: eac@eacnyc.org
http://www.eacnyc.org

A network of concerned citizens who devote time and money to spreading information on ecological and environmental clean-up efforts. Offers environmental education and sponsors programs for professionals in the field, citizenactivists, volunteers, teachers, students and labor leaders.
Paul C Berizzi, Executive Director

1241 Environmental Technology Seminar

PO Box 391   516-931-3200
Bethpage, NY 11714   http://app.nea.gov.sg
Jean Wood, President

1242 Federation of New York State Bird Clubs

New York Birders
PO Box 440
Loch Sheldrake, NY 12759   E-mail: mkoeneke@a-znet.com
http://www.fnysbc.com

The objectives are to document the ornithology of New York State; to foster interest in and appreciation of birds; and to protect birds and their habitats.
Sue Adadair, Treasurer

1243 Great Lakes United

State University College at Buffalo
1300 Elmwood Avenue   716-886-0142
Cassety Hall   Fax: 716-886-0303
Buffalo, NY 14222   E-mail: glu@glu.org
http://www.glu.org

Founded: 1982
Terry Yonker, Executive Director

1244 Hudsonia Limited

PO Box 5000   845-758-0600
Annandale, NY 12504-5000   Fax: 845-758-7033
E-mail: kiviat@bard.edu
http://www.hudsonia.org

Conducts research in the environmental sciences and provides information to decision makers and the public. We specialize in ecology and field biology in the northeastern states. Nonadvocacy nonprofit institute.
Founded: 1981
Gretchen Stevens, Staff Botanist
Carol Cadmus, Administrator

1245 INFORM

120 Wall Street   212-361-2400
14th Floor   Fax: 212-361-2412
New York, NY 10005-4001   http://www.informinc.org

INFORM is an independent research organization that examines the effects of business practices on the environment and on human health. Our goal is to identify ways of doing business that ensure environmentally sustainable economicgrowth. Our reports are used by government, industry, and environmental leaders around the world.
Founded: 1974
Joanna Underwood, President

1246 Montefiore Medical Center Lead Poisoning Prevention Program

111 E 210th Street   718-920-5016
Bronx, NY 10467   Fax: 718-920-4377

The Montefiore Lead Poisoning Prevention Program addresses all aspects of childhood lead poisoning from diagnosis and treatment to education and research. Their mission is to treat lead-poisoned children and their families and toeducate families at risk, other medical providers and local, state and national legislators and policy makers.
Nancy Redkey, Project Coordinator

1247 National Flyway Council: Atlantic Division of Fish and Wildlife Department

50 Wolf Road   518-402-8995
Albany, NY 12233   Fax: 518-402-9027
http://www.dec.state.ny.us

Gerald Barnhart, Chairman

1248 Nature Conservancy: Eastern New York Chapter

200 Broadway   518-272-0195
3rd Floor   Fax: 518-272-0298
Troy, NY 12180   E-mail: kmichasiow@tnc.org
http://http://nature.org/wherewework/northamerica/states/newyork/

Founded: 1951
Kristina Schvejda, Executive Director

1249 Nature Conservancy: New York Long Island Chapter

The Nature Conservancy
250 Lawerence Hill Road   631-367-3225
Cold Spring Harbor, NY 11724   800-628-6860
Fax: 516-367-4715
E-mail: comment@tnc.org
http://www.nature.org

Founded: 1951
Henry M Paulson Jr, Chairman
Steven J McCormick, President/CEO

1250 New York Association of Conservation Districts

HRC 335E   518-629-7645
HVCC, 80 Vandenburgh Avenue   Fax: 518-629-7646
Troy, NY 12180   E-mail: nyacd@AOL.com
http://www.nyacd.org

Founded: 1958
Linda Coffin, President
Gregory Bell, Executive Director

1251 New York Coalition for Alternatives to Pesticides

353 Hamilton Street   518-426-8246
Albany, NY 12210-1709   Fax: 518-426-3052
E-mail: nycap@crisn.or
http://www.altpest.org

The mission of the New York Coalition for Alternatives to Pesticides is to eliminate the use of hazardous chemicals through education and outreach.NYCAP seeks to improve public and environmental health by promoting the use of saferalternatives to pesticides, cleaning supplies and other chemicals, by advocating for the reduction of risks in the manufacture, transportation, use and disposal of toxic chemicals and by campaiging for environmentally sound public policy.
Founded: 1989
Pamela Hadad-Hurst, Executive Director

1252 New York Forest Owners Association

PO Box 180   716-377-6060
Fairport, NY 14450   800-836-3566
Fax: 716-388-7592
E-mail: nyfoainc@hotmail.com
http://www.nyfoa.org

Founded: 1905
Ronald Pederson, President
Daniel Palm, Executive Director

1253 New York Healthy Schools Network

773 Madison Ave   518-462-0632
Albany, NY 12208   Fax: 518-462-0433
E-mail: info@healthyschools.org
http://www.healthyschools.org

The New York Healthy Schools Network is a state wide coalition of parent, environment, health and education organizations working to assure every child and school employee an environmentally healthy school which is clean and in goodrepair, through shared advocacy and information and referral.
Founded: 1995
Claire L Barnett, Executive Director

1254 New York Parks and Conservation Association

29 Elk Street   518-434-1583
Albany, NY 12207   Fax: 518-427-0067
E-mail: nypca@nypca.org
http://www.nypca.org

Founded: 1985
Douglas R McCuen, Chairman
Jeffrey P Swain, Vice Chairman

1255 New York State Council of Landscape Architects

235 Lark Street   212-431-3609
Albany, NY 12210   E-mail: kmatthews@mnlandscape.com
http://www.nyscla.org

Associations & Organizations/New York

W Dean Gomolka, President
Alexander F Kurnicki, President-Elect

1256 New York State Department of Environmental Conservation
625 Broadway   518-402-8540
Albany, NY 12233   Fax: 518-402-9016
       http://www.dec.state.ny.us
DEC protects, improves and conserves the state's land, water air, fish, wildlife and other resources to enhance the health, safety and welfare of the people and their overall economic and social well-being.
Denise M Sheehan, Commissioner

1257 New York Turtle and Tortoise Society
PO Box 878   212-459-4803
Orange, NJ 07051-0878   http://www.nytts.org
Suzanne Dohm, President

1258 New York Water Environment Association
126 N Salina Street   315-422-7811
Suite 200
Syracuse, NY 13202   Fax: 315-422-3851
       E-mail: pcr@nywea.org
       http://www.nywea.org
The New York Water Environment Association is a nonprofit educational association dedicated to the development and dissemination of information concerning water quality management and the nature, collection, treatment, and disposal ofwastewater. Founded in 1929, the Association has over 2,100 members. The NYWEA is a member association of the Water Environment Federation.
Founded: 1929
Patricia Cerro-Reehil, Executive Director

1259 New York-New Jersey Trail Conference
232madison Avenue   212-685-9699
New York, NY 10016   Fax: 212-779-8102
       E-mail: office@nynjtc.org
       http://www.nynjtc.org
Gary Haugland, President

1260 Radioactive Waste Campaign
118 N 11th Street   718-387-8786
Brooklyn, NY 11211
Jean Fazzino, Executive Director

1261 Rene Dubos Center for Human Environments
81 Pondfield Road   914-337-1636
Suite 387   Fax: 914-771-5206
Bronxville, NY 10708   http://www.dubos.org
The Rene Dubos Center for Human Environments is a non-profit education and research organization focused on the social and humanistic aspects of environmental problems.
Noel Brown, President

1262 Sagamore Institute
PO Box 146   315-354-5311
Raquette Lake, NY 13436   Fax: 315-354-5851
       E-mail: sagamore@telnet.com
       http://www.sagamore.org
Educational programs on history, ecology and culture of Adirondack Park.
Michael Wilson

1263 Scenic Hudson
1 Civic Center Plaza   845-473-4440
Suite 200   Fax: 845-473-2648
Poughkeepsie, NY 12601   E-mail: info@scenichudson.org
       http://www.scenichudson.org
Founded in 1963, Scenic Hudson is a nonprofit environmental organization and separately incorporated land trust that works from Manhattan to the foothills of the Adirondacks. Headquartered in downtown Poughkeepsie, the group works withcitizens, government and business to preserve landscapes and create public parks, reclaim urban waterfronts and foster community planning, improve air and water quality, and promote environmental awareness and access to the Hudson River.

Christopher Davis, Vice Chairman

1264 Selikoff Clinical Center for Occupational /Environmental Medicine
Mount Sinai School of Medicine
Department of Community Medicine   212-241-8689
1 Gustave Levy Place, Box 1043   Fax: 212-360-6965
New York, NY 10029-6574   http://www.mssm.edu/cpm
Internationally respected diagnostic referral center and an important interface between the research programs of the Division of Environmental Health Science and populations exposed to environmental hazards.

1265 Sierra Club
85 Washington Street   518-587-9166
Saratoga Springs, NY 12866   Fax: 518-583-9062
       E-mail: ne.field@sierraclub.org
       http://www.sierraclub.org
To advance the preservation and protection of the natural environment by empowering the citizenry, especially democratically based grassroots organizations with charitable resources to further the cause of environmental protection, thevehicle through which The Sierra Club Foundation generally fulfills its charitable mission.
Founded: 1892
Mark Bettinger, Staff Director
Chris Ballantyne, Senior Regional Representati

1266 Sierra Club: New York City Office
National Sierra Club
31st Floor 116 John Street   212-791-9291
Suite 3100   Fax: 212-791-0839
New York, NY 10038   E-mail: ne.field@sierraclub.org
       http://www.sierraclub.org or
       http://newyork.sierraclub.org/nyc/
Part of the national Sierra Club with over 700,000 members, headquarters in San Francisco, offices in Washington DC, and staff in state capitals around the nation. Our mission is to enjoy and protect the earth's ecosystems andresources, and to enlist others to do the same. The Atlantic Chapter consists of 32,000 members in 11 local groups throughout New York State. We use the media, grassroots education and personal contact to bring our issues to our communities and ourpublic officials.
Founded: 1892   Quarterly
Lisa Renstrom, President
Bernard Zaleha, Vice President

1267 Trout Unlimited
111 High Ponit Mountain Road   914-892-8630
West Shokan, NY 12494-5337   E-mail: karwac@ibm.org
Chester Karwatowski, Chairman

1268 Tug Hill Tomorrow Land Trust
PO Box 6063   315-779-8240
Watertown, NY 13601   Fax: 315-785-2574
       E-mail: thtomorr@northnet.org
       http://www.tughilltomorrowlandtrust.org
A regional, nonprofit land trust and education organization working to retain the farm, forest, recreation and wild lands of northern New York's Tug Hill region. Offers conservation planning, land registry and conservation easementprograms for landownwers and educational resources and programs.
Linda M Garrett, Executive Director

1269 West Harlem Environmental Action
271 W 125th Street   212-961-1000
Suite 308   Fax: 212-961-1015
New York, NY 10027   E-mail: wheact@igc.apc.org
       http://www.weact.org
West Harlem Environmental Action, is a nonprofit organization working to improve environmental quality and to secure environmental justice in predominately African-American and Latino communities.
Founded: 1988
Dennis A Derryck, Chairman
Peggy M Shepard, Executive Director

1270 Women's Environment and Development Organization
355 Lexington Avenue   212-973-0325
3rd Floor   Fax: 212-973-0335
New York, NY 10017   E-mail: wedo@wedo.org
       http://www.wedo.org

WEDO is an international advocacy network that seeks to increase the power of women worldwide as policymakers in government and in policy making institutions, forums and processes, at all levels, to achive economic and social justice,a peaceful and healthy planet and human rights for all.
Founded: 1990
June Zeitlin, Executive Director
Betsy Apple, Deputy Director

North Carolina

1271 ASLA: North Carolina Chapter
PO Box 221256
Charlotte, NC 28222-1256   704-367-1012
                                            888-999-2752
                                            Fax: 704-367-1013
                                 E-mail: howard@ncat.edu
                                       http://www.asla.org
Founded: 1899
Lahill Aylward, Director
Perry Howard, President-Elect

1272 Acid Rain Foundation
1410 Varsity Drive   919-828-9443
Raleigh, NC 27606-2010   Fax: 919-515-3593
Dr. Harriett S Stubbs, Executive Director

1273 American Fisheries Society-Early Life History
NOAA-National Marine Fisheries Service
Beaufort Laboratory
101 Pivers Island Road   206-526-4108
Beaufort, NC 28526   Fax: 252-728-8747
                             E-mail: art.kendal@noaa.org
Art Kendal, President

1274 American Lung Association of North Carolina
American Lung Association
3801 Lake Boone Trail, Suite 190   919-832-8326
PO Box 27985   800-892-5650
Raleigh, NC 27611   Fax: 919-856-8530
                                  E-mail: info@lungnc.org
                                     http://www.lungnc.org
Founded: 1904
Deborah C Bryan, President/CEO
Kathryn Rawlings, VP Development

1275 Carolina Bird Club
5009 Crown Point Lane   910-791-9034
Wilmington, NC 28409   Fax: 910-791-7228
                         E-mail: hq@carolinabirdclub.org
                              http://www.carolinabirdclub.org
A nonprofit educational and scientific association, open to anyone interested in the study and conservation of wildlife, particularly birds. Meets each winter, spring and fall. Meeting sites are selected to give participants anopportunity to see many different kinds of birds. Guided field trips, informative programs and business sessions are combined for an exciting weekend of meeting with people who share an enthusiasm and concern for birds.
Founded: 1937
Stephen Harris, President
Dana Harris, Headquarters Secretary

1276 Carolina Recycling Association
274 Pittsboro Elementary School Rd.   919-545-9050
Suite #6   Fax: 919-545-9060
Pittsboro, NC 27312   http://www.cra-recycle.org
Founded: 1989
Kerry Krumsiek, Executive Director
Karen Hales, Program Manager

1277 Environmental Educators of North Carolina
PO Box 4901   919-250-1050
Chapel Hill, NC 27515-4901   Fax: 919-250-1058
Sheila Jones, President

1278 Forest History Society
701 Vickers Avenue   919-682-9319
Durham, NC 27701   Fax: 919-682-2349

E-mail: stevena@duke.edu
http://www.lib.duke.edu/forest/
Brings lessons of forest and conservation history to bear on current issues in natural resource management. Identifies, collects, preserves, interprets and disseminates information on forest and conservation history with the goals ofimproving the public's understanding of the forest industry and forest products; providing original resource material to researchers writing forest history, to assure accuracy on facts and interpretation; and facilitating development of naturalresource policy.
Founded: 1946
Steven Anderson, President
Andrea Anderson, Secretary

1279 North Carolina Association of Soil & Water Conservation Districts
Department of Environment and Natural Resources
1614 Mail Service Center   919-733-2302
Releigh, NC 27699-1614   Fax: 919-715-3559
                            http://www.enr.state.nc.us/dswc/
David Williams, Acting Director
Lynn Sprague, Chief, District Programs

1280 North Carolina Coastal Federation
3609 Highway 24   252-393-8185
Newport, NC 28570   800-232-6210
                                    Fax: 252-393-7508
                                  E-mail: nccf@nccoast.org
                                    http://www.nccoast.org/
Founded: 1982
Todd Miller, Executive Director
Melvin Shepard, Jr, President

1281 North Carolina Museum of Natural Sciences
11 W Jones Street   919-733-7450
Raleigh, NC 27601-1029   877-4NA-TSCI
                                    Fax: 919-733-1573
                       E-mail: museum@naturalsciences.org
                              http://www.naturalsciences.org
Founded: 1985
Fran Nolan, Executive Director

1282 North Carolina Wild Flower Preservation Society
Botanical Center
CB #3775 Totten Center UNC-CH   336-370-8172
Chapel Hill, NC 27599-3375   http://www.ncwfps.org
Charlotte Patterson, President

1283 Scenic North Carolina
19 W Hargett Street   919-832-3687
Raleigh, NC 27601   Fax: 919-832-3299
                          E-mail: scenic.nc@worldnet.att.net
                              http://www.mcis.duke.edu/snc.htm
The only national nonprofit organization dedicated to protecting natural beauty and distinctive community character. We provide technical assistance across the nation and through affiliates on scenic byways, billboard and sign control,context sensitive highway design, wireless telecommunications tower location, transportation enhancements, and other scenic conservation issues.
Dale McKeel, Executive Director

1284 Trout Unlimited
135 Tacoma Circle   704-684-5178
Ashville, NC 28801-1625   Fax: 704-687-1689
Founded: 1959
Kirk Otey, Chairman

1285 Wildlife Society
5410 Grosvenor Lane   301-897-9770
Suite 200   Fax: 301-530-2471
Bethesda, MD 20814-2144   E-mail: TWS@Wildlife.org
                                       http://www.wildlife.org
A nonprofit scientific and educational organization that serves professionals such as government agencies, academia, industry, and non-government organizations in all areas related to the conservation of wildlife and natural resourcesmanagement.
Founded: 1937
Sandra Staples-Bortner, Program Director
William Rooney, Managing Editor

North Dakota

1286 American Lung Association of North Dakota
American Lung Association
PO Box 5004   701-223-5613
Bismarck, ND 58502-5004   800-252-6325
                                     Fax: 701-223-5727
           E-mail: amerlungnd@gcentral.com
Founded: 1909

1287 International Association for Impact Assessment
1330 23rd Street S   701-297-7908
Suite C   Fax: 701-297-7917
Fargo, ND 58103   E-mail: info@iaia.org
                                     http://www.iaia.org
The International Association for Impact Assessment is an inter-disciplinary society dedicated to developing international capacity to anticipate, plan and manage the consequences of development. The Association has over 2,500 membersin over 100 nations. IAIA seeks to ensure that political, environmental, social and technological dimensions of decisions are understood by those making them.
           Founded: 1980
Rita Hamm, CEO

1288 North Dakota Association of Soil Conservation Districts
Lincoln Oakes Nurseries
3310 University Drive PO Box 1601   701-223-8575
Bismarck, ND 58502-1601   Fax: 701-223-1291
           E-mail: lincolnoakes@tic.bisman.com
                   http://www.lincolnoakes.com
           Founded: 1947
Gary L Puppe, Executive Vice President

Ohio

1289 Akron Zoological Park
500 Edgewood Avenue   330-375-2550
Akron, OH 44307   Fax: 330-375-2575
                                     E-mail: info@akronzoo.org
                                     http://www.akronzoo.org
           Founded: 1953

1290 American Lung Association of Ohio
1950 Arlingate Lane   614-279-1700
Columbus, OH 43228   800-586-4872
                                     Fax: 614-279-4940
                                     E-mail: tracy@ohiolung.org
                                     http://www.ohiolung.org
           Founded: 1904
Tracy Ross, Interim President
Tracy Ross, CEO

1291 American Lung Association of Ohio, Northeast Region
American Lung Association
6100 Rockside Woods Boulevard   216-524-5864
#260   800-586-4872
Independence, OH 44131   Fax: 216-524-7647
                                     E-mail: bgalvin@ohiolung.org
                                     http://www.ohiolung.org
           Founded: 1901
Charles A Heinrich, Chairman
James M Anderson, Secretary

1292 American Lung Association of Ohio, Northwest Region
226 State Route 61 East   419-663-5864
Norwalk, OH 44857   Fax: 419-668-2575
                                     E-mail: pjvolz@accnorwalk.com

1293 American Lung Association of Ohio, Southwest Region
American Lung Association of Ohio
11113 Kenwood Road   513-985-3990
Cincinatti, OH 45242   Fax: 513-985-3995
                                     E-mail: jkaplan@ohiolung.org
                                     http://www.ohiolung.org
           Founded: 1904
Joel B kaplan, Executive Director

1294 Association for Facilities Engineering
Crooked Tree Golf Course   515-536-2508
5171 Sentinel Oak Drive   Fax: 513-558-1739
Mason, OH 45040   E-mail: webmaster@afe2.org
                                     http://www.afe.org
AFE provides education, certification, technical information and other relevant resources for plant and facility engineering, operations and maintenance professionals worldwide.
           Founded: 1915
Byron Denny, Chapter President

1295 Central Ohio Anglers and Hunters Club
1773 Huy Road   614-447-0116
PO Box 28224   http://www.dnr.state.oh.us
Columbus, OH 43224
Doug Eakens, President

1296 Cincinnati Nature Center
4949 Tealtown Road   513-831-1711
Milford, OH 45150   Fax: 513-831-8052
                                     E-mail: cnc@cincynature.org
                                     http://www.cincynature.org
           Founded: 1965
Rhonda Barnes-Kloth, Communications Manager
Bill Hopple, Executive Director

1297 Environmental Enterprises
Environmental Enterprises Inc.
10163 Cincinnati Dayton Road   513-772-2818
Cincinnati, OH 45241   800-722-2818
                                     Fax: 513-782-8950
                                     http://www.eeienv.com
Provides analytical testing for drinking and ground water and waste analysis. Part B treatment facility offers waste disposal including: aerosols, consumer return goods, acids, caustics, oxidizers, cyanides, flammables, heavy metal,wastewaters, sludges, lab chemicals and household hazardous waste. Field Div. provides emergency response, Phase I & II, mold investigation and RCRA and OSHA training. Lab Pack Div. offers on-site ID, profiling, packaging, manifesting and removal oflab chemicals.
Daniel McCabe, President
Jerry Trumpey, Director Sales/Marketing

1298 Holden Arboretum
9500 Sperry Road   440-946-4400
Kirtland, OH 44094   Fax: 440-602-3857
                                     E-mail: holden@holdenarb.org
                                     http://www.holdenarb.org
           Founded: 1913
Elaine Price, Interim Executive Director

1299 League of Ohio Sportsmen
642 West Broad Street   614-224-8970
Columbus, OH 43215   Fax: 614-224-8971
           E-mail: info@leagueofohipsportsmen.org
                   http://www.leagueofohiosportsmen.org
           Founded: 1978
Larry Mitchell Sr., President
Dan Olszak, Vice President

1300 Nature Conservancy: Ohio Chapter
6375 Riverside Drive   614-717-2770
Suite 50   800-628-6860
Dublin, OH 43017   Fax: 614-717-2777
                                     E-mail: ohio@tnc.org
                                     http://nature.org
A global conservation organization dedicated to preserving plants, animals and natural communities that represent the diversity of life on Earth by protecting the lands and water they need to survive. Since its inception in 1951, TheNature Conservancy has protected more than 12 million acres in the US and helped through partnerships preserve more tan 80 million acres in Latin America, the Caribbean, Canada, Asia and the Pacific.
           Founded: 1915
Steven J McCormick, President

1301 Ohio Alliance for the Environment
14 Beck St   614-833-4223
Canal Winchester, OH 43110   Fax: 614-833-4223
                                     E-mail: probasco@ohioalliance.org
                                     http://www.ohioalliance.org

Founded: 1978
Peggy Smith, Executive Director
Mike Parkes, President

1302 Ohio BASS Chapter Federation
43 Portsmouth Rd.   740-446-9810
Gallipolis, OH 45631   Fax: 740-446-9819
                            E-mail: jdoss@ohiobass.org
                                http://www.ohiobass.org

Jim Doss, President
Karl Guegold, Vice President

1303 Ohio Energy Project
670 Enterprise Drive   614-785-1717
Suite A   Fax: 614-785-1731
Lewis Center, OH 43035   E-mail: rsmith@ohioenergy.org
                                http://www.ohioenergy.org
An organization providing energy and energy efficiency education using current complete and unbrased information, as well as
hands-on, engaging and innovative techniques. OEP's kids
teching kids approach also helps develop leadershipteam work
and presentation skills. An affiliate of NEED, OEP has been
named a National Energy Champion and one of the Top 12 Environmental Education Programs in Ohio.
Rich Smith, Executive Director
Deb Yerkes, Director

1304 Ohio Environmental Council
1207 Grandview Avenue Suite 201   614-487-7506
Columbus, OH 43212-3449   Fax: 614-487-7510
                            E-mail: oec@theOEC.org
                                http://www.theoec.org

Vicki L Deisner, Executive Director
Keith Dimoff, Deputy Director

**1305 Ohio Federation of Soil and Water Conservation Dis-
tricts**
PO Box 24518   614-784-1900
Columbus, OH 43224   Fax: 614-784-9181
                        E-mail: jeaneen.hooks@ofswcd.org
                                http://www.ofswcd.org

Ken Reidlinger, President
Clark Sheets, Jr, Vice President

1306 Ohio Native Plant Society
6 Louise Drive   440-338-6622
Chagrin Falls, OH 44022   E-mail: inky5@juno.com
          http://www.dir.gardenweb.com/directory/onps1/
A Malmquist, Executive Secretary

1307 Ohio Parks and Recreation Association
1069A W Main Street   614-895-2222
Westerville, OH 43081-1181   800-238-1108
                                 Fax: 614-895-3050
                            E-mail: opra@opraonline.org
                                http://www.opraonline.org
            Founded: 1938
Michelle Park, Executive Director
Julia Collins, Office Manager

1308 Sierra Club
36 W Gay Street   614-461-0734
Suite 314   Fax: 614-461-0730
Columbus, OH 43215   E-mail: Enid.Nagel@thomson.com
          http://www.sierraclub.org/chapters/oh/
To advance the preservation and protection of the natural environment by empowering the citizenry, especially democratically-based grassroots organizations, with charitable resources
to further the cause of environmental protection.The vehicle
through which The Sierra Club Foundation generally fulfills its
charitable mission.
Enid Nagel, Chairman
Earl Clausson, Vice Chairman

1309 Wildlife Society
952 Lima Avenue   419-424-5000
Box A   Fax: 419-422-4875
Findlay, OH 45840

A nonprofit scientific and educational organization that serves
professionals such as government agencies, academia, industry,
and non-government organizations in all areas related to the
conservation of wildlife and natural resourcesmanagement.
J Butterworth, President

Oklahoma

**1310 American Fisheries Society: Fisheries Management Sec-
tion**
OK Fish RS Laboratory
500 E Consellation   405-325-7288
Norman, OK 73072   Fax: 405-325-7631
                            E-mail: main@fisheries.org
                                http://www.fisheries.org
        Founded: 1870
Kurk Kuklinski, President
Brent Gordon, Seretary

1311 American Lung Association of Oklahoma
2805 East Skelly Drive   918-747-3441
Suite 806   800-654-2452
Tulsa, OK 74105   Fax: 918-747-4629
                            E-mail: ktodd@oklung.org
                                http://www.oklung.org
Kay Todd, Chief Executive Officer

1312 American Lung Association— Oklahoma City Office
3160 West Britton Road   405-748-4674
Suite E   800-586-4872
Oklahoma City, OK 73120-2037   Fax: 405-748-6274
                    E-mail: karenstoltz@earthlink.net
                                http://www.oklung.org
        Founded: 1904

1313 Nature Conservancy: Oklahoma Chapter
23 W 4th Street   918-585-1117
Tulsa, OK 74103   Fax: 918-585-2383
                            http://www.tnc-oklahoma.org
Mary Collins, State Director

1314 Oklahoma Association of Conservation Districts
PO Box 107   405-340-8884
Chelsea, OK 74016-0107   Fax: 405-842-8744
Mark Moehle, VP

1315 Oklahoma BASS Chapter Federation
BASS Federation
2300 E Coleman Road   580-765-0165
Ponca City, OK 74604   Fax: 580-765-7002
                        E-mail: president@okbass.org
                                http://www.okbass.org
        Founded: 1972
Gary Gunter, President
Hollis Spear, Vice President

1316 Oklahoma Ornithological Society
1701 W Will Rogers   918-343-7706
Claremore, OK 74017   Fax: 918-343-7563
                            E-mail: info@okbirds.org
                                http://www.okbirds.org
        Founded: 1982
Suzy Harris, President

1317 Trout Unlimited
9528 E 55th Street   918-822-6633
Siute A   Fax: 918-627-2383
Tulsa, OK 74145
Dale Deuvall, Chairman

1318 Underground Injection Practices Council
Underground Injection Practices Research Found.
827 NW 63rd Street   405-848-0690
Oklahoma City, OK 73116   Fax: 405-848-0722
Michel Paque, Executive Director

Oregon

1319 ASLA: Oregon Chapter
PO Box 40709   503-297-1005
Portland, OR 97240   E-mail: dwalters@migcom.com
           http://www.aslaoregon.org

Dave Walters, President
Rodney Lamb, Vice President

1320 American Fisheries Society: Marine Fisheries
Hatfield Marine Science Center
2030 Marine Sciences Drive   541-867-0135
Newport, OR 97365   Fax: 541-867-0138
    E-mail: steve.berkeley@hmsc.orst.edu
Steven Berkeley, President

1321 American Fisheries Society: Water Quality Section
324 25th Street   801-625-5358
Ogden, UT 84401   Fax: 801-625-5756
    E-mail: glampman@fs.fed.us
Section objectives are to: maintain an association of persons involved in the protection of watersheds, water quality, and aquatic habitat, and the abatement of water pollution and aquatic habitat and water deterioration.
Gina Lampman, President

1322 American Lung Association of Oregon
THE American Lung Association
7420 Southwest Bridgeport Road   503-924-4094
Suite 200   Fax: 503-924-4120
Tigard, OR 97224   E-mail: admin@lungoregon.org
           http://www.lungoregon.org

Founded: 1972
Sue Fratt, CEO
Sandy Jimnez, Assistant Executive

1323 Columbia Basin Fish and Wildlife Authority
851 SW Sixth Avenue, Suite 260   503-229-0191
Pacific First Building   Fax: 503-229-0443
Portland, OR 97204   E-mail: Jann.Eckman@cbfwa.org
           http://www.cbfwa.org

Founded: 1993
Brian Lipscomb, Executive Director
Jann Eckman, President

1324 Natural Resources Information Council
StreamNet Library
c/o Beth Thomsett-Scott, University   503-736-3581
P.O. Box 305190   Fax: 503-731-1260
Denton, TX 76203-5190   E-mail: bscott@library.unt.edu
           http://www.nric.info
NRIC is made up of federal, state, academic, and special natural resources librarians and information specialists from the US and Canada. The mission of NRIC is to facilitate the exchange of information on sustainable naturalresources. Its goals are to build a network of resource people to collect and dissemate information on sustainable natural resources and to provide continuing education in the technology of delivering information and to provide service to users.
Founded: 1993
Inev Hopkins, Chairperson
Beth Thomsett-Scott

1325 Northwest Coalition for Alternatives to Pesticides
PO Box 1393   541-344-5044
Eugene, OR 97440-1393   Fax: 541-344-6923
    E-mail: info@pesticide.org
           http://www.pesticide.org
NCAP works to protect people and the environment from the hazards of pesticide chemicals.
Founded: 1977
Jean Cameron, President
Norma Grier, Executive Director

1326 Oregon BASS Chapter Federation
4775 Gardner Road SE   503-588-1920
Salem, OR 97302   http://www.orbass.oregonbass.net

Orville Alleman, President
Rod McKenzie, Vice President

1327 Oregon Chapter Sierra Club
2950 Stark Street SE   503-238-0442
Suite 110   Fax: 503-238-6281
Portland, OR 97214   E-mail: oregon.chapter@sierraclub.org
           http://www.oregon.sierraclub.org
To advance the preservation and protection of the natural environment by empowering the citizenry, especially democratically based grassroots organizations, with charitable resources to further the cause of environmental protection,the vehicle through which The Sierra Club Foundation generally fulfills its charitable mission.
Robyn Conroy, Office Manager

1328 Oregon Natural Resources Council
5825 N Greeley Avenue   503-283-6343
Portland, OR 97217   Fax: 503-283-0756
    E-mail: info@onrc.org
           http://www.onrc.org
ONRC works to aggressively protect and restore Oregon's wildlands, wildlife and waters as an enduring legacy.
Founded: 1974
Regna Merritt, Executive Director

1329 Oregon Refuse and Recycling Association
PO Box 2186   503-588-1837
Salem, OR 97308-2186   800-527-7624
           Fax: 503-399-7784
    E-mail: orrainfo@orra.net
           http://www.orra.net

Founded: 1965
Max Brittingham, Executive Director
Kimera Coady, Executive Assistant

1330 Oregon State Public Interest Research Group
Oregon Student Public Interest Research Group
1536 SE 11th Avenue   503-231-4181
Portland, OR 97214   Fax: 503-231-4007
    E-mail: info@ospirgstudents.org
           http://www.ospirg.org
Founded: 1983
Rhett Lawrence, Environmental Advocate
Maureen Kirk, Executive Director

1331 Oregon Trout
117 SW Naito Parkway   503-222-9091
Portland, OR 97204   Fax: 503-222-9187
    E-mail: info@ortrout.org
           http://www.ortrout.org
Dedicated to the protection and restoration of native fish and the ecosystems upon which they depend. Oregon Trout is a science-based conservation group that achieves its mission through policy monitoring, on-the-ground restoration,and its Salmon Watch education program.
Founded: 1983
Joe S Whitworth, Executive Director

1332 Oregon Water Resources Congress
1201 Court Street NE   503-363-0121
Suite 303   Fax: 503-371-4926
Salem, OR 97301   E-mail: owrc@owrc.org
           http://www.owrc.org

Founded: 1912
Anita Winkler, Executive Director
Carol Zielinski, Office Manager

1333 Pacific Rivers Council
PO Box 10798   541-345-0119
Eugene, OR 97440   Fax: 541-345-0710
    E-mail: info@pacrivers.org
           http://www.pacrivers.org
Born from the commitment to protect the free flowing rivers of Oregon, it began as a statewide nonprofit organization in 1987 with a volunteer staff. By 1993, we had developed into one of the most influential river conservation groupsin the country. PRC now has 11 professionals on staff, with a budget of more than $1 million and projects throughout the west. Our mission is to protect and restore rivers, their watersheds and native aquatic species.
Founded: 1987
David Bayles, Executive Director

1334 Pacific States Marine Fisheries Commission
45 82nd Drive SE   503-650-5400
Suite 100   Fax: 503-650-5426
Gladstone, OR 97027-2522   http://www.psmfc.org
Randy Fisher, Executive Director

1335 University of Oregon Environmental Studies Center
5223 University of Oregon   541-346-5000
Eugene, OR 97403-5223   Fax: 541-346-5954
E-mail: ecostudy@uoregon.edu
http://darkwing.uoregon.edu/~ecostudy/
Environmental Studies crosses the boundaries of traditional disciplines, challenging faculty and students to look at the relationship between humans and their environment from a new perspective. They are dedicated to gaining greaterunderstanding of the natural world from an ecological perspective; devising policy and behavior that address contemporary environmental problems; and promoting a rethinking of basic cultural premises, ways of structuring knowledge and the rootmetaphors of society.
Mary Ann Larkin, Office Manager
Daniel Udovic, Program Director

1336 Wildlife Society
2280 Riverview   541-937-2131
Eugene, OR 97403   Fax: 541-937-3401
E-mail: kat.beal@sauce.army.mil
A nonprofit scientific and educational organization that serves professionals such as government agencies, academia, industry, and non-government organizations in all areas related to the conservation of wildlife and natural resourcesmanagement.
Katherine Beal, Secretary

Pennsylvania

1337 ASLA: Pennsylvania/Delaware Chapter
American Society Of Landscape Architects
908 North Second Street   717-441-6041
Harrisburg, PA 17102 E-mail: info@landscapearchitects.org
http://www.landscapearchitects.org
Founded: 1899
Lisa A Kunst Vavro, President
Richard Rauso, Vice President

1338 Air and Waste Management Association
1 Gateway Center   412-232-3444
3rd Floor   800-270-3444
Pittsburgh, PA 15222   Fax: 412-232-3450
E-mail: info@awma.org
http://www.awma.org
The Air & Waste Management Association (A&WMA) is a nonprofit, nonpartisan professional organization that provides training, information, and networking opportunities to thousands of environmental professionals in 65 countries.
Founded: 1907
Adrianne Corolla, Executive Director

1339 Alliance for the Chesapeake Bay Harrisburg Office
3310 Market Street   717- 73- 862
Suite A   Fax: 717- 73- 865
Camp Hill, PA 17011   E-mail: acbpa@acb-online.org
http://www.acbpa.acb-online.org
Founded: 1971
Charles Conklin, Chairman
Anne Crawford White, Vice Chairman

1340 Audubon Society of Western Pennsylvania Beechwood Farms Nature Reserve
614 Dorseyville Road   412-963-6100
Pittsburgh, PA 15238-1618   Fax: 412-963-6761
E-mail: aswp@aswp.org
http://www.aswp.org/
Carolyn Sanford, President

1341 Brandywine Conservancy
PO Box 141   610-388-2700
Chadds Ford, PA 19317   Fax: 610-388-1197
E-mail: emc@brandywine.org
http://www.brandywinemuseum.org

The Conservancy is a nonprofit land and water conservation organization protecting natural resources in southeastern PA and northern DE. It provides conservation solutions to landowners, farmers and municipalities by taking acomprehensive approach to cutting-edge environmental planning and management. Through conservation easements, historic preservation, and water protection efforts, the Conservancy has been instrumental in permanently protecting more than 33,0 acres ofland.
Founded: 1972
George Weymuth, Chairman
James Duss, Executive Director

1342 GemNet Global Education Motivators
Global Education Motivators
9601 Germantown Avenue   215-248-1150
Philadelphia, PA 19128   877-451-7925
Fax: 215-248-7056
E-mail: gem@chc.edu
http://www.gem-ngo.org
Founded: 1981
Wayne Jacoby, President
Sabrina Cusimano, Director

1343 Hawk Mountain Sanctuary Association
RR 2   215-756-6961
PO Box 191   Fax: 215-756-4468
Kempton, PA 19529 E-mail: webmaster@hawkmountain.org
http://www.dep.state.pa.us/dep/deputate/enved/hawk
Cynthia Lenhart, Executive Director

1344 Nature Conservancy: Pennsylvania Chapter
Nature Conservancy
15 East Ridge Pike   610-834-1323
Suite 500   800-628-6860
Conshohocken, PA 19428   Fax: 610-834-6533
E-mail: pa_chapter@tnc.org
http://http://nature.org/wherewework/northamerica/states/pennsylvan
Founded: 1951
Steven J McCormick, President
Philip J James, Senior Vice President

1345 Northeast Conservation Law Enforcement Chiefs' Association
83 Park Street   401-277-2284
Providence, PA 02908
Ronald Alie, President

1346 Penn State Institutes of the Environment
Pennsylvania State University
Land and Water Research Building   814-863-0291
University Park, PA 16802   Fax: 814-865-3378
E-mail: emb7@psn.edu
http://www.psie.psu.edu
Eva Brownawell, Info Specialist

1347 Pennsylvania Association of Accredited Environmental Laboratories
128 Roosevelt Street   570-888-4768
Sayre, PA 18840-2934   Fax: 570-882-8538
E-mail: judygraves@paael.org
http://www.paael.org
Founded: 1987
David Barrett, President
Anita Martin, Secretary

1348 Pennsylvania Association of Conservation Districts
25 N Front Street   717-238-7223
Harrisburg, PA 17101   Fax: 717-238-7201
E-mail: pacd@pacd.org
http://www.pacd.org
Founded: 1950
Victor Cappucci, President
Larry Kehl, Vice President

1349 Pennsylvania BASS Chapter Federation
769 N Cottage Road
Mercer, PA 16137   http://www.pabass.com
Mike Dunkerley, President

1350 Pennsylvania Environmental Council
117 S 17th Street
Suite 2300
Philadephia, PA 19103
215-563-0250
800-322-9214
Fax: 215-563-0528
E-mail: amcelwaine@pecpa.org
http://www.pecpa.org
The Pennsylvania Environmental Council is an independent, nonprofit advocacy organization founded in 1969. The Council promotes public participation in environmental policy and focuses on land use, watershed protection and innovation.
Founded: 1970
Andrew McElwaine, President
Brian Hill, Executive Vice President

1351 Pennsylvania Forestry Association
56 E Main Street
Mechanicsburg, PA 17055
717-766-5371
800-835-8065
Fax: 610-338-2909
E-mail: lms28@psu.edu
http://pfa.cas.psu.edu

Earl Higgins, President
Harry V Wiant Jr, Vice President

1352 Pennsylvania Resources Council
3606 Providence Road
Newtown Square, PA 19073
610-353-1555
800-468-6772
Fax: 610-353-6257
E-mail: mlweller@aol.com
http://www.prc.org
The purpose of the Pennsylvania Resources Council is to inform the public through education and other appropriate means of the need for sound conservation practices to promote preservation of natural resources and protection of theenvironment.
Founded: 1939
Bryan Clark, President
Larry Myers, Executive Director

1353 Pocono Environmental Education Center
Rr 2
PO Box 1010
Dingsman Ferry, PA 18328
570-828-2319
Fax: 570-828-9695
E-mail: peec@ptd.net
http://www.peec.org

Founded: 1972
John Padalino, President

1354 Rodale Institute
222 Main Street
Emmaus, PA 18098
610-683-1400
Fax: 215-967-8959
E-mail: info@rodaleinstitute.org
The Rodale Institute offers creative opportunities and solutions that contribute to regenerating environmental and human health worldwide. Our mission statement is clear: The Rodale Institute works worldwide to achieve a regenerativefood system that improves environmental and human health.
John Haberern, President

1355 Sierra Club: Pennsylvania Chapter
PO Box 663
Harrisburg, PA 17108
717-232-0101
Fax: 717-238-6330
E-mail: sierraclub.pa@paonline.com
http://www.sierraclub.org/chapter/pa/
Includes 11 local Sierra Club groups. Emphasis is on state environmental policy advocacy, outings, education and local environmental protection efforts.
Jeff Schmidt, Sr. Chapter Director

1356 Western Pennsylvania Conservancy
209 4th Avenue
Pittsburgh, PA 15222
412-288-2777
866-JOI-NWPC
Fax: 412-281-1792
E-mail: wpc@paconserve.org
http://www.paconserve.org
WPC, working together to save the places we care about, protects natural lands, promotes healthy and attractive communities, and preserves Fallingwater, Frank Lloyd Wright's masterwork in Mill Run, which was entrusted to theConservancy in 1963. Since its inception in 1932, the Conservancy has protected more than 280,000 acres of natural lands in Pennnsylvania. We continue to work to secure lands of ecological significance that frequently offer recreational and scenicvalues.
Larry Schweiger, President

Rhode Island

1357 American Lung Association of Rhode Island
298 West Exchange Street
Providence, RI 02903-3700
401-421-6487
Fax: 401-331-5266
E-mail: ALARI@lungri.org

1358 Audubon Society of Rhode Island
12 Sanderson Road
Smithfield, RI 02917-2600
401-949-5454
Fax: 401-949-5788
E-mail: audubon_ri@ids.net
http://www.asri.org
The purposes of the Audubon Society of RI are to foster conservation of wild birds and other animal and plant life; to conserve wildlife habitat and unique natural areas through acquisition or other means; to carry out a broad programof public conservation and education; to focus public attention on natural resource problems; provide leadership in action on natural resource problems;and to do all other things necessary to foster better management of the environment for thebenefit of all life.
Lee C Schisler Jr, Executive Director

1359 Nature Conservancy: Northeast Division Office
159 Waterman Street
Providence, RI 02906
401-751-2521
Fax: 401-751-7596
John Cook, VP

1360 Nature Conservancy: Rhode Island Chapter
159 Waterman Street
Providence, RI 02906
401-331-7110
Fax: 401-273-4902
http://www.nature.org
An international nonprofit organization dedicated to preserving the plants, animals and natural communities that represent the diversity of life on Earth by protecting the lands and waters they need to survive.
Terry Sullivan, State Director

1361 Rhode Island BASS Chapter Federation
64 Oakton Street
Woonsocket, RI 02895
Lucien Bibeault, President

1362 Rhode Island State Association of Conservation
35 Abbott Run Valley Road
Cumberland, RI 02864
Robert Swanson, Secretary

South Carolina

1363 ASLA: South Carolina Chapter
South Carolina Department of Natural Resources
1000 Assembly Street
PO Box 167
Columbia, SC 29202-0167
803-734-9131
E-mail: rem@lowcountry.com
http://www.scasla.org
Jessica McClung, President
Andrea Almond, President-Elect

1364 American Lung Association of South Carolina
1817 Gadsden Street
Columbia, SC 29201-2392
803-779-5864
800-849-5864
Fax: 803-254-2711
E-mail: alasc@lungsc.org
http://www.lungsc.org
Gabrielle Steele, Regional Director

1365 American Lung Association of South Carolina- Coastal Region
1941 Savage Road
Suite 200-A
Charleston, SC 29407-4789
843-556-8451
Fax: 843-556-3332
E-mail: scatlin@lungsc.org
Sally Catlin, Regional Director

1366 American Lung Association of South Carolina— Up-state Region
11 Brendan Way 864-233-0517
B-2 Fax: 864-233-2124
Greenville, SC 29615 E-mail: altompkins@lungsc.org
Al Tompkins, Regional Director

1367 Friends of the Reedy River
PO Box 9351 864-255-5009
Greensville, SC 29604 Fax: 864-288-9262
E-mail: reedyrvr@aol.com
http://www.reedyrvr.org
Dave Hargett, Executive Director

1368 Nature Conservancy: South Carolina Chapter
PO Box 5475 803-254-9049
Columbia, SC 29250 Fax: 803-252-7134
http://www.nature.org/states/southcarolina/
Nonprofit conservation organization dedicated to preserving the plants, animals and natural communities that represent the diversity of life on Earth by protecting the lands and waters they need to survive. Buys significant tracts of land in its project areas and later re-sells the tracts to a public agency partner such as US Fish and Wildlife Service, US Forest Service and the SC Department of Natural Resources. Also supports and encourages conservation easements.
Mark Robertson, Executive Director
Mary Choate, Director Development

1369 South Atlantic Fishery Management Council
S Park Building 843-571-4366
Summerville, SC 29483 Fax: 843-769-4520
E-mail: kim.iverson@noaa.gov
http://www.safmc.noaa.gov
Kim Iverson, Public Information Officer

1370 South Carolina BASS Chapter Federation
1469 Schurlknight Road 803-567-4680
St. Stephen, SC 29479
Tony Bennett, President

1371 South Carolina Native Plant Society
PO Box 597 864-898-1221
Pickens, SC 29671 Fax: 864-898-0191
http://www.clemson.edu/scnativeplants
Rick Huffman, President

1372 Southern Appalachian Botanical Society
Newberry College
2100 College Street 803-321-5257
Newberry, SC 29108 Fax: 803-321-5636
E-mail: chron@newberry.edu
http://www.newberrynet.com/sabs/
This is a professional organization for those interested in botanical research, especially in the areas of ecology, floristics and systematics. To this end, we publish a journal, CASTANEA, and a newsletter, CHINQUAPIN.
Zack Murrell, President
Charles Horn, Treasurer

South Dakota

1373 ASLA: Great Plains Chapter
c/o Wyss Associates
728 Sixth Street 605-348-2268
Rapid City, SD 57701 Fax: 605-348-6506
E-mail: PatW@wyssassociates.com
http://www.asla.org/chapters/greatplains
Pat Wyss, President
Eirik Heikes, President-Elect

1374 American Lung Association of South Dakota
American Lung Association
1212 West Elkhorn 605-336-7222
Suite 1 800-873-5864
Sioux Falls, SD 57104-0233 Fax: 605-336-7227
E-mail: lung@americanlungsd.org
http://www.lungusa.org
Founded: 1904
Kathleen Sweere, Executive Director

1375 Great Plains Native Plant Society
PO Box 641 605-745-3397
Hot Springs, SD 57747 Fax: 605-745-3397
E-mail: cascade@gwtc.net
http://www.gnps.org
A membership organization founded to promote the study and appreciation of Great Plains Native Plants. A major project is the Great Plains Botanic Garden, to be located near Rapid City, South Dakota.
Cynthia Reed, President

1376 Nature Conservancy: South Dakota Chapter
The Nature Conservancy
2601 South Minnesota Avenue 612-331-0700
Suite 105-319 Fax: 612-331-0770
Sioux Falls, SD 57105-4730 E-mail: comment@tnc.org
http://http://nature.org/wherewework/northamerica/states/southdakot
Founded: 1951
Bob Paulson, Black Hills Program Director
Mary Miller, Manager

1377 South Dakota Association of Conservation Districts
14321 465th Avenue 605-895-4099
Marvin, SD 57251 800-729-4099
Fax: 605-895-9424
E-mail: conserpe@wcenet.com
http://www.sd.nacdnet.org
Angela Elhers, Executive Director
Duane Murphy, Project Coordinator

1378 South Dakota Ornithologists Union
NSU Box 740 605-677-6175
Aberdeen, SD 57401 Fax: 605-677-6557
E-mail: tallmand@northern.edu
Founded: 1949
Dan Tallman, President
Rosemary Draeger, Vice President

1379 South Dakota Wildlife Federation
PO Box 7075 605-224-7524
Pierre, SD 57501 Fax: 605-224-7524
E-mail: sdwf@pie.midco.net
http://www.sdwf.org
The Federation as a 501 nonprofit organization, seeks to attain its conservation goals through educational means. The federation is the leading advocate enforcing the legislative sessions and at the Game, Fish & Parks commission meetings. For preserving quality outdoor opportunities for South Dakota's outdoor enthusiasts.
Founded: 1945
Chris Hesla, Executive Director
Jeff Albrecht, President

1380 Wildlife Society
Box 218 605-854-9105
DeSmet, SD 57231-0218 E-mail: paul.coughlin@state.sd.us
A nonprofit scientific and educational organization that serves professionals such as government agencies, academia, industry, and non-government organizations in all areas related to the conservation of wildlife and natural resources management.
Founded: 1937
Will Morlock, South Dakota State President

Tennessee

1381 American Lung Association of Tennessee
One Vantage Way 615-329-1151
Suite B130 800-432-5864
Nashville, TN 37228 Fax: 615-329-1723
E-mail: alastaff@alatn.org
http://www.lungtn.org
Founded: 1904
Doug Halleen, Executive Director

1382 American Lung Association of Tennessee- Southeast Office

American Lung Association
3335 Ringgold Road
Suite 105
Chattanooga, TN 37412-4305

423-629-1098
800-LUN- USA
Fax: 423-629-0054
E-mail: scudabacalatn@bellsouth.net
http://www.lungtn.org

Founded: 1904
Doug Halleen, Executive Director

1383 Kentucky-Tennessee Society of American Foresters

PO Box 4188
Oak Ridge, TN 37831

865-483-3572
Fax: 865-483-3572
E-mail: jmyoung@utk.edu
http://www.ktsaf.org

Founded: 1900
JR Anderson, Chairman
James Anderson, Secretary

1384 Kids for a Clean Environment

PO Box 158254
Nashville, TN 37215

615-331-7381
800-952-3223
Fax: 615-333-9879
E-mail: kidsface@mindspring.com
http://www.kidsface.org

Established in 1989 to help children who wanted to learn more about the world in which they live, provide a way for children to be involved in the protection of nature and connect children with other children who share their concernsabout global environmental issues.
Allen Spalt, President
Katherine M Shea, Vice President

1385 Nature Conservancy: Tennessee Chapter

Nature Conservancy
2021 21st Avenue South
Suite C-400
Nashville, TN 37212

615-383-9909
800-628-6860
Fax: 615-383-9717
E-mail: tennessee@tnc.org
http://www.nature.org

Founded: 1951
Steven J McCormick, President
Henry M Paulson Jr., Chairman

1386 Scenic Tennessee

71 Lakewood Drive
Winchester, TN 37398

931-962-1813
E-mail: sanders@cafes.net
http://www.scenictn.zfx.com

Scenic America is the only national nonprofit organization dedicated to protecting natural beauty and destinctive community character. We provide technical assistance across the nation and through affiliates on scenic byways, billboardand sign control, context sensitive highway design, wireless telecommunications tower location, transportation enhancements, and other scenic conservation issues.
Suzanna Askew, VP

1387 Tennessee Association of Conservation Districts

National Conservation Association
144,South East
Parkway Suite 210
Franklin, TN 37064

615-595-9979
Fax: 615-595-9982
E-mail: gillisr@k12tn.net
http://www.nacdnet.org

Founded: 1982
Roy L Gills, President
Barry Lake, Vice President

1388 Tennessee Citizens for Wilderness Planning

130 Tabor Road
Oak Ridge, TN 37830

865-481-0286
E-mail: chewonthis@pcwp.org
http://www.pcwp.org

Founded: 1966
Cindy Kendrick, President
Sandra Goss, Executive Director

1389 Tennessee Environmental Council

1 Vantage Way
Suite D 105
Nashville, TN 37228

615-248-6500
Fax: 615-248-6545

E-mail: tec@tectn.org
http://www.tectn.org

Founded: 1970
Will Callaway, Executive Director
Claudia Schenck, Office Director

1390 Tennessee Woodland Owners Association

PO Box 1400
Crossville, TN 38557

615-484-5535
Fax: 915-484-1924
E-mail: reharrison@multipro.com

Robert Harrison, Secretary/Treasurer

1391 Toxicology Information Response Center

1060 Commerce Park
MS 6480
Oak Ridge, TN 37830-6480

865-576-1746
Fax: 865-574-9888
E-mail: slusherkg@ornl.gov
http://www.ornl.gov/TechResources/tirc/hmepg.html

Makes available extensive information and reference services on individual chemicals, chemical classes and a wide variety of toxicology-related topics.
Kim Slusher, Administrator

1392 Trout Unlimited

1300 North 17th St
Suite 500
Arlington, VA 22209-2404

703-522-0200
Fax: 703-284-9400
E-mail: trout@tu.org
http://www.tu.org

Founded: 1960
Charles Gauvin, President/CEO
Kenneth Mendez, Vice President/COO

1393 Wildlife Society

PO Box 1071
Knoxville, TN 37901

423-974-7981
Fax: 423-974-4714
E-mail: imuller@utk.edu

A nonprofit scientific and educational organization that serves professionals such as government agencies, academia, industry, and non-government organizations in all areas related to the conservation of wildlife and natural resourcesmanagement.
Founded: 1937
Craig Harper, President
David Lingo, Secretary

Texas

1394 American Lung Association of Texas

PO Box 26460
Austin, TX 78755-0460

512-467-6753
800-252-5864
Fax: 512-467-7621
E-mail: info@texaslung.org
http://www.texaslung.org

Founded: 1908
Lewis Brown, Chairman
Bennie McWilliams, Vice Chairman

1395 American Lung Association of Texas, Dallas/ Fort Worth Region

American Lung Association
8150 Brookriver Drive
S-102, LB-151
Dallas, TX 75247

214-631-5864
800-LUN- USA
Fax: 214-630-8092
E-mail: dallas@texaslung.org
http://www.texaslung.org

Founded: 1904
Tessie Holloway, Executive Director

1396 American Lung Association of Texas, Rio Grande Valley Region

3827 North Tenth Street
Suite 203
McAllen, TX 78501

956-631-0514
800-548-8252
Fax: 956-668-9615
E-mail: kyliehood@sbcglobal.net
http://www.lungusa.org

Founded: 1904
Fjack Sutter, Chairman

1397 American Lung Association of Texas, Western Region
American Lung Association of Texas
4141 Pinnacle Street   915-532-6776
Suite 212   800-252-5864
El Paso, TX 79902   Fax: 915-532-7231
E-mail: info@texaslung.org
http://www.texaslung.org

Founded: 1904
Elizabeth Terrazas, Executive Director
Tony Baird, Chairman

1398 American Lung Association of Texas- Alamo and Southern Region
8207 Callaghan Road   210-308-8978
Suite 140   Fax: 210-308-8992
San Antonio, TX 78230   E-mail: alasoutx@texaslung.org
http://www.texaslung.org

Senator Van de Putte, Chair Person

1399 American Lung Association of Texas- Houston and Southeast Region
American Lung Association
2030 North Loop West   713-629-5864
Suite 250   800-586-4872
Houston, TX 77018   Fax: 713-629-5828
http://www.lungusa.org

Founded: 1904
Charles A Heinrich, Chairman
Robert A Green, Vice Chairman

1400 Association of Texas Soil and Water Conservation Districts
PO Box 658   254-773-8741
Temple, TX 76503   800-792-3485
Fax: 254-773-3311

Robert G Buckley, Executive Advisor

1401 Big Bend Natural History Association: Big Bend National Park
PO Box 196   432-477-2236
Big Bend National Park, TX 79834   Fax: 432-477-2234
E-mail: bbnha@nps.gov
http://www.nps.bigbendbookstore.org
Serves Big Bend National Park which is situated on the boundary with Mexico along the Rio Grande, a place where countries and cultures meet. A place that merges natural environments, from desert to mountains. A place where south meetsnorth and east meets west, creating a great diversity of plants and animals. The park covers over 801,000 acres of west Texas in the place where the Rio Grande makes a sharp turn-the Big Bend.
Mike Boren, Executive Director

1402 National Wildlife Federation
44 East Avenue   512-476-9805
Suite 200   Fax: 512-476-9810
Austin, TX 78701   http://www.nwf.org

Founded: 1933
Larry Schweiger, President
Dulce Gomez-Zormelo, Chief Financial Officer

1403 Scenic Texas
3015 Richmond   713-533-9149
Suite 220   Fax: 713-629-0485
Houston, TX 77098   E-mail: scenic@scenictexas.org
http://www.scenictexas.org
Scenic America is the only national nonprofit organization dedicated to protecting natural beauty and distinctive community character. We provide technical assistance across the nation and through affiliates on scenic byways, billboardand sign control, context sensitive highway design, wireless telecommunications tower location, transportation enhancements, and other scenic conservation issues.
Cece Fowler, Executive Director

1404 Sierra Club
54 Chicon Street   512-477-1729
Austin, TX 78702-5451   Fax: 512-477-8526
E-mail: scls@igc.org
http://www.sierraclub.org/chapters/tx/

To advance the preservation and protection of the natural environment by empowering the citizenry, especially democratically-based grassroots organizations, with charitable resources to further the cause of environmental protection.The vehicle through which The Sierra Club Foundation generally fulfills its charitable mission.
Jennifer Walker, Adminstrative Assistant

1405 Texas Association of Soil and Water Conservation Districts
279 Hostetter   254-778-8741
New Waverly, TX 77358   Fax: 254-773-3311
Jose Dodier, President

1406 Texas Committee on Natural Resources
3532 Bee Caves Road   512-441-1122
Suite 110   Fax: 512-327-2115
Austin, TX 78746   E-mail: tconr@texas.net
http://www.eden.com/~bezanson
Founded: 1970
Janice Bezanson, Executive Director
Ned Fritz, TCONR Founder

1407 Texas Environmental Health Association
PO Box 8453   940-322-3232
Wichita Falls, TX 76307-8453   Fax: 940-322-3232
Thomas L Edmonson Jr, Exec. Secretary-Treasurer

1408 Texas Solar Energy Society
PO Box 1447   512-326-3391
Austin, TX 78767-1447   800-465-5049
Fax: 512-326-1785
E-mail: info@txses.org
http://www.txses.org

Lorin Hull, Chairman
Jaya Jackson, Vice Chairman

1409 Texas Water Conservation Association
221 E 9th Street   512-472-7216
Suite 206   Fax: 512-472-0537
Austin, TX 78701   E-mail: robbins@twca.org
http://www.twca.org

Founded: 1944
Leroy Goodson, General Manager
Dean Robbins, Assistant General Manager

Utah

1410 ASLA: Utah Chapter
PO Box 511125
Salt Lake City, UT 84151   E-mail: kelly@crsarchitects.com
http://host.asla.org/chapters/utahasla
J Kelly Gillman, President
Mark Vlasic, President-Elect

1411 American Lung Association of Utah
American Lung Association
1930 South 1100 East   801-484-4456
Salt Lake City, UT 84106-2317   800-586-4872
Fax: 801-484-5461
E-mail: info@utahlung.org
http://www.utahlung.org

Founded: 1904
Craig Cutright, Executive Director

1412 Arizona Strip District Advisory Council
345 E Riverside Drive   435-688-3222
St. George, UT 84790
Rodger Taylor

1413 Grand Canyon Trust
Utah Office MOAB
PO Box 1801   435-259-5284
Maob, UT 84532-9610   Fax: 435-259-5348
http://www.grandcanyontrust.org

1414 Jack H Berryman Institute for Wildlife Management: Department of Fisheries & Wildlife
Utah University 435-797-2436
Logan, UT 84322-5210 Fax: 435-797-1871
E-mail: director@berrymaninstitute.org
http://www.berrymaninstitute.org
There is no other organization in the world like the Berryman Institue. Based at Utah State University, the Berryman Institute is a functional component of the Department of Fisheries and Wildliufe and the College of Natural Resources.Its faculty members hold academic appontments in various departments throughout Utah State University and at other universitoes. This multidisciplinary approach to speed the discovery and development of innovative methods to solve wildlife conflicts.

1415 Nature Conservancy: Utah Chapter
Nature Conservancy
559 E South Temple 801-531-0999
Salt Lake City, UT 84102 800-628-6860
Fax: 801-531-1003
E-mail: utah@tnc.org
http://www.tnc.org/utah
Founded: 1951
Steven J McCormick, Chief Executive Director
Dave Livermore, Executive Director

1416 Sierra Club
2120 South 1300 East 801-467-9294
Suite 204 Fax: 801-467-9296
Salt Lake City, UT 84106 http://www.sierraclub.org/ut
To advance the preservation and protection of the natural environment by empowering the citizenry, especially democratically-based grassroots organizations, with charitable resources to further the cause of environmental protection.The vehicle through which The Sierra Club Foundation generally fulfills its charitable mission.
Founded: 1892
Carl Pope, Executive Director

1417 Southern Utah Wilderness Alliance
122 C Street NW 202-546-2215
Suite 240 Fax: 202-544-5197
Washington, DC 20001 E-mail: heidi@suwa.org
http://www.suwa.org
Founded: 2003
Hansjorg Wyss, Chairman
Ted Wilson, Vice Chairman

1418 Utah Association of Conservation Districts
1860 N 100 E 435-753-6029
Logan, UT 84341-1784 Fax: 435-755-2117
Gordon L Younker, Executive VP

1419 Utah Association of Soil Conservation Districts
1860 N 100 E 435-753-6029
Logan, UT 84341 Fax: 435-753-4037
William Rigby, Executive Board Member

1420 Utah Wildlife Federation
PO Box 526367 801-487-1946
Salt Lake City, UT 84152-6367 Fax: 801-846-0611
E-mail: rdiamond@sisna.com
http://www.wildlife.utah.gov
Founded: 1999
Max Morgan, Chairman

Vermont

1421 ASLA: Vermont Chapter
ASLA Group
PO Box 1263 802-862-0098
Montpelier, VT 05601 E-mail: pmclean@dufresne-henry.com
http://host.asla.org/chapters/vermont/index.cfm
Stephen Plunkard, President
Patrick T McLean, President-Elect

1422 American Lung Association of Vermont
30 Farrell Street 802-863-6817
South Burlington, VT 05403-6196 Fax: 802-863-6818
E-mail: info@vtlung.org
http://www.lungusa2.org/vermont/
Danielle Hunt, Contact
John J Cronin, Chief Executive Officer

1423 Bluebirds Across Vermont Project
The Birdhouse Network
255 Sherman Hollow Road 802-434-3068
Green Mountain Abdubon Society Fax: 802-434-4686
Huntington, VT 05462 E-mail: bluebirdhousing@ellijay.com
http://www.cornell.edu
Jim Shallow, Executive Director

1424 National Wildlife Federation
11100 Wildlife Center Drive 802-229-0650
Reston, VA 20190-5362 800-822-9919
Fax: 802-229-4532
E-mail: info@nwf.org
http://www.nwf.org
Founded: 1936
Jerome C Ringo, Chairman
Stephen E Petron, Vice Chairman

1425 Noise Pollution Clearinghouse
PO Box 1137 888-200-8332
Montpelier, VT 05601-1137 888-200-8332
E-mail: freepnc@nonoise.org
http://www.nonoise.org
The Noise Pollution Clearinghouse is a nonprofit organization with extensive online noise related resources. The mission is to create more civil cities and more natural rural and wilderness areas by reducing noise pollution and itssources.

1426 Northeast Recycling Council
Northeast Recycling Council
139 Main Street 802-254-3636
Suite 401 Fax: 802-254-5870
Brattleboro, VT 05301 E-mail: info@nerc.org
http://www.nerc.org
John Trevor, President
Jeff Bednar, Vice President

1427 Trout Unlimited
PO Box 163 802-334-1674
Island Pond, VT 05846-0163 Fax: 802-334-2991
Francis Smith, Chairman

1428 Vermont Association of Conservation Districts
487 Rowell Hill Road 802-229-9250
Berlin, VT 05602 Fax: 802-229-6920
E-mail: mdomi15978@aol.com
http://www.vacd.org
Mike Domingue, President
Christine Kaiser, Vice President

1429 Vermont BASS Chapter Federation
Bassin' USA
19 Pinewood Road 802-223-7793
Montpelier, VT 05602 E-mail: nsk@together.net
http://www.bassinusa.com
Founded: 1968
Brendan Cucinello, President
Bob Crino, Secretary

1430 Vermont Haulers and Recyclers Association
PO Box 976 802-864-3615
Williston, VT 05495 Fax: 802-660-8553
http://www.zella.com

1431 Vermont Land Trust
8 Bailey Avenue 802-223-5234
Montpelier, VT 05602 800-639-1709
Fax: 802-223-4223
E-mail: info@vlt.org
http://www.vlt.org

Founded: 1977
Darby Bradley, President
Dawn Lee Minter, Executive Assistant

1432 Vermont Public Interest Research Group
141 Main Street
Montpelier, VT 05602
802-223-5221
Fax: 802-223-6855
E-mail: vpirg@vpirg.org
http://www.vpirg.org
Watchdog and advocacy group that promotes and protects the health of Vermont's environment, people, and locally based economy. By informing and mobilizig individuals across the state, VPIRG brings the voices of citizens to publicpolicy debates that shape the future of Vermont.
Founded: 1972
Paul Borns, Executive Director
Duane Peterson, President

1433 Vermont State-Wide Environmental Education Programs
9 Bailey Avenue
Montpelier, VT 05602
802-985-8686
Susan Clark, Chair

1434 Wildlife Society
RD1 Box 2161
Pittsford, VT 05763
E-mail: scott.darling@anr.state.vt.us
A nonprofit scientific and educational organization that serves professionals such as government agencies, academia, industry, and non-government organizations in all areas related to the conservation of wildlife and natural resourcesmanagement.
Scott Darling, President

Virginia

1435 ASLA: Virginia Chapter
American Society of Landscape Architects
11712C Jefferson Avenue
Suite 249
Newport News, VA 23606
757-412-2664
Fax: 757-412-4637
E-mail: info@vaasla.org
http://www.vaasla.org
Founded: 1899
Luigi Mignardi, President
Lynn Crump, President-Elect

1436 American Lung Association of Virginia
American Lung Association
9221 Forest Hill Avenue
Richmond, VA 23235
804-267-1900
800-586-4872
Fax: 804-267-5634
E-mail: resourcecenter@lungva.org
http://www.lungusa.org/
Founded: 1909
Catherine Hamm, Executive Director
Richard Pierson, Dir Finance/Administration

1437 Arlington Outdoor Education Association
Phoebe Hall Knipling Outdoor Laboratory
PO Box 5646
Arlington, VA 22205
703-228-7650
Fax: 540-349-3336
E-mail: roffice@arlington.k12.va.us
Founded: 1967
Mike Nardolli, President
Teresa Rusnak, Vice President

1438 Center for Health, Environment and Justice
PO Box 6806
Falls Chruch, VA 22040-6806
703-237-2249
Fax: 703-237-8389
E-mail: chej@chej.org
http://www.chej.org
The Center for Health, Environment nad Justice trains and assists local people to fight for justice, become empowered to protect their communities from environmental threats and build strong locally controlled organizations. CHEJconnects these strong local groups with each other to build a movement from the bottom up.
Founded: 1981
Lois Marie Gibbs, Executive Director
Maryll Kleibrink, Director of Development

1439 Center for the Evaluation of Risks to Human Reproduction
National Institute of Environmental Health Science
1800 Diagonal Road
Suite 500
Alexandria, VA 22314
703-838-9440
Fax: 703-684-2223
E-mail: shelby@niehs.nih.gov
http://http://cerhr.niehs.nih.gov/
The Center's mission includes the following: to provide timely and unbiased, scientifically sound assessments of reproductive health hazards associated with human exposure to naturally occurring and man-made chemicals; to make theseassessments readily available to the public, to state and federal agencies and to the scientific community; and to build an electronic resource for providing, or directing one to, information of public interest concerning human reproductive health.
Founded: 1998
Michael D Shelby, Director

1440 Chesapeake Bay Foundation
1108 E Main Street
Suite 1600
Richmond, VA 23219
804-780-1392
Fax: 804-648-4011
E-mail: chesapeake@cbf.org
http://www.savethebay.cbf.org
Founded: 1964
Ann Jennings, Executive Director
Jeff Corvin, Deputy Director

1441 Citizens Clearinghouse for Hazardous Waste
PO Box 6806
Falls Church, VA 22040-6806
703-237-2249
Fax: 703-237-8389
E-mail: info@chej.org
http://www.chej.org
Nonprofit organization serves citizens' groups, individuals and small municipalities working to solve hazardous and solid waste problems. Supplies information needed to understand, prevent, reduce or eliminate exposure to toxicchemicals through customized assistance, both in-house and on referral, a research library and service, publications and newsletters.
Founded: 1981
Louis Marie Gibbs, Executive Director
Stephen Lester, Science Director

1442 Environmental Safety
1700 N Moore Street
Arlington, VA 22209-1921
703-527-8300
Fax: 703-527-8383
William Drayton, Executive Director

1443 Fairfax Audubon Society
Potomac Flier
4022 Hummer Road
Annandale, VA 22003
703-256-6895
800-659-2622
Fax: 703-256-2060
E-mail: fairfax@erols.com
http://www.fairfaxaudubon.org
The Fairfax Audubon Society (FAS), is committed to the Audubon mission, which is to conserve and restore natural ecosystems, focusing on birds and other wildlife, and their habitats. FAS carries out this mission through a three-prongedstrategy of education, advocacy, and conservation programs in order to foster a culture of conservation and environmental ethic throughout the communities of northern Virginia.
Founded: 1980
Stephanie Farleyw, Executive Director
Pat Patterson, President

1444 Institute of Scrap Recycling Industries: Seaboard Chapter
1325 G Street NW
Suite 1000
Washington, DC 20005-3104
202-737-1770
Fax: 202-626-0900
E-mail: robinwiener@isri.org
http://www.isri.org/
Founded: 1987
Robin Wiener, President
John Sacco, Secretary

1445 National Wildlife Federation
11100 Wildlife Center Drive
Reston, VA 20190-5362
703-438-6000
800-822-9919
Fax: 907-258-4811
E-mail: info@nwf.org
http://www.nwf.org/

The National Wildlife Federation is the largest member-supported conservation group, uniting, individuals, organizations, businesses and government to protect wildlife, wild places and the environment.
Founded: 1938
Larry J Schweiger, President/CEO
Eileen Morgan Johnson, Secretary

1446 Nature Conservancy
490 Westfield Road
Charlottesville, VA 22901
434-295-6106
800-628-6860
E-mail: dwhite@tnc.orgg
http://www.tnc.org

To preserve the plants, animals and natural communities that represent the diversity of life on Earth by protecting the lands and waters they need to survive.

1447 Potomac Appalachian Trail Club
118 Park Street SE
Vienna, VA 22180-4609
703-242-0693
Fax: 703-242-0968
E-mail: info@patc.net
http://www.patc.net

The PATC, through volunteer efforts, education and advocacy, acquires, maintains and protects the trail and lands of the Appalachian Trail, other trails and related facilities in the Mid-Atlantic Region for the enjoyment of present andfuture hikers. PATC publishes hiking guides, maps and history books of the Appalachian Trail and other trails in our area of responsibility. A monthly newsletter is sent to members and upon request.
Founded: 1927
Thomas Johnson, President
Wilson Riley, Director Administration

1448 Scenic Virginia
1904 Byrd Avenue
Suite 108
Richmond, VA 23230
804-282-5522
Fax: 804-282-5506
E-mail: email@scenicvirginia.org
http://www.scenicva.org

Scenic America is the only national nonprofit organization dedicated to protecting natural beauty and distinctive community character. We provide technical assistance across the nation and through affiliates on scenic byways, billboardand sign control, context sensitive highway design, wireless telecommunications tower location, transportation enhancements, and other scenic conservation issues.
Hylah Boyd, Chairman
David Kenerson, President

1449 Sierra Club
6 N 6th Street
Suite 401
Richmond, VA 23219-241E9mail: information@sierraclub.org
804-225-9113
Fax: 804-225-9114
http://www.sierraclub.org

To advance the preservation and protection of the natural environment by empowering the citizenry, especially democratically-based grassroots organizations, with charitable resources to further the cause of environmental protection.The vehicle through which The Sierra Club Foundation generally fulfills its charitable mission.
Sen Gaylord Nelson, President
Brock Evans, Executive Director

1450 Society for Occupational and Environmental Health
6728 Old McLean Village Drive
McLean, VA 22101
703-556-9222
Fax: 703-556-8729
E-mail: soeh@degnon.org
http://www.soeh.org

Provides a neutral forum where occupational safety and health and environmental issues can be discussed and resolved. Actively seeks to improve the quality of both working and living places.
Founded: 1972
Ronald Denny Dobbin, President
George K Degnon, Executive Director

1451 Student Conservation Association
1800 N Kent Street
Suite 102
Arlington, VA 22209
703-524-2441
Fax: 703-524-2451
E-mail: ask-us@thesca.org
http://www.thesca.org

To build the next generation of conservation leaders and inspire lifelong stewardship of our environment and communities by engaging young people in hands-on service to the land.

Founded: 1957
Dale Penny, President

1452 Teratology Society
1821 Michael Faraday Drive
Suite 300
Reston, VA 20190-5332
703-438-3104
Fax: 703-438-3113
E-mail: tshq@teratology.org
http://http://teratology.org/

A multidisciplinary scientific society founded in 1960, the members of which study the causes and biological processes leading to abnormal development and birth defects at the fundamental and clinical level, and appropriate measuresfor prevention.
Founded: 1973
Shawn Lamb, President
Tonia Masson, Vice President

1453 Virginia Association of Conservation Districts
Virginia Association of Conservation Districts
7308, Hanover Green Drive
Mechanicsville, VA 23111
804-559-0324
Fax: 804-559-0325
http://www.vacd.org

Founded: late
James G Byrne, President

1454 Virginia Association of Soil and Water Conservation Districts
7308 Hanover Green Drive
suite 100
Mechanicsville, VA 23111
804-559-0324
Fax: 804-559-0325
E-mail: info@vaswcd.org
http://www.vaswcd.org

Founded: 1930
James G Byrne, President
Richard O Rash, Vice President

1455 Virginia Conservation Network
1001 E Broad Street
Suite LL 35-C
Richmond, VA 23219
804-644-0283
Fax: 804-644-0286
E-mail: vcngeneral@aol.com
http://www.vcnva.org

Founded: 1990
Martha Wingfield, President
Skip Stiles, Vice President

1456 Virginia Forestry Association
3808 Augusta Avenue
Richmond, VA 23230-3910
804-278-8733
Fax: 804-320-1447
E-mail: vfa@verizon.net
http://www.vaforestry.org

Founded: 1943
John Burke, Chairman
Frank Sherwood, Vice Chairman

1457 Virginia Native Plant Society
400 Blandy Farm Lane
Unit 2
Boyce, VA 22620
540-837-1600
Fax: 540-837-1523
E-mail: rccsca@visuallink.com
http://www.vnps.org

Founded: 1982
Sally Anderson, President
Nicky Staunton, Vice President

1458 Virginia Waste Industries Association
508 SOMERSET AVENUE
Richmond, VA 23226
757-686-5960
Fax: 757-686-0010
E-mail: mdobson@envasns.org
http://www.vwia.com/index.html

Members include waste haulers, recyclers and landfill operators.
Bud Ross, Chairman
Robert Dick, Vice Chairman

1459 Water Environment Federation
601 Wythe Street
Alexandria, VA 22314-1994
703-684-2400
800-666-0206
Fax: 703-684-2492
http://www.wef.org

Nonprofit international membership organization that develops and disseminates technical information on the nature, collection, treatment and disposal of domestic and industrial wastewater.
Founded: 1928
William Bertera, Executive Director
Tim Ricker, Vice President

Washington

1460 ASLA: Washington Chapter
603 Stewart Street   206-443-9484
Seattle, WA 98101   Fax: 425-450-9077
E-mail: office@asla.org
http://www.asla.org

Founded: 1899
Andrew L Estep, Executive Director
Maureen C Colaizzi, President-Elect

1461 American Lung Association of Washington
2625 Third Avenue   206-441-5100
Seattle, WA 98121   800-586-4872
Fax: 206-441-3277
E-mail: alaw@alaw.org
http://www.alaw.org

Founded: 1906
Marina Cofer-Wildsmith, CEO
Paul Payton, Director Communications

1462 American Lung Association of Washington- Western Region
American Lung Association
223 Tacoma Avenue South   253-272-8777
Tacoma, WA 98402   Fax: 253-593-8827
E-mail: lnoren@alaw.org
http://www.ala.org

Founded: 1905
marina Cofer-Wildsmith, Chief Executive Director
leanne Noren, Operations Officer

1463 American Lung Association of Washington— Eastern Region
110 South 9th Avenue   509-248-4384
Yakima, WA 98902   Fax: 509-248-4943
E-mail: alaw@alaw.org
http://www.alaw.org

Founded: 1906
Marina Cofer-Wildsmith, Chief Executive Officer

1464 American Lung Association of Washington: Spokane Branch
American Lung Association
1817 East Springfield   509-325-6516
Suite E   800-732-9339
Spokane, WA 99202   Fax: 509-323-5380
E-mail: alaw@alaw.org
http://www.alaw.org

Founded: 1904
Marina Cofer-Wildsmith, Chief Executive Officer
Leanne Noren, Chief Operations Officer

1465 American Rivers Northwest Regional Office
4005 20th Ave West   206-213-0330
Suite 221   877-4RI-ERS
Seattle, WA 98199   Fax: 206-213-0334
E-mail: arnw@amrivers.org
http://www.americanrivers.org

Founded: 1992
Rebecca R Wooder, President
Eric Eckl, Director of Media Affairs

1466 Environmental Education Association of Washington
Environmental Education Association of Washington
EEAW   360-943-6643
P.O. Box 6277   Fax: 360-497-7132
Olympia, WA 98507   E-mail: eeaw@eeaw.org
http://www.eeaw.org/index.htm

Founded: 1991
Beverly Walker, President
Martin Fortin, Treasurer

1467 Friends of Discovery Park
PO Box 99662   206-285-6862
Seattle, WA 98199-0662   888-291-6104
Fax: 253-872-6668
E-mail: info@discoveryparkfriends.org
http://www.discoveryparkfriends.com

Founded: 1970
Valerie Cholvin, President

1468 Friends of the San Juans
PO Box 1344   360-378-2319
Friday Harbor, WA 98250   877-757-3629
Fax: 360-378-2324
E-mail: friends@sanjuans.org
http://www.sanjuans.org

Founded: 1979
Stephanie Buffum, Executive Director
Tina Whitman, Science Director

1469 Hood Canal Land Trust
3721, Kitsap Way   360-373-3500
Suite 5   866-373-3504
Bremerton, WA 98312   Fax: 360-377-0239
E-mail: info@greatpeninsula.org
http://www.greatpeninsula.org

Founded: 2000
Ann D Haines, Executive Director
Kate Kuhlman, Director of Development

1470 International Bicycle Fund
4887 Columbia Drive South   206-767-0848
Seatle, Wa 98108-1919   Fax: 206-767-0848
E-mail: ibike@ibike.org
http://www.ibike.org

Advocate for sustainable transportation in Africa, Asia, North and South America.
Founded: 1983
David Mozer, President
John Dowlin, Executive Director

1471 Issaquah Alps Trails Club
PO Box 351   425-392-4432
Issaquah, WA 98027   E-mail: IATCDrew@aol.com
http://www.issaquahalps.org

Founded: 1985
Doug Simpson, President
Steve Drew, Vice President

1472 Mountaineers Conservation Division
The Mountaineers
300 3rd Avenue West   206-284-6310
Seattle, WA 98119   800-573-8484
Fax: 206-284-4977
E-mail: clubmail@mountaineers.org
http://www.mountaineers.org

Founded: 1906
Ron Eng, President
Steve Costie, Executive Director

1473 National Wildlife Federation
6 NICKERSON ST.   206-285-8707
SUITE 200   Fax: 206-285-8698
Seattle, WA 98109   E-mail: nwnrc@nwf.org
http://www.nwf.org

Founded: 1936
Paula Delgiudice, Director

1474 Nature Conservancy: Washington Chapter
217 Pine Street   206-343-4344
Suite 1100   800-628-6860
Seattle, WA 98101   Fax: 206-343-5608
E-mail: washington@tnc.org
http://www.tnc-washington.org

Founded: 1951
John Rose, Chairman

1475 North Cascades Conservation Council
PO Box 95980   206-282-1644
Seattle, WA 98145-2980   Fax: 206-684-1379
E-mail: steveb@premier1.net
http://www.northcascades.org

Founded: 1957
Marc Bardsley, President

1476 Northwest Ecosystem Alliance
1208 Bay Street   360-671-9950
Bellingham, WA 98225-4301   Fax: 360-671-8429
E-mail: wild@conservationnw.org
http://www.conservationnw.org

Founded: 1988
Alex Loeb, President
Bill Donnelly, Vice President

1477 Olympic Park Associates
2433 Del Campo Drive   206-364-3933
Everett, WA 98208   Fax: 206-364-6379
E-mail: pollytdyer@juno.com
http://www.drizzle.com/~rdpayne/opa.html/
Founded: 1948
Polly Dyer, President
Bruce Babbit, Secretary

1478 People for Puget Sound
911,Western Avenue   206-382-7007
Suite 580   Fax: 206-382-7006
Seattle, WA 98104   E-mail: people@pugetsound.org
http://www.pugetsound.org
People for Puget Sound is a regional citizen's organization founded in 1991 to educate and involve ordinary - and extraordinary - people in protecting and restoring the land and waters of Puget Sound. People for Puget Sound's programsare based on partnership and collaborations, scientific credibility, creative use of communications and technology, and a hands-on-style. People for Puget Sound publishes a quarterly newsletter, and many scientific publications.
Founded: 1991
Kathy Fletcher, Executive Director
Stacey Jurgensen, Communications Director

1479 Rivers Council of Washington
509 10th Avenue E   206-568-1380
Suite 200   Fax: 206-568-1381
Seattle, WA 98102   E-mail: RIVERSWA@BRIadoon.com
Doug North, President

1480 Sierra Club: Cascade Chapter
8511 15th Avenue NE   206-523-2147
Room 201   Fax: 206-729-2468
Seattle, WA 98115   E-mail: cascade.chaptersierraclub.org
http://www.cascadechapter.org
They focus on protecting our quality of life and natural heritage. Members participate in: outings-hikes and restoration projects; committees-to protect Washington's salmon, lakes and rivers, forests and open spaces, watershed andnative habitat; electing-proenvironment candidates to public office; communicating-with legislators and other public officals; meetings-seminars and other special events; and socializing-with others who share their enviromental interests.
Roy D Goodman, Chapter Coordinator

1481 Student Conservation Association Northwest
1265 S Main Street   206-324-4649
Suite 210   Fax: 206-324-4998
Seattle, WA 98144   E-mail: webmaster@thesca.org
http://www.thesca.org/
SCA is a national organization with regional offices in Seattle, Oakland, Pittsburg, Washington DC and headquartered in Charlestown NH. Our mission is to build the next generation of conservation leaders and inspire lifelongstewardship of our environment and communities by engaging young people in hands-on service to the land. We offer a wide range of internships and crew based programs for ages 16 years and up.
Founded: 1957
Dale M Penny, President/CEO
Mark Bodin, Executive VP/COO

1482 Trout Unlimited
Trout Unlimited
2401 Bristol Court SW   360-754-2131
suite A18   800-533-0852
Olympia, WA 98502   Fax: 360-754-4240
E-mail: troutunlimited@localaccess.com
http://www.troutunlimitedwashington.org
Terry Turner, President
Rick Abbett, Corporate Executive

1483 Washington Association of Conservation Districts
185 Beebe Road   509-773-5065
Goldendale, WA 98620   Fax: 509-773-5600
E-mail: wacd@ncia.com
http://www.wacd.org
Mike Bailey, Finance Director

1484 Washington Environmental Council
615 2nd Avenue   206-622-8103
Suite 380   800-561-8294
Seattle, WA 98104   Fax: 206-622-8113
E-mail: wec@wecprotects.org
http://www.wecprotects.org
Founded: 1967
Joan Crooks, Executive Director

1485 Washington Public Interest Research Group
3240 Eastlake Avenue E   206-568-2850
Suite 100   800-213-7383
Seattle, WA 98102   Fax: 206-568-2858
E-mail: washpirg@pirg.org
http://www.washpirg.org
Robert Pregulman, Executive Director

1486 Washington Recreation and Park Association
350 S 333rd Street   253-874-1283
Suite 103   Fax: 254-661-3929
Federal Way, WA 98003   E-mail: wrpa@seanet.com
Tracy Thomas, Secretary

1487 Washington Refuse and Recycling Association
4160 6th Avenue SE   360-943-8859
Suite 205   866-788-9772
Lacey, WA 98503   Fax: 360-357-6958
E-mail: office@wrra.org
http://www.wrra.org
Steve Wheatley, President
Nancy LeMay, Vice-President

1488 Washington Toxics Coalition
4649 Sunnyside Avenue N   206-632-1545
Suite 540   800-844-7233
Seattle, WA 98103   Fax: 206-632-8661
E-mail: info@watoxics.org
http://www.watoxics.org
Founded: 1981
Jon Stier, President
Gregg Small, Executive Director

1489 Washington Wilderness Coalition
4649 Sunnyside Avenue N   206-633-1992
Suite 520   800-627-0062
Seattle, WA 98103   Fax: 206-633-1996
E-mail: info@wawild.org
http://www.wawild.org
Founded: 1979
Tom Geiger, President

1490 Washington Wildlife Federation
PO Box 1966   360-705-1903
Olympia, WA 98507   E-mail: www.washingtonwildlife.org
Thea Levkovitz, Representative

1491 Wildlife Society
3526 103rd Place SE   509-997-2131
Everett, WA 98208-4613   Fax: 509-997-9770
E-mail: tillzinger@communitynet.org
http://http://www.washingtonwildlifesoc.org

A nonprofit scientific and educational organization that serves professionals such as government agencies, academia, industry, and non-government organizations in all areas related to the conservation of wildlife and natural resourcesmanagement.
Bill Vogel, President
Peter Singleton, President-Elect

West Virginia

1492 ASLA: West Virginia Chapter
5088 Washington Street West   304-776-7473
Cross Lanes, WV 25313   Fax: 304-776-6426
E-mail: tschoolcraft@elrobinson.com
http://www.wvasla.org

Peter J Williams, President
Laura L Cox, Secretary

1493 American Lung Association of West Virginia
American Lung Association
PO Box 3980   304-342-6600
Charleston, WV 25339-3980   800-LUN- USA
Fax: 304-342-6096
E-mail: tatty@alawv.org
http://www.alawv.org

Founded: 1904
Sara Crickenberger, Executive Director
Chantal Fields, Assistant Executive Director

1494 West Virginia Bureau for Public Health
West Virginia Departmwent of Health and Human Reso
350 Capitol Street   304-558-2971
Suite 702   Fax: 304-558-1035
Charleston, WV 25301-37   E-mail: ASOsupport@wvdhhr.org
http://www.wvdhhr.org
Organizational activities not directed specifically toward children are education, regulation and research.
Chris Curtis, Commisioner
Ronald Forren, Deputy Commisioner

1495 West Virginia Forestry Association
PO Box 718   304-372-1955
Ripley, WV 25271   888-372-9663
Fax: 304-372-1957
E-mail: wvfa@wvadventures.net
http://www.wvfa.org
Nonprofit association.
Dick Waybright, Executive Director

1496 West Virginia Highlands Conservancy
1525 Hampton Road   304-342-8989
Charleston, WV 25314   E-mail: peter@mountain.net
http://www.wvhighlands.org/
Founded: 1967
Hugh Rogers, President
Peter Shoenfeld, Senior Vice President

1497 West Virginia Woodland Owners Association
PO Box 13695   304-594-3648
Sissonville, WV 25360   Fax: 304-594-3648
Russ Richardson, VP

Wisconsin

1498 ASLA: Wisconsin Chapter
1320 Pewaukee Road   262-548-7805
Room 230   800-394-4309
Waukesha, WI 53188   E-mail: stkelly1@wisc.edu
http://www.asla.org
Founded: 1899
Gerald Beaulieu, Managing Director
Daniel Williams, President-Elect

1499 Botanical Club of Wisconsin
Wisconsin Academy of Science, Arts, & Letters
1922 University Avenue
Madison, WI 53705   608-262-5489
Fax: 608-265-2993

Emmet Judziewicz, President

1500 Center for Alternative Mining Development Policy
210 Avon Street   608-784-4399
Suite 4   Fax: 608-785-8486
La Crosse, WI 54603   E-mail: gedicks.al@uwlax.edu
http://www.wrpc.net
Provides technical information on mining impacts to facilitate public and tribal involvement in the decision-making process for metallic sulfide mining in Wisconsin. Distributes educational videos about mining and about native andenvironmental cooperation to resist ecologically destructive mining projects.
Founded: 1974
Will David, Chair Person
Richard Nuna, Director

1501 Central Wisconsin Environmental Station (CWES)
10186 County Road MM   715-824-2428
Amherst Junction, WI 54407   Fax: 715-824-3201
E-mail: sjohnson@uwsp.edu
http://www.uwsp.edu/cnr/cwes/
Founded: 1975
Scott Johnson, Director

1502 Citizens for Animals: Resources and Environment
PO Box 18772   414-466-1250
Milwaukee, WI 53218
Debi Zweifel, Director

1503 River Alliance of Wisconsin
306 E Wilson   608-257-2424
Suite 2W   Fax: 608-260-9799
Madison, WI 53703   E-mail: wisrivers@wisconinrivers.org
http://www.wisconsinrivers.org
Jake Barnes, Treasurer

1504 Sierra Club-John Muir Chapter (Wisconsin)
222 S Hamilton Street   608-256-0565
Suite 1   Fax: 608-256-4562
Madison, WI 53703   http://www.sierraclub.org/chapters/wi/
Preserve and protect the natural environment by empowering citizens, especially democratically based grassroots organizations with charitable resources to further the cause of environmental protection, the vehicle through which TheSierra Club Foundation generally fulfills its charitable mission.
Caryl Terrell, Executive Director
Chris Nehrbass, Chairman

1505 Sixteenth Street Community Health Center
1032 S 16th Street   414-672-1353
S Cesar E Chavez Drive   Fax: 414-672-9190
Milwaukee, WI 53204   http://www.sschc.org
The mission of the Sixteenth Street Community Health Center is to improve the health and well-being of Milwaukee's Near South Side residents by providing quality, family-based health care, health education and social services, freefrom linguistic, cultural and economic barriers.
Founded: 1969
John Bartkowski, Chief Executive Officer

1506 Trees for Tomorrow Natural Resources Educational Center
519 Sheridan Street   715-479-6456
PO Box 609   800-838-9472
Eagle River, WI 54521   Fax: 715-479-2318
E-mail: learning@treesfortomorrow.com
http://www.treesfortomorrow.com
Accredited natural resource specialty school. Hosts three day workshops for middle/high school students during the school year. Workshops emphasize conservation, proper land management and environmental basics.
Founded: 1944
Gail Gilson Pierce, Director
Sheri Buller, Assistant Director

1507 Trout Unlimited
PO Box 228   703-522-0200
Eau Claire, WI 54702-0228   Fax: 715-831-9568
E-mail: jwelter@discover-net.net
John Welter, Chairman

1508 Wildlife Society
6315 Clovernock Road   608-221-6344
Middletown, WI 53562-3824   Fax: 608-221-6353
                        E-mail: barteg@dnr.state.wi.us
A nonprofit scientific and educational organization that serves
professionals such as government agencies, academia, industry,
and non-government organizations in all areas related to the
conservation of wildlife and natural resourcesmanagement.
Gerald Adrian Bartelt Wydeven, President

1509 Wisconsin Association for Environmental Education
8 Nelson Hall   715-346-3835
University of Wisconsin   E-mail: waee@uwsp.edu
Stevens Point, WI 54481   http://www.uwsp.edu/cnr/waee/
        Founded: 1975
Carrie Hembree, Administrative Assistant
Carol Weston, Administrative Assistant

1510 Wisconsin Association of Lakes
PO Box 126   715-364-3604
Stevens Point, WI 54481-0126   Fax: 715-346-3624
                            E-mail: wal@coredcs.com

Jim Burgess, President

1511 Wisconsin Land and Water Conservation Association
One Point Place   608-833-1833
Suite 101   Fax: 608-833-7179
Madison, WI 53719   E-mail: ginakaminski@wlwca.org
                                http://www.wlwca.org

Wilbur Petroskey, President
Rebecca Baumann, Executive Director

1512 Wisconsin Society for Ornithology
2022 Sherryl Lane   262-547-6128
Waukesha, WI 53188-3142   E-mail: dcreel@execpc.com
                        http://www.uwgb.edu/birds/wso/
        Founded: 1939
Jeffrey L Baughman, President
Dave Sample, Vice President

1513 Wisconsin Wildlife Federation
242 Keoller Avenue   412-235-9136
Oshkosh, WI 54901-9136   Fax: 414-235-6030
                            E-mail: wiwf@execpc.com
                    http://www.easy-axcess.com/wwf
Russell Hitz, Representative

1514 Wisconsin Woodland Owners Association
PO Box 285   715-346-4798
Stevens Point, WI 54481-0285   800-838-9472
                                    Fax: 715-346-4821
                            E-mail: nbozek@uwsp.edu
                    http://www.wisconsinwoodlands.org
        Founded: 1979
Nancy Livingston, Chairman

Wyoming

1515 Jackson Hole Conservation Alliance
685 S Cache   307-733-9417
PO Box 2728   Fax: 307-733-9008
Jackson, WY 83001-2728   E-mail: info@jhalliance.org
                                http://www.jhalliance.org
An organization dedicated to responsible land stewardship in
Jackson Hole, Wyoming to ensure that human activities are in
harmony with the area's irreplaceable wildlife, scenery and
other natural resources.
        Founded: 1979
Franz J Camenzind PhD, Executive Director
Margie Lynch, Program Director

1516 Nature Conservancy: Wyoming Chapter
Nature Conservancy
258 Main Street   307-332-2971
Suite 200   Fax: 307-332-2974
Lander, WY 82520   http://www.nature.org

Founded: 1950
Andrea Erickson, State Director
Paula Hunker, Associate State Director

1517 Powder River Basin Resource Council
934 North Main   307-672-5809
Sheridan, WY 82801   Fax: 307-672-5800
                    E-mail: resources@powderriverbasin.org
                        http://www.powderriverbasin.org
Committed to the preservation and enrichment of Wyoming's
agricultural heritage and rural lifestyle; the conservation of
Wyomings unique land, mineral, water and clean air resources,
consistent with responsible use of those resourcesto sustain the
vitality of present and future generations; the education and empowerment of Wyoming's citizens to raise a coherent voice in
decisions. They are the only group in Wyoming that addresses
both agricultural and conservation issues.
Bernie Barlow, Chairman
Clay Rowley, Vice Chairman

1518 Sierra Club
Sierra CLub
45 East Loucks   307-672-0425
Suite 109   Fax: 307-674-6187
Sheridan, WY 82801   http://www.wyoming.sierraclub.org
To advance the preservation and protection of the natural environment by empowering the citizenry, especially democratically-based grassroots organizations, with charitable resources
to further the cause of environmental protection.The vehicle
through which The Sierra Club Foundation generally fulfills its
charitable mission.
        Founded: 1892
Todd Herreid, Chairman
Mary Byrnes, Treasurer

1519 Trout Unlimited
PO Box 4069   307-733-6991
Jackson, WY 83001   Fax: 307-733-9678
                        E-mail: jpiotrowski@hpllp.net
                                http://www.tu.org

        Founded: 1959
James Piotrowski, President

1520 Western Association of Fish and Wildlife Agencies
5400 Bishop Boulevard   307-777-4569
Cheyenne, WY 82006   Fax: 307-777-4699
                        E-mail: ikruck@state.wy.us
                                http://www.wafwa.org

Ken Ambrock, President

1521 Wyoming Association of Conservation Districts
517 E 19th Street   307-632-5716
Cheyenne, WY 82001   Fax: 307-638-4099
                            E-mail: waocd@trib.com
                        http://www.conservewy.com
        Founded: 1945
Ralph Brokaw, President
Jean Dickinson, Vice President

1522 Wyoming Native Plant Society
PO Box 2500   307-766-3020
Laramie, WY 82073   E-mail: wndd@uwyo.edu
        http://uwadmnweb.uwyo.edu/wyndd/wnps/wnps_home.htm
                Founded: 1981
Bonnie Heidel, President
Laura Hudson, Vice President

1523 Wyoming Wildlife Federation
PO Box 106   307-637-5433
Cheyenne, WY 82003   800-786-5434
                                    Fax: 307-637-6629
                    E-mail: admin@wyomingwildlife.org
                        http://www.wyomingwildlife.org
The Wyoming Wildlife Federation, established in 1937, is
Wyomings oldest and largest conservation group advocating
sportsmen and sportswomen. The Federation's mission is to
work for hunters, anglers and other wildlife enthusiasts
toprotect and enhance habitat; propetuate quality hunting and
fishing; protect citizens rights to use public lands and waters;
and promote ethical hunting and fishing.

Founded: 1937
Dave Gowdey, Executive Director
Mark Winland, President of the Board

Environmental

1524 Adirondack Council Conservationist of the Year

103 Hand Avenue
Suite 3
Elizabethtown, NY 12932

518-873-2240
877-873-2240
Fax: 518-873-6675
E-mail: adkcouncil@aol.com
http://www.adirondackcouncil.org

The Adirondac Council is a not-for-profit environmental group that has been working since 1975 to protect the open-space resources of New York State's six million acre Adirondack Park and to help sustain the natural and humancommunities of the region. Based in the Adirondacks with a second office in Albany, the Adirondack Council has a staff of 15 and a strong and vocal membership of 18,000.
John F Sheehan, Communications Director

1525 Aerospace Medical Association

320 S Henry Street
Alexandria, VA 22314

703-739-2240
Fax: 703-739-9652
E-mail: pday@asma.org
http://www.asma.org

AsMA is dedicated to uniting the world's professionals in aviation, space and environmental medicine: advancing the frontiers of aerospace medicine by dissemination of knowledge throughout industry, the general public, and governmentalagencies worldwide. Ensuring the highest level of safety, health and performance of those involved in aerospace, AsMA is recognized as the international authority in aerospace medicine.
Daniel B Lestage, Chair

1526 Air Force Association

1501 Lee Highway
Arlington, VA 22209-1198

703-247-5810
Fax: 703-247-5853
E-mail: service@afa.org
http://www.afa.org

The Air Force Association's mission is to advocate aerospace power and a strong national defense; to support the United States Air Force and Air Force Family; and to promote aerospace education to the American people.

James A McDonnell Jr

1527 Air and Waste Management Association

1 Gateway Center
420 Fort Duquesne Boulevard
Pittsburgh, PA 15222

412-232-3444
Fax: 412-232-3450
E-mail: info@awma.org
http://www.awma.org/

The Air and Waste Management Association is a nonprofit, nonpartisan professional organization that provides training, information and networking opportunites to thousands of environmental professionals in 65 countries.
Edith Mijares Ardiente, President

1528 American Association of Engineering Societies

1828 L Street NW
Suite 906
Washington, DC 20036-5110

202-296-2237
888-400-2237
Fax: 202-296-1151
E-mail: info@aaes.org
http://www.aaes.org

Multidisciplinary organization dedicated to advancing the knowledge, understanding and practice of engineering in the public interest. Its members represent the mainstream of US engineering-affecting over 1,000,000 engineers inindustry, government and education. Through its councils, commissions, committees and task forces, the AAES addresses questions relating to the engineering profession.
Paul J Kostek, Chair
John R Parker, Chair-Elect

1529 American Chemical Society

1155 16th Street NW
Washington, DC 20036

860-872-4408
Fax: 202-872-6317

The American Chemical Society is a self-governed individual membership organization that consists of more than 158,000 members in the field of chemistry. The organizations provides a broad range of opportunities for peer intereactionand career development, regardless of professional or scientific interests.
William J Carroll Jr, President

1530 American Conference of Governmental Industrial Hygienists

1330 Kemper Meadow Drive
Cincinnati, OH 45240

513-742-6163
Fax: 513-742-3355
E-mail: mail@acgih.org
http://www.acgih.org

The American Conference of Governmental INdustrial Hygeienists (ACGIH) is a member-based organization and community of professionals that advances worker health and safety through education and the development and dissemination ofscientific and technical knowledge.
Cindy Coe Laseter, Chair
Robert D Soule, Vice Chair

1531 American Forest and Paper Association

111 Nineteenth Street, NW
Suite 800
Washington DC 20036

800-878-8878
E-mail: info@afandpa.org
http://http://afandpa.org

The American Forest and Paper Association (AF&PA) is the national trade association of the forest, pulp, paperboard wood products industry. We represent member companies engaged i ngrowning, harvesting and processing wood and woodfiber, manufacturing pulp, paper and paperboard products from both virgin and recycled fiber, and producing engineered and traditional wood products.

1532 American Institute of Chemical Engineers

3 Park Avenue
New York, NY 10016-5911

212-591-8100
800-242-4363
Fax: 212-591-8888
E-mail: xpress@aiche.org
http://www.aiche.org

Founded in 1908, a professional association of more than 50,000 members that provides leadership in advancing the chemical engineering profession. Fosters and disseminates chemical engineering knowledge, supports the professional andpersonal growth of its members, and applies the expertise of its member to address societal needs through the world
Scott Berger, Director
Bette Lawler, Sr. Director, Operations

1533 American Institute of Mining, Metallurgical and Petroleum Engineers

3 Park Avenue
New York, NY 10016-5998

212-419-7676
Fax: 212-419-7671
E-mail: aimeny@aimeny.org
http://www.idis.com/aime

Organized and operated exclusively to advance, record and disseminate significant knowledge of engineering and the arts and sciences involved in the production and use of minerals, metals, energy sources and materials for the benefitof humankind, both directly as AIME and through Member Services.
Nellie Guernsey, Executive Director

1534 American Nuclear Society

55 N Kensington Avenue
La Grange Park, IL 60526

708-352-6611
Fax: 708-352-0499
E-mail: nucleus@ans.org
http://www.ans.org

The American Nuclear Society is a not-for-profit, international, scientific and educational organization. It was established by a group of individuals who recognized the need to unify the professional activities within the diversefields of nuclear science and technology.
E James Reinsch, President
Harry Bradley, Executive Direcotr

1535 American Society of Civil Engineers

1801 Alexander Bell Drive
Reston, VA 20191

800-548-2723

Founded in 1852, the American Society of Civil Engineers represents more than 137,500 members of the civil engineering profession worldwide, and is America's oldest national engineering society. ASCE's vision is to position engineersas global leaders building a better quality of life.
Dennis R Martenson, President Elect
Patrick J Natale, Executive Director

1536 American Society of Heating, Refrigerating and Air-Conditioning

1791 Tullie Circle NE   404-636-8400
Atlanta, GA 30329   Fax: 404-321-5478
http://www.ashrae.org

Organized for the sole purpose of advancing the arts and sciences of heating, ventilation, air conditioning and refrigeration for the public's benefit through research, standards writing, continuing education and publications for thepublic's benefit.
Lee Burgett, President
Terry Townsend, President Elect

1537 American Sportfishing Association

225 Reinekers Lane   703-519-9691
Suite 420   Fax: 703-519-1872
Alexandria, VA 22314   E-mail: info@asafishing.org
http://www.asafishing.org

Mike Nussman, President/CEO
Gordon Robertson, VP

1538 American Water Resources Association

4 West Federal Street   540-687-8390
PO Box 1626   Fax: 540-687-8395
Middleburg, VA 20118-2192   E-mail: info@awra.org
http://www.awra.org

Founded in 1964, the American Water Resources Association is a non-profit professional association dedicated to the advancement of men and women in water resources management, research, and education. AWRA's membership ismultidisciplinary; its diversity is its hallmark. It is the professional home of a wide variety of water resources experts including engineers, educators, foresters, biologists, ecologists, geographers, managers, regulators, hydrologists andattorneys.
Mindy Lalor, President
David R DeWalle, President Elect

1539 Association for Conservation Information

Montana Department of Fish, Wildlife and Parks
1420 E 6th Avenue   406-444-2535
PO Box 200701   Fax: 406-444-4952
Helena, MT 59620   http://www.aci-net.org

ACI, the Association for Conservation Information, is a non-profit association of information and education professionals representing state, federal and Canadian agencies and private conservation organizations. ACI memberprofessionals play a major role in providing natural resource, environmental, wildlife and other information and education to the public through a variety of means, many of which are continental in scope.
David Chanda, President

1540 Association of Conservation Engineers

2901 w. Truman Boulevard   573-751-4115
PO Box 180   Fax: 573-751-4467
Jefferson City, MO 65109   http://www.conservation.state.mo.us

Organization of conservation engineers and technicians who are working to conserve and improve our nation's natural heritage. Brings together engineers and allied personnel employed by conservation and recreation agencies andconsultants who have a community of specialized interests in the areas of fish, wildlife, parks, forests and related conservation/recreation fields. Members pool experience and information pertaining to conservation engineering to make naturalresources more accessible.
Anita B Gorman, Chairman
Lowell Mohler, Vice-Chairman

1541 Association of Consulting Foresters of America, Inc

312 Montgomery Street   703-548-0990
Suite 208   Fax: 703-548-6395
Alexandria, VA 22314   E-mail: director@acf-foresters.com
http://www.acf-foresters.com

To protect the public welfare and property in the practice of forestry, to raise the professional standards and work of ACF Consultants and all other consulting foresters. To develop and expand the services of ACF Consultants. To serveas a forum for the exchange of information. To certify qualified foresters for competency in specific areas of professional practice.
Lynn C Wilson, Executive Director
Dick Courter, President

1542 Audubon Naturalist Society of Central Atlantic

8940 Jones Mill Road   301-652-9188
Chevy Chase, MD 20815   Fax: 301-951-7179

E-mail: contact@audubonnaturalist.com
http://www.audubonnaturalist.com

The Audubon Naturalist Society is an independent environmental education and conservation organization with over 10,000 members in the Washington DC area. The society offers a wide variety of natural history classes and campaigns forthe protection and renewal of the Mid-Atlantic regions natural resources.

Joseph Coleman, President
Francis J O'Donnell, VP

1543 Audubon Society of New Hampshire

3 Silk Farm Road   603-224-9909
Concord, NH 03301   Fax: 603-226-0902
E-mail: asnh@nhaudubon.org
http://www.nhaudubon.org

The Audubon Society of New Hampshire, a nonprofit statewide membership organization, is dedicated to the conservation of wildlife and habitat throughout the state. Independent of the National Audubon Society, ASNH has offered programsin wildlife conservation, land protection, environmental policy and environmental education since 1914.

Julian Zelazny, Environmental Policy Directo

1544 Big Thicket Conservation Association

PO Box 198   409-892-8976
Saratoga, TX 77585   http://www.btatx.org

Ellen Buchanan, President
Judy Arnow, Vice President

1545 Botanical Society of America

PO Box 299   314-577-9566
St. Louis, MO 63166-0299   Fax: 314-577-9515
E-mail: bsa-manager@botany.org
http://www.botany.org

Promote botany, the field of basic science dealing with the study and inquiry into the form, function, diversity, reproduction, evolution, and uses of plants and their interactions within the biosphere.
Allison A Snow, President
Edward L Schneider, President Elect

1546 Chicago Community Trust

111 East Wacker Drive   312-616-7955
Suite 1400   Fax: 312-856-1703
Chicago, IL 60601   E-mail: info@cct.org
http://www.cct.org

Terry Mazany, President/CEO
Anne Blanton, Executive VP Administration

1547 Connecticut River Watershed Council

15 Bank Row   413-772-2020
Greenfield, MA 01301   Fax: 413-772-2090
E-mail: crwc@ctriver.org
http://www.ctriver.org

The Connecticut River Watershed Council (CRWC) is the only broad-based citizen advocate for the environmental well-being of the entire Connecticut River. Our primary mission is to promote improvement of water quality and therestoration, conservation, wise development and use of the natural resources of the Connecticut River watershed.
Chelsea Gwyther, Executive Director

1548 Conservation Essay and Poster Contest

Department of Natural Resources
663 Teton Trail   502-564-3080
Frankfort, KY 40601   Fax: 505-564-9195
E-mail: hduncan@ca.uky.edu

Stanley Head, Director

1549 Conservationist of the Year

Adirondack Council
103 Hand Avenue   877-873-2240
Suite 3   Fax: 518-873-6675
Elizabethtown, NY 12932   http://www.adirondackcouncil.org

John F Sheehan, Communications Director

1550 Ecological Society of America
1707 H Street NW   202-833-8773
Suite 400   Fax: 202-833-8775
Washington, DC 20006   E-mail: esahq@esa.org
     http://www.esa.org
The Ecological Society of America is a non-partisan, nonprofit organization of scientists founded in 1915 to : promote ecological science by improving communications among ecologists; raise the public's level of awareness of theimportance of ecological science; increase the resource available for the conduct of ecological science.
Katherine S McCarter, Executive Director

1551 Federal Aviation Administration
Office of Public Affairs
800 Independence Avenue SW   202-267-3883
Room 908   Fax: 202-267-5047
Washington, DC 20591   http://www.faa.gov
The major roles of the Federal Aviation Administration (a part of the Department of Transportation) include regulation, development, and research in the areas of civil aviation, civil aeronautics, and U.S. commercial spacetransportation.
Marion C Blakey, Administrator
Robert A Sturgell, Deputy Administrator

1552 Federation of Fly Fishers
215 E. Lewis   406-222-9369
Livingston, MT 59047   Fax: 406-222-5823
     http://www.fedflyfishers.org
The Federation of Fly Fishers seeks to cultivate and advance the art science and sport of flyfishing as the most sporting and enjoyable method of angling and the way of fishing most consistent with the preservation and use of game fishresources; to be the voice for organized fly fishing; to promote conservation of recreational resources; to facilitate and improve the knowledge of fly fishing; and to elevate the standard of integrity, honor and courtesy of anglers.
RP van Gytenbeek, Chief Executive Officer
Bob Wiltshire, Chief Operating Officer

1553 Florida Audubon Society
1101 Audubon Way   407-644-0190
Maitland, FL 32751   Fax: 407-644-8940
     http://www.audubonofflorida.org
To conserve, protect and restore Florida's natural resources and to create a conservation ethic among all Floridians. The mission is to conserve and restore natural ecosystems, focusing on birds and other wildlife for the benefit ofhumanity and the earth's biological diversity.

1554 Frank A Chambers Award
Air and Waste Management Association
One Gateway Center, 3rd Floor   412-232-3444
420 Fort Duquesne Boulevard   Fax: 412-232-3450
Pittsburgh, PA 15222   http://www.awma.org
Award for outstanding achievement in the science and art of air pollution control. It requires accomplishment of a technical nature on the part of the recipient which is considered to be a major contribution to the science and art ofair pollution control, the merit of which has been widely recognized by persons in the field.

1555 German Marshall Fund of the United States
1744 R Street NW   202-745-3950
Washington, DC 20009   Fax: 202-265-1662
     E-mail: info@amfus.org
     http://www.gmfus.org
To stimulate the exchange of ideas and promote cooperation between the United States and Europe in the spirit of the postwar Marshall Plan. GMF was created in 1972 by a gift from the German people as a permanent memorial to MarshallPlan aid.

1556 Global Tomorrow Coalition
Capital Research Center, Green Watch
1513 16th Street NW   202-483-6900
Washington, DC 20036-3104   Fax: 202-483-6990
     E-mail: contact@capitalresearch.org
     http://www.capitalresearch.org/gw

Green Watch is an online database and information clearinghouse providing factual information on over 500 nonprofit environmental groups. This free service identifies the location, leadership and membership of each profiled group.Green Watch also produces timely news reports and analyses of the environmental movement.
Terrence Scanlon, President

1557 Golden Gate Audubon Society
2530 San Pablo Avenue   510-843-2222
Suite G   Fax: 510-843-5351
Berkeley, CA 94702   E-mail: ggas@goldengateaudubon.org
     http://www.goldengateaudobon.org
A conservation and education organization that has birds as its key component. We seek to protect and enjoy wildlife and their natural habitat in San Francisco and East Bay through interaction between our members and the community.
Elizabeth Murdock, Executive Director

1558 Goldman Environmental Foundation
One Lombard Street   415-788-9090
Suite 303   Fax: 415-788-7890
San Francisco, CA 94111   E-mail: info@goldmanprize.org
     http://www.goldman-prize.org
Goldman Environmental Prizes are awarded for sustained and important efforts to preserve the natural environment, including, but not limited to:Æprotecting endangered ecosystems and species, combatting destructive development projects,promoting sustainability, influencing environmental policies and striving for environmental justice.

1559 Great Lakes Commission
2805 S Industrial Highway   734-971-9135
Suite 100   Fax: 734-971-9150
Ann Arbor, MI 48104-4816   http://www.glc.org
Binational agency that promotes the orderly, integrated and comprehensive development, use and conservation of the water and related natural resources of the Great Lakes basin and St Lawrence River.
Thomas R Crane, Interim Executive Director

1560 Honorary Membership
Air and Waste Management
1 Gateway Center, 3rd Floor   412-232-3444
420 Fort Duquesne Blvd   Fax: 412-232-3450
Pittsburgh, PA 15222   http://www.awma.org
May be conferred upon persons who have attained eminence in some field related to the mission and objectives of the Association who have rendered valuable service to the Association.

1561 Institute of Environmental Sciences and Techno logy
5005 Newport Drive   847-255-1561
Suite 506   Fax: 847-255-1699
Rolling Meadows, IL 60008-3841   E-mail: iest@iest.org
     http://www.iest.org

Julie Kendrick, Executive Director

1562 International Desalination Association
PO Box 387   978-887-0410
Topsfield, MA 01983   Fax: 978-887-0411
     E-mail: info@idadesal.org
     http://www.idadesal.org
IDA is committed to the development and promotion of the appropriate use of desalination and desalination technology worldwide. We endeavor to carry out these goals by encouraging research and development, exchanging, promotingcommunication and disseminating information.
Abdulhamid A Al-Mansour, President
Lisa R Henthorne, 1st Vice President

1563 International Studies Association
University of South Carolina   803-777-2933
Columbia, SC 29208   http://www.csf.colorado.edu/isa/
William A Welsh, Executive Director

1564 International Wildlife Film Festival: Media Center
718 S Higgins Avenue   406-728-9380
Missoula, MT 59801   Fax: 406-728-2881
     E-mail: iwff@wildlifefilms.org
     http://www.wildlifefilms.org

Goal is to be the preeminent wildlife film, television and media organization, showcasing the world's best wildlife films and television programs, providing educational resources and events seminars, workshops, field classes, filmtours and many hands-on activities, that emphasize the most up-to-date, factual and ethical scienced based information.
Janet Rose, Executive Director

1565 Irrigation Association

6540 Arlington Boulevard   703-536-7080
Falls Church, VA 22042   Fax: 703-536-7019
E-mail: webmaster@irrigation.org
http://www.irrigation.org
To improve the products and practices used to manage water resources and to held shape the worldwide business environment of the irrigation industry.
Thomas H Kimmell, Executive Director

1566 John Burroughs Association

79th Street   212-769-5469
New York, NY 10024-5192   Fax: 212-769-5495
E-mail: peling@amnh.org
http://www.nimidi.amnh.org/burroughs/jba
Each year a medal is awarded to the author of a distinguished book of natural history, a list of exceptional national history books for young readers is selected. and an outstanding nature essay is identified.

1567 Keep America Beautiful

1010 Washington Boulevard   203-323-8987
Stamford, CT 06901   Fax: 203-325-9199
E-mail: info@kab.org
http://www.kab.org
Nonprofit organization whose network of local, statewide and international affiliate progams educates individuals about litter prevention and ways to reduce, reuse, recycle and properly manage wase materials.
G Raymon Empson, President

1568 Keep North Carolina Beautiful

2720 Kilgore Avenue   919-820-4940
Raleigh, NC 27607   Fax: 919-834-9869
North Carolina Keep America Beautiful is a nonprofit public education organization dedicated to enhancing the natural beauty of North Carolina communities, improving waste handling practices and empowering individuals to take greaterresponsibility for improving community environments.
Brenda Barger, Chairwoman

1569 Lawrence K Cecil Award

American Institute of Chemical Engineers
3 Park Avenue   212-591-8100
New York, NY 10016   800-242-4363
Fax: 212-591-8888
E-mail: awards@aiche.org
http://www.aiche.org
Recognizes an individual's outstanding chemical engineering contribution and achievement in the preservation or improvement of the environment.

1570 Lyman A Ripperton Award

Air and Waste Management Association
One Gateway Center, 3rd Floor   412-232-3444
420 Fort Duquesne Boulevard   Fax: 412-232-3450
Pittsburgh, PA 15222   E-mail: info@awma.org
http://www.awma.org
Awarded for distinguished achievement as an educator in some field of air pollution control. Awarded to an individual, who by precept and example, has inspired students to achieve excellence in all their professional and socialendeavors.
Edith Mijares Ardiente, President

1571 National Association for Environmental Education

PO Box 400   937-698-6493
Troy, OH 45373   Fax: 937-335-5623
The National Association for Environmental Education is a network of professionals, students and volunteers working in the field of environmental education throughout North America and in over 55 countries around the world. NAAEE takesa cooperative, nonconfrontational, scientifically-based approach to promoting education about environmental issues.

Joseph Baust, President
Martha Monroe, President Elect

1572 National Association of Conservation Districts

408 E Main Street   281-332-3402
PO Box 855   Fax: 281-332-5259
League City, TX 77574-0855
The National Association of Conservation Districts has a membership consisting of local subdivisions of state governements which work to conserve and develop land, water, forests, wildlife and related natural resources. It was startedin 1946 and now has about 3,000 member districts.
Bill Wilson, President
Tim Reich, Vice President

1573 National Association of Environmental Professionals

PO Box 2086   301-860-1140
Bowie, MD 20718   888-251-9902
Fax: 301-860-1141
http://www.naep.org
NAEP is the multidisciplianry association dedicated to the advancement of the environmental professions in the US and abroad. A forum for state-art-the-art information on environmental planning, research and mangement. A network ofprofessional contacts and exchange on information, among colleagues in industry, government, academic and the private sector.
Gary Kelman, President

1574 National Audubon Society

700 Broadway   212-979-3000
New York, NY 10003   Fax: 212-353-0377
http://www.audubon.org
The mission of the National Audubon Society is to conserve and restore natural ecosystems, focusing on birds and other wildlife for the benefit of humanity and the earth's biological diversity. Founded in 1905, the National AudubonSociety is named for John James Audubon, famed orithologist, explorer, and wildlife artist.
Erica Barton, Contact

1575 National Bison Association

1400 West 122nd Avenue   303-292-2833
Suite 106   Fax: 303-292-2564
Westminster, CO 80234   http://www.bisoncentral.com
Dave Carter, Executive Director

1576 National Environmental Training Association

5320 N 16th Street   602-956-6099
Suite 114   Fax: 602-956-0399
Phoenix, AZ 85018-3241   E-mail: neta@ehs-training.org
http://ehs-training.org
The National Environmental Training Assoication, is a nonprofit international organization of enviromental, health and safety, other technical training professionals. Activities centeral to NETA's educational services include itssupport for trainer networking, professional development and competency certification for its members, EH&S training information and programs for industry, and development of training competency standards.
Charles L Richardson, Executive Director

1577 National Ocean Industries Association

1120 G Street NW   202-347-6900
Suite 900   Fax: 202-347-8650
Washington, DC 2005-3801   E-mail: tom@noia.org
http://www.noia.org
The National Ocean Industries Assoication, founded in 1972 with 35 members, represents all facets of the domestic offshore and related industries. Today, our more than 300 member companies are dedicated to the development of offshoreoil and natural gas for the coninued growth and secirity of the United States. NOIA members are engaged in many business activities, in addition to those listed below, including enviromental safeguards, equipment supply, gas transmission,naviogation,ect.

1578 National Press Club

529 14th Street NW   202-662-7500
Suite 1300   Fax: 202-662-7569
Washington, DC 20045   E-mail: info@press.org
http://www.press.org

Professional organization of reporters, writers and newspeople employed by newspapers, wire services, magazines, radio and television stations, and other forms of news media; and former newspeople and associates of newspeople. Sponsorsprofessional, sports, travel and cultural events; book rap sessions with news figures and authors; and newsmaker and luncheon speaker sessions. Houses reference library and archives. Offers computer training. Publishes a weekly newsletter.
John Bloom, General Manager

1579 National Recreation and Park Association

2775 S Quincey Street   703-820-4940
Suite 300   Fax: 703-671-6772
Alexandria, VA 22206   http://www.nrpa.org

The mission of the National Recreation and Park Association is to a advance parks, recreation and environmental conservation efforts that enhance the quality of life for all people.
Laurie Kusek

1580 National Recycling Coalition

1727 King Street   703-683-9025
Suite 105   Fax: 703-683-9026
Alexandria, VA 22314   http://www.nrc-recycle.org

NRC is a not-for-profit organization dedicated to the advancement and improvement of recycling, source reduction, composting, and reuse by providing technical information, education, training, outreach, and advocacy services to itsmembers in order to conserve resources and benefits the environment.
Ben Walker, President
Stephen Bantillo, Vice President

1581 National Sanitation Foundation

789 N Disboro Road   734-769-8010
PO Box 130140   Fax: 734-769-0109
Ann Arbor, MI 48113-0140   E-mail: info@nsf.org
     http://www.nsf.org

NSF International, The Public Health and Safety Company, is an independent, not-for-profit organization providing a wide range of services around the world. For more than 55 years, NSF has been committed to public health, safety andprotection of the enviroment.

1582 National Water Resources Association

3800 N Fairfax Drive   703-524-1544
Suite 4   Fax: 703-524-1548
Arlington, VA 22203-1703   http://www.nwra.org/

The National Water Resources Association is a nonprofit federation of state organizations whose membership includes rural water districts, municipal water entities, commerical companies and individuals. As an Association we areconcerned with the appropriate management, conservation, and use of water and land resources on a national scope.
Norm M Semanko, President

1583 National Wild Turkey Federation

770 Augusta Road   803-637-3106
PO Box 530   800-THE-NWTF
Edgefield, SC 29824   Fax: 803-637-0034
     E-mail: nwtf@nwtf.net
     http://www.nwtf.org/

The NWTF, an international nonprofit conservation and education organization dedicated to conserving wild turkeys and preserving hunting traditions. Growth and progress define the NWTF as it has expanded from 1,300 members in 1973 tonearly a half million today.
Tammy Bristow Sapp, President Communications

1584 National Wildlife Federation

11100 Wildlife Center Drive   202-797-6800
Reston, VA 20190   Fax: 202-797-6646
     http://www.nwf.org

National Wildlife Federation inspires Americans to protect wildlife for our children's future. We represent the power and commitment of four million members and supporters joined by affiliated wildlife organizations in 47 states andterritories. We channel the energy of thousands of volunteers from all walks of life to take action because they care about wildlife.
Larry J Schweiger, President\CEO

1585 Natural Resources Defense Council

40 W 20th Street   212-727-2700
New York, NY 10011   Fax: 212-727-1773
     E-mail: nrdcinfo@nrdc.org
     http://www.nrdc.org

We work to foster the fundamental right of all people to have a voice in decisions that affect their environment. We seek to break down the pattern of disproportionate environmental burdens borne by people of color and others who facesocial or economic inequities. Ultimately,NRDC strives to create a new way of life for humankind, one that can be sustained indefinitely without fouling or depleting the resources that support all life on earth.
Francis Beinecke, Executive Director

1586 Nature Conservancy

4245 N Fairfax Drive   703-841-8744
Suite 100   800-628-6860
Arlington, VA 22203   Fax: 703-841-4850

The Nature Conservancy is a leading international, nonprofit organization dedicated to rpeserving the diversity of life on Earth. The mission of The Nature Conservancy is to preserve teh plants, animals and natural communities thatrepresent the diversity of life on Earth by protecting the lands and waters they need to survive.
Henry M Paulson, Jr, Chairman
Steven J McCormick, President\CEO

1587 New England Wild Flower Society

180 Hemenway Road   508-877-7630
Framingham, MA 01701   Fax: 508-877-3658
     E-mail: newfs@newfs.org
     http://www.newfs.org

The Wild Flower Society is a recognized leader in native plant conservation. Founded in 1900, the Society is the oldest plant conservation organization in the US. Its purpose is to promote the conservation of temperate North Americanplants through key programs-conservation and research, education, horticulture and habitat preservation. They publish three magazines annually, a seed and book catalog, nursery catalog and brochures and pamphlets about native plant conservation andhorticulture.
Gwen Stauffer, Executive Director
Francis H Clark, President

1588 New York Botanical Garden

200th Street & Kazimiroff Boulevard   718-817-8700
Bronx, NY 10458-5126   Fax: 718-562-8474
     http://www.nybg.org

Founded in 1891, the Garden is one of the world's great collections of plants, the region's leading educational center for gardening and horticulture, and an international center for plant research. The New York Botanical Garden is anadvocate for teh plant kingdom.
Wilson Nolen, Chairman
Gregory Long, President

1589 Outdoor Writers Association of America

121 Hickory Street   406-728-7434
Suite 1   Fax: 406-728-7445
Missoula, MT 59801   http://www.owaa.org

The mission of Outdoor Writers Assciation of America is to improve the professional skills of our members, set the highest ethical and communications standards, encourage public enjoyment and conservation of natural resources, and bementors for the next generation of professional outdoor communicators.
Marty Malin, Chairman
Spence Turner, President

1590 Ozark Society

PO Box 2914   501-225-1795
Little Rock, AR 72203   http://www.ozarksociety.net

The Ozark Society, was founded in 1962 by Dr. Neil Compton of BEntonville, an Ozark native, and group of associates for the immediate purpose of saving the Buffalo River from dams proposed by the US Army Corps of Engineers. Societyfounders, working with Sen. JW Fullbright, helped get the National Park Service to survey the Buffalo River area and then began to campiagn for the creation of the Buffalo National River as an alternative to the dams.
Steve Eyler, Chairman
Rick Larned, Vice Chairman

1591 Pennsylvania Association of Environmental Professionals
PO Box 7202   717-582-8540
Mechanicsburg, PA 17050   E-mail: paep@earthlink.net
                                          http://www.paep.org
A nonpolitical interdisciplinary organization of individuals working in environmental management, planning, impact assessment, environmental protection, compliance, research, engineering, design and education.
Eric H Buncher, President
Mark A Fedosick, Vice President

1592 Power
McGraw-Hill
11 W 19th Street   212-337-4060
New York, NY 10011   Fax: 212-627-3811

1593 Sea Grant Association
Center for Wetlande Resources
Louisiana State University   225-388-6710
Baton Rouge, LA 70803-7507   Fax: 225-388-6331
The Sea Grant Association (SGA) is a non-profit organization dedicated to furthering the Sea Grant program concept. SGA provides the mechanism for academic institutions to coordinate their activities, to set program priorities at boththe regional and national level, and to proved a unified voice for the institutions on issues of importance to the oceans and coasts.
Jonathan Kramer, President

1594 Sierra Club
85 Second Street   415-977-5500
2nd Floor   Fax: 415-977-5799
San Francisco, CA 94105   http://www.sierraclub.org
To advance the preservation and protection of the natural environment by empowering the citizenry, especially democratically-based grassroots organizations, with charitable resources to further the cause of environmental protection.The vehicle through which The Sierra Club Foundation generally fulfills its charitable mission.
Lisa Renstrom, President
Bernard Zaleha, Vice President

1595 Society of American Foresters
5400 Grosvenor Lane   301-897-8720
Bethesda, MD 20814-2198   Fax: 301-897-3690
                                          http://www.safnet.org
The mission of the Society of American Foresters is to advance the science, education, technology, and practice of forestry; to enhance the competency of its membersw; to establish professional excellence; and, to use the knowledge,skills, and conservation ethic of the profession to ensure the continued health and use of forest ecosystems and the present and future availability of forest resources to benefit society.
Michael T Goergen, Jr, Executive Vice-President\CEO

1596 Society of American Travel Writers
1500 Sunday Drive   202-429-6639
Suite 102   Fax: 202-775-4674
Raleigh, NC 27607   E-mail: satw@satw.org
                                          http://www.satw.org
SATW is a tax-exempt professional association whose purpose is to promote responsible journalism, provide professional support and development for our members. and encourage the conservation and preservation of travel resourcesworldwide.
Milton Fullman, President
Edwin Malone, President-Elect

1597 Society of Petroleum Engineers
PO Box 833836   972-952-9393
Richardson, TX 75083-3836   Fax: 972-952-9434
                                       E-mail: spedal@spe.org
                                          http://www.spe.org

Eve Sprunt, President

1598 Soil and Water Conservation Society
7515 NE Ankeny Road   515-289-2331
Ankeny, IA 50021-9764   Fax: 515-289-1227
                                          http://www.swcs.org/

Foster the science and the art of soil, water and related natural resource management to achieve sustainability. To promote and practice an ethic recognizing the interdependence of the people in the environment.
Deborah Cavanaugh-Grant, President
Jeffrey Vonk, Vice President

1599 Solar Energy Industries Association
805 15th Street NW   202-628-0556
Suite 510   Fax: 202-628-0559
Washington, DC 20005   E-mail: info@seia.org
                                          http://www.seia.org/
SEIA's primary mission is to expand the use of solar technologies in the global marketplace. National members combined with chapter members in 22 states exceed 500 compines providing solar thermal and solar electric products andservices.
Chris O'Brien, Chairman
Jeffrey D Wolfe, Division Chair

1600 Trout Unlimited
1300 North 17th Street   703-522-0200
Suite 500   800-834-2419
Arlington, VA 22209   Fax: 703-284-9400
                                       E-mail: trout@tu.org
                                          http://www.tu.org
Our mission is to conserve, protect and restore North America's trout and salmon fisheries and their watersheds. We accomplish this mission on local, state and national levels with an extensive and dedicated volunteer network.
Charles Gauvin, President\CEO
Kenneth Mendez, Executive Vice President

1601 US Army Corps of Engineers
20 Massachusetts Avenue NW   202-272-0011
Washington, DC 20314-1000   Fax: 202-504-4032
                                    http://www.usace.army.mil/
Our mission is to provide quality, responsive engineering services to the nation including: planning, desiging, building and operating water resources and other civila works projects.
LTG Carl A Stock, Chief of Engineers\Commander
Ronald Hawthorne, Deputy Director

1602 US Department of Energy
1000 Independence Avenue SW   202-586-5000
James Forrestal Building   Fax: 202-586-4403
Washington, DC 20585   http://www.energy.gov
The Department of ENergy's mission is to advance the national, economic and energy security of the US; to promote scientific and technological innovation; and to ensure the environmental cleanup of the national nuclear weapons complex.
Samuel W Bodman, Secretary Of Energy

1603 US Department of the Interior
1849 C Street NW   202-208-3100
Washington, DC 20240   Fax: 202-208-6956
                                          http://www.doi.gov

Villere C Reggio

1604 US Environmental Protection Agency
Office of External Liaison and Education
401 M Street SW   202-260-4454
Washington, DC 20460   Fax: 202-260-0130
                                          http://www.epa.gov
The mission of the EPA is to protect human health and the environment.
Stephen L Johnson, Administrator

1605 Underwater Society of America
PO Box 628   650-583-8492
Daly City, CA 94017   Fax: 408-294-3496
                              http://www.underwater-society.org/
The Underwater Society of America was founded in 1959 by the existing skin-diving councils; it was composed of and represented all divers in North America. It is the public diving organization of the United States. It is controlled byits Executive committee, board of directors and delegates of the member councils and clubs meeting annually.
Carol Rose, President

1606 Washington Journalism Center

2600 Virginia Avenue NW   202-662-7351
Suite 502
Washington, DC 20037
Murielle Nagl, Administrative Assistant

1607 Water Environment Federation

601 Wythe Street   703-684-2400
Alexandria, VA 22314-1994   800-666-0206
     Fax: 703-684-2492
     E-mail: webfeedback@wef.org
     http://www.wef.org
Founded in 1928, the Water Environment Federation (WEF) is a not-for-profit technical and educational organization with members from varied disciplines who work toward the WEF vision of preservation and enhancement of the global waterenvironment. The WEF network includes more than 100,000 water quality professionals from 77 member associations in 31 countries.
William J Bertera, Executive Director

1608 Western Forestry and Conservation Association

4033 SW Canyon Road   503-226-4562
Portland, OR 97221   Fax: 503-228-3624
     E-mail: richard@westernforestry.org
     http://www.westernforestry.org
Offers continuing education workshops and seminars for professional foresters.

1609 Whooping Crane Conservation Association

1007 Carmel Avenue   337-234-6339
Lafayette, LA 70501   Fax: 361-364-2650
     E-mail: wcca@excelonline.com
     http://www.whoopingcrane.com
Works to conserve and protect the Whooping Crane and prevent its extinction. Collects and disseminates knowledge of this species to advocate and encourage public appreciation and understanding of the Whooping Crane.

1610 Wilderness Society

1615 M Street NW   202-833-2300
Washington, DC 20036   800-THE-WILD
     E-mail: tws@wilderness.org
     http://www.wilderness.org
Our goal is to ensure that future generations will enjoy the clean air and water, wildlife, beauty and opportunites for recreation and renewal thatpristine forests, rivers, deserts and mountains provide.
Rebecca Rom, Chair
Brenda Davis, Vice Chair

1611 Wildlife Society

6550 Rock Springs Drive   301-897-9700
Suite 600   800-368-2806
Bethesda, MD 20817   Fax: 301-897-9794
A nonprofit scientific and educational organization that serves professionals such as government agencies, academia, industry, and non-government organizations in all areas related to the conservation of wildlife and natural resourcesmanagement.

1612 Willowbrook Wildlife Haven Preservation

National Wildlife Rehabilitation Association
2625 Clearwater Road   320-230-9920
Suite 110   http://www.nwrawildlife.org
St. Cloud, MN 56301
The National Wildlife Rehabilitators Association is a nonprofit international membership organization committed to promoting and improving the integrity and professionalism of wildlife rehabilitation and contributing to thepreservation of natural ecosystems.
Daniel R Ludwig PhD, Awards/Grants

1613 World Environment Center

1300 Pennsylvania Avenue   212-683-4700
Suite 550   Fax: 212-683-5053
Washington, DC 20004   http://www.wec.org
The World Environment Center contributes to sustainable development worldwide by strengthening industrial and urban environment, health, and safety policy and practices.
John Mizroch, President\CEO
Cecilia Ho, Vice President

1614 World Wildlife Fund

1250 24th Street NW   202-778-9555
Washington, DC 20037   Fax: 202-293-9211
     http://www.worldwildlife.org/
World Wildlife Fund is dedicated to protecting the world's wildlife and wildlands. The largest privately supported international conservation organization in the world, WWF has more than 1 million members in the US alone. Since itsinception in 1961, WWF has invested in over 13,100 projects in 157 countries.
Carter S Roberts, President

Environmental

1615 Air and Waste Management Association Annual Conference and Exhibition

1 Gateway Center, 3rd Floor	412-232-3444
420 Fort Duquesne Boulevard	800-270-3444
Pittsburgh, PA 15222-1435	Fax: 412-232-3450
	E-mail: info@awma.org
	http://www.awma.org

Environmental professionals from all sectors of the economy including colleges, universities, natural resource manufacturing and process industries, consultants, local state, provincial, regional and federal governments, construction,utilities industries.
Edith M Ardiente, President
Richard C Scherr, Secretary

1616 American Academy of Environmental Medicine Conference

7701 East Kellogg	316-684-5500
Suite 625	Fax: 316-684-5709
Wichita, KS 67207	E-mail: administrator@aaem.com
	http://www.aaem.com

Aims to support physicians and other professionals in serving the public through education about the interaction between humans and their environment, and to promote optimal health through prevention and safe, effective treatment ofthe causes, not the illness.

35+ booths with 200 attendees and 35+ exhibits
James W Willoughby II, President
James F Coy, President Elect

1617 American Industrial Hygiene Association Conference and Exposition

2700 Prosperity Avenue	703-849-8888
Suite 250	Fax: 703-207-3561
Fairfax, VA 22031	E-mail: infonet@aiha.org
	http://www.aiha.org

AIHA promotes, protects and enhances industrial hygienists and other occupational health, safety and environmental professionals in their efforts to improve the health and well-being of workers, the community and the environment.
Frank Renshaw, President Elect
Beverly S Cohen, Vice Chair

1618 American Solar Energy Society Conference

2400 Central Avenue	303-443-3130
Suite A	Fax: 303-442-3212
Boulder, CO 80301	E-mail: ases@ases.org
	http://www.ases.org

The American Solar Energy Society Conference (ASES) is the United States section of the International Solar Energy Society. ASES is a nonprofit organization dedicated to the development and adoption of renewal energy in all its forms.

750 attendees and 75-100 exhibits
Thomas Starrs, Chair
Ronal W Larson, Chair-Elect

1619 American Water Resources Association Conference

4 West Federal Street	540-687-8390
PO Box 1626	Fax: 540-687-8395
Middleburg, VA 20118-1626	E-mail: info@awra.org
	http://www.awra.org

Founded in 1964, the American Water Resources Association is a non-profit professional association dedicated to the advancement of men and women in water resources management, research and education.
Kenneth D Reid, Executive VP

1620 Arkansas Association of Conservation Districts Annual Conference

101 East Capitol	501-682-2915
Suite 350	Fax: 501-682-3991
Little Rock, AR 72201	http://www.aracd.org

Affiliated with the National Association of Conservation Districts. Membership is $25.00 per year for individuals and $1,200.00 per year for organizations and companies. November.
Debbie Moreland, Program Administrator
R D Sonny Jones, President

1621 Atlantic States Marine Fisheries Commission Annual Meeting

1444 Eye Street Northwest	202-289-6400
6th Floor	Fax: 202-289-6051
Washington, DC 20005-2210	E-mail: info@asmfc.org
	http://www.asmfc.org

The mission of the Atlantic States Marine Fisheries Commission Annual Meeting is to promote the better utilization of the fisheries, marine, shell, and anadramous, of the Atlantic seaboard by the development of a joint program for thepromotion and protection of such fisheries, and by the prevention of physical waste of the fisheries from any cause.
John V O'Shea, Executive Director
Preston P Pate, Jr., Chair

1622 Children's Environmental Health: Research, Practice, Prevention and Policy

110 Maryland Avenue Northeast	202-543-4033
Suite 505	Fax: 202-543-8797
Washington, DC 20002	E-mail: cehn@cehn.org
	http://www.cehn.org

Children's Environmental Health Network holds an annual conference to focus on children's environmental health research. The framework for the the conference, Children's Environmental Health: Research, Practice, Prevention, and Policy,emerged from the four priority health outcomes identifed by the Networks Research Committee.
J Routt Reigart, MD, Chair
Cynthia Bearer, MD, Vice Chair

1623 Coastal Society Conference

PO Box 25408	703-933-1599
Alexandria, VA 22313-5408	Fax: 703-933-1596
	E-mail: coastalsoc@aol.com
	http://www.thecoastalsociety.org

Conference dedicated to addressing coastal issues by fostering dialogue, forging partnerships, promoting communication, and education. Conference in May.
Judy Tucker, CAE, Executive Director

1624 Colorado Water Congress Annual Meeting

1580 Logan Street	303-837-0812
Suite 400	Fax: 303-837-1607
Denver, CO 80203	E-mail: macravey@cowatercongress.org
	http://www.cowatercongress.org

Protects and conserves Colorado's water resources by means of advocacy and education.

January
350 attendees and 9 exhibits
David Merritt, President
David Robbins, Vice President

1625 Connecticut Forest and Park Association Annual Meeting

16 Meriden Road	860-346-2372
Rockfall, CT 06481-2961	Fax: 860-347-7463
	E-mail: conn.forest.assoc@snet.net
	http://www.ctwoodlands.org

An organization for forest and wildlife conservation. Develops outdoor recreation and natural resources. Provides forest management, construction of hiking trails and consultation in the areas of forestry and environment.

Spring
Richard Whitehouse, President
Gordon Anderson, VP

1626 ESTECH

Institute of Environmental Sciences & Technology

5005 Newport Drive	847-255-1561
Suite 506	Fax: 847-255-1699
Rolling Meadows, IL 60008-3841	E-mail: iest@iest.org
	http://www.iest.org

An international professional society that serves the environmental sciences in the areas of contamination control in electronics manufacturing and pharmaceutical processes, design, test and evaluation of commercial and militaryequipment and product reliability issues.

April
50 booths with 500 attendees and 48 exhibits
Julie Kendrick, Executive Director
Carolyn Chapman, Publications Sales Coord.

1627 Earth Technologies Forum: Conference on Climate Change and Ozone Protection
2111 Wilson Boulevard   703-807-4052
8th Floor   Fax: 703-528-1734
Arlington, VA 22201   E-mail: earthforum@alcalde-fay.com
http://www.earthforum.com
The Earth Technologies Forum Annual Conference evolved from the International Conference on Ozone Protection Technologies and the International Climate Change Conference & Technologies Exhibition. The merger of these twointernationally-recognized conferences preserves the best of each, providing the most comprehensive educational program and exhibition featuring the latest ozone protection and climate change technologies and policies.
Tonya Hunt

1628 Environmental Technology Expo
Association of Energy Engineers
4025 Pleasantdale Road   770-447-5083
Suite 420   Fax: 770-446-3969
Atlanta, GA 30340-4264   E-mail: info@aeecenter.org
http://www.aeecenter.org
AEE is a source of information in the field of energy efficiency, utility deregulation, facility management, plant engineering, and environmental compliance. Outreach programs include technical seminars, conferences, books, joblistings and certification programs.
Stephen A Roosa, President
Timothy B Janos, President Elect

1629 Federation of Environmental Technologists
PO Box 624   262-644-0700
Slinger, WI 53086   Fax: 262-644-7106
E-mail: info@fetinc.org
http://www.fetinc.org
A nonprofit organization formed to assist industry in interpretation of and compliance with environmental regulations. Membership is open to all industries, municipalities, organizations and individuals concerned about environmentalregulations. Currently there are approximately 1000 members and 125 patron companies.

March
200 attendees and 70 exhibits
Triese Haase, Administrator

1630 Forestry Conservation Communications Association Annual Meeting
PO Box 3217   717-338-1505
Gettysburg, PA 17325   Fax: 717-334-5656
E-mail: nfc@fcca-usa.org
http://www.fcca-usa.org
The FCCA is a national organization. Its main function is to assist federal, state and local governments in public safety two-way radio operations by locating suitable frequencies within specified operating areas, recommending theirassignment to the FCC for licensing, and protecting them once licensed.
Ralph Haller, Executive Manager
Sgt. John McIntosh, President

1631 Global Warming International Conference and Expo
22W381, 75th Street   630-910-1551
Naperville, IL 60565-9245   Fax: 630-910-1561
http://www.globalwarming.net
The GWIC is the international body disseminating information on global warming science and policy, serving both governmental,and non-governmental organizations and industries in more than 145 countries. It sponsors unbiased researchsupporting the understanding of global warming and its mitigation.
Sinyan Shen

1632 GlobalCon
Association of Energy Engineers
4025 Pleasantdale Road   404-761-0509
Suite 420   Fax: 770-446-3969
Atlanta, GA 30340-4264   E-mail: info@aeecenter.org
http://www.globalconevent.com
Energy/environmental technological equipment.
Ruth Bennett, Information Services Dir.

1633 Greenprints: Sustainable Communities by Design
Southface Energy Institute
241 Pine Street Northeast   404-872-3549
Atlanta, GA 30308   Fax: 404-872-5009
E-mail: info@southface.org
http://www.southface.org
Southface promotes sustainable homes, workplaces and communities through education, research, advocacy and technical assistance.Greenprints is a conference and trade show produced by the Southface Energy Institute.

March 18-19, 2004
100 booths with 1200 attendees
Dennis Creech, Executive Director
Howard Katzman, Technical Project Manager

1634 HydroVision
HCI Publications
410 Archibald Street   816-753-4830
Kansas City, MO 64111-3001   Fax: 816-931-2015
E-mail: hci@aol.com.us
http://www.hcipub.com
Serves the hydroelectric industry.
Leslie Eden, Manager

1635 Institute of Scrap Recycling Industries Convention
1325 G Street Northwest   202-737-1770
Suite 1000   Fax: 202-626-0900
Washington, DC 20005-3104   http://www.isri.org
Equipment for the recycling industries.
Robin Weiner, President
Joel Denbo, Chair

1636 International Association for Energy Economics Conference
International Association for Energy Economics
28790 Chagrin Boulevard   216-464-5365
Cleveland, OH 44122-4630   Fax: 216-464-2737
E-mail: iaee@iaee.org
http://www.iaee.org
A nonprofit professional organization to provide an interdisciplinary forum for the exchange of ideas and experiences among energy professionals and others interested in shaping opinions and preparing for the events which effect energyindustries.
David L Williams, Executive Director

1637 International Conference on Solid Waste
Widener University
One University Place   610-499-4042
Chester, PA 19013-5792   Fax: 610-499-4059
E-mail: solid.waste@widener.edu
http://www.widener.edu/solid.waste
An annual conference on solid waste technology and management. Over 100 speakers from 25 countries present their work. Proceedings available.
Ronald L Mersky, Director

1638 Maryland Recyclers Coalition Annual Conference
2105 Laurel Bush Road   443-640-1050
Suite 200   Fax: 443-640-1031
Bel Air, MD 21015-6185 E-mail: info@marylandrecyclers.org
http://www.marylandrecyclers.org
MRC's mission is to promote sustainable reduction, reuse and recycling of materials otherwise destined for disposal and promote and increase buying products made with recycled material content.

June
Brian Ryerson, President
Virginia Lipscomb, Vice President

1639 Massachusetts Association of Conservation Commissions Conference

10 Juniper Road
Belmont, MA 02478
617-489-3930
Fax: 617-489-3935
E-mail: staff@maccweb.org
http://www.maccweb.org

We host the MACC Annual Environmental Conference, the largest such event in New England, with over 40 workshops and nearly 50 exhibitors.

March
1100 attendees and 50 exhibits
Patrick Garner, Board President
Kenneth Pruitt, Executive Director

1640 Massachusetts Water Pollution Control Association Annual Conference

PO Box 221
Groveland, MA 01834-0221
978-374-0170
Fax: 978-521-4083
E-mail: mwpca1965@verizon.net
http://www.mwpca.org

September
John Connor, Secretary/Treasurer
Michael Burke, President

1641 Michigan Association of Conservation Districts Annual Meeting

201 North Mitchell Street
Suite 203
Cadillac, MI 49601
231-876-0328
Fax: 231-876-0372
E-mail: macd@macd.org
http://www.macd.org

The Michigan Association of Conservation Districts is a non-governmental, non-profit organization, established to represent and provide services to Michigan's 80 Conservation Districts. The Association represents its members at the state level by working with legislators, cooperating agencies, and special interest groups whose programs affect the care and management of Michigan's natural resources, especially on private lands.

November
Marilyn Shy, Executive Director
Tom Middleton, President

1642 Michigan Forest Association Annual Meeting

1558 Barrington
Ann Arbor, MI 48103-5603
734-665-8279
Fax: 734-913-9167
E-mail: info@michiganforests.com
http://www.michiganforests.com

An organization composed mainly of private owners of small woodlands. Our purpose is to promote good forest management and stewardship of all forest lands.

Summer
McClain B Smith Jr, Executive Director

1643 Minnesota Association of Soil and Water Conservation Districts Annual Meeting

790 Cleveland Avenue S
Suite 201
St. Paul, MN 55116
651-690-9028
Fax: 651-690-9065
E-mail: lbuck@pioneerplanet.infi.net

MASWCD is a nonprofit organization which exists to provide a common voice for Minnesota's soil and water conservation districts and to maintain an positive, results-oriented relationship with rule-making agencies, partners and legislators; expanding education opportunities to the districts so they may carry out effective conservation programs.

December
Carol Berg, President
Loyal Fisher, Vice President

1644 Montana Association of Conservation Districts Annual Meeting

501 North Sanders
Helena, MT 59601
406-443-5711
Fax: 406-443-0174
E-mail: mail@macdnet.org
http://www.macdnet.org

Montana's 58 Conservation Districts utilize locally-led and largely non-regulatory approaches to successfully address general natural resource issues. CD's have a decades-long history of conserving our state's resources by helping local people match their needs with technical and financial resources, thereby getting good conservation practices on the ground to benefit all of Montanans.

November
Bob Fossum, President
Buzz Mattelin, Vice President

1645 Montana Water Environment Association Annual Meeting

516 N Park Street
Suite A
Helena, MT 59601
406-449-7913
Fax: 406-449-6350

Spring

1646 NEHA Annual Education Conference and Exhibition

National Environmental Health Association
720 South Colorado Boulevard
Suite 970, South Tower
Denver, CO 80246-1925
303-756-9090
Fax: 303-691-9490
E-mail: staff@neha.org
http://www.neha.org

Containing 100 booths and 100 exhibits.
Larry Marcum, Manager
Tabby Bernardo, Executive Coordinator

1647 National Association Civilian Conservation Corps Alumni

16 Hancock Avenue
Saint Louis, MO 63125
314-487-8666
http://www.cccalumni.org

The NACCCA was established as a non-profit organization in 1977 in California. The NACCCA offers annual national reunions, and a scholarship to a descendent of a NACCCA member.

10 booths
John Moscinski, Executive Director

1648 National Environmental Balancing Bureau Meeting

National Environmental Balancing Bureau
8575 Grovemont Circle
Gaithersburg, MD 20877-4121
301-977-3698
Fax: 301-977-9589

The NEBB is a nonprofit organization founded by contractors in the heating, ventilating and air conditioning (HVAC) industry. NEBB exists to help architects, engineers, building owners and contractors produce great buildings with HVAC systems that perform in ways they have visualized and designed.
Joseh Miller, President
John G Capell, Vice President

1649 National Environmental Health Association Annual Education Conference

National Environmental Health Association
720 South Colorado Boulevard
Suite 970, South Tower
Denver, CO 80246-1925
303-756-9090
Fax: 303-691-9490
E-mail: staff@neha.org
http://www.neha.org

The NEHA AEC and Exhibition is a six-day educational event consisting of nine different environmental health and protection conferences and highlighting a two-day exhibition. It is the only conference that emcompasses all areas of environmental health and protection, including, but not limited to: food protection, onsite wastewater, chemical and bioterrorism preparedness, indoor air quality, hazardous waste, and drinking water.

Late June-Early July
120 booths with 1300 attendees and 120 exhibits
Larry Marcum, Manager
Tabby Bernardo, Executive Coordinator

1650 National Environmental, Safety and Health Trai ning Association Show

2720 E Thomas Road, Suite 253C
PO Box 10321
Phoenix, AZ 85016-4202
602-956-6099
Fax: 602-956-6399
http://www.neshta.org

The National Environmental, Safety and Health Training Association is a non-profit international society for environmental, safety, health and other technical training and adult education professionals. NESHTA promotes trainercompetency through training and education standards, voluntary certification, and peer networking.
CL Richardson, Show Manager

1651 National Real Estate Environmental Conference
National Society of Environmental Consultants
PO Box 12528   210-225-2897
San Antonio, TX 78212-0528   800-486-3676
                                       Fax: 956-225-8450
Environmentally responsible management of real estate.

1652 National Recycling Congress Show
1325 G Street Northwest   202-347-0450
Suite 1025   Fax: 202-347-0449
Washington, DC 10005   E-mail: info@nrc-recycle.org
                              http://www.nrc-recycle.org
Founded in 1978, the National Recycling Coalition, Inc. provides technical education, disseminates public information on selected recycling issues, shapes public and private policy on recycling and operates programs that encouragerecycling markets and economic development.
Ben Walker, President
Ronald Kolbash, Chair

1653 National Water Resources Association Annual Conference
3800 North Fairfax Drive   703-524-1544
Suite 4   Fax: 703-524-1548
Arlington, VA 22203   E-mail: nwra@nwra.org
                                    http://www.nwra.org
Conservation of water resources in the 17 western reclamation states.
Norm M Semanko, President

1654 Nebraska Association of Resources Districts Annual Meeting
601 South 12th Street   402-471-7670
Suite 201   Fax: 402-471-7677
Lincoln, NE 68508   E-mail: nard@nrdnet.org
                                    http://www.nrdnet.org
Our mission is to assist NRDs in a coordinated effort to accomplish collectively what may not be accomplished individually to conserve, sustain, and improve our natural resources and environment.

September
Dean E Edson, Executive Director
Jeanne Dryburgh, Office Manager

1655 New England Enviro Expo
Industrial Shows Northeast
330 North Wabash   312-628-5870
Suite 3201   Fax: 312-628-5878
Chicago, IL 60611   E-mail: kstromberg@zweigwhite.com
                                 http://www.enviroexpo.com
Environmental products/services for industrial, municipal, and government uses.

May
400 booths with 5000 attendees
Kristen Stromberg, Conference Manager

1656 New England Water Environment Association Annual Meeting
NEWEA
100 Tower Office Park   781-939-0908
Suite K   Fax: 781-939-0907
Woburn, MA 01801   E-mail: mail@newea.org
                                    http://www.newea.org
We are a regional member association of the Water Environmental Federation. We provide technical and education for the waste water industry.

150+ booths with 1500 attendees and 90 exhibits
Elizabeth Cutone, Executive Director
Janice Moran, Program Coordinator

1657 New Hampshire Association of Conservation Commissions Annual Meeting
54 Portsmouth Street   603-224-7867
Concord, NH 03301-5486   Fax: 603-228-0423
                                   E-mail: info@nhacc.org
                                    http://www.nhacc.org
The New Hampshire Association of Conservation Commissions is authorized to research its water and land areas, keep an index of all open space or natural areas, and recommend programs for the protection or betterment of all areas in theindex.

November
Marjory M Swope, Executive Director
Mason Westfall, President

1658 New Jersey Society for Environmental Economic Development Annual Conference
30 W Lafayette Street   609-392-8899
Trenton, NJ 08608-0000   Fax: 609-396-0891

October
James C Morford, President

1659 New Jersey Water Environment Association Conference
PO Box 1212   201-670-5576
Fair Lawn, NJ 07410   Fax: 201-251-4573
                              E-mail: JLagrosa@NJWEA.org
                                    http://www.njwea.org
The New Jersey Water Environment Association is a nonprofit educational organization dedicated to preserving and enhancing the water environment.
Henry Penley, President
Blake Maloney, Vice President

1660 New Mexico Association of Soil and Water Conservation Annual Conference
New Mexico Association Of Conservation Districts
163 Trail Canyon Road   505-981-2400
Carlsbad, NM 88220   Fax: 505-981-2422
                           E-mail: conserve@dellcity.com
                               http://http://nm.nacdnet.org
The mission of NMACD is to facilitate conservation of the natural resources in New Mexico by providing opportunities and quality support to local conservation districts through representation and leadership.

Fall
Debbie Hughes, Executive Director

1661 New York Water Environment Association Semi- Annual Conferences
126 E Salina Street   315-422-7811
Suite 200   Fax: 315-422-3851
Syracuse, NY 13202   http://www.nywea.org
The New York Water Environment Association is a nonprofit educational association dedicated to the development and dissemination of information concerning water quality management and the nature, collection, treatment, and disposal ofwastewater. Founded in 1929, the Association has over 2,500 members. The NYWEA is a member association of the Water Environment Federation.

Winter and Summer
Robert Hennigan, Executive Director

1662 North American Lake Management Society International Symposium
4513 Vernon Boulevard, Suite 103   608-233-2836
PO Box 5443   Fax: 608-233-3186
Madison, WI 53705-0443   E-mail: nalms@nalms.org
                                    http://www.nalms.org
The North American Lake Management Society's mission is to forge partnerships among citizens, scientists and professionals to foster the management and protection of lakes and reservoirs for today and tomorrow.

Oct and Nov
100 booths with 850 attendees and 50 exhibits
Gene Medley, President

1663 North Carolina Association of Soil and Water Conservation Districts Annual Conference

1614 Mail Service Center   919-733-2302
Raleigh, NC 27699   Fax: 919-715-3559
       http://www.ncaswcd.org

The association is an indepented, nonpartisan conservation organization created in 1944 to represent the interests of the 96 local soil and water conservation districts and the 492 district supervisors who direct their local district'sconservation programs.

January
Tom Davidson, President
Don Rawls, First Vice President

1664 North Dakota Association of Soil Conservation Districts Annual Conference

PO Box 1601   701-223-8518
Bismarck, ND 58502   Fax: 701-223-1291
       http://www.lincolnoakes.com

November
Gary L Puppe, Executive VP

1665 Northeast Recycling Council Conference

139 Main Street   802-254-3636
Suite 401   Fax: 802-254-5870
Brattleboro, VT 05301 E-mail: info@nerc.org
       http://www.nerc.org

NERC's mission is to leverage the strengths and resources of its member states to advance an environmentally stable economy in the Northeast by promoting source reduction, recycling, and the purchasing of environmentally preferableproducts and services.

March and October
Lynn Rubenstein, Executive Director
Mary Ann Remolador, Assistant Director

1666 Northeast Resource Recovery Association Annual Conference

9 Bailey Road   603-798-5777
Chichester, NH 03258   Fax: 603-798-5744
      E-mail: nrra@conknet.com
      http://www.recyclewithus.org

The Northeast Resource Recovery Association is a pro-active nonprofit working with its membership to make their recycling programs strong, efficient, and financially sucessful by providing cooperative marketing, cooperative purcashing,education and networking opportunities; developing innovative recycling programs; creating sustainable alternatives to reduce the volume and toxicity of the waste, and educating and informing local officials about recycling and solid waste issues.

June
Elizabeth Bedard, Executive Director

1667 Northeast Sustainable Energy Association Conferences

50 Miles Street   413-774-6051
Greenfield, MA 01301   Fax: 413-774-6053
      E-mail: nesea@nesea.org
      http://www.nesea.org

NESEA is the nation's leading regional membership organization focused on promoting the understanding, development, and adoption of energy conservation and non-polluting renewable energy technologies. We work to bring cleanelectricity, green transportation, and healthy, efficient buildings into everyday use.

March and May
Nancy Hazard, Executive Director
Janet Nokes, Business Manager

1668 Pacific Fishery Management Council Conferences

7700 Northeast Ambassador Place 503-820-2286
Suite 200   866-806-7204
Portland, OR 97220-1384 Fax: 503-820-2299
    E-mail: Donald.McIsaac@noaa.gov
      http://www.pcouncil.org

The Pacific Council has developed fishery management plans for salmon, groundfish and coastal species in the US Exclusive Economic Zone off the coast of Washington, Oregon and California, and recommends Pacific halibut harvestregulations to the International Pacific Halibut Commission.

5x year
Donald McIsaac, Executive Director
John Coon, Deputy Director

1669 Pacific States Marine Fisheries Commission Annual Conference

205 Southeast Spokane Street 503-595-3100
Portland, OR 97202   Fax: 503-595-3232
   E-mail: front_office@psmfc.org
      http://www.psmfc.org

The goal of Pacific States Marine Fisheries Commission is to promote and support policies and actions directed at the conservation, development and management of fishery resources through mutual concern to member states through acoordinated regional approach to research, monitoring and utilization.

Fall
Randy Fisher, Executive Director
Stephen Phillips, Program Manager

1670 Parks and Trails Council of Minnesota Annual Meeting

275 East 4th Street   651-726-2457
#642   800-944-0707
Saint Paul, MN 55101-1651 E-mail: info@parksandtrails.org
      http://www.parksandtrails.org

The Parks and Trails Council of Minnesota exists to acquire, protect and enhance critical lands for the public's enjoyment now and in the future.

March
Dorian Grilley, Executive Director
Judith Erickson, Government Relations

1671 Plant and Facilities Expo

Association of Energy Engineers
4025 Pleasantdale Road   404-761-0509
Suite 420   Fax: 770-446-3969
Atlanta, GA 30340-4264 E-mail: info@aeecenter.org
      http://www.aeecenter.org

Occupational health/safety systems.
Ruth Bennett, Information Services Dir.

1672 Professional Conference on Industrial Hygiene

American Board of Industrial Hygiene
6015 West St Joseph   517-321-2638
Suite 102   Fax: 517-321-4624
Lansing, MI 48917-3980 E-mail: abih@abih.org
      http://www.abih.org

Industrial hygiene equipment.
Lynn C O'Donnell, CIH, Executive Director
Barbara A Saalfeld, Certification Maintenance

1673 Solar Cookers International World Conference

1919 21st Street   916-444-6616
Suite 101   Fax: 916-444-5379
Sacramento, CA 95814-6827 E-mail: sbci@igc.apc.org

Equipment for solar cooking and pasteurization of drinking water.

1674 South Dakota Association of Conservation Districts Conference

PO Box 515   605-895-4099
Presho, SD 57568   Fax: 605-895-9424
     E-mail: conserve@wcenet.com
      http://http://sd.nacdnet.com

The association's mission is to assist, lead and coordinate conservation districts in their efforts to promote sensible, voluntary, self-governed conservation management and development of South Dakota's natural resources for ourselvesand our posterity.

September
Angela Ehlers, Executive Secretary

1675 South Dakota Environmental Health Association Annual Conference

State Department of Health
600 East Capitol Avenue   605-773-3361
Pierre, SD 57501-2536   800-738-2301

E-mail: DOH.INFO@state.sd.us
http://www.state.sd.us/doh

April
Doneen Hollingsworth, Secretary of Health

1676 Southeastern Association of Fish and Wildlife Agencies Annual Meeting
8005 Freshwater Farms Road — 850-893-1204
Tallahassee, FL 32308 — Fax: 850-893-1024
The SEAFWA conducts an annual conference each fall to provide a forum for presentation of information and exchange of ideas regarding the management and protection of fish and wildlife resources throughout the nation but with emphasis on the southeast.

October/November
John Frampton, President
Robert Cook, Vice President

1677 Take it Back
Raymond Communications
5111 Berwin Road — 301-345-4237
Suite 115 — Fax: 301-345-4768
College Park, MD 20740 — E-mail: michele@raymond.com
http://www.raymond.com
Top recycling experts and practical sessions.
250 pages
Michele Raymond, Publisher/Editor

1678 Texas Environmental Health Association Annual Conference
PO Box 10 — 903-572-7278
Leesburg, TX 75451 — http://www.myteha.org
The mission of TEHA is to work for the betterment of the health and welfare of people through the improvement of the environment.

March
Richard Briley, President
Maggie Earl, Vice President

1679 Texas Solar Energy Society Annual Conference
PO Box 1447 — 512-326-1785
Austin, TX 78767-1447 — 800-465-5049
Fax: 512-326-1285
E-mail: info@txses.org
http://www.txses.org
The Texas Solar Energy Society is dedicated to educating the public about the use of solar and other renewable energy technologies. Our membership includes educators, researchers, students, bankers, electrical contractors, architects, builders, building inspectors, home owners and solar enthusiasts.

September
Lorin Vant-Hull, Chair
Jaya Pichumani Jackson, Vice Chair

1680 Texas Water Conservation Association Annual Conference
221 E 9th Street — 512-472-7216
Suite 206 — Fax: 512-472-0537
Austin, TX 78701 — http://www.twca.org

March
Dr Peggy Glass, President
Greg Rothe, President Elect

1681 Utah Association of Conservation Districts Annual Conference
1860 N Street — 435-753-6029
Suite 100 — Fax: 435-755-2117
Logan, UT 84341-1784 — http://http://uacd.org
The Utah Association of Conservation Districts is a nonprofit corporation representing Utah's 38 soil conservation districts. By working with landowners, organizations and government, the conservation districts work through voluntary, incentive-based programs to protect soil, water quality and other natural resources.

November
Larry Johnson, President
Bill Rasmussen, Vice President

1682 Virginia Association of Soil and Water Conservation Districts Annual Conference
7308 Hanover Green Drive — 804-559-0324
Suite 100 — Fax: 804-559-0325
Mechanicsville, VA 23111 — E-mail: info@vaswcd.org
http://www.vaswcd.org
The Virginia Association of Soil and Water Conservation Districts (VASWCD) is a private nonprofit association of 47 soil and water conservation districts in Virginia. It is a voluntary, nongovernmental association of Virginia's districts that provides and promotes leadership in the conservation of natural resources through stewardship and education programs.

December
Stephen Carlos, Executive Director
James Byrne, President

1683 Virginia Forestry Association Annual Conference
3308 Augusta Avenue — 804-278-8733
Richmond, VA 23230-3910 — E-mail: vafa@verizon.net
http://www.vaforestry.org
VFA promotes stewardship and wise use of the Commonwealth's forest resources for the economic and environmental benefits of all Virginians. Membership consists of forest landowners, forest product businesses, forestry professionals, and a variety of individuals and groups who are concerned about the future and well-being of Virginia's forest resources.

Late Spring
John Burke, Chair
Frank Sherwood, Vice Chair

1684 WEFTEC Show
Water Environment Federation
601 Wythe Street — 703-684-2400
Alexandria, VA 22314-1994 — 800-666-0206
Fax: 703-684-2471
E-mail: confinfo@wef.org
http://www.wef.org
North America's largest annual water quality conference and exposition. Covers a wide spectrum of critical water quality issues.

Fall
16,000 attendees and 800+ exhibits
Nina Maxberry

1685 Waste Expo
Environmental Industries Asociation
4301 Connecticut Avenue NW — 202-244-4700
Suite 300 — 800-424-2869
Washington, DC 20008 — Fax: 202-966-4841
Waste/recycling equipment and technology.
Bruce Parkert, President

1686 WasteExpo
Primedia Business Exhibitions
11 Riverbend Drive S — 203-358-4134
Stamford, CT 06907 — 800-927-5007
Fax: 203-358-5815
E-mail: lmagliola@primediabusiness.com
http://www.wasteexpo.com
WasteExpo is the largest tradeshow in North America serving the $43 billion solid waste and recycling industries.

11,500 attendees and 450 exhibits
Catherine Picketta, Show Coordinator
Rita Ugianskis-Fishman, Show Director

1687 West Virginia Forestry Association Annual Conference
PO Box 718 — 304-372-1955
Ripley, WV 25271 — 888-372-9663
Fax: 304-372-1957
E-mail: wvfa@wvadventures.net
http://www.wvfa.org

Conferences & Trade Shows/Environmental

The West Virginia Forestry Association is a non-profit organization funded by its membership. Our members include individuals and businesses involved in forest management, timber production and wood product manufacturing. Our members are concerned with protecting the environment, as well as enhancing the future of West Virginia's forests through multiple-use management.

Summer
Richard Waybright, Executive Director

1688 Western Association of Fish and Wildlife Agencies Annual Meeting
Alberta Sustained Resource Development
9920-108 Street
Great West Life Building, 2nd Floor
Edmonton, AB

July
Ken Ambrock, President

1689 Western Forestry and Conservation Association Conference
4033 Canyon Road SW   503-226-4562
Portland, OR 97221   Fax: 503-228-3624
http://www.westernforestry.org
Offers continuing education workshops and seminars for professional foresters throughout the west.

January/February

1690 Western Society of Naturalists Annual Meeting
San Diego State University Department Of Biology
5500 Campanile Drive   818-677-3256
San Diego, CA 92182   Fax: 818-677-2034
http://www.wsn-online.org
Members include researchers, educators, academics and others with an interest in the area's biology, particularly its marine life. Membership is $15.00 per year for individuals and $7.00 for students.
Mark Carr, President
Ralph Larson, President Elect

1691 Wildlife Habitat Council Annual Symposium
Wildlife Habitat Council
8737 Colesville Road   301-588-8994
Suite 800   Fax: 301-588-4629
Silver Spring, MD 20910-5600   E-mail: whc@wildifehc.org
http://www.wildlifehc.org
Develops solutions for developing lands for wildlife.
David W Carroll, Chairman
Lawrence A Selzer, Vice Chairman

1692 Wildlife Society Annual Conference
Wildlife Society
5410 Grosvenor Lane   301-897-9770
Suite 200   Fax: 301-530-2471
Bethesda, MD 20814   E-mail: tws@wildlife.org
http://www.wildlife.org
Annual conference of wildlife professionals, organized by the Wildlife Society.

50 booths with 1400 attendees and 15 exhibits
Richard A Lancia, President
John F Organ, Vice President

1693 Wisconsin Association for Environmental Education Annual Conference
8 Nelson Hall   715-346-2796
University of Wisconsin http://www.uwsp.edu/waee
Stevens Point, WI 54481
WAEE is a statewide non-profit organization composed of people interested in learning about and helping others learn about environmental issues. Our goal is to promote responsible environmental action through education in the classroom and in the community.

Fall
Carrie Hembree, Chair
Jeremy Higgins, Chair Elect

1694 Wisconsin Land and Water Conservation Association Annual Conference
One Point Place   608-833-1833
Suite 101   Fax: 608-833-7179
Madison, WI 53719   E-mail: wlwca@exectc.com
http://www.wlwca.org
The Wisconsin Land and Water Conservation Association (WLWCA) is a nonprofit organization representing Wisconsin's 72 County Board Land Conservation committees and departments. The mission of the WLWCA is "to assist Land Conservation committees and departments with the protection, enhancement and sustainable use of Wisconsin's natural resources and represent Land COnservation Committees and Departments through education and governmental interaction.

December
Hugh Mulliken, President
Marvin Fox, Vice President

1695 Wisconsin Woodland Owners Association Annual Conference
PO Box 285   715-346-4798
Stevens Point, WI 54481-0285   Fax: 715-346-4821
E-mail: bardenalb@nnex.net
http://www.wisconsinwoodlands.org
The Wisconsin Woodland Owners Association, a nonprofit educational organization, was established in 1979 to advance the interests of woodland owners and the cause of forestry; develop public appreciation for the value of Wisconsin's woodlands and their importance in the economy and overall welfare of the state; foster and encourage wise use and management of Wisconsin's woodlands for timber production, wildlife habitat and recreation; and to educate those interested in managing the woodlands.

October
Alvin Barden, President
Dale Zaug, President Elect

1696 World Energy Engineering Congress
Association of Energy Engineers
4025 Pleasantdale Road   770-447-5083
Suite 420   Fax: 770-446-3969
Atlanta, GA 30340-4264   E-mail: info@aeecenter.org
http://www.aeecenter.org
Equipment and services.
Ruth Bennett, Information Services Dir.

Environmental

1697 3D/International
1900 West Loop South
Suite 400
Houston, TX 77027

713-871-7000
Fax: 713-871-7171
E-mail: contact@3di.com
http://www.3di.com

Enrvironmental compliance and consulting services

John Murph, PE, President/CEO
Gary Boyd, AIA, Executive VP

1698 AAA Lead Consultants and Inspections
1307 West 6th Street
Suite 134
Corona, CA 92882

951-582-9071

Offers quality consulting, inspections, monitoring and project design for lead based paint.

1699 AB2MT Consultants
9400 South Dadeland Boulevard
Suite 370
Miami, FL 33156

305-670-1011
Fax: 305-670-1016
E-mail: ab2mt@aol.com

Environmental and engineering consulting.
Paula H Church, President

1700 ABB Environmental Services
511 Congress Street
PO Box 7050
Portland, ME 04112

207-775-5401
Fax: 207-772-4762
http://www.abb.com

1701 ACC Environmental Consultants
7977 Capwell Drive
Suite 100
Oakland, CA 94621

510-638-8400
800-525-8838
Fax: 510-638-8404
E-mail: info@accenv.com
http://www.accenv.com

An employee owned environmental and energy consulting firm. Helps companies and public agencies throughout California identify and manage environmental hazards, comply with their OSHA and EPA requirements.
Kenneth R Churchill, CEO
James Wilson, President

1702 ACRT Environmental Specialists
1333 Home Avenue
Akron, OH 44310

800-622-2562
E-mail: askacrt@acrtinc.com
http://www.acrt.com

Appraisal, Research and Training is an international consulting service and training organization in the utility and urban forestry, arboricultural, environmental, natural resource and horticultural services.

1703 ADS Corporation
5030 Bradford Boulevard
Suite 210, Building 1
Huntsville, AL 35805

800-633-7246
Fax: 256-430-6633
E-mail: info@adsenv.com
http://www.adsenv.com

ADS Evnironmental Services offers water and wastewater system diagnostic products and services, combining innovative technology, professional field services and expertise to obtain accurate information about and undergroundinfrastructure system's condition, inventory, and performance.
Karl Boone, President/CEO
Joseph Goustin, Cheif Financial Officer

1704 AECOS
45-939 Kamehameha Highway
Room 104
Kaneohe, HI 96744

808-234-7770
Fax: 808-234-7775
E-mail: aecos@aecos.com
http://www.aecos.com

Environmental counseling firm providing the services of scientists and facilities in the environmental sciences to clients throughout the Pacific area. Specializes in aquatic (both fresh water and marine) biology and water quality,with practiced expertise in analytical chemistry, oceanography, water pollution, and marine and fresh water ecology.
Eric Guinther, President

1705 AKT Peerless Environmental Services
22725 Orchard Lake Road
Farmington, MI 48336

248-615-1333
Fax: 248-615-1334
http://www.akt.com

Providing environmental services to facilitate real estate transfer, development, and redevelopment. Services include phase I ESA, subsurface investigation, remediation, Brownfield's redevelopment, Brownfield's financial incentives.

1706 APEC-AM Environmental Consultants
2525 Northwest Expressway
Suite 301D
Oklahoma City, OK 73112

405-840-9327
Fax: 405-840-9328

Environmental assessment
Charlie Bowlin, Principal
Saleem Nizami, Principal

1707 ATC Associates
600 W. Cummings Park
Suite 5500
Woburn, MA 01801-6350

781-932-9400
Fax: 781-952-6211
http://www.atc-enviro.com

Environmental consulting firm with 1,600 experts in 65 offices throughout the United States, including engineers, scientists, technicians, and regulatory specialists.
Pam O'Deen, Business Development

1708 ATC Associates: Omaha
3712 S 132nd Street
Omaha, NE 68144

402-697-9747
Fax: 402-697-9170
http://www.atcassociates.com

Environmental and worker exposure consulting firm for EPA and OSHA compliance.

1709 ATS-Chester Engineers
260 Airside Drive
Moon Township, PA 15108

412-809-6600
Fax: 412-809-6611
http://www.atsengineers.com

Provides services in waste water treatment and air pollution control.
Robert Agbede, President

1710 Aarcher
910 Commerce Road
Annapolis, MD 21401

410-897-9100
Fax: 410-897-9104
E-mail: cschwartz@aarcherinc.com
http://www.aarcherinc.com

Aarcher is a small business providing environmental management, assessment and planning services nationwide from its headquarters and regional offices. Aarcher provides environmental compliance audits; NEPA analysis and documentation;natural and cultural resource management planning; site assessment and investigation; plans and permits; and environmental liability assessment and control. Our consulting services are guided by a comprehensive understanding of current environmentalregulations.
Craig J Schwartz, President

1711 Abacus Environmental
123 Pinney Street
PO Box 365
Ellington, CT 06029

860-871-6216
800-343-9970
Fax: 860-872-8044
E-mail: dweeks@abacusenvironmental.com
http://www.abacusenvironmental.com

Full service environmental and occupational safety and health consulting firm. Services include project management, indoor air quality and industrial hygiene, asbestos and lead project management, expert witness testimony and allaspects of workplace safety. Goals are to help reduce client's operating costs, minimize liability for environmental and occupational safety reguations, guidelines, and requirements.
Donald M Weekes Jr, President

1712 Abco Engineering Corporation

1601 South Yosemite Street   303-220-8220
Suite 205   Fax: 303-796-0810
Englewood, CO 80112-1413   E-mail: abco@abco-corp.com
    http://www.abco-corp.com
Provides a full spectrum of engineering and environmental services pertaining to both new construction and existing buildings, including Property Condition Assessment Reports, Phase I Environmental Assessments, Quality Control Reportsand other technical support services related to buildings and building systems including feasibility reports, construction observation and cost eliminating.
Joe Johnson, Director
Michael R Dannecker, Director

1713 Abonmarche Environmental

95 West Main Street   269-927-2295
PO Box 1088   Fax: 269-927-1017
Benton Harbor, MI 49023   E-mail: aci@abonmarche.com
    http://www.abonmarche.com
Full-service architectural, engineering, land surveying and planning firm.

1714 AccuTech Environmental Services

3 Cass Street   732-739-6444
Keyport, NJ 07735-1425   800-644-ISRA
    Fax: 732-739-0451
   E-mail: info@accutechenvironmental.com
   http://www.accutechenvironmental.com
Aims to meet the environmental consulting needs generated by New Jersey's Environmental Cleanup Responsibility Act by preparing and managing complete environmental sampling and cleanup programs.
Harry Moscatello, President

1715 Accutest Laboratories

2235 Route 130   732-329-0200
Building B   Fax: 732-329-3499
Dayton, NJ 08810   E-mail: info@accutest.com
    http://www.accutest.com
Privately held, independent testing laboratory delivering legally defensible data, providing a full range of environmental analytical services to industrial, engineering/consulting and government clients throughout the United States.Operating from coordinated laboratories in New Jersey, Massachusetts, Florida and Texas, resources include a staff of over 200, five million dollars worth of laboratory instrumentation and equipment, and more than 80,000 square feet of laboratoryspace.

1716 Acheron Engineering Services

147 Main Street   207-368-5700
Newport, ME 04953-1118   Fax: 207-368-5120
   E-mail: WBall@AcheronEngineering.com
    http://www.acheronengineering.com
Provides solutions to the most challenging engineering, environmental and geologic issues.
William B Ball, President
Kirk Ball, Engineering Field Technician

1717 Acumen Industrial Hygiene

PO Box 423570   415-642-6050
San Francisco, CA 94142   Fax: 415-642-6051
    E-mail: info@acumen-ih.com
    http://www.acumen-ih.com
Industrial hygiene consultation.
Michael Connor, Principal

1718 Advanced Chemistry Labs

PO Box 88610   770-409-1444
Atlanta, GA 30356   800-277-0520
    Fax: 770-409-1844
    E-mail: acl@mindspring.com
   http://www.advancedchemistrylabs.com
Laboratory analysis firm.

1719 Advanced Resources International

4501 Fairfax Drive   703-528-8420
Suite 910   Fax: 703-528-0439
Arlington, VA 22203-1661   E-mail: ari-info@adv-res.com
    http://www.adv-res.com
Independent consulting firm focused on providing technical services to the international energy industry.
Vello A Kuuskraa, President
Jonathan Kelafant, Senior Vice President

1720 Advanced Waste Management Systems

6430 Hixson Pike   423-843-2206
Hixson, TN 37343   Fax: 423-843-2310
    E-mail: info@awm.net
    http://www.awm.net/
Provides a wide range of environmental and engineering services to domestic and international clients, including governments, corporations, and provate citizens.
Richard Ellis PhD, CEO
James Mullican, PE, President

1721 Aerosol Monitoring and Analysis

4475 Forbes Boulevard   301-459-2640
Lanham, MD 20706   Fax: 301-459-2643
    E-mail: amalab@aol.com
Aerosol Monitoring & Analysis, provides Industrial Hygiene, Environmental and Health & Safety to government agencies, institutions, building owners, property managers, architects and engineers.

1722 Aguirre Engineers

13276 E Fremont Place   303-799-8378
PO Box 3814   Fax: 303-799-8392
Englewood, CO 80112-3909   E-mail: infoteam@aquirre1.com
Aguirre Engineers is an environmental engineering firm based in Englewwod, CO. Over the past 20 years we have augmented our service offering, from a commercial geotechnical leader, into providing full-scale government contractingservices for radioactive and hazardous waste remediation. The continued growth or our company throughout the years is a solid indicator of high-quality performance and client satisfaction.

1723 Air Consulting and Engineering

2106 NW 67th Place   352-335-1889
Suite 4   Fax: 352-335-1891
Gainesville, FL 32653   http://www.airconsulting.com
Air Consulting and Engineers, provides air pollution testing services utilizing United States Environmental Protection Agency. The company was founded in April 1984 to provide prefessional source emission testing and engineering, andair permitting to industries located in Florida and throughout the world.
Charles Simon, Sr. Scientist

1724 Air Sciences

1301 Washington Avenue   303-988-2960
Suite 200   Fax: 303-988-2968
Golden, CO 80401   E-mail: air@airsci.com
    http://www.airsci.com
Air Sciences was founded in the Denver-metro area in 1980 with the purpose of providing superior air pollution consulting services. Air Sciences attained this goal and presently enjoys a unique reputation as a firm that provides bothindustry and government a high quality service in air quality consulting. Our future is focused on emerging disciplines in the air quality arena driven by new air quality standards and regional haze regulations. Air Sciences is an employee ownedfirm.

1725 Aires Consulting Group

1550 Hubbard Avenue   630-879-3006
Batavia, IL 60510-4400   800-247-3799
    Fax: 630-879-3014
    E-mail: info@airesconsulting.com
    http://www.airesconsulting.com
National full-service industrial hygiene, environmental and occupational health consulting firm. Assists clients in the control of liability through the application of risk management principals.
Rich Rapacki, Contact

1726 Airtek Environmental Corporation

39 West 38th Street   212-768-0516
12th Floor   Fax: 212-768-0759
New York, NY 10018-5502   E-mail: info@airtekenv.com
    http://www.airtekenv.com
Environmental investigation and mangement professionals specializing in multi-jurisdictional regulatory climates.

Mike S Zouak, President

1727 Alan Plummer and Associates

7524 Mosier View Court   817-806-1700
Suite 200   Fax: 817-589-0072
Fort Worth, TX 76118   E-mail: aplummer@apaienv.com
     http://www.apaienv.com

Provides civil and environmental engineering services primarly in the areas of water, wastewater, and water resources. Clients include municipalities, local, regional, state and federal agencies, water districts, water authorities, andprivate-sector entities.
Richard Smith, Director
Alan Plummer, PE, Office Contact

1728 Allee, King, Rosen and Fleming

440 Park Avenue South   212-696-0670
7th Floor   Fax: 212-779-9721
New York, NY 10016-8022   E-mail: nycinfo@akrf.com
     http://www.akrf.com

Environmental consulting firm.
Debra C Allee, AICP, Principal
Edward A Applebome, ITE, Principal

1729 Allied Engineers

2420 Camino Ramon   925-867-4646
Suite 220   Fax: 925-867-0736
San Ramon, CA 94583   http://www.alliedengineersinc.com

Consulting services in wastewater and industrial waste, including emissions testing.
Robert Dawyat, President

1730 Allstate Power Vac

928 East Hazelwood Avenue   732-815-0220
Rahway, NJ 07065   Fax: 732-815-9892
     http://www.aspvac.com

Industrial and environmental waste management.

1731 Allwest Environmental

530 Howard Street   415-391-2510
Suite 300   Fax: 415-391-2008
San Francisco, CA 94105   E-mail: info@allwest1.com
     http://www.allwest1.com

Practical, business-oriented consulting firm specializing in Environmental and Engineering Due Diligence offering expertise to the real estate industry. Helps clients to understand and manage potential environmental and buildingliabilites, and to advocate their interests through the discovery and mitigation process.
Marc Cunningham, President
Chris Marinescu, Vice President

1732 Alpha-Omega Environmental Services

933 Northwest 31 Avenue   954-969-5906
Pompano Beach, FL 33069   866-969-6653
     Fax: 954-969-5232
     E-mail: dave@aomegagroup.com
     http://www.aomegagroup.com

Environmental engineering consulting

1733 Alternative Resources

1732 Main Street   978-371-2054
Concord, MA 01742-2842   Fax: 978-371-7269
     E-mail: info@alt-res.com
     http://www.alt-res.com

Alternative Resources is an independent consulting firm providing management, engineering, environmental, economic and financial advisory in the fields of water and wastewater treatment, solid waste management, residuals management,environmental compliance, and energy production.
Gretchen Karlson, Personnel

1734 Ambient Engineering

100 Main Street
Suite 330   888-262-6232
Concord, MA 01742   Fax: 978-369-8380
     E-mail: info@ambient-engineering.com
     http://ambient-engineering.com

An environmental engineering and consulting firm incorporated in 1994. We provide environmental services and solutions to our clients throughout the Northeastern US.
T J Stevenson, President
Steve Boynton, PE, Director of Engineering

1735 American Archaeology Group

PO Box 1017   512-556-4100
Lampasas, TX 76550-1017   Fax: 512-556-3373
     E-mail: info@american-archaeology.com
     http://www.american-archaeology.com

Archaeological and historical consulting specializing in energy, development, and municipal related archaeological permitting throughout the United States.

1736 American Engineering Testing

550 Cleveland Avenue North   651-659-9001
St Paul, MN 55114-1804   800-972-6364
     Fax: 651-659-1379
     E-mail: aet@amengtest.com
     http://www.amengtest.com

America's people, technology, innovation and quality commited to fulfilling client tequirements.
Terrence E Swor, President

1737 American Services Associates

18154 41st Place SE   425-641-5130
Issaquah, WA 98027   Fax: 425-641-5138
     E-mail: airsampler@aol.com

Consultants in emission testing, permitting, emission control system design, training and continuous emission moniters (CEMs). Producer of video training programs on E{A emission sampling methods in CD and VHS formats. Offices are inIssaquah, Washington.
Wes Snowden, President
John Vareski, Vice President

1738 Ana-Lab Corporation

2600 Dudley Road   903-984-0551
PO Box 9000   Fax: 903-984-5914
Kilgore, TX 75663   E-mail: corp@ana-lab.com
     http://www.ana-lab.com

Ana-Lab Corporation, a Texas corporation, is privately owned small business established in 1965 specializing in environmental chemistry. Hundreds of satisfied clients have made Ana Lab one of the oldest and largest environmentallaboratories in the Southwest. Ana Lab is an industry leader providing customers with improved turn around time, reduced cost and highest analysis quality, making it the clear choice among analytical chemistry laboratories.
C H Whiteside, President
Bill Peery, Jr., Executive VP

1739 Andco Environmental Processes

595 Commerce Drive   716-691-2100
Buffalo, NY 14228   Fax: 716-691-2880
     E-mail: Andco@Localnet.com

Manufacturers of waste disposal treatment systems.
Jack I Reich, Sales Manager

1740 Andersen 2000 Inc/Crown Andersen

306 Dividend Drive   770-486-2000
Peachtree City, GA 30269   800-241-5424
     Fax: 770-487-5066
     http://www.crownandersen.com

Supplying World Industry with Incineration and Air Pollution Control Systems

1741 Anderson Consulting Group

PO Box 407   610-918-7461
Downingtown, PA 19335   Fax: 610-918-9469
     E-mail: info@andersonconsultinggroup.com
     http://www.andersonconsultinggroup.com

Anderson Consulting Group has helped companies and publics agencies manage their project development risk, drive down construction cost, and improve schedules. Anderson Consulting Group's environmental and geotechnical services areuniquely designed to address client objectives. Our engineering solutions have earned engineering leadeship and innovation awards.

1742 Applied Ecological Services

17921 Smith Road 608-897-8641
PO Box 256 Fax: 608-897-8486
Brodhead, WI 53520-9803 E-mail: Info@AppliedEco.com
http://www.appliedeco.com

A broad-based ecological consulting, contracting, and restoration firm providing services to foundations, governmental units, corporations, and commercial/ residential developers nationwide scientist, prairie and consisting of amultidisciplinary team of botanists, wildlife biologist, wetland scientist, prairie and ecosystem restoration specialists and scientists, manage approximately 100 projects per year.
Steve Apfelbaum, President
Carl V Korfmacher, MLA, Vice President

1743 Applied Geoscience and Engineering

1300 New Holland Road 610-777-5027
Reading, PA 19607 Fax: 610-777-4276
E-mail: office@appliedgeoscience.com
http://www.appliedgeoscience.com

Environmental engineering, site assessments, and testing services through subcontractors.

1744 Applied Marine Ecology

1359 SW 22nd Terrace 305-757-0018
Miami, FL 33145 Fax: 305-858-6697

Marine ecology research firm.
Anitra Thorhaug, President
Andrew Oerke, CEO

1745 Applied Science Associates

70 Dean Knauss Drive 401-789-6224
Narragansett, RI 02882 Fax: 401-789-1932
E-mail: asa@appsci.com
http://www.appsci.com

ASA's mission is to create cost-effective solutions to environmental problems in surface waters; To supply supporting, specialized consulting services for engineering projects; To manage information with users-friendly computer models, integrating geographical information systems (GIS) display capabilities, environmental data monitoring and modeling.
J Craig Swanson, Principal
Eric Anderson, President

1746 Aqua Sierra

8350 South Mariposa Drive 303-697-5486
Morrison, CO 80465-2418 800-524-FISH
Fax: 303-697-5069
E-mail: inforequest@aqua-sierra.com
http://www.aqua-sierra.com

Aqua Sierra is a complete company servicing fisheries, aquaculture, water quality, wastewater, and database management interest. Aqua Sierra can assure an efficient, effective cost-conscious approach to managing aquatic resources bycombining a broad base of experience in all aspects of fisheries, aquatic ecology, and water quality management.

William J Logan, President

1747 Aqua Survey

469 Point Breeze Road 908-788-8700
Flemington, NJ 08822-4720 Fax: 908-788-9165
E-mail: Mail@AquaSurvey.com
http://www.aquasurvey.com

Aqua Survey is a full service ecotoxicology company founded in 1975. Aqua Survey provides laboratory testing, field sampling and consulting services to a wide variety of clients throughout the world including many of the largest UScorporations, internationally reconized environmental consulting firms, and the public sector.
Kenneth R Hayes, President

1748 Aqualogic

30 Devine Street 203-248-8959
North Haven, CT 06473 800-989-8959
Fax: 203-288-4308
E-mail: rheller@aqualogic.com
http://www.aqualogic.com

Water pollution control. Recovery and zero discharge.
Dick Heller

1749 Arcadis

630 Plaza Drive 720-344-3500
Suite 200 800-225-8419
Highlands Ranch, CO 80129 Fax: 720-344-3535
E-mail: info@arcadis-us.com
http://www.arcadis-us.com

Complete environmental services and remediation.
Alan Hurley, Area Manager

1750 Arcadis, Geraghty and Miller

630 Plaza Drive 720-344-3500
Suite 200 Fax: 720-344-3535
Highlands Ranch, CO 80129 http://www.arcadis-us.com

Arcadis G & M is a leading, global, knowledge-driven service provider, active in the fields of infrastructure, environment, buildings, and communications. Know for our innovation and full service capabilities, we develop and implementintegrated solutions based on personal service, open communication, and commitment to quality.
Steve Blake, Chairman/CEO
Michael Myers, President/COO

1751 Architectural Energy Corporation

2540 Frontier Avenue 303-444-4149
Suite 201 800-450-4454
Boulder, CO 80301 Fax: 303-444-4304
E-mail: aecinfo@archenergy.com
http://www.archenergy.com

Michael J Holtz, FAIA, President

1752 Arctech

14100 Park Meadow Drive 703-222-0280
Chantilly, VA 20151-2217 Fax: 703-222-0299
E-mail: bdrc@arctech.com
http://www.arctech.com

Arctech a diverse American Corporation is providing cost-effective solutions for energy, environmental, and agriculture market sectors. Arctech group through 25 years of experience in energy, energetics, environment and agriculture,has created holistic solutions in these interrelated market sectors. The enterpreneurial scientist and engineers at Arctech have pioneered the use of our vast resources of coal to make coal-derived humic acid products.
Daman S Walia, President/CEO
Madhu Walia, Administration Director

1753 Ardea Consulting

10 1st Street 530-669-1645
Woodland, CA 95695 Fax: 530-669-1674
E-mail: birdtox1@ardeacon.com
http://www.ardeacon.com

Ardea Consulting provides avian and wildlife toxicology guidance to engineering and environmental firms, government agencies, business and non-governmental organizations.
Joseph P Sullivan PhD, Sr. Consultant

1754 Argus/King Environmental Limited

7271 Wurzbach 210-493-2560
Suite 202 800-698-6018
San Antonio, TX 78240 Fax: 210-342-9027

Industrial hygiene and indoor air quality management. Mold and bacteria sampling, asbestos and lead testing and forensic.
Robert W Miller, CIH
Henry King, Consultant

1755 Arro Consulting

270 Granite Run Drive 717-569-7021
Lancaster, PA 17601-6822 800-229-6009
Fax: 717-560-0577
http://www.thearrogroup.com

Environmental engineering.
GM Brown, President

1756 Arro Laboratory

PO Box 686 815-727-5436
Caton Farm Road Fax: 815-740-3238
Joliet, IL 60434

Testing and analysis laboratory specializing in resource conservation and recovery act sampling. Research results published in confidential reports to clients.

1757 Artemel and Associates

218 North Lee Street
Suite 316
Alexandria, VA 22314

703-683-3838
Fax: 703-836-1370
E-mail: aiusa@artemel.com
http://www.artemel.com

Artemel & Associates is the technical arm of the Artemel Group of companies, with planning, engineering and analytical capabilities. The firm's ares of professional expertise directly complement Artemel International's areas ofspecialization. Artemel & Associates has been serving clients since 1984, and has benn credited with a variety of successful technical accomplishments both in the United States and around the globe.
Engin Artemel, President

1758 Associates in Rural Development

159 Bank Street
Suite 300
Burlington, VT 05401

802-658-3890
Fax: 802-658-4247
E-mail: ard@ardinc.com
http://www.ardinc.com

ARD was founded in 1977 as a Vermont corporation. Vermont's reputation for leadership in environmental affairs and its heritage of local participatory government embody ARD's ideals. Services include: watershed management,resource/sector assessments, EIA, urban environmental magagement, policy and action planning, natural resource assessment and evaluation, NR-based enterprise development, biodiversity conservation and finance, integrated water resource planning andmanagement.
George Burrill, President
Jim Talbot, Sr. VP

1759 Astbury Environmental Engineering

5645 West 79th Street
Indianapolis, IN 46278-1729

317-472-0999
Fax: 317-472-0993
E-mail: info@aeeindy.com
http://www.aeeindy.com

Astbury Environmental Engineering is a privately owned Indianapolis company that provides a full range of environmental services that include invironemntal management, site investigation and corrective action, health and safety, aircompliance, solid and hazardous waste management, and wastewater.
Steve Wilcox, President/CEO
Fred Nichols, VP, Business Development

1760 Astorino Branch Environmental

227 For Pitt Boulevard
Pittsburg, PA 15222

412-765-1700
800-518-0464
Fax: 412-765-1711
E-mail: marketing@ldastorino.com

Astorino Companies is an architectural, engineering and environmental consulting firm headquartered in downtown Pittsburg, PA.
Louis Astorino, President & CEO

1761 Atkins Environmental HELP

PO Box 222320
Santa Clarita, CA 91322-2320

661-260-2260
800-750-0622
Fax: 661-253-3555
E-mail: info@atkinsenvironmental.com
http://www.atkinsenvironmental.com

Environmental, health and safety compliance. Support services.
BJ Atkins, President

1762 Atlantic Testing Laboratories

6431 US Highway 11
Canton, NY 13617

315-386-4578
Fax: 315-386-1012
E-mail: info@AtlanticTesting.com
http://www.atlantictest.com

ATL is a full-service engineering support firm offering environmental services, subsurface investigations, geoprobe services, water-based investigations, geotechnical engineering, construction materials testing and engineering, specialinspection services, pavement engineering, nondestructive testing, and surveying from our ten offices. The firm currently has extensive capabilities in the areas of underground and aboveground storage tank testing and management and other relatedareas.
Marijean Remington, President
Spencer Thew, CEO

1763 Atlas Environmental Engineering

15701 Chemical Lane
Huntington Beach, CA 92649

714-890-7129
Fax: 714-890-7149
E-mail: info@aeei.com
http://www.aeei.com

Atlas Environmental Engineering provides very cost effective site assessments, investigations, corrective and remedial action plans and risk-based corrective action for low risk sites, along with groundwater monitoring, sampling, freeproduct removal activities and reporting. We also provide complete groundwater and soil remediation and cleanup activities, including all necessary equipment. Our goal is to provide clients with site closure in a minimal time period.
Karl H Kerner, VP

1764 Ayres Associates

3433 Oakwood Hills Parkway
PO Box 1590
Eau Claire, WI 5470

715-834-3161
Fax: 715-831-7500
E-mail: pearsont@AyresAssociates.com
http://www.ayresassociates.com

Ayres Associates has assisted both public and private clients with their engineering and architectural design challenges. We design roads and buildings. We evaluate and develop solutions for environmental problems. We design dams andstorm water solutions.
Patrick Quinn PE, President
Thomas Pearson, VP Marketing

1765 BBS Corporation

1103 Schrock Road
Suite 400
Columbus, OH 43229

614-888-3100
Fax: 614-888-0043
E-mail: email@bbsengineers.com
http://www.bbsengineers.com

A full service multi-disciplinary engineering firm specializing in the planning, design and construction administration of water and wastewater treatment, distribution and collection systems. Other services include data conversion anddatabase design for geographical information systems projects.
Edward Vance, Chairman

1766 BCI Engineers and Scientists

2000 East Edgewood Drive, Suite 215
Lakeland, FL 33803

863-667-2345
Fax: 863-667-2662
E-mail: info@bcieng.com
http://www.bcieng.com

BCI was founded in the early 1970s by former MIT professor, Dr. L.G. Bromwell. At that time, the majority of our clients were in the phosphate and Geotechnical industry. Since its beginning, BCI has grown into a multidisciplinary firmwith over 100 employees. We are proud to offer our valued clients a diverse team of professionals with a unique and complimentary blend of expertise. Our unusual blend of experience allows us to develop solutions to complex engineering andenvironmental challenges.
Richard M Powers, President

1767 BE and K/Terranext

155 South Madison Street
Suite 311
Denver, CO 80209

303-399-6145
Fax: 303-399-6146
E-mail: kmartin@terranext.net
http://www.bektnxt.com

We have been nationally recognized for excellence in environmental services since 1985. The goal of management is to develop and execute appropriate solutions to complex issues and act as strong advocates for our clients.
Kim Martin, President

1768 BHE Environmental

11733 Chesterdale Road
Cincinnati, OH 45246

513-326-1500
Fax: 513-326-1550
http://www.bheenv.com

BHE's mission is to provide a full range of environmental consulting and remediation services that set the standard for quality and responsibility. We strive to serve the total needs of clients and to create a challenging andsupportive work environment for our employees.

1769 Bac-Ground

3216 Georgetown
Houston, TX 77005-2906

713-664-8452
Fax: 713-664-2629
E-mail: ebaca@bac-ground.com
http://www.bac-ground.com

Environmental consultant
Ernesto Baca

1770 Badger Laboratories and Engineering Company

501 West Bell Street   920-729-1100
Neenah, WI 54956   800-776-7196
     Fax: 920-729-4945
E-mail: information@badgerlabs.com
http://www.badgerlabs.com
Badger Laboratories and Engineering provides customers with analytical, engineering and technical services focusing on the environmental field.
Steve Taylor, President
Jeff Wagner, Laboratory Services

1771 Baltec Associates

69 Fields Lane   845-279-7448
Brewster, NY 10509   Fax: 845-279-7467
E-mail: info@baltecusa.com
http://www.baltecusa.com
Baltec Associates is an international environmental consulting firm specializing in groundwater and soil remediation. We offers professional expertise and technical services for environmental management and planning, assessment,engineering, and remediation projects around the world.

1772 Barco Enterprises

11200 Pulaski Highway   410-335-0660
PO Box 0074   800-832-7538
White Marsh, MD 21162   Fax: 410-335-0790
E-mail: barco.enterprises@verizon.net
http://www.barcoenterprises.com
Hazardous materials handling

1773 Barer Engineering

PO Box 290   518-236-7070
Mooers, NY 12958   800-878-2806
     Fax: 518-236-5796
E-mail: info@barer.com
http://www.barer.com
Environmental engineering; pollution control

1774 Baron Consulting Company

273 Pepes Farm Road   203-874-5678
Milford, CT 06460-3671   Fax: 203-874-7863
E-mail: analyze@baronconsulting.com
http://www.baronconsulting.com
Chemical, environmental and biological testing firm.
Harry Agahigian, Technical Director
Barbara Obert, Lab Manager

1775 Barr Engineering Company

450 South Wagner Road   734-327-1200
Ann Arbor, MI 48103   800-632-2277
     Fax: 734-327-1212
E-mail: askbarr@barr.com
http://www.barr.com
Barr provides engineering, environmental, and information technology services to clients across the nation and around the world. We were incorporated as an employee-owned firm in 1966 and trace our orgins back to the early 1900s.Today, our more than 300 engineers, scientists, and technical support staff in Minnesota, Michigan, and Missouri work with clients in numerous industries, as well as at all levels of government.
Karin Clemon, Contact

1776 Batta Environmental Associates

6 Garfield Way   302-737-3376
Deleware Industrial Park   800-543-4807
Newark, DE 19713-5817   Fax: 302-737-5764
E-mail: bcbatta@battaenv.com
http://www.battaenv.com
BATTA was establihed in 1982 and is a Deleware Corporation registered to conduct work in the states of the Mid-Atlantic Region. BATTA has the in-house expertise in the scientific disciplines of geology, hydrogeology, civil andenvironmental engineering, chemistry, toxicology, health and safety, project design and construction management to adequately perform work without the use of outside consultants.
Naresh C Batta

1777 Baxter and Woodman

8678 Ridgefield Road   815-459-1260
Crystal Lake, IL 60012   Fax: 815-455-0450
E-mail: info@baxwood.net
http://www.baxterwoodman.com
Municipal waste, water, transportation, control systems, and mapping services. Our mission statement is: we will be the leader in consulting engineering based on our reputation for trust, integrity, and client service.
Darrel R Gavle, PE, DEE, President/CEO
Steve A Larson, PE, DEE, VP

1778 Baystate Environmental Consultants, Inc

296 North Main Street   413-525-3822
East Longmeadow, MA 01028   Fax: 413-525-8348
E-mail: ccarranza@b-e-c.com
http://www.b-e-c.com
BEC offers a wide range of civil engineering, water resources and environmental expertise. BEC was incorporated in 1972 and specializes in lake and pond restoration services, environmental assessment under MEPA/NEPA and wetlandscience. The staff at BEC is exceptionally diverse, having had formal training and long-term experience in multi-disiplinary projects involving site development options, conceptual layout planning, environmental permitting and civil engineeringservices.
Carlos Carranza, President

1779 Beak Consultants

4600 Northgate Boulevard   916-565-7929
Suite 215   Fax: 916-565-7900
Sacramento, CA 95834
Beak provides a fully integrated approach to environmental planning, assessment and problem solving. Our professional and technical specialists include ecologists, environmental auditors, risk assessors, geochemists, ecotoxicologists,contaminant hydrogeologists, modellers and environmental engineers. We integrate these specialties to provide our clients with a broad range of services.

Rick Swift
Amy Stuhr

1780 Beals and Thomas

144 Turnpike Road   508-366-0560
Route 9   Fax: 508-366-4391
Southborough, MA 01772-2104   E-mail: info@btiweb.com
http://www.btiweb.com
Beals and Thomas is a multidisciplinary consulting firm providing services to support the development and conservation of land and water resources throughout New England and the northeastern United States. Founded in 1984, BTI islocated in Southborough MA. Our mission is to advocate and assist in the attainment of our clients' project goals. We strive to provide creative and solution-oriented land eplanning and design services that are balanced with an environmental ethic.
John E Thomas, President
John E Bensley, Principal

1781 Bear West Company

145 S 400 E   801-355-8816
Salt Lake City, UT 84111-2104   Fax: 801-355-2090
E-mail: bearwest@burgoyne.com
Consultants on enviornmental issues
Ralph Becker, President

1782 Beaumont Environmental Systems

108 Lintel Drive   724-941-1743
McMurray, PA 15317   Fax: 561-382-6455
http://www.besmp.com
Beaumont company provides Particulate and Gaseous Air Pollution Control Equipment, Systems and Services for Power Generation, Waste Incineration, Utility, Steel, Mining, Cement, Foundry and Pulp and Paper Industries. They design andfurnish systems that include Fabric Filters, Electrostatic Precipitators, Wet Scrubbers, Semi Dry Scrubbers and Evaporative Cooloers along with the other necessary systems components.
Will Goss, President

1783 Becher-Hoppe Associates

PO Box 8000   715-845-8000
Wausau, WI 54402-8000   800-845-8009
     Fax: 715-845-8008

E-mail: mailbox@becherhoppe.com
http://www.bhassoc.com

Becher-Hoppe Associates is a firm of consulting engineers, architects, scientists, real estate specialists and surveyors. We provide a spectrum of professional services to governmental, industry and the private sector for airport,highway, municipal, facilities maintainance, water/wastewater, solid waste and environmental projects. From our location in central Wisconsin, we provide upper Midwest clients with neighborly promptness and efficiency.
Randy W Van Natta, PE, President
Phil Valitchka, Business Development

1784 Benchmark Environmental Consultants

6116 North Central Expressway   214-363-5996
Suite 808   Fax: 214-363-5994
Dallas, TX 75206   E-mail: info@benchmarkenviro.com
http://www.benchmarkenviro.com/index2.htm

A progressive environmental consulting firm which specialized in solving environmental problems by using a practical business and technical approach.

1785 Bendix Environmental Research

1950 Addison Street   415-861-8484
Suite 202   Fax: 510-845-8484
Berkeley, CA 94704   http://home.earthlink.net/~bendix/

Specializes in toxicology, hazardous materials management, and preparation of environmental documents, or appropriate parts of environmental documents dealing with hazardous materials. Provides expert witness serrvices and litigationresearch and support for toxic tort cases, including workplace and environmental exposures and chemical cancer causation.

1786 Beta Associates

915 Greenville Avenue   513-772-9296
Cincinnati, OH 45246   Fax: 513-772-9296
E-mail: BetaBob@BetaAssociates.com
http://www.betaassociates.com/

A professional project management organization which provides complete problem analysis, feasibility study and design services. These include specification and construction management for control and monitoring systems.

1787 Better Management Corporation of Ohio

41738 Esterly Drive
PO Box 130   877-293-4300
Columbiana, OH 44408   Fax: 330-482-9242
E-mail: bmc@bmcohio.com
http://www.bmcohio.com

A focus of providing transportation and disposal services of baled, compacted and loose municipal solid waste as well as C&D material to both public and private waste transfer station companies located mainly along the East Coast fromMaine to Florida.
Jerry Stoneburner, President
Bob Ruggeri, VP

1788 Beyaz and Patel

800 South Broadway   925-934-0707
Suite 200   888-431-0707
Walnut Creek, CA 94596-5218   Fax: 925-934-0318
E-mail: info@beyazpatel.com
http://www.beyazpatel.com

Structural engineering firm specializing in the structural design and construction management of public works infrastructure projects.
Yogesh B Patel, President
Subhash Patel, VP

1789 Bhate Associates

1608 13th Avenue South   205-918-4000
Suite 300   800-806-4001
Birmingham, AL 35205   Fax: 205-918-4050
E-mail: kgallant@bhate.com
http://www.bhate.com

Consulting environmental engineers
Kathleen Gallant, CPA, Human Resources Director

1790 Bioengineering Group

18 Commercial Street   978-740-0096
Salem, MA 01970   Fax: 978-740-0097
E-mail: mail@bioengineering.com
http://www.bioengineering.com

Provides a full range of consulting services in the field of bioengineering for erosion control, water quality, habitat restoration and stormwater management.

1791 Bioenvironmental Associates

4117 Sumpter Square   970-225-9549
Fort Collins, CO 80525   Fax: 970-484-4147
http://www.toolcity.net/~richreen/Bio.htm

Management plans, environmental permitting and compliance monitoring. We specialize in biological inventories for threated and endangered species.
Rex E Thomas PhD, Principal

1792 Biological Frontiers Institute

PO Box 313   707-996-2863
Sonoma, CA 95476   http://www.zoogenetics.com

Genetics, preservation of endangered species
Fred T Shultz, President

1793 Biological Monitoring

1800 Kraft Drive   540-953-2821
Suite 101   877-953-2821
Blacksburg, VA 24060   Fax: 540-951-1481
E-mail: bmi@biomon.com
http://www.biomon.com/

Environmental consulting group

1794 Biological Research Associates

3910 US Highway 301 N   813-664-4500
Suite 180   Fax: 813-664-0440
Tampa, FL 33619 E-mail: callahan@BiologicalResearch.com
http://www.biolresearch.com/

BRA professionals act as proponents of our clients' interest, helping them through the maze of environmental regulations.
Richard Callahan, CEO
J Steve Godley, President

1795 Bioremediation Consulting Incorporated

39 Clarendon Street   617-923-0976
Watertown, MA 02472   Fax: 617-923-0959
E-mail: bioremediation@bcilabs.com
http://www.bcilabs.com

Bioremediation Consulting responds to clients needs by providing high quality, cost effective planning and management for treatment of contaminants in soil and groundwater.
Margaret Findlay, President
Sam Fogel, VP

1796 Biospec

PO Box 788   918-336-3363
Bartlesville, OK 74005-0788   800-617-3363
Fax: 918-336-6060
E-mail: info@biospec.com
http://www.biospec.com

Laboratory scientific equipment.

1797 Bison Engineering

3423 Rivers Edge Trail   281-359-2476
Kingwood, TX 77339   888-502-4766
Fax: 281-359-2591
E-mail: inforequestz@bisonengineering.com
http://www.bisonengineering.com

Environmental services company.

Patsy Permenter

1798 Bjaam Environmental

472 Elm Ridge Avenue, Suite B   330-854-5300
PO Box 523   800-666-5331
Canal Futon, OH 44614   Fax: 330-854-5340
E-mail: bjaam@bjaam.com
http://www.riskassessment.com/

Provide one source for reliable, affordable environmental consulting and contracting services, as well as industrial wastewater pre-treatment systems and service

1799 Black and Veatch Engineers: Architects

11401 Lamar Avenue 913-458-2000
Overland Park, KS 66211 Fax: 913-458-2934
E-mail: info@bv.com
http://www.bv.com

Linda Heil, Communications Specialist
Corrine Smith, Vice President

1800 Blackhawk GeoSciences

301 Commercial Road 303-278-8700
Suite B Fax: 303-278-0789
Golden, CO 80401-5613 E-mail: geoinfo@blackhawkgeo.com
http://www.blackhawkgeo.com
High quality geophysical contracting and consulting services over the full spectrum of geophysical technologies, and to apply the geophysical technologies to several cross-cutting areas of engineering and exploration.

1801 Blasland, Bouck and Lee

6723 Towpath Road 315-446-9120
PO Box 66 Fax: 315-449-0017
Syracuse, NY 13214-0066 E-mail: info@bbl-inc.com
http://www.bbl-inc.com/bblinc/
Hazardous waste, environmental compliance, air quality; engineering, solid waste, water, wastewater engineering
Robert K Goldman, PE, CEO/Chairman

1802 Block Environmental Services

2451 Estand Way
Pleasant Hill, CA 94523 800-682-7255
Fax: 925-686-0399
E-mail: dblock@blockenviron.com
http://www.blockenviron.com
A environmental consulting firm specializing in indoor air quality and toxicology. We are a Certified Aquatic Bioassay Laboratory.
Ronald Block, President

1803 Boelter and Yates

1300 Higgins Road 847-692-4700
Suite 301 Fax: 847-692-3127
Park Ridge, IL 60068-5772 E-mail: info@boelter-yates.com
http://www.boelter-yates.com/
Provide environmental engineering, occupational health and safety management, design engineering, and consulting services.

Fred Boelter, Chairperson
Thomas Kowalski, President/CEO

1804 Bollyky Associates

31 Strawberry Hill Avenue 203-967-4223
Stamford, CT 06902-2608 Fax: 203-967-4845
E-mail: ljbbai@bai-ozone.com
http://www.bai-ozone.com
Engineering firm specializing in Ozone technology, water and wastewater treatment, treatability studies.

L Joseph Bollyky, President
Thomas Kleiber, Office Manager

1805 Bottom Line Consulting

27248 Twin Pond Road 847-381-0597
Barrington, IL 06010-1125 Fax: 847-381-0598
Plastics recycling.
John Fearncombe, President

1806 Braun Intertec Corporation

11001 Hampshire Avenue South 952-995-2000
Minneapolis, MN 55438 800-279-6100
Fax: 952-995-2020
E-mail: info@braunintertec.com
http://www.braunintertec.com

An engineering firm providing consulting, management and testing services to clients in the commercial, industrial and residential real estate, institutional, retail, financial and government markets.
George D Kluempke, PE, President/CEO

1807 Bregman and Company

5272 River Road 301-652-4818
Suite 550 Fax: 301-652-4819
Bethesda, MD 20816 E-mail: bob@bregmanandcompany.com
http://www.bregmanandcompany.com
Environmental consulting firm.
Robert Edell, President/CEO

1808 Brinkerhoff Environmental Services

1913 Atlantic Avenue 732-223-2225
Suite R 5 800-246-7358
Manasquan, NJ 08736 Fax: 732-223-3666
E-mail: lbrinkerhoff@brinkenv.com
http://www.brinkenv.com
Groundwater remediation, environmental site assessments and sensitive area mapping; hazardous material management.
Laura A Brinkerhoff, President/CEO
Eileen Della Volle, Consultant

1809 Brooks Laboratories

9 Issac Street 203-853-9792
Norwalk, CT 06850 800-843-1631
Fax: 203-853-0273
E-mail: brooklabs@aol.com
http://www.brookslabs.com
Consulting and testing air, soil and water for contamination. Accident and disease prevention.
Michael Zubarev, President
Kalonji Diyoka, VP

1810 Brown, Vence and Associates

115 Sansome Street 415-434-0900
Suite 800 Fax: 415-956-6220
San Francisco, CA 94104 http://www.brownvence.com
Waste management energy consulting firm.

1811 Buck, Seifert and Jost

65 Oak Street 201-767-3111
PO Box 415 Fax: 201-767-3178
Norwood, NJ 07648 E-mail: bsjinc@bsjinc.com
http://www.bsjinc.com
Consultancy for the water and wastewater industries.
Ronald von Autenried, PE, President
Guido von Autenried, PE, Director/Chief Engineer

1812 Burk-Kleinpeter

4176 Canal Street 504-486-5901
New Orleans, LA 70119 Fax: 504-488-1714
E-mail: mjackson@bkiusa.com
http://www.bkiusa.com
A full service firm bringing together resources from our Engineering, Architecture, Planning and Environmental Science Divisions. Our Divisions, which may function independently also work as a team, providing our clients withassistance from the first conceptual idea through final construction. We also provide services through our professional support groups which include landscape architecture, construction management and inspection, graphic design, aerial photographyand marketing.
Michael G Jackson, PE, Executive VP

1813 Burns and McDonnell

9400 Ward Parkway 816-333-9400
Kansas City, MO 64114 Fax: 816-333-3690
E-mail: busdev@burnsmcd.com
http://www.burnsmcd.com/index.html
A multidisciplinary engineering, architectural, construction and environmental service firm. More than 1,400 engineers, architects, scientists and other specialists plan, design and build quality projects around the world.

1814 C&H Environmental

216 Stiger Street 908-852-4855
PO Box 188 Fax: 908-852-5275
Hackettstown, NJ 07840 http://www.candhenvironmental.com

Environmental consulting firm.

John H Crow, PhD
Timir B Hore, PhD

1815 CA Rich
17 DuPont Street
Plainview, NY 11803

516-576-8844
Fax: 516-576-0093
E-mail: info@carichinc.com
http://www.carichinc.com

An independently owned, private consulting firm providing targeted, solution oriented hydrogeologic and environmental engineering services. Assists in the conception, development, design, implementation, documentation and defense of site evaluations and remedial action.
Charles A Rich, Founder/President
Richard J Izzo, Human Resources Director

1816 CBA Environmental Services
57 Park Lane
Hegins, PA 17938

570-682-8742
Fax: 570-682-8915
E-mail: info@cbaenvironmental.com
http://www.cbaenvironmental.com

Provides environmenta soulutions from general plant maintenance and cleaning to large-scale soil remediation projects.
Bruce L Bruso, Principal

1817 CDS Laboratories
75 Suttle Street
PO Box 2605
Durango, CO 81302

303-247-4220
800-553-6266
Fax: 303-247-4227

Specializes in analytical analysis and testing, consulting, QA, environmental, and dyes.

1818 CEDA
3519 Old Red Trail
PO Box 787
Mandan, ND 58554

701-663-0307
Fax: 701-667-2090

Provides services in hazardous waste, spill response, and asbestos abatement.

WF Mowatt, President

1819 CIH Environmental
1044 Victory Circle
Reading, PA 19605

610-372-6692
Fax: 610-372-0862
E-mail: cihenv@fast.net
http://www.cihenv.com

Provides services in indoor air quality and industrial hygiene and mold/bacterial contaminations.
James E Detwiler, President

1820 CIH Services
7148 Creekside Lane
Indianapolis, IN 46250

317-797-7768
Fax: 317-913-1895
E-mail: cihservices@juno.com
http://www.cih-services.com

Services in indoor air quality.
John Beltz

1821 CII Engineered Systems
6767 Forrest Hill Avenue
Richmond, VA 23225-1856

804-320-1405
800-768-2545
Fax: 804-320-9625

Services in energy conservation.
Jack Thacker

1822 CK Environmental
1020 Turnpike Street
#8
Canton, MA 02021

781-828-5200
888-253-0303
Fax: 781-828-5380
E-mail: info@ckenvironmental.com
http://www.eco-web.com

Serivces in regulatory compliance.

1823 CRB Geological and Environmental Services
4573 Ponce De Leon Boulevard
Coral Gables, FL 33146

305-447-9777
Fax: 305-567-2853

E-mail: blivieri@crbgeo.net
http://www.crbgeo.net/

Environmental consulting.
Frederick R Baddour, President

1824 CTE Engineers
303 East Wacker Drive
Suite 600
Chicago, IL 60601

312-938-0300
Fax: 312-938-1109
E-mail: tony.bouchard@cte.aecom.com
http://www.cte-eng.com

Consulting services in the use of computers and electronics to solve pollution problems.

Tony Bouchard, Environmental Services
Carl Mahr, Environmental Services

1825 CTI and Associates, Inc
12482 Emerson Drive
Brighton, MI 48116

248-486-5100
800-468-7499
Fax: 248-486-5050
E-mail: robertanderson@cti-assoc.com
http://www.cti-assoc.com/

Environmental engineering firm.
Robert Anderson, Environmental Services

1826 CTL Environmental Services
24404 South Vermont Avenue
Suite 307
Harbor City, CA 90710

310-530-5006
800-777-0605
Fax: 310-530-0792
E-mail: info@ctles.com
http://www.ctles.com

Industrial hygiene and safety, asbestos/lead based paint surveys, environmental site assessments, risk mangement, indoor air quality, radon testing and mold investigation.

1827 CZR
1061 East Indiantown Road
Suite 100
Jupiter, FL 33477-5143

561-747-7455
Fax: 561-747-7576
E-mail: czrjup@aol.com
http://members.aol.com/czrwilm

Environmental impact studies, wetlands delineation and threatened species surveys.

C A Mullen, Office Staff

1828 Cabe Associates
144 South Governors Avenue
PO Box 877
Dover, DE 19903

302-674-9280
800-542-7979
Fax: 302-674-1099
E-mail: jpj@cabe.com
http://www.cabe.com

Environmental and pollution control.
Robert Kerr, Contact

1829 California Environmental
1161 Calla Suerte
Suite C
Camario, CA 91362

818-991-1542
Fax: 818-991-0793

Industrial safety and hygiene.
Michael R Tiffany, CIH

1830 California Geo-Systems
1545 Victory Boulevard
2nd Floor
Glendale, CA 91201

818-500-9533
Fax: 818-500-0134
http://www.geosys1.com

Geotechnical environmental services.
Vince Carnegie, President
Rachel Fischer, Sr. Environmental Geologist

1831 Callidus Consulting
1094 Quince Avenue
Boulder, CO 80304

303-443-2316
Fax: 303-938-9420
E-mail: frank.aldrich@atsuchsc.edu
http://www.callidusconsulting.com

Expert witness in toxicology, environmental health, indoor air quality.
Franklin D Aldrich MD, President

1832 Cambridge Environmental
58 Charles Street
Cambridge, MA 02141-2179
617-225-0810
Fax: 607-225-0813
E-mail: info@cambridgeenvironmental.com
http://www.cambridgeenvironmental.com
Consulting and research firm that assesses and helps to minimize risks to health and the environment.
Laura Green, Senior Scientist
Michael Ames, Senior Engineer

1833 Camiros Limited
411 South Wells
Suite 400
Chicago, IL 60607
312-922-9211
Fax: 312-922-9689
E-mail: sbland@camiros.com
http://www.camiros.com
Camiros is an active proponent of Sustainable Growth as well as other environmentally sensitive aspects of urban planning and design. In particular, the firm has drafted land use plans and zoning ordinances that pay careful attentionto environmental issues.

Shirelle Brand, Administrative

1834 Camo Pollution Control
1610 State Route 376
Wappingers Falls, NY 12590-6126
845-463-7310
Environmental consultants.
Michael Tremper, Vice President

1835 Camtech
4550 McKnight Road
Suite 202
Pittsburg, PA 15237-3162
412-931-1210
Fax: 412-931-1304
Environmental consulting.

1836 Canin Associates
500 Delaney Avenue
Suite 404
Orlando, FL 32801-3850
407-422-4040
Fax: 407-425-7427
E-mail: edoran@canin.com
http://www.canin.com
Environmental services.
Elizabeth Doran, Business Development
Myrna Canin, VP

1837 Cape Environmental Management
2302 Parklake Drive NE
Suite 200
Atlanta, GA 30345-2902
770-908-7200
800-488-4372
Fax: 770-908-7219
Environmental consulting firm.
Fernando Rios, President

1838 Capital Environmental Enterprises
2244 Profit Drive
Indianapolis, IN 46241-5019
317-240-8085
888-376-4315
Fax: 317-241-4180
http://www.capitalenvironmental.net
Environmental consulting.

1839 Cardinal Environmental
3303 Paine Avenue
Sheboygan, WI 53081
920-459-2500
800-413-7225
Fax: 920-459-2503
E-mail: info@cardinalenvironmental.com
http://www.cardinalenvironmental.com/
Environmental consulting.

1840 Carpenter Environmental Associates
PO Box 656
Monroe, NY 10949
845-781-4844
Fax: 845-782-5591
E-mail: b.bell@cea-enviro.com
http://www.ceaenviro.com/
Environmental engineering and assessment services, including wastewater and storm water management, wetlands and ecological investigations, site assessments, environmental compliance and contingency planning, permitting services, andlitigation support.

Bruce A Bell, PhD, PE, DEE, President

1841 Carr Research Laboratory
17 Waban Street
Wellesley, MA 02482-6310
508-651-7027
Fax: 508-647-4737
E-mail: cann@carr-research.lab.com
http://www.carr-research.lab.com
Environmental consulting research laboratory.
Jerome B Carr PhD, President
Jessica S Veysey, Environmental Biologist

1842 Catlin Engineers and Scientists
220 Old Dairy Road
PO Box 10279
Wilmington, NC 28405
910-452-5861
800-346-730
Fax: 910-452-7563
E-mail: info@catlinusa.com
http://www.catlinusa.com
Specializes in providing quality service in the fields of environmental, civil, and geotechnical engineering. Services include soil and ground water remediation, wastewater treatment system design, public infrastructure,environmentally secure landfills, and safe, clean, water supplies.
Richard Catlin, President

1843 Center for Energy and Environmental Analysis Oak Ridge Laboratory
Energy and Environmental Analysis
PO Box 2008
Oak Ridge, TN 37831-6205
865-576-4160
Fax: 865-574-8884
E-mail: sdb@ornl.gov
http://www.ornl.gov/ceea/
In the Center for Energy and Environmental Analysis, a part of the Energy Division at Oak Ridge National Laboratory, we provide our customers with analysis of energy and environmental issues of local, regional, national, and globalimportance so as to provide decision makers with information on which to base major policy, program, and project decisions.
Dr. Michael O Lerner, President

1844 Central States Environmental Services
1079 Copple Road
Centralia, IL 62801-6587
618-532-4784
800-367-3090
Fax: 618-532-5615
Environmental clean-up contractor.
Elvin Copple, President

1845 Challenge Environmental Laboratories
2270 Worth Lane
Suite D
Springdale, AR 72764
479-927-1008
Fax: 479-927-1000
E-mail: kent@challenge-sys.com
http://www.challenge-sys.com
Challenge Environmental Laboratories specializes in conducting highly technical test to determine the biological and physical chemical treatability of industrial wastes.
Mark L Kuss, Founder

1846 Chapman Environmental Control
PO Box 288
Osceola, IN 46561-0288
800-675-8706
Air pollution control.
Frank X Chapman, President

1847 Chelsea Group
1 Pierce Place
Suite 275e
Itasca, IL 60143-2618
800-626-6722
Fax: 630-775-9231
E-mail: info@chelsea-grp.com
http://www.chelsea-grp.com
Specializes in strategic, technical, and marketing consulting to major corporations for enhanced positioning of products and services relating to the indoor environment.
George Benda, Senior Princpial
David Munn, PE, Principal

1848 Chemical Data Management Systems
6515 Trinity Court
Suite 201
Dublin, CA 94568
925-551-7300
Fax: 925-829-3886
E-mail: info@cdms.com
http://www.cdms.com

Provides a full range of hazardous material and OSHA regulatory compliance services to industries using hazardous materials, including implementing compliance programs, submitting necessary reports to all regulatory agencies, andproviding a full range of training services to industry, agencies, and industrial groups.

1849 Chicago Chem Consultants Corporation

14 North Peoria Street
Suite 2C
Chicago, IL 60607-2609

312-226-2436
Fax: 312-226-8886
E-mail: info@chichem.com
http://www.chichem.com

Provides innovative, technologically sophisticated, cost-effective, and risk protective environmental and engineering services.
Jeffrey P Perl, President
Stanley Yoslov, Senior Associate

1850 Cigna Loss Control Services

Oklahoma City, OK 73150-1703

405-524-2127
http://www.cigna.com

Provides services in the field of industrial hygiene.
H Edward Hanway, Chairman/CEO
Michael W Bell, Executive VP/CFO

1851 Clayton Group Services

45525 Grand River Avenue
Suite 200
Novi, MI 48374

248-344-8550
Fax: 248-344-0229
E-mail: info@claytongrp.com
http://www.claytongrp.com

A full service environmental, occupational health and safety, and laboratory services consulting firm serving both public and private clients.
Lisa Barnes, PE CIH, Sr VP/COO
MJ Haught, Marketing Director

1852 Clean Air Engineering

500 West Wood Street
Palatine, IL 60067-4975

847-991-3300
800-627-0033
Fax: 847-991-3385
E-mail: contact@cleanair.com
http://www.cleanair.com

Environmentally responsible consulting and permitting, process engineering, equipment rental and manufacture, measurement and analytical services.
William I Walker, President
Frank Kilvinger, VP

1853 Clean Environments

10803 Gulfdale
Suite 210
San Antonio, TX 78216

210-349-7242
Fax: 210-349-1132
E-mail: cei@cleanenvironments.com
http://www.cleanenvironments.com

Environmental consulting firm.

1854 Clean Technologies

2700 Capitol Trail
Newark, DE 19711-6814

302-999-0924
Fax: 302-999-0925

Environmental engineering firm.
Deborah A Buniski, President

1855 Clean World Engineering

1737 S Naperville Road
Suite 200
Wheaton, IL 60187-8132

630-260-0200
800-761-9603
Fax: 630-260-0797
E-mail: cwe@clean-world.com
http://www.clean-world.com

A woman-owned environmental engineering firm established in 1985. They specialize in meeting the needs of small businesses to FORTUNE 500 companies to large government agencies. They develop practical and cost-efficient soulutions toincreasingly stringent and complex environmental reguations.
Rita Kapur, President/CEO

1856 Coastal Lawyer

173 E Blithedale Avenue
Suite 3
Mill Valley, CA 94941

415-383-3715
Fax: 415-383-3718
E-mail: del@greendogcampaigns.com

Consultation for environmental causes and initiatives. Legal representation, public relations and campaign consulting.

Dotty E LeMieux, Principal

1857 Coastal Planning and Engineering

2481 Northwest Boca Raton Boulevard
Boca Raton, FL 33431

561-391-8102
Fax: 561-391-9116
E-mail: mail@coastalplanning.net
http://www.coastalplanning.net

Environmental consulting firm providing services in coastal engineering, coastal planning, coastal surveying, environmental science, and regulatory permitting.

1858 Cohen Group

3 Waters Park Drive
San Mateo, CA 94403

650-349-9737
Fax: 650-349-3378
E-mail: jcohen@thecohengroup.com
http://www.thecohengroup.com

Provides a complete range of environmental health and safety services to business and government including indoor air quality, asbestos, respiratory protection, and industrial hygiene safety.
Joel M Cohen, President

1859 Cohrssen Environmental

1990 Lombard Street
Suite 200
San Francisco, CA 94123

415-775-1105

Industrial hygiene services.
Barbara Cohrssen, Principal Executive

1860 Columbia Analytical Services

1317 South 13th Avenue
Kelso, WA 98626

360-577-7222
800-695-7222
Fax: 360-425-9096
http://www.caslab.com

Areas of expertise and services include environmental testing of air, water, soil, hazardous waste, sediments and tissues; process and quality control testing ; analytical method development; sampling and mobile laboratory services;and consulting and data management services.

Stephen W Vincent, President/CEO
Jerry Watega, Employee Representative

1861 Combustion Unlimited

PO Box 8856
Philadelphia, PA 19117

215-537-0871
Fax: 215-884-3074

Engineering consulting services in air pollution control and combustion.

John F Straitz III, President

1862 Committee for Environmentally Effective Packaging

601 13th Street NW
Suite 900S
Washington, DC 20005

202-783-5594
Fax: 203-783-5595
http://www.epa.gov/epaoswer

Monitors legislation and regulations affecting packaging in the food service industry and educates decision makers on packaging.

1863 Commonwealth Engineering and Technology CET Engineering Services

1240 North Mountain Road
Harrisburg, PA 17112

717-541-0622
Fax: 717-541-8004
E-mail: contact@cet-inc.com
http://http://cet-inc.com

Pollution control utilizing the ability of Geographic Informaiton Systems as an analytical tool to be used to analyze water and wastewater systems.

Jeffrey Wendle, President

1864 Commonwealth Laboratory

1602 Skipwith Road
Richmond, VA 23229-5205

804-289-5608
Fax: 804-644-5820

1865 Community Conservation Consultants Howlers Forever

50542 One Quiet Lane
Gays Mills, WI 54631

608-735-4717
Fax: 512-519-8494

E-mail: communityconservation@mwt.net
http://www.communityconservation.org
Works together with local rural people to aid in the protection of their wildlife and forests. Projects undertaken have mainly been in Belize and Wisconsin with an emphasis on primates and other species. Additional projects are evolving in Central America and abroad.

Robert Horwich, Director

1866 Compass Environmental

1751 McCollum Parkway   770-499-7127
Kennesaw, GA 30144   Fax: 770-423-7402
E-mail: staff@compassenv.com
http://www.compassenv.com
Asbestos indoor air quality and industrial hygiene services. Established as an alternative to large companuies that are typically structured to provide routine testing and consulting services.
Eva M Ewing, Vice President
William M Ewing, President

1867 Comprehensive Environmental

21 Depot Street   603-424-8444
Merrimack, NH 03054-3423   800-725-2550
Fax: 800-331-0892
E-mail: webmaster@ceiengineers.com
http://www.ceiengineers.com
Provides water, wastewater and hazardous waste services with a mission to protect the client, public health and the environment, to be client advocates, to build trust, and to provide objectivity.
Eileen Pannetier, President

1868 Comprehensive Environmental Strategies

11950 Rocky Brook Court   703-791-7700
Manassas, VA 20112   Fax: 703-791-7700
Asbestos, indoor air quality, industriel hygiene services.
Reginald B Simmons, Principal Executive

1869 Conestoga-Rovers and Associates

2055 Niagra Falls Boulevard   716-297-6150
Suite 3   Fax: 716-297-2265
Niagra Falls, NY 14304   E-mail: info@cra.com
http://www.craworld.com
Family of companies that provide a full-service engineering, environmental, construction and information technology services worldwide.

Frank A Rovers, Principal Executive

1870 Conservtech

Division of Delphey, Gerdes Engineering
3655 South Soto Street   323-583-6897
Vernon, CA 90058-1783   Fax: 323-587-8132
E-mail: bob@conservtechgroup.com
http://www.conservtechgroup.com
Pollution control systems, site assessment, waste problems, water and waste systems.

Robert J MacDonald, President/CEO

1871 Consoer Townsend Envirodyne Engineers

303 East Wacker Drive   312-938-0300
Suite 600   Fax: 312-938-1109
Chicago, IL 60601-5276 E-mail: tony.bouchard@cte.aecom.com
http://www.cte-eng.com
Environmental engineering, hazardous and toxic waste, environmental and energy products, cost-effective soulutions to modern, urban infrastructure problems.

Tony Bouchard, Environmental Services
Carl Mahr, Environmental Services

1872 Consultox

PO Box 51210   504-529-7500
New Orleans, LA 70125   Fax: 504-926-0638
E-mail: info@consultox.com
http://www.consultox.com
Toxicology consulting firm.

Richard A Parent, President

1873 Converse Consultants

222 East Huntington Drive   626-930-1200
Suite 211   Fax: 626-930-1212
Monrovia, CA 91016-3500 E-mail: converse@converseconsultants.com
http://www.converseconsultants.com
Environmental consulting and engineering firm with offices throughout the United States.

1874 Cook Flatt and Strobel Engineers

6111 29th Street Southwest   785-272-4706
Topeka, KS 66614   Fax: 785-272-4736
E-mail: cfsengr@cfse.com
http://www.cfse.com
Environmental engineering.

1875 Cornerstone Environmental, Health and Safety

880 Lennox Court   317-733-2637
Zionsville, IN 46077   800-285-2568
Fax: 317-577-2481
E-mail: info@corner-enviro.com
http://www.corner-enviro.com
Environmental, health, and safety services.
Jill Para, Contact

1876 Cox Environmental Engineering

82 Dresser Hill Road   508-248-5185
Charlton, MA 01507-5133   Fax: 508-248-5003
Environmental engineering consulting.

1877 Crouse & Company

912 Greengate North Plaza   724-832-3114
Greensburgh, PA 15601   Fax: 724-832-3627
http://www.crouse.com
Natural resources management company dedicated to ushering traditional engineering and natural sciences consulting services into the age of modern information technology.
Jeffery P Evers, Director

1878 Cultural Resource Consultants International Archaeology & Ecology

7400 Jones Drive   832-592-9549
Galveston, TX 77551   Fax: 713-468-4263
E-mail: postmaster@culturalresource.com
http://www.culturalresource.com
Historic land use and environmentally sensitive projects.
Robert P d'Aigle, Owner
Nataliya Hryshechko, Laboratory Services Director

1879 Curt B Beck Consulting Engineer

408 W Kingsmill Street   806-665-9281
PO Box 2442   888-665-9281
Pampa, TX 79006-2442   Fax: 806-665-1965
E-mail: curtbbeck@cableone.net
Pollution control services.
Curt B Beck, Owner

1880 Custom Environmental Services

233 Forest Drive   805-968-2112
Santa Barbara, CA 93117-1108   Fax: 805-968-2137
E-mail: info@Custom-env.com
http://www.custom-env.com
Small environmental consulting business offering services that help individuals and companies of all sizes comply with environmental laws, including the Clean Air Act, Clean Water Act, and Medical Waste Management Act.
Rosalie A Skefich, Founder

1881 D'Appolonia

275 Center Road   412-856-9440
Monroeville, PA 15146-1451   Fax: 412-856-9535
E-mail: info@dappolonia.com
http://www.dappolonia.com
Provides engineering, scientific and construction management services for projects involving large civil works and special earth/structure interaction issues.

1882 D/E3

18234 S Miles Road
Suite 44
Cleveland, OH 44128-4232

216-663-1500
Fax: 216-663-1501

Environmental impact statements, corrective process, pollution abatement, hazardous waste, radon and asbestos hazards.

Harold N Danto, President

1883 DPRA

200 Research Drive
Manhattan, KS 66503

785-539-3565
Fax: 785-539-5353
E-mail: info@dpra.com
http://www.dpra.com

Environmental, economic, regulatory and technical research company. Research results published by information services.
Richard Seltzer, President

1884 Datanet Engineering

6334 Dogwood Road
Baltimore, MD 21207-5227

410-944-3600
888-896-7133
Fax: 410-944-5154
E-mail: info@datanetengineering.com
http://www.datanetengineering.com

Failure investigation causation studies for fuel systems corrosion control designs and testing for water and fuel systems, EPA training programs for fuel tank installation and management, NACE, API, ABIH, and AWS accreditations. Corrosion and environmental engineering as well as industrial hygiene services.

1885 DeVany Industrial Consultants

14507 NW 19th Avenue
Vancouver, WA 98685-8003

360-546-0999
Fax: 360-546-0777
E-mail: mdevany@earthlink.net

Strive to provide a full range of safety and industrial hygiene services customized for your particular environment.

1886 Dell Engineering

3352 128th Avenue
Holland, MI 49424-9263

616-399-3500
Fax: 616-399-3777

1887 Dennis Breedlove and Associates

330 West Canton Avenue
Winter Park, FL 32789

407-677-1882
800-304-1882
Fax: 407-657-7008

Environmental and natural resources consulting firm.

1888 Detail Associates: Environmental Engineering Consultants

300 Grand Avenue
Englewood, NJ 07631

201-569-6708
Fax: 201-569-4378
E-mail: dainfo@daienviro.com
http://www.daienviro.com

Asbestos management programs. Indoor air quality. Analytical services. Lead surveys. Phase I, II and III environmental audits.

1889 Donald Friedlander

1091 Willowbrook Road
Staten Island, NY 10314-6514

718-698-7545

1890 Dunn Corporation

12 Metro Park Road
Albany, NY 12205

518-458-1313
Fax: 518-458-2472

1891 ENSR Consulting and Engineering

2 Technology Park Drive
Westford, MA 01886-3140

978-589-3000
800-722-2440
Fax: 978-589-3100
http://www.ensr.com

Provides consulting, engineering, remediation, and related services to industrial and commercial companies, municipalities, and regulated government agencies throughout the United States, Europe, Latina America, and Asia.
Robert C Weber, President/CEO

1892 ENTRIX

5252 Westchester
Suite 250
Houston, TX 77005

713-666-6223
Fax: 713-666-5227
E-mail: webmaster@entrix.com
http://www.entrix.com

Provides environmental management services.
Dan Taylor, President
Richard Firth, Executive VP

1893 ETS

1401 Municipal Road Northwest
Roanoke, VA 24012-1309

540-265-0004
Fax: 540-265-0131
E-mail: jmck@esti-inc.com
http://www.esti-inc.com

ETS is a full-service environmental consulting and training firm specializing in air emissions control, measurement, engineering and consulting services.
John McKenna, Contact
Jack Mycock, Contact

1894 Earth Science Associates

444 West Ocean Boulevard
Suite 1510
Long Beach, CA 90802

562-437-7373
Fax: 562-437-7722
E-mail: contactESA@earthsci.com
http://www.earthsci.com

Earth Science Associates is a consultancy serving the international oil and gas industry. ESA specializes in resource assessment, economic evaluation and risk studies and the development of custom geographic information systems. Oiland gas companies use our assessment studies in evaluating the geologic potential of prospects, plays and basins.

1895 Earth Science Associates/ESA Consultants

PO Box 12067
Knoxville, TN 37912-0067

800-467-6380
E-mail: esa@halos.com
http://www.halos.com

1896 EcoLogic Systems

7977 Capwell Drive
Suite 150
Oakland, CA 94621

800-223-0609
Fax: 510-634-7402
E-mail: gjames@accenv.com
http://www.ecologicsystems.com

The developer of leading suite of hazardous material management and environmental health and safety compliance software.
Geoff James, Business Development

1897 Ecology and Environment

368 Pleasant View Drive
Lancaster, NY 14086

716-684-8060
Fax: 716-684-0844
E-mail: ckarpowicz@ene.com
http://www.ene.com

Ecology and Environment is a multidisciplinary environmental science and engineering company with more than 25 offices in the US and offices and partners in more than 35 countries. We are a world leader in providing environmentalconsulting services and litigation support.
Cheryl Karpowicz, VP
Janet Steinbruckner, Director of Human Resources

1898 Ed Caicedo Engineers/Consultants

PO Box 22256
Lexington, KY 40522

859-259-0042
E-mail: info@eciengineers.com
http://www.eciengineers.com

We are a Kentucky-based Consulting Engineering firm which provides professional and technical services to the public and private sectors. These services cover all facets of the construction process, including the preparation offeasibility studies, conceptual designs, environmental impact studies, cost-effectiveness analysis, final project design, construction administration, operation and maintenance management.
Eduardo Caicedo, President
William H Meadows, Principal Engineer

1899 Elinor Schwartz

318 South Abingdon Street
Arlington, VA 22204

703-920-5389
Fax: 703-920-5402

1900 En Safe
5724 Summer Trees Drive
Memphis, TN 38134-7309

901-372-7962
800-588-7962
Fax: 901-372-2452
http://www.ensafe.com

Services include implementation of occupational health and safety programs, worker's compensation programs, and environmental management systems for clients ranging from large corporations to small businesses to decrease accidentfrequency and accident costs.
Phillip G Coop, President/CEO
Michael Wood, Vice President/CFO

1901 Energy Technology Consultants
2020 E 1st Street
Santa Ana, CA 92705-4015

714-835-6886
Fax: 714-667-7147

1902 Enviro Equipment Sales
PO Box 20760
Floral Park, NY 11001

516-354-1212
800-346-5926
Fax: 516-354-2434
E-mail: enviroeq@ix.netcom.com

Bag breakers-openers, balers, carts, can recycling equipment, compactors, glass crushers, odor control systems, sherrers/parts and supplies.
Harvey Podolsky

1903 EnviroTest Laboratories
315 Fillerton Avenue
Newbrugh, NY 12550

845-562-0890
Fax: 845-562-0841

Test soil and water.
Scott Morris, President

1904 Envirocorp
7020 Portwest Drive
Suite 100
Houston, TX 77024

713-880-4640
800-535-4105
Fax: 713-880-3248
E-mail: pfh@subsurfacegroup.com
http://www.envirocorpinc.com

Envirocorp, Inc. is a consultant firm with over 25 years of expeoence around the world involving site assessments and remediations. Envirocorp's expertise includes UST removals and remediations, hazardous waste reporting,environmental audits, oil & gas property assessments, and asbestos inspection. Envirocorp has offices in South Bend, Indiana; Baton Rouge, Louisiana and its's corporate office in Houston, Texas.

1905 Environmental Compliance Consulting
PO Box 11417
Green Bay, WI 54307-1417

920-434-6380
888-ECC-INOW

Environmental assessments.

1906 Environmental Consultants
391 Newman Avenue
Clarksville, IN 47129-3247

812-282-8481
Fax: 812-282-8554

Environmental consulting firm.
Robert E Fuchs, President

1907 Environmental Resource Associates
6000 West 64th Avenue
Arvada, CO 80002

303-431-8454
800-372-0122
Fax: 303-421-0159
E-mail: info@eraqc.com
http://www.eraqc.com

1908 Environmental Risk Limited
120 Mountain Avenue
Bloomfield, CT 06002

860-242-9933
Fax: 860-243-9055
E-mail: info@erl.com
http://www.erl.com

Environmental consulting and engineering firm offers environmental permitting and compliance assistance, site investigation and remediation services, air quality impact analyses, pollution prevention planning, aquatic toxicitylaboratory, hazardous waste management and chemical accident prevention program assistance.
David I Brandwein, Esq., Principal

1909 Environmental Risk Management
3109 N McColl Road
PO Box 3213
McAllen, TX 78502-3213

956-686-6569
800-880-9582
Fax: 956-668-7227
E-mail: office@enrisk.com
http://www.enrisk.com

Provides experienced environmental consulting to South Texas. Services includes Phase I, II and III environmental site assessments, leaking petroleum storage tank assessments and project management, asbestos inspections and managementplans, remedial services and non process waste management.
Mark Barron, President

1910 Environmental Science Associates
225 Bush Street
Suite 1700
San Francisco, CA 94107

415-896-5900
Fax: 415-896-0332
E-mail: mabell@esassoc.com
http://www.esassoc.com

Environmental Science Associates is an environmental consulting firm committed to helping clients meet the environmental challenges of tommorrow today.
Marty Abell, VP

1911 Environmental Science Services
401 Wampanoag Trail
Suite 400
Providence, RI 02915

401-434-5560
Fax: 401-434-8158
E-mail: questions@essgroup.com
http://www.essgroup.com

Environmental Science Services is an multi-disciplinary environmental consulting and engineering firm with offices located in Wellesley, Massachusetts and Providence, Rhode Island. ESS was established in 1979, and has experiencedsteady growth and market diversification to become a recognized leader in the environmental consulting and engineering services business.
Charles J Natale, Jr., President
Robert V Bibbo, CEO

1912 Environmental Strategies Corporation
11911 Freedom Drive
Suite 900
Reston, VA 20190

703-709-6500
Fax: 703-709-8505
E-mail: jchizzonite@escva.com
http://www.environmental-strategies.com

ECS is a complete environmental consulting, management, and engineering firm specializing in identifying potential or actual environmental liabilities and preventing or remediating them.
John Simon, Executive Partner
Jan Chizzonite, Managing Executive Partner

1913 Environmental Testing and Consulting
2790 Whitten Road
Memphis, TN 38133

901-213-2400
Fax: 901-213-2440
E-mail: nathan.pera@etcmemphis.com
http://www.etcmemphis.com

Environmental laboratory
Nathan Pera, President

1914 Enviroplan
81 Two Bridges Road
Fairfield, NJ 07004

973-575-2555
Fax: 973-575-6617
E-mail: contact@enviroplan.com
http://www.enviroplan.com/

Enviroplan Consulting is one of the nation's leading environmental consulting compaines. Since 1972, our staff has conducted over 2,000 studies and monitoring programs for over 350 industrial and governmental clients.
Howard Ellis, Contact

1915 Epcon Industrial Systems NV, Ltd
17777 I-45 South
Conroe, TX 77385

936-273-3300
800-447-7872
Fax: 936-273-4600
E-mail: epcon@epconinc.com
http://www.epconind.com

Provides a broad line of technology advanced, yet user friendly air pollution control products, finishing systems and heat processing equipment.

1916 Floyd Browne Associates
107 North Main Street   740-383-2187
Suite 200   800-325-7647
Marion, OH 43302-3029   http://www.floydbrowne.com

Services include water supply, treatment, transmission, distrobution, wastewater treatment and collection, solid waste management, environmental compliance, hydrological/geophysical assessments, investigations and remediation, civiland structural engineering, contract operations, and land development.
Jay W Shutt, PE, President
Jay M Shoup, PE, Office Director

1917 Foothill Engineering Consultants
350 Indiana Street   303-278-0622
Suite 415   Fax: 303-278-0624
Golden, CO 80401   E-mail: dpunceles@foothillmc.com
http://www.foothillmc.com

Offers services in environmental studies and design, decontamination and decommissioning support, mining services, radiological engineering, waste management, civil engineering, waste management, water resources planning andengineering, and cultural, natural, and physical resources evaluation.
Darrin Punceles

1918 Franklin D Aldrich MD, PhD
1094 Quince Avenue   303-443-2316
Boulder, CO 80304-0704   Fax: 303-938-9420
E-mail: w1@fa@hotmail.com

Environmental/ clinical toxicology and consulting.

1919 GEO/Plan Associates
30 Mann Street   617-740-1340
Hingham, MA 02043-1316

Michu Tcheng, Partner
Peter Rosen, Partner

1920 Gabbard Environmental Services
7611 Hope Farm Road   260-493-2982
Fort Wayne, IN 46815-6541   Fax: 219-493-4043

Consulting services for environmental affairs such as permitting, compliance, plans and programs.
William D Gabbard, President

1921 Galson Corporation
6601 Kirkville Road   315-432-0506
East Syracuse, NY 13057   800-950-0506
Fax: 315-437-0509

Environmental consulting and engineering and analytical services, specializing in the air management of indoor and outdoor environments.
Cindy Kuiper

1922 Geo-Marine Technology
899 Windridge Circle   760-736-4805
San Marcos, CA 92078   Fax: 760-744-2306
E-mail: jr@geomarinetech.com
http://www.geomarinetech.com

Provides geological, geophysical, and hydrographic survey consultancy services to offshore oil and gas industries, offshore survey industries, and governments.

1923 GeoResearch
7806 MacArthur Boulevard   301-229-8111
Cabin John, MD 20818-2043   Fax: 301-229-7980
http://www.georesearch.com

Provides grography-related services from forestry to telecommunications, real-time mobile interactive geographic technologies and databases.

1924 Geomet Technologies
20251 Century Boulevard   301-428-9898
Suite 300   Fax: 301-428-9482
Germantown, MD 20874   E-mail: marketing@geomet.com
http://www.geomet.com

Provides consultant, technical, and material evaluation services in the areas of personal protective systems, indooor and ambient air quality, energy, chemical testing, and environmental services to government agencies, private andcommercial clients.

1925 Geospec
17912 Sotile Drive   225-753-8811
Baton Rouge, LA 70809   877-503-5618
Fax: 225-753-8877
E-mail: info@geospec-llc.com
http://www.geospec-llc.com

Geophysical services. Gound penetrating radar, EM, conductivity and resistivity surveys to detect and identify potential environmental hazards or contamination. Borehole and excavation utility clearance.

1926 Gradient Corporation
20 University Road   617-395-5000
Cambridge, MA 02138   Fax: 617-395-5001
E-mail: info@gradientcorp.com
http://www.gradientcorp.com

A consulting firm with nationally recognized specialities in risk and environmental sciences.

1927 Granville Composite Products Corporation
6 North Parkway Court   717-247-2879
Lewistown, PA 17044   800-350-4660
Fax: 412-291-3291
E-mail: infosales@granville.cc
http://www.granville.cc

Plastic recycling. Molder of recycled pastics.

1928 Great Lakes Educational Consultants
4109 Apple Bluff Drive   269-382-2314
Kalamazoo, MI 49006-1953   Fax: 616-382-6495
E-mail: rjonaiti@kresanet.org

A consulting firm which develops safety/security/emergency plans to protect the educational/business environment including buildings, grounds, personnel and students. We conduct a hazard analysis to determine planning requirements andprovide a proposal for your consideration.
Robert Jonaitis, President

1929 Greeley-Polhemus Group
105 South High Street   215-692-2224
West Chester, PA 19382-1008   Fax: 215-692-4052

Specializes in providing consulting services to the United States Army Corps of Engineers and to non-Federal local sponsors of proposed Federal projects. Provides services in the areas of project planning, economics, finance,institutional strategy development, and environmental studies related to flood control, land uses, recreation, water supply, and navigation.

1930 Greystone Environmental Consultants
5231 South Quebec Street   303-850-0930
Greenwood Village, CO 80111-1809   Fax: 303-721-9298
E-mail: greystone@greystoneus.com
http://www.greystoneus.com

Environmental consultants
Randy Schroeder, President

1931 Groundwater Technology
100 River Ridge Drive   781-769-7600
Norwood, MA 02062   800-635-0053
Fax: 781-769-7992

Groundwater Technology has been a leader in the development and application of advanced technologies for environmental restoration of contaminated sites. One of the largest environmental consulting, engineering and remediation firms.GTI is widely recognized for its innovative, bioremedial technology for rapid cleanup of soil and groundwater, both above-ground and in situ.

1932 HC Nutting Company
611 Lunken Park Drive   513-321-5816
Cincinnati, OH 45226-1813   Fax: 513-321-0294
E-mail: cincinnati@hcnutting.com
http://www.hcnutting.com

Materials tesing company, geotechnical and environmental engineering firm.
Jack Scott, CEO
Jim Cahill, CFO

1933 HE Cramer Company
8249 Shangrila Circle 801-561-4964
Sandy, UT 84094-1322 Fax: 801-562-4964
E-mail: checo1@qwest.net
H.E. Cramer Company does air pollution consulting, computer software development, and environmental consulting. Research results are published in project reports and professional journals.

Jay R Bjorklund, President

1934 HYGIENETICS Environmental Services
432 Columbia Street 617-621-0363
Suite 16A Fax: 617-621-1609
Cambridge, MA 02141 http://www.hygienetics.com
Hygienetics provides comprehensive analysis, design, and program management services to a diverse group of private sector customers. Primary areas of expertise include environmental site assessments for property transaction, soil andgroundwater investigation and remediation, air resource management, industrial hygiene and asbestos/lead management.
Carmen Pombiero, General Information Contact

1935 Harold I Zeliger PhD
1270 Sacandaga Road 518-882-6800
West Charlton, NY 12010 Fax: 518-882-6926
E-mail: hiz@zeliger.com
http://www.zeliger.com
Areas of expertise include occupational and environmental exposure to toxic chemicals, hazard communication, chemical formulating and processing impact and toxic waste.
Dr Harold I Zeliger, Principal

1936 Hart Crowser
1910 Fairview Avenue East 206-324-9530
Seattle, WA 98102-3699 Fax: 206-328-5581
E-mail: rick.moore@hartcrowser.com
http://www.hartcrowser.com

Rick Moore, Principal, Env. Svce. Mgr.

1937 Hasbrouck Geophysics
2473 North Leah Lane 928-778-6320
Prescott, AZ 86301 Fax: 928-925-4424
E-mail: jim@hasgeo.com
http://www.hasgeo.com
Over 30 years experience in all major surface, airbone, and borehole geophisical methods plus strong geological background.
Jim Hasbrouck, Principal

1938 HazMat Environmental Group
60 Commerce Drive 716-827-7200
Buffalo, NY 14218 Fax: 716-827-7217
E-mail: rwickham@hazmatinc.com
http://www.hazmatinc.com
Transportation services - specializing in hazardous materials and hazardous waste transportation.

Ricky Wickham, General Manager of Operation

1939 Heritage Environmental Services
7901 West Morris Street 317-243-7475
Indianpolis, IN 46231 877-436-8778
Fax: 317-486-5095
E-mail: webmaster@heritage-enviro.com
http://www.heritage-enviro.com
Provides environmental management, integrated environment remediation services, product recovery and recycling services, waste services, analytical services, consulting and engineering services, and plant and industrial services.
Mike Karpinski, Quality Manager

1940 Hermann Associates
117 Church Road 847-446-7640
Winnetka, IL 60093-3903

1941 Huff and Huff
512 West Burlington 708-579-5940
Suite 100 Fax: 708-579-3526
La Grange, IL 60525

E-mail: jhuff@huffnhuff.com
http://www.huffnhuff.com
Multi-diciplined firm providing environmental, civil, and chemical engineering and consulting services.
James Huff, PE
Richard Trzupek

1942 Hydrogeologic
1165 Herndon Parkway 703-478-5186
Suite 900 Fax: 703-471-4180
Herndon, VA 21070
Jack Robertson, President

1943 In-Flight Radiation Protection Services
211 E 70th Street 212-288-7201
Suite 12G E-mail: robbarish@aol.com
New York, NY 10021 http://robbarish.tripod.com
Our mission is to educate flight crew members and business frequent flyers about the risks of cosmic radiation exposure during air travel.

1944 Integrated Chemistries Inc.
Po Box 10558 651-426-3224
White Bear Lake, MN 55110 Fax: 651-426-3114
E-mail: info@integratedchemistries.com
http://www.capsuleinc.com
Environmental management.
James E Nash, President

1945 Integrated Environmental Management
9040 Executive Park Drive 865-531-9140
Suite 205 Fax: 865-531-9130
Knoxville, TN 37923
IEM is a women-owned small business that provides strategic consulting and services in the areas of radiation, radioactivity, and the environment.

1946 International Certification Accreditation Board
PO Box 2099 847-724-6631
Glenview, IL 60025 Fax: 847-724-4223
E-mail: icab@icab.cc
http://www.icab.cc
A legally recognized, nonprofit, accreditation organization. Established in 2000, ICAB accredits credentialing programs of environmental, safety, medical, pharaceutical, information technology, educational, industrial and trainingorganizations.
James Wade, President

1947 Interpoll Laboratories
4500 Ball Road NE 763-786-6020
Circle Pines, MN 55014 Fax: 763-786-7854
E-mail: interpoll@interpoll-labs.com
http://www.interpoll-labs.com

Dan Despen, President
Timothy MacDonald, Manager Field Services

1948 Jack J Bulloff
8140 Township Line Road 317-824-0014
Indianapolis, IN 46260-5832 E-mail: jbulloff@ind.net
Environmental consultant
Jack J Bulloff

1949 James Anderson and Associates
2123 University Park Drive 517-349-8066
Suite 130 Fax: 517-349-7870
Okemos, MI 48864 E-mail: info@jaa-hlp.cpm
http://www.safe-at-work.com
A leading worldwide provider of noise control, sound exposure, and hearing loss prevention services for the industry.
Lee D Hagev, Executive VP

1950 James W Sewall Company
136 Center Street 207-827-4456
PO Box 433 Fax: 207-827-3641
Old Town, ME 04468-0433 E-mail: info@jws.com
http://www.jws.com

Founded in 1880, Sewall provides comphrehensive services in forestry appraisal and inventory, aerial imagery, GIS consulting and engineering. Sewall's expertise in GIS project implementation is supported by 50 years' experience inaerial photography and photogrammetry and 30 years' experience in data conversion, database design and application development.
Scott E Graham, PE, Vice President
Aaron Shaw, PE, Project Manager

1951 John Zink Company

11920 East Apache
Tulsa, OK 74116

918-234-1800
800-421-9242
Fax: 918-234-2700
E-mail: jzinfo@kochind.com
http://www.johnzink.com

John Zink offers technologically advanced equipment and systems for the clean and efficient combustion of fossil fuels and for the removal of contaminates from process affluents entering the atmosphere.

1952 Kemstar Corporation

3456 Wade Street
Los Angeles, CA 90066

310-390-0180
Fax: 310-391-8143

1953 Kimre

16201 Southwest 95 Avenue
Suite 303
Miami, FL 33157

305-233-4249
Fax: 305-233-8687
E-mail: sales@kimre.com
http://www.kimre.com

Manufacturer of air/water pollution control equipment mainly used in the fertilizer and most other chemical applications.

1954 LA Weaver Company

308 E Jones Street
Releigh, NC 27601-1028

919-832-6242
Fax: 919-831-1130

1955 LSI Adapt Engineering

615 8th Avenue South
Seattle, WA 98104

206-654-7045
800-643-9932
Fax: 360-674-7048
E-mail: seattle@lsi-industries.com
http://www.lsiadapt.com

Consulting and engineering firm specializing in petroleum engineering, geotechnical and environmental issues. Team of professional engineers, licensed environmental site assessors, hydrogeologiststs and geotechnical engineers offer avariety of strengths and services to clients.

1956 Landau Associates

10 N Post Street
Suite 218
Spokane, WA 99201

509-327-9737
Fax: 509-327-9691
E-mail: information@landauinc.com
http://www.landauinc.com

Provided environmental and geotechnical services on nearly 2,000 projects for more than 500 private and public clients in the Pacific Northwest and western US since we opened our doors in 1982. A valued resource in waterfrontdevelopment for public and private ports, also expanded beyond the waterfront to serve clients in some of the best known industries, municipal government, and site development.

Steve Johnston, Principal and CEO
Dennis Stettler, Principal

1957 Law Environmental

3200 Town Point Drive NW
#100
Kennesaw, GA 30144-7088

770-421-3400
Fax: 770-421-3526

1958 Lawler, Matusky and Skelly Engineers

1 Blue Hill Plaza
Pearl River, NY 10965

845-735-8300
Fax: 845-735-7466
E-mail: cnevel@lmseng.com
http://www.lmseng.com

Environmental engineering and consulting firm. Research results published in client reports and professional journals.
Christy Nevel, Director Marketing

1959 Lenox Institute of Water Technology

101 Yokun Avenue
PO Box 1639
Lenox, MA 01240

413-637-3025
Fax: 413-637-3362

Provides services in the area of municipal and industrial water and wastewater treatment systems.
Charles L Smith, Executive VP

1960 Les A Cartier and Associates

191 Main Street
PO Box 559
Candia, NH 03034-2409

603-483-2180
800-639-7703
Fax: 603-483-8986

Environmental service company; Health and Safety Course in their Hazardous Materials Management Series
Leslie A Cartier, President

1961 Louis Berger Group

100 Halsted Street
East Orange, NJ 07018

973-678-1960
Fax: 973-672-4284
http://www.louisberger.com

Offers professional services in the areas of civil, structural, mechanical, electrical and environmental engineering; program management; planning; environmental sciences; cultural resources; information services; economics; policy andmanagement analysis; and construction management and support.
Derish Wolff, Chairman

1962 Louis Defilippi

208 Edgewood Lane
Palatine, IL 60067

847-925-8524
Fax: 847-303-1731
E-mail: defilip1@flash.net
http://www.flash.net/~defilip1/Default.htm

We offer consulting services in three broad areas: industrial biotechnology, bioprocessing, and proteomics; environmental and regulatory compliance; biotechnology and applied engineering for the microbiological treatment of hazardouswaste. We are especially valuable when you don't really need a full time in-house expert, but need someone to take an important load off your shoulders, to review your processes and procedures, or just to get things moving faster, on an as-neededbasis.

Dr. Louis DeFilippi, President

1963 Marc Boogay Consulting Engineer

326 Main Street
Vista, CA 92084

760-407-4000
Fax: 760-407-4004
E-mail: boogay@sdnc.quik.com
http://www.boogay.com

Marc Boogay and staff have completed more than 600 site assessments. This work has include all varieties of developed and undeveloped properties. Projects have range from Phase I investigations through sampling surveys, remediationdesigns, abatement monitoring, risk assessment, and expert witness testimony. Projects have benn conducted in several states, meeting standards of government agencies as well as many leading/investment institutions.
Marc Boogay, PE, Principal
Todd Jacquay, Soils Engineer

1964 McVehil-Monnett Associates

44 Inverness Drive East
Building C
Englewood, CO 80112

303-790-1332
Fax: 303-790-7820
http://www.mcvehil-monnett.com

MMA is a experienced consulting firm of atmospheric scientists, engineers and environmental specialists providing air quality and environmental management system (EMS) services worldwide. Serves the mining, oil and gas, electric powerand manufacturin industries, as well as government agencies and engineering and law firms. Leader in air quality permitting, modeling, monitoring and litigation supoort services as well as environmental management and planning services.

William R Monnett, President
George McVehil, Principal

1965 Mercury Technology Services

23014 Lutheran Church Road
Tomball, TX 77377

281-255-3775
Fax: 281-357-0721
E-mail: smw@htech.com
http://www.hgtech.com

Specialists in solving problems related to mercury pollution and contamination. MTS provides technical services to companies having metals contamination and expert testimony on mercury pollution and remediation.
Mark Wilhelm, President

1966 Meteorological Evaluation Services Company

165 Broadway   516-691-3395
Amityville, NY 11701   Fax: 516-691-3550
http://www.mesamity.com

Consultants in Applied Meteorology, Air Quality and the Environment.

1967 Miceli Kulik Williams and Associates

39 Park Avenue   201-933-7809
Rutherford, NJ 07070   Fax: 201-933-8702
E-mail: jwilliams@mkwla.com
http://www.mkwla.com

Offers complete services covering the various aspects of landscape architecture, site planning, and urban design. Present scope of work includes neighborhood rehabilitation, housing and community development, park, recreational andopen space planning, landscape architecture, impact assessment, educational, municipal, commercial and industrial commissions. Project involvement extends throughout the Eastern States.

1968 Michael Baker Corporation

Airside Business Park
100 Airside Drive   412-269-6300
Moon Township, PA 15108   800-MIB-AKER
Fax: 412-375-3980
E-mail: CorpCom@mbakercorp.com
http://www.mbakercorp.com

Michael Baker Corporation has evolved into one of the leading engineering and energy management firms by consistently solving complex problems for its clients. A project's challenges are not viewed as obstacles, but as invitations toinovate.
David Higie, Corporate Communications

1969 Michael Brandman Associates

220 Commerce   714-508-4100
Suite 200   Fax: 714-508-4110
Irvine, CA 92602   http://www.brandman.com

Michael Brandman Associates is a comprehensive environmental planning services firm specializing in environmental documentation, planning, and natural resources management.
Erika Bennett, Marketing Manager
Michael Brandman, President/CEO

1970 Micro-Bac

3200 North Interstate Highway 35   512-310-9000
Round Rock, TX 78681-2410   877-559-1800
Fax: 512-310-8800
E-mail: mail@micro-bac.com
http://www.micro-bac.com

Delvelops and manufactures biological products for remediation of contaminated substances; reduction of waste and odor in food processing, agriculture, and sewage; and control of paraffin in oil production.

1971 Midstream Farm

3117 Stone Arbor Lane   804-262-6552
Suite 1133   877-641-2536
Glen Allen, VA 23059   Fax: 804-262-6555
http://www.usaclem.com

Clement Mesavage Jr, Proprietor

1972 Milton R Beychok: Consulting Engineer

1126 Colony Plaza   949-718-1360
Newport Beach, CA 92660   Fax: 949-718-1360
E-mail: milt@air-dispersion.com
http://www.air-dispersion.com

Environmental assessments.
Milton R Beychok, Principal

1973 Mostardi Platt Environmental

1520 Kensington Road   630-993-2100
Suite 204   Fax: 630-993-9017
Oak Brook, IL 60523-2139 http://www.mostardiplattenv.com

Mostardi Platt Environmental-your full service environmental management partner. Offers innovative solutions and strategies to assist our clients comply with environmental, health and safety regulations and develop environmentalprograms that save long-term costs. We understand our clients need the best possible compliance options. We evaluate a wide variety of technical and economic concerns and work with our clients to establish the best path towards compliance.
Joseph J Macak III, President
Robert A Gere, Engineering Consultant

1974 NTH Consultants

38955 Hills Tech Drive   248-553-6300
PO Box 9173   Fax: 248-324-5179
Farmington Hills, MI 48333   http://www.nthconsultants.com

NTH Consultants has provided consulting engineering services to clients throughout the United States since 1968. Headquartered in Farmington Hills, MI, NTH has maintained an office in downtown Detroit since 1980, a regional,full-service office in Exton, PA and offices in Lansing and Grand Rapids, MI since 1992.
Jerome C Neyer, Chairman

1975 National Environment Management Group

PO Box 5131   708-771-7350
River Forest, IL 60305   Fax: 312-733-2478

Environmental consulting.
Jack Hughes, Chairman

1976 National Environmental

1019 W Manchester Boulevard   310-645-4516
Suite 102   800-870-1719
Inglewood, CA 90301   Fax: 310-645-0148
E-mail: customerservice@natlenviro.com

Training school. Training in use of lead, asbestos and hzardous materials.
James McFarland, President
David P Fuller, VP

1977 National Sanitation Foundation

789 North Dixboro Road   734-769-8010
PO Box 130140   800-NSF-MARK
Ann Arbor, MI 48113-0140   Fax: 734-769-0109
E-mail: info@nsf.org
http://www.nsf.org

NSF International, The Public Health and Safety Company, is an independent, not for profit organization providing a wide range of services around the world. For more than 55 years, NSF has been committes to public health, safety andprotection of the environment.
Robert Ferguson, Vice President
Tim Bruursema, General Manager

1978 National Society of Environmental Consultants

303 West Cypress Street   210-271-0781
PO Box 12528   800-486-3676
San Antonio, TX 78212-0528   Fax: 210-225-8450
E-mail: lincolncenter@worldnet.att.net
http://nsec.lincoln-grad.org/

The mission is to encourage an awareness of environmental risk and the regulations regarding their impact on real property value, to advocate reponsible use and development of real estate resources in harmony with the environment, toelevate the competency of the membership through information and education and to promote the development of ethics and standards of professional practice for the speciality of environmental consultants
Gary T Deane, Executive Director

1979 Natural Resources Consulting Engineers

131 Lincoln Avenue   970-224-1851
Suite 300   Fax: 970-224-1885
Fort Collins, CO 80524   E-mail: office@nrce.com
http://www.nrce.com

Water supply investigations. Native American water rights expert witness testimony.

1980 Network Environmental Systems

1141 Sibley Street   916-353-2360
Folsom, CA 95630-4701   800-637-2384
Fax: 916-353-2375

E-mail: office@networkenvironmental.com
http://www.networkenvironmental.com

Network Environmental Systems was incorporated in 1988 to privide high quality professional industrial hygiene and environmental management services through customer service excellence.
Jerry Bucklin, President/CEO
Donald Rothenbaum, Senior Vice President

1981 Ninyo and Moore

5710 Ruffin Road 858-576-1000
San Diego, CA 92123 Fax: 858-576-9600
E-mail: nminquiries@ninyoandmoore.com
http://www.ninyoandmoore.com

As a leading geotechnical and environmental scieces engineering and consulting firm, Ninyo & Moore provides specialized services to clients in both public and private sectors.
Avram Ninyo, Principal Engineer

1982 Nordlund and Associates

813 East Ludington Avenue 231-843-3485
Ludington, MI 49431 Fax: 231-843-7676
E-mail: Nordlund@T-one.net

Water systems, wastewater treatment, sanitary landfills and hydrogeological studies.
James T Nordlund Sr, President

1983 Normandeau Associates

25 Nashua Road 603-472-5191
Bedford, NH 03110-5500 Fax: 603-472-7052
E-mail: nai@normandeau.com or pkinner@normandeau.com
http://www.normandeau.com

Normandeau Associates is an employee owned natural resources management consulting and testing services firm that provides: permit assistance, water quality studies, aquatic and terrestrial ecology, environmental impact assessments,property transfer site assessments, wetlands services, contamination studies and biological laboratory services.
Pamela Hall, President
Peter Kinner, Senior VP

1984 Norton Associates

46 Leland Road 508-528-3357
Norfolk, MA 02056 E-mail: norton@designofmachinery.com
http://www.designofmachinery.com

Professor Norton and his associates have been providing engineering consulting services since 1970. Areas of expertise include: cam design and analysis, linkage design and analysis, street analysis, vibrations in machinery, dynamicsignal analysis, machinery monitoring, and machine dynamic analysis. We also can provide short courses and seminars on site in cam design, dynamic signal analysis and machinery vibrations.
Robert L Norton, President

1985 NuChemCo

5765-F Burke Centre Parkway 703-548-3200
#149 800-682-4362
Burke, VA 22015-2233 Fax: 703-978-0642
E-mail: info@nuchemco.com
http://www.nuchemco.com

Neil B Jurinski
Joseph B Jurinski

1986 OCCU-TECH

6501 East Commerce Avenue 816-231-5580
Suite 230 800-950-1953
Kansas City, MO 64120-2141 Fax: 816-231-5641
E-mail: service@occutec.com
http://www.occutec.com

OCCU-TECH is a leading safety, health and environmental services company. From OSHA to EPA issues, safety assessments to program development, asbestos inspections to environmental management, our expertise has been relied on for over16 years.

1987 Oak Creek

60 Oak Creek 207-929-6375
Buxton, ME 04093-6616 Fax: 207-929-6374
E-mail: oak-creek@oak-creek.net
http://www.oak-creek.net

James S Smith Jr, PhD, President/Toxicologist
Brad House, Senior Scientist

1988 Occupational Health and Safety Management

117 La Farge 303-665-8528
Louisville, CO 80027-1715 Fax: 303-673-0785

Industrial hygiene/safety consulting.
Mary Ann Heaney

1989 Occupational Safety and Health Consultants

12000 6th Street East 727-345-1552
Saint Petersburg, FL 33706-5100 Fax: 727-363-8151
E-mail: oshc@oshc.com
http://www.oshc.com

Air pollution control/industrial hygiene.

1990 Occupational and Environmental Health Consulting Services

635 Harding Road 630-325-2083
Hinsdale, IL 60521-4814 Fax: 630-325-2098
E-mail: bobb@safety-epa.com
http://www.safety-epa.com

A full service regulatory, safety, industrial hygiene, and environmental engineering consulting firm. Specialize in assisting all sizes of companies and corporations. Clients include very small businesses up to Fortune 100corporations.
Bob Brandys PhD,MPH,PE,CIH, President

1991 Occusafe

135 Mountain View Drive 413-323-1036
Belchertown, MA 01007 Fax: 413-323-1039
E-mail: occusafe@map.com
http://www.occusafe.net

OCUSAFE is a full service consulting firm specializing in assistance to management in the areas of occupational safety, industrial hygiene, and environment.

1992 Ocean City Research

50 Tennessee Avenue 609-399-2417
Ocean City, NJ 08226 Fax: 609-399-5233
E-mail: PAult@corrpro.com
http://www.corrpro.com

Ocean City Research Corporation, incorporated in 1963, is a wholly owned subsidiary of Corrpro Companies, Collectively, the Corrpro affiliated companies represent the largest, independent consulting corrosion engineering organizationin the world.
J Peter Ault, PE

1993 Omega Waste Management

1900 Highway 90 W 985-399-5100
Patterson, LA 70392-5506 Fax: 985-399-7963

A consulting firm, whose unique and innovative approach to waste removal and recycling has made it one of the largest volume purchasers of waste services in the nation.

1994 Owen Engineering and Management Consultants

5353 West Dartmouth Avenue 303-969-9393
Suite 509 Fax: 303-969-9394
Denver, CO 80227

Water/Wastewater design systems.
Webster J Owen, President

1995 PACE Analytical Services

1700 Elm Street 612-607-1700
Suite 200 Fax: 612-607-6444
Minneapolis, MN 55414 E-mail: info@pacelabs.com
http://www.pacelabs.com

Provider of air, water, soil and environmental testing services.
Bruce E Warden, General Manager
Jeff Smith, Marketing Contact

1996 PAR Environmental

1906 21st Street 916-739-8356
PO Box 160756 Fax: 916-739-0626
Sacramento, CA 95816 E-mail: mlmaniery@aol.com
http://www.parenvironmental.com

PAR Environmantal Services mission is to provide technical reports on time, within budget, and with meticulous attention to detail.
Mary L Maniery, President

1997 PBR HAWAII

1001 Bishop Street, ASB Tower   808-521-5631
Suite 650   Fax: 808-523-1402
Honolulu, HI 96813-3484 E-mail: sysadmin@pbrhawaii.com
    http://www.pbrhawaii.com

Consulting services in environmental studies, land planning and landscape architecture.
Thomas S Witten ASLA, President
Frank Brandt Falsa, Chairman

1998 PBS Environmental Building Consultants

4412 SW Corbett Ave   503-248-1939
Portland, OR 97239   888-248-1939
    Fax: 503-248-0223
    http://www.pbsenv.com

PBS specializes in program development, identification, assessment, testing and corrective action consultation in the areas of: Environmental Engineering, Geotechnical Engineering, Hazardous Materials Management, Industrial HygieneServices, Natural Resources Studies, Training and Laboratory.
Stephen Smiley, President

1999 PE LaMoreaux and Associates

PO Box 2310   205-752-5543
Tuscaloosa, AL 35405   Fax: 205-752-4043
    E-mail: info@pela.com
    http://www.pela.com

For over three decades, PELA's integration of qualified personnel, up-to-date technology, and sound management has established PELA as an international leader in the environmental consulting field. PELA's expertise in hydrogeology,geotechnical analysis, design and construction management, remediation, computer graphics and models, and permitting can get your project on two feet quicker than you might think.
James W Moreaux, President
Bashir A Memon, Executive VP

2000 PEER Consultants

12300 Twinbrook Parkway   301-816-0700
Suite 410   Fax: 301-816-9291
Rockville, MD 20852   E-mail: peercpc@peercpc.com
    http://www.peercpc.com

For nearly a quarter of a century, PEER Consultants has provided civil, sanitary, and environmental engineering consulting services for public and private sector clients nationwide.
Lilia Abron, President

2001 Pacific Soils Engineering

10653 Progress Way   714-220-0770
Cypress, CA 90630   Fax: 714-220-9589

Services include: Geotechnical Services, Laboratory Testing, Field Observation and Testing, Consultation and Review of Geotechnical Reports.

Daniel Martinez, President

2002 Parish, Weiner and Maffia

101 Executive Boulevard   914-345-9230
Elmsford, NY 10523   Fax: 914-345-8972
    E-mail: pwsioi@aol.com

Environmental impact studies, traffic studies, land planning, zoning studies, community planning, feasibility studies.
Nat Parish

2003 Pavia-Byrne Engineering Corporation

6305 Elysian Fields Avenue   504-283-5080
New Orleans, LA 70184   Fax: 504-283-4090

Provides services for environmental control and water treatment including definition, process development, and start up services.
Edgar H Pavia, President

2004 Perry-Carrington Engineering Corporation

214 West Second Street   715-384-2133
Marshfield, WI 54449   Fax: 715-384-9797
    E-mail: 2perryear@temet.com

Water pollution control systems.
David L LaFontaine, President

2005 Petra Environmental

10550 North 6th Avenue   715-536-7870
Merrill, WI 54452   800-458-3772
    Fax: 715-536-7890
    E-mail: info@petraenvironmental.net
    http://www.petraenvironmental.net

PETRA Environmental Consultants, is an environmental engineering firm specializing in environmental compliance, hydrogeological investigations, and environmental assessments.
Anthony M Ungerer, PG, Principal Hydrogeologist
Christopher J Rog, PG, Senior Hydrogeologist

2006 Phase One

2680 Walnut Avenue   714-669-8055
Suite B   800-524-8877
Tustin, CA 92780   Fax: 714-669-8025
    E-mail: info@phasei.com
    http://www.phasei.com

A focused environmental consulting practice that specializes in real property assessments for any type of property transfer, leasing development, special uses, and/or financing purposes. Founded in response to the business community'sneed for affordable, standardized and consistently high quality assessment reports that provide recommendations for sound real estate decisions.
Eric D Kieselbach, President/CEO

2007 Planning Resources

402 W Liberty Drive   630-668-3788
Wheaton, IL 60187-4937   Fax: 630-668-4125
    http://www.planres.com

Land use and environmental planning.
Keven Graham, Managing Principal

2008 Post, Buckley, Schuh and Jernigan

2001 107th Avenue Northwest   305-592-7275
Miami, FL 33172-2507   800-597-7275
    Fax: 305-599-3809
    http://www.pbsj.com

PBS&J was founded in 1960 by four respected engineers who joined forces to help develop Florida's first planned community. Their tenacity in meeting production schedules, commitment to client service, and ability to provide innovativesolutions to difficult challenges quickly earned our firm a reputation for excellence and laid the foundation for future grouth.
Todd J Kenner, President

2009 Presnell Associates

815 West Market   502-585-2222
Louisville, KY 40202   Fax: 502-581-0406
    E-mail: presnell@thepoint.net
    http://www.qk4.com

The professional practice of Prenell encompasses a variety of services directly related to preserving the environment, including potable water system planning and design, municipal and industrial wastewater treatment, solid wastemanagement, landfill siting, asbestos management, contamination screening assessments, indoos air quality, and lead paint abatement.
Wendell Wright, President

2010 Priester and Associates

1345 Garner Lane   803-798-4377
Suite 105   877-798-4377
Columbia, SC 29210   Fax: 803-798-4378
    E-mail: priester@conterra.com

Provides personalized environmental services ranging from short-term consulting to extensive remediation and management activities.
LE Priester, President

2011 Process Applications

2627 Redwing Road   970-223-5787
Suite 340   Fax: 970-223-5786
Fort Collins, CO 80526-6310

Environmental engineering consultants.
Bob A Hegg, President

2012 Professional Analytical and Consulting Services (PACS)
409 Meade Drive   724-457-6576
Coraopolis, PA 15108   800-367-2587
Fax: 724-457-1214
E-mail: hnpacs@aol.com

Training courses and conferences. Provides short courses in spectrocopy, chomatography, quality, safety, environmental, and management. Provides professional manuals and software products. Provides laboratory testing and consultingservices. Company also goes by the following names: Activated Carbon Services, PACS Testing and Consulting, PACS Courses and Conferences.
Henry G Nowicki PhD, President
Barbara Sherman, Manager of Operations

2013 Psomas and Associates
4540 California Avenue   661-631-2311
Suite 210   866-9PS-OMAS
Bakersfield, CA 93309   Fax: 661-631-2782
E-mail: info@psomas.com
http://www.psomas.com

Psomas is a leading consulting engineering firm offering services in land development, water and natural resources, transportaion, public works, survey and information systems to public and private sector clients.
George Psomas, Chairman

2014 QORE
4201 Pleasant Hill Road Northwest   770-232-0235
Suite A   877-767-3462
Duluth, GA 30096   Fax: 770-232-0238
E-mail: corporate@qore.net
http://www.qore.net

Consultants in property science, in fields of geology, geotechnical and environmental engineering.
Richard D Heckel, PE, President
Ed Heustess, Chief Financial Officer

2015 RDG Geoscience and Engineering
10360 Sapp Brothers Drive   402-894-2678
Omaha, NE 68138   888-260-0893
Fax: 402-894-9043
E-mail: info@rdgge.com
http://www.rdgge.com

Is an earth science and engineering consulting firm that has completed over 1200 projects throughout the mid-west and mountain west of US.
Jon Gross, President
Robert Kalinski, Vice President

2016 RGA Environmental
1466 66th Street   510-547-7771
Emeryville, CA 94608-2907   800-776-5696
Fax: 510-547-1983
E-mail: rga@rgaenv.com
http://www.rgaenv.com

Founded in 1985, RGA Environmental is a specialty consultant in the environmental sciences. Our mission is to provide high-quality environmental engineering, health & safety consulting services to meet the special needs of our clients.
Steven C Rosas, COO, Director of Business

2017 RMT
744 Heartland Trail   608-831-4444
PO Box 8923   800-283-3443
Madison, WI 53717-1934   Fax: 608-831-3334
E-mail: info@rmtinc.com
http://www.rmtinc.com

We serve industrail compaines throughout the world who value environmental and engineering solutions that improve productivity and profitability. RMT's diverified staff of over 550 engineers, scientists and technicians takesresponsibility for managing environmental issues so our clients can concentrate on their core business.
Jodi Burmester, Corporate Communications

2018 Raterman Group
75 East Wacker Drive   312-345-0111
Suite 500   Fax: 312-345-9950
Chicago, IL 60601   E-mail: susan@ratermangroup.com
http://www.ratermangroup.com

Industrial hygiene and environmental assessments.

Susan M Raterman, President

2019 React Environmental Engineers
1120 South 6th Street   314-678-1398
St. Louis, MO 63104   800-325-1398
Fax: 314-678-6610
E-mail: react@mvp.net
http://www.react-env.com

Turnkey environmental services including site assessments, remedial action and emergency response.
Stewart Ryckman, President
Henry Stremlav, VP Operations

2020 Reclamation Services Unlimited
701 Temple Street   270-754-3976
Central City, KY 42330   Fax: 270-754-4374

Environmental consulting services.
Sue Poole Cardwell, President

2021 Redniss and Mead
22 1st Street   203-327-0500
PO Box 3247   800-404-2060
Stamford, CT 06905-0247   Fax: 203-357-1118
http://www.redniss-mead.com

Over the years the firm has expanded to become leading land use consultants, producing many innovative designs by combing traditional skills with creative planning and environmental senitivity.
Aubrey E Mead, Jr., PE, VP
Raymond L Redniss, PLS, Senior Vice President

2022 Refuse Management Systems
99 Tulip Avenue   516-354-1212
#303   800-346-5926
Floral Park, NY 11001-1974   Fax: 516-354-2434
E-mail: enviroeq@ix.netcom.com

Environmental consultants.
Harvey Podolsky, President

2023 Regional Services Corporation
3200 Sycamore Court   812-372-9511
Suite 2C   Fax: 812-372-9520
Columbus, IN 47203

Solid waste disposal.
Mark Richards, President

2024 Regulatory Management
6190 Lehman Drive   719-531-6883
Suite 106   Fax: 719-599-4410
Colorado Springs, CO 80918-3445   E-mail: maxlab@usa.net

Environmental consulting group.
James T Egan, President

2025 Resource Applications
9291 Old Keene Mill Road   703-644-0401
Burke, VA 22015-4202   Fax: 703-644-7143

Hazardous waste management, pollution prevention/site remediation.
Paul Singh, Director, Corp. Development

2026 Resource Concepts
340 North Minnesota Street   775-883-1600
Carson City, NV 89703   Fax: 775-883-1656
E-mail: john@rci-nv.com
http://www.rci-nv.com

RCI has years of experience and demonstrated accomplisjment working with environmentally sensitive projects. Combining technical abilities and excellent working relationships with regulatory agencies results in highly effective projectplanning and permitting services.
Bruce R Scott, Principal
John McLain, Principal

2027 Resource Conversion Systems
10190 Old Katy Road   713-461-9484
Suite 430   800-706-4070
Houston, TX 77043   Fax: 713-461-9482
E-mail: hefty@granville.cc
http://www.granville.cc

Solid waste separation/recycling.
Henry O Hefty, VP

2028 Resource Decisions

934 Diamond Street   415-282-5330
San Francisco, CA 94114   E-mail: marvin@resourcedecisions.net
     http://www.resourcedecisions.net
Assisting clients to evaluate trade-offs which foster the wise allocation of resources is primary mission of Resource Decisions. To accomplish this mission we apply a wide range of economic and decision-making tools.
Marvin Feldman, PhD, Principal

2029 Resource Management

625 Chapin Road   803-345-0200
Chapin, SC 29036-8066   Fax: 803-345-6520
     E-mail: resourc9@winusa.com
Hazardous waste management.
Don Dicus, President

2030 Resource Technology Corporation

2931 Soldier Springs Road   307-745-7936
PO Box 1346   800-576-5690
Laramie, WY 82070   Fax: 307-742-5452
     E-mail: RTC@RT-corp.com
     http://www.rt-corp.com
They offer Laboratory Proficiency Testing for drinking water, waste water and USEPA RCCRA Program. Certified analytical standards and Certified Reference Materials.

2031 Respec Engineering

3824 Jet Drive   605-394-6400
Rapid City, SD 57703-4757   877-4RE-SPEC
     Fax: 605-394-6456
     http://www.respec.com
Since our founding in 1969, RESPEC has remained committed to its original purpose of providing clients with high-quality technical and advisory services.

2032 Reston Consulting Group (RCG)

462 Herndon Parkway   703-834-1155
Herndon, VA 20170-5233   Fax: 703-834-3086
     E-mail: information@rcg.com
     http://www.rcg.com

2033 Rich Tech

2410 Devonshire Drive   815-229-1122
Rockford, IL 61107-1534   Fax: 815-229-1525
Water pollution control.
Gail Rivitts, President
Rich Rivitts, Vice President

2034 Rizzo Associates

1 Grant Street   508-903-2000
Framingham, MA 01701-9005   Fax: 508-903-2001
     E-mail: rmoore@rizzo.com
     http://www.rizzo.com
A leading engineering, transportation, and environmental engineering firm. We work with you throughout the development process to reslove the challenges that arise in planning, permitting, design, and construction phases of complexprojects.
Rick Moore, President

2035 Robert B Balter Company

18 Music Fair Road   410-363-1555
Owings Mills, MD 21117   Fax: 410-363-8073
Environmental consultation.
Sharon Ames-Burgess, Marketing Director

2036 Rockwood Environmental Services Corporation

50 Kearney Road   781-449-8740
Needham, MA 02494-2508   Fax: 781-449-8741
     E-mail: bwhite@rockwood-enviro.com
     http://www.rockwood-enviro.com
Rockwood specializes in solving the problems of hazardous waste management and disposal for New England generators. By shipping wastes directly to ultimate disposal sites on a regular basis, Rockwood reduces current disposal costs andreduces long-term liability exposure.

William A White III, President

2037 Rodriguez, Villacorta and Weiss

16750 Hedgecroft   281-447-1726
Suite 500   Fax: 281-447-2299
Houston, TX 77060   E-mail: sweiss@rvw.net
     http://rvw.net/
Our mission is to provide cost-effective and thorough work product for claims services and loss control. Maximum integration of all in-house and affiliated expertise will guarantee prompt service, nurturing strong client relationshipsbased on dependability, trust and competence.
Steve Weiss, Principal

2038 Roux Associates

209 Shafter Street   631-232-2600
Islandia, NY 11749   Fax: 631-232-9898
     E-mail: sisadiker@rouxinc.com
     http://www.rouxinc.com
Environmental Consulting and Management.

Steve Sadiker, Vice President

2039 SLC Consultants/Constructors

6362 Robinson Road   716-433-0776
Lockport, NY 14094   800-932-0157
     Fax: 716-433-0802

2040 Safina

953 N Plum Grove Road   847-956-8617
Suite A-1   Fax: 847-956-8619
Schaumburg, IL 60173-5190
Environmental due diligence.

Sanjiv Pillai, General Manager

2041 Schneider Instrument Company

8115 Camargo Road   513-561-6803
Cincinnati, OH 45243   Fax: 513-527-4375
     E-mail: schneidxcompany@aol.com
Gary Schneider, Vice President

2042 Schoell and Madson

10580 Wayzata Boulevard   952-546-7601
Suite 1   Fax: 952-546-9065
Minneotonka, MN 55305   E-mail: mail@schoellmadson.com
     http://www.schoellmadson.com
We are dedicated to creatively serving our clients by meeting or exceeding their needs in a responsive and cost-effective manner while providing an interesting and rewarding experience for our employees.

2043 SciComm

7735 Old Georgetown Road   301-652-1900
12th Floor   E-mail: info@scicomm.com
Bethesda, MD 20814   http://www.scicomm.com
A professional services firm specializing in communications, engineering, environmental, and information management services. Organized to carry out the interest, expertise, and vision of co-founder Laura Chen and Dan Lewis.

Laura Chen, Founder

2044 SevernTrent Laboratories

4101 Shuffel Drive NW   330-497-9396
North Canton, OH 44720   Fax: 330-497-0772
     http://www.stl-inc.com
The two compaines merged as Wadsworth/Alert Laboratories in early 1980's and the core business focused on environmental testing, with a specialization in on-site and emergency response projects. Mobile Labs were placed as far north asMichigan, and south to Florida, east to New York, and west to Missouri.
Rachel Jannetta, President

2045 Shaw Environmental

200 Horizon Center Boulevard   609-584-8900
Trenton, NJ 08691   Fax: 609-588-6300
     E-mail: general@shawgrp.com
     http://www.shawgrp.com

Hazardous waste remediation.
Ron Prann, Division Manager

2046 Shell Engineering and Associates

2403 West Ash Street   573-445-0106
Columbia, MO 65203-0045   Fax: 573-445-0137
E-mail: Charles@shellengr.com
http://www.shellengr.com

Shell Engineering provides services firm specializing in communications, engineering, environmental monitoring and engineering. Shell Engineering has completed hundreds of projects since 1975 throughout the United States, Canada.Central America, South America, Asia and Africa.
Harvey D Shell, CEO/Chairman
Charles A Shell, President

2047 Sierra Geological and Environmental Consultants

91 South Main Street   616-678-5157
PO Box 136   Fax: 616-678-5149
Kent City, MI 49330   E-mail: info@sierraconsultants.net
http://www.sierraconsultants.net

A full service environmental consulting firm providing assassment, investigation, and cleanup services throughout Michigan and the Great Lakes States.

2048 Slakey and Associates

375 Village Square   925-254-4164
PO Box 944   Fax: 925-254-0679
Orinda, CA 94563

Consulting, civil, mechanical, environmental engineers with 40 years experience in indoor air quality, air pollution control. Design of systems and equipment for collection abatement of fugitive and source missions of dusts, odor andfumes. Industrial clients only.
Philip Slakey, President

2049 Slosky & Company

1675 Broadway   303-825-1911
Suite 1400   Fax: 303-892-3882
Denver, CO 80202   E-mail: Lslosky@slosky.com

Full service environmental consulting firm.
Leonard Slosky, President

2050 Snyder Research Company

509 Miller Creek Road   415-499-3463
San Rafael, CA 94903   http://www.sdforum.org

2051 Staunton-Chow Engineers

100 W 32nd Street   212-683-8865
7th Floor   Fax: 201-798-4992
New York, NY 10001-3120   http://www.stauntonchow.com

Known widely as a small premiere multidisciplined engineering/architectural consulting firm providing professional services for new construction, repair, alterations, and maintenance for nearly 50 years.
Kin Chow, President

2052 Strata Environmental Services

110 Perimeter Park   865-539-2077
Suite E   Fax: 865-539-3970
Knoxville, TN 37922   E-mail: info@strataenv.com
http://www.strataenv.com

Founded to provide consulting services in geosciences, engineering, air quality, water quality, regulatory compliance, and environmental due diligence.

2053 TECHRAD Environmental Services

4619 North Santa Fe Avenue   405-528-7016
Oklahoma City, OK 73118   800-375-7016
Fax: 405-528-3346

Analytical laboratory, environmental site assessments, underground storage tank management and remediation, industrial hygiene, stormwater and hazardous waste management, asbestos consulting and analysis and regulatory compliance.
Edward M Wall, President/CEO

2054 THP

40 Brunswick Woods Drive   732-257-4040
East Brunswick, NJ 08816   Fax: 732-257-7953

Engineering traffic and engineering planning consulting firm.
Lester J Nebenzahl, President

2055 Technos

10430 Northwest 31st Terrace   305-718-9594
Miami, FL 33172   Fax: 305-718-9621
E-mail: info@technos-inc.com
http://www.technos-inc.com

A geologic and geophysical consulting firm specializing in subsurface site characterization for geotechnical, environmental, and groundwater projects.

Richard Benson, Senior Scientist
Lynn Yuhr, President

2056 Terryn Barill

301 N Harrison
Suite 484   800-718-6690
Princeton, NJ 08540   Fax: 609-243-8703
E-mail: terryn1@mail.com
http://www.terryn.com

Audits/assessments, training, implementation and facilitation.

2057 Tetra Tech

3475 East Foothill Boulevard   626-351-4664
Pasadena, CA 91107   Fax: 626-351-5291
E-mail: info@tetratech.com
http://www.tetratech.com

We provide comprehensive resource management, infraestructure and communications services, including, research and development, applied science, management consulting, engineering and architectural design, construction management, andoperation and maintenance.
Sam W Box, President
Mark A Walsh, Senior VP

2058 Theil Consulting

1136 South Fort Thomas Avenue   859-781-2651
Fort Thomas, KY 41075-2440   Fax: 859-781-2356
E-mail: larry@theilair.com
http://www.theilair.com

Experts in industrial process exhausts—especially submicron particles created by heat or other high energy in a process.
Greg Theil, Technical Director
Larry Olson, Sales Manager

2059 Titan Corp. Ship and Aviation Engineering Group

3033 Science Park Road   858-552-9500
San Diego, CA 92121   Fax: 858-552-9645
E-mail: corpcomm@titan.com
http://www.titan.com

TITAN provides a wide range of engineering and environmental services. Experience includes ISO 14001 and ISO 9000 series and its implementation, pollution prevention planning, hazardous materials/waste management, database management.
Gene W Ray, Chairman of the Board
Lawrence J Delaney, VP of Operations

2060 Tradet Laboratories

Battle Run Road   304-547-9094
Triadelphia, WV 26059   Fax: 304-547-9097

Coal, analytical and environmental services.
G William Kald, President

2061 Transviron

1624 York Road   410-321-6961
Lutherville, MD 21093   Fax: 410-494-9321

2062 Trinity Consultants

12770 Merit Drive   972-661-8100
Suite 900   800-229-6655
Dallas, TX 75251   Fax: 972-385-9203
E-mail: information@trinityconsultants.com
http://www.trinityconsultants.com

An environmental consulting company that assists industrial facilities with issues related to regulatory compliance and environmental management. Founded in 1974, this nationwide firm
has particular expertise in air quality issues. Trinity also sells environmental software and professional education. T3, a Trinity
Consultants Company, provides EH&S management information systems (EMIS) implementation and integration services.
Jay Hofmann, President
Richard H Schulze, CEO

2063 Troppe Environmental Consulting

17 South Main Street   330-375-1900
Akron, OH 44224

Provides level I and level II assessments, water and oil testing,
and amtm standards.
Fred Troppe, President

2064 Turner Technologies

560 S Zimmer Road   574-267-3305
Box 1096   Fax: 574-269-6569
Warsaw, IN 46581-1096

Environmental laboratory also provides water treatment systems.
David Turner, President

2065 Versar

6850 Versar Center   703-750-3000
Springfield, VA 22151   800-283-7727
          Fax: 703-642-6807
          E-mail: info@versar.com
          http://www.versar.com

Versar is a public-held, international professional services firm
that applies technology, science, and management skills to enhance its customers' performance.
Dennis Rankin, VP

2066 Water and Air Research

6821 SW Archer Road   352-372-1500
Gainesville, FL 32608   800-242-4927
          Fax: 352-378-1500
          E-mail: services@waterandair.com
          http://www.waterandair.com

Mission is to be an international environmental consulting firm
that achieves extraordinary results by partnering with clients
that to make informed and responsible decisions regarding the
environment.

William C Zegel, President
William Kinser, Director/Manager

2067 Weavertown Group Optimal Technologies

2 Dorrington Road   724-746-4850
Carnegie, PA 15106   800-746-4850
          Fax: 724-746-9024
          E-mail: optimal@optimaltech.com
          http://www.weavertown.com

We are an environmental engineering and consulting firm.
Dawn Fuchs, President

2068 Wenck Associates

1800 Pioneer Creek Center   763-479-4200
PO Box 249   800-472-2232
Maple Plain, MN 55359   Fax: 763-479-4242
          E-mail: wenckmp@wenck.com
          http://www.wenck.com

Our mission is to provide our customers strategic advice and
technical excellence.

2069 Westinghouse Remediation Services

675 Park N Boulevard   770-469-6522
Building F, Suite 100   888-937-7101
Clarkston, GA 30021   Fax: 770-469-3225

2070 Weston Solutions, Inc

1400 Weston Way   610-701-3000
Box 2653   800-7WE-STON
West Chester, PA 19380   Fax: 610-701-3186
          E-mail: info@westonsolutions.com
          http://www.westonsolutions.com

Weston is a leading infrastructure redevelopment services firm
delivering integrated environmental engineering solutions to industry and government worldwide. With an emphasis on creating lasting economic value for its clients, the company provides
services in site remediation, redevelopment, infrastructure operations and knowledge management.
Edmund B Pettiss, Jr., Senior VP
Patrick G McCann, CEO/President

Associations

2071 Acid Rain Foundation
1410 Varsity Drive   919-828-9443
Raleigh, NC 27606-2010   Fax: 919-515-3593
Dr. Harriett S Stubbs, Executive Director

2072 Action on Smoking and Health
2013 H Street NW   202-659-4310
Washington, DC 20016   http://http://ash.org
Organized to use the power of the law to protect the rights of non-smokers. Emphasis is placed on legal efforts to protect non-smokers and to get courts to support the rights of nonsmokers. Also conducts educational and awarenesscampaigns regarding the problem of smoking and the rights of nonsmokers.
John Banzhaf, Executive Director

2073 Agency for Toxic Substances and Disease Registry
1600 Clifton Road NE   404-498-0110
Mail Stop E-29   888-422-8737
Atlanta, GA 30333   Fax: 404-498-0093
    http://www.atsdr.cdc.gov
The mission of the agency is to prevent exposure and adverse human health effects and diminished quality of life associated with exposure to hazardous substances from waste sites, unplanned releases, and other sources of pollutionpresent in the enviroment. ATSDR is an operating division of the US Department of Health and Human Services. It divides its activities between those related to a particular site and those related to a specific hazardous substance.
Julie L. Gerberding, MD, MPH, Administrator
Tom Sinks, PhD, Acting Director

2074 Air and Waste Management Association
1 Gateway Center   412-232-3444
3rd Floor   Fax: 412-232-3450
Pittsburgh, PA 15222   E-mail: info@awma.org
    http://www.awma.org
The Air & Waste Management Association (A&WMA) is a nonprofit, nonpartisan professional organization that provides training, information, and networking opportunities to thousands of environmental professionals in 65 countries.
John A Thorner, Executive Director

2075 Alliance for Acid Rain Control and Energy Policy
444 N Capitol Street   202-624-5475
Suite 602   Fax: 202-508-3829
Washington, DC 20001

2076 Alternatives for Community and Environment Roxbury Environmental Empowerment Project
2181 Washington Street   617-442-3343
Suite 301   Fax: 617-442-2425
Roxbury, MA 02119   E-mail: info@ace-ej.org
    http://www.ace-ej.org
ACE is a community-based, nonprofit, environmental justice, law and education center. ACE works in partnership with community groups from low income communities and communities of color to help them address their environmental andenvironmental heath issues by providing free legal, educational and organizing services.
Penn Loh, Executive Director
Warren Goldstein-Gelb, Program Director

2077 American Academy of Environmental Medicine
7701 E Kellog   316-684-5000
Suite 625   Fax: 316-684-5709
Wichita, KS 67207   E-mail: centraloffice@aaem.com
    http://www.aaem.com
The Academy is interested in expanding the knowledge of interactions between human individuals and their environment, as these may be demonstrated to be reflected in their total health. The Academy is comprised primarily of medicalprofessionals who sponsor publications, seminars and courses. A newsletter and journal are among the organization's publications.
Bobbie J Hinshaw, Executive Director

2078 American Association in Support of Ecological Initiatives
150 Coleman Road   860-364-2967
Middletown, CT 06457   Fax: 860-347-8459
    E-mail: Wwasch@wesleyan.edu
AASEI is a US 501 nonprofit organization which suports international environmental initiatives in Russian Nature Reserves. In cooperation sith Russia and foreign scientists, students, and universities, AASEI organizes scientificresearch projects, academic internships, work camps, environmental exchanges, and eco-tourism. Our aim is to provide practical support to Russina Reserves, expand opportunities for international scientific research, and promote internationalunderstanding.
William K Walsh Jr, Executive Officer

2079 American Association of Poison Control Centers
3201 New Mexico Avenue NW   202-362-7217
Suite 310   Fax: 202-362-8377
Washington, DC 20016   E-mail: ras@poison.org
    http://www.poison.org
Nationwide organization of poison centers and interested individuals. Provides a forum for poison centers and interested individuals to promote the reduction of morbidity and mortality from poisonings through public and professionaleducation and scientific research.

2080 American Board of Environmental Medicine
4205 McAuley Boulevard   405-749-0193
Suite 385   Fax: 405-751-5168
Oklahoma City, OK 73120
Clifton R Brooks MD, Executive Director

2081 American Board of Industrial Hygiene
6015 W St Joseph Highway   517-321-2638
Suite 102   Fax: 517-321-4624
Lansing, MI 48917   E-mail: abih@abih.org
    http://www.abih.org
The American Board of Industrial Hygiene, a nonprofit corporation, was organized to improve the practice and educational standards of industrial hygiene.
Lynn O'Donnell, CIH, Executive Director

2082 American Cancer Society
1599 Clifton Road NW   404-329-7686
Atlanta, GA 30329   Fax: 404-321-4669
    http://www.cancer.org
The American Cancer Society is the nationwide, community-based, voluntary health organization dedicated to eliminating cancer as a major health problem by preventing cancer, saving lives from cancer, and diminishing suffering fromcancer through research, education and service.
Clark W Heath Jr MD, VP

2083 American College of Occupational and Environmental Medicine
1114 North Arlington Heights Road   847-818-1800
Arlington Heights, IL 60004   Fax: 847-818-9266
    E-mail: mdreger@acoem.org
    http://www.acoem.org
Made up of physicians in industry, government, academia, private practice and the military, who promote the health of workers through preventive medicine, clinical care, research and education.
Marianne Dreger, Director Of Communications
Darleene Shah, Communications Assistant

2084 American Conference of Governmental Industrial Hygienists
1330 Kemper Meadow Drive   513-742-2020
Cincinnati, OH 45240   Fax: 513-742-3355
    E-mail: pubs@acgih.org
    http://www.acgih.org
The American Conference of Governmental Industrial hygienists is a member-based organization and community of professionals that advances worker health and saftey through education and the development and dissemination of scientificand technical knowledge. Examples include annual editions of the TLVs and BELs and work practice guides in ACGIH's Signature Series of Publications.

90 pages Magazine
Cindy Coe, Chair
Robert D Soule, Vice Chair

2085 American Council on Science and Health
1995 Broadway   212-362-7044
2nd Floor   Fax: 212-362-4919
New York, NY 10023-5860   E-mail: acsh@acsh.org
                                       http://www.acsh.org
A consumer education organization based in New York City.
The council is an independent association that promotes scientifically balanced evaluations of food, chemicals and the environment, and their relationship to human health.
Eliabeth M Whelen, President

2086 American Federation of Teachers
555 New Jersey Avenue NW   202-879-4463
Washington, DC 20001-2079   Fax: 202-879-4597
                                       http://www.aft.org
The American Federation of Teachers is a union that represents
K-12 teachers and other school emploees, health care profes-
sionals and public employees. The union considers itself an ad-
vocacy organization for children and the public.
Darryl Alexander, Occ and Env Health Coord

2087 American Indian Environmental Office
1200 Pennsylvania Avenue (4104M)   202-564-0303
Washington, DC 20460   http://www.epa.gov/indian
Coordinates the US environmental Protection Agency-wide ef-
fort to strengthen public health and environmental protection in
Indian Country, with a special emphasis on building Tribal ca-
pacity to administer their own environmentalprograms.
Carol Jorgensen, Director
Gary Hudiburgh, Deputy Director

2088 American Industrial Hygiene Association
2700 Prosperity Avenue   703-849-8888
Suite 250   Fax: 703-207-3561
Fairfax, VA 22031   E-mail: infonet@aiha.org
                                       http://www.aiha.org
To promote the highest quality of occupational and environmental health and safety within the workplace and the community
through advocacy and the provision of services and tools to enhance the professional practice of our members.
Donna M Doganiero, Chair

2089 American Lung Association
1150 18th Street NW   202-785-3355
Suite 900   Fax: 202-452-1805
Washington, DC 20036   http://www.lungusa.org
The American Lung Association has been fighting lung disease
for nearly 100 years. The work continues as we strive to make
breathing easier for everyone through education, community
service, advocacy and research programs. Our missionis to prevent lung disease and promote lung health.
Paul Billings, Assistant VP

2090 American Medical Association
515 N State Street   312-464-5000
Chicago, IL 60610   800-621-8335
                                  Fax: 312-464-4184
                                   http://www.ama-assn.org
Medical doctors concerned with environmentally related health
issues.
John C Nelson, President

2091 American Public Health Association
800 I Street Northwest   202-777-APHA
Washington, DC 20001-3710   Fax: 202-777-2532
                              E-mail: comments@apha.org
                                       http://www.apha.org
The APHA brings together researchers, health service providers, administrators, teachers and other health workers in a
unique, multidisciplinary environment of professional exchange, study and action. APHA is concerned with a broad setof
issues affecting personal and environmental health, including
federal and state funding for health programs, pollution control,
programs and policies related to chronic and infectious diseases,
a smoke-free society and a professional education inpublic
health.
George C Benjamin, Executive Director
Patricia D Mail, President Elect

2092 American Society for Microbiology
1752 N Street NW   202-737-3600
Washington, DC 20036-2804   Fax: 202-942-9329
                                       http://www.asm.org
The mission of the American Society for Microbiology is to promote the microbiological sciences and their applications for the
common good. To meet the needs of our members, our activities
include: publishing journals and books,convening meetings;
conducting and supporting education, training and public information programs to facilitate the dissemination and application
of new microbiological knowledge.
James Tiedje, President

2093 American Society of Safety Engineers
1800 East Oakton Street   847-699-2929
Des Plaines, IL 60018-2187   Fax: 847-768-3434
                         E-mail: customerservice@asse.org
                                       http://www.asse.org
ASSE is a global association providing professional development and representation for those engaged in the practice of
safety, health and environmental issues. Provides services to the
private and public sectors to protect people,property and the environment.
Jack H Dobson, President
Fred J Fortman, Executive Director

2094 Asbestos Information Association of North America
1235 Jefferson Davis Highway   703-560-2980
PMB 114   Fax: 703-560-2981
Arlington, VA 22202   E-mail: aiabjpigg@aol.com
The Asbestos Information Association/North America was
founded in 1970 to represent the interest of the asbestos industry
and to collect and disseminate information about asbestos and
asbestos products, with emphasis on safety, health,and environmental issues. The Association appears before Federal regulatory bodies and works with Government agencies to develop and
implement standards for worker protection.
Bob Pigg, President

2095 Asian Pacific Environmental Network
310 8th Street   510-834-8920
Suite 309   Fax: 510-834-8926
Oakland, CA 94607-4253   E-mail: apen@apen4ej.org
                                       http://www.apen4ej.org
The Asian Pacific Environmental Network (APEN) empowers
low-income Asian Pacific Islander (API) communities to take
action on environmental and social justice issues. APEN builds
organizations in dis-empowered API communities todevelop
lasting capacity of the community to achieve solutions to problems affecting people's lives.
Pronita Gupta, Chair
Helen Chen, Vice Chair

2096 Association for Environmental Health of Soils
150 Fearing Street   413-549-5170
Amherst, MA 01002   Fax: 413-549-0579
                                       http://www.aehs.com
The AEHS was created to facilitate communications and foster
cooperation among professionals concerned with the challenge
of soil protection and cleanup. Members represent the many disciplines involved in making decisions and solvingproblems affecting soils. AEHS recognizes that widely acceptable solutions
to the problem can be found only through the integration of scientific and technological discovery, social and political judgement and hands on practice.
Paul Kostecki, Executive Director
Marc A Nascarella, Managing Ed/Conference Coor

2097 Association of American Pesticide Control Officials
PO Box 1249   802-472-6956
Hardwick, VT 05843   Fax: 802-472-6957
                              E-mail: aapco@vtlink.net
                          http://www.aapco.ceris.purdue.edu/
To encourage uniformity among the states in their pesticide reg-
ulatory programs.
Paul Liemandt, President
Steven Rutz, President-Elect

2098 Association of Battery Recyclers
PO Box 290286   813-626-6151
Tampa, FL 33687   Fax: 813-622-8388
                                  http://www.batteryrecyclers.com

To keep members abreast of environmental, health, and safety requirements that affect our industry.
Joyce Moralis, Contact

2099 Asthma and Allergy Foundation of America
1233 20th Street, NW   202-466-7643
Suite 402   Fax: 202-466-8940
Washington, DC 20036   E-mail: info@aafa.org
        http://www.aafa.org
Dedicated to improving the quality of life for people with asthma and allergies through education, advocacy and research.

2100 Beyond Pesticides/National Coalition Against the Misuse of Pesticides
701 E Street SE   202-543-5450
Suite 200   Fax: 202-543-4791
Washington, DC 20003   E-mail: info@beyondpesticides.org
        http://www.beyondpesticides.org
Nonprofit organization formed in February 1981 to bring together health, environmental, labor, farm, consumer and church groups, as well as concerned individuals, to focus attention on the potential hazards associated with pesticideuse.
Gregg Small, President
Audrey Their, Vice President

2101 Bio Integral Resource Center
PO Box 7414   510-524-2567
Berkeley, CA 94707   Fax: 510-524-1758
        E-mail: birc@igc.org
        http://www.birc.org
The goal of the Bio Integral Resources Center is to reduce pesticide use by educating the public about effective, less toxic alternatives for pest problems.
Dr. William Quarles, Executive Director

2102 Bison World
National Bison Association
1400 W 122nd Avenue   303-292-2833
Suite 106   Fax: 303-292-2564
Westminster, CO 80234   E-mail: info@bisoncentral.com
        http://www.bisoncentral.com
An organization of bison producers dedicated to awareness of the healthy properties of bison meat and bison production.
Laurie Dineen, Editor

2103 Center for Health, Environment and Justice
PO Box 6806   703-237-2249
Falls Chruch, VA 22040-6806   Fax: 703-237-8389
        E-mail: chej@chej.org
        http://www.chej.org
The Center for Health, Environment nad Justice trains and assists local people to fight for justice, become empowered to protect their communities from environmental threats and build strong locally controlled organizations. CHEJconnects these strong local groups with each other to build a movement from the bottom up.
Lois Marie Gibbs, Executive Director

2104 Center for Holistic Resource Management
The Savory Center
1010 Tijeras NW   505-842-5252
Albuquerque, NM 87102   Fax: 505-843-7900
        http://www.holisticmanagement.org
Seeks to improve the human environment and quality of life through holistic management.
Shannon Horst, Interim Executive Director
Ann Adams, Managing Editor

2105 Center for Science in the Public Interest
1875 Connecticut Avenue NW   202-332-9110
Suite 300   Fax: 202-265-4954
Washington, DC 20009-5728   E-mail: cspi@cspinet.org
        http://www.cspinet.org
A nonprofit education and advocacy organization that focuses on improving the safety and nutritional quality of our food supply and on reducing the carnage caused by alcoholic beverages. CSPI seeks to promote health through educatingthe public about nutrition and alcohol; it represents citizen's interests before legislative, regulatory and judicial bodies; and it works to ensure advances in science are used for the public good.
Kathleen O'Reilly, President

2106 Center for the Evaluation of Risks to Human Reproduction
Po Box 12233   703-838-9440
Research Triangle Park, NC 27709   Fax: 703-684-2223
        E-mail: shelby@niehs.nih.gov
        http://http://cerhr.niehs.nih.gov
The Center's mission includes the following: to provide timely and unbiased, scientifically sound assessments of reproductive health hazards associated with human exposure to naturally occurring and man-made chemicals; to make theseassessments readily available to the public, to state and federal agencies and to the scientific community; and to build an electronic resource for providing, or directing one to, information of public interest concerning human reproductive health.
Dr. Michael D Shelby

2107 Centers for Disease Control & Prevention
National Center for Environmental Health
4770 Buford Highway NE   770-488-7020
Suite F-29   888-232-6789
Atlanta, GA 30341   Fax: 770-488-7024
        http://www.cdc.gov/nceh
To provide national leadership, through science and service, to promote health and quality of life by preventing and controlling disease and death resulting from interactions between people and their environment.
Dr. Howard Frumkin, Director

2108 Centers for Disease Control and Prevention
1600 Clifton Road NE   404-639-3311
Building 16   800-311-3435
Atlanta, GA 30333   Fax: 404-639-2657
        http://www.cdc.gov
The Centers for Disease Control and Prevention protect the public health of the nation by providing leadership and direction in the prevention and control of diseases and other preventable conditions, and responding to public healthemergencies.

Julie Gerberding, Director
William Gimson, Chief Operating Officer

2109 Chemical Injury Information Network
PO Box 301   406-547-2255
White Sulphur Springs, MT 59645   Fax: 406-547-2455
        E-mail: chemicalinjury@ciin.org
        http://www.ciin.org
Nonprofit tax-exempt support and advocacy organization run by the chemically injured for the benefit of the chemically injured. CIIN serves an international membership, and focuses primarily on eductaion, credible multiple sensitivityresearch and the empowerment of the chemically injured.
Cindy Wilson, Executive Director

2110 Communities for a Better Environment
1440 Broadway   510-302-0430
Suite 701   Fax: 510-302-0437
Oakland, CA 94612   E-mail: cbeca@mail.com
        http://www.cbecal.org
Nonprofit statewide, multiracial, urban environmental health and justice organization that works with urban communities and grassroots organizations, using science based research, legal tactics and organizing strategies to prevent airand water pollution, eliminate toxic hazards and improve public health. Long-term goals are to develop an environmentally sustainable manufacturing base, minimize the use of toxins, expand pollution prevention strategies and involve people most atrisk.
Jose T Bravo, President
Luke Cole, Vice President

2111 Council of State and Territorial Epidemiologists
2872 Woodcock Boulevard   770-458-3811
Suite 303   Fax: 770-458-8516
Atlanta, GA 30341-4015   http://www.cste.org
The Council of State and Territorial Epidemiologists is an organization of epidemiologists working together to establish effective relationships amoung state and other epidemiologists, to consult and advise with appropriate disciplinesin other health agencies, and to provide technical assistance to the Association of State and Territorial Health Officials.
Mack Sewell, President
Christine Hahn, Vice President

2112 Dangerous Goods Advisory Council

1100 H Street NW   202-289-4450
Suite 740   Fax: 202-289-4074
Washington, DC 20005   E-mail: info@dgac.org
                                http://www.hmac.org

HMAC promotes improvement in the safe transportation of hazardous materials/dangerous goods globally by: providing education, assistance, and information to the private and public sectors; through our unique status with regulatorybodies; and the adversity and technical strengths of our membership.
Alan I Roberts, President
Mike Morrissette, Vice President

2113 Environmental Defense Fund

257 Park Avenue S   212-505-2100
New York, NY 10010   http://www.environmentaldefense.org

To prevent environmentally induced harm to human populations.
Fred Krupp, President
David Yarnold, Executive Vice President

2114 Environmental Health Coalition

401 Mile of Cars Way   619-474-0220
Suite 310   Fax: 619-474-1210
National City, CA 92101 http://www.environmentalhealth.org

One of the oldest and most effective grassroots organizations in the US, using social change strategies to achive environmental and social justice. We believe that justice is accomplished by empowered communities acting together tomake social change. We organize and advocate to protect public health and the environment threatened by toxic pollution. EHC supports broad efforts that create a just society which foster a healthy and sustainable quality of life.
Diane Takvorian, Executive Director
Sonya Holmquist, Associate Director

2115 Environmental Health Education Center

655 West Lombard Street   410-706-1849
Baltimore, MD 21230   Fax: 410-706-0295
                                http://www.envirn.umaryland.edu

The overall mission of the Center is to engage in research and provide training and education programs on topics related to occupational and environmental health and safety. Our focus is broad and the workplace, community and home areall included in our defination of environment. The audiences for our training and education programs include professionals, labor, industry and community members. Through our efforts we hope to prevent occupation and/or environment related injuriesand illnesses.
Barbara Sattler RN DrPH

2116 Environmental Health Network

PO Box 1155   415-541-5075
Larkspur, CA 94977-0074   http://www.chnca.org

Nonprofit, volunteer organization, whose main goal is to promote public awareness of evironmental sensitivities and causative factors. EHN's focus is on issues of access and developments relating to the health and welfare of theenvironmentally sensitive.

2117 Environmental Justice Resource Center

223 James Brawley Drive SW   404-880-6911
Atlanta, GA 30314   Fax: 404-880-6909
                                E-mail: ejrc@cau.edu
                                http://www.ejrc.cau.edu

Since 1994, a research, policy and information clearinghouse on issues related to environmental justice, race and the environment, civil rights, facility siting, land use planning, brownfields, transportation equity, suburban sprawland Smart Growth. The overall goal of the center is to assist, support, train and educate people of color, students, professionals and grassroots community leaders with the goal of facilitating their inclusion into the mainstream of environmentaldecision-making.
Robert D Bullard PhD, Director

2118 Environmental Safety

1700 North Moore Street   703-527-8300
Arlington, VA 22209-2793   Fax: 703-527-8383
                                http://www.ashoka.org

Ashoka's mission is to shape a citizen sector that is entrepreneurial, productive and globally integrated, and to develop the profession of social entrepreneurship around the world. Ashoka identifies and invests in leading socialentrepreneurs-extraordinary individuals and unprecedented ideas for change in their communities-supporting the individual, idea and institution through all phases of their career. Once elected to Ashoka, Fellows benefit from being part of the globalFellowship for life.
William Drayton, Executive Director

2119 Environmental Working Group

1718 Connecticut Avenue NW   202-667-6982
Suite 600   Fax: 202-232-2592
Washington, DC 20009   http://www.ewg.org

The Environmental Working Group is a small, computer powered research organization dedicated to improving environmental protection through the analysis of federal and state regulatory policies and performance and through technicalassistance and education.
Kelsey Wirth, Chair

2120 Environmental and Occupational Health Science Institute

170 Frelinghuysen Road   732-445-0200
PO Box 1179   Fax: 732-445-0131
Piscataway, NJ 08855   http://www.cohsi.rutgers.edu

Environmental and Occupational Health Sciences Institute sponsors research, education and service programs in a setting that fosters interaction among experts in environmental health, toxicology, occupational health, exposureassessment, public policy and health education. The Institute also serves as an unbiased source of expertise about environmental problems for communities, employers and government in all areas of occupational and environmental health, toxicology andrisk assessment.
Dr. Cory Slechta, Director

2121 Food Safety and Inspection Service US Department of Agriculture

Food Safety Education Office
1400 Independence Avenue SW   301-504-9605
Washington, DC 20250   Fax: 301-504-0203
                                E-mail: fsis.webmaster@usda.gov
                                http://www.fsis.usda.gov

The Food Safety and Inspection Services (FSIS) is the public health agency in the U.S. Department of Agriculture responsible for ensuring that the nation's commercial supply of meat, poultry, and egg products is safe, wholesome, andcorrectly labeled and packaged.

2122 Food and Drug Administration

US Department of Health and Human Services
Office of Public Affairs   301-827-3666
5600 Fishers Lane   888-463-6332
Rockville, MD 20857   Fax: 301-443-0017
                                http://www.fda.gov

FDA is determined to keep its public health protections what they have have always been: an effective armor against public health hazards, and one of the greatest bargains that the US government delivers to all citizens.
Lester M Crawford, DVM, PhD, Commissioner

2123 Friends of the River

915 20th Street   916-442-3155
Sacramento, CA 95814   Fax: 916-442-3396
                                E-mail: info@friendsoftheriver.org
                                http://www.friendsoftheriver.org

Friends of the River educates, organizaes, and advocates to protect and restore California rivers, streams, and watersheds.
Pete Ferenbach, Executive Director

2124 Genesis Fund/National Birth Defects Center: Pregnancy Environmental Hotline

40 2nd Avenue   781-466-8474
Suite 520   800-322-5014
Waltham, MA 02451   Fax: 781-487-2361
                                E-mail: peh@thegenesisfund.org
                                http://www.thegenesisfund.org/peh.htm

General information service that provides information regarding exposure to environmental factors during pregnancy and the effects on the developing fetus.

2125 Hazardous Waste Resource Center Environmental Technology Council

734 15th Street NW   202-783-0870
Suite 720   Fax: 202-737-2038
Washington, DC 20005   E-mail: mail@etc.org
      http://www.etc.org

The Environmental Technology Council (ETC) is a trade association of commercial firms that recycle, treat and dispose of industrial and hazardous wastes; and firms involved in cleanup of contaminated sites.

2126 Healthy Mothers, Healthy Babies

121 North Washington Street   703-836-6110
Suite 300   Fax: 703-836-3470
Alexandria, VA 22314   E-mail: info@hmhb.org
      http://www.hmhb.org

Healthy Mothers, Healthy Babies is a coalition of national, state and local providers, advocates and administrators concerned about health of pregnant women, infants and families. The coalition serves as a forum for informationexchange and as a catalyst to encourage collaborative partnerships amoung its members and colleagues.
George Guido, Chairperson
Leith Merrow Mullaly, Vice Chair

2127 Human Ecology Action League (HEAL)

PO Box 29629   404-248-1898
Atlanta, GA 30359   Fax: 404-248-0162
      E-mail: HEALNatnl@aol.com
      http://members.aol.com/HEALNatnl/index/html

The Human Ecology Action League Inc (HEAL) is a nonprofit organization founded in 1977 to serve those whose health has been adversely affected by environment exposures; to provide information to those who are concerned about the healtheffects of chemicals; and to alert the general public about the potential dangers of chemicals. Referrals to local HEAL chapters and other support groups are available from the League.
Katherine P Collier, Contact

2128 INFORM

120 Wall Street   212-361-2400
14th Floor   Fax: 212-361-2412
New York, NY 10005-4001   http://www.informinc.org

INFORM is an independent research organization that examines the effects of business practices on the environment and on human health. Our goal is to identify ways of doing business that ensure environmentally sustainable economicgrowth. Our reports are used by government, industry, and environmental leaders around the world.
Joanna Underwood, President

2129 Institute for Agriculture and Trade Policy

2105 First Avenue South   612-870-0453
Minneapolis, MN 55404   Fax: 612-870-4846
      http://www.sustain.org

The mission of the Institute for Agriculture and Trade Policy is to foster socially, economically and environmentally sustainable rural communities and regions.
Dr Arie van den Brand, Board Chair
Mark Ritchie, President

2130 Institute of Hazardous Materials Management

11900 Parklawn Drive   301-984-8969
Suite 450   Fax: 301-984-1516
Rockville, MD 20852   E-mail: ihmminfo@ihmm.org
      http://www.ihmm.org

Mission is to provide recognition for professionals engaged in the managment and engineering control of hazardous materials who have attained the required level of education, experience and competence; foster continued professionaldevelopment of Certified Hazardous Materials Managers (CHMM).
John H Frick, PhD, CHMM, Executive Director

2131 Laotian Organizing Project

220 25th Street   510-236-4616
Richmond, CA 94804-1808   Fax: 510-236-4572
      E-mail: apen@apen4ej.org
      http://www.apen4ej.org

The Laotian Organization Project (LOP), a project of the Asian Pacific Environmental Network, is a membership-based organization of Laotian residents in Richmond and San Pablo, California. LOP works to bring people from the Laotiancommunity together to identify problems, develop solutions, and take action for a more healthy, safe, and just community.
Vivian Chang, Executive Director
Pronita Gupta, Chair

2132 MCS Referral and Resources

618 Wyndhurst Avenue #2   410-889-6666
Baltimore, MD 21210   Fax: 410-889-4944
      E-mail: adonnaya@mcsrr.org
      http://www.mcsrr.org

The mission of MCS Refferal and Resources is to further the diagnosis, treatment, accomodation and prevention of multiple chemical sensitivity (MCS) disorders.
Dr. Anne McCampbell

2133 Midwest Center for Environmental Science and Public Policy

1845 North Farewell Avenue   414-271-7280
Suite 100   Fax: 414-273-7293
Milwaukee, IL 53202   E-mail: mcespp@mcespp.org
      http://www.mcespp.org

Protecting human health and the environment through research, advocacy, public education and citizen empowerment.
Jeffrey Foran, President

2134 Montefiore Medical Center Lead Poisoning Prevention Program

111 E 210th Street   718-920-5016
Bronx, NY 10467   Fax: 718-920-4377

The Montefiore Lead Poisoning Prevention Program addresses all aspects of childhood lead poisoning from diagnosis and treatment to education and research. Their mission is to treat lead-poisoned children and their families and toeducate families at risk, other medical providers and local, state and national legislators and policy makers.
Nancy Redkey, Project Coordinator

2135 Mount Sinai School of Medicine: Division of Environmental Health Science

Department of Community and Preventive Medicine
1 Gustave Levy Place   212-241-8689
Box 1043   Fax: 212-360-6965
New York, NY 10029-6574   http://www.mssm.edu/cpm

The Division's ultimate goal is the protection of the public's health by understanding, elucidating and preventing diseases that arise from environmental exposures.
Philip J Landrigan M.D.

2136 National Accreditation Council for Environmental Health Science and Protection

2632 Southeast 25th Avenue   503-235-6047
Suite D   Fax: 503-235-7300
Portland, OR 97202   E-mail: ehacinfo@aehap.org
      http://www.ehaoffice.org

The National Environmental Health Science and Protection Accreditation Council was established in 1967 to enhance the education and training of students who intend to become environmental health science and protection practitioners/professionals. The council, originally known as the Health Curricula, is composed of highly qualified proessionals.
Sharon LaFollette, Chairperson
Richard Rowe, Director

2137 National Alliance for Hispanic Health

1501 16th Street NW   202-387-5000
Washington, DC 20036   Fax: 202-265-8027
      E-mail: alliance@hispanichealth.org
      http://www.hispanichealth.org

The mission of the National Alliance for Hispanic Health is to improve the health and well-being of Hispanics. Issues covered include the full range of health and human services issues, including environmental health.
Jane L Delgado PhD, MS, President/CEO

2138 National Association of City and County Health Officials
1100 17th Street NW   202-783-5550
2nd Floor   Fax: 202-783-1583
Washington, DC 20036   http://www.naccho.org

The National Assoication of County and City Health Officials is a nonprofit, membership organization serving all 3,000 local health departments nationwide. NACCHO is dedicated to improving the health of people and communities byassuring an effective local public health system. As the Voice of local public health officials at the national level, NACCHO is able to promote the local perspective on national health programs and policies.
Michael C Caldwell, President
Poki S Namkung, Vice President

2139 National Association of Noise Control Officials
53 Cubberly Road   609-586-2684
West Windsor, NJ 08690   Fax: 609-799-2616
                                         http://www.arcat.com

The National Association of Noise Control Officials consists of employees of the federal and state governments, consultants, scientists, and students concerned with acoustical control in the environment. It was started in 1978 and nowhas about 70 members.
Edward J DiPolzere, Executive Director

2140 National Association of Physicians for the Environment
1643 Prince Street   703-684-1760
Alexandria, VA 22314-2818   Fax: 703-549-2772
                                         E-mail: nape@napenet.org
                                         http://www.greenlink.org

The National Association of Physicians for the Environment works to involve physicians and other health care professionals, particularly through their geographic and medical specialty organizations, to deal with the impact ofpollutants on organs and systems of the human body.
Jerome Goldsmith, Founder

2141 National Association of School Nurses
1416 Park Street   303-663-2329
Suite A   866-627-6767
Castle Rock, CO 80109   Fax: 303-663-0403
                                         E-mail: nasn@nasn.org
                                         http://www.nasn.org

The mission of The National Association of School Nurses is to advance the practice of school nursing and provide leadership in the delivery of quality health programs to school communities.
Wanda Miller, Executive Director

2142 National Cancer Institute
National Institutes of Health
9000 Rockville Pike
Health Building 10, Room 2A33   800-422-6237
Bethesda, MD 20892   E-mail: cancergovstaff@mail.nih.gov
                                         http://www.cancer.gov

Leads the Nation's fight against cancer by supporting and conducting ground-breaking research in cancer biology, causation, prevention, detection, treatment and survivorship.
Andrew Eschenbach, Director

2143 National Center for Environmental Health Strategies
1100 Rural Avenue   856-429-5358
Voorhees, NJ 08043   Fax: 856-429-5358
                                         E-mail: mary@ncehs.org
                                         http://www.ncehs.org/

Fosters the development of creative solutions to environmental health problems with a focus on indoor air quality, chemical sensitivities and environmental disabilities.
Mary Lamielle, Executive Director

2144 National Center for Lead-Safe Housing
10227 Wincopin Circle   410-992-0712
Suite 205   Fax: 410-715-2310
Columbia, MD 21044   http://www.leadsafehousing.org

The National Center for Lead-Safe Housing was founded to bring the public health, housing and environmental communities together to combat our nation's epidemic of childhood lead poisoning.
Don Ryan, Executive Director

2145 National Conference of Local Environmental Health Administrators
University of Washington
Department of Environmental Health   206-616-2097
Box 357234   Fax: 206-543-8123
Seattle, WA 98195-7234   E-mail: webmaster@ncleha.org
                                         http://www.ncleha.org

The NCLEHA's purpose is to provide a forum for local administrators to share common concerns and solutions to mutual problems, and to provide a professional organization for environmental health administrators, focused on the issuesand problems of local environmental health programs.
Mel Knight, Chairperson

2146 National Conference of State Legislatures
7700 East First Place   303-364-7700
Denver, CO 80230   Fax: 303-364-7800
                                         http://www.ncsl.org

The National Conference of State Legislatures serves the legislators and staffs of the nation's 50 states and its commonwealths and territories. NCSL is a bipartisan organization with three objectives: to improve the quality andeffectiveness of state legislatures; to foster interstate communications and cooperation; and to ensure states a strong cohesive voice in the federal system.
Douglas Farquhar, Program Principal

2147 National Education Association Health Information Network
1201 16th Street NW   202-822-7570
Suite 521   800-718-8387
Washington, DC 20036   Fax: 202-822-7775
                                         E-mail: info@neahin.org
                                         http://www.neahin.org

The National Education Association Health Information Network believes that sound public education must begin with school employees and students who are healthy and free of preventable diseases and supported with information, materialsand training opportunities that reaffirm these values.
Jerald Newberry, Executive Director

2148 National Environmental Coalition of Native Americans
Po Box 988   405-567-4297
Claremore, OK 74018   E-mail: norteno_84@hotmail.com
                                     http://http://oraibi.alphacdc.com

Nonprofit organization formed to educate Indians and Non-Indians about the health dangers of radioactivity and the transport of nuclear waste on America's rails and roads. Networks with environmentalists to develop grassrootscounter-movement to the efforts of the nuclear industry and develop Tribal nuclear free zones across the nation.
Grace Thorpe, President

2149 National Environmental Health Association (NEHA)
720 S Colorado Boulevard   303-756-9090
Suite 970, S Tower   Fax: 303-691-9490
Denver, CO 80246-1925   E-mail: staff@NEHA.org
                                         http://www.neha.org

NEHA is the only national association that represents all of environmental health and protection from terrorism and all-hazards preparedness, to food safety and protection and on site wastewater systems. Over 4500 members and theprofession are served by the association through its Journal of Environmental Health, Annual Education Conference and Exhibition credentialing programs, research and development activities and other services.

Nelson Fabian, Executive Director

2150 National Environmental Trust
1200 18th Street NW   202-887-8800
5th Floor   Fax: 202-887-8877
Washington, DC 20036   http://www.environet.org

Manages comprehensive media and public policy campaigns around national environmental issues.
Phil Clapp, President

2151 National Institute for Global Environmental Change
University of California, Davis
1419 Drew Avenue   530-757-3350
Suite 140   Fax: 530-756-6499
Davis, CA 95616-8756

E-mail: wlgates@ucdavis.edu
http://nigec.ucdavis.edu

Sponsored by the Office of Health and Environmental Research and the Office of Environmental Analysis of the US Department of Energy at the University of California, Davis. Offers congressional briefings and education materials forschools, universities, and public education programs.
W. Lawrence Gates, Sc.D. Director

2152 National Institute for Occupational Safety and Health

Humphrey Boulevard Room 715 H   202-401-0721
200 Indiana Avenue   800-356-4674
Washington, DC 20201   http://www.cdc.gov

The National Institute for Occupational Safety and Health was created to conduct research and training and make recommendations for the prevention of work-related illnesses and injuries. NIOSH provides scientific information andrecommendations that may be used in setting workplace regulations.
Bill Halpherin, Director

2153 National Institute of Child Health and Human Development

Po Box 3006
Rockville, MD 20847   800-370-2943
Fax: 301-984-1473
E-mail: NICHDinformationresourcecenter@mail.nih.gov
http://www.nih.gov/nichd

National Institute of Child Health and Human Development supports and conducts basic, clinical and epidemiological research on the reproductive, neurobiological, developmental and behavioral processes that determine and maintain thehealth of children, adults, families and populations.
Duane Alexander, Director

2154 National Institute of Environmental Health Sciences

Po Box 12233   301-402-3378
Research Triangle Park, MD 27709   Fax: 301-496-0563
http://www.niehs.nih.gov

The mission of the National Institute of Environmental Health Sciences is to reduce the burden of environmentally associated diseases and dysfunctions.
Kenneth Olden, Director

2155 National Oceanic & Atmospheric Administration

14th Street/Constitution Avenue NW   202-482-6090
Room 6217   Fax: 202-482-3154
Washington, DC 20230   E-mail: d.james.baker@noaa.gov
http://www.noaa.gov

Describes and predicts changes in the Earth's environment and conserves and wisely manages the nation's coastal and marine resources. Goals and objectives include advance short-term warning and forecast services.
Conrad C Lautenbacher Jr, NOAA Administrator
James R Mahoney PhD, NOAA Deputy Adminstrator

2156 National Parent Teachers Association

330 N Wabash Avenue   312-670-6782
Suite 2100   800-307-4782
Chicago, IL 60611   Fax: 312-670-6783
E-mail: info@pta.org
http://www.pta.org

The mission of the National PTA is to support and speak on behalf of children and youth in the school, in the community and before governmental bodies and other organizations; to assist parents in developing the skills they need toraise and protect their children and to encourage parent and public involvement in the public schools.
Gabriella Hayes, Program Manager

2157 National Pesticide Information Center

Oregon State University
333 Weniger Hall
Corvallis, OR 97331-6502   800-858-7378
Fax: 541-737-0761
E-mail: npic@ace.orst.edu
http://http://npic.orst.edu

NPIC provides objective science based information about a wide variety of pesticide-related topics, including: pesticide product information, toxicology, environmental chemistry, referrals for lab analyses, investigation of pesticideincidents, emergency treatment information, safety information, health and environmental effects and clean-up/disposal procedures. NPIC is a cooperative effort between Oregon State University and the US EPA.
Terry Miller, Director

2158 National Religious Partnership for the Environment

1047 Amsterdam Avenue   212-316-7441
New York, NY 10025   Fax: 212-316-7547
http://www.nrpe.org

The mission of the National Religious Partnership for the Environment is to permanently integrate issues of environmental sustainability and justice across all aspects of organized religious life.
Paul Gorman, Executive Director

2159 National Safety Council

Environmental Health Center
1025 Connecticut Avenue NW   202-293-2270
Suite 1200   Fax: 202-293-0032
Washington, DC 20036   E-mail: ehc@nsc.org
http://www.nsc.org

The Mission of the Environmental Health Center is to foster improved public understanding of significant health risk and challenges facing modern society. This goal reinforces the National Safety Council's commitment to increased andmore effective citizen involvement in safety, health and environmental decision-making.
Jeffrey Shavelson, Policy Analyst

2160 Natural Resources Defense Council

40 West 20th Street   212-727-2700
New York, NY 10011   Fax: 212-727-1773
E-mail: nrdcinfo@nrdc.org
http://www.nrdc.org

His organization is dedicated to protecting endangered natural resources and improving the quality of the human environment. Areas of concentration are air and water pollution, nuclear safety, land use, urban environment, toxicsubstances control, resource management, wilderness and wildlife protection, Alaska, coastal zone management, energy conservation, soil erosion, and forestry. Founded in 1970, it has a membership of more than 500,000.
Francis Beinecke, Executive Director

2161 Navy Environmental Health Center

620 John Paul Jones Circle   757-953-0700
Suite 1100   http://www.-nehc.med.navy.mil
Portsmouth, VA 23708-2103

Ensures Navy and Marine Corps readiness through leadership in prevention of disease and promotion of health.
Captain David Hiland, Commanding Officer

2162 Noise Pollution Clearinghouse

PO Box 1137
Montpelier, VT 05601-1137   888-200-8332
E-mail: freenpc@nonoise.org
http://www.nonoise.org

The Noise Pollution Clearinghouse is a nonprofit organization with extensive online noise related resources. The mission is to create more civil cities and more natural rural and wilderness areas by reducing noise pollution and itssources.

2163 North American Association for Environmental Education

2000 P Street NW   202-419-0412
Suite 540   Fax: 202-419-0415
Washington, DC 20036   E-mail: ashotkin@smtp.aed.org
http://www.naaee.org

The NAAEE, originally the National Association for Environmental Education, is a professional association of educators, communicators, and managers established to assist and support the work of individuals and groups engaged inenvironmental education, research, and service. In addition, they specifically promote environmental education at the post-secondary, secondary, and elementary levels. Founded in 1971, the NAAEE has over 1,800 members.

William H Dent Jr, Executive Director
Joseph Baust, President

2164 Northwest Coalition for Alternatives to Pesticides
PO Box 1393   541-344-5044
Eugene, OR 97440-1393   Fax: 541-344-6923
                                     E-mail: info@pesticide.org
                                     http://www.pesticide.org
NCAP works to protect people and the environment from the
hazards of pesticide chemicals.
Jean Cameron, President
Norma Grier, Executive Director

2165 Occupational Safety and Health Administration: US Department of Labor
Office of Administrative Services
200 Constitution Avenue NW   202-693-1999
Room N-310   800-321-6742
Washington, DC 20210   http://www.osha.gov
OSHA's mission is to send every worker home whole and
healthy every day. Since the agency was created in 1971, workplace fatalities have been cut in half and occupational injury and
illness rates have declined 40 percent. At the sametime, US employment has nearly doubled from 56 million workers at 3.5 million worksites to 105 milion workers at nearly 6.9 million sites.

2166 Pesticide Action Network North America
49 Powell Street   415-981-1771
Sutie 500   Fax: 415-981-1991
San Francisco,]A 94102   E-mail: panna@panna.org
                                     http://www.panna.org
Pesticide Action Network North American advocates the adoption of ecologically sound practices in place of pestices in place
of pesticide use. PANNA works with more than 100 affiliated
orgarnizations in Canada, Mexico and US, as well aswith Pesticide Action Network partners around the world to demand that
development agencies and governments redirect support from
pesticides to safe alternatives.
Monica Moore, Co-Director & Program Dir
Stephen Scholl-Buckwald, Co-Director & Managing Dir

2167 Pesticide Education Center
Box 225279   415-665-4722
San Francisco, CA 94122-5279   Fax: 415-665-2693
                                     E-mail: pec@igc.org
                                     http://www.pesticides.org/pesticides
The mission of the Pesticide Education Center is to educate the
public about the adverse health effects of exposure to pesticides
in the home, community and at work.
Marion Moses, President

2168 Physicians for Social Responsibility
1875 Connecticut Avenue, Northwest   202-667-4260
Suite 1012   Fax: 202-667-4201
Washington, DC 20005   E-mail: psrnatl@psr.org
                                     http://www.psr.org/crh.htm
Physicians for Social Responsibility-the US affiliate of the International Physicians for the Prevention of Nuclear War is committed to the elimination of nuclear weapons of mass
destruction, the achievement of a sustainableenvironment, and
the reduction of violence and its causes. The active conscience
of American medicine.
Robert Musil, Executive Director

2169 Public Citizen
1600 20th Street NW   202-588-1000
Washington, DC 20009   Fax: 202-588-7796
                                     http://www.citizen.org
Founded by Ralph Nader in 1971, Public Citizen is the consumer's eyes in Washington. With the support of more than
15,000 people like you, we fight for safer drugs and medical devices, cleaner and safer energy sources, a cleanerenvironment,
fair trade and a more open and democratic government.
Joan Claybrook, President

2170 Rachel Carson Council
8940 Jones Mill Road   301-652-1877
Chevy Chase, MD 20815   Fax: 301-951-7179
                                     E-mail: rccouncil@aol.com
                                     http://www.members.aol.com/rccouncil

Independent nonprofit scientific organization dedicated to protecting the environment against toxic and chemical threats, particularly those of pesticides.
Dr. Diana Post, Executive Director
Martha Collins, VP Marketing

2171 Rene Dubos Center for Human Environments
81 Pondfield Road   914-337-1636
Suite 387   Fax: 914-771-5206
Bronxville, NY 10708   http://www.dubos.org
The Rene Dubos Center for Human Environments is a non-profit
education and research organization focused on the social and
humanistic aspects of environmental problems.
Noel Brown, President

2172 Rodale Institute
222 Main Street   610-683-1400
Emmaus, PA 18098   Fax: 215-967-8959
                                     E-mail: info@rodaleinstitute.org
The Rodale Institute offers creative opportunities and solutions
that contribute to regenerating environmental and human health
worldwide. Our mission statement is clear: The Rodale Institute
works worldwide to achieve a regenerativefood system that improves environmental and human health.
John Haberern, President

2173 Safer Pest Control Project
25 E Washington   312-641-5575
Suite 1515   Fax: 312-641-5454
Chicago, IL 60602
Rachel Rosenberg, Program Associate
Julie Dick, Program Associate

2174 Silicone Health Council
1850 M Street Northwest   202-721-4100
Suite 700   Fax: 202-296-8120
Washington, DC 20036   E-mail: info@socma.com
                                     http://www.socma.com
Coordinates health, environmental and safety programs. Con-
veys scientifically sound information about silicones.
Joseph Acker, President

2175 Silicones Environmental Health and Safety Council of North America
2325 Dulles Center Boulevard   703-788-6570
Suite 500   Fax: 703-788-6345
Herndon, VA 20171   E-mail: sehsc@sehsc.com
                                     http://www.sehsc.com
An organization of organosilicones manufacturers. Formed to
coordinate programs dealing with health, environmental and
safety issues.
Reo Menning, Executive Director

2176 Society for Occupational and Environmental Health
6728 Old McLean Village Drive   703-556-9222
McLean, VA 22101   Fax: 703-556-8729
                                     E-mail: soeh@degnon.org
                                     http://www.soeh.org
Provides a neutral forum where occupational safety and health
and environmental issues can be discussed and resolved. Actively seeks to improve the quality of both working and living
places.

2177 Society of Environmental Toxicology and Chemistry
SETAC N America Office
1010 N 12th Avenue   850-469-1500
Pensacola, FL 32501   Fax: 850-469-9778
                                     E-mail: setac@setac.org
                                     http://www.setac.org
Provides a forum for individuals and institutions engaged in
study of environmental issues, management and conservation of
natural resources, environmental education and environmental
research and development.
Mimi Meredith, Sr. Manager

2178 Teratology Society
1821 Michael Farday Drive   703-438-3104
Suite 300   Fax: 703-483-3113
Reston, VA 20190   E-mail: tshq@teratology.org
                                     http://www.teratology.org

A multidisciplinary scientific society founded in 1960, the members of which study the causes and biological processes leading to abnormal development and birth defects at the fundamental and clinical level, and appropriate measuresfor prevention.
Kenneth Jones, President
Mellisa Tassinari, Vice President

2179 Toxicology Information Response Center

1060 Commerce Park   865-576-1746
MS 6480   Fax: 865-574-9888
Oak Ridge, TN 37830-6480   E-mail: slusherkg@ornl.gov
http://www.ornl.gov/TechResources/tirc/hmepg.html
Kim Slusher, Administrator

2180 US Consumer Product Safety Commission

4340 East West Highway
Bethesda, MD 20814   800-638-2772
Fax: 301-504-0124
http://www.cpsc.gov
An independent federal regulatory agency. Helps keep American families safe by reducing the risk of injury or death from consumer products.
Hal Stratton, Chairman
Thomas Moore, Vice Chair

2181 US Nuclear Regulatory Commission

Reference Librarian   301-415-4737
Public Document Room (01F-13)   800-368-5642
Washington, DC 20555   Fax: 301-415-8200
E-mail: pdr@nrc.gov
http://www.nrc.gov
Ensures adequate protection of the public health and safety, the common defense and security, and the environment in the use of nuclear materials in the United States.
Nils J Diaz, Chairman

2182 Water Environment Federation

601 Wythe Street   703-684-2400
Alexandria, VA 22314-1994   800-666-0206
Fax: 703-684-2492
http://www.wef.org
Nonprofit international membership organization that develops and disseminates technical information on the nature, collection, treatment and disposal of domestic and industrial wastewater.
Janet Blatt, Director
Mincaiee Brown, Executive Director

2183 Western Occupational and Environmental Medical Association

74 New Montgomery   415-764-4803
Suite 230   Fax: 415-927-5726
San Francisco, CA 94105   E-mail: woema@hp-asso.com
http://www.woema.org
The mission is to represent and be a resource to members in the profession and practice of occupational and environmental medicine and to enhance their efforts to promote and improve health in the workplace.
Robert Orford, President

2184 White Lung Association

PO Box 1483   410-243-5864
Baltimore, MD 21203-1483   Fax: 410-243-5234
E-mail: jfite@whitelung.org
http://www.whitelung.org
National nonprofit organization dedicated to the education of the public to the hazards of asbestos exposure. Has developed programs of public education and consults with victims of asbestos exposure, school boards, building owners,government agencies and others interested in identifying asbestos hazards and developing control programs.
James Fite, Executive Director

Pediatric Health: Associations

2185 Allergy and Asthma Network: Mothers of Asthmatics

2751 Prosperity Avenue
Suite 150   800-878-4403
Fairfax, VA 22031   Fax: 703-573-7794
http://www.aanma.org
Our mission is to eliminate suffering and death due to asthma and allergies through education, advocacy, community outreach and research.
Nancy Sander, President
Bradley Chipps MD, Chairman

2186 Alliance to End Childhood Lead Poisoning

227 Massachusetts Avenue NE   202-543-1147
Suite 200   Fax: 202-543-4466
Washington, DC 20002   E-mail: aeclp@aeclp.org
http://www.aeclp.org
The Alliance is a national, nonprofit, public interest organization created to launch a comprehensive attack on the epidemic of childhood lead poisoning. Its mission is to frame the national agenda, formulate innovative approaches, andbring critical resources to bear to prevent childhood lead poisoning.
Bailus Walker Jr, Chair

2187 Ambulatory Pediatric Association

6728 Old McLean Village Drive   703-556-9222
McLean, VA 22101   Fax: 703-556-8729
http://www.ambpeds.org
The mission of the APA is to foster the health and well-being of children and their families by: promoting health services, education and research in general pediatrics; affecting public and governmental policies regarding issues vitalto child health and to education and research in general pediatrics; and supporting the professional growth and development of faculty in general pediatrics.
Diane Kittredge M.D., President

2188 American Academy of Pediatrics: Committee on Environment Health

141 Point Boulevard NW   847-434-4000
PO Box 927   Fax: 847-434-8000
Elk Grove Villiage, IL 60009-0927   http://www.aap.org
The AAP is commited to the attainment of optimal physical, mental and social health for all infants, children, and young adults. This mission will be accomplished by engaging in the following activities: professional education,advocacy for children and youth, advocacy for pediatricians, public education, membership service and research.
Michael W Shannon M.D., Chairman

2189 Association of Maternal and Child Health Program

1220 19th Street N.W.   202-775-0436
Suite 801   Fax: 202-775-0061
Washington, DC 20036   http://www.amchp1.org
The mission of Maternal and Child Health Programs is to assure the health of all women, children and families.
Peter Sybinsky PhD, Chief Executive Officer

2190 Benton Foundation: Kids Campaigns

1625 K Street NW   202-638-5770
11th Floor   Fax: 202-638-5771
Washington, DC 20006   http://www.kidscampaigns.org
The Benton Foundation deals with communications, strategies and products in the pubilc interest.
Susan Nall Bales, Program Director

2191 Childhood Lead Poisoning Prevention Program

Ohio Department of Health
246 N High Street   614-728-9454
Columbus, OH 43215   Fax: 614-728-6793
E-mail: BCFHS@odh.ohio.gov
http://www.odh.state.oh.us
The mission of Ohio's Childhood Lead Poisoning Prevention Program is to eliminate childhood lead poisoning through screening, environmental inspection, abatement, education and case management.
J. Nick Baird M.D., Director

2192 Childhood Lead Poisoning Program
Oklahoma Department of Health
1000 Northeast 10th Street   405-271-5600
Oklahoma, OK 73117   Fax: 405-271-6199
http://www.health.state.ok.us
The Oklahoma State Department of Health is to protect and promote the health of citizens of Oklahoma and to prevent disease and injury, and to assure the conditions by which our citizens can be healthy.
Pamela G Rollins, Director Pediatrics

2193 Children's Defense Fund
25 E Street NW   202-628-8787
Washington, DC 20001   Fax: 202-622-3560
http://www.childrensdefense.org
To provide a strong and effective voice for all children of America who cannot vote, lobby or speak out for themselves. Pays particular attention to the needs of poor, minority and disabled children. To educate the nation about theneeds of children and encourage preventive investments in children before they get sick, drop out of school, suffer family breakdown or get into trouble.
Marti Teitelbaum, Senior Health Analyst

2194 Children's Environmental Health Network
110 Maryland Avenue NE   202-543-4033
Suite 505   Fax: 202-543-8797
Washington, DC 20002   E-mail: cehn@cehn.org
http://www.cehn.org/
Mission is to promote a healthy environment and to protect the fetus and child from environmental hazards. Three areas of concentration for the Network are education, research and policy. Network's goals are: to promote the developmentof sound public health and child-focused national policy; to stimulate prevention-oriented research; to educate health professionals, policymakers and community members in preventive strategies; and to elevate public awareness of environmentalhazards to children.
Nsedu Obot Witherspoon, Executive Director
Lynn Goldman, Board Chair

2195 Children's Health Environmental Coalition
12300 Wilshire Blvd   310-820-2030
Suite 410   Fax: 310-820-0207
Los Angeles, CA 90025   E-mail: chec@checnet.org
http://www.checnet.org
National grassroots nonprofit organization, researches the casues of childhood cancers and their relation to hazards in the environment. Dedicated to educating and organizing parents and grassroots groups around the need to protectchildren from the threat of environmental toxins.
Nancy Chuda, Co-Founder & President
Tessa Hill, Board Chair

2196 Coalition for Clean Air
523 W. 6th Street   213-630-1192
10th Floor   Fax: 213-630-1158
Los Angeles, CA 90014   http://www.coalitionforcleanair.org
To advocate policies and strategies that remedy air quality problems; to provide effective education and outreach programs to the community about the causes and health effects of air pollution and ways to help clean our air; and toadvocate for responsible public health policy and to serve as a source of technical and educational expertise.
Tim Carmichael, President & CEO
Janine Hamner, VP & COO

2197 Community-Based Hazard Management Program
George Perkins Marsh Institute
Clark University   508-751-4604
950 Main Street   Fax: 508-751-4615
Worcester, MA 01610   E-mail: otaylor@clarku.edu
http://www.clarku.edu/departments/marsh/projects/community/
The Community-Based Hazardous Management Program (formerly the Childhood Cancer Research Institute) is engaged in capacity building in communities affected by nuclear weapons production and testing and also specializes in radiationhealth risk assessment and management.
Octavia Taylor, Program Manager

2198 Healthy Schools Network
773 Madison Avenue   518-462-0632
Albany, NY 12208   Fax: 518-462-0433
E-mail: healthyschools@aol.com
http://www.healthyschools.org
HSN is a nationally recognized state-based advocate for the protection of children's environmental health in schools. Engages in research, education, outreach, technical assistance and coalition building to create schols that areenvironmentally responsible to children, and to their communities. Publishes a quarterly newsletter and maintains an Information Clearinghouse and Referral Service.

Leyla Erk McCurdy, President
John Shaw, Vice-President

2199 Institute of Medicine: Board of Children, Youth and Families
500 Fifth Street, NW   202-334-1935
Washington, DC 20001   Fax: 202-334-3584
E-mail: bocyf@nas.edu
http://www.bocyf.org
The mission of the Institute of Medicine is to advance scientific knowledge and the health and well-being of all people of this nation and the world, consistent with the role conferred by its congressional authority. It accomplishesthis mission by providing objective, timely and authoritative information to goverment, the professions and the public through its elected membership and access to the best expertise.
Rosemary Chalk, Director

2200 Kids Against Pollution
311 Main Street   315-266-0185
3rd Floor   Fax: 315-266-0186
Utica, NY 13501   E-mail: kap@kidsagainstpollution.org
http://www.kidsagainstpollution.org
Nonprofit organization of active youth dedicated to solving and preventing pollution problems through educational projects and events in order to protect children's health and the planet.

2201 Kids for Conservation: Today and Tomorrow
Illinois Department of Conservation
524 S 2nd Street   217-524-4126
Room 510   http://www.dnr.state.il.us
Springfield, IL 62701-1787
Provides a variety of programs for young Illinois hunters.
Kathy Andrews, Director Education Programs

2202 Kids for Saving Earth Worldwide
PO Box 421118   763-559-1234
Minneapolis, MN 55442   Fax: 763-559-6980
E-mail: kseww@aol.com
http://www.kidsforsavingearth.org/
To help protect the Earth through kids and adults. To educate and inspire them to participate in Earth-saving actions.
Tessa Hill, President and Director

2203 Kids for a Clean Environment
PO Box 158254   615-331-7381
Nashville, TN 37215   Fax: 615-333-9879
E-mail: kidsface@mindspring.com
http://www.kidsface.org
Established in 1989 to help children who wanted to learn more about the world in which they live, provide a way for children to be involved in the protection of nature and connect children with other children who share their concernsabout global environmental issues.
Trish Poe

2204 March of Dimes Birth Defects Foundation
1275 Mamaroneck Avenue   914-428-7100
White Plains, NY 10605   Fax: 914-428-8203
http://www.modimes.org
The mission of the March of Dimes Birth Defects Foundation is to improve the health of babies by preventing birth defects and reducing infant mortality. The March of Dimes carries out the mission through research, community service,education and advocacy.
Ann Umemoto, Associate Director

2205 US Environmental Protection Agency
Office of Children's Health Protection
1200 Pennsylvania Avenue, NW   202-564-2188
Mail Code 1107A   Fax: 202-564-2733
Washington, DC 20004   http://www.epa.gov
The mission of the Office of Children's Health Protection is to make the protection of children's health a fundamental goal of public health and environmental protection in the United States.
William H Sandersrn, Acting Director
Joanne Rodman, Associate Director

2206 US Public Interest Research Group
218 D Street SE   202-546-9707
Washington, DC 20003   Fax: 202-546-2461
                                      E-mail: uspirg@pirg.org
                                      http://www.pirg.org
The United States Interest Research Group is the national lobbying office for state PIRGS-nonpartisan, nonprofit environmental and consumer advocacy organizations active in more than 30 states. USPIRG combines the expertise ofprofessional staff with the energy, political voice and financial support of more than one million members to win reforms for a clean environment, consumer rights and a government accountable to the people.
Gene Karpinski, Executive Director

2207 US-Mexico Border Health Association
5400 Suncrest Dr   915-833-6450
Suite C-5   Fax: 915-833-7840
El Paso, TX 79912   http://www.usmba.org
Promotes public and individual health along the United States-Mexico border through reciprocal technical cooperation.
Dr Federico Saracho Weber, President
Dr Hector Gonzalez, First Vice President

Pediatric Health: Publications

2208 Case Studies in Environmental Medicine
1600 Clifton Road NE   404-639-0727
Atlanta, GA 30333   888-422-8737
                                      E-mail: astdric@cdc.gov
                                      http://www.astdr.cdc.gov/HEC/csem.html
Case Studies in Environmental Medicine is a series of self-instructional publications designed to increase primary care providers's knowledge of hazardous substances in the environment and to aid in the evaluation of potentiallyexposed patients. Continuing medical education credits and continuing education units are offered by the Agency for Toxic Substances and Disease Registry in support of this series.

2209 Child Health and the Environment
Oxford University Press
198 Madison Avenue   212-726-6000
New York, NY 10016   Fax: 212-726-6440
                                      http://www.oup-usa.org
Focus on environmental threats to child health. The first three chapters provide overviews of key children's environmental health issues as well as the role of environmental epidemiology and risk assessment in child health protection.Later chapters address the health affects of metal, PCBs, dioxins, pesticides, hormonally active agents, radiation, indoor and outdoor air pollution, and water contaminants.
416 pages
ISBN: 0-195135-59-8
Donald T Wigle, Author

2210 Children's Defense Fund
PO Box 90500   202-662-3652
Washington, DC 20090-0500   Fax: 202-628-8333
                                      http://www.childerndefense.org/
Annual report on the status of US children. To understand what is happening, The State of America's Children Yearbook 2000 examines cause and effect and offers thoughtful analyses on a variety of topics: child poverty, health care,child care, education, families in crisis and violence.

2211 Handbook of Pediatric Environmental Health
American Academy of Pediatrics
141 NW Point Boulevard   847-434-4000
Elk Grove Village, IL 60007-1098   Fax: 847-434-8000
                                      http://www.aap.org
Comprehensive guide to the identification, prevention of pediatric environmental health problems. Environmental hazards are among parent's top health concerns for their children, very little time is spent training physicians torecognize and treat ailments resulting from exposure to harmful substances and environments. This book provides insights and training to physicians in the field of pediatric environmental health.
723 pages
ISBN: 1-581100-29-9
Ruth A Etzel PhD, Editor

2212 Handle with Care: Children and Environmental Carcinogens
Natural Resources Defense Council
40 West 20th Street   212-727-2700
New York, NY 10011
This booklet discusses the exposure routes and effects of 10 pesticides and 11 other synthetic chemicals. The booklet raises concern about multipleexposures and discusses the current regulatory status of each chemical. It is written ina concsie, easy-to-understand language and provides an appendix of contacts for additional information.
50 pages
Lawrie Mott, Author

2213 Kids Count Data Book: State Profiles of Child Well-Being
701 St. Paul Street   410-223-2890
Baltimore, MD 21202   http://www.aecf.org/kidscount
Kids Count is national and state-by-state effort to provide makers and citizens with benchmarks of child well-being. It includes variables such as percentage of low birth-weight babies, child death rates, percentage of childern inpoverty, and percentage of childern without health insurance.

2214 Pediatric Annals: A Journal of Continuing Pediatric Education
515 North State Street   312-464-5192
Chicago, IL 60610   http://www.slackinc.com
Pediatric Annals is America's most widely read journal for continuing education in the field of pediatrics. Each monthly issue provides a through, practical review of a single topic in pediatrics, carefully chosen by an editorial boardcomposed of the leading pediatricians in the United States. Three credit hours in Category 1 for the physician's Recognition Award of the AMA are offered for the successful completion of a CME quiz.

2215 Pesticides and the Immune System: Public Health Risks
10 G Street NE   202-729-7600
Suite 800   Fax: 202-729-7610
Washington, DC 2002   E-mail: lauralee@wri.org
                                      http://www.wri.org
Brings together for the frist time an extensive body of experimental and epidemiological research from around the world documenting the the effects of widely used pesticides on the immune system and the attendent health risks. In sodoing, it documents that pesticide-related health risks are much more serious than genrally known, especially in developing countries where exposure is widespread and infectious diseases take a heavy toll.
100 pages
ISBN: 1-569730-87-3
Robert Repetto, Author
Sanjay S Baliga, Author

2216 Pesticides in Children's Food
Environmental Working Group
1718 Connecticut Avenue NW   202-667-6982
Suite 600
Washington, DC 20009
This report from Washington-based Environmental Working Group found widespread, general low pesticide levels in food. Based on data from nearly 20,000 food samples tested between 1990 and 1992 by the Food and Drug Administration andprivate labs, the EWG reported that more than half of the food samples had detectable pesticide resides of at least one pesticide. Of the samples that had dectectable residues, there were few violations of current tolerance limits.

2217 Preventing Child Exposures to Environmental Hazards: Research and Policy Issues

5900 Hollis Street   510-597-1393
Suite R3   Fax: 510-597-1399
Emeryville, CA 94608   E-mail: cehn@cehn.org
                    http://www.cchn.org/cchn/cehnsymposium.pdf

Preventing Children's Exposures to Environmental Hazards: Research and Policy Issues, the first symposium to address a national research and policy agenda for pediatric environmental health, generated over one hundred specificrecommendations. Two hundred recognized experts on research, policy, clinical practice, and advocacy contributed their expertise and vison for protecting childern from exposures to environmental hazards.

2218 Resource Guide on Children's Environmental Health

5900 Hollis Street   510-597-1393
Suite R3   http://www.cchn.org
Emeryville, CA 94608-2008

The Children's Environmental Health Network has developed the Resource Guide on Childern's Environmental Health to assist community leaders, policy makers, health and environmental specialists, members of the advocacy community andmedia, and the general public in identifying and accessing key resources in childern's environmental health.

Associations

2219 Atlantic States Legal Foundation
658 West Onondaga Street   315-475-1170
Syracuse, NY 13204-3757   Fax: 315-475-6719
E-mail: atlantic.states@aslf.org
http://www.aslf.org
Atlantic States Legal Foundation was established in 1982 to provide legal, technical, and organizational assistance on environment issues to citizen organizations (NGOs), individuals, local governments, and others.

Samuel H Sage, President
Daisy Hollis, Chair

2220 Business & Legals Reports
141 Mill Rock Road East   860-510-0100
Old Saybrook, CT 06475-6011   800-727-5257
Fax: 860-510-7220
E-mail: service@blr.com
http://www.blr.com
Provides essential tools for safety and environmental compliance and training needs.
Brian Gurnham, Chief Operating Officer
Peggy Carter-Ward, Editor-in-Chief

2221 Center for Community Action and Environmental Justice
PO Box 33124   909-360-8451
Riverside, CA 92519   Fax: 909-360-5950
E-mail: admin@ccaej.org
http://www.ccaej.org
The Center for Community Action and Environmental Justice serves as a resource center for community groups working on environmental justice issues.
Cindy Lopez-Elwell, President
Deenaz Coachbuilder, President-Elect

2222 Center for Health, Environment, and Justice
150 S Washington Street, Ste 300   703-237-2249
PO Box 6806   Fax: 703-237-8389
Falls Church, VA 22040   E-mail: chej@chej.org
http://www.chej.org
Supports environmental justice, the principle that people have the right to a clean and healthy environment. Helps local citizens and organizations hold industry and government accountable and work toward a healthy, environmentallysustainable future.
Lois Marie Gibbs, Executive Director

2223 Center for International Environmental Law
1367 Connecticut Avenue NW   202-785-8700
Suite 300   Fax: 202-785-8701
Washington, DC 20036-1860   E-mail: info@ciel.org
http://www.ciel.org
Public interest not-for-profit environmental law firm founded in 1989 to strengthen international and comparative environmental law and policy around the world. CIEL provides a full range of environmental legal services in bothinternational and comparative national law, including policy research and publication, advice, advocacy and education.
Daniel B Magraw Jr, President
Frederick R Anderson, Chair

2224 Center for Investigative Reporting
131 Steuart Street   415-543-1200
Suite 600   Fax: 415-543-8311
San Francisco, CA 94105   E-mail: center@cironline.org
http://www.muckraker.org
The only independent, nonprofit organization in the country dedicated to investigative reporting in the public interest on a broad range of issues.
Tom Golstein, President
Burton Glass, Executive Director

2225 Communities for a Better Environment
1400 Broadway   510-302-0430
Suite 701   Fax: 510-302-0437
Oakland, CA 94612

E-mail: cbeca@mail.com
http://www.cbecal.org
Nonprofit statewide, multiracial, urban environmental health and justice organization that works with urban communities and grassroots organizations, using science based research, legal tactics and organizing strategies to prevent airand water pollution, eliminate toxic hazards and improve public health. Long-term goals are to develop an environmentally sustainable manufacturing base, minimize the use of toxins, expand pollution prevention strategies and involve people most atrisk.
Jose T Bravo, President
Luke Cole, Vice President

2226 Community Environmental Council
930 Miramonte Drive   805-963-0583
Santa Barbara, CA 93109-1303   Fax: 805-962-9080
E-mail: admin@cecmail.org
http://www.communityenvironmentalcouncil.org
The Community Environmental Council is a nonprofit environmental organization headquartered in Santa Barbara, California. Our community involvement includes managing two recycling centers, a household hazardous waste facility, anurban farm, and three community gardens as well as the environmental education program Art From Scrap. In addition CEC provides research, technical assistance and education on local and statewide land use planning, and solid waste and integrated pestmanagement.
Bob Ferris, Executive Director

2227 Community Rights Counsel
1301 Connecticut Avenue NW   202-296-6889
Suite 502   Fax: 202-296-6895
Washington, DC 20036   E-mail: crc@communityrights.org
http://www.communityrights.org
A nonprofit public interest law firm that was formed in 1997 to assist communities in protecting their health and welfare by regulating permissible land uses, and that provides strategic assistance to state and local governmentattorneys in defending land use laws.
Douglas T Kendall, Founder/Executive Director
Timothy J Dowling, Chief Counsel

2228 Conservation Law Foundation
62 Summer Street   617-350-0990
Boston, MA 02110-1016   Fax: 617-350-4030
E-mail: ihartclf.org
http://www.clf.org
A nonprofit, member-supported organization that works to solve the environmental problems that threaten the people, natural resources and communities of New England. CLF's advocates use law, economics and science to design andimplement strategies that conserve natural resources, protect public health and promote vital communities in our region.

Philip Warburg, President
Rob Gardiner, Vice President

2229 Earthjustice
426 17th Street   510-550-6700
6th Floor   Fax: 510-550-6740
Oakland, CA 94612-2820   E-mail: eajus@earthjustice.org
http://www.earthjustice.org
Nonprofit public interest law firm dedicated to protecting the magnificent places, natural resources, and wildlife of this earth and to defending the right of all people to a healthy environment. It enforces and strengthensenvironmental laws on behalf of hundreds of organizations and communities.
Vawter (Buck) Parker, Executive Director

2230 Environmental Compliance in Your State
141 Mill Rock Road East   860-510-0100
Old Saybrook, CT 06475   800-727-3257
Fax: 860-510-7220
E-mail: service@blr.com
http://www.blr.com
Guide to state and federal environmental compliance regulations. Plain english analysis of regulations make them easy to understand and state vs. federal comparison tell you what rule to follow. Monthly state newsletter includedfree.
Robert Brady, Author
Tom Picinich, Marketing Director

2231 Environmental Defense Fund

257 Park Avenue S   212-505-2100
17th Floor   Fax: 212-505-2375
New York, NY 10010   http://www.environmentaldefense.org

Dedicated to protecting the environmental rights of all people, including future generations. Among these rights are clean air, clean water, healthy, nourishing food and a flourishing ecosystem. The solutions we advocate will be basedon science, even when it leads in unfamiliar directions. We will work to create solutions that win lasting political, economic and social support.
Frederic D Krupp, Executive Director
David Yarnold, Executive Vice President

2232 Environmental Law Alliance Worldwide

1877 Garden Avenue   541-687-8454
Eugene, OR 97403   Fax: 541-687-0535
       E-mail: elawus@elaw.org
       http://www.elaw.org

E-LAW advocates serve low income communities around the world, helping citizens strengthen and enforce laws to protect communities from toxic pollution and environmental degradation.
Bern Johnson, Executive Director
Lori Maddox, Associate Director

2233 Environmental Law Institute

2000 L Street NW   202-939-3800
Suite 620   Fax: 202-939-3868
Washington, DC 20036   E-mail: law@eli.org
       http://www.eli.org

Community Education and Training Program provides citizens and grassroots groups with information on environmental law and policy that can help them participate effectively in the decisions that impact public health and the environmentin their communities. Program's activities have included training courses on right-to-know laws and a series of workshops in demystifying the law, which focus on using the tools of public participation to address issues ranging from hazardous wasteto land use.
Leslie Carothers, President
Kenneth Berlin, Chairman

2234 Environmental Law and Policy Center of the Midwest

35 E Wacker Drive   312-673-6500
Suite 1300   Fax: 312-795-3730
Chicago, IL 60601   E-mail: elpc@elpc.org
       http://www.elpc.org

Howard Learner, Executive Director

2235 Environmental Support Center

1500 Massachusetts Avenue NW   202-331-9700
Suite 25   Fax: 202-331-8592
Washington, DC 20005   http://www.envsc.org/

The mission of the Environmental Support Center is to promote the quality of the natural environment, human health, and community sustainability by increasing the organizational effectiveness of local, state, and regional organizationsworking on environmental issues and for environmental justice. To be eligible for assistance, your organization must be a local, state or regional nonprofit organization with a portion of its resources devoted to environmental issues.
James Abernathy, Executive Director

2236 Federation of Environmental Technologists

9451 N 107th Street   414-354-0700
Milwaukee, WI 53224   Fax: 414-354-0073
       E-mail: info@fetinc.org
       http://www.fetinc.org

A nonprofit organization formed to assist industry in interpretation of and compliance with environmental regulations. Membership is open to all industries, municipalities, organizations and individuals concerned about environmentalregulations. Currently there are approximately 1000 members and 125 patron companies.
Triese Haase, Administrator
Julie Jansett, Assistant Administrator

2237 Fund for Animals

200 West 57th Street   212-246-2096
New York, NY 10019   Fax: 212-246-2633
       E-mail: hdquarters@fund.org
       http://www.fund.org

The Fund for Animals was founded in 1967 by author/humanitarian Cleveland Amory, and works in advocacy fields, using legislation, litigation and education as tools to advance the cause of animals.

2238 Greenhouse Action

PO Box 16743   206-935-8314
Seattle, WA 98116

Citizen's action group with focus on legislation, lobbying and media attention, inspired by the very successful hunger lobbying group results.
Patti Lowe, Executive Officer

2239 Greenpeace

702 H Street NW   202-462-1177
Suite 300   Fax: 202-462-4507
Washington, DC 20001   E-mail: info@WDC.greenpeace.org
       http://www.Greenpeaceusa.org

Greenpeace is an independent campaigning organization which uses non-violent creative confrontation to expose global environmental problems, and to force solutions that are essential to a green and peaceful future.
John Passacantaudo, Executive Director

2240 Harvard Environmental Law Society

Harvard Law School
Austin 201   617-495-3125
Cambridge, MA 02138   E-mail: jbaran@law.harvard.edu
       http://www.law.harvard.edu/studorgsslels/

HELS has been one of the most active student organizations at the Law School. The central mission of the society has remained essentially unchanged: to strengthen the capacity of its members to make real contributions to the field ofenvironmental law.
Miriam Seifter, President

2241 LandWatch Monterey County

PO Box 1876   831-422-9390
Salinas, CA 93902-1876   Fax: 831-422-9391
       E-mail: landwatchg@mclw.org
       http://www.landwatch.org

LandWatch is a nonprofit membership organization, founded in 1997. LandWatch works to promote and inspire sound land use legislation at the city, country, and regional lands, through grassroots community action.
Michael DeLapa, President
Chris Fitz, Executive Director

2242 League for Ecological Democracy

PO Box 1858   310-833-2633
San Pedro, CA 90042-1858

Publishes a newsletter that deals with the evolution of Green politics in the United States.
Robert W Long, Executive Officer

2243 League of Conservation Voters

1920 L Street NW   202-785-8683
Suite 800   Fax: 202-835-0491
Washington, DC 20036   http://www.lcv.org

Works to create a Congress more responsive to your environmental concerns. As the nonpartisan political voice for over nine million members of environmental and conservation groups, LCV is the only national environmental organizationdedicated full-time to educating citizens about the environmental voting records of Members of Congress.
Bill Roberts, Board Chair
Deb Callahan, President

2244 League of Women Voters of the United States

1730 M Street NW   202-429-1965
Suite 1000   Fax: 202-429-0854
Washington, DC 20036-4508   E-mail: lwv@lwv.org
       http://www.lwv.org

The League of Women Voters, a nonpartisan political organization, encourages the informed and active participation of citizens in government, works to increase understanding of major public policy issues and influences public policythrough education and advocacy.
Kay J Maxwell, President
Nancy Tate, Executive Director

2245 Legacy International

1020 Legacy Drive
Bedford, VA 24523

540-297-5982
Fax: 540-297-1860
E-mail: mail@legacyintl.org
http://www.legacyintl.org

Creates environments where people can address personal, community, and global needs while developing skills and effective responses to change.Whether working with youths, corporate leaders, educational professionals, entrepreneurs, orindividuals on opposing sides of a conflict, our goal is the same. Programs provide experiences, skills, and strategies that enable people to build better lives for themselves and others around them.
J E Rash, President

2246 Legal Environmental Assistance Foundation (LEAF)

1114 Thomasville Road
Suite E
Tallahassee, FL 32303

850-681-2591
Fax: 850-224-1275
E-mail: cvalencic@leaflaw.org
http://www.leaflaw.org

To protect human health and life-sustaining natural resources from pollution in Florida, Georgia and Alabama.
David Ludder, President/General Counsel

2247 National Association of Conservation Districts League City Office

PO Box 855
League City, TX 77574-0855

281-332-3402
Fax: 281-332-5259
E-mail: ron-francis@nacdnet.org
http://www.nacdnet.org

The nonprofit organization that represents the nation's 3,000 conservation districts and 17,000 men and women who serve on their governing boards. Conservation districts, local units of government established under state law to carryout natural resource management programs at the local level, work with more than 2.5 million cooperating landowners and operators to help them manage and protect land and water resources on nearly 98% of the private lands in the United States.
Bill Wilson, President
Tim Reich, Vice President

2248 Natural Resources Defense Council

40 West 20th Street
New York, NY 10011

212-727-2700
Fax: 212-727-1773
http://www.nrdc.org

NRDC is a national nonprofit organization with 500,000 members and a staff lawyers, scientists, and other environmental specialists. NRDC's mission is to safeguard the earth, its people, its plants and animals, and the natural systemson which all life depends.
John H Adams, President
Frances G Beinecke, Vice President

2249 Natural Resources Law Center

University of Colorado
School of Law
Box 401
Boulder, CO 80309-0401

303-492-1280
Fax: 303-492-1297
E-mail: peavyd@spot.colorado.edu
http://www.colorado.edu/Law/centers/nrlc/index.htm

Founded in 1981 as a function of the University of Colorado School of Law, the Center conducts research on environmental and natural resources law and policy, including water, public lands, minerals, and Native American law. The Centersponsors conferences and workshops, hosts visiting scholars, and publishes books, research papers, and newsletters.
Mark Squillace, Director

2250 New Mexico Environmental Law Center

1405 Luisa Street
Suite 5
Santa Fe, NM 87505

505-989-9022
Fax: 505-989-3769
E-mail: nmelc@nmelc.org
http://www.nmenvirolaw.org

An organization dedicated to protecting New Mexico's natural environment and communities from pollution and degradation. Over 80 percent of our clients are indigenous Native American or Hispanic and low income. Cases often includemining issues, growth impacts, water protection, air pollution, public lands protection or indigenous land claims. The organization is supported by grants from foundations, contributions from individuals and fees.
Douglas Meiklejohn, Executive Director
Earl James, Development Director

2251 Southern Environmental Law Center

201 W Main Street
Charlottesville, VA 22902-5065

434-977-4090
Fax: 434-977-1483
E-mail: selcva@selcva.org
http://www.southernenvironment.org

Dedicated to protecting the natural resources of Alabama, Georgia, North Carolina, South Carolina, Tennessee and Virginia. Works with more than 100 partner groups to safeguard southern forests, wetlands, coastal resources, rivers, airand water quality, wildlife habitat and rural landscapes through policy reform, public education, and direct legal action.
Rick Middleton, Executive Director

2252 Stanford Environmental Law Society

559 Nathan Abbott Way
Stanford, CA 94305-8610

650-723-4421
E-mail: sranchod@stanford.edu
http://www.elj.stanford.edu

Provides students with a unique set of opportunities to tap into structured programs or to create and pursue their own projects. Both organizations complement the Stanford Environmental and Natural Reources Law and Policy Program.
Zachary Fabish, Co-President
Craig Segall, Co-President

2253 Student Environmental Action Coalition

PO Box 31909
Philadelphia, PA 19104-0609

215-222-4711
Fax: 215-222-2896
E-mail: seac@seac.org
http://www.seac.org

Student and youth run national network of progressive organizations and individuals whose aim is to uproot environmental injustices through action and education. Works to create progressive social change on both the local and globallevels.
Maren Cummings, National Council Coordinator
Alison Brooks, Working Comte Coordinator

2254 US Public Interest Research Group

218 D Street SE
Washington, DC 20003

202-546-9707
Fax: 202-546-2461
E-mail: uspirg@pirg.org
http://www.uspirg.org

US PIRG is an advocate for the public interest. When consumers are cheated, or our natural environment is threatened, or the voices of ordinary citizens are drowned out by special interest lobbyists, US PIRG speaks up and takes action.We uncover threats to public health and well-being and fight to end them, using the time-tested tools of investigative research, media exposes, grassroots organizing, advocacy and litigation.
Anna Aurilio, Legislative Director
Gene Karpinski, Executive Director

2255 Western Environmental Law Center

1216 Lincoln Street
Eugene, OR 97401

541-485-2471
Fax: 541-485-2457
E-mail: eugene@westernlaw.org
http://www.westernlaw.org

The Western Environmental Law Center was established to meet the need for a regional public interest law firm to protect and restore the West's forests, grasslands, rivers, wildlife and human communities.
Corrie Yackulic, President
Lori Maddox, Vice President

Publications

2256 A Guide to Environmental Law in Washington DC

Environmental Law Institute
2000 L Street NW
Suite 620
Washington, DC 20036

202-939-3800
Fax: 202-939-3868
E-mail: law@eli.org
http://www.eli.org

Leslie Carothers, President
Kenneth Berlin, Chairman

2257 Buying Green: Federal Purchasing Practices and the Environment

Government Printing Office
732 North Capitol Street NW
Washington, DC 20401

202-512-0000
E-mail: www.gpo.gov

2258 Clean Water Act Twenty Years Later
Island Press
1718 Connecticut Avenue NW
Suite 300
Washington, DC 20009-1148
202-232-7933
Fax: 202-234-1328
E-mail: info@islandpress.org
333 pages
ISBN: 1-559632-65-8
Richard W Alder; Jessica C Landman; Diane Cameron, Author

2259 Comparative Environmental Law and Regulation
Oceana Publications, Inc
75 Main Street
Dobbs Ferry, NY 10522
914-693-8100
800-831-0758
Fax: 914-693-0402
E-mail: info@oceanalaw.com
http://www.oceanalaw.com
Key environmental laws, regulations and implementation systems and agencies of 37 countries from around the world.
2 vol pages Semi-Annual
ISBN: 0-379012-51-0
Nicholas A Robinson, Editor

2260 Compendium of Superfund Program Publications
National Technical Information Service
5285 Port Royal Road
Springfield, VA 22161
800-553-6847
Fax: 703-321-8547

2261 Environmental Defense
257 Park Avenue S
17th Floor
New York, NY 10010-7304
212-505-2100
800-684-3322
Fax: 212-505-2375
http://www.environmentaldefense.org
Recycled products and gifts.
Fred Krupp, President

2262 Environmental Defense Newsletter
257 Park Avenue S
17th Floor
New York, NY 10010
212-505-2100
Fax: 212-505-2375
http://www.environmentaldefense.org
Dedicated to protecting the environmental rights of all people, including future generations. Among these rights are clean air, clean water, healthy, nourishing food and a flourishing ecosystem. The solutions we advocate will be basedon science, even when it leads in unfamiliar directions. We will work to create solutions that win lasting political, economic and social support because they are bipartisan, efficient and fair.
Frederic D Krupp, Executive Director
Marcia Aronoff, Deputy Direrector Programs

2263 Environmental Law Lexicon
Law Journal Seminars-Press
111 8th Ave.
New York, NY 10011
212-741-8300
800-888-8300
Fax: 212-463-5526
A unique quick-reference guide specifically prepared to help lawyers understand the most frequently encountered scientific terminology concerning the environment, ecology and many other scientific disciplines, and the specific legalmeanings that these terms have acquired through their use in laws, regulations, and judicial decisions.
400 pages
ISBN: 1-588520-57-9
Nicholas A Robinson, Author

2264 Environmental Law and Compliance Methods
Oceana Publications, Inc
75 Main Street
Dobbs Ferry, NY 10522
914-693-8100
800-831-0758
Fax: 914-693-0402
E-mail: info@oceanalaw.com
http://www.oceanalaw.com
Presents practical information tailored to professionals responsible for day-to-day compliance with the environmental laws of the US.
678 pages One Time
ISBN: 0-379214-26-1
Edward E Shea, Author

2265 Environmental Politics and Policy
Congressional Quarterly Press
1255 22nd Street NW
Suite 400
Washington, DC 20037
202-729-1800
366 pages
ISBN: 1-568028-78-4
Walker A Rosenbaum, Author

2266 Environmental Regulatory Glossary
Government Institutes
4 Research Place
Suite 200
Rockville, MD 20850
301-921-2300
Fax: 301-921-0373
623 pages
Thomas F P Sullivan, Author

2267 How Wet is a Wetland?: The Impacts of the Proposed Revisions to the Federal Wetlands Manual
Environmental Defense Fund
257 Park Avenue South
New York, NY 10010
212-505-2100
800-684-3322
Fax: 212-505-0892
E-mail: media@environmentaldefense.org
http://www.environmentaldefense.org

Fred Krupp, President

2268 Insider's Guide to Environmental Negotiation
Lewis Publishers
2719 Dover Lane
Albany, GA 31721
229-438-1080
http://www.lewispub.com
242 pages
ISBN: 0-873715-09-8
Dale M Gorczynski, Author

2269 International Environmental Policy: From the Twentieth to the Twenty-First Century
Duke University Press
905 W Main Street, Ste 18-B
PO Box 90660
Durham, NC 27708-0660
919-687-3600
Fax: 919-688-4574
E-mail: orders@dukepress.edu
http://www.dukespress.com
496 pages
Lynton Keith Caldwell, Author

2270 Making Development Sustainable: Redefining Institutions, Policy, and Economics
Island Press
Box 7
Covelo, CA
800-828-1302
Fax: 707-983-6432
362 pages
ISBN: 1-559632-13-5
Johan Holmberg, Author

2271 Managing Planet Earth: Perspectives on Population, Ecology and the Law
Bergin & Garvey Publishers
88 Post Road W
Box 5007
Westport, CT 06881
203-226-3571
800-225-5800
http://www.greenwood.com
184 pages
ISBN: 0-897892-16-X
Miguel A Santos, Author

2272 Natural Resources Policy and Law: Trends and Directions
Island Press; Natural Resources Law Center
1718 Connecticut Avenue NW
Suite 300
Washington, DC 20009

202-232-7933
800-828-1302
Fax: 202-234-1328
E-mail: info@islandpress.org
http://www.islandpress.org

255 pages
ISBN: 1-559632-46-1
Lawrence J MacDonnell; Sarah F Bates, Author

2273 New Mexico Environmental Law Center: Green Fire Report
1405 Luisa Street
Suite 5
Santa Fe, NM 87505

505-989-9022
Fax: 505-989-3769
E-mail: nmelc@nmelc.org
http://www.nmenvirolaw.org

A publication from the organization dedicated to protecting New Mexico's natural environment and communities from pollution and degradation. Over 80 percent of our clients are indigenous Native American or Hispanic and low income.Cases often include mining issues, growth impacts, water protection, air pollution, public lands protection or indigenous land claims. The organization is supported by grants from foundations, contributions from individuals and fees.
12 pages Quarterly
Douglas Meiklejohn, Executive Director
Earl James, Development Director

2274 Oversight of Implementation of the Clean Air Act Amendments of 1990
Government Printing Office
N Capitol & H Streets NW
Washington, DC 20401

202-512-0000

ISBN: 0-160388-26-0

2275 People for the Ethical Treatment of Animals
501 Front Street
Norfolk, VA 23510

757-622-7382
Fax: 757-628-0457
E-mail: info@peta.org
http://www.peta.org

People for the Ethical Treatment of Animals, with more than seven hundred members, is the largest animal rights organization in the world. Founded in 1980, PETA is dedicated to establishing and protecting the rights of all animals.PETA operates under the simple principle that animals are not ours to eat, wear, experiment on, or use for entertainment.

2276 Renewable Resource Policy: The Legal-Institutional Foundation
Island Press
1718 Connecticut Avenue NW
Suite 300
Washington, DC 20009

202-232-7933
800-828-1302
Fax: 202-234-1328

572 pages
ISBN: 1-559632-25-9
David A Adams, Editor

2277 Saving All the Parts: Reconciling Economics and the Endangered Species Act
Island Press
Box 7
Covelo, CA

707-983-6432

280 pages
ISBN: 1-559632-02-X
Rocky Barker, Author

2278 Searching Out the Headwaters: Change and Rediscovery in Western Policy
Island Press
Box 7
Covelo, CA

800-828-1302
Fax: 707-983-6432

253 pages
ISBN: 1-559632-17-8
Sarah F Bates, et al, Author

2279 Setting National Priorities: Policy for the Nineties
Brookings Institution
1775 Massachusetts Ave., NW
Washington, DC 20036-2188

202-797-6000
800-275-1447
Fax: 202-797-6004
http://www.brookings.edu

2280 Trade and the Environment: Law, Economics and Policy
Island Press
Box 7
Covelo, CA

800-828-1302
Fax: 707-983-6432

333 pages
ISBN: 1-559632-67-4
Durwood Zaelke et al, Editors

2281 Understanding Environmental Administration and Law
Island Press
Box 7
Covelo, CA 95428

800-828-1302

239 pages
ISBN: 1-559634-74-X
Susan J Buck, Author

Foundations & Charities

2282 AMETEK Foundation
37 N Valley Road
PO Box 1764   610-647-2121
Paoli, PA 19301   800-473-1286
Fax: 215-323-9337
E-mail: www.ametek.com
http://webmaster@ametek.com
Robert W Yannarell, Assistant Secretary

2283 ARCO Foundation
515 South Flower Street   213-486-3342
Los Angeles, CA 90071
Russell Sakaguchi, Program Officer

2284 Abelard Foundation
2530 San Pable Avenue   510-644-1904
Suite B
Berkeley, CA 94702-2013
Leah Brumer, Executive Director

2285 Acid Rain Foundation
1410 Varsity Drive   919-828-9443
Raleigh, NC 27606   Fax: 919-515-3593
Dr. Harriet S Stubbs, Executive Director

2286 Acorn Foundation
2530 San Pablo Avenue   510-644-1904
Suite B
Berkeley, CA 94702
Leah Brummer, Executive Director

2287 African Wildlife Foundation
1717 Massachusetts Avenue NW   202-939-3333
Suite 602   Fax: 202-939-3332
Washington, DC 20036
Paul T Schindler, President

2288 Amax Foundation
200 Park Avenue   212-856-4250
New York, NY 10166
Sonja Michaud, President

2289 American Association of Petroleum Geologists Foundation
Po Box 979   918-584-2555
Tulsa, OK 74101   Fax: 918-560-2636
E-mail: info@aapg.org
http://www.aapg.org
News for explorationists of oil, gas and minerals as well as for geologists with environmental and water well concerns.

2290 American Electric Power
1 Riverside Plaza   614-223-1000
Columbus, OH 43215-2373   E-mail: corpcomm@aep.com
http://www.aep.com
Richard M McMorrow, Assistant VP

2291 American Rivers
801 Pennsylvania Avenue SE   202-347-7550
Suite 400   Fax: 202-347-9240
Washington, DC 20003
Kevin J Coyle, President

2292 Amoco Foundation
200 East Randolph Drive   312-856-6306
Chicago, IL 60601   E-mail: foundation@amoco.com
http://www.amoco.com
Patricia Wright, Executive Director

2293 Andrew W. Mellon Foundation
140 East 62nd Street   212-838-8400
New York, NY 10021   Fax: 212-223-2778
William Robertson IV, Cons. and Envir. Pro. Dir.

2294 Asthma and Allergy Foundation of America
1233 20th Street, NW   202-466-7643
Suite 402   Fax: 202-466-8940
Washington, DC 20036   E-mail: info@aafa.org
http://www.aafa.org
Dedicated to improving the quality of life for people with asthma and allergies through education, advocacy and research.

2295 Atherton Family Foundation
1164 Bishop Street   808-566-5524
Suite 800   Fax: 808-521-6286
Honolulu, HI 96813 http://www.athertonfamilyfoundation.org
Lisa Schiff, Priv Found Services Officer

2296 Audubon Naturalist Society of the Central Atlantic States
8940 Jones Mill Road   301-652-9188
Chevy Chase, MD 20815   Fax: 301-951-7179
Mike Nelson, Executive Director

2297 BP America
200 Public Square 36-A   216-586-8625
Cleveland, OH 44114   http://www.bp.com
Lance C Buhl, Dir. Corporate Contributions

2298 Baltimore Gas & Electric Foundation
PO Box 1475   410-685-0123
Baltimore, MD 21203   800-685-0123
E-mail: corporate.communications@bge.com
http://www.bge.com
Gary R Fuhrman, Asst. Secretary/Treasurer

2299 Bauman Foundation
1731 Connecticut Avenue   202-234-8547
Suite 400   Fax: 202-234-8584
Washington, DC 20009-1146
Patricia Bauman, Co-Director

2300 Bay Foundation
17 West 94th Street   212-663-1115
New York, NY 10025
Robert W Ashton, Executive Director

2301 Blandin Foundation
100 North Pokegama Avenue   218-326-0523
Grand Rapids, MI 55744   Fax: 218-327-1949
Paul M Olson, President

2302 Boise Cascade Corporation
PO Box 50   208-384-7189
Boise, ID 83728-0001   Fax: 208-384-7189
E-mail: bcweb@bc.com
http://www.bc.com
Connie E Weaver, Contributions Administrator

2303 Cape Branch Foundation
5 Independence Way   609-987-0300
Princeton, NJ 08540   Fax: 609-452-1024
Dorothy Frank, Partner

2304 Cargill Foundation
15407 McGinty Road   952-742-2546
PA-50   Fax: 952-742-7224
West Wayzata, MN 55391-2399   http://www.cargill.com
James S Hield, Secretary, Contributions

2305 Caribbean Conservation Corporation
4424 NW 13th Street 352-373-6441
Suite A1 800-678-7853
Gainesville, FL 32609 Fax: 352-375-2449
E-mail: ccc@cccturtle.org
http://cccturtle.org
Caribbean Conservation Corporation is a nonprofit membership organization based in Gainesville. CCC was the first marine turtle conservation organization in the world, and has more than 40 years of experience in national andinternational sea turtle conservation, research and educational endeavors.
David Godfrey, Executive Director

2306 Carolyn Foundation
901 Marquette Avenue 612-596-3266
Suite 2630 Fax: 612-338-2084
Minneapolis, MN 55402 E-mail: carolyn@winternet.com
http://www.carolynfoundation.org
The Carolyn Foundation is a general foundation. Please check out our website www.carolynfoundation.org for the most up-to-date information regarding funding priorities, guidelines and application process.
Rebecca Erdahl, Executive Director
Cindy Mellin, Foundation Administrator

2307 Caterpillar Foundation
100 Northeast Adams Street 309-675-4464
Peoria, IL 61629-1480 http://www.caterpillar.com
Edward W Siebert, Vice President/Manager

2308 Charles Engelhard Foundation
645 5th Avenue 212-935-2430
7th Floor
New York, NY 10022
Elaine Catterall, Secretary

2309 Chesapeake Bay Foundation
6 Herndon Avenue 410-268-8816
Annapolis, MD 21403 Fax: 410-268-6687
William C Baker, President

2310 Chevron Corporation
575 Market Street 415-894-7700
San Francisco, CA 94105 E-mail: chevweb@chevron.com
http://www.chevron.com
Skip Rhodes, Mgr. Corporate Contributions

2311 Clean Water Action
4455 Connecticut Avenue 202-895-0420
Suite A300 Fax: 202-895-0438
Washington, DC 20008 http://www.cleanwateraction.org
David Zwick, Executive Director

2312 Collins Foundation
1618 SW 1st Avenue 503-227-7171
Suite 305 http://members.tripod.com/collinsfoundation
Portland, OR 97201-5708
Brian Coakley, President

2313 Compton Foundation
525 Middlefield Road 650-328-0101
Suite 115 Fax: 650-328-0171
Menlo Park, CA 94025
Edith T Eddy, Administrative Director

2314 Conservation International
1919 M Street NW 202-912-2306
Suite 600 800-406-2306
Washington, DC 20036 Fax: 202-887-5188
http://www.conservation.org
Our mission is to conserve the Earth's living natural heritage, our global biodiversity, and to demonstrate that human societies are able to live harmoniously with nature.
Russell A Mittermeier, President

2315 Conservation Treaty Support Fund
3705 Cardiff Road 301-654-3150
Chevy Chase, MD 20815-5943 800-654-3150
Fax: 301-652-6390
E-mail: ctsf@conservationtreaty.org
http://www.conservationtreaty.org
Supports projects of international wildlife and habitat treaties.
George A Furness Jr, President

2316 Conservation and Research Foundation
PO Box 909 913-268-0076
Shelburne, VT 05482 Fax: 913-268-0076
Dr Mary Wetzel, President

2317 Cooper Industries Foundation
PO Box 4446 713-209-8800
Houston, TX 77210-4446 Fax: 713-209-8982
E-mail: info@cooperindustries.com
http://www.cooperindustries.com
Cooper was primarily a one-market company, manufactguring power and compression equipment for the transmission of natural gas. Eventually broadening its product lines to include petroleum and industrial equipment, electrical powerequipment, automotive products tools and hardware.
Victoria B Guennewig, President

2318 Cricket Foundation
Exchange Place 617-570-1130
Suite 2200 Fax: 617-523-1231
Boston, MA 02109-2881
George W Butterworth III, Counsel

2319 Crystal Channel Foundation
PO Drawer 329 254-897-2960
Glen Rose, TX 76043 Fax: 254-897-3785
Christine Jurzykowski, President

2320 Curtis and Edith Munson Foundation
321 N Clark Street 312-527-5545
Suite 950 Fax: 312-527-9064
Chicago, IL 60610
C Wolcott Henry III, President

2321 Deer Creek Foundation
720 Olive Street 314-241-3228
Suite 1975
St. Louis, MO 63101
Mary Stake Hawker, Administrator

2322 Defenders of Wildlife
1011 14th Street NW 202-682-9400
Suite 1400 Fax: 202-682-1331
Washington, DC 20005 E-mail: info@defenders.org
http://www.defenders.org
Dedicated to the protection of all native wild animals and plants in their natural communities. We focus on what scientists consider two of the most serious environmental threats to the planet: the accelerating rate of extinction ofspecies and the associated loss of biological diversity, and habitat alteration and destruction. Long known for our leadership on endangered species issues.
Roger Schlickeisen, President

2323 Digital Equipment Corporation
111 Powder Mill Road 978-493-9210
Unit B14 http://www.digitalcentury.com
Maynard, MA 01754
Jane Hamel, Mgr. Corporate Contributions

2324 Dunspaugh-Dalton Foundation
9040 SW 72nd Street 305-668-4192
Suite 30 Fax: 305-668-4247
Miami, FL 33173
William A Lane Jr, President

2325 Earth Share
3007 Tilden Street NW
Washington, DC 20008   800-875-3863

2326 Earth Society Foundation
238 E 58th Street   212-832-3659
Suite 2400   Fax: 212-826-6213
New York, NY 10022   E-mail: trusteeone@aol.com
                                http://www.earthsite.org
Started the original Earth Day, which is devoted to peace, justice
and the care of earth. It invites everyone to think and act as trust-
ees of earth.
Mary Carlin, Secretary
John McConnell, Chairman Emeritus

2327 Echoing Green
60 East 42nd Street   212-689-1165
Suite 520   Fax: 212-689-9010
New York, NY 10165   E-mail: info@echoinggreen.org
                              http://www.echoinggreen.org
Echoing Green is a global science venture fund that provides
seed funding and support to visionary leaders with bold new
ideas for social change.
Cheryl Dorsey, President

2328 Edward John Noble Foundation
32 East 57th Street   212-055-4212
New York, NY 10022-2513
EJ Noble Smith, Executive Director

2329 Elmina Sewall Foundation
245 Commercial Street
Portland, ME 04101
Elmina B Sewall, President

2330 Energy Foundation
75 Federal Street   415-561-6700
San Francisco, CA 94107   Fax: 415-561-6709
Hal Harvey, Executive Director

2331 Environmental Law Institute
2000 L Street NW   202-939-3800
Suite 620   Fax: 202-939-3868
Washington, DC 20036   E-mail: law@eli.org
                                http://www.eli.org
Leslie Carothers, President
Kenneth Berlin, Chairman

2332 Exxon Education Foundation
225 E John W Carpenter Freeway   972-444-1000
Room 1429   http://www.exxon.com
Irving, TX 75062-2298
Leonard Fleischer, Mgr. Corporate Contributions

2333 First Hawaiian Foundation
999 S Bishop Street   808-525-7000
29th Floor   http://www.fhb.com
Honolulu, HI 96813
Herbert E Wolff, Secretary

2334 First Interstate Bank of Nevada Foundation
PO Box 11007   775-784-3844
Reno, NV 89520
Kevin Day, President

2335 FishAmerica Foundation
6101 E Apache Street   918-836-5581
Tulsa, OK 74115-3370   800-444-5581
                                                  Fax: 918-836-3542

2336 FishAmerica Foundation.
Grant Guidelines
225 Reinekers Lane
Suite 420   703-519-9691
Alexandria, VA 22314   Fax: 703-519-1872

E-mail: fishamerica@asafishing.org
http://www.fishamerica.org
The FishAmerica Foundation provides funding for local,
hands-on projects to enhance fish populations, restore fisheries
habitat, improve water quality, and advancing fisheries research
in North America; thereby increasing theopportunity for
sportfishing success.

Johanna Laderman, Managing Director
Jeff Bloem, Grants Coordinator

2337 Frank Weeden Foundation
11 Broadway
New York, NY 10004
2125090579
Alan N Weeden, President

2338 Friends of the Earth Foundation
1717 Massachusetts Avenue NW 600
Washington, DC 20036-2002   877-843-8687
                                                  Fax: 202-783-0444
                                             E-mail: foe@foe.org
                                              http://www.foe.org
Defends the environment and champions a healthy and just
world.
Brent Blackwelder, President

2339 Frost Foundation
511 Armijo   505-986-0208
Suite A   E-mail: info@frostfound.org
Santa Fe, NM 87501   http://www.frostfound.org
Dr. Theodore Kauss, Executive Director

2340 Fund for Animals
200 West 57th Street   212-246-2096
New York, NY 10019   Fax: 212-246-2633
                                        E-mail: hdquarters@fund.org
                                http://www.fundforanimals.org/about/
The Fund for Animals was founded in 1967 by author/humani-
tarian Cleveland Amory to speak for those who can't. We work
in advocacy fields, using legislation, litigation, and education as
tools to advance the cause of animal.
Marian Probst, President

2341 Fund for Preservation of Wildlife and Natural Areas
Boston Safe Deposit and Trust Company
1 Boston Place   617-722-7340
Boston, MA 02108   Fax: 617-722-7129
Sylvia Salas, Director

2342 George B Storer Foundation
PO Box 1270
Saratoga, WY 82331   307-326-8308
Peter Storer, President

2343 George W Perkins Memorial Foundation
660 Madison Avenue
New York, NY 10021

2344 Georgia Pacific Foundation
133 Peachtree Street NE   404-652-4000
Atlanta, GA 30303   http://www.gp.com
Wayne Tamblyn, President

2345 Geraldine R. Dodge Foundation
163 Madison Avenue   973-540-8442
PO Box 1239   Fax: 973-540-1211
Morristown, NJ 07962
David Grant, Executive Director
Robert Perry, Program Director

2346 Greensward Foundation
104 Prospect Park West
Brooklyn, NY 11215   E-mail: info@greenswardsparks.org
                          http://www.greenswardparks.org
The Greensward Foundation, through its local branches, celebrates and suppports our communities's public parks.

Robert M Makla, Director

2347 HKH Foundation
33 Irving Place   212-682-7522
10th Floor
New York, NY 10003
Harriet Barlow, Adv.

2348 Helen Clay Frick Foundation
7227 Reynolds Street   412-371-0600
PO Box 86190
Pittsburgh, PA 15208
DeCoursey E McIntosh, Executive Director

2349 Henry L and Consuelo S Wenger Foundation
100 Renaissance Center   313-567-1212
Detroit, MI 48226
Shelly Raines, Principal Manager

2350 Hoffman-La Roche Foundation
PO Box 278   973-235-3797
Nutley, NJ 07110
Rosemary Bruner, Administrative Director

2351 INFORM
120 Wall Street   212-361-2400
14th Floor   Fax: 212-361-2412
New York, NY 10005-4001   http://www.informinc.org
INFORM is an independent research organization that examines the effects of business practices on the environment and on human health. Our goal is to identify ways of doing business that ensure environmentally sustainable economicgrowth. Our reports are used by government, industry, and environmental leaders around the world.
Joanna Underwood, President

2352 International Primate Protection League
PO Box 776   843-871-2280
Summerville, SC 29484   Fax: 843-871-7988
    E-mail: info@ippl.org
    http://www.ippl.org
An organization that works worldwide for the conservation and protection of apes and monkeys.
Dr Shirley McGreal, Chairperson
Marjorie Doggett, Secretary

2353 International Wildlife Coalition
634 N Falmouth Highway   508-564-9980
Box 388   Fax: 508-563-2843
North Falmouth, MA 02556-0388
Daniel J Morast, President

2354 International Wildlife Conservation Society
Grants Management Association
230 Congress Street   617-426-7172
Boston, MA 02110   http://www.wcs.org
Ala H Reid, Administrator

2355 Jessie Smith Noyes Foundation
6 E 39th Street   212-684-6577
12th Floor   Fax: 212-689-6549
New York, NY 10016   E-mail: noyes@noyes.org
    http://www.noyes.org
Promotes a sustainable and just social and natural system by supporting grassroots organizations and movements committed to this goal.

Victor DeLuca, President

2356 John D and Catherine T MacArthur Foundation
140 S Dearborn Street
Suite 1100   Fax: 312-917-0334
Chicago, IL 60603   E-mail: 4answers@macfound.org
    http://www.macfound.org
A private independent grantmaking institution dedicated to helping groups and individuals foster lasting improvement in the human condition.

2357 Jules and Doris Stein Foundation
PO Box 30   323-276-2101
Beverly Hills, CA 90213
Linda L Valliant, Secretary

2358 Kangaroo Protection Foundation
1900 L Street NW   202-452-1100
Suite 526
Washington, DC 20036
Marian Newman, Program Director

2359 Keep America Beautiful
1010 Washington Boulevard   203-323-8987
Stamford, CT 06901   Fax: 203-325-9199
    E-mail: blyons@kab.org
    http://www.kab.org
Nonprofit organization whose network of local, statewide and international affiliate programs educates individuals about litter prevention and ways to reduce, reuse, recycle and properly manage waste materials. Through partnershipsand strategic alliances with citizens, businesses and government, Keep America Beautiful's programs motivate millions of volunteers annually to clean up, beautify and improve their neighborhoods, thereby creating healthier and safer communityenvironments.
Becky Lyons, VP Training/Affiliate

2360 Kraft General Foods Foundation
Kraft Court   847-998-7031
Unit 2W   800-543-5335
Glenview, IL 60025   http://www.kraftfoods.com
Pamela Hollie, Dir. Corporate Contributions

2361 Kroger Company Foundation
1014 Vine Street   513-762-4443
PO Box 1199   866-221-4141
Cincinnati, OH 45202   http://www.kroger.com
Paul Bernish, VP/Secretary

2362 Liz Claiborne Foundation
1441 Broadway Avenue
New York, NY 10018   E-mail: corporate.secretary@liz.com
    http://http://www.lizclaiborneinc.com/foundation/default.asp
Established to serve as the Company's center for charitable activiteis. Works to meet the needs of the communities where the major facilities of Liz Claiborne, Inc. are located. Projects focus primarily on helping disadvantaged womengain their self-sufficiency through job training and microenterprise development. The Foundation also provides ongoing support to many artistic and cultural institutions which enhance the livability of our communities.

Paul R Charron, Chairman of the Board/CEO
Angela J Ahrendts, Executive Vice President

2363 Liz Claiborne and Art Ortenberg Foundation
650 5th Avenue   212-333-2536
15th Floor   Fax: 212-956-3531
New York, NY 10019   E-mail: lcaof@fcc.net
The Foundation has 2 primary program interests: mitigation of conflict between the land and resources needs of local communities and conservation of biological diversity in rural landscapes outside of parks and reserves; implementationof relevant, field based scientific, technical and practical training programs for local people. The Foundation typically funds modest, carefully designed field activities-primarily in developing countries and the Northern Rockies.
James Murtaugh, Director
Jeffery T Olson, Director

2364 Louis and Anne Abrons Foundation
437 Madison Avenue   212-756-3376
New York, NY 10017
Richard Abrons, President and Director

2365 Louisiana Land and Exploration Company
PO Box 60350   504-566-6500
New Orleans, LA 70160

Karen A Overson, Contributions Coordinator

2366 MNC Financial Foundation
10 Light Street   301-244-5000
PO Box 987-MS251001
Baltimore, MD 21203
Geeorge BP Ward Jr, Secretary/Treasurer
Alfred Lerner, Chairman

2367 Mark and Catherine Winkler Foundation
4900 Seminary Road   703-998-0400
Alexandria, VA 22311   Fax: 703-578-7899
Lynne Ball, Executive Director

2368 Mars Foundation
6885 Elm Street   703-821-4900
McLean, VA 22101   E-mail: ifoundmars@aol.com
                http://www.multiple-sclerosis-mf.org
Roger G Best, Secretary

2369 Marshall and Ilsley Foundation
770 North Water Street   414-765-7835
Milwaukee, WI 53201   http://www.micorp.com
Diana L Sebion, Secretary

2370 Mary Reynolds Babcock Foundation
2920 Reynolda Village   336-748-9222
Winston-Salem, NC 27106   Fax: 336-777-0095
                        E-mail: info@mrbr.org
                        http://www.mrbf.org
Our mission is to help people and places to move out of poverty
and achieve greater social and economic justice.
Gayle Williams, Executive Director
Sandra Mikush, Assistant Director

2371 Max McGraw Wildlife Foundation
PO Box 9   847-741-8000
Dundee, IL 60118   Fax: 847-741-8157
                E-mail: mcgrawwild@AOL.COM
The foundation's mission: education, research, and land man-
agement. Currently, the Foundation is invovled in over 15 re-
search and land management projects through the Chicago
region, including participation in the Chicago
Wildernessinitiative. Situated on 1,225 acres, the Foundation
property is managed by professional land management staff.
Stanley W Koenig, Executive Director
John Thompson, Director Research

2372 Max and Victoria Dreyfus Foundation
575 Madison Avenue   212-605-0354
New York, NY 10022
Lucy Gioia, Administrative Assistant

2373 May Stores Foundation
611 Olive Street   314-342-6300
St. Louis, MO 63101   Fax: 314-342-4461
                http://www.maycompany.com
James Abrams, VP Corporate Communications

2374 McIntosh Foundation
1730 M Street NW   202-338-8055
Washington, DC 20036   Fax: 202-234-0745
                        E-mail: mcf@aol.com
Michael A McIntosh, President

2375 Nathan Cummings Foundation
475 Tenth Avenue   212-787-7300
Fourteenth Floor   Fax: 212-787-7377
New York, NY 10018   E-mail: info@nathancummings.org
                http://http://www.nathancummings.org
The Nathan Cummings Foundation is rooted in the Jewish tradition and committed to democratic values and social justice, including fairness, diversity, and community. They seek to build a
socially and economically just society thatvalues nature and
protects the ecological balance for future generations; promotes
humane health care; and fosters arts and cultures that enriches
communities.
, Environment Program Director

2376 National Audubon Society
700 Broadway   212-979-3000
New York, NY 10003   Fax: 212-979-3188
                E-mail: jbianchi@audubon.org
The mission of the National Audubon Society is to conserve and
restore natural ecosystems, focusing on birds and other wildlife
for the benefit of humanity and the earth's biological diversity.
Founded in 1905, the National AudubonSociety is named for
John James Audubon, famed orinthologist, exployer, and wild-
life artist.
John Flicker, President

2377 National Fish and Wildlife Foundation
1120 Connecticut Avenue NW   202-857-0166
Suite 900   Fax: 202-857-0162
Washington, DC 20036   http://www.nfwf.org
John Berry, Executive Director
Karen Sprecher-Keating, General Counsel

2378 National Geographic Society Education Foundation
1145 17th Street NW   202-828-6672
Washington, DC 20036   Fax: 202-775-6141
Dori Jacobson, Program Development Officer

2379 National Parks Conservation Association
1300 19th Street   202-223-6722
#300   Fax: 202-659-0650
Washington, DC 20036   http://www.eparks.org
Thomas Kiernan, President

2380 National Wildlife Federation
1400 16th Street NW   202-797-6800
Washington, DC 20036   Fax: 202-797-6646
                                http://www.nwf.org
Jim Lyon, Director

2381 Nature Conservancy
1815 North Lynn Street   703-841-5300
Arlington, VA 22209   Fax: 703-841-1283
John C Sawhill, President

2382 New England Biolabs Foundation
8 Enon st #2B   978-927-2404
Beverly, MA 01915-1116   Fax: 978-998-6837
                        E-mail: fosters@nebf.org
                        http://www.nebf.org
NEBF funds grass roots organizations in developing countries
that focus on environmental issues and education.

2383 New York Times Company Foundation
229 W 43rd Street   212-556-1234
New York, NY 10036   Fax: 212-556-3690
                                http://www.nytco.com
Arthur Gelb, President

2384 New-Land Foundation
1114 Avenue of the Americas   212-479-6162
46th Floor   Fax: 212-841-6275
New York, NY 10036-7798
Seeks to foster positive change throughout the global community through its grant making.
Robert Wolf, President

2385 Norcross Wildlife Foundation
250 W 88th Street   212-362-4831
New York, NY 10024   Fax: 212-812-4299
                E-mail: norcross_wf_po@prodigy.net
                        http://http://norcrossws.org
Richard Reagan, President

2386 Northwest Area Foundation
332 Minnesota Street   651-224-9635
Suite E-1201   Fax: 651-225-3881
St. Paul, MN 55101-1373
Terry Tinson Saario, President

2387 Oliver S and Jennie R Donaldson Charitable Trust
US Trust Company of New York
114 W 47th Street   212-852-3683
New York, NY 10036   Fax: 212-852-3377
Anne L Smith-Ganey, Secretary

2388 Overbrook Foundation
Overbrook Management Corporation   212-661-8710
521 5th Avenue
New York, NY 10175
Ms Sheila McGoldrick

2389 Pacific Whale Foundation
101 N Kihei Road   808-879-8811
Suite 25   800-942-5311
Kihei, HI 96753   Fax: 808-879-2615
Gregory D Kaufman, President

2390 Patrick and Anna Cudahy Fund
PO Box 11978   708-866-0760
Milwaukee, WI 53211
Types of support: general/operating support; continuing support; annual campaigns; building/renovation; equipment; program development; seed money; technical assistance; matching funds.
Judith Borchers, OSB, Executive Director

2391 Pew Charitable Trusts
2005 Market Street   215-575-9050
Suite 1700   Fax: 215-575-4939
Philadelphia, PA 19103-7017
Joshua S Reichert, Dir. Environmental Programs

2392 Providence Journal Charitable Foundation
75 Fountain Street   401-277-7000
Providence, RI 02902   http://www.projo.com

2393 Public Welfare Foundation
1200 U Street NW   202-965-1800
Washington, DC 20009-4443   Fax: 202-265-8851
E-mail: general@publicwelfare.org
http://www.publicwelfare.org
Larry Kressley, Acting Executive Director

2394 Puget Sound Energy Consumer Affairs Department
Bellevue, WA 98004   425-454-2000

2395 RARE Center for Tropical Bird Conservation
1529 Walnut Street   215-568-0420
Philadelphia, PA 19102
John Guarnaccia, Executive Director

2396 Rainforest Action Network
221 Pine Street   415-398-4404
Suite 500   Fax: 415-398-2732
San Francisco, CA 94104   E-mail: rainforest@ran.org
http://www.ran.org
Rainforest Action Network works to protect the Earth's rainforests and support the rights of their inhabitants through education, grassroots organizing and non-violent direct action.
Michael Brune, Executive Director

2397 Rainforest Alliance
65 Bleecker Street   212-677-1900
New York, NY 10012   Fax: 212-677-2187
Daniel R Katz, President/Executive Director

2398 Raytheon Company
141 Spring Street   617-862-6600
Lexington, MA 02173   http://www.raytheon.com
Janet Taylor, Admin. Corp. Contributions

2399 Richard King Mellon Foundation
PO Box 2930   412-392-2800
Pittsburg, PA 15230-2930   Fax: 412-392-2837
George H Taber, Vice President and Director

2400 Richard Lounsberry Foundation
159-A E 61st Street   212-319-7033
New York, NY 10021
Alan F McHenry, President

2401 Rockefeller Brothers Foundation
437 Madison Avenue   212-812-4200
37th Floor   Fax: 212-812-4299
New York, NY 10022   http://www.rbf.org
A philantrophic organization working to promote social change that contributes to a more just, sustainable and peaceful world. The Fund's programs are intended to develop leaders, strenghten institutions, engage citizens, buildcommunity, and foster partnerships that include government, business, and civil society.
Benjamin R Shute Jr, Secretary

2402 Rockefeller Family Fund
437 Madison Avenue   212-812-4252
37th Floor   Fax: 212-812-4299
New York, NY 10022   E-mail: mmccarthy@rffund.org
http://www.rffund.org
For thirty years, the Rockefeller Family Fund has worked at the cutting edge of advocacy in such areas as environmental protection, advancing the economic rights of women, and holding public and private institutions accountable fortheir actions.
, Director

2403 Safari Club International Conservation Fund
4800 West Gates Pass Road   602-620-1220
Tucson, AZ 85745   Fax: 602-622-1205
Doug Yajko, President

2404 Samuel Roberts Noble Foundation
2510 Sam Noble Parkway   508-223-5810
Ardmore, OK 73401   Fax: 508-224-6217
http://www.noble.org
One of the largest international offshore drilling contractors in the world.

Michael Cawley, President/CEO
Patrick Jones, VP/CFO/Treasurer

2405 Save the Redwoods League
114 Sansome Street   415-362-2352
Room 605   Fax: 415-362-7017
San Francisco, CA 94104
John B Dewitt, Executive Director

2406 Scherman Foundation
16 E 52nd Street   212-832-3086
#601   Fax: 212-838-0154
New York, NY 10022   http://www.scherman.org
Mr Mike Pratt, Program Officer/Treasurer

2407 Sequoia Foundation
820 A Street   206-627-1634
Suite 345
Tacoma, WA 98402
Mr Frank D Underwood, Executive Director

2408 Sierra Club Foundation
220 Sansome Street   415-291-1800
Suite 1100   Fax: 415-291-1791
San Francisco, CA 94104
To advance the preservation and protection of the natural environment by empowering the citizenry, especially democratically-based grassroots organizations, with charitable resources to further the cause of environmental protection.The vehicle through which The Sierra Club Foundation generally fulfills its charitable mission.
Stephen M Stevick, Executive Director

2409 Switzer Foundation New Hampshire Charitable Foundation
37 Pleasont Street
PO Box 1335
Concord, NH 03301-4005
603-225-6641
800-464-6641
Fax: 603-225-1700
E-mail: info@nhcf.org

Judith Burrows, Director Student Aid

2410 Territory Resource
603 Stewart Street
Suite 1007
Seattle, WA 98101
206-624-4081
Fax: 206-382-2640

Progressive foundation dedicated to creating a morejust society. Funds grass-roots community-based organizations in Idoho, Montana, Wyoming, Washington, and Oregon.
Stephanie Austin, Administrative Assistant

2411 Texaco Foundation
2000 Westchester Avenue
White Plains, NY 10650
914-701-0320
http://www.texaco.com

Maria Mike-Mayer, Secretary

2412 The National Arbor Day Foundation
211 North 12th Street
Lincoln, NE 68508
402-474-5655
888-448-7337
Fax: 402-474-0820
E-mail: info@arborday.org
http://www.arborday.org

A nonprofit educational, environmental organization that helps people plant and care for trees. We are committed to tree-planting and environmental stewardsip. Newsletter free with $10.00 annual membership.
8 pages Bi-Monthly
John Rosenow, President
Gary Brienzo, Info. Coordinator

2413 Threshold Foundation
1388 Sutter Street
10th Floor
San Francisco, CA 94109
415-771-4308
Fax: 415-771-0535

Drummond Pike, Foundation Manager

2414 Times Mirror Foundation
202 W First Street
Los Angeles, CA 90012
213-237-3945
Fax: 213-237-2116

The Times Mirror Foundation, an affiliate of Tribune Company, is dedicated to supporting nonprofit organizations that measurably improve the quality of life in communities we serve. The Foundation focuses its support on programs thatimprove the quality of journalism, education and literacy, strengthen the fabric of the community, and enhance cultural appreciation and understanding.
Cassandra Malry, Treasurer

2415 Tinker Foundation
55 E 59th Street
New York, NY 10022
212-421-6858
Fax: 212-223-3326
E-mail: tinker@tinker.org
http://fdncenter.org/grantmaker/tinker

Endeavors to promote better understanding among the peoples of the US, Latin America, and Iberia. In the environmental policy program area, grants are awarded to 501(c)(3) or equivalent organizations for projects addressingenvironmental issues, particularly incentive-based environmental activities, projects supporting the collaboration of NGOs and corporate interests, and those activities supporting the formulation of the necessary regulatory framework essential togood environmental goverance

Margaret Cushing, Program Officer

2416 Town Creek Foundation
PO Box 159
Oxford, MD 21654
410-226-5315
Fax: 410-226-5468
E-mail: towncrk@dmn.com
http://www.towncreekfdn.org

Christine B Shelton, Executive Director

2417 TreePeople
12601 Mulholland Drive
Beverly Hills, CA 90210
818-753-4600
Fax: 818-753-4625

Andy Lipkis, Executive Director

2418 Trout Unlimited
800 Follin Lane SE
Suite 250
Vienna, VA 22180
703-281-1100
Fax: 703-281-1825
E-mail: trout@tu.org
http://www.tu.org

Our mission is to conserve, protect a nd restore North America's trout and salmon fisheries and their network. We accomplish this mission on local, state and national levels with an extensive and dedicated volunteer network.
Charles Gauvin, Executive Director

2419 True North Foundation
508 Westwood Drive
Fort Collins, CO 80524
970-223-5285
Fax: 970-495-0892

Kerry K Anderson, President

2420 Turner Foundation
133 Luckie Street NW
2nd Floor
Atlanta, GA 30303
404-681-9900
Fax: 404-681-0172
http://www.turnerfoundation.org

This Foundation is committed to preventing damage to the natural systems, water, air and land, on which all life depends.

Micheal Finley, President

2421 US-Japan Foundation
145 East 32nd Street
New York, NY 10016
212-481-8753
Fax: 212-481-8762

Tom Foran, Program Officer

2422 USF and G Foundation
100 Light Street
Baltimore, MD 21202-1036
410-685-3047

2423 Union of Concerned Scientists
2 Brattle Square
Cambridge, MA 02238-9105
617-547-5552
Fax: 617-864-9405
E-mail: ucs@ucsusa.org
http://www.ucsusa.org

A partnership of citizens and scientists working to preserve our health, protect our safety, and enhance our quality of life. Since 1969, we've used rigorous scientific analysis, innovative policy development, and tenacious citizenadvocacy to advance practical solutions for the environment.

Kevin Knobloch, President
Kathleen Rest, Executive Director

2424 Unitarian Universalist Veatch Program at Shelter Rock
48 Shelter Rock Road
Manhasset, NY 11030
516-627-6576
Fax: 516-627-6596
E-mail: jan@veatch.org

Marjorie Fine, Executive Director

2425 Victoria Foundation
946 Bloomfield Avenue
Glen Ridge, NJ 07028
973-748-5300
Fax: 973-748-0016
E-mail: info@victoriafoundation.org
http://www.victoriafoundation.org

A private grantmaking institution. Since the early 1960's the Foundation's trustees have targeted giving to programs that impact the cycle of poverty in Newark, New Jersey.

Catherine McFarland, Executive Officer

2426 Vidda Foundation
10 E 40th Street
New York, NY 10016
212-696-4052
Fax: 212-889-7791

Gerald E Rupp, Manager

2427 Virginia Environmental Endowment
Three James Center
1051 East Cary Street   804-644-5000
PO Box 790   http://www.freenet.vcu.edu/vee
Richmond, VA 23206-0790
Gerald P McCarthy, Executive Director

2428 W Alton Jones Foundation
232 East High Street   804-295-2134
Charlottesville, VA 22902-5178   Fax: 804-295-1648
    E-mail: earth@wajones.org
    http://www.wajones.org
Dr. JP Meyers, Director

2429 Wallace Genetic Foundation
4801 Massachusetts Avenue NW   202-966-2932
Suite 400   http://www.voideinternational.org
Washington, DC 20016

2430 Waste Management
3003 Butterfield Road   708-572-8800
Oak Brook, IL 60521
Paul Pyrcik, Director of Programs

2431 Wilderness Society
1615 M Street NW   202-833-2300
Washington, DC 20036   800-THE-WILD
    Fax: 202-429-3958
    E-mail: tws@wilderness.org
    http://www.wilderness.org
Our goal is to ensure that future generations will enjoy the clean
air and water, wildlife, beauty and opportunities for recreation
and renewal that pristine forests, rivers, deserts and mountains
provide.
Karin Sheldon, President

2432 Wildfowl Foundation
1101 14th Street NW   202-371-1808
Suite 725
Washington, DC 20005
Richard McCabe, Secretary

2433 Wildlife Conservation Fund of America
801 Kingsmill Parkway   614-888-4868
Columbus, OH 43229-1137   Fax: 614-888-0326
Rick Stony, Executive Director

2434 Wildlife Preservation Trust International
3400 West Girard Avenue   215-222-3636
Philadelphia, PA 19104   Fax: 215-222-2191
Empowers local conservation scientists worldwide to protect
nature and safeguard ecosystem and human health.

Dr. Mary Pearl

2435 William Bingham Foundation
1250 Leader Building   216-781-3275
Cleveland, OH 44114
Ms Laura C Hitchcox, Director

2436 William H Donner Foundation
60 East 42nd Street   212-949-0404
Suite 1651   Fax: 212-949-6022
New York, NY 10165
When we build let us...build forever. Let it not be for present de-
light nor for present alone. Let it be such work as our descen-
dants will thank us for.
Joseph W Donner, Jr., President
Deborah Donner, Vice President

2437 William Penn Foundation
Two Logan Square 11th Floor   215-988-1830
100 North 18th Street   Fax: 215-988-1823
Philadelphia, PA 19103   E-mail: moreinfo@williampennfoundation.org
    http://www.williampennfoundation.org

To improve the quality of life in the greater Philadelphia region
through efforts that foster rich cultural expression, strenghten
children's futures, and deepen connections to nature and community. In partnerships with others, we workto advance a vital,
just and caring community.
Feather O Houstoun, President
Olive Mosier, Director

2438 William and Flora Hewlett Foundation
525 Middlefield Road   415-329-1070
Suite 200   Fax: 415-329-9342
Menlo Park, CA 94025-3495
Stephen Toben, Program Officer

2439 Winston Foundation for World Peace
2040 S Street, NW   202-483-4215
Washington, DC 20009   Fax: 202-483-4219
    E-mail: winstonfoun@igc.apc.org
    http://www.wf.org
John H. Adams, Director

2440 Wisconsin Energy Corporation Foundation
1035 W Canal Street   414-221-2345
Milwaukee, WI 53233   Fax: 414-221-2554
    http://www.wisconsinenergy.com
Jerry G Remmel, Treasurer

2441 World Parks Endowment
1616 Place Street NW   202-939-3808
Suite 200   Fax: 202-939-3868
Washington, DC 20036 E-mail: worldparks@worldparks.org
    http://www.worldparks.org
The World Parks Endowment provides the opportunity to buy
rainforest land and establish new protected areas that conserve
rainforests and other sites of high biodiversity value. Our projects target lands that conserve rare orendangered species, and
are low price, so the minimum amount of the funds protect high
priority areas.
Daniel Katz, President

2442 World Research Foundation
PO Box 10187   310-827-0070
Marina del Rey, CA 90295   Fax: 310-827-5010
    http://www.wrf.org
The purpose of the foundation is to locate, gather, codify, evaluate, classify and disseminate information dealing with health
and the environment. All countries are contacted to collect the
best information in an unbiased, neutral andindependent manner.
LaVerne Boeckmann, Vice President/Founder

2443 World Resources Institute
10 G Street, NW   202-729-7600
Suite 800   Fax: 202-729-7610
Washington, DC 20002   http://www.wri.org
An independent nonprofit organization with a staff of more than
100 scientists, economists, policy experts, business analysts,
statistic analysists, mapmakers and communicators working to
protect the Earth and improve people's lives.Our four goals are:
protect the Earth's living systems, increase access to information, create sustainable enterprise and opportunity and reverse
global warming.
Jonathan Lash, President

2444 World Society for the Protection of Animals
34 Deloss Street   508-879-8350
Framingham, MA 01702   800-883-9772
    Fax: 508-620-0786
    E-mail: wspa@wspausa.com
    http://www.wspa.org.uk or www.wspa-americas.org
Laura Salter, USA Regional Manager
Susan Sherwin, Campaigns Manager

2445 World Wildlife Fund
1250 24th Street NW   202-293-4800
Washington, DC 20037-1175   Fax: 202-293-9211
Jennifer A Zadwick, Program Information Coord.

2446 Xerces Society
10 Ash Street SW
Portland, OR 97204
Melody Allen, Executive Director

503-222-2788
Fax: 503-222-2763

Scholarships

2447 AGI Minority Geoscience Scholarship
American Geological Institute
4220 King Street
Alexandria, VA 22302-1507

703-379-2480
800-336-4764
Fax: 703-379-7563

2448 Abundant Life Seed Foundation
930 Lawrence Street
PO Box 772
Port Townsend, WA 98368   E-mail: abundant@olypen.com
http://www.abundantlifeseed.org
A nonprofit organization dedicated to the preservation of rare, heirloom and native seeds. ALSF grows and distributes open-pollinated seeds and offers them for sale in an annual catalog. Seeds are also sent to people in need throughthe World Seed Fund. ALSF teaches seed saving through workshops, appreticeships and school programs.
Matthew Dillon, Executive Director
Elsa Golts, Board President

425-385-5660
Fax: 360-385-7455

2449 Alexander Hollaender Distinguished Postdoctoral Fellowships
PO Box 117
Oak Ridge, TN 37831-0117   E-mail: holmesl@orau.gov
http://www.orau.gov/orise/contacts.htm
Barbara Dorsey
Linda Holmes

423-576-9975

2450 American Association for the Advancement of Science
1333 H Street NW
Washington, DC 20005

202-326-6600
Fax: 202-289-4950

2451 American Fishing Tackle Manufacturers Association
1250 Grove Avenue
Suite 300
Barrington, IL 60010

847-381-9490
Fax: 847-381-9518

2452 American Geophysical Union Member Programs Division
2000 Florida Avenue NW
Washington, DC 20009

202-462-6900
Fax: 202-328-0566

2453 American Indian Science and Engineering Society
2305 Renard SE
PO Box 9828
Albuquerque, NM 87119-9828

505-765-1052
Fax: 505-765-5608
E-mail: info@aises.org
http://www.aises.org

Pam Silas, Executive Director
Teresa Gomez, Deputy Director

2454 American Museum of Natural History
Central Park West & 79th Street
New York, NY 10024-5192

2455 American Nuclear Society
555 North Kensington Avenue
La Grange Park, IL 60525

708-352-6611
Fax: 708-352-0499

2456 American Society of Naturalists
Queens College - CUNY
Department of Biology
Flushing, NY 11367

718-997-3426

2457 Apple Computer Earth Grants: Community Affairs Department
1 Infinite Loop
Cupertino, CA 95014

408-996-1010
800-692-7753
http://www.apple.com

Beverly Long, Program Manager

2458 Beldon Fund
99 Madison Avenue
8th Floor
New York, NY 10016

212-616-5600
800-591-9595
Fax: 212-616-5656
E-mail: info@beldon.org
http://www.beldon.org

Judy Donald, Executive Director

2459 Beldon II Fund: Old Kent Bank and Trust Company
Old Kent Bank
300 Old Kent Bank Building
Grand Rapids, MI 49503
John R Hunting, President and Director

616-771-5326

2460 CS Fund
469 Bohemian Highway
Freestone, CA 95472
Martin Teitel, Executive Director

707-874-2942
Fax: 707-874-1734

2461 Charles A Lindbergh Fund
2150 Third Avenue North
Suite 310
Anoka, MN 55303-2208

763-576-1596
Fax: 763-576-1664
E-mail: info@lindberghfoundation.org
http://www.lindberghfoundation.org/grants/
Each year, the Charles A. and Anne Morrow Lindbergh Foundation provides grants to men and women whose individual initiative and work in a wide spectrum of disciplines furthers the Lindbergh's vision of a balance between the advance oftechnology and the perservation of the natural/human environment.
Steven R. Whitley, Chairman/CEO

2462 Cousteau Society
710 Settler's Landing Road
Hampton, VA 23669

757-523-9335
Fax: 757-722-8185
E-mail: cousteau@cousteausociety.org
http://www.cousteausociety.org

Francie Cousteau, President

2463 DRB Communications
1234 Summer Street
Stamford, CT 06905

800-323-1550
Fax: 203-324-7175

Robyn DeWolf

2464 Delmar Publishers Scholarship
National FAA Foundation
6060 FFA Drive
PO Box 68960
Indianapolis, IN 46268-0960
Carrie Powers, Contact

317-802-6060
Fax: 317-802-6061
http://www.ffa.org

2465 Du Pont de Nemours and Company
1007 Market Street
Room 8065
Wilmington, DE 19898

302-774-2036
800-441-7515
E-mail: info@dupont.com
http://www.dupont.com
Peter C Morrow, Mgr. Corporate Contributions

2466 Earth Island Institute
300 Broadway
Suite 28
San Francisco, CA 94133

415-788-3666
Fax: 415-788-7324
E-mail: earthisland@earthisland.org
http://www.earthisland.org
Incubates and supports over 35 projects working on environmental issues worldwide. Publishes quarterly Earth Island Journal. Our project support programs halps aspiring and veteran activists alike put ideas into action. Our youthprogram in creases the visibility, effectiveness and influence of youth leadership in the environmental movement, inspiring other young people to work for the Earth.

John A Knox, Executive Director

2467 Environmental Defense
257 Park Avenue S   212-505-2100
17th Floor   800-684-3322
New York, NY 10010-7304   Fax: 212-505-2375
E-mail: members@environmentaldefense.org
http://www.environmentaldefense.org
Dedicated to protecting the environmental rights of all people, including future generations. Among these rights are clean air, clean water, healthy, nourishing food and a flourishing ecosystem. Advocates solutions based on science,even when it leads in unfamiliar directions. Works to create solutions that win lasting political, economic and social support because they are bipartisan, efficient and fair.
Frederick D Krupp, Executive Director
Marcia Aronoff, Deputy Director Programs

2468 Environmental Grantmakers Association
437 Madison Avenue   212-812-4260
37th Floor   Fax: 212-812-4299
New York, NY 10022

2469 Environmental Protection Agency: Grants Administration Division
Grants Operation Branch   202-260-5260
401 M Street SW
Washington, DC 20460

2470 Environmental Science Amoco Foundation Scholarships
Future Farmers of America
6060 FFA Drive   317-802-6060
PO Box 68960   Fax: 317-802-6061
Indianapolis, IN 46268-0960   http://www.ffa.org
Carrie Powers, Contact

2471 Environmental and Engineering Fellowship
American Association for the Advancement
1333 H Street NW   202-326-6600
Washington, DC 20005   Fax: 202-289-4950

2472 Financial Support for Graduate Work
Women's Seamen's Friend Society of Connecticut
291 Whitney Avenue   203-777-2165
New Haven, CT 06511

2473 Ford Motor Company Fund
American Road   313-845-8711
PO Box 6248   800-392-3673
Dearborn, MI 48126   http://www.fordfound.com
Leo J Brennan Jr, Executive Director

2474 Forest History Society
701 William Vickers Avenue   919-682-9319
Durham, NC 27701-3162   Fax: 919-682-2349
E-mail: coakes@duke.edu
http://http://www.lib.duke.edu/forest
The Forest History Society is a non-profit educational institution that links the past to the future by identifying, collecting, preserving, interpreting, and disseminating information on the history of people, forests, and theirrelated resources.

Cheryl Oakes, Librarian

2475 Garden Club of America
598 Madison Avenue   212-753-8287
New York, NY 10022   Fax: 212-753-0134
Sellers Thomas Jr, President

2476 Georgia M. Hellberg Memorial Scholarships
National Future Federation of America
6060 FFA Drive   317-802-6060
PO Box 68960   Fax: 317-802-6061
Indianapolis, IN 46268-0960   http://www.ffa.org
Carrie Powers, Contact

2477 German Marshall Fund of the United States
11 Dupont Circle NW   202-745-3950
Suite 750   Fax: 202-265-1662
Washington, DC 20036   http://www.gmfus.org
Marianne L Gindburg, Program Officer

2478 Great Lakes Protection Fund
1560 Sherman Avenue   847-425-8150
Suite 880   Fax: 847-424-9832
Evanston, IL 60201   http://www.glpf.org

A private non profit organization formed by the Govenors of the Great Lakes States. It is a permanent environmental endowment that supports collaborative actions to improve the health of the Great Lakes ecosystem.

Russell Van Herik, Executive Director

2479 Hawk Mountain Sanctuary Association
Route 2   215-756-6961
Kempton, PA 19529   Fax: 215-756-4468
Cynthia R Lenhart, Executive Director

2480 Hazardous Waste Reduction Loan Program
California Department of Commerce
1001 I Street   916-341-6181
PO Box 4025   E-mail: grants@ciwmb.ca.gov
Sacramento, CA 95812-4025   http://www.ciwmb.ca.gov
Merri Stevenson

2481 Heller Charitable and Educational Fund
PO Box 336   415-434-3160
Kentfield, CA 94914   Fax: 415-434-3807
Ruth B Heller, Correspondence Secretary

2482 JM Kaplan Fund
30 Rockefeller Plaza   212-767-0630
Suite 4250   Fax: 212-767-0639
New York, NY 10112
Anthony C Wood, Program Officer

2483 Jessie Ball duPont Religious, Charitable and Educational Fund
1 Independent Drive   904-353-0890
Suite 1400   800-252-3452
Jacksonville, FL 32202   Fax: 904-353-3870
E-mail: jbennett@dupontfund.org
http://www.dupontfund.org
A national grantmaking foundation limited in its giving to 325 eligible organizations to which Mrs. duPont personally contributed to in a five year period 1960-1964. The duPont fund accomplishes its mission by working creatively withthese organizations and their partners.

Sherry Magill, PhD, President
JoAnn P Bennett, Director of Administration

2484 Johnson's Wax Fund
1525 Howe Street   262-260-2000
Racine, WI 53403   http://www.scjohnsonwax.com

2485 Joint Oceanographic Institutions
1201 New York Avenue NW
Suite 400   202-232-3900
Washington, DC 20005   Fax: 202-265-4409
E-mail: info@joiscience.org
http://www.joiscience.org
A consortium of 20 premier oceanographic research institutions that serves the US scientific community through management of large scale, global research programs in the fields of marine geology and geophysics and oceanography.
Steven Bohlen, President
Amy Castner, Executive Program Associate

2486 LSB Leakey Foundation
P.O. Box 29346   415-561-4646
1002A O'Reilly Aveanue   Fax: 415-561-4647
San Francisco, CA 94129-3750

E-mail: info@leakeyfoundation.org
http://www.leakeyfoundation.org/foundation.f4jsp
The mission of the Leakey Foundation is to increase scientific knowledge and public understanding of human origins and evolution.

2487 MJ Murdock Charitable Trust
PO Box 1618   360-694-8415
Vancouver, WA 98668   Fax: 360-649-1819
http://www.murdock-trust.org
The mission is to enrich the quality of life in the Pacific Northwest by providing grants organizations that seek to strenghten the region's educational and cultural base in creative and sustainable ways.
Neal O Thorpe, Executive Director

2488 Mary Flagler Cary Charitable Trust
122 East 42nd Street   212-953-7700
Room 3505   Fax: 212-953-7720
New York, NY 10168   http://http://www.carytrust.org/
The Trust was established as a testamentary, charitable trust by the will of the late Mary Flagler Cary. The trustees have worked to use the assets of the Trust to carry forward Mrs. Cary's interests, and to elaborate on them in lightof new circumstances and needs. A major part of the Trust's assets continue to be devoted to special commitments relating to the origins of the Trust, especially the Institute of Ecosystem Studies at the Mary Flagler Cary Arboretum in Millbrook, NewYork.

, Trustee

2489 Maryland Sea Grant
University of Maryland
4321 Hardwick Road   301-403-4220
Suite 300   Fax: 301-403-4250
College Park, MD 20740 E-mail: mdsg@mdsg.umd.edu
http://www.mdsg.umd.edu
Supports innovative marine research and education, with a special focus on the Chesapeake Bay.
Dr Jonathan Kramer, Director

2490 National Academy of Sciences
2001 Constitution Avenue NW   202-334-2000
Washington, DC 20418   Fax: 202-334-2158
http://www.nas.edu/legal.html

2491 National Center for Atmospheric Research
PO Box 3000   303-497-1601
Boulder, CO 80307   Fax: 303-497-1314

2492 National Environmental Health Association NEHA/AAS Scholarships
720 S Colorado Boulevard   303-756-9090
Suite 970-S   Fax: 303-691-9490
Denver, CO 80246-1925 E-mail: credentialing@neha.org
http://www.neha.org
V Potter, Member Liaison

2493 Needmor Fund
42 South Street   419-255-5560
Clair Street   http://www.fdncenter.org/grantmaker/needmor
Toledo, OH 43602
Lynn Gisi, Coordinator

2494 Nixon Griffis Fund for Zoological Research: New York Zoological Society
Bronx Zoo
185th Street & Southern Boulevard   212-220-5152
Bronx, NY 10460   Fax: 212-220-7114
Supports research in zoology, conservation, and marine science. Grants lilmited to $3,000. Grants made four times a year. Applications reviewed by selected US zoo personnel.
John Behler, Contact

2495 North American Loon Fund Grants
PO Box 329   603-528-4711
Holderness, NH 03245   800-462-5666
http://www.facstaff.uww.edu/wentzl/nalf

THe North American Loon Fund's mission is to promote the preservation of loons and their lake habitats through research, public education, and the involvement of people who share their lakes with loons.
Linda O'Bara, Director

2496 Oak Ridge Institute Science/Engineering Education Division
MS 36 120 Badger Avenue   865-576-3424
PO Box 117   Fax: 865-241-2923
Oak Ridge, TN 37831-0117 E-mail: westm@orau.gov
http://www.orau.org

2497 Oak Ridge Institute for Science and Education
120 Badger Avenue   865-576-3424
PO Box 117, MS 36   Fax: 865-241-2923
Oak Ridge, TN 37831-0117 E-mail: westm@orau.gov
http://www.orau.org

2498 Oklahoma State University
003 Life Sciences E   405-744-9995
Stillwater, OK 74078   Fax: 405-744-7673
Dr. Norman N Durham, Director

2499 Resources for the Future
1616 P Street NW   202-328-5000
Washington, DC 20036   Fax: 202-939-3460
E-mail: info@rff.org
RFF is a nonprofit and nonpartisan organization, or think tank, that conducts independent research-rooted primarily in economics and other social sciences-on environmental and natural resource issues. Its research scope comprisesprograms in nations around the world.

Lesli Creedon, Vice Pres., External Aff.
Stan Wellborn, Director of Communications

2500 University of Colorado: Boulder
Campus Box 216   303-492-1143
Boulder, CO 80309-0216   Fax: 303-492-1149
Dr. Robert Sievers, Director

2501 University of Tennessee at Knoxville
Energy, Environment and Resource Center
676 Danbney Hall   865-974-8080
Knoxville, TN 37996   Fax: 865-974-8086
E-mail: sayler@utk.edu
http://www.web.utk.edu/~enr/contact.htm
Dr. William Colglazier, Director

2502 WERC Undergraduate Fellowships
New Mexico State University   505-646-2038
Box 30001, MSC WERC   800-523-5996
Las Cruces, NM 88003-0001   Fax: 505-646-4149
E-mail: bdelrio@nmsu.edu
http://www.werc.net
WERC a consortium for environmental education and technology development.
Barbara Valdez, Contact

2503 Water Environment Federation
601 Wythe Street   703-684-2400
Alexandria, VA 22314-1994   800-666-0206
Fax: 703-684-2492
Founded in 1928, the Water Environment Federation (WEF) is a not-for-profit technical and educational organization with members from varied disciplines who work toward the WEF vision of preservation and enhancement of the global waterenvironment. The WEF network includes more than 100,000 water quality professionals from 77 member associations in 31 countries.
Janet Blatt, Director
Mincaiee Brown, Executive Director

2504 Weston Institute
Weston Way   215-430-3100
West Chester, PA 19380   Fax: 215-692-6503
William Gaither, Director

2505 **Wildlife Conservation Society**
New York Zoological Society   718-220-6864
Bronx, NY 10460   Fax: 718-364-4275
Dr. Mary Pearl

2506 **Women's Seamen's Friend Society of Connecticut**
291 Whitney Avenue   203-467-3887
New Haven, CT 06511

2507 **Yale Institute for Biospheric Studies (YIBS)**
21 Sachem Street   203-432-9856
PO Box 208105   Fax: 203-432-9927
New Haven, CT 06520   E-mail: roserita.riccitelli@yale.edu
      http://www.yale.edu/yibs
The Yale Institute for Biospheric Studies (YIBS) serves as a principal focus for Yale University's research and training efforts in the environmental sciences, and is committed to the teaching of environmental studies to futuregenerations. It provides physical and intellectual centers for research and education that address fundamental questions that will inform the ability to generate solutions to the biosphere's most critical environmental solutions.

Rose Rita Riccitelli, Administrator
Derek Briggs, Director

Federal

2508 Advisory Committee on Nuclear Waste
US Nuclear Regulatory Commission
Office Of Public Affairs   301-415-8200
Washington, DC 20555   800-368-5642
                                            Fax: 301-415-5575
                                            E-mail: ram2@nrc.gov

Dr John T Larkins, Acting Executive Director

2509 Advisory Council on Historic Preservation
1100 Pennsylvania Avenue NW   202-606-8503
Suite 809, Old Post Office Building E-mail: achp@achp.gov
Washington, DC 20004   http://www.achp.gov
John Fowler, Executive Director

2510 Advisory Panel Ecosystem Studies
National Science Foundation
1800 G Street NW   202-357-9596
Washington, DC 20550

Dr Richard Dame, Designated Officer

2511 Advisory Panel for Ecology
National Science Foundation
4201 Wilson Boulevard   703-292-8480
Arlington, VA 22230   Fax: 202-357-1191

2512 Agency for Toxic Substances and Disease Registry
Centers for Disease Control and Prevention
1600 Clifton Road NE   404-639-7000
Building 16, Room 5125   Fax: 404-639-7111
Atlanta, GA 30333   E-mail: ATSDRIC@cdc.gov
                                    http://www.atsdr1.cdc.gov:8080
This agency provides leadership and direction to programs and
activities designed to protect both the public and workers from
exposure or adverse health effects of hazardous substances in
storage sites or released in fires, explosions,or transportation accidents. The agency also collects, maintains, analyzes and disseminates information relating to serious diseases, mortality
and human exposure to toxic or hazardous substances.
Clarie V Broome, Director

2513 Air and Radiation Research Committee
Environmental Protection Agency
1200 Pennsylvania Avenue   202-564-7400
Washington, DC 20004   Fax: 202-501-0826
                                            http://www.epa.gov/air

2514 American Farmland Trust
1200 18th Street NW   202-331-7300
Washington, DC 20036   E-mail: info@farmland.org
                                http://www.farmland.org
Works to prevent industrial development on agricultural lands.

2515 American Indian Environmental Office
1200 Pennsylvania Avenue (4104M)   202-564-0303
Washington, DC 20460   http://www.epa.gov/indian/
Coordinates the US environmental Protection Agency-wide ef-
fort to strengthen public health and environmental protection in
Indian Country, with a special emphasis on building Tribal ca-
pacity to administer their own environmentalprograms.
Carol Jorgensen, Director
Gary Hindburgh, Deputy Director

**2516 Animal and Plant Health Inspection Service Protection
Quarantine**
1400 Independence Avenue SW   202-720-5601
Whitten Building, Room 302-E   Fax: 202-690-0472
Washington, DC 20250-3431 E-mail: aelder@aphis.usda.gov
                                http://www.aphis.usda.gov

This division is responsible for the control or eradication of
plant pests and diseases. Programs are carried out in cooperation
with the states involved, other federal agencies, farmers and private organizations.
Alfred Elder, Deputy Administrator

2517 Antarctica Project
1630 Connecticut Avenue NW   202-234-2480
3rd Floor   Fax: 202-387-4823
Washington, DC 20009   E-mail: antarctica@igc.org
                                http://www.asoc.org
Concerned with educating the public about environmental prob-
lems in the arctic regions. Conducts research pertaining to
Antarctica.
Beth Clark, Director
Josh Stevens, Campaign Associate

2518 Army Corps of Engineers
441 G Street NW   202-761-0000
Washington, DC 20314-1000   Fax: 202-761-1683
    http://www.usace.army.mil/inet/organization/execoffc.htm
The Army Corps of Engineers serves as the Army's real property
manager; manages and executes civil works programs, including research and development, planning, design, construction,
operation and maintenance and real estate activitiesrelated to
rivers, harbors and waterways; administers laws for protection
and preservation of navigable waters and related resources such
as wetlands, and assists in recovery from natural disasters.
Joe Ballard, Commander/Chief Engineers

2519 Aspen Institute
One Dupont Circle NW   202-736-5800
Suite700   Fax: 202-466-4568
Washington, DC 20036   http://www.aspeninst.org
A forum that addresses critical environmental issues. Interests
include energy, the environment and economics. Partner insti-
tutes located in Japan, Italy, Germany and France.

2520 Atlantic States Marine Fisheries Commission
1444 I Street NW   202-289-6400
6th Floor   Fax: 202-289-6051
Washington, DC 20005-2210   E-mail: info@asmfc.org
                                http://www.asmfc.org

John H Dunnigan, Executive Director

2521 Biodiversity Support Program
1250 24th Street NW   202-861-8347
Washington, DC 20037   Fax: 202-861-8324
                                http://www.bsponline.org
Concerned with environmental, social and economic development in developing countries. Assists partner programs with
training and research.

2522 Bioengineering & Environmental Systems Section
National Science Foundation
1800 G Street NW   202-357-7955
Room 1132   http://www.eng.nsf.gov/bes/
Washington, DC 20550

The Bioengineering and Environmental Systems (BES) Division supports research that expands the knowledge base of bioengineering, applies engineering principles to the
understanding of living systems, improves our ability to
applyengineering principles to avoid and/or correct problems
that impair the usefulness of land, air and water and advances
fundamental engineering knowledge of the ocean environment.
Norman Caplan, Head

**2523 Blue Mountain Natural Resource Institute Advisory
Board**
US Department of Agriculture
1400 Independence Avenue SW   202-720-3291
Room 240W   Fax: 202-720-2191
Washington, DC 20250   http://www.fs.fed.us

Charles R Hilty, Management Office

2524 Board of Scientific Counselors: National Institute of Environmental Health
National Institute of Health
111 Alexander Drive   919-541-3201
Room B242   Fax: 919-541-2260
Research Triangle Park, NC 27709

Dr John McLachlan, Designated Federal Employee

2525 Bureau of Economic Analysis
Economics and Statistics Administration
1441 L Street NW   202-606-9600
Room 6006   Fax: 202-606-5311
Washington, DC 20230   E-mail: john.landefeld@bea.doc.gov
                                        http://www.bea.doc.gov
The BEA provides a clear picture of the US economy through the preparation, development, and interpretation of the national income and product accounts, which show the business and other components of national wealth; the input-outputaccounts, which trace the interrelationships among industrial markets; personal income and related economic series by geographic area; the US balance of payments accounts and associated foreign investment accounts, and the measures relating toenvironmental change.
John Landefeld, Director

2526 Bureau of Land Management, Land & Renewable Resources
1849 C Street NW Room #240LS   202-208-4896
MS-5650-MIB   Fax: 202-452-5124
Washington, DC 20240   http://www.blm.gov
The mission of the Bureau of Land Management is to sustain the health, diversity, and productivity of the public lands for the use and enjoyment of present and future generations. Offers environmental education, news about theactivities of the Bureau, events, and regulations. In addition, there is information about ALMRS (Automated Land and Management Record System). This is an information system that contains more than one billion land and mineral records.
Michael J Penfold, Assistant Director

2527 Bureau of Oceans International Environmental & Scientific Affairs
US Department of State
2201 C Street NW   202-647-4000
Washington, DC 20520
OES coordinates US international oceans, environmental and health policy, integrating US domestic concerns with geopolitical concerns. OES promotes the full range of US interests in the oceans to advance our national security,facilitate commerce, manage fish resources, foster scientific understanding and protect the marine environment through bilateral, regional and multilateral fora.
John F Turner, Assistant Secretary

2528 Center for Disease Control and Prevention
National Center for Environmental Health
4770 Buford Highway NE   770-488-7020
Suite F-29   Fax: 770-488-7042
Atlanta, GA 30341-3724   http://www.cdc.gov
CDC'c National Center for Environmental Health (NCEH) strives to promote health and quality of life by preventing or controlling those diseases or deaths that result from interactions between people and their environment.
Alison Kelly, Public Health Analyst

2529 Center for Sustainable Development in the Americas
33 Livingston Avenue   732-932-5680
New Brunswick, NJ 08901   Fax: 732-932-0934
                                        http://www.csdanet.org
Supports sustainable development in Latin America.
Nora Lovrien, Contact

2530 Centers for Disease Control and Prevention
1600 Clifton Road NE   404-639-3311
Building 16, Road 5125   800-311-3435
Atlanta, GA 30333   Fax: 404-639-2657
                                        http://www.cdc.gov

The Centers for Disease Control and Prevention protect the public health of the nation by providing leadership and direction in the prevention and control of diseases and other preventable conditions, and responding to public healthemergencies.
Julie Gerberding, Director
William Gimson, Chief Operating Officer

2531 Chesapeake Bay Critical Areas Commission
1804 W Street   410-260-3460
Suite 100   http://www.dnr.state.md.us/criticalarea/
Annapolis, MD 21401
Develops criteria used by local jurisdictions to develop individual Critical Area programs and amend local comprehensive plans, zoning ordinances and subdivision regulations. Programs are designed to address the unique characteristicsand needs of each county and municipality and together they represent a comprehensive land use strategy for preserving and protecting Maryland's most important natural resource, the Chesapeake Bay.
LeeAnne Chandler, Natural Resources Planner

2532 Chief of Engineers Environmental Advisory Board
441 G Street NW   202-761-0008
Washington, DC 20314   Fax: 202-761-1683
                            http://www.hq.usace.army.mil/hqhome/

2533 Civil Division: Consumer Litigation Office
550 11th Street NW   202-514-6786
The Todd Building, Room 6114
Washington, DC 20530
Represents the interests of the United States, including its officers and agents, in suits where monetary judgments are sought for damages resulting from negligent or wrongful acts. Also handles actions involving injury or damage togovernment property. Approximately 130 attorneys employed by the Torts Branch are divided among four different sections: Aviation and Admiralty; Federal Tort Claims Act; Environmental Torts; and Constitutional and Specialized Tort Litigations.

2534 Clean Air Scientific Advisory Committee
US Environmental Protection Agency
1200 Pennsylvania Avenue NW   202-260-6552
Washington, DC 20460
The Clean Air Scientific Advisory Committee (CASAC) has a statutorily mandated responsibility to review and offer scientific advice on the air quality criteria and regulatory documents which form the basis for the National Ambient AirQuality Standards (NAAQS), which are currently lead, particulate matter (PM), ozone and other photochemical oxidants (O3), carbon monoxide (CO), nitrogen oxides (NOx) and sulfur oxides (SOx).

2535 Coast Guard
2100 2nd Street SW   202-267-2229
Washington, DC 20593   http://www.uscg.mil

2536 Coastal States Organization
444 N Capitol Street NW   202-508-3860
Hall of the States, Suite 322   Fax: 202-508-3843
Washington, DC 20001   E-mail: cso@sso.org
                                  http://www.coastalstate.org
Since 1970, the Coastal States Organization (CSO) has represented the governors of the US coastal states, territories and commonwealths as an advocate for improved management of the nation's coasts, oceans and Great Lakes.
Jena Carter, Dir/Government Affairs

2537 Committee of State Foresters
NASF-444 N Capitol Street NW   202-624-5415
Suite 540   E-mail: nasf@sso.org
Washington, DC 20001   http://www.stateforesters.org
Richard Wright, Designated Fed. Empl.

2538 Committee on Agriculture
1301 House Office Building   202-225-2171
Washington, DC 20515   E-mail: agriculture@mail.house.gov
                               http://www.house.gov/agriculture/

2539 Committee on Agriculture, Nutrition, and Forestry
United State Senate   202-224-2035
Washington, DC 20510-6000   Fax: 202-224-1725
http://www.senate.gov.-agriculture/

2540 Committee on Appropriations
The Capitol Building   202-224-3471
Room S-128   Fax: 202-228-0587
Washington, DC 20510http://www.senate.gov-appropriations/
Ted Stevens, R-AK

2541 Committee on Commerce
Committee on Energy and Commerce
2125 Rayburn House Office Building   202-225-2927
Washington, DC 20515   E-mail: commerce@mail.house.gov
http://www.houise.gov/commerce

2542 Committee on Commerce, Science, and Transportation
Room SD-508   202-224-5115
Washington, DC 20510-6125   Fax: 202-224-1259
E-mail: webmaster@commerce.senate.gov
http://www.senate.gov/-commerce/

2543 Committee on Education and the Workforce
US House of Representatives
2181 Rayburn House Office Building   202-225-4527
Washington, DC 20515   http://www.house.gov/eeo/
The committee is composed of 49 Members of the House of Representatives, 27 Members are from the Republican party and 22 Members are from the Democrat party. The Members are chosen to serve on the committee by their respective partycaucus.

2544 Committee on Energy and Natural Resources
364 Dirksen Building   202-224-4971
Washington, DC 20510   Fax: 202-224-6163
E-mail: committee@energy.senate.gov
http://energy.senate.gov/general/main_header.htm

2545 Committee on Environment and Public Works Republicans
Senate Dirksen Office Building   202-224-6176
Room 410   Fax: 202-224-5167
Washington, DC 20510   http://www.senate.gov/-epw/

2546 Committee on Glen Canyon Environmental Studies Review
National Research Council
2101 Constitution Avenue   202-334-3422
Washington, DC 20418   Fax: 202-334-1961
http://ww4.nas.edu/cger/wstb.nsf/
Sheila D Davis, Staff Officer

2547 Committee on Government Reform and Oversight
2157 Rayburn House Office Building   202-225-5074
Washington, DC 20515   http://www.house.gov/reform/

2548 Committee on Labor and Human Resources
428 Dirksen Senaate Office Building   202-224-5375
Washington, DC 20510-6300   http://www.senate.gov/-labor

2549 Committee on Resources
US House of Representatives
1324 Longworth House Office Bldg.   202-225-2761
Washington, DC 20515   http://www.house.gov/resources/

2550 Committee on Science
2320 Rayburn House Office Building   202-225-8772
Washington, DC 20515   http://www.house.gov
Sherwood L Boehlert, Chairman

2551 Committee on Small Business: House of Representatives
Small Business Committee
2361 Rayburn House Office Bldg.   202-225-5821
Washington, DC 20515   Fax: 202-225-3587

E-mail: smbiz@mail.house.gov
http://www.house.gov/smbiz/

2552 Committee on Small Business: US Senate
428A Russell Senate Office Building   202-224-5175
Washington, DC 20510   Fax: 202-224-4885
http://www.senate.gov
Weston J Coulam, Staff Director

2553 Committee on Transportation and Infrastructure
2165 Rayburn HOB   202-225-9446
Washington, DC 20510   http://www.house.gov/transporation/
US Rep. Don Young (R-Alaska), Chairman

2554 Community Greens
1700 N Moore Street   703-527-8300
Suite 2000   Fax: 703-527-8383
Arlington, VA 22209   E-mail: info@communitygreens.org
http://www.communitygreens.org
The mission of Community Greens is to improve social and environmental conditions in cities and suburbs across the United States by transforming the interiors of urban blocks into resplendent shared parks and gardens that are owned andmanaged by the residents who live around them. Community greens are not community gardens, nor public parks, nor private backyards but they function as all three.
Kae Herrod, Director

2555 Community Planning and Development Planning and Community Viability
451 7th Street SW   202-708-1112
Washington, DC 20410   http://www.hud.gov
The Community Planning and Developmental office administers programs to help communities plan and finance their growth and development. The Planning and Community Viability Division is responsible for efforts to improve theenvironment, pursuant to the National Environmental Policy Act of 1969 and related statutes and executive orders.
Richard Broun, Director

2556 Cooperative Forestry Research Advisory Council
Department of Agriculture
Cooperative State Research Service   202-720-4318
Washington, DC 20250   http://www.reeusda.gov
Peter A Muscato

2557 Council on Environmental Quality
722 Jackson Place, Northwest   202-395-5750
Washington, DC 20503   Fax: 202-456-6546
http://www.whitehouse.gov/ceq/contact.html
Formulates and recommends national environmental policies.

Larry Flick, Designated Fed. Empl.

2558 Dangerous Goods Advisory Council
1100 H Street, Northwest   202-289-4550
Suite 740   Fax: 202-289-4074
Washington, DC 20005   E-mail: info@dgac.org
http://www.dgac.org
DGAC is an international, nonprofit, educational organization that promotes safety in the transportation of hazardous materials and dangerous goods, including hazardous substances and hazardous wastes.

Al Roberts, President

2559 Department of Agriculture
USDA Forest Service
PO Box 96090
Washington, DC 20090-6090   202-205-1657
Fax: 202-205-1174
http://www.fs.fed.us

2560 Department of Agriculture: Research Department, Forest Environment Research
USDA Forest Service, Research   202-205-1524
PO Box 96090   http://www.fs.fed.us/research
Washington, DC 20090-6090
Dr. Robert Lewis, Deputy Chief

2561 Department of Agriculture: Forest Inventory, Economics
Research Department
USDA Forest Service, Research 202-205-1747
PO Box 96090 http://www.fs.fed.us/research
Washington, DC 20090-6090
Dr. Robert Lewis, Deputy Chief

2562 Department of Agriculture: Forest Service Public Affairs
Sidney R Yates Federal Building 202-205-1006
201 14th Street SW at Independence Ave http://www.fs.fed.us
Washington, DC 20250
Chris Holmes, Assistant Director

2563 Department of Agriculture: National Forest Watershed and Air Management
US Consilidated Farm Service Agency
14th Street & Independence Avenue 202-720-3467
Washington, DC 20250-0001
Donna V Lamb, Manager

2564 Department of Agriculture: National Forest Watershed and Soil Resource
USDA Forest Service 202-205-1657
PO Box 96090 Fax: 202-205-1174
Washington, DC 20090-6090 http://www.fs.fed.us/spf/
Peter E Avers, Manager

2565 Department of Agriculture: Natural Resources State and Private Forestry Division
USDA Forest Service 202-205-1657
PO Box 96090 Fax: 202-205-1174
Washington, DC 20090-6090 http://www.fs.fed.us/spf
Links forestry and conservation with people from the inner city to the rural countryside. Connects people to resources, ideas and to one another so we can all care for the forests and sustain our communities.
Allan J West, Dep Chief

2566 Department of Agriculture: Research Department
USDA Forest Service, Research 202-453-9575
PO Box 96090 http://www.fs.fed.us/research
Washington, DC 20090-6090
Dr. Robert Lewis, Deputy Chief

2567 Department of Agriculture: Research Department Fire Sciences Program
USDA Forest Service 201 202-205-1561
PO Box 96090 http://www.fs.fed.us/recreation
Washington, DC 20090-6090
Linda R Donoghue, Specialist

2568 Department of Agriculture: Soil Conservation
411 S Fort St. & Indepen. Ave. SW 202-720-4525
Washington, DC 20250 Fax: 202-720-7690
Galen S Bridge, Acting Cheif

2569 Department of Agriculture: State & Private Forestry
USDA Forest Service 201, 4th Floor 202-205-1657
PO Box 96090 Fax: 202-205-1174
Washington, DC 20090-6090 http://www.fs.fed.us/spf/
Tony Dorrell, Director

2570 Department of Commerce: National Oceanic & Atmospheric Administration
1100 Wayne Avenue 301-427-2089
Suite 1210 Fax: 301-427-2073
Silver Spring, MD 20910
J Hall, Director

2571 Department of Commerce: National Marine
1315 E West Highway 301-713-2322
SSMC3 Fax: 301-713-0376
Silver Spring, MD 20910 http://ww.nmfs.noaa.gov
Charles Karnella, Acting Director

2572 Department of Commerce: National Ocean Service
Office of Ocean Resources Conservation/Assessment
1305 East-West Highway 301-713-2989
Silver Spring, MD 20910 Fax: 301-713-4389
Charles N Ehler, Director

2573 Department of Commerce: Ocean Observation Division
National Ocean Services
1315 East-West Highway 301-713-2780
Silver Spring, MD 20910-3282 Fax: 301-713-4019
E-mail: erich.frey@noaa.gov
http://chartmaker.ned.noaa.gov
Erich Frey, Office Coast Survey

2574 Department of Energy: Office of Alcohol Fuels
1000 Independence Avenue SW 202-586-9220
EE-1 Forrestal Building, Room 6A-016 Fax: 202-586-9260
Washington, DC 20585 http://www.eren.doe.gov/organization
The Department of Energy and its predecessor agencies have been involved in the research, production and testing of nuclear weapons since the 1940s. These facilities generated large quantities of radioactive and hazardous materialswhich resulted in contamination of many facilities and surrounding areas. The Office of Environmental Restoration was created to consolidate, centralize and promote the cleanup of contaminated waste sites and surplus facilities within the DOEComplex.

2575 Department of Energy: Office of NEPA Policy and Compliance
1000 Independence Avenue SW 202-586-4600
EH-42 800-472-2756
Washington, DC 20585-0119 Fax: 202-586-7031
E-mail: tis.eh.doe.gov/nepa
http://www.eh.doe.gov/nepa
This office serves as the contact point for NEPA matters for the US Department of Energy.
Carol M Borgstrom, Director

2576 Department of Energy: Transportation and Alternative Fuels
1000 Independence Avenue SW 202-586-1723
Washington, DC 20585 Fax: 202-586-9811
E-mail: john.garbak@hq.doe.gov
http://www.ott.doe.gov/oaat/afv.html
John Bgarbak

2577 Department of Energy: Utility Technologies Department
US Department of Energy
1000 Independence Avenue SW 202-586-1720
Washingtion, DC 20585 Fax: 202-586-8148
Robert H Annan, Director

2578 Department of Justice: Environment and Natural Resources Division
10th Street/Constitution Ave 2143 202-514-2701
Washington, DC 20530 Fax: 202-514-0557

2579 Department of Justice: Environment and Resources, Environmental Defense
10th Street/Constitution Ave 7110 202-514-2219
Washington, DC 20530 Fax: 202-514-0557
Letitia J Grishaw, Chairman

2580 Department of State: Bureau of Economic and Business
2201 C Street NW Room 3529 202-647-1498
Washington, DC 20520 Fax: 202-647-8758
William C Ramsay, Assistant Deputy Secretary

2581 Department of State: Bureau of Oceans and International Environmental and Scientific Affairs
2201 C Street NW Room 7831   202-647-2232
Washington, DC 20520   Fax: 202-647-0217
E-mail: www.state.gov/g/oes
Anthony Rock, Principal Dep Asst Secretary

2582 Department of State: Energy Resources and Food Policy Department
Bureau of Economic and Business Affairs
2201 C Street NW   202-647-2875
Washington, DC 20520   Fax: 202-647-4037
David Brown, Director

2583 Department of State: National Office of Environmental Protection
US Department of State
2201 C Street NW   202-647-9266
Washington, DC 20520   Fax: 202-647-5947
Arnold Schifferdecker, Director

2584 Department of State: Ocean and Fisheries Affairs
US Department of State
Office of Marine Conservation   202-647-2335
2201 C Street NW, Room 5806   Fax: 202-736-7350
Washington, DC 20520   http://www.foia.state.gov/records.asp
Larry Snead, Director

2585 Department of State: Office of Ecology, Health, and Conservation State Department
2201 C Street NW   202-647-2418
OES/ETC, Room 4333   Fax: 202-736-7351
Washington, DC 20520
Paul Blakeburn, Director

2586 Department of State: Office of Global Change
Office of Global Change   202-647-4069
OES/EGC, US Department State   Fax: 202-647-0191
Washington, DC 20520
Daniel Reifsnyder, Director

2587 Department of Transportation: Administrator for Pipeine Safety, Alaska
400 7th Street SW   202-366-4595
Room 2335   Fax: 202-366-4566
Washington, DC 20590-0001   http://ops.dot.gov/request.htm
Lloyd W Ulrich, Director

2588 Department of Transportation: Associate Administrator for Pipeline Safety
400 7th Street SW Room 2335   202-366-0656
Washington, DC 20590
George W Tenley Jr, Associate Administrator

2589 Department of Transportation: Office of Marine Safety and Environmental
US Coast Guard
USCG Headquarters, Room 2100   202-267-0518
Washington, DC 20593   Fax: 202-267-4065
Captain Michael Donohue, Chief

2590 Department of Transportation: Office of Marine Safety, Security & Environmental
U.S. Coast Guard
2100 2nd Street SW Room 2408   202-267-2200
Washington, DC 20593   Fax: 202-426-1405
Rear Adm Arthur Henn, Cheif

2591 Department of the Interior
1849 C Street Room 6151   202-208-7351
Washington, DC 20240   Fax: 202-208-5048

2592 Department of the Interior, U.S. Fish & Wildlife Service
1849 C Street NW Room 3242   202-208-4646
Washington, DC 20240   Fax: 202-208-6916
Renne Lohoefener, Assistant Director
Elizabeth Stevens, Deputy Assistant Director

2593 Department of the Interior: National Parks Service
1849 C Street NW Room 3104   202-208-4621
Washington, DC 20240   Fax: 202-208-7520
Roger G Kennedy, Director

2594 Department of the Interior: Bureau of Land Management
1849 C Street NW   202-208-3801
Interior Building, Room 5660   Fax: 202-208-5902
Washington, DC 20240   E-mail: pshea@blm.gov
http://www.blm.gov
Patrick Shea, Director

2595 Department of the Interior: Division of Parks and Wildlife
Office of the Solicitor
1849 C Street NW Room 6557   202-208-4344
Washington, DC 20240   Fax: 202-208-3877
David Watts, Acting Associate Solicitor

2596 Department of the Interior: Land & Renewable Wildlife & Fisheries Division
Special Status Program
1620 L Street NW Room 204   202-653-9210
Washington, DC 20036
Ken Berg, Manager

2597 Department of the Interior: Land and Renewable Forestry Division
Division of Forestry Bureau of Land Management
1849 C Streets NW   202-653-8864
Washington, DC 20240
Mel Berg, Dhief

2598 Department of the Interior: National Park Serv ices
1849 C Street NW   202-208-3100
Washington, DC 20240   E-mail: patti_reilly_@nps.gov
http://www.nps.gov
Alan A Rubin, President

2599 Department of the Interior: National Resources Department
National Park Service
Interior Main Interior Building   202-208-5391
1849 C Street NW, Room 3127   Fax: 202-208-4620
Washington, DC 20240   E-mail: gmachlis@uidaho.edu
F Eugene Hester, Associatate Director

2600 Department of the Interior: Office of the Solicitor
Division of Land and Water
1849 C Street NW, MS 6412   202-208-5757
Washington, DC 20240   Fax: 202-219-1792
Vacant , Associate Solicitor

2601 Department of the Interior: Policy, Management Office of Environmental Affairs
Willie Taylor Office of Env. Pol. & Compliance
Main Interior Building   202-208-3891
Mail Stop 3891   Fax: 202-208-6970
Washington, DC 20240   E-mail: willie_taylor@os.doi.gov
Jonathan P Deason, Director

2602 Department of the Interior: Soil, Water & Air
Land and Renewable Rangeland Resources Division
1620 L Street NW Room 204   202-208-4621
Washington, DC 20036   Fax: 202-208-7520
Donald D White, Chief

2603 Department of the Interior: US Fish & Wildlife, Fisheries Department
1849 C Street NW   202-358-1718
Room 3248
Washington, DC 20240
James G Geiger, Chief

2604 Department of the Interior: US Fish & Wildlife Services
1849 C Street NW, Room 3251   202-208-5333
Washington, DC 20240   Fax: 202-208-3082
William F Hartwig, Natl. Wildlife Refuge Sys.

2605 Department of the Interior: US Fish and Wildlife
US Department of Interior Fish & Wildlife Service
4401 N Fairfax Drive   703-358-1700
Arlington, VA 22203
Cathleen Short, Chief

2606 Department of the Interior: Water and Science, Water Resources Division
Water Resources Division   703-648-5215
12201 Sunrise Valley Drive   Fax: 703-648-7031
Reston, VA 20192   E-mail: rhirsch@usgs.gov
                                               http://water.usgs.gov
Provides reliable, impartial, timely information to understand the water resources of the United States.
Robert M Hirsch, Chief Hydrologist

2607 Department of the Interior: Water and Science Bureau of Reclamation
1849 C Street NW   202-513-0501
Washington, DC 20240-0001   Fax: 202-513-0315
                                               http://www.usbr.gov
To manage, develop, and protect water and related resources in an environmentally and economically sound manner in the interest of the American public.

John Keys, Commissioner

2608 Department of the Interior: Wild Horses and Burros
Land and Renewable Resources
1620 L Street NW Room 204   202-653-5258
Washington, DC 20036
Bob Bainbridge, Advisor

2609 Department of the Interior: Wildlife and Vegetation
Wildlife & Veg. Div. Nat. Park Ser.   202-343-8100
PO Box 37127
Washington, DC 20013-7127
Michael Ruggiero, Chief

2610 Dept. of Agriculture: National Forest Watershed and Hydrology
14th Street SW Building 201   202-205-0886
3rd Floor
Washington, DC 20250
Keith McLaughlin, Water Quality/Hydrology Prog

2611 Dialogue Committee on Phosphoric Acid Product Consensus and Dispute Resolution
Environmental Protection Agency
400 M Street SW   202-260-5495
Washington, DC 20460
Deborah Dalton

2612 EPA: Border Environmental Plan
Public Advisory Office of International Activities
401 M Street SW A-106   202-260-4890
Washington, DC 20460   http://www.epa.org
EPA works with other countries on the entire range of international environmental issues such as climate change, protection of marine environments, lead phase-out, and international transport of hazardous waste.
Sylvia Correa, Designated Federal Officer

2613 EPA: Department of Solid Chemical Emergency Preparedness & Prevention
401 M Street SW Room M3103   202-260-8600
Washington, DC 20460   Fax: 202-260-0927
James L Makris, Director

2614 EPA: Office of Solid Waste, Municipal & & Industrial Solid Waste
401 M Street SW Room M2105   703-308-7267
Washington, DC 20460   Fax: 703-308-8686
                              E-mail: levy.steve@epamail.epa.gov
Bruce R Weddle, Director

2615 Emission Standards Office
National Air Pollution Control Techniques
Office of Air Quality   919-541-5571
Mail Drop 13   Fax: 919-541-0072
Research Triangle Park, NC 27711
The Emission Standards Division (ESD) is responsible for establishing emission standards and managing federal programs for nationwide control of hazardous and criteria pollutant emissions from stationary sources. The Division developsand implements emission standards for hazardous and criteria air pollutants, new source performance standards, control technique guidelines, hazardous waste standards, alternative control techniques documents, and guidance for implementing standards.
Bruce C Jordon, Designated Federal Officer
Sally Shaver, Director

2616 Endangered Species Committee
Endangered Species Committee
1849 C Street NW   202-208-4077
1 Dept. of Interior, Room 4426
Washington, DC 20240
Jon H Goldtstein, Staff Economist

2617 Energy Research Office
1000 Independence Avenue SW   202-586-5430
Room 7B-058   Fax: 202-586-4120
Washington, DC 20585   E-mail: martha.krebs@oer.doe.gov
                                               http://www.oer.doe.gov
This office manages a comprehensive program of basic and applied research and development, and reviews the effectiveness of all departmental programs related to energy research and development.
Martha Krebs, Director

2618 Environment Safety and Health Office
Department of Energy
1000 Independence Avenue SW GA-076   202-586-2481
Washington, DC 20585   Fax: 202-586-3915
The division coordinates and develops the Departmental position on emerging Resources Conservation and Recovery Act, Comprehensive Environmetal Response Compensation and Liability Act, Emergency Planning and Community Right-toKnow Act,Pollution Prevention Act, and Toxic Substance Control Act requirements.
Thomas T Traceski, Director

2619 Environment and Natural Resources: Environmental Crimes Section
Department of Justice
10th Street/Constitution Ave 6101   202-272-9877
Washington, DC 20530   Fax: 202-514-0557
Neil Cartusciello, Chairman

2620 Environment, Safety and Health
1000 Independence Avenue SW   202-586-6151
Room 7A-097   Fax: 202-586-0956
Washington, DC 20585   E-mail: peter.brush@hq.doe.gov
                                               http://www.hq.doe.gov
This office assures the development of comprehensive departmental policy and procedures pertaining to environment, safety, and health laws and regulations, and assures execution of the policy through independent oversight.
Peter Brush, Assistant Secretary

2621 Environmental Change and Security Program:
Woodrow Wilson International Center for Scholars
1300 Pennsylvania Avenue NW   202-691-4130
Washington, DC 20004-3027   Fax: 202-691-4001
E-mail: ecsp@wwic.si.edu
http://www.wilsoncenter.org/ecsp
The ECSP provides specialists and interested individuals with a
road-map to the myriad conceptions, activities and policy initiatives related to environment, population and security. The project pursues three basic activities:gathering information on
related international academic and policy initiatives; organizing
meetings of experts and public seminars; and publishing the
ECSP Report, The China Environment Series and related pa-
pers.ECSP explores a wide range of environment related issues.

Geoffrey D Dabelko, Director
Meaghan Parker, Writer/Editor

2622 Environmental Enforcement
Dept. of Justice/Environment/Natural Resources
1425 New York Avenue NW   202-514-1604
Washington, DC 20005   Fax: 202-514-0557
John Cruden, Chair

2623 Environmental Health Sciences Review Committee
National Institute/Environmental Health Science
PO Box 12233   919-541-7508
Research Triangle Park, NC 27709   Fax: 919-541-2503
E-mail: malone@niehs.nih.gov
Dr Donald McRee PHD, Designated Federal Employee

2624 Environmental Management
1000 Independence Avenue SW   202-586-7710
Room 5A-014   Fax: 202-586-7757
Washington, DC 20585 E-mail: james.ownedoff@em.doe.gov
http://www.em.doe.gov
The Assistant Secretary for EM provides program policy guidance, manages the assessment and cleanup of departmental inactive waste sites and facilities in compliance with federal, state
legal and regulatory requirements. The AS alsodirects a program of safe and effective waste management operations, develops and implements an aggressive applied waste research and
development program to provide innovative environmental
technologies to yield permanent and cost-effective
disposalsolutions.
James Owendoff, Assistant Secretary

2625 Environmental Management Advisory Board
1000 Independence Avenue SW   202-586-7709
Room 180671   Fax: 202-586-0293
Washington, DC 20585   E-mail: terri.lamb@em.doe.gov
http://www.em.doe.gov
The mission of the Environmental Management Advisory Board
is to provide advice, information and recommendations to the
Assistant Secretary for Environmental Management regarding
environmental restoration and waste management issues.
Terri Lamb, Executive Director

2626 Environmental Monitoring Management Council
Environmental Protection Agency
401 M Street SW   202-260-7077
Washington, DC 20460
E Ramona Trovato, Executive Secretary

2627 Environmental Protection Agency
Ariel Rios Building   202-272-0167
1200 Pennsylvania Avenue NW   http://www.epa.gov
Washington, DC 20460
EPA's mission is to protect human health and to safeguard the
natural environment- air, awter and land- upon which life depends. For 30 years, EPA has been working for a cleaner, healthier environment for the American people.
Carol M Browner, Adminsitrator

2628 Environmental Protection Agency Emergency Hazard-
ous Site Control
1235 Jefferson Davis Highway 14th   703-603-8800
Arlington, VA 22202
David Bennett, Acting Director

2629 Environmental Protection Agency Global Change Divi-
sion
401 M Street SW Room 739   202-233-9190
Washington, DC 20460
John Hoffman, Director

2630 Environmental Protection Agency Ground Water and
Drinking Water
1200 Pennsylvania Avenue NW   202-260-5543
Ariel Rios Building   Fax: 202-260-4656
Washington, DC 20460-0003   http://www.epa.gov
James R Elder, Director

2631 Environmental Protection Agency Resource Conserva-
tion and Recovery Act
US Environmental Protection Agency
401 M Street SW, Room M2714   202-260-4808
Washington, DC 20460   Fax: 202-260-1400
RCRA gave EPA the authority to control hazardous waste from
the cradle-to-grave. This includes the generation, transportation, treatment, storage, and disposal of hazardous waste.
Susan Bromm, Director

2632 Environmental Protection Agency: Health & Environ-
mental Review Division
401 M Street SW Room 617   202-260-4241
Washington, DC 20460
J Merenda, Director

2633 Environmental Protection Agency: Air & Radiation
Office of Radiation
401 M Street SW   202-233-9320
Washington, DC 20460   Fax: 202-233-9651
Margo T Oge, Director

2634 Environmental Protection Agency: Health Effects Divi-
sion
401 M Street SW   703-305-7351
Washington, DC 20460   Fax: 703-305-5147
HED is responsible for reviewing and validating data on properties and effects of pesticides, as well as, characterizing and assessing exposure and risks to humans and domestic animals.
Margaret J Stasikowski, Director
Edward Zager, Associate Director

2635 Environmental Protection Agency: Indoor Air Division
Office of Radiation
401 M Street SW   202-233-9315
Washington, DC 20460
Robert Axelrad, Sr. Policy Advisor

2636 Environmental Protection Agency: Office of Pollution
Prevention
Pollution Prevention Division
401 M Street S   202-564-3810
EPA OPPT
Washington, DC 20460
David Kling, Acting Director

2637 Environmental Protection Agency: Office of Health &
Environmental Assessment
401 M Street SW Room M3700   202-260-7315
Washington, DC 20460   Fax: 202-260-0393
William H Farland, Director

2638 Environmental Protection Agency: Office of Air Emis-
sion Standards Division
Research Triangle Park, NC 27709   919-541-5571
Fax: 919-541-0072
E-mail: shaver.sally@epa.gov
Sally Shaver, Director
Jack Edward, Associate Director

2639 Environmental Protection Agency: Water
Environmental Protection Agency
Ariel Rios Building   202-272-0167
1200 Pennsylvania Avenue NW   http://www.epa.gov
Washington, DC 20004
Jonathan Fox, Assistant Administror(Actg)

2640 Environmental Security
Environmental Security US Dept. of Defense
3400 Defense Pentagon   703-695-6639
Room 3E792   Fax: 703-693-7011
Washington, DC 20301-3400 E-mail: sgoodman@acq.osd.mil
          http://www.dtic.mil/defenselink/faq/comment.html
Responsible to the Under Secretary of Defense [Aquisition & Technology] for global Environmental Security policy, oversight, advocacy, representation, and implementation of environmental, safety, occupational health, and fire andemergency services programs for Defense activities including the relationship between the environment and the military missions of the Department of Defense.
Curits Bowlingan Goodman, Force Protection
Patrick Meehan, Program Integration

2641 Federal Aviation Administration
800 Independence Avenue, SW   202-267-3111
Washington, DC 20591   http://www.faa.gov
The FAA operates the world's biggest and best aviation system. The FAA is responsable for the safety and certification of aircraft and pilots, security of our airports, and for the around-the-clock operation of our nations's airtraffic control system. FAA is also responsable for guiding and helping develop commercial space transportation.
Jane Garvey, Administrator

2642 Federal Energy Regulatory Commission
888 1st Street NE   202-502-8004
Room 11A-1   Fax: 202-208-2106
Washington, DC 20426   http://www.ferc.gov
FERC is an independent commission within the department which has retained many of the functions of the Federal Power Commission, such as setting rates and charges for the transporation and scale of natural gas and for the transmissionand sale of electricity and the licensing of hydroelectric power projects. In addition, the commission establishes rates or charges for the transportation of oil by pipeline, as well as the valuation of such pipelines.
Joseph T Kelliher, Chair

2643 Federal Facilities Environmental Restoration
Office of Federal Facilities
401 M Street SW   202-260-1270
Washington, DC 20460   http://es.epa.gov/aoeca/fedbac
The Federal Facilities Enforcement Office, as part of the Office of Enforcement and Compliance Assurance, has responsibility for ensuring that Federal facilities take actions necessary to prevent, control, and abate environmentalpollution. FFEO's functions include policy and guidance development, regional program, enforcement, and information support, interagency agreement, negotiation support, and technical assistance along with capacity building.
Nicholas Morgan, Designated Federal Officer

2644 Federal Highway Administration
400 Seventh Street Sw   202-366-4000
Washington, DC 20590   E-mail: douwes@fhws.dot.gov
                              http://www.fhwa.dot.gov
Our services are designed to meet the present-day needs of our partners and customers while laying the foundation to address the future transportation needs of our Nation.
Kenneth Wykle, Adiministrator

2645 Federal Highway Administration Recreational Trails Programs
400 7th Street SW   202-366-5013
Room 3240   Fax: 202-366-3409
Washington, DC 20590E-mail: christopher.douwes@fhwa.dot.gov
          http://www.fhwa.dot.gov/environment/rectrail
The Recreational Trails Program provides funds to the states to develop and maintain recreational trails for motorized and nonmotorized uses.
Christopher B Douwes, Trails/Rec Program Manager

2646 Federal Highway Administration Transportation Enhancement Activities
400 7th Street SW   202-366-5013
Room 3240   Fax: 202-366-3409
Washington, DC 20590il: christopher.douwes@fhwa.dot.gov.
     http://www.fhwa.dot.gov/environment/te.htm and www.enhancements.org
Transportation Enhancement funds may be used for pedestrian and bicycle facilities, scenic and historic highway programs, landscaping, historic preservation, and environmental mitigation. Projects must relate to surfacetransportation.
Christopher B Douwes, Trails/Enhancments Pro. Mgr.

2647 Federal Railroad Administration
1120 Vermont Avenue NW   202-493-6304
Mail Stop 25   E-mail: Jolene.Molitoris@FRA.DOT.GOV
Washington, DC 20590   http://www.fra.dot.gov
Office of Acquisition and Grants Services is a centralized procurement Office that negotiates, awards, and administers contracts, purchases grants, and cooperative agreemnts in support of the Federal Railroad Administration. The Officeprocures supplies, services, research development, architecture-engineering, information technology and services, and other requirements related to FRA's mission.
Joseph H Boardman, Administrator

2648 Federal Task Force on Environmental Education
Office of Environmental Education Code 111
1849 C Street NW   202-452-5078
Washington, DC 20240   Fax: 202-452-5199
          http://www.epa.gov/enviroed/FTFmembr.html
The Federal Task Force on Environmental Education facilities communication and collaboration among federal agencies and departments that have common interests in supporting and implementing EE programs. The task force places emphasison supporting joint interagency EE projects that leverage both federal and non-federal dollars.
Kathleen MacKinnon

2649 Federal Transit Administration
US Department of Transportation
400 7th Street SW   202-366-4043
TPA-1/Room9400   Fax: 202-366-3472
Washington, DC 20590
Gordon Linton, Administrator Office

2650 Food Safety and Inspection Service
Technical Service Center
1299 Farnam Street   402-221-7400
Landmark Center, Suite 300   800-233-3935
Omaha, NE 68102   Fax: 402-221-7438
               E-mail: TechCenter@fsis.usda.gov
                    http://www.fsis.usda.gov
The Food Safety and Inspection Service (FSIS) is thepublic health agency in the U.S. Department of Agriculture responsible for ensuring that the nation's commercial supply of meat, poultry, and egg products is safe, wholesome, andcorrectly labeled and packaged.
Lynvel Johnson, Director

2651 Food and Drug Administration
US Department of Health and Human Services
Office of Public Affairs   301-827-3666
5600 Fishers Lane   888-463-6332
Rockville, MD 20857   Fax: 301-443-4915
                              http://www.cfsan.fda.gov
FDA is determined to keep its public health protections what they have always been: an effective armor against public health hazards, and one of the greatest bargains that the US government delivers to all citizens.

2652 General Services Administration
1800 F Street NW   202-501-1231
GSA Building   Fax: 202-501-1300
Washington, DC 20405   E-mail: sharon.lighton@gsa.gov
                              http://www.gsa.gov

Our mission is to provide other federal agencies the workspace, products, services, technology, and policy they need to accomplish their missions. The mission statement contained in our Strategic Plan reflects our recognition that wemust provide Federal agencies with the highest quality service at a competitive cost.
Bill Bearden, Contact

2653 Global Learning and Observations to Benefit the Environment
Uchar-The Globe Program
PO Box 3000   800-858-9947
Boulder, CO 80307   Fax: 970-491-8768
http://www.globe.gov
A large network composed of students, scientists and teachers working to heighten environmental consciousness.

2654 House Committee on Agriculture Operations, Oversight, Nutrition
House Agriculture Committee
1301 Longworth House Office Buildg.   202-225-2171
Washington, DC 20515   Fax: 202-225-0917
http://agriculture.house.gov

Bob Goodlatte, Chair

2655 House Committee on Commerce Telecommunications, Trade and Consumer Protection
2125 Rayburn House Office Building   202-225-2927
Washington, DC 20515   Fax: 202-225-1919
Thomas Bliley, Chair

2656 House Committee on International Relations
2170 Rayburn House Office Building   202-225-5021
Washington, DC 20515   Fax: 202-225-2035
E-mail: HIRC@mail.house.gov
http://www.house.gov/international_relations
Our Committee is charged with overseeing US foreign policy programs and agencies. We manage legislation regarding foreign policy, State Department management, foreign assistance, trade promotion, export controls, foreign arms sales,student exchanges, international broadcasting and many other issues.
Henry J Hyde, Chair

2657 House Committee on Transportation and Infrastructure
Room 216, Rayburn House Office Bldg   202-225-4472
Washington, DC 20515   Fax: 202-226-0921
Bud Shuster, Chairman

2658 Indian Affairs Office of Management Support Services
1849 C Street NW   202-208-6364
MS-3657-MIB   Fax: 202-208-1605
Washington, DC 20240   http://www.ios.doi.gov
Responsible for activities pertaining to Indian affairs, including resources and the environment.
Debbie McBride, Chief, Div. of Env. Mgt.

2659 Installation Management
600 Army Pentagon   703-693-3233
Room 1E668   Fax: 703-693-3507
Washington, DC 20310-0600   http://www.army.mil
This office is responsible for policy and oversight of construction, utilization, improvement, alteration, maintenance, repair and disposal of real estate and facilities.
Jan Menig, Deputy Assistant Chief

2660 Installations and Environment
Secretary of the Navy Installation & Environment
1000 Navy Pentagon   703-693-4530
Room 4E729   Fax: 703-693-1165
Washington, DC 20350-1000   http://www.ncts.navy.mil
This office is responsible for policy and oversight of construction, utilization, improvement, alteration, maintenance, repair and disposal of real estate and facilities, inslucing capital equipment, utilities and quarters; baseconversion and redevelopment; force basing and infrastructure requirement analysis; environmental compliance, planning, analysis and technology, pollution prevention, and contamination cleanup; natural and cultural resources conservation, andoccupational safety.

Robert Pirie, Assistant Secretary

2661 Inter-American Foundation
901 N Stuart Street   703-306-4301
10th Floor   Fax: 703-306-4365
Arlington, VA 22203   http://www.iaf.gov
Supports grassroots organizations and NGOs in Latin America and the Caribbean.

2662 International Center for Development Policy
731 8th Street SE   202-547-3800
Washington, DC 20003   Fax: 202-546-4784
E-mail: etoledo@newforestsproject.com
http://www.newforestsproject.com
Erick Toledo, Director

2663 International Joint Commission
1250 23rd Street NW   202-736-9024
Suite 100   Fax: 202-467-0746
Washington, DC 20440   http://www.ijc.org
Many rivers and some of the largest lakes in the world lie along, or flow across, the border between the United States and Canada. The International Joint Commission assists governments in finding solutions to problems in these waters.

Irene B Brooks, Commissioner
Allen I Olson, Commissioner

2664 International Trade Administration Trade Development
Herbet C Hoover Building, Room 3832   202-482-1461
14th Street & Construction Avenue NW   Fax: 202-482-5697
Washington, DC 20230   E-mail: emottur@ita.doc.gov
http://www.ita.doc.gov
The Assistant Secretary for Trade Development is responsible for an integrated program of policy, promotion and analysis to assist key US industrial sources, including basic industries and technology and aerospace.
Ellis Mottur, Assistant Secretary

2665 Land & Renewable Recreation & Wilderness Resources: Wilderness Branch
Department of the Interior
1620 L Street NW Room 302   202-653-8828
Wahington, DC 20036
Keith H Corrigall, Chief

2666 Land and Minerals Management
1849 C Street NW   202-208-5676
Interior Building, Room 6613   Fax: 202-208-3144
Washington, DC 20240   E-mail: bob.armstrong@mms.gov
http://www.mms.gov
Rebecca Watson, Assistant Secretary

2667 Land and Minerals Office of Surface Mining Reclamation & Enforcement
Department of the Interior
1951 Constitution Avenue NW   202-208-4006
Washington, DC 20240   Fax: 202-219-3106
E-mail: jjarrett@osmre.gov
http://www.osmre.gov
Our mission is to carry out the requirments of the Surface Mining Control and Reclamation Act in cooperation with States and Tribes. Our primary objectives are to ensure that coal mines are operated in a manner that protects citizensand the environment during mining and assures that the land is restored to beneficial use following mining, and to mitigate the effects of past mining byaggressively pursuing reclamation of abandoned coal mines.
Jeffrey D Jarrett, Director

2668 Management Advisory Group to the Assistant Administrator
Communications and Information Management
401 M Street SW   202-260-5554
Washington, DC 20460

Mission: To assist owners and managers of businesses to reach their full potential in a cost effective, reliable and collaborative manner by utilizing the strengths of all members of the client's team and providing resources to offsetweaknesses that may be impeding such success.
Michelle Hiller, Designated Federal Officer

2669 Management and Budget Office: Natural Resources, Energy and Science

Old Executive Office Building   202-395-4561
Washington, DC 20503   Fax: 202-395-4639

This division of the Management and Budget Office assists the President by clearing and coordinating advice on proposed legislation and by making recommendations as to presidential action on legislative enactments related to naturalresources, energy and science.
TJ Glauthier, Associate Director

2670 Manpower, Reserve Affairs, Installations and Environment

Environment, Safety and Occupational Health
1665 Air Force Pentagon   703-697-9297
Washington, DC 20330-1665   Fax: 703-614-2884
    http://www.af.mil

This office provides guidance, direction, and oversight for the department on all matters pertaining to the environment, safety, and occupational health.
Thomas McCall Jr, Deputy Assistant Secretary

2671 Marine Mammal Commission

4340 East West Highway   301-504-0087
Suite 905   Fax: 301-504-0099
Bethesda, MD 20814   E-mail: mmc@mmc.gov
    http://www.mmc.gov

Developing, reviewing and making recommendations on domestic and international actions and policies with respect to marine mammal protection, conservation and with carrying out a research program. Primary objective is to ensure thatfederal programs are being administered in ways that maintain the health and stability of marine ecosystems and do no disadvantage marine mammal populations or species.
David Cottingham, Executive Director
Timothy J Ragen, PhD, Scientific Program Director

2672 Maritime Administration

Maritime Administration (MARAD)
400 7th Street SW   202-366-5823
Room 7206   Fax: 202-366-3890
Washington, DC 20590   E-mail: pao.marad@marad.dot.gov
    http://www.dot.gov

The overall mission of the Maritime Administration is to promote the development and maintenance of an adequate, well-balanced, United States merchant marine, sufficient to carry the Nation's domestic waterborne commerce and asubstantial portion of its waterborn foreign commerce, and capable of serving as a naval and military auxiliary in time of war or national emergency.
Bruce Carlton, Acting Deputy Administrator

2673 Marketing and Regulatory Programs and Support Services Agricultural Marketing

1400 Independence Avenue SW   202-720-4276
South Building, Room 3069   Fax: 202-720-8477
Washington, DC 20250   http://www.usda.gov

The Marketing Programs Support Services provide numerous agricultural services including the administering of the Pesticide Data Program which, in cooperation with states, samples and analyzes fresh fruits and vegetables for pesticideresidues. It shares residue test results with the Environmental Protection Agency and other public agencies.
Kenneth Clayton, Associate Administrator

2674 Migratory Bird Conservation Commission

4401 N Fairfax Drive   703-358-1716
Suite 622   Fax: 703-358-2223
Arlington, VA 22203   E-mail: mbcc@fws.gov

The Migratory Bird Conservation Commission (MBCC) is responsible for considering and approving for acquistion areas of migratory bird habitat (other than waterfowl production areas) that have been submitted by regional offices andrecommended by the Secretary. The MBCC is composed of representatives from the Legislative and Executive Branches of government, fixes the price at which such areas may be purchased or rented, and meets three times a year.

Eric Alvarez, Secretary

2675 Migratory Bird Regulations Committee Office of Migratory Bird Management

4401 N Fairfax Drive   703-358-1714
MBSP 4107   Fax: 703-358-2217
Arlington, VA 22203   E-mail: brian_a_millsap@fws.gov
    http://policy.fws.gov/723fw2.html

The Service Regulations Committee will review information provided to it each year on regulatory issues and submit recommendations to the Director. In this regard, the committe receives guidance fromt he office of Migratory BirdManagement, the Division of Law Enforcement, and from the Regional migratory Bird Coordinators. In addition, Flyway Consultants, from the four Flyway Councils, may provide technical data and certain advice as limited by memoranda of understanding.
Brian A Millsap
Hope Grey

2676 Mine Safety and Health Administration

1100 Wilson Boulevard   202-693-9899
Arlington, VA 22203   Fax: 202-693-9873
    http://www.msha.gov

The mission of the Mine Safety and Health Administration is to administer the provisions of the Federal Mine Safety and Health Act of 1977 and to enforce compliance with mandatory safety and health standards as means to eliminate fatalaccidents; to reduce the frequency and severity of nonfatal accidents; to minimize health hazards; and to promote improved safety and health conditions in the Nation's mines. MSHA carries out the mandates of the Mine Act at all mining and mineralprocessing areas.
J Davitt McAteer

2677 NOAA Sanctuaries and Reserves Management Divisions

Department of Commerce
NOAA's National Marine Sanctuaries   301-713-3125
1305 E West Highway, 11th Floor   Fax: 301-713-0404
Silver Spring, MD 20910   E-mail: nmscomments@noaa.gov
    http://www.sanctuaries.nos.noaa.gov

2678 National Acid Precipitation Assessment Program

NOAA, Mailcode R/E   301-713-0460
1315 E West Highway   Fax: 301-713-1459
Silver Spring, MD 20910   E-mail: napap@noaa.gov
    http://www.oar.noaa.gov/organization/napap.html

The National Acid Precipitation Assessment Program (NAPAP) is an interagency scientific research, monitoring and assessment program on the effects of sulfur and nitrogen oxides on the environment and human health. NAPAP acts as acoordinating office between six Federal agencies, which also fosters cooperation among its members, other governments, States, universities, and the private sector.
Dr Derek Winstanley

2679 National Aeronautics and Space Administration

NASA
300 E Street SW   202-358-0000
Washington, DC 20024-3210   Fax: 202-358-3251
    E-mail: info-center@hq.nasa.gov
    http://www.nasa.gov

Research includes a variety of global environmental conditions.

2680 National Animal Damage Control Advisory Committee

1400 Independence Avenue SW   202-720-2054
Washington, DC 20250-3402   Fax: 202-690-0053
Bobby R Acord, Executive Secretary

2681 National Cancer Institute: Cancer Epidemiology and Genetics Division

6130 Executive Board   301-496-1611
Executive Plaza N, Room 543   Fax: 301-402-3256
Rockville, MD 20852   E-mail: fraumeni@nih.gov
     http://www.nih.gov
The National Cancer Institute expands existing scientific knowledge on cancer cause and prevention as well as on the diagnosis, treatment, and rehabilitation of cancer patients. This division conducts research on cancer epdiemiologyand genetics.
Joseph Fraumeni, Director

2682 National Center for Health Statistics

Nat. Cen. for Health Stats. Division of Data Serv.
6525 Belcrest Road   301-458-4636
Hyattsville, MD 20782-2003   E-mail: nchsquery@cdc.gov
     http://www.cdc.gov/nchs/
The mission of the National Center for Health Statistics (NCHS) is to provide statistical information that will guide actions and policies to improve the health of the American people. As the Nation's principal health statisticsagency, NCHS leads the way with accurate, relevant, and timely data.
Edward Sondik, Director

2683 National Climatic Data Center

Federal Building   828-271-4800
151 Patton Avenue   Fax: 828-271-4876
Asheville, NC 28801-5001   http://www.ncdc.noaa.gov
The collection center and custodian of all US weather records. It is the central source of historical weather information, some of which it publishes as serial or periodic publications. The center responds to inquiries for statisticalweather data, such as temperature, precipitation and wind.

2684 National Council on Radiation Protection and Measurements

7910 Woodment Avenue   301-657-2652
Suite 400   800-229-2652
Bethesda, MD 20814   Fax: 301-907-8768
     E-mail: ncrp@ncrp.com
     http://www.ncrp.com
Nonprofit federally chartered corporation. Its objectives are to collect, analyze and disseminate information and recommendations about radiation protection and measurement and to stimulate the exchange of ideas and promote cooperationamong organizations concerned with radiation problems throughout the world.
William Beckner, Executive Director
Thomas S Tenforde, President

2685 National Environmental Justice Advisory Council

401 M Street Southwest (MC 2201A)   202-564-2515
Washington, DC 20460   800-962-6215
     Fax: 202-501-0740
     E-mail: environmental-justice-epa@epamail.epa.gov
     http://es.epa.gov/oeca/main/ej/nejac
A federal advisory committee established by charter to provide independent advice, consultation and recommendations to the Administrator of the US Environmental Protection Agency on matters related to environmental justice.

2686 National Environmental Satellite Data & Information Service

National Oceanic and Atmospheric Administration
Suitland & Silver Hill Roads   757-824-3446
Federal Office Building, Room 2069   Fax: 757-824-7300
Suitland, MD 20233   E-mail: rwinokur@nesdis.noaa.gov
     http://www.nesdis.noaa.gov
The National Environmental Satellite, Data and Information Service operates a national environmental satellite system. It acquires, stores and disseminates worldwide environmental data through its data centers.
Gregory W Withee, Assistant Administrator

2687 National Guard Bureau: Army and Air Force

2500 Army Pentagon   703-695-6987
Room 2E394   http://www.guardnet.ngb.army.mil
Washington, DC 20310-2500

Today's National Guard provides to the states units trained and equipped to protect life and property, while providing to the nation units trained, equipped and ready to defend the United States and its interests around the globe.
Edward Baca, Chief

2688 National Health Information Center (NHIC) Services

US Department of Health and Human Services
PO Box 1133   301-565-4167
Washington, DC 20013-1133   800-336-4797
     Fax: 301-984-4256
     E-mail: healthfinder@nhic.org / info@nhic.org
     http://www.health.gov / www.healthfinder.gov
A health information referral service which links consumers and health professionals who have health questions with organizations best able to provide answers. Spanish language information specialists are available. We do notdiagnose medical conditions or offer medical advice.

2689 National Institute for Occupational Safety and Health

Humphrey Boulevard Room 715 H   202-401-0721
200 Indiana Avenue   800-356-4674
Washington, DC 20201   http://www.cdc.gov
To ensure safe and healthful working conditions for all working people, occupational safety and health standards are developed, and research and other activities are carried out, through the National Institute for Occupational Safetyand Health.
Bill Halpherin, Director

2690 National Institute of Environmental Health Sciences

PO Box 12233   919-541-3201
Research Triangle Park, NC 27709   Fax: 919-541-2260
     E-mail: olden@niehs.nih.gov
     http://www.niehs.nih.gov
Environmental research.
Kenneth Olden PhD, Director

2691 National Institutes of Health

1 Center Drive, Building 1   301-496-2433
Room 126   Fax: 301-496-8276
Bethesda, MD 20892   E-mail: execsec1@od.nih.gov
     http://www.nih.gov
The National Institutes of Health conducts, supports and promotes biomedical research to improve the health of the American people by increasing the understanding of processes underlying human health, disability and disease. Theinstitutes advance knowledge concerning the health effects of interactions between humans and the environment.
Dr Elias A Zerhouni, Director

2692 National Lead Information Center

8601 Georgia Avenue
Suite 503   800-424-LEAD
Silver Spring, MD 20910   Fax: 301-585-7976
     E-mail: hotline.lead@epamail.epa.gov
     http://www.epa.gov/lead/nlic.htm
Provides the general public and professionals with information about lead hazards and their prevention.

2693 National Marine Fisheries Service

National Oceanic & Atmospheric Administration
1315 E West Highway   301-713-2239
Silver Spring Metro Center, Room 14555 Fax: 301-713-2258
Silver Spring, MD 20910E-mail: roland.schmitten@noaa.gov
     http://www.noaa.gov
NMFS conducts an integrated program of management, research and services for the protection and rational use of living marine resources. It also is responsible for the protection of marine mammals.
Rolland Schmitten, Assistant Administrator

2694 National Oceanic & Atmospheric Administration

14th Street & Constitution Ave NW   202-482-6090
Room 6013   Fax: 202-482-3154
Washington, DC 20230   E-mail: d.james.baker@noaa.gov
     http://www.rdc.noaa.gov

The Administration's mission is to explore, map, and chart the global ocean and its living resources and to manage, use, and conserve those resources; to describe, monitor, and predict conditions in the atmosphere, ocean, sun, andspace environment; to issue warnings against impending destructive natural events; to assess the consequences of inadvertent environmental modification over several scales of time, and to manage and disseminate long-term environmental information.
Conrad C Lautenbacher Jr, NOAA Administrator
James R Mahoney PhD, NOAA Deputy Adminsitrator

2695 National Organic Standards Board Agricultural Marketing Service
USDA/AMS
1400 Independence Avenue SW   **202-720-2704**
Washington, DC 20250-0264   **Fax: 202-690-4948**
                    E-mail: NOSB.Materials@usda.gov
Dr Ahrold S Ricker, Staff Director

2696 National Park Service: Fish, Wildlife and Parks
National Park Service
1849 C Street NW   **202-208-4621**
PO Box 37127   **Fax: 202-208-7889**
Washington, DC 20240   **E-mail: nps_director@nps.gov**
                    http://www.nps.gov
Fran Mainella, Director

2697 National Park System Advisory Board
Department of the Interior
1849 C Street NW   **202-343-8163**
Washington, DC 20013-7127
Herb S Cables Jr, Designated Federal Officer
Patricia Henry

2698 National Petroleum Council
1625 K Street NW   **202-393-6100**
Suite 600   **Fax: 202-331-8539**
Washington, DC 20006   **E-mail: info@npc.org**
The purpose of the NPC is solely to represent the views of the oil and natural gas industries in advising, informing, and making recommendations to the Secretary of Energy with respect to any matter relating to oil and natural gas, orto the oil and gas industries submitted to it or approved by the Secretary. The NPC does not concern itself with trade practices, nor does it engage in any of the usual trade association activities.
Carla Byrd

2699 National Science Foundation
4201 Wilson Boulevard   **703-292-5111**
Arlington, VA 22230   **800-877-8339**
                    http://www.nsf.gov
It is the National Science Foundation's mission to promote the progress of science; to advance the national health, prosperity, and welfare; and to secure the national defense.
Rita Colwell, Director

2700 National Science Foundation Directorate for Geosciences
4201 Wilson Boulevard   **703-292-8500**
Room 705   **Fax: 703-292-9042**
Arlington, VA 22230
The Directorate for Geosciences (GEO)'s mission is to support research in the atmospheric, earth, and ocean sciences. As the principal source of federal funding for university-based fundamental research in the geosciences, GEOaddresses the nation's need to understand, predict, and respond to environmental events and changes to use Earth's resources wisely.
Robert W Corell, Assistant Director

2701 National Science Foundation Office of Polar Programs
4201 Wilson Boulevard   **703-292-8030**
Room 755 S   **Fax: 703-292-9081**
Arlington, VA 22230   **http://www.nsf.gov/od/opp/**
OPP shares in the vision and goals expressed in NSF's strategic plan: enable world leadership in science and engineering; promote discovery, dissemination and employment of new knowledge; and support excellence in science mathematical,engineering and technology education. Polar research and the associated logistics activities make a recognized and visible contribution to these goals.

Karl A Erb, Director

2702 Natural Resources Conservation Service
PO Box 2890   **202-720-3210**
Washington, DC 20013   **Fax: 202-720-1564**
                    http://www.nrcs.usda.gov
Provides leadership in a partnership effort to help people conserve, maintain and improve our natural resources and environment. NRCS is the technical delivery arm of USDA and provides conservation information, incentive programs andtechnical assistance at the state and county levels. Contact the Conservation Communications Staff.

2703 Natural Resources Conservation Service (NRCA)
US Agriculture Department
14th and Independence Avenue SW   **202-720-7246**
Room 5105-A   **Fax: 202-720-7690**
Washington, DC 20250 E-mail: bruce.knight@wdc.usda.gov
                    http://www.nrcs.usda.gov
The Under Secretary for Natural Resources and Environment is responsible for fostering sound stewardship of 75 percent of the nation's total land area. Ecosystems are the underpinning for the department's operating philosophy in thearea, in order to maximize stewardship of our natural resources. This approach ensures that products, values, services, and uses desired by people are produced in ways that sustain healthy, productive ecosystems.
Bruce Knight, Chief
Dana York, Associate Chief

2704 Natural Resources and Environment Forest Service
Forest Service   **202-205-1661**
PO Box 96090   **Fax: 202-205-1765**
Washington, DC 20090-6090 E-mail: mdombeck@fsweb.gov
                    http://www.fsweb.gov
The Forest Service is a federal agency that manages public lands in national forests and grasslands. It is the largest forestry research organization in the world, and provides technical and financial assistance to state and privateforestry agencies.
Dale Bosworth, Chief

2705 Naval Sea Systems Command
1333 Isaac Hull Avenue SE   **202-781-4124**
Washington Navy Yard, DC 20376   **Fax: 202-781-4713**
                    http://www.ncts.navy.mil
The Naval Sea Systems Command provides material support to the Navy and Marine Corps, and for mobilization purposes to the Department of Defense and Department of Transportation, for ships, submarines, and other sea platforms,shipboard combat systems and components, other surface and undersea warfare and weapons systems, and ordinance expendables not specifically assigned to other system commands.
George Sterner, Commander

2706 Navy Environmental Health Center
620 John Paul Jones Circle   **757-953-0700**
Suite 1100   **http://www.-nehc.med.navy.mil**
Portsmouth, VA 23708-2103
Ensures Navy and Marine Corps readiness through leadership in prevention of disease and promotion of health.
Captain David Hiland, Commanding Officer

2707 North American Wetlands Conservation Council
4401 N Fairfax Drive   **703-358-1784**
Mailstop 4075   **Fax: 703-358-2282**
Arlington, VA 22203   **E-mail: DBHC@fws.gov**
                    http://www.fws.gov/birdhabitat
The North American Wetlands Conservation Council reviews and recommends wetlands conservation project proposals for funding under the North American Wetlands Conservation Act.

David A Smith, Coordinator

2708 Nuclear Materials, Safety, Safeguards & Operations
Nuclear Regulatory Commission
One White Flint N Building   **301-492-1713**
11555 Rockville Pike   **http://www.nrc.gov**
Rockville, MD 20852

The Division of Industrial and medical Nuclear safety within the Office of Nuclear Materials Safety and safeguards at the NRC has the responsibility for NRC's principal rulemaking and guidance development, licensing, inspection, eventresponse and regulatory activities for material licensed under the Atomic Energy Act of 1954, as amended, to ensure safety and quality associated with the possession, processing, and handling of nuclear material.
Hugh L Thompson Jr, Department Executive Direct

2709 Occupational Safety and Health Administration: US Department of Labor
Office of Administrative Services
200 Constitution Avenue NW   202-693-1999
Room N-310   800-321-6742
Washington, DC 20210   http://www.osha.gov
OSHA's mission is to send every worker home whole and healthy every day. Since the agency was created in 1971, workplace fatalities have been cut in half and occupational injury and illness rates have declined 40 percent. At the sametime, US employment has nearly doubled from 56 million workers at 3.5 million worksites to 105 milion workers at nearly 6.9 million sites.

2710 Oceanic and Atmospheric Research Office
National Oceanic and Atmospheric Administration
1315 E West Highway   301-713-2458
Room 11627   Fax: 301-713-0163
Silver Spring, MD 20910   E-mail: david.evans@noaa.gov
                                        http://www.oar.noaa.gov
The Office of Oceanic and Atmospheric Research is where much of the work is done that results in better weather forecasts, longer warning lead times for natural disasters, new products from the sea, and greater understanding of ourclimate, atmosphere, and oceans.
David Evans, Assistant Administrator

2711 Office of Civil Water Enforcement Division
Environmental Protection Agency
US Env. Protect. Agency, Water Enf.   202-564-2240
1200 Pennsylvania Avenue   Fax: 202-564-0018
Washington, DC 20460   http://www.epa.gov
Mark Pollins, Director
Kate Anderson, Associate Director

2712 Office of Environmental Restoration & Waste Management
1000 Independence Avenue SW 5A-014   202-586-5430
Washington, DC 20585   Fax: 202-586-4120
                                http://www.envirosources.com
In 1989, the Office of Environmental Restoration was created within the newly established Office of Environmental Management to consolidate, centralize and promote the cleanup of contaminated waste sites and surplus facilities withinthe DOE Complex.
Thomas P Grumbly, Assistant Secretary

2713 Office of Research & Engineering Hazardous Materials
National Transportation Safety Board
490 L'Enfant Plaza East SW   202-382-6585
Room 5131
Washington, DC 20594
Robert J Chipkevich, Chief

2714 Office of Solid Waste Management & Emergency Response
401 M Street SW   202-382-7486
Room 308 NE Mail   http://www.epa.gov/swerrims/about.htm
Washington, DC 20460
Develops guidelines and standards for the land disposal of hazardous wastes and for underground storage tanks. Furnishes techincal assistance in the development, management and operation of solids waste activities and analyzes therecovery of useful energy from solid waste. The Office has undertaken the development and implementatin of a program to respond to abandoned and active hazardous waste sites and accidental release as well as the encouragement of new technology.
Michael H Shapiro, Acting Assistant Admin.

2715 Office of Surface Mining
1951 Constitution Avenue NW   202-208-2719
Washington, DC 20240   http://www.osmre.gov
Aids in maintaining proper safety precautions during coal reclamation.

2716 Office of Surface Mining Reclamation & Enforcement
US Department of the Interior
1951 Constitution Avenue NW   202-208-2719
Washington, DC 20240   E-mail: getinfo@osmre.gov
The Office of Surface Mining is the bureau of the US Department of the Interior with responsability, in cooperation with the state and indian Tribes, to protect citizens and the environment during coal mining and reclamation, and toreclaim mines abandoned before 1977.
Margy White, Chief of Staff

2717 Office of the Chief Economist
Jamie L Whitten Federal Building   202-720-5447
Room 112-A   http://www.usda.gov
Washington, DC 20250-3800
The Office of the Chief Economist advises the Secretary on the economic implications of policies and programs affecting the US food and fiber systems and rural areas. The Chief Economist coordinates, reviews, and approves theDepartment's commodity and farm sector forecast.
Keith Collins, Chief Economist

2718 Office of the Executive Clerk
17th Street & Pennsylvania Ave NW   202-456-2226
Washington, DC 20500   Fax: 202-456-2569
                                        http://www.whitehouse.gov
This office provides information on when a bill was signed or vetoed, the dates of presidential messages, executive orders, and dates of other presidential actions.
Timothy Saunders, Executive Clerk

2719 Office of the General Counsel
1400 Independence Avenue NW   202-720-3351
Jamie L Whitten Federal Building   http://www.usda.gov
Washington, DC 20250
The Office of the General Counsel provides legal services for all programs, operations and activities of the department.
Bonnie Luken, General Counsel

2720 Office of the Secretary of Agriculture
1400 Independence Avenue NW   202-720-3631
Jamie L Whitten Federal Building   Fax: 202-720-2119
Washington, DC 20250   http://www.web.fie.com/web/fed/agi
The responsibility of the Secretary of Agriculture is to improve and maintain farm income; develop and expand markets abroad for agricultural products; help curb and cure poverty, hunger and malnutrition; enhance the environment andmaintain production capacity; protect the soil, water forests and other natural resources; provide rural development, credit and conservation programs; carry out agricultural research and provide food inspection and grading services.
Ann M Veneman, Secretary of Agriculture

2721 Office of the Secretary of Energy
1000 Independence Avenue SW   202-586-6210
Forrestal Building, Room 7A-257   Fax: 202-586-4403
Washington, DC 20585   http://www.doe.gov
The Secretary of Energy provides the framework for a comprehensive and balanced national energy plan through the coordination and administration of the energy functions of the federal government.
Federico Pena, Secretary of Energy

2722 Office of the Secretary of Health and Human Services
200 Independence Avenue SW   202-690-7000
Hubert H Humphery Building   Fax: 202-690-7203
Washington, DC 20201   E-mail: donna.shalala@os.dhhs.gov
                                        http://www.os.dhhs.gov
The Secretary of Health and Human Services advises the President on health, welfare, and income security plans, policies, and programs of the federal government.
Donna Shalala, Sect'y Health/Human Service

2723 Office of the Secretary of the Interior

1849 C Street NW
Interior Building, Room 6156
Washington, DC 20240

202-208-7351
Fax: 202-208-7545
E-mail: exsec@doe.gov
http://www.doi.gov

The Secretary of the Interior is responsible for the administration of over 500 million acres of federal land, and holds in trust approximately 50 million acres of land, mostly Indian reservations; the conservation and development of mineral and water resources; the conservation, development, and utilization of fish and wildlife resources; the coordination of federal and state recreational programs; the preservation and administration of the nation's scenic and historic areas.
Bruce Babbitt, Secretary of the Interior

2724 Office of the Solicitor

1849 C Street NW
Interior Building, Room 6352
Washington, DC 20240

202-208-4423
Fax: 202-208-5584
http://www.doi.gov

The Solicitor performs all of the legal work of the department and is the principal legal adviser to the secretary and the chief law officer of the department.
John Leshy, Solicitor

2725 Office of the US Trade Representative Environment and Natural Resources

600 17th Street NW
Representative Winder Building
Washington, DC 20506

202-395-7320
http://www.ustr.gov

This division of the Office of the US Trade Representative is responsible for the direction of all trade negotiations and the formulation of trade policy for the United States as related to the environment and natural resources.
Jennifer Haverkamp, Assistant US Trade

2726 Outer Continental Shelf Advisory Board Minerals Management Service

Department of the Interior
Room 4230
Washington, DC 20240

703-787-1211
http://www.mms.gov/eppd/scicom/

The purpose of the Outer Continental Shelf Scientific Committe of the Minerals Management Advisory Board is to advise the Director, Minerals Management Service, on the feasibility, appropriateness, and scientific value of the MMS/OCS.Environmental Studies Program main function is to obtain environmental information through research to support the decision process at all stages of the oil and gas leasing program.
Terry Holman

2727 Peace Corps

1111 20th Street NW
Washington, DC 20526

202-692-2100
800-424-8580
http://www.peacecorps.gov

The goal of the Peace Corps is to help people of interested countries in meeting their need for trained men and women, to promote a better understanding of Americans on the part of peolple served and to help promote a betterunderstanding of other peoples on the part of Americans.
Charles Baquet, Deputy Director

2728 President's Council on Sustainable Development

730 Jackson Place NW
Washington, DC 20503

202-408-5331
Fax: 202-408-6839
E-mail: infopcsd@aol.com
http://www.whitehouse.gov/PSCD

The council is responsible for the beginning of the implementation of the recommendation written for the President for a national sustainable development action strategy that will encourage economic development, job creation andprotection of natural and cultural resources. Council membership includes representatives from industry, government and environmental, labor and civil rights organization.
Martin Spitzer, Executive Director

2729 Public Affairs Directorate

1690 Air Force Pentagon
Room 4D922
Washington, DC 20330-1690

703-694-6061
http://www.af.mil

The Public Affairs Directorate advises the secretary and chief of staff on aspects of Air Force missions impacting upon the public, and establishes programs for communicating with Air Force personnel and the public, including newsmedia worldwide, to gain informed public support.
Ronald Sconyers, Director

2730 Research and Special Programs Administration

400 7th Street SW
Nassif Building, Room 8410
Washington, DC 20590

202-366-4433
http://www.rspa.dot.gov

As one of the nine major agencies of the US Department of Transportation the RSPA provides vital services to America's dynamic multimodel transportation system. Its safety and research programs strengthen the nation's industrialcompetitiveness, especially in a global economy where intermodel transportation is essential.
Kelley Coyner, Administrator

2731 Research, Education and Economics Cooperative State Research/Extension Service

1400 Independence Avenue SW
Jamie L Whitten Federal Building
Washington, DC 20250

202-720-4423
Fax: 202-720-8987
E-mail: brobinson@reeusda.gov
http://www.reeusda.gov

The CSREES expands the research and higher education functions of the former Cooperative State Research Service and the education and outreach functions of the former Extension Service. In cooperation with its partners and customers,CSREES provides the focus to advance a global system of research, extension, and higher education in the food and agricultural sciences and related environmental and human sciences to benefit people, communities, and the nation.
BH Robinson, Administrator

2732 Research, Education and Economics Economic Research Service

Economics Division
1800 M Street NW
Washington, DC 20036-5831

202-694-5500
Fax: 202-694-5773
http://www.usda.gov/insiders

This division of the Economic Research Service provides economic and other social science information and analysis pertaining to resources for improving the performance of agriculture and rural America. Information is made available tothe public through research monographs, situation and outlook reports, standardized data products in electronic media, and professional and trade jounrals.
Margot Anderson, Director

2733 Research, Education, and Economics Economic Research Service Energy Office

300 7th Street SW
Washington, DC 20036-5831

202-401-0461
Fax: 202-401-0533
E-mail: rconway@oce.usda.gov
http://www.usda.gov/oec/oepnu

The Energy Office serves as the focal point for all energy-related matters within the department. The office is responsible for developing and coordinating all USDA energy policies, reviewing and evaluating all USDA energy andenergy-related programs, and providing liaison with the Department of Energy and other federal agencies and departments on energy activities that may effect agriculture and rural America.
Roger Conway, Director

2734 Research, Education, and Economics National Agricultural Statistics Service

1400 Independence Avenue SW
S Building, Room 4117
Washington, DC 20250

202-720-2707
Fax: 202-720-9013
http://www.usda.gov/nass

This service prepares estimates and reports on production, supply, price and other items necessary for the orderly operation of the US agricultural economy. These reports include statistics on field crops, fruits and vegetables,cattle, hogs, sheep, poultry, and related commodities or processed products. Other estimates concern pesticide use, farm numbers, prices recieved by farmers for products sold, prices paid for commodities and services, indexes of prices paid forcommodities and services.
Donald Bay, Administrator

2735 Risk Assessment and Cost Benefit Analysis Office
Office of the Chief Economist
1400 Independence Avenue SW   202-720-8022
S Building, Room 5248   E-mail: jcallahan@oce.usda.gov
Washington, DC 20250   http://www.usda.gov

ORACBA's primary role is to ensure that major regulations proposed by USDA are based on sound scientific and economic analysis.
Alwynelle Ahl, Director

2736 Rural Utilities Service
1400 Independence Avenue SW   202-720-9540
S Building, Room 4055   E-mail: bstockto@rus.usda.gov
Washington, DC 20250   http://www.rus.usda.gov

Modern utilities came to rural America through some of the most successful goverment initiatives in American history, carried out through the USDA working with rural cooperatives, non-profit associations, public bodies, and for-profitutilities. Today, they carry on this tradition helping rural utilities expand and keep their technology up to date, helping establish new and vital services such as distance learning and telemedicine.
Blaine D Stockton, Acting Administrator

2737 Saint Lawrence Seaway Development Corporation
US Department of Transportation
400 7th Street   202-366-0091
Suite 5424
Washington, DC 20590

Saint Lawrence Seaway Development Corporation operates and maintains the Great Lakes/St. Lawrence System, which encompasses the St. Lawrence River and the five Great lakes.
Albert Jacquez, Administrator

2738 Science Advisory Board Environmental Protection Agency
401 M Street SW   202-382-4126
Room 1145   E-mail: thomas.patrical@epamail.epa.gov
Washington, DC 20460   http://www.epa.gov/sab

The SAB was established by Congress to provide independent scientific and engineering advice to the EPA Administrator on the technical basis for EPA regulations. Expressed in terms of the current parlance of the risk assessment/riskmanagement paradigm of decision making, they deal with risk assessment issues and only that portion of risk management that deals strictly with the technical issues associated with various contorol options.
Dr Donald G Barnes, Staff Director

2739 Science and Technology Policy Office
Old Executive Office Bldg, Rm 424   202-456-7116
17th Street & Pennsylvania Avenue NW Fax: 202-456-6021
Washington, DC 20502   http://www.whitehouse.gov.ostp

The Science and Technology Policy Office serves as a source of scientific, engineering, and technological analysis and judgement for the President with respect to major policies, plans, and programs of the federal government. Incarrying out this mission, the office advises the President of scientific and technological considerations involved in areas of national concern, including the economy, national security, health, foreign relations, and the environment.
John Gibbons, Director

2740 Senate Committee on Appropriations
SD-128, Capitol Building   202-224-3471
Washington, DC 20510
Ted Stevens, Chair

2741 Senate Committee on Energy and Natural Resources
Dirksen Duilding   202-224-4971
Room SD-364
Washington, DC 20510
Frank Murkowski, Chair

2742 Senate Committee on Foreign Relations
439 Dirksen Senate Office Builing   202-224-4651
Washington, DC 20510-6225   Fax: 202-224-0836
E-mail: senator@biden.senate.gov
http://foreign.senate.gov

No foreign policy can be sustained without the informed consent of the American people. Our government works best when citizens care enough to become involved.
Richard Lugar, Chairman

2743 Smithsonian Institution Office of Environmental Awareness
S Dillon Ripley Center   202-357-4797
Room 3123   http://www.si.edu/newstart.htm
Washington, DC 20560

The Environmental Awareness Program (EAP), active at the Smithsonian Institution from April 1990 to September 1996, sought to address growing concern about the world's environment by gathering and disseminating information about awide range of issues. EAP worked with Smithsonian bureaus and outside groups to reach the general public and professional audiences nationally, through exhibitions, publications, conferences, and workshops.
Judith Gradwohl, Director

2744 Smithsonian Tropical Research Institute
Smithsonian Institution Research Department
900 Jefferson Drive SW   202-786-2817
Room 2207   http://www.stri.org
Washington, DC 20560
George Angehr, Liaison Officer

2745 State Energy Advisory Board: Office of the Secretary Department of Energy
1000 Independence Avenue SW 7A-257   202-586-3279
Washington, DC 20585   http://steab.org

STEAB subscribes to the concept of sustainable energy development, by which the nation seeks to meet current energy, economic, and environmental needs while preserving the means to meet the needs of future generations. Federal andstate investments in deploying developed energy efficient technologies have helped to free American capital for more productive uses. Further, the ongoing manufacture of new efficient and renewable technologies has lead us to capture additionalmarkets worldwide.
Rachael T Murphy

2746 Take Pride in America Advisory Board Department of the Interior
1849 C Street NW   202-208-3726
Washington, DC 20240   http://resources.lawlinks.com

The purposes of the program include the following: 1-to establish and maintain a public awareness campaign in cooperation with public and private organizations and individuals; 2- To conduct a national awards program to honor thoseindividuals and entities which, in the opinion of the Secretary of the Interior, have distinguished themselves in the above mentioned activities.
Clifton White, Director

2747 Task Force on Environmental Cancer & Heart & Lung Disease
401 M Street SW RD-683   202-260-5900
Washington, DC 20460   http://envirotext.eh.doe.gov

The Task Force shall: recommend a comprehensive research program to determine and qualify the relationship between environmental pollution and human cancer, heart, and lung disease; recommend comprehensive strategies to reduce oreliminate the risks of cancer or such other deseases associated with environmental pollution; recommend research and such other measures as may be appropriate to prevent or reduce the incidence of environmentally related cancer, heart, and lungdiseases.
Ken Sexton

2748 Technology Administration: National Institute of Standards & Technology
Technology Policy Office
Administration Building   301-975-2300
Room A1134   Fax: 301-869-8972
Gaithersburg, MD 20899 E-mail: raymond.kammer@nist.gov
http://www.nist.gov

The National Institute of Standards and Technology assists industry in the development of technology needed to improve product quality, modernize manufacturing processes, ensure product reliability, and facilitate rapidcommercialization of products based on new scientific discoveries. NIST's primary mission is to promote US economic growth by working with industry to develop and apply technology, measurements, and standards.
Raymond Kammer, Director

2749 US Agency for International Development Information Center
Ronald Reagan Building
Washington, DC 20523
202-712-4810
Fax: 202-216-3524
http://www.usaid.gov/environment
Aids in the development of urban environmental programs, forest conservation teams, watershed management and promotes improved pollution control.

2750 US Consumer Product Safety Commission
4340 E West Highway
Bethesda, MD 20814
800-638-2772
Fax: 301-504-0124
http://www.cpsc.gov
An independent federal regulatory agency. Helps keep American families safe by reducing the risk of injury or death from consumer products.
Hal Stratton, Chairman
Thomas Moore, Vice Chairman

2751 US Customs Service
Washington, DC 20229
202-927-1000
http://www.customs.ustreas.gov
Works to prevent illegal acts against endangered wildlife species.

2752 US Department of Agriculture
14th Street & Independence Avenue
Washington, DC 20250
202-720-8732
http://www.usda.gov
USDA Mission: Enhance the quality of life for the American people by supporting production of agriculture: ensuring a safe, affordable, nutritious, and accessible food supply; caring for agricultural, forest, and range lands;supporting sound development of rural communities; providing economic opportunities for farm and rural residents; expanding global markets for agricultural and forest products and services; and working to reduce hunger in America and throughout theworld.
Dan Glickman, Secretary

2753 US Department of Education
400 Maryland Avenue SW
Washington, DC 20202-0498
800-USA-LEAR
http://www.ed.gov
Richard Riley, Secretary

2754 US Department of Housing and Urban Development
451 7th Street SW
HUD Building
Washington, DC 20410
202-708-1112
http://www.hud.gov
Xavier Briggs, Deputy Assistant Secretary

2755 US Department of Housing and Urban Development
Office of Lead Hazard Control
451 7th Street SW
Washington, DC 20410
202-755-1785
Fax: 202-755-1000
http://www.hud.gov/lea/leahome.html
The office works to ensue that hazard controls are conducted in the safest, most cost-effective and efficient way possible to preserve our nation's stock of affordable housing while still ensuring that our children are properlyprotected.
Dolline Hatchett, COO

2756 US Department of Labor
200 Constitution Avenue NW
Washington, DC 20210
202-693-5000
http://www.dol.gov

The US Department of Labor is charged with preparing the American workforce for new and better jobs, and ensuring the adequacy of America's workplaces. It is responsible for the administration and enforcement of over 180 federalstatutes. These legislative mandates and the regulations produced to implement them cover a wide variety of workplace activities including protecting workers' wages, health and safety, employment and pension rights, promoting equal employmentopportunity.
Alexis Herman, Secretary

2757 US Department of Treasury
1500 Pennsylvania Avenue NW
Washington, DC 20220
202-622-2000
Fax: 202-622-6415
http://www.ustreas.gov
Stuart Eizenstat, Deputy Secretary

2758 US Department of the Air Force Major Air Commands
1700 N Moore Street
Suite 2300
Arlington, VA 22209-2802
703-696-5536
Jerry Cleaver, Conservation Manager

2759 US Department of the Army: Office of Public Affairs
1500 Army Pentagon
Washington, DC 20310
703-693-0677
http://www.army.mil
Vic Diersing, Conservation Team Leader

2760 US Department of the Interior: Bureau of Indian Affairs
1849 C Street NW
Washington, DC 20240
202-208-5116
http://www.doi.gov
The Bureau of Indian Affairs' mission is to enhance the quality of life, promote economic opportunity, and to carry out the responsibility to protect and improve the trust assets of American Indians, Indian tribes and Alaska Natives.We will accomplish this through the delivery of quality services, maintaining government-to-government relationships within the spirit of Indian self-determination.
Kevin Gover, Assistant Secretary

2761 US Department of the Navy: US Marine Corps
2 Navy Annex
Washington, DC 20380-4775
703-695-8332
Jim Omans, Head of Natural Resources

2762 US Environmental Protection Agency
Ariel Rios Building
1200 Pennsylvania Avenu NW
Washington, DC 20024
202-272-0167
http://www.epa.gov
The mission of the US Environmental Protection Agency is to protect human health and to safeguard the natural environment; air, water and land upon which life depends.

2763 US Environmental Protection Agency Office of Children's Health Protection
1200 Pennsylvania Avenue NW
Mail Code 1107A
Washington, DC 20004
202-564-2188
Fax: 202-564-2733
http://www.epa.gov
The mission of this office is to make the protection of children's environmental health a fundamental goal of public health and environmental protection in the US.
William H Sanderson, Acting Director
Joanne Rodman, Associate Director

2764 US Environmental Protection Agency: Information Resource Center
1200 Pennsylvania Avenue NW
3404T
Washington, DC 20460
202-566-0556
800-438-4318
Fax: 202-566-0562
E-mail: library-hq@epa.gov
http://www.epa.gov/natlibra/hqirc
Offers a range of professional services to help you find the information you need to get your job done. When appropriate, the IRC refers inquiries to the proper technical program or regional office.

2765 US Environmental Protection Agency: Clean Air Markets Division
1200 Pennsylvania Avenue NW   202-233-9150
Mail Code 6204N
Washington, DC 20460
Brian J McLean, Director

2766 US Environmental Protection Agency: Office of Air and Radiation
1200 12th Street NW   202-564-7433
Washington, DC 20004   Fax: 202-501-0826
                                http://www.epa.gov/oar/
Protects human health and the environment by preventing air
pollution and exposure to radiation through effective management
of public and private resources.

2767 US Environmental Protection Agency: Office of Environmental Justice
1200 Pennsylvania Avenue NW   202-564-2515
Washington, DC 20460   800-962-6215
                                Fax: 202-501-0740
E-mail: environmental-justice-epa@epamail.epa.org
            http://es.epa.gov/oeca/main/ej/index.html
Serves as a focal point for ensuring that communities comprised
mainly of people of color or low income receive protection under
environmental laws.

2768 US Forest Service
U.S. Dept. of Agriculture
PO Box 96090   202-205-1650
Washington, DC 20090   Fax: 202-205-1603
                                http://www.fs.fed.us.
Manages National Forests and Grasslands. Provides assistance
to private forest operators.

2769 US National Committee for Man and the Biosphere
PO Box 96090   202-205-0908
Washington, DC 20090   Fax: 202-205-1530
                                http://www.usmab.org
A network of agencies committed to solving social,
environmental and economic issues.

2770 US Nuclear Regulatory Commission
Reference Librarian   301-415-4737
Public Document Room (01F-13)   800-397-4209
Washington, DC 20555   Fax: 301-415-3548
                                E-mail: pdr@nrc.gov
                                http://www.nrc.gov
Ensures adequate protection of the public health and safety, the
common defense and security, and the environment in the use of
nuclear materials in the United States.

2771 USDA Forest Service: Watershed and Air Management
14th Street Building 201   202-205-1473
3rd Floor   E-mail: dlamb@fs.fed.us
Washington, DC 20250   http://www.fs.fed.us/clean
William McCleese, Director

2772 Wildlife and Marine Resources Section
Dept. of Justice/Environment/Natural Resources
Ben Franklin Station   202-272-4421
PO Box 7369
Washington, DC 20044-7369
James C Kilbourne, Chief

Alabama

2773 Agriculture and Industries Department
Pesticide Laboratory
1445 Federal Drive   334-240-7100
Montgomery, AL 36107-1100   Fax: 334-240-7190
                                http://www.agi.stste.al.us/
Danny Lecompte, Director

2774 Alabama Cooperative Extension System
109 Duncan Hall   334-844-4444
Auburn, AL 36849-5612   http://www.aces.edu
Operates as the primary outreach organization for the land-grant
function of Alabama A&M University and Auburn University.
Identifies statewide educational needs, audiences, and optimal
educational programs that are delivered through anetwork of
public and private partners supported by county, state, and federal
governments.
James Armstrong, Extension Wildlife Specialis

2775 Alabama Department of Agriculture and Industries
Richard Beard Building   334-240-7100
PO Box 3336   Fax: 334-240-7190
Montgomery, AL 36109   E-mail: commone@agi.state.al.us
Charles Bishop, Commissioner

2776 Alabama Department of Environmental Management
1400 Coliseum Boulevard   334-271-7700
PO Box 301463   Fax: 334-271-7950
Montgomery, AL 36130-1463
James W Warr, Director

2777 Alabama Forestry Commission
513 Madison Avenue   334-240-9300
Montgomery, AL 36104   Fax: 334-240-9390

2778 Conservation and Natural Resources Departmenet: Lands Division
N Union Street   334-242-3484
Montgomery, AL 36130   Fax: 334-242-0999
                        http://www.dcnr.state.al.us/Lands.htm
James Griggs, Director

2779 Conservation and Natural Resources Department: Game & Fish Division
64 N Union Street   334-242-3465
Room 702   http://www.dcnr.state.al.us/agfd
Montgomery, AL 36130
MN Corky Pughey, Director
Fred Harders, Assistant Director

2780 EPA: National Air and Radiation Environmental Laboratory
540 S Morris Avenue   334-270-3400
Montgomery, AL 36115-2601   Fax: 334-270-3454
                                http://www.epa.gov/narel/
The National Air and Radiation Environmental Laboratory is a
comprehensive environmental laboratory, and provides services to a wide range of clients, including other EPA offices,
Federal agencies, and, in somes cases, the privatesector. The
mission is the commitment to developing and applying the most
advanced methods for measuring environmental radioactivity
and evaluating its risk to the public.
Charles M Petko, Program Analyst

2781 United States Department of the Air Force Major US Installations
Maxwell AFB, AL   334-953-3892
Dennis Tates, Natural Resources Manager

Alaska

2782 Alaska Cooperative Fish and Wildlife Research Unit
University of Alaska
209 Irving I Building   907-474-7661
PO Box 757020   Fax: 907-474-7872
Fairbanks, AK 99775-7020

The Alaska Unit is a part of a nationwide program created to foster college-level research and graduate student training in support of science-based management of fish and wildlife, and their habitats. The Unit exists by cooperativeagreement between the AK Department of Fish and Game, University of Alaska Fairbanks, US Geological Survey, Wildlife Management Institute, and US Fish and Wildlife Service. The Unit is staffed by 5 USGS scientists, who are also research faculty.
F Margraf, Leader

2783 Alaska Department of Fish and Game
PO Box 25526   907-465-4100
Juneau, AK 99802-5526

Aims to manage, protect, maintain and improve the fish, game and aquatic plant resources of Alaska. The primary goals are to ensure that Alaska's renwable fish and wildlife resources and their habitats are conserved and managed on thesustained yield prinicpal, and the use of development of these resources are in the best interest of the economy and well-being of the people of the state.
Frank Rue, Commissioner
Rob Bosworth, Deputy Commissioner

2784 Alaska Department of Public Safety
PO Box 111200   907-465-4322
Juneau, AK 99811   Fax: 907-465-4362
     http://www.dps.state.ak.us

Bill Tandeske, Commissioner

2785 Alaska Division of Forestry: Central Office
550 W 7th Avenue   907-269-8463
Suite 1450   Fax: 907-269-8931
Anchorage, AK 99501-3566

Jeff Jahnke, State Forester
Dean Brown, Deputy State Forester

2786 Alaska Division of Forestry: Coastal Region Office
101 Airport Road   907-465-2491
Palmer, AK 99645   Fax: 907-586-3113

Jim Eleazer, Regional Forester

2787 Alaska Division of Forestry: Delta Area Office Office
Mi. 267.5 Richardson Highway   907-895-4225
PO Box 1149   Fax: 907-895-4934
Delta Junction, AK 99737

Al Edgren, Area Forester

2788 Alaska Division of Forestry: Fairbanks Area Office
3700 Airport Way   907-451-2600
Fairbanks, AK 99709-4699   Fax: 907-451-2633

Marc Lee, Area Forester

2789 Alaska Division of Forestry: Haines Area Office
Gateway Building   907-766-2120
PO Box 263   Fax: 907-766-3225
Haines, AK 99827

Roy Josephson, Area Forester

2790 Alaska Division of Forestry: Kenai/Kodiak Area Office
42499 Sterling Highway   907-262-4124
Soldotna, AK 99669   Fax: 907-262-6390

Jim Peterson, Area Forester

2791 Alaska Division of Forestry: Ketchikan Area Office
2030 Sea Level Drive   907-225-3070
#217   Fax: 907-247-3070
Ketchikan, AK 99801

Mike Curran, Area Forester

2792 Alaska Division of Forestry: MacGrath Field Office
Box 130   907-524-3010
MacGrath, AK 99627   Fax: 907-524-3932

Judy Reese, Fire Management Officer

2793 Alaska Division of Forestry: Mat-Su/Southwest Area Office
101 Airport Road   907-761-6300
Palmer, AK 99645   Fax: 907-761-6319

Ken Bullman, Area Forester

2794 Alaska Division of Forestry: Northern Region Office
3700 Airport Way   907-451-2660
Fairbanks, AK 99709-4699   Fax: 907-451-2690

Chris Maisch, Regional Forester

2795 Alaska Division of Forestry: State Forester's Office
550 W 7th Avenue   907-269-8474
Suite 1450   Fax: 907-269-8931
Anchorage, AK 99501-3566

Jeff Jahnke, State Forester

2796 Alaska Division of Forestry: Tok Area Office
Mile 123 Glenn Highway   907-883-5134
Box 22   Fax: 907-883-5135
Tok, AK 99780

2797 Alaska Division of Forestry: Valdez/Copper River Area Office
Mile 110 Richardson Highway   907-822-5534
Box 185   Fax: 907-822-5539
Glennallen, AK 99588

Martin Maricle, Area Forester

2798 Alaska Health Project
218 E 4th Avenue   907-276-2864
Anchorage, AK 99501   Fax: 907-279-3089

Daniel Middaugh, Executive Director

2799 Alaska Oil and Gas Conservation Commission
333 W 7th Avenue   907-279-1433
Suite 100   Fax: 907-276-7542
Anchorage, AK 99501 http://www. state.ak us./local/ak pages/
     ADMIN/ogc/homeogc.html

Protecting the oil and gas of Alaska.
John K ' Norman, Chairman

2800 Alaska Resource Advisory Council
Bureau of Land Management
822 W 7th Avenue   907-271-5555
Suite 13   Fax: 907-272-3430
Anchorage, AK 99513   http://www.ak.blm.gov/rac/

A statewide resource advisory council which advises BLM on land management issues for 90 million acres of public lands in Alaska. Membership is comprised of representatives from industry, conservation, recreation, Alaska Nativeorganizations and the public at large. The council meets quarterly.
Teresa McPherson, Council Coordinator

2801 Anchorage Office: Alaska Department of Environmental Conservation
555 Cordova Street   907-269-7500
Anchorage, AK 99501-2617   Fax: 907-269-7600
     E-mail: website@dec.state.ak.us
     http://www.state.ak.us/dec/dsps

The people and industries that operate in our state have both the corporate conscience and the technical ability to work with us on constuctive solutions to basic environmental management and public health issues. We anticipate,collaborate, negotiate, educate and communicate to address the most important environmental and public health risks to Alaska and Alaskans. Investigation, legislation, regulation and litigation are available tools, but not the first tools of choice.
Sally Smith, Environmental Tech

2802 Bethel Office: Alaska Department of Environmental Conservation
PO Box 228   907-543-3215
Bethel, AK 99559   Fax: 907-543-3216
     E-mail: randy_dowd@dec.state.ak.us
     http://state.ak.us/dec

Randall Dowd, Environmental Specialist

2803 Bristol Bay Office: Alaska Department of Environmental Conservation
PO Box 438   907-246-6636
King Salmon, AK 99613   Fax: 907-246-6610

2804 Cooperative Extension Service: University of Alaska Fairbanks
2221 E Northern Lights Boulevard   907-786-6300
Suite 118   Fax: 907-786-6312
Anchorage, AK 99508-4143   E-mail: dffes@uaa.alaska.edu
                            http://www.uaf.edu/coop-ext
Fred Sorensen, Land Resources Program Chair

2805 Department of Justice, Environment and Natural Resources Division
Alaska Field Office
801 B Street   907-271-5452
Suite 504   Fax: 907-271-5827
Anchorage, AK 99501
This office conducts affirmative and defensive litigation on behalf of the United States in judicial actions concerning enforcement of federal environmental laws and management of federal lands within Alaska.

2806 Fairbanks Office: Alaska Department of Environmental Conservation
610 University Avenue   907-451-2360
Fairbanks, AK 99709   Fax: 907-451-2188

2807 Food Safety Laboratory: Alaska Department of Environmental Conservation
500 S Alaska Street   907-745-3236
Suite A   Fax: 907-745-8125
Palmer, AK 99645

2808 Juneau Office: Alaska Department of Environmental Conservation
410 Willoughby Avenue   907-465-5010
Suite 303   Fax: 907-465-5097
Juneau, AK 99801-1795

2809 Kenai Office: Alaska Department of Environmental Conservation
43335 Kalifornsky Beach Road   907-262-5210
Suite 11, Red Diamond Center   Fax: 907-262-2294
Soldotna, AK 99669

2810 Ketchikan Office: Alaska Department of Environmental Conservation
540 Water Street   907-225-6200
Ketchikan, AK 99901   Fax: 907-225-0620

2811 Kodiak Office: Alaska Department of Environmental Conservation
PO Box 515   907-486-3350
Kodiak, AK 99615   Fax: 907-486-5032

2812 Mat-Su Office: Alaska Department of Environmental Conservation
PO Box 871064   907-376-5038
Wasilla, AK 99687   Fax: 907-376-2382

2813 Natural Resources Department Public Affairs Information Office
3601 C Street Suite 1210   907-762-2452
PO Box 107005http://www.dnr.state.ak.us/pic/dnrdirectory.htm
Anchorage, AK 99510-7005
Carol D Larsen, Public Affairs Info Officer

2814 Nome Office: Alaska Department of Environmental Conservation
PO Box 966   907-443-3294
Nome, AK 99762   Fax: 907-443-7498

2815 North Slope Office: Alaska Department of Environmental Conservation
Box 140   907-569-2215
Deadhorse, AK 99734   Fax: 907-569-0231

2816 Northern Alaska Advisory Council Bureau of Land Management
Department of the Interior
1150 University Avenue   907-474-2200
Fairbanks, AK 99709   http://aurora.ak.blm.gov
William J Robertson, Designated Federal Employee

2817 Palmer Office: Alaska Department of Environmental Conservation
500 S Alaska Street   907-745-3236
Suite A   Fax: 907-745-8125
Palmer, AK 99645

2818 Sitka Office: Alaska Department of Environmental Conservation
901 Halibut Point Road   907-747-8614
Sitka, AK 99835   Fax: 907-747-7419

2819 State Chemistry Laboratory: Alaska Department of Environmental Conservation
10107 Brentwood Street   907-790-2169
Juneau, AK 99801   Fax: 907-790-2451

2820 Subsistance Resource Commission Cape Krusenstern National Monument
National Park Service
Box 1029   907-683-2294
Kotzebue, AK 99752   http://www.nps.gov/cakr
Bob Gerhard, Designated Fedeeral Employee

2821 Subsistence Resource Gates of the Artic National Park
National Park Service
National Park Service-Fairbanks HQ 907-457-5752
201 1st Avenue   Fax: 907-455-0601
Fairbanks, AK 99701   http://www.nps.gov/gaar
Dave Mills, Superintendent
Fred Anderson, Designated Federal Employee

2822 Subsistence Resource Commission Kobuk Valley National Park
National Park Service
373 A 2nd Ave   907-442-3890
Kotzebue, AK 99752
Bob Gerhard, Designated Federal Employee

2823 United States Department of the Army US Army Corps of Engineers
PO Box 898   907-753-2520
Anchorage, AK 99506-0898
Design and constructs military projects for the Army, Air Force, civil works and water resources development projects for coastal communities. Conducts military Real Estate transactions, is responsible for Emergency Operationsinvolving national emergency and natural disaster, and regulates development in navigable waters, and placement of fill material in waters and wetlands.

2824 Valdez Office: Alaska Department of Environmental Conservation
213 Meals Avenue, Room 17   907-835-4698
Po Box 1709   Fax: 907-835-2429
Valdez, AK 99686

Arizona

2825 Arizona Department of Agriculture: Animal Services Division
1688 West Adams — 602-542-6309
Phoenix, AZ 85007 — Fax: 602-542-3244
http://www.agric.state.az.us

Sheldon Jones, Director

2826 Arizona Department of Environmental Protection
University of Arizona
Biological Sciences E — 520-621-1959
Room 104 — Fax: 520-621-8801
Tucson, AZ 85721
Scott Bonar, Leader

2827 Arizona Department of Environmental Quality
1110 West Washington Street — 602-771-2300
Phoenix, AZ 85007 — 800-234-5677
http://www.azdeq.gov/index.html
The Arizona Department of Environmental Quality was established in 1987 to preserve, protect and enhance the environmental and public health through the maintenance of air, land and water resources. The department oversees compliance with state and federal environmental regulations and works with industry and local governments.
Steve Owens, Director

2828 Arizona Department of Health Services: Office of Environmental Health
1740 W Adams Street — 602-542-1025
Phoenix, AZ 85007 — Fax: 602-542-1062
http://www.hs.state.az.us
Childhood Lead Poisoning Program. Geographic focus: Arizona No membership Other organizational activities (not directed specifically toward children): advocacy, education, organizing research.
Kristina Schaller, Epidemiolgist

2829 Arizona Environmental Quality Department
1110 West Washington Street — 602-771-2300
Phoenix, AZ 85007 — 800-234-5677
http://www.azdeq.gov/index.html
Preserves, protects and enhances the environment and public health through the maintenance of air, land and water resources. The department oversees compliance with state and federal environmental regulations and works with industry and local governments. ADEQ issues permits, administers air quality improvement programs and regulates vehicular emissions. ADEQ offers services to enhance compliance, educate the public and develop innovative solutions to potential enviromental hazards.

Steve Owens, Director

2830 Arizona Game & Fish Department
2221 West Greenway Road — 602-942-3000
Phoenix, AZ 85023-4399 — http://www.gf.state.az.us
Aims to conserve, enhance and restore Arizona's diverse wildlife resources and habitats through aggressive protection and management programs, and to provide wildlife resources and safe watercraft and off-highway vehicle recreation for the enjoyment, appreciation, and use by present and future generations.

2831 Arizona Game & Fish Department: Region I
HC 66 — 520-367-4281
Box 57201
Pinetop, AZ 85935

2832 Arizona Game & Fish Department: Region II
3500 South Lake Mary Road — 520-774-5045
Flagstaff, AZ 85935

2833 Arizona Game & Fish Department: Region III
5325 North Stockton Hill Road — 520-692-7700
Kingman, AZ 86401

2834 Arizona Game & Fish Department: Region IV
9140 East 28th Street — 520-342-0091
Yuma, AZ 85365

2835 Arizona Game & Fish Department: Region V
555 North Greasewood Road — 520-628-5376
Tucson, AZ 85745

2836 Arizona Game & Fish Department: Region VI
7200 East University Road — 602-981-9400
Mesa, AZ 85207

2837 Arizona Geological Survey
416 W Congress Street — 520-770-3500
Suite 100 — Fax: 520-770-3505
Tucson, AZ 85701 — http://www.azgs.state.az.us
Our mission is to inform and advise the public about the geologic character of Arizona in order to foster understanding and prudent development of the State's land, water, mineral and energy resources.
Rose Ellen McDonnell, Assistant Director Geologist

2838 Arizona State Parks
1300 West Washington — 602-542-4174
Phoenix, AZ 85007 — 800-285-3703
Fax: 602-542-4180
E-mail: feedback@pr.state.az.us
http://www.pr.state.az.us

Ken Travous, Executive Director

2839 Arizona State Parks Board
1300 West Washington Avenue — 602-542-4174
Phoenix, AZ 85007 — 800-285-3703
Fax: 602-542-4180
http://www.azstateparks.com

Ken Travous, Director

2840 Environmental and Analytical Chemistry Laboratory
Arizona Department of Health — 602-542-6108
1520 W Adams — Fax: 602-364-0281
Phoenix, AZ 85007 — http://www.hs.state.az.us
State public health laboratory both in chemistry and microbiology. Supports investigations into environmental contamination by analyzing water, soil, air, hazardous materials, food and miscellaneous items for the presence of hazardous and toxic chemicals. Microbiology tests for pathogens and/or indicator organisms.
Patricia A Adler, Chemistry Manager
Wesley B Press, Bureau Chief

2841 Gila Box Riparian National Conservation Area BLM Safford District Office
425 East 4th Street — 520-428-4040
Safford, AZ 85546 — http://www.az.blm.gov
There are more than 14 million acres of public lands in Arizona that people have put in our trust. It's an awesome responsibility, and one that we take very seriously. We don't try to do it alone. Every day, we work with people to help make sure we are doing what is right for Arizona's environment, wildlife, culture, and history... for the people who rely upon the land to earn a living or to manufacture the things which make our lives a little easier...and most importantly, for Arizona's future.

2842 Phoenix District Advisory Council Bureau of Land Management
21605n 7th Avenue — 623-580-5500
Phoenix, AZ 85027-2099 — Fax: 623-580-5580
http://www.az.blm.gov
Manages public lands and resources in central Arizona, and supports related statewide initiatives and functions to sustain their health, diversity and productivity while providing for customer service and meeting public demand resulting from the expanding Phoenix metropolitan area and growth of adjoining communities.

Arkansas

2843 Arkansas Department of Parks and Tourism
One Capitol Mall
Little Rock, AR 72201
501-682-7777
Fax: 501-682-1364
http://www.arkansas.com

Richard Davies, Executive Director

2844 Arkansas Fish and Game Commission
#2 Natural Resources Drive
Little Rock, AR 72205
501-223-6300
800-364-4263
Fax: 501-223-6447
http://www.agfc.com

2845 Arkansas Natural Heritage Commission
1500 Tower Building
323 Center Street
Little Rock, AR 72201
501-324-9619
Fax: 501-324-9618
E-mail: info@arkansasheritage.org
http://www.naturalheritage.org

Raymond Abramson, Chairman

2846 Arkansas State Plant Board
1 Natural Resources Drive
Po Box 1069
Little Rock, AR 72203
501-225-1598
Fax: 501-225-3590
http://www.natuallyarkansas.org

Gerald King, Director
Darryl Little, Assistant Director

2847 Department of Environment Quality
8001 National Drive
Po Box 8913
Little Rock, AR 72219-8913
501-682-0744
Fax: 501-682-0798
http://www.adeq,state.ar.us

Richard Weiss, Director

2848 Game and Fish Commission Wildlife Management Division
2 Natural Resources Drive
Little Rock, AR 72205
501-223-6359
Fax: 501-223-6452
http://www.agfc.com

Dedicated to managing wildlife in the state of Arkansas.
Doyle Shook, Chief

2849 Pine Bluff Cooperative Fishery Research Project
University of Arkansas at Pine Bluf
1200 N University Street
PO Box 4005
Pine Bluff, AR 71611-2799
501-543-8165
http://www.nwrc.usgs.gov

Agricultural Experiment Station that provides research services.
Steve Lochmann, Project Leader

California

2850 American Cetacean Society
PO Box 1391
San Pedro, CA 90733-1391
310-548-6270
Fax: 310-548-6950
E-mail: info@acsonline.org
http://www.acsonline.org

A non profit organization that is the oldest whale conservation group in the world. Founded to protect whales, dolphins, porpoises, and their habitats through public education, research grants, and conservation actions.

Patty Geary, President
Mason Weinrich, Vice President

2851 Bakersfield Field Office
Bakersfield Field Office
3801 Pegasus Drive
Bakersfield, CA 93308
661-391-6000
Fax: 661-391-6041
http://www.ca.blm.gov

Ron Huntsinger, Field Manager

2852 California Department of Education Office of Environmental Education
721 Capitol Mall
PO Box 944272
Sacramento, CA 94244-2720
916-322-9503
Fax: 916-322-9360
E-mail: www.cde.ca.gov
Bill Andrews, Consultant

2853 California Department of Fish and Game
1416 9th Street
Sacramento, CA 95814
916-653-7664
Fax: 916-653-1856
http://www.dfg.ca.gov

Manages California's diverse fish, wildlife, and plant resources, and the habitats upon which they depend, for their ecological values and for their use and enjoyment by the public.

2854 California Department of Water Resources
1416-9th Street
Room 1104-1
Sacramento, CA 95814
916-653-6192
Fax: 916-653-4684
http://www.dwr.water.ca.gov

2855 California Desert District Advisory Council Bureau of Land Management
California State Office
2800 Cottage Way
Suite W-1834
Sacramento, CA 95825-1886
916-978-4400
http://www.ca.blm.gov

Henri Bisson, Designated Fed. Empl.

2856 California Environmental Protection Agency
PO Box 2815
Sacramento, CA 95812
916-445-3846
Fax: 916-445-6401
http://www.calepa.ca.gov

Winston Hickox, Secretary

2857 California Institute of Public Affairs
PO Box 189040
Sacramento, CA 95818
916-442-2472
Fax: 916-442-2478
http://www.interenvironment.org

A forum for policy dialogue and research on California and international environmental issues. Publishes the online World Directory of Environmental Organizations.
Thaddeus C Trzyna, President

2858 California Pollution Control Financing Authority
915 Capitol Mall, Room 457
PO Box 942809
Sacramento, CA 95814-0001
916-654-5610
Fax: 916-657-4821

Kirsten Snow Spalding, Executive Director

2859 Department of Agriculture: Forest Service, Pacific Southwest Region
630 Sansome Street
San Francisco, CA 94111
415-705-2884
Fax: 415-705-2836

2860 Department of Justice Environment and Natural Resources
San Francisco Office
301 Howard Street
Room 870
San Francisco, CA 94105-6605
415-744-6491

2861 Department of the Interior: National Parks Pacific West Region
600 Harrison Street
Suite 600
San Francisco, CA 94107-1372
415-427-1300
http://www.nps.gov/legacy/regions.html

2862 Department of the Interior: Water Resources Western
Placer Hall
6000 J Street
Sacramento, CA 95819
916-278-3000
800-833-3661
Fax: 916-278-3070
E-mail: dc_ca@usgs.gov
http://http://water.wr.usgs.gov

The US Geological Survey has the principal responsibility within the Federal Government to provide the hydrologic information and understanding needed by others to achieve the best use and management of the nation's water resources. Toaccomplish this mission, the Water Resources Division cooperates with state, local and other federal agencies.

2863 Energy Resources Conservation & Development Commission
1516 9th Street
MS-29
Sacramento, CA 95814-5504
916-654-5000
Fax: 916-654-4420
http://www.energy.ca.gov
Charles R Imbrecht, Chairman

2864 Environmental Protection Agency Region IX
75 Hawthorne Street
San Francisco, CA 94105-000
415-744-1702
E-mail: mailto:9.info@epa.gov
http://www.epa.gov/region09/
Region 9 covers Arizona, California, Hawaii, Nevada, the Pacific Islands subject to US law, and approximately 140 Tribal Nations. We work together with state, local, and tribal governments in the region to carry out the nationsenvironmental laws.
Felicia Marcus, Regional Administrator

2865 Environmental Protection Office: Hazard Identification
Environmental Health Hazard Assessment Office
2151 Berkely Way Annex 11
Berkeley, CA 94704
510-540-3063
The mission of the Office of Environmental Health Hazard Assessment (OEHHA) is to protect and enhance public health and the environment by objective scientific evaluation of risks posed by hazardous substances.

2866 Environmental Protection Office: Toxic Substance Control Department
400 P Street 4th Floor
PO Box 806
Sacramento, CA 95812-0806
916-323-9723
Fax: 916-323-3215
The Department's mission is to restore, protect and enhance the environment, to ensure public health, environmental quality and economic vitality by regulating hazardous waste, conducting and overseeing cleanups, and developing andpromoting pollution prevention.
William F Soo Hoo, Director

2867 Golden Gate National Recreation Area
Fort Mason
Building 201
San Francisco, CA 94123-0022
415-556-0560
Fax: 415-561-4750
E-mail: George_Su@nps.gov
The Golden Gate National Recreation Area (GGNRA) is the largest urban national park in the world. The total park area is 74,000 acres of land and water. Approximately 28 miles of coastline line within its boundaries. It is nearly twoand one-half times the size of San Francisco.

2868 Inter-American Tropical Tuna Commission
8604 La Jolla Shores Drive
La Jolla, CA 92037-1508
858-546-7100
Fax: 858-546-7133
E-mail: info@iattc.org
http://www.iattc.org/homeeng.htm
The IATTC, established by international convention in 1950, is responsible for the conservation and management of fisheries for tunas and other species taken by tuna-fishing vessels in the eastern Pacific Ocean. The IATTC also hassignificant responsibilities for the implementation of the International Dolphin Conservation Program (IDCP), and provides the Secretariat for that program.
Robin Allen, Director

2869 Klamath Fishery Management Council US Fish & Wildlife Service
1829 South Oregon Street
Yreka, CA 96097-1006
530-842-5763
Fax: 530-842-4517
E-mail: www.pacific.fws.gov/yreka
Phil Detrich, Project Leader

2870 Mono Basin National Forest Scenic Area Advisor Forest Service
PO Box 429
Lee Vining, CA 93541
760-647-3044
Located to the immediate east of Yosemite National Park, Mono is the westernmost basin of the Basin and Range province, which stretches across western North America between the Rocky Mountains and Sierra Nevada. In the heart of theBasin lies the strange and majestic Mono Lake, a vast inland sea nestled amidst the 13,000-foot peaks of the High Sierras to the west, the ancient volcanic Bodie Hills to the north, rolling oceans of sagebrush to the east, and the Mono Crater to thesouth.
JoEllen Keil, Acting District Ranger

2871 Native American Heritage Commission
915 Capitol Mall
Room 364
Sacramento, CA 95814
916-653-4082
Fax: 916-657-5390
E-mail: nahc@pacbell.net
http://www.nahc.ca.gov
The mission of the Native American Heritage Comm. is to provide protection to Native American burials from vandalism and inadvertent destruction, provide a procedure for the notification of most likely descendents regarding thediscovery of Native American human remains and associated grave goods, bring legal action to prevent severe and irreparable damage to sacred shrines, ceremonial sites, sanctified cemeteries and place of worship on pub. property, and maintain aninventory of sacred places.

Larry Myers, Executive Secretary
Carol Gaubatz, Program Analyst

2872 Pesticide Regulation, Environmental Monitoring and Pesticide Management
Environmental Protection Office
1001 I Street
PO Box 4015
Sacramento, CA 95812
916-445-4300
Fax: 916-324-1452
http://www.cdpr.ca.gov
Cal/EPA's Department of Pesticide Regulation (DPR) protects human health and the environment by regulating pesticide sales and use, and by fostering reduced-risk pest management.

2873 Resources Agency: California Coastal Commission
45 Fremont Street
Suite 2000
San Francisco, CA 94105-2219
415-904-5250
Fax: 415-904-5400
http://ceres.ca.gov/cra/
The California Resources Agency is responsible for the conservation, enhancement, and management of California's natural and cultural resorces, including land, water, wildlife, parks, minerals and historic sites. The Agency is composedof departments, boards, conservancies, commissions, and programs.
John Dixon

2874 Resources Agency: California Conservation Corps
1719 24th Street
Sacramento, CA 95816
916-341-3100
800-952-5627
Fax: 916-323-8922
http://www.ccc.ca.gov
Engages young men and women in meaningful work, public service educational activities that assist them in becoming more responsible citizens, while protecting and enhancing California's environment, human resources and communities.
Al Arambury, Director

2875 Resources Agency: State Coastal Conservancy
1330 Broadway Suite 1100
Oakland, CA 94612-2530
510-286-1015
Fax: 510-286-0470
E-mail: dwayman@scc.ca.gov
http://www.coastalconservancy.ca.gov
The coastal conservancy is a state agency that works with the people of california to protect and improve the coast and San Francisco Bay. The conservancy has opened over 70 miles of coast and bay shores to the public and has helpedto preserve over 60,000 acres of wetlands, wildlife habitat, parks, and farmland.
Quarterly Magazine
Samuel Schuchat, Executive Officer
Dick Wayman

2876 Southwestern Low-Level Radioactive Waste Commission

PO Box 277727   916-448-2390
Sacramento, CA 95827-7727   Fax: 815-361-3848
E-mail: mailto:swllrwcc@swllrwcc.org
http://www.swllrwcc.org

The Southwestern Low-Level Radioactive Waste Commission is the governing body for the Southwestern Low-Level Radioactive Waste Disposal Compact, consisting of Arizona, California, North Dakota, and South Dakota. Created by public law100-712 in 1988, its key duties include controlling the importation and exportation of low-level waste into and out of the region. The Commission has no authority over disposal facility siting, but can make recommendations and comments to ensure safedisposal.

Donald Womeldorf, Executive Director

2877 Susanville District Advisory Council Bureau of Land Management

2950 Riverside Drive   530-257-5381
Susanville, CA 96130   Fax: 530-256-3539
Herrick E Hanls, Designated Federal Employee

2878 Toxic Substances: Hazardous Waste Management Division

Environmental Protection Office
1001 I Street   916-445-3846
Sacramento, CA 95814   Fax: 916-324-0908

Regulates hazardous waste through its permitting, enforcement and unified program activities. HWMP maintains the US Environmental Protection Agency authorization to implement the Resource Conservation and Recovery Act program inCalifornia and develops regulations, policies, guidance and technical assistance/training to assure the safe storage, treatment, transportation and disposal of hazardous wastes.
Ed Lowry, Director

2879 Toxic Substances: Pollution Prevention and Environmental Technology

Environmental Protection Office
1001 I Street   916-445-3846
Sacramento, CA 95814   Fax: 916-324-0908

Responsible for promoting pollution prevention and environmental technologies through statewide leadership, demonstration projects, technology certification, assistance to technology developers, local government and regulatoryagencies.

2880 Trinity River Basin Fish & Wildlife Task Force Mid-Pacific Region

2800 Cottage Way   916-979-2482
MP-150
Sacramento, CA 95825-1893
Chip Bruss, Secretary

2881 United States Department of Agriculture Research Education and Economics

800 Buchanan Street   510-559-6060
ARS Pacific West Office: USDA   Fax: 510-559-5779
Albany, CA 94710
Dwayne R Buxton, Area Director

2882 United States Department of the Army US Army Corps of Engineers

333 Market Street   415-977-8600
US Army Engineer District   Fax: 415-977-8316
San Francisco, CA 64105-2195

2883 United States Department of the Interior United States Fish and Wildlife Service

2800 Cottage Way   916-414-6464
Room W-2606   Fax: 916-414-6486
Sacramento, CA 95825
Steve Thompson, Manager
Ken McDermond, Deputy Manager

Colorado

2884 Bureau of Land Management

Department of the Interior
2815 H Road   970-244-3000
Grand Junction, CO 81506   Fax: 970-244-3083

Manages 8.3 million acres of public lands in Colorado. These lands are managed for a multitude of uses including, but not limited to, recreation, mining, wildlife habitat and grazing. Along with these 8.3 million acres, BLM oversees27.3 million subsurface acres for mineral development.
Catherine Robertson, Field Manager

2885 Bureau of Land Management: Little Snake Field Office

Little Snake Field Office   970-826-5000
455 Emerson Street   Fax: 970-826-5002
Craig, CO 81625-1129   http://www.blm.com
John Husband, Director

2886 Canon City District Advisory Council

3170 E Main Street   719-269-8500
Canon City, CO 81212   Fax: 719-269-8599
http://www.blm.gov/education/state/colorado.html
Ken Smith, Contact

2887 Cheyenne Mountain Zoological Park

4250 Cheyenne Mountain Zoo Road   719-633-9925
Colorado Springs, CO 80906   Fax: 719-633-2254
http://www.cmzoo.org

2888 Colorado Department of Agriculture

700 Kipling Street   303-239-4100
Suite 4000   Fax: 303-239-4125
Lakewood, CO 80215   http://www.ag.state.co.us
Don Ament, Commisioner

2889 Colorado Department of Natural Resources

1313 Sherman Street   303-866-3311
Room 718   800-536-5308
Denver, CO 80203   Fax: 303-866-2115
http://www.dnr.state.co.us
Greg Walcher, Executive Director

2890 Colorado Department of Natural Resources- Division of Water Resources

1313 Sherman Street   303-866-3581
Room 818   Fax: 303-866-3589
Denver, CO 80203   http://www.water.state.co.us

The Colorado Division of Water Resources is an agency of the State of Colorado, Department of Natural Resources, operating under the direction of specific state stautes, court decrees, and interstate compacts. The DWR is empowered toadminister all surface and ground water rights throughout the state and ensure that the doctrine of prior appropiation is enforced.
Harold Simpson, State Engineer

2891 Colorado Department of Public Health Environment Consumer Protection Division

4300 Cherry Creek Drive South   303-692-3620
Denver, CO 80246-1530   Fax: 303-753-6809
http://www.cdphe.state.co.us/cp/about.html
The Consumer Protection Division assumes the responsiblity for protecting Colorado residents and visitors by prevention of a wide array of health hazards.
Barbara Hruska, Division Director

2892 Colorado Department of Public Health and Environment

4300 Cherry Creek Drive South   303-692-2000
Denver, CO 80246-1530   800-886-7689
http://www.cdphe.state.co.us

2893 Colorado State Forest Service
Colorado State University
5060 Campus Delivery   970-491-6303
Fort Collins, CO 80523-5060   Fax: 970-491-7736
http://www.nationalwoodlands.org/nwoa/state.asp
Jeff Jahnke, State Forester

2894 Department of Commerce National Oceanic & Atmospheric Laboratories
325 Broadway   303-497-6000
Boulder, CO 80303   Fax: 303-497-6951
E-mail: hhorton@erl.noaa.gov

2895 Environmental Protection Agency Region VIII (CO, MT, ND, SD, UT, WY)
999 18th Street, Suite 300   303-312-6312
Denver, CO 80202-2466   800-227-8917
http://www.epa.gov/region8/
To restore and protect the ecological integrity of the mountains, plains and deserts and to protect the health of their inhabitants.
William Yellowtail, Regional Administrator

2896 Governors Office of Energy, Management and Conservation: Colorado
225 E 16th Avenue   303-866-2100
Suite 650   Fax: 303-866-2930
Denver, CO 80202-4613   E-mail: oemc@state.co.us
http://www.state.co.us/oemc
Supports cost-effective programs, grants and partnerships that benefit Colorado's economic and natural environment.
Rick Grice, Director

2897 Minerals Management Service/Minerals Revenue Management
PO Box 25165 Ms 300B3   303-231-3899
Denver, CO 80225   Fax: 303-231-3194
http://www.mms.gov
Donald T Sant, Deputy Associate Director

2898 National Park Service Natural Resources Department
1201 Oak Ridge Drive Room 250   970-225-3500
Fort Collins, CO 80525
Dan Kimbah, Acting Director

2899 Natural Resources Department: Air Quality Division
Department of the Interior
PO Box 25287   303-969-2070
Denver, CO 80225   Fax: 303-969-2822
E-mail: christine_shaver@nps.gov
Christine Shaver, Chief

2900 Natural Resources Department: Oil & Gas Conservation Commission
1120 Lincoln Street Suite 801   303-894-2100
Denver, CO 80203   Fax: 303-866-2115
E-mail: dnr.ogcc@state.co.us
http://www.oil-gas.state.co.us
Rich Griebling, Director

2901 Natural Resources Department: Wildlife Division
6060 Broadway   303-291-7208
Denver, CO 80216   Fax: 303-294-0874
Manages the state's 960 wildlife species. Regulates hunting and fishing activities by issuing licenses and enforcing regulations. Conducts research to improve wildlife management activities, provides technical assistance to private andother land owners concerning wildlife and habitat management and develops programs to protect and recover threatened and endangered species.
Perry Olson, Director

2902 Office of Surface Mining Reclamation & Enforcement
1999 Broadway   303-844-1400
Suite 3320   Fax: 303-844-1546
Denver, CO 80202   http://www.wrcc.osmre.gov/

2903 Rocky Mountain Low-Level Waste Radioactive Waste Board
1675 Broadway   303-825-1912
Suite 1400
Denver, CO 80202
Leonard Slosky, Executive Director

2904 State Forest Service
Colorado University   970-491-6303
Fort Collins, CO 80532   Fax: 970-491-7736
http://www.nationalwoodlands.org/nwoa/state.asp
James Hubbard, State Forester

2905 United States Department of the Air Force Major Air Commands
150 Vandenberg Street   719-554-9915
Suite 1105   Fax: 719-554-3849
Peterson AFB, CO 80914
Stan Rogers, Natural Resources Manager

2906 United States Department of the Interior United States Fish and Wildlife Service
134 Union Boulevard
PO Box 25486
Denver, CO 80225
Ralph Morgenweck, Regional Director

2907 United States Forest Service: United States Department of Agriculture
240 W Prospect Street   970-498-1126
Rocky Mountain Research Station   Fax: 970-498-1010
Fort Collins, CO 80526-2098
Denver Burns, Director

Connecticut

2908 Connecticut Department of Agriculture
765 Asylum Avenue   860-713-2503
Hartford, CT 06105   Fax: 860-713-2516
E-mail: ctdeptag@po.state.ct.us
http://www.state.ct.us/doag
Shirley Ferris, Commissioner

2909 Connecticut Department of Environmental Protection
Department of Environmental Protection
79 Elm Street   860-424-4180
Hartford, CT 06106-5127   http://dep.state.ct.us

2910 Connecticut Department of Public Health
410 Capitol Avenue   860-509-7745
PO Box 34308   Fax: 860-509-7785
Hartford, CT 06134-0308   http://www.state.ct.us
Has long recognized the adverse public health impact of environmental sources of lead in many of Connecticut's childern. Established dedicated staff to evaluate these environmental sources and began funding local programs in the1970's. The Childhood Lead Posioning Prevention Program has continued to be active in addressing this issue by implementing additional state and community programs, especially in towns that have been identified as high risk.
Renee Coleman-Mitchell, Program Director
Eileen Boulay, Case Manager

District of Columbia

2911 District of Columbia State Extension Services
4200 Connecticut Avenue   202-274-7115
Building 352, Suite 322   Fax: 202-274-7130
Washington, DC 20008

Delaware

2912 Delaware Association of Conservation Districts
PO Box 242   302-739-4411
Dover, DE 19903-0242   Fax: 302-739-6724
Coordinates the three state conservation districts.
Terry L Pepper, President

2913 Delaware Cooperative Extension
University of Delaware
Townsend Hall   302-831-2501
Newark, DE 19716-2103   Fax: 302-831-6758
E-mail: jseitz@udel.edu
http://ag.udel.edu/extension/
Contact person Janice A Seitz's title is Associate Dean for Extension and Outreach Director of Extension College of Agriculture and Natural Resources.
Janice A Seitz, Associate Dean

2914 Delaware Department of Agriculture
232 S DuPont Highway   302-693-4500
Dover, DE 19901-4811   Fax: 302-697-4463
E-mail: mscuse@dda.state.edu.us
http://www.state.de.us/deptagri/
As part of the state government, the department's mission is to sustain and promote the viability of food, fiber and agricultural industries in Delaware through quality services that protect and enhance the environment, health andwelfare of the general public.
Michael T Scuse, Secretary

2915 Delaware Department of Natural Resources and Environmental Control
DNREC
89 Kings Highway   302-739-9902
Dover, DE 19901   Fax: 302-739-6242
http://www.dnrec.state.de.us
Protects and manages the state's vital natural resources, protects public health and safety, provides quality outdoor recreation, and serves and educates the citizens of the First State about the wise use, conservation and enhancementof Delaware's environment.

2916 Delaware Sea Grant Program
University of Delaware   302-831-2841
Newark, DE 19716   Fax: 302-831-4389
E-mail: ntargett@udel.edu
http://www.ocean.udel.edu/seagrant

Dr. Nancy Targett, Interim Director

2917 Mid-Atlantic Fishery Management Council
300 S New Street   302-674-2331
Room 2115   Fax: 302-674-5399
Dover, DE 19904   E-mail: dfurlong@mafmc.org
http://www.mafmc.org

Daniel T Furlong, Executive Director

2918 United States Department of the Interior United States Fish and Wildlife Service
Delaware Bay Estuary Project
2610 Whitehall Neck Road   302-653-9152
Smyrna, DE 19977-2910   Fax: 302-653-9421

Florida

2919 Department of Commerce National Oceanic & Atlantic Oceanographic & Meteorological Laboratory
4301 Rickenbacker Causeway   305-361-4300
Miami, FL 33149   Fax: 305-361-4449
E-mail: webmaster@aoml.noaa.gov
http://www.aoml.noaa.gov
Kristina B Katsaros, Director

2920 Florida Cooperative Extension Service
University of Florida
1308 McCarthy Hall   352-392-1761
PO Box 110210   Fax: 352-392-3583
Gainsville, FL 32611-0210   http://www.ifas.ufl.edu
Christine Waddill, Dean of Extension

2921 Florida Department of Agriculture & Consumer Service
The Capitol, Pl 10   850-488-3022
400 South Monroe Street   Fax: 850-488-7585
Tallahassee, FL 32399
Charles H Bronson, Commissioner

2922 Florida Department of Environmental Protection
3900 Commonwealth Boulevard, M.S.49
Tallahassee, FL 32399   http://www.dep.state.fl.us

2923 Florida Department of Health and Rehabilitative Services
1317 Winewood Boulevard   850-245-4250
Tallahassee, FL 32399-0700   Fax: 850-410-1375
To promote and protect the health and safety of all residents in this state through the establishment and maintenance of high quality standards for the public health environment and the delivery of public health services.
Lyle Jerret, Div. Environmental Health

2924 Florida State Department of Health
2020 Capital Circle SE   904-487-2945
Bin #A00   Fax: 850-410-1375
Tallahassee, FL 32399-1701

2925 Game & Fresh Water Fish Commission Informational Services
20 South Meridian Street   850-488-3796
Tallahassee, FL 32399-1600   Fax: 850-488-6988
http://www.stste.fl.us/fwc

Allan Egbert, Executive Director

2926 Gulf of Mexico Fishery Management Council
3018 US Highway 301 N   813-228-2815
Suite 1000   Fax: 813-225-7015
Tampa, FL 33619-2266 E-mail: gulfcouncil@gulfcouncil.org
http://www.gulfcouncil.org
The Gulf of Mexico Fishery management Council is one of eight regional Fishery Management Councils which were established by the Fishery conservation and Management Act in 1976 (now called the Magnuson-Stevens Fishery Conservation andMagnuson Act). The Council prepares fishery plans which are designed to manage fishery resources from where state waters end to the 200 mile limit of the Gulf of mexico.
Wayne E Swingle, Executive Director

2927 Lee County Parks & Recreation
3410 Palm Beach Boulevard   239-461-7400
Fort Meyers, FL 33916   Fax: 239-461-7450
http://www.leeparks.org/
Our mission is to provide safe, clean and functional Parks & Recreation facilities; to provide programs and services that add to the quality of life for all Lee County residents and visitors; to enhance tourism through special eventsand attractions. We are committed to fulfilling this mission through visionary leadership, individual dedication and the trustworthy use of available resources.
John Yarbrough, Director
Barbara Manzo, Deputy Director

2928 Natural Resources Department: Beaches & Division
3900 Commenwealth Boulevard   850-487-4469
Tallahassee, FL 32399-3000   Fax: 850-487-1469
Kirby Green, Director

2929 Natural Resources Department: Recreation & Parks Division
3900 Commonwealth Boulevard   850-488-6131
Tallahassee, FL 32399-3000   Fax: 850-487-1469

Fran Mainella, Director

2930 Southwest Florida Water Management District
2379 Broad Street, US 41 South   352-796-7211
Brookville, FL 34604-6899   Fax: 352-754-6885
     http://www.watermatters.org
Manages the water and water-related resources within its boundaries. Maintains balance between the water needs of current and future users while protecting and maintaining the natural systems that provide the District with its existingand future water supply. The Conservation Projects Section, in the Resource Conservation and Development Department, is reponsible for managing water conservation, reclaimed water and other alternative source projects, and estimating future waterdemands.
Kathy F Scott, Cons. Projects Section Mgr.

2931 United States Department of Treasury US Customs Service
909 Southeast 1est Avenue   305-810-5120
Miami, FL 33131   Fax: 305-810-5143
Howard Cooperman, Director

Georgia

2932 Blackbeard Island National Wildlife Refuge
1000 Business Center Drive   912-652-4415
Suite 10   Fax: 912-652-4385
Savannah, GA 31405   http://savannahcoastal.fws.gov
This Georgia barrier island's 5,618 acres includes maritime forest, saltmarsh, freshwater marsh, and beach habitat, 3,000 of which has been set aside as National wilderness of variety of recreational activities are availableyear-round, including wildlife observation, birdwatching, hiking and beachcombing.
James D Browning, Refuge Manager

2933 Board of Scientific Counselors: Agency for Toxic Substance and Disease Registry
1600 Clifton Road NE   404-498-0004
Atlanta, GA 30333   888-422-8737
     Fax: 404-498-0083
     E-mail: ATSDRIC@edc.gov
     http://www.atsdr.cdc.gov

Robert W Amler MD, Chief Medical Officer
Henry Falk MD, MPH, Assistant Administrator

2934 Chattahoochee River National Recreation Area: Advisory Commission
Department of Interior
1978 Island Ford Parkway   770-399-8070
Atlanta, GA 30350-3400   Fax: 770-392-7045
     E-mail: CHAT_Superintendent@rips.gov
     http://www.nps.gov/chat/pphtml/contacts.html
Marvin Madry

2935 Georgia Department of Agriculture
19 Martin Luther King Drive   404-656-3600
Atlanta, GA 30334   Fax: 404-656-9380
     http://www.agr.state.ga.us
Tommy Irvin, Commissioner

2936 Georgia Department of Education
2066 Twin Towers East   404-656-2800
Atlanta, GA 30334   Fax: 404-651-8737
     http://www.doe.k12.ga.us
To function as a service oriented and policy driven agency that meets the needs of local school systems as they go about the business of preparing all students for college or a career in a safe and drug free environment where we ensurethat no chils is left behind.
Kathy Cox, CFO

2937 Georgia Department of Natural Resources: Historic Preservation Division
34 Peachtree Street NW   404-656-2840
Suite 1600   Fax: 404-651-8739
Atlanta, GA 30303-2316

     E-mail: ray luce@dnr.state.ga.us
     http://www.gashpo.org
To promote the preservation and use of historic places for a better Georgia.
Ray Luce, Director

2938 Georgia Department of Natural Resources: Pollution Prevention Assistance Division
7 Martin Luther King Jr Drive SW   404-651-5120
Suite 450   Fax: 404-651-5130
Atlantic, GA 30334-9004   http://www.p2ad.org
G Kerr, Director
Venessa Freeman, Information Manager

2939 Georgia Sea Grant College Program
University of Georgia
Marine Science Bulding Room 220   706-542-5954
Room220   Fax: 706-542-3652
Athens, GA 30602   http://alpha.marsci.uga.edu/gaseagrant.html
Goal is to better understand the complex interactions between the physical, chemical, biological and geological processes that are manifested in the area where land and sea come together, and to make that knowledge available and usefulto Georgia's citizens. Sea Grant strives to deepen our understanding of coastal and estuarine ecology, the critical role of fresh water interaction and to expand our knowledge of action beyond the marshes and estuaries and into the life of the riversand streams.
Mac Rawson, Director

2940 National Center for Environmental Health
4770 Buford Highway NE   770-488-7030
Mail Stop F-29   888-232-6789
Atlanta, GA 30341-3724   Fax: 770-488-7042
     E-mail: ncehinfo@cehod1.em.cdc.gov
     http://www.cdc.gov/nceh/ncehhome.htm
Provides national leadership, through science and service, to promote health and quality of life by preventing and controlling disease, birth defects, disability and death resulting from interactions between people and theirenvironment.

2941 Natural Resource Department
205 Butler Street Suite 1353   404-656-2770
Atlanta, GA 30334   Fax: 404-651-5871
     http://www.state.ga.us/dnr/
The mission of the Department of Natural Resources is to sustain, enhance, protect and conserve Georgia's natural, historic and cultural resources for present and future generations, while recognizing the importance of promoting thedevelopment of commerce and industry that utilize sound environmental practices.
Lonice Barrett, Director

2942 Natural Resources Department: Air Protection
4244 International Parkway   404-363-7000
Suite 120   Fax: 404-363-7100
Atlanta, GA 30354
Ron Methier, Branch Chief

2943 Natural Resources Department: Coastal Division
1 Conservation Way   912-264-7218
Brunswick, GA 31520   Fax: 912-262-3143
     http://www.dnr.state.ga.us
Duane Harris, Director

2944 Natural Resources Department: Environmental Protection Division
205 Butler Street SE Suite   404-657-5947
Suite 1152, E Tower   888-373-5947
Atlanta, GA 30334   Fax: 404-651-5778
     http://www.state.ga.us/dnr/environ/
Provides Georgia's citizens with clean air, clean water, healthy lives and productive land by assuring compliance with environmental laws and by assisting others to do their part for a better environment.
Harold Reheis, Director

2945 Natural Resources Department: Land Protection
4244 International Parkway   404-362-2537
Atlanta, GA 30354   Fax: 404-362-2654

John Taylor, Chief

2946 United States Department of the Army US Army Corps of Engineers

US Army Engineer Distric   912-652-5625
PO Box 889   Fax: 912-652-6012
Savannah, GA 31402

2947 United States Fish and Wildlife Service

1875 Century Boulevard
Atlanta, GA 30345

Sam Hamilton, Regional Director

2948 Wassaw National Wildlife Refuge

1000 Business Center Drive   912-652-4415
Suite 10   Fax: 912-652-4385
Savannah, GA 31405   http://savannahcoastal.fws.gov

The most primitive of Georgia's barrier islands, the 10,053 acre refuge, includes beaches with rolling dunes, live oak and slash pine woodlands, and vast salt marshes. The island supports rookeries for egrets and herons, and a varietyof leading birds are abundant in the summer months. Wassaw also provides prime nesting habitat for the loggerhead seaturtles. Refuge visitors may enjoy recreational activities such as birdwatching, beachocombing, hiking, and general nature studies.

James D Browning, Refuge Manager

Hawaii

2949 Agriculture Department

Plant Industry Division, Pesticides Branch
PO Box 22159   808-973-9401
Honolulu, HI 96822   Fax: 808-973-9418
E-mail: hdoainfo@exec.state.hi.vs
http://www.hawaiiag.org/hdoa/

Robert Boesch, Manager

2950 College of Tropical Agriculture and Human Resources

University of Hawaii
3050 Maile Way   808-956-8131
Honolulu, HI 96822   Fax: 808-956-9105
E-mail: research@ctahr.hawaii.edu
http://www2.ctarhr.hawaii.edu/extout.asp

The vision of the college is to be the premier resource for tropical agricultural systems and resource management in the Asia-Pacific region. Its mission outlines a commitment to the preparation of students and all citizens of Hawaiifor life in a global community through research and education programs supporting tropical agricultural systems that foster viable communities, a diversified economy and a healthy environment.

Catherine Chanhalbrendt, Associate Dean

2951 Department of Land and Natural Resources Division of Water Resource Management

PO Box 621   808-587-0214
Honolulu, HI 96809   Fax: 808-587-0219
http://www.state.hi.us/dlnr/cwrm

Linnel Nichioka, Deputy

2952 Environmental Center: University of Hawaii

2500 Dole Street   808-956-7361
Krauss Annex 19   Fax: 808-956-3980
Honolulu, HI 96822   E-mail: envctr@hawaii.edu
http://www2.hawaii.edu/~envctr/

The Center's three areas of focus are education, research and service. The education function of the Center includes the administration of the Environmental Studies Major Equivalent and Certificate program. It fulfills its researchfunction by identifying and addressing environmentally related research needs, particularly those pertinent to Hawaii. The service function primarily involves the coordination and transfer of technical information from the University community togovernment agencies.

Dr John T Harrison, Coordinator

2953 Hawaii Department of Agriculture

PO Box 22159   808-973-9560
Honolulu, HI 96823-2159   Fax: 808-973-9613
http://www.hawaiiag.org/hdoa

Contains devisions such as: Administrative; Animal Industry; Marketing; Measurement Standards; and Plant Industry. Carries out programs to conserve, develop and utilize the agricultural resources of the state. Enforces laws, andformulates and enforces rules and regulation to further control the management of these resources.

James Nakatani, Chairperson

2954 Hawaii Institute of Marine Biology University of Hawaii

PO Box 1346   808-236-7401
Kaneohe, HI 96744   Fax: 808-236-7443
http://www.hawaii.edu/HIMB

Paul Nachtigall, Interm Director

2955 Health Department: Environmental Quality Control

235 S Beretania Street   808-586-4185
Room 702   Fax: 808-586-4186
Honolulu, HI 96813   E-mail: oeqc@mail.health.state.hi.us
http://www.state.hi.us/health/oeqc/index.html

The office is tasked to implement Chapter 343, Hawaii Revised Statues and Title 11, Chapter 200. This is a systematic process to ensure consideration is given to the environmental consequences of actions proposed within our state. Thereview process offers many opportunities to prevent environmental degradation and protect human communities through decreased citizen involvement and informed decision making.

Genevieve Salmonson, Director

2956 Health Department: Noise & Radiation Branch

591 Ala Moana Boulevard   808-586-4700
Honolulu, HI 96813   Fax: 808-586-4729

Russell Pakata, Chief

2957 State of Hawaii: Department of Land and Natural Resources

1151 Punchbowl Street   808-587-0320
Kalanimoku Building, Room 130   Fax: 808-587-0390
Honolulu, HI 96813   E-mail: dlnr@hawaii.gov
http://www.hawaii.gov/dlnr

State agency.

2958 United States Department of the Air Force Major Air Commands

800 Scott Circle   808-449-2490
Hickam AFB, HI 96853-5328   Fax: 808-449-3017
http://www2.hickam.af.mil/wingpa/fact_sht/hickam.htm

Hickam AFB now consists of 2,850 acres of land and facilities. Sharing its runways with adjacent Honolulu International Airport, Hickman and the HIA constitute a single airport complex operated under a joint-use agreement.

Arthur Buckman, Natural Resources Manager

2959 Water Resources Research Center University of Hawaii

2540 Dole Street   808-956-7847
Holmes Hall 283   Fax: 808-956-5044
Honolulu, HI 96822   http://www.wrrc.hawaii.edu

The Water Resources Research Center was organized under the federal Water Resources Research Act of 1964. The Center is supported by university funds, external grants, contracts and a small annual federal grant. WRRC faculty coverthe areas of engineering, hydrology, microbiology, ecology, economics and zoology. Cooperating faculty come from numerous other disciplines. WRRC is open to consideration of any question related to water supply or water quality.

James Moncur, Director
Philip Moravcik, Technology Transfer Spec.

Idaho

2960 Coeur d'Alene District Advisory Council
Bureau of Land Management
1808 North Third Street   208-769-5030
Coeur d'Alene, ID 83814   Fax: 208-769-5050
                                 http://www.id.blm.gov
Fritz Rennebaum, Designated Fed. Empl.

2961 Department of Lands
954 W Jefferson Street   208-334-0200
Po Box 83720   Fax: 208-334-2339
Boise, ID 83720-0050   E-mail: boise@idl.state.id.us
                                http://www.state.id.us.lands
Dirk Kempthrone, State Board Commissioner

2962 Health & Welfare Department: Legal Services, Environmental Quality
1410 N Hilton   208-373-0502
Boise, ID 83706-9990   Fax: 208-373-0417
                                E-mail: webmastr@deq.state.id.us
To protect human health and preserve the quality of Idaho's air, land and water for use and enjoyment today and in the future.
Curt A Fransen, Senior Deputy Attorney Genr

2963 Idaho Association of Soil Conservation Districts
6003 Overland Road Suite 204   208-338-5900
PO Box 2637   Fax: 208-338-9537
Boise, ID 83709   E-mail: kfoster@idahoag.us
                                http://www.iascd.state.id.us
Our mission is to keep the Soil Conservation District as the leading organization providing action at the local level for promoting wise and beneficial conservation of natural resources with emphasis on soil and water. Providesdistricts with information and educational opportunities, financial and technical assistance and provide a formum for discussing conservation issues and concerns.

J Kent Foster, Executive Director

2964 Idaho Cooperative Extension
PO Box 443163   208-885-6639
Moscow, ID 83844-3163   Fax: 208-885-5050
                                http://www.uidaho.edu/extension/
Leroy Luft, Extension Director

2965 Idaho Department of Environemtal Quality: Idaho Falls Regional Office
900 N Skyline   208-528-2650
Suite B   Fax: 208-528-2695
Idaho Falls, ID 83402   http://www.state.id.us/deq

2966 Idaho Department of Environmental Quality: Pocatello Regional Office
444 Hospital Way #300   208-236-6160
Pocatello, ID 83201   Fax: 208-236-6168
                                http://www.state.id.us/deq

2967 Idaho Department of Environmental Quality: State Office
1410 North Hilton Street   208-373-0502
Boise, ID 83706-1255   Fax: 208-373-0417
                                http://www.state.id.us/deq
J Sandoval, Chief Staff

2968 Idaho Department of Fish & Game: Clearwater Region
1540 Warner Avenue   208-799-5010
Lewiston, ID 83501   Fax: 208-799-5012
                                http://wwwfishandgame.idaho.gov

2969 Idaho Department of Fish & Game: Headquarters
600 S Walnut Street   208-334-3700
PO Box 25   Fax: 208-334-2114
Boise, ID 83707   http://www.state.id.us/fishgame
Stephen Barton, Bureau Chief Administration

2970 Idaho Department of Fish & Game: Magic Valley Region
868 E Main Street   208-324-4350
PO Box 428   Fax: 208-324-1160
Jerome, ID 83338

2971 Idaho Department of Fish & Game: McCall
555 Deinhard Lane   208-634-8137
McCall, ID 83638   Fax: 208-634-4320

2972 Idaho Department of Fish & Game: Panhandle Region
2750 Kathleen Avenue   208-769-1414
Coeur d'Alene, ID 83815   Fax: 208-769-1418

2973 Idaho Department of Fish & Game: Salmon Region
99 Highway 93 N   208-756-2271
PO Box 1336   Fax: 208-756-6274
Salmon, ID 83467

2974 Idaho Department of Fish & Game: Southeast Region
1345 Barton Road   208-232-4703
Pocatello, ID 83204   Fax: 208-233-6430

2975 Idaho Department of Fish & Game: Southwest Region
3101 S Powerline Road   208-465-8465
Nampa, ID 83686   Fax: 208-465-8467
                                http://www.fishandgame.idaho.gov
Mission statement is to preserve, protect and perpetuate and manage all wild animals, wild birds and fish within the state of Idaho.

2976 Idaho Department of Fish & Game: Upper Snake Region
1515 Lincoln Road   208-525-7290
Idaho Falls, ID 83401   Fax: 208-523-7604

2977 Idaho Department of State Parks and Recreation
PO Box 83720   208-344-4199
Boise, ID 83720-0065   E-mail: rjust@idpr.state.id.us
                                http://www.idahoparks.org
Manages 27 state parks. We also run the registration program for snowmobiles, boats and off-highway vehicles. Money from registrations and other sources goes to develop and maintain trails, facilities and programs statewide for thepeople who use those vehicles.

2978 Idaho Department of Water Resources
322 E Front Street   208-287-4800
Boise, ID 83720-0098   Fax: 208-287-6700
                                E-mail: IDWRInfo@idwr.idaho.gov
                                http://www.idwr.idaho.gov/about/
Working for a controlled development and wise management of Idaho's resources. Documents and reports on topics of public interest such as drought, salmon, wilderness and the Snake River Basin.

, Director

2979 Idaho Falls District Advisory Council Bureau of Land Management
Department of the Interior
1405 Hollipark Drive   208-524-7500
Idaho Falls, ID 83401   Fax: 208-524-7505
Jim May, District Manager

2980 Idaho Geological Survey
University of Idaho
Morrill Hall   208-885-7991
3rd Floor   Fax: 208-885-5826
Moscow, ID 83844-3014   E-mail: igs@uidaho.edu
                                http://www.idahogeology.org
Earl Bennett, Director

2981 Idaho State Department of Agriculture
Agricultural Resources
623 11th Avenue S   208-465-8442
Nampa, ID 83651   Fax: 208-465-8471
Victor Mason, Program Manager

2982 Lands Department: Soil Conservation Commission
1215 West State Street   208-334-0214
Boise, ID 83720   208-334-2339
          http://www.scc.state.id.us/scc_facts.htm
Responsibilities of the Commission are: organize Districts and
provide assistance, coordination, information and training to
Disrict supervisors; ensure that Districts function legally and
properly as local subdivisions of stategovernment; administer
general funds appropriated by the Idaho Legislature to Districts
so they can install resource conservation practices and provide
technical assistance personnel to Districts administering water
quality projects and conductingsoil surveys.
Wayne R Faude, Adminsitrative Officer

2983 Salmon District Advisory Council Bureau of Land Management
50 Highway 93 S   208-756-5400
Salmon Field Office   Fax: 208-756-5436
Salmon, ID 83467
Dave Krosting, Field Manager

2984 United States Department of the Air Force Major US Installations
Mountain Home AFB, ID   208-828-4297
          http://www.mountainhome.af.mil/
Nathan Rowland, Natural Resources Manager

2985 United States Department of the Interior Bureau of Land Management
3833 South Development Avenue   208-387-5512
Boise, ID 83705   Fax: 208-387-5797
Lester Rosenkrance, Director

Illinois

2986 Association for Conservation Information Illinois Department of Conservation
One Natural Resources Way   217-782-7454
Springfield, IL 62702-1271   800-720-0298
          E-mail: gthomas@dnrmail.state.il.us
                   http://www.dnr.state.il.us
Gary Thoams, President

2987 Association of Illinois Soil and Water Conservation Districts
2520 Main Street   217-744-3414
Springfield, IL 62702   Fax: 217-744-3420
          E-mail: aiswcdep@aol.com
          http://www.ilconservation.com
To foster and promote charitable and educational purposes designed to further the principles of soil conservation and stewardship, water conservation and energy conservation. Provides,
conducts and sponsors programs to aidindividuals, groups, organizations, government bodies, association and all entities in
combating soil erosion and energy water waste.
Richard Nichols, Executive Director

2988 Central Midwest Interstate Low-Level Radioactive Waste Commission
1035 Outer Park Drive   217-785-9982
Springfield, IL 62704   Fax: 217-785-9977
          http://www.state.il.us/iema/about
Gary N Wright, Secretary/Treasurer
Marcia S Marr, Executive Director

2989 Conservation Deartment: Resource Marketing and Education
Illinois Department of Conservation
One Natural Resources Way   217-782-6302
Springfield, IL 62702-1271   Fax: 217-524-5612
          E-mail: parksadmin@dnrmail.state.il.us
                   http://www.dnr.state.il.us
Joel Brunsvold, Director
Brice Sheriff, Deputy Director

2990 Conservation Department
Illinois Conservation Foundation
524 S 2nd Street   217-524-4126
Springfield, IL 62701-1787   Fax: 217-782-5177
          E-mail: Kandrews@dnrmail.state.il.us
                   http://www.dnr.state.il.us
John Schmitt, Executive Director

2991 Construction Engineering Research Laboratory
US Army Engineer Research and Development Center
PO Box 9005   217-352-6511
Champaign, IL 61826-9005   Fax: 217-373-7222
          E-mail: Dana.L.Finney@erdc.usace.mil
                   http://www.cecer.army.mil
CERL conducts research to support sustainable military installations. Research is directed toward increasing the Army's ability to more efficiently construct, operate and maintain its
installations and ensure environmental quality andsafety at a reduced life-cycle cost. Excellent facilities support the Army's
training, readiness, mobilization and sustainability missions.
Dana Finney, Public Affairs Officer

2992 Environmental Protection Agency: Region 5
77 West Jackson Boulevard   312-353-2000
Chicago, IL 60604   800-621-8431
                               Fax: 312-353-1120
          http://www.epa.gov/region5
Elissa Speizman, Director Public Affairs

2993 Environmental Protection Agency: Water Pollution Control
Water Bureau
4500 South Sixth Street Road   217-786-6892
Springfield, IL 62706
The Bureau of Water conducts the following programs in striving to protect and enchance the quality of the State's surface waters: permitting programs, compliance/enforcement, surface
water quality monitoring and assessment,
watershedmanagement, operator certification and field operations.
James Park, Director

2994 Illinois Department of Agriculture Bureau of Land and Water Resources
PO Box 19281   217-782-6297
Springfield, IL 62794-9281   Fax: 217-557-0993
          http://www.agr.state.il.us
Steve Chard, Acting Bureau Chief

2995 Illinois Department of Transportation
2300 S Dirksen Parkway   217-782-6953
Springfield, IL 62764   Fax: 217-782-8714
          E-mail: Monseurmj@nt.dot.state.il.us
Provides cost-effective, safe and efficient transportation for the
people who live, work, visit and do business in Illinois, and ensures that the system supports the state's economic growth.
Mike Monseur

2996 Illinois Dept. Commerce & Econimic Opportunity Division of Recycling & Waste Reduction
620 East Adams Street   217-785-1997
Springfield, IL 62701-1615   Fax: 217-785-2618
          http://http://www.illinois.biz.biz/com/recycling/index.html
Provides technical assistance and grants to advance recycling
and waste reduction in Illinois. Grant projects must occur in Illi-
nois.
David Ross, Manager
David E Smith, Recycling Specialist

2997 Illinois Nature Preserves Commission
One Natural Resources Way   217-785-8686
Springfield, IL 62702-1271   Fax: 217-785-2438
E-mail: dmcfall@dnrmail.state.il.us
http://www.dnr.state.il.us/inpc/index/htm
To assist private and public landowners in protecting high quality natral areas and habitats of endangered and threatened species in perpetuity, through voluntary dedication or registration of such lands into the Illinois NaturePreserves Systems.
Don McFall, Acting Director

2998 United States Department of Agriculture Research Education and Economics
1815 N University Street   309-681-6602
Peoria, IL 61604   Fax: 309-681-6684
E-mail: Sbuxton@mail.ncaur.usda.gov
Adrianna Hewings, Area Director

2999 United States Department of the Army US Army Corps of Engineers
US Army Engineer Distric
Clock Tower Building   309-794-5224
PO Box 2004   Fax: 309-794-5181
Rock Island, IL 61204-2004

Indiana

3000 Indiana Department of Natural Resources
402 West Washington Street   317-232-4020
Indianapolis, IN 46204   Fax: 317-232-8036
http://www.ai.org/dnr
Protects, enhances, preserves, and wisely uses natural, cultural, and recreational resources for the benefit of Indiana's citizens through professional leadership, management and education.

3001 Indiana Department of Natural Resources, Mine Reclamation
RR2 Box 129   812-665-2207
Jasonville, IN 47438-9517   800-722-6463
Fax: 812-665-5041
E-mail: cbaughma@reclmation.dnr.state.in.us
The Indiana Department of Natural Resources' Division of Reclamation is agency responsible for the regulation of coal mining and reclamation and for abandoned mine land reclamation in the state of Indiana.
Timothy Taylor, Regulatory Program Director
Steve Herbert, Abandoned Land Mine Dir.

3002 Indiana State Department of Health
2 N Meridan Street   317-233-7400
Indianapolis, IN 46204   Fax: 317-233-7387
E-mail: gwilson@isdh.state.in.us
http://www.state.in.us/isdh
Greg Wilson, State Health Commissioner

3003 Natural Resources Department: Fish & Wildlife
402 West Washington Street RMW273   317-232-4080
Indianapolis, IN 46204   Fax: 317-232-8150
http://www.state.in.us/dnr/fishwild/index.htm
VACANT , Director

3004 Natural Resources Department: Soil Conservation
402 West Washington Street RoomW265   317-233-3870
Indianapolis, IN 46204-2739   Fax: 317-233-3882
http://www.state.in.us/dnr/soilcons
Harry Nikides, Director

Iowa

3005 Iowa Association of County Conservation Boards
405 SW 3rd Street   515-963-9582
Suite 1   Fax: 515-963-9582
Ankeny, IA 50021   E-mail: iaccb@ecity.net

IACCB is a nonprofit organization assisting member county conservation boards in areas of board member education, public relations and legislation. The association's main purposes are to promote the objectives and supplement theactivities of conservation boards, exchange information, assist boards and members in program development and provide a unified voice in the legislature. IACCB is governed by a nine-member board elected by member counties.
Don Brazelton, President

3006 Iowa Department of Agriculture & Land Bureau of Field Services
E 9th & Grand Avenue   515-281-5258
Wallace Building   E-mail: jim.gillespie@idals.state.ia.us
Des Moines, IA 50319
James Gillespie

3007 Iowa Department of Agriculture & Land Stewards Bureau of Mines & Minerals
East 9th & Grand Avenue   515-281-6142
Wallace Building   Fax: 515-281-6170
Des Moines, IA 50319-0034 http://www.state.ia.us/agriculture
Kenneth Tow

3008 Iowa Department of Agriculture, and Land Stewardship division of Soil Conservation
502 E 9th   515-281-5851
Wallace State Office Building   Fax: 515-281-6170
Des Moines, IA 50319   http://www.agriculture.state.ia.us
Kenneth R Tow, Director

3009 Iowa Department of Natural Resources Administrative Services Division
502 E 9th Street   515-281-5918
Wallace Office Building   Fax: 515-281-8895
Des Moines, IA 50319   http://www.state.ia.us
Aims to manage, protect, conserve and develop Iowa's natural resources in cooperation with other public and private organizations and individuals, so that the quality of life for Iowans is significantly enhanced by the use, enjoymentand understanding of those resources.
Linda Hanson, Administrator

3010 Iowa State Extension Services
1032 Wallace Road Office Building   515-294-6192
Ames, IA 55011   Fax: 515-294-4715
E-mail: jlpease@iastate.edu
http://www.extension.iastate.edu/
Dr James L Pease, Specialist

3011 Natural Resource Department
Wallace Building   515-281-5385
Des Moines, IA 50319-0034   Fax: 515-281-8895
Jeffrey R Venk, Director

Kansas

3012 Emporia Research and Survey Office Kansas Department of Wildlife & Parks
PO Box 1525   316-342-0658
Emporia, KS 66801-1525   Fax: 316-342-6248
E-mail: randys@wp.state.ks.us
Randall D Schultz, Contact

3013 Environmental Affairs Office
502 Landon State Office Building   785-296-2281
900 SW Jackson Street   Fax: 785-296-6953
Topeka, KS 66612-1233
Steve Adams, Officer

Government Agencies & Programs/Kansas

3014 Environmental Protection Agency: Region 7, Air & Toxics Division
901 North 5th Street 913-551-7020
Kansas City, KS 66101 Fax: 913-551-7065
http://www.epa.gov/region7/
Responsible for management of programs for air, hazardous waste, toxic substances, radiation and pollution prevention in Iowa, Kansa, Missouri and Nebraska as required by the following legistlation: The Clean Air Act, The ResourceConservation and Recovery Act, the Toxic Substances Control Act and the Emergency planning and Community Right-to Know Act.
Carol A Kather, Deputy Director
William A Spratlin, Director

3015 Health & Environment Department
1000 SW Jackson Street 785-296-0461
Topeka, KS 66612 Fax: 785-368-6368
http://www.kdhe.state.ks.us
An organization dedicated to optimizing the promotion and protection of the health of Kansas through efficient and effective public health programs and services and through preservation, protection and remediation of natural resourcesof the environment.
Roderick L Bremby, Secretary

3016 Health & Environment Department: Air & Radiation
100 SW Jackson 785-296-1593
Suite 310 Fax: 785-296-7455
Topeka, KS 66612
Mission: To protect the public from the harmful effects of radiation and air pollution and conserve the natural resources of the state by preventing damage to the environment from releases of radioactive materials or air contaminants.
Clark Duffy, Director

3017 Health & Environment Department: Environment Division
1000 SW Jackson Street 785-296-1535
Topeka, KS 66612-1367 Fax: 785-296-8464
E-mail: rhammers@kdhe.state.ks.us
The mission of the Division of Environment is the protection of the public health and environment. The division conducts regulatory programs involving public water supplies, industrial discharges, wastewater treatment systems, solidswaste landfills, hazardous waste, air emissions, radioactive materials, asbestos removal, refined petroleum storage tanks and other sources which impact the environment.
Ronald Hammerschmidt, Director

3018 Health & Environment Department: Waste Management
1000 Sw Jackson 785-296-1600
Suite 320 Fax: 785-296-1592
Topeka, KS 66612-1366 http://www.kdhe.state.ks.us/waste
Regulates landfills, HHW, Hazardous Waste Permitting, Solid Waste Permitting, Public Outreach, Illegal Dumps.
William Bider, Director

3019 Information Systems Division Public Information Services
900 Jackson Street 785-296-3343
London State Office Building, Room 751 Fax: 785-296-1168
Topeka, KS 66620-1275 E-mail: discweb@state.ks.us
http://da.state.ks.us/disk/
Provides cost-effective data processing and telecommunication services to all state agencies and Regents.
Greg Crawford, Director

3020 Kansas Association of Soil Conservation Districts
522 Winn Road 785-827-2547
Salina, KS 67401-3668 Fax: 785-827-7784
E-mail: ngjjaj@midusa.net

3021 Kansas Cooperative Fish & Wildlife Research
Kansas State University
205 Leasure Hall 785-532-6070
Manhattan, KS 66506-3501 Fax: 785-532-7159
Philip Gipson, Leader

3022 Kansas Corporation Commission Conservation Division
Finney State Office Building 316-337-6200
130 S Market, Room 2078 Fax: 316-337-6211
Wichita, KS 67202 http://www.kcc.state.ks.us
State of Kansas oilfield regulatory agency. The KCC is responsible for the preservation of Kansas' hydro and carbo resources, protection of corrullative right and the prevention and remediation of oil field pollution.
ML Korphage, Director

3023 Kansas Department of Wildlife & Parks
900 SW Jackson Street 785-296-2281
Suite 502 Fax: 785-296-6953
Topeka, KS 66612-1233 http://www.kdwp.state.ks.us
Steven Williams, Secretary

3024 Kansas Department of Wildlife & Parks Region 1
1426 Highway 183 Alt 785-628-8614
PO Box 338 Fax: 785-623-2945
Hays, KS 67601 http://www.kdwp.state.ks.us
Mary Jane Pfannenstiel, Secretary

3025 Kansas Department of Wildlife & Parks Region 2
3300 SW 29th 785-273-6740
Topeka, KS 66614 Fax: 785-273-6757
http://www.dkwp.state.ks.us
Manages and promotes the wildlife and natural resources of Kansas. Administered by a secetary of Wildlife and Parks and is advised by a seven-member Wildlife and Parks Commission.

3026 Kansas Department of Wildlife & Parks Region 3
1001 McArtor Drive 620-227-8609
Dodge City, KS 67801 Fax: 620-227-8600
http://www.kdwp.state.ks.us
Cindy Konda, Region Director

3027 Kansas Department of Wildlife & Parks Region 4
6232 E 29th Street N 316-683-8069
Wichita, KS 67220 http://www.dkwp.state.ks.us

3028 Kansas Department of Wildlife & Parks Region 5
1500 W 7th Street 620-431-0380
Po Box 777 Fax: 620-431-0381
Chanute, KS 66720-0777 http://www.kdwp.state.ks.us
This region is made up of 18 counties in the southeastern corner of the state. This area is dominated by the Osage Questas physiographic region, which is characterized by rolling grasslands, limestone bluffs, and heavily timberedbottomlands.

3029 Kansas Geological Survey
Kansas University
1930 Constant Avenue 785-864-3965
W Campus Fax: 785-864-5317
Lawrence, KS 66047 http://www.kgs.ku.edu
Conducts geological studies and research and collects, correlates, preserves and disseminates information leading to a better understanding of the geology of Kansas, with special emphasis on nautral resources of economic value, waterquality and quantity and geologic hazards. This information is published in books and maps both technical and educational and also provides computer programs and data bases derived from geologic investigations.
William E Harrison, Interior Dir/Stat Geologist

3030 Kansas Health & Environmental Laboratories
Forbes Field 785-296-1620
Suilding 740 Fax: 785-296-1641
Topeka, KS 66620-0001 http://www.kdhe.state.ks.us/labs
Provides timely and accurate analytical information for public health benefit in Kansas and assures the quality of statewide laboratory sevices though certification and improvement programs.
Roobert H Carlson PhD, Manager

3031 Kansas Water Office
901 S Kansas Avenue 913-296-3185
Topeka, KS 66612 888-KAN-WATE

204

Fax: 785-296-0878
http://www.kwo.org

Works to achieve proactive solutions for resource issues of the state and to ensure good quality water to meet the needs of the people and the environment of Kansas. Evaluates and develops public policies, coordinating the waterresource operations of agencies at all levels of government.
Joe Harkins, Director

3032 Pratt Operations Office Kansas Department of Wildlife & Parks
512 SE 25th Avenue   316-672-5911
Pratt, KS 67124   E-mail: kenb@wp.state.ks.us

3033 Wildlife and Parks Department Environmental Affairs Office
502 Landon State Office Building   785-296-2281
900 SW Jackson Street   Fax: 785-296-6953
Topeka, KS 66612-1233
Theodore D Ensley, Secretary

Kentucky

3034 Agriculture Department Consumer Safety: Office Pesticides Division
500 Metro Street   502-564-7274
Capital Plaza Tower, 7th Floor
Frankfort, KY 40601
Ronald Egnew, Director

3035 Attorney General's Office Civil and Environmental Law Division
700 Capitol Avenue   502-696-5300
Capitol Building, Suite 118   Fax: 502-564-2894
Frankfort, KY 40601-3449http://www.kyattorneygeneral.com
Robert S Jones, Acting Director
Ryan M Halloran, Assistant Director

3036 Breaks Interstate Park Commission
131 Summit Drive   606-432-1447
Pikeville, KY 41501   Fax: 606-432-1440
E-mail: cbyers@summit-engr.com

3037 Department for Environmental Protection
14 Reilly Road   502-564-2150
Fort Boone Plaza
Frankfort, KY 40601
Lloyd Cress, Commisioner

3038 Economic Development Cabinet: Community Development Department Brokerage Division
500 Metro Street   502-564-7140
Capital Plaza Tower, 5th Floor   Fax: 502-564-3256
Frankfort, KY 40601   http://www.thinkkentucky.com
Richard Cirre, Director

3039 Environmental Education Department
500 Metro Street   502-564-5525
Capital Plaza Tower, 5th Floor   Fax: 502-564-3354
Frankford, KY 40601   E-mail: environment@ky.gov
http://www.environment.ky.gov/education
Julie B Smither, Coordinator

3040 Environmental Protection Department: Management Services Branch
Fort Boone Plaza   502-564-2150
14 Reilly Road
Frankfort, KY 40601
Lloyd Cress, Commisioner

3041 Environmental Protection Cabinet: Law Department
500 Metro Street   502-564-2150
Capital Plaza Tower, 5th Floor   Fax: 502-564-4245
Frankfort, KY 40601   E-mail: dep@ky.gov
http://www.dep.ky.gov
Judith Villines, Commisioner

3042 Environmental Protection Department: Environmental Services Division
Fort Boone Plaza   502-564-2150
14 Reilly Road
Frankfort, KY 40601
William Davis, Director

3043 Environmental Protection Department: Waste Management Division
Fort Boone Plaza   502-564-6716
14 Reilly Road   Fax: 502-564-4049
Frankfort, KY 40601   E-mail: waste@ky.gov
http://www.waste.ky.gov
Bruce Scott, Director

3044 Environmental Protection Department: Water Division
Fort Boone Plaza   502-564-3410
14 Reilly Road   Fax: 502-564-0111
Frankfort, KY 40601   E-mail: water@ky.gov
http://www.water.ky.gov
David Morgan, Director

3045 Fish and Wildlife Resources Department: Fisheries Division
One Game Farm Road   502-564-3596
Frankfort, KY 40601   800-858-1549
Fax: 502-564-6501
E-mail: info.center@ky.gov
http://www.fw.ky.gov
John Gassett, Commissioner

3046 Fish and Wildlife Resources Department: Conservation Education Division
One Game Farm Road   502-564-4762
Frankfort, KY 40601   800-858-1549
Fax: 502-564-6501
John Gassett, Commissioner

3047 Kentucky Department for Public Health
275 E Main Street   502-564-2154
Franfort, KY 40621   Fax: 502-564-8389
Ann Johnson, Program Coordinator

3048 Kentucky Environmental and Public Protection Cabinet
500 Metro Street   502-564-3350
Capital Plaza Tower, 5th Floor   Fax: 502-564-3354
Frankfort, KY 40601   E-mail: environment@ky.gov
http://www.environment.ky.gov

Provides a safe, clean environment in the Commonwealth, while working with business and industry to help ensure adequate jobs and a strong economy.
LaJuana Wilcher, Secretary

3049 Kentucky Natural Resources and Environmental Protection Cabinet
Capitol Plaza Tower   502-564-5525
4th Floor   Fax: 502-564-2043
Frankfort, KY 40601   http://www.nr.state.ky.us/nrhome

3050 Kentucky State Cooperative Extension Services
University of Kentucky
S-107 Agricultural Science Bldg N   859-257-4302
Lexington, KY 40546-0091   Fax: 859-323-1031
http://www.ca.uky.edu/coopext/

3051 Kentucky State Nature Preserves Commission
801 Schenkel Lane 502-573-2886
Frankfort, KY 40601 Fax: 502-573-2355
Aims to protect Kentucky's natural heritage by identifying, acquiring and managing natural areas that represent the best known occurrences of rare native species and natural communities and working together to protect biologicaldiversity.
Judy Cunningham, Nature Officer

3052 Natural Resources & Environmental Project Cabinet: Department for Natural Resources
#2 Hudson Hollow 502-564-6940
Frankfort, KY 40601 Fax: 502-564-5698
E-mail: naturalresources@ky.gov
http://www.naturalresources.ky.gov
The Department is responsible for the conservation, preservation, protection, and perpetuation of natural resources in the Commonwwealth and consists of three divisions and a Commissioner's Office. The Office provides policy directionand coordination for them and provides administrative support to the Kentucky Heritage Land Conservation Fund Board. The Division protects and enhances the forest resources through a public informed of the environmental and economic importance of theresources.
Susan Busch, Commissioner

3053 Natural Resources & Environmental Protection Cabinet
Capital Plaza Tower 502-564-3350
5th Floor Fax: 502-564-3354
Frankfort, KY 40601 http://www.kyenvironmet.org
Henry C List, Secretary
Mark York, Deputy Secretary

3054 Natural Resources Department: Conservation Division
375 Versailles Road 502-573-3080
Frankfort, KY 40601 Fax: 502-573-1692
E-mail: conservation@ky.gov
http://www.conservation.ky.gov
Assists Kentucky's local conservation districts in the development and implementation of sound soil and water conservation programs to manage, enhance, and promote the wise use of the Commonwealth's natural resources.
Stephen A Coleman, Director

3055 Natural Resources Department: Division of Fore stry
627 Comanche Trail 502-564-4496
Frankfort, KY 40601 Fax: 502-564-6553
E-mail: forestry@ky.gov
http://www.forestry.ky.gov

Leah W MacSwords, Director

3056 Natural Resources Department: Energy Division
663 Teton Trail 502-564-7192
Frankfort, KY 40601 Fax: 502-564-7484
E-mail: kyenergy@mail.state.ky.us
http://www.energy.ky.gov/dnrdoe.html
Provides leadership to maximize the benefits of energy effeciency and alternate energy through awareness, technology development, energy preparedness and new partnerships and resources.
John Stapleton, Director

3057 Natural Resources and Environment Protection Cabinet: Environmental Quality Commission
14 Reilly Road 505-564-2150
Frankfort, KY 40601 Fax: 502-567-9634
http://www.eqc.ky.gov
Aloma Dew, Chairperson
Leslie Cole, Executive Director

3058 Natural Resources and Environmental Protection Cabinet
500 Metro Street 502-564-2150
Capitol Plaza Tower, 5th Floor Fax: 502-564-3354
Frankfort, KY 40601
LaJuana Wilcher, Secretary

3059 Surface Mining Reclamation and Enforcement Department
Two Hudson Hollow 502-564-6940
Frankfort, KY 40601 Fax: 502-564-5848
Dave Rosenbaum, Commissioner

3060 Tourism Cabinet: Parks Department
2400 Capital Plaza Tower 502-564-2172
500 Metro Street Fax: 502-564-6100
Frankfort, KY 40601
Mark Lovely, Commissioner

3061 United States Department of the Army: US Army Corps Engineers
441 G Street NW 202-761-0001
Washington, DC 20314-1000 Fax: 202-761-1683
http://www.usace.army.mil/
LTG Carl Strock, Commander

Louisiana

3062 Agriculture & Forestry: Soil & Water Conservation
Louisiana Department of Agriculture and Forestry
PO Box 3554 225-922-1269
Baton Rouge, LA 70821-3554 Fax: 225-922-2577
Bradley Spicer, Executive Director

3063 Central Interstate Low-Level Radioactive Waste Commission
1033 "O" Street 402-476-8247
Suite 530 Fax: 402-476-8205
Lincoln, NE 68508 http://www.cillrwcc.org
Rita Houskie, Administrator

3064 Culture, Recreation and Tourism Department State Parks Office
Office of State Parks 225-342-8111
PO Box 44426 888-677-1400
Baton Rouge, LA 70804-4426 E-mail: parks@crt.state.la.us
http://www.crt.state.la.us
Jim Ball, Assistant Secretary

3065 Delta Region Preservation Commission
365 Canal Street 504-589-3882
Suite 2400 Fax: 504-589-3851
New Orleans, LA 70130-1142
Robert Belous, Superintendent

3066 Department of Natural Resources: Office of Mineral Resources
PO Box 2827 225-342-4543
Baton Rouge, LA 70821 Fax: 225-342-3885
E-mail: OMR@dnr.state.la.us.us
http://www.dnr.louisianna.gov/MIN/min.asp
Provides staff support to the State Mineral Board in granting and administering leases on state-owned lands and waterbottoms for the production and development of minerals, primarily oil and gas, for the purpose of optimizing revenueto the state from the royalties, bonuses and rentals generated therefrom.
Gus C Rodemacher, Assistant Secretary

3067 Louisana Department of Natural Resources
625 N Fourth Street 504-342-1375
PO Box 94396 Fax: 225-342-2707
Baton Rouge, LA http://www.dnr.state.la.us
James R Hanchey, Assistant Secretary

3068 Louisiana Department of Health and Hospitals: Office of Public Health
PO Box 3214 504-925-7209
Baton Rouge, LA 70821
To protect the public health and the people of Louisiana.
Louis Trachtman MD, MPH, Asst State Health Officier

3069 Louisiana Department of Natural Resources Office of Coastal Restoration and Management

625 N 4th Street 225-342-7308
Baton Rouge, LA 70802 Fax: 225-342-9417
E-mail: philp@dnr.state.la.us
http://www.savelawetlands.org

Develops, implements and monitors costal vegetated wetland restoration, creation and conservation measures. Preforms engineering, planning and monitoring functions essential to successful development and implementation of wetlandconservation and restoration plans and projects as directed by the Costal Wetlands Conservation and Restoration Plan.
Bill Good, Director

3070 Louisiana State Extension Services

PO Box 25100 225-578-6083
Baton Rouge, LA 70894-5100 Fax: 225-578-4225
http://www.lsu.edu.center.edu

Jack Bagent, Director Extension Services

3071 Natural Resources: Coastal Management Division

PO Box 44487 225-342-7591
Baton Rouge, LA 70804 800-267-4019
http://www.savelawetlands.org

Uses its regulatory authority to protect, restore and enhance coastal resources so that loss of coastal wetlands resulting from activites regulated by the division is offset by activities which provide equivalent wetland functionalvalue. Also supports and encourages multiple use of coastal resources to allow for adequate economic growth and minimizing adverse effects of one resource use upon another.
Terry Howey, Director

3072 Natural Resources: Conservation Office

617 N 3rd Street 225-342-5540
PO Box 94275 Fax: 225-342-3705
Baton Rouge, LA 70802-9275 http://wwww.dnr.state.la.us

Regulatory oil and gas agency, State of Louisiana.
James H Welsh, Commissioner
Felix J Boudreaux, Assistant Commissioner

3073 Natural Resources: Geological Oil & Gas Division

625 N 4th Street 225-342-5515
PO Box 94275 Fax: 225-342-4438
Baton Rouge, LA 70804-9275

Administers a regulatory program to prevent waste of oil and gas and the drilling of unnecessary wells. Protects individual property rights and conserves the state's natural resources in a geologically-approved manner.
Charles King, Director

3074 Natural Resources: Injection & Mining Division

PO Box 94275 225-342-5515
Baton Rouge, LA 70804 Fax: 225-342-3094

Has the responsibility for implementation of major environmental programs statutorily charged to the Office of Conservation. Administers a regulatory and permit program to protect underground sources of drinking water fromendangerment; is responsible for regulating exploration, development and surface mining operations for coal and lignite; and protection of state and private lands.
Joseph S Ball Jr, Director

Maine

3075 Conservation Department: Waste Reduction and Recycling

286 Water Street 207-287-2211
Key Bank Plaza Fax: 207-287-2400
Augusta, ME 04333 http://www.state.me.us/doc/offices.html

3076 Maine Cooperative Fish & Wildlife Research Unit

University of Maine
USGS Biological Resources Division 207-581-2870
5755 Nutting Hall Fax: 207-581-2858
Orono, ME 04469-5755

Dr. William B Krohn, Leader

3077 Maine Department of Environmental Protection: Augusta

17 State House Station 207-287-7688
Augusta, ME 04333-0017 800-452-1942
http://janus.state.me.us/dep/home.html

Responsible for environmental protection and regulation in the state of Maine. Engages in a wide range of activities, makes reccomendations to the Legistlature regarding measures to minimize and eliminate environmental pollution,grants licenses, initiates enforcement actions, and provides information and technical assistance.

3078 Maine Department of Conservation

22 State House Station 207-287-2211
Augusta, ME 04333 Fax: 207-287-2216
http://www.state.me.us

Ronald Lovaglio, Commissioner

3079 Maine Department of Conservation: Ashland Regional Office

45 Radar Road 207-435-7963
Ashland, ME 04732 Fax: 207-435-7184
http://www.state.me.us

3080 Maine Department of Conservation: Bangor Regional Office

106 Hogan Road 207-941-4014
BMHI Complex Fax: 207-941-4222
Bangor, ME 04401 http://www.state.me.us

3081 Maine Department of Conservation: Bolton Hill Regional Office

2870 North Belfast Avenue 207-624-3700
Augusta, ME 04330 Fax: 207-287-8534
http://www.state.me.us

Alan Johnston, Regional Supervisor

3082 Maine Department of Conservation: Bureau of Parks & Lands

22 State House Station 207-287-2211
Augusta, ME 04333-0022 Fax: 207-287-2216
http://www.state.me.us/doc

Patrick K McGowan, Commissioner

3083 Maine Department of Conservation: Entomology Laboratory

48 Hospital Street 207-287-2431
Augusta, ME 04330 Fax: 207-287-2432
http://www.state.me.us

David Struble, Contact

3084 Maine Department of Conservation: Farmington Regional Office

114 Corn Shop Lane 207-778-8211
Farmington, ME 04938 800-442-6382
Fax: 207-778-8210
http://www.state.me.us

3085 Maine Department of Conservation: Greenville Regional Office

Department of Inland Fisheries and Wildlife
PO Box 551 207-695-3756
Greenville, ME 04441 Fax: 207-695-2022
http://www.state.me.us

3086 Maine Department of Conservation: Hallowell Regional Office

1 Beech Street 287-624-6080
Stevens Complex Fax: 287-624-6081
Hallowell, ME 04330 http://www.maine.gov/doc

Patrick K McGowan, Commissioner

3087 Maine Department of Conservation: Jonesboro Regional Office
PO Box 220   207-434-5927
Jonesboro, ME 04648-0220   Fax: 207-624-6081
http://www.state.me.us

3088 Maine Department of Conservation: Land Use Regulation Commission
18 Elkins Lane, Harlow Building   207-287-2631
22 State House Station   Fax: 207-287-7439
Augusta, ME 04333 E-mail: Catherine.M.Carroll@maine.gov
http://www.maine.gov/doc/lurc/offices/augusta.shtml
The Maine Land Use Regulation Commission meets monthly in various locations throughout the state to discuss jurisdiction-related issues and to act upon pending cases.
E Bart Harvey, III, Chair
Stephen W Wight, Co-Chair

3089 Maine Department of Conservation: Millinocket Regional Office
191 Main Street   287-746-2244
East Millinocket, ME 04430   Fax: 207-512-1003
http://www.state.me.us

3090 Maine Department of Conservation: Old Town Regional Office
Airport Road   207-827-1800
PO Box 415   Fax: 207-827-8441
Old Town, ME 04468   http://www.state.me.us

3091 Maine Department of Conservation: Rangeley Regional Office
2352 Main Street   207-864-5064
PO Box 887   Fax: 207-512-1004
Rangeley, ME 04970   http://www.state.me.us

3092 Maine Department of Environmental Protection: Presque Isle
1235 Central Drive   207-764-0477
Skyway Park   888-769-1053
Presque Isle, ME 04769-2094   Fax: 207-764-1507

3093 Maine Department of Environmental Protection: Portland
312 Canco Road   207-822-6300
Portland, ME 04103   888-769-1036
Fax: 207-822-6303

3094 Maine Inland Fisheries & Wildlife Department
284 State Street Station 41SHS   207-287-8000
Augusta, ME 04333-0041   Fax: 207-287-6395
E-mail: ifw.online@maine.gov
http://www.mefishwildlife.com

Roland Martin, Commissioner

3095 Maine Natural Areas Program
93 State House Station   207-287-8044
Augusta, ME 04333   Fax: 207-287-8040
E-mail: maine.nap@maine.gov
http://www.state.me.us/doc/nrimc/mnap/home.html
The most comprehensive source of information of the State's important nautral features. The program maintains a cross-referenced date management system containing current and historic information about natural features from acrossMaine. This information is shared with other State agencies, town planners, land trusts and other groups interested in natural resource and conservation planning.
Molly Docherty, Director

3096 Maine Sea Grant College Program
University of Maine
5715 Coburn Hall, 14   207-581-1435
Orono, ME 04469-5715   Fax: 207-581-1426
E-mail: davison@maine.maine.edu
http://www.seagrant.umaine.edu

Ian Davison, Director

3097 St Croix International Waterway Commission
PO Box 610   506-466-7550
Calais, ME 04619   Fax: 506-466-7551
E-mail: staff@stcroix.org
http://www.stcroix.org
The State of Maine and the Province of New Brunswick established the St. Croix International Waterway Commission to develop a transboundary management plan for the St. Croix River corridor and to facilitate its longterm, voluntaryimplementation by governmental and local interests, on an international basis.

Lee Sochasky, Executive Director

3098 University of Maine Cooperative Extension Forestry & Wildlife Office
5755 Nutting Hall   207-581-2892
Orono, ME 04469-5755   Fax: 207-581-3466
http://www.umext.maine.edu/

Jim Philp, Forestry Specialist
Les Hyde, Education Educator

Massachusetts

3099 Connecticut River Atlantic Salmon Commission
103 E Plumtree Road   413-548-8002
Sunderland, MA 01375   Fax: 413-548-9746
E-mail: info@ctriversalmon.org
http://www.ctriversalmon.org
To support the effort to restore Atlantic salmon in the Connecticut River basin. An invitation is extended to explore the site to learn more about our organization and the work being done to re-establish a species that had been extinctin this region since about 1800.
Jan Rowan, Executive

3100 Department of Homeland Security Customs and Border Patrol
10 Causeway Street   617-565-6210
Room 801   Fax: 617-565-6277
Boston, MA 02222   http://www.customs.treas.gov
Philip Spayd, Director

3101 EPA: Region 1, Air Management Division
JFK Federal Building   617-565-3800
Boston, MA 02203   http://www.epa.gov/region01
The region's Air Enforcement program is responsible for conducting various compliance monitoring and enforcement activities under the Clean Air Act that govern the operation of stationary pollution sources. In addition, we are chargedwith the responsibility of overseeing the performance of state air pollution enforcement program.

3102 Environmental Affairs Agricultural Development
251 Causeway Street   617-727-9800
Boston, MA 02114   Fax: 617-727-2754
E-mail: Valerie.Mahoney@state.ma.us
http://www.mass.gov/bb/fy97h1/EOENV.HTM#ENA
Robert W. Gollege Jr., Commissioner

3103 Environmental Affairs Bureau of Markets
251 Causeway Street   617-626-1750
Suite 500   Fax: 617-626-1850
Boston, MA 02114-2151http://www.massdfa.org/agricult.html
Susan Black, Chief

3104 Environmental Affairs Metropolitan District
Executive Office of Environmental Affairs
100 Cambridge Street   617-626-1000
Suite 900   Fax: 617-626-1181
Boston, MA 02114   E-mail: env.internet@state.ma.us
http://www.mass.gov/envir/eoea.htm
Stephen R. Pritchard, Secretary

3105 Environmental Affairs: Environmental Protection
251 Causeway Street   617-626-1000
9th Floor   Fax: 617-626-1181
Boston, MA 02114

Daniel Greenbaum, Commissioner

3106 Environmental Affairs: Animal Health
100 Cambridge Street   617-727-3018
Boston, MA 02202   Fax: 617-727-7235
          E-mail: dwebber@state.ma.us

**3107 Environmental Affairs: Environmental Management
Department**
251 Causeway Street Suite 700   617-727-9800
Boston, MA 02114   Fax: 617-727-2754
     E-mail: Valerie.Mahoney@state.ma.us
    http://www.mass.gov/bb/fy97h1/EOENV.HTM#ENA
Robert W. Gollege Jr., Commisioner

**3108 Environmental Affairs: Environmental Law Fisheries &
Wildlife Division**
251 Causeway Street   617-727-3151
Suite 900   Fax: 617-727-7288
Boston, MA 02114

The DFWELE is responsible for protecting and managing Massachusetts's various natural habitats, native plants, wildlife, freshwater fish, and marine species. The divisions consist of: Fisheries; Wildlife; Heritage and EndangeredSpecies; Realty; and Publ;ic Information and education.
David Peters, Comm. of Fisheries/Wildlife

3109 Environmental Affairs: Fisheries, Wildlife and Environmental Law Enforcement Department
175 Portland Street   617-727-3190
Boston, MA 02114   Fax: 617-727-8551
William McKeon, Deputy Director

**3110 Environmental Affairs: Hazardous Waste Facilities Site
Safety Council**
100 Cambridge Street   617-727-6629
Boston, MA 02202

Regina McCarthy, Director

**3111 Environmental Protection Agency Region 1 (CT, ME,
MA, NH, RI, VT)**
1 Congress Street   617-918-1111
Suite 1100   Fax: 617-918-1112
Boston, MA 02114-2023   http://www.epa.gov/region1

The mission of the US Environmental Protection Agency is to protect human health and to safeguard the natural environment-air, water and land- upon which life depends.

3112 Massachusetts Department of Environmental Protection
1 Winter Street   617-292-5500
Boston, MA 02108   800-462-0444
          http://www.state.ma.us/dep

The Masachusetts Department of Environmental Protection is a state agency responsible for protecting human health and the environment by ensuring clean air and water, the safe management and disposal of solid and hazardous wastes, thetimely cleanup of hazardous waste sites and spills, and the preservation of wetlands and coastal resources.

3113 Massachusetts Highway Department
10 Park Plaza   617-973-7484
Boston, MA 02116   Fax: 617-973-8879
          http://www.state.ma.us/mhd
Gregory Prendergast, Deputy Chief Engineer

3114 New England Interstate Water Pollution Control Commission
Boot Hills S   978-323-7929
100 Foot of John Street   Fax: 978-323-7919
Lowell, MA 01852   E-mail: mail@neiwpcc.org

The New England Interstate Water Pollution Control Commission, a nonprofit interstate agency established by an Act of Congress, serves and assist its members states individually and collectively by providing coordination, publiceducation, training and leadership in the management and protection of water quality in the New England Region and New York.
Ronald F Poltak, Executive Director

**3115 United States Department of the Army US Army Corps
of Engineers**
696 Virginia Road   978-318-8220
Concord, MA 01742-2751   Fax: 978-318-8821

3116 Waquoit Bay National Estuarine Research Program
149 Waquoit Highway   508-457-0495
PO Box 3092   Fax: 617-727-5537
Waquoit Bay, MA 02536

In 1979, The Commonwealth of Massachusetts designated Waquoit Bay as an Area of Critical Environment Concern (ACEC) in recognition of its significant natural resources. The designation provides a state-wide umbrella of protection andoversight under the exisiting regulations of different state agencies which include higher standards of protection for ACEC's. The Waquoit Bay ACEC includes parts of the Bay proper. Although the ACEC covers 2522 acres, including Washburn Island andSouth Cape Beach.

Maryland

3117 Chesapeake Bay Executive Council
410 Severn Avenue, Suite 109   410-267-5700
Chesapeake Bay Program Office   800-908-0229
Annapolis, MD 21403   Fax: 410-267-5777
       http://www.chesapeakebay.net/exec.htm

The Chesapeake Bay Executive Council establishes the policy direction for the restoration and protection of the Chesapeake Bay and its living resources. A series of Directives, Agreements, and Amendments signed by the ExecutiveCouncil set goals and guide policy for the Bay restoration. The Council meets annually.

Edward Rendell, Chairman
Michael Burke, Acting Assoc. Dir. of Comm.

**3118 Department of Commerce: National Marine Office of
Habitat Protection Chesapeake Bay**
410 Severn Avenue 107   410-267-5660
Chesapeake Bay Office   Fax: 410-267-5666
Annapolis, MD 21403   http://noaa.chesapeakebay.net
Mary E Gillelan, Chief

3119 Health & Environmental Research Advisory Committee: AC-16
Department of Energy, Office of Energy Research
19901 Germantown Road   301-903-2987
Germanton, MD 20874-1290
Jean Hummer

3120 Interstate Commission on the Potomac River Basin
51 Monroe Street   301-984-1908
PE-08   Fax: 301-984-5841
Rockville, MD 20850   E-mail: info@icprb.org
          http://www.potomacriver.org

The Interstate Commission on the Potomac River Basin is an interstate compact agency established to help protect the Potomac River and its 14,670-square-mile watershed. Its mission is to enhance, protect, and conserve the water andassociated land resources of the Potomac River and its tributaries through regional and interstate cooperation.

Joseph Hoffman, Executive Director

**3121 Maryland Department of the Environment: Southern
Maryland Field Office**
200 Duke Street   410-414-3400
Prince Frederick, MD 20678
Robert Cole, Manager

3122 Maryland Department of Agriculture
50 Harry S Truman Parkway   410-841-5700
Annapolis, MD 21401   800-492-5590
          Fax: 410-841-5914
          http://www.mda.state.md.us

Established on the basis of agriculture's growing importance and impact to the economy of the state. Many activities are regulatory in nature, others are assigned to a category of public service and some are educational or promotional in scope. All are intended to provide the maximum protection possible for the consumer as well as promote the economic well-being of farmers, food and fiber processors and businesses engage in agricultural related operations.
Lewis R Riley, Secretary

3123 Maryland Department of Agriculture: State Soil Conservation Committee
50 Harry S Truman Parkway   410-841-5863
Annapolis, MD 21401   Fax: 410-841-5736
Dave Thomas, Chairman

3124 Maryland Department of Health and Mental Hygiene
201 W Preston Street   410-767-6500
Baltimore, MD 21201   877-463-3464
    Fax: 410-767-6489
    http://www.dhmh.state.md.us/
The Maryland Department of Health and Mental Hygiene's mission is to protect and promote health and prevent disease and injury. This is accomplished through the provision of population-based health services and core public health;assessment, assurance and policy development.
S Anthony McCann, Secretary
Lisa Ellis, Administration

3125 Maryland Department of Natural Resources Chesapeake Bay & Watershed
Tawes State Office Building   410-260-8116
Annapolis, MD 21401   http://www.dnrstate.md.us/bay
Verna Harrison, Assistant Secretary

3126 Maryland Department of Natural Resources: Resource Management Services
580 Taylor Avenue Room C-4   410-260-8100
Annapolis, MD 21401   Fax: 410-260-8111
    http://www.dnr@state.md.us
Sarah Taylor-Rogers, Secretary

3127 Maryland Department of the Environment: Water Management Field Office
1800 Washington Boulevard   301-689-8494
Suite 505   Fax: 301-689-6543
Baltimore, MD 21230-1718 E-mail: mdecambr@intercom.net
To restore and maintain the quality of the State's ground and surface waters, protect wetland habitats throughout the State, and manage the utilization of Maryland's mineral resources.
Marley Lewis, Manager

3128 Maryland Department of the Environment/ Emergency Response Division: Westport Field Office
2500 Broening Highway   410-333-2950
Baltimore, MD 21224   Fax: 410-333-3728
    E-mail: awilliams@mde.state.md.us
Alan Williams, Manager

3129 Maryland Department of the Environment/Water Management: Nontidal Wetlands & Waterways
District Court/Multi Service Bldg   410-543-6703
201 Baptist Street, Suite 22   Fax: 410-543-6740
Salisbury, MD 21801   E-mail: salisb@shore.intercom.net
Steve Dawson, Manager

3130 Maryland Department of the Environment: Air & Radiation Management Field Office
2500 Broening Highway   410-333-2950
Baltimore, MD 21224   Fax: 410-333-3728
    E-mail: awilliams@mde.state.md.us

3131 Maryland Department of the Environment: Air and Radiation Management Main Field Office
160 South Water Street   301-689-5756
Frostburg, MD 21532   Fax: 301-689-6544
    E-mail: frostbur@hereintown.net
Laurie K Bucher, Manager

3132 Maryland Department of the Environment: Field Operations Office
416 Chinquapin Round Road   410-974-3238
Annapolis, MD 21401
John Steinfort, Manager

3133 Maryland Department of the Environment: Main Office
2500 Broening Highway   410-631-3000
Baltimore, MD 21224   800-633-6101
    E-mail: mdeprf@olg.com
To protect and restore the quality of Maryland's air, land, and water resources, while fostering economic development, healthy and safe communities, and quality environmental education for the benefit of the environment, public health,and future generations.

3134 Maryland: National Capital Park & Planning Commission
6611 Kenilworth Avenue   301-454-1740
Riverdale, MD 20737   Fax: 301-454-1750
    E-mail: webmanager@mncppc.org
    http://www.mncppc.org
Is a bi-county agency empowered by the State of Maryland to acquire, develop, maintain and administer a regional system of parks withing Montgomery and Prince George's Counties, and to prepare and administer a general plan for thephysical development of the two counties.
Trudye Morgan Johnson, Executive Director

3135 Research, Education and Economics Agricultural Research Service
1400 Independence Avenue SW   202-720-3656
Jamie L Whitten Federal Building   http://www.ars.usda.gov
Washington, DC 20250
The USDAs Research, Education and Economics mission agencies and their university partners lack a central integrated, user friendly electronic information system capable of providing a knowledge base of the thousands of programs andprojects for which they are responsible that focus on food, agriculture, natural resources, and rural development. Such a system is increasingly needed to enable the Department and its partners to readily conduct both comprehensive baseline andongoing assessments.
Floyd Horn, Administrator

Michigan

3136 Great Lakes Environmental Research Laboratory
2205 Commonwealth Boulevard   734-741-2393
Ann Arbor, MI 48105-2945   Fax: 734-741-2055
    http://www.glerl.noaa.gov
Conducts integrated, interdisciplinary environmental research in support of resource management and environmental services in coastal and estuarine waters with a special emphasis on the Great Lakes. Laboratory performs field,analytical, and laboratory investigations to improve understanding and prediction of coastal and estaurine processes and the interdependencies with the atmosphere and sediments.
Michael Quigley, Public Affairs Officer

3137 Great Lakes Fishery Commission
2100 Commonwealth Boulevard   734-662-3209
Ann Arbor, MI 48105   Fax: 734-741-2010
    http://www.glfc.org
The Great Lakes Fishery Commision was established by the Convention on Great Lakes Fisheries between Canada and the US in 1955. It has two major responsibilities: to develop coordinated programs of research on the Great Lakes and, onthe basis of the findings, to recommend measures which will permit the maximum sustained productivity in stocks of fish of common concern; and to formulate and implement a program to eradicate or minimize sea lampry populations in the Great Lakes.

3138 Michigan Department of Community Health: Bureau of Child and Family Services
3423 N Martin Luther King Jr Blvd. 517-335-8885
PO Box 301 866-691-5323
Lansing, MI 48909 Fax: 517-335-8509
http://www.michigan.gov/MDCH
A state agency which continually and diligently endeavors to prevent disease, prolong life and promote the public health.
Sharon Hudson RN, MSN, Coordinator
Carol Hinkle RN, Nurse Consultant

3139 Michigan Department of Environmental Quality
525 W Allegan Street, 6th Floor 517-373-7917
PO Box 30473 Fax: 517-241-7401
Lansing, MI 48909 http://www.michigan.gov/DEQ
Steven Chester, Director

3140 Michigan Department of Natural Resources
Box 30444 517-241-7427
Lansing, MI 48909 Fax: 517-241-7428
E-mail: dnr-wld-webpages@state.mi.us
http://www.dnr.states.mi.us
K Cool, Director

3141 Michigan State University Extension
11 Agriculture Hall 517-355-0108
East Lansing, MI 48824 Fax: 517-432-1048
E-mail: loverids@msue.msu.edu
http://www.msue.msu.edu/msue/
Michigan State University Extension helps people improve their lives through an educational process that applies knowledge to critical issues, needs and opportunities.
Scott Loveridge, Professor

3142 Natural Resources: Waste Management Division
PO Box 30028 517-373-2730
Lansing, MI 48909 517-373-1547
Jim Sygo, Chief

3143 Natural Resources: Wildlife Division
PO Box 30028 517-373-1263
Lansing, MI 48909 Fax: 517-373-1547
http://www.dnr.state.mi.us
George Burgoyne, Chief

Minnesota

3144 Midwest Interstate Low-Level Radioactive Waste Commission
336 Robert Street N 614-292-1868
Room 130 http://www.ag.ohio-state.edu/~rer/rerhtml/re_61.html
St Paul, MN 55101

3145 Minnesota Board of Water & Soil Resources
520 Lafayette Road North 651-296-3767
St Paul, MN 55115 Fax: 651-297-5615
http://www.bwsr.state.mn.us
The Minnesota Board of Water and Soil Resources assists local governments to manage and conserve their irreplaceable water and soil resources.
Ron Harnack, Executive Director

3146 Minnesota Department of Agriculture
90 West Plato Boulevard 651-297-2200
St Paul, MN 55107-2094 800-967-2474
Fax: 651-297-5522
http://www.mda.state.mn.us
The MDA's mission is to work toward a diverse ag industry that is profitable as well as environmentally sound; to protect the public health safety regarding food and ag products; and to ensure orderly commerce in agricultural and foodproducts. We have two major branches of the department to accomplish this mission: regulatory divisions and non-regulatory divisions.
Gene Hugoson, Commissioner

3147 Minnesota Department of Natural Resources
500 Lafayette Road 651-296-6157
St. Paul, MN 55155-4040 888-646-6367
E-mail: info@dnr.state.mn.us
http://www.dnr.state.mn.us
The DNR vision hinges on the concept of sustainability. To DNR, sustainability means protecting and restoring the natural environment while enhancing economic opportunity and community well-being. DNR endorsed ecosystem-basedmanagement as its method to achieve sustainability goals. Sustainability addresses three related elements: the environment, the economy and the community. The goal is to maintain all three elements in a healthy state indefinitely.

3148 Minnesota Environmental Quality Board
658 Cedar Street 651-296-9535
Centennial Building E-mail: eqb@mnplan.state.mn.us
St Paul, MN 55155 http://www.mnplan.state.mn.us/eqb
The mission of the Environmental Quality Board is to lead Minnesota environmental policy by responding to key issues, providing appropriate review and coordination, serving as a public forum and developing long-range strategies toenhance Minnesota's environmental quality. The Environmental Quality Board consists of 10 state agency commissioners or directors and five citizen members. It was established by the Minnesota Legislature in 1973.

3149 Minnesota Pollution Control Agency
520 Lafayette Road North 651-296-6300
St. Paul, MN 55155 800-657-3864
Fax: 651-296-7923
http://www.pca.state.mn.us
Established in 1967 to protect Minnesota's environment through monitoring environmental quality and enforcing environmental regulations.
Karen Studders, Comm./Chair Citizens' Board

3150 Minnesota Pollution Control Agency: Duluth
525 S Lake Avenue Suite 400 218-723-4660
Duluth, MN 55802 800-657-3864
Fax: 218-723-4727
http://www.pca.state.mn.us
MPCA staff at Duluth and the other six agency offices: identify environmental problems through testing, monitoring, inspections and research; develop environmental priorities; set standards and propose rules to protect people and theenvironment; develop permits; provide technical assistance; respond to emergencies and encourage pollution prevention and sustainability.
Suzanne Hanson, Manager

3151 Minnesota Sea Grant College Program
University of Minnesota
208 Washburn Hall 218-726-8106
2305 E 5th Street Fax: 218-726-6556
Duluth, MN 55812 E-mail: seagr@d.umn.edu
http://www.seagrant.umn.edu

3152 Mississippi River Coordinating Commission: National Park Service
175 5th Street E 612-290-4160
Suite 418
St Paul, MN 55101

3153 United States Department of Agriculture United States Forest Service
1992 Folwell Avenue 651-649-5249
St. Paul, MN 55108

3154 United States Department of the Army US Army Corps of Engineers
190 5th Street East 651-290-5300
St. Paul, MN 55101-1638 Fax: 651-290-5478

Mississippi

3155 Army Coastal Engineering Research Board
Waterways Experiment Station   601-634-2000
3909 Halls Ferry Road   Fax: 601-634-2818
Vicksburg, MS 39180-6199

Leonard G Hassel, Executive Secretary

3156 Gulf Coast Research Laboratory
703 E Beach Drive (39564)   228-872-4200
PO Box 7000   Fax: 228-872-4204
Ocean Springs, MS 39566-7000   http://www.ims.usm.edu
Robert Vanaller, Interm Director

3157 Gulf of Mexico Fisheries Commission
PO Box 726   228-875-5912
Ocean Springs, MS 39564   Fax: 228-875-6604
E-mail: mailto:webmaster@gsmfc.org
http://gsmfc.html
The Gulf States Marine Fisheries commission is an organization of the five states whose coastal waters are the Gulf of Mexico. It has as its principal objective the conservation, development, and full utilization of the fisheryresources of the Gulf of Mexico, to provide food, employment, income, and recreation to the people of these United States.
Thomas Gollott, Chairman

3158 Land & Water Resources Bureau
2380 Highway 80 W   601-961-5200
Jackson, MS 39204   Fax: 601-354-6290
E-mail: Jamie_Crawford@deq.state.ms

3159 Mississippi Alabama Sea Grant Consortium
703 E Beach Dr Caylor Bldg   228-818-8836
Suite 200 PO Box 7000   Fax: 228-818-8841
Ocean Springs, MS 39566-7000   http://www.masgc.org
A federal state partnership that is dedicated to activities that foster the conservation and sustainable development of coastal and marine resources in Mississippi and Alabama.

Dr Edward R Richardson, Acting President
Dr Shelby F Thames, President

3160 Mississippi Department Agriculture & Commerce
PO Box 1609   601-354-7050
Jackson, MS 39215-1609   Fax: 601-354-6290
E-mail: Spell@mdac.state.ms.us
http://www.mdac.state.ms.us/Index.asp
The mission of the Mississippi Department of Agriculture and Commerce is to regulate and promote agricultural-related businesses within the state and to promote Mississippi's products throughout both the state and the rest of the worldfor the benefit of all Mississippi citizens.
Umesh Sanjanwala, Director of Informational Sy

3161 Mississippi Department of Wildlife, Fisheries and Parks
1505 Eastover Drive   601-432-2400
Jackson, MS 39211
It is the mission of the Mississippi Department of Wildlife, Fisheries and Parks to conserve and enhance Mississippi's natural resources, to provide continuing outdoor recreational opportunities, to maintain the ecological integrityand aesthetic quality of the resources and to ensure socioeconomic and educational opportunities for present and future generations.

3162 Mississippi Forestry Commission
301 North Lamar Street Suite 300   601-359-1386
Jackson, MS 39201   Fax: 601-359-1349
http://www.mfc.state.ms.us

James L Sledge Jr, State Forester

3163 Mississippi Sea Grant Program
Kinard Hall Wing E Room 262   662-915-7775
PO Box 1848   Fax: 662-915-5267
University, MS 38677-1848   E-mail: sealaw@olemiss.edu
http://www.olemiss.edu

The legal arm of the Mississippi-Alabama Sea Grant Consortium, the legal program serves the legal needs of Gulf of Mexico region. Program attorney work closely with the Consortium, extension agents and state agencies to provide timelylegal research and advice on Gulf issues.
Stephanie Shawalter, Director

3164 Mississippi State Department of Health Bureau of Child/Adolescent Health
PO Box 1700   601-960-7634
Jackson, MS 39215-1700   Fax: 601-576-7364
http://www.msdh.state.ms.us
The Mississippi Department of Health's mission is to promote and protect the health of the citizens of Missippi. Geographic focus: Missippi other organizational activies, not directed specifically toward children; advocacy, directservice delivery, education, organizing, regulation, social services
Sam Valentine, Director

3165 Mississippi State Health Department Environmental Health Bureau
PO Box 1700   601-576-7680
Jackson, MS 39215-1700   Fax: 601-576-7364
http://www.msdh.state.ms.us

Rick Harrington, Director

3166 Mississippi: Environmental Quality Pollution Control Department
PO Box 10385   601-961-5171
Jackson, MS 39289-0385   Fax: 601-354-6612
E-mail: charles-chisolm@deq.state.ms.us
http://www.deq.state.ms.us
Charles Chisolm, Director

3167 Policy Review Board of the Gulf of Mexico: Gulf of Mexico Program
Building 1103   228-688-3726
Room 202   Fax: 228-688-2709
Stennis Space Center, MS 39529   http://www.gmpo.gov
The Gulf of Mexico Program has identified 12 priority coastal areas. We believe that through assitance to the States agencies and communities that are dealing with the critical issues in these 12 local areas, the Gulf Program will makea vital contribution to improving the quality of life for the citizens in those areas as well as to restoring and protecting the health of the entire Gulf ecosystem.
Dr Douglas A Lipka

3168 United States Department of the Army US Army Corps of Engineers
4155 Clay Street   601-631-5010
Vicksburg, MS 39180-3435   Fax: 601-631-5296

Missouri

3169 Missouri Conservation Department
2901 W Truman Boulevard   573-751-4115
PO Box 180   Fax: 573-751-4467
Jefferson City, MO 65102
John D Hoskins, Director

3170 Missouri Department of Natural Resources
PO Box 176   573-751-4732
Jefferson City, MO 65102   800-361-4827
Fax: 573-751-7627
E-mail: oac@dnr.mo.gov
http://www.dnr.mo.gov
The Department of Natural Resources preserves, protects and enhances Missouri's natural, cultural and energy resources and works to inspire their enjoyment and responsible use for present and future generations. Our staff work toensure that our state enjoys clean air to breathe, clean water for drinking and recreation and land that sustains a diversity of life.
Doyle Childers, Director
Connie Patterson, Director Communications

3171 Natural Resources Department: Pollution Control
PO Box 176
Jefferson City, MO 65102   800-334-6946
                                        Fax: 573-751-8656
        E-mail: cleanair@mail.dnr.state.mo.us
Maintains the purity of Missouri's air to protect the health, general welfare and property of the people.
Charles H Chisolm, Director

3172 Natural Resources: Energy Division
PO Box 176   573-751-2254
Jefferson City, MO 65102   Fax: 573-751-7627
Anita Randolph, Director

3173 Natural Resources: Environmental Improvement and Energy Resources
PO Box 744   573-751-4919
Jefferson City, MO 65102-0176   Fax: 573-751-7627
Thomas Welch, Director

3174 United States Department of the Army US Army Corps of Engineers
US Army Engineer District
601 E 12th Street   816-983-3201
Kansas City, MO 64106-2896   Fax: 816-983-5575

Montana

3175 Butte District Advisory Council
106 N Parkmont   406-494-5059
PO Box 3388   Fax: 406-494-3474
Butte, MT 59701   http://www.mt.blm.gov/bdo/

Rick Hotaling, Field Manager

3176 Crown of the Continent Research Learning Center - Glacier National Park
PO Box 128   406-888-5827
West Glacier, MT 59936   Fax: 406-888-7903
        E-mail: leigh_welling@nps.gov
http://www.nps.gov/glac/learningcenter/learningcenter.htm
The center initiates and facilitates research and learning in Glacier National Park and the Crown of the Continent Ecosystem so that communities, both regional and global, can make informed decisions as stewards of the region's vastrepositories of cultural and natural resources.
Leigh Welling, Director
Sallie Hejl PhD, Resource Ed Specialist

3177 Environmental Quality Council
State Capitol, Room 171   406-444-3742
PO Box 201704   Fax: 406-444-3971
Helena, MT 59620-1704   E-mail: teverts@state.mt.us
The EQC is a state legislative committee create by the 1971 Montana Environmental Policy Act. As outlined in MEPA, the EQC'S purpose is to encourage conditions under which people can coexist with nature in productive harmony. TheCouncil fulfills this purpose by assisting the Legislature in the development of natural resource and environmental policy, by conducting studies on related issues and by serving in an advisory capacity to the state's natural resource programs.
Todd Events, Environment Analyst

3178 Interagency Grizzly Bear Committee USFWS NS 312
University of Montana
University Hall 309   406-329-3223
Missoula, MT 59812   Fax: 406-329-3212
The Rocky Mountains are one of the few places in the lower 48 states that are home to both black bears and grizzly bears. By practicing the adequate safety techniques, we can reduce the chances of an encounter.
Skip Ladd, Director

3179 Lewiston District Advisory Council Bureau of Land Management
Box 1160 Airport Road   406-538-7461
Lewistown, MT 59457   Fax: 406-538-1904
                            http://www.nps.gov/lecl/fedsites.htm
David L Mari, Designated Federal Employee

3180 Montana Department of Agriculture
PO Box 200201   406-444-3144
Helena, MT 59620-0201   Fax: 406-444-5109
                                   E-mail: agr@state.mt.us
                                   http://www.agr.state.mt.us
W Peck, Director

3181 Montana Natural Heritage Program
1515 East 6th Avenue   406-444-3009
Helena, MT 59620-1800   Fax: 406-444-0581
                            E-mail: mtnhp@nris.state.mt.us
                            http://www.nris.state.mt.us/mtnhp
Susan Crispin, Director

3182 Natural Resources & Conservation Department
1625 11th Avenue   406-444-2074
Po Box 201601   Fax: 406-444-2684
Helena, MT 59620-1601   http://www.dncr.state.mt.us
Bob Clinch, Director

3183 Natural Resources & Conservation Department: Conservation & Resource Development Division
1520 East 6th Avenue   406-444-6667
Helena, MT 59620-2301   Fax: 406-444-6721
                                   E-mail: rbeck@mt.gov
                                   http://www.dnrc.state.mt.us

Ray Beck, Administrator

3184 Natural Resources & Conservation: Oil & Gas
2535 St Johns Avenue   406-656-0040
Billings, MT 59102   Fax: 406-657-1604
A quasi-judicial body that is attached to the DNRC for administrative purposes only. The board's regulatory actions serve three primary purposes: 1) to prevent waste of oil and gas resources, 2) to conserve oil and gas by encouragingmaximum efficient recovery of the resource, and 3) to protect the correlative rights of the mineral owners, i.e., the right of each owner to recover its fair share of the oil and gas underlying its lands.
Tom Richmond, Administrator

3185 Natural Resources & Conservation: Water Resource
1520 East 6th Avenue   406-444-6605
Helena, MT 59620-2301   Fax: 406-444-6721
Jack Stults, Administrator

Nebraska

3186 Department of Agriculture: Natural Resources Conservation Service
National Soil Survey Center
100 Centennial Mall N Room 152   402-437-5499
Lincoln, NE 68508   Fax: 402-437-5336
                                   http://http://soils.usda.gov

3187 Department of Natural Resources
301 Centennial Mall S   402-471-2363
PO Box 94676   Fax: 402-471-2900
Lincoln, NE 68509-4676   http://www.dnr.state.ne
State Natural Resources Agency
Roger K Patterson, Director

3188 Games & Parks Commission: Nebraska
Ak-Sar-Ben Aquarium   402-332-3901
21502 W Highway 31   Fax: 402-332-5853
Gretna, NE 68028   http://www.ngpc.state.ne.us
Darrell Fiet, Director

3189 Health Department: Drinking Water and Environmental Sanitation Division
301 Centennial Mall S
PO Box 95007
Lincoln, NE 68509-5007
402-471-2541
Fax: 402-471-6436

3190 Mississippi River Corridor Study Commission Planning & Resources Preservation Natl. Park Svc.
1709 Jackson Street
Omaha, NE 68102-2571
402-221-3431
Fax: 402-221-3461
William W Schenk, Regional Director

3191 Nebraska Department of Agriculture
301 Centennial Mall S
PO Box 94947
Lincoln, NE 68509
402-471-2341
Fax: 402-471-6876
http://www.agr.state.ne.us
Regulatory state agency.

Greg Ibach, Director
Jamie Karl, Assistant Director

3192 Nebraska Department of Environmental Quality
1200 N Street Suite 400
PO Box 98922
Lincoln, NE 68509   E-mail: MoreInfo@NDEQ.state.NE.US
402-471-2186
Fax: 402-471-2909
http://www.deq.state.ne.us
To protect Nebraska's air, land and water resources
Mike Linder, Director

3193 Nebraska Department of Natural Resources
301 Centennial Mall S
PO Box 94676
Lincoln, NE 68509-4876
402-471-2363
Fax: 402-471-2900
http://www.dnr.state.ne.us
State Natural Resources Agency.
Roger Patterson, Director

3194 Nebraska Ethanol Board
301 Centennial Mall S
PO Box 94922
Lincoln, NE 68509   E-mail: tsneller@ethanol.state.ne.us
402-471-2941
Fax: 402-471-2470
http://www.ne-ethanol.org
The Nebraska Ethanol Board assists ethanol producers with programs and strategies for marketing ethanol and related co-products. The Board supports organizations and policies that advocate the increased use of ethanol fuels, andadministers public information, education and ethanol research projects. The Board also assists companies and organizations in the development of ethanol production facilities in Nebraska.

Todd C Sneller, Administrator

3195 Nebraska Game & Parks Commission: Fisheries Division
2200 N 33rd Street
PO Box 30370
Lincoln, NE 68503   E-mail: dgabel@ngpc.state.ne.us
402-471-5515
Fax: 402-471-5528
http://www.ngpc.state.ne.us
Don Gabelhouse, Division Administrator

3196 Nebraska Games & Parks Commission
2200 N 33rd Street
Lincoln, NE 68503
402-471-0641
Fax: 402-471-5528
E-mail: ngpc@ngpc.state.ne.us
http://www.ngpc.state.ne.us
Rex Amack, Director

3197 Nebraska Games & Parks: Wildlife Division
2200 N 33rd Street
PO Box 30370
Lincoln, NE 68503-0370   E-mail: jim.douglas@ngpc.ne.gov
402-471-5411
Fax: 402-471-5528
http://www.ngpc.state.ne.us

3198 United States Department of the Army US Army Corps of Engineers
US Army Engineer District
215 N 17th Street
Omaha, NE 68102-4978
402-221-3900
Fax: 402-221-3128

Nevada

3199 Bureau of Land Management
Department of the Interior
HC 33 Box 33500
Ely, NV 89301
775-289-1800
Fax: 775-289-1910
http://www.nv.blm.gov/Ely.
Gene A Kolkman, Designated Federal Employee

3200 Carson City Field Office Advisory Council
Bureau of Land Management
5665 Morgan Mill Road
Carson City, NV 89701-1448
775-885-6000
Fax: 775-885-6147
http://www.nv.blm.gov/carson/default.htm
A Federal Land Management Agency.
John Singlaub, Field Manager

3201 Colorado River Basin Salinity Control Program
US Bureau of Land Reclamation
125 S State Street
Room 6335
Salt Lake City, UT 84138-1147
801-524-3753
Fax: 801-524-5499
http://www.usbr.gov
The Colorado River and its tributaries provide municipal and industrial water to about 27 million people and irrigation to nearly four million acres of land in the US. The river also serves about 2.3 million people and 500,000 acres inMexico. The threat of salinity is a major concern in both the US and Mexico. Salinity affects agricultural, municipal and industrial water users. We work to control the salinity of the Colorado river and thereby to protect the land and people.
Kib Jacobs, Program Manager

3202 Conservation and Natural Resources Department
Wildlife Division, Conservation Education
1100 Valley Road
Reno, NV 89512
775-688-1558
Fax: 775-688-1595
E-mail: ndowinfo@govmail.state.nv.us
Nevada Division of Wildlife is the state's fish and game agency, responsible for the management of the state's wildlife and boating on state waters.
Maureen Angel, Publications Writer

3203 Conservation and Natural Resources: Water Resources Division
123 West Nye Lane
Carson City, NV 89706-0818
775-687-4380
Fax: 775-687-6972
E-mail: hricci@wr.state.nv.us
http://ndwr.state.nv.us
Ricci Hugh, State Engineer

3204 Department of the Interior: Bureau of Reclamation
Lower Colorado Regional Office
PO Box 61470
Boulder City, NV 89006
702-293-8000
Fax: 702-293-8418
http://www.usbr.gov/lc
Manage a number of environmental managment programs related to the lower Colorado River.
Deanna Miker, Dir., Resource Mgt. Office

3205 Elko District Advisory Council: Bureau of Land Management
3900 E Idaho Street
Elko, NV 89801
775-753-0200
http://www.nv.blm.gov/elko

3206 Las Vegas Bureau of Land Management
4701 N Torrey Pines Drive
Las Vegas, NV 89130
702-515-5000
http://www.nv.blm.gov
The Bureau of Land Management's mission is to help sustain the health, diversity and productivity of public lands so they can be used and enjoyed by both present and future generations.

3207 Nevada Bureau of Mines & Geology
University of Nevada
Mail Stop 178   775-784-6691
Reno, NV 89557-0088   Fax: 775-784-1709
http://www.nbmg.unr.edu
Jonathan Price, Director/State Geologist

3208 Nevada Department of Conservation and Natural Resources
901 South Stewart Street   775-684-2700
Suite 5001   Fax: 775-684-2715
Carson City, NV 89701-0818   http://www.state.nv.us
Protecting the natural resources of Neveda
Allen Biaggi, Director
Kay Scherer, Assistant Director

3209 Nevada Department of Conservation: Wildlife Division
PO Box 10678   775-688-1500
Reno, NV 89520   Fax: 775-688-1595
http://www.state.nv.us/cnr
Terry Crawforth, Administrator

3210 Nevada Natural Heritage Program
1500 East College Parkway Suite 145   775-687-4245
Carson City, NV 89706-7921   Fax: 775-687-1288
http://www.stat.nv.us/nvnhp
Glen Clemmer, Program Manager

3211 Tahoe Regional Planning Agency (TRPA) Advisory Planning Commission
308 Dorla Court   775-588-4547
PO Box 1038   Fax: 775-588-4527
Zephyr Cove, NV 89448-1038   E-mail: mrhoades@trpa.org
http://www.trpa.org
The TRPA leads the cooperative effort to preserve, restore and enhance the natural and human environment of the Lake Tahoe Region. The Code of Ordinances regulates, among other things, land use, density, rate of growth, land coverage,excavation and scenic impacts. These regulations are designed to bring the region into compliance with the threshold standards established for water quality, air quality, soil conservation, wildlife habitat, vegetation, noise, recreation and scenicresources.

New Hampshire

3212 New Hampshire Agriculture Markets & Foods: Department Conservation Committee
PO Box 2042   603-271-3551
Concord, NH 03302-2042   Fax: 603-271-1109
Coordinates the work of the ten county conservation districts in the state of New Hampshire.
Samuel Doyle, Chairperson
Joanna Pellerin, Coordinator

3213 New Hampshire Department of Environmental Services
6 Hazen Drive   603-271-3503
PO Box 95   800-735-2964
Concord, NH 03302   http://www.des.state.nh.us

3214 New Hampshire Fish and Game Department
2 Hazen Drive   603-271-3422
Concord, NH 03301   Fax: 603-271-1438
E-mail: info@wildlife.state.nh.us
http://www.wildlife.state.nh.us
As the guardian of the states fish, wildlife and marine resources, the department works with the public to: conserve, manage and protect these resources and their habitats; inform and advise the public about these resources; providethe public with opportunities to use and appreciate these resources.
Richard Moquin, Commission Chairman
Jim Jones, Commission Secretary

3215 Northeastern Forest Fire Protection Commission
P.O. Box 6192   207-968-3782
21 Parmenter Terrace   Fax: 207-968-3782
China Village, ME 04926
E-mail: necompact@pivot.net
http://www.nffpc.org/html
The mandate of the Northeastern Forest Fire Protection Commission is to provide the means for its member states and provinces to cope with fires that might be beyond the capabilities of a singler member through infromation, technologyand resources sharing activities.
Thomas G. Parent, Executive Director

3216 Resources & Development Council: State Planning
Department of Resources & Economic Development
172 Pembroke Road   603-271-2411
P.O. Box 1856   Fax: 603-271-2629
Concord, NH 03302-1856   http://www.state.nh.us
Sean O'Kane, Commissioner

3217 University of New Hampshire Cooperative Extension
131 Main Street   603-862-1520
215 Nesmith Hall   Fax: 603-862-1585
Durham, NH 03824   http://ceinfo.unh.edu
Robert Edmonds, Program Ldr Forest/Wildlife

3218 Wildcat River Advisory Commission White Mountain National Forest
Carter Notch Road   603-383-6547
Jackson, NH 03846   http://www.irn.org

New Jersey

3219 Environmental Protection and Energy: Fish and Wildlife
501 E State Street, 3rd Fl, Bldg 5   609-292-9410
PO Box 400   Fax: 609-984-1414
Trenton, NJ 08625-0400   http://www.nfishandwildlife.com
Our mission is to protect and manage the state's fish and wildlife to maximize their long-term biological, recreation and economic values for all New Jerseyans.
Robert McDowell, Director

3220 New Jersey Department of Agriculture State Soil & Conservation Committee
PO Box 330   609-292-3976
Trenton, NJ 08625   Fax: 609-292-3978
Arthur Brown, Chairman

3221 New Jersey Department of Environmental Protection
401 E State Street   609-292-2885
7th Floor, E Wing   Fax: 609-292-7695
Trenton, NJ 08625   http://www.state.nj.us.dep
A state department dedicated to protecting New Jersey's air, land, water and natural resources.

Bradley M Campbell, Commissioner

3222 New Jersey Dept. Environmental Protection: Division of Publicly Funded Site Remediation
401 E State Street Cn 028   609-984-2902
PO Box 413   Fax: 609-777-0756
Trenton, NJ 08625-0402   http://www.state.nj.us/edt
Anthony Farro, Director

3223 New Jersey Geological Survey Department Environmental Protection
PO Box 427   609-292-1185
Trenton, NJ 08625-0427   Fax: 609-633-1004
E-mail: karl.muessig@dep.state.nj.us
http://www.njgeology.org
State agency that maps, interprets and provides geoscience information to the public on geology and ground water resources.

Karl Muessig, State Geologist

3224 New Jersey Pinelands Commission
PO Box 7   609-894-7300
New Lisbon, NJ 08064   Fax: 609-894-7330
                            E-mail: info@njpines.state.nj.us
                            http://www.state.nj.us/pinelands/
The 15-member Pinelands Commission is a volunteer citizen
body composed of one member appointed by each of the seven
counties in the Pinelands, one member appointed by the US Secretary of the Interior, and seven members appointed by the Governor. The Commission reviews applications for development,
works with local officials to implement planning strategies, conducts scientific research, and educates the public.
Annette M Barbaccia, Executive Director

**3225 New Jersey: Department Environmental Protection
Law Enforcement**
401 E State Street Cn 422   609-292-9430
PO Box 400   Fax: 609-984-1414
Trenton, NJ 08625-0400   http://www.state.nj.us/dep/fgw
Rob Winkel, Chief

New Mexico

3226 Albuquerque Bureau of Land Management
435 Montano Road NE   505-761-8700
Albuquerque, NM 87107-4935   Fax: 505-761-8911

3227 Attorney General
Environmental Enforcement
PO Drawer 1508
Santa Fe, NM 87504-1508   505-827-6000
                                    Fax: 505-827-4440

3228 Canadian River Commission
New Mexico Interstate Stream Commission
PO Box 25102   505-827-6160
Santa Fe, NM 87504   Fax: 505-827-6188
                                http://www.ose.state.nm.us
Philip B Mutz, Secretary

**3229 Energy, Minerals & Natural Resources: Energy Conser-
vation Management**
2040 South Pacheco   505-827-5900
Santa Fe, NM 87505   Fax: 505-438-3855
                        http://www.emnrd.state.nm.us/ecmd/
Encourages efficient energy use in New Mexico by offerin programs and information for state agencies, companies and
induviduals.
Dianne Caron, Director

3230 Energy, Minerals and Natural Resources Department
1220 S St. Francis Drive   505-476-3200
PO Box 6429   Fax: 505-476-3220
Santa Fe, NM 87505   http://www.emnrd.state.nm.us
Mission is to provide leadership in the protection, conservation,
management and responsible use of New Mexico's natural re-
sources.
Jennifer Salisbury, Cabinet Secretary

**3231 Las Cruces District Advisory Bureau of Land Manage-
ment**
Department of the Interior
1800 Marquess Street   505-525-4300
Las Cruces, NM 88005   Fax: 505-525-4412
Linda SC Rundell, Designated Ferderal Employee

3232 New Mexico Bureau of Geology & Mineral Resources
801 Leroy Place   505-835-5420
Socorro, NM 87801   Fax: 505-835-6333
                            E-mail: scholle1@nmt.edu
                            http://www.geoinfo.nmt.edu
A service and research division of the New Mexico Institute of
Mining and Technology. Acts as the geological survey of New
Mexico.
Peter Scholle, Director & State Geologist

3233 New Mexico Cooperative Fish & Wildlife Research Unit
New Mexico University MSC 4901   505-646-6053
PO Box 30003   Fax: 505-646-1281
Las Cruces, NM 88003-0003
Bruce Thompson, Leader
Lewis Bender, Assisstant Unit Leader

3234 New Mexico Department of Game & Fish
566 N Telshor Boulevard   505-522-9796
Las Cruces, NM 88011   Fax: 505-522-8382
                            E-mail: charthorn@state.nm.us
                            http://www.gmfsh.state.nm.us

3235 New Mexico Department of Game and Fish
One Wildlife Way   505-476-8000
PO Box 25112   E-mail: ispa@state.nm.us
Santa Fe, NM 87504   http://www.gmfsh.state.nm.us
Larry Bell, Director

3236 New Mexico Environment Department
1190 St. Francis Drive   505-827-2855
Harold S Runnels Building   800-219-6157
Santa Fe, NM 87502-0110   http://www.nmenv.state.nm.us
The New Mexico Environment Department's mission is to pro-
vide the highest quality of life throughout the state by promoting
a safe, clean, and productive environment.
Peter Maggiore, Cabinet Secretary

3237 New Mexico Soil & Water Conservation Commission
MSC APR PO Box 30005   505-646-2642
Las Cruces, NM 88003-8005   Fax: 505-646-1540
                        E-mail: acoleman@nmda.nmsu.edu
Julie Maitland, Division Director
Anne Coleman, Administrative Secretary

**3238 Roswell District Advisory Council: Bureau of Land
Management**
2909 W 2nd Street   505-627-0272
Roswell, NM 88201   Fax: 505-627-0276

**3239 United States Department of the Interior: United States
Fish and Wildlife Service**
500 Gold Avenue SW   505-248-6282
PO Box 1306   Fax: 505-248-6845
Albuquerque, NM 87102   E-mail: diane_knudson@fws.gov
                            http://southwest.fws.gov
Diane Knudson

3240 United States Forest Service: Southwestern R-3
517 Gold Avenue SW   505-476-3300
Albuquerque, NM 87102

New York

3241 Adirondack Park Agency
State Route 86   518-891-4050
PO Box 99   Fax: 518-891-3938
Ray Brook, NY 12977   http://www.apa.state.ny.us
Daniel T Fitts, Executive Director

3242 Environmental Conservation
625 Broadway   518-402-8540
Albany, NY 12233   Fax: 518-402-8541
Denise Sheehan, Acting Commissioner

**3243 Environmental Conservation Department: Air Re-
sources**
50 Wolf Road   518-457-7230
Albany, NY 12233
Thomas Allen, Director

3244 Environmental Conservation: Marine Resources
205 N Belle Mead Road   631-444-0430
Suite 1   Fax: 631-444-0434
East Setauket, NY 11733
Gordon Colvin, Director

3245 Environmental Financial Advisory Board (EFAB)
US EPA, Office of Enterprise Technology & Innovat.
Environmental Finance Program (Mail   202-564-4994
Code) 2731R 1200 Pennsylvania Ave. NWFax: 202-565-2587
Washington, DC 20460   E-mail: bowie.vanessa@epa.gov
                http://http://www.epa.gov/efinpage/efab.htm
The EFAB provides advice to the Environmental Protection
Agency's Administrator and Program Offices on "how to pay"
questions for environmental protection. They are a federally
chartered advisory committee operating under the
FederalAdvisory Committee Act.

Vanessa Bowie, Contact
Alecia Crichlow, Contact

3246 Interstate Sanitation Commission
311 West 43rd Street Room 201   212-582-0380
New York, NY 10036 E-mail: OW-General@epamail.epa.gov
                http://www.epa.gov/305b/98report/interstate.html
Established in 1936 by the Tri-State Compact, which has approved by its member states and the US Congress, the Interstate
Sanitation Commission (ISC) is a tri-state environmental
agency formed by the states of New York, New Jersey,
andConnecticut. The Interstate Sanitation District encompasses
approximatly 797 square miles of estuarine waters in the Metropolitan Area shared by the states.
Howard Golub, Director/Cheif Engineer

3247 NYS Tug Hill Commission
PO Box 6063   315-779-8240
Watertown, NY 13601   Fax: 315-785-2574
                E-mail: thtomorr@northnet.org
                http://www.tughilltomorrowlandtrust.org
Assists local governments and organizations in the 2,100 square
mile Tug Hill region in natural resource management and rural
economic development.

Linda Garrett, Executive Director

3248 Natural Resources and Environmental Programs
Agriculture and Markets Department
10 B Airline Drive   518-457-0752
Albany, NY 12235   Fax: 518-457-3087
                http://www.agmkt.state.ny.us
Rick Zimmerman, Deputy Commissioner

3249 New York Cooperative Fish & Wildlife Research Unit
Cornell University, Fernow Hall   607-255-2151
Natural Resources Department   Fax: 607-255-1895
Ithaca, NY 14853   E-mail: dnrcru-mailbox@cornell.edu
                http://www.dnr.cornell.edu/f@wres/nycf@wru.htm

Dr. Milo Richmond, Unit Leader

3250 New York Department of Environmental Conservation
SUNY - Building 40   631-444-0354
Stony Brook, NY 11790   Fax: 631-444-0349
                http://www.dec.state.ny.us
The New York Department of Environmental Conservation's
mission is to conserve, improve, and protect its natural resources
and environment, and control water, land and air pollution, in or-
der to enhance the health, safety, and welfareof the people of the
state and their overall economic and social well being. Serving
Nassau and Suffolk counties
Peter A Scully, Regional Director

3251 New York Department of Health
Flanigan Square   518-402-7500
547 River Street   Fax: 518-402-7509
Troy, NY 12180-2216   http://www.health.state.ny.us
Working together and committed to excellence, we protect and
promote the health of New Yorkers through prevention, science
and the assurance of quality health care delivery.
Ronald Tramontano, Health Program Administrator

**3252 New York Department of Health: Center for Environ-
mental Health**
2 University Place
Albany, NY 12203   800-458-1158
                                        Fax: 518-458-6732
                        http://www.health.state.ny.us
The Center for Environmental Health applies scientific, medical, engineering and public health expertise to identify, understand, prevent and mitigate risks to human health from New
York's living and working environments.
Richard Svenson

**3253 New York State Office of Parks, Recreation and His-
toric Preservation**
Empire State Plaza   518-474-0456
Agency Building 1   Fax: 518-486-2924
Albany, NY 12238   http://www.nysparks.com
The agency operates 168 parks offering a wide variety of recreational, cultural and education activities, and 35 state historic
sites; sponsors boating and snowmobiling, nature study and outreach programs; manages grant programs forboating and snowmobiling enforcement and aid to zoos, botanical gardens,
aquariums; and administers funds for federal historical preservation and parks programs, the Environmental Protection Fund
and the 1996 Clean Water/Clean Air Bond act.
Bernadette Castro, Commissioner

3254 New York State Soil and Conservation Committee
10B Airline Drive   518-457-3738
Albany, NY 12235   Fax: 518-457-3412
                E-mail: Barb.Silvestri@agmkt.state.ny.us
                        http://www.nys-soilandwater.org
The New York State Soil and Water Conservation Committee is
composed of voting and advisory members who represent a wide
range of agricultural, environmental and other interests. The
Committee operates through a network of partnershipsbetween
state, federal and local agencies, as well as citizen interests and
the private sector. The mission of the Committee is to develop
and oversee and agricultural nonpoint source water quality
program for New York State.
Dennis Hill, Chair
Ronald Kaplewicz, Director

**3255 United States Department of the Army US Army Corps
of Engineers**
1766 Niagara Street   716-879-4200
Buffalo, NY 14207-3199   Fax: 716-879-4195
                        http://www.lrb.usace.army.mil

North Carolina

3256 Agriculture Department
Food & Drug Administration
2109 Blue Ridge Road   919-733-3556
PO Box 27647   Fax: 919-733-9796
Raleigh, NC 27611   http://www.ncagr.com
John L Smith Jr, Director

3257 Carnivore Preservation Trust
1940 Hanks Chapel Road   919-542-4684
Pittsboro, NC 27312   Fax: 919-542-4454
                                        E-mail: info@cptigers.org
                                        http://www.cptigers.org
Is a wildlife sanctuary, offering unique opportunities to learn
about these animals and their critical importance to our quality
of life on Earth.
Pam Fulk, Executive Director

**3258 Environment, Health & Natural Resources: Marine
Fisheries**
3441 Arendell Street   252-726-7021
Morehead City, NC 28557   Fax: 252-726-0254

3259 North Carolina Board of Science and Technology
301 North Wilmington Street   919-733-6500
1326 Mail Service Center   Fax: 919-733-8356
Raleigh, NC 27699-1326

E-mail: ncbst@nccommerce.com
http://www.ncscienceandtechnology.com
Encourages, promotes, and supports scientific, engineering, and industrial research applications in North Carolina.
Robert McMahan, PhD, Executive Director

3260 North Carolina Department of Agriculture
1001 Mail Service Center   919-733-7125
Raleigh, NC 27699-1001   Fax: 919-733-1141
http://www.agr.state.nc.us
Steve Trexler, Commissioner

3261 North Carolina Department of Environment and Natural Resources
1601 Mail Service Center   919-733-4984
Raleigh, NC 27699-1601   Fax: 919-715-3060
http://www.enr.state.nc.us

3262 United States Department of the Army US Army Corps of Engineers
69 Darlington Avenue   910-251-4501
PO Box 1890   910-251-4185
Wilmington, NC 28402-1890 http://www.saw.usace.army.mil
Provides quality planning, design, construction, and operations products and services to meet the needs of civilian and military customers.

North Dakota

3263 Dakotas Resource Advisory Council: Department of the Interior
Bureau of Land Management
2933 3rd Avenue W
Dickinson, ND 58601   701-227-7700
          Fax: 701-227-8510
E-mail: mramsey@blm.gov
http://www.blm.gov/nhpl/facts/index.htm
The Dakotas Council currently has 15 members. It is structured to provide a balance of membership by area of expertise, training, and experience. It consists of five individuals in each of three categories.

Lonny Bagley, Field Manager
Mary Ramsey, RAC Coordinator

3264 Institute for Ecological Studies
University of North Dakota
PO Box 7110
Grand Forks, ND 58203   701-777-2851
Richard Crawford, Director

3265 ND Game and Fish Department
100 N Bismarck Expressway   701-328-6300
Bismarck, ND 58501-5095   Fax: 701-328-6352
E-mail: ndgf@state.nd.us
http://www.discovernd.com/gnf
To protect, conserve and enhance fish and wildlife populations and their habitats for sustained public consumptive and nonconsumptive use.
Dean Hildebrand, Director
Roger Rostvet, Deputy Director

3266 North Dakota Forest Service
307 First Street E   701-228-5422
Bottineau, ND 58318-1100   Fax: 701-228-5448
E-mail: Larry.Kotchman@ndsu.edu
http://www.state.nd.us/forest
The ND Forest Service administers forestry programs statewide. The agency operates a nursery at Towner specializing in the production of conifer tree stock. The nursery is the sole supplier of evergreen seedlings in North Dakota.Technial assistance relating to the management of private forest lands, state forest lands, urban and community forests, tree planting and wildland fire protection is provided by the agency. The ND Forest Service also owns and manages app. 13,278acres of state lands.

Larry Kotchman, State Forester

3267 North Dakota Parks and Recreation Department
1600 E Century Ave.   701-328-5357
Suite 3   Fax: 701-328-5363
Bismarck, ND 58503   E-mail: parkrec@state.nd.us
http://www.ndparks.com
The state government agency charged with managing North Dakota's state parks and recreation areas; the state's nature preserves and natural area programs; motorized and non-motorized trail programs; recreational grants and state-widerecreation planning; and state scenic byways program.

Donna Schouweiler, Public Info Coordinator

Ohio

3268 Childhood Lead Poisoning Prevention Program
Ohio Department of Health
246 N High Street   937-285-6250
6th Floor   Fax: 937-285-6306
Columbus, OH 43266   http://www.odh.state.oh.us/data
The mission of Ohio's Childhood Lead Poisoning Prevention Program is to eliminate childhood lead poisoning through screening, environmental inspection, abatement, education and case management.
Pamela G Rollins, Director of Pediatrics

3269 Environmental Protection Agency: Center for Environmental Research Information
Office of Research
26 West Martin Luther King Drive   513-569-7562
Cincinnati, OH 45268   Fax: 513-569-7585
Calvin Lawrence, Director

3270 Environmental Protection Agency: Environmental Services Division
1571 Perry Street   614-644-4247
Columbus, OH 43201   Fax: 614-644-4272

3271 Environmental Protection Agency: Ohio Division of Surface Water
PO Box 1049   614-644-2001
Columbis, OH 43216-1049   Fax: 614-644-2745
http://www.epa.state.oh.us
To protect, enhance and restore all waters of the state for the health, safety and welfare of present and future generations. We accomplish this mission by monitoring the aquatic environment, permitting, enforcing environmental laws,using and refining scientifically sound methods and regulations, planning, coordinating, educating, providing technical assistance and encouraging pollution prevention practices.
Lisa Morris, Chief

3272 Ohio Department of Natural Resources
Division of Geological Survey   614-265-6576
2045 Morse Road C-L   Fax: 614-447-1918
Columbus, OH 43229   E-mail: geo.survey@dnr.state.oh.us
http://www.ohiodnr.com/geosurvey
Provides geologic information and services for responsible managemtn of Ohio's natural resources. Geologic maps, reports and data files developed by the division can be used by individuals, educators, industry, business andgovernment.

3273 Ohio Environmental Protection Agency
122 S Front Street   614-644-3020
PO Box 1049   Fax: 614-644-2329
Columbus, OH 43216 E-mail: request@www.epa.state.oh.us
http://www.epa.state.oh.us
Christopher Jones, Director

3274 Ohio River Valley Water Sanitation Commission
181 Renslar Avenue   513-231-7719
Cincinnati, OH 45228   Fax: 513-231-7761
E-mail: info@orsanco.org
http://www.orsanco.com

ORSANCO operates programs to improve water quality in the Ohio River and its tributaries, including: setting waste water discharge standards; performing biological assessment; monitoring for the chemical and physical properties of the waterways; and conducting special surveys and studies. Also coordinates emergency response activities for spills or accidental discharges to the river and promotes public participating programs.

3275 Ohio Water Development Authority
88 East Broad Street Suite 1300   614-466-5822
Columbus, OH 43215   Fax: 614-644-9964
                                         http://owda.org
Provides financial assistance for environmental infrastructure from the sale of municipal revenue bonds through loans to local governments in Ohio and issuing Industrial Revenue Bonds for qualified projects. The vision of OWDA is to continue to provide assistance for environmental infrastructure by being responsive to the needs of local government agencies, enhancing the provision of financial and technical assistance and developing new financial assistance products for the private sector.
Steven J Grossman, Executive Director

3276 United States Department of the Air Force Major Air Commands
4225 Logistics Avenue   937-656-1409
Suite 8   Fax: 937-257-5875
Wright Patterson, OH 45433
Michael Cornelius, Natural Resource Manager

3277 United States Department of the Army: US Army Corps of Engineers
PO Box 1159   513-684-3002
Cincinnati, OH 45201-1159   Fax: 513-684-2085
                                E-mail: CELRD-DE@usace.army.mil

Oklahoma

3278 Childhood Lead Poisoning Program
Oklahoma Department of Health
1000 Northeast 10th Street   405-271-4471
Oklahoma, OK 73117   Fax: 405-271-6199
                                         http://www.health.state.ok.us
The Oklahoma State Department of Health is to protect and promote the health of citizens of Oklahoma and to prevent disease and injury, and to assure the conditions by which our citizens can be healthy.
Pamela G Rollins, Director Pediatrics

3279 Conservation Commission: Water Quality Division
5225 N Shurtel   405-810-1009
Oklahoma City, OK 73118   Fax: 405-810-1046

Larry Edmison, Director

3280 Oklahoma Department of Environmental Quality
707 N Robinson   405-702-1000
PO Box 1677   Fax: 405-702-1001
Oklahoma City, OK 73101-1677
Administers environmental laws considering both the economy of today and the environment of tomorrow.

3281 Salt Plains National Wildlife Refuge
Route 1 Box 76   405-626-4794
Jet, OK 73749   Fax: 405-626-4793
                                         http://www.gorp.com

Rodney F Krey, Manager

Oregon

3282 Burns District Advisory Council
Bureau of Land Management
12533 Highway 20 West   508-808-6001
Hines, OR 97738-9409   E-mail: or020mb@or.blm.gov
                                http://www.or.blm.gov/Burns/

Mike Green, Designated Fed. Empl.

3283 Department of Transportation
355 Capitol St. NE
Salem, OR 97301-3871   888-275-6368
                                         Fax: 503-986-3432
Lori Sundstrom, Manager

3284 Eugene District Advisory Council: Bureau of Land Management
Department of the Interior
PO Box 10226   541-683-6600
Eugene, OR 97440-2226   888-442-3061
                                         Fax: 541-683-6981
                                E-mail: ro090mb@or.blm.gov
                                http://www.edo.or.blm.gov
The Eugene District manages several ecosystems ranging from coastal inlands to dense Douglas-fir, hemlock, and cedar forests. The wide variation in the lands managed by the District offers the perfect compromise between the urban parks in the cities and the high elevation recreation opportunities in the adjacent Willamette, Siuslaw and Umpqua National Forest.
Doug Huntington

3285 Governor's Office: Natural Resources and Environment
254 State Capitol   503-378-6827
Salem, OR 97301-4047   Fax: 503-378-4863
Louise Solliday, Senior Policy Advisor

3286 Klamath River Compact Commission
280 Main Street   541-882-4436
Klamath Falls, OR 97601
Created by the Klamath River Compact in 1957, KRCC is a three member commission whose purpose, with respect to the water of the Klamath River Basin, is to faciliate and promote the orderly, integrated, and comprehensive developement, use, conservation and control of water for development of lands by irrigation, protection of fish and wildlife, domestic and industrial use, hydropower, navigation, and flood protection.
Richard Fairclo, Consultant

3287 Lakeview District Advisory Council: Bureau of Land Management
Department of the Interior
Highway 395 North   541-947-2177
Hc 64 Box 60   Fax: 541-947-6399
Lakeview, OR 97630   http://www.fs.fed.us
Renee Snyder

3288 Medford District Advisory Council: Bureau of Land Management
Withcombe Hall   503-808-6001
Corvallis, OR 97331-6704
The Bureau of Land Management's Medford District oversees approximately 862,000 acres of scattered public lands between the Cascade and Siskiyou mountain ranges and from the Oregon/California border to Canyon Creek and southern Douglas County. This large land base is divided into four Resource Areas: Ashland, Butte Falls, Grants Pass and Glendale.

3289 Oregon Department of Environmental Quality
811 SW Sixth Avenue   503-229-5983
Portland, OR 97204-1390   800-452-4011
                                         Fax: 503-229-6924
                                E-mail: deq.info@deq.state.or.us
                                http://www.deq.state.or.us

3290 Oregon Department of Fish and Wildlife
3406 Cherry Avenue NE   503-947-6000
Salem, OR 97303   800-720-6339
                                E-mail: odfw.info@state.or.us
                                http://www.dfw.state.or.us
Commissioners formulate general state programs and policies concerning management and conservation of fish wildlife resources and establishes seasons, methods and bag limits for recreational and commercial take.

3291 Oregon Department of Forestry
2600 State Street
Salem, OR 97310
503-945-7200
800-437-4490
Fax: 503-945-7212
http://www.odf.state.or.us

3292 Oregon Department of Land Conservation and Development
635 Capitol Street NE
Suite 150
Salem, OR 97301-2540
503-373-0050
Fax: 503-378-5518
http://www.lcd.state.or.us
Shelia Preston, Executive Assistant

3293 Oregon Water Resource Department
158-12th S & NE
Salem, OR 97301-4172
503-378-8455
Fax: 503-378-2496

3294 Pacific States Marine Fisheries Commission
205 SE Spokane Street
Suite 100
Portland, OR 97215
503-595-3100
Fax: 503-595-3232
E-mail: front.office@psmfc.org
http://www.psmfc.org
PSMFC is one of three Insterstate Commissions dedicated to resolving fishery issues. It is composed of five member states: Alaska, California, Idaho, Oregon & Washington. While it has no regulatory or management authority, it wascreated to provide collective participation by states to work on mutual problems of the fisheries resource.

Randy Fisher, Executive Director
Sharon Perkins, Executive Assistant

3295 Policy Management Office of Environmental Affairs: Pacific Northwest
Department of the Interior
500 NE Multnomah Street
Suite 600
Portland, OR 97232-2036
503-231-6157
Fax: 503-231-2361

3296 Prineville District Advisory Council: Bureau of Land Management
Department of the Interior
3050 NE Third Street
Prineville, OR 97754
541-416-6700
Fax: 541-416-6798
E-mail: or050mb@or.blm.gov
http://www.or.blm.gov
A Barron Bail, Designated Federal Employee

3297 Roseburg District Advisory Council Bureau of Land Management
777 NW Garden Valley Boulevard
Roseburg, OR 97470
541-440-4930
Fax: 541-440-4948
E-mail: joseph_ross@or.blm.gov
Public lands of the Roseburg District, located in southwestern Oregon, contain some of the most productive forests in the world. An important mainstay of the local economy, which acquires timber from both private and federal lands inthe region. The district is criss-crossed with streams and rivers that support sport fishing. With Interstate 5 running through the middle of the district, and east-west state highways connecting Crater Lake to the Pacfic coast, the district drawsmany tourists.

3298 Salem District Advisory Council: Bureau of Land Management
Department of the Interior
1717 Fabry Road Southeast
Salem, OR 97306
503-375-5646
Fax: 503-375-5622
E-mail: salem_mail@blm.gov
BLM Mission: to sustain the health, diversity and productivity of the public lands for the use and enjoyment of present and future generations. Salem District manages 400,000 acres scattered across 13 counties. Seventy three percent ofOregon's population live within the boundries of this district. Their major focus is an ecosystem management approach involving many different disciplines. Salem employs 200 full-time employees working in forestry, land surveying, wildlife biology,hydrology, etc.
Denis Williamson, District Manager
Paul Jeske, Supervisor/Natural Resources

3299 United States Department of the Army: US Army Corps of Engineers
US Army Engineer Division
PO Box 2946
Portland, OR 97208-2946
503-808-5150

3300 Vale District Advisory Council: Bureau of Land Management
100 Oregon Street
Vale, OR 97918-9630
541-473-3144
Fax: 541-473-6213
E-mail: vale_mail@or.blm.gov
http://www.or.blm.gov
The Vale District of the Bureau of Land Management manages 4.9 million acres of public land in eastern Oregon. The mission of the BLM is to sustain the health, diversity, and productivity of the public lands for the use and enjoymentof present and future generations.
Jim May, Designated Federal Employee

Pennsylvania

3301 Childhood Lead Poisoning Prevention Program
Pennsylvania Department of Health
PO Box 90
Harrisburg, PA 17108
717-783-8451
Fax: 717-772-0323
The mission of the Pennsylvania Department of Health, Childhood Lead Poisoning Prevention Program is to make the citizens of the Commonwealth aware of the dangers of lead poisoning and to reduce the number of children who becomelead-poisoned.

3302 Citizens Advisory Council
Pennsylvania Department Environmental Protection
13th Floor, RCSOB
PO Box 8459
Harrisburg, PA 17105-8459
717-787-4527
Fax: 717-787-2878
E-mail: stmioff@state.pa.us
http://www.cacdep.state.pa.us
Susan Wilson, Executive Director
Stephanie Mioff, Administrative Assistant

3303 Delaware Water Gap National Recreation Area Citizen Advisory Commission
1 River Road
Bushkill, PA 18324
570-588-2418

3304 Department of the Interior: National Parks
200 Chestnut Street
5th Floor
Philadelphia, PA 19106
215-597-7013
Fax: 215-597-8015

3305 Environmental Protection Agency: Region 3 Hazardous Waste Management
1650 Arch Street (3CG00)
Philadelphia, PA 19103-2029
215-814-5000
The Waste Minimization National Plan emphasizes source reduction and environmentally sound recycling over waste treatment and disposal. The goal is to prevent transfers chemical releases from one medium (air, water, or land) toanother.

3306 Environmental Protection Agency: Region III
1650 Arch Street
Philadelphia, PA 19103-2029
215-814-2900
Fax: 215-814-2901
http://www.epa.gov/region03
Region III is responsible for federal environmental programs in Delaware, Maryland, Pennsylvania, Virginia, West Virginia and District of Columbia. Programs include air and water pollution control; toxic substances, pesticides, anddrinking water regulation; wetlands protection; hazardous waste management, hazardous waste dump site cleanup; and some aspects of radioactive materials regulation.

3307 Northern Allegheny National Wild & Scenic River/Allegheny National Forest

Forest Service
222 Liberty Street
Warren, PA 16365
814-723-5150
Fax: 814-726-1465
E-mail: anf/r9_allegheny@fs.fed.us
http://www.fs.fed.us

An organization dedicated to providing advice for development of the corridor management plan for the northern section of the Allegheny River that has been designated as a National Wild and Scenic River.
Don Clymer, Staff Contact

3308 Pennsylvania Department of Conservation and Natural Resources

Rachel Carson State Office Building
7th Floor, PO Box 8767
Harrisburg, PA 17105-8767
717-787-2869
Fax: 717-772-9106
http://www.dcnr.state.pa.us

3309 Pennsylvania Fish and Boat Commission: Northeast Region

PO Box 88
Sweet Valley, PA 18656
570-477-5717
Fax: 570-477-3221
E-mail: wdietz@state.pa.us
http://www.fish.state.pa.us

Our mission is to provide fishing and boating opportinities through the protection and management of Pennsylvania's aquatic resources.
Sally A Corl, Law Enforcement Manager
Walter M Dietz, Aquatic Resources Specialist

3310 Pennsylvania Forest Stewardship Program

PO Box 8552
Harrisburg, PA 17105-8552
717-787-2106

Forest stewardship is a US Forest Service program with the goal of helping private landowners manage their lands for various objectives. Landowners participating in the Forest Stewardship Program work with a private forestry consultantto depelop a customized plan for their land and objetives. Studies show that landowners that work with professionals and follow their customized plan are more likely to engage in practices that sustain forest values.

3311 Pennsylvania Game Commission

2001 Elmerton Avenue
Harrisburg, PA 17110-9797
717-787-4250
Fax: 717-772-0542
http://www.pgc.state.pa.us

The Pennsylvania Game Commission has the specific responsibility of acting as steward of the Commonwealth's wild birds and wild animals for the benefit of present and future generations. In carrying out this state constitutionalmandate, the Pennsylvania Game Commission will: Protect, conserve and manage the diversity of wildlife and their habitats; Provide wildlife related education, services, and recreational opportunities for both consumptive and non-consumptive uses ofwildlife.

Dr. Nicholas Spook, President

3312 Regional Center for Environmental Information US

EPA Region 3
1650 Arch Street (3PM52)
Philadelphia, PA 19103
215-814-5254
Fax: 215-814-5253
E-mail: library-reg3@epa.gov
http://www.epa.gov/reg3rcei

The Regional Center for Environmental Information provides information services to EPA staff and the general public.
Diane McCreary, Task Order Project Officer

3313 Susquehanna River Basin Commission

1721 N Front Street
Harrisburg, PA 17102
717-238-0423
Fax: 717-238-2436
E-mail: srbc@srbc.net
http://www.srbc.net

The responsibility of SRBC is to enhance public welfare through comprehensive planning, water supply allocation & management of the water resources of the Susquehanna River Basin. The SRBC works to reduce damages caused by floods;provide for the reasonable & sustained development & use of surface & ground water for municipal, agricultural, recreational, commercial & industrial purposes; protect & restore fisheries, wetlands & aquatic habitat; protect water quality & instreamuses.
Paul O Swartz, Executive Director

3314 United States Department of the Army US Army Corps of Engineers

US Army Engineer District
1000 Liberty Avenue
Pittsburgh, PA 15222
412-395-7103
Fax: 412-644-4093

Jean Yeager, Administrative Assistant

Rhode Island

3315 Environmental Management: Division of Fish and Wildlife

4808 Tower Hill Road
Wakefield, RI 02879
401-789-3094
Fax: 401-783-4460
http://www.state.ri.us/dems/programs/

Agency manages the fish and wildlife resources of the State of Rhode Island inluding marine fisheries. The division has 60 employees and is located in 4 stations statwide.

John OBrien, Acting Chief
Michael Lapisky, Deputy Chief

3316 Environmental Management: Parks and Recreation

2321 Hartford Avenue
Johnston, RI 02919
401-277-2632
Fax: 401-934-0610
E-mail: riparks@earthlink.net
http://www.riparks.com

William Hawkins Jr, Chief

3317 Rhode Island Department of Environmental Management

235 Promenade Street
Providence, RI 02908-5767
401-222-6800
http://www.state.ri.us/dem

We are committed to preserving the quality of Rhode Island's environment, maintaining and safety of its residents and protecting the natural systems upon which life depends. Together with many partners, we offer assistance toindividuals, business and municipalities, conduct research, find solutions, and enforce laws created to protect the environment.

3318 Rhode Island Department of Evironmental Management: Forest Environment

1037 Hartford Pike
North Scituate, RI 02857
401-647-3367
Fax: 401-647-3590
http://www.state.ri.us/dem.programs.bnates/forest/index.htm

Coordinates a statewide forest fire protection plan, provides forest fire protection on state lands, assists rural volunteer fire departments, and develops forest and wildlife management plans for private landowners who choose tomanage their property in ways that will protect these resources on their land. The program promotes public understanding of environmental conservation, enforces Department rules and regulations on DEM lands.

3319 Rhode Island State Water Resources Board

100 North Main Street
Providence, RI 08903
401-222-3961
Fax: 401-222-4707
http://www.wrb.state.ri.us

An executive agency of state government charged with managingthe proper development, utilization and conservation of water resources. The primary responsibility is to ensure that sufficient water supply is available for present andfuture generations, apportioning available water to all areas of the state.
Daniel W Varin, Chairman
Wiliiam J Penn, Vice Chairman

3320 URI Cooperative Extension Education Center
University of Rhode Island
3 E Alumni Avenue   401-874-2900
Kingston, RI 02881   Fax: 401-874-2259
E-mail: ceec@etal.uri.edu
http://www.uri.edu/ce/ceec
The URI Cooperative Extension Education Center is a one-stop facility for the latest research-based, environmentally-sound home and garden practices. Situated in the beautiful URI Learning Landscape Gardens, the Center is home to theURI Master Gardener Program, the URI GreenShare Program and the URI Environmental Education School Program. Through this facility one can also access the URI Plant Protection Clinic, the URI Food Safety Education Program and the URI Master GardenerAssociation.
Marion Gold, Center Director

South Carolina

3321 Department of Interior: South Carolina Cooperative Fish and Wildlife
Clemson University
G27 Lehotsky Hall   864-656-0168
Clemson, SC 29634-0168   Fax: 864-656-1034
David Otis, Unit Leader

3322 Department of Parks: Recreation and Tourism
1205 Pendleton Street   803-734-1700
Edgar A Brown Building   Fax: 803-734-1409
Columbia, SC 29201   http://www.travelsc.com
John Durst, Director

3323 Office of Environmental Laboratory Certification
PO Box 72   803-896-0970
State Park, SC 29147   Fax: 803-896-0850
http://www.scdhec.net/envserv/html
We offer certification to any environmental laboratory wishing to analyze samples for South Carolina's Department of Health and Environmental Control [DHEC]. This scope of certification covers the Safe Drinking Water Act (SDWA), theClean Water Act (NPDES), and solid & hazardous wastes including RCRA and CERCLA requirements (SW846 methodologies).

3324 South Atlantic Fishery Management Council
S Park Building   843-571-4366
Summerville, SC 29483   Fax: 843-769-4520
E-mail: kim.iverson@noaa.gov
http://www.safmc.noaa.gov
The South Atlantic Fishery Management Council is headquartered in Charleston, SC, and is responsible for the conservation and management of fish stocks within the 200-mile limit of the Atlantic off the coasts of North Carolina, SouthCarolina, Georgia, and east florida to Key West.
Kim Iverson, Public Information Officer

3325 South Carolina Department of Health and Environmental Control
2600 Bull Street   803-898-8940
Columbia, SC 29201   Fax: 803-898-8941
http://www.scdhec.net
The Office of Environmental Quality Control is the environmental regulatory arm of the South Carolina Department of Health and Environmental Control. EQC is responsible for the enforcement of federal and state environmental laws andregulations, and for the issuing of permits, licenses and certifications for activities wich may affect the environment. EQC is composed of four program areas, each concerned with specific aspect of environmental protection.

3326 South Carolina Department of Natural Resources
1000 Assembly Street, PO Box 167   803-734-4007
Rembert C Dennis Building   Fax: 803-734-4300
Columbia, SC 29202   http://www.dnr.state.sc.us

John E Frampton, Director

3327 South Carolina Forestry Commission
PO Box 21707   803-896-8800
Columbia, SC 29221   Fax: 803-798-8097
http://www.state.sc.us
The mission of the Forestry Commission is to protect, promote, enhance and nurture the forest lands of South Carolina in a manner consistent with achieving the greatest good for its citizens. Responsibilities extend to all forestlands, both rural and urban, and to all associated forest values and amenities including, but not limited to: timber, wildlife, water quality, air quality, soil protection, recreation and aesthetics.
Bob Schowalter, State Forester
Jennie Morris, Division Director

3328 United States Department of the Army US Army Corps of Engineers
PO Box 919
Charleston, SC 29401-0919

South Dakota

3329 Attorney General's Office Natural Resources Division
500 East Capitol Avenue   605-773-3215
Pierre, SD 57501   Fax: 605-773-4106
E-mail: atghelp@state.sd.us
http://www.state.sd.us/attorney/
The Natural Resources Division of the South Dakota Attorney General's Office provides specialized legal counsel to state agencies in environmental, agricultural, financial, Indian law and natural resource matters. It focuses on (1)state boards and agencies which issue environmental, water, and agricultural permits and the lease of state mineral lands; (2) environmental litigation before boards and agencies and in the courts; and (3) jurisdictional disputes.
John Guhin, Assistant Attorney General
Roxanne Giedd, Deputy Attorney General

3330 Department of Environment & Natural Resources
523 E Capitol Avenue   605-773-3151
Pierre, SD 57501   Fax: 605-773-6035
E-mail: DENRINTERNET@state.sd.us
http://www.state.sd.us/denr/denr.html
Our mission is to provide environmental and natural resources assessment, financial assistance, and regulation in a customer service manner that protects the public health, conserves natural resources, preserves the environment andpromotes economic development.
Steve Pirner, Secretary of the Department
Kim Smith, Public Information

3331 Department of Wildlife and Fishery Sciences
Soutn Dakota State University
Box 2140b   605-688-6121
Brookings, SD 57007   Fax: 605-688-4515
Charles Berry, Leader

3332 South Dakota Department of Game, Fish & Parks
523 E Capitol Avenue   605-773-3485
Pierre, SD 57501-3182   Fax: 605-773-5842
E-mail: wildinfo@state.sd.us
http://www.state.sd.us/gfp
The purpose of the Department of Game, Fish and Parks is to perpetuate, conserve, manage, protect and enhance South Dakota's wildlife resources, parks and outdoor recreational opportunities for the use, benefit and enjoyment of thepeople of this state and its visitors, and to give the highest priority to the welfare of this states's wildlife and parks, and their environment, in planning and decisions.

3333 South Dakota Environmental Health Association
State Department of Health
600 E Capitol Avenue   605-773-3364
Pierre, SD 57501-2536   Fax: 605-773-5904
http://www.state.sd.us/doh
Dave Micklof, Secretary-Treasurer

3334 South Dakota State Extension Services
South Dakota State University
Box 2207D   605-688-4792
Brookings, SD 57007-0093   Fax: 605-688-6347

Tennessee

3335 Carbon Dioxide Information Analysis Center
Oak Ridge National Laboratory
Building 1509   865-574-0390
PO Box 2008   Fax: 865-574-2232
Oak Ridge, TN 37831-6335   E-mail: cdiac@ornl.gov
http://cdiac.esd.ornl.gov/
The primary global-change data and information analysis center of the US Department of Energy. Responds to data and information requests from users from all over the world who are concerned with the greenhouse effect and global climatechange.
Robert M Cushman, Director

3336 Obed Wild & Scenic River
PO Box 429   423-346-6294
Warburg, TN 37887   Fax: 423-346-3362
E-mail: krisstoehr@nps.gov
http://www.nps.gov/obed/
Approximately 45 miles of wild and scenic river are comprised of the Obed River, Clear Creek, Daddy's Creek and Emory River. These water courses have cut rugged gorges leaving exciting whitewater gorges with bluffs as high as 500 feetabove the water.
Kris Stoehr, Unit Manager

3337 Tennessee Agricultural Extension Service
2621 Morgan Circle   865-974-7114
Knoxville, TN 37996   Fax: 865-974-1068
E-mail: clnorman@utk.edu
http://www.utextension.utk.edu/
Statewide educational organization that brings research-based information about agriculture, family and consumer sciences, and resource development to the people of Tennessee where they live and work.
Charles Norman, Dean of Extension Service

3338 Tennessee Department of Agriculture
Ellington Agricultural Center   615-837-5103
Melrose Station, PO Box 40627   Fax: 615-837-5333
Nashville, TN 37204   E-mail: twomack@mail.state.tn.us
http://www.state.tn.us
Ken Givens, Commissioner
Mike Countess, Assistant Commissioner

3339 Tennessee Department of Environment and Conservation
401 Church Street
21st Floor, L&C Tower   888-891-8332
Nashville, TN 37243
Milton H Hamilton Jr, Commissioner

3340 Tennessee Valley Authority
400 W Summit Hill Drive   865-632-2101
Knoxville, TN 37092-1499   E-mail: tvainfo@tva.gov
http://www.tva.gov
TVA generates prosperity in the Tennessee Valley by promoting economic development, supplying low-cost, reliable power and supporting a thriving river system.
Mark Medford, Executive VP

3341 United States Army Engineer District: Memphis
167 N Main Street   901-544-3221
Room B202   800-317-4156
Memphis, TN 38103-1894   http://www.mvm.usace.army.mil
Provides flood control, navigation, environmental stewardship, emergency operations, and other authorized civil works to benefit the region and the Nation.

3342 Wildlife Resources Agency
PO Box 40747   615-781-6500
Nashville, TN 37204   Fax: 615-741-4606

The Tennessee Wildlife Resources Agency develops, manages and maintains sound programs of hunting, fishing, trapping, boating, and other wildlife related outdoor recreational activities.
Gary Myers, Executive Director
William W Cox, Commission Chairman

3343 Wildlife Resources Agency: Fisheries Division
Ellington Agricultural Center   615-781-6575
PO Box 40747   Fax: 615-781-6667
Nashville, TN 37204
Bill Reeves, Chief

3344 Wildlife Resources Agency: Wildlife Management Division
PO Box 40747   615-781-6610
Nashville, TN 37204   Fax: 615-781-6654
Larry Marcum, Chief

Texas

3345 Animal and Plant Health Inspection Service
501 W Felix Street, Building 11   817-885-6923
Animal Care Regional Central Office   Fax: 817-885-6917
Fort Worth, TX 76115
Walter Christensen, Regional Director

3346 Attorney General of Texas Natural Resources Division (NRD)
300 W 15th Street, 10th Floor   512-463-2012
PO Box 12548   Fax: 512-320-0052
Austin, TX 78711-2548   http://www.oag.state.tx.us
The NRD represents the enviromental and energy agencies of the State of Texas in court. NRD's primary activity is the prosecution of lawsuits, referred by state agencies, that involve violations of the state's enviromental and naturalresources protection laws. NRD also defends permits issued by agencies uder those laws and defends challenges to the statues and regulations themselves.NRD also has primary enforcement responsibility for protecting the public's access to Texasbeaches.
Karen Kornell, Chief NRD/Asst Attn General

3347 Bureau of Economic Geology
University of Texas
University Station   512-471-1534
Box X   Fax: 512-471-0140
Austin, TX 78713-7508   E-mail: beg@utexas.edu
W Fisher, Director

3348 Chihuahuan Desert Research Institute
PO Box 1334   915-837-8370
Alpine, TX 79831
Conducts research on the Chihuahuan Desert.

3349 Environmental Protection Agency: Region VI
1445 Ross Avenue, Fountain Place   214-665-6444
12th Floor, Suite 1200   Fax: 214-665-2146
Dallas, TX 75202-2733   http://www.epa.gov/region06
Region 6 encompasses the ecologically, demographically and economically diverse of states of Arkansas, Louisiana, New Mexico, Oklahoma and Texas. The regional vision is to meet the environmental needs of a changing world.
Gregg Cooke, Regional Administrator

3350 Guadalupe: Blanco River Authority
933 E Court   830-379-5822
Seguin, TX 78155   Fax: 830-379-9718
http://www.gbra.org
Aims to conserve and protect the water resources of the Guadalupe River basin and make them available for beneficial use. Services include water and wastewater treatment, water quality testing, the management of water rights anddelivery of stored water, the production of electricity from seven hydroelectric plants and engineering design support.
WE West Jr, General Manager

3351 Parks & Wildlife: Conservation Communication
4200 Smith School Road   512-389-4994
Austin, TX 78744   Fax: 512-389-4448

3352 Parks & Wildlife: Public Lands Division
4200 Smith School Road   512-389-4866
Austin, TX 78744   Fax: 512-389-4960
In 1963 the Parks Board merged with the game and Fish Commission to form the Texas Parks and Wildlife Department. The merger created the Parks Division, currently the Public Lands Division. In 1967 park acquisition and developmentincreased with the passage of a $75 million parks bond authorization and the dedication of a portion of the state's cigarette tax to the development of state and local parks.
Ron Holliday, Director

3353 Parks & Wildlife: Resource Protection Division
4200 Smith School Road   512-389-4864
Austin, TX 78744
The Resource Protection Division protects Tezas fich, wildlife, plant and mineral resources from degradation or depletion. The division investigates any environmental contamination that may cause loss of fish or wildlife. It providesinformation and recommendations to other government agencies and participates in administrative and judicial proceedings concerning pollution incidents, development, development projects and other actions that may affect fish and wildlife.
Larry McKinney, Director

3354 Pecos River Commission
3600 S Stockton   915-943-5171
Monahans, TX 79756   Fax: 915-943-3267

3355 Rio Grande Compact Commission
PO Box 1917   915-532-0196
El Paso, TX 79950-1917   Fax: 915-532-6891
Joe G Hanson, Commisioner

3356 Sabine River Compact Administration
Honorable Wayne Reeh   409-882-0354
3720 N Highway 87   Fax: 409-882-0563
Orange, TX 77632   E-mail: webmaster@www.state.tx.us
http://www.state.tx.us
Wayne Reeh, Commissioner

3357 Texas Agricultural Extension Service
Texas A&M University   409-845-4747
College Station, TX 77843-7101   Fax: 409-862-1637
http://agextension.tamu.edu/
Edward Hiler, Director

3358 Texas Department of Agriculture
PO Box 12847   512-463-7476
Austin, TX 78711   Fax: 512-463-1104
http://www.agr.state.tx.us
Susan Combs, Commissioner
Martin Hubert, Deputy Commissioner

3359 Texas Department of Health
1100 W 49th Street   512-458-7375
Austin, TX 78756   888-983-7111
Fax: 512-458-7686
E-mail: web.master@dshs.state.tx.us
http://www.tdh.state.tx.us
The Texas Department of Health is the state government agency charged with protecting and promoting the health of the public.

3360 Texas Forest Service
301 Tarrow Suite 364   979-458-6600
College Station, TX 77840-7896   Fax: 979-458-6610
http://http://texasforestservice.tamu.edu
The mission is to provide statewide leadership and professional assistance to assure the states's forest, tree and related natural resources are wisely used, nurtured, protected and perpetuated for the benefit of all Texans.
James Hull, Director

3361 Texas Natural Resource Conservation Commission
12100 Park 35 Circle   512-239-1000
Austin, TX 78753   Fax: 512-239-4430
http://www.tnrcc.state.tx.us
The Texas Natural Resource Conservation Commission strives to keep our state's human and natural resources consistent with sustainable economic development. Our goal is clean air, clean water and the safe management of waste.

3362 Texas Parks & Wildlife Department
4200 Smith School Road   512-389-4800
Austin, TX 78744   800-792-1112
Fax: 512-389-4814
E-mail: webcomments@tpwd.state.tx.us
http://www.tpwd.state.tx.us

3363 Texas State Soil & Water Conservation Board
PO Box 658   254-773-2250
Temple, TX 76503   Fax: 254-773-3311
http://www.tsswcb.state.tx.us
The Texas State Soil & Water Conservation Board (TSSWCB) is a state agency that administers Texas' soil and water conservation law and coordinates conservation and pollution abatement programs throughout the state.
Rex Isom, Executive Director

3364 United States Department of the Army: US Army Corps of Engineers
PO Box 17300   817-978-2300
Fort Worth, TX 76102-0300   Fax: 817-978-3311

Utah

3365 Cedar City District Advisory Council
Bureau of Land Management
176 East D.L. Sargent Drive   435-586-2401
Cedar City, UT 84720   Fax: 435-865-3058
http://www.ut.blm.gov/cedarcity_fo/index.htm
Todd S Christiansen, Field Manager

3366 Colorado River Basin Salinity Control Advisory Council: Upper Colorado Region
US Bureau of Reclamation
125 S State Street   801-524-3774
Room 6107   Fax: 801-524-3856
Salt Lake City, UT 84138-110 E-mail: Regional Public Affairs
http://dataweb.usbr.gov
Stan Grappa, Designated Fed. Empl.

3367 Moab District Advisory Council Bureau of Land Management
Department of the Interior
82 East Dogwood   435-259-2100
Moab, UT 84532   Fax: 435-259-2106
Roger Zortman, Designated Federal Employee

3368 Richfield Field Office: Bureau of Land Management
150 E 900 N   435-896-1500
Richfield, UT 84701   Fax: 435-896-1550
Jerry W Goodman, Manager

3369 Salt Lake District Multiple Use Advisory Council
324 S State Street   801-539-4001
PO Box 45155   Fax: 801-539-4013
Salt Lake City, UT 84145-0155   http://www.ut.blm.gov/
The BLM administers public lands within a framework of numerous laws. The most comprehensive of these is the Federal Land Policy and Management Act of 1976. All Bureau policies, procedures and management actions must be consistent withFLPMA and the other laws that govern use of the public lands. It is their mission to sustain the health, diversity and productivity of the public lands for the use and enjoyment of present and future generations.
David Zeller, Designated Federal Employee

3370 Solid Hazardous Waste Bureau
288 North 1460 West   801-538-6170
Salt Lake City, UT 84116   http://www.eq.state.ut.us
To protect public health and the environment by ensuring proper
management of solid and hazardous wastes within the State of
Utah.
Don Verbica, Contact

3371 Upper Colorado River Commission
355 S 400 E Street   801-531-1150
Salt Lake City, UT 84111   Fax: 801-789-4883
The Upper Colorado River Commission is an interstate compact
administration agency created by the Upper Colorado River Ba-
sin Compact of 1948. Since its inception, the Commission (made
up of Commissioners appointed by the Governor of each Upper
Division State and one appointed by the President of the United
States) has actively participated in the development, utilization
and conservation of the water resources of the Colorado River
Basin.
PJ Magura, Administrative Secretary

3372 Utah Department of Agriculture and Food
350 N Redwood Road   801-538-7100
Salt Lake City, UT 84114-6500 Fax: 801-538-7126
                             http://www.ag.state.ut.us
Leonard M Blackham, Commissioner
Kyle Stephens, Deputy Commissioner

3373 Utah Geological Survey
1594 W North Temple, Suite 3110 801-537-3300
PO Box 146100   Fax: 801-537-3400
Salt Lake City, UT 84114   http://www.ugs.utah.gov
The Utah Geological Survey is an applied scientific agency that
creates, interprets and provides information about Utah's geologic environment, resources and hazards to promote safe, beneficial and wise use of land.
Richard Allis, Director

3374 Utah Natural Resources: Water Resources Section
1594 W North Temple, Room 310 801-538-7230
PO Box 146201
Salt Lake City, UT 84114-6201
D Larry Anderson, Director

3375 Utah Natural Resources: Wildlife Resource Division
1594 W N Temple, Suite 2100 801-538-4700
PO Box 146301   Fax: 801-538-4709
Salt Lake City, UT 84114   http://www.nr.utah.gov
Kevin Conway, Director
Max Morgan, Chairman UT Wildlife Board

**3376 Utah State Department of Natural Resources: Office of
Energy & Resource Planning**
1594 W North Temple, Suite 3610 801-538-5428
PO Box 146480   Fax: 801-521-0657
Salt Lake City, UT 84114-6480 E-mail: nroerp@stateut.us
Jeffrey Burks, Director

**3377 Utah State Department of Natural Resources: Division
of Forestry, Fire, & State Lands**
1594 W North Temple, Suite 3520 801-538-5555
PO Box 145703   Fax: 801-533-4111
Salt Lake City, UT 84114-5703
Arthur Dufault, State Forester/Director
Karl Kappe, Strategic Planner

3378 Utah State Soil Conservation Commission
350 North Redwood Road   801-538-7120
Salt Lake City, UT 84116   Fax: 801-538-4949
                             E-mail: vwatkins@state.ut.us
Miles Ferry, Chairman
K Jacobson, Executive

**3379 Vernal District Advisory Council Bureau of Land Man-
agement**
170 South 500 East   435-781-4400
Vernal, UT 84078   Fax: 435-781-4410
          http://www.blm.gov/utah/vernal/whatwedo.html
David E Little, Designated Federal Employee

Vermont

**3380 Agency of Natural Resources Department of Forests,
Park, and Recreation**
103 South Main Street   802-241-3670
Waterbury, VT 05671   Fax: 802-244-1481
Jonathan Wood, Commissioner

**3381 University of Vermont Extension Communication &
Technology Resources**
The University of Vermont   802-656-0385
Burlington, VT 05405   E-mail: kim.parker@uvm.edu
                              http://www.uvm.edu/extension/ctr
Kim Parker, Director

3382 Vermont Agency of Natural Resources
103 S Main Street   802-241-3800
Waterbury, VT 05671   Fax: 802-241-3287
                             http://www.anr.state.vt.us

**3383 Vermont Department of Agriculture, Food and Mar-
kets: Natural Resources Conservation Council**
116 State Street   802-828-3529
Montpelier, VT 05620   Fax: 802-828-2361
                             E-mail: jwa@agr.state.vt.us
Fred Humphrey, Chairperson
Jon Anderson, Executive Secretary

3384 Vermont Department of Health
108 Cherry Street   802-863-7280
PO Box 70   800-464-4343
Burlington, VT 05402   Fax: 802-863-7475
                             http://www.state.vt.us/health
The Vermont Department of Health, the state's public health
agency, works to protect and improve the health of our population through core public health functions. Core public health
functions are those activities that lay the groundwork for healthy
communities.
Jan Carney, Commissioner
Larry Crist, Director Health Protection

Virginia

3385 Commerce and Trade: Energy Division
1300 E Main Street   804-371-9611
Richmond, VA 23219   800-552-7945

3386 Commerce and Trade: Gas and Oil Division
PO Box 1416   540-676-5423
Abingdon, VA 24212-1416   Fax: 540-676-5459
                             E-mail: mdd@mme.state.va.us

**3387 Commerce and Trade: Mines, Minerals and Energy De-
partment**
Depart. Mines, Minerals & Energy 804-692-3200
9th St Office Bldg, 202 North 9th Street Fax: 804-692-3200
Richmond, VA 23219-3402   http://www.mme.state.va.us/
George P. Willis, Chief Deputy Director

3388 Conservation & Development of Public Beaches Board
Virginia Department of Conservation & Recreation
203 Governor Street   804-786-1712
Suite 213   http://www.dcr.state.va.us/sw/pubbeach.htm
Richmond, VA 23219-2094

3389 Department of Conservation & Recreation: Division of Dam Safety
203 Governor Street
Suite 402
Richmond, VA 23219
804-371-6095
Fax: 804-786-0536
E-mail: dam@dcr.state.va.us
Regulates dams larger than 50 acre-feet in a maximum capacity or over 25 feet in height.
Chandra McPherson

3390 Department of the Interior: Water & Science US Geological Survey
Water Resources Northeastern
1730 East Parham Road
Richmond, VA 23228
804-261-2600
Fax: 804-261-2600

3391 Division of Planning & Recreation Resources
Virginia Department of Conservation & Recreation
203 Governor Street Suite 326
Richmond, VA 23219
804-786-5046
Fax: 804-371-7899
http://gwwebdcr.state.va.us
John Davy, Director

3392 Division of State Parks
Virginia Department of Conservation & Recreation
203 Governor Street Suite 306
Richmond, VA 23219
804-692-0403
800-923-7275
Fax: 804-786-9294
E-mail: joeelton@dcr.virginia.gov
http://www.dcr.virginia.gov

Joseph Elton, Director/State Parks

3393 Economic Development Secretariat: Forestry Department
PO Box 3758
Charlottesville, VA 22903
804-977-6555
Fax: 804-977-7749
E-mail: garnerj@hq.forestry.state.va

3394 Economic Development: Mineral Resources Division
PO Box 3667
Charlottesville, VA 22903
804-951-6340
Fax: 804-951-6365
Stanley S Johnson, State Geologist

3395 Geological Survey
12201 Sun Rise Valley Drive
Reston, VA 20192
703-648-4000
Fax: 703-648-6693
http://www.usgs.gov
Services encompass topography, geology, mineral and water resources.

3396 Smithsonian Institution Conservation Programs: Conservation & Research
1500 Remount Road
Front Royal, VA 22630
540-635-6522
Fax: 540-635-6551
E-mail: nzpcrc04@sivm.si.edu
http://www.si.edu/crc/
The Conservation and Research Center's mission is to advance conservation of biological diversity. In meeting the Smithsonian Institution's mandate, CRC increases knowledge through investigations of threatened species, habitats andcommunities, and disseminates knowledge through advanced studies, professional training and public outreach.

3397 United States Department of the Army
US Army Training & Doctrine Command ATBO-BPS
5 N Gate Road
Fort Monroe, VA 23651
757-727-2077
Bob Anderson, Natural Resources Specialist

3398 Virginia Cooperative Fish & Wildlife Research Unit
Virginia Polytechnic Institute & State Unversity
106 Cheatham Hall
Blacksburg, VA 24061-0321
540-231-5927
Fax: 540-231-7580
E-mail: mussel@vt.edu

A field station of the US Geological Survey, dedicated to research and management of fish and wildlife resources in Virginia and surrounding states. Expertise includes freshwater fish and mollusks and large game mammals. The unitincludes 3 research scientists.
Dr. Richard Neves, Unit Leader

3399 Virginia Department of Environmental Quality
629 E Main Street
PO Box 10009
Richmond, VA 23219
804-698-4000
800-592-5482
Fax: 804-698-4500
E-mail: vanaturally@deq.virginia.gov
http://www.deq.virginia.gov
Virginia's regulatory state agency for air, water waste management and coastal resources. The department is also the coordinating clearinghouse for environmental education and information; and maintains the state's gateway.
Bob Burnley, Director
Ann Regn, Environmental Education Mgr

3400 Virginia Department of Game & Inland Fisheries: Region II
1132 Thomas Jefferson Road
Forest, VA 24551-9223
804-525-7522
Fax: 804-525-7720
E-mail: dgifwe@dgif.state.va.us
http://www.dgif.state.va.us
To manage Virginia's wildlife and inland fish to maintain optimum population of all species to serve the needs of the commonwealth; to provide opportunity for all to enjoy wildlife, inland fish, boating and related outdoor recreation;to promote safety for persons and property in connection with boating, hunting and fishing.

3401 Virginia Department of Game & Inland Fisheries: Region IV
4010 W Broad Street
Richmond, VA 23230
804-367-1000
E-mail: dgifweb@dgif.state.va.us
http://www.dgif.state.va.us

3402 Virginia Department of Game & Inland Fisheries: Wildlife Division
4010 West Broad Street
Richmond, VA 23230
804-367-9588
Fax: 804-367-9147
Robert W Duncan, Director

3403 Virginia Department of Game & Inland Fisheries Fisheries Division
4010 West Broad Street
Richmond, VA 23230
804-367-8704
Fax: 804-367-9147
Gary Martel, Director

3404 Virginia Department of Health Commissioners Office
1500 East Main Street Suite 214
Richmond, VA 23219
804-786-3561
Fax: 804-786-4616
Anne Peterson MD MPH, Commissioner
Helen Tarantino, Deputy Commissioner of Admin

3405 Virginia Department of Mines, Minerals & Energy: Division of Mined Land Reclamation
3405 Mountain Empire Road
P.O. Drawer 900
Big Stone Gap, VA 24219
276-523-8100
Fax: 276-523-8148
http://www.mme.state.va.us
Ernie Barker, Manager
Gerald Collins, Manager

3406 Virginia Department of Mines, Minerals and Energy: Division of Mineral Resources
PO Box 3667
Charlottesville, VA 22903
804-951-6340
Fax: 804-951-6365
http://www.geology.state.va.us
State Geological Survey with reports and maps on Geology and Mineral Resources.
Stanley Johnson, State Geologist

3407 Virginia Museum of Natural History
1001 Douglas Avenue 540-666-8600
Martinsville, VA 24112 Fax: 540-632-6487
http://www.vmnh.net
We are a state museum of natural history with research scientists in marine biology, vertebrate and invertebrate paleontology, archaeology, earth sciences, entomology, and mammalogy. Creates education programs, exhibits, and fieldtrips focused on natural history and environmental issues. Its publishing division specializes in works by natural scientists and environmental educators in the US and abroad. Writing, editorial, and design services available for books, reports, textbooks, etc..
Susan Felker, Managing Editor\Outreach

3408 Virginia Natural Resources: Game & Inland Fisheries
4010 W Broad Street 804-367-1000
PO Box 11104 Fax: 804-367-9147
Richmond, VA 23230 E-mail: dgifweb@dgif.state.va.us
http://www.dgif.state.va.us
William Woodfin, Director
Julia Smith, Media Relations Coordinator

3409 Virginia Sea Grant Program
170 Rugby Road, Madison House 804-924-5965
Charlottesville, VA 22903 Fax: 804-982-3694
E-mail: rickards@virginia.edu
http://www.virginia.edu/virginia-sea-grant/
William Rickards, Director

Washington

3410 Department of Commerce: Pacific Marine Environmental Laboratory
7600 San Dpoint Way NE 206-526-6239
Seattle, WA 98115-6239 Fax: 206-526-6815
http://www.pmel.noaa.gov
Eddie N Bernard, Director

3411 Department of Fish and Wildlife
600 Capitol Way North 360-902-2200
Olympia, WA 98501-1091 Fax: 360-902-2947
http://www.wa.gof/wdfw
Jeff Tayer, Regional Director

3412 Environmental Protection Agency: Region 10 Environmental Services
1200 6th Avenue 206-553-1200
Seattle, WA 98101 800-424-4372
Fax: 206-553-1809
E-mail: philip.jeff@epa.gov
http://www.epa.gov/r10earth
L Michael Bogart, Administrator

3413 Julia Butler Hansen Refuge for the Columbian White-Tailed Deer
PO Box 566 360-795-3915
Cathlamet, WA 98612 Fax: 360-795-0803
Offers critical habitat for the endangered Columbian white-tailed deer. The refuge also provides a wintering area for tundra swans, Canada geese, mallards, American wigeon and pintails. Deer and elk are easily observed and photographedfrom the country road that circles the mainland portion of the refuge. Evenings and mornings are the best time to spot animals. Open year-round. No fees charged.

3414 Washington Cooperative Fish & Wildlife Research Unit
University of Washington
Box 355020 206-543-6475
Seattle, WA 98195-5020

3415 Washington Department of Fish and Wildlife- Wildlife Program
600 Capitol Way N 360-902-2515
Olympia, WA 98501-1091 Fax: 360-902-2162
E-mail: wildthing@dfw.wa.gov

3416 Washington Department of Fish and Wildlife: Public Affairs
600 Capitol Way N 360-902-2250
Olympia, WA 98501-1091 Fax: 360-902-2171
E-mail: turcocmt@dfw.wa.gov

3417 Washington Department of Ecology
PO Box 47600 360-407-6000
Olympia, WA 98504-7600 Fax: 360-459-6007
http://www.ecy.wa.gov
The mission is to protect, preserve and enhance Washington's environment and to promote the wise management of our air, land and water.
Jay Manning, Director
Polly Zehm, Deputy Director

3418 Washington Department of Ecology: Southwestern Office
PO Box 4775 360-407-6300
Olympia, WA 98504-7775 Fax: 360-407-6305
E-mail: smay461@ecy.wa.gov
http://www.ecy.wa.gov
Sue Mauermann, Regional Director

3419 Washington Department of Ecology: Eastern Region Office
4601 North Monroe 508-456-2926
Spokane, WA 99205 Fax: 509-456-6175
http://www.ecy.wa.gov
Tony Grover, Staff

3420 Washington Department of Ecology: Public Information & Education
PO Box 47600 360-407-6000
Olympia, WA 98504-7600 Fax: 360-459-6007
http://www.ecy.wa.gov
Sandy Howard, Officer

3421 Washington Department of Fish & Wildlife: Business Services
600 Capitol Way N 360-902-2200
Olympia, WA 98501-1091 Fax: 360-902-2230
E-mail: director@dfw.wa.gov
Business Services includes the Director's Office, Hunter Education, the License Division, Personnel and Volunteer Services.

3422 Washington Department of Fish & Wildlife: Fish and Wildlife Commission
600 Capitol Way N 360-902-2267
Olympia, WA 98501-1091 Fax: 360-902-2448
E-mail: commission@dfw.wa.gov

3423 Washington Department of Fish & Wildlife: Habitat Program
600 Capitol Way N 360-902-2534
Olympia, WA 98501 Fax: 360-902-2946
E-mail: habitatprogram@dfw.wa.gov
Includes Maps and Digital Info Requests.

3424 Washington Department of Natural Resources: Southwest Region
PO Box 280 360-577-2025
Capital Rock, WA 98611-0280 Fax: 360-274-4196
Rick Cooper, Staff

3425 Washington Department of Natural Resources: Central Region
1405 Rush Road 360-748-2383
Chelias, WA 98532 Fax: 360-748-2387
Vicky Christenson, Staff

Government Agencies & Programs/West Virginia

3426 Washington Dept. of Natural Resources: Northwest Division
919 North Township Street
Sedro Woolley, WA 98284
360-856-3500
Fax: 360-856-2150
http://www.wadnr.gov/par

Bill Walace, Regional Manager

3427 Washington Dept. of Natural Resources: South Puget Sound Region
PO Box 68
Enumclaw, WA 98022
360-577-2025
Fax: 360-825-1672
Bonnie Bunning, Staff

3428 Washington Sea Grant Program
University of Washington
3716 Brooklyn Avenue NE
Seattle, WA 98105-6716
206-543-6600
Fax: 206-685-0380
E-mail: seagrant@u.washington.edu
http://www.wsg.washington.edu/

Louie S Echols, Director
Andrea Copping, Assistant Director

3429 Washington State Parks & Recreation Commission: Eastern Region
2201 North Duncan Drive
Wenatchee, WA 98801-1007
509-662-0420
Fax: 509-663-9754
http://www.parks.wa.gov

Jim Harris, Regional Manager

3430 Washington State Parks & Recreation Commission Puget Sound Region
2840 Riverwalk Drive
Auburn, WA 98002
253-931-3902
Fax: 253-931-3963
E-mail: infocent@parks.wa.gov
http://www.parks.wa.gov
The Washington State Parks and Recreation Commision acquires, operates, enhances and protects a diverse system of recreational, cultural, historical and natural sites. The commission fosters outdoor recreation and education statewide to provide enjoyment and enrichment for all, and a valued legacy to future generations.

West Virginia

3431 Federal Energy Technology Center
3610 Collins Ferry Road
PO Box 880
Morgantown, WV 26507-0880
304-285-4511
Fax: 301-285-4292
E-mail: rbajur@fetc.doe.gov
http://www.fetc.doe.gov
The Federal Energy Technology Center is an international leader in solving energy and environmental problems. FETC implements a broad range of energy and environmental programs for the department.
Rita Bajura, Director

3432 Gauley River National Recreation Area Advisory National Park Service
PO Box 246
Glen Jean, WV 25846
304-465-0508
Fax: 304-465-0591
E-mail: katy_miller@nps.gov
http://www.nps.gov/gar www.nps.gov/neri
www.nps.gov/blue
Located in the southern West Virginia, New River Gorge National River was established in 1978 to conserve and protect 53 miles of the New River as a free-flowing waterway. This unit of the National Park System encompasses over 70,000acres of land along the New River between the towns of Hinton and Fayetteville. New River Gorge National River and Bluestone National Scenic River are both managed by our same office in Glen Jean, WV.
Lorrie Sprague, Public Information Officer

3433 West Virginia Cooperative Fish & Wildlife Research Unit USGS
W Virginia University
PO Box 6125
Morgantown, WV 26506
304-293-3794
Fax: 304-293-4826
E-mail: wvcoop@wvu.edu

Patricia Mazik, Unit Leader Fisheries
Petra Wood, Assistant Leader Wildlife

3434 West Virginia Department of Agriculture Soil Conservation Agency
1900 Kanawha Boulevard East
Charlestown, WV 25305-0193
304-558-2204
Fax: 304-558-1635
E-mail: taborl@wvic.wvnet.edu
http://www.wvsca.org

Lance Tabor, Executive Director

3435 West Virginia Department of Environmental Protection
WV Dpt of Environmental Protection
1356 Hansford Street
Charleston, WV 25301
304-926-3647
Fax: 304-926-3637
http://www.dep.state.wv.us

3436 West Virginia Division of Natural Resources
1900 Kanawha Boulevard
State Capitol Complex, Building 3 Rm 669
Charleston, WV 25305-0060
304-558-2754
Fax: 304-558-2768
http://www.wvdnr.gov
Frank Jezioro, Director

3437 West Virginia Geological & Economic Survey
1 Mont Chateau Road
Morgantown, WV 26508
304-594-2331
Fax: 304-594-2575
E-mail: info@geosrv.wvnet.edu
http://www.wvgs.wvnet.edu
Carl J Smith, Director/State Geologist

Wisconsin

3438 Great Lakes Indian Fish and Wildlife Commission
PO Box 9
Odanah, WI 54861
715-682-6619
Fax: 715-682-9294

James H Schlender, Executive Director

3439 Natural Resources Department
PO Box 7921
Madison, WI 53707
608-266-2121
Fax: 608-266-6983
Darrell Baggel, Secretary

3440 United States Department of Agriculture United States Forest Service
310 W Wisconsin Avenue
Suite 500
Milwaukee, WI 53203
414-297-3600
Fax: 414-297-3808

3441 Wisconsin Cooperative Fishery Research Unit
University of Wisconsin
College of Natural Resources
Stevens Point, WI 54481
715-346-2178
Fax: 715-346-3624
E-mail: coopfish@uwsp.edu
http://www.uwsp.edu/cnr/wicfru

Michael Bozek, Leader

3442 Wisconsin Department of Agriculture Trade and Consumer Protection: Land & Water Resources Bureau
2811 Agriculture Drive
PO Box 8911
Madison, WI 53708-8911
608-224-4622
Fax: 608-224-4615

David Jelinski, Bureau Director
Rod Nilsestuen, Secretary

3443 Wisconsin Geological & Natural History Survey
University of Wisconsin Extension
3817 Mineral Point Road
Madison, WI 53705
608-262-1705
Fax: 608-262-8086
http://www.uwex.edu/wgnhs

228

Provides scientific information about the geology, mineral resources, water resources, soil and biology of Wisconsin. Communicates the results of activities through publications, talks and responses to inquiries from the public. Activities support informed decision making by government, industry, liasons and individual citizens.
James Robertson, State Geologist & Director
Thomas J Evans, Geology Program Leader

3444 Wisconsin State Extension Services Community Natural Resources & Economic Development
University of Wisconsin Extension
432 N Lake Street 608-262-1748
Madison, WI 53706 Fax: 606-262-9166
http://www.uwex.edu/ces/
Carl O'Conner, Dean/Director Coop Ext
Patrick Walsh, Statewide Program Leader

Wyoming

3445 Casper District Advisory Council
Bureau of Land Management
2987 Prospector Drive 307-261-7600
Casper, WY 82604-2698 Fax: 307-261-7587
E-mail: casper_wymail@blm.gov
http://www.blm.gov/Director/fo_map/casper_fo.lhtml
Jim Murkin, Field Manager

3446 Environmental Quality Department
122 W 25th StreetHerschler Building 307-777-7937
Herschler Building, 4th Floor Fax: 307-777-7682
Cheyenne, WY 82002 E-mail: deqwyo@state.wy.us
John Corra, Director

3447 Rock Springs Field Office Bureau of Land Management
280 Highway 191 N 307-352-0256
Rock Springs, WY 82901 Fax: 307-352-0329
http://www.wy.blm.gov
BLM's Rock Springs Field Office is a federal agency in the USA that manages over 3.6 million acres of public land surface and 3.5 million acres of public sub-surface minerals in the southwestern part of the great State of Wyoming. For these public lands, BLM administers a variety of programs including mineral exploration and development, wildlife habitat, outdoor recreation, wild horses, lifestock grazing and historic trails.

3448 State Parks & Cultural Resources Division of State Parks & Historical Sites
1st Floor Herschler Building 307-777-6323
Cheyenne, WY 82002 Fax: 307-777-6474
E-mail: sphs@state.wy.us
http://www.wyo-park.com
Bill Gentle, Director

3449 Wyoming Board of Land Commissioners
Herschler Building 307-777-7331
Cheyenne, WY 82002 Fax: 307-777-5400
E-mail: slfmail@state.wy.us
http://www.state.wy.us
Ron Arnold, Director
Stephen Reynolds, Secretary

3450 Wyoming Cooperative Fishery and Wildlife Research Unit
University of Wyoming
Biological Sciences Building 307-766-5415
Box 3166 Fax: 307-766-5400
Laramie, WY 82071
Unit conducts fish and wildlife research for the state of Wyoming conservation, fish department and federal agencies.
Dr. Stanley H Anderson, Leader

3451 Wyoming State Forestry Division
1100 W 22nd Street 307-777-7586
Cheyenne, WY 82002 Fax: 307-777-5986
E-mail: forest@state.wy.us

The Forestry Division's general reposnsibility and objectives are to promote and assist the multiple use management and protection of Wyoming's 270,000 acres of state and 1.9 million acres of private forest lands; to provide forestryassistance and information to landowners, industry, communities and public agencies; and to help provide rural, range, and forest land fire protection, equipment and training.
Thomas Ostermann, State Forester
Daniel Perko, State Forester

3452 Wyoming State Geological Survey
PO Box 1347 307-766-2286
Laramie, WY 82073 Fax: 307-766-2605
E-mail: wsgs-info@uwyo.edu
http://www.wsgs.uwyo.edu
The Wyoming State Geological Survey's mission is to promote the beneficial and environmentally sound use of Wyoming's vast geologic, mineral, and energy resources while helping to protect the public from geologic hazards.

Ronald C Surdam, State Geologist

Alabama: US Forests, Parks, Refuges

3453 Bon Secour National Wildlife Refuge
12295 State Highway 180 251-540-7720
Gulf Shores, AL 36542 Fax: 251-540-7301
E-mail: bonsecour@fws.gov
http://http://bonsecour.fws.gov/index_files/slide0001.htm
Gulf Shores offer nature enthusiasts much to explore. The Bon Secour NWR, which lies just 6 miles west of Gulf Shores, caters equally to the angler, the hiker and the birder. The refuge encourages guests to enjoy a leisurely hikethrough the grounds or a fishing excursion on the 40-acre fresh water Gator Lake. Pack a picnic lunch and your binoculars, park your blanket on one of the many secluded beaches and savor the scenery. Call 1-866-SEA TURTLE to report sea turtleactivity.
Jerome T Carroll, Manager

3454 Choctaw National Wildlife Refuge
PO Box 808 251-246-3583
Jackson, AL 36545 Fax: 251-246-5414
E-mail: choctaw@fws.gov
http://www.choctaw.fws.gov//index.html
The objectives of the Refuge are: to manage habitat for wintering waterfowl, maintain habitat and provide protection for threatened and endangered species, manage wood duck nest boxes and brood rearing habitat, maintain wildlifediversity, manage forest to be productive bottomland hardwoods, and to provide wildlife dependent recreation.

Robert Dailey, Refuge Manager

3455 Conecuh National Forest
Rt 5 Box 157 334-222-2555
Andalusia, AL 36420
Blue Lake provides focal points for outdoor recreation in the Conecuh National Forest. Other areas of recreational interest in the Forest include the Conecuh Trail, which winds for 20 miles through Alabama's coastal plain, and BrookHines, offering picnicking and fishing.

3456 Little River Canyon National Preserve
2141 Gault Avenue N 256-845-9605
Fort Payne, AL 35967 Fax: 256-997-9129
E-mail: LIRI_Superintendent@nps.gov
Little River flows for most of its length atop Lookout Mountain in northeast Alabama. The river and canyon systems are spectacular Appalachian Plateau landscapes any season of the year. Forested uplands, waterfalls, canyon rims andbluffs, stream riffles and pools, boulders and sandstone cliffs offer settings for a variety of recreational activities. Natural resources and cultural heritage come together to tell the story of the preserve, a special place in the SouthernAppalachians.

John Bundy, Supt

3457 Wheeler National Wildlife Refuge Complex
2700 Refuge Headquarters Road 256-353-7243
Decatur, AL 35603-5202 Fax: 256-340-9728
E-mail: wheeler@fws.gov

The 35,000 acre wildlife refuge was established in 1938 The refuge is located between Decatur and Huntsville in the Tennessee River Valley of northern Alabama.
C Dwight Cooley, Manager

3458 William B Bankhead National Forest

Bankhead Ranger District   205-489-5111
PO Box 278   Fax: 205-489-3427
Double Springs, AL 35553
The William B Bankhead national Forest covers 180,000 acres in Franklin, Winston and Lawrence counties. Within the forest are the 26,000 acre Sipsey Wilkderness and the Sipsey Wild and Scenic River, offering 61.4 miles of seasonalcanoeing.

Alaska: US Forests, Parks, Refuges

3459 Alagnak Wild River Katmai National Park

PO Box 7   907-246-3305
King Salmon, AK 99611   Fax: 907-246-4286
The Alagnak river offers 69 miles of outstanding white-water floating. The river is also noted for abundant wildlife and sport fishing for five species of salmon.

3460 Alaska Maritime National Wildlife Refuge

95 Sterling Highway   907-235-6546
Suite 101   Fax: 907-235-7783
Homer, AK 99603
To administer a national network of lands and waters for the conservation management and where appropiate, restoration of the fish, wildlife, and plant resources and their habitats within the US for the benefit of present and futuregeneration of Americans.
Gregory Siekaniec, Refuge Manager

3461 Alaska Penninsula National Wildlife Refuge

PO Box 277   907-246-3339
King Salmon, AK 99613   Fax: 907-246-6696
http://www.gorp.com/gorp/resource/us_nwr/ak_ak_pe.htm
Ronald Hood, Manager

3462 Becharof National Wildlife Refuge

PO Box 277   907-246-3339
King Salmon, AK 99613   Fax: 907-246-6696

3463 Chugach National Forest

3301 C Street   907-271-2773
Suite 300   Fax: 907-271-3992
Anchorage, AK 99503   http://www.fs.fed.us/r10/chugach

3464 Denali National Park and Preserve

PO Box 9   907-683-2294
Denali Park, AK 99755   Fax: 907-683-9617
E-mail: denali_info@nps.gov
http://www.nps.gov

3465 Innoko National Wildlife Refuge

PO Box 69   907-524-3251
McGrath, AK 99627   Fax: 907-524-3141
http://www.gorp.com/gorp/resource/us_nwr/ak_innok.htm
The Innoko National Wildlife Refuge was established December 2, 1980, with the passage of the Alaska National Interest Lands Conservation Act. This 3.85 million acre refuge supports a large nesting waterfowl population, and is wellpopulated with moose, bear, and other animals, as well as a variety of game birds and neotropical bird species and is a relatively flat plain covering much of the drainage area of the Innoko and Iditarod rivers. The vegetation of the reguse is atransition zone.
William H Schaff, Refuge Manager

3466 Izembek National Wildlife Refuge

Box 127   907-532-2445
Cold Bay, AK 99571   877-837-6332
Fax: 907-532-2549
E-mail: izembek@fws.gov
http://www/R7/fws.gov/nwr/izembek/index.htm

Established to conserve fish, wildlife and habitats in their natural diversity including, waterfowl, shorebirds, other migratory birds, brown bears and salmon; to fulfill treaty obligations; to provide the opportunity for continuedsubsistence uses by local residents consistent with the purposes previously mentioned; and to ensure necessary water quality and quantity.

Sandra Siekaniec, Refuge Manager

3467 Kanuti National Wildlife Refuge

Ferderal Building & Courthouse   907-456-0329
Box 20 101-12th Avenue   Fax: 907-456-0506
Fairbanks, AK 99701
Tom Early, Manager

3468 Katmai National Park and Preserve

#1 King Salmon Mall   907-271-3751
PO Box 7   Fax: 907-246-4286
King Salmon, AK 99613   E-mail: Kristi_Bergeron@nps.gov
http://www.nps.gov

3469 Kenai Fjords National Park

1212 4th Avenue   907-224-3175
PO Box 1727   Fax: 907-224-2144
Seward, AK 99664   E-mail: KEFJ_Superintendant@nps.gov
http://www.nps.gov/kefj
This park encompasses over 600,000 acres of wild coastal Alaska. The Harding Icefield dominates most of the park. This 300-square mile bowl of ice spills out into numerous glaciers at its edges. Tidewater glaciers and the amazingmarine wildlife of the park can be viewed from boat or air. Humpback whales, orca, many species of sea birds, Steller sea lions and other marine wildlife come here because of the rich variety of foods. Sea birds and sea lions also raise their youngon the rocky sites.

Jeff Mow, Superintendent
Jim Ireland, Chief Interpretation

3470 Kenai National Wildlife Refuge

PO Box 2139   907-262-7021
Soldotna, AK 99669-2139   Fax: 907-487-2144
http://www.gorp.com/gorp/resource/us_nwr/ak_kenai.htm
Jay Bellinger, Manager

3471 Kobuk Valley National Park

PO Box 1029   907-442-3760
Kotzebue, AK 99752   Fax: 907-442-8316
E-mail: NWAK_superintendant@nps.gov
http://www.nps.gov

3472 Kodiak National Wildlife Refuge

1390 Buskin River Road   907-487-2600
Kodiak, AK 99615   888-408-3514
Fax: 907-487-2144
E-mail: kodiak@fws.gov
http://www.gorp.com/gorp/resource/us_nwr/ak_kodia.htm
Leslie Kerr, Manager

3473 Koyukuk and Nowitna National Wildlife Refuge

PO Box 287   907-656-1231
Galena, AK 99741   800-656-1231
Fax: 907-656-1708
E-mail: r7kynwr@fws.gov
http://http://nowitna.fws.gov/index.htm or
http://koyukik.fws.gov/
Approximately 200 miles west of Fairbanks, the refuge lies within a solar basin encircled by rolling hills capped by alpine tundra. The Nowitna River, a nationally designated Wild River, bisects the refuge and forms a broad meanderingfloodplain. The river passes through a scenic 15 mile canyon with peaks up to 2,100 feet.
Vacant , Refuge Manager
Greg McClellan, Refuge Manager Deputy

3474 Lake Clark National Park and Preserve

1 Park Place   907-781-2218
Port Alsworth, AK 99653   Fax: 907-781-2119
E-mail: jeelan_eastlack@nmps.gov
http://www.nps.gov

Jeelan Eastlack, Office Automation Assistant

3475 Selawik National Wildlife Refuge
PO Box 270 907-442-3799
Kotzebue, AK 99752 Fax: 907-442-3124
http://www.gorp.com

Mike Rearden, Manager

3476 Tetlin National Wildlife Refuge
PO Box 115 907-883-5312
Tok, AK 99780 Fax: 907-883-5747
http://www.gorp.com

Steve Bresser, Manager

3477 Tongass National Forest: Chatham Area
204 Siginaka Way 907-747-6671
Sitka, AK 99835-7316 Fax: 907-747-4331
The Tongass National Forest is a forest of islands and trees and rain. It also abounds in animals and birds and fish, with unsurpassed scenery. It's a place where eagles are commonplace, most every road is a deer crossing, and bearsuse the trails too. The spirituality and scenery demands respect.
Fred Salinas, Forest Supervisor

3478 Tongass National Forest: Ketchikan Area
648 Mission Street 907-225-3101
Federal Building Fax: 907-228-6292
Ketchikan, AK 99901-6591
The Tongass National Forest is a forest of islands and trees and rain. It also abounds in animals and birds and fish, with unsurpassed scenery. It's a place where eagles are commonplace, most every road is a deer crossing, and bearsuse the trails too. The Tongass is a wild place, where the natural world is a strong presence that nurtures spirituality and materially demands respect.
Tom Puchlerz, Forest Supervisor

3479 Tongass National Forest: Stikine Area
15 N 12th Street 907-772-3841
PO Box 309 Fax: 907-772-5895
Petersburg, AK 99833-0309
The Tongass National Forest is a forest of islands and trees and rain. It also abounds in animals and birds and fish, with unsurpassed scenery. It's a place where eagles are commonplace, most every road is a deer crossing, and bearsuse the trails too. The Tongass is a wild place, where the natural world is a strong presence that nurtures spirituality and materially demands respect.

3480 US Fish & Wildlife Service Togiak National Wildlife Refuge
PO Box 270 907-842-1063
Dillingham, AK 99576 800-817-2538
Fax: 907-842-5402
http://togiak.fws.gov
Established to conserve fish and wildlife populations and habitats in their natural diversity including salmon, marine birds, mammals, migrating birds and large mammals, to fulfill international treaty obligations; to provide forcontinued subsistence uses; and to ensure necessary water quality and quantity.
Paul Liedberg, Refuge Manager

3481 Yukon Delta National Wildlife Refuge
PO Box 346 907-543-3151
Bethel, AK 99559-0346 Fax: 907-456-0447
http://www.gorp.com

Ted Huer, Manager

3482 Yukon Flats National Wildlife Refuge
101 12th Avenue 907-456-0440
Room 264 Fax: 907-456-0447
Fairbanks, AK 99701 E-mail: yukonflats@fws.gov
http://www.r7.fws.gov/nwr/yf/
Located about 100 air miles north of Fairbanks, encompassing about 12 million acres along the Yukon River. In the spring, millions of migrating birds converge on the refuge. With its 40,000 lakes and other wetlands, it has one of thehighest waterfowl nesting densities in North America for ducks, geese, sandhill cranes, loons, grebes and songbirds. Each year, the Yukon Flats is a major contributor to the agencies that occur along the North American flyways.

3483 Yukon-Charley Rivers National Preserve
PO Box 167 907-547-2234
Eagle, AK 99738 Fax: 907-547-2247
E-mail: yuch_eagle_cheifofoperation
http://www.nps.gov/yuch
Kevin Fex, Cheif Operations

Arizona: US Forests, Parks, Refuges

3484 Apache-Sitgreaves National Forest
National Forests 520-333-4301
PO Box 640 Fax: 520-333-6357
Springerville, AZ 85938
Taking care of the land while making the forest resources available to all shareholders. Resources include: high quality water, wilderness, and outdoor recreation; quality habitat for many plants and animals; wood for paper, homes, andhundreds of other uses; forage for wildlife and livestock; a source of minerals.
Donna Sherwood, Personnel Management

3485 Bill Williams National Wildlife Refuge
60911 Highway 95 520-667-4144
Parker, AZ 85344 E-mail: r2rw_bw@mail.fws.gov
http://www.gorp.com
Bill Williams River National Wildlife Refuge is located along the Bill Willaims River in La Paz and Mojave Counties, Arizona, with the river as the dividing line between the two counties. The refuge was established in 1941 as part ofHavasu NWR as mitigation for the Boulder and Parker Dam projects. In 1993, the two refuges were seperated and the Bill W Unit became the Bill Williams River NWR.

3486 Buenos Aires National Wildlife Refuge
PO Box 109 520-823-4251
Sasabe, AZ 85633 Fax: 520-823-4247
http://ww.gorp.com

Wayne A Shifflett, Manager

3487 Cabeze Prieta National Wildlife Refuge
1611 N 2nd Avenue 520-387-6483
Ajo, AZ 85321 Fax: 520-387-5359
http://southwest.fws.gov/refuges/arizona/cabeze.htm
Roge Dikesa, Manager
Michael Luck, Assistant Refuge Manager

3488 Chiricahua National Monument
13063 E Bonita Canyon Road 520-824-3560
Wilcox, AZ 85643-9737 Fax: 520-824-3421
The monument is mecca for hikers and birders. At the intersection of the Chiricahuan and Sonoran deserts, and the southern Rocky Mountains and northern Sierra Madre in Mexico, Chiricahua plants and animals represents one of the premiereareas for biological diversity in the northern hemisphere.
Neil Mangum, Supt

3489 Coconino National Forest
2323 East Greenlaw Lane 520-527-3600
Flagstaff, AZ 86004 Fax: 520-527-3620
http://www.fs.fed.us/r3/coconino/

3490 Coronado National Forest
300 W Congress Street 520-670-4552
Tucson, AZ 85701 Fax: 520-670-4584
The Coronado National Forest covers 1,780,000 acres of southeastern Arizona and southwestern New Mexico. Elevations range from 3,000 feet to 10,720 feet in 12 widely scattered mountain ranges or sky islands that rise dramatically fromthe desert floor, supporting plant communities as biologically diverse as those encountered on a trip from Mexico to Canada. The views are spectacular from these mountains, and visitors may experience all four seasons during a single day's journey.

3491 Glen Canyon National Recreation Area
Glen Canyon NRA 520-645-8200
PO Box 1507 E-mail: GLCA_CHVC@nps.gov
Page, AZ 86040-1507

Glen Canyon National Recreation Area offers unparalleled opportunities for water-based & backcountry recreation. The recreation area stretches for hundreds of miles from Lees Ferry in Arizona to the Orange Cliffs of southern Utah, encompassing scenic vistas, geologic wonders, and a panorama of human history. Additionally, the controversy surrounding the construction of Glen Cayon Dam and the creation of Lake Powell contributed to the birth of modern day environmental movement.

John Lancaster, Supt

3492 Grand Canyon National Park

PO Box 129   520-638-7888
Grand Canyon, AZ 86023   Fax: 520-638-7797
E-mail: grca_superintendant@nps.gov
http://www.nps.gov

3493 Imperial National Wildlife Refuge

PO Box 72217   602-783-3371
Martinez Lake, AZ 85365   Fax: 602-783-0682
E-mail: FW2_RW_Imperial@fws.gov
http://www.fws.gov/southwest/refuges/arizona/imperial.html

3494 Kaibab National Forest

800 South Sixth Street   520-635-2681
Williams, AZ 86046   Fax: 520-635-8208
http://www.fs.fed.us/r3/kai/

3495 Kofa National Wildlife Refuge

356 West 1st Street   602-783-7861
Yuma, AZ 85364   Fax: 602-783-8611
E-mail: FW2_RW_Kofa@fws.gov
http://www.fws.gov/southwest/refuges/arizona/kofa.html
J. Paul Cornes, Manager

3496 Organ Pipe Cactus National Monument

10 Organ Pipe Drive   520-387-6849
Ajo, AZ 85321-9626   Fax: 520-387-7144
E-mail: orpi_information@nps.gov
Organ Pipe Cactus National Monument celebrates the life and landscape of the Sonoran Desert. Here, in this desert wilderness of plants and animals and dramatic mountains and plains scenery, you can drive a lonely road, hike abackcountry trail, camp beneath a clear desert sky, or just soak in the warmth and beauty of Southwest.

Harold J Smith

3497 Petrified Forest National Park

PO Box 2217   520-524-6228
Petrified Forest National Park, AZ 86028 Fax: 520-524-3567
E-mail: PEFO_superintendant@nps.gov
http://www.nps.gov
Petrified Forest is a surprising land of scenic wonders and fascinating science. The park is located in northeast Arizona and features one of the world's largest and most colorful concentrations of petrified wood. Also included in thepark's 93,533 acres are the multihued badlands of the Chinle Formation known as the Painted Desert, historic structures, archeological sites and displays of 225 millio-year-old fossils.

3498 Prescott National Forest

344 S Cortez Street   520-771-4700
Prescott, AZ 86303   Fax: 520-771-4708
This involves taking care of the land while making the forest resources available to all shareholders. Resources include high quality water, wilderness and outdoor recreation; quality habitat for many plants and animals; wood forpaper, homes and hundreds of other uses; forage for wildlife and livestock; and minerals.

3499 Saguaro National Park

3693 S Old Spanish Trail   520-733-5100
Tuscon, AZ 85730   Fax: 520-733-5183
E-mail: mailto:sagu_information@nps.gov
http://www.nps.gov/sagu
This unique desert is home to the most recognizable cactus in the world, the majestic saguaro. Visitors of all ages are fascinated and enchanted by these desert gaints, especially their many interesting and complex interrelationshipswith other desert life. With the average life span of 150 years, a mature saguaro may grow to the height of 50 feet and weigh over 10 tons.

Sarah Craighead, Superintendent
Chip Littlefield, Interpretive Ranger

3500 San Bernardino/Leslie Canyon National Refuge

1408 10th Street   602-364-2104
Douglas, AZ 85607   Fax: 602-364-2130
http://www.fws.gov/southwest/refuges/arizona/sanb.html
Bill Radke, Manager

3501 Sunset Crater Volcano National Monument

Route 3 Box 149   520-556-7042
Flagstaff, AZ 86004   Fax: 520-526-0502
Here, past meets present. Pueblos and cliff dwellings are so well presevec that it's hard to believee their builders moved on 700 years ago. Amid lava and cinders, one can imagine a landscape still hot to the touch. Welcome to theFlagstaff Area NAtional Monuments! There is something for everyone: prehistoic cliff dwellings at Walnut Canyon, the mountain scenery and geology of Sunset Crater Volcano, and the painted desert landscape and masonry pueblos of Wupatki NationalMonument.

3502 Tonto National Forest

2324 East McDowell Road   602-225-5200
Phoenix, AZ 85010   Fax: 602-225-5295
The Tonto National Forest occupies about 2.8 million acres which generally lie northeast of Phoenix, Ariz., to the Mogollon Rim and east to the San Carlos and Fort Apache Indain Reservations. The west side approximately interstate 17which stretches north of Phoenix to Flagstaff. The lower elevations are of the Sonoran Desert type while the northern portion of the Forest is generally Pinon, Juniper, and Ponderosa Pine types.

3503 Walnut Canyon National Monument

Walnut Canyon Road #3   520-526-3367
Flagstaff, AZ 86004-9705   Fax: 520-526-4259
Hike down into Walnut Canyon and walk in the footsteps of the people that lived here over 900 years ago. Under limestone overhanges, were occupied from about 1100 to 1250. Look down into the canyon and imagine the creek runningthrough. Visualize a woman hiking up from the bottom with a pot of water on her back. Imagine the men on the rim farming corn or hunting deer. Think of a cold winter night with your family huddled around the fire.

Arkansas: US Forests, Parks, Refuges

3504 Bald Knob National Wildlife Refuge

26320 Highway 33 South   870-347-2614
Augusta, AR 72006   Fax: 870-347-2908
E-mail: r4rw_ar_nea@fws.gov
http://www.gorp.com
The refuge facts established in 1993 has 14,800 acres and the location is in White County, Ar approximately two miles south of Bald Knob, AR on Coal Chute Road. It provides habitat for migratory waterfowl and other birds. And also forendangered species recreational and environmental education opportunities.

3505 Big Lake National Wildlife Refuge Northeast Arkansas Refuges

PO Box 279   501-343-2595
Turrell, AR 72442
The Northeast Arkansas and Southeast Missouri area is extremely rich in archeological history. This area was covered with a spruce forest prior to the Ice Age, but this gave way to an oak/hickory environment which was inhabited byPaleo Indians and many now extinct animals including the mastadon, llama, tapir, horse, camel, and sloth. The earliest documented human occupation of the Big Lake Area was in the eighth century by the horticultural based Woodland Indians.

Luke F Eggering, Manager

3506 Buffalo National River

402 N Walnut   870-741-5443
Suite 136   Fax: 870-741-7286
Harrison, AR 72601   E-mail: buff_information@nps.gov
http://www.nps.gov/buff

One of the few remaining unpolluted, free-flowing rivers in the lower 48 states offering both swift-running and placid stretches. The river encompasses 135 miles of the 150 mile long river. It begins as a trickle in the BostonMountains 15 miles above the park boundary. Following what is likely an ancient riverbed, the Buffalo cuts its way through massive limestone bluffs traveling eastward throug the Ozarks and into the Whtie River.

3507 Cache River National Wildlife Refuge Northeast Arkansas Refuges

Route 2 Box 126-T
Augusta, AR 72006

501-347-2614
E-mail: r4rw_ar_nea@fws.gov

3508 Felsenthal National Wildlife Refuge

PO Box 1157
Crossett, AR 71635

870-364-3167
Fax: 870-364-7800
E-mail: r4rw_ar.fsl@fws.gov
http://www.gorp.com

Robert J Bridges, Manager

3509 Greers Ferry National Fish Hatchery

349 Hatchery Road
Heber Springs, AR 72543

501-362-3615
Fax: 501-362-4007
E-mail: greersferry@fws.gov
http://www.gorp.com

Through self-guided tours at the hatchery, visitors can observe techniques of trout production, view information exhibits in the aquarium, and see trout in the outdoor raceways. Adjacent to the hatchery, visitors can camp at JFK Parkand trout fish the Little Red River. Nearby, the US Army Corps of Engineers has a visitor center, two mini-hiking trails and an overlook of Greers Ferry Dam. Greers Ferry Lake on the other side of the dam offers camping, swimming, fishing and otherwater sports.
Sherri Shoults, Hatchery Manager

3510 Holla Bend National Wildlife Refuge

10448 Holla Bend Road
Dardanelle, AR 72834

479-229-4300
Fax: 479-229-4302
E-mail: hollabend@fws.gov
http://www.fws.gov/southeast/HollaBend

Part of a system of over 475 national wildlife refuges located across the country. Administered by US Fish and Wildlife Service, this system of refuges, the finest in the world, protects important habitat needed to provide a home for awide variety of wildlife. These refuges also provide the public with valuable opportunities to see and learn about wildlife and to enjoy outdoor activities such as hunting and fishing. Holla Bend's main purpose is to provide a winter home for ducksand geese.

Ben Mense, Manager

3511 Hot Springs National Park

PO Box 1860
Hot Springs, AR 71902

501-624-3383
Fax: 501-624-3458
E-mail: HOSP_Interpretation@nps.gov
http://www.nps.gov/hosp

The park protects eight historic bathhouses with the former luxurious Fordyce Bathhouse housing the park visitor center. The entire "Bathhouse Row" area is national Historic Landmark District that contains the grandest collection ofbathhouses of its kind in North America. By protecting the 47 hot springs and their watershed, the National Park Service continues to provide visitors with historic leisure activities such as hiking, picnicking and scenic drives.

Josie Fernandez, Superintendent

3512 Ouachita National Forest

PO Box 1270
Hot Springs, AR 71902

501-321-5202
Fax: 501-321-5353

3513 Overflow National Wildlife Refuge

5531 Highway 82 W
PO Box 1157
Crossett, AR 71635

870-364-3167
Fax: 870-364-3757
E-mail: fw4_rw_felsenthal@fws.gov
http://southeast.fws.gov/Overflow/index.html

Refuge objectives are to: provide a diversity of habitat types for migratory waterfowl and other birds; provide habitat and protection for the threatened bald eagle; provide opportunities for environmental and ecological research;provide a variety of recreational opportunities consistent with primary wildlife objectives; and expand the public's understanding of and appreciation for the environment with special emphasis on natural resources.
Jim C Johnson, Project Leader

3514 Ozark-St. Francis National Forest

605 W Main Street
PO Box 1008
Russleville, AR 72801

501-968-2354

The Ozark-St. Francis National Forests are really two separate Forests with many differences. They are distinct in their own topographical, geological, biological, cultural and social differences, yet each makes up a part of theoverall National Forest system.

3515 Pond Creek National Wildlife Refuge

Highway 82 W
PO Box 1157
Crossett, AR 71635

870-364-3167
Fax: 870-364-3757
E-mail: r4rw_ar.fsl@fws.gov
http://www.gorp.com

3516 Wapanocca National Wildlife Refuge

178 Hammond Avenue
PO Box 279
Turrell, AR 72384

870-343-2595
Fax: 870-343-2416
E-mail: glen_miller@fws.gov

The 5,484 acre refuge is an important stopover for waterfowl traveling the Mississippi Flyway and for songbirds as they migrate to and from Central and South America. The refuge is open to limited small and big game hunting. Auto tourroutes offers excellent wildlife observation, photography and hiking opportunities. An observation platform is located on the east side of the 600 acre Wapanocca Lake.

Glen Miller, Manager

3517 White River National Wildlife Refuge

321 W 7th Street
PO Box 308
DeWitt, AR 72042

870-946-1468
Fax: 870-946-2591
E-mail: fw4_rw_white_river@fws.gov
http://southeast.fws.gov/whiteriver/

Refuge objectives are to provide: optimum habitat for migratory birds; habitat and protection for endangered species: a natural diversity of wildlife common to the White River bottoms; opportunities and facilities for wildlife orientedrecreation and environmental education; cooperation with other water and land managing agencies and private interests to foster proper management of the White River Basin's resources; and preservation of appropriate wooded areas in their naturalcondition.
Larry E Mallard, Refuge Manager

California: US Forests, Parks, Refuges

3518 Angeles National Forest

701 North Santa Anita Avenue
Arcadia, CA 91006

626-574-1613
Fax: 626-574-5233
http://www.dgif.state.va.us

The Angeles National Forest covers 650,000 acres and is the backyard playground to the huge metropolitan area of Los Angeles. The Los Angeles National Forest manages the watersheds eithin its boundaries to provide valuable water tosouthern California and to protect surrounding communities from catastrophics floods.

3519 Antioch Dunes National Wildlife Refuge

San Francisco Bay NWR Complex
Po Box 524
Newark, CA 94560-0524

415-792-0222
http://www.gorp.com

Primary wildlife protects critical habitat for three endangered species: Lange's metalmark butterfly, Contra Costa wallflower, and Antioch Dunes evening primrose. Isolated dunes ecosystems with a unique assemblage of plants, insectsand reptiles. The habitat consists of 67 acres of sand dunes and former dune areas. A major restoration project is restoring and improving dune habitat.

3520 Bear Valley National Wildlife Refuge

Klamath Basin NWR Complex 916-667-2231
Route 1 Box 74 Fax: 916-667-3299
Tulelake, CA 96134-9715 E-mail: tom_stewart@fws.gov
http://www.gorp.com

Refuge was established to protect a major winter night roost site for bald egales. The acquisition program was completed in 1991. Klamath Basin hosts the largest wintering popluation of blad eagles in the contiguous United States, withnumbers some years approaching 1,000. Refuge serves as one of serveral eagle roots in the Basin. It consists of large stands of old-growth tinmber, which protects the birds at night from the harsh winter weather.

3521 Bitter Creek National Wildlife Refuge

Hopper Mountain NWR 805-644-5185
Po Box 5839 Fax: 805-644-1732
Ventura, CA 93005-0839 E-mail: marc_weitzel@fws.gov
http://www.gorp.com

The primary wildlife traditional feeding and roosting habitat for the California condor. Also provides habitat for the San Joaquin kit fox, golden eagle, Southern bald eagle and American peregrine falcon. 14,000 contiguous acres,mostly annual grasslands with some juniper and scrub oak with grass understory. Public use is severely limited because of the sensitive situation of the California condor. The refuge can be viewed from the Cerro Noroeste Road.

3522 Blue Ridge National Wildlife Refuge

Kern NWR 661-725-2767
Po Box 670 http://www.gorp.com
Delano, CA 93216

Primary wildlife area is a traditional summer roosting site for the endangered Califorina condor. The habitat includes 897 acres of rugged mountains, rock outcroppings, chaparral and coniferous trees. The refuge is closed to publicaccess due to the sensitivity of California condors and its isolation and difficulty in access.

3523 Castle Rock National Wildlife Refuge

Humboldt Bay NWR 707-733-5406
PO Box 576 Fax: 707-733-1946
Loleta, CA 95551 http://www.gorp.com

The Refuge is a 14-acre offshore rock with steep cliffs and sparse vegetation. It was established in 1981 to protect an important migration staging area of the threatened Aleutian Canada goose. Over 21,000 of these roost on the island,which contains the second largest seabird breeding colony in California. Haul-out for a variety of marine mammals, including California sea lion, Stellar sea lion and northern elephant seal. Not open to the public, but wildlife can be observed fromshore.

3524 Channel Islands National Park

1901 Spinnaker Drive 805-658-5730
Ventura, CA 93001 Fax: 805-658-5799
E-mail: chis_interpretation@nps.gov
http://www.channel.island.national-park.com

Encompasses five of the eight California Channel Islands and their ocean environment, preserving and protecting a wealth of natural and cultural resources. Marine life ranges from microscopic plankton to the blue whale, the largestanimal to live on Earth. Archeological and cultural resources span a period of more than 10,000 years of human habitation.

3525 Cibola National Wildlife Refuge

Route 2 928-857-3253
Box 138 Fax: 928-387-3420
Cibola, AZ 85328 E-mail: FW2_RW_Cibola
http://www.fws.gov/southwest/refuges/arizona/cibola.html
Tom Alexander, Manager

3526 Clear Lake National Wildlife Refuge

Klamath Basin NWR Complex 916-667-2231
Route 1 Box 74 Fax: 916-667-3299
Tulelake, CA 96134 http://www.gorp.com

3527 Cleveland National Forest

10845 Rancho Bernardo Road 858-673-6130
Suite 200 Fax: 858-673-6192
San Diego, CA 92127 http://www.fs.fed.us/r5/cleveland

Pete Gomben, NEPA Coordinator

3528 Coachella Valley National Wildlife Refuge

906 W Siclair Road 760-348-5278
PO Box 120 Fax: 760-348-7245
Calipatria, CA 92233-0120 E-mail: clark_bloom@fws.gov
http://www.gorp.com

Contains 13,000 acres consisting of palm oasis woodlands, perennial desert pools and blow-sand habitat. This habitat is critical for the Coachella Valley fringe-toed lizard (Uma inornata) and flat-tailed horned lizard. These threatenedspecies are restricted to the refuge dune system and a few other small areas. Also has the state's second largest grove of native fan palms and the Coachella milk-vetch, a species of special concern.

3529 Death Valley National Park

PO Box 579 760-786-3200
Death Valley, CA 92328 Fax: 760-786-3283
E-mail: deva_superintendant@nps.gov
http://www.nps.gov/deva

Death Valley National Park has more than 3.3 million acres of spectacular desert scenery, interesting and rare desert wildlife, complex geology, undisturbed wilderness and sites of historical and cultural interest. The National ParkService is dedicated to the protection and preservation of this park's unique resources for everyone to enjoy now and for future generations.

3530 Delevan National Wildlife Refuge

Sacramento NWR Complex 530-934-2801
752 County Road 99W E-mail: gary_w_kramer@fws.gov
Willows, CA 95988 http://www.gorp.com

Delevan NWR is part of the Sacramento NWR Complex and is located in the Sacramento Valley of north-central California. The refuge consists of nearly 5,800 acres comprised of seasonal marsh, permanent ponds, watergrass and uplands inColusa County.

3531 Devil's Postpile National Monument

Devil's Postpile National Monument 760-934-2289
Minaret Road
Mammoth Lakes, CA 93546

The geologic formation that is the Postpile is the world's finest example of unusual columnar basalt. Its columns of lava, with their four to seven sides, display of honeycomb pattern of order and harmony. Another jewel in the Monumentis the San Joaquin River.

3532 Eldorado National Forest

100 Forni Road 530-622-5061
Placeville, CA 95667 Fax: 530-621-5297

Situated near the California gold discovery site on the American River at Coloma, this forest still boast numerous gold-bearing rivers and streams.Fishing opportunities are abundant. Only 34 miles of waterways are stocked, theremainder contain resident trout. Winter sports are cross country ski, sonowmobile and snowshoe. Backcountry exploration takes place year round in the Desolation and Mokelumme wildernesses.

3533 Ellicott Slough National Wildlife Refuge

San Francisco Bay NWR Complex 415-792-0222
Po Box 524 http://www.gorp.com
Newark, CA 94560-0524

3534 Farallon National Wildlife Refuge

San Francisco NWR Complex 415-792-0222
Po Box 524 http://www.gorp.com
Newark, CA 94560-0524

3535 Havasu National Wildlife Refuge

317 Mesquite Avenue 760-326-3853
PO Box 3009 Fax: 760-326-5745
Needles, CA 92363 E-mail: aimee_haskew@fws.gov
http://www.fws.gov/southwest/refuges/arizona/havasu

John Earle, Manager

3536 Hopper Mountain National Wildlife Refuge
PO Box 5839   805-644-1585
Ventura, CA 93005   http://www.gorp.com
The area is a traditional feeding site for the endangered California condor. Condors use the area frequently from October through May. A variety of other birds occur during migration and year round. The habitat includes 2,471 acres ofgrassland, chaparral and coastal sage scrub. There is a small, 350 acre area of intact California black walnut groves, some of the last remaining in southern California.
Marc Weitzel, Manager

3537 Humboldt Bay National Wildlife Refuge San Francisco Bay National Wildlife Refuge
PO Box 576   707-733-5406
Loleta, CA 95551   Fax: 707-733-1946
                               http://www.gorp.com

3538 Inyo National Forest
873 North Main Street   760-873-5841
Bishop, CA 93514   http://www.r5.fs.fed.us/inyo/
The Inyo National Forest is a unique and special area of public land located along the aestern edge of California and Sierra Nevada. Extending 165 miles along the California/Nevada border between Los Angeles and Reno, the Inyo Nationalforest includes 1.9 million acres of pristine lakes, fragile meadows, winding streams, rugged Sierra Nevada peaks, and arid Great Baisn Mountains. Elevations range frome 4,000 to 14,495 feet, providing diverse habitats that support vegetationpatterns ranging.

3539 Joshua Tree National Park
74485 National Park Drive   760-367-5500
Twentynine Palms, CA 92277   Fax: 760-367-6392
                               E-mail: JOTR_info@nps.gov
                               http://www.nps.gov/jotr
Joshua Tree National Park's 794,000 acres span the transition between the Mojave and Colorado deserts of Southern California. Proclaimed a National Monument in 1936 and a Biosphere Reserve in 1984, Joshua Tree was designated a NationalPark In 1994. The area possesses a rich human history and a pristine natural environment.

3540 Joshua Tree National Park Association
74485 National Park Drive   760-367-5525
Twentynine Palms, CA 92277-3597 Fax: 760-367-5583
                                 E-mail: mail@joshuatree.org
                                 http://www.joshuatree.org
The Joshua Tree National Park Association is a not-for-profit organization formed to assist with preservation, education, historical and scientific programs for the benefit of Joshua Tree National Park and its visitors.

3541 Kern National Wildlife Refuge
PO Box 670   661-725-2767
Delano, CA 93216   Fax: 661-725-6041
                               http://www.gorp.com

Thomas J Charmley, Manager

3542 Kesterson National Wildlife Refuge
PO Box 2176   209-826-3508
Los Banos, CA 93635   Fax: 209-826-1445
                               http://www.gorp.com
Habitat includes 5,900 acres of marshland and native grasslands superimposed on Bureau of Reclamation land. Seasonal marsh is being restored and enlarged and some permanent marsh is being created. Wildlife includes pintails,green-winged teal, shovelers, mallards, gadwalls, geese, phalaropes, yellowlegs, dowitchers, sandpipers, long-billed curlews, cinnamon teal, avocets, black-necked stilts and several species of herons and Krets. The endangered San Joaquin kit fox ispresent in small numbers.

3543 Kings Canyon National Park
Sequoia & Kings Canyon Ntl Park 559-565-3134
Three Rivers, CA 93271
Kings Canyon National Park, located in California's Sierra Nevada Mountains, is a park most famous for its pristine stands of Giant Sequoia trees. Visitors explore Grant Grove along a network of easy trails. For the more adventurous,longer trails including those in the remote Kings Canyon Backcountry provide a greater challenge.

Thomas Ritter, Supt

3544 Klamath Basin National Wildlife Refuges
4009 Hill Road   530-667-2231
Tulelake, CA 96134   Fax: 530-667-3299
                               E-mail: Dave_Menke@fws.gov
                               http://http://klamathbasinrefuges.fws.gov
The Klamath Basin National Wildlife Refuges Complex consists of six refuges in Northern California and Southern Oregon. The refuges support the largest concentration of migratory water fowl on the west coast and the largest winteringnumbers of bald eagles in the lower 48 states.
Ron Cole, Refuge Manager
Francis Maiss, Assistant Refuge Manager

3545 Klamath National Forest
1312 Fairlane Road   530-842-6131
Yreka, CA 96097

3546 Lake Tahoe Basin Management Unit
870 Emerald Bay Road   530-573-2600
S Lake Tahoe, CA 96150   Fax: 530-573-2693
                               http://www.fs.fed.us/r5/ltbmu/
Majestic sceenery and diverse recreation oppotunities draw millions of visitors to the Lake Tahoe Basin annually. Changing colors throughout the year afford a brilliant backdrop to the many available activities in all seasons. TheBasin is home to a rich diversity of plants and animals that can be viewed during walks at interpertive sites and on many forest trails.

3547 Lassen National Forest
2550 Riverside Drive   530-257-2151
Susanville, CA 96130   Fax: 530-257-8282
                               http://www.fs.fed.us/r5/lassen/contact
Lassen National Forest lies at the heart of a fascinating part of California, a crossroads of people and nature. This is where the Sierra Nevada, the Cascades, the Modoc Plateau and the Great Basin meet. Within Lassen National Forest,you can explore a lava tube or the land of Ishi, the last survivor of the Yahi Tana Native American tribe: watch prong-horn antelope glide across sage flats; drive four-wheel trails into granite country appointed with sapphire lakes or discoverwildflowers on foot.

3548 Lava Beds National Monument
1 Indian Well Headquarters   530-667-2282
Tulelake, CA 96134   Fax: 530-667-2737
                               E-mail: LABE_SUperintendent@nps.gov
Volcanic eruptions on the Medicine Lake shield volcano have created an incredibly rugged landscape punctuated by cinder cones, lava flows, spatter cones, lava tube caves and pit craters. During the Modoc War of 1872-1873, the ModocIndains used these tortuous lava flows to their advantage. Under the leadership of Captain Jack, the Modocs took refuge in "Captain Jack's Stronghold," a natural lava fortress.

3549 Los Padres National Forest
6755 Hollister Avenue   805-968-6640
Suite 150   Fax: 805-961-5729
Goleta, CA 93117   http://www.fs.fed.us/r5/lospadres
Los Padres National Forest encompasses nearly 2 million acres of the central coastal mountains of California.
Kathy Good, Public Affairs Officer
Maeton Freel, Ecosystem Staff Officer

3550 Lower Klamath National Wildlife Refuge
Hill Road, Route 1   530-667-2231
Box 74   E-mail: tom_stewart@mail.fws.gov
Tulelake, CA 96134   http://www.klamathnwr.org
The objectives of the refuge are to: maintain habitat for endangered, threatened and sensitive species; provide and enhance habitat for fall and spring migrant waterfowl; protect native habitats and wildlife representative of thenatural biological diversity of the Klamath Basin; integrate the maintenance of productive wetland habitats and sustainable agriculture; provide high quality wildlife-dependent visitor services.

3551 Mendocino National Forest
825 Humboldt Avenue   530-934-3316
Willows, CA 95988   Fax: 530-934-7384
                               E-mail: mailroom_r5_mendocino@fs.fed.us

The Mendocino national Forest was set aside by President Roosevelt in 1907. It was frist named the Stony Creek Reserve and then the Stony Creek National Forest. It was later named the California National Forest and in 1932 became theMendocino National Forest. The MNF straddles the eastern spur of the Coastal Mountain Range in northernwestern Califonia, just a three hour drive north of San Francisco and Sacramento.
Jim Fenwood, Forest Supervisor

3552 Merced National Wildlife Refuge
San Luis NWR Complex
947 W Pacheco Boulevard, Suite C   209-826-3508
PO Box 2176   Fax: 209-826-1445
Los Banos, CA 93635   http://www.pacific.fws.gov
7,035 acres of native grasslands, agricultural fields, and wetlands provide the primary wintering area for the largest flock of lesser sandhill cranes and Ross' geese in the Pacific Flyway, as well as important habitat for northernpintails, cackling Canada geese, and a wide variety of shorebirds.
, Project Manager

3553 Modoc National Forest
441 North Main Street   530-233-5811
Alturas, CA 96101   E-mail: %20rhaggard@fs.fed.us
A land of contrasts and unspoiled vaction-hideaway settings. Nestled in the extreme northeastern corner of California, The Modoc National Forest is 140 miles east of Redding on Highway 299, and 169 miles north of Reno, Nevada, viahighway 395. The Modoc National Forest features several mountain areas. The Warner Mountains, on its east, are the western edge of Great Basin Province, the Medicine Lake Highlands, to northwest, are a couthern spur of the Cascade Range.

3554 Modoc National Wildlife Refuge
PO Box 1610   530-233-3572
Alturas, CA 96101   Fax: 530-233-4143
    http://modoc.fws.gov
A 7,000+ acre refuge established to manage and protect migratory waterfowl.

Steve Clay, Manager

3555 Pinnacles National Monument: National Park Service, Department of the Interior
5000 Highway 146   831-389-4485
Paicines, CA 95043   Fax: 831-389-4489
    E-mail: pinnvisitorinformation@nps.gov
    http://www.nps.gov/pinn
Rising out of the chaparral-covered Gabilan Mountains, east of central California's Salinas Valley, are the spectacular remains of an ancient volcano. Massive monoliths, spires, sheer-walled canyons and talus passages define millionsof years of erosion, faulting and tectonic plate movement is reowned for the beauty and variety of its spring wildflowers. Hiking, rock climbing, picnicing and sildlife observation can be enjoyed throughout the year.
Steve Shackelton, Superintendent

3556 Pixley National Wildlife Refuge
Kern NWR
Po Box 670   805-725-2767
Delano, CA 93216-0670   http://www.gorp.com

3557 Plumas National Forest
159 Lawrence Street   530-283-2050
PO Box 11500   Fax: 530-283-7746
Quincy, CA 95971-6025
The Plumas National Forest has fresh conifer forests, rugged cayons, crystal clear lakes, grassy meadows, trout filled streams and brilliant star-filled skies. Located where the Sierra Nevada and Cascade Mountain ranges meet, thisforest has more than 100 lakes, 1,000 miles of rivers and streams, and over a million acres of National Forest.

3558 Redwood National Park
1111 2nd Street   707-464-6101
Crescent City, CA 95531   Fax: 707-464-1812
    http://www.nps.gov/redw

The world's tallest living trees can found along the northern California coast. Of the coast redwood forests still around today, almost one half of them can be found within the projected boundaries of Redwood National and State Parks.In 1994, the National Park Service and the Caifornia State Parks joined forces to manage four parks: Redwood National, Jebediah Smith, Del Norte Coast, and Prairie Creek Redwoods state Parks collectively known as Redwood National and State Parks.
Bill Pierce, Supt

3559 Sacramento River National Wildlife Refuge
Sacremento NWR Complex   530-934-2801
752 County Road 99W   Fax: 530-934-7814
Willows, CA 95988-96E9mail: gary_w_kramer@mail.fws.gov
    http://www.gorp.com
Gary W Kramer, Manager

3560 Salinas River Wildlife Refuge
San Francisco Bay NWR Complex   415-792-0222
Po Box 524   http://www.gorp.com
Newark, CA 94560-0524

3561 Salton Sea National Wildlife Refuge
906 W Sinclair Road   619-348-5278
PO Box 120   http://www.gorp.com
Calipatria, CA 92233-0120
Kenneth Voget, Manager

3562 San Bernardino National Forest
1824 S Commercenter Circle   909-383-5588
San Bernardino, CA 92408   Fax: 909-383-5770
In the San Bernardino Mountains, the forest service has developed an extensive network of campgrounds and dozens of picnic areas for families and groups who want to enjoy a day in the mountains. The forest offers camping, picnicking,fishing, boating, swimming, hiking, horseback riding and more. During the winter, visitors come to the forest to cross-contry and down ski, snowboard and snowmobile.

3563 San Francisco Bay National Wildlife Refuge
PO Box 524   415-792-0222
Newark, CA 94560-0524   Fax: 415-792-5828
    http://www.gorp.com
Richard A Coleman, Manager

3564 San Joaquin River National Wildlife Refuge
PO Box 2176   209-826-3508
Los Banos, CA 93635   Fax: 209-826-1445
    http://www.gorp.com

3565 San Luis National Wildlife Refuge
947 W Pacheco Boulevard, Suite C   209-826-3508
PO Box 2176   Fax: 209-826-1445
Los Banos, CA 93635http://http://www.fws.gov/pacific/sanluis/
This National Wildlife Refuge Complex is comprised of 3 National Wildlife Refuges and the Grasslands Wildlife Management Area located in the San Joaquin Valley of California. They are part of the US Fish and Wildlife Service, a bureauwithin the Department of the Interior. Their mission: Working with others in the Pacific Region to conserve, protect, and enhance fish, wildlife and their habitats for the continuing benefit of the American people.
Forrest, Refuge Manager

3566 San Pablo Bay National Wildlife Refuge
San Francisco Bay NWR Complex   415-792-0222
Po Box 524   http://www.gorp.com
Newark, CA 94560-0524

3567 Santa Monica Mountains National Recreation
401 West Hillcrest Drive   805-370-2301
Thousand Oaks, CA 91360   Fax: 805-370-1850
Santa Monica Mountains rise above Los Angeles, widen to meet the curve of Santa Monica Bay and reach their highest peaks facing the ocean, forming a beautiful and multi-faceted landscape. Santa Monica Mountains National Recreation Areais a cooperative effort that joins federal, state and local park agencies with private preserves and landowners to protect the natural and cultural resources of this transverse mountain range and seashore.

Arthur E Eck, Superintendent

3568 Seal Beach National Wildlife Refuge
Kern NWR   661-725-2767
Po Box 670   http://www.gorp.com
Delano, CA 93216

3569 Sequoia National Forest
900 W Grand Avenue   559-784-1500
Porterville, CA 93257   Fax: 559-781-4744

The Sequoia National Forest is at the southern tip of the Sierra Nevada range. Its highest point is 12,432 foot Florence Peak in the Golden Trout Wilderness. The forest has five wildernesses, a scenic byway and four wild and scenic rivers. About 10,000 cows graze on the forest land. Camping, water sports, hiking, downhill and cross-country skiing and horseback riding are amoung the forest's many recreational activities.

3570 Sequoia and Kings Canyon National Parks
47050 Generals Highway   559-565-3341
Three Rivers, CA 93271-9651   Fax: 559-565-3730
              E-mail: SEKI_interpretation@nps.gov
              http://www.nps.gov

3571 Shasta-Trinity National Forests
2400 Washington Avenue   530-244-2978
Redding, CA 96001   Fax: 530-242-2233
              http://www.fs.fed.us/r5/

Sharon Heywood, Forest Supervisor

3572 Sierra National Forest
1600 Tollhouse Road   559-297-0706
Clovis, CA 93611-0532   Fax: 559-294-4809
              E-mail: dkohut@fs.fed.us

3573 Six Rivers National Forest
1330 Bayshore Way   707-442-1721
Eureka, CA 95501   Fax: 707-442-9242

3574 Stanislaus National Forest
19777 Greenley Road   909-532-3671
Sonora, CA 95370

The Stanislaus National Forest, created on February 22, 1897, is among the oldest of the National Forests. It is named for the Stanislaus River whose headwaters rise within Forest boundaries. The Spanish explorer Gabriel Moraga named the river "Our Lady of Guadalupe" during an 1806 expedition. Later, the river was renamed in honor of Estanislao, an Indian leader.

3575 Sutter National Wildlife Refuge
Sacramento NWR Complex   530-934-2801
752 County Road 99W   Fax: 530-934-7814
Willows, CA 95988   http://www.gorp.com
Kevin Forester, Project Manager

3576 Sweetwater Marsh National Wildlife Refuge
301 Caspian Way   619-575-2704
Imperial Beach, CA 91932   Fax: 619-575-6913
              E-mail: rebecca_young@fws.gov
              http://www.gorp.com

The refuge has 316 acres of salt marsh and coastal uplands. It includes the largest emergent wetlands area remaining in San Diego Bay.

3577 Tahoe National Forest
631 Coyote Street   530-265-4531
Nevada City, CA 95959   Fax: 530-478-6107
              http://www.fs.fed.us/r5/tahoe/

3578 Tijuana Slough National Wildlife Refuge
Tijuana River NERR   619-575-2704
301 Caspian Way   Fax: 619-575-6913
Imperial Beach, CA 91932   E-mail: rebecca_young@fws.gov

Established in 1980 to conserve and protect endangered and threatened fish, wildlife and plant species. Conservation of the light-footed clapper rail was the primary impetus for the creation of the refuge. The refuge is part of a larger unit called the Tijuana River National Estuarine Research Reserve, which is administered by the National Oceanographic and Atmospheric Administration.
Phil Jenkins, Reserve Manager

3579 Tule Lake National Wildlife Refuge
4009 Hill Road   530-667-2231
Tulelake, CA 96134   Fax: 530-667-3299
              http://www.gorp.cim

3580 Upper Klamath National Wildlife Refuge
Klamath Basin NWR Complex   530-667-2231
Route 1, Box 74   Fax: 530-667-3299
Tulelake, CA 96134 http://http://klamathbasinrefuges.fws.gov/

3581 Yosemite National Park
PO Box 577   209-372-0200
Yosemite National Park, CA 95389   Fax: 209-372-0220
              E-mail: yose_web_manager@nps.gov
              http://www.nps.gov/yose/home.htm

Yosemite National Park embraces a spectacular tract of mountain- and-valley scenery in the Sierra Nevada, which was set aside as a national park in 1890. The park harbors a grand collection of waterfalls, meadows, and forests that include groves of giant sequoias, the world's largest living things. Highlights of the park include Yosemite Valley, and its high cliffs and waterfalls; Wawona's history center and historic hotel; the Mariposa Grove, which contains hundreds of ancient giant sequoias.
David A Mihalic, Superintendent

Colorado: US Forests, Parks, Refuges

3582 Alamosa/Monte Vista National Wildlife Refuge
9383 El Rancho Lane   719-589-4021
Alamosa, CO 81144   Fax: 719-587-0595
              E-mail: alamosa@fws.gov
              http://http://alamosa.fws.gov

The Valley extends over 100 miles from north to south and 50 miles from east to west, with dwarfing mountains in three directions. The surrounding mountains feed the arid valley with precious surface water, as well as replenish an expansive underground reservoir. This liquid wealth has made two National Wildlife Refuges possible in the San Luis Valley: Alamosa and Monte Vista. These wetland gems near the heart and on the edge of the Valley are places for wildlife and people.
, Manager

3583 Arapaho National Forest
2995 Baseline Road Road   970-498-2770
Boulder, CO 80303   877-444-6777
        http://www.llbean.com/parksearch/parks/html/15238gd.htm
Camping, hiking, hunting, fishing, skiing, mountain biking and enjoying scenic drives are popular activities in the Arapaho National Forest. More than 400 species of wildlife inhabit the forest. The Continental Divide and the Frotn Range of the Rocky Mountains form the backbone of the forest. Along their flanks are high plateaus, rolling foothills and open stretches of high prarie.

3584 Arapaho National Wildlife Refuge
953 JC Rd 32   970-723-8202
Walden, CO 80480   Fax: 970-723-8528
              E-mail: r6rw_arp@fws.gov
              http://www.gorp.com

Arapaho National Wildlife Refuge supports diverse wildlife habitats including sagebrush grassland uplands, grassland meadows, willow riparian areas, wetlands and mixed conifer and aspen woodland. This refuge is one in a system of over 500 National Wildlife Refuges, a network of lands set aside and managed specifically for wildlife. It is administered by the US Fish and Wildlife Service.

Ann M Timberman, Manager

Government Agencies & Programs/Connecticut: US Forests, Parks, Refuges

3585 Browns Park National Wildlife Refuge
1318 Highway 318 970-365-3613
Maybell, CO 81640 Fax: 970-365-3614
http://www.brownspark.fws.gov
The primary purpose of Browns Park Refuge is to provide high quality nesting and migration habitat for the Great Basin Canada Goose, ducks and other migratory birds. Before Flaming Gorge Dam was constructed in 1962, the Green River flooded annually, creating excellent waterfowl nesting, feeding and resting marshes in the backwater sloughs and old stream meanders. The dam stopped the flooding, eliminating much of this waterfowl habitat.
Jerry Rodriguez, Refuge Manager

3586 Colorado National Monument
Fruita, CO 81521 970-858-3617
Fax: 970-858-0372
E-mail: COLM_Info@nps.gov
Colorado National Monument consists of geological features including: towering red sandstone monoliths, deep sheer-walled canyons and a variety of wildlife.
Palma Wilson, Supt

3587 Curecanti National Recreation Area
102 Elk Creek 970-641-2337
Gunnison, CO 81230 Fax: 970-641-3127
E-mail: CURE_Vis_Mail@nps.gov
Three reservoirs, named for corresponding dams on the Gunnison River, form the heart of Curecanti National Recreation Area; Blue Mesa Reservoir, Morrow Point Reservoir and the Crystal Reservoir.
William Wellman, Superintendent

3588 Dinosaur National Monument
4545 Highway 40 970-374-3000
Dinosaur, CO 81610 Fax: 970-374-3003
http://www.nps.gov/dino/dinos.htm
Dinosaur Monument protects a large deposit of fossil dinosaur bones that lived millons of years ago.
Dennis K Huffman

3589 Florissant Fossil Beds National Monument
15807 Teller County 1 719-748-3253
Florissant, CO 80816 Fax: 719-748-3164
E-mail: FLFO_Information@nps.gov
Huge petrified redwoods and incredibly detailed fossils of ancient insects and plants reveal a very different Colorado of long ago. A lake formed in the valley and the fine-grained sediments at its bottom became the final resting-place for thousands of insects and plants. These sediments compacted into layers of shale and preserved the delicated details of these organisms as fossils.
Dale Ditmanson, Supt

3590 Grand Mesa National Forest
2250 US Highway 50 970-874-6600
Delta, CO 81416 Fax: 970-874-6698
E-mail: fjreed@fs.fed.us
The Grand Mesa national Forest encompasses the Grand Mesa, which is one of the world's largest flattop mountains and has an average elevation of 10,000 feet. It is dotted with over 300 alpine lakes and reservoirs. There are campgrounds views from elevations af more then 11,000 feet. The Grand Mesa Scenic Byway runs from Cedaredge to Plateau Valley, with Land End Road effering breathtaking views.

3591 Great Sand Dunes National Monument
11999 Highway 150 719-378-6399
Mosca, CO 81146
These dunes are the tallest in North America, rising 750 feet from the valley floor. The dunes are home to some unique and spectacular species of flora and fauna. Besides a large variety of birds, there are quite a few species of mammals that visit or reside within the dunes. Few reptiles are found here due to the high altitude.
Bill Wellman, Supt

3592 Pike National Forest
1920 Valley Road 719-545-8737
Pueblo, CO 81008 Fax: 719-543-8926
http://www.fs.fes.us/r2/psicc/

Perhaps as embedded in American folklore as the Brooklyn Bridge, Pike's Peak is America's easternmost 14'er (a mountain with an elevation of 14,000 feet or higher). It is atop Pike's Peak that the song "America the Beautiful" was written. This ecologically conscious suburbia is snaked with numerous green belts and effectively serves as the eastern boundary of Pike National Forest.
, Ranger

3593 Rio Grande National Forest
1803 W US Highway 160 719-852-5941
Monte Vista, CO 81144 Fax: 719-852-6250
The Rio Grande National Forest is 1.86 million acres located in southwestern Colorado and remains one of the true undiscovered jewels of Colorado. The Continental Divide runs for 236 miles along most of the western border of the Forest. The Forest present myraid ecosystems; from 7600-ft alpine desert to over 14,300-ft in the majestic Sangre de Cristo Wilderness on the eastern side.

3594 Rocky Mountain National Park
1000 Highway 36 970-586-1206
Estes Park, CO 80517 Fax: 970-586-1256
E-mail: ROMO_informatin@nps.gov
http://www.nps.gov
Dick Putney, Information Office Manager

3595 Routt National Forest
925 Weiss Drive 970-879-1870
Steamboat Springs, CO 80487 Fax: 970-870-2284
http://www.dgif.state.va.us
The varied terrain of the forest provides habitat for a range of wildlife, including threatened and endangered species such as sandhill cranes and green-backed cutthroat trout. At Fish Creek Falls, cascading water tumbles 263 feet over shear rock walls. Most of the mountain areas of the Routt are covered with lodgepole pine, Engelmann spruce, subalpine fir and Douglas-fir, interspersed with large extensions of pine parks and aspen stands.

3596 San Juan National Forest
15 Burnett Court 970-247-4874
Durango, CO 81301 Fax: 970-385-1243
San Jaun National Forest, a region of forested mountains, 14,000-foot peaks, scenic roads, geological wonders, hisoric and prehistoric communities, and a narrow-gauge railroad.

3597 White River National Forest
900 Grand Avenue 970-945-2521
PO Box 948 Fax: 970-945-3266
Glenwood Springs, CO 81602-0948
The two and one quarter million acre White River National Forest is located in the heart of the Colorado Rocky Mountains, approximately two to four hours west of Denver on Interstate 70. The scenic beauty of the area, along with ample developed and undeveloped recreation opportunities on the forest, accounts for the fact the White River consistently ranks as one of the top five Forests nationwide for total recreation use.

Connecticut: US Forests, Parks, Refuges

3598 Stewart B McKinney National Wildlife Refuge: Ninigret National Wildlife Refuge
PO Box 1030 860-399-2513
Westbrook, CT 06498 Fax: 860-399-2515
E-mail: r5rw_sbmnwr@fws.mail.gov
http://www.gorp.com
Paul Casey, Manager

Delaware: US Forests, Parks, Refuges

3599 Bombay Hook National Wildlife Refuge
2591 Whitehall Neck Road 302-653-9345
Smyrna, DE 19977 Fax: 302-653-0684
E-mail: fw5rw_bhnwr@fws.gov
http://bombayhook.fws.gov

238

Stretching about eight miles along Delaware Bay and covering nearly 16,000 acres, Bombay Hook NWR was established as a refuge for migratory waterfowl. Today, the Refuge provides habitat for a diversity of wildlife. The refuge offersauto tours, walking trails, observation towers, and interpretive displays for the visiting public.

Terry Villanueva, Refuge Manager
Tina Watson, Outdoor Recreation Planner

3600 Prime Hook National Wildlife Refuge

11978 Turkle Pond Road   302-684-8419
Milton, DE 19968   Fax: 302-684-8504
          E-mail: FW5RW_PHNWR@fws.gov
              http://primehook.fws.gov
The Prime Hook National Wlidlife Refuge spans about 10,000 acres along the western Delaware Bay. The marshes of the refuge are ideal habitat for thousands of migrating ducks, geese, and shorebirds. The refuge is also home to woodlandand grassland birds, reptiles, amphibians, and mammals, including the endangered Delmarva Peninsula Fox Squirrel. Avid photographers can enjoy the beauty of wildlife from a photography blind and wheel-chair accessible observation platform.
Jonathan Schafler, Project Leader
Brian Brandis, Deputy Project Manager

District of Columbia: US Forests, Parks, Refuges

3601 Battleground National Cemetery

6625 Georgia Avenue, NW   202-426-6924
Washington, DC 20240   Fax: 202-426-1845
                  http://www.nps.gov
Battleground National Cemetery, located at 6625 Georgia Avenue NW, was established shortly after the Battle of Fort Stevens in the summer of 1864. The battle, which lasted two days (July 11-12, 1864) marked the defeat of General JubalA Early's Confederate campaign to launch an offensive action against the poorly defended Nation's Capital. Near the entrance are monuments commemorating those units which fought at Fort Stevens.

Florida: US Forests, Parks, Refuges

3602 Apalachicola National Forest

325 John Knox Road   850-523-8500
Suite F-100   http://www.dgif.state.va.us
Tallahassee, FL 32303
The Apalachicola National Forest contains 564,000 acres and is the largest forest in the state. Much of the forest is wet lowland dressed with cypress, oaks and magnolias. On the upland flatwoods, slash and longleaf pine. The abundanceand diversity of water features like rivers, creeks, lakes, sink holes ans savannahs provide outstanding recreation opportunities and excellent habitat for rare and endangered plants and animals.

3603 Arthur R Marshall Loxahatchee National Wildlife Refuge

10216 Lee Road   401-732-3684
Boynton Beach, FL 33437-4796   Fax: 561-732-3684
        http://www.loxahatchee.fws.gov/news-n-happenings/coe-200503.asp
Welcome to the Arthur R. Marshall Loxahatchee National Wildlife Refuge, the last northernmost portion of the unique Everglades. With over 221 square miles of Everglades habitat, A.R.M. Loxahatchee National Wildlife Refuge is home tothe American alligator and the endangered Everglades snail kite. In any given year, as many as 257 species of birds may use the refuge's diverse wetland habitats.
Burkett S Neely Jr, Manager

3604 Big Cypress National Preserve

HCR 61, Box 110   239-695-2000
Ochopee, FL 33141   Fax: 941-695-3901
        E-mail: sandy_snell-dobert@nps.gov

The 729,000 acre Big Cypress National Preserve was set aside to ensure the preservation, conservation, and protection of the natural scenic, floral and faunal, and recreational values of the Big Cypress Watershed. The importance ofthis watershed to the Everglades National Park was a major consideration for its establishment. The name Big Cypress refers to the large size of this area. Vast expanses of cypress strands span this unique landscape.

Wally Hibbard, Superintendent

3605 Biscayne National Park

9700 SW 328 Street   305-230-7275
Homestead, FL 33033   Fax: 305-230-1190
            E-mail: BISC_Information@nps.gov
                http://www.nps.gov/bisc
Turquoise waters, emeral islands and fish-bejeweled reefs make Biscayne National Park a paradise for wildlife-watching, snorkeling, diving, boating, fishing and other activities. Within the park boundaries are the longest stretch ofmangrove forest left on Florida's east coast, the clear shallow waters of Biscayne Bay, over 40 of the northernmost Florida Keys, and a spectacular living coral reef. Superimposed on all of this natural beauty is 10,000 years of human history.
Linda Canzanelli, Supt

3606 Canaveral National Seashore

308 Julia Street   321-267-1110
Titusville, FL 32796   Fax: 321-264-2906
        E-mail: CANA_Superintendant@nps.gov
            http://www.nbbd.com/godo/cns/
Canaveral National Seashore is a step into the past, protection for the present and a doorway into the future. The 100 Timucuan Mounds that are within our boundaries are evidence of past generations of people that lived here. CanaveralNational Seashore covers 57,000 acres and is the longest stretch (24 miles) of undeveloped beach on Florida's east coast. Fourteen endangered species make their home within Canaveral's boundaries.
Robert Newkirk, Superintendent

3607 Chassahowitzka National Wildlife Refuge Complex

1502 SE Kings Bay Drive   352-563-2088
Crystal River, FL 34429   Fax: 352-795-7961
        E-mail: fw4rwchassahowitzka/r4/fws/dol@fws
              http://chassahowitzka.fws.gov
A complex of 5 National Wildlife Refuges on the Gulf Coast of Florida including Crystal River NWR. Established in 1983 for the protection of the endangered West Indian manatee. Office hours are from 7:30 AM to 4:00 PM Monday thruFriday. The office is also open Saturdays and Sundays during the winter months (November 15 - March 31) please call the office at 352-563-2088 for more information.
Jim Kraus, Project Leader

3608 Dry Torgus National Park/Everglades National Park

40001 State Road 9336   305-242-7700
Key West, FL 33041   Fax: 305-242-7711
Richard Ring, Supt

3609 Egmont National Wildlife Refuge

1502 SE Kings Bay Drive   352-563-2088
Crystal River, FL 34429   Fax: 352-795-7961
                  http://www.gorp.com
This barrier island refuge is approximately 350 acres and was established to provide nesting, feeding and resting habitat for brown pelicans and other migratory birds. The combined resources of the US Fish and Wildlife Service andFlorida Park Service provide protection for Egmont Key and its wildlife, as well as an enjoyable experience for the visitor.

3610 Everglades National Park

4001 State Road 9336   315-242-7700
Homestead, FL 33034   Fax: 305-242-7728
            E-mail: EVER_Information@nps.gov
Spanning the southern tip of the Florida peninsula and most of Florida Bay, Everglades National Park is the only subtropical preserve in North America. It contains both temperature and tropical plant communities, as well as marine andmarsh environments and is known for its rich bird life. It is also the only place in the world where alligators and crocodiles exist side by side.
Maureen Finnerty, Supt

3611 Florida Panther National Wildlife Refuge
3860 Tollgate Boulevard   813-353-8442
Suite 300
Naples, FL 33942-5444
Todd Logan, Manager

3612 Hobe Sound National Wildlife Refuge
PO Box 645   561-546-6141
Hobe Sound, FL 33475-0645   Fax: 561-545-7572
E-mail: fw4_rw_hobe_sound@fws.gov
http://www.mrcirl.org/mapirl.hobe.html
Refuge objectives are to: maintain and restore diverse habitats designed to achieve refuge purposes and wildlife population objectives; maintain viable diverse populations of native flora and fauna consistent with sound biologicalprinciples; manage natural and cultural resources through land protection and partnership; and develop and implement wildlife dependent recreation and environmental education that leads to enjoyable recreation experiences and a greater understandingof resources.
Margo Stahl, Refuge Manager

3613 JN Darling National Wildlife Refuge
1 Wildlife Drive   239-472-1100
Sanibel, FL 33957   Fax: 239-472-4061
The refuge on Sanibel Island is a subtropical barrier island in the Gulf of Mexico hemmed by mangrove trees, shallow bays and white sandy beaches. The 6,300-acre refuge is connected to the mainland by a three-mile causeway. Named in1967 for Jay Norwood (Ding) Darling, an editorial cartoonist, pioneer conservationist and originator of the federal Duck Stamp Program. Darling, who was the first director of what is now the US Fish and Wildlife Service, wintered on neighboringCaptiva Island.
Louis Hinds, Manager

3614 Lake Woodruff National Wildlife Refuge
4490 Grand Avenue   904-985-4673
PO Box 488   Fax: 904-985-0926
DeLeon Springs, FL 32130-0488
Encompasses two large lakes offering sights of diverse habitats and a variety of wildlife. Fishing and boating are the primary recreational activities.
Leon I Rhodes, Manager

3615 National Key Deer Refuge
PO Box 430510   305-872-2239
Big Pine Key, FL 33043-0510   Fax: 305-872-3675
E-mail: fw4_rw_key_deer@fws.gov
http://nationalkeydeer.fws.gov/
Refuge objectives are: to protect and preserve Key deer and other wildlife resources in the Florida Keys; to conserve endangered and threatened fish, wildlife and plants; to provide habitat and protection for migratory birds; and toprovide opportunities for environmental education and public viewing of refuge wildlife and habitats.
Jim Halpin, Project Leader

3616 Pelican Island National Wildlife Refuge
1339 20th Street   772-562-3909
Vero Beach, FL 32960-3559   Fax: 772-299-3101
E-mail: pelicanisland@fws.gov
http://www.fws.gov/pelicanisland

Paul Tritaik, Refuge Manager
Fran Breedlove, Administrative Assistant

3617 St. Marks National Wildlife Refuge
PO Box 68   850-925-6121
St Marks, FL 32355
Saint Marks National Wildlife Refuge is in Wakulla, Jefferson and Taylor counties along the Gulf coast of north Florida. The refuge is approximately 25 miles south of Tallahassee. The refuge encompasses 65,000 acres of divided tidalflats; and freshwater impoundments harbor a large variety of wildlife, including 434 verebrate species, excluding fish. Over quarter million vistors enjoy a variety of outdoor recreation opportunities annually.
Joe D White, Manager

3618 St. Vincent National Wildlife Refuge
479 Market Street   850-653-8808
Apalachicola, FL 32320   Fax: 850-653-9893
E-mail: terry_peacock@fws.gov
The historic St Vincent National Wildlife Refuge is a large barrier island, four miles wide and nine miles long. It was inhabited as early as 240A.D. and is known to have been visited by Franciscan friars in the early 1600s. Over itshistory, private landowners developed the island into a preserve housing Asian and African wildlife and an assortment in between. The US Fish and Wildlife Service purchased the island in 1968 bringing an end to the exotic jungle.

3619 Timucuan Ecological & Historic Preserve
13165 Mount Pleasant Road   904-221-5568
Jacksonville, FL 32225   Fax: 904-221-5248
http://www.nps.gov/timu
The 46,000 acre Timucan Ecological and Historic Preserve was established to protect one of the last unspoiled coastal wetlands on the Atlantic Coast and to preserve historic and prehistoric sites within the area. The estuarineecosystem includes aslt marsh, coastal dunes, hardwood hammock, as well as salt, fresh, and brackish waters, all rich in native vegetation and animal life.
Barbara Goodman, Supt

Georgia: US Forests, Parks, Refuges

3620 Chattahoochee-Ocnee National Forest
1755 Cleveland Highway   770-297-3000
Gainesville, GA 30501   Fax: 770-297-3011
http://www.dgif.state.va.us
Together, these forest offer more than ten wilderness areas, six beaches, and thousands of acres of lakes and stream. The forest is real draw for history buffs, who can get absorbed in tracing events such as the Trail of Tears and manya Civil War battle.

3621 Cumberland Island National Seashore
PO Box 806   912-882-4335
Saint Marys, GA 31558   Fax: 912-673-7747
E-mail: cuis_information@nps.gov
Cumberland Island is 17.5 miles long and totals 36,415 acres of which 16,850 are marsh, mud flats, and todal crreks. Well know for its sea turtles, abundant shore bierds, dune fields, maritime forest, salt marshes, and historicstructures.
Rolland Swain, Supt

3622 Okefenokee National Wildlife Refuge
Highway 121   912-496-7366
Folkston, GA 31537   Fax: 912-496-3332
Okefenokee NWR was established in 1937 to preserve the 438,000 acre Okefenokee Swamp. Presently the refuge encompasses approximately 396,000 acres of which 353,000 are designated as Wilderness. Habitats include open wet prairies,cypress forests, scrub-shrub, oak hammocks and longleaf pine forests. The prosperity and survival of the swamp, and the species dependent on it, is directly tied with maintaining the integrity of complex ecological processes, including hydrology andfire.
M Skippy Reeves, Refuge Manager

3623 Piedmont National Wildlife Refuge
718 Juliette Road   478-986-5441
Round Oak, GA 31038   Fax: 478-986-9646
E-mail: fw4rwpiedmont.fws.gov
http://piedmont.fws.gov/
35,000 acre national wildlife refuge with hiking trails, gravel roads throughout, wildlife drive, hunting and fishing opportunties, bird watching.
Clarke Dirks, Refuge Manager
Carolyn Johnson, Assistant Refuge Manager

3624 Pinckney Island National Wild Refuge
Savannah Coastal Refuges
1000 Business Center Drive   912-652-4415
Suite 10   Fax: 912-652-4385
Savannah, GA 31405   http://www.savannahcoastal.fws.gov

A group of islands and small hammocks, the 4,053-acre refuge includes a variety of land types: saltmarsh, forestland, brushland, fallow fields and freshwater ponds. Pinckney, the largest of the refuge islands, is 3.8 miles long and 1.75 miles across at its greatest width. Boaters who navigate the refuge's estuarine waters may view shore and wading birds, including the endangered wood stork, that feeds on mudflats, oysterbeds and shores.

James D Browning, Manager

3625 Savannah National Wildlife Refuge: Savannah Coastal Refuges

1000 Business Center Drive   912-652-4415
Parkway Business Center, Suite 10   Fax: 912-652-4385
Savannah, GA 31405   E-mail: savannacoastal.fws.gov
          http://fwy_rw_savannah_coastal@gws.govnah/
Refuge objectives are to provide: a refuge and feeding ground for native birds and wild animals; habitat and protection for threatened and endangered plants and animals; habitat and sanctuary for migratory birds consistent with the objectives of the Atlantic Flyway; habitats for other species of indigenous wildlife and fishery resources; management of furbearers, deer and other upland animals; and opportunities for environmental education, interpretation and recreation for the visiting public.
James D Browning, Manager

Hawaii: US Forests, Parks, Refuges

3626 Baker, Howland, & Jarvis Islands & Johnston Atoll National Wildlife Refuge

300 Ala Moana Boulevard Room 5231   808-541-1201
Po Box 50167   Fax: 808-541-1216
Honolulu, HI 96850   http://www.gorp.com
Baker and Howland Island lie about 40 miles apart just north of the equator in the central Pacific Ocean and about 1,600 miles southwest of Honolulu. Jarvis Island lies a few miles south of the equator about 1,300 miles south of Honolulu and 1,000 miles east of Baker and Howland. Jarvis consists of a 1,086-acre island surrounded by 35,397 acres of submerged land.

3627 Hakalau Forest National Wildlife Refuge Hawaiian & Pacfic Islands Complex

154 Waianuenue Avenue Room 219   808-969-9909
Hilo, HI 96720   Fax: 808-934-7473
                                 http://www.gorp.com

Richard C Wass, Manager

3628 Haleakala National Park

PO Box 369   808-572-9306
Makawao, HI 96768   Fax: 808-572-9306
The Park preserves the outstanding volcanic landscape of the upper slopes of Haleakala on the island of Maui and protects the unique and fragile ecosystems of Kipahulu Valley, the scenic pools along Oheo gulch, and many rare and endangered species. Haleakala National Park was designated an International Biosphere Reserve.
Don Reeser, Supt

3629 Hanalei National Wildlife Refuge: Kauai

PO Box 1128   808-828-1413
Kilauea   Fax: 808-828-6634
Kauai, HI 96754-1128 http://www.gorp.com
The refuge was established to protect the endangered Koloa (Hawaiian duck), Hawaiian gallinule, Hawaiian coot and the Hawaiian stik. The refuge also provides habitat for migratory shorebirds and waterfowl. The habitat includes 917 acres of river bottom land, taro farms and wooded slopes in the Hanalei River Valley on the northern coast of Kauai. Recreational activities include wildlife observation and photography and fishing from the riverbank.

3630 Hawaii Volcanoes National Park

PO Box 52   808-985-6000
Hawaii National Park, HI 96718   Fax: 808-985-6004
                               http://www.nps.gov/havo
Established in 1916, this 333,000 acre National Park encompasses coastal lava plains, rain forests and deserts. It preserves and protects active volcanoes, rare and endangered plants and animals, and Hawaiian archeological sites.

3631 Hawaiian Island & Midway Atoll National Wildlife Refuge

PO Box 50167   808-541-1201
Honolulu, HI 96850-5167 Fax: 808-541-1216
                        http://www.gorp.com

3632 Huleia National Wildlife Refuge: Kauai

PO Box 1128   808-828-1413
Kilauea   Fax: 808-828-6634
Kauai, HI 96754 http://www.gorp.com

3633 James Campbell National Wildlife Refuge

Oahu Refuge Complex   808-637-6330
66-590 Kam Highway, Rm 2C   Fax: 808-637-3578
Haleiwa, HI 96712   E-mail: sylvia_pelissa@fws.gov
                    http://www.fws.gov/pacific/pacificislands/wnwr/ojamesnwr.html
James Campbell NWR lies at the northernmost tip of Oahu near the community of Kahuku and serves as a strategic landfall for native and migratory birds coming from as far away as Alaska, Siberia, and Asia. The specific purpose of the Refuge is to provide habitat for Hawaii's four endemic, endangered waterbirds and other native wildlife, as well as migratory waterfowl and shorebirds. A total of 102 bird species have been documented on the Refuge since its creation.

Sylvia R Pelizza, Refuge Manager

3634 Kakahaua National Wildlife Refuge

66-590 Kamehameha   808-637-6330
Room 2C   http://www.gorp.com
Haleiwa, HI 96712-1484

3635 Kealia Pond National Wildlife Refuge

PO Box 1042   808-875-1582
Kihei, HI 96753-1042 Fax: 808-875-2945
                      E-mail: katherine_smith@fws.gov
                      http://www.gorp.com

3636 Kilauea Point National Wildlife Refuge

PO Box 1128   808-828-1413
Kilauea, HI 96754 Fax: 808-828-6634
This National Wildlife Refuge provides nesting habitat for seabirds; notably red-footed boobies, Laysan albatross and wedge-tailed shearwaters. The Refuge is also home to the historic Kilauea Point Lighthouse and native marine coastal plant communities. In winter, the 180 foot high precipice provides an ideal site for viewing humpback whales in offshore waters.
Mike Hawkes, Refuge Manager

3637 Oahu National Wildlife Refuge

66-590 Kamehameha Highway   808-637-6330
Room 2C
Haleiwa, HI 96712-1484

3638 Pearl Harbor National Wildlife Refuge

66-590 Kamehameha Highway   808-637-6330
Room 2C
Haleiwa, HI 96712-1484

Idaho: US Forests, Parks, Refuges

3639 Bear Lake National Wildlife Refuge Southeast Idaho Complex

1450 E Fairview Avenue   208-847-1757
Montpelier, ID 83254   Fax: 208-847-1319
                         http://www.gorp.com

Richard Sjostrom, Manager

3640 Boise National Forest

1249 S Vinnell Way   208-373-4100
Suite 200   Fax: 208-373-4111
Boise, ID 83709

The predominantly Ponderosa pine and Douglas fir ecosystem provides homes for fish and wildlife; fiber for wood and paper products; forage for cows and sheep; precious metals for industrial and personal use; and an unlimited menu of year round opportunities. The Boise National Forest also contains a number of unique sites, including the Experimental Forest, the Lucky Peak Nursery and Bogus Basin Ski Area.

3641 Camas National Wildlife Refuge Southeast Idaho Refuge Complex

2150 E 2350 N
Hamer, ID 83425

208-662-5423
Fax: 208-662-5525
http://www.gorp.com

Cama NWR is 36 miles north of Idaho Falls in southeast Idaho, in the Cama Creek floodplain. Elevation is 4,800 feet. About half of its acreage consists of lakes, ponds and marshlands; the remainder is grass sagebrush uplands, meadows and farm fields. Camas Creek flows for 8 miles through the length of the refuge and is the source of water for many of the lakes and ponds. Tall cottonwood trees along the creek attract a wide variety of songbirds.
Jack L Richardson, Manager

3642 Caribou-Targhee National Forest

1405 Hollipark Drive
Idaho Falls, ID 83401

208-524-7500
E-mail: jjbennett@fs.fed.us
http://www.fs.fed.us/r4/caribou-targhee

The Caribou National Forest was created to help preserve wilderness land in an area marked by mining activity and westward migration. The forest now covers more than 1 million acres in southeast Idaho, with small portions in Utah and Wyoming. Several north-south mountain ranges of the Overthrust Belt dominate the landscape. Caribou National Park offers a wide variety of outdoor activities, including camping, hiking, fishing, climbing, skiing and horseback riding.
Jerry Reese, Supervisor

3643 Clearwater National Forest

12730 Highway 12
Orofino, ID 83544

208-476-4541
Fax: 208-476-8329
E-mail: elozar@fs.fed.us.com
http://www.fs.fed.us/r1/clearwater/

The Clearwater National Forest covers 1.8 million acres from the jagged peaks of the Bitterroot Mountains in the east to the river canyons and rolling hills of the Palouse Prairie in the west. The North Fork of the Clearwater & the Lochsa rivers provide miles of tumbling white water interspersed with quiet pools for migratory and resident fish. The mountains provide habitat for elk, moose, whitetail & mule deer, gray wolf, cougar, mountain goats, and many smaller mammals.
Tom Reilly, Forest Supervisor
Kimberly Nelson, Public Affairs

3644 Craters of the Moon National Monument

Highway 20
PO Box 29
Arco, ID 83213

208-527-3257
Fax: 208-527-3073
http://http://www.nps.gov/crmo/

A sea of lava flows with scattered islands of cinder cones and sagebrush characterizes this "wierd and scenic landscape" known as Craters of the Moon. Craters of the Moon National Monument and Preserve contains three young lava fields covering almost half a million acres. These remarkably well preserved volcanic features resulted from geologic events that appear to have happened yesterday and will likely continue tomorrow...

, Superintendent

3645 Deer Flat National Wildlife Refuge

13751 Upper Embankment Road
Nampa, ID 83686

208-467-9278
Fax: 208-467-1019
E-mail: deerflat@fws.gov
http://http://deerflat.fws.gov

One of the nation's oldest refuges. It includes the Lake Lowell sector and the Snake River Islands sector. The refuge provides a mix of wildlife habitats, including open waters, wetland edges around the lake, sagebrush uplands, grasslands and riparian forests. More than 250 birds and 30 mammals have been seen on the refuge, providing excellent wildlife observation and photography opportunities.

Elaine Johnson, Manager

3646 Grays Lake National Wildlife Refuge Southeast Idaho Refuge Complex

74 Grays Lake Road
Wayan, ID 83285

208-574-2755
Fax: 208-662-5525
E-mail: mike_fisher@fws.gov
http://www.gorp.com

Twenty-seven miles north of Soda Springs in southeast Idaho, the refuge lies in a high mountain valley at 6,400 feet. Grays Lake is actually a large, shallow marsh with dense vegetation and little open water. Most of the marsh vegetation is bulrush and cattail. Adjacent lands are primarily wet meadows and grasslands.
Mike Fisher, Manager

3647 Idaho Panhandle National Forest: Kaniksu

3815 Schreiber Way
Coeur d'Alene, ID 83815-8363

208-765-7223
Fax: 208-765-7307

3648 Kootenai National Wildlife Refuge

HCR 60 Box 283
Bonners Ferry, ID 83805

208-267-3888
Fax: 208-267-5570
E-mail: Dan_Pennington@r1.fws.gov
http://kootenai.fws.gov/

The 2,774 acre refuge is located in the northern panhandle of Idaho, and serves as a resting and feeding area for migratory birds.
Larry D Napier, Manager

3649 Minidoka National Wildlife Refuge Southeast Idaho Refuge Complex

961 E Minidoka Dam
Rupert, ID 83350-9414

208-436-3589
E-mail: mike_r_johnson@fws.gov
http://www.gorp.com

Lying 12 miles northeast of Rupert in the Snake River Valley in south-central Idaho, the refuge extends upstream about 25 miles from the Minidoka Dam along both shores of the Snake River and includes all of Lake Walcott. Over half the refuge is open water, with some small marsh area.
Martha Collins, Manager

3650 Nez Perce National Forest

Route 2 E US Highway 13
PO Box 475
Grangeville, ID 83530

208-983-1950
Fax: 208-983-4099

3651 Payette National Forest

800 W Lakeside Avenue
PO Box 1026
McCall, ID 83638

208-634-0700
Fax: 208-634-0744
E-mail: saschmitz@fs.fed.us
http://www.fs.fed.us/r4/payette

Payette National Forest spans over 2.3 million acres of some of west-central Idaho's most beautiful and diverse country. In one day you can travel from hot desert grasslands through cool conifer forests to snow-capped peaks. The specacular land is bordered by two of the deepest canyons in North America— the Salmon River Canyon On the north and Hells Canyon of the Sake River on the west. To the east lies 2.4 million-acre the largest Congressionally designated wilderness in the lower 48 states.
Kelly Woods, Public Affairs Specialist

3652 Salmon-Challis National Forest: Forest Service Building

US Highway 93 N
HC 63 Box 1669
Challis, ID 83226

208-879-4100
Fax: 208-879-2224

Ralph Rau, District Ranger

3653 Sawtooth National Forest

2647 Kimberly Road E
Twins Falls, ID 83301

208-737-3243

The Sawtooth National Forest encompasses 2.1 million acres of some of the nation's most magnificent country. Managed and protected by the US Department of Agriculture's Forest Service, the Sawtooth National Forest is working, producing forest that has been providing goods and services to the american people since its establishment in 1905.

3654 Southeast Idaho National Wildlife Refuge
4425 Durley Drive
Suite A
Chubbuck, ID 83202

208-237-6616
Fax: 208-237-8213
http://www.gorp.com

Dick Munoz, Project Leader

3655 Targhee National Forest
420 N Bridge Street
PO Box 208
St Anthony, ID 83445

208-624-3151

The Forest lies almost entirely within the Greater Yellowstone Area or the Greater Yellowstone Ecosystem. An area of 12 million acres and the largest remaining block of relatively undisturbed plant and animal habitat in the contiguousUnited States. The Forest lies along the Continental Divide.

Illinois: US Forests, Parks, Refuges

3656 Mark Twain National Wildlife Refuge Complex
1704 N 24th Street
Quincy, IL 62301

217-224-8580
Fax: 217-224-8583
E-mail: Dick_Steinbach@fws.gov
http://midwest.fws.gov/marktwain

Administrative office over five refuges; Port Louisa NWR, Great River NWR, Clarence Common NWR, Two Rivers NWR, and Middle Mississippi River NWR.
Richard Steinbach, Manager

3657 Shawnee National Forest
50 Highway 145 South
Harrisburg, IL 62946

618-253-7114
800-699-6637

3658 Upper Mississippi River National Wildlife & Fish Refuge: Savanna District
7071 Riverview Road
Thomson, IL 61285

815-273-2732
Fax: 815-273-2960

The Upper Mississippi River National Wildlife and Fish Refuge covers nearly 240,000 acres and extends along 281 miles of the Mississippi River. This Refuge is home to a diverse collection of wildlife, including bald eagles, great blueherons, sandhill cranes and spectacular concentrations of waterfowl. Local residents and visitors enjoy a wide array of opportunties throughout the year such as fishing, hunting, wildlife observation, photography, interpretation and environmentaleducation.

Ed Britton, District Manager
Pam Steinhaus, Visitor Service Manager

Indiana: US Forests, Parks, Refuges

3659 Indiana Dunes National Lakeshore
1100 N Mineral Springs Road
Porter, IN 46304

219-926-7561

Indiana Dunes National Lakeshore, authorized by Congress in 1996, is located approximately 50 miles southeast of Chicago, Illinois in the counties of Lake, Porter and LaPorte in Northwest Indiana. The national lakeshore runs for nearly25 miles along southern Lake Michigan, bordered by Michigan City, Indiana on the east and Gary on the west. The park contains approximately 15,000 acres, 2,182 of which are located in Indiana Dunes State Park managed by the Indiana Department ofNatural Resources.
Dale B Engquist, Supt

3660 North Carolina Department of Environment
PO Box 29534
Raleigh, NC 27626-0534

919-715-5381
Fax: 919-715-3227

The mission of the North Carolina Department of Environment, health and Natural Resources is to work on environment protection and public health issues including childhood lead-poisoning prevention and child care sanitation.
Ed Norman, Environmental Epidemiologist

Iowa: US Forests, Parks, Refuges

3661 DeSoto National Wildlife Refuge US Fish & Wildlife Service
1434 316th Lane
Missouri Valley, IA 51555

712-642-4121
Fax: 712-642-2877
E-mail: desoto@fws.gov
http://http://midwest.fws.gov.desoto/

DeSoto Refuge is located along the Missouri River, 25 miles north of Omaha, NE. Popular activities include fishing, picnicking, mushroom-picking, hiking, boating, and wildlife observation. Peak viewing of 500,000 waterfowl inmid-November. The DeSoto Visitor Center houses over 200,000 artifacts from the 1860's steamboat Bertrand.

Larry Klimek, Refuge Manager
Mindy Sheets, Assistant Refuge Manager

3662 McGregor District Upper: Mississippi River National Wildlife & Fish Refuge
PO Box 460
McGregor, IA 52157

563-873-3423
Fax: 563-873-3803

J Lindell, Manager

Kansas: US Forests, Parks, Refuges

3663 Kirwin National Wildlife Refuge
702 East Xavier Road
Kirwin, KS 67644

785-543-6673
Fax: 785-543-5464
E-mail: erich_gilbert@fws.gov
http://mountain-prairie.fws.gov/refuges/ks

Erich Gilbert, Manager

Kentucky: US Forests, Parks, Refuges

3664 Daniel Boone National Forest
1700 Bypass Road
Winchester, KY 40391

859-745-3100
Fax: 853-744-1568
E-mail: bluhn@fs.fed.us
http://www.r8web.com/boone/

The mission is to achive quality land management under the sustainable multiple-use management concept to meer the diverse need of people.

3665 Mammoth Cave National Park
PO Box 7
Mammoth Cave, KY 42259

270-758-2180
Fax: 270-758-2349
E-mail: MACA_Park_information@nps.gov
http://www.nps.gov

Established to preserve the cave system, including Mammoth Cave, the scenic river valleys of the Green and Nolin Rivers, and a section of the hilly country of south central Kentucky. This is the longest recorded cave system in theworld with more than 360 miles explored and mapped. Established July 1, 1941. Designated a World Heritage Site October 27, 1981. Designated a Biosphere Reserve in 1990.
Vickie Carson, Public Information Officer

Louisiana: US Forests, Parks, Refuges

3666 Atchafalaya National Wildlife Refuge Sherburne Wildlife Management Area
PO Box 127
Krotz Springs, LA 70750

318-566-2251
http://www.gorp.com

The State of Louisiana's Sherburne Wildlife Management Area is located in the upper third of the Atchafalaya River Basin between Interstate Highway 10 and US Highway 190. It covers approximately 11,780 acres and was established in 1983by the Louisiana Department of Wildlife and Fisheries. The area supervisor's headquaters is located east of Krotz Springs, Louisiana, on LA 975 approximately three miles south of US Highway 190.

3667 Barataria Preserve

7400 Highway 45   504-589-2330
Marrero, LA 70072   http://www.gorp.com

The Barataria Preserve is one of six sites comprising Jean Lafitte National Historical Park and Preserve. The Barataria Preserve interprets the culture of peoples, past and present, who settled the delta and the unique ecosystem whichsustained them. It preserves a representative example of the delta's environment, containing natural levee forests, bayous, swamps and marshes. Though wild and teeming with wildlife, this is not a pristine wilderness.

3668 Cameron Prairie National Wildlife Refuge

1428 Highway 27   318-598-2216
Bell City, LA 70630   E-mail: r4rw_la.cam@mail.fws.gov
                                   http://www.gorp.com

The refuge contains 9,621 acres of fresh marsh, coastal prairie and old rice fields. East Cove Unit of the refuge contains 14,927 acres of brackish and salt marsh. Seasonal visitors include the Peregrine falcon, alligators, whitetailed deer, wading and shorebirds, ducks and geese and various migratory birds.
Paul Yakupzack, Manager

3669 Catahoula National Wildlife Refuge

PO Drawer Z   318-992-5261
Rhinehart, LA 71363-0201   Fax: 318-992-6023
                        E-mail: catahoula@fws.gov
                        http://catahoula.fws.gov

The objectives of the refuge are: to provide wintering habitat for migratory waterfowl consistent with Mississippi Flyway objectives; to provide habitat and protection for endangered species; preserve bottomland hardwoods and providehabitat necessary for wildlife diversity; provide opportunities for environmental education, interpretation and wildlife oriented recreation.
Andrew Hammond, Refuge Manager

3670 D'Arbonne National Wildlife Refuge

11372 Highway 143   318-726-4400
Farmerville, LA 71363-0201   Fax: 318-726-4667
                        E-mail: northlarefuges@fws.gov
                        http://darbonne.fws.gov//index.html

The Refuge provides habitat for a diversity of migratory birds and resident wildlife species, provides habitat and protection for endangered species such as the bald eagle, wood stork and red-cockaded woodpecker and providesopportunities for wildlife-oriented recreation, environmental education and interpretation.

Kelby Ouchley, Manager

3671 Kisatchie National Forest

2500 Shreveport Highway   318-473-7160
Pineville, LA 71360   Fax: 318-473-7117
            E-mail: kisatchie@www.southernregion.fs.fed.us

The Kisatchie National Forest has a lot to offer vistors, such as 355 miles of trails for hiking, camping, mountain biking, horseback riding, ORV riding. Other recreational opportunites include four lakes, an 8700 acre Wilderness anddozens of caming sites. The forest also provides opportunities to hunt and fish.

3672 Lacassine National Wildlife Refuge

209 Nature Road   337-774-5923
Lake Arthur, LA 70549   Fax: 337-774-9913
                    E-mail: FW4_RW_lacassine@fws.gov
                    http://lacassine.fws.gov

The nearly 35,000 acre refuge is mostly freshwater marsh habitat. It preserves one of the major wintering grounds for waterfowl in the US. Wintering populations of ducks and geese at Lacassine are among the largest in the NationalWildlife Refuge System. Portions of the refuge are open year-round from one hour before sunrise until one hour after sunset. Please consult refuge brochures or contact the refuge office for more details. A vicinity map and refuge map are available.

3673 Sabine National Wildlife Refuge

3000 Holly Beach Highway   337-762-3816
Hackberry, LA 70645   Fax: 337-762-3780
                    E-mail: sabine@fws.gov
                    http://www.fws.gov/sabine/

The objectives of the refuge is to provide habitat for migratory waterfowl and other birds, to preserve and enhance coastal marshes for fish and wildlife, and to provide outdoor recreation and environmental education for the public.

Don Voros, Project Manager
Terence Delaine, Refuge Manager

Maine: US Forests, Parks, Refuges

3674 Acadia National Park Advisory Commission

PO Box 177   207-288-8800
Bar Harbor, ME 04609-0177   Fax: 208-288-8813
                E-mail: Acadia_Information@nps.gov
                http://www.nps.gov/acad/faqs.htm

The purpose of the commission is to consult with the Secretary of the Interior, or his designee, on matters relating to the management and development of the park, including but not liited to the acquisition of lands and interests inlands and termination of rights of use and occupancy.
Len Babinchock, Acting Superintendent

3675 Cross Island National Wildlife Refuge

PO Box 279   207-546-2124
Milbridge, ME 04658-0279   Fax: 207-546-7805
                E-mail: r5rw_pmnwr@fws.gov
                http://www.gorp.com

Cross Island is a forested land of islands with stands of yellow birch and inhabited by white-tailed deer accomanying eagles and osprey. Waterfowl, shorebirds and raptors stop here on their migrations.

3676 Pond Island National Wildlife Refuge

PO Box 279   207-546-2124
Milbridge, ME 04658   Fax: 207-546-7805
                E-mail: r5rw_pmnwr@fws.gov
                http://www.gorp.com

3677 Sunkhaze Meadows National Wildlife Refuge

1168 Main Street   207-827-6138
Old Town, ME 04468   Fax: 207-827-6099
                E-mail: Tom_Comish@fws.gov
                http://www.sunkhaze.org

Sunkhaze Meadows NWR was established to protect a large peat bog and its associated wildlife. There are three divisions of the refuge, totalling 11,772 acres. The areas are open to wildlife-dependent recreation.

Tom Comish, Refuge Manager

Maryland: US Forests, Parks, Refuges

3678 Antietam National Battlefield

PO Box 158   301-432-5124
Sharpsburg, MD 21782-0158   Fax: 301-432-4590
                E-mail: anti_superintendent@nps.gov
                http://www.nps.gov

3679 Assateague Island National Seashore

7206 National Seashore Lane   410-641-1441
Berlin, MD 21811   Fax: 410-641-1099
                E-mail: christopher_seymour@nps.gov
                http://www.nps.gov/asis/home.htm

A windswept barrier island that offers many opportunities for seashore recreation and nature study along its thirty-seven miles. Ocean swimming, camping, bayside canoeing, crabbing, clamming, hunting, surf fishing and off-road vehicleuse are all popular. The famous wild horses roam throughout and share beach to bay habitats with a host of other animals, including over 300 species of birds. The park's land and water boundaries encompass over 48,000 acres in the states of MD andVA.
Roger Rector, Supt

3680 Patuxent Research Refuge

10901 Scarlet Tanager Loop   301-497-5580
Laurel, MD 20708-4027   Fax: 301-497-5515
                E-mail: susan_mcmahon@fws.gov
                http://www.gorp.com

One of over 500 refuges in the National Wildlife Refuge System, a network of lands and waters specifically for the protection of wildlife and wildlife habitat.

Massachusetts: US Forests, Parks, Refuges

3681 Boston National Historical Park
Charleston Navy Yard   617-242-5642
Boston, MA 02129-4543   Fax: 617-242-6006
E-mail: bost_email@nps.gov
http://www.nps.gov

3682 Cape Cod National Seashore
99 Marconi Station Site Road   508-349-3785
Wellfleet, MA 02667   Fax: 508-349-9052
E-mail: CACO_Superintendent@nps.gov
http://www.nps.gov/caco/home.html
Cape Cod National Seashore comprises 43,604 acres of shoreline and upland landscape features, including a forty-mile long stretch of pristine sandy beach, dozens of clear, deep, freshwater kettle ponds, and upland scenes that depictevidence of how people have used the land. A variety of historic structures are within the boundary of the Seashore, including lighthouses, a lifesaving station, and numerous Cape Cod style houses.
Andrew Ringgold, Supt

3683 Great Meadows National Wildlife Refuge
73 Weir Hill Road   978-443-4661
Sudbury, MA 01776   Fax: 978-443-2898
http://easternmanwrcomplex.fws.gov
The Sudbury office serves as the headquarters for the eight refuge Eastern Massachusetts National Wildlife Refuge Complex. Public use opportunities are available at a number of the complex's refuges.
Elizabeth Herland, Refuge Manager
Tim Prior, Deputy Refuge Manager

3684 Silvio O Conte National Wildlife Refuge
52 Avenue A   413-863-0209
Turners Falls, MA 01376   Fax: 413-863-3070
E-mail: boardman@k12.oit.umass.edu
http://www.fws.gov/r5soc
The Connecticut River Watershed, 7.2 million acres in four states, is larger and more heavily populated than areas usually considered when creating a refuge. The refuge's purposes are also much broader than usual. The new scientificand social challenge of protecting natural diversity cannot be met by land acquistion alone. The refuge's primary action is to involve people of the watershed, especially landowners and land managers, in environmental education programs andcooperative management.

3685 Wapack National Wildlife Refuge
73 Weir Hill Road   978-443-4661
Sudbury, MA 01776   Fax: 978-443-2898
E-mail: r5rw_gmnwr@fws.gov
http://www.gorp.com

Michigan: US Forests, Parks, Refuges

3686 Huron-Manistee National Forest
1755 S Mitchell Street   231-775-2421
Cadillac, MI 49601   800-821-6263
http://www.fs.fed.us/r9/hmnf/
The Huron-Manistee National Forest comprise almost a million acres of public lands extending acroee the northern lower peninsula of Michigan. The Huron-Manistee national Forest provide recreation opportunities for visitors, habitat forfish and wildlife, and resources for local industry.

3687 Ottawa National Forest
E6248 US Highway 2   906-932-1330
Ironwood, MI 49938

The almost one million acres of the Ottawa National Forest are located in the western Upper Peninsula of Michigan. It extends from the south shore of Lake Superior down to Wisconsin and the Nicolet National Forest. The area is rich inwildlife viewing opportunities; topography in the northern portion is the most dramatic with breathtaking views of rolling hills dotted with lakes, rivers and spectacular waterfalls.

3688 Pictured Rocks National Lakeshore
N8391 Sand Point Road   906-387-2607
PO Box 40   Fax: 906-387-4025
Munising, MI 49862-0040   E-mail: pamela_baker@nps.gov
http://www.nps.gov/piro/home.htm
Multicolored sandstone cliffs, beaches, and dunes, waterfalls, inland lakes, wildlife and the forest of Lake Superior shoreline beckon visitors to explore this 73,000+ acre park. Attractions include a lighthouse and former Coast Guardlife-saving stations along with old farmsteads and orchards. The park is a four season recreational destination where hiking, camping, hunting, nature study, and winter activities abound.
Grant Petersen, Supt

3689 Seney National Wildlife Refuge
Seney, MI 49883   906-586-9851
Fax: 906-586-3800
http://midwest.fws.govt

Trudy Casselman, Manager

3690 Shiawassee National Wildlife Refuge
6975 Mower Road   989-777-5930
Saginaw, MI 48601   Fax: 989-777-9200
http://http://www.fws.gov/midwest/shiawassee
The refuge is comprised of over 9400 acres of wetlands, uplands, and bottomland hardwood forests. Four rivers flow through. Over 12 miles of hiking trails, bank fishing sites, and two photography blinds are available. Environmentaleducation progrms are offered at Green Point Environmental Learning Center in Saginaw.

Becky Goche, Park Ranger

3691 Sleeping Bear Dunes National Lakeshore
9922 Front Street   231-326-5137
Empire, MI 49630   Fax: 231-326-5382
Sleeping Bear Dunes National Lakeshore encompasses a 60 km stretch of Lake Michigan's eastern coastline, as well as North and South Manitou Islands. The park was established primarily for its outstanding natural features, includingforests, beaches, dune formations and ancient glacial phenomena. The Lakeshore also contains many cultural features including an 1871 lighthouse, three former Life-Saving Service (Coast Guard) Stations and an extensive rural historic farm district.
Dusty Shultz, Superintendent

Minnesota: US Forests, Parks, Refuges

3692 Agassiz National Wildlife Refuge
22996 290th Street Northeast   218-449-4115
Middle River, MN 56737-9653   Fax: 218-449-3241
E-mail: margaret_anderson@fws.gov
http://midwest.fws.gov/agassiz/
The National Refuge provides resting, nesting and feeding habitat for waterfowl and other migratory birds. Agassiz NWR is designated a Globally Important Bird Area by the American Bird Conservancy. It protects endangered and threatenedspecies. It also provides for biodiversity, public opportunities for outdoor recreation and environmental education.
Margaret Anderson, Manager

3693 Big Stone National Wildlife Refuge
44843 County Road 19   320-273-2191
Odessa, MN 56276   Fax: 320-273-2231
E-mail: BigStone@fws.gov
http://http://midwest.fws.gov/bigstone

Fig Stone NWR is one of more than 545 National Wildlife Refuges administered as part of the National Wildlife Refuge System by the US Fish and Wildlife Service. The Refuge now overlays 11,585.8 acres of the Minnesota River Valley in western Minnesota. A unique visual and geological feature of the Refuge is the red, lichen covered granite outcrops for which the Refuge was named. The Refuge offers an auto tour route, nature trails, wildlife observation, hunting and fishing opportunities.

Alice Hanley, Manager

3694 Chippewa National Forest

200 Ash Avenue	218-335-8600
Cass Lake, MN 56633	Fax: 281-335-8637
	http://www.fs.fed.us/r9/chippewa/

The Chippewa was the first National Forest established east of the Mississippi. The Forest boundary encompasses 1.6 millon, of which over 666,325 acres are managed by the USDA Forest Service. Aspen, birch, pines, balsam fir and maples blanket the uplands. Water is abundant, with over 1300 lakes, 923 miles of rovers and streams, and 400,000 acres of wetlands.

3695 Crane Meadows National Refuge

19502 Iris Road	320-632-1575
Little Falls, MN 56345	Fax: 320-632-5471
	E-mail: r3rw-cmd@fws.gov
	http://www.gorp.com

3696 Detroit Lakes Wetland Management District

26624 N Tower Road	218-844-3406
Detroit Lakes, MN 56501	Fax: 218-847-4156
	E-mail: DetroitLakes@fws.gov
	http://www.fws.gov/Midwest.detroitlakes/

is divided into three general landscape areas covering approximately 6000 square miles. From east to west, these are: the Red River Valley floodplain, the glacial moraine/prairie pothole region, and the hardwood/coniferous forest. Land acquisition and management efforts are focused in the prairie pothole region of the WMD, with a goal of providing habitat for nesting waterfowl.

Mark Chase, Manager

3697 Hamden Slough National Wildlife Refuge

21212 210th Street	218-439-6319
Audubon, MN 56511	Fax: 218-439-6907
	E-mail: mike_t_murphy@fws.gov
	http://www.gorp.com

Mike Murphy, Manager

3698 Hiawatha National Forest

2727 North Lincoln Road	906-786-4062
Escanaba, MI 49829	Fax: 906-789-3311
	http://www.fs.fed.us/r9/hiawatha

Located in the central and easter Upper Peninsula of Michigan, the firest affords visitors access to white sand, scenic beaches and relatively undeveloped shorelines along three of Americas's Inland Seas, Lake Superior, Michigan and Huron.

3699 Mille Lacs National Wildlife Refuge

Rice Lake NWR	218-768-2402
Route 2 Box 67	Fax: 218-768-3040
McGregor, MN 55760	E-mail: ralph_lloyd@fws.gov
	http://www.gorp.com

Comprised of two small islands in Mille Lacs Lake in central Minnesota. The islands are boulder and gravel outcrops important for colonial nesting birds including common terns and ring billed gulls.

3700 Minnesota Valley National Wildlife Refuge

3815 East 80th Street	612-854-5900
Bloomington, MN 55425-1600	Fax: 612-725-3279
	E-mail: r3rw_mnv@fws.gov
	http://www.gorp.com

Thomas Larson, Manager

3701 Rice Lake National Wildlife Refuge

Route 2 Box 67	218-768-2402
McGregor, MN 55760	Fax: 218-768-3040
	http://www.fws.gov/~r3pao/rice_lk

Ralph Lloyd, Manager

3702 Rydell National Wildlife Refuge

Route 3 Box 105	218-687-2222
Erskine, MN 56535	Fax: 218-687-2225
	E-mail: Paul_Soler@fws.gov
	http://www.gorp.com

3703 Sherburne National Wildlife Refuge

17076 293rd Avenue	763-389-3323
Zimmerman, MN 55398	Fax: 612-389-3493
	E-mail: sherburne@fws.gov
	http://www.fws.gov/midwest/sherburne/index.htm

Jay Hammernick, Manager

3704 Superior National Forest

8901 Grand Avenue Place	218-626-4300
Duluth, MN 55808	Fax: 218-626-4398
	E-mail: r9 superior NF@fs.fed.us
	http://www.fs.fed.us/r9/forests/superior

Located in northeastern tip of Minnesota, the Superior National Forest stretches 150 miles along the US-Canadian border, encompassing 3.85 million acres. This hilly deep pine forest is home to moose, wolves, black bears, loons and migratory birds. More than 2,250 miles of stream flow within the forest, including the renowned Boundary Waters Canoe Area, where you can canoe, portage and camp in the spirit of the French Canadian voyagers of 200 years ago.

Jim Sanders, Forest Supervisor

3705 Tamarac National Wildlife Refuge

35704 County Highway 26	218-847-2641
Rochert, MN 56578-9638	Fax: 218-847-9141
	E-mail: jay_m_johnson@fws.gov
	http://midwest.fws.gov

Established in 1938, Tamarac Refuge is dedicated to providing a breeding ground and sanctuary for migratory birds and other wildlife. Situated at a unique transitional zone where hardwood forest, boreal forest and tallgrass prairie meets. Tamarac provides boundless opportunities for visitors to observe wildlife in their natural surroundings. Spring and fall migrations of songbirds and waterfowl can be spectacular. Their visitor center offers interpretive displays and programs.

Jay Johnson, Manager

3706 Voyageurs National Park

3131 Highway 53 S	218-283-9821
International Falls, MN 56649	Fax: 218-285-7407
	E-mail: VOYA_superintendent@nps.gov
	http://www.nps.gov

The park lies in the southern part of the Canadian Shield, representing some of the oldest exposed rock formations in the world. This bedrock has been shaped and carved by at least four periods of glaciation. The topography of the park is rugged and varied; rolling hills are interspersed between bogs, beaver ponds, swamps, islands, small lakes and four large lakes.

3707 Winona District National Wildlife Refuge Upper Mississippi River National Wildlife and Fish

51 E 4th Street	507-454-7351
Room 203	Fax: 507-452-0851
Winona, MN 55987	E-mail: bob_drieslein@fws.gov
	http://www.umesc.usgs.gov/umr_refuge.html

Refuge objectives are to: protect and preserve one of America's premier fish and wildlife areas; provide habitat for migratory birds, fish, plants and resident wildlife; protect and enhance habitat for endangered species; provide interpretation, environmental education and wildlife-oriented recreational public use opportunities; and conserve a diversity of plant life.

Mary Stefanski, District Manager

Mississippi: US Forests, Parks, Refuges

3708 Bienville National Forest

3873 Highway 35 S	601-469-3811
Route 2 Box 1239	Fax: 601-469-2513
Forest, MS 39074	

The forest offers camping, picnicking, swimming, hiking, fishing and historic sites. Bienville boasts the largest known cluster of old growth pine forest in Mississippi in the 180-acre Bienville Pines Scenic Area. Here visitors canwander among towering loblolly and shortleaf pines, many more than two centuries old. The 23-mile Shockaloe Horse Trail, a national recreation trail, starts near the town of Forest.

3709 Mississippi Sandhill Crane National Wildlife Refuge: US Fish & Wildlife Service

7200 Crane Lane   228-497-6322
Gautier, MS 39553   Fax: 228-497-5407
http://mississippisandhillcrane.fws.gov
The Refuge is located in southeast Mississippi in Jackson County a few miles north of the Gulf of Mexico. This 20,000 acre refuge was established in 1975 to protect the sandhill cranes and the wet pine savanna habitat they prefer.
Alan Schriver, Refuge Manager

3710 Noxubee National Wildlife Refuge

Route 1   662-323-5548
Box 142   Fax: 662-323-6390
Brooksville, MS 39739   E-mail: noxubee@fws.gov
http://noxubee.fws.gov
Established in 1940 to protect and enhance habitat for the conservation of migratory birds, endangered species and other wildlife. The recreational and educational opportunities provided on the refuge help the public experience natureand learn how sound management ensures that future generations continue to enjoy fish and wildlife and their habitats.
Jimmie L Tisdale, Manager

3711 Panther Swamp National Wildlife Refuge

13695 River Road   662-746-5060
Yazoo City, MS 39194   Fax: 662-746-5055
E-mail: yazoo@fws.gov
http://pantherswamp.fws.gov//index.html
Refuge objectives are: to provide resting, nesting and feeding habitat for waterfowl and other migratory birds; to provide habitat for resident wildlife; to protect endangered and threatened species; and to provide public useopportunities for outdoor recreation and environmental education.
Bo Sloan, Refuge Manager

Missouri: US Forests, Parks, Refuges

3712 Mark Twain National Forest

401 Fairgrounds Road   573-364-4621
Rollo, MO 65401   Fax: 573-341-7475
The Mark Twain National forest is located in southern and centeral Missouri, and extends from the St Francois Mountains in the southeast to glades in the southwest, from the southwest, from the prairie lands the Missouri River to thenation's most ancient mountains in the south.
Ronnie Raum, Manager

3713 Mingo National Wildlife Refuge

24279 State Highway 51   573-222-3589
Puxico, MO 63960   Fax: 573-222-6343
E-mail: mingo@fws.gov
http://www.fws.gov/midwest/mingo
Vergial Harp, Park Ranger

3714 Ozark National Scenic Riverways

404 Watercress Drive   573-323-4236
PO Box 490   Fax: 573-323-4140
Van Buren, MO 63965 E-mail: ozar_superintendent@nps.gov
Missouri's largest National Park and America's first to preserve a free flowing river in its wild state. Covers some 80,000 acres along 134 miles of the Current and Jacks Fork Rivers. Staff provide elementary level environmentaleducation programs on natural history, with an emphasis on karst and water issues. Publishes "More Than Skin Deep, a Teacher's Guide to Caves and Groundwater," a curriculum guide suitable for grades K-12.

Noel Poe, Supt

3715 Swan Lake National Wildlife Refuge

Route 1 Box 29A   660-856-3323
Sumner, MO 64681   Fax: 660-856-3687
John Guthrie, Manager

Montana: US Forests, Parks, Refuges

3716 Beaverhead National Forest

420 Barrett Street   406-683-3900
Dillon, MT 59725-3572 Fax: 406-683-3855
http://www.fs.fed.us/fl/bdnf/
The Forest Service is a federal agency that manages public land in the National Forests and Grasslands. The Forest Service is also the largest forestry research organization in the world, providing technical and financial assistance tostate forestry agencies.

3717 Benton Lake National Wildlife Refuge

922 Bootlegger Trail   406-727-7400
Gerat Falls, MT 59404 Fax: 406-727-7432
http://www.gorp.com

Jim McCollum, Manager

3718 Bighorn Canyon National Recreation Area

T Avenue B   406-666-2412
PO Box 7458   Fax: 406-666-2415
Fort Smith, MT 59035
This dam, named after the famous Crow chairman Robert Yellowtail, harnessed the waters of the Bighorn River and turned this variable stream into a magnificent lake. The Afterbay Lake below the Yellowtail Dam is a good spot for troutfishing and wildlife viewing for ducks, geese and other animals. The Bighorn River below the Afterbay Dam is a world class trout fishing area. Bighorn Canyon National Recreation Area boasts breath-takign scenery, countless varieties of wildlife.

Bill Binnewies, Supt

3719 Bitterroot National Forest

1801 N First Street   406-363-7161
Hamilton, MT 59840-3114 Fax: 406-363-7106
E-mail: mailroom/rl_bitterroot@fs.fed.us
http://www.fs.fed.us/rl/bitterroot/
The 1.6 million acre Bitterroot National Forest, in west central Montana and east central Idaho, is part of the Norther Rocky Mountains. National Forest land begins above the foothills of the Bitterroot River Valley in two mountainranges—the Bitterroot Mountains on the west and the Sapphire Mountains on the east side of the valley.

3720 Bowdoin National Wildlife Refuge: Refuge Manager

Malta, MT 59538   406-654-2863
Fax: 406-654-2866

3721 Charles M Russell National Wildlife Refuge: San Creek Wildlife Station

Highway 191 North   406-464-5181
River Route, MT 59471 Fax: 406-464-5182
http://www.ohwy.com/mt/s/sancrkws.htm
Gene Williams, Manager

3722 Charles M Russell National Wildlife Refuge: Jordan Wildlife Station

Highway 200   406-557-6145
PO Box 63
Jordan, MT 59337
Part of the Charles M Russell National Wildlife Refuge which is a 1.3 million-acre wildlife refuge surrounding Fort Peck reservoir. The refuge is classified as a Missouri River breaks habitat which is typified by deep coulees and largegumbo buttes. Throughout this untamed wilderness wildlife abounds. A plethora of species is located here including elk, mule and white-tailed deer, bighorn sheep, pronghorn antelope, mountain lion, bobcat, coyote and badger. Bald and golden eaglesare common.
Chad Karges, Manager

3723 Charles M Russell National Wildlife Refuge
PO Box 110   406-538-8706
Lewistown, MT 59457   E-mail: r6rw_cmr@mail.fws.gov
                                              http://http://cmr.fws.gov
Located in Central Montana, the 1.1 million-acre refuge contains native prairies, forested coulees, river bottoms, badlands and the 250,000-acre Fort Peck Reservoir. Refuge wildlife includes mule and white-tailed deer, elk, bighornsheep, antelope, coyote, bobcat, beaver, sharp-tailed grouse and numerous other species.
John Foster, Manager

3724 Charles M Russell National Wildlife Refuge: Fort Peck Wildlife Station
PO Box 166   406-526-3464
Fort Peck, MT 59223   Fax: 406-526-3464
Field unit containing 1.2 million acres. Natural setting with mostly unimproved roads and no modern campgrounds. Roads are rough and travel by street-type automobile is impossible. Four-wheel drive vehicles are recommended. Winterweather is severe.
Genes Sipe, Manager

3725 Custer National Forest
1310 Main Street   406-657-2000
PO Box 2556   Fax: 406-657-6222
Billings, MT 59105   http://www.fs.fed.us/r1/custer/
The Custer national Forest is made up of 1.2 million acres of high alpine mountain country, and small pockets of timbered buttes and grasslands scattered across two states, Montana and South Dakota.

3726 Deerlodge National Forest
420 Barrett Street   406-683-3900
Dillon, MT 59725-3572   http://www.fs.fed.us/r1/bdnf

3727 Flathead National Forest
1935 3rd Avenue E   406-758-5200
Kalispell, MT 59901   Fax: 406-758-5367
                             http://www.fs.fed.us/r1/flathead/
The 2.3 million-acre Flathead national Forest is bordered by Canada to the north, Glacier National Park to the north and east and Clark National Forest to the east, the Lolo national Forest to the south, and the Kootenai NationalForest to the west.

3728 Gallatin National Forest
10 E Babcock Avenue   406-587-6701
PO Box 130   Fax: 406-587-6758
Bozeman, MT 59771   http://www.fs.fed.us

3729 Helena National Forest
2880 Skyway Drive   406-449-5201
Helena, MT 59601   Fax: 406-449-5436
                             http://www.fs.fed.us/r1/helena/
The Helena National Forest offers close to one million acres of diverse landscapes and wildland opportunities. Located in west central Montana, the Helena National Forest boasts some of the most vivid glimpses into the past of thishistorically rich area.

3730 Kootenai National Forest
1101 Highway 2 West   406-293-6211
Libby, MT 59923   Fax: 406-293-6139
The Kootenai National Forest, conataining 2.2 million acres, is located in the extreme northwest corner of Montana, Bordered on the north by Canada and on the west by Idaho. Of the total acres, 50,384 are in the state of Idaho. Accessinto the forest is available from US highways 2 and 93, and Montana State Highways 37, 56, 200, and 508.

3731 Lee Metcalf National Wildlife Refuge
PO Box 247   406-777-5552
Stevensville, MT 59870   Fax: 406-777-4344
                                              http://www.gorp.com
Mission is to manage habitat for a diversity of wildlife species with emphasis on migratory birds and endangered and threatened species.

3732 Lewis & Clark National Forest
1101 15th Street N   406-791-7700
PO Box 869   Fax: 406-731-5302
Great Falls, MT 59405   http://www.fs.fed.us/rl/lewisclark/
The 1.8 million acres of the Lewis and clark National Forest are scatteered into seven separate mountain ranges. The Forest is situated i west central Montana. The boundaries spread eastward from the rugged, mountainous ContinentalDivide onto the plains. When looking at a map, the National Forest System lands appear as islands of forest within oceans of prairie. Because of its wide-ranging land pattern, the forest is separated into two divisions: the Rocky Mountain and theJefferson.
Leslie Thompson, Administrator

3733 Lolo National Forest
Building 24-A Fort Missoula   406-329-3750
Missoula, MT 59804   Fax: 406-329-1049
                http://www.fs.fed.us/r1/lolo/general-info/general.htm
Of the 15 National Forests in the Northern Region of the USDA Forest Service, the Lolo National Forest is estimated to be the third largest. It is located in western Montana. Several major tributaries to the Clark Fork River of theColumbia River Basin flow through the Forest. Its 2.1 million acres of diverse and spectacular mountainous country extend into seven counties.

3734 Medicine Lake National Wildlife Refuge Refuge Manager
Medicine Lake, MT 59247   406-789-2305
                                  Fax: 406-789-2350
                                              http://www.gorp.com

Ted Gutzke, Manager

3735 National Bison Range National Wildlife Refuge
132 Bison Range Road   406-644-2211
Moiese, MT 59824   Fax: 406-644-2661
                                              http://www.gorp.com

David Wiseman, Manager

3736 Pablo National Wildlife Refuge
132 Bison Range Road   406-644-2211
Moiese, MT 59824   E-mail: r6rw_nbr@fws.gov
                                              http://www.gorp.com
Located on tribal trust lands of the Confederated Salish and Kootenai Tribes. It is superimposed on the irrigation reservoir managed by the Bureau of Indian Affairs Flathead Irrigation Project.

3737 Red Rocks Lakes National Wildlife Refuge
27820 Southside Centennial Road   406-276-3536
Lima, MT 59739   Fax: 406-276-3538
                             E-mail: redrocks@fws.gov
                                  http://redrocks.fws.gov
Primarily a high elevation mountain wetland-riparian area. Red rock creek flows through the upper end of the Centennial Valley, within which the Refuge lies, creating the impressive Upper Red Rock Lake, River Marsh and Lower Red RockLake marshlands. The rugged Centennial Mountains border the Refuge on the south, catching the snows of winter that replenish the Refuge's lakes and marshes.
Mike Parker, Refuge Manager

3738 Swan River National Wildlife Refuge
132 Bison Range Road   406-858-2216
Moiese, MT 59824   E-mail: R6FFA_CRE@fws.gov
                                              http://www.gorp.com
The Swan River National Wildlife Refuge is located in northwest Montana, 32 miles southeast of the town of Creston, in the serene and picturesque Swan Valley Mountain Range. The refuge boundary lies within the floodplain of the SwanRiver above Swan Lake and between the Swan Mountain Range to the east and the Mission Mountain Range to the west.

Nebraska: US Forests, Parks, Refuges

3739 Agate Fossil Beds National Monument/Scotts Bluff National Monument
PO Box 27   308-668-2211
Gering, NE 69341-0027   Fax: 308-668-2318

Well-known quarries, one open for the public display, contain numerous, concentrated and well preserved Miocene mammal fossils, representing an important chapter in mammalian evolution. Visitor center, fossil exhibits and a selfguiding trail to area of exposed fossils.
Larry Reed, Supt

3740 Crescent Lake National Wildlife Refuge
10630 Road 181   308-762-4893
Ellsworth, NE 69340   Fax: 308-762-7606
                              E-mail: r6rw_crl@fws.gov
                              http://crescentlake.fws.gov
Located in the panhandle of Western Nebraska, Crescent Lake consists of 45,818 acres of rolling sandhills interspersed with numerous shallow wetlands and lakes. Plant and animal species call Crescent Lake home, while visitors canparticipate in a variety of public use activities.
Neil Powers, Refuge Manager

3741 Fort Niobrara National Wildlife Refuge: Valentine
Hidden Timber Route   402-376-3789
HC 14, Box 67   Fax: 402-376-3217
Valentine, NE 69201   http://www.gorp.com
Robert M Ellis, Manager

3742 Missouri National Recreation River/Niobrara Missouri National Scenic Riverways
PO Box 591   402-336-3970
O'Neill, NE 68763   Fax: 402-336-3981
Warren H Hill, Contact

3743 Nebraska National Forest
125 N Main Street   308-432-0300
Chadron, NE 69337   Fax: 308-432-0309
                         E-mail: r2nebraskannfinfo@fs.fed.us
                              http://www.fs.fed.us/r2/nebraska
The nearly 1.1 million acres administered by the Nebraska National Forest Supervisor are scattered across a large arc extending from central Nebraska west to the northern Panhandle, into southwestern South Dakota and east to thestate's center. Representing a cross section of the northern Great Plains ecosystems are three National Grasslands, the Buffalo Gap and Fort Pierre in South Dakota and the Oglala in Nebraska, which along with two National Forests, the Nebraska andSamuel R McKelvie.
Jerry Schumacher, Public Affairs Specialist

3744 North Platte National Wildlife Refuge Crescent Lake National Wildlife Refuge
115 Railway Street   308-635-7851
Suite C109   Fax: 308-635-7841
Scottsbluff, NE 69361   E-mail: mckinney@fws.gov
                              http://www.r6.fws.gov/crescentlake/
The refuge lies in the panhandle of Western Nebraska. Established in 1916, the refuge is superimposed over four Bureau of Reclamation irrigation projects. For interested educators, http://panesu.esu14.k12.ne.us/~epa, several of theunits of the North Platte Refuge can be accessed through this link.
Bradley McKinney, Manager

Nevada: US Forests, Parks, Refuges

3745 Anaho Islands National Wildlife Refuge
Stillwater Wildlife Management
9604 Auction Road   702-423-5128
PO Box 1236   http://www.gorp.com
Fallon, NV 89406-1236
Primary wildlife Anaho Island has one of the largest white pelican nesting colonies in North America, as well as cormorant, great blue heron and gull nesting colonies. An island in Pyramid Lake in Great Basin. Closed to public entry toprotect wildlife. Contact refuge manager for information.

3746 Ash Meadows National Wildlife Refuge
HCR 70   775-372-5435
Box 610-Z   http://www.gorp.com
Amargosa Valley, NV 89020

Refuge staff are responsible for managing the 22,117 acre refuge, most of wich is spring-fed wetland and alkaline desert upland The refuge area habitat for at least 24 plants and animals found nowhere else in the world. Four fishes andone plant are currently listed as endangered. Species found on the refuge include numerous endemic species, the greatest concentration in the US and the second greatest in all of North America.
Suzanne Cottman, Manager

3747 Desert National Wildlife Range
Desert National NWR Complex   702-646-3401
1500 N Decatur Boulevard E-mail: ken_voget@mail.fws.gov
Las Vegas, NV 89108-1218   http://www.gorp.com
Established in 1936 for perpetuating the desert bighorn sheep. Two threatened, and twenty-nine species of concern can be found at the refuge. Wildlife observation is one of the most popular refuge activities. Big game hunting is verylimited, but also very popular. Bird watching is another popular activity. A growing program provides additional opportunities and students are able to earn college credits through an internship at the refuge.

3748 Great Basin National Park
Highway 488   775-234-7331
Baker, NV 89311   Fax: 775-234-7269
                         E-mail: grba_interpretation@nsp.gov
                         http://www.great.basin.national-park.com
Great Basin National Park includes streams, lakes, alpine plants, abundant wildlife, a variety of forest types including groves of ancient bristlecone pines, and numerous limestone caverns, including beatiful Lehman Caves.
Albert J Hendricks, Supt

3749 Humboldt National Forest
2035 Last Chance Road   702-778-0209
Elko, NV 89801
The Humboldt National Forest is scattered throughout northern and eastern Nevada.

3750 Lake Mead National Recreation Area (NRA)
601 Nevada Way   702-293-8990
Boulder City, NV 89005   Fax: 702-293-8936
                         E-mail: LAME_Interpretation@nps.gov
                              http://www.nps.gov/lame
Lake Mead NRA, which includes Lake Mohave, offers a wealth of things to do and places to go year-round. Its huge lakes cater to boaters, swimmers, sunbathers and fishermen while its desert rewards hikers, wildlife photographers androadside sightseers. Three of America's four desert ecosystems: the Mojave, the Great Basin and the Sonoran Desert meet in here. As a result, this seemingly barren area contains a surprising variety of plants animals, some of which may be foundnowhere else.
William K Dickinson, Superintendent
Kent Turner, Chief Resources Management

3751 Moapa Valley National Wildlife Refuge
Desert National Refuge Complex   702-879-6110
HCR 38, Box 700   http://www.gorp.com
Las Vegas, NV 89124

3752 Pahranagat National Wildlife Refuge
PO Box 510   775-725-3417
Alamo, NV 89001   Fax: 775-725-3389
                         E-mail: jim-docktor@fws.gov
                         http://www.r1.fws.gov/desert/pah-noframe.htm
Refuge staff are responsible for managing the 5,380 acre refuge, a mixture of desert, open water, native grass meadows, cropland and marsh. The refuge provides habitat for migratory birds of the Pacific Flyway and several speciesabound at this desert oasis. Wildlife observation is one of the most popular refuge activities. Waterfowl and small game hunting are also very popular, as is bird watching. A growing program provides additional opportunities, including internshipsfor college students.

Jim Docktor, Contact
Linda Miller, Deputy Project Leader

3753 Ruby Lake National Wildlife Refuge
HC 60 Box 860   775-779-2237
Ruby Valley, NV 89933   Fax: 775-779-0416
                                      http://www.gorp.com
Daniel L Pennington, Manager

3754 Toiyabe National Forest
1200 Franklin Way 775-331-6444
Sparks, NV 89431 Fax: 775-355-5398
The Toiyabe National Forest is located mostly in Nevada, with a portion also in California. Sections of the Forest stretch from Lake Tahoe and Reno, south to the Las Vegas area.

New Hampshire: US Forests, Parks, Refuges

3755 Great Bay National Wildlife Refuge
100 Merrimac Drive 603-431-7511
Newington, NH 03801-2903 Fax: 603-431-6014
E-mail: FWSRW_GBNWR@fws.gov or
jimmie_reynolds@fws.gov
http://www.gorp.com
Contains a variety of habitat including forested uplnads, open grasslands, brushy areas and freshwater/forested wetlands that support a wide variety of resident wildlife and migratory birds.
Jimmie J Reynolds, Refuge Manager

3756 John Hay National Wildlife Refuge
Great Bay NWR 603-431-7511
100 Merrimac Drive Fax: 603-431-6014
Newington, NH 03801-2903 E-mail: rsrw_gbnwr@fws.gow

3757 Lake Umbagog National Wildlife Refuge
PO Box 240 603-482-3415
Errol, NH 03579 Fax: 603-482-3308
E-mail: r5rw_lunwr@fws.gov
http://lakeumbagog.fws.gov/
Northern Forest refuge of New Hampshire and Maine provides long-term conservation of important wetland/upland habitats for wildlife, migratory birds and protected species.
Betty Chamagne, Administrative Assistant

3758 White Mountain National Forest
719 N Main Street 603-528-8721
Laconia, NH 03246 Fax: 603-528-8783

New Jersey: US Forests, Parks, Refuges

3759 Cape May National Wildlife Refuge
24 Kimbles Beach Road 609-463-0994
Cape May Courthouse, NJ 08210-4207 Fax: 609-463-1667
E-mail: howard_schlegel@fws.gov
http://www.fws.gov/northeast/capemay/
Howard Schlegel, Refuge Manager

3760 Great Swamp National Wildlife Refuge
152 Pleasant Plains Road 973-425-1222
Basking Ridge, NJ 07920 Fax: 973-425-7309
Swamp woodland, hardwood ridges, cattail marsh and grassland are typical of this approximately 7,500 acre refuge. The Swamp contains many large old oak and beach trees, stands of mountain laurel, mosses, ferns and species of many otherplants of both Nortern and Southern botanical zones.
William Koch, Refuge Manager

3761 Supawna Meadows National Wildlife Refuge
197 Lighthouse Road 856-935-1487
Pennsville, NJ 08070 Fax: 856-935-1198
E-mail: FW5RW_SMNWR@fws.gov
http://northeast.fws.gov/nj/spm.htm
The refuge lies along the Delaware River, north of the Salem River in Salem County, New Jersey. Approximately 75 percent of the current acreage is brackish tidal marsh. The refuge provides wintering and migrating waterfowl with animportant feeding and resting area.
Tom Walker, Refuge Manager

New Mexico: US Forests, Parks, Refuges

3762 Bitter Lake National Wildlife Refuge
4065 Bitter Lake Road 505-622-6755
Roswell, NM 88201 Fax: 505-623-9039
http://http://southwest.fws.gov/refuges/newmex/bitter.html
Native grasses, sand dunes, brushy bottomlands, seven lakes and a red-rimmed plateau make up Bitter Lake National Wildlife Refuge, winter home for thousands of migratory birds. The Lakes on the refuge were formed within the ancientriver beds of the Pecos River. These lakes store about 1,000 acres of water at their highest levels, while nearby marshland, mudflats and the Pecos Rriver provide an additional 24,500 acres of habitat.
Ken Butts, Manager

3763 Bosque del Apache National Wildlife Refuge
PO Box 1246 505-835-1828
Socorro, NM 87801 Fax: 505-835-0314
E-mail: daniel_perry@fws.gov
http://http://southwest.fws.gov/refuges/newmex/bosque
Bosque del Apache NWR is located in south-central New Mexico, along the Rio Grande in the northern reach of the Chihuahuan Desert. Habitats include cottonwood forests, seasonally managed wetlands, farm fields, saltgrass meadows, anddesert uplands. The refuge features large concentrations of sandhill cranes, light geese, and migrating waterfowl in fall and winter, shorebirds and songbirds travel through in spring and fall, and hummingbirds are abundant in summer.
Maggie O'Connell, Visitor Services Chief

3764 Capulin Volcano National Monument National Park Service
PO Box 40 505-278-2201
Capulin, NM 88414 Fax: 505-278-2211
http://www.nps.gov/cavo
Capulin Volcano is long extinct, and today the forested slopes provide habitat for mule deer, wild turkey, black bear and other wildlife. Abundant displays of wildflowers bloom on the mountain each summer. A two mile paved roadspiraling to the volcano rim makes Capulin Volcano one of the most accessible volcanos in the world. Trails leading around the rim allow exploration of this classic cinder cone.
, Superintendent, Park Ranger

3765 Carlsbad Caverns National Park
3225 National Parks Highway 505-785-2232
Carlsbad, NM 88220 Fax: 505-785-2133
E-mail: cave_park_information@nps.gov
http://www.nps.gov/cave
Established to preserve Carlsbad Cavern and numerous other caves within a Permian-age fossil reef, the park contains over 100 known caves, including Lechuguilla Cave-the nation's deepest limestone cave and third longest. CarlsbadCavern, with one of the world's largest underground chambers and countless formations, is highly accessible, with a variety of tours offered year-round.
John Benjamin, Superintendent
Chuck Barat, Deputy Superintendent

3766 Carson National Park
208 Cruz Alta Road 505-758-6200
Taos, NM 87571 Fax: 505-758-6213
Some of the finest mountain scenery in the southwest is found in the 1.5 million acres covered by the Carson National Forest. Elevations rise from 6,000 feet to 13,161 feet. The scenic Sangre de Cristo Mountains include Wheeler Peak,the highest peak in New Mexico.

3767 Cibola National Forest
2113 Osuna Road Northeast Suite A 505-346-3900
Albuquerque, NM 87113-1001 Fax: 505-346-3901
http://www.fs.fed.us/r3/cibola
Cibola, pronounced See'-bo-lah, is thought to be the original Zuni Indian name for their group of pueblos or tribal lands. Later, the Spanish interpreted the word to mean, "buffalo." Valued for its recreation opportunities, naturalbeauty, timber, watersheds, water, forage, and wilderness resources, the forest is managed to give the American people the greatest benefits that can be produced on a permanent basis.

3768 El Malpais National Monument
123 E Roosevelt Avenue   505-876-2783
201 E Roosevelt   Fax: 505-285-5661
Grants, NM 87020   E-mail: Leslie_DeLong@nps.gov
This monument preserves 114,277 acres of which 109,260 acres
are federal and 5,017 acres are private. Volcanic features such as
lava flows, cinder cones, pressure ridges and complex lava tube
systems dominate the landscape. Sandstone bluffs and mesas
border the eastern side, providing to vast wilderness.
Douglas E Eury

3769 Las Vegas National Wildlife Refuge
Route 1 Box 399   505-425-3581
Las Vegas, NM 87701   http://www.gorp.com
Joe B Rodriguez, Manager

3770 Lincoln National Forest
LNF Federal Building   505-437-6030
11th and New York   http://www.fs.fed.us/r3/lincoln/
Alamogordo, NM 88310

3771 Maxwell National Wildlife Refuge
PO Box 276   505-375-2331
Maxwell, NM 87728   Fax: 505-375-2332
                    E-mail: fw2_rw_maxwell@fws.gov
        http://southwest.fws.gov/refuges/newmex/maxwell.html
At an altitude of 6,050 feet, the refuge is made up of more than
3,000 acres of gently rolling prairie, playa lakes and farmland
for waterfowl. Rangeland and reclaimed farmland on the refuge
are made up of a variety of grasses including blue grama, galleta,
sand dropseed, threeawn and buffalo grass, as well as fourwing
saltbush and cactus. Several lakes provide approximately 700
acres of roosting and feeding habitat for waterfowl. Supports
waterfowl nesting and is also beneficial to shore birds.
Patty Hoban, Refuge Manager

3772 San Andres National Wildlife Refuge
PO Box 756   505-382-5047
Las Cruces, NM 88004   Fax: 505-382-5454
                  E-mail: fw2_rw_sanandres@fws.gov
        http://www.fws.gov/southwest/refuges/newmex/sanandres/in-
        dex.html
Refuge not open to the public due to its location within the
boundaries of U.S. Department of Army, White Sands Missile
Range. Primary emphasis has been focused on restoring a rem-
nant population of desert bighorn sheet (Ovis Canadensis
Mexicana)
Kevin Cobble, Refuge Manager

3773 Sevilleta National Wildlife Refuge
PO Box 1248   505-864-4021
Socorro, NM 87801   Fax: 505-864-7761
                   E-mail: fw2_rw_sevilleta@fws.gov
        http://southwest.fws.gov/refuges/newmex/sevill.html
Home to over 1200 species of plants, 89 species of mammals,
225 species of birds and 15 species of amphibians. More commonly seen species include mule deer, coyotes, pronghorns,
red-tailed hawks, northern harriers, western
diamondback rattlesnakes, roadrunners, sandhill cranes and
many different types of waterfowl and migrating shorebirds.
Bobcats, elk, bighorn sheep and an occasional mountain lion
also roam the hillsides.
Terry Tadano, Manager

3774 White Sands National Monument
PO Box 1086   505-479-6124
Holloman Air Force Base, NM 88330   Fax: 505-479-4333
White Sands National Mounment preserves a major portion of
this gypsum dune field, along with the plants and animals that
have successfully adapted to this constantly changing environ-
ment.
Dennis L Ditmanson

New York: US Forests, Parks, Refuges

3775 Fire Island National Seashore
120 Laurel Street   631-289-4810
Patchogue, NY 11772   Fax: 631-289-4898

E-mail: fiis_interpretation@nps.gov
http://www.nps.gov/fiis/home.htm
There are 32 miles of sandy beaches and saltwater marshes, a
sunken forest of 300 year old holly trees, hiking trails, a wilderness area and many other sites on the Fire Island National Seashore.
John Hauptman, Supt

3776 Gateway National Recreation Area
Building 69 Floyd Bennett Field   718-338-3575
Brooklyn, NY 11234   E-mail: carole_silano@nps.gov
Gateway NRA is a 26,000 acre recreation area located in the
heart of the New York metropolitan area. The park extends
through three New York City boroughs and into northern New
Jersey. Parks sites offer a variety of recreation opportunities,
along with a chance to explore many significant cultural re-
sources.
Kevin Buckley, Supt

3777 Iroquois National Wildlife Refuge
1101 Casey Road   585-948-5445
Basom, NY 14013-9730   Fax: 585-948-9538
                  http://http://iroquoisnwr.fws.gov
Iroquois National Wildlife Refuge lies within the rural township
of Alabama, New York, midway between Buffalo and Rochester. Part of what the locals call the "Alabama Swamps", its
10,818 acres of freshwater marshes, and hardwood swamps
bounded by woods, forest, pastures and wet meadows, serve the
habitat needs of many animals as a major stopover for migrating
birds and as a year-round residence.

Robert Lamayr, Refuge Manager

3778 Montezuma National Wildlife Refuge
3395 Routes 5 & 20 East   315-568-5987
Seneca Falls, NY 13148   Fax: 315-568-8835
                   E-mail: R5RW_MZNWR@fws.gov
                   http://www.gorp.com
Grady E Hocutt, Manager

**3779 Seatuck National Wildlife Refuge: Long Island National
Wildlife Refuge Complex**
Long Island NWR Complex   516-581-1538
PO Box 21   Fax: 516-581-2003
Shirley, NY 11967   E-mail: R5RW_STKNWR@fws.gov
Located along the southern shore of Long Island, the Refuge
consists of half salt marsh and half freshwater wetlands, ponds
and sparsely wooded areas. It is part of the larger Great South
Bay, which is a significant coastal habitat for migrating birds.
Limited recreation includes viewing wildlife and bird watching.
Charles Stenvall, Manager

**3780 Target Rock National Wildlife Refuge Long Island Na-
tional Wildlife Refuge Complex**
PO Box 21   516-286-0485
Shirley, NY 11967   Fax: 516-286-4003
The refuge was established in 1967. It consists of mixed upland
forest, a half mile of rocky beach, a brackish and several vernal
ponds. The offshore, beach and pond habitats provide foraging
areas for piping plover, wintering waterfowl and fish species.

North Carolina: US Forests, Parks, Refuges

3781 Alligator River National Wildlife Refuge
PO Box 1969   252-473-1131
Manteo, NC 27954   Fax: 252-473-1668
                   E-mail: alligatorriver@fws.gov
                   http://http://www.fws.gov/alligatorriver

Bonnie Strawser, Wildlife Interpretive Spec

3782 Cape Hatteras National Seashore
1401 National Park Drive   252-473-2111
Manteo, NC 27954   Fax: 252-473-2595
                   E-mail: CAHA_Information@nps.gov
                   http://www.nps.gov/caha/

A thin broken strand of islands curves out into the Atlantic Ocean and then back again in a sheltering embrace of North Carolina's mainland coast and its offshore sounds. These are the Outer Banks of North Carolina. Today their longstretches of beach, sand dunes, marshes and woodlands are set aside as Cape Hatteras National Seashore.
Lawrence A Belli, Superintendent

3783 Cape Lookout National Seashore

131 Charles Street   252-728-2250
Harkers Island, NC 28531   Fax: 252-728-2160
                    E-mail: CALO_information@nps.gov
                                   http://www.nps.gov/calo

The seashore is a 56 mile long section of the Outer Banks of North Carolina running from Ocracoke Inlet on the northeast to Beafort Inlet on the southeast. The four undeveloped barrier island, make up the seashore- North Core Banks,South Core Banks, Middle Core Banks and Shackleford Banks- may seem barren and isolated but they offer many natural and historical features that can make a visit very rewarding.
Robert A Vogel, Superintendent
Donna O Tiptor, Administrative Officer

3784 Cedar Island National Wildlife Refuge

38 Mattamuskeet Road   252-926-4021
Swan Quarter, NC 27885   Fax: 252-926-1743

3785 Currituck National Wildlife Refuge

Mackay Island NWR   919-429-3100
PO Box 39   Fax: 919-429-3185
Knotts Island, NC 27950-0039   http://www.gorp.com

3786 Mattamuskeet National Wildlife Refuge

Route 1   919-926-4021
Box N-2   E-mail: r4rw_nc.mtk@fws.gov
Swan Quarter, NC 27885   http://www.gorp.com

Located in eastern North Carolina in Hyde County, the Mattamuskeet Refuge consists of more than 50,000 acres of water, marsh, timber and crop lands. The refuge's most significant feature is lake Mattamuskeet, the largest natural lakein North Carolina. The lake is 18 miles long and five to 6 miles wide, encompassing approximately 40,000 acres, but averages 2 feet in depth.
Don Temple, Manager

3787 Nantahala National Forest

160-A Zillicoa Street   828-257-4200
Asheville, NC 28802   Fax: 828-257-4263

The National Forests of North Carolina include four national forests covering 1.2 million acres from the mountains to the sea. The Nantahala is located in the Appalachians of southwest North Carolina. The Nantahala is the largest ofthe four forests, totaling 528,541 acres. The Nantahala sits adjacent to Great Smokey Mountains National Park.

3788 Pea Island River National Wildlife Refuge

PO Box 1969   252-473-1131
Manteo, NC 27954   Fax: 252-473-1668
                    E-mail: alligatorriver@fws.gov
                                   http://http://www.fws.gov/peaisland
Bonnie Strawser, Wildlife Interprtive Spec

3789 Pee Dee National Wildlife Refuge

Route 1, Box 92   704-694-4424
Wadesboro, NC 28170   Fax: 704-694-6570
                    E-mail: fw4_rw_pee_dee@fws.gov
                                   http://peedee.fws.gov//index.html

Refuge objectives are to: provide habitat for migratory waterfowl and song birds; to provide habitat and protection for an endangered species, the red-cockaded woodpecker; to provide recreation, environmental education andinterpretation for the public; to engage in dynamic partnering.
Dan Frisk, Refuge Manager

3790 Pocosin Lakes National Wildlife Refuge

205 S Ludington Drive   252-796-3004
PO Box 329   Fax: 252-796-3010
Columbia, NC 27925   E-mail: pocosinlakes@fws.gov
                                   http://www.fws.gov/pocosinlakes/

The 112,000 acre refuge was established to protect and ehnhance a unique habitat called a "pocosin" and contains a variety of wildlife including endangered species such as the red wolf, bald eagle, peregrine falcon and red-cockadedwoodpecker as well as natural vegetation and scenic areas.

Howard Philips, Refuge Manager
David Kitts, Deputy Manager

3791 Roanoke River National Wildlife Refuge

114 W Water Street   252-794-3808
PO Box 430   Fax: 252-794-3780
Windsor, NC 27983   E-mail: roanokeriver@fws.gov
                                   http://roanokeriver.fws.gov/

Refuge objectives are: to provide habitat for migratory waterfowl, neo-tropical migrants and other birds; to provide migrating, spawning and nursery habitat for anadromous fish; (i.e. blueback herring, alewife, hickory shad and stripedbass); to enhance and protect forested wetlands consisting of bottomland hardwoods and swamps; to protect and manage for endangered and threatened wildlife; and to provide recreation and environmental education for the public

Harvey Hill, Refuge Manager

3792 Swanquarter National Wildlife Refuge

Route 1, Box N-2   252-926-4021
Swan Quarter, NC 27885   Fax: 252-926-1743

North Dakota: US Forests, Parks, Refuges

3793 Des Lacs National Wildlife Refuge Complex

PO Box 578   701-385-4046
Kenmare, ND 58746   Fax: 701-385-3214
                    E-mail: r6rw_dsl@fws.gov
             http://http://mountain-prairie.fws.gov/dslcomplex

Des Lacs NWR encompasses more than 19,500 acres along the Des Lacs River from the Canadian border to a point eight miles south of Kenmare, ND. A mix of natural lakes and managed wetlands in the valley provide a haven for migrating andnesting waterfowl and marsh birds.
Tim K Kessler, Manager

3794 Devils Lake Wetland Management District

Devils Lake WMD   701-662-8611
PO Box 908   Fax: 701-662-8612
Devils Lake, ND 58301

Located in the heart of the Prairie Pothole Region of the US. The northeastern North Dakota counties of Towner, Cavalier, Pembina, Benson, Ramsey, Walsh, Nelson and Grand Forks are included in the District. Managed by the US Fish andWildlife Service, the district provides wetland areas needed by waterfowl in the spring and summer for nesting and feeding. Hundreds of thousands of waterfowl also use these wetlands in the spring and fall for feeding and resting during longmigratory flights.

3795 Lake Ilo National Wildlife Refuge Audubon Complex

489 102 Avenue SW   701-548-8110
Dunn Center, ND 58626-0127   Fax: 701-548-8108

Located near the center of Dunn County in west central North Dakota, the refuge habitat is made up of native prairie, planted grasslands and wetlands. The uplands are characterized by gently sloping hills and terraces with creeks andan occasional slough. The average rainfall of 16.8 inches supports a prairie environment with a climate of hot dry summers, occasional thunderstorms and cold winters.

3796 Long Lake National Wildlife Refuge

12000 353rd Street 56   701-387-4397
Moffit, ND 58560   Fax: 701-387-4767
                    E-mail: r6rw-llk@fws.gov
                                   http://www.r6.fws.gov/refuge

The Refuge is about 18 miles long and contains 22,300 acres. The Refuge attracts a diversity and abundance of animals and waterfowl, both resident and migratory. Over 200 species of birds use the Refuge for breeding, rearing theiryoung and as a migratory stop. Long Lake Refuge is open for birdwatching, fishing, photography, boating, hiking and regulated hunting.

3797 Lostwood National Wildlife Refuge Des Lacs Complex
8315 Highway 8   701-848-2722
Kenmare, ND 58746-9046   Fax: 701-848-2702

Lies in the highly productive pothole region that produces more ducks than any other region in the lower 48 states. The refuge is a land of rolling hills mantled in short-grass and mixed with grass prairie interspersed with numerouswetlands. Established to preserve a unique wildlife habitat, Lostwood is an important link in our nation's system of more than 410 wildlife refuges.
Bob Barrett, Manager

3798 Theodore Roosevelt National Park
Box 7   701-623-4466
Medora, ND 58645-0007   Fax: 701-623-4840
                         E-mail: susan_recce@nps.gov
                                    http://www.nps.gov

Here in the North Dakota badlands, where many of his personal concerns first gave rise to his later environmental efforts, Roosevelt is remembered in a national park that bears his name and honors the memory of this greatconservationist. Theodore Roosevelt NAtional Park is colorful North Dakota badlands and is home to a variety of plants and animals, including bison, prairie dogs, and elk.

Ohio: US Forests, Parks, Refuges

3799 Cuyahoga Valley National Park
15610 Vaughn Road   216-524-1497
Brecksville, OH 44141   800-445-9667
            E-mail: cuva_canal_visitor_center@nps.gov
       http://www.nps.gov/cuva or www.dayinthevalley.com

Cuyahoga Valley National Park protects 33,000 acres along the Cuyahoga River between Cleveland and Akron, Ohio. Managed by the National Park Service, CVNP combines cultural, historical, recreational and natural activities in onesetting. Visitors can hike, bike, birdwatch, golf, fish, ski, ride Cuyahoga Valley Scenic Railroad, explore the history of the Ohio and Erie Canal on a 20 mile section of the Towpath Trail, and attend national park ranger-guided programs, concerts,art exhibits and more.
Colleen Brown, Visitor Use Assistant

3800 Ottawa National Wildlife Refuge
14000 W State Route 2   419-898-0014
Oak Harbor, OH 43449   Fax: 419-898-7895
                         E-mail: dan_frisk@fws.gov
                      http://midwest.fws.gov/ottawa.html

Refuge objectives are: to restore optimum acreage to a natural floodplain condition; to improve and restore wetland habitat, to improve fishery and wildlife resources, to provide for biodiversity; and to provide public opportunitiesfor outdoor recreation and environmental education.
Dan Frisk, Refuge Manager

Oklahoma: US Forests, Parks, Refuges

3801 Deep Fork National Wildlife Refuge
PO Box 816   918-756-0815
Okmulgee, OK 74447   Fax: 918-756-0275
                      E-mail: fw2_rw_deepfork@fws.gov
       http://southwest.fws.gov/refuges/oklahoma/deepfrk.html

Protecting important wetlands along the Deep Fork River, Deep Fork National Wildlife Refuge in eastern Oklahoma is a newcomer to the National Wildlife Refuge System. Established in 1993, the 9,000 acre refuge is subject to flooding atleast once a year. This flooding results in excellent conditions for waterfowl, including mallard, blue-winged teal, shoveler, pintail and wood ducks.
Darrin B Unruh, Manager

3802 Little River National Wildlife Refuge
PO Box 340   405-584-6211
Broken Bow, OK 74728   Fax: 405-584-2034
                                    http://www.gorp.com

Berlin A Heck, Manager

3803 Oklahoma Bat Caves National Wildlife Refuge
Route 1 Box 18-A   918-773-5251
Vian, OK 74962   Fax: 918-773-5252
                                    http://www.gorp.com

3804 Optima National Wildlife Refuge
Route 1 Box 68   580-664-2205
Butler, OK 73625   Fax: 580-664-2206
                         E-mail: r2rw_wa@fws.gov
       http://southwest.fws.gov/refuges/oklahoma/optima.html

Located in the middle of the Oklahoma panhandle, the 4,333-acre refuge is made up of grasslands and wooded bottomland on the Coldwater Creek arm of the Army Corps of Engineers Optima Reservoir Project.
David Maple, Manager

3805 Sequoyah National Wildlife Refuge
Route 1 Box 18 A   918-773-5251
Vian, OK 74962   Fax: 918-773-5598
                    E-mail: FW2_RW_Sequoyah@fws.gov
       http://http://southwest.fws.gov/refuges/oklahoma/sequoy.html

Henry R Sansing, Refuge Manager
Craig Heflebower, Refuge Operations Specialist

3806 US Fish & Wildlife Service Tishomingo National Wildlife Refuge
12000 S Refuge Road   580-371-2402
Tishomingo, OK 73460   Fax: 580-371-9312
                    E-mail: fw2_rw_tishomingo@fws.gov
       http://http://southwest.fws.gov/refuges/oklahoma/tishomingo

The 16,464-acre Refuge lies at the upper Washita arm of Lake Texoma and is administered for the benefit of migratory waterfowl in the Central Flyway. It offers a variety of aquatic habitats for wildlife. The murky water of theCumberland Pool provides abundant nutrients for innumerable microscopic plants and animals. Seasonally flooded flats and willow shallows lying at the Pool's edge also provide excellent wildlife habitat.
Kris Patton, Refuge Manager

3807 Washita National Wildlife Refuge
Route 1 Box 685   405-664-2205
Butler, OK 73625   E-mail: r2rw_wa@fws.gov
                                    http://www.gorp.com

Jon M Brock, Manager

Oregon: US Forests, Parks, Refuges

3808 Ankeny National Wildlife Refuge
Western Oregon NWR Complex   503-588-2071
20301 Wintel Road   Fax: 541-757-4450
Jefferson, OR 97352-9758   http://www.gorp.com

Refuge's primary management goal is to provide vital wintering habitat for dusky Canada geese. The refuge includes flat to gently rolling land near the confluence of the Willamette and Sanitiam rivers. The refuge's fertile farmedfields, hedgerows, forests, and wetlands provide a variety of wildlife habitats. the refuge is open to limited opportunites for wildlife-oriented education and recreation. Ducks, geese, and swans are commonly seen in refuge fields and ponds throughthe fall and winter.

3809 Bandon Marsh National Wildlife Refuge
Western Oregon NWR Complex   503-757-7236
26208 Finley Refuge Road   http://www.gorp.com
Corvallis, OR 97333

3810 Baskett Slough National Wildlife Refuge
Western Oregon NWR Complex   503-757-7236
26208 Finley Refuge Road   http://www.gorp.com
Corvallis, OR 97333

Includes 2,492 acres typical of Willamette Vallley's irrigated hillsides, oak-covered knolls and grass fields. Wetlands include Morgan Lake and Baskett Slough. The refuge's objective is the protection and management of winteringhabitat for dusky Canada geese. Several species of waterfowl, herons, hawks, quail, shorebirds, mourning doves, woodpeckers and a variety of songbirds frequent the area, as well as mammals, amphibians and reptiles. Recreation includes observation,study and photography.
Richard Guadagno, Manager

3811 Cape Meares National Wildlife Refuge
Western Oregon NWR Complex   503-757-7236
26208 Finley Refuge Road   http://www.gorp.com
Corvallis, OR 97333

3812 Cold Springs National Wildlife Refuge
PO Box 700   541-992-3232
Umatilla, OR 97882   E-mail: gary_hagedorn@fws.gov
                                    http://www.gorp.com
Cold Springs NWR lies in sharp contrast with the arid desert surroundings of northeastern Oregon. The refuge, a tree-lined reservoir, lies 7 miles east of the agricultural community of Hermiston. The variety of refuge habitats attractsan abundance of wildlife. Cold Springs supports peak populations of over 45,000 winter waterfowl comprised mainly of mallards and Canada geese.

3813 Columbia River Gorge National Scenic Area
902 Wasco Avenue   541-386-2333
Waucoma Center, Suite 200http://www.fs.fed.us/r6/columbia/
Hood River, OR 97031
The Columbia River Gorge is a espaectacular river canyon cutting the only sea-level route through the Cascade Mountain Range. It's 80 miles long and to 4,00 feet deep with the north canyon walls in Washington State and the south canyonwalls in Oregon State.

3814 Crater Lake National Park
Highway 62   541-594-2211
PO Box 7   Fax: 541-594-2261
Crater Lake, OR 97604   http://www.nps.gov/crla/home.html
During the summer, visitors may navigate the Rim Drive around the lake, enjoy boat tours, stay in the historic Crater Lake Lodge Camp or hike some of the park's various trails. The winter brings some of the heaviest snowfall in thecountry, averaging 533 inches per year. Although park facilities mostly close for the snow season, visitors may view the lake during fair weather, enjoy cross-country skiing, and participate in weekend snowshoe hikes.
Dave Morris, Supt

3815 Deschutes National Forest
1645 US Highway 20 NE   541-388-2715
Bend, OR 97701
Scenic backdrop of volcanic attraction, evergreen forest, mountain lakes, caves, desert areas and alpine meadows.

3816 Fremont National Forest
1300 S G Street   541-947-2151
HC 10, Box 337   Fax: 541-947-6399
Lakeview, OR 97630   http://www.fs.fed.us/r6/fremont/
Located in Oregon's Outback, the forest provides the self reliant recreationist the opportunity to discover nature in a rustic environment.

3817 Malheur National Forest
431 Patterson Bridge Road   541-575-1731
PO Box 909
John Day, OR 97845
The 1,460,000 acre Malheur National Forest is located in the blue Mountains of Eastern Oregon. The diverse and beautiful scenery of the forest includes high desert grasslands, sage and juniper, pine, fir and other tree species, and thehidden gem of alpine lakes and meadows. Elevations vary from about 4000feet (1200 meters) to the 9038 foot (2754 meters) top Strawberry Mountain. The Strawberry Mountain range extends east to west through the center of the forest.
Bonnie J Wood, Forest Supervisor

3818 Malheur National Wildlife Refuge
36391 Sodhouse Lane   541-493-2612
Princeton, OR 97721   Fax: 541-493-2405
                  http://http://pacific.fws.gov/malheur
Forrest Cameron, Manager

3819 McKay Creek National Wildlife Refuge
Umatilla NWR Complex   503-922-3232
PO Box 239   http://www.gorp.com
Umatilla, OR 97882

3820 Mount Hood National Forest
16400 Champion Way   503-668-1700
Sandy, OR 97055   Fax: 503-668-1641
Located 20 miles east of the city of Portland and the northern Willamette River valley, Mt Hood National Forest extends south from the strikingly beautiful Columbia River Gorge across more than sixty miles of forested mountains, lakesand streams to OlAllie Scenic Area, a high lake basin under the slopes of Mt Jefferson. Our many visitors enjoy fishing, caming, boating and hiking in the summer, hunting in the fall, skiing and other snow sports in the winter.

3821 National Park Service: John Day Fossil Beds National Monument
32651 Highway 19   541-987-2333
Kimberly, OR 97845   Fax: 541-987-2336
              E-mail: joda_interpretation@nps.gov
                      http://www.nps.gov/joda
Within the heavily eroded volcanic deposits of the scenic John Day Fossil Basin is a great diversity of well-preserved plant and animal fossils. This remarkably complete record spans more than 40 of the 65 million years of the CenozoicEra (the Age of Mammals). The monument was established in 1975.
John Fiedor, Chief Visitor Services

3822 Ochoco National Forest
3061 NE 3rd Street   541-416-6500
Po Box 490   Fax: 541-416-6695
Prineville, OR 97754
With a total of almost 1,500 square miles, the Ochoco National Forest is endowed with vast natural resources, scenic grandeur and tremendous recreation opportunities. People are drawn to the Ochoco for its majestic ponderosa pinestands, picturesque rimrock vantage points, deep canyons, unique geologic formations, abundant wildlife and plentiful sunshine.

3823 Oregon Caves National Monument
19000 Caves Highway   541-592-2100
Cave Junction, OR 97523   Fax: 541-592-3981
              E-mail: roger_brandt_@nps.gov
Oregan Caves National Monument is small in size, 480 acres, but rich in diversity. Above ground, the monument encompasses a remnant old-growth coniferous forest. It harbors a fantastic array of plants, and a Douglas-fir tree with thewildest known girth in Oregon. Three hiking trails access this forest. Below ground is an active marble cave created by natural forces over hundreds of thousands of years in one of the world's most diverse geologic realms.
Craig Ackerman

3824 Oregon Coastal Refuges
2127 SE OSU Drive   541-867-4550
Newport, OR 97365   Fax: 541-867-4551
Nancy Morrissey, Manager

3825 Oregon Islands National Wildlife Refuge
Western Oregon NWR Complex   503-757-7236
26208 Finley Refuge Road   http://www.gorp.com
Corvallis, OR 97333

3826 Rogue River National Forest
333 W 8th Street   541-858-2200
PO Box 520   Fax: 541-858-2220
Medford, OR 97501   http://www.fs.fed.us/r6/rogue

The Rogue River National Forest encompasses roghly 630,000 acres of Southern Oregon's most beautiful territory. Staddling the Siskiyou and Cascade mountain ranges, the forest provides vistors and local residents with an array of resources and recreation opportunities.

3827 Sheldon National Wildlife Refuge
US Fish and Wildlife Service- Pacific Region
PO Box 111 503-947-3315
18 South G Fax: 503-947-4414
Lakeview, OR 97630 http://www.fws.gov
Brian Day, Manager

3828 Siskiyou National Forest
200 NE Greenfield Road 541-471-6500
Po Box 440 Fax: 541-471-6514
Grants Pass, OR 97528

The Siskiyou National Fforest embodies the most complex soils, geology, landscape, and plant communities in the Pacific Northwest. World-class rivers, biological diversity, fisheries, and complex watersheds rank the Siskiyou high inthe Nation as an outstanding resource.

3829 Siuslaw National Forest
4077 SE Research Way 541-750-7000
PO Box 1148 Fax: 541-750-7234
Corville, OR 97339

The Siuslaw National Forest encompasses one of the most productive and diverse landscapes in the world from fertile soils, which support tall stands of Douglas fir, western hemlock and Sitka spruce forests laced with miles of riversand streams, to miles of open sand dunes. These rich settings from habitats for a broad array of plants and animals and provide endless opportunities for learning.

3830 Three Arch Rocks National Wildlife Refuge
Western Oregon NWR Complex 503-757-7236
26208 Finley Refuge Road http://www.gorp.com
Corvallis, OR 97333

3831 Umatilla National Forest
2517 SW Hailey Avenue 541-278-3716
Pendleton, OR 97801

The Umatilla National Forest, located in the Blue Mountains of southeast Wasington and northeast Oregon, covers 1.4 million acres of diverse landscapes and plant communities. The forest has some mountainous terrain, but most of theforest consists of v-shaped valleys separated by narrow ridges or plateaus.

3832 Umatilla National Wildlife Refuge
Mid-Columbian River Refuges
2805 St. Andrews Loop 509-545-8588
PO Box 2527 Fax: 509-545-8670
Pasco, WA 99301 http://http://midcolumbiariver.fws.gov
Morris C LeFever, Manager

3833 Umpqua National Forest
2900 NW Stewart Parkway 541-672-6601
PO Box 1008 Fax: 541-957-3495
Roseburg, OR 97470 E-mail: jcaplan@fs.fed.us

The Umpqua National Forest covers nearly one million acres and is located in the western slopes of the Cascades in Southwest Oregon. The forest encommpasses a diverse area of rugged peaks, high rolling meadows, sparkling rivers andlakes and deep canyons producing a wealth of water resources, timber, forage, minerals, wildlife and outdoor recreation opportunities.

3834 Wallowa-Whitman National Forest
1500 Dewey Avenue 541-523-6391
PO Bos 907 Fax: 541-523-1315
Baker, OR 97814

The Wallowa-Whitman National Forest contains 2.3 million acres ranging in elevation from 875 feet in Hells Canyon, to 9845 feet in the Eagle Cap Wilderness. Our varied forests are managed as sustainable ecosystems providing cleanwater, wildlife habitat and valuable forest products. For things to do and places to be, the Wallowa-Whitman is the setting for a variety of year-round recreation. You are welcome at the Wallowa-Whitman National Forest.

3835 William L Finley National Wildlife Refuge
26208 Finley Refuge Road 503-757-7236
Corvallis, OR 97333 http://www.gorp.com

3836 Winema National Forest
2819 Dahlia Street 541-883-6714
Klamath Falls, OR 97601 Fax: 541-883-6709

The 1.1 million acre Winema National Forest lies on the eastern slopes of the Cascade Mountain Range in South Central Oregon, an area noted for its year-round sunshine. The Forest borders Crater Lake National Park near the crest of theCascades and stretches eastward into the Klamath River Basin. Near the floor of the Basin the Forest gives way to vast marshes and meadows assoicated with Upper Klamath Lake and the Williamson River.

Pennsylvania: US Forests, Parks, Refuges

3837 Allegheny National Forest
222 Liberty Street 814-723-5150
Warren, PA 16365 Fax: 814-726-1465
E-mail: anf/r9_allegheny@fs.fed.us
http://www.fs.fed.us/r9/allegheny
An organization dedicated to providing advice for development of the corridor management plan for the northern section of the Allegheny River that has been designated as a National Wild and Scenic River.
Don Clymer, Staff Contact

3838 Delaware National Scenic River/Delaware Water Gap National Recreation Area
Delaware Water Gap National Recreation Area
HQ River Road - Route 209 570-588-2435
Bushkill, PA 18324 Fax: 570-588-2780
E-mail: dewa_interpretation@nps
http://www.nationalparksgallery.com/

3839 Erie National Wildlife Refuge
11296 Wood Duck Lane 814-789-3585
Guys Mills, PA 16327 Fax: 814-789-2909
http://www.fws.gov/northwest/erie
A haven for migratory birds consisting of two divisions: the Sugar Lake Division and the Seneca Division. Refuge management objectives include: providing waterfowl and other migratory birds with nesting, feeding, brooding, and restinghabitat; providing habitat to support a diversity of other wildlife species; and enhancing opportunities for wildlife-oriented public recreation and environmental education.

Patty Nagel, Deputy Refuge Manager

3840 Gettysburg National Military Park
97 Taneytown Road 717-334-1124
Gettysburg, PA 17325-2804 Fax: 717-334-1891
E-mail: gett_superintendant@nps.gov
http://www.nps.gov
A unit of the national park service preserving 6000 acres of Gettysburg battlefield, and the Soldiers' National Cemetery, site of Lincoln's Gettysburg Address.

3841 Upper Delaware Scenic & Recreational River
Rural Route 2 570-729-7134
Box 2428 Fax: 570-729-7918
Beach Lake, PA 18405-9737 http://www.npa.gov/upde
As a part of the National Wild and Scenic Rivers System, upper Delaware Scenic and Recreational River streches 73.4 miles (118.3 km) along the New York/Pennsylvania border. The longest free flowing river in the Northeast, it includesriffles and Class I and II rapids between placid pools eddies. Public fishing and boating accesses are provided, although most land along the river is privately owned. Wintering bald eagles are among the wildlife that may be seen here.
Dave Forney, Superintendent
Michael Reubet, Chief Resource Management

Rhode Island: US Forests, Parks, Refuges

3842 Rhode Island National Wildlife Refuge Complex
3769 D Old Post Road 401-364-0170
PO Box 307 Fax: 401-364-0170
Charlestown, RI 02813-0307 E-mail: fw5rw_rinwr@fws.gov
http://www.northeast.fws./ri.htm
Charles Vandemoer, Complex Refuge Manager
Gary M Andres, Deputy Refuge Manager

South Carolina: US Forests, Parks, Refuges

3843 Ace Basin National Wildlife Refuge
PO Box 848 843-889-3084
Hollywood, SC 29449 Fax: 843-889-3282
http://acebasin.fws.gov
The Ace Basin National Wildlife Refuge was established in 1990 to assist in preserving the nationally significant wildlife and related habitats within the 350,000-acre Ashepoo, Combahee and South Edisto (ACE) rivers basin. The wetlandshabitat of the area has been preserved during the last several centuries through careful management by private landowners. An antebellum mansion that survived the Civil War now serves in part as office space for the refuge.
Jane Griess, Refuge Manager

3844 Cape Romain National Wildlife Refuge
5801 Highway 17 North 843-928-3264
Awendaw, SC 29429 Fax: 843-928-3803
http://caperomain.fws.gov/DonnyBrowning
Refuge objectives are to: provide habitat for waterfowl, shorebirds, wading birds and resident species; provide habitat and management of endangered and threatened species; provide protection of Class I Wilderness Area; and provideenvironmental education and recreation for the public.
James "Donny" Browning, Refuge Manager

3845 Carolina Sandhills National Wildlife Refuge
Route 2 Box 330 803-335-8401
McBee, SC 29101 Fax: 803-335-8406
http://www.gorp.com

Richard P Ingram, Manager

3846 Francis Marion-Sumter National Forest
4931 Broad River Road 803-561-4000
Columbia, SC 29212-3530 http://www.fs.fed.us/r8/fms/
Headquaters in the capital city of Colombia, both forests are managed for many uses; including timber and wood production, watershed protection and improvement, habitat for wildlife and fish species, wilderness area management,minerals leasing and outdoor recreation.

3847 Santee National Wildlife Refuge
2125 Fort Watson Road 803-478-2217
Summerton, SC 29148 Fax: 803-478-2314
E-mail: fw4_rw_santee@fws.gov
http://southeast.fws.gov/santee/
Santee NWR is located in Clarendon County, SC. The 15,000 acre refuge lies within the Atlantic Coastal Plain and consists of mixed hardwoods, mixed pine-hardwoods, pine plantations, marsh, cropland, old fields, ponds, impoundments andopen water. This tremendous diversity of habitat supports many kinds of wildlife. The four management units of the refuge stretch over 18 miles along the northern side of Santee Cooper's Lake Marion.

Mark Purcell, Refuge Manager
Kay McCutcheen, Park Ranger

South Dakota: US Forests, Parks, Refuges

3848 Badlands National Park
PO Box 6 605-433-5361
Interior, SD 57750 Fax: 605-433-5404

E-mail: badl_information@nps.gov
http://www.nps.gov/badl/
Consists of acres of sharply eroded buttes, pinnacles and spires blended with the largest protected mixed grass prarie in the US. The Badlands Wilderness Area covers 64,000 acres and is the site of the reintroduction of theblack-footed ferret, the most endangered land mammal in North America. The Stronghold Unit is co-managed with the Oglala Sioux Tribe and includes site of 1890s Glost Dances. Over 11,000 years of human history pale to the ages old paleontologicalresources.

William R Supernaugh, Supt

3849 Black Hills National Forest
25041 N Highway 16 605-673-9200
Custer, SD 57730 Fax: 605-673-9350
http://www.fs.fed.us/r2/blackhills/
Eleven reservoirs, 30 campgrounds, 2 scenic byways, 1300 miles of streams, 13,000 acres of wilderness, 353 miles of trails, and much more. The forest is managed for multiple use so don't be surprised to see mining, logging, cattlegrazing, and summer homes on your travel.

3850 Huron Wetland Management District
200 4th Street SW 605-352-5894
Federal Building, Room 309 Fax: 605-352-6709
Huron, SD 57350 E-mail: huronwetlands@fws.gov
http://huron wetlands.fws.gov
The public lands of the HWMD, called Waterfowl Production Areas, are part of the National Wildlife Refuge System. The refuges and WPAs are vitally important to wildlife and people. They provide food, water, cover and space for hundredsof species of birds, mammals, reptiles, amphibians, fish and plants. Managed to benefit endangered species, migratory birds and other wildlife and provide places to learn about and enjoy wildlife. HWMD's mission is to preserve wetlands and managehabitat.
harris Hoisted, Project Leader

3851 Jewel Cave National Monument
Rural Route 1 Box 60-AA 605-673-2288
Custer, SD 57730 Fax: 605-673-3294
E-mail: mailto:JECA_Interpretation@nps.gov
http://www.dgif.state.va.us
With more than 125 miles sureyed, jewel cave is recognized as the third longest cave in the world. Airflow within its passages indicates a vast area yet to be explored. Cave tours provide opportunities for viewing this pristine cavesystem and its wide varitey of speleothems including stalactites, stalagmites, draperies, frostwork, flowstone, boxwork and hydromagnesite balloons. The cave is an important hibernaculum for several species of bats.
Kate Cannon, Supt

3852 Sand Lake National Wildlife Refuge
39650 Sand Lake Drive 605-885-6320
Columbia, SD 57433 Fax: 605-885-6333
E-mail: sandlake@fws.gov
http://sandlake.fws.gov
Sand Lake Refuge is haven for wildlife and those who enjoy it. Home to more than 266 species of birds, 40 mammal species and a variety of fish, reptiles and amphibians, this 22,000 acre refuge is a mosaic of wildlife and the wildplaces they need. Sand Lake is also a very popular recreation spot. Wildlife observation, fishing, hunting, photography, interpretation and environmental education are all popular activities at the refuge.
Refuge Manager

3853 Wind Cave National Park
26611 US Highway 385 605-745-4600
Hot Springs, SD 57747-9430 Fax: 605-745-4207
http://www.nps.gov
One of the world's longest and most complex caves and 28,295 acres of mixed-grass prairie, ponderosa pine forest, and associated display of boxwork, an unusual cave formation composed of thin calcite fins resembling honeycombs. Thepark's mixed grass prairie is one of the few remaining and is home to native wildlife such as bison, elk, pronghorn, mule deer, coyotes, and prairie dogs.

Tennessee: US Forests, Parks, Refuges

3854 Big South Fork National River Recreation Area
4564 Leatherwood Road   931-879-3625
Onieda, TN 37841   Fax: 423-569-5505
The free-flowing Big South Fork of the Cumberland River and its tributaries pass through 90 miles of scenic gorges and valleys containing a wide range of natural and historic features. The area offers a broad range of recreationalopportunities including camping, whitewater rafting, kayaking, canoeing, hiking, horseback riding, mountain biking, hunting and fishing, The US Army Corps of Engineers, with its experience in managing river basins, was charged with land acquisition,planning and deve
William K Dickinson, Supt

3855 Cherokee National Forest
2800 N Ocoee Street NW   423-476-9700
Po Box 2010   Fax: 423-476-9721
Cleveland, TN 37320
The Cherokee is steeped in colorful history and rich in the grandeur of the Appalachian Mountains. The forest is separated into two sections by Great Smoky Mountains Park and shares other boundaries with national forest in Georgia,North Carolina and Virginia.

3856 Chickasaw National Wildlife Refuge
1505 Sand Bluff Road   731-635-7621
Ripley, TN 38063   Fax: 731-635-0178
E-mail: fw4_rw_chickasaw.fws.gov
http://www.fws.gov
Established to provide essentail habitat for migratory birds in the Lower Mississippi Valley. The refuge supports a variety of wildlife. Visitors can see large numbers of migratory wasterfowl in the winter. Neotropical migratory birdsand shorebirds are a common site yearround. The refuge is open to hunting and fishing-special regulations apply. Please contact the refuge manager for current regulations.
Curt McMurl, Refuge Manager

3857 Cross Creeks National Wildlife Refuge
643 Wildlife Road   931-232-7477
Dover, TN 37058   Fax: 931-232-5958
E-mail: fw4_web_manager@fws.gov
http://www.fws.gov/crosscreeks/

Vicki C Grafe, Manager

3858 Great Smokey Mountains National Park
107 Park Headquarters Road   865-436-1200
Gatlinburg, TN 37738   Fax: 865-436-1220
E-mail: grsm_smokies_information@nps.gov
http://www.nps.gov
The national park, in the state of North Carolina is world renowned for thr diversity of its plant and animal resources, the beaty of its ancient mountains, the quality of its remmants of Southern Appalachin mountain culture, and thedepth and integrity of the wildernees sanctuary within its boundaries, it is one the largest protected areas in the east.
Randall R Pope, Supt

3859 Hatchie National Wildlife Refuge
4172 Highway 76 South   901-772-0501
Brownsville, TN 38012-8332   Fax: 901-772-7839
E-mail: r4rw_tn.htc@fws.gov
http://www.gorp.com

Marvin L Nichols, Manager

3860 Lower Hatchie National Wildlife Refuge
1505 Sand Bluff Road   901-635-7621
Ripley, TN 38063   Fax: 901-635-7621
E-mail: r4rw_tn.rlf@fws.gov
http://www.gorp.com

3861 Reelfoot National Wildlife Refuge
4343 Highway 157   901-538-2481
Union City, TN 38261   Fax: 901-538-9760
E-mail: r4rw_tn.rlf@fws.gov
http://www.gorp.com

Randy Cook, Manager

3862 Tennessee National Wildlife Refuge
PO Box 849   901-642-2091
Paris, TN 38242   Fax: 901-644-3351
E-mail: r4rw_tn.tns@fws.gov
http://www.gorp.com

John Taylor, Manager

Texas: US Forests, Parks, Refuges

3863 Alibates Flint Quarries National Monument/Lake Meredith National Recreation Area
PO Box 1460   806-857-3151
Fritch, TX 79036   Fax: 806-857-2319
http://www.nps.fov/alfl
ALIBATES: The only national monument in Texas. Preserves over 700 archeological sites. The monument can only be viewed by ranger-led guided tours. LAKE MEREDITH: A 45,000 acre recreation area that includes a 10,000 acre reservoirwhere visitors can enjoy water and land recreational activities such as hunting, fishing, boating, horseback riding, off-road vehicles, jetskies and the like.

Karren Brown, Supt

3864 Amistad National Recreation Area
PO Box 420367   830-775-7491
Del Rio, TX 78840-9350   Fax: 830-775-7299
E-mail: interpretation@nps.gov
Situated on the United States-Mexico border, is know primarily for excellent year round, water-based recreation including: fishing , boating, swimming, suba diving. Also provides opportunities for picnicking, camping and hinting. Thereservoir, at the confluence of the Rio Grande, Devils and Pecos rivers, was created by Amistad Dam in 1969, This area is reach in technology and rock art, and contains a wide variety of plant and animal life.
Robert Reyes, Supt

3865 Angelina National Forest
701 N 1st Street   409-639-8620
Room 100   Fax: 409-639-8624
Lufkin, TX 75901
The Angelina National Forest is located in the heart of Texas. The reservoir, a 114,500 acre lake on the Angelina River is noted for its fishing, boating and water skiing.

3866 Big Bend National Park
PO Box 129   915-477-2251
Big Bend National Park, TX 79834   Fax: 915-477-1175
The Big Bend National Park is situated on the boundry with Mexico along the Rio Grande. It is a place where countries and cultures meet, also a place that merges natural environments, from desert to mountains. It's a place where southmeets north and east meets west, creating a great diversity of plants and animals. The park covers over 801,000 acres of west Texas in the place where the Rio Grande makes a sharp turn - the Big Bend.
Robert Arnberger, Supt

3867 Big Thicket National Preserve
3785 Milam Street   409-951-6800
Beaumont, TX 77701-4724   Fax: 409-951-6868
E-mail: BITH_Administration@nps.gov
http://www.nps.gov/bith/default.htm
The Preserve consists of nine land units and six water corridors encompassing more than 97,000 acres. Big Thicket was the first Preserve in the National Park System and protects and area of rich biological diversity. A convergence ofecosystems occured here during the last Ice Age. It brought together, in one geographical location, the eastern hardwood forests, the Gulf coastal plains and the midwest praries.

Ronald Switzer, Supt

3868 Grulla National Wildlife Refuge
Muleshoe NWR   806-946-3341
PO Box 549   Fax: 806-946-3317
Muleshoe, TX 79347   http://southwest.fws.gov

Located in Roosevelt County, New Mexico, near the small town of Arch, approximately 25 miles northwest of Muleshoe National Wildlife Refuge. Grulla NWR, which is managed by the staff at Muleshoe NWR, has 3,236 acres, more than 2,000 of which make up the saline lake bed of Salt Lake. The rest of the refuge is grassland. When the lake holds sufficient water, Grulla NWR is a wintering area for lesser sandhill cranes. Ring-necked pheasant, scaled quail and lesser prairie chickens maybe seen.

Harold Beierman, Manager

3869 Guadalupe Mountains National Park

HC 60 Box 400   915-828-3251
Salt Flat, TX 79847-9400   Fax: 915-828-3269
            E-mail: GUMO_Superintendent@nps.gov
            http://www.nps.gov/gumo/gumo/home.html
This mountain mass contains portions of the world's most extensive and significant Permian limestone fossil reef, earth fault peaks, unusual flora and fauna. Guadalupe Peak, highest point in Texas at 8,749 feet.
Larry Henderson, Supt

3870 Hagerman National Wildlife Refuge

6465 Refuge Road   903-786-2826
Sherman, TX 75092-5817   Fax: 903-786-3327
            E-mail: fw2_rw_hagerman@fws.gov
            http://southwest.fws.gov/refuges/texas/hagermn.html
Hagerman NWR lies on the Big Mineral Arm of Lake Texoma, on the Red River between Oklahoma and Texas. Established in 1946, the refuge includes 3,000 acres of marsh and water and 8,000 acres of upland and farmland. During fall, winter and spring, the marshes and waters are in constant use by migrating and wintering waterfowl.
James M Williams, Manager

3871 Padre Island National Seashore

PO Box 181300   361-949-8173
Corpus Christi, TX 78480-1300   Fax: 361-949-8023
            http://www.nps.gov/pais
Encompassing 130,434 acres, the longest remaining undeveloped stretch of barrier island in the world, and offers a wide variety of flora and fauna as well as recreation.

3872 Santa Ana National Wildlife Refuge

Route 2   956-784-7500
Box 202A   Fax: 956-787-8338
Alamo, TX 78516   E-mail: r2rw_sta@fws.gov
            http://southwest.fws.gov/refuges/texas/santana.html
The 2,088 acre refuge along the banks of the lower Rio Grande was established in 1943 for the protection of migratory birds. Considered the jewel of the refuge system, this essential island of thorn forest habitat is host or home to nearly 400 different types of birds and a myriad of other species, including the indigo snake, malachite butterfly and the endangered ocelot. Provides habitat for thousands of migrating birds and about one half of all butterfly species found in North America.
Jeff Howland, Refuge Manager

Utah: US Forests, Parks, Refuges

3873 Arches National Park

PO Box 907   435-719-2299
Moab, UT 84532-0907   Fax: 435-719-2305
            E-mail: archinfo@nps.gov
Arches National Park preserves over two thousand natural sandstone arches and a variety of other unique geological resources. The extraordinary features of the park are highlighted by a striking environment of contrasting colors, landforms and textures.
            ] pages
Laura Jess, Supt

3874 Ashley National Forest

PO Box 279   435-784-3445
Manila, UT 84046   Fax: 435-781-5295

3875 Bear River Migratory Bird Refuge

58 S   801-723-5887
Brigham City, UT 84302   E-mail: R6RW_BRR@fws.gov

To date, close to 1 million cubic yards of earth has been moved to restore and enhance the refuge. Forty-seven primary water control structures have been restored along with over forty-seven miles of dikes. Through volunteer efforts, debris has been removed from the old headquaters site and a new pavilion, restroom, demonstration pond, and kiosk have been built on the site. The 12-mile auto tour route has been reopened to the public.
Alan K Trout, Manager

3876 Bryce Canyon National Park

PO Box 170001   435-834-5322
Bryce Canyon, UT 84717   Fax: 435-834-4102
            http://www.nps.gov/brca
Consists of 37,277 acres of scenic colorful rock formations and desert wonderland. Bryce Canyon National Park is named for one of a series of horseshoe-shaped amphitheaters carved from the eastern edge of the Paunsaugunt Plateau in southern Utah. Erosion has shaped colorful Claron limestones, sandstones and mudstones into thousands of spires, fins, pinnacles and mazes. Collectively called hoodoos, these unique formations are whimsically arranged and tinted with colors too numerous to name.

Craig C Axtell

3877 Canyonlands National Park

2282 S West Resource Boulevard   435-719-2313
Moab, UT 84532-3298   Fax: 435-719-2300
            E-mail: canyinfo@nps.gov
            http://www.nps.gov/cany/home.htm
Canyonlands National Park preserves a stunning landscape of sedimentary sandstones eroded into countless canyons, mesas and buttes by the Colorado River and its tributaries. Largely undeveloped, the park is a popular backcountry destination and scientific research site.

3878 Capitol Reef National Park

HC 70 Box 15   435-425-3791
Torrey, UT 84775-9602   Fax: 435-425-3026
            E-mail: care_administration@nps.gov
            http://www.capitol.reef.national-park.com
The Waterpocket Fold, a 100 mile long wrinkle in the earth's know as a monoclide, extends from nearby Thousand Lakes Mountain to the Colorado River. Capitol Reef National Park was established to protect this grand and colorful geologic feature, as well as the unique historical and cultural history found in the area.

Albert J Hendricks, Supt

3879 Cedar Breaks National Monument

2390 W Highway 56   435-586-9451
Suite 11   Fax: 435-586-3813
Cedar City, UT 84720-4151
Millons of years of sedimentation, uplift and erosion continue to create a deep canyon of rock walls, fins, spires and columms, that spans some three miles, and over 2,000 feet deep. The rim of the canyon is over 10,000 feet above sealevel, and is forested with islands of Englemann spruce, subalpine fir and aspen: separated by broad meadows of brillant summertime wild flowers.
Denny Davies, Supt
Ateve Robinson, Chief Ranger

3880 Dixie National Forest

82 N 100 E   435-865-3700
Cedar City, UT 84720   Fax: 435-865-3791
The Dixie is located adjacent to three National Parks, Bryce Canyon, Zion and Capitol Reef. The red sandstone formations of Red Canyon rival those of Bryce Canyon National park. From the top of Powell Point, it is possible to see far into three different states. Boulder Mountain and the many different lakes provide opportunities for hiking, fishing and viewing outstanding scenery.
Mary Wagner, Forest Supervisor

3881 Fish Springs National Wildlife Refuge

PO Box 568   435-831-5353
Dugway, UT 84022   Fax: 435-831-5354

Located at the southern end of the Great Salt Lake Desert in western Utah, Fish Springs National Wildlife Refuge encompasses 17,992 acres between two small mountain ranges. Five major springs and several lesser springs and seeps flowfrom a faultline at the base of the eastern front of the Fish Springs Mountain Range, These warm, saline springs provide virtually all of the water for the Refuge's 10,000-acre marsh system.
Jerry Bana, Manager

3882 Fishlake National Forest
115 E 900 N   435-896-9233
Richfield, UT 84701   Fax: 435-896-9347
Fishlake National Forest is known for its beatiful aspen forest, scenic byways, motorized and non-motorized trails, elk hunting and mackinaw and trout fishing. recreational opportunities include scenic drives, mountain biking,snowmobiling, ATV use, hiking, and camping.
Mary Erickson, Forest Supervisor

3883 Manti-LaSai National Forest
599 W Price River Drive   435-637-2817
Price, UT 84501   Fax: 435-637-4940
Elaine Zieroth, Forest Supervisor

3884 Natural Bridges National Monument
HC 60 Box 1   435-259-5174
Lake Powell, UT 84533-0101   Fax: 435-692-1111
                                          E-mail: narino@nps.gov
Natural Bridges protects some of the finest examples of ancient stone architecture in the southwest. The monument is located in the southeast Utah on a pinyon-juniper covered mesa bisected by deep canyons of Permian age Ceder MesaSandstone. Where meandering streams cut through the cayon walls, three natural bridges formed: Kachina, Owachomo and Sipapu.
Steve Chaney

3885 Ouray National Wildlife Refuge
266 West 100 North #2   801-789-0351
Vernal, UT 84078   Fax: 801-789-4805
                                          http://www.gorp.com

Gary Montoya, Manager

3886 Timpanogos Cave National Monument
Rural Route 3 Box 200   801-756-5239
American Fork, UT 84003-9803 Fax: 801-756-5661
Timpanogos Cave Natioanl Monument sits high in the Wasatch Mountains. The cave system consists of three spectacularly decorated caverns. Each cavern has unique colors and formations. Helicitites and anthodites are just a few of themany dazzling formations to be found in the many chambers. As visitors climb to the cavern entrance, on a hike gaining over 1,000 feet in elevation, they are offered incredible views of American Fork Canyon.
Sue McGill, Supt

3887 Uinta National Forest
88 W 100 N   801-342-5100
Provo, UT 84601   Fax: 801-342-5144
The Uinta National Forest ranges from high western desert at Vernon to lofty mountain peaks such as Mount Nebo (elevation 11,877 feet, the highest peak in the Wasatch Range) and Mount Timpanogos (elevation 11,750 feet). The forestcontains three wilderness areas: the Lone Peak, the Mount Timpanogos and the Mount Nebo Wildernesses. The Forest surrounds the Timpanogos Cave National Mounment.
Peter Karp, Forest Supervisor

3888 Wasatch-Cache National Forest
8236 Federal Building   801-524-3900
125 S State Street   Fax: 801-524-3172
Salt Lake City, UT 84138
Wasatch-Cache National Forest lands are located in three major areas: the northern and western slopes of the Uinta Mountains. The Wasatch Front from Lone Peak north to the Idaho border including the Wasatch, Monte cristo, and BearRiver Ranges. The Stansbury Range, in the Great Basin.
Tom Tidwell, Forest Supervisor

3889 Zion National Park
Star Route 9   435-772-3256
Springdale, UT 84767   Fax: 435-772-3426

E-mail: zion_park_information@nps.gov
                http://www.ups.gov/zion
Protected within Zion National Park's 229 square miles (593.1 km) is a spectacular cliff-and-cayon landscape and wilderness full of the unexpected including the world's largest arch -Kolob Arch- with a span that measures 310 feet(94.5m). Wildlife such as mule deer, golden eagles, and mountain lions also inhabit the park. Mukuntuweap National Monument proclaimed July 31, 1909; incorporated in Zion National Monument March 18, 1918; established as national park Nov. 19, 1919.
Ron Terry, Public Information Officer

Vermont: US Forests, Parks, Refuges

3890 Green Mountains & Finger Lakes National Forest
231 North Main Street   802-747-0300
Rutland, VT 05701   Fax: 802-747-6766
    http://www.fs.fed.us/r9/gmfl/Green Mountain National Forest

3891 Missisquoi National Wildlife Refuge
29 Tabor Road   802-868-4781
Swanton, VT 05488   Fax: 802-868-2379
                              E-mail: missisquoi@fws.gov
                    http://http://missisquoi.fws.gov
The 6,592-acre refuge includes most of the Missisquoi River delta where it flows into Missisquoi Bay. The refuge consists of quiet waters and wetlands which attract large flocks of migratory birds.

Mark Sweeny, Manager

Virginia: US Forests, Parks, Refuges

3892 George Washington National Forest
5162 Valleypointe Parkway   540-265-5100
PO Box 233   Fax: 540-265-5145
Roanoke, VA 24019
Outstanding hiking trails, campsites, fishing and canoeing are the hallmarks of George Washington Forest in Virginia and West Virginia, part of the George Washington and Jefferson National Forest.

3893 Jefferson National Forest
5162 Valleypointe Parkway   540-265-5100
Roanoke, VA 24019   888-265-0019
The Jefferson National Forest is prize Appalachia country: tumbling waterfalls, rare wildflowers, vividly colored hills and Virginia's highest peak. Jefferson National Forest spreads 690,000 acres of hardwood and conifer forest acrosswest-central Virginia, West Virginia and Kentucky, including the ridge province of the Blue Ridge mountains.

3894 Mason Neck National Wildlife Refuge
14344 Jefferson Davis Highway   703-690-1297
Woodbridge, VA 22191
The refuge, the Mason Neck State Park, the Northern Virginia Park Authority, the Gunston Hall Plantation and the Virginia Department of Game and Inland Fisheries are cooperating in the management of their combined 5,000+ acres on theMason Neck peninsula. This cooperation provides a wide variety of recreational activities while protecting natural resources. The primary objective of the refuge is to protect essential nesting, feeding and roosting habitat for bald eagles.
J Frederick Milton, Manager

3895 Shenandoah National Park
3655 US Highway 211 East   540-999-2243
Luray, VA 22835-9036   Fax: 540-999-3500
Shenandoah National park lies astride a beautiful section of the blue rige mountains, which from the eastern rampart of the Appalachian Mountains between Pennsylvania and Georgia. The Shenandoah River flows through the valley to thewest, with Massanutten Mountain, 40 miles long, standing between the river's north and south forks. The rolling Piedmont country lies to the east of the park. Skyline Drive, a 105- mile road that winds along the crest of the mountainsthrough thelength of the park.
William Wade, Supt

Washington: US Forests, Parks, Refuges

3896 Colville National Forest
765 South Main Street   509-684-3711
Coleville, WA 99114   Fax: 509-684-7280
                               http://www.fed.us/cvnf
The Colville National Forest encompasses over one million acres in northeastern Washington. The Sherman Pass National Scenic Byway leads through a portion of the Forest, with camping, fishing, hiking, picnicking, mountain biking,cross-country skiing, sonowmobiling and other recreational activities. 49 Degrees North, a full service ski resort, is located east of Chewelah, about one hour north of Spokane. The Salmo-Prist Wilderness Area sits in the northeast corner of theforest.

3897 Conboy Lake National Wildlife Refuge
PO Box 5   509-364-3410
Glenwood, WA 98619-0005   Fax: 509-364-3667
                               E-mail: harold_cole@fws.gov
Located in the northwest corner of Klickitat County, Washington, the refuge was established primarily for waterfowl. The broad range of habitat diversity provides for a broad diversity of resident wildlife species.
Harold E Cole, Manager

3898 Gifford Pinchot National Forest
6926 E 4th Plain Boulevard   360-891-5000
PO Box 8944   Fax: 360-891-5045
Vancouver, WA 98668-8944   http://www.fs.fed.us/gpnf/
The Gifford Pinchot National Forst is one of the oldest National Forests in the United States. Include as part of The Mount Rainier Forest Reserve in 1897, this area was set aside as the columbia National Forest in 1908, and renamedthe Gifford Pinchot National Forest in 1949. The Forest, located in southwest Washington State, now contains 1,312,000 acreas and includes the 110,000- acre Mount St. Helens National Volcanic Mounument established by congress in 1982.

3899 Lewis & Clark National Wildlife Refuge
Julia Butler Hansen NWR   206-795-3915
PO Box 566   http://www.gorp.com
Cathlamet, WA 98612

3900 McNary National Wildlife Refuge
64 Maple Road   509-547-4942
PO Box 544
Burbank, WA 99723
A resting and feeding area for up to 100,000 migrating waterfowl. It includes 3,629 acres of water and marsh, croplands, grasslands, trees and shrubs.
David Linehan, Manager

3901 Mount Baker-Snoqualmie National Forests
21905 64th Avenue West   425-744-3200
Mountlake Terrace, WA 98043   800-627-0062

3902 Mount Rainier National Park
Tahoma Woods, Star Route   253-569-2211
Ashford, WA 98304-9751   Fax: 360-569-2170
                               E-mail: MORAInfo@nps.gov
Established in 1899. 235,625 acres (97% is designated Wilderness). Includes mount Rainier (14,410'), an active volcano encased in over 35 square miles of snow and ice. The park contains outstanding examples of old growth forests andsubalpine meadows. Designated a National Historic landmark Distract in 1997 as a showcase for the NPS Rustic style architecture of the 1920s and 1930s.
William Briggle, Supt

3903 North Cascades National Park Service Complex
810 State Route 20   360-856-5700
Sedro-Woolley, WA 98284-1239   Fax: 360-856-1934
                               E-mail: NOCA_Interpretation@nps.gov
                               http://www.nps.gov/noca

3904 Okanogan National Forest
1240 S 2nd Avenue   509-826-3068
PO Box 950
Okanogan, WA 98840
There is a variety of country from craggy peaks to rolling meadows, to rich old growth forest and classic groves of ponderosa pine. We're called the Sunny Okanogan and for good reason: summers here are hot and dry, and our winters arefamous for brilliant clear skies and plenty of snow.

3905 Okanogan and Wenatchee National Forests Headquarters
215 Melody Lane   509-664-9200
Wenatchee, WA 98801   Fax: 509-664-9280
                               http://www.fs.fed.us/r6/wenatchee/forest/formain.htm
The 2.2 million acre Wenatchee National Forest extends about 135 miles along the east side of the crest of the Cascade Mountains in Washington State. This National Forest is most noted for its wide range of recreation opportunities.There truly is "something for everyone" who likes to have fun in the outdoors.

3906 Olympic National Forest
1835 Black Lake Boulevard SW   360-956-2402
Olympia, WA 98512-5623   http://www.fs.fed.us/r6/olympic
The National Forests are part of America's great outdoors and are public lands. They are managed for the multiple uses of recreation, wildlife, timber, gazing, mining, oil and gas, watershed and wilderness. The Olympic National Forestis over 632,000 acres in size and is divided into two Ranger Districts: Hood Canal and Pacific.

3907 Olympic National Park
600 East Park Avenue   360-565-3000
Port Angeles, WA 98362   Fax: 360-565-3015
                               E-mail: olym_visitor_center@nps.gov
Often referred to as three parks in one, Olympic National Park encompasses three distinctly different ecosystems-rugged glacier capped mountains, over 60 miles of wild Pacific coast and magnificent are still largely pristine incharacter and are Olympic's gift to you.
Bill Laitner, Superintendent

3908 Ross Lake National Recreation Area North Cascades National Park
810 State Route 20   360-856-5700
Sedro Woolley, WA 98284-1239   Fax: 360-856-1934
                               E-mail: mailto:NOCA_Interp@nps.gov
                               http://www.nps.gov/rola
The most accessible part of the North Cascades National Park Service Complex. Is also the corridor for scenic Washington State Route 20, the North Cascades Highway, and includes three reservoirs.

3909 Toppenish National Wildlife Refuge
21 Pumphouse Road
Toppenish, WA 98948-9791   800-344-9453
                               E-mail: refuges@fws.gov
                               http://www.fws.gov/refuges/
An important migration and wintering area for waterfowl in the Yakima Valley of eastern Washington. Wetland impoundments along Toppenish Creek provide natural foods for wintering mallards and other ducks. Ducks and other water birdsbreed in the wetland impoundments during the summer. Native shrub-steppe communities and riparian areas along Toppenish and Snake creeks provide habitat for many other species of birds. The refuge has active hunting and wildlife-viewing programs.
George J Fenn, Manager

3910 Willapa National Wildlife Refuge
HC 01   206-484-3482
Box 910
Llwaco, WA 98624-9707
Located on Willapa Bay in Pacific County, the southernmost coastal county in Washington. The upland forest varies in successional stages from recently logged areas to a unique remnant of virgin, coastal cedar-hemlock forest home todeer, bear, elk, grouse, beaver and numerous songbirds and small mammals.
James A Hidy, Manager

West Virginia: US Forests, Parks, Refuges

3911 Monongahela National Forest
200 Sycamore Street   304-636-1800
Elkins, WV 26241   Fax: 304-636-1875

The Monongahela National Forest was established following passage of the 1911 Weeks Act. This act authorized the purchase of land for long-term watershed protection and natural resource management following massive cutting of theEastern forests in the late 1800's and at the turn of the century.

3912 Ohio River Islands National Wildlife Refuge
PO Box 1811   304-422-0752
Parkersburg, WV 26102-1811   Fax: 304-422-0754
E-mail: fw5rw_ohrinwr@fws.gov

The refuge extends 362 river miles from Shippingport, Pennsylvania to Manchester, Ohio along one of the nation's busiest waterways. Ohio River Islands and their back channels have long been recognized for high quality fish andwildlife, recreation, scientific and natural heritage values.
Jerry L Wilson, Manager

Wisconsin: US Forests, Parks, Refuges

3913 Apostle Islands National Lakeshore
415 Washington Street   715-779-3397
Route 1, Box 4   Fax: 715-779-3049
Bayfield, WI 54814   E-mail: APIS_Webmaster@nps.gov

The national lakeshore includes 21 islands and 12 miles of mainland Lake Superior shoreline, featuring pristine stretches of sand beach, spectacular sea caves, remnant old growth forests, resident bald eagles and black bears and thelargest collection of lighthouses anywhere in the National Park System.
Robert J Krumenaker, Superintendant

3914 Chequamegon National Forest
1170 4th Avenue S   715-762-2461
Park Falls, WI 54552   Fax: 715-762-5179

Shaped principally by glacial action some 10,000 years ago, the forest offers a variety of hiking, ATV, and cross-country ski trails at different levels of difficulty. These campgrounds are located on either a lake or a river and offerfishing and boating.

3915 Ice Age National Scientific Reserve
PO Box 7921   608-266-2183
Madison, WI 53707   Fax: 608-267-7474
E-mail: brigit.brown@dnr.state.wi.us

This first national scientific reserve contains nationally significant features of continental glaciation. State parks in the area are open to the public.
Tom Gilbert, Supt

3916 Nicolet National Forest
68 S Stevens Street   715-362-1300
Rhinelander, WI 54501   Fax: 715-362-1359

Located in Wisconsin's Northwoods where towering pine and hardwood forests are interspersed with hundreds of crystal clear lakes and streams, the Nicolet offers you many opportunities to enjoy the outdoors. Within a day's drive of theChicago, Milwaukee, St. Paul and Minneapolis metropolitan areas, the forest is a place where urban dwellers can truly get away from it all in the scenic beauty of the northwoods.

3917 St Croix National Scenic Riverway
401 N Hamilton Street   715-483-3284
St Croix Falls, WI 54024   Fax: 715-483-3288
http://www.nps.gov

The St. Croix National Scenic Riverway is home to the endangered Higgins Eye and Winged Mapleleaf mussels, bald eagles, gray wolves, and the prehistoric paddlefish. The 252 miles of Riverway provide numerous recreational opportunitiesfor boaters, canoeists, kayakers and others.

Tom Bradley, Superintendent
Ron Erickson, Education Team Manager

Wyoming: US Forests, Parks, Refuges

3918 Bighorn National Forest
2013 Eastside 2nd Street   307-674-2600
Sheridan, WY 82801   Fax: 307-674-2668

The forest has 32 campgrounds, 14 picnic areas, 2 visitor centers, 2 ski areas, 7 lodges, 2 recreation lakes, 3 scenic byways and over 1500 miles of trails. The Bighorn National Forest is 80 miles long and 30 miles wide. The mostcommon tree is the lodgepole pine. The Bighorn River, flowing along the west side of the forest was first named by American Indians due to the great herds of bighorn sheep at its mouth.

3919 Bridger-Teton National Forest
340 N Cache-Forest Service Bldg   307-739-5500
PO Box 1888   Fax: 307-739-5010
Jackson, WY 83001   E-mail: tbills@fs.fed.us
http://www.fs.fed.us/btnf/

With it's 3.4 million acres, it is the second largest National Forest outside Alaska. Included are more than 1.2 million acres of wilderness in the Bridger, Gros Ventre, and Teton Wildernesses. The Bridger-Teton is a land of variedrecreational opportunities, beautiful vistas, and abundant wildlife. Its crystal blue skies are puctuated by awesome mountain ranges which include the Gros Ventre, Teton, Salt River, Wind River, and Wyoming Mountain Ranges.

3920 Devils Tower National Monument
PO Box 10   307-467-5283
Devils Tower, WY 82714-0010   Fax: 307-467-5350
E-mail: deto_interpretation@nps.gov
http://www.nps.gov/deto/home.html

This unit of the National Park Service protects the nearly vertical monolith known as Devil's Tower. The rolling hills of this 1347 acre park are covered with pine forests, deciduous woodlands, and prairie grasslands. Known byseveral northern plains tribes as Bear Lodge, it is sacred to many American Indians. Devil's Tower was proclaimed in September, 1906 as the nation's first national monument by President Theodore Roosevelt.

Lisa Eckert, Superintendent

3921 Fossil Butte National Monument
PO Box 592   307-877-4455
Kemmerer, WY 83101-0592   Fax: 307-877-4457
E-mail: FOBU_Superintendent@nps.gov
http://www.nps.gov/fobu

Located in southwest Wyoming, Fossil Butte National Monument represents one of the richest fossil localities in the world. Fifty million-year-old fish, insects, birds, reptiles, and plants are nearly perfectly preserved in limestone.
David E McGinnis, Superintendent
Marcia Fagnant, Park Ranger

3922 Grand Teton National Park
PO Box 170   307-739-3300
Moose, WY 83012   Fax: 307-739-3438
E-mail: GRTE_info@nps.gov
http://www.nps.gov/grte

Established in 1929 and enlarged in 1950 to protect a rugged, awe-inspiring mountain range with numerous piedmont lakes nestled amoung its flanks, and a wide sagebrush-covered valley called Jackson Hole. Administered by the NationalPark Service under the Department of the Interior, Grand Teton is one of 384 units within the national park system. It encompasses approximately 310,000 acres or 485 square miles of northwestern Wyoming, just south of Yellowstone National Park.

Mary Gibson Scott, Supt

3923 Medicine Bow National Forest
2468 Jackson Street   307-745-2300
Laramie, WY 82070-6535   Fax: 307-745-2398
http://www.fs.fed.us/r2/mbr/about/districts/laramie.shtml/

The Medicine Bow National Forest dates back to May 22, 1902, with the establishment of the Medicine Bow Forest Reserve by President Theodore Roosevelt. In 1929, the former Hayden National Forest along the Continental Divide wasformerly a War Department target and maneuver reservation under joint administration by the Forest Service and the War Department. In 1959, the area formerly used by the military was added to the Medicine Bow National Forest.

3924 National Elk Refuge

PO Box 510   307-733-9212
Jackson, WY 83001   Fax: 307-733-9729
E-mail: don_delong@fws.gov
http://nationalelkrefuge.fws.gov/
More than 7,500 elk make the winter range of National Elk Refuge their home from October until May. Adjacent to the north side of Jackson, Wyoming, the 25,000-acre refuge includes nearly 1600 acres of open water and marsh lands, 47different mammals and nearly 175 species of birds.
Mike Hedrick, Manager

3925 Seedskadee National Wildlife Refuge

PO Box 700   307-875-2187
Green River, WY 82935-0700   Fax: 307-875-4425
E-mail: Seedskadee@fws.gov
http://http://seedskadee.fws.gov/
Fishery resource is managed cooperatively with the state G&F and includes a special regulations area to promote catch and release fishing for trophy trout (brown, Snake River cutthroat and rainbow trout). Refuge lands are rich inhistorical and cultural resources as the area was utilized by nomadic Indian tribes, fur trappers, early pioneers and travelers heading for the better life of California and Oregon. Many of the old campsites, river crossings and early structuresstill exist.
Francis Maiss, Manager

3926 Shoshone National Forest

808 Meadow Lane   307-527-6241
Cody, WY 82414-4516   Fax: 307-578-1212
The Shoshone consists of 2.4 million acres of varied terrain ranging from sagebrush flats to rugged mountain peaks and includes portions of the Absaroka, Wind River, and Beartooth Mountain Ranges. Elevations on the Shoshone range from4,600 feet at the mouth of the spectacular Clarks Fork Canyon to 13,804 feet on ganneett Peak, Wyoming's highest point. Geologists delightedly call the Shoshone's varied topography an open book.

3927 Yellowstone National Park

PO Box 168   307-344-7381
Yellowstone National Park, WY 82190   Fax: 307-344-2005
E-mail: yell_visitor_services@nps.gov
http://www.nps.gov
Established on March 1, 1872, Yellowstone National Park is the first and oldest national park in the world. Preserved within Yellowstone are Old Faithful Geyser and some 10,000 hot springs and geysers, the majority of the plant'stotal. These geothermal wonders are evidence of one of the world's largest active volcanoes; its last eruption created a crater or caldera that spans almost half of the park.
Suzanne Lewis, Supt

Directories & Handbooks: Air & Climate

3928 Acid Rain
Watts, Franklin
90 Sherman Turnpike
Danbury, CT 06816
203-797-3500
800-621-1115
Fax: 203-797-3657

Lists over 4,000 citations, with abstracts, to the worldwide literature on the sources of acid rain and its effects on the environment.

3929 Weather America
Grey House Publishing
185 Millerton Road
PO Box 860
Millerton, NY 12546
518-789-8700
800-562-2139
Fax: 518-789-0545
E-mail: books@greyhouse.com
http://www.greyhouse.com

Provides extensive climatological data for over 4,000 national and cooperative weather stations throughout the US. Includes a new major storms section and a nationwide ranking section that provides rankings for maximum and minimumtemperatures, precipitation, snowfall, fog, humidity and wind speed. Each of 50 state sections contains a city index for locating the nearest weather station to the city/county being researched and a narrative description of the state's climaticconditions.
2,013 pages
ISBN: 1-891482-29-7
Leslie Mackenzie, Publisher
David Garoogian, Editor

Directories & Handbooks: Business

3930 American Caves
American Caves Conservation Association
119 E Main Street
PO Box 409
Horse Cave, KY 42749
270-786-1466
Fax: 270-786-1467
E-mail: debraheavers@caven.org
http://www.cavern.org

A bi-annual membership publication. Published by the American Caves Conservation Association and available by subscription to nonmembers.
David Foster, Executive Director/Author
Debra Heavers, Editor

3931 Associations Canada Directory
20 Victoria Street
Toronto, Ontario M5C 2N8
416-362-5211
Fax: 416-362-6161
E-mail: info@micromedia.ca
http://www.micromedia.ca

Directory published by Micromedia.
Publication Date: 1991   1200 pages
ISBN: 1-895021-95-2
Patricia Petruga, Author
Peter Asselstine, Marketing Manager
Bryan Moore, Sales and Marketing Manager

3932 Business and the Environment: A Resource Guide
Island Press
1718 Connecticut Avenue NW
Suite 300
Washington, DC 20009-1148
202-232-7933
800-828-1302
Fax: 202-234-1328
E-mail: info@islandpress.org
http://www.islandpress.org

Includes more than 1,000 references to material from scholarly journals, government agencies, case clearing-houses, research organizations, trade magazines and the popular press. It was the most current (1992) listing of research onself-monitoring and compliance programs and environmental performance strategies for corporate competitiveness.
Publication Date: 1992   382 pages
ISBN: 1-559631-59-7
Barbara Dean, Executive Editor
Jonathan Cobb, Executive Editor

3933 California Certified Organic Farmers: Membership Directory
1115 Mission Street
Santa Cruz, CA 95060
831-423-2263
Fax: 831-423-4528
E-mail: ccof@ccof.org
http://www.ccof.org

Annual
Brian Leahy, Executive Director
Helge Hellberg, Marketing Director

3934 Directory of Environmental Information Sources
Government Institutes
4 Research Place
Suite 200A
Rockville, MD 20850-6209
301-907-1000
Fax: 301-921-2362

Over 1,400 federal and state government agencies, professional and scientific organizations and trade associations are profiled.
322 pages

3935 Directory of New York City Environmental Organizations
Council on the Environment of New York City
51 Chambers Street
Room 228
New York, NY 10007
212-788-7900
Fax: 212-788-7913
E-mail: conyc@cenyc.org
http://www.cenyc.org

Promotes environmental awareness and solutions to environmental problems.

3936 Directory of Professional Services
Professional Services Institute
1730 Rhode Island Avenue NW
Suite 1000
Washington, DC 20036
202-659-4613
800-424-2869
Fax: 202-775-5917

3937 Directory of Socially Responsible Investments
Funding Exchange
666 Broadway
Suite 500
New York, NY 10012 E-mail: mailto:%20FEXEXC@aol.com
212-529-5300
Fax: 212-982-9272
http://www.udc.edu/index-b.htm

Network of 15 community foundations around the country with a national office in New York City. Staff at the national office are responsible for three main program areas: grantmaking, donor programs and member fund saervices.

3938 EH&S Compliance Auditing & Teaching Software Report
Donley Technology
220 Garfield Avenue
PO Box 152
Colonial Beach, VA 22443-0152
804-224-9427
800-201-1595
Fax: 804-224-7958
E-mail: donleytech@donleytech.com
http://www.donleytech.com

Profiles 25 software packages for achieving and maintaining compliance, including detailed product descriptions, tables comparing system features, and contact information.
240 pages
ISBN: 1-891682-08-3
Elizabeth Donley, Editor

3939 EPA Information Resources Directory
National Technical Information Service
5285 Port Royal Road
Springfield, VA 22161
703-605-6000
http://www.ntis.gov

Supports the nation's economic growth and job creation by providing access to information that stimulates innovation and discovery. NTIS accomplishes this mission through two major programs: information collection and dissemination tothe public and production and other services to federal agencies.

3940 EnviroSafety Directory
IEI Publishing Division
1635 W Alabama St
Houston, TX 77006
713-529-1616
800-654-1480
Fax: 281-529-0936
E-mail: iei@mail.infohwy.com
http://www.oilonline.com

Approximately 6,000 environmental services, state agencies and EPA/Superfund sites within the EPA regions 4, 6 and 9.
James W Self, Editor
Janis Johnson, Managing Editor

3941 Environment: Books by Small Presses of the General Society of Mechanics & Tradesmen
Small Press Center
20 West 44th Street   212-764-7021
New York, NY 10036   Fax: 212-354-5365
E-mail: smallpress@aol.com
http://www.smallpress.org
Publication Date: 1992   250 pages
ISBN: 0-962276-93-6
Paula Matta, Author

3942 Environmental Address Book: The Environment's Greatest Champions and Worst Offenders
Perigee Books
200 Madison Avenue   212-951-8400
New York, NY 10016   http://www.penguinputnam.com

3943 Environmental Cost Estimating Software Report
Donley Technology
220 Garfield Ave   804-224-9427
PO Box 152   800-201-1595
Colonial Beach, VA 22443-0152   Fax: 804-224-7958
E-mail: donleytech@donleytech.com
http://www.donleytech.com
Profiles 20 software packages for estimating the cost of environmental projects, including detailed product descriptions, tables comparing system features, and contact information.
162 pages
ISBN: 1-891682-05-9
Elizabeth Donley, Managing Editor/Publisher
John Donley, Editor

3944 Fibre Market News: Paper Recycling Markets Directory
Recycling Media Group GIE Publishers
4012 Bridge Avenue   216-961-4130
Cleveland, OH 44113   800-456-0707
Fax: 216-961-0364
A list of over 2,000 dealers, brokers, packers and graders of paper stock in the US and Canada.
Dan Sandoval, Internet/Senior Editor
Jim Keefe, Group Publisher

3945 Greenpeace Guide to Anti-Environmental Organizations
Odonian Press
PO Box 776   510-486-0313
Berkeley, CA 94701-0776   800-326-0959
Fax: 415-512-8699
Corporations, foundations and public relations firms determined to be anti-environmental despite their attempts to project the green image.
Publication Date: 1993   112 pages
ISBN: 1-878825-05-3
Carl Deal, Author

3946 Guide to Curriculum Planning in Environmental Education
125 S Webster Street   608-266-2188
PO Box 7841   800-441-4563
Madison, WI 53707-7841   Fax: 608-267-9110
E-mail: sandi.mcnamer@dpi.state.wi.us
http://www.dpi.state.wi.us/pubsales
Provides a direction in planning a comprehensive environmental education program based on perceptual awareness knowledge, environmental ethics, citizen action skills and citizen action experience.
Publication Date: 1994   167 pages Book
Sandi McNamer, Publications Director

3947 Handbook on Air Filtration
IEST
5005 Newport Drive   847-255-1561
Suite 506   Fax: 847-255-1699
Rolling Meadows, IL 60008-3841   E-mail: iest@iest.org
http://www.iest.org
Covers a broad range of applications for users who require removal of airborn particulate contamination for maximum air cleanliness.
ISBN: 1-877862-60-6
Julie Kendrick, Executive Director

3948 Harbinger File
Harbinger Communications, Inc
5 N Union Street   847-622-0905
Elgin, IL 60123   800-320-7206
Fax: 847-622-0830
E-mail: info@harbingeronline.com
http://www.harbingeronline.com

3949 National Environmental Data Referral Service
US National Environmental Data Referral Service
1825 Connecticut Avenue NW   202-606-4089
Washington, DC 20235
More than 22,200 data resources that have available data on climatology and meteorology, ecology and pollution, geography, geophysics and geology, hydrology and limnology, oceanography and transmissions from remote sensing satellites.

3950 National Environmental Organizations
US Environmental Directories
PO Box 65156
St Paul, MN 55165

3951 New Jersey Environmental Directory
Youth Environmental Society
PO Box 441   609-655-8030
Cranbury, NJ 08512
Environmental education and leadership programs for high school students in New Jersey.

3952 Opportunities in Environmental Careers
VGM Career Books
4255 W Touphy Avenue   847-679-5500
Lincolnwood, IL 60646-1975   800-323-4900
Fax: 847-679-2494
Odom Fanning, Author

3953 Research Services Directory
Grey House Publishing
185 Millerton Road   518-789-8700
PO Box 860   800-562-2139
Millerton, NY 12546   Fax: 518-789-0545
E-mail: books@greyhouse.com
http://www.greyhouse.com
This Ninth Edition provides access to well over 8,000 corporate and independent commercial research firms and laboratories offering contract services for hands-on, basic or applied research in environmental and other areas. Providesthe company's name and addresses, as well as a company description and research and technical fields served.
1,200 pages
ISBN: 1-891482-30-0
Leslie Mackenzie, Publisher
Richard Gottlieb, Editor

3954 State Environmental Agencies on the Internet
Government Institutes
4 Research Place   301-907-1000
Suite 200A   Fax: 301-921-2362
Rockville, MD 20850   http://www.govinst.com
Provides a concise profile of each state agency's requirements and resources-including hard-to-find online laws, rules, and regulations-in one quicl-reference guide.

3955 Water Environment & Technology Buyer's Guide and Yearbook

Water Environment Federation
601 Wythe Street
Alexandria, VA 22314-1994

703-684-2400
800-666-0206
Fax: 703-684-2492
E-mail: webfeedback@wef.org
http://www.wef.org

Founded in 1928, the Water Environment Federation (WEF) is a not for profit technical and educational organization with members from varied disciplines who work toward the WEF vision of preservation and enhancement of the global waterenvironment. The WEF network includes water quality professionals from 76 Member Associations in 30 countries.
William J Bertera, Executive Director

3956 World Directory of Environmental Organizations On-line

California Institute of Public Affairs
PO Box 189040
Sacramento, CA 95814

916-442-2472
Fax: 916-442-2478
http://www.interenvironment.org

A guide to governmental and nongovernmental organizations and programs concerned with protecting the earth's resources. It also covers national and international organizations throughout the world. Only available online.

Directories & Handbooks: Design & Architecture

3957 Directory of International Periodicals and Newsletters on Built Environments

Van Nostrand Reinhold

More than 1,400 international periodicals and newsletters that cover architectural design and the building industry and the aspects of the environment that deal with the industry are covered.
Publication Date: 1992   175 pages
ISBN: 0-442230-03-6
Frances C Gretes, Author

3958 Land Improvement Contractors of America Membership Buyer's Guide

3080 Ogden Avenue
Lisle, IL 60532

630-548-1984
Fax: 630-548-9189
E-mail: nlica@aol.com
http://www.licanational.com

90 pages
Wayne March, Executive VP

Directories & Handbooks: Disaster Peparedness & Response

3959 Association of State Floodplain Managers

Association of State Floodplain Managers
2609 Fish Hatchery Road
Suite 204
Madison, WI 53713

608-255-8599
Fax: 608-274-0696
E-mail: asfpm@floods.org
http://www.floods.org

A complete name/address/phone listing for all key floodplain managers in the nation, comprehensive summary of ASFPM's activities of past year and planned future directions, key federal agency programs, much more. Free to currentmembers.
Diane Watson, Editor

3960 El Environmental Services Directory Online

Environmental Information Limited
PO Box 390266
Edina, MN 55439

952-831-2473
Fax: 952-831-6550
E-mail: ei@enviro-information.com
http://www.envirobiz.com

The most comprehensive, largest directory of environmental services in the United States. Coverage includes asbestos & lead abatement, consulting, laboratories, transportation, industrial cleaning, municipal solid waste facilities,hazardous waste facilities, indsutrial waste facilities, well drilling, soil boring, drum reconditioning, spill response, and remediation services.
Cary Perket

3961 EnviroSafety Directory

IEI Publishing Division
1635 W Alabama St
Houston, TX 77006-4101

713-529-1616
800-654-1480
Fax: 281-529-0936
E-mail: iei@mail.infohwy.com
http://www.oilonline.com

Approximately 6,000 environmental services, state agencies and EPA/Superfund sites within the EPA regions 4, 6 and 9.
James W Self, Editor
Janis Johnson, Managing Editor

3962 Floodplain Management: State & Local Programs

Association of State Floodplain Managers
2809 Fish Hatchery Road
Madison, WI 53713

608-274-0123
Fax: 608-274-0696
E-mail: asfpm@floods.org
http://www.floods.org

The most comprehensive source assembled to date, this report summarizes and analyzes various state and local programs and activities.
Publication Date: 2005

3963 Hazardous Materials Regulations Guide

JJ Keller
3003 West Breezewood Lane
Neenah, WI 54956

920-722-2848
877-564-2333
Fax: 800-727-7516
E-mail: sales@jjkeller.com
http://www.jjkeller.com

A complete reference guide of hazardous materials regulations.
May/Novemeber
ISBN: 0-934674-94-9
Tom Ziebell, Editor

3964 Institute of Chemical Waste Management Directory of Hazardous Waste Treatment

National Solid Wastes Management Assn.
1730 Rhode Island Avenue NW
Suite 1000
Washington, DC 20036

202-659-4613
800-424-2869
Fax: 202-775-5917
http://www.nswma.org

3965 Pesticide Directory: A Guide to Producers and Products, Regulators, and Researchers

Thomson Publications
Box 9335
Fresno, CA 93791

559-266-2964
Fax: 559-266-0189
http://www.agbook.com

This directory is for the person who needs to know anything about the US pesticide industry. It includes basic manufacturers and formulators along with their products, key personnel, managers, district/regional offices and otherpertinent information. Other sections include Universities, State Extension Centers, USDA, EPA, National Organizations, US Forest Service, Poison Control Centers and much more.
Publication Date: 1987   153 pages   Biannual
ISBN: 0-913702-45-5
WT Thomson, Author
Susan Heflin, President/Owner

3966 SEEK

525 S Lake Avenue
Suite 400
Duluth, MN 55802

218-529-6258
888-668-3224
E-mail: seek@moea.state.mn.us
http://www.seek.state.mn.us/comment.cfm

Minnesotas's interactive directory of environmetal education resources.

3967 Tracking Toxic Wastes in CA: A Guide to Federal and State Government Information Sources
INFORM
120 Wall Street   212-361-2400
14th Floor   Fax: 212-361-2412
New York, NY 10005 http://www.informinc.org/INFORM.html

Directories & Handbooks: Energy & Transportation

3968 Alternative Energy Network Online
Environmental Information Networks
119 South Fairfax Street   703-683-0774
Alexandria, VA 22314-3301   Fax: 703-683-3893
E-mail: sales@eintoday.com
http://www.eintoday.com
Reports on news of all energy sources designed as alternatives to conventional fossil fuels, including wind, solar and alcohol fuels.

3969 Current Alternative Energy Research and Development in Illinois
Department of Energy & Natural Resources
325 W Adams   217-785-2800
Room 300   800-252-8955
Springfield, IL 62704-1892   Fax: 217-785-2618

3970 Department of Energy Annual Procurement and Financial Assistance Report
US Department of Energy
Mail Stop 142
Washington, DC 20585-0001   800-342-5303
Fax: 202-586-4403
Offers a list of universities, research centers and laboratories that represent the Department of Energy.

3971 Directory of Solar-Terrestrial Physics Monitoring Stations
Air Force Geophysics Laboratory
Department of Defense   781-377-3977
Hanscom Air Force Base, MA 01731   Fax: 781-377-4498

3972 Energy Science and Technology
US Department of Energy
175 Oak Ridge Turnpike   865-576-1188
PO Box 62   Fax: 865-576-2865
Oak Ridge, TN 37831   E-mail: ISTIWebmaster@osti.gov
http://www.osti.gov/resource.html
To collect, preserve, disseminate, and leverage the scientific and technical information (STI) resources of the Department of Energy to provide access to national and global STI for use by DOE, the scientific research community, academia, US industry, and the public to expand the knowledge base of science and technology.

3973 Energy Statistics Spreadsheets
Institute of Gas Technology
3424 S State St   312-842-4100
Chicago, IL 60616-3834   Fax: 773-567-5209
The coverage of this database encompasses worldwide energy industry statistics, including production, consumption, reserves, imports and prices.

3974 Interstate Oil Compact Commission and State Oil and Gas Agencies Directory
Interstate Oil & Gas Compact Commission
Box 53127   405-525-3556
Oklahoma City, OK 73152-3127   800-822-4015
Fax: 405-525-3592
E-mail: iogcc@iogcc.state.ok.us
http://www.iogcc.oklaosf.state.ok.us
A directory of members and in the back is a list of state oil and gas agencies

3975 Women's Council on Energy and the Environment Membership Directory
PO Box 33211   703-351-7850
Washington, DC 20033   Fax: 202-318-2506
Offers valuable information on over 300 members representing consulting firms, private industry and the environmental community.
4 pages

Directories & Handbooks: Environmental Engineering

3976 Association of Conservation Engineers: Membership Directory
Engineering Section Alabama Dept. of Conservation
573-522-2323
Fax: 573-522-2324
E-mail: mihalg@mail.conservation.state.mo.us
http://www.conservation.state.mo.us/engineering/ace

3977 Energy Engineering: Directory of Software for Energy Managers and Engineers
Fairmont Press
700 Indian Trail   770-925-9388
Liburn, GA 30347   Fax: 770-381-9865
E-mail: linda@fairmontpress.com
http://www.fairmontpress.com
Directory of services and supplies to the industry.
Publication Date: 1904   80 pages   Bimonthly
ISSN: 0199-8895
Wayne C Turner, Author
Wayne C Turner, Editor

3978 NEPA Lessons Learned
Office of NEPA Policy & Compliance
1000 Independence Avenue SW   202-586-4600
EH-42   800-472-2756
Washington, DC 20585-0119   Fax: 202-586-7031
E-mail: denise.freeman@eh.doe.gov
http://www.eh.doe.gov/nepa
Publication Date: 1994   Quarterly
Carol M Borgstrom, Director

3979 New York State Conservationist
NYS Department of Environmental Conservation
625 Broadway   518-402-8047
2nd Floor   Fax: 518-402-9036
Albany, NY 12233-4502 E-mail: dinnelson@gw.dec.state.ny.us
http://www.dec.state.ny.us/
An informative and entertaining full-color bi-monthly magazine featuring New York State's natural resources and peoples' enjoyment of those resources.
Bi-Monthly
ISSN: 0010-650X
David Nelson, Editor
Alex Hyatt, Assistant Editor

3980 Pollution Abstracts
Cambridge Scientific Abstracts
7200 Wisconsin Avenue   301-961-6700
Suite 601   800-843-7751
Bethesda, MD 20814-4823   Fax: 301-961-6720
E-mail: sales@csa.com
http://www.csa.com
This database provides fast access to the environmental information necessary to resolve day to day problems, ensure ongoing compliance, and handle emergency situations more effectively.
James P McGinty, President
Ted Caris, Publisher

Directories & Handbooks: Environmental Health

3981 American Academy of Environmental Medicine Directory

American Academy of Environmental Medicine
7701 E Kellogg
316-684-5500
Suite 625
Fax: 316-684-5709
Wichita, KS 67207
E-mail: administrator@aaem.com
http://www.aaem.com
To suppoprt physicians and other professionals in serving the publi through education about the interaction between humans and their environment. Also to promote optimal health through prevention, and safe and effective treatment ofthe causes, not the illness.
Dee Rogers, Contact

3982 Carcinogenicity Information Database of Environmental Substances

Technical Database Services
10 Columbus Cir
212-246-3629
New York, NY 10019-1203
Fax: 212-247-0587
This database contains test results on the carcinogenic and mutagenic effects of approximately 1000 substances of environmental or health concerns.

3983 Directory of NEHA Credentialed Professionals

720 S Colorado Boulevard
303-691-9490
Suite 970 S Tower
Fax: 303-691-9490
Denver, CO 80246-1925
E-mail: staff@neha.org
http://www.neha.org
This is a directory of all NEHA credentialed professionals. It is available to NEHA credentialed professionals only.
Catalog 569

3984 Ecosystem Change and Public Health: A Global Perspective

Johns Hopkins University Press
2715 N Charles Street
410-516-6900
Baltimore, MD 21218-4363
800-537-5487
Fax: 410-516-6968
http://www.press.jhu.edu/books
The strength of the John Hopkins University Press' publications in medicine is in part a reflection of the university and medical institution's excellence and long term tradition of exceptional research and clinical care. Joan Aron'sEcosystem Change and Public Health is the first textbook devoted to this emerging field. The book covers such topics as global climate change, stratospheric ozone depletion, water resources management, ecology and infectious disease.
Publication Date: 2001   526 pages
ISBN: 0-801865-82-4
Joan Aron, Author
Joan Aron, Editor
Jonathan Pratz, Editor

3985 Environmental Encyclopedia

Thomson Gale
27500 Drake Road
248-699-4253
Farmington Hills, MI 48331
800-877-4253
Fax: 800-414-5043
http://www.galegroup.com
Consisting of nearly 1,300 signed articles and term definitions. The encyclopedia provides in-depth, worldwide coverage of environmental issues. Each article written in a nontechnical style and provides current status, analysis andsuggests solutions whenever possible.
Publication Date: 2002   2000 pages
ISBN: 0-787654-86-8
Virginia Regish, Contact

3986 Environmental Guide to the Internet

Government Institutes
4 Research Place
301-907-9100
Suite 200A
Fax: 301-921-2362
Rockville, MD 20850
http://www.govinst.com

Provides information for the best sites in the internet dealing with the preservation and protection of the environment, ecology, and conservation and offers over 320 new listings and addresses. Writin for environmental consultants,industry professionals, researchers, lawyers, educators, and students, contains the top 1,200 environmental internet resources, including, newletters and journals, and world wide web sites
Publication Date: 1997   384 pages
ISBN: 0-865875-78-2
Carol Briggs-Erickson and Toni Murphy, Author

3987 Environmental Guidebook: A Selective Guide to Environmental Organizations and Related Entities

Environmental Frontlines
PO Box 43
650-323-8452
Menlo Park, CA 94026
E-mail: info@envirofront.org
http://www.envirofront.org
Designed to serve as an essential reference book profiling nearly 500 national organizations and other entities actively engaged in environmental issues in the US and beyond.
312 pages
ISBN: 0-972068-50-3
Jeff Staudinger, Author

3988 Environmental Key Contacts and Information Sources

Government Institutes
4 Research Place
301-907-9100
Suite 200A
Fax: 301-921-2362
Rockville, MD 20850
http://www.govinst.com
An updated and revised compilation of Government Institutes' two previous directories, this reference contains more than 400 pages of contact information for more than 2,700 federal, state, and local environmental agencies andorganizations. This directory also includes contacts for information concerning environmental protection, hazardous waste materials, clean water and air, environmental assessment and management, pesticides, pollution control, recycling, naturalresources and conservation.
Publication Date: 1998   424 pages
ISBN: 0-865876-39-8
Charlene Ikonomou and Diane Pacchione, Author

3989 Pesticide Directory: A Guide to Producers and Products, Regulators, and Researchers

Thomson Publications
Box 9335
559-266-2964
Fresno, CA 93791
Fax: 559-266-0189
http://www.agbook.com
This directory is for the person who needs to know anything about the US pesticide industry. It includes basic manufacturers and formulators along with their products, key personnel, managers, district/regional offices and otherpertinent information. Other sections include Universities, State Extension Centers, USDA, EPA, National Organizations, US Forest Service, Poison Control Centers and much more.
Publication Date: 1987   153 pages Biannual
ISBN: 0-913702-45-5
WT Thomson, Author
Susan Heflin, President/Owner

Directories & Handbooks: Habitat Preservation & Land Use

3990 Alliance for Wildlife Rehabilitation and Education

Wildlife Care Directory
1912 Harbor Boulevard
949-722-0606
Costa Mesa, CA 92627

3991 Biodiversity Action Network

1630 Connecticut Avenue
202-547-8902
3rd Floor
Fax: 202-265-0222
Washington, DC 20009
http://www.bionet-us.org
An information exchange network launched by the Center for International Environmental Law.

3992 Conservation Directory 2004: The Guide to Worldwide Environmental Organizations
National Wildlife Federation
11100 Wild Life Center Drive   703-438-6000
Reston, VA 20190-5362   800-822-9919
http://www.nwf.org
Comprehensive listing of conservation and environmental organizations, with information on nearly 4,000 government agencies, nongovernmental organizations, and colleges and universities, as more than 18,000 officials concernedwith environmental conservation, education, and natural resource use management.
768 pages
ISBN: 1-559634-15-4
Robin Assa, Sales Assistant

3993 County Conservation Board Outdoor Adventure Guide
Iowa Association of County Conservation Boards
405 SW 3rd Street   515-963-9582
Suite 1   Fax: 515-963-9582
Ankeny, IA 50021   E-mail: iaccb@ecity.net
http://http://george.ecity.net/iaccb/guide.htm
The 2002 issue has information on 1,614 areas covering approximately 159,899 acres managed by County Conservation Boards. This guide also includes a map of each county with the area to be shaded in or a pinpoint of the location, andhas information on cabin rentals, camping, shelters, playgrounds, swimming, fishing, boating, boat rental, sports and fields, hunting, nature centers, praires, historic sites, wildlife exhibits and more.
184 pages
Don Brazelton, Contact

3994 DOCKET
US Environmental Protection Agency
US EPA Region 3   215-814-2993
1650 Arch Street (3PM52)   Fax: 215-814-5102
Philadelphia, PA 19103   E-mail: teller.lawrence@epa.gov
http://www.epa.gov
This database offers the complete text of summaries of all justice cases filed by the US Department of Justice on behalf of the US Environmental Protection Agency.

3995 Directory of Resource Recovery Projects and Services
Institute of Resource Recovery
1730 Rhode Island Avenue NW   202-659-4613
Suite 1000   Fax: 202-775-5917
Washington, DC 20036

3996 Ecology Abstracts
Cambridge Scientific Abstracts
7200 Wisconsin Avenue   301-961-6700
Suite 601   800-843-7751
Bethesda, MD 20814-4823   Fax: 301-961-6720
E-mail: sales@csa.com
http://www.csa.com
This large database updated continuously, offers over 150,000 citations, with abstracts, to the worldwide literature available on ecology and the environment.
James P McGinty, President
Theodore Caris, Publisher

3997 Environmental Bibliography
International Academy at Santa Barbara
5385 Hollister Avenue   805-683-8889
#210   Fax: 805-965-6071
Santa Barbara, CA 93111   E-mail: info@iasb.org
http://www.iasb.org
Over 615,000 citations are offered in this database, aimed at scientific, technical and popular periodical literature dealing with the environment.

ISSN: 1053-1440

3998 Environmental Concerns: Wetland Planning Guide for the Northeastern United States
Business Research Division-Univ. of Colorado
PO Box 420   303-492-8227
Boulder, CO 80309-0420   Fax: 303-492-3620
Approximately 1,300 private businesses, government organizations and corporations in Colorado that contribute to environmental protection and rehabilitation.

Publication Date: 1993
ISBN: 1-883226-02-3
Gin Hayden, Editor
Sean Shepherd, Editor

3999 Environmental Encyclopedia
Thomson Gale
27500 Drake Road   248-699-4253
Farmington Hills, MI 48331   800-877-4253
Fax: 800-414-5043
http://www.galegroup.com
Consisting of nearly 1,300 signed articles and term definitions. The encyclopedia provides in depth, worldwide coverage of environmental issues. Each article written in a nontechnical style and provides current status, analysis andsuggests solutions whenever possible.
Publication Date: 2002   2000 pages
ISBN: 0-787654-86-8
Virgina Regish, Contact

4000 Environmental Guide to the Internet
Government Institutes
4 Research Place   301-907-9100
Suite 200A   Fax: 301-921-2362
Rockville, MD 20850   http://www.govinst.com
Provides information for the best sites in the internet dealing with the preservation and protection of the environment, ecology, and conservation and offers over 320 new listings and addresses. Writin for environmental consultants,industry professionals, researchers, lawyers, educators, and students, contains the top 1,200 environmental internet resources, including, newletters and journals, and world wide web sites
Publication Date: 1997   384 pages
ISBN: 0-865875-78-2
Carol Briggs-Erickson and Toni Murphy, Author

4001 Environmental Guidebook: A Selective Guide to Environmental Organizations and Related Entities
Environmental Frontlines
PO Box 43   650-323-8452
Menlo Park, CA 94026   E-mail: info@envirofront.org
http://www.envirofront.org
Designed to serve as an essential reference book profiling nearly 500 national organizations and other entities actively engaged in environmental issues in the US and beyond.
312 pages
ISBN: 0-972068-50-3
Jeff Staudinger, Author

4002 Environmental Key Contacts and Information Sources
Government Institutes
4 Research Place   301-907-9100
Suite 200A   Fax: 301-921-2362
Rockville, MD 20850   http://www.govinst.com
An updated and revised compilation of Government Institutes' two previous directories, this reference contains more than 400 pages of contact information for more than 2,700 federal, state and local environmental agencies andorganizations. Also includes contacts for information concerning environmental protection, hazardous waste materials, clean water and air, environmental assessment and management, pesticides, pollution control, recycling, natural resources andconservation.
Publication Date: 1998   424 pages
ISBN: 0-865876-39-8
Charlene Ikonomou and Diane Pacchione, Author

4003 Helping Out in the Outdoors: A Directory of Volunteer Opportunities on Public Lands
American Hiking Society
1422 Fenwick Lane   301-565-6704
Silver Spring, MD 20910   Fax: 301-565-6714
E-mail: info@americanhiking.org
http://www.americanhiking.org
Mary Margaret Sloan, President

4004 Hospitality Directory
Human Ecology Action League (HEAL)
PO Box 29629   404-248-1898
Atlanta, GA 30359-0629   Fax: 404-248-0162

E-mail: HEALNatnl@aol.com
http://members.aol.com/HEALNatnl/index.html
Nonprofit organization founded in 1977 to serve those whose health has been adversely affected by environment exposures; to provide information to those who are concerned about the health effects of chemicals; and to alert the generalpublic about the potential dangers of chemicals.
Katherine P Collier, Contact

4005 Human Ecology Action League Directory
Human Ecology Action League
PO Box 29629   404-248-1898
Atlanta, GA 30359-0629   Fax: 404-248-0162
E-mail: HEALNatnl@aol.com
http://members.aol.com/HEALNatnl/index.html
The Human Ecology Action League Inc (HEAL) is a nonprofit organization founded in 1977 to serve those whose health has been affected by environmental exposures; to provide information to those who are concerned about the health effectsof chemicals; and to alert the general public about the potential dangers of chemicals. Referrals to local HEAL chapters and other support groups are available from the League.
Katherine P Collier, Contact

4006 Hummingbird Connection
249 E Main Street Suite 9
PO Box 394   302-369-3699
Newark, DE 19715   800-529-3699
Fax: 302-369-1816
E-mail: info@hummingbird.org
http://www.hummingbirdsociety.org
Published by the Hummingbird Society.
Publication Date: 1992   16 pages   Quarterly
ISSN: 1097-3427
H Ross Hawkins, Author/Editor

4007 International Society of Tropical Foresters: Membership Directory
5400 Grosvenor Lane   301-897-8720
Bethesda, MD 20814   Fax: 301-897-3690
E-mail: istfi@igc.apc.org
The International Society of Tropical Foresters, Inc. (ISTF) is a nonprofit organization committed to the protection, wise management and rational use of the world's tropical forests. Established in 1950, ISTF was reactivated in 1979.It has about 1500 members in more than 100 countries. Financial support comes from membership dues, donations and grants. ISTF sponsors meetings, promotes chapters in other countries, maintains a web site and has chapters at universities.
Warren T Doolittle, President

4008 Journal of Wildlife Rehabilitation
International Wildlife Rehabilitation Council
PO Box 8187   408-271-2685
San Jose, CA 95155   Fax: 408-271-9285
E-mail: office@iwrc-online.org
http://www.iwrc-online.org
A peer reviewed scientific journal that has served as a primary reference for wildlife rehabilitators and others involved in the care and conservation of wildlife. Features articles, columns and reviews, with topics ranging from allaspects of wildlife care to administration, fundraising, education programs, case studies, environmental issues, legalities, ethics and more. And is also a benefit of membership to IWRC.
Publication Date: 1978   40 pages   Quarterly
Jennifer Gursu, Executive Director

4009 LEXIS Environmental Law Library
Lexis Nexis Group
PO Box 933   937-865-6800
Dayton, OH 45401-0933   800-227-9597
Fax: 937-865-6909
http://www.lexis-nexis.com
This database contains decisions related to environmental law from the Supreme Court and other legislative bodies.

4010 Managed Area Basic Record
The Nature Conservancy
490 Westfield Road   804-295-6106
Charlottesville, VA 22901   Fax: 804-979-0370
E-mail: cmullen@tnc.org
http://http://nature.org

4011 Minienvironments
IEST
5005 Newport Drive   847-255-1561
Suite 506   Fax: 847-255-1699
Rolling Meadows, IL 60008-3841   E-mail: iest@iest.org
http://www.iest.org
The purpose of this document is to provide a framework for describing minienvironments for microelectronics and similar applications.
Publication Date: 2002   28 pages
ISBN: 1-877862-83-5
Julie Kendrick, Executive Director

4012 Morrison Environmental Directory
PO Box 2312   316-262-0100
Wichita, KS 67201

ISSN: 1060-488

4013 National Directory of Conservation Land Trusts
Land Trust Alliance
1331 H Street NW   202-638-4725
Suite 400   Fax: 202-638-4730
Washington, DC 20005-4734   E-mail: lta@lta.org
http://www.lta.org
More than 1,200 nonprofit land conservation organizations at the local and regional levels are profiled.
210 pages

4014 New York State Department of Environmental Conservation Personnel Directory: Internet Only
NYS Department of Environmental Conservation
625 Broadway   518-402-8013
2nd Floor   Fax: 518-402-9036
Albany, NY 12233   E-mail: dinnelson@gw.dec.state.ny.us
http://www.dec.state.ny.us
Internet only, this directory includes DEC's executive management and division directors. Executive managers are appointed by the Governor to carry out the policies of the state. Division directors have direct management responsibilityfor the department's programs.
Mary A Kadlecek, Chief Internet Publications

4015 Nonprofit Sample and Core Repositories Open to the Public in the United States
Branch of Sedimentary Processes
MS 939 Federal Center   303-236-5760
Denver, CO 80225   Fax: 303-236-0459
Walter E Dean, Contact

4016 Range and Land Management Handbook
Wyoming Association of Conservation Districts
517 E 19th Street   307-632-5716
Cheyenne, WY 82001   Fax: 307-638-4099
http://www.conservewy.com
This publication is intended for people from all walks of life who want to gain an appreciation of rangelands. This publication is also an introduction to the various fields of range management.
Annual

4017 Takings Litigation Handbook: Defending Takings Challenges to Land Use Regulations
American Legal Publishing Corporation
432 Walnut Street
Suite 1200   800-445-5588
Cincinnati, OH 45202   Fax: 513-763-3562
E-mail: customerservice@amlegal.com
http://www.amlegal.com
No government attorney, land use planner or other local official can effectively protect their community from harmful land use without a working knowledge of takings law. Developers and other landowners increasingly are attempting touse takings litigation, or the mere threat of takings litigation, to convince government agencies to relax or abandon vital protections for our neighborhoods and natural environment.
Publication Date: 2000   404 pages
Kendall, Dowling and Schwartz, Author
Douglas Kendall, Executive Director

4018 Trout Unlimited Chapter and Council Handbook

Trout Unlimited
1300 North 17th Street   703-522-0200
Suite 500   Fax: 703-284-9400
Arlington, VA 22209   E-mail: trout@tu.org
                               http://www.tu.org

Charles Gauvin, President/CEO
Kenneth Mendez, Executive VP/COO

4019 Turtle Help Network

New York Turtle and Tortoise Society
PO Box 878
Orange, NJ 07051-0878   212-459-4803

4020 Wisconsin Department of Public Instruction

125 S Webster Street   608-266-2188
PO Box 7841   800-441-4563
Madison, WI 53707-7841   Fax: 608-267-9110
               E-mail: sandi.mcnamer@dpi.state.wi.us
                       http://www.dpi.state.wi.us/pubsales
State education department publisher of K-12 curriculum plan-
ning guides in 25 subject areas including environmental educa-
tion and science.
Sandi McNamer, Publications Director

Directories & Handbooks: Recycling & Pollution Prevention

4021 A Glossary of Terms and Definitions Relating to Contamination Control

IEST
5005 Newport Drive   847-255-1561
Suite 506   Fax: 847-255-1699
Rolling Meadows, IL 60008-3841 E-mail: iest@iest.org
                               http://www.iest.org
A publication from the Institute of Environmental Science and
Technology.
        Publication Date: 1995   32 pages
        ISBN: 1-877862-28-2
Julie Kendrick, Executive Director

4022 American Recycling Market Directory: Reference Manual

Recycling Data Management Corp.
PO Box 577   315-471-0707
Ogdensburg, NY 13669   800-267-0707
                               Fax: 613-471-3258
Comprehensive directory/reference manual to materials recycling markets. Helps individuals locate buyers and sellers of recyclable materials on a regional basis throughout North
America. Contains 20,000 cross-referenced company
andagency listings. Sections include: scrap metals, waste paper,
paper mills, auto dismantlers, demolition, glass, oil, rubber and
textiles recyclers, recycling centers, MRF's composting, equipment and consulting services, industry references UBC
specsand more.

4023 Analysis of the Stockholm Convention on Persistent Organic Pollutants

Oceana Publications, Inc
75 Main Street   914-693-8100
Dobbs Ferry, NY 10522   800-831-0758
                               Fax: 914-693-0402
                       E-mail: info@oceanalaw.com
                       http://www.oceanalaw.com
This book analyzes the Stockholm Convention on Persistent Organic Pollutants. Prepared under the auspices of the UN Environment Programme Chemical Division.
        Publication Date: 2003   200 pages One Time
        ISBN: 0-379215-06-3
Mario Antonio Olsen, Author

4024 Criteria Pollutant Point Source Directory

North American Water Office
PO Box 174   651-770-3861
Lake Elmo, MN 55042   Fax: 651-770-3976

                               E-mail: gwillc@mtn.org
                               http://www.mtn.org

4025 Directory of Key Recycling Contacts

Resource Recycling
PO Box 42270   503-233-1305
Portland, OR 97242   Fax: 503-233-1356
               E-mail: info@resource-recycling.com
                       http://www.resource-recycling.com
Provides nearly 1,000 updated listings and email addresses and
web sites. This is the best directory for recycling, composting,
and solid waste professionals.
        64 pages
        ISSN: 0744-4710
Andrew Santosusso, Managing Editor
Mary Lynch, Circulation

4026 Directory of Municipal Solid Waste Management Facilities

The Institute of Solid Waste Disposal
1730 Rhode Island Avenue NW   202-659-4613
Suite 1000   Fax: 202-296-7915
Washington, DC 20036

4027 EI Environmental Services Directory

Environmental Information Networks
7301 Ohms Lane   952-831-2473
Suite 460   Fax: 952-831-6550
Eding, MN 55439   E-mail: ei@mr.net
                               http://www.envirobiz.com
Waste-handling facilities, transportation and spill response
firms, laboratories and the broad scope of environmental services. Online versions are also available.

        ISSN: 1053-475N
Cary Perket

4028 Environmental Encyclopedia

Thomson Gale
27500 Drake Road   248-699-4253
Farmington Hills, MI 48331   800-877-4253
                               Fax: 800-414-5043
                               http://www.galegroup.com
Consisting of nearly 1,300 signed articles and definitions. The
encyclopedia provides in depth, worldwide coverage of environmental issues. Each article written in a non technical style
and provides current status, analysis andsuggests solutions
whenever possible.
        Publication Date: 2002   2000 pages
        ISBN: 0-787654-86-8
Virginia Regish, Contact

4029 Environmental Guide to the Internet

Government Institutes
4 Research Place   301-907-9100
Suite 200A   Fax: 301-921-2362
Rockville, MD 20850   http://www.govinst.com
Provides information for the best sites in the internet dealing
with the preservation and protection of the environment, ecology, and conservation and offers over 320 new listings and addresses. Writin for environmental consultants,industry
professionals, researchers, lawyers, educators, and students,
contains the top 1,200 environmental internet resources, including, newletters and journals, and world wide web sites
        Publication Date: 1997   384 pages
        ISBN: 0-865875-78-2
Carol Briggs-Erickson and Toni Murphy, Author

4030 Environmental Guidebook: A Selective Guide to Environmental Organizations and Related Entities

Environmental Frontlines
PO Box 43   650-323-8452
Menlo Park, CA 94026   E-mail: info@envirofront.org
                               http://www.envirofront.org
Designed to serve as an essential reference book profiling nearly
500 national organizations and other entities actively engaged
in environmental issues in the US and beyond.

312 pages
ISBN: 0-972068-50-3
Jeff Staudinger, Author

4031 Environmental Key Contacts and Information Sources
Government Institutes
4 Research Place   301-907-9100
Suite 200A   Fax: 301-921-2362
Rockville, MD 20850   http://www.govinst.com

An updated and revised compilation of Government Institutes' two previous directories, this reference contains more than 400 pages of contact information for more than 2,700 federal, state and local environmental agencies andorganizations. Also includes contacts for information concerning environmental protection, hazardous waste materials, clean water and air, environmental assessment and management, pesticides, pollution control, recycling, natural resources andconservation.
Publication Date: 1998   424 pages
ISBN: 0-865876-39-8
Charlene Ikonomou and Diane Pacchione, Author

4032 Fibre Market News: Paper Recycling Markets Directory
Recycling Media Group GIE Publishers
4012 Bridge Avenue   216-961-4130
Cleveland, OH 44113   800-456-0707
                                Fax: 216-961-0364

A list of over 2,000 dealers, brokers, packers and graders of paper stock in the US and Canada.
Dan Sandoval, Editor

4033 Hazardous Materials Regulations Guide
JJ Keller
3003 West Breezewood Lane   920-722-2848
Neenah, WI 54956   877-564-2333
                                 Fax: 800-727-7516
                                 E-mail: sales@jjkeller.com
                                 http://www.jjkeller.com
A complete reference guide of hazardous materials regulations.
May/November
ISBN: 0-934674-94-9
Tom Ziebell, Editor

4034 How-To: 1,400 Best Books on Doing Almost Everything
R.R. Bowker Company
121 Chanlon Road   908-464-6800
New Providence, NJ 07974   888-269-5372
                                 E-mail: info@bowker.com
                                 http://www.bowker.com/bowkerweb/

4035 Institute of Chemical Waste Management Directory of Hazardous Waste Treatment and Dispos.
National Solid Wastes Management Assn.
1730 Rhode Island Avenue NW   202-659-4613
Suite 1000   800-424-2869
Washington, DC 20036   Fax: 202-775-5917
                                 http://www.nswma.org

4036 International Handbook of Pollution Control
Greenwood Publishing Group
88 Post Road W   203-226-3571
Westport, CT 06881   E-mail: webmaster@greenwood.com
                                 http://www.greenwood.com
Publication Date: 1989   482 pages
ISBN: 0-313240-17-5
Edward J Kormondy, Author

4037 KIND News
National Assn for Humane & Environmental Education
67 Norwich Essex Turnpike   860-434-8666
East Haddam, CT 06423-1736   Fax: 860-434-6282
                                 E-mail: nahee@nahee.org
                                 http://www.nahee.org
Classroom newspaper for kids in grades K-6. It features articles, puzzles and celebrity interviews that teach children the value of showing kindness and respect to animals, the environment, and one another.

Publication Date: 1985   4 pages 9 per year
William DeRosa, Executive Director
Dorothy Weller, Director Outreach & Fulf.

4038 List of Water Pollution Control Administrators
Assn. of State and Interstate Water Pollution Con.
750 1st Street NE   202-898-0905
Suite 1010   Fax: 202-898-0929
Washington, DC 20002   E-mail: admin1@aswipca.org
                                 http://www.asiwpca.org

Roberta Savage, Executive Director
Linda Eichmiller, Deputy Director

4039 Nebraska Recycling Resource Directory
Nebraska Dept of Environmental Quality
12200 N Street Suite 400   402-471-2186
PO Box 98922   877-253-2603
Lincoln, NE 68509   Fax: 402-471-2909
                    E-mail: MoreInfo@NDEQ.state.NE.US
                                 http://www.deq.state.ne.us
Publication Date: 1986   139 pages Bi-Annually
Steve Danahy, Unit Supervisor

4040 Pesticide Directory: A Guide to Producers and Products, Regulators, and Researchers
Thomson Publications
Box 9335   559-266-2964
Fresno, CA 93791   Fax: 559-266-0189
                                 http://www.agbook.com
This directory is for the person who needs to know anything about the US pesticide industry. It includes basic manufacturers and formulators along with their products, key personnel, managers, district/regional offices and otherpertinent information. Other sections include Universities, State Extension Centers, USDA, EPA, National Organizations, US Forest Service, Poison Control Centers and much more.
Publication Date: 1987   153 pages Biannual
ISBN: 0-913702-45-5
WT Thomson, Author
Susan Heflin, President/Owner

4041 Pollution Abstracts
Cambridge Scientific Abstracts
7200 Wisconsin Avenue   301-961-6700
Suite 601   800-843-7751
Bethesda, MD 20814-4823   Fax: 301-961-6720
                                 E-mail: sales@csa.com
                                 http://www.csa.com
This database provides fast access to the environmental information necessary to resolve day to day problems, ensure ongoing compliance, and handle emergency situations more effectively.
James P McGinty, President
Ted Caris, Publisher

4042 Product Cleanliness Levels and Contamination Control Program
IEST
5005 Newport Drive   847-255-1561
Suite 506   Fax: 847-255-1699
Rolling Meadows, IL 60008-3841   E-mail: iest@iest.org
                                 http://www.iest.org
Intended to provide a basis for specifying product cleanliness levels and contamination control program requirments with emphasis on contaminants that affect product performance.
Publication Date: 2002   20 pages
ISBN: 1-877862-82-7
Julie Kendrick, Executive Director

4043 Recycling Related Newsletters, Publications and Periodicals
Continnuus
PO Box 416   303-575-5676
Denver, CO 80201   Fax: 970-292-2136

4044 Recycling Today: Recycling Products & Services Buyers Guide
Recycling Today GIE Publishers
4012 Bridge Ave   216-961-4130
Cleveland, OH 44113-3320   800-456-0707

Fax: 216-961-0364
E-mail: jkeefe@gie.net
http://www.recyclingtoday.com
Directory of services and supplies to the industry.
Dan Sandoval, Internet/Senior Editor
James Keefe, Group Publisher

4045 Scholastic Environmental Atlas of the United States
Scholastic
730 Broadway   212-505-3000
New York, NY 10003   E-mail: uwpress@washinton.edu
http://www.washington.edu/uwpress/

4046 Scrap Plastics Markets Directory
Resource Recycling
PO Box 42270   503-233-1305
Portland, OR 97242-0270   Fax: 503-233-1356
E-mail: info@resource-recycling.com
http://www.resource-recycling.com
Offers listings of more than 400 firms that purchase scrap plastics, including buyers of plastic containers, film, sheet and mixed thermoplastics.
64 pages Yearly
ISSN: 0744-4710
Mary Lynch, Circulation Manager

4047 Tracking Toxic Wastes in CA: A Guide to Federal and State Government Information Sources
INFORM
120 Wall Street   212-361-2400
New York, NY 10005   Fax: 212-361-2412
http://www.informinc.org/INFORM.html

4048 Waste Age: Resource Recovery Acitivities Update Issue
National Solid Wastes Management Assn.
1730 Rhode Island Avenue NW
Suite 1000   202-659-4613
Washington, DC 20036

4049 Waste Age: Waste Industry Buyer Guide
National Solid Wastes Management
1730 Rhode Island Avenue NW   202-659-4613
Suite 1000   800-424-2869
Washington, DC 20036   Fax: 202-659-0925

4050 Waste Manifest Software Report
Donley Technology
200 Garfield Avenue   804-224-9427
PO Box 152   800-201-1595
Colonial Beach, VA 22443-0152   Fax: 804-224-7958
E-mail: donleytech@donleytech.com
http://www.donleytech.com
Profiles 30 software packages for solid and hazardous waste management, including detailed product descriptions, tables comparing system features, and contact information.
118 pages
ISBN: 1-891682-01-6
Elizabeth Donley, Editor

4051 Wastes to Resources: Appropriate Technologies for Sewage Treatment and Conversion
National Center for Appropriate Techology
3040 Continental Drive   406-494-4572
Butte, MT 59701   800-275-6228
Fax: 406-494-2905
E-mail: info@ncat.org
http://www.ncat.org
Kathy Hadley, Executive Director

Directories & Handbooks: Travel & Tourism

4052 Access America: An Atlas and Guide to the National Parks for Visitors with Disabilities
Northern Cartographic
4050 Williston Road   802-860-2886
South Burlington, VT 05403   Fax: 802-865-4912
E-mail: info@ncarto.com
http://www.ncarto.com
Publication Date: 1988
ISBN: 0-944187-00-5

4053 American Shore and Beach Preservation Association Directory
1724 Indian Way   510-339-2818
Oakland, CA 94611   Fax: 510-339-6710
E-mail: business@asbpa.org
http://www.asbpa.org
Federal, state and local government agencies and individuals interested in conservation, development and restoration of beaches and shorefronts.
Gregory Woodell, President

4054 Audubon Society Field Guide to the Natural Places of the Northeast
National Audubon Society
700 Broadway   212-979-3000
New York, NY 10003   Fax: 212-979-3188
http://www.audubon.org

4055 Audubon Society Field Guide to the Natural Places of the Mid-Atlantic States
National Audubon Society
700 Broadway   212-979-3000
New York, NY 10003   Fax: 212-979-3188
http://www.audubon.org

4056 Camper's Guide to California Parks, Lakes, Forests and Beaches
Gulf Publishing Company
PO Box 2608   713-529-4301
Houston, TX 77252   Fax: 713-520-4433
http://www.gpcbooks.com/
A goldmine of information about exploring Southern California, with hundreds of photographs, campsite maps and information about park fees and rules.
Publication Date: 1997   180 pages
ISBN: 0-884152-45-6

4057 Complete Guide to America's National Parks: The Official Visitor's Guide
National Park Foundation
11 Dupont Circle NW   202-238-4200
Suite 600   800-285-2448
Washington, DC 20036   Fax: 202-234-3103
E-mail: ask-npf@nationalparks.org

4058 Field Guide to American Windmills
University of Oklahoma Press
2800 Venture Drive   405-325-2000
Norman, OK 73069-8216   800-627-7377
Fax: 405-364-5798
This guide to America's windmills is both a complete general history of turbine wheel mills and an identification guide to the 112 most common models, which still dot landscapes today.
Publication Date: 1985   528 pages
T Lindsay Baker, Author

4059 Guide to the National Wildlife Refuges
Macmillan Publishing Company
National Wildlife Guide
590 Madison Avenue   http://www.nationalwildlifeguide.com   212-832-2101
New York, NY 10022

More than 500 National Wildlife Refuges and satellite refuges are listed.
684 pages

4060 National Parks Visitor Facilities and Services

Conference of National Park Concessioners
PO Box 29041   480-967-6006
Phoenix, AZ 85038   http://www.nps.gov/legacy/business.html
Within the parks, private businesses provide accommodations and services for visitors under concession contracts.
Rex G Maughan, Chairman

4061 National Parks: National Park Campgrounds Issue

National Parks Conservation Association
1300 19th Street Northwest
Suite 300   800-628-7275
Washington, DC 20036   Fax: 202-659-0650
E-mail: npca@npca.org
http://www.npca.org
To safeguard the scenic beauty, wildlife, and historical and cultural treasures of the largest and most diverse park system in the world.
Thomas C Kiernan, President
Tom Martin, Executive Vice President

4062 National Wildlife Refuges: A Visitor's Guide

Fish and Wildlife Services, Interior Department
1849 C Street NW   703-358-2043
Washington, DC 20242   E-mail: webteam@ios.doi.gov
http://www.fws.gov
Contains a map showing national wildlife refuges that provide recreational and educational opportunities. Provides tips for visiting national wildlife refuges. Also list refuges in all 50 States, Puerto Rico and the Virgin Islans, withthe best wildlife viewing season and the features of each refuge.

ISBN: 0-160617-00-6

4063 Nature Center Directory

Wisconsin Association for Environmental Education
8 Nelson Hall   715-346-2796
University of Wisconsin-Stevens Point   Fax: 715-346-3835
Stevens Point, WI 54481   E-mail: waee@uwsp.edu
http://www.uwsp.edu/waee

Annual

4064 Recreation Sites in Southwestern National Forests

USDA Forest Service
Public Affairs Office   505-842-3292
333 Broadway Blvd Se   Fax: 505-842-3106
Albuqurque, NM 87102   http://www.fs.fed.us/r3
Listings for all recreation sites for Arizona and New Mexico.
72 pages

4065 Sierra Club Guide to the Natural Areas of California

Sierra Club
85 2nd Street   415-977-5500
2nd Floor   Fax: 415-977-5799
San Francisco, CA 94105 E-mail: information@sierraclub.org
http://www.sierraclub.org
Revised and updated, this comprehensive guide makes more than 200 wilderness areas in California, including many lesser known natural areas, accessible to the outdoor enthusiast.
352 pages
ISBN: 0-871568-50-0
John Perry and Jane Greverus Perry, Author

4066 Thermal Springs of Wyoming

Wyoming State Geological Survey
PO Box 1347   307-766-2286
Laramie, WY 82073   Fax: 307-766-2605
E-mail: wsgs-info@uwyo.edu
http://www.wsgs.uwyo.edu

Ronald C Surdam, Agency Director
Richard W Jones, Editor

4067 Traveler's Guide to the Smoky Mountains Region

Harvard Common Press
535 Albany Street   617-423-5803
Boston, MA 02118   Fax: 619-695-9794
E-mail: orders@harvardcommonpress.com
http://www.harvardcommonpress.com
Features museums, events of the South Appalachians of Tennessee, North Carolilna, Virginia and Georgia
Publication Date: 1985   288 pages
ISBN: 0-916782-64-6
Valerie Cimino, Executive Editor
Christine Alaimo, Associate Publisher

4068 Wild Places & Open Spaces Map

Division of Fish and Wildlife
PO Box 400   609-292-9450
Trenton, NJ 08625-0400   Fax: 609-984-1414
http://www.njfishandwildlife.com
Designed similar to a road map, offers the outdoors person a welath of information on locating and exploring New Jersey's open spaces in compact and easy to read format. Showcasing a full color map of New Jersey, with more than 700,000acres of public open space.
Carol Nash, Customer Service

Directories & Handbooks: Water Resources

4069 Citizen's Directory for Water Quality Abuses

Izaak Walton League of America
707 Conservation Lane   301-548-0150
Gaithersburg, MD 20878   800-453-5463
Fax: 301-548-0146
E-mail: general@iwla.org
http://www.iwla.org

Paul Hansen, Executive Director

4070 Coordination Directory of State and Federal Agency Water Resources Officials: Missouri Basin

Department of Water Resources
301 Centennial Mall South   402-471-2363
Lincoln, NE 68508-2529   http://ne.water.usgs.gov

4071 Directory of Water Resources Expertise

California Water Resources Center
University of California   714-787-4327
Riverside, CA 92521   http://www.nceas.ucsb.edu/exp/
A searchable, comprehensive database of specialists, their affiliations, and areas of expertise.

4072 How Wet is a Wetland?: The Impacts of the Proposed Revisions to the Federal Wetlands Manual

Environmental Defense Fund
257 Park Avenue South   212-505-2100
New York, NY 10010   800-684-3322
Fax: 212-505-0892
E-mail: media@environmentaldefense.org
http://www.environmentaldefense.org
Publication Date: 1992
Fred Krupp, President

4073 Hydro Review: Industry Source Book Issue

HCI Publications
410 Archibald Street   816-931-1311
Kansas City, MO 64111   Fax: 816-931-2015
E-mail: info@hcipub.com
http://www.hcipub.com
List of over 800 manufacturers and suppliers of products and services to the hydroelectric industry in the US and Canada.
January
Carl Vansant, Editor-In-Chief

4074 List of Water Pollution Control Administrators

Assn. of State and Interstate Water Pollution Con.
750 1st Street NE   202-898-0905
Suite 1010   Fax: 202-898-0929
Washington, DC 20002

E-mail: admin1@aswipca.org
http://www.asiwpca.org

Roberta Savage, Executive Director
Linda Eichmiller, Deputy Director

4075 Thermal Springs of Wyoming
Wyoming State Geological Survey
PO Box 1347   307-766-2286
Laramie, WY 82073   Fax: 307-766-2605
E-mail: wsgs-info@uwyo.edu
http://www.wsgs.uwyo.edu

Ronald C Surdam, Agency Director
Richard W Jones, Editor

4076 Water Environment & Technology Buyer's Guide and Yearbook
Water Environment Federation
601 Wythe Street   703-684-2400
Alexandria, VA 22314-1994   800-666-0206
Fax: 703-684-2492
E-mail: webfeedback@wef.org
http://www.wef.org

Founded in 1928, the Water Environment Foundation (WEF) is a not for profit technical and educational organization with members from varied disiplines who work toward the WEF vision of preservation and enhancement of the globalwaterenvironment. The WEF network includes water quality professionals from 76 Member Associations in 30 countries.
William J Bertera, Executive Director

Periodicals: Air & Climate

4077 Air/Water Pollution Report
Business Publishers
8737 Colesville Road   301-589-5103
10th Floor   800-274-6737
Silver Spring, MD 20910   Fax: 301-589-8493
E-mail: custserv@bpines.com
http://www.bpinews.com

Regulatory activities and governmental legislation and litigation are covered in this pulication.
Publication Date: 1963   Monthly
Leonard Eiserer, Publisher

4078 Bulletin of the American Meteorological Society
45 Beacon Street   617-227-2425
Boston, MA 02108-3693   Fax: 617-742-8718
E-mail: amsinfo@ametsoc.org
http://www.ametsoc.org

The American Meteorological Society promotes the development and dissemination of information and education on the atmospheric and related oceanic and hydrologic sciences and the advancement of their professional applications.
Publication Date: 1919   Monthly
Ronald D McPherson, Executive Director

4079 Climate Institute: Climate Alert
Coping with Climate Change
1785 Massachusetts Avenue NW   202-547-0104
Washington, DC 20036   Fax: 202-547-0111
E-mail: info@climate.org
http://www.climate.org

The Climate Institute works to protect the balance between climate and life on earth by facilitating dialogue among scientists, policy makers, business executives and citizens. In all its efforts, the institute strives to be a sourceof objective, reliable information.
Publication Date: 1988   8-12 pages Quarterly
ISSN: 1071-3271
John Topping, President

4080 Earth Share of Georgia Newsletter
1447 Peachtree Street   404-873-3173
Suite 214   Fax: 404-873-3135
Atlanta, GA 30309   E-mail: info@earthsharega.org
http://www.earthsharega.org

Nonprofit federation of local, national and global environmental groups addressing the critical environmental issues. ESGA raises funds for these groups through workplace giving campaigns, special events and individual contributions.

Publication Date: 1992   Bi-Monthly
Richard Judy, Executive Director

4081 Environmental Policy Alert
Inside Washington Publishers
1225 Jefferson Davis Highway   703-416-8500
Suite 1400   800-424-9068
Arlington, VA 22202-4301   Fax: 703-416-8543
E-mail: iwp@iwpnews.com
http://www.iwpnews.com

Addresses the legislative news and provides reports on the federal environmental policy process.
Publication Date: 1980
Paul Finger, Publisher

4082 Global Environmental Change Report
Cutter Information Corporation
37 Broadway   781-648-8700
Suite 1   800-888-8939
Arlington, MA 02474   Fax: 781-648-1950
E-mail: service@cutter.com
http://www.cutter.com

News and analysis of policy, science and industry developments in the areas of global warming and acid rain.
Karen Fine Coburn, Publisher
Brad Hurley, Editor

4083 Journal of the Air Pollution Control Association
Air Pollution Control Association
420 Fort Duquesne   412-232-3444
Boulevard #3
Pittsburgh, PA 15222-1420

A comprehensive journal offering information to the environment and conservation industry.

4084 Journal of the Air and Waste Management Association
Air and Waste Management Association
One Gateway Center   412-232-3444
3rd Floor   Fax: 412-232-3450
Pittsburgh, PA 15222-1435   E-mail: info@awma.org
http://www.awma.org

Published for the working environmental professional and carries peer-reviewed technical papers on a variety of topics form control technology to science.
Publication Date: 1951   Monthly
Maura Moktar, Managing Editor

4085 Population Reference Bureau: Household Transportation Use and Urban Pollution
1875 Connecticut Avenue NW   202-483-1100
Suite 520   800-877-9881
Washington, DC 20009-5728   Fax: 202-328-3937
E-mail: popref@prb.org
http://www.prb.org

Publication Date: 1929
Peter Donaldson, President
Mary Mederios Kent, Editor

4086 Population Reference Bureau: Population & Environment Dynamics
1875 Connecticut Avenue NW   202-483-1100
Suite 520   800-877-9881
Washington, DC 20009-1435   Fax: 202-328-3937
E-mail: popref@prb.org
http://www.prb.org

PRB publishes the quarterly Population Bulletin, the annual World Population Data Sheet, and PRB Reports on America, as well as specialized publications covering population and public policy issues in the U.S. and abroad, particularlyin developing countries.
Publication Date: 1929
Peter Donaldson, President
Mary Mederios Kent, Editor

4087 Population Reference Bureau: Water
1875 Connecticut Avenue NW   202-483-1100
Suite 520   800-877-9887
Washington, DC 20009   Fax: 202-328-3937
E-mail: popref@prb.org
http://www.prb.org

Publication Date: 1929
Peter Donaldson, President
Mary Mederios Kent, Editor

4088 Trinity Consultants Air Issues Review

12770 Merit Drive　　　　　　　**972-661-8100**
Suite 900　　　　　　　　　　　**800-229-6655**
Dallas, TX 75251　　　　　**Fax: 972-385-9203**
　　　　E-mail: information@trinityconsultants.com
　　　　　http://www.trinityconsultants.com
An environmental consulting company that assists industrial facilities with issues related to regulatory compliance and environmental management. Founded in 1974, this nationwide firm has particular expertise in air quality issues. Trinity also sells environmental software and professional education. T3, a Trinity Consultants Company, provides EH&S management information systems (EMIS) implementation and integration services.
　　　Publication Date: 1990　8 pages　Quarterly
John Hofmann, VP
Patrick Delamater, VP

4089 Weather & Climate Report

Nautilus Press
1054 National Press Building　　**202-347-6643**
Washington, DC 20045-2001
Reports on federal actions which impact weather, climate research and global changes in climate.
　　　Monthly
　　　ISSN: 0730-8256
John R Botzum, Editor

4090 World Resource Review

SUPCON International
22w381 75th Street　　　　　　**630-910-1551**
Naperville, IL 60565　　　**Fax: 630-910-1561**
　　　　E-mail: syshen@megsinet.net
　　　　http://www.globalwarming.net
For business and government readers, provides expert worldwide reviews of global warming and extreme events in relation to the management of natural, mineral and material resources. Subjects include global warming impacts on agriculture, energy, and infrastructure, monitoring of changes in resources using remote sensing, actions of national and international bodies, global carbon budget, greenhouse budget and more.
　　　Publication Date: 1990　Quarterly
　　　ISSN: 1042-8011
Dr. Sinya Shen, Production Manager

4091 World Watch

Worldwatch Institute
1776 Massachusetts Avenue NW　　**202-452-1999**
Suite 800　　　　　　　　　**Fax: 202-296-7365**
Washington, DC 20036E-mail: worldwatch@worldwatch.org
　　　　　http://www.worldwatch.org
Magazine on global environmental issues.
　　　Publication Date: 1975　40 pages
　　　ISSN: 0896-0615
Ed Ayres, Author
Lester Brown, Founding Publisher
Lisa Mastny, Senior Editor

Periodicals: Business

4092 AFE Journal

8160 Corporate Park Drive　　　**513-489-2473**
Suite 125　　　　　　　　　**Fax: 513-247-7422**
Cincinnati, OH 45242　　**E-mail: mail@afe.org**
　　　　　　　　　http://www.afe.org
AFE Journal is a bimonthly publication from the Association for Facilities Engineering.
　　　48 pages　Bimonthly
　　　ISSN: 1088-5900
Gabriella Jacobs, Author
Gabriella Jacobs, Editor

4093 ALBC News

American Livestock Breeds Conservancy
PO Box 477　　　　　　　　　**919-542-5704**
Pittsboro, NC 27312　　　　**Fax: 919-545-0022**

E-mail: albc@albc-usa.org
http://www.albc-usa.org
ALBC News is a bi-monthly newsletter published by the American Livestock Breeds Conservancy.
　　　Publication Date: 1987　20 pages　Bi-Monthly
　　　ISSN: 1064-1599
Cindy Rubel, Author
Cindy Rubel, Editor

4094 Abstracts of Presentations

Wildlife Society
5410 Grosvenor Lane　　　　　**301-897-9770**
Suite 200　　　　　　　　　**Fax: 301-530-2471**
Bethesda, MD 20814-2144　　**E-mail: tws@wildlife.org**
　　　　　　　　http://www.wildlife.org
A yearly publication of the Wildlife Society.
　　　Publication Date: 1994　300 pages　Yearly
Gene Pozniak, Production Editor

4095 Advisor

Great Lakes Commission
Eisenhower Corporate Park　　　**734-665-9135**
2805 South Industrial Hwy, Suite 100　**Fax: 734-665-4370**
Ann Arbor, MI 48104-6791　**E-mail: glc@great-lakes.net**
　　　　　　　　http://www.glc.org
Covers economic and environmental issues of the Great Lakes region with a special focus on activities of the Great Lakes Commission.
　　　Publication Date: 1955　12 pages　Bi-Monthly
Julie Wagemakers, Production Manager

4096 Agribusiness Fieldman

Western Agricultural Publishing Company
4969 E Clinton Way　　　　　　**559-252-7000**
Suite 104　　　　　　　　　**Fax: 559-252-7387**
Fresno, CA 93727-1549
Aimed at keeping the professional agriculture consultant posted on changes in the agricultural-chemical industry. Provides news about pests and control measures for all segments of the agricultural-chemical industry.
Paul Baltimore, Publisher
Margi Katz, Editor

4097 American Environmental Laboratory

International Scientific Communications
PO Box 870　　　　　　　　　**203-926-9300**
Shelton, CT 06484-0870　　**Fax: 203-926-9310**
　　　　　E-mail: iscpubs@iscpubs.com
　　　　　http://www.iscpubs.com
Laboratory activities, new equipment, and analysis and collection of samples are the main topics.
　　　Bi-Monthly
Brian Howard, Publisher/Editor-in-Chief

4098 Annual Newsletter and Report

The Peregrine Fund
5668 W Flying Hawk Lane　　　**208-362-3716**
Boise, ID 83709　　　　　　**Fax: 208-362-2376**
　　　　　E-mail: tpf@peregrinefund.org
　　　　　http://www.peregrinefund.org
Yearly publication from The Peregrine Fund. Free with $25 membership fee.
　　　Publication Date: 1970　Yearly
William Burnham, Author
Dr William Burnham, President

4099 Bison World

National Bison Association
1400 West 122nd Avenue　　　**303-292-2833**
Suite 106　　　　　　　　　**Fax: 303-292-2564**
Westminster, CO 80234　**E-mail: info@bisoncentral.com**
　　　　　　http://www.bisoncentral.com
Published by the NBA, an organization of bison producers dedicated to awareness of the healthy properties of bison meat and bison production.
　　　Publication Date: 1975　100 pages　Quarterly
Sam Albrecht, Publisher
Laurie Dineen, Editor

4100 Business and the Environment

Cutter Information Corporation
37 Broadway   781-648-8700
Suite 1   800-888-8939
Arlington, MA 02474   Fax: 781-648-1950
                    E-mail: service@cutter.com
                        http://www.cutter.com
Environmental investment trends, deals and market developments.
Karen Fine Coburn, Publisher
Kathleen Victory, Editor

4101 CAC Annual Reports

Citizens Advisory Council
13th Floor, RCSOB   717-787-4527
PO Box 8459   Fax: 717-787-2878
Harrisburg, PA 17105   E-mail: mioff.stephanie@state.pa.us
                     http://www.cacdep.state.pa.us
Publisher by the Citizens Advisory Council.
        Publication Date: 1977   20-40 pages   Annual
Susan Wilson, Executive Director
Stephanie Mioff, Administrative Assistant

4102 Chemosphere

Pergamon Press
660 White Plains Road   914-524-9200
Tarrytown, NY 10591-5104   Fax: 914-592-3625
Related to environmental affairs. Accepts advertising.
        100 pages
T Stephen, Editor
Rosemarie Fazzolari, Advertising

4103 Connecticut Sea Grant

1080 Shennecossett Road   860-405-9110
Groton, CT 06340   Fax: 860-405-9109
                   http://www.seagrant.uconn.edu
Based at the University of Connecticut, CT Sea Grant is part of
the National Sea Grant network, whose mission is the conservation and wise use of coastal and marine resources through research, education and outreach.
Peg Van Patten, Communications Director

4104 Earth First! Journal

PO Box 3023   520-620-6900
Tucson, AZ 85702   Fax: 413-254-0057
         E-mail: collective@earthfirstjournal.org
              http://www.earthfirstjournal.org/
Earth First! Journal was founded in 1979 in response to a lethargic, compromising and increasingly corporate environmental community. Earth First! takes a decidedly different tack toward environmental issues. We believe in using allthe tools in the toolbox, ranging from grassroots organizing and involvement in the legal process to civil disobedience and monkeywrenching.
        Publication Date: 1979   64 pages   Bimonthly
        ISSN: 1055-8411

4105 Earth Island Journal

300 Broadway   415-788-3666
Suite 28   Fax: 415-788-7324
San Francisco, CA 94133   http://www.earthisland.org
Quarterly publication from the Earth Island Institute.
        Publication Date: 1982
        ISSN: 1041-0406
Dave Phillips, Executive Director
Chris Clarke, Editor

4106 Economic Opportunity Report

Business Publishers
8737 Colesville Road   301-587-6300
10th Floor   800-274-6737
Silver Spring, MD 20910-3928   Fax: 301-587-4530
                   E-mail: custserv@bpinews.com
                       http://www.bpinews.com
Antipoverty news coverage and analysis which gives insight into developments that affect social programs.
        Publication Date: 1963
Leonard A Eiserer, Publisher
Beth Early, Operations Director

4107 Environmental Business Journal

Environmental Business International
4452 Park Boulevard   619-295-7685
Suite 306   Fax: 619-295-5743
San Diego, CA 92116   E-mail: ebi@ebiusa.com
                            http://www.ebiusa.com
EBJ provides industry data and market analysis for the environmental industry. Each issue features one of the fourteen industry segments (consulting and engineering, site remediation, water and wastewater. etc.) and the results ofmarket surveys as well as analysis of emerging trends and business strategies. Annual subscription to EBJ as well as other market reports and industry data are available on EBJ's web site.
        Publication Date: 1988   16 pages   Annual
Grant Ferrier, President

4108 Environmental Compliance Update

High Tech Publishing Company
PO Box 1275   413-534-4500
Amherst, MA 01004-1275   Fax: 413-256-6378
Identifies and analyzes the issues and business and economic impact of environmental compliance laws and regulations. Monitors the relevant changes due to legislation, court decisions, private rulings and technology.
Lori Reilly, Editor

4109 Environmental Ethics

Center for Environmental Philosophy
PO Box 310980   940-565-2727
Denton, TX 76203-0980   Fax: 940-565-4439
                          E-mail: cep@unt.edu.
                       http://www.cep.unt.edu
Published by the Center for Environmental Philosophy.
        Publication Date: 1979   112 pages   Quarterly
        ISSN: 0163-4275
Eugene C Hargrove, Author/Editor

4110 Environmental News

CA Business Publications
PO Box 3359   817-924-5301
Fort Worth, TX 76113-3359   Fax: 817-922-8893
                          E-mail: txenv@aol.com
Follows the progress of public environmental stock companies, provides updates on environmental contract opportunities, news of international environmental opportunities, and profiles innovative new companies.
Carolyn Ashford, Publisher/Editor

4111 Environmental Packaging

Thompson Publishing Group
1725 K Street NW   202-872-4000
Suite 700   800-444-8741
Washington, DC 20006-1401   Fax: 202-739-9578
                        http://www.thompson.com
A newletter aimed at the environmental regulatory specialist, product development managers, purchasing managers, legal counsel and package designers covering state-by-state regulations and the FCA guidelines and enforcement.
        Publication Date: 1972
Daphne Musselwhite, Publisher

4112 Florida Forests Magazine

Florida Forestry Association
PO Box 1696   850-222-5646
Tallahassee, FL 32302   Fax: 850-222-6179
                      E-mail: info@forestfla.org
                    http://www.floridaforest.org
A publication of the Florida Forestry Association.
        Publication Date: 1997   26-32 pages   Quarterly
J Doran, Executive VP

4113 George Miksch Sutton Avian Research Center Sutton Newsletter

PO Box 2007   918-336-7778
Bartlesville, OK 74005-2007   Fax: 918-336-7783
                       E-mail: gmsarc@aol.com
                    http://www.suttoncenter.org
Newsletter published by George Miksch Sutton Avian Research Center.
        Publication Date: 1990   8-10 pages   Semiannual
Steve Sherrod, Executive Director

4114 Green Business Letter
Tilden Press
6 Hillwood Place   202-332-1700
Oakland, CA 94610-1810   Fax: 202-332-3028
E-mail: gbl@greenbiz.com
http://www.greenbiz.com
Hands-on journal for environmentally conscious companies, covering management strategies, facilities management, personnel policies and procurement with environmental consciousness. Emphasis on products, resources and how-toinformation.
8 pages
ISSN: 1056-490X
Joel Makower, Editor

4115 In Business: The Magazine for Sustainable Enterprises and Communities
JG Press, Inc
419 State Avenue   610-967-4135
Emmaus, PA 18049   E-mail: advert@jgpress.com
http://www.jgpress.com
Jerome Goldstein, Editor

4116 International Environment Reporter
Bureau of National Affairs
1231 25th Street NW   202-452-4200
Washington, DC 20037-1197   800-372-1033
Fax: 202-822-8092
http://www.bna.com
A four-binder information and reference service covering international environmental law and developing policy in the major industrial nations.
William A Beltz, Publisher

4117 International Environmental Systems Update
BSI Management Systems
12110 Sunset Hills Road   703-437-9000
Suite 140   800-862-4977
Reston, VA 20190-3231   Fax: 703-435-7979
E-mail: solutions@bsiamencas.com
http://www.bsiamencas.com
Provides accurate, up-to-date and useful information for environmental professionals around the globe. Monthly publication brings current environmental events into the limelight, dissecting complex issues, helping hundreds oforganizations improve their environmental and business preformance.
Publication Date: 1994 24 pages Monthly
ISSN: 1079-0837
Marcus Darby, Publisher

4118 McCoy's RCRA Unraveled
McCoy & Associates
25107 Genesee Trail Road   303-526-2674
Suite 200   Fax: 303-526-5471
Golden, CO 80401-5708   E-mail: info@mccoyseminars.com
http://www.mccoyseminars.com
This book addresses the most troublesome areas in 40 CFR Parts 261 and 262 of the federal regulations. Our engineers have researched every scrap of guidance EPA has ever issued on these troublesome topics, studied the Federal Registerpreamble language, and talked to thousands of people who attented our RCRA seminars and shared their real-world experiences. It includes a keyword index with more than 1,300 entries, 200 probing examples from EPA's own guidance documents and ahelpful acronym list.
Publication Date: 2005 752 pages Yearly
ISBN: 0-930469-25-9
Paul Gallagher, President
Nancy Pribble, Marketing Manager

4119 NFPA Journal
One Batterymarch Park   617-770-3000
Quincy, MA 02169   Fax: 617-770-0200
E-mail: nfpa@nfpa.org
http://www.nfpa.org
A bi-monthly journal published by the National Fire Protection Association.
Bi-Monthly

4120 NSS News
National Speleological Society
2813 Cave Avenue   256-852-1300
Huntsville, AL 35810-4413 Fax: 256-851-9241

E-mail: nss@caves.org
http://www.caves.org
Published by the National Speleological Society.
Publication Date: 1942 Monthly
ISSN: 0027-7010
Dave Bunnell, Editor

4121 Newsleaf
4949 Tealtown Road   513-831-1711
Milford, OH 45150   Fax: 513-831-8052
E-mail: cnc@cincynature.org
http://www.cincynature.org
Newsleaf is a quarterly publication for Cincinatti Nature Center members. This publication provides informative articles that teach readers about native flora and fauna.
Publication Date: 1965 20-24 pages Quarterly
Rhonda Barnes-Kloth, Communications Manager

4122 Proceedings of the Desert Fishes Council
Desert Fishes Council
PO Box 337   760-872-8751
Bishop, CA 93515   http://www.desertfishes.org
Yearly publication from Desert Fishes Council.
Publication Date: 1969 Yearly
ISSN: 1068-0381
E P Pister, Executive Secretary

4123 Proceedings of the Southeastern Association of Fish and Wildlife Agencies
8005 Freshwater Farms Road 850-893-1204
Tallahassee, FL 32308   Fax: 850-893-6204
E-mail: SEAFWA@aol.com
http://www.seafwa.org
Proceedings of the SEAFWA - an annual publication.
Publication Date: 1947 4-900 pages Yearly
Robert M Brantly, Executive Secretary

4124 Pumper
COLE Publishing
PO Box 220   715-546-3346
Three Lakes, WI 54562-0220 Fax: 715-546-3786
E-mail: cole@pumper.com
http://www.pumper.com
Emphasis on companies, individuals and industry events while focusing on customer service, environmental issues and employment trends.
Publication Date: 1947
Ken Lowther, Editor

4125 Regulatory Update
Arkansas Environmental Federation
1400 W Markham Street   501-374-0263
Suite 250   Fax: 501-374-8752
Little Rock, AR 72201   http://www.environmentark.org
The AEF focuses on practical, common-sense laws and regulations based on sound science; a teamwork approach to compliance; waste minimization and pollution prevention. The AEF enables information to be exchanged on a daily basisbetween its members, government regulators, and policy makers.
Publication Date: 1967 Semi-Annual
Randy Thurman, Executive Director

4126 Risk Policy Report
Inside Washington Publishers
1225 South Clark Street   703-416-8500
Suite 1400   800-424-9068
Arlington, VA 22202   E-mail: iwp@iwpnews.com
http://www.iwnews.com
Contains analysis, great perspectives, industry news, policymaking profiles and a calendar of events.
Monthly
David P Clarke, Publisher/Editor

4127 Semillero
731 8th Street SE   202-547-3800
Washington, DC 20003   Fax: 202-546-4784
E-mail: etoledo@newforestsproject.com
http://www.newforestsproject.com

Our electronic publication has more than 2500 subscribers in the US and Latin America. It provides useful information and references to specifice resources regarding agro-forestry, rural development and grant information.
Publication Date: 1982
Erick Toledo and Catalina Serna, Author
Erick Toledo, Director

Periodicals: Design & Architecture

4128 MSW Management
Forester Communications
PO Box 3100   805-682-1300
Santa Barbara, CA 93130   Fax: 805-682-0200
E-mail: erosion@ix.netcom.com
http://www.mswmanagement.com
Provides general news on facility construction, financing, new equipment and revenue issues.
Publication Date: 1991   7 Times Yearly
Daniel Waldman, Publisher

Periodicals: Disaster Preparedness & Response

4129 Hazard Technology
EIS International
555 Herndon Parkway   703-478-9808
Herndon, VA 20170   Fax: 703-787-6720
Application of technology to the field of emergency and environmental management to save lives and protect property.
James W Morentz PhD, Publisher
Leslie Atkin, Managing Editor

4130 Hazardous Materials Newsletter
Hazardous Materials Publishing
243 West Main Street   610-683-6721
Kutztown, PA 19530   Fax: 610-683-3171
E-mail: lheffner@hazmat-tsp.com
http://www.hazmatpublishing.com
Focuses on response to and control of hazardous materials emergencies. Particularly appropriate tools, equipment, materials, methods, procedures, strategies and lessons learned. Addresses leak, fore and spill control for incidentcommanders and experienced responders, including incident clauses, prevention and remedial actions; decisionmaking; scene management; control and containment; response teams; and product identification and hazards.
12 pages

4131 Natural Hazards Research and Applications Information Center Newsletter
University of Colorado
482UCB   303-492-6818
Boulder, CO 80309-0482   Fax: 303-492-2151
E-mail: hazctr@colorado.edu
http://www.colorado.edu/hazards
The Natural Hazards Research and Applications Information Center serves as a national clearinghouse for social science research data on natural disasters, related technological events, and programs to reduce damage from naturaldisasters.
Publication Date: 1976   28 pages Bi Monthly
Dennis Mileti, Director

Periodicals: Energy & Transportation

4132 Alternative Energy Resources Organization Newsletter
432 North Last Chance Gulch   406-443-7272
Helena, MT 59601   Fax: 406-442-9120
E-mail: aero@aeromt.org
AERO is a statewide grassroots group whose members work together to strengthen communities through promoting sustainable agriculture, local food production and citizen-based Smart Growth community planning.
Publication Date: 1978   4-24 pages Quarterly
Kelly Rusoff, Office Manager

4133 Butane-Propane News
Butane-Propane News, Inc
PO Box 660698   626-357-2168
Arcadia, CA 91066   Fax: 626-303-2854
E-mail: arey@bpnews.com
http://www.bpnews.com
Offers information to professionals that are involved in the distribution, production, shipping and sales of butane and propane in the US and internationally. $32 for US one year subscription; $60 for international one yearsubscription.
Publication Date: 1939   48-96 pages Monthly
ISSN: 0007-7259
Ann Rey, Editor/Director

4134 Defense Cleanup
Pasha Publications
1600 Wilson Boulevard   703-528-1244
Suite 600   800-424-2908
Arlington, VA 22209-2509   Fax: 703-528-1253
Covers contracting opportunities with the Energy Department and Defense Department in Environmental remediation and waste management at current and former DOE/DOD facilities.
Quarterly
Harry Baisden, Group Publisher
Michael Hopps, Editor

4135 E&P Environment
Pasha Publications
1600 Wilson Boulevard   703-528-1244
Suite 600   800-424-2908
Arlington, VA 22209-2510   Fax: 703-528-1253
Reports on environmental regulations, advances in technology and litigation aimed specifically at the exploration and production segments of the oil and gas industry.
Harry Baisden, Group Publisher
Michael Hopps, Editor

4136 Energy Engineering
Association of Energy Engineers
4025 Pleasantdale Road   770-447-5083
Suite 420   Fax: 770-446-3969
Atlanta, GA 30340   E-mail: webmaster@aeecenter.org
http://www.aeecenter.org
Engineering solutions to cost efficiency problems and mechanical contractors who design, specify, install, maintain, and purchase non-residential heating, ventilating, air conditioning and refrigeration equipment and components.
Wayne Turner, Editor-in-Chief

4137 Getting Around without Gasoline
Northeast Sustainable Energy Association
50 Miles Street   413-774-6051
Greenfield, MA 01301   Fax: 413-774-6053
E-mail: nesea@nesea.org
http://www.nesea.org
This interdisciplinary science/social studies curriculum allows students to explore the transportation and environmental issues in their own lives. Activities cover: transportation systems, health impacts, environmental andtransportation histories, carpooling, and mass transit.
Bi-Annually
Chris Mason, Education Director

4138 Heliographs
Illinois Solar Energy Association
PO Box 634   630-260-0424
Wheaton, IL 60189-0634   E-mail: info@illinoissolar.org
http://www.illinoissolar.org
Publication Date: 1975   12 pages Quarterly

4139 International Journal of Hydrogen Energy
PO Box 248266   305-284-4666
Coral Gables, FL 33124   Fax: 305-284-4792
E-mail: ayfer@iahe.org
http://www.iahe.org
A monthly publication serving to inform scientists and the public of advances made in hydrogen energy research and development.

Publication Date: 1976   120 pages   Monthly
ISSN: 0360-3199
T Nejat Veziroglu, Author
T Neja Veziroglu, Editor-in-Chief

4140 Midwest Renewable Energy Association Newsletter
7558 Deer Road   715-592-6595
Custer, WI 54423   Fax: 715-592-6596
                                      E-mail: info@the-mrea.org
                                            http://www.the-mrea.org
ReNews includes articles on energy issues, book reviews, case studies, and other general information about renewable energy.
Quarterly
Tehri Parker, Executive Director

4141 Northeast Sun
Northeast Sustainable Energy Association
50 Miles Street   413-774-6051
Greenfield, MA 01301   Fax: 413-774-6053
                                      E-mail: nesea@nesea.org
                                            http://www.nesea.org
Promotes responsible use of energy for a stronger economy and cleaner environment.
Publication Date: 1982   Bi-Annually
Nancy Hazard, Executive Director

4142 Nuclear Monitor
Nuclear Information & Resource Services
1424 16th Street NW   202-328-0002
Suite 404   Fax: 202-462-2183
Washington, DC 20036-2239   E-mail: nirsnet@nirs.org
                                            http://www.nirs.org
Nuclear power, radioactive waste and sustainable energy news for environmental activities, state and local officials and investment communities.
12 pages   Monthly
Michael Mariotte, Editor

4143 Nuclear Waste News
Business Publishers
8737 Colesville Road   301-589-5103
10th Floor   800-274-6737
Silver Spring, MD 20910   Fax: 301-589-8493
                                      E-mail: custserv@bpinews.com
                                            http://www.bpinews.com
Worldwide coverage of the nuclear waste management industry, including waste generation, radiological environmental remediation, packaging, transport, processing and disposal.
Weekly
Leonard A Eiserer, Publisher
Beth Early, Associate Publisher

4144 Radwaste Magazine
American Nuclear Society
555 North Kensington Avenue   708-352-6611
LaGrange Park, IL 60526-5535   Fax: 708-352-0499
                                      E-mail: radwaste@ans.org
                                            http://www.ans.org
Addresses issues in all fields of radioactive waste management, removal, handling, disposal, treatment, cleanup and environmental restoration.
6x per Year
Nancy Zacha, Publisher

4145 ReNews
Midwest Renewable Energy Association
PO Box 249   715-592-6595
Amherst, WI 54406-0249   Fax: 715-592-6596
                                      E-mail: info@the-mrea.org
                                            http://www.the-mrea.org
ReNews is a quarterly newsletter that includes articles on energy issues, book reviews, case studies, and other general information about renewable energy.
Quarterly

4146 Solar Energy
Elsevier Science
360 Park Avenue South   212-989-5800
11th Floor   Fax: 212-633-3680
New York, NY 10010   E-mail: usinfo-f@elsevier.com
                                            http://www.elsevier.com

John A Duffie, Editor

4147 Solar Energy Report
PO Box 782   949-837-7430
Rio Vista, CA 94571   Fax: 949-709-8043
                                            http://www.calseia.org
Bi-Monthly

4148 Solar Today
American Solar Energy Society
2400 Central Avenue   303-443-3130
Suite A   Fax: 303-443-3212
Boulder, CO 80301   E-mail: ases@ases.org
                                            http://www.ases.org
Provides information and case histories and reviews of a variety of renewable energy technologies, including solar, wind, biomass and geothermal.
Donna McClane, Publications

4149 TXSES Newsletter
Texas Solar Energy Society
PO Box 1447   512-326-3391
Austin, TX 78767-1447   800-465-5049
                                          Fax: 512-326-1785
                                      E-mail: info@txses.org
                                            http://www.txses.org
The Texas Solar Energy Society is a non-profit organization open to anyone with an interest in renewable energy. It is dedicated to increasing awareness of the uses and benefits of solar and other renewable energy resources. TheSociety publishes a quarterly newsletter and engages in educational projects and conferences.
Quarterly

Periodicals: Environmental Engineering

4150 Air and Waste Management Association's Magazine for Environmental Managers
Air and Waste Management Association
One Gateway Center   412-232-3444
420 Fort Duquesne Blvd., 3rd Floor   Fax: 412-232-3450
Pittsburgh, PA 15222-1416   E-mail: info@awma.org
                                            http://www.awma.org
A magazine that contains sections of Washington and Canadian reports, a calendar of events, government affairs, news focus, campus research, business briefs, district control news, porfessional development programs, professionalservices and other issues facing the environmental professionals.
Todd Zahniser, Publisher/Editor

4151 Asbestos & Lead Abatement Report
Business Publishers
8737 Colesville Road   301-589-5103
Suite 1100   800-274-6737
Silver Spring, MD 20910   Fax: 301-589-8493
                                      E-mail: bpinews@bpinews.com
                                            http://www.bpinews.com
Tracks the major legal, legislative, regulatory, business and technological developments in the asbestos and lead abatement industries.
Weekly
Leonard A Eiserer, Publisher
Beth Early, Associate Publisher

4152 Association for Environmental Health of Soils (AEHS):
Magazine, Newsletter
150 Fearing Street   413-549-5170
Amherst, MA 01002   Fax: 413-549-0579
                                            http://www.aehs.com
The AEHS was created to facilitate communications and foster cooperation among professionals concerned with the challenge of soil protection and cleanup. AEHS provides the network. Members represent the many disciplines involved inmaking decisions and solving problems affecting soils. AEHS recognizes that widely acceptable solutions to the problem can be found only through the intergration of scientific and tecnological discovery, social and political judgement and hands onpractice.
James Dragun, Editor-in-Chief
Denise Leonard, Managing Editor

4153 Curt B Beck Consulting Engineer: Newsletter

408 W Kingsmill Street   806-665-9281
PO Box 2442   888-665-9281
Pampa, TX 79006-2442   Fax: 806-665-1965
E-mail: curtbbeck@cableone.net
http://www.pan-tex.net/usr/b/beck
Pollution control services.
Publication Date: 1984   Monthly
Curt B Beck, Owner

4154 Defense Cleanup

Business Publishers
8737 Colesville Road   301-587-6300
10th Floor   800-274-6737
Silver Spring, MD 20910   Fax: 301-587-4530
E-mail: bpinews@bpinews.com
http://www.bpinews.com
Covers the latest news and analysis of defense cleanup activity.
Including base remediation and closure, contract awards and site
cleanups.
Weekly
Leonard A Eiserer, Publisher
Beth Early, Associate Publisher

4155 EI Digest

Environmental Information
PO Box 390266   952-831-2473
Minneapolis, MN 55439-0266   Fax: 952-831-6550
E-mail: ei@enviro-information.com
http://www.envirobiz.com
Contains market studies of commercial hazardous waste man-
agement companies with in-depth analysis of trends in policy,
regulations, technology and business.
Publication Date: 1983
ISSN: 1042-251X
Cary Perket, Editor

4156 Environment

Helen Dwight Reid Educational Foundation
1319 18th Street NW   202-296-6267
Washington, DC 20036-1802   800-365-9753
Fax: 202-296-5149
E-mail: subscribe@heldref.org
http://www.heldref.org
Provides environment professionals and concerned citizens
with authoritative yet accessible articles that provide critical
analysis of environmental science and policy issues, book rec-
ommendations, commentaries, news briefs and reviewson envi-
ronmental websites and major governmental and institutional
reports.
Publication Date: 1958   48 pages Monthly
ISSN: 0013-9157
Douglas Kirkpatrick, Executive Director
Barbara T Richman, Managing Editor

4157 Environment 21

Florida Environments Publishing
4010 Newberry Road   352-373-1401
#F   Fax: 352-373-1405
Gainesville, FL 32607-2368   E-mail: info@enviroworld.com
http://www.enviroworld.com
Regulations, wildlife, hazard waste/materials, ground/sur-
face/drinking water and other issues concerning Florida's environment are emphasized in this publication for the
environmental management team.
Dave Newport, Publisher

4158 Environmental Engineering Science

Mary Ann Liebert
140 Huguenot Street   914-740-2100
3rd Floor   800-MLI-EBER
New Rochelle, NY 10801-5215   Fax: 914-740-2101
E-mail: info@liebertpub.com
http://www.liebertpub.com
The focus is on pollution control of the suface, ground, and
drinking water, and highlight research news and product devel-
opments that aid in the fight against pollution.
Bi-Monthly
ISSN: 1092-8758
Mary Ann Liebert, Publisher
Dumpnico Grosso PhD, Editor-in-Chief

4159 Environmental Manager

Air and Waste Management Association
1 Gateway Center   412-232-3444
3rd Floor   800-270-3444
Pittsburgh, PA 15222-1435   Fax: 412-232-3450
E-mail: info@awma.org
http://www.awma.org
Features timely articles on business, regulatory, and technical
issues of interest to the environmental industry.
Tim Keener, Editor
Lisa Bucher, Managing Editor

4160 Environmental and Energy Study Institute

122 C Street NW   202-628-1400
Suite 630   Fax: 202-628-1825
Washington, DC 20001   E-mail: eesi@eesi.org
http://www.eesi.org
A nonprofit organization dedicated to promoting environmentally sustainable societies. EESI believes meeting this goal requires transitions to social and economic patterns that sustain
people, the environment and the natural resourcesupon which
present and future generations depend. EESI produces credible,
timely information and innovative public policy initiatives that
lead to these transitions. These products take the form of publications, briefings, work shops and taskforces.
Publication Date: 1984
Carol Werner, Executive Director

4161 Federation of Environmental Technologists

9451 North 107th Street   262-644-0700
Milwaukee, WI 53224   Fax: 262-644-7106
E-mail: info@fetinc.org
http://www.fetinc.org
A nonprofit organization formed to assist industry in interpreta-
tion of and compliance with environmental regulations. Mem-
bership is open to all industries, municipalities, organizations
and individuals concerned about environmentalregulations.
Currently there are approximately 1000 members and 125 pa-
tron companies.
Publication Date: 1982   Monthly
Triese Haase, Administrator

4162 Food Protection Trends

International Association for Food Protection
6200 Aurora Avenue   515-276-3344
Suite 200W   800-369-6337
Des Moines, IA 50322-2864   Fax: 515-276-8655
E-mail: info@foodprotection.org
http://www.foodprotection.org
Published as the general membership publication by the International Association for Food Protection, each issue contains ref-
ereed articles on applied research, applications of current
technology and general interest subjects for foodsafety professionals. Regular features include industry and association news,
an industry related product section and a calendar of meetings,
seminars and workshops.Updates of government regulations
and sanitary design is also featured. All membersreceive FPT.
Publication Date: 1981   80+ pages Monthly
ISSN: 1043-3546
David W Tharp, Executive Director
Lisa K Hovey, Managing Editor

4163 Hazardous Materials Intelligence Report

World Information Systems
129 Mount Auburn   617-491-5100
Cambridge, MA 02238-0535   Fax: 617-492-3312
Provides news analysis on environmental business, hazardous
materials, waste management, pollution prevention and control.
Covers regulations, legislation and court decisions, new tech-
nology, contract opportunities and awards andconference no-
tices.
Richard S Golob, Publisher
Roger B Wilson Jr, Editor

4164 Hazmat Transport News

Business Publishers
8737 Colesville Road   301-587-6300
10th Floor   800-274-6737
Silver Spring, MD 20910   Fax: 301-587-4530
E-mail: bpinews@bpinews.com
http://www.bpinews.com

Reports on the regulatory, enforcement, legislative and litigation developments affecting hazardous materials transportation.
Monthly
Leonard A Eiserer, Publisher
Beth Early, Associate Publisher

4165 Integrated Waste Management
McGraw Hill
1221 Avenue Of The Americas   212-904-2000
Suite C3A   800-372-1033
New York, NY 10020-1095   Fax: 212-512-2723
                                         http://www.mcgraw-hill.com
Articles geared toward integration of solid waste management.
8 pages
Kevin Hamilton, Publisher

4166 Iowa Academy of Science Journal
Iowa Academy of Science
174 Baker Hall   319-273-2581
Cedar Falls, IA 50614-0508   Fax: 319-273-2807
                                         E-mail: craig.johnson@uni.edu
                                         http://www.iacad.org

Quarterly

Craig Johnson, Executive Director

4167 Journal of Air and Waste Management Association
Air and Waste Management Association
1 Gateway Center, 3rd Floor   412-232-3444
420 Fort Duquesne Boulevard   800-270-3444
Pittsburgh, PA 15222-1435   Fax: 412-232-3450
                                         E-mail: info@awma.org
                                         http://www.awma.org
Publishes original, peer-reviewed research on a range of environmental topics.

ISSN: 1047-3289
Todd E Zahniser, Editor
Andy Knopes, Production Manager

4168 Journal of Environmental Engineering
American Society of Civil Engineers
1801 Alexandria Bell Drive   703-295-6257
Reston, VA 20191-4400   800-548-2723
                                         Fax: 703-295-6333
                                         http://www.asce.org
The journal of Environmental Engineering presents a collection of broad interdisciplinary information on the practice and status of research in environmental engineering science, systems engineering, and sanitation.
Publication Date: 2005
Mark Rood, Editor

4169 Journal of Environmental Quality
American Society of Agronomy
677 S Segoe Road   608-273-8080
Madison, WI 53711   Fax: 608-273-2021
                                         http://www.agronomy.org
Written for university, government and industry scientists interested in the impacts of environmental perturbations on the biological and physical sciences.
Bi-Monthly
G M Pierzynski, Editor
N T Basta, Technical Editor

4170 Journal of IEST
Institute of Environmental Sciences & Technology
5005 Newport Drive   847-255-1561
Suite 506   Fax: 847-255-1699
Rolling Meadows, IL 60004-3841   E-mail: iest@iest.org
                                         http://www.iest.org
An annual journal published by the Institute of Environmental Sciences & Technology.
Publication Date: 1957   185 pages   Annual
ISSN: 1098-4321
Charles W Berndt, VP Communications

4171 Kennedy-Jenks Consultants: Alert Newsletter
622 Folsom Street   415-243-2150
San Francisco, CA 94107   Fax: 415-896-0999
                                         http://www.kennedyjenks.com

Environmental engineering consulting company.
Publication Date: 1990   6 pages   3 times/year
Gordon Morris, Editor
Rina Chin, Editor

4172 Kennedy-Jenks Consultants: Spotlights
622 Folsom Street   415-243-2150
San Francisco, CA 94107   Fax: 415-896-0999
                                         http://www.kennedyjenks.com
Environmental engineering consulting company.
Publication Date: 1981   12 pages   3 times/year
Gordon Morris, Editor
Rina Chin, Edito

4173 Lead Detection & Abatement Contractor
IAQ Publications
7920 Norfolk Avenue   301-913-0115
Suite 900   Fax: 301-913-0119
Bethesda, MD 20814   E-mail: iaqpubs@aol.com
                                         http://www.iaqpubs.com
Feature articles include new on legislation, operational and safety issues that affect the removal of lead and lead by-products from paint, water, soil, and air.
Susan Valenti, Editor

4174 Leading Edge
Society of Exploration Geophysicists
8801 South Yale Avenue   918-497-5500
Tulsa, OK 74137   Fax: 918-497-5557
                                         E-mail: jlawnick@seg.org
                                         http://www.seg.org
Addresses a broad spectrum of topics related to applied geophysics. Material immediately accessible to a broad audience.
116 pages
ISSN: 1070-485X
Dean Clark, Editor
Dolores Proubasta, Associate Editor

4175 McCoy's Hazardous Waste Regulatory Update Service
McCoy & Associates
25107 Genessee Trail Road   303-526-2674
Suite 200   Fax: 303-526-5471
Golden, CO 80228
Offers a complete text of the federal hazardous waste regulations, summaries, interpretations and indexes.

4176 McCoy's Regulatory Analysis Service
McCoy & Associates
25107 Genessee Trail Road   303-526-2674
Suite 200   Fax: 303-526-5471
Golden, CO 80401
Provides timely, in-depth analyses of hazardous waste regulations within 10 working days after their publication in the Federal Register.

4177 Medical Waste News
Business Publishers
8737 Colesville Road   301-589-5103
10th Floor   800-274-6737
Silver Spring, MD 20910-3928   Fax: 301-589-8493
                                         E-mail: custserv@bpinews.com
                                         http://www.bpinews.com
Reports on the rapidly evolving legislative and regulatory actions in medical waste management. Includes coverage of incineration, laboratory wastes, infection control, liability and legal issues and waste transport.
Publication Date: 1963
Leonard A Eiserer, Publisher
Beth Early, Operations Director

4178 Noise Regulation Report
Business Publishers
8737 Colesville Road   301-589-5103
10th Floor   800-274-6737
Silver Spring, MD 20910-3928   Fax: 301-589-8493
                                         E-mail: custserv@bpinews.com
                                         http://www.bpinews.com
Exclusive coverage of airport, highway, occupational and open space noise, noise control and mitigation issues.

Publication Date: 1963
Leonard A Eiserer, Publisher
Beth Early, Operations Director

4179 Pollution Engineering

Cahners Business Information
2000 Clearwater Drive 630-320-7000
Oak Brook, IL 60523 Fax: 630-288-8282
http://www.pollutionengineering.com
Serves the field of pollution control in manufacturing industries, utilities, consulting engineers and constructors. Also serves government agencies including administration of federal, state and local environmental programs.

ISSN: 0032-3640
Barbara Olsen, Publisher
Roy Bigham, Managing Editor

4180 RMT Newsletter

744 Heartland Trail 608-831-4444
PO Box 8923 800-283-3443
Madison, WI 53717-1934 Fax: 608-831-3334
E-mail: info@rmtinc.com
http://www.rmtinc.com
Global engineering and management consulting firm that develops environmental solutions for industry. With a 600 person staff and 20 offices throughout the US and Europe, we help clients sustain the environment while meeting theirbusiness objectives. Engineers, scientists and construction managers can take a project from conception through successful completion. Expertise includes air, water and waste permitting, remediation, hazardous/solid waste management, air pollutioncontrol and more.
8 pages Quarterly
Jodi Burmester, Marketing Communications

4181 SPAC Newsletter

Soil and Plant Analysis Council
621 Rose St 402-437-4944
Lincoln, NE 68502-2040 Fax: 402-476-7598
E-mail: spcouncil@aol.com
http://www.spcouncil.com
Quarterly newsletter.
Quarterly
Mark Flock, President
Bryon Vaughan, Secretary/Treasurer

4182 Sludge

Business Publishers
8737 Colesville Road 301-589-5103
10th Floor 800-274-6737
Silver Spring, MD 20910-3928 Fax: 301-589-8493
E-mail: custserv@bpinews.com
http://www.bpinews.com
Premier insider guide to the biosolids industry. Follows developments in and management of beneficial use and wastewater residuals, with practical information about industrial sludge, incineration, special wastes, permits andlandfills.
BiWeekly
Leonard A Eiserer, Publisher
Beth Early, Operations Director

4183 Solid Waste Report

Business Publishers
8737 Colesville Road 301-589-5103
10th Floor 800-274-6737
Silver Spring, MD 20910-3928 Fax: 301-589-8493
E-mail: custserv@bpinews.com
http://www.bpinews.com
Comprehensive news and analysis of legislation, regulation and litigation in solid waste management. Regularly features federal rules, congressional actions, state updates and business trends.
Weekly
Leonard A Eiserer, Publisher
Beth Early, Associate Publisher

4184 Waste News

Crain Communications
1725 Merriman Road 330-836-9180
#300 Fax: 330-836-1692
Akron, OH 44313 E-mail: editorial@wastenews.com
http://www.wastenews.com

Editorial content focuses on waste management and recycling issues, primarily how businesses deal with the waste they generate. Covers waste management service providers, legistlative and regulatory environmental issues, emergingtechnologies, municipal recycling and waste issues, commodity market price, mergers, aquisitions and expansions.
Publication Date: 1995 Bi-Weekly
ISSN: 1091-699
Allan Gerlat, Editor
Brennan Lafferty, Managing Editor

4185 Widener University: International Conference on Solid Waste Magazine

One University Place 610-499-4042
Chester, PA 19013-5792 Fax: 610-499-4059
E-mail: solid.waste@widener.edu
http://www.widener.edu/solid.waste
Publication of the annual conference on solid waste technology and management. Over 100 speakers from 25 countries present their work. Proceedings available.
Publication Date: 1983 Quarterly
ISSN: 1088-1697
Ronald L Mersky, Editor
Wen K Shieh, Associate Editor

Periodicals: Environmental Health

4186 American College of Toxicology

9650 Rockville Pike 301-571-1840
Bethesda, MD 20814 Fax: 301-571-1852
E-mail: ekagan@actox.org
http://www.actox.org
The American College of Toxicology is a 501-0-3 nonprofit organization. It is not a degree-granting organization. The American College of Toxicology is dedicated to providing an interactive forum for the advancement and exchange oftoxicologic information between industry, goverment, and academia. There is an annual meeting in November each year. The ACT publiches a journal, International Journal of Toxicology on a bi-monthly basis.

ISSN: 1091-5818
Carol L Lemire, Executive Director
Eve Gamzu Kagan, Assistant Executive Director

4187 Asbestos & Lead Abatement Report

Business Publishers
8737 Colesville Road 301-589-5103
10th Floor 800-274-6737
Silver Spring, MD 20910-3928 Fax: 301-589-8493
E-mail: custserv@bpinews.com
http://www.bpinews.com
Contains articles on regulation compliance, environmental trends and business opportunities.
Leonard A Eiserer, Publisher

4188 Aviation, Space and Environmental Medicine

Aerospace Medical Association
320 S Henry Street 703-739-2240
Alexandria, VA 22314-3579 Fax: 703-739-9652
E-mail: pday@asma.org
http://www.asma.org
Provides contact with physicians, life scientists, bioengineers, and medical specialists working in both basic medical research and in its clinical applications.
Publication Date: 1929 112 pages Monthly
ISSN: 0095-5562
Sarah A Nunneley MD, Author
Sarah A Nunneley, Editor-in-Chief
Sarah A Pierce-Rubio, Editor Assistant

4189 Bio Integral Resource Center: Common Sense Pest Control

PO Box 7414 510-524-2567
Berkeley, CA 94707 Fax: 510-524-1758
E-mail: birc@igc.org
http://www.birc.org
Features least toxic solutions to pest problems of the home and garden. Those who are chemically sensitive and looking for alternatives may find what they need in the Quarterly.

Publication Date: 1984   24 pages   Quarterly
ISSN: 8756-7881
Dr. William Quarles, Executive Director

4190 Bio Integral Resource Center: IPM Practitioner
PO Box 7414   510-524-2567
Berkeley, CA 94707   Fax: 510-524-1758
                                     E-mail: birc@igc.org
                                     http://www.birc.org
Focuses on management alternatives for pests such as insects,
mites, ticks, vertebrates, weeds and plant pathogens.
Publication Date: 1979   24 pages   10 Times a Year
ISSN: 0738-969x
Dr. William Quarles, Executive Director

4191 Center for Statistical Ecology & Environmental Statistics: Environmental & Ecological Statistics
Kluwer Academic Publishers
Pennsylvania State University   814-865-9442
421 Thomas Building, Dept of Statistics   Fax: 814-865-1278
University Park, PA 16802   E-mail: gpp@stat.psu.edu
                                     http://www.stat.psu.edu
An international forum for cross-disciplinary discussion for sta-
tistical ecology and environmental statistics
Publication Date: 1994   110 pages   Quarterly
ISSN: 1352-8505
Ganapati P Patil, Editor-in-Chief
Wolfgang Urfer, Associate Editor

4192 EH&S Software News Online
Donley Technology
220 Garfield Avenue   804-224-9427
PO Box 152   Fax: 804-224-7958
Colonial Beach, VA 22443   E-mail: donleytech@donleytech.com
                                     http://www.donleytech.com
Reports on news and upgraded software products, database, and
on-line systems from commercial developers and government
resources.
Quarterly
John Donley, Editor
Elizabeth Donley, Managing Editor/Publisher

4193 Enterprise Software: Essential EH&S
Essential Technologies
1401 Rockville Pike   301-284-3000
#500   Fax: 301-284-3001
Rockville, MD 20852-1446   E-mail: info@essentech.com
                                     http://www.essential-technologies.com
Integrated solutions for emissions management, hazard commu-
nication, compliance management, occupational health and
safety and contingency management.
James Morentz, Publisher

4194 Environmental Connections
Connecticut College
270 Mohegan Avenue   860-439-5417
Box 5293   Fax: 860-439-2418
New London, CT 06320-4196   E-mail: ccbes@conncoll.edu
                                     http://www.ccbes.conncoll.edu
Publication Date: 1998   10 pages   2 times per year
Robert Askins, Director
Diane Whitelaw, Assistant Director

4195 Environmental Dimensions
Trine Publishers
28 Kilbarry Crescent   613-749-3735
Ottawa, Ontario K1K-0G8   Fax: 613-749-6807
                                     E-mail: trine@istar.ca
                                     http://www.envirodim.com
Provides the environmental professionals with news and information on current environmental health issues, solution, and
hazards, as well as examining governmental policies, legal news
and Canadian's environmental preformance.
22 times per year
Roland Blassnig, Editor

4196 Environmental Health Information Service
1001 Winstead Drive   866-541-3841
Suite 355   Fax: 919-678-8696
Cary, NC 27513   E-mail: ehponline@niehs.nih.gov
                                     http://www.ehis.niehs.nih.gov

Monthly
Thomas J Goehl, Editor-in-Chief

4197 Environmental Health Letter
Business Publishers
8737 Colesville Road   301-589-5103
10th Floor   800-274-6737
Silver Spring, MD 20910-3928   Fax: 301-589-8493
                                     E-mail: custserv@bpinews.com
                                     http://www.bpinews.com
Comprehensive coverage of the latest policies and
ground-breaking research that explores the potential links between environmental factors and human health.
8 pages
Leonard A Eiserer, Publisher
Beth Early, Operations Director

4198 Florida Journal of Environmental Health
Florida Environmental Health Association
PO Box 271823   813-962-0176
Tampa, FL 33688-1823   Fax: 813-962-0176
                                     E-mail: Scexedir@aol.com
                                     http://www.feha.org
Promotes public health by means of advanced environmental
control.
Quarterly
ISSN: 0897-1823
Lu Grimm, Editor
Jennifer M Willimas, Executive Director

4199 Healthy Schools Network Newsletter
773 Madison Avenue   518-462-0632
Albany, NY 12208   Fax: 518-462-0433
                                     E-mail: info@healthyschools.rog
                                     http://www.healthyschools.org
HSN is a nationally recognized state-based advocate for the protection of children's environmental health in schools. Engages
in research, education, outreach, technical assistance and coalition building to create schols that areenvironmentally responsible to children, and to their communities. Publishes a quarterly
newsletter and maintains an Information Clearinghouse and Referral Service.
Quarterly
Claire L Barnett, Executive Director

4200 Human Ecology Action League Magazine
2250 N Druid Hills Road NE   404-248-1898
Atlanta, GA 30329   Fax: 404-248-0162
                                     E-mail: HEALNatnl@aol.com
                                     http://members.aol.com/HEALNatnl/index/html
The Human Ecology Action League Inc (HEAL) is a nonprofit
organization founded in 1977 to serve those whose health has
been adversely affected by environment exposures; to provide
information to those who are concerned about the healtheffects
of chemicals; and to alert the general public about the potential
dangers of chemicals. Referrals to local HEAL chapters and
other support groups are available from the League.
Publication Date: 1977   35 pages   Quarterly
ISSN: 8755-7878
Diane Thomas, Editor

4201 Indoor Environment Review
IAQ Publications
7920 Norfolk Avenue   301-913-0115
Suite 900   Fax: 301-913-0119
Bethesda, MD 20814   E-mail: iaqpubs@aol.com
                                     http://www.iaqpubs.com
New technology, research and legislation concerning all indoor
air and water quality issues.
Robert Morrow, Editor

4202 Industrial Health and Hazards Update
InfoTeam
PO Box 15640   954-473-9560
Plantation, FL 33318-5640   Fax: 954-473-0544
                                     E-mail: infoteamma@aol.com
Covers occupational safety, health hazards, and disease; mitiga-
tion and control of hazardous situations; waste recycling and
treatment.
20 pages
Dr. David Allen, Associate Editor

4203 Journal of Environmental Health
National Environmental Health Association
720 S Colorado Boulevard   303-756-9090
Suite 970-S   Fax: 303-691-9490
Denver, CO 80246-1925   E-mail: staff@neha.org
        http://www.neha.org
A practical journal containing information on a variety of environmental health issues.
        Publication Date: 1937   70-76 pages   10 Times a Year
        ISSN: 0022-0892
Nelson E Fabain, Executive Director
Julie Collins, Content Editor

4204 Journal of Medical Entomology
Journal of Entomology
10001 Derkwood Land   301-731-4535
Suite 100   Fax: 301-731-4538
Lanham, MD 20706-4876   E-mail: esa@entsoc.org
        http://www.entsoc.org
Contributions report on all phases of medical entomology and medical acarology, including the systematics and biology of insects, acarines, and other arthropods of public health and veterinary significance.
        Bi=Monthly
John Edman, Editor-in-Chief

4205 Journal of Pesticide Reform
PO Box 1393   541-344-5044
Eugene, OR 97440-1393   Fax: 541-344-6923
        E-mail: info@pesticide.org
        http://www.pesticide.org
Pesticide factsheets, alternatives factsheets for common pest problems, and helpful information on how to take action for change are featured in this journal. Each issue also includes updates on NCAP's work, news on pesticide issues,and reviews of books and videos.
        Publication Date: 1984   24 pages   Quarterly
Caroline Cox, Editor
Norma Grier, Executive Director

4206 National Institute of Environmental Health Sciences Journal
PO Box 12233   919-496-2433
Research Triangle Park, NC 27709   Fax: 919-496-8276
        E-mail: olden@NIEHS.nih.gov
        http://www.niehs.nih.gov
The National Institute of Environmental Health Sciences is the principal federal agency for basic biomedical research on the health effects of environmental agents. It is the headquarters for the National Toxicology Program whichcoordinates toxicology studies within the Department of Health and Human Services.
        Publication Date: 1972   150 pages   Monthly
        ISSN: 0091-6765
Kenneth Olden PhD, Director

4207 Natural Resources Council of America: Environmental Resource Handbook
Universal Reference Publications
1025 Thomas Jefferson Street NW   518-789-0545
Suite 109   Fax: 202-333-0412
Washington, DC 20001http://www.NaturalResourcesCouncil.org
Environmental Resource Handbook updates.
        4 pages
Laura Seal, Membership Coordinator

4208 Natural Resources Council of America: Conservation Voice
1025 Thomas Jefferson Street NW   202-333-0411
Suite 109   Fax: 202-333-0412
Washington, DC 20001http://www.naturalresourcescouncil.org
Charts the news, events, and personnel that shape the face of the conservation movement and includes the quarterly supplemental publication NEPA news.
        Publication Date: 1958   6-8 pages   Bi-Monthly
Andrea Yank, Executive Director

4209 Natural Resources Council of America: NEPA News
1025 Thomas Jefferson Street NW   202-333-0411
Suite 109   Fax: 202-333-0412
Washington, DC 20001http://www.naturalresourcescouncil.org

        Publication Date: 1994   8 pages   Quarterly
Andrea Yank, Executive Director

4210 Noise Regulation Report
Business Publishers
8737 Colesville Road   301-587-6300
Suite 1100   800-274-6737
Silver Spring, MD 20910   Fax: 301-587-4530
        E-mail: bpinews@bpinews.com
        http://www.bpinews.com
Exclusive coverage of airport, highway, occupational and open space noise, legislation and mitigation.
        Monthly
Leonard A Eiserer, Publisher
Beth Early, Associate Publisher

Periodicals: Gaming & Hunting

4211 American Bass Association Newsletter
402 N Prospect Avenue   310-376-1026
Redondo Beach, CA 90277   Fax: 310-376-5072
        E-mail: feedback@americanbass.com
        http://www.americanbass.com
        Publication Date: 1989   24 pages   Quarterly
Craig Sutherland, Editor

4212 Caller
770 Augusta Road   803-637-3106
PO Box 530   800-THE-NWTF
Edgefield, SC 29824   Fax: 803-637-0034
        E-mail: nwtf@nwtf.net
        http://www.nwtf.org
News magazine is the NWTF's member publication that allows NWTF volunteers to share information about their successful conservation projects, Wild Turkey Super Fund Banquets and educational events for youth, women and disabledindividuals. Free with membership.
        Quarterly
Jason Gilbertson, Editor

4213 IPPL News
PO Box 776   843-871-2280
Summerville, SC 29484   Fax: 843-871-7988
        E-mail: info@ippl.org
        http://www.ippl.org
Educates readers in more than 50 countries about action that can be taken to protect primates.
        Publication Date: 1974   32 pages   3 Times a Year
        ISSN: 1040-3027
Dr Shirley McGreal, Chairperson
Marjorie Doggett, Secretary

4214 International Game Fish Association Newsletter
300 Gulf Stream Way   954-927-2628
Dania Beach, FL 33004   Fax: 954-924-4299
        E-mail: hq@igfa.org
        http://www.igfa.org
Founded as record-keeper and to maintain fishing rules. Today, emphasis is on conservation and education. Newsletters published are World Record Game Fish, annually and The International Angler, bi-monthly.
        Bi-Monhtly
Rob Kramer, President

4215 JAKES Magazine
Wild Turkey Center   803-637-3106
PO Box 530   800-THE-NWTF
Edgefield, SC 29824   Fax: 803-637-0034
        E-mail: nwtf@nwtf.net
        http://www.nwtf.org
JAKES (Juniors Acquiring Knowledge, Ethics and Sportsmanship) is a magazine which provides fun and educational articles focusing on young hunters, outdoor activities, the environment and other items of interest to readers 17 years oldand younger. Free with membership.
        Quarterly
Matt Lindler, Editor

4216 Mid-Atlantic Fishery Management Council Newsletter

300 S New Street
Federal Building, Room 2115
Dover, DE 19904

302-674-2331
Fax: 302-674-5399
http://www.mafmc.org

Publication Date: 1998 8 pages Quarterly
Daniel T Furlong, Executive Director

4217 Turkey Call

770 Augusta Road
PO Box 530
Edgefield, SC 29824

803-637-3106
800-THE-NWTF
Fax: 803-637-0034
E-mail: nwtf@nwtf.net
http://www.nwtf.org

A magazine for turkey hunting enthusiasts, provides articles to help you improve your hunting skills and learn how to enhance your land for wildlife. Free with membership.
Bi-Monthly
Doug Howlett, Editor

4218 Wheelin' Sportsmen

770 Augusta Road
PO Box 530
Edgefield, SC 29824

803-637-3106
800-THE-NWTF
Fax: 803-637-0034
E-mail: nwtf@nwtf.net
http://www.nwtf.org

Magazine for all disabled people and their able-bodied partners who are interested in the outdoors, especially recreational shooting, hunting and fishing. Free with membership.
Quarterly
Karen Roop, Editor

4219 Women in the Outdoors

770 Augusta Road
PO Box 530
Edgefield, SC 29824

803-637-3106
800-THE-NWTF
Fax: 803-637-0034
E-mail: nwtf@nwtf.net
http://www.nwtf.org

Magazine that delivers features on a variety of outdoor topics of interest to the novice and experienced outdoorswoman. Free with membership.
Quarterly
Karen Roop, Editor

Periodicals: Habitat Preservation & Land Use

4220 ANJEC Report

Association/New Jersey Environmental Commissions
PO Box 157
Mendham, NJ 07945

973-539-7547
Fax: 973-539-7713
E-mail: info@anjec.org
http://www.anjec.org

Nonprofit organization promoting public interest in natural resorce protection and supporting municipal environmental commissions throughout New Jersey.
Publication Date: 1970 20 pages Quarterly
Sandy Batly, Executive Director

4221 Afield

4705 University Drive
Suite 290
Durham, NC 27707

919-403-8558
Fax: 919-403-0379
E-mail: northcarolina@tnc.org
http://www.nature.org/northcarolina

Published by The Nature Conservancy, saving the last great places of North Carolina.
Quarterly Newsletter
Katherine Skinner, Executive Director

4222 Aldo Leopold Foundation: The Leopold Outdoor

400 Oak Street
Baraboo, WI 53913-0077

608-355-0279
Fax: 608-356-7309
E-mail: mail@aldoleopold.org
http://www.aldoleopold.org

A nonprofit organization founded in 1982, works to promote the philosophy of Aldo Leopold and the land ethic he so eloquently defined in his writing. The foundation actively integrates programs on land stewardship, environmentaleducation and scientific research to promote care of natural resources and have an ethical relationship between people and land.

Publication Date: 1999 8 pages Quarterly
Steve Swenson, Ecologist
Buddy Huffaker, Executive Director

4223 American Association for Advancement of Science: Animal Keeper's Forum

1200 New York Avenue Northwest
Washington, DC 20005

202-326-6400
Fax: 202-289-4985
E-mail: webster@aaas.org
http://www.aaas.org

The American Association for the Advancement of Science is the world's largest general science and publisher of the peer-reviewed journal. With more than 138,000 members and 275, AAAS serves as an authoritative source for informationon the latest developments in science and bridges gaps among scientists, policy-makers and the public to advance science and science education.
Publication Date: 1947 54 pages Monthly
ISSN: 0164-9531
Alan Leshner, Executive Director

4224 American Birding Association: Birding

PO Box 6599
Colorado Springs, CO 80934

719-578-9703
800-850-2473
Fax: 719-578-1480
E-mail: member@aba.org
http://www.americanbirding.org

The American Birding Association represents the interests of birdwatchers in various arenas, and helps birders increase their knowledge, skills, and enjoyment of birding. ABA also contributes to bird conservation by linking the skillsof its members to on-the-ground projects. ABA promotes field-birding skills through meetings, workshops, equipment, and guided involvement in birding, promoting national and international birders networks and publications.
Publication Date: 1972 8 pages Bi Monthly
Steve Runnels, Executive Director
Ted Floyd, Editor

4225 American Birding Association: Winging it

PO Box 6599
Colorado Springs, CO 80934-6599

719-578-9703
800-850-2473
Fax: 719-578-1480
E-mail: member@aba.org
http://www.americanbirding.org

The American Birding Association represents the interests of birdwatchers in various arenas, and helps birders increase their knowledge, skills, and enjoyment of birding. ABA also contributes to bird conservation by linking the skillsof its members to on-the-ground projects. ABA promotes field-birding skills through meetings, workshops, equipment, and guided involvement in birding, promoting national and international birders networks and publications.
Publication Date: 1989 24 pages Bi Monthly
Steve Runnels, Executive Director
Rick Wright, Editor

4226 American Entomologist

Journal of Entomology
9301 Annapolis Road
Lanham, MD 20706-3115

301-731-4535
Fax: 301-731-4538
E-mail: esa@entsoc.org
http://www.entsoc.org

American Entomologist is a quarterly, general interest entomology magazine written for both scientists and nonscientists. It publishes colorful, illustrated feature articles, peer-reviewed scientific reports, provocative and humorouscolumns, letters, book reviews, and obituaries.
Publication Date: 1955 64 pages Quarterly
ISSN: 1046-2821
Gene Kristky, Editor-in-Chief

4227 American Forests Magazine

910 17th Street NW, Suite 600
PO Box 2000
Washington, DC 20013

202-955-4500
800-368-5748
Fax: 202-955-4588
E-mail: info@amfor.org
http://www.americanforests.org

Published by American Forests, the oldest national citizens conservation organization in the US.
Quarterly
Gerald Gray, VP Forest Policy Center
Deborah Gangloff, Executive Director

4228 Annals of the Entomological Society of America
Journal of Entomology
10001 Derekwood Lane 301-731-4535
Suite 100 Fax: 301-731-4538
Lanham, MD 20706-4876 E-mail: esa@entsoc.org
http://www.entsoc.org
Contributions report on the basic aspects of the biology of anthropods and are divided into categories by subject matter; systematics; ecology and population biology; arthropods in relation to plant disease; conservation biology andbiodiversity; physiology, biochemistry, and toxicology; morphology, history, and fine sructure; genetics and behavior.
Bi-Monthly
Joe B Keiper, Editor-in-Chief

4229 Annual Review of Entomology
Annual Reviews
10001 Derekwood Lane 301-731-4535
Suite 100 Fax: 301-731-4538
Lanham, MD 20706-4876 E-mail: esa@entsoc.org
http://www.entsoc.org
This is published in Januray and made available through ESA on a regular subscription basis. The series occupies a special place within the field of entomology. Authoritative critical reviews by eminent scientists provide a valuableresource for students, teachers, and researchers: specialists and nonspecialists.
Yearly
Alan Kahan, Director Communications

4230 Appalachian Mountain Club
Appalachian Mountain Club Books
5 Joy Street 617-523-0655
Boston, MA 02108 800-262-4455
Fax: 617-523-0722
E-mail: information@outdoors.org
http://www.outdoors.org
The AMC, founded in 1876, promotes the protection, enjoyment, and wise use of the mountains, rivers and trails of the Northeast. We encourage people to enjoy and protect the natural world because we believe that successful conservationdepends on this experience. The AMC publishes an award-winning magazine and more than 60 guide books to the Northeast.
Andrew Falender, Executive Director
Clair O'Connell, Director Development

4231 Arthropod Management Tests
Journal of Entomology
10001 Derekwood Lane 301-731-4535
Suite 100 Fax: 301-731-4538
Lanham, MD 20706-4876 E-mail: esa@entsoc.org
http://www.entsoc.org
This is published in late spring. The purpose is to promote timely dissemination of information on preliminary and routine screening tests on management of arthropods, both beneficial and harmful. Pest management methods tested andreported may include the use of chemical pesticides as well as other materials or agents, such as insect growth regulators, pheromones, natural enemies for biological control, or pest-resistant plants/animals. Reports are based on tests conducted byreseachers.
Yearly
David L Kerns, Editor-in-Chief

4232 Blowhole
3625 Brigantine Boulevard 609-266-0538
PO Box 773 Fax: 609-266-6300
Brigantine, NJ 08203 E-mail: mmsc@verizon.net
http://marinemammalstrandingcenter.org
A quarterly newsletter published by the Marine Mammal Stranding Center.
Publication Date: 1978 8 pages Quarterly
Robert C Schoelkopf, Director
Sheila M Dean, Co-Director

4233 Carolina Bird Club
5009 Crown Point Lane 910-791-5726
Wilmington, NC 28409 Fax: 910-791-7228
E-mail: hq@carolinabirdclub.org
http://www.carolinabirdclub.org

A nonprofit educational and scientific association, open to anyone interested in the study and conservation of wildlife, particularly birds. Meets each winter, spring and fall. Meeting sites are selected to give participants anopportunity to see many different kinds of birds. Guided field trips, informative programs and business sessions are combined for an exciting weekend of meeting with people who share an enthusiasm and concern for birds.
Publication Date: 1937 Quarterly
ISSN: 0009-1987
Dana Harris, HQ Secretary
Ken Fiala, Editor

4234 Coalition for Education in the Outdoors
Outdoor Education Research Education
SUNY at Cortland 607-753-4971
Box 2000 Fax: 607-753-5982
Cortland, NY 13045 E-mail: rts@nycorva.cortland.edu
http://www.outdooredcoalition.org
The Coalition is a nonprofit network of outdoor and environmental education centers, nature centers, conservation and recreation organizations, outdoor education and experiential education associations, institutions of higher learning,public and private schools and fish and wildlife agencies that share a mission-the support and furtherance of environmental and outdoor education and its goals.
Publication Date: 1993 48-52 pages Bi-Annual
ISSN: 1065-5204
Charles Yaple, Executive Director

4235 College of Tropical Agriculture and Human Resources:
Impact Report
University of Hawaii
3050 Maile Way 808-956-7056
Gilmore 119 Fax: 808-956-5966
Honolulu, HI 96822 E-mail: ocs@ctahr.hawaii.edu
http://www.ctahr.hawaii.edu
The vision of the college is to be the premier resource for tropical agricultural systems and resource management in the Asia-Pacific region. Its mission outlines a commitment to the preparation of students and all citizens of Hawaiifor life in a global community through research and education programs supporting tropical agricultural systems that foster viable communities, a diversified economy and a healthy environment.
Annual
Andrew G Hashimoto, Dean/Director

4236 Connecticut Woodlands
Middlefield 860-346-2372
16 Meriden Road Fax: 860-347-7463
Rockfall, CT 06481-2961 E-mail: conn.forest.assoc@snet.net
http://www.ctwoodlands.org
Quarterly publication of the Connecticut Forest and Park Association, an organization for forest and wildlife conservation. Develops outdoor recreation and natural resources. Provides forest management, construction of hiking trailsand consultation in the areas of forestry and environment.
Quarterly
Chris Woodside, Editor

4237 Conservancy of Southwest Florida
Eye On The Issues
1450 Merrihue Drive 232-262-0304
Naples, FL 34102 Fax: 232-262-0672
E-mail: info@conservancy.org
http://www.conservancy.org
8 pages Quarterly
Kathy Prosser, President/CEO
Sheila Etalamaki, Director Finance

4238 Conservation & Natural Resources: Water Resources
Division, Nevada Wildlife Almanac
123 West Nye Lane 775-687-4380
Room 246 Fax: 775-687-6972
Carson City, NV 89706-0818 E-mail: hricci@wr.state.nv.us
http://ndwr.state.nv.us
Publication Date: 1996 12 pages Twice/year
Hugh Ricci, State Engineer

4239 Conservation Commission News
New Hampshire Association of Conservation Comm.
54 Portsmouth Street   603-224-7867
Concord, NH 03301-5486   E-mail: info@nhacc.org
              http://www.nhacc.org
Encourage conservation and appropriate use of New Hampshire's natural resources by providing assistance to New Hampshire's municipal conservation commissions and by facilitating communication among commissions and between commissionsand other public and private agencies involved in conservation.
8 pages
Marjory Swope, Publisher

4240 Conservation Communique
Wyoming Association of Conservation Districts
517 East 19th Street   307-632-5716
Cheyenne, WY 82001   Fax: 307-638-4088
            http://www.conservewy.com
Quarterly
Kelly Brown, Editor

4241 Conservation Conversation
Montana Association of Conservation Districts
201 North Mitchell Street   231-876-0328
Suite 203   Fax: 231-876-0312
Cadillac, MI 49601   E-mail: macd@macd.org
             http://www.macd.org
Quarterly

4242 Conservation Leader
Utah Association of Conservation Districts
1860 North 100 East   435-753-6029
Logan, UT 84341-1784   Fax: 435-755-2117
Quarterly

4243 Conservation Notes: New England Wildflower Society
180 Hemenway Road   508-877-7630
Framingham, MA 01701   508-877-6553
           Fax: 508-877-3658
       E-mail: newfs@newfs.org
          http://www.newfs.org
36 pages   Yearly

4244 Conservation Partner
Arkansas Association of Conservation Districts
101 E Capitol   501-682-2915
Suite 350   Fax: 501-682-3991
Little Rock, AR 72201   http://www.aracd.org
Quarterly
Debbie Moreland, Program Administrator

4245 Conservation Visions
Nebraska Association of Resources Districts
601 S 12th Street   402-471-7670
Suite 201   Fax: 402-471-7677
Lincoln, NE 68508   E-mail: nard@nrdnet.org
           http://www.nrdnet.org
Bi-Monthly
Dean E Edson, Executive Director

4246 Conservogram
Soil and Water Conservation Society
945 Southwest Ankeny Road   515-289-2331
Ankeny, IA 50023   Fax: 515-289-1227
       E-mail: pubs@swcs.org
          http://www.swcs.org
Published for the professionals in the natural resource fields, and contains highlights on the news and ideas in the preservation of natural resources.
Monthly
Suzi Case, Editor

4247 Consultant
Assoc of Consutling Foresters of America, Inc
312 Montgomery Street   703-548-0990
Suite 208   888-540-8733
Alexandria, VA 22314   Fax: 703-548-6395
     E-mail: director@acf-foresters.com
       http://www.acf-foresters.com
Publication from Association of Consulting Foresters of America, Inc.
Lynn C Wilson, Executive Director

4248 Cornell Lab of Ornithology: Birdscope
Birdscope
159 Sapsucker Woods Road   607-254-2475
Ithaca, NY 14850   Fax: 607-254-2415
      E-mail: cornellbirds@cornell.edu
        http://www.birds.cornell.edu
Quarterly,Newsletter
Allison Wills, Communications Director

4249 Cornell Lab of Ornithology: Living Bird
Birdscope
159 Sapsucker Woods Road   607-254-2475
Ithaca, NY 14850   Fax: 607-254-2415
      E-mail: cornellbirds@cornell.edu
        http://www.birds.cornell.edu
Quarterly, Magazine
Allison Wills, Communications Director

4250 Department of Natural Resources
301 Centennial Mall South   402-471-2363
PO Box 94676   Fax: 402-471-2900
Lincoln, NE 68509-4676   E-mail: webmaster@dnr.state.ne.us
          http://www.dnr.state.ne
Quarterly
Roger K Patterson, Director

4251 District Connection Newsletter
North Carolina Assoc. of Soil/Water Cons. Dist.
512 North Salisbury Street   919-733-2302
Raleigh, NC 27604   Fax: 919-715-3559
       http://www.enr.state.nc.us\DSWC
Monthly
David Williams, Acting Director

4252 E Magazine
Doug Moss
28 Knight St   203-852-0773
Norwalk, CT 06851   Fax: 203-866-0602
      E-mail: jessica@emagazine.com
        http://www.emagazine.com
A comprehensive magazine dealing with environmental issues and national conservation concerns.
Publication Date: 1990   64 pages   Bimonthly
ISSN: 1046-8021
Doug Moss, Publisher/Executive Director

4253 ESA Newsletter
Journal of Entomology
10001 Derekwood Lane   301-731-4535
Suite 100   Fax: 301-731-4538
Lanham, MD 20706-4876   E-mail: esa@entsoc.org
        http://www.entsoc.org
ESA Newsletter is a monthly publication presenting timely information of interest to ESA members. In addition to feature articles, it contains meeting announcements, society business, listing of employment oppurtunities, notice ofgrants and awards, member profiles, and branch and section news.
Monthly
Lisa Spurlock, Editor

4254 Eagle Nature Foundation Ltd: Bald Eagle News
Bald Eagle News
300 E Hickory   815-594-2306
Apple River, IL 61001   Fax: 815-594-2305
      E-mail: eaglenature.tni@juno.com
        http://eaglenature.org
A quarterly publication from the Eagle Nature Foundation.
Publication Date: 1992   20 pages   Quarterly
Terrence N Ingram, Executive Director

4255 Earth Steward
Michigan Association of Conservation Districts
201 North Mitchell Street   231-876-0328
Suite 203   Fax: 231-876-0372
Cadillac, MI 49601   E-mail: macd@macd.org
                                         http://macd.org

Quarterly
Marilyn Shy, Executive Director

4256 Ecosphere
Forum International
91 Gregory Lane   925-671-2900
Suite 21   800-252-4475
Pleasant Hill, CA 94523-4914   Fax: 925-671-2993
                    E-mail: fti@foruminternational.com
                    http://www.foruminternational.com
Accepts advertising. The first ever environmental/ecological
magazine. It is dedicated to the interrelations of man in nature
and a balanced approach of its biological, economic, socio-po-
litical and spiritual components. Since 1965.
        Publication Date: 1965   16-48 pages Bi Monthly
Dr. Nicolas Hetzer, Production Manager
J McCormack, Circulation Director

4257 Elm Leaves
Elm Research Institute
11 Kit Street   603-358-6198
Keene, NH 03431   800-367-3567
                                         Fax: 603-358-6305
                    E-mail: libertyelm@webryders.com
                    http://www.libertyelm.com
A semi annual publication published by the Elm Research Insti-
tute. Free with membership.
        Publication Date: 1967   4-9 pages Semi-Annually
John Hansel, Editor/Executive Director
Yvonne Spalthoff, Assistant Director

**4258 Environmental Concern: The Wonders of Wetlands , an
Educators Guide**
POW-The Planning of Wetlands
201 Boundary Lane   410-745-9620
PO Box P   Fax: 410-745-3517
St Michaels, MD 21663   E-mail: order@wetland.org
                                         http://www.wetland.org
Since its founding in 1972, EC has been specializing in consulting, planning design, education services, construction services
and research related to all aspects of wetlands. As wetlands and
contiguous upland forests and meadows areinteracting ecosystems EC specializes in consulting, planning, design, and project
supervision services for such upland ecosystem constructions
and restorations for the purpose of wetland buffers, reforestation, wildlife habitat and critical areas ofpreservation.
        Publication Date: 1995   Quarterly
        ISSN: 1095-2063
Edgar W Garbisch, President

4259 Environmental Entomology
Journal of Entomology
10001 Derekwood Lane   301-731-4535
Suite 100   Fax: 301-731-4538
Lanham, MD 20706-4876   E-mail: esa@entsoc.org
                                         http://www.entsoc.org
Contributions report on the interaction of insects with biological, chemical, and physiological and chemical ecology (abiotic
effects, pheromonea, effects of miscellaneous pollutants), com-
munity/ecosystem ecology (trophic-levelsstudies, associations), population ecology (mating, reproduction, movement,
behavior, parasitism, predation, microbial ecology, insect-plant
relations), pest management and sampling (integrated pest management, sampling, distribution), and biologicalcontrol.
        Bi-Monthly
E Alan Cameron, Editor-in-Chief

4260 Everglades Reporter
Friends of the Everglades
7800 Red Road   305-669-0858
Suite 215K   Fax: 305-669-4108
South Miami, FL 33143   E-mail: info@everglades.org
                                         http://www.everglades.org
Protecting the Everglades. A bi-annual publication from Friends
of the Everglades.

        Publication Date: 1971   8 pages Quarterly
David Reiner, President

4261 Fisheries
American Fisheries Society
5410 Grosvenor Lane   301-897-8616
#100   Fax: 307-897-8096
Bethesda, MD 20814-2144   E-mail: main@fisheries.org
                                         http://www.fisheries.org
Peer-reviewed articles that address contemporary issues and
problems, techniques, philosophies and other areas of interest to
the general fisheries profession. Monthly features include letters, meeting notices, book listings andreviews, environmental
essays and organization profiles.
Kristin Merriman-Clarke, Editor

4262 Fisheries Focus: Habitat Hotline Atlantic
Atlantic States Marine Fisheries Commission
1444 I Street Northwest   202-289-6400
6th Floor   Fax: 202-289-6051
Washington, DC 20005-2210   E-mail: comments@asmfc.org
                                         http://www.asmfc.org
        Monthly
John V O'Shea, Executive Director
Tina L Berger, Public Affairs/Resource Spec

4263 Forest History Society
701 William Vickers Avenue   919-682-9319
Durham, NC 27701-3162   Fax: 919-682-2349
                                E-mail: coakes@duke.edu
                    http://www.lib.duke.edu/forest
The Forest History Society is a non-profit educational institution that links the past to the future by identifying, collecting,
preserving, interpreting, and disseminating information on the
history of people, forests, and theirrelated resources.
        Publication Date: 1946
Cheryl Oakes, Librarian

4264 Forest Voice
Native Forest Council
PO Box 2190   541-688-2600
Eugene, OR 97402   Fax: 541-461-2156
                                E-mail: info@forestcouncil.org
                                http://www.forestcouncil.org
Quarterly publication from Native Forest Council.
        Publication Date: 1989   16 pages Quarterly
        ISSN: 1069-2002
Timothy Hermach, President
Debbie Shivers, Office Manager

4265 Forestry Source
Society of American Foresters
5400 Grosvernor Lane   301-897-8720
Bethesada, MD 20814-2198   Fax: 301-897-3690
                                E-mail: safweb@safnet.org
                                http://www.safnet.org
Tabloid newsletter covering important information regarding
critical issues in forestry research and technology, legislative
updates and news about SAF programs and activities on a national and local level.
        20 pages Monthly
Matt Walls, Editor

4266 Friday Letter
Michigan Association of Conservation Districts
201 North Mitchell Street   231-876-0328
Suite 203   Fax: 231-876-0372
Cadillac, MI 49601   E-mail: macd@macd.org
                                         http://www.macd.org
        Monthly
Marilyn Shy, Executive Director

**4267 Game & Fish Commission Wildlife Management Divi-
sion Newsletter**
2 Natural Resources Drive   501-223-6300
Little Rock, AR 72205   800-364-4263
                                         Fax: 501-223-6452
                                         http://www.agfc.com
Dedicated to managing wildlife in the state of Arkansas.
        Publication Date: 1920   33-35 pages 5x year
Donny Harris, Chief

4268 Golden Gate Audubon Society
2530 San Pablo Avenue
Suite G
510-843-2222
Fax: 510-843-5351
Berkeley, CA 94702   E-mail: ggas@goldengateaudubon.org
http://www.goldengateaudubon.org
Monthly publication from Golden Gate Audubon Society. Published 10 times yearly.
Publication Date: 1917   12 pages Monthly
ISSN: 0164-971x
Elizabeth Murdock, Executive Director

4269 Great Plains Native Plant Society
PO Box 641
Hot Springs, SD 57747
605-745-3397
Fax: 605-745-3397
E-mail: cascade@gwtc.net
Publication Date: 1990   4-8 pages Intermittant
Cynthia Reed, President

4270 Green Space
New York Parks and Conservation Association
29 Elk Street
Albany, NY 12207
518-434-1583
Fax: 518-427-0067
E-mail: ptny@ptny.org
http://www.nypca.org
Semi-Annual
Robin Dropkin, Executive Director

4271 Habitat Hotline
Atlantic States Marine Fisheries Commission
1444 I Street Northwest
6th Floor
Washington, DC 20005-2210
202-289-6400
Fax: 202-289-6051
E-mail: info@asmfc.org
http://www.asmfc.org
6 pages Quarterly
John V O'Shea, Executive Director

4272 Illinois Audubon Society
425B N Gilbert Street
PO Box 2418
Danville, IL 61832
217-446-5085
Fax: 217-446-6375
E-mail: director@pdnt.com
http://www.illinoisaudubon.org
A membership organization dedicated to the preservation of Illinois Wildlife and the habitats which support them. Has sanctuaries, conservation education and land acquisition programs and publishes quarterly magazines and newsletters.
Publication Date: 1916   28 pages Quarterly
ISSN: 1061-9801
Marilyn F Campbell, Executive Director

4273 Illinois Environmental Council: IEC Bulletin
107 W Cook
Suite E
Springfield, IL 62704
217-544-5954
Fax: 217-544-5958
E-mail: iec@ilenviro.org
http://www.ilenviro.org
The IEC is a coalition of over 70 environmental, conservation and health groups.
Bi-Monthly
Jonathan Goldman, Executive Director
Jennifer Sublett, Outreach Coordinator

4274 In Brief
223 S King Street
Suite 400
Honolulu, HI 96813-4501
808-599-2436
Fax: 808-521-6841
E-mail: eajushi@earthjustice.org
http://www.earthjustice.org
Newsletter of Earthjustice, a nonprofit public interest law firm dedicated to protecting the magnificent places, natural resources and wildlife of this earth and to defending the right of all people to a healthy environment. We bringabout far-reaching change by enforcing and strengthening environmental laws on behalf of hundreds of organizations and communities.
Quarterly
Douglas Hannold, Managing Attorney

4275 Iowa Cooperative Fish & Wildlife Research Unit: Annual Report
Iowa State University
Science Hall II
Ames, IA 50011
515-294-3056
Fax: 515-294-5468
Publication Date: 1932   50 pages Annual

4276 Iowa Native Plant Society Newsletter
Iowa State University
Botany Department
Ames, IA 50011-1020
515-294-9499
Fax: 515-294-1337
E-mail: dlewis@iastate.edu
An organization of amateur and professional botanists and native plant enthusiasts who are interested in the scientific, educational and cultural aspects, as well as the preservation and conservation of the native plants of Iowa. TheSociety was organized in 1995 to create a forum where plant enthusiasts, gardners and professional botanists could exchange ideas and coordinate activities such as field trips, work shops, and restoration of natural areas.
Publication Date: 1995   12 pages 3/4 times X year
Tom Rosburg, President
Deb Lewis, Contact Person

4277 Journal of Caves & Karst Studies
National Speleological Society
2813 Cave Avenue
Huntsville, AL 35810-4413
256-852-1300
Fax: 256-851-9241
E-mail: nss@caves.org
http://www.caves.org
A quarterly journal published by the National Speleological Society.
Quarterly
ISSN: 1090-6924
Stephanie Searles, Operations Manager
Dave Bunnell, Editor

4278 Journal of Economic Entomology
Journal of Entomology
10001 Derekwood Lane
Suite 100
Lanham, MD 20706-4876
301-731-4535
Fax: 301-731-4538
E-mail: esa@entsoc.org
http://www.entsoc.org
Contributions report on the economic significance of insects and are divided into categories by subject matter: apiculture and social insects; arthropods in relation to plant disease; biological and microbial disease; ecology andbehavior; ecotoxicology; extension; field and forage crops; forest entomology; horticultural entomology; household and structural insects; insecticide resistance and resistance management; medical entomology; plant resistance; sampling andbiostatistics.
Bi-Monthly
John T Trumble, Editor-in-Chief

4279 Land Use Law Report
Business Publishers
8737 Colesville Road
10th Floor
Silver Spring, MD 20910-3928
301-587-6300
800-274-6737
Fax: 301-587-4530
E-mail: custserv@bpinews.com
http://www.bpinews.com
Provides timely news on court decisions, legislation and regulations that impact today's most pressing land-use policy planning and legal issues.
Biweekly
James D Lawlor, Author
Adam Goldstein, Publisher
James D Lawlor, Editor

4280 Land and Water Magazine
Land and Water
320 A Street
Fort Dodge, IA 50501
515-576-3191
Fax: 515-576-2606
E-mail: landandwater@frontiernet.net
http://www.landandwater.com
Edited for contractors, engineers, architects, government officials and those working in the field of natural resource management and restoration from idea stage through project completion and maintenance.
Publication Date: 1974   72 pages Bimonthly
ISSN: 0192-9453
Amy Dencklau, Publisher/Editor

4281 Leaves Newsletter
Michigan Forest Association
1558 Barrington
Ann Arbor, MI 48103-5603
734-665-8279
Fax: 734-913-9167
E-mail: mfa@i-star.com
http://www.michiganforests.com
A monthly publication from the Michigan Forest Association.

Monthly
McClain B Smith Jr, Executive Director

4282 MACC Newsletter
Alba Press
10 Juniper Road   617-489-3930
Belmont, MA 02478   Fax: 617-489-3935
                                E-mail: staff@maccweb.org
                                http://www.maccweb.org
Published six times a year, each issue features carefully chosen
technical and interpreative articles, updates on government actions and policies, notices of workshops and meetings, publications, listings and a professional directory.
16 pages Bi-Monthly
Ingeborg Hegamann, President
Kenneth Pruitt, Executive Director

4283 Michigan Forests Magazine
Michigan Forest Association
1558 Barrington   734-665-8279
Ann Arbor, MI 48103-5603   Fax: 734-913-9167
                                E-mail: mfa@i-star.com
                                http://www.michiganforests.com
A quarterly magazine published by the Michigan Forest Associ-
ation.
Quarterly
McClain B Smith Jr, Executive Director

4284 Minnesota Plant Press
Minnesota Native Plant Society
220 Biological Science Center
1445 Gortner Avenue   E-mail: president@mnnps.org
Saint Paul, MN 55108   http://www.mnnps.org
A nonprofit organization dedicated to the conservation of the native plants of Minnesota through public education and advocacy. Offered are monthly meetings, field trips, symposia and a
regular newsletter.
4 per year
Jason Husveth, President
Gerry Drewry, Editor

4285 Monitor
Florida Defenders of the Environment
4424 NW 13 Street   352-378-8465
Suite C-8   Fax: 352-377-0869
Gainesville, FL 32609-1885   E-mail: fde@fladefenders.org
                                http://www.fladefenders.org
Newsletter of Florida Defenders of the Environment, one of the
oldest and most accomplished conservation organizations in
Florida with a network of scientists, economists and other professionals dedicated to preserving and protecting the state's natural resources. FDE's top priority is currently the restoration of
a 16-mile stretch of the Ocklawaha River and its 9,000-acre
floodplain forest by removal of Rodman Dam- the last vestige of
the Cross-Florida Barge Canal.
Publication Date: 1982 6-8 pages Bi-Annually
Nick Williams, Executive Director

4286 Montana Land Reliance Newsletter
324 Fuller Avenue   406-443-7027
PO Box 355   Fax: 406-443-7061
Helena, MT 59624-0355   E-mail: info@mtlandreliance.org
                                http://www.mtlandreliance.org
Montana's only private, statewide land trust, an apolitical, nonprofit corporation. Our mission is to provide permanent protection for private lands that are ecologically significant for
agricultural production, fish and wildlife habitat and scenic open
space. We publish a newsletter twice per year.
8 pages Spring/Fall
Jay Erickson, Managing Director
Rock Ringling, Managing Director

4287 NACD News & Views
National Association of Conservation Districts
509 Capitol Court, NE   202-547-6223
Washington, DC 20002-4937   Fax: 202-547-6450
                                http://www.nacdnet.org

Newsletter of the nonprofit organization that represents the nation's 3,000 conservation districts and 17,000 men and women
who serve on their governing boards. Conservation districts, local units of government established under statelaw to carry out
natural resource management programs at the local level, work
with more than 2.5 million cooperating landowners and operators to help them amange and protect land and water resources
on nearly 98% of the private lands in the UnitedStates.
Publication Date: 1952 8 pages Bi-Monthly
Maxine Mathis, Production Manager

4288 National Gardener Magazine
National Garden Clubs, Inc
4401 Magnolia Avenue   314-776-7574
St Louis, MO 63110-3492   800-550-6007
                                            Fax: 314-776-5108
                                E-mail: headquarters@gardenclub.org
                                http://www.gardenclub.org
National Garden Clubs, Inc publishes The National Gardener
Magazinequarterly.
Publication Date: 1970 48 pages Quarterly
ISSN: 0027-9331
Susan Davidson, Author
Susan Davidson, Editor

4289 National Grange Newsletter
1616 H Street Northwest   202-628-3507
Washington, DC 20006-4999   888-447-2643
                                            Fax: 202-347-1091
                                E-mail: rweiss@nationalgrange.org
                                http://www.nationalgrange.org
The Grange is a family based community organization with a
special interest in agriculture and rural america as well as in legislative efforts regarding these issues.
6 pages Bi-Monthly
William Steele, President
Richard Weiss, COO

4290 National Recreation and Park Association, (NRPA):
Parks & Recreation Magazine
Parks and Recreation Magazine
22377 Belmont Ridge Road   703-858-0784
Ashburn, VA 20148   Fax: 703-858-0794
                                E-mail: info@nrpa.org
                                http://www.nrpa.org
The NRPA, headquartered in Ashburn Virginia, is a national
nonprofit organization devoted to advancing park, recreation
and conservation efforts that enhance the quality of life for all
Americans. The Association works to extendsocial, health, cultural and economic benefits of parks and recreation, through its
network of 23,000 recreation and park professionals and civic
leaders. NRPA encourages recreation initiatives for youth in
high-risk environments.
100 pages Monthly
ISSN: 0031-2215
John Thorner, Executive Director
Rachel Roberts, Editor

4291 National Wildlife Federation, Western Natural Re-
source Center: National Wildlife Magazine
Western Natural Resource Center
6 Nickerson Street   206-285-8707
Suite 200   Fax: 206-285-8693
Seattle, WA 98109   E-mail: wnrc@nwf.org
                                http://www.nwf.org
The Western Natural Resource Center of NWF connects the peo-
ple and issues of the West-educating, inspiring and empowering
individuals from all walks of life to protect and conserve the re-
gions diverse wildlife and natural resources.
Monthly
Jeannie Bowen, Office Manager
Paula Delgadies, Center Director

4292 Natural Resources Department: Fish & Wildlife News-
letter
402 West Washington Street RMW273   317-232-4080
Indianapolis, IN 46204   Fax: 317-232-8150
                                http://www.state.in.us/dnr/fishwild/index.htm
Publication Date: 1985 12 pages Quarterly

Publications/Periodicals: Habitat Preservation & Land Use

4293 Nature Conservancy: Nebraska Chapter Newsletter
1025 Leavenworth Street 402-342-0282
Omaha, NE 68102 Fax: 402-342-0474
E-mail: nebraska@tnc.org
http://nature.org
The mission of the Nature Conservancy is to preserve the plants, animals and natural communities that represent the diversity of life on Earth by protecting the lands and waters they need to survive.
Publication Date: 1987 12 pages Quarterly
Jill Jeffrey, Donor Relations Manager

4294 New England WildFlower: New England Wildflower Society
180 Hemenway Road 508-877-7630
Framingham, MA 01701 Fax: 508-877-3658
E-mail: newfs@newfs.org
http://www.newfs.org
36 pages Twice a Year

4295 New Jersey Environmental Lobby News
204 W State Street 609-396-3774
Trenton, NJ 08608 Fax: 609-396-4521
E-mail: njelcurtis@aol.org
http://www.njenvironment.org
Quarterly publication from New Jersey Environmental Lobby.
Publication Date: 1971 8 pages Quarterly
ISSN: 1535-2021
Anne Poole, President
Marie A Curtis, Executive Director

4296 North Dakota Association of Soil Conservation Districts Newsletter
3310 University Drive 701-223-8518
Bismarck, ND 58504 Fax: 701-223-1291
E-mail: gpuppe@tic.bisman.com
http://www.ndascd.org
Quarterly
Gary Puppe, Executive VP

4297 Outdoors Unlimited
121 Hickory Street 406-728-7434
Suite 1 Fax: 406-728-7445
Missoula, MT 59801 E-mail: owaa@montana.com
http://www.owaa.org
Magazine of the Outdoor Writers Association of America. Membership fees are $175.00 (individual), $325.00 (supporting), $40.00 (student)
Publication Date: 1962 35 pages Monthly
Kevin Rhoades, Executive Director

4298 Pacific Fishery Management Council Newsletter
7700 NE Ambassador Place 503-820-2280
Suite 200 866-806-7204
Portland, OR 97220-1384 Fax: 503-820-2299
E-mail: Donald.McIsaac@noaa.gov
http://www.pcouncil.org
24 pages 5x Year
Donald McIsaac, Executive Director
John Coon, Deputy Director

4299 Pacific States Marine Fisheries Commission Newsletter
205 SE Spokane Street 503-595-3100
Suite 100 Fax: 503-595-3232
Gladstone, OR 97202 E-mail: front_office@psmfc.org
http://www.psmfc.org
Quarterly
Randy Fisher, Executive Director

4300 Parks and Trails Council of Minnesota: Newsletter
275 East 4th Street 651-726-2457
#642 800-944-0707
Saint Paul, MN 55101-1651 E-mail: info@parksandtrails.org
http://www.parksandtrails.org
Quarterly
Judy Erickson, Editor

4301 Powder River Basin Resource Council: Powder River Breaks Newsletter
934 North Main 307-672-5809
Sheridan, WY 82801 Fax: 307-672-5800
E-mail: resources@powderriverbasin.org
http://www.powderriverbasin.org
Committed to the preservation and enrichment of Wyoming's agricultural heritage and rural lifestyle; the conservation of Wyomings unique land, mineral, water and clean air resources, consistent with responsible use of those resourcesto sustain the vitality of present and future generations; the education and empowerment of Wyoming's citizens to raise a coherent voice in decisions. They are the only group in Wyoming that addresses both agricultural and conservation issues.
Publication Date: 1973 8-12 pages 6x year
Jillian Malone, Editor
Stephanie Avey, Assistant Editor

4302 Prairie Naturalist Magazine
600 Park Street 785-628-4214
Department of Biological Sciences Fax: 316-341-5607
Hays, KS 67601-4099 E-mail: efinck@fhsu.edu
http://www.fhsu.edu
Published by the North Dakota Natural Science Society, a regional organization with interests in the natural history of grasslands and the Great Plains.
Publication Date: 1968 260 pages Quarterly
ISSN: 0091-0376
Elmer J Finck, Editor

4303 Reef Line
Reef Relief Environmental Center
201 Williams Street 305-294-3100
#5 Fax: 305-293-9515
Key West, FL 33040 E-mail: reef@bellsouth.net
http://www.reefrelief.org
Reef Line is a quarterly publication from Reef Relief.
Publication Date: 1986 16 pages Quarterly
Michael Blades, Project Director

4304 Roger Troy Peterson Institute of Natural History Newsletter
311 Curtis Street 716-665-2473
Jamestown, NY 14701 800-758-6841
Fax: 716-665-3794
E-mail: mail@rtpi.org
http://www.rtpi.org
RPTI is a national nature education organization with a mission to create passion for and knowledge of the natural world in the hearts and minds of childern by inspiring and guiding the study of nature in our schools and communities.To accomplish its mission RTPI offers programs and workshops for teachers, preserves and exhibits the works of Roger Troy Peterson, creates and hosts educational exhibits and natural history programs, develops and produces nature education materials.
15-30 pages Quarterly
Jim Berry, President

4305 SWOAM News
Small Woodland Owners Association of Maine
153 Hospital Street 207-626-0005
PO Box 836 877-467-9626
Augusta, ME 04332-0836 Fax: 207-626-7992
E-mail: info@swoam.com
http://www.swoam.com
Monthly
Tom Doak, Executive Director

4306 Save San Francisco Bay Association: Watershed Newsletter
350 Frank H Ogawa Plaza 510-452-9261
Suite 900 Fax: 510-452-9266
Oakland, CA 94612 E-mail: SAVEBAY@savesfbay.org
http://www.savesfbay.org
Save the Bay has worked for over 40 years to protect the San Francisco Bay-Delta from pollution, fill, shoreline destruction and fresh water diversion. We have launched a century of renewal to restore bay fish and wildlife, reclaimtidal wetlands and make the bay safe and accessible to all.
8-10 pages 3-4 times/year
David Lewis, Executive Director

4307 Scenic America Newsletter

Scenic America
1634 I Street NW   202-638-0550
Suite 510   Fax: 202-638-3171
Washington, DC 20006   E-mail: Scenic@scenic.org
                                  http://www.scenic.org

12 pages 3 Times Per Year
Kevin Fry, President
Peggy Lint, Office Manager

4308 Shore and Beach

5460 Beaujolais Lane   239-489-2616
Fort Myers, FL 33919   Fax: 239-489-9917
                            E-mail: ExDir@asbpa.org
                                  http://www.asbpa.org
A quarterly publication from the American Shore and Beach
Preservation Association.
24 pages Quarterly
ISSN: 0037-4237
Ken Gooderham, Executive Director/Editor
Kate Gooderham, Executive Director/Editor

4309 Sierra Club, NJ Chapter: The Jersey Sierran

139 W Hanover Street   609-656-7612
Trenton, NJ 08618   Fax: 609-656-7618
                E-mail: webmaster@sierraactivist.org
                            http://www.sierraactivist.org
The Sierra Club is our country's oldest and most effective grass-
roots environmental organization. Hikes and outings are sched-
uled throughout the year. We are dedicated to fighting sprawl
and over-development.
Publication Date: 1992 14 pages Quarterly
Jeff Tittle, Director
Dennis Schvejda, Conservation Coordinator

4310 Sierra Club: Pennsylvania Chapter Newsletter

600 North 2nd Street   717-232-0101
Suite 300A   Fax: 717-238-6330
Harrisburg, PA 17101   E-mail: sierraclub.pa@paonline.com
                        http://www.sierraclub.org/chapter/pa/
The Pennsylvania chapter includes 11 local Sierra Club groups.
Emphasis is on state environmental policy advocacy, outings,
education and local environmental protection efforts.
18 pages Quarterly
Jeff Schmidt, Sr. Chapter Director

4311 Southern Appalachian Botanical Society: Gastanea

Newberry College
2100 College Street   803-321-5257
Newberry, SC 29108   Fax: 803-321-5636
                        E-mail: chorn@newberry.edu
                        http://www.newberrynet.com/sabs/
This is a professional organization for those interested in botani-
cal research, especially in the areas of ecology, floristics and
systematics. To this end, we publish a journal, CASTANEA, and
a newsletter, CHINQUAPIN.
Publication Date: 1936 350 pages Quarterly
ISSN: 0008-7475
Michael E Held, PhD, President
Charles Horn, Treasurer

4312 Systematics Museums Newsletter

University of Kansas
Dyche Hall   785-864-4540
Lawrence, KS 66045-7561   Fax: 785-864-5335

A comprehensive research, graduate education and public service institution dedicated to biodiversity science and collections. Collections of more than 7 million plant and animal
specimens, with particular strengths in neotropical amphibians,
great plains, flora, bees and antarctic plant fossils.
Publication Date: 1978 6 pages Quarterly
Dr Leonard Krishtalka, Director

4313 Terrain Magazine

2530 San Pablo Avenue   510-548-2220
Berkeley, CA 94702   Fax: 510-548-2240
                        E-mail: info@ecologycenter.org
                            http://www.ecologycenter.org
A quarterly magazine published by the Ecology Center of Berke-
ley, CA.

Publication Date: 1971 39 pages Quarterly
ISSN: 1526-8322
Linnea Due, Editor-in-Chief

4314 Tidbits Newsletter

Minnesota Association/Soil and Water Cons. Dist.
790 Cleveland Avenue S   651-690-9028
Suite 216   Fax: 651-690-9065
St. Paul, MN 55075   E-mail: leann.buck@maswcd.org
                                  http://www.maswcd.org

Quarterly
Le Ann Buck, Executive Director
Sheila Vanney, Editor

4315 Tide

Coastal Conservation Association
4801 Woodway Drive   713-626-4234
Houston, TX 77056   800-201-FISH
                        E-mail: ccantl@joincca.org
TIDE is the official bimonthly magazine of the Coastal Conservation Association. It has received local, state and national acclaim for writing, photography and layout and currently boasts a
circulation of more than 70,000. TIDE is available only to members of the Coastal Conservation Association.

4316 Upper Mississippi River Conservation Committee Newsletter

4469 48th Avenue Court   309-793-5800
Rock Island, IL 61201   Fax: 309-793-5804
                E-mail: umrcc@mississippi-river.com/umrcc
                        http://mississippi-river.com/umrcc
A bimonthly newsletter published by the Upper Mississippi
River Conservation Committee.
10 pages Bi monthly
Mike McGhee, Chairman

4317 Urban Land Magazine

Urban Land Institute
1025 Thomas Jefferson Street NW   202-624-7000
Suite 500 W   800-321-5011
Washington, DC 20007-5201   Fax: 202-624-7140
                                  http://www.uli.org
Nonprofit research and education organization dedicated to im-
proving land use policy and development practice. Publishes a
monthly magazine, several quarterly publications and books.
Topics relate to real estate development including government
sensitive development, smart growth, sustainable development
and city parks.
Monthly
Kristina Kessler, Chief Editor

4318 Utah Geological Survey: Survey Notes

1594 W N Temple Suite 3110   801-537-3300
PO Box 146100   Fax: 801-537-3400
Salt Lake City, UT 84114   http://www.ugs.state.ut.us
The Utah Geological Survey is an applied scientific agency that
creates, interprets and provides information about Utah's geo-
logic environment, resources and hazards to promote safe, bene-
ficial and wise use of land. This is their publication, which is
issued three times yearly.
Publication Date: 1964 3 Times Yearly
ISSN: 1061-7930
Richard Allis, Director

4319 Virginia Forests Magazine

Virginia Forestry Association
3808 Augusta Avenue   804-278-8733
Richmond, VA 23255-2080   Fax: 804-320-1447
                        E-mail: vafa@erols.com
                            http://www.vaforestry.org
Quarterly magazine published by the Virginia Forestry Associa-
tion.
Quarterly
Paul Howe, VFA Executive Vice President

4320 WAEE Bulletin

Wisconsin Association for Environmental Education
8 Nelson Hall   715-346-2796
University of Wisconsin   E-mail: waee@uwsp.edu
Stevens Point, WI 54481

Quarterly
Carol Weston, Administrative Assistant

4321 West Virginia Forestry Association Newsletter
PO Box 718
Ripley, WV 25271
304-372-1955
888-372-9663
Fax: 304-372-1957
E-mail: wvfa@wvadventures.net
http://www.wvfa.org
Monthly
Richard Waybright, Executive Director

4322 Western Pennsylvania Conservancy Newsletter
209 4th Avenue
Pittsburgh, PA 15222
412-288-2777
866-JOI-NWPC
Fax: 412-281-1792
E-mail: wpc@paconserve.org
http://www.paconserve.org
WPC, working together to save the places we care about, protects natural lands, promotes healthy and attractive communities, and preserves Fallingwater, Frank Lloyd Wright's masterwork in Mill Run, which was entrusted to theConservancy in 1963. Since its inception in 1932, the Conservancy has protected more than 280,000 acres of natural lands in Pennnsylvania. We continue to work to secure lands of ecological significance that frequently offer recreational and scenicvalues.
16 pages Quarterly
Larry Schweiger, President

4323 Western Proceedings Newsletter
Western Association of Fish and Wildlife Agencies
5400 Bishop Boulevard
Cheyenne, WY 82006
307-777-4569
Fax: 307-777-4699
E-mail: ikruck@state.wy.us
Annual

4324 Western Society of Naturalists: Newsletter
San Diego State University Biology Department
5500 Campanile Street
San Diego, CA 92182
818-677-3256
Fax: 818-677-2034
http://www.wsn-online.org
Mark Carr, President
Ralph Larson, President Elect

4325 Whalewatcher
PO Box 1391
San Pedro, CA 90733-1391
310-548-6279
Fax: 310-548-6950
E-mail: info@acsonline.org
http://www.acsonline.org
A bi-annual publication from the American Cetacean Society. Cost included with membership fees.
Publication Date: 1967 30 pages Bi-Annual
Diane Alps, Administrative Assistant

4326 Wilderness Education Association
900 E 7th Street
Bloomington, IN 47405
812-855-4095
Fax: 812-855-8697
E-mail: wea@indiana.edu
http://www.weainfo.org/
Publication Date: 1976 4-6 pages 3 Times Per Year

4327 Wildfowl Trust of North America: Newsletter
Wildfowl Trust of North America
600 Discovery Lane
PO Box 519
Grasonville, MD 21638
410-827-6694
Fax: 410-827-6713
E-mail: cbec@cbec-wtna.org
http://www.cbec-wtna.org
Published by the Wildfowl Trust of North America.
Publication Date: 1995 8 pages Qaurterly
Judy Wink, Executive Director
Sharyn B Harlow, Executive Admin. Assistant

4328 Wildlife Law News Quarterly
University of New Mexico School of Law
1117 Stanford NE
Albuquerque, NM 87131
505-277-5006
Fax: 505-277-7064
E-mail: dfarris@unm.edu
http://wildlifenews.unm.edu

A quarterly publication from the New Mexico Center for Wildlife Law.
Publication Date: 1993 16 pages Quarterly
ISSN: 1085-7338
R Musgrave, Editor
D Macke, Editor

4329 Wildlife Society Bulletin
Wildlife Society
5410 Grosvenor Lane
Suite 200
Bethesda, MD 20814-2144
301-897-9770
Fax: 301-530-2471
E-mail: tws@wildlife.org
http://www.wildlife.org
A quarterly publication from the Wildlife Society.
Publication Date: 1973 Quarterly
ISSN: 0091-7648
Warren Ballard, Editor

4330 Woodland Management Newsletter
Wisconsin Woodland Owners Association
PO Box 285
Stevens Point, WI 54481-0285
715-346-4798
Fax: 715-346-4821
E-mail: nbozek@uwsp.edu
http://www.wisconsinwoodlands.org
Quarterly
Tim Eisele, Editor

4331 Woodland Report
National Woodland Owners Association
374 Maple Avenue E
Suite 310
Vienna, VA 22180
703-255-2700
800-476-8733
Fax: 703-281-9200
E-mail: argow@nwoa.net
http://www.nationalwoodlands.org
Provides timely information about forestry and forest practices with news from Washington, DC and state capitals. Written for non-industrial, private woodland owners. Includes state landowner association news.
2 pages
Keith A Argow, Editor
Eric Johnson, Editor

4332 World Wildlife Fund: US Focus
1250 24th Street NW
Suite 500
Washington, DC 20037-1124
202-293-4800
http://www.worldwildlife.org
WWF projects.
8 pages
Pat Sullivan, Publisher

Periodicals: Recycling & Pollution Prevention

4333 AARA Newsletter
Arizona Automotive Recyclers Association
PO Box 20662
Phoenix, AZ 85036
480-391-2888
Fax: 480-391-2390
E-mail: info@aara.com
http://www.aara.com
Quarterly newsletter of the AARA, a select group of recyclers providing quality recycled parts for the benefit of our customers, communities and environment. There are 90 member companies. AARA is affiliated with the AutomotiveDismantlers and Recyclers Association.
Quarterly
Robert Sibbio, President

4334 Air/Water Pollution Report
Business Publishers
8737 Colesville Road
10th Floor
Silver Spring, MD 20910-3928
301-589-5103
800-274-6737
Fax: 301-589-8493
E-mail: custserv@bpinews.com
http://www.bpinews.com
Provides comprehensive coverage of economic, political, legislative, regulatory and domestic and international implications of air and water pollution.

Weekly
Leonard A Eiserer, Publisher
Beth Early, Associate Publisher

4335 American Waste Digest

Charles G Moody
226 King Street   610-326-9480
Pottstown, PA 19464-9150   800-442-4215
                                                    Fax: 610-326-9752
E-mail: awd@americanwastedigest.com
http://www.americanwastedigest.com
Provides reviews on new products, profiles on sucessful waste
removal businesses, and provides discussion on legislation on
municipal regulations on recycling.
86 pages Monthly
Charles G Moody, III, Publisher/Editor

4336 BNA's Environmental Compliance Bulletin

Bureau of National Affairs
1231 25th Street NW   202-452-4200
Washington, DC 20037-1157   800-372-1033
                                                    Fax: 202-452-5331
                                                    http://www.bna.com
Cover the water and air pollution, waste management and regulatory updates, as well as a summary of selected regulatory actions and a list of key environmental compliance dates.
Kevin Fepherston, Managing Editor

4337 Bio-Integral Resource Center: IPM Practitioner

PO Box 7414   510-524-2567
Berkeley, CA 94707   Fax: 510-524-1758
                                                    E-mail: birc@igc.org
                                                    http://www.birc.org
The goal of the Bio Integral Resources Center is to reduce pesti-
cide use by educating the public about effective, least-toxic al-
ternatives for pest problems.
Publication Date: 1979 6-12 pages
ISSN: 0738-968x
Dr. William Quarles, Executive Director

4338 C&D Recycler

Gie Publishing
4012 Bridge Avenue   216-961-4130
Cleveland, OH 44113   800-456-0707
                                                    Fax: 216-961-0364
                                                    E-mail: btaylor@gie.net
                                                    http://www.cdrecycler.com
Brian Taylor, Editor

4339 Common Sense Pest Control Quarterly

PO Box 7414   510-524-2567
Berkeley, CA 94707   Fax: 510-524-1758
                                                    E-mail: birc@igc.org
                                                    http://www.birc.org
A quarterly publication published by the Bio Integral Resource
Center
Publication Date: 1984 24 pages Quarterly
ISSN: 8756-7881
Dr. William Quarles, Executive Director

4340 Composting News

McEntee Media Corporation
9815 Hazelwood Avenue   440-238-6603
Cleveland, OH 44149   Fax: 440-238-6712
                                                    E-mail: ken@recycle.cc
                                                    http://www.recycle.cc
New composting projects, research, regulations and legislation,
as well as the latest news in the composting industry.
Publication Date: 1992 Monthly
Ken McEntee, Publisher

4341 Daily Environment Report

Bureau of National Affairs
1231 25th Street NW   202-452-4200
Washington, DC 20037-1197   800-372-1033
                                                    Fax: 202-822-8092
                                                    http://www.bna.com

A 40-page daily report providing comprehensive, in-depth coverage of national and international environmental news. Each
issue contains summaries of the top news stories, articles, and
in-brief items, and a journal of meetings, agency activities, hearings and legal proceedings. Coverage includes air and water
pollution, hazardous substances, and hazardous waste, solid
waste, oil spills, gas drilling, pollution prevention, impact statements and budget matters.
40 pages
ISSN: 1060-2976
William A Beltz, Publisher

4342 E-Scrap News

Resource Recycling
PO Box 42270   503-233-1305
Portland, OR 97242-0270   Fax: 503-233-1356
                                                    E-mail: info@resource-recycling.com
                                                    http://www.resource-recycling.com
64 pages
ISSN: 0744-4710
Jerry Powell, Editor

4343 Earth Preservers

PO Box 6   908-654-9293
Westfield, NJ 07091   E-mail: earthpreservers@att.net
                                                    http://www.earthpreserves.com
Award winning monthly environmental newspaper for school
children aged 7 to 15.
4 pages
Bill Paul, Publisher

4344 Environment Reporter

Bureau of National Affairs
1231 25th Street NW   202-452-4200
Washington, DC 20037-1197   800-372-1033
                                                    Fax: 202-822-8092
                                                    http://www.bna.com
A weekly notification and reference service covering the
full-spectrum of legislative, administrative, judicial, industrial
and technological developments affecting pollution control and
environmental protection.

ISSN: 0013-9211
William A Beltz, Publisher
Patricia Spencer, Managing Editor

4345 Environmental Engineering Science

Mary Ann Liebert
140 Huguenot Street   914-740-2100
3rd Floor   800-MLI-EBER
New Rochelle, NY 10801-5215   Fax: 914-740-2101
                                                    E-mail: info@liebertpub.com
                                                    http://www.liebertpub.com
The focus is on pollution control of the suface, ground, and
drinking water, and highlight research news and product devel-
opments that aid in the fight against pollution.
Bi-Monthly
ISSN: 1092-8758
Mary Ann Liebert, Publisher
Dumpnico Grosso PhD, Editor-in-Chief

4346 Environmental Notice

235 S Beretania Street   808-586-4185
Room 702   Fax: 808-586-4186
Honolulu, HI 96813   E-mail: oeqc@mail.health.state.hi.us
                                    http://www.state.hi.us/health/oeqc/index.html
A bi-monthly publication from the Health Department Environ-
mental Quality Control division.
Publication Date: 1978 24 pages Bi monthly
Genevieve Salmonson, Director

4347 Environmental Protection

Stevens Publishing Corporation
5151 Beltline Road   972-687-6700
Suite 1010   Fax: 972-687-6770
Dallas, TX 75240   E-mail: jlaws@stevenspublishing.com
                                    http://www.stevenspublishing.com
Angela Neville, Editor-in-Chief
Jason Goodman, Managing Editor

4348 Environmental Regulation
State Capitals Newsletters
PO Box 7376   703-768-9600
Alexandria, VA 22307-7376   Fax: 703-768-9690
E-mail: newsletters@statecapitals.com
http://www.statecapitals.com
Weekly news from the state capitals keeps you informed on state programs, recycling, wetlands, ground water protection, beach renourishment, land management, greenspace laws, brownfields, livestock regulation, wilderness preservation, urban sprawl and solid waste.
Publication Date: 1946   4-10 pages   Newsletter 48x/Yr
ISSN: 1061-9682
Ellen Klein, Editor

4349 Environmental Regulatory Advisor
JJ Keller
3003 W Breezewood Lane   920-722-2848
PO Box 368   877-564-2333
Neenah, WI 54957-0368   Fax: 800-727-7516
E-mail: sales@jjkeller.com
http://www.jjkeller.com
Covers developments at the EPA.
12 pages
ISSN: 1056-3164
Tom Ziebell, Editor

4350 Environmental Science and Technology
American Chemical Society
1155 16th Street NW   202-872-4582
Washington, DC 20036   Fax: 202-872-4403
E-mail: service@acs.org
http://www.acs.org
Articles on pollution control, waste treatment, climate changes and various other environmental interests.
110 pages   Semi-Monthly
Bruce Poorman, Ad Manager
Steve Cole, Managing Editor

4351 Environmental Systems Corporation Newsletter
200 Tech Center Drive   865-688-7900
Knoxville, TN 37912-2799   Fax: 865-687-8977
E-mail: esccorp@envirosys.com
http://www.envirosys.com
Data acquisition and reporting systems for electric power producers and industrial sources, ESC is the leading supplier of CEM and ambient data systems in the US Newsletter is free.
Publication Date: 1994   4 pages   Quarterly
Steve Drevik, Sr. Marketing Manager

4352 Environmental Times
Environmental Assessment Association
1224 N Nokomis NE   320-763-4320
Alexandria, MN 56308-5078   Fax: 320-763-9290
E-mail: eaa@iami.org
http://www.iami.org/eaa.cfm
This publication contains environmental conferences and expos, industry trends, federal regulations related to the environment and industry assessments.
Robert Johnson, Publisher/Editor

4353 From the Ground Up
Ecology Center
117 Division Street   734-761-3186
Ann Arbor, MI 48104-1523   Fax: 734-663-2414
E-mail: info@crocenter.org
http://www.crocenter.org
Progressive environmental news from southeast Michigan.
32 pages
Michael Garfield, Editor

4354 Full Circle
Northeast Resource Recovery Association
PO Box 721   603-798-5777
Concord, NH 03302-0721   Fax: 603-798-5744
E-mail: nrra@tds.net
http://www.recyclewithus.org
Bi-Monthly
Elizabeth Bedard, Executive Director

4355 Hauler
Hauler Magazine
166 South Main Street   215-997-3622
PO Box 508   Fax: 215-997-3623
New Hope, PA 18938   E-mail: mag@thehauler.com
http://www.thehauler.com
This magazine serves as an advertising guide to new products in the waste management, recycling, and environmental industries.
Publication Date: 1978   Monthly
Thomas N Smith, Publisher/Editor

4356 HazMat Management
Business Information Group
1450 Don Mills Road   416-442-2223
Don Mills, Ontario M3B-2X7   888-702-1111
Fax: 416-442-2917
E-mail: sales@hazmatmag.com
http://www.hazmatmag.com
Solutions for the environment.
Publication Date: 1989   Bi-Annual
ISSN: 0843-9303
Thea Papadakis, Publisher
Connie Vitello, Editor

4357 Hazardous Waste News
Business Publishers
8737 Colesville Road   301-589-5103
10th Floor   800-274-6737
Silver Spring, MD 20910-3928   Fax: 301-589-8493
E-mail: custserv@bpinews.com
http://www.bpinews.com
Comprehensive federal, state and local coverage of legislation and regulation affecting all aspects of the hazardous waste industry including Superfund, Resource Conservation and Recovery Act, US EPA, incineration, land disposal and more.
8 pages
Leonard A Eiserer, Publisher
Beth Early, Operations Director

4358 Hazardous Waste/Superfund Week
Business Publishers
8737 Colesville Road   301-589-5103
Suite 1100   800-274-6737
Silver Spring, MD 20910-3928   Fax: 301-589-8493
E-mail: custserv@bpinews.com
http://www.bpinews.com
Provides comprehensive coverage on hazardous waste disposal and cleanup, behind-the-scenes coverage of congressional action, EPA initiatives, Superfund sites, regulatory changes, court cases, enforcement news, contract opportunities, new technologies, research findings and business developments.
Weekly
Leonard A Eiserer, Publisher
Beth Early, Associate Publisher

4359 Hazmat Transportation News
Bureau of National Affairs
1231 25th Street NW   202-452-4200
Washington, DC 20037-1197   800-372-1033
Fax: 202-822-8092
http://www.bna.com
A two-binder service containing the full-text of rules and regulations governing shipment of hazardous material by rail, air, ship, highway and pipeline, including DOT's Hazardous Materials Tables and EPA's rules for its hazardous waste tracking system.
Stan Pond, Managing Editor

4360 Inside EPA
Inside Washington Publishers
1225 South Clark Street   703-416-8500
Suite 1400   Fax: 703-415-8543
Arlington, VA 22202   E-mail: iwp@iwpnews.com
http://www.iwpnews.com
Gives timely information on all facets of waste, water, air, and other environmental regulatory programs.
Publication Date: 1980   Weekly
Al Sosenko, Publisher

4361 Institute of Scrap Recycling Industries
1325 G Street Northwest
Suite 1000   202-737-1770
Washington, DC 20005-3104   Fax: 202-626-0900
E-mail: kentkiser@scrap.org
http://www.isri.org
Publication Date: 1988   148 pages   Bi monthly
ISSN: 0036-9527
Frank Cozzi, President
Kent Kiser, Publisher/Editor-in-Chief

4362 Journal of Environmental Education
Heldref Publications
1319 18th Street NW   202-296-6267
Washington, DC 20036-1802   Fax: 202-296-5149
E-mail: webmaster@heldref.org
http://www.heldref.org
The issues featured are case studies, environmental philosophy and policy discussions, new research evaluations and information on environmental education.
Douglas Kirkpatrick, Publisher/Executive Director

4363 Legislative Bulletin
Arkansas Environmental Federation
1400 W Markham Street   501-374-0263
Suite 250   Fax: 501-374-8752
Little Rock, AR 72201   http://www.environmentark.org
Publication Date: 1967
Randy Thurman, Executive Director

4364 Minnesota Pollution Control Agency Minnesota Environment Magazine
520 Lafayette Road North   651-296-6300
St. Paul, MN 55155-4194   800-657-3864
Fax: 651-296-7923
E-mail: darlene.sigstad@pca.state.mn.us
http://www.pca.state.mn.us
Established in 1967 to protect Minnesota's environment through monitoring environmental quality and enforcing environmental regulations.
Publication Date: 2000   16-20 pages   Quarterly
Fheryl Corrigan, Comm./Chair Citizens' Board

4365 Northeast Recycling Council Bulletin
139 Main Street   802-254-3636
Suite 401   Fax: 802-254-5870
Brattleboro, VT 05301   E-mail: info@nerc.org
http://www.nerc.org
Monthly
Lynn Rubinstein, Executive Director

4366 Northeast Recycling Council News
139 Main Street   802-254-3636
Suite 401   Fax: 802-254-5870
Brattleboro, VT 05301   E-mail: info@nerc.org
http://www.nerc.org
3x Year
Lynn Rubenstein, Executive Director

4367 Oregon Refuse and Recycling Association Newsletter
PO Box 2186   503-588-1837
Salem, OR 97308-2186   800-527-7624
Fax: 503-399-7784
E-mail: orrainfo@orra.net
http://www.orra.net
Monthly
Max Brittingham, Executive Director
Kristin Mitchell, Editor

4368 Plastics Recycling Update
Resource Recycling
PO Box 42270   503-233-1305
Portland, OR 97242-0270   Fax: 503-233-1356
E-mail: subscriptions@resource-recycling.com
http://www.resource-recycling.com
Monthly
Jerry Powell, Editor

4369 Pollution Equipment News
Rimbach Publishing
8650 Babcock Boulevard   412-364-5366
Suite 1   800-245-3182
Pittsburgh, PA 15237   Fax: 412-369-9720
E-mail: info@rimbach.com
http://www.rimbach.com
Provides information to those responsible for selecting products and services for air, water, wastewater and hazardous waste pollution abatement.
Publication Date: 1967   64 pages   Bi-Monthly
ISSN: 0032-3659
Raquel Rimbach, Editor
Norberta Rimbach, Publisher/President

4370 Pollution Prevention News
US EPA
1200 Pennsylvania Avenue, NW   202-272-0167
Washington, DC 20460   Fax: 202-260-2219
http://www.epa.gov
Articles include recent information on source reduction and sustainable technologies in industry, transportation, consumer, agriculture, energy, and the international sector.
Maureen Eichelberger, Editor

4371 Recharger Magazine
Recharger Magazine
2800 West Sahara Avenue   702-438-5557
Suite S-C   Fax: 702-438-4025
Las Vegas, NV 89102   E-mail: info@rechargermag.com
http://www.rechargermag.com
Information including articles that cover business and marketing, technical updates, association and industry news, and company profiles. On the remanufactured imaging supplies industry, related features focus on the importance of recycling, government legislation, and product comparisons. Annual trade event in Las Vegas.
Publication Date: 1989   300+ pages   Monthly
ISSN: 1053-7503
Julie Kerrane, Author
Phyllis Gurgevich, Publisher
Amy Turner, Managing Editor

4372 Recycled News
Maryland Recyclers Coalition
2105 Laurel Bush Road   443-640-1050
Suite 200   Fax: 443-640-1031
Bel Air, MD 21015-6185 E-mail: info@marylandrecyclers.org
http://www.marylandrecyclers.org
2 pages   Bi-Monthly
Jackie King, Executive Director

4373 Recycling Laws International
Raymond Communications
5111 Berwin Road   301-345-4237
Suite 115   Fax: 301-345-4768
College Park, MD 20740   E-mail: circulation@raymond.com
http://www.raymond.com
Covers recycling, takeback and green labeling policy for business in 38 countries. Available online.
Publication Date: 1995   Bi-Monthly
Bruce Popka, Vice President

4374 Recycling Markets
NV Business Publishers Corporation
43 Main Street   732-502-0500
Avon by the Sea, NJ 07717   Fax: 732-502-9606
E-mail: recycling@nvpublications.com
http://www.nvpublications.com
Contains profiles on recycling mills, as well as large users and generators of recycled materials for the broker, dealers and processors of paper stock, scrap metal, plastics and glass.
Jim Curley, Editor
Anna Dutco, Managing Editor

4375 Recycling Product News
Baum Publications
201-2323 Boundary Road   604-291-9900
Vancouver, Can, BC V5M-4V8   Fax: 604-291-1906
E-mail: webadmin@baumpub.com
http://www.baumpub.com

Published for the recycling center operators and other waste mangers, articles discuss technology and new products.
Engelbert J Baum, Publisher
Keith Barker, Editor

4376 Recycling Today

GIE Media
4012 Bridge Ave   216-961-4130
Cleveland, OH 44113-3320   800-456-0707
                                       Fax: 216-961-0364
                            E-mail: dtoto@gie.net
                            http://www.recyclingtoday.com
Published for the secondary commodity processing/recycling market.
James R Keefe, Group Publisher
Brian Taylor, Editor

4377 Resource Recovery Report

5313 38th Street NW   540-347-4500
PO Box 3356   800-627-8913
Warrenton, VA 20188-1956   Fax: 540-348-4540
                            E-mail: rwill@coordgrp.com
                            http://www.coordgrp.com
Covers all alternatives to landfills, i.e., recycling, energy recovery, composting in North America, Government, industry, associations, universities, etc. are included.
12 pages
Richard Will, Production Manager

4378 Resource Recycling Magazine

Resource Recycling
PO Box 42270   503-233-1305
Portland, OR 97242-0270   Fax: 503-233-1356
                            E-mail: info@resource-recycling.com
                            http://www.resource-recycling.com
Monthly
Rick Downing, Advertising Director

4379 Resource Recycling: North America's Recycling & Composting Journal

Resource Recycling
PO Box 42270   503-233-1305
Portland, OR 97242-0270   Fax: 503-233-1356
                            E-mail: info@resource-recycling.com
                            http://www.resource-recycling.com
The nation's leading recycling and composting magazine. This monthly journal focuses on efforts in the US and Canada to recover materials from homes and businesses for recycling. Accepts advertising.
64 pages Monthly
ISSN: 0744-4710
Judy Roumpf, Publisher
Rick Downing, Advertising Director

4380 Reuse/Recycle Newsletter

Technomic Publishing Company
PO Box 3535   717-291-5609
Lancaster, PA 17601   800-233-9936
                                       Fax: 717-295-4538
                            E-mail: aflannery@techpub.com
                            http://www.techpub.com
Provides news and information on important developments in both industrial and municipal recycling, and focuses on large-scale post-consumer, post-commercial, and post-industrial waste recycling.
8 pages
ISSN: 0048-7457
Susan E Selke, Author
Amy Flannery, Marketing

4381 Scrap

Institute of Scrap Recycling Industries
1325 G Street Northwest Suite 1000   202-662-8545
Washington, DC 20005-3104   Fax: 202-626-0945
                            E-mail: ellenross@scrap.org
                            http://www.scrap.org
Serves the scrap processing and recycling industry. Subscription: $32.95 (US), $38.95 (Canada/Mexico) & $104.95 (all other international)
Bi-Monthly
Kent Kiser, Publisher/Editor-in-Chief
Ellen Ross, Production Director

4382 Solid Waste & Recycling

Southam Environment Group
1450 Don Mills Road   416-442-5600
Don Mills, Ontario M3B-2X7   800-387-0273
                                       Fax: 416-510-5130
                            E-mail: bobrien@solidwastemag.com
                            http://www.solidwastemag.com
Published to emphasize on municipal and commercial aspects of collection, handling, transportation, hauling, disposal and treatment of solid waste , including incineration, recycling and landfill technology.
Brad O'Brien, Publisher
Guy Crittenden, Editor-in-Chief

4383 Solid Waste Report

Business Publishers
8737 Colesville Road   301-589-5103
10th Floor   800-274-6737
Silver Spring, MD 20910-3928   Fax: 301-589-8493
                            E-mail: custserv@bpinews.com
                            http://www.bpinews.com
Comprehensive news and analysis of legislation, regulation and litigation in solid waste management including resource recovery, recycling, collection and disposal. Regularly features international news, state updates and businesstrends.
Bi-Weekly
Leonard A Eiserer, Publisher
Beth Early, Operations Director

4384 State Recycling Laws Update

Raymond Communications
5111 Berwin Road   301-345-4237
Suite 115   Fax: 301-345-4768
College Park, MD 20740   E-mail: michele@raymond.com
                            http://www.raymond.com
Provides coverage of recycling legislation affecting business, as well as the outlook on future legislation across the US and Canada.
Bruce Popka, Vice President

4385 Waste Age

Environmental Industry Associations
4301 Connecticut Avenue NW   202-244-4700
#300   800-424-2869
Washington, DC 20008-2304   Fax: 202-966-4868
                            E-mail: ptom@primediabusiness.com
                            http://www.wasteage.com
Contents focus on new system technologies, recycling, resource recovery and sanitary landfills with regular features on updates in the status of government regulations, new products, guides, company profiles, exclusive surveyinformation, legislative implications and news.
Patricia-Anne Tom, Editor
Stephen Ursery, Managing Editor

4386 Waste Age's Recycling Times

Environmental Industry Associations
4301 Connecticut Avenue NW   202-244-4700
#300   800-424-2869
Washington, DC 20008-2304   Fax: 202-966-4868
                            E-mail: ptom@primediabusiness.com
                            http://www.wasteage.com
Features municipalities, recycling goals and rates, program innovations, waste habits, and new materials being recycled.
Patricia-Anne Tom, Editor
Stephen Ursery, Managing Editor

4387 Waste Handling Equipment News

Lee Publications
6113 State Highway 5   518-673-3237
PO Box 121   800-218-5586
Palatine Bridge, NY 13428   Fax: 518-673-2381
                            E-mail: rbrown@leepub.com
                            http://www.leepub.com
Dicusses the latest developments in woodwaste, C&D, scrapmetal, concrete, asphalt, recycling and composting with the emphasis on equipment.
Publication Date: 1993 50 pages Monthly
Coyle Rockwell, Author
Holly Reiser, Editor
Richard Brown, Production Coordinator

4388 Waste Recovery Report
Icon
211 S 45th Street   215-349-6500
Philadelphia, PA 19104   Fax: 215-349-6502
E-mail: wasterec@aol.com
http://www.iconworldwide.com/recovery
Contains information on waste-to-energy, recycling, compost-
ing and other technologies.
Publication Date: 1975   6 pages   Monthly
ISSN: 0889-0072
Alan Krigman, Publisher/Editor

Periodicals: Sustainable Development

4389 Alternative Energy Resources Organization Newsletter
432 North Last Chance Gulch   406-443-7272
Helena, MT 59601   Fax: 406-442-9120
E-mail: aero@aeromt.org
AERO is a statewide grassroots group whose members work to-
gether to strengthen communities through promoting sustain-
able agriculture, local food production and citizen-based Smart
Growth community planning.
Publication Date: 1978   4-24 pages   Quarterly
Jennifer Rasmuson, Office Manager

**4390 California Association of Resource Conservation Dis-
tricts- CCP News**
3823 V Street   916-457-7094
Suite 3   Fax: 916-457-7934
Sacramento, CA 95817   http://www.carcd.org
Quarterly
Patrick Truman, President
Brian Leahy, Executive Director

4391 Californians for Population Stabilization: CAPS News
1129 State Street   805-564-6626
Suite 3-D   Fax: 805-564-6636
Santa Barbara, CA 93101   E-mail: info@capsweb.org
http://www.capsweb.org
A nonprofit, public interest organization that works to protect
California's environment and quality of life by turning the tide
of population growth.
Publication Date: 1986   8 pages   3x year
Diana Hull PhD, President
Ben Zuckerman PhD, Vice President

4392 Cultivar
Center for Agoecology
1156 High Street
Santa Cruz, CA 95064   831-459-2506
Fax: 831-459-2867
http://www.agroecology.org
Publication Date: 1985   9-12 pages   Bi-Yearly
Steven Gliessman, Professor Agroecology
Martha Brown, Editor

4393 Ecosphere
Forum International
91 Gregory Lane   925-671-2900
Suite 21   800-252-4475
Pleasant Hill, CA 94523   Fax: 925-671-2993
E-mail: fti@foruminternational.com
http://www.foruminternational.com
Accepts advertising. The first ever environmental/ecological
magazine. It is dedicated to the interrelations of man in nature
and a balanced approach of its biological, economic, socio-po-
litical and spiritual components. Since 1965.
Publication Date: 1965   16-48 pages   Bi-monthly
Dr. Nicolas Hetzer, Production Manager
J McCormack, Circulation Director

4394 EnviroNews
600 Forbes Avenue   412-396-6000
331 Fisher Hall   800-456-0590
Pittsburgh, PA 15282   Fax: 412-396-4092
E-mail: bembic@duq.edu
http://www.science.duq.edu

Educating environmental professionals for the twenty-first century is the focus of the Duquesne University Environmental Science and Management (ESM) Masters Degree Program. The
program grew out of the perceived need to combine depth of
knowledge in environmental science with a comprehensive understanding of the business, legal and policy implications surrounding environmental issues.
4 pages   Semester Newsletter
Sonia Bembic, Program Advisor

4395 Environmental News
Arkansas Environmental Federation
1400 W Markham Street   501-374-0263
Suite 250   Fax: 501-374-8752
Little Rock, AR 7220   E-mail: rthurman@environmentark.org
http://www.environmentark.org
Randy Thurman, Executive Director

4396 Forest Magazine
Forest Service Employees for Environmental Ethics
PO Box 11615   541-484-2692
Eugene, OR 97440-3815   Fax: 541-484-3004
E-mail: fseee@fseee.org
http://www.fseee.org
FSEEE is the largest forest watchdog organization in the nation.
Since 1989, FSEEE has defended the rights and responsibilities
of brave scientists and resource professionals working to assure
the long-term health and vitality of ournational forests. FSEEE
publishes Forest Magazine every other month to educate the
public on forest issues.
Publication Date: 1989   50 pages   Bi-Monthly
ISSN: 1534-9284
Andy Stahl, Executive Director
Patricia Marshall, Editor

4397 Forest Service Employees for Environmental Ethics
PO Box 11615   541-484-2692
Eugene, OR 97440-3815   Fax: 541-484-3004
E-mail: fseee@fseee.org
http://www.fseee.org
FSEEE is the largest forest watchdog organization in the nation.
Since 1989, FSEEE has defended the rights and responsibilities
of brave scientists and resource professionals working to assure
the long-term health and vitality of ournational forests. FSEEE
publishes Forest Magazine every other month to educate the
public on forest issues.
Andy Stahl, Executive Director

4398 Greens/Green Party USA Green Politics Green Politics
PO Box 3568   978-682-4353
Eureka, CA 95502   866-GRE-ENS2
Fax: 978-682-4318
E-mail: info@greenparty.org
http://www.greenparty.org
Is a national non-profit membership organization dedicated to
advancing the Green Ten Key Values as a guiding force in Amer-
ican society and politics.
12 pages   Quarterly
Don Fitz, Editor

4399 International Boreal Forest Newsletter
Institute for World Resource Research
PO Box 5275   630-910-1551
Woodridge, IL 60517-0275   Fax: 630-910-1561
http://www.globalwarming.net
Covers all phases of developments in forestry and reforestation
of northern nations including the US, Canada, Russia, Sweden,
Finland, Norway, China, Japan and others. Its goal is to increase
the worldwide understanding of theecological and economic
roles of the northern forest regions of the world.
Dr. Yuan Lee, Editor-in-Chief
BJ Jefferson, Advertising/Sales

**4400 International Society of Tropical Foresters: ISTF No-
tices**
5400 Grosvenor Lane   301-897-8720
Bethesda, MD 20814   Fax: 301-897-3690
E-mail: istfi@igc.apc.org

The International Society of Tropical Foresters, Inc. (ISTF) is a nonprofit organization committed to the protection, wise management and rational use of the world's tropical forests. Established in 1950, ISTF was reactivated in 1979.It has about 1500 members in more than 100 countries. Financial support comes from membership dues, donations and grants. ISTF sponsors meetings, promotes chapters in other countries, maintains a web site and has chapters at universities.
Warren T Doolittle, President

4401 Jackson Hole Conservation Alliance: Alliance News

685 S Cache   307-733-9417
PO Box 2728   Fax: 307-733-9008
Jackson, WY 83001   E-mail: info@jhalliance.org
                                        http://www.jhalliance.org
An organization dedicated to responsible land stewardship in Jackson Hole, Wyoming to ensure that human activities are in harmony with the area's irreplaceable wildlife, scenery and other natural resources.
          20 pages Quarterly
Franz Camenzind PhD, Executive Director

4402 Leopold Letter

Leopold Center for Sustainable Agriculture
209 Curtiss Hall   515-294-3711
Ames, IA 50011-1050   Fax: 515-294-9696
                                        E-mail: leopold@iastate.edu
                                        http://www.leopold.iastate.edu
To inform diverse audiences about Leopold Center programs and activities; to encourage increased interest in and use of sustainable farming practicies; and to stimulate public discussion about sustainable agriculture in Iowa.
          Publication Date: 1989 12 pages Quarterly
          ISSN: 1065-2116
Frederick L Kirschenmann, Director
Laura Miller, Editor

4403 Minnesota Department of Agriculture: MDA Quarterly

90 West Plato Boulevard   651-297-2200
St Paul, MN 55107   800-967-2474
                                        Fax: 651-297-5522
                                        E-mail: webinfo@mda.state.mn.us
                                        http://www.mda.state.mn.us
The MDA's mission is to work toward a diverse ag industry that is profitable as well as environmentally sound; to protect the public health safety regarding food and ag products; and to ensure orderly commerce in agricultural and foodproducts. We have two major branches of the department to accomplish this mission: regulatory divisions and non-regulatory divisions.
          Publication Date: 2000
Gene Hugoson, Commissioner
Michael Schommer, Editor

4404 Mountain Research and Development

PO Box 1978   530-752-8330
Davis, CA 95617 http://www.mrd-journal.org/about_mrd.htm
The leading journal specifically devoted to the world's mountains. It has been published since 1981 and has established itself as a renowned international publication containing well-researched, peer-reviewed scientific articles byauthors from around the world.
Professor Hans Hurni, Editor-in-Chief

4405 Northeast Sun

Northeast Sustainable Energy Association
50 Miles Street   413-774-6051
Greenfield, MA 01301   Fax: 413-774-6053
                                        E-mail: nesea@nesea.org
                                        http://www.nesea.org
Regional magazine promoting responsible use of energy for a stronger economy and cleaner environment.
          2x Year
Nancy Hazard, Executive Director

4406 Pinchot Letter

1616 P Street NW   202-797-6580
Suite 100   Fax: 202-797-6583
Washington, DC 20036   E-mail: pinchot@pinchot.org
                                        http://www.pinchot.org

A tri-annual newsletter published by the Pinchot Institute for Conservation, an independent nonprofit organization that works collaboratively with all Americans-from federal and state policymakers to citizens in rural communities-tostrengthen forest conservation by advancing sustainable forest management, developing conservation leaders and providing science-based solutions to natural resource issues.
          Publication Date: 1995 20 pages Tri-Annual
Dr V Alaric Sample, President

4407 Population Institute Newsletter

107 2nd Street NE   202-544-3300
Suite 207   188-787-0038
Washington, DC 20002   Fax: 202-544-0068
                                        E-mail: web@populationinstitute.org
                                        http://www.populationinstitute.org
The Population Institute is the World's largest independent non-profit, educational organization dedicated exclusively to achieving a more equitable balance between the worlds population, environment, and resources. Established in 1969,the Institute, with members in 172 countries, is headquartered on Capitol Hill in Washington DC. The Institute uses a variety of resources and programs to bring its concerns about the consequences of rapid poulation growth to the forefront of thenational agenda.
          Publication Date: 1988 8 pages Bi-Monthly
Werner Fornos, President
Hal Burdett, Executive Editor

4408 Population Reference Bureau: World Population Data Sheet

1875 Connecticut Avenue NW   202-483-1100
Suite 520   800-877-9881
Washington, DC 20009-5728   Fax: 202-328-3937
                                        E-mail: popref@prb.org
                                        http://www.prb.org

Up-to-date demographic data and estimates for all the countries and major regions of the world.
William P Butz, President/CEO

4409 Reporter

Population Connection
1400 16th Street NW   202-332-2200
Suite 320   Fax: 202-332-2302
Washington, DC 20036   E-mail: info@popconnect.org
                                        http://www.popconnect.org
Looks at the connections between overpopulation and the environment around the world and features reports from our activists on Capitol Hill Days 2005. This publication is included in your $25.00 memberhsip fee.
          Publication Date: 1972 24 pages Quarterly
          ISSN: 0199-0071
John Seager, President/CEO
Mara Nelson Grynavinski, Editor

4410 Resource Development Newsletter

University of Tennessee
PO Box 1071   865-974-7448
Knoxville, TN 37901-1071   Fax: 423-974-7448
Community development information.
          4 pages
Dr Alan Barefield, Publisher

4411 Restoration Ecology Magazine

Blackwell Science
350 Main Street   781-388-8250
Malden, MA 02148   Fax: 781-388-8270
                                        http://www.blackwellpublishing.com
Provides the most recent developments in the ecological and biological restoration field for both the fundamental and practical implications of restorations.
Richard Hobbs, Editor

4412 Society of American Foresters Information Center Newsletter

5400 Grosvenor Lane   301-897-8720
Bethesda, MD 20814   Fax: 301-897-3690
                                        E-mail: safweb@safnet.org
                                        http://www.safnet.org
An organization that represents the forestry profession in the United States. Its mission is to advance the science, education, technology and practice of forestry.

Publication Date: 1996   24 pages Monthly
Michael T Goergen, Jr, EVP/CEO

4413 Solar Energy Magazine

Elsevier Science
360 Park Avenue S   212-989-5800
11th Floor   Fax: 212-633-3680
New York, NY 10010   **E-mail: usinfo-f@elseview.com**
                    http://www.elseview.com
Devoted exclusively to the science and technology of solar energy applications.
Publication Date: 1957
ISSN: 0380-92X
D Yogi Goswami, Editor-in-Chief

4414 Solar Energy Report

California Solar Energy Industries Association
PO Box 782   949-837-7430
Rio Vista, CA 94571   Fax: 949-709-8043
                    E-mail: info@calseia.org
                    http://www.calseia.org
Bi-Monthly
Les Nelson, President

4415 Southface Journal of Sustainable Building

Southface Energy Institute
241 Pine Street   404-872-3549
Atlanta, GA 30308   Fax: 404-872-5009
                    E-mail: info@southface.org
                    http://www.southface.org
Contains articles on numerous sustainable building topics. Free to members and available online.
Publication Date: 1978   24 pages Quarterly
Dennis Creech, Executive Director/Editor

4416 Tall Timbers Research Station: Bulletin Series

13093 Henry Beadel Drive   850-893-4153
Tallahassee, FL 32312-0918   Fax: 850-668-7781
                    E-mail: rose@ttrs.org
                    http://www.talltimbers.org
Dedicated to protecting wildlands and preserving natural habitats. Promotes public education on the importance of natural disturbances to the environment and the subsequent need for wildlife and land management. Conducts fire ecologyresearch and other biological research programs through the Tall Timbers Research Station. Operates museum.
Publication Date: 1962
ISSN: 0496-7631
Lane Green, Executive Director
R Todd Engstrom, Editor

4417 Tall Timbers Research Station: Fire Ecology Conference Proceedings

13093 Henry Beadel Drive   850-893-4153
Tallahassee, FL 32312   Fax: 850-668-7781
                    E-mail: rose@ttrs.org
                    http://www.talltimbers.org
Dedicated to protecting wildlands and preserving natural habitats. Promotes public education on the importance of natural disturbances to the environment and the subsequent need for wildlife and land management. Conducts fire ecologyresearch and other biological research programs through the Tall Timbers Research Station. Operates museum.
Publication Date: 1962
ISSN: 0082-1527
Lane Green, Executive Director
R Todd Engstrom, Editor

4418 Tall Timbers Research Station: Game Bird Seminar Proceedings

13093 Henry Beadel Drive   850-893-4153
Tallahassee, FL 32312-0918   Fax: 850-668-7781
                    E-mail: rose@ttrs.org
                    http://www.talltimbers.org
Dedicated to protecting wildlands and preserving natural habitats. Promotes public education on the importance of natural disturbances to the environment and the subsequent need for wildlife and land management. Conducts fire ecologyresearch and other biological research programs through the Tall Timbers Research Station. Operates museum.

Publication Date: 1980
ISSN: 1087-4372
Lane Green, Executive Director
R Todd Engstrom, Editor

4419 Tall Timbers Research Station: Miscellaneous Series

13093 Henry Beadel Drive   850-893-4153
Tallahassee, FL 32312-0918   Fax: 850-668-7781
                    E-mail: rose@ttrs.org
                    http://www.talltimbers.org
Dedicated to protecting wildlands and preserving natural habitats. Promotes public education on the importance of natural disturbances to the environment and the subsequent need for wildlife and land management. Conducts fire ecologyresearch and other biological research programs through the Tall Timbers Research Station. Operates museum.
Publication Date: 1961
ISSN: 0494-764x
Lane Green, Executive Director
R Todd Engstrom, Editor

4420 Totally Tremendous Activities

Northeast Sustainable Energy Association
50 Miles Street   413-774-6051
Greenfield, MA 01301   Fax: 413-774-6053
                    E-mail: nesea@nesea.org
                    http://www.nesea.org
Nancy Hazard, Executive Director

4421 Woodland Steward

Massachusetts Forestry Association
270 Jackson Street   413-323-7326
Belchertown, MA 01007   Fax: 413-323-9594
This publication is full of information about Massachusett's forest and ways that landowners can manage their woodlands to achieve their goals in an environmentally sustainable manner. Free with membership.
Bi-Monthly
Gregory Cox, Executive Director

4422 Worldwatch Institute: State of the World

1776 Massachusetts Avenue NW   202-452-1999
Suite 800   Fax: 202-296-7365
Washington, DC 20036E-mail: worldwatch@worldwatch.org
                    http://www.worldwatch.org
The most authorative go-to resource for those who understand the importance of nuturing a safe, sane and healthy global environment through both policy and action.
Annual
ISBN: 0-393326-66-7
Christopher Flavin, President
Tom Prugh, Editor

4423 Worldwatch Institute: Vital Signs

1776 Massachusetts Avenue NW   202-452-1999
Suite 800   Fax: 202-296-7365
Washington, DC 20036E-mail: worldwatch@worldwatch.org
                    http://www.worldwatch.org
Provides comprehensive, user-friendly information on key trends and includes tables and graphs that help readers access the developments that are changing their lives for better or for worse.
Annual
ISBN: 0-393326-89-6
Christopher Flavin, President
Tom Prugh, Editor

4424 Worldwatch Institute: World Watch

1776 Massachusetts Avenue NW   202-452-1999
Suite 800   Fax: 202-296-7365
Washington, DC 20036E-mail: worldwatch@worldwatch.org
                    http://www.worldwatch.org
The Worldwatch Institute is an independent, nonprofit environmental research organization in Washington DC. Its mission is to foster a sustainable society in which human needs are met in ways that do not threaten the health of thenatural environment or future generations. To this end, this Institute conducts interdisciplinary research on emerging global issues, the results of which are published and disseminated to decision-makers and the media.
Bi-Monthly
Christopher Flavin, President
Tom Prugh, Editor

4425 Worldwatch Institute: Worldwatch Papers

1776 Massachusetts Avenue NW 202-452-1999
Suite 800 Fax: 202-296-7365
Washington, DC 20036 E-mail: worldwatch@worldwatch.org
http://www.worldwatch.org

Provides cutting-edge analysis on an environmental topic that is making - or is about to make - headlines worldwide.
50-70 pages 5x times year
Christopher Flavin, President
Tom Prugh, Editor

Periodicals: Travel & Tourism

4426 New York State Parks, Recreation and Historic Preservation

Empire State Plaza 518-474-0456
Agency Building 1 Fax: 518-486-2924
Albany, NY 12238 http://www.nysparks.com

Publishes New York State Boat Launching Guide, Camping/Cabin Reservation Info, Snowmobiling Guide and Preservation Magazine.
Bernadette Castro, Commissioner

4427 Noxubee National Wildlife Refuge Newsletter

2970 Bluff Lake Road 662-323-5548
Brooksville, MS 39739 Fax: 662-323-6390
E-mail: noxubee@fws.gov
http://noxubee.fws.gov

Noxubee National Wildlife Refuge was established in 1940 to protect and enhance habitat for the conservation of migratory birds, endangered species and other wildlife. The recreational and educational opportunities provided on therefuge help the public experience nature and learn how sound management ensures that future generations continue to enjoy fish and wildlife and their habitats.
2 pages Bi-Annual
Andrea Duncan, Editor

4428 Parks and Recreation Magazine

National Recreation and Park Association
22377 Belmont Ridge Road 703-858-0784
Ashburn, VA 20148 Fax: 703-858-0794
E-mail: info@nrpa.org
http://www.activeparks.comornrpa.org

Informs, motivates and inspires professionals, civic leaders and citizens to elevate the value of parks and recreation as a public service.
Monthly
John Thorner, Exectutive Director
Rachel Roberts, Editor

4429 Potomac Appalachian

118 Park Street SE 703-242-0693
Vienna, VA 22180-4609 Fax: 703-242-0968
E-mail: info@patc.net
http://www.patc.net

Published by the Potomac Appalachian Trail Club, which through volunteer efforts, education and advocacy, acquires, maintains and protects the trail and lands of the Appalachian Trail, other trails and related facilities in theMid-Atlantic Region for the enjoyment of present and future hikers. PATC publishes hiking guides, maps and history books of the Appalachian Trail and other trails in our area of responsibility. The monthly newsletter is sent to members and uponrequest. Free to members.
16-20 pages Monthly
ISSN: 098 -8154
Thomas R Johnson, President

Periodicals: Water Resources

4430 Air Water Pollution Report's Environment Week

Business Publishers
8737 Colesville Road 301-589-5103
10th Floor 800-274-6737
Silver Spring, MD 20910 Fax: 301-589-8493
E-mail: custserv@bpinews.com
http://www.bpinews.com

Provides a balanced, insightful update on the week's most important environmental news from Washington, DC.
Leonard A Eiserer, Publisher
David Goeller, Editor

4431 Air/Water Pollution Report

Business Publishers
8737 Colesville Road 301-589-5103
10th Floor 800-274-6737
Silver Spring, MD 20910 Fax: 301-589-8493
E-mail: custserv@bpinews.com
http://www.bpinews.com

Regulatory activities and governmental legislation, in addition to litigation are covered in this pulication.
Leonard A Eiserer, Publisher
David Goeller, Editor

4432 American Fisheries Society: Water Quality Matters

324 25th Street 801-625-5358
Ogden, UT 84401 Fax: 801-625-5756
E-mail: glampman@fs.fed.us

8 pages 1-2 per year
Georgina Lampman, President
Gregg Lomincky, Editor

4433 American Shore & Beach Preservation Association

1724 Indian Way 310-305-9537
Oakland, CA 94561 Fax: 310-821-6345
E-mail: president@asbpa.org
http://www.asbpa.org

This Association is formed in recognition of the fact that the shores of our oceans, lakes and rivers constitute assets for promoting the health and physical well-being of the people of this nation, and that their contiguity to outgreat centers of population affords an opportunity for wholesome and necessary rest and recreation not equally available in any other form. The purpose of the Association is to bring together for cooperation and mutual helpfulness.
24 pages Quarterly
ISSN: 0037-4237
Honorable Harry Simmons, President

4434 American Water Resources Association: Journal of the American Water Resources Association

4 W Federal Street 540-687-8390
PO Box 1626 Fax: 540-687-8395
Middleburg, VA 20118 E-mail: info@awra.org
http://www.awra.org

AWRA is a nonprofit, scientific educational association for individuals and organizations involved in all aspects of water resources. Its goal is to advance multidisciplinary water resources management and research through itsconferences, publications, technical commettees, state sections and student chapters.
Publication Date: 1964 Bi-Monthly
ISSN: 1093-474X
Mindy Lalor, President
John J Warwick, Editor

4435 American Water Resources Association: Water Resources IMPACT

4 W Federal Street 540-687-8390
PO Box 1626 Fax: 540-687-8395
Middleburg, VA 20118-1626 E-mail: info@awra.org
http://www.awra.org

AWRA is a nonprofit, scientific educational association for individuals and organizations involved in all aspects of water resources. Its goal is to advance multidisciplinary water resources management and research through itsconferences, publications, technical commettees, state sections and student chapters.
Bi-Monthly
ISSN: 1093-474X
Mindy Lalor, President
Earl Spangenberg, Editor-in-Chief

4436 American Well Owner

American Ground Water Trust
16 Centre Street 603-228-5444
Concord, NH 03301 Fax: 603-228-6557
E-mail: trustinfo@agwt.org
http://www.agwt.org

This publication helps citizens, communities, businesses and farms maintain safe, reliable water supplies from their wells.

Publication Date: 1998 4 pages Quarterly
Andrew Stone, Director

4437 Blue Planet Magazine
The Ocean Conservancy
2029 K Street   202-429-5609
Washington, DC 20006   800-519-1541
Fax: 202-429-0056
E-mail: info@oceanconservancy.org
http://www.oceanconservancy.org
To educate peoeple about ocean issues; inspire readers with the
beauty and wonder of oceans; encourage dedication to appreci-
ating and protecting marine resources; and enlist new volunteers
in the ocean community. Free with membershipfee of $25.00
46 pages Quarterly
Roger Rufe,Jr, President/CEO
Sara Bennington, Editor

4438 Clean Water Network: CWN Status Water Report
Spills and Kills
1200 New York Avenue, NW   202-289-2521
Suite 400   Fax: 202-289-1060
Washington, DC 20005   E-mail: info@cwn.org
http://www.cwn.org
A nonprofit network of over 1,000 organizations that deal with
clean water issues covered by the Clean Water Act. Our member
organizations consist of a variety of organizations representing
environmentalists, family farmers, recreationanglers, commercial fishermen, surfers, boaters, faith communities, labor unions
and civic associates. We publish a monthly newsletter and various reports.
8-12 pages Monthly
Katherine Smitherman, Executive Director

4439 Clean Water Report Newsletter
Business Publishers
8737 Colesville Road   301-589-5103
10th Floor   800-274-6737
Silver Spring, MD 20910-3928   Fax: 301-589-8493
E-mail: custserv@bpinews.com
http://www.bpinews.com
Follows the latest news from the EPA, Congress, the states, the
courts, and private industry. A key information source for environmental professionals, covering the important issues of
ground and drinking water, wastewater treatment,wetlands,
drought, coastal protection, non-point source pollution,
agrichemical contamination and more.
8 pages Bi-Weekly
ISSN: 0009-8620
Leonard A Eiserer, Publisher
Louise Harris, Editor

4440 Clearwaters Magazine
New York Water Environment Association
126 East Salina Street   315-422-7811
Suite 200   Fax: 315-422-3851
Syracuse, NY 13202   E-mail: pcr@nywea.org
http://www.nywea.org
Published by The New York Water Environment Association, a
nonprofit educational association dedicated to the development
and dissemination of information concerning water quality man-
agement and the nature, collection, treatment, anddisposal of
wastewater. Founded in 1929, the Association has over 2,500
members. The NYWEA is a member association of the Water
Environment Federation.
Quarterly
Patricia Cerro-Reehil, Executive Director
Robert D Hennigan, Executive Editor

**4441 Colorado Department of Natural Resources: Division of
Water Resources: StreamLines**
1313 Sherman Street   303-866-3581
Room 818   Fax: 303-866-3589
Denver, CO 80203   http://www.water.state.co.us
The Colorado Division of Water Resources is an agency of the
State of Colorado, Department of Natural Resources, operating
under the direction of specific state stautes, court decrees, and
interstate compacts. The DWR is empowered toadminister all
surface and ground water rights throughout the state and ensure
that the doctrine of prior appropiation is enforced.
Publication Date: 1988 4-8 pages Quearterly
Hal D Simpson, Director Water Resources
Russell George, Executive Director

4442 Colorado Water Rights
1580 Logan Street   303-837-0812
Suite 400   Fax: 303-837-1607
Denver, CO 80203   E-mail: macravey@cowatercongress.org
http://www.cowatercongress.org
This newsletter helps the Colorado Water Congress protect and
conserve Colorado's water resouces by educating its readers.
Publication Date: 1982 4-16 pages Quarterly
David Merritt, President

4443 Environmental Policy Alert
Inside Washington Publishers
1225 Jefferson Davis Highway   703-416-8500
Suite 400   800-424-9068
Arlington, VA 22202   Fax: 703-416-8543
E-mail: iwp@iwpnews.com
http://www.iwpnews.com
Is a reliable resource for all regulatory, congressional and litiga-
tion developments in air quality, waste cleanup, clean water and
other environmental quality issues. Also provides a special fo-
cus on efforts to reinvent environmentalpolicies.
Publication Date: 1984 Bi-Weekly
Jeremy Bernstein, Editor

**4444 Georgia Water and Pollution Control Association: Op-
erator**
2221 New Market Parkway   770-618-8690
Suite 134   Fax: 770-618-8695
Marietta, GA 30067   E-mail: info@gwpca.org
http://www.gawponline.org
The GW+PCA is dedicated to education, dissemination of technical and scientific information, increased public understanding
and promotion of sound public laws and programs in the water
resources and related environmental fields.Founded in 1932.
Publication Date: 1970 56-68 pages Quarterly
Jack C Dozier, PE, Executive Director

**4445 Georgia Water and Pollution Control Association: News
& Notes**
2221 New Market Parkway   770-618-8690
Suite 134   Fax: 770-618-8695
Marietta, GA 30067   E-mail: info@gwpca.org
http://www.gawponline.org
The GW+PCA is dedicated to education, dissemination of tech-
nical and scientific information, increased public understanding
and promotion of sound public laws and programs in the water
resources and related environmental fields.Founded in 1932.
Publication Date: 1970 20-28 pages Monthly
Jack C Dozier, PE, Executive Director

4446 Gulf of Mexico Science
Marine Envir. Sci. Consort.
Dauphin Island Sea Lab   251-861-2141
101 Bienville Boulevard   Fax: 251-861-4646
Dauphin Island, AL 36528   http://www.disl.org
Journal devoted to disemminating knowledge of the Gulf of
Mexico and adjacent areas. Appropriate topics of consideration
for publication include all areas of marine science.
2x Year
ISSN: 1087-688X
Dr. George F Crozier, Executive Director
Dr William W Schroeder, Editor

4447 International Desalination and Water Reuse Quarterly
Lineal Publishing Company
306 Eagle Dr   561-451-9429
Jupiter, FL 33477-4066   Fax: 561-451-9435
Disseminates technical information, reviews and analyzes regional developments in the field, as well as new products and
processes. The publication provides, on a continuing basis, a
major vehicle in which to promote desalination andwater reuse
technologies, equipment, and design to potential users.
Irv Lineal, Publisher

4448 Journal of Soil and Water Conservation
Soil & Water Conservation Society
945 SW Ankeny Road   515-289-2331
Ankeny, IA 50023-9723   Fax: 515-289-1227
E-mail: swcs@swcs.org
http://www.swcs.org

Publication includes a variety of conservation subjects, as well as international conservation issues.
Craig Cox, Executive Director
Deb Happe, Editor/Communications Dir

4449 Journal of the New England Water Environment
New England Water Environment Association
100 Tower Office Park   781-939-0908
Suite K   Fax: 781-939-0907
Woburn, MA 01801   E-mail: main@newea.org
                               http://www.newea.org
Bi-annual publication from New England Water Environment Association.
Publication Date: 1929 150 pages Bi-Annual
ISSN: 1077-3002
Elizabeth Cutone, Executive Director

4450 Mass Waters
Massachusetts Water Pollution Control Association
PO Box 221   978-374-0170
Groveland, MA 01834   Fax: 978-521-4083
                               E-mail: mwpca@mwpca.org
                               http://www.mwpca.org

Quarterly
John Connor, Secretary/Treasurer

4451 Mono Lake Committee Newsletter
Corner of Hwy 395 & 3rd Street   760-647-6595
Lee Vining, CA 93541   Fax: 760-647-6377
                                   E-mail: info@monolake.org
                                   http://www.monolake.org
Nonprofit citizen's group dedicated to: protecting and restoring the Mono Basin ecosystem; educating the public about Mono Lake and the impacts on the environment of excessive water use; promoting cooperative solutions that protectMono Lake and meet real water needs without transferring environmental problems to other areas.
Publication Date: 1978 28 pages Quarterly
Francis Spivy-Weber, Executive Director
Arya Degenhardt, Editor

4452 Montana Environmental Training Center Newsletter
2100 16th Avenue S   406-771-4433
PO Box 6010   Fax: 406-771-4317
Great Falls, MT 59406   E-mail: boylej@msun.edu
                        http://www.msun.edu/grants/metc/gary.asp
METC is a cooperative effort between Montana State University-Northern and the Montana Department of Environmental Quality. Basic, advance training, and continuing education in the areas of water and wastewater operation, maintenance,safety, process control, cross connection and backflow prevention along with courses in basic water science and watershed awareness define the training activities of METC. A newsletter and Training Announcement are published quarterly.
Quarterly
Jan Boyle, Director

4453 Montana Environmental Training Center Training Announcement
2100 16th Avenue S   406-771-4433
Great Falls, MT 59406-6010   Fax: 406-771-4317
                                   E-mail: boylej@msun.edu
                        http://www.msun.edu/grants/metc/gary.asp
METC is a cooperative effort between Montana State University-Northern and the Montana Department of Environmental Quality. Basic, advance training, and continuing education in the areas of water and wastewater operation, maintenance,safety, process control, cross connection and backflow prevention along with courses in basic water science and watershed awareness define the training activities of METC. A newsletter and Training Announcement are published quarterly.
Quarterly
Jan Boyle, Director

4454 Montana Water Environment Association: Newsletter
516 N Park Street   406-449-7913
Suite A   Fax: 406-449-6350
Helena, MT 59601-2773
        Semi-Annual
Carl Anderson, President

4455 New Mexico Rural Water Association Newsletter
3413 Carlisle Boulevard NE   505-884-1031
Albuquerque, NM 87110   Fax: 505-884-1032
                               E-mail: contact @nmrwa.org
                               http://www.nmrwa.org
To provide top quality, responsive technical assistance and training for rural water and wastewater systems in New Mexico.
        Quarterly
Matthew Holmes, Executive Director
Robert Matthews, Co-Editor

4456 New York Water Environment Association Clearwaters
126 N Salina Street   315-422-7811
Suite 200   Fax: 315-422-3851
Syracuse, NY 13202   E-mail: pcr@nywea.org
                               http://www.nywea.org
Contains articles on environmental issues, regulatory changes, technological advances as well as, updates on members and activities.
        50 pages Quarterly
Patricia Cerro-Reehil, Executive Director
Hope Dodge, Editor

4457 Oregon Water Resources Congress Newsletter
1201 Court Street NE   503-363-0121
Suite 303   Fax: 503-371-4926
Salem, OR 97301   E-mail: owrc@owrc.org
                               http://www.owrc.org
Is to promote the protection and use of water rights and the wise stewardship of water resources.
        Quarterly
Anita Winkler, Executive Director
Carol Zielinski, Editor

4458 Ozark National Scenic Riverways
Ozark National Scenic Riverways
404 Watercress Drive   573-323-4236
PO Box 490   Fax: 573-323-4140
Van Buren, MO 63965E-mail: ozar_superintendent@nps.gov
                               http://www.nps.gov/ozar
Missouri's largest National Park and America's first to preserve a free flowing river in its wild state. Covers some 80,000 acres along 134 miles of the Current and Jacks Fork Rivers. Staff provide elementary level environmentaleducation programs on natural history, with an emphasis on karst and water issues. Publishes "More Than Skin Deep, a Teacher's Guide to Caves and Groundwater," a curriculum guide suitable for grades K-12.
        Publication Date: 1964 Annual
Noel Poe, Superintendent

4459 Pacific Rivers Council: Freeflow
PO Box 10798   541-345-0119
Eugene, OR 97440   Fax: 541-345-0710
                               E-mail: info@pacrivers.org
                               http://www.pacrivers.org
Promoting the protection and restoration of rivers, their watersheds, and native aquatic species.
        Quarterly
David Bayles, Executive Director
Holly Spencer, Editor

4460 Pipeline
National Evironmental Services Center
NRCCE Building, Evandale Drive   304-293-4191
258 Stewart Street   800-624-8301
Morgantown, WV 26505   Fax: 304-293-3161
                        E-mail: nsfc_orders@mail.nesc.wvu.edu
                               http://www.nsfc.wvu.edu
Newsletter of the National Small Flows Clearinghouse, a nonprofit national source of information about small flows technologies-those systems that have fewer than one million gallons of wastewater flowing through them per day-rangingfrom individual septic systems to small sewage treatment plants. Free to US residents.
        Publication Date: 1990 8 pages Quarterly
        ISSN: 1060-0043
Sanjay Saxena, NSFC Program Coordinator
Jen Hause, NSFC Engineering Scientist

4461 Puerto Rico Water Resources and Environmental Research Institute Newsletter
University of Puerto Rico
College of Engineering   787-265-3826
PO Box 9040   Fax: 787-832-0119
Mayaguez, PR 00681-9040   E-mail: wrri_rum@uprm.edu
                    http://www.ece.uprm.edu/rumhp/prwrri
Its objectives are to: conduct research aimed at resolving local and national water resources problems; train scientists and engineers through hands-on participation in research; and to facilitate the incorporation of research resultsin the knowledge base of water resources professionals.
        Publication Date: 1990   4 pages Quarterly/thru Email
Jose R Cedeno, Associate Director

4462 Runoff Rundown
Center for Watershed Protection
8391 Main Street   410-461-8323
Ellicott, MD 21043   Fax: 410-461-8324
                        E-mail: center@cwp.org
        http://www.cwp.org or www.stormwatercenter.net
Electronic newsletter published by the Center for Watershed Protection, a nonprofit 501(c)3 organization dedicated to finding new ways to protect and restore our nation's streams, lakes, rivers and estuaries. The center publishesnumerous technical publications on all aspects of watershed protection, including stormwater management, watershed planning and better site design. All of our publications are available oneline at www.cwp.org.
        Publication Date: 2000   Quarterly
Hye Yeong Kwon, Executive Director
Lauren Lasher, Editor

4463 Save San Francisco Bay Association: Watershed Newsletter
350 Frank H Ogawa Plaza   510-452-9261
Suite 900   Fax: 510-452-9266
Oakland, CA 94612-2016 E-mail: SAVEBAY@savesfbay.org
                        http://www.savesfbay.org
Save the Bay has worked for over 40 years to protect the San Francisco Bay-Delta from pollution, fill, shoreline destruction and fresh water diversion. We have launched a century of renewal to restore bay fish and wildlife, reclaimtidal wetlands and make the bay safe and accessible to all.
        8-10 pages 3-4 times/year
David Lewis, Executive Director
Paul Revier, Editor

4464 Small Flow Quarterly
NRCCE Building, Evandale Drive   304-293-4191
258 Stewart Street   800-624-8301
Morgantown, WV 26505   Fax: 304-293-3161
                E-mail: nsfc_orders@mail.nesc.wvu.edu
                        http://www.nsfc.wvu.edu
Magazine of the National Small Flows Clearinghouse, a nonprofit national source of information about small flows technologies-those systems that have fewer than one million gallons of wastewater flowing through them per day-rangingfrom individual septic systems to small sewage treatment plants. Free to US residents.
        Publication Date: 2000   50 pages Quarterly
            ISSN: 1528-6827
Sanjay Saxena, NSFC Program Coordinator
Jen Hause, NSFC Engineering Scientist

4465 South Carolina Sea Grant Consortium
287 Meeting Street   843-727-2078
Charleston, SC 29401   Fax: 843-727-2080
                        http://www.scseagrant.org
A state agency that supports coastal and marine research, education, outreach, and one technical assistance program that fosters sustainable economic development and resource conservation. The consortium represents eight university andstate research organizations and induces a number of information products on coastal and marine resource topics.
        Publication Date: 1982   16 pages
M Richard DeVoe, Executive Director

4466 TCS Bulletin
PO Box 25408   703-768-1599
Alexandria, VA 22313-5408   Fax: 703-768-1596
                    E-mail: coastalsoc@aol.com
                    http://www.thecoastalsociety.org

Organization of private sector, academic, government professionals and students dedicated to actively addressing emerging coastal issues by fostering dialogue, forging partnerships and promoting communication and education. Thispublication covers issues of aquaculture-related law and coastal management research.
        Publication Date: 1975   24 pages Yearly
Paul Ticco, President
John Duff, Editor

4467 Tide
Coastal Conservation Association
4801 Woodway Drive   713-626-4234
Houston, TX 77056   800-201-FISH
                    E-mail: ccantl@joincca.org
                        http://www.joincca.org
TIDE is the official bimonthly magazine of the Coastal Conservation Association. It has received local, state and national acclaim for writing, photography and layout and currently boasts a circulation of more than 70,000. TIDE isavailable only to members of the Coastal Conservation Association.
        Bi-Monthly
Pat Murray, Executive Director
Ted Venker, Editor

4468 Utah Watershed Review
Utah Association of Conservation Districts
1860 North 100 East   435-753-6029
Logan, UT 84341-1784   Fax: 435-755-2117
                        http://www.uacd.org
Provides information about what's new in Utah and watershed volunteer work and management.
        Bi-Monthly
Gordon Younker, EVP
Jack Wilbur, Editor

4469 Water & Wastes Digest
Scranton Gillette Communications
380 E Northwest Highway   847-298-6622
Suite 200   Fax: 847-390-0408
Des Plaines, IL 60016   http://www.scrantongillette.com
This serves readers in the water and/or wastewater industries. These people work for municipalities, in industry, or as engineers. They design, specify, buy, operate and maintain equipment, chemicals, software and wastewater treatmentservices.
        128 pages
            ISSN: 0043-1181
Dennis Martyka, Publisher
Tim Gregorski, Editorial Director

4470 Water Quality Products
Scranton Gillette Communications
380 E Northwest Highway   847-298-6622
Suite 200   Fax: 847-390-0408
Des Plaines, IL 60016   http://www.scrantongillette.com
Provides balanced editorial content including developments in water conditioning, filtration and disinfection for residential, commercial and industrial systems.
        68 pages
            ISSN: 1092-0978
Dennis Martyka, Publisher
Tracy Fabre, Editor

4471 Water Resource Center: Minnegram
University of Minnesota
173 McNeal 1985 Buford Avenue   612-624-9282
St Paul, MN 55108   Fax: 612-625-1263
                                        Fax: '
                    E-mail: ander045@umn.edu
                    http://www.wrc.coafes.umn.edu
Four University water programs, Extension Water Quality Program, Center for Hydrocultural Impacts on Water Quality, Water Resources Research Center and Water Resources Science Graduate Program make up the Water Resources Center. Thecenter sponsors and coordinates programs in research, graduate education, outreach and service to address water resource management issues.
        Publication Date: 1986   Quarterly
Jim Anderson, Co-Director
Debra Swackhamer, Co-Director

4472 WaterMatters
Southwest Florida Water Management District
2379 Broad Street   352-796-7211
Brookville, FL 34604-6899   Fax: 352-754-6885
                              http://www.watermatters.org
Newletter of the Southwest Florida Water Management District,
which manages the water and water-related resources within its
boundaries. Maintains balance between the water needs of current and future users while protecting andmaintaining the natural systems that provide the District with its existing and future
water supply. The Conservation Projects Section, is reponsible
for managing water conservation, reclaimed water, other alternative source projects, and estimatingfuture water demands.
           2 pages Monthly
Dave Moore, Executive Director
Rebecca Bray, Editor

Books: Air & Climate

**4473 Air Pollution Control and the German Experience: Les-
sons for the United States**
Center for Clean Air Policy
750 1st Street NE   202-408-9260
Suite 940   Fax: 202-408-8896
Washington, DC 20002   E-mail: communications@ccap.org
                                     http://www.ccap.org

4474 Caring for Our Air
Enslow Publishers
40 Industrial Road   908-771-9400
PO Box 398   800-398-2504
Berkeley Heights, NJ 07922-0398   Fax: 908-771-0925
                     E-mail: customerservice@enslow.com
                                     http://www.enslow.com

4475 Center for Resource Economics
1718 Connecticut Avenue NW   202-232-7933
Suite 300   Fax: 202-234-1328
Wasington, DC 20009-1148   E-mail: info@islandpress.org
                                     http://www.islandpress.org
Works to educate the public about global environmental issues.
Methods include publishing literature on environmental con-
cerns.

**4476 Confronting Climate Change: Strategies for Energy Re-
search and Development**
National Academy Press
500 5th Street NW   202-334-3313
Lockbox 285   888-624-8373
Washington, DC 20055   Fax: 202-334-2451
                                     http://www.nap.edu
       Publication Date: 1990   144 pages
       ISBN: 0-309043-47-6

4477 Fight Global Warming: 29 Things You Can Do
Environmental Defense Fund
257 Park Avenue South   212-505-2100
New York, NY 10010-7304   800-684-3322
                                       Fax: 212-505-2375
                                       http://www.edf.org

4478 Fundamentals of Stack Gas Dispersion
Milton R. Beychok Consulting
1126 Colony Plaza   949-718-1360
Newport Beach, CA 92660   Fax: 949-718-1360
                  E-mail: mbeychok@air-dispersion.com
                               http://www.air-dispersion.com
The most comprehensive single-source reference book on
dispertion modeling of continuous buoyant pollution plumes.
Milton R Beychok, Principal

**4479 Healing the Planet: Strategies for Resolving the Envi-
ronmental Crisis**
Addison-Wesley Publishing Company
75 Arlington Street   617-848-7500
Suite 300   http://www.aw.com
Boston, MA 02116

4480 Indoor Air Quality: Design Guide Book
Fairmont Press
700 Indian Trail   770-279-4386
Liburn, GA 30047   Fax: 770-381-9865
                        E-mail: linda@fairmountpress.com
                                     http://www.fairmontpress.com

**4481 To Breath Free: Eastern Europe's Environmental Cri-
sis**
John Hopkins University Press
3400 N Charles Street   410-516-6900
Baltimore, MD 21218   800-537-5487
                                     http://www.jhubookis.com
Adam Glazer, Promotions Manager

Books: Business

4482 Butterflies of Delmarva
Delware Nature Society
PO Box 700   301-239-2334
Hockessin, DE 19707   Fax: 302-239-2473
                           E-mail: e-mail@dnashland.org
                       http://www.delawarenaturesociety.org
       Publication Date: 1994 138 pages
       ISBN: 0-870334-53-0
Peter Flint, President
Dr Elton N Woodbury, Author/Editor

**4483 Environmental Career Guide: Job Opportunities with
the Earth in Mind**
J Wiley & Sons
605 3rd Avenue   212-850-6000
6th Floor   Fax: 212-850-6088
New York, NY 10158   http://www.wiley.com
       Publication Date: 1991   208 pages
       ISBN: 0-471534-13-7
Nicholas Basta, Author

**4484 Environmental Disputes: Community Involvement in
Conflict Resolution**
Island Press
PO Box 7   707-983-6432
Covelo, CA 95428   800-828-1302
                                       Fax: 707-983-6414
A book published by Island Press which helps citizen groups,
business and government understand how Enviornmental Dis-
pute Settlement-a set of procedures for settling disputes over en-
vironmental policies without litigation-can work forthem.
       Publication Date: 1990   295 pages
       ISBN: 0-933280-74-2
James E Crowfoot, Julia Wondolleck, Author

**4485 In the US Interest: Resources, Growth, and Security in
the Developing World**
State University of New York Press
90 State Street   518-472-5000
Albany, NY 12246   518-472-5038
                                     http://www.sunypress.edu

**4486 Shopping for a Better Environment: Brand Name
Guide to Environmentally Responsible Shopping**
Meadowbrook Press
5451 Smetana Drive
Minnetonka, MN 55343   800-338-2232
                                       Fax: 952-930-1940
                        E-mail: info@meadowbrookpress.com
                                     http://www.meadowbrookpress.com

Books: Design & Architecture

4487 Designing Healthy Cities
Krieger Publishing Co.
PO Box 9542   321-724-9542
Melbourne, FL 32902-9542   800-724-0025
                                                          Fax: 321-951-3671
E-mail: info@krieger-publishing.com
http://www.krieger-publishing.com
Krieger Publishing Company produces quality books in various
fields of interest. We have an extensive Natural Science listing.
Publication Date: 1998   158 pages
ISBN: 0-894649-27-2
Cheryl Stanton, Advertising

4488 Indoor Air Quality: Design Guide Book
Fairmont Press
700 Indian Trail   770-279-4386
Liburn, GA 30047   Fax: 770-381-9865
E-mail: linda@fairmountpress.com
http://www.fairmontpress.com

Books: Disaster Preparedness & Response

4489 Acceptable Risk?: Making Decisions in a Toxic Environment
University of California Press
2120 Berkeley Way   510-642-4247
Berkeley, CA 94720-1012   Fax: 510-643-7127
E-mail: askucp@ucpress.edu
http://www.ucpress.edu

4490 Borrowed Earth, Borrowed Time: Healing America's Chemical Wounds
Plenum Publishers
233 Spring Street   212-620-8000
7th Floor   Fax: 212-463-0742
New York, NY 10013-1578   http://www.plenum.com
Publication Date: 1991

Books: Energy & Transportation

4491 Confronting Climate Change: Strategies for Energy Research and Development
National Academy Press
500 5th Street NW   202-334-3313
Washington, DC 20055   888-624-8373
                                                       Fax: 202-334-2451
                                               http://www.nap.edu
Publication Date: 1990   144 pages
ISBN: 0-309043-47-6

4492 Energy & Environmental Strategies for the 1990's
Fairmont Press
700 Indian Tr.   770-279-4386
Lilburn, GA 30247   Fax: 770-381-9865
E-mail: linda@fairmontpress.com
http://www.fairmontpress.com

4493 Energy Management and Conservation
National Conference of State Legislatures
7700 E First Place 700   303-364-7700
Denver, CO 80230   Fax: 303-364-7800
                                               http://www.ncsl.org

4494 Getting Around Without Gasoline
Northeast Sustainable Energy Association
50 Miles Street   413-774-6051
Greenfield, MA 01301   Fax: 413-774-6053
E-mail: nesea@nesea.org
http://www.nesea.org

This interdisciplinary science/social studies curriculum allows
students to explore the transportation and environmental issues
in their own lives. Activities cover: transportation systems,
health impacts, environmental andtransportation histories,
carpooling, and mass transit.
54 pages Curriculum
Chris Mason, Education Director

4495 Global Science: Energy, Resources, Environment
Kendall-Hunt Publishing Company
4050 Westmark Drive   319-589-1000
PO Box 1840   800-772-9165
Dubuque, IA 52004-1840   http://www.kendallhunt.com

Books: Environmental Engineering

4496 Global Science: Energy, Resources, Environment
Kendall-Hunt Publishing Company
4050 Westmark Drive   319-589-1000
PO Box 1840   800-772-9165
Dubuque, IA 52004-1840   http://www.kendallhunt.com

4497 Principles of Environmental Science and Technology
Elsevier Science Publishers
655 Avenue of the Americas   212-633-3650
New York, NY 10010   Fax: 212-633-3990
                                               http://www.elsevier.com

Books: Environmental Health

4498 Ecologue: The Environmental Catalogue and Consumer's Guide for a Safe Earth
Prentice Hall Press (Simon & Schuster Division)
1 Gulf & Western Plaza   212-373-8500
New York, NY 10023   800-223-1360
                                               http://www.prenhall.com

Books: Gaming & Hunting

4499 Better Trout Habitat: A Guide to Stream Restoration
Island Press
1718 Connecticut Avenue NW   202-232-7933
Suite 300   Fax: 202-234-1328
Washington, DC 20009-1148   E-mail: info@islandpress.org
                                               http://www.islandpress.org

Books: Habitat Preservation & Land Use

4500 50 Simple Things Kids Can Do to Save the Earth
Andrews and McMeel
4520 Main Street   816-932-6700
Suite 700   Fax: 816-932-6706
Kansas City, MO 64111

4501 Access EPA: Clearinghouses and Hotlines
National Technical Information Service
5285 Port Royal Road   703-487-4650
Springfield, VA 22161   E-mail: info@ntis.gov
                                               http://www.ntis.gov
Publication Date: 1991   57 pages

4502 Access EPA: Library and Information Services
National Technical Information Service
5285 Port Royal Road   703-487-4650
Springfield, VA 22161   E-mail: info@ntis.gov
                                               http://www.ntis.gov
Publication Date: 1990   110 pages

4503 After Earth Day: Continuing the Conservation Effort
University of North Texas Press
PO Box 311336   940-565-2142
Denton, TX 76203-1336   800-826-8911
                                       Fax: 940-565-4590
                          E-mail: rchrisman@unt.edu
                          http://www.unt.edu/untpress
          Publication Date: 1992   241 pages
                ISBN: 1-574414-44-0
Karen DeVinney, Managing Editor

**4504 Agatha's Feather Bed: Not Just Another Wild Goose
Story**
Peachtree Publishers
1700 Chattahoochee Avenue   404-876-8761
Atlanta, GA 30318-2112   Fax: 404-875-2578
                      E-mail: hello@peachtree-online.com
                      http://www.peachtree-online.com
          32 pages
          ISBN: 1-561450-08-1

4505 America in the 21st Century: Environmental Concerns
Population Reference Bureau
1875 Connecticut Avenue   202-483-1100
Suite 520   800-877-9881
Washington, DC 20009-5728   Fax: 202-328-3937
                          E-mail: popref@prb.org
               http://www.prb.org; www.popplanet.org

4506 Association of State Wetland Managers Symposium
Association of State Wetland Managers
2 Basin Road   207-892-3399
Windham, ME 04062   Fax: 207-892-3089
                          E-mail: aswm@aswm.org
                          http://www.aswm.org

4507 At Odds with Progress: Americans and Conservation
University of Arizona Press
355 S Euclid Avenue   520-621-1441
Tucson, AZ 85719   Fax: 520-621-8899
                      http://www.uapress.arizona.edu
          Publication Date: 1991   255 pages
                ISBN: 0-816509-17-4
Bret Wallach, Author

4508 Balancing on the Brink of Extinction
Island Press
1718 Connecticut Avenue NW   202-232-7933
Suite 300   Fax: 202-232-1328
Washington, DC 20009-1148   E-mail: info@islandpress.org
                          http://www.islandpress.org
          Publication Date: 1991   329 pages
Kathryn A Kohm, Author

**4509 Beyond the Beauty Strip: Saving What's Left of Our
Forests**
Tilbury House Publishers
132 Water Street   207-582-1899
Gardiner, ME 04345   Fax: 202-582-8227
                      E-mail: tilbury@tilburyhouse.com
                      http://www.tilburyhouse.com

4510 Biodiversity and Ecosystem Function
Springer-Verlag
233 Spring Street   212-460-1500
New York, NY 10013   800-777-4643
                                       Fax: 212-460-1575
                      E-mail: service-ny@springer-sbm.com
                      http://www.springeronline.com
          Publication Date: 1994   528 pages

4511 Bioemediation
McGraw-Hill
1221 Avenue of the Americas   212-512-2000
New York, NY 10020   800-722-4726
                      http://www.magraw-hill.com

4512 Bluebird Bibliography
North American Bluebird Society
PO Box 244   330-359-5511
Wilmot, OH 44689-0244   Fax: 330-359-5455
                      E-mail: info@nabluebirdsociety.org
                      http://www.nabluebirdsociety.org

4513 Clean Sites Annual Report
Clean Sites
228 S Washington Street   703-519-2140
Suite B30   Fax: 703-519-2141
Alexandria, VA 22314   E-mail: cses@cleansites.com
                              http://www.cleansites.com
We apply sound project management principles, real-world ex-
perience, and cost control measures to find creative solutions to
environmental remediation and land reuse problems.
Douglas Ammon, Contact

**4514 Connections: Linking Population and the Environment
Teaching Kit**
Population Reference Bureau
1875 Connecticut Avenue   202-483-1100
Suite 520   800-877-9881
Washington, DC 20009-5728   Fax: 202-328-3937
                          E-mail: popref@prb.org
                          http://www.prb.org

**4515 Conservation and Research Foundation Five Year Re-
port**
Conservation and Research Foundation
PO Box 909   913-268-0076
Shelburne, VT 05482   Fax: 913-268-0076
Publication from Conservation and Research Foundation which
is published every five years and distributed to contributors.
This publication is also available upon request. Next edition to
be published in Fall 2003. Please, call beforefaxing!
          Publication Date: 1998   43 pages
Dr Mary Wetzel, President

**4516 Decade of Destruction: The Crusade to Save the Ama-
zon Rain Forest**
Henry Holt and Company
175 Fifth Avenue   646-307-5095
New York, NY 10010   Fax: 212-633-0748
                          E-mail: publicity@hholt.com
                          http://www.henryholt.com
          215 pages
Adrian Cowell, Author

**4517 Discordant Harmonies: A New Ecology for the
Twenty-first Century**
Oxford University Press
198 Madison Avenue   212-679-7300
New York, NY 10016   Fax: 212-725-2972
                          http://www.oup.co.uk
          Publication Date: 1992   254 pages
                ISBN: 0-195074-69-6
Daniel B Botkin, Author

4518 Earth Keeping
Zondervan Publishing House
5300 Patterson Avenue SE   616-698-6900
Grand Rapids, MI 49530   Fax: 616-698-3439
                          http://www.zondervan.com

4519 Earthright
Prima Publishing & Communications
PO Box 1260BK   916-786-0426
Rocklin, CA 95677   800-632-8676
                                       Fax: 916-632-4405
                      http://www.primapublishing.com

4520 Ecology of Greenways: Design and Function of Linear Conservation
University of Minnesota Press
2037 University Ave SE   612-624-2516
Minneapolis, MN 55414   800-388-3863
                    E-mail: lfreeman@epx.cis.umn.edu
                    http://www.upress.umn.edu
            Publication Date: 1994   238 pages
            ISBN: 0-816621-57-8
Daniel S Smith, Paul Cawood Hellmund, Author

4521 Eli's Songs
MacMillan Publishing Company
866 3rd Avenue   212-702-2000
New York, NY 10022   800-257-5755
                    http://www.macmillian.com

4522 Endangered Kingdom: The Struggle to Save America's Wildlife
John Wiley & Sons
605 3rd Avenue   212-850-6890
New York, NY 10158-0012   800-825-7550
                                        Fax: 212-850-8800
                    Publication Date: 1991   241 pages
                    ISBN: 0-471528-22-6
Roger L DiSilvestro, Author

4523 Environment in Peril
Smithsonian Institution Press
PO Box 960   202-287-3738
Herndon, VA 20172-0960   800-782-4612
                                        Fax: 202-287-3184
                                        http://www.si.edu

4524 Environmental Concern in Florida and the Nation
University of Florida Press
15 NW 15th Street   352-392-1351
Gainesville, FL 32611   800-226-3822
                                        Fax: 352-392-7302
                                        http://www.upf.com
                    Publication Date: 1997   144 pages
                    ISBN: 0-813010-56-X
Lance Dehaven-Smith, Author

4525 Environmental Concern: A Comprehensive Review of Wetlands Assessment Producers
POW-The Planning of Wetlands
201 Boundary Lane   410-745-9260
PO Box P   Fax: 410-745-3517
St Michaels, MD 21663   E-mail: order@wetland.org
                                        http://www.wetland.org
Since its founding in 1972, EC has been specializing in consulting, planning design, education services, construction services and research related to all aspects of wetlands. As wetlands and contiguous upland forests and meadows areinteracting ecosystems EC specializes in consulting, planning, design, and project supervision services for such upland ecosystem constructions and restorations for the purpose of wetland buffers, reforestation, wildlife habitat and critical areas ofpreservation.
                    Publication Date: 1999   Quarterly
                    ISBN: 1-883226-04-x
Edgar W Garbisch, President

4526 Environmental Concern: Evaluation for Planned Wetlands
POW-The Planning of Wetlands
201 Boundary Lane   410-745-9260
PO Box P   Fax: 410-745-3517
St Michaels, MD 21663   E-mail: order@wetland.org
                                        http://www.wetland.org

Since its founding in 1972, EC has been specializing in consulting, planning design, education services, construction services and research related to all aspects of wetlands. As wetlands and contiguous upland forests and meadows areinteracting ecosystems EC specializes in consulting, planning, design, and project supervision services for such upland ecosystem constructions and restorations for the purpose of wetland buffers, reforestation, wildlife habitat and critical areas ofpreservation.
                    Publication Date: 1994   Quarterly
                    ISBN: 1-883226-03-1
Edgar W Garbisch, President

4527 Environmental Concern: The Planning of Wetlands
POW-The Planning of Wetlands
201 Boundary Lane   410-745-9260
PO Box P   Fax: 410-745-3517
St Michaels, MD 21663   E-mail: order@wetland.org
                                        http://www.wetland.org
Since its founding in 1972, EC has been specializing in consulting, planning design, education services, construction services and research related to all aspects of wetlands. As wetlands and contiguous upland forests and meadows areinteracting ecosystems EC specializes in consulting, planning, design, and project supervision services for such upland ecosystem constructions and restorations for the purpose of wetland buffers, reforestation, wildlife habitat and critical areas ofpreservation.
                    Publication Date: 2000
                    ISBN: 1-883226-05-8
Edgar W Garbisch, President

4528 Environmental Concern: The Wonders of Wetlands
POW-The Planning of Wetlands
201 Boundary Lane   410-745-9260
PO Box P   Fax: 410-745-3517
St Michaels, MD 21663   E-mail: order@wetland.org
                                        http://www.wetland.org
Since its founding in 1972, EC has been specializing in consulting, planning design, education services, construction services and research related to all aspects of wetlands. As wetlands and contiguous upland forests and meadows areinteracting ecosystems EC specializes in consulting, planning, design, and project supervision services for such upland ecosystem constructions and restorations for the purpose of wetland buffers, reforestation, wildlife habitat and critical areas ofpreservation.
                    Publication Date: 1995
                    ISBN: 1-888631-00-7
Edgar W Garbisch, President

4529 Environmental Crisis: Opposing Viewpoints
Greenhaven Press
PO Box 289009   858-485-9549
San Diego, CA 92128-9009   800-231-5163
                                        Fax: 800-550-5448
                    E-mail: info@grennhaven.com
                    http://www.greenhaven.com

4530 Environmental Profiles: A Global Guide to Projects and People
Garland Publishing
717 5th Avenue   212-751-7447
25th Floor   Fax: 212-308-9399
New York, NY 10022   http://www.garlandpub.com
                    Publication Date: 1993   1112 pages
                    ISBN: 0-815300-63-8

4531 Forgotten Forest
Sierra Club Books
730 Polk Street   415-776-2211
San Francisco, CA 94109   Fax: 415-776-4868
                    http://www.sierraclub.org/books

4532 Friends of the Earth Foundation Annual Report
Friends of the Earth Found.
218 D Street SE   202-544-2600
Washington, DC 20003   Fax: 202-543-4710
                    http://www.oceanic-society.org

4533 Future Primitive
University of North Texas Press
PO Box 311336   940-565-2142
Denton, TX 76203-1336   800-826-8911

Fax: 940-565-4590
E-mail: rchrisman@unt.edu
http://www.unt.edu/untpress
Publication Date: 1996   223 pages
ISBN: 1-574410-07-5
Ronald Chrisman, Director
Karen DeVinney, Managing Editor

4534 Going Green: A Kid's Handbook to Saving the Planet
Puffin Books
375 Hudson Street   212-366-2403
New York, NY 10014   http://www.puffin.co.uk
Out of print—limited availability.
Publication Date: 1990
ISBN: 0-140345-97-3

4535 Guide to Spring Wildflower Areas
Minnesota Native Plant Society
220 Biological Science Center
1445 Gortner Avenue   E-mail: MNPS@HotPOP.com
Saint Paul, MN 55108   http://www.stolaf.edu
Updated its guide to over 40 wildflower sites in the Twin Cities area. The guide contains a description and location for each.
4 per year

4536 Guide to Urban Wildlife Management
National Institute for Urban Wildlife
10921 Trotting Ridge Way   301-596-3311
Columbia, MD 21044

4537 Hawaii Undersea Research Laboratory: In Deep Waters
University of Hawaii at Manoa
1000 Pope Road   808-956-6335
MSB 303   Fax: 808-956-9772
Honolulu, HI 96822   E-mail: malahoff@soest.hawaii.edu
                        http://www.soeast.hawaii.edu/hurl
One of six research centers funded by NOAA's National Undersea Research Program. HURL operates two 2000-meter Pisces submersibles and a remotely operated vehicle. Research projects include fisheries research, geology and biology of thedeepsea around the Hawaiian Islands.
Publication Date: 1998   25 pages One Time
ISBN: 0-824819-46-2
Dr. Alexander Malahoff, Director

4538 How to Save Your Neighborhood, City, or Town: The Sierra Club Guide to Community Organizing
Sierra Club Books
100 Bush Street   415-291-1196
13th Floor   http://www.sieraaclub.org/books/
San Francisco, CA 94104

4539 Information Please Environmental Almanac
Houghton Mifflin Company
Beacon Street   617-725-5000
30th Floor   http://www.hmco.com
Boston, MA 02108
Publication Date: 1992   704 pages
ISBN: 0-395637-67-8

4540 International Protection of the Environment
Oceana Publications, Inc
75 Main Street   914-693-8100
Dobbs Ferry, NY 10522   800-831-0758
                                    Fax: 914-693-0402
                        E-mail: info@oceanalaw.com
                        http://www.oceanalaw.com
This set provides the documents which form the framework of "softlaw" administrative instruments for the implementation of international environment treaties under Agenda 21.
Publication Date: 1995   7 vol pages Bi-Monthly
ISBN: 0-379102-95-1
Nicholas A Robinson & Wolfgang Burhenne, Author

4541 International Society for Endangered Cats
3070 Riverside Drive Suite 160   614-487-8760
Columbus, OH 43221   Fax: 614-487-8769

Publication Date: 1990   237 pages
ISBN: 0-816019-44-4
Bill Simpson, President
Patricia Currie, Executive Director

4542 Just A Dream
Houghton Mifflin Company
Beacon Street   617-725-5000
30th Floor
Boston, MA 02108

4543 Krieger Publishing Company: Wildlife Habitat Management of Wetlands
Krieger Publishing Co.
PO Box 9542   321-724-9542
Melbourne, FL 32902-9542   800-724-0025
                                    Fax: 321-951-3671
                E-mail: info@krieger-publishing.com
                http://www.krieger-publishing.com
Krieger Publishing Company produces quality books in various fields of interest. We have an extensive Natural Science listing.
Publication Date: 1992   572 pages
ISBN: 1-575240-89-0
Cheryl Stanton, Advertising

4544 Krieger Publishing Company: Wildlife Habitat Management of Forestlands/Rangelands/Farmlands
Krieger Publishing Co.
PO Box 9542   321-724-9542
Melbourne, FL 32902-9542   800-724-0025
                                    Fax: 321-951-3671
                E-mail: info@krieger-publishing.com
                http://www.krieger-publishing.com
Krieger Publishing Company produces quality books in various fields of interest. We have an extensive Natural Science listing.
Publication Date: 1994   868 pages
ISBN: 1-575240-93-9
Cheryl Stanton, Advertising

4545 Last Extinction
MIT Press
5 Cambridge Center   617-253-5646
Cambridge, MA 02142-1493   Fax: 617-258-6779
Today there is a new and more widespread awareness of what some consider to be the great tragedy of our time - organisms which took many thousands or even millions of years to evolve are being snuffed out permanently owing to humanactivity.
Publication Date: 1993
Les Kaufman, Kenneth Mallory, Author

4546 Lessons of the Rainforest
Sierra Club Books
85 2nd Street   415-977-5500
2nd Floor   Fax: 415-977-5799
San Francisco, CA 94105   http://www.sierraclub.org/books/
256 pages
ISBN: 0-871566-82-6
Suzanne Head, Editor
Robert Heinzman, Editor

4547 Mastering Nepa: A Step-By-Step Approach
Solano Press
46201 Iversen Drive   707-884-4508
PO Box 773   Fax: 707-884-4109
Point Arena, CA 95468   http://www.solano.com
Publication Date: 1993   250 pages
ISBN: 0-923956-14-x
Ronald E Bass, Albert I Herson, Author

4548 National Wildlife Rehabilitators Association Annual Report
National Wildlife Rehabilitators Association
14 N 7th Avenue   320-259-4086
St Cloud, MN 56303-4766   Fax: 320-259-4086
                        E-mail: nwra@nwrawildlife.org
                        http://www.nwrawildlife.org
Please, call before you fax!

4549 Nature and the American: Three Centuries of Changing Attitudes
Unviersity of Nebraska Press
901 N 17th Street   402-472-3581
Room 327   http://www.nebraskapress.unl.edu
Lincoln, NE 68588-0520

4550 Ordinance Information Packet
Scenic America
21 Dupont Circle NW   202-833-4300
Washington, DC 20036   Fax: 202-833-4304
            http://www.scenic.org

4551 Ozone Diplomacy: New Directions in Safeguarding the Planet
Harvard University Press
79 Garden Street   617-495-2600
Cambridge, MA 02138-1499   Fax: 617-495-5898
            http://www.hup.harvard.edu
Offers an insider's view of the politics, economics, science and diplomacy involved in creating the precedent-setting treaty to protect the Earth: the 1987 Montreal Protocol on Substances That Deplete the Ozone Layer.
Richard Elliot Benedick, Author

4552 Practical Guide to Environmental Management
1616 P Street NW   202-939-3800
Suite 200   Fax: 202-939-3868
Washington, DC 20036

4553 Preserving the World Ecology
H W Wilson Company
950 University Avenue   718-588-8400
Bronx, NY 10452-4224   Fax: 718-588-6365
            http://www.hwwilson.com

4554 Quill's Adventures in Grozzieland
John Muir Publications
PO Box 613   505-982-4078
Santa Fe, NM 87504   800-888-7504
            Fax: 505-988-1680

4555 RARE Center for Tropical Conservation Annual Report
Rare Center for Tropical Conservation
1616 Walnut Street   215-735-3510
Suite 911   Fax: 215-735-3615
Philadelphia, PA 19103   http://www.rarecenter.org

4556 Resource Conservation and Management
Wadsworth Publishing Company
10 Davis Drive   415-595-2350
Belmont, CA 94002   Fax: 415-637-7544
            http://www.wadsworth.com

4557 Revolution for Nature: From the Environment to the Connatural World
University of North Texas Press
PO Box 311336   940-565-2142
Denton, TX 76203-1336   800-826-8911
            Fax: 940-565-4590
            E-mail: rchrisman@unt.edu
            http://www.unt.edu/untpress
145 pages
ISBN: 1-574417-0X-
Ronald Chrisman, Director
Karen DeVinney, Managing Editor

4558 Seed Listing
Native Seeds/SEARCH
2509 N Campbell Ave #325   602-327-9123
Tucson, AZ 85719   Fax: 602-883-2500
            http://www.nativeseeds.org

4559 Statement of Policy and Practices for Protection of Wetlands
National Wildlife Fed. Corporate Conservation Coun
1400 16th Street NW   202-797-6870
Washington, DC 20036   Fax: 202-797-6871

4560 Student Conservation Association Northwest: Lightly on the Land
1265 S Main Street   206-324-4649
Suite 210   Fax: 206-324-4998
Seattle, WA 98144   http://www.sca-inc.org
SCA is a national organization with regional offices in Seattle, Oakland, Pittsburg, Washington DC and headquartered in Charlestown NH. Our mission is to build the next generation of conservation leaders and inspire lifelongstewardship of our environment and communities by engaging young people in hands-on service to the land. We offer a wide range of internships and crew based programs for ages 16 years and up.
Publication Date: 1996   267 pages
ISBN: 0-898869-91-7
Su Thieds, Director of Regional Progams

4561 Transactions of Annual North American Wildlife and Natural Resources Conference
Wildlife Management
1101 14th Street NW   202-371-1808
Suite 801   Fax: 202-408-5059
Washington, DC 20005

4562 Urban Wildlife Manager's Notebook
National Institute for Urban Wildlife
10921 Trotting Ridge Way   301-596-3311
Columbia, MD 21044

4563 Wetlands Protection: The Role of Economics
Environmental Law Institute
1616 P Street NW   202-939-3800
Suite 200   Fax: 202-939-3868
Washington, DC 20036

4564 Wilderness Society Annual Report
The Wilderness Soc.
1615 M Street Northwest   202-833-2300
Washington, DC 20036   800-THE-WILD
            Fax: 202-429-3958
            E-mail: tws@wilderness.org
            http://www.wilderness.org

4565 Wildlife Conservation in Metropolitan Environments
National Institute for Urban Wildlife
10921 Trotting Ridge Way   301-596-3311
Columbia, MD 21044

4566 Wildlife Habitat Relationships in Forested Ecosystems
Timber Press
9999 SW Wilshire   503-292-0745
Portland, OR 97225   http://www.timber-press.com
Available by special order.
Publication Date: 1997
David R Patton, Author

4567 Wildlife Research and Management in the National Parks
University of Illinois Press
1325 S Oak Street   217-333-0950
Champaign, IL 61820-6903   Fax: 217-244-8082
            http://www.press.uillinois.edu
Publication Date: 1992   240 pages
ISBN: 0-252018-24-9
Gerald R Wright, Author

4568 Wildlife Reserves and Corridors in the Urban Environment: A Guide to Ecological Landscape
National Institute for Urban Wildlife
10921 Trotting Ridge Way   301-596-3311
Columbia, MD 21044
Out of print—limited availability.
Publication Date: 1989   91 pages
ISBN: 0-942015-02-9
Lowell W Adams, Louise E Dove, Author

4569 Wildlife-Habitat Relationships: Concepts and Applications
University of Wisconsin Press
114 N Murray Street   608-262-8782
Madison, WI 53715-1199
Anyone working with wildlife must be concerned with its habitat identification, measurement and analysis. Wildlife-Habitat Relationships goes beyond introductory wildlife biology texts and specialized studies of single species toprovide a broad but advanced understanding of habitat relationships applicable to all terrestrial species.
Publication Date: 1998   416 pages
ISBN: 0-299156-40-0
Michael L Morrison, Bruce G Marcot, Author

Books: Recycling & Pollution Prevention

4570 Aunt Ipp's Museum of Junk
HarperCollins
10 E 53rd Street   212-207-7000
New York, NY 10022   800-424-6234
Fax: 212-207-7433
http://www.harpercollins.com

4571 Beyond 40 Percent: Record-Setting Recycling and Composting Programs
Island Press
1718 Connecticut Avenue NW   202-232-7933
Suite 300   Fax: 202-234-1328
Washington, DC 20009-1148   E-mail: info@islandpress.org
http://www.islandpress.org
Publication Date: 1991   280 pages
ISBN: 1-559630-73-6

4572 Borrowed Earth, Borrowed Time: Healing America's Chemical Wounds
Plenum Publishers
233 Spring Street   212-620-8000
7th Floor   Fax: 212-463-0742
New York, NY 10013-1578   http://www.plenum.com
Publication Date: 1991

4573 Caring for Our Air
Enslow Publishers
PO Box 777
Hillside, NJ 07205   800-398-2504
Fax: 908-964-4116
http://www.enslow.com

4574 Community Recycling: System Design to Management
Prentice Hall
Route 9W   201-592-2000
Englewood Cliffs, NJ 07632   800-947-7700
E-mail: orders@prenhall.com
http://www.prenhall.com
A guide for getting into the growing business of community recycling, for those with little or no previous experience with the technical details of recycling. Discusses marketing, management, equipment, profit comparisons of variousprocessing methods and legal considerations.
Publication Date: 1992   240 pages
ISBN: 0-131557-89-0
Nyles V Reinfeld, Carl M Layman, Author

4575 Garbage and Recycling
Kingfisher Publications
http://www.kingfisherpub.com
Publication Date: 1995   32 pages
ISBN: 1-856976-15-7
Rosie Harlow, Sally Morgan, Author

4576 How On Earth Do We Recycle Glass?
Millbrook Press
2 Old New Milford Road   203-740-2220
PO Box 335   800-462-4703
Brookfield, CT 06804   Fax: 203-740-2526
http://www.millbrookpress.com

4577 Let's Talk Trash: The Kids' Book About Recycling
Waterfront Books
85 Crescent Road   802-658-7477
Burlington, VT 05401-4126http://www.waterfrontsbooks.com

4578 Plastic: America's Packaging Dilemma
Island Press
1718 Connecticut Avenue NW   202-232-7933
Suite 300   Fax: 202-234-1328
Washington, DC 20009-1148   E-mail: info@islandpress.org
http://www.islandpress.org

4579 Pollution Knows No Frontiers
Paragon House of Publishers
90 Fifth Avenue   212-725-3380
New York, NY 10011-+E-mail: paragon@paragonhouse.com
http://www.paragonhouse.com

4580 Recycle!: A Handbook for Kids
Little, Brown & Company
1271 Avenue of the Americas
New York, NY 10020   800-759-0190
Fax: 212-522-0885
http://www.twbookmark.com
Publication Date: 1996   32 pages
Gail Gibbons, Editor

4581 Recycling Paper: From Fiber to Finished Product
TAPPI Press
PO Box 105113   404-394-6130
Atlantic, GA 30348-5113   http://www.tappi.org

4582 Reducing Toxics
Island Press
1718 Connecticut Avenue NW   202-232-7933
Suite 300   Fax: 202-234-1328
Washington, DC 20009-1148   E-mail: info@islandpress.org
http://www.islandpress.org
Publication Date: 1995   460 pages
Robert Gottlieb, Editor

Books: Sustainable Development

4583 Biodiversity Prospecting: Using Genetic Resources for Sustainable Development
World Resources Institue
1709 New York Ave., NW   202-638-6300
Washington, DC 20006   http://www.wri.org

4584 Building Sustainable Communities: An Environmental Guide for Local Government
Global Cities Project
2926 Philmore Street   415-775-0791
San Francisco, CA 94123   http://www.globalcities.org

4585 Center for Ecoliteracy
2528 San Peblo Avenue   510-845-4595
Berkeley, CA 94702   Fax: 510-845-1439

E-mail: info@ecoliteracy.org
http://www.ecoliteracy.org

The Center for Ecoliteracy is dedicated to fostering a profound understanding of the natural world, grounded in direct experience that leads to sustainable patterns of living.
Publication Date: 2000   90 pages   Paperback
ISBN: 0-967565-23-5
Zenobia Barlow, Executive Director

4586 Constructing Sustainable Development
State University of New York Press
194 Washington Avenue   518-472-5000
Suite 305   Fax: 518-472-5038
Albany, NY 12210-2384
Publication Date: 2000   175 pages
Neil Harrison, Editor

4587 Environmental Defense Annual Report
Environmental Defense
257 Park Avenue S   212-505-2100
17th Floor   800-684-3322
New York, NY 10010-7304   Fax: 212-505-2375
http://www.environmentaldefense.org
Environmental Defense believes that a sustainable environment will require economic and social systems that are equitable and just.
Fred Krupp, Executive Director

4588 Environmental Profiles: A Global Guide to Projects and People
Garland Publishing
717 5th Avenue   212-751-7447
25th Floor   Fax: 212-308-9399
New York, NY 10022   http://www.garlandpub.com
Publication Date: 1993   1112 pages
ISBN: 0-815300-63-8

4589 Global Environment
Jones and Bartlett Publishers
1 Extor Plaza   617-859-3900
Boston, MA 02116   http://www.jbpub.com

4590 Gnat is Older than Man: Global Environment and Human Agenda
Princeton University Press
41 William Street   609-258-4900
Princeton, NJ 08540   800-777-4726
http://www.pup.princeton.edu

4591 Managing Sustainable Development
Earthscan Publications
8-12 Camden High Street   207-387-8558
London   Fax: 207-387-8998
Publication Date: 2001   304 pages
Michael Carley, Editor

4592 Practice of Sustainable Development (The)
Urban Land Institute
1025 Thomas Jefferson Street NW   202-624-7000
Suite 500 W   Fax: 202-624-7140
Washington, DC 20007   E-mail: customerservice@uli.org
http://www.uli.org
Publication Date: 2000   160 pages
ISBN: 0-874208-31-9
Douglas R Porter, Author

4593 Sustainable Planning and Development
WIT Press
                               978-667-5841
                               Fax: 978-667-7582
                     E-mail: salesUSA@witpress.com
                               http://www.witpress.com
Publication Date: 2003   1048 pages
ISBN: 1-853129-85-2
Linda Ouellette, Customer Service Manager

4594 Worldwatch Paper 101: Discarding the Throwaway Society
Worldwatch Intitutes
Massachusetts   202-452-1999
Washington, DC 20036   http://www.worldwatch.org

Books: Travel & Tourism

4595 Appalachian Mountain Club
Appalachian Mountain Club Books
5 Joy Street   617-523-0636
Boston, MA 02108   800-262-4455
                               Fax: 617-523-0722
                     E-mail: information@outdoors.org
                               http://www.outdoors.org
The AMC, founded in 1876, promotes the protection, enjoyment, and wise use of the mountains, rivers and trails of the Northeast. We encourage people to enjoy and protect the natural world because we believe that successful conservationdepends on this experience. The AMC publishes an award-winning magazine and more than 60 guide books to the Northeast.
Andrew Falender, Executive Director
Chase O'Connell, Director Development

4596 Prospect Park Handbook
Greensward Found
Lenox Hill Station   212-473-6283
PO Box 610   http://www.greenswardparks.org
New York, NY 10021

Books: Water Resources

4597 And Two if By Sea: Fighting the Attack on America's Coasts
Coast Alliance
                               202-546-9609
This book is the benchmark in the effort to save the coasts.
Publication Date: 1986

4598 Comparative Health Effects Assessment of Drinking Water Treatment Technologies
Government Institutes Division
4 Research Place   301-921-2300
Suite 200   Fax: 301-921-0373
Rockville, MD 20850   http://www.govinst.com
The report evaluates the public health impact of the most widespread drinking water treatment technologies, with particular emphasis on disinfection.
Publication Date: 1988   20 pages

4599 Dying Oceans
Gareth Stevens, Inc
330 W Olive Street   414-332-3520
Suite 100   800-542-2595
Milwaukee, WI 53212   Fax: 414-332-3567
                               http://www.garethstevens.com
Publication Date: 1991
ISBN: 0-836804-76-7

4600 Hawaii Undersea Research Laboratory: In Deep Waters
University of Hawaii at Manoa
1000 Pope Road, MSB 303   808-956-6335
Honolulu, HI 96822   Fax: 808-956-9772
                     E-mail: malahoff@soest.hawaii.edu
                               http://www.soeast.hawaii.edu/hurl
One of six research centers funded by NOAA's National Undersea Research Program. HURL operates two 2000-meter Pisces submersibles and a remotely operated vehicle. Research projects include fisheries research, geology and biology of thedeepsea around the Hawaiian Islands.
Publication Date: 1998   25 pages   One Time
ISBN: 0-824819-46-2
Dr. Alexander Malahoff, Director

4601 Managing Troubled Water: The Role of Marine Environmental Monitoring
Duke University Press
6697 College Station   919-684-2173
Durham, NC 27708   Fax: 919-684-8644
http://www.dukepress.edu
Publication Date: 1990

4602 Turning the Tide: Saving the Chesapeake Bay
Island Press
Box 7   707-983-6432
Covelo, CA 95428   800-828-1302
E-mail: info@islandpress.org
http://www.islandpress.org
The Chesapeake Bay is one of the most productive and important ecosystems on earth, and as such is a model for other estuaries facing the demands of commerce, tourism, transportation, recreation and other uses. Turning the Tide presents a comprehensive look at two decades of efforts to save the bay, outlining which methods have worked and which have not.
Publication Date: 2003   352 pages
ISBN: 1-559635-48-7
Tom Horton, Author

4603 Using Common Sense to Protect the Coasts: The Need to Expand Coastal Barrier Resources
Coast Alliance
600 Pennsylvania Avenue SE   202-546-9554
Suite 340   E-mail: coast@coastalliance.org
Washington, DC 20003   http://www.coastalliance.org
This report gives a common sense approach to protecting coastal areas from unwise development that would benefit American taxpayers.
Publication Date: 1990

Library Collections

4604 Acres International Library
140 John James Audubon Parkway   716-689-3737
Amherst, NY 14228-1180   Fax: 716-689-3749
E-mail: amherst@acres.com
http://www.acres.com
Serves clients in the hydroelectric power, highways and bridges, mining, heavy industrial, civil/geotechnical and environmental and hazardous waste sectors.
Marion D'Amboise, Librarian

4605 Alaska Department of Fish and Game Habitat Library
333 Raspberry Road   907-267-2314
Anchorage, AK 99518-1599   Fax: 907-349-1723
Celia Rozen, Contact

4606 Alaska Resources Library and Information Services
3150 C Street   907-272-7547
Suite 100   Fax: 907-271-4742
Anchorage, AK 99503   E-mail: reference@arlis.org
http://www.arlis.org
Carrie Holba, Reference Services Coord

4607 American Academy of Pediatrics
141 NW Point Boulevard
PO Box 747   888-227-1710
Elk Grove Village, IL 60009-0747   Fax: 847-228-1281
http://www.aap.org
Dedicated to the health of all children.

4608 American Water Works Association
6666 W Quincy Avenue   303-794-7711
Denver, CO 80235   Fax: 303-795-1440
http://www.awwa.org
Dedicated to the promotion of public health and welfare in the provision of drinking water of unquestionable quality and sufficient quantity. AWWA must be proactive and effective in advancing the technology, science, management and government policies relative to the stewardship of water.
Jack W Hoffbuhr, Executive Director

4609 Aquatic Research Institute Aquatic Sciences and Technology Archive
2242 Davis Court   510-782-4058
Hayward, CA 94545   Fax: 510-784-0945
Library and data base in aquatic sciences and technologies also research faculty in aquatic sciences.
V Parker, Archv

4610 Arizona State Energy Office Information Center
3800 N Central   602-280-1402
Suite 1200   Fax: 602-280-1445
Phoenix, AZ 85012   E-mail: energy@azcommerce.com
Maxine Robertson, Assistant Director

4611 Arizona State University Architecture and Environmental Design Library
College of Architecture and Environmental Design
4300   480-965-6400
Tempe, AZ 85287-1705   Fax: 480-727-6965
E-mail: deborah.koshinsky@asu.edu
http://www.asu.edu/caed/AEDlibrary
Deborah H Koshinsky, Director

4612 Arkansas Energy Office Library
1 State Capitol Mall   501-682-1370
Little Rock, AR 72201   Fax: 501-682-2703
E-mail: cbenson@1800arkansas.com

4613 Atmospheric Sciences Model Division Library
US Environmental Protection Agency
79 TW Alexander Drive   919-541-4536
4201 Building, Room 308   Fax: 919-541-1379
Research Triangle Park, NC 27711   http://www.epa.gov
Serves the NOAA Division assigned to support the EPA National Exposure Laboratory and Office of Air Quality Planning and Standards. The major field of interest is the meteorological aspects of air pollution, including numerical and physical model development and application.
Evelyn M Poole-Kober, Tech. Pubns.

4614 Belle W Baruch Institute for Marine Biology and Coastal Research Library
607 EWS Building   803-777-5288
Columbia, SC 29208   Fax: 803-777-3935

4615 Bickelhaupt Arboretum Education Center
340 S 14th Street   319-242-4771
Clinton, IA 52742

4616 Brown University Center for Environmental Studies Library
135 Angel Street   401-863-3449
Box 1943   Fax: 401-863-3503
Providence, RI 02912   E-mail: UEL_DCC@brown.edu
http://www.brown.edu/Department/Environmental_Studies/
The Center for Environmental Studies at Brown University was established with the primary aim of educating individuals to solve challenging environmental problems both at the local and global levels. It also works directly to improve human well-being and environmental quality through community, city, and state partnerships in service and research.
Patti Caton, Administrative Manager

4617 Burroughs Audubon Center and Library
Burroughs Audubon Society
21905 SW Woods Chapel Road   816-795-8177
Independence, MO 64050

4618 CH2M Hill
Corvallis Regional Office Library
2300 NW Walnut Boulevard   541-752-4271
PO Box 428   Fax: 541-752-0276
Corvallis, OR 97339-0428   http://www.ch2m.com
Tony Salmon, Lib. Supv.

4619 California Energy Commission Library
1516 9th Street 916-654-4292
MS 10 Fax: 916-654-4046
Sacramento, CA 95814 E-mail: library@energy.state.ca.us
http://www.energy.ca.gov/asd/library
Serves Energy Commission staff, California state government
agencies, the Legislature and its staff, and members of the public.
Karen Hamilton, Librarian

4620 California State Resources Agency Library
1416 9th Street 916-653-2225
Room 117 Fax: 916-653-1856
Sacramento, CA 95814

Contains books, documents and subscriptions on topics including: flood control; natural resources (in California); endangered
species (in California); soil conservation; water; water pollution; water quality; water resources;conservation and water supply.

4621 Center for Coastal Fisheries and Habitat Research:
Rice Library
101 Pivers Island Road 252-728-8713
Beaufort, NC 28516-9722 Fax: 252-728-8784
E-mail: patti.marraro@noaa.gov
http://www.shrimp.ccfhrb.noaa.gov/library/index.html
Ensures the delivery of scientific, technical, and legistlative information to library users including NOAA staff, general public,
academia, industry, and governmental agencies. Houses comprehensive coverage of marine fisheries,fisheries statistics, habitat restoration, mapping and remote sensing, marine chemistry,
pollution and toxicology, living marine resources, protected
species, and oceanography.
Patti M Marraro, Technical Info Specialist

4622 Center for Health, Environment and Justice Library
PO Box 6806 703-237-2249
Falls Church, VA 22040 Fax: 703-237-8389
E-mail: chej@chej.org
http://www.essential.org/cchw
Believes in environmental justice, the principle that people have
the right to a clean and healthy environment. Believes the most
effective way to win environmental justice is from the bottom up
through community organizing andempowerment.

4623 Clinton River Watershed Council Library
1970 East Auburn Road 248-853-9580
Rochester Hils, MI 48307 Fax: 248-853-0486
http://www.crwc.org

4624 Colorado River Board of California
770 Fairmont Avenue 818-543-4676
Suite 100 Fax: 818-543-4685
Glendale, CA 91203-1035 http://www.crb.ca.gov

4625 Columbia River Inter-Tribal Fish Commission
StreamNet Library
729 NE Oregon Street 503-731-1304
Suite 190 Fax: 503-731-1260
Portland, OR 97232 E-mail: fishlib@critfc.org
http://www.fishlib.org
Serving the scientific and environmental community of the Pacific Northwest, The StreamNet Library works in cooperation
with the region's fish and wildlife recovery efforts. The library
provides access to technical information on theColumbia Basin
fisheries, ecosystem and other relevant subjects for states in the
Pacific Northwest. The library collections emphasize less commonly available grey literature, such as consultant's reports,
state documents and nonprofit organizations'reports.
Lenora Oftedahl, Librarian
Todd Hannon, Assistant Librarian

4626 DER Research Library
Pennsylvania Department of Environmental Resources
Box 8458 717-787-9647
Harrisburg, PA 17105-8458 Fax: 717-772-0288

4627 Dawes Arboretum Library
7770 Jacksontown Road SE 740-323-2355
Newark, OH 43056 800-44D-AWES
http://www.dawesarb.org

4628 Delaware River Basin Commission Library
25 State Police Drive 609-883-9500
Box 7360 Fax: 609-883-9522
West Trenton, NJ 08628 http://www.drbc.net
The Commission is a federal/interstate agency responsible for
managing the water resources at the Delaware River Basin.
Carol R Collier, Executive Director
Christopher Roberts, Staff Contact PIO

4629 Division of Water Resources Library
Kansas Department of Agriculture
109 SW 9th Street 785-296-3717
2nd Floor Fax: 785-296-1176
Topeka, KS 66612-1283

4630 Duke University Biology: Forestry Library
Duke University
Perkins Library 919-660-5880
Durham, NC 27708 Fax: 919-684-2855
http://www.lib.duke.edu

David M Talbert

4631 Earth Island Institute Library
300 Broadway 415-788-3666
Suite 28 Fax: 415-788-7324
San Francisco, CA 9413 E-mail: earthisland@earthisland.org
http://www.earthisland.org
John A Knox, Executive Director

4632 Earthworm Recycling Information Center
35 Medford Street 617-628-1844
Somerville, MA 02143 Fax: 617-628-2773
John Perkins, Contact

4633 Eastern States Office Library
US Bureau of Land Management
7450 Boston Boulevard 703-440-1561
Springfield, VA 22153-3121 Fax: 703-440-1599
Terry Lewis, Contact

4634 Eastern Technical Associates Library
PO Box 1009 919-878-3188
Garner, NC 27529 Fax: 919-872-5199
E-mail: tomrose@eta-is-opacity.com
http://www.eta-is-opacity.com
Environmental consulting firm. Research results published in
government reports.
Publication Date: 1979
Thomas H Rose, President

4635 Ecology Center Library
2530 San Pablo Avenue 510-548-2220
Berkeley, CA 94702 Fax: 510-548-2240
E-mail: info@ecologycenter.org
http://www.ecologycenter.org.

4636 Environment and Natural Resources Branch Library
US Department of Justice
One Congress Street 617-918-1807
Suite 1100 Fax: 617-918-1810
Boston, MA 02114-2023 E-mail: friedman.fred@epa.gov
http://www.eoa.gov
Research library for Solid Wasteto conduct research and answer
questions in the subject fields of nonhazarodus solid waste and
recycling.
Leola Decker, Librarian

4637 Environmental Action Coalition Library: Resource Center
625 Broadway   212-677-1601
2nd Floor   Fax: 212-505-8613
New York, NY 10012   http://www.enviro-action.org
Paul Berizzi, Executive Director

4638 Environmental Coalition on Nuclear Power Library
433 Orlando Avenue   814-237-3900
State College, PA 16803   Fax: 814-237-3900
Dr. J H Johnsrud, Executive Officer

4639 Environmental Contracting Center Library
ENSR Consulting and Engineering
Box 2105   970-493-8878
Fort Collins, CO 80522   800-722-2440
                            E-mail: faq/default.asp
                            http://www.ensr.com

Beth Mullan, Librarian

4640 Environmental Research Associates Library
414 Mill Road   610-449-7400
Havertown, PA 19083-3740   Fax: 610-449-7404

4641 Federated Conservationists of Westchester County (FCWC) Office Resource Library
78 N Broadway   914-422-4053
White Plains, NY 10603   Fax: 914-289-0539
                            E-mail: fcwc@genesis.law.pace.edu
                            http://www.fcwc.org

4642 Fish and Wildlife Reference Service
5430 Grosvenor Lane   301-492-6403
Suite 110   800-582-3421
Bethesda, MD 20814   Fax: 301-564-4059
To provide policy guidance regarding the operation and use of the Fish and Wildlife Reference Service.
Paul E Wilson, Project Manager

4643 Florida Conservation Foundation
1191 Orange Avenue   407-644-5377
Winter Park, FL 32789

4644 Forest History Society Library and Archives
701 William Vickers Avenue 919-682-9319
Durham, NC 27701-3162   Fax: 919-682-2349
                            E-mail: coakes@duke.edu
                            http://http://www.lib.duke.edu/forest
The Forest History Society is a non-profit educational institution that links the past to the future by identifying, collecting, preserving, interpreting, and disseminating information on the history of people, forests, and theirrelated resources.
        Publication Date: 1946
Cheryl Oakes, Librarian

4645 Galveston District Library
US Army Corps of Engineers
Box 1229   409-766-3196
Galveston, TX 77553   Fax: 409-766-3905
                            E-mail: clark.bartee@usace.army.mil
                            http://www.swg.usace.army.mil/library.htm
Clark Bartee

4646 Georgia State Forestry Commission Library
PO Box 819   912-751-3480
Macon, GA 31202-0819   Fax: 912-751-3465
Fred Allen, Director

4647 Glen Helen Association Library
405 Corry Street   937-767-7375
Yellow Springs, OH 45387

4648 Great Lakes Environmental Research Laboratory
2205 Commonwealth Boulevard 734-741-2235
Ann Arbor, MI 48105-2945   Fax: 734-741-2055
                             http://www.glerl.com
Conducts integrated interdiciplinary environmental research in support of resource management and environmental services in costal and esturine water with special emphasis on the Great Lakes.

4649 Huxley College of Environmental Studies
Western Washington University   360-650-3000
Bellingham, WA 98225   E-mail: huxley@cc.wwu.edu
One of the oldest environmental studies colleges in the nation. Innovative and indisciplinary academic programs reflect a broad view of the physical, biological, social and cultural world.
Hailey Outzs, Coordinater

4650 Illinois State Water Survey Library
208 Water Survey Research Center 217-333-4956
2204 Griffith Drive   Fax: 217-333-6540
Champaign, IL 61820-7495   E-mail: library@sws.uiuc.edu
                           http://www.sws.uiuc.edu/chief
The Illinois State Water Survey, a division of the office of Scientific Research and Analysis of the Illinois Department of Natural Resources and affiliated with the University of Illinois, is the primary agency in Illinois concernedwith water and atmosheric resources.
Patricia G Morse, Librarian

4651 Institute of Ecosystem Studies
65 Sharon Turnpike   845-677-5343
Box AB   Fax: 845-677-5976
Millbrook, NY 12545   E-mail: Cadwalladerj@ecostudies.org
                           http://www.ecostudies.org
Ecology research and education institution; independent; international.
Jill Cadwallader, Public Information Officer

4652 International Academy at Santa Barbara Library
800 Garden Street   805-965-5010
Suite D   Fax: 805-965-6071
Santa Barbara, CA 93101
Susan J Shaffer, Office Manager

4653 International Game Fish Association
300 Gulf Stream Way   954-927-2628
Dania Beach, FL 33004   Fax: 954-924-4299
                            E-mail: igfahq@aol.com
                            http://www.igfa.org
Founded as record-keeper and to maintain fishing rules. Today, emphasis is on conservation and education. Encourages youngsters to enter the sport and maintains a huge library on the subject of fishing. Has a network of well over 300representatives around the world, many of whom are conservation leaders in their communities.
Rob Kramer, President

4654 Interstate Oil and Gas Compact Commission Library
900 NE 23rd Street   405-525-3556
Box 53127   Fax: 405-525-3592
Oklahoma City, OK 73152   E-mail: iogcc@iogcc.state.ok.us
                          http://www.iogcc.oklaosf.state.ok.us
W Timothy Dowd, Executive Director

4655 Lake Michigan Federation
220 S State Street   312-939-0838
Suite 1900   Fax: 312-939-2708
Chicago, IL 60604   http://www.lakemichigan.org
Works to restore fish and wildlife habitat, conserve land and water, and eliminate pollution in the watershed of America's largest lake. We achieve these through education, research, law, science, economics and strategic partnerships.

4656 Lionael A Walford Library
74 Magruder Road   732-872-3035
Highlands, NJ 07732   Fax: 732-872-3088
Claire L Steimle, Librarian

4657 Los Angeles County Sanitation District Technical Library
PO Box 4998
Whittier, CA 90607-4998
562-699-7411
Fax: 562-699-5422
http://www.lacsd.org

4658 Louisiana Department of Environmental Quality Information Resource Center
7290 Bluebonnet Boulevard
2nd Floor
Baton Rouge, LA 70810
225-765-0169
Fax: 225-765-0222
E-mail: pattyb@deq.state.la.us
To promote a healthy environment by providing a specialized environmental library to meet the informational and educational needs of the DEQ employees and the citizens of Louisiana.
Patty Birkett, Tech. Librarian

4659 Madison Water Resources Institute Library
University of Wisconsin
1975 Willow Drive, Floor 2
Madison, WI 53706-1177
608-262-3069
Fax: 608-262-0591
E-mail: AskWater@wri.wisc.edu
http://www.wri.wisc.edu/library
The Water Resources Institute (WRI) Library collection covers al major topics in water resources, but is particularly strong in Wisconsin and Great Lakes water issues, groundwater protection, wetlands issues, and the impacts ofagricultural chemicals. The collection consists of over 21,000 hard copy and microfiche documents, over 35 journals and 130 newsletters. The collection may be searched at http://madcat.library.wisc.edu
JoAnn M Savoy, Librarian

4660 Marine Environmental Sciences Consortium
Dauphin Island Sea Lab
101 Bienville Boulevard
Dauphin Island, AL 36528
251-861-2141
Fax: 251-861-4646
http://www.disl.org

Connie Mallon, Librarian

4661 Massachusetts Audubon Society Berkshire Sanctuaries Library
Pleasant Valley Wildlife Sanctuary
472 W Mountain Road
Lenox, MA 01240
413-637-0320
Fax: 413-637-0499
E-mail: berkshires@massaudubon.org
http://www.massaudubon.org
The Massachusetts Audubon Society is an environmental organization with emphases in conservation, advocacy and education. The advocacy effort is statewide and features a legislative team in Boston.

4662 Matt Cole Memorial Library
Williams College Center for Environmental Studies
Kellogg House
Box 632
Williamstown, MA 01267
413-597-2500
Fax: 413-597-3489
E-mail: nparker@williams.edu
http://www.williams.edu/ces
Norm Parker, Information Specialist

4663 Minneapolis Public Library and Information Center
Technology and Science Department
300 Nicolet Mall
Minneapolis, MN 55401
612-372-6570
Fax: 312-372-6546
The varied collection in the Technology/Science/Government Documents department runs from agriculture to zoology, computers to cooking, engineering to handicrafts, medicine to motorcycle repair to military science. Special resourcesinclude a complete US Patent and Trademark collection and the CASSIS Patent Trademark Databases, a collection of US industrial standards, including publications from ANSI (American National Standards Institutes).

4664 Minnesota Department of Natural Resources DNR Library
500 Lafayette Road
St. Paul, MN 55155-4021
651-297-4929
Fax: 651-297-4946
E-mail: dnr.library@dnr.state.mn.us
15,000 titles on natural resource subjects available on interlibrary loan.
Char Feist, Librarian

4665 Minnesota Department of Trade and Economic Development Library
500 Metro Square
121 7th Place E
St. Paul, MN 55101-2112
651-296-8902
Fax: 651-296-1290
Pat Fenton, Sr. Librarian

4666 Minot State University Bottineau Library
105 Simrall Boulevard
Bottineau, ND 58318
701-228-5454
Fax: 701-228-5468
http://www.misu-b.nodak.edu
Jan Wysocki, Library Director

4667 Mississippi Department of Environmental Quality Library
PO Box 20307
Jackson, MS 39289-1307
601-961-5024
Fax: 601-354-6965
E-mail: ronnie_sanders@deq.state.ms.us
http://www.deq.state.ms.ud
Geology and environmental reference library. Holdings in geosciences, hydrology, pollution control, paleontology, petroleum geology, land and water resources. Special collections; Topographic maps, United States, State andInternational Geoloical survey publications.
Ronnie Sanders, Librarian

4668 Missouri Department of Natural Resources Geological Survey & Resource Assessment Division
Box 250
Rolla, MO 65401
573-368-2101
Fax: 573-368-2111
Mimi Garstang, Director/State Geologist

4669 National Audubon Society: Aullwood Audubon Center and Farm Library
1000 Aullwood Road
Dayton, OH 45414
937-890-7360
Fax: 937-890-2382
E-mail: aullwood@gemair.com
Known as the Miami Valley's first educational farm, here visitors will discover a variety of native grasses and flowers, 300 year old oak trees and threatened bird species.

4670 National Institute for Urban Wildlife Library
10921 Trotting Ridge Way
Columbia, MD 21004
301-596-3311
http://www.webdirectory.com/wildlife/
Louise E Dove, Wildlife Biology

4671 Native Americans for a Clean Environment Resource Office
Box 1671
Tahlequah, OK 74465
918-458-4322
Fax: 918-458-0322
NACE is to raise the consciousness of Indian people and the general public about environment hazards, with an emphasis on the nuclear industry.
Lance Hughes, Executive Director

4672 Nature Conservancy Long Island Chapter
Uplands Farm Environmental Center
250 Lawrence Hill Road
Cold Spring Harbor, NY 11724
516-367-3225
Fax: 516-367-4715

4673 Nebraska Natural Resources Commission Planning Library
301 Centennial Mall S
Box 94876
Lincoln, NE 68509-4876
402-471-2081
Fax: 402-471-3132
E-mail: mosaic@nrcdec.nrc.state.ne.us
http://www.nrc.state.ne.us/

4674 New England Coalition on Nuclear Pollution Library
PO Box 545
Brattleboro, VT 05302
802-257-0336
E-mail: energy@necnp.org
http://www.necnp.org

4675 New England Governors' Conference Reference Library

76 Summer Street   617-423-6900
Boston, MA 02110   Fax: 617-423-7327
E-mail: info@negc.org
http://www.negc.org

4676 Occupational Safety and Health Library

1111 3rd Avenue   206-553-5930
Suite 715   Fax: 206-553-6499
Seattle, WA 98101-3212

4677 Ohio Environmental Protection Agency Library

122 South Front Street   614-644-3024
Columbus, OH 43215   Fax: 614-728-9500
http://www.epa.state.oh.us

Ruth Ann Evans, Librarian

4678 Peninsula Conservation Foundation Library of the Environment

3921 E Bayshore Road   650-962-9876
Palo Alto, CA 94303   Fax: 650-962-8234

4679 People, Food and Land Foundation Library

35751 Oak Springs   559-855-3710
Tollhouse, CA 93667   E-mail: sunmt@sunmt.org
http://www.sunmt.org

4680 Rainforest Action Network Library

221 Pine Street   415-398-4404
Suite 500   Fax: 415-398-2732
San Francisco, CA 94104   E-mail: rainforest @ran.org

Rainforest Action Network works to protect the Earth's rainforests and support the rights of their inhabitants through education, grassroots organizing and non-violent direct action.
Michael Brune, Executive Director

4681 Region 2 Library

US Environmental Protection Agency
290 Broadway   212-637-3185
16th Floor   Fax: 212-637-3086
New York, NY 10007   http://www.epa.gov/region02/library

Is a research and reference library for use by EPA staff, EPA contractors, other government agencies, and the public. The library contains or has access to scientific and technical materials in paper and electronic media related to awide variety of environmental issues, with and emphasis on EPA's Region 2.
Eveline M Goodman, Head Librarian

4682 Region 9 Library

US Environmental Protection Agency
75 Hawthorne Street 13th Floor   415-774-1510
San Francisco, CA 94105   Fax: 415-744-1474
E-mail: libaray-reg9@epa.gov
http://www.epa.gov/region9/library
Deborra Samuels, Hd. Libn./Coord.

4683 Rob and Bessie Welder Wildlife Foundation Library

Walker Wildlife Foundation   361-364-2643
PO Box 1400   Fax: 361-364-2650
Sinton, TX 78387   E-mail: welderwf@aol.com
http://www.members.aol.com/welderwf/welderhome
Private, nonprofit operation foundation which conducts research and education in wildlife management and related fields. Funds graduate fellowships and conducts its reserach and education program on its 7,800 acre wildlife refuge inthe surrounding South Texas region and throughout the United States.
Dr. D Lynn Drawe, Director
Vandra Davis, Librarian

4684 Schuylkill Center for Environmental Education

8480 Hagy's Mill Road   215-482-7300
Philadelphia, PA 19128-9978   Fax: 215-482-8158
http://www.schuylkillcenter.org

Karin James, Resource Librarian

4685 Society of American Foresters Information Center

5400 Gosvenor Lane   301-897-8720
Bethesda, MD 20814   Fax: 301-897-3690
http://www.safnet.org
An organization that represents the forestry profession in the United States. Its mission is to advance the science, education, technology and practice of forestry.
Jeff Ghannam, Director Media Relations

4686 Solartherm Library

1315 Apple Avenue   301-587-8686
Silver Spring, MD 20910   Fax: 301-587-8688
http://www.solartherm.com

4687 Solid Waste Association of North America

PO Box 7219   301-585-2898
Silver Spring, MD 20910   800-467-9262
Fax: 301-589-7068
E-mail: info@swana.org
http://www.swana.org
Nonprofit trade association designed to serve the municipal solid waste industry in cutting-edge informational and technilogical practices.
Dr John Skinner, Executive Director

4688 Southeast Fisheries Laboratory Library

75 Virginia Beach Drive   305-361-4229
Miami, FL 33149-1099   Fax: 305-361-4499

4689 Southwest Research and Information Center

105 Stanford SE   505-262-1862
PO Box 4524   Fax: 505-262-1864
Albuquerque, NM 87106   E-mail: sricdon@earthlink.net
http://www.sric.org
SRIC exists to provide timely, accurate information to the public on matters that affect the environment, human health, and communities in order to protect natural resources, promote citizen participation, and ensure environmental andsocial justice now and for future generations.
Dan Hancock, Administrator
Annette Aguayo, Information Specialist

4690 St Paul Plant Pathology Library

395 Borlaug Hall   612-625-9777
St Paul Campus
St Paul, MN 55108

Subject oriented library, specializing in plant diseases, plant virology, mycology, mycotoxicology and the effects of air pollution on vegetation. The collection contains approximately 8000 volumes of books, periodicals and PlantPathology theses. Over 50 current periodicals are recieved.

4691 St. Paul Forestry Library

University of Minnesota
B-50 Skok Hall   612-624-3222
2003 Upper Buford Circle   Fax: 612-624-3733
St. Paul, MN 55108   E-mail: heroL228@umn.edu
http://http://forestry.lib.umn.edu
Houses a general collection of books, journals, government documents, maps, and pamphlets relating to the subjects of forestry, forest products, outdoor recreation, range management, and remote sensing. There is also a small generalrefernce section. Also compiles and maintains four databases focused on aspects of forestry: Social Sciences in Forestry, Urban Forestry, Tropical Conservation and Development, Trails Planning Construction and Maintenance Planning.
Philip Herold, Librarian

4692 State University of New York

College of Environmental Science and Forestry
Environamental Science and Forestry   315-470-6715
Syracuse, NY 13210   Fax: 315-470-6512
E-mail: eaelkins@esf.edu
http://esf.edu
Moon Library supports the SUNY College of Environmental Science and Forestry where students major in Engineering, Chemistry, Biology, Landscape Architecture, Forest Resources Management and Environmental Studies.
Elizabeth A Elkins, Director/College Libraries

4693 Staten Island Institute of Arts and Sciences
William T Davis Education Center
75 Stuyvesant Place   718-987-6233
State Island, NY 10301   Fax: 718-273-5683
Patricia Salmon, Curator of History

4694 Texas Water Commission Library
PO Box 13087   512-463-7834
Austin, TX 78711-3087

4695 Turner, Collie and Braden Library
Box 130089   713-267-2826
Houston, TX 77219-0089   Fax: 713-780-0838
                            E-mail: rushbrookd@tcbhou.com
                                      http://www.tcbhou.com

David Rushbrook, Librarian
Renee Miller, Library Assistant

4696 US Bureau of Land Management
California State Office
2135 Butano Drive   916-978-4400
Sacramento, CA 95825   Fax: 916-978-4305
It is the mission of the Bureau of Land management to sustain the health, diversity and productivity of the public lands for the use an employment of present and future generations.

4697 US Bureau of Land Management Library
Denver Federal Center Building 50   303-236-6648
Box 25047   Fax: 303-236-4810
Denver, CO 80225-0047   E-mail: blm_library@blm.gov
                    http://www.blm.gov/nstc/library/library.html
The BLM Library serves the information and research needs of BLM personnel. The library also serves as the point of contract for bureau publications and information with other federal agencies and the public. The collection covers allaspects of land management and natural resources.
Barbara Campbell, Director

4698 US Department of Agriculture: National Agricultural Library, Water Quality Info Center
10301 Baltimore Boulevard   301-504-6077
Beltsville, MD 20705   Fax: 301-504-7098
                            E-mail: wqic@nal.usda.gov/wqic
                                   http://www.nal.usda.gov/wqic
Collects, organizes and communicates the scientific findings, educational methologies and public policy issues related to water and agriculture.
Joseph R Makuch, Coord. WQIC

4699 US Geological Survey: Great Lakes Science Center
1451 Green Road   734-994-3331
Ann Arbor, MI 48105   Fax: 734-994-8780
                     E-mail: GS-B-GLSC-Webmaster@usgs.gov
                                       http://www.glsc.usgs.gov
The USGS Great Lakes Science Center exists to meet the Nation's need for scientific information for restoring, enhancing, managing, and protecting living resources and their habitats in the Great Lakes. The center is headquartered inAnn Arbor, Michigan, and has biological research stations and vessels located throughout the Great Lakes basin.
Leon Carl, Center Director

4700 US Geological Survey: National Wetlands Research Center
700 Cajundome Boulevard   337-266-8545
Lafayette, LA 70506   Fax: 337-266-8513
                              E-mail: judy_buys@usgs.gov
                                    http://www.nwrc.gov/library.html
The National Wetlands Research Center is a source and clearinghouse of science information about wetlands in the United States and the world for fellow agencies, private entities, academia, and the public at large. Staff members obtainand provide this information by performing original scientific research and developing research results into literature and technological tools. They then disseminate that information through a variety of means.
Judy Buys, Librarian

4701 US Geological Survey: Upper Midwest Environmental Sciences Center Library
2630 Fanta Reed Road   608-781-6215
La Crosse, WI 54603   Fax: 608-783-6066
A federal library with technical holdings mainly in aquatic sciences, bird and amphibean materials.
Kathy Mannstedt, Librarian

4702 US Geological Survey: Water Resources Division Library
375 S Euclid Avenue   520-670-6201
Tucson, AZ 85719

4703 Unexpected Wildlife Refuge Library
110 Unexpected Road   856-697-3541
Newfield, NJ 08344   http://www.animalplace.org

4704 University of California
1 Shields Avenue   530-752-1011
Davis, CA 95616

4705 University of Florida Coastal Engineering Archives
209 Yon Hall   352-392-2710
Gainesville, FL 32611   Fax: 352-392-2710
                            E-mail: Twedell@coastal.ufl.edu
Helen Twedell, Archivist
Kimberly Hunt, Sr. Library Technical Asst

4706 University of Hawaii at Manoa Water Resources Center
2540 Dole Street   808-956-7847
Homes Hall 283   Fax: 808-956-5044
Honolulu, HI 96822   E-mail: morav@hawaii.edu
                                      http://www.wrrc.hawaii.edu
Coordinates and conducts research to identify, characterize and quantify water/environmental related problems in the state of Hawaii. Based on the research WRRC makes recommendations to all agencies and organizations withresponsibilities to manage the water/ environmental resources in Hawaii.
Phillip Morakik, Technology Transfer Spec
James Moncur, Director

4707 University of Illinois at Chicago
Energy Resource Center
851 S Morgan Street   312-996-4490
12th Floor   Fax: 312-996-5420
Chicago, IL 60607-7054   E-mail: rsanka1@uic.edu
                              http://h008.erc.uic.edu/welcome.htm
The Energy Resources center is an interdiciplinary public service, research, and special projects organization dedicated to improving energy efficiency and the environment. Conducts studies in the fields of energy and environment andprovides industry, utilities, government agencies and the public with assistance, information, and advice on new technologies, public policy, and professional development training.
James Hartnett, Director

4708 University of Maryland: Center for Environmental Science Chesapeake Biological Lab
1 Willams Street   410-326-7287
Box 38   Fax: 410-326-7302
Solomons, MD 20688   http://www.cbl.umces.edu
Kathleen A Heil, Librarian

4709 University of Montana Wilderness Institute Library
Forestry Building   406-243-5361
Room 207   Fax: 406-243-4845
Missoula, MT 59812   E-mail: wi@forestry.umt.edu
                                      http://www.forestry.umt.edu/wi

4710 Vermont Institute of Natural Sciences Library
27023 Church Hill Road   802-457-2779
Woodstock, VT 05091   Fax: 802-457-2779

4711 Voices from the Earth
Box 4524
Albuquerque, NM 87106

505-262-1862
E-mail: sricdon@earthlink.net
http://www.sric.org

SRIC exists to provide timely, accurate information to the public on matters that affect the environment, human health, and communities in order to protect natural resources, promote citizen participation, and ensure environmental andsocial justice now and for future generations.
Dan Hancock, Administrator
Annette Aguayo, Information Specialist

4712 Wasserman Public Affairs Library
University of Texas at Austin
General Libraries
Sid Richardson Hall 3243
Austin, TX 78712

512-495-4400
Fax: 512-495-4347
E-mail: pal@lib.utexas.edu
http://www.lib.utexas.edu/pal

Stephen Littrell, Head Librarian

4713 Western Ecology Division Library
US Environmental Protection Agency
200 SW 35th Street
Corvallis, OR 97333

541-754-4731
Fax: 541-754-4799

Mary OBrien, Librarian

4714 Wildlife Management Institute Library
1101 14th Street NW
Suite 725
Washington, DC 20005

202-371-1808
Fax: 202-408-5059

Richard E McCabe, Sec./Dir., Pubns.

4715 Wisconsin Department of Natural Resources Library
Box 7921
Madison, WI 53707-7921

608-266-8933
Fax: 608-266-5226

Contains books, journals, and EPA reports on air pollution, geology, hazardous waste, natural resources management, recycling, soil pollution, solid waste, toxic substances, waste minimization, wastewater, water pollution, andwetlands.
Erin Matiszik, Librarian

4716 Yale University School of Forestry and Environmental Studies Library
205 Prospect Street
New Haven, CT 06511

203-432-5132
Fax: 203-432-5942
http://www.library.yale.edu/scilib/forestl.html

A part of the Yale University Library System, the library serves the resource needs of the graduate students and faculty of Yale's 100 year old school of Forestry and Environmental Studies.
Carla Heister, Librarian

Publishers

4717 Academic Press: New York
Academic Press
15 E 26th Street
15th Floor
New York, NY 10010

212-592-1000
E-mail: ap@acad.com
http://customerservice.apnet.com

4718 Adison Wesley Longman
Pearson
26 Prince Andrew Place
Toronto ONT M3C-2T8

905-853-7888
800-563-9196
Fax: 800-263-7733
E-mail: webinfo.pubcanada@pearson.com
http://pearson.com

One integrated and diverse company offering learning resources on an extraordinary level. Pearson has an estblished reputation for producing market-leading educational products and services as well as a comprehensive range ofbest-selling consumer, environmental technical and professional titles.

4719 Blackwell Publishers
Blackwell Publishers
350 Main Street
Malden, MA 02148

781-388-8200
Fax: 781-388-8210

Blackwell Publishers are dedicated to serving the global academic community. We recognize that publishing is about making connections. Knowledge is not constrained by national or liguistic boundries. Many academics are engaged in bothteaching and research. Our readers are often our authors as well. We develop books for students which take account of the latest research and we aim to make the journals we publish as acessible as possible.

4720 Boxwood Press
183 Ocean View Boulevard
Pacific Grove, CA 93950

408-375-9110
Fax: 408-375-0430
E-mail: boxwood@boxwoodpress.com

Publishes significant titles in the areas of Natural History, Area Studies, General Sciences and Local and Special Interest. Founded in 1952, it first published lab manuals, then expanded to include a variety of mainly biologicaltitles.

4721 CABI Publishing
CABI Publishing
CAB International
44 Brattle Street, 4th Floor
Cambridge, MA 02138

617-395-4056
800-528-4841
Fax: 617-354-6875
E-mail: cabi-nao@cabi.org
http://www.cabi-publishing.com

CABI publishing is a leading international, nonprofit publisher in applied life sciences, including animal science, nutrition, integrated crop management and forestry. Our products have a global reputation for quality, relevance andauthority, and are used in over 100 countries. Our long-established print publishing activities include a substantial book and reference work list, and an expanding primary and review journal program.

4722 CRC Press
CRC Press
2000 NW Corporate Boulevard
Boca Raton, FL 33431

561-994-0555
800-272-7737
Fax: 800-374-3401
http://www.crcpress.com

CRC Press LLC is recognized as a leader in scientific, medical, environmental science, engineering, business, technical, mathamatical, and statistics publishing. CRC Press LLC publishes books, journals, newsletters and databases.Customers have access to publications through individual purchases, bookstores, libraries and on-line acess at www.crcpress.com

4723 Chelsea Green Publishing Company
85 N Main Street
PO Box 428
White River Junction, VT 05001

802-295-6300
800-639-4099
Fax: 802-295-6444
E-mail: publicity@chelseagreen.com
http://www.chelseagreen.com

Chelsea Green publishes information that helps us lead pleasurable lives on a planet where human activities are in harmony and balance with nature.Free catolog listing over 250 titles on sustainable living, innovative shelter andorganic gardening.
Alice Blackmer, Publicity Director

4724 DK Publishing
DK Publishing
95 Madison Avenue
New York, NY 10016

212-213-4800
Fax: 212-213-5240

Dorling Kindersley is an international publishing company specialising in the creation of high quality, illustrated information books, interactive software, TV programs and online resources for childern and adults. Founded in London1974, DK now has offices in the UK, USA, Australia, South Africa, India France, Germany and Russia.
Publication Date: 1974

4725 Elsevier Science
Elsevier Sciences
655 Avenue of the Americas
New York, NY 10010-5107

212-633-3730
Fax: 212-633-3680
E-mail: usinfo-f@elsevier.com

Our focus will be entirely on scientific, technical and medical publishing. Together we can offer customers choice across our portfolio, with outstanding platforms for the delivery of electronic services and a high level of investmentto ensure the development of leading electronic products.

Publications/Publishers

4726 Institute for Food and Development Policy
398 60th Street 510-654-4400
Oakland, CA 94618 Fax: 510-654-4551
http://www.foodfirst.org
Publishes numerous health related books.

4727 Island Press
Distribution Center
PO Box 7 707-983-6432
Covelo, CA 95428 800-828-1302
Fax: 707-983-6414
E-mail: service@islandpress.com
http://www.islandpress.com
Mission-oriented nonprofit publisher organized in 1984 to help meet the need for accessible, solutions-oriented information through a unique approach that addresses the multidisciplinary nature of environmental problems. Our program isdesigned to translate technical information from a range of disciplines into a book format that is accessible and informative to citizen activists, educators, students and professionals involved in the study or management of environmental problems.
Bernice Hiatt, Customer Service

4728 It's Academic
29 West 35th Street 212-216-7800
New York, NY 10001 Fax: 212-564-7854
A tool for teachers who use Routledge books in their classes. To aid in finding the books best suited for your needs, we offer: pages which highlight books designed specifically for your courses, a list of conferences at which wedisplay our books, journal information, a forum for instructors to send us their comments, supplements available on line and a subject search menu.

4729 Kluwer Academic Publishers
101 Philip Drive 781-871-6600
Assinippi Park Fax: 781-871-6528
Norwell, MA 02061
A sector of the Wolters Kluwer publishing group. Operates world-wide from offices in Dordrecht, Boston, New York and London. All over the world, scientists and professionals hold our publications in high esteem.

4730 Krieger Publishing Company
Krieger Publishing Co.
PO Box 9542 321-724-9542
Melbourne, FL 32902-9542 800-724-0025
Fax: 321-951-3671
E-mail: info@krieger-publishing.com
http://www.krieger-publishing.com
Krieger Publishing Company produces quality books in various fields of interest. We have an extensive Natural Science listing.
Cheryl Stanton, Advertising

4731 MIT Press
5 Cambridge Center 617-253-5646
Cambridge, MA 02142 Fax: 617-258-6779
http://www.press.mit.edu
The only university press in the US whose list is based in science and technology. Our environment list is also strong in policies and the social sciences. We are committed to the edges and frontiers of the world - to exploring newfields and new modes of inquiry. We publish about 200 new books a year and over 40 journals including Global Environmental Politics. We have a long-term commitment to both design excellence and the efficient and creative use of new technologies.
Clay Morgan, Senior Acquisition Editor

4732 McGraw-Hill Education
The McGraw-Hill Companies
1221 Avenue of the Americas 212-512-2000
40th Floor Fax: 212-512-6111
New York, NY 10020-1095
A global leader in educational materials and professional information, with offices in more than 30 countries and publications in more than 40 languages, we develop products that influence people's lives from preschool through career.The scope of our operations, the quality of our editorial product and the pace at which we are developing new media to fulfill our customers' information requirements are increasing.

4733 Micromedia
20 Victoria Street 416-362-5211
Toronto, Ontario M5C 2N8 Fax: 416-362-6161
E-mail: info@micromedia.ca
http://www.micromedia.ca
Canada's largest developer, publisher and distributor of value-added reference information for the academic, library, government and corporate markets. Our mission is to be Canada's one stop shop for information products and services.We license content from media, government and other sources and organize, abstract and compile this content into databases. Through a combination of technology expertise and a full service approach, our solutions provide access to a wide range ofinformation.
Peter Asselstine, Marketing Manager
Bryan Moore, Sales and Marketing Manager

4734 National Information Service Corporation
NISC USA, Wyman Towers 410-243-0797
3100 St. Paul Street Fax: 410-243-0982
Baltimore, MD 21218
Publishes information products for access through BiblioLine, our Web search service, or on CD-ROM. Some of our abstract and index services are available in print. NISC's bibliographic and full-text databases cover a wide range oftopics in the natural and social sciences, arts and humanities. Some titles provide comprehensive coverage of particular geographic regions, such as Latin America, Africa, South-East Asia or the Arctic and Antarctic.

4735 O'Reilly & Associates
101 Morris Street
Sebastopol, CA 95472 800-998-9938
Fax: 707-829-0104
Premier information source for leading-edge computer technologies. We offer the knowledge of experts through our books, conferences and web sites. Our books, known for their animals on the covers, occupy a treasured place on theshelves of developers building the next generation of software. Conferences and summits bring innovators together to shape the ideas that spark new industries. From the Internet to the web, Linux, Open Source and peer-to-peer networking, we puttechnologies on the map.

4736 Random House
1540 Broadway 212-782-9000
New York, NY 10036 Fax: 212-302-7985
The world's largest English-language general trade book publisher. It is a division of the Bertelsmann Book Group of Bertelsmann AG, one of the foremost media companies in the world.

4737 Simon & Schuster
1230 Avenue of the Americas 212-698-7000
New York, NY 10020 Fax: 212-698-2359
E-mail: ken.riel@simonandschuster.com.
Ken Riel

4738 Springer-Verlag New York
175 Fifth Avenue 212-460-1500
New York, NY 10010 Fax: 212-473-6272
Founded in 1964 and maintained its position last year as the Springer Group's largest foreign subsidiary. In 1999, 426 new titles were released. In addition, 50 journals were published, most of them available in electronic form as wellas via the Springer information system LINK. The number of license agreements in the North American market has increased fivefold due to the increasing demand for this leading Online Library.

4739 Virginia Museum of Natural History
1001 Douglas Avenue 540-666-8600
Martinsville, VA 24112 Fax: 540-632-6487
http://www.vmnh.org
Our publishing division specializes in works by natural scientists and environmental educators in the US and abroad. Writing, editorial, and design services available for books, reports, manuals, text books, presentations, fieldguides, etc.. A catalogue is available on request.
24 pages Quarterly
ISSN: 1085-5084
Susan Felker, Managing Editor\Outreach

4740 WW Norton & Company

500 5th Avenue   212-354-5500
New York, NY 10110   Fax: 212-869-0856

The oldest and largest publishing house owned wholly by its employees, strives to carry out the imperative of its founder to publish books of long-term value in the areas of fiction, nonfiction and poetry. The roots of the company dateback to 1923, when William Warder Norton and his wife, M.D. Herter Norton, began publishing lectures delivered at the People's Institute, the adult education division of New York City's Cooper Union.

4741 Wiley North America

605 3rd Avenue   212-850-6000
New York, NY 10158-0012   Fax: 212-850-6088

The company was founded in 1807, during the Jefferson presidency. In the early years, Wiley was best known for the works of Washington Irving, Edgar Allen Poe, Herman Melville and other 19th century American literary giants. By theturn of the century, Wiley was established as a leading publisher of scientific and technical information.

Corporate & Commercial Centers

4742 3M: 201 Technical Library
3M Center
St. Paul, MN 55133
651-575-1300
Fax: 651-736-3940
http://www.3M.com
High-tech library that manages its collection with 3M digital identification.

4743 AAA & Associates
28 West Adams
Suite 1511
Detroit, MI 48226
313-961-4122
Fax: 313-588-6232

Katherine Banicki, President

4744 AB Gurda Company
6061 Whitnall Way
Hales Corners, WI 53130-2271
414-529-3116
Environmental testing and analysis firm.

4745 ABC Research Corporation
3437 SW 24th Avenue
Gainesville, FL 32607-4599
352-372-0436
Fax: 352-378-6483
E-mail: info@abcr.com
http://www.abcr.com
Research and analysis laboratory. Research results published in scientific journals.
Dr William L Brown, President
Dr Peter Bodnaruck, VP

4746 ABS Consulting
16800 Greenspoint Park Drive
Suire 300 South
Houston, TX 77060-2393
281-673-2800
Fax: 281-673-2812
E-mail: info@absconsulting.com
http://www.plg.com
ABS Consulting provides rational engineering, science and technology-based solutions that blend effective management controls, state-of-the-art engineering analyses, practical loss-control measures and innovative risk-transfer options.
Frank Iarossi, Chairman/CEO

4747 ACRES Research
6621 W Ridgeway Avenue
Waterloo, IA 50701
319-277-6661
Fax: 319-266-7569
E-mail: acresres@aol.com
Environmental research and testing.
Bert Schou PhD, President

4748 ACZ Laboratories, Inc
2773 Downhill Drive
Steamboat Springs, CO 80487
970-879-6590
800-334-5493
Fax: 970-879-2216
E-mail: sales@acz.com
http://www.acz.com
A full service environmental analytical lab with inorganic, organic and radiochemical capabilities. We perform analysis on a wide variety of matrices including water, wastewater, waste, soil, plant and animal tissue as well as fishtissue.

Tim VanWyngarden, Manager Business Development
Brad Craig, President/CEO

4749 ADA Technologies
8100 Shaffer Parkway
Suite 130
Littleton, CO 80127
303-792-5615
800-232-0296
Fax: 303-792-5633
E-mail: ada@adatech.com
http://www.adatech.com
Product development and testing of environmental technologies.
Judith Armstrong PhD, President

4750 AER
840 Memorial Drive
Cambridge, MA 02139
617-547-6207
Fax: 617-661-6479
E-mail: ross@aer.com
http://www.aer.com

4751 AF Meyer and Associates
9060 Meadowood Street
Baton Rouge, LA 70815
225-925-0630
Fax: 225-928-7848
E-mail: afmal@webtv.net
http://www.erols.com/afma
Environmental consulting firm.
AF Meyer, President

4752 AM Kinney
2900 Vernon Place
Cincinnati, OH 45219
513-281-2900
Fax: 513-281-1123
E-mail: nielseng@amkinney.com
http://www.amkinney.com
Provides creative and cost effective solutions in the planning, design and delivery of clients' projects.
George Finch, President

4753 AMA Analytical Services
4475 Forbes Boulevard
Lanham, MD 20706
301-459-2640
Fax: 301-459-2643
http://www.amalab.com
Environmental research. Asbestos, lead and explosives analysis.
David P Hood, CEO

4754 ANA-Lab Corporation
2600 Dudley Road
PO Box 9000
Kilgore, TX 75663
903-984-0551
Fax: 903-984-5914
E-mail: corp@ana-lab.com
http://www.ana-lab.com
Environmental laboratory. Offers ICP-MS which allows Ana-Lab to offer improved turn around time, reduce costs, and achieve better quantitation of regulated parameters. Tests are performed by methods specified by the EPA. Specializes inenvironmental chemistry.
C H Whiteside, President

4755 APC Lab
13760 Manolia Avenue
Chino, CA 91710
909-590-1828
Fax: 909-590-1498
E-mail: apcl@apclab.com
Environmental and industrial testing laboratory. Research results published in journals and conference reports.
Irene Huang, Public Relations

4756 APS Technology
800 Corporate Row
Cromwell, CT 06416-2072
E-mail: info@aps-tech.com
http://www.aps-tech.com
Product development, conceptual design, engineering, prototype manufacture and test analysis.
William E Turner, President
Denis Bigin, VP

4757 ARCADIS
630 Plaza Drive
Suite 200
Highlands Ranch, CO 80129
720-344-3500
Fax: 720-344-3535
http://www.arcadis-us.com
A leading, global, knowledge-driven service provider. Active in the fields of infrastructure, environment and buildings. Feasibility studies, design, engineering, project management, implementation and facility management, plusrelated legal and financial services.
John Boyette, President
Steven B Blake, CEO

4758 ARDL
400 Aviation Drive
Mount Vernon, IL 62864
618-244-3235
Fax: 618-244-1149
Environmental sampling and testing laboratory; Research Development Engineering. Alternate Name: Applied Research and Development Laboratories, Inc.
Larry Gibbons PhD, President
Don Gillespie, Marketing Manager

4759 ASTB Analytical Services
4027 New Castle Avenue   302-571-8881
New Castle, DE 19720   800-221-5170
    Fax: 302-571-0582
E-mail: astbanalytical@aol.com
State of the art analytical and testing services for the chemical, cosmetics, food, forensics, industrial, pharmaceutical and consumer product sectors. Complete consulting services in legal and regulatory compliance matters, R&D,IP/Product development, trade verification and worldwide technical representation.

Dr. Frank Coleman

4760 ASW Environmental Consultants
20 N Plains Industrial Road   203-265-0509
PO Box 495   Fax: 203-265-1476
Wallingford, CT 06492
Jason J Sarojak, PE

4761 ATC Associates
7988 Centerpoint Drive   317-849-4990
Suite 100   Fax: 317-849-4278
Indianapolis, IN 46256   http://www.atc-enviro.com
Technical engineering research and environmental consulting firm.
John Mundell, Director/Manager

4762 ATC Environmental
720 E Benson Road   605-338-0555
Sioux Falls, SD 57103
Donald Beck

4763 ATL
2912 W Clarendon Avenue   602-241-1097
Phoenix, AZ 85017-4609   Fax: 602-277-1306
Technical engineering evaluation firm. Research results published in test summaries and project reports.
Frank C Rivera, President
David P Hayes, VP

4764 AW Research Laboratories
16326 Airport Road   218-829-7974
Brainerd, MN 56401-5804   Fax: 218-829-1316
    http://www.awlab.com
A.W. Research Laboratories, Inc. (AWRL) provides environmental consulting services and water quality analysis. AWRL specializes in the use of remote sensing techniques for lake analysis and management.
Alan W Cibuzar, CEO

4765 AZTEC Laboratories
6402 Stadium Drive   816-921-3922
PO Box 7953
Kansas City, MO 64129
Data collection and analysis, systems design and product development firm.

4766 Aaron Environmental
189 Atwater Street   860-276-1201
Plantsville, CT 06479   800-372-1233
    Fax: 860-276-1233
E-mail: info@aaronenvironmental.com
http://www.aaronenvironmental.com
Joyce Kogut, President

4767 Accurate Engineering Laboratories
2707 W Chicago Avenue   773-384-4522
Chicago, IL 60622   Fax: 773-384-8681
Environmental engineering laboratory.
Noel Buczkowski, President

4768 Accutest Laboratories
2235 Route 130 S   732-329-0200
Building B   Fax: 732-329-3499
Dayton, NJ 08810   http://www.accutest.com
Environmental testing firm.
Vincent Pagrissi, President

4769 Acts Testing Labs
100 Northpointe Parkway   716-505-3300
Buffalo, NY 14228   Fax: 716-505-3301
Global consumer products testing organization providing quality assurance testing, inspections and consulting services.
Tom Fatta, Contact

4770 Adelaide Associates
690 N Broadway G13   914-949-3109
White Plains, NY 10603   Fax: 914-949-8103
E-mail: adelaide@bestweb.net
Environmental health consulting and testing firm. Additional offices: White Plains, NY, Poughkeepsie, NY and Perth Amboy, NJ.
Ron Birlinski, CEO

4771 Adelaide Environmental Health Associates
111-115 Court Street   607-722-6839
Binghamton, NY 13901   Fax: 607-771-0752
Roland E Bielinski, President

4772 Adirondack Environmental Services
314 N Pearl Street   518-434-4546
Albany, NY 12207   Fax: 518-434-0891
E-mail: aes@adirondackenvironmental.com
http://www.adirondackenvironmental.com
Analytical medical laboratory.

4773 Advance Pump and Filter Company
10 Cales Highway   603-868-3212
Lee, NH 03824   Fax: 603-868-3230

Cathleen Pleadwell, Business Manager

4774 Advanced Civil Engineering Materials Research Lab
2350 Hayward Street   734-764-3368
Ann Arbor, MI 48109   Fax: 734-764-4292
E-mail: vcli@umich.edu
http://www.ace-mrl.engin.umich.edu.acemrl
A research laboratory affiliated with the Department of Civil and Environmental Engineering, University of Michigan. Activities include research in sustainable infrastructure development, hazard mitigation and environmental improvementthrough materials technology.
Victor Li, Professor

4775 Advanced Terra Testing
833 Parfet Street   303-232-8308
Unit A   888-859-8378
Lakewood, CO 80215   Fax: 303-232-1579
E-mail: terratest@aol.com
http://www.terratesting.com
Geotechnical and geosynthetic testing firm.
Chris Wienecke, Director/Manager

4776 AeroVironment
825 S Myrtle Avenue   626-357-9983
Monrovia, CA 91016-3524   Fax: 626-359-9628
http://www.aerovironment.com
Research, service and consulting firm specializing in the environment, alternative energy and aerodynamic design. Research results published in project reports and technical journals.

4777 Aerosol Monitoring & Analysis
PO Box 646   410-684-3327
Hanover, MD 21076-0646   Fax: 410-684-3384
http://www.amatraining.com
Environmental services firm.

4778 AgriDyne Technologies
417 Wakara Way, 3rd Floor   801-583-3500
University Research Park   Fax: 801-583-2945
Salt Lake City, UT 84108
Willem SO Spiegel, VP/CFO

4779 Agvise Laboratories
PO Box 510
Northwood, ND 58267-0510
701-587-6010
Fax: 701-587-6013
E-mail: agvise@polarcomm.com
Applied and product research in environmental applications.
Robert Wallace, CEO

4780 Aigner McLaughlin Associates
1813 Northwest 27th
Portland, OR 97210
503-274-2922
Fax: 503-274-1429
Phyllis A McLaughlin, Partner

4781 Alan Plummer and Associates
7524 Mosier View Court
Suite 200
Fort Worth, TX 76118
817-806-1700
Fax: 817-589-0072
http://www.apaienv.com
Civil and environmental engineering consulting.
Alan H Plummer Jr, President

4782 Alar Engineering Corporation
9651 West 196th Street
Mokena, IL 60448
708-479-6100
Fax: 708-479-9059
Alex Doncer, President

4783 Alden Research Laboratory
30 Shrewsbury Street
Holden, MA 01520-1843
508-829-6000
Fax: 508-829-5939
E-mail: arlmail@aldenlab.com
http://www.aldenlab.com
Hydraulic engineering firm solving air and water flow problems using physical and CFD models and field testing, for areas such as fish passage/protection systems, free surface and closed conduit flow, pump/turbine performance, hydraulic structures, environmental hydraulics, fluid equipment, 3D air flow, and flow meter calibration.
Edward P Taft II, President

4784 Allied Laboratories
716 North Iowa
Villa Park, IL 60181
630-279-0390
Fax: 630-279-3114
Irving I Domsky, Director

4785 Alloway Testing
1101 N Cole Street
Lima, OH 45805-2003
419-223-1362
Fax: 419-227-3792
http://www.alloway.com
Environmental sampling and analysis laboratory.
John R Hoffman, President

4786 Alpha Manufacturing Company
100 Old Barnwell Road
PO Box 2809
West Columbia, SC 29171
803-739-4500
Fax: 803-739-0517
E-mail: mail@alphamfg.com
http://www.alphamfg.com
Physical testing of environmental testing and repair services, and instrument design.
William L Cowley, President

4787 Alpha Solarco
11534 Gondola
Cincinnati, OH 45241
513-771-8086
Fax: 513-771-5902
Edward Schmidt, President

4788 Alton Geoscience
21 Technology Drive
Irvine, CA 92618-2302
949-753-0101
Fax: 949-753-0111
http://www.trcsolutions.com
Environmental remediation and consulting firm.
Jenny Rue, Public Relations
Larry Farrington, Contact

4789 Amalgamated Technologies
13901 N 73rd Street
Suite 208
Scottsdale, AZ 85260-3125
480-991-2901

Firm providing metals and materials development, processing and testing.
Roy E Beal, President

4790 American Analytical Laboratories
840 South Main Street
Akron, OH 44311-1516
350-535-1300
800-837-2251
Fax: 350-535-7246
American Analytical Laboratories is a full service environmental laboratory specializing in air toxics, groundwater, wastewater, drinking water, and industrial hygiene analyses and consulting.
Richard E. Moore, CIH., President

4791 American Environmental Network
9151 Rumsey Road
Suite 150
Columbia, MD 21045
410-730-8525
Fax: 410-997-2586
Paul Jackson, Marketing Manager

4792 American Laboratory for Environmental Excellence
21 Ash Street
Suite 2
Frankfort, IL 60423-1591
815-469-7500
Fax: 815-469-2260
Barbara A Olsen, President

4793 American Research & Testing
14934 S Figueroa Street
Gardena, CA 90248-1711
310-538-9709
800-538-1655
Fax: 310-538-9965
E-mail: chemist@americanresearch.com
http://www.americanresearch.com
Company specializing in chemical analysis and physical testing.
Rita R Boggs, President
Barbara Belmont, Director/Manager

4794 American Resources Group
Signet Bank Building
Vienna, VA 22180
703-255-2700
Fax: 703-281-9200
Dr Keith A Argow, President

4795 American Testing Laboratory
1230 8th Avenue N
PO Box 731
Bessemer, AL 35020-5526
205-424-1390
Fax: 205-425-8118
Multidisciplinary testing laboratory.

4796 American Waste Processing
2100 W Madison Street
Maywood, IL 60153
708-681-3999
800-841-6900
Fax: 708-681-5583
E-mail: american@american-waste.com
http://www.american.waste.com
Hazardous waste management, remediation and dust removal.
William Vajdik, President

4797 Ana-Lab Corporation
PO Box 9000
Kilgore, TX 75663-9000
903-984-0551
Fax: 903-984-5914
E-mail: chw@ana-lab.com
http://www.ana-lab.com
Ana-Lab Corporation is an environmental testing laboratory which includes a wide variety of analyses including air quality, hazardous waste classification, water and wastewater, soils and sludge, and petroleum testing. A2LA accredited.
CH Whiteside, President

4798 Analab
630 Heron Drive
PO Box 336
Bridgeport, NJ 08014
856-467-4555
800-262-5229
Fax: 856-467-1212
E-mail: info@analab1.com
http://www.analab1.com
Compliance lab services for EMC/EMI, safety and ESD.
Jason Smith, Director/Manager

Research Centers/Corporate & Commercial Centers

4799 Analyte Laboratories
6630 Baltimore National Pike
Route 40 West
Baltimore, MD 21228
410-747-3844
Fax: 410-747-4007

4800 Analytical Laboratories and Consulting
361 West 5th Avenue
Eugene, OR 97401
541-485-8404
Fax: 541-484-5995
Rory E White, Sr. Analyst

4801 Analytical Process Laboratories
8222 W Calumet Road
Milwaukee, WI 53223-3845
414-355-3909
Fax: 414-355-3099
http://www.apl-lab.com
Materials analysis laboratory and environmental engineering.
Jitendra Shah, President
Dr. Taxla Shah, CEO

4802 Analytical Resources
4611 S 134th Place
Tukwila, WA 98168-3240
206-695-6200
Fax: 206-695-6201
http://www.arilabs.com
Environmental testing and analysis laboratory.
Mark Weidner, President
Stephanie Lucas, Project Manager

4803 Analytical Services
110 Technology Parkway
Nocross, GA 30092
770-734-4200
Fax: 770-734-4201
http://www.asi-lab.com
Environmental testing and analysis firm.
Robert G Owens, President

4804 Anametrix
1961 Concourse Drive
Suite E
San Jose, CA 95131
408-432-8192
Fax: 408-432-8198
Doug Robbins, President

4805 AndCare
PO Box 14566 Parkway
Research Triangle Park, NC 27709
919-544-8220
Fax: 919-544-9808
Development and commercialization of low cost, simple to use diagnostic devices and tests for medical, environmental, and laboratory markets.
Dr. Steven Wagner, PhD, President

4806 Anderson Engineering Consultants
10205 W Rockwood Road
Little Rock, AR 72204-8100
501-455-4545
Fax: 501-455-4552
Firm providing engineering, inspection, and testing services specializing in geotechnology and materials, and environmental sciences. Services include site studies, soil testing, engineering surveys, specification evaluation, andfailure investigation.

4807 Andrea Aromatics
PO Box 3091
Princeton, NJ 08543
609-695-7710
Fax: 609-392-8914
E-mail: Orders@andreasaromatics.com
Natural essential oils, fragrances, deodorants, odor neutralizers.
Michael D'Andrea, VP

4808 Andrew D Sauter Consulting
1576 Sweet Home Road
Amherst, NY 14228
702-896-5413
Fax: 702-896-5413
E-mail: nanogenesys@email.com
A D Sauter

4809 Anlab
1910 S Street
Sacramento, CA 95814
916-447-2946

4810 Anoroc Scientific
PO Box 2001
South Hackensack, NJ 07606-0601
201-891-8919
Fax: 201-641-0789
Chemical manufacturing and testing company.

4811 Anteon Corporation
3211 Jermantown Road
Suite 200
Fairfax, VA 22030
703-246-0200
Fax: 703-246-0797
Don MacDougall, President Services

4812 Apollo Energy Systems
PO Box 238
Navasota, TX 77868
936-825-8188
Fax: 936-825-2625
Gary Conway, VP

4813 Applied Biomathematics
100 N Country Road
Setauket, NY 11733-1300
631-751-4350
Fax: 631-751-3435
E-mail: info@ramas.com
http://www.ramas.com
Environmental and ecological software development firm.
Lev Ginzburg, President

4814 Applied Coastal Research & Engineering
766 Falmouth Road
Building A, Unit 1-C
Mashpee, MA 02649-3340
508-539-3737
Fax: 508-539-3739
E-mail: info@appliedcoastal.com
http://www.appliedcoastal.com
Environmental analysis.
Mark Bynes PhD, President

4815 Applied P&C Laboratory
4066 East Mission Boulevard
Pomona, CA 91766
909-622-5148
Fax: 909-622-3199
Yongfeng Zhang, PhD, Director

4816 Applied Physics
8100 Mountain Road NE
Sutie 217
Albuquerque, NM 8711
505-265-2464
Fax: 505-265-2237
E-mail: 74407.1604@compuserve.com
Environmental applications.
Lori Thueson, President

4817 Applied Technical Services
1190 Atlanta Industrial Drive
Marietta, GA 30066-6603
770-423-1400
Fax: 770-514-3299
http://www.atslab.com
Environmental, chemical, and mechanical testing and consulting company.
Jim F Hill, President

4818 Aptus
PO Box 55021750 Cedar Avenue
Lakeville, MN 55044
877-240-1022
Fax: 952-469-5140
Environmental services firm specializing in hazardous and toxic waste transportation, storage, incineration, and decontamination.

4819 Aqua Tech Environmental Laboratory
181 S Main Street
Marion, OH 43302-3964
740-382-5991
800-783-5991
Fax: 740-389-1481
Environmental testing laboratory.

4820 Aquatec Chemical International
408 Auburn Avenue
Pontiac, MI 48342
313-334-4747
Douglas Schwartz, President

4821 Architectural Energy Corporation
2540 Frontier Avenue
Suite 201
Boulder, CO 80301
303-444-4149
Fax: 303-444-4304

E-mail: aecinfo@archenergy.com
http://www.archenergy.com
Energy, daylighting and sustainable design and analysis; LEED certification services; building commissioning, energy auditing and diagnostic testing; home energy rating software (RFOM/Rate); commercial energy analysis software(VisualDOE); and data acquisition equipment.

Michael J Holtz, FAIA, President

4822 Ardaman & Associates

8007 S Orange Avenue	407-855-3860
PO Box 593003	800-688-SOIL
Orlando, FL 32859	Fax: 407-859-8121

E-mail: mmongeau@ardaman.com
http://www.ardaman.com
Geotechnical, environmental and materials consultants.

Mark L Mongeau, VP

4823 Arete Associates

5000 Van Nuys Boulevard	818-501-2880
PO Box 6024	Fax: 818-501-2905
Sherman Oaks, CA 91413-6024	http://www.arete.com

Environmental research.
Dr Stephen C Lubard, President

4824 Aroostook Testing & Consulting Laboratory

160 Airport Drive	207-762-5771
Presque Isle, ME 04769-2044	Fax: 207-764-8123

E-mail: atclabs@ainop.com
Toxiological and environmental laboratory.

G Noel Currie III, President

4825 Arro

Caton Farm Road	815-727-5436
PO Box 686	Fax: 815-740-3234
Joliet, IL 60434	

Robert J Rolih, President

4826 Artesian Laboratories

630 Churchmans Road	302-266-9121
Newark, DE 19702-1900	Fax: 302-454-8720

Environmental sampling and testing laboratory. Research results published in methods reports to clients.

4827 Arthur L Conn & Associates

1469 East Park Place	312-842-6388
Chicago, IL 60637	Fax: 773-667-1527

Arthur L Conn, President

4828 Asbestos Management

PO Box 347	313-961-6135
Grosse Ile, MI 48138	Fax: 734-692-3883

LaDonna Slifco, President

4829 Ascension Technology

PO Box 527	802-893-6657
Burlington, VT 05402	Fax: 802-893-6659

E-mail: ascension@ascension-tech.com
http://www.ascension-tech.com
Manufactures motion tracking equipment and provides a full year's warranty, free telephonic support, and on-site support.
Edward C Kern Jr, President

4830 Association of Ecosystem Research Centers

730 11th Street NW	202-628-1500
Washington, DC 20001-4521	Fax: 202-628-1509

E-mail: aerc@culter.colorado.edu

Brings together 39 US research programs in universities and private, state and federal laboratories that conduct research, provide training and analyze policy at the ecosystem level of environmental science and natural resourcesmanagement. Although AERC is an association of professional scientists rather than environmental activists, its goals and interest complement those of conservation organizations.
John E Hobbie, President

4831 Astro-Chem Services

4102 2nd Avenue W	701-572-7355
PO Box 972	Fax: 701-774-3907
Williston, ND 58802-0972	

David Vander, President

4832 Astro-Pure Water Purifiers

1441 SW 1st Way	954-422-8966
Deerfield Beach, FL 33441	Fax: 954-422-8966

Manufacturers complete line of water treatment equipment, purifiers, De Calcifiers, R.O., V.V., iron filters, chemical feed equipment. Sizes for portable, point of use, central commercial and industrial. Manufacturers and privatelabels counter top units.

RL Stefl, President

4833 Athena Environmental Sciences

1450 S Rolling Road	410-455-6319
Baltimore, MD 21227	888-892-8408
	Fax: 410-455-1155

E-mail: athenaes@athenaenvironmental.com
http://www.athenaenvironmental.com
Designs and develops novel products that represent environmentally responsible and economically sound solutions to environmental problems. Contract services are provided to clients for product development. Products developed includethe company's own Spill Pill (TM), a proprietary cleaning agent for petroleum contamination on concrete and other building surfaces, Bilge Tech, Inc.'s Bilge Pill (TM), a cleaning agent for removing oil and dirt buildup in boat bilges, and expertisein biotechnology.
Sheldon Broedel PhD, CEO

4834 Atlantic Testing Laboratories

6431 US Highway 11	315-386-4578
PO Box 29	Fax: 315-386-1012
Canton, NY 13617	E-mail: atl-test@northweb.com
	http://www.atlantictest.com

Atlantic Testing Laboratories, Limited, is a technical services firm that provides engineering, construction and industrial clients; contractors; and governmental entities throughout the Northeastern US. Services provided by ATLinclude: Construction Materials Testing, Aggregate Testing, Geotechnical Engineering, Subsurface Investigation, Geoprobe Services, Surveying, UST/AST Tank Testing, Environmental Consulting and Geomembrane and landfill Services.
Spencer F Thew PE/LS, CEO
Marijean B Remington, President

4835 Atlas Weathering Services Group

45601 N 47th Avenue	623-465-7356
Phoenix, AZ 85087-7042	800-255-3738
	Fax: 623-465-9409

E-mail: info@atlaswsg.com
http://www.atlaswsg.com
Technical research firm specializing in environmental testing. Research results published in reports to clients and in archival journals.
Jack Martin, President

4836 Atmospheric & Environmental Research

131 Hartwell Avenue	781-761-2288
Lexington, MA 02421-3126	Fax: 781-761-2299
	E-mail: aer@aer.com
	http://www.aer.com

Firm providing research, consulting, and assessment on atmospheric chemistry, meteorology, climate, and air quality.
Nien Dak Sze, President
Cecilia Sze, CEO

4837 Axiom Laboratories
24 Tobey Road   860-242-6291
Bloomfield, CT 06002-3522   Fax: 860-286-0634
E-mail: mackeyw@worldnet.att.net
Environmental and materials analytical testing services.
William AG Macke, President

4838 B&P Laboratories
5635 Delridge Way SW   206-937-3644
Seattle, WA 98106-1445   Fax: 206-937-1348
E-mail: bplabor@qwest.net
Environmental testing and chemical laboratory water analyses (ICP);Mercury analyzer sulfates, fluorides, chlorides (DIONEX); cyandies (RONTES); storm waters; fats, oil & greas (FOG); Karl Fischer corrosion testing

Victor Broto, President

4839 BC Analytical
4100 Atlas Court   661-327-4911
Bakersfield, CA 93308   Fax: 661-327-1918
http://www.bclabs.com/
Lab capabilities include diversified sample matrices for drinkg waters, ground water monitoring and waste acceptance. Diversified analytical methods include general chemistry and field services including field analysis, sampling andcourier service.

4840 BC Laboratories
4100 Atlas Court   661-327-4911
Bakersfield, CA 93308-4510   Fax: 661-327-1918
Chemical analysis and environmental monitoring of hazardous waste. Research results published in project reports.

4841 BC Research
3650 Westbrook Mall   604-224-4331
Vancouver, BC V6S2L   Fax: 604-224-0540
Research results published in scientific journals and trade magazines.
Hugh Wynne-Edwards, President
Marion Webber, Director/Manager

4842 BCM Engineers
One Plymouth Meeting Mall   215-825-3800
Plymouth Meeting, PA 19462   Fax: 215-834-8236
Mark Z Hanlon, President/CEO

4843 BCR National Laboratory
500 William Pitt Way   412-826-3030
Pittsburgh, PA 15238-1331   Fax: 412-826-3353
Environmental research laboratory.

4844 Babcock & Wilcox Company
1562 Beeson Street   330-753-4511
Alliance, OH 44601   Fax: 330-823-0639

Robert E Howson, President

4845 Badger Laboratories & Engineering Company
1110 South Oneida Street   262-739-9213
Appleton, WI 54915   Fax: 262-739-5399
Stephen Taylor, President

4846 Baird Scientific
532 Oak Street   417-358-5567
Carthage, MO 64836-1814
Gary Baird, President/Owner

4847 Baker Environmental
420 Rouser Road   412-269-6000
Coraopolis, PA 15108-2722   Fax: 412-269-2534
Environmental engineering company.
Andrew P Paja, President

4848 Baker-Shiflett
5701 East Loop 820 South   817-478-8254
Fort Worth, TX 76119   Fax: 817-478-8874

Larry Gardner, Administrative Manager

4849 Barnebey & Sutcliffe Corporation
PO Box 2526   614-258-9501
Columbus, OH 43216   Fax: 614-258-3464

Amanda L Fisher, Marketing Coordinator

4850 Barton & Loguidice
290 Elwood Davis Road   315-457-5200
Box 3107   Fax: 315-451-0052
Syracuse, NY 13220 E-mail: B&L@BartonandLoguidice.com
http://www.BartonandLoguidice.com
Since 1961,Barton and Loguidice, P.C. has assisted a wide variety of clients in meeting their engineering requirements. As a full service, multi-disciplinary firm, B and L has the expertise and capacity to perform a wide array of highquality engineering services including bridge and highway, facilities, water, wastewater, environmental, and solid waste engineering. The firm continues to serve as an engineering services leader in Upstate New York.
Tom Aiston, President

4851 Baxter and Woodman
8678 Ridgefield Road   815-459-1260
Crystal Lake, IL 60012-2797   Fax: 847-948-2887
Environmental engineering firm.

4852 Bell Evaluation Laboratory
17300 Mercury Drive   281-488-3701
Houston, TX 77058-2732   Fax: 281-488-8543
E-mail: bellabs@bellabs.com
http://www.bellabs.com
Coating testing and evaluation firm.
Robert T Bell, Contact

4853 Beltran Associates
1133 East 35th Street   718-338-3311
Brooklyn, NY 11210   Fax: 718-253-9028

Mike Beltran, President

4854 Benchmark Analytics
4777 Saucon Creek Road   215-974-8100
Center Valley, PA 18034-9004   Fax: 610-974-8104

4855 Bendix Environmental Research
1950 Addison Street   415-861-8484
Suite 202   Fax: 510-845-8484
Berkeley, CA 94704   http://home.earthlink.net/~bendix/#A1
Selina Bendix, PhD, President

4856 Benson C Stone & Associates
PO Box 28658   858-485-7779
San Diego, CA 92198-0658   Fax: 858-487-0367
Environmental conservation research firm.
Benson C Stone, President

4857 Best Environmental, Safety and Industrial Products
10 Charlestowne Boulevard   504-467-9640
Suite C, PO Box 777   Fax: 504-469-7420
St. Rose, LA 70087-4034
Sales and distribution of environmental spill response products. Containment booms, skimmers, oil and chemical sorbents, custom spill response kits, response trailers and safety supplies.
Isidore J Prescia, President

4858 Bhate Environmental Associates
5115 Maryland Way   615-377-0725
Brentwood, TN 37027-7512   Fax: 615-661-4226
Environmental consulting.

4859 Bio-Chem Analysts
PO Box 3270   256-859-2161
Huntsville, AL 35810-0270   Fax: 256-859-9222

Environmental testing of water, wastewater, air, soil, and hazardous waste.

Vijay Thakore, President

4860 Bio-Science Research Institute
4813 Cheyenne Way 909-628-3007
Chino, CA 91710-5510 Fax: 909-590-8948
Independent environmental testing laboratory.

4861 Bio/West
1063 West 1400 North 435-752-4202
Logan, UT 84321 Fax: 435-752-0507
Paul Holden, Principal

4862 BioTest
2682 Welsh Road 215-671-8493
Philadelphia, PA 19152-1548
Environmental laboratory.
Karl L Gabriel, President

4863 BioTrol
10300 Valley View Road 952-942-8032
Suite 107 Fax: 952-942-8526
Eden Prairie, MN 55344

Michael S Gratz, President/CEO

4864 Bioengineering Resources
1650 E Emmaus Road 479-521-2745
Fayetteville, AR 72701 Fax: 479-521-2749
Environmental research.
JL Gaddy, President

4865 Biological Research Associates
3910 N US Highway 301 813-664-4500
Suite 180 Fax: 813-664-0440
Tampa, FL 33619-1283
Environmental research and consulting firm.

4866 Biomarine
16 E Main Street 978-281-0222
PO Box 1153 Fax: 978-283-6296
Gloucester, MA 01930 E-mail: info@biomarinelab.com
http://www.biomarinelab.com
Provides water and seafood analysis and consulting for the public, private companies and government.

John Marletta, Laboratory Director

4867 Bionetics Corporation Analytical Laboratories
20 Research Drive 757-865-0880
Hampton, VA 23666-1396 Fax: 757-865-8014
Joseph A Stern, President

4868 Bioscience
1550 Valley Center Parkway 610-974-9693
Suite 140 800-627-3069
Bethlehem, PA 18017-2263 Fax: 610-691-2170
E-mail: bioscience@bioscienceinc.com
http://www.bioscienceinc.com
Specialized microbes for wastewater and hazardous waste, biological treatment. BOD and COD monitoring instruments and test kits for water and wastewater analysis.

Thomas G Zitrides, President
Richard Bleam, Director of Technical Svc.

4869 Biospherics
12051 Indian Creek Court 301-419-3900
Beltsville, MD 20705 Fax: 301-210-4909

Gilbert V Levin, PhD, President

4870 Bison Engineering
1400 11th Avenue 406-442-5768
Helena, MT 59601-5744 Fax: 406-449-6653
E-mail: hrobbins@bison-eng.com
http://www.bison-eng.com
Bison Engineering, Inc. is a full-service environmental consulting firm with extensive experience in air quality permitting, air emissions testing and ambient air monitoring. Our knowledge of the water quality regulatory environmentis substantial and expanding. We also provide an array of other environmental services through our staff and associates to complement our air and water quality expertise.

Hal Robbins, President

4871 Black Rock Test Lab
5 Eastgate Plaza 304-296-8347
Morgantown, WV 26501

4872 BlazeTech Corporation
24 Thorndike Street 617-661-0700
Cambridge, MA 02141 Fax: 617-661-9242
E-mail: office@blazetech.com
http://www.blazetech.com
An engineering consulting firm specializing in fire, explosion, environmental safety and homeland defense. They have developed specialized software for the chemical, petroleum, aerospace and power industries: ADORA, BLAZETANK andothers.
Albert Moussa, President

4873 Bollyky Associates
31 Strawberry Hill Avenue 203-967-4223
Stamford, CT 06902 Fax: 203-967-4845
E-mail: ljbbai@bai-ozone.com
http://www.bai-ozone.com
Engineering firm specializing in Ozone technology, water and wastewater treatment, treatability studies.

L Joseph Bollyky, President
Thomas Kleiber, Office Manager

4874 Bolt Technology Corporation
4 Duke Place 203-853-0700
Norwalk, CT 06854 Fax: 203-854-9601

Raymond M Soto, President

4875 Braun Intertec Corporation
11001 Hampshire Avenue South 952-995-2000
Minneapolis, MN 55438 800-279-6100
Fax: 952-995-2020
E-mail: info@braunintertec.com
http://www.braunintertec.com
Full-service engineering, environmental and infrastructure consulting and testing organization.

George D Kluempke, President
Scott C Barnard, CFO

4876 Braun Intertec Northwest
PO Box 17126 503-289-1778
Portland, OR 97217
Testing and quality control monitoring laboratory specializing in construction inspections, materials testing, soils engineering and geological services.
John Ulmer, Public Relations
Jack Braun, Contact

4877 Bregman and Company
5272 River Road 301-652-4818
Suite 550 Fax: 301-652-4819
Bethesda, MD 20814E-mail: info@bregmanandcompany.com
http://www.bregmanandcompany.com
Environmental consulting firm.
Robert Edell, President/CEO

4878 Briggs Associates
100 Weymouth 781-871-6040
Rockland, MA 02370-1236 Fax: 781-871-7982

Environmental engineering and testing facility.

4879 Brighton Analytical
2105 Pless Drive
Brighton, MI 48116-1238
810-229-7575
Fax: 810-229-8650
J Shawn Letwin, Laboratory Director

4880 Bromwell & Carrier
PO Box 5467
Lakeland, FL 33807-5467
863-607-4292
Fax: 863-644-5920
Environmental and civil engineering, geotechnical processes, and chemical research and development firm.

4881 Brooks Companies
9 Isaac Street
Norwalk, CT 06850
203-853-9792
Fax: 203-853-0273
Margaret Y Brooks PhD, President

4882 Brooks Laboratories
9 Isaac Street
Norwalk, CT 06850
203-853-9792
800-843-1631
Fax: 203-853-0273
E-mail: brookslabs@aol.com
http://www.brookslabs.com
Consulting and testing air, soil and water for contamination. Accident and disease prevention.

4883 Brotcke Engineering Company
140 E Division Road, Suite A-6
PO Box 1168
Fenton, MO 63026
426-483-2060
Paul Brotcke, President

4884 Buchart-Horn
445 W Philadelphia Street
York, PA 17404
717-852-1400
800-274-2224
E-mail: pace40@aol.com
This company provides environmental engineering, consulting, civil engineering, facility design and planning as well as laboratory and testing services.
Dennis Miner, VP Business Development
Gene Schenck, Public Relations Director

4885 Buffalo Testing Laboratories
902 Kenmore Avenue
Buffalo, NY 14216
716-873-2302
Fax: 716-873-9914
Edward J Kris, President

4886 Burt Hill Kosar Rittelmann Associates
40 Morgan Center
Butler, PA 16001
724-285-4761
Fax: 724-285-6815
P Richard Rittelmann, FAIA, Sr. VP

4887 Business Health Environmental Lab
33 E 7th Street
300 Doctors Building
Covington, KY 41011
859-431-6224
Fax: 859-431-6228
Dan Moos, President

4888 C L Technology
280 N Smith Avenue
Corona, CA 91720
909-734-9600
Fax: 909-734-2803

4889 CDS Laboratories
RR 2
Box 234
Loganton, PA 17747
570-725-3411

4890 CENSOL
582 Hawthorne
L9T-4N8
Milton, ON
416-219-6950
Fax: 905-878-8775
E-mail: info2003@censol.ca
http://www.censol.ca

Environmental consulting and testing firm. Research results published in technical association papers.

4891 CET Environmental Services
7032 South Revere Parkway
Englewood, CO 80112
720-875-9115
Fax: 720-875-9114
http://www.cetenvironmental.com
Provides env. consulting, engineering, remediation & construction servies. The Group has three primary segments: Industrial services, environmental remediation & gvmt. programs. The industrial services include water & wastewatertreatment, facility cleaning, operation, construction & emergency response. The environmental rem. services include environmental assessment & remedial investigation studies. The government programs include emergency response, remediation, &water-related services.
Steven H Davis, President & CEO
Dale W Bleck, CFO

4892 CONSAD Research Corporation
121 N Highland Avenue
Pittsburgh, PA 15206-3050
412-363-5500
Fax: 412-363-5509
E-mail: info@consad.com
http://www.consad.com
Social science research and consulting firm. Research results published in journals and project and government reports.
Wilbur A Steger, President
Frederick H Rueter, VP

4893 CPAC
2364 Leicester Road
Leicester, NY 14481
585-382-3223
Fax: 585-382-3031
E-mail: cpacinfo@cpac.com
http://www.cpac-fuller.com
CPAC, Inc. manages holdings in two industries: Cleaning and Personal Care and Imaging. The Fuller Brands segment develops, manufactures, and markets over 2799 branded and private lavel products for commercial cleaning, householdcleaning, and personal care. CPAC Imaging manufactures, packages, and distrubtes branded and private label chemicals for photographic, health care, and graphic arts markets as well as associated imaging equipment and silver refining services.

Thomas N Hendrickson, President

4894 Cascadia Research
218 1/2 W 4th Avenue
Olympia, WA 98501
206-943-7325
http://www.cascadiaresearch.org
Nonprofit tax-exempt scientific and educational organization founded to conduct research needed to manage and protect threatened marine mammals.

4895 Cedar Grove Environmental Laboratories
100 Gallagherville Road
Downingtown, PA 19335-3640
610-269-6977
Fax: 610-269-6965
http://www.cgelab.com
Environmental analysis firm serving agriculture and industry.

4896 Ceimic Corporation
10 Dean Knauss Drive
Narragansett, RI 02882
401-782-8900
Fax: 401-782-8905
Environmental testing for water and soil.
Margaret Marple, Marketing

4897 Center for Technology, Policy & Industrial Development
1 Amherst Street
Cambridge, MA 02139-4301
617-253-8973
Fax: 617-253-7140
http://web.mit.edu/ctpid
Environmental research.
Fred Moavenzadeh, Director/Manager

4898 Central Virginia Laboratories and Consultants
3109 Odd Fellows Road
PO Box 10938
Lynchburg, VA 24506
804-847-2852
Fax: 804-847-2830
Adrian K Mood, President

4899 Centre Analytical Laboratories

3048 Research Drive 814-231-8032
State College, PA 16801-2782 Fax: 814-231-1253
  http://www.centrelab.com
Environmental and chemical testing research laboratory.
M Michael Arjmand, President

4900 Century Testing Laboratories

825 NE Multnomah Street
Suite 425
Portland, OR 97232 800-231-6482
  Fax: 503-231-6482
Tests drinking water and provides waste analysis services.
Sheena Fischer, Director

4901 Century West Engineering Corporation

1444 NW College Way 541-388-3500
Bend, OR 97709

4902 Chas. T Main: Environamental Division

Prudential Center 617-262-3200
Boston, MA 02199 Fax: 781-401-2575
Environmental consulting firm specializing in site assessments, surveys, and tests and analysis.

4903 Chemical Resource Processing

2525 Battleground Road 281-930-2525
PO Box 1914 Fax: 281-930-2535
Deer Park, TX 77536-1914
Environmental consulting and chemical processing firm.

4904 Chemical Waste Disposal Corporation

4214 19th Avenue 718-274-3339
Astoria, NY 11105-1018 Fax: 718-726-7917
Environmental waste disposal and consulting company.

4905 Chemir Analytical Services

2672 Metro Boulevard 314-291-6620
Maryland Heights, MO 63043 800-659-7659
  Fax: 314-291-6630
  E-mail: info@chemir.com
  http://www.chemir.com
Provides a wide range of chemical analysis and chemical testing services. Experienced at solving difficult problems including product failure analysis, materials identification, plastic testing or reverse engineering. An independenttesting lab with scientists that can provide litigation support such as expert witness testimony in intellectual property or products liability cases.

Dr Shri Thanedar, President

4906 Chesner Engineering

2171 Jericho Turnpike 516-431-4031
Commack, NY 11725 E-mail: mail@chesnerengineering.com
  http://www.chesnerengineering.com
Civil, environmental, and waste management firm that provides professional services to industry and government. Specializes in the areas of waste and by-product material recycling and stabliization, marine and dredge environmentalmanagement, risk assessment and environmental modeling, environmental database program development, remedial site investigations and cleanup management, and water and wastewater treatment.
Warren Chesner, President

4907 Chihuahuan Desert Research Institute

PO Box 1334 915-837-8370
Alpine, TX 79831
Conducts research on the Chihuahuan Desert.

4908 Chopra-Lee

1815 Love Road 716-773-7625
Grand Island, NY 14072-2207 800-508-5419
  Fax: 716-773-7624
  E-mail: pburger@mailexcite.com
Environmental services, assessments, sampling, and data collection, testing, and analysis firm. Industrial hygiene/indoor air quality; materials testing/failure analysis and hazardous waste analysis.

4909 ChromatoChem

2837 Fort Missoula Road 406-728-5897
Missoula, MT 59801 Fax: 406-728-5924
Biotechnology firm. Research results published in proposals to the Environmental Protection Agency.

4910 Chyun Associates

267 Wall Street 609-924-5151
Princeton, NJ 08540

4911 Clark Engineering Corporation

621 Lilac Drive N 763-545-9196
Minneapolis, MN 55422 E-mail: mheller@clark-eng.com
  http://www.clark-eng.com

Stephen E Clark, President

4912 Clean Air Engineering

500 W Wood Street 847-991-3300
Palatine, IL 60067-4975 800-627-0033
  Fax: 847-991-3385
  E-mail: contact@cleanair.com
  http://www.cleanair.com
Environmental consulting and permitting, process engineering, equipment rental and manufacture, measurement and analytical services.
William I Walker, President
Frank Kilvinger, VP

4913 Clean Harbors

1200 Crown Colony Drive 781-849-1800
PO Box 9137 Fax: 781-848-2141
Quincy, MA 02169-0938
Environmental consulting firm specializing in soil analysis, site assessments, and water sample testing.

4914 Clean Water Systems

2322 Marina Drive 541-882-9993
Klamath Falls, OR 97601 Fax: 541-882-9994
  E-mail: cws@internetcds.com
  http://www.cleanwatersysintl.com
Environmental research firm. Designs, develops and manufactures ultra-violet electronic water purification units and systems and electronic measuring systems. The Company has developed lines of proprietary electronic monitoring andcontrol systems and electronic ballast.
Charles Romary, President

4915 Coast to Coast Analytical Services

2400 Cumberland Drive 800-688-6522
Valparaiso, IN 46383-2502 Fax: 219-462-2953
Environmental engineering research and consulting firm.
A Sami El-Naggar, President

4916 Coastal Resources

2988 Environment Soloman Road 410-956-9000
Annapolis, MD 21037 Fax: 410-956-0566
Environmental impact assessments and nontidal wetlands identification expert testimony.

4917 Colorado Analytical

240 S Main Avenue 303-659-2313
PO Box 507 Fax: 303-659-2315
Brighton, CO 80601-0507
Agricultural consulting and testing laboratory. Research results published in project reports and test summaries.

4918 Colorado Research Associates

3380 Mitchell Lane 303-415-9701
Boulder, CO 80301-5410 Fax: 303-415-9702
  http://www.co-ra.com
Environmental research.
David C Fritts, VP

4919 Columbus Instruments International

950 N Hague Avenue 614-276-0861
Columbus, OH 43204-2121 Fax: 614-276-0529

E-mail: sales@colinst.com
http://www.colinst.com

Manufacturer of biomedical and environmental research equipment which includes respirometers and gas analysis monitoring systems.

4920 Columbus Water and Chemical Testing Laboratory
4628 Indianola Avenue   614-262-4372
Columbus, OH 43214

4921 Commonwealth Laboratory
2209 E Broad Street   804-648-8358
Richmond, VA 23223

4922 Commonwealth Technology
2526 Regency Road   859-276-3091
Lexington, KY 40503   800-467-3091
                             Fax: 859-276-4374
                             E-mail: fyi@ctienv.com
                             http://www.ctienv.com
Environmental engineering and analysis firm.

4923 CompuChem Environmental Corporation
501 Madison Avenue   919-379-4000
Cary, NC 27513   800-833-5097
                             Fax: 919-379-4050
                             http://www.compuchemlabs.com
Environmental testing laboratory.
Gerard Verkerk, Contact

4924 Conjun Laboratories
9283 Highway 15   606-633-8027
Isom, KY 41824

4925 Conservation Foundation
1919 M Street NW   202-912-1000
Suite 600   Fax: 202-912-0765
Washington, DC 20036   http://www.conservation.org
Conducts research and develops knowledge and techniques to improve the quality of the environment.

4926 Consumer Testing Laboratories
430 S Congress Avenue   561-330-3081
Suite 1B   Fax: 561-330-7712
Delray Beach, FL 33445
Research in textiles, safety wear.
Stewart Satter, President

4927 Container Testing Laboratory
607 Fayette Avenue   914-381-2600
Mamaroneck, NY 10543   Fax: 914-381-0143
                             E-mail: contestlab@hotmail.com
Independent third party testing laboratory.
Anton Cotaj, Laboratory Director/Manager

4928 Conti Testing Laboratories
3190 Industrial Boulevard   412-854-3751
Bethel Park, PA 15102   Fax: 412-854-0373

Patricia A Otroba, President

4929 Continental Shelf Associates
759 Parkway Street   561-746-7946
Jupiter, FL 33477-4567   Fax: 561-747-2954
                             E-mail: csa@gate.net

Environmental consulting firm.
Dr David A Gettleson, President
Robert C Stevens Jr, CEO

4930 Continental Systems
7870 Deering Avenue   818-340-3217
Canoga Park, CA 91304-5005   Fax: 818-340-2405
Environmental laboratory specializing in soil and water analysis.
Janis Butler, President

4931 Controlled Environment Corporation
29 Sanford Drive   207-854-9126
Gorham, ME 04038-2647   Fax: 207-854-4357
Firm providing research, design, and development services relating to clean rooms and contamination control.
Matthew F Pec, President

4932 Controls for Environmental Pollution
1925 Rosina Street   505-982-9841
Box 5351   800-545-2188
Santa Fe, NM 87502   Fax: 505-982-9289

James J Mueller, President

4933 Converse Consultants
3 Century Drive   973-605-5200
Parsippany, NJ 07054-4610   Fax: 973-605-8145
                             E-mail: convers@mailidt.net
Applied and product research in environmental studies.
R Brian Ellwood, VP

4934 Copper State Analytical Lab
710 East Evans Boulevard   520-388-4922
Tucson, AZ 85713   Fax: 520-884-5133
                             E-mail: csalinc@aol.com
                             http://www.csalinc.com
Hazardous waste characterization, organic and inorganic waste oil characterization, waste water analysis, potable water analysis, microbiology, general waters and soil chemistry.

DA Shah, President

4935 Corning
1 Riverfront Plaza   607-974-9000
Corning, NY 14831   Fax: 607-974-8091
                             http://www.corning.com

4936 Corporate Environmental Advisors
127 Hartwell Street   508-835-8822
West Boylston, MA 01583   Fax: 508-835-8812
Environmental engineering and consulting firm.

4937 Corrosion Testing Laboratories
60 Blue Hen Drive   302-454-8200
Newark, DE 19711   Fax: 302-454-8204
                             E-mail: ctl@corrosionlab.com
                             http://www.corrosionlab.com
Corrosion testing laboratory. Research results published in technical journals and conference proceedings.
Richard A Corbett, President

4938 Coshocton Environmental Testing Service
709 Main Street   740-622-3328
Coshocton, OH 43812   Fax: 740-622-3368
Environmental testing service.

4939 Crane Environmental
2600 Eisenhower Avenue   610-631-7700
Trooper, PA 19043   Fax: 610-631-6800
                             http://www.craneenv.com
Industrial water treatment equipment.

4940 Crosby & Overton
1610 W 17th Street   562-432-5445
Long Beach, CA 90813   800-827-6729
                             Fax: 562-436-7540
                             http://www.crosbyoverton.com
Fully permitted RCRA Part B TSD facility located in Southern California. The Facility can process both bulk and drummed waste, including lab-packs. Crosby & Overton can process a wide variety of D,F,K,P and U listed RCRA waste.
Bob Ritter, Sales Manager
Michelle Dalot, Sales

4941 Curtis & Tompkins
2323 5th Street   510-486-0900
Berkeley, CA 94710   Fax: 510-486-0532

C Bruce Godfrey, President

4942 Cutter Environment
37 Broadway 781-678-8702
Arlington, MA 02474 800-964-8702
Fax: 781-648-1950
E-mail: environment@cutter.com
http://www.cutter.com/environment
Environmental research.
Karen Coburn, President

4943 Cyberchron Corporation
US Route 9 845-265-3700
PO Box 160 Fax: 845-265-3752
Cold Spring, NY 10516-0160
Computer manufacturing firm specializing in the development of computers designed to withstand extreme travel, environmental, and work conditions.

4944 Cyrus Rice Consulting Group
200 High Tower Road 412-788-2468
Suite 302 Fax: 412-788-1797
Pittsburgh, PA 15205
Al Owens, VP

4945 DE3
18234 S Miles Road 216-663-1500
Cleveland, OH 44128 Fax: 216-663-1501
Environmental engineering research and testing firm.
Harold N Danto, President

4946 DPRA
200 Research Drive 785-539-3565
Manhattan, KS 66503 Fax: 785-539-5353
http://www.dpra.com
Environmental, economic, regulatory and technical research company. Research results published by information services.
Richard Seltzer, President

4947 DW Ryckman and Associates
2208 Welsh Industrial Court 314-569-0991
St. Louis, MO 63146-4222 Fax: 314-432-2845

DW Ryckman, Chairman

4948 Daily Analytical Laboratories
1621 West Candletree Drive 309-692-5252
Peoria, IL 61614 Fax: 309-692-0488
Kurt Stepping, Chief Chemist

4949 Dan Raviv Associates
57 E Willow Street 973-564-6006
Millburn, NJ 07041-1416 Fax: 973-564-6442
E-mail: ddrai@ix.netcom.com
http://www.danraviv.com
Environmental consulting firm specializing in environmental impact studies, waste management, site assessment and litigation support.
Dan D Raviv PhD, President
John J Trela PhD, Director/Manager

4950 Danaher Corporation
2099 Pennsylvania Avenue NW 202-828-0850
Washington, DC 20006 Fax: 202-828-0860
http://www.danaher.com
Development of process and environmental controls, tools and components.

4951 Data Testing
3434 Country Club Avenue 479-649-8378
Fort Smith, AR 72903 Fax: 479-649-8486
Patrick J Mickle, PE, Chief Engineer

4952 Datachem Laboratories
960 W Levoy Drive 801-266-7700
Salt Lake City, UT 84123-2502 Fax: 801-268-9992

Analytical laboratory provides lab analysis of soil, water, air and asbestos samples.
Brent Stephens, VP/Laboratory Director

4953 Davis Research
134 Hobart Road 662-332-1943
Avon, MS 38723 Fax: 662-332-0081
Agricultural, food and environmental testing and research firm.
R G Davis, President
Diane Barnham, Director/Manager

4954 Dellavalle Laboratory
1910 W McKinley 559-233-6129
Suite 110 800-228-9896
Fresno, CA 93728 Fax: 559-268-8174
http://www.dellavallelab.com
Agricultural laboatory analyzes plant, soil, fertilizer, manure and water (ag, domestic, wastewater). Certified Professional Soil Scientits/ Agronomists/Crop Advisors and others provide consultation on nutrient and fertilizermanagement, crop feasibility, regulatory compliance, troubleshooting and related areas.
Nat B Dellavalle, President

4955 Dobbin Milus International
304 Springwood Court NE 703-255-1170
Suite 100 Fax: 703-255-0754
Vienna, VA 22180-3579
Coastal land use planning, zone managements and environmental assessment firm.

4956 Douglass Environmental Services
8649 Bash Street 317-595-9108
Indianapolis, IN 46256-1202 Fax: 317-822-8362
Environmental engineering and testing firm.

4957 Dowl Engineers
4040 B Street 907-562-2000
Anchorage, AK 99503 Fax: 907-563-3953
http://www.dowl.com
Serves clients' needs in the areas of environmental planning, National Environmental Policy Act (NEPA) documentation, permitting, engineering and public involvement. Environmental studies and analyses include wetland delineation andfunction and values assessment, vegetation mapping, GIS mapping and analysis, environmental site assessment, air and noise impact analysis. Section 106 consultation, hydrology studies, and secondary and cumulative impact analysis.

Linda Hulleen, Marketing Manager
Kristen J Hansen, Service Environ. Spec.

4958 Duke Solutions
1 Winthrop Square 617-482-8228
Boston, MA 02110 Fax: 617-482-3784

4959 Dynamac Corporation
22575 Research Boulevard 301-417-9800
Suite 500 Fax: 301-417-9801
Rockville, MD 20850 E-mail: ibaumel@dynamac.com
http://www.dynamac.com
A scientific research, engineering, and information technology company, conducting state of the art field and laboratory research, and providing scientific and technical support to federal and state environmental programs.
William J Silver, President
Diana Mac Arthur, CEO

4960 E&S Environmental Chemistry
PO Box 609 541-758-5777
Corvallis, OR 97339-0609 Fax: 541-758-4413
Environmental research.
Tim Sullivan, President

4961 EA Engineering Science and Technology
11019 McCormick Road 410-584-7000
Hunt Valley, MD 21031-1412 Fax: 410-771-1625
E-mail: ea@eaest.com
http://www.eaest.com

Loren Jensen, President
James Gift, R&D

4962 EADS Group
1126 8th Avenue   814-944-5035
Altoona, PA 16602   800-626-0904
     Fax: 814-944-4862

Charles F Welker, President

4963 EAI Corporation
1308 Continental Drive   410-676-1449
Suite J   Fax: 410-671-7241
Abingdon, MD 21009-2334   E-mail: info@eaicorp.com
     http://www.eaicorp.com
Environmental engineering and scientific firm. Research results published in private reports to clients.
Charles Speranzella, President
Tom Albro, VP

4964 EETCO
17117 W 9 Mile Road   248-569-8604
Suite 537   Fax: 248-594-1739
Southfield, MI 48075-4502
Environmental engineering firm.

4965 EJS Consulting
9594 First Avenue NE
Seattle, WA 98115
Environmental consulting service specializing in radiological monitoring and analysis, laboratory services, health physics services, radiation and hazardous waste analysis and monitoring, regulatory compliance, water treatment andmonitoring, sewage treatment, and industrial waste outfall monitoring.

4966 EMCO Testing & Engineering
PO Box 644   860-886-0697
Norwich, CT 06360   Fax: 860-886-0697
     E-mail: emco@99main.com
Water treatment and environmental research.
Dr Ernie Cohen, President/R&D

4967 EMCON Alaska
201 E 56th Avenue   907-562-3452
Suite 300   Fax: 907-563-2814
Anchorage, AK 99518-1283
Environmental engineering firm.

4968 EMMES Corporation
401 North Washington Street   301-251-1161
Suite 700   Fax: 301-251-1355
Rockville, MD 20850   E-mail: info@emmes.com
     http://www.emmes.com
Firm providing medical data management and statistical support services. Research results published in scientific literature.
Donald M Stablein, PhD, President

4969 EMS Laboratories
117 W Bellevue Drive   626-568-4065
Pasadena, CA 91105   800-675-5777
     Fax: 626-796-5282
     E-mail: contact@emslabs.com
     http://www.emslabs.com
Environmental testing lab services. Asbestos/Lead/Mold testing, I.H. testing. Fully accreditted AIHA lab
Bernadine Kolk, President
Anthony Kolk, CEO

4970 EN-CAS Analytical Laboratories
2359 Farrington Point Drive   336-785-3252
Winston-Salem, NC 27107-2453   Fax: 336-785-3262
     http://www.en-cas.com
Chemical and environmental testing and analysis company.

4971 ENSR Consulting and Engineering
1601 Prospect Parkway   970-493-8878
Fort Collins, CO 80525   Fax: 970-493-0213
Environmental engineering and consulting firm.
Will Wright, Director/Manager

4972 ENTRIX
5252 Westchester Street   713-666-6223
Suite 250   Fax: 713-666-5227
Houston, TX 77005-4102   http://www.entrix.com
Environmental engineering firm.

4973 ENVIRO Tech Services
361 N Ohio Street   785-827-1682
Salina, KS 67401-2433   Fax: 785-827-8765
Jack Esidck, President

4974 ENVIRON Corporation
4350 N Fairfax Drive   703-516-2300
Arlington, VA 22203-1695   Fax: 703-516-2345
Health and environmental sciences consultants.
Grover Wrenn, President
Cindy Holloman, Contact

4975 ERC Research and Development Company
PO Box 10107   703-246-0200
Fairfax, VA 22030-8007   Fax: 703-246-0409
Environmental and technical services firm.

4976 ESA Laboratories
Laboratories, 22 Alpha Road   978-250-7150
Chelmsford, MA 01824   Fax: 978-250-7171
Environmental and biological testing laboratory.

4977 ESS Group
888 Worcester Street   781-431-0500
Suite 240   Fax: 781-431-7434
Wellesley, MA 02482   E-mail: cnatale@essgroup.com
     http://www.essgroup.com
A multi-disciplinary team of engineers, scientits, and regulatory consultants. Staff are experienced and certified in varied disciplines including ecological services, water resources management, hydrogeology, site investigations,risk assessment and remedial action design, air quality consulting, industrial hygiene, and civil, coastal and wastewater engineering.

4978 ETS
1401 Municipal Road NW   540-265-0004
Roanoke, VA 24012-1309   Fax: 540-265-0131
     E-mail: jmck@etsi-inc.com
     http://www.etsi-inc.com
Research results published by The Environmental Protection Agency, Department of Energy, and Air Pollution Control Association in papers and government reports. ETS is a full-service environmental consulting and training firmspecializing in air emissions control, measurement, engineering and consulting services.
John McKenna, CEO

4979 ETTI Engineers and Consultants
1717 E Erwin Street   903-595-4421
Tyler, TX 75702-6346   Fax: 909-595-6113
     E-mail: ettinc@ettinc.com
     http://www.ettinc.com
Engineering research and testing laboratory specializing in construction services and environmental needs. Research results published in project reports and test summaries.

4980 Eaglebrook Environmental Laboratories
1152 Junction Avenue   219-322-0450
Schererville, IN 46375-1305   Fax: 219-322-0440
Environmental testing service.

4981 Earth Dimensions
1091 Jamison Road   716-655-1717
Elma, NY 14059   Fax: 716-655-2915
Geotechnical soil investigations and wetland delineations.

Don Owens, President
Brian Bartron, Geologist/Drilling Manager

4982 Earth Regeneration Society

1442A Walnut Street 510-849-4155
Number 57 Fax: 510-849-0183
Berkeley, CA 94709 E-mail: csiri@igc.apc.org
http://www.imaja.com

The Earth Regeneration Society does research and education on climate change, ozone, and pollution, and calls for full employment and full social support based on surival programs and national and international networking.
Alden Bryant, President

4983 Earth Tech

4135 Technology Parkway 920-458-8711
PO Box 1067 Fax: 920-458-0537
Sheboygan, WI 53083-1825 E-mail: earthtech.com
http://www.earthtech.com

Specializes in the planning, design, and construction management and observation of environmental and infrastructure projects including water/wastewater; solid, hazardous and process waste facilities; environmental restoration;transportation; and architecture. The company's staff size, multiple office locations and comprehensive mix of expertise and experience combined iwht an in-depth knowledge of technical and regulatory issues, provide our clients with a valuableresource for solutions.
Diane Creel, President

4984 Earth Technology Corporation

300 Oceangate 562-951-2000
Suite 700 Fax: 562-951-2100
Long Beach, CA 90802 http://www.earthtech.com/

Alan P. Krusi, President

4985 EarthNet Laboratories

414 W California Avenue 318-255-0060
Ruston, LA 71270-4904 Fax: 318-251-5625

Environmental testing and research consulting laboratories. Research results published in test summaries and project reports.
Rollie Thompson, Contact

4986 East Texas Testing Laboratory

1717 E Erwin Street 903-595-4421
Tyler, TX 75702-6346 Fax: 903-595-6113

Engineering research and testing laboratory specializing in construction services and environmental needs. Research results published in project reports and test summaries.

4987 Eastern Laboratory Service Associates

517 North George Street 717-846-4953
York, PA 17404 Fax: 717-846-4986

KG Rao, PhD, President

4988 Eastern Technical Associates

PO Box 1009 919-878-3188
Garner, NC 27529 Fax: 919-872-5199
E-mail: tomrose@eta-is-opacity.com
http://www.eta-is-opacity.com

Environmental consulting firm. Research results published in government reports.

Thomas H Rose, President

4989 Eberline Services

4501 Indian School Road NE 505-262-2694
Suite 105 Fax: 505-262-2698
Albuquerque, NM 8E-mail: marketing@eberlineservices.com
http://www.eberlineservices.com

Environmental and radiological services provider specializing in ES&H management; hazardous and radioactive waste management; radiological characterization and analysis.
Nelson R Johnson, President

4990 Eco-Analysts

105 E 2nd Street 208-882-2588
Suite 1 Fax: 208-883-4288
Moscow, ID 83843 E-mail: eco@ecoanalysts.com
http://http://www.ecoanalysts.com/contact.html

Specializes in Aquatic Taxonomy and Bioassessment. An independent environmental consulting firm located in Moscow, Idaho. Experienced in the identification of freshwater organisms; macroinvertebrates, periphyton, plankton, and fish.Offer aquatic bioassessment and biological monitoring services.

Gary Lester

4991 EcoTest Laboratories

377 Sheffield Avenue 631-422-5777
North Babylon, NY 11703 Fax: 631-422-5770

Environmental testing laboratory.

4992 Ecological Engineering Associates

13 Marconi Lane 508-748-3224
Marion, MA 02738 Fax: 508-748-9740

Bruce Strong, Operations Manager

4993 Ecology and Environment

Buffalo Corporate Center 716-684-8060
368 Pleasant View Drive Fax: 716-684-0844
Lancaster, NY 14086-1397 E-mail: nsiekmann@ene.com
http://www.ene.com

E and E is a multidisciplinary environmental science and engineering company with more than 25 offices in the US and offices and partners in more than 35 countries. We are a world leader in providing environmental consulting servicesand litigation support.

Ronald J Skare, Sr. VP Marketing/Sales

4994 Economists

1200 New Hampshire Avenue NW 202-223-4700
Suite 400 Fax: 202-296-7138
Washington, DC 20036-6806 http://www.ei.com

Firm providing economic analysis and public policy evaluation, with emphasis on private antitrust litigation, communications regulation, and the Environment.
Bruce M Owen, President

4995 Ecotope

4056 9th Avenue NE 206-322-3753
Seattle, WA 98105 Fax: 206-325-7270
http://www.ecotope.com

Energy efficiency research, architecture, and engineering.
David Baylon, President

4996 Eder Associates

480 Forest Avenue 516-671-8440
Locust Valley, NY 11560-2118 Fax: 516-671-3349

Environmental engineering and consulting firm.
Leonard J Eden, President

4997 Eichleay Corporation of Illinois

11919 S Avenue O 773-731-7010
Chicago, IL 60617

Environmental consulting firm.

4998 El Dorado Engineering

2964 W 4700 South 801-966-8288
Suite 109 Fax: 801-966-8499
Salt Lake City, UT 84118-2558 E-mail: eldorado50@aol.com

Environmental applications.
Ralph W Haye, President

4999 Electron Microprobe Laboratory

Northern Arizona University 928-523-0388
NAU Box 5640 Fax: 928-523-7500
Flagstaff, AZ 86011 http://www.jan.ucc.edu

James R Kessler, Director

5000 Elm Research Institute

11 Kit Street 603-358-6198
Keene, NH 03431 800-367-3567
Fax: 603-358-6305

E-mail: libertyelm@webryders.com
http://www.libertyelm.com

A nonprofit organization dedicated to the restoration and preservation of the American Elm. Provides disease-resistant American Liberty Elms to municipalities, colleges and volunteer nonprofit groups for public planting. Distribution of Elm Fungicide for treatment of Dutch Elm Disease.
John Hansel, Executive Director
Yvonne Spalthoff, Assistant Director

5001 Emcon Baker-Shiflett

5701 E Loop S　　　　　　　817-478-8254
Fort Worth, TX 76119　　　　Fax: 817-478-8874

Consulting engineers providing research services to the construction industry.
Larry Gardner, Contact

5002 Energetics

7164 Columbia Gateway Drive　410-290-0370
Columbia, MD 21046-2979　　Fax: 410-290-0377

Environmental engineering firm.

5003 Energy & Environmental Technology

110 Daventry Lane　　　　　502-458-0600
Louisville, KY 40223

Environmental research.
Shirish Phulgaonkar, President

5004 Energy Conversion Devices

2956 Waterview Drive　　　　248-293-0440
Rochester Hills, MI 48309　　800-528-0617
　　　　　　　　　　　　　Fax: 248-844-1214
　　　　　　　　　　　　　http://www.ovonic.com

Maintained a strong core competence in materials research and advanced product development throughout its forty plus year history. The company protects the results of these efforts through an extensive patent collection.

Stanford Ovshinsky, President/CTO
Iris Ovshinsky, Vice President

5005 Energy Laboratories

2393 Salt Creek Highway　　307-235-0515
PO Box 3258　　　　　　　　888-235-0515
Casper, WY 82602-3258　　　Fax: 307-234-1639
　　　　　　　　　　E-mail: casper@energylab.com
　　　　　　　　　　http://www.energylab.com

Environmental data collection, testing and analysis firm.

5006 Energy and Environmental Analysis

1655 Fort Myer Drive　　　　703-528-1900
Suite 600　　　　　　　　　Fax: 703-528-5106
Arlington, VA 22209-3117　　http://www.eea-imc.com

Consulting firm offering technical, analytical, and managerial services in the energy/environmental field.
Michael O Lerner, President

5007 Energy and Environmental Engineering

35 Medford Street　　　　　617-666-5500
Somerville, MA 02143　　　　Fax: 617-666-5802

James H Porter, PhD, President

5008 Engineering & Environmental Management Group

11251 Roger Bacon Drive　　703-318-4522
Reston, VA 20910　　　　　Fax: 703-318-4729

Environmental applications.

5009 Engineering Analysis

715 Arcadia Circle　　　　　256-533-9391
Huntsville, AL 35801-5941　　Fax: 256-533-9325
　　　　　　　　　　E-mail: eai@mindspring.com
　　　　　　　　　　http://eai.home.mindspring.com

Environmental and safety research and analysis organization. Research results published in client and technical reports and professional journals.
Frank B Tato, President

5010 Engineering Research

1400 Kings Drive　　　　　501-575-4108
Fayetteville, AR 72701

5011 Entek Environmental & Technical Services

1724 5th Avenue　　　　　518-271-2000
Troy, NY 12180-3320　　　800-888-9200
　　　　　　　　　　　　Fax: 518-273-6595
　　　　　　　E-mail: McDonough@entek-env.com
　　　　　　　http://www.entek-env.com/

Environmental consulting and engineering firm.
Patrick J McDonough, Director/Manager

5012 Entropy

PO Box 90067　　　　　　　919-781-3550
Raleigh, NC 27675-0067　　　800-486-3550
　　　　　　　　　　　　Fax: 919-787-8442
　　　　　　E-mail: sales@entropyinc.com
　　　　　　http://www.entropyinc.com

Provides air emission testing services.
Robert Drew, President

5013 Enviro Dynamics

7923 Jones Branch Drive # 3201　703-760-0023
Mc Lean, VA 22102　　　　　Fax: 703-760-9382
　　　　　　　　　　　　E-mail: ian@2edi.com
　　　　　　　　　　　　http://www.2edi.com

Occupational and environmental health analysis and consulting firm.
William J Keanet, President

5014 Enviro Systems

1 Lafayette Road　　　　　603-926-3345
PO Box 778　　　　　　　　Fax: 603-926-3521
Hampton, NH 03842

Petra Karbe, VP Marketing

5015 Enviro-Bio-Tech

140 Sharsville Road　　　　610-488-7664
Bernville, PA 19506　　　　Fax: 610-488-9185
　　　　　　　　　　　　http://www.harpi@aol.com

Chemical analysis firm.
Harpal Singh, President

5016 Enviro-Lab

45-10 Court Square　　　　718-392-0185
Long Island City, NY 11101　Fax: 718-392-8654
　　　　　　　　　E-mail: info@envirolab.com
　　　　　　　　　http://www.envirolab.com

Environmental Toxicology Laboratory (ETL) is a research, development and testing laboratory; concentrating its efforts on new approaches to toxicity testing. The mission of ETL is the further advancement of the Tetramitis Assay aswell as the promotion and commercialization of the test.

Dr. Robert L Jaffe, Ph.D, Director of Lab. Science

5017 Enviro-Sciences

111 Howard Boulevard　　　973-398-8183
Suite 108　　　　　　　　Fax: 973-398-8037
Mount Arlington, NJ 07856　http://www.enviro-sciences.com

Environmental sciences firm.
Irving D Cohen, CEO
Glenn Lechner, Marketing Manager

5018 EnviroAnalytical

627 Main Street　　　　　203-459-1800
Monroe, CT 06468　　　　800-459-5060
　　　　　　　　　　　　Fax: 203-459-1466
　　　　　　E-mail: dsethi@enviroanalytical.com

Environmental compliance analysis, R&D, personnel training and analytical method development.
Dr S Sethi, Job Director

5019 EnviroMed Laboratories

414 West California Avenue　318-251-5621
Ruston, LA 71270　　　　　Fax: 318-251-5624

Patricia Flournoy, President

5020 EnviroTech Southeast
1819 Albert Street   904-346-3532
Jacksonville, FL 32202-1103   Fax: 904-346-3576
Firm specializes in technology of industrial wastewater treatment and used oil recycling.

5021 Envirochem Environmental Services
1005 Investment Boulevard   919-362-9010
Apex, NC 27502   Fax: 919-362-9005
Jerry Deakle, President

5022 Enviroclean Technology
13015 SW 89th Place   305-232-8249
Suite 212   Fax: 305-232-1011
Miami, FL 33176   E-mail: bill lorenz@erm.com
    http://www.erm.com
Environmental laboratory service company.

5023 Envirodyne Engineers
303 E Wacker Drive   312-938-0300
Suite 600   Fax: 312-938-1109
Chicago, IL 60601
Environmental science research firm.

5024 Environ Laboratories
9725 Girard Avenue S   952-888-7795
Minneapolis, MN 55431-2621   800-826-3710
    Fax: 952-888-6345
    http://www.environlab.com
Laboratory providing environmental and physical testing of products and materials to commercial and military specifications.
Alan G Thompson, President

5025 Environment Associates
9604 Variel Avenue   818-709-0568
Chatsworth, CA 91311-4913   800-354-1522
    Fax: 818-709-8914
    E-mail: info@eatest.com
    http://www.eatest.com
Provides a full spectrum of environmental test services to Aerospace, Military and commercial manufacturers including temperature, humidity, altitude, thermal vacuum, shock, vibration, corrosive atmosphere, a DSCC approved connectortest lab, hydraulic and pneumatic test capabilities, flow testing firewall testing, EMMI/EMC testing and more. Services also include test procedure development formal test reports, certification and test fixture modifications and adaptations.

William Spaulding, President
Andrew Spaulding, Sales

5026 Environment/One Corporation
PO Box 773   518-346-6161
Schenectady, NY 12301-0773   Fax: 518-346-6188
Environmental analysis and instrumentation firm. Research results published in technical journals.

5027 Environmental Consulting Laboratories
1005 Boston Post Road   203-245-0568
Madison, CT 06443-3335   800-246-9624
    Fax: 203-318-0830
    E-mail: eclinc@aol.com
Environmental testing and consulting firm.

5028 Environmental Abatement
1434 Sadler Circle East Drive   317-359-2898
Indianapolis, IN 46239   Fax: 317-359-2899
Environmental analysis firm.

5029 Environmental Acoustical Research
PO Box 2146   303-447-2619
Boulder, CO 80306-2146   866-327-7584
    Fax: 303-447-2637
    http://www.freehearingtest.com
Acoustics research firm.

5030 Environmental Analysis
3278 N Highway 67   314-921-4488
Florissant, MO 63033   Fax: 314-921-4494
Environmental analytical laboratory.
R M Ferris, President

5031 Environmental Analytical Laboratory
95 Beaver Street   781-893-3124
Waltham, MA 02453   Fax: 781-893-4414
    E-mail: sboyle@hubtesting.net
    http://www.hubtesting.net
Environmental testing services company specializing in asbestos abatement, waste water, soils and surveys. Also mold remediation and screening.
Frederick Boyle, President

5032 Environmental Audits
120 Bishops Way   262-785-9322
Suite 130   Fax: 262-785-9323
Brookfield, WI 53005-6214   E-mail: consult@execpc.com
    http://www.environmentalaudits.net
Consulting firm specializing in environmental science.
John R Ruetz, President

5033 Environmental Chemical
6954 Cornell Road   513-489-2001
Suite 300   http://www.ecc.net/Offices/about_office.html
Cincinnati, OH 45242
Research and testing firm.

5034 Environmental Company
1721 W Plano Parkway   972-516-1169
Suite 210   Fax: 972-516-1170
Plano, TX 75075
Robert Wright, President

5035 Environmental Consultants
391 Newman Avenue   812-282-8481
Clarksville, IN 47129-3247   Fax: 812-282-8554
Environmental consulting firm.
Robert E Fuchs, President

5036 Environmental Consulting Laboratories
1005 Boston Post Road   203-245-0568
Madison, CT 06443-3335   800-246-9624
    E-mail: eclinc@aol.com
Environmental testing and consulting firm.

5037 Environmental Control
5 Baca Lane   505-473-0982
Santa Fe, NM 87507   Fax: 505-438-3801
Environmental consulting and analytical laboratory. Research results published in project reports and test summaries.
James J Meuller, President
Lisa Ann Wilburn, Marketing Manager

5038 Environmental Control Laboratories
38818 Talyor Industrial Parkway   440-353-3700
North Ridgeville, OH 44039   800-962-0118
    Fax: 440-353-3773
    E-mail: eclabs@compuserve.com
E.C. Labs does environmental testing, such as: waste water, soil, asbestos, remediation and constrution projects.
Ron Schiedel, Marketing Manager
Phyllis Conley, Lab Manager

5039 Environmental Control Technology
3985 Research Park Drive   734-761-1389
Ann Arbor, MI 48108   Fax: 734-761-1034
    E-mail: remead@aol.com
    http://www.aiche.org

5040 Environmental Data Resources

440 Wheelers Farms Road   203-255-6606
Milford, CT 06460   800-352-0050
      Fax: 800-231-6802
      http://www.edrnet.com
Applied and product research in environmental applications.
Robert D Barber, CEO
Mark Cerino, COO

5041 Environmental Elements Corporation

3700 Koppers Street   410-368-7000
Baltimore, MD 21227-1020   800-333-4331
      Fax: 410-368-7252
      http://www.EEC1.com
Environmental Elements Corporation, the leading supplier of air pollution control systems for over 50 years, designs, installs, and maintains electrostatic precipitators, fabric filters, gas, and particulate scrubbing andAmmonia-on-Demand (AOD) Systems. EEC technologies enable customers in a broad range of power and generation, pulp, and paper, waste-to-energy, rock products, metals and petrochemical industries to operate their facilities in compliance withparticulate and gaseous emissions.
John L Sans, President/CEO
Neil R Davis, Sr. VP Operations

5042 Environmental Enterprises

10163 Cincinnati Dayton Road   513-772-2818
Cincinnati, OH 45241   800-722-2818
      Fax: 513-782-8950
      E-mail: dbledsoe@eeienv.com
      http://www.eeienv.com
A full service environmental service company that can sample, analize, package, transport and dispose of materials without using subcontractors. RCRA, TOSCA, ATF treatment and storage facility,state certified laboratory, and licensedhazardous waste transportation.

Daniel McCube, President
Jerry Trumpey, VP Sales/Marketing

5043 Environmental Innovations

9600 West Flag Avenue   414-358-7760
Milwaukee, WI 53225   Fax: 414-358-7770
Environmental engineering and consulting firm.

5044 Environmental Laboratories

142 Temple Street   203-789-1260
New Haven, CT 06510   Fax: 203-789-8261
Ray Macaluso, VP/General Manager

5045 Environmental Management: Guthrie

PO Box 700   405-282-8510
Guthrie, OK 73044-0700   Fax: 405-282-8533
      http://www.emiok.com
Environmental assessment and management firm. Provides emergency response remediation and routie waste management.
Terry Bobo, President

5046 Environmental Management: Waltham

95 Beaver Street   781-891-4750
Waltham, MA 02453   Fax: 781-893-4414
      E-mail: sboyle@hubtesting.net
      http://www.hubtesting.net
Environmental testing services, consulting services, environmental abatement services, water quality/chemical testing, asbestos remediation monitoring and inspections, mold testing/mold remediation, industrial hygiene services.
Susan Boyle, VP

5047 Environmental Measurements

2660 California Street   415-567-8089
San Francisco, CA 94115   Fax: 415-398-7664
      E-mail: sales@langan.net
      http://http://lpi.langan.net

5048 Environmental Monitoring Laboratory

59 N Plains Industrial Road   203-284-0555
Suite A   Fax: 203-284-2064
Wallingford, CT 06492

Laboratory providing environmental chemistry services including analysis, bioassays, product efficacy and research and development in the areas of water and wastewater, agricultural chemicals, protective coatings, petroleum products,metals and chemicals.
Jan D Dunn PhD, Director/Manager

5049 Environmental Quality Protection Systems Company

5150 Keele Street   601-961-5650
Jackson, MS 39206-4313   Fax: 601-354-6612
Environmental engineering and science firm.

5050 Environmental Research Associates

414 Mill Road   610-449-7400
Havertown, PA 19083-3740   Fax: 610-449-7404
Research and consulting ecologists and testing firm. Research results published in professional journals.

5051 Environmental Resource Associates

5540 Marshall Street   800-372-0122
Arvada, CO 80002   Fax: 303-421-0159
Engineering consultant.

5052 Environmental Risk Limited

120 Mountain Avenue   860-242-9933
Bloomfield, CT 06002   Fax: 860-243-9055
      E-mail: info@erl.com
      http://www.erl.com
Environmental consulting and engineering firm offers environmental permitting and compliance assistance, site investigation and remediation services, air quality impact analyses, pollution prevention planning, aquatic toxicitylaboratory, hazardous waste management and chemical accident prevention program assistance.
Gordon T Brookman, President

5053 Environmental Risk: Clifton Division

1373 Broad Street   973-773-8322
Suite 301   Fax: 973-243-9055
Clifton, NJ 07013-4200
Environmental engineering and consulting services.

5054 Environmental Science & Engineering

8901 N Industrial Road   309-692-4422
Peoria, IL 61615-1509   Fax: 309-692-9364
Comprehensive environmental and engineering consulting firm.
Richard Holm, Director/Manager

5055 Environmental Services International

6404 Maccorkle Avenue   304-768-2233
Saint Albans, WV 25177-2328   Fax: 304-768-9988
      E-mail: esi@citynet.net
Consulting, engineering, and analytical firm. Research results published in reports to clients.

5056 Environmental Systems Corporation

200 Tech Center Drive   865-688-7900
Knoxville, TN 37912-2799   Fax: 865-687-8977
      E-mail: esccorp@envirosys.com
      http://www.envirosys.com
Data acquisition and reporting systems for electric power producers and industrial sources, ESC is the leading supplier of CEM and ambient data systems in the US.
Steve Drevik, Senior Marketing Manager

5057 Environmental Technical Services

834 Castle Ridge Road   512-327-6672
Austin, TX 78746   Fax: 512-327-1974
      http://www.wetlands.com/
Firm conducts sewer rehabilitation and tank testing.

5058 Environmental Technology & Education Center

4500 Hawkins Street NE   505-345-2707
Suite B   E-mail: jnimitz@etec-nm.com
Albuquerque, NM 87109-4541   http://www.etec-nm.com
Environmental technology firm.

Dr Jon Nimitz, President/CEO
Dr Patrick Dhooge, Marketing Manager

5059 Environmental Testing Laboratories
PO Box 137   609-693-3100
Lanoka Harbor, NJ 08734

Walter Holm Jr, Laboratory Director

5060 Environmental Testing Services
95 Beaver Street   781-893-8339
Waltham, MA 02453   Fax: 781-893-4414
                    E-mail: sboyle@hubtesting.net
                    http://www.hubtesting.net
Environmental testing services company specializing in asbestos abatement, waste water, soils and surveys. Also mold remediation and screening.
Frederick Boyle, President

5061 Environmental Testing and Consulting
2924 Walnut Grove Road   901-327-2750
Memphis, TN 38111-2714   Fax: 901-327-6334
Environmental testing service. Research results published in proprietary reports.

5062 Environmental Working Group
1718 Connecticut Avenue NW   202-667-6982
Suite 600   Fax: 202-232-2592
Washington, DC 20009   http://www.ewg.org
Cutting-edge research on health and the environment.

5063 Environmetrics
11401 Moog Drive   314-432-0550
St Louis, MO 63146   800-333-3278
                          Fax: 314-432-4977
                 E-mail: lab@environmetrics.net
                    http://www.environmetrics.net
Environmental testing laboratory.
Barbara Earhart, President/CEO
Keith Earhart, VP

5064 Enviropro
9765 Eton Avenue   818-998-7197
Chatsworth, CA 91311-4306 Fax: 818-998-7258
Environmental engineering services. Spcialize in: Site investigation/remediation; Real Estate transfers: Phase I and II assessments; feasibility studies; clean-up of contaminated property, soil and groundwater; Methane gasinvestigations.
Zvia Uziel, President
Dr Michael Uziel, Director/Manager

5065 Enviroscan Inc
1051 Columbia Avenue   717-396-8922
Lancaster, PA 17603   Fax: 717-396-8746
                 E-mail: email@enviroscan.com
                    http://www.enviroscan.com
Specializes in non-intrusive, non-deestructive land marine and borhole geophysics for engineers, environmental consultants, architects, industry, government and others. Geophysics is the earth science equivalent of medical radiology,and is used to locate subsurface objects such as utilities, underground storage tanks, drums, bedrock depths, sinkholes, contaminant plumes, fractures, graves, downed aircraft(in oceans, lakes) and submerged items.
Mary Christie, Technical Accounts Manager

5066 Envisage Environmental
PO Box 152   440-526-0990
Richfield, OH 44286-0152   Fax: 440-526-8555
Environmental engineering firm.

5067 Eppley Laboratory
12 Sheffield Avenue   401-847-1020
PO Box 419   Fax: 401-847-1031
Newport, RI 02840   E-mail: eplab@mail.bbsnet.com
Produces radiometer, pyranometers, pyrheliometers and pyrgeometers that measure solar and terrestrial radiation.
George L Kirk, President

5068 Era Laboratories
24 N 21st Avenue W   218-727-6380
Duluth, MN 55806-2017   Fax: 218-727-3049
Environmental laboratory serving the agricultural industry through chemical analysis and sampling.
Robert D Manuson, President

5069 Ernaco
PO Box 6522   301-598-5025
Silver Spring, MD 20906
Firm offers biomedical, health and environmental research services.
Dr Muriel M Lippman, President

5070 Eureka Laboratories
6794 Florin Perkins Road   916-381-7953
Sacramento, CA 95828
Shao-Pin Yo, Laboratory Director

5071 Eustis Engineering Company
3011 28th Street   504-834-0157
Metairie, LA 70002-6019   800-966-0157
                          Fax: 504-834-0354
                 E-mail: info@eustiseng.com
                    http://www.eustiseng.com
Geotechnical firm performing complete investigations, dynamic pile testing, cone penetrometer testing, CQC and materials testing and environmental services.
Lloyd A Held Jr, President
William W Gwyn, VP

5072 Evans Cooling Systems
255 Route 41 North   860-364-5130
Sharon, CT 06069   Fax: 860-364-0888
Environmental applications.

5073 Everglades Laboratories
1602 Clare Avenue   561-833-4200
West Palm Beach, FL 33401 Fax: 561-833-7280
                 E-mail: info@everglades.com
                    http://www.evergladeslabs.com/
Dr. Ben Martin, Director

5074 Excel Environmental Resources
825 Georges Road   732-545-9525
2nd Floor   Fax: 732-545-9425
New Brunswick, NJ 08902   http://www.excelenv.com
Environmental research.
Laura Dodge, President

5075 First Coast Environmental Laboratory
8818 Arlington Expressway   904-725-4847
Jacksonville, FL 32211   Fax: 904-725-2215
Analytical laboratory.
Adolph W Wollitz, Director/Manager

5076 Fishbeck, Thompson, Carr and Huber
1515 Arboretum Drive, SE   616-575-3824
Grand Rapids, MI 49546   Fax: 616-464-3993
                          E-mail: info@ftch.com
                          http://www.ftch.com
Environmental consulting and engineering firm.
James D. Townley, P.E., President

5077 Flowers Chemical Laboratories
481 Newburyport Avenue   407-339-5984
PO Box 150597   800-669-5227
Altamonte Springs, FL 32715-0597   Fax: 407-260-6110
                 E-mail: jeff@flowerslabs.com
                    http://www.flowerslabs.com
Analytical consulting firm. Research results published in reports to clients. Displays report in a pdf or html format.

Dr Jefferson Flowers, President

5078 Ford Chemical Laboratory
40 West Louise Avenue   801-466-8761
Salt Lake City, UT 84115   Fax: 801-466-8763
Ralph Price, President

5079 Forensic Engineering
PO Box 102   775-359-4692
Sparks, NV 89432   Fax: 775-358-6090

Joe M Beard, President

5080 Fredericktowne Labs
3020 Ventrie Court   301-293-3340
PO Box 245   800-332-3340
Myersville, MD 21773   Fax: 301-293-2366
E-mail: info@Fredericktownelabs.com
http://www.Fredericktownelabs.com
Environmental testing lab performing analyses on drinking water, waste water and natural waters for microbiological, inorganic, metal and organic contaminants. State certified
laboratory. State certified sample collectors.Consulting services.

Mary Miller, PhD, Laboratory Director
Kathy Ryan, Special Projects Coordinator

5081 Free-Col Laboratories: a Division of Modern Industries
11618 Cotton Road   814-724-6242
Meadville, PA 16335   Fax: 814-333-1466
E-mail: johnp@modernind.com
http://www.modernind.com
Full service environmental laboratory - drinking water, waste
water, solid waste, industrial hygiene testing; materials testing
& engineering; non-destructive testing; mechanical testing;
chemical analysis; failure analysis;consulting.
John Paraska, Director

5082 Froehling & Robertson
3015 Dumbarton Road   804-264-2701
Richmond, VA 23228-5831   Fax: 804-264-1202
Environmental and construction materials testing lab.

5083 FuelCell Energy
3 Great Pasture Road   203-825-6000
Danbury, CT 06813   E-mail: dferenz@fce.com
http://www.ercc.com
Developer and manufacturer of clean and efficient electric
power generators. Products are designed for distributed generation users including schools, data centers, hospitals, buildings,
waste water treatment plants and othercommercial and industrial applications.

5084 Fugro McClelland
1107 W Gibson Street   512-443-6551
Austin, TX 78704-2375   Fax: 512-444-3996
http://www.fmmg.fugro.com/
Environmental, geotechnical, marine geoscience and environmental engineering firm.
Frank Marshall, President

5085 G&E Engineering
5601 NW 72nd Street   405-840-0301
Suite 290   Fax: 405-840-4307
Oklahoma City, OK 73132-5903
Environmental impact assessment firm.
Richard Adams, President

5086 GEO-CENTERS
7 Wells Avenue   617-964-7070
Newton, MA 02459-3212   Fax: 617-527-7592
http://www.geo-centers.com
Provider of WMD homeland security preparedness services and
products with major strengths in chemical and biological research.
Edward P Marram, CEO

5087 GEO/Plan Associates
30 Mann Street   781-740-1340
Hingham, MA 02043-1316   Fax: 781-740-1340

Environmental planning and analysis firm.

5088 GFG Environmental
1850 William Penn Way   717-293-0173
Suite 201   Fax: 717-293-0178
Lancaster, PA 17601-6737
Geological and environmental research and consulting firm.

5089 GKY and Associates
5411 Backlick Road   703-642-5080
Suite E   Fax: 703-642-5367
Springfield, VA 22151-3933
Civil and environmental systems engineering consulting orga-
nization. Research results published in project reports, govern-
ment publications, and professional journals.
GK Young, President

5090 GL Applied Research
142 Hawley Street   847-223-2220
PO Box 187   Fax: 847-223-2287
Grayslake, IL 60030-0187   E-mail: glapplied@aol.com
Analytical and process control instrumentation development
and manufacture; photometric analyzers.

Edgar Watson Jr, President

5091 GZA GeoEnvironmental
One Edgewater Drive   781-278-3700
Norwood, MA 02062   Fax: 781-278-5701
E-mail: info@gza.com
http://www.gza.com
Geotechnical and geohydrological testing and analysis firm.

William R Beloff, President/CEO
Joseph P Hehir, CFO

5092 Gabriel Laboratories
1421 North Elston   773-486-2123
Chicago, IL 60622   Fax: 773-486-0004
Donna Panek, Laboratory Director

5093 Galson Laboratories
6601 Kirkville Road   315-432-5227
PO Box 369   888-432-5227
East Syracuse, NY 13057-0369   Fax: 315-437-0509
http://www.galsonlabs.com
Environmental industrial hygiene and biological testing service.

5094 Gas Desulfurization Corporation
820 Harden Drive   412-364-1822
Pittsburgh, PA 15229-1109   Fax: 412-367-1254
E-mail: wgwg@aol.com
Environmental studies.
William G Wilson, President

5095 Gas Monitoring
PO Box 13666   919-469-9461
Rtp, NC 27709-3666   Fax: 919-469-5427
Environmental monitoring firm. Research results published in
professional journals.

Russ Scales, President

5096 Gas Technology Institute
1700 S Mount Prospect Road   847-768-0500
Des Plaines, IL 60018-1804   Fax: 847-768-0501
E-mail: info@gastechnology.org
http://www.gastechnology.org
Energy and environmental research.
John F Riordan, President
Robert A Stokes, R&D

5097 Gaynes Labs
9708 Industrial Drive   708-223-6655
Bridgeview, IL 60455   Fax: 708-233-6985
E-mail: gayneslabs@aol.com

Research laboratories. Research results published in confidential test summaries and project reports.

5098 General Engineering Labs

2040 Savage Road   843-556-8171
Charleston, SC 29407-4731   Fax: 843-766-1178
                                    http://www.gel.com

Environmental testing lab.

5099 General Sciences Corporation

4600 Powder Mill Road   301-931-2900
Suite 400   Fax: 301-931-3797
Beltsville, MD 20705-2651   http://www.saic-gsc.com

Consulting and research firm specializing in environmental sciences.
Jeffrey Chen, President

5100 General Systems Division

1025 West Nursery Road   410-636-8700
Suite 120   Fax: 410-636-8708
Linthicum Heights, MD 21090-1205 http://www.nct-active.com
Environmental studies.
Michael Parella, President

5101 Geo Environmental Technologies

140 Broadway   401-421-4140
Providence, RI 02903-3003   Fax: 401-751-8613
Environmental testing and consulting company.

5102 Geo-Con

4075 Monroeville Boulevard   412-856-7700
Suite 400   Fax: 412-373-3357
Monroeville, PA 15146-2533
Environmental services firm provides soil remediation by mixing soil with chemicals designed to eliminate the contaminants.

5103 Geo-Research/Geo-Test

3960 E Gilman Street   562-597-3977
Long Beach, CA 90815-1753   Fax: 562-597-8459
Environmental testing firm. Research results published in papers presented at meetings.

5104 GeoPotential

22323 E Wild Fern Lane   503-622-0154
Brightwood, OR 97011   Fax: 503-492-4404
                        E-mail: geopotential@aol.com
            http://www.members.aol.com/resiii/geomain.htm
GeoPotential provides subsurface mapping surveys to locate underground objects such as underground storage tanks, utilities, geology, etc. They use geophysical methods consisting of ground penetrating radar, magnetics, electromagnetics and gravity.

Ralph Soule, President

5105 GeoResearch

2070 Chain Bridge Road   703-404-2229
Suite 350   Fax: 703-448-5611
Vienna, VA 22182-2573
Environmental sciences and computerized mapping firm. Research results published in reports, journals, and professional society papers. Publishes material on environmental monitoring data, computer programs, regulatory analyses, and special studies.
Douglas Richardson, President

5106 Geological Sciences & Laboratories

3133 N Main Street   606-439-3373
Hazard, KY 41701

5107 Geomatrix

2443 Sidney Avenue # A   410-752-5388
Baltimore, MD 21230
Environmental engineering and consulting firm.

5108 Geomet Technologies

20251 Century Boulevard   301-428-9898
Germantown, MD 20874   Fax: 301-428-9482

Robert L Durfee, President

5109 Geophex

605 Mercury Street   919-839-8515
Raleigh, NC 27603-2343   Fax: 919-839-8528
                        E-mail: geophex@geophex.com
                             http://www.geophex.com
Environmental services firm.
IJ Won, President

5110 George Miksch Sutton Avian Research Center

PO Box 2007   918-336-7778
Bartlesville, OK 74005   Fax: 918-336-7783
                        E-mail: gmsarc@aol.com
                           http://www.suttoncenter.org
Finding cooperative conservation solutions for birds and the natural world through science and education.
Steve Sherrod, Executive Director

5111 Geosyntec Consultants

621 NW 53rd Street   561-731-3440
Suite 650   Fax: 561-995-0925
Boca Raton, FL 33487-8220
Environmental engineering and consulting firm investigates contaminated sites and designs systems necessary for cleanup. Also conducts soil remediation, landfill design, site assessments, and component quality assurance monitoring.

5112 Geotechnical and Materials Testing

22446 Davis Drive   703-406-8702
Suite 127   Fax: 703-406-8708
Sterling, VA 20164-7111
Environmental engineering and consulting firm.
Ahmed N Elrefai, President

5113 Geoteknika

127 Old Farms Road   860-496-0332
Torrington, CT 06790-2240   Fax: 860-496-0332
                        E-mail: geotek@mail.snet.net
Environmental studies.

5114 Geraghty & Miller

125 East Bethpage Road   631-249-7600
Plainview, NY 11803   Fax: 631-249-7610
David Miller, CEO/President

5115 Gerhart Laboratories

Route 219   814-634-0820
Garrett, PA 15542
Environmental testing laboratory.
Michael Gerhart, President

5116 Giblin Associates

PO Box 6172   707-528-3078
Santa Rosa, CA 95406-0172   Fax: 707-528-2837
Environmental and geotechnical engineering firm.
Jere A Giblin, President

5117 Gilbert/Commonwealth

Box 1498   215-775-2600
Reading, PA 19603   Fax: 215-775-2670
Timothy S Cobb, President

5118 Global Geochemistry Corporation

6919 Eton Avenue   818-992-4103
Canoga Park, CA 91303-2110   Fax: 818-992-8940
Consulting firm in the fields of geochemistry and environmental sciences. Research results published by the firm's scientists in journals.
Isaac Kaplan, President

5119 Globetrotters Engineering Corporation
300 S Wacker Drive   312-922-6400
Suite 200   Fax: 312-922-2953
Chicago, IL 60606

Niranjan S Shah, Chair

5120 GoodKind & O'Dea
31 Saint James Avenue   617-695-3400
Suite 1601   Fax: 617-695-3310
Boston, MA 02116-4103
Architectural and engineering consulting firm.
David K Blake, Contact

5121 Gordon & Associates
6975 North County Road   765-478-4801
550 West PO Æ Box 25   Fax: 765-478-9073
Benstonville, IN 47331
Environmental consulting firm.

Paul W Gordon, President

5122 Gordon Piatt Energy Group
Box 650   316-221-4770
Winfield, KS 67156   Fax: 316-221-6289

Jim Salomon, President

5123 Gradient Corporation
1343 Main Streett   941-366-7797
Sarasota, FL 39236   Fax: 941-366-8389
Environmental analysis firm.

5124 Grand Junction Laboratories
435 North Avenue   970-242-7618
Grand Junction, CO 81501   Fax: 970-243-7235

Brian Bauer, Director

5125 Graseby Nutech
4022 Stirrup Creek Drive   919-544-5775
Durham, NC 27703-9000   Fax: 919-544-3770

Gale Woods, Director Marketing

5126 Greeley-Polhemus Group
1310 Birmingham Road   610-793-9440
West Chester, PA 19382
Environmental engineering and economic analysis firm.

5127 Green Environmental Services
3950 River Ridge Drive Northeast   319-395-0578
PO Box 9007   Fax: 319-395-9410
Cedar Rapids, IA 52409
Ralph J Russell, PE, President

5128 Ground Technology
14227 Fern Drive   281-597-8866
Houston, TX 77079   Fax: 281-597-8308
E-mail: ground@groundtechinc.com
http://http://www.groundtechinc.com
A multi-disciplinary engineering firm specializing in environmental services, geotechnical engineering, and construction materials and inspection services. GTI is a woman owned business enterprise as well as minority/disadvantagedenterprise certified by TxDOT, METRO, the State of Texas, and City of Houston, HISD, and the Houston Minority Business Council.

Ruma Acharya, President

5129 Ground Water Associates
771 Brooksedge Plaza Drive   614-882-3136
Westerville, OH 43081   Fax: 614-882-0676
Environmental engineering firm.

5130 Groundwater Specialists
3806 Telluride Place   303-494-8122
Boulder, CO 80305   Fax: 303-494-5443
E-mail: gws@qwest.net
Groundwater exploration; dewatering; waterwell design; mitigation of high groundwater problems.

William H Bellis, Hydrologist

5131 Gruen, Gruen & Associates
564 Howard Street   415-433-7598
San Francisco, CA 94105-3071
Nina J Gruen, Principle Sociologist

5132 Guanterra Environmental Services
1721 S Grand Avenue   714-258-8610
Santa Ana, CA 92705-4808   Fax: 714-258-0921
Chemical analysis technical and consulting research firm. Research results published in project reports and professional journals.

5133 Guardian Systems
1108 Ashville Road NE   205-699-6647
PO Box 190   866-729-7211
Leeds, AL 35094   Fax: 205-699-3882
E-mail: gsilab@gsilab.com
http://www.gsilab.com
Laboratory division provides a full range of analysis for inorganic, organicand physical testing of drinking water, wastewater, groundwater sediments, sludge, waste materials and soils. Industrial Hygiene division provides equipmentand analysis to meet OSHA requirements. Bio-Assay division can accommodate Aquatic Toxicity monitoring requirement.

Gerald Miller, President
Linda Miller, Executive Vice President

5134 Gulf Coast Laboratories
2417 Bond Street   708-534-5200
University Park, IL 60466   Fax: 708-534-5211

5135 Gutierrez, Smouse, Wilmut and Associates
11117 Shady Trl   972-620-1255
Dallas, TX 75229-4646   Fax: 972-620-8028
Environmental engineering consulting firm. Research results published in project reports and technical journals.
Charles G Wilmut, President

5136 H John Heinz III Center for Science Economics and the Environment
1001 Pennsylvania Avenue NW   202-737-6307
Suite 735 S   Fax: 202-737-6410
Washington, DC 20004   E-mail: info@heinzctr.org
http://www.heinzctr.org
The Heinz Center is a nonpartisan, nonprofit institution dedicated to improving the scientific and economic foundation for environmental policy through multisectoral collaboration. The Heinz Center fosters collaboration amongindustry, environmental organizations, academia, and all levels of government in each of its program areas.

Thomas E Lovejoy, President

5137 H2M Group: Holzmacher, McLendon & Murrell
575 Broad Hollow Road   631-756-8000
Melville, NY 11747   Fax: 631-694-4122
E-mail: labs@h2m.com
http://www.h2mlabs.com
Engineers, hydrogeologists, geologists and scientists strive to balance society's dynamic industrial and commercial growth with appropriate development and conservation of natural and man-made resources.

John J Molloy, President

5138 HC Nutting Company
611 Lunken Park Drive   513-321-5816
Cincinnati, OH 45226-1813   Fax: 513-321-0294

E-mail: cincinnati@hcnutting.com
http://www.hcnutting.com

A materials testing company, geotechnical and environmental engineering firm.
R Jackson Scott, President
Jim Cahill, CEO

5139 HTS
416 Pickering Street   713-692-8373
Houston, TX 77091

Ron Langston, President

5140 HWS Consulting Group
PO Box 80358   402-479-2200
Lincoln, NE 68501   Fax: 402-479-2276

James Linderholm, President

5141 Hach Company
Box 389   970-669-3050
Loveland, CO 80539   Fax: 970-669-2932

Bruce Hach, President

5142 Haley & Aldrich
465 Medford Street   617-886-7400
Boston, MA 02129   Fax: 617-886-7600
E-mail: info@haleyaldrich.com
http://www.haleyaldrich.com

Geotechnical engineering and environmental consulting firm, operating in remedial solutions for industry and redevelopment of contaminated sites.
Bruce E Beverly, CEO

5143 Halliburton Company
500 North Akard   214-978-2600
Dallas, TX 75201   Fax: 214-978-2611

One of the world's largest providers of products and services to the oil and gas industries.

Thomas H Cruikshank, Chairman/CEO

5144 Hamilton Research, Ltd.
80 Grove Street   914-631-9194
Tarrytown, NY 10591   Fax: 914-631-6134
E-mail: rwh@rwhamilton.com
http://www.rwhamilton.com

Hamilton Research is a consulting firm specializing in environmental physiology. Our focus is mainly on exposure to people to pressures less and greater than atmospheric, and involves dealing with different breathing gases, especiallyhigh and low levels of oxygen, and the consequences of changes in pressure. Another important area of interest is hyperbaric oxygen therapy.
R W Hamilton, President

5145 Hampton Roads Testing Laboratories
611 Howmet Drive   757-826-5310
Hampton, VA 23661

Independent third party testing laboratory. Performs sampling and analysis in accordance with the required ASTM or ISO Standards.

5146 Handex Environmental Recovery
500 Campus Drive   732-536-8500
PO Box 451   Fax: 732-536-7751
Morganville, NJ 07751-1257

Environmental management and analysis firm.
CL Smith, CEO

5147 Harris Laboratories
624 Peach Street   402-476-2811
PO Box 80837   Fax: 402-476-7598
Lincoln, NE 68501

Provides air quality, soil and water analysis, radiation detection and formaldehyde testing services.

Frederick Boyle, President

5148 Hart Crowser
1910 Fairview Avenue E   206-324-9530
Suite 100   Fax: 206-328-5581
Seattle, WA 98102-369E-mail: rick.moore@hartcrowser.com
http://www.hartcrowser.com

Hart Crowser, Inc. provides a full range of services from initial site studies through regulatory permitting, design, and construction. They integrate thses services as required by each project. They know what kind of informaiton isimportant, how to collect it and apply it to the selection of viable solutions, and how actions are perceived by regulatory agencies and the public. Consequently, they design an appraoch that is practical, cost-effective, and client-oriented.

Rick Moore, Principal, Env. Svce. Mgr.

5149 Hatch Mott MacDonald
PO Box 1008   973-379-3400
Millburn, NJ 07041   800-832-3272
Fax: 973-376-1072
E-mail: info@hatchmott.com
http://www.hatchmott.com

Hatch Mott MacDonald is a client-focused consulting firm providing planning, investigation, design and management capabilities in engineering disciplines and environmental sciences. Areas of expertise include industrial wastewater,site utilities engineering, hazardous and solid waste management and environmental site assessments.
Russell Shallieu, Associate
Dennis Suler, Executive VP

5150 Hatcher-Sayre
905 Southlake Boulevard   804-794-0216
Richmond, VA 23236-3943 Fax: 804-379-8934

Environmental consulting and engineering services firm.

5151 Havens & Emerson
700 Bond Court Building   216-621-2407
1300 E 9th Street   Fax: 216-621-4972
Cleveland, OH 44114

Environmental engineering firm.
Gary Siegel, President/CEO

5152 Hayden Environmental Group
6015 Manning Road   937-439-3764
Dayton, OH   Fax: 937-439-3767

Testing, sampling, and analysis service.

5153 Hayes, Seay, Mattern & Mattern
PO Box 13446   540-857-3100
Roanoke, VA 24034   Fax: 540-857-3296

Troy S Kincer, PE, Principal Associate
Guy E Slagle, PE, LS, Vice President

5154 HazMat Environmental Group
60 Commerce Drive   716-827-7200
Buffalo, NY 14218   Fax: 716-827-7217
http://www.hazmatinc.com

Transportation services - specializing in hazardous materials and hazardous waste transportation.

Ricky F Wickham, General Manager of Operation

5155 Heinrichs Geoexploration Company
810 W Grant Road   520-623-0578
PO Box 5964   Fax: 520-326-4019
Tucson, AZ 85703-0964

Walter E Heinrichs Jr, President

5156 Henry Souther Laboratories
24 Tobey Road   860-242-6291
Bloomfield, CT 06002   Fax: 860-286-0634

Richard J Lombardi, VP

5157 Heritage Laboratories
7901 W Morris Street   317-243-8304
Indianapolis, IN 46231-1366   Fax: 317-486-5085
Environmental testing laboratory.

5158 Heritage Remediation Engineering
4925 Heller Street   502-473-0638
Louisville, KY 40218-3481   Fax: 502-459-4988
Environemtal management and remediation company.

5159 Hess Environmental Services
6057 Executive Centre Drive   901-377-9139
Suite 6   Fax: 901-377-9150
Memphis, TN 38134-7631   E-mail: HES@hessenv.com
http://www.hessenv.com
Hess Environmental Services, Inc. (HES) is an environmental consulting/engineering firm. Their primary activities are: Indoor air quality (IAQ) (Mold and Bacterial) Investigations; Title V Air and Other Permit Applications; Phase I,II, III Property Assessments; Remedial Investigations, Audits, Enviromental Health and Safety, Storm Water, Wastewater, Air Monitoring, Asbestos Inspection and Sampling.

Connie Hess, President
Gary Siebenschuh, VP

5160 Hidell-Eyster Technical Services
PO Box 325   781-749-8040
Accord, MA 02018-0325   Fax: 781-749-2304
E-mail: hidell@hidelleyster.com
Environmental and bottled water assessment and consulting company.
Henry R Hidell, President

5161 Hillmann Environmental Company
1080 Cedar Avenue   908-686-3335
Union, NJ 07083   Fax: 908-686-2636
Joseph Hillmann, Executive VP

5162 Honeywell Technology Center
3660 Technology Drive   612-951-1000
Minneapolis, MN 55418-1096   800-328-5111
Fax: 612-951-7438
E-mail: info@htc.honeywell.com
http://www.htc.honeywell.com
Parent holding company with numerous high-tech units involved in environmental, energy, computer hardware and industrial automation research and development.

5163 Hoosier Microbiological Laboratory
912 West McGalliard   765-288-1124
Muncie, IN 47303   Fax: 765-288-8378
Donald A Hendrickson, Owner

5164 Horner & Shifrin
5200 Oakland Avenue   314-531-4321
Saint Louis, MO 63110-1490   Fax: 314-531-6966
Civil, structural, and environmental engineering firm.

5165 Houston Advanced Research Center
4800 Research Forest Drive   281-367-1348
The Woodlands, TX 77381-4142   Fax: 281-363-7914
http://www.harc.edu
Environmental studies.
John Butler, CEO

5166 Howard R Green
3851 River Ridge Drive NE   319-395-0578
PO Box 9007   Fax: 319-395-9410
Cedar Rapids, IA 52409-9007
Environmental engineering and consulting firm.

5167 Hub Testing Laboratory
95 Beaver Street   781-893-8330
Waltham, MA 02453   Fax: 781-893-4414
E-mail: sboyle@hubtesting.net
http://www.hubtesting.net

Environmental testing services company specializing in asbestos abatement, waste water, soils and surveys. Also mold remediation and screening.

Frederick Boyle, President

5168 Humphrey Energy Enterprises
216 7th Avenue South   406-538-3132
Lewistown, MT 59457
Dr. John P Humphrey, Consultant

5169 Huntingdon Analytical Services
140 Telegraph Road   716-735-3400
Middleport, NY 14105-1330   Fax: 716-735-3653
Environmental engineering consulting firm.

5170 Huntingdon Chen-Northern
96 S Zuni Street   303-744-7105
Denver, CO 80223-1209   Fax: 303-744-0210
Environmental analysis and consulting firm.

5171 Huntingdon Engineering & Environmental
1940 Orange Tree Lane   909-793-2691
Redlands, CA 92374-4552   Fax: 909-793-1704
http://www.dell.com/outlet
Offers the following service(s): Environmental remediation, engineering services, environmental consultant, environmental research, petroleum, mining, and chemical engineers and sanitary engineers.

5172 Hydro Science Laboratories
320 West Water Street   732-349-9692
P.O. Box 4978   800-624-3100
Toms River, NJ 08753-0137   Fax: 609-693-4682
E-mail: info@HydroscienceInc.com
http://www.hydroscienceinc.com
Environmental testing and analysis laboratory.
Robert Salt, Director of Business Dvlpmt

5173 Hydro-logic
1927 North 1275 Road   785-542-2518
Eudora, KS 66025   Fax: 785-542-3971
E-mail: Logic913@aol.com
http://www.hydro-logic.com
Offers professional environmental services, specializing in hydraulic soil and groundwater sampling. Maintains a multidisiplinary team of geologists, hydrologists, chemists, and regulatory compliance specialists.
Thomas Barr, President

5174 Hydrocomp
2386 Branner Drive   650-561-9030
Menlo Park, CA 94025   Fax: 650-561-9031
Dr. N Crawford, President

5175 Hydrologic
122 Lyman Street   828-258-3973
Asheville, NC 28801-4372   Fax: 828-258-3973
Environmental laboratory services firm.
Thomas Barr, President

5176 IC Laboratories
PO Box 721   914-962-2477
Amawalk, NY 10501-0721   Fax: 914-962-5564
Firm providing qualitative and quantitative materials analysis through X-ray diffraction. Studies focus on powders, metals, fibers, and clays, including analysis of crystallinity, thin films, environmental dusts, geological materials,and fabrics. Also provides limited research and development and consulting.

5177 ICS Radiation Technologies
8416 Florence Avenue   562-923-1837
Suite 207   Fax: 562-923-3609
Downey, CA 90240-3949   E-mail: mike@icsrad.com
http://www.icsrad.com
Testing, engineering and consulting firm specializing in radiation effects in semiconductor devices.
Dr Michael K Gauthier, President

5178 IHI Environmental
4527 N 16th Street
Suite 105
Phoenix, AZ 85016-5353
602-776-0300
Fax: 602-776-0301
E-mail: phoenix@ihi-env.com
http://www.ihi-env.com/

5179 INFORM
120 Wall Street
14th Floor
New York, NY 10005-4001
212-361-2400
Fax: 212-361-2412
E-mail: inform@informinc.org
http://www.informinc.org
INFORM is an independent research organization that examines the effects of business practices on the environment and on human health. Our goal is to identify ways of doing business that ensure environmentally sustainable economicgrowth. Our reports are used by government, industry, and environmental leaders around the world.
Joanna Underwood, President

5180 Ike Yen Associates
867 Marymount Lane
Claremont, CA 91711
714-621-2302

5181 Image
4525 Kingston Street
Denver, CO 80239-3016
303-371-3338
Fax: 303-371-3299
Biochemistry and environmental research firm. Research results published in professional journals.

5182 ImmuCell Corporation
56 Evergreen Drive
Portland, ME 04103
207-878-2770
Fax: 207-878-2117
http://www.immucell.com
Biotechnology testing kits, animal health products and environment water testing.
Michael F Brigham, President

5183 Immunotox
PO Box 21705
Carson City, NV 89721-1705
775-885-9400
800-854-1903
Fax: 775-885-9721
Research and consultation firm specializing in the establishment of drug-free work environments.

5184 Inchcape Testing Services
7979 GSRI Avenue
Baton Rouge, LA 70820
504-392-7961
Fax: 225-767-5717
Environmental and industrial testing laboratory.
Kenny Duke, Marketing Manager

5185 Inchcape Testing/NDRC Laboratories
11155 South Main Street
Houston, TX 77025
713-661-8150
Raj Naran, General Manager

5186 Industrial Laboratories
1450 East 62nd Avenue
Denver, CO 80216
303-287-9691
800-456-5288
Fax: 303-287-0964
http://www.industriallabs.net
Provides quality laboratory analysis and consultation. ICP Mineral Analysis.

Larisa Moore, Business Development Manager

5187 Informatics Division of Bio-Rad
3316 Spring Garden Street
Philadelphia, PA 19104
215-382-7800
Fax: 215-382-7800
Richard Shaps, Division Manager

5188 Innovative Biotechnologies International
335 Lang Boulevard
Grand Island, NY 14702
716-773-4232
Fax: 716-773-4257
E-mail: info@ibi.cc
http://www.ibi.cc
Manufacturing technology of biosensing technology.

5189 Inprimis
500 West Cypress Creek Road
Suite 1
Fort Lauderdale, FL 33309-2732
954-556-4020
Fax: 954-556-4031
E-mail: info@ener1.com
http://www.inprimis.com/
Provides hardware and software technology, communications solutions that enbale data transmission, connectivity of devices, and access to applications and information via the Internet, personal computers, and/or server-basedenvironments. Also designs, manufactures, markets, and supports quality, innovative products that have a cost, performance, and time-to-market advantage.
Kevin P. Fitzgerald, Chairman/CEO
Ronald N. Stewart, Executive VP

5190 Integral System
5000 Philadelphia Way
Suite A
Lanham, MD 20706-4419
301-731-4233
Fax: 301-731-9606
http://www.integ.com
Custom computer systems for satellite control; environmental monitoring.

5191 Inter Ag Services IAS Laboratories
2515 E University Drive
Phoenix, AZ 85034
602-273-7248
Fax: 602-275-3836
E-mail: vickin@iaslab.com
http://www.iaslab.com

5192 Inter-Mountain Laboratories
1673 Terra Avenue
Sheridan, WY 82801-6116
307-672-8945
Fax: 307-672-6053
Provides high-quality analytical, engineering and field services to industry and governmental agencies.
Duane Madsen, President

5193 International Asbestos Testing Laboratories
16000 Horizon Way
Unit 100
Mount Laurel, NJ 08054
856-231-9449
Fax: 856-231-9818
E-mail: info@iatl.com
An environmental laboratory specializing in asbestos, lead and mold analysis. Provides environmental laboratory services to environmental consultants, engineers, building owners and govt. agencies throughout the US, Canada and othercountries. Accredited by numerous agencies including the National Voluntary Laboratory Accreditation Program (NVLAP) and the American Industrial Hygiene Association (AIHA).

Emil M Ondra, President
Shirley Clark, Business Development

5194 International Maritime, Inc
110 Pine Avenue
Suite 1070
Long Beach, CA 90802
562-624-4343

5195 International Science and Technology Institute
1820 North Fort Myer Drive
Suite 600
Arlington, VA 22209
703-807-2080
Fax: 703-807-1126
E-mail: isti@istiinc.com
http://www.istiinc.com
Provides technical assistance in project design, implementation, and evaluation; database development and maintenance; institutional and human resource development; policy and economic analysis, methodological research and analysis;strategic planning; and workshop and conference design and organization.
BK Wesley Copeland, Vice Chair

5196 International Society of Chemical Ecology
University of California
Department of Entomology
Riverside, CA 92521
909-787-5821
Fax: 909-787-3086
E-mail: jocelyn.millar@ucr.edu
http://www.isce.ucr.edu/Society/
ISCE is organized specifically to promote the understanding of interactions between organisms and their environment. Research areas include the chemistry, biochemistry and function of natural products, their importance at all levels ofecological organization, their evolutionary origin and their practical application.

John Hildebrand, President

5197 Interpoll Laboratories
4500 Ball Road NE 763-786-6020
Circle Pines, MN 55014 Fax: 763-786-7854
E-mail: interpoll@interpoll-labs.com
http://www.interpoll-labs.com
Interpoll is a full service environmental laboratory with a multidisciplinary staff. They provide their clients with responsive and accurate solutions to their environmental needs. Interpoll offers a full range of environmentaltesting services including stationary source testing, laboratory analysis, groundwater monitoring, ambient air monitoring and pharmaceutical analysis.

Dan Despen, President
Timothy MacDonald, Manager Field Services

5198 Invensys Climate Controls
191 East North Avneue 630-260-3402
Carol Stream, IL 60188 http://www.icca.invensys.com
Formely the Robertshaw Controls Company. Founded after a successfully designing and manufacturing a line of top quality smoke alarms for the residential smoke alarm market.

5199 J Dallon and Associates
16 Fox Hollow Road 201-825-4574
Ramsey, NJ 07446
Research and consulting firm specializing in hortoculture.
Dr Joseph Dallon Jr, President

5200 J Phillip Keathley: Agricultural Services Division
25330 Ruess Avenue 209-599-2800
Ripon, CA 95366-9635
Dr. J Phillip Keathley, President

5201 JABA
2766 North Country Club Road 520-327-7440
Tucson, AZ 85716 Fax: 520-327-7450
E-mail: jbriscoe@jaba.com
http://www.jaba.com
Mining exploration and environmental analysis firm.
James A Briscoe, President

5202 JH Kleinfelder & Associates
7133 Koll Center Parkway 925-484-1700
Suite 100 Fax: 925-484-5838
Pleasanton, CA 94566-3183
Geotechnical and environmental Engineering firm.

5203 JH Stuard Associates
22 Tanglewood Drive 802-878-5171
Woodstock, VT 05091
Environmental Consulting firm.
Joe Shockcor, President

5204 JJ Keller and Associates
3003 Breezewood Lane 920-722-2848
Neenah, WI 54956-9611 Fax: 920-727-7455
E-mail: sales@jjkeller.com
http://www.jjkeller.com
Regulatory compliance, best practices, and training for environmental, safety, and transportation issues.

, *Sales*

5205 JK Research Associates
86 Gold Hill Road 970-453-1760
Breckenridge, CO 80424

5206 JL Rogers & Callcott Engineers
PO Box 5655 864-232-1556
Greenville, SC 29606-5655 Fax: 864-233-9058
Environmental engineering research firm.

5207 JM Best
119 S College Street 724-222-2102
Washington, PA 15301
Performs geologic, economic and engineering evaluations for oil and gas well drilling. Also provides completion operations, environmental studies, map preparations and investigative studies.

5208 JR Henderson Labs
123 Seaman Avenue 732-341-1211
Beachwood, NJ 08722 Fax: 732-505-1658
Environmental laboratory.
Elmer Hemphill, President

5209 JWS Delavau Company
2140 Germantown Avenue 215-235-1100
Philadelphia, PA 19122-1499 Fax: 215-671-1401
International environmental and technical company.
David L Sokol, President

5210 James M Montgomery Consulting Engineers
250 N Madison Avenue 626-796-9141
Pasadena, CA 91101-1639 Fax: 626-568-6308
Environmental engineering and consulting firm.

5211 James R Reed and Associates
770 Pilot House Drive 757-873-4703
Newport News, VA 23606 800-873-4703
Fax: 757-873-1498
E-mail: claiborne@jrreed.com
http://www.jrreed.com
Full service environmental testing facility offering quality analysis and reliable technical services to industry, local and federal government, engineers and private citizens. Areas of expertise include organic and inorganic chemicalanalyses, microbiological testing, aquatic toxicity monitoring and industrial hygiene.
Han Ping Huang, President
Elaine Claiborne, Laboratory Director

5212 James W Bunger and Associates
PO Box 520037 801-975-1456
Salt Lake City, UT 84152-0037 Fax: 801-975-1530
Energy research and development firm specializing in environmental and oil remediation.

5213 Jane Goodall Institute for Wildlife Research, Education and Conservation
8700 Georgia Avenue 240-645-4000
Suite 500 800-99C-HIMP
Silver Spring, MD 20910 Fax: 301-565-3188
http://www.janegoodall.org
A tax-exempt, nonprofit corporation, founded in 1977 focusing on Jane Goodall.
William Johnson, President & CEO

5214 Johnson Company
100 State Street 802-229-4600
Suite 600 Fax: 802-229-5876
Montpelier, VT 05602 E-mail: info@jcomail.com
http://www.johnsonco.com
Physical and product testing of environmental products.

5215 Johnson Controls
5757 N Green Bay Avenue 414-228-1200
Milwaukee, WI 53209-4408 Fax: 414-228-2446
Research in environmental controls.
James Keyes, President

5216 Johnson Research Center
University of Alabama at Huntsville 256-890-6343
Huntsville, AL 35899-0001 Fax: 256-890-6848
Environmental research.
Dr. Michael Eley, CEO

5217 Jones & Henry Laboratories
2567 Tracy Road 419-666-0411
Northwood, OH 43619 Fax: 419-666-1657
E-mail: jhlabs@glasscity.net
Environmental sampling and testing laboratory.
Fred W Doering, President
David Collins, Marketing Manager

5218 Joyce Environmental Consultants
5051 North Lane 407-297-7980
Orlando, FL 32808 Fax: 407-290-0388

Connie Morrison, VP

5219 KAI Technologies
16 Marin Way 603-778-1888
Stratham, NH 03885 Fax: 603-778-0700
E-mail: cliff@kaitech.com
http://www.kaitech.com
Applied and product research in the environment.
Bruce L. Cliff, Director

5220 KCM
1917 1st Avenue 206-443-5300
Seattle, WA 98101 Fax: 206-443-5372
http://www.tetratech.com
Applied and product research in the environment.
Stephen Wagner, President

5221 KE Sorrells Research Associates
8002 Stanton Road 501-562-8139
Little Rock, AR 72209 Fax: 501-562-7025
Analytical chemistry and applied research company providing consulting services in water technology and stream ecology.
KE Sorrells, President

5222 KLM Engineering
3394 Lake Elmo Avenue N 651-773-5111
PO Box 897 888-959-5111
Lake Elmo, MN 55042 Fax: 651-773-5222
E-mail: jkollmer@klmengineering.com
Structural engineering and inspection firm specializing in the industry of steel and concrete plate structures.

Jack R Kollmer, President/Principal
Shawn A Mulhern, Vice President-Sales/Mktg.

5223 Kag Laboratories International
2323 Jackson Street 920-426-2222
Oshkosh, WI 54903 800-356-6045
Fax: 920-426-2664
E-mail: info@kaglab.com
http://www.kaglab.com/
An independent agricultural testing and consulting laboratory. Professional scientific services for agriculture, soil, feed, plant, water and other fields. Total farm management services including high value crops such as cranberry,stevia, blueberry, ginseng, strawberry, herbs, etc. Consultation and recommendation to increase net yield. Available for contractual applied research for all agribusiness industries in Wisconsin, North America and world-wide.
Dr. Akhtar Khwaja, President
Ruma Roy, Vice President/Chemist

5224 Kansas City Testing Laboratory
2012 W 104th Street 913-648-2303
Shawnee Mission, KS 66206-2647 Fax: 913-321-8181
Consulting engineering firm employed in geotechnical, materials, and environmental engineering. Research results published in Project reports.
Donald Cesso, President

5225 Kar Laboratories
4425 Manchester Road 269-381-9666
Kalamazoo, MI 49001-0802 Fax: 269-381-9698
E-mail: info@karlabs.com
http://www.karlabs.com
Environmental testing laboratory, wastewater, drinking water, hay waste, soil and air.

William Rauch, President
Jayne Rauch, Marketing Manager

5226 Karl A Riggs PhD
PO Box Kr 662-323-8889
Mississippi State, MS 39762-5857 Fax: 662-325-2907
Geological services: petroleum, mining, environment and engineering geology. Consulting and research. Published results in technical journals.
Karl A Riggs Jr, President

5227 Kemron Environmental Services
109 Starlite Park 740-373-4071
Marietta, OH 45750 Fax: 740-373-4835
Environmental testing and analysis firm.
Cindy Arnold, Contact

5228 Kennedy-Jenks Consultants
622 Folsom Street 415-243-2150
San Francisco, CA 94107 Fax: 415-896-0999
http://www.KennedyJenks.com
Environmental engineering consulting company.
Blaine L Harrison, CEO

5229 Kentucky Resource Laboratory
Highway 421 606-598-2605
Manchester, KY 40962 Fax: 606-598-1544
Environmental testing firm.
Roy Rice, President

5230 Kenvirons
452 Versailles Road 502-698-4357
Frankfort, KY 40601-3886 Fax: 502-695-4363
E-mail: dgriffin@kenvirons.com
A multi-disciplined environmental and civil engineering firm. Offers engineering services in a range of areas to include water and wastewater related studies and system design, dam design, hydrological studies, environmentalassessments, air and water quality studies, urban and industrial planning, solid waste management, energy-environment interface, computer technology and laboratory services.

Douglas Griffen, President

5231 Keystone Labs
600 East 17th Street South 800-858-5227
Newton, IA 50208 Fax: 641-792-7989
http://http://www.keystonelabs.com
Keystone Laboroatories, Inc. is a full service environmental laboratory committeed to providing the highest quality services at competitive prives.
, President

5232 Kinnetic Laboratories
307 Washington Street 831-457-3950
Santa Cruz, CA 95060 Fax: 831-426-0405
Environmental marine, physical, toxicological, water quality, biological research and scientific consulting and services.
Mary Lee Kinney, President
Mark Savoie, VP

5233 Kleinfelder
981 Garcia Avenue 925-427-6477
Suite A Fax: 925-427-6478
Pittsburg, CA 94565
Laboratories testing.
Gerry Salontai, President/CEO

5234 Konheim & Ketcham
175 Pacific Street 718-330-0550
Brooklyn, NY 11201-6298 Fax: 718-330-0582
E-mail: csk@konheimketcham.com
http://www.konheimketcham.com
Environmental and transportation planning.
Carolyn S Konheim, President

5235 Kraim Environmental Engineering Services
11437 Etiwanda Avenue 818-363-0952
Northridge, CA 91326 Fax: 818-363-0492
E-mail: luftmench@aol.com
Environmental engineering firm.
Jerry Kraim, President

5236 Kramer & Associates
4501 Bogan Avenue NE 505-881-0243
Suite A1 Fax: 505-881-7738
Albuquerque, NM 87109-2225
Environmental monitoring firm. Research results published in conference proceedings.
Gary Kramer, Contact

5237 LRA Environmental
4221 Northgate Boulevard 916-929-9267
Suite 6 Fax: 916-929-9269
Sacramento, CA 95834

5238 LaBella Associates
300 State Street 585-454-6110
Suite 201 Fax: 585-454-3066
Rochester, NY 14614 E-mail: info@labellapc.com
http://www.labellapc.com
Civil and environmental engineering firm.

Salvatore A LaBella, President

5239 LaQue Center for Corrosion Technology
702 Causeway Drive 910-256-2271
Wrightsville Beach, NC 28480-0656 Fax: 910-256-9816
E-mail: info@laque.com
http://www.laque.com
Corrosion technology firm. Research results published in trade journals and presented at technical association meetings.
W T Raines, President
D G Melton, VP

5240 Laboratory Corporation of America Holdings
1904 Alexander Drive 919-572-6900
Research Triangle Park, NC 27709 800-533-0567

5241 Laboratory Resources
100 Hollister Road 201-288-3700
Teterboro, NJ 07608

5242 Laboratory Services Division of Consumers Energy
135 W Trail Street 517-788-2238
Jackson, MI 49201-1314 800-736-4147
Fax: 517-788-1104
E-mail: naserafin@cmsenergy.com
http://www.laboratoryservices.com
Laboratory Services is a full-service testing laboratory. Services include: calibration, nondestructive testing, metallurgy, materials testing and chemistry. They are A2LA accredited (ISO/IEC 17025)-request scope-and 10CFR50 AppendixB authorized.

Nick Serafin, Marketing Manager

5243 Lancaster Laboratories
2425 New Holland Pike 717-656-2300
Lancaster, PA 17601-5994 Fax: 717-656-2681
http://www.lancasterlabs.com
Premier contract testing laboratory serving environmental, pharmaceutical and biophamaceutical clients worldwide. Offers a broad range of high quality analytical services in full compliance with EPA and FDA regulations and clientrequirements.

J Wilson Hershey PhD, President
Anne Osborn, Dir/Marketing/Communications

5244 Lancy Environmental
181 Thorn Hill Road 724-772-0044
Warrendale, PA 15086 Fax: 724-772-1360
Gerald Rogers, President

5245 Land Management Decisions
3048 Research Drive 814-231-1248
State College, PA 16801 Fax: 814-231-1253
Dr. Dale E Baker, President

5246 Land Management Group
3805 Wrightsville Avenue 910-452-0001
Suite 15 Fax: 910-452-0060
Wilmington, NC 28403 E-mail: info@lmgroup.net
http://www.lmgroup.net
Specializes in providing environmental services within the coastal plain of the Carolina's and Virginia and has demonstrated experience in the pedimont and mountain regions as well.
Robert L Moul, President

5247 Land Research Management
1300 N Congress Avenue 561-686-2481
Suite C Fax: 561-684-8709
West Palm Beach, FL 33409 E-mail: lrmi@bellsouth.net
Land planning and zoning, environmental assessments and market analysis firm. Research results published in reports.

Kevin McGinley, President

5248 Lark Enterprises
16 Sunset Drive 508-949-2672
Dudley, MA 01571 Fax: 508-943-8833
E-mail: rjlark@aol.com
Environmental research.
Lother Frank, President

5249 Laticrete International
1 Laticrete Park N 203-393-0010
Bethany, CT 06524 Fax: 203-393-1684
http://www.laticrete.com
Firm providing chemical, mechanical, and environmental simulation testing of concrete and aggregrate building materials.

5250 Law & Company
1763 Montreal Circle 770-934-8200
Tucker, GA 30084
RE Kieffer, Partner

5251 Law & Company of Wilmington
1711 Castle Street 910-762-7082
Wilmington, NC 28403 Fax: 910-762-8785

Richard W Spivey, President

5252 Lawler, Matusky and Skelly Engineers
1 Blue Hill Plaza 845-735-8300
Pearl River, NY 10965 Fax: 845-735-7466
E-mail: cnevel@lmseng.com
http://www.lmseng.com
We anticipate the environmental and engineering needs of our clients and contribute to their success by providing creative solutions.
Christy Nevel, Director Marketing

5253 Lawrence G Spielvogel
31343 Valley Forge Road 610-783-6350
King of Prussia, PA 19406-3785 Fax: 610-783-6349
A consulting engineer who specializes in energy management and procurement and problem solving in buildings.
Lawrence G Spielvogel, President

5254 Ledoux and Company
359 Alfred Avenue 201-837-7160
Teaneck, NJ 07666 Fax: 201-837-1235

LA Ledoux, President

5255 Lee Wilson and Associates
105 Cienega Street 505-988-9811
Santa Fe, NM 87501 Fax: 505-986-0092

Environmental consulting firm. Research results published in project reports.
Lee Wilson, President

5256 Leighton & Associates
17781 Cowan Street 949-250-1421
Irvine, CA 92614 Fax: 949-250-1114
http://www.leightongeo.com
Geotechnical and environmental engineering firm.
Bruce Clark, Contact

5257 Life Science Resources
2 Fremontia Street 650-851-0225
Portola Valley, CA 94028-8032
Biomedical and environmental sciences research firm.

5258 Life Systems
24755 Highpoint Road 216-464-3291
Suite 1 Fax: 216-464-8146
Cleveland, OH 44122-6098
Environmental engineering research and consulting organization. Research results published In project reports and in technical journals.
R Wynveen, President

5259 Los Alamos Technical Associates
2400 Louisiana Boulevard NE 505-665-8616
Building 1, Suite 400 Fax: 505-880-3560
Albuquerque, NM 87110 http://www.lata.com
Environmental studies.
LP Reinig, CEO

5260 Louisville Testing Laboratory
1401 West Chestnut Street 502-584-5914
Louisville, KY 40203 Fax: 502-584-5914
Kenneth Smith Jr, President

5261 Lowry Engineering
219 Measow Street 207-374-3502
Rockport, ME 04856 Fax: 207-374-3503
Environmental engineering consulting company.
Sylvia Lowry, President

5262 Lowry Systems
PO Box 1239
Blue Hill, ME 04614-1239 800-434-9080
Fax: 207-374-3503
E-mail: info@lowryh2o.com
http://www.lowryh2o.com
Environmental research.
Sylvia Lowry, President

5263 Lycott Environmental Research
600 Charlton Street 508-765-0101
Southbridge, MA 01550 800-462-8211
Fax: 508-765-1352
E-mail: lycottine@aol.com
http://www.lycott.com
Environmental science and ecological planning consultant and research firm. Research results published in project reports.
Lee D Lyman, President

5264 Lyle Environmental Management
1507 Chambers Road 614-488-1022
Columbus, OH 43212-1568 Fax: 614-488-1198
Chemical research and consulting service.

5265 Lyle Laboratories
1507 Chambers Road 614-488-1022
Columbus, OH 43212 Fax: 614-488-1198
Dr. Thomas Eggers, Director

5266 Lynntech
7610 Eastmark Drive 979-693-0017
Suite 105 Fax: 979-764-7479
College Station, TX 77840

Oliver J Murphy, President

5267 MBA Labs
340 South 66th Street 713-928-2701
Houston, TX 77261 800-472-1485
Fax: 281-292-7492
E-mail: mbalabs@mbalabs.com
http://www.mbalabs.com
Independently owned and operated since 1968, mba Labs serves industry, government agencies and private citizens in Houston, the continental US and even across the globe. Conform to standards established by the EPA, the TNRCC and meetthe equivalent of ISO 9000 requirements for laboratories through their accreditation by NELAC.

Herman J Kresse

5268 MBA Polymers
500 West Ohio Avenue 521-231-9031
Richmond, CA 94804-2040 Fax: 521-231-0320
E-mail: info@mbapolymers.com
http://www.mbapolymers.com/
Environmental research.
Mike Biddle, President

5269 MBC Applied Environmental Sciences
3000 Redhill Avenue 714-850-4830
Costa Mesa, CA 92626-4524 Fax: 714-850-4840
E-mail: info@mbcnet.net
http://www.mbcnet.net
Environmenatl consultants since 1969. Specializing in marine biology and ecology, oceanography, EIR, EIS, EA, toxicity testing, technical meetings, expert witnesses. MBE/DBE certified.

5270 MIRCON
17 W Pennsylvania Avenue 410-296-7971
Suite 400 Fax: 410-296-3419
Towson, MD 21204-5020
Environmental research firm.

5271 MSE
PO Box 4078 406-782-0463
Butte, MT 59702-4078 Fax: 406-723-8328
Technology testing and evaluation, engineering and environmental consulting, and analytical testing laboratory services.

Donald Peoples, President

5272 MWH Laboratories
750 Royal Oaks Drive 626-386-1100
#100 800-566-LABS
Monrovia, CA 91016 Fax: 626-386-1101
E-mail: mwhlabs@mwhlabs.com
http://www.mwhlabs.com
Environmental testing laboratory focused on drinking water testing.
Dr. Andrew Eaton, Lab Director
Rick Zimmer, Sales Manager

5273 Mabbett & Associates: Environmental Consultants and Engineers
5 Alfred Circle 781-275-6050
Bedford, MA 01730 800-877-6050
Fax: 781-275-5651
E-mail: info@mabbett.com
http://www.mabbett.com
Mabbett & Associates (M&A) provides multi-disciplinary environmental, health and safety services to manufacturing and commercial industry, institutions and public agencies. M&A's services include pollution prevention and wasteminimization, site assessment and remediation, environmental pollution control, environmental management systems and auditing, training and occupational safety and health.
Arthur N Mabbett, President
Joyce Cote, Administrative Assistant

5274 Mack Laboratories
2199 Dartmore Street 412-885-2900
Pittsburgh, PA 15210-4037

5275 Magma-Seal
10116 Aspen Street   512-836-4936
Austin, TX 78758-5102   Fax: 512-836-4936
E-mail: tkdw39a@prodigy.com
Develops materials (plastic and rubber) to withstand severe environmental conditions.
Earl Dumitro, President

5276 Malcolm Pirnie
104 Corporate Park Drive   914-694-2100
White Plains, NY 10602   800-478-6870
Fax: 914-694-9286
E-mail: webmaster@pirnie.com
http://www.pirnie.com
Provides environmental engineering, science and consulting services to over 3,000 public and private clients.
Paul L Busch, PhD, President

5277 Marine Environmental Research Institute
772 West End Avenue   212-864-6285
New York, NY 10025   Fax: 212-864-1470
E-mail: meri@interport.net
The Marine Environmental Research Institute is a 501 (c)3 non-profit charitable organization dedicated to protecting the health and biodiversity of the marine environment for the future generations. MERI accomplishes its goals through multidisciplinary ecotoxicological research, research dissemination, environmental education, and international programs.
Lemuel Evans, Chairman
Joan Koven, Secretary

5278 Martin Water Laboratories
Box 1468   915-943-3234
Monahans, TX 79756   Fax: 915-563-1040
Waylan C Martin, Owner

5279 Maryland Spectral Services
1500 Caton Center Drive   410-247-7600
Suite G   Fax: 410-247-7602
Baltimore, MD 21227
Samuel Hamner, VP

5280 Massachusetts Technological Lab
330 Pleasant Street   617-484-7314
Belmont, MA 02178   E-mail: masstechlab@juno.com
Applies research in the following areas: telecommunications and Internet.
Dr Ta-Ming Fang, President

5281 Mateson Chemical Corporation
1025 E Montgomery Avenue   215-423-3200
Philadelphia, PA 19125-3491   Fax: 215-423-1164
Environmental, toxic, materials, hazardous waste research.

5282 Mayhew Environmental Training Associates (META)
PO Box 786   785-842-6382
Lawrence, KS 66044   800-444-6382
Fax: 785-842-6993
E-mail: salesmeta@cs.com
http://www.metaworld.org
Environmental testing lab offering site assessments.
Thomas Bradford Mayhew, President

5283 McCoy & McCoy Laboratories
1800 Kentucky Avenue   270-444-6547
Paducah, KY 42003-2809   Fax: 270-444-6572
Environmental assessment laboratory.

5284 McIlvaine Company
191 Waukegan Road   847-784-0012
Suite 208   Fax: 847-784-0061
Northfield, IL 60093 E-mail: editor@mcilvainecompany.com
http://www.mcilvainecompany.com
Environmental research and consulting firm. Research results published in manuals updated by newsletters and abstracts.
Robert W McIlvaine, President
Marilyn McIlvaine, Marketing Manager

5285 McLaren-Hart
3039 Kilgore Road   916-638-3696
Rancho Cordova, CA 95670-6041   Fax: 916-638-6840
http://www.mclaren-hart.com
Environmental research.

5286 McNamee Advanced Technology
3135 S State Street   734-665-5553
Suite 301   Fax: 734-665-2570
Ann Arbor, MI 48108-1653
Environmental engineering firm, offering environmental consulting and environmental testing services.

5287 McVehil-Monnett Associates
44 Inverness Drive E   303-790-1332
Suite C   Fax: 303-790-7820
Englewood, CO 80112   http://www.mcvehil-monnett.com
Experienced consulting firm of atmospheric scientsits, engineers and environmental specialists providing air quality and environmental management system (EMS) services worldwide. Serves the mining, oil and gas, electric power and manufacturing industries as well as government agencies and engineering and law firms.

William R Monnett, President/CEO

5288 McWhorter and Associates
33 Bull Street   912-234-8891
Box 9419   Fax: 912-234-8892
Savannah, GA 31412

Thomas McWhorter, President

5289 Mega Engineering
10800 Lockwood Drive   301-681-4778
Silver Spring, MD 20901   Fax: 301-681-5683

Richard E Dame, PE

5290 Membrane Technology & Research Corporate Headquarters
1360 Willow Road   650-328-2228
Suite 103   Fax: 650-328-6580
Menlo Park, CA 94025-1516   E-mail: sales@mtrinc.com
http://www.mtrinc.com
Supplier of membrane-based hydrocarbon recovery systems. Company capabilities include membrane and module manufacturing, process and system design, project engineering and commissioning services.
Dr. Hans Wijmans, Director Research

5291 Merck & Company
126 East Lincoln Avenue   732-574-4000
Rahway, NJ 07065   Fax: 732-594-3810

Bill Hamilton, Dir/NJ Environmental Affairs

5292 Merrimack Engineering Services
66 Park Street   978-475-3555
Andover, MA 01810-3644   Fax: 978-475-1448
E-mail: merreng@aol.com
Research of all forms of environmental studies.
Stephen Stapinski, President

5293 Metlab Testing Services
10835 E Independence   800-324-5757
Suite 102   Fax: 918-234-7152
Tulsa, OK 74116
George Dust, VP

5294 Metro Services Laboratories
6309 Fern Valley Pass   502-964-0865
Louisville, KY 40228-1059   Fax: 502-241-4347
Environmental testing laboratory offering air, water and soil testing services.

5295 Miami Marine Research & Testing Station
3141 SW 112 Avenue
Miami, FL 33165   305-534-0100
   Fax: 305-227-9606
E-mail: mmats@bellsouth.net
Research and testing firm. Research results published in magazines or submitted to clients per customer's instructions.
Carlos L Perez, President

5296 Michael Baker Jr: Civil and Water Division
4301 Dutch Ridge Road
Beaver, PA 15009   724-495-7711
   Fax: 724-495-4017
E-mail: hchakrav@mbakercorp.com
http://www.mbakercorp.com

5297 Michael Baker Jr: Environmental Division
420 Rouser Road
Coraopolis, PA 15108   412-269-6000
   Fax: 412-269-6097
Andrew P Pajak, President

5298 Micro Controls Systems
9046 North 51st Street
Brown Deer, WI 53223   414-362-7880
   Fax: 414-362-7887
GJ Walloch, President

5299 Microseeps, Inc
220 William Pitt Way
Pittsburgh, PA 15238   412-826-5245
   800-659-2887
   Fax: 412-826-3433
E-mail: rpirkle@microseeps.com
http://www.microseeps.com
A full service, NELAP certified environmental library which specializes in elucidation of groundwater geochemistry for use in in-situ remediation processes.

Robert J Pirkle, President
Frank Phillips, VP Sales

5300 Microspec Analytical
3352 128th Avenue
Holland, MI 49424-9263   616-399-6070
   Fax: 616-399-6185
E-mail: info@mspec.com
http://www.mspec.com
Environmental research and resting firm. Research results published in journals and client reports.
Tom Beamish, President

5301 Midwest Environmental Assistance Center
6561 N Seeley Avenue
Chicago, IL 60645-5511   773-973-4850
   Fax: 773-973-4851
E-mail: meac2@aol.com
Noise pollution research firm.
Howard R Schechter, President

5302 Midwest Laboratories
13611 B Street
Omaha, NE 68144   402-334-7770
   Fax: 402-334-9121
E-mail: pohlman@midwestlabs.com
http://www.midwestlabs.com
Midwest Laboratories, Inc. offers analytical services to agriculture, industry and municipal entities throughout the US and Canada. Using wet chemistry methods, they have the capability of testing soil, water, feed, food, plants,fertilizers and residues. Their quality assurance program (QA/QC) provides consistent production of reliable data with high accuracy and precision.

Ken Pohlman, President
John DeBoer, VP

5303 Midwest Research Institute
425 Volker Boulevard
Kansas City, MO 64110-2299   816-753-7600
   Fax: 816-753-8420
E-mail: bduncan@mriresearch.org
http://www.mriresearch.org
Midwest Research Institute is an independent, not-for-profit organization that performs contract research for clients in business, industry and government. MRI conducts programs in the areas of environment, health, engineering,technology development and energy research.
James Spigarelli, President

5304 Mikropul Environmental Systems
20 Chatham Road
Summit, NJ 07901-1302   908-598-1100
   Fax: 908-598-1302
Dust collecting systems.

5305 Miller Engineers
5308 S 12th Street
Sheboygan, WI 53081-8015   920-458-6164
   Fax: 920-458-0369
Civil and environmental engineering firm.
Roger G Miller, President

5306 Minnesota Valley Testing Laboratories
1126 N Front Street
New Ulm, MN 56073-1176   507-354-8517
   Fax: 507-359-1231
Independent bacteriological and chemical analysis firm, with services in environmental, agricultural, and energy fields. Research results published in project reports.
Henry Nupson, President

5307 Mirage Systems
PO Box 820
DeLand, FL 32721   386-740-9222
   Fax: 386-740-9444
Environmental research.
Robert S Ziernicki, President

5308 Miranda Associates
2000 L Street NW
Washington, DC 20036   202-857-0430

5309 Montgomery Watson
300 North Lake Avenue
Pasadena, CA 91101   626-796-9141
   Fax: 626-568-6619
http://www.mwhglobal.com
Engineering company that has designed, constructed, financed and managed many of the largest and most technologically significant infrastructures in the world.
Robert B Uhler, COO

5310 Montgomery Watson Mining Group
1475 Pine Grove Road
Suite 109
Steamboat Springs, CO 80477   970-879-6260
   Fax: 970-879-9048
   http://www.mw.com
Mine engineering and environmental services firm.
Alan Krause, SVP/COO

5311 Mycotech
630 S Utah
PO Box 4109
Butte, MT 59702   406-782-2386
   Fax: 406-782-9912
Clifford Bradley, Director R&D

5312 Myra L Frank & Associates
811 W 7th Street
Suite 800
Los Angeles, CA 90017   213-627-5376
   Fax: 213-627-6853
E-mail: fwilliams@myrafrank.com
http://www.myrafrank.com
Environmental impact analysis firm. Architectural historic surveys.
Florence Williams

5313 Mystic Air Quality Consultants
1204 North Road
Route 117
Groton, CT 06340   860-449-8903
   Fax: 860-449-8860
Rich Haffey, VP/General Manager

5314 NET Pacific
11135 Rush Street
Suite Q
South El Monte, CA 91733   626-350-4241

5315 National Loss Control Service Corporation
1 Kemper Drive
Long Grove, IL 60049-0001   847-320-2488
   Fax: 847-320-4331
Environmental science laboratory.

Joan Wronski, Laboratory Manager

5316 National Oceanic & Atmospheric Administration Environmental Research Laboratories
325 Broadway 303-497-3000
Boulder, CO 80303-3337 Fax: 303-497-6554
http://www.noaa.gov

Environmental research.

5317 Nebraska Testing Corporation
4453 South 67th Street 402-331-4453
Omaha, NE 68117 Fax: 402-331-5961

Marc D Shannon, President

5318 Neilson Research Corporation
245 South Grape Street 541-770-5678
Medford, OR 97501-3123 Fax: 541-770-2901
E-mail: clientservices@nrclabs.com
http://www.nrclabs.com
Provides analytical services to support environmental projects
including testing of drinking water, wastewater, ground and surface water, foods soils, sediments, sludges, filters, air, and hazardous waste samples.
John WT Neils, CEO

5319 Neponset Valley Engineering Company
378 Page Street 781-297-7040
Suite 10 Fax: 781-297-7050
Stoughton, MA 02072-1124
Environmental engineering analysis and consulting firm.

5320 New England Testing Laboratory
1254 Douglas Avenue 401-353-3420
North Providence, RI 02904 Fax: 401-354-8951
Mark Bishop, VP Operations

5321 New York Testing Laboratories
143-05 Emery Avenue 718-658-7300
Jamaica, NY 11432 800-281-3329
Fax: 718-657-3902
http://www.nytesting.com
Consulting on a range of disciplines including environmental.

Charles Realmuto, Director Marketing

5322 Newport Electronics
2229 South Yale Street 714-540-4914
Santa Ana, CA 92704-4401 800-639-7678
Fax: 714-546-3022
E-mail: info@newportinc.com
http://www.newportinc.com
Manufacturer of industrial and environmental instrumentation
including signal conditioners, digital panel meters, PID controllers and temperature sensors.
Milton Hollander, President

5323 Nobis Engineering
18 Chenell Drive 603-224-4182
Concord, NH 03301-8537 Fax: 603-224-2507
Environmental engineering consulting firm.

5324 Normandeau Associates
102 South Boundary 803-652-2206
New Ellenton, SC 29809 Fax: 803-652-7428
Jean Eidson

5325 North American Environmental Services
PO Box 26521 512-264-2828
Austin, TX 78755-0521
Environmental science research firm.
D Craig Kissock, President

5326 North American Science Association
Box 1305 770-427-3101
Marietta, GA 30061 Fax: 770-426-5692

George Kordares, Manager Chemistry

5327 Northeast Test Consultants
587 Spring Street 207-854-3939
Westbrook, ME 04092 Fax: 207-854-3658
Asbestos and lead testing/industrial hygiene.

Stephen Broadhead, Laboratory Manager

5328 Northeastern Analytical Corporation
4 E Stow Road 856-985-8000
Marlton, NJ 08053-3150 Fax: 856-985-8000
Environmental sampling and analysis firm.

5329 Northern Power Systems
1 North Wind Road 802-496-2955
Moretown, VT 05660 Fax: 802-496-2953
John Kueffner, VP Operations

5330 Nuclear Consulting Services
7000 Huntley Road 614-846-5710
Columbus, OH 43229 Fax: 614-431-0858
Joseph C Enneking, VP

5331 O'Brien & Gere Engineers
Box 4873 315-437-6100
Syracuse, NY 13221 Fax: 315-463-7554

Cornelius B Murphy, President

5332 OA Laboratories and Research
1430 N Stadium Drive 317-639-2626
Indianapolis, IN 46202-2151 Fax: 317-636-6760
OA Laboratories and Research, Inc. serves customers in Indiana
and throughout the United States by meeting their Analytical
needs.

Bela Jones, Senior Chemist

5333 OEA Research
PO Box 1209 406-443-5560
Helena, MT 59624-1209

5334 Oak Ridge Institute for Science and Education
120 Badger Avenue 865-576-3424
PO Box 117, MS 36 Fax: 865-241-2923
Oak Ridge, TN 37831-0117 E-mail: westm@orau.gov
http://www.orau.org

Dr. Nathaniel W Revis, Director

5335 Occupational Health Conservation
5118 N 56th Street 813-626-8156
Tampa, FL 33610-5416 800-229-8156
Fax: 813-623-6702

Environmental impact assessment firm.
James F Rizk, President

5336 Occusafe
240 East Lake Street 630-941-3001
Addison, IL 60101-2890 800-323-7597
Fax: 630-941-3865
E-mail: info@occusafe-inc.com
http://www.occusafe-inc.com
Employee safety, industrial hygiene and environmental consulting firm.
Bob McKinley, President

5337 Ogden Environment & Energy Services Company
4455 Brookfield Corporate Drive 703-488-3700
Suite 100 Fax: 703-488-3701
Chantilly, VA 20151
Environmental engineering and consulting company.
J Mark Elliot, President

5338 Ogden Environment & Energy Services Company
5510 Morehouse Drive   858-458-9044
San Diego, CA 92121-3720   Fax: 858-458-0943
Scientific and environmental engineering; analytical chemistry.
Mike Nienberg, Executive VP

5339 Ogden Environment & Energy Services Company
33254 Perimeter Hill Drive   615-333-0630
Nashville, TN 37211   Fax: 615-781-0655
Environmental engineering and consulting company.

5340 Oil-Dri Corporation of America
410 N Michigan Avenue # 400   312-321-1515
Chicago, IL 60611-4213   Fax: 312-321-1271
Absorbents for consumers, industrial, agricultural, environmental and fluid purification.

5341 Olver
1116 S Main Street   540-552-5548
Blacksburg, VA 24060-5548   Fax: 540-552-5577
E-mail: info@olver.com
http://www.olver.com
Engineering research and consulting firm specializing in environmental design and analysis. Research results published in project reports.

5342 Omega Thermal Technologies
21 Elbo Lane   856-232-1399
Mount Laurel, NJ 08054-9624   Fax: 856-232-1772
E-mail: email@ottusa.com
Environmental studies.
Kenneth Hladun, President

5343 Oneil M Banks
336 S Main Street   410-879-4676
Suite 2D   Fax: 410-836-8685
Bel Air, MD 21014-3978
Industrial and environmental hygiene and toxicology consulting company.

5344 Online Environs
201 Broadway   617-577-0202
Suite 7   Fax: 617-577-0772
Cambridge, MA 02139-1955   http://www.environs.com
Telecommunications and Internet research.
Anrew Yu, President

5345 Operational Technology Corporation
4100 NW Loop 410 Street   210-731-0000
Suite 230   800-677-8072
San Antonio, TX 78229-4253   Fax: 210-731-0008
E-mail: webmaster@otcorp.com
http://www.otcorp.com
Employment research firm providing information technologies, computer sales and service and environmental services.
John Fernandez, CEO

5346 Orlando Laboratories
PO Box 149127   407-896-6645
Orlando, FL 32814-9127   Fax: 407-898-6588
Independent environmental testing and analysis laboratory.

5347 Osmonics
5951 Clearwater Drive   952-933-2277
Minnetonka, MN 55343-8990   Fax: 952-933-0141

David J Paulson, Research Director

5348 Ostergaard Acoustical Associates
200 Executive Drive   973-731-7002
West Orange, NJ 07052   Fax: 973-731-6680
E-mail: kherbert@acousticalconsultant.com
http://acousticalconsultant.com
Environmental, acoustic and noise control testing and analysis firm. Research results published in project reports.

R Kring Herbert, Principal
John Erdriech PhD, Principal

5349 Ozark Environmental Laboratories
PO Box 806   573-364-8900
Rolla, MO 65402-0806   Fax: 573-341-2040
Firm providing construction materials testing on soils, aggregates, and asphaltic and portland cement concrete; water and wastewater physical and chemical analysis; and quality control studies encompassing physical measurements andchemical analysis.

5350 P&P Laboratories
2025 Woodlynne Avenue   856-962-6188
Oaklyn, NJ 08107-2236
Environmental testing and chemical toxicology laboratory.

5351 PACE
100 Marshall Drive   724-772-0610
Warrendale, PA 15086-7554   Fax: 724-772-1686
Environmental testing and analysis firm.

5352 PACE Analytical Services
1700 Elm Street   612-607-1700
Suite 200   E-mail: info@pacelabs.com
Minneapolis, MN 55414   http://www.pacelabs.com
Provider of air, water, soil and environmental testing services.
Steve A Vanderboom, CEO
Gabe LeBrun, Director/Manager

5353 PACE Environmental Products
5240 W Coplay Road   610-262-3818
Whitehall, PA 18052-2212   800-303-4532
Fax: 610-262-4445
E-mail: emission@pacecems.com
http://www.pacecems.com
Manufacturer and Integrator of continuous emissions monitoring systems (EMS). Regulatory, process, and certification stack testing. In-shop analyzer repair, CEMS field service. Parts, sales, rentals, repairs and service.

Bernard Sinkiewicz, Vice President
David DeSantis, Sales Manager

5354 PACE Resources, Incorporated
40 S Richland Avenue   717-852-1300
York, PA 17404-3470   800-711-8075
Fax: 717-852-1301
E-mail: pace40@aol.com
This company is the parent of units involved in environmental engineering and consulting, civil engineering, architectural planning, data processing, printing and other services.
Russell E Horn, Jr, President

5355 PARS Environmental
6A S Gold Drive   609-890-7277
Robbinsville, NJ 08691   Fax: 609-890-9116
E-mail: hgill@parsenviro.com
http://www.parsenviro.com
Environmental consulting company.
HS Gill, President

5356 PCCI
300 N Lee Street   703-684-2060
Suite 201   Fax: 703-684-5343
Alexandria, VA 22314   E-mail: all@pccii.com
http://www.pccii.com
Provides sensible solutions to difficult engineering and environmental problems in coastal, ocean and inland environments. Specialties include: environmental compliance; all hazards emergency response planning, trainings, drills andexercises; and marine engineering.
Robert W Urban, President
Alan R Becker, VP

5357 PDC Laboratories
2231 West Altorser Drive   309-692-9688
Peoria, IL 61615   Fax: 309-692-9689
Environmental laboratory performs air sample analysis, soil
analysis, and potential toxic waste analysis.

5358 PE LaMoreaux & Associates
PO Box 2310   205-752-5543
Tuscaloosa, AL 35403-2310   Fax: 205-752-4043
Consulting hydrologists, geologists, engineers, and environ-
mental scientists. Research results published in brochures, pam-
phlets, news releases, speeches, seminars, studies, and reports.
James W Lamoreaux, President

5359 PEI Associates
11499 Chester Road   513-782-4700
Suite 200   Fax: 513-782-4807
Cincinnati, OH 45246-4098
Environmental consulting firm. Research results published in
government publications.

5360 PELA
PO Box 2310   205-752-5543
Tuscaloosa, AL 35403   Fax: 205-752-4043
                                        E-mail: pela@dbtech.net
                                        http://www.pela.com
For over three decades, PELA's integration of qualified personnel, up to date technology, and sound management has established PELA as an international leader in the environmental
consulting field. PELA's expertise in hydrology, geotechnical
analysis, design and construction management, remediation,
computer graphics and models, and permitting can get your project on the two feet quicker than you might think.

James W LaMoreaux, President

5361 PRC Environmental Management
233 N Michigan Avenue   312-938-0300
Suite 1621   Fax: 312-931-1109
Chicago, IL 60601
Robert Banosten, VP

5362 PRD Tech
7103 Turfway Road   859-525-2350
Suite 305-B   Fax: 859-525-2373
Florence, KY 41042-2061   E-mail: rsmprdt@aol.com
                                        http://www.prdtechinc.com
Biological and chemical research and commercial technology
development firm, serving primarily the baking, brewing, and
other food industry segments with their environmental control
needs - odor and volatile organic compound (VOC) control applications.
Ramesh Melarkode, President

5363 PSC Environmental Services
550 Pinetown Road   215-643-5466
Suite 166   800-292-2510
Fort Washington, PA 19034   Fax: 215-643-2772
                                        http://www.contactpsc.com

Environmental services
Paul Butsavage, Area Operations Manager

5364 PSI
1901 South Meyers Road   630-691-1587
Suite 400   800-548-7901
Oakbrook Terrace, IL 60181   Fax: 630-691-1587
                                        E-mail: info@psiusa.com
                                        http://www.psiusa.com
Distinguished as a leader in environmental consulting,
geotechnical engineering, and construction testing services, PSI
is nationally recognized in several disciplines, including: construction services, materials testing, roof consulting and asbestos management.

5365 Pace
2400 Cumberland Drive   219-464-2389
Valparaiso, IN 46383-2502   Fax: 219-462-2953
Environmental testing laboratory.
Les Arnold, President

5366 Pace Laboratory
9893 Brewers Court   301-490-9860
Laurel, MD 20723-1905
Environmental testing laboratory.

5367 Pace New Jersey
284 Raritan Center Parkway   973-257-9300
Edison, NJ 08837-3610   Fax: 973-257-0777
Environmental analytical laboratory and data management
firm.

5368 Pacific Materials Laboratory
35 S La Patera Lane   805-688-7587
Suite A
Goleta, CA 93117

5369 Pacific Northwest National Lab
902 Battelle Boulevard   509-375-2121
PO Box 999   888-375-7665
Richland, WA 99352   Fax: 509-375-2491
                                        E-mail: inquiry@pnl.gov
                                        http://www.pnl.gov
Contract research and development for the government environ-
mental restoration, energy, national security and health.

5370 Pacific Nuclear
1010 South 336th Street   253-874-2235
Federal Way, WA 98003   Fax: 253-874-2401

5371 Package Research Laboratory
41 Pine Street   973-627-4405
Rockaway, NJ 07866   Fax: 973-627-4407
                                        E-mail: info@package-testing.com
                                        http://www.package-testing.com
Packaged product testing facility. Research results published in
reports, videos and pictures. Custom tests designed. DOT/UN
certification on hazardous materials. Extreme environment testing. Pallet load and pallet merchandizing testing. Design and
packaging development, consulting, package analysis, project
management, and vendor audits.
David Dixon, VP
Brian Berg, R&D

5372 Pan American Laboratories
4099 Highway 190   985-893-4097
Covington, LA 70433   Fax: 985-893-6195
                                        E-mail: pamlab@pamlab.com
                                        http://www.pamlab.com/

Pharmaceutical manufacturer.
Mary L Lipps, President

5373 Pan Earth Designs
307 E Yelm Avenue   360-458-9173
PO Box 1928   Fax: 360-458-9123
Yelm, WA 98597-7677
Environmental research firm.

5374 Par Enterprises
12601 Clifton Hunt Lane   703-818-9274
Clifton, VA 20124

5375 Par Environmental Services
1906 21st Street   916-739-8356
PO Box 160756   Fax: 916-739-8356
Sacramento, CA 95816-0756 http://www.parenvironmental.com
Environmental research firm.

5376 Parsons Engineering Science
100 W Walnut Street   626-440-2000
Pasadena, CA 91124   Fax: 626-440-2630
                                        E-mail: erin.kuhlman@parsons.com
                                        http://www.parsons.com
Environmental engineering testing and consulting company
with expertise in advanced wastewater treatment.

Frank A DeMartino, President
Erin Kuhlman, VP Corporate Relations

5377 Penniman & Browne

PO Box 65309   410-825-4131
Baltimore, MD 21209   Fax: 410-321-7384
E-mail: clientservices@pandbinc.com
http://www.pandbinc.com
Independent testing laboratory whose mission is to provide excellent client service with its scope of both engineering and chemical services.

Hans V Steer, Client Services Manager

5378 Peoria Disposal Company

4700 N Sterling Avenue   309-688-0760
Suite 2   Fax: 309-688-0881
Peoria, IL 61615-3600
Environmental services firm, especially hazardous waste testing.

5379 Pharmaco LSR

Mettlers Road   732-873-2550
Box 2360   Fax: 732-873-3992
East Millstone, NJ 08875-2360

Dr. Geoffrey K Hogan, President

5380 Philip Environmental Services

210 W Sand Bank Road   618-281-7173
PO Box 230   Fax: 618-281-5120
Columbia, IL 62236-1044   http://www.philipinc.com
Environmental research and analysis firm.
Jenny Penland, President

5381 Physical Sciences

20 New England Business Center   978-689-0003
Andover, MA 01810   Fax: 978-689-3232
http://www.psicorp.com
PSI focuses on providing contract research and development services in a variety of technical areas to both government and commercial customers. Our interests range from basic research to technology development, with an amphasis onapplied research.
George Caledonia, President & CEO
David Green, President, R & D Operations

5382 Pittsburgh Mineral & Environmental Technology

700 5th Avenue   724-843-5000
New Brighton, PA 15066   Fax: 724-843-5353
E-mail: pmet@pmet-inc.com
http://www.pmet-inc.com
A full service company specializing in metals and mineral processing, coal ash utilization, waste stream management, and precision analysis. Also develops technologies dedicated to waste minimibation, treatment, and conversion tosafe,usable, profitable products.

Thomas E Weyand, President
William F Sutton, CEO

5383 Planning Concepts

309 Commercial Street   530-265-8068
Nevada City, CA 95959-2409   Fax: 916-265-5042
Environmental impact assessment firm.

5384 Planning Design & Research Engineers

2000 Lindell Avenue   615-298-2065
Nashville, TN 37203   Fax: 615-269-4119
E-mail: ttichenor@pdre.net
http://PDRE.net
Environmental engineers, asbestos, lead paint design, testing underground tanks, hazardous waste projects, Phase I and II site assessments.

Teresa Tichenor, Office Manager

5385 Planning Resources

402 W Liberty Drive   630-668-3788
Wheaton, IL 60187-4937   Fax: 630-668-4125
E-mail: dri@sprintmail.com
Land use and environmental planning.
Lan R Richart, President
Pamela J Richart, VP

5386 Plant Research Technologies

525 Del Rey Avenue Unit C   408-245-4423
PO Box 6008   Fax: 408-245-8043
Sunnyvale, CA 94086
Contact research organization which provides agricultural and analytical applied services.
Basil Burke PhD, President

5387 Plasma Science & Fusion Center

167 Albany Street   617-253-8100
Cambridge, MA 02139-4301   Fax: 617-253-0570
E-mail: info@psfc.mit.edu
http://www.psfc.mit.edu
Plasma science and technology and plasma fusion energy research.
Miklos Porkolab, Director/Manager
Paul Rivenberg, Communications Manager

5388 Polaroid Corporation

549 Technology Square   617-577-2000
Cambridge, MA 02139   Fax: 617-577-5618
http://www.polaroid.com

5389 Polyengineering

1935 Headland Avenue   334-793-4700
Dothan, AL 36303   http://www.polyengineering.com
Offers a broad range of professional engineering and architectural services as well as financial services and administrative support.

AE Parsons, President

5390 Polytechnic

3740 W Morse Avenue   847-677-0450
Lincolnwood, IL 60712   Fax: 847-677-0480

5391 Porter Consultants

4400 Old William Penn Highway   412-380-7500
Suite 200   Fax: 214-689-9
Monroeville, PA 15146   http://www.porter-consulting.com
Executive recruiting firm specializing in national and international placement of Sales, Marketing, Management, Executive-level, and Technical Support professionals within a wide rang of industries including High Tech, Exhibit.Telecommunications, Medical, and Pharmaceutical.
SW Porter Jr, COO

5392 Powell Labs Limited

1915 Aliceanna Street   410-558-3540
Baltimore, MD 21231
Provides services in the specialty fields of metallurgical investigations, failure analysis, metal overheating and corrosion failures, remaining life assessments of high temperature components, identification of casting andmanufacturing defects, microbiological investigations, alloy identification, cycle water, cooling water, drinking water, high purity water, industrial process water, waste water, water and stream formed deposits, field examinations and training.

5393 Precision Environmental

180 Canada Larga Road   805-641-9333
Ventura, CA 93001   800-375-7786
Fax: 805-648-6999
http://http://www.precisionenv.com
Precision Environmental, Inc. was founded at Stanford University with the purpose of providing quality environmental contracting services to clients with asbestos contamination problems. State licensed and registered.

5394 Princeton Energy Resources International
1700 Rockville Pike 301-881-0650
Suite 550 Fax: 301-230-1232
Rockville, MD 20852-1695 http://www.perihq.com
Engineering and consulting firm: engineering and environmental technology, environmental management and global climate change issues, economic research, aviation economics, and human factors
Thomas Schweizer, President

5395 Priorities Institute
3233 Vallejo Street 303-477-3792
#3B Fax: 303-838-8105
Denver, CO 80211-0089 E-mail: mail@priorities.org
http://www.priorities.org
Nonprofit, educational research organization that explores issues of critical importance that are not adequately researched by existing educational, media, research, governmental or other organizations.
Logan Perkins, Director/Founder

5396 Professional Service Industries
8936 Nieman Road 913-310-1600
Overland Park, KS 66214 Fax: 913-310-1601
Milda Cooper, Department Manager

5397 Professional Service Industries Laboratory
4106 NW Riverside Drive 816-741-9466
Riverside, MO 64150 Fax: 816-587-2996
Engineering test laboratory.
Stephen Fitzer, President

5398 Professional Service Industries/Jammal & Associates Division
1675 Lee Road 407-645-5560
Winter Park, FL 32789 Fax: 407-645-1320
William N Phillips, Executive VP

5399 Purewater Corporation
PO Box 597 913-342-9436
Shawnee Mission, KS 66201 Fax: 913-342-9457

Daniel Katz

5400 Puricons
101 Quaker Avenue 215-644-5488
Malvern, PA 19355 Fax: 215-644-5545
Dr. Sallie A Fisher, President

5401 Q-Lab
1005 SW 18th Avenue 305-245-5600
PO Box 349490 Fax: 305-245-5656
Homestead, FL 33034 E-mail: mcrewdson@q-panel.com
http://www.q-panel.com
Firm providing environmental simulation testing.
Michael J Crewdson, Director/Manager

5402 QC
1205 Industrial Highway 215-355-3900
Southampton, PA 18966-4010 Fax: 215-355-7231
Environmental testing lab.

5403 Quantum Environmental
167 Little Lake Drive 734-930-2600
Ann Arbor, MI 48104-1041 Fax: 734-930-2798
Environmental remediation firm.

5404 R&R Visual
1828 W Olson Road 219-223-5426
Rochester, IN 46975 800-656-4225
Fax: 219-223-7953
E-mail: info@rapidview.com
http://www.rapidview.com
Developing and providing unique inspection solutions to the nuclear, petrochemical, industrial and municipal sewer industries.

Rex Robinson, President

5405 RE/SPEC
Box 725 605-394-6400
Rapid City, SD 57709 Fax: 605-394-6456
Tom Zeller, VP Finance

5406 RMC Corporation Laboratories
214 W Main Plaza 417-256-1101
West Plains, MO 65775-2726 Fax: 417-256-1103
Environmental waste studies. Research results published in journals.
Joseph Cooke, President
Dr R Soundararajan, Director R&D

5407 RMT
744 Heartland Trail 608-831-4444
PO Box 8923 800-283-3443
Madison, WI 53717-1934 Fax: 608-831-3334
E-mail: info@rmtinc.com
http://www.rmtinc.com
Global engineering and management consulting firm that develops environmental solutions for industry. With a 600 person staff and 20 offices throughout the US and Europe, we help clients sustain the environment while meeting theirbusiness objectives. Engineers, scientists and construction managers can take a project from conception through successful completion. Expertise includes air, water and waste permitting, remediation, hazardous/solid waste management, air pollutioncontrol and more.
8 pages Quarterly
Jodi Burmester, Corporate Communications

5408 RPM Systems
938 Chapel Street 203-776-2358
New Haven, CT 06510-2515 Fax: 203-773-3657
Environmental and energy conservation firm.

5409 RV Fitzsimmons & Associates
1860 Arthur Road 630-231-0680
West Chicago, IL 60185-1602 Fax: 630-231-0811
Environmental testing and consulting firm.
Robert Fitzsimmons, President

5410 RVSI Acuity CiMatrix
5 Shawmut Road 781-821-0830
Canton, MA 02021-1408 800-646-6664
Fax: 781-828-8942
E-mail: info@rvsi.com
http://www.rvsi.com
Machine vision systems, bar code and advanced two-dimensional symbology readers for manufacturing and distribution environments.
Katie Catalogna, Marketing Specialist

5411 Radian Corporation
PO Box 201088 512-244-0100
Austin, TX 78720-1088 Fax: 512-388-0966
Environmental science and industrial safety research and consulting firm. Research results published in project reports and in professional journals.

5412 Ralph Stone and Company
10954 Santa Monica Boulevard 310-478-1501
Los Angeles, CA 90025 Fax: 310-478-7359
Richard Kahle, President

5413 Ramco
6362 Ferris Square 858-452-5963
Suite C Fax: 858-453-0625
San Diego, CA 92121
Richard A McCormack, President

5414 Ray W Hawksley Company
220 Cutting Boulevard 510-235-5780
Richmond, CA 94804

5415 Raytheon Engineers and Constructors
13105 NW Freeway 281-488-5510
Suite 200 Fax: 281-280-4083
Houston, TX 77040-6343
Environmental testing firm.

5416 Recon Environmental Corporation
5 Johnson Drive 908-526-1000
PO Box 130 Fax: 908-526-7886
Raritan, NJ 08869-1651
Environmental engineering, consulting, and laboratory services. Research results published in project reports and government publications.
Norman J Weinstein, President

5417 Recon Systems
5 Johnson Drive 908-526-1000
PO Box 130 Fax: 908-526-7886
Raritan, NJ 08869-0130

Dr. Norman J Weinstein, President

5418 Recra Environmental
10 Hazelwood Drive
Suite 110 800-527-3272
Amherst, NY 14228-2298 Fax: 716-691-2617
http://www.clu-in.org/products/site/complete/rcraenvi.htm
Research and development chemical and environmental measurement information.

Kenneth Kinecki, Technology Developer Contact

5419 Reed and Associates
2430 South Arlington Heights Road 847-718-0101
Arlington Heights, IL 60005 Fax: 847-718-0202
Environmental testing laboratory.

5420 Reid, Quebe, Allison, Wilcox & Associates
4755 Kingsway Drive 317-255-6060
Suite 400 Fax: 317-255-8354
Indianapolis, IN 46205-1570
Architectural and environmental engineering research firm.
J Edward Doyle, President

5421 Reliance Laboratories
Benedum Airport Industrial Park 304-842-5285
PO Box 625 Fax: 304-842-5351
Bridgeport, WV 26330
William F Kirk Jr, President

5422 Remtech
110 12th Street NW 205-682-7900
Suite E 106 Fax: 205-682-7953
Birmingham, AL 35203
Systems design and engineering firm specializing in energy and environmental control applications. Research results published in project reports and are presented in papers at conferences.
Gene Fuller, President

5423 Research Planning
1121 Park Street 803-256-7322
Columbia, SC 29201-3140 Fax: 803-254-6445
Scientific consulting firm specializing in the environment and natural resource assessment. Extensive experience in field surveys, EIS, spatial data analysis, and international work in Central America, West Africa and the Middle East. Research results published in professional journals, proceedings, and project reports. Woman-owned, small business concern.
Jacqueline Michel, President

5424 Resource Technologies Corporation
248 E Calder Way 814-237-4009
Suite 300 Fax: 814-237-1769
State College, PA 16801 E-mail: clients@resourcetec.com
http://www.resourcetec.com

An independent research, development and technical services firm located in central Pennsylvania. Specializes in appraisal and assessment services, information system development, assessment appeals and digitalmapping, web basedapplications, geotechnical services, environmental and ecological analysis and planning and management services.
Jeffrey R Stern, President
Ronald W Stingelin, Contact

5425 Resource Technologies Group
9210 Sky Park Court 858-637-7410
San Diego, CA 92123 Fax: 858-637-7411
Research and consulting firm specializing in process development, biosensors, detection, and environmental toxicology.

5426 Resources for the Future
1616 P Street NW 202-328-5000
Washington, DC 20036 Fax: 202-939-3460
E-mail: info@rff.org
http://www.rff.org
RFF is a nonprofit and nonpartisan think tank located in Washington DC that conducts independent research-rooted primarily in economics and other social sciences on environmental and natural resource issues. RFF was founded in 1952.

Lesli Creedon, Vice Pres., External Aff.
Stan Wellborn, Diretor of Communications

5427 Responsive Management
130 Fraklin Street 540-432-1888
Harrisonburg, VA 22801 Fax: 540-432-1892
E-mail: mdduda@rica.net
http://www.responsivemanagement.com
Responsive Management is a Virginia-based public opinion polling and survey research firm specializing in fisheries, wildlife, natural resource, outdoor recreation and environmental issues.
Mark Duda, Executive Director

5428 Revet Environmental and Analytical Lab
181 Cedar Hill Street 508-460-7600
Marlborough, MA 01752-3035 Fax: 508-460-7777
Environmental analysis and consulting laboratory.
V Taylor, President

5429 Ricerca Biosciences LLC
7528 Auburn Road 440-357-3300
PO Box 1000 Fax: 440-354-6276
Concord, OH 44077-1000 E-mail: info@ricerca.com
http://www.ricerca.com
Ricerca, a premier solution provider, offers expertise in both biology and chemistry to enable life sciences companies to fully leverage integrated, cost-effective, best practices approach to lead optimization and drug development. Services include in-vitro/in-vivo ADME, pharmacology, toxicology, medicinal, process, analytical chemistry, cGMP API scale-up production, regulatory support.

Thomas Bradshaw, President & CEO
Detlef Rethage, Executive VP - Operations

5430 Rich Technology
2410 Devonshire Drive 815-229-1122
Rockford, IL 61107-1534 Fax: 815-229-1525
Environmental engineering research firm.

5431 Riviana Foods: RVR Package Testing Center
1702 Taylor Street 713-861-8221
Houston, TX 77007 Fax: 713-861-9939
Lejo C Brana, Director Packaging

5432 Robert Bosch Corporation
32104 State Road 2 574-237-2100
New Carlisle, IN 46552-9605 Fax: 219-654-8755
Controlled-road environmental testing of automotive components for passenger cars, trucks, buses, tractor-trailers and off-road vehicles; certification to federal brake commission and fuel economy requirements.

5433 Robert D Niehaus
5951 Encina Road
Suite 105
Santa Barbara, CA 93117-6248
805-962-0611
Fax: 805-962-0097
Socioeconomic and environmental planning organization. Research results published in reports.

5434 Robert H Wilder
6706 Maxalea Road
Baltimore, MD 21239
410-377-6533
Fax: 410-377-0825

Charles E Sells, President

5435 Rone Engineers
8908 Ambassador Row
Dallas, TX 75247
214-630-9745
Fax: 214-630-9819
http://www.roneengineers.com
Provider of Geotechnical, Construction Materials Testing and Environmental Consulting services throughout Texas and the Southwest.
Mark Gray, PE, Geotechnical & Engineering

5436 Roux Associates
209 Shafter Street
Islandia, NY 11749
631-232-2600
Fax: 631-232-9898
E-mail: sisadiker@rouxinc.com
http://www.rouxinc.com
Environmental Consulting and Management.

Steve Sadiker, Vice President

5437 Rummel, Klepper & Kahl
81 Mosher Street
Baltimore, MD 21217-4243
410-728-2900
800-787-3755
Fax: 410-728-2992
http://www.rkkengineers.com
Civil, site, transpotation, environmental, structural engineering services.

5438 S-F Analytical Laboratories
6125 W National Avenue
PO Box 14513
Milwaukee, WI 53214
414-475-6700
800-300-6700
Fax: 414-475-7216
E-mail: dkliber@sflabs.com
http://www.sflabs.com
Environmental and materials testing laboratory.

David L Kliber, President/CEO

5439 SCS Engineers
3900kilroy Airport Way
Suite 100
Long Beach, CA 90806
562-426-9544
Fax: 562-427-0805
E-mail: service@scsengineers.com
http://www.scsengineers.com
Delivers economically and environmentally sound solutions for solid waste management and site remediation projects throughout the world. Provides engineering, construction, and contract operations services to private and public sectorclients through a network of more than 40 offices and 500 professional staff working in the US and abroad.

5440 SGI International
1200 Prospect Street
Suite 325
La Jolla, CA 92037-3660
858-551-1090
Fax: 858-551-0247
E-mail: info@sgiinternational.com
http://www.sgiinternational.com
Environmental applications.
Michael L Rose, President

5441 SGS Environmental Services Inc
200 W Potter Drive
Anchorage, AK 99518
907-562-2343
Fax: 907-562-0119
E-mail: george.wolters@sgs.com
http://www.us.sgs.com
Environmental laboratory services.

Chuck Homestead, General Manager
George Wolters, Business Development

5442 SHB AGRA
3232 W Virginia Avenue
Phoenix, AZ 85009-1502
602-995-3916
Fax: 602-995-3921
Geotechnical and environmental research firm.

5443 SJS Archeological Services
Continental Business Center, A-10
Front and Ford Streets
Bridgeport, PA 19405
215-272-3144
Fax: 215-272-3144
Glenn W Sheehan, Treasurer

5444 SMC Martin
501 Allendale
King of Prussia, PA 19406
215-265-2700
Fax: 215-337-1875
Daniel Shoemaker, President

5445 SP
45 Congress Street Suite 4
PO Box 848
Salem, MA 01970-0948
978-745-4569
Fax: 978-745-4881
Bruce Poole, Contact

5446 SPECTROGRAM Corporation
287 Boston Post Road
Madison, CT 06443
203-318-0535
Fax: 203-318-0535
E-mail: spectrogram@msn.com
http://www.spectrogram.com
Reseach, development and manufacturing firm which produces analytical instrumentation and systems in the fields of analytical chemistry (environmental) and elastomeric physical testing (rubber and plastics).

HR Gram, President

5447 STL Denver
4955 Yarrow Street
Arvada, CO 80002
303-736-0100
800-572-8958
Fax: 303-431-7171
Testing and analysis services.

5448 STS Consultants
750 Corporate Woods Parkway
Vernon Hills, IL 60061-3153
847-279-2500
800-859-7871
Fax: 847-279-2510
http://stsltd.com
Consulting engineering firm offering an integrated package of services in geotechnical engineering, waste management, environmental management, and construction technology.
Thomas W Wolf, CEO

5449 STS Consultants
111 Pfingsten Road
Northbrook, IL 60062
630-272-6520
Fax: 847-498-2721

Mike Russell, President

5450 Samtest
3730 James Savage Road
Midland, MI 48642-6517
989-496-3610
Fax: 989-496-3190
Geotechnical and environmental services firm.

5451 Sari Sommarstrom
PO Box 219
Etna, CA 96027
530-467-5783
Fax: 530-467-3623
E-mail: sari@sisqtel.net

5452 Savannah Laboratories
PO Box 13548
Savannah, GA 31416-0548
912-354-7854
Fax: 912-352-0165
Environmental and biological research and testing laboratory with expertise in fish farming technology.

5453 Science Applications International Corporation
221 Third Street
Newport, RI 02840-1087
401-847-4210
Fax: 401-849-1585
http://www.saic.com

Environmental applications.
Donald Jagoe, CEO

5454 Scientific & Technical Resources
11-D Union Valley Road
Oak Ridge, TN 37830-8045
865-481-6088
Fax: 865-481-6057
E-mail: nrevis@str.nxs.net
http://www.scitechresources.com
Research laboratory for environmental, medical, and toxicology studies. Data storage and retrieval, technical staff augmentation. Research results published in professional journals.
Nathaniel W Revis, President

5455 Scientific Associates
1639 12th Street
Santa Monica, CA 90404
310-450-1334
Fax: 310-450-1364

Dr. Eugene B Nebeker, President

5456 Scitest
Route 66 Professional Center
PO Box 339
Randolph, VT 05060-0339
802-728-6313
Fax: 802-728-6044

Environmental testing and analysis laboratory.
Roderick J Lamothe, President

5457 Separation Systems Technology
4901 Morena Boulevard
Suite 809
San Diego, CA 92117-7325
858-581-3765
Fax: 858-581-1211
E-mail: riley1034@aol.com
Environmental research.
Robert L Riley, President

5458 Shannon & Wilson
PO Box 300303
Seattle, WA 98103-8636
206-632-8020
Fax: 206-695-6777
http://www.shannonwilson.com

Environmental research.

5459 Sheladia Associates
15825 Shady Grove Road
Suite 100
Rockville, MD 20850-4023
301-590-3939
Fax: 301-948-7174

Consulting firm specializing in environmental studies. Research results published in research reports for the government.

A Moytayek, President

5460 Shell Engineering and Associates
2403 West Ash
Columbia, MO 65203
573-445-0106
Fax: 573-445-0137

Harvey D Shell, COO

5461 Sherry Laboratories
PO Box 9-2662316 Mecca Drive
Lafayette, LA 70509
337-235-0483
Fax: 337-233-6540

Analytical environmental laboratory.
Mel Burnell, President

5462 Shive-Hattery Engineers & Architects
800 1st Street Northwest
PO Box 1803
Cedar Rapids, IA 52406
319-364-0227
Fax: 319-364-4251

Donald P Hattery, Chairman

5463 Sidney R Frank Group
229 El Monte Drive
Santa Barbara, CA 93109

SR Frank, President

5464 Siebe Appliance Controls
2809 Emerywood Parkway
Richmond, VA 23294-3743
804-756-6500
Fax: 804-756-6563
Automatic temperature, environmental, electronic appliance, heating, cooling and gas safety controls and valves; thermostats and oven burners.

5465 Simon Hydro-Search
401 W Main Street
#222
Norman, OK 73069
405-329-8300
Fax: 405-366-8722

Craig Eisen, President

5466 Simons and Associates
2601 S Lemay Avenue
Fort Collins, CO 80525-2247
970-223-3957
Fax: 970-223-9958
Civil, water resources, and environmental engineering consulting firm. Research results published in internal reports. Publishes 120 reports and publications per year.

5467 Simpson Electric Company
853 Dundee Avenue # 859
Elgin, IL 60120-3090
847-697-2260
Fax: 847-697-2272
Analog and digital panel meters, meter relays, controllers, volt-ohm-milliammeters, scopes and industrial and environmental test instruments.

5468 Skinner and Sherman Laboratories
1st Avenue
Waltham, MA 02451
781-890-7200
Fax: 781-890-7003

5469 Smith & Mahoney
540 Broadway
PO 22047
Albany, NY 12201-2047
518-463-4107
Fax: 518-463-3823

Michael W McNarney, President

5470 Snell Environmental Group
1425 Keystone Avenue
PO Box 22127
Lansing, MI 48909-2127
517-393-6800
Fax: 517-272-7390
E-mail: seg-adm@ix.netcom.com
http://www.dlzcorp.com

Consulting structural engineers.
John O'Mallia, President

5471 Socio Technical Research Applications
1101 Wilson Boulevard
Suite 1950
Arlington, VA 22209-2277
703-243-9100
Fax: 703-243-9455
Environmental research firm.
Manuel Gallardo, President

5472 Soil Engineering Testing
9301 Bryant Avenue S
Suite 107
Bloomington, MN 55420
952-884-6833
Fax: 952-884-6923
http://www.soilengineeringtesting.com/
A comprehensive soil mechanics laboratory facility for engineering disciplines, environmental and hydrological applications.
Gordon R Eischens, President

5473 Solar Energy Research Institute
United States Department of Energy
1617 Cole Boulevard
Golden, CO 80401-3305
303-275-4700
Fax: 303-275-4788
Created by the Solar Energy Research, Development and Demonstration Act of 1974, which authorized a federal program aimed at developing solar energy as a viable source of the nation's future energy needs. SERI conducts and coordinatessolar research, technology development and testing functions as developed by the US Department of Energy.

5474 Solar Power Engineering Company
PO Box 91
Morrison, CO 80465
303-697-8144
Fax: 303-781-8568

Harrison C Wroton, President

5475 Solar Resources
623 Delores Road
Taos, NM 87571-1337
505-758-9788
Fax: 505-751-7125
E-mail: skenin@newmex.com

Stephen Kenin, President

5476 Solar Testing Laboratories
5399 Lancaster Drive
Brooklyn Heights, OH 44131-1848
216-741-7007
Fax: 216-741-7011
Geotechnical, environmental engineering, materials testing, and construction inspection laboratory.
George J Ata, President

5477 Solonex
331 Cornelia Street
Plattsburgh, NY 12901
518-561-3160
Fax: 514-747-3906

Alex Kalil, President

5478 Solsorce, Inc
1700 W Cloverdale Drive
Appleton, WI 54911
920-585-2888

5479 Southeastern Engineering & Testing Laboratories
4761 SW 51st Street
Davie, FL 33314-5525
954-584-4322
E-mail: jack@seetl.com
http://www.seetl.com
Geotechnical and environmental engineering consulting firm and construction materials engineering laboratory.

5480 Southern Petroleum Laboratory
PO Box 20807
Houston, TX 77225-0807
713-660-0901
800-969-6773
Fax: 713-660-8975
E-mail: HRBrown@spl-inc.com
National environmental and petrochemical lab network with focus on customer satisfaction and adherence to highest quality standards.
HR Brown, President

5481 Southern Research Institute
PO Box 55305
Birmingham, AL 35255-5305
205-581-2000
800-967-6774
Fax: 205-581-2568
E-mail: southern@sri.org
http://www.southernresearch.com
Southern Research Institute is an independent research corporation with established capabilities in pharmaceutical discovery and development, engineering, chemical and biological defense, environmental and energy-related sciences.Research is conducted through contracts and grants with government and commerical clients.
Robert C Lonergan, President

5482 Southern Testing & Research Laboratories
3809 Airport Drive
Wilson, NC 27896-8653
252-237-4175
Fax: 252-237-9341
http://www.southerntesting.com
Full-service laboratory with over 100 chemists, microbiologists and support personnel that provides personalized service to clients. Capabilities include pharmaceutical, foods and feeds, environmental, industrial hygiene, agriculturaland microbiological sciences. Laboratory is FDA-inspected GLP/cGMP laboratory utilizing AOAC, USP, EPA, USDA, AACC, AOCS, ISO, client and in-house validated methods.
Kim Bauchman PhD, Managing Director
Walter Hogg, Business Development

5483 Southwestern Laboratories
PO Box 8768
Houston, TX 77249-8768
713-692-9151
Fax: 713-696-6307
Materials analysis and testing laboratory offering environmental control services.

5484 Spears Professional Environmental & Archeological Research Service
13858 South Hwy 170
West Fork, AR 72774-9491
479-839-3663
Fax: 479-839-2575
Archeological research service.
Carol S Spears, President

5485 Spectrochem Laboratories
545 Commerce Street
Franklin Lakes, NJ 07417-1309
201-337-4774
Fax: 201-337-1255
Research and development firm specializing in environmental sciences and inorganic chemistry. Research results published in proceedings at technical conferences.
Irene Van Dren, President

5486 Spectrum Research
PO Box 122
Montpelier, VT 05602
802-223-7088
Fax: 802-223-7088
Fred Kent, President

5487 Spectrum Sciences and Software
PO Box 8
Mary Esther, FL 32569-0008
850-796-0909
Fax: 850-244-9650
E-mail: lcars@specsci.com
http://www.specsci.com
Environmental research.
Donal R Myrick, President

5488 Spotts, Stevens and McCoy
1047 North Park Road
Reading, PA 19610
610-621-2000
Fax: 610-621-2001
E-mail: information@ssmgroup.com
http://www.ssmgroup.com
An engineering and consulting firm, serving business, industry, and government for more than 70 years.

5489 St. Louis Testing Laboratories
2810 Clark Avenue
Saint Louis, MO 63103-2506
314-531-8080
Fax: 314-531-8085
Research and testing laboratory specializing in chemical, metallurgical, nondestructive and environmental testing and field services. Research results published in project reports.
W Trowbridge, President

5490 Standard Testing and Engineering
3400 Lincoln Boulevard
Oklahoma City, OK 73105
405-528-0541
Fax: 405-528-0559
Thomas J Kelly, President

5491 Stanford Technology Corporation
57 Poplar Street
PO Box 2100D
Glenbrook, CT 06906-0100
203-348-4080
Fax: 203-327-5225
E-mail: stctestlab@aol.com
High technology research firm. Research results published in confidential project reports.
Charles C Cullari, President
Gerald T Ciccone, VP

5492 Stantech: Division of Standard Testing
4300 Lincoln Boulevard
Oklahoma City, OK 73105
405-424-8378
Fax: 405-528-0559
Environmental research firm.
Thomas J Kelly, President

5493 Steven Winter Associates
50 Washington
Norwalk, CT 06854
203-852-0110
Fax: 203-852-0741

Steven Winter, President

5494 Stillwell & Gladding
130 Cedar Street
New York, NY 10006
212-732-4033
Fax: 212-732-4034

Leonard Maltese, President

5495 Stilson Laboratories
6121 Huntley Road 614-848-4333
Columbus, OH 43229-1003 Fax: 614-841-0818
Environmental testing laboratory. Research results published in professional journals and project reports.

5496 Stone Environmental
58 East State Street 802-229-4541
Montpelier, VT 05602-3043 800-959-9987
Fax: 802-229-5417
E-mail: sei@stone-env.com
http://www.stone-env.com
Environmental applications.
Christopher Stone, President

5497 Suburban Laboratories
4140 Litt Drive 708-544-3260
Hillside, IL 60162-1120 Fax: 708-544-8587
Environmental laboratory providing chemical, chromatographic, and spectrographic analysis of biological materials, including water and groundwater, soil, and hazardous materials for priority pollutants, metals, and pesticide residues.

5498 Sunsearch
PO Box 590 203-453-6591
Guilford, CT 06437 Fax: 203-458-9011
E-mail: ebarber@sunsearchinc.com
http://www.sunsearchinc.com

Everett M Barber Jr, President

5499 Swanson Environmental
24156 Haggerty Road 313-478-2700
Farmington Hills, MI 48335 Fax: 313-478-3819

5500 Sylvanus Environmental
45 West Street 919-545-0552
PO Box 848 Fax: 919-545-0553
Pittsbro, NC 27312-0848
Environmental studies.
Steve Freedman, President

5501 Systech Corporation
245 North Valley Road 937-372-8077
Xenia, OH 45385 Fax: 937-372-8099
Michael J Balchunas, President

5502 Systech Environmental Corporation
3085 Woodman Drive 937-643-1240
Suite 300 800-888-8011
Dayton, OH 45420-1159 Fax: 937-643-1203
E-mail: Erica.Hawk@lafarge-na.com
http://www.sysenv.com
Provider of alternative fuels to cement kilns.
Erica Hawk, Corporate Mktg Specialist

5503 Systems Applications International
101 Lucas Valley Road 415-507-7100
San Rafael, CA 94903-1791 Fax: 415-507-7177
Environmental management consulting firm.

C Shepherd Burton, President

5504 TECH
333 E Main Street 505-327-3311
Farmington, NM 87401-2740 Fax: 505-325-3311
Firm providing environmental and quality control for the construction industry.

5505 TMA/Norcal
2030 Wright Avenue 510-235-2633
Richmond, CA 94804 Fax: 510-235-0438

Dan Stremer, Marketing Director

5506 TRAC Laboratories
113 Cedar Street 940-566-3359
PO Box 215 Fax: 940-566-2698
Denton, TX 76201 http://www.traclaboratories.com
Provide multi-disciplinary problem-solving approaches to environmental and public health issues.
Dr. Barney J Venables, Research Driector
Stephen B Junot, Laboratory Director

5507 TRC Environmental Corporation
21 Griffin Road North 860-298-9692
Windsor, CT 06095 Fax: 860-298-6399
E-mail: czoephel@tresolutions.com
http://www.trcsolutions.com
A leading provider of engineering, financial, risk managemnet and construction services to large industrial and government customers throughout the United States. TRC's 2,700 employees provide customer focused solutions in threeprimary markets: environmental, energy and infrastructure.

Carl Zoephel, Dir. Business Development

5508 TRC Garrow Associates
3772 Pleasantdale Road 770-270-1192
Suite 200 Fax: 770-270-1392
Atlanta, GA 30340-4270 E-mail: bgarrow@trcgarrow.com
http://www.trcgarrow.com
Environmental analysis and planning firm.
Barbara Garrow, President

5509 Taka Asbestos Analytical Services
8 Pine Hill Court 631-261-2117
Northport, NY 11768-3441
Environmental testing and analysis firm.

5510 Talos
460 Herndon Parkway 703-715-3500
Herndon, VA 20170-5278 Fax: 703-471-1228
Environmental computer company.

5511 Taylor Engineering
9000 Cypress Green Drive 904-731-7040
Suite 200 Fax: 904-731-9847
Jacksonville, FL 32256 http://www.taylorengineering.com
Services in coastal engineering consulting, dredging and dredged material management, hydrology and hydraulics, environmental services, and construction support services.
Bruce Taylor, President

5512 Tech Reps
5000 Marble Avenue Northeast 505-266-5678
Albuquerque, NM 87110 Fax: 505-260-1163

Donald E Tiano, CEO/President

5513 Tech-Art
400 Pacific Avenue 415-362-1110
Ground Floor Fax: 415-362-2811
San Francisco, CA 94133 E-mail: info@techart.com
http://www.techart.com

5514 Techrad
4619 North Santa Fe 405-528-7016
Oklahoma City, OK 73118 Fax: 405-528-3346
Dawn Rickard, President

5515 Teledyne Isotopes
50 Van Buren Avenue 201-664-7070
Westwood, NJ 07675-3193 Fax: 201-664-5586
Research firm specializing in environmental monitoring of radioactive materials.
DF Schutz, President

5516 Tellus Institute
11 Arlington Street 617-266-5400
Boston, MA 02116-3406 Fax: 617-266-8303

E-mail: info@telllus.org
http://www.tellus.org

Environmental research.
Paul D Raskin PhD, President

5517 Tenco Laboratories
1152 Junction Avenue 219-322-0450
Schererville, IN 46375 Fax: 219-322-0440

Robert W Hoole, Marketing Director

5518 Tepra Tech
11629 Central Street 781-344-6446
Stoughton, MA 02072 Fax: 781-575-8915
Environmental consulting firm. Research results published in
presentations, journals, and newsletters.
Eliot Epstein, Chief Environmental Scient.

5519 Testing Engineers & Consultants
1343 Rochester Road 248-588-6200
Troy, MI 48083-6022 Fax: 248-588-6232
E-mail: tec@tectest.com
http://www.testingengineers.com
TEC is a certified Woman Business Enterprise specializing in
environmental and geotechnical engineering, materials testing,
roof systems management, facility asset management, and indoor air quality. Headquartered in Troy with branchoffices in
Ann Arbor and Detroit, Michigan.

Katherine Banicki, President
Duncan R Mein PE, Mgr, Environ. Assesment

5520 Tetra Tech
56 W Main Street 302-738-7551
Suite 400 800-462-0910
Christiana, DE 19702 Fax: 302-454-5980
http://www.tetratech.com
Tetra Tech is a leading provider of specialized management consulting and technical services in three principal business areas:
resource management, infrastructure and communication. The
company's clients include a diverse base ofpublic and private
sector organizations serviced through more than 200 offices located in the US and internationally.

Arkan Say, PE, VP

5521 Thermo Analytical
160 Taylor Street 626-357-3247
Monrovia, CA 91016 Fax: 626-359-5036

Michelle Miller, President

5522 Thermo Electron Corporation
81 Wyman Street 781-622-1000
Waltham, MA 02254 Fax: 781-622-1207

Dr. George Hatsopoulos, President

5523 ThermoRetec Corporation
1011 SW Klickitat Way 206-624-9349
Suite 207 Fax: 206-624-2839
Seattle, WA 98134-1162 http://www.thermoretec.com
Testing laboratory; environmental consultants.

5524 Thermoenergy Corporation
323 Center Street 501-376-6477
Suite 1300 Fax: 501-375-5249
Little Rock, AR 72201-2628 http://www.thermoenergy.com
Environmental research.
Dennis C Cossey, CEO

5525 Thermotron Industries
291 Kollen Park Drive 616-392-1491
Holland, MI 49423-3487 Fax: 616-392-5643
E-mail: info@thermotron.com
http://www.thermotron.com
Environmental test chambers, instruments, electrodynamic vibration systems and test equipment systems integration

5526 Thompson Engineering Testing
3707 Cottage Hill Road 251-666-2443
Mobile, AL 36609 Fax: 251-666-6422
E-mail: info@thompsonengineering.com
http://www.thompsonengineering.com
A multi disciplined engineering design, environmental consulting, construction management, construction inspection and materials testing firm.

Henry R Seawell III PE, Chairman/CEO
James H Shumock CPA, CFO

5527 Thornton Laboratories
1145 E Cass Street 813-223-9702
PO Box 2880 Fax: 813-223-9332
Tampa, FL 33601 E-mail: randy.cigarran@thorntonlab.com
http://www.thorntonlab.com
Environmental sampling and testing laboratory and general ana-
lytical testing lab.
Randy Cigarran, Project Manager

5528 Tighe & Bond
53 Southampton Road 413-562-1600
Suite 3 Fax: 413-562-5317
Westfield, MA 01085-5308 E-mail: info@tighebond.com
http://www.tighebond.com
Civil/environmental engineering.
David G Healey, President
John W Power, CEO

5529 Timber Products Inspection and Testing
1641 Sigman Road 770-922-8000
PO Box 919 Fax: 770-922-1290
Conyers, GA 30012 E-mail: cbarber@tpinspection.com
http://www.tpinspection.com
Chris Barber, Laboratory Manager

5530 Total Design Four
5700 South Staples 361-993-6980
#F-5 Fax: 361-993-6426
Corpus Christi, TX 78413-3703
Clarence Upchurch, President

5531 Touch Vision Interactive Kiosks
330 Main Street 562-626-8200
Suite 201 Fax: 562-626-8203
Seal Beach, CA 90740-6352
Environmental assessment firm.
Larry Mahar, President

5532 Towne, Richards and Chaudiere
105 Northeast 56th Street 206-523-3350
Seattle, WA 98105
Robin M Towne, President

5533 Townley Laboratories
1750 West Front Street 908-757-1137
Plainfield, NJ 07063 888-TOW-NLEY
Fax: 908-757-0335
E-mail: townley@eclipse.net
Environmental testing laboratory analyzing and sampling of
drinking water, wastewater, soil and hazardous waste. NJDEP
certified.

TR Komline, President

5534 Tox Scan
42 Hangar Way 831-724-4522
Watsonville, CA 95076 Fax: 831-724-3188
E-mail: pcarpenter@toxscan.com
http://www.toxscan.com
Environmental and hazardous waste analytical chemistry;
bioassay and bioaccumulation testing, specializing in marine
applications and low-detection limit analyses.

Philip D Carpenter, CEO

5535 Trac Laboratories

113 Cedar Street
Denton, TX 76201-4101

940-566-3359
Fax: 940-566-2698
E-mail: traclabs@gte.net

Environmental research.
Barney J Venables, President

5536 Trace Minerals International

6545 Gun Park Drive
Suite 240
Boulder, CO 80301

303-530-5135
Fax: 303-530-5296

E Blaurock-Busch, PhD, President

5537 Transviron

1624 York Road
Lutherville, MD 21093-5603

410-321-6961
Fax: 410-494-9321

Civil and environmental engineering firm.

5538 Tri-State Laboratories

2870 Salt Springs Road
Youngstown, OH 44509

330-797-8844
Fax: 330-797-3264
E-mail: trislabs@aol.com
http://www.tristatelabs.net

Environmental testing labs.

A Bari Lateef PhD, CEO
Wendy Hanna, COO

5539 Tri-Tech Laboratories

599 Waldron Road
La Vergne, TN 37086-4109

615-793-7547
Fax: 615-793-5070
E-mail: TTLINC@home.com

Environmental laboratory specializing in waste water studies.
Research results published in conference proceedings.
Wendy Ingram, Lab Director

5540 Tribble & Richardson

4875 Riverside Drive
Suite 1
Macon, GA 31210-1198

912-474-6100
Fax: 912-474-8933

Consulting engineers, surveyors, planners, and laboratory services in environmentsl protection.

5541 Turner Laboratories

2445 North Coyote Drive
Suite 104
Tucson, AZ 85745

520-882-5880
Fax: 520-882-9788
E-mail: nturner@turnerlabs.com
http://www.turnerlabs.com

Water, wastewater, soil, and environmental services.

Nancy D Turner, President
Shari Bauman, Lab Director

5542 UEC Industrial Hygiene and Environmental Health Laboratories

4000 Tech Center Drive
Monroeville, PA 15146-3057

412-825-2400
888-LAB-SUCC
Fax: 412-825-2407
E-mail: ratheys@uss.com
http://www.uec.com/labs/ih

General industrial hygiene consulting laboratory and field services.

5543 URS

911 Wilshire Boulevard
Suite 800
Los Angeles, CA 90017-3436

213-996-2200
Fax: 213-996-2290

Environmental analysis and comprehensive engineering services.

5544 US Filter/Control Systems

1239 Willow Lake Boulevard
Vadnais Heights, MN 55110-5145

651-766-2700
800-224-9474
Fax: 651-766-2701
http://www.controlsystems.usfilter.com

Environmental devices and controls.

5545 US Public Interest Research Group

218 D Street SE
Washington, DC 20003

202-546-9707
Fax: 202-546-2461
E-mail: uspirg@pirg.org
http://www.pirg.org

US PIRG is an advocate for the public interest. We uncover threats to public health and well-being and fight to end them, using the time-tested tools of investigative research, media exposes, grassroots organizing, advocacy andlitigation.
Anna Aurilio, Legislative Director

5546 USDA Forest Service: Pacific Southwest Researc Station

PO Box 245
Berkeley, CA 94701

510-559-6300
Fax: 510-559-6440
http://www.psw.fs.fed.us

A Governmental Research Organization specializing in research on forest ecosystems, including fire, watersheds, forest genetics and diversity, wildlife, forest diseases, and urban forestry.

5547 Umpqua Research Company

125 Volunteer Way
PO Box 609
Myrtle Creek, OR 97457-0102

541-863-7770
Fax: 541-863-7775
E-mail: umpqua@urcmail.net
http://www.umpqua-research.com

Environmental research.
James E Atwater, President

5548 Underground Tank Testing & Service

PO Box 8148
Greenville, NC 27835-8148

252-758-0001
Fax: 252-758-9652

Environmental and geological testing and analysis firm.

5549 United Environment Systems

PO Box 524
Chester, NJ 07930-0524

973-927-1488
Fax: 973-927-5598

Environmental testing and analysis firm.

5550 United States Land, Air, and Water Environment Industries

PO Box 351
Cape May, NJ 08204-0351

609-884-7595
Fax: 609-884-7595

Research and development service concentrating on environmental conservation.
Joseph Poole, President

5551 Universal Environmental Technologies

87 Technology Way
Nashua, NH 03060-4220

603-883-9312
Fax: 603-883-9314
E-mail: uetm@aol.com
http://www.techexpo.com/firms/univenvi.html

Environmental research firm.
Sharon McMillin, VP/Remedial Services

5552 Upstate Laboratories

Eastwood Station
PO Box 289
Syracuse, NY

315-437-0255
Fax: 315-437-1209

Testing laboratory specializing in environmental and organic/synthetic analysis.

5553 Vapex Environmental Technologies

480 Neponset Street
Canton, MA 02021-1971

781-821-5560
Fax: 781-821-4967

Environmental technology firm.

5554 Vara International: Division of Calgon Corporation

1201 19th Place
Suite 400
Vero Beach, FL 32960-3599

561-567-1320
Fax: 561-567-4108
E-mail: vara@calgoncarbon.com
http://www.varatechnologies.com

Environmental and industrial process research firm.
Michael S Thomas, President

5555 Versar
6850 Versar Center
PO Box 1549
Springfield, VA 22151-0549
703-750-3000
800-283-7727
Fax: 703-642-6825
E-mail: whitelaw@versar.com
http://www.versar.com
Engineering and environmental research organization. Research results published in project reports, government publications, books, articles, and technical reports.
Theodore Prosciv, President

5556 Villanueva Associates
299 Alhambra Circle
Suite 406
Coral Gables, FL 33134-5114
305-448-7274
Fax: 305-448-7274
Civil and environmental engineering firm.
Plinio M Villanueva, President

5557 Vista Research
755 N Mary Avenue
Sunnyvale, CA 94086-2909
408-830-3300
Fax: 408-830-3399
E-mail: info@vrinc.com
http://www.vistaleakdetection.com
Applied and product research in environmental studies.
Harold Guthard, President

5558 Volumetric Techniques
317 Bernice Drive
Bayport, NY 11705
631-472-4848
Fax: 631-472-4991
http://www.strux.com
Laboratory providing environmental sampling and analysis services.
Sander Sternig, President

5559 WCH Industries
PO Box 441040
Fort Washington, MD 20749-1040
301-292-6460
Fax: 301-292-6282
Environmental engineering and consulting firm.

Bill Clark Harrison, President

5560 WERC: Consortium for Environmental Education & Technology Development
New Mexico State University
Box 30001, MSC WERC
Las Cruces, NM 88003
505-646-2038
800-523-5996
Fax: 505-646-5474
E-mail: aghassem@nmsu.edu
http://www.werc.net
A consortium focusing on Environmental Education including health, energy, environment, sustainability and Food Safety as well as Research and outreach in the listed areas.

Dr Abbas Ghassemi, Executive Director

5561 WQS Environmental Laboratory
17459 Village Green Drive
Houston, TX 77040
713-466-0958
Fax: 713-466-9882
Chemical and environmental analysis laboratory.

5562 Wadsworth/Alert Laboratories
PO Box 29124101 Shuffel Drive NW
North Canton, OH 44720
330-497-9396
Fax: 330-497-0772
Environmental testing and analysis laboratory.

5563 Waid & Associates
14205 N Mo Pac Expressway
Suite 500
Austin, TX 78728-6530
512-255-9999
Fax: 512-255-8780
E-mail: waid@waid.com
http://www.waid.com
Environmental engineering firm.
Patrick Murin, PE, President

5564 Wallgren Environmental Services
435 Isam Road
Suite 228
San Antonio, TX 78216
210-340-0343
Fax: 210-344-5407

Environmental testing and analysis firm.

5565 Waste Compliance Services
12255 Wormer
Redford, MI 48239
313-255-9601
Fax: 313-255-9205
Vishnu Peketi, Laboratory Manager

5566 Waste Water Engineers
5751 Old Hickory Boulevard
Suite 207
Hermitage, TN 37076-2046
615-883-7100
Fax: 615-889-6101
Environmental science research consultant. Research results published in project reports. Environmental civil engineering consultant.
Thomas H Patton Jr, President

5567 Water Quality Services
17459 Village Green Drive
Houston, TX 77040
713-466-0958
Fax: 713-466-9882
Dr. R Kevin Gibben, Acting Manager

5568 Water Resources Associates
4041 N Central
Suite 1050
Phoenix, AZ 85012-3393
602-248-8808
Fax: 602-248-7722
Michael Noble, President

5569 Water and Air Research
6821 SW Archer Road
Gainesville, FL 32608-4748
352-372-1500
Fax: 352-378-1500
E-mail: info@waterandair.com
http://www.waterandair.com
Environmental research and consulting firm. Research results published in client reports.
William C Zegel, President
Connie Bieber, Director/Manager

5570 Watkins Environmental Sciences
7000 E Genesee Street
Suite A7
Fayetteville, NY 13066-1156
315-446-4763
Fax: 315-446-4764
Environmental assessments, septic system designs, residential water sampling and testing services, home inspections, radon.
Andrew A Watkins PE, President

5571 Weather Services Corporation
131A Great Road
Bedford, MA 01730
781-275-8860
Fax: 781-271-0178
Michael Leavitt, President

5572 West Coast Analytical Service
9840 Alburtis Avenue
Santa Fe Springs, CA 90670
562-948-2225
Fax: 562-948-5850
DJ Northington, PhD, President

5573 West Michigan Testing
815 E Ludington Avenue
Ludington, MI 49431
231-843-3353
Fax: 231-843-7676
We provide soil borings, geotechnical services, environmental assessments, construction materials testing and asbestos inspection.

James T Nordlund Jr, Vice President

5574 West More Mechanical Testing and Research
PO Box 388
Youngstown, PA 15696
724-537-8686
Fax: 724-537-3151
James Dague, Laboratory Manager

5575 Western Environmental Services
913 N Foster Road
Casper, WY 82601
307-234-5511
Fax: 307-234-8324
E-mail: aroylance@testair.com
http://www.testair.com
Air emission testing.

Alan Roylance, President
J Scott Mortimer, VP

5576 Western Michigan Environmental Services
3552 128th Avenue   616-399-6070
Holland, MI 49424   Fax: 616-399-6185

Cheryl A Dell, President

5577 Westinghouse Electric Company
11 Stanwix Street   412-244-2000
Pittsburgh, PA 15222   Fax: 412-642-4985

5578 Westinghouse Remediation Services
675 Park N Boulevard   404-298-7101
Suite F-100   Fax: 404-296-9752
Clarkston, GA 30021-1987
Environmental remediation firm.

5579 Weston Solutions, Inc
1400 Weston Way   610-701-3000
Box 2653   800-7WE-STON
West Chester, PA 19380   Fax: 610-701-3124
          E-mail: info@westonsolutions.com
             http://www.westonsolutions.com
Weston is a leading infrastructure redevelopment services firm delivering integrated environmental engineering solutions to industry and government worldwide. With an emphasis on creating lasting economic value for its clients, thecompany provides services in site remediation, redevelopment, infrastructure operations and knowledge management.
Patrick McCann, President/Chief Executive
William Robertson, Chairman

5580 Whale Scientific
4945 Monaco Street   303-289-4781
Commerce City, CO 80022-4609   Fax: 303-289-7957

Tony Leonardo, Production Manager

5581 Whibco
River Road   856-455-9200
PO Box 259   Fax: 856-455-8884
Leesburg, NJ 08327   http://www.whibco.com
Andrew R Strelczyk, Director Quality Control

5582 Wik Associates
PO Box 230   302-322-2558
New Castle, DE 19720-0230   Fax: 302-322-8921
Environmental testing and analysis firm.

5583 William T Lorenz & Company
3541 Norwegian Hollow Road   608-935-9285
Dodgeville, WI 53533-8702   Fax: 608-935-2010
Environmental and water resources marketing, consulting, and product research firm.

5584 William W Walker Jr
1127 Lowell Road   978-369-8061
Concord, MA 01742-5522   Fax: 978-369-4230
          E-mail: wwwalker@wwwalker.net
            http://www.wwwalker.net
William W Walker, Jr, Environmental Engineer

5585 Woods End Research Laboratory
20 Old Rome Road   207-293-2457
Mount Vernon, ME 04352   Fax: 207-293-2488
          E-mail: info@woodsend.org
            http://www.woodsend.org
Compost analysis; bioremediation design; solvita test kits for soil and compost. Quality Seal of Approval Program for Compost Products.

William Brinton, President

5586 World Resources Company
1600 Anderson Road   703-734-9800
Suite 200   Fax: 703-790-7245
Mc Lean, VA 22102 http://www.worldresourcescompany.com
World Resources Company (WRC) is a highly specialized environmental risk management company that designs, implements and manages recycling activities and provides environmental services for non-ferrous metal industries nationally andinternationally. This support includes regulatory, environmental, transportation, production, and all other aspects of business inherent to recycling services.
Peter T Halpin, CEO

5587 Yellowstone Environmental Science
320 South Wilson Avenue   406-586-3905
Bozeman, MT 59715   Fax: 406-587-5109

Mary M Hunter, Managing Partner

5588 Yes Technologies
320 S Willson Avenue   406-586-2002
Bozeman, MT 59715-4633   Fax: 406-586-8818
            E-mail: yes@yestech.com
             http://www.yestech.com
Environmental and public health research and development. Patent consulting.
Mary M Hunter, President

5589 Zimpro Environmental
301 West Military Road   715-359-7211
Rothschild, WI 54474   Fax: 715-355-3219
William Copa, VP Technical Services

5590 Zurn Industries
5900 Elwin Buchanan Drive   919-775-2255
Sanford, NC 27330-2767   800-997-3876
            Fax: 919-775-3541
Environmental systems including air, land, thermal and water; energy systems including steam and heat; mechanical systems.

University Centers

5591 Adirondack Ecological Center: SUNY College of Environmental Science & Forestry
6312 State Route 28N   518-582-4551
Science & Forestry, Huntington Forest   Fax: 518-582-2181
Newcomb, NY 12852   E-mail: aechwf@esf.edu
             http://www.esf.edu/aec/
Provides the organizational framework for research, instructional, and public service activities thoughout the Adriondack region.
Dr. William F Porter, Director

5592 Adirondack Lakes Survey Corporation
Route 86   518-897-1354
PO Box 296   Fax: 518-897-1364
Ray Brook, NY 12977 E-mail: admin@adirondacklakessurvey.org
          http://www.adirondacllakessurvey.org
Determines the extent and magnitude of acidification of lakes and ponds in the Adironack region.

5593 Agricultural Research and Development Center
University of Nebraska
1071 County Road G   402-624-8000
Ithaca, NE 68033   Fax: 402-624-8010
            http://ardc.unl.edu
Serves as the primary site for field based reseach with 5,000 acres of row crops and 5,000 domestic farm animals used for teaching and research.
Daniel J Duncan, Director

5594 Agromedicine Program Medical University of South Carolina
19 Hagwood Avenue   843-792-2281
Suite 305   800-922-5250
Charleston, SC 29425   Fax: 843-792-4702

E-mail: Simpsowm@musc.edu
http://www.musc.edu/oem/

Information, consultation, referral service for professional and lay persons involved in or in contact with agriculture or agricultural products.
Dr. William M Simpson, Jr., Medical Director
JoAnn Stokes, Administative Assistant

5595 Akron Center for Environmental Studies

University of Akron
215 Crouse Hall   330-972-5389
Akron, OH 44325   Fax: 330-972-7611
E-mail: ids@uakron.edu
http://www.uakron.edu/envstudies/

Ira D Sasowsky, Director

5596 Albrook Hydraulics Laboratory

Washington State University
PO Box 642910   509-335-2576
Pullman, WA 99164-2910   Fax: 509-335-7632
E-mail: rhh@wsu.edu
http://www.wsu.edu

Research laboratory capable of performing projects with physically scaled hydraulic models. 15,000 square feet of floor space, discharge capacity up to 70 cubic feet per second, modern instrumentaion and shop facilities.
Rollin H Hotchkiss, Director

5597 Alternative Energy Institute

Texas A&M University   806-656-2296
PO Box 248 WT   Fax: 806-656-2733
Canyon, TX 79016

Dr. Vaughn Nelson, Director

5598 American Petroleum Institute

1220 L Street Northwest   202-682-8000
Washington, DC 20005   Fax: 202-682-8232

Charles J DiBona, President

5599 American Society of Primatologists

University of Washington
PO Box 357330   206-543-0440
Seattle, WA 98195   Fax: 206-685-0305
http://www.asp.org

Conducts research on primates.

5600 Applied Energy Research Laboratory

North Carolina State University   919-515-5236
Raleigh, NC 27695   http://www.mae.ncsu.edu/centers/aerl/

Dr. John A Edwards, Director

5601 Aquatic Research Laboratory: Lake Superior State University

650 W Easterday Avenue   906-635-1949
Sault Sainte Marie, MI 49783   888-800-LSSU
Fax: 906-635-2266

Administered through the college of Arts and Sciences.
Prof Alex Litvinov, Director

5602 Architecture Research Laboratory

University of Arizona
College of Architecture   520-621-6751
Tucson, AZ 85721   Fax: 520-621-8700
E-mail: Whampton@ccit.arizona.edu

Provides assistance in the areas of education, applied research and public service.

5603 Argonne National Laboratory

9700 S Cass Avenue   630-252-2000
Argonne, IL 60439   http://www.anl.gov/

One of the US Department of Energy's largest research centers.

5604 Aspen Global Change Institute

100 East Francis Street   970-925-7376
Aspen, CO 81611   Fax: 970-925-7097
E-mail: agcimail@agci.org
http://www.agci.org

A Colorado nonprofit dedicated to furthering the understanding of Earth systems through interdisciplinary science meetings, publications, and educational programs about global environmental change.
John Katzenberger, Director
Sue Bookhout, Director Operations

5605 Auburn University Environmental Institue and Water Resources Research Institute

Auburn University
101 Comer Hall   334-844-4132
Auburn, AL 36849   Fax: 334-844-4462
E-mail: hatchlu@auburn.edu
http://www.ave.auburn.edu

Dr Upton Hatch, Director

5606 Belle W Baruch Institute for Marine Biology and Coastal Research

607 EWS Building   803-777-5288
Columbia, SC 29208   Fax: 803-777-3935

Conducts basic and applied research in marine and coastal environments.

5607 Bio-Integral Resource Center

PO Box 7414   510-524-2567
Berkeley, CA 94707   Fax: 510-524-1758
E-mail: birc@igc.org
http://www.birc.org

The goal of the Bio Integral Resources Center is to reduce pesticide use by educating the public about effective, least-toxic alternatives for pest problems.
Dr. William Quarles, Executive Director

5608 Biological Reserve

Denison University   740-587-6261
Granville, OH 43023   E-mail: stocker@denison.edu
http://webby.cc.denison.edu/biology/bioreserve/DUBR.shml

Enhances the education of students in Biology and the Environmental Sciences by providing opportunities for field studies.
Dr. John E Fauth, Contact

5609 Caesar Kleberg Wildlife Research Institute: Texas A&M University

700 University Boulevard   361-595-3922
MSC 218
Kingsville, TX 78363-8202

Facilitates complex wildlife-related research studies. Includes modern high-tech facilities, specially designed wildlife pens, and rangeland tracts.
Dr. Sam L Beasom, Director

5610 California Sea Grant College Program

University of California
9500 Gilman Drive   858-534-4440
Deptartment 0232   Fax: 858-534-2231
La Jolla, CA 92093-0232   http://www.csgc.ucsd.edu

Russell Moll, Director
Marsha Gear, Communications Director

5611 California State University: Monterey Bay Watershed Project

100 Campus Center   831-582-4120
Seaside, CA 93955   Fax: 831-582-4122
E-mail: essp_comments@csumb.edu
http://watershed.csumb.edu

The Watershed Institute is a direct action, community based coalition of researchers, restoration ecologists, educators, students, planners and area volunteers dedicated to restoring the watersheds of the Monterey Bay region throughrestoration, education and research. Their policy is to work with state and federal agencies, private landowers and local planners to gain access to critical lands. Institute staff are involved in local land and water planning.
Robert Curry, Adjunxt Professor

5612 Cedar Creek Natural History Area

University of Minnesota
2660 Fawn Lake Drive NE   763-434-5131
Bethel, MN 55005   Fax: 763-434-7361

Twenty-two hundred hectare experimental ecological reserve.

5613 Center for Applied Energy Research
University of Kentucky
2540 Research Park Drive   859-257-0305
Lexington, KY 40511-8410   Fax: 859-257-0220
     http://www.gaer.uky.edu
An applied research and development center with an international reputation, focusing on the optimal use of Kentucky's energy resources for the benefit of its people.
Ari Geertsema, Director

5614 Center for Applied Environmental Research
432 N Saginaw Street   810-767-7373
Suite 805   Fax: 810-767-7183
Flint, MI 48502-1950   E-mail: hblecker@umich.edu
     http://www.umf-outreach.edu/caer/index.htm
Harry S Blecker, Interim Director

5615 Center for Aquatic Research and Resource Management
Florida State University
Conradi Building, 136-B   850-644-4887
Tallahassee, FL 32306   Fax: 850-644-9829
     E-mail: livingston@bio.fsu.edu
     http://www.bio.fsu.edu/carrma.htm
Conducts research designed to answer aquatic resource-management questions posed by government agencies and private concerns. Research is conducted in lakes, rivers, and near-shore coastal systems throughout the southeastern UnitedStates with a multi-disciplinary approach to topics such as light, nutrients, primary productivity, fate and effects of storm water pollutants, sediment-water interactions, community assemblages of fish and invertebrates in various habitats andtrophic dynamics.
Dr. Robert J Livingston, Director

5616 Center for Cave and Karst Studies Western Kentucky University: Geography & Geology
1906 College Heights Blvd. #31066   270-745-3252
Bowling Green, KY 42101-1066   Fax: 270-745-3961
     E-mail: caveandkarst@wku.edu
     http://caveandkarst.wku.edu
Promotes research on all aspects of cave and karst studies with amphasis upon solving environmental problems associated with karst.
Dr. Nicholas Crawford, Director

5617 Center for Crops Utilization Research
Iowa State University of Science & Technology
Dairy Industries Building   515-294-0160
Ames, IA 50011   Fax: 515-294-6261
     E-mail: ljohnson@iastate.edu
Incorporates various aspects of new product and product research, applications development, and technology transfer. Activities focus on developing technologies for producing food and industrial products from agricultural materials,developing agricultural substitutes for petrochemicals, and exploring and modifying the functional properties of crop-derived materials.
Dr. Lawrence A Johnson, Professor-In-Charge

5618 Center for Earth & Environmental Science
State University of NY College   518-564-2028
Plattsburgh, NY 12901   877-554-1041
     Fax: 518-564-5267
     E-mail: cees@plattsburgh.edu
     http://www.plattsburgh.edu/cees
Undergraduate degree programs in environmental science, geology, planning and geography, with special emphasis on watershed science, remote sensing and geographic information systems, and aquatic and terrestrial ecology. The center isone of the oldest and largest environmental programs in the US, with 16 full-time interdisciplinary faculty and diverse field sites.
Robert Fuller, Director

5619 Center for Environmental Communications (CEC)
Rutgers University
31 Pine Street   732-932-1966
New Brunswick, NJ 08901-2883   Fax: 732-932-9544
     E-mail: cec@aesop.rutgers.edu

The CEC, located on the Cook College Campus, brings together university investigators to provide a social science perspective to environmental problem solving. CEC has gained international recognition for responding to environmentalcommunication dilemmas with research, training, and public service. Established in 1986, CEC is now jointly sponsored by the New Jersey Agricultural Experiment Station and the Edward J. Bloustein School of Planning and Public Policy.

Caron Chess, Director

5620 Center for Environmental Health Sciences
77 Massachusetts Avenue   617-253-6220
Building 16-743   Fax: 617-258-5424
Cambridge, MA 02139   http://www.mit.edu
William G Thilly, Director

5621 Center for Environmental Medicine Asthma & Lung Biology: University of North Carolina
104 Mason Farm Road, CB #7310   919-962-0126
Room 552 EPA Human Studies Facility   Fax: 919-966-9863
Chapel Hill, NC 27599-7310   E-mail: tmg@med.unc.edu
     http://www.med.unc.edu/envlung/welcome1.htm
The CEMALB are a group of investigators with diverse research interests that include cardiopulmonary medicine, immunology, lung physiology, cell biology, cell and molecular immunology, molecular toxicology and epidemiology. We conductresearch studies involving human volunteers that are aimed at understanding the negative health effects of air pollution on the lung and heart.
Philip A Bromberg, MD, Scientific Director
David B Peden, MD, MS, Center Director

5622 Center for Environmental Research Education
State University of New york
1300 Elmwood Avenue   716-878-4329
Upton Hall, Room 314   Fax: 716-878-6644
Buffalo, NY 14222-1095   E-mail: zolnowsa@buffalostate.edu
     http://www.buffalostate.edu/~glc/
Dr. Charles Beasley, Contact

5623 Center for Environmental Studies
Kellogg House   413-597-2346
PO Box 632   Fax: 413-597-3489
Williamstown, MA 01267   http://www.williams.edu/CES
Provides students with the opportunity to learn how environmental issues are interconnected with many traditional fields of study. Offered as a concentration, the program encourages students to become well grounded in a single field bypursuing a major in a traditional discipline or department, while focusing several of their elective courses on the interdisciplinary study of the environment.

Karen Merrill, Director
Sarah Gardner, Associate Director

5624 Center for Environmental Toxicology and Technology
Colorado State University   970-491-8522
Foothills Campus   Fax: 970-491-8304
Fort Collins, CO 80521   http://www.cvmbs.colostate.edu
Raymond Yang, Director

5625 Center for Environmental, Earth and Space Studies
Bemidji State University
1500 Brischmont Drive NE   218-755-2040
Bemidji, MN 56601   888-345-1721
     Fax: 218-755-2074
     E-mail: admissions@vax1.bemidji.msus.edu
     http://www.bemidji.msus.edu
The Center for Environment, Earth and Space Studies offers a unique variety of interdisciplinary degree programs that a student may choose to pursue. Degrees in Environmental Studies include both BS and MS and a BS or BA with a geologyminor is also availible.
Fu-Hsian Chang, Interim Director

5626 Center for Field Biology
Austin Pay State University
PO Box 4718   931-221-7019
Clarksville, TN 37044   Fax: 931-221-6372

E-mail: fieldbiology@apsu.edu
http://www.apsu.edu/field_biology
The Center of Excellence for Field Biology at Austin Peay State University brings together scholars and students from various biological disciplines to conduct research on topics in field biology and ecology, including toxicology,population and community ecology, and the ecology and biology of rare, threatened and endangered species. Major research efforts have focused on the ecology and biology of the flora and fauna of the Land Between the Lakes.
Dr. Andrew N Barrass, Director

5627 Center for Global & Regional Environmental Research
University of Iowa
204 IATL   319-355-3333
Iowa City, IA 52242   Fax: 319-335-3337
E-mail: jfrank@cgrer.uiowa.edu
http://www.cgrer.uiowa.edu
Jane Frank, Administrative Assistant

5628 Center for Global Change Science (MIT)
Massachusetts Institute of Technology
77 Massachusetts Avenue   617-253-4902
Room 54-1312   Fax: 617-253-0354
Cambridge, MA 02139   E-mail: cgcs@mit.edu
http://mit.edu/cgcs/
Addresses long-standing scientific problems whose solution is necessary for accurate prediction of changes in the global environment. The CGCS is interdisciplinary and interdepartmental, and builds on research and educational programsin earth sciences and engineering. The Center is also involved in substantial cooperative efforts focused on climate modeling, and on climate-policy research.

Dr. Ronald Prinn, Director
Anne Slinn, Administrator

5629 Center for Hazardous Substance Research
Kansas State University   785-532-6519
Ward Hall 104   Fax: 785-532-5985
Manhattan, KS 66506-2502   E-mail: hsre@ksu.edu
Dr. Larry Erickson, Director

5630 Center for International Development Research
Duke University
Sanford Institute of Public Policy   919-684-8894
PO Box 90237   Fax: 919-684-2861
Durham, NC 27708-0237

5631 Center for International Food and Agricultural Policy
University of Minnesota
1994 Buford Avenue   612-625-9208
332k Classroom Office   Fax: 612-625-2729
St. Paul, MN 55108   E-mail: frunge@apec.umn.edu
http://www.apec.umn.edu/cifap
With its interdisciplinary approach, CIFAP uses its research and education activities to increase international understanding about food, agriculture, nutrition, natural and human resources, and the environment, and to positivelyaffect the policies of both developed and developing countries.

C Ford Runge, Director

5632 Center for International Studies
292 Main Street   617-253-8093
Building E38-651   Fax: 617-253-9330
Cambridge, MA 02139-4307   E-mail: lauries@mit.edu
http://web.mit.edu/cis

5633 Center for Lake Superior Ecosystem Research: Michigan Technological University
1400 Townsend Drive   906-487-2769
Houghton, MI 49931-1295   E-mail: wkerfoot@mtu.edu
http://www.mtu.edu/level3/centers.html
An interdisiplinary center with goals to promote and strengthen ecological research and graduate programs at MTU through developing and applying technological advances to ecological problems, to advocate an ecosystem perspective forstudying aquatic and terrestrial portions of the Lake Superior watershed and to become a resource center for basic information on watershed and lake properties.

5634 Center for Marine Biology
85 Adams Point Road   603-862-2175
Durham, NH 03824   Fax: 603-862-1101
E-mail: ray.grizzle@unh.edu
http://http://marine.unh.edu/jacksonlab.htm
The Center for Marine Biology (CMB) fosters excellence in marine biological research and education. Its primary goals are to strengthen and focus research and graduate education in modern marine biology and to encourage the developmentof high-quality undergraduate programs in all aspects of marine biology. The center helps faculty members compete for external grant funds and fosters coordination of marine research efforts, both with the life sciences and in other disciplines.
Ray Grizzle, Research Scientist

5635 Center for Population Biology
University of Aclifornia
One Shields Avenue   530-752-1274
Davis, CA 95616   Fax: 530-752-1449
Founded in 1989, the Center for Population Biology unites UC Davis' population biologists. The center's membership comprises graduate students enrolled in the http://www-eve.ucdavis.edu/popbio.htm, graduate students interested inpopulation biology who are earning their degrees in graduate programs such as ecology or entomology, postdoctoral researchers, from nine academic departments and sections, 17 of whom have faculty appointments in the division.

Dr H Bradley Shaffer, Director

5636 Center for Resource Management
1104 Ashton Avenue   801-466-3600
Suite 210   Fax: 801-328-3457
Salt Lake City, UT 84106-2374
Facilitates communication between environmental, corporate and government agencies.
Terrell Minger, President
Paul Parker, Vice President

5637 Center for Resource Policy Studies
1450 Linden Drive   608-262-8254
Room 240
Madison, WI 53706
The Center for Resource Policy Studies and Programs uses interdisciplinary research, teaching and extension efforts to analyze resource policies and development programs. This center gives particular emphasis to the social scienceaspects of natural resource policy issues.

5638 Center for Solid & Hazardous Waste Management
2207 NW 13th Street   352-392-6264
Suite D   Fax: 352-846-0183
Gainesville, FL 32609   E-mail: center@floridacenter.org
http://www.floridacenter.org
The center serves the citizens of Florida by providing leadership in the field of waste management research and by supporting the Florida Department of Environmental Protection in its mission to preserve and protect the state's naturalresources.
John D Schert, Executive Director

5639 Center for Streamside Studies
University of Washington
Box 352100   206-543-6920
Seattle, WA 98195-2100   Fax: 206-543-3254
http://dept.washington.edu/cssuw/
The mission of the Center for Streamside Studies is to provide scientific information necessary for the resolution of management issues related to the production and protection of forest, fish, wildlife, and water resources associatedwith the streams and rivers in the Pacific Northwest.
Robert J Naiman, Director

5640 Center for Tropical Agriculture
University of Florida
3081 McCarty Hall   352-392-2643
Box 110286   Fax: 352-846-0816
Gainesville, FL 32611   E-mail: hlp@ufl.edu
Enhances research and education on tropical agriculture between University of Florida and tropical countries.
Dr. Hugh L Popenoe, Director

5641 Center for the Management, Utilization and Protection of Water Resources
Box 5033
Cookeville, TN 38505
931-372-3507
Fax: 931-372-6346
E-mail: dgeorge#tntech.edu
http://www.tntech.edu/wrc
The Center for the Management, Utilization and Protection of Water Resources at Tennessee Technological University is dedicated to the vision of enhancing environmental education through research. Using interdisciplinary teams ofresearchers, the Center focuses its work in the core areas of environmental resource management and protection, environmental hazards, and environmental information.
Dr. Dennis George, Director

5642 Center for the Study of Earth from Space
University of Colorado
CB 216
Boulder, CO 80309
303-492-5086
Fax: 303-492-5070
E-mail: cses@cires.colorado.edu
http://cires.colorado.edu/cves
CSES provides a focus for the development and application of modern remote sensing techniques used in the research of all aspects of earth sciences at the University of Colorado. Although measurements from space are emphasized,aircraft and field measurements are integral to any remote sensing project.
Dr. Alexander Goetz, Director

5643 Clean Energy Research Institute
University of Miami
219 McArthur Building
Coral Gables, FL 33124
305-284-4666
Fax: 305-284-4792
E-mail: veziroglu@miami.edu
Acts as the focal point of energy and environmental related activities in the College of Engineering. Its goals are to conduct research and to generate research proposals to investigate energy and environmental problems; to organizeseminars, workshops and conferences using researchers within and without the University; to assemble, compile, publish and disseminate information on every aspect of energy and environmental problems; and to cooperate with other organs of theUniversity.
Dr T Nejat Veziroglu, Director

5644 Cobbs Creek Community Environment Educational Center (CCCEEC)
Penn State Cooperative Extension
4601 Market Street, 2nd Floor
Philadelphia, PA 19139
215-471-2223
Fax: 215-471-2231
E-mail: ccceec@cobbscreek.org
CCCEEC is designated to institutionalize the practice of Urban Environmental Education. Their mission is to preserve the quality for residents living in the Cobbs Creek area of Philadelphia through the establishment of a center foreducating and informing people about the issues affecting their environment.
Alan G Fastman, Executive Director
Jasa M Porciello, Programing Coordinator

5645 College of Forest Resources
University of Washington
College of Forest Resources
107 Anderson Hall
Seattle, WA 98195-2100
206-543-2730
Fax: 206-685-0790
E-mail: cfruw@u.washington.edu
http://www.cfr.washington.edu
The University of Washington College of Forest Resources is dedicated to generating and disseminating knowledge for the stewardship of natural and managed environments and the sustainable use of their products and services throughteaching, research and outreach.

5646 Colorado Cooperative Fish & Wildlife Research Unit
Colorado State University
1484 Campus Delivery
Fort Collins, CO 80523
970-491-5396
Fax: 970-491-1413
http://www.colostate.edu/depts/coopunit
The Colorado Cooperative Wildlife Research Unit was founded in 1947, and the Colorado Cooperative Fishery Research Unit was established in 1963. The two Units were combined in 1984 into the Colorado Cooperative Fish and WildlifeResearch Unit. This unit is staffed, supported, and coordinated by the Colorado Division of Wildlife, Colorado State University , the United states Geological Survey , and the Wildlife Management Insistute.
Dana L Winkelman, Unit Leader

5647 Cornell Waste Management Institute
Cornell University
101 Rice Hall Hall, Envrn'l Rsch. Center
Ithaca, NY 24853
607-255-1187
Fax: 607-255-8207
E-mail: cwmi@cornell.edu
http://cwmi.css.cornell.edu
Conduct applied research and outreach focused on composting and land application of sewage sludges.

Lauri Wellin, Administrative Assistant

5648 Eagle Lake Biological Field Station
California State University
Department of Biology Sciences
Chico, CA 95929-0515
530-898-4490
E-mail: rbogiatto@oavax.csuchico.edu
http://www.csuchico.edu/biol/eaglelakehtml
The Eagle Lake Biological Field Station, located 26 miles northwest of Susanville in Lassen County, California is a ten building facility on the eastern shore of Eagle Lake. The field station is administered by California StateUniversity, Chico and the CSUC Foundation with support from the University of California Natural Reserve System and UC Davis. The ELBFS is open to any individual or group whose purpose is primarily academic and whose activities are consistent withthe isolation.
Raymond J Bogiatto, Director

5649 Earthwatch International
3 Clock Tower Place, Suite 100
Box 75
Maynard, MA 01754
978-461-0081
800-776-0188
Fax: 978-461-2332
E-mail: info@earthwatch.org
http://www.earthwatch.org
Engages people worldwide in scientific research and education to promote the understanding and action necessary for a sustainable environment.
Ed Wilson, Acting President

5650 Ecology Center
Utah State University
UMC 5205
Logan, UT 84322-5200
435-797-2555
E-mail: ecol@cc.usu.edu
The Utah State University Ecology Center is an administrative structure in the University that supports and coordinates ecological research and graduate education in the science of ecology, and provides professional information andadvice for decision makers considering actions that affect the environment.
Frederic H Wagner, Director

5651 Energy Resources Center
Univeristy of Illinois at Chicago
851 S Morgan Street
Chicago, IL 60607-7054
312-996-4490
Fax: 312-996-5620
E-mail: erc@uic.edu
http://www.erc.uic.edu
The Center is a University of Illinois at Chicago interdiciplinary research and public service organization. It was established in 1973 by the University's Board of Trustees to conduct studies in the field of energy and to providelocal, state and federal governments and the public with current information on energy technology and policy.
William M Worek, Director

5652 Energy, Environment & Resource Center
University of Tennesse at Knoxville
676 Dabney Hall
Knoxville, TN 37996
865-974-8080
Fax: 865-974-8086
E-mail: sayler@utk.edu
http://www.web.utk.edu/~enr/contact.htm
Dr. Jack Barkenbus, Director

5653 Environmental & Water Resources Engineering Area
Texas A&M University
Civil Engineering Department
College Station, TX 77843-3136
979-845-3011
Fax: 979-862-1542
Bill Batchelor, Area Leader

5654 Environmental Center: University of Hawaii
2500 Dole Street
Krauss Annex 19
Honolulu, HI 96822
808-956-7361
Fax: 808-956-3980
E-mail: envctr@hawaii.edu
http://www2.hawaii.edu/~envctr/

Research Centers/University Centers

The Center's three areas of focus are education, research and service. The education function of the Center includes the administration of the Environmental Studies Major Equivalent and Certificate program. It fulfills its research function by identifying and addressing environmentally related research needs, particularly those pertinent to Hawaii. The service function primarily involves the coordination and transfer of technical information from the University community to government agencies.
Dr John T Harrison, Coordinator

5655 Environmental Chemistry and Technology Program
University of Wisconsin at Madison
660 N Park Street 608-263-3264
Room 122 Fax: 608-262-0454
Madison, WI 53706 http://www.engr.wisc.edu/interd/wcp
D E Armstrong, Chair

5656 Environmental Exposure Laboratory
University of California
1000 Veterans Avenue 310-825-2739
Rehabilitation Center, Room A163
Los Angeles, CA 90024
Dr. Henry Gong Jr, Director

5657 Environmental Health Sciences Center: University of Rochester Medical Center
601 Elmwood Avenue 716-275-3911
Rochester, NY 14642 Fax: 716-256-2591
E-mail: slechta@ehsct7.envmed.rochester.edu
The Community Outreach and Education Program of the Rochester Environmental Health Sciences Center serve as a local, national and international sources of environmental health information. They are designed to educate students, teachers, health care professionals, public officials and other members of the community. Our programs are bassed on identified community needs and on issues that mirror ous EHS Center research.
Deborah Cory-Slechta PhD

5658 Environmental Health Sciences Research Laboratory
127 New Market Street 504-394-2233
PO Box 379 Fax: 504-394-7982
Belle Chasse, LA 70037

5659 Environmental Human Toxicology
University of Florida
PO Box 110885 352-362-4700
Gainesville, FL 32611 Fax: 352-392-4707
http://www.floridatox.org
The Center serves as an interface between basic research and its applications for evaluation of human health and environmental risk. The research and teaching activities of the Center provide a resource to identify and reduce risk associated with environmental pollution, food contamination, and workplace hazards. The center provides a forum for the discussion of specific and general problems concerning the potential adverse human health effects associated with chemical exposure.
Dr. Stephen Roberts

5660 Environmental Institute of Houston
University ofg Houston
Environmental Institute of Houston 281-283-3950
Box 540 Fax: 281-283-3044
Houston, TX 77058-1098 E-mail: eih@cl.uh.edu
The mission of EIH is to help people in the Houston region participate more effectively in environmental improvement. Information and technology will be obtained and disseminated from research supported by EIH in critical areas including pollution prevention, natural resource conservation, public policy and societal issues. EIH will seek to expand balanced environmental education based on objective scholarship to empower the community to make sound decisions on environmental issues.
Candy Allison, Senior Secretary
Heather Biggs, GIS Analyst

5661 Environmental Law Institute
1616 P Street NW 202-939-3800
Suite 20036 Fax: 202-939-3868
Washington, DC 20036
J William Futrell, President

5662 Environmental Remote Sensing Center
University of Wisconsin
1225 W Dayton Street 608-263-3251
Madison, WI 53706 Fax: 608-262-5964
http://www.ersc.wisc.edu
A university research center focused on application of remote sensing and attending geospatial technologies in government, business and science. Particular heritage in the application of remote sensing in natural resource management and environmental monitoring. A NASA-sponsored Affiliated Research Center.
Prof Thomas Lillesand, Director

5663 Environmental Research Institute
University of Idaho
Food Research Center 103 208-885-6580
Moscow, ID 83844 Fax: 208-885-5741
E-mail: crawford@uidaho.edu
http://image.fs.uidaho.edu/
The faculty, associated with the institute, perform multidisciplinary research in environmental molecular ecology, restoration of contaminated soils and waters, and microbial genomics related to environmental processes.
Dr Ronald L Crawford, Director

5664 Environmental Resource Center
One Washington Square #1 408-924-5467
San Jose, CA 95192-0116 E-mail: ncowan@email.sjsu.edu
The Environmental Resource Center is a nonprofit information and outreach organization within the http://www.sjsu.edu/depts/envstudies/ at http://www.sjsu.edu/ (SJSU), serving the San Jose community since 1971. Several of our projects are funded by the http://as.sjsu.edu of SJSU.
Annemarie Vallesteros, Executive Director

5665 Environmental Science & Engineering Program
Clarkson University
210 Clarkson Hall 315-268-3786
Potsdam, NY 13699-5715 Fax: 315-268-4266
E-mail: feitelsb@clarkson.edu
Samuel B Feitelberg PT, MA, FAPTA, Professor

5666 Environmental Studies Institute
256 Smith Walkity 215-573-3164
Philadelphia, PA 19104 E-mail: ies_penn@sas.upenn.edu
The Institute for Environmental Studies is dedicated to improving the understanding of key scientific, economic, and political issues that underlie environmental problems and their management. The mission of the Institute is to bring scholars together from across the University in order to promote collaborations in education and research endeavors in the area of environmental issues. These collaborative endeavors span basic and applied sciences, engineering and the social and human sciences.
Dr. Bernard Hamel, Director

5667 Environmental Systems Application Center
107 S Indiana Avenue 812-855-4848
Bloomington, IN 47405-7000
The goals of the Center are to promote excellence in environmental science research and to foster increased interdisciplinary collaboration among environmental science faculty on the Indiana University-Bloomington campus. The Center has no degree programs. The Center can be listed as an affiliation of the associated faculty in publications and in correspondence.

5668 Environmental Systems Engineering Institute
University of Central Florida
4000 Central Florida Boulevard 407-823-2000
Orlando, FL 32816

5669 Environmental Toxicology Center
1710 University Avenue 608-263-4580
Room 290 Fax: 608-262-5245
Madison, WI 53706-4098 http://www.wisc.edu/etc/
Environmental Toxicology is the study of the adverse effects on individual life forms and ecosystems of environmental agents (chemical, physcial, biological) whether of natural origin or released through human activity, and origins and control of these harmful agents.
Prof. Colin R Jefcoate, Director

369

5670 Environmental and Occupational Health Science Institute

170 Frelinghuysen Road   732-445-0200
PO Box 1179   Fax: 732-445-0131
Piscataway, NJ 08855   http://www.cohsi-rutgers.edu

The major objectives of the institute are to: improve understanding of the impact of environmental chemicals on human health; to find ways to quantify and prevent exposure to hazardous substances; and develop methods to identify andtreat people adversely affected by environmental agents. Devises approaches for educating the public about the relative risks from chemical exposure. Trains professionals to accomplish these tasks.
Dr. Cory Slechta, Director

5671 Feed and Fertilizer Laboratory

Louisiana State University
Department Agriculture and Forestry   225-388-2755
Baton Rouge, LA 78021

David Wall, Contact

5672 Fitch Natural History Reservation

University of Kansas
2060 E 1600 Road   785-843-3612
Lawrence, KS 66044

5673 Florida Cooperative Fish and Wildlife Research Unit

University of Florida
Box 110450   352-392-1861
Gainesville, FL 32611-0450   Fax: 352-846-0841
     http://biology.usgs.gov/coop/unitpages/fl_cfwru.html

The Cooperative Research Unit has three facets to its mission: education—Cooperative Unit scientists teach university courses at the graduate level, provide academic guidence to graduate students, and serve on academic committees;research—Cooperative Unit scientists conduct research that is designed to meet the information needs expressed by unit cooperators; technical Assistance— unit provides technical assistance and training to State and federal personnel and othernatural resources.
Dr. Wiley M Kitchens, Leader

5674 Florida Museum of Natural History

University of Florida
Museum Drive   352-392-1721
PO Box 117800   Fax: 352-392-8783
Gainesville, FL 32611   E-mail: gdshaak@flmnh.ufl.edu
     http://www.flmnh.ufl.edu

The Florida Museum of Natural History, on the University of Florida Campus, is one of the leading university natural history museums in the nation. With over 23 million specimens and artifacts in its permanent collections, it is thelargest collection-based museum in the southeastern US. The museum was established by the Legislature in 1917 at the University of Florida where it functions in a dual capacity as the official state museum of Florida and the University Museum.

Dr Douglas S Jones, Director
Dr Graid D Shaak, Associate Director

5675 Forestry Experiment Stations and Arboretum

University of Tennessee
901 S Illinois Avenue   865-483-3571
Oak Ridge, TN 37830   Fax: 865-483-3572
     E-mail: utforest@utk.edu
     http://web.utk.edu/-utforest/arb/index.htm

The UT Forestry Experiment Station mission is to:(1) provide the land and supporting resources necessary for conducting modern and effective forestry, wildlife, and associated social, biological and ecological research programs;(2)demonstrate the application of optimal forest and wildlife management technologies; and (3) assist with transfer of new technology to forest land owners and industries.
Richard M Evans, Contact

5676 Formaldehyde Institute

1330 Connecticut Avenue NW   202-833-2131
Washington, DC 20036   Fax: 202-659-1699
     http://www.ainc.org

John F Murray, Executive Director

5677 Foundation for Research on Economics and the Environment (FREE)

662 Ferguson Avenue   406-585-1776
Bozeman, MT 59718   Fax: 406-585-3000
     http://www.free-eco.org

FREE mission is to advance conservation and environmental values consistent with individuals freedom and responsibility. The Foundation's intellectual entrepreneurs develop environmental policies featuring private property rights,market incentives, and voluntary organizations. FREE achieves its mission by working with leaders in universities, businesses, environmental groups, government, the media, and think tanks.
John Baden, Chairman

5678 Gannett Energy Laboratory

Florida Institute of Technology
150 W University Boulevard   321-768-8000
Melbourne, FL 32901

5679 Geological Survey of Alabama

University of Alabama
420 Hackberry Lane   205-349-2852
Tuscaloosa, AL 35486-9780   Fax: 205-349-2861
     E-mail: info@gsa.state.al.us
     http://www.gsa.state.al.us

Dr. EA Mancini, Director

5680 Global Change & Environmental Quality Program

University of Colorado
Campus Box 214   303-492-7943
Boulder, CO 80309-0214   Fax: 303-492-1414

In addition to addressing CU's overall objectives, the Global Change and Environmental Quality Program is pursuing three main goals; studying environmental issues at the local level, including the cleanup and restoration of toxicsites, such as Rocky Flats, the Rocky Mountain Arsenal, and mine tailing sites; waste treatment and water quality; and land use.
Dr. Robert Sievers, Director

5681 Graduate Program in Community and Regional Planning

University of Texas
Main Building   512-471-0134
Austin, TX 78701   Fax: 512-471-0716

The Graduate Program in Community and Regional Planning educates and prepares students for professional practice in planning in the context of urban growth. The program is designed to expose students to full range of issues, but alsoto develop skills in areas of specialization. This approach respects the knowledge shared by all professional planners while recognizing roles involved in comtemporary planning practice.

5682 Great Lakes Coastal Research Laboratory

Purdue University
School of Civil Engineering   765-494-4600
West Lafayette, IN 47907

5683 Great Lakes Tomorrow

9315 Glenwood Trail   440-838-4176
Brecksville, OH 44141   Fax: 440-838-4176
     E-mail: jcowdeni@ibm.net

James W Cowden, Director

5684 Great Lakes/Mid-Atlantic Hazardous Substance Research Center

University of Michigan
EWRE Building   734-763-2274
Suite 181
Ann Arbor, MI 48109

The mission of the Great Lakes Mid- Atlantic Center for Hazardous Substances Research is to foster and support integrated, interdisciplinary, and collaborative efforts that advance the science and technology of hazardous substancemanagement to benefit human and environmental health and well-being.
Dr. Walter J Weber Jr, Director

5685 Great Plains: Rocky Mountain Hazardous Substance Research Center

Ward Hall 101   785-532-6519
Manhattan, KS 66506-2502   Fax: 785-532-6952
E-mail: hsrc@ksu.edu
http://www.engg.ksu.edu/HSRC

Conducts research and transfers technology on hazardous substance management, and remediation of contaminated soil and water.
Dr. Larry Erickson, Director

5686 Greenley Memorial Research Center

Greenley Memorial Center   660-739-4410
Box 126   Fax: 660-739-4500
Novelty, MO 63460   E-mail: smootr@missouri.edu
Randall Smoot, Supervisor

5687 HT Peters Aquatic Biology Laboratory

Bemidji State University   218-755-2877
Bemidji, MN 56601   E-mail: dcloutman@bemidjistate.edu

5688 Harry Reid Center for Environmental Studies

4505 Maryland Parkway   702-895-3382
Box 454009   Fax: 702-895-3094
Las Vegas, NV 89154-4009

The HRC was started in 1981 under UNLV's Marjorie Barrick Museum of Natural History. HRC currently includes 65 staff members and a 65,000 square foot building with four laboratories.

5689 Hawaii Cooperative Fishery Research Unit

University of Hawaii at Manoa
2538 The Mall   808-956-8350
Honolulu, HI 96822   Fax: 808-956-4238
E-mail: parrishj@hawaii.edu
Dr. James D Parrish, Unit Leader

5690 Hawaii Undersea Research Laboratory

University of Hawaii at Manoa
1000 Pope Road   808-956-6335
MSB 303
Honolulu, HI 96822

One of six research centers funded by NOAA's National Undersea Research Program. HURL operates two 2000-meter Pisces submersibles and a remotely operated vehicle. Research projects include fisheries research, geology and biology of thedeepsea around the Hawaiian Islands.
Dr. Alexander Malahoff, Director

5691 Henry S Conrad Environmental Research Area

1210 Grinnell College   641-269-4457
Grinnell, IA 50112   Fax: 641-269-4984

5692 Highlands Biological Station

265 Sixth Street   828-526-2602
Highlands, NC 28741   Fax: 828-526-2797
E-mail: hbs@email.wcu.edu
http://www.wcu.edu/hbs

The Station is an interinstitutional center of the University of North Carolina and includes the Highlands Nature Center and Botanical Gardens, as well as the Biological Laboratory. Our mission, for more than 75 years has been tofoster education and research focused on the rich natural heritage of the Highlands Plateau.
Tom Martin, Interim Director
Peggy Cowart, Office Assistant IV

5693 Hudsonia

Bard College Field Station   845-758-7053
PO Box 5000   Fax: 845-758-7033
Annandale, NY 12504-5000   E-mail: kiviat@bard.edu
http://www.hudsonia.org

Since 1981 Hudsonia has conducted environmental research, education, training and technical assistance to protect the Hudson River Valley's natural heritage. Nonpartisan and non-ideological, Hudsonia serves as a neutral voice in thechallenging process of land conservation.
Erik Kiviat, Executive Director

5694 Huntsman Environmental Research Center

Utah State University   435-750-1418
UMC 6005   Fax: 435-797-2567
Logan, UT 84322-6000   E-mail: herc@cc.usu.edu

The establishment of the Hunts man Environmental Research Center recognized the fundamental interdependence of the health of man and the health of the environment. The HERC's mission is to engage in research in the key areas ofrecycling, degradability, improvement of air and water quality and conservation of trees. The center purpose is to solve environmental problems and to provide realistic and comprehensive research solutions for our environment.
Dr. Maurice Thomas, Director

5695 INFORM

120 Wall Street   212-361-2400
14th Floor   Fax: 212-361-2412
New York, NY 10005-4001   http://www.informinc.org

INFORM is an independent research organization that examines the effects of business practices on the environment and on human health. Our goal is to identify ways of doing business that ensure environmentally sustainable economicgrowth. Our reports are used by government, industry, and environmental leaders around the world.
Joanna Underwood, President

5696 Idaho Cooperative Fish & Wildlife Research Unit

PO Box 44-1141   208-885-2750
Moscow, ID 83844-1141   Fax: 208-885-9080
E-mail: sarahm@uidaho.edu
http://www.cnrhome.uidaho.edu/coop

Personnel of the Idaho Cooperative Fish and Wildlife Research Unit will: (1) conduct research on problems of state, regional and national interest; (2) train graduate students for careers in the fish and wildlife professions;(3)provide technical assistance to state and federal managers and researchers.

Dr J Michael Scott, Unit Leader
Sarah Martinez, Administrative Asst. II

5697 Institute for Alternative Agriculture

9200 Edmonston Road   301-441-8777
Suite 117   Fax: 301-220-0164
Greenbelt, MD 20770-1551   E-mail: hawiaa@access.digex.net

The Wallace Institute advances this goal by providing the leadership, and policy research and analysis necessary to influence national agriculturalpolicy. It is a contributing member of a growing national alternaative agriculturenetwork, and works directly with government agencies, educational and research institutions, producer groups, farmers, scientists, advocates, and other organizations that provide agricultural research, education, and information services.
Dr. I Garth Youngberg, Executive Director

5698 Institute for Applied Research

840 La Goleta Way   916-482-3120
Sacramento, CA 95864

5699 Institute for Biopsychological Studies of Color, Light, Radiation, Health

One Washington Square   408-924-1000
Psychology Department   Fax: 408-924-1018
San Jose, CA 95192   http://www.sjsu.edu
San Jose State University, Author

5700 Institute for Crustal Studies

University of California
1140 Girvetz Hall   805-893-8231
Santa Barbara, CA 93106-1100   Fax: 805-893-8649
http://www.crustal.ucsb.edu/ics/

Purpose is to increase our understanding of the geological processes and evolution of the earth's crust and lithosphere, and the impact these processes have on society.
Douglas W Burbank, Director

5701 Institute for Ecological Infrastructure Engineering
Losuisiana State University
College of Engineering   225-578-1399
102 ELAB   Fax: 225-578-8662
Baton Rouge, LA 70803   E-mail: eielab@eiel.lsu.edu
      http://www.eiel.lsu.edu
Institute for Ecological Infrastructure Engineering integrates engineering with science (physical, chemical, life & social) for the co-development of society and nature (ecosystems).
Lily A Rusch, Director

5702 Institute for Environmental Education
16 Upton Drive   978-658-5272
Wilmington, MA 01887   800-823-6239
      Fax: 978-658-5435
      E-mail: courseinfo@ieetrains.com
      http://www.ieetrains.com
IEE is New England's largest environmental training provider with over 57 classes in asbestos, lead and environmental health and safety.
Martin Wood, President
Ed Crowell, VP Marketing

5703 Institute for Environmental Policy & Planning
One Forestry Drive   315-470-6636
107 Marshall Hall   Fax: 315-470-6915
Syracuse, NY 13210

5704 Institute for Environmental Science
University of Texas at Dallas
PO Box 830688   972-794-4000
Richardson, TX 75083-0688
Dr. John Ward, Director

5705 Institute for Lake Superior Research
University of Minnesota
135 Medical Building   218-726-8000
Duluth, MN 55812   http://www.uwsuper.edu
Dr. Carlson, Director

5706 Institute for Regional and Community Studies
Western Illinois University   309-298-1566
Tillman Hall 413B
Macomb, IL 61455

5707 Institute for Urban Ports and Harbors
State University of New York
Marine Sciences Research Center   631-632-8700
Stony Brook, NY 11794   Fax: 631-632-8820
      http://www.msrc.sunysb.edu

5708 Institute of Analytical and Environmental Chemistry
University of New Haven
300 Orange Avenue   203-932-7171
West Haven, CT 06515   800-342-5864
      http://www.newhaven.edu
Prof. George Wheeler, Director

5709 Institute of Chemical Toxicology
Wayne State University
2727 2nd Avenue   313-577-0100
Room 4000   Fax: 313-577-0082
Detroit, MI 48201-2654   http://www.wayne.edu

5710 Institute of Ecology
University of California
1 Shields Avenue   530-752-3026
Davis, CA 95616   Fax: 530-752-3350
      E-mail: aking@ucdavis.edu
      http://www.ucdavis.edu

5711 Institute of Marine and Atmospheric Sciences
138th Street and Convent Avenue   212-650-7000
New York, NY 10031   http://www.ccny.cuny.edu

5712 Interdisciplinary Center for Aeronomy & Other Atmospheric Sciences
University of Florida
311 Space Sciences Research Bldg   352-392-2001
Gainesville, FL 32611   Fax: 352-392-2003
Prof. Alex ES Green, Director

5713 International Center for the Solution of Environmental Problems
2345 Navigation, Suite 310   713-527-8711
Houston, TX 77003-1017   Fax: 713-527-8025
      E-mail: icsp@neosoft.com

5714 Iowa Cooperative Fish & Wildlife Research Unit
Iowa State University Science Hall   515-294-3056
II   Fax: 515-294-5468
Ames, IA 50011
VACANT , Leader

5715 Iowa Waste Reduction Center
1005 Technology Parkway   319-273-8905
Cedar Falls, IA 50613   800-422-3109
      Fax: 319-268-3733
      http://www.iwrc.org
A service of the University of Northern Iowa, provides free and confidential environmental regulatory assistance to Iowa small businesses. At the request of a small business, an IWRC specialist will tour the facility and offer specificsolutions to regulatory issues.
John Konefes, Director

5716 James H Barrow Field Station
Hiram College Biological Station   330-527-2141
Garrettsville, OH 44231   Fax: 330-527-3187
The James H Barrow Field Station was established in 1967 to provide Hiram College students the opportunity to supplement classroom activities with hands-on learning experiences. Over the Last 32 years the Station has grown anddeveloped into an active research and educational facility that not only echances the College's science and environmental studies programs, but also provides a means for the general public to increase their understanding and appreciation of Ohio'snatural history.

5717 John D MacArthur Agro Ecology Research Center
300 Buck Island Ranch Road   727-669-0242
Lake Placid, FL 33852   Fax: 863-699-2217
      E-mail: maerc@archbold-station.org
      http://www.archbold-station.org/maerc/maerc.htm
The MacArthur Agro Ecology Research Center at Buck Island Ranch is dedicated to a mission of long-term research, education and outreach related to the ecological and social value of subtropical grazing lands. The Center is at a 10,300acre cattle ranch on a long-term lease to Archbold Biological Station from the John D and Catherine T MacArthur Foundation. Provides researchers the opportunity to evaluate the relationship between economic and ecological factors and how these changeover time.

5718 John F Kennedy School of Government Environmental and Natural Resources Program
Harvard University
79 John F Kennedy Street   617-495-1351
Cambridge, MA 02138   Fax: 617-495-1635
      E-mail: enrp@ksg.harvard.edu
      http://http://bcsia.ksg.harvard.edu/?program+ENRP
Henry Lee, Director

5719 John F Kennedy School of Government - Harvard University
Harvard University
79 John F Kennedy Street   617-495-1100
Cambridge, MA 02138
We would like to indroduce you to our teaching degree and executive programs; to our research centers where faculty, fellows and students concentrate around particular policy issues - from international affairs to state and localgovernment; as well as to many research projects and programs and to work of individual faculty and students.
David Ellwood, Dean

5720 John Heinz National Wildlife Refuge at Tinicum
86th Street & Lindbergh Boulevard   215-365-3118
Philadelphia, PA 19153   Fax: 215-365-2846
E-mail: FW5RW_JHTNWR@fws.gov
http://http://Heinz.fws.gov
The John Heinz National Wildlife Refuge at Tinicum is administered by the Department of Interior's U.S. Fish and Wildlife Service and is located in Philadelphia and Delaware Counties, Pennsylvania. The refuge protects the last 200acres of freshwater tidal marsh in Pennsylvania. The refuge has become a resting and feeding area for more than 20 species of birds, 80 of which nest here. Fox, deer, muskrat, turtles, fish, frogs and a wide variety of wildflowers and plants callthe refuge "home".

, Refuge Manager

5721 Juneau Center School of Fisheries & Ocean Sciences
University of Alaska Fairbanks
1120 Glacier Highway   907-465-6441
Juneau, AK 99801   Fax: 907-465-6447
E-mail: fsfosj@uaf.edu
http://www.sfos.uaf.edu/
JCSFOS has the primary responsibility within the University for education, research and public service in support of fisheries related areas of oceanography, marine biology and limnology with emphasis on Alaskan waters and the Arctic.The school's goal is to maintain and develop the broad expertise among its faculty and students needed to contribute to the wise use of Alaska's natural resources.

5722 Kansas Rural Center
PO Box 133   913-873-3431
Whiting, KS 66552   Fax: 913-873-3432
E-mail: ksruralctr@aol.com
http://www.kansasruralcenter.org

5723 Kemper Research Foundation
122 Main Street   513-249-2489
Milford, OH 45150
Richard Kemper, Director

5724 Kresge Center for Environmental Health
Harvard University
665 Huntington Avenue   617-732-1272
Boston, MA 02115   E-mail: brain@hsph.harvard.edu
http://www.hsph.harvard.edu/kresge
The Kresge Center serves as the focus for research and training activities in environmental health at the Harvard School of Public Health and elsewhere in the University. The Center was established in 1958 to promote interactions amongbiological scientists, physical scientists and engineers working on environmental problems of human health concern.
Joseph D Brain, Director

5725 Lacawac Sanctuary Foundation
RR 1   570-689-9494
Box 1683   Fax: 570-689-9494
Lake Ariel, PA 18436   E-mail: director@lacawac.org
http://www.lacawac.org/

5726 Large Experimental Aquifer Program
PO Box 91000   503-690-1193
Portland, OR 97291   Fax: 503-690-1273
E-mail: rjohnson@ese.ogi.edu
Richard Johnson, Director

5727 Lawrence Berkeley Laboratory: Structural Biology Division
1 Cyclotron Road   510-486-4311
Mail Stop 3-0226   Fax: 510-486-6059
Berkeley, CA 94720   http://www.lbl.gov/sbdiv/

5728 Leopold Center for Sustainable Agriculture
Iowa State University of Science &   515-294-3711
Technology, 209 Curtiss Hall   Fax: 515-294-9696
Ames, IA 50011-1050   E-mail: leocenter@iastate.edu
http://www.leopold.iastate.edu
The center was created by the 1987 Iowa Ground Water Protection Act with a three fold mission: 1) to identify and reduce adverse environmental impacts of farming practices, 2) develop profitable farming systems that conserve naturalresources, and 3) create educational programs with the ISU Extension Service. The center opertes a competitive grant program and supports several muti-desciplinary research teams and initatives. It is named after internationally acclaimed and Iowaborn Aldo Leopold.
Frederick L Kirschenmann, Director

5729 Limnological Research Unit
109 University Square   814-871-7641
Erie, PA 16541-0001   Fax: 814-871-7633

5730 Living Marine Resources Institute
Stony Brook University   631-632-8656
Stony Brook, NY 11794-5000   Fax: 631-632-9441
E-mail: wwise@notes.cc.sunysb.edu
http://www.sunysb.edu
LIMRI is one of several specialized institutes subsumed within the Marine Sciences Research Center of Stony Brook University. LIMRI's program of research includes investigations on marine fisheries, harmful algal blooms, marine law &policy, and aquaculture. The Institute operates the Flax Pond Marine Laboratory, a seaside, seawater-equipped facility for experimental work, located 5 miles north of the main campus on a tidal pond adjacent to Long Island Sound.
William Wise, Director

5731 Long-Term Ecological Research Project
University of Colorado
1560 30th Street   303-492-3302
Campus Box 450   Fax: 303-492-0434
Boulder, CO 80309-0450   E-mail: tims@culter.colorado.edu
Dr T R Seastedt, Director

5732 Louisiana Sea Grant College Program
1315 East West Highway   301-713-2431
Silver Spring, MD 70803   Fax: 301-713-0799
http://www.laseagrant.org
Jack. R Van Lopik, Executive Director

5733 MIT Center for Advanced Educational Services
77 Massachusetts Avenue   617-253-4962
Building 9, Room 9-234   Fax: 617-253-8301
Cambridge, MA 02139-4307   E-mail: caes-info@mit.edu
http://www-caes.mit.edu
CAES is home to several independent research groups aiming to advance the use of information technology in academia. The Center for Educational Computing Initiatives (CECI) was created to advance the state-of-the-art and practical useof computation and communication technologies for learning and teaching. The Hypermedia Teaching Facility (HTF) is a functioning academic environment whose objective is to deliver a quality hypermedia curriculum via the Internet.

5734 MIT Energy Laboratory
77 Massachusetts Avenue   617-253-3400
Room E40-455   Fax: 617-253-8013
Cambridge, MA 02139-4301   E-mail: elabweb@mit.edu
http://web.mit.edu/energylab/www
Conducts research, educating students, and performing public service in support of economically sound, globally conscious, and environmentally responsible energy technologies and policies a variety of research opportunities forstudents at all levels.

5735 MIT Sea Grant College Program
Massachusetts Institute of Technology
292 Main Street   617-253-7131
E 38-300   Fax: 617-258-5730
Cambridge, MA 02139-9910   E-mail: chrys@deslab.mit.edu
http://web.mit.edu/sargent/
Chrys Chryssostomidis, Director

5736 Marine Science Institute
University of Texas
750 Channel View Drive   361-749-6711
Port Aransas, TX 78373-5015   Fax: 361-749-6777

The Marine Institute is an organized research unit of The University of Texas at Austin. Institute scientists are engaged in both multi-investigator, multi-disciplinary studies and individual research projects in the local area and throughout the world. Many of these projects are combinations of field and laboratory investigations. The Institute receives an operating budget annually that is based on a two-year advanced budget approval by the state legislature.

5737 Marine and Freshwater Biomedical Sciences Center
University of Miami
4600 Rickenbacker Cswy   305-361-4736
Rosenstiel School   888-232-8635
Miami, FL 33149   Fax: 305-421-4001
E-mail: toxmaster@rsmas.miami.edu
http://www.rsmas.miami.edu/groups/niehs
The Environmental Health Science Center is an integral part of the University, with 20 faculty postdoctoral fellows and outreach personnel. Supported by the US government our research focus is on human health applications for disease prevention. Research includes neurotoxicology, potent marine metabolites present in seafood, environmental intoxicants, fisheries models for hepatic metabolism and development of sentinel species for xenobiotic evaluation. Courses, conferences, seminars, outreach.
Dr Pat Walsh, Center Director
Lora Fleming, Associate Director

5738 Massachusetts Cooperative Fish & Wildlife Unit
University of Massachusetts
Holdsworth Natural Resources Center   413-545-0080
Amherst, MA 01003   Fax: 413-545-4358

5739 Massachusetts Public Interest Research Group
29 Temple Place   617-292-4800
Boston, MA 02111   Fax: 617-292-8057
E-mail: pburns@igc.apc.org

5740 Masschusetts Water Resources Research Center
University of Massachusetts
Blaisdell House   413-253-5686
Amherst, MA 01003   Fax: 413-253-1309
E-mail: godfrey@tei.umass.edu
http://www.umass.edu/tei/wrrc
The Center has three objectives: 1) to develop, through research, new technology and more efficient methods for resolving local, state and national water resources problems; 2) to train water scientists and engineers through on-the-job participation in water resources research and outreach; 3) to facilitate water research coordination and the application of research results by means of information dissemination, technology transfer and outreach.
Paul Godfrey, Director

5741 Michigan Atmospheric Deposition Laboratory
University of Michigan
1539 Space Research Building   734-763-6213
Ann Arbor, MI 48109-2143   Fax: 734-763-5376
E-mail: samson@engin.umich.edu
Prof Perry J Samson, Director

5742 Millar Wilson Laboratory for Chemical Research
Jacksonville University
2800 University Bolevard N   904-744-3950
Jacksonville, FL 32211   Fax: 904-744-0101
http://dept.ju.edu/mwl/

5743 Mining and Mineral Resources Research Center
MTU 512 M&ME Building   906-487-2630
1400 Townsend Drive   Fax: 906-487-2934
Houghton, MI 49931-1295   http://www.mtu.edu

5744 Mississippi Cooperative Fish & Wildlife Research Unit
Mississippi State University
Thompson Hall, Room 271   662-325-3174
Box 9690   Fax: 662-325-8750
Mississippi State, MS 39762-9690

5745 Mississippi State Chemical Laboratory
Mississippi State University   662-325-3584
PO Box CR   Fax: 662-325-1618
Mississippi State, MS 39762 www.msstate.edu/dept/chemistry/
Dr. Earl G Allen, State Chemist

5746 Missouri Cooperative Fish & Wildlife Research Unit
University of Missouri
302 Anheuser-Busch Natural Resource   573-882-3634
Columbia, MO 65211-7240   Fax: 573-884-5070
Dr. Charles S Rabeni, Leader

5747 Montana Cooperative Fishery Research Unit
Montana St University Dept Ecology   406-994-3491
Po Box 173460   Fax: 406-994-7479
Bozeman, MT 59717-3460   E-mail: bobwhite@montana.edu
Dr. Robert G White, Leader

5748 Museum of Zoology
University of Massachusetts
Zoology Department   413-545-2287
Amherst, MA 01003   http://www.umass.edu
Dr. DJ Klingener, Director

5749 National Center for Ground Water Research
University of Oklahoma
660 Parrington Oval   405-325-0311
Norman, OK 73019-0390   Fax: 405-325-7596
E-mail: canter@ou.edu
Dr Larry Canter, Director

5750 National Center for Vehicle Emissions Control & Safety
Colrado State University
Old Industrial Sciences Building   970-491-7240
Room 100   Fax: 970-491-7801
Fort Collins, CO 80523   E-mail: ncvecs@cahs.colostate.edu
http://www.NCVECS.colostate.edu
NCVECS is a nationally and internationally recognized university based research and training center devoted to motor vehicle emission issues. NCVECS primarily assists states with research and training related to their local vehicle inspection program. In addition, research is conducted on OBDII systems, alternative fuels and diesel vehicle issues.
Dr Lenora Bohren, Director

5751 National Institute for Global Environmental Change: South Central Regional Center
Tulane University
605 Lindy Boogs Center   504-865-5250
New Orleans, LA 70118-5674   Fax: 504-865-6745
E-mail: nigec@tulane.edu
Dr Stathis Michaelides, Director

5752 National Institute for Urban Wildlife
10921 Trotting Ridge Way   301-596-3311
Columbia, MD 21044   http://www.webdirectory.com/wildlife/
Gomer E Jones, President

5753 National Mine Land Reclamation Center: Eastern Region
106 Land & Water Resources Building   814-863-0291
University Park, PA 16802   Fax: 814-865-3378
E-mail: ajm2@psu.edu

5754 National Mine Land Reclamation Center: Midwest Region
Southern Illinois University
1201 W Gregory   618-453-2496
Carbondale, IL 62901   Fax: 217-333-8816

5755 National Mine Land Reclamation Center: Western Region
Highway 6 S   701-777-5217
Mandan, ND 58554   E-mail: jsolc@eerc.und.nodak.edu

5756 National Park Service Cooperative Unit: Athens
University of Georgia
Institute of Ecology   706-542-8301
Athens, GA 30602
Dr. Stephen Cover-Shabica, Contact

5757 National Research Center for Coal and Energy NRCCE
West Virginia University   304-293-2867
PO Box 6064   Fax: 304-293-3749
Morgantown, WV 26506   E-mail: NRCCE@mail.wvu.edu
http://www.nrcce.wvu.edu
The NRCCE identifies, develops, promotes, coordinates and
supports multidisciplinary energy and environmental research
and service programs that are important to industry, the state and
the nation in collaboration with facultythroughout West Virginia University (WVU) schools and colleges. In addition to involving WVU faculty in the development and execution of
NECCE programs, the center develops partnerships with other
academic and industrial research institutionsnationally.
Richard A Bajura, Director
Trina K Wofle, Associate Director

5758 National Undersea Research Center
University of Hawaii at Manoa
1000 Pope Road   808-956-6802
Honolulu, HI 96822   Fax: 808-956-2136
E-mail: malahoff@soest.hawaii.edu
Alexander Malahoff, Director

5759 Natural Energy Laboratory of Hawaii Authority
73-4460 Queen Kachumanu Highway   808-329-7341
101 Keahole Point   Fax: 808-326-3262
Kailua Kona, HI 96740   E-mail: inquires@nelha.org
http://nelha.org
NELHI, an agency of the State of Hawaii, operates facilities at
Keahole Point on Hawaii Island that pumps ashore cold deep and
warm surface seawater for commercial and research tenants
from the private and public sectors. Tenantsutilize the seawater
and NELHA's high sunlight and consistant temperatures in a
wide range of aquaculture and energy projects.
Jeff Smith, Executive Director

5760 New Hampshire Sea Grant Program
University of New Hampshire
Kingman Farm   603-749-1565
Durham, NH 03824-3512   Fax: 603-743-3997
E-mail: abc@cisunix.unh.edu
http://www.seagrant.unh.edu
A component of the National Sea Grant College Program, NH
Sea Grant works toward the conservation, wise use and develop-
ment of marine resources in the state and region.
Ann Bucklin, Director

5761 New York Cooperative Fish & Wildlife Research Unit
Cornell University
Natural Resource Department   607-255-2151
Fernow Hall   Fax: 607-255-1895
Ithaca, NY 14853   E-mail: dnrcru-mailbox@cornell.edu
http://www.dnr.cornell.edu/f@wres/nycf@wru.htm

Dr. Milo Richmond, Unit Leader

5762 New York State Water Resources Institute
Cornell University
204A Rice Hall   607-255-5941
Ithaca, NY 14853-5601   Fax: 607-255-5945
E-mail: nyswri@cornell.edu
http://www.cfe.cornell.edu/wri/
Keith S Porter, Director

5763 Northern Lights Institute
210 N Higgins #326   406-721-7415
PO Box 8084   Fax: 406-721-7415
Missoula, MT 59807
Donald Snow, Program Director

5764 Northwoods Field Station
Hiram College
PO Box 123   216-569-3211
Hiram, OH 44234   http://www.hiram.edu
Dr WR Knight, Director

5765 Occupational & Environmental Health Laboratory
Chemistry & Industrial Hygiene Dept   256-765-4622
Box 5049   Fax: 256-765-4958
Florence, AL 35632-0001   http://www2.una.edu/chemdept/

5766 Ocean & Coastal Policy Center
University of California
Woolley-5134   805-893-8393
Santa Barbara, CA 93106-6105   Fax: 805-893-8062

5767 Ocean Engineering Center
33 College Road   603-862-1898
Kingsbury Hall   Fax: 603-862-0241
Durham, NH 03824   E-mail: kcb@cisunix.unh.edu
Kenneth Baldwin, Director

5768 Oceanic Institute
Makapuu Point   808-259-3102
41-2020 Kalanianole Highway   Fax: 808-259-5971
Waimanalo, HI 96795   E-mail: oi@oceanicinstitute.org
http://www.oceanicinstitute.org
Oceanic Institute is a not-for-profit organization dedicated to
research, development and transfer of oceanographic, marine
Environmental, and aquaculture technologies. Oceanic Institute is a world leader in conducting appliedresearch in
aquaculture production and marine resource conservation. Its
mission is to develop and transfer environmentally responsible
technologies to increase aquatic food production while promoting the sustainable use of ocean resources.

Dr Bruce S Anderson, President
Gary E Karr, Director Of Comm & Education

5769 Oregon Cooperative Fishery Research Unit
Oregon State University
104 Nash Hall   541-737-1938
Corvallis, OR 97331-3803   Fax: 541-737-3590
E-mail: or_cfwru@orst.edu
http://www.orst.edu/

5770 Oregon Cooperative Park Studies Unit
Oregon State University
3200 Jefferson Way   541-737-2056
Corvallis, OR 97331   E-mail: starkeye@ccmail.orst.edu
http://www.cof.orst.edu/
Dr Edward E Starkey, Codirector

5771 Oregon Sea Grant College Program
Oregon State University
322 Kerr Administrative Building   541-737-2714
Corvallis, OR 97331   Fax: 541-737-7958
http://seagrant.oregonstate.edu/
Robert E Malouf, Director

5772 Pacific Gamefish Research Foundation
47-381 Kealakehe Parkway   808-329-6105
PO Box 4800   Fax: 808-329-1148
Kailua Kona, HI 96740

5773 Pacific Northwest Research Institute
720 Broadway   206-726-1200
Seattle, WA 98122-4327   Fax: 206-726-1217
http://www.pnri.org
Established as Pacific Northwest Research Foundation in 1956
by Dr. William B Hutchinson, Sr. as the first private nonprofit
biomedical and clinical research institute in the Northwest. As
founder and first director, Dr. Hutchinson'sprimary objective
was to provide a facility for basic and clinical research dedicated
to the improvement of patient care. Sponsors basic science efforts in biochemistry, molecular biology and immunology as
they pertain to the clinical areas of cancerand diabetes.
R. Paul Robertson, MD, CEO/Scientific Director

5774 Pennsylvania Cooperative Fish & Wildlife Research Unit
Pennsylvania State University
Merkle Building   814-865-4511
University Park, PA 16802   Fax: 814-863-4710
                                             E-mail: klc2@psu.edu
                                             http://pacfwru.cas.psu.edu

Dr Robert F Carline, Leader

5775 Permaculture Gap Mountain
9 Old County Road   603-532-6877
Jaffrey, NH 03452

5776 Pesticide Research Center
Michigan State University
107 Pesticide Research Center   517-353-9430
East Lansing, MI 48824   E-mail: whalon@pilot.msu.edu
                                             http://www.msu.edu

5777 Pesticide Research Laboratory & Graduate Study Center
Pennsylvania State University
Entomology Department   814-863-7789
University Park, PA 16802   Fax: 814-865-3048
                                             E-mail: gwf10@psu.edu
                                             http://www.ento.psu.edu

Gary W Felton, Director

5778 Planning Institute
University of Southern California
Von KleinSmid Center 351   213-740-6842
Los Angeles, CA 90089-0004   Fax: 213-740-1801
                                             E-mail: sppd@usc.edu
                                             http://www.usc.edu

Michael Horst, Contact

5779 Practical Farmers of Iowa
2035 190th Street   515-432-1560
Boone, IA 50036   http://www.practicalfarmers.org

5780 Program for International Collaboration in Agroecology: University of California Santa Cruz
Center for Agroecology   831-459-2506
1156 High Street   Fax: 831-459-2867
Santa Cruz, CA 95064   http://www.agroecology.org
Researches, develops, and advances sustainable food and agriultural systems that are environmentally sound, economically viable, socially responsible, nonexploitive, and that serve as a foundation for future generations. A specialfocus on promoting an international network of training programs in agroecology.
Steven Gliessman, Professor of Agroecology

5781 Program in Freshwater Biology
University of Mississippi
214 Shoemaker Hall   662-232-7203
Biology Department   Fax: 662-915-5144
University, MS 38677   E-mail: biology@olemiss.edu
                                             http://www.olemiss.edu

Dr. James Kushlan, Chairman

5782 Rare and/or Endangered Species Research Center
215 Mitchell Street   256-760-4429
Florence, AL 35630

5783 Red Butte Garden and Arboretum
300 Wakara Way   801-581-4747
Salt Lake City, UT 84108   http://www.redbuttegarden.org

5784 Remote Sensing/Geographic Information Systems Facility
Indiana State University   812-237-2444
Science Building, Room 159   Fax: 812-237-8029
Terre Haute, IN 47809

E-mail: gga@baby.indstate.edu
http://http://baby.indstate.edu/geo/ggafolder/ggafacil.html
Dr Susan Berta, Interim Chairperson

5785 Renew America
1200 18th Street Northwest   202-721-1545
Suite 1100   Fax: 202-467-5780
Washington, DC 20036   http://www.solstice.crest.org
Tina Hobson, President

5786 Research Triangle Institute
3040 Cornwallis Road   919-541-6000
PO Box 12194   Fax: 919-541-7155
Research Triangle Park, NC 27709   E-mail: listen@rti.org
                                             http://www.rti.org
Clients around the world rely on RTI to conduct innovative, multidisciplinary research to meet their R and D challenges. RTI's staff of more then 1,850 people represents a diverse set of technical capabilities in health and medicine,environmental protection, technology commercialization, decision support systems and education and training.
Victoria F Haynes, President
Dennis F Naugle, VP Enviro and Engineering

5787 Resources for the Future
1616 P Street Northwest   202-328-5000
Washington, DC 20036   Fax: 202-939-3460
                                             E-mail: info@rff.org
                                             http://www.rff.org
RFF is a nonprofit and nonpartisan think tank located in Washington DC that conducts independent research-rooted primarily in economics and other social sciences on environmental and natural resource issues. RFF was founded in 1952.

Lesli Creedon, Vice Pres., External Affairs
Stan Wellborn, Director of Communications

5788 Resources for the Future: Energy & Natural Resources Division
1616 P Street, Northwest   202-328-5000
Washington, DC 20036   Fax: 202-939-3460
                                             http://www.rff.org

Douglas Bohi, Director

5789 Resources for the Future: Quality of the Environment Division
1616 P Street, Northwest   202-328-5000
Washington, DC 20036   Fax: 202-939-3460
                                             http://www.rff.org

Raymond J Kopp, Director

5790 Rice Creek Field Station
Oswego State University
Rice Creek Field Station   315-342-0961
Oswego, NY 13126   Fax: 315-342-0347
                                             E-mail: anelson@oswego.edo
Dedicated to the support of academic instruction, research, and public service in all aspects of natural history, especially the natural sciences and the environmental education.

5791 River Studies Center
University of Wisconsin
Department of Microbiology   608-785-8238
La Crosse, WI 54601   Fax: 608-785-6460

5792 Robert J Bernard Biological Field Station
1400 N Amherst Avenue   909-624-6661
Claremont, CA 91711   E-mail: Stephen.Dreher@cgu.edu
                                             http://www.bfs.claremont.edu/

5793 Rocky Mountain Biological Laboratory
PO Box 519   970-349-7231
Crested Butte, CO 81224   Fax: 970-349-7481
                                             E-mail: admin@rmbl.org
                                             http://www.rmbl.org

High-altitude field station whose principal purpose is to provide quality research and teaching facilities for biologists and biology students of all diciplines who can benefit personally and intellectually from studying at thislocation. An important further purpose of the Laboratory is to promote the understanding and protection of the high altitude ecosystems of Colorado and the watershed of the Qunnison River through through the professional activuty of its members.
Ian Billick, Director
Michele Simpson, Assistant Director

5794 Rocky Mountain Mineral Law Foundation

9191 Sheridan Boulevard #203   303-321-8100
Westminster, CO 80031   Fax: 303-321-7657
E-mail: info@rmmlf.org
http://www.rmmlf.org

David P Phillips, Executive Director
Mark Holland, Senior Attorney

5795 Romberg Tiburon Centers

San Francisco State University
3150 Paradise Drive   415-435-7100
PO Box 855   Fax: 415-435-7120
Tiburon, CA 94920

5796 Roosevelt Wildlife Station

1 Forestry Drive   315-470-6798
Syracuse, NY 13210   Fax: 315-470-6934
E-mail: wfporter@esf.edu

5797 Salt Institute

700 N Fairfax Street   703-549-4648
Suite 600   Fax: 703-548-2194
Alexandria, VA 22314   E-mail: info@saltinstitute.org
http://www.saltinstitute.org
The Salt Institute is an international trade association of salt producers. It has information about the environmental impacts of salt production and use.

Richard L Hanneman, President

5798 School for Field Studies

10 Federal Street   978-741-3544
Suite 24   800-989-4435
Salem, MA 01970-3876   Fax: 978-741-3551
E-mail: admissions@fieldstudies.org
http://www.fieldstudies.org
Students conduct hands-on, community-focused environmental field work around the world. Addresses critical environmental issues including preserving entire ecosystems or individual species, balancing economic development andconservation, and finding ways to manage and maintain wildlife, marine and agricultural resources.
Paul Houlihan, President

5799 School of Marine Affairs (SMA)

University of Washington
3707 Brooklyn Avenue NE   206-543-7004
Seattle, WA 98105   Fax: 206-543-1417
SMA is a masters-level, professional school within the University of Washington specializing in the interdisciplinary teaching and research on contemporary coastal and ocean resources, environmental and developmental problems.
Marc J Hershman, Professor

5800 Science and Public Policy

Rockefeller University   212-327-7917
1230 York Avenue, Box 234   Fax: 212-327-7519
New York, NY 10021-6399   http://www.rockefeller.edu

5801 Seatuck Foundation: Seatuck Research Program

500 Saint Marks Lane   631-581-6908
Islip, NY 11751

5802 Society for Ecological Restoration

1955 West Grant Road   520-622-5485
Suite 150   Fax: 520-622-5491
Tucsonn, AZ 85745   E-mail: info@ser.org
http://www.ser.org

John Rieger, President

5803 Society for the Application of Free Energy

1315 Apple Avenue   301-587-8686
Silver Spring, MD 20910   Fax: 301-587-8688
E-mail: uv@uvbi.com
http://www.solarthem.com

5804 Soil and Water Research

4115 Gourrier Avenue   225-757-7726
Baton Rouge, LA 70808   Fax: 225-757-7728
http://mse.ars.usda.gov/la/btn/swr/
Mission is to characterize and quantify the transport and fate of agrochemicals in high water table soils, develop integrated soil, water, and agrochemical management systems that provide profitable yields and improve water table soilsin the humid, warm temperature areas of the US and develop improved soil and water management systems and operational procedures that enhance crop production conditions and increase the efficiency of conducting farming operationsin a timely manner.
Dr. James Fouss, Research Leader

5805 Solar Energy Group

University of Chicago
5640 S Ellis Avenue   773-702-7756
Chicago, IL 60637   Fax: 773-702-6317
E-mail: winston@rainbow.uchicago.edu

5806 Solar Energy and Energy Conversion Laboratory

University of Florida
Mechanical Engineering Department   352-392-0812
Gainesville, FL 32611   Fax: 352-392-1071
E-mail: solar@mae.ufl.edu
http://http://seecl.mae.ufl.edu/solar/
Has uniquely influenced the development of solar energy and renewable energy conversion systems all over the world through its research, education and training. Has pioneered research in many areas of solar energy, energy conversionand conservation. The lab has been designated as an ASME National Landmark.

Dr. D Yogi Goswami, Professor & Director

5807 South Carolina Sea Grant Consortium

287 Meeting Street   843-727-2078
Charleston, SC 29401   Fax: 843-727-2080
http://www.scseagrant.org
A state agency that supports coastal and marine research, education, outreach, one technical assistance program that foster sustainable economic development and resource conservation. The consortium represents eight university andstate research organizations and induces a number of information products on coastal and marine resource topics.
16 pages
M Richard DeVoie, Executive Director

5808 Southern California Pacific Coastal Water Research Project

7171 Fenwick Lane   714-894-2222
Westminster, CA 92683   Fax: 714-894-9699
http://www.sccwrp.org/#
SCCWRP is a joint powers agency focusing on marine environmental research. A joint powers agency is one that is formed when several government agencies have a common mission that can be better addressed by pooling resources andknowledge. In our case, the common mission is to gather the necessary scientific information so that our member agencies can effectively and cost-efficiently protect the Southern California marine environment.

5809 Southwest Consortium on Plant Genetics & Water Resources

New Mexico State University   505-646-6553
Box 3GL   http://www.nmsu.edu
Las Cruces, NM 88003
Dr. John D Kemp, Chairman

5810 Strom Thurmond Institute of Government & Public Affairs, Regional Development Group
200 Excelsior Mill Road   864-646-4700
PO Box 158
Pendleton, SC 29670

5811 Stroud Water Research Center
970 Spencer Road   610-268-2153
Avondale, PA 19311   Fax: 610-268-0490
E-mail: webmaster@stroudcenter.org
http://www.stroudcenter.org
The Stroud Water Research Center seeks to understand streams and rivers and to use the knowledge gained from its research to promote environmental stewardship and resolve freshwater challenges throughout the world.
B W Sweeney, Laboratory Director

5812 Sustainable Agriculture Research & Education Program
University of California
1 Shields Avenue   530-752-7556
Davis, CA 95616   Fax: 530-754-8550
E-mail: sarep@ucdavis.edu
http://www.sarep.ucdavis.edu
Sean Swezey, Director
Linda Fugitt, Office Manager

5813 Systematics Museums
University of Kansas
Dyche Hall   785-864-4540
Lawrence, KS 66045-7561   Fax: 785-864-5335
A comprehensive research, graduate education and public service institution dedicated to biodiversity science and collections. Collections of more than 7 million plant and animal specimens, with particular strengths in neotropicalamphibians, great plains, flora, bees and antarctic plant fossils.
Dr Leonard Krishtalka, Director

5814 Tennessee Cooperative Fishery Research Unit
Tennessee Technological University
Biology Department   931-372-3094
PO Box 5114   Fax: 931-372-6257
Cookeville, TN 38505   E-mail: jim_layzer@tntech.edu
James Layzer, Leader

5815 Texas Center for Policy Studies
44 East Avenue   512-474-0811
Suite 306   Fax: 512-474-7846
Austin, TX 78768   http://www.texascenter.org
Mary Kelly, Executive Director

5816 Texas Water Resources Institute
Texas A&M University   979-845-1851
College Station, TX 77843-2118   Fax: 979-845-8554
E-mail: twri@tamu.edu
http://twri.tamu.edu
C Allan Jones, Director

5817 Throckmorton-Purdue Agricultural Center
8343 US231 Southy   765-538-3422
Lafayette, IN 47909-9049   Fax: 765-538-3423
E-mail: jjf@aes.purdue.edu
John Trott, Farm Director

5818 Toxic Chemicals Laboratory
Cornell University
Cornell University, Tower Road   607-255-4538
New York State College of Agriculture
Ithaca, NY 14853-4538

5819 US Forest Service: Wildlife Habitat & Silviculture Laboratory
506 Hayter Street   936-569-7981
Nacogdoches, TX 75965   Fax: 936-569-9681
http://www.srs.fs.fed.us/wildlife

We assess impacts of forest management practices on wildlife populations and their habitats and provide guide lines to land managers for improving their management to accommodate wildlife.
Ronald E Thill, Project Leader

5820 USDA Forest Service: North Central Research Station
1407 South Harrison Road   517-355-7740
Suite 220   Fax: 517-355-5121
East Lansing, MI 48823   E-mail: rhaack@fs.fed.us
Dr. Robert Haack, Project Leader

5821 USDA Forest Service: Rocky Mountain Research Station
2150 Centre Avenue   970-295-5926
Building A   Fax: 970-295-5927
Fort Collins, CO 80526-1891
We seek to be an unbiased source of scientific information; provide tools that consider the multidisciplinary nature of natural resource decisions and recognize that resource managers need scientific information that is integrated anddeveloped for application. Research results are made available through a variety of technical reports, seminars, demonstrations, exhibits and personal consultations. These help resource managers and planners balance economic and environmental demandsworldwide.
Marcia Patton-Mallory, Station Director

5822 UVA Institute for Environmental Negotiation
University of Virginia
104 Emmet Street   434-924-1970
Charlottesville, VA 22903   Fax: 434-924-0231
E-mail: envneg@virgina.edu/
http://www.virginia.edu/ien
The Institute for Environmental Negotiation is committed to building a sustainable future for Virginia's communities and beyond by: bringing people together to develop sustainable solutions; providing people with learning opportunitiesto be creative and collaborative leaders; and building understanding of best collaborative practices.
Dr E Franklin Dukes, Director

5823 University Forest
University of Missouri
Route 2   573-222-8373
Box 139   Fax: 573-222-8829
Williamsville, MO 63967
Albert R Vogt, Director

5824 Vantuna Research Group
Moore Laboratory of Zoology
Occidental College   323-259-2675
1600 Campus Road   http://www.oxy.edu
Los Angeles, CA 90041
John S Stephens, Director

5825 Virginia Center for Coal & Energy Research
Virginia Tech   540-231-5458
Blacksburg, VA 24061   Fax: 540-231-4078
E-mail: vccer@vt.edu
http://www.energy.vt.edu
Created by an Act of the VA General Assembly in 1977 as a study, research, information and resource facility for the commonwealth of VA, and is located at VA Tech. The mission involves four primary functions: research in energy andcoal related issues of interest to the Commonweatlth; coordination of coal and energy research at VA Tech; dissemination of coal and energy data to users in the Commonwealth; examination of socio-economic implications and environmental impacts ofcoal and energy.
Dr Michael Karmis, Director
Margaret Radcliffe, Manager

5826 Washington Cooperative Fishery Research Unit
University of Washington
Box 355020   206-543-6475
Seattle, WA 98195-5020

5827 Waste Management Education & Research Consortium

New Mexico State University
Box 30001, MSC WERC
Las Cruces, NM 88003-0001

505-646-2038
800-523-5996
Fax: 505-646-4149
E-mail: bdelrio@nmsu.edu
http://www.werc.net

A key component of WERC is higher education degree programs. To support this component, WERC administers a Fellowship Program at each academic partner institution. The primary objective for the WERC Fellowship program is to helpstudents develop a program which will lead to environmental related career opportunities upon graduation.

Barbara Valdez, Program Facilitator

5828 Waste Management Research & Education Institute

University of Tennessee
676 Dabney Hall
Knoxville, TN 37996-0710

865-974-8080
Fax: 865-974-8086
E-mail: sayler@utk.edu
http://www.web.utk.edu/~enr/contact.htm

Dr. Gary Sayler, Acting Director

5829 Water Quality Laboratory

Western Wyoming Community College
PO Box 428
Rock Springs, WY 82901

307-382-1662
http://www.wwcc.cc.wy.us/

Craig Thompson, Director

5830 Water Resource Center

University of Minnesota
1985 Buford Avenue
St Paul, MN 55108

612-625-2282
Fax: 612-625-1263
E-mail: ander045@umn.edu
http://wrc.coafes.umn.edu

The center coordinatoes research, education and extension programs on water resource issues. Administrative responsibility for Water Resource Sciences Graduate Program. Is the Water Resources Institute for Minnesota.

Jim Anderson, Co-Director
Deb Swackhamer, Co-Director

5831 Water Resources Center

University of Kansas, Geology Dept
1475 Jayhawk Boulevard, Room 120
Lawrence, KS 66045

785-864-4974
Fax: 785-864-5276
http://www.ku.edu/~geology/index.html

5832 Water Resources Institute

University of Wisconsin
1975 Willow Drive
Floor 2
Madison, WI 53706-1177

608-262-3577
Fax: 608-262-0591
http://www.wri.wisc.edu

The University of Wisconsin Water Resources Institute's primary mission is to plan, develop and coordinate research programs that address present and emerging water-and land-related issues. It has developed a broadly based statewideprogram of basic and applied research that has effectively confronted a spectrum of societal concerns. It is one of 54 institutes or centers located at the Land Grant College in each state.

Anders W Andren, Director

5833 Water Resources Research Institute at Kent University

Kent State University
136 Science Research Building
PO Box 5190
Kent, OH 44242-0001

330-672-2529
Fax: 330-672-4834
E-mail: rheath@kent.edu
http://www.dept.kent.edu/wrri

The institute fosters a broad-based approach to the evaluation and analysis of environmental problems related to water use. WRRI is a resouce for citizens, governmental agencies and policy makers, providing reliable scientificinformation on which to base decisions related to the wise use and management of water and land management, water policy decisions and environmental conservation.

Dr. Robert T Heath, Director

5834 Water Resources Research of the University of North Carolina

North Carolina State University
1131 Jordan Hall, Box 7912
Raleigh, NC 27695-7912

919-515-2815
Fax: 919-515-2839
E-mail: water_resources@nesu.edu
http://www.ncsu.edu/wrri/

One of 54 state water institutes authorized to administer and promote federal/state partnerships in research and information transfer on water-related issues. Identifies and supports research needed to help solve water quality andwater resources problems in NC. Publishes peer-reviewed reports on completed research projects. Sponsors educational seminars and conferences and provides public information on water issues through publication of a newsletter.

Dr David H Moreau, Director
Kelly Porter, Env. Ed. & Comm. Coordinator

5835 Water Testing Laboratory

Morehead State University
Box 804
Morehead, KY 40351

606-783-2961
http://www.morehead-st.edu

Rita Wright, Contact

5836 Weather Analysis Center

University of Michigan
Space Research Building
2455 Hayward Street
Ann Arbor, MI 48109-2143

734-936-0482
Fax: 734-763-0437
E-mail: dbaker@umich.edu
http://www.aoss.engin.umich.edu

Atmospheric, planetary and space science engineering.
Dr Dennis Baker, Director

5837 West Virginia Water Research Institute

West Virginia University
Room 202 NRCCE
Box 6064
Morgantown, WV 26506

304-293-2867
Fax: 304-293-7822
E-mail: pziemkie@wvu.edu

The West Virginia Water Research Institute (WVWRI) has served as a statewide vehicle for performing research related to water issues. WVWRI serves as the coordinating body for the following programs: the National Mine LandReclamation Center, Appalachian Clean Streams Initiatve, Acid Drainage Technology Initiative, Combustion Byproducts Recycling Consortium, National Environmental Education Training Center, Hydrogeology Research Center, State Water Institutes, andmore.

Paul Ziemkiewicz, PhD, Director

5838 Western Region Hazardous Substance Research Center

Oregon State Univtersity
204 Apperson Hall
Corvallis, OR 97331-2302

541-737-2751
Fax: 541-737-3099
E-mail: wrhsrc@engr.orst.edu
http://http://wrhsrc.oregonstate.edu

Perry L McCarty, Director

5839 Western Research Farm

36515 Highway 34 E
Castana, IA 51010

712-885-2802
E-mail: wroush@iastate.edu

Wayne B Roush, Ag Specialist

5840 Wetland Biogeochemistry Institute

Louisiana State University
Baton Rouge, LA 70803-7511

225-388-8806
Fax: 225-388-6423

William H Patrick Jr, Director

5841 Wilderness Institute: University of Montana

School of Forestry
Missola, MT 59812

406-243-5361
800-462-8636
Fax: 406-243-4845
E-mail: wi@forestry.umt.edu
http://www.forestry.umt.edu.si

Mission is to further understand wilderness and its stewardship through education, outreach, and scholarship. Activity is guided by the philosophy that wildlands are increasingly significant, ecologically and socially, and educateddialogue about the role of wild places in our nation's future should be promoted. Engaged in undergraduate education, graduate student research, the dissemination of wilderness information and the promotion of scholarship on wilderness issues.
Wayne Freimund, Director
Laurie Ashley, Outreach Coordinator

5842 Wilderness Research Center
University of Idaho
College of Natural Resources 208-885-7911
Room 18a Fax: 208-885-6226
Moscow, ID 83844-1144 E-mail: rrt@uidaho.edu
http://http://www.cnr.uidaho.edu/wrc/
The mission of the WRC is to study the human dimensions of wilderness ecosystems. The WRC conducts research and teaches courses on the use of wilderness for personal growth, therapy, education, and leadership development.
Steve Hollenhorst, Ph.D, Director
Lilly Steinhorst, Administrative Assistant

5843 Wisconsin Applied Water Pollution Research Consortium: University of Wisconsin-Madison
University of Wisconsin
3232 Engineering Hall 608-263-7773
Madison, WI 53706-1691 Fax: 608-262-5199
E-mail: harringt@engr.wisc.edu
http://www.engr.wisc.edu/consortial/cawper
This consortium seeks effective and economical solutions to water supply problems and pollution control in Wisconsin. It conducts innovative practical research that cannot be carried out effectively by individual organizations.
Greg Harrington, Director

5844 Wisconsin Rural Development Center
USDA Rural Development-WI
4949 Kirschling Ct. 715-345-7615
Stevens Point, WI 54481 Fax: 715-345-7614
E-mail: RD.Webmaster@wi.usda.gov
http://www.rurdev.usda.gov/wi/index.htm
Frank Frassetto, State Director

5845 Wisconsin Sea Grant Institute
University of Wisconsin
1975 Williw Drive 608-262-0905
Floor 2 Fax: 608-262-0591
Madison, WI 53706-1177 http://www.seagrant.wisc.edu
The University of Wisconsin Sea Grant Institute is a statewide program of basic and applied research, education, and technology transfer dedicated to the wise stewardship and sustainable use of Great Lakes and ocean resources. It ispart of a national network of 30 university-based programs.

Dr. Anders W Andren, Director

5846 Yale Institute for Biospheric Studies (YIBS)
21 Sachem Street 203-432-9856
PO Box 208105 Fax: 203-432-9927
New Haven, CT 06520 E-mail: roserita.riccitelli@yale.edu
http://www.yale.edu/yibs

Rose Rita Riccitelli, Administrator
Derek Briggs, Director

Universities

5847 20/20 Vision
8403 Colesville Road   301-587-1782
Suite 860   Fax: 301-587-1848
Silver Spring, MD 20910   E-mail: vision@2020vision.org
http://www.2020vision.org
A national grassroots nonprofit organization that works to increase citizen participation in public policy related to peace and the environment. Each month 20/20 Vision produces an action-alert postcard so that our members can quicklyand easily contact policymakers and weigh in on a timely issue. Acts to dog government and corporations and collaborate with dozens of groups and experts.
Tom Z Collina, Executive Director

5848 Agronomy and Environmental Science
Delaware Valley College
700 East Butler Avenue   215-489-2334
Doylestown, PA 18901   Fax: 215-489-2404
E-mail: Hepner@devalcol.edu
http://www.devalcol.edu/agronomy/index.htm
The Department of Agronomy and Environmental Science offers courses designed to give you a broad, workable background in the plant, soil, turf or environmental sciences. We focus on the environmental issues facing society today, and wegive you the knowledge and training necessary to be successful. And, our programs provide you with excellent preparation for grad school.
Larry Hepner, Contact

5849 Allegheny College
520 North Main Street   814-332-3100
Meadville, PA 16335   800-521-5293
Fax: 814-337-0431
E-mail: info@allegheny.edu
http://www.allegheny.edu
Environmental Science is the study of interrelationships between human activities and the environment. Two major programs: 1) Environmental Science. Core courses include biology, chemistry, geology, and mathematics. Upper levelcourses synthesize, integrate and apply basic sciences toward solving real environmental problems; 2) Environmental Studies. Objective is to study the concept of sustainability in an integrated way.
Dr. Eric Pallant, Program Director
Gayle Pollock, Contact

5850 Alliance for Environmental Education
51 Main Street   540-253-5812
PO Box 368   Fax: 540-253-5811
The Plains, VA 22171
Steven C Kussmann, Chairman

5851 Amazon Center for Environmental Education Research
Ten Environs Park   205-428-1700
Helena, AL 35080   Fax: 205-428-1711
E-mail: aceer@jetravel.com
http://www.erri.psu.edu/web/aceer.htm
Opened in 1993, the ACEER, Amazon Center for Environmental Education and Research, is a joint effort with other private companies and individuals to provide a research station for scientists in the rainforest and opportunity for thelayman to share in their discovery. The large, thatched buildings are identical in construction to those found in Explorama Lodge.

5852 American Chemical Society
1155 16th Street Northwest   202-872-4600
Washington, DC 20036   800-227-5558
Fax: 202-776-8258
E-mail: webmaster@acs.org
http://www.chemistry.org
Promotes the public perception and understanding of chemistry and the chemical sciences through public outreach programs and public awareness campaigns.
John K Crumb, Executive Director

5853 American Geological Institute
4220 King Street   703-379-2480
Alexandria, VA 22302-1502   Fax: 703-379-7563

E-mail: agi@agiweb.org
http://www.agiweb.org
A 501(c)(3) nonprofit federation of 43 scientific and professional associations that represent more than 120,000 geologists, geophysicists, and other earth scientists. Provides information services to geoscientists, serves as a voiceof shared interests in our profession, plays a major role in strengthening geoscience education, and strives to increase public awareness of the vital role the geosciences play in society's use of resources and interaction with the environment.
Marcus Milling, Executive Director

5854 American Insitute of Biological Sciences
1444 I Street   202-628-1500
Suite 200   Fax: 202-628-1509
Washington, DC 20005   http://www.aibs.org
AIBS facilities communication and interactions among biologists, biological societies, and biological disciplines in order to serve and advance the interests of organismal and integrative biology in the broader scientific community andother components of society on issues related to research, education, and public policy.
Marvalee Wake, President

5855 American Nature Study Society
Poconos Environmental Education Center
Rural Route 2   507-828-9692
Box 1010   Fax: 507-828-9695
Dinghams Ferry, PA 18328   http://http://hometown.aol.com
The American Nature Study Society, founded in 1908, is America's oldest environmental education organization. Affiliated with the American Association for the Advancement of Science, the ANSS is focusing on environmental literacy andnature writing.
Steve Melcher, President
Betty McKnight, Secretary

5856 Antioch College
Glen Helen Outoor Education Center
1075 State Route 343   937-767-7648
Yellow Springs, OH 45387   Fax: 937-767-6655
E-mail: mdavidson@antioch-college.edu
http://www.glenhelen.org
Training in residential naturalist instruction for upper elementary aged students. Classes and field experience in outdoor education methods and natural history. Care for hawk or owl in Raptor Center.
Motyka Davidson, Day Programs Director

5857 Antioch New England Graduate School
40 Avon Street   603-357-6265
Keene, NH 03431-3552   800-553-8920
Fax: 603-357-0718
E-mail: admissions@antiochne.edu
http://www.antiochne.edu
Doctoral Program in Environmental Studies intended for people who are committed to scholarly excellence and wish to design, implement and evaluate research regarding crucial environmental issues. Program cultivates a dynamic learningcommunity of environmental scholars/practioners who combine scope and vision with breadth and precision, conceptualizing and implementing research strategies and designs that contribute to solving and anticpating regional, national and globalenvironmental issues.
Steven P Guerriero, Chairperson

5858 Antioch University Seattle
2326 6th Avenue   206-268-4202
Seattle, WA 98121-1814   Fax: 206-441-3307
E-mail: admissions@antiochsea.edu
http://www.seattleantioch.edu
Program approaches environmental concerns by emphasizing social science perspectives and natural science literacy. Program is part of the Center for Creative Change, an integrated professional studies center.
Ormond Smythe, Dean

5859 Arizona State University
Box 877705   480-965-8972
Tempe, AZ 85287-7705   Fax: 480-965-0865
http://www.asu.edu

The general studies program requirement consists of courses intended to provide every student pursuing an undergraduate degree basic skills and familiarity areas of knowledge. Introductory and advanced courses are required and studentsmay select from an approved menu of courses. The general studies program consists of five core areas and three awareness areas. The core areas are literacy and critical inquiry, numeracy, humanities and fine arts, social and behavioral sciences.
Michael Crow, President

5860 Arkansas Tech University

1509 North Boulder Avenue	479-968-0294
Russellville, AR 72801	800-582-6953
	Fax: 479-964-0837
	http://www.atu.edu

A two-year preparatory program in Wildlife Conservation with an outlined Wildlife Curriculum was developed at Arkansas Tech University in 1956. Two years later, plans were made to elevate this program to a four-year program. During the1959-1960 academic year, a full slate of courses was developed that provided the foudation for degree that specialized in fisheries and wildlife management.
Deborah Wilson, MLA Program Director

5861 Association of American Geographers

1710 16th Street NW	202-234-1450
Washington, DC 20009-3198	Fax: 202-234-2744
	E-mail: gaia@aag.org
	http://www.aag.org

The Association of American Geographers (AAG) is a scientific and educational society founded in 1904. Its 6,500 members share interests in the theory, methods, and practice of geography, which they cultivate through the AAG's AnnualMeeting, two scholarly journals (the Annals of the Association of American Geographers and The Professional Geographer), the monthly AAG Newsletter, and the activities of its two affinity groups, nine regional divisions and 53 specialty groups.

Dr Douglas Richardson, Executive Director

5862 Association of Environmental and Resource Economists

1616 P Street NW	202-328-5077
Room 510	Fax: 202-939-3460
Washington, DC 20036	E-mail: voigt@rff.org

A means of exchanging ideas, stimulating research, and promoting graduate training in resource and environmental economics. Members come from academic institutions, the public sector, and private industry.
Marilyn Voigt, Secretary

5863 Auburn University

Environmental Institute	
59 Duggar Drive	334-844-4000
Extension Cottage	Fax: 334-844-5748
Auburn Unviersity, AL 36849	

Serves faculty, governments, and the general public in a coordinating role to bring together teams to develop acceptable and economically feasible means of enhancing the environmental quality of the state and nation.
John Heilman, Provost & VP Academic Affair

5864 Audubon Expedition Institute

29 Everett Street	207-338-5859
Cambridge, MA 02138	888-287-2234
	Fax: 207-338-1037
	E-mail: ldoubleday@aei-audubon.org
	http://www.lesley.edu/gsass/audubon

Students challenge themselves and their assumptions through experimental learning and direct contact with social, natural, historical and urban environments. Subjects are studied and integrated through real life experiences.
Wendy Watson, Development Director

5865 BLM Environmental Education Program

1620 L Street NW	202-452-0365
Suite 406	Fax: 202-452-5199
Washington, DC 20036	E-mail: Mary_Tisdale@blm.gov
	http://http://www.blm.gov/education

Mission: To provide public education about public lands resources and management issues. Identify educational needs and resource gaps; collaborate with partner groups, volunteers, schools, and other agencies; make products andprograms available to BLM field offices, communities and schools.
Mary E Tisdale, Manager Environmental Edu.

5866 Ball State University

Natural Resources and Environmental Management	
2000 West University	765-285-5780
Muncie, IN 47306-0495	Fax: 765-285-2606
	E-mail: nrem@bsu.edu
	http://www.bsu.edu/nrem

The Natural Resources and Environmental Management Department enhances students' scientific competence and prepares them for a variety of environmental careers. Their students can focus on topics such as air, energy, land,occupational hugiene, parks, recreation, soil, waste management, and water. They offer B.A., B.S., M.A., M.S., and Ed.D. degree programs.

Hugh Brown, Chairperson

5867 Bothell Environmental Science

University of Washington-Bothell	
18115 Campus Way NE	425-352-5000
Bothell, WA 98011	E-mail: uwbothell@u.washington.edu
	http://www.bothell.washington.edu/IAS/BSES

Primary goal of this degree program is to train a new generation of interdisciplinary scientists who are able to work in both the public and private sectors to address some of the pressing environmental issues that face our society.
Linda Watts, Professor/Director IAS
Martha Groom, Contact

5868 Bradley University

Geological Sciences Program	
1501 West Bradley Avenue	309-677-3018
Peoria, IL 61625	Fax: 309-677-3558
	E-mail: kdm@bradley.edu
	http://www.bradley.edu/academics/las/bio/

Aims to develop an awareness of the Earth as a dynamic and unified system in time and space. Curriculum is preparatory for careers in geology, engineering geology, geophysics, hydrogeology, oceanography or secondary Earth scienceteaching.
Kelly McConnaughay, Plant Ecology

5869 Brooklyn College Environmental Studies Program

2900 Bedford Avenue	718-951-4159
143A New Ingersoll	Fax: 718-951-4546
Brooklyn, NY 11210-2889	E-mail: yklein@brooklyn.cuny.edu
	http://http://academic.brooklyn.cuny.edu/environ/maintext.html

Liberal arts program aimed at educating students to be fluent in the languages of the social and physical sciences in the range of areas related to the environment, broadly construed. In addition to the requirements drawn from existingcourses in the humanities, social sciences, mathematics, and sciences, two new interdisciplinary courses, designed both to introduce the field of environmental studies and to apply this knowledge in a capstone seminar emphasizing specific cases, arerequired.
Micha Tomkiewicz, Program Director
Yehuda L Klein, Deputy Director

5870 Brown University Center for Environmental Studies

135 Angell Street	401-863-3449
Providence, RI 02912	Fax: 401-863-3503
	E-mail: enustu@brown.edu
	http://www.envstudies.brown.edu

Assists students to gain the skills, confidence and judgement to work effectively on environmental issues of social significance.
Osvaldo E Sala, Director

5871 CO2 Science

Po Box 25697	480-966-3715
Tempe, AZ 85285-5697	Fax: 480-966-0758
	E-mail: staff@co2science.org
	http://www.co2science.org

The Center accomplishes its goals through reviews on articles, books, and other educational materials. It attemps to separate reality from rhetoric in the emotionally-charged debate that swirls around the subject of carbon dioxide andglobal change. The Center maintains on-line isntructions on how to conduct CO_2 enrichment and depletion experiments in its Global Change Laboratory.
Sherwood B Idso, President

5872 California Polytechnic State University City and Regional Planning Department
One Grand Avenue 805-756-1315
Dexter Building 34-251 Fax: 805-756-1340
San Luis Obispo, CA 93407 E-mail: crp@calopoly.edu
http://www.planning.calpoly.edu
The Institute was developed to coordinate interdisciplinary projects and research relating to the management of watersheds, urban areas, marine environments and related natural and human resources. Specialists in various areas such asbiological science, business administation, city and regional planning, civil and environmental engineering, economics, geology, landscape architecture, natural resources management, political science and soil science participate in order to make theinstitute.
William Siembieda, Department Head

5873 California State University at Fullerton
PO Box 34080 714-278-2011
Fullerton, CA 92834 http://www.fullerton.edu
The Environmental Studies program is an interdisciplinary program that broadens environmental knowledge and awareness. The program will prepare the individual student to work as a professional in the environmental field by providing anopportunity to learn applicable skills and to develop an appropriate body of knowledge.
Milton Gordon, President

5874 California State University: Monterey Bay Capstone Project
100 Campus Center 831-582-4120
Seaside, CA 93955 Fax: 831-582-4122
E-mail: essp_comments@csumb.edu
http://watershed.csumb.edu
Capstone Projects encompass an extremely broad array of student interests. Earth Systems Science and Policy students complete a Capstone Project, which is similar to a senior thesis project at other universites. These projects showcasestudents mastery of ESSP skills, and fulfill CSUMB graduation requirements. All Projects follow a set of outcome based, interdisciplinary criteria used to determined the quality of a student's work.

5875 California University of Pennsylvania
Biological and Environmental Sciences
Biological & Environmental Sciences 724-938-4200
250 University Avenue Fax: 724-938-1514
California, PA 15419 E-mail: webteam@cup.edu
http://www.cup.edu
Department includes intensive scientific curricula that prepare students for graduate work in the biological and environmental sciences and career work in many related areas.
William Kimmel, Professor
David Argent, Assistant Professor

5876 Center for Ecoliteracy
2528 San Peblo Avenue 510-845-4595
Berkeley, CA 94702 Fax: 510-845-1439
E-mail: info@ecoliteracy.org
http://www.ecoliteracy.org
The Center for Ecoliteracy is dedicated to fostering a profound understanding of the natural world, grounded in direct experience that leads to sustainable patterns of living.
Zenobia Barlow, Executive Director

5877 Center for Environmental Education
40 Avon Street 603-355-3251
Keene, NH 03431-3552 Fax: 603-357-0718
E-mail: cee@antiochne.edu
http://www.schoolsgogreen.org

The center continues to integrate environmental education into all aspects of the K-12 experience through: developing and replicating model community-based EE programs; connecting communities and schools to useful resources; providingenvironmental curricula literature and other materials directly to educators and students.
David Sobel, Co-Executive Director

5878 Center for Environmental Information
46 Prince Street 716-271-3550
Rochester, NY 14607-1016 Fax: 716-271-0606
Elizabeth Thorndike, Executive Officer

5879 Center for Environmental Philosophy
University of North Texas
PO Box 310980 940-565-2727
Denton, TX 76203-0980 Fax: 940-565-4439
E-mail: cep@unt.edu.
http://www.cep.unt.edu
Conducts environmental research. Supports environmental education.
Eugene C Hargrove, President

5880 Center for Environmental Science and Policy
Stanford University
Encina Hall East 650-723-5924
Suite 400 Fax: 650-723-5920
Stanford, CA 94305-502 E-mail: lmcvay@leland.stanford.edu
http://http://cesp.stanford.edu
Focuses on important environmental problems and draws methods and analyses from multiple diciplines.
Lori McVay, Director

5881 Center for Environmental Strategies: Academy for Educational Development
1825 Connecticut Avenue NW 202-884-8898
Washington, DC 20009-5721 Fax: 202-884-8997
E-mail: rbossi@aed.org
http://www.aed.org
To find sustainable solutions to global environmental protection and natural resource management problems through individual and institutional behavior change, education, training and communication strategies. such efforts are drivenby a strong commitment to improve or maintain environmental quality as well as the quality of life for diverse communities and groups through the provision of technical assistance, guided practice and capacity building support.
Stephen F Moseley, President & CEO

5882 Center for Environmental Study
143 Bostwick Avenue NE 616-234-3935
Grand Rapids, MI 49503 Fax: 616-234-3936
E-mail: ces1@iserv.net
http://www.cesmi.org
Creating awareness for a sustainable future.
Rick Sullivan, Executive Director

5883 Center for Environmental: Earth and Space Study
Bemidji State University
1500 Birchmont Drive NE 218-755-4104
Bemidji, MN 56601 888-345-1721
Fax: 218-755-4107
http://www.bemidjistate.edu/
The Center for Environmental, Earth and Space Studies offer a unique variety of interdisciplinary degree programs that a student may choose to pursue. Degrees in Environmental Studies include both B.S. and M.S. and a B.S. or B.A. withgeology minor is also available.
Fu-Hsian Chang, Director

5884 Center for Great Lakes Environmental Education
PO Box 56 716-878-3175
Buffalo, NY 14205-0056 Fax: 716-885-5292
E-mail: info@greatlakesed.org
http://www.greatlakesed.org
Our Mission leads to awareness of and access to information about Great Lakes environmental subjects by promoting learning links for teachers, students and other stakeholders in the international Great Lakes-St Lawrence basinecosystem.
Lauren Makeyenko, Education/Program Specialist

5885 Center for Northern Studies

Po Box 72   802-586-7711
Craftsbury Common, VT 05827   Fax: 802-586-2596
E-mail: north@sterlingcollege.edu
http://www.sterlingcollege.edu/CNS

The Center for Northern Studies is a small, not for profit, undergraduate teaching and research institution located in Wolcott, Vermont. Its program is interdisciplinary in nature, integrating social and natural sciences, humanities and resource issues in the Circumpolar North.

50 pages
Dr. Steven B Young, Founder

5886 Center for Statistical Ecology & Environmental Statistics

Pennsylvania State University
Department of Statistics   814-865-9442
421 Thomas Building   Fax: 814-865-1278
University Park, PA 16802   E-mail: gpp@stat.psu.edu
http://www.stat.psu.edu/~gpp/

The Center is the first of its kind in the world and enjoys national and international reputation. They have an ongoing program of research that integrates statistics, ecology and the environment. The emphasis is on the environment and collaborative research, training and exposition on improving the quantification and communication of man's impact on the environment. Major interest also lies in statistical investigations of the impact of the environment on man. Contact them for full listings.
Ganapati Patil, Director

5887 Chelonia Institute

3330 Washington Boulevard   703-516-2600
Arlington, VA 22201   Fax: 703-522-1427
http://www.truland.com

Robert Truland, Director and Trustee

5888 City University of New York

Convent Avenue at 138th Street   212-650-8099
New York, NY 10031   Fax: 212-650-8097
http://www.library.csi.cuny.edu/dept/as/ces/escpgm.htm

In 1993, City College's Schools of Engineering opened the Center for Water Resources and Environmental Research under the direction of Dr. Reza Khanbilvardi. This center was established to bring together multidisciplinary teams of scientists and engineers to help tackle the diverse problems of water resources and environmental issues.
Alfred Levine, Director

5889 Clemson University

PO Box 340919   864-656-5567
Clemson, SC 29634   Fax: 864-656-0672
http://www.cesclemson.edu/ees/schintro.html

Alan Elzerman, School Director

5890 Coalition for Education in the Outdoors

SUNY at Cortland   607-753-4971
Box 2000   Fax: 607-753-5982
Cortland, NY 13045   E-mail: info@outdooredcoalition.org
http://www.outdooredcoalition.org

The Coalition is a nonprofit network of outdoor and environmental education centers, nature centers, conservation and recreation organizations, outdoor education and experiential education associations, institutions of higher learning, public and private schools and fish and wildlife agencies that share a mission-the support and furtherance of environmental and outdoor education and its goals.
Charles Yaple, Executive Director

5891 College of William and Mary

PO Box 1346   804-684-7000
Gloucester Point, VA 23062-1346   Fax: 804-684-7097
E-mail: dsbrad@facstaff.wm.edu
http://www.vims.edu/

The Center for Conservation Biology is an organization dedicated to discovering innovative solutions to environmental problems that are both scientifically sound and practical within today's social context. Established in the fall of 1991, the Center has been a leader in conservation issues throughout the mid-Atlantic region. Our philosophy has been to use general systems approach to locate critical information needs and to plot a deliberate course of action to reach what what we believe.
M Roberts, Chair

5892 Colorado Mountain College Central Administration and Admissions Office

831 Grand Avenue   970-945-8691
PO Box 10001   800-621-8559
Glenwood Springs, CO 81601   Fax: 970-947-8385
E-mail: joinus@coloradomtn.edu
http://www.coloradomtn.edu

The Natural Resource Management program developed from our former Environmental Technology program which began in 1974. It was one of the most well established programs of its kind in the country. The degree specializes in helping students graduate with entry-level skills in a variety of environmental fields. It combines aquatic and terrestrial resource management. Students will be trained in career fields of environmental site assessment, hydrology, soil science, environmental law and others.

5893 Colorado State University

College of Natural Resources   970-491-6909
Fort Collins, CO 80523   Fax: 970-491-7799
E-mail: info@cnr.colostate.edu
http://www.cnr.colostate.edu

The College of Natural Resources is one of the most comprehensive environment and natural resources programs in the nation. We have four departments and offer eight undergraduate majors, 9 minors and 13 concentrations. Our students address the most current issues in environment and natural resources such as: endangered species; water quality; biological diversity; parks, forests and wildlife management; recreation; and environmental and ecosystem sciences.
Joyce Berry, Associate Dean

5894 Conservation Education Association

3914 Foxdale Road   573-751-4115
New Bloomfield, MO 65063

5895 Conservation Education Center

2473 160th Road   641-747-8383
Guthrie Center, IA 50115   Fax: 641-747-3951
E-mail: ajay.winter@dnr.state.ia.us

We are an educational facility serving 20,000 participants each year. We focus on hands-on natural resource topics.
AJay Winter, Specialist

5896 Conservation Leadership School

306 Ag Administration Building   814-865-8301
University Park, PA 16802-2601   Fax: 814-865-7050
E-mail: cam174@psu.edu
http://www.cas.psu.edu/docs/casconf/CLS/CLS.html

The curriculum focuses on the application of principles of forestry, wildlife management, soil management, watershed management, and other areas of conservation. Students develop a multiple use master plan for more than 700 acres of field, forest, and water resources. They then present the plan, which integrates most of the principles of conservation resource management, to a faculty panel.

5897 Conservation Management Institute

College of Natural Resources
1900 Kraft Drive, Suite 250   540-231-7348
Blacksburg, VA 24061   Fax: 540-231-7019
E-mail: fwiexchg@vt.edu
http://fwie.fw.vt.edu/

To better address multi-disciplinary research questions that affect conservation management effectiveness in Virginia, North American and the world. Faculty from research institutions work collaboratively on projects ranging from endangered species propagation to natural resource-based satellite imagery interpretation.

5898 Consortium for Environmental Education in Medicine

99 Chauncy Street   617-292-7771
Sixth Floor   Fax: 617-292-0150
Boston, MA 02111-1703   E-mail: ceem@secondnature.org
http://www.secondnature.org/programs/ceem.nsf

The Consortium is dedicated to advancing human's quality of life by demostrating the close links between human health and the environment. CEEM's goal is to make the relationship of environment to human health an integral part of medical education.
Kimberly Shakins, Acting Director

5899 Cornell Center for the Environment
Cornell University
201 Rice Hall
Ithaca, NY 14853-5601
607-255-7535
Fax: 607-255-0238
E-mail: cucfe@cornell.edu
http://www.environment.cornell.edu/
The Graduate Field of Natural Resources offers opportunities for graduate study in the ecology, management, and policy of fishery, forest, wetland, wildlife, and other environmental resources. There also are opportunities to focus on conservation biology, agroforestry, environmetnal change, and conservation and sustainable development.
Michele Van De Walle, Executive Director
Mark B. Bain, Director

5900 Cornell University
118 Fernow Hall
Ithaca, NY 14853
607-253-3354
Fax: 607-253-3317
http://www.cals.cornell.edu/
The graduate program in environmental toxicology stresses a broad interdisciplinary education based on strong discipline-based skills, amplified by specific areas of competence acquired through course work and research in one of three major areas of concentration.
Charles Smith

5901 Dartmouth College
6105 Sherman Fairchild Hall
Hanover, NH 03755-3571
603-646-2373
Fax: 603-646-3922
E-mail: earth.sciences@dartmouth.edu
http://www.dartmouth.edu/
The department of Earth Sciences offers opportunities for learning and research in all major disciplines devoted to the study of the earth, including its structure and development, the oceans and atmosphere, weather and climate. Teaching and research at a more advanced level emphasize watershed processes, environmental biogeochemistry, geophysics and mechanics, sedimentology, paleontology, economic geology, end remote sensing of the earth from aircraft and satellites.
Suzanne Auerbach, Contact

5902 Department of Environmental Sciences
291 McCormick Road
PO Box 400123
Charlottesville, VA 22904-4123
434-924-7761
Fax: 434-982-2137
http://www.evsc.virginia.edu
The Department of Environmental Sciences is an academic department in the College of Arts and Sciences, offering instrution and conducting research in the areas of Ecology, Geosciences, Hydrology, and Atmospheric Sciences. The research endeavors of both faculty and graduate students, whether disciplinary or interdisciplinary, deal largely with problems of fundamental scientific interest and to a lesser extent with applied sciences, management or policy making.
Jay Zieman, Chair

5903 Department of Geography and Environmental Engineering
Johns Hopkins University
3400 N Charles Street
313 Ames Hall
Baltimore, MD 21218
410-516-7092
Fax: 410-516-8996
E-mail: dogee@jhu.edu
http://www.jhu.edu/~dogee
Concerned with understanding the nature and dynamics of ecosystems, engineered systems, and societies. Offers a broad range of graduate programs including the natural, social and engineering sciences.
Edward J Bouwer, Director Admissions

5904 Department of Geological Sciences
New Mexico State University
Box 30001
MSC 3AB
Las Cruces, NM 88003
505-646-2708
Fax: 505-646-1056
E-mail: geology@nmsu.edu
http://www.nmsu.edu/~geology
The Department offers both undergraduate and graduate study leading to advanced degrees in geological science. Advanced training offered through the program of study can qualify students for employment in such branches of geological science as mining, petroleum, environmental and engineering geology, for gov't service or may serve in preparation for further graduate study. The education experience may include sedimentology, geochemistry, volcanology, stratigraphy, geotectonics and paleontology.

Dr Timothy F Lawton, Department Head

5905 Department of Oceanography
University of Hawaii at Manoa
1000 Pope Road MSB 205
Honolulu, HI 96822
808-956-9937
Fax: 808-956-9225
E-mail: ges@soest.hawaii.edu
http://www.soest.hawaii.edu/oceanography/GES/
The degree program in Global Environmental Science is a holistic, scientific approach to the study of Earth and Earth's physical, chemical, biological, and human systems. The ultimate objective of the program is to produce a student in the environmental sciences at a rigorous level who is able to go on to graduate or professional school; enter the work force in environmental science positions, and become a wise environmental steward of the planet.
Jane Schumaker, Director

5906 Donald Bren School of Environmental Science
University of California
Bren Hall
Santa Barbara, CA 93106-5131
805-893-7611
Fax: 805-893-7612
E-mail: info@bren.ucsb.edu
http://www.bren.ucsb.edu
A professional degree intended for students who are going to enter or re-enter the workforce after graduation. This degree is designed to produce working professionals with training beyond the bachelor's degree. It is not designed as an intermediate degree for the PhD, although MESM graduates will be well prepared for PhD study. Some may later apply to a PhD program either in the Bren School or elsewhere.
Jill Richardson, Mgr. Academic & Student Aff.

5907 Duquesne University
600 Forbes Avenue
331 Fisher Hall
Pittsburgh, PA 15282
412-396-4749
Fax: 412-396-4092
E-mail: bembic@duq.edu
http://www.science.duq.edu
Educating environmental professionals for the twenty-first century is the focus of the Duquesne University Environmental Science and Management (ESM) Masters Degree Program. The program grew out of the perceived need to combine depth of knowledge in environmental science with a comprehensive understanding of the business, legal and policy implications surrounding environmental issues.
David W Seybert, Dean

5908 E2: Environment and Education Developers of Environmental ACTION
PO Box 20515
Boulder, CO 80308
303-442-3339
Fax: 303-442-6633
E-mail: e2ee@enviroaction.org
http://www.enviroaction.org
E2: Environment & Education, formed in 1994, is a nonprofit organization dedicated to helping young people develop an awareness of environmental issues, and the understanding and skill to take informed and effective action. Through education, E2 promotes the development of knowledge about the environment and encourages personal responsibility and postive action to improve environmental quality.
Leslie Crawford

5909 Earth Force
1908 Mount Vernon Avenue
2nd Floor
Alexandria, VA 22301
703-299-9400
Fax: 703-299-9485
E-mail: earthforce@earthforce.org
Earth Force is youth for a change! Earth Force youth create lasting solutions to environmental problems in their communities. Earth Force offers educators innovative programs and resources to help youth become environmental problemsolvers.

Vince Meldrum, President
Jerry Pharr, Vice President

5910 Earth Regeneration Society
1442A Walnut Street
Number 57
Berkeley, CA 94709
510-849-4155
Fax: 510-849-0183
E-mail: csiri@igc.apc.org
http://www.imaja.com
The Earth Regeneration Society does research and education on climate change, ozone, and pollution, and calls for full employment and full social support based on survial programs and national and international networking.

Alden Bryant, President

5911 Earth System Science Community Curriculum
Ecology Corporation
19 Eye Street NW   202-218-4100
Washington, DC 20001   800-NET-4001
E-mail: askeco@ecologic.net
http://www.ecologic.net
The challenge to Earth System Science is to develop the capability to understand those changes that will occur in the next decade to century, both naturally and in response to human activity.

5912 Eastern Environmental Program
Eastern Nazarene College
23 East Elm Avenue   617-745-3546
Quincy, MA 02170   800-88E-NC88
                                    Fax: 617-745-3907
E-mail: jonathan.e.twining@enc.edu
http://www.enc.edu
Cross-disciplinary program which provides for the student strong pereparation in the several scientific disciplines involved in the study of environmental issues. Jointly sponsored by the Department of Biology and Chemistry in orderto provide the appropriate basis in all the sciences for students wishing to pursue environmental careers or graduate school.
Jonathan E Twining, Instructor, Env. Science

5913 Eastern Illinois University
600 Lincoln Avenue   217-581-2817
Charleston, IL 61920-3099   Fax: 217-581-7141
E-mail: rufischer@eiu.edu
http://www.ux1.eiu.edu/~cfruf
Dr. Ficher's general interests include aquatic ecology, fisheries biology, and physiological ecology. His special interests include community aaanalysis of stream fishes, life history and demographics of fish, bioenergetics ofdevelopment and life history phenomena, and lipid storage and utilization patterns of fish.
Robert Fischer, Coordinator

5914 Eastern Michigan University
2816 Fish Lake Road   810-667-2350
Lapeer, MI 48446   Fax: 810-677-2350
http://www.emich.edu
The Kresge Environmental Education Center, is located in Mayfield Township of Lapeer County, about six miles north east of the city of Lapeer. The main buildings are located a quarter of a mile off Fish Lake Road in the middle of thecenter's 240 acres. In addition to the centers 240 acres, there are 7,000acres of state land contiguous to or nearby the property.
Ben Czinski, Director, EMU's KEEC
Robert Winning, Biology Department Head

5915 Ecological Society of America
1707 H Street NW   202-833-8773
Suite 400   Fax: 202-833-8775
Washington, DC 20067   E-mail: esahq@esa.org
http://www.esa.sdsc.edu
The Ecological Society of America is a non-partisan, nonprofit organization of scientists founded in 1915 to : promote ecological science by improving communications among ecologists; raise the public's level of awareness of theimportance of ecological science; increase the resource available for the conduct of ecological science.

5916 Educational Communications
1800 North Stonelake Drive   812-335-7675
Suite 2   877-677-AECT
Bloomington, IN 47404   Fax: 812-355-7678
E-mail: aect@aect.org
http://www.aect.org
The mission of the Association for Educational Communications and Technology is to provide leadership in educational communications and technology by linking professionals holding a common interest in the use of educational technologyand its application to the learning process.
Phillip Harris, Executive Director

5917 Emporia State University
1200 Commerical Street   620-341-1200
Emporia, KS 66801   Fax: 620-341-5589

E-mail: ugadmiss@emporia.edu
http://www.emporia.edu/
A comprehensive Regents university primarily serving residents of Kansas by providing leadership in quality instruction, related scholarship and service.
Dr Kay Schallenkamp, President

5918 Environmental Education
Office of Environmental Affairs   617-626-1000
100 Cambridge, 9th Floor   Fax: 617-626-1181
Boston, MA 02114   http://www.state.ma.us/envir
The Executive Office of Environmental Affairs is a resource for your environmental literacy. Our goal is to reconnect people to the natural world, and to inspire a sense of responasbility in the public. Each one of us is an importantlink to our collective health. Through our decisions and choices, we can make a difference.
Stephen R Pritchard, Secretary

5919 Environmental Education Program
Southern Oregon University
1250 Siskiyou Boulevard   541-552-6876
Siskiyou Environmental Center   Fax: 541-552-6415
Ashland, OR 97520   E-mail: seec@sou.edu
http://www.sou.edu/biology/enved/mainpage.htm
Designed to promote a better understanding of the environment and environmental issues and to develop awarenesss and knowledge of biodiversity and ecosystem complexity and to inspire a sense of wonder in the natural world. Seeks toprepare students for active roles in education and social change related to resolution of environmental problems and conflicts affecting present and future generations.
Roger Christianson, Executive Director
James Stewart, Professor, Department Chair

5920 Environmental Engineering University of Nevada, Reno
University of Nevada
Department of Civil & Env. Engin.   775-784-1474
Mail Stop 258   Fax: 775-784-4466
Reno, NV 89557   E-mail: vdadams@unr.neveda.edu
http://www.coeweb.engr.unr.edu
We provide an educational program focused at the undergraduate and graduate level in environmental engineering. Environmental engineers have taken an increasingly important role in the application on engineering and scientificprinciples to protect and preserve human health and environment. The cirriculum is designed with the goal of providing each student with the necessary fundamentals and background in engineering science and engineering science design to address manydifferent challenges.
Dr. Ted E Batchman, Dean
Dr. V Dean Adams, Professor/Assoc. Dean

5921 Environmental Engineering Department
1300 W Park Street   406-496-4115
Butte, MT 59701   Fax: 406-496-4650
E-mail: KPeterson@mtech.edu
http://www.mtech.edu/ee
The Environmental Engineering Department of Montana Tech of The University of Montana Tech is one of the best of its kind in the country. The EE degree is accredited by the Engineering Accreditation Commission of the AccreditationBoard of Engineering and Technology (ABET/EAC).
Kumar Gamesan, Professor/Head EE Dept

5922 Environmental Institute
Carnegie Mellon University   412-268-2940
5000 Forbes Avenue   Fax: 412-268-7813
Pittsburgh, PA 12513-3890   http://www.ce.cmu.edu
A major function of the Environmental Institute is to enable Carnegie Mellon to play a leadership role in developing educational programs on environmental issues. These include initiatives at both undergraduate and graduate levels.
Michael Balderson, Administrative Assistant

5923 Environmental Management
University of California-Berkeley   510-643-9927
1995 University Avenue, Suite 300   Fax: 510-643-8290
Berkeley, CA 94720   E-mail: course@unex.berkeley.edu
http://www.unex.berkeley.edu/em/
Roxanne Hernandez, Program Representative

5924 Environmental Management Program

Lake Erie College
391 West Washington Street   216-639-4708
Painesville, OH 44077   Fax: 216-352-3533
          E-mail: pbelange@lakeerie.edu

Interdisciplinary major, grounded in the sciences and liberal arts, designed for those who want to pursue career paths utilizing environmental science in decision making.   Students take courses in Environmental Management, Biology, Chemistry, Mathematics, and Business. Program is designed so students will build a solid knowledge base in regional and global environmental issues.
Dr. Paul Belanger, Director

5925 Environmental Management and Policy Program

Rensselaer Polytechnic Institute
110 8th Street   518-276-6565
Troy, NY 12180-3590   Fax: 518-276-6783
          E-mail: mendef@rpi.edu
          http://www.rpi.edu/

Masters of Science degree program dedicated to providing leaders in business and the environment. Incorporates technical training and a high level practicum assignment in degree requirements. Graduates are employed in environmentalmanagement, government relations, marketing communications, management of information systems, investor relations, accounting and other functions that must incorporate environmental considerations.
Frank J Mendelson, Director

5926 Environmental Media Association

10780 Santa Monica Boulevard   310-446-6244
Suite 210   Fax: 310-446-6255
Los Angles, CA 90025   E-mail: ema@ema-online.org
                        http://www.ema-online.org

501(c)3 non-profit organization.
Patie Maloney, Vice President

5927 Environmental Mutagen Society

1821 Michael Faraday Drive   703-438-8220
Suite 300   Fax: 703-438-3113
Reston, VA 20190   E-mail: emshq@ems-us.org
                        http://www.ems-us.org

The Environmental Mutagen Society is the primary scientific society fostering research on the basic mechanisms of mutagensis as well as on the application of this knowledge in the field of genetic toxicology. EMS has seven corescientific content areas.
David Eastman, President

5928 Environmental Policy Programs

Munching Hall   301-405-6362
College Park, MD 20742   Fax: 301-403-3737
          E-mail: spaoep@umd.edu
          http://www.puaf.umd.edu

This part-time degree program is intented for highly ambitious mid-career professionals who are ready to advance within the field, understand the importance and value of a professional degree and able to attend one or twoo classes onelate afternoon/evening per week for two years. A minimum of 5 years of policy related work experience; for-profit, nonprofit and public sectro can be considered. A minimum udergraduate GPA of 3.0 is required.
Tom Kennedy, Director

5929 Environmental Resource Center

Local Dirt
411 East Sixth Street   208-726-4333
PO Box 819   Fax: 208-726-1531
Ketchum, ID 83340   E-mail: erc@ercsv.org
                        http://www.ercsv.org

An oraganization offering environmental education for the community.
Craig Barry, Executive Director

5930 Environmental Science

University of Florida
College Natural Resources & Envir.   352-846-1634
Box 116455   Fax: 352-392-9748
Gainesville, FL 32611-6455 E-mail: mwade@ufl.edu
                        http://http://snre.ufl.edu

Science based, multidisciplinary and academically rigorous. The College of Natural Resources and Environment has students and a curriculum, with 200 courses taught in 56 departments of other colleges. The 290 affiliate faculty havetheir primary appointments in discipline centered deparments of other colleges.
Dr. Stephen Humphrey, Director, Academic Programs
Meisha Wade, Coord. Academic Programs

5931 Environmental Science Program

Texas Christian University
PO Box 298830   817-921-7271
Fort Worth, TX 76129   Fax: 817-921-7789
          E-mail: environmentalscience@tcu.edu
          http://www.ensc.tcu.edu

A program helping students to understand the connection between science and the earth.
Dr. Leo Newland, Director

5932 Environmental Science and Policy Program

Drake University
2507 University Avenue   515-271-3812
Olin Hall   Fax: 515-271-3702
Des Moines, IA 50311   E-mail: thomas.rosburg@drake.edu

Environmental Science and Policy Program at Drake is an interdisciplinary program that awards BS and BA degrees in two majors: Environmental Science and Environmental Policy. There are 40 to 45 students in either of these majors.

Thomas Rosburg, Director

5933 Environmental Sciences

Oklahoma State University
Stillwater, OK 74078   405-744-9229
                                    Fax: 405-744-7673
          E-mail: tmh9620@okway.okstate.edu

Environmental Science Graduate Program's mission is to broaden the scope of scientific and technological study through a multidisciplinary approach encompassing social and legal aspects of environmental concerns and based on ecologicalfoundations.
Dr. Edward T Knobbe, Program Director
Talya M Henderson, Contact

5934 Environmental Sciences Program

State University of New York College at Fredonia
SUNY Fredonia   716-673-3817
130 Jewett Hall   Fax: 716-673-3493
Fredonia, NY 14063   E-mail: fuenteap@fredonia.edu
                        http://www.fredonia.edu/

Rigorous, interdisciplinary program in environmental science with 68 semester hours of core courses in mathematics, biology, chemistry, environmental sciences, and geosciences. Students are prepared to pursue graduate studies, professional certifications, or employment in the private or public sector.
Dr Alicia Perez-Fuentetaja, Coordinator Program

5935 Environmental Studies

Middlebury College
Emma Willard House   802-443-3000
RD 2, Box 2005   Fax: 802-443-2056
Middlebury, VT 05753 E-mail: admissions@middlebury.edu

Explores the relationship between humans and their environment. Students pursuing the ES major work in a variety of disciplines, including biology, chemistry, economics, geography, geology, literature, the performing arts, philosophy, political science, religion and sociology.
Dr. Christopher McGrory Klyza, Program Director
John Hanson, Director of Admissions

5936 Environmental Studies Department

St. Lawrence University
Canton, NY 13617   315-229-5814
                                    Fax: 315-229-5802
          E-mail: cjohns@stlawu.edu

B.A. and B.S. degree programs in environmental studies/science. ten options for combining environmental studies with traditional disciplines (eg. biology, economics) plus B.A. program in Environmental Studies.
Carolyn Johns, Environmental Studies Prog.

5937 Environmental Studies Institute

800 Garden Street 805-965-5010
Suite D
Santa Barbara, CA 93101

5938 Environmental Studies Program

5223 University of Oregon 541-346-5000
Eugene, OR 97403 Fax: 541-346-5954
E-mail: ecostudy@oregon.uoregon.edu
http://darkwing.uoregon.edu/~ecostudy/

Environmental Studies crosses the boundaries of traditional disciplines, challenging faculty and students to look at the relationship between humans and their environment from a new perspective. They are dedicated to gaining greaterunderstanding of the natural world from an ecological perspective; devising policy and behavior that address contemporary environmental problems; and promoting a rethinking of basic cultural premises, ways of structuring knowledge and the rootmetaphors of society.
Daniel Udovic, Program Director

5939 Environmental Studies and Environmental Science Programs

Brown University 401-863-3449
Providence, RI 02912 Fax: 401-863-3503
http://www.brown.edu/departments/

The Center for environmental Studies (CES) at Brown University was established with the primary aim of educating individuals to solve challenging environmental problems, both at the local and global levels. CES also works directly toimprove human well-being and environmental quality through community, city and state partnerships in service and research.

5940 Environmental Studies and Planning Department

Sonoma State University
1801 E Cotati Avenue 707-664-2306
Rohnert Park, CA 94928 Fax: 707-664-4202
E-mail: ensp@sonoma.edu
http://www.sonoma.edu/ensp/

ENSP was founded as an interdisciplinary program in 1972 during a period of growing environmental concern. The department has evolved and matured, and today stresses: the development of a global prespective by synthesizing knowledgefrom a variety of scientific and academic disciplines; the acquisition of specific professional skills through a focused course of study; and the application of knowledge and skills through effective strategies for environmental management.

Steve Orlick, Department Chair

5941 Environmental Working Group

1718 Connecticut Avenue NW 202-667-6982
Suite 600 Fax: 202-232-2592
Washington, DC 20009 http://www.ewg.org

The Environmental Working Group is a leading content provider for public interest groups and concerned citizens who are campaigning to protect the environment. Offers reports, articles, technical assistance and the development ofcomputer databases and Internet resources.
Kelsey Wirth, Chair

5942 Environmental and Energy Study Institute

122 C Street NW 202-628-1400
Suite 630 Fax: 202-628-1825
Washington, DC 20001 E-mail: eesi@eesi.org
http://www.eesi.org

The Environmental and Energy Study Institute is a nonprofit organization dedicated to promoting environmentally sustainable societies. EESI believes meeting this goal requires transitions to social and economic patterns that sustainpeople, the environment and the natural resources upon which present and future generations depend.

Carol Werner, Executive Director

5943 Ferrum College

Garber Hall, Room 309 540-365-2121
PO Box 1000 800-868-9797
Ferrum, VA 24088 Fax: 540-365-4203
E-mail: webmaster@ferrum.edu
http://www.ferrum.edu

Professor of chemistry and environmental science, received both MS and PhD in environmental chemistry from the University of Michigan. His research has included air pollutant deposition in the Great Lakes and the formulation ofmembrances for ion-selective electrodes. He manages a long-term water quality monitoring project on Smith Mountain Lake and is using a geographical information system to model soil loss in its watershed.
Carolyn Thomas, Biology Coordnator

5944 Florida Center for Environmental Studies

3932 RCA Boulevard 561-799-8554
PO Box 3210 Fax: 561-626-1404
Palm Beach Gardens, FL 33410-3208 E-mail: tdodge@ces.fau.edu
http://www.ces.fau.edu/

Represents the ten state universities and the major private universities in regard to environmental studies and research.
Terry Dodge, Information Services
Jo Ann Jolley, Associate Director

5945 Florida State University

Tallahassee, FL 32306 850-644-2525
Fax: 850-644-0197
E-mail: admissions@admin.fsu.edu
http://www.fsu.edu

Environmental issues as they relate to geological phenomena, which include volcanic and earthquake hazards, resource and land- use planning, air and water pollution, waste disposal, glaciation and sea-level change, landslides,flooding, shoreline erosion, and global change issues.
T.K. Wetherell, President

5946 George Washington University

2023 G Street NW 202-994-7123
Lisner Hall Room 341 Fax: 202-994-8113
Washington, DC 20052 E-mail: hmerchnt@awu.edu
http://www.gwu.edu/

Int'l Environmental Policy and Management course and Marketing Manageent Course to take place in Sydney and Brisbane Australia during July 1998. Total of 6 credits (undergraduate or graduate level) maximum. NO degree granted, butcredit potentially transferrable to a degree program student is currently attending. Total cost of program (all inclusive) is between $3800-$5000.
Henry Merchant

5947 Georgetown University

600 New Jersey Avenue NW 202-662-9000
Washington, DC 20001 Fax: 202-662-9444

The Environmental Studies minor is an interdisciplinary program designed to allow an undergraduate of the college majoring in any discipline to focus on environmental issues. Environmental Studies provides a framework for the study offundamental mechanisms of ecosystems and human interaction with the Earth. Environmental studies encompasses the humanities, social sciences and natural sciences as they relate to environmental questions.
Bill Butler, Interim Director

5948 Global Action and Information Network

740 Front Street 408-457-0130
Suite 355 Fax: 408-457-0133
Santa Cruz, CA 95060 E-mail: info@gain.org
http://www.gain.org

Bill Leland, Executive Director

5949 Global Environmental Assessments: Information, Institutions and Influence

Cambridge, MA: MIT Press
Department of Political Science 617-495-3981
University of Oregon Fax: 617-495-8963
Eugene, OR 97403 E-mail: rmitchel@oregon.uoregon.edu

The influence of global environmental assessments on policy requires institutional designs that allow them to be salient, credible and legitimate to international negotiators, national policymakers and local decisionmakers.
Mitchell, Ronald, William Clark, David Cash, Author
William C Clark, Director

5950 Global Environmental Institute

4949 Tamiami Trail North 941-643-1444
Suite 203 Fax: 941-434-2469
Naples, FL 34103 E-mail: btraub@gei.org
http://www.internetserver.com/~smaes/GEI/PUBLIC/geifr.htm
The program primary goal is to break down barriers that exist between disciplines, professions and nations. A policy-maker and an industrial engineer will come to better understand the constraints under which the other operates.Students from other countries in different stages of economic develompment will better understand each other's needs. We strongly encourage students from developing countries to apply.

5951 Global and Environmental Education Resources

61 Route 9W 845-365-8930
PO Box 1000 Fax: 845-365-8922
Palisades, NY 10964-+ E-mail: help@gcrio.org
http://www.gcrio.org/edu.html
Multidisciplinary and international in scope, this collection of resources was selected for its relevance to global change and environmental education. Included is a wide range of resources in a variety of formats for educators andstudents at all levels (K-12 and higher education), librarians, citizens and community groups.

5952 Goodwin-Niering Ctr for Conservation Biology and Environmental Studies: Connecticut College

270 Mohegan Avenue 860-439-5417
Box 5293 Fax: 860-439-2418
New London, CT 06320-4196 E-mail: ccbes@conncoll.edu
http://ccbes.conncoll.edu
This study is a comprehensive, interdisciplinary program that builds on one of the nation's leading undergraduate environmental studies programs. The Center fosters research and education aimed at understanding contemporary ecologicalchallenges. Its Certificate Program offers students the opportunity to blend thier interest in the environment with a non-science major and is of particular interest to those planning careers in environmental policy, law, economies or education.
Diane Whitelaw, Assistant Director

5953 Gore Natural Science School

400 Pine Street 970-827-9725
PO Box 250 Fax: 970-827-9730
Red Cliff, CO 81649 E-mail: markianf@gorerange.org
http://www.gorerange.org
During the Summer Gore Range Natural Science School offers day and overnight programs for students in 3rd grade up. During academic school year GRNSS provides integrated field science education to local and visiting schools. Ourmission is to raise environmental awareness and inspire stewardship of the Eagle River watershed. GRNSS is a non-advocacy organization.
Markian Feduschak, Executive Director

5954 Green Media Toolshed

1200 New York Avenue NW 202-326-6200
Suite 300 Fax: 202-682-2154
Washington, DC 20005E-mail: info@greenmediatoolshed.org
http://www.greenmediatoolshed.org
Green Media Toolshed is a nonprofit organization that provides the environmental community with access to a host of high quality communications tools for an affordable cost.

Martin Kearns, Executive Director
Bobbi Russell, Media Svcs./Marketing Dir.

5955 Green TV

1125 Hayes Street 415-255-4797
San Francisco, CA 94117 Fax: 415-255-4664
E-mail: fgreen@greentv.org
http://www.greentv.org
Green TV is a San Francisco based nonprofit video production company. These writers and filmakers combine environmental journalism with dramatic wildlife and natural history footage.
Frank Green, Owner/President

5956 Harvard Medical School Center for Health and the Global Environment

401 Park Drive 617-384-8530
2nd Floor E Fax: 617-384-8585
Boston, MA 02215 http://www.med.harvard.edu/chge

Conducts research on human health and its relationship to the global environment. Informs medical staff, scientists and policy makers about health concerns.
Eric Chivian, Director
Tracy Sachs, Director of Programs

5957 Hocking College

3301 Hocking Parkway 740-753-3591
Nelsonville, OH 45764 Fax: 740-753-2021
E-mail: admissions@hocking.edu
http://www.hocking.edu
Growing concern for the environment has increased the need for technicians qualified in the restoration of environmentally unstable land, water, and air. Hocking College's Environmental Restoration technology prepares you to meet thisneed.
Russell Tippett, Dean

5958 Idaho State University

741 South Seventh Avenue 208-282-3785
Pocatello, ID 83209-8007 Fax: 208-282-4570
E-mail: bios@isu.edu
http://www.isu.edu/departments/bios/
Our department has recently developed an emphasis in environmental geochemistry and hydrogeology. This specialty is ideal in southern Idaho where problems of nuclear and toxic waste clean-up at the Idaho National EnvironmentalEngineering Laboratory will require study and generate research monies for years to come.
Rod Seeley, Chairman

5959 Institute for Community and Environment

Colby-Sawyer College
541 Main Street 603-526-3793
New London, NH 03257 Fax: 603-526-3429
E-mail: jcallewaert@colby-sawyer.edu
Bachelor of Science degree in Community and Environmental Studies. A minor in CES is also available.
John H Callewaert, Director Institute

5960 Institute for Earth Education

Cedar Cove 304-832-6404
Greenville, WV 24945 Fax: 304-832-6077
E-mail: iee1@aol.com
http://www.eartheducation.org
The Institute for Earth Education is the world's alternative agency- and industry- sponsored supplemental environmental education. IEE develops and disseminates instructional programs aimed at helping people live more lightly on theearth. Earth education is the process of helping people live more harmoniously and joyously with the natural world.
Steve Van Matre, Chair

5961 Institute for Environmental Studies

Louisiana State University
1285 Energy, Coast and Environment 225-578-8521
Building Fax: 225-578-4286
Baton Rouge, LA 70803 E-mail: evstev@unix1.sncc.lsu.edu
http://http://info.envs.lsu.edu
LSU Master of Science in Environmental Sciences program is designed to provide a broad-based graduate education to prepare students for careers in industrial, government, and academia. The program builds on a strong undergraduatebackground in the sciences.
Michael Wascom, Program Director
Craig Stevens, Contact

5962 Institute for Resource and Security Studies

27 Ellsworth Avenue 617-491-5177
Cambridge, MA 02139 Fax: 617-491-6904
E-mail: info@irss-usa.org
http://www.irss-usa.org
To conduct technical and policy analysis and public education, with the objective of promoting peace and international security.
Gordon D Thompson, Executive Director
Paula Gutlove, Deputy Director

5963 Institute of Environmental Sciences and Technology

5005 Newport Drive 847-255-1561
Suite 506 Fax: 847-255-1699
Rolling Meadows, IL 60008-3841 E-mail: iest@iest.org
http://www.iest.org

International professional society that serves members and the industries they represent through education and the development of recommended practices and standards.
Julie Kendrick, Director

5964 International Association for the Advancement of Earth and Environmental Sciences
Northeastern Illinois University
5500 North St. Louis
Chicago, IL 60625   773-794-2628
                          Fax: 847-824-8436
Dr. Musa Qutub, President

5965 International Center for Earth Concerns
D162 Baldwin Road   805-649-3535
Ojai, CA 93023   Fax: 805-649-1757
          E-mail: information@earthconcerns.org
          http://www.earthconcerns.org
Dedicated to providing for public use, a world class botanic garden, outdoor learning-ecology center and a 50-passenger all-electric floating classroom on Lake Casitas. These facilities are used to promote a better understanding ofman's place in the environment, as well as to help develop a sense of respect, responsibility and compassion for animals and nature.
John Taft, Chairman

5966 International Center for Tropical Ecology
B216 Benton Hall   314-516-6203
One University Boulevard   Fax: 314-516-6233
St Louis, MO 63121-4499   E-mail: icte@umsl.edu
          http://icte.umsl.edu
The International Center for Tropical Ecology at the University of Missiour provides a focal point for interdisciplinary research and graduate education in all aspects of the conservation of tropical ecosystems. The Center, formed in1990 in collaboration with the Missouri Botanical Garden, supports a network of students, scientists, and conservationists from tropical countries in the United States at which to study issues related to biodiversity conservation.
Dr. Patrick Osborne, Executive Director

5967 International Center for the Solution of Environmental Problems
5120 Woodway Drive   713-527-8711
Suite 8009   Fax: 713-961-5157
Houston, TX 77056-1746   E-mail: icsep@airmail.net
          http://http://icsep.com
To anticipate or detect environmental problems and either solve the problems or design and demonstrate their solutions using scientific methods in concert with nature. To provide environmental information.
Dr. Joseph Goldman, Director

5968 Iowa State University
131 Bessey Hall   515-294-3651
Ames, IA 50011-2011   800-262-3810
          Fax: 515-294-2592
          E-mail: ssprong@iastste.edu
          http://www.envs.ag.iastate.edu/
The Environmental Studies Program deals with the relationship between humans and nature, or between humans and natural systems. The curriculum is designed to give students an understanding of regional and global environmental issuesand an appreciation of different perspectives regarding these issues. Courses are provided for students pursuing careers related to the environment and for others who simply want to know more about environmental issues.
Dr. Stephen Schmidt, Contact

5969 Jane Goodall Institute for Wildlife Research, Education and Conservation
8700 Georgia Avenue   240-645-2000
Suite 500   Fax: 301-565-3188
Silver Spring, MD 20910   http://www.janegoodall.org
A tax-exempt, nonprofit corporation, founded in 1977 focusing on Jane Goodall.
William Johnston, President/CEO

5970 Kansas State University
117 Waters Hall   785-532-6151
Manhatten, KS 66506-5506   800-432-8270

          Fax: 785-532-6897
          E-mail: jax1@ksu.edu
          http://www.ag.ksu.edu/
Physcial and chemical qualities of natural environments and health from a geologic perspective detecion and prediction of environmental changes, identification of sources of pollutants and their movements in soils, rocks, and waters.
Thomas Warner

5971 Keene State College
229 Main Street   603-358-2276
Keene, NH 03435-2301   800-572-1909
          Fax: 603-358-2767
          E-mail: admissions@keene.edu
          http://www.keene.edu
Environmental Studies is an interdisciplinary program comprised of courses in Biology, Chemistry, Economics, Geography, Geology, and Political Science. The major is designed with two options, Environmental Policy and EnviromentalScience, to prepare students for a wide range of environment-related career opportunities. Students intending to major in Environmental Studies should select an advisor and formally declare their major as early as possible, preferably by the end oftheir first year.
Tim Allen, Associate Professor

5972 Lally School of Management and Technology
110 8th Street, Pittsburgh Building   518-276-6565
Rensselaer Polytechnic Institute   Fax: 518-276-2665
Troy, NY 12180-3590   E-mail: lallymba@rpi.edu
          http://http://lallyschool.rpi.edu
Rensselaer's Lally School of Management and Technology is committed to integrating green business strategy throughout our management curriculum. An MBA with an Environmental Management and Policy Concentration is truly designed as aninterdisciplinary degree. We expect that an MBA graduate of the Lally School will be employable in a "traditional" business function, with the knowledge and skill-sets to help the organization realize environmental, health and safety strategy.
Frank J Mendelson, Director

5973 Lost Valley Educational Center
81868 Lost Valley Lane   541-937-3351
Dexter, OR 97431   E-mail: info@lostvalley.org
          http://www.lostvalley.org
We offer a wide variety of programs that express this approach, including residential interships, educational workshops that emphasize hands on, experiential learning and personal/spiritual growth workshops. We also provide asupportive and nourishing place to hold individuals and organizations who share our vision to hold their own conferences, retreats and workshops.

5974 Louisiana Tech University
PO Box 10138   318-257-4985
Ruston, LA 71272   Fax: 318-257-5061
          E-mail: jadams@lans.latech.edu
          http://www.ans.latech.edu/forestry-index.html
The Wildlife Conservation degree program meets the certification requirements of the Wildlife Society, and graduates may apply for certification as an Associate Wildlife Biologist.
James Adams, Director

5975 MacBride Raptor Project
6301 Kirkwood Boulevard SW   319-398-5495
Cedar Rapids, IA 52406   Fax: 319-398-5611
          E-mail: jcancil@kirkwood.edu
          http://www.macbrideraptorproject.org
A nonprofit organization jointly sponsored by the University of Iowa and Kirkwood Community College. The project has two main facilities: the educational display facility and rehabilitation flight cage at the Macbride NatureRecreational Area, and the Raptor Clinic and educational display at KCC. The project is dedicated to the preservation of birds of prey and their habitats through rehabilitation of injured raptors, public education programs and raptor reseach.
Jodeane Cancilla, Director
Kristene Lake, Assistant Director

5976 Marine/Freshwater Biomedical Center
Nicholas School of the Environment
Marine Laboratory   252-504-7503
135 Duke Marine Lab Road   Fax: 252-504-7648
Beaufort, NC 28516-9721

The Marine/Freshwater Center of Duke University is a problem-oriented center that is nationally and internationally recognized for its contributions to environmental health. It integrates unique facilities and faculty expertiseavailable on the Beaufort and Durham campuses of Duke University and applies this powerful collective strength to challenging problems of human and environmental health significance, with a focus on adverse effects associated with the toxicity ofmetals and free radicals.
Michael K Orbach, Director

5977 Miami University: Institute of Environmental Sciences

102 Boyd Hall	513-529-5811
Oxford, OH 45056	Fax: 513-529-5814
	E-mail: boardman@muohio.edu
	http://www.muohio.edu/ies

Offers a professional Master of Environmental Science degree. This interdisciplinary program stresses problem solving and community service. The curriculum provides practical experience in an area of concentration, preparing studentsfor a variety of practical careers in public and private sector jobs.

Mark Boardman, Director

5978 Michigan Technological University

1400 Townsend Drive	906-487-2454
Houghton, MI 49913-1295	Fax: 906-487-2915
	E-mail: forest@mtu.edu
	http://forestry.mtu.edu/

A.B.S. in Applied Ecology and Environmental Sciences from Michigan Tech prepares students to address complex environmental problems posed by the use of natural resources. Students learn how to protect the integrity of ecosystems andhelp assure that natural resources will be managed wisely for generations of sustainable use.
Margaret Gale, Dean

5979 Minority Science & Engineering Program

University of Washington

207 Loew Hall	206-685-8688
College of Engineering Box 352180	Fax: 206-685-0666
Seattle, WA 98195	E-mail: shpe@u.washington.edu

Focused on providing assistance and opportunities to minority students, the program pursues a strategic plan that starts with community partnerships and ends with help with job placement. Specifically, MSEP provides the followingservices: scholarships and financial assistance, counseling/advising, problem solving workshops, tutoring, pre-freshman summer interships, Study Center, Computer Learning Center, motivational talks to pre college students, career planning and studentsocieties.
Scott Pinkam, Academic Advisor

5980 Montana Environmental Training Center

2100 16th Avenue South	406-771-4433
PO Box 6010	Fax: 406-771-4317
Great Falls, MT 59406-6010	E-mail: boylej@msun.edu
	http://www.msun.edu/stuaffairs/grants/metc

METC is a cooperative effort between Montana State University-Northern and the Montana Department of Environmental Quality. Basic, advance training, and continuing education in the areas of water and wastewater operation, maintenance,safety, process control, cross connection and backflow prevention along with courses in basic water science and watershed awareness define the training activities of METC. A newsletter and Training Announcement are published quarterly.
Jan Boyle, Training Coordinator
Michelle Jackman, Training/Development Spec.

5981 Natural Resource Management

Colorado Mountain College

831 Grand Avenue	970-945-8691
Po Box 10001	800-621-8559
Glenwood Springs, CO 81601	Fax: 970-947-8385
	E-mail: joinus@coloradomtn.edu
	http://www.coloradomtn.edu

Rocky Mountains are the perfect place to learn how to keep our land and water pure. Study in many ecosystems, from high alpine to river and semi-desert. Natural Resource Management program will teach you the skills to assess problemsand recommend solutions for a healthy environment.

5982 Natural Resources and Environmental Management

PO Box 6305	601-325-2224
Mississippi State, MS 39762	Fax: 601-325-7360

E-mail: admit@admissions.msstate.edu
http://www.msstate.edu

The NREM emphasis option provides an enhanced background in geology, hydrogeology, resource conservation, and water quality for students pursuing careers that require environmental training. Enhanced knowledge of ground and surfacewater supplies, water transport, and water pollution provides graduates with the opportunity to excel in positions requiring environmental management skills.
Cynthia West, Head

5983 Natural Resources and Environmental Science

Purdue University

3440 Lilly Hall	765-494-8060
West Lafayette, IN 47907-1150	Fax: 765-496-2926
	E-mail: jgraveel@purdue.edu
	http://www.agry.purdue.edu/nres/nres.htm

NRES is an interdisciplinary program at the Purdue School of Agriculture designed to prepare students to work with environmental problems which impact our basic natural resources: land, air and water. Faculty from all departments inthe school contribute to the curriculum. NRES is a flexible program which allows students, working closely with an academic advisor, to develop their personal curriculum to meet individual career goals.
Dr John Graveel, Program Director

5984 North Dakota State University

Department of Biological Sciences	701-231-7087
Fargo, ND 58105	Fax: 701-231-7149
	E-mail: williamblier@ndsu.edu
	http://www.biology.ndsu.nodak.edu/

Department offers undergraduate and graduate degrees in biological disciplines, including environmental and conservation sciences.

Will Bleier, Professor
Craig Stockwell, Associate Professor

5985 Northern Arizona University

NAU Box 5640	928-523-0388
Flagstaff, AZ 86011-5621	Fax: 928-523-7500
	E-mail: biology@nau.edu
	http://www.nau.edu/

The Bachelor of Science in Environmental Sciences is designed to offer students a technically rigorous foundation and broad exposure to the environmental science. The core courses in environmental sciences are interdisciplinary,usually team taught by scientists with different backgrounds and specialties, providing multiple perspectives and rich learning experiences.
Lee Drickamer, Chair

5986 Northern Michigan University

1401 Presque Isle Avenue	906-227-2310
Marquette, MI 49855	800-682-9797
	Fax: 906-227-1063
	http://www.nmu.edu

Research focuses on the Upper Peninsula environment, ethnic groups, economy, politics, folklore and literature.
Neil Cumberlidge, Department Biology

5987 Northland College

1411 Ellis Avenue	715-682-1699
Ashland, WI 54806-3999	800-753-1840
	Fax: 715-682-1308
	E-mail: info@northland.edu
	http://www.northland.edu

Northland College is a private four year environmental liberal arts college with a comprehensive range of programs that integrate traditional study with a keen eye toward problem-solving and environmental impact.
Karen Halberslabin, President

5988 Ohio State University School of Natural Resources

2021 Coffey Road	614-292-2265
210 Kottman Hall	Fax: 614-292-7432
Columbus, OH 43210-1085	E-mail: giese.1@osu.edu
	http://snr.osu.edu

An interdisciplinary degree program focusing on the science and management of natural resources and the environment. Four integrated undergraduate programs of study provide the foundation to a variety of career paths dealing with natural resources and the environment. Graduates are employed as environmental and ecosystem scientists; forest, wildlife and fisheries researchers and biologists; environmental educators, communicators and naturalists; park, forest and wildlife managers and rangers.
Dr Jerry M Bigham, Program Director

5989 Oregon State University
Corvallis, OR 97331-4501   541-737-1000
                                 Fax: 541-737-3590
                                 http://www.osu.orst.edu
Oregon State University is offering MS, MA, and PhD degrees in Environmental Sciences. Environmental sciences are central to the mission of Oregon State University—a university with extensive programs related to the environment and wise use of natural resources. Recognized as a Land, Sea, and Space Grant institution, OSU has exceptional strength in many of the disciplines that are required to provide a high-quality interdisciplinary education for future environmental scientists.
Micheal Unsworth, Director

5990 P.I.N.E.S.
120-13 Whitebog Road   609-893-4646
Browns Mill, NJ 08015   Fax: 609-893-4646
Program emphasis on the ecological relationships of humans in their environment.

5991 Portland State University
PO Box 751   503-725-4982
Portland, OR 97207-0751 Fax: 503-725-9040
                         E-mail: envir@pdx.edu
                         http://www.esr.pdx.edu
Environmental studies are central to the mission of Portland State University, which serves the state's major urban center. The Environmental Sciences and Resources program offers both undergraduate and graduate degrees.
John Reuter, Program Director

5992 Prescott College
Department of Environmental Studies
220 Grove Avenue   928-778-2090
Prescott, AZ 86301   800-628-6364
                          Fax: 928-776-5137
          http://www.prescott.edu/rdp/rdp_es.html
Small groups of students work actively on real-world projects with faculty who are leaders in the field of environmental studies. Offers dynamic and active laboratories for students and gives them the opportunity to be on the cutting edge of environmental and sustainabilty research.
Tim Crews, Program Coordinator

5993 Program in Environmental Science & Policy: Clark University
IDCE Department
950 Main Street   508-793-7102
Worcester, MA 01610-1477 Fax: 508-793-8821
                          E-mail: idce@clarku.edu
          http://www.clarku.edu/departments/idce
An interdisciplinary graduate and undergraduate program that emphasizes policy questions involving the environment and the use and misuse of science and technology. Its goal is to enable individuals to deal with technical and environmental issues in social and political areas. Urgent topics, such as assessment and management of environmental risks to humans and ecosystems; capacity for sustainable development in the third worlds; integrated watershed management and others, are addressed.
William Fisher, Director

5994 Roger Tory Peterson Institute of Natural History
311 Curtis Street   716-665-2473
Jamestown, NY 14701   800-758-6841
                           Fax: 716-665-3794
                           E-mail: mail@rtpi.org
                           http://www.rtpi.org
Nature education organization that works on the national level. Its audience is primarily adults who are interested in gaining skills for educating young people about the natural world. RTPI also houses the life's work of Roger Tory Peterson.

Jim Berry, President

5995 Roger Williams University
Center for Environmental Development
One Old Ferry Road   401-254-9087
Bristol, RI 02809   Fax: 401-254-3310
                            E-mail: mdg@alpha.rwu.edu
                            http://www.rwu.edu/
Undergraduate program in marine biology combining chemistry, biology, physics, and mathmatics. Designed to keep and develop interest in the sciences by using field research and laboratory experimentation.

5996 Rosenstiel School of Marine and Atmospheric Science
University of Miami
4600 Rickenbacker Causeway   305-421-4000
Miami, FL 33149   Fax: 305-421-4711
          E-mail: fmillero@rsmas.miami.edu
          http://www.rsmas.miami.edu/
Established in 1943 as the Marine Laboratory of the University of Miami. It has grown from its modest beginnings in a boathouse to be one of the nations leading institutions for oceanographic research and education.
Otis B Brown, Dean

5997 Santa Barbara Environmental Studies Program
University of California
Girvetz Hall   805-893-2968
2nd Floor   Fax: 805-893-8686
Santa Barbara, CA 93106-4160 E-mail: esinfo@ucsb.edu
                              http://www.es.ucsb.edu
An undergraduate program designed to provide students with the scholarly background and intellectual skills necessary to understand complex environmental problems and formulate decisions that are environmentally sound. Academic process is interdisciplinary, drawing upon not only ES faculty, but also the resources of a variety of environmentally related departments and disciplines at UCSB. Offers two Bachelor of Science degrees and one Bachelor of Arts degree.
Amanda Grundman, Program Assistant
Eric Zimmerman, Academic Advisor

5998 School for Field Studies
10 Federal Street   978-741-3544
Salem, MA 01970   800-989-4418
                           Fax: 978-741-3551
          E-mail: admissions@fieldstudies.org
          http://www.fieldstudies.org
Teaching students to address critical environmental problems using an interdisciplinary, experiential approach to education.
Nigel Barton, Center Director

5999 School of Forest Resources and Conservation
University of Florida
PO Box 110410   352-846-0850
Gainesville, FL 32611-0410 Fax: 352-392-1707
                            E-mail: sfrc@ifas.ufl.edu
                            http://www.sfrc.ufl.edu
Offers baccalaureate (BSFRC) and graduate (MFRC, MS and PhD, including a joint JD with College of Law) programs, conducts fundamental and applied research and provides public service through extension education. Programs include the forest sciences, management and policy as well as related programs in natural resources education, ecotourism and agroforestry.
Timothy L White, Director
George Blakeslee, Associate Director

6000 Slippery Rock University
Department of Environmental Geosc.   724-738-2495
1 Morrow Way   800-SRU-9111
Slippery Rock, PA 16057-1326   Fax: 724-738-4807
                            E-mail: EGeoInfo@SRU.edu
                            http://www.sru.edu/
Prepares students for ocupations with industrial laboratories concerned with air, water and soil pollution control, engineering firms that study industrial pollution and prepare environmental impact statements, and state and federal agencies charged with monitoring the environment
James Hathaway, Chair

6001 Smithsonian Institution

Smithsonian Information 202-357-2700
SI Building, Room 153 E-mail: info@info.si.edu
Washington, DC 20560-0010 http://www.si.edu

Independent trust instrumentality of the United States holding more than 140 million artifacts and specimens in its trust for the increase and diffusion of knowledge. Also a center for research dedicated to public education, nationalservice, and scholarship in the arts, sciences, and history.
Lawrence Small, Secretary

6002 Southern Connecticut State University

501 Crescent Avenue 203-392-6600
Jennings Hall, Room 342 888-500-SCSU
New Haven, CT 06515 Fax: 203-392-6614

Environmental Education program focuses on practicality and application of thoery bringing about environmental change through educational processes. The objective is to prepare well informed people who are dedicated to improvingenvironmental conditions.
Cheryl J Horton, President

6003 Southwest Environmental Health Sciences Center: University of Arizona

PO Box 210207 520-626-5594
Tucson, AZ 85721-0207 Fax: 520-626-6944
E-mail: swehsc-info@pharmacy.arizona.edu

The Southwest Environmental Health Sciences Center serves as a platform to promote the study of health effects of environmental agents. The SWEHSC promotes interdisciplinary research collaborations driven by cutting-edge technologies.Research in the SWEHSC is focused on mechanisms of action of environmetnal agents in living systems.
Serrine S Lau, Director

6004 State University of New York College of Environmental Science and Forestry

One Forestry Drive 315-470-6633
312 Bray Hall Fax: 315-470-6958
Syracuse, NY 13210-2778 E-mail: esfengg@mailbox.syr.edu
http://www.esf.edu

As part of it's education mission, the Faculty offers an accredited engineering undergraduate program in Forest Engineering and graduate programs at both the masters and doctoral levels. The Faculty also conducts research and publicservice programs.
Robert H Brock Jr, Professor

6005 State University of New York-College at Plattsburgh

101 Broad Street 518-564-2000
Plattsburgh, NY 12901 http://www.plattsburgh.edu

One of the largest and most established environmental science programs in the US, with 16 full time interdisciplinary faculty (4 part time) and nearly 300 majors among five degree programs. Outstanding opportunities for hands-on workand practical experience are provide by close proximity to the huge Adirondack Mountains State Forest Preserve, Plattsburgh's location on the banks of Lake Champlain, and our affiliations with the WH Miner Agricultural Research Institute and more.
David Franzi, Professor of Geology

6006 Sterling College

PO Box 72 802-586-7711
Craftsbury Common, VT 05827 Fax: 802-586-2596
E-mail: admissions@sterlingcollege.edu
http://www.sterlingcollege.edu

Sterling College is a community-oriented, environmentally focused, four-year, private college that combines traditional academics with experiential learning. They offer an integrated curriculum, a comprehensive internship program, andglobal field studies. Academic majors: Conservation Ecology, Sustainable Agriculture, Outdoor Education and Leadership, Northern Studies. Financial Aid is available.
Gwyn Harris, Director Admissions

6007 Tennessee Technological University

Tennessee Technical University 931-372-3888
P.O. Box 5006 866-733-8324
Cookeville, TN 38505-0001 E-mail: visit@tntech.ecu
http://www.tntech.edu

Prepares graduates for high-level careers in various areas of biology.

Robert Bell, President
Dr. Marvin Barker, Provost/VP Academic Affairs

6008 Texas A and M University at College Station

Center for Natural Resource Information Technology
113 Administration Building 409-845-5548
College Station, TX 77843-2142 Fax: 409-845-6430
E-mail: j-stuth@tamu.eud
http://cnrit.tamu.eud/cnrit

Serves as a point of contact for external organizations seeking cooperative efforts to assemble and disseminate information, create information technologies, research critical concepts limiting application of information technilogy andfacilitate technology transfer through training of end users and establishing necessary information infrastructures.
Bob Brown, Director

6009 The Department of Ecology and Organismal Biolo gy

Indiana State University 812-237-2400
Terre Haute, IN 47809 Fax: 812-237-4480
E-mail: Isamlan@scifac.indstate.edu
http://www.indstate.edu/ecology

The Department of Ecology and Organismal Biology of Indiana State University consists of dedicated researchers in the areas of ecology, evolution, and conservation. EWE offer graduate studies leading to the M.S. and Ph.D degrees. Ourstate-of-the-art laboratories and local field staaaations enable students to conduct innovative research under the guidance of our nationally-recognized faculty.
Dr. Charles Amianer, Director

6010 Thorne Ecological Institute

Po Box 19107 303-499-3647
Boulder, CO 80308-2107 Fax: 720-565-3873
E-mail: info@thorne-eco.org
http://www.thorne-eco.org

Offers hands-on environmental education for young people along the Front Range of Colorado.
Jessica Feld, Executive Director

6011 Three Circles Center for Multicultural Environmental Education

PO Box 1946 415-331-4540
Sausalito, CA 94965 Fax: 415-331-4540

To introduce, encourage and cultivate multicultural perspectives and values in environmental and outdoor education, recreation and interpretation. The three circles refers to the dynamic interacting systems of culture, ecology andcommunity.

6012 Treasure Valley Community College

Biology Department
650 College Boulevard 541-881-TVCC
Ontario, OR 97914 Fax: 541-881-2721
E-mail: russell@tvcc.cc
http://www.tvcc.cc.or.us/natres/

Offers several courses for those seeking careers in natural resource management including range management, wildland fire management, and forest management.
James Sorenson, President

6013 Trees for Tomorrow

519 Sheridan Street 715-479-6456
PO Box 609 800-838-9472
Eagle River, WI 54521 Fax: 715-479-2318
E-mail: learning@treesfortomorrow.com
http://www.treesfortomorrow.com

Private, nonprofit natural resource education center which uses a combination of field studies and classroom presentations to teach conservation values, as well as demonstrate the benefits of modern resource management.
Gail Gilson Pierce, Executive Director

6014 Tropical Forest Foundation

2121 Eisenhower Avenue 703-518-8834
Suite 200 Fax: 703-518-8974
Alexandria, VA 22314 E-mail: tff@igc.org
http://www.tropicalforestfoundation.org

Keister Evans, Executive Director
Wendy Beer, Staff Executive

6015 Tulane University

6823 St. Charles Avenue   504-865-5731
New Orleans, LA 70118   800-873-9283
    Fax: 504-862-8715
    E-mail: pr@tulane.edu
    http://www.tulane.edu

Environmental Health Sciences offers several graduate degree programs: the Master of Public Health, the Master of Science in Public health, and the Doctor of Science. The MPH is designed to meet the needs of public health professionalssuch as environmental health and health officers. The MSPH id designed to prepare students to undertake responsible positions in government, industrail facilities or whom will continue their education at the doctoral level. The SCD is a researchdegree.
Scott Cowen, President

6016 UIS Department of Environmental Studies

Univeristy of Illinois at Springfield
PAC 322   217-206-6720
One University Plaza   Fax: 217-206-7279
Springfield, IL 62703   E-mail: ens@uis.edu
    http://www.uis.edu/ens/main.html

Goal of the environmental studies program is to enhance society's ability to create an environmentally acceptable future. Program faculty with diverse backgrouns in the social and natural sciences and in the humanities are committedto developing interdisciplinary approaches to environmental problem solving. The primary objective is to educate citizens and professionals who are aware of environmental issues and their origins, causes, effects, and resolutions.
Nancy Rachelle, Secretary

6017 Unity College in Maine

90 Quaker Hill Road   207-948-3131
Unity, ME 04988   Fax: 207-948-6277
    E-mail: admissions@unity.edu
    http://www.unity.edu/

Unity College is a small independent college, in a rural setting, that exists to educate students in liber arts context for professional preparation in fields of environmental science, natural resource management, wilderness-basedoutdoor recreation leadership, and related fields.
Robert Pollis, Chair
Joan Amory, Vice Chair

6018 University of California at Los Angeles

595 Charles Young Drive East   310-825-3880
Box 951567   Fax: 310-825-2779
Los Angeles, CA 90095-1567   E-mail: info@ess.ulca.edu
    http://www.ulca.edu

Principles and practice of soil mechanics and foundation engineering in light of geologic conditions, recognition, prediction, and control or abatement of subsidence, landslides, earthquakes, and other geologic aspects of urbanplanning and subsurface disposal of liduids and solid wastes.
Roger Wakimoto, Chair

6019 University of California at Riverside

900 University Avenue   909-787-4531
Riverside, CA 92521   Fax: 909-787-6344
    E-mail: marylynn.yates@ucr.edu
    http://www.ucr.edu

The University of California, Riverside combines the opportunities of major research university and its state-of-the-art facilities with a friendly campus environment that features small classes and close faculty-student interaction.
Marylynn Yates, Chair

6020 University of California: Santa Cruz

Sciences Building   831-459-2634
405 Interdisciplinary   Fax: 831-459-4015
Santa Cruz, CA 95064   E-mail: envstudies@ucsc.edu
    http://envs.ucsc.edu

BA and Ph.D. programs in Environmental Studies offered at University of California, Santa Cruz.
Daniel Pressan, Chairperson

6021 University of Colorado

Environmental Engineering Program
428 UCB   303-495-0253
College of Engineering & Applied ScienceFax: 303-492-7115
Boulder, CO 80309-0428

    E-mail: apply@colorado.edu
    http://www.colorado.edu

Environmental Engineering Progam in the Department of Civil, Environmental, and Architectural Engineering at the University of Colorado in Boulder.We welcome prospective students, alumni of the program, and colleagues in environmentalengineering and invite you to explore our areas of emphasis, B.S., M.S., and Ph.D programs, research, and facilities.
Bob Sievers, Director
Rosella Chavez, Administrative Assistant

6022 University of Georgia

Savannah River Ecology Laboratory   803-725-2472
Drawer E   Fax: 803-725-3309
Aiken, SC 29803   http://www.uga.edu/srel

Our goal is to attract collaborating scientists from across the DOE complex and the nation for collaborative work at the interface of fundamental and applied environmental research in order to improve the management of contaminatedstites. The staff and facilities of AACES are available to researchers in environmental science and engineering and to practitoneers from industry, government, academia and private foundations.
Rosemary Forrest, Public Relations Coordinator

6023 University of Idaho

College of Natural Resources
PO Box 441138   208-885-2397
Moscow, ID 83844   Fax: 208-885-5534
    E-mail: cnr@uidaho.edu/cnr
    http://www.cnrhome.edu

Consists of 5 departments which together form a comprehensive educational program on the study and management of natural resources. Each department has several degree options to provide students with a flexible curriculum for theirdegree. We educate resource professionals with truly integrated resource management skills using innovative instructional programs. Our education occurs in a residential setting and provides a balance between theoretical and pratical experiences.
Steven B Daley Laursen, Dean

6024 University of Illinois at Urbana

Department of Natural Resources and Environment
1102 South Goodwin Avenue   217-333-2770
W-503 Turner Hall   Fax: 217-244-3219
Urbana, IL 61801   E-mail: nres@uiuc.edu
    http://www.nres.uiuc.edu

Establishes and implements research and educational programs that enhance environmental stewardship in the management and use of natural, agricultural, and urban systems in a socially responsible manner.
Wesley Jarrell, Professor & Head

6025 University of Kansas Field Stations

Kansas Biological Survey
2101 Constant Avenue   785-864-1505
Lawrence, KS 66047-3759   Fax: 785-864-5093
    http://www.ksr.ku.edu

The Kansas Ecological Reseves is the environmental field station operated by the University of Kansas; it consists of seven tracts of land with attendant research, teaching, and support facilities. The Reserves were establishedthroughthe efforts of concerned citizens, faculty, and others dedicated to establishing a regional field station to promote research in the natural sciences and to provide expanded educational opportunities for people of all ages.
Edward Martinko, KSR Director

6026 University of Maine at Fort Kent

23 University Street   207-834-7600
Fort Kent, ME 04743   888-try-umfk
    Fax: 207-834-7609
    E-mail: umfkadm@maine.edu
    http://www.umfk.maine.edu

Students develop a broad knowledge of the natural and social sciences, and focus on an aspect of this diverse area of study that is of personal interest. Students learn to critically identify environmental problems, collect andinterpret data, communicate today's complex environmental issues, and explore creative solutions. Students work closely with an interdisciplinary group of faculty with expertise in biology, chemistry, forestry, the social sciences and thehumanities.
Steven Selva, Professor

6027 University of Maryland at Baltimore

1000 Hilltop Circle   410-455-2291
Baltimore, MD 21250   800-UMB-C4US
     Fax: 410-455-1094
E-mail: adnmissions@umbc.edu
http://www.umbc.edu

The goal of the MEES program is to train students with career interests in some aspect of environmental science involving terrestrial, freshwater, marine, or estuarine systems. The program is university-wide and interdisciplinary,allowing students to use facilities and interact with faculty of the entire university system in order to plan a program best suited to their particular interests.
Lasse Lindahl, Professor

6028 University of Maryland at College Park

0107 Chemistry Building   301-405-1788
University of Maryland   800-422-5867
College Park, MD 20742-4454   Fax: 301-314-9121
E-mail: um-admit@uga.umd.eduu
http://www.chem.umd.edu

Since 1972, the Chemistry and Biochemistry Depasrtment has offered specialized training at the graduate level in environmental chemistry. In addition to course work in traditional chemistry subjects, students in this specialty takespecific environmental course offerings and do research under the guildance of faculty members specializing in this area.
Michael Doyle, Professor & Chair

6029 University of Minnesota at St. Paul

277 Coffey Hall   612-624-3009
1420 Eckles Avenue   Fax: 612-625-1260
St.Paul, MN 55108 E-mail: college-info@mail.coafes.umn.edu
http://www.coafes.umn.edu

Welcome to the College of Agricultural, Food, and Environmental Sciences. We are a landgrant college whose mission is to teach, conduct research, and share our findings through outreach efforts.
Dave Lime, Senior Research Associate

6030 University of Montana

Missoula, MT 59812   406-243-5209
     Fax: 406-243-6090
E-mail: len.broberg@umontana.edu

Interdisciplinary graduate and undergraduate program in environmental studies. See www.umt.edu/evst.

Len Broberg, Program Director

6031 University of Nebraska

Environmental Studies Program   402-472-7527
103 Natural Resources Hall   800-742-8800
Lincoln, NE 68583-0844   Fax: 402-472-3574
E-mail: nuhusker@unl.edu
http://www.unl.edu

The Environmental Studies Program is designed to serve a variety of students concerned about environmental issues and change. The program provides a thorough, holistic view of the environment and human-environmental interaction and thetechnical skills for active participation in an environmental career.
Robert Kuzelka, Program Director
Meghan Sittler, Program Advisor

6032 University of Nevada at Las Vegas

4505 S Maryland Parkway   702-895-3701
Box 454015   Fax: 702-895-3936
Las Vegas, NV 89154-4015   E-mail: ce-info@ce.unlv.edu
http://www.ce.unlv.edu

Department of Civil and Environmental Engineering in the Howards R. Hughes College of Engineering offers programs leading to a Master of Science in Engineering and Doctor of Philosophy, with concentration in six areas: environmentalengineering, fluid mechanics and hydraulics, geotechnical engineering, structural engineering, construction engineering, and transportation systems.
Jodie Sabedra, Management Assistant
Allen Sampson, Sr Development Tech

6033 University of New Haven

300 Orange Avenue   203-932-7319
Westhaven, CT 06516   Fax: 203-931-6093

E-mail: adminfo@charger.newhaven.edu
http://www.newhaven

The bachelor of science program in environmental science is designed to give the student a strong foundation in the fudamental sciences (biology, chemistry, physics, and geology).
Dr. Laurence Davis, B.S. Environ Science Biology

6034 University of North Carolina at Chapel Hill

Campus Box 2200   919-966-3621
Jackson Hall   Fax: 919-962-3045
Chapel Hill, NC 27599   E-mail: uadm@email.unc.edu
http://www.admissions.unc.edu

The Environmental Science and Studies program leads to a Bachelor of Science degree in Environmental Science or a Bachelor of Arts degree in Environmental Studies. Students investigate the relationship between the environment andsociety, focusing on environmental management, law and business. The program combines traditional classroom teaching with extensive use of interdisciplinary, team-based field projects, internships, study abroad and research.

6035 University of Pennsylvania

119 Meyerson Hall   215-898-7507
210 S 34th Street   Fax: 215-573-3770
Philadelphia, PA 19104   http://www.upenn.edu

The mission of our department is to bring the time perspective of the Earth scientist/historian to bear on contemporary problems of natural resource conservation and environmental quality.
James Corner, Chairman

6036 University of Pittsburgh Department of Geology and Planetary Science

4107 O'Hara Street   412-624-8780
200 SRCC Building   Fax: 412-624-3914
Pittsburgh, PA 15260-3332   E-mail: geology@pitt.edu
http://www.geology.pitt.edu

The Bachelor of Arts in Environmental Studies equips students with an understanding of earth systems and the impact of humans on the biosphere, atmosphere and hydrosphere. Courses in the natural and social sciences, humanities, andschools of law, business, and public health provide a comprehensive, interdisciplinary bachground in environmental issues and public policy.
Mark Collins, Environmental Studies Coordi

6037 University of Redlands

1200 E Colton Avenue   909-793-2121
Po Box 3080   Fax: 909-793-2029
Redlands, CA 92373-0999   http://www.redlands.edu

To promote a new way of thinking and acting about our relationship to the world. We believe students should be environmentally literate, sensitive to competing demands and conflicting values of each issue and finally, they should havethe creativity, confidence and conviction to begin effecting change.
Dr James Appleton, President

6038 University of South Carolina Baruch Marine Field Lab

Hobgaw Barony   843-546-3623
Po Box 1630   Fax: 843-546-1632
Georgetown, SC 29442   E-mail: dallen@belle.baruch.sc.edu
http://www.cas.sc.edu/baruch/

Environmental research and education programs at the BMFL are focused on estuarine systems and their associated watersheds. More than 160 investigators representing 30 academic institutes and agencies are affiliated with 89 projects in2005-06. The lab provides support for undergraduate classes, graduate students, and senior scientists. Long-term environmental monitoring, training programs & outreach activities are sponsored by the North-inlet-Winyah Bay National Estuarine researchreserve prog.

Dennis Allen, Resident Director

6039 University of Southern California

Environmental Studies Program   213-740-7770
Hancock Building, Room 232   Fax: 213-740-8566
Los Angeles, CA 90089   http://www.usc.edu

Educational Resources & Programs/Universities

The USC Environmental Sciences, Policy and Engineering Sustainable Cities Program, a multidisciplinary doctoral training program funded by the National Science Foundation, prepares doctoral students to confront, analyze and resolve thechallenges posed by problems of urban sustainablilty. The Program allows doctoral students to transcend disciplinary, policy-relevant research on major environmental problems should be conducted.
Albert A Herrara, Director & Professor
Angel Grigorian, Advisor & Program Coord.

6040 University of Southern Mississippi

University of Southern Mississippi 601-266-4748
118 College Drive #5018 Fax: 601-266-5797
Hattiesburg, MS 39406-0001 E-mail: admissions@usm.edu
http://www.biology.usm.edu
The Environmental Science concentration focuses on industrail problems related to the working environment, pollution control, and safety. Courses address major industrail issues, including environmental impact statements, industrailhygiene and environmental laws and reulations.
Frank Moore, Chair
Patricia Brewer, Administrative Assistant

6041 University of Tennessee at Martin

University Street 901-587-7020
Martin, TN 38238 Fax: 901-587-7029
E-mail: admitme@utm.edu
http://www.utm.edu
The application of geology to the interaction between man and the environment. Topics include geohazards, chemical and nuclear contamination of soils and water, remediation of environmental problems and governmental environmentalagencies and laws.
Jim Byford, Dean

6042 University of West Florida

11000 University Parkway 850-474-2748
Pensacola, FL 32514-5750 Fax: 850-474-2749
E-mail: biology@uwf.edu
http://www.uwf.edu
The B.S. program in Environmental Studies consists of a multi-disciplinary approach that combines natural science and resource management. Students learn to analyze physical and socioeconomic environments and to reach decisionsconcerning environmental use and protection. It offers a core curriculum that is designed to provide the student with a solid foundation in the earth sciences, the life sciences, and the basic sciences as wellas in modern methods and techniques.
George Stewart, Department Of Biology

6043 University of Wisconsin at Green Bay

2420 Nicolet Drive 920-465-2111
Green Bay, WI 54311-7001 888-367-8942
Fax: 920-465-5754
E-mail: admissns@uwgb.edu
http://www.uwgb.edu
The program of study in the environmental science major is interdisciplinary, emphasizing an integrated approach to knowledge in the field. Because the study of environmental science major is grounded in the natural sciences andmathematics. The curriculum includes a social science component, enabling students to gain an understanding of environmental economic and policy issues. Field experiences, internships and practicums are emphasized.
Denise Scheberle, Asst Prof Political Science

6044 University of Wisconsin at Madison

Experiment Station 608-262-4930
140 Agricultural Hall Fax: 608-262-1429
Madison, WI 53706 E-mail: eaberle@cals.wisc.edu
http://www.cals.wisc.edu
Remote sensing and geographic information systems offer sophisticated and powerful tools for monitoring the environment on large geographic scales over time. Students in the Environmental Monitoring Program learn to employ thesetechnologies in fields of their choice, from forestry to urban planning to environmental engineering.
Elton Aberle, Dean

6045 University of Wisconsin at Stevens Point

Stevens Point, WI 54481 715-346-4185
Fax: 715-346-3624
http://www.uwsp.edu/acad/cnr

The environmental Task Force Program is a water lab which was initiated in 1972. The Program includes two water chemistry labs which tests for organics and inorganics; we have five full time staff, plus a part time faculty director. Wehire and/ or train 35 to 40 students per year. The program has near state-of-the-art field sampling and laboratory analytical equipment nutrients, pesticides, polynuclear aromatic hydrocarbons, polychlorinated biphenyls, and volatile organiccompounds.
Christine Thomas, Dean

6046 University of the South

735 University Avenue 931-598-1271
Sewanee, TN 37383-1000 800-522-2234
Fax: 931-598-1667
E-mail: collegeadmission@sewanee.edu
http://www.sewanee.edu/forestry_gelology/forestrygeology.html
The Environmental Studies concentration is an interdisciplinary program that offers students an informed and broad understanding of the environment. The concentration is open to students majoring in any discipline. By adding theconcentration to their chosen field of study, students will develop skills of inquiry, analysis, and stewardship that enable them to evaluate and address complex environmetnal issues from multiple perspectives.
Scott Torreano, Associate/Professor/Chair

6047 Urban and Environmental Policy Program

Rice University
Rice University MS-24 713-527-8101
PO Box 1892 Fax: 713-348-5277
Huston, TX 77251 E-mail: dho@rice.edu
http://www.ruf.rice.edu/
Gives students a background in how environmental policies are developed and how science and engineering issues are included in effective policy.
Dr. Donald Ostdiek, Director

6048 Utah State University

College of Natural Resources 435-797-2459
5210 Old Main Hill Fax: 435-797-1871
Logan, UT 84322-5210 E-mail: awerinfo@cc.usu.edu
http://www.cnr.usu.edu
There is no other organization in the world like Berryman Institute. Based at Utah State University, the Berryman Institute is a functional component of the Department of Fisheries and Wildlife and the College of Natural Resources. ItsFaculty members hold academics appointments in various departments throughout Utah State University and other universities. This multidisciplinary approach is calculated to speed the discovery and development of innovative methods to solve humanwildlife conflict.
Micheal Conover, Director

6049 Vanderbilt University

Dept. Earth, Environmental Sciences
2301 Vanderbilt Place 615-322-2976
VU Station B, #351805 800-288-0432
Nashville, TN 32235-1805 Fax: 615-322-2138
E-mail: admissions@vanderbilt.edu
http://www@vuse.vanderbilt.edu
VCEMS promotes and develops partnerships among industry, government and academia to study the relationship of environmental policy to business management and operations. The Center is a Vanderbilt University system-wide initiativejointly led by the School of Engineering, the Graduate School, the Owen Graduate School of Management, the Law School, and the Insititute for Public Policy Studies.
David Jon Furbish, Chair

6050 Vermont Law School

Chelsea Street
PO Box 96 800-227-1395
South Royalton, VT 05068 E-mail: jloftus@vermontlaw.edu
http://www.vermontlaw.edu
The Environmental Law Center administers three degrees in environmental law: the Master of Studies in Environmental Law (MSEL) and JD/MSEL (joint-degree) and the LL.M. in Environmental Law. Each degree is adaptable to career objectivesin both public and private sectors. The mission is to educate for stewardship, to teach an awareness of underlining environmental issues and values, to provide a solid knowledge of environmental law and to develop skills to administer and improvepolicies.

Jennifer Loftus, Admissions Coordinator

6051 Virginia Polytechnic Institute Center
CSES Department   540-231-6300
240 Smyth Hall, W Campus Drive   Fax: 540-231-3431
Blacksburg, VA 24061-0404   E-mail: sybil@mail.vt.edu
Designed to prepare environmental professionals needed in the private and public sector and by nonprofit organizaitons. It is built on a rigorous interdisciplinary curriculum that stresses the basic sciences, engineering andenvironmental technologies, soils and geology, computational and analytical skills, communication skills, as well as social sciences such as economics and environmental law.
Sybil Phoenix, Administrative Assistant

6052 Virginia Polytechnic Institute and State University
330 Smyth Hall (0404)   540-231-6305
Blacksburg, VA 24061   Fax: 540-231-3431
E-mail: vtadmiss@vt.edu
http://www.vt.edu
The major is about people and their environment. It deals with crop production, soil utilization, and environmental stewardship. Its professionals are concerned with helping to feed the world and protect the environment; their ranksinclude women and men who work to grow crucial commodities, improve water quality, develop environmentally acceptable methods for protecting crops from pests, advise municipalities on use of the land resource.
Steven C Hodges, Department Head

6053 Warren Wilson College
PO Box 9000   828-298-3325
Asheville, NC 28815   Fax: 828-299-4841
E-mail: lweber@warren-wilson.edu
http://www.warren-wilson.edu
Combines rigorous courses in the natural and social sciences with abundant natural resources near the classrooom. Courses and work crews give students a balance of theory, first hand knowledge and field experience. Successful programsmost often result when students, with the help of an advisor, begin planning course work and indentifying goals during their first year.
Dr Louise Weber, Chair

6054 Washington State University
305 Troy Hall   509-335-8538
Pullman, WA 99164   Fax: 509-335-7636
E-mail: budd@wsu.edu
http://www.esrp.wsu.edu
Because of the diversity in the field of environmental science, the course of study for each student is flexible. Majors can specialize in agricultural ecology, biological science, environmental education, environmental quality (airand water), natural resource management, systems, environmental/land use planning or hazardous waste management.
Dr William Budd, Chair

6055 Washington State University Extension
Washington State University   509-335-2933
PO Box 646230   Fax: 509-335-2926
Pullman, WA 99164   E-mail: lkfox@wsu.edu
http://www.ext.wsu.edu/
Linda Kirk Fox, Dean/Director

6056 Welder Wildlife Foundation
PO Box 1400   361-364-2643
Sinton, TX 78387   Fax: 361-364-2650
E-mail: welderwf@aol.com
http://http://hometown.aol.com/welderwf/welderweb.html
A private non profit organization that has gained international recognition through its graduate student research program. The mission of the Foundation is to conduct research and education in the field of wildlife management andconservation and other closely related fields.
Lynn Drawe, Director

6057 West Virginia University
425 White Hall   304-293-5603
Po Box 6300   800-344-WVU1
Morgantown, WV 26506   Fax: 304-293-6522
E-mail: cafcs@wvu.edu
http://www.caf.wvu.edu

The Environmental Geosciences undergraduate degree features an interdisciplinary approach to environmental issues. Graduates will be well prepared to face the environmental challenges of the future.
Hope Stewart, General Info Contact
Dr. Trevor M. Harris, Department Chair

6058 Western Montana College Environmental Science
710 S Atlantic Street   406-683-7331
Dillon, MT 59725   866-UMN-MONT
Fax: 406-683-7331
E-mail: admissions@umwestern.edu
http://www.umwestern.edu
The mission of the environmental sciences programs is to provide students with an in-depth understanding of the natural processes which create and shape our environment. Students will become informed, critical thinkers capable ofscientifically evaluating complex issues involving the environment. Student development will occur through interdisciplinary, field-based research projects that have societal relevance.
Craig Zaspel, Professor Environ. Sciences

6059 Wilderness Education Association
900 E 7th Street   812-855-4095
Bloomington, IN 47405   Fax: 812-855-8697
E-mail: wea@indiana.edu
http://www.weainfo.org/
WEA provides professional instruction, leadership training and wilderness travel. Promotes safe, ethical and professional wilderness leaders.

6060 Wildlife Management Institute
1146 19th Street NW   202-371-1808
Suite 700   Fax: 202-408-5059
Washington, DC 20036   http://www.wildlifemgt.org
WMI is a private, nonprofit, scientific and educational organization. It is committed to the conservation, anhacement and professional management of North America's wildlife and other natural resources.
Steve Williams, President

6061 Williams College
Center for Environmental Studies
PO Box 632   413-597-2346
Kellogg House   Fax: 412-597-3489
Williamstown, MA 01267   E-mail: szepke@williams.edu
http://www.williams.edu
The Environmental Studies program provides students with tools, ideas, and opportunities to engage constructively with the environmental and social issues brought about by changes in population, economic activity, and values.Environmental studies in interdisciplinary, and it is broad, including the coditions of inner-city poverty as well as the magnificent scenery of wildlands, ecompassing the view of planet earth from near space as well as the rich nuance of theculteral anthropologists.
Hank Art, Director

6062 World Resources Institute
10 G Street, NW   202-729-7600
Suite 800   Fax: 202-729-7610
Washington, DC 20002   http://www.wri.org
WRI provides information, ideas, and solutions to global environmental problems. Our mission is to move human society to live in ways that protect the environment for current and future generations. Our program meets global challengesby using knowledge to catalyze public and private action. Goals are to: safeguard earth's climate from further harm; protect the ecosystems; reduce the use of materials and generation of wastes in the production of goods and services.
Jonathan Lash, President

6063 Yale University
PO Box 208234   203-432-9316
New Haven, CT 06520   Fax: 203-432-9392
E-mail: undergraduate.admission@yale.edu
http://www.yale.edu

Our mission is to provide the new leadership and new knowledge neede to restore and sustain both the health of the biosphere and the well-being of its people. We believe that human enterprise can and must be conducted in harmony withthe environment, using natural resources in ways that sustain both resources and ourselves. We believe that solving environmental problems must incorporate human values and motivations and a deep respect for both human and natural communities.

Richard Charles Levin, President

Workshops & Camps

6064 A Closer Look at Plant Life
Educational Images
PO Box 3456   607-732-1090
Elmira, NY 14905   800-527-4264
                                   Fax: 607-732-1183
                        E-mail: info@edimages.com
                   http://www.educationalimages.com
Access a wealth of information on every major group of vascular and nonvascular plants. Includes details on plant microanatomy; external and internal structures; life cycles; and processes such as growth transpiration andphotosynthesis.

Charles R Belinky, Ph.D, CEO

6065 A Closer Look at Pondlife - CD-ROM
Educational Images
PO Box 3456   607-732-1090
Elmira, NY 14905   800-527-4264
                                   Fax: 607-732-1183
                        E-mail: info@edimages.com
                   http://www.educationalimages.com
Through the wonders of close-up photography, this unique CD-ROM brings students face-to-face with the inner workings of a freshwater pond, the myriad creatures and plants that reside there, and the dynamic interactions that go onbeneath the surface. This disk features a library of reference information, images, illustrations, clip art, video clips and more!

Charles R Belinky, Ph.D, CEO

6066 Air and Waste Management Association
One Gateway Center   412-232-3444
3rd Floor   Fax: 412-232-3450
Pittsburg, PA 15222

The Air & Waste Management Association is a non profit, nonpatism professional organization that provides training, information, and networking opportunities to 12,000 environmental professionals in 65 countries. THe Association'sgoals are to stengthen the environmental professionals in critical environmental decision making to benefit society.

6067 American Museum of Natural History
Center for Biodiversity and Conservation
Central Park West   212-769-5742
79th Street   Fax: 212-769-5292
New York, NY 10024   http://www.research.amnh.org
Conducts research and field projects based on information provided by Museum departments.

6068 Animal Tracks and Signs
Educational Images
PO Box 3456   607-732-1090
Elmira, NY 14905   800-527-4264
                                   Fax: 607-732-1183
                        E-mail: info@edimages.com
                   http://www.educationalimages.com
Presents various animal tracks and signs throughout the seasons, and provides useful information about the special characteristics and natural history of the animals that left the signs. Footprints, scratch marks, nesting places,wallows, scats and signs of food gathering are all detailed. Coverage includes deer, fox, porcupine, rabbit, bear, mink, otter, owl, woodpecker, killdeer, wild turkey, sapsucker and grouse.

Charles R Belinky, Ph.D, CEO

6069 Annotated Invertebrate Clipart CD-ROM
Educational Images
POÆBox 3456   607-732-1090
Elmira, NY 14905   800-527-4264
                                   Fax: 607-732-1183
                        E-mail: info@edimages.com
                   http://www.educationalimages.com
780 colorful graphics of invertebrates from protists through urochordates, supported by extensive written annotations in addition to traditional labels. Includes presentation graphics and page after page of supplemental information onclassification, anatomy, evolution, development, reproduction, etc.

Charles R Belinky, Ph.D, CEO

6070 Annotated Vertebrate Clipart CD-ROM
Educational Images
PO Box 3456   607-732-1090
Elmira, NY 14905   800-527-4264
                                   Fax: 607-732-1183
                        E-mail: info@edimages.com
                   http://www.educationalimages.com
792 colorful graphics of vertebrates from urochordates and tunicates through mammals, supported by extensive written annotations in addition to traditional labels. Includes presentation graphics and page after page of supplementalinformation on classification, organ systems, anatomy, evolution, development, reproduction, etc.

Charles R Belinky, Ph.D, CEO

6071 Argonne National Laboratory
9700 S Cass Avenue   630-252-2000
Argonne, IL 60439   http://www.anl.gov/
We focus on four broad strategic environmental areas under which specific programs and projects are conducted. Our staff of over 100 multidisciplinary professionals are organized in a matrix fashion to undertake programs and projectswith technical managers and staff from seven sections, each specializing in specific technical disciplines.

6072 Ashland Nature Center
Delaware Nature Society   302-239-2334
PO Box 700   Fax: 302-239-2473
Hockessin, DE 19707   E-mail: webpage@dnsashland.org
                        http://delawarenaturesociety.org
Open year round, seven days a week, Ashland is headquarters of the Delaware Nature Society. Any time of year, you can enjoy four self-guiding trails, about 4 miles in total length, traversing 131 acres of rolling hills, flood plain andmarsh.

6073 Association for Environmental Health and Sciences
150 Fearing Street   413-549-5170
Amherst, MA 01002   Fax: 413-549-0579
                        E-mail: info@aehs.com
Created to facilitate communication and foster cooperation among professionals concerned with the challenge of soil protection and cleanup. Experience over the past decades has revealed the need for a consistent and reliable networkfor the exchange of information derived from multiple sources and disciplines among people who, because of different disciplinary affiliations and interests, may not have easy access to significant portions of the information map.

6074 Biosystems and Agricultural Engineering
Univerity of Kentucky
218 Agricultural Engineering Bldg   606-257-3000
Lexington, KY 40546-0276   Fax: 606-257-5671
                        E-mail: lturner@bae.uky.edu
                   http://www.bae.uky.edu/
Biosystems and Agricultural Engineering provides an essential link between the biological sciences and the engineering profession. The linkage is necessary for the development of food and fiber production and processing systems whichpreserves our natural resources base.

6075 Bishop Resource Area
785 N Main Street   760-872-4881
Suite E
Bishop, CA 93514

The Bishop Resource Area has facilitated aerial photo interpretation and remote sensing programs in local schools through corporate and public partnerships. The program incorporates aerial photo interpretation, its relationship to mapping and land use history.

6076 Bog Ecology

Educational Images
PO Box 3456
Elmira, NY 14905
607-732-1090
800-527-4264
Fax: 607-732-1183
E-mail: info@edimages.com
http://www.educationalimages.com

A comprehensive program that explores the origin and formation of bogs, common plants and animals, and compares bogs to other types of wetlands. Bog succession is illustrated by use of diagrams and photographs. 74 frames and guide.

Charles R Belinky, Ph.D, CEO

6077 Camp Fire USA

4601 Madison Avenue
Kansas City, MO 64112-1278
816-756-1950
Fax: 816-756-0258
E-mail: info@campfireusa.org
http://www.campfireusa.org

Not-for-profit, youth development organization, Camp Fire USA provides fun, coeducational programs for approximately 650,000 youth from birth to age 21. Helps boys and girls learn and play side by side in comfortable, informal settings.

Gregg Hibbeler, Director Market/Communicate

6078 Camp Habitat Northern Alaska Environmental Center

830 College Road
Fairbanks, AK 99701
907-452-5021
Fax: 907-452-3100
E-mail: camphabitat@northern.org
http://northern.org

Camp Habitat is a nature education program for young people ages 4-17 sponsored by the Northern Alaska Environmental Center, Friends of Creamer's Field, and Alaska Department of Fish & Game. The mission of Camp Habitat is to provide young people with guided explorations of their natural surroundings through interactive, hands-on activities. Skilled instructors and resource specialists lead small groups through new outdoor activities focusing on the habitats of Interior Alaska.

6079 Center for Environmental Research and Conservation

1200 Amsterdam Avenue
New York, NY 10027
212-845-8179
Fax: 212-854-8188
E-mail: cerc@columbia.edu

CERC, a consortium of five education and research institutions, was created in response to critical environmental concerns facing the Earth. Within the next fifty years human influence will affect every place on the planet. That impact will almost certainly result in species extinctions, ecosystem degradation and a loss of the benefits those species and ecosystems provide to people.

6080 Center for Geography and Environmental Education

311 Conference Center Building
Knoxville, TN 37996-4134
865-974-4251
Fax: 865-974-1838

Focuses on research and development related to geography and environmental education. Geography is the chosen disciplinary vehicle for environmental education because it integrates both the natural and social sciences in studying environmental problems and issues. CGEE designs environmental literacy and citizenship-assessment instruments, defines the socio-political-cultural foundations of environmental education, and surveys the environmental education component of teacher-preparation programs.

Dr. Rosalyn McKeown-Ice, Director

6081 Center for Mathematical Services

4202 East Fowler Avenue
Tampa, FL 33620
813-974-9568
Fax: 974-974-2700

Mission is to help prepare students of all levels to effectively use mathematics as a tool to analyze situations and resolve problems. In the field of mathematical sciences it serves as an interface for the University with the secondary schools in the area served by the University of South Florida. By means of this interface special programs in the mathematics, science, and engineering are offered at the University of South Florida for secondary students.

Kenneth L Pothoven, Director

6082 Cetacean Society International

PO Box 953
Georgetown, CT 06829
203-431-1606
Fax: 203-431-1606
E-mail: rossiter@csiwhalesalive.org
http://www.csiwhalesalive.org

All volunteer, nonprofit conservation, educational and research organization to benefit whales, dolphins, porpoises and the marine environment. Promotes education and conservation programs, including whale and dolphin watching, and noninvasive, benign research. Advocates for laws and treaties to prevent commercial whaling, habitat destruction and other harmful or destructive human interactions. CSI's world goal is to minimize cetacean killing and captures and to enhance public awareness.

William W Rossiter, President
Barbara Kilpatrick, VP

6083 Chicago Botanic Garden

1000 Lake Cook Road
Glencoe, IL 60022
847-835-5440

The Chicago Horticultural Society has been promoting gardens and gardening since 1890. Generations of Chicagoans have been touched by the Society's flower shows, victory gardens, horticultural lectures and more. The mission encompasses three important components: collections, programs and research. A living museum, the Chicago Botanic Garden serves both a public and a scientific community.

6084 Clean Ocean Action

PO Box 505
Sandy Hook
Highlands, NJ 07732
732-872-0111
Fax: 732-872-8041
E-mail: sandyhook@cleanoceanaction.org
http://www.cleanoceanaction.org

Clean Ocean Action is a broad-based coalition of over 150 conservation, community, diving, fishing, environmental, surfing, women's and business groups that works to improve and protect the waters off the New York and New Jersey coast.

Kari Jermansen, Outreach Director

6085 Cleaner and Greener Environment

1526 Chandler Street
Madison, WI 53711
608-280-0256
877-977-9277
Fax: 608-255-7202
E-mail: info@cleanerandgreener.org

Cleaner and Greener Environment is a program of Leonardo Academy, a 501 environmental nonprofit organization. Leonardo Academy reports reductions in emissions, and promotes the development of markets for the emission reductions that result from energy efficiency, renewable energy, and other emission reduction action.

6086 Coastal Resources Center

University of Rhode Island
Narragansett, RI 02882
401-874-6224
Fax: 401-789-4670
E-mail: info@crc.uri.edu
http://www.crc.uri.edu

Mobilizes governments, business and communities around the world to work together as stewards of coastal ecosystems. With partners we strive to define and achieve the health, equitable allocation of wealth, and sustainable intensities of human activity at the transition between the land and sea.

Stephen Olsen, Director
Chip Young, Communications Liasion

6087 Comet Halley: Once in a Lifetime!

Educational Images
PO Box 3456
Elmira, NY 14905
607-732-1090
800-527-4264
Fax: 607-732-1183
E-mail: info@edimages.com
http://www.educationalimages.com

Particularly relevant because of the recent appearance of Hale-Bopp, this program presents the reactions to comets in ancient, historic and relatively modern times, press coverage of the 1910 Halley return, superstitions and beliefs,current scientific knowledge and research, and much more.

Charles R Belinky, Ph.D, CEO

6088 Cooperative Institute for Research in Environmental Sciences: K-12 and Public Outreach
University of Colorado
Campus Box 216   303-492-5431
Boulder, CO 80309   Fax: 303-492-1149
We educate people about Earth and environmental science issues that are relevant to our everyday lives, through outreach to the public and to the K-12 education community.

6089 Deep Portage Conservation Reserve
2197 Nature Center Drive NW   218-682-2325
Hackensack, MN 56452   Fax: 218-682-3121
E-mail: portage@uslink.net
http://www.deep-portage.org
Deep Portage serves schools, groups, organizations, research teams, area residents and visitor with resident environmental education programs, weekly classes, interpretive programs, wildflower garden displays, land use demonstrations,summer youth camps and recreation opportunities of birding, hiking, hunting and skiing.

6090 Department of Energy and Geo-Environmental Engineering
110 Hosler Building   814-865-3437
University Park, PA 16802-5000   Fax: 814-865-3248
E-mail: egee@ems.psu.edu
Through education, research and service, EGEE aspires to insure that socisty is provided with an affordable supply of energy and minerals, concomitant with protecting the environment.

6091 Earth Day Network
1616 P Street NW   202-518-0044
Suite 340   Fax: 202-518-8794
Washington, DC 20036   E-mail: earthday@earthday.net
http://www.earthday.net
This nonprofit network was created to be a vehicle for increased awareness & responsibility through the promotion of Earth Day. Offers workshops.
Kathleen Rogers, President
Mary Minette, Senior VP Programs

6092 Ecological & Environmental Learning Services
46 Back Bone Hill Road   732-577-5599
Clarksburg, NJ 08510   800-206-6672
Fax: 732-577-5598
Ecological & Environmental Learning Services provides K-12 education consulting for teacher professional development, curriculum development and education programs and assemblies. EELS has the expertise and experience to providesolutions for enhancing the academic excellence of students in the following areas: Science (particularly in science research); Ecology; Environmental Science; and Environmental Education.

6093 Ecology and Environmental Sciences
206 Winslow Hall   207-581-3198
Orono, ME 04469   E-mail: mark.anderson@umit.maine.edu
Faculty from five different academic departments, covering biological, physical and social sciences, work together to offer a broad educational experience for our students. Since these faculty have active research programs, studentsnot only get access to the most up-to-date information, but also get employment opportunities in their fields of study during the academic year and the summer months.

6094 Economic Development/Marketing California Environmental Business Council
UC Extention
3120 De La Cruz   408-748-2170
Santa Clara, CA 95054   Fax: 408-748-2189
E-mail: br1027@aol.com

The CEBC is a nonprofit trade and business assoication that promotes and assists California's environmental technology and services industry at the state, national, and international levels. Founded in 1994, the CEBC currently has morethan 100 member compines and other organizations throughout the state that represent all segments of the environmental industry.

6095 Environmental Data Resources
440 Wheelers Farms Road   203-255-6606
Milford, CT 06460   800-352-0050
Fax: 800-231-6802
http://www.edrnet.com
Applied and product research in environmental applications.
Robert D Barber, CEO
Mark Cerino, COO

6096 Environmental Education Council of Ohio
1972 Clark Avenue   330-823-3655
Alliance, OH 44601   Fax: 330-823-8531
E-mail: mcclauca@muc.edu
http://www.muc.edu
EECO believes that: we are all learners interacting with others in lifelong process, education is vital for individuals to reach their full potential as members of our global community, a healthy and sustainable environment isessential to the survival of the planet. It is the mission of EECO to lead in facilitating and promoting environmental education which nurtures knowledge, attitudes and behaviors that foster global stewardship.

6097 Environmental Education K-12
PO Box 2057   863-465-2571
Lake Placid, FL 33862   Fax: 863-699-1927
E-mail: archbold@archbold-station.org
Archbold Biological Station provides environmental education programs to help people af all ages discover and understand the unique and endangered Florida scrib. Several programs for children Grades K-12 are offered each year. Theprogram goals are; promote a sound foundation in ecological proinciples, nurture a sense of stewardship for Florida scrub habitat, demostrate the value of scientific research, and develop a deeper understanding of the importance of natural habitatsfor investigation.

6098 Environmental Forum of Marin
PO Box 74   415-479-7814
Larkspur, CA 94977   E-mail: forum@MarinEFM.org
Dedicated to protecting and enhancing the environment by educating its members and the Marin citizenry on environmental issues. In futherance of this goal, the Environmental Forum of Marin conducts annual training programs onenvironmental matters, provides continuing education for its members and public, and supports citizen action to influence environmental decision-making and public policy.

6099 Environmental Resources
W275 N1990 Cabin Creek Ct   262-691-7413
Pewaukee, WI 53072-5349   Fax: 262-691-1579
Owner and co-founder of Moraine Multimedia, and Environmental Resources developes continuing education programs.

6100 Environmental Sciences
2507 Laclade Avenue   314-977-3131
St. Louis, MO 63103   Fax: 314-977-3117
E-mail: enviro@eas.slu.edu
Environmental Sciences is concerned with the near-surface realm of Earth and the way humans interact with that environment. Environmental scientist are concerned with water availability and equal, waste disposal, the use of Earth'slimited resources, and natural hazards such as earthquakes, landslides, and floods. Environmental scientists use the principles of geology, physics, chemistry, and biology to understand these phenomena and solve environmental problems.

6101 Environmental Technologies Exports
International Trade Administration   202-482-5225
HCHB Room 1003   Fax: 202-482-5665
Washington, DC 20230   E-mail: OETE1@ita.doc.gov
The Environmental Technologies Exports office is the principal resource and key contact point within the US Department of Commerce for American environmental technology compines. ETE's goal is to facilitate and increase exports ofenvironmental technologies-goods and services-by providing support and guidance to US exporters.

Anne Alonzo, Deputy Assistant Seceretary

6102 Exploring Animal Life - CD-ROM
Educational Images
PO Box 3456   607-732-1090
Elmira, NY 14905   800-527-4264
                              Fax: 607-732-1183
                    E-mail: info@edimages.com
              http://www.educationalimages.com
A curriculum oriented presentation and an instant encyclopedia, filled with superb photographs, informative text, exciting video clips, printable diagrams and illustrations, and lab activities. Provides a fascinating survey of themajor divisions of animal life and their characteristics: sponges, molluscs, insects, arthropods, fish, reptiles, birds and mammals are fully presented in the order in which you teach them.

Charles R Belinky, Ph.D, CEO

6103 Exploring Environmental Science Topics CD-ROM
Educational Images
PO Box 3456   607-732-1090
Elmira, NY 14905   800-527-4264
                              Fax: 607-732-1183
                    E-mail: info@edimages.com
              http://www.educationalimages.com
Provides a curriculum oriented presentation, an instant encyclopedia, superb photographs, video clips, informative text, printable diagrams & illustrations, & lab activities. This program offers a fascinating survey of environmentaltopics & concerns such as the environmental costs of energy; acid rain; energy flow and the greenhouse effect; oil spills; tundra, chaparral, desert, grassland and forest biomes; the hydrological cycle and water pollution; and the recycling elementsin the biosphere.

Charles R Belinky, Ph.D, CEO

6104 Exploring Freshwater Communities - CD-ROM
Educational Images
PO Box 3456   607-732-1090
Elmira, NY 14905   800-527-4264
                              Fax: 607-732-1183
                    E-mail: info@edimages.com
              http://www.educationalimages.com
A complete resource for studying freshwater biomes. It provides a fascinating survey of the ecology of swamps, bogs, marshes, wetlands, streams, ponds, lakes and the Everglades. There is even an introduction to fish restoration andwater pollution.

Charles R Belinky, Ph.D, CEO

6105 Five Winds International
626 Meadow Drive   610-431-5782
Westchester, PA 19380   Fax: 610-431-5783
                    E-mail: j.fava@fivewinds.com
                      http://www.fivewinds.com
The goal of this union is to create a global consultancy that helps clients improve their business and environmental performance. Our mission is to help organizations integrate product sustainability into their core business practices.

6106 Fossil Rim Wildlife Center
2155 County Road 2008   254-897-2960
PO Box 2189   888-775-6742
Glen Rose, TX 76043   Fax: 254-897-3785
              E-mail: vistor-services@fossilrim.org
                      http://www.fossilrim.org
Fossil Rim Wildlife Center is dedicated to conservation of species in peril, scientific research, training of professionals, creative management of natural resources, and impactful public education. Through these activities we providea diversity of compelling learning experiences which invoke positive change in the way people think, feel and act environmentally. Also provides scenic drives and lodgings for visitors, and is open all seasons.

Cheryl Joslin, Reservations

6107 GLOBE
Mailstop T28H
Moffett Field, CA 94035   800-858-9947
                              Fax: 650-604-1913

E-mail: help@globe.gov
http://www.globe.gov
GLOBE is a worldwide hands-on, primary and secondary school based science and education program.

6108 Glacier Institute
137 Main Street   406-755-1211
PO Box 1887   Fax: 406-755-7154
Kalispell, MT 59903   E-mail: register@glacierinstitute.org
                      http://www.glacierinstitute.org
The Glacier Institute serves adults and children as an educational leader in the Crown of the Continent ecosystem with Glacier National Park at its center. Emphasizing field based learning experiences, the Institute provides anobjective and science based understanding of the area's ecology and its interaction with people. Through this non advocacy approach to outdoor education, participants can be better prepared to make informed and constructive decisions which impactthis & other ecosystems.

Joyce Baltz, Executive Director

6109 Global Nest
Michigan State University
Institute of International Health   517-353-8992
B-301 W Fee Hall   Fax: 517-355-1894
East Lansing, MI 48824-1315   E-mail: secretary@gnest.org
                      http://www.NESTHOME.com
The Global Nest constitutes an international association of scientists, technologists, engineers and other interested groups involved in all scientific and technological aspects of the environment as well as in application techniquesaiming at the development of sustainable solutions. Its main target is to support and assist the dissemination of information regarding the most contemporary methods for improving quality of life through the development and application oftechnologies.

6110 Groundwater Foundation
PO Box 22558   402-434-2740
Lincoln, NE 68542-2558   800-858-4844
                              Fax: 402-434-2742
                    E-mail: info@groundwater.org
The Groundwater Foundation is a nonprofit organization that is dedicated to informing the public about one of our greatest hidden resources, groundwater. Since 1985, our programs and publications present the benefits everyone receivesfrom groundwater and the risks that threaten groundwater quality. We make learning about groundwater fun and understandable for kids and adults alike.

6111 Hazardous Chemicals: Handle With Care
Educational Images
PO Box 3456   607-732-1090
Elmira, NY 14905   800-527-4264
                              Fax: 607-732-1183
                    E-mail: info@edimages.com
              http://www.educationalimages.com
Shows the importance of hazardous chemicals in our daily lives and problems caused by ignorance, mistakes, accidents and occasionally, recklessness in their use. Four case studies show how toxic chemicals were introduced into theenvironment causing serious health and environmental effects. Video, 56 page guide with lesson plans, projects, reproducible handouts.

Charles R Belinky, CEO

6112 Hidden Villa
26870 Moody Road   650-949-8650
Los Altos Hills, CA 94022   Fax: 650-948-4159
                    E-mail: info@hiddenvilla.org
A non-profit 1600 acre organic farm and wilderness preserve serving approx. 50,000 visitors. Programs and offerings include: Hidden Villa Environmental Education Program, Summer Camp, a resident intern program, Community SupportedAgriculture, a Hostel for domestic and international travelers, meeting/retreat rental space, Community Programs and eight miles of hiking trails and picnic areas.
Beth Ross, Executive Director

6113 Ice Age Relics: Living Glaciers and Signs of Ancient Ice Sheets

Educational Images
PO Box 3456
Elmira, NY 14905

607-732-1090
800-527-4264
Fax: 607-732-1183
E-mail: info@edimages.com
http://www.educationalimages.com

Glaciers, living relics of the ice age, are still important today. They hold much of the earth's fresh water, sculpted much of North America, and promise an early return to finish their work. Provides a coherent picture of howglaciers work, and what they did. 26 page guide. Video or filmstrips.

Charles R Belinky, Ph.D, CEO

6114 Institute for Environmental Studies

UW Madison, 550 N Park Street
70 Science Hall
Madison, WI 53706-1491

608-263-1796
Fax: 608-262-2273

Few institutions can match the University of Wisconsin-Madison's expertise in environmental studies. Literally hundreds of professors teach and conduct research in environmentally related subjects ranging from agriculture to zoology.Their scholarship and achievement are widely recognized. In dozens of academic fields, the university is consistently rated the nation's best and most prolific.

6115 International Center for Earth Concerns

D162 Baldwin Road
Ojai, CA 93023

805-649-3535
Fax: 805-649-1757
E-mail: information@earthconcerns.org
http://www.earthconcerns.org

Dedicated to providing for public use, a world class botanic garden, outdoor learning-ecology center and a 50-passenger all-electric floating classroom on Lake Casitas. These facilities are used to promote a better understanding ofman's place in the environment, as well as to help develop a sense of respect, responsibility and compassion for animals and nature.

John Taft, Chairman

6116 Invertebrate Animal Videos

Educational Images
PO Box 3456
Elmira, NY 14905

607-732-1090
800-527-4264
Fax: 607-732-1183
E-mail: info@edimages.com
http://www.educationalimages.com

Four part series. Each a 40-minute multimedia presentation with easy going narration and hundreds of interactive links. Part I: sponges, anemones, corals and flatworms. Part II: molluscs, segmented worms and minor phyla. Part III:the insects. Part IV: noninsect arthropods and echinoderms.

Charles R Belinky, Ph.D, CEO

6117 Jones & Stokes

11820 Northup Way
Suite E300
Bellevue, WA 98005-1946

425-822-1077
Fax: 425-822-1079

An employee-owned company, Jones & Stokes is the best consulting source for integrated environmental planning and natural resources management services in the western United States.
Grant Bailey

6118 Killer Whales: Lords of the Sea

Educational Images
PO Boc 3456
Elmira, NY 14905

607-732-1090
800-527-4264
Fax: 607-732-1183
E-mail: info@edimages.com
http://www.educationalimages.com

Separates facts from myth about these majestic, maligned and usually misrepresented scagoing mammals: both wild and captive killer whales, their mental and physical powers, their feeding and reproductive behavior, physiology,sociology and echolocation. Information on other cetaceans is presented for comparison and better understanding. Provides scientific information and reports on ongoing research.

Charles R Belinky, Ph.D, CEO

6119 Legacy International

1020 Legacy Drive
Bedford, VA 24523

540-297-5982
Fax: 540-297-1860
E-mail: mail@legacyintl.org
http://www.legacyintl.org

Creates environments where people can address personal, community, and global needs while developing skills and effective responses to change.Whether working with youths, corporate leaders, educational professionals, entrepreneurs, orindividuals on opposing sides of a conflict, our goal is the same. Programs provide experiences, skills, and strategies that enable people to build better lives for themselves and others around them.
J Rash, President

6120 Lesley/Audubon Environmental Education Programs

Lesley University
29 Everett Street
Cambridge, MA 02138

617-349-8320
E-mail: info@lesley.edu
http://www.lesley.edu/gsass/75audubon.html

In partership with Audubon Expedition Institute in Belfast, Lesley University offers a Bachelor of Science degree in Environmental Studies, a Master of Science degree in Environmental Education and a Master of Science in EcologicalTeaching and Learning. Students travel throughout the US earning academic credit and gaining first-hand experience of environmental issues.

6121 Let's Grow Houseplants

Educational Images
Po Box 3456
Elmira, NY 14905

607-732-1090
800-527-4264
Fax: 607-732-1183
E-mail: info@edimages.com
http://www.educationalimages.com

Details different kinds of plants and their needs, when to water, selection, how to start from seeds and cuttings, how to make inexpensive pots, etc. Perfect to initiate an elementary classroom gardening project. 74 frames and guide.For elementary and preschool.

Charles R Belinky, Ph.D, CEO

6122 MacKenzie Environmental Education Center

Wisconsin Department of Natural Resources
W7303 County Highway CS
Poynette, WI 53955

608-635-8110
Fax: 608-635-8107
http://www.naturenet.com/mackenzie/

Located only 20 miles north of Madison, the MacKenzie Center offers a wide array of outdoor experiences. Five themed nature trails, prairie restorations, picnic area, nature study, three museums and a wildlife exhibit containing liveanimals that are native to Wisconsin, are here to help you gain a better understanding of our natural resources.

Derek A Duane, Director

6123 Marine Biological Laboratory

167 Water Street
Woods Hole, MA 02543

508-548-3705
Fax: 508-457-1924
E-mail: comm@mbl.edu

For more than a century, scientists from around the world have been gathering in Woods Hole. The best students from the best universities, the brightest young faculty, the most succesful scientists working at the pinnacle of theprofession, an unmatched collection of researchers and educators congregates every year in the seaside village whose name has become synonimous with science.

6124 Microscopic Pond

Educational Images
PO Box 3456
Elmira, NY 14905

607-732-1090
800-527-4264
Fax: 607-732-1183

THE MICROSCOPIC POND: This exquisitely photographed program introduces students to both the micrscopic plant and animal life of a pond. Various groups of algae are discussed and illustrated, including desmids, Pediastrum, Pithophora,Spyrogyra, Volvox, Nostac, calothrix, Bacillariophyseae, Dinophyseae, and amoebas (includeing Amoeba proteus), Arcella, the testaceans, and many others. With only a few exceptions, all of the organisms in this program were photographed live.

Charles R Belinky, Ph.D, CEO

6125 Mote Environmental Services

1600 Ken Thompson Parkway   941-388-4441
Sarasota, FL 34236   Fax: 941-388-4312

Mote Environmental Services offers consulting services focused on marine and coastal issues, where our expertise is strongest. We provide superior, results-oriented investigations and management planning service within our areas oftechnical and policy specialty. MESI is a wholly owned subsidary of Mote Marine Laboratory, an independent, nonprofit research and public education institution dedicated to excellence in marine and environmental sciences.

6126 Narrow Ridge Earth Literacy Center

1936 Liberty Hill Road   865-497-2753
Washburn, TN 37888   Fax: 865-497-2297
    E-mail: narrowr@korrnet.org

Provides experiential learning of Earth Literacy based on the cornerstones of spirituality, sustainability and community. The vision of Narrow Ridge is one of justice for human beings, animals and Earth, and sustainability in the waywe live our lives and create our institutions and technologies.

6127 National Environmental Health Association (NEHA)

720 South Colorado Boulevard   303-756-9090
Suite 970-S   Fax: 303-691-9490
Denver, CO 80246-1925   E-mail: staff@neha.org
    http://www.neha.org

NEHA is the only national association that represents all of environmental health and protection from terrorism and all-hazards preparedness, to food safety and protection and onsite wastewater systems. Over 4500 members and theprofession are served by the association through its Journal of Environmental Health, Annual Educational Conference & Exhibition, credentialing programs, research and development activities and other services.

Nelson Fabian, Executive Director

6128 National Institute of Environmental Health Sciences

111 TW Alexander Drive   919-541-3345
PO Box 12233   Fax: 919-541-4395
Research Triangle Park, NC 27709 http://www.niehs.nih.gov

The National Institute of Environmental Health Sciences is the principal federal agency for basic biomedical research on the health effects of environmental agents. It is the headquarters for the National Toxicology Program whichcoordinates toxicology studies within the Department of Health and Human Services.

Dr David A Schwartz, Director

6129 National PTA: Environmental Project

330 N Wabash Avenue   312-670-6782
Suite 2100   800-307-4782
Chicago, IL 60611   Fax: 312-670-6783
    E-mail: info@pta.org
    http://www.pta.org

The mission of the National PTA is to support and speak on behalf of children and youth in the school, in the community and before governmental bodies and other organizations; to assist parents in developing the skills they need toraise and protect their children and to encourage parent and public involvement in the public schools. Engages in advocacy and education, including workshops and lobbying.

Gabriella Hayes, Program Manager

6130 Natural Resources Conservation and Management

708 Garrigus Building   606-257-3275
Rural Sociology Department   E-mail: greider@uky.edu
Lexington, KY 40546-0215

As a trained professional, you will have a variety of challenging employment opportunities in public agencies and industry to contribute to sustained productivity and equality of all of our natural resources. In addition, some studentsfind that the Natural Resource Conservation and Management program satisfies their desire for a career in environmental education or environmental journalism.

6131 Nielsen Environmental Field School

4686 State Route 605 S   740-965-5026
Galena, OH 43021-9652   Fax: 740-965-5027

In 1990, the Nielsen Environmental Field School was created in reponse to a demand from the environmental industry for practically oriented, hands-on environmental field training.

6132 Northwest Environmental Education Council

3213 W Wheeler Street   206-352-1510
PMB 366   E-mail: emcwayne@nweec.org
Seattle, WA 98199   http://www.nweec.org

The Northwest Environmental Education Council increases environmental awareness, appreciation and stewardship by providing environmental education and science training opportunities for youth and adults.

6133 Northwest Interpretive Association

909 1st Avenue   206-220-4140
Suite 630   Fax: 206-220-4143
Seattle, WA 98104-3627   http://www.nwpubliclands.com

Works with public and management agencies to operate educational bookstores. Our mission is to provide visitors with information they need to learn about the nature and natural history of public lands so they can make wise choicesabout the lands use, preservation and protection. We accomplish our mission by selling educational and interpretive materials directly to visitors as well as returning net proceeds to the site where they were guarenteed to help fund other programs.

Mary Quackenbush, Executive Director

6134 Perkiomen Watershed Conservancy

1 Skippack Pike   610-287-9383
Schwenksville, PA 19473   Fax: 610-287-9237
    E-mail: pwc@pvwatershed.org
    http://www.perkiomenwatershed.org/

A nonprofit organization founded in 1964 by local citizens that works to protect the watershed of the Perkiomen Creek and its tributaries. This is accomplished through environmental education, conservation programs and watershedstewardship activities.

6135 Pollution Prevention Trust Fund

Dade County Dept. of Env. Res. Mgm.   305-372-6825
33 SW 2nd Avenue, #1200   Fax: 305-372-6760
Miami, FL 33130-1540

Receives monies from sources such as fees from pollution prevention events, grants, allocations, appropriations and workshop fees. These funds are then used in developing, promoting and conducting environmental workshops, expositions,symposia, conferences and other forms of public information for the purpose of educating industry, government and the public about pollution prevention.

Nichole Hefty

6136 Primary Ecological Succession

Educational Images
PO Box 3456   607-732-1090
Elmira, NY 14905   800-527-4264
    Fax: 607-732-1183
    E-mail: info@edimages.com
    http://www.educationalimages.com

An illustrated explanation of basic concepts of primary succession: the pioneer community; tolerant vs. intolerant species; stabilization; stratification and the climax community. Concise overview followed by classic, specificexamples of succession - on bare rock, on the sand dunes of Lake Michigan, on the outer banks of North Carolina - all explored in detail. 72 frames and guide.

Charles R Belinky, Ph.D, CEO

6137 Project Oceanology

1084 Shennecossett Road   860-445-9007
Avery Point   800-364-8472
Groton, CT 06340   Fax: 860-449-8008
    E-mail: info@oceanology.org
    http://www.oceanology.org

Project Oceanology is owned and operated by Interdistrict Committee for Project Oceanology and association of 25 educational institutions in Massachusetts, Rodhe Island, Connecticut and New York. Members of this associations includepublic school districts, private schools, states university, public and private colleges, a maritime museum and an aquarium. The project is governed by an assembly of delegates representing the member institutions.

Educational Resources & Programs/Workshops & Camps

6138 Project WILD
5555 Morningside Drive   713-520-1936
Suite 212   Fax: 713-520-8008
Houston, TX 77005   E-mail: info@projectwild.org
Project WILD is one of the most widely-used conservation and environmental education programs among educators of students in kindergarten through high school. Project WILD is based on the premise that young people and educators have a vital interest in learning about natural world.
Bill Andrews, Director
Amina Abdel-Haum, Program Assistant

6139 Resource-Use Education Council
The Virginia Natural Resources Education Guide
PO Box 11104   804-698-4442
Richmond, VA 23230   Fax: 804-698-4522
http://WWW.VRUEC.ORG/
In the mid 1950s, representatives from Virginia and federal natural resource agencies, along with professors in the colleges of education and resource management, came together as the Virginia Resource Use Education Council. For 35years, the VRUEC sponsored a summer conservation course for teachers at four of Virginia's colleges.
Ann Regn, Chairman

6140 Risk Management Internet Services
Managerial Technologies Corporation
900 Ogden Avenue   630-221-9116
#337   Fax: 630-964-2571
Downers Grove, IL 60515   E-mail: info@rmis.com
http://rmis.com
The rmFamily of sites and services is dedicated to bringing risk management related professions together with the consultants, developers and providers who service them.

6141 Ross & Associates Environmental Consulting,Ltd
1218 Third Avenue   206-447-1805
Suite 1207   Fax: 206-447-0956
Seattle, WA 98101   E-mail: rossmail@ross-assoc.com
Ross and Associates environmental consulting, is a small group of highly motivated professionals committed to helping environmental and natural resources agencies improve management programs and achieve better environmental results.

6142 SEEK
525 South Lake Avenue   218-529-6258
Suite 400   888-668-3224
Duluth, MT 55802
The SEEK directory works as a clearinghouse for all types of environmental education resources, from articles to lesson plans, from performances to displays, and many more. These resources come a variety of organizations throughoutMinnesota, including schools and colleges, government agencies, libraries and businesses.

6143 Sacramento River Discovery Center
1000 Sale Lane   530-527-1196
Red Bluff, CA 96080   Fax: 530-527-1312
E-mail: lgreen@tehama.k12.ca.us
http://www.srdc.tehama.k12.ca.us
The mission of the Sacramento River Discovery Center is to educate the public's school programs. Teacher professional development, camping, rafting and tourist events are available.
Lupe Green, Executive Director
Anna Draper, Program Manager

6144 Save the Dolphins Project Earth Island Institute
300 Broadway   415-788-3666
Suite 28   800-DOL-PHIN
San Francisco, CA 94133   http://www.earthisland.org
David Phillips

6145 Schlitz Audubon Nature Center
1111 E Brown Deer Road   414-352-2880
Milwaukee, WI 53217   Fax: 414-352-6091

A unique urban area of green just 15 minutes north of downtown Milwaukee, we are located along the shore of Lake Michigan. Escape from the world of concrete to hike seven miles of trails, walk along the beach and feel far away from thecity or view forests and wildlife from the 60-foot observation tower. Remember to bring your binoculars, you never know what you may want to take a closer look at while visiting the Center. New Sustainable Environmental Learning Center opened in2003!

6146 School of Public & Environmental Affairs
Indiana University
SPEA 240   812-855-0635
Bloomington, IN 47405   800-765-7755
The School of Public Environmental Affairs offer environmental science summer programs to high school students and middle/high school teachers who want answers to environmental questions.

6147 Science House
909 Capability Drive   919-515-6118
Suite 1200   Fax: 919-515-7545
Raleigh, NC 27695-8211   E-mail: science_house@ncsu.edu
http://www.science-house.org
The activities of The Science House is itself a partership of facultu and stuff from science and education departments across the NC State campus, and collaborates with many other k-12 support organization in North Carolina.

6148 Seacamp Association, Inc
1300 Big Pine Avenue   305-872-2331
Big Pine Key, FL 33043   Fax: 305-872-2555
E-mail: info@nhmi.org
http://www.nhmi.org
To create awareness of the complex and fragile marine world and to foster critical thinking and informed decision making about man's use of natural resources. One of the few organizations in the US providing experiential education inmarine studies to students aged 8 to 21 years.
Irene Hooper, Executive Director
Chuck Brand, NHMI Director at Seacamp

6149 Setting Up a Small Aquarium
Educational Images
PO Box 3456   607-732-1090
Elmira, NY 14905   800-527-4264
Fax: 607-732-1183
E-mail: info@edimages.com
http://www.educationalimages.com
Details the exact procedure to be followed in setting up either a marine or freshwater aquarium successfully. Methods are scientifically sound, well documented, and up-to-date. A scaled-down version of the methods used in largepublic aquaria, the system works! 90 frames and guide.

Charles R Belinky, Ph.D, CEO

6150 Sierra Club
85 Second Street   415-977-5500
2nd Floor   Fax: 415-977-5799
San Francisco, CA 94105   http://www.sierraclub.org
Aims to explore, enjoy, and protect the wild places of the earth, to practice and promote the responsible use of the earth's ecosystems and resources, to educate and enlist humanity to protect and restore the quality of the naturaland human environment, and to use all lawful means to carry out these objectives.

6151 Smithsonian Environmental Research Center
Edgewater, MD 21037   443-482-2218
Fax: 301-261-3415
The Smithsonian Environmental Research Center advances stewardship of the biosphere through interdisciplinary research and educational outreach. SERC's scientists study a variety of interconnected ecosystems at the Center's primaryresearch site here in Maryland, and at affiliated sites around the world.

6152 Society of Environmental Toxicology and Chemistry
SETAC N America Office
1010 N 12th Avenue   850-469-1500
Pensacola, FL 32501   Fax: 850-469-9778
E-mail: setac@setac.org
http://www.setac.org

The Society of Environmental Toxicology and Chemistry provides a forum for the examination of environmental issues by environmental professionals from industry, academia, government, and public-interest groups.
Mimi Meredith, Sr. Manager

6153 Southwest Environmental Health Sciences: Community Outreach and Education Program

PO Box 210207	520-626-3692
Tucson, AZ 85721-0207	Fax: 520-626-4468

E-mail: coep-info@pharmacy.arizona.edu

The COEP goals are to review, develop, and disseminate quality environmental health science curricula. Develop and host K-12 teacher training workshops, communicate with the general public about local and common environmental healthscience concerns, share research results from SWEHSC investigators with the COEP target audiences.

6154 Spiders in Perspective: Their Webs, Ways and Worth

Educational Images

PO Box 3456	607-732-1090
Elmira, NY 14905	800-527-4264
	Fax: 607-732-1183

Comprehensive coverage of the nature of spiders, their diversity of structures, and their remarkable behavior patterns. Presents the unique world of creatures you may have ignored before, but probably never will again. Coverslocomotion, the various perceptual senses, silk production, camouflage and mimicry, webs, hunting, predation by wasps, kleptoparasitism, courtship and reproduction, population densities, impact on humans, etc. 2 parts, 76 & 78 frames. Video, slidesor filmstrip.

Charles R Belinky, Ph.D., CEO

6155 Student Conservation Association

689 River Road	603-543-1700
PO Box 550	Fax: 603-543-1828
Charlestown, NH 03603	E-mail: internships@thesca.org
	http://www.thesca.org

America's largest and oldest provider of conservation service opportunities, outdoor education and career training for youth. SCA is building the next generation of conservation leaders and inspire lifelong stewardship of ourenvironment and communities.
Dale Penny, President
Shaundrea Kenyon, Operations Director/Recruitm

6156 Triumvirate Environmental

61 Inner Belt Road	617-628-8098
Somerville, MA 02143	Fax: 617-628-8099

E-mail: contactus@triumvirate.com

Triumvirate Environmental is a full-service environmental management firm headquartered in eastern Massachusetts. Serving the environmental and hazardous waste needs of clients throughout the northeast in the areas of biotechnology andpharmaceuticals, education, health care, metal platers and finishers, manufacturing, and utilities, Triumvirate Environmental is the industry leader in personalized service.

6157 US Environmental Protection Agency: Great Lakes National Program Office

77 West Jackson Boulevard	312-886-4040
Chicago, IL 60604	Fax: 312-353-2018

The focus for the State of the Lakes Ecosystem Conference (SOLEC) 1996 is the nearshore zone of the Great Lakes. Nearshore ecosystems are complex and dynamic with many measurable parameters. The nearshore area is extremely important tooverall ecosystem function. It is the most productive zone within each of the Great Lakes and is the area most affected by human activity. Nearshore zones include embayments, tributaries and tributary mouths, marshes and other wetlands, and dunes.

6158 Water Resources Management

550 N Park Street	608-263-2273
70 Science Hall	Fax: 608-262-2273
Madison, WI 53706-1491	E-mail: iesgrad@mail.ies.wisc.edu

The program addresses the complex, interdisciplinary aspects of managing resources by helping students integrate the biological and phisical sciences with engineering and law and the social sciences. The workshop provides anopportunity for students to work outside of the textbook environment and tackle a rea-world problem.

6159 Wilderness Education Association

Colorado University

Fort Collins, CO 80523	970-223-6252
	E-mail: wea@lamar.colostate.edu
	http://www.prienet.org/

David Cockrell, President

6160 Windows on the Wild

World Wildlife Fund

1250 24th Street Northwest	202-293-4800
Washington, DC 20037-1175	800-225-5993
	Fax: 202-293-9211

Provides educators with interdisciplinary curriculum materials and training programs. By using biodiversity as its organizing theme, WOW provides students with a unique window for exploring a range of topics including science,economics, social studies, language arts, geography and civics.
Jennifer A Zadwick, Program Information Coord.

6161 Young Entomologists Society

6907 West Grand River Avenue	517-886-0630
Lansing, MI 48906-9131	Fax: 517-886-0630
	E-mail: yesbugs@aol.com

http://www.members.@aol.com/YESbugs/mainmenu.html

To provide young people with a combination of programs, publications, and educational materials that enrich their insect and spider studies through dynamic, innovative, and enjoyable learning experience.
Gary Dunn, Director Of Education

Environmental

6162 ABS Consulting Training Services

http://www.absconsulting.com/svc_training.cfm
Government Institutes Division provides continuing education and practical information on government regulatory topics. We recognize that you face unique challenges presented by the ever-increasing number of new regulations and theresulting rapid evolution of new technologies.

6163 ASCE Geotechnical Engineering Seepage: Groundwater Modelling

http://www.ggsd.com/ggsd/index.cfm
This site contains a summary of links to sites describing software used for seepage/groundwater flow modeling.

6164 Academy of Natural Sciences

http://www.acnatsci.org
Our mission is to create the basis for a healthy and sustainable planet through exploration, research and education.

6165 ActiveSet.org

http://www.activeset.org
ActiveSet.org was created for the benefit of professionals involved with all aspects of environmental air quality testing, monitoring and management. Environmental Managers can easily research and contact emissions testing firms,services and products either by state, region, or using the site's built-in search engine. Facility owners and managers use ActiveSet.org's free online request for proposals form to reach stack testing firms all over the country for their nextproject.

6166 Adirondack Council

http://www.adirondackcouncil.org
The Adirondack Council is a nonprofit environmental group working to protect the open space resources of New York's six million acre Adirondack Park and to help sustain the natural and human communities of the region. It monitorsdevelopment on private lands and ensures the mandated constitutional protection of public lands.

6167 Advanced Forest Technologies Program

http://www.pfc.forestry.ca/index_e.html
Advanced Forest Technologies Program purpose is to develop new approaches in the areas of remote sensing, geographic information systems, artifical intelligence, expert systems and decision support systems to: assist the resource andenvironmental manager with integrated resource planning; provide the land manager with tools for sustainable development.

6168 Advanced Recovery

http://www.advancedrecovery.com
Advanced Recovery, based in New Jersey, is involved in the recycling industry. They promote the proper disposal of all scrap, but are particularly interested in the disposal of electronic equipment, such as computer monitors consistingof lead.

6169 Advanced Technologies for Commercial Buildings

http://www.advancedbuildings.org
A building professional's guide to more than 90 environmentally-appropriate technologies and practices. Architects, engineers and buildings managers can improve the energy and resource efficiency of commercial, industrial andmulti-unit residential buildings through the use of the technologies and practices described in this web site.

6170 Advanced Technology Environmental Education Center

http://www.ateec.org
ATEEC's mission is the advancement of environmental technology education through curriculum development, professional development and program improvement in the nation's community colleges and high schools. ATEEC is funded by theNational Science Foundation and is a partnership of the Hazardous Materials Training and Research Institute, the National Partnership for Environmental Technology Education, and the University of Northern Iowa.

6171 African Environmental Research and Consulting Group

http://www.africaenviro.org
The AERCG is a US nonprofit organization with offices in Africa and Western countries. The group focuses on improving the quality of life, mitigating environmental hazards and the protection of human health in Africa. They are involvedin promoting sustainable development in African communities, their site contains information about their efforts.

6172 Agency for Toxic Substances and Disease Registry

http://www.atsdr.cdc.gov/
The mission of the Agency for Toxic Substances and Disease Registry, as an agency of the US Department of Health and Human Services, is to prevent exposure and adverse human health effects and diminished quality of life associated withexposure to hazardous substances from waste sites, unplanned releases and other sources of pollution present in the environment.

6173 Agriculture Network Information Center

http://www.agnic.org
The Agriculture Network Information Center is a voluntary alliance of the National Agricultural Library, land-grant universities and other agricultural organizations in cooperation with citizen groups and government agencies. AgNICfocuses on providing agricultural information in electronic format over the World Wide Web via the Internet.

6174 Air Force Center for Environmental Excellence

http://www.afcee.brooks.af.mil
The Air Force Center for Environmental Excellence provides our customers with a complete range of world class enviromental, architectural and landscape design, planning and construction management services and products.

6175 Air and Waste Management Association

http://www.awma.org
The Air and Waste Management Association is a nonprofit, nonpartisan professional organization that provides training, information and networking opportunities to thousands of environmental professionals in 65 countries.

6176 Alabama Department of Environmental Management

http://www.adem.state.al.us
Provides environmental stewardship through the implementation of authorized environmental statutes, advocating statutory change as needed.

6177 Alaska Chilkat Bald Eagle Preserve

http://www.dnr.state.ak.us/parks/units/eagleprv.htm
The Alaska Chilkat Bald Eagle Preserve was created by the State of Alaska in June 1982. The Preserve was established to protect and perpetuate the world's largest concentration of bald eagles and their critical habitat. It alsosustains and protects the natural salmon runs.

6178 Albany Research Center

http://www.alrc.doe.gov
Our mission is to provide stewardship for the Nation's mineral resources by conserving materials produced from minerals.

6179 Alfred Wegener Institute for Polar and Marine Research

http://www.awi-bremerhaven.de/index-e.html
Polar and Marine research are central themes of Global system and Environmental Science. The Alfred Wegener Institute conducts research in the Arctic, the Antarctic and at temperate latitudes. It coordinates Polar research in Germanyand provides both the necessary equipment and the essential logistic back up for polar expeditions. Website is in the German language.

6180 Alliance for Environmental Technology

http://www.aet.org
The Alliance for Environmental Technology is an international association of chemical manufacturers and forest products companies dedicated to improving the environmental performance of the pulp and paper industry. AET supports the useof Elemental Chlorine-Free technology based on chlorine dioxide.

6181 American Academy of Environmental Engineers

http://www.aaee.net/
The American Academy of Environmental Engineers is dedicated to excellence in the practice of environmental engineering to ensure the public health, safety and welfare to enable humankind to co-exist in harmony with nature.

6182 American Chemical Society

http://www.chemistry.org/portal/a/c/s/l/home.html
The mission is to promote the public perception and understanding of chemistry and the chemical sciences through public outreach programs and public awareness campaigns; involve the Society's more than 163,000 member's in improving thepublic's perception of chemistry.

6183 American Conference of Governmental Industrial Hygienists

http://www.acgih.org/home.htm
Member-based organization and community of professionals that advances worker health and safety through education and the development and dissemination of scientific and technical knowledge.

6184 American Council for an Energy-Efficient Economy

http://www.aceee.org
The American Council for an Energy-Efficient Economy is a nonprofit organization dedicated to advancing energy efficiency as a means of promoting both economic prosperity and environmental protection.

6185 American Farmland Trust

http://www.farmland.org
American Farmland Trust is a private nonprofit organization founded in 1980 to protect our nation's farmland. AFT works to stop the loss of productive farmland and to promote farming practices that lead to a healthy environment.

6186 American Forests

http://www.americanforests.org
American Forests is a world leader in planting trees for environmental restoration, a pioneer in the science and practice of urban forestry and a primary communicator of the benefits of trees and forests.

6187 American Geophysical Union

http://earth.agu.org
AGU's mission is to promote the scientific study of Earth and its environment in space and to disseminate the results to the public, to promote cooperation among scientific organizations involved in geophysics and related disciplinesand to initiate and participate in geophysical research programs.

6188 American Hydrogen Association

http://www.clean-air.org
The Mission of AHA is to facilitate achievements of prosperity without pollution and to close the information gap between researchers, industry and the public, drawing on world-wide developments concerning hydrogen, solar, wind, hydro,ocean and biomass resource materials, energy conversion, wealth-addition economics and the environment.

6189 American Rivers

http://www.americanrivers.org/site/PageServer
American Rivers is a national nonprofit conservation organization dedicated to protecting and restoring America's rivers and to fostering a river stewardship ethic.

6190 American Solar Energy Society

http://www.ases.org
The American Solar Energy Society is a national organization dedicated to advancing the use of solar energy for the benefit of US citizens and the global environment. ASES promotes the widespread near-term and long-term use of solarenergy.

6191 Ames Laboratory: Environment Technology Department

http://www.etd.ameslab.gov
The Ames Laboratory's Environmental & Protection Sciences Program is playing an important role in the US Department of Energy's initiative to cleanup hazardous waste, responding to remediation problems that need faster, safer, betteror cheaper technological solutions. You'll find information here about those technologies, the scientists behind them and our efforts to move these technologies into the marketplace.

6192 Antarctic and Southern Ocean Coalition

http://www.asoc.org
The Antartic Project is the Secretariat of the Antartic and Southern Ocean Coalition which contains nearly 230 organizations in 49 countries and leads the national and international campaigns to protect the biological diversity andpristine wilderness of Antartica, including its oceans and marine life. We work for passage of strong measures which protect the marine ecosystem from the harmful effects of overfishing, and work to ensure that the integrity of the land and animalsis maintained.

6193 Appropriate Technology for Community and Environment

http://www.pactok.net.au/docs/apace/home.htm
The mission is to undertake activities in developing countries for the purposes of ecologically sustainable development and the relief of proverty. To undertake research into and development of technologies appropriate to particularenvironmental and social circumstances.

6194 Argonne National Laboratory

http://www.anl.gov
Argonne National Laboratory is one of the US Department of Energy's largest research centers. It is also the nation's first national laboratory, chartered in 1946.

6195 Arizona Geological Survey

http://www.azgs.state.az.us/
To inform and advise the public about the geologic character of Arizona in order to foster understanding and prudent development of the State's land, water, mineral and energy resources.

6196 Arkansas Natural Heritage Commission

http://www.naturalheritage.org

The Arkansas Natural Heritage Commission (ANHC) is responsible for maintaining the most up-to-date and comprehensive source of information concerning the rare plant and animal species, and high-quality natural communities of Arkansas. Systematic analysis of this natural heritage data can be used to identify locations that hold exceptional importance for the state's natural diversity, but that lack formal protection.

6197 Asia-Pacific Centre for Environmental Law

http://http://law.nus.edu.sg/apcel

APCEL was established in response to the need for capacity-building in environmental legal education and the need for promotion of awareness in environmental issues. It is currently working closely with IUCN's Commission onEnvironmental Law.

6198 Association of Energy Engineers

http://www.aeecenter.org

The administration's energy plan slashes R&D funding for energy efficiency technologies by $160 million. In addition, the US Department of Energy Federal Energy Management Program's budget has been reduced by 48%. This plan does notrecognize the valuable contributions energy efficiency technologies have made in the last decade.

6199 Association of State Flood Plain Managers

http://www.floods.org/home/default.asp

Promotes common interest in flood damage abatement, supports environmental protection for floodplain areas, provides education on floodplain management practices and policy and urges incorporating multi-objective management, approaches to solve local flooding problems.

6200 Associations of University Leaders for a Sustainable Future

http://www.ulsf.org

The mission of the Association of University Leaders for Sustainable Future is to make sustainability a major focus of teaching, research, operations and outreach at colleges and universities worldwide. ULSF pursues this missionthrough advocacy, education, research, assessment, membership support and international partnerships to advance education for sustainability.

6201 Atlantic Salmon Federation

http://www.asf.ca

The Atlantic Salmon Federation is an international nonprofit organization which promotes the conservation and wise management of the wild Atlantic salmon and its environment.

6202 Australian Cooperative Research Centres

http://www.crc.gov.au/Information/default.aspx

The Cooperative Research Centers, generally known as CRCs, bring together researchers from universities, CSIRO and other government laboratories and private industry or public sector agencies in long-term collaborative arrangementswhich support research and development and education activities that achieve real outcomes of national economic and social significance.

6203 Australian Oceanographic Data Centre

http://www.aodc.gov.au/about.html

The mission of the Australian Oceanographic Data Centre is to acquire, manage and distribute oceanographic information and provide specialist oceanographic advice to; enable the Australian Defence Force to exploit the above and belowwater physical operating environments for strategic, operational and tactical advantage meet national and international obligations to manage oceanographic information.

6204 Bat Conservation International

http://www.batcon.org

The mission of Bat Conservation International is to protect and restore bats and their habitats worldwide.

6205 Battelle Seattle Research Group

http://www.seattle.battelle.org/

Battelle Memorial Institute is a multidimensional organization dedicated to making the future better for everyone. Although that may sound a bit grand, it really is Battelle's mission. From 1929 to the present, putting innovation andtechnology to practical use has been our goal.

6206 Bellona Foundation

http://www.bellona.no

Bellona Foundation on the web brings you news and background on important environmental issues.

6207 Benton Foundation

http://www.benton.org/

Since 1981, the Foundation has worked to articulate a public interest vision for the digital age and to demonstrate the value of communications for solving social problems. Through its projects, the foundation bridges the worlds ofphilanthropy, public policy and community action to promote the use of digital media to engage, equip and connect people for social change.

6208 Best Manufacturing Practices Center of Excellence

http://www.bmpcoe.org

The Office of Naval Research's Best Manufacturing Practices Program is a unique, innovative technology transfer effort that improves the competitiveness of the US industrial base both here and abroad. The main goal at BMP is toincrease the quality, reliability and maintainability of goods produced by American firms.

6209 Biocatalysis/Biodegradation Database

http://www.umbbd.ahc.umn.edu/

This database contains information on microbial biocatalytic reations and biodegradation pathways for primarily xenobiotic, chemical compounds. The goal of the UM-BBD is to provide information on microbial enzyme catalyzed reactionsthat are important for biotechnology.

6210 Bioelectromagnetics Society

http://www.bioelectromagnetics.org/

The Bioeletromagnetics Society was established in 1978 as an independent organization of biological and physical scientists, physicians and engineers interested in the interactions of non-ionizing radiation with biological systems.BEMS is incorporated as a nonprofit organization in the District of Columbia and is registered with the Internal Revenue Service as a educational and training organization.

6211 Biology: Careers and Jobs

http://ublib.buffalo.edu/libraries/asl/guides/bio/bio_careers.html

SEL librarians have developed a comprehensive inventory of career and job-finding resources. The SEL Career Information Site provides access to general and subject-specific resources for assisting you in finding resources related tojobs and careers.

6212 Birding on the Web

http://www.birder.com

Birding is the most extensive section of this site. You will find checklists which span the globe, birding Hot Spots, and rare bird alert phone numbers. In the Backyard Birders section you can also find information on seeds to attractbirds, building bird houses, and links to home pages of other bird watchers.

6213 British Atmospheric Data Centre

http://www.badc.rl.ac.uk/

The role of the BADC is to assist UK atmosperic researchers to locate, access and interpret atmospheric data to ensure the long-term integrity of atmospheric data produced by NERC projects. The BADC has substantial data holdings of itsown and also provides information and links to data held by other data centres.

6214 Brookhaven National Laboratory

http://www.bnl.gov/world/Default.asp

The department of Energy's Brookhaven National Laboratory conducts research in physical, biomedical and environmental sciences, as well as in energy technologies. Brookhaven also builds and operates major facilities available touniversity, industrial and government scientists.

6215 Brown is Green

http://www.brown.edu/Departments/Brown_Is_Green/

The listserv dicussion lists provide for some good conservation, conjecture and occasional facts and figures, but are not reliable sources of information for students, faculty and staff working on campus environmental programs. The WebServer is our attempt at making current information on Brown's and other University campus' environmental programs freely available over the internet.

6216 Bureau of International Recycling

http://www.bir.org

BIR is an international trade federation representing the world's recycling industry, covering in particular ferrous and non-ferrous metals, paper and textiles. Plastics, rubber, tires and glass are also studied and traded by some BIRmembers.

6217 Bureau of Reclamation

http://www.usbr.gov

The mission of the Bureau of Reclamation is to manage, develop and protect water and related resources in an environmentally and economically sound manner in the interest of the American public.

6218 Bureau of Transportation Statistics

http://www.bts.gov

The 1991 Intermodal Surface Transportation Efficiency Act established the Bureau of Transportation Statistics for data collection, analysis and reporting and to ensure the most cost-effective use of transportation-monitoring resources.We strive to increase public awareness of the nation's transportation system and its implications and improve the transportation knowledge base of decision makers.

6219 Business & Legal Reports

http://www.blr.com/

Business & Legal Reports has been helping employers avoid legal problems for 25 years. Human Resources, Compensation, Environmental and Safe managers know that they can count on our compliance and training products to keep them out oftrouble. BLR's attorneys are constantly researching federal and state legislation, best practices, industry trends and impending changes that can affect your organization.

6220 California Conservation Corps

http://www.ccc.ca.gov/cccweb/index.htm

The CCC is the oldest, largest and longest-running youth conservation corps in the world! Nearly 90,000 young men and women have worked more than 50 million hours to protect and enhance California's environment and communities and haveprovided six million hours of assistance with emergencies like fires, floods and earthquakes. We're proud of our accomplishments and hope you are too!

6221 California Energy Commission

http://www.energy.ca.gov

The California Energy Commission is the state's primary energy policy and planning agency. It was created by the Legislature in 1974 and located in Sacramento.

6222 California Environmental Protection Agency

http://www.calepa.ca.gov

The mission of the Cailfornia Environmental Protection Agency is to restore, protect and enhance the environment, to ensure public health, environmental quality and economic vitality.

6223 California Environmental Resources Evaluation System (CERES)

http://http://resources.ca.gov/

The goal of CERES is to improve environmental analysis and planning by integrating natural and cultural resource information from multiple contributors and by making it available and useful to wide variety of users. CERES collects andintegrates data and information and distributes it via the World Wide Web.

6224 California League of Conservation Voters

http://www.ecovote.org/

The California League of Conservation Voters is the nation's largest and oldest state political action organization for the environment. Founded in 1972, the League mobilizes California voters to support environmentally responsiblecandidates and issues and serves as a watchdog to hold elected officials accountable for their environmental votes.

6225 California Resources Agency

http://http://resources.ca.gov/

The California Resources Agency is responsible for the conservation, enhancement and management of California's natural and cultural resources, including land, water, wildlife, parks, minerals and historic sites. The Agency is composedof departments, boards, conservations, commissions and programs.

6226 Canadian Chlorine Chemistry Council

http://www.cfour.org/cms/

To facilitate dialogue and promote coordinated action in Canada among key stakeholders in order to bring about a balanced view of chlorine chemistry to enable society to make informed, science based decisions on issues involvingchlorine.

6227 Canadian Council of Ministers of the Environment (CCME)

http://www.ccme.ca

CCME works to promote cooperation on and coordination of interjurisdictional issues such as waste management, air pollution and toxic chemicals. CCME members propose nationally-consistent environmental standards and objectives so as toachieve a high level of environmental quality across the country.

6228 Canadian Environmental Assessment Agency

http://www.ceaa.gc.ca/index_e.htm

To provide Canadians with high quality environmental assessments that contribute to informed decision-making in support of sustainable development.

6229 Canadian Institute for Environmental Law and Policy (CIELAP)

http://www.cielap.org

To provide leadership in the research and development of environmental law and policy that promotes the public interest and principles of sustainability.

6230 Carbon Dioxide Information Analysis Center (CDIAC)

http://www.cdiac.esd.ornl.gov/home.html

CDIAC responds to data and information requests from users all over the world who are concerned with the greenhouse effect and global climate change. CDIAC's data holdings include records of the concentrations of carbon dioxide andother radiatively active gases in the atmosphere.

6231 Carnegie Institute of Technology, Department of Civil & Environmental Engineering

http://www.ce.cmu.edu

Carnegie Mellon's Department of Civil and Environmental Engineering is a part of the engineering college, Carnegie Institute of Technology. The department maintains a commitment to excellence and innovation in education and research.

6232 Center for Disease Control

http://www.cdc.gov

To promote health and quality of life by preventing and controlling disease, injury and disability. CDC seeks to accomplish its mission by working with partners throughout the nation and world to monitor health, detect and investigatehealth problems, conduct research to enhance prevention, develop and advocate sound public health policies, implement prevention strategies, promote healthy behaviors, foster safe and healthful environments and provide leadership and training.

6233 Center for Environmental Biotechnology

http:////www.ceb.utk.edu/

The Center for Environmental Biotechnology at the University of Tennessee, Knoxville was established in 1986 to foster a multidisciplinary approach for training the next generation of environmental scientists and solving environmentalproblems through biotechnology.

6234 Center for Environmental Citizenship

http://www.envirocitizen.org/

The Center for Environmental Citizenship is a national nonpartisan 501 organization. We were founded by young activists in 1992 to encourage college students to be environmental citizens. CEC is dedicated to educating, training andorganizing a diverse national network of young leaders to protect the environment.

6235 Center for Environmental Design Research College of Environmental Design

http://www.cedr.berkeley.edu/

The Center for Environmental Design Research is an Organized Research Unit of the University of California at Berkeley. The Center's mission is to encourage research in environmental planning and design, in order to increase thefactual content of design decisions and to promote systematic approaches to design decision making.

6236 Center for Health Effects of Environmental Contamination

http:////www.cheec.uiowa.edu/

The University of Iowa Center for Health Effects of Environmental Contamination supports and conducts research to identify, measure and prevent adverse health outcomes related to exposure to environmental toxins. CHEEC organizes andparticipates in educational and outreach programs, provides environmental health expertise to local, state and federal entities and serves as a resource to Iowans in the field of environmental health.

6237 Center for International Earth Science Information Network (CIESIN)

http://www.ciesin.org

CIESIN works at the intersection of the social, natural and information services, specializing in on-line data and information management, spatial data integration and training and interdisciplinary research related to humaninteractions in the environment.

6238 Center for International Environmental Law

http://www.ciel.org

The Center for International Environmental Law is a public interest nonprofit environmental law firm founded in 1989 to strengthen international and comparative environmental law and policy around the world. CIEL provides a full rangeof environmental legal services in both international and comparative national law.

6239 Center for Plant Conservation

http://www.centerforplantconservation.org/

The CPC is a consortium of 28 American botanical gardens and arboreta whose mission is to conserve and restore the rare native plants of the US. To meet this end, they are involved in plant conservation, research and education. Thissite includes information about the National Collection of Endangered Plants which is maintained by the group.

6240 Center for Renewable Energy and Sustainable Technology

http://www.crest.org

CREST's goal is to accelerate the use of renewable energy by providing credible information, insightful analysis and innovative strategies amid changing energy markets and mounting environmental needs. The combined CREST organizationboasts a strong platform for research, publication and dissemination of timely information regarding sustainable energy.

6241 Center for Resourceful Building Technology

http://www.crbt.org/

CRBT is a project of the National Center for Appropriate Technology, NCAT. CRBT is dedicated promoting environmentally responsible practices in construction. It works to serve as both catalyst and facilitator in encouraging buildingtechnologies which realize a sustainable and efficient use of resources.

6242 Central European Environmental Data Request Facility

http://www.cedar.at/sitemap/htm

CEDAR was created to provide computing and Internetworking facilities to support international data exchange with the Central and Eastern European environmental community. Focusing at first on mainly Central and Eastern Europeancountries, CEDAR's activities expanded quickly to an audience all over the world.

6243 Centre for Alternative Transportation Fuels

http://http://catf.bcresearch.com/catf/catf.nsf

The CATF is operated by BC Research, its staff have maintained a database of technical papers since 1995 to support the tenchical and business community in the rapidly moving area of alternative transportation fules.

6244 Centre for the Analysis and Dissemination of Demonstrated Energy Technologies

http://www.caddet.org

A unique source of global information on proven commercial applications covering the full range of energy-saving technologies.

6245 Cetacean Society International

http://csiwhalesalive.org/

CSI is all volunteer nonprofit conservation, education and research organization based in the USA, with volunteer representatives in 26 countries around the world. The goal of the Cetacean Society International is to achieve on aglobal basis the optimum utilization of cetacean resources through benign utilization and the elimination of all killing and captive display of whales, dolphins, and porpoises. Our ultimate aim is peaceful coexistence and mutual enrichment for humansand cetaceans.

6246 Chanslor Wetlands Wildlife Project

http://www.sonomawetlands.org

Chanslor Wetlands Wildlife Project protects 250 acres of crucial habitat adjoining the historic fishing community of Bodega Bay just 1.25 hours north of San Francisco. Dedicated in 1973, the Chanslor Wetlands encompasses a rarebrackish marsh as well as freshwater marshes, vernal pools and ponds bordered by Salmon Creek.

6247 Charles Darwin Research Station

http://www.darwinfoundation.org/

Our mission is to conduct scientific research and environmental education about conservation and natural resource management in the Galapagos archipelago and its surrounding Marine Reserve. Scientific research and monitoring projects are conducted at the CDRS in conjunction and cooperation with our chief partner, the Galapagos National Park Service, which functions as the principal government authority in charge of conservation and natural resource issues in the Galapagos.

6248 Chemcyclopedia

http://www.mediabrains.com/client/chemcyclop/BGI/search.asp

Chemcyclopedia, an annual supplement to Chemical & Engineering News, provides users and purchasers of chemicals with a unique opportunity to obtain information that aids in making buying decisions. In addition to the listing of a given chemical, suppliers have been asked to provide trade names, packaging, special shipping requirements, potential applications and CAS Registry Numbers, if available.

6249 Chemical Industry Institute of Toxicology

http://www.ciit.org

Founded in 1974, CIT is a nonprofit toxicology research institute dedicated to providing an improved scientific basis for understanding and assessing the potential adverse effects of chemicals, pharmaceuticals and consumer products on human health.

6250 Chicago Wilderness

http://www.chiwild.org

The lands stretching south and west from the shores of Lake Michigan hold one of North America's great metropolises. More than nine million people live in northwestern Indiana, northeastern Illinois and southeastern Wisconsin. Living among them, on islands of green, are thousands of species of native plants and animals-species that make up some of the rarest natural communities on earth. We call these communities and the lands and waters that are their homes Chicago Wilderness.

6251 China Council for International Cooperation on Environment and Development

http://www.iisd.org/trade/cciced/

The China Council for International Cooperation on Environment & Development will continue to act as a bridge between China and other countries in cooperation on environment and development by introducing useful experiences of other countries to China and communicating to the world the determination and aspiration of the Chinese Government and people for sustainable development.

6252 Chlorine Chemistry Council

http://www.c3.org

The Chlorine Chemistry Council, a business council of the American Chemistry Council, is a national trade association based in Arlington, VA representing the manufacturers and users of chlorine and chlorine-related products. Chlorine is widely used as a diease-fighting disinfection agent, as a basic component in pharmaceuticals and myriad other products that are essential to modern life.

6253 City Farmer

http://www.cityfarmer.org/

Our nonprofit society promotes urban food production and environmental conservation from a small office in downtown Vancouver, British Columbia and from our demonstration food garden in nearby Kitsilano, a residential neighborhood.

6254 Climate Change and Human Health

http://www.jhu.edu/~climate/

The Climate Change and Human Health Integrated Assessment Web provodes recent and relevant information about the potential impacts of climate change through integrated assessment.

6255 Coastal Conservation Association

http://www.joincca.org

CCA is a national organization dedicated to the conservation and preservation of marine resources.

6256 Code of Federal Regulations

http://www.gpoaccess.gov/cfr/

The Code of Federal Regulations is a codification of the general and permanent rules published in the Federal Register by the Executive departments and agencies of the Federal Government.

6257 Colorado Department of Natural Resources

http://www.dnr.state.co.us/index.asp

The Colorado Department of Natural Resources was created to develop, protect and enhance Colorado natural resources for the use and enjoyment of the state's present and future residents, as well as for visitors to the state.

6258 Colorado School of Mines

http://www.mines.edu/index_js.shtml

The Colorado School of Mines shall be a specialized baccalaureate and graduate research institution with high admission standards. The Colorado School of Mines shall have a unique mission in energy, mineral and materials science and engineering and associated engineering and science fields.

6259 Columbia Earth Institute: Columbia University

http://www.earthinstitute.columbia.edu

The Earth Institute at Columbia University is the world's leading academic center for the integrated study of Earth, its environment and society. The Earth Institute builds upon excellence in the core disciplines-earth sciences, biological sciences, engineering sciences, social sciences and health sciences-and stresses cross-disciplinary approaches to complex problems.

6260 Connecticut Department of Environmental Protection

http://http://dep.state.ct.us/

The mission of the Department of Environmental Protection is to conserve, improve and protect the natural resources and environment of the State of Connecticut while preserving the natural environment and the life forms it supports in a delicate, interrelated and complex balance, to the end that the state may fulfill its responsibility as trustee of the environment for present and future generations.

6261 Conservation Fund

http://www.conservationfund.org

Works with private and public agencies and organizations to protect wildlife habitats, historic sites and parks.

6262 Conservation International

http://www.conservation.org/

Our mission is to conserve the Earth's living natural heritage, our global biodiversity and to demonstrate that human societies are able to live harmoniously with nature.

6263 Conservation Treaty Support Fund

http://www.conservationtreaty.org

The unique mission of the Conservation Treaty Support Fund is to support major inter governmental treaties which conserve wild natural resources for their own sake and the benefit of people. The fund believes these undertakings have the best potential for global conservation, because they stem from the will of the nations of the world, are premised on the goal of sustaining living natural resources and have created a framework for effective conservation supported by many agencies.

6264 Consortium on Green Design and Manufacturing (CGDM)

http://http://cgdm.berkeley.edu/

The Consortium on Green Design and Manfacturing (CGDM) is an interdisciplinary research initiative at the University of California, Berkeley and an industry/government/university partnership to develop linkages between manufacturingand design and their environmental effects and to integrate engineering information, management practices and government policy-making.

6265 Consultative Group on International Agricultural Research (CGIAR)

http://www.cgiar.org

To contribute to food security and poverty eradication in developing countries through research, partnerships, capacity building and policy support. Promotes sustainable agricultural development based on the environmentally soundmanagement of natural resources.

6266 Coral Health and Monitoring Program (CHAMP)

http://coral.aoml.noaa.gov

The mission of the Coral Health and Monitoring Program is to provide services to help improve and sustain coral reef health throughout the world.

6267 Coral Reef Alliance

http://www.coralreefalliance.org/

The Coral Reef Alliance (CORAL) is a member supported, non-profit organization dedicated to keeping coral reefs alive around the world. Coral reefs are one of nature's most magnificent creations-filled with thousands of unique andvaluable plants and animals. CORAL works with marine park managers, businesses and communities to help increase their capacity to protect their local coral reefs.

6268 Cornell University Center for the Environment

http://www.cfe.cornell.edu

The Cornell Center for the Environment is committed to research, teaching and outreach focused on environmental issues, with the goals of enhancing the quality of life, encouraging economic vitality and promoting the conservation ofnatural resources for sustainable future.

6269 Council for Agricultural Science & Technology (CAST)

http://www.cast-science.ogr/

CAST is a nonprofit organization composed of scientific societies and many individual, student, company, nonprofit and associate society members. CAST's Board of Directors is composed of representatives of the scientific societies andindividual members as well as an Executive Committee. CAST assembles, interprets and communicates science based information regionally, nationally and internationally on food, fiber, agricultural, natural resources and related environmental issues tostakeholders.

6270 Council on Environmental Quality (CEQ)

http://ceq.eh.doe.gov or www.whitehouse.gov/CEQ/

The Council on Environmental Quality coordinates federal environmental efforts and works closely with agenices and other White House offices in the development of environmental policies and initiatives.

6271 Coweeta LTER Site

http://coweeta.ecology.uga.edu/

The program was developed to support research of ecological phenomena that occur on time scales of decades or centuries, periods of time normally investigated with research support from National Science Foundation.

6272 CropLife Canada

http://www.cropro.org/

To support sustainable agriculture in Canada, in cooperation with others, by building trust and appreciation for plant life science technologies.

6273 Declining Amphibian Populations Task Force

http://www.open.ac.uk/daptf/

Established in 1991, the DAPTF consists of a network of over 3,000 scientists and conservationists belonging to national and regional working groups which now cover more than 90 countries around the world. The mission of the DAPTF isto determine the nature, extent and causes of declines of amphibians throughout the world and to promote means by which declines can be halted or reversed.

6274 Defenders of Wildlife

http://www.defenders.org

Defenders of Wildlife is dedicated to the protection of all native wild animals and plants in their natural communities. We focus our programs on what scientis consider two of the most serious environmental threats to the planet:the accelerating rate of extinction of species and the associated loss of biological diversity and habitat alteration and destruction.

6275 Defense Technical Information Center (DTIC)

http://www.dtic.mil

To improve the productivity of those who use scientific and technical information to accomplish a Defense mission objective, DTIC manages 13 Information Analysis Centers staffed by experienced information specialists, scientists andengineers who help customers locate, analyze and use scientific and technical information in a specialized subject area.

6276 Delaware Department of Natural Resources and Environmental Control
DNREC

http://www.dnrec.state.de.us

Delaware Department Of Natural Resources and Environmental Control's mission is to protect Delaware's environment for future generations.

6277 Department of Conservation

http://www.conservation.state.mo.us

The mission is to protect and manage the fish, forest and wildlife resources of the state; to serve the public and facilitate their participation in resource management activities; to provide opportunity for all citizens to use, enjoyand learn about fish, forest and wildlife resources.

6278 Department of Defense Environmental Cleanup Home Page

http://www.dtic.mil/envirodod/

The mission of the Environmental Cleanup program is to protect the environment while reducing risks to US troops, their families and local communities from pollutants due to past practices.

6279 Department of Energy

http://www.doe.gov/engine/content.do

The Department of Energy's mission is to foster a secure and reliable energy system that is environmentally and economically sustainable, to be a responsible steward of the nation's nuclear weapons, to cleanup our own facilities and tosupport continued US leadership in science and technology.

6280 Department of the Interior

http://www.doi.gov

The Interior Department has had a wide range of responsibilities enstrusted to it: the construction of the national capital's water system, the colonization of freed slaves in Haiti, exploration of western wilderness, oversight of theDistrict of Columbia jail, regulation of territorial governments, management of hospitals and universities, management of public parks, the basic responsibilities for Indians, public lands, patents and pensions.

6281 DiveWeb

http://www.sandiegodiving.com/resources/links/items/277.html

DiveWeb is the original comprehensive online resource for information about commerical diving, ROV, marine technology, offshore/telecommunications and inland/coastal underwater industries.

6282 EDIE: Environmental Data Interactive Exchange

http://www.edie.net

EDIE is a free, personalized, interactive news, information and communications service for water, waste and environmental professionals around the world. With comprehensive independent coverage, powerful search facilities, e-mailalerts and discussion forums, EDIE provides a one-stop-shop for the exchange of specialized information on the Web.

6283 EE-Link (Environmental Education-Link)

http://http://eelink.net/pages/EE-Link+Introduction

EE-Link is a participant in the Environmental Education and Training Partnership of the North American Association for Environmental Education.

6284 Earth Council

http://www.ecouncil.ac.cr

Earth Council is an international NGO that was created in 1992 to promote and advance the implementation of the Earth Summit agreements. Three fundamental objectives have guided the work of the Earth Council since its inception: topromote awareness for the needed transistion to more sustainable and equitable patterns of development; to encourage public participation in decision making processes; to build bridges of understanding and cooperation between civil society andgovernments worldwide.

6285 Earth Day Network

http://www.earthday.net/

Earth Day Network is the nonprofit coordinating body of worldwide Earth Day activities. Our goal is to promote a healthy environment and a peaceful, just, sustainable world by sending environmental awareness through educationalmaterials and publications and by organizing events, activities and annual campaigns. Our network includes more than 5,000 organizations in 184 countries.

6286 Earth Observing System Amazon Project

http://boto.ocean.washington.edu/eos/

This project is a NASA Earth Observing System Interdisciplinary Investigation. The purpose of this research project is to understand the biogeochemistry, hydrology and sedimentation of the Amazon River and its drainage basin.

6287 Earth Preservers

http://www.earthpreservers.com

The web site for the environmental newspaper-for kids and adults. Features of environmental work being done by kids all over the world. Contains articles and interesting facts.

6288 Earth Resources Laboratory at MIT

http://http://eaps.mit.edu/erl/

The Earth Resources Laboratory, formed in 1982, brings together faculty, staff and students dedicated to research in applied geophysics that will further our understanding of the Earth, its resources and the environment.

6289 EarthVote.com

http://www.earthvote.com

Twenty-four hour global resource for domestic and global environmental and social issues. Input your own voting topic or vote on current issues.

6290 Earthlink

http://www.earthlink.org.au/earthlink/listings.php

Earthlink is dedicated to creating a just sustainable world by harnessing economic power for positive change. Earthlink is designed to educate and encourage people to use their spending and investing power to bring about increasedsocial justice and environmental responsibility.

6291 Earthwatch Institute

http://www.earthwatch.org

The mission of the Earthwatch Institute is to promote sustainable conservation of our natural resources and cultural heritage by creating partnerships between scientists, educators and the general public.

6292 EcoTradeNet

http://www.ecosecretariat.org/ECOTradeNet/

This Site provides trade information in respect of Member States of the Economic Cooperation Organziation. It's objective is not only to facilitate intra-regional trade cooperation but also provide relevant and updated information tothe prospective trader from outside the ECO region.

6293 Ecologia

http://www.ecologia.org

Ecologia is a private nonprofit organization providing information, training and technical support for grassroots environmental groups. Ecologia offers technical and humanitarian assistance to individuals and organizations working tosolve ecological problems at the local, regional, national and global levels.

6294 Ecology Action Centre

http://www.ecologyaction.ca

The Ecology Action Centre has been an active advocate protecting the environment since 1972. The Centre's earliest projects included recycling, composting and energy conservation and these are now widely recognized environmentalissues.

6295 Edison Electric Institute

http://www.eei.org/

Edison Electric Institute is the association of US shareholder owned electric compaines, international affiliates and industry associates worldwide. Our US members serve over 90 percent of all customers served by the shareholder ownedsegment of the industry.

6296 Edwards Aquifer Research and Data Center

http://www.eardc.txstate.edu/

The Edwards Aquifer Research and Data Center was established in 1979 by special funding for Southwest Texas State University to provide a public service in the study, understanding and use of the very fragile natural resource known asthe Edwards Aquifer.

6297 Electric Power Research Institute (EPRI)

http://http://my.epri.com/portal/server.pt?

EPRI is a nonprofit organization committed to providing science and technology-based solutions of indispensable value to our global energy customers. To carry out our mission, we manage a far-reaching program of scientific research,technology development and product implementation.

6298 Elsevier Science Tables of Contents

http://www.elsevier.com/wps/find/homepage.cws_home

Elsevier Science has become the undisputed market leader in the publication and dissemination of literature covering the broad spectrum of scientific endeavors.

6299 Endangered Species Recovery Program

http://http://esrp.csustan.edu/

The Endangered Species Recovery Program's mission is to facilitate endangered species recovery and resolve conservation conflicts through scientifically based recovery planning and implementation.

6300 Energy & Environmental Research Center (EERC)

http://www.eerc.und.nodak.edu/

The EERC is dedicated to moving promising technologies out of the laboratory and into the marketplace to produce energy cleanly and efficiently, minimizing enviromental impacts and conserving precious natural resources.

6301 Energy Ideas Clearinghouse

http://www.energyideas.org/

EnergyIdeas is the most comprehensive technical resource that Northwest businesses, industy, government and utilities use to implement energy technologies and practices.

6302 Energy Technology Data Exchange

http://www.etde.org/

ETDE through its member countries provides an extensive bibliographic database announcing published energy research and technology information.

6303 Enviroene

http://http://es.epa.gov/

Enviroene is a free public environmental information system resident on the Internet's World Wide Web. This Web provides users with pollution prevention/cleaner production solutions, compliance and enforcement assistance informationand innovative technology and policy options.

6304 Enviro-Access

http://www.enviroaccess.ca

Enviro-Access is a business partner investing in the development of environmental technologies by supplying compaines in the environemtal sector with the professional services required during the various steps of bringing theirproducts and services to the market-place.

6305 EnviroOne.com

http://enviroone.com

EnviroOne.com is your one stop center for everything environmental. We have created propriety technology that combines a vast array of resources, timely content, innovative tools and e-commerce functionality that closes the loop onsearching for information and having the ability to immediately act upon that information.

6306 Envirolink Library

http://library.envirolink.org

EnviroLink is a nonprofit organization, a grassroots online community that unites hundreds of organizations and volunteers around the world with millions of people in more than 150 countries. EnviroLink is dedicated to providingcomprehensive, up-to-date environmental information and news. We recognize that our technologies are just tools, and that the solutions to our ecological challenges lie within our communities and their connection to the Earth itself.

6307 Envirolink Network

http://envirolink.org

EnviroLink is a nonprofit organization, a grassroots online community that unites hundreds of organizations and volunteers around the world with millions of people in more than 150 countries. EnviroLink is dedicated to providingcomprehensive, up-to-date environmental information and news. We recognize that our technologies are just tools, and that the solutions to our ecological challenges lie within our communities and their connection to the Earth itself.

6308 Environment Council UK

http://www.the-environment-council.org.uk/

The Environment Council is an independent UK charity which brings together people from all sectors of business, non-governmental organizations, government and the community to develop long term solutions to environmental issues.

6309 Environment in Asia

http://www.asianenviro.com/

AsianEnviro is a collaboration between AET Ltd and ERM Japan. This site draws on the established regional networks of environmental professionals in both organizations to deliver targeted environmental business intelligence. Pleasenote that translation is necessary to view this page.

6310 Environmental Alliance for Senior Involvement

http://www.easi.org/links.html

The mission of the Environmental Alliance for Senior Involvement is to build, promote and utilize the environmental ethic, expertise and commitment of older persons to expand citizen involvement in protecting and caring for ourenvironment for present and future generations.

6311 Environmental Assessment Association

http://www.iami.org/eaa.html

The Environmental Assessment Association is an international organization dedicated to providing members with information and education in the Real Estate Industry in respect to Environmentaal Inspections, Testing and HazardousMaterial Removal.

6312 Environmental Careers Organization

http://www.eco.org

ECO's mission is to protect and enhance the environment through the development of diverse leaders, the promotion of environmental careers and the inspiration of individual action. This is accomplished through internships, careeradvice, career products and research and consulting.

6313 Environmental Change Network (ECN)

http://www.ecn.ac.uk/

To establish and maintain a selected network of sites within the UK from which to obtain comparable long-term datasets through the monitoring ofraange of variables identified as being of major environmental importance.

6314 Environmental Compliance Assistance Center

http://www.epa.gov/compliance/

The purpose of this site is to bring complicated environmental laws and regulations into everyday language that the normal business person can understand, and to give that person associated information on educational opportunities.

6315 Environmental Contaminants Encyclopedia

http://www.nature.nps.gov/hazardssafety/toxic/index.cfm

This product differs from existing databases in that it has an environmental toxicology emphasis and it summarizes information on these issues into a single, easily searchable source.

6316 Environmental Defense

http://www.environmentaldefense.org/home.cfm

Environmental Defense is dedicated to protecting the environmental rights of all people, including future generations. Among these rights are clean air, clean water, healthy, nourishing food and flourishing ecosystem.

6317 Environmental Measurements Laboratory

http://www.eml.doe.gov

EML's current mission is to conduct scientific investigations and develop technologies related to environmental restoration, site and facility characterization and environmental surveillance and monitoring.

6318 Environmental News Network

http://www.enn.com/

Since 1993, the Environmental News Network has been working to educate the world about environmental issues facing our Earth. We began as a monthly print publication called Environmental News Briefing, and two years later discoveredthe Internet as an effective means of reaching a broader, more diverse audience.

6319 Environmental Organization Web Directory

http://www.webdirectory.com/

Our goal is simple: we strive to make it easy for people from around the world to find your web page. We currently have 23 terrific staff members all dedicated to provide a free service to the environmental community.

6320 Environmental Protection Agency, US

http://www.epa.gov/

The mission of the United States Environmental Protection Agency is to protect human health and to safeguard the natural environment air, water, and land upon which life depends.

6321 Environmental Research Institute of Michigan-A ltarum

http://www.altarum.org/

ERIM promotes sustainable societal wellbeing by helping our customers - and through them, society - employ new knowledge and decision support tools to solve complex systems problems in the healthcare, national security and energy,environment and transportation sectors.

6322 Environmental Resource Center

http://www.ercweb.com

Environmental Resource Center is a full-service environmental consulting firm that has been serving the needs of private industry and government for over seventeen years with the highest standards of quality at a competitive price.

6323 Environmental Resources Information Network/ Environment Australia Online

http://kaos.erin.gov.au

ERIN, the Environmental Resources Information Network, provides environmental information for policy developers and decision makers. ERIN is a National facility, using the latest computing technology to provide access to a vastreservoir of information on the Australian environment, and the analytical tools to interpret it.

6324 Environmental Resources Management

http://www.erm.com

Environmental Resources Management has well-established reputation as one of the world's largest providers of environmental management consulting services. We have 25 year's experience working with both the public and private sectorsacross a broad spectrum of industries.

6325 Environmental Routenet

http://www.jsu.edu/depart/library/graphic/routnot.htm

Environmental RouteNet provides a single gateway to the world's foremost databases and information sources available via the Internet. The service includes searchable links to hundreds of carefully-screened environmentally-relatedresources, selected and indexed by the editors at Cambridge Scientific Abstracts.

6326 Environmental Simulations

http://www.groundwatermodels.com

Our mission is to provide our software clients with superior and technical support. To provide cost-effective in house training courses, state-of-the-art groundwater modeling services and high quality independent hydrogeologicalconsulting.

6327 Environmental Treaties and Resource Indicators

http://sedac.ciesin.columbia.edu/entri/

ENTRI, a database of searchable treaties and resource indicators, is provided by the CIESIN organization with the assistance of many other groups. They use nine specific issue areas to search the database. Those areas are: 1) landuse/land cover change and deseration, 2)global climate changr, 3)stratospheric ozone depletion, 4)transboundary air pollution, 5)conservation of biologicaldiversity, 6)deforestation, 7)oceans, 8)trade and environment, and 9) population.

6328 Environmental Working Group

http://www.ewg.org

The Environmental Working Group is a leading content provider for public interest groups and concerned citizens who are campaiging to protect the environment.

6329 Environmental and Societal Impacts Groups

http://www.isse.ucar.edu/

ESIG studies environmental change and responses to such change inorder to gain insights into how decision makers, from individuals to governements to international coalitions, might better understand and cope with impacts associatedwith the complex relationship of the atmosphere, environment, and society.

6330 Envirotext

http://tamora.cs.umass.edu/info/envirotext/index.html

EnviroText is an on-line searchable library that provides you with easy access to environmental laws, regulations, and gudidance as well as Native American Treaties and Constitutions. The purpose of EnviroText is to provide you with a"one-stop-shop" for all of your environmental regulatory needs.

6331 Essential Information

http://www.essential.org/about.html

Founded by Ralph Nader. A nonprofit, tax-exempt organization. We are involved in a variety of projects to encourage citizens to become active and engaged in their communities. We provide provocative information to the public onimportant topics neglected by the mass media and policy makers. Publishes a monthly magazine, books and reports, sponsors investigative journalism conferences, provides writers with grants to persue investigations and operates information clearinghouses.

6332 European Centre for Nature Conservation

http://www.ecnc.nl

ECNC actively promotes, by bringing the gap between science and policy, the conservation of nature and especially of biodiversity in Europe, because of their intrinsic values and their relevance to economy and European culture; therebyECNC seeks the integration of nature conservation considerations into other policies.

6333 European Forest Institute (EFI)

http://www.efi.fi

EFI's mission is to promote, conduct and co-operate in research of forests, forestry and forest products at the pan-European level; and to make the results of the research known to all interested parties, notably in the areas of policyformulation and implementation, in order to promote the conservation and sustainable management of forests in Europe.

6334 Everglades Information Network

http://everglades.fiu.edu/

The Everglades Information Network is a program of library and information services in support of research, restoration, and resource management of the south Florida environment. The EIN serves researchers, resource managers,educators, students, researchers, decision makers, and concerned citizens both within south Florida and around the world.

6335 Extension Toxicology Network ETOXNET

http://http://extoxnet.orst.edu/
Some of the goals of EXTOXNET are to stimulate dialog on toxicology issues, develop and make available information relevant to extension toxicology, and facilitate the exchange of toxicology-related information in electronic form,accessible to all with access to the Internet.

6336 Federal Emergency Management Agency FEMA

http://www.fema.gov/
Advising on building codes and flood plain management, teaching people how to get through a disaster, helping equip local and state emergency preparedness, the range of FEMA's activities is broad indeed.

6337 Federal Geographic Data Committee

http://www.fgdc.gov/
The Federal Geographic Data Committee is a 19 member interagency committee composed of representatives from the Executive Office of the President, Cabinet-level and independent agencies. The FGDC is developing the National Spatial DataInfrasturcture (NSDI) in cooperation with organizations from State, local and tribal governments, the academic community, and the private sector.

6338 Fedworld Information Network

http://www.fedworld.gov
We here at FedWorld have enjoyed thinking outside the box to offer multiple distribution channels to disseminate information to the public and to the Federal Governement. The modes of access, the variety of documents available, and thetechnological expertise at FedWorld are expanding with technology.

6339 Finnish Forest Research Institute: METLA

http://www.metla.fi
The Finnish Forest Research Institute builds the future of the forest sector through research. METLA social task is to promote through research economically, ecologically and socially sustainable management and utilisation of theforests.

6340 Fish and Wildlife Information Exchange Homepage

http://http://gis.lib.vt.edu/vt_ogis/
The FWIE is a technical assistance center and information clearinghouse for fish, wildlife, and land management agencies and organizations. The FWIE also assists with the planning, development, implementation, and maintenance ofinformation management and delivery systems.

6341 Florida Center for Environmental Studies

http://www.ces.fau.edu
The Mission of the Center is to collect, analyze, and promote the use of scientifically-sound information concerning tropical and suptropical freshwater ecosystems.

6342 Florida Cooperative Extension Service

http://www.ifas.ufl.edu/extension.ces.htm
Part of the mission of the Florida Cooperative Extension Service is to disseminate and provide access to science-based information that will contribute to the solution of natural resource problems of concern to the people of Florida.Wildlife Extension specialists in the Department of Wildlife Ecology and Conservation serve , advice, and develop educayional programs for Florida citizens in conjunction with county extension agents and other state , county and local organizations.

6343 Florida Department of Environmental Protection

http://www.dep.state.fl.us
The mission of the Department of Environmental Protection is: More Protection, Less Process. The Department accomplishes its mission in a manner that provides stewardship of Florida's ecosystems so that the State's unique quality oflife may be preserved for present and future generations.

6344 Forest History Society

http://www.lib.duke.edu/forest
The Forest History Society is a non-profit educational institution that links the past to the future by identifying, collecting, preserving, interpreting, and disseminating information on the history of people, forests, and theirrelated resources.

6345 Forest Service Employees for Environmental Ethics

http://www.fseee.org/
Forest Service Employess for Environmental Ethics is a 501 nonprofit organization. Our mission is to forge a socially responsible value system for the Forest Service based on a land ethic that ensures ecologically and economicallysustainable resource management.

6346 Friends of the Earth International

http://www.foe.org/international/
Friends of the Earth International is a worldwide federation of national environmental organizations. This federation aims to: protect the earth against futher deterioration and repair damage inflicted upon the environment by humanactivities and negligence; preserve the earth's ecological, cultural and ethic diversity.

6347 GAP (Gap Analysis Program) National

http://www.gap.uidaho.edu/
The mission of the GAP Analysis Program is to provide regional assessments of the conservation status of native vertebrate species and natural land cover types and to facilitate the application of this information to land managementactivities.

6348 GLOBE Program

http://www.globe.gov
GLOBE is a worldwide hands-on, primary and secondary school based science and education program.

6349 Galapagos Coalition

http://www.law.emory.edu/PI/GALAPAGOS/
The Galapagos Coalition is a group of biologists, other scientists, and lawyers with expertise in environemntal and international law, many of whom have done research in the Galapagos and all of whom are interested in the understandingthe relationship between the conservation of the Galapagos and human activities.

6350 General Accounting Office

http://www.gao.gov
The General Accounting Office is the audit, evalution and investigative arm of Congress. GAO exists to support the congress in meeting its Constitutional responsibilities and to help improve the performance and ensure accountability ofthe federal government for the American people. GAO examines the use of public funds, evaluates federal programs and activities and provides analysis, options, recommendations and other assistance to help the Congress make effective oversite, policyand funding.

6351 Georgia Department of Natural Resources

http://www.gadnr.org/
The mission of the Department of Natural Resources is to sustain, enhance, protect and conserve Georgia's natural, historic and cultural resources for present and future generations, while recognizing the importance of promoting thedevelopment of commerce and industry that utilize sound environmental practices.

6352 Germinal Project

http://http://lasig.epfl.ch/projets/germinal/Germinal.html

GERMINAL is an interdissiplinary project established by the institutes of the Department of Rural Engineeing and by the directorship of the Swiss Federal Institute of Technology at Lausanne. Its goal is the development of a global, integrated approach for land use planning and environmental management based on the use of Geographic Information Systems.

6353 Global Change Master Directory (GCMD)

http://gcmd.gsfc.nasa.gov/
The mission of the Global Change Master Directory is to assist the scientific community in the discovery of and linkage to Earth science data, as well as to provide data holders a means to advertise their data to the Earth ScienceCommunity.

6354 Global Change Research Information Office

http://gcrio.org
The US Global Change Research Information Office provides access to data and information on global change research, adaption/mitigation strategies and technologies, and global change-related educational resources on behalf of thevarious US Federal Agenices and Organizations that are involved in the US Global Change Research Program

6355 Global Ecovillage Network

http://http://gen.ecovillage.org/
GEN's main aim is to support and encourage the evolution of sustainable settlements across the world through: Internal and External Communications services; facilitating information exchange and flow about ecovillages anddemonstrations sites.

6356 Global Environmental Options (GEO)

http://www.genonetwork.org
GEO was created to bring attention to the impact that buildings and planning have on the environment and biodiversity. To reduce these impacts, GEO develops integrated high performance green building stragies for professionals andother decision makers worldwide, coordinating hands-on building and planningprojects that showcase green design.

6357 Global Futures Foundation (GFF)

http://www.gffltd.com/services/index.asp
Global Futures Foundation is to resolve conflicts and unify business, social, and environmental interest. We do so by helping discover true common ground, forging productive partnerships, and together developing programs and policiesdesigned for the next century, not the last one.

6358 Global Network of Environment & Technology

http://www.gnet.org/
GNET provides worldwide access to timely information on environmental news, products and services, marketing opportunities, contracts, government programs, policy and law, and business assitance resources via the World Wide Web.

6359 Global Research Information Database (GRID)

http://www.grida.no/
GRID-Arendal provides environmental information, communications and capacity buildings services for information management and assessment. Established to strengthen the United Nations through its Environmental Programme, our focus isto make credible, science-based knowledge understandable to the public and to decision making for sustainable development.

6360 Great Lakes Fishery Commission
2100 Commonwealth Boulevard **734-662-3209**
Ann Arbor, MI 48105 **Fax: 734-741-2010**
http://www.glfc.org
To develop coordinated programs of research on the Great Lakes, and, on the basis of the findings, to recommend measures which will permit the maximum sustained productivity in stocks of fish of common concern.

6361 Great Lakes Information Network (GLIN)

http://www.great-lakes.net
The Great Lakes Information Network is a partnership that provides one place to find information relating to the binational Great Lakes-St Lawrence region of North America. GLIN offers a wealth of data and information about theregion's environment, economy, tourism, education and more.

6362 Green Mountain Institute for Environmental Democracy

http://www.gmied.org
The Green Mountain Institute for Environmental Democracy seeks to reinvigorate the essential connections among the public, government and information necessary for effective improvements in environmental quality.

6363 Green Seal

http://www.greenseal.org
Green Seal is the independent nonprofit organization dedicated to protecting the environment by promoting the manufacture and sale of enviromentally responsible consumer products. It sets environmental standards and awards a Green Sealof Approval to products that cause less harm to the environment than other similar products.

6364 Green University Initiative

http://www.gwu.edu/~greenu/
The George Washington Green University Initiative began as a grassroot movement to implement sustainable practices into all aspects of life at GW. At Green University we work towards ecosystem protection, incorporating environmentaljustice into daily activities and decisions.

6365 Greenbelt Alliance

http://www.greenbelt.org/
Our mission is to make the nine-country San Francisco Bay Area a better place to live by protecting the region's Greenbelt and improving the livability of its cities and towns. Scince 1958 we have worked in partnership with diversecoalitions on public policy development, advocacy and education.

6366 Greenhouse Gas Technology Information Exchange GREENTIE

http://www.greentie.org
IEA GREENTIE is an international information network that distributes details of suppliers whose technologies help to reduce greenhouse gas emissions. GREENTIE also provides information on leading international organizations and IEAprograms whose R&D and information activities center around clean energy technologies.

6367 Ground-Water Remediation Technologies Analysis Center

http://www.gwrtac.org
The Groundwater Remediation Technologies Analysis Center compiles, analyzes and disseminates information on innovative ground-water remediation technologies. GWRTAC prepares reports by technical teams selectively chosen from ConcurrentTechnologies Corporation, the University of Pittsburgh and other supporting institutions, also maintaining an active outreach program.

6368 Harbor Branch Oceanographic Institution

http://www.hboi.edu
Harbor Branch Oceanographic Institution is dedicated to exploring the world's oceans, integrating the science and technology of the sea with the needs of humankind.

6369 Harvard Forest

http://http://harvardforest.fas.harvard.edu/

Through the years researchers at the Forest have focused on silviculture and forest management, soils and the development of forest site concepts, the biology of temperate and tropical trees, forest ecology and economics and ecosystemdynamics.

6370 Hawaii Biological Survey

http://http://hbs.bishopmuseum.org/hbs1.html

It was created to locate, identify and evaluate all native and non-native species of flora and fauna within the State and maintain the reference collections of that flora and fauna for a wide range of uses.

6371 Hawaiian Ecosystems at Risk (HEAR)

http://www.hear.org

The mission of the Hawaiian Ecosystem at Risk project is to promote technology, methods and information to decision-makers, resource managers and the general public to aid in the fight against harmful alien species in Hawaii and thePacific Basin.

6372 Hawk Mountain Sanctuary

http://www.hawkmountain.org/default.shtml

Hawk Mountain's mission is to foster the conservation of birds of prey worldwide and to create a better understanding of and further the conservation of the natural environment, particularly the Central Appalachian region.

6373 Hawkwatch International

http://www.hawkwatch.org/

Our mission is to protect hawks, eagles, other birds of prey and their environment through research, education, and conservation.

6374 HazDat-Hazardous Substance Release: Health Effects Database

http://www.atsdr.cdc.gov/hazdat.html

HazDat is the scientific and administrative database developed to provide access to information on the release of hazardous substances from Superfund sites or from emergency events and on the effects of hazardous substances on thehealth of human populations.

6375 Hazardous Substance Research Centers

http://www.hsrc.org

Hazardous Substance Research Cente is a national organization that carries out an active program of basic and applied research, technology transfer and training. Our activities are conducted regionally by five multi-university centers,which focus on different aspects of hazardous substance management.

6376 Hazardous Waste Clean-Up Information (CLU-IN)

http://www.clu-in.com

Providing information about innovative treatment and site characterization technologies while acting as a forum for all waste remediation stakeholders.

6377 Headwaters Science Center

http://www.hscbemidji.org/

Headwaters Science Center is dedicated to science education and environmental awareness. It features hands-on exhibits, a live animal collection and special events and science-related programs and demonstrations.

6378 Heartwood

http://www.heartwood.org/

An association of groups, individuals and businesses dedicated to the health and well being of the native forest of the Central Hardwood region and its interdependent plant, animal and human communities.

6379 Holland Island Preservation Foundation

http://www.intercom.net/local/holland/index.html

Our goal is to stabilize and preserve this beautiful island, not only for the people that once lived there, but also for the wildlife that thrives there still. With modern tools and techniques, this fragile ecosystem can be saved fromultimate destruction.

6380 Horned Lizard Conservation Society (HLCS)

http://www.hornedlizards.com

The Texas Chapter of the Horned Lizard Conservation Society (HLCS) is devoted to discovering why the Texas Horned Lizard has declined in numbers so dramatically in recent years and what can be done to reverse the process.

6381 Houston Audubon Society

http://www.houstonaudubon.org/

The Houston Audubon Society works for the thoughtful conservation of the earth's natural resources by educating people to the value of the natural world; protecting, preserving and enhancing wildlife habitat and encouraging the passageof legislation to protect the environemnt.

6382 Howl: The PAWS Wildlife Center

http://www.paws.org/wildlife/index.htm

The PAWS Wildlife Center is a world reowned wildlife rehabiliation facility. Formerly known as HOWL, the PAWS Wildlife Center receives over 5,000 injured or displaced wild animals every year. The center houses and rehabilitateswildanimals, and prepares them for eventual release back into the wild.

6383 Hydrographic Survey Data

http://www.ngdc.noaa.gov/mgg/geodas/bathymetry/hydro.html

The NOS hydrographic data base, maintained by NGDC in conjunction with NOS, comprises the majority of NGDC's area survey holdings and provide extensive survey coverage of the coastal waters and Exclusive Economic Zone of the US and itsterritories.

6384 IFAW: International Fund for Animal Welfare

http://www.ifaw.org

IFAW's mission is to improve the welfare of wild and domestic animals throughout the world by reducing commercial exploitation of animals in distress. We seek to motivate the public to prevent cruelty to animals and promote animalwelfare and conservation policies that advance the well-being of both animals and people.

6385 IISDnet: International Institute for Sustainable Development

http://www.iisd.org/

Our mission is to champion innovation, enabling societies to live sustainably.

6386 Illinois Recycling Association

http://www.illinoisrecycles.org/

The Illinois Recycling Association's mission is to encourage the responsible use of resources by promoting waste reduction, re-use and recycling.

6387 Indiana Department of Natural Resources

http://www.in.gov/dnr/

The mission of the Indiana Department of Natural Resources is to protect, enhance, preserve and wisely use natural, cultural and recreational resources for the benefit of Indiana's citizens through professional leadership, managementand education.

6388 Information Center for the Environment (ICE)

http://ice.ucdavis.edu/

The Foundation wishes to participate actively in all activities and efforts to protection, preservation and exploitation of mangrove ecosystem for the prosperity of all people.

6389 Inland Seas Education Association

http://www.schoolship.org/

Inland Seas Education Association is a nonprofit organization whose mission is to provide a floating classroom where people of all ages can gain first-hand training and experience in the Great Lakes ecosystem. The knowledge gainedthrough these experiences will provide the leadership, understanding and commitment needed for the long-term stewardship of the Great Lakes.

6390 International Arid Lands Consortium

http://http://ag.arizona.edu/OALS/IALC/Home.html

The International Arid Lands Consortium works to achive research and development educational and traning initiahves, demonstration projects, workshops, and other technology-transfer activities applied to the development, management,restoration, and reciamation of and semiand lands in the US the MIddle East, and elsewhere in the world.

6391 International Association for Energy Economics

http://www.iaee.org

The International Association for Energy Economics provides a forum for the exchange of ideas, experience and issues among professionals interested in energy economics. Its scope is worldwide, as are its members who come from diversebackgrounds-corporate, academic, scientific and government.

6392 International Association for Environmental Hydrology

http://www.hydroweb.com

Worldwide association of environmental hydrologists dedicated to the protection and cleanup of fresh water resources.

6393 International Bee Research Association (IBRA)

http://www.ibra.org/uk

The International Bee Research Association is the world information specialist on bees. Founded in 1949, IRBA is a nonprofit organization with members in almost every country in the world. It exists to increase peoples' awareness ofthe vital role of bees in agriculture and the natural environment.

6394 International Canopy Network (ICAN)

http://www.evergreen.edu/ican/main/ican.html

The International Canopy Network is devoted to facilitating the continuing interaction of people concerned with forest canopies and forest ecosystems around the world. ICAN is a nonprofit organization supported by a global community ofscientists, conservation advocates, canopy educators and environmental professionals. The organization is funded by subscriber dues, donations and grants.

6395 International Centre for Gas Technology Information

http://www.etde.org/abtetde/icgti.html

ICGTI benefits us by: Offering gas technology decisionmakers a competitive information edge; affording immediate connectionn with the world's natural gas technology leaders; and by allowing easy access to provider services andequipment that serve the natural gas industry.

6396 International Council for the Exploration of the Sea(ICES)

http://www.ices.dk/

ICES is a leading forum for the promotion, coordination and gas dissemination of research on the physical, chemical and biological systems in the North Atlantic and advice on human impact on its environment, in particular fisherieseffects in the Northeast Atlantic.

6397 International Crane Foundation

http://www.savingcranes.org/

The International Crane Foundation works worldwide to conserve cranes and the wetland and grasslands communities on which they depend. ICF is dedicated to providing experience, knowledge, and inspiration to involve people in resolvingthreats to these ecosystems.

6398 International Energy Agency Solar Heating and Cooling Programme

http://www.iea-shc.org

The Solar Heating and Cooling Programme was one of the frist IEA Implementing Agreements to be established. Since 1977, its 21 members have collaborating to advance active solar, passive solar and photovoltaic thecnologies and theirapplication in buildings.

6399 International Geosphere-Biosphere Programme

http://www.igbp.kva.se

Our scientific objective is to describe and understand the interactive physical, chemical and biological processes that regulate the total Earth System, the unique environment that it provides for life, the changes that are occurringin this system, and the manner in which they are influenced by human actions.

6400 International Ground Source Heat Pump Association(IGSHPA)

http://www.igshpa.okstate.edu/default.htm

As an organization, IGSHPA pursues these goals: Supporting GHP industry research and development; promoting the GHP-related current events internationally. Developing and distuting internationally recognized training materials.

6401 International Institute for Industrial Environmental Economics

http://www.iiee.lu.se/

The mission of the Institute is to the international advancement of sustainable development by conducting at the forefront of issues pretaining to cleaner production, and to educate present and future decision makers within all sectorsof society in the formulation and implemantation of preventive environmental strategies. The Institute is founded on the firm conviction that a preventive approach to environmental problems is necessary for the perpetuation of life on this planet.

6402 International Marine Mammal Association

http://www.imma.org

The International Marine Mammal Association is a non-for-profit organization dedicated to promoting the conservation of marine mammals and their habitats worldwide, through research and education.

6403 International Marinelife Alliance

http://www.marine.org/

The IMA condems destructive fishing practices, and advocates environmental defense, popularize alternative livelihood for surplus fisherfolk, promotes clean harvesting technology, nurtures environmental activism, propagatesenvironmental causes, supports environmental education, adheres to the principles of sustainable development and safeguards marine biodiversity.

6404 International Otter Survival Fund

http://www.otter.org/

The International Otter Survival Fund is a global organization working to conserve all 13 species of otter by helping to support scientists and other workers in practical conservation, education, research and rescue and rehabilitation.

6405 International Primate Protection League (IPPL)

http://www.ippl.org

The International Primate Protection League was funded in 1973, and since this time has been working continuously for the well-being of primates.

6406 International Research Institute for Climate Prediction

http://http://in.columbia.edu/

The vision for the IRI is that of an innovative science institution working to accelerate the ability of societies worldwide to cope with climate fluctuations, especially those that cause devastating impacts on humans and theenvironment, thereby reaping the benefits of decades of research on the predictability of El Nino-Southern Oscillation phenomenon and other climate variations.

6407 International Rivers Network

http://www.irn.org/index.html

IRN's mission is to halt and reverse the degradation of river systems; to support local communities in protecting and restoring the well-being of the people, cultures and ecosystems that depend on rivers; to promote sustainable,environmentally sound alternatives to damming and channeling rivers.

6408 International Satellite Land Surface Climatology Project (ISLSCP)

http://www.gewex.org

ISLSCP Objective is to demosnstrate the type of surface and near-surface satellite measurements that are relevant to climate and global change studies. Develop and improve algorithms for the interpretation of satellite measurements ofland-surface features.

6409 International Snow Leopard Trust

http://www.snowleopard.org

The International Snow Leopard Trust is dedicated to the conservation of the endangered snow leopard and its mountain ecosystem through a balanced approach that considers the needs of the people and the environment.

6410 International Society for Ecological Modelling (ISEM)

http://www.isemna.org

The International Society for Ecological Modelling promotes the international exchange of ideas, scientific results, and general knowledge in the area of the application of systems analysis and simulation in ecology and naturalresource management.

6411 International Society for Environmental Ethics

http://www.cep.unt.edu

ISEE now maintains this website, which includes the largest bibliography in the world on environmental ethics, over 7,000 entries. Newsletters over the last ten years also available here, and by consulting these a full historicalrecord may be obtained.

6412 International Society of Arboriculture

http://www.isa-arbor.com

The mission is through research, technology, and education promote the professional practice of arboriculture and foster a greater public awareness of the benefits of trees.

6413 International Solar Energy Society

http://www.ises.org/

The mission is to encourage the use and acceptance of Renewable Energy technologies; to realise a global community of industry, individuals and institutions in support of renewable energy and to create a structure to faciliatecooperation and exchange.

6414 International Union of Forestry Research Organizations (IUFRO)

http://www.iufro.org

IUFRO is a nonprofit, non-governmental international network of forest scientists. Its objective are to promote international cooperation in forestry and forest products research. IUFRO's activities are organized primarily through its268 specialized Units in 8 technical Divisions.

6415 International Wildlife Coalition

http://www.iwc.org

The International Wildlife Coalition is a federally recognized, nonprofit taxexempt charitable organization. Founded in 1984, the Coalition is dedicated to public education, research, rescue, rebilitation, litigation and internationaltreaty negotiations concerning global wildlife and natural habitat protection issues.

6416 International Wolf Center

http://www.wolf.org

The mission of the International Wolf Center is profoundly simple. We support the survival of the wolf around thr world by teaching about its life, its association with other species and dynamic relationships to humans.

6417 International Year of the Ocean -1998

http://www.yoto98.noaa.gov

The overall objective is to focus and reinforce the attention of the public, governments and decision makers at large on the importance of the oceans and the marine environment as resources for sustainable development.

6418 Iowa Department of Natural Resources

http://www.iowa.gov

The department's mission is to manage, protect, conserve, and develop Iowa's natural reources in cooperation with other public and private organizations and individuals, so that the quality of life for Iowans is significantly enchancedby the use, enjoyment and understanding of those resources.

6419 Irish Peatland Conservation Council

http://www.ipcc.ie

The Irish Peatland Conservation Council is an independent conservation charity. We were established in 1982 to campaign for the conservation of a representative sample of living intact Irish bogs and peatlands and we need your support.

6420 Island Wildlife Natural Care Centre

http://www.islandstrust.bc.ca

Its mission is to function in a twofold manner. Frist, by rehabilitating North American wildlife, including marine mammals, with emphasis on alternative, non-toxic, non-invasive treatments. Second, educationally, by furtheringknowledge of treatments available to professionals in the field, and by educationg the public on both rehabilitation and the interaction of man and wild animals.

6421 Izaak Walton League

http://www.iwla.org

The mission is to conserve, maintain, protect and restore the soil, forest, water and other natural resources of the United States and other lands; to promote means and opportunities for education of the public with respect to suchresources and their enjoyment and wholesome utilization.

6422 Jane Goodall Institute

http://www.gsn.org

The Jane Goodall Institute advances the power of individuals to take informed and compassionate action to improve the environment of all living things.

6423 Jefferson Land Trust

http://www.saveland.org

Jefferson Land Trust is a private, nonprofit, grass-roots organization with a mission to conserve property and natural resources. Landowners may work with a Land Trust when they wish to permanently protect the ecological, scenic, historic, or recreational qualities of land they own from inappropriate development.

6424 John M Judy Environmental Education Consortium

http://www.utm.edu/departments/ed/cece/john.html

The purpose of the Consortium is to promote and enhance environmental education through systemic change toward infusing a collective environmental consciousness and individual environmental ethic into the learning and teaching process.

6425 Joint Center for Energy Management (JCEM)

http://bechtel.colorado.edu/Graduate Programs/Jcem/jcemmain.html

The Joint Center for Energy Management is a research center in the Department of Civil, Environmental, and Architectural Engineering at the University of Colorado at Boulder. It is dedicated to excellence in energy-related research, development, education, and technical assistance.

6426 Journey North

http://www.learner.org/jnorth/index.html

Journey North engages students in a global study of wildlife migration and seasonal change. K-12 students share their own field observations with classmates across North America. Widely considered a best-practices model for education, we are the nation's "citizen science" project for children.

6427 Kansas Environmental Almanac

http://www.idir.net/-chsjones/

Kansas is a magnificent place. We owe it to ourselves, to earlier generations and those yet to come, to take good care of our home. With that goal in mind, the Kansas Environmental Almanac will collect and house information pertaining to the Kansas environment and its protection.

6428 Kentucky Department of Fish and Wildlife Resources

http://www.kdfwr.state.ky.us/

We are stewards of Kentucky's fish and wildlife resources and their habitats. We manage for the perpetuation of these resources and their use by present and future generations. Through partnerships, we will enhance wildlife diversity and promote sustainable use, including hunting, fishing, boating and other nature-related recreation.

6429 Kentucky Water Resources Research Institute

http://www.uky.edu/WaterResources/

The institute mission is to stimulate water resources and water-related environmental research. To assist and stimulate academic units in the conduct of undergraduate and graduate education in water resources and water-related environmental issues.

6430 Kola Ecogeochemistry

http://www.ngu.no/kola

The primary aims are to map the extent of contamination by inorganic elements in various media around industrial centres. To map the content of raionuclides in topsoil throughout the Project area and to shed light on the process and dynamics of trace element cycling in catchments.

6431 LIFE

http://life.csu.edu.au/

Founded in 1992, the LIFE Site is Australia's frist information service on the World Wide Web. The main focus is on biological information, especially the environment and biodiversity.

6432 LTER (US Long-Term Ecological Research)

http://www.lternet.edu

The mission of the LTER Network is to facilitate and conduct ecological research through: Understanding ecological pheonmena over long temporal and large spatial scales. Creating a legacy of well-designed and documented long-term experiments and observations for future generations.

6433 Lake Pontchartrain Basin Foundation

http://www.saveourlake.org

The Lake Pontchartrain Basin Foundation, a membership-basede citizens organization, is the public's independent voice dedicated to restoring and preserving the Lake Pontchartrain Basin.

6434 Land Conservancy of San Luis Obispo County

http://www.special-places.org/members.htm

We at the Land Conservancy take pride in our active approach to land conservation. We pursue the protection of open space through land aquisition, conservation easements, restoration, and stewardship.

6435 Land Trust Alliance

http://www.lta.org

Founded in 1982, the Land Trust Alliance is the national leader of the private land conservation movement, promoting voluntary land conservation across the country and providing resources, leadership and training to the nation's 1,200 plus nonprofit, grassroots land trusts, helping them to protect important open spaces.

6436 League of Conservation Voters

http://www.lcv.org

The League of Conservation Voters works to create a Congress more responsive to your environmental concerns. As the nonpartisan political voice for over nine million members of environmental and conservation groups LCV is the only national environmental organization dedicated full-time to educating citizens about the environmental voting records of Members of Congress.

6437 Leave No Trace

http://www.leavenotrace.com

The mission of the Leave No Trace program is to promote and inspire responsible outdoor recreation through education, research, and partnership. The program is managed by LNT, a nonprofit organization located in Boulder, Colorado.

6438 Living on Earth

http://www.loe.org

Living on Earth with Steve Curwood is the weekly environmental news and information program distributed by National Public Radio. This Year marks Living on Earth's eighth anniversary.

6439 Lloyd Center for Environment Studies

http://www.thelloydcenter.org

The Lloyd Center for Environmental Studies is a nonprofit organization that provides education programs and conducts research to develop a scientific and public understanding of coastal, estaurine, and watershed environments in southeastern New England.

6440 Louisiana Department of Agriculture & Forestry

http://www.ldaf.state.la.us

The Louisiana Department of Agriculture & Forestry was created in accordance with the provisions of Article IV, Section 10 of the Constitution of Louisiana. The commissioner of agriculture and forestry heads the department and exercises all functions of the state relating to the promotion, protection, and advancement of agriculture and forestry, expert research and educational functions expressly allocated by the constitution or by law to other state agencies.

6441 Louisiana Energy & Environmental Resources & Information Center

http://www.leeric.lsu.edu/

A primary LEERIC objective is to serve information needs of LSU's faculty, staff, and researchers. LEERIC also provides energy and environmental educational programs for consumers and non-college educators and students.

6442 Lower Rio Grande Ecosystem Initiative

http://www.cerc.usgs.gov/lrgrei.htm

The Lower Rio Grande Ecosystem Initiative was established by the Biological Division of the USGS to address research and information needs pertinent to the biotic resources of the river and its adjacent terrestrial habitats.

6443 Macaw Landing Foundation

http://www.cybernw.com/-mlf/index.html

The foundation is operated solely by volunteers, which allows us to spend all the donated money entirely on Macaws. The Foundation is dedicated to preservation of Macaws.

6444 Maine Department of Conservation

http://www.state.me.us/doc/dochome.htm

Created in 1973, the Department of Conservation's Mission is to benefit the citizens, landowners, and users of the state's natural resources by promoting and performing stewardship and ensuring responsible balanced use of Marine'sland, forest, water, and mineral resources.

6445 Maine Department of Environmental Protection

http://www.maine.gov/dep/indes.shtml

The Maine Department of Environmental Protecion engages in a wide range of activities. It makes recomendations to the Legislature regarding measures to minimize and climate environmental pollution; grants licenses; initiatesenforcement actions; and provides information and technical assistance.

6446 Maine Department of Inland Fisheries & Wildlife

http://www.state.me.us/ifw

The Vision is of an IF&W that: conserves, protects, and enhances the inland fisheries and wildlife resources and promotes efficiency in program management through employee involvement, intitiative, innovation, and teamwork.

6447 Mangrove Replenishment Initiative

http://www.mangrove.org

The Mangrove Replenishment Initiative began as a local project along the central east coast of Florida; however, in the last few years it has contributed to wide range of habitat creation and restoration programs that are internationalin scope.

6448 Manomet Center for Conservation Science

http://www.manomet.org

Manomet's mission is to conserve natural resources for the benefit of wildlife and human populations. Through research and collaboration, Manomet builds science-based, cooperative solutions to environmental problems.

6449 Marine Biological Association

http://www.chm.org.uk

The Marine Biological Association of the United Kingdom is a professional body for marine biologists with some 1200 members world-wide. The current programme encourages fundamental research in marine biology by resident Fellows,interwoven with that of visting scientists.

6450 Marine Environmental Research Institute

http://www.downeast.net

The Marine Environmental Research Institute is a 501 nonprofit charitable organization dedicated to scientific research and education on the impacts of pollution on marine life, and to protecting the health and biodiversity of themarine environment for furture generations.

6451 Marine Mammal Center

http://www.tmmc.org

We recognize our interdependence with marine mammals, their improtance as sentinels of the ocean environment, and our responsibility to use our awareness, compassion and intelligence to ensure their survival and the conservation oftheir habitat.

6452 Marine Mammal Stranding Center

http://www.mmsc.org

The Marine Mammal Stranding Center is a private nonprofit organization fully licensed by the state and federal governments to rescue, rehabilitate, and release stranded or otherwise distressted marine mammals whales, dolphins, sealsand turtles that comes ashore along the New Jersey coast.

6453 Marine Technology Society

http://www.mtsociety.org

Our mission is to disseminate marine science and technical knowledge, to promote and support education for marine scientists, engineers and technicians, advance the development of tools and procedures required to explore, study andfurther the responsible and sustainable use of the oceans. Provide servicesthat create a broader understanding of the relevance of marine sciences to other technologies, arts and human affairs.

6454 Maryland Department of Natural Resources

http://www.dnr.state.md.us

The Maryland Department of Natural Resources is the state agency which overseas the management and wise use of the living and natural resources of the Chesapeake Bay and its tributaries. The resources of Maryland portion of thewatershed include its state forests and parks, fisheries, wildlife and the recreation of citizens.

6455 Maryland Forests Association

http://www.mdforests.org

Incorporated in 1976, the Maryland Forests Association is a nonprofit 501 citizens organization whose membership includes more than 500 individuals and companies from throughout Maryland and the tri-state area.

6456 Massachusetts Department of Fisheries, Wildlife and Environmental Law Enforcement

http://www.state.ma.us/dfwele

Massachusetts state agency responsible for the management and conservation of the state's fisheries and wildlife, including rare and endangered species.

6457 Mediterranean Oceanic Data Base

http://modbg.oce.ulg.ac.be/

The general objective of the MODB is to deliver advanced data products to mediterranean research projects supported by the MAST programme of the European Union. All products are however freely distributed to the whole scientificcommunity.

6458 Medomak Valley Land Trust

http://www.mltn.org

The Medomak Valley Land Trust is a local, private nonprofit oragnization edstablished in 1991 to preserve the natural, recreational, scenic and productive values of the Medomak River watershed. Our goals are to foster a regionalperspective of the watershed and to encourage valley residents to work together to ensure that the resources they value will remain for future generations.

6459 Messinger Woods Wildlife Care and Education Center

http://www.wildlifecare.org

The mission is to promote community awareness, education and instruction, involvement, understanding, appre4ciation, and acceptance of our wildlife in order to conserve it. To co-exist and protect each other, our natural surroundings,and all the inhabitants of our earth by education & example.

6460 Michigan Department of Environment Quality

http://www.michigan.gov/deq

Our mission is to drive improvements in environmental quality for the protection of public health and natural resources to benefit current and future generations. This will be accomplished through effective administration of agencyprograms, providing for the use of innovative strategies, while helping to foster a strong and sustainable economy.

6461 Michigan Department of Natural Resources

http://www.michigan.gov/deq

The Department of Natural Resources is responsible for the stewardship of Michigan's natural resources and for the provision of outdoor recreational opportunities; a role is has relished since creation of the original ConservationDepartment in 1921.

6462 Michigan Environmental Science Board

http://www.mesb.org

The MESB is an independent state egency established to provide scientific and technical advice to the Governor of Michigan and to state departments, on matters affecting the protection and management of Michigan's environmental andnatural resources.

6463 Michigan Forest Association

http://www.michiganforests.com

The mission of the Michigan Forest Association is to promote good management on all forest land, to educate our members about good forest practices and stewardship of the land. To inform the general public about forestry issues and thebenefits of good forest management.

6464 Michigan Pulp and Paper Pollution Preservation Program

http://www.deq.state.mi.us/ead/p2sect/p5/

The Pulp and Paper Pollution Program (P5) is a voluntary pollution prevention (P2) partnership between the Michigan Pulp and Paper Environmental Council and the MDEQ. P5 promotes environmental improvement in concert with economicgrowth and security. P2 goals are set annually by mills to reduce the generation, discharge, and emissions of substances of concern. P5 membership is open to all forest products companies in Michigan.

6465 Michigan United Conservation Clubs

http://www.mucc.org

MUCC is the largest statewide conservation in the nation. The mission of Uniting Citizens to Conserve Michigan's Natural Resources and Protect our outdoor heritage. MUCC works to conserve Michigan's wildlife, fisheries, waters, forest,air, and soils by providing information, education and advocacy.

6466 Midwest Center for Environmental Science and Public Policy

http://www.mcespp.org

Protecting human health and the environment through research, advocacy, public education and citizen empowerment. We make it a top priority to safeguard human health by decreasing pollution; we focus on enhancing the environment wherepeople live, work and play and we strive to involve citizens in decision-making that affects the health of their environment.

6467 Midwest Renewable Energy Association (MREA)

http://www.the-mrea.org

The Midwest Renewable Energy Association is a nonprofit network for sharing ideas, resources, and information with individuals, business, and communities to promote a sustainable future through renewable energy and energy efficiency.

6468 Milton Keynes Wildlife Hospital

http://technology.open.ac.uk/staff/robert/mkwh.htm

We are one of the few establishments in the UK to be licensed by the Department of the Environment to care for certain species of birds. A purely voluntary organization and rely solely on donations from the public and local companiesin order to keep the hospital open.

6469 Mineral Policy Center

http://www.mineralpolicy.org

Mineral Policy Center is a nonprofit environmental organization dedicated to protecting communities and the environment by preventing the environmental impacts associated with irresponsible mining and mineral development, and bycleaning up pollution caused by past mining.

6470 Minnesota Department of Natural Resources

http://www.dnr.state.mn.us

The DNR vision hinges on the concept of sustainability. Protecting and restoring the natural environment while enhancing economic opportunity and community well-being. DNR endorsed ecosystem-based management as its methods to achievesustainability, and uses the concept of ecosystems integrity as a benchmark to measure progress toward sustainability goals. The goal is to maintain environment, econony and the community in a healthy state indefinitely.

6471 Minnesota Pollution Control Agency

http://www.pca.state.mn.us

This site includes a great deal of information on air, water, and waste pollution in Minnesota. It also contains data on regulations and permits, clean-up techniques, prevention, publications, and programs to protect Minnesota'senvironment. The MPCA site has a calender of events, information for childern, news releases, training opportunities, and conference information.

6472 Minnesotans for An Energy-Efficient Economy (ME3)

http://www.me3.org

ME3 is a coalition working to improve the quality of life, the environment and the economy of Minnesota by promoting energy efficiency and the sound use of renewable energy. Through a program research, public education, and theintervention in the decision making process. ME3 seeks to develop and build consensus for energy vision that will ensure the well being of future generations.

6473 Missouri Audubon Council

http://www.audubon.org/chapter/mo

The Audubon Society's mission is to conserve and restore natural ecosystems, focusing on birds and other wildlife for the benefit of humanity and earth's biological diversity. The purpose of the Missouri Audubon Council is to representthe interests of the 14 chapters of the National Audubon Society on a state level.

6474 Missouri Department of Conservation

http://www.conservation.state.mo.us

The mission of the Missouri Department of Conservation is to protect and manage the fish, forest, and wildlife resources of the state, to serve the public and facilitate their participation in resources management activities, toprovide opportunity fos all citizens to use, enjoy, and learn about fish, forest and wildlife resources.

6475 Missouri Prairie Foundation

http://www.moprairie.org

The Missouri Prairie Foundation mission is to preserve the Greater Prairie Chicken. To restore the vegetative and faunal balance to the grassland ecosystem, not just to one species, even through the praire chicken is on the verge ofextirpation from Missouri.

6476 Mmarie

http://www.kuleuven.ac.be/mmarie

The general objectives of MMARIE are to create an interdisciplinary forum for the exchange of information and experience, related to projects carried out by the participants, and to facilitate collaboration between the partners.

6477 Monarch Watch

http://www.monarchwatch.org

Our goals are to further science education, particularly in primary and secondary school systems, to promote the conservation of Monarch butterflies; and to invlove thousands of students and adults in a cooperative study of theMonarch's spectacular fall migration.

6478 Mountain Lion Foundation

http://www.mountainlion.org

The Mountain Lion Foudation is a nonprofit conservation and education organization dedicated to protecting the mountain lion, its wild habitat, and the wildlif that shares that habitat. The foundation is dedicated to the propositionthat much can be done to preserve the cougar as a viable species and that the success of this effort can assure the survival of other species.

6479 NEMO: Oceanographic Data Server

http://nemo.ucsd.edu

Nemo is a collection of data useful for physical oceanographers here at Scripps Institutions of Oceanography.

6480 NIREX

http://www.nirex.co.uk

NIREX, with the agreement of the Government, to examine safe, environmental and exconomic aspects of deep geological disposal of radioactive waste. We deal with intermediate level waste, which accounts for the majority of radioactivewaste currently in storage, and also with some low-level waste.

6481 NOAA (National Oceanic and Atmospheric Administration)

http://www.noaa.gov

NOAA's mission is to describe and predict changes in the Earth's environment, and conserve and wisely manage the Nation's coastal and marine resources. NOAA's strategy consist of seven interrelated Strategic Goals for environmentassessment,, predictions and stewardship.

6482 Napa County Resource Conservation

http://www.naparcd.org

Napa County Resource Conservation mission is to encourage and assist acceptance of individual responsibility for watershed management; the goals are enhacement of wildlife habitat, reduction of soil erosion, protection and enhacementof water quality, and promotion of land stewardship and sustainable agriculture.

6483 National Agricultural Pest Information System

http://www.ceris.purdue.edu

NAPIS is the database for the Cooperative Agricultural Pest Survey and is maintained by the Center for Environmental and Regulatory Information Systems (CERIS). This site contains pest information, the NAPIS User Guide, the CAPSProgram Guidebook, and the APHIS Environmental Manual. There is also a list of government certified nurseries.

6484 National Arborist Association

http://www.natlarb.com

NAA is a trade association of commercial tree care firms that develops that safety and education programs, satndards of tree care practice, and management information for arboriculture firms around the world.

6485 National Association for Environmental Management

http://www.naem.org

Dedicated to advancing the profession of environmental management and supports the professional corporate and facility environmental manager.

6486 National Association for Pet Container Resources

http://www.napcor.com

The National Association for Pet Container Resources is the trade association for the PET plastic industry in the United States and Canada. Its mission is to facilitate PET plastic recycling and to promote the usage of PET packaging.

6487 National Association of Conservation Districts

http://www.nacdnet.org

Association works with landowners, organizations and agency partners in the district helping to protect the soil, water, forest wildlife and other resources.

6488 National Association of Environmental Professionals

http://www.enfo.com/NAEP

NAEP is a multi-disciplinary association dedicated for the advancement of persons in the environmental profession in the US and abroad, a forum for state of the art information on environmental planning, research and management, anetwork of professional contacts and exchange on information among colleagues in industry, government, academic, and the private sector.

6489 National Association of State Foresters

http://www.stateforesters.org

State Foresters provide management assistance and protection services for over-two-thirds of the nation's forests.

6490 National Audubon Society

http://www.audubon.org

The mission of the National Audubon Society is to conserve and restor natural ecosystems, focusing on birds and other wildlife for the benefit of humanity and the earth's biological diversity.

6491 National BioEnergy Industries Association

http://solstice.crest.org

Bioenergy, renewable energy, energy efficiency, sustainable living

6492 National Center for Appropriate Technology

http://www.ncat.org

Helping people by championing small-scale, local, and sustainable solutions to reduce poverty, promote healthy communities, and protect natural resources.

6493 National Center for Atmospheric Research

http://www.ncar.ucar.edu

NCAR's mission to plan, organize, and conduct atmospheric and related research programs in collaboration with the universities and other institutions, to provide state of the art research tools and facilities to the atmosphericsciences community, to support and enhance university atmospheric science education, and to facilitate the transfer of technology to both the public and private sectors.

6494 National Center for Ecological Analysis and Synthesis

http://www.nceas.ucsb.edu/

The mission is to advance the state of ecological knowledge through the search for general patterns and principles, to organize and synthesize ecological information in a manner useful to researchers, resource managers, and policymakers addressing important environmental issues.

6495 National Councils for Sustainable Development

http://www.ncsdnetwork.org/

National Councils for Sustainable Development (NCSD) embraces the ideal of civil societies becoming partners with government in making policies for the impllementation os sustainable development.

6496 National Energy Foundation

http://www.nefl.org

The NEF is a nonprofit educational organization that has become a national leader in teacher training and student programs, instructional materials development and distribution and has done so in cooperation with sponsors frombusiness, government agencies and the education community.

6497 National Environmental Health Association (NEHA), CO

http://www.neha.org

NEHA is the only national association that represents all of environmental health and protection from terrorism and all-hazards preparedness, to food safety and protection and onsite wastewater systems. Over 4500 members and theprofession are served by the association through it Journal of Environmental Health, Annual Educational Conference and Exhibition, credentialing programs, research and development activities and other services.

6498 National Estuary Program

http://www.epa.gov/nep/nep.html

The National Estuary Program was established in 1987 by amendments to Clean Water to identify, restore, and protect nationally significant estuaries of the United States. NEP targets a broad range of issues and engages localcommunities in the process. The program focuses not just on improving water quality in an estuary, but on maintaining the integrity of the whole systems-its chemical, physical and biological properties, as well as its economic, recreational, andaesthetic values.

6499 National Ground Water Association

http://www.ngwa.org/

Providing and protecting the world's ground water resource. Enhance the skills and credibility of all ground water professionals, develop and exchange industry knowledge and promote the ground water industry and understanding ofground water resources.

6500 National Institute for Environmental Studies

http://www.nies.go.jp

The National Institute for Environmental Studies was established in 1974 at Tsukuba Science City, about 60 Km northeast of Tokyo, as the main research branch of the Environment Agency of the Government of Japan. NIES is the solenational institute for comprehensive research in the environmental sciences.

6501 National Institute of Environmental Health Science (NIEHS)

http://www.niehs.nih.gov/

The mission is to reduce the burden of human illness and dysfunction from environmental causes by understanding each of these elements and how they interrelate. The NIEHS achives its mission through multidisciplinary biomedicalresearch programs, prevention and intervention efforts, and communication strategies that encompass training, education, technology transfer, and community outreach.

6502 National Library for the Environment

http://www.cnie.org/nle

The National Library for the Environment is a universal, timely, easy to use,single point entry to quality environmental data and information for the use of all participants in the environmental enterprise. This online library icludesdirectories of academic environmental programs, journals, funding sources, meetings, job opportunities, news sources, laws and treaties, reports, reference materials, and more.

6503 National Oceanic and Atmospheric Administration

http://www.noaa.gov

NOAA mission is to describe and predict changes in the Earth's environment, and conserve and wisely manage the Nation's coastal and marine resources.

6504 National Outdoor Leadership School

http://www.nols.edu/nols.html

Over the past 35 years NOLS has become the leader in wilderness education. NOLS is now the largest backcountry permit holder in the United States and runs courses on four continents. NOLS has gone from 100 students in 1965 toapproximately 3,070 students in 1999. As NOLS enters the 21st century, it remainscommited to the quality of courses and programs that it offers, as well as to the wilderness environment that serves as our classroom.

6505 National Park and Conservation Association

http://www.npca.org/

The National Parks & Conservation Association has been the sole voice of the American people in the fight to safeguard the scenic beauty, wildlife, and historical and cultural treasures of the largest and most diverse park system inthe world.

6506 National Pollutant Inventory

http://www.npi.gov.au

Provides Australians with free access to information on the types and amounts of pollutants being emitted in their community.

6507 National Pollution Prevention Center for Higher Education (NPPC)

http://www.snre.umich.edu/nppc/

The National Pollutant Release Inventory was established under the Canadian Environmental Protection Act, to provide information on the type and quality of pollutantas being released into Canada's environment. The inventory is aNationwide, publicly accessible database of releases and transfers of 178 specified substances to air, water and land.

6508 National Renewable Energy Laboratory (NREL)

http://www.nrel.gov/

As the nation's leading center for renewable energy research, NREL is developing new energy technologies to benefit both the environment and the economy.

6509 National Sea Grant Library

http://nsgd.gso.uri.edu

The National Sea Grant Library was established as an archive and lending library for Sea Grant funded documents. These documents cover a wide variety of subjects, including oceanography, marine education, aquaculture, fisheries,limonology, coastal zone management, marine recreation and law. NSGL staff lends documents all over the world to aid scientists, teachers, students, fishermen and many other individuals in their reseacrh and studies.

6510 National Society for Clean Air

http://www.nsca.org.uk

NCSA is a nonprofit group made up of organizations and individuals who promote clean air through the reduction of air, water, and land pollution, noise and other contaminats, while having due regard for other aspects of theenvironment. The society exmines environmental policy issues from air quality perspective and aims to place them in a broader social and economic context.

Industry Web Sites/Environmental

6511 National Wildlife Health Center

http://www.nwhc.usgs.gov
The National Wildlife Health Center's mission is to improve information, technical assistance, and research on national and international wildlife health issues. To fulfill the NWHC mission, the Center monitors disease and assesses the impact od disease on wildlife populations; defines ecological relationships leading to occirrence of disease; transfers technology for disease prevention and control; and provides guidance, training and on site assistance for reducing wildlifelosses.

6512 National Wildlife Rehabilitators Association

http://www.nwrawildlife.org
The National Wildlife Rehabilitation Association is a nonprofit international membership organization committed to promoting and improving the integrity and professionalism of wildlife rehabilitation and contributing to thepreservation of natural ecosystems. Please, call before you fax!

6513 National Woodland Owners Association

http://www.nationalwoodlands.org
Provides timely information about forestry and forest practices with news from washington,Dc and state capitals. written for non-industrial land owners. Includes state landowner association news.

6514 Native Americans and the Environment

http://www.cnie.org/NAE
The mission is to educate the public on environmental problems in Native American Communities; to explore the values and historical experiences that Native Americans bring to bear on environmental issues and to promote conservationmeasures that respect Native American land and resource rights.

6515 Native Forest Council

http://www.forestcouncil.org
The mission is to provide visionary leadership and to ensure the integrity of public land ecosystems, without compromising people or forests.

6516 Native Forest Network

http://www.native forest.org/
The Native Forest Network's mission is to protect the world's remaining native forest by they temperature or otherwise, to ensure they can survive, flourish and maintain their evolutionary potential.

6517 Natural Energy Laboratory of Hawaii

E-mail: inquires@nelha.org
http://www.nelha.org
NELHA mangers comprehensive environmental monitoring of seawater. This site describes the participants in this program and the projects in which they are involved. NELHA's goal is to help businesses utilize Hawaii's natural resources.Information available here includes water quality data, NELHA's Seawater Delivery System information, Ocean Thermal Energy Converstion material, and bibliographical information.
Thomas H Daniel, Scientific Manager

6518 Natural Environmental Research Council

http://www.nerc.ac.uk/
The mission is to promote and support, by any means, high quality basic, strategic and applied research, survey, long-term environmental monitoring and related postgraduate training in terrestrial, marine and freshwater biology andEarth, atmospheric, hydrological, oceanographic and polar sciences and Earth observation. To provide advice on, dieeminate knowledge and promote public understanding of the fields aforesaid.

6519 Natural Resources Defense Council

http://www.nrdc.org/nrdc/

The Natural Resources Defense Council's purpose is to safaguard the Earth. We work to restore the integrity of the elements that sustain life, and to defend endangered natural places. We seek to establish sustainability and goodstewardship of the Earth as central ethical imperatives oh human society. NRCDaffirms the integral place of human beings in the environment. We strive to protect nature in ways that advance the long-term welfare of present and future generations.

6520 Nature Conservancy

http://www.tnc.org/
The mission of The Nature Conservancy is to preserve the plants, animals and natural communities that represent the diversity of life on Earth by protecting the lands and waters they need to survive.

6521 Nature Conservancy of Texas

http://www.tnc.org/texas/
The Nature Conservancy of Texas conserves habitat for native wildlife, using science based research and a cooperative approach to protect the animals and plants that represent Texa's precious natural heritage.

6522 Nature Node

http://www.naturenode.com
Online community celebrating the diversity of life on our planet. It is a place to build a virtual community of like-minded individuals.

6523 Nature Node's Animal Interviews

http://www.naturenode.com/rebop/rebop.html
Altrough these animal interviews are aimed at kids, adults love them too. How animals interact with and adapt to their natural environments is stressed. The natural history of each animal interviewed is also covered in detail. Theinterviews are presented in play form with the intention of having them used with finger puppets that kids make.

6524 Nature Saskatchewan

http://www.naturesask.ca
Our mission is to promote appreciation and understanding of our natural environment through education, and through conservation and research, to protec and preserve natural ecosystems and their biodiversity.

6525 NatureNet

http://naturenet.net/index.html
NatureNet is a voluntary enterprise to provide a good resource for practical nature conservation and countryside management on the Web. Based in UK, and most of the information available on Naturenet relates to the UK, particularyEngland.

6526 New England Wild Flower Society

http://www.newfs.org
The New England Wild Flower Society is the oldest plant conservation organization in the United States, promoting the conservation of temperature North American plants through key programs.

6527 New England Wildlife Center

http://www.newildlife.com
The New England Wildlife Center is a native wildlife preservation, rehabilitation , animal habitat and environmetal protection, educational organization. We afford humane care to native and naturalized wild animals through our wildlifemedicine hospital. We conduct no research on patients. The center is a nonprofit environmental native habitat protection and preservation organization.

6528 New Forest Project

http://www.newforestsproject.com/

To protect conserve, and enhance the health of the Earth's ecosystems by supporting integrated grassrroots efforts to maintain and rebuild the world's forest through the promotion of agroforestry, reforestation, the protection ofwatersheds, and the initiation of renewable energy products.

6529 New Hampshire Department of Environmental Sciences (DES)

http://www.state .nh.us/des/descover.htm

The mission of the Department of Environmental Services is to protect, maintain and anhance environmental quality public health in New Hampshire.

6530 New Hampshire Fish and Game Department

http://www.wildlife.state.nh.us/

The mission is to conserve, manage and protect these resources and their habitats; inform and educate the public about these resources and provide the public with opportunities to use and appreciate these resources.

6531 New Jersey Department of Environmental Protection

http://www.state.nj.us/dep/

The mission is to assist the residents of New Jersey in preserving, retoring, sustaining, protecting and enhancing the environment to ensure the integration of high environmental quality, public health and economic vitality.

6532 New Jersey Division of Fish, Game and Wildlife

http://www.state.nj.us/dep/fgw/

The mission of the Divison of Fish, Game, and Wildlife is to protect and manage the State's fish and wildlife resources to maximize their long term economic, recreational and biological values for the citizens of New Jersey.

6533 New Mexico Wilderness Alliance

http://www.nmwild.org

The New Mexico Wilderness Allince is dedicated to the protection, restoration, and rewilding of New Mexico's Wilderness areas. We focus on forward-looking measures to develop an active and educates Wilderness constituency throughoutthe state.

6534 New Mexico Wildlife Association

http://www.wildlifewest.org

The New Mexico Wildlife Association is a nonprofit corporation dedicated to yhe preservation of the rich heritage of native New Mexico wildlife and its habitat through education, scientific research, and the sponsorship of wildlifepark.

6535 New York Association for Reduction, Reuse and Recycling

http://www.recyclexchange.com

The mission is to provide state-wide leadership on waste reduction, reuse and recycling issues and practices.

6536 New York State Department of Environmental Conservation

http://www.dec.state.ny.us

The mission of the department is to conserve, improve, and protect its natural resources and environment, and control water, land and air ollution, in order to enhance the health, safety and welfare of the peolple of the state andtheir overall economic and social well being.

6537 North American Commission for Environmental Co operation

http://www.cec.org

The Commission for Environmental Cooperation is an international organization created by Canada, Mexico and the US under the North American Agreement on Environmental Cooperation. The CEC was established to address regionalenvironmental concerns, help prevent potential trade and environmental conflicts and to promote the effective enforcement of environmental law.

6538 North American Lake Management Society

http://www.nalms.org

Members are academics, lake managers and others interested in furthering the understanding of lake ecology.

6539 North Carolina Coastal Federation

http://www.nccoast.org/

NNCF is a nonprofit tax exempt organization which seeks to protect and restore the states's coastal environment, culture and economy through citizen involvement in the management of coastal resources.

6540 North Carolina Department of Environment and Natural Resources

http://www.enr.state.nc.us

The mission is to lead stewardship agency for preservation and protection of North Carolina's outstanding natural resources. The organization administers regulatory programs designed to protect air quality, water quality, and thepublic's health.

6541 North Cascades Conservation Council

http://www.northcascades.org

The NCCC keeps government officials, environmental organizations, and the general public informed about issues affecting the Greater North Cascade Ecosystem. Action is pursued through legislative, legal, and public participationchannels to protect the lands, waters, plants and wildlife.

6542 Northeast Advanced Vehicle Consortium

http://www.navc.org/home.html

The Northeast Advanced Vehicle Consortium is a nonprofit association of private and public sector firms and agencies workimg together to promote advanced vehicle technologies in the Northeast US. NAVC is now the principal multi-state,nonprofit funding mechanism for advanced transportation research, technology development and demostration in the region.

6543 Northeast Sustainable Energy Association

http://www.nesa.org/neshome.htm

NESEA is a regional membership organization comprised of engineers, educators, builders, students, energy experts, environmental activist, transportation planners, architects, and other citizens interested in responsible energy use.Ourgoal is to bring clean electricity, green transportation, and healthy, efficient buildings into everyday use in order to strengthen the economy and improve the environment.

6544 Northern Prairie Wildlife Research Center

http://www.npwrc.usgs.gov/index.htm

Our mission is to develop research information on the quantitative requirements for sustainable wildlife populations. To design and conduct studies of numbers and distribution of flora and fauna including identification of changeresulting from habitat loss and modification. To disseminate the latest in technical information and research findings such that interested audiences benefit to the maximum extent possible.

6545 Ocean Voice International

http://www.ovi.ca/

Ocean Voice International is a nonprofit membership based marine environmental organization.

6546 Oceania Project

http://nornet.nor.com.au/users/oceania/index.html

The Oceania Project mission is to promotr awareness and co-operation to instigate and maintain the process of rehabilitation, preservation and conservation of Cetacea and the Oceans; the promotion and undertaking of scientific researchof Cetacea and the Oceans for the benefit of the community. To provide environmentally sensitive Ocean platforms for non-manipulative research, education and experiential programmes of Cetacea and the Oceans using sensitive vessels.

6547 Oceanic Resource Foundation

http://www.orf.org/

The Oceanic Resource Foundation is a nonprofit, scientific researchorganization dedicated to the preservation of the global marine environment and marine biological diversity.

6548 Office of Energy Efficiency

http://oee.nrcan.gc.ca/oee_e.cfm

The office of Energy Efficiency, Canada's centre of excellence for energy efficiency and alternative fuels information, is pplaying a dynamic leadership role in helping Canadians save millions of dollars in energy cost while addressingthe challenges of climate change.

6549 Office of Protected Resources

http://www.nmfs.noaa.gov

The Office of Protected Resources provides program oversight, national policy direction and guidance on the conservation of those marine mammals and endangered species, and their habitats, under the jurisdiction of the Secretary ofCommerce; develops national guidelines and policies for relevant protected resources programs, and provides oversight, advice and guidance on scientific aspects of managing protected species and marine protected areas.

6550 Ohio Environmental Protection Agency

http://www.epa.state.oh.us

The Ohio EPA has authority to implement laws and regulations regarding air and water quality standards; solid hazardous and infectious waste disposal standards; quality planning, supervision of sewage treatment and public drinkingwater supplies; and cleanup of unregulated hazardous waste sites. The Ohio EPA cooperates with government and private agencies, manages some federally funded pollution control projects, obtains technical and laboratory services, investigateenvironment problems, etc.

6551 Ohio Wildlife Center

http://www.ohiowildlifecenter.org/

The Ohio Wildlife Center is a nonprofit educational organization that promotes increased appreciation and understanding of the natural environment, with particular emphasis on wildlife. OWC is supported by individuals from all walks oflife who wish to improve their own understanding of native wild species and local wildlife issues.

6552 Oklahoma Department of Wildlife Conservation

http://www.wildlifedepartment.com

The mission is to manage Oklahoma's wildlife resources and habitat to provide scientific, educational, aesthetic, economic and recreational benefits for present and future generations of hunters, anglers and others who appreciatewildlife.

6553 Ontario Environment Network

http://www.oen.ca

The Ontario Environment Network is a nonprofit, nongovernmental network serving Ontario's environmental nonprofit, nongovernmental community. The OENÆseeks to increase awareness of these organizations and encourage discussions aboutmeans to protect the environment.

6554 Organization of American States: Department of Regional Development and Environment

http://www.oas.org

Conducts technical cooperation and training programs to assist the member States in their efforts to preserve natural resources. It works with the countries on planning sustainable development, managing the environment and preparinginvestment programs and projects.

6555 PureZone

http://www.purezone.com

PureZone is a group effort among manufacturers, suppliers, and installers of Heating and Ventilating systems designed to maximize the quality of aour air in our buildings.

6556 Rachel Carson Council

http://www.rachelcarsoncouncil.com

A clearinghouse and library with information at both scientific and layperson levels on pesticides related issues. Rachel Carson Council develops its knowledge from literature searches and conservations with experts. It then providesanswers to the public and also products various publications clarifying pestcide dangers and bringing alternative pest controls to the public's attention.

6557 Renewable Fuels Association

http://www.ethenolrfa.org

Members are companies and individuals involved in the production and use of ethanol. Ethanol is sold nationwide as a high-octane fuel that delivers improved vehicle performance while reducing emmissions and improving air quality.

6558 Renewable Natural Resources Foundation

http://www.rnrf.org

Consortium of professional and scientific organizations with an interest in natural rsources. Established to advance sciences and education in renewable natural resources; promote the application of sound scientific practices inmanaging and conserving renewable natural resources; foster coordination and cooperation among professional, scientific and educational organizations having leadership responsibilities for renewable natural resources; and develop a Renewable NaturalResources Center.

6559 Silicone Health Council

http://www.socma.com

Coordinates health, environmental and safety programs. Conveys scientifically sound information about silicones.

6560 Society for Conservation Biology (SCB)

http://www.conbio.org/

The mission of the Center for Conservation Biology Network is to help develop the technical means for the protection, maintenance and restoration of life on this planet-its species, its ecological and evolutionary processes and itsparticular and total environment; to help raise awareness, educate and encourage personal involvement of the public and academics alike.

6561 Steel Recycling Institute

http://www.recycle-steel.org

Promotes steel recycling and works to forge a coalition of steelmakers, can manufacturers, legislators, government officials, solid waste managers, business and consumer groups.

6562 Student Conservation Association

http://www.sca-inc.org

To build the next generation of conservation leaders and inspire lifelong stewardship of our environment and communities by engaging young people in hands on service to the land.

6563 TechKnow

http://www.techknow.org

TechKnow is an interactive database that is available over the Internet. The database is a springboard for people interested in Environmentally Sustainable technologies. Originally, TechKnow only contained environmental remediationtechnologies but has now branched off to other forms of sustainable technologies. One of the main goals of the database is to remain current and topical.

6564 The Environment Directory

http://www.webdirectory.com

Includes agriculture, animals, arts, business, databases, design, disasters, education, employment, energy, forestry, general environmental interest, government, health, land conservation, parks and recreation, pollution, products andservices, publications, recycling, science, social science, sustainable development, transportation, usenet newsgroups, vegetarianism, water resources, weather and wildlife.

6565 US Environmental Protection Agency: Environmental Monitoring and Assessment Program

http://www.epa.gov

A research program to develop the tools necessary to monitor and assess the status and trends of national ecological resources.

6566 United States Geological Survey: National Earthquake Information Centre

http://www.earthquake.usgs.gov

Determine location and size of all destructive eartquakes worldwide and immediatly disseminate this information to concerned national and agencies, scientistsm, and the general public

6567 Vertical Net

http://www.verticalnet.com

Vertical Net is the Internet's leading bussines e-commerce enabler, providing ent to end e-commerce salutions that are targeted at district business segments through two strategic business units- Vertical Markets and Vertical NetSolutions. While both units focus on a core area of eexpertise, each also leverages the strengths, resources and experience of the other.

6568 Wisconsin Sea Grant Program

http://www.seagrant.wisc.edu

Wisconsin Sea grant is a statewide program of basic and applied research, education, and outreach and technology transfer dedicated to the stewardship and sustainable use of the nation's Great Lakes and ocean resources.

6569 World Data Centre: National Geophysical Data Centre

http://www.ngdc.noaa.gov

Data management in the broadest sense. Play an integral role in the nation's research into the environment, and at the same tome provide public domain data to a wide group of users.

6570 World Fish Center

http://www.worldfishcenter.org/

Our mission is to promote sustainable development and use of living aquatic based on environmentally sound management.

6571 World Women in Environment and Development Organization

http://www.wedo.org

International advocacy organization that seeks to increase the power of women worldwide as policymakers at all levels in governments, institutions and forums to achive economic and social justice, healthy and peaceful planet, and humanrights for all.

Environmental

6572 Acid Rain Information Clearinghouse
46 Prince Street   706-271-3550
Rochester, NY 14607   Fax: 716-271-0606

6573 Air Resources Information Clearinghouse Center
46 Prince Street   716-271-3550
Rochester, NY 14607   Fax: 716-271-0606
Elizabeth Thorndike, President

6574 Air Risk Information Support Center Hotline: US Office of Air Quality Planning and Standards
Mail Drop 13   919-541-0888
Research Triangle Park, NC 27709   Fax: 919-541-4028
Holly Reid, Co-chair

6575 Alternative Treatment Technology Information Center
4 Research Place   301-670-6294
Suite 210   Fax: 301-670-3815
Rockville, MD 20850
Gary Turner, System Operator

6576 Asbestos Ombudsman Clearinghouse/Hotline
401 M Street SW   703-305-5938
A-149 C   800-368-5888
Washington, DC 20460   Fax: 703-305-6462
Karen V Brown, Ombudsman

6577 Bureau of Explosives Hotline
50 F Street NW   202-639-2222
Washington, DC 20001   Fax: 412-741-0609
                    http://www.aar.org/aarhome.nsf

6578 CARIERS US Department of Energy
PO Box 3048
Merrifield, VA 22116   800-523-2929
Paul Hesse, Project Manager

6579 CQS Health and Environmental
2192 Massachusetts Avenue   617-491-7646
Cambridge, MA 02140   E-mail: jon@cqs.com
                              http://www.cqs.com

6580 Carbon Dioxide Information Analysis Center: Environmental Services Division
Oak Ridge National Lab Bldg 1000   423-574-0390
PO Box 2008
Oak Ridge, TN 37830
RI Vanhook, Dir. Opns. & User Servs. Mgr

6581 Center for Environmental Research Information ORD Research Information Unit
26 West Martin Luther King Drive   513-569-7562
MS G-72
Cincinnati, OH 45268
Dorothy Williams, Executive Officer

6582 Center for Environmental and Regulatory Information Systems: Purdue University
1231 Cumberland Avenue   765-494-6614
Suite A   Fax: 765-494-9727
West Layfayette, IN 47906-13E7mail: info@ceris.purdue.edu
                http://www.ceris.purdue.edu/ceris/main.html
Virginia Walters, Environmental Contact

6583 Center for Health, Environment and Justice
PO Box 6806   703-237-2249
Falls Church, VA 22040

The Center for Heath, Environment and Justice trains and assists local people to fight for justice, become empowered to protect their communities from environmental threats and build strong locally controlled organizations. CHEJconnects these strong local groups with each other to build a movement from the bottom up.
Lois Marie Gibbs, Executive Director

6584 Center for Sustainable Systems University of Michigan
440 Church Street   734-764-1412
Dana Building   Fax: 734-647-5841
Ann Arbor, MI 48109-1115   E-mail: css.info@umich.edu
                              http://http://css.snre.umich.edu
CSS advances concepts of sustainability through interdisciplinary research and education. Collaborates with diverse stakeholders to develop and apply life cycle based models and sustainability metrics for systems that meet societalneeds. Promotes tools and knowledge that support the design, evaluation, and improvement of complex systems.

Jonathan W Bulkley, Co-Director
Gregory A Keoleian, Co-Director

6585 Chemtrec Center
2501 M Street NW
Washington, DC 20037   800-CMA-8200

6586 Chemtrec Hotline
2501 M Street NW
Washington, DC 20037   800-424-9300

6587 Clean Ocean Action
PO Box 505   732-872-0111
Sandy Hook   Fax: 732-872-8041
Highlands, NJ 077E2mail: SandyHook@cleanoceanaction.org
                              http://www.cleanoceanaction.org
Clean Ocean Action is a broad-based coalition of over 150 conservation, community, diving, fishing, environmental, surfing, women's and business groups that works to improve and protect the waters off the New York and New Jersey coast.

Kari Jermansen, Outreach Director

6588 Clean-Up Information Bulletin Board System: US EPA Technology Innovation Office
401 M Street   301-589-8368
OS-10W   Fax: 301-589-8487
Washington, DC 20460
Beth Ann Kyle, System Operator

6589 Congressional Clearinghouse on the Future
H2-555 House Annex 2   202-226-3434
Washington, DC 20515

6590 Conservation Locator
NYS Department of Environmental Conservation
625 Broadway   518-402-8013
2nd Floor   Fax: 518-402-9036
Albany, NY 12233 http://www.dec.state.ny.us/website/locator
A web-based index of current DEC publications, with listings by subject, links to publications available on the web and directions for obtaing copies of print-only publications.
Helen Paruolo, EnCon Program Assistant

6591 Consumer Energy Council of America Research Foundation
2000 L Street NW   202-659-0404
Suite 802   Fax: 202-659-0407
Washington, DC 20036   E-mail: eberman@cecarf.org
                              http://www.cecarf.org
Founded in 1973, CECA is the senior public interest organization in the US focusing on the energy, telecommunications and other network industries that provide essential services to consumers.
Ellen Berman, President

6592 Consumer Product Safety Commission Hotline
Office of the Secretary
Washington, DC 20207   800-638-2882

6593 Control Technology Center Emission Standards Division
US EPA, MD 13   919-541-0800
Research Triangle Park, NC 27709   Fax: 919-541-0072
Bob Blaszczak, ESD/ QAQPS

6594 DOT Hotline US Department of Transportation
400 7th Street, Room 8100   202-366-4488
Office of Hazardous Material Standards 888-327-4236
Washington, DC 20530-0001 E-mail: hotline@nhtsa.dot.gov
                           http://www.nhtsa.dot.gov/hotline

6595 EPA Model Clearinghouse
Office of Air Quality Planning   919-541-5683
Research Triangle Park, NC 27711   Fax: 919-541-2464
Dean A Wilson

6596 EPA Public Information Center: US EPA
401 M Street   202-260-7751
PM- 211B
Washington, DC 20460

6597 EPCRA (SARA Title III) Hotline: US EPA
401 M Street SW   202-479-2449
OS120   800-535-0202
Washington, DC 20460   Fax: 703-412-3333

6598 ERIC Clearinghouse for Science, Mathematics and Environmental Education: Ohio State University
1929 Kenney Road   614-292-6717
Room 200   Fax: 614-292-0263
Columbus, OH 43210   E-mail: ericse@osu.edu
                           http://www.stemworks.org/bhm.html
ERIC, the Educational Resources Information Center clearinghouses are federally funded, nation-wide information networks designed to provide you with ready access to education literature. Each clearinghouse is subject specific. ERIC/CSMEE covers science, mathematics, and environmental education at all levels, including the development of curriculum and instruction materials; teachers and teacher education; learning theory and outcomes; research and evaluative studies, etc.
Dr David L Haury, Director
Susan B Eshbaugh, User Services Coordinator

6599 Educational Resource Information Center/ Clearinghouse for Science, Math and Env. Education
Ohio State University
1929 Kenny Road   614-292-6717
Coulumbus, OH 43210-1080   E-mail: ericse@osu.edu
                                       http://www.ericse.org
Dr. David L Haury, Director

6600 Emergency Planning and Community Right-to-Know Information Hotline- Booz-Allen and Hamilton
1725 Jefferson Davis Highway   703-920-8977
Arlington, VA 22202   800-535-0202
                                       Fax: 703-486-3333
Dan Kovacs, Contractor

6601 Emission Factor Clearinghouse: US EPA
MD-14   919-541-5477
Research Triangle Park, NC 27709   Fax: 919-541-0684
Dennis Shipman

6602 Enviro Village
                           http://www.envirovillage.com
The #1 Indoor Air Quality and Indoor Environment Resource on the Web.

6603 Environmental Connection: Austin
PO Box 27437   512-463-5381
Austin, TX 78755-2437
Jo Ann Farrell

6604 Environmental Financing Information Network: US EPA
EFN, WH-547   202-260-0420
401 M Street, East Tower, Room 1117   Fax: 202-260-1827
Washington, DC 20460
June Lobit

6605 Green Committees of Correspondence Clearinghouse
PO Box 30208   816-931-9366
Kansas City, MO 64112
Amy Belanger, Coordinator

6606 Green Lights Program: The Bruce Company
1850 K Street 290   202-775-6650
Washington, DC 20006   Fax: 202-775-6680
Maria Theesen

6607 Hazardous Waste Ombudsman Program: US EPA
401 M Street SW   202-260-9361
OS-130, Room SE 315   800-262-7937
Washington, DC 20460   Fax: 202-260-8929
                           http://www.epa.gov/earth100/records/000154.html
Bob Knox, Headquarters Contact

6608 Indoor Air Quality Information Center
PO Box 37133   703-356-4020
Washington, DC 20013-7133   800-438-4318
                                       Fax: 703-356-5386
                           E-mail: iaqinfo@aol.com
                           http://www.epa.gov/iaq/iaqinfo.html
Susan Dolgin

6609 Inspector General Hotline: US Environmental Protection Agency
401 M Street SW
Room 307 NE   800-424-4000
Washington, DC 20460   Fax: 202-260-6976
Ed Maddox

6610 International Clearinghouse on the Military and the Environment
PO Box 150753   718-788-6071
Brooklyn, NY 11215   Fax: 718-788-6071
                           E-mail: fbp@igc.apc.org
John Miller

6611 International Ground Water Modeling Center
Colorado School of Mines   303-273-3103
Golden, CO 80401   Fax: 303-273-3278
Paul van der Hijde, Director

6612 Kentucky Partners: State Waste Reduction Center
University of Louisville
Ernst Hall   502-588-7260
Room 312
Louisville, KY 40292
Joyce St. Clair, Executive Director

6613 Learning about Backyard Birds
                           http://www.birdfeeding.org
Free resource available from the National Bird-feeding Society. Download this big, loose-leaf collection of resource material perfect for kids. It features background and guidance for instructors and facilitators as well as working materials and projects for group use.

6614 Life on a Rocky Shore
Earthwise Media
PO Box 1223   360-271-1584
Poulsbo, WA 98370   Fax: 360-394-2168
E-mail: info@earthwisevideos.com
http://www.earthwisevideos.com
A colorful Powerpoint slide show, provides an introduction to
the plants and animals of the Pacific Northwest's rich Intertidal
Zone. Produced in cooperation with the Marine Science Society
of the Pacific Northwest.

Nancy Sefton, Creative Director
Wes Nicholson, Technical Director

6615 Methods Information Communications Exchange
Reston, VA 20190   703-676-4690
Fax: 703-318-4682
E-mail: mice@cpmx.saic.com
http://www.epa.gov/sw-846/mice.htm
Sara Hartwell, Contractor

**6616 Minority Energy Information Clearinghouse: Office of
Minority Economic Inpact, US Dept Energy**
100 Independence Ave SW   202-586-5876
Forrestal Building, Room 5B-110   800-543-2325
Washington, DC 20585
Effie A Young, Officer

**6617 Montana State Library: Montana Natural Resource In-
formation System**
1515 East 6th Avenue   406-444-3115
Helena, MT 59620   Fax: 406-444-5612
E-mail: msl.state.mt.us
Alan Cox

**6618 National Air Toxics Information Clearinghouse: US
EPA**
Mail Drop 13   919-541-0850
Office of Air Quality and Standards Fax: 919-541-4028
Research Triangle Park, NC 27709
Vasu Kilaru, Database Administrator

**6619 National Capital Poison Center Georgetown University
Hospital**
3800 Resevior Road NW   202-625-3333
Washington, DC 20007

6620 National Center for Biotechnology Information
National Library of Medicine   301-496-2475
Building 38A, Room 8N805   Fax: 301-480-9241
Bethseda, MD 20894   E-mail: info@ncbi.nkm.nih.gov
http://www.ncbi.nlm.nih.gov
The National center for Biotechnology Information creates a
public databases, conducts research in computational biology,
develops software tools for analyzing genome data, and
dissemanates biomediacl information, all for the
betterunderstanding of molecular processes affecting human
health and disease.

6621 National Ground Water Information Center
6375 Riverside Drive   614-761-0700
Dublin, OH 43017   800-242-4965
Fax: 614-761-3446
Kevin McCray, Assistant Executive Director

6622 National Pesticide Information Retrieval System
1231 Cumberland Avenue   765-494-6614
Suite A   Fax: 765-494-9727
West Lafayette, IN 4706-1317
Virginia Walters

**6623 National Pesticide Telecommunications Network Dept.
of Preventive Medicine & Community Health**
Texast Tech University Health
School of Medicine   806-858-7378
Sciences Center   800-858-PEST
Lubbock, TX 79430   Fax: 806-743-3094
Frank L Davido

6624 National Radon Hotline National Safety Council
1025 Connecticut Avenue NW   202-293-2270
#1200   800-767-7236
Washington, DC 20036   Fax: 202-293-0032
E-mail: airqual@nsc.org
http://www.nsu.org/issues/radon/index.htm
For twelve years, the National Safety Council has been providing information on Radon and other indoor air quality issues
through various toll-free hotlines and publications.
Kristin Marstiller, Senior Program Manager

**6625 National Renewable Energy Laboratory Technical In-
quiry Service**
1617 Cole Boulevard   303-275-4099
Golden, CO 80401-3393
Steve Rubin, Manager

**6626 National Response Center: US Coast Guard Headquar-
ters**
2100 2nd Street SW   202-267-2675
Room 2611   800-424-8802
Washington, DC 20593-0001   Fax: 202-267-2181
Commander David Beach

**6627 National Small Flows Clearinghouse- West Virginia
University**
NRCCE Building, Evandale Drive   304-293-4191
PO Box 6064   800-624-8301
Morgantown, WV 26506   Fax: 304-293-3161
E-mail: nsfc_orders@mail.nesc.wvu.edu
http://www.nsfc.wvu.edu
Nonprofit national source of information about small flows
technologies-those systems that have fewer than one million
gallons of wastewater flowing through them per day-ranging
from individual septic systems to small sewage treatmentplants.
Offers more than 450 free and low-cost educational products, a
toll-free technical assistance hotline, five computer databases,
two free publications and an online discussion group.
Peter Casey, NSFC Program Coordinator
Jen Hause, NSFC Engineering Scientist

**6628 New York State Department of Environmental Conser-
vation**
625 Broadway   518-402-8540
Pollution Prevention Unit
Albany, NY 12233-1011
John E Iannotti, PE

6629 Northeast Industrial Waste Exchange
720 Erie Boulevard West   315-422-6572
Suite 211   Fax: 315-422-4005
Syracuse, NY 13204
Carrie Mauhs-Pugh, President

**6630 Northeast Waste Management Officials' Association:
NE Multi-Media Pollution Prevention**
85 Merrimac Street   617-367-8558
Boston, MA 02114
Terri Goldberg, Program Manager

6631 Nuclear Information and Resource Service
1424 16th Street NW   202-328-0002
Suite 404   Fax: 202-462-2183
Washington, DC 20036   E-mail: nirsnet@nirs.org
http://www.nirs.org
NIRS is the information and networking center for citizens and
environmental organizations concerned about nuclear power,
radioactive waste, radiation, and sustainable energy issues.

Michael Mariotte, Executive Director

6632 OTS Chemical Assessment Desk
401 M Street SW   202-260-3583
(TS-778)
Washington, DC 20460
Terry O'Bryan, Executive Officer

6633 Office of Research and Development Electronic Bulletin Board System: US EPA
25 West Martin Luther King Drive   513-569-7610
Cincinnati, OH-4526   800-258-9605
                              Fax: 513-569-7566

Charles W Guion

6634 Pesticide Action Network North America
116 New Montgomery   415-541-9140
Suite 810   Fax: 415-541-9253
San Francisco, CA 94105
Monica Moore, Program Director

6635 Powder River Basin Resource Council Energy Conservation Education
23 North Scott   307-672-5809
Sheridan, WY 82801
Jill Morrison, Organizer

6636 Public Information Center: US EPA
401 M Street SW   202-260-7751
PM-211B
Washington, DC 20460
Alison Cook, Director

6637 QAQPS TNN Bulletin Board System
Office of Air Quality Planning and   919-541-5742
Standards Technology Transfer Network
Research Triangle Park, NC 27709
Hersch Rorex, System Manager

6638 RACT/BACT/LAER Clearinghouse Office of Air Quality Planning and Standards
Emissions Standards Division   919-541-0800
MD-13   Fax: 919-541-0072
Research Triangle Park, NC 27709
Bob Blaszczak, ESD

6639 Rachel Carson Council
8940 Jones Mill Road   301-652-1877
Chevy Chase, MD 20815
Dr. Diana Post, Executive Director

6640 Records of Decision System Hotline Computer Sciences Corporation
401 M Street SW   202-260-3770
Room L101
Washington, DC 20460
Thomas Batts

6641 Risk Communication Hotline: US EPA
401 M Street SW   202-260-5606
W Tower, Room 425   Fax: 202-260-9757
Washington, DC 20460
Ernestine Thomas

6642 Safe Drinking Water Hotline
LaBat-Anderson, Inc.

                              800-426-4791

6643 Small Business Ombudsman Clearinghouse/Hotline US EPA Office of Small Business
401 M Street SW   703-305-5938
Washington, DC 20460   800-368-5888
                              Fax: 703-305-6462

Karen V Brown, Manager

6644 Solid Waste Information Clearinghouse and Hotline
1100 Wayne Avenue, Suite 700
PO Box 7219   800-467-9262
Silver Spring, MD 20907-7219   Fax: 301-589-7068
                      E-mail: tvondeak@swana.org
                              http://www.swana.org
Todd von Deak, Dir Marketing/Member Service

6645 Stratospheric Ozone Information Hotline/The Bruce Company
US Environmental Protection Agency
1200 Pennsylvania Avenue NW   202-343-9410
Washington, DC 20460-0001   800-296-1996
                          Fax: 202-343-2363
                  E-mail: green.vanessa@epa.gov
                          http://www.epa.gov\ozone

Eric Firstenberg, Contractor

6646 Sustainable Buildings Industry Council
1331 H Street NW   202-628-7400
Suite 600   Fax: 202-393-5043
Washington, DC 20005   E-mail: SBIC@SBICouncil.org
                          http://www.sbicouncil.org
Helen English, Executive Director

6647 TSCA Assistance Information Service Hotline Environmental Assistance Division
1200 Pennsylvania Avenue NW   202-554-1404
Mail Code 74080   Fax: 202-554-5603
Washington, DC 20460 E-mail: tsca-hotline@epamail.epa.gov
The information service furnishes TSCA regulation information to the chemical industry, labor and trade organization, environmental groups and the general public. Technical as well as general information is available.
John Alter, Primary EPA

6648 Toxicology Information Response Center Oak Ridge National Laboratory
Building 2001   865-574-4160
PO Box 2008   Fax: 865-574-0595
Oak Ridge, TN 37831   E-mail: stairbr@ornl.gov
                              http://www.ornl.gov

Jeffrey Wadsworth, Director

6649 Toxnet

                      http://toxnet.nlm.nih.gov
TOXNET offers a high quality database, some scientifically peer-reviewed, in an easy to use interface, and includes links to additional sources of dat related toxicology and environmental health. An array of information to inform thepublic and the scientific community about environmental hazards. TOXNET is a product of the National Library of Medicine's Toxicology and Environmental Infromation Program.

6650 US Global Change Data and Information System

                      http://globalchange.gov
Gateway to Global Change Data.

6651 US Global Research Information Office
61 Route 9W   845-365-8930
PO Box 1000   Fax: 845-365-8922
Palisades, NY 10964   E-mail: help@gcrio.org
                          http://gcrio.ciesin.org
The US Global Research Information Office, provides access to data and information on global environmetal change research, adaptation/mitigation strategies and technologies, and global change related educational resources.

6652 Waste Exchange Clearinghouse Waste Systems Institute
of Michigan
400 Ann Street NW   616-363-3262
Number 201-A
Grand Rapids, MI 49504-2054
Jeffery L Duphin

6653 Wastewater Treatment Information Exhange National
Small Flows Clearinghouse
West Virginia University
PO Box 6064   800-544-1936
Morgantown, WV 56506-6064   Fax: 304-293-3161
Loukis Kissonergis

6654 Wetlands Protection Hotline Geo/Resource Consultants
1555 Wilson Boulevard   703-527-5190
Suite 500   800-832-7828
Arlington, VA 22209
John Ruffing

6655 White Lung Association
PO Box 1483   410-243-5864
Baltimore, MD 21203-1483   Fax: 410-243-5892
James Fite, Executive Director

6656 Wisconsin Energy Information Clearinghouse Wiscon-
sin Division of Energy
Department of Administration
101 East Wilson Street   608-266-8234
PO Box 7868   608-266-8870
Madison, WI 53707-7868   Fax: 608-267-6931
E-mail: patmeier@doa.state.wi.us
http://www.doa.state.wi.us/energy/
Patrick Meier, State Energy Office Director

Environmental

6657 AIMS Multimedia
9710 DeSoto Avenue   818-773-4300
Chatsworth, CA 91311-4409   800-367-2467
E-mail: info@aimsmultimedia.com
http://www.aimsmultimedia.com
AIMS is a leader in the production and distribution of educational videos, DVDs and interactive CD-ROMS. Our library of award-winning programs, targeted at the K-12 and adult markets, features many environmental and ecological titles.
Wynn Sherman, President
Don Lukenbill, Communicators Director

6658 Acid Rain: A North American Challenge
National Film Board of Canada
1123 Broadway Suite 307   212-629-8890
New York, NY 10010   Fax: 212-629-8502
E-mail: j.sirabella@nfb.ca
http://www.nfb.ca
Summerizes what we know today about the causes and effect of the menace of acid rain.

6659 Adventures of the Little Koala & Friends
Family Home Entertainment
15400 Sherman Way   818-499-5827
PO Box 10124   http://www.familyhome ent.com
Van Nuys, CA 91410-0124

6660 Air Pollution: A First Film
2349 Chaffee Drive   314-569-0211
St Louis, MO 63146

6661 Black Waters
Green Mountain Post Films   413-863-4754
PO Box 229   Fax: 413-863-8248
Turner Falls, MA 01376   E-mail: info@gmpfilms.com
http://www.gmpfilms.com

6662 Bog Ecology
Educational Images
PO Box 3456   607-732-1090
Elmira, NY 14905   800-527-4264
Fax: 607-732-1183
E-mail: info@edimages.com
http://www.educationalimages.com
Charles R Belinky PhD, CEO

6663 Captain Planet & the Planeteers: Toxic Terror
Turner Home Entertainment Company
1 CNN N Tower   404-827-2000
12th Floor   http://www.turner.com
Atlanta, GA 30348

6664 Carnivores
Walt Disney Home Video
500 S Buena Vista Street   818-562-3560
500 South Buena Vista Street   http://www.disneyvideos.com
, CA 91521

6665 Chelyabinsk: The Most Contaminated Spot on the Planet
Filmakers Library
124 E 40th Street   212-808-4980
New York, NY 10016   Fax: 212-808-4983
E-mail: info@filmakers.com
http://www.filmakers.com
Until now Chelyabinsk, where an atomic weapons complex was located, was shrouded in secrecy. In 1957, there was an explosion of the cooling system. Ten years later, a storm spread radioactive dust. For six years, the complex dumpedradioactive waste into the river that supplied water for villages. In 1992 when the area was opened up to foreigners, the filmmaker began recording the details of ordinary life in this deadly environment.

Linda Gottesman, Co-President

6666 Chemical Kids
Filmakers Library
124 E 40th Street   212-808-4980
New York, NY 10016   Fax: 212-808-4983
E-mail: info@filmakers.com
http://www.filmakers.com
Most people want the conveniences of modern life - the cell phones, computers, and perfect-looking food. But it seems we are paying a high price: being poisoned by chemicals. This excellent film contains many alarming facts aboutthe man-made chemicals we inject in our food and water. Children are the most affected - damaged physically and mentally by the fish their mothers ate during pregnancy, by eating too much fish or meat themselves, or by drinking water drawn too closeto waste dumps.

Linda Gottesman, Co-President

6667 Children of Chernobyl
Filmakers Library
124 E 40th Street   212-808-4980
New York, NY 10016   Fax: 212-808-4983
E-mail: info@filmakers.com
http://www.filmakers.com
This disturbing film reveals for the first time the true depths of the tragedy at Chernobyl through exclusive archival film and eyewitness accounts. It is a story of deception and cover-up on a grand scale. The true legacy of thedisaster can be seen on the children's wards of hospitals. Total hair loss is one of the more visible signs. Sadly for the children, the number of cancer diagnoses is growing alarmingly.

Linda Gottesman, Co-President

6668 City of the Future
University of California
2176 Shattuck Avenue   510-642-0460
Media Center   http://www.berkley.edu
Berkeley, CA 94704

6669 Clouds of Doubt
Electronics Arts Intermix
536 Braodway   212-966-4605
9th Floor   http://www.eci.org
New York, NY 10012

6670 Cocos Island: Treasure Island
ESPN Home Video
ESPN Plaza   860-585-2000
Bristol, CT 06010   http://www.espn.com

6671 Coral Cities of the Caribbean
Nancy Sefton/Triton Productions (Earthwise Media)
Po Box 1223   360-271-1584
Poulsbo, WA 98370   Fax: 360-394-2168
E-mail: info@earthwisevideos.com
http://www.earthwisevideos.com
A journey of discovery among the coral reefs of the Caribbean. Watch the various fishes and invertebrates that reside in these coral cities as they follow their clever strategies for survival. Filmed beneath the waters of LittleCayman Island, a diving Mecca with rich, colorful coral reefs, Coral Cities of the Carribean opens with a typical morning among the corals.
Nancy Sefton, Creative Director
Wes Nicholson, Technical Director

6672 Cousteau Collection, Vol. 1: Alaska, Outrage at Valdez
Facets Multimedia
1517 E Fullertion Avenue   773-281-4114
Chicago, IL 60614   http://www.centerstage.net/film/cinemas/

6673 Cousteau Odyssey, Vol. 10: Warm-Blooded Sea, Mammals of the Deep
Warner Home Video
4000 Warner Boulevard   818-954-6000
Burbank, CA 91522   http://www.store.warnervideo.com

6674 Cousteau's Undersea World, Vol. 10: Whales
Churchill Media
12210 Nebraska Avenue   310-207-6600
Los Angeles, CA 90025   Fax: 310-207-1330

6675 Cut Your Home Heating Cost
Rodale Press
33 E Minor Street   215-967-5171
Emmaus, PA 18098-0099   Fax: 610-967-8963
E-mail: info@rodate.com
http://www.rodate.com

6676 Designing the Environment
NETCHE
1800 N 33rd Street   402-472-3611
Lincoln, NE 68503   Fax: 402-472-1785
E-mail: netche@unl.edu
http://www.netdb.unl.edu/netchevideo

6677 Disappearance of the Great Rainforest
Arthur Mokin Productions   707-542-4868
PO Box 1866   800-238-4868
Santa Rosa, CA 95402-1866

6678 Dolphin
Media Guild   858-755-9191
PO Box 910534   Fax: 858-755-4931
San Diego, CA 92191-0534

6679 Double Envelope Solar House: Living with Tomorrow
Willow Mixed Media   845-657-2914
PO Box 194   E-mail: willowmx@ulster.net
Glenford, NY 12433   http://www.hudsonvalley.com/willow/

6680 Earth Science Video Library
Great Plains National (GPN)
1800 N 33rd Street   402-472-2007
Box 80669   800-228-4630
Lincoln, NE 68501-0669   Fax: 402-472-1785
E-mail: gpn.unl.edu
http://gpn.unl.edu
GPN is a producer and distributor of educational media - video, CD-ROM, DVD, Internet-for-16. Free previews available on line and on video.
100 pages Annual Catalog
Steve Lenzen, Director

6681 Earth Summit: What Next?
EcuFilm
810 12th Avenue S   615-242-6277
Nashville, TN 37203   800-251-4091
http://www.ecufilm.com

6682 Earth at Risk Environmental Series
Library Video Company   610-645-4000
PO Box 580   800-843-3620
Wunnewood, PA 19096   Fax: 610-645-4040
E-mail: comments@libraryvideo.com
http://www.libraryvideo.com

6683 Earth's Physical Resources
Media Guild   858-755-9191
PO Box 910534   Fax: 858-755-4931
San Diego, CA 92191-0534

6684 Earthwise Media
PO Box 1223   360-271-1584
Poulsbo, WA 98370   Fax: 360-394-2168
E-mail: info@earthwisevideos.com
http://www.earthwisevideos.com
Earthwise Media, LLC is a full service video production company specializing in producing environmental educational videos, multi-media presentations and computer aided learning tools.
Nancy Sefton, Creative Director
Wes Nicholson, Technical Director

6685 Ecological Realities: Natural Laws at Work
University of California
2176 Shattuck Avenue   510-642-0460
Berkeley, CA 94704

6686 Educational Images
PO Box 3456   607-732-1090
Elmira, NY 14905   800-527-4264
Fax: 607-732-1183
E-mail: info@edimages.com
http://www.educationalimages.com
Publish a very wide range of educational CD ROMs, slide sets and videos on ecology, geology, aquatic life and related biological science and medical topics. Provide stock photos for digital and print use.
Charles R Belinky PhD, CEO

6687 Effluents of Affluence
University of Michigan
Film Video Library   734-764-5360
919 S University Avenue, Room 207   Fax: 734-764-6849
Ann Arbor, MI 48109-1185   E-mail: ful.office@umich.edu
http://www.lib.umich.edu

6688 Elephant Boy
HBO Home Video
1114 6th Avenue   212-512-7400
New York, NY 10036   http://www.hbohomevideo.com

6689 Empire of the Red Bear
Discovery Home Entertainmnet
7700 Wisconsin Avenue   301-986-1999
Betseda, MD 20814-3579   E-mail: letters@discovery.com
http://www.discovery.com

6690 Endangered Species: Massasauga Rattler and Bog Turtle
Educational Images
PO Box 3456   607-732-1090
Elmira, NY 14905   800-527-4264
Fax: 607-732-1183
E-mail: info@edimages.com
http://www.educationalimages.com
Charles R Belinky PhD, CEO

6691 Enemies of the Oak
Carolina Biological Supply Company
2700 York Road   919-584-0381
Burlington, NC 27215   800-334-5551
E-mail: carolina@carolina.com
http://www.carolina.com

6692 Energetics of Life
The Media Guild   858-755-9191
PO Box 910534   Fax: 858-755-4931
San Diego, CA 92191-0534

6693 Energy Now
Educational Images
PO Box 3456   607-732-1090
Elmira, NY 14905   800-527-4264
Fax: 607-732-1183
E-mail: info@edimages.com
http://www.educationalimages.com
Charles R Belinky PhD, CEO

6694 Energy to Go Around
The Media Guild   858-755-9191
PO Box 910534   Fax: 858-755-4931
San Diego, CA 92191-0534

6695 Energy: The Alternatives
The Media Guild   858-755-9191
PO Box 910534   Fax: 858-755-4931
San Diego, CA 92191-0534

6696 Everglades Region: An Ecological Study
The Media Guild   858-755-9191
PO Box 910534   Fax: 858-755-4931
San Diego, CA 92191-0534

6697 Exploring the Forest
Alfred Higgins Productions
6350 Laurel Canyon Boulevard   818-762-3300
North Hollywood, CA 91606   Fax: 818-762-8223
http://www.alfredhigginsprod.com

6698 Fascinating World of Forestry
Educational Images
PO Box 3456   607-732-1090
Elmira, NY 14905   800-527-4264
Fax: 607-732-1183
E-mail: info@edimages.com
http://www.educationalimages.com
Charles R Belinky PhD, CEO

6699 Flag: The Story of the White White-Tailed Deer
Educational Images
PO Box 3456   607-732-1090
Elmira, NY 14905   800-527-4264
Fax: 607-732-1183
E-mail: info@edimages.com
http://www.educationalimages.com
Charles R Belinky PhD, CEO

6700 Flooding River
The Media Guild   858-755-9191
PO Box 910534   800-527-4264
San Diego, CA 92191-0534   Fax: 760-732-1183

6701 Florida Bay and the Everglades
Educational Images
PO Box 3456   607-732-1090
Elmira, NY 14905   800-527-4264
Fax: 607-732-1183
E-mail: info@edimages.com
http://www.educationalimages.com
Charles R Belinky PhD, CEO

6702 Food from the Rainforest
The Media Guild   858-755-9191
PO Box 910534   Fax: 858-755-4931
San Diego, CA 92191-0534

6703 Forms of Energy
Educational Images
PO Box 3456   607-732-1090
Elmira, NY 14905   800-527-4264
Fax: 607-732-1183
E-mail: info@edimages.com
http://www.educationalimages.com
Charles R Belinky PhD, CEO

6704 Freshwater and Saltwater Marshes
Educational Images
PO Box 3456   607-732-1090
Elmira, NY 14905   800-527-4264
Fax: 607-732-1183
E-mail: info@edimages.com
http://www.educationalimages.com
Charles R Belinky PhD, CEO

6705 Great American Woodlots
Cornell University
8 Business & Technology Park   607-255-2091
Itahaca, NY 14850   607-255-9946
http://www.cornell.edu

6706 Guardians of the Cliff: The Peregrine Falcon Story
Educational Images
PO Box 3456   607-732-1090
Elmira, NY 14905   800-527-4264

Fax: 607-732-1183
E-mail: info@edimages.com
http://www.educationalimages.com
Charles R Belinky PhD, CEO

6707 Happy Campers with Miss Shirley & Friends
Kids Express   417-889-2234
PO Box 11176   888-492-5437
Kspringfield, MO 65808   Fax: 417-883-9157
E-mail: missshirley@kids-express.com
http://www.kids-express.com

6708 I Walk in the Desert
Educational Images
PO Box 3456   607-732-1090
Elmira, NY 14905   800-527-4264
Fax: 607-732-1183
E-mail: info@edimages.com
http://www.educationalimages.com
Charles R Belinky PhD, CEO

6709 Joe Albert's Fox Hunt
Education Development Center
55 Chapel Street   617-969-7100
Newton, MA 02160   800-225-4276
E-mail: www@edc.org
http://www.edc.org

6710 John Muir
Educational Images
PO Box 3456   607-732-1090
Elmira, NY 14905   800-527-4264
Fax: 607-732-1183
E-mail: info@edimages.com
http://www.educationalimages.com
Charles R Belinky PhD, CEO

6711 Legacy of an Oil Spill: Ten Years After Exxon Valdez
Media Guild   858-755-9191
PO Box 910534   800-886-9191
San Diego, CA 92191-0534   Fax: 858-755-4931
E-mail: info@mediaguild.com
http://www.mediaguild.com
In 1989 the super tanker spilled over 11 million gallons of crude oil into Alaska's Prince William Sound, damaging over 1,000 miles of shoreline. An estimated 250,000 sea birds, 2,800 sea otters, 300 harbor seals and up to 22 killerwhales died as a result. A decade later, only two of the 28 species injured have recovered. This award-winning video allows viewers to join scientists in the field on an ecosystem research project to determine the long-term effects of the spill onseveral species.
Ruth Pipitone

6712 Life on the Edge: A Guide to Pacific Coastal Habitats
Earthwise Media
PO Box 1223   360-271-1584
Poulsbo, WA 98370   Fax: 360-394-2168
E-mail: info@earthwisevideos.com
http://www.earthwisevideos.com
Explore key Pacific coastal habitats from Alaska to California. Meet the residents of sandy beaches, eelgrass beds, rocky shores, estuaries and kelp forests. Includes discussions of: coastal geology, tides, intertidal zonation, keyelements of each habitat, and their residences. Produced in conjunction with the Seattle Aquarium.
Nancy Sefton, Creative Director
Wes Nicholson, Technical Director

6713 Manatees: A Living Resource
Educational Images
PO Box 3456   607-732-1090
Elmira, NY 14905   800-527-4264
Fax: 607-732-1183
E-mail: info@edimages.com
http://www.educationalimages.com
Charles R Belinky PhD, CEO

6714 Mitzi A Da Si: A Visit to Yellowstone National Park
Educational Images
PO Box 3456
Elmira, NY 14905

607-732-1090
800-527-4264
Fax: 607-732-1183
E-mail: info@edimages.com
http://www.educationalimages.com

Charles R Belinky PhD, CEO

6715 Modeling Photosynthesis
Media Guild
PO Box 910534
San Diego, CA 92191-0534

858-755-9191
Fax: 858-755-4931

6716 Our Precious Environment
Educational Images
PO Box 3456
Elmira, NY 14905

607-732-1090
800-527-4264
Fax: 607-732-1183
E-mail: info@edimages.com
http://www.educationalimages.com

Charles R Belinky PhD, CEO

6717 RMC Medical Inc
3019 Darnell Road
Philadephia, PA 19154-3201

215-824-4100
800-332-0672
Fax: 215-824-1371
E-mail: rmcmedical@cs.com
http://www.rmcmedical.com

Manufacturer of decontamination emergency response equipment.

Lois White, Sales/Marketing Manager

6718 Rainbows in the Sea: A Guide to Earth's Coral Reefs
Earthwise Media
PO Box 1223
Poulsbo, WA 98370

360-271-1584
Fax: 360-394-2168
E-mail: info@earthwisevideos.com
http://www.earthwisevideos.com

"Rainbows in the Sea" is a journey of discovery across our planet's magnificent coral reefs, from the Caribbean to the far reaches of the Indo-Pacific. It showcases the wide variety of colorful fishes and invertebrates that survive inthese complex ecosystems.
Nancy Sefton, Creative Director
Wes Nicholson, Technical Director

6719 Return of the Dragon
Educational Images
PO Box 3456
Elmira, NY 14905

607-732-1090
800-527-4264
Fax: 607-732-1183
E-mail: info@edimages.com
http://www.educationalimages.com

Charles R Belinky PhD, CEO

6720 Salt Marshes-A Special Resource
Educational Images
PO Box 3456
Elmira, NY 14905

607-732-1090
800-527-4264
Fax: 607-732-1183
E-mail: info@edimages.com
http://www.educationalimages.com

Charles R Belinky PhD, CEO

6721 Sand Dune Ecology and Formation
Educational Images
PO Box 3456
Elmira, NY 14905

607-732-1090
800-527-4264
Fax: 607-732-1183
E-mail: info@edimages.com
http://www.educationalimages.com

Charles R Belinky PhD, CEO

6722 Seals
Media Guild
PO Box 910534
San Diego, CA 92191-0534

858-755-9191
Fax: 858-755-4931

6723 Survey of Environment
Educational Images
PO Box 3456
Elmira, NY 14905

607-732-1090
800-527-4264
Fax: 607-732-1183
E-mail: info@edimages.com
http://www.educationalimages.com

Charles R Belinky PhD, CEO

6724 The World Between the Tides: A Guide to Pacific Rocky Shores
Earthwise Media
PO Box 1223
Poulsbo, WA 98370

360-271-1584
Fax: 360-394-2168
E-mail: info@earthwisevideos.com
http://www.earthwisevideos.com

Low tide along the Pacific coast's rocky shores uncovers a kaleidoscope of marine creatures glowing like scattered jewels in shallow pools. This narrated journey along the rich intertidal area of Pacific coast rocky shores looks atanimals and plants exquisitely adapted to live in one of Nature's toughest neighborhoods.
Nancy Sefton, Creative Director
Wes Nicholson, Technical Director

6725 Tropical Rainforest
The Media Guild
PO Box 910534
San Diego, CA 92191-0534

858-755-9191
Fax: 858-755-4931

6726 Tropical Rainforests Under Fire
Educational Images
PO Box 3456
Elmira, NY 14905

607-732-1090
800-527-4264
Fax: 607-732-1183
E-mail: info@edimages.com
http://www.educationalimages.com

Charles R Belinky PhD, CEO

6727 Warming Warning
Media Guild
PO Box 910534
San Diego, CA 92191-0534

858-755-9191
Fax: 858-755-4931

6728 Water Resources Videos
Educational Images
PO Box 3456
Elmira, NY 14905

607-732-1090
800-527-4264
Fax: 607-732-1183
E-mail: info@edimages.com
http://www.educationalimages.com

Charles R Belinky PhD, CEO

6729 Watershed: Canada's Threatened Rainforest
Media Guild
PO Box 910534
San Diego, CA 92191-0534

858-755-9191
800-886-9191
Fax: 858-755-4931
E-mail: info@mediaguild.com
http://www.mediaguild.com

Chronicles the explorations of the Rainforest Conservation Society's Summer Expedition to the temperate rainforest of Canada's Central Pacific Coast. Viewers are introduced to environmental issues, the complex ecosystem of therainforest and the efforts of conservationists and indigenous people who are trying to preserve this wilderness. Powerful images of pristine wilderness contrasted with decimated forests dramatically portray the struggle to preserve this fragileenvironment.
Ruth Pipitone

6730 Wetlands: Development, Progress, and Environmental Protection under the Changing Law
American Law Institute
4025 Chestnut Street
Philadelphia, PA 19104

215-243-1600
800-CLE-NEWS
Fax: 215-243-1664
E-mail: jmendicino@ali.org
http://www.ali.org

6731 Wilderness Video: Yellowstone, Yosemite, Grand Canyon, Big Sur Coastline, New Nature in Motion

PO Box 3150   541-488-9363
Ashland, OR 97520   Fax: 541-488-9363
E-mail: Bob@Wildernessvideo.com
http://www.wildernessvideo.com

Wilderness Video has a beautiful collection of high definition stock footage which includes Timelapse, Nature, National Parks, and Cities.

Bob Glusic, Owner

6732 Windrifters: The Bald Eagle Story

Educational Images
PO Box 3456   607-732-1090
Elmira, NY 14905   800-527-4264
Fax: 607-732-1183
E-mail: info@edimages.com
http://www.educationalimages.com

Charles R Belinky PhD, CEO

General

6733 Acorn Designs
5066 Mott Evans Road   607-387-3424
Trumansburg, NY 14886   800-299-3997
                              Fax: 607-387-5609
                    E-mail: acorndes@lightlink.com
                    http://www.acorndesigns.org
Totes and other items with wildlife and nature themes, recycled
paper, tree-free kenaf paper and hemp paper stationery,
notecards, journals and organic cotton tee shirts.
Steve Sierigk, Owner

6734 Alexandra Avery Purely Natural Body Care
4717 SE Belmont Street   503-236-5926
Portland, OR 97215-1736   800-669-1863
                              Fax: 503-234-7272
                    E-mail: aavery42@earthlink.net
100% natural and cruelty free aromatherapy products for face
and body care.
Alexandra Avery, President

6735 Artistic Video
87 Tyler Avenue   631-744-5999
Sound Beach, NY 11789   888-982-4244
                              Fax: 631-744-5993
                    E-mail: bobklien@movementsofmagic.com
                    http://www.movementsofmagic.com
Instructional videos and DVDs on health and fitness, alternative
healing, children's programs about animals, free interactive section with articles, discussion forums, video clips, classes, instructions, Tai-Chi and other natureoriented cultures and
practices.

Bob Klein, President

6736 BDM Holdings
7915 Jones Branch Drive   703-848-5000
McLean, VA 22102
Bennie Dibona, VP Engineering/Environment

6737 Balance of Nature
Unviersity of Chicago Press
1427 East 60th Street   773-702-7700
Chicago, IL 60637   Fax: 773-702-9756
                    http://www.press.chicago.edu

6738 Bio-Sun Systems
RR 2 Box 134A   570-527-2200
Millerton, PA 16936   800-847-8840
                              Fax: 570-537-6200
                    E-mail: biosun@npacc.net
                    http://www.bio-sun.com
Composting toilets and modular restrooms.
Donna White, President
Al White, VP

6739 Cotton Clouds
5176 S 14th Avenue   520-428-7000
Safford, AZ 85546-9252   Fax: 520-428-6630
                    E-mail: cottonclouds@az.org
Yarn, patterns, books, video's, kit for weaving crochet, knitting
and spinning.

6740 Earth Options
Solar Electric Engineering
117 Morris Street   707-824-4150
Sebastopol, CA 95472-3858   882-198-1986
                              Fax: 707-542-4358
                    http://882-198-1986
Environmental retail products.

6741 Earth Science
PO Box 1925   909-371-7565
Corona, CA 91718   800-222-6720
                              Fax: 909-371-0509
                    http://800-222-6720
All-natural, environmentally sound skin and hair care products.
Kristine Schoenauer, President

6742 Eco-Store
2441 Edgewater Drive   407-426-9949
Orlando, FL 32804   800-556-9949
                              Fax: 407-649-3148
                    E-mail: beth@eco-store.com
                    http://800-556-9949
Environmental home products, gifts, etc.

6743 Ecology Store
6928 Queens Boulevard   718-446-4444
Flushing, NY 11377-5102   800-548-9660
                              Fax: 718-446-9860
                    http://800-548-9660
Range of environmental products.

6744 Energy Efficient Environments
2119 Inverness Lane   847-475-3005
Glen View, IL 60025   800-336-3749
                    E-mail: info@eeenvironments.com
                    http://www.eeenvironments.com
Efficient and earth-friendly devices for energy, water and light
use.

6745 Environment Friendly Papers
Cherry Paper
13520 Liberty Avenue   718-297-3000
Jamaica, NY 11419   Fax: 718-297-2986
                    E-mail: cherryop@AOL.com
100% recycled gift stationary, with original designs depicting
flora and fauna of South Africa and environmental themes.

6746 Erlander's Natural Products
Nature's Department Store
2279 Lake Avenue   626-797-7004
Altadena, CA 91001-2414   800-562-8873
                              Fax: 626-798-2663
                    E-mail: erlander@webtv.net
                    http://800-562-8873
Natural olive oil-wine soap; bar and liquid roach killer from
herbs, organic red zinfandel wine, 100% organic cotton pillows
and mattresses, jewelry - semi-precious and costume, washing
compound/non-detergent, sodium sesquicarbonate.
Leatrice Erlander, Co-Owner
Stig Erlander, Co-Owner

6747 GAIA Clean Earth Products
PO Box 1906   717-840-1638
York, PA 17405-1906   800-726-5496
                              Fax: 800-726-5496
                    E-mail: gaia@blazenet.net
                    http://800-726-5496
Environmentally-compatible products.
Brian N Hartman, President

6748 Greenpeace
564 Mission Street Box 416   510-538-7842
San Francisco, CA 94105-3008 800-326-0959
                              Fax: 202-462-4517
                    E-mail: greenpeace@npgear.com
                    http://800-326-0959
Environmentally and socially responsible apparel, accessories
and gifts.

6749 Jason Natural Cosmetics
8468 Warern Drive
Culver City, CA 90232   877-JAS-ON01
                              Fax: 310-838-9274
                    E-mail: jnp@jason-natural.com
                    http://877-JAS-ON01
All natural cosmetics.
Jeffrey Light, President

6750 Look Alive!
Rice Lake Products
100 27th Street NE
Minot, ND 58703-5164

701-857-6357
800-998-7450
Fax: 701-857-6300
E-mail: ricelake@dalotah.com
http://800-998-7450

Environmentally safe movement devices for hunting decoys, as well as bird and pest deterring owls for home garden, boats and businesses.
Virgil Farstad, President

6751 Real Goods Trading Company
PO Box 8507
Ukiah, CA 95482-8507

707-744-2100
800-347-0070
Fax: 707-468-9394
http://800-347-0070

Recycled papers, environmental gifts and household goods.

6752 Second Renaissance Books
17 George Washington Plaza
Gaylordsville, CT 06755-1500

800-729-6149
Fax: 860-355-7160
E-mail: inquiries@secondrenaissance.com
http://800-729-6149

Editorially selected books and audio tapes. Complete selection of writing and letters by Ayn Rand.

6753 Sparky Boy Enterprises
1512 Gold Avenue
Bozeman, MT 59715-2471

406-587-5891
800-289-6656
Fax: 406-587-0223
E-mail: ecostoke@mcn.net
http://800-289-6656

Natural products for the home, lawn and garden.
Wayne Vinje, President

6754 Sunrise Lane Products
780 Greenwich Street
New York, NY 10014-5919

212-243-4745

Environmentally safe and cruelty-free products for home and personal care.
Rossella Mocerino, President

6755 Williams Distributors
1801 S Cardinal Lane
New Berlin, WI 53151-1543

262-597-9865

All natural products; nutrition, health, home care and personal care, plus water purification systems.
GL Williams, President

Maximum air quality concentrations by metropolitan statistical area, 2003

Metropolitan Statistical Area	2000 Population	CO 8-hr (ppm)	NO_2 AM (ppm)	O_3 1-hr (ppm)	O_3 8-hr (ppm)	PM_{10} Wtd AM (ug/m^3)	PM_{10} 24-hr (ug/m^3)	$PM_{2.5}$ Wtd AM (ug/m^3)	$PM_{2.5}$ 24-hr (ug/m^3)	SO_2 AM (ppm)	SO_2 24-hr (ppm)
Akron, OH	694,960	3	ND	0.13	0.09	21	62	15.4	37	0.009	0.054
Albany, GA	120,822	ND	ND	ND	ND	20	43	13.4	27	ND	ND
Albany—Schenectady—Troy, NY	875,583	3	ND	0.11	0.09	ND	ND	12.2	34	0.004	0.024
Albuquerque, NM	712,738	4	0.018	0.1	0.08	41	262	10.6	26	ND	ND
Alexandria, LA	126,337	ND	ND	ND	ND	ND	ND	IN	IN	ND	ND
Allentown—Bethlehem—Easton, PA	637,958	1	0.015	0.11	0.09	32	114	14.3	37	0.009	0.038
Altoona, PA	129,144	1	0.013	0.1	0.08	20	95	ND	ND	0.007	0.03
Amarillo, TX	217,858	ND	ND	ND	ND	ND	ND	6.7	16	ND	ND
Anchorage, AK	260,283	7	ND	ND	ND	32	187	5.9	16	ND	ND
Ann Arbor, MI	578,736	ND	ND	0.11	0.09	ND	ND	14.6	39	ND	ND
Appleton—Oshkosh—Neenah, WI	358,365	ND	ND	0.09	0.08	ND	ND	10.4	26	ND	ND
Asheville, NC	225,965	ND	ND	0.09	0.07	ND	ND	12.6	31	ND	ND
Athens, GA	153,444	ND	ND	0.08	0.07	ND	ND	14.3	33	ND	ND
Atlanta, GA	4,112,198	3	0.017	0.12	0.09	28	91	17.7	44	0.003	0.022
Atlantic—Cape May, NJ	354,878	ND	ND	0.12	0.09	21	35	IN	IN	0.003	0.01
Augusta—Aiken, GA—SC	477,441	ND	0.004	0.09	0.08	IN	60	14.8	31	ND	ND
Austin—San Marcos, TX	1,249,763	1	0.004	0.1	0.08	22	53	IN	IN	IN	IN
Bakersfield, CA	661,645	2	0.023	0.15	0.12	52	128	19.6	52	ND	ND
Baltimore, MD	2,552,994	3	0.026	0.13	0.09	27	62	16.8	42	IN	IN
Bangor, ME	90,864	ND	ND	ND	ND	22	48	10.6	31	ND	ND
Baton Rouge, LA	602,894	3	0.016	0.14	0.1	31	53	13.2	28	0.005	0.03
Beaumont—Port Arthur, TX	385,090	ND	0.01	0.11	0.08	ND	ND	12	26	0.004	0.034
Bellingham, WA	166,814	ND	ND	0.07	0.06	12	24	7	18	ND	ND
Benton Harbor, MI	162,453	ND	ND	0.11	0.09	ND	ND	12.5	34	ND	ND
Bergen—Passaic, NJ	1,373,167	2	0.02	0.11	0.09	37	75	13.3	40	0.004	0.022
Billings, MT	129,352	4	ND	ND	ND	17	45	IN	IN	0.006	0.032
Biloxi—Gulfport—Pascagoula, MS	363,988	ND	0.004	0.11	0.09	IN	33	12	27	0.002	0.01
Binghamton, NY	252,320	ND	ND	ND	ND	ND	ND	IN	IN	ND	ND
Birmingham, AL	921,106	5	ND	0.1	0.08	57	178	17.4	39	0.006	0.049
Bismarck, ND	94,719	ND	ND	ND	ND	IN	51	7.2	16	0.006	0.042
Bloomington—Normal, IL	150,433	ND	ND	0.08	0.07	ND	ND	13.2	34	ND	ND
Boise City, ID	432,345	4	ND	0.1	0.07	22	88	8	22	ND	ND
Boston, MA—NH	3,406,829	2	0.023	0.12	0.09	IN	64	IN	IN	0.006	0.022
Boulder—Longmont, CO	291,288	3	ND	0.1	0.08	17	42	9.2	29	ND	ND
Brazoria, TX	241,767	ND	0.009	0.12	0.1	ND	ND	IN	IN	ND	ND
Bremerton, WA	231,969	ND	ND	ND	ND	ND	ND	IN	IN	ND	ND
Bridgeport, CT	459,479	2	ND	0.14	0.1	21	49	12.3	40	0.005	0.032
Brockton, MA	255,459	ND	ND	ND	ND	ND	ND	9.8	35	ND	ND
Brownsville—Harlingen—San Benito, TX	335,227	2	ND	0.08	0.07	38	110	10.7	40	ND	ND
Buffalo—Niagara Falls, NY	1,170,111	2	0.02	0.11	0.09	17	42	12.3	37	0.009	0.078
Burlington, VT	169,391	2	0.015	ND	ND	19	49	10.1	29	ND	ND
Canton—Massillon, OH	406,934	2	ND	0.12	0.09	23	63	16.8	35	0.007	0.032
Casper, WY	66,533	ND	ND	ND	ND	24	68	ND	ND	ND	ND
Cedar Rapids, IA	191,701	2	IN	0.08	0.07	22	50	11	35	0.004	0.042
Champaign—Urbana, IL	179,669	ND	ND	0.08	0.08	ND	ND	13	33	ND	ND
Charleston—North Charleston, SC	549,033	3	0.01	0.09	0.07	19	43	11	23	0.003	0.011
Charleston, WV	251,662	ND	ND	0.11	0.09	25	60	16.1	36	0.01	0.04
Charlotte—Gastonia—Rock Hill, NC—SC	1,499,293	3	0.015	0.13	0.1	24	52	14.6	33	0.003	0.013
Charlottesville, VA	159,576	ND	ND	ND	ND	17	30	ND	ND	ND	ND
Chattanooga, TN—GA	465,161	ND	ND	0.11	0.08	22	57	16.5	38	ND	ND
Cheyenne, WY	81,607	ND	ND	ND	ND	14	27	4.9	13	ND	ND
Chicago, IL	8,272,768	3	0.031	0.1	0.08	33	103	16.8	45	0.006	0.023
Chico—Paradise, CA	203,171	2	0.011	0.1	0.09	21	41	10.5	32	ND	ND
Cincinnati, OH—KY—IN	1,646,395	2	0.022	0.12	0.1	29	57	17.3	38	0.009	0.04
Clarksville—Hopkinsville, TN—KY	207,033	ND	ND	0.09	0.08	19	39	IN	IN	0.006	0.022
Cleveland—Lorain—Elyria, OH	2,250,871	3	0.022	0.12	0.1	36	177	17.6	46	0.01	0.052
Colorado Springs, CO	516,929	4	ND	0.09	0.08	22	61	8	19	ND	ND

443

Maximum air quality concentrations by metropolitan statistical area, 2003 - *continued*

Metropolitan Statistical Area	2000 Population	CO 8-hr (ppm)	NO$_2$ AM (ppm)	O$_3$ 1-hr (ppm)	O$_3$ 8-hr (ppm)	PM$_{10}$ Wtd AM (ug/m^3)	PM$_{10}$ 24-hr (ug/m^3)	PM$_{2.5}$ Wtd AM (ug/m^3)	PM$_{2.5}$ 24-hr (ug/m^3)	SO$_2$ AM (ppm)	SO$_2$ 24-hr (ppm)
Columbia, MO	135,454	ND	ND	ND	ND	ND	ND	12.5	31	ND	ND
Columbia, SC	536,691	2	0.003	0.1	0.08	35	155	13.1	30	0.004	0.024
Columbus, GA—AL	274,624	ND	ND	0.09	0.07	24	52	15.4	32	ND	ND
Columbus, OH	1,540,157	3	ND	0.12	0.09	33	89	16.4	39	0.004	0.022
Corpus Christi, TX	380,783	ND	ND	0.09	0.08	IN	48	10.6	27	0.002	0.019
Dallas, TX	3,519,176	2	0.018	0.13	0.09	32	74	14.2	33	0.002	0.028
Danbury, CT	217,980	ND	ND	0.11	0.09	ND	ND	IN	IN	0.003	0.021
Davenport—Moline—Rock Island, IA—IL	359,062	ND	0.007	0.09	0.08	39	137	12.7	32	0.002	0.012
Dayton—Springfield, OH	950,558	3	ND	0.13	0.09	27	46	15.9	43	0.004	0.023
Daytona Beach, FL	493,175	ND	ND	0.09	0.07	19	53	8.5	18	ND	ND
Decatur, AL	145,867	ND	ND	0.09	0.08	ND	ND	13.7	30	ND	ND
Decatur, IL	114,706	ND	ND	0.08	0.07	ND	ND	13.6	34	0.003	0.026
Denver, CO	2,109,282	5	0.022	0.14	0.1	38	103	10.6	26	0.003	0.013
Des Moines, IA	456,022	2	0.012	0.08	0.06	30	83	10.7	31	ND	ND
Detroit, MI	4,441,551	3	0.022	0.13	0.1	45	203	19.1	46	0.007	0.054
Dothan, AL	137,916	ND	ND	ND	ND	IN	69	IN	IN	ND	ND
Dover, DE	126,697	ND	ND	0.12	0.08	ND	ND	12.7	35	ND	ND
Duluth—Superior, MN—WI	243,815	2	ND	0.08	0.07	27	63	8.3	23	ND	ND
Dutchess County, NY	280,150	ND	ND	0.09	0.08	ND	ND	IN	IN	ND	ND
Eau Claire, WI	148,337	ND	ND	ND	ND	16	40	ND	ND	ND	ND
El Paso, TX	679,622	6	0.02	0.11	0.08	63	719	IN	43	0.002	0.007
Elkhart—Goshen, IN	182,791	ND	ND	0.1	0.09	ND	ND	14.9	37	ND	ND
Elmira, NY	91,070	ND	ND	0.09	0.08	ND	ND	ND	ND	0.004	0.014
Enid, OK	57,813	ND	ND	ND	ND	ND	ND	9.4	21	ND	ND
Erie, PA	280,843	3	0.012	0.11	0.09	16	54	IN	IN	0.011	0.038
Eugene—Springfield, OR	322,959	3	ND	0.1	0.08	19	60	12.3	53	ND	ND
Evansville—Henderson, IN—KY	296,195	3	0.012	0.1	0.08	25	50	15.3	36	0.012	0.065
Fargo—Moorhead, ND—MN	174,367	ND	0.006	0.07	0.07	20	60	7.9	18	0.001	0.002
Fayetteville, NC	302,963	3	ND	0.1	0.09	19	35	13.4	28	ND	ND
Fayetteville—Springdale—Rogers, AR	311,121	ND	ND	ND	ND	ND	ND	10.5	23	ND	ND
Flagstaff, AZ—UT	122,366	ND	ND	0.08	0.07	IN	51	IN	IN	ND	ND
Flint, MI	436,141	IN	ND	0.1	0.09	20	44	12	32	0.002	0.011
Florence, AL	142,950	ND	IN	0.09	0.08	ND	ND	12.9	29	0.002	0.016
Florence, SC	125,761	ND	ND	ND	ND	ND	ND	12.1	25	ND	ND
Fort Collins—Loveland, CO	251,494	3	ND	0.1	0.09	17	36	7.3	16	ND	ND
Fort Lauderdale, FL	1,623,018	4	0.009	0.08	0.07	19	41	8.2	17	0.002	0.011
Fort Myers—Cape Coral, FL	440,888	ND	ND	0.08	0.07	18	48	8	16	ND	ND
Fort Pierce—Port St. Lucie, FL	319,426	ND	0.009	0.08	0.07	17	43	8.4	18	ND	ND
Fort Smith, AR—OK	207,290	ND	ND	ND	ND	ND	ND	11.5	23	ND	ND
Fort Wayne, IN	502,141	3	ND	0.1	0.09	30	65	14.1	35	ND	ND
Fort Worth—Arlington, TX	1,702,625	2	0.013	0.12	0.1	20	56	13.4	32	ND	ND
Fresno, CA	922,516	4	0.02	0.14	0.11	43	89	17.8	56	IN	IN
Gadsden, AL	103,459	ND	ND	0.09	0.08	IN	44	14.3	29	ND	ND
Gainesville, FL	217,955	ND	ND	0.09	0.07	16	44	9.7	19	ND	ND
Galveston—Texas City, TX	250,158	ND	0.006	0.16	0.09	23	62	10.2	20	0.003	0.02
Gary, IN	631,362	3	0.02	0.1	0.08	34	166	17.4	46	0.004	0.029
Goldsboro, NC	113,329	ND	ND	ND	ND	18	36	13	26	ND	ND
Grand Junction, CO	116,255	3	ND	ND	ND	24	82	8.8	20	ND	ND
Grand Rapids—Muskegon—Holland, MI	1,088,514	2	0.018	0.12	0.1	21	60	13.5	36	0.002	0.007
Great Falls, MT	80,357	3	ND	ND	ND	ND	ND	IN	IN	0.003	0.025
Greeley, CO	180,936	IN	ND	0.11	0.08	21	41	9	21	ND	ND
Green Bay, WI	226,778	ND	ND	0.1	0.08	IN	IN	11.3	34	0.003	0.013
Greensboro—Winston-Salem—High Point, NC	1,251,509	4	0.015	0.12	0.09	19	59	15.2	36	0.006	0.017
Greenville, NC	133,798	ND	ND	0.1	0.08	ND	ND	12.2	27	ND	ND
Greenville—Spartanburg—Anderson, SC	962,441	3	0.014	0.1	0.08	IN	65	15.1	35	0.003	0.014

Maximum air quality concentrations by metropolitan statistical area, 2003 - *continued*

Metropolitan Statistical Area	2000 Population	CO 8-hr (ppm)	NO₂ AM (ppm)	O₃ 1-hr (ppm)	O₃ 8-hr (ppm)	PM₁₀ Wtd AM (ug/m³)	PM₁₀ 24-hr (ug/m³)	PM₂.₅ Wtd AM (ug/m³)	PM₂.₅ 24-hr (ug/m³)	SO₂ AM (ppm)	SO₂ 24-hr (ppm)
Hagerstown, MD	131,923	ND	ND	0.1	0.08	ND	ND	13	34	ND	ND
Hamilton—Middletown, OH	332,807	ND	ND	0.12	0.09	26	63	15.8	35	0.006	0.029
Harrisburg—Lebanon—Carlisle, PA	629,401	2	0.016	0.1	0.08	21	66	16.2	42	0.005	0.017
Hartford, CT	1,183,110	5	0.017	0.11	0.09	IN	38	11.7	35	0.004	0.022
Hattiesburg, MS	111,674	ND	ND	ND	ND	ND	ND	13.6	27	ND	ND
Hickory—Morganton—Lenoir, NC	341,851	ND	ND	0.1	0.08	27	45	15	36	0.004	0.012
Honolulu, HI	876,156	2	0.005	0.05	0.04	17	72	5.8	12	0.001	0.007
Houma, LA	194,477	ND	ND	0.1	0.08	ND	ND	10.9	24	ND	ND
Houston, TX	4,177,646	4	0.019	0.19	0.11	39	95	14.7	29	0.006	0.033
Huntington—Ashland, WV—KY—OH	315,538	1	0.01	0.11	0.09	31	56	15.5	34	0.009	0.042
Huntsville, AL	342,376	ND	IN	0.09	0.08	19	49	13.8	27	0.002	0.005
Indianapolis, IN	1,607,486	3	0.016	0.12	0.1	26	55	16.3	41	0.005	0.024
Iowa City, IA	111,006	ND	ND	ND	ND	ND	ND	11.8	27	ND	ND
Jackson, MS	440,801	3	ND	0.08	0.08	ND	ND	IN	IN	0.002	0.007
Jackson, TN	107,377	ND	ND	ND	ND	18	35	IN	IN	ND	ND
Jacksonville, FL	1,100,491	2	0.014	0.09	0.08	32	185	9.8	21	0.003	0.031
Jacksonville, NC	150,355	ND	ND	ND	ND	14	30	10.8	21	ND	ND
Jamestown, NY	139,750	ND	ND	0.12	0.09	12	28	IN	IN	0.007	0.063
Janesville—Beloit, WI	152,307	ND	ND	0.09	0.08	ND	ND	IN	IN	ND	ND
Jersey City, NJ	608,975	3	0.025	0.11	0.08	IN	61	14.8	46	0.009	0.034
Johnson City—Kingsport—Bristol, TN—VA	480,091	2	0.015	0.1	0.08	ND	ND	13.9	30	0.009	0.044
Johnstown, PA	232,621	2	0.013	0.1	0.08	22	67	15.5	37	0.008	0.028
Jonesboro, AR	82,148	ND	ND	ND	ND	ND	ND	12.1	29	ND	ND
Joplin, MO	157,322	ND	ND	ND	ND	33	156	13.4	31	ND	ND
Kalamazoo—Battle Creek, MI	452,851	ND	ND	0.1	0.09	ND	ND	13.9	37	ND	ND
Kansas City, MO—KS	1,776,062	3	0.017	0.11	0.09	IN	68	14.4	35	0.003	0.015
Kenosha, WI	149,577	ND	ND	0.11	0.09	ND	ND	IN	IN	ND	ND
Knoxville, TN	687,249	ND	ND	0.11	0.09	40	148	15.4	39	0.012	0.077
Kokomo, IN	101,541	ND	ND	ND	ND	ND	ND	14.3	33	ND	ND
Lafayette, LA	385,647	ND	ND	0.1	0.08	ND	ND	11.2	25	ND	ND
Lafayette, IN	182,821	ND	ND	ND	ND	ND	ND	14	35	ND	ND
Lake Charles, LA	183,577	ND	0.006	0.1	0.08	ND	ND	11.2	23	0.006	0.017
Lakeland—Winter Haven, FL	483,924	ND	ND	0.09	0.08	20	49	9.2	16	0.005	0.015
Lancaster, PA	470,658	2	0.015	0.12	0.08	20	53	17.6	52	0.005	0.018
Lansing—East Lansing, MI	447,728	ND	ND	0.1	0.09	ND	ND	13	29	ND	ND
Laredo, TX	193,117	4	ND	0.08	0.07	25	60	IN	IN	ND	ND
Las Cruces, NM	174,682	3	0.011	0.11	0.08	63	268	11.2	35	0.001	0.004
Las Vegas, NV—AZ	1,563,282	5	0.022	0.11	0.09	46	211	10.7	32	0.002	0.007
Lawrence, KS	99,962	ND	ND	0.1	0.08	ND	ND	ND	ND	ND	ND
Lawrence, MA—NH	396,230	ND	ND	ND	ND	ND	ND	IN	IN	ND	ND
Lawton, OK	114,996	2	ND	0.09	0.08	ND	ND	IN	IN	ND	ND
Lewiston—Auburn, ME	90,830	ND	ND	ND	ND	22	63	11.4	33	ND	ND
Lexington, KY	479,198	1	0.012	0.09	0.07	23	43	13.8	28	0.004	0.016
Lima, OH	155,084	ND	ND	0.1	0.09	IN	IN	ND	ND	0.003	0.015
Lincoln, NE	250,291	3	ND	0.07	0.06	ND	ND	9.3	23	ND	ND
Little Rock—North Little Rock, AR	583,845	2	0.014	0.1	0.08	ND	ND	13.4	29	0.002	0.007
Longview—Marshall, TX	208,780	ND	0.006	0.11	0.08	ND	ND	12.5	27	0.002	0.022
Los Angeles—Long Beach, CA	9,519,338	7	0.035	0.19	0.14	44	96	22.1	61	0.003	0.006
Louisville, KY—IN	1,025,598	3	0.018	0.12	0.09	27	78	16	38	0.006	0.035
Lowell, MA—NH	301,686	3	ND	ND	ND	ND	ND	ND	ND	ND	ND
Lubbock, TX	242,628	ND	ND	ND	ND	ND	ND	8.2	21	ND	ND
Lynchburg, VA	214,911	ND	ND	ND	ND	ND	ND	IN	IN	ND	ND
Macon, GA	322,549	ND	ND	0.09	0.08	30	75	14.8	30	0.002	0.009
Madison, WI	426,526	ND	ND	0.08	0.08	IN	IN	11.9	32	ND	ND
Manchester, NH	198,378	3	0.012	0.09	0.07	20	48	10.3	29	0.004	0.021
Mayaguez, PR	253,347	ND	ND	ND	ND	ND	ND	IN	IN	ND	ND
McAllen—Edinburg—Mission, TX	569,463	ND	ND	0.09	0.07	33	109	12.1	38	ND	ND
Medford—Ashland, OR	181,269	5	ND	0.09	0.07	22	59	11.3	39	ND	ND

Maximum air quality concentrations by metropolitan statistical area, 2003 - *continued*

Metropolitan Statistical Area	2000 Population	CO 8-hr (ppm)	NO₂ AM (ppm)	O₃ 1-hr (ppm)	O₃ 8-hr (ppm)	PM₁₀ Wtd AM (ug/m³)	PM₁₀ 24-hr (ug/m³)	PM₂.₅ Wtd AM (ug/m³)	PM₂.₅ 24-hr (ug/m³)	SO₂ AM (ppm)	SO₂ 24-hr (ppm)
Melbourne—Titusville—Palm Bay, FL	476,230	ND	ND	0.09	0.07	19	79	7.4	16	ND	ND
Memphis, TN—AR—MS	1,135,614	3	0.02	0.12	0.09	27	51	13.8	34	0.006	0.038
Merced, CA	210,554	ND	0.012	0.12	0.11	32	73	15.7	44	ND	ND
Miami, FL	2,253,362	4	0.013	0.09	0.07	27	50	9.4	18	0.001	0.004
Middlesex—Somerset—Hunterdon, NJ	1,169,641	2	0.018	0.12	0.09	ND	ND	13	45	0.005	0.02
Milwaukee—Waukesha, WI	1,500,741	3	IN	0.12	0.1	27	59	12.7	34	0.003	0.02
Minneapolis—St. Paul, MN—WI	2,968,806	4	0.012	0.09	0.08	37	75	12	29	0.003	0.058
Missoula, MT	95,802	4	IN	ND	ND	24	116	IN	IN	ND	ND
Mobile, AL	540,258	ND	ND	0.1	0.08	23	49	12.8	31	0.002	0.009
Modesto, CA	446,997	3	0.017	0.12	0.09	31	75	14.5	47	ND	ND
Monmouth—Ocean, NJ	1,126,217	2	ND	0.14	0.1	ND	ND	11.6	39	ND	ND
Monroe, LA	147,250	ND	ND	0.09	0.08	ND	ND	11.7	26	0.002	0.01
Montgomery, AL	333,055	ND	ND	0.09	0.07	20	39	14	32	ND	ND
Muncie, IN	118,769	ND	ND	0.1	0.09	ND	ND	IN	IN	ND	ND
Myrtle Beach, SC	196,629	ND	ND	ND	ND	ND	ND	10.8	20	ND	ND
Naples, FL	251,377	ND	ND	0.08	0.07	ND	ND	ND	ND	ND	ND
Nashua, NH	190,949	4	ND	0.1	0.08	18	39	9.7	35	IN	IN
Nashville, TN	1,231,311	4	0.009	0.1	0.09	26	56	14.3	33	0.004	0.011
Nassau—Suffolk, NY	2,753,913	2	0.021	0.13	0.1	18	37	12.4	39	0.006	0.03
New Bedford, MA	175,198	ND	ND	0.11	0.1	ND	ND	ND	ND	ND	ND
New Haven—Meriden, CT	542,149	2	0.025	0.13	0.1	41	130	16.9	44	0.005	0.031
New London—Norwich, CT—RI	293,566	ND	ND	0.12	0.1	IN	IN	11.7	38	ND	ND
New Orleans, LA	1,337,726	4	0.015	0.1	0.09	26	53	12	25	0.003	0.019
New York, NY	9,314,235	3	0.03	0.12	0.09	22	75	15.8	46	0.011	0.051
Newark, NJ	2,032,989	3	0.032	0.12	0.09	IN	IN	16	48	0.008	0.025
Newburgh, NY—PA	387,669	ND	ND	0.11	0.09	ND	ND	11.8	31	ND	ND
Norfolk—Virginia Beach—Newport News, VA—NC	1,569,541	3	0.016	0.1	0.08	17	35	12.8	33	0.004	0.017
Oakland, CA	2,392,557	3	0.017	0.12	0.08	20	46	9.7	34	0.002	0.007
Ocala, FL	258,916	ND	ND	0.09	0.08	ND	ND	9.3	18	ND	ND
Odessa—Midland, TX	237,132	ND	ND	ND	ND	ND	ND	7.8	17	ND	ND
Oklahoma City, OK	1,083,346	3	0.011	0.1	0.08	24	64	10.2	25	IN	IN
Olympia, WA	207,355	ND	ND	0.09	0.07	13	40	7.8	26	ND	ND
Omaha, NE—IA	716,998	3	ND	0.08	0.07	44	133	11	30	0.002	0.018
Orange County, CA	2,846,289	4	0.028	0.14	0.1	33	77	17.3	52	0.002	0.01
Orlando, FL	1,644,561	2	0.011	0.09	0.08	23	47	9.4	19	0.001	0.004
Owensboro, KY	91,545	ND	0.01	0.09	0.07	20	47	IN	IN	0.004	0.027
Panama City, FL	148,217	ND	ND	0.11	0.08	IN	53	10.8	28	ND	ND
Parkersburg—Marietta, WV—OH	151,237	ND	ND	0.12	0.08	27	75	14.9	32	0.01	0.04
Pensacola, FL	412,153	ND	0.007	0.1	0.08	20	40	11.2	31	0.003	0.021
Peoria—Pekin, IL	347,387	3	ND	0.09	0.08	25	55	13.7	35	0.005	0.051
Philadelphia, PA—NJ	5,100,931	4	0.025	0.13	0.1	31	74	16.1	40	0.008	0.034
Phoenix—Mesa, AZ	3,251,876	6	0.034	0.12	0.09	62	197	11.3	27	0.003	0.007
Pine Bluff, AR	84,278	ND	ND	ND	ND	ND	ND	12.4	25	ND	ND
Pittsburgh, PA	2,358,695	2	0.021	0.12	0.09	37	145	20.2	67	0.012	0.083
Pittsfield, MA	84,699	ND	ND	0.11	0.08	ND	ND	IN	IN	ND	ND
Pocatello, ID	75,565	ND	ND	ND	ND	22	61	5.9	17	0.005	0.02
Ponce, PR	361,094	ND	ND	ND	ND	39	74	6.6	13	ND	ND
Portland, ME	243,537	1	0.017	0.1	0.07	26	57	12.7	36	0.003	0.016
Portland—Vancouver, OR—WA	1,918,009	5	IN	0.1	0.08	18	32	9.1	27	ND	ND
Portsmouth—Rochester, NH—ME	240,698	ND	0.009	0.1	0.08	IN	IN	9.6	31	IN	IN
Providence—Fall River—Warwick, RI—MA	1,188,613	2	0.019	0.11	0.09	25	57	13	39	0.006	0.027
Provo—Orem, UT	368,536	4	0.022	0.11	0.08	25	118	9.2	29	ND	ND
Pueblo, CO	141,472	ND	ND	ND	ND	25	64	7.6	17	ND	ND
Racine, WI	188,831	IN	ND	0.1	0.08	ND	ND	ND	ND	ND	ND
Raleigh—Durham—Chapel Hill, NC	1,187,941	3	ND	0.12	0.09	21	51	13.7	33	0.003	0.01
Rapid City, SD	88,565	ND	ND	0.08	0.07	36	107	7.4	20	ND	ND
Reading, PA	373,638	2	0.018	0.09	0.08	25	54	16.1	45	0.008	0.023

Maximum air quality concentrations by metropolitan statistical area, 2003 - *continued*

Metropolitan Statistical Area	2000 Population	CO 8-hr (ppm)	NO₂ AM (ppm)	O₃ 1-hr (ppm)	O₃ 8-hr (ppm)	PM₁₀ Wtd AM (ug/m³)	PM₁₀ 24-hr (ug/m³)	PM₂.₅ Wtd AM (ug/m³)	PM₂.₅ 24-hr (ug/m³)	SO₂ AM (ppm)	SO₂ 24-hr (ppm)
Redding, CA	163,256	ND	ND	0.11	0.09	21	52	7.5	16	ND	ND
Reno, NV	339,486	4	ND	0.09	0.08	37	84	7.3	19	ND	ND
Richland—Kennewick—Pasco, WA	191,822	ND	ND	ND	ND	25	138	IN	IN	ND	ND
Richmond—Petersburg, VA	996,512	2	0.016	0.12	0.09	17	44	14	38	0.005	0.021
Riverside—San Bernardino, CA	3,254,821	4	0.034	0.17	0.14	56	227	24.8	77	0.003	0.01
Roanoke, VA	235,932	2	0.013	0.09	0.08	30	64	13.8	32	0.003	0.009
Rochester, MN	124,277	ND	ND	0.08	0.07	ND	ND	IN	IN	ND	ND
Rochester, NY	1,098,201	2	ND	0.1	0.08	ND	ND	11.1	28	0.006	0.02
Rockford, IL	371,236	2	ND	0.09	0.08	ND	ND	12.2	27	ND	ND
Rocky Mount, NC	143,026	ND	ND	0.1	0.09	ND	ND	IN	IN	ND	ND
Sacramento, CA	1,628,197	4	0.018	0.14	0.11	28	70	12.2	43	0.001	0.004
Saginaw—Bay City—Midland, MI	403,070	ND	ND	ND	ND	ND	ND	10.9	27	ND	ND
St. Cloud, MN	167,392	2	ND	0.08	0.07	ND	ND	8.9	23	ND	ND
St. Joseph, MO	102,490	ND	ND	ND	ND	32	63	11.9	28	IN	IN
St. Louis, MO—IL	2,603,607	3	0.02	0.13	0.09	44	180	17.5	41	0.006	0.044
Salem, OR	347,214	5	ND	0.1	0.07	ND	ND	IN	IN	ND	ND
Salinas, CA	401,762	1	0.007	0.09	0.07	30	71	7.3	14	ND	ND
Salt Lake City—Ogden, UT	1,333,914	4	0.026	0.11	0.08	40	209	12.3	46	0.005	0.014
San Antonio, TX	1,592,383	3	0.017	0.11	0.09	26	79	15.5	47	0.005	0.01
San Diego, CA	2,813,833	7	0.021	0.12	0.09	52	115	15.5	47	0.002	0.006
San Francisco, CA	1,731,183	3	0.018	0.09	0.07	22	48	10.1	33	ND	ND
San Jose, CA	1,682,585	4	0.021	0.11	0.09	24	55	11.7	37	0.003	0.009
San Juan—Bayamon, PR	1,967,627	4	IN	ND	ND	37	127	IN	IN	0.003	0.009
San Luis Obispo—Atascadero—Paso Robles, CA	246,681	2	0.009	0.09	0.07	25	77	8.2	21	0.004	0.019
Santa Barbara—Santa Maria—Lompoc, CA	399,347	2	0.011	0.1	0.09	24	49	8.6	16	0.001	0.006
Santa Cruz—Watsonville, CA	255,602	1	IN	0.08	0.07	27	65	7.4	14	0.001	0.005
Santa Fe, NM	147,635	2	ND	ND	ND	IN	42	5.2	11	ND	ND
Santa Rosa, CA	458,614	2	0.012	0.09	0.06	17	34	8.8	30	ND	ND
Sarasota—Bradenton, FL	589,959	3	0.006	0.11	0.08	21	41	8.6	18	0.002	0.009
Savannah, GA	293,000	ND	ND	0.08	0.07	22	58	13.3	26	0.003	0.022
Scranton—Wilkes-Barre—Hazleton, PA	624,776	2	0.014	0.1	0.08	21	77	13.1	35	0.005	0.021
Seattle—Bellevue—Everett, WA	2,414,616	6	0.018	0.1	0.08	22	52	10.7	40	0.003	0.009
Sharon, PA	120,293	ND	ND	0.12	0.09	ND	ND	13.8	35	0.006	0.025
Sheboygan, WI	112,646	ND	ND	0.12	0.09	ND	ND	ND	ND	ND	ND
Shreveport—Bossier City, LA	392,302	ND	ND	0.1	0.08	24	42	12.3	25	0.002	0.007
Sioux City, IA—NE	124,130	ND	ND	ND	ND	26	81	10.8	29	ND	ND
Sioux Falls, SD	172,412	ND	0.008	0.07	0.07	27	101	10.4	27	0.001	0.006
South Bend, IN	265,559	ND	0.014	0.1	0.09	IN	37	13.8	35	ND	ND
Spokane, WA	417,939	5	ND	0.09	0.08	27	89	8.2	25	ND	ND
Springfield, IL	201,437	2	ND	0.09	0.08	ND	ND	13	34	0.003	0.031
Springfield, MO	325,721	2	0.011	0.09	0.07	17	39	11.7	29	0.003	0.028
Springfield, MA	591,932	3	0.02	0.11	0.08	21	55	IN	IN	0.006	0.024
Stamford—Norwalk, CT	353,556	3	0.016	0.14	0.1	15	75	13.5	45	0.005	0.034
State College, PA	135,758	ND	0.009	0.11	0.09	ND	ND	13.6	35	0.006	0.019
Steubenville—Weirton, OH—WV	132,008	9	ND	0.11	0.08	33	110	17.7	43	0.013	0.063
Stockton—Lodi, CA	563,598	3	0.018	0.1	0.08	28	112	13.6	41	ND	ND
Sumter, SC	104,646	ND	ND	ND	ND	ND	ND	ND	ND	ND	ND
Syracuse, NY	732,117	2	ND	0.11	0.09	ND	ND	IN	IN	0.003	0.013
Tacoma, WA	700,820	5	ND	0.1	0.08	20	56	9.9	37	ND	ND
Tallahassee, FL	284,539	ND	ND	0.09	0.07	IN	47	11.8	24	ND	ND
Tampa—St. Petersburg—Clearwater, FL	2,395,997	3	0.01	0.11	0.08	27	64	10.5	22	0.006	0.047
Terre Haute, IN	149,192	ND	ND	0.1	0.08	24	54	14.1	35	0.008	0.035
Texarkana, TX—Texarkana, AR	129,749	ND	ND	ND	ND	ND	ND	13.3	27	ND	ND
Toledo, OH	618,203	ND	ND	0.1	0.09	22	60	14.5	37	0.007	0.026
Topeka, KS	169,871	ND	ND	ND	ND	22	39	11.1	32	ND	ND

Maximum air quality concentrations by metropolitan statistical area, 2003 - *continued*

Metropolitan Statistical Area	2000 Population	CO 8-hr (ppm)	NO$_2$ AM (ppm)	O$_3$ 1-hr (ppm)	O$_3$ 8-hr (ppm)	PM$_{10}$ Wtd AM (ug/m^3)	PM$_{10}$ 24-hr (ug/m^3)	PM$_{2.5}$ Wtd AM (ug/m^3)	PM$_{2.5}$ 24-hr (ug/m^3)	SO$_2$ AM (ppm)	SO$_2$ 24-hr (ppm)
Trenton, NJ	350,761	ND	0.015	0.12	0.09	23	70	13.5	41	ND	ND
Tucson, AZ	843,746	3	0.017	0.09	0.08	40	129	6.5	16	0.001	0.004
Tulsa, OK	803,235	3	0.01	0.11	0.09	27	69	11.9	28	0.006	0.039
Tuscaloosa, AL	164,875	ND	ND	0.08	0.07	ND	ND	IN	IN	ND	ND
Tyler, TX	174,706	ND	0.005	0.09	0.08	ND	ND	ND	ND	ND	ND
Utica—Rome, NY	299,896	ND	ND	0.11	0.08	IN	IN	IN	IN	0.001	0.006
Vallejo—Fairfield—Napa, CA	518,821	3	0.012	0.1	0.07	21	37	9.4	25	0.001	0.003
Ventura, CA	753,197	3	0.015	0.12	0.09	31	122	14.2	33	0.001	0.002
Victoria, TX	84,088	ND	ND	0.09	0.08	ND	ND	ND	ND	ND	ND
Vineland—Millville—Bridgeton, NJ	146,438	ND	ND	0.11	0.09	ND	ND	ND	ND	0.004	0.015
Visalia—Tulare—Porterville, CA	368,021	2	0.018	0.13	0.11	43	89	18.2	47	ND	ND
Waco, TX	213,517	ND	ND	ND	ND	ND	ND	IN	IN	ND	ND
Washington, DC—MD—VA—WV	4,923,153	4	0.026	0.14	0.1	23	55	16.2	39	0.008	0.035
Waterbury, CT	228,984	ND	ND	ND	ND	21	45	12.6	38	0.004	0.02
Waterloo—Cedar Falls, IA	128,012	ND	ND	ND	ND	27	53	IN	IN	ND	ND
Wausau, WI	125,834	ND	ND	0.08	0.07	ND	ND	ND	ND	ND	ND
West Palm Beach—Boca Raton, FL	1,131,184	2	0.014	0.08	0.07	30	53	7.5	16	0.001	0.002
Wheeling, WV—OH	153,172	2	ND	0.11	0.08	25	51	15.4	33	0.011	0.044
Wichita, KS	545,220	3	0.01	0.1	0.08	26	57	11.2	26	ND	ND
Williamsport, PA	120,044	ND	ND	0.1	0.08	20	41	ND	ND	0.005	0.017
Wilmington—Newark, DE—MD	586,216	2	0.019	0.12	0.09	21	56	15.3	37	0.008	0.052
Wilmington, NC	233,450	3	ND	0.09	0.08	ND	ND	9.2	18	0.005	0.046
Worcester, MA—CT	511,389	2	IN	0.1	0.08	IN	IN	10.8	36	IN	IN
Yakima, WA	222,581	3	ND	ND	ND	25	64	10.2	35	ND	ND
Yolo, CA	168,660	1	0.011	0.1	0.08	23	52	8.4	28	ND	ND
York, PA	381,751	2	0.017	0.11	0.08	24	77	17.4	47	0.004	0.012
Youngstown—Warren, OH	594,746	ND	ND	0.12	0.09	25	71	15	36	0.007	0.039
Yuba City, CA	139,149	2	0.014	0.11	0.09	26	77	9.4	29	ND	ND
Yuma, AZ	160,026	ND	ND	0.09	0.08	38	93	ND	ND	ND	ND

Notes: *Data from exceptional events are not included. The reader is cautioned that this summary is not adequate in itself to numerically rank MSAs according to their air quality. The monitoring data represent the quality of air in the vicinity of the monitoring site but may not necessarily represent urban-wide air quality; ND - Indicates data not available; IN - Indicates insufficient data to calculate summary statistic; Wtd AM - Weighted Annual Mean; AM - Annual mean; ug/m^3 - micrograms per cubic meter; ppm - parts per million; NAAQS - National Ambient Air Quality Standards*

Pollutants:

CO (Carbon Monoxide)- Highest second maximum non-overlapping 8-hour concentration (Applicable NAAQS is 9 ppm)

NO$_2$ (Nitrogen Dioxide) - Highest arithmetic mean concentration (Applicable NAAQS is 0.053 ppm)

O$_3$ 1-hr (Ozone) - Highest second daily maximum 1-hour concentration (Applicable NAAQS is 0.12 ppm)

O$_3$ 8-hr (Ozone) - Highest fourth daily maximum 8-hour concentration (Applicable NAAQS is 0.08 ppm)

PM$_{10}$ Wtd Am (Particulate Matter 10 - particles with diameters of 10 micrometers or less) - Highest weighted annual mean concentration (Applicable NAAQS is 50 ug/m^3)

PM$_{10}$ 24-hr (Particulate Matter 10 - particles with diameters of 10 micrometers or less) - Highest second maximum 24-hour concentration (Applicable NAAQS is 150 ug/m^3)

PM$_{2.5}$ Wtd AM (Particulate Matter 2.5 - particles with diameters of 2.5 micrometers or less) - Highest weighted annual mean concentration (Applicable NAAQS is 15 ug/m^3)

PM$_{2.5}$ 24-hr (Particulate Matter 2.5 - particles with diameters of 2.5 micrometers or less)- Highest second maximum 24-hour concentration (Applicable NAAQS is 65 ug/m^3)

SO$_2$ AM (Sulfur Dioxide) - Highest annual mean concentration (Applicable NAAQS is 0.03 ppm)

SO$_2$ 24-hr (Sulfur Dioxide) - Highest second maximum 24-hour concentration (Applicable NAAQS is 0.14 ppm)

Source: *U.S. Environmental Protection Agency, National Air Quality and Emissions Trends Report, 2003*

Number of days with AQI¹ values greater than 100 at trend sites, 1994-2003, and all sites in 2003

Metropolitan Statistical Area	No. of Trend Sites	1994	1995	1996	1997	1998	1999	2000	2001	2002	2003	All Sites Active in 2003	AQI¹>100 in 2003 Using All Sites
Akron, OH	8	8	12	11	6	14	25	9	22	24	7	9	7
Albany-Schenectady-Troy, NY	5	6	3	4	3	3	6	1	11	8	5	12	6
Albuquerque, NM	15	1	0	0	0	0	1	0	1	0	2	32	6
Allentown-Bethlehem-Easton, PA	5	3	7	6	12	18	19	10	20	25	8	11	8
Atlanta, GA	16	12	33	21	26	43	61	35	24	15	11	38	16
Austin-San Marcos, TX	1	4	10	0	0	5	8	6	0	5	3	13	4
Bakersfield, CA	14	98	105	109	55	76	139	132	124	149	141	29	142
Baltimore, MD	20	40	36	28	30	51	41	23	33	42	20	34	20
Baton Rouge, LA	10	7	15	7	8	14	17	29	5	5	15	22	17
Bergen-Passaic, NJ	6	0	0	0	0	1	2	1	1	1	2	9	7
Birmingham, AL	22	6	32	15	8	23	51	49	36	17	10	28	11
Boston, MA-NH	16	0	0	0	0	0	3	0	3	7	6	28	8
Buffalo-Niagara Falls, NY	11	4	6	3	1	13	8	5	13	22	8	14	8
Charleston-North Charleston, SC	10	2	1	3	3	3	5	7	0	3	0	14	0
Charlotte-Gastonia-Rock Hill, NC-SC	9	9	11	16	24	47	34	22	14	27	4	22	9
Chicago, IL	46	7	23	6	9	10	19	13	33	23	10	64	10
Cincinnati, OH-KY-IN	17	16	19	10	11	13	18	15	15	31	10	33	13
Cleveland-Lorain-Elyria, OH	28	23	24	16	10	20	36	21	29	30	17	48	19
Columbus, OH	6	4	6	5	3	7	17	10	10	18	8	15	11
Dallas, TX	9	0	0	0	0	0	0	0	0	3	0	41	21
Dayton-Springfield, OH	11	14	11	18	9	19	19	9	11	27	7	16	12
Denver, CO	19	1	2	0	0	5	2	2	7	4	5	29	20
Detroit, MI	32	11	14	13	11	17	23	16	31	28	19	33	19
El Paso, TX	15	5	0	5	2	0	5	4	8	9	7	38	13
Fort Lauderdale, FL	9	1	1	1	0	1	4	3	3	3	0	20	0
Fort Worth-Arlington, TX	7	31	28	14	14	17	19	17	17	23	25	19	31
Fresno, CA	16	55	61	70	75	67	133	131	139	152	127	30	128
Gary, IN	17	6	17	11	12	9	21	16	29	21	7	25	10
Grand Rapids-Muskegon-Holland, MI	10	6	9	5	6	5	11	6	14	14	10	14	13
Greensboro-Winston Salem-High Point, NC	11	7	6	6	13	25	22	14	11	24	4	19	7
Greenville-Spartanburg-Anderson, SC	7	4	5	7	8	19	14	14	15	25	3	11	4
Harrisburg-Lebanon-Carlisle, PA	9	12	13	3	9	22	19	16	22	21	10	11	12
Hartford, CT	11	18	14	5	16	10	18	7	18	23	9	11	9
Honolulu, HI	7	0	0	0	0	0	2	2	2	2	2	22	2
Houston, TX	23	38	65	26	47	38	51	42	27	21	31	62	39
Indianapolis, IN	26	22	19	13	12	19	23	8	14	24	11	33	16
Jacksonville, FL	13	0	0	0	0	3	2	1	3	0	0	16	0
Jersey City, NJ	7	12	16	5	9	7	20	4	7	8	5	7	5
Kansas City, MO-KS	24	10	21	7	16	14	3	11	7	7	11	39	15
Knoxville, TN	14	13	22	19	36	52	61	31	21	42	14	21	15
Las Vegas, NV-AZ	10	2	0	2	0	0	0	1	4	2	5	65	16
Little Rock-North Little Rock, AR	6	2	7	1	1	2	6	16	4	11	1	10	1
Los Angeles-Long Beach, CA	55	139	113	94	60	56	60	87	89	81	88	62	112
Louisville, KY-IN	22	27	21	10	13	24	42	17	17	24	12	28	13
Memphis, TN-AR-MS	13	10	18	18	14	27	29	27	10	13	8	19	9
Miami, FL	14	1	2	1	3	8	7	2	1	1	1	16	1
Middlesex-Somerset-Hunterdon, NJ	4	9	16	8	18	21	23	9	14	21	9	6	10
Milwaukee-Waukesha, WI	20	8	13	5	4	10	18	5	21	9	9	29	9
Minneapolis-St. Paul, MN-WI	21	2	4	0	0	0	0	5	5	1	1	46	1
Monmouth-Ocean, NJ	3	3	6	12	12	19	12	5	8	18	11	4	13
Nashville, TN	20	21	26	23	20	30	37	20	7	16	7	19	7
Nassau-Suffolk, NY	5	10	9	6	8	10	11	4	4	10	7	13	7
New Haven-Meriden, CT	8	13	14	8	19	9	19	9	15	25	16	14	17
New Orleans, LA	15	8	18	8	6	7	18	20	5	2	8	19	8
New York, NY	15	13	17	11	22	14	22	19	21	26	14	44	14
Newark, NJ	16	12	20	11	13	22	25	13	19	30	12	22	12

Number of days with AQI[1] values greater than 100 at trend sites, 1994-2003, and all sites in 2003 - *continued*

Metropolitan Statistical Area	No. of Trend Sites	1994	1995	1996	1997	1998	1999	2000	2001	2002	2003	All Sites Active in 2003	AQI[1]>100 in 2003 Using All Sites
Norfolk-Virginia Beach-Newport News, VA-NC	13	4	3	4	15	13	16	2	8	14	4	17	4
Oakland, CA	18	1	8	4	0	6	17	10	10	19	7	46	9
Oklahoma City, OK	10	5	13	2	4	7	4	7	2	3	3	19	4
Omaha, NE-IA	14	1	1	1	0	5	5	2	4	0	2	20	2
Orange County, CA	11	14	8	6	3	5	15	31	31	20	19	16	26
Orlando, FL	12	3	1	1	4	11	4	3	4	1	0	16	1
Philadelphia, PA-NJ	39	26	30	22	32	37	32	21	35	35	20	57	22
Phoenix-Mesa, AZ	24	10	22	15	12	14	10	11	7	9	9	86	15
Pittsburgh, PA	41	19	25	11	20	39	40	32	55	50	37	69	39
Portland-Vancouver, OR-WA	17	2	2	6	0	3	6	7	5	6	0	22	0
Providence-Fall River-Warwick, RI-MA	8	0	0	0	0	0	3	1	1	3	3	14	10
Raleigh-Durham-Chapel Hill, NC	7	2	1	1	13	21	27	8	4	18	5	19	8
Richmond-Petersburg, VA	10	9	14	5	19	22	21	6	15	22	5	16	7
Riverside-San Bernardino, CA	43	149	124	118	104	95	121	144	156	146	138	67	141
Rochester, NY	7	1	6	0	6	4	9	1	8	13	3	8	3
Sacramento, CA	16	28	32	27	2	20	47	34	39	46	22	53	52
St. Louis, MO-IL	48	32	34	20	15	23	31	20	20	34	13	65	15
Salt Lake City-Ogden, UT	16	11	4	8	1	12	13	20	27	33	10	42	16
San Antonio, TX	2	3	17	2	3	6	9	0	0	17	4	12	8
San Diego, CA	28	46	48	31	14	33	33	31	31	19	20	36	20
San Francisco, CA	11	0	2	0	0	0	2	2	5	1	0	16	1
San Jose, CA	6	2	10	7	0	5	1	0	3	4	4	12	11
SanJuan-Bayamon, PR	12	0	0	1	1	0	0	0	1	0	0	27	0
Scranton-Wilkes Barre-Hazleton, PA	11	6	10	4	9	7	12	3	11	18	6	12	6
Seattle-Bellevue-Everett, WA	12	1	0	0	0	3	6	8	6	7	2	29	4
Springfield, MA	15	12	9	5	10	7	14	4	17	17	9	14	9
Syracuse, NY	2	0	0	0	0	0	1	0	1	2	1	9	5
Tacoma, WA	6	2	0	1	0	4	4	14	11	7	2	8	3
Tampa-St. Petersburg-Clearwater, FL	22	3	2	3	4	11	12	8	4	0	2	44	6
Toledo, OH	4	7	5	10	3	5	8	4	12	13	6	10	9
Tucson, AZ	20	0	3	0	1	0	7	0	0	3	1	30	1
Tulsa, OK	10	11	19	12	5	7	13	10	6	5	7	18	9
Ventura, CA	15	63	65	62	44	29	24	31	25	11	20	23	33
Washington, DC-MD-VA-WV	32	20	27	18	28	45	42	22	27	31	12	64	13
West Palm Beach-Boca Raton, FL	5	0	0	0	0	2	1	1	1	0	0	11	0
Wilmington-Newark, DE-MD	6	5	12	3	6	8	12	8	17	15	5	18	11
Youngstown-Warren, OH	5	0	0	0	0	0	7	3	16	5	3	15	8

Note: (1) Air Quality Index - The AQI runs from 0 to 500. The higher the AQI value, the greater the level of air pollution and the greater the health danger. For example, an AQI value of 50 represents good air quality and little potential to affect public health, while an AQI value over 300 represents hazardous air quality. An AQI value of 100 generally corresponds to the national air quality standard for the pollutant, which is the level EPA has set to protect public health. So, AQI values below 100 are generally thought of as satisfactory. When AQI values are above 100, air quality is considered to be unhealthy—at first for certain sensitive groups of people, then for everyone as AQI values get higher.

Source: U.S. Environmental Protection Agency, National Air Quality and Emissions Trends Report, 2003

Brownfields

City	State	Population	Estimated Number of Sites	Estimated Number of Acres	Estimated Tax Revenue Gained (conservative)	Estimated Tax Revenue Gained (optimistic)	Potential Number of Jobs Created
Akron	OH	217,074	36	210	1,900,000	2,500,000	2,200
Alameda	CA	72,259	38	2,000	*	*	18,000
Albany	NY	95,658	75	50	*	*	*
Albuquerque	NM	448,607	50	*	*	*	*
Alton	IL	30,496	11	1,000	1,000,000	5,000,000	1,500
Amesbury	MA	16,450	20	200	300,000	400,000	500
Anaheim	CA	328,014	11	78	*	500,000	1,000
Arlington Heights	IL	76,031	5	5	*	*	*
Atlanta	GA	416,474	685	513	411,000	1,682,000	465
Attleboro	MA	42,068	5	35	800,000	1,000,000	1,016
Augusta	GA	199,775	30	*	*	*	100
Aurora	CO	276,393	121	*	*	*	*
Austin	TX	656,652	450	1,170	5,000,000	75,000,000	1,000
Bangor	ME	31,473	17	66	2,805,460	9,517,275	960
Bartlett	IL	36,706	1	10	200,000	1,000,000	30
Baton Rouge	LA	227,818	400	1,725	335,111	670,222	*
Bayonne	NJ	61,842	12	160	*	*	*
Bedford Heights	OH	11,375	3	14	350,000	500,000	*
Bell Gardens	CA	44,054	8	16	20,000	40,000	100
Billings	MT	89,847	*	*	*	*	*
Binghamton	NY	47,380	12	60	500,000	2,000,000	500
Birmingham	AL	242,820	*	*	*	*	*
Bloomfield	NJ	47,683	1	3	10,000	20,000	30
Bloomington	IN	69,291	8	100	200,000	700,000	560
Bossier City	LA	56,461	10	20	100,000	50,000	150
Boston	MA	589,141	50	50	3,000,000	10,000,000	300
Bridgeport	CT	139,529	100	600	50,000,000	100,000,000	*
Brookfield	WI	38,649	4	6	1,500	3,000	*
Brooklyn	OH	11,586	*	*	500,000	2,000,000	600
Brooklyn Center	MN	29,172	1	50	*	*	*
Broomfield	CO	38,272	4	50	*	*	*
Buffalo	NY	292,648	300	2,000	*	*	5,000
Burbank	CA	100,316	10	65	1,000,000	2,000,000	2,000
Burlington	VT	38,889	30	40	50,000	100,000	150
Caguas	PR	140,502	6	42	15,000	25,000	70
Calumet City	IL	39,071	*	*	*	*	*
Canton	OH	80,806	8	300	1,900,000	3,000,000	6,000
Carbondale	IL	20,681	10	40	250,000	1,000,000	100
Carolina	PR	186,076	12	34	*	*	90
Cathedral City	CA	42,647	*	*	*	*	*
Charleston	SC	96,650	30	*	1,000,000	2,500,000	150
Charlotte[1]	NC	540,828	800	43,780	*	*	*
Chesapeake	VA	199,184	5	250	800,000	3,000,000	1,600
Chicago	IL	2,896,016	*	*	*	78,000,000	34,000
Chicago Heights	IL	32,776	30	400	500,000	750,000	500
Clearwater	FL	108,787	188	1,842	*	*	1,200
Clifton	NJ	78,672	65	80	1,800,000	2,800,000	1,000
Concord	NC	55,977	300	*	500,000	2,000,000	*
Coon Rapids	MN	61,607	6	20	50,000	100,000	*
Costa Mesa	CA	108,724	*	*	*	*	*
Dallas	TX	1,188,580	*	*	25,000,000	52,000,000	*
Danbury	CT	74,848	30	60	*	*	*
Davenport	IA	98,359	25	500	1,000,000	4,000,000	300
Dayton	OH	166,179	20	250	4,000,000	12,000,000	2,500
Dearborn	MI	97,775	200	500	25,000,000	50,000,000	1,000
Delray Beach	FL	60,020	100	45	*	*	*
Denver	CO	554,636	100	500	*	*	100
Des Moines	IA	198,682	250	1,400	200,000,000	600,000,000	7,000
Detroit	MI	951,270	500	2,500	15,000,000	90,000,000	7,000

Brownfields - *continued*

City	State	Population	Estimated Number of Sites	Estimated Number of Acres	Estimated Tax Revenue Gained (conservative)	Estimated Tax Revenue Gained (optimistic)	Potential Number of Jobs Created
Dover	DE	32,135	25	30	6,000	10,000	40
Durham	NC	187,035	100	150	130,000	500,000	300
East Lansing	MI	46,525	5	20	1,000,000	20,000,000	50
East St. Louis	IL	31,542	90	600	3,000,000	8,000,000	7,000
Easthampton	MA	15,994	12	42	18,000	35,000	5
El Paso	TX	563,662	400	4,000	1,000,000	10,000,000	2,000
Elizabeth	NJ	120,568	50	625	6,000,000	15,000,000	15,000
Erie	PA	103,717	14	200	950,000	110,000	1,950
Evansville	IN	121,582	350	1,000	2,300,000	4,285,000	*
Everett	MA	38,037	*	380	500,000	1,000,000	11,000
Fairfield	CA	96,178	39	946	*	*	*
Fajardo	PR	40,712	5	*	5,000	25,000	50
Fayetteville	NC	121,015	140	500	10,000,000	25,000,000	250
Fitchburg	MA	39,102	25	50	50,000	250,000	600
Flint	MI	124,943	20	300	*	*	*
Fort Collins	CO	118,652	1	352	*	*	*
Fort Myers	FL	48,208	65	750	500,000	1,500,000	500
Fort Wayne	IN	205,727	40	500	1,000,000	5,000,000	400
Fort Worth	TX	534,694	263	650	*	*	*
Gadsen	AL	38,978	5	800	*	*	750
Gainesville	FL	95,447	100	200	500,000	2,000,000	150
Galesburg	IL	33,706	10	30	50,000	150,000	200
Galveston	TX	57,247	35	650	*	*	*
Gardena	CA	57,746	47	100	1,000,000	1,500,000	75
Garfield Heights	OH	30,734	11	280	12,000,000	22,000,000	3,000
Gary	IN	102,746	120	*	30,000,000	40,000,000	33,146
Glencoe	IL	8,762	1	6	400,000	800,000	*
Green Bay	WI	102,313	20	150	1,500,000	5,000,000	300
Greenfield	WI	35,476	1	30	75,000	150,000	*
Greensboro	NC	223,891	*	*	*	*	*
Hagerstown	MD	36,687	55	383	1,070,000	250,000	4,000
Hamilton	OH	60,690	30	150	400,000	600,000	75
Holly Hill	FL	12,119	7	18	50,000	150,000	100
Hollywood	FL	139,357	103	*	*	*	60
Hormigueros	PR	16,614	2	15	10,000	50,000	100
Houston	TX	1,953,631	*	*	500,000	1,000,000	5,000
Huntington Park	CA	61,348	2	55	600,000	1,200,000	1,500
Indianapolis	IN	791,926	175	400	*	*	1,200
Inkster	MI	30,115	3	32	1,000,000	5,000,000	*
Irvington	NJ	60,695	40	*	*	*	500
Jackson	MS	184,256	180	720	7,000,000	21,000,000	25,000
Jackson	MI	36,316	250	2,000	5,000,000	25,000,000	7,000
Jacksonville	FL	735,617	300	11,520	35,000,000	75,000,000	2,500
Jersey City	NJ	240,055	*	*	*	*	*
Kansas City	MO	441,545	*	*	*	*	*
Kenner	LA	70,517	*	*	*	*	*
Kettering	OH	57,502	2	20	500,000	1,000,000	400
Knoxville	TN	173,890	11	1,000	350,000	2,500,000	1,800
Kokomo	IN	46,113	15	150	*	*	*
La Crosse	WI	51,818	20	100	1,000,000	2,225,000	450
La Habra	CA	58,974	10	2	50,000	100,000	*
La Verne	CA	31,638	*	*	*	*	*
Lafayette	IN	56,397	4	10	*	*	*
Lakeland	FL	78,452	1	99	*	*	*
Lakewood	CO	144,126	150	500	*	*	*
Lancaster	OH	35,335	5	40	50,000	75,000	50
Lansing	MI	119,128	100	100	15,000,000	30,000,000	1,000
Lansing	IL	28,332	12	50	100,000	250,000	200
Las Vegas	NV	478,434	20	40	20,000	150,000	150

Brownfields - *continued*

City	State	Population	Estimated Number of Sites	Estimated Number of Acres	Estimated Tax Revenue Gained (conservative)	Estimated Tax Revenue Gained (optimistic)	Potential Number of Jobs Created
Lawrence	MA	72,043	25	100	1,000,000	1,500,000	500
Leigh Valley	PA	106,632	35	150	*	*	1,000
Lenexa	KS	40,238	*	*	*	*	*
Lewiston	ME	35,690	*	*	3,000,000	6,000,000	6,000
Lima	OH	40,081	200	300	400,000	1,100,000	3,000
Lincoln	NE	225,581	12	100	*	*	*
Livonia	MI	100,545	3	250	220,000	440,000	750
Long Beach	CA	461,522	100	15	*	*	433
Louisville	KY	256,231	650	5,080	10,000,000	18,750,000	*
Loveland	OH	11,677	3	15	105,000	150,000	150
Lowell	MA	105,167	365	1,000	3,000,000	15,000,000	*
Lubbock	TX	199,564	5	*	*	*	*
Lynn	MA	89,050	10	235	1,500,000	3,500,000	15,000
Lynwood	CA	69,845	7	58	1,000,000	1,500,000	50
Madison	WI	208,054	10	20	280,000	540,000	*
Malden	MA	56,340	141	70	1,500,000	6,000,000	3,750
Manchester	NH	107,006	1	20	500,000	600,000	300
Marlborough	MA	36,255	6	*	*	*	*
McKinney	TX	54,369	*	*	*	*	*
Medford	MA	55,765	5	10	*	*	*
Memphis	TN	650,100	212	4,230	4,000,000	6,000,000	15,000
Meriden	CT	58,244	14	27	10,000,000	20,000,000	100
Meridian	MS	39,968	5	200	100,000	1,000,000	250
Mesquite	TX	124,523	*	*	*	*	*
Miami-Dade	FL	2,057,000	300	250	150,000	2,000,000	5,500
Middleton	CT	43,167	10	30	*	*	200
Midland	MI	41,685	12	25	10,000	250,000	50
Minneapolis	MN	382,618	1,300	3,000	*	*	*
Missoula	MT	57,053	12	200	1,000,000	2,000,000	500
Modesto	CA	188,856	10	300	250,000	750,000	1,500
Moline	IL	43,768	3	2	*	*	*
Monroe	LA	53,107	15	225	2,500,000	10,000,000	8
Montebello	CA	62,150	1	55	2,500,000	5,000,000	1,000
Montgomery	AL	201,568	20	200	1,000,000	6,000,000	5,000
Mount Vernon	NY	68,381	10	3	500,000	1,000,000	400
Mountain View	CA	70,708	4	210	9,900,000	13,900,000	8,000
Murfreesboro	TN	68,816	10	100	20,000	300,000	50
Murray	UT	34,024	3	150	500,000	750,000	2,000
Muskegon Heights	MI	40,105	92	105	*	*	*
Nacogdoches	TX	29,914	6	60	45,000	100,000	70
Nashville	TN	569,891	13	170	3,000,000	10,000,000	150
New Bedford	MA	93,768	44	228	*	*	*
New Brunswick	NJ	48,573	*	*	500,000	1,000,000	*
New Castle	PA	28,334	6	75	*	*	300
New Haven	CT	123,626	78	92	1,250,000	2,500,000	1,150
New Orleans	LA	484,674	500	2,500	5,000,000	20,000,000	7,500
New York	NY	8,008,278	6,000	3,500	*	*	200,000
Newark	CA	42,471	5	100	2,000,000	3,500,000	1,500
Niagara Falls	NY	55,593	100	300	*	*	1,000
North Chicago	IL	35,918	10	50	10,000,000	15,000,000	250
North Providence	RI	32,411	3	*	40,000	100,000	*
North Tonawanda	NY	33,262	2	30	*	400,000	200
Norwalk	CA	103,298	19	82	180,000	540,000	*
Norwich	CT	36,117	20	200	2,000,000	5,000,000	*
Oak Brook	IL	8,702	*	*	*	*	*
Oakland	CA	399,484	1,000	1,000	*	*	*
Oakley	CA	25,619	3	120	1,500,000	4,000,000	5,000
Ocala	FL	45,943	9	108	*	*	150
Orland Park	IL	51,077	*	*	*	*	*

Brownfields - *continued*

City	State	Population	Estimated Number of Sites	Estimated Number of Acres	Estimated Tax Revenue Gained (conservative)	Estimated Tax Revenue Gained (optimistic)	Potential Number of Jobs Created
Palatine	IL	65,479	20	25	*	*	*
Park Ridge	IL	37,775	5	3	*	*	*
Pembroke Pines	FL	137,427	*	*	*	*	*
Pensacola	FL	56,255	30	60	*	*	*
Perth Amboy	NJ	47,303	72	900	1,400,000	3,200,000	2,500
Philadelphia	PA	1,517,550	*	*	*	*	*
Pine Bluff	AR	55,085	4	12	33,000	70,000	90
Plantation	FL	82,934	*	*	*	*	*
Pocatello	ID	51,466	*	*	*	*	*
Pompano Beach	FL	78,191	*	3,028	200,000	300,000	238
Port Huron	MI	32,338	*	*	*	*	*
Portland	OR	529,121	650	806	*	*	*
Providence	RI	173,618	30	*	*	*	*
Rantoul	IL	12,857	5	15	200,000	500,000	40
Rapid City	SD	59,607	*	*	*	*	*
Redondo Beach	CA	63,261	5	50	*	*	*
Reno	NV	180,480	*	*	*	*	*
Richardson	TX	91,802	*	*	*	*	*
Richmond	VA	197,790	121	190	100,000,000	150,000,000	1,000
Rochester Hills	MI	68,825	15	300	500,000	8,000,000	*
Rock Island	IL	39,684	40	20	125,000	250,000	400
Rome	NY	34,950	1	200	250,000	500,000	500
Rosemead	CA	53,505	*	*	*	*	*
St. Joseph	MO	73,990	2	500	10,000,000	15,000,000	950
St. Louis	MO	348,189	1,000	2,000	*	*	6,411
St. Paul	MN	287,151	56	1,200	20,000,000	30,000,000	18,000
St. Petersburg	FL	248,232	150	500	*	*	200
Salt Lake City	UT	181,743	*	650	1,000,000	3,000,000	*
Santa Rosa	CA	147,595	6	22	*	*	*
Savannah	GA	131,510	3	30	500,000	800,000	260
Schaumburg	IL	75,386	*	*	*	*	*
Shaker Heights	OH	29,405	1	*	25,000	45,000	*
South Salt Lake	UT	22,038	3	25	500,000	1,500,000	150
Southgate	MI	30,136	5	30	15,000,000	30,000,000	*
Springfield	OH	65,358	15	155	1,455,000	5,500,000	2,962
Springfield	MO	151,580	600	250	*	*	*
Stamford	CT	117,083	30	60	5,000,000	*	1,000
Stockton	CA	243,771	60	175	50,000	500,000	500
Strongsville	OH	43,858	*	*	*	*	*
Tallahassee	FL	150,624	40	80	37,000,000	45,000,000	1,300
Tampa	FL	303,447	1,000	1,000	*	*	*
Taylor	MI	65,868	35	465	2,200,000	3,500,000	459
Tempe	AZ	158,625	20	500	2,000,000	5,000,000	3,000
Trenton	NJ	85,403	98	250	*	*	*
Tucson	AZ	486,699	*	*	*	*	*
Tupelo	MS	34,211	1	50	*	*	*
University Heights	OH	14,146	*	*	*	*	*
Villa Park	IL	22,075	3	4	1,000,000	*	50
Visalia	CA	91,565	1	1	*	*	*
Waco	TX	113,726	11	152	637,500	900,000	1,800
Walnut Creek	CA	64,296	10	6	1,000,000	1,500,000	200
Walton Hills	OH	2,400	3	22	*	*	200
Warrensville Heights	OH	15,109	12	65	1,500,000	3,500,000	500
Washington	DC	572,059	*	*	*	*	*
West Hollywood	CA	35,716	9	9	500,000	2,000,000	450
West Jordan	UT	68,336	1	120	*	*	*
West Valley City	UT	108,896	1	40	500,000	1,250,000	150
West Warwick	RI	29,268	3	15	500,000	1,000,000	50
Westland	MI	86,602	3	72	300,000	1,000,000	*

Brownfields - *continued*

City	State	Population	Estimated Number of Sites	Estimated Number of Acres	Estimated Tax Revenue Gained (conservative)	Estimated Tax Revenue Gained (optimistic)	Potential Number of Jobs Created
Wheeling	WV	31,419	15	*	*	*	*
Whittier	CA	83,680	9	12	150,000	500,000	50
Wichita	KS	344,284	285	4,400	250,000	1,000,000	200
Winston-Salem	NC	185,776	100	*	55,000	220,000	5,250
Woodbridge	NJ	97,203	20	525	500,000	1,000,000	5,000
Worcester	MA	172,648	250	200	20,000,000	30,000,000	8,100
Wyandotte	MI	28,006	15	1,300	845,000	1,000,000	30

Notes: *Brownfields are defined as abandoned, idled, or under-used industrial and commercial facilities where expansion or redevelopment is complicated by real or perceived environmental contamination; (*) City unable to provide estimates; (1) Charlotte's 43,780 estimated brownfield acres were identified based on tax data and a categorization of the uses of such properties.*
Source: *The United States Conference of Mayors, Recycling America's Land: A National Report on Brownfields Redevelopment-Volume IV, 2003*

Statistics & Rankings / Children's Environmental Health

Percentage of children living in counties in which air quality standards were exceeded

Gas	1990	1991	1992	1993	1994	1995	1996	1997	1998	1999	2000	2001
Ozone one-hour standard	22.7	25.2	16.9	21.1	19.3	27.8	16.5	18.6	20.8	21.7	13.3	15.0
Ozone eight-hour standard	na	na	na	na	na	na	39.1	38.9	48.5	46.9	27.9	39.8
PM-10	8.0	6.3	9.6	2.6	2.3	10.0	1.5	2.4	2.0	2.1	2.4	3.2
PM-2.5	na	na	na	na	na	na	na	na	na	na	27.2	25.4
Carbon monoxide	9.4	8.4	6.1	5.0	6.4	4.9	5.6	3.7	4.3	3.7	3.8	0.2
Lead	2.2	6.0	1.8	2.1	1.7	1.8	1.6	1.4	1.6	0.2	0.5	1.0
Sulfur dioxide	0.5	2.1	0.1	0.5	0.1	0.1	0.1	0.1	0.1	0.1	0.1	0.0
Nitrogen dioxide	3.7	3.7	0.0	0.0	0.0	0.0	0.0	0.0	0.0	0.0	0.0	0.0
Any standard[1]	28.0	31.8	20.9	24.2	23.5	30.8	19.8	21.9	23.7	24.0	15.5	18.5

Note: *(1) Does not include ozone eight-hour or PM-2.5 standards; na not available*
Source: *U.S. Environmental Protection Agency, America's Children and the Environment: Measures of Contaminants, Body Burdens, and Illness, Second Edition, February 2003*

Percentage of children living in areas served by public water systems that exceeded a drinking water standard or violated treatment requirements

Type of standard violated	1993	1994	1995	1996	1997	1998	1999
Lead and copper[1]	2.2	0.9	1.4	1.6	1.7	1.6	1.5
Microbial contaminants	8.3	7.5	4.1	4.3	3.6	2.8	2.5
Chemical and radiation	4.7	4.7	2.2	1.8	2.4	1.2	1.0
Nitrate/nitrite	0.23	0.12	0.25	0.20	0.37	0.17	0.21
Treatment and filtration	10.7	8.1	4.5	3.7	3.6	3.4	3.0
Any health-based violations	20.2	15.5	12.0	10.7	10.7	8.6	8.0

Notes: *(1) Lead and copper represents the lead and copper rule, which is a set of standards and implementation measures*
Source: *U.S. Environmental Protection Agency, America's Children and the Environment: Measures of Contaminants, Body Burdens, and Illness, Second Edition, February 2003*

Percentage of children with asthma[1]

1980	1981	1982	1983	1984	1985	1986	1987	1988	1989	1990	1991	1992	1993	1994	1995	1996	1997	1998	1999	2000	2001
3.6	3.7	4.1	4.5	4.3	4.8	5.1	5.3	5.0	6.1	5.8	6.4	6.3	7.2	6.9	7.5	6.2	5.4	5.3	5.3	5.5	5.7
																	11.4	12.1	10.8	12.3	12.6

Note: *(1) The survey questions for asthma changed in 1997. From 1980 -1996 the figures show the percentage of children with asthma in the past 12 months. From 1997 -2001 the top figure shows the percentage of children ever diagnosed with asthma and having an asthma attack in the past 12 months. The bottom figure shows the percentage of children ever diagnosed with asthma. Therefore data before 1997 cannot be directly compared to later data.*
Source: *U.S. Environmental Protection Agency, America's Children and the Environment: Measures of Contaminants, Body Burdens, and Illness, Second Edition, February 2003*

Cancer incidence for children under 20 by type

Cancer type	1974-78	1979-83	1984-88	1989-93	1994-98
Acute lymphoblastic leukemia	23.7	24.9	27.6	28.2	28.3
Acute myeloid leukemia	5.2	4.9	3.8	5.1	4.8
Central nervous system tumors	23.2	22.2	27.9	30.1	27.3
Ewing's sarcoma	2.4	3.5	3.4	3.0	3.1
Germ cell tumors	8.3	9.9	9.7	11.7	11.7
Hepatoblastoma	0.7	0.7	1.1	1.2	1.4
Hodgkin's lymphoma	13.9	14.2	14.1	14.1	12.8
Malignant melanoma	3.9	4.3	5.4	6.5	6.2
Neuroblastoma	7.3	7.2	7.9	7.7	8.0
Non-Hodgkin's lymphoma	9.1	9.5	10.4	10.2	11.2
Osteosarcoma	3.8	4.8	5.0	5.0	5.5
Soft tissue sarcomas	10.0	10.7	10.9	11.4	11.6
Thyroid carcinoma	4.8	4.8	5.1	5.1	5.4
Wilms' tumor	5.4	6.5	5.6	6.3	6.5

Note: *All figures are rates per million children*
Source: *U.S. Environmental Protection Agency, America's Children and the Environment: Measures of Contaminants, Body Burdens, and Illness, Second Edition, February 2003*

Cancer incidence for children under 20 by age and type, 1994-1998

Cancer Type	Ages 0-4	Ages 5-9	Ages 10-14	Ages 15-19
Acute non-lymphocytic leukemia	9.7	4.3	6.5	7.8
CNS and misc. intracranial and intraspinal neoplasms	34.5	29.7	25.0	19.2
Epithelial and unspecified	3.1	3.3	11.6	40.5
Ewing's sarcoma	0.6	2.2	4.2	4.7
Germ cell, trophoblastic, other gonadal neoplasms	7.2	2.1	7.2	28.1
Hepatic tumors	4.6	0.6	0.5	1.2
Hodgkin's lymphoma	0.9	3.7	11.8	32.0
Lymphocytic leukemia	61.0	30.6	18.4	14.9
Neuroblastoma and ganglioneuroblastoma	25.7	3.2	0.8	0.4
Non-Hodgkin's lymphoma	3.6	5.9	7.7	13.2
Osteosarcoma	0.2	2.6	7.9	8.9
Soft tissue sarcomas	11.1	7.9	10.5	15.0
Wilms' tumor	18.8	4.8	0.8	0.3

Note: *All figures are rates per million children*

Source: *U.S. Environmental Protection Agency, America's Children and the Environment: Measures of Contaminants, Body Burdens, and Illness, Second Edition, February 2003*

Projected number of high risk units and associated change in lead poisoning prevalence

	1993	1994	1995	1996	1997	1998	1999	2000
High Risk Housing Units								
Percent High Risk (%)	41.5	40.1	38.8	37.5	36.2	34.9	33.7	32.5
Change in High Risk Percent (%)	-3.2	-3.3	-3.3	-3.4	-3.4	-3.4	-3.5	-3.5
Lead Poisoning Prevalence								
High Risk, PIR>1.3 pre-1940 (%)	4.19	4.1	3.9	3.8	3.7	3.5	3.4	3.3
High Risk, PIR>1.3 1940-74 (%)	3.96	3.8	3.7	3.6	3.5	3.3	3.2	3.1
Low Risk, PIR>1.3 (%)	0.22	0.2	0.2	0.2	0.2	0.2	0.2	0.2
High Risk, PIR<1.3 pre-1940 (%)	20.38	19.7	19.1	18.4	17.8	17.2	16.6	16.0
High Risk, PIR<1.3 1940-74 (%)	9.84	9.5	9.2	8.9	8.6	8.3	8.0	7.7
Low Risk, PIR<1.3 (%)	4.33	4.2	4.0	3.9	3.8	3.6	3.5	3.4
Pre-1940, PIR<1.3 (%)	16.6	15.8	15.1	14.4	13.7	13.0	12.4	11.8
1940-74, PIR<1.3 (%)	7.3	7.0	6.7	6.5	6.2	5.9	5.7	5.4

Projected Number of children under 6 (in thousands) with blood lead levels above 10 ug/dL with PIR > 1.3

Housing Category	Children<6/unit	%PIR>1.3	1993	1994	1995	1996	1997	1998	1999	2000
High-Risk										
Pre-1940	0.214	67.0	89	85	81	77	72	68	64	60
1940-59	0.216	66.0	71	68	64	61	57	54	51	48
1960-74	0.199	67.3	64	60	56	53	49	46	44	41
Low Risk										
Pre-1940	0.214	67.0	1	1	2	2	2	2	2	2
1940-59	0.216	66.0	2	2	2	2	2	2	2	2
1960-74	0.199	67.3	4	4	4	4	4	3	3	3
Post 1974	0.249	77.7	14	14	14	13	13	13	13	13

Projected Number of children under 6 (in thousands) with blood lead levels above 10 ug/dL with PIR < 1.3

Housing Category	Children<6/unit	%PIR>1.3	1993	1994	1995	1996	1997	1998	1999	2000
High-Risk										
Pre-1940	0.214	33.0	313	278	243	208	173	162	152	143
1940-59	0.216	34.0	104	96	89	81	73	69	65	61
1960-74	0.199	32.7	88	81	73	66	59	56	53	50
Low Risk										
Pre-1940	0.214	33.0	20	19	18	17	15	15	15	15
1940-59	0.216	34.0	26	25	24	24	23	22	22	22
1960-74	0.199	32.7	45	43	40	37	34	33	32	31
Post 1974	0.249	22.3	82	80	78	76	74	75	75	75
All Children < 6 with blood lead levels > 10 ug/dL			925	857	788	720	651	621	593	565

Notes: PIR = Poverty Income Ratio (e.g. PIR<1.3 covers households with incomes less than 1.3 times the poverty level)
Source: President's Task Force on Environmental Health Risks and Safety Risks to Children, "Eliminating Childhood Lead Poisoning: A Federal Strategy Targeting Lead Paint Hazards," February 2000

System size and reported health-based violations, by state

State	Number of Systems			Population Served			CWSs with Reported Health-Based Violations			
	CWS	NTNCWS	TNCWS	CWS	NTNCWS	TNCWS	No. of Systems	Pct. of Systems	Population Served	Percent of Pop. Served
Alaska	436	220	969	465,008	96,784	47,418	74	17.0	48,636	10.5
Alabama	619	31	70	5,177,165	7,374	18,518	11	1.8	41,031	0.8
Arkansas	728	40	341	2,504,010	17,021	10,268	96	13.2	225,605	9.0
Arizona	789	197	605	4,912,776	120,380	128,937	106	13.4	1,533,665	31.2
California	3,123	1,418	3,055	32,877,694	6,095,389	422,576	178	5.7	4,276,549	13.0
Colorado	830	163	944	4,986,055	270,601	72,630	56	6.7	597,835	12.0
Connecticut	586	658	1,737	2,707,276	60,611	128,777	68	11.6	45,332	1.7
District of Colum.	3	3	0	606,000	0	4,670	1	33.3	595,000	98.2
Delaware	226	105	177	791,675	57,325	24,531	24	10.6	159,034	20.1
Florida	1,881	1,009	3,341	17,205,824	296,166	264,565	116	6.2	1,719,617	10.0
Georgia	1,689	242	550	7,363,830	88,178	70,645	46	2.7	124,556	1.7
Hawaii	115	13	3	1,280,327	400	11,043	3	2.6	18,001	1.4
Iowa	1,142	144	713	2,559,611	87,600	47,609	58	5.1	162,431	6.3
Idaho	752	244	1,029	953,210	113,744	51,614	89	11.8	124,206	13.0
Illinois	1,792	405	3,700	11,568,029	390,982	144,382	202	11.3	899,345	7.8
Indiana	840	686	2,897	4,667,741	408,140	196,992	95	11.3	223,762	4.8
Kansas	911	52	101	2,562,950	3,788	19,107	97	10.6	181,965	7.1
Kentucky	417	50	65	4,594,369	8,566	18,482	40	9.6	500,522	10.9
Louisiana	1,111	179	312	4,886,463	72,822	69,719	99	8.9	441,520	9.0
Massachusetts	523	250	939	8,975,932	174,008	68,704	41	7.8	690,253	7.7
Maryland	502	573	2,621	4,845,205	171,787	169,401	25	5.0	18,850	0.4
Maine	399	369	1,221	616,267	196,857	71,923	76	19.0	125,060	20.3
Michigan	1,438	1,631	8,841	7,278,655	1,038,931	354,802	69	4.8	116,492	1.6
Minnesota	965	563	6,273	3,963,562	510,591	89,455	21	2.2	36,774	0.9
Missouri	1,463	241	1,016	4,909,561	128,360	73,620	186	12.7	240,894	4.9
Mississippi	1,170	95	118	3,040,537	23,596	74,394	19	1.6	67,171	2.2
Montana	676	227	1,157	681,326	173,913	60,483	78	11.5	38,618	5.7
North Carolina	2,174	566	4,322	6,435,706	355,293	156,004	264	12.1	598,465	9.3
North Dakota	320	29	174	551,910	15,423	4,167	22	6.9	27,360	5.0
Nebraska	606	188	583	1,416,485	87,732	43,547	134	22.1	387,901	27.4
New Hampshire	698	444	1,137	803,172	232,690	90,925	101	14.5	62,471	7.8
New Jersey	607	870	2,644	7,891,239	381,246	217,056	16	2.6	173,806	2.2
New Mexico	645	147	483	1,590,051	131,448	38,900	82	12.7	135,400	8.5
Nevada	253	106	251	1,894,696	28,109	38,253	23	9.1	49,989	2.6
New York	2,816	757	6,432	17,900,676	2,916,321	326,776	115	4.1	1,673,911	9.4
Ohio	1,318	970	3,191	10,037,355	482,841	236,836	79	6.0	229,759	2.3
Oklahoma	1,135	120	355	3,442,683	26,734	19,699	161	14.2	1,012,659	29.4
Oregon	874	337	1,437	2,924,951	223,858	75,556	133	15.2	110,513	3.8
Pennsylvania	2,135	1,213	6,544	10,453,283	841,775	513,299	124	5.8	2,234,932	21.4
Rhode Island	83	78	321	977,398	55,041	27,160	5	6.0	22,208	2.3
South Carolina	659	171	576	3,425,186	35,613	54,632	66	10.0	203,420	5.9
South Dakota	467	28	194	659,281	26,868	12,109	44	9.4	10,652	1.6
Tennessee	681	46	424	5,346,416	59,529	28,795	25	3.7	197,510	3.7
Texas	4,489	785	1,225	22,674,849	245,421	321,559	280	6.2	872,721	3.8
Utah	451	63	433	2,490,321	68,476	27,582	39	8.6	120,478	4.8
Virginia	1,263	602	1,293	6,083,683	216,406	305,751	162	12.8	658,592	10.8
Vermont	435	234	678	514,473	164,736	43,024	64	14.7	34,977	6.8
Washington	2,274	315	1,541	5,597,194	415,122	138,684	264	11.6	377,592	6.7
Wisconsin	1,086	907	9,376	3,858,783	689,414	190,514	54	5.0	291,652	7.6
West Virginia	536	154	534	1,442,234	34,899	44,510	34	6.3	70,367	4.9
Wyoming	276	87	385	439,026	76,068	18,947	15	5.4	5,140	1.2
United States	51,407	19,025	87,328	265,832,109	5,689,550	18,424,977	4,280	8.3	22,815,199	8.6

Notes: Data as of 4th Quarter, 2004; CWS=Community Water System - water systems that serve the same people year-round (e.g. in homes or businesses); NTNCWS=Non-Transient Non-Community Water System - water systems that serve the same people, but not year-round (e.g. schools that have their own water system); TNCWS=Transient Non-Community Water System - water systems that do not consistently serve the same people (e.g. rest stops, campgrounds, gas stations)

Source: U.S. Environmental Protection Agency, Office of Ground Water and Drinking Water, Factoids: Drinking Water and Ground Water Statistics for 2004

Community water system (CWS) violations reported, 1993–2004

Fiscal Year	MCL[1]	MRDL[2]	TT[3]	M/R[4]	Other[5]	Total[6]
Number of Violations						
2004	5,562	6	2,168	n/a	14,459	n/a
2003	4,884	1	2,299	68,840	12,671	88,695
2002	4,693	-	2,408	80,635	11,759	99,495
2001	5,231	-	2,493	63,476	11,455	82,655
2000	4,754	-	3,053	51,912	10,634	70,353
1999	5,497	-	2,202	56,307	1,025	65,031
1998	6,315	-	2,503	52,998	1,614	63,430
1997	5,785	-	2,707	63,432	1,473	73,397
1996	7,351	-	3,022	118,577	1,138	130,088
1995	7,126	-	3,705	133,378	1,796	146,005
1994	7,804	-	4,788	155,264	3,645	171,501
1993	9,604	-	2,128	104,649	3,193	119,574
Number of Systems in Violation						
2004	3,455	4	1,409	n/a	8,947	n/a
2003	3,069	1	1,458	12,880	8,109	20,343
2002	3,193	-	1,482	12,430	8,386	20,232
2001	3,241	-	1,572	13,639	8,631	20,996
2000	3,161	-	1,682	13,736	8,937	21,308
1999	3,298	-	1,046	9,259	618	11,946
1998	3,733	-	1,095	9,853	893	12,868
1997	3,710	-	1,094	10,809	861	13,864
1996	4,391	-	1,310	12,870	731	16,229
1995	4,638	-	1,782	14,397	950	18,032
1994	4,945	-	2,225	21,563	1,483	25,095
1993	5,713	-	1,460	21,666	1,535	25,235
Population Affected						
2004	15,060,424	3,617	16,588,301	n/a	17,833,120	n/a
2003	16,082,938	2,250	16,969,087	49,812,213	12,726,332	81,672,086
2002	8,800,095	-	9,245,467	32,503,639	17,809,478	56,644,512
2001	11,165,733	-	15,474,893	52,618,908	25,780,359	77,845,089
2000	11,947,655	-	15,394,154	34,186,550	13,881,076	62,921,077
1999	11,078,268	-	15,854,547	19,829,544	2,528,920	38,124,231
1998	10,384,541	-	18,483,671	28,227,720	2,366,440	48,858,948
1997	13,843,906	-	22,056,053	27,179,042	3,879,561	58,038,069
1996	16,039,066	-	22,678,716	32,526,904	4,670,817	63,680,826
1995	23,559,606	-	26,405,992	40,631,838	8,707,474	73,163,910
1994	27,720,949	-	28,469,122	51,367,424	9,729,524	82,581,010
1993	32,759,178	-	29,785,722	73,961,496	10,871,425	106,353,974

Notes: *FY2003 and 2004 compliance information includes new Disinfectant and Disinfection Byproduct (D-DBP) and Interim Enhanced Surface Water Treatment (IESWTR) rules; (1) Maximum Contaminant Level (MCL) - The highest level of a contaminant that is allowed in drinking water. MCLs are set as close to Maximum Contaminant Level Goals (MCLGs) as feasible using the best available treatment technology and taking cost into consideration. MCLs are enforceable standards; (2) Maximum Residual Disinfectant Level (MRDL) - The highest level of a disinfectant allowed in drinking water. There is convincing evidence that addition of a disinfectant is necessary for control of microbial contaminants; (3) Treatment Technique - A required process intended to reduce the level of a contaminant in drinking water; (4) Monitoring and Reporting - EPA sets minimum monitoring schedules that drinking water systems must follow. These minimum reporting schedules (systems may monitor more frequently) vary by system size as well as by contaminant. Some contaminants are monitored for daily, others need to be checked far less frequently (every nine years is the longest monitoring cycle). For instance, at a minimum, drinking water systems will monitor continuously for turbidity, monthly for bacteria, and once every four years for radio nuclides. A Monitoring and Reporting Violation means the system did not perform the required testing, take adequate samples, or report a violation as required; (5) Jump in FY2000 due to new violations for failing to issue, or issuing an insufficient, Consumer Confidence report; (6) Totals for the number of systems in violation, and for population affected, should be lower than the sum in each row. This is because some systems will have incurred more than one type of violation; n/a not available.*

Source: *U.S. Environmental Protection Agency, Office of Ground Water and Drinking Water, Factoids: Drinking Water and Ground Water Statistics for 2004*

Non-transient non-community water system (NTNCWS) violations reported, 1993–2004

Fiscal Year	MCL[1]	MRDL[2]	TT[3]	M/R[4]	Other[5]	Total[6]
Number of Violations						
2004	1,381	0	681	n/a	538	n/a
2003	1,391	0	680	n/a	538	n/a
2002	1,336	-	720	38,805	227	41,088
2001	1,258	-	696	32,847	305	35,106
2000	1,243	-	834	23,138	210	25,425
1999	1,369	-	272	27,487	145	29,273
1998	1,324	-	179	26,612	256	28,371
1997	1,389	-	227	27,913	156	29,685
1996	1,829	-	437	54,147	198	56,611
1995	1,689	-	683	58,089	257	60,718
1994	2,082	-	1,029	57,515	373	60,999
1993	2,146	-	219	30,518	360	33,243
Number of Systems in Violation						
2004	1,033	0	474	n/a	356	n/a
2003	1,029	0	495	n/a	317	n/a
2002	1,005	-	529	5,322	155	6,267
2001	921	-	515	5,368	213	6,241
2000	925	-	580	5,513	169	6,445
1999	995	-	190	3,852	101	4,650
1998	959	-	109	4,067	164	4,810
1997	1,023	-	107	4,772	108	5,531
1996	1,317	-	252	6,233	138	7,132
1995	1,262	-	403	6,416	170	7,421
1994	1,460	-	583	9,569	245	10,532
1993	1,439	-	151	8,443	216	9,262
Population Affected						
2004	288,280	0	130,242	n/a	88,003	n/a
2003	266,522	0	140,053	n/a	85,026	n/a
2002	302,322	-	165,522	1,566,034	57,016	1,897,686
2001	359,556	-	222,120	1,616,708	82,910	2,012,850
2000	276,758	-	216,701	1,759,909	37,941	2,075,184
1999	373,845	-	56,710	883,464	36,223	1,229,143
1998	244,960	-	26,063	1,108,715	55,336	1,306,710
1997	285,722	-	34,087	1,311,222	39,417	1,560,659
1996	455,990	-	93,866	1,592,965	35,782	1,892,307
1995	462,317	-	145,607	1,834,011	53,585	2,158,323
1994	668,123	-	178,835	2,878,698	85,328	3,254,116
1993	500,914	-	122,861	2,619,837	93,685	2,911,225

Notes: *FY2003 and 2004 compliance information includes new Disinfectant and Disinfection Byproduct (D-DBP) and Interim Enhanced Surface Water Treatment (IESWTR) rules; (1) Maximum Contaminant Level (MCL) - The highest level of a contaminant that is allowed in drinking water. MCLs are set as close to Maximum Contaminant Level Goals (MCLGs) as feasible using the best available treatment technology and taking cost into consideration. MCLs are enforceable standards; (2) Maximum Residual Disinfectant Level (MRDL) - The highest level of a disinfectant allowed in drinking water. There is convincing evidence that addition of a disinfectant is necessary for control of microbial contaminants; (3) Treatment Technique - A required process intended to reduce the level of a contaminant in drinking water; (4) Monitoring and Reporting - EPA sets minimum monitoring schedules that drinking water systems must follow. These minimum reporting schedules (systems may monitor more frequently) vary by system size as well as by contaminant. Some contaminants are monitored for daily, others need to be checked far less frequently (every nine years is the longest monitoring cycle). For instance, at a minimum, drinking water systems will monitor continuously for turbidity, monthly for bacteria, and once every four years for radio nuclides. A Monitoring and Reporting Violation means the system did not perform the required testing, take adequate samples, or report a violation as required; (5) Jump in FY2000 due to new violations for failing to issue, or issuing an insufficient, Consumer Confidence report; (6) Totals for the number of systems in violation, and for population affected, should be lower than the sum in each row. This is because some systems will have incurred more than one type of violation; n/a not available.*

Source: *U.S. Environmental Protection Agency, Office of Ground Water and Drinking Water, Factoids: Drinking Water and Ground Water Statistics for 2004*

Transient non-community water system (TNCWS) violations reported, 1993–2004

Fiscal Year	MCL[1]	MRDL[2]	TT[3]	M/R[4]	Other[5]	Total[6]
Number of Violations						
2004	5,032	0	415	n/a	3,028	n/a
2003	5,082	0	573	n/a	8,640	n/a
2002	4,837	-	626	39,451	2,339	47,253
2001	4,596	-	677	38,794	1,467	45,534
2000	4,408	-	309	35,328	915	40,960
1999	4,967	-	223	36,651	740	42,581
1998	5,224	-	287	28,988	2,157	36,656
1997	5,382	-	390	33,851	1,795	41,418
1996	5,935	-	588	57,284	1,814	65,621
1995	4,883	-	529	50,987	2,267	58,666
1994	4,114	-	573	48,631	3,226	56,544
1993	4,352	-	245	43,485	2,756	50,838
Number of Systems in Violation						
2004	3,974	0	195	n/a	1,697	n/a
2003	4,071	0	245	n/a	3,685	n/a
2002	3,858	-	249	19,191	1,142	22,193
2001	3,687	-	256	19,623	826	22,358
2000	3,632	-	177	17,725	656	20,325
1999	3,862	-	139	19,658	494	22,185
1998	3,798	-	94	16,006	1,113	18,773
1997	4,041	-	134	17,485	971	20,505
1996	4,521	-	169	25,611	1,067	28,516
1995	3,853	-	179	25,118	1,166	27,735
1994	3,271	-	240	22,660	1,283	25,117
1993	3,264	-	178	20,439	1,325	22,733
Population Affected						
2004	559,673	0	51,915	n/a	234,291	n/a
2003	541,950	0	40,518	n/a	340,950	n/a
2002	504,967	-	60,692	2,154,654	137,795	2,552,563
2001	558,780	-	46,987	2,385,309	118,847	2,792,810
2000	506,759	-	44,406	2,568,632	78,759	2,979,498
1999	566,066	-	39,547	2,220,821	62,786	2,628,376
1998	477,911	-	58,575	2,071,470	138,161	2,423,132
1997	537,539	-	64,156	2,205,834	122,530	2,622,932
1996	550,463	-	51,969	3,255,986	110,644	3,581,030
1995	589,816	-	72,894	3,109,977	124,258	3,512,069
1994	472,556	-	86,207	2,669,415	146,978	3,020,884
1993	682,030	-	48,309	2,661,227	174,674	3,156,825

Notes: FY2003 and 2004 compliance information includes new Disinfectant and Disinfection Byproduct (D-DBP) and Interim Enhanced Surface Water Treatment (IESWTR) rules; (1) Maximum Contaminant Level (MCL) - The highest level of a contaminant that is allowed in drinking water. MCLs are set as close to Maximum Contaminant Level Goals (MCLGs) as feasible using the best available treatment technology and taking cost into consideration. MCLs are enforceable standards; (2) Maximum Residual Disinfectant Level (MRDL) - The highest level of a disinfectant allowed in drinking water. There is convincing evidence that addition of a disinfectant is necessary for control of microbial contaminants; (3) Treatment Technique - A required process intended to reduce the level of a contaminant in drinking water; (4) Monitoring and Reporting - EPA sets minimum monitoring schedules that drinking water systems must follow. These minimum reporting schedules (systems may monitor more frequently) vary by system size as well as by contaminant. Some contaminants are monitored for daily, others need to be checked far less frequently (every nine years is the longest monitoring cycle). For instance, at a minimum, drinking water systems will monitor continuously for turbidity, monthly for bacteria, and once every four years for radio nuclides. A Monitoring and Reporting Violation means the system did not perform the required testing, take adequate samples, or report a violation as required; (5) Jump in FY2000 due to new violations for failing to issue, or issuing an insufficient, Consumer Confidence report; (6) Totals for the number of systems in violation, and for population affected, should be lower than the sum in each row. This is because some systems will have incurred more than one type of violation; n/a not available.

Source: U.S. Environmental Protection Agency, Office of Ground Water and Drinking Water, Factoids: Drinking Water and Ground Water Statistics for 2004

Maximum Contaminant Level (MCL), Maximum Residual Disinfectant Level (MRDL), and Treatment Technique (TT) violations reported

Contaminant/ Rule	Key	System Size (population served)					Total
		Very Small (25-500)	Small (501-3,300)	Medium (3,301-10,000)	Large (10,001-100,000)	Very Large (>100,000)	
Microbials:							
TCR/T[1,8]	No. of violations	7,629	942	268	242	18	9,099
	No. of sys. in vio.	5,939	754	241	194	11	7,139
	Population affected	658,437	965,775	1,434,325	4,335,013	3,989,416	11,382,966
Stage 1 DBP[4,9]	No. of violations	110	273	124	223	11	741
	No. of sys. in vio.	75	162	64	97	5	403
	Population affected	13,613	239,795	359,708	2,355,202	4,629,814	7,598,132
SWTR[2,10]	No. of violations	945	394	175	29	8	1,551
	No. of sys. in vio.	516	222	75	25	6	844
	Population affected	85,889	311,309	432,884	703,988	6,715,034	8,249,104
IESWTR[3,11]	No. of violations	2	5	0	110	13	130
	No. of sys. in vio.	2	4	0	46	6	58
	Population affected	129	4,093	0	1,460,032	2,519,510	3,983,764
Organics:							
VOC[5,12]	No. of violations	80	3	7	8	0	98
	No. of sys. in vio.	44	3	3	6	0	56
	Population affected	3,698	6,339	17,633	169,877	0	197,547
SOC[6,12]	No. of violations	18	6	2	3	0	29
	No. of sys. in vio.	13	3	2	2	0	20
	Population affected	1,655	4,593	8,835	100,000	0	115,083
Inorganics:							
Nitrates[12]	No. of violations	722	136	8	4	0	870
	No. of sys. in vio.	469	69	5	3	0	546
	Population affected	49,376	73,185	28,556	60,402	0	211,519
Arsenic[12]	No. of violations	51	11	6	4	0	72
	No. of sys. in vio.	28	7	2	3	0	40
	Population affected	4,018	10,497	8,940	48,324	0	71,779
Other IOC[7,12]	No. of violations	262	38	12	4	0	316
	No. of sys. in vio.	129	23	6	4	0	162
	Population affected	21,368	33,135	41,350	214,333	0	310,186
Radionuclides[13]	No. of violations	311	193	50	60	4	618
	No. of sys. in vio.	122	71	18	18	1	230
	Population affected	23,381	94,217	95,959	372,153	106,221	691,931
Lead & Copper[12]	No. of violations	1,073	256	38	25	6	1,398
	No. of sys. in vio.	830	209	31	22	3	1,095
	Population affected	123,037	276,427	189,019	589,483	1,040,814	2,218,780

Notes: *Data as of 4th Quarter, 2004; (1) Total Coliform Rule/Turbidity; (2) Surface Water Treatment Rule; (3) Interim Enhanced Surface Water Treatment Rule; (4) Disinfectants By-Products Rule; (5) Regulated volatile organic contaminants (e.g., Benzene) other than TTHMs (which are listed in DBP Rule); (6) Synthetic organic chemicals (e.g., Atrazine); (7) Regulated inorganic contaminants (e.g., arsenic) other than Copper, Lead, Nitrate, and Nitrite; (8) Applies to all water systems; (9) Applies to surface water systems that disinfect (TNCWSs that use ClO2); (10) Applies to surface water systems; (11) Applies to surface water systems serving more than 9,999 people; (12) Applies to CWS and NTNCWS; (13) Applies to CWS.*

Source: *U.S. Environmental Protection Agency, Office of Ground Water and Drinking Water, Factoids: Drinking Water and Ground Water Statistics for 2004*

Monitoring and Reporting (M/R) violations reported

Contaminant/ Rule	Key	System Size (population served)					Total
		Very Small (25-500)	Small (501-3,300)	Medium (3,301-10,000)	Large (10,001-100,000)	Very Large (>100,000)	
Microbials:							
TCR/T[1,8]	No. of violations	30,311	2,915	383	233	20	33,862
	No. of sys. in vio.	17,851	1,767	289	166	18	20,091
	Population affected	1,819,497	2,167,938	1,637,496	4,754,595	7,955,803	18,335,329
Stage 1 DBP[4,9]	No. of violations	2,284	885	250	237	21	3,677
	No. of sys. in vio.	760	386	120	140	7	1,413
	Population affected	137,481	519,075	726,687	3,500,020	3,770,667	8,653,930
SWTR[2,10]	No. of violations	2,227	620	123	57	9	3,036
	No. of sys. in vio.	427	165	71	33	5	701
	Population affected	72,497	216,099	405,911	1,169,644	3,867,870	5,732,021
IESWTR[3,11]	No. of violations	0	0	3	61	7	71
	No. of sys. in vio.	0	0	2	24	4	30
	Population affected	0	0	14,273	674,162	2,391,250	3,079,685
Organics:							
VOC[5,12]	No. of violations	n/a	n/a	n/a	n/a	n/a	n/a
	No. of sys. in vio.	n/a	n/a	n/a	n/a	n/a	n/a
	Population affected	n/a	n/a	n/a	n/a	n/a	n/a
SOC[6,12]	No. of violations	n/a	n/a	n/a	n/a	n/a	n/a
	No. of sys. in vio.	n/a	n/a	n/a	n/a	n/a	n/a
	Population affected	n/a	n/a	n/a	n/a	n/a	n/a
Inorganics:							
Nitrates[12]	No. of violations	n/a	n/a	n/a	n/a	n/a	n/a
	No. of sys. in vio.	n/a	n/a	n/a	n/a	n/a	n/a
	Population affected	n/a	n/a	n/a	n/a	n/a	n/a
Arsenic[12]	No. of violations	n/a	n/a	n/a	n/a	n/a	n/a
	No. of sys. in vio.	n/a	n/a	n/a	n/a	n/a	n/a
	Population affected	n/a	n/a	n/a	n/a	n/a	n/a
Other IOC[7,12]	No. of violations	n/a	n/a	n/a	n/a	n/a	n/a
	No. of sys. in vio.	n/a	n/a	n/a	n/a	n/a	n/a
	Population affected	n/a	n/a	n/a	n/a	n/a	n/a
Radionuclides[13]	No. of violations	n/a	n/a	n/a	n/a	n/a	n/a
	No. of sys. in vio.	n/a	n/a	n/a	n/a	n/a	n/a
	Population affected	n/a	n/a	n/a	n/a	n/a	n/a
Lead & Copper[12]	No. of violations	9,773	1,857	331	193	21	12,175
	No. of sys. in vio.	6,667	1,297	254	138	14	8,370
	Population affected	921,199	1,606,104	1,429,897	3,689,863	7,468,972	15,116,035

Notes: Data as of 4th Quarter, 2004 except for Chem M/Rs which are from 1st Quarter, 2005; (1) Total Coliform Rule/Turbidity; (2) Surface Water Treatment Rule; (3) Interim Enhanced Surface Water Treatment Rule; (4) Disinfectants By-Products Rule; (5) Regulated volatile organic contaminants (e.g., Benzene) other than TTHMs (which are listed in DBP Rule); (6) Synthetic organic chemicals (e.g., Atrazine); (7) Regulated inorganic contaminants (e.g., arsenic) other than Copper, Lead, Nitrate, and Nitrite; (8) Applies to all water systems; (9) Applies to surface water systems that disinfect (TNCWSs that use ClO2); (10) Applies to surface water systems; (11) Applies to surface water systems serving more than 9,999 people; (12) Applies to CWS and NTNCWS; (13) Applies to CWS; n/a not available.
Source: U.S. Environmental Protection Agency, Office of Ground Water and Drinking Water, Factoids: Drinking Water and Ground Water Statistics for 2004

National primary drinking water regulations

Contaminant	MCLG[1] (mg/L)[2]	MCL or TT[1] (mg/L)[2]	Potential Health Effects from Ingestion of Water	Sources of Contaminant in Drinking Water
Microorganisms				
Cryptosporidium	zero	TT[3]	Gastrointestinal illness (e.g., diarrhea, vomiting, cramps)	Human and animal fecal waste
Giardia lamblia	zero	TT[3]	Gastrointestinal illness (e.g., diarrhea, vomiting, cramps)	Human and animal fecal waste
Heterotrophic plate count	n/a	TT[3]	HPC has no health effects; it is an analytic method used to measure the variety of bacteria that are common in water. The lower the concentration of bacteria in drinking water, the better maintained the water system is.	HPC measures a range of bacteria that are naturally present in the environment
Legionella	zero	TT[3]	Legionnaire's Disease, commonly known as pneumonia	Found naturally in water; multiplies in heating systems
Total Coliforms (including fecal coliform and *E. Coli*)	zero	5.0%[4]	Not a health threat in itself; used as an indicator that other potentially harmful bacteria may be present[5]	Coliforms are naturally present in the environment; fecal coliforms and E. coli come from human and animal fecal waste.
Turbidity	n/a	TT[3]	Turbidity is a measure of the cloudiness of water. It is used to indicate water quality and filtration effectiveness (e.g., whether disease-causing organisms are present). Higher turbidity levels are often associated with higher levels of disease-causing microorganisms such as viruses, parasites and some bacteria. These organisms can cause symptoms such as nausea, cramps, diarrhea, and associated headaches.	Soil runoff
Viruses (enteric)	zero	TT[3]	Gastrointestinal illness (e.g., diarrhea, vomiting, cramps)	Human and animal fecal waste
Disinfectants	**MRDLG[1] (mg/L)[2]**	**MRDL[1] (mg/L)[2]**		
Chloramines (as Cl$_2$)	MRDLG=4[1]	MRDL=4.0[1]	Eye/nose irritation; stomach discomfort, anemia	Water additive used to control microbes
Chlorine (as Cl$_2$)	MRDLG=4[1]	MRDL=4.0[1]	Eye/nose irritation; stomach discomfort	Water additive used to control microbes
Chlorine dioxide (as ClO$_2$)	MRDLG=0.8[1]	MRDL=0.8[1]	Anemia; infants & young children: nervous system effects	Water additive used to control microbes
Disinfection Byproducts				
Bromate	zero	0.010	Increased risk of cancer	Byproduct of drinking water disinfection
Chlorite	0.8	1.0	Anemia; infants & young children: nervous system effects	Byproduct of drinking water disinfection
Haloacetic acids (HAA5)	n/a[6]	0.060	Increased risk of cancer	Byproduct of drinking water disinfection
Total Trihalomethanes (TTHMs)	none[7]; n/a[6]	0.10; 0.080	Liver, kidney or central nervous system problems; increased risk of cancer	Byproduct of drinking water disinfection

National primary drinking water regulations - *continued*

Contaminant	MCLG[1] (mg/L)[2]	MCL or TT[1] (mg/L)[2]	Potential Health Effects from Ingestion of Water	Sources of Contaminant in Drinking Water
Inorganic Chemicals				
Antimony	0.006	0.006	Increase in blood cholesterol; decrease in blood glucose	Discharge from petroleum refineries; fire retardants; ceramics; electronics; solder
Arsenic	0[7]	0.010 as of 1/23/06	Skin damage; circulatory system problems; increased risk of cancer	Erosion of natural deposits; runoff from glass & electronics production wastes
Asbestos (fiber >10 micrometers)	7 million fibers per liter	7 MFL	Increased risk of developing benign intestinal polyps	Decay of asbestos cement in water mains; erosion of natural deposits
Barium	2	2	Increase in blood pressure	Discharge of drilling wastes; discharge from metal refineries; erosion of natural deposits
Beryllium	0.004	0.004	Intestinal lesions	Discharge from metal refineries and coal-burning factories; discharge from electrical, aerospace, and defense industries
Cadmium	0.005	0.005	Kidney damage	Corrosion of galvanized pipes; erosion of natural deposits; discharge from metal refineries; runoff from waste batteries and paints
Chromium (total)	0.1	0.1	Some people who use water containing chromium well in excess of the MCL over many years could experience allergic dermatitis	Discharge from steel and pulp mills; erosion of natural deposits
Copper	1.3	TT[8]; Action Level=1.3	Short term exposure: Gastrointestinal distress. Long term exposure: Liver or kidney damage. People with Wilson's Disease should consult their personal doctor if their water systems exceed the copper action level.	Corrosion of household plumbing systems; erosion of natural deposits
Cyanide (as free cyanide)	0.2	0.2	Nerve damage or thyroid problems	Discharge from steel/metal factories; discharge from plastic and fertilizer factories
Fluoride	4.0	4.0	Bone disease (pain and tenderness of the bones); Children may get mottled teeth.	Water additive which promotes strong teeth; erosion of natural deposits; discharge from fertilizer and aluminum factories
Lead	zero	TT[8]; Action Level=0.015	Infants and children: Delays in physical or mental development. Adults: Kidney problems; high blood pressure	Corrosion of household plumbing systems; erosion of natural deposits
Mercury (inorganic)	0.002	0.002	Kidney damage	Erosion of natural deposits; discharge from refineries and factories; runoff from landfills and cropland
Nitrate (measured as Nitrogen)	10	10	Infants below the age of six months who drink water containing nitrate in excess of the MCL could become seriously ill and, if untreated, may die. Symptoms include shortness of breath and blue-baby syndrome.	Runoff from fertilizer use; leaching from septic tanks, sewage; erosion of natural deposits

National primary drinking water regulations - *continued*

Contaminant	MCLG[1] (mg/L)[2]	MCL or TT[1] (mg/L)[2]	Potential Health Effects from Ingestion of Water	Sources of Contaminant in Drinking Water
Nitrite (measured as Nitrogen)	1	1	Infants below the age of six months who drink water containing nitrite in excess of the MCL could become seriously ill and, if untreated, may die. Symptoms include shortness of breath and blue-baby syndrome.	Runoff from fertilizer use; leaching from septic tanks, sewage; erosion of natural deposits
Selenium	0.05	0.05	Hair or fingernail loss; numbness in fingers or toes; circulatory problems	Discharge from petroleum refineries; erosion of natural deposits; discharge from mines
Thallium	0.0005	0.002	Hair loss; changes in blood; kidney, intestine, or liver problems	Leaching from ore-processing sites; discharge from electronics, glass, and pharmaceutical companies
Organic Chemicals				
Acrylamide	zero	TT[9]	Nervous system or blood problems; increased risk of cancer	Added to water during sewage/wastewater treatment
Alachlor	zero	0.002	Eye, liver, kidney or spleen problems; anemia; increased risk of cancer	Runoff from herbicide used on row crops
Atrazine	0.003	0.003	Cardiovascular system problems; reproductive difficulties	Runoff from herbicide used on row crops
Benzene	zero	0.005	Anemia; decrease in blood platelets; increased risk of cancer	Discharge from factories; leaching from gas storage tanks and landfills
Benzo(a)pyrene (PAHs)	zero	0.0002	Reproductive difficulties; increased risk of cancer	Leaching from linings of water storage tanks and distribution lines
Carbofuran	0.04	0.04	Problems with blood or nervous system; reproductive difficulties.	Leaching of soil fumigant used on rice and alfalfa
Carbon tetrachloride	zero	0.005	Liver problems; increased risk of cancer	Discharge from chemical plants and other industrial activities
Chlordane	zero	0.002	Liver or nervous system problems; increased risk of cancer	Residue of banned termiticide
Chlorobenzene	0.1	0.1	Liver or kidney problems	Discharge from chemical and agricultural chemical factories
2,4-D	0.07	0.07	Kidney, liver, or adrenal gland problems	Runoff from herbicide used on row crops
Dalapon	0.2	0.2	Minor kidney changes	Runoff from herbicide used on rights of way
1,2-Dibromo-3-chloropropane (DBCP)	zero	0.0002	Reproductive difficulties; increased risk of cancer	Runoff/leaching from soil fumigant used on soybeans, cotton, pineapples, and orchards
o-Dichlorobenzene	0.6	0.6	Liver, kidney, or circulatory system problems	Discharge from industrial chemical factories
p-Dichlorobenzene	0.075	0.075	Anemia; liver, kidney or spleen damage; changes in blood	Discharge from industrial chemical factories
1,2-Dichloroethane	zero	0.005	Increased risk of cancer	Discharge from industrial chemical factories
1,1-Dichloroethylene	0.007	0.007	Liver problems	Discharge from industrial chemical factories

National primary drinking water regulations - *continued*

Contaminant	MCLG[1] (mg/L)[2]	MCL or TT[1] (mg/L)[2]	Potential Health Effects from Ingestion of Water	Sources of Contaminant in Drinking Water
cis-1,2-Dichloroethylene	0.07	0.07	Liver problems	Discharge from industrial chemical factories
trans-1,2-Dichloroethylene	0.1	0.1	Liver problems	Discharge from industrial chemical factories
Dichloromethane	zero	0.005	Liver problems; increased risk of cancer	Discharge from pharmaceutical and chemical factories
1,2-Dichloropropane	zero	0.005	Increased risk of cancer	Discharge from industrial chemical factories
Di(2-ethylhexyl) adipate	0.4	0.4	General toxic effects or reproductive difficulties	Leaching from PVC plumbing systems; discharge from chemical factories
Di(2-ethylhexyl) phthalate	zero	0.006	Reproductive difficulties; liver problems; increased risk of cancer	Discharge from rubber and chemical factories
Dinoseb	0.007	0.007	Reproductive difficulties	Runoff from herbicide used on soybeans and vegetables
Dioxin (2,3,7,8-TCDD)	zero	0.00000003	Reproductive difficulties; increased risk of cancer	Emissions from waste incineration and other combustion; discharge from chemical factories
Diquat	0.02	0.02	Cataracts	Runoff from herbicide use
Endothall	0.1	0.1	Stomach and intestinal problems	Runoff from herbicide use
Endrin	0.002	0.002	Nervous system effects	Residue of banned insecticide
Epichlorohydrin	zero	TT[9]	Stomach problems; reproductive difficulties; increased risk of cancer	Discharge from industrial chemical factories; added to water during treatment process
Ethylbenzene	0.7	0.7	Liver or kidney problems	Discharge from petroleum refineries
Ethelyne dibromide	zero	0.00005	Stomach problems; reproductive difficulties; increased risk of cancer	Discharge from petroleum refineries
Glyphosate	0.7	0.7	Kidney problems; reproductive difficulties	Runoff from herbicide use
Heptachlor	zero	0.0004	Liver damage; increased risk of cancer	Residue of banned termiticide
Heptachlor epoxide	zero	0.0002	Liver damage; increased risk of cancer	Breakdown of hepatachlor
Hexachlorobenzene	zero	0.001	Liver or kidney problems; reproductive difficulties; increased risk of cancer	Discharge from metal refineries and agricultural chemical factories
Hexachlorocyclopentadiene	0.05	0.05	Kidney or stomach problems	Discharge from chemical factories
Lindane	0.0002	0.0002	Liver or kidney problems	Runoff/leaching from insecticide used on catttle, lumber, gardens
Methoxychlor	0.04	0.04	Reproductive difficulties	Runoff/leaching from insecticide used on fruits, vegetables, alfalfa, livestock
Oxamyl (Vydate)	0.2	0.2	Slight nervous system effects	Runoff/leaching from insecticide used on apples, potatoes, and tomatoes

National primary drinking water regulations - continued

Contaminant	MCLG[1] (mg/L)[2]	MCL or TT[1] (mg/L)[2]	Potential Health Effects from Ingestion of Water	Sources of Contaminant in Drinking Water
Polychlorinated biphenyls (PCBs)	zero	0.0005	Skin changes; thymus gland problems; immune deficiencies; reproductive or nervous system difficulties; increased risk of cancer	Runoff from landfills; discharge of waste chemicals
Pentachlorophenol	zero	0.001	Liver or kidney problems; increased risk of cancer	Discharge from wood preserving factories
Picloram	0.5	0.5	Liver problems	Herbicide runoff
Simazine	0.004	0.004	Problems with blood	Herbicide runoff
Styrene	0.1	0.1	Liver, kidney, and circulatory problems	Discharge from rubber and plastic factories; leaching from landfills
Tetrachloroethylene	zero	0.005	Liver problems; increased risk of cancer	Discharge from factories and dry cleaners
Toluene	1	1	Nervous system, kidney, or liver problems	Discharge from petroleum factories
Toxaphene	zero	0.003	Kidney, liver, or thyroid problems; increased risk of cancer	Runoff/leaching from insecticide used on cotton and cattle
2,4,5-TP (Silvex)	0.05	0.05	Liver problems	Residue of banned herbicide
1,2,4-Trichlorobenzene	0.07	0.07	Changes in adrenal glands	Discharge from textile finishing factories
1,1,1-Trichloroethane	0.20	0.20	Liver, nervous system, or circulatory problems	Discharge from metal degreasing sites and other factories
1,1,2-Trichloroethane	0.003	0.005	Liver, kidney, or immune system problems	Discharge from industrial chemical factories
Trichloroethylene	zero	0.005	Liver problems; increased risk of cancer	Discharge from petroleum refineries
Vinyl chloride	zero	0.002	Increased risk of cancer	Leaching from PVC pipes; discharge from plastic factories
Xylenes (total)	10	10	Nervous system damage	Discharge from petroleum factories; discharge from chemical factories
Radionuclides				
Alpha particles	none[7]; zero	15 picocuries per Liter (pCi/L)	Increased risk of cancer	Erosion of natural deposits
Beta particles and photon emitters	none[7]; zero	4 millirems per year	Increased risk of cancer	Decay of natural and man-made deposits
Radium 226 and Radium 228 (combined)	none[7]; zero	5 pCi/L	Increased risk of cancer	Erosion of natural deposits
Uranium	zero	30 ug/L	Increased risk of cancer, kidney toxicity	Erosion of natural deposits

Notes: *(1) Definitions: Maximum Contaminant Level (MCL) - The highest level of a contaminant that is allowed in drinking water. MCLs are set as close to MCLGs as feasible using the best available treatment technology and taking cost into consideration. MCLs are enforceable standards. Maximum Contaminant Level Goal (MCLG) - The level of a contaminant in drinking water below which there is no known or expected risk to health. MCLGs allow for a margin of safety and are non-enforceable public health goals. Maximum Residual Disinfectant Level (MRDL) - The highest level of a disinfectant allowed in drinking water. There is convincing evidence that addition of a disinfectant is necessary for control of microbial contaminants. Maximum Residual Disinfectant Level Goal (MRDLG) - The level of a drinking water disinfectant below which there is no known or expected risk to health. MRDLGs do not reflect the benefits of the use of disinfectants to control microbial contaminants. Treatment Technique - A required process intended to reduce the level of a contaminant in drinking water; (2) Units are in milligrams per liter (mg/L) unless otherwise noted. Milligrams per liter are equivalent to parts per million; (3) EPA's surface water treatment rules* require systems using surface water or ground water under the direct influence of surface water to one, disinfect their water, and two, filter their water or meet criteria for avoiding filtration so that the following contaminants are controlled at the following levels: Cryptosporidium: (as of January 1, 2002 for systems serving >10,000 and 1/14/05 for systems serving <10,000) 99% removal; Giardia lamblia: 99.9% removal/ inactivation; Viruses: 99.99% removal/inactivation; Legionella: No limit, but EPA believes that if Giardia and viruses are removed/inactivated, Legionella will also be controlled. Turbidity: At no time can turbidity (cloudiness of water) go above 5 nephelolometric turbidity units (NTU); systems that filter must ensure that the turbidity go no higher than 1 NTU (0.5 NTU for conventional or direct filtration) in at least 95% of the daily samples in any month. As of January 1, 2002, turbidity may never exceed 1 NTU, and must not exceed 0.3 NTU in 95% of daily samples in any month; HPC: No more than 500 bacterial colonies per milliliter; Long Term 1 Enhanced Surface Water Treatment (Effective Date: January 14, 2005): Surface water systems or (GWUDI) systems serving fewer than 10,000 people must comply with the applicable Long Term 1 Enhanced Surface Water Treatment Rule provisions (e.g. turbidity standards, individual filter monitoring, Cryptosporidium removal requirements, updated watershed control requirements for unfiltered systems); Filter Backwash Recycling: The Filter Backwash Recycling Rule requires systems that recycle to return specific recycle flows through all processes of the system's existing conventional or direct filtration system or at an alternate location approved by the state; (4) No more than 5.0% samples total coliform-positive in a month. (For water systems that collect fewer than 40 routine samples per month, no more than one sample can be total coliform-positive). Every sample that has total coliforms must be analyzed for fecal coliforms. There may not be any fecal coliforms or E. coli; (5) Fecal coliform and E. coli are bacteria whose presence indicates that the water may be contaminated with human or animal wastes. Disease-causing microbes (pathogens) in these wastes can cause diarrhea, cramps, nausea, headaches, or other symptoms. These pathogens may pose a special health risk for infants, young children, and people with severely compromised immune systems; (6) Although there is no collective MCLG for this contaminant group, there are individual MCLGs for some of the individual contaminants: Trihalomethanes: bromodichloromethane (zero); bromoform (zero); dibromochloromethane (0.06 mg/L); Chloroform is regulated with this group but has no MCLG; Haloacetic acids: dichloroacetic acid (zero); trichloroacetic acid (0.3 mg/L); Monochloroacetic acid, bromoacetic acid, and dibromoacetic acid are regulated with this group but have no MCLGs; (7) MCLGs were not established before the 1986 Amendments to the Safe Drinking Water Act. Therefore, there is no MCLG for this contaminant; (8) Lead and copper are regulated by a Treatment Technique that requires systems to control the corrosiveness of their water. If more than 10% of tap water samples exceed the action level, water systems must take additional steps. For copper, the action level is 1.3 mg/L, and for lead is 0.015 mg/L; (9) Each water system must certify, in writing, to the state (using third-party or manufacturer's certification) that when acrylamide and epichlorohydrin are used in drinking water systems, the combination (or product) of dose and monomer level does not exceed the levels specified, as follows: Acrylamide = 0.05% dosed at 1 mg/L (or equivalent); Epichlorohydrin = 0.01% dosed at 20 mg/L (or equivalent)

Source: U.S. Environmental Protection Agency, Office of Ground Water and Drinking Water, Current Drinking Water Standards, June 20, 2005

National secondary
drinking water regulations

Contaminant	Secondary Standard
Aluminum	0.05 to 0.2 mg/L
Chloride	250 mg/L
Color	15 (color units)
Copper	1.0 mg/L
Corrosivity	noncorrosive
Fluoride	2.0 mg/L
Foaming Agents	0.5 mg/L
Iron	0.3 mg/L
Manganese	0.05 mg/L
Odor	3 threshold odor number
pH	6.5-8.5
Silver	0.10 mg/L
Sulfate	250 mg/L
Total Dissolved Solids	500 mg/L
Zinc	5 mg/L

Notes: *National Secondary Drinking Water Regulations (NSDWRs or secondary standards) are non-enforceable guidelines regulating contaminants that may cause cosmetic effects (such as skin or tooth discoloration) or aesthetic effects (such as taste, odor, or color) in drinking water. EPA recommends secondary standards to water systems but does not require systems to comply. However, states may choose to adopt them as enforceable standards.*
Source: *U.S. Environmental Protection Agency, Office of Ground Water and Drinking Water, Current Drinking Water Standards, June 20, 2005*

List of threatened and endangered species

Common name	Scientific name	Group	Historic range	Vertebrate population where endangered or threatened	Status
acornshell, southern	Epioblasma othcaloogensis	Clams	U.S.A. (AL, GA, TN)	Entire	E
agave, Arizona	Agave arizonica	Flowering Plants	U.S.A. (AZ)	Entire	E
'Ahinahina	Argyroxiphium sandwicense ssp. macrocephalum	Flowering Plants	U.S.A. (HI)	Entire	T
'Ahinahina	Argyroxiphium sandwicense ssp. sandwicense	Flowering Plants	U.S.A. (HI)	Entire	E
'aiakeakua, popolo	Solanum sandwicense	Flowering Plants	U.S.A. (HI)	Entire	E
'aiea	Nothocestrum breviflorum	Flowering Plants	U.S.A. (HI)	Entire	E
'aiea	Nothocestrum peltatum	Flowering Plants	U.S.A. (HI)	Entire	E
akepa, Hawaii (honeycreeper)	Loxops coccineus coccineus	Birds	U.S.A. (HI)	Entire	E
akepa, Maui (honeycreeper)	Loxops coccineus ochraceus	Birds	U.S.A. (HI)	Entire	E
akialoa, Kauai (honeycreeper)	Hemignathus procerus	Birds	U.S.A. (HI)	Entire	E
akiapola'au (honeycreeper)	Hemignathus munroi	Birds	U.S.A. (HI)	Entire	E
'akoko	Chamaesyce celastroides var. kaenana	Flowering Plants	U.S.A. (HI)	Entire	E
'akoko	Chamaesyce deppeana	Flowering Plants	U.S.A. (HI)	Entire	E
'akoko	Chamaesyce herbstii	Flowering Plants	U.S.A. (HI)	Entire	E
'akoko	Chamaesyce kuwaleana	Flowering Plants	U.S.A. (HI)	Entire	E
'akoko	Chamaesyce rockii	Flowering Plants	U.S.A. (HI)	Entire	E
'akoko	Euphorbia haeleeleana	Flowering Plants	U.S.A. (HI)	Entire	E
'akoko, Ewa Plains	Chamaesyce skottsbergii var. kalaeloana	Flowering Plants	U.S.A. (HI)	Entire	E
ala balik (trout)	Salmo platycephalus	Fishes	Turkey	Entire	E
alani	Melicope adscendens	Flowering Plants	U.S.A. (HI)	Entire	E
alani	Melicope balloui	Flowering Plants	U.S.A. (HI)	Entire	E
alani	Melicope haupuensis	Flowering Plants	U.S.A. (HI)	Entire	E
alani	Melicope knudsenii	Flowering Plants	U.S.A. (HI)	Entire	E
alani	Melicope lydgatei	Flowering Plants	U.S.A. (HI)	Entire	E
alani	Melicope mucronulata	Flowering Plants	U.S.A. (HI)	Entire	E
alani	Melicope munroi	Flowering Plants	U.S.A. (HI)	Entire	E
alani	Melicope ovalis	Flowering Plants	U.S.A. (HI)	Entire	E
alani	Melicope pallida	Flowering Plants	U.S.A. (HI)	Entire	E
alani	Melicope quadrangularis	Flowering Plants	U.S.A. (HI)	Entire	E
alani	Melicope reflexa	Flowering Plants	U.S.A. (HI)	Entire	E
alani	Melicope saint-johnii	Flowering Plants	U.S.A. (HI)	Entire	E
alani	Melicope zahlbruckneri	Flowering Plants	U.S.A. (HI)	Entire	E
Albatross, Amsterdam	Diomedia amsterdamensis	Birds	Indian Ocean_Amsterdam Island	Entire	E
albatross, short-tailed	Phoebastria (=Diomedea) albatrus	Birds	North Pacific Ocean and Bering Sea_Canada, China, Japan, Mexico, Russia, Taiwan, U.S.A. (AK, CA, HI, OR, WA)	Entire	E
alethe, Thyolo	Alethe choloensis	Birds	Malawi, Mozambique	Entire	E
aleus, Varecia	Lemuridae	Mammals	Malagasy Republic (=Madagascar)	Entire	E
alligator, American	Alligator mississippiensis	Reptiles	Southeastern U.S.A.	Entire	SAT
alligator, Chinese	Alligator sinensis	Reptiles	China	Entire	E
allocarya, Calistoga	Plagiobothrys strictus	Flowering Plants	U.S.A. (CA)	Entire	E
alopecurus, Sonoma	Alopecurus aequalis var. sonomensis	Flowering Plants	U.S.A. (CA)	Entire	E
amaranth, seabeach	Amaranthus pumilus	Flowering Plants	U.S.A. (DE, MA, MD, NC, NJ, NY, RI, SC, VA)	Entire	T
ambersnail, Kanab	Oxyloma haydeni kanabensis	Snails	U.S.A. (AZ, UT)	Entire	E
ambrosia, San Diego	Ambrosia pumila	Flowering Plants	U.S.A. (CA), Mexico	Entire	E
ambrosia, south Texas	Ambrosia cheiranthifolia	Flowering Plants	U.S.A. (TX)	Entire	E

List of threatened and endangered species - *continued*

Common name	Scientific name	Group	Historic range	Vertebrate population where endangered or threatened	Status
amole, purple	Chlorogalum purpureum	Flowering Plants	U.S.A. (CA)	Entire	T
amphianthus, little	Amphianthus pusillus	Flowering Plants	U.S.A. (AL, GA, SC)	Entire	T
amphipod, Hay's Spring	Stygobromus hayi	Crustaceans	U.S.A. (DC)	Entire	E
amphipod, Illinois cave	Gammarus acherondytes	Crustaceans	U.S.A. (IL)	Entire	E
amphipod, Kauai cave	Spelaeorchestia koloana	Crustaceans	U.S.A. (HI)	Entire	E
amphipod, Peck's cave	Stygobromus (=Stygonectes) pecki	Crustaceans	U.S.A. (TX)	Entire	E
'anaunau	Lepidium arbuscula	Flowering Plants	U.S.A. (HI)	Entire	E
anoa, lowland	Bubalus depressicornis	Mammals	Indonesia	Entire	E
anoa, mountain	Bubalus quarlesi	Mammals	Indonesia	Entire	E
anole, Culebra Island giant	Anolis roosevelti	Reptiles	U.S.A. (PR_Culebra Island)	Entire	E
antelope, giant sable	Hippotragus niger variani	Mammals	Angola	Entire	E
'anunu	Sicyos alba	Flowering Plants	U.S.A. (HI)	Entire	E
argali	Ovis ammon	Mammals	Afganistan, China, India, Kazakhstan, Kyrgyzstan, Mongolia, Nepal, Pakistan, Russia, Tajikistan, Uzbekistan	Entire except Kyrgyzstan, Mongolia, and Tajikistan	E
argali	Ovis ammon	Mammals	Afganistan, China, India, Kazakhstan, Kyrgyzstan, Mongolia, Nepal, Pakistan, Russia, Tajikistan, Uzbekistan	Kyrgyzstan, Mongolia, and Tajikistan	T
armadillo, giant	Priodontes maximus	Mammals	Venezuela and Guyana to Argentina	Entire	E
armadillo, pink fairy	Chlamyphorus truncatus	Mammals	Argentina	Entire	E
arrowhead, bunched	Sagittaria fasciculata	Flowering Plants	U.S.A. (NC, SC)	Entire	E
ass, African wild	Equus asinus	Mammals	Somalia, Sudan, Ethiopia	Entire	E
ass, Asian wild	Equus hemionus	Mammals	Southwestern and Central Asia	Entire	E
aster, decurrent false	Boltonia decurrens	Flowering Plants	U.S.A. (IL, MO)	Entire	T
aster, Florida golden	Chrysopsis floridana	Flowering Plants	U.S.A. (FL)	Entire	E
aster, Ruth's golden	Pityopsis ruthii	Flowering Plants	U.S.A. (TN)	Entire	E
aupaka	Isodendrion hosakae	Flowering Plants	U.S.A. (HI)	Entire	E
aupaka	Isodendrion laurifolium	Flowering Plants	U.S.A. (HI)	Entire	E
aupaka	Isodendrion longifolium	Flowering Plants	U.S.A. (HI)	Entire	T
avahi	Avahi laniger (entire genus)	Mammals	Malagasy Republic (=Madagascar)	Entire	E
avens, spreading	Geum radiatum	Flowering Plants	U.S.A. (NC, TN)	Entire	E
'awikiwiki	Canavalia molokaiensis	Flowering Plants	U.S.A. (HI)	Entire	E
awiwi	Centaurium sebaeoides	Flowering Plants	U.S.A. (HI)	Entire	E
awiwi	Hedyotis cookiana	Flowering Plants	U.S.A. (HI)	Entire	E
aye-aye	Daubentonia madagascariensis	Mammals	Malagasy Republic (=Madagascar)	Entire	E
ayenia, Texas	Ayenia limitaris	Flowering Plants	U.S.A. (TX), Mexico	Entire	E
ayumodoki (loach)	Hymenophysa curta	Fishes	Japan	Entire	E
a'e	Zanthoxylum dipetalum var. tomentosum	Flowering Plants	U.S.A. (HI)	Entire	E
a'e	Zanthoxylum hawaiiense	Flowering Plants	U.S.A. (HI)	Entire	E
babirusa	Babyrousa babyrussa	Mammals	Indonesia	Entire	E
baboon, gelada	Theropithecus gelada	Mammals	Ethiopia	Entire	T
baccharis, Encinitas	Baccharis vanessae	Flowering Plants	U.S.A. (CA)	Entire	T
bandicoot, barred	Perameles bougainville	Mammals	Australia	Entire	E
bandicoot, desert	Perameles eremiana	Mammals	Australia	Entire	E
bandicoot, lesser rabbit	Macrotis leucura	Mammals	Australia	Entire	E
bandicoot, pig-footed	Chaeropus ecaudatus	Mammals	Australia	Entire	E
bandicoot, rabbit	Macrotis lagotis	Mammals	Australia	Entire	E
bankclimber, purple (mussel)	Elliptoideus sloatianus	Clams	U.S.A. (AL, GA, FL)	Entire	T
banteng	Bos javanicus	Mammals	Southeast Asia	Entire	E
Barberry, island	Berberis pinnata ssp. insularis	Flowering Plants	U.S.A. (CA)	Entire	E
barberry, Nevin's	Berberis nevinii	Flowering Plants	U.S.A. (CA)	Entire	E
bariaco	Trichilia triacantha	Flowering Plants	U.S.A. (PR)	Entire	E
bat, Bulmer's fruit (=flying fox)	Aproteles bulmerae	Mammals	Papua New Guinea	Entire	E

List of threatened and endangered species - *continued*

Common name	Scientific name	Group	Historic range	Vertebrate population where endangered or threatened	Status
bat, bumblebee	Craseonycteris thonglongyai	Mammals	Thailand	Entire	E
bat, gray	Myotis grisescens	Mammals	Central and southeastern U.S.A.	Entire	E
bat, Hawaiian hoary	Lasiurus cinereus semotus	Mammals	U.S.A. (HI)	Entire	E
bat, Indiana	Myotis sodalis	Mammals	Eastern and Midwestern U.S.A	Entire	E
bat, lesser long-nosed	Leptonycteris curasoae yerbabuenae	Mammals	U.S.A. (AZ, NM), Mexico, Central America	Entire	E
Bat, little Mariana fruit	Pteropus tokudae	Mammals	Western Pacific Ocean_U.S.A. (Guam)	Entire	E
Bat, Mariana fruit (=Mariana flying fox)	Pteropus mariannus mariannus	Mammals	Western Pacific Ocean,U.S.A. (GU, MP)	Guam	T
Bat, Mariana fruit (=Mariana flying fox)	Pteropus mariannus mariannus	Mammals	Western Pacific Ocean,U.S.A. (GU, MP)	Aguijan, Tinian, Saipan pops	T
Bat, Mariana fruit (=Mariana flying fox)	Pteropus mariannus mariannus	Mammals	Western Pacific Ocean,U.S.A. (GU, MP)	(Rota, northern island populations)	T
bat, Mexican long-nosed	Leptonycteris nivalis	Mammals	U.S.A. (NM, TX), Mexico, Central America	Entire	E
bat, Ozark big-eared	Corynorhinus (=Plecotus) townsendii ingens	Mammals	U.S.A. (MO, OK, AR)	Entire	E
bat, Rodrigues fruit (=flying fox)	Pteropus rodricensis	Mammals	Indian Ocean_Rodrigues Island	Entire	E
bat, Singapore roundleaf horseshoe	Hipposideros ridleyi	Mammals	Malaysia	Entire	E
bat, Virginia big-eared	Corynorhinus (=Plecotus) townsendii virginianus	Mammals	U.S.A. (KY, NC, WV, VA)	Entire	E
Beaked-rush, Knieskern's	Rhynchospora knieskernii	Flowering Plants	U.S.A. (NJ)	Entire	T
bean, Cumberland (pearlymussel)	Villosa trabalis	Clams	U.S.A. (AL, KY, TN, VA)	Entire Range; Except where listed as Experimental Populations	E
bean, Cumberland (pearlymussel)	Villosa trabalis	Clams	U.S.A. (AL, KY, TN, VA)	U.S.A. (AL;The free-flowing reach of the Tennessee R. from the base of Wilson Dam downstream to the backwaters of Pickwick Reservoir [about 12 RM (19 km)] and the lower 5 RM [8 km] of all tributaries to this reach in Colbert and Lauderdale Cos., see 17.85(a))	EXPN
bean, purple	Villosa perpurpurea	Clams	U.S.A. (TN, VA)	Entire	E
bear, American black	Ursus americanus	Mammals	North America	U.S.A. (LA, all counties; MS, TX, only within the historic county range of the Louisiana black bear)	SAT
bear, Baluchistan	Ursus thibetanus gedrosianus	Mammals	Iran, Pakistan	Entire	E
bear, brown	Ursus arctos arctos	Mammals	Palearctic	Italy	E
bear, brown	Ursus arctos pruinosus	Mammals	China (Tibet)	Entire	E
bear, grizzly	Ursus arctos horribilis	Mammals	Holarctic	U.S.A. experimental non-essential (portions of ID and MT, see 17.84(I)	EXPN
bear, grizzly	Ursus arctos horribilis	Mammals	Holarctic	U.S.A., conterminous (lower 48) States, except where listed as an experimental population	T

List of threatened and endangered species - *continued*

Common name	Scientific name	Group	Historic range	Vertebrate population where endangered or threatened	Status
bear, Louisiana black	Ursus americanus luteolus	Mammals	U.S.A. (LA_all counties; MS_all counties south of or touching a line from Greenville, Washington County, to Meridian, Lauderdale County; TX_all counties east of or touching a line from Linden, Cass County, SW to Bryan, Brazos County, thence SS	Entire	T
bear, Mexican grizzly	Ursus arctos	Mammals	Holarctic	Mexico	E
bear-poppy, dwarf	Arctomecon humilis	Flowering Plants	U.S.A. (UT)	Entire	E
beardtongue, Penland	Penstemon penlandii	Flowering Plants	U.S.A. (CO)	Entire	E
beargrass, Britton's	Nolina brittoniana	Flowering Plants	U.S.A. (FL)	Entire	E
beauty, Harper's	Harperocallis flava	Flowering Plants	U.S.A. (FL)	Entire	E
beaver	Castor fiber birulai	Mammals	Mongolia	Entire	E
bedstraw, El Dorado	Galium californicum ssp. sierrae	Flowering Plants	U.S.A. (CA)	Entire	E
bedstraw, island	Galium buxifolium	Flowering Plants	U.S.A. (CA)	Entire	E
beetle, American burying	Nicrophorus americanus	Insects	U.S.A. (eastern States south to FL, west to SD and TX), eastern Canada	Entire	E
beetle, Coffin Cave mold	Batrisodes texanus	Insects	U.S.A. (TX)	Entire	E
beetle, Comal Springs dryopid	Stygoparnus comalensis	Insects	U.S.A. (TX)	Entire	E
beetle, Comal Springs riffle	Heterelmis comalensis	Insects	U.S.A. (TX)	Entire	E
beetle, delta green ground	Elaphrus viridis	Insects	U.S.A. (CA)	Entire	T
beetle, Helotes mold	Batrisodes venyivi	Insects	U.S.A. (TX)	Entire	E
Beetle, Hungerford's crawling water	Brychius hungerfordi	Insects	U.S.A. (MI, Canada)	Entire	E
beetle, Kretschmarr Cave mold	Texamaurops reddelli	Insects	U.S.A. (TX)	Entire	E
beetle, Mount Hermon June	Polyphylla barbata	Insects	U.S.A. (CA)	Entire	E
beetle, Tooth Cave ground	Rhadine persephone	Insects	U.S.A. (TX)	Entire	E
beetle, valley elderberry longhorn	Desmocerus californicus dimorphus	Insects	U.S.A. (CA)	Entire	T
bellflower, Brooksville	Campanula robinsiae	Flowering Plants	U.S.A. (FL)	Entire	E
birch, Virginia round-leaf	Betula uber	Flowering Plants	U.S.A. (VA)	Entire	T
bird's beak, palmate-bracted	Cordylanthus palmatus	Flowering Plants	U.S.A. (CA)	Entire	E
bird's-beak, Pennell's	Cordylanthus tenuis ssp. capillaris	Flowering Plants	U.S.A. (CA)	Entire	E
bird's-beak, salt marsh	Cordylanthus maritimus ssp. maritimus	Flowering Plants	U.S.A. (CA), Mexico (Baja California)	Entire	E
bird's-beak, soft	Cordylanthus mollis ssp. mollis	Flowering Plants	U.S.A. (CA)	Entire	E
birds-in-a-nest, white	Macbridea alba	Flowering Plants	U.S.A. (FL)	Entire	T
Bison, wood	Bison bison athabascae	Mammals	Canada, northwestern U.S.A	Canada	E
bittercress, small-anthered	Cardamine micranthera	Flowering Plants	U.S.A. (NC, VA)	Entire	E
blackbird, yellow-shouldered	Agelaius xanthomus	Birds	U.S.A. (PR)	Entire	E
bladderpod, Dudley Bluffs	Lesquerella congesta	Flowering Plants	U.S.A. (CO)	Entire	T
bladderpod, kodachrome	Lesquerella tumulosa	Flowering Plants	U.S.A. (UT)	Entire	E
bladderpod, lyrate	Lesquerella lyrata	Flowering Plants	U.S.A. (AL)	Entire	T
bladderpod, Missouri	Lesquerella filiformis	Flowering Plants	U.S.A. (AR, MO)	Entire	T
bladderpod, San Bernardino Mountains	Lesquerella kingii ssp. bernardina	Flowering Plants	U.S.A. (CA)	Entire	E
bladderpod, Spring Creek	Lesquerella perforata	Flowering Plants	U.S.A. (TN)	Entire	E
bladderpod, white	Lesquerella pallida	Flowering Plants	U.S.A. (TX)	Entire	E
bladderpod, Zapata	Lesquerella thamnophila	Flowering Plants	U.S.A. (TX)	Entire	E
blazingstar, Ash Meadows	Mentzelia leucophylla	Flowering Plants	U.S.A. (NV)	Entire	T
blazingstar, Heller's	Liatris helleri	Flowering Plants	U.S.A. (NC)	Entire	T
blazingstar, scrub	Liatris ohlingerae	Flowering Plants	U.S.A. (FL)	Entire	E

List of threatened and endangered species - *continued*

Common name	Scientific name	Group	Historic range	Vertebrate population where endangered or threatened	Status
blindcat, Mexican (catfish)	Prietella phreatophila	Fishes	Mexico	Entire	E
blossom, green (pearlymussel)	Epioblasma torulosa gubernaculum	Clams	U.S.A. (TN, VA)	Entire	E
blossom, tubercled (pearlymussel)	Epioblasma torulosa torulosa	Clams	U.S.A. (AL, IL, IN, KY, TN, WV)	Entire Range; Except where listed as Experimental Populations	E
blossom, tubercled (pearlymussel)	Epioblasma torulosa torulosa	Clams	U.S.A. (AL, IL, IN, KY, TN, WV)	U.S.A. (AL;The free-flowing reach of the Tennessee R. from the base of Wilson Dam downstream to the backwaters of Pickwick Reservoir [about 12 RM (19 km)] and the lower 5 RM [8 km] of all tributaries to this reach in Colbert and Lauderdale Cos., see 17.85(a))	EXPN
blossom, turgid (pearlymussel)	Epioblasma turgidula	Clams	U.S.A. (AL, TN)	Entire Range; Except where listed as Experimental Populations	E
blossom, turgid (pearlymussel)	Epioblasma turgidula	Clams	U.S.A. (AL, TN)	U.S.A. (AL;The free-flowing reach of the Tennessee R. from the base of Wilson Dam downstream to the backwaters of Pickwick Reservoir [about 12 RM (19 km)] and the lower 5 RM [8 km] of all tributaries to this reach in Colbert and Lauderdale Cos., see 17.85(a))	EXPN
blossom, yellow (pearlymussel)	Epioblasma florentina florentina	Clams	U.S.A. (AL, TN)	Entire Range; Except where listed as Experimental Populations	E
blossom, yellow (pearlymussel)	Epioblasma florentina florentina	Clams	U.S.A. (AL, TN)	U.S.A. (AL;The free-flowing reach of the Tennessee R. from the base of Wilson Dam downstream to the backwaters of Pickwick Reservoir [about 12 RM (19 km)] and the lower 5 RM [8 km] of all tributaries to this reach in Colbert and Lauderdale Cos., see 17.85(a))	EXPN
blue-star, Kearney's	Amsonia kearneyana	Flowering Plants	U.S.A. (AZ)	Entire	E
bluecurls, Hidden Lake	Trichostema austromontanum ssp. compactum	Flowering Plants	U.S.A. (CA)	Entire	T
bluegrass, Hawaiian	Poa sandvicensis	Flowering Plants	U.S.A. (HI)	Entire	E
bluegrass, Mann's	Poa mannii	Flowering Plants	U.S.A. (HI)	Entire	E
bluegrass, Napa	Poa napensis	Flowering Plants	U.S.A. (CA)	Entire	E
bluegrass, San Bernardino	Poa atropurpurea	Flowering Plants	U.S.A. (CA)	Entire	E
bluet, Roan Mountain	Hedyotis purpurea var. montana	Flowering Plants	U.S.A. (NC, TN)	Entire	E
boa, Jamaican	Epicrates subflavus	Reptiles	Jamaica	Entire	E
boa, Mona	Epicrates monensis monensis	Reptiles	U.S.A. (PR)	Entire	T
boa, Puerto Rican	Epicrates inornatus	Reptiles	U.S.A. (PR)	Entire	E
boa, Round Island bolyeria	Bolyeria multocarinata	Reptiles	Indian Ocean_Mauritius	Entire	E
boa, Round Island casarea	Casarea dussumieri	Reptiles	Indian Ocean_Mauritius	Entire	E
boa, Virgin Islands tree	Epicrates monensis granti	Reptiles	U.S.A. (PR), British Virgin Islands	Entire	E
bobcat, Mexican	Lynx (=Felis) rufus escuinapae	Mammals	Central Mexico	Entire	E

List of threatened and endangered species - *continued*

Common name	Scientific name	Group	Historic range	Vertebrate population where endangered or threatened	Status
bobwhite, masked (quail)	Colinus virginianus ridgwayi	Birds	U.S.A. (AZ), Mexico (Sonora)	Entire	E
bonamia, Florida	Bonamia grandiflora	Flowering Plants	U.S.A. (FL)	Entire	T
Bontebok	Damaliscus pygarus (=dorcas) dorcas	Mammals	South Africa	Entire	E
bonytongue, Asian	Scleropages formosus	Fishes	Thailand, Indonesia, Malaysia	Entire	E
booby, Abbott's	Papasula (=Sula) abbotti	Birds	Indian Ocean_Christmas Island	Entire	E
boxwood, Vahl's	Buxus vahlii	Flowering Plants	U.S.A. (PR, VI)	Entire	E
bristlebird, western	Dasyornis longirostris (=brachypterus l.)	Birds	Australia	Entire	E
bristlebird, western rufous	Dasyornis broadbenti littoralis	Birds	Australia	Entire	E
brodiaea, Chinese Camp	Brodiaea pallida	Flowering Plants	U.S.A. (CA)	Entire	T
brodiaea, thread-leaved	Brodiaea filifolia	Flowering Plants	U.S.A. (CA)	Entire	T
broom, San Clemente Island	Lotus dendroideus ssp. traskiae	Flowering Plants	U.S.A. (CA)	Entire	E
buckwheat, cushenbury	Eriogonum ovalifolium var. vineum	Flowering Plants	U.S.A. (CA)	Entire	E
buckwheat, Ione (incl. Irish Hill)	Eriogonum apricum (incl. var. prostratum)	Flowering Plants	U.S.A. (CA)	Entire	E
buckwheat, scrub	Eriogonum longifolium var. gnaphalifolium	Flowering Plants	U.S.A. (FL)	Entire	T
buckwheat, steamboat	Eriogonum ovalifolium var. williamsiae	Flowering Plants	U.S.A. (NV)	Entire	E
bulbul, Mauritius olivaceous	Hypsipetes borbonicus olivaceus	Birds	Indian Ocean_Mauritius	Entire	E
bullfinch, Sao Miguel (finch)	Pyrrhula pyrrhula murina	Birds	Eastern Atlantic Ocean_Azores	Entire	E
bulrush, Northeastern	Scirpus ancistrochaetus	Flowering Plants	U.S.A. (MA, MD, NH, NY, PA, VA, VT, WV)	Entire	E
bush-clover, prairie	Lespedeza leptostachya	Flowering Plants	U.S.A. (IA, IL, MN, WI)	Entire	T
bush-mallow, San Clemente Island	Malacothamnus clementinus	Flowering Plants	U.S.A. (CA)	Entire	E
bush-mallow, Santa Cruz Island	Malacothamnus fasciculatus var. nesioticus	Flowering Plants	U.S.A. (CA)	Entire	E
bush-shrike, Ulugura	Malaconotus alius	Birds	Tanzania	Entire	T
bushwren, New Zealand	Xenicus longipes	Birds	New Zealand	Entire	E
bustard, great Indian	Ardeotis (=Choriotis) nigriceps	Birds	India, Pakistan	Entire	E
Buttercup, autumn	Ranunculus aestivalis (=acriformis)	Flowering Plants	U.S.A. (UT)	Entire	E
Butterfly plant, Colorado	Gaura neomexicana var. coloradensis	Flowering Plants	U.S.A. (CO, NE, WY)	Entire	T
butterfly, bay checkerspot	Euphydryas editha bayensis	Insects	U.S.A. (CA)	Entire	T
butterfly, Behren's silverspot	Speyeria zerene behrensii	Insects	U.S.A. (CA)	Entire	E
butterfly, callippe silverspot	Speyeria callippe callippe	Insects	U.S.A. (CA)	Entire	E
butterfly, Corsican swallowtail	Papilio hospiton	Insects	Corsica, Sardinia	Entire	E
butterfly, El Segundo blue	Euphilotes battoides allyni	Insects	U.S.A. (CA)	Entire	E
butterfly, Fender's blue	Icaricia icarioides fenderi	Insects	U.S.A. (OR)	Entire	E
butterfly, Homerus swallowtail	Papilio homerus	Insects	Jamaica	Entire	E
butterfly, Karner blue	Lycaeides melissa samuelis	Insects	U.S.A. (IL, IN, MA, MI, MN, NH, NY, OH, PA, WI), Canada (Ont.)	Entire	E
butterfly, Lange's metalmark	Apodemia mormo langei	Insects	U.S.A. (CA)	Entire	E
butterfly, lotis blue	Lycaeides argyrognomon lotis	Insects	U.S.A. (CA)	Entire	E
butterfly, Luzon peacock swallowtail	Papilio chikae	Insects	Philippines	Entire	E

List of threatened and endangered species - *continued*

Common name	Scientific name	Group	Historic range	Vertebrate population where endangered or threatened	Status
butterfly, mission blue	Icaricia icarioides missionensis	Insects	U.S.A. (CA)	Entire	E
Butterfly, Mitchell's satyr	Neonympha mitchellii mitchellii	Insects	U.S.A. (IN, MI, NJ)	Entire	E
butterfly, Myrtle's silverspot	Speyeria zerene myrtleae	Insects	U.S.A. (CA)	Entire	E
butterfly, Oregon silverspot	Speyeria zerene hippolyta	Insects	U.S.A. (CA, OR, WA)	Entire	T
butterfly, Palos Verdes blue	Glaucopsyche lygdamus palosverdesensis	Insects	U.S.A. (CA)	Entire	E
butterfly, Queen Alexandra's birdwing	Troides alexandrae	Insects	Papua New Guinea	Entire	E
butterfly, Quino checkerspot Insects	Euphydryas editha quino (=E. e. wrighti) U.S.A. (CA), Mexico)	Entire	E		
butterfly, Saint Francis' satyr	Neonympha mitchellii francisci	Insects	U.S.A. (NC)	Entire	E
butterfly, San Bruno elfin	Callophrys mossii bayensis	Insects	U.S.A. (CA)	Entire	E
butterfly, Schaus swallowtail	Heraclides aristodemus ponceanus	Insects	U.S.A. (FL)	Entire	E
butterfly, Smith's blue	Euphilotes enoptes smithi	Insects	U.S.A. (CA)	Entire	E
butterfly, Uncompahgre fritillary	Boloria acrocnema	Insects	U.S.A. (CO)	Entire	E
butterweed, Layne's	Senecio layneae	Flowering Plants	U.S.A. (CA)	Entire	T
butterwort, Godfrey's	Pinguicula ionantha	Flowering Plants	U.S.A. (FL)	Entire	T
button, Mohr's Barbara	Marshallia mohrii	Flowering Plants	U.S.A. (AL, GA)	Entire	T
button-celery, San Diego	Eryngium aristulatum var. parishii	Flowering Plants	U.S.A. (CA)	Entire	E
cactus, Arizona hedgehog	Echinocereus triglochidiatus var. arizonicus	Flowering Plants	U.S.A. (AZ)	Entire	E
cactus, Bakersfield	Opuntia treleasei	Flowering Plants	U.S.A. (CA)	Entire	E
cactus, black lace	Echinocereus reichenbachii var. albertii	Flowering Plants	U.S.A. (TX)	Entire	E
cactus, Brady pincushion	Pediocactus bradyi	Flowering Plants	U.S.A. (AZ)	Entire	E
Cactus, Chisos Mountain hedgehog	Echinocereus chisoensis var. chisoensis	Flowering Plants	U.S.A. (TX)	Entire	T
cactus, Cochise pincushion	Coryphantha robbinsorum	Flowering Plants	U.S.A. (AZ), Mexico (Sonora)	Entire	T
cactus, Key tree	Pilosocereus robinii	Flowering Plants	U.S.A. (FL), Cuba	Entire	E
cactus, Knowlton	Pediocactus knowltonii	Flowering Plants	U.S.A. (CO, NM)	Entire	E
cactus, Kuenzler hedgehog	Echinocereus fendleri var. kuenzleri	Flowering Plants	U.S.A. (NM)	Entire	E
cactus, Lee pincushion	Coryphantha sneedii var. leei	Flowering Plants	U.S.A. (NM)	Entire	T
cactus, Lloyd's Mariposa	Echinomastus mariposensis	Flowering Plants	U.S.A. (TX), Mexico (Coahuila)	Entire	T
cactus, Mesa Verde	Sclerocactus mesae-verdae	Flowering Plants	U.S.A. (CO, NM)	Entire	T
cactus, Nellie cory	Coryphantha minima	Flowering Plants	U.S.A. (TX)	Entire	E
cactus, Nichol's Turk's head	Echinocactus horizonthalonius var. nicholii	Flowering Plants	U.S.A. (AZ)	Entire	E
cactus, Peebles Navajo	Pediocactus peeblesianus peeblesianus	Flowering Plants	U.S.A. (AZ)	Entire	E
cactus, Pima pineapple	Coryphantha scheeri var. robustispina	Flowering Plants	U.S.A. (AZ), Mexico (Sonora)	Entire	E
cactus, San Rafael	Pediocactus despainii	Flowering Plants	U.S.A. (UT)	Entire	E
cactus, Siler pincushion	Pediocactus (=Echinocactus,=Utahia) sileri	Flowering Plants	U.S.A. (AZ, UT)	Entire	T
cactus, Sneed pincushion	Coryphantha sneedii var. sneedii	Flowering Plants	U.S.A. (NM, TX)	Entire	E
cactus, star	Astrophytum asterias	Flowering Plants	U.S.A. (TX), Mexico	Entire	E
cactus, Tobusch fishhook	Ancistrocactus tobuschii	Flowering Plants	U.S.A. (TX)	Entire	E

List of threatened and endangered species - *continued*

Common name	Scientific name	Group	Historic range	Vertebrate population where endangered or threatened	Status
Cactus, Uinta Basin hookless	Sclerocactus glaucus	Flowering Plants	U.S.A. (CO, UT)	Entire	T
cactus, Winkler	Pediocactus winkleri	Flowering Plants	U.S.A. (UT)	Entire	T
cactus, Wright fishhook	Sclerocactus wrightiae	Flowering Plants	U.S.A. (UT)	Entire	E
cahow	Pterodroma cahow	Birds	North Atlantic Ocean_Bermuda	Entire	E
caiman, Apaporis River	Caiman crocodilus apaporiensis	Reptiles	Colombia	Entire	E
caiman, black	Melanosuchus niger	Reptiles	Amazon basin	Entire	E
caiman, broad-snouted	Caiman latirostris	Reptiles	Brazil, Argentina, Paraguay, Uruguay	Entire	E
Caiman, brown	Caiman crocodilus fuscus (includes Caiman crocodilus chiapas	Reptiles	Mexico, Central America, Colombia, Ecuador, Venezuela, Peru	Entire	SAT
caiman, common	Caiman crocodilus crocodilus	Reptiles	Brazil, Colombia, Ecuador, French Guiana, Guyana, Suriname, Venezuela, Bolivia, Peru	Entire	SAT
Caiman, Yacare	Caiman yacare	Reptiles	Bolivia, Argentina, Peru, Brazil	Entire	T
camel, Bactrian	Camelus bactrianus	Mammals	Mongolia, China	Entire	E
campeloma, slender	Campeloma decampi	Snails	U.S.A. (AL)	Entire	E
campion, fringed	Silene polypetala	Flowering Plants	U.S.A. (FL, GA)	Entire	E
capa rosa	Callicarpa ampla	Flowering Plants	U.S.A. (PR)	Entire	E
caracara, Audubon's crested	Polyborus plancus audubonii	Birds	U.S.A. (AZ, FL, LA, NM, TX) south to Panama; Cuba	U.S.A. (FL)	T
caribou, woodland	Rangifer tarandus caribou	Mammals	U.S.A. (AK, ID, ME, MI, MN, MT, NH, VT, WA, WI), Canada	U.S.A. (ID, WA), Canada (that part of S.E. British Columbia bounded by the U.S.-Can. border, Columbia R., Kootenay R., Kootenay L., and Kootenai R.)	E
cat's-eye, Terlingua Creek	Cryptantha crassipes	Flowering Plants	U.S.A. (TX)	Entire	E
cat, Andean	Felis jacobita	Mammals	Chile, Peru, Bolivia, Argentina	Entire	E
cat, Asian golden (=Temmnick's)	Catopuma (=Felis) temminckii	Mammals	Nepal, China, Southeast Asia, Indonesia (Sumatra)	Entire	E
cat, black-footed	Felis nigripes	Mammals	Southern Africa	Entire	E
cat, flat-headed	Prionailurus (=Felis) planiceps	Mammals	Malaysia, Indonesia	Entire	E
cat, Iriomote	Prionailurus (=Felis) bengalensis iriomotensis	Mammals	Japan (Iriomote Island, Ryukyu Islands)	Entire	E
cat, leopard	Prionailurus (=Felis) bengalensis bengalensis	Mammals	India, Southeast Asia	Entire	E
cat, marbled	Pardofelis (=Felis) marmorata	Mammals	Nepal, Southeast Asia, Indonesia	Entire	E
cat, Pakistan sand	Felis margarita scheffeli	Mammals	Pakistan	Entire	E
cat, tiger	Leopardus (=Felis) tigrinus	Mammals	Costa Rica to northern Argentina	Entire	E
Catchfly, Spalding's	Silene spaldingii	Flowering Plants	U.S.A. (ID, MT, OR, WA)	Entire	T
catfish	Pangasius sanitwongsei	Fishes	Thailand	Entire	E
catfish, Thailand giant	Pangasianodon gigas	Fishes	Thailand	Entire	E
catfish, Yaqui	Ictalurus pricei	Fishes	U.S.A. (AZ), Mexico	Entire	T
catspaw (=purple cat's paw pearlymussel)	Epioblasma obliquata obliquata	Clams	U.S.A. (AL, IL, IN, KY, OH, TN)	Entire Range; Except where listed as Experimental Populations	E

List of threatened and endangered species - *continued*

Common name	Scientific name	Group	Historic range	Vertebrate population where endangered or threatened	Status
catspaw (=purple cat's paw pearlymussel)	Epioblasma obliquata obliquata	Clams	U.S.A. (AL, IL, IN, KY, OH, TN)	U.S.A. (AL;The free-flowing reach of the Tennessee R. from the base of Wilson Dam downstream to the backwaters of Pickwick Reservoir [about 12 RM (19 km)] and the lower 5 RM [8 km] of all tributaries to this reach in Colbert and Lauderdale Cos., see 17.85(a))	EXPN
catspaw, white (pearlymussel)	Epioblasma obliquata perobliqua	Clams	U.S.A. (IN, MI, OH)	Entire	E
cavefish, Alabama	Speoplatyrhinus poulsoni	Fishes	U.S.A. (AL)	Entire	E
cavefish, Ozark	Amblyopsis rosae	Fishes	U.S.A. (AR, MO, OK)	Entire	T
cavesnail, Tumbling Creek	Antrobia culveri	Snails	U.S.A. (MO)	Entire	E
ceanothus, coyote	Ceanothus ferrisae	Flowering Plants	U.S.A. (CA)	Entire	E
ceanothus, Pine Hill	Ceanothus roderickii	Flowering Plants	U.S.A. (CA)	Entire	E
ceanothus, Vail Lake	Ceanothus ophiochilus	Flowering Plants	U.S.A. (CA)	Entire	T
centaury, spring-loving	Centaurium namophilum	Flowering Plants	U.S.A. (CA, NV)	Entire	T
chaff-flower, round-leaved	Achyranthes splendens var. rotundata	Flowering Plants	U.S.A. (HI)	Entire	E
chaffseed, American	Schwalbea americana	Flowering Plants	U.S.A. (AL, CT, DE, FL, GA, LA, MA, MD, MI, MS, NC, NJ, NY, SC, TN, VA)	Entire	E
chamois, Apennine	Rupicapra rupicapra ornata	Mammals	Italy	Entire	E
Checker-mallow, Keck's	Sidalcea keckii	Flowering Plants	U.S.A. (CA)	Entire	E
checker-mallow, Kenwood Marsh	Sidalcea oregana ssp. valida	Flowering Plants	U.S.A. (CA)	Entire	E
checker-mallow, Nelson's	Sidalcea nelsoniana	Flowering Plants	U.S.A. (OR, WA)	Entire	T
checker-mallow, pedate	Sidalcea pedata	Flowering Plants	U.S.A. (CA)	Entire	E
checkermallow, Wenatchee Mountains	Sidalcea oregana var. calva	Flowering Plants	U.S.A. (WA)	Entire	E
cheetah	Acinonyx jubatus	Mammals	Africa to India	Entire	E
chimpanzee Africa-see 17.40(c)(3)	Pan troglodytes Wherever found in the wild	Mammals E			
chimpanzee	Pan troglodytes	Mammals	Africa-see 17.40(c)(3)	Wherever found in captivity	T
chimpanzee, pygmy	Pan paniscus	Mammals	Zaire	Entire	E
chinchilla	Chinchilla brevicaudata boliviana	Mammals	Bolivia	Entire	E
chub, bonytail	Gila elegans	Fishes	U.S.A. (AZ, CA, CO, NV, UT, WY)	Entire	E
chub, Borax Lake	Gila boraxobius	Fishes	U.S.A. (OR)	Entire	E
chub, Chihuahua	Gila nigrescens	Fishes	U.S.A. (NM), Mexico (Chihuahua)	Entire	T
chub, humpback	Gila cypha	Fishes	U.S.A. (AZ, CO, UT, WY)	Entire	E
chub, Hutton tui	Gila bicolor ssp.	Fishes	U.S.A. (OR)	Hutton	T
chub, Mohave tui	Gila bicolor mohavensis	Fishes	U.S.A. (CA)	Entire	E
chub, Oregon	Oregonichthys crameri	Fishes	U.S.A. (OR)	Entire	E
chub, Owens tui Fishes	Gila bicolor snyderi U.S.A. (CA)	Entire	E		
chub, Pahranagat roundtail	Gila robusta jordani	Fishes	U.S.A. (NV)	Entire	E
chub, slender	Erimystax cahni	Fishes	U.S.A. (TN, VA)	Entire	T
chub, Sonora	Gila ditaenia	Fishes	U.S.A. (AZ), Mexico	Entire	T
chub, spotfin	Cyprinella monacha	Fishes	U.S.A. (AL, GA, NC, TN, VA)	Tellico River, between the backwaters of the Tellico Reservoir and the Tellico Ranger Station, in Monroe County, Tennessee	EXPN
chub, spotfin	Cyprinella monacha	Fishes	U.S.A. (AL, GA, NC, TN, VA)	Entire	T
Chub, Virgin River	Gila seminuda (=robusta)	Fishes	U.S.A. (AZ, NV, UT)	Entire	E
chub, Yaqui	Gila purpurea	Fishes	U.S.A. (AZ), Mexico	Entire	E

483

List of threatened and endangered species - *continued*

Common name	Scientific name	Group	Historic range	Vertebrate population where endangered or threatened	Status
chuckwalla, San Esteban Island	Sauromalus varius	Reptiles	Mexico	Entire	E
Chumbo, Higo	Harrisia portoricensis	Flowering Plants	U.S.A. (PR)	Entire	T
chupacallos	Pleodendron macranthum	Flowering Plants	U.S.A. (PR)	Entire	E
cicek (minnow)	Acanthorutilus handlirschi	Fishes	Turkey	Entire	E
civet, Malabar large-spotted	Viverra civettina (=megaspila c.)	Mammals	India	Entire	E
cladonia, Florida perforate	Cladonia perforata	Lichens	U.S.A. (FL)	Entire	E
clarkia, Pismo	Clarkia speciosa ssp. immaculata	Flowering Plants	U.S.A. (CA)	Entire	E
clarkia, Presidio	Clarkia franciscana	Flowering Plants	U.S.A. (CA)	Entire	E
clarkia, Springville	Clarkia springvillensis	Flowering Plants	U.S.A. (CA)	Entire	T
clarkia, Vine Hill	Clarkia imbricata	Flowering Plants	U.S.A. (CA)	Entire	E
Cliff-rose, Arizona	Purshia (=Cowania) subintegra	Flowering Plants	U.S.A. (AZ)	Entire	E
clover, Monterey	Trifolium trichocalyx	Flowering Plants	U.S.A. (CA)	Entire	E
clover, running buffalo	Trifolium stoloniferum	Flowering Plants	U.S.A. (AR, IL, IN, KS, KY, MO, OH, WV)	Entire	E
clover, showy Indian	Trifolium amoenum	Flowering Plants	U.S.A. (CA)	Entire	E
clubshell	Pleurobema clava	Clams	U.S.A. (AL, IL, IN, KY, MI, OH, PA, TN, WV)	Entire Range; Except where listed as Experimental Populations	E
clubshell	Pleurobema clava	Clams	U.S.A. (AL, IL, IN, KY, MI, OH, PA, TN, WV)	U.S.A. (AL;The free-flowing reach of the Tennessee R. from the base of Wilson Dam downstream to the backwaters of Pickwick Reservoir [about 12 RM (19 km)] and the lower 5 RM [8 km] of all tributaries to this reach in Colbert and Lauderdale Cos., see 17.85(a))	EXPN
clubshell, black	Pleurobema curtum	Clams	U.S.A. (AL, MS)	Entire	E
clubshell, ovate	Pleurobema perovatum	Clams	U.S.A. (AL, GA, MS, TN)	Entire	E
clubshell, southern	Pleurobema decisum	Clams	U.S.A. (AL, GA, MS, TN)	Entire	E
cobana negra	Stahlia monosperma	Flowering Plants	U.S.A. (PR), Dominican Republic	Entire	T
cochito	Phocoena sinus	Mammals	Mexico (Gulf of California)	Entire	E
combshell, Cumberlandian	Epioblasma brevidens	Clams	U.S.A. (AL, KY, MS, TN, VA)	Entire Range; Except where listed as Experimental Populations	E
combshell, Cumberlandian	Epioblasma brevidens	Clams	U.S.A. (AL, KY, MS, TN, VA)	U.S.A. (AL;The free-flowing reach of the Tennessee R. from the base of Wilson Dam downstream to the backwaters of Pickwick Reservoir [about 12 RM (19 km)] and the lower 5 RM [8 km] of all tributaries to this reach in Colbert and Lauderdale Cos., see 17.85(a))	EXPN
combshell, southern	Epioblasma penita	Clams	U.S.A. (AL, MS)	Entire	E
combshell, upland	Epioblasma metastriata	Clams	U.S.A. (AL, GA, TN)	Entire	E
condor, Andean	Vultur gryphus	Birds	Colombia to Chile and Argentina	Entire	E
condor, California	Gymnogyps californianus	Birds	U.S.A. (AZ, CA, OR), Mexico (Baja California)	U.S.A. only, except where listed as an experimental population below	E
condor, California	Gymnogyps californianus	Birds	U.S.A. (AZ, CA, OR), Mexico (Baja California)	U.S.A. (specific portions of Arizona, Nevada, and Utah)	EXPN
coneflower, smooth	Echinacea laevigata	Flowering Plants	U.S.A. (GA, MD, NC, PA, SC, VA)	Entire	E

List of threatened and endangered species - *continued*

Common name	Scientific name	Group	Historic range	Vertebrate population where endangered or threatened	Status
coneflower, Tennessee purple	Echinacea tennesseensis	Flowering Plants	U.S.A. (TN)	Entire	E
coot, Hawaiian	Fulica americana alai	Birds	U.S.A. (HI)	Entire	E
cooter (=turtle), northern redbelly (=Plymouth)	Pseudemys rubriventris bangsi	Reptiles	U.S.A. (MA)	Entire	E
coqui, golden	Eleutherodactylus jasperi	Amphibians	U.S.A. (PR)	Entire	T
cory cactus, bunched	Coryphantha ramillosa	Flowering Plants	U.S.A. (TX), Mexico (Coahuila)	Entire	T
cotinga, banded	Cotinga maculata	Birds	Brazil	Entire	E
cotinga, white-winged	Xipholena atropurpurea	Birds	Brazil	Entire	E
Crane, black-necked	Grus nigricollis	Birds	China (Tibet)	Entire	E
crane, Cuba sandhill	Grus canadensis nesiotes	Birds	West Indies_Cuba	Entire	E
Crane, hooded	Grus monacha	Birds	Japan, Russia	Entire	E
Crane, Japanese	Grus japonensis	Birds	China, Japan, Korea, Russia	Entire	E
crane, Mississippi sandhill	Grus canadensis pulla	Birds	U.S.A. (MS)	Entire	E
Crane, Siberian white	Grus leucogeranus	Birds	C.I.S. (Siberia) to India, including Iran and China	Entire	E
Crane, white-naped	Grus vipio	Birds	Mongolia	Entire	E
crane, whooping	Grus americana	Birds	Canada, U.S.A. (Rocky Mountains east to Carolinas), Mexico	Entire, except where listed as an experimental population	E
crane, whooping	Grus americana	Birds	Canada, U.S.A. (Rocky Mountains east to Carolinas), Mexico	U.S.A. (CO, ID, FL, NM, UT, and the western half of Wyoming)	EXPN
crane, whooping	Grus americana	Birds	Canada, U.S.A. (Rocky Mountains east to Carolinas), Mexico	U.S.A.(AL, AR, GA, IL, IN, IA, KY, LA, MI, MN, MS, MO, NC, OH, SC, TN, VA, WI, WV)	EXPN
crayfish, cave	Cambarus aculabrum	Crustaceans	U.S.A. (AR)	Entire	E
crayfish, cave	Cambarus zophonastes	Crustaceans	U.S.A. (AR)	Entire	E
crayfish, Nashville	Orconectes shoupi	Crustaceans	U.S.A. (TN)	Entire	E
crayfish, Shasta	Pacifastacus fortis	Crustaceans	U.S.A. (CA)	Entire	E
creeper, Hawaii	Oreomystis mana	Birds	U.S.A. (HI)	Entire	E
creeper, Molokai	Paroreomyza flammea	Birds	U.S.A. (HI)	Entire	E
creeper, Oahu	Paroreomyza maculata	Birds	U.S.A. (HI)	Entire	E
crocodile, African dwarf	Osteolaemus tetraspis tetraspis	Reptiles	West Africa	Entire	E
crocodile, African slender-snouted	Crocodylus cataphractus	Reptiles	Western and central Africa	Entire	E
crocodile, American	Crocodylus acutus	Reptiles	U.S.A. (FL), Mexico, Caribbean, Central and South America	Entire	E
crocodile, Ceylon mugger	Crocodylus palustris kimbula	Reptiles	Sri Lanka	Entire	E
crocodile, Congo dwarf	Osteolaemus tetraspis osborni	Reptiles	Congo R. drainage	Entire	E
crocodile, Cuban	Crocodylus rhombifer	Reptiles	Cuba	Entire	E
crocodile, Morelet's	Crocodylus moreletii	Reptiles	Mexico, Belize, Guatemala	Entire	E
crocodile, mugger	Crocodylus palustris palustris	Reptiles	India, Pakistan, Iran, Bangladesh	Entire	E
crocodile, Nile	Crocodylus niloticus	Reptiles	Africa, Middle East	Entire	T
crocodile, Orinoco	Crocodylus intermedius	Reptiles	South America_Orinoco R. basin	Entire	E
crocodile, Philippine	Crocodylus novaeguineae mindorensis	Reptiles	Philippine Islands	Entire	E
crocodile, saltwater	Crocodylus porosus	Reptiles	Southeast Asia, Australia, Papua New Guinea, Islands of the West Pacific Ocean	Entire, except Papua New Guinea and Australia	E
crocodile, saltwater	Crocodylus porosus	Reptiles	Southeast Asia, Australia, Papua New Guinea, Islands of the West Pacific Ocean	Australia	T
crocodile, Siamese	Crocodylus siamensis	Reptiles	Southeast Asia, Malay Peninsula	Entire	E

List of threatened and endangered species - *continued*

Common name	Scientific name	Group	Historic range	Vertebrate population where endangered or threatened	Status
Crow, Hawaiian (='alala)	Corvus hawaiiensis	Birds	U.S.A. (HI)	Entire	E
Crow, Mariana (=aga)	Corvus kubaryi	Birds	Western Pacific Ocean_U.S.A. (Guam, Rota)	Entire	E
crow, white-necked	Corvus leucognaphalus	Birds	U.S.A. (PR), Dominican Republic, Haiti	Entire	E
crownbeard, big-leaved	Verbesina dissita	Flowering Plants	U.S.A. (CA), Mexico.	Entire	T
crownscale, San Jacinto Valley	Atriplex coronata var. notatior	Flowering Plants	U.S.A. (CA)	Entire	E
cuckoo-shrike, Mauritius	Coquus typicus	Birds	Indian Ocean_Mauritius	Entire	E
cuckoo-shrike, Reunion	Coquus newtoni	Birds	Indian Ocean_Reunion	Entire	E
cui-ui	Chasmistes cujus	Fishes	U.S.A. (NV)	Entire	E
curassow, razor-billed	Mitu mitu mitu	Birds	Brazil (Eastern)	Entire	E
curassow, red-billed	Crax blumenbachii	Birds	Brazil	Entire	E
curassow, Trinidad white-headed	Pipile pipile pipile	Birds	West Indies_Trinidad	Entire	E
curlew, Eskimo	Numenius borealis	Birds	Alaska and northern Canada to Argentina	Entire	E
Cycladenia, Jones	Cycladenia jonesii (=humilis)	Flowering Plants	U.S.A. (AZ, UT)	Entire	T
cypress, Gowen	Cupressus goveniana ssp. goveniana	Conifers and Cycads	U.S.A. (CA)	Entire	T
cypress, Santa Cruz	Cupressus abramsiana	Conifers and Cycads	U.S.A. (CA)	Entire	E
dace, Ash Meadows speckled	Rhinichthys osculus nevadensis	Fishes	U.S.A. (NV)	Entire	E
dace, blackside	Phoxinus cumberlandensis	Fishes	U.S.A. (KY, TN)	Entire	T
dace, Clover Valley speckled	Rhinichthys osculus oligoporus	Fishes	U.S.A. (NV)	Entire	E
dace, desert	Eremichthys acros	Fishes	U.S.A. (NV)	Entire	T
dace, Foskett speckled	Rhinichthys osculus ssp.	Fishes	U.S.A. (OR)	Foskett	T
dace, Independence Valley speckled	Rhinichthys osculus lethoporus	Fishes	U.S.A. (NV)	Entire	E
dace, Kendall Warm Springs	Rhinichthys osculus thermalis	Fishes	U.S.A. (WY)	Entire	E
dace, Moapa	Moapa coriacea	Fishes	U.S.A. (NV)	Entire	E
daisy, lakeside	Hymenoxys herbacea	Flowering Plants	U.S.A. (IL, MI, OH), Canada (Ont.)	Entire	T
daisy, Maguire	Erigeron maguirei	Flowering Plants	U.S.A. (UT)	Entire	T
daisy, Parish's	Erigeron parishii	Flowering Plants	U.S.A. (CA)	Entire	T
daisy, Willamette	Erigeron decumbens var. decumbens	Flowering Plants	U.S.A. (OR)	Entire	E
darter, amber	Percina antesella	Fishes	U.S.A. (AL, GA, TN)	Entire	E
darter, bayou	Etheostoma rubrum	Fishes	U.S.A. (MS)	Entire	T
Darter, bluemask (=jewel)	Etheostoma /	Fishes	U.S.A. (TN)	Entire	E
darter, boulder	Etheostoma wapiti	Fishes	U.S.A. (AL, TN)	Entire	E
darter, boulder	Etheostoma wapiti	Fishes	U.S.A. (AL, TN)	Shoal Creek	EXPN
darter, Cherokee	Etheostoma scotti	Fishes	U.S.A. (GA)	Entire	T
darter, duskytail	Etheostoma percnurum	Fishes	U.S.A. (TN, VA)	Entire	E
darter, duskytail	Etheostoma percnurum	Fishes	U.S.A. (TN, VA)	Tellico River, between the backwaters of the Tellico Reservoir and the Tellico Ranger Station, in Monroe County, Tennessee	EXPN
darter, Etowah	Etheostoma etowahae	Fishes	U.S.A. (GA)	Entire	E
darter, fountain	Etheostoma fonticola	Fishes	U.S.A. (TX)	Entire	E
darter, goldline	Percina aurolineata	Fishes	U.S.A. (AL, GA, TN)	Entire	T
darter, leopard	Percina pantherina	Fishes	U.S.A. (AR, OK)	Entire	T
darter, Maryland	Etheostoma sellare	Fishes	U.S.A. (MD)	Entire	E
darter, Niangua	Etheostoma nianguae	Fishes	U.S.A. (MO)	Entire	T
darter, Okaloosa	Etheostoma okaloosae	Fishes	U.S.A. (FL)	Entire	E
darter, relict	Etheostoma chienense	Fishes	U.S.A. (KY)	Entire	E
darter, slackwater	Etheostoma boschungi	Fishes	U.S.A. (AL, TN)	Entire	T
darter, snail	Percina tanasi	Fishes	U.S.A. (AL, GA, TN)	Entire	T

List of threatened and endangered species - *continued*

Common name	Scientific name	Group	Historic range	Vertebrate population where endangered or threatened	Status
darter, vermilion	Etheostoma chermocki	Fishes	U.S.A. (AL).	Entire	E
darter, watercress	Etheostoma nuchale	Fishes	U.S.A. (AL)	Entire	E
dawn-flower, Texas prairie	Hymenoxys texana	Flowering Plants	U.S.A. (TX)	Entire	E
deer, Bactrian	Cervus elaphus bactrianus	Mammals	Tajikistan, Uzbekistan, Afghanistan	Entire	E
deer, Barbary	Cervus elaphus barbarus	Mammals	Morocco, Tunisia, Algeria	Entire	E
deer, Calamianes (=Philippine)	Axis porcinus calamianensis	Mammals	Philippines (Calamian Islands)	Entire	E
Deer, Cedros Island mule	Odocoileus hemionus cerrosensis	Mammals	Mexico (Cedros Island)	Entire	E
deer, Columbian white-tailed	Odocoileus virginianus leucurus	Mammals	U.S.A. (WA, OR)	Columbia River (Clark, Cowliz, Pacific, Skamania, and Wahkiakum Counties, WA., and Clatsop, Columbia, and Multnomah Counties, OR.)	E
deer, Corsican red	Cervus elaphus corsicanus	Mammals	Corsica, Sardinia	Entire	E
Deer, Eld's brow-antlered	Cervus eldi	Mammals	India to Southeast Asia	Entire	E
deer, Formosan sika	Cervus nippon taiouanus	Mammals	Taiwan	Entire	E
deer, Indochina hog	Axis porcinus annamiticus	Mammals	Thailand, Indochina	Entire	E
deer, key	Odocoileus virginianus clavium	Mammals	U.S.A. (FL)	Entire	E
deer, Kuhl's (=Bawean)	Axis porcinus kuhli	Mammals	Indonesia	Entire	E
deer, marsh	Blastocerus dichotomus	Mammals	Argentina, Uruguay, Paraguay, Bolivia, Brazil	Entire	E
deer, McNeill's	Cervus elaphus macneilii	Mammals	China (Sinkiang, Tibet)	Entire	E
deer, musk	Moschus spp. (all species)	Mammals	Central and eastern Asia	Afghanistan, Bhutan, Burma, China (Tibet, Yunnan), India, Nepal, Pakistan, Sikkim	E
deer, North China sika	Cervus nippon mandarinus	Mammals	China (Shantung and Chihli Provinces)	Entire	E
deer, pampas	Ozotoceros bezoarticus	Mammals	Brazil, Argentina, Uruguay, Bolivia, Paraguay	Entire	E
deer, Persian fallow	Dama mesopotamica (=dama m.)	Mammals	Iraq, Iran	Entire	E
deer, Ryukyu sika	Cervus nippon keramae	Mammals	Japan (Ryukyu Islands)	Entire	E
deer, Shansi sika	Cervus nippon grassianus	Mammals	China (Shansi Province)	Entire	E
deer, South China sika	Cervus nippon kopschi	Mammals	Southern China	Entire	E
deer, swamp	Cervus duvauceli	Mammals	India, Nepal	Entire	E
deer, Visayan	Cervus alfredi	Mammals	Philippines	Entire	E
deer, Yarkand	Cervus elaphus yarkandensis	Mammals	China (Sinkiang)	Entire	E
desert-parsley, Bradshaw's	Lomatium bradshawii	Flowering Plants	U.S.A. (OR,WA)	Entire	E
dhole	Cuon alpinus	Mammals	C.I.S., Korea, China, India, Southeast Asia	Entire	E
dibbler	Antechinus apicalis	Mammals	Australia	Entire	E
diellia, asplenium-leaved	Diellia erecta	Ferns and Allies	U.S.A. (HI)	Entire	E
dog, African wild	Lycaon pictus	Mammals	Sub-Saharan Africa	Entire	E
dogweed, ashy	Thymophylla tephroleuca	Flowering Plants	U.S.A. (TX)	Entire	E
Dolphin, Chinese River	Lipotes vexillifer	Mammals	China	Entire	E
dolphin, Indus River	Platanista minor	Mammals	Pakistan (Indus R. and tributaries)	Entire	E
dove, cloven-feathered	Drepanoptila holosericea	Birds	Southwest Pacific Ocean_New Caledonia	Entire	E
dove, Grenada gray-fronted	Leptotila rufaxilla wellsi	Birds	West Indies_Grenada	Entire	E
dragonfly, Hine's emerald	Somatochlora hineana	Insects	U.S.A. (IL, IN, OH, WI)	Entire	E
drill	Mandrillus (=Papio) leucophaeus	Mammals	Equatorial West Africa	Entire	E
dropwort, Canby's	Oxypolis canbyi	Flowering Plants	U.S.A. (DE, GA, MD, NC, SC)	Entire	E

List of threatened and endangered species - *continued*

Common name	Scientific name	Group	Historic range	Vertebrate population where endangered or threatened	Status
Duck, Hawaiian (=koloa)	Anas wyvilliana	Birds	U.S.A. (HI)	Entire	E
duck, Laysan	Anas laysanensis	Birds	U.S.A. (HI)	Entire	E
duck, pink-headed	Rhodonessa caryophyllacea	Birds	India	Entire	E
duck, white-winged wood	Cairina scutulata	Birds	India, Malaysia, Indonesia, Thailand	Entire	E
dudleya, Conejo	Dudleya abramsii ssp. parva	Flowering Plants	U.S.A. (CA)	Entire	T
dudleya, marcescent	Dudleya cymosa ssp. marcescens	Flowering Plants	U.S.A. (CA)	Entire	T
dudleya, Santa Clara Valley	Dudleya setchellii	Flowering Plants	U.S.A. (CA)	Entire	E
dudleya, Santa Cruz Island	Dudleya nesiotica	Flowering Plants	U.S.A. (CA)	Entire	T
dudleya, Verity's	Dudleya verityi	Flowering Plants	U.S.A. (CA)	Entire	T
dudleyea, Santa Monica Mountains	Dudleya cymosa ssp. ovatifolia	Flowering Plants	U.S.A. (CA)	Entire	T
dugong	Dugong dugon	Mammals	East Africa to southern Japan, including U.S.A. (Trust Territories)	Entire, except Palau	E
dugong	Dugong dugon	Mammals	East Africa to southern Japan, including U.S.A. (Trust Territories)	Palau	E
duiker, Jentink's	Cephalophus jentinki	Mammals	Sierra Leone, Liberia, Ivory Coast	Entire	E
dwarf-flax, Marin	Hesperolinon congestum	Flowering Plants	U.S.A. (CA)	Entire	T
eagle, bald	Haliaeetus leucocephalus	Birds	North America south to northern Mexico	U.S.A., conterminous (lower 48) States.	T
eagle, Greenland white-tailed	Haliaeetus albicilla groenlandicus	Birds	Greenland and adjacent Atlantic islands	Entire	E
eagle, harpy	Harpia harpyja	Birds	Mexico south to Argentina	Entire	E
eagle, Madagascar sea	Haliaeetus vociferoides	Birds	Madagascar	Entire	E
eagle, Madagascar serpent	Eutriorchis astur	Birds	Madagascar	Entire	E
eagle, Philippine	Pithecophaga jefferyi	Birds	Philippines	Entire	E
eagle, Spanish imperial	Aquila heliaca adalberti	Birds	Spain, Morocco, Algeria	Entire	E
egret, Chinese	Egretta eulophotes	Birds	China, Korea	Entire	E
eider, spectacled	Somateria fischeri	Birds	U.S.A. (AK), Russia	Entire	T
Eider, Steller's	Polysticta stelleri	Birds	U.S.A. (AK), Russia, winters to Scandanavia	U.S.A. (AK breeding population only)	T
eland, western giant	Taurotragus derbianus derbianus	Mammals	Senegal to Ivory Coast	Entire	E
Elepaio, Oahu	Chasiempis sandwichensis ibidis	Birds	U.S.A. (HI)	Entire	E
elephant, African	Loxodonta africana	Mammals	Africa	Entire	T
elephant, Asian	Elephas maximus	Mammals	South-central and southeastern Asia	Entire	E
elimia, lacy (snail)	Elimia crenatella	Snails	U.S.A. (AL)	Entire	T
elktoe, Appalachian	Alasmidonta raveneliana	Clams	U.S.A. (NC, TN)	Entire	E
elktoe, Cumberland	Alasmidonta atropurpurea	Clams	U.S.A. (KY, TN)		
Entire	E				
erubia	Solanum drymophilum	Flowering Plants	U.S.A. (PR)	Entire	E
evening-primrose, Antioch Dunes	Oenothera deltoides ssp. howellii	Flowering Plants	U.S.A. (CA)	Entire	E
evening-primrose, Eureka Valley	Oenothera avita ssp. eurekensis	Flowering Plants	U.S.A. (CA)	Entire	E
evening-primrose, San Benito	Camissonia benitensis	Flowering Plants	U.S.A. (CA)	Entire	T
fairy shrimp, Conservancy	Branchinecta conservatio	Crustaceans	U.S.A. (CA)	Entire	E
fairy shrimp, longhorn	Branchinecta longiantenna	Crustaceans	U.S.A. (CA)	Entire	E
fairy shrimp, Riverside	Streptocephalus woottoni	Crustaceans	U.S.A. (CA)	Entire	E
fairy shrimp, San Diego	Branchinecta sandiegonensis	Crustaceans	U.S.A. (CA)	Entire	E
fairy shrimp, vernal pool	Branchinecta lynchi	Crustaceans	U.S.A. (CA, OR)	Entire	T

List of threatened and endangered species - *continued*

Common name	Scientific name	Group	Historic range	Vertebrate population where endangered or threatened	Status
falcon, Eurasian peregrine	Falco peregrinus peregrinus	Birds	Europe, Eurasia south to Africa and Mideast	Entire	E
falcon, northern aplomado	Falco femoralis septentrionalis	Birds	U.S.A. (AZ, NM, TX), Mexico, Guatemala	Entire	E
fanshell	Cyprogenia stegaria	Clams	U.S.A. (AL, IL, IN, KY, OH, PA, TN, VA, WV)	Entire	E
fatmucket, Arkansas	Lampsilis powelli	Clams	U.S.A. (AR)	Entire	T
fern, Alabama streak-sorus	Thelypteris pilosa var. alabamensis	Ferns and Allies	U.S.A. (AL)	Entire	T
fern, Aleutian shield	Polystichum aleuticum	Ferns and Allies	U.S.A. (AK)	Entire	E
fern, American hart's-tongue	Asplenium scolopendrium var. americanum	Ferns and Allies	U.S.A. (AL, MI, NY, TN), Canada (Ont.)	Entire	T
fern, Elfin tree	Cyathea dryopteroides	Ferns and Allies	U.S.A. (PR)	Entire	E
fern, pendant kihi	Adenophorus periens	Ferns and Allies	U.S.A. (HI)	Entire	E
ferret, black-footed	Mustela nigripes	Mammals	Western U.S.A., western Canada	Entire, except where listed as an experimental population below	E
ferret, black-footed	Mustela nigripes	Mammals	Western U.S.A., western Canada	U.S.A. (specific portions of AZ, CO, MT, SD, UT, and WY, see 17.84(g))	EXPN
fiddleneck, large-flowered	Amsinckia grandiflora	Flowering Plants	U.S.A. (CA)	Entire	E
finch, Laysan (honeycreeper)	Telespyza cantans	Birds	U.S.A. (HI)	Entire	E
finch, Nihoa (honeycreeper)	Telespyza ultima	Birds	U.S.A. (HI)	Entire	E
Fir, Guatemalan (=pinabete)	Abies guatemalensis	Conifers and Cycads	Mexico, Guatemala, Honduras, El Salvador	Entire	T
flannelbush, Mexican	Fremontodendron mexicanum	Flowering Plants	U.S.A. (CA), Mexico	Entire	E
flannelbush, Pine Hill	Fremontodendron californicum ssp. decumbens	Flowering Plants	U.S.A. (CA)		
Entire	E				
fleabane, Zuni	Erigeron rhizomatus	Flowering Plants	U.S.A. (NM)	Entire	T
fly, Delhi Sands flower-loving	Rhaphiomidas terminatus abdominalis	Insects	U.S.A. (CA)	Entire	E
flycatcher, Euler's	Empidonax euleri johnstonei	Birds	West Indies_Grenada	Entire	E
flycatcher, Seychelles paradise	Terpsiphone corvina	Birds	Indian Ocean_Seychelles	Entire	E
flycatcher, southwestern willow	Empidonax traillii extimus	Birds	U.S.A. (AZ, CA, CO, NM, TX, UT), Mexico	Entire	E
flycatcher, Tahiti	Pomarea nigra	Birds	South Pacific Ocean_Tahiti	Entire	E
fody, Mauritius	Foudia rubra	Birds	Indian Ocean_Mauritius	Entire	E
fody, Rodrigues	Foudia flavicans	Birds	Indian Ocean_Rodrigues Island (Mauritius)	Entire	E
fody, Seychelles (weaver-finch)	Foudia sechellarum	Birds	Indian Ocean_Seychelles	Entire	E
four-o'clock, MacFarlane's	Mirabilis macfarlanei	Flowering Plants	U.S.A. (ID, OR)	Entire	T
fox, northern swift	Vulpes velox hebes	Mammals	U.S.A. (northern plains), Canada	Canada	E
fox, San Joaquin kit	Vulpes macrotis mutica	Mammals	U.S.A. (CA)	Entire	E
Fox, San Miguel Island	Urocyon littoralis littoralis	Mammals	U.S.A. (CA)	Entire	E
Fox, Santa Catalina Island	Urocyon littoralis catalinae	Mammals	U.S.A. (CA)	Entire	E
Fox, Santa Cruz Island	Urocyon littoralis santacruzae	Mammals	U.S.A. (CA)	Entire	E
Fox, Santa Rosa Island	Urocyon littoralis santarosae	Mammals	U.S.A. (CA)	Entire	E
fox, Simien	Canis simensis	Mammals	Ethiopia	Entire	E
francolin, Djibouti	Francolinus ochropectus	Birds	Djibouti	Entire	E
frankenia, Johnston's	Frankenia johnstonii	Flowering Plants	U.S.A. (TX), Mexico (Nuevo Leon)	Entire	E
freira	Pterodroma madeira	Birds	Atlantic Ocean_Madeira Island	Entire	E
frigatebird, Andrew's	Fregata andrewsi	Birds	East Indian Ocean	Entire	E
fringe-tree, pygmy	Chionanthus pygmaeus	Flowering Plants	U.S.A. (FL)	Entire	E

List of threatened and endangered species - *continued*

Common name	Scientific name	Group	Historic range	Vertebrate population where endangered or threatened	Status
fringepod, Santa Cruz Island	Thysanocarpus conchuliferus	Flowering Plants	U.S.A. (CA)	Entire	E
Fritillary, Gentner's	Fritillaria gentneri	Flowering Plants	U.S.A. (OR)	Entire	E
frog, California red-legged	Rana aurora draytonii	Amphibians	U.S.A. (CA), Mexico.	Entire (excluding Del Norte, Humboldt, Trinity, & Mendocino Cos., CA; Glenn, Lake, & Sonoma Cos., CA, west of the Central Valley Hydrologic Basin; Sonoma & Marin Cos., CA, west & north of San Francisco Bay drainages and Walker Creek drainage; and NV	T
frog, Chiricahua leopard	Rana chiricahuensis	Amphibians	U.S.A. (AZ, NM), Mexico	Entire	T
Frog, Goliath	Conraua goliath	Amphibians	Cameroon, Equatorial Guinea, Gabon	Entire	T
frog, Israel painted	Discoglossus nigriventer	Amphibians	Israel	Entire	E
Frog, Mississippi gopher	Rana capito sevosa	Amphibians	U.S.A. (AL, FL, LA, MS)	Wherever found west of Mobile and Tombigbee Rivers in AL, MS, and LA	E
frog, mountain yellow-legged	Rana muscosa	Amphibians	U.S.A. (CA, NV) including San Diego, Orange, Riverside, San Bernardino, and Los Angeles Counties	U.S.A., southern California DPS	E
frog, Panamanian golden	Atelopus varius zeteki	Amphibians	Panama	Entire	E
frog, Stephen Island	Leiopelma hamiltoni	Amphibians	New Zealand	Entire	E
gambusia, Big Bend	Gambusia gaigei	Fishes	U.S.A. (TX)	Entire	E
gambusia, Clear Creek	Gambusia heterochir	Fishes	U.S.A. (TX)	Entire	E
gambusia, Pecos	Gambusia nobilis	Fishes	U.S.A. (NM, TX)	Entire	E
gambusia, San Marcos	Gambusia georgei	Fishes	U.S.A. (TX)	Entire	E
gardenia (=Na'u), Hawaiian	Gardenia brighamii	Flowering Plants	U.S.A. (HI)	Entire	E
gavial	Gavialis gangeticus	Reptiles	Pakistan, Burma, Bangladesh, India, Nepal	Entire	E
gazelle, Arabian	Gazella gazella	Mammals	Arabian Peninsula, Palestine, Sinai	Entire	E
gazelle, Clark's	Ammodorcas clarkei	Mammals	Somalia, Ethiopia	Entire	E
gazelle, Mhorr	Gazella dama mhorr	Mammals	Morocco	Entire	E
gazelle, Moroccan	Gazella dorcas massaesyla	Mammals	Morocco, Algeria, Tunisia	Entire	E
gazelle, mountain (=Cuvier's)	Gazella cuvieri	Mammals	Morocco, Algeria, Tunisia	Entire	E
gazelle, Pelzeln's	Gazella dorcas pelzelni	Mammals	Somalia	Entire	E
gazelle, Rio de Oro Dama	Gazella dama lozanoi	Mammals	Western Sahara	Entire	E
gazelle, sand	Gazella subgutturosa marica	Mammals	Jordan, Arabian Peninsula	Entire	E
gazelle, Saudi Arabian	Gazella dorcas saudiya	Mammals	Israel, Iraq, Jordan, Syria, Arabian Peninsula	Entire	E
gazelle, slender-horned	Gazella leptoceros	Mammals	Sudan, Egypt, Algeria, Libya	Entire	E
gecko, day	Phelsuma edwardnewtoni	Reptiles	Indian Ocean_Mauritius	Entire	E
gecko, Monito	Sphaerodactylus micropithecus	Reptiles	U.S.A. (PR)	Entire	E
gecko, Round Island day	Phelsuma guentheri	Reptiles	Indian Ocean_Mauritius	Entire	E
gecko, Serpent Island	Cyrtodactylus serpensinsula	Reptiles	Indian Ocean_Mauritius	Entire	T
geranium, Hawaiian red-flowered	Geranium arboreum	Flowering Plants	U.S.A. (HI)	Entire	E
gerardia, sandplain	Agalinis acuta	Flowering Plants	U.S.A. (CT, MA, MD, NY, RI)	Entire	E
gibbons	Hylobates spp. (including Nomascus)	Mammals	China, India, Southeast Asia	Entire	E
gilia, Hoffmann's slender-flowered	Gilia tenuiflora ssp. hoffmannii	Flowering Plants	U.S.A. (CA)	Entire	E

List of threatened and endangered species - *continued*

Common name	Scientific name	Group	Historic range	Vertebrate population where endangered or threatened	Status
gilia, Monterey	Gilia tenuiflora ssp. arenaria	Flowering Plants	U.S.A. (CA)	Entire	E
gnatcatcher, coastal California	Polioptila californica californica	Birds	U.S.A. (CA), Mexico	Entire	T
goby, tidewater	Eucyclogobius newberryi	Fishes	U.S.A. (CA)	Entire	E
goetzea, beautiful	Goetzea elegans	Flowering Plants	U.S.A. (PR)	Entire	E
goldenrod, Blue Ridge	Solidago spithamaea	Flowering Plants	U.S.A. (NC, TN)	Entire	T
goldenrod, Houghton's	Solidago houghtonii	Flowering Plants	U.S.A. (MI), Canada (Ont.)	Entire	T
goldenrod, Short's	Solidago shortii	Flowering Plants	U.S.A. (KY)	Entire	E
goldenrod, white-haired	Solidago albopilosa	Flowering Plants	U.S.A. (KY)	Entire	T
goldfields, Burke's	Lasthenia burkei	Flowering Plants	U.S.A. (CA)	Entire	E
goldfields, Contra Costa	Lasthenia conjugens	Flowering Plants	U.S.A. (CA)	Entire	E
goose, Hawaiian	Branta (=Nesochen) sandvicensis	Birds	U.S.A. (HI)	Entire	E
gooseberry, Miccosukee	Ribes echinellum	Flowering Plants	U.S.A. (FL, SC)	Entire	T
Goral	Naemorhedus goral	Mammals	East Asia	Entire	E
gorilla	Gorilla gorilla	Mammals	Central and western Africa	Entire	E
goshawk, Christmas Island	Accipiter fasciatus natalis	Birds	Indian Ocean_Christmas Island	Entire	E
gourd, Okeechobee	Cucurbita okeechobeensis ssp. okeechobeensis	Flowering Plants	U.S.A. (FL)	Entire	E
Grackle, slender-billed	Quiscalus palustris	Birds	Mexico	Entire	E
grass, Colusa	Neostapfia colusana	Flowering Plants	U.S.A. (CA)	Entire	T
grass, Eureka Dune	Swallenia alexandrae	Flowering Plants	U.S.A. (CA)	Entire	E
grass, Solano	Tuctoria mucronata	Flowering Plants	U.S.A. (CA)	Entire	E
grass, Tennessee yellow-eyed	Xyris tennesseensis	Flowering Plants	U.S.A. (AL, GA, TN)	Entire	E
grasshopper, Zayante band-winged	Trimerotropis infantilis	Insects	U.S.A. (CA)	Entire	E
grasswren, Eyrean (flycatcher)	Amytornis goyderi	Birds	Australia	Entire	E
Grebe, Alaotra	Tachybaptus rufolavatus	Birds	Madagascar	Entire	E
grebe, Atitlan	Podilymbus gigas	Birds	Guatemala	Entire	E
greenshank, Nordmann's	Tringa guttifer	Birds	Russia, Japan, south to Malaya, Borneo	Entire	E
ground beetle, [unnamed]	Rhadine exilis	Insects	U.S.A. (TX)	Entire	E
ground beetle, [unnamed]	Rhadine infernalis	Insects	U.S.A. (TX)	Entire	E
ground-plum, Guthrie's (=Pyne's)	Astragalus bibullatus	Flowering Plants	U.S.A. (TN)	Entire	E
groundsel, San Francisco Peaks	Senecio franciscanus	Flowering Plants	U.S.A. (AZ)	Entire	T
guajon	Eleutherodactylus cooki	Amphibians	U.S.A. (PR)	Entire	T
guan, horned	Oreophasis derbianus	Birds	Guatemala, Mexico	Entire	E
guan, white-winged	Penelope albipennis	Birds	Peru	Entire	E
guineafowl, white-breasted	Agelastes meleagrides	Birds	West Africa	Entire	T
gull, Audouin's	Larus audouinii	Birds	Mediterranean Sea	Entire	E
gull, relict	Larus relictus	Birds	India, China	Entire	E
gumplant, Ash Meadows	Grindelia fraxino-pratensis	Flowering Plants	U.S.A. (CA, NV)	Entire	T
haha	Cyanea acuminata	Flowering Plants	U.S.A. (HI)	Entire	E
haha	Cyanea asarifolia	Flowering Plants	U.S.A. (HI)	Entire	E
haha	Cyanea copelandii ssp. copelandii	Flowering Plants	U.S.A. (HI)	Entire	E
haha	Cyanea copelandii ssp. haleakalaensis	Flowering Plants	U.S.A. (HI)	Entire	E
haha	Cyanea dunbarii	Flowering Plants	U.S.A. (HI)	Entire	E
haha	Cyanea glabra	Flowering Plants	U.S.A. (HI)	Entire	E
haha	Cyanea grimesiana ssp. grimesiana	Flowering Plants	U.S.A. (HI)	Entire	E
haha	Cyanea grimesiana ssp. obatae	Flowering Plants	U.S.A. (HI)	Entire	E
haha	Cyanea hamatiflora carlsonii	Flowering Plants	U.S.A. (HI)	Entire	E

List of threatened and endangered species - *continued*

Common name	Scientific name	Group	Historic range	Vertebrate population where endangered or threatened	Status
haha	Cyanea hamatiflora ssp. hamatiflora	Flowering Plants	U.S.A. (HI)	Entire	E
haha	Cyanea humboldtiana	Flowering Plants	U.S.A. (HI)	Entire	E
haha	Cyanea koolauensis	Flowering Plants	U.S.A. (HI)	Entire	E
haha	Cyanea lobata	Flowering Plants	U.S.A. (HI)	Entire	E
haha	Cyanea longiflora	Flowering Plants	U.S.A. (HI)	Entire	E
haha	Cyanea macrostegia ssp. gibsonii	Flowering Plants	U.S.A. (HI)	Entire	E
haha	Cyanea mannii	Flowering Plants	U.S.A. (HI)	Entire	E
haha	Cyanea mceldowneyi	Flowering Plants	U.S.A. (HI)	Entire	E
haha	Cyanea pinnatifida	Flowering Plants	U.S.A. (HI)	Entire	E
haha	Cyanea platyphylla	Flowering Plants	U.S.A. (HI)	Entire	E
haha	Cyanea procera	Flowering Plants	U.S.A. (HI)	Entire	E
haha	Cyanea recta	Flowering Plants	U.S.A. (HI)	Entire	T
haha	Cyanea remyi	Flowering Plants	U.S.A. (HI)	Entire	E
haha	Cyanea shipmannii	Flowering Plants	U.S.A. (HI)	Entire	E
haha	Cyanea st-johnii	Flowering Plants	U.S.A. (HI)	Entire	E
haha	Cyanea stictophylla	Flowering Plants	U.S.A. (HI)	Entire	E
haha	Cyanea superba	Flowering Plants	U.S.A. (HI)	Entire	E
haha	Cyanea truncata	Flowering Plants	U.S.A. (HI)	Entire	E
haha	Cyanea undulata	Flowering Plants	U.S.A. (HI)	Entire	E
hala pepe	Pleomele hawaiiensis	Flowering Plants	U.S.A. (HI)	Entire	E
hare, hispid	Caprolagus hispidus	Mammals	India, Nepal, Bhutan	Entire	E
harebells, Avon Park	Crotalaria avonensis	Flowering Plants	U.S.A. (FL)	Entire	E
harperella	Ptilimnium nodosum	Flowering Plants	U.S.A. (AL, AR, GA, MD, NC, SC, WV)	Entire	E
hartebeest, Swayne's	Alcelaphus buselaphus swaynei	Mammals	Ethiopia, Somalia	Entire	E
hartebeest, Tora	Alcelaphus buselaphus tora	Mammals	Ethiopia, Sudan, Egypt	Entire	E
harvestman, Bee Creek Cave	Texella reddelli	Arachnids	U.S.A. (TX)	Entire	E
harvestman, Bone Cave	Texella reyesi	Arachnids	U.S.A. (TX)	Entire	E
Harvestman, Cokendolpher Cave	Texella cokendolpheri	Arachnids	U.S.A. (TX).	Entire	E
hau kuahiwi	Hibiscadelphus giffardianus	Flowering Plants	U.S.A. (HI)	Entire	E
hau kuahiwi	Hibiscadelphus hualalaiensis	Flowering Plants	U.S.A. (HI)	Entire	E
hau kuahiwi	Hibiscadelphus woodii	Flowering Plants	U.S.A. (HI)	Entire	E
hawk, Galapagos	Buteo galapagoensis	Birds	Ecuador (Galapagos Islands)	Entire	E
Hawk, Hawaiian (='Io)	Buteo solitarius	Birds	U.S.A. (HI)	Entire	E
hawk, Puerto Rican broad-winged	Buteo platypterus brunnescens	Birds	U.S.A. (PR)	Entire	E
hawk, Puerto Rican sharp-shinned	Accipiter striatus venator	Birds	U.S.A. (PR)	Entire	E
ha'iwale	Cyrtandra crenata	Flowering Plants	U.S.A. (HI)	Entire	E
ha'iwale	Cyrtandra dentata	Flowering Plants	U.S.A. (HI)	Entire	E
ha'iwale	Cyrtandra giffardii	Flowering Plants	U.S.A. (HI)	Entire	E
ha'iwale	Cyrtandra limahuliensis	Flowering Plants	U.S.A. (HI)	Entire	T
ha'iwale	Cyrtandra munroi	Flowering Plants	U.S.A. (HI)	Entire	E
ha'iwale	Cyrtandra polyantha	Flowering Plants	U.S.A. (HI)	Entire	E
ha'iwale	Cyrtandra subumbellata	Flowering Plants	U.S.A. (HI)	Entire	E
ha'iwale	Cyrtandra tintinnabula	Flowering Plants	U.S.A. (HI)	Entire	E
ha'iwale	Cyrtandra viridiflora	Flowering Plants	U.S.A. (HI)	Entire	E
heartleaf, dwarf-flowered	Hexastylis naniflora	Flowering Plants	U.S.A. (NC, SC)	Entire	T
heather, mountain golden	Hudsonia montana	Flowering Plants	U.S.A. (NC)	Entire	T
heau	Exocarpos luteolus	Flowering Plants	U.S.A. (HI)	Entire	E
hedyotis, Na Pali beach	Hedyotis st.-johnii	Flowering Plants	U.S.A. (HI)	Entire	E
heelsplitter, Alabama (=inflated)	Potamilus inflatus	Clams	U.S.A. (AL, LA, MS)	Entire	T
heelsplitter, Carolina	Lasmigona decorata	Clams	U.S.A. (NC, SC)	Entire	E
hermit, hook-billed (hummingbird)	Ramphodon (=Glaucis) dohrnii	Birds	Brazil	Entire	E
hibiscus, Clay's	Hibiscus clayi	Flowering Plants	U.S.A. (HI)	Entire	E

List of threatened and endangered species - *continued*

Common name	Scientific name	Group	Historic range	Vertebrate population where endangered or threatened	Status
higgins eye (pearlymussel)	Lampsilis higginsii	Clams	U.S.A. (IA, IL, MN, MO, NE, WI)	Entire	E
higuero de sierra	Crescentia portoricensis	Flowering Plants	U.S.A. (PR)	Entire	E
hog, pygmy	Sus salvanius	Mammals	India, Nepal, Bhutan, Sikkim	Entire	E
holei	Ochrosia kilaueaensis	Flowering Plants	U.S.A. (HI)	Entire	E
holly, Cook's	Ilex cookii	Flowering Plants	U.S.A. (PR)	Entire	E
honeycreeper, crested	Palmeria dolei	Birds	U.S.A. (HI)	Entire	E
honeyeater, helmeted	Lichenostomus melanops cassidix (=Meliphaga c.)	Birds	Australia	Entire	E
Honohono	Haplostachys haplostachya	Flowering Plants	U.S.A. (HI)	Entire	E
hornbill, helmeted	Buceros (=Rhinoplax) vigil	Birds	Thailand, Malaysia	Entire	E
horse, Przewalski's	Equus przewalskii	Mammals	Mongolia, China	Entire	E
howellia, water	Howellia aquatilis	Flowering Plants	U.S.A. (CA, ID, MT, OR, WA)	Entire	T
huemul, north Andean	Hippocamelus antisensis	Mammals	Ecuador, Peru, Chile, Bolivia, Argentina	Entire	E
huemul, south Andean	Hippocamelus bisulcus	Mammals	Chile, Argentina	Entire	E
hutia, Cabrera's	Capromys angelcabrerai	Mammals	Cuba	Entire	E
hutia, dwarf	Capromys nana	Mammals	Cuba	Entire	E
hutia, large-eared	Capromys auritus	Mammals	Cuba	Entire	E
hutia, little earth	Capromys sanfelipensis	Mammals	Cuba	Entire	E
hyena, Barbary	Hyaena hyaena barbara	Mammals	Morocco, Algeria, Tunisia	Entire	E
Hyena, brown	Parahyaena (=Hyaena) brunnea	Mammals	Southern Africa	Entire	E
hypericum, highlands scrub	Hypericum cumulicola	Flowering Plants	U.S.A. (FL)	Entire	E
Iagu, Hayun (=(Guam), Tronkon guafi (Rota))	Serianthes nelsonii	Flowering Plants	Western Pacific Ocean-U.S.A. (GU, MP-Rota)	Entire	E
ibex, Pyrenean	Capra pyrenaica pyrenaica	Mammals	Spain	Entire	E
ibex, Walia	Capra walie	Mammals	Ethiopia	Entire	E
ibis, Japanese crested	Nipponia nippon	Birds	China, Japan, Russia, Korea	Entire	E
ibis, northern bald	Geronticus eremita	Birds	Southern Europe, southwestern Asia, northern Africa	Entire	E
iguana, Acklins ground	Cyclura rileyi nuchalis	Reptiles	West Indies_Bahamas	Entire	T
iguana, Allen's Cay	Cyclura cychlura inornata	Reptiles	West Indies_Bahamas	Entire	T
iguana, Andros Island ground	Cyclura cychlura cychlura	Reptiles	West Indies_Bahamas	Entire	T
iguana, Anegada ground	Cyclura pinguis	Reptiles	West Indies_British Virgin Islands (Anegada Island)	Entire	E
iguana, Barrington land	Conolophus pallidus	Reptiles	Ecuador (Galapagos Islands)	Entire	E
iguana, Cayman Brac ground	Cyclura nubila caymanensis	Reptiles	West Indies_Cayman Islands	Entire	T
iguana, Cuban ground	Cyclura nubila nubila	Reptiles	Cuba	Entire	T
iguana, Exuma Island	Cyclura cychlura figginsi	Reptiles	West Indies_Bahamas	Entire	T
iguana, Fiji banded	Brachylophus fasciatus	Reptiles	Pacific_Fiji, Tonga	Entire	E
iguana, Fiji crested	Brachylophus vitiensis	Reptiles	Pacific_Fiji	Entire	E
iguana, Grand Cayman ground	Cyclura nubila lewisi	Reptiles	West Indies_Cayman Islands	Entire	E
iguana, Jamaican	Cyclura collei	Reptiles	West Indies_Jamaica	Entire	E
iguana, Mayaguana	Cyclura carinata bartschi	Reptiles	West Indies_Bahamas	Entire	T
Iguana, Mona ground	Cyclura cornuta stejnegeri	Reptiles	U.S.A. (PR_Mona Island)	Entire	T
iguana, Turks and Caicos	Cyclura carinata carinata	Reptiles	West Indies_Turks and Caicos Islands	Entire	T
iguana, Watling Island ground	Cyclura rileyi rileyi	Reptiles	West Indies_Bahamas	Entire	E

List of threatened and endangered species - *continued*

Common name	Scientific name	Group	Historic range	Vertebrate population where endangered or threatened	Status
iguana, White Cay ground	Cyclura rileyi cristata	Reptiles	West Indies_Bahamas	Entire	T
ihi'ihi	Marsilea villosa	Ferns and Allies	U.S.A. (HI)	Entire	E
iliau, dwarf	Wilkesia hobdyi	Flowering Plants	U.S.A. (HI)	Entire	E
impala, black-faced	Aepyceros melampus petersi	Mammals	Namibia, Angola	Entire	E
indian paintbrush, San Clemente Island	Castilleja grisea	Flowering Plants	U.S.A. (CA)	Entire	E
indri	Indri indri (entire genus)	Mammals	Malagasy Republic (=Madagascar)	Entire	E
ipomopsis, Holy Ghost	Ipomopsis sancti-spiritus	Flowering Plants	U.S.A. (NM)	Entire	E
iris, dwarf lake	Iris lacustris	Flowering Plants	U.S.A. (MI, WI), Canada (Ont.)	Entire	T
irisette, white	Sisyrinchium dichotomum	Flowering Plants	U.S.A. (NC)	Entire	E
ischaemum, Hilo	Ischaemum byrone	Flowering Plants	U.S.A. (HI)	Entire	E
isopod, Lee County cave	Lirceus usdagalun	Crustaceans	U.S.A. (VA)	Entire	E
isopod, Madison Cave	Antrolana lira	Crustaceans	U.S.A. (VA)	Entire	T
isopod, Socorro	Thermosphaeroma thermophilus	Crustaceans	U.S.A. (NM)	Entire	E
ivesia, Ash Meadows	Ivesia kingii var. eremica	Flowering Plants	U.S.A. (NV)	Entire	T
jacquemontia, beach	Jacquemontia reclinata	Flowering Plants	U.S.A. (FL)	Entire	E
jaguar	Panthera onca	Mammals	U.S.A. (AZ, CA, LA, NM, TX), Mexico, Central and South America	Entire	E
jaguarundi, Guatemalan	Herpailurus (=Felis) yagouaroundi fossata	Mammals	Mexico, Nicaragua	Entire	E
jaguarundi, Gulf Coast	Herpailurus (=Felis) yagouaroundi cacomitli	Mammals	U.S.A. (TX), Mexico	Entire	E
jaguarundi, Panamanian	Herpailurus (=Felis) yagouaroundi panamensis	Mammals	Nicaragua, Costa Rica, Panama	Entire	E
Jaguarundi, Sinaloan	Herpailurus (=Felis) yagouaroundi tolteca	Mammals	U.S.A. (AZ), Mexico	Entire	E
jatropha, Costa Rican	Jatropha costaricensis	Flowering Plants	Costa Rica	Entire	E
jay, Florida scrub	Aphelocoma coerulescens	Birds	U.S.A. (FL)	Entire	T
jewelflower, California	Caulanthus californicus	Flowering Plants	U.S.A. (CA)	Entire	E
jewelflower, Metcalf Canyon	Streptanthus albidus ssp. albidus	Flowering Plants	U.S.A. (CA)	Entire	E
jewelflower, Tiburon	Streptanthus niger	Flowering Plants	U.S.A. (CA)	Entire	E
joint-vetch, sensitive	Aeschynomene virginica	Flowering Plants	U.S.A. (DE, MD, NC, NJ, PA, VA)	Entire	T
kagu	Rhynochetos jubatus	Birds	South Pacific Ocean_New Caledonia	Entire	E
kakapo	Strigops habroptilus	Birds	New Zealand	Entire	E
kamakahala	Labordia cyrtandrae	Flowering Plants	U.S.A. (HI)	Entire	E
kamakahala	Labordia lydgatei	Flowering Plants	U.S.A. (HI)	Entire	E
kamakahala	Labordia tinifolia var. lanaiensis	Flowering Plants	U.S.A. (HI)	Entire	E
kamakahala	Labordia tinifolia var. wahiawaensis	Flowering Plants	U.S.A. (HI)	Entire	E
kamakahala	Labordia triflora	Flowering Plants	U.S.A. (HI)	Entire	E
Kamanomano	Cenchrus agrimonioides	Flowering Plants	U.S.A. (HI)	Entire	E
kangaroo rat, Fresno	Dipodomys nitratoides exilis	Mammals	U.S.A. (CA)	Entire	E
kangaroo rat, giant	Dipodomys ingens	Mammals	U.S.A. (CA)	Entire	E
kangaroo rat, Morro Bay	Dipodomys heermanni morroensis	Mammals	U.S.A. (CA)	Entire	E
kangaroo rat, San Bernardino Merriam's	Dipodomys merriami parvus	Mammals	U.S.A. (CA)	Entire	E
kangaroo rat, Stephens'	Dipodomys stephensi (incl. D. cascus)	Mammals	U.S.A. (CA)	Entire	E
kangaroo rat, Tipton	Dipodomys nitratoides nitratoides	Mammals	U.S.A. (CA)	Entire	E
kangaroo, Tasmanian forester	Macropus giganteus tasmaniensis	Mammals	Australia (Tasmania)	Entire	E
kauai hau kuahiwi	Hibiscadelphus distans	Flowering Plants	U.S.A. (HI)	Entire	E
kauila	Colubrina oppositifolia	Flowering Plants	U.S.A. (HI)	Entire	E
kaulu	Pteralyxia kauaiensis	Flowering Plants	U.S.A. (HI)	Entire	E

List of threatened and endangered species - *continued*

Common name	Scientific name	Group	Historic range	Vertebrate population where endangered or threatened	Status
kestrel, Mauritius	Falco punctatus	Birds	Indian Ocean_Mauritius	Entire	E
kestrel, Seychelles	Falco araea	Birds	Indian Ocean_Seychelles Islands	Entire	E
Kidneyshell, triangular	Ptychobranchus greenii	Clams	U.S.A. (AL, GA, TN)	Entire	E
kingfisher, Guam Micronesian	Halcyon cinnamomina cinnamomina	Birds	West Pacific Ocean_U.S.A. (Guam)	Entire	E
kio'ele	Hedyotis coriacea	Flowering Plants	U.S.A. (HI)	Entire	E
kiponapona	Phyllostegia racemosa	Flowering Plants	U.S.A. (HI)	Entire	E
kite, Cuba hook-billed	Chondrohierax uncinatus wilsonii	Birds	West Indies_Cuba	Entire	E
kite, Everglade snail	Rostrhamus sociabilis plumbeus	Birds	U.S.A. (FL), Cuba	U.S.A. (FL)	E
kite, Grenada hook-billed	Chondrohierax uncinatus mirus	Birds	West Indies_Grenada	Entire	E
Koala	Phascolarctos cinereus	Mammals	Australia	Entire	T
kohe malama malama o kanaloa	Kanaloa kahoolawensis	Flowering Plants	U.S.A. (HI)	Entire	E
kokako (wattlebird)	Callaeas cinerea	Birds	New Zealand	Entire	E
koki'o	Kokia drynarioides	Flowering Plants	U.S.A. (HI)	Entire	E
Koki'o	Kokia kauaiensis	Flowering Plants	U.S.A. (HI)	Entire	E
koki'o ke'oke'o	Hibiscus arnottianus ssp. immaculatus	Flowering Plants	U.S.A. (HI)	Entire	E
koki'o ke'oke'o	Hibiscus waimeae ssp. hannerae	Flowering Plants	U.S.A. (HI)	Entire	E
koki'o, Cooke's	Kokia cookei	Flowering Plants	U.S.A. (HI)	Entire	E
kolea	Myrsine juddii	Flowering Plants	U.S.A. (HI)	Entire	E
kolea	Myrsine linearifolia	Flowering Plants	U.S.A. (HI)	Entire	T
Kopa	Hedyotis schlechtendahliana var. remyi	Flowering Plants	U.S.A. (HI)	Entire	E
kouprey	Bos sauveli	Mammals	Vietnam, Laos, Cambodia, Thailand	Entire	E
ko'oko'olau	Bidens micrantha ssp. kalealaha	Flowering Plants	U.S.A. (HI)	Entire	E
ko'oko'olau	Bidens wiebkei	Flowering Plants	U.S.A. (HI)	Entire	E
ko'oloa'ula	Abutilon menziesii	Flowering Plants	U.S.A. (HI)	Entire	E
kuahiwi laukahi	Plantago hawaiensis	Flowering Plants	U.S.A. (HI)	Entire	E
kuahiwi laukahi	Plantago princeps	Flowering Plants	U.S.A. (HI)	Entire	E
kuawawaenohu	Alsinidendron lychnoides	Flowering Plants	U.S.A. (HI)	Entire	E
kula wahine noho	Isodendrion pyrifolium	Flowering Plants	U.S.A. (HI)	Entire	E
kulu'i	Nototrichium humile	Flowering Plants	U.S.A. (HI)	Entire	E
ladies'-tresses, Canelo Hills	Spiranthes delitescens	Flowering Plants	U.S.A. (AZ)	Entire	E
ladies'-tresses, Navasota	Spiranthes parksii	Flowering Plants	U.S.A. (TX)	Entire	E
ladies'-tresses, Ute	Spiranthes diluvialis	Flowering Plants	U.S.A. (CO, ID, MT, NE, NV, UT, WA, WY)	Entire	T
lampmussel, Alabama	Lampsilis virescens	Clams	U.S.A. (AL, TN)	Entire Range; Except where listed as Experimental Populations	E
lampmussel, Alabama	Lampsilis virescens	Clams	U.S.A. (AL, TN)	U.S.A. (AL;The free-flowing reach of the Tennessee R. from the base of Wilson Dam downstream to the backwaters of Pickwick Reservoir [about 12 RM (19 km)] and the lower 5 RM [8 km] of all tributaries to this reach in Colbert and Lauderdale Cos., see 17.85(a))	EXPN
Langur, capped	Trachypithecus (=Presbytis) pileatus	Mammals	India, Burma, Bangladesh	Entire	E
langur, Douc	Pygathrix nemaeus	Mammals	Cambodia, Laos, Vietnam	Entire	E
langur, Francois'	Trachypithecus (=Presbytis) francoisi	Mammals	China (Kwangsi), Indochina	Entire	E
langur, golden	Trachypithecus (=Presbytis) geei	Mammals	India (Assam), Bhutan	Entire	E

List of threatened and endangered species - continued

Common name	Scientific name	Group	Historic range	Vertebrate population where endangered or threatened	Status
langur, gray (=entellus)	Semnopithecus (=Presbytis) entellus	Mammals	China (Tibet), India, Pakistan, Kashmir, Sri Lanka, Sikkim, Bangladesh	Entire	E
langur, long-tailed	Presbytis potenziani	Mammals	Indonesia	Entire	T
langur, Pagi Island	Nasalis concolor	Mammals	Indonesia	Entire	T
langur, purple-faced	Presbytis senex	Mammals	Sri Lanka	Entire	T
larch, Chilean false	Fitzroya cupressoides	Conifers and Cycads	Chile, Argentina	Entire	T
lark, Raso	Alauda razae	Birds	Atlantic Ocean_Raso Island (Cape Verde)	Entire	E
larkspur, Baker's	Delphinium bakeri	Flowering Plants	U.S.A. (CA)	Entire	E
larkspur, San Clemente Island	Delphinium variegatum ssp. kinkiense	Flowering Plants	U.S.A. (CA)	Entire	E
larkspur, yellow	Delphinium luteum	Flowering Plants	U.S.A. (CA)	Entire	E
lau 'ehu	Panicum niihauense	Flowering Plants	U.S.A. (HI)	Entire	E
Laulihilihi	Schiedea stellarioides	Flowering Plants	U.S.A. (HI)	Entire	E
layia, beach	Layia carnosa	Flowering Plants	U.S.A. (CA)	Entire	E
lead-plant, Crenulate	Amorpha crenulata	Flowering Plants	U.S.A. (FL)	Entire	E
leather flower, Alabama	Clematis socialis	Flowering Plants	U.S.A. (AL)	Entire	E
leather flower, Morefield's	Clematis morefieldii	Flowering Plants	U.S.A. (AL)	Entire	E
lechwe, red	Kobus leche	Mammals	Southern Africa	Entire	T
lemurs	Lemuridae (incl. genera Lemur, Phaner, Hapalemur, Lepilem				
leopard	Panthera pardus	Mammals	Africa, Asia	Wherever found, except where it is listed as Threatened as set forth below	E
leopard	Panthera pardus	Mammals	Africa, Asia	In Africa, in the wild, south of, and including, the following countries: Gabon, Congo, Zaire, Uganda, Kenya	T
leopard, clouded	Neofelis nebulosa	Mammals	Southeastern and south-central Asia, Taiwan	Entire	E
leopard, snow	Uncia (=Panthera) uncia	Mammals	Central Asia	Entire	E
lessingia, San Francisco	Lessingia germanorum (=L.g. var. germanorum)	Flowering Plants	U.S.A. (CA)	Entire	E
lichen, rock gnome	Gymnoderma lineare	Lichens	U.S.A. (NC,TN)	Entire	E
liliwai	Acaena exigua	Flowering Plants	U.S.A. (HI)	Entire	E
lilliput, pale (pearlymussel)	Toxolasma cylindrellus	Clams	U.S.A. (AL, TN)	Entire	E
lily, Minnesota dwarf trout	Erythronium propullans	Flowering Plants	U.S.A. (MN)	Entire	E
lily, Pitkin Marsh	Lilium pardalinum ssp. pitkinense	Flowering Plants	U.S.A. (CA)	Entire	E
lily, Western	Lilium occidentale	Flowering Plants	U.S.A. (OR, CA)	Entire	E
limpet, Banbury Springs	Lanx sp.	Snails	U.S.A. (ID)	Entire	E
linsang, spotted	Prionodon pardicolor	Mammals	Nepal, Assam, Vietnam, Cambodia, Laos, Burma	Entire	E
lion, Asiatic	Panthera leo persica	Mammals	Turkey to India	Entire	E
lioplax, cylindrical (snail)	Lioplax cyclostomaformis	Snails	U.S.A. (AL, GA)	Entire	E
liveforever, Laguna Beach	Dudleya stolonifera	Flowering Plants	U.S.A. (CA)	Entire	T
liveforever, Santa Barbara Island	Dudleya traskiae	Flowering Plants	U.S.A. (CA)	Entire	E
lizard, blunt-nosed leopard	Gambelia silus	Reptiles	U.S.A. (CA)	Entire	E
lizard, Coachella Valley fringe-toed	Uma inornata	Reptiles	U.S.A. (CA)	Entire	T
lizard, Hierro giant	Gallotia simonyi simonyi	Reptiles	Spain (Canary Islands)	Entire	E
lizard, Ibiza wall	Podarcis pityusensis	Reptiles	Spain (Balearic Islands)	Entire	T
lizard, Island night	Xantusia riversiana	Reptiles	U.S.A. (CA)	Entire	T
lizard, Maria Island ground	Cnemidophorus vanzoi	Reptiles	West Indies_St. Lucia (Maria Islands)	Entire	E
lizard, St. Croix ground	Ameiva polops	Reptiles	U.S.A. (VI)	Entire	E
locoweed, Fassett's	Oxytropis campestris var. chartacea	Flowering Plants	U.S.A. (WI)	Entire	T
logperch, Conasauga	Percina jenkinsi	Fishes	U.S.A. (GA, TN)	Entire	E

List of threatened and endangered species - *continued*

Common name	Scientific name	Group	Historic range	Vertebrate population where endangered or threatened	Status
logperch, Roanoke	Percina rex	Fishes	U.S.A. (VA)	Entire	E
lomatium, Cook's	Lomatium cookii	Flowering Plants	U.S.A. (OR)	Entire	E
loosestrife, rough-leaved	Lysimachia asperulaefolia	Flowering Plants	U.S.A. (NC, SC)	Entire	E
loris, lesser slow	Nycticebus pygmaeus	Mammals	Indochina	Entire	T
lousewort, Furbish	Pedicularis furbishiae	Flowering Plants	U.S.A. (ME), Canada (N.B.)	Entire	E
love grass, Fosberg's	Eragrostis fosbergii	Flowering Plants	U.S.A. (HI)	Entire	E
lo'ulu	Pritchardia affinis	Flowering Plants	U.S.A. (HI)	Entire	E
lo'ulu	Pritchardia kaalae	Flowering Plants	U.S.A. (HI)	Entire	E
lo'ulu	Pritchardia munroi	Flowering Plants	U.S.A. (HI)	Entire	E
lo'ulu	Pritchardia napaliensis	Flowering Plants	U.S.A. (HI)	Entire	E
lo'ulu	Pritchardia remota	Flowering Plants	U.S.A. (HI)	Entire	E
lo'ulu	Pritchardia schattaueri	Flowering Plants	U.S.A. (HI)	Entire	E
lo'ulu	Pritchardia viscosa	Flowering Plants	U.S.A. (HI)	Entire	E
lupine, clover	Lupinus tidestromii	Flowering Plants	U.S.A. (CA)	Entire	E
Lupine, Kincaid's	Lupinus sulphureus (=oreganus) ssp. kincaidii (=var. kincaid	Flowering Plants	U.S.A. (OR, WA)	Entire	T
lupine, Nipomo Mesa	Lupinus nipomensis	Flowering Plants	U.S.A. (CA)	Entire	E
lupine, scrub	Lupinus aridorum	Flowering Plants	U.S.A. (FL)	Entire	E
Lynx, Canada	Lynx canadensis	Mammals	U.S.A. (AK, CO, ID, ME, MI, MN, MT, NH, NY, OR, UT, VT, WA, WI, WY), Canada	Entire	T
lynx, Spanish	Felis pardina	Mammals	Spain, Portugal	Entire	E
macaque, Formosan rock	Macaca cyclopis	Mammals	Taiwan	Entire	T
macaque, Japanese	Macaca fuscata	Mammals	Japan (Shikoku, Kyushu and Honshu Islands)	Entire	T
macaque, lion-tailed	Macaca silenus	Mammals	India	Entire	E
macaque, stump-tailed	Macaca arctoides	Mammals	India (Assam) to southern China	Entire	T
macaque, Toque	Macaca sinica	Mammals	Sri Lanka	Entire	T
macaw, glaucous	Anodorhynchus glaucus	Birds	Paraguay, Uruguay, Brazil	Entire	E
macaw, indigo	Anodorhynchus leari	Birds	Brazil	Entire	E
macaw, little blue	Cyanopsitta spixii	Birds	Brazil	Entire	E
madtom, Neosho	Noturus placidus	Fishes	U.S.A. (KS, MO, OK)	Entire	T
madtom, pygmy	Noturus stanauli	Fishes	U.S.A. (TN)	Entire	E
madtom, Scioto	Noturus trautmani	Fishes	U.S.A. (OH)	Entire	E
madtom, smoky	Noturus baileyi	Fishes	U.S.A. (TN)	Entire	E
madtom, smoky	Noturus baileyi	Fishes	U.S.A. (TN)	Tellico River, between the backwaters of the Tellico Reservoir and the Tellico Ranger Station, in Monroe County, Tennessee	EXPN
madtom, yellowfin	Noturus flavipinnis	Fishes	U.S.A. (TN, VA)	N. Fork Holston R., VA, TN; S. Fork Holston R., upstream to Ft. Patrick Henry Dam, TN; Holston R., downstream to John Sevier Detention Lake Dam, TN; and all tributaries thereto	EXPN
madtom, yellowfin	Noturus flavipinnis	Fishes	U.S.A. (TN, VA)	Entire, except where listed as an experimental population below	T
madtom, yellowfin	Noturus flavipinnis	Fishes	U.S.A. (TN, VA)	Tellico River between the backwaters of the Tellico Reservoir and the Tellico Ranger Station, in Monroe County, Tennessee	EXPN
magpie-robin, Seychelles (thrush)	Copsychus sechellarum	Birds	Indian Ocean_Seychelles Islands	Entire	E
mahoe	Alectryon macrococcus	Flowering Plants	U.S.A. (HI)	Entire	E

List of threatened and endangered species - continued

Common name	Scientific name	Group	Historic range	Vertebrate population where endangered or threatened	Status
makou	Peucedanum sandwicense	Flowering Plants	U.S.A. (HI)	Entire	T
malacothrix, island	Malacothrix squalida	Flowering Plants	U.S.A. (CA)	Entire	E
malacothrix, Santa Cruz Island	Malacothrix indecora	Flowering Plants	U.S.A. (CA)	Entire	E
malimbe, Ibadan	Malimbus ibadanensis	Birds	Nigeria	Entire	E
malkoha, red-faced (cuckoo)	Phaenicophaeus pyrrhocephalus	Birds	Sri Lanka (=Ceylon)	Entire	E
mallow, Kern	Eremalche kernensis	Flowering Plants	U.S.A. (CA)	Entire	E
mallow, Peter's Mountain	Iliamna corei	Flowering Plants	U.S.A. (VA)	Entire	E
manaca, palma de	Calyptronoma rivalis	Flowering Plants	U.S.A. (PR)	Entire	T
manatee, Amazonian	Trichechus inunguis	Mammals	South America (Amazon R. basin)	Entire	E
manatee, West African	Trichechus senegalensis	Mammals	West Coast of Africa from Senegal R. to Cuanza R	Entire	T
manatee, West Indian	Trichechus manatus	Mammals	U.S.A. (southeastern), Caribbean Sea, South America	Entire	E
mandrill	Mandrillus (=Papio) sphinx	Mammals	Equatorial West Africa	Entire	E
mangabey, Tana River	Cercocebus galeritus galeritus	Mammals	Kenya	Entire	E
mangabey, white-collared	Cercocebus torquatus	Mammals	Senegal to Ghana; Nigeria to Gabon	Entire	E
manioc, Walker's	Manihot walkerae	Flowering Plants	U.S.A. (TX), Mexico	Entire	E
manzanita, Del Mar	Arctostaphylos glandulosa ssp. crassifolia	Flowering Plants	U.S.A. (CA), Mexico.	Entire	E
manzanita, Ione	Arctostaphylos myrtifolia	Flowering Plants	U.S.A. (CA)	Entire	T
manzanita, Morro	Arctostaphylos morroensis	Flowering Plants	U.S.A. (CA)	Entire	T
manzanita, pallid	Arctostaphylos pallida	Flowering Plants	U.S.A. (CA)	Entire	T
Manzanita, Presidio	Arctostaphylos hookeri var. ravenii	Flowering Plants	U.S.A. (CA)	Entire	E
manzanita, Santa Rosa Island	Arctostaphylos confertiflora	Flowering Plants	U.S.A. (CA)	Entire	E
mapele	Cyrtandra cyaneoides	Flowering Plants	U.S.A. (HI)	Entire	E
Mapleleaf, winged (mussel)	Quadrula fragosa	Clams	U.S.A. (AL, IA, IL, IN, KY, MN, MO, NE, OH, OK, TN, WI)	Entire; except where listed as experimental populations	E
Mapleleaf, winged (mussel)	Quadrula fragosa	Clams	U.S.A. (AL, IA, IL, IN, KY, MN, MO, NE, OH, OK, TN, WI)	U.S.A. (AL	EXPN
Mapleleaf, winged (mussel) Quadrula fragosa	Clams	U.S.A. (AL, IA, IL, IN, KY, MN, MO, NE, OH, OK, TN, WI)	U.S.A. (ALThe free-flowing reach of the Tennessee R. from the base of Wilson Dam downstream to the backwaters of Reservoir [about 12 RM (19 km)] and the lower 5 RM [8 km] of all tributaries to this reach in Colbert and Lauderdale .,see 17.85(a))	EXPN	
margay	Leopardus (=Felis) wiedii	Mammals	U.S.A. (TX), Central and South America	Mexico southward	E
mariposa lily, Tiburon	Calochortus tiburonensis	Flowering Plants	U.S.A. (CA)	Entire	T
markhor, chiltan (=wild goat)	Capra falconeri (=aegragrus) chiltanensis	Mammals	Chiltan Range of west-central Pakistan	Entire	E
markhor, Kabul	Capra falconeri megaceros	Mammals	Afghanistan, Pakistan	Entire	E
markhor, straight-horned	Capra falconeri jerdoni	Mammals	Afghanistan, Pakistan	Entire	E
marmoset, buff-headed	Callithrix flaviceps	Mammals	Brazil	Entire	E
marmoset, cotton-top	Saguinus oedipus	Mammals	Costa Rica to Colombia	Entire	E

List of threatened and endangered species - *continued*

Common name	Scientific name	Group	Historic range	Vertebrate population where endangered or threatened	Status
marmoset, Goeldi's	Callimico goeldii	Mammals	Brazil, Colombia, Ecuador, Peru, Bolivia	Entire	E
marmoset, white-eared (=buffy tufted-ear)	Callithrix aurita (=jacchus a.)	Mammals	Brazil	Entire	E
marmot, Vancouver Island	Marmota vancouverensis	Mammals	Canada (Vancouver Island)	Entire	E
marstonia, royal (snail)	Pyrgulopsis ogmorhaphe	Snails	U.S.A. (TN)	Entire	E
marsupial, eastern jerboa	Antechinomys laniger	Mammals	Australia	Entire	E
marsupial-mouse, large desert	Sminthopsis psammophila	Mammals	Australia	Entire	E
marsupial-mouse, long-tailed	Sminthopsis longicaudata	Mammals	Australia	Entire	E
marten, Formosan yellow-throated	Martes flavigula chrysospila	Mammals	Taiwan	Entire	E
ma'o hau hele, (=native yellow hibiscus)	Hibiscus brackenridgei	Flowering Plants	U.S.A. (HI)	Entire	E
ma'oli'oli	Schiedea apokremnos	Flowering Plants	U.S.A. (HI)	Entire	E
ma'oli'oli	Schiedea kealiae	Flowering Plants	U.S.A. (HI)	Entire	E
meadowfoam, Butte County	Limnanthes floccosa ssp. californica	Flowering Plants	U.S.A. (CA)	Entire	E
Meadowfoam, large-flowered woolly	Limnanthes floccosa ssp. grandiflora	Flowering Plants	U.S.A. (OR)	Entire	E
meadowfoam, Sebastopol	Limnanthes vinculans	Flowering Plants	U.S.A. (CA)	Entire	E
meadowrue, Cooley's	Thalictrum cooleyi	Flowering Plants	U.S.A. (FL, NC)	Entire	E
megapode, Maleo	Macrocephalon maleo	Birds	Indonesia (Celebes)	Entire	E
megapode, Micronesian	Megapodius laperouse	Birds	West Pacific Ocean_Palau Islands, U.S.A. (Mariana Islands)	Entire	E
mehamehame	Flueggea neowawraea	Flowering Plants	U.S.A. (HI)	Entire	E
mesa-mint, Otay	Pogogyne nudiuscula	Flowering Plants	U.S.A. (CA), Mexico (Baja California)	Entire	E
mesa-mint, San Diego	Pogogyne abramsii	Flowering Plants	U.S.A. (CA)	Entire	E
Meshweaver, Braken Bat Cave	Cicurina venii	Arachnids	U.S.A. (TX).	Entire	E
Meshweaver, Government Canyon Bat Cave	Cicurina vespera	Arachnids	U.S.A. (TX).	Entire	E
Meshweaver, Madla's Cave	Cicurina madla	Arachnids	U.S.A. (TX).	Entire	E
Meshweaver, Robber Baron Cave	Cicurina baronia	Arachnids	U.S.A. (TX).	Entire	E
milk-vetch, Applegate's	Astragalus applegatei	Flowering Plants	U.S.A. (OR)	Entire	E
milk-vetch, Ash meadows	Astragalus phoenix	Flowering Plants	U.S.A. (NV)	Entire	T
milk-vetch, Braunton's	Astragalus brauntonii	Flowering Plants	U.S.A. (CA)	Entire	E
milk-vetch, Clara Hunt's	Astragalus clarianus	Flowering Plants	U.S.A. (CA)	Entire	E
milk-vetch, Coachella Valley	Astragalus lentiginosus var. coachellae	Flowering Plants	U.S.A. (CA)	Entire	E
milk-vetch, coastal dunes	Astragalus tener var. titi	Flowering Plants	U.S.A. (CA)	Entire	E
milk-vetch, Cushenbury	Astragalus albens	Flowering Plants	U.S.A. (CA)	Entire	E
milk-vetch, Deseret	Astragalus desereticus	Flowering Plants	U.S.A. (UT)	Entire	T
milk-vetch, Fish Slough	Astragalus lentiginosus var. piscinensis	Flowering Plants	U.S.A. (CA)	Entire	T
milk-vetch, heliotrope	Astragalus montii	Flowering Plants	U.S.A. (UT)	Entire	T
milk-vetch, Holmgren	Astragalus holmgreniorum	Flowering Plants	U.S.A. (AZ, UT)	Entire	E
milk-vetch, Jesup's	Astragalus robbinsii var. jesupi	Flowering Plants	U.S.A. (NH, VT)	Entire	E
milk-vetch, Lane Mountain	Astragalus jaegerianus	Flowering Plants	U.S.A. (CA)	Entire	E
milk-vetch, Mancos	Astragalus humillimus	Flowering Plants	U.S.A. (CO, NM)	Entire	E
milk-vetch, Osterhout	Astragalus osterhoutii	Flowering Plants	U.S.A. (CO)	Entire	E
milk-vetch, Peirson's	Astragalus magdalenae var. peirsonii	Flowering Plants	U.S.A. (CA)	Entire	T
milk-vetch, Sentry	Astragalus cremnophylax var. cremnophylax	Flowering Plants	U.S.A. (AZ)	Entire	E
milk-vetch, Shivwitz	Astragalus ampullarioides	Flowering Plants	U.S.A. (UT)	Entire	E
milk-vetch, triple-ribbed	Astragalus tricarinatus	Flowering Plants	U.S.A. (CA)	Entire	E

List of threatened and endangered species - *continued*

Common name	Scientific name	Group	Historic range	Vertebrate population where endangered or threatened	Status
Milk-vetch, Ventura Marsh	Astragalus pycnostachyus var. lanosissimus				
Flowering Plants	U.S.A. (CA)	Entire	E		
milkpea, Small's	Galactia smallii	Flowering Plants	U.S.A. (FL)	Entire	E
milkweed, Mead's	Asclepias meadii	Flowering Plants	U.S.A. (IA, IL, IN, KS, MO, WI)	Entire	T
milkweed, Welsh's	Asclepias welshii	Flowering Plants	U.S.A. (AZ, UT)	Entire	T
millerbird, Nihoa (old world warbler)	Acrocephalus familiaris kingi	Birds	U.S.A. (HI)	Entire	E
minnow, Devils River	Dionda diaboli	Fishes	U.S.A. (TX), Mexico	Entire	T
minnow, loach	Tiaroga cobitis	Fishes	U.S.A. (AZ, NM), Mexico	Entire	T
minnow, Rio Grande silvery	Hybognathus amarus	Fishes	U.S.A. (NM, TX), Mexico	Entire	E
mint, Garrett's	Dicerandra christmanii	Flowering Plants	U.S.A. (FL)	Entire	E
mint, Lakela's	Dicerandra immaculata	Flowering Plants	U.S.A. (FL)	Entire	E
mint, longspurred	Dicerandra cornutissima	Flowering Plants	U.S.A. (FL)	Entire	E
mint, scrub	Dicerandra frutescens	Flowering Plants	U.S.A. (FL)	Entire	E
moccasinshell, Alabama	Medionidus acutissimus	Clams	U.S.A. (AL, GA, MS)	Entire	T
moccasinshell, Coosa	Medionidus parvulus	Clams	U.S.A. (AL, GA, TN)	Entire	E
moccasinshell, Gulf	Medionidus penicillatus	Clams	U.S.A. (AL, FL, GA)	Entire	E
moccasinshell, Ochlockonee	Medionidus simpsonianus	Clams	U.S.A. (FL, GA)	Entire	E
monardella, willowy	Monardella linoides ssp. viminea	Flowering Plants	U.S.A. (CA), Mexico	Entire	E
monitor, desert	Varanus griseus	Reptiles	North Africa to Aral Sea, through Central Asia to Pakistan, Northwest India	Entire	E
monitor, Indian (=Bengal)	Varanus bengalensis	Reptiles	Iran, Iraq, India, Sri Lanka, Malaysia, Afghanistan, Burma, Vietnam, Thailand	Entire	E
monitor, Komodo Island	Varanus komodoensis	Reptiles	Indonesia (Komodo, Rintja, Padar, and western Flores Island)	Entire	E
monitor, yellow	Varanus flavescens	Reptiles	West Pakistan through India to Bangladesh	Entire	E
monkey, black colobus	Colobus satanas	Mammals	Equatorial Guinea, People's Republic of Congo, Cameroon, Gabon	Entire	E
monkey, black howler	Alouatta pigra	Mammals	Mexico, Guatemala, Belize	Entire	T
monkey, Diana	Cercopithecus diana	Mammals	Coastal West Africa	Entire	E
monkey, Guizhou snub-nosed	Rhinopithecus brelichi	Mammals	China	Entire	E
monkey, L'hoest's	Cercopithecus lhoesti	Mammals	Upper eastern Congo R. Basin, Cameroon	Entire	E
monkey, mantled howler	Alouatta palliata	Mammals	Mexico to South America	Entire	E
monkey, Preuss' red colobus	Procolobus (=Colobus) preussi (=badius p.)	Mammals	Cameroon	Entire	E
monkey, proboscis	Nasalis larvatus	Mammals	Borneo	Entire	E
monkey, red-backed squirrel	Saimiri oerstedii	Mammals	Costa Rica, Panama	Entire	E
monkey, red-bellied	Cercopithecus erythrogaster	Mammals	Western Nigeria	Entire	E
monkey, red-eared nose-spotted	Cercopithecus erythrotis	Mammals	Nigeria, Cameroon, Fernando Po	Entire	E
monkey, Sichuan snub-nosed	Rhinopithecus roxellana	Mammals	China	Entire	E
monkey, spider	Ateles geoffroyi frontatus	Mammals	Costa Rica, Nicaragua	Entire	E
monkey, spider	Ateles geoffroyl panamensis	Mammals	Costa Rica, Panama	Entire	E
monkey, Tana River red colobus	Procolobus (=Colobus) rufomitratus (=badius r.)	Mammals	Kenya	Entire	E
monkey, Tonkin snub-nosed	Rhinopithecus avunculus	Mammals	Vietnam	Entire	E
monkey, woolly spider	Brachyteles arachnoides	Mammals	Brazil	Entire	E

List of threatened and endangered species - *continued*

Common name	Scientific name	Group	Historic range	Vertebrate population where endangered or threatened	Status
monkey, yellow-tailed woolly	Lagothrix flavicauda	Mammals	Andes of northern Peru	Entire	E
monkey, Yunnan snub-nosed	Rhinopithecus bieti	Mammals	China	Entire	E
monkey, Zanzibar red colobus	Procolobus (=Colobus) pennantii (=kirki) kirki	Mammals	Tanzania	Entire	E
monkey-flower, Michigan	Mimulus glabratus var. michiganensis	Flowering Plants	U.S.A. (MI)	Entire	E
monkeyface, Appalachian (pearlymussel)	Quadrula sparsa	Clams	U.S.A. (TN, VA)	Entire	E
monkeyface, Cumberland (pearlymussel)	Quadrula intermedia	Clams	U.S.A. (AL, TN, VA)	Entire Range; Except where listed as Experimental Populations	E
monkeyface, Cumberland (pearlymussel)	Quadrula intermedia	Clams	U.S.A. (AL, TN, VA)	U.S.A. (AL;The free-flowing reach of the Tennessee R. from the base of Wilson Dam downstream to the backwaters of Pickwick Reservoir [about 12 RM (19 km)] and the lower 5 RM [8 km] of all tributaries to this reach in Colbert and Lauderdale Cos., see 17.85(a))	EXPN
monkshood, northern wild	Aconitum noveboracense	Flowering Plants	U.S.A. (IA, NY, OH, WI)	Entire	T
moorhen, Hawaiian common	Gallinula chloropus sandvicensis	Birds	U.S.A. (HI)	Entire	E
moorhen, Mariana common	Gallinula chloropus guami	Birds	West Pacific Ocean_U.S.A. (Guam, Tinian, Saipan, Pagan)	Entire	E
morning-glory, Stebbins'	Calystegia stebbinsii	Flowering Plants	U.S.A. (CA)	Entire	E
moth, Blackburn's sphinx	Manduca blackburni	Insects	U.S.A. (HI)	Entire	E
moth, Kern primrose sphinx	Euproserpinus euterpe	Insects	U.S.A. (CA)	Entire	T
mountain balm, Indian Knob	Eriodictyon altissimum	Flowering Plants	U.S.A. (CA)	Entire	E
mountain beaver, Point Arena	Aplodontia rufa nigra	Mammals	U.S.A. (CA)	Entire	E
mountain-mahogany, Catalina Island	Cercocarpus traskiae	Flowering Plants	U.S.A. (CA)	Entire	E
mouse, Alabama beach	Peromyscus polionotus ammobates	Mammals	U.S.A. (AL)	Entire	E
mouse, Anastasia Island beach	Peromyscus polionotus phasma	Mammals			
U.S.A. (FL)	Entire	E			
mouse, Australian native	Notomys aquilo	Mammals	Australia	Entire	E
mouse, Australian native	Zyzomys pedunculatus	Mammals	Australia	Entire	E
mouse, Choctawhatchee beach	Peromyscus polionotus allophrys	Mammals	U.S.A. (FL)	Entire	E
mouse, Field's	Pseudomys fieldi	Mammals	Australia	Entire	E
mouse, Gould's	Pseudomys gouldii	Mammals	Australia	Entire	E
mouse, Key Largo cotton	Peromyscus gossypinus allapaticola	Mammals	U.S.A. (FL)	Entire	E
mouse, New Holland	Pseudomys novaehollandiae	Mammals	Australia	Entire	E
mouse, Pacific pocket	Perognathus longimembris pacificus	Mammals	U.S.A. (CA)	Entire	E
mouse, Perdido Key beach	Peromyscus polionotus trissyllepsis	Mammals	U.S.A. (AL, FL)	Entire	E
mouse, Preble's meadow jumping	Zapus hudsonius preblei	Mammals	U.S.A. (CO, WY)	Entire	T
mouse, salt marsh harvest	Reithrodontomys raviventris	Mammals	U.S.A. (CA)	Entire	E
mouse, Shark Bay	Pseudomys praeconis	Mammals	Australia	Entire	E
mouse, Shortridge's	Pseudomys shortridgei	Mammals	Australia	Entire	E
mouse, smoky	Pseudomys fumeus	Mammals	Australia	Entire	E
mouse, southeastern beach	Peromyscus polionotus niveiventris	Mammals	U.S.A. (FL)	Entire	T

List of threatened and endangered species - *continued*

Common name	Scientific name	Group	Historic range	Vertebrate population where endangered or threatened	Status
mouse, St. Andrew beach	Peromyscus polionotus peninsularis	Mammals	U.S.A. (FL)	Entire	E
mouse, western	Pseudomys occidentalis	Mammals	Australia	Entire	E
mucket, orangenacre	Lampsilis perovalis	Clams	U.S.A. (AL, MS)	Entire	T
mucket, pink (pearlymussel)	Lampsilis abrupta	Clams	U.S.A. (AL, AR, IL, IN, KY, LA, MO, OH, PA, TN, VA, WV)	Entire	E
muntjac, Fea's	Muntiacus feae	Mammals	Northern Thailand, Burma	Entire	E
murrelet, marbled	Brachyramphus marmoratus marmoratus	Birds	U.S.A. (AK, CA, OR, WA), Canada (B.C.)	U.S.A. (CA, OR, WA)	T
mussel, oyster	Epioblasma capsaeformis	Clams	U.S.A. (AL, GA, KY, NC, TN, VA)	Entire Range; Except where listed as Experimental Populations	E
mussel, oyster	Epioblasma capsaeformis	Clams	U.S.A. (AL, GA, KY, NC, TN, VA)	U.S.A. (AL;The free-flowing reach of the Tennessee R. from the base of Wilson Dam downstream to the backwaters of Pickwick Reservoir [about 12 RM (19 km)] and the lower 5 RM [8 km] of all tributaries to this reach in Colbert and Lauderdale Cos., see 17.85(a)]	EXPN
mussel, scaleshell	Leptodea leptodon	Clams	U.S.A. (AL, AR, IL, IN, IA, KY, MN, MO, OH, OK, SD, TN, WI)	Entire	E
mustard, Carter's	Warea carteri	Flowering Plants	U.S.A. (FL)	Entire	E
mustard, Penland alpine fen	Eutrema penlandii	Flowering Plants	U.S.A. (CO)	Entire	T
mustard, slender-petaled	Thelypodium stenopetalum	Flowering Plants	U.S.A. (CA)	Entire	E
nani wai'ale'ale	Viola kauaiensis var. wahiawaensis	Flowering Plants	U.S.A. (HI)	Entire	E
nanu	Gardenia mannii	Flowering Plants	U.S.A. (HI)	Entire	E
native-cat, eastern	Dasyurus viverrinus	Mammals	Australia	Entire	E
naucorid, Ash Meadows	Ambrysus amargosus	Insects	U.S.A. (NV)	Entire	T
naupaka, dwarf	Scaevola coriacea	Flowering Plants	U.S.A. (HI)	Entire	E
navarretia, few-flowered	Navarretia leucocephala ssp. pauciflora (=N. pauciflora)	Flowering Plants	U.S.A. (CA)	Entire	E
navarretia, many-flowered	Navarretia leucocephala ssp. plieantha	Flowering Plants	U.S.A. (CA)	Entire	E
navarretia, spreading	Navarretia fossalis	Flowering Plants	U.S.A. (CA), Mexico (Baja California)	Entire	T
na'ena'e	Dubautia herbstobatae	Flowering Plants	U.S.A. (HI)	Entire	E
na'ena'e	Dubautia latifolia	Flowering Plants	U.S.A. (HI)	Entire	E
na'ena'e	Dubautia pauciflorula	Flowering Plants	U.S.A. (HI)	Entire	E
na'ena'e	Dubautia plantaginea ssp. humilis	Flowering Plants	U.S.A. (HI)	Entire	E
nehe	Lipochaeta fauriei	Flowering Plants	U.S.A. (HI)	Entire	E
nehe	Lipochaeta kamolensis	Flowering Plants	U.S.A. (HI)	Entire	E
nehe	Lipochaeta lobata var. leptophylla	Flowering Plants	U.S.A. (HI)	Entire	E
nehe	Lipochaeta micrantha	Flowering Plants	U.S.A. (HI)	Entire	E
nehe	Lipochaeta tenuifolia	Flowering Plants	U.S.A. (HI)	Entire	E
nehe	Lipochaeta waimeaensis	Flowering Plants	U.S.A. (HI)	Entire	E
nekogigi (catfish)	Coreobagrus ichikawai	Fishes	Japan	Entire	E
nightjar, Puerto Rican	Caprimulgus noctitherus	Birds	U.S.A. (PR)	Entire	E
nioi	Eugenia koolauensis	Flowering Plants	U.S.A. (HI)	Entire	E
niterwort, Amargosa	Nitrophila mohavensis	Flowering Plants	U.S.A. (CA, NV)	Entire	E
No common name	Abutilon eremitopetalum	Flowering Plants	U.S.A. (HI)	Entire	E
No common name	Abutilon sandwicense	Flowering Plants	U.S.A. (HI)	Entire	E
No common name	Achyranthes mutica	Flowering Plants	U.S.A. (HI)	Entire	E
No common name	Adiantum vivesii	Ferns and Allies	U.S.A. (PR)	Entire	E
No common name	Alsinidendron obovatum	Flowering Plants	U.S.A. (HI)	Entire	E
No common name	Alsinidendron trinerve	Flowering Plants	U.S.A. (HI)	Entire	E

List of threatened and endangered species - *continued*

Common name	Scientific name	Group	Historic range	Vertebrate population where endangered or threatened	Status
No common name	Alsinidendron viscosum	Flowering Plants	U.S.A. (HI)	Entire	E
No common name	Amaranthus brownii	Flowering Plants	U.S.A. (HI)	Entire	E
No common name	Aristida chaseae	Flowering Plants	U.S.A. (PR)	Entire	E
No common name	Asplenium fragile var. insulare	Ferns and Allies	U.S.A. (HI)	Entire	E
No common name	Auerodendron pauciflorum	Flowering Plants	U.S.A. (PR)	Entire	E
No common name	Bonamia menziesii	Flowering Plants	U.S.A. (HI)	Entire	E
No common name	Calyptranthes thomasiana	Flowering Plants	U.S.A. (PR, VI) British VI	Entire	E
No common name	Catesbaea melanocarpa	Flowering Plants	U.S.A. (PR, VI), Antigua, Barbuda, Guadalupe	Entire	E
No common name	Chamaecrista glandulosa var. mirabilis	Flowering Plants	U.S.A. (PR)	Entire	E
No common name	Chamaesyce halemanui	Flowering Plants	U.S.A. (HI)	Entire	E
No common name	Cordia bellonis	Flowering Plants	U.S.A. (PR)	Entire	E
No common name	Cranichis ricartii	Flowering Plants	U.S.A. (PR)	Entire	E
No common name	Cyanea (=Rollandia) crispa	Flowering Plants	U.S.A. (HI)	Entire	E
No common name	Daphnopsis hellerana	Flowering Plants	U.S.A. (PR)	Entire	E
No common name	Delissea rhytidosperma	Flowering Plants	U.S.A. (HI)	Entire	E
No common name	Delissea undulata	Flowering Plants	U.S.A. (HI)	Entire	E
No common name	Diellia falcata	Ferns and Allies	U.S.A. (HI)	Entire	E
No common name	Diellia pallida	Ferns and Allies	U.S.A. (HI)	Entire	E
No common name	Diellia unisora	Ferns and Allies	U.S.A. (HI)	Entire	E
No common name	Diplazium molokaiense	Ferns and Allies	U.S.A. (HI)	Entire	E
No common name	Elaphoglossum serpens	Ferns and Allies	U.S.A. (PR)	Entire	E
No common name	Eugenia woodburyana	Flowering Plants	U.S.A. (PR)	Entire	E
No common name	Gahnia lanaiensis	Flowering Plants	U.S.A. (HI)	Entire	E
No common name	Geocarpon minimum	Flowering Plants	U.S.A. (AR, LA, MO)	Entire	T
No common name	Gesneria pauciflora	Flowering Plants	U.S.A. (PR)	Entire	T
No common name	Gouania hillebrandii	Flowering Plants	U.S.A. (HI)	Entire	E
No common name	Gouania meyenii	Flowering Plants	U.S.A. (HI)	Entire	E
No common name	Gouania vitifolia	Flowering Plants	U.S.A. (HI)	Entire	E
No common name	Hedyotis degeneri	Flowering Plants	U.S.A. (HI)	Entire	E
No common name	Hedyotis parvula	Flowering Plants	U.S.A. (HI)	Entire	E
No common name	Hesperomannia arborescens	Flowering Plants	U.S.A. (HI)	Entire	E
No common name	Hesperomannia arbuscula	Flowering Plants	U.S.A. (HI)	Entire	E
No common name	Hesperomannia lydgatei	Flowering Plants	U.S.A. (HI)	Entire	E
No common name	Ilex sintenisii	Flowering Plants	U.S.A. (PR)	Entire	E
No common name	Lepanthes eltoroensis	Flowering Plants	U.S.A. (PR)	Entire	E
No common name	Leptocereus grantianus	Flowering Plants	U.S.A. (PR)	Entire	E
No common name	Lipochaeta venosa	Flowering Plants	U.S.A. (HI)	Entire	E
No common name	Lobelia gaudichaudii ssp. koolauensis	Flowering Plants	U.S.A. (HI)	Entire	E
No common name	Lobelia monostachya	Flowering Plants	U.S.A. (HI)	Entire	E
No common name	Lobelia niihauensis	Flowering Plants	U.S.A. (HI)	Entire	E
No common name	Lobelia oahuensis	Flowering Plants	U.S.A. (HI)	Entire	E
No common name	Lyonia truncata var. proctorii	Flowering Plants	U.S.A. (PR)	Entire	E
No common name	Lysimachia filifolia	Flowering Plants	U.S.A. (HI)	Entire	E
No common name	Lysimachia lydgatei	Flowering Plants	U.S.A. (HI)	Entire	E
No common name	Lysimachia maxima	Flowering Plants	U.S.A. (HI)	Entire	E
No common name	Mariscus fauriei	Flowering Plants	U.S.A. (HI)	Entire	E
No common name	Mariscus pennatiformis	Flowering Plants	U.S.A. (HI)	Entire	E
No common name	Mitracarpus maxwelliae	Flowering Plants	U.S.A. (PR)	Entire	E
No common name	Mitracarpus polycladus	Flowering Plants	U.S.A. (PR), Saba	Entire	E
No common name	Munroidendron racemosum	Flowering Plants	U.S.A. (HI)	Entire	E
No common name	Myrcia paganii	Flowering Plants	U.S.A. (PR)	Entire	E
No common name	Neraudia angulata	Flowering Plants	U.S.A. (HI)	Entire	E
No common name	Neraudia ovata	Flowering Plants	U.S.A. (HI)	Entire	E
No common name	Neraudia sericea	Flowering Plants	U.S.A. (HI)	Entire	E
No common name	Nesogenes rotensis	Flowering Plants	U.S.A. (MP)	Entire	E
No common name	Osmoxylon mariannense	Flowering Plants	U.S.A. (MP)	Entire	E

List of threatened and endangered species - *continued*

Common name	Scientific name	Group	Historic range	Vertebrate population where endangered or threatened	Status
No common name	Phyllostegia glabra var. lanaiensis	Flowering Plants	U.S.A. (HI)	Entire	E
No common name	Phyllostegia hirsuta	Flowering Plants	U.S.A. (HI)	Entire	E
No common name	Phyllostegia kaalaensis	Flowering Plants	U.S.A. (HI)	Entire	E
No common name	Phyllostegia knudsenii	Flowering Plants	U.S.A. (HI)	Entire	E
No common name	Phyllostegia mannii	Flowering Plants	U.S.A. (HI)	Entire	E
No common name	Phyllostegia mollis	Flowering Plants	U.S.A. (HI)	Entire	E
No common name	Phyllostegia parviflora	Flowering Plants	U.S.A. (HI)	Entire	E
No common name	Phyllostegia velutina	Flowering Plants	U.S.A. (HI)	Entire	E
No common name	Phyllostegia waimeae	Flowering Plants	U.S.A. (HI)	Entire	E
No common name	Phyllostegia warshaueri	Flowering Plants	U.S.A. (HI)	Entire	E
No common name	Phyllostegia wawrana	Flowering Plants	U.S.A. (HI)	Entire	E
No common name	Platanthera holochila	Flowering Plants	U.S.A. (HI)	Entire	E
No common name	Poa siphonoglossa	Flowering Plants	U.S.A. (HI)	Entire	E
No common name	Polystichum calderonense	Ferns and Allies	U.S.A. (PR)	Entire	E
No common name	Pteris lidgatei	Ferns and Allies	U.S.A. (HI)	Entire	E
No common name	Remya kauaiensis	Flowering Plants	U.S.A. (HI)	Entire	E
No common name	Remya montgomeryi	Flowering Plants	U.S.A. (HI)	Entire	E
No common name	Sanicula mariversa	Flowering Plants	U.S.A. (HI)	Entire	E
No common name	Sanicula purpurea	Flowering Plants	U.S.A. (HI)	Entire	E
No common name	Schiedea haleakalensis	Flowering Plants	U.S.A. (HI)	Entire	E
No common name	Schiedea helleri	Flowering Plants	U.S.A. (HI)	Entire	E
No common name	Schiedea hookeri	Flowering Plants	U.S.A. (HI)	Entire	E
No common name	Schiedea kaalae	Flowering Plants	U.S.A. (HI)	Entire	E
No common name	Schiedea kauaiensis	Flowering Plants	U.S.A. (HI)	Entire	E
No common name	Schiedea lydgatei	Flowering Plants	U.S.A. (HI)	Entire	E
No common name	Schiedea membranacea	Flowering Plants	U.S.A. (HI)	Entire	E
No common name	Schiedea nuttallii	Flowering Plants	U.S.A. (HI)	Entire	E
No common name	Schiedea sarmentosa	Flowering Plants	U.S.A. (HI)	Entire	E
No common name	Schiedea spergulina var. leiopoda	Flowering Plants	U.S.A. (HI)	Entire	E
No common name	Schiedea spergulina var. spergulina	Flowering Plants	U.S.A. (HI)	Entire	T
No common name	Schiedea verticillata	Flowering Plants	U.S.A. (HI)	Entire	E
No common name	Schoepfia arenaria	Flowering Plants	U.S.A. (PR)	Entire	T
No common name	Silene alexandri	Flowering Plants	U.S.A. (HI)	Entire	E
No common name	Silene hawaiiensis	Flowering Plants	U.S.A. (HI)	Entire	T
No common name	Silene lanceolata	Flowering Plants	U.S.A. (HI)	Entire	E
No common name	Silene perlmanii	Flowering Plants	U.S.A. (HI)	Entire	E
No common name	Spermolepis hawaiiensis	Flowering Plants	U.S.A. (HI)	Entire	E
No common name	Stenogyne angustifolia var. angustifolia	Flowering Plants	U.S.A. (HI)	Entire	E
No common name	Stenogyne bifida	Flowering Plants	U.S.A. (HI)	Entire	E
No common name	Stenogyne campanulata	Flowering Plants	U.S.A. (HI)	Entire	E
No common name	Stenogyne kanehoana	Flowering Plants	U.S.A. (HI)	Entire	E
No common name	Tectaria estremerana	Ferns and Allies	U.S.A. (PR)	Entire	E
No common name	Ternstroemia subsessilis	Flowering Plants	U.S.A. (PR)	Entire	E
No common name	Tetramolopium arenarium	Flowering Plants	U.S.A. (HI)	Entire	E
No common name	Tetramolopium filiforme	Flowering Plants	U.S.A. (HI)	Entire	E
No common name	Tetramolopium lepidotum ssp. lepidotum	Flowering Plants	U.S.A. (HI)	Entire	E
No common name	Tetramolopium remyi	Flowering Plants	U.S.A. (HI)	Entire	E
No common name	Tetramolopium rockii	Flowering Plants	U.S.A. (HI)	Entire	T
No common name	Thelypteris inabonensis	Ferns and Allies	U.S.A. (PR)	Entire	E
No common name	Thelypteris verecunda	Ferns and Allies	U.S.A. (PR)	Entire	E
No common name	Thelypteris yaucoensis	Ferns and Allies	U.S.A. (PR)	Entire	E
No common name	Trematolobelia singularis	Flowering Plants	U.S.A. (HI)	Entire	E
No common name	Vernonia proctorii	Flowering Plants	U.S.A. (PR)	Entire	E
No common name	Vigna o-wahuensis	Flowering Plants	U.S.A. (HI)	Entire	E
No common name	Viola helenae	Flowering Plants	U.S.A. (HI)	Entire	E
No common name	Viola lanaiensis	Flowering Plants	U.S.A. (HI)	Entire	E
No common name	Viola oahuensis	Flowering Plants	U.S.A. (HI)	Entire	E
No common name	Xylosma crenatum	Flowering Plants	U.S.A. (HI)	Entire	E
nohoanu	Geranium multiflorum	Flowering Plants	U.S.A. (HI)	Entire	E
nukupu'u (honeycreeper)	Hemignathus lucidus	Birds	U.S.A. (HI)	Entire	E

List of threatened and endangered species - *continued*

Common name	Scientific name	Group	Historic range	Vertebrate population where endangered or threatened	Status
numbat	Myrmecobius fasciatus	Mammals	Australia	Entire	E
nuthatch, Algerian	Sitta ledanti	Birds	Algeria	Entire	E
'o'o, Kauai (honeyeater)	Moho braccatus	Birds	U.S.A. (HI)	Entire	E
'o'u (honeycreeper)	Psittirostra psittacea	Birds	U.S.A. (HI)	Entire	E
oak, Hinckley	Quercus hinckleyi	Flowering Plants	U.S.A. (TX)	Entire	T
ocelot	Leopardus (=Felis) pardalis	Mammals	U.S.A. (AZ, TX) to Central and South America	Entire	E
oha	Delissea rivularis	Flowering Plants	U.S.A. (HI)	Entire	E
oha	Delissea subcordata	Flowering Plants	U.S.A. (HI)	Entire	E
'oha wai	Clermontia drepanomorpha	Flowering Plants	U.S.A. (HI)	Entire	E
'oha wai	Clermontia lindseyana	Flowering Plants	U.S.A. (HI)	Entire	E
'oha wai	Clermontia oblongifolia ssp. brevipes	Flowering Plants	U.S.A. (HI)	Entire	E
'oha wai	Clermontia oblongifolia ssp. mauiensis	Flowering Plants	U.S.A. (HI)	Entire	E
'oha wai	Clermontia peleana	Flowering Plants	U.S.A. (HI)	Entire	E
'oha wai	Clermontia pyrularia	Flowering Plants	U.S.A. (HI)	Entire	E
'oha wai	Clermontia samuelii	Flowering Plants	U.S.A. (HI)	Entire	E
'ohe'ohe	Tetraplasandra gymnocarpa	Flowering Plants	U.S.A. (HI)	Entire	E
ohai	Sesbania tomentosa	Flowering Plants	U.S.A. (HI)	Entire	E
olulu	Brighamia insignis	Flowering Plants	U.S.A. (HI)	Entire	E
onion, Munz's	Allium munzii	Flowering Plants	U.S.A. (CA)	Entire	E
opuhe	Urera kaalae	Flowering Plants	U.S.A. (HI)	Entire	E
orangutan	Pongo pygmaeus	Mammals	Borneo, Sumatra	Entire	T
orchid, eastern prairie fringed	Platanthera leucophaea	Flowering Plants	U.S.A. (AR, IA, IL, IN, ME, MI, MO, NE, NJ, NY, OH, OK, PA, VA, WI), Canada (Ont., N.B.)	Entire	T
Orchid, western prairie fringed	Platanthera praeclara	Flowering Plants	U.S.A. (IA, KS, MN, MO, ND, NE, SD), Canada (Man.)	Entire	T
Orcutt grass, California	Orcuttia californica	Flowering Plants	U.S.A. (CA)	Entire	E
Orcutt grass, hairy	Orcuttia pilosa	Flowering Plants	U.S.A. (CA)	Entire	E
Orcutt grass, Sacramento	Orcuttia viscida	Flowering Plants	U.S.A. (CA)	Entire	E
Orcutt grass, San Joaquin	Orcuttia inaequalis	Flowering Plants	U.S.A. (CA)	Entire	T
Orcutt grass, slender	Orcuttia tenuis	Flowering Plants	U.S.A. (CA)	Entire	T
oryx, Arabian	Oryx leucoryx	Mammals	Arabian Peninsula	Entire	E
ostrich, Arabian	Struthio camelus syriacus	Birds	Jordan, Saudi Arabia	Entire	E
ostrich, West African	Struthio camelus spatzi	Birds	Spanish Sahara	Entire	E
otter, Cameroon clawless	Aonyx congicus (=congica) microdon	Mammals	Cameroon, Nigeria	Entire	E
otter, giant	Pteronura brasiliensis	Mammals	South America	Entire	E
otter, long-tailed	Lontra (=Lutra) longicaudis (incl. platensis)	Mammals	South America	Entire	E
otter, marine	Lontra (=Lutra) felina	Mammals	Peru south to Straits of Magellan	Entire	E
otter, southern river	Lontra (=Lutra) provocax	Mammals	Chile, Argentina	Entire	E
otter, southern sea	Enhydra lutris nereis	Mammals	West Coast, U.S.A. (CA, OR, WA) south to Mexico (Baja California)	All areas subject to U.S. jurisdiction south of Pt. Conception, CA (34026.9' N. Lat.) [Note—status governed by Pub. L. 99-625, 100 Stat. 3500.]	EXPN
otter, southern sea	Enhydra lutris nereis	Mammals	West Coast, U.S.A. (CA, OR, WA) south to Mexico (Baja California)	Entire, except where listed below	T
owl's-clover, fleshy	Castilleja campestris ssp. succulenta	Flowering Plants	U.S.A. (CA)	Entire	T
owl, Anjouan scops	Otus rutilus capnodes	Birds	Indian Ocean_Comoro Island	Entire	E
owl, giant scops	Mimizuku (=Otus) gurneyi	Birds	Philippines_Marinduque and Mindanao Island	Entire	E

List of threatened and endangered species - *continued*

Common name	Scientific name	Group	Historic range	Vertebrate population where endangered or threatened	Status
owl, Madagascar red	Tyto soumagnei	Birds	Madagascar	Entire	E
owl, Mexican spotted	Strix occidentalis lucida	Birds	U.S.A. (AZ, CO, NM, TX, UT), Mexico	Entire	T
owl, northern spotted	Strix occidentalis caurina	Birds	U.S.A. (CA, OR, WA), Canada (B.C.)	Entire	T
owl, Seychelles scops	Otus magicus (=insularis) insularis	Birds	Indian Ocean_Seychelles Islands	Entire	E
owlet, Morden's	Otus ireneae	Birds	Kenya	Entire	E
oxytheca, cushenbury	Oxytheca parishii var. goodmaniana	Flowering Plants	U.S.A. (CA)	Entire	E
oystercatcher, Canarian black	Haematopus meadewaldoi	Birds	Atlantic Ocean_Canary Islands	Entire	E
paintbrush, ash-grey	Castilleja cinerea	Flowering Plants	U.S.A. (CA)	Entire	T
Paintbrush, golden	Castilleja levisecta	Flowering Plants	U.S.A. (WA), Canada (B.C.)	Entire	T
paintbrush, soft-leaved	Castilleja mollis	Flowering Plants	U.S.A. (CA)	Entire	E
paintbrush, Tiburon	Castilleja affinis ssp. neglecta	Flowering Plants	U.S.A. (CA)	Entire	E
palila (honeycreeper)	Loxioides bailleui	Birds	U.S.A. (HI)	Entire	E
palo colorado	Ternstroemia luquillensis	Flowering Plants	U.S.A. (PR)	Entire	E
palo de jazmin	Styrax portoricensis	Flowering Plants	U.S.A. (PR)	Entire	E
palo de nigua	Cornutia obovata	Flowering Plants	U.S.A. (PR)	Entire	E
palo de ramon	Banara vanderbiltii	Flowering Plants	U.S.A. (PR)	Entire	E
palo de rosa	Ottoschulzia rhodoxylon	Flowering Plants	U.S.A. (PR), Dominican Republic	Entire	E
pamakani	Tetramolopium capillare	Flowering Plants	U.S.A. (HI)	Entire	E
pamakani	Viola chamissoniana ssp. chamissoniana	Flowering Plants	U.S.A. (HI)	Entire	E
panda, giant	Ailuropoda melanoleuca	Mammals	China	Entire	E
Pangolin, Temnick's ground	Manis temminckii	Mammals	Africa	Entire	E
panicgrass, Carter's	Panicum fauriei var. carteri	Flowering Plants	U.S.A. (HI)	Entire	E
panther, Florida	Puma (=Felis) concolor coryi	Mammals	U.S.A. (LA and AR east to SC and FL)	Entire	E
parakeet, blue-throated (=ochre-marked)	Pyrrhura cruentata	Birds	Brazil	Entire	E
parakeet, Forbes'	Cyanoramphus auriceps forbesi	Birds	New Zealand	Entire	E
parakeet, golden	Aratinga guarouba	Birds	Brazil	Entire	E
parakeet, golden-shouldered	Psephotus chrysopterygius	Birds	Australia	Entire	E
parakeet, Mauritius	Psittacula echo	Birds	Indian Ocean_Mauritius	Entire	E
parakeet, Norfolk Island	Cyanoramphus cookii (=novaezelandiae c.)	Birds	Australia (Norfolk Island)	Entire	E
parakeet, orange-bellied	Neophema chrysogaster	Birds	Australia	Entire	E
parakeet, paradise	Psephotus pulcherrimus	Birds	Australia	Entire	E
parakeet, scarlet-chested	Neophema splendida	Birds	Australia	Entire	E
parakeet, turquoise	Neophema pulchella	Birds	Australia	Entire	E
parrot, Bahaman or Cuban	Amazona leucocephala	Birds	West Indies_Cuba, Bahamas, Caymans	Entire	E
parrot, ground	Pezoporus wallicus	Birds	Australia	Entire	E
parrot, imperial	Amazona imperialis	Birds	West Indies_Dominica	Entire	E
parrot, night (=Australian)	Geopsittacus occidentalis	Birds	Australia	Entire	E
parrot, Puerto Rican	Amazona vittata	Birds	U.S.A. (PR)	Entire	E
parrot, red-browed	Amazona rhodocorytha	Birds	Brazil	Entire	E
parrot, red-capped	Pionopsitta pileata	Birds	Brazil	Entire	E
parrot, red-necked	Amazona arausiaca	Birds	West Indies_Dominica	Entire	E
parrot, red-spectacled	Amazona pretrei pretrei	Birds	Brazil, Argentina	Entire	E
parrot, red-tailed	Amazona brasiliensis	Birds	Brazil	Entire	E
parrot, Seychelles lesser vasa	Coracopsis nigra barklyi	Birds	Indian Ocean_Seychelles (Praslin Island)	Entire	E
parrot, St Vincent	Amazona guildingii	Birds	West Indies_St. Vincent	Entire	E
parrot, St. Lucia	Amazona versicolor	Birds	West Indies_St. Lucia	Entire	E
parrot, thick-billed	Rhynchopsitta pachyrhyncha	Birds	Mexico, U.S.A. (AZ, NM)	Mexico	E

List of threatened and endangered species - *continued*

Common name	Scientific name	Group	Historic range	Vertebrate population where endangered or threatened	Status
parrot, vinaceous-breasted	Amazona vinacea	Birds	Brazil	Entire	E
parrotbill, Maui (honeycreeper)	Pseudonestor xanthophrys	Birds	U.S.A. (HI)	Entire	E
pauoa	Ctenitis squamigera	Ferns and Allies	U.S.A. (HI)	Entire	E
pawpaw, beautiful	Deeringothamnus pulchellus	Flowering Plants	U.S.A. (FL)	Entire	E
pawpaw, four-petal	Asimina tetramera	Flowering Plants	U.S.A. (FL)	Entire	E
pawpaw, Rugel's	Deeringothamnus rugelii	Flowering Plants	U.S.A. (FL)	Entire	E
pearlshell, Louisiana	Margaritifera hembeli	Clams	U.S.A. (LA)	Entire	T
pearlymussel, birdwing	Conradilla caelata	Clams	U.S.A. (TN, VA)	Entire Range; Except where listed as Experimental Populations	E
pearlymussel, birdwing	Conradilla caelata	Clams	U.S.A. (TN, VA)	U.S.A. (AL;The free-flowing reach of the Tennessee R. from the base of Wilson Dam downstream to the backwaters of Pickwick Reservoir [about 12 RM (19 km)] and the lower 5 RM [8 km] of all tributaries to this reach in Colbert and Lauderdale Cos., see 17.85(a))	EXPN
pearlymussel, cracking	Hemistena lata	Clams	U.S.A. (AL, IL, IN, KY, OH, TN, VA)	Entire Range; Except where listed as Experimental Populations	E
pearlymussel, cracking	Hemistena lata	Clams	U.S.A. (AL, IL, IN, KY, OH, TN, VA)	U.S.A. (AL;The free-flowing reach of the Tennessee R. from the base of Wilson Dam downstream to the backwaters of Pickwick Reservoir [about 12 RM (19 km)] and the lower 5 RM [8 km] of all tributaries to this reach in Colbert and Lauderdale Cos., see 17.85(a))	EXPN
pearlymussel, Curtis	Epioblasma florentina curtisii	Clams	U.S.A. (AR, MO)	Entire	E
pearlymussel, dromedary	Dromus dromas	Clams	U.S.A. (AL, KY, TN, VA)	Entire Range; Except where listed as Experimental Populations	E
pearlymussel, dromedary	Dromus dromas	Clams	U.S.A. (AL, KY, TN, VA)	U.S.A. (AL;The free-flowing reach of the Tennessee R. from the base of Wilson Dam downstream to the backwaters of Pickwick Reservoir [about 12 RM (19 km)] and the lower 5 RM [8 km] of all tributaries to this reach in Colbert and Lauderdale Cos., see 17.85(a))	EXPN
pearlymussel, littlewing	Pegias fabula	Clams	U.S.A. (AL, KY, NC, TN, VA)	Entire	E
pearlymussel, Nicklin's	Megalonaias nicklineana	Clams	Mexico	Entire	E
pearlymussel, Tampico	Cyrtonaias tampicoensis tecomatensis	Clams	Mexico	Entire	E
pebblesnail, flat	Lepyrium showalteri	Snails	U.S.A. (AL)	Entire	E
pelican, brown	Pelecanus occidentalis	Birds	U.S.A (Carolinas to TX, CA, OR, WA), West Indies, coastal Central and South America	Entire U.S., except U.S. Atlantic coast, FL, AL	E
pelos del diablo	Aristida portoricensis	Flowering Plants	U.S.A. (PR)	Entire	E
penguin, Galapagos	Spheniscus mendiculus	Birds	Ecuador (Galapagos Islands)	Entire	E
penny-cress, Kneeland Prairie	Thlaspi californicum	Flowering Plants	U.S.A. (CA)	Entire	E

List of threatened and endangered species - *continued*

Common name	Scientific name	Group	Historic range	Vertebrate population where endangered or threatened	Status
pennyroyal, Todsen's	Hedeoma todsenii	Flowering Plants	U.S.A. (NM)	Entire	E
penstemon, blowout	Penstemon haydenii	Flowering Plants	U.S.A. (NE)	Entire	E
pentachaeta, Lyon's	Pentachaeta lyonii	Flowering Plants	U.S.A. (CA)	Entire	E
pentachaeta, white-rayed	Pentachaeta bellidiflora	Flowering Plants	U.S.A. (CA)	Entire	E
peperomia, Wheeler's	Peperomia wheeleri	Flowering Plants	U.S.A. (PR)	Entire	E
petrel, Hawaiian dark-rumped	Pterodroma phaeopygia sandwichensis	Birds	U.S.A. (HI)	Entire	E
petrel, Mascarene black	Pterodroma aterrima	Birds	Indian Ocean_Mauritius (Reunion Island)	Entire	E
phacelia, clay	Phacelia argillacea	Flowering Plants	U.S.A. (UT)	Entire	E
phacelia, island	Phacelia insularis ssp. insularis	Flowering Plants	U.S.A. (CA)	Entire	E
phacelia, North Park	Phacelia formosula	Flowering Plants	U.S.A. (CO)	Entire	E
pheasant, bar-tailed	Syrmaticus humaie	Birds	Burma, China	Entire	E
pheasant, Blyth's tragopan	Tragopan blythii	Birds	Burma, China, India	Entire	E
pheasant, brown eared	Crossoptilon mantchuricum	Birds	China	Entire	E
pheasant, Cabot's tragopan	Tragopan caboti	Birds	China	Entire	E
pheasant, cheer	Catreus wallichii	Birds	India, Nepal, Pakistan	Entire	E
pheasant, Chinese monal	Lophophorus lhuysii	Birds	China	Entire	E
pheasant, Edward's	Lophura edwardsi	Birds	Vietnam	Entire	E
pheasant, Elliot's	Syrmaticus ellioti	Birds	China	Entire	E
pheasant, imperial	Lophura imperialis	Birds	Vietnam	Entire	E
pheasant, Mikado	Syrmaticus mikado	Birds	Taiwan	Entire	E
pheasant, Palawan peacock	Polyplectron emphanum	Birds	Philippines	Entire	E
pheasant, Sclater's monal	Lophophorus sclateri	Birds	Burma, China, India	Entire	E
pheasant, Swinhoe's	Lophura swinhoii	Birds	Taiwan	Entire	E
pheasant, western tragopan	Tragopan melanocephalus	Birds	India, Pakistan	Entire	E
pheasant, white eared	Crossoptilon crossoptilon	Birds	China (Tibet), India	Entire	E
phlox, Texas trailing	Phlox nivalis ssp. texensis	Flowering Plants	U.S.A. (TX)	Entire	E
phlox, Yreka	Phlox hirsuta	Flowering Plants	U.S.A. (CA)	Entire	E
pigeon wings	Clitoria fragrans	Flowering Plants	U.S.A. (FL)	Entire	T
pigeon, Azores wood	Columba palumbus azorica	Birds	East Atlantic Ocean_Azores	Entire	E
pigeon, Chatham Island	Hemiphaga novaeseelandiae chathamensis	Birds	New Zealand	Entire	E
pigeon, Mindoro imperial (=zone-tailed)	Ducula mindorensis	Birds	Philippines	Entire	E
pigeon, pink	Columba mayeri	Birds	Indian Ocean_Mauritius	Entire	E
Pigeon, Puerto Rican plain	Columba inornata wetmorei	Birds	U.S.A. (PR)	Entire	E
Pigeon, white-tailed laurel	Columba junoniae	Birds	Atlantic Ocean_Canary Islands	Entire	T
pigtoe, Cumberland	Pleurobema gibberum	Clams	U.S.A. (TN)	Entire	E
pigtoe, dark	Pleurobema furvum	Clams	U.S.A. (AL)	Entire	E
pigtoe, finerayed	Fusconaia cuneolus	Clams	U.S.A. (AL, TN, VA)	Entire Range; Except where listed as Experimental Populations	E
pigtoe, finerayed	Fusconaia cuneolus	Clams	U.S.A. (AL, TN, VA)	U.S.A. (AL;The free-flowing reach of the Tennessee R. from the base of Wilson Dam downstream to the backwaters of Pickwick Reservoir [about 12 RM (19 km)] and the lower 5 RM [8 km] of all tributaries to this reach in Colbert and Lauderdale Cos., see 17.85(a))	EXPN
pigtoe, flat	Pleurobema marshalli	Clams	U.S.A. (AL, MS)	Entire	E
pigtoe, heavy	Pleurobema taitianum	Clams	U.S.A. (AL, MS)	Entire	E
pigtoe, oval	Pleurobema pyriforme	Clams	U.S.A. (AL, FL, GA)	Entire	E
pigtoe, rough	Pleurobema plenum	Clams	U.S.A. (AL, IN, KY, PA, TN, VA)	Entire	E

List of threatened and endangered species - *continued*

Common name	Scientific name	Group	Historic range	Vertebrate population where endangered or threatened	Status
pigtoe, shiny	Fusconaia cor	Clams	U.S.A. (AL, TN, VA)	Entire Range; Except where listed as Experimental Populations	E
pigtoe, shiny	Fusconaia cor	Clams	U.S.A. (AL, TN, VA)	U.S.A. (AL;The free-flowing reach of the Tennessee R. from the base of Wilson Dam downstream to the backwaters of Pickwick Reservoir [about 12 RM (19 km)] and the lower 5 RM [8 km] of all tributaries to this reach in Colbert and Lauderdale Cos., see 17.85(a))	EXPN
pigtoe, southern	Pleurobema georgianum	Clams	U.S.A. (AL, GA, TN)	Entire	E
pikeminnow (=squawfish), Colorado	Ptychocheilus lucius	Fishes	U.S.A. (AZ, CA, CO, NM, NV, UT, WY), Mexico	Entire, except Salt and Verde R. drainages, AZ	E
pikeminnow (=squawfish), Colorado	Ptychocheilus lucius	Fishes	U.S.A. (AZ, CA, CO, NM, NV, UT, WY), Mexico	Salt and Verde R. drainages, AZ	EXPN
pilo	Hedyotis mannii	Flowering Plants	U.S.A. (HI)	Entire	E
pimpleback, orangefoot (pearlymussel)	Plethobasus cooperianus	Clams	U.S.A. (AL, IA, IL, IN, KY, OH, PA, TN)	Entire	E
pink, swamp	Helonias bullata	Flowering Plants	U.S.A. (DE, GA, MD, NC, NJ, NY, SC, VA)	Entire	T
pinkroot, gentian	Spigelia gentianoides	Flowering Plants	U.S.A. (AL, FL)	Entire	E
piperia, Yadon's	Piperia yadonii	Flowering Plants	U.S.A. (CA)	Entire	E
piping-guan, black-fronted	Pipile jacutinga	Birds	Argentina	Entire	E
pitaya, Davis' green	Echinocereus viridiflorus var. davisii	Flowering Plants	U.S.A. (TX)	Entire	E
pitcher-plant, Alabama canebrake	Sarracenia rubra alabamensis	Flowering Plants	U.S.A. (AL)	Entire	E
pitcher-plant, green	Sarracenia oreophila	Flowering Plants	U.S.A. (AL, GA, NC, TN)	Entire	E
pitcher-plant, mountain sweet	Sarracenia rubra ssp. jonesii	Flowering Plants	U.S.A. (NC, SC)	Entire	E
pitta, Koch's	Pitta kochi	Birds	Philippines	Entire	E
planigale, little	Planigale ingrami subtilissima	Mammals	Australia	Entire	E
planigale, southern	Planigale tenuirostris	Mammals	Australia	Entire	E
plover, New Zealand shore	Thinornis novaeseelandiae	Birds	New Zealand	Entire	E
Plover, piping	Charadrius melodus	Birds	U.S.A. (Great Lakes, northern Great Plains, Atlantic and Gulf coasts, PR, VI), Canada, Mexico, Bahamas, West Indies	Great Lakes watershed in States of IL, IN, MI, MN, NY, OH, PA, and WI and Canada (Ont.)	E
Plover, piping	Charadrius melodus	Birds	U.S.A. (Great Lakes, northern Great Plains, Atlantic and Gulf coasts, PR, VI), Canada, Mexico, Bahamas, West Indies	Entire, except those areas where listed as endangered above	T
plover, western snowy	Charadrius alexandrinus nivosus	Birds	U.S.A. (AZ, CA, CO, KS, NM, NV, OK, OR, TX, UT, WA), Mexico	U.S.A. (CA, OR, WA), Mexico (within 50 miles of Pacific coast)	T
plum, scrub	Prunus geniculata	Flowering Plants	U.S.A. (FL)	Entire	E
pochard, Madagascar	Aythya innotata	Birds	Madagascar	Entire	E
pocketbook, fat	Potamilus capax	Clams	U.S.A. (AR, IA, IL, IN, KY, MO, MS, OH)	Entire	E
pocketbook, finelined	Lampsilis altilis	Clams	U.S.A. (AL, GA)	Entire	T
pocketbook, Ouachita rock	Arkansia wheeleri	Clams	U.S.A. (AR, OK)	Entire	E
pocketbook, shinyrayed	Lampsilis subangulata	Clams	U.S.A. (AL, FL, GA)	Entire	E
pocketbook, speckled	Lampsilis streckeri	Clams	U.S.A. (AR)	Entire	E

List of threatened and endangered species - *continued*

Common name	Scientific name	Group	Historic range	Vertebrate population where endangered or threatened	Status
pogonia, small whorled	Isotria medeoloides	Flowering Plants	U.S.A. (CT, DC, DE, GA, IL, MA, MD, ME, MI, MO, NC, NH, NJ, NY, PA, RI, SC, TN, VA, VT,WV), Canada (Ont.)	Entire	T
polygala, Lewton's	Polygala lewtonii	Flowering Plants	U.S.A. (FL)	Entire	E
polygala, tiny	Polygala smallii	Flowering Plants	U.S.A. (FL)	Entire	E
Polygonum, Scotts Valley	Polygonum hickmanii	Flowering Plants	U.S.A. (CA)	Entire	E
pondberry	Lindera melissifolia	Flowering Plants	U.S.A. (AL, AR, FL, GA, LA, MO, MS, NC, SC)	Entire	E
Pondweed, Little Aguja (=Creek)	Potamogeton clystocarpus	Flowering Plants	U.S.A. (TX)	Entire	E
poolfish, Pahrump	Empetrichthys latos	Fishes	U.S.A. (NV)	Entire	E
popcornflower, rough	Plagiobothrys hirtus	Flowering Plants	U.S.A. (OR)	Entire	E
popolo ku mai	Solanum incompletum	Flowering Plants	U.S.A. (HI)	Entire	E
poppy, Sacramento prickly	Argemone pleiacantha ssp. pinnatisecta	Flowering Plants	U.S.A. (NM)	Entire	E
poppy-mallow, Texas	Callirhoe scabriuscula	Flowering Plants	U.S.A. (TX)	Entire	E
porcupine, thin-spined	Chaetomys subspinosus	Mammals	Brazil	Entire	E
Possum, Leadbeater's	Gymnobelideus leadbeateri	Mammals	Australia	Entire	E
possum, mountain pygmy	Burramys parvus	Mammals	Australia	Entire	E
possum, scaly-tailed	Wyulda squamicaudata	Mammals	Australia	Entire	E
potato-bean, Price's	Apios priceana	Flowering Plants	U.S.A. (AL, IL, KY, MS, TN)	Entire	T
potentilla, Hickman's	Potentilla hickmanii	Flowering Plants	U.S.A. (CA)	Entire	E
po'e	Portulaca sclerocarpa	Flowering Plants	U.S.A. (HI)	Entire	E
po'ouli (honeycreeper)	Melamprosops phaeosoma	Birds	U.S.A. (HI)	Entire	E
prairie dog, Mexican	Cynomys mexicanus	Mammals	Mexico	Entire	E
prairie dog, Utah	Cynomys parvidens	Mammals	U.S.A. (UT)	Entire	T
prairie-chicken, Attwater's greater	Tympanuchus cupido attwateri	Birds	U.S.A. (TX)	Entire	E
prairie-clover, leafy	Dalea foliosa	Flowering Plants	U.S.A. (AL, IL, TN)	Entire	E
prickly-apple, fragrant	Cereus eriophorus var. fragrans	Flowering Plants	U.S.A. (FL)	Entire	E
prickly-ash, St. Thomas	Zanthoxylum thomasianum	Flowering Plants	U.S.A. (PR, VI)	Entire	E
primrose, Maguire	Primula maguirei	Flowering Plants	U.S.A. (UT)	Entire	T
pronghorn, peninsular	Antilocapra americana peninsularis	Mammals	Mexico (Baja California)	Entire	E
pronghorn, Sonoran	Antilocapra americana sonoriensis	Mammals	U.S.A. (AZ), Mexico	Entire	E
pseudoscorpion, Tooth Cave	Tartarocreagris texana	Arachnids	U.S.A. (TX)	Entire	E
Pua 'ala	Brighamia rockii	Flowering Plants	U.S.A. (HI)	Entire	E
Pudu	Pudu pudu	Mammals	Southern South America	Entire	E
puma (=cougar), eastern	Puma (=Felis) concolor couguar	Mammals	Eastern North America	Entire	E
puma (=mountain lion)	Puma (=Felis) concolor (all subsp. except coryi)	Mammals	Canada to South America	U.S.A. (FL)	SAT
puma, Costa Rican	Puma (=Felis) concolor costaricensis	Mammals	Nicaragua, Panama, Costa Rica	Entire	E
pupfish, Ash Meadows Amargosa	Cyprinodon nevadensis mionectes	Fishes	U.S.A. (NV)	Entire	E
pupfish, Comanche Springs	Cyprinodon elegans	Fishes	U.S.A. (TX)	Entire	E
pupfish, desert	Cyprinodon macularius	Fishes	U.S.A. (AZ, CA) Mexico	Entire	E
pupfish, Devils Hole	Cyprinodon diabolis	Fishes	U.S.A. (NV)	Entire	E
pupfish, Leon Springs	Cyprinodon bovinus	Fishes	U.S.A. (TX)	Entire	E
pupfish, Owens	Cyprinodon radiosus	Fishes	U.S.A. (CA)	Entire	E
pupfish, Warm Springs	Cyprinodon nevadensis pectoralis	Fishes	U.S.A. (NV)	Entire	E
pussypaws, Mariposa	Calyptridium pulchellum	Flowering Plants	U.S.A. (CA)	Entire	T
pu'uka'a	Cyperus trachysanthos	Flowering Plants	U.S.A. (HI)	Entire	E
pygmy-owl, cactus ferruginous	Glaucidium brasilianum cactorum	Birds	U.S.A. (AZ, TX), Mexico	U.S.A. (AZ)	E
python, Indian	Python molurus molurus	Reptiles	Sri Lanka and India	Entire	E

List of threatened and endangered species - *continued*

Common name	Scientific name	Group	Historic range	Vertebrate population where endangered or threatened	Status
quail, Merriam's Montezuma	Cyrtonyx montezumae merriami	Birds	Mexico (Vera Cruz)	Entire	E
quetzel, resplendent	Pharomachrus mocinno	Birds	Mexico to Panama	Entire	E
quillwort, black spored	Isoetes melanospora	Ferns and Allies	U.S.A. (GA, SC)	Entire	E
quillwort, Louisiana	Isoetes louisianensis	Ferns and Allies	U.S.A. (LA, MS)	Entire	E
quillwort, mat-forming	Isoetes tegetiformans	Ferns and Allies	U.S.A. (GA)	Entire	E
quokka	Setonix brachyurus	Mammals	Australia	Entire	E
rabbit, Lower Keys marsh	Sylvilagus palustris hefneri	Mammals	U.S.A. (FL)	Entire	E
Rabbit, pygmy	Brachylagus idahoensis	Mammals	U.S.A. (CA, ID, MT, NV, OR, UT, WA, WY)	Columbia Basin DPS	E
rabbit, riparian brush	Sylvilagus bachmani riparius	Mammals	U.S.A. (CA)	Entire	E
rabbit, Ryukyu	Pentalagus furnessi	Mammals	Japan (Ryukyu Islands)	Entire	E
rabbit, volcano	Romerolagus diazi	Mammals	Mexico	Entire	E
rabbitsfoot, rough	Quadrula cylindrica strigillata	Clams	U.S.A. (TN, VA)	Entire	E
rail, Aukland Island	Rallus pectoralis muelleri	Birds	New Zealand	Entire	E
rail, California clapper	Rallus longirostris obsoletus	Birds	U.S.A. (CA)	Entire	E
rail, Guam	Rallus owstoni	Birds	Western Pacific Ocean_U.S.A. (Guam)	Entire, except Rota	E
rail, Guam	Rallus owstoni	Birds	Western Pacific Ocean_U.S.A. (Guam)	Rota	EXPN
rail, light-footed clapper	Rallus longirostris levipes	Birds	U.S.A. (CA), Mexico (Baja California)	U.S.A. only	E
rail, Lord Howe wood	Gallirallus (=Tricholimnas) sylvestris	Birds	Australia (Lord Howe Island)	Entire	E
rail, Yuma clapper	Rallus longirostris yumanensis	Birds	Mexico, U.S.A. (AZ, CA)	U.S.A. only	E
rat, false water Australia	Xeromys myoides Entire	Mammals E			
rat, stick-nest	Leporillus conditor	Mammals	Australia	Entire	E
rat-kangaroo, brush-tailed	Bettongia penicillata	Mammals	Australia	Entire	E
rat-kangaroo, desert (=plain)	Caloprymnus campestris	Mammals	Australia	Entire	E
rat-kangaroo, Gaimard's	Bettongia gaimardi	Mammals	Australia	Entire	E
rat-kangaroo, Lesuer's	Bettongia lesueur	Mammals	Australia	Entire	E
rat-kangaroo, Queensland	Bettongia tropica	Mammals	Australia	Entire	T
rattlesnake, Aruba Island	Crotalus unicolor	Reptiles	Aruba Island (Netherland Antilles)	Entire	T
rattlesnake, New Mexican ridge-nosed	Crotalus willardi obscurus	Reptiles	U.S.A. (AZ, NM), Mexico	Entire	T
rattleweed, hairy	Baptisia arachnifera	Flowering Plants	U.S.A. (GA)	Entire	E
reed-mustard, Barneby	Schoenocrambe barnebyi	Flowering Plants	U.S.A. (UT)	Entire	E
reed-mustard, clay	Schoenocrambe argillacea	Flowering Plants	U.S.A. (UT)	Entire	T
reed-mustard, shrubby	Schoenocrambe suffrutescens	Flowering Plants	U.S.A. (UT)	Entire	E
remya, Maui	Remya mauiensis	Flowering Plants	U.S.A. (HI)	Entire	E
rhea, lesser (incl. Darwin's)	Rhea (=Pterocnemia) pennata	Birds	Argentina, Bolivia, Peru, Uruguay	Entire	E
rhinoceros, black	Diceros bicornis	Mammals	Sub-Saharan Africa	Entire	E
rhinoceros, great Indian	Rhinoceros unicornis	Mammals	India, Nepal	Entire	E
rhinoceros, Javan	Rhinoceros sondaicus	Mammals	Indonesia, Indochina, Burma, Thailand, Sikkim, Bangladesh, Malaysia	Entire	E
rhinoceros, northern white	Ceratotherium simum cottoni	Mammals	Zaire, Sudan, Uganda, Central African Republic	Entire	E
rhinoceros, Sumatran	Dicerorhinus sumatrensis	Mammals	Bangladesh to Vietnam to Indonesia (Borneo)	Entire	E
rhododendron, Chapman	Rhododendron chapmanii	Flowering Plants	U.S.A. (FL)	Entire	E

List of threatened and endangered species - *continued*

Common name	Scientific name	Group	Historic range	Vertebrate population where endangered or threatened	Status
rice rat	Oryzomys palustris natator	Mammals	U.S.A. (FL)	Lower FL Keys (west of Seven Mile Bridge)	E
ridge-cress, Barneby	Lepidium barnebyanum	Flowering Plants	U.S.A. (UT)	Entire	E
riffleshell, northern	Epioblasma torulosa rangiana	Clams	U.S.A. (IL, IN, KY, MI, OH, PA, WV), Canada (Ont.)	Entire	E
riffleshell, tan	Epioblasma florentina walkeri (=E. walkeri)	Clams	U.S.A. (AL, KY, NC, TN, VA)	Entire	E
ring pink (mussel)	Obovaria retusa	Clams	U.S.A. (AL, IL, IN, KY, OH, PA, TN, WV)	Entire	E
riversnail, Anthony's	Athearnia anthonyi	Snails	U.S.A. (AL, GA, TN)	Entire Range; Except where listed as Experimental Populations	E
riversnail, Anthony's	Athearnia anthonyi	Snails	U.S.A. (AL, GA, TN)	U.S.A. (AL;The free-flowing reach of the Tennessee R. from the base of Wilson Dam downstream to the backwaters of Pickwick Reservoir [about 12 RM (19 km)] and the lower 5 RM [8 km] of all tributaries to this reach in Colbert and Lauderdale Cos., see 17.85(a))	EXPN
robin, Chatham Island	Petroica traversi	Birds	New Zealand	Entire	E
robin, dappled mountain	Arcanator orostruthus	Birds	Mozambique, Tanzania	Entire	T
robin, scarlet-breasted (flycatcher)	Petroica multicolor multicolor	Birds	Australia (Norfolk Island)	Entire	E
rock-cress, Braun's	Arabis perstellata	Flowering Plants	U.S.A. (KY,TN)	Entire	E
rock-cress, Hoffmann's	Arabis hoffmannii	Flowering Plants	U.S.A. (CA)	Entire	E
rock-cress, McDonald's	Arabis mcdonaldiana	Flowering Plants	U.S.A. (CA)	Entire	E
rock-cress, shale barren	Arabis serotina	Flowering Plants	U.S.A. (VA, WV)	Entire	E
rockcress, Santa Cruz Island	Sibara filifolia	Flowering Plants	U.S.A. (CA)	Entire	E
rockfowl, grey-necked	Picathartes oreas	Birds	Cameroon, Gabon	Entire	E
rockfowl, white-necked	Picathartes gymnocephalus	Birds	Africa_Togo to Sierra Leone	Entire	E
rocksnail, painted	Leptoxis taeniata	Snails	U.S.A. (AL)	Entire	T
rocksnail, plicate	Leptoxis plicata	Snails	U.S.A. (AL)	Entire	E
rocksnail, round	Leptoxis ampla	Snails	U.S.A. (AL)	Entire	T
roller, long-tailed ground	Uratelornis chimaera	Birds	Malagasy Republic (=Madagascar)	Entire	E
rosemary, Apalachicola	Conradina glabra	Flowering Plants	U.S.A. (FL)	Entire	E
rosemary, Cumberland	Conradina verticillata	Flowering Plants	U.S.A. (KY, TN)	Entire	T
rosemary, Etonia	Conradina etonia	Flowering Plants	U.S.A. (FL)	Entire	E
rosemary, short-leaved	Conradina brevifolia	Flowering Plants	U.S.A. (FL)	Entire	E
roseroot, Leedy's	Sedum integrifolium ssp. leedyi	Flowering Plants	U.S.A. (MN, NY)	Entire	T
rush-pea, slender	Hoffmannseggia tenella	Flowering Plants	U.S.A. (TX)	Entire	E
rush-rose, island	Helianthemum greenei	Flowering Plants	U.S.A. (CA)	Entire	T
saiga, Mongolian (antelope)	Saiga tatarica mongolica	Mammals	Mongolia	Entire	E
saki, southern bearded	Chiropotes satanas satanas	Mammals	Brazil	Entire	E
saki, white-nosed	Chiropotes albinasus	Mammals	Brazil	Entire	E
salamander, Barton Springs	Eurycea sosorum	Amphibians	U.S.A. (TX)	Entire	E
Salamander, California tiger	Ambystoma californiense	Amphibians	U.S.A. (CA)	Entire	T
salamander, Cheat Mountain	Plethodon nettingi	Amphibians	U.S.A. (WV)	Entire	T
salamander, Chinese giant	Andrias davidianus (=davidianus d.)	Amphibians	Western China	Entire	E
salamander, desert slender	Batrachoseps aridus	Amphibians	U.S.A. (CA)	Entire	E
salamander, flatwoods	Ambystoma cingulatum	Amphibians	U.S.A. (AL, FL, GA, SC)	Entire	T
salamander, Japanese giant	Andrias japonicus (=davidianus j.)	Amphibians	Japan	Entire	E
salamander, Red Hills	Phaeognathus hubrichti	Amphibians	U.S.A. (AL)	Entire	T

List of threatened and endangered species - *continued*

Common name	Scientific name	Group	Historic range	Vertebrate population where endangered or threatened	Status
salamander, San Marcos	Eurycea nana	Amphibians	U.S.A. (TX)	Entire	T
salamander, Santa Cruz long-toed	Ambystoma macrodactylum croceum	Amphibians	U.S.A. (CA)	Entire	E
salamander, Shenandoah	Plethodon shenandoah	Amphibians	U.S.A. (VA)	Entire	E
Salamander, Sonora tiger	Ambystoma tigrinum stebbinsi	Amphibians	U.S.A. (AZ), Mexico	Entire	E
salamander, Texas blind	Typhlomolge rathbuni	Amphibians	U.S.A. (TX)	Entire	E
salmon, Atlantic	Salmo salar	Fishes	U.S.A., Canada, Greenland, western Europe	U.S.A. ME Gulf of Maine Atlantic Salmon Distinct Population Segment, which includes all naturally reproducing wild populations and those river-specific hatchery populations of Atlantic salmon having historical, river-specific characteristics found north of and including tributaries of the lower Kennebec River to, but not including, the mouth of the St. Croix River at the U.S.-Canda border. To date, the Services have determined that these populations are found in the Dennys, East Machias,, E Sacramento R.	E
salmon, chinook	Oncorhynchus (=Salmo) tshawytscha	Fishes	North America from Ventura R. in California to Point Hope, Alaska, and the Mackenzie R. area in Canada; Northeast Asia from Hokkaido, Japan, to the Anadyr R., Russia	(U.S.A._CA) winter run, wherever found	E
salmon, chinook	Oncorhynchus (=Salmo) tshawytscha	Fishes	North America from Ventura R. in California to Point Hope, Alaska, and the Mackenzie R. area in Canada; Northeast Asia from Hokkaido, Japan, to the Anadyr R., Russia	U.S.A.(WA) all naturally spawned populations in the Columbia R. tributaries upstream of Rock Island Dam and downstream of Chief Joseph Dam, excluding the Okanogan R., and the Columbia R. from a line between the west end of Clatsop jetty, OR, and the west end of Peacock jetty, WA, upstream to Chief Joseph Dam, including spring-run hatchery stocks (and their progeny) in Chiwawa R., Methow R., Twisp R., Chewuch R., White R., and Nason Creek	E
salmon, chinook	Oncorhynchus (=Salmo) tshawytscha	Fishes	North America from Ventura R. in California to Point Hope, Alaska, and the Mackenzie R. area in Canada; Northeast Asia from Hokkaido, Japan, to the Anadyr R., Russia	U.S.A.(WA) all naturally spawned populations from rivers and streams flowing into Puget Sound, including the Straits of Juan De Fuca from the Elwha R. eastward, and Hood Canal, South Sound, North Sound and the Strait of Georgia	T

List of threatened and endangered species - *continued*

Common name	Scientific name	Group	Historic range	Vertebrate population where endangered or threatened	Status
salmon, chinook	Oncorhynchus (=Salmo) tshawytscha	Fishes	North America from Ventura R. in California to Point Hope, Alaska, and the Mackenzie R. area in Canada; Northeast Asia from Hokkaido, Japan, to the Anadyr R., Russia	Snake R. (U.S.A._ID,OR,WA) mainstem and the following subbasins_Tucannon R., Grande Ronde R., Imnaha R., and Salmon R.; spring/summer run, natural population(s), wherever found	T
salmon, chinook	Oncorhynchus (=Salmo) tshawytscha	Fishes	North America from Ventura R. in California to Point Hope, Alaska, and the Mackenzie R. area in Canada; Northeast Asia from Hokkaido, Japan, to the Anadyr R., Russia	Snake R. (U.S.A._ID, OR, WA) mainstem and the following subbasins_Tucannon R., Grande Ronde R., Imnaha R., Salmon R, and Clearwater R.; fall run, natural population(s), wherever found	T
salmon, chinook	Oncorhynchus (=Salmo) tshawytscha	Fishes	North America from Ventura R. in California to Point Hope, Alaska, and the Mackenzie R. area in Canada; Northeast Asia from Hokkaido, Japan, to the Anadyr R., Russia	U.S.A.(OR, WA) all naturally spawned populations from the Columbia R. and its tributaries upstream from its mouth to a point east of the Hood R. and White Salmon R. to Willamette Falls in Oregon, excluding the spring run in the Clackamas R.	T
salmon, chinook	Oncorhynchus (=Salmo) tshawytscha	Fishes	North America from Ventura R. in California to Point Hope, Alaska, and the Mackenzie R. area in Canada; Northeast Asia from Hokkaido, Japan, to the Anadyr R., Russia	U.S.A.(CA) all naturally spawned spring-run populations from the Sacramento San Joaquin R. mainstem and its tributaries	T
salmon, chinook	Oncorhynchus (=Salmo) tshawytscha	Fishes	North America from Ventura R. in California to Point Hope, Alaska, and the Mackenzie R. area in Canada; Northeast Asia from Hokkaido, Japan, to the Anadyr R., Russia	U.S.A.(CA) from Redwood Creek south to Russian R., inclusive, all naturally spawned populations in mainstems and tributaries	T
salmon, chinook	Oncorhynchus (=Salmo) tshawytscha	Fishes	North America from Ventura R. in California to Point Hope, Alaska, and the Mackenzie R. area in Canada; Northeast Asia from Hokkaido, Japan, to the Anadyr R., Russia	U.S.A.(OR) all naturally spawned populations in the Clackamas R. and the Willamette R. and its tributaries above Willamette Falls	T
salmon, chum	Oncorhynchus (=Salmo) keta	Fishes	North Pacific Rim from Korea and the Japanese Island of Honshu east to Monterey Bay California; Arctic Ocean from the Laptev Sea in Russia to Mackenzie R. in Canada	U.S.A. (OR, WA) all naturally spawned populations in the Columbia R. and its tributaries	T

List of threatened and endangered species - *continued*

Common name	Scientific name	Group	Historic range	Vertebrate population where endangered or threatened	Status
salmon, chum	Oncorhynchus (=Salmo) keta	Fishes	North Pacific Rim from Korea and the Japanese Island of Honshu east to Monterey Bay California; Arctic Ocean from the Laptev Sea in Russia to Mackenzie R. in Canada	U.S.A. (WA) all naturally spawned summer-run populations in Hood Canal and its tributaries and Olympic Penninsula rivers between Hood Canal and Dungeness Bay	T
salmon, coho	Oncorhynchus (=Salmo) kisutch	Fishes	North Pacific Basin from U.S.A. (CA to AK) to Russia and Japan	U.S.A. (CA), naturally spawning populations in streams between Punta Gorda, Humboldt Co., CA and the San Lorenzo River, Santa Cruz, Co., CA	T
salmon, coho	Oncorhynchus (=Salmo) kisutch	Fishes	North Pacific Basin from U.S.A. (CA to AK) to Russia and Japan	U.S.A. (natural populations in river basins between Cape Blanco in Curry County, OR and Punta Gorda in Humboldt Co., CA)	T
salmon, sockeye	Oncorhynchus (=Salmo) nerka	Fishes	North Pacific Basin from U.S.A. (CA) to Russia	U.S.A. (Snake River, ID stock wherever found.)	E
salmon, sockeye	Oncorhynchus (=Salmo) nerka	Fishes	North Pacific Basin from U.S.A. (CA) to Russia	U.S.A. (WA) all naturally spawned population in Ozette Lake and its tributary streams	T
sand-verbena, large-fruited	Abronia macrocarpa	Flowering Plants	U.S.A. (TX)	Entire	E
sandalwood, Lanai (='iliahi)	Santalum freycinetianum var. lanaiense	Flowering Plants	U.S.A. (HI)	Entire	E
sandlace	Polygonella myriophylla	Flowering Plants	U.S.A. (FL)	Entire	E
sandwort, Bear Valley	Arenaria ursina	Flowering Plants	U.S.A. (CA)	Entire	T
sandwort, Cumberland	Arenaria cumberlandensis	Flowering Plants	U.S.A. (KY, TN)	Entire	E
Sandwort, Marsh	Arenaria paludicola	Flowering Plants	U.S.A. (CA)	Entire	E
schiedea, Diamond Head	Schiedea adamantis	Flowering Plants	U.S.A. (HI)	Entire	E
scrub-bird, noisy	Atrichornis clamosus	Birds	Australia	Entire	E
Sculpin, pygmy	Cottus paulus (=pygmaeus)	Fishes	U.S.A. (AL)	Entire	T
sea turtle, green	Chelonia mydas	Reptiles	Circumglobal in tropical and temperate seas and oceans	Breeding colony populations in FL and on Pacific coast of Mexico	E
sea turtle, green	Chelonia mydas	Reptiles	Circumglobal in tropical and temperate seas and oceans	Wherever found except where listed as endangered	T
sea turtle, hawksbill	Eretmochelys imbricata	Reptiles	Tropical seas	Entire	E
sea turtle, Kemp's ridley	Lepidochelys kempii	Reptiles	Tropical and temperate seas in Atlantic Basin, incl. Gulf of Mexico	Entire	E
sea turtle, leatherback	Dermochelys coriacea	Reptiles	Tropical, temperate, and subpolar seas	Entire	E
sea turtle, loggerhead	Caretta caretta	Reptiles	Circumglobal in tropical and temperate seas and oceans	Entire	T
sea turtle, olive ridley	Lepidochelys olivacea	Reptiles	Circumglobal in tropical and temperate seas	Breeding colony populations on Pacific coast of Mexico	E
sea turtle, olive ridley	Lepidochelys olivacea	Reptiles	Circumglobal in tropical and temperate seas	Wherever found except where listed as endangered below	T
sea-lion, Steller	Eumetopias jubatus	Mammals	U.S.A. (AK, CA, OR, WA), Canada, Russia; North Pacific Ocean	Population segment west of 1440 W. Long	E
sea-lion, Steller	Eumetopias jubatus	Mammals	U.S.A. (AK, CA, OR, WA), Canada, Russia; North Pacific Ocean	Entire, except the population segment west of 1440 W. Long	T
seablite, California	Suaeda californica	Flowering Plants	U.S.A. (CA)	Entire	E
seagrass, Johnson's	Halophila johnsonii	Flowering Plants	U.S.A. (FL)	Entire	T

List of threatened and endangered species - *continued*

Common name	Scientific name	Group	Historic range	Vertebrate population where endangered or threatened	Status
seal, Caribbean monk	Monachus tropicalis	Mammals	Caribbean Sea, Gulf of Mexico	Entire	E
seal, Guadalupe fur	Arctocephalus townsendi	Mammals	U.S.A. (Farallon Islands of CA) south to Mexico (Islas Revillagigedo)	Entire	T
seal, Hawaiian monk	Monachus schauinslandi	Mammals	U.S.A. (HI)	Entire	E
seal, Mediterranean monk	Monachus monachus	Mammals	Mediterranean, Northwest African Coast and Black Sea	Entire	E
seal, Saimaa	Phoca hispida saimensis	Mammals	Finland (Lake Saimaa)	Entire	E
sedge, golden	Carex lutea	Flowering Plants	U.S.A. (NC)	Entire	E
sedge, Navajo	Carex specuicola	Flowering Plants	U.S.A. (AZ, UT)	Entire	T
sedge, white	Carex albida	Flowering Plants	U.S.A. (CA)	Entire	E
seledang	Bos gaurus	Mammals	Bangladesh, Southeast Asia, India	Entire	E
serow	Naemorhedus (=Capricornis) sumatraensis	Mammals	East Asia, Sumatra	Entire	E
serval, Barbary	Leptailurus (=Felis) serval constantina	Mammals	Algeria	Entire	E
shagreen, Magazine Mountain	Mesodon magazinensis	Snails	U.S.A. (AR)	Entire	T
shama, Cebu black (thrush)	Copsychus niger cebuensis	Birds	Philippines	Entire	E
shapo	Ovis vignei vignei	Mammals	Kashmir	Entire	E
shearwater, Newell's Townsend's	Puffinus auricularis newelli	Birds	U.S.A. (HI)	Entire	T
sheep, bighorn	Ovis canadensis	Mammals	U.S.A. (Western conterminous states), Canada (southwestern), Mexico (northern)	U.S.A. (CA) Peninsular Ranges	E
sheep, bighorn	Ovis canadensis californiana	Mammals	U.S.A. (Western conterminous states), Canada (southwestern), Mexico (northern)	U.S.A. (CA) Sierra Nevada	E
shiner, Arkansas River	Notropis girardi	Fishes	U.S.A. (AR, KS, NM, OK, TX)	Arkansas River Basin (AR, KS, NM, OK, TX)	T
shiner, beautiful	Cyprinella formosa	Fishes	U.S.A. (AZ, NM), Mexico	Entire	T
shiner, blue	Cyprinella caerulea	Fishes	U.S.A. (AL, GA, TN)	Entire	T
shiner, Cahaba	Notropis cahabae	Fishes	U.S.A. (AL)	Entire	E
shiner, Cape Fear	Notropis mekistocholas	Fishes	U.S.A. (NC)	Entire	E
shiner, palezone	Notropis albizonatus	Fishes	U.S.A. (AL, KY, TN)	Entire	E
shiner, Pecos bluntnose	Notropis simus pecosensis	Fishes	U.S.A. (NM)	Entire	T
shiner, Topeka	Notropis topeka (=tristis)	Fishes	U.S.A. (IA, KS, MN, MO, NE, SD)	Entire	E
shou	Cervus elaphus wallichi	Mammals	Tibet, Bhutan	Entire	E
Shrew, Buena Vista Lake ornate	Sorex ornatus relictus	Mammals	U.S.A. (CA)	Entire	E
shrike, San Clemente loggerhead	Lanius ludovicianus mearnsi	Birds	U.S.A. (CA)	Entire	E
shrimp, Alabama cave	Palaemonias alabamae	Crustaceans	U.S.A. (AL)	Entire	E
shrimp, California freshwater	Syncaris pacifica	Crustaceans	U.S.A. (CA)	Entire	E
shrimp, Kentucky cave	Palaemonias ganteri	Crustaceans	U.S.A. (KY)	Entire	E
shrimp, Squirrel Chimney Cave	Palaemonetes cummingi	Crustaceans	U.S.A. (FL)	Entire	T
siamang	Symphalangus syndactylus	Mammals	Malaysia, Indonesia	Entire	E
sifakas	Propithecus spp.	Mammals	Malagasy Republic (=Madagascar)	Entire	E
silverside, Waccamaw	Menidia extensa	Fishes	U.S.A. (NC)	Entire	T
silversword, Mauna Loa (=Ka'u)	Argyroxiphium kauense	Flowering Plants	U.S.A. (HI)	Entire	E
siskin, red	Carduelis cucullata	Birds	South America	Entire	E
skink, bluetail mole	Eumeces egregius lividus	Reptiles	U.S.A. (FL)	Entire	T
skink, Round Island	Leiolopisma telfairi	Reptiles	Indian Ocean_Mauritius	Entire	T
skink, sand	Neoseps reynoldsi	Reptiles	U.S.A. (FL)	Entire	T

List of threatened and endangered species - *continued*

Common name	Scientific name	Group	Historic range	Vertebrate population where endangered or threatened	Status
skipper, Carson wandering	Pseudocopaeodes eunus obscurus	Insects	U.S.A. (CA, NV)	Entire	E
skipper, Laguna Mountains	Pyrgus ruralis lagunae	Insects	U.S.A. (CA)	Entire	E
skipper, Pawnee montane	Hesperia leonardus montana	Insects	U.S.A. (CO)	Entire	T
skullcap, Florida	Scutellaria floridana	Flowering Plants	U.S.A. (FL)	Entire	T
skullcap, large-flowered	Scutellaria montana	Flowering Plants	U.S.A. (GA, TN)	Entire	T
slabshell, Chipola	Elliptio chipolaensis	Clams	U.S.A. (AL, FL)	Entire	T
sloth, Brazilian three-toed	Bradypus torquatus	Mammals	Brazil	Entire	E
smelt, delta	Hypomesus transpacificus	Fishes	U.S.A. (CA)	Entire	T
snail, armored	Pyrgulopsis (=Marstonia) pachyta	Snails	U.S.A. (AL)	Entire	E
snail, Bliss Rapids	Taylorconcha serpenticola	Snails	U.S.A. (ID)	Entire	T
snail, Chittenango ovate amber	Succinea chittenangoensis	Snails	U.S.A. (NY)	Entire	T
Snail, flat-spired three-toothed	Triodopsis platysayoides	Snails	U.S.A. (WV)	Entire	T
snail, Iowa Pleistocene	Discus macclintocki	Snails	U.S.A. (IA, IL)	Entire	E
snail, Manus Island tree	Papustyla pulcherrima	Snails	Pacific Ocean_Admiralty Is. (Manus Is.)	Entire	E
snail, Morro shoulderband (=Banded dune)	Helminthoglypta walkeriana	Snails	U.S.A. (CA)	Entire	E
snail, Newcomb's	Erinna newcombi	Snails	U.S.A. (HI)	Entire	T
snail, noonday	Mesodon clarki nantahala	Snails	U.S.A. (NC)	Entire	T
snail, painted snake coiled forest	Anguispira picta	Snails	U.S.A. (TN)	Entire	T
snail, Snake River physa	Physa natricina	Snails	U.S.A. (ID)	Entire	E
snail, Stock Island tree	Orthalicus reses (not incl. nesodryas)	Snails	U.S.A. (FL)	Entire	T
snail, tulotoma	Tulotoma magnifica	Snails	U.S.A. (AL)	Entire	E
snail, Utah valvata	Valvata utahensis	Snails	U.S.A. (ID)	Entire	E
snail, Virginia fringed mountain	Polygyriscus virginianus	Snails	U.S.A. (VA)	Entire	E
snails, Oahu tree	Achatinella spp.	Snails	U.S.A. (HI)	Entire	E
snake, Atlantic salt marsh	Nerodia clarkii taeniata	Reptiles	U.S.A. (FL)	Entire	T
snake, Concho water	Nerodia paucimaculata	Reptiles	U.S.A. (TX)	Entire	T
snake, copperbelly water	Nerodia erythrogaster neglecta	Reptiles	U.S.A. (IL, IN, KY, MI, OH)	IN north of 400 N. Lat., MI, OH	T
snake, eastern indigo	Drymarchon corais couperi	Reptiles	U.S.A. (AL, FL, GA, MS, SC)	Entire	T
snake, giant garter	Thamnophis gigas	Reptiles	U.S.A. (CA)	Entire	T
snake, Lake Erie water	Nerodia sipedon insularum	Reptiles	U.S.A. (OH), Canada (Ont.)	Lake Erie offshore islands and their adjacent waters (located more than 1 mile from mainland)_U.S.A. (OH), Canada (Ont.)	T
snake, Maria Island	Liophus ornatus	Reptiles	West Indies_St. Lucia (Maria Islands)	Entire	E
snake, San Francisco garter	Thamnophis sirtalis tetrataenia	Reptiles	U.S.A. (CA)	Entire	E
snakeroot	Eryngium cuneifolium	Flowering Plants	U.S.A. (FL)	Entire	E
sneezeweed, Virginia	Helenium virginicum	Flowering Plants	U.S.A. (MO, VA)	Entire	T
snowbells, Texas	Styrax texanus	Flowering Plants	U.S.A. (TX)	Entire	E
solenodon, Cuban	Solenodon cubanus	Mammals	Cuba	Entire	E
solenodon, Haitian	Solenodon paradoxus	Mammals	Dominican Republic, Haiti	Entire	E
sparrow, Cape Sable seaside	Ammodramus maritimus mirabilis	Birds	U.S.A. (FL)	Entire	E
sparrow, Florida grasshopper	Ammodramus savannarum floridanus	Birds	U.S.A. (FL)	Entire	E
sparrow, San Clemente sage	Amphispiza belli clementeae	Birds	U.S.A. (CA)	Entire	T
sparrowhawk, Anjouan Island	Accipiter francesii pusillus	Birds	Indian Ocean_Comoro Islands	Entire	E
Spider, Government Canyon Bat Cave	Neoleptoneta microps	Arachnids	U.S.A. (TX).	Entire	E

List of threatened and endangered species - *continued*

Common name	Scientific name	Group	Historic range	Vertebrate population where endangered or threatened	Status
spider, Kauai cave wolf or pe'e pe'e maka 'ole	Adelocosa anops	Arachnids	U.S.A. (HI)	Entire	E
spider, spruce-fir moss	Microhexura montivaga	Arachnids	U.S.A. (NC,TN)	Entire	E
Spider, Tooth Cave	Leptoneta myopica	Arachnids	U.S.A. (TX)	Entire	E
spikedace	Meda fulgida	Fishes	U.S.A. (AZ, NM), Mexico	Entire	T
spinedace, Big Spring	Lepidomeda mollispinis pratensis	Fishes	U.S.A. (NV)	Entire	T
spinedace, Little Colorado	Lepidomeda vittata	Fishes	U.S.A. (AZ)	Entire	T
spinedace, White River	Lepidomeda albivallis	Fishes	U.S.A. (NV)	Entire	E
spineflower, Ben Lomond	Chorizanthe pungens var. hartwegiana	Flowering Plants	U.S.A. (CA)	Entire	E
spineflower, Howell's	Chorizanthe howellii	Flowering Plants	U.S.A. (CA)	Entire	E
spineflower, Monterey	Chorizanthe pungens var. pungens	Flowering Plants	U.S.A. (CA)	Entire	T
spineflower, Orcutt's	Chorizanthe orcuttiana	Flowering Plants	U.S.A. (CA)	Entire	E
spineflower, Robust (incl. Scotts Valley)	Chorizanthe robusta (incl. vars. robusta and hartwegii)	Flowering Plants	U.S.A. (CA)	Entire	E
spineflower, slender-horned	Dodecahema leptoceras	Flowering Plants	U.S.A. (CA)	Entire	E
spineflower, Sonoma	Chorizanthe valida	Flowering Plants	U.S.A. (CA)	Entire	E
spinymussel, James	Pleurobema collina	Clams	U.S.A. (VA, WV)	Entire	E
spinymussel, Tar River	Elliptio steinstansana	Clams	U.S.A. (NC)	Entire	E
spiraea, Virginia	Spiraea virginiana	Flowering Plants	U.S.A. (GA, KY, NC, OH, PA, TN, VA, WV)	Entire	T
springfish, Hiko White River	Crenichthys baileyi grandis	Fishes	U.S.A. (NV)	Entire	E
springfish, Railroad Valley	Crenichthys nevadae	Fishes	U.S.A. (NV)	Entire	T
springfish, White River	Crenichthys baileyi baileyi	Fishes	U.S.A. (NV)	Entire	E
springsnail, Alamosa	Tryonia alamosae	Snails	U.S.A. (NM)	Entire	E
springsnail, Bruneau Hot	Pyrgulopsis bruneauensis	Snails	U.S.A. (ID)	Entire	E
springsnail, Idaho	Fontelicella idahoensis	Snails	U.S.A. (ID)	Entire	E
springsnail, Socorro	Pyrgulopsis neomexicana	Snails	U.S.A. (NM)	Entire	E
spurge, deltoid	Chamaesyce deltoidea ssp. deltoidea	Flowering Plants	U.S.A. (FL)	Entire	E
spurge, Garber's	Chamaesyce garberi	Flowering Plants	U.S.A. (FL)	Entire	T
spurge, Hoover's	Chamaesyce hooveri	Flowering Plants	U.S.A. (CA)	Entire	T
spurge, telephus	Euphorbia telephioides	Flowering Plants	U.S.A. (FL)	Entire	T
squirrel, Carolina northern flying	Glaucomys sabrinus coloratus	Mammals	U.S.A. (NC, TN)	Entire	E
squirrel, Delmarva Peninsula fox	Sciurus niger cinereus	Mammals	U.S.A. (Delmarva Peninsula to southeastern PA)	Entire, except Sussex Co., DE	E
squirrel, Delmarva Peninsula fox	Sciurus niger cinereus	Mammals	U.S.A. (Delmarva Peninsula to southeastern PA)	U.S.A. (DE_Sussex Co.)	EXPN
squirrel, Mount Graham red	Tamiasciurus hudsonicus grahamensis	Mammals	U.S.A. (AZ)	Entire	E
squirrel, northern Idaho ground	Spermophilus brunneus brunneus	Mammals	U.S.A. (ID)	Entire	T
squirrel, Virginia northern flying	Glaucomys sabrinus fuscus	Mammals	U.S.A. (VA, WV)	Entire	E
stag, Barbary	Cervus elaphus barbarus	Mammals	Tunisia, Algeria	Entire	E
stag, Kashmir	Cervus elaphus hanglu	Mammals	Kashmir	Entire	E
starling, Ponape mountain	Aplonis pelzelni	Birds	West Pacific Ocean_Federated States of Micronesia	Entire	E
starling, Rothschild's (myna)	Leucopsar rothschildi	Birds	Indonesia (Bali)	Entire	E

List of threatened and endangered species - *continued*

Common name	Scientific name	Group	Historic range	Vertebrate population where endangered or threatened	Status
steelhead	Oncorhynchus (=Salmo) mykiss	Fishes	North Pacific Ocean from the Kamchatka Peninsula in Asia to the northern Baja Peninsula	All naturally spawned populations (and their progeny) in rivers from the Santa Maria R., San Luis Obispo County, CA (inclusive) to Malibu Cr., Los Angeles County, CA (inclusive)	E
steelhead	Oncorhynchus (=Salmo) mykiss	Fishes	North Pacific Ocean from the Kamchatka Peninsula in Asia to the northern Baja Peninsula	All naturally spawned populations (and their progeny) in the Sacramento and San Joaquin Rivers and their tributaries, excluding San Francisco and San Pablo Bays and their tributaries	T
steelhead	Oncorhynchus (=Salmo) mykiss	Fishes	North Pacific Ocean from the Kamchatka Peninsula in Asia to the northern Baja Peninsula	All naturally spawned populations and their progeny in river basins from Redwood Creek in Humboldt County, CA to the Gualala River in Mendocino County, CA (inclusive)	T
steelhead	Oncorhynchus (=Salmo) mykiss	Fishes	North Pacific Ocean from the Kamchatka Peninsula in Asia to the northern Baja Peninsula	U.S.A. (OR, WA) All naturally spawned populations in streams above and excluding the Wind R. in Washington, and the Hood R. in Oregon, upstream to, and including, the Yakima R. Excluded are steelhead from the Snake R. Basin	T
steelhead	Oncorhynchus (=Salmo) mykiss	Fishes	North Pacific Ocean from the Kamchatka Peninsula in Asia to the northern Baja Peninsula	U.S.A. (OR) All naturally spawned winter-run populations in the Willamette R. and its tributaries from Willamette Falls to the Calapooia R., inclusive	T
steelhead	Oncorhynchus (=Salmo) mykiss	Fishes	North Pacific Ocean from the Kamchatka Peninsula in Asia to the northern Baja Peninsula	All naturally spawned populations (and their progeny) in streams from the Pajaro R. (inclusive) located in Santa Cruz County, CA, to (but not including) the Santa Maria R	T
steelhead	Oncorhynchus (=Salmo) mykiss	Fishes	North Pacific Ocean from the Kamchatka Peninsula in Asia to the northern Baja Peninsula	All naturally spawned populations (and their progeny) in streams and tributaries to the Columbia R. between the Cowlitz and Wind Rivers, WA, inclusive, and the Willamette and Hood Rivers, OR, inclusive, excluding the Upper Willamette River Basin above Willamette Falls and excluding the Little and Big White Salmon Rivers in WA	T

List of threatened and endangered species - *continued*

Common name	Scientific name	Group	Historic range	Vertebrate population where endangered or threatened	Status
steelhead	Oncorhynchus (=Salmo) mykiss	Fishes	North Pacific Ocean from the Kamchatka Peninsula in Asia to the northern Baja Peninsula	All naturally spawned populations (and their progeny) in streams from the Russian R. to Aptos Cr., Santa Cruz County, CA (inclusive), and the drainages of San Francisco and San Pablo Bays eastward to the Napa R. (inclusive), Napa County, CA, excluding the Sacramento-San Joaquin R. Basin of the Central Valley of CA	T
steelhead	Oncorhynchus (=Salmo) mykiss	Fishes	North Pacific Ocean from the Kamchatka Peninsula in Asia to the northern Baja Peninsula	All naturally spawned populations (and their progeny) in streams in the Snake R. Basin of southeast WA, northeast OR, and ID	T
steelhead	Oncorhynchus (=Salmo) mykiss	Fishes	North Pacific Ocean from the Kamchatka Peninsula in Asia to the northern Baja Peninsula	All naturally spawned populations (and their progeny) in the Upper Columbia R. Basin upstream from the Yakima R., WA, to the U.S./Canada border, and also including the Wells Hatchery stock	E
stickleback, unarmored threespine	Gasterosteus aculeatus williamsoni	Fishes	U.S.A. (CA)	Entire	E
stickseed, showy	Hackelia venusta	Flowering Plants	U.S.A. (WA)	Entire	E
stilt, Hawaiian	Himantopus mexicanus knudseni	Birds	U.S.A. (HI)	Entire	E
stirrupshell	Quadrula stapes	Clams	U.S.A. (AL, MS)	Entire	E
stonecrop, Lake County	Parvisedum leiocarpum	Flowering Plants	U.S.A. (CA)	Entire	E
stork, oriental white	Ciconia boyciana (=ciconia b.)	Birds	China, Japan, Korea, Russia	Entire	E
stork, wood	Mycteria americana	Birds	U.S.A., (CA, AZ, TX, to Carolinas), Mexico, C. and S. America	U.S.A. (AL, FL, GA, SC)	E
sturgeon, Alabama	Scaphirhynchus suttkusi	Fishes	U.S.A. (AL, MS)	Entire	E
sturgeon, Beluga	Huso huso	Fishes	Black Sea, Caspian sea, Adriatic Sea and Sea of Azov.	Entire	T
sturgeon, gulf	Acipenser oxyrinchus desotoi	Fishes	U.S.A. (AL, FL, GA, LA, MS)	Entire	T
sturgeon, pallid	Scaphirhynchus albus	Fishes	U.S.A. (AR, IA, IL, KS, KY, LA, MO, MS, MT, ND, NE, SD, TN)	Entire	E
sturgeon, shortnose	Acipenser brevirostrum	Fishes	U.S.A. and Canada (Atlantic Coast)	Entire	E
sturgeon, white	Acipenser transmontanus	Fishes	U.S.A. (ID, MT), Canada (B.C.)	U.S.A. (ID, MT), Canada (B.C.), (Kootenai R. system)	E
sucker, June	Chasmistes liorus	Fishes	U.S.A. (UT)	Entire	E
sucker, Lost River	Deltistes luxatus	Fishes	U.S.A. (CA, OR)	Entire	E
Sucker, Modoc	Catostomus microps	Fishes	U.S.A. (CA)	Entire	E
sucker, razorback	Xyrauchen texanus	Fishes	U.S.A. (AZ, CA, CO, NM, NV, UT, WY), Mexico	Entire	E
sucker, Santa Ana	Catostomus santaanae	Fishes	U.S.A. (CA)	Los Angeles River basin, San Gabriel River basin, Santa Ana River basin	T
Sucker, shortnose	Chasmistes brevirostris	Fishes	U.S.A. (CA, OR)	Entire	E
sucker, Warner	Catostomus warnerensis	Fishes	U.S.A. (OR)	Entire	T
sumac, Michaux's	Rhus michauxii	Flowering Plants	U.S.A. (GA, NC, SC, VA)	Entire	E
sunbird, Marungu	Nectarinia prigoginei	Birds	Zaire	Entire	E
sunburst, Hartweg's golden	Pseudobahia bahiifolia	Flowering Plants	U.S.A. (CA)	Entire	E

List of threatened and endangered species - *continued*

Common name	Scientific name	Group	Historic range	Vertebrate population where endangered or threatened	Status
sunburst, San Joaquin adobe	Pseudobahia peirsonii	Flowering Plants	U.S.A. (CA)	Entire	T
sunflower, Eggert's	Helianthus eggertii	Flowering Plants	U.S.A. (AL, KY, TN)	Entire	T
sunflower, Pecos (=puzzle, =paradox)	Helianthus paradoxus	Flowering Plants	U.S.A. (NM, TX)	Entire	T
sunflower, San Mateo woolly	Eriophyllum latilobum	Flowering Plants	U.S.A. (CA)	Entire	E
sunflower, Schweinitz's	Helianthus schweinitzii	Flowering Plants	U.S.A. (NC, SC)	Entire	E
suni, Zanzibar	Neotragus moschatus moschatus	Mammals	Zanzibar (and nearby islands)	Entire	E
sunray, Ash Meadows	Enceliopsis nudicaulis var. corrugata	Flowering Plants	U.S.A. (NV)	Entire	T
sunshine, Sonoma	Blennosperma bakeri	Flowering Plants	U.S.A. (CA)	Entire	E
swiftlet, Mariana gray	Aerodramus vanikorensis bartschi	Birds	Western Pacific Ocean_U.S.A. (Guam, Rota, Tinian, Saipan, Agiguan)	Entire	E
tadpole shrimp, vernal pool	Lepidurus packardi	Crustaceans	U.S.A. (CA)	Entire	E
tahr, Arabian	Hemitragus jayakari	Mammals	Oman	Entire	E
tamaraw	Bubalus mindorensis	Mammals	Philippines	Entire	E
tamarin, golden-rumped	Leontopithecus spp.	Mammals	Brazil	Entire	E
tamarin, pied	Saguinus bicolor	Mammals	Brazil	Entire	E
tamarin, white-footed	Saguinus leucopus	Mammals	Colombia	Entire	T
tango, Miyako (=Toyko bitterling)	Tanakia tanago	Fishes	Japan	Entire	E
tapir, Asian	Tapirus indicus	Mammals	Burma, Laos, Cambodia, Vietnam, Malaysia, Indonesia, Thailand	Entire	E
tapir, Central American	Tapirus bairdii	Mammals	Southern Mexico to Colombia and Ecuador	Entire	E
tapir, mountain	Tapirus pinchaque	Mammals	Colombia, Ecuador and possibly Peru and Venezuela	Entire	E
tapir, South American (=Brazilian)	Tapirus terrestris	Mammals	Colombia and Venezuela south to Paraguay and Argentina	Entire	E
taraxacum, California	Taraxacum californicum	Flowering Plants	U.S.A. (CA)	Entire	E
Tarplant, Gaviota	Deinandra increscens ssp. villosa	Flowering Plants	U.S.A. (CA)	Entire	E
tarplant, Otay	Deinandra (=Hemizonia) conjugens	Flowering Plants	U.S.A. (CA), Mexico	Entire	T
tarplant, Santa Cruz	Holocarpha macradenia	Flowering Plants	U.S.A. (CA)	Entire	T
tarsier, Philippine	Tarsius syrichta	Mammals	Philippines	Entire	T
tartaruga	Podocnemis expansa	Reptiles	South America_Orinoco R. and Amazon R. basins	Entire	E
teal, Campbell Island flightless	Anas aucklandica nesiotis	Birds	New Zealand (Campbell Island)	Entire	E
temoleh, Ikan (minnow)	Probarbus jullieni	Fishes	Thailand, Cambodia, Vietnam, Malaysia, Laos	Entire	E
tern, California least	Sterna antillarum browni	Birds	Mexico, U.S.A. (CA)	Entire	E
tern, least	Sterna antillarum	Birds	U.S.A. (Atlantic and Gulf coasts, Miss. R. Basin, CA), Greater and Lesser Antilles, Bahamas, Mexico; winters Central America, northern South America	U.S.A. (AR, CO, IA, IL, IN, KS, KY, LA_Miss. R. and tribs. N of Baton Rouge, MS_Miss. R., MO, MT, ND, NE, NM, OK, SD, TN, TX_except within 50 miles of coast)	E
tern, roseate	Sterna dougallii dougallii	Birds	Tropical and temperate coasts of Atlantic Basin and East Africa	U.S.A. (Atlantic Coast south to NC), Canada (Newf., N.S, Que.), Bermuda	E

List of threatened and endangered species - *continued*

Common name	Scientific name	Group	Historic range	Vertebrate population where endangered or threatened	Status
tern, roseate	Sterna dougallii dougallii	Birds	Tropical and temperate coasts of Atlantic Basin and East Africa	Western Hemisphere and adjacent oceans, incl. U.S.A. (FL, PR, VI), where not listed as endangered	T
terrapin, river	Batagur baska	Reptiles	Malaysia, Bangladesh, Burma, India, Indonesia	Entire	E
thelypody, Howell's spectacular	Thelypodium howellii spectabilis	Flowering Plants	U.S.A. (OR)	Entire	T
thistle, Chorro Creek bog	Cirsium fontinale var. obispoense	Flowering Plants	U.S.A. (CA)	Entire	E
thistle, fountain	Cirsium fontinale var. fontinale	Flowering Plants	U.S.A. (CA)	Entire	E
thistle, La Graciosa	Cirsium loncholepis	Flowering Plants	U.S.A. (CA)	Entire	E
thistle, Loch Lomond coyote	Eryngium constancei	Flowering Plants	U.S.A. (CA)	Entire	E
thistle, Pitcher's	Cirsium pitcheri	Flowering Plants	U.S.A. (IL, IN, MI, WI), Canada (Ont.)	Entire	T
thistle, Sacramento Mountains	Cirsium vinaceum	Flowering Plants	U.S.A. (NM)	Entire	T
thistle, Suisun	Cirsium hydrophilum var. hydrophilum	Flowering Plants	U.S.A. (CA)	Entire	E
thornmint, San Diego	Acanthomintha ilicifolia	Flowering Plants	U.S.A. (CA), Mexico	Entire	T
thornmint, San Mateo	Acanthomintha obovata ssp. duttonii	Flowering Plants	U.S.A. (CA)	Entire	E
thrasher, white-breasted	Ramphocinclus brachyurus	Birds	West Indies_St. Lucia, Martinique	Entire	E
three-ridge, fat (mussel)	Amblema neislerii	Clams	U.S.A. (FL, GA)	Entire	E
Thrush, large Kauai (=kamao)	Myadestes myadestinus	Birds	U.S.A. (HI)	Entire	E
thrush, Molokai	Myadestes lanaiensis rutha	Birds	U.S.A. (HI)	Entire	E
thrush, New Zealand (wattlebird)	Turnagra capensis	Birds	New Zealand	Entire	E
Thrush, small Kauai (=puaiohi)	Myadestes palmeri	Birds	U.S.A. (HI)	Entire	E
thrush, Taita	Turdus olivaceus helleri	Birds	Kenya	Entire	E
tiger	Panthera tigris	Mammals	Temperate and tropical Asia	Entire	E
tiger beetle, northeastern beach	Cicindela dorsalis dorsalis	Insects	U.S.A. (CT, MA, MD, NJ, NY, PA, RI, VA)	Entire	T
tiger beetle, Ohlone	Cicindela ohlone	Insects	U.S.A. (CA)	Entire	E
tiger beetle, Puritan	Cicindela puritana	Insects	U.S.A. (CT, MA, MD, NH, VT)	Entire	T
tiger, Tasmanian	Thylacinus cynocephalus	Mammals	Australia	Entire	E
tinamou, solitary	Tinamus solitarius	Birds	Brazil, Paraguay, Argentina	Entire	E
toad, arroyo (=arroyo southwestern)	Bufo californicus (=microscaphus)	Amphibians	U.S.A. (CA), Mexico.	Entire	E
toad, Cameroon	Bufo superciliaris	Amphibians	Equatorial Africa	Entire	E
toad, Houston	Bufo houstonensis	Amphibians	U.S.A. (TX)	Entire	E
toad, Monte Verde golden	Bufo periglenes	Amphibians	Costa Rica	Entire	E
toad, Puerto Rican crested	Peltophryne lemur	Amphibians	U.S.A. (PR), British Virgin Islands	Entire	T
Toad, Wyoming	Bufo baxteri (=hemiophrys)	Amphibians	U.S.A. (WY)	Entire	E
toads, African viviparous	Nectophrynoides spp.	Amphibians	Tanzania, Guinea, Ivory Coast, Cameroon, Liberia, Ethiopia	Entire	E
tomistoma	Tomistoma schlegelii	Reptiles	Malaysia, Indonesia	Entire	E
topminnow, Gila (incl. Yaqui)	Poeciliopsis occidentalis	Fishes	U.S.A. (AZ, NM), Mexico	U.S.A. only	E
torreya, Florida	Torreya taxifolia	Conifers and Cycads	U.S.A. (FL, GA)	Entire	E
tortoise, angulated	Geochelone yniphora	Reptiles	Malagasy Republic (=Madagascar)	Entire	E
tortoise, Bolson	Gopherus flavomarginatus	Reptiles	Mexico	Entire	E

List of threatened and endangered species - *continued*

Common name	Scientific name	Group	Historic range	Vertebrate population where endangered or threatened	Status
tortoise, desert	Gopherus agassizii	Reptiles	U.S.A. (AZ, CA, NV, UT), Mexico	AZ south and east of Colorado R., and Mexico, when found outside of Mexico or said range in AZ	SAT
tortoise, desert	Gopherus agassizii	Reptiles	U.S.A. (AZ, CA, NV, UT), Mexico	Entire, except AZ south and east of Colorado R., and Mexico	T
tortoise, Galapagos	Geochelone nigra (=elephantopus)	Reptiles	Ecuador (Galapagos Islands)	Entire	E
tortoise, gopher	Gopherus polyphemus	Reptiles	U.S.A. (AL, FL, GA, LA, MS, SC)	Wherever found west of Mobile and Tombigbee Rivers in AL, MS, and LA	T
tortoise, Madagascar radiated	Geochelone radiata	Reptiles	Malagasy Republic (=Madagascar)	Entire	E
totoaba (seatrout or weakfish)	Cynoscion macdonaldi	Fishes	Mexico (Gulf of California)	Entire	E
towhee, Inyo California	Pipilo crissalis eremophilus	Birds	U.S.A. (CA)	Entire	T
townsendia, Last Chance	Townsendia aprica	Flowering Plants	U.S.A. (UT)	Entire	T
tracaja	Podocnemis unifilis	Reptiles	South America_Orinoco R. and Amazon R. basins	Entire	E
trembler, Martinique (thrasher)	Cinclocerthia ruficauda gutturalis	Birds	West Indies_Martinique	Entire	E
trillium, persistent	Trillium persistens	Flowering Plants	U.S.A. (GA, SC)	Entire	E
trillium, relict	Trillium reliquum	Flowering Plants	U.S.A. (AL, GA, SC)	Entire	E
trout, Apache	Oncorhynchus apache	Fishes	U.S.A. (AZ)	Entire	T
Trout, bull	Salvelinus confluentus	Fishes	U.S.A. (AK, Pacific NW into CA, ID, NV, MT), Canada (NW Territories)	U.S.A., conterminous, (lower 48 states)	T
trout, Gila	Oncorhynchus gilae	Fishes	U.S.A. (AZ, NM)	Entire	E
trout, greenback cutthroat	Oncorhynchus clarki stomias	Fishes	U.S.A. (CO)	Entire	T
trout, Lahontan cutthroat	Oncorhynchus clarki henshawi	Fishes	U.S.A. (CA, NV, OR, UT)	Entire	T
trout, Little Kern golden	Oncorhynchus aguabonita whitei	Fishes	U.S.A. (CA)	Entire	T
trout, Paiute cutthroat	Oncorhynchus clarki seleniris	Fishes	U.S.A. (CA)	Entire	T
tuatara	Sphenodon punctatus	Reptiles	New Zealand	Entire	E
tuatara, Brother's Island	Sphenodon guntheri	Reptiles	New Zealand (N. Brother's Island)	Entire	E
tuctoria, Greene's	Tuctoria greenei	Flowering Plants	U.S.A. (CA	Entire	E
turaco, Bannerman's	Tauraco bannermani	Birds	Cameroon	Entire	E
turtle dove, Seychelles	Streptopelia picturata rostrata	Birds	Indian Ocean_Seychelles	Entire	E
turtle, Alabama red-belly	Pseudemys alabamensis	Reptiles	U.S.A. (AL)	Entire	E
turtle, aquatic box	Terrapene coahuila	Reptiles	Mexico	Entire	E
turtle, black softshell	Trionyx nigricans	Reptiles	Bangladesh	Entire	E
turtle, bog (=Muhlenberg)	Clemmys muhlenbergii	Reptiles	U.S.A. (CT, DE, GA, MA, MD, NC, NJ, NY, PA, SC, TN, VA)	U.S.A. (GA, NC, SC, TN, VA)	SAT
turtle, bog (=Muhlenberg)	Clemmys muhlenbergii	Reptiles	U.S.A. (CT, DE, GA, MA, MD, NC, NJ, NY, PA, SC, TN, VA)	Entire, except GA, NC, SC, TN, VA	T
turtle, Brazilian sideneck	Phrynops hogei	Reptiles	Brazil	Entire	E
turtle, Burmese peacock	Morenia ocellata	Reptiles	Burma	Entire	E
turtle, Cat Island	Trachemys terrapen	Reptiles	West Indies_Jamaica, Bahamas	Cat Island in the Bahamas	E
turtle, Central American river	Dermatemys mawii	Reptiles	Mexico, Belize, Guatemala	Entire	E
turtle, Cuatro Cienegas softshell	Trionyx ater	Reptiles	Mexico	Entire	E
turtle, flattened musk	Sternotherus depressus	Reptiles	U.S.A. (AL)	Black Warrior R. system upstream from Bankhead Dam	T

List of threatened and endangered species - *continued*

Common name	Scientific name	Group	Historic range	Vertebrate population where endangered or threatened	Status
turtle, geometric	Psammobates geometricus	Reptiles	South Africa	Entire	E
turtle, Inagua Island	Trachemys stejnegeri malonei	Reptiles	West Indies_Bahamas (Great Inagua Island)	Entire	E
turtle, Indian sawback	Kachuga tecta tecta	Reptiles	India	Entire	E
turtle, Indian softshell	Trionyx gangeticus	Reptiles	Pakistan, India	Entire	E
turtle, peacock softshell	Trionyx hurum	Reptiles	India, Bangladesh	Entire	E
turtle, ringed map	Graptemys oculifera	Reptiles	U.S.A. (LA, MS)	Entire	T
turtle, short-necked or western swamp	Pseudemydura umbrina	Reptiles	Australia	Entire	E
turtle, South American red-lined	Trachemys scripta callirostris	Reptiles	Colombia, Venezuela	Entire	E
turtle, spotted pond	Geoclemys hamiltonii	Reptiles	North India, Pakistan	Entire	E
turtle, three-keeled Asian	Melanochelys tricarinata	Reptiles	Central India to Bangladesh and Burma	Entire	E
turtle, yellow-blotched map	Graptemys flavimaculata	Reptiles	U.S.A. (MS)	Entire	T
twinpod, Dudley Bluffs	Physaria obcordata	Flowering Plants	U.S.A. (CO)	Entire	T
uakari (all species)	Cacajao spp.	Mammals	Peru, Brazil, Ecuador, Colombia, Venezuela	Entire	E
uhiuhi	Caesalpinia kavaiense	Flowering Plants	U.S.A. (HI)	Entire	E
urial	Ovis musimon ophion	Mammals	Cyprus	Entire	E
uvillo	Eugenia haematocarpa	Flowering Plants	U.S.A. (PR)	Entire	E
vanga, Pollen's	Xenopirostris polleni	Birds	Madagascar	Entire	T
vanga, Van Dam's	Xenopirostris damii	Birds	Madagascar	Entire	T
vervain, Red Hills	Verbena californica	Flowering Plants	U.S.A. (CA)	Entire	T
vetch, Hawaiian	Vicia menziesii	Flowering Plants	U.S.A. (HI)	Entire	E
vicuna	Vicugna vicugna	Mammals	South America (Andes)	Entire	E
viper, Lar Valley	Vipera latifii	Reptiles	Iran	Entire	E
Vireo, black-capped	Vireo atricapilla	Birds	U.S.A. (KS, LA, NE, OK, TX), Mexico.	Entire	E
vireo, least Bell's	Vireo bellii pusillus	Birds	U.S.A. (CA), Mexico	Entire	E
vole, Amargosa	Microtus californicus scirpensis	Mammals	U.S.A. (CA)	Entire	E
vole, Florida salt marsh	Microtus pennsylvanicus dukecampbelli	Mammals	U.S.A. (FL)	Entire	E
vole, Hualapai Mexican	Microtus mexicanus hualpaiensis	Mammals	U.S.A. (AZ)	Entire	E
wahane	Pritchardia aylmer-robinsonii	Flowering Plants	U.S.A. (HI)	Entire	E
wallaby, banded hare	Lagostrophus fasciatus	Mammals	Australia	Entire	E
wallaby, brindled nail-tailed	Onychogalea fraenata	Mammals	Australia	Entire	E
wallaby, crescent nail-tailed	Onychogalea lunata	Mammals	Australia	Entire	E
wallaby, Parma	Macropus parma	Mammals	Australia	Entire	E
wallaby, western hare	Lagorchestes hirsutus	Mammals	Australia	Entire	E
wallaby, yellow-footed rock	Petrogale xanthopus	Mammals	Australia	Entire	E
wallflower, Ben Lomond	Erysimum teretifolium	Flowering Plants	U.S.A. (CA)	Entire	E
wallflower, Contra Costa	Erysimum capitatum var. angustatum	Flowering Plants	U.S.A. (CA)	Entire	E
wallflower, Menzies'	Erysimum menziesii	Flowering Plants	U.S.A. (CA)	Entire	E
Walnut, West Indian	Juglans jamaicensis	Flowering Plants	U.S.A. (PR), Cuba, Hispaniola	Entire	E
wanderer, plain (=collared-hemipode)	Pedionomus torquatus	Birds	Australia	Entire	E
warbler (=wood), Bachman's	Vermivora bachmanii	Birds	U.S.A. (Southeastern), Cuba	Entire	E
warbler (=wood), Barbados yellow	Dendroica petechia petechia	Birds	West Indies_Barbados	Entire	E
warbler (=wood), golden-cheeked	Dendroica chrysoparia	Birds	U.S.A. (TX), Mexico, Guatemala, Honduras, Nicaragua, Belize	Entire	E
warbler (=wood), Kirtland's	Dendroica kirtlandii	Birds	U.S.A. (principally MI), Canada, West Indies_Bahama Islands	Entire	E
warbler (=wood), Semper's	Leucopeza semperi	Birds	West Indies_St. Lucia	Entire	E

List of threatened and endangered species - *continued*

Common name	Scientific name	Group	Historic range	Vertebrate population where endangered or threatened	Status
Warbler, Aldabra (old world warbler)	Nesillas aldabranus	Birds	Indian Ocean_ Seychelles (Aldabra Island)	Entire	E
warbler, nightingale reed (old world warbler)	Acrocephalus luscinia	Birds	West Pacific Ocean_U.S.A. (Guam, Alamagan, Saipan)	Entire	E
warbler, Rodrigues (old world warbler)	Bebrornis rodericanus	Birds	Mauritius (Rodrigues Islands)	Entire	E
warbler, Seychelles (old world warbler)	Bebrornis sechellensis	Birds	Indian Ocean_Seychelles Island	Entire	E
warea, wide-leaf	Warea amplexifolia	Flowering Plants	U.S.A. (FL)	Entire	E
wartyback, white (pearlymussel)	Plethobasus cicatricosus	Clams	U.S.A. (AL, IL, IN, KY, TN)	Entire	E
water-plantain, Kral's	Sagittaria secundifolia	Flowering Plants	U.S.A. (AL, GA)	Entire	T
water-umbel, Huachuca	Lilaeopsis schaffneriana var. recurva	Flowering Plants	U.S.A. (AZ), Mexico	Entire	E
water-willow, Cooley's	Justicia cooleyi	Flowering Plants	U.S.A. (FL)	Entire	E
watercress, Gambel's	Rorippa gambellii	Flowering Plants	U.S.A. (CA)	Entire	E
wattle-eye, banded	Platysteira laticincta	Birds	Cameroon	Entire	E
wawae'iole	Huperzia mannii	Ferns and Allies	U.S.A. (HI)	Entire	E
wawae'iole	Lycopodium (=Phlegmariurus) nutans	Ferns and Allies	U.S.A. (HI)	Entire	E
weaver, Clarke's	Ploceus golandi	Birds	Kenya	Entire	E
wedgemussel, dwarf	Alasmidonta heterodon	Clams	U.S.A. (CT, DC, DE, MA, MD, NC, NH, NJ, NY, PA, VA, VT), Canada (N.B.)	Entire	E
whale, blue	Balaenoptera musculus	Mammals	Oceanic	Entire	E
whale, bowhead	Balaena mysticetus	Mammals	Oceanic (north latitudes only)	Entire	E
whale, finback	Balaenoptera physalus	Mammals	Oceanic	Entire	E
whale, gray	Eschrichtius robustus	Mammals	North Pacific Ocean—coastal and Bering Sea, formerly North Atlantic Ocean	Entire, except eastern North Pacific Ocean—coastal and Bering, Beaufort, and Chukchi Seas	E
whale, humpback	Megaptera novaeangliae	Mammals	Oceanic	Entire	E
whale, right	Balaena glacialis (incl. australis)	Mammals	Oceanic	Entire	E
whale, Sei	Balaenoptera borealis	Mammals	Oceanic	Entire	E
whale, sperm	Physeter catodon (=macrocephalus)	Mammals	Oceanic	Entire	E
whipbird, western	Psophodes nigrogularis	Birds	Australia	Entire	E
whipsnake (=striped racer), Alameda	Masticophis lateralis euryxanthus	Reptiles	U.S.A. (CA)	Entire	T
white-eye, bridled	Zosterops conspicillatus conspicillatus	Birds	Western Pacific Ocean_U.S.A. (Guam)	Entire	E
white-eye, Norfolk Island	Zosterops albogularis	Birds	Indian Ocean_Norfolk Islands	Entire	E
white-eye, Ponape greater	Rukia longirostra	Birds	West Pacific Ocean_Federated States of Micronesia	Entire	E
White-eye, Rota bridled	Zosterops rotensis	Birds	U.S.A. (MP)	Entire	E
white-eye, Seychelles	Zosterops modesta	Birds	Indian Ocean_Seychelles	Entire	E
whitlow-wort, papery	Paronychia chartacea	Flowering Plants	U.S.A. (FL)	Entire	T
wild-buckwheat, clay-loving	Eriogonum pelinophilum	Flowering Plants	U.S.A. (CO)	Entire	E
wild-buckwheat, gypsum	Eriogonum gypsophilum	Flowering Plants	U.S.A. (NM)	Entire	T
wild-buckwheat, southern mountain	Eriogonum kennedyi var. austromontanum	Flowering Plants	U.S.A. (CA)	Entire	T
wild-rice, Texas	Zizania texana	Flowering Plants	U.S.A. (TX)	Entire	E
wire-lettuce, Malheur	Stephanomeria malheurensis	Flowering Plants	U.S.A. (OR)	Entire	E
wireweed	Polygonella basiramia	Flowering Plants	U.S.A. (FL)	Entire	E

List of threatened and endangered species - continued

Common name	Scientific name	Group	Historic range	Vertebrate population where endangered or threatened	Status
wolf, gray	Canis lupus	Mammals	Holarctic	Southwestern Distinct Population Segment U.S.A. (AZ, NM, CO south of Interstate Highway 70, UT south of U.S. Highway 50, OK and TX, except those parts of OK and TX east of Interstate Highway 35; except where listed as an experimental population); Mexico	E
wolf, gray	Canis lupus	Mammals	Holarctic	U.S.A. (WY and portions of ID and MT_see 17.84(i))	EXPN
wolf, gray	Canis lupus	Mammals	Holarctic	U.S.A. (portions of AZ, NM, and TX - see section 17.84(k))	EXPN
wolf, gray	Canis lupus	Mammals	Holarctic	Western Distinct Population Segment U.S.A. (CA, ID, MT, NV, OR, WA, WY, UT north of U.S. Highway 50, and CO north of Interstate Highway 70, except where listed as an experimental population)	T
wolf, gray	Canis lupus	Mammals	Holarctic	Eastern Distinct Population Segment U.S.A. (CT, IA, IL, IN, KS, MA, ME, MI, MN, MO, ND, NE, NH, NJ, NY, OH, PA, RI, SD, VT, and WI)	T
wolf, maned	Chrysocyon brachyurus	Mammals	Argentina, Bolivia, Brazil, Paraguay, Uruguay	Entire	E
wolf, red	Canis rufus	Mammals	U.S.A. (SE U.S.A., west to central TX)	Entire, except where listed as experimental populations below	E
wolf, red	Canis rufus	Mammals	U.S.A. (SE U.S.A., west to central TX)	U.S.A. (portions of NC and TN_see 17.84(c)(9))	EXPN
wombat, Queensland hairy-nosed (incl. Barnard's)	Lasiorhinus krefftii (formerly L. barnardi and L. gillespiei	Mammals	Australia	Entire	E
woodland-star, San Clemente Island	Lithophragma maximum	Flowering Plants	U.S.A. (CA)	Entire	E
woodpecker, imperial Entire	Campephilus imperialis E	Birds	Mexico		
woodpecker, ivory-billed	Campephilus principalis	Birds	U.S.A. (southcentral and southeastern), Cuba	Entire	E
woodpecker, red-cockaded	Picoides borealis	Birds	U.S.A. (southcentral and southeastern)	Entire	E
woodpecker, Tristam's	Dryocopus javensis richardsi	Birds	Korea	Entire	E
woodrat, Key Largo	Neotoma floridana smalli	Mammals	U.S.A. (FL)	Entire	E
woodrat, riparian (=San Joaquin Valley)	Neotoma fuscipes riparia	Mammals	U.S.A. (CA)	Entire	E
woolly-star, Santa Ana River	Eriastrum densifolium ssp. sanctorum	Flowering Plants	U.S.A. (CA)	Entire	E
wooly-threads, San Joaquin	Monolopia (=Lembertia) congdonii	Flowering Plants	U.S.A. (CA)	Entire	E
woundfin	Plagopterus argentissimus	Fishes	U.S.A. (AZ, NV, UT)	Entire, except Gila R. drainage, AZ, NM	E
woundfin	Plagopterus argentissimus	Fishes	U.S.A. (AZ, NV, UT)	Gila R. drainage, AZ, NM	EXPN
wren, Guadeloupe house	Troglodytes aedon guadeloupensis	Birds	West Indies_Guadeloupe	Entire	E
wren, St. Lucia house	Troglodytes aedon mesoleucus	Birds	West Indies_St. Lucia	Entire	E
yak, wild	Bos mutus (=grunniens m.)	Mammals	China (Tibet), India	Entire	E
yellowhead, desert	Yermo xanthocephalus	Flowering Plants	U.S.A. (WY)	Entire	T
yerba santa, Lompoc	Eriodictyon capitatum	Flowering Plants	U.S.A. (CA)	Entire	E

List of threatened and endangered species - *continued*

Common name	Scientific name	Group	Historic range	Vertebrate population where endangered or threatened	Status
zebra, Grevy's	Equus grevyi	Mammals	Kenya, Ethiopia, Somalia	Entire	T
zebra, Hartmann's mountain	Equus zebra hartmannae	Mammals	Namibia, Angola	Entire	T
zebra, mountain	Equus zebra zebra	Mammals	South Africa	Entire	E
ziziphus, Florida	Ziziphus celata	Flowering Plants	U.S.A. (FL)	Entire	E

Notes: *Data as of July 19, 2005*
Status: E=Endangered; T=Threatened; E or T prefaced by SA=Similarity of appearance species; EXPN=Experimental populations
Source: *U.S. Fish & Wildlife Service, Threatened and Endangered Species System (TESS)*

Delisted Species

Species name	Date species first listed	Date species delisted	Reason delisted
Alligator, American (Alligator mississippiensis)	3/11/1967	6/4/1987	Recovered
Barberry, Truckee (Berberis (=Mahonia) sonnei)	11/6/1979	10/1/2003	Original data in error (Taxonomic rev.)
Bidens, cuneate (Bidens cuneata)	2/17/1984	2/6/1996	Original data in error (Taxonomic rev.)
Broadbill, Guam (Myiagra freycineti)	8/27/1984	2/23/2004	Extinct
Butterfly, Bahama swallowtail (Heraclides andraemon bonhotei)	4/28/1976	8/31/1984	Original data in error (Act amendment)
Cactus, Lloyd's hedgehog (Echinocereus lloydii)	10/26/1979	6/24/1999	Original data in error (Taxonomic rev.)
Cactus, spineless hedgehog (Echinocereus triglochidiatus var. inermis)	11/7/1979	9/22/1993	Original data in error (Not a listable entity)
Cinquefoil, Robbins' (Potentilla robbinsiana)	9/17/1980	8/27/2002	Recovered
Cisco, longjaw (Coregonus alpenae)	3/11/1967	9/2/1983	Extinct
Deer, Columbian white-tailed Douglas County DPS (Odocoileus virginianus leucurus)	7/24/2003	7/24/2003	Recovered
Dove, Palau ground (Gallicolumba canifrons)	6/2/1970	9/12/1985	Recovered
Duck, Mexican U.S.A. only (Anas 'diazi')	3/11/1967	7/25/1978	Original data in error (Taxonomic rev.)
Falcon, American peregrine (Falco peregrinus anatum)	6/2/1970	8/25/1999	Recovered
Falcon, Arctic peregrine (Falco peregrinus tundrius)	6/2/1970	10/5/1994	Recovered
Flycatcher, Palau fantail (Rhipidura lepida)	6/2/1970	9/12/1985	Recovered
Gambusia, Amistad (Gambusia amistadensis)	4/30/1980	12/4/1987	Extinct
Globeberry, Tumamoc (Tumamoca macdougalii)	4/29/1986	6/18/1993	Original data in error (New info. discovered)
Goose, Aleutian Canada (Branta canadensis leucopareia)	3/11/1967	3/20/2001	Recovered
Hedgehog cactus, purple-spined (Echinocereus engelmannii var. purpureus)	10/11/1979	11/27/1989	Original data in error (Taxonomic rev.)
Kangaroo, eastern gray (Macropus giganteus)	12/30/1974	3/9/1995	Recovered
Kangaroo, red (Macropus rufus)	12/30/1974	3/9/1995	Recovered
Kangaroo, western gray (Macropus fuliginosus)	12/30/1974	3/9/1995	Recovered
Mallard, Mariana (Anas oustaleti)	6/2/1977	2/23/2004	Extinct
Milk-vetch, Rydberg (Astragalus perianus)	4/26/1978	9/14/1989	Original data in error (New info. discovered)
Monarch, Tinian (old world flycatcher) (Monarcha takatsukasae)	6/2/1970	9/21/2004	Recovered
Owl, Palau (Pyrroglaux podargina)	6/2/1970	9/12/1985	Recovered
Pearlymussel, Sampson's (Epioblasma sampsoni)	6/14/1976	1/9/1984	Extinct
Pelican, brown U.S. Atlantic coast, FL, AL (Pelecanus occidentalis)	n/a	2/4/1985	Recovered
Pennyroyal, Mckittrick (Hedeoma apiculatum)	7/13/1982	9/22/1993	Original data in error (New info. discovered)
Pike, blue (Stizostedion vitreum glaucum)	3/11/1967	9/2/1983	Extinct
Pupfish, Tecopa (Cyprinodon nevadensis calidae)	10/13/1970	1/15/1982	Extinct
Shrew, Dismal Swamp southeastern (Sorex longirostris fisheri)	9/26/1986	2/28/2000	Original data in error (New info. discovered)
Sparrow, Santa Barbara song (Melospiza melodia graminea)	6/4/1973	10/12/1983	Extinct
Sparrow, dusky seaside (Ammodramus maritimus nigrescens)	3/11/1967	12/12/1990	Extinct
Treefrog, pine barrens FL pop. (Hyla andersonii)	11/11/1977	11/22/1983	Original data in error (New info. discovered)
Trout, coastal cutthroat Umpqua R. (Oncorhynchus clarki clarki)	9/13/1996	4/26/2000	Original data in error (Taxonomic rev.)
Turtle, Indian flap-shelled (Lissemys punctata punctata)	6/14/1976	2/29/1984	Original data in error (Erroneous data)
Whale, gray except where listed (Eschrichtius robustus)	6/16/1994	6/16/1994	Recovered
Wolf, gray U.S.A. (delisting of all other lower 48 states or portions of lower 48 states not otherwise included in the 3 distinct population segments) (Canis lupus)	n/a	4/1/2003	Original data in error (Taxonomic rev.)
Woolly-star, Hoover's (Eriastrum hooveri)	7/19/1990	10/7/2003	Recovered and original data in error (New information discovered)

Notes: Data as of July 19, 2005
Source: U.S. Fish & Wildlife Service, Threatened and Endangered Species System (TESS)

Summary of listed species by group

Group	Endangered		Threatened		Total species	U.S. species with recovery plans[1]
	U.S.	Foreign	U.S.	Foreign		
Animals						
Amphibians	11	8	10	1	30	15
Arachnids	12	0	0	0	12	5
Birds	77	175	13	6	271	78
Clams	62	2	8	0	72	69
Crustaceans	18	0	3	0	21	13
Fishes	71	11	43	1	126	95
Insects	35	4	9	0	48	31
Mammals	68	251	10	20	349	55
Reptiles	14	64	22	16	116	33
Snails	21	1	11	0	33	22
Animal Subtotal	389	516	129	44	1,078	416
Plants						
Conifers and Cycads	2	0	1	2	5	3
Ferns and Allies	24	0	2	0	26	26
Flowering Plants	571	1	144	0	716	584
Lichens	2	0	0	0	2	2
Plant Subtotal	599	1	147	2	749	615
Grand Total	988	517	276	46	1,827[a]	1,031

Notes: Data as of July 19, 2005; (1) There are 546 distinct approved recovery plans. Some recovery plans cover more than one species, and a few species have separate plans covering different parts of their ranges. This count includes only plans generated by the USFWS or jointly by the USFWS and NMFS, and includes only listed species that occur in the United States; (a) There are 1,857 total listings (1,292 U.S.). A listing is an E or a T in the "status" column of the table titled "The List of Endangered and Threatened Species," also found in this chapter. The following types of listings are combined as single counts in the table above: species listed both as threatened and endangered (9 animal species have dual status in the U.S.), and subunits of a single species listed as distinct population segments. Only the endangered population is tallied for dual status populations (except for the following: Olive ridley sea turtle; for which only the threatened U.S. population is tallied). The dual status U.S. species that are tallied as endangered are: Chinook salmon, Gray wolf, Green sea turtle, Piping Plover, Roseate tern, Sockeye salmon, Steelhead, Steller sea-lion. The dual status foreign species that are tallied as endangered are: Argali, Chimpanzee, Leopard, Saltwater crocodile. Distinct population segments tallied as one include: Chinook salmon, Chum salmon, Coho salmon, Dugong, Gray wolf, Mariana fruit bat (=Mariana flying fox), Steelhead. Entries that represent entire genera or families include: African viviparous toads, Gibbons, Lemurs, Musk deer, Oahu tree snails, Sifakas, Uakari (all species).
Source: U.S. Fish & Wildlife Service, Threatened and Endangered Species System (TESS)

Number of U.S. listed species per year (1980-2002)

Calendar Year	Amphi-bians	Arach-nids	Birds	Clams	Crusta-ceans	Fishes	Insects	Mam-mals	Plants	Reptiles	Snails	Total[1]
Endangered												
1980	5	0	58	23	1	33	7	32	50	13	2	224
1981	5	0	58	23	1	33	7	32	51	13	3	226
1982	5	0	58	23	2	35	7	32	57	14	3	236
1983	5	0	58	23	3	34	7	35	58	14	3	240
1984	5	0	66	23	3	33	8	37	71	14	3	263
1985	5	0	68	23	3	40	8	43	93	14	3	300
1986	5	0	71	23	4	42	8	43	114	14	3	327
1987	5	0	73	28	5	42	8	46	139	15	3	364
1988	5	4	72	31	8	46	11	50	153	15	3	398
1989	6	4	72	34	8	50	12	51	166	15	3	421
1990	6	4	72	37	8	53	12	53	179	15	3	442
1991	6	4	72	40	8	54	14	56	238	15	7	514
1992	6	4	72	40	9	55	16	56	295	15	11	579
1993	6	4	72	50	11	61	17	56	323	14	12	626
1994	7	4	74	51	14	66	19	57	420	14	15	741
1995	7	5	75	51	14	66	20	57	432	14	15	756
1996	7	5	74	51	14	67	20	57	513	14	15	837
1997	9	5	76	56	16	67	28	57	553	14	15	896
1998	9	5	76	61	17	70	28	59	567	14	18	924
1999	9	5	74	61	17	69	28	61	581	16	18	939
2000	10	12	78	61	18	70	33	63	592	14	20	971
2001	11	12	78	62	18	71	35	64	595	14	21	981
2002	13	12	78	62	18	71	35	65	599	14	21	988
Threatened												
1980	3	0	3	0	0	14	7	4	9	12	5	57
1981	3	0	3	0	0	14	6	4	10	12	5	57
1982	3	0	3	0	1	14	6	4	10	12	5	58
1983	3	0	3	0	1	15	6	4	11	12	5	60
1984	3	0	3	0	1	18	5	5	11	12	5	63
1985	3	0	4	0	1	24	5	5	25	12	5	84
1986	3	0	4	0	1	28	7	6	27	14	5	95
1987	4	0	9	0	1	32	7	6	35	17	5	116
1988	4	0	9	0	1	31	7	6	48	17	5	128
1989	5	0	9	0	1	32	7	7	51	17	6	135
1990	5	0	11	2	2	33	9	8	61	17	6	154
1991	5	0	11	2	2	34	9	8	64	17	6	158
1992	5	0	12	2	2	36	9	9	74	18	7	174
1993	5	0	16	3	2	37	9	9	80	19	7	187
1994	5	0	16	3	3	39	9	9	90	19	7	200
1995	5	0	16	6	3	39	9	9	93	19	7	206
1996	6	0	16	6	3	40	9	9	101	19	7	216
1997	7	0	17	6	3	41	9	9	115	22	7	236
1998	7	0	17	8	3	49	9	10	135	22	10	270
1999	8	0	15	8	3	43	9	8	140	22	10	266
2000	8	0	15	8	3	44	9	9	144	22	11	273
2001	8	0	14	8	3	44	9	9	145	22	11	273
2002	9	0	14	8	3	44	9	9	146	22	11	275

Notes: *Data as of July 19, 2005; (1) Totals are not additive. Number of species listed fluctuate between years because of new listings, reclassifications, delistings, new information on taxonomy, and other reasons. For the nine species that have dual status, only the endangered population is tallied except for one species for which only the threatened population is tallied.*
Source: *U.S. Fish & Wildlife Service, Threatened and Endangered Species System (TESS)*

Energy overview (quadrillion Btu)

	Production	Imports	Exports	Adjustments[a]	Consumption
1973 Total	63.585	14.613	2.033	-0.456	75.708
1975 Total	61.357	14.032	2.323	-1.067	71.999
1980 Total	67.241	15.796	3.695	-1.054	78.289
1985 Total	67.647	11.781	4.196	1.238	76.469
1990 Total	70.729	18.817	4.752	-.126	84.668
1995 Total	71.156	22.260	4.511	2.315	91.221
1996 Total	72.472	23.702	4.633	2.683	94.224
1997 Total	72.389	25.215	4.514	1.637	94.727
1998 Total	72.787	26.581	4.299	.078	95.146
1999 Total	71.652	27.252	3.715	1.585	96.774
2000 Total	71.218	28.973	4.006	2.720	98.905
2001 Total	71.793	30.157	3.770	-1.799	96.380
2002 Total	70.673	29.406	3.661	1.370	97.788
2003 January	6.002	2.429	.377	1.262	9.316
February	5.406	2.180	.300	1.201	8.487
March	5.917	2.585	.316	.230	8.416
April	5.771	2.613	.333	-.358	7.693
May	5.952	2.747	.357	-.667	7.674
June	5.823	2.661	.351	-.494	7.639
July	5.893	2.752	.339	-.030	8.275
August	5.936	2.731	.335	.069	8.401
September	5.761	2.666	.325	-.474	7.627
October	5.898	2.668	.349	-.402	7.815
November	5.581	2.458	.338	.178	7.879
December	5.983	2.624	.345	.739	9.001
Total	69.921	31.115	4.066	1.253	98.223
2004 January	6.069	2.572	.299	1.048	9.390
February	5.608	2.506	.312	.924	8.727
March	5.995	2.793	.388	-.009	8.391
April	5.747	2.615	.410	-.196	7.755
May	5.824	2.805	.390	-.279	7.959
June	5.908	2.785	.390	-.389	7.915
July	5.991	2.906	.372	-.174	8.351
August	5.998	2.907	.375	-.223	8.307
September	5.697	2.638	.362	-.158	7.815
October	5.748	2.850	.351	-.297	7.950
November	R 5.702	2.776	.350	R -.076	8.052
December	R 6.056	2.852	.434	R .679	R 9.153
Total	R 70.343	33.004	4.433	R .850	R 99.764
2005 January	R 5.980	2.784	.340	R .918	R 9.343
February	R 5.557	R 2.665	.354	R .441	R 8.308
March	6.091	2.828	.393	.194	8.721
3-Month Total	17.628	8.277	1.086	1.553	26.372
2004 3-Month Total	17.673	7.871	.999	1.963	26.508
2003 3-Month Total	17.324	7.195	.994	2.693	26.218

Notes: (a) A balancing item. Includes stock changes, losses, gains, miscellaneous blending components, and unaccounted-for supply; R=Revised; E=Estimate; Totals may not equal sum of components due to independent rounding; Geographic coverage is the 50 States and the District of Columbia.

Definitions: ENERGY PRODUCTION-Includes production of fossil fuels (coal, dry natural gas, crude oil and lease condensate, and natural gas plant liquids), nuclear electric power, pumped-storage hydroelectric power, and renewable energy. Renewable energy production is assumed to be equivalent to: end-use consumption of wood, waste, alcohol fuels, geothermal heat pump and direct use energy, and solar thermal direct use and photovoltaic energy; and electric utility and nonutility net electricity generation from conventional hydroelectric power, wood, waste, geothermal, solar, and wind; ENERGY CONSUMPTION-Includes consumption of fossil fuels (coal, natural gas, and petroleum), some secondary energy derived from fossil fuels (supplemental gaseous fuels, coal coke net imports, and electricity net imports from fossil fuels), nuclear electric power, pumped-storage hydroelectric power, and renewable energy. Renewable energy consumption includes: end-use consumption of wood, waste, alcohol fuels, geothermal heat pump and direct use energy, and solar thermal direct use and photovoltaic energy; electric utility and nonutility net electricity generation from conventional hydroelectric power, wood, waste, geothermal, solar, and wind; and net imports of electricity from hydroelectric power and geothermal energy; ENERGY IMPORTS-Includes imports of fossil fuels (coal, natural gas, and petroleum, including crude oil imported for the Strategic Petroleum Reserve), some secondary energy derived from fossil fuels (coal coke imports, and electricity imports from fossil fuels), and renewable energy (electricity imports derived from hydroelectric power and geothermal energy); ENERGY EXPORTS-Includes exports of fossil fuels (coal, natural gas, and petroleum), some secondary energy derived from fossil fuels (coal coke exports, and electricity exports from fossil fuels), and renewable energy (electricity exports derived from hydroelectric power).

Source: U.S. Department of Energy, Energy Information Administration, Monthly Energy Review, June 2005

Energy production by source (quadrillion Btu)

	Fossil Fuels					Nuclear Electric Power	Hydro-electric Pumped Storage[c]	Renewable Energy[a]					Total
	Coal	Natural Gas (Dry)	Crude Oil[b]	Natural Gas Plant Liquids	Total			Conventional Hydroelectric Power	Wood, Waste, Alcohol[d]	Geo-thermal	Solar and Wind	Total	
1973 Total	13.992	22.187	19.493	2.569	58.241	0.910	(e)	2.861	1.529	0.043	NA	4.433	63.585
1975 Total	14.989	19.640	17.729	2.374	54.733	1.900	(e)	3.155	1.499	.070	NA	4.723	61.357
1980 Total	18.598	19.908	18.249	2.254	59.008	2.739	(e)	2.900	2.485	.110	NA	5.494	67.241
1985 Total	19.325	16.980	18.992	2.241	57.539	4.076	(e)	2.970	2.864	.198	(s)	6.033	67.647
1990 Total	22.456	18.326	15.571	2.175	58.529	6.104	-.036	3.046	2.662	.336	.089	6.133	70.729
1995 Total	22.029	19.082	13.887	2.442	57.440	7.075	-.028	3.205	3.068	.294	.102	6.669	71.156
1996 Total	22.684	19.344	13.723	2.530	58.281	7.087	-.032	3.590	3.127	.316	.104	7.137	72.472
1997 Total	23.211	19.394	13.658	2.495	58.758	6.597	-.041	3.640	3.006	.325	.104	7.075	72.389
1998 Total	23.935	19.613	13.235	2.420	59.204	7.068	-.046	3.297	2.835	.328	.101	6.561	72.787
1999 Total	23.186	19.341	12.451	2.528	57.505	7.610	-.062	3.268	2.885	.331	.115	6.599	71.652
2000 Total	22.623	19.662	12.358	2.611	57.254	7.862	-.057	2.811	2.907	.317	.123	6.158	71.218
2001 Total	23.490	20.205	12.282	2.547	58.523	8.033	-.091	2.242	2.640	.311	.135	5.328	71.793
2002 Total	22.622	19.439	12.163	2.559	56.783	8.143	-.089	2.689	2.648	.328	.170	5.835	70.673
2003 January	1.902	1.661	1.040	.204	4.807	.721	-.008	.211	.229	.029	.012	.481	6.002
February	1.686	1.510	.940	.190	4.327	.635	-.008	.203	.210	.027	.012	.452	5.406
March	1.827	1.709	1.046	.200	4.782	.625	-.008	.248	.226	.029	.016	.518	5.917
April	1.832	1.636	1.005	.191	4.664	.592	-.006	.254	.224	.027	.017	.521	5.771
May	1.857	1.671	1.031	.181	4.740	.648	-.006	.301	.225	.028	.016	.569	5.952
June	1.814	1.618	.992	.177	4.602	.669	-.008	.293	.223	.029	.016	.560	5.823
July	1.815	1.639	.994	.191	4.638	.726	-.008	.254	.238	.029	.015	.536	5.893
August	1.836	1.671	1.006	.197	4.711	.719	-.008	.235	.236	.029	.014	.514	5.936
September	1.854	1.610	.989	.198	4.651	.663	-.008	.189	.223	.028	.015	.455	5.761
October	1.928	1.665	1.013	.211	4.817	.625	-.006	.189	.230	.028	.014	.462	5.898
November	1.727	1.592	.968	.206	4.493	.621	-.007	.202	.230	.027	.015	.474	5.581
December	1.889	1.644	1.003	.200	4.736	.715	-.007	.246	.247	.030	.016	.539	5.983
Total	21.970	19.626	12.026	2.346	55.968	7.959	-.087	2.825	2.740	.339	.178	6.082	69.921
2004 January	1.912	E 1.679	E 1.015	.208	4.814	.739	-.007	.235	.243	.030	.016	.523	6.069
February	1.771	E 1.560	E .939	.194	4.465	.669	-.007	.213	.224	.028	.015	.481	5.608
March	1.940	E 1.667	E 1.011	.211	4.829	.660	-.006	.231	.234	.028	.019	.513	5.995
April	1.875	E 1.605	E .969	.199	4.648	.612	-.006	.212	.236	.027	.018	.493	5.747
May	1.782	E 1.627	E 1.009	.207	4.626	.678	-.007	.242	.234	.028	.023	.527	5.824
June	1.940	E 1.596	E .940	.194	4.670	.708	-.007	.255	.235	.028	.019	.537	5.908
July	1.886	E 1.653	E .972	.209	4.720	.751	-.007	.235	.246	.029	.017	.527	5.991
August	1.946	E 1.649	E .949	.215	4.759	.742	-.008	.220	.241	.029	.016	.505	5.998
September	1.911	E 1.536	E .886	.201	4.534	.688	-.007	.208	.230	.027	.016	.482	5.697
October	1.891	E 1.604	E .919	.210	4.625	.653	-.007	.193	.239	.029	.016	.477	5.748
November	1.884	RE 1.574	E .939	.209	R 4.606	.615	-.006	.213	.232	.028	.015	.488	R 5.702
December	1.949	RE 1.644	E .980	.210	R 4.782	.716	-.006	.267	.251	.029	.017	.564	R 6.056
Total	22.686	RE 19.395	E 11.528	2.468	R 56.077	8.232	-.082	2.725	2.845	.340	.206	6.116	R 70.343
2005 January	R 1.897	RE 1.645	E .970	.209	R 4.721	.728	-.007	.248	.247	.029	.015	.539	R 5.980
February	R 1.820	E 1.530	E .888	.194	R 4.432	.635	-.004	.221	.234	.025	.013	.493	R 5.557
March	2.067	E 1.654	E .988	.215	4.924	.641	-.005	.234	.248	.029	.019	.530	6.091
3-Month Total	5.784	E 4.830	E 2.846	.617	14.078	2.004	-.016	.703	.729	.083	.048	1.562	17.628
2004 3-Month Total	5.623	E 4.907	E 2.965	.613	14.108	2.069	-.020	.680	.701	.086	.050	1.517	17.673
2003 3-Month Total	5.416	4.880	3.026	.594	13.916	1.981	-.024	.661	.665	.085	.040	1.451	17.324

Notes: (a) End-use consumption and electricity net generation; (b) Includes lease condensate; (c) Pumped storage facility production minus energy used for pumping; (d) Alcohol is ethanol blended into motor gasoline; (e) Included in conventional hydroelectric power; R=Revised; NA=Not Available. E=Estimate; F=Forecast; (s)=Less than +0.5 trillion Btu and greater than -0.5 trillion Btu; Totals may not equal sum of components due to independent rounding; Geographic coverage is the 50 States and the District of Columbia.

Definitions: ENERGY PRODUCTION-Includes production of fossil fuels (coal, dry natural gas, crude oil and lease condensate, and natural gas plant liquids), nuclear electric power, pumped-storage hydroelectric power, and renewable energy. Renewable energy production is assumed to be equivalent to: end-use consumption of wood, waste, alcohol fuels, geothermal heat pump and direct use energy, and solar thermal direct use and photovoltaic energy; and electric utility and nonutility net electricity generation from conventional hydroelectric power, wood, waste, geothermal, solar, and wind.

Source: U.S. Department of Energy, Energy Information Administration, Monthly Energy Review, June 2005

Energy consumption by source (quadrillion Btu)

	Fossil Fuels				Nuclear Electric Power	Hydro-electric Pumped Storage[f]	Renewable Energy[a]					Total[d,h]
	Coal	Natural Gas[b]	Petro-leum[c,d]	Total[e]			Conventional Hydroelectric Power	Wood, Waste, Alcohol[d,g]	Geo-thermal	Solar and Wind	Total	
1973 Total	12.971	22.512	34.840	70.316	0.910	([i])	2.861	1.529	0.043	NA	4.433	75.708
1975 Total	12.663	19.948	32.731	65.355	1.900	([i])	3.155	1.499	.070	NA	4.723	71.999
1980 Total	15.423	20.394	34.202	69.984	2.739	([i])	2.900	2.485	.110	NA	5.494	78.289
1985 Total	17.478	17.834	30.922	66.221	4.076	([i])	2.970	2.864	.198	(s)	6.033	76.469
1990 Total	19.173	19.730	33.553	72.460	6.104	-.036	3.046	2.662	.336	.089	6.133	84.668
1995 Total	20.089	22.784	34.553	77.488	7.075	-.028	3.205	3.068	.294	.102	6.669	91.221
1996 Total	21.002	23.197	35.757	79.979	7.087	-.032	3.590	3.127	.316	.104	7.137	94.224
1997 Total	21.445	23.328	36.266	81.086	6.597	-.041	3.640	3.006	.325	.104	7.075	94.727
1998 Total	21.656	22.936	36.934	81.592	7.068	-.046	3.297	2.835	.328	.101	6.561	95.146
1999 Total	21.623	23.010	37.960	82.650	7.610	-.062	3.268	2.885	.331	.115	6.599	96.774
2000 Total	22.580	23.916	38.404	84.965	7.862	-.057	2.811	2.907	.317	.123	6.158	98.905
2001 Total	21.914	22.906	38.333	83.182	8.033	-.091	2.242	2.640	.311	.135	5.328	96.380
2002 Total	21.904	23.628	38.401	83.994	8.143	-.089	2.689	2.648	.328	.170	5.835	97.788
2003 January	2.019	2.800	3.314	8.134	.721	-.008	.211	.229	.029	.012	.481	9.316
February	1.774	2.589	3.046	7.423	.635	-.008	.203	.210	.027	.012	.452	8.487
March	1.757	2.276	3.262	7.299	.625	-.008	.248	.226	.029	.016	.518	8.416
April	1.617	1.805	3.177	6.602	.592	-.006	.254	.224	.027	.017	.521	7.693
May	1.710	1.567	3.202	6.481	.648	-.006	.301	.225	.028	.016	.569	7.674
June	1.845	1.415	3.171	6.435	.669	-.008	.293	.223	.029	.016	.560	7.639
July	2.046	1.653	3.326	7.031	.726	-.008	.254	.238	.029	.015	.536	8.275
August	2.077	1.704	3.408	7.190	.719	-.008	.235	.236	.029	.014	.514	8.401
September	1.866	1.475	3.193	6.537	.663	-.008	.189	.223	.028	.015	.455	7.627
October	1.802	1.615	3.341	6.762	.625	-.006	.189	.230	.028	.014	.462	7.815
November	1.813	1.817	3.184	6.817	.621	-.007	.202	.230	.027	.015	.474	7.879
December	1.994	2.355	3.423	7.778	.715	-.007	.246	.247	.030	.016	.539	9.001
Total	22.321	23.069	39.047	84.487	7.959	-.087	2.825	2.740	.339	.178	6.082	98.223
2004 January	2.020	2.758	3.378	8.160	.739	-.007	.235	.243	.030	.016	.523	9.390
February	1.827	2.585	3.185	7.606	.669	-.007	.213	.224	.028	.015	.481	8.727
March	1.736	2.166	3.340	7.251	.660	-.006	.231	.234	.028	.019	.513	8.391
April	1.612	1.805	3.240	6.680	.612	-.006	.212	.236	.027	.018	.493	7.755
May	1.779	1.622	3.348	6.786	.678	-.007	.242	.234	.028	.023	.527	7.959
June	1.887	1.533	3.260	6.699	.708	-.007	.255	.235	.028	.019	.537	7.915
July	2.036	1.636	3.413	7.094	.751	-.007	.235	.246	.029	.017	.527	8.351
August	2.015	1.624	3.435	7.081	.742	-.008	.220	.241	.029	.016	.505	8.307
September	1.875	1.529	3.272	6.675	.688	-.007	.208	.230	.027	.016	.482	7.815
October	1.801	1.605	3.436	6.848	.653	-.007	.193	.239	.029	.016	.477	7.950
November	1.801	1.837	3.332	6.975	.615	-.006	.213	.232	.028	.015	.488	8.052
December	2.003	2.396	3.492	7.899	.716	-.006	.267	.251	.029	.017	.564	R 9.153
Total	22.390	23.096	40.130	85.754	8.232	-.082	2.725	2.845	.340	.206	6.116	R 99.764
2005 January	R 2.015	R 2.679	3.400	R 8.105	.728	-.007	.248	.247	.029	.015	.539	R 9.343
February	R 1.775	R 2.328	3.090	R 7.206	.635	-.004	.221	.234	.025	.013	.493	R 8.308
March	1.851	2.284	3.435	7.579	.641	-.005	.234	.248	.029	.019	.530	8.721
3-Month Total	5.641	7.292	9.924	22.890	2.004	-.016	.703	.729	.083	.048	1.562	26.372
2004 3-Month Total	5.583	7.508	9.903	23.017	2.069	-.020	.680	.701	.086	.050	1.517	26.508
2003 3-Month Total	5.550	7.664	9.622	22.855	1.981	-.024	.661	.665	.085	.040	1.451	26.218

Notes: *(a) End-use consumption and electricity net generation; (b) Natural gas, plus a small amount of supplemental gaseous fuels that cannot be identified separately; (c) Petroleum products supplied, including natural gas plant liquids and crude oil burned as fuel. Beginning in 1993, also includes ethanol blended into motor gasoline; (d) Beginning in 1993, ethanol blended into motor gasoline is included in both "Petroleum" and "Wood, Waste, Alcohol," but is counted only once in total consumption; (e) Includes coal coke net imports; (f) Pumped storage facility production minus energy used for pumping; (g) "Alcohol" is ethanol blended into motor gasoline; (h) Includes coal coke net imports and electricity net imports, which are not separately displayed; (i) Included in conventional hydroelectric power; R=Revised; NA=Not available; E=Estimate; F=Forecast; (s)=Less than +0.5 trillion Btu and greater than -0.5 trillion Btu; Totals may not equal sum of components due to independent rounding. Geographic coverage is the 50 States and the District of Columbia.*

Definitions: *ENERGY CONSUMPTION-Includes consumption of fossil fuels (coal, natural gas, and petroleum), some secondary energy derived from fossil fuels (supplemental gaseous fuels, coal coke net imports, and electricity net imports from fossil fuels), nuclear electric power, pumped-storage hydroelectric power, and renewable energy. Renewable energy consumption includes: end-use consumption of wood, waste, alcohol fuels, geothermal heat pump and direct use energy, and solar thermal direct use and photovoltaic energy; electric utility and nonutility net electricity generation from conventional hydroelectric power, wood, waste, geothermal, solar, and wind; and net imports of electricity from hydroelectric power and geothermal energy.*

Source: *U.S. Department of Energy, Energy Information Administration, Monthly Energy Review, June 2005*

Cost of fuels to end users in constant (1982-84) dollars

	Consumer Price Index (Urban)[a]	Motor Gasoline[b]		Residential Heating Oil[c]		Residential Natural Gas[b]		Residential Electricity[b]	
	Index 1982-1984=100	Cents per Gallon	Dollars per Million Btu	Cents per Gallon	Dollars per Million Btu	Cents per Thousand Cubic Feet	Dollars per Million Btu	Cents per Kilowatthour	Dollars per Million Btu
1973 Average	44.4	NA	NA	NA	NA	290.5	2.85	5.6	16.50
1975 Average	53.8	NA	NA	NA	NA	317.8	3.12	6.5	19.07
1980 Average	82.4	148.2	11.85	118.2	8.52	446.6	4.36	6.6	19.21
1985 Average	107.6	111.2	8.89	97.9	7.06	568.8	5.52	6.87	20.13
1990 Average	130.7	93.1	7.44	81.3	5.86	443.8	4.31	5.99	17.56
1995 Average	152.4	79.1	6.37	56.9	4.10	397.6	3.87	5.51	16.15
1996 Average	156.9	82.1	6.61	63.0	4.54	404.1	3.93	5.33	15.62
1997 Average	160.5	80.4	6.48	61.3	4.42	432.4	4.21	5.25	15.39
1998 Average	163.0	68.4	5.51	52.3	3.77	418.4	4.05	5.07	14.85
1999 Average	166.6	73.3	5.91	52.6	3.79	401.6	3.91	4.90	14.36
2000 Average	172.2	90.8	7.32	76.1	5.49	450.6	4.39	4.79	14.02
2001 Average	177.1	86.4	6.97	70.6	5.09	543.8	5.27	4.87	14.27
2002 Average	179.9	80.1	6.46	62.8	4.52	438.6	4.26	4.70	13.78
2003 January	181.7	85.7	6.91	73.3	5.29	444.7	4.30	4.39	12.87
February	183.1	92.1	7.43	82.4	5.94	462.0	4.47	4.36	12.79
March	184.2	97.2	7.84	83.6	6.02	523.3	5.07	4.51	13.21
April	183.8	92.7	7.48	73.2	5.28	546.8	5.29	4.79	14.05
May	183.5	86.5	6.98	69.0	4.98	581.5	5.63	4.90	14.36
June	183.7	84.8	6.84	66.2	4.78	651.1	6.30	5.01	14.68
July	183.9	85.2	6.87	63.3	4.56	686.2	6.64	4.97	14.57
August	184.6	90.5	7.30	63.7	4.59	689.1	6.67	4.97	14.57
September	185.2	95.6	7.71	64.1	4.63	658.2	6.37	4.81	14.08
October	185.0	89.0	7.18	66.8	4.82	568.6	5.50	4.81	14.08
November	184.5	85.5	6.90	69.5	5.01	523.6	5.07	4.74	13.88
December	184.3	83.5	6.73	72.8	5.25	509.5	4.93	4.52	13.25
Average	**184.0**	**89.0**	**7.18**	**73.6**	**5.31**	**517.4**	**5.01**	**4.73**	**13.86**
2004 January	185.2	88.3	7.11	76.5	5.52	523.8	5.08	4.45	13.04
February	186.2	92.1	7.42	76.9	5.55	528.5	5.13	4.47	13.10
March	187.4	96.5	7.77	75.4	5.44	533.6	5.18	4.60	13.48
April	188.0	99.7	8.03	75.1	5.41	559.6	5.43	4.75	13.92
May	189.1	108.4	8.73	75.1	5.41	614.0	5.96	4.80	14.07
June	189.7	109.8	8.84	74.2	5.35	687.9	6.67	4.88	14.29
July	189.4	104.6	8.43	75.4	5.44	710.1	6.89	4.93	14.45
August	189.5	102.4	8.25	79.1	5.70	727.7	7.06	5.00	14.65
September	189.9	101.8	8.20	84.1	6.07	699.8	6.79	4.93	14.46
October	190.9	108.5	8.74	94.6	6.82	611.3	5.93	4.77	13.97
November	191.0	107.5	8.66	95.6	6.89	599.0	5.81	4.69	13.75
December	190.3	101.2	8.15	94.2	6.79	R 583.8	R 5.66	4.51	13.21
Average	**188.9**	**101.8**	**8.20**	**81.8**	**5.90**	**568.6**	**5.51**	**4.73**	**13.87**
2005 January	190.7	97.9	7.88	94.8	6.83	R 577.9	R 5.60	4.45	13.05
February	191.8	102.2	8.23	96.1	6.93	568.3	5.51	4.55	13.32
March	193.3	109.0	8.78	R 100.4	R 7.24	R 567.0	R 5.50	R 4.58	R 13.42
April	194.6	119.5	9.62	NA	NA	NA	NA	NA	NA

Notes: (a) Consumer Price Index, All Urban Consumers, All Items, 1982-1984 = 100.0; (b) Includes taxes; (c) Excludes taxes; R=Revised; NA=Not available; Fuel costs are calculated by using the Urban Consumer Price Index (CPI) developed by the Bureau of Labor Statistics; Annual averages may not equal average of months due to independent rounding; Geographic coverage is the 50 States and the District of Columbia.

Source: U.S. Department of Energy, Energy Information Administration, Monthly Energy Review, June 2005

Motor vehicle mileage, fuel consumption, and fuel rates

	Passenger Cars[a]			Vans, Pickup Trucks, and Sport Utility Vehicles[b]			Trucks[c]			All Motor Vehicles[d]		
	Mileage (miles per vehicle)	Fuel Consumption (gallons per vehicle)	Fuel Rate (miles per gallon)	Mileage (miles per vehicle)	Fuel Consumption (gallons per vehicle)	Fuel Rate (miles per gallon)	Mileage (miles per vehicle)	Fuel Consumption (gallons per vehicle)	Fuel Rate (miles per gallon)	Mileage (miles per vehicle)	Fuel Consumption (gallons per vehicle)	Fuel Rate (miles per gallon)
1973	9,884	737	13.4	9,779	931	10.5	15,370	2,775	5.5	10,099	850	11.9
1974	9,221	677	13.6	9,452	862	11.0	14,995	2,708	5.5	9,493	788	12.0
1975	9,309	665	14.0	9,829	934	10.5	15,167	2,722	5.6	9,627	790	12.2
1976	9,418	681	13.8	10,127	934	10.8	15,438	2,764	5.6	9,774	806	12.1
1977	9,517	676	14.1	10,607	947	11.2	16,700	3,002	5.6	9,978	814	12.3
1978	9,500	665	14.3	10,968	948	11.6	18,045	3,263	5.5	10,077	816	12.4
1979	9,062	620	14.6	10,802	905	11.9	18,502	3,380	5.5	9,722	776	12.5
1980	8,813	551	16.0	10,437	854	12.2	18,736	3,447	5.4	9,458	712	13.3
1981	8,873	538	16.5	10,244	819	12.5	19,016	3,565	5.3	9,477	697	13.6
1982	9,050	535	16.9	10,276	762	13.5	19,931	3,647	5.5	9,644	686	14.1
1983	9,118	534	17.1	10,497	767	13.7	21,083	3,769	5.6	9,760	686	14.2
1984	9,248	530	17.4	11,151	797	14.0	22,550	3,967	5.7	10,017	691	14.5
1985	9,419	538	17.5	10,506	735	14.3	20,597	3,570	5.8	10,020	685	14.6
1986	9,464	543	17.4	10,764	738	14.6	22,143	3,821	5.8	10,143	692	14.7
1987	9,720	539	18.0	11,114	744	14.9	23,349	3,937	5.9	10,453	694	15.1
1988	9,972	531	18.8	11,465	745	15.4	22,485	3,736	6.0	10,721	688	15.6
1989	[a]10,157	[a]533	[a]19.0	11,676	724	16.1	22,926	3,776	6.1	10,932	688	15.9
1990	10,504	520	20.2	11,902	738	16.1	23,603	3,953	6.0	11,107	677	16.4
1991	10,571	501	21.1	12,245	721	17.0	24,229	4,047	6.0	11,294	669	16.9
1992	10,857	517	21.0	12,381	717	17.3	25,373	4,210	6.0	11,558	683	16.9
1993	10,804	527	20.5	12,430	714	17.4	26,262	4,309	6.1	11,595	693	16.7
1994	10,992	531	20.7	12,156	701	17.3	25,838	4,202	6.1	11,683	698	16.7
1995	11,203	530	21.1	12,018	694	17.3	26,514	4,315	6.1	11,793	700	16.8
1996	11,330	534	21.2	11,811	685	17.2	26,092	4,221	6.2	11,813	700	16.9
1997	11,581	539	21.5	12,115	703	17.2	27,032	4,218	6.4	12,107	711	17.0
1998	11,754	544	21.6	12,173	707	17.2	25,397	4,135	6.1	12,211	721	16.9
1999	11,848	553	21.4	11,957	701	17.0	26,014	4,352	6.0	12,206	732	16.7
2000	11,976	547	21.9	11,672	669	17.4	25,617	4,391	5.8	12,164	720	16.9
2001	11,831	534	22.1	11,204	636	17.6	26,602	4,477	5.9	11,887	695	17.1
2002	12,202	555	22.0	11,364	650	17.5	27,071	4,642	5.8	12,171	719	16.9
2003[P]	12,242	550	22.3	11,467	647	17.7	27,286	4,750	5.7	12,210	716	17.0

Notes: (a) Motorcycles are included through 1989; (b) Includes a small number of trucks with 2 axles and 4 tires, such as step vans; (c) Single-unit trucks with 2 axles and 6 or more tires, and combination trucks; (d) Includes buses and motorcycles, which are not shown separately; P=Preliminary; Geographic coverage is the 50 States and the District of Columbia.

Source: U.S. Department of Energy, Energy Information Administration, Monthly Energy Review, June 2005

Residential sector energy consumption (quadrillion Btu)

| | Primary Consumption | | | | | | | | | | | |
| | Fossil Fuels | | | | Renewable Energy[a] | | | | | | Electrical System | |
	Coal	Natural Gas[b]	Petroleum	Total	Wood	Geo-thermal[c]	Solar[d]	Total	Total Primary	Electricity Retail Sales[e]	Energy Losses[f]	Total
1973 Total	0.094	4.977	2.825	7.896	0.354	NA	NA	0.354	8.250	1.976	4.703	14.930
1975 Total	.063	5.023	2.495	7.580	.425	NA	NA	.425	8.006	2.007	4.829	14.842
1980 Total	.031	4.866	1.748	6.645	.859	NA	NA	.859	7.504	2.448	5.897	15.848
1985 Total	.039	4.571	1.483	6.093	.899	NA	NA	.899	6.992	2.709	6.227	15.928
1990 Total	.031	4.523	1.263	5.817	.581	.006	.056	.642	6.460	3.153	7.287	16.900
1995 Total	.017	4.981	1.356	6.355	.596	.007	.065	.667	7.022	3.557	8.073	18.653
1996 Total	.017	5.383	1.489	6.888	.595	.007	.065	.667	7.556	3.694	8.393	19.643
1997 Total	.016	5.118	1.448	6.582	.433	.008	.065	.506	7.088	3.671	8.308	19.067
1998 Total	.012	4.669	1.322	6.003	.387	.008	.065	.459	6.462	3.856	8.733	19.052
1999 Total	.014	4.858	1.452	6.324	.414	.009	.064	.486	6.810	3.906	8.917	19.634
2000 Total	.011	5.126	1.506	6.643	.433	.009	.061	.503	7.147	4.069	9.238	20.453
2001 Total	.012	4.919	1.539	6.470	.370	.009	.060	.439	6.909	4.103	9.248	20.261
2002 Total	.011	5.031	R 1.463	R 6.504	.313	.010	.059	.382	R 6.886	4.323	9.670	R 20.879
2003 January	.001	.977	R .181	R 1.159	.030	.001	.005	.037	R 1.196	.425	.936	R 2.557
February	.001	.913	R .155	R 1.069	.028	.001	.004	.033	R 1.102	.380	.784	R 2.267
March	.001	.697	R .136	R .833	.030	.001	.005	.037	R .870	.340	.751	R 1.961
April	.001	.428	R .109	R .537	.030	.001	.005	.036	R .573	.286	.635	R 1.494
May	.001	.256	R .097	R .354	.030	.001	.005	.037	R .391	.300	.698	R 1.388
June	.001	.162	R .088	R .251	.030	.001	.005	.036	R .287	.343	.793	R 1.423
July	.001	.131	R .096	R .227	.030	.001	.005	.037	R .264	.442	1.000	R 1.706
August	.001	.120	.105	R .225	.030	.001	.005	.037	.262	.455	1.017	1.734
September	.001	.133	.110	R .244	.030	.001	.005	.036	.279	.385	.787	1.452
October	.001	.239	R .123	R .363	.030	.001	.005	.037	R .399	.306	.665	R 1.370
November	.001	.427	R .124	R .552	.030	.001	.005	.036	R .588	.297	.675	R 1.559
December	.002	.763	R .171	R .936	.030	.001	.005	.037	R .973	.387	.877	R 2.237
Total	.010	5.246	R 1.494	R 6.750	.359	.017	.058	.434	R 7.184	4.345	9.627	R 21.157
2004 January	.001	.997	R .182	R 1.181	.028	.002	.005	.035	R 1.215	.433	.964	R 2.612
February	.001	.888	R .161	R 1.050	.026	.001	.005	.032	R 1.082	.386	.818	R 2.286
March	.001	.612	R .141	R .754	.028	.002	.005	.035	R .788	.338	.725	R 1.851
April	.001	.393	R .116	R .510	.027	.001	.005	.033	R .544	.292	.625	R 1.460
May	.001	.220	R .108	R .329	.028	.002	.005	.035	R .364	.309	.726	R 1.399
June	.001	.149	R .100	R .250	.027	.001	.005	.033	R .283	.383	.847	R 1.513
July	.001	.129	R .105	R .236	.028	.002	.005	.035	R .270	.443	.986	R 1.699
August	.001	.123	.109	.232	.028	.002	.005	.035	.267	.432	.947	1.646
September	.001	.129	.109	R .239	.027	.001	.005	.033	.272	.384	.815	1.471
October	.001	.223	R .129	R .352	.028	.002	.005	.035	R .387	.319	.694	R 1.400
November	.001	.420	R .129	R .550	.027	.001	.005	.033	R .583	.306	.675	R 1.564
December	.002	.746	R .172	R .920	.028	.002	.005	.035	R .954	.388	.887	R 2.230
Total	.011	R 5.030	R 1.561	R 6.602	.332	.018	.057	.408	R 7.009	4.413	9.710	R 21.133
2005 January	.001	R .917	R .177	R 1.096	.028	.002	.005	.035	R 1.130	.429	.948	R 2.507
February	.001	.780	R .150	R .931	.025	.001	.004	.031	R .962	.366	.752	R 2.080
March	.001	.698	.153	.852	.028	.002	.005	.035	.886	.356	.770	2.012
3-Month Total	.003	2.395	.480	2.878	.082	.004	.014	.101	2.979	1.150	2.470	6.599
2004 3-Month Total	.003	2.497	.484	2.984	.083	.004	.014	.101	3.085	1.157	2.507	6.749
2003 3-Month Total	.003	2.587	.471	3.061	.089	.004	.014	.107	3.168	1.146	2.471	6.785

Notes: RESIDENTIAL SECTOR-An energy-consuming sector that consists of living quarters for private households. Common uses of energy associated with this sector include space heating, water heating, air conditioning, lighting, refrigeration, cooking, and running a variety of other appliances. The residential sector excludes institutional living quarters; (a) All values are estimated; (b) Natural gas, plus a small amount of supplemental gaseous fuels that cannot be identified separately; (c) Geothermal heat pump and direct use energy; (d) Solar thermal direct use and photovoltaic electricity generation. Includes small amounts of commercial sector use; (e) Electricity retail sales to ultimate customers reported by electric utilities and other energy service providers; (f) Electrical system energy losses are calculated as the difference between total primary consumption by the electric power sector and the total energy content of electricity retail sales. Most of these losses occur at steam-electric power plants (conventional and nuclear) in the conversion of heat energy into mechanical energy to turn electric generators; R=Revised; NA=Not available; Totals may not equal sum of components due to independent rounding; Geographic coverage is the 50 States and the District of Columbia

Source: U.S. Department of Energy, Energy Information Administration, Monthly Energy Review, June 2005

Commercial sector energy consumption (quadrillion Btu)

| | Primary Consumption | | | | | | | | | | | |
| | Fossil Fuels | | | | Renewable Energy[a] | | | | | | | |
	Coal	Natural Gas[b]	Petroleum	Total	Hydro-power[c]	Wood and Waste	Geo-thermal[d]	Total	Total Primary	Electricity Retail Sales[e]	Electrical System Energy Losses[f]	Total
1973 Total	0.160	2.649	1.565	4.374	NA	0.007	NA	0.007	4.381	1.517	3.609	9.507
1975 Total	.147	2.558	1.310	4.015	NA	.008	NA	.008	4.023	1.598	3.845	9.466
1980 Total	.115	2.674	1.288	4.076	NA	.021	NA	.021	4.097	1.906	4.591	10.594
1985 Total	.137	2.508	1.039	3.684	NA	.024	NA	.024	3.708	2.351	5.405	11.465
1990 Total	.124	2.701	.913	3.739	.001	.067	.003	.071	3.810	2.860	6.611	13.281
1995 Total	.117	3.113	.710	3.940	.001	.086	.005	.092	4.032	3.252	7.381	14.665
1996 Total	.122	3.244	.743	4.108	.001	.103	.005	.110	4.218	3.344	7.599	15.161
1997 Total	.129	3.302	.704	4.135	.001	.107	.006	.113	4.248	3.503	7.928	15.679
1998 Total	.093	3.098	.653	3.845	.001	.102	.007	.111	3.956	3.678	8.330	15.964
1999 Total	.103	3.130	.637	3.870	.001	.106	.007	.114	3.984	3.766	8.597	16.347
2000 Total	.092	3.265	.726	4.083	.001	.100	.008	.109	4.192	3.956	8.982	17.129
2001 Total	.097	3.116	.742	3.955	.001	.080	.008	.089	4.044	4.086	9.208	17.337
2002 Total	.091	3.235	R .732	R 4.058	(s)	.081	.009	.090	R 4.148	R 4.155	R 9.296	R 17.599
2003 January	.010	.540	R .080	R .630	(s)	.007	.001	.009	R .639	.343	.754	R 1.736
February	.009	.503	R .072	R .584	(s)	.007	.001	.008	R .592	.310	.640	R 1.542
March	.006	.404	R .065	R .475	(s)	.007	.001	.009	R .484	.316	.699	R 1.499
April	.007	.272	R .050	R .329	(s)	.007	.001	.008	R .338	.305	.679	R 1.322
May	.005	.187	R .047	R .239	(s)	.007	.001	.009	R .247	.327	.761	R 1.335
June	.004	.142	R .042	R .188	(s)	.007	.001	.009	R .197	.347	.804	R 1.347
July	.006	.137	R .047	R .190	(s)	.008	.001	.009	R .199	.391	.885	R 1.474
August	.006	.135	R .052	R .193	(s)	.008	.001	.009	R .202	.396	.886	R 1.484
September	.004	.141	R .050	R .195	(s)	.007	.001	.008	.204	.364	.743	1.310
October	.005	.187	R .058	R .251	(s)	.007	.001	.009	R .259	.342	.743	R 1.345
November	.008	.268	R .057	R .333	(s)	.007	.001	.008	R .341	.317	.721	R 1.379
December	.012	.407	R .078	R .498	(s)	.008	.001	.009	R .507	.335	.760	R 1.602
Total	.084	3.323	R .698	R 4.105	.001	.087	.014	.102	R 4.207	4.093	9.070	R 17.371
2004 January	.011	.504	R .082	R .597	(s)	.007	.001	.009	R .605	.339	.753	R 1.697
February	.009	.473	R .076	R .557	(s)	.007	.001	.008	R .565	.320	.679	R 1.565
March	.006	.353	R .121	R .480	(s)	.008	.001	.009	R .489	.325	.697	R 1.511
April	.007	.249	R .105	R .362	(s)	.007	.001	.009	R .370	.318	.680	R 1.368
May	.005	.169	R .087	R .262	(s)	.008	.001	.009	R .271	.343	.805	R 1.419
June	.005	.136	R .074	R .215	(s)	.008	.001	.009	R .224	.368	.813	R 1.405
July	.007	.126	R .051	R .184	(s)	.008	.001	.009	R .193	.395	.879	R 1.466
August	.006	.126	R .095	R .227	(s)	.008	.001	.009	R .236	.391	.856	R 1.482
September	.005	.128	.053	.186	(s)	.007	.001	.008	.195	.374	.792	1.360
October	.005	.171	R .064	R .240	(s)	.007	.001	.009	R .249	.348	.759	R 1.356
November	.008	.254	R .061	R .323	(s)	.007	.001	.009	R .331	.326	.721	R 1.379
December	.013	R .399	R .081	R .493	(s)	.008	.001	.009	R .502	.345	.790	R 1.637
Total	.087	R 3.088	R .951	R 4.125	.001	.089	.015	.106	R 4.230	4.192	9.223	R 17.645
2005 January	R .010	.480	R .084	R .574	(s)	.008	.001	.009	R .583	.346	.766	R 1.695
February	R .008	.429	R .074	R .511	(s)	.007	.001	.008	R .519	.319	.655	R 1.493
March	.008	.391	.071	.470	(s)	.008	.001	.009	.479	.337	.729	1.545
3-Month Total	.026	1.300	.230	1.555	(s)	.023	.004	.027	1.582	1.002	2.150	4.734
2004 3-Month Total	.026	1.330	.278	1.634	(s)	.022	.004	.026	1.660	.984	2.129	4.772
2003 3-Month Total	.025	1.447	.217	1.689	(s)	.021	.004	.025	1.714	.969	2.093	4.777

Notes: *COMMERCIAL SECTOR-An energy-consuming sector that consists of service-providing facilities and equipment of: businesses; Federal, State, and local governments; and other private and public organizations, such as religious, social, or fraternal groups. The commercial sector includes institutional living quarters. Common uses of energy associated with this sector include space heating, water heating, air conditioning, lighting, refrigeration, cooking, and running a wide variety of other equipment; (a) All values are estimated; (b) Natural gas, plus a small amount of supplemental gaseous fuels that cannot be identified separately; (c) Conventional hydroelectric power; (d) Geothermal heat pump and direct use energy; (e) Electricity retail sales to ultimate customers reported by electric utilities and other energy service providers; (f) Electrical system energy losses are calculated as the difference between total primary consumption by the electric power sector and the total energy content of electricity retail sales. Most of these losses occur at steam-electric power plants (conventional and nuclear) in the conversion of heat energy into mechanical energy to turn electric generators; (s) Less than 0.5 trillion Btu.; R=Revised; NA=Not available; E=Estimate; F=Forecast; Totals may not equal sum of components due to independent rounding; Geographic coverage is the 50 States and the District of Columbia.*
Source: *U.S. Department of Energy, Energy Information Administration, Monthly Energy Review, June 2005*

Industrial sector energy consumption (quadrillion Btu)

	Primary Consumption									Electricity Retail Sales[h]	Electrical System Energy Losses[i]	Total[c]
	Fossil Fuels				Renewable Energy[a]							
	Coal	Natural Gas[b]	Petroleum	Total[c]	Hydro-power[d]	Wood[e] and Waste[f]	Geo-thermal[g]	Total	Total Primary			
1973 Total	4.057	10.388	9.104	23.541	0.035	1.165	NA	1.200	24.741	2.341	5.571	32.653
1975 Total	3.667	8.532	8.146	20.359	.032	1.063	NA	1.096	21.454	2.346	5.647	29.447
1980 Total	3.155	8.395	9.525	21.040	.033	1.600	NA	1.633	22.673	2.781	6.698	32.152
1985 Total	2.760	7.080	7.805	17.632	.033	1.875	NA	1.908	19.540	2.855	6.563	28.958
1990 Total	2.756	8.502	8.305	19.568	.031	1.634	.002	1.667	21.235	3.226	7.457	31.918
1995 Total	2.488	9.637	8.552	20.738	.055	1.847	.003	1.905	22.643	3.455	7.842	33.941
1996 Total	2.434	9.947	8.989	21.393	.061	1.907	.003	1.971	23.364	3.527	8.014	34.905
1997 Total	2.395	9.976	9.214	21.632	.058	1.915	.003	1.976	23.608	3.542	8.017	35.167
1998 Total	2.335	9.806	9.017	21.226	.055	1.784	.003	1.841	23.067	3.587	8.124	34.777
1999 Total	2.227	9.415	9.284	20.983	.049	1.791	.004	1.843	22.826	3.611	8.242	34.679
2000 Total	2.256	9.535	9.055	20.912	.042	1.781	.004	1.828	22.740	3.631	8.245	34.616
2001 Total	2.192	8.725	9.220	20.166	.033	1.593	.005	1.630	21.796	3.290	7.415	32.501
2002 Total	2.019	8.870	R 9.162	R 20.112	.039	1.565	.005	1.608	R 21.720	3.317	7.420	R 32.457
2003 January	.170	.807	R .835	R 1.813	.004	.129	(s)	.133	R 1.946	.279	.613	R 2.838
February	.170	.751	R .737	R 1.672	.003	.118	(s)	.121	R 1.794	.270	.557	R 2.621
March	.175	.737	R .783	R 1.698	.004	.127	(s)	.131	R 1.829	.274	.605	R 2.708
April	.166	.690	R .785	R 1.645	.002	.126	(s)	.129	R 1.774	.279	.622	R 2.675
May	.164	.672	R .749	R 1.587	.004	.126	(s)	.130	R 1.717	.286	.666	R 2.669
June	.167	.620	R .712	R 1.503	.004	.124	(s)	.128	R 1.631	.292	.677	R 2.600
July	.169	.688	R .762	R 1.624	.004	.133	(s)	.138	R 1.761	.299	.675	R 2.735
August	.167	.695	R .758	R 1.621	.004	.130	(s)	.135	R 1.756	.308	.690	R 2.754
September	.168	.675	.763	1.609	.003	.125	(s)	.129	1.738	.293	.599	2.630
October	.174	.720	R .783	R 1.681	.003	.130	(s)	.133	R 1.814	.296	.644	R 2.755
November	.175	.710	R .806	R 1.694	.004	.127	(s)	.131	R 1.825	.282	.641	R 2.749
December	.177	.770	R .845	R 1.799	.005	.137	(s)	.142	R 1.941	.280	.635	R 2.855
Total	2.041	8.534	R 9.318	R 19.944	.043	1.533	.005	1.581	R 21.525	3.439	7.620	R 32.584
2004 January	.175	.812	R .834	R 1.826	.005	.141	(s)	.146	R 1.972	.274	.610	R 2.857
February	.171	.771	R .763	R 1.715	.005	.129	(s)	.134	R 1.849	.272	.576	R 2.696
March	.179	.758	R .747	R 1.692	.004	.132	(s)	.137	R 1.829	.284	.610	R 2.724
April	.165	.714	R .713	R 1.615	.004	.137	(s)	.141	R 1.755	.285	.611	R 2.651
May	.164	.696	R .761	R 1.658	.004	.131	(s)	.135	R 1.793	.299	.702	R 2.795
June	.163	.686	R .716	R 1.585	.003	.133	(s)	.137	R 1.722	.298	.658	R 2.678
July	.162	.697	R .779	R 1.648	.003	.139	(s)	.143	R 1.791	.302	.673	R 2.767
August	.165	.710	R .780	R 1.661	.004	.136	(s)	.140	R 1.801	.306	.670	R 2.777
September	.164	.691	R .782	R 1.635	.005	.129	(s)	.135	R 1.770	.294	.623	R 2.687
October	.173	.717	R .812	R 1.708	.004	.138	(s)	.142	R 1.850	.293	.639	R 2.783
November	.171	R .732	R .874	R 1.783	.005	.130	(s)	.135	R 1.918	.289	.638	R 2.845
December	.174	.792	R .853	R 1.826	.006	.144	(s)	.150	R 1.975	.286	.655	R 2.916
Total	2.025	8.776	R 9.412	R 20.351	.051	1.620	.005	1.676	R 22.027	3.483	7.664	R 33.174
2005 January	R .167	R .805	R .852	R 1.834	.004	.139	(s)	.143	R 1.977	.281	.621	R 2.880
February	R .164	.710	R .747	R 1.634	.003	.133	(s)	.136	R 1.771	.274	.564	R 2.609
March	.169	.726	.842	1.746	.004	.134	(s)	.138	1.884	.289	.625	2.798
3-Month Total	.500	2.241	2.440	5.214	.012	.406	.001	.418	5.632	.844	1.811	8.287
2004 3-Month Total	.525	2.341	2.343	5.233	.014	.403	.001	.418	5.651	.830	1.796	8.277
2003 3-Month Total	.515	2.295	2.355	5.183	.010	.374	.001	.386	5.569	.823	1.775	8.167

Notes: *INDUSTRIAL SECTOR-An energy-consuming sector that consists of all facilities and equipment used for producing, processing, or assembling goods. The industrial sector encompasses the following types of activity: manufacturing; agriculture, forestry, and fisheries; mining; and construction. Overall energy use in this sector is largely for process heat and cooling and powering machinery, with lesser amounts used for facility heating, air conditioning, and lighting. Fossil fuels are also used as raw material inputs to manufactured products; (a) All values are estimated; (b) Natural gas, plus a small amount of supplemental gaseous fuels that cannot be identified separately; (c) Includes coal coke net imports, which are not separately displayed; (d) Conventional hydroelectric power; (e) Wood, black liquor, and other wood waste; (f) Municipal solid waste, landfill gas, sludge waste, tires, agricultural byproducts, and other biomass; (g) Geothermal heat pump and direct use energy; (h) Electricity retail sales to ultimate customers reported by electric utilities and other energy service providers; (i) Electrical system energy losses are calculated as the difference between total primary consumption by the electric power sector and the total energy content of electricity retail sales. Most of these losses occur at steam-electric power plants (conventional and nuclear) in the conversion of heat energy into mechanical energy to turn electric generators; (s) Less than 0.5 trillion Btu.; R=Revised; NA=Not available; E=Estimate; F=Forecast; Totals may not equal sum of components due to independent rounding; Geographic coverage is the 50 States and the District of Columbia.*

Source: *U.S. Department of Energy, Energy Information Administration, Monthly Energy Review, June 2005*

Transportation sector energy consumption (quadrillion Btu)

	Primary Consumption						Electricity Retail Sales[f]	Electrical System Energy Losses[g]	Total[d]
	Fossil Fuels				Renewable Energy[a]	Total Primary[d]			
	Coal	Natural Gas[b]	Petroleum[c,d]	Total	Alcohol Fuels[d,e]				
1973 Total	0.003	0.743	17.831	18.576	NA	18.576	0.011	0.025	18.612
1975 Total	.001	.595	17.614	18.209	NA	18.209	.010	.024	18.244
1980 Total	(h)	.650	19.008	19.658	NA	19.658	.011	.027	19.696
1985 Total	(h)	.519	19.504	20.023	.052	20.075	.014	.033	20.122
1990 Total	(h)	.680	21.792	22.472	.063	22.535	.016	.037	22.589
1995 Total	(h)	.724	23.181	23.905	.117	23.905	.017	.039	23.960
1996 Total	(h)	.737	23.719	24.456	.084	24.456	.017	.038	24.511
1997 Total	(h)	.780	23.973	24.753	.106	24.753	.017	.038	24.808
1998 Total	(h)	.666	24.635	25.301	.117	25.301	.017	.038	25.357
1999 Total	(h)	.675	25.375	26.050	.122	26.050	.017	.040	26.108
2000 Total	(h)	.672	25.973	26.645	.139	26.645	.018	.042	26.705
2001 Total	(h)	.659	25.556	26.215	.147	26.215	.019	.042	26.276
2002 Total	(h)	.702	R 26.084	R 26.786	.174	R 26.786	R .019	R .042	R 26.847
2003 January	(h)	.086	R 2.092	R 2.178	.017	R 2.178	.002	.005	R 2.185
February	(h)	.080	R 1.974	R 2.054	.020	R 2.054	.002	.004	R 2.060
March	(h)	.070	R 2.176	R 2.246	.017	R 2.246	.002	.004	R 2.252
April	(h)	.055	R 2.145	R 2.200	.020	R 2.200	.002	.004	R 2.206
May	(h)	.048	R 2.229	R 2.277	.019	R 2.277	.002	.004	R 2.283
June	(h)	.043	R 2.219	R 2.261	.019	R 2.261	.002	.005	R 2.268
July	(h)	.050	R 2.298	R 2.348	.020	R 2.348	.002	.005	R 2.355
August	(h)	.052	R 2.365	R 2.417	.021	R 2.417	.002	.005	R 2.424
September	(h)	.045	2.182	2.227	.018	2.227	.002	.004	2.233
October	(h)	.049	R 2.292	R 2.341	.021	R 2.341	.002	.004	R 2.347
November	(h)	.056	R 2.131	R 2.187	.024	R 2.187	.002	.004	R 2.193
December	(h)	.072	R 2.230	R 2.303	.025	R 2.303	.002	.004	R 2.309
Total	(h)	.706	R 26.332	R 27.038	.239	R 27.038	.024	.053	R 27.115
2004 January	(h)	.084	R 2.132	R 2.216	.024	R 2.216	.002	.005	R 2.223
February	(h)	.079	R 2.095	R 2.173	.022	R 2.173	.002	.005	R 2.180
March	(h)	.066	R 2.237	R 2.303	.024	R 2.303	.002	.004	R 2.309
April	(h)	.055	R 2.217	R 2.272	.024	R 2.272	.002	.004	R 2.279
May	(h)	.050	R 2.289	R 2.339	.025	R 2.339	.002	.005	R 2.346
June	(h)	.047	R 2.262	R 2.309	.025	R 2.309	.002	.005	R 2.316
July	(h)	.050	R 2.356	R 2.407	.025	R 2.407	.002	.005	R 2.414
August	(h)	.050	R 2.339	R 2.389	.024	R 2.389	.002	.005	R 2.396
September	(h)	.047	R 2.240	R 2.287	.026	R 2.287	.002	.005	R 2.294
October	(h)	.050	R 2.355	R 2.404	.025	R 2.404	.002	.005	R 2.411
November	(h)	.056	R 2.202	R 2.258	.025	R 2.258	.002	.005	R 2.265
December	(h)	.073	R 2.288	R 2.361	.026	R 2.361	.002	.005	R 2.369
Total	(h)	.708	R 27.011	R 27.720	.296	R 27.720	.026	.058	R 27.803
2005 January	(h)	.082	R 2.169	R 2.250	.026	R 2.250	.003	.006	R 2.259
February	(h)	.071	R 2.048	R 2.119	.028	R 2.119	.002	.005	R 2.127
March	(h)	.070	2.288	2.358	.033	2.358	.002	.005	2.365
3-Month Total	(h)	.223	6.505	6.727	.087	6.727	.007	.016	6.751
2004 3-Month Total	(h)	.229	6.463	6.692	.070	6.692	.007	.014	6.713
2003 3-Month Total	(h)	.236	6.242	6.478	.054	6.478	.006	.013	6.497

Notes: *TRANSPORTATION SECTOR-An energy-consuming sector that consists of all vehicles whose primary purpose is transporting people and/or goods from one physical location to another. Included are automobiles; trucks; buses; motorcycles; trains, subways, and other rail vehicles; aircraft; and ships, barges, and other waterborne vehicles. Vehicles whose primary purpose is not transportation (e.g., construction cranes and bulldozers, farming vehicles, and warehouse tractors and forklifts) are classified in the sector of their primary use; (a) All values are estimated; (b) Natural gas consumed in the operation of pipelines (primarily in compressors) and small amounts consumed as vehicle fuel; (c) Beginning in 1993, includes ethanol blended into motor gasoline; (d) Beginning in 1993, ethanol blended into motor gasoline is included in both "Petroleum" and "Alcohol Fuels," but is counted only once in both total primary consumption and total consumption; (e) "Alcohol Fuels" is ethanol blended into motor gasoline; (f) Electricity retail sales to ultimate customers reported by electric utilities and, beginning in 1996, other energy service providers; (g) Electrical system energy losses are calculated as the difference between total primary consumption by the electric power sector and the total energy content of electricity retail sales. Most of these losses occur at steam-electric power plants (conventional and nuclear) in the conversion of heat energy into mechanical energy to turn electric generators; (h) Since 1978, the small amounts of coal consumed for transportation are reported as industrial sector consumption; (s) Less than 0.5 trillion Btu.; R=Revised; NA=Not available; E=Estimate; F=Forecast; Totals may not equal sum of components due to independent rounding; Geographic coverage is the 50 States and the District of Columbia.*
Source: *U.S. Department of Energy, Energy Information Administration, Monthly Energy Review, June 2005*

Electric power sector energy consumption (quadrillion Btu)

	Primary Consumption												
	Fossil Fuels						Renewable Energy						
	Coal	Natural Gas[a]	Petroleum	Total	Nuclear Electric Power	Hydro-electric Pumped Storage[b]	Conventional Hydroelectric Power	Wood[c] and Waste[d]	Geo-thermal[e]	Solar[f] and Wind[g]	Total	Electricity Net Imports	Total Primary
1973 Total	8.658	3.748	3.515	15.921	0.910	(h)	2.827	0.003	0.043	NA	2.873	0.049	19.753
1975 Total	8.786	3.240	3.166	15.191	1.900	(h)	3.122	.002	.070	NA	3.194	.021	20.307
1980 Total	12.123	3.810	2.634	18.567	2.739	(h)	2.867	.005	.110	NA	2.982	.071	24.359
1985 Total	14.542	3.160	1.090	18.792	4.076	(h)	2.937	.014	.198	(s)	3.150	.140	26.158
1990 Total[i]	16.261	3.332	1.289	20.883	6.104	-.036	3.014	.317	.326	.033	3.689	.008	30.647
1995 Total	17.466	4.325	.755	22.546	7.075	-.028	3.149	.422	.280	.038	3.889	.134	33.616
1996 Total	18.429	3.883	.817	23.129	7.087	-.032	3.528	.438	.300	.039	4.305	.137	34.626
1997 Total	18.905	4.146	.927	23.977	6.597	-.041	3.581	.446	.309	.039	4.375	.116	35.024
1998 Total	19.216	4.698	1.306	25.220	7.068	-.046	3.241	.444	.311	.036	4.032	.088	36.363
1999 Total	19.279	4.926	1.211	25.416	7.610	-.062	3.218	.453	.312	.051	4.034	.099	37.097
2000 Total	20.220	5.316	1.144	26.680	7.862	-.057	2.768	.453	.296	.062	3.579	.115	38.180
2001 Total	19.614	5.481	1.277	26.371	8.033	-.091	2.209	.450	.289	.075	3.023	.075	37.411
2002 Total	19.783	5.785	.961	26.529	8.143	-.089	2.650	.516	.305	.111	3.581	.078	38.243
2003 January	1.835	.392	.126	2.353	.721	-.008	.207	.045	.026	.007	.286	.005	3.357
February	1.595	.343	.109	2.047	.635	-.008	.199	.039	.024	.008	.270	.004	2.949
March	1.578	.370	.103	2.051	.625	-.008	.244	.044	.025	.011	.324	-.001	2.991
April	1.446	.361	.089	1.896	.592	-.006	.251	.041	.025	.012	.329	.003	2.813
May	1.542	.404	.081	2.026	.648	-.006	.297	.042	.025	.011	.374	.001	3.044
June	1.673	.446	.111	2.230	.669	-.008	.289	.043	.026	.012	.370	.001	3.262
July	1.868	.646	.124	2.637	.726	-.008	.251	.046	.026	.010	.333	.010	3.698
August	1.899	.701	.128	2.727	.719	-.008	.231	.047	.026	.009	.313	.008	3.758
September	1.693	.480	.088	2.261	.663	-.008	.186	.043	.025	.010	.264	-.002	3.178
October	1.624	.419	.085	2.128	.625	-.006	.185	.042	.025	.010	.262	-.006	3.003
November	1.631	.357	.065	2.053	.621	-.007	.198	.043	.024	.010	.275	-.003	2.940
December	1.802	.344	.098	2.245	.715	-.007	.241	.046	.027	.011	.326	.001	3.280
Total	20.185	5.264	1.205	26.653	7.959	-.087	2.781	.522	.303	.120	3.725	.022	38.272
2004 January	1.831	.361	.148	2.340	.739	-.007	.230	.042	.026	.011	.309	(s)	3.381
February	1.646	.375	.091	2.112	.669	-.007	.209	.040	.025	.011	.284	.000	3.058
March	1.554	.377	.095	2.026	.660	-.006	.227	.042	.025	.014	.308	-.003	2.985
April	1.443	.393	.089	1.924	.612	-.006	.209	.040	.024	.014	.286	(s)	2.816
May	1.610	.485	.103	2.197	.678	-.007	.238	.042	.025	.018	.323	.001	3.192
June	1.717	.512	.108	2.338	.708	-.007	.252	.042	.025	.015	.333	.002	3.374
July	1.862	.631	.121	2.615	.751	-.007	.231	.046	.026	.012	.315	.010	3.685
August	1.841	.614	.112	2.567	.742	-.008	.216	.045	.026	.011	.297	.012	3.610
September	1.705	.532	.088	2.324	.688	-.007	.203	.041	.024	.012	.280	.003	3.288
October	1.623	.443	.077	2.143	.653	-.007	.188	.041	.026	.011	.266	.004	3.059
November	1.622	.375	.066	2.062	.615	-.006	.209	.042	.025	.010	.285	.005	2.961
December	1.815	.387	.098	2.300	.716	-.006	.261	.045	.026	.012	.344	.005	3.359
Total	20.268	5.486	1.195	26.948	8.232	-.082	2.673	.508	.302	.149	3.632	.039	38.769
2005 January	1.834	.396	.119	2.349	.728	-.007	.243	.045	.025	.011	.325	.005	3.399
February	1.602	.339	.071	2.011	.635	-.004	.217	.041	.022	.009	.289	.006	2.938
March	1.674	.399	.081	2.153	.641	-.005	.230	.045	.025	.014	.315	.008	3.113
3-Month Total	5.110	1.134	.270	6.513	2.004	-.016	.691	.131	.073	.034	.929	.019	9.450
2004 3-Month Total	5.031	1.113	.334	6.478	2.069	-.020	.666	.124	.076	.035	.901	-.003	9.424
2003 3-Month Total	5.008	1.105	.338	6.451	1.981	-.024	.651	.128	.076	.026	.880	.009	9.297

Notes: *ELECTRIC POWER SECTOR-An energy-consuming sector that consists of all utility and nonutility facilities and equipment used to generate, transmit, and/or distribute electricity. Although the energy-use allocations are made according to these aggregations as closely as possible, some data are collected by using different classifications. For example, electric utilities may classify commercial and industrial users by the quantity of electricity purchased rather than by the business activity of the purchaser; Data are for fuels consumed to produce electricity and useful thermal output. The electric power sector comprises electricity-only and combined-heat-and-power (CHP) plants within the NAICS 22 category whose primary business is to sell electricity, or electricity and heat, to the public; (a) Natural gas, plus a small amount of supplemental gaseous fuels that cannot be identified separately; (b) Pumped storage facility production minus energy used for pumping; (c) Wood, black liquor, and other wood waste; (d) Municipal solid waste, landfill gas, sludge waste, tires, agricultural byproducts, and other biomass; (e) Geothermal electricity net generation; (f) Solar thermal and photovoltaic electricity net generation; (g) Wind electricity net generation; (h) Included in conventional hydroelectric power; (i) Through 1988, data are for consumption at electric utilities only. Beginning in 1989, data also include consumption at independent power producers; (s) Less than 0.5 trillion Btu.; R=Revised; NA=Not available; E=Estimate; F=Forecast; Totals may not equal sum of components due to independent rounding; Geographic coverage is the 50 States and the District of Columbia.*
Source: *U.S. Department of Energy, Energy Information Administration, Monthly Energy Review, June 2005*

Petroleum overview: supply

	Supply							
	Field Production[a]			Refinery and Blender Net Production	Imports			Adjust-ments[d]
	Crude Oil	Natural Gas Plant Liquids[b]	Total		Crude Oil[c]	Petroleum Products	Total	
	Thousand Barrels per Day							
1973 Average	9,208	1,738	10,946	13,854	3,244	3,012	6,256	18
1975 Average	8,375	1,633	10,007	13,685	4,105	1,951	6,056	41
1980 Average	8,597	1,573	10,170	14,622	5,263	1,646	6,909	64
1985 Average	8,971	1,609	10,581	13,750	3,201	1,866	5,067	200
1990 Average	7,355	1,559	8,914	15,272	5,894	2,123	8,018	338
1995 Average	6,560	1,762	8,322	15,994	7,230	1,605	8,835	496
1996 Average	6,465	1,830	8,295	16,324	7,508	1,971	9,478	528
1997 Average	6,452	1,817	8,269	16,759	8,225	1,936	10,162	487
1998 Average	6,252	1,759	8,011	17,030	8,706	2,002	10,708	495
1999 Average	5,881	1,850	7,731	16,989	8,731	2,122	10,852	567
2000 Average	5,822	1,911	7,733	17,243	9,071	2,389	11,459	532
2001 Average	5,801	1,868	7,670	17,285	9,328	2,543	11,871	501
2002 Average	5,746	1,880	7,626	17,273	9,140	2,390	11,530	527
2003 January	5,785	1,758	7,543	16,405	8,633	2,471	11,104	245
February	5,791	1,812	7,603	16,363	8,474	2,447	10,921	427
March	5,817	1,729	7,545	16,914	9,226	2,819	12,044	656
April	5,774	1,701	7,475	17,601	9,928	2,671	12,599	592
May	5,733	1,564	7,297	18,146	10,153	2,765	12,918	458
June	5,701	1,582	7,283	17,739	10,038	2,962	13,001	485
July	5,526	1,649	7,175	17,811	10,034	2,702	12,736	568
August	5,595	1,703	7,299	18,053	10,023	2,746	12,769	505
September	5,683	1,761	7,445	17,650	10,287	2,581	12,868	431
October	5,635	1,818	7,453	17,461	10,063	2,310	12,373	526
November	5,560	1,839	7,399	17,660	9,351	2,361	11,712	581
December	5,579	1,723	7,302	17,957	9,684	2,349	12,033	257
Average	**5,681**	**1,719**	**7,400**	**17,487**	**9,665**	**2,599**	**12,264**	**478**
2004 January	E 5,644	1,803	E 7,447	16,766	9,322	2,405	11,727	462
February	E 5,584	1,798	E 7,382	16,623	9,258	3,071	12,329	673
March	E 5,622	1,829	E 7,451	17,184	10,073	3,000	13,073	287
April	E 5,568	1,784	E 7,351	18,032	10,062	2,389	12,450	765
May	E 5,612	1,795	E 7,408	18,299	10,324	2,665	12,989	671
June	E 5,403	1,737	E 7,140	18,294	10,505	2,796	13,301	947
July	E 5,404	1,810	E 7,214	18,368	10,302	3,087	13,389	681
August	E 5,280	1,859	E 7,139	18,414	10,447	3,042	13,489	499
September	E 5,091	1,797	E 6,888	17,248	9,669	2,863	12,532	539
October	E 5,112	1,822	E 6,934	17,588	10,328	2,995	13,323	427
November	E 5,397	1,873	E 7,270	17,940	10,108	3,111	13,219	813
December	E 5,448	1,818	E 7,266	18,467	10,018	2,913	12,931	623
Average	**E 5,430**	**1,811**	**E 7,241**	**17,774**	**10,038**	**2,861**	**12,899**	**614**
2005 January	E 5,394	1,809	E 7,203	17,137	9,844	2,818	12,661	657
February	E 5,469	1,859	E 7,327	17,504	10,158	3,378	13,536	532
March	E 5,498	1,858	E 7,356	17,442	10,144	2,776	12,919	657
April	RE 5,488	R 1,830	RE 7,318	R 18,508	R 10,314	R 3,062	R 13,376	R 730
May	E 5,526	E 1,858	E 7,384	NA	E 10,456	E 3,093	E 13,549	NA
5-Month Average	**E 5,475**	**E 1,843**	**E 7,317**	**NA**	**E 10,183**	**E 3,018**	**E 13,201**	**NA**
2004 5-Month Average	**E 5,606**	**1,802**	**E 7,408**	**17,387**	**9,813**	**2,703**	**12,516**	**569**
2003 5-Month Average	**5,780**	**1,711**	**7,491**	**17,097**	**9,295**	**2,638**	**11,933**	**476**

Notes: Crude oil includes lease condensate; (a) Crude oil production on leases, and natural gas liquids (liquefied petroleum gases, pentanes plus, and a small amount of finished petroleum products) production at natural gas processing plants. Excludes what was previously classified as "Field Production" of finished motor gasoline, motor gasoline blending components, and other hydrocarbons and oxygenates; these are now included in "Adjustments"; (b) Due to differences internal to EIA data processing systems, some small discrepancies exist between data in the Monthly Energy Review (MER) and the Petroleum Supply Annual (PSA) and Petroleum Supply Monthly (PSM); (c) Includes Strategic Petroleum Reserve imports; (d) An adjustment for crude oil, and for motor gasoline blending components and fuel ethanol. Through 1988, also includes a small amount of distillate fuel oil production at natural gas processing plants; NA=Not Available; R=Revised; E=Estimate; Geographic coverage is the 50 States and the District of Columbia.
Source: U.S. Department of Energy, Energy Information Administration, Monthly Energy Review, June 2005

Petroleum overview: disposition and stocks

	Disposition								Stocks[a]		
	Stock Change[b]			Refinery and Blender Net Inputs	Exports			Petroleum Products Supplied	Crude Oil[c,d]	Petroleum Products[d,e]	Total[d]
	Crude Oil[c,d]	Petroleum Products[d,e]	Total[d]		Crude Oil	Petroleum Products[f]	Total[f]				
	Thousand Barrels per Day								Million Barrels		
1973 Average	-11	146	135	13,401	2	229	231	17,308	242	766	1,008
1975 Average	17	[d]15	[d]32	13,225	6	204	209	16,322	271	862	1,133
1980 Average	98	42	140	14,025	287	258	544	17,056	466	[d]926	[d]1,392
1985 Average	50	-153	-103	13,192	204	577	781	15,726	814	705	1,519
1990 Average	-35	142	107	14,589	109	748	857	16,988	908	712	1,621
1995 Average	-93	-153	-246	15,220	95	855	949	17,725	895	668	1,563
1996 Average	-124	-28	-151	15,487	110	871	981	18,309	850	658	1,507
1997 Average	51	93	143	15,909	108	896	1,003	18,620	868	692	1,560
1998 Average	74	165	239	16,144	110	835	945	18,917	895	752	1,647
1999 Average	-118	-304	-422	16,103	118	822	940	19,519	852	641	1,493
2000 Average	-70	(s)	-69	16,295	50	990	1,040	19,701	826	641	1,468
2001 Average	99	227	325	16,382	20	951	971	19,649	862	724	1,586
2002 Average	40	-145	-105	16,316	9	975	984	19,761	877	671	1,548
2003 January	-110	-1,293	-1,403	15,472	10	1,202	1,212	20,017	873	631	1,504
February	-106	-1,464	-1,570	15,441	5	1,062	1,067	20,375	870	590	1,460
March	339	114	452	15,949	10	1,042	1,051	19,708	881	594	1,474
April	338	383	720	16,664	12	1,041	1,053	19,830	891	605	1,496
May	-75	1,263	1,188	17,190	15	1,082	1,097	19,344	889	644	1,533
June	150	745	895	16,755	45	1,020	1,065	19,793	893	667	1,560
July	135	209	344	16,876	7	969	976	20,094	897	673	1,570
August	15	35	50	17,044	4	943	947	20,586	898	674	1,572
September	441	426	867	16,635	3	956	960	19,933	911	687	1,598
October	468	-348	120	16,540	14	956	970	20,182	926	676	1,602
November	-356	241	-116	16,663	21	911	933	19,873	915	683	1,598
December	-244	-721	-965	16,845	4	986	990	20,679	907	661	1,568
Average	84	-28	56	16,513	12	1,014	1,027	20,034	907	661	1,568
2004 January	199	-692	-493	15,753	6	742	748	20,393	913	639	1,552
February	380	-549	-170	15,582	8	1,038	1,046	20,549	924	623	1,547
March	720	-91	629	16,181	19	1,005	1,024	20,161	946	620	1,566
April	379	-111	268	16,970	55	1,099	1,153	20,207	957	617	1,574
May	186	646	831	17,275	26	1,026	1,052	20,209	963	637	1,600
June	130	831	961	17,320	45	1,025	1,070	20,333	967	662	1,629
July	-186	782	596	17,376	18	1,062	1,080	20,601	961	686	1,647
August	-381	695	314	17,405	13	1,078	1,091	20,732	949	708	1,657
September	-151	-307	-458	16,294	35	927	961	20,411	945	699	1,643
October	450	-576	-126	16,577	25	1,052	1,078	20,743	959	681	1,639
November	187	407	594	16,874	42	950	992	20,782	964	693	1,657
December	-79	-327	-406	17,330	30	1,253	1,284	21,080	962	683	1,645
Average	152	61	212	16,750	27	1,021	1,048	20,517	962	683	1,645
2005 January	207	-136	71	16,147	40	877	917	20,524	968	679	1,647
February	619	-98	521	16,470	22	1,237	1,259	20,650	986	676	1,661
March	686	-836	-150	16,485	36	1,272	1,308	20,732	1,007	650	1,657
April	[R]518	[R]393	[R]912	[R]17,459	[R]97	[R]1,285	[R]1,382	[R]20,179	[R]1,022	[R]662	[R]1,684
May	[E]198	[E]664	[E]863	NA	[E]19	[E]1,073	[E]1,092	[E]20,449	[E]1,025	[E]667	[E]1,692
5-Month Average	[E]442	[E]-3	[E]439	NA	[E]43	[E]1,146	[E]1,189	[E]20,506	[E]1,025	[E]667	[E]1,692
2004 5-Month Average	373	-155	218	16,358	23	980	1,003	20,301	963	637	1,600
2003 5-Month Average	79	-178	-99	16,154	10	1,086	1,097	19,844	889	644	1,533

Notes: *Crude oil includes lease condensate; (a) Stocks are at end of period; (b) A negative value indicates a decrease in stocks and a positive value indicates an increase. Current month stock change estimates are based on the change from the previous month's stocks estimates, rather than the actual stocks values shown in this table; (c) Includes Strategic Petroleum Reserve stocks; (d) New Stock Basis: In January 1975, 1979, 1981, and 1983, numerous respondents were added to bulk terminal and pipeline surveys, affecting subsequent stocks reported and stock change calculations. Using the expanded coverage (new basis), the end-of-year stocks, in million barrels, would have been: Crude Oil: 1982 - 645 (Total) and 351 (Other Primary). Crude Oil and Petroleum Products: 1974 - 1,121; 1980 - 1,425; and 1982 - 1,461. Motor Gasoline: 1974 - 225; 1980 - 263 (Total) and 214 (Finished); 1982 - 244 (Total) and 202 (Finished). Distillate Fuel Oil: 1974 - 224; 1980 - 205; and 1982 - 186. Residual Fuel Oil: 1974 - 75; 1980 - 91; and 1982 - 69. Jet Fuel: 1974 - 30 (Total) and 24 (Kerosene Type); 1980 - 42 (Total) and 36 (Kerosene Type); and 1982 - 39 (Total) and 32 (Kerosene Type). Liquefied Petroleum Gases: 1974 - 113; 1978 - 136; 1980 - 128; and 1982 - 102. Propane and Propylene: 1978 - 86; 1980 - 69; and 1982 - 57. Other Petroleum Products: 1974 - 190; 1980 - 207; and 1982 - 219; (e) Does not include distillate stocks in the Northeast Heating Oil Reserve; (f) Due to differences internal to EIA data processing systems, some small discrepancies exist between data in the Monthly Energy Review (MER) and the Petroleum Supply Annual (PSA) and Petroleum Supply Monthly (PSM); (s) Less than +500 barrels per day and greater than -500 barrels per day; R=Revised; NA=Not available; E=Estimate.*
Source: *U.S. Department of Energy, Energy Information Administration, Monthly Energy Review, June 2005*

Natural gas overview (billion cubic feet)

	Dry Gas Production[a]	Supplemental Gaseous Fuels[b]	Trade			Net Storage Withdrawals[c]	Balancing Item[d]	Consumption[e]
			Imports	Exports	Net Imports			
1973 Total	[f]21,731	NA	1,033	77	956	-442	-196	22,049
1975 Total	[f]19,236	NA	953	73	880	-344	-235	19,538
1980 Total	19,403	155	985	49	936	23	-640	19,877
1985 Total	16,454	126	950	55	894	235	-428	17,281
1990 Total	17,810	123	1,532	86	1,447	-513	307	[g]19,174
1995 Total	18,599	110	2,841	154	2,687	415	396	22,207
1996 Total	18,854	109	2,937	153	2,784	2	860	22,610
1997 Total	18,902	103	2,994	157	2,837	24	871	22,737
1998 Total	19,024	102	3,152	159	2,993	-530	657	22,246
1999 Total	18,832	98	3,586	163	3,422	172	-119	22,405
2000 Total	19,182	90	3,782	244	3,538	829	-305	23,333
2001 Total	19,616	86	3,977	373	3,604	-1,166	99	22,239
2002 Total	18,928	68	4,015	516	3,499	468	44	23,007
2003 January	1,611	6	365	60	305	865	-72	2,716
February	1,465	6	314	59	255	698	87	2,511
March	1,658	5	329	55	275	139	130	2,207
April	1,587	5	317	52	266	-162	55	1,750
May	1,621	6	328	50	277	-424	40	1,520
June	1,569	5	310	54	256	-483	25	1,372
July	1,589	6	345	50	296	-372	84	1,603
August	1,621	6	337	51	286	-319	60	1,653
September	1,562	5	326	55	271	-423	15	1,430
October	1,615	5	336	61	275	-292	-37	1,566
November	1,544	6	322	71	251	89	-128	1,763
December	1,594	7	367	76	291	489	-97	2,284
Total	19,036	68	3,996	692	3,305	-194	161	22,375
2004 January	E 1,631	6	373	67	306	811	-76	2,677
February	E 1,515	6	346	70	276	600	114	2,510
March	E 1,618	5	349	91	258	103	117	2,103
April	E 1,558	5	325	62	263	-198	124	1,753
May	E 1,580	6	327	61	266	-379	103	1,575
June	E 1,549	1	342	64	278	-397	57	1,488
July	E 1,605	2	375	67	308	-366	39	1,588
August	E 1,601	5	360	67	293	-345	24	1,577
September	E 1,491	5	345	74	270	-325	44	1,485
October	E 1,558	5	336	61	274	-248	-32	1,558
November	RE 1,528	5	369	86	282	65	R -97	R 1,784
December	RE 1,596	E 5	413	83	330	567	R -172	R 2,327
Total	RE 18,830	E 55	4,259	854	3,404	-110	R 245	R 22,424
2005 January	RE 1,597	E 4	E 402	E 63	E 339	713	R -53	R 2,601
February	E 1,486	E 5	RE 356	E 65	RE 291	429	R 49	R 2,260
March	E 1,606	E 6	E 364	E 72	E 292	284	29	2,218
3-Month Total	E 4,689	E 16	E 1,122	E 200	E 922	1,426	26	7,079
2004 3-Month Total	E 4,764	17	1,068	228	840	1,514	155	7,290
2003 3-Month Total	4,734	17	1,009	173	836	1,702	145	7,434

Notes: *(a) Marketed production (wet) minus extraction loss; (b) Supplemental Gaseous Fuels: Any gaseous substance that, introduced into or commingled with natural gas, increases the volume available for disposition. Such substances include, but are not limited to, propane-air, refinery gas, coke oven gas, still gas, manufactured gas, biomass gas, or air or inert gases added for Btu stabilization. Annual data beginning with 1980 are from the Energy Information Administration (EIA) Natural Gas Annual (NGA). Unknown quantities of supplemental gaseous fuels are included in consumption data for 1979 and earlier years. Monthly data are considered preliminary until after the publication of the EIA NGA. Monthly estimates are based on the annual ratio of supplemental gaseous fuels to the sum of dry gas production, net imports, and net withdrawals from storage. The ratio is applied to the monthly sum of the three elements to compute a monthly supplemental gaseous fuels figure; (c) Net withdrawals from underground storage. For 1980-2003, also includes net withdrawals of liquefied natural gas in above-ground tanks. Natural gas in storage at the end of a reporting period may not equal the quantity derived by adding or subtracting net injections or withdrawals from the quantity in storage at the end of the previous period. The difference is due to changes in the quantity of native gas included in the base gas and/or losses in base gas due to migration from storage reservoirs. Monthly underground storage data are collected from the Federal Energy Regulatory Commission (FERC) Form FERC-8 (interstate data) and EIA Form EIA-191 (intrastate data). Beginning in January 1991, all data are collected on the revised Form EIA-191. Injection and withdrawal data from the FERC-8/EIA-191 survey are adjusted to correspond to data from Form EIA-176 following publication of the EIA NGA. The final monthly and annual storage and withdrawal data for 1980-2003 include both underground and liquefied natural gas (LNG) storage. Annual data on LNG additions and withdrawals are from Form EIA-176. Monthly data are estimated by computing the ratio of each month's underground storage additions and withdrawals to annual underground storage additions and withdrawals and applying the ratio to the annual LNG data; (d) Since 1980, excludes transit shipments that cross the U.S.-Canada border (i.e., natural gas delivered to its destination via the other country). The balancing item for natural gas represents the difference between the sum of the components of natural gas supply and the sum of components of natural gas disposition. The differences may be due to quantities lost or to the effects of data reporting problems. Reporting problems include differences due to the net result of conversions of flow data metered at varying temperature and pressure bases and converted to a standard temperature and pressure base; the effect of variations in company accounting and billing practices; differences between billing cycle and calendar period time frames; and imbalances resulting from the merger of data reporting systems which vary in scope, format, definitions, and type of respondents; (e) Consumption includes use for lease and plant fuel, pipelines and distribution, vehicle fuel, and electric power plants, as well as deliveries to residential, commercial, and other industrial customers. Final data for series other than "Other Industrial CHP" and "Electric Power Sector" are from the EIA NGA. Monthly data are considered preliminary until after publication of the EIA NGA. For more detailed information on the methods of estimating preliminary and final monthly data, see the EIA NGM; (f) May include unknown quantities of nonhydrocarbon gases; (g) For 1989-1992, a small amount of consumption at independent power producers may be counted in both "Other Industrial" and "Electric Power Sector" on Table 4.4. Consumption, 1989-1992: Prior to 1993, deliveries to nonutility generators were not separately collected from natural gas companies on Form EIA-176, "Annual Report of Natural and Supplemental Gas Supply and Disposition." As a result, for 1989 through 1992, those volumes are probably included in both the industrial and electric power sectors and double-counted in total consumption. In 1993, 0.28 trillion cubic feet was reported as delivered to nonutility generators; R=Revised; E=Estimate; NA=Not available; Totals may not equal sum of components due to independent rounding; Geographic coverage is the 50 States and the District of Columbia.*
Source: *U.S. Department of Energy, Energy Information Administration, Monthly Energy Review, June 2005*

Crude oil and natural gas drilling activity measurements

	Rotary Rigs in Operation[a]					Total Footage Drilled[c]	Active Well Service Rig Count[d]
	By Site		By Type				
	Onshore	Offshore	Crude Oil	Natural Gas	Total[b]	Thousand Feet	Number
	Average						
1973 Average	1,110	84	NA	NA	1,194	138,223	NA
1975 Average	1,554	106	NA	NA	1,660	180,494	NA
1980 Average	2,678	231	NA	NA	2,909	314,654	NA
1985 Average	1,774	206	NA	NA	1,980	313,045	NA
1990 Average	902	108	532	464	1,010	153,701	NA
1995 Average	622	101	323	385	723	117,832	NA
1996 Average	671	108	306	464	779	129,045	NA
1997 Average	821	122	376	564	943	156,661	NA
1998 Average	703	123	264	560	827	143,454	NA
1999 Average	519	106	128	496	625	99,410	NA
2000 Average	778	140	197	720	918	141,392	NA
2001 Average	1,003	153	217	939	1,156	187,616	NA
2002 Average	717	113	137	691	830	138,310	1,830
2003 January	743	111	132	718	854	12,962	1,898
February	797	110	153	750	907	10,866	1,928
March	836	105	171	767	941	13,269	1,950
April	877	106	185	795	983	14,409	1,954
May	921	113	167	864	1,034	14,515	1,927
June	958	109	152	910	1,067	15,080	1,957
July	974	107	153	924	1,081	15,637	2,016
August	979	111	153	932	1,090	15,776	2,026
September	984	109	154	936	1,093	15,796	1,966
October	997	105	158	941	1,102	16,156	2,064
November	1,005	106	158	952	1,111	16,307	1,973
December	1,010	104	153	959	1,114	16,301	1,946
Average	924	108	157	872	1,032	177,074	1,967
2004 January	1,001	100	143	955	1,101	15,957	2,019
February	1,020	99	153	961	1,119	13,531	2,043
March	1,041	94	164	968	1,135	16,508	2,047
April	1,058	93	154	996	1,151	16,642	2,050
May	1,068	96	156	1,007	1,164	16,687	2,095
June	1,080	96	164	1,011	1,176	16,905	2,067
July	1,116	97	170	1,041	1,213	17,174	2,068
August	1,139	95	170	1,063	1,234	17,462	2,106
September	1,148	92	166	1,073	1,240	17,485	2,078
October	1,145	95	171	1,068	1,240	17,543	2,111
November	1,160	102	183	1,077	1,262	17,936	2,024
December	1,140	106	180	1,064	1,246	17,693	2,063
Average	1,095	97	165	1,025	1,192	201,523	2,064
2005 January	1,153	102	178	1,075	1,255	17,791	2,091
February	1,170	106	192	1,083	1,276	18,218	2,144
March	1,209	97	186	1,118	1,306	18,622	2,143
April	1,241	93	171	1,163	1,334	18,776	2,216
May	1,229	91	150	1,170	1,320	18,138	2,242
5-Month Average	1,201	98	175	1,124	1,300	91,545	2,167
2004 5-Month Average	1,039	96	154	978	1,135	79,325	2,051
2003 5-Month Average	835	109	161	780	944	66,021	1,931

Notes: (a) Rotary rigs in operation are reported weekly. Monthly data are averages of 4- or 5-week reporting periods, not calendar months. Multi-month data are averages of the reported data over the covered months, not averages of the weekly data. Annual data are averages over 52 or 53 weeks, not calendar years. Published data are rounded to the nearest whole number; (b) Sum of rigs drilling for crude oil, rigs drilling for natural gas, and other rigs (not shown) drilling for miscellaneous purposes, such as service wells, injection wells, and stratigraphic tests; (c) Values shown are totals. The number of rigs doing true workovers (where tubing is pulled from the well), or doing rod string and pump repair operations, and that are, on average, crewed and working every day of the month; NA=Not available; Geographic coverage is the 50 States and the District of Columbia.
Source: U.S. Department of Energy, Energy Information Administration, Monthly Energy Review, June 2005

Coal overview (thousand short tons)

	Production[a]	Waste Coal[b,c]	Imports	Exports	Stock Change[d]	Losses and Unaccounted for[e]	Consumption
1973 Total	598,568	NA	127	53,587	([f])	[g] -17,476	562,584
1975 Total	654,641	NA	940	66,309	32,154	-5,522	562,640
1980 Total	829,700	NA	1,194	91,742	25,595	10,827	702,730
1985 Total	883,638	NA	1,952	92,680	-27,934	2,796	818,049
1990 Total	1,029,076	3,339	2,699	105,804	26,542	-1,730	904,498
1995 Total	1,032,974	8,561	9,473	88,547	-275	632	962,104
1996 Total	1,063,856	8,778	8,115	90,473	-17,456	1,411	1,006,321
1997 Total	1,089,932	8,096	7,487	83,545	-11,253	3,678	1,029,544
1998 Total	1,117,535	8,690	8,724	78,048	24,228	-4,430	1,037,103
1999 Total	1,100,431	8,683	9,089	58,476	23,988	-2,906	1,038,647
2000 Total	1,073,612	9,089	12,513	58,489	-48,309	938	1,084,095
2001 Total	1,127,689	([c])	19,787	48,666	41,630	-2,966	1,060,146
2002 Total	1,094,283	([c])	16,875	39,601	10,215	-5,012	1,066,355
2003 January	92,804	([c])	1,134	3,680	-6,051	-2,718	99,026
February	82,264	([c])	1,804	2,428	-3,488	-1,904	87,032
March	89,134	([c])	2,017	2,410	4,064	-1,505	86,182
April	89,378	([c])	2,390	3,571	6,634	2,251	79,312
May	90,610	([c])	2,109	3,875	4,490	464	83,889
June	88,511	([c])	1,894	4,003	-2,803	-1,302	90,508
July	88,534	([c])	2,619	4,223	-11,519	-1,932	100,381
August	89,586	([c])	2,133	4,164	-10,204	-4,113	101,872
September	90,444	([c])	2,300	3,707	-4,539	2,067	91,510
October	94,058	([c])	2,545	3,997	2,134	2,078	88,395
November	84,266	([c])	2,358	3,737	-433	-5,627	88,947
December	92,163	([c])	1,742	3,219	-4,945	-2,176	97,808
Total	1,071,753	([c])	25,044	43,014	-26,659	-14,419	1,094,861
2004 January	93,681	([c])	1,748	3,447	-13,475	5,855	99,602
February	86,767	([c])	1,789	2,276	-3,288	-537	90,105
March	95,023	([c])	1,788	3,965	6,336	891	85,620
April	91,850	([c])	2,157	5,359	9,357	-191	79,482
May	87,311	([c])	2,232	4,910	-263	-2,837	87,732
June	95,048	([c])	2,464	4,987	-2,508	1,976	93,058
July	92,401	([c])	2,531	3,957	-5,627	-3,816	100,418
August	95,354	([c])	2,494	4,067	-6,015	430	99,367
September	93,647	([c])	2,779	4,178	-5,072	4,867	92,453
October	92,635	([c])	2,678	3,358	7,162	-4,017	88,810
November	92,288	([c])	2,258	3,144	3,121	-527	88,809
December	95,472	([c])	2,361	4,350	-7,948	2,620	98,811
Total	1,111,479	([c])	27,280	47,998	-18,221	4,715	1,104,267
2005 January	R 92,935	([c])	2,014	4,075	R -9,585	R 1,071	R 99,389
February	R 89,166	([c])	2,315	3,008	R 2,227	R -1,291	R 87,537
March	R 101,278	([c])	3,277	3,046	R 6,922	R 3,304	R 91,283
April	91,614	([c])	R 2,376	R 4,294	NA	NA	NA
May	88,514	([c])	NA	NA	NA	NA	NA
5-Month Total	463,507	([c])	NA	NA	NA	NA	NA
2004 5-Month Total	454,633	([c])	9,715	19,957	-1,332	3,182	442,541
2003 5-Month Total	444,190	([c])	9,453	15,964	5,650	-3,412	435,441

Notes: (a) Beginning in 2001, includes bituminous refuse; (b) Waste coal (including anthracite culm, bituminous gob, fine coal, and lignite waste) consumed by independent power producers. For 1989-2000, waste coal is counted as a supply-side item to balance the same amount of waste coal included in "Consumption"; (c) Beginning in 2001, bituminous refuse is included in "Production"; to avoid double counting, waste coal is not counted as a separate supply-side item for 2001 forward. d A negative value indicates a decrease in stocks; a positive value indicates an increase; (e) "Losses and Unaccounted for" is calculated as the sum of production, imports, and waste coal, minus exports, stock change, and consumption; (f) Included in "Losses and Unaccounted for"; (g) Includes stock change; R=Revised; NA=Not available; Totals may not equal sum of components due to independent rounding. Geographic coverage is the 50 States and the District of Columbia.

Source: U.S. Department of Energy, Energy Information Administration, Monthly Energy Review, June 2005

Electricity overview (billion kilowatthours)

	Net Generation				Imports[d]	Exports[d]	T&D Losses[e] and Unaccounted for[f]	End Use		
	Electric Power Sector[a]	Commercial Sector[b]	Industrial Sector[c]	Total				Retail Sales[g]	Direct Use[h]	Total
1973 Total	1,861	NA	3	1,864	17	3	165	1,713	NA	1,713
1975 Total	1,918	NA	3	1,921	11	5	180	1,747	NA	1,747
1980 Total	2,286	NA	3	2,290	25	4	216	2,094	NA	2,094
1985 Total	2,470	NA	3	2,473	46	5	190	2,324	NA	2,324
1990 Total	2,901	6	131	3,038	18	16	203	2,713	125	2,837
1995 Total	3,194	8	151	3,353	43	4	229	3,013	151	3,164
1996 Total	3,284	9	151	3,444	43	3	231	3,101	153	3,254
1997 Total	3,329	9	154	3,492	43	9	224	3,146	156	3,302
1998 Total	3,457	9	154	3,620	40	14	221	3,264	161	3,425
1999 Total	3,530	9	156	3,695	43	14	240	3,312	172	3,484
2000 Total	3,638	8	157	3,802	49	15	244	3,421	171	3,592
2001 Total	3,580	7	149	3,737	39	16	226	3,370	163	3,532
2002 Total	3,698	7	153	3,858	36	14	253	3,463	166	3,629
2003 January	327	1	14	342	3	1	21	307	E 15	323
February	287	1	12	299	3	2	5	282	E 13	295
March	291	1	13	304	3	3	17	273	E 14	287
April	273	1	12	286	3	2	18	256	E 13	269
May	294	1	13	308	3	2	26	268	E 14	282
June	315	1	13	329	3	2	27	288	E 14	302
July	360	1	14	374	4	1	30	332	E 15	347
August	367	1	14	382	4	1	29	340	E 15	355
September	310	1	13	323	2	2	3	306	E 14	320
October	293	1	13	307	1	3	14	277	E 14	291
November	285	1	12	298	1	2	20	263	E 13	277
December	318	1	13	332	2	2	24	294	E 14	308
Total	3,721	7	155	3,883	30	24	233	3,488	168	3,656
2004 January	331	1	13	345	2	2	24	307	E 14	322
February	300	1	12	313	2	2	12	287	E 13	301
March	293	1	13	307	2	3	14	278	E 14	292
April	277	1	12	290	2	2	14	263	E 13	276
May	313	1	13	326	2	2	33	280	E 14	293
June	331	1	13	344	3	2	23	308	E 14	322
July	361	1	14	376	4	1	29	335	E 15	350
August	353	1	13	367	5	1	25	332	E 15	346
September	321	1	13	335	3	2	13	309	E 14	323
October	299	1	12	311	3	2	17	282	E 13	295
November	287	1	12	300	3	2	18	270	E 13	283
December	326	1	13	340	3	2	28	300	E 14	313
Total	3,794	7	152	3,953	34	23	248	3,551	E 166	3,717
2005 January	330	1	13	343	3	2	20	310	E 14	324
February	286	1	12	298	3	1	5	282	E 13	295
March	305	1	13	318	3	1	18	288	E 14	302
3-Month Total	920	2	37	959	10	4	44	880	E 41	921
2004 3-Month Total	925	2	38	965	6	7	50	873	E 42	914
2003 3-Month Total	905	2	39	946	8	6	43	863	E 42	905

Notes: *(a) Electricity-only and combined-heat-and-power (CHP) plants within the NAICS 22 category whose primary business is to sell electricity, or electricity and heat, to the public. Through 1988, data are for electric utilities only; beginning in 1989, data are for electric utilities and independent power producers; (b) Commercial combined-heat-and-power (CHP) and commercial electricity-only plants; (c) Industrial combined-heat-and-power (CHP) and industrial electricity-only plants. Through 1988, data are for industrial hydroelectric power only; (d) Electricity transmitted across U.S. borders with Canada and Mexico; (e) Transmission and distribution losses (electricity losses that occur between the point of generation and delivery to the customer); (f) Data collection frame differences and nonsampling error; (g) Electricity retail sales to ultimate customers by electric utilities and, beginning in 1996, other energy service providers; (h) Use of electricity that is 1) self-generated, 2) produced by either the same entity that consumes the power or an affiliate, and 3) used in direct support of a service or industrial process located within the same facility or group of facilities that house the generating equipment. Direct use is exclusive of station use; E=Estimate; NA=Not available; Totals may not equal sum of components due to independent rounding; Geographic coverage is the 50 States and the District of Columbia.*
Source: *U.S. Department of Energy, Energy Information Administration, Monthly Energy Review, June 2005*

Nuclear energy overview

	Total Operable Units[a,b]	Net Summer Capacity of Operable Units[b,c]	Nuclear Electricity Net Generation	Nuclear Share of Electricity Net Generation	Capacity Factor[d]
	Number	Million Kilowatts	Million Kilowatthours	Percent	
1973 Year	42	22.683	83,479	4.5	53.5
1975 Year	57	37.267	172,505	9.0	55.9
1980 Year	71	51.810	251,116	11.0	56.3
1985 Year	96	79.397	383,691	15.5	58.0
1990 Year	112	99.624	576,862	19.0	66.0
1995 Year	109	99.515	673,402	20.1	77.4
1996 Year	109	100.784	674,729	19.6	76.2
1997 Year	107	99.716	628,644	18.0	71.1
1998 Year	104	97.070	673,702	18.6	78.2
1999 Year	104	97.411	728,254	19.7	85.3
2000 Year	104	97.860	753,893	19.8	88.1
2001 Year	104	98.159	768,826	20.6	89.4
2002 Total	104	98.657	780,064	20.2	90.3
2003 January	104	99.209	69,211	20.2	93.8
February	104	99.209	60,942	20.4	91.4
March	104	99.209	59,933	19.7	81.2
April	104	99.209	56,776	19.9	79.5
May	104	99.209	62,202	20.2	84.3
June	104	99.209	64,181	19.5	89.9
July	104	99.209	69,653	18.6	94.4
August	104	99.209	69,024	18.1	93.5
September	104	99.209	63,584	19.7	89.0
October	104	99.209	60,016	19.6	81.3
November	104	99.209	59,600	20.0	83.4
December	104	99.209	68,612	20.7	93.0
Total	104	99.209	763,733	19.7	87.9
2004 January	104	R 99.615	70,806	20.5	R 95.5
February	104	R 99.615	64,102	20.5	R 92.5
March	104	R 99.615	63,263	20.6	R 85.4
April	104	R 99.615	58,620	20.2	R 81.7
May	104	R 99.615	64,917	19.9	R 87.6
June	104	R 99.615	67,787	19.7	R 94.5
July	104	R 99.615	71,975	19.2	R 97.1
August	104	R 99.615	71,064	19.3	R 95.9
September	104	R 99.615	65,932	19.7	R 91.9
October	104	R 99.615	62,530	20.1	R 84.4
November	104	R 99.615	58,941	19.7	R 82.2
December	104	99.615	68,617	20.2	92.6
Total	104	99.615	788,556	19.9	R 90.1
2005 January	104	99.615	69,828	20.3	94.2
February	104	99.615	60,947	20.4	R 91.1
March	104	99.615	61,539	19.4	83.0
3-Month Total	104	99.615	192,314	20.0	89.4
2004 3-Month Total	104	99.615	198,171	20.5	91.1
2003 3-Month Total	104	99.209	190,086	20.1	88.7

Notes: (a) Total of nuclear generating units holding full-power licenses, or equivalent permission to operate, at the end of the period. Although Browns Ferry 1 was shut down in 1985, the unit has remained fully licensed and thus has continued to be counted as operable during the shutdown; in May 2002, the Tennessee Valley Authority announced its intenton to have the unit resume operation in 2007; (b) At end of period; (c) Net Summer Capacity: The steady hourly output that generating equipment is expected to supply to system load, exclusive of auxiliary power, as demonstrated by test at the time of summer peak demand. Auxiliary power of a typical nuclear power plant is about 5 percent of gross generation; (d) Capacity: Nuclear generating units may have more than one type of net capacity rating, including the following: 1) Net Summer Capacity - The steady hourly output that generating equipment is expected to supply to system load, exclusive of auxiliary power, as demonstrated by test at the time of summer peak demand. Auxiliary power of a typical nuclear power plant is about 5 percent of gross generation; 2) Net Design Capacity or Net Design Electrical Rating (DER)—the nominal net electrical output of a unit, specified by the utility and used for plant design. The monthly capacity factors are computed as the actual monthly generation divided by the maximum possible generation for that month. The maximum possible generation is the number of hours in the month multiplied by the net summer capacity at the end of the month. That fraction is then multiplied by 100 to obtain a percentage. Annual capacity factors are averages of the monthly values for that year; R=Revised; Nuclear electricity net generation totals may not equal sum of components due to independent rounding; Geographic coverage is the 50 States and the District of Columbia.
Source: U.S. Department of Energy, Energy Information Administration, Monthly Energy Review, June 2005

Renewable energy consumption by source

	Conventional Hydroelectric Power[a]	Wood[b]	Waste[c]	Alcohol Fuels[d]	Geothermal[e]	Solar[f]	Wind[g]	Total
1973 Total	2,861	1,527	2	NA	43	NA	NA	4,433
1975 Total	3,155	1,497	2	NA	70	NA	NA	4,723
1980 Total	2,900	2,483	2	NA	110	NA	NA	5,494
1985 Total	2,970	2,576	236	52	198	(s)	(s)	6,033
1990 Total	3,046	2,191	408	63	336	60	29	6,133
1995 Total	3,205	2,420	531	117	294	70	33	6,669
1996 Total	3,590	2,467	577	84	316	71	33	7,137
1997 Total	3,640	2,350	551	106	325	70	34	7,075
1998 Total	3,297	2,175	542	117	328	70	31	6,561
1999 Total	3,268	2,224	540	122	331	69	46	6,599
2000 Total	2,811	2,257	511	139	317	66	57	6,158
2001 Total	2,242	1,980	514	147	311	65	70	5,328
2002 Total	2,689	1,899	576	174	328	64	105	5,835
2003 January	211	163	49	17	29	5	6	481
February	203	148	43	20	27	5	8	452
March	248	160	49	17	29	5	11	518
April	254	157	47	20	27	5	11	521
May	301	158	48	19	28	6	10	569
June	293	157	47	19	29	6	11	560
July	254	168	50	20	29	6	10	536
August	235	166	49	21	29	6	8	514
September	189	158	47	18	28	5	9	455
October	189	163	47	21	28	5	9	462
November	202	160	46	24	27	5	10	474
December	246	171	50	25	30	5	11	539
Total	2,825	1,929	571	239	339	64	115	6,082
2004 January	235	173	46	24	30	5	11	523
February	213	159	43	22	28	5	11	481
March	231	164	46	24	28	5	13	513
April	212	166	46	24	27	5	13	493
May	242	159	50	25	28	6	17	527
June	255	161	49	25	28	6	14	537
July	235	173	49	25	29	6	11	527
August	220	168	49	24	29	6	10	505
September	208	160	45	26	27	5	11	482
October	193	169	45	25	29	5	10	477
November	213	161	45	25	28	5	10	488
December	267	177	48	26	29	5	12	564
Total	2,725	1,989	560	296	340	63	143	6,116
2005 January	248	171	49	26	29	5	10	539
February	221	162	43	28	25	5	9	493
March	234	166	49	33	29	5	14	530
3-Month Total	703	500	141	87	83	15	33	1,562
2004 3-Month Total	680	496	135	70	86	15	33	1,517
2003 3-Month Total	661	472	140	54	85	15	25	1,451

Notes: *(a) Hydroelectricity generated by pumped storage is not included in renewable energy; (b) Wood, black liquor, and other wood waste; (c) Municipal solid waste, landfill gas, sludge waste, tires, agricultural byproducts, and other biomass; (d) Ethanol blended into motor gasoline; (e) Geothermal electricity net generation, heat pump, and direct use energy; (f) Solar thermal and photovoltaic electricity net generation, and solar thermal direct use energy; (g) Wind electricity net generation; (s) Less than 0.5 trillion Btu; NA=Not Available; Totals may not equal sum of components due to independent rounding; Geographic coverage is the 50 states and the District of Columbia.*

Source: *U.S. Department of Energy, Energy Information Administration, Monthly Energy Review, June 2005*

Estimated number of alternative-fueled vehicles in use, by state, 2001-2003

State	2001	2002	2003
Alabama	7,501	8,979	9,870
Alaska	1,288	1,277	1,433
Arizona	11,046	11,771	13,303
Arkansas	2,873	2,839	3,014
California	66,366	71,501	77,761
Colorado	11,120	11,925	12,447
Connecticut	3,981	5,147	5,606
Delaware	679	1,378	1,492
District of Columbia	3,105	3,243	3,674
Florida	15,959	16,542	17,829
Georgia	12,959	15,567	17,912
Hawaii	2,487	2,513	2,707
Idaho	2,759	5,233	5,821
Illinois	12,912	15,401	16,521
Indiana	5,515	6,584	7,405
Iowa	3,163	4,139	4,823
Kansas	5,633	5,985	6,332
Kentucky	4,676	5,718	6,298
Louisiana	3,154	3,325	3,582
Maine	376	390	417
Maryland	9,031	9,157	9,791
Massachusetts	2,478	2,700	2,946
Michigan	10,675	12,307	14,335
Minnesota	4,403	6,032	6,482
Mississippi	1,908	1,876	1,990
Missouri	6,302	7,102	7,540
Montana	3,812	3,557	4,228
Nebraska	5,142	5,814	6,303
Nevada	5,318	5,571	5,968
New Hampshire	935	1,096	1,218
New Jersey	5,854	5,956	6,569
New Mexico	9,643	10,624	11,042
New York	26,890	32,423	37,559
North Carolina	8,661	9,770	10,695
North Dakota	1,818	1,819	2,133
Ohio	8,296	9,939	11,097
Oklahoma	21,440	22,283	23,336
Oregon	5,769	5,878	6,568
Pennsylvania	7,326	7,611	8,351
Rhode Island	745	844	936
South Carolina	6,018	7,460	7,992
South Dakota	1,765	1,802	1,906
Tennessee	5,430	6,654	7,343
Texas	54,254	56,190	55,820
Utah	6,583	7,162	7,621
Vermont	675	748	844
Virginia	9,686	10,495	11,706
Washington	9,122	9,166	9,764
West Virginia	1,022	1,012	1,098
Wisconsin	4,168	5,813	6,457
Wyoming	2,737	2,780	2,924
U.S. Total	425,457	471,098	510,805

Notes: *Excludes gasoline-electric hybrids. Estimates for 2003 are based on plans or projections. Estimates for historical years may be revised in future reports if new information becomes available. Total may not equal sum of components due to independent rounding.*
Source: *Energy Information Administration, Office of Coal, Nuclear, Electric, and Alternate Fuels and the DOE/GSA Federal Automotive Statistical Tool (FAST).*

Statistics & Rankings / Energy

Estimated number of alternative-fueled vehicles in use, by state and fuel type, 2002

State	Liquified Petroleum Gases	Natural Gas	Methanol	Ethanol	Electricity	Total
Alabama	4,289	1,341	0	2,713	636	8,979
Alaska	145	401	0	720	11	1,277
Arizona	1,082	7,243	201	1,583	1,662	11,771
Arkansas	2,199	340	0	300	0	2,839
California	21,537	24,990	4,787	9,517	10,670	71,501
Colorado	5,611	2,694	3	3,491	126	11,925
Connecticut	379	2,762	1	1,849	156	5,147
Delaware	85	489	10	783	11	1,378
Dist. of Columbia	7	1,462	50	1,408	316	3,243
Florida	4,171	4,152	6	7,856	357	16,542
Georgia	4,418	4,484	39	2,076	4,550	15,567
Hawaii	842	0	0	1,467	204	2,513
Idaho	1,581	3,412	0	240	0	5,233
Illinois	5,259	3,120	17	6,916	89	15,401
Indiana	1,426	3,397	0	1,670	91	6,584
Iowa	2,179	18	27	1,903	12	4,139
Kansas	3,565	748	1	1,649	22	5,985
Kentucky	2,214	1,191	0	2,313	0	5,718
Louisiana	1,117	896	3	1,309	0	3,325
Maine	158	77	0	134	21	390
Maryland	2,570	3,634	7	2,901	45	9,157
Massachusetts	249	1,006	36	1,331	78	2,700
Michigan	4,822	991	48	4,840	1,606	12,307
Minnesota	2,162	509	0	3,361	0	6,032
Mississippi	1,193	140	0	543	0	1,876
Missouri	2,642	476	95	3,878	11	7,102
Montana	2,980	268	0	309	0	3,557
Nebraska	4,338	370	0	1,095	11	5,814
Nevada	1,487	3,111	0	973	0	5,571
New Hampshire	718	42	0	169	167	1,096
New Jersey	358	2,723	4	2,681	190	5,956
New Mexico	6,069	1,969	11	2,140	435	10,624
New York	6,213	13,100	88	3,723	9,299	32,423
North Carolina	4,560	559	0	4,539	112	9,770
North Dakota	1,310	155	0	354	0	1,819
Ohio	2,487	2,647	26	4,537	242	9,939
Oklahoma	17,839	3,322	0	1,122	0	22,283
Oregon	3,084	1,034	20	1,528	212	5,878
Pennsylvania	1,107	2,299	108	4,008	89	7,611
Rhode Island	122	331	0	391	0	844
South Carolina	3,047	362	0	4,051	0	7,460
South Dakota	1,374	44	0	384	0	1,802
Tennessee	2,623	763	0	3,068	200	6,654
Texas	39,279	9,961	162	6,706	82	56,190
Utah	3,227	1,961	8	1,966	0	7,162
Vermont	366	5	0	199	178	748
Virginia	927	4,735	7	3,740	1,086	10,495
Washington	4,397	1,925	73	2,760	11	9,166
West Virginia	39	378	0	595	0	1,012
Wisconsin	1,459	1,207	35	3,075	37	5,813
Wyoming	2,368	303	0	87	22	2,780
U.S. Total	187,680	123,547	5,873	120,951	33,047	471,098

Notes: *Natural gas includes compressed natural gas (CNG) and liquefied natural gas (LNG). Methanol includes M85 and M100. Ethanol includes E85 and E95. Excludes gasoline-electric hybrids. Totals may not equal sum of components due to independent rounding.*
Source: *Energy Information Administration, Office of Coal, Nuclear, Electric, and Alternate Fuels and the DOE/GSA Federal Automotive Statistical Tool (FAST).*

Number of onroad alternative-fueled buses made available, by vehicle type, and fuel type, 2000-2003

Bus Type/Fuel	2000	2001	2002	2003
School Bus	234	333	530	319
Liquefied Petroleum Gas (LPG)	84	39	71	71
Compressed Natural Gas (CNG)	148	292	457	248
Liquefied Natural Gas (LNG)	0	0	0	0
Electricity (EVC)	2	2	2	0
Transit Bus[1]	1,432	1,819	1,840	1,200
Liquefied Petroleum Gas (LPG)	101	106	113	181
Compressed Natural Gas (CNG)	973	1,273	1,447	799
Liquefied Natural Gas (LNG)	343	245	84	44
Electricity (EVC)[2]	15	195	196	176
Intercity Bus	79	32	0	0
Liquefied Petroleum Gas (LPG)	0	0	0	0
Compressed Natural Gas (CNG)	79	32	0	0
Liquefied Natural Gas (LNG)	0	0	0	0
Electricity	0	0	0	0
Total Buses	1,745	2,184	2,370	1,519

Note: *(1) Includes shuttle busses and trolley replicas; (2) Includes gasoline/diesel-electric hybrid vehicles which are outside EPACT's definition of alternative fueled vehicle*
Source: *U.S. Department of Energy, Energy Information Administration, Form EIA-886, "Annual Survey of Alternative Fuel Vehicle Suppliers and Users."*

Number of onroad alternative fuel and gasoline/diesel-electric hybrid vehicles planned to be made available, by vehicle type and fuel type, 2004

Fuel Type	Cars	Vans	Pickup Trucks	Light Duty Trucks	Medium Duty Trucks	Heavy Duty Trucks	Buses	Other Onroad Vehicles	Total
Compressed Natural Gas (CNG)	2,706	w	3,040	0	290	190	1,353	w	10,439
Electric (EVC)[1]	w	0	w	w	w	w	w	w	2,465
Ethanol, 85 Percent (E85)[2]	w	w	w	w	0	0	0	0	598,757
Gasoline/Diesel-Electric Hybrid[3]	w	0	0	w	0	4	462	0	81,584
Hydrogen (HYD)	0	0	0	0	0	0	w	0	w
Liquefied Natural Gas (LNG)	0	0	0	0	0	w	w	0	w
Liquefied Petroleum Gas (LPG)	w	w	w	w	w	w	295	w	6,789
Total	195,909	2,933	15,886	475,251	5,415	397	2,183	2,232	700,206

Notes: (1) Does not include gasoline/diesel-electric hybrids; (2) The remaining portion of 85-percent methanol is gasoline; (3) Gasoline/diesel-electric hybrids are no longer grouped under the Electric fuel category because the input fuel is gasoline or diesel rather than an alternative transportation fuel. DOE has ruled that gasoline/diesel-electric hybrids are not "alternative fuel vehicles;" Vans include light- and medium-duty passenger and cargo vans; Light Duty Trucks include pickups and other trucks less than or equal to 8,500 GVWR; Medium Duty Trucks include pickups and other trucks 8,501 to 26,000 GVWR; Heavy Duty Trucks include vehicles 26,001 GVWR and over; Other Onroad Vehicles includes motorcycles and low speed vehicles such as neighborhood electric vehicles; w = withheld to avoid disclosure of individual company data.

Source: U.S. Department of Energy, Energy Information Administration, Form EIA-886, "Annual Survey of Alternative Fuel Vehicle Suppliers and Users."

Estimated number of alternative-fueled vehicles in use in the United States, by fuel, 1995-2004

Fuel	1995	1996	1997	1998	1999	2000	2001	2002	2003	2004
Liquefied Petroleum Gases (LPG)	172,806	175,585	175,679	177,183	178,610	181,994	185,053	187,680	190,438	194,389
Compressed Natural Gas (CNG)	50,218	60,144	68,571	78,782	91,267	100,750	111,851	120,839	132,988	143,742
Liquefied Natural Gas (LNG)	603	663	813	1,172	1,681	2,090	2,576	2,708	3,030	3,134
Methanol, 85 Percent (M85)[1]	18,319	20,265	21,040	19,648	18,964	10,426	7,827	5,873	4,917	4,592
Methanol, Neat (M100)	386	172	172	200	198	0	0	0	0	0
Ethanol, 85 Percent (E85)[1,2]	1,527	4,536	9,130	12,788	24,604	87,570	100,303	120,951	133,776	146,195
Ethanol, 95 Percent (E95)[1]	136	361	347	14	14	4	0	0	0	0
Electricity[3]	2,860	3,280	4,453	5,243	6,964	11,830	17,847	33,047	45,656	55,852
Non-LPG Subtotal	74,049	89,421	104,526	117,847	143,692	212,670	240,404	283,418	320,367	353,515
Total	246,855	265,006	280,205	295,030	322,302	394,664	425,457	471,098	510,805	547,904

Notes: (1) The remaining portion of 85-percent methanol and both ethanol fuels is gasoline; (2) In 1997, some vehicle manufacturers began including E85-fueling capability in certain model lines of vehicles. For 2002, the EIA estimated that the number of E-85 vehicles that are capable of operating on E85, gasoline, or both, is about 4.1 million. Many of these alternative-fueled vehicles (AFVs) are sold and used as traditional gasoline-powered vehicles. In this table, AFVs in use include only those E85 vehicles believed to be intended for use as AFVs. These are primarily fleet-operated vehicles; (3) Excludes gasoline-electric hybrids; Estimates for 2003 and 2004 are based on plans or projections. Estimates for historical years may be revised in future reports if new information becomes available.
Sources: 1995: Science Applications International Corporation, "Alternative Transportation Fuels and Vehicles Data Development," unpublished final report prepared for the Energy Information Administration (McLean, VA, July 1996) and U.S. Department of Energy, Office of Energy Efficiency and Renewable Energy. 1996-2004: Energy Information Administration, Office of Coal, Nuclear, Electric, and Alternate Fuels. Beginning in 2000, Federal data were derived from the DOE/GSA Federal Automotive Statistical Tool (FAST).

Statistics & Rankings / Energy

Estimated consumption of vehicle fuels in the United States, 1998-2004

Fuel	1998	1999	2000	2001	2002	2003	2004
Alternative Fuels							
Liquefied Petroleum Gases (LPG)[1]	241,386	209,817	212,576	215,876	223,143	230,486	242,368
Compressed Natural Gas (CNG)[1]	72,412	79,620	86,745	104,496	120,670	141,726	159,464
Liquefied Natural Gas (LNG)[1]	5,343	5,828	7,259	8,921	9,382	10,514	10,868
Methanol, 85 Percent (M85)[1,2]	1,212	1,073	585	439	337	274	257
Methanol, Neat (M100)	449	447	0	0	0	0	0
Ethanol, 85 Percent (E85)[1,2]	1,727	3,916	12,071	14,623	17,783	20,092	22,405
Ethanol, 95 Percent (E95)[1,2]	59	62	13	0	0	0	0
Electricity[3]	1,202	1,524	3,058	4,066	7,274	9,633	11,836
Subtotal[1]	323,790	302,287	322,307	348,421	378,589	412,725	447,198
Oxygenates							
Methyl Tertiary Butyl Ether (MTBE)[4]	2,903,400	3,402,600	3,296,100	3,352,200	2,383,000	na	na
Ethanol in Gasohol[1]	889,500	950,300	1,085,800	1,143,300	1,413,600	1,792,900	2,052,000
Biodiesel	na	na	6,816	7,076	16,917	26,758	36,599
Total Alternative and Replacement Fuels	4,116,690	4,655,187	4,711,023	4,850,997	4,192,106	2,232,383	2,535,797
Traditional Fuels							
Gasoline[5]	122,849,000	125,111,000	125,720,000	127,768,000	131,299,000	132,961,000	136,374,000
Diesel	33,665,360	35,796,800	36,990,370	37,085,270	38,305,630	39,930,170	40,740,760
Total Fuel Consumption[1]	151,597,859	156,838,150	161,210,087	163,032,677	165,201,691	169,983,219	173,303,895

Notes: Fuel quantities are expressed in a common base unit of gasoline-equivalent gallons to allow comparisons of different fuel types. Gasoline-equivalent gallons do not represent gasoline displacement. Gasoline equivalent is computed by dividing the lower heating value of the alternative fuel by the lower heating value of gasoline and multiplying this result by the alternative fuel consumption value. Lower heating value refers to the Btu content per unit of fuel excluding the heat produced by condensation of water vapor in the fuel. Totals may not equal sum of components due to independent rounding. Estimates for 2003 and 2004, in italics, are based on plans or projections. Estimates for historical years may be revised in future reports if new information becomes available; (1) 1999 and 2002 estimates have been revised; (2) The remaining portion of 85-percent methanol and both ethanol fuels is gasoline. Consumption data include the gasoline portion of the fuel; (3) Excludes gasoline-electric hybrids; (4) Includes a very small amount of other ethers, primarily Tertiary Amyl Methyl Ether (TAME) and Ethyl Tertiary Butyl Ether (ETBE); (5) Gasoline consumption includes ethanol in gasohol and MTBE; (6) Total fuel consumption is the sum of alternative fuel, gasoline, and diesel consumption. Oxygenate consumption is included in gasoline consumption.

Sources: *1993-2002 Oxygenate Consumption: Energy Information Administration, Petroleum Supply Monthly. 1993-2004 Traditional Fuel Consumption: Energy Information Administration, Petroleum Supply Annual, Volume 1 (June 2000). Highway use of gasoline was estimated as 97.1 percent of consumption, based on data in the Transportation Energy Data Book: Edition 16, prepared by Oak Ridge National Laboratory for the U.S. Department of Energy (July 1996). Diesel consumption was adjusted for highway use by multiplying by .568 derived from Energy Information Administration, Fuel Oil and Kerosene Sales 1999. 2002-2004 Oxygenate and Traditional Fuel Consumption: Energy Information Administration, Short Term Energy Outlook, September 2002. Alternative Fuel Consumption: Science Applications International Corporation 1992-1995, "Alternative Transportation Fuels and Vehicles Data Development." unpublished final report prepared for the Energy Information Administration (McLean, VA, July 1996) and Energy Information Administration, Office of Coal, Nuclear, Electric and Alternate Fuels for 1996-2004.*

Effective dates of appliance efficiency standards, 1988-2005

Product	1988	1990	1992	1993	1994	1995	2000	2001	2005
Clothes dryers[1]	X				X				
Clothes washers[1]	X				X				
Dishwashers[1]	X				X				
Refrigerators and freezers[1]		X		X				X	
Kitchen ranges and ovens		X							
Room air conditioners[1]		X					X		
Direct heating equipment		X							
Fluorescent lamp ballasts		X							X
Water heaters		X							
Pool heaters		X							
Central air conditioners and heat pumps			X						
Furnaces									
Central (>45,000 Btu per hour)			X						
Small (<45,000 Btu per hour)			X						
Mobile home		X							
Boilers			X						
Fluorescent lamps, 8 foot							X		
Fluorescent lamps, 2 and 4 foot (U tube)								X	

Notes: *(1) Multiple revisions were made to the product's efficiency standard from 1988-2005*
Source: *U.S. Department of Energy, Office of Codes and Standards; and Electric Power Research Institute, "Energy Conservation Standards for Consumer Products."*

United States refueling site counts by state and fuel type

State	CNG[1]	E85[2]	LPG[3]	Electric	BD[4]	HY[5]	LNG[6]	All
Alabama	1	0	79	0	0	0	0	80
Alaska	0	0	12	0	0	0	0	12
Arizona	31	2	67	18	3	1	6	128
Arkansas	4	0	61	0	0	0	0	65
California	180	3	259	468	17	8	30	966
Colorado	21	11	74	4	16	0	0	126
Connecticut	12	0	24	4	1	0	0	41
Delaware	2	0	5	0	3	0	0	10
Dist. of Columbia	1	0	0	0	0	1	0	2
Florida	27	3	111	6	4	0	0	151
Georgia	21	1	55	0	4	0	0	81
Hawaii	0	0	6	11	3	0	0	20
Idaho	8	1	30	0	1	0	1	41
Illinois	13	27	83	0	6	0	0	129
Indiana	11	0	42	0	10	0	0	63
Iowa	0	24	35	0	1	0	0	60
Kansas	3	4	52	0	5	0	0	64
Kentucky	0	4	36	0	4	0	0	44
Louisiana	12	0	25	0	0	0	0	37
Maine	0	0	12	0	3	0	0	15
Maryland	16	3	23	0	4	0	0	46
Massachusetts	9	0	33	31	1	0	0	74
Michigan	14	2	94	0	12	2	0	124
Minnesota	3	108	40	0	1	0	0	152
Mississippi	0	0	42	0	1	0	0	43
Missouri	6	11	106	0	2	0	0	125
Montana	2	3	32	0	5	0	0	42
Nebraska	1	9	24	0	1	0	0	35
Nevada	16	0	25	0	9	1	0	51
New Hampshire	1	0	17	9	6	0	0	33
New Jersey	18	0	13	0	1	0	0	32
New Mexico	7	3	60	0	2	0	0	72
New York	33	6	47	1	0	0	0	87
North Carolina	9	3	69	0	31	0	0	112
North Dakota	4	5	18	0	0	0	0	27
Ohio	13	4	77	0	10	0	0	104
Oklahoma	54	3	78	1	1	0	0	137
Oregon	16	0	41	0	6	0	0	63
Pennsylvania	40	0	75	0	3	0	1	119
Rhode Island	6	0	6	1	0	0	0	13
South Carolina	4	14	41	0	22	0	0	81
South Dakota	0	11	25	0	0	0	0	36
Tennessee	4	5	59	0	8	0	0	76
Texas	29	3	708	2	5	0	2	749
Utah	63	3	27	0	1	0	0	94
Vermont	1	0	12	10	0	0	0	23
Virginia	16	2	39	0	9	0	2	68
Washington	19	2	70	2	15	0	0	108
West Virginia	2	2	8	0	0	0	0	12
Wisconsin	20	9	59	0	1	0	0	89
Wyoming	12	2	33	0	12	0	0	59
Totals by Fuel	785	293	3069	568	250	13	42	5021

Notes: (1) Compressed Natural Gas; (2) 85% Ethanol; (3) Liquified Petroleum Gas; (4) Biodiesel; (5) Hydrogen; (6) Liquified Natural Gas
Source: U.S. Department of Energy, Alternative Fuels Data Center, Refueling Sites, July 25, 2005

Environmental revenue and expenditures, by state

State	Revenue ($/per capita)						Current Operational Expenses ($/per capita)						
	NRA[1]	NRF[3]	NRO[4]	PR[5]	S[6]	SWM[7]	NRA[1]	NRFG[2]	NRF[3]	NRO[4]	PR[5]	S[6]	SWM[7]
Alabama	0.66	1.12	0.13	5.40	0.00	0.00	26.42	4.20	5.25	10.05	4.23	0.00	1.00
Alaska	0.85	6.79	6.54	0.00	0.00	0.00	21.71	190.90	75.14	65.55	12.41	0.00	0.00
Arizona	1.06	0.00	1.42	2.34	0.00	0.04	18.04	7.37	2.95	11.36	7.66	0.00	4.40
Arkansas	3.09	0.31	0.91	4.32	0.00	0.00	27.65	14.76	7.11	3.93	20.18	4.81	5.22
California	8.16	0.18	16.80	2.00	0.00	0.00	29.57	6.41	5.52	22.02	7.51	3.98	24.28
Colorado	1.86	0.00	1.55	0.46	0.00	0.76	9.51	17.47	0.00	10.86	10.79	0.00	3.21
Connecticut	0.01	0.13	0.23	1.19	0.00	44.76	5.67	4.30	2.95	15.89	28.70	0.11	50.37
Delaware	0.11	0.45	5.46	12.59	0.00	59.97	8.56	9.80	3.40	46.94	39.35	0.00	31.43
Florida	0.95	0.57	1.05	1.47	0.00	1.57	28.84	8.41	2.79	35.93	9.19	0.00	11.09
Georgia	2.76	0.00	0.02	9.11	0.00	3.23	23.05	4.69	4.61	16.76	16.31	0.00	0.05
Hawaii	4.07	0.00	9.55	6.13	0.00	2.87	48.67	4.67	8.26	14.73	37.17	0.00	2.52
Idaho	0.51	27.22	1.42	0.75	0.00	0.00	39.59	37.42	21.16	11.67	13.26	22.32	0.12
Illinois	0.46	0.00	0.00	0.72	0.00	0.00	14.40	3.42	0.50	5.88	8.75	2.04	3.78
Indiana	4.38	0.50	1.14	2.42	0.00	1.03	24.38	4.69	1.45	9.92	5.93	1.48	2.34
Iowa	7.65	0.00	1.51	1.00	0.00	1.28	44.88	5.61	1.35	18.02	5.61	0.00	1.08
Kansas	5.15	0.00	0.20	1.34	0.00	0.00	44.15	8.45	0.00	10.11	2.09	0.00	0.00
Kentucky	9.03	0.34	0.15	12.97	0.02	0.05	40.23	7.21	3.33	12.68	22.82	4.44	9.94
Louisiana	0.92	0.45	0.97	21.06	0.00	0.00	35.80	9.44	3.44	24.45	31.48	0.00	4.08
Maine	4.11	0.05	6.26	3.28	0.00	17.87	32.16	32.28	1.17	42.94	7.80	0.02	9.03
Maryland	1.32	0.70	1.00	3.34	0.22	0.00	15.04	15.40	0.00	20.52	15.82	10.55	3.53
Massachusetts	0.09	0.00	3.13	3.79	0.00	0.71	3.01	5.78	0.01	18.01	11.93	22.26	5.75
Michigan	0.73	2.39	0.56	0.47	0.00	0.00	15.90	7.67	2.82	17.98	8.71	0.00	5.44
Minnesota	4.99	3.65	2.55	3.47	0.00	0.00	37.92	15.79	9.80	16.49	19.04	0.00	7.66
Mississippi	3.71	0.81	0.59	2.73	0.00	0.00	40.47	12.50	8.63	3.69	12.26	0.00	0.00
Missouri	1.07	0.47	1.41	0.53	0.00	0.01	15.94	2.39	1.24	19.95	6.57	0.00	4.95
Montana	7.21	17.64	11.17	2.65	0.00	0.73	44.40	49.85	28.68	51.48	6.17	0.00	0.73
Nebraska	9.01	0.00	0.44	1.66	0.00	0.00	59.87	14.33	0.00	7.62	13.57	0.00	0.00
Nevada	0.65	0.85	0.53	1.21	0.00	1.11	18.49	8.02	3.43	8.48	6.68	0.00	2.51
New Hampshire	1.22	0.69	0.42	10.02	0.05	0.05	7.47	15.68	2.05	15.00	6.18	3.63	9.02
New Jersey	0.29	0.92	5.23	28.89	2.18	5.15	6.41	3.44	1.10	7.24	38.47	1.87	12.96
New Mexico	9.18	0.07	1.96	1.34	0.00	0.00	33.03	15.94	2.56	40.79	27.17	0.00	2.31
New York	0.07	0.00	1.63	10.11	0.17	0.00	5.43	2.17	1.57	7.55	18.66	1.85	3.78
North Carolina	1.76	0.53	2.54	0.79	0.00	0.04	22.20	7.41	4.62	15.52	11.96	0.18	9.47
North Dakota	12.44	0.91	0.27	3.69	0.00	0.00	130.14	27.33	2.03	17.43	16.83	0.00	0.00
Ohio	1.28	0.17	0.29	1.31	0.19	0.00	12.79	3.32	0.86	7.80	5.73	0.98	2.38
Oklahoma	0.97	0.07	0.01	6.36	0.00	0.31	27.18	5.49	0.82	13.53	16.92	0.04	1.49
Oregon	4.19	32.46	2.50	5.38	0.00	3.03	28.64	22.58	36.26	12.15	14.47	0.00	7.31
Pennsylvania	0.32	2.63	1.49	1.11	0.01	0.03	6.08	7.56	4.17	19.94	9.77	0.26	6.24
Rhode Island	0.00	0.00	4.24	3.59	1.19	4.99	4.14	5.34	1.91	24.17	21.67	27.01	40.64
South Carolina	0.76	0.75	1.41	3.52	0.00	1.67	20.02	14.11	4.69	1.62	14.03	0.00	0.81
South Dakota	10.83	0.04	1.49	1.77	0.00	0.00	66.95	31.58	1.75	11.12	28.68	0.00	0.00
Tennessee	0.91	0.32	0.26	4.74	0.00	0.00	15.40	8.72	5.60	2.97	16.79	0.00	3.43
Texas	1.10	0.11	0.38	1.46	0.00	0.69	13.57	6.37	1.86	8.21	2.70	0.00	6.84
Utah	1.83	0.29	1.01	3.31	0.00	1.31	26.73	17.37	10.17	19.77	10.90	0.00	5.56
Vermont	1.38	0.63	1.82	11.58	0.00	0.06	47.87	20.81	7.73	65.38	16.09	0.00	15.20
Virginia	0.00	0.21	0.02	1.71	0.00	0.00	6.70	6.34	2.90	6.27	9.76	7.74	5.25
Washington	1.02	20.85	4.14	0.52	0.00	0.91	27.03	20.70	12.25	22.20	9.59	0.00	9.19
West Virginia	2.25	0.47	18.78	11.73	0.00	7.01	19.06	13.07	3.46	54.71	36.98	0.00	7.34
Wisconsin	4.67	0.93	3.60	2.72	0.00	0.20	20.04	13.17	11.05	17.13	8.63	0.00	15.66
Wyoming	3.26	1.52	4.58	2.31	0.00	0.01	55.81	79.01	21.32	83.61	44.79	0.00	4.38

Notes: (1) Natural Resources, Agriculture; (2) Natural Resources, Fish & Game; (3) Natural Resources, Forestry; (4) Natural Resources, Other; (5) Parks & Recreation; (6) Sewerage; (7) Solid Waste Management
Source: U.S. Census Bureau, Government Finances, 2003

Environmental revenue and expenditures for U.S counties with populations of 100,000+

County	St.	Revenue ($/per capita)						Current Operational Expenses ($/per capita)						
		NRA[1]	NRF[3]	NRO[4]	PR[5]	S[6]	SWM[7]	NRA[1]	NRFG[2]	NRF[3]	NRO[4]	PR[5]	S[6]	SWM[7]
Baldwin	AL	(a)	(a)	(a)	0.09	0.00	46.35	(a)	(a)	(a)	0.00	3.09	0.00	32.01
Calhoun	AL	(a)	(a)	(a)	0.53	0.00	13.04	(a)	(a)	(a)	0.59	1.61	0.00	15.03
Etowah	AL	(a)	(a)	(a)	0.00	0.00	11.93	(a)	(a)	(a)	0.00	0.00	0.00	0.14
Jefferson	AL	(a)	(a)	(a)	0.00	136.83	5.72	(a)	(a)	(a)	0.00	22.21	61.34	6.43
Lee	AL	(a)	(a)	(a)	0.00	0.00	0.00	(a)	(a)	(a)	0.00	0.00	0.00	0.00
Madison	AL	(a)	(a)	(a)	0.15	0.00	21.49	(a)	(a)	(a)	0.00	1.26	16.23	20.08
Mobile	AL	(a)	(a)	(a)	0.53	0.00	0.00	(a)	(a)	(a)	1.37	9.72	0.00	6.21
Montgomery	AL	(a)	(a)	(a)	0.00	0.00	0.00	(a)	(a)	(a)	0.00	7.69	0.00	0.64
Morgan	AL	(a)	(a)	(a)	0.00	0.00	17.11	(a)	(a)	(a)	0.00	7.07	0.00	16.13
Shelby	AL	(a)	(a)	(a)	0.00	18.83	23.71	(a)	(a)	(a)	0.51	8.55	12.75	10.31
Tuscaloosa	AL	(a)	(a)	(a)	0.00	0.00	0.79	(a)	(a)	(a)	0.27	5.00	0.00	1.81
Benton	AR	(a)	(a)	(a)	0.00	0.00	0.00	(a)	(a)	(a)	0.60	0.00	0.00	2.33
Pulaski	AR	(a)	(a)	(a)	0.00	0.00	5.52	(a)	(a)	(a)	0.60	0.00	0.00	5.49
Sebastian	AR	(a)	(a)	(a)	5.55	0.00	0.00	(a)	(a)	(a)	0.00	12.03	0.00	0.00
Washington	AR	(a)	(a)	(a)	0.00	0.00	0.00	(a)	(a)	(a)	0.00	0.00	0.00	0.85
Cochise	AZ	(a)	(a)	(a)	0.00	0.00	0.00	(a)	(a)	(a)	0.00	0.00	0.00	0.00
Coconino	AZ	(a)	(a)	(a)	7.16	0.00	4.47	(a)	(a)	(a)	0.00	19.67	0.00	20.30
Maricopa	AZ	(a)	(a)	(a)	1.09	0.00	0.02	(a)	(a)	(a)	9.92	2.47	0.00	1.01
Mohave	AZ	(a)	(a)	(a)	5.71	0.00	6.65	(a)	(a)	(a)	26.96	6.35	0.00	2.79
Navajo	AZ	(a)	(a)	(a)	0.00	0.00	0.00	(a)	(a)	(a)	0.00	0.00	0.00	0.00
Pima	AZ	(a)	(a)	(a)	1.02	51.43	29.49	(a)	(a)	(a)	2.29	14.09	70.04	7.60
Pinal	AZ	(a)	(a)	(a)	0.00	0.00	0.06	(a)	(a)	(a)	0.00	12.83	0.00	6.74
Yavapai	AZ	(a)	(a)	(a)	0.00	0.00	0.00	(a)	(a)	(a)	0.56	0.00	0.31	8.47
Yuma	AZ	(a)	(a)	(a)	0.00	0.00	0.16	(a)	(a)	(a)	2.56	0.10	0.00	4.49
Alameda	CA	(a)	(a)	(a)	0.12	0.00	0.00	(a)	(a)	(a)	51.28	0.28	0.00	0.00
Butte	CA	(a)	(a)	(a)	0.00	0.00	24.30	(a)	(a)	(a)	7.18	0.05	0.00	32.74
Contra Costa	CA	(a)	(a)	(a)	0.00	22.60	1.25	(a)	(a)	(a)	20.57	0.00	14.67	0.25
El Dorado	CA	(a)	(a)	(a)	0.00	0.00	0.00	(a)	(a)	(a)	18.15	6.00	0.00	0.00
Fresno	CA	(a)	(a)	(a)	0.31	5.90	18.64	(a)	(a)	(a)	0.53	3.05	5.24	9.40
Humboldt	CA	(a)	(a)	(a)	1.93	0.00	3.13	(a)	(a)	(a)	1.89	4.26	0.00	2.25
Imperial	CA	(a)	(a)	(a)	0.44	0.00	20.58	(a)	(a)	(a)	1.96	0.44	0.13	17.48
Kern	CA	(a)	(a)	(a)	9.55	1.10	42.93	(a)	(a)	(a)	7.77	20.32	3.31	48.68
Kings	CA	(a)	(a)	(a)	0.44	0.00	0.00	(a)	(a)	(a)	0.98	5.95	0.00	0.00
Los Angeles	CA	(a)	(a)	(a)	6.87	1.81	2.50	(a)	(a)	(a)	19.37	16.95	1.87	3.43
Madera	CA	(a)	(a)	(a)	0.26	0.00	23.48	(a)	(a)	(a)	6.71	0.37	0.00	21.19
Marin	CA	(a)	(a)	(a)	8.37	22.51	0.00	(a)	(a)	(a)	12.27	31.47	18.86	0.00
Merced	CA	(a)	(a)	(a)	0.98	0.00	35.62	(a)	(a)	(a)	12.31	5.28	0.00	25.04
Monterey	CA	(a)	(a)	(a)	10.88	3.95	0.11	(a)	(a)	(a)	36.63	16.81	5.42	0.47
Napa	CA	(a)	(a)	(a)	1.86	104.47	0.00	(a)	(a)	(a)	327.85	5.21	56.81	0.00
Orange	CA	(a)	(a)	(a)	8.83	0.00	36.69	(a)	(a)	(a)	8.08	22.24	0.00	18.48
Placer	CA	(a)	(a)	(a)	1.82	19.73	15.98	(a)	(a)	(a)	2.02	10.11	16.20	7.67
Riverside	CA	(a)	(a)	(a)	1.59	0.00	26.92	(a)	(a)	(a)	39.90	6.64	0.07	19.96
Sacramento	CA	(a)	(a)	(a)	12.51	0.22	196.58	(a)	(a)	(a)	0.29	17.70	77.18	39.56
San Bernardino	CA	(a)	(a)	(a)	5.02	2.19	19.26	(a)	(a)	(a)	19.91	8.05	2.05	18.00
San Diego	CA	(a)	(a)	(a)	1.23	5.43	0.26	(a)	(a)	(a)	4.89	6.59	4.89	4.34
San Joaquin	CA	(a)	(a)	(a)	1.88	0.00	25.44	(a)	(a)	(a)	8.08	5.16	1.12	22.68
San Luis Obispo	CA	(a)	(a)	(a)	22.17	7.47	4.75	(a)	(a)	(a)	42.84	31.24	5.15	1.52
San Mateo	CA	(a)	(a)	(a)	1.63	21.29	6.50	(a)	(a)	(a)	1.62	10.61	18.25	6.97
Santa Barbara	CA	(a)	(a)	(a)	9.27	8.79	47.47	(a)	(a)	(a)	21.20	22.13	7.95	44.97
Santa Clara	CA	(a)	(a)	(a)	1.80	0.00	0.05	(a)	(a)	(a)	3.12	11.80	1.49	3.58
Santa Cruz	CA	(a)	(a)	(a)	5.30	63.42	36.79	(a)	(a)	(a)	10.80	21.91	47.45	35.93
Shasta	CA	(a)	(a)	(a)	0.00	2.45	17.49	(a)	(a)	(a)	0.91	0.63	2.38	25.69
Solano	CA	(a)	(a)	(a)	0.83	0.00	0.00	(a)	(a)	(a)	0.53	2.14	0.00	0.00
Sonoma	CA	(a)	(a)	(a)	3.12	29.05	67.83	(a)	(a)	(a)	1.20	22.17	19.18	57.93
Stanislaus	CA	(a)	(a)	(a)	3.42	0.00	30.68	(a)	(a)	(a)	0.77	16.70	0.00	31.57
Tulare	CA	(a)	(a)	(a)	0.79	0.37	23.31	(a)	(a)	(a)	2.32	3.56	0.28	26.76
Ventura	CA	(a)	(a)	(a)	10.61	30.94	0.79	(a)	(a)	(a)	26.54	10.16	28.77	3.21
Yolo	CA	(a)	(a)	(a)	0.70	0.00	43.13	(a)	(a)	(a)	1.48	3.31	0.09	40.54
Adams	CO	(a)	(a)	(a)	8.34	0.00	0.52	(a)	(a)	(a)	17.09	10.84	0.00	0.76
Arapahoe	CO	(a)	(a)	(a)	0.06	0.00	0.00	(a)	(a)	(a)	0.00	0.94	0.00	0.00

Environmental revenue and expenditures for U.S counties with populations of 100,000+ - *continued*

County	St.	Revenue ($/per capita)						Current Operational Expenses ($/per capita)						
		NRA[1]	NRF[3]	NRO[4]	PR[5]	S[6]	SWM[7]	NRA[1]	NRFG[2]	NRF[3]	NRO[4]	PR[5]	S[6]	SWM[7]
Boulder	CO	(a)	(a)	(a)	0.78	0.00	10.10	(a)	(a)	(a)	6.05	23.40	0.00	9.91
Douglas	CO	(a)	(a)	(a)	1.16	0.00	0.03	(a)	(a)	(a)	9.53	11.56	0.00	0.50
El Paso	CO	(a)	(a)	(a)	2.74	0.00	1.25	(a)	(a)	(a)	1.75	8.07	0.00	1.15
Jefferson	CO	(a)	(a)	(a)	0.00	0.00	0.00	(a)	(a)	(a)	1.07	38.84	0.00	0.82
Larimer	CO	(a)	(a)	(a)	2.15	0.00	22.73	(a)	(a)	(a)	6.40	14.67	0.00	14.83
Mesa	CO	(a)	(a)	(a)	0.12	0.00	23.12	(a)	(a)	(a)	0.86	10.11	0.00	11.88
Pueblo	CO	(a)	(a)	(a)	5.30	0.00	0.00	(a)	(a)	(a)	3.33	8.48	0.00	0.00
Weld	CO	(a)	(a)	(a)	0.00	0.00	6.28	(a)	(a)	(a)	4.07	1.58	0.00	1.51
Kent	DE	(a)	(a)	(a)	0.00	75.30	9.52	(a)	(a)	(a)	0.00	10.84	61.28	8.87
New Castle	DE	(a)	(a)	(a)	2.62	89.90	0.00	(a)	(a)	(a)	0.59	8.85	95.66	0.00
Sussex	DE	(a)	(a)	(a)	0.00	139.41	0.00	(a)	(a)	(a)	1.79	0.10	42.36	0.73
Alachua	FL	(a)	(a)	(a)	0.00	0.00	31.51	(a)	(a)	(a)	7.70	11.50	0.00	59.07
Bay	FL	(a)	(a)	(a)	20.39	48.91	62.91	(a)	(a)	(a)	1.42	31.18	44.42	60.32
Brevard	FL	(a)	(a)	(a)	11.03	0.00	57.21	(a)	(a)	(a)	22.75	40.93	0.00	55.78
Broward	FL	(a)	(a)	(a)	9.92	26.15	55.73	(a)	(a)	(a)	11.62	29.74	13.11	51.07
Charlotte	FL	(a)	(a)	(a)	5.32	105.68	104.48	(a)	(a)	(a)	25.42	40.71	33.02	97.71
Citrus	FL	(a)	(a)	(a)	0.00	8.12	27.25	(a)	(a)	(a)	21.32	7.98	7.01	20.81
Clay	FL	(a)	(a)	(a)	0.00	0.00	26.40	(a)	(a)	(a)	3.45	5.21	0.00	59.58
Collier	FL	(a)	(a)	(a)	1.48	0.00	76.34	(a)	(a)	(a)	9.78	64.14	0.00	74.90
Escambia	FL	(a)	(a)	(a)	10.21	0.00	32.83	(a)	(a)	(a)	3.64	21.65	0.00	18.28
Hernando	FL	(a)	(a)	(a)	0.00	0.00	46.67	(a)	(a)	(a)	3.35	19.47	0.00	31.15
Hillsborough	FL	(a)	(a)	(a)	1.55	0.00	62.71	(a)	(a)	(a)	22.94	31.15	0.00	50.86
Indian River	FL	(a)	(a)	(a)	29.29	76.56	83.87	(a)	(a)	(a)	4.84	56.21	31.63	63.48
Lake	FL	(a)	(a)	(a)	0.00	0.00	59.45	(a)	(a)	(a)	8.75	2.54	1.54	52.41
Lee	FL	(a)	(a)	(a)	5.42	57.29	115.20	(a)	(a)	(a)	15.05	41.20	27.51	63.59
Leon	FL	(a)	(a)	(a)	0.00	0.00	24.25	(a)	(a)	(a)	11.58	6.43	0.00	16.32
Manatee	FL	(a)	(a)	(a)	17.88	127.13	98.45	(a)	(a)	(a)	18.19	44.86	81.82	106.97
Marion	FL	(a)	(a)	(a)	3.15	11.28	23.69	(a)	(a)	(a)	2.29	9.77	4.32	39.59
Martin	FL	(a)	(a)	(a)	0.00	62.57	96.55	(a)	(a)	(a)	66.42	64.39	32.61	80.06
Miami-Dade	FL	(a)	(a)	(a)	11.90	99.11	88.37	(a)	(a)	(a)	13.75	51.64	53.44	91.52
Okaloosa	FL	(a)	(a)	(a)	0.00	55.22	44.07	(a)	(a)	(a)	3.52	13.70	35.94	48.91
Orange	FL	(a)	(a)	(a)	41.90	0.00	51.66	(a)	(a)	(a)	12.14	27.19	0.00	38.17
Osceola	FL	(a)	(a)	(a)	7.62	0.00	16.70	(a)	(a)	(a)	0.00	23.63	0.00	25.75
Palm Beach	FL	(a)	(a)	(a)	7.10	27.93	41.06	(a)	(a)	(a)	10.12	36.14	20.19	82.77
Pasco	FL	(a)	(a)	(a)	1.55	66.66	76.43	(a)	(a)	(a)	1.07	21.82	42.53	43.95
Pinellas	FL	(a)	(a)	(a)	0.00	47.67	70.77	(a)	(a)	(a)	15.12	18.30	25.45	41.92
Polk	FL	(a)	(a)	(a)	0.00	0.00	56.98	(a)	(a)	(a)	14.28	13.66	0.00	51.30
Santa Rosa	FL	(a)	(a)	(a)	0.00	0.00	15.44	(a)	(a)	(a)	2.71	6.91	0.70	6.50
Sarasota	FL	(a)	(a)	(a)	3.82	67.17	65.97	(a)	(a)	(a)	55.75	57.59	36.91	97.29
Seminole	FL	(a)	(a)	(a)	1.69	39.85	33.41	(a)	(a)	(a)	11.18	6.95	24.78	43.93
St Johns	FL	(a)	(a)	(a)	28.95	34.67	65.11	(a)	(a)	(a)	0.00	72.08	22.22	69.33
St Lucie	FL	(a)	(a)	(a)	9.63	5.98	32.96	(a)	(a)	(a)	12.77	43.21	8.05	18.47
Volusia	FL	(a)	(a)	(a)	15.09	7.35	46.00	(a)	(a)	(a)	21.44	32.98	5.09	42.49
Bibb	GA	(a)	(a)	(a)	3.54	0.00	15.20	(a)	(a)	(a)	1.31	12.46	0.00	16.22
Chatham	GA	(a)	(a)	(a)	5.09	2.68	1.90	(a)	(a)	(a)	0.03	10.81	3.20	9.47
Cherokee	GA	(a)	(a)	(a)	0.00	0.00	9.28	(a)	(a)	(a)	0.95	7.87	0.00	5.00
Clayton	GA	(a)	(a)	(a)	6.01	0.00	8.93	(a)	(a)	(a)	0.98	22.24	0.00	6.20
Cobb	GA	(a)	(a)	(a)	3.61	83.56	7.34	(a)	(a)	(a)	0.00	22.31	26.13	18.90
Dekalb	GA	(a)	(a)	(a)	4.79	54.47	80.39	(a)	(a)	(a)	1.84	26.04	37.27	34.88
Forsyth	GA	(a)	(a)	(a)	5.50	51.97	2.93	(a)	(a)	(a)	2.45	31.56	54.67	9.15
Fulton	GA	(a)	(a)	(a)	1.63	48.78	0.30	(a)	(a)	(a)	0.91	14.30	34.54	2.98
Gwinnett	GA	(a)	(a)	(a)	3.95	95.11	0.64	(a)	(a)	(a)	0.82	28.66	65.44	0.61
Hall	GA	(a)	(a)	(a)	3.97	0.07	30.35	(a)	(a)	(a)	2.34	16.24	0.01	26.03
Henry	GA	(a)	(a)	(a)	1.57	0.00	0.00	(a)	(a)	(a)	1.12	10.48	0.00	0.00
Houston	GA	(a)	(a)	(a)	0.00	15.84	48.83	(a)	(a)	(a)	0.64	0.03	12.51	44.58
Hawaii	HI	(a)	(a)	(a)	10.29	39.85	22.00	(a)	(a)	(a)	1.17	86.95	32.43	76.68
Maui	HI	(a)	(a)	(a)	12.78	158.55	47.22	(a)	(a)	(a)	0.00	150.65	102.58	62.36
Black Hawk	IA	(a)	(a)	(a)	1.95	0.00	0.00	(a)	(a)	(a)	0.76	13.73	0.00	0.64
Johnson	IA	(a)	(a)	(a)	0.00	0.00	0.00	(a)	(a)	(a)	1.03	8.32	0.00	3.62
Linn	IA	(a)	(a)	(a)	2.08	0.00	0.00	(a)	(a)	(a)	2.64	16.99	0.00	0.13

Environmental revenue and expenditures for U.S counties with populations of 100,000+ - *continued*

County	St.	Revenue ($/per capita)						Current Operational Expenses ($/per capita)						
		NRA[1]	NRF[3]	NRO[4]	PR[5]	S[6]	SWM[7]	NRA[1]	NRFG[2]	NRF[3]	NRO[4]	PR[5]	S[6]	SWM[7]
Polk	IA	(a)	(a)	(a)	10.68	1.65	0.00	(a)	(a)	(a)	9.42	9.87	0.55	1.99
Scott	IA	(a)	(a)	(a)	6.48	0.00	0.00	(a)	(a)	(a)	9.53	11.36	0.00	3.59
Woodbury	IA	(a)	(a)	(a)	0.00	0.00	0.00	(a)	(a)	(a)	7.23	7.84	0.00	0.00
Ada	ID	(a)	(a)	(a)	0.53	0.00	20.07	(a)	(a)	(a)	11.14	1.34	0.00	13.59
Canyon	ID	(a)	(a)	(a)	0.00	0.00	16.06	(a)	(a)	(a)	8.06	0.00	0.00	6.11
Kootenai	ID	(a)	(a)	(a)	0.00	0.00	65.82	(a)	(a)	(a)	8.90	2.19	0.00	37.39
Champaign	IL	(a)	(a)	(a)	0.00	0.00	0.00	(a)	(a)	(a)	15.45	0.00	0.00	0.01
Cook	IL	(a)	(a)	(a)	8.70	0.00	0.00	(a)	(a)	(a)	0.00	22.23	0.00	0.00
Du Page	IL	(a)	(a)	(a)	5.32	10.03	0.44	(a)	(a)	(a)	38.95	24.40	14.08	1.17
Kane	IL	(a)	(a)	(a)	0.00	0.00	12.62	(a)	(a)	(a)	8.39	0.00	0.00	2.78
Kankakee	IL	(a)	(a)	(a)	0.00	0.00	0.00	(a)	(a)	(a)	0.73	0.00	0.00	4.31
La Salle	IL	(a)	(a)	(a)	0.00	0.00	0.00	(a)	(a)	(a)	0.00	0.96	0.00	0.00
Lake	IL	(a)	(a)	(a)	0.00	31.68	0.00	(a)	(a)	(a)	28.83	0.00	27.55	4.01
Macon	IL	(a)	(a)	(a)	0.00	0.00	0.00	(a)	(a)	(a)	0.00	0.00	0.00	0.00
Madison	IL	(a)	(a)	(a)	0.00	6.48	2.77	(a)	(a)	(a)	0.00	0.62	6.53	0.89
Mchenry	IL	(a)	(a)	(a)	0.00	0.00	0.00	(a)	(a)	(a)	0.00	0.00	0.06	0.00
Mclean	IL	(a)	(a)	(a)	0.00	0.00	0.00	(a)	(a)	(a)	0.00	2.63	0.00	0.00
Peoria	IL	(a)	(a)	(a)	0.00	0.00	0.00	(a)	(a)	(a)	0.00	0.00	0.00	0.00
Rock Island	IL	(a)	(a)	(a)	5.49	0.00	0.00	(a)	(a)	(a)	13.34	14.83	0.00	0.00
Sangamon	IL	(a)	(a)	(a)	0.00	0.00	0.00	(a)	(a)	(a)	0.00	0.00	0.00	0.00
St Clair	IL	(a)	(a)	(a)	0.00	0.00	0.00	(a)	(a)	(a)	0.00	0.00	0.00	0.00
Tazewell	IL	(a)	(a)	(a)	0.00	0.00	0.00	(a)	(a)	(a)	0.00	0.00	0.00	3.08
Will	IL	(a)	(a)	(a)	0.00	0.00	0.43	(a)	(a)	(a)	10.67	0.00	0.00	2.12
Winnebago	IL	(a)	(a)	(a)	7.83	0.00	0.00	(a)	(a)	(a)	13.28	5.37	0.00	0.00
Allen	IN	(a)	(a)	(a)	15.27	0.00	0.00	(a)	(a)	(a)	2.85	29.34	1.76	0.00
Delaware	IN	(a)	(a)	(a)	0.00	0.00	0.00	(a)	(a)	(a)	12.72	15.91	0.03	0.00
Elkhart	IN	(a)	(a)	(a)	0.02	0.00	15.87	(a)	(a)	(a)	2.51	7.48	0.00	8.61
Hamilton	IN	(a)	(a)	(a)	0.75	0.00	0.00	(a)	(a)	(a)	6.19	8.45	0.00	0.00
Johnson	IN	(a)	(a)	(a)	1.50	0.00	0.00	(a)	(a)	(a)	3.23	7.03	0.00	0.00
La Porte	IN	(a)	(a)	(a)	0.00	0.00	0.20	(a)	(a)	(a)	4.48	13.82	0.00	6.03
Lake	IN	(a)	(a)	(a)	8.84	0.30	0.00	(a)	(a)	(a)	1.07	25.54	0.32	0.00
Madison	IN	(a)	(a)	(a)	0.00	0.00	1.33	(a)	(a)	(a)	3.98	3.08	0.10	0.42
Monroe	IN	(a)	(a)	(a)	1.59	0.00	0.00	(a)	(a)	(a)	1.37	13.17	0.00	0.00
Porter	IN	(a)	(a)	(a)	0.00	0.00	0.00	(a)	(a)	(a)	3.05	1.41	0.00	4.00
St Joseph	IN	(a)	(a)	(a)	0.00	0.00	0.00	(a)	(a)	(a)	0.00	0.00	0.00	0.00
Tippecanoe	IN	(a)	(a)	(a)	0.66	0.00	0.00	(a)	(a)	(a)	4.66	12.76	0.00	16.68
Vanderburgh	IN	(a)	(a)	(a)	35.22	0.52	0.00	(a)	(a)	(a)	2.60	98.12	0.61	0.00
Vigo	IN	(a)	(a)	(a)	1.15	0.00	0.00	(a)	(a)	(a)	0.81	15.66	0.00	0.00
Johnson	KS	(a)	(a)	(a)	28.84	55.66	0.00	(a)	(a)	(a)	1.58	52.21	43.79	0.00
Sedgwick	KS	(a)	(a)	(a)	4.48	0.00	2.73	(a)	(a)	(a)	14.57	13.00	0.00	2.49
Shawnee	KS	(a)	(a)	(a)	4.67	0.00	30.51	(a)	(a)	(a)	2.69	25.31	0.00	30.47
Jefferson	KY	(a)	(a)	(a)	0.00	0.00	0.00	(a)	(a)	(a)	0.00	0.00	0.00	0.00
Kenton	KY	(a)	(a)	(a)	16.31	159.93	0.00	(a)	(a)	(a)	0.01	21.35	95.66	0.00
Bossier	LA	(a)	(a)	(a)	0.00	0.00	0.00	(a)	(a)	(a)	2.67	0.33	0.00	0.01
Caddo	LA	(a)	(a)	(a)	0.05	2.71	0.00	(a)	(a)	(a)	1.15	3.27	2.68	6.92
Calcasieu	LA	(a)	(a)	(a)	5.25	0.58	0.00	(a)	(a)	(a)	21.02	30.77	0.76	18.74
Jefferson	LA	(a)	(a)	(a)	10.38	34.11	23.71	(a)	(a)	(a)	54.34	36.57	50.81	37.56
Ouachita	LA	(a)	(a)	(a)	0.81	15.77	0.00	(a)	(a)	(a)	0.10	5.08	7.83	0.00
Rapides	LA	(a)	(a)	(a)	1.60	3.07	0.00	(a)	(a)	(a)	1.67	3.18	2.67	0.00
St Tammany	LA	(a)	(a)	(a)	0.64	3.18	0.00	(a)	(a)	(a)	23.70	25.99	2.66	0.00
Barnstable	MA	(a)	(a)	(a)	0.00	0.00	0.00	(a)	(a)	(a)	7.38	0.00	0.00	0.00
Bristol	MA	(a)	(a)	(a)	0.00	0.00	0.00	(a)	(a)	(a)	0.00	0.00	0.00	0.00
Norfolk	MA	(a)	(a)	(a)	0.00	0.00	0.00	(a)	(a)	(a)	0.00	0.00	0.00	0.00
Plymouth	MA	(a)	(a)	(a)	0.00	0.00	0.00	(a)	(a)	(a)	0.39	0.00	0.00	0.00
Anne Arundel	MD	(a)	(a)	(a)	10.10	108.77	72.55	(a)	(a)	(a)	1.12	27.16	37.32	64.17
Baltimore	MD	(a)	(a)	(a)	10.71	163.72	1.21	(a)	(a)	(a)	7.15	28.37	52.52	45.26
Carroll	MD	(a)	(a)	(a)	8.69	18.07	33.42	(a)	(a)	(a)	4.38	25.34	21.63	39.19
Charles	MD	(a)	(a)	(a)	16.12	85.19	46.17	(a)	(a)	(a)	4.31	43.16	49.34	34.00
Frederick	MD	(a)	(a)	(a)	1.86	42.31	42.55	(a)	(a)	(a)	16.06	13.44	29.45	25.42
Harford	MD	(a)	(a)	(a)	2.67	43.12	51.79	(a)	(a)	(a)	8.34	25.13	21.96	43.10

Environmental revenue and expenditures for U.S counties with populations of 100,000+ - *continued*

County	St.	Revenue ($/per capita)						Current Operational Expenses ($/per capita)						
		NRA[1]	NRF[3]	NRO[4]	PR[5]	S[6]	SWM[7]	NRA[1]	NRFG[2]	NRF[3]	NRO[4]	PR[5]	S[6]	SWM[7]
Howard	MD	(a)	(a)	(a)	45.94	56.97	41.05	(a)	(a)	(a)	4.03	66.77	84.66	42.48
Montgomery	MD	(a)	(a)	(a)	41.49	105.65	100.29	(a)	(a)	(a)	1.20	125.72	43.97	96.93
Prince Georges	MD	(a)	(a)	(a)	17.74	113.95	88.23	(a)	(a)	(a)	1.40	147.74	70.11	73.23
Washington	MD	(a)	(a)	(a)	9.69	68.20	36.88	(a)	(a)	(a)	2.29	14.91	42.34	21.91
Androscoggin	ME	(a)	(a)	(a)	0.00	0.00	0.00	(a)	(a)	(a)	0.00	0.00	0.00	0.00
Cumberland	ME	(a)	(a)	(a)	0.00	0.00	0.00	(a)	(a)	(a)	0.00	0.00	0.00	0.00
Kennebec	ME	(a)	(a)	(a)	0.00	0.00	0.00	(a)	(a)	(a)	0.00	0.00	0.00	0.00
Penobscot	ME	(a)	(a)	(a)	0.00	0.00	0.00	(a)	(a)	(a)	0.58	0.00	0.00	0.00
York	ME	(a)	(a)	(a)	0.00	0.00	0.00	(a)	(a)	(a)	0.00	0.00	0.00	0.00
Bay	MI	(a)	(a)	(a)	10.59	17.99	0.00	(a)	(a)	(a)	6.09	24.64	21.47	0.00
Berrien	MI	(a)	(a)	(a)	2.17	0.00	18.45	(a)	(a)	(a)	8.85	8.40	0.00	0.93
Calhoun	MI	(a)	(a)	(a)	0.00	0.00	0.00	(a)	(a)	(a)	2.85	0.00	0.00	0.00
Eaton	MI	(a)	(a)	(a)	0.38	0.00	2.23	(a)	(a)	(a)	8.98	3.05	0.00	2.46
Genesee	MI	(a)	(a)	(a)	5.00	0.00	0.00	(a)	(a)	(a)	3.39	17.24	0.00	0.00
Ingham	MI	(a)	(a)	(a)	0.71	0.00	0.00	(a)	(a)	(a)	4.01	18.58	0.00	0.00
Jackson	MI	(a)	(a)	(a)	5.72	0.00	42.63	(a)	(a)	(a)	24.46	9.86	0.01	36.40
Kalamazoo	MI	(a)	(a)	(a)	2.03	0.00	0.00	(a)	(a)	(a)	2.89	8.64	0.00	0.00
Kent	MI	(a)	(a)	(a)	2.23	0.00	69.57	(a)	(a)	(a)	0.00	13.73	4.24	67.16
Livingston	MI	(a)	(a)	(a)	0.00	0.00	0.00	(a)	(a)	(a)	0.00	0.00	0.00	0.00
Macomb	MI	(a)	(a)	(a)	0.28	0.00	0.00	(a)	(a)	(a)	1.75	1.19	4.85	0.00
Monroe	MI	(a)	(a)	(a)	0.00	0.00	1.59	(a)	(a)	(a)	30.83	2.77	17.92	1.25
Muskegon	MI	(a)	(a)	(a)	3.07	59.18	11.36	(a)	(a)	(a)	0.00	3.70	59.81	6.05
Oakland	MI	(a)	(a)	(a)	6.65	20.50	0.00	(a)	(a)	(a)	5.92	13.44	69.04	0.00
Ottawa	MI	(a)	(a)	(a)	0.98	0.00	0.00	(a)	(a)	(a)	4.97	9.90	0.00	1.98
Saginaw	MI	(a)	(a)	(a)	7.36	0.00	0.00	(a)	(a)	(a)	6.07	18.03	0.00	1.07
St Clair	MI	(a)	(a)	(a)	0.17	3.63	25.81	(a)	(a)	(a)	4.46	7.96	5.18	25.70
Washtenaw	MI	(a)	(a)	(a)	7.63	0.00	0.00	(a)	(a)	(a)	0.00	26.13	0.00	2.24
Wayne	MI	(a)	(a)	(a)	7.88	39.89	0.00	(a)	(a)	(a)	6.40	22.96	31.13	0.00
Anoka	MN	(a)	(a)	(a)	4.69	0.00	16.64	(a)	(a)	(a)	2.46	17.82	0.00	23.83
Dakota	MN	(a)	(a)	(a)	1.22	0.00	12.17	(a)	(a)	(a)	5.19	23.11	0.00	11.98
Hennepin	MN	(a)	(a)	(a)	0.00	0.00	50.37	(a)	(a)	(a)	5.22	0.56	0.00	47.04
Olmsted	MN	(a)	(a)	(a)	0.00	0.00	77.46	(a)	(a)	(a)	5.95	9.52	0.00	55.00
Ramsey	MN	(a)	(a)	(a)	8.67	0.00	0.00	(a)	(a)	(a)	2.72	18.93	0.00	30.99
Scott	MN	(a)	(a)	(a)	0.00	0.00	0.00	(a)	(a)	(a)	5.96	3.24	0.00	0.00
St Louis	MN	(a)	(a)	(a)	0.00	0.00	25.88	(a)	(a)	(a)	33.01	3.71	0.00	24.22
Stearns	MN	(a)	(a)	(a)	0.00	0.00	0.96	(a)	(a)	(a)	11.92	5.49	0.00	1.77
Washington	MN	(a)	(a)	(a)	0.10	0.00	0.00	(a)	(a)	(a)	1.76	8.79	0.00	0.00
Boone	MO	(a)	(a)	(a)	0.00	0.00	0.00	(a)	(a)	(a)	0.00	0.24	0.67	0.16
Clay	MO	(a)	(a)	(a)	8.96	0.00	0.00	(a)	(a)	(a)	0.57	17.41	0.00	0.00
Greene	MO	(a)	(a)	(a)	0.00	0.00	0.00	(a)	(a)	(a)	0.00	0.00	0.00	0.00
Jackson	MO	(a)	(a)	(a)	7.27	20.04	0.00	(a)	(a)	(a)	0.03	17.15	11.65	0.00
Jasper	MO	(a)	(a)	(a)	0.00	0.00	0.00	(a)	(a)	(a)	0.00	0.00	0.00	0.00
Jefferson	MO	(a)	(a)	(a)	0.43	0.00	0.00	(a)	(a)	(a)	0.00	3.05	0.00	0.00
St Charles	MO	(a)	(a)	(a)	13.39	0.00	0.00	(a)	(a)	(a)	1.22	28.89	0.00	0.00
St Louis	MO	(a)	(a)	(a)	1.71	3.01	0.00	(a)	(a)	(a)	0.00	23.05	2.66	2.14
Harrison	MS	(a)	(a)	(a)	0.00	0.34	0.00	(a)	(a)	(a)	9.39	10.60	2.65	10.95
Hinds	MS	(a)	(a)	(a)	0.00	0.00	0.00	(a)	(a)	(a)	2.97	5.50	0.00	0.00
Jackson	MS	(a)	(a)	(a)	3.00	0.00	0.00	(a)	(a)	(a)	1.74	22.08	0.00	15.79
Rankin	MS	(a)	(a)	(a)	0.00	0.00	0.00	(a)	(a)	(a)	4.43	9.45	0.00	0.00
Yellowstone	MT	(a)	(a)	(a)	34.12	0.00	0.00	(a)	(a)	(a)	3.62	48.53	0.00	0.99
Alamance	NC	(a)	(a)	(a)	0.07	0.00	28.12	(a)	(a)	(a)	2.09	7.33	0.00	12.26
Buncombe	NC	(a)	(a)	(a)	7.41	0.00	26.81	(a)	(a)	(a)	1.56	22.29	0.00	23.84
Cabarrus	NC	(a)	(a)	(a)	10.24	0.00	7.23	(a)	(a)	(a)	4.53	21.49	0.00	6.16
Catawba	NC	(a)	(a)	(a)	37.13	0.00	39.99	(a)	(a)	(a)	1.41	3.42	0.25	19.95
Cumberland	NC	(a)	(a)	(a)	6.58	0.91	7.17	(a)	(a)	(a)	1.42	24.13	0.69	12.21
Davidson	NC	(a)	(a)	(a)	0.15	0.03	21.97	(a)	(a)	(a)	3.02	3.32	0.58	22.86
Durham	NC	(a)	(a)	(a)	0.25	14.93	3.82	(a)	(a)	(a)	4.36	0.00	7.84	6.45
Forsyth	NC	(a)	(a)	(a)	12.09	0.00	0.00	(a)	(a)	(a)	1.90	21.81	0.00	0.00
Gaston	NC	(a)	(a)	(a)	0.22	0.00	15.15	(a)	(a)	(a)	2.49	4.18	0.00	13.89
Guilford	NC	(a)	(a)	(a)	0.00	0.95	0.00	(a)	(a)	(a)	0.87	2.07	0.95	1.93

Environmental revenue and expenditures for U.S counties with populations of 100,000+ - *continued*

County	St.	Revenue ($/per capita)						Current Operational Expenses ($/per capita)						
		NRA[1]	NRF[3]	NRO[4]	PR[5]	S[6]	SWM[7]	NRA[1]	NRFG[2]	NRF[3]	NRO[4]	PR[5]	S[6]	SWM[7]
Iredell	NC	(a)	(a)	(a)	1.51	0.00	39.03	(a)	(a)	(a)	2.37	5.77	0.00	23.17
Johnston	NC	(a)	(a)	(a)	0.00	76.49	33.67	(a)	(a)	(a)	3.01	0.76	71.67	21.13
Mecklenburg	NC	(a)	(a)	(a)	3.42	0.00	14.02	(a)	(a)	(a)	7.38	35.46	0.00	14.53
New Hanover	NC	(a)	(a)	(a)	0.62	57.53	72.57	(a)	(a)	(a)	2.59	28.78	30.47	55.91
Onslow	NC	(a)	(a)	(a)	0.47	36.12	29.33	(a)	(a)	(a)	1.35	7.48	21.25	12.32
Orange	NC	(a)	(a)	(a)	0.86	0.39	37.93	(a)	(a)	(a)	2.99	7.78	0.72	61.36
Pitt	NC	(a)	(a)	(a)	0.00	0.00	43.70	(a)	(a)	(a)	3.51	0.33	0.00	49.99
Randolph	NC	(a)	(a)	(a)	0.00	0.00	22.23	(a)	(a)	(a)	2.24	0.00	0.00	23.90
Robeson	NC	(a)	(a)	(a)	0.00	42.50	18.94	(a)	(a)	(a)	3.05	5.20	19.93	12.73
Rowan	NC	(a)	(a)	(a)	5.80	2.15	18.06	(a)	(a)	(a)	1.84	12.14	0.56	13.02
Union	NC	(a)	(a)	(a)	2.54	91.61	18.32	(a)	(a)	(a)	0.54	8.86	43.19	20.46
Wake	NC	(a)	(a)	(a)	0.17	0.37	26.92	(a)	(a)	(a)	0.64	2.78	0.22	23.66
Wayne	NC	(a)	(a)	(a)	0.00	0.78	29.53	(a)	(a)	(a)	3.91	0.22	2.14	14.92
Cass	ND	(a)	(a)	(a)	0.00	0.00	0.00	(a)	(a)	(a)	14.43	5.71	0.00	0.00
Douglas	NE	(a)	(a)	(a)	0.00	0.00	29.35	(a)	(a)	(a)	0.00	1.41	0.00	24.23
Lancaster	NE	(a)	(a)	(a)	0.00	0.00	0.00	(a)	(a)	(a)	4.11	3.86	0.00	0.00
Sarpy	NE	(a)	(a)	(a)	0.00	14.01	37.97	(a)	(a)	(a)	1.74	1.91	10.98	25.80
Hillsborough	NH	(a)	(a)	(a)	0.00	0.00	0.00	(a)	(a)	(a)	1.21	0.00	0.00	0.00
Merrimack	NH	(a)	(a)	(a)	0.00	0.00	0.00	(a)	(a)	(a)	2.22	0.00	0.00	0.00
Rockingham	NH	(a)	(a)	(a)	0.00	0.00	0.00	(a)	(a)	(a)	1.54	0.00	0.00	0.00
Strafford	NH	(a)	(a)	(a)	0.00	0.00	0.00	(a)	(a)	(a)	1.84	0.00	0.00	0.00
Atlantic	NJ	(a)	(a)	(a)	3.13	0.00	0.00	(a)	(a)	(a)	1.62	6.14	0.00	0.26
Bergen	NJ	(a)	(a)	(a)	5.90	0.00	0.00	(a)	(a)	(a)	0.00	9.91	0.00	0.00
Burlington	NJ	(a)	(a)	(a)	0.00	0.00	45.24	(a)	(a)	(a)	0.03	0.18	0.00	31.64
Camden	NJ	(a)	(a)	(a)	0.38	0.00	0.00	(a)	(a)	(a)	0.00	7.75	0.02	2.06
Cumberland	NJ	(a)	(a)	(a)	0.00	0.00	0.00	(a)	(a)	(a)	2.50	0.96	0.00	0.00
Essex	NJ	(a)	(a)	(a)	6.65	0.00	0.00	(a)	(a)	(a)	0.00	12.95	0.00	0.09
Gloucester	NJ	(a)	(a)	(a)	4.55	0.00	0.00	(a)	(a)	(a)	1.42	11.79	0.00	0.00
Hudson	NJ	(a)	(a)	(a)	0.00	0.00	0.00	(a)	(a)	(a)	0.02	8.63	0.00	0.00
Hunterdon	NJ	(a)	(a)	(a)	5.11	0.00	0.00	(a)	(a)	(a)	3.71	16.06	0.00	2.25
Mercer	NJ	(a)	(a)	(a)	13.78	0.00	0.00	(a)	(a)	(a)	0.76	26.71	0.00	0.00
Middlesex	NJ	(a)	(a)	(a)	0.20	0.00	0.42	(a)	(a)	(a)	0.37	8.60	0.00	1.76
Monmouth	NJ	(a)	(a)	(a)	9.81	0.00	44.89	(a)	(a)	(a)	0.59	29.09	0.00	40.99
Morris	NJ	(a)	(a)	(a)	19.01	0.00	0.00	(a)	(a)	(a)	0.76	20.81	0.00	0.00
Ocean	NJ	(a)	(a)	(a)	3.49	0.00	0.00	(a)	(a)	(a)	2.07	8.94	0.00	3.81
Passaic	NJ	(a)	(a)	(a)	3.42	0.00	0.48	(a)	(a)	(a)	0.34	4.62	0.00	1.13
Somerset	NJ	(a)	(a)	(a)	10.95	0.00	1.61	(a)	(a)	(a)	2.04	18.47	0.00	10.25
Sussex	NJ	(a)	(a)	(a)	0.00	0.00	0.00	(a)	(a)	(a)	1.58	0.87	0.00	0.59
Union	NJ	(a)	(a)	(a)	9.33	0.00	0.00	(a)	(a)	(a)	0.78	16.62	0.00	0.57
Warren	NJ	(a)	(a)	(a)	0.00	0.00	107.65	(a)	(a)	(a)	3.41	1.90	0.00	76.38
Bernalillo	NM	(a)	(a)	(a)	0.68	0.00	5.13	(a)	(a)	(a)	7.49	12.91	0.00	5.68
Dona Ana	NM	(a)	(a)	(a)	0.00	0.81	8.36	(a)	(a)	(a)	4.44	0.24	0.95	10.68
San Juan	NM	(a)	(a)	(a)	7.21	0.00	1.52	(a)	(a)	(a)	6.06	1.40	0.00	15.74
Santa Fe	NM	(a)	(a)	(a)	0.00	0.00	0.00	(a)	(a)	(a)	0.34	2.75	0.00	0.00
Clark	NV	(a)	(a)	(a)	10.89	52.03	0.00	(a)	(a)	(a)	10.54	24.32	24.29	0.00
Washoe	NV	(a)	(a)	(a)	34.35	10.99	0.00	(a)	(a)	(a)	5.96	90.21	7.15	0.00
Albany	NY	(a)	(a)	(a)	21.32	1.72	0.00	(a)	(a)	(a)	0.00	19.09	24.18	0.00
Broome	NY	(a)	(a)	(a)	3.94	0.00	27.09	(a)	(a)	(a)	2.53	24.17	0.00	24.79
Chautauqua	NY	(a)	(a)	(a)	0.00	23.79	44.22	(a)	(a)	(a)	0.18	5.63	16.35	31.82
Dutchess	NY	(a)	(a)	(a)	0.29	0.00	0.02	(a)	(a)	(a)	5.07	7.44	0.00	5.00
Erie	NY	(a)	(a)	(a)	0.97	4.27	0.00	(a)	(a)	(a)	0.08	18.01	22.45	0.00
Jefferson	NY	(a)	(a)	(a)	0.00	0.00	14.54	(a)	(a)	(a)	5.17	3.52	0.00	17.30
Monroe	NY	(a)	(a)	(a)	1.69	20.31	10.06	(a)	(a)	(a)	0.00	22.00	49.12	17.87
Nassau	NY	(a)	(a)	(a)	14.90	6.93	0.00	(a)	(a)	(a)	0.15	45.76	48.94	3.12
Niagara	NY	(a)	(a)	(a)	4.52	1.78	2.10	(a)	(a)	(a)	0.00	13.69	9.97	4.02
Oneida	NY	(a)	(a)	(a)	0.00	27.28	0.00	(a)	(a)	(a)	9.76	4.66	23.48	0.00
Onondaga	NY	(a)	(a)	(a)	4.14	105.39	0.00	(a)	(a)	(a)	4.62	30.98	73.78	0.00
Ontario	NY	(a)	(a)	(a)	0.19	21.63	73.88	(a)	(a)	(a)	5.83	3.44	12.57	43.57
Orange	NY	(a)	(a)	(a)	5.00	9.73	25.89	(a)	(a)	(a)	3.80	14.18	12.23	27.79
Oswego	NY	(a)	(a)	(a)	0.35	0.00	34.96	(a)	(a)	(a)	1.64	13.29	0.00	49.89

Environmental revenue and expenditures for U.S counties with populations of 100,000+ - *continued*

County	St.	Revenue ($/per capita)						Current Operational Expenses ($/per capita)						
		NRA[1]	NRF[3]	NRO[4]	PR[5]	S[6]	SWM[7]	NRA[1]	NRFG[2]	NRF[3]	NRO[4]	PR[5]	S[6]	SWM[7]
Rensselaer	NY	(a)	(a)	(a)	0.00	25.00	0.00	(a)	(a)	(a)	0.69	2.77	21.69	0.00
Rockland	NY	(a)	(a)	(a)	0.00	0.06	0.00	(a)	(a)	(a)	0.00	6.90	41.61	0.00
Saratoga	NY	(a)	(a)	(a)	0.00	45.21	0.00	(a)	(a)	(a)	4.86	4.63	26.42	0.00
Schenectady	NY	(a)	(a)	(a)	1.89	0.00	2.66	(a)	(a)	(a)	3.92	5.26	0.00	4.15
St Lawrence	NY	(a)	(a)	(a)	0.00	0.00	27.40	(a)	(a)	(a)	5.08	2.52	0.00	25.49
Suffolk	NY	(a)	(a)	(a)	7.57	14.63	0.00	(a)	(a)	(a)	2.34	11.17	41.70	0.00
Ulster	NY	(a)	(a)	(a)	0.64	0.00	45.79	(a)	(a)	(a)	3.11	5.47	0.00	75.72
Westchester	NY	(a)	(a)	(a)	30.25	2.94	13.75	(a)	(a)	(a)	1.10	47.90	47.84	63.86
Allen	OH	(a)	(a)	(a)	0.00	37.01	0.00	(a)	(a)	(a)	10.84	0.00	24.67	0.00
Ashtabula	OH	(a)	(a)	(a)	0.00	14.97	6.07	(a)	(a)	(a)	4.08	0.00	16.29	9.84
Butler	OH	(a)	(a)	(a)	0.00	67.93	0.00	(a)	(a)	(a)	1.36	0.15	47.11	0.14
Clark	OH	(a)	(a)	(a)	0.00	20.69	5.01	(a)	(a)	(a)	3.10	0.00	16.32	5.99
Clermont	OH	(a)	(a)	(a)	0.00	74.23	1.70	(a)	(a)	(a)	0.00	0.00	33.65	2.77
Columbiana	OH	(a)	(a)	(a)	0.00	11.18	0.00	(a)	(a)	(a)	0.00	0.00	10.78	0.00
Cuyahoga	OH	(a)	(a)	(a)	0.00	0.86	1.33	(a)	(a)	(a)	0.12	0.00	6.28	1.78
Fairfield	OH	(a)	(a)	(a)	0.00	22.26	0.00	(a)	(a)	(a)	3.20	0.94	17.78	0.40
Franklin	OH	(a)	(a)	(a)	3.30	2.52	0.00	(a)	(a)	(a)	0.00	4.84	1.31	0.00
Greene	OH	(a)	(a)	(a)	0.39	97.61	5.01	(a)	(a)	(a)	3.87	13.93	50.27	4.30
Hamilton	OH	(a)	(a)	(a)	6.84	145.67	4.74	(a)	(a)	(a)	0.00	22.76	0.00	1.50
Lake	OH	(a)	(a)	(a)	0.00	52.69	25.00	(a)	(a)	(a)	0.00	0.00	35.96	20.06
Licking	OH	(a)	(a)	(a)	0.00	12.45	0.00	(a)	(a)	(a)	3.40	3.60	9.38	14.17
Lorain	OH	(a)	(a)	(a)	0.00	3.66	9.46	(a)	(a)	(a)	1.01	0.00	2.86	9.83
Lucas	OH	(a)	(a)	(a)	4.50	13.85	4.28	(a)	(a)	(a)	0.05	20.65	7.85	3.66
Mahoning	OH	(a)	(a)	(a)	0.00	67.42	15.12	(a)	(a)	(a)	0.00	0.00	51.02	15.22
Medina	OH	(a)	(a)	(a)	0.00	59.30	41.40	(a)	(a)	(a)	0.04	0.40	60.54	28.33
Montgomery	OH	(a)	(a)	(a)	1.20	64.31	54.51	(a)	(a)	(a)	0.00	2.74	45.81	30.27
Portage	OH	(a)	(a)	(a)	0.00	47.66	17.78	(a)	(a)	(a)	0.00	0.00	23.17	15.28
Richland	OH	(a)	(a)	(a)	0.00	14.74	0.00	(a)	(a)	(a)	10.04	1.18	4.09	0.00
Stark	OH	(a)	(a)	(a)	0.00	41.96	0.00	(a)	(a)	(a)	0.00	0.00	33.10	0.37
Summit	OH	(a)	(a)	(a)	4.27	45.73	0.00	(a)	(a)	(a)	0.76	15.52	37.16	0.00
Trumbull	OH	(a)	(a)	(a)	0.00	33.89	0.00	(a)	(a)	(a)	0.00	0.00	17.41	0.00
Warren	OH	(a)	(a)	(a)	0.00	34.39	0.00	(a)	(a)	(a)	0.00	0.00	25.44	1.47
Wayne	OH	(a)	(a)	(a)	0.00	2.32	0.00	(a)	(a)	(a)	4.32	0.27	5.08	3.00
Wood	OH	(a)	(a)	(a)	0.00	0.00	21.45	(a)	(a)	(a)	4.22	1.01	0.00	20.99
Cleveland	OK	(a)	(a)	(a)	0.00	0.00	0.00	(a)	(a)	(a)	2.24	0.00	0.00	0.00
Comanche	OK	(a)	(a)	(a)	0.00	0.00	0.00	(a)	(a)	(a)	1.98	1.10	0.00	0.00
Oklahoma	OK	(a)	(a)	(a)	0.00	0.00	0.00	(a)	(a)	(a)	0.61	0.00	0.00	0.00
Tulsa	OK	(a)	(a)	(a)	2.85	0.00	0.00	(a)	(a)	(a)	1.35	11.76	0.00	0.00
Clackamas	OR	(a)	(a)	(a)	7.46	48.55	0.00	(a)	(a)	(a)	3.55	30.33	36.59	3.88
Deschutes	OR	(a)	(a)	(a)	0.00	0.00	47.08	(a)	(a)	(a)	1.86	15.56	0.00	28.57
Douglas	OR	(a)	(a)	(a)	29.99	3.26	1.91	(a)	(a)	(a)	19.53	40.82	2.67	19.23
Jackson	OR	(a)	(a)	(a)	10.78	0.00	0.00	(a)	(a)	(a)	10.04	16.90	0.00	0.00
Lane	OR	(a)	(a)	(a)	0.00	0.00	33.27	(a)	(a)	(a)	12.19	6.47	0.00	26.01
Linn	OR	(a)	(a)	(a)	11.83	0.00	0.00	(a)	(a)	(a)	3.84	14.37	0.00	0.00
Marion	OR	(a)	(a)	(a)	0.20	1.20	61.68	(a)	(a)	(a)	0.00	0.70	2.06	61.19
Multnomah	OR	(a)	(a)	(a)	0.00	0.01	0.00	(a)	(a)	(a)	0.84	0.00	0.42	0.00
Washington	OR	(a)	(a)	(a)	0.00	137.01	0.00	(a)	(a)	(a)	2.92	1.88	72.14	0.00
Allegheny	PA	(a)	(a)	(a)	2.95	0.00	0.00	(a)	(a)	(a)	0.20	27.48	0.00	0.28
Beaver	PA	(a)	(a)	(a)	2.30	0.00	2.31	(a)	(a)	(a)	1.50	5.60	0.00	2.37
Berks	PA	(a)	(a)	(a)	0.00	0.00	0.00	(a)	(a)	(a)	2.33	3.38	0.00	0.00
Blair	PA	(a)	(a)	(a)	0.00	0.00	0.00	(a)	(a)	(a)	0.48	0.52	0.00	1.37
Bucks	PA	(a)	(a)	(a)	2.51	0.00	0.00	(a)	(a)	(a)	0.47	6.13	0.01	0.51
Butler	PA	(a)	(a)	(a)	1.28	0.00	0.00	(a)	(a)	(a)	9.65	1.71	0.00	0.00
Cambria	PA	(a)	(a)	(a)	0.00	0.00	0.00	(a)	(a)	(a)	6.61	7.38	0.00	0.00
Centre	PA	(a)	(a)	(a)	0.00	0.00	0.13	(a)	(a)	(a)	4.69	0.25	0.00	0.25
Chester	PA	(a)	(a)	(a)	0.50	0.00	0.00	(a)	(a)	(a)	0.90	14.18	0.00	0.00
Cumberland	PA	(a)	(a)	(a)	0.00	0.00	2.49	(a)	(a)	(a)	4.54	2.54	0.00	3.75
Dauphin	PA	(a)	(a)	(a)	0.12	0.00	2.42	(a)	(a)	(a)	7.01	2.98	0.00	2.87
Delaware	PA	(a)	(a)	(a)	0.00	0.00	0.00	(a)	(a)	(a)	0.35	2.22	0.00	19.99
Erie	PA	(a)	(a)	(a)	0.00	0.00	0.00	(a)	(a)	(a)	0.00	0.00	0.00	0.00

Environmental revenue and expenditures for U.S counties with populations of 100,000+ - *continued*

County	St.	Revenue ($/per capita)						Current Operational Expenses ($/per capita)						
		NRA[1]	NRF[3]	NRO[4]	PR[5]	S[6]	SWM[7]	NRA[1]	NRFG[2]	NRF[3]	NRO[4]	PR[5]	S[6]	SWM[7]
Fayette	PA	(a)	(a)	(a)	0.00	0.00	0.00	(a)	(a)	(a)	0.79	0.27	0.00	0.00
Franklin	PA	(a)	(a)	(a)	0.00	0.00	0.00	(a)	(a)	(a)	0.00	0.00	0.00	3.34
Lackawanna	PA	(a)	(a)	(a)	33.12	0.00	0.00	(a)	(a)	(a)	1.90	29.84	0.00	6.53
Lancaster	PA	(a)	(a)	(a)	0.45	0.00	0.00	(a)	(a)	(a)	19.21	4.34	0.00	0.00
Lebanon	PA	(a)	(a)	(a)	0.00	0.00	0.00	(a)	(a)	(a)	6.24	0.00	0.00	0.00
Lehigh	PA	(a)	(a)	(a)	2.86	10.46	1.69	(a)	(a)	(a)	2.07	9.85	11.80	2.61
Luzerne	PA	(a)	(a)	(a)	0.31	0.00	0.18	(a)	(a)	(a)	2.90	2.24	0.00	0.57
Lycoming	PA	(a)	(a)	(a)	0.00	0.00	88.00	(a)	(a)	(a)	10.61	2.01	3.20	63.56
Mercer	PA	(a)	(a)	(a)	0.00	0.00	0.00	(a)	(a)	(a)	2.32	0.25	0.00	0.25
Monroe	PA	(a)	(a)	(a)	2.42	1.70	0.00	(a)	(a)	(a)	5.22	4.33	1.02	0.00
Montgomery	PA	(a)	(a)	(a)	0.00	0.00	0.59	(a)	(a)	(a)	0.00	5.41	0.00	0.66
Northampton	PA	(a)	(a)	(a)	0.07	0.00	0.00	(a)	(a)	(a)	2.41	7.64	0.00	0.00
Schuylkill	PA	(a)	(a)	(a)	0.00	0.00	0.00	(a)	(a)	(a)	15.92	2.53	0.00	2.72
Washington	PA	(a)	(a)	(a)	0.36	0.00	0.00	(a)	(a)	(a)	0.48	2.06	0.00	0.00
Westmoreland	PA	(a)	(a)	(a)	0.68	0.00	0.00	(a)	(a)	(a)	3.80	6.09	0.00	0.00
York	PA	(a)	(a)	(a)	0.62	0.00	0.00	(a)	(a)	(a)	6.87	8.28	0.00	0.00
Aiken	SC	(a)	(a)	(a)	0.53	31.22	10.73	(a)	(a)	(a)	0.00	9.20	30.82	26.16
Anderson	SC	(a)	(a)	(a)	2.87	12.02	19.32	(a)	(a)	(a)	0.00	13.93	20.71	25.38
Beaufort	SC	(a)	(a)	(a)	0.00	34.86	0.27	(a)	(a)	(a)	0.00	32.65	30.19	52.92
Berkeley	SC	(a)	(a)	(a)	2.55	0.00	0.00	(a)	(a)	(a)	0.00	13.42	0.00	0.00
Charleston	SC	(a)	(a)	(a)	0.00	0.00	88.56	(a)	(a)	(a)	0.00	12.77	0.00	74.62
Florence	SC	(a)	(a)	(a)	3.73	0.00	8.57	(a)	(a)	(a)	0.00	14.26	0.00	23.69
Greenville	SC	(a)	(a)	(a)	0.00	0.00	7.91	(a)	(a)	(a)	5.11	0.00	0.00	16.65
Horry	SC	(a)	(a)	(a)	0.00	0.00	65.06	(a)	(a)	(a)	8.99	28.25	0.00	58.92
Lexington	SC	(a)	(a)	(a)	0.01	0.00	6.83	(a)	(a)	(a)	2.11	7.71	0.00	29.24
Pickens	SC	(a)	(a)	(a)	1.22	0.00	8.26	(a)	(a)	(a)	0.00	4.60	0.00	26.51
Richland	SC	(a)	(a)	(a)	0.00	0.00	29.21	(a)	(a)	(a)	0.32	0.00	4.07	38.35
Spartanburg	SC	(a)	(a)	(a)	1.75	0.00	24.99	(a)	(a)	(a)	0.00	20.28	0.00	11.72
Sumter	SC	(a)	(a)	(a)	5.04	0.00	13.81	(a)	(a)	(a)	0.00	23.92	0.00	37.33
York	SC	(a)	(a)	(a)	0.96	0.00	23.82	(a)	(a)	(a)	0.00	21.01	0.00	34.72
Minnehaha	SD	(a)	(a)	(a)	0.00	0.00	0.00	(a)	(a)	(a)	0.98	5.51	0.00	0.00
Blount	TN	(a)	(a)	(a)	0.00	0.00	0.00	(a)	(a)	(a)	1.77	0.00	0.00	0.66
Hamilton	TN	(a)	(a)	(a)	1.13	0.71	1.47	(a)	(a)	(a)	1.48	23.40	0.46	1.60
Knox	TN	(a)	(a)	(a)	1.09	0.00	0.00	(a)	(a)	(a)	0.85	23.11	0.00	10.39
Montgomery	TN	(a)	(a)	(a)	0.10	0.00	0.00	(a)	(a)	(a)	1.98	0.59	0.00	0.75
Rutherford	TN	(a)	(a)	(a)	0.05	0.00	6.40	(a)	(a)	(a)	1.83	1.44	0.00	10.81
Shelby	TN	(a)	(a)	(a)	0.49	0.00	0.05	(a)	(a)	(a)	1.10	1.04	0.00	0.27
Sullivan	TN	(a)	(a)	(a)	1.24	0.00	7.50	(a)	(a)	(a)	1.05	3.77	0.00	13.03
Sumner	TN	(a)	(a)	(a)	0.00	0.00	0.00	(a)	(a)	(a)	1.83	0.00	0.00	0.00
Washington	TN	(a)	(a)	(a)	0.00	0.00	1.31	(a)	(a)	(a)	2.72	1.25	0.00	12.14
Williamson	TN	(a)	(a)	(a)	17.09	0.00	13.27	(a)	(a)	(a)	2.02	42.00	0.00	33.15
Bell	TX	(a)	(a)	(a)	0.97	0.00	0.00	(a)	(a)	(a)	1.55	6.55	0.00	0.00
Bexar	TX	(a)	(a)	(a)	0.00	0.00	0.00	(a)	(a)	(a)	0.30	1.65	0.00	0.00
Brazoria	TX	(a)	(a)	(a)	0.09	0.00	0.00	(a)	(a)	(a)	1.85	7.28	0.00	0.02
Brazos	TX	(a)	(a)	(a)	0.00	0.00	0.17	(a)	(a)	(a)	1.31	0.00	0.00	0.17
Cameron	TX	(a)	(a)	(a)	9.88	0.00	0.00	(a)	(a)	(a)	0.59	8.94	0.60	0.00
Collin	TX	(a)	(a)	(a)	0.11	0.00	0.00	(a)	(a)	(a)	0.37	0.92	0.00	0.00
Dallas	TX	(a)	(a)	(a)	0.00	0.00	0.00	(a)	(a)	(a)	0.10	0.05	0.07	0.02
Denton	TX	(a)	(a)	(a)	0.00	0.00	0.00	(a)	(a)	(a)	0.70	0.18	0.00	0.10
Ector	TX	(a)	(a)	(a)	5.44	0.00	0.00	(a)	(a)	(a)	0.83	13.23	0.00	0.00
El Paso	TX	(a)	(a)	(a)	1.87	0.14	0.00	(a)	(a)	(a)	1.45	4.27	0.24	0.78
Ellis	TX	(a)	(a)	(a)	0.00	0.00	0.00	(a)	(a)	(a)	1.80	0.00	0.00	0.00
Fort Bend	TX	(a)	(a)	(a)	0.24	0.00	0.00	(a)	(a)	(a)	11.69	3.08	0.00	0.00
Galveston	TX	(a)	(a)	(a)	0.00	0.00	0.00	(a)	(a)	(a)	6.98	7.91	0.00	0.00
Grayson	TX	(a)	(a)	(a)	0.00	0.00	0.00	(a)	(a)	(a)	1.60	0.69	0.00	0.40
Gregg	TX	(a)	(a)	(a)	0.00	0.00	0.00	(a)	(a)	(a)	1.36	0.00	0.00	0.33
Harris	TX	(a)	(a)	(a)	0.15	0.00	0.00	(a)	(a)	(a)	13.15	9.76	0.00	0.00
Hidalgo	TX	(a)	(a)	(a)	0.12	0.00	0.00	(a)	(a)	(a)	7.60	3.11	0.00	5.49
Jefferson	TX	(a)	(a)	(a)	0.00	0.00	0.00	(a)	(a)	(a)	0.00	0.19	0.00	0.00
Johnson	TX	(a)	(a)	(a)	0.00	0.00	0.00	(a)	(a)	(a)	0.70	0.00	0.00	0.02

Environmental revenue and expenditures for U.S counties with populations of 100,000+ - *continued*

County	St.	Revenue ($/per capita)						Current Operational Expenses ($/per capita)						
		NRA[1]	NRF[3]	NRO[4]	PR[5]	S[6]	SWM[7]	NRA[1]	NRFG[2]	NRF[3]	NRO[4]	PR[5]	S[6]	SWM[7]
Lubbock	TX	(a)	(a)	(a)	0.00	0.00	0.00	(a)	(a)	(a)	0.00	0.63	0.00	0.00
Mclennan	TX	(a)	(a)	(a)	0.00	0.00	0.00	(a)	(a)	(a)	1.66	0.05	0.00	0.14
Midland	TX	(a)	(a)	(a)	0.00	0.00	0.00	(a)	(a)	(a)	1.57	3.03	0.00	0.20
Montgomery	TX	(a)	(a)	(a)	0.81	0.00	0.00	(a)	(a)	(a)	1.82	2.20	0.00	0.00
Nueces	TX	(a)	(a)	(a)	0.00	0.00	0.00	(a)	(a)	(a)	0.64	8.57	0.00	0.36
Potter	TX	(a)	(a)	(a)	0.00	0.00	0.00	(a)	(a)	(a)	0.69	0.00	0.00	0.00
Randall	TX	(a)	(a)	(a)	0.00	0.00	0.00	(a)	(a)	(a)	1.44	0.00	0.00	0.00
Smith	TX	(a)	(a)	(a)	0.00	0.00	0.00	(a)	(a)	(a)	1.00	0.00	0.00	0.00
Tarrant	TX	(a)	(a)	(a)	0.02	0.00	0.00	(a)	(a)	(a)	0.01	0.01	0.00	0.00
Taylor	TX	(a)	(a)	(a)	0.00	0.00	0.00	(a)	(a)	(a)	1.31	0.00	0.00	1.20
Tom Green	TX	(a)	(a)	(a)	0.00	0.00	0.00	(a)	(a)	(a)	1.33	0.68	0.00	0.84
Travis	TX	(a)	(a)	(a)	0.50	0.00	0.00	(a)	(a)	(a)	0.72	5.24	0.00	0.00
Webb	TX	(a)	(a)	(a)	2.08	0.00	0.00	(a)	(a)	(a)	0.00	3.84	0.00	0.00
Wichita	TX	(a)	(a)	(a)	0.00	0.00	0.00	(a)	(a)	(a)	0.89	0.00	0.00	0.00
Williamson	TX	(a)	(a)	(a)	0.00	0.00	0.00	(a)	(a)	(a)	0.85	0.00	0.00	0.00
Davis	UT	(a)	(a)	(a)	0.00	0.00	62.05	(a)	(a)	(a)	6.19	10.18	0.00	29.50
Salt Lake	UT	(a)	(a)	(a)	7.95	0.00	11.71	(a)	(a)	(a)	5.28	62.73	0.00	8.60
Utah	UT	(a)	(a)	(a)	0.00	0.21	14.58	(a)	(a)	(a)	0.00	3.77	7.26	14.72
Weber	UT	(a)	(a)	(a)	0.00	0.00	28.07	(a)	(a)	(a)	0.05	20.39	0.00	28.31
Arlington	VA	(a)	(a)	(a)	19.35	27.70	46.42	(a)	(a)	(a)	31.10	134.46	68.50	54.63
Chesterfield	VA	(a)	(a)	(a)	0.91	82.66	6.81	(a)	(a)	(a)	1.01	32.68	46.96	22.02
Fairfax	VA	(a)	(a)	(a)	33.98	108.41	90.27	(a)	(a)	(a)	3.48	95.61	70.75	96.02
Henrico	VA	(a)	(a)	(a)	4.70	125.39	24.89	(a)	(a)	(a)	1.13	41.72	72.10	31.48
Loudoun	VA	(a)	(a)	(a)	33.78	0.00	4.62	(a)	(a)	(a)	6.07	83.14	0.00	10.26
Prince William	VA	(a)	(a)	(a)	38.22	102.07	37.98	(a)	(a)	(a)	0.00	76.96	36.15	29.22
Spotsylvania	VA	(a)	(a)	(a)	5.15	144.74	3.20	(a)	(a)	(a)	1.26	16.84	47.92	28.31
Stafford	VA	(a)	(a)	(a)	10.92	68.01	0.00	(a)	(a)	(a)	1.54	30.84	35.52	1.96
Chittenden	VT	(a)	(a)	(a)	0.00	0.00	0.00	(a)	(a)	(a)	0.00	0.00	0.00	0.00
Benton	WA	(a)	(a)	(a)	0.04	0.00	0.74	(a)	(a)	(a)	2.95	4.12	0.00	0.77
Clark	WA	(a)	(a)	(a)	7.31	17.22	3.86	(a)	(a)	(a)	2.06	11.47	4.24	5.43
King	WA	(a)	(a)	(a)	2.66	121.18	64.07	(a)	(a)	(a)	20.02	20.98	45.26	36.30
Kitsap	WA	(a)	(a)	(a)	3.09	45.51	28.39	(a)	(a)	(a)	3.27	15.88	37.10	26.52
Pierce	WA	(a)	(a)	(a)	2.74	29.83	3.04	(a)	(a)	(a)	6.76	14.19	19.84	0.25
Skagit	WA	(a)	(a)	(a)	20.35	0.00	70.86	(a)	(a)	(a)	26.14	15.01	2.53	52.98
Snohomish	WA	(a)	(a)	(a)	4.02	0.00	61.57	(a)	(a)	(a)	2.06	7.42	17.01	30.40
Spokane	WA	(a)	(a)	(a)	8.29	31.83	0.00	(a)	(a)	(a)	0.92	14.02	17.09	2.60
Thurston	WA	(a)	(a)	(a)	2.99	1.67	55.78	(a)	(a)	(a)	5.12	4.85	1.68	55.80
Whatcom	WA	(a)	(a)	(a)	1.73	0.00	4.31	(a)	(a)	(a)	15.77	14.71	0.00	3.58
Yakima	WA	(a)	(a)	(a)	0.62	0.40	21.76	(a)	(a)	(a)	6.39	2.71	0.32	19.50
Brown	WI	(a)	(a)	(a)	7.97	0.00	23.33	(a)	(a)	(a)	14.59	33.42	0.00	20.22
Dane	WI	(a)	(a)	(a)	19.01	0.00	16.07	(a)	(a)	(a)	6.92	25.37	0.00	14.15
Kenosha	WI	(a)	(a)	(a)	18.50	0.00	1.74	(a)	(a)	(a)	12.69	28.48	0.00	1.30
La Crosse	WI	(a)	(a)	(a)	2.45	0.00	71.39	(a)	(a)	(a)	11.39	6.22	0.00	95.13
Marathon	WI	(a)	(a)	(a)	4.23	0.00	16.40	(a)	(a)	(a)	11.57	17.95	0.00	22.00
Milwaukee	WI	(a)	(a)	(a)	41.57	0.00	0.00	(a)	(a)	(a)	0.99	101.40	0.00	0.00
Outagamie	WI	(a)	(a)	(a)	0.40	0.00	23.64	(a)	(a)	(a)	11.66	6.49	0.00	47.49
Racine	WI	(a)	(a)	(a)	3.63	0.00	0.91	(a)	(a)	(a)	8.32	12.54	0.00	0.52
Rock	WI	(a)	(a)	(a)	0.09	0.00	0.00	(a)	(a)	(a)	2.71	3.95	0.00	0.00
Sheboygan	WI	(a)	(a)	(a)	0.00	0.00	0.00	(a)	(a)	(a)	15.50	1.83	0.00	0.71
Washington	WI	(a)	(a)	(a)	14.44	0.00	0.00	(a)	(a)	(a)	10.33	39.84	0.00	0.30
Waukesha	WI	(a)	(a)	(a)	14.86	0.00	2.15	(a)	(a)	(a)	1.50	31.83	0.00	5.42
Winnebago	WI	(a)	(a)	(a)	1.34	0.00	26.09	(a)	(a)	(a)	10.07	9.41	0.00	45.10
Kanawha	WV	(a)	(a)	(a)	8.94	0.00	0.00	(a)	(a)	(a)	0.00	12.60	0.00	0.00

Notes: *(1) Natural Resources, Agriculture; (2) Natural Resources, Fish & Game; (3) Natural Resources, Forestry; (4) Natural Resources, Other; (5) Parks & Recreation; (6) Sewerage; (7) Solid Waste Management; (a) Data not collected at the county level*
Source: *U.S. Census Bureau, Governments Finances, 2003*

Environmental revenue and expenditures for U.S cities with populations of 50,000 or more

City	St.	Revenue ($/per capita)						Current Operational Expenses ($/per capita)						
		NRA[1]	NRF[3]	NRO[4]	PR[5]	S[6]	SWM[7]	NRA[1]	NRFG[2]	NRF[3]	NRO[4]	PR[5]	S[6]	SWM[7]
Abilene	TX	(a)	(a)	(a)	1.49	61.66	75.70	(a)	(a)	(a)	0.00	43.37	41.42	50.48
Abington	PA	(a)	(a)	(a)	11.17	132.14	43.11	(a)	(a)	(a)	0.00	47.42	24.98	94.27
Akron	OH	(a)	(a)	(a)	5.72	156.52	47.30	(a)	(a)	(a)	0.00	26.97	75.79	44.77
Albany	GA	(a)	(a)	(a)	13.21	94.81	83.98	(a)	(a)	(a)	0.00	95.97	88.29	80.08
Albuquerque	NM	(a)	(a)	(a)	26.81	101.42	87.86	(a)	(a)	(a)	0.00	90.24	35.24	63.40
Alexandria	VA	(a)	(a)	(a)	13.19	144.42	25.60	(a)	(a)	(a)	0.00	130.57	138.85	44.91
Algonquin	IL	(a)	(a)	(a)	0.00	0.00	0.00	(a)	(a)	(a)	0.00	0.00	0.00	0.00
Alhambra	CA	(a)	(a)	(a)	31.93	11.04	77.65	(a)	(a)	(a)	0.00	70.06	8.66	73.89
Allentown	PA	(a)	(a)	(a)	13.46	111.14	67.89	(a)	(a)	(a)	0.00	37.73	78.09	84.39
Amarillo	TX	(a)	(a)	(a)	16.64	66.93	69.61	(a)	(a)	(a)	0.00	51.48	41.17	42.37
Amherst	NY	(a)	(a)	(a)	19.72	0.07	2.49	(a)	(a)	(a)	13.63	82.06	104.65	58.80
Anaheim	CA	(a)	(a)	(a)	89.12	0.00	129.57	(a)	(a)	(a)	0.00	128.38	0.00	119.36
Anchorage	AK	(a)	(a)	(a)	10.67	94.32	85.32	(a)	(a)	(a)	0.00	72.54	57.18	58.44
Ann Arbor	MI	(a)	(a)	(a)	11.20	139.24	0.00	(a)	(a)	(a)	9.96	91.03	87.11	54.00
Antioch	CA	(a)	(a)	(a)	19.19	18.84	0.00	(a)	(a)	(a)	0.00	25.72	19.85	1.59
Arlington	TX	(a)	(a)	(a)	37.29	99.77	22.57	(a)	(a)	(a)	0.00	70.01	86.46	18.44
Arlington Heights	IL	(a)	(a)	(a)	0.00	0.00	27.11	(a)	(a)	(a)	0.00	0.00	0.00	20.37
Arvada	CO	(a)	(a)	(a)	88.84	75.12	0.00	(a)	(a)	(a)	0.00	162.00	60.92	0.00
Athens-Clarke County	GA	(a)	(a)	(a)	6.79	97.73	49.63	(a)	(a)	(a)	1.04	55.22	66.96	34.53
Atlanta	GA	(a)	(a)	(a)	64.00	201.67	118.51	(a)	(a)	(a)	0.00	66.98	184.47	89.91
Augusta-Richmond Co.	GA	(a)	(a)	(a)	5.37	90.37	75.99	(a)	(a)	(a)	1.34	54.12	64.02	61.13
Aurora	CO	(a)	(a)	(a)	32.08	84.79	0.00	(a)	(a)	(a)	0.00	101.53	67.79	0.00
Aurora	IL	(a)	(a)	(a)	11.80	18.73	0.03	(a)	(a)	(a)	0.00	39.25	16.85	2.38
Austin	TX	(a)	(a)	(a)	33.33	160.00	56.79	(a)	(a)	(a)	0.00	94.26	77.91	51.87
Babylon	NY	(a)	(a)	(a)	4.64	0.00	128.79	(a)	(a)	(a)	2.31	41.47	0.72	216.90
Bakersfield	CA	(a)	(a)	(a)	20.27	83.74	98.80	(a)	(a)	(a)	0.00	38.58	28.93	89.59
Baton Rouge-East Baton Rouge	LA	(a)	(a)	(a)	18.51	116.02	55.28	(a)	(a)	(a)	0.27	63.36	69.30	82.76
Battle Creek	MI	(a)	(a)	(a)	71.28	190.10	39.87	(a)	(a)	(a)	0.00	134.26	176.07	38.53
Beaumont	TX	(a)	(a)	(a)	7.13	87.34	56.21	(a)	(a)	(a)	0.00	30.68	38.32	53.18
Bellevue	WA	(a)	(a)	(a)	28.34	222.29	6.46	(a)	(a)	(a)	10.78	127.27	179.83	1.46
Bensalem	PA	(a)	(a)	(a)	3.37	0.00	0.00	(a)	(a)	(a)	0.00	38.71	0.00	0.00
Berkeley	CA	(a)	(a)	(a)	76.61	130.96	210.96	(a)	(a)	(a)	0.00	138.77	70.29	211.31
Bethlehem	PA	(a)	(a)	(a)	25.10	96.36	0.00	(a)	(a)	(a)	0.00	42.37	62.52	12.99
Billings	MT	(a)	(a)	(a)	17.25	103.05	75.94	(a)	(a)	(a)	0.00	41.78	49.68	57.63
Birmingham	AL	(a)	(a)	(a)	4.38	9.83	1.83	(a)	(a)	(a)	0.00	100.63	5.99	173.65
Bismarck	ND	(a)	(a)	(a)	58.43	81.43	66.99	(a)	(a)	(a)	0.00	34.52	19.42	53.51
Bloomington	MN	(a)	(a)	(a)	36.28	83.29	4.39	(a)	(a)	(a)	0.00	100.57	36.86	8.25
Boise	ID	(a)	(a)	(a)	20.41	102.91	67.28	(a)	(a)	(a)	0.00	62.55	86.85	71.95
Bolingbrook	IL	(a)	(a)	(a)	24.33	51.16	3.98	(a)	(a)	(a)	0.00	67.25	35.37	46.96
Boston	MA	(a)	(a)	(a)	0.02	177.52	0.00	(a)	(a)	(a)	1.66	25.49	42.55	90.80
Brick	NJ	(a)	(a)	(a)	6.50	184.37	1.12	(a)	(a)	(a)	0.00	25.54	142.75	73.59
Bridgeport	CT	(a)	(a)	(a)	17.26	140.95	4.30	(a)	(a)	(a)	0.00	40.36	108.75	51.16
Bristol	CT	(a)	(a)	(a)	2.73	63.53	23.32	(a)	(a)	(a)	0.00	33.32	52.69	75.25
Bristol	PA	(a)	(a)	(a)	2.59	0.00	0.00	(a)	(a)	(a)	0.00	0.81	0.00	58.69
Brockton	MA	(a)	(a)	(a)	9.35	94.12	31.91	(a)	(a)	(a)	0.44	54.53	47.06	70.29
Brookhaven	NY	(a)	(a)	(a)	5.65	0.00	151.55	(a)	(a)	(a)	2.00	32.42	0.08	170.30
Brookline	MA	(a)	(a)	(a)	24.90	174.78	38.36	(a)	(a)	(a)	1.93	56.88	11.59	44.52
Brooklyn Park	MN	(a)	(a)	(a)	51.21	73.07	11.02	(a)	(a)	(a)	0.00	104.55	50.76	12.49
Brownsville	TX	(a)	(a)	(a)	5.18	92.60	55.39	(a)	(a)	(a)	0.00	40.11	49.13	26.49
Bryan	TX	(a)	(a)	(a)	7.14	139.00	87.39	(a)	(a)	(a)	0.00	44.04	98.79	67.66
Buffalo	NY	(a)	(a)	(a)	3.84	188.53	52.39	(a)	(a)	(a)	0.00	19.36	163.07	50.48
Burbank	CA	(a)	(a)	(a)	43.85	98.83	95.48	(a)	(a)	(a)	0.00	76.35	89.88	75.43
Calumet	IN	(a)	(a)	(a)	0.00	0.00	0.00	(a)	(a)	(a)	0.00	0.00	0.00	0.00
Cambridge	MA	(a)	(a)	(a)	20.53	265.81	0.27	(a)	(a)	(a)	0.42	114.80	12.16	50.34
Canton	OH	(a)	(a)	(a)	0.00	144.41	54.23	(a)	(a)	(a)	0.00	29.46	88.89	52.02
Cape Coral	FL	(a)	(a)	(a)	32.85	103.01	0.00	(a)	(a)	(a)	0.00	89.84	45.15	0.00
Carrollton	TX	(a)	(a)	(a)	22.32	83.28	57.42	(a)	(a)	(a)	0.00	65.71	6.06	42.07

Environmental revenue and expenditures for U.S cities/townships with populations of 50,000+ - *continued*

City	St.	Revenue ($/per capita) NRA[1]	NRF[3]	NRO[4]	PR[5]	S[6]	SWM[7]	Current Operational Expenses ($/per capita) NRA[1]	NRFG[2]	NRF[3]	NRO[4]	PR[5]	S[6]	SWM[7]
Carson	CA	(a)	(a)	(a)	33.08	0.00	2.03	(a)	(a)	(a)	0.00	75.62	0.00	1.58
Cary	NC	(a)	(a)	(a)	13.86	190.28	28.14	(a)	(a)	(a)	0.00	53.99	115.80	46.90
Cedar Rapids	IA	(a)	(a)	(a)	24.65	164.10	118.83	(a)	(a)	(a)	0.00	109.72	110.93	102.63
Centennial	CO	(a)	(a)	(a)	0.00	0.00	0.00	(a)	(a)	(a)	0.00	0.07	0.00	0.00
Center	IN	(a)	(a)	(a)	0.21	0.00	0.00	(a)	(a)	(a)	0.00	3.03	0.00	0.00
Cerritos	CA	(a)	(a)	(a)	157.54	0.00	39.05	(a)	(a)	(a)	0.00	105.97	0.00	85.63
Chandler	AZ	(a)	(a)	(a)	0.00	95.32	45.82	(a)	(a)	(a)	0.00	2.89	57.92	41.11
Chapel Hill	NC	(a)	(a)	(a)	6.95	0.00	4.55	(a)	(a)	(a)	13.67	37.36	0.00	81.14
Charlotte	NC	(a)	(a)	(a)	0.05	182.25	16.15	(a)	(a)	(a)	18.75	17.48	76.29	59.00
Chattanooga	TN	(a)	(a)	(a)	17.48	210.79	4.29	(a)	(a)	(a)	0.00	84.36	212.68	59.25
Cheektowaga	NY	(a)	(a)	(a)	8.19	0.03	1.73	(a)	(a)	(a)	3.78	42.55	88.87	59.29
Cherry Hill	NJ	(a)	(a)	(a)	0.00	0.00	0.00	(a)	(a)	(a)	0.00	0.00	0.00	0.00
Chesapeake	VA	(a)	(a)	(a)	3.29	37.43	0.00	(a)	(a)	(a)	0.00	27.23	23.50	68.17
Chicago	IL	(a)	(a)	(a)	0.00	48.97	0.00	(a)	(a)	(a)	0.00	19.83	25.68	65.09
Chino	CA	(a)	(a)	(a)	23.78	71.17	137.72	(a)	(a)	(a)	0.00	66.41	68.05	141.61
Chula Vista	CA	(a)	(a)	(a)	4.46	93.43	1.66	(a)	(a)	(a)	0.00	78.82	81.61	2.29
Cicero	IL	(a)	(a)	(a)	0.00	22.73	13.82	(a)	(a)	(a)	0.00	0.00	21.98	24.34
Cincinnati	OH	(a)	(a)	(a)	40.80	0.00	1.07	(a)	(a)	(a)	0.00	116.28	232.91	53.52
Citrus Heights	CA	(a)	(a)	(a)	0.00	0.00	2.92	(a)	(a)	(a)	0.00	0.00	0.00	2.92
Clarkstown	NY	(a)	(a)	(a)	21.92	0.00	125.63	(a)	(a)	(a)	20.69	54.69	6.02	237.10
Clarksville	TN	(a)	(a)	(a)	0.00	101.55	0.00	(a)	(a)	(a)	0.00	31.94	70.05	0.00
Clay	IN	(a)	(a)	(a)	33.81	0.00	0.00	(a)	(a)	(a)	0.00	17.68	0.00	0.00
Clay	NY	(a)	(a)	(a)	4.25	0.00	0.00	(a)	(a)	(a)	13.35	9.76	7.39	32.96
Clearwater	FL	(a)	(a)	(a)	42.64	139.04	167.04	(a)	(a)	(a)	0.00	192.72	105.44	143.81
Cleveland	OH	(a)	(a)	(a)	18.39	39.49	4.89	(a)	(a)	(a)	0.00	111.82	47.32	56.86
College Station	TX	(a)	(a)	(a)	11.01	116.27	61.05	(a)	(a)	(a)	0.00	83.88	57.11	62.07
Colonie	NY	(a)	(a)	(a)	24.64	66.24	99.17	(a)	(a)	(a)	1.00	40.21	54.38	41.46
Colorado Springs	CO	(a)	(a)	(a)	7.93	68.35	0.00	(a)	(a)	(a)	0.00	57.20	60.33	0.00
Columbia	MO	(a)	(a)	(a)	28.27	79.13	110.22	(a)	(a)	(a)	0.00	89.40	111.43	109.47
Columbia	SC	(a)	(a)	(a)	2.85	295.30	7.27	(a)	(a)	(a)	12.22	63.15	96.55	63.50
Columbus	GA	(a)	(a)	(a)	10.66	75.77	49.82	(a)	(a)	(a)	0.66	60.48	45.06	31.02
Columbus	OH	(a)	(a)	(a)	10.86	204.56	0.07	(a)	(a)	(a)	0.00	87.78	103.47	49.49
Compton	CA	(a)	(a)	(a)	1.36	9.89	74.24	(a)	(a)	(a)	0.00	16.29	6.89	75.20
Concord	CA	(a)	(a)	(a)	32.00	112.84	0.00	(a)	(a)	(a)	0.00	78.59	93.58	0.00
Coral Springs	FL	(a)	(a)	(a)	19.05	60.39	0.00	(a)	(a)	(a)	0.00	81.11	0.00	0.00
Corona	CA	(a)	(a)	(a)	7.58	96.74	39.38	(a)	(a)	(a)	0.00	49.54	101.11	39.18
Corpus Christi	TX	(a)	(a)	(a)	26.49	103.78	68.97	(a)	(a)	(a)	0.00	65.41	56.94	59.96
Costa Mesa	CA	(a)	(a)	(a)	16.10	8.24	1.69	(a)	(a)	(a)	0.00	33.33	6.55	1.38
Council Bluffs	IA	(a)	(a)	(a)	18.47	81.94	37.47	(a)	(a)	(a)	0.00	57.90	62.35	49.73
Daly City	CA	(a)	(a)	(a)	40.21	143.17	16.17	(a)	(a)	(a)	0.00	79.37	109.07	12.04
Davenport	IA	(a)	(a)	(a)	46.87	109.05	12.84	(a)	(a)	(a)	0.00	98.20	97.29	35.96
Dayton	OH	(a)	(a)	(a)	19.33	266.33	0.00	(a)	(a)	(a)	0.00	74.13	144.59	0.00
Dearborn	MI	(a)	(a)	(a)	40.77	162.07	0.56	(a)	(a)	(a)	0.00	136.32	28.75	49.98
Decatur	IL	(a)	(a)	(a)	7.36	25.54	5.92	(a)	(a)	(a)	0.00	11.20	15.34	5.94
Denver	CO	(a)	(a)	(a)	69.38	114.40	0.00	(a)	(a)	(a)	0.00	316.38	48.86	0.00
Des Moines	IA	(a)	(a)	(a)	15.31	144.84	41.25	(a)	(a)	(a)	0.00	73.73	55.26	57.95
Des Plaines	IL	(a)	(a)	(a)	0.15	22.61	38.82	(a)	(a)	(a)	0.00	1.69	12.70	38.41
Detroit	MI	(a)	(a)	(a)	13.66	311.45	0.73	(a)	(a)	(a)	0.00	111.92	181.34	38.95
Dover	NJ	(a)	(a)	(a)	22.13	134.69	0.00	(a)	(a)	(a)	0.00	30.29	33.15	76.73
Downey	CA	(a)	(a)	(a)	43.65	1.16	1.81	(a)	(a)	(a)	0.00	38.97	4.99	3.60
Duluth	MN	(a)	(a)	(a)	148.30	190.91	0.00	(a)	(a)	(a)	0.00	182.86	159.87	0.00
East Orange	NJ	(a)	(a)	(a)	11.48	128.36	0.00	(a)	(a)	(a)	0.00	28.00	112.59	68.59
Edinburg	TX	(a)	(a)	(a)	28.88	55.32	153.06	(a)	(a)	(a)	0.00	86.35	46.51	95.12
Edison	NJ	(a)	(a)	(a)	1.06	87.48	0.63	(a)	(a)	(a)	0.00	30.86	78.96	78.75
Edmond	OK	(a)	(a)	(a)	30.75	68.95	74.79	(a)	(a)	(a)	0.00	53.96	35.03	62.52
El Cajon	CA	(a)	(a)	(a)	4.68	127.82	0.00	(a)	(a)	(a)	0.00	48.78	112.02	0.00
El Monte	CA	(a)	(a)	(a)	2.69	0.00	0.00	(a)	(a)	(a)	0.00	31.93	0.00	0.00
Elgin	IL	(a)	(a)	(a)	23.52	45.26	2.48	(a)	(a)	(a)	0.00	95.60	15.23	38.52
Elizabeth	NJ	(a)	(a)	(a)	0.76	139.00	8.49	(a)	(a)	(a)	0.00	62.29	91.82	64.26
Elk Grove	CA	(a)	(a)	(a)	0.00	0.00	0.00	(a)	(a)	(a)	0.00	0.00	0.00	0.00

Environmental revenue and expenditures for U.S cities/townships with populations of 50,000+ - *continued*

City	St.	Revenue ($/per capita)						Current Operational Expenses ($/per capita)						
		NRA[1]	NRF[3]	NRO[4]	PR[5]	S[6]	SWM[7]	NRA[1]	NRFG[2]	NRF[3]	NRO[4]	PR[5]	S[6]	SWM[7]
Elyria	OH	(a)	(a)	(a)	3.84	163.94	44.45	(a)	(a)	(a)	0.00	45.18	113.11	50.41
Escondido	CA	(a)	(a)	(a)	20.47	114.53	0.00	(a)	(a)	(a)	0.00	30.23	51.40	0.00
Eugene	OR	(a)	(a)	(a)	36.43	224.57	0.00	(a)	(a)	(a)	0.00	107.93	113.10	9.04
Evanston	IL	(a)	(a)	(a)	49.96	215.28	0.00	(a)	(a)	(a)	0.00	131.32	24.16	64.04
Evansville	IN	(a)	(a)	(a)	34.21	192.03	2.99	(a)	(a)	(a)	11.16	94.87	104.41	3.97
Everett	WA	(a)	(a)	(a)	35.76	188.66	8.22	(a)	(a)	(a)	0.34	110.80	123.13	4.34
Fairfield	CA	(a)	(a)	(a)	64.06	0.00	0.00	(a)	(a)	(a)	0.00	90.96	0.00	0.71
Fairfield	CT	(a)	(a)	(a)	50.85	52.67	46.69	(a)	(a)	(a)	0.00	57.32	59.07	56.05
Fall River	MA	(a)	(a)	(a)	0.10	88.15	0.00	(a)	(a)	(a)	0.05	10.36	74.02	11.26
Fargo	ND	(a)	(a)	(a)	38.19	79.16	70.63	(a)	(a)	(a)	0.00	69.80	28.96	58.47
Farmington Hills	MI	(a)	(a)	(a)	43.52	117.17	13.35	(a)	(a)	(a)	0.00	86.41	29.13	39.15
Fayetteville	AR	(a)	(a)	(a)	0.00	147.04	90.15	(a)	(a)	(a)	0.00	44.03	109.74	94.83
Fayetteville	NC	(a)	(a)	(a)	2.53	253.61	0.04	(a)	(a)	(a)	23.80	43.09	152.80	36.74
Flagstaff	AZ	(a)	(a)	(a)	0.00	114.44	139.20	(a)	(a)	(a)	0.00	85.71	74.82	122.54
Flint	MI	(a)	(a)	(a)	0.00	0.00	0.00	(a)	(a)	(a)	0.00	0.00	0.00	0.00
Folsom	CA	(a)	(a)	(a)	32.85	30.92	119.07	(a)	(a)	(a)	0.00	123.43	36.44	109.43
Fontana	CA	(a)	(a)	(a)	7.21	62.80	0.00	(a)	(a)	(a)	0.00	25.98	45.65	0.00
Fort Collins	CO	(a)	(a)	(a)	75.48	99.62	0.00	(a)	(a)	(a)	0.00	135.32	72.37	0.00
Fort Lauderdale	FL	(a)	(a)	(a)	42.85	0.00	120.55	(a)	(a)	(a)	0.00	172.35	0.00	118.85
Fort Smith	AR	(a)	(a)	(a)	0.00	109.74	118.29	(a)	(a)	(a)	0.00	45.61	72.27	86.53
Fort Wayne	IN	(a)	(a)	(a)	25.00	164.68	29.61	(a)	(a)	(a)	0.00	95.84	72.91	28.99
Fort Worth	TX	(a)	(a)	(a)	1.57	107.17	45.93	(a)	(a)	(a)	0.00	65.42	65.75	47.05
Framingham	MA	(a)	(a)	(a)	5.28	177.52	1.80	(a)	(a)	(a)	1.80	30.06	32.41	46.01
Franklin	NJ	(a)	(a)	(a)	0.00	151.26	0.00	(a)	(a)	(a)	0.00	16.55	94.92	10.74
Fremont	CA	(a)	(a)	(a)	14.22	0.00	11.96	(a)	(a)	(a)	0.00	53.94	0.00	10.09
Fullerton	CA	(a)	(a)	(a)	10.11	30.67	61.03	(a)	(a)	(a)	0.00	38.19	2.98	59.25
Gainesville	FL	(a)	(a)	(a)	55.53	170.78	48.15	(a)	(a)	(a)	0.00	87.31	180.36	54.69
Garden Grove	CA	(a)	(a)	(a)	7.69	0.00	0.00	(a)	(a)	(a)	0.00	15.99	0.00	0.00
Garland	TX	(a)	(a)	(a)	13.76	104.24	93.39	(a)	(a)	(a)	0.00	51.90	48.73	61.09
Gary	IN	(a)	(a)	(a)	0.00	199.79	0.00	(a)	(a)	(a)	0.00	74.14	274.67	8.42
Gilbert	AZ	(a)	(a)	(a)	10.96	76.72	64.83	(a)	(a)	(a)	0.00	49.41	63.21	56.86
Glendale	AZ	(a)	(a)	(a)	4.01	93.06	87.78	(a)	(a)	(a)	0.00	43.12	38.30	73.26
Glendale	CA	(a)	(a)	(a)	7.96	95.48	65.35	(a)	(a)	(a)	0.00	54.14	9.68	64.99
Grand Prairie	TX	(a)	(a)	(a)	32.70	94.32	57.02	(a)	(a)	(a)	0.00	136.41	57.32	65.28
Grand Rapids	MI	(a)	(a)	(a)	33.70	176.36	19.82	(a)	(a)	(a)	0.00	89.86	72.37	49.30
Greece	NY	(a)	(a)	(a)	4.83	1.61	0.00	(a)	(a)	(a)	3.34	15.39	13.98	5.93
Green Bay	WI	(a)	(a)	(a)	62.42	124.67	0.77	(a)	(a)	(a)	42.89	64.61	115.34	56.20
Greenburgh	NY	(a)	(a)	(a)	11.11	0.00	0.37	(a)	(a)	(a)	0.00	72.36	5.19	32.83
Greensboro	NC	(a)	(a)	(a)	76.46	149.79	65.81	(a)	(a)	(a)	30.37	154.06	112.44	84.74
Greenwich	CT	(a)	(a)	(a)	43.86	0.16	10.29	(a)	(a)	(a)	0.00	168.23	94.77	99.78
Gresham	OR	(a)	(a)	(a)	0.00	106.60	0.00	(a)	(a)	(a)	0.00	20.39	77.80	4.88
Hamburg	NY	(a)	(a)	(a)	17.46	1.15	0.00	(a)	(a)	(a)	0.28	54.66	46.28	3.54
Hamden	CT	(a)	(a)	(a)	0.00	73.11	0.00	(a)	(a)	(a)	0.00	42.17	66.53	0.00
Hamilton	NJ	(a)	(a)	(a)	0.17	132.19	0.00	(a)	(a)	(a)	0.00	24.05	84.17	95.26
Hammond	IN	(a)	(a)	(a)	5.40	133.90	0.09	(a)	(a)	(a)	0.65	40.46	129.98	49.07
Hampton	VA	(a)	(a)	(a)	70.95	58.31	44.41	(a)	(a)	(a)	0.00	121.46	19.15	59.09
Hartford	CT	(a)	(a)	(a)	0.09	0.00	0.14	(a)	(a)	(a)	0.00	23.31	0.00	118.98
Haverhill	MA	(a)	(a)	(a)	0.00	58.07	43.92	(a)	(a)	(a)	3.71	11.40	67.98	42.12
Hayward	CA	(a)	(a)	(a)	6.14	106.38	0.00	(a)	(a)	(a)	0.00	25.48	67.87	0.00
Hempstead	NY	(a)	(a)	(a)	7.64	0.00	14.39	(a)	(a)	(a)	8.40	74.65	0.00	124.88
Henderson	NV	(a)	(a)	(a)	22.60	111.40	0.00	(a)	(a)	(a)	0.00	136.86	74.46	0.00
Hialeah	FL	(a)	(a)	(a)	8.30	121.43	53.29	(a)	(a)	(a)	0.00	40.57	86.16	53.78
Hollywood	FL	(a)	(a)	(a)	28.53	189.76	103.91	(a)	(a)	(a)	0.00	52.16	112.82	94.47
Honolulu	HI	(a)	(a)	(a)	18.30	129.55	91.03	(a)	(a)	(a)	0.00	72.75	73.57	135.78
Houston	TX	(a)	(a)	(a)	11.00	135.50	0.47	(a)	(a)	(a)	0.00	50.09	63.93	29.23
Howell	NJ	(a)	(a)	(a)	0.00	99.08	0.00	(a)	(a)	(a)	0.00	22.78	87.35	10.35
Huntington	NY	(a)	(a)	(a)	25.34	3.73	114.62	(a)	(a)	(a)	5.04	50.37	12.03	196.85
Huntington Beach	CA	(a)	(a)	(a)	16.76	30.95	47.00	(a)	(a)	(a)	0.00	54.38	8.81	49.75
Huntsville	AL	(a)	(a)	(a)	41.99	124.78	56.75	(a)	(a)	(a)	0.00	135.43	50.96	57.95
Idaho Falls	ID	(a)	(a)	(a)	48.59	110.50	43.53	(a)	(a)	(a)	0.00	102.67	70.30	42.59

Environmental revenue and expenditures for U.S cities/townships with populations of 50,000+ - *continued*

City	St.	Revenue ($/per capita)						Current Operational Expenses ($/per capita)						
		NRA[1]	NRF[3]	NRO[4]	PR[5]	S[6]	SWM[7]	NRA[1]	NRFG[2]	NRF[3]	NRO[4]	PR[5]	S[6]	SWM[7]
Independence	MO	(a)	(a)	(a)	1.03	112.50	0.00	(a)	(a)	(a)	0.00	33.92	89.93	0.00
Indianapolis	IN	(a)	(a)	(a)	28.36	74.06	14.08	(a)	(a)	(a)	2.86	65.57	81.14	42.55
Inglewood	CA	(a)	(a)	(a)	2.52	19.38	85.21	(a)	(a)	(a)	0.00	68.89	13.08	76.06
Irondequoit	NY	(a)	(a)	(a)	7.06	0.46	0.00	(a)	(a)	(a)	0.00	27.54	33.54	4.49
Irvine	CA	(a)	(a)	(a)	26.95	0.00	0.00	(a)	(a)	(a)	0.00	130.04	0.00	0.00
Irving	TX	(a)	(a)	(a)	15.07	92.47	36.64	(a)	(a)	(a)	0.00	56.92	92.62	35.99
Irvington	NJ	(a)	(a)	(a)	0.00	50.55	1.64	(a)	(a)	(a)	0.00	14.36	47.97	40.01
Islip	NY	(a)	(a)	(a)	18.93	0.00	0.02	(a)	(a)	(a)	2.83	30.64	0.00	100.07
Jackson	TN	(a)	(a)	(a)	18.72	163.60	176.86	(a)	(a)	(a)	0.00	108.93	159.17	168.58
Jacksonville	FL	(a)	(a)	(a)	5.82	134.56	48.84	(a)	(a)	(a)	1.42	48.01	60.95	77.49
Jacksonville	NC	(a)	(a)	(a)	3.11	84.33	23.79	(a)	(a)	(a)	0.00	25.65	89.59	52.07
Janesville	WI	(a)	(a)	(a)	12.64	81.19	35.73	(a)	(a)	(a)	8.16	42.66	82.25	50.72
Jersey City	NJ	(a)	(a)	(a)	0.00	145.62	5.36	(a)	(a)	(a)	0.00	22.28	75.99	127.39
Joliet	IL	(a)	(a)	(a)	13.79	97.00	7.11	(a)	(a)	(a)	0.00	20.64	48.20	36.77
Jonesboro	AR	(a)	(a)	(a)	0.00	58.33	16.79	(a)	(a)	(a)	0.00	45.25	34.72	60.21
Kansas City	MO	(a)	(a)	(a)	29.53	126.58	0.00	(a)	(a)	(a)	0.00	106.06	74.15	53.55
Kennewick	WA	(a)	(a)	(a)	34.34	85.04	0.00	(a)	(a)	(a)	0.31	92.51	21.93	1.90
Killeen	TX	(a)	(a)	(a)	10.98	109.52	87.34	(a)	(a)	(a)	0.00	32.17	59.88	56.49
Knoxville	TN	(a)	(a)	(a)	1.61	166.04	4.17	(a)	(a)	(a)	0.00	39.54	114.37	55.71
Lafayette	LA	(a)	(a)	(a)	67.20	96.06	63.12	(a)	(a)	(a)	31.20	137.26	63.22	64.91
Lakeland	FL	(a)	(a)	(a)	63.96	160.21	94.32	(a)	(a)	(a)	0.00	168.49	93.15	84.84
Lakewood	CA	(a)	(a)	(a)	12.08	0.00	43.39	(a)	(a)	(a)	0.00	81.37	0.00	38.06
Lakewood	CO	(a)	(a)	(a)	46.77	18.03	0.00	(a)	(a)	(a)	0.00	126.38	14.21	3.10
Lakewood	NJ	(a)	(a)	(a)	0.00	71.59	0.00	(a)	(a)	(a)	0.00	19.32	50.75	22.51
Lakewood	OH	(a)	(a)	(a)	23.10	63.05	0.00	(a)	(a)	(a)	0.00	46.68	55.08	75.48
Lancaster	CA	(a)	(a)	(a)	9.90	0.40	0.00	(a)	(a)	(a)	0.00	74.25	0.00	0.00
Lansing	MI	(a)	(a)	(a)	18.56	219.63	28.06	(a)	(a)	(a)	0.00	83.58	97.35	32.95
Laredo	TX	(a)	(a)	(a)	2.13	64.86	66.22	(a)	(a)	(a)	0.00	28.23	33.36	44.72
Las Vegas	NV	(a)	(a)	(a)	13.09	120.68	0.00	(a)	(a)	(a)	4.11	66.42	60.96	8.08
Lawton	OK	(a)	(a)	(a)	3.80	210.43	69.84	(a)	(a)	(a)	0.00	33.35	34.78	33.37
League City	TX	(a)	(a)	(a)	3.52	123.82	0.00	(a)	(a)	(a)	0.00	35.25	53.76	0.00
Lees Summit	MO	(a)	(a)	(a)	10.37	133.15	33.58	(a)	(a)	(a)	0.00	41.78	101.12	24.22
Lewisville	TX	(a)	(a)	(a)	12.83	98.50	13.77	(a)	(a)	(a)	0.00	46.94	27.88	5.96
Lexington-Fayette	KY	(a)	(a)	(a)	16.83	100.65	29.55	(a)	(a)	(a)	0.00	52.69	50.86	69.81
Libertyville	IL	(a)	(a)	(a)	0.00	0.00	0.00	(a)	(a)	(a)	0.00	0.00	0.00	0.00
Lincoln	NE	(a)	(a)	(a)	26.41	64.67	24.15	(a)	(a)	(a)	0.00	60.75	37.15	16.82
Little Rock	AR	(a)	(a)	(a)	25.49	114.94	66.11	(a)	(a)	(a)	0.00	56.05	88.07	50.19
Livermore	CA	(a)	(a)	(a)	33.39	283.59	0.90	(a)	(a)	(a)	0.00	30.68	127.18	7.02
Livonia	MI	(a)	(a)	(a)	24.26	122.92	0.99	(a)	(a)	(a)	0.00	90.69	130.12	146.05
Long Beach	CA	(a)	(a)	(a)	111.86	19.08	146.99	(a)	(a)	(a)	0.00	148.61	15.49	135.95
Louisville-Jefferson County	KY	(a)	(a)	(a)	37.78	0.00	0.87	(a)	(a)	(a)	0.00	52.63	0.00	28.34
Loveland	CO	(a)	(a)	(a)	94.42	141.75	58.44	(a)	(a)	(a)	0.00	164.73	45.61	39.13
Lowell	MA	(a)	(a)	(a)	0.96	96.89	17.69	(a)	(a)	(a)	0.00	27.11	73.77	55.76
Lower Merion	PA	(a)	(a)	(a)	1.52	105.96	83.33	(a)	(a)	(a)	0.00	36.42	50.04	104.91
Lynn	MA	(a)	(a)	(a)	10.10	132.74	0.00	(a)	(a)	(a)	0.01	4.59	61.22	53.09
Lyons	IL	(a)	(a)	(a)	0.00	0.00	0.00	(a)	(a)	(a)	0.00	0.00	0.00	0.00
Madison	WI	(a)	(a)	(a)	50.81	75.17	16.36	(a)	(a)	(a)	44.58	124.91	84.59	60.29
Maine	IL	(a)	(a)	(a)	0.00	0.00	0.00	(a)	(a)	(a)	0.00	0.00	0.00	0.00
Manchester	CT	(a)	(a)	(a)	2.70	91.71	98.14	(a)	(a)	(a)	0.00	43.73	61.14	75.23
Manchester	NH	(a)	(a)	(a)	47.57	103.60	0.30	(a)	(a)	(a)	1.29	49.31	57.51	17.05
Maple Grove	MN	(a)	(a)	(a)	50.38	89.35	6.65	(a)	(a)	(a)	0.00	99.19	73.39	10.23
Marietta	GA	(a)	(a)	(a)	2.45	153.18	48.94	(a)	(a)	(a)	0.00	15.74	99.05	43.94
Mcallen	TX	(a)	(a)	(a)	10.52	71.71	75.58	(a)	(a)	(a)	0.00	80.12	40.69	64.75
Medford	MA	(a)	(a)	(a)	0.80	170.21	0.00	(a)	(a)	(a)	0.09	14.98	22.33	67.40
Memphis	TN	(a)	(a)	(a)	75.82	59.80	34.96	(a)	(a)	(a)	2.85	123.13	61.80	65.63
Mesa	AZ	(a)	(a)	(a)	22.53	108.77	75.67	(a)	(a)	(a)	0.00	79.20	42.13	46.63
Mesquite	TX	(a)	(a)	(a)	6.63	95.97	41.17	(a)	(a)	(a)	0.00	44.25	58.92	35.23
Miami	FL	(a)	(a)	(a)	29.78	25.79	137.08	(a)	(a)	(a)	0.00	86.37	40.08	54.88
Miami Beach	FL	(a)	(a)	(a)	117.24	372.77	61.43	(a)	(a)	(a)	21.75	389.93	271.52	58.71

Environmental revenue and expenditures for U.S cities/townships with populations of 50,000+ - *continued*

City	St.	Revenue ($/per capita)						Current Operational Expenses ($/per capita)						
		NRA[1]	NRF[3]	NRO[4]	PR[5]	S[6]	SWM[7]	NRA[1]	NRFG[2]	NRF[3]	NRO[4]	PR[5]	S[6]	SWM[7]
Middletown	NJ	(a)	(a)	(a)	6.05	117.70	0.00	(a)	(a)	(a)	0.00	32.66	75.87	54.76
Milford	CT	(a)	(a)	(a)	2.32	0.00	2.80	(a)	(a)	(a)	0.00	15.46	80.91	80.66
Millcreek	PA	(a)	(a)	(a)	0.00	107.56	0.00	(a)	(a)	(a)	0.00	10.13	87.01	0.00
Milpitas	CA	(a)	(a)	(a)	19.61	194.27	4.85	(a)	(a)	(a)	0.00	59.75	101.52	7.35
Milton	IL	(a)	(a)	(a)	0.00	0.00	0.00	(a)	(a)	(a)	0.00	0.00	0.00	0.00
Minneapolis	MN	(a)	(a)	(a)	46.61	172.39	71.17	(a)	(a)	(a)	0.00	263.06	25.57	62.99
Mission Viejo	CA	(a)	(a)	(a)	8.34	0.00	0.00	(a)	(a)	(a)	0.00	91.15	0.00	0.00
Missouri City	TX	(a)	(a)	(a)	3.35	6.32	0.00	(a)	(a)	(a)	0.00	32.66	5.80	0.00
Modesto	CA	(a)	(a)	(a)	12.48	115.72	0.47	(a)	(a)	(a)	0.00	62.56	86.07	4.62
Moreno Valley	CA	(a)	(a)	(a)	4.80	0.00	0.82	(a)	(a)	(a)	0.00	53.91	0.00	0.30
Mount Prospect	IL	(a)	(a)	(a)	0.00	9.80	16.20	(a)	(a)	(a)	0.00	4.96	7.58	57.04
Muncie	IN	(a)	(a)	(a)	5.95	0.03	0.00	(a)	(a)	(a)	0.00	18.30	113.22	102.84
Naperville	IL	(a)	(a)	(a)	4.22	57.28	0.00	(a)	(a)	(a)	0.00	18.06	42.71	38.32
Nashua	NH	(a)	(a)	(a)	0.00	116.64	33.80	(a)	(a)	(a)	0.00	36.53	84.19	57.01
Nashville and Davidson County	TN	(a)	(a)	(a)	21.60	134.03	6.82	(a)	(a)	(a)	0.00	127.17	59.90	40.39
New Bedford	MA	(a)	(a)	(a)	4.91	158.64	0.00	(a)	(a)	(a)	0.82	20.13	124.13	36.47
New Haven	CT	(a)	(a)	(a)	11.38	115.37	9.87	(a)	(a)	(a)	0.00	84.84	94.17	1.10
New Orleans	LA	(a)	(a)	(a)	1.28	113.12	47.26	(a)	(a)	(a)	0.00	30.29	95.17	69.02
New Trier	IL	(a)	(a)	(a)	0.00	0.00	0.00	(a)	(a)	(a)	0.00	0.00	0.00	0.00
New York	NY	(a)	(a)	(a)	7.21	128.06	1.40	(a)	(a)	(a)	0.00	57.78	31.42	116.41
Newark	NJ	(a)	(a)	(a)	0.00	135.15	0.00	(a)	(a)	(a)	0.00	40.60	22.41	92.93
Newport Beach	CA	(a)	(a)	(a)	26.35	35.46	9.65	(a)	(a)	(a)	0.00	292.74	30.85	62.99
Newport News	VA	(a)	(a)	(a)	29.39	72.27	49.05	(a)	(a)	(a)	0.00	95.50	19.31	45.40
Newton	MA	(a)	(a)	(a)	2.21	217.62	0.00	(a)	(a)	(a)	1.12	51.31	26.19	66.94
Norfolk	VA	(a)	(a)	(a)	22.21	91.25	31.00	(a)	(a)	(a)	0.00	130.52	50.14	54.32
Norman	OK	(a)	(a)	(a)	6.91	107.64	76.88	(a)	(a)	(a)	0.00	50.00	61.45	63.74
North	IN	(a)	(a)	(a)	0.00	0.00	0.00	(a)	(a)	(a)	0.00	4.70	0.00	0.00
North Hempstead	NY	(a)	(a)	(a)	37.32	4.06	69.95	(a)	(a)	(a)	0.00	64.56	26.79	112.66
North Las Vegas	NV	(a)	(a)	(a)	31.91	91.89	0.00	(a)	(a)	(a)	0.00	82.17	66.22	0.00
North Miami	FL	(a)	(a)	(a)	11.31	154.01	80.04	(a)	(a)	(a)	0.00	88.90	121.70	90.63
Norwalk	CA	(a)	(a)	(a)	3.11	0.00	0.62	(a)	(a)	(a)	0.00	81.83	0.00	4.49
Norwalk	CT	(a)	(a)	(a)	6.37	109.52	0.00	(a)	(a)	(a)	0.00	127.25	55.67	0.00
Oakland	CA	(a)	(a)	(a)	20.04	49.33	25.79	(a)	(a)	(a)	0.00	89.30	39.65	2.52
Oceanside	CA	(a)	(a)	(a)	8.54	118.53	103.54	(a)	(a)	(a)	0.00	35.16	109.44	99.11
Odessa	TX	(a)	(a)	(a)	0.56	97.87	73.58	(a)	(a)	(a)	0.00	40.67	26.53	60.80
Olathe	KS	(a)	(a)	(a)	9.72	84.43	69.49	(a)	(a)	(a)	0.00	19.13	39.63	62.03
Old Bridge	NJ	(a)	(a)	(a)	15.49	135.53	0.10	(a)	(a)	(a)	0.00	30.91	58.62	7.58
Omaha	NE	(a)	(a)	(a)	20.33	85.86	0.58	(a)	(a)	(a)	0.00	76.06	52.10	36.97
Ontario	CA	(a)	(a)	(a)	19.32	60.70	128.67	(a)	(a)	(a)	0.00	42.90	46.19	100.95
Orem	UT	(a)	(a)	(a)	14.30	87.26	31.40	(a)	(a)	(a)	0.00	40.28	57.85	16.18
Orland	IL	(a)	(a)	(a)	2.96	0.00	0.00	(a)	(a)	(a)	0.00	1.52	0.00	0.00
Orlando	FL	(a)	(a)	(a)	83.28	254.42	81.51	(a)	(a)	(a)	0.00	203.73	199.85	84.00
Overland Park	KS	(a)	(a)	(a)	31.78	0.00	0.00	(a)	(a)	(a)	0.00	52.79	0.00	0.00
Owensboro	KY	(a)	(a)	(a)	22.00	126.02	89.54	(a)	(a)	(a)	0.00	61.87	65.91	67.19
Oxnard	CA	(a)	(a)	(a)	19.75	124.34	166.78	(a)	(a)	(a)	0.00	55.32	83.41	144.29
Oyster Bay	NY	(a)	(a)	(a)	13.62	1.13	37.68	(a)	(a)	(a)	8.83	61.07	6.22	156.00
Palmdale	CA	(a)	(a)	(a)	26.44	0.31	0.00	(a)	(a)	(a)	0.00	43.99	0.00	0.00
Palo Alto	CA	(a)	(a)	(a)	107.07	421.11	376.95	(a)	(a)	(a)	0.00	257.23	330.05	410.09
Palos	IL	(a)	(a)	(a)	0.00	0.00	0.00	(a)	(a)	(a)	0.00	0.24	0.00	0.00
Parma	OH	(a)	(a)	(a)	13.20	4.02	0.00	(a)	(a)	(a)	0.00	40.40	8.80	35.20
Parsippany-Troy Hills	NJ	(a)	(a)	(a)	74.31	235.97	29.64	(a)	(a)	(a)	0.00	89.72	117.65	59.45
Pasadena	CA	(a)	(a)	(a)	133.60	31.00	66.65	(a)	(a)	(a)	0.00	132.20	19.10	54.10
Pasadena	TX	(a)	(a)	(a)	7.87	56.15	33.70	(a)	(a)	(a)	0.00	47.99	44.88	43.49
Paterson	NJ	(a)	(a)	(a)	0.00	39.86	1.61	(a)	(a)	(a)	0.00	15.79	0.00	55.87
Pembroke Pines	FL	(a)	(a)	(a)	44.04	76.42	0.00	(a)	(a)	(a)	0.00	106.54	89.10	0.00
Penn	IN	(a)	(a)	(a)	0.00	0.00	0.00	(a)	(a)	(a)	0.00	0.65	0.00	0.00
Peoria	AZ	(a)	(a)	(a)	36.34	125.04	65.17	(a)	(a)	(a)	0.00	99.90	60.47	40.26

Environmental revenue and expenditures for U.S cities/townships with populations of 50,000+ - continued

City	St.	Revenue ($/per capita)						Current Operational Expenses ($/per capita)						
		NRA[1]	NRF[3]	NRO[4]	PR[5]	S[6]	SWM[7]	NRA[1]	NRFG[2]	NRF[3]	NRO[4]	PR[5]	S[6]	SWM[7]
Philadelphia	PA	(a)	(a)	(a)	10.39	142.75	0.76	(a)	(a)	(a)	0.00	46.05	92.25	61.33
Phoenix	AZ	(a)	(a)	(a)	20.19	60.00	68.11	(a)	(a)	(a)	0.00	80.90	38.09	47.38
Pico Rivera	CA	(a)	(a)	(a)	21.97	0.99	0.00	(a)	(a)	(a)	0.00	87.13	0.00	0.00
Pine Bluff	AR	(a)	(a)	(a)	23.22	108.97	39.73	(a)	(a)	(a)	0.00	91.77	93.56	41.41
Piscataway	NJ	(a)	(a)	(a)	0.33	105.30	0.02	(a)	(a)	(a)	0.00	24.39	38.89	18.93
Pittsburg	CA	(a)	(a)	(a)	29.31	42.71	0.00	(a)	(a)	(a)	0.00	58.36	11.91	0.00
Pittsburgh	PA	(a)	(a)	(a)	4.33	0.00	0.00	(a)	(a)	(a)	0.00	28.25	0.00	73.35
Plainfield	IL	(a)	(a)	(a)	0.00	0.00	0.00	(a)	(a)	(a)	0.00	0.00	0.00	0.00
Plano	TX	(a)	(a)	(a)	25.61	142.17	60.62	(a)	(a)	(a)	0.00	88.60	22.20	62.84
Plantation	FL	(a)	(a)	(a)	13.98	53.16	3.28	(a)	(a)	(a)	0.00	86.57	36.01	5.89
Plymouth	MN	(a)	(a)	(a)	46.97	75.84	10.50	(a)	(a)	(a)	0.00	87.47	27.34	11.38
Pomona	CA	(a)	(a)	(a)	3.07	16.42	45.34	(a)	(a)	(a)	0.00	24.71	8.43	39.72
Pompano Beach	FL	(a)	(a)	(a)	41.25	131.91	41.67	(a)	(a)	(a)	0.00	96.65	112.84	33.87
Port St Lucie	FL	(a)	(a)	(a)	11.56	181.04	0.00	(a)	(a)	(a)	0.00	50.93	62.07	0.00
Portland	ME	(a)	(a)	(a)	22.65	241.63	0.00	(a)	(a)	(a)	0.00	96.62	59.06	27.57
Portland	OR	(a)	(a)	(a)	46.05	292.49	2.68	(a)	(a)	(a)	0.00	102.96	152.63	4.98
Portsmouth	VA	(a)	(a)	(a)	14.67	110.95	68.99	(a)	(a)	(a)	0.00	92.38	77.58	77.21
Providence	RI	(a)	(a)	(a)	29.52	0.00	0.00	(a)	(a)	(a)	0.00	104.96	2.18	43.91
Provo	UT	(a)	(a)	(a)	11.74	41.83	23.18	(a)	(a)	(a)	0.00	47.52	36.89	22.81
Quincy	MA	(a)	(a)	(a)	0.25	154.39	0.00	(a)	(a)	(a)	0.00	25.08	35.54	57.63
Raleigh	NC	(a)	(a)	(a)	32.87	131.92	20.43	(a)	(a)	(a)	8.61	108.27	105.32	59.95
Ramapo	NY	(a)	(a)	(a)	22.40	0.22	2.18	(a)	(a)	(a)	0.00	42.26	10.42	19.80
Rancho Cordova	CA	(a)	(a)	(a)	0.00	0.00	0.00	(a)	(a)	(a)	0.00	0.00	0.00	0.00
Rancho Cucamonga	CA	(a)	(a)	(a)	13.12	10.44	0.00	(a)	(a)	(a)	0.00	26.29	0.29	0.23
Reading	PA	(a)	(a)	(a)	19.77	241.30	27.69	(a)	(a)	(a)	0.00	16.36	59.41	27.84
Redding	CA	(a)	(a)	(a)	5.03	110.81	172.94	(a)	(a)	(a)	0.00	33.49	65.99	114.30
Reno	NV	(a)	(a)	(a)	26.70	125.19	0.00	(a)	(a)	(a)	0.00	81.64	101.51	0.00
Renton	WA	(a)	(a)	(a)	51.36	210.26	161.69	(a)	(a)	(a)	5.60	128.63	186.34	132.80
Rialto	CA	(a)	(a)	(a)	9.57	100.60	11.10	(a)	(a)	(a)	0.00	22.79	70.10	5.03
Richardson	TX	(a)	(a)	(a)	50.11	108.85	91.33	(a)	(a)	(a)	0.00	137.11	111.06	69.59
Richmond	CA	(a)	(a)	(a)	13.59	78.21	0.00	(a)	(a)	(a)	0.00	73.87	129.53	0.00
Richmond	VA	(a)	(a)	(a)	13.50	222.73	44.09	(a)	(a)	(a)	0.00	249.09	127.30	92.27
Rochester	MN	(a)	(a)	(a)	51.55	99.27	0.00	(a)	(a)	(a)	0.00	106.13	56.08	0.00
Rochester	NY	(a)	(a)	(a)	13.25	0.00	94.49	(a)	(a)	(a)	0.00	63.29	2.75	85.30
Rockford	IL	(a)	(a)	(a)	0.00	0.00	40.38	(a)	(a)	(a)	0.00	0.00	0.00	48.55
Rocky Mount	NC	(a)	(a)	(a)	4.46	219.08	100.74	(a)	(a)	(a)	0.61	60.48	156.14	90.42
Salem	OR	(a)	(a)	(a)	0.00	206.93	0.00	(a)	(a)	(a)	0.00	59.39	184.75	0.00
Salinas	CA	(a)	(a)	(a)	14.78	25.63	0.00	(a)	(a)	(a)	0.00	50.64	14.94	0.00
San Angelo	TX	(a)	(a)	(a)	5.99	83.47	9.00	(a)	(a)	(a)	0.00	47.80	45.90	4.85
San Antonio	TX	(a)	(a)	(a)	26.23	138.57	38.86	(a)	(a)	(a)	4.99	76.07	85.16	37.69
San Buenaventura	CA	(a)	(a)	(a)	32.23	122.34	0.00	(a)	(a)	(a)	0.00	76.03	88.63	0.00
San Clemente	CA	(a)	(a)	(a)	50.80	112.83	1.61	(a)	(a)	(a)	0.00	129.85	74.86	2.39
San Francisco	CA	(a)	(a)	(a)	23.64	177.03	2.78	(a)	(a)	(a)	0.00	252.05	120.34	0.00
San Jose	CA	(a)	(a)	(a)	24.52	99.83	67.77	(a)	(a)	(a)	0.00	110.11	79.06	71.49
San Rafael	CA	(a)	(a)	(a)	61.98	0.00	0.00	(a)	(a)	(a)	0.00	132.69	0.00	0.00
Sandy	UT	(a)	(a)	(a)	18.04	33.85	39.02	(a)	(a)	(a)	0.00	34.67	125.21	37.47
Santa Ana	CA	(a)	(a)	(a)	10.90	8.92	34.44	(a)	(a)	(a)	0.00	51.69	4.70	31.54
Santa Clarita	CA	(a)	(a)	(a)	18.54	0.00	0.00	(a)	(a)	(a)	0.00	71.92	0.00	19.05
Santa Monica	CA	(a)	(a)	(a)	107.05	125.67	168.46	(a)	(a)	(a)	0.00	313.62	133.96	164.97
Santa Rosa	CA	(a)	(a)	(a)	31.28	296.48	0.00	(a)	(a)	(a)	0.00	76.17	178.56	0.00
Savannah	GA	(a)	(a)	(a)	4.24	133.64	268.56	(a)	(a)	(a)	0.00	122.12	138.66	269.09
Schaumburg	IL	(a)	(a)	(a)	0.00	0.00	0.00	(a)	(a)	(a)	0.00	0.00	0.00	0.00
Scottsdale	AZ	(a)	(a)	(a)	21.04	128.12	75.31	(a)	(a)	(a)	0.00	137.06	84.20	55.72
Scranton	PA	(a)	(a)	(a)	0.00	0.00	57.03	(a)	(a)	(a)	0.00	9.29	0.00	18.95
Seattle	WA	(a)	(a)	(a)	76.73	253.29	196.50	(a)	(a)	(a)	0.00	275.49	71.17	183.11
Shelby	MI	(a)	(a)	(a)	3.56	54.80	0.00	(a)	(a)	(a)	0.00	22.75	31.14	0.00
Shoreline	WA	(a)	(a)	(a)	12.09	0.00	0.00	(a)	(a)	(a)	4.02	38.40	0.00	0.00
Shreveport	LA	(a)	(a)	(a)	8.73	81.56	27.14	(a)	(a)	(a)	0.00	61.96	56.77	75.43
Simi Valley	CA	(a)	(a)	(a)	0.00	78.79	3.17	(a)	(a)	(a)	0.00	2.36	73.10	9.87

Environmental revenue and expenditures for U.S cities/townships with populations of 50,000+ - continued

City	St.	Revenue ($/per capita)						Current Operational Expenses ($/per capita)						
		NRA[1]	NRF[3]	NRO[4]	PR[5]	S[6]	SWM[7]	NRA[1]	NRFG[2]	NRF[3]	NRO[4]	PR[5]	S[6]	SWM[7]
Sioux City	IA	(a)	(a)	(a)	35.93	124.73	38.94	(a)	(a)	(a)	0.00	104.72	81.74	37.55
Sioux Falls	SD	(a)	(a)	(a)	2.53	65.54	25.52	(a)	(a)	(a)	0.00	95.72	42.72	15.13
Skokie	IL	(a)	(a)	(a)	17.60	0.00	0.00	(a)	(a)	(a)	0.00	17.84	0.00	53.89
Smithtown	NY	(a)	(a)	(a)	10.53	0.00	55.06	(a)	(a)	(a)	4.75	33.82	0.00	129.05
Somerville	MA	(a)	(a)	(a)	0.00	0.00	0.00	(a)	(a)	(a)	0.52	7.88	0.00	47.75
South Bend	IN	(a)	(a)	(a)	68.78	196.17	37.72	(a)	(a)	(a)	0.00	182.55	201.38	37.67
South Gate	CA	(a)	(a)	(a)	7.01	13.49	26.68	(a)	(a)	(a)	0.00	43.06	8.71	36.20
Southampton	NY	(a)	(a)	(a)	16.55	0.00	20.58	(a)	(a)	(a)	0.00	404.36	0.00	36.68
Southfield	MI	(a)	(a)	(a)	38.97	0.00	29.78	(a)	(a)	(a)	0.00	103.55	0.00	42.00
Sparks	NV	(a)	(a)	(a)	36.40	166.73	0.00	(a)	(a)	(a)	0.00	96.42	50.93	0.00
Spokane Valley	WA	(a)	(a)	(a)	0.00	0.00	0.00	(a)	(a)	(a)	0.00	0.00	0.00	0.00
Springdale	AR	(a)	(a)	(a)	10.46	155.20	0.00	(a)	(a)	(a)	0.00	38.95	104.36	0.00
Springfield	IL	(a)	(a)	(a)	5.82	53.30	1.56	(a)	(a)	(a)	0.00	29.70	32.99	3.69
Springfield	MA	(a)	(a)	(a)	8.56	177.96	0.00	(a)	(a)	(a)	0.00	32.24	117.44	32.50
Springfield	MO	(a)	(a)	(a)	41.93	135.89	23.44	(a)	(a)	(a)	0.55	32.24	117.44	32.50
St Cloud	MN	(a)	(a)	(a)	47.03	86.19	35.81	(a)	(a)	(a)	0.00	93.70	62.25	17.47
St George	UT	(a)	(a)	(a)	102.78	130.27	41.89	(a)	(a)	(a)	0.00	60.85	47.40	32.27
St Paul	MN	(a)	(a)	(a)	50.29	133.77	0.00	(a)	(a)	(a)	0.00	170.46	83.79	41.22
St Petersburg	FL	(a)	(a)	(a)	55.52	146.04	121.58	(a)	(a)	(a)	0.00	139.20	23.94	8.58
Stamford	CT	(a)	(a)	(a)	23.35	96.85	24.51	(a)	(a)	(a)	0.00	168.39	131.86	121.61
Sterling Heights	MI	(a)	(a)	(a)	0.00	80.76	0.00	(a)	(a)	(a)	0.00	120.64	50.56	86.39
Stockton	CA	(a)	(a)	(a)	13.52	185.57	10.68	(a)	(a)	(a)	0.00	27.21	2.26	36.93
Stratford	CT	(a)	(a)	(a)	12.40	0.00	94.72	(a)	(a)	(a)	0.00	40.60	83.25	9.61
Suffolk	VA	(a)	(a)	(a)	14.21	94.99	0.00	(a)	(a)	(a)	0.00	25.05	0.00	141.26
Sunnyvale	CA	(a)	(a)	(a)	59.03	116.03	346.95	(a)	(a)	(a)	0.00	57.00	52.88	24.77
Sunrise	FL	(a)	(a)	(a)	16.43	285.31	123.41	(a)	(a)	(a)	0.00	121.14	95.56	325.26
Syracuse	NY	(a)	(a)	(a)	2.45	32.13	0.00	(a)	(a)	(a)	0.00	93.06	173.13	126.63
Tacoma	WA	(a)	(a)	(a)	10.05	263.34	207.47	(a)	(a)	(a)	14.01	43.76	19.83	41.44
Tallahassee	FL	(a)	(a)	(a)	21.57	214.47	109.21	(a)	(a)	(a)	0.00	32.72	134.09	123.55
Tampa	FL	(a)	(a)	(a)	20.88	200.42	176.53	(a)	(a)	(a)	0.00	86.98	154.50	102.82
Taylor	MI	(a)	(a)	(a)	83.27	70.36	2.52	(a)	(a)	(a)	0.00	106.73	118.71	123.64
Taylorsville	UT	(a)	(a)	(a)	0.00	0.00	0.00	(a)	(a)	(a)	0.00	147.44	28.15	20.81
Tempe	AZ	(a)	(a)	(a)	39.50	108.08	65.81	(a)	(a)	(a)	0.00	2.00	0.00	0.00
Terrebonne Parish	LA	(a)	(a)	(a)	11.66	40.86	49.50	(a)	(a)	(a)	57.90	101.26	32.55	56.77
Thousand Oaks	CA	(a)	(a)	(a)	19.54	153.28	9.05	(a)	(a)	(a)	0.00	64.87	34.33	119.33
Toledo	OH	(a)	(a)	(a)	0.34	86.63	24.87	(a)	(a)	(a)	0.00	49.28	81.64	9.16
Tonawanda	NY	(a)	(a)	(a)	49.02	50.25	5.22	(a)	(a)	(a)	3.22	19.90	80.66	43.84
Topeka	KS	(a)	(a)	(a)	23.19	124.26	0.00	(a)	(a)	(a)	0.00	97.88	107.93	58.54
Torrance	CA	(a)	(a)	(a)	29.95	10.23	57.95	(a)	(a)	(a)	0.00	83.36	81.92	0.00
Tracy	CA	(a)	(a)	(a)	38.71	77.07	116.38	(a)	(a)	(a)	0.00	90.24	1.19	54.80
Trenton	NJ	(a)	(a)	(a)	0.00	112.26	0.00	(a)	(a)	(a)	0.00	53.02	64.57	107.35
Troy	MI	(a)	(a)	(a)	48.25	110.90	2.55	(a)	(a)	(a)	0.00	41.70	93.53	93.03
Tucson	AZ	(a)	(a)	(a)	33.05	0.00	20.48	(a)	(a)	(a)	4.28	119.23	11.42	34.48
Tyler	TX	(a)	(a)	(a)	4.86	74.38	79.29	(a)	(a)	(a)	0.00	111.41	0.00	110.56
Union	NY	(a)	(a)	(a)	2.88	6.75	0.00	(a)	(a)	(a)	0.00	41.85	48.03	72.83
Union City	NJ	(a)	(a)	(a)	0.00	0.00	0.00	(a)	(a)	(a)	0.63	17.06	17.65	22.88
Upper Darby	PA	(a)	(a)	(a)	4.51	69.44	40.42	(a)	(a)	(a)	0.00	20.18	0.00	87.31
Vacaville	CA	(a)	(a)	(a)	34.67	188.46	0.00	(a)	(a)	(a)	0.00	29.85	0.00	21.99
Vallejo	CA	(a)	(a)	(a)	512.57	141.91	6.86	(a)	(a)	(a)	0.00	77.76	118.58	2.69
Vineland	NJ	(a)	(a)	(a)	0.00	131.26	0.00	(a)	(a)	(a)	0.00	469.35	100.17	12.51
Virginia Beach	VA	(a)	(a)	(a)	35.97	78.13	2.35	(a)	(a)	(a)	0.00	11.16	96.79	1.85
Visalia	CA	(a)	(a)	(a)	43.45	115.54	107.11	(a)	(a)	(a)	0.00	102.91	58.80	61.69
Vista	CA	(a)	(a)	(a)	37.98	67.90	6.60	(a)	(a)	(a)	0.00	57.65	86.70	95.54
Waco	TX	(a)	(a)	(a)	25.95	130.43	107.13	(a)	(a)	(a)	0.00	73.72	49.94	2.40
Warren	IL	(a)	(a)	(a)	0.00	0.00	0.00	(a)	(a)	(a)	0.00	125.52	45.76	69.27
Warren	IN	(a)	(a)	(a)	0.00	0.00	0.00	(a)	(a)	(a)	0.00	22.61	0.00	0.00
Warren	MI	(a)	(a)	(a)	9.53	94.96	0.00	(a)	(a)	(a)	0.00	0.00	0.00	0.00
Warwick	RI	(a)	(a)	(a)	7.40	114.22	0.00	(a)	(a)	(a)	0.00	42.43	65.95	49.07
Washington D C	DC	(a)	(a)	(a)	20.11	167.55	1.72	(a)	(a)	(a)	0.00	26.59	38.24	39.65
Waterbury	CT	(a)	(a)	(a)	12.88	195.48	0.00	(a)	(a)	(a)	0.00	155.63	167.71	77.13
											0.00	47.86	58.67	49.44

Environmental revenue and expenditures for U.S cities/townships with populations of 50,000+ - *continued*

City	St.	Revenue ($/per capita)						Current Operational Expenses ($/per capita)						
		NRA[1]	NRF[3]	NRO[4]	PR[5]	S[6]	SWM[7]	NRA[1]	NRFG[2]	NRF[3]	NRO[4]	PR[5]	S[6]	SWM[7]
Waterloo	IA	(a)	(a)	(a)	31.10	134.42	40.60	(a)	(a)	(a)	0.00	55.68	67.59	38.71
Waukegan	IL	(a)	(a)	(a)	0.69	20.94	0.00	(a)	(a)	(a)	0.00	6.13	14.71	30.21
Wayne	IL	(a)	(a)	(a)	0.00	0.00	0.00	(a)	(a)	(a)	0.00	0.00	0.00	0.00
Wayne	NJ	(a)	(a)	(a)	10.23	182.37	0.00	(a)	(a)	(a)	0.00	46.07	92.49	63.59
West Bloomfield	MI	(a)	(a)	(a)	7.58	108.68	0.00	(a)	(a)	(a)	0.00	54.67	70.41	0.00
West Chester	OH	(a)	(a)	(a)	0.00	0.00	0.00	(a)	(a)	(a)	0.00	16.08	0.00	0.00
West Covina	CA	(a)	(a)	(a)	9.68	0.00	0.00	(a)	(a)	(a)	0.00	35.69	0.00	0.00
West Hartford	CT	(a)	(a)	(a)	56.71	0.00	0.00	(a)	(a)	(a)	0.00	69.62	0.00	0.00
West Palm Beach	FL	(a)	(a)	(a)	7.55	227.67	126.25	(a)	(a)	(a)	0.00	110.33	140.61	86.82
West Valley	UT	(a)	(a)	(a)	26.31	0.00	14.15	(a)	(a)	(a)	0.00	108.64	0.00	15.44
Westland	MI	(a)	(a)	(a)	12.27	108.40	0.00	(a)	(a)	(a)	0.00	33.15	95.63	46.36
Westminster	CA	(a)	(a)	(a)	5.72	0.00	0.00	(a)	(a)	(a)	0.00	25.64	0.00	0.00
Westminster	CO	(a)	(a)	(a)	81.87	111.94	0.00	(a)	(a)	(a)	0.00	113.15	34.17	0.00
Weymouth	MA	(a)	(a)	(a)	0.24	214.14	0.00	(a)	(a)	(a)	0.95	10.67	31.05	87.14
Wheaton	IL	(a)	(a)	(a)	0.00	0.00	0.00	(a)	(a)	(a)	0.00	0.00	0.00	0.00
Whittier	CA	(a)	(a)	(a)	25.61	11.55	76.88	(a)	(a)	(a)	0.11	74.95	11.84	69.32
Wichita	KS	(a)	(a)	(a)	17.52	76.40	1.03	(a)	(a)	(a)	0.00	68.57	44.04	7.25
Wichita Falls	TX	(a)	(a)	(a)	10.68	48.90	74.01	(a)	(a)	(a)	0.00	40.71	28.17	78.65
Wilmington	DE	(a)	(a)	(a)	14.44	209.43	0.00	(a)	(a)	(a)	0.00	141.72	166.81	0.00
Wilmington	NC	(a)	(a)	(a)	12.85	201.79	67.97	(a)	(a)	(a)	28.21	57.25	122.26	70.10
Winston Salem	NC	(a)	(a)	(a)	39.30	162.17	73.16	(a)	(a)	(a)	13.55	69.89	113.57	105.53
Woodbridge	NJ	(a)	(a)	(a)	0.75	122.12	3.46	(a)	(a)	(a)	0.00	26.69	86.26	92.62
Worcester	MA	(a)	(a)	(a)	6.12	79.43	17.34	(a)	(a)	(a)	0.00	23.16	60.57	24.62
Wyandotte County and Kansas City	KS	(a)	(a)	(a)	8.79	113.47	36.28	(a)	(a)	(a)	0.31	44.41	81.10	19.62
Wyoming	MI	(a)	(a)	(a)	3.80	124.15	0.00	(a)	(a)	(a)	0.00	50.21	100.20	0.13
Yonkers	NY	(a)	(a)	(a)	6.36	12.93	0.00	(a)	(a)	(a)	0.00	36.06	7.02	85.02
Youngstown	OH	(a)	(a)	(a)	1.75	168.83	0.00	(a)	(a)	(a)	0.00	32.59	135.52	0.00
Ypsilanti	MI	(a)	(a)	(a)	17.63	0.00	5.16	(a)	(a)	(a)	0.00	47.26	0.00	47.32

Notes: *(1) Natural Resources, Agriculture; (2) Natural Resources, Fish & Game; (3) Natural Resources, Forestry; (4) Natural Resources, Other; (5) Parks & Recreation; (6) Sewerage; (7) Solid Waste Management; (a) Data not collected at the city level*
Source: *U.S. Census Bureau, Government Finances, 2003*

Recent trends in U.S. greenhouse gas emissions and sinks (Tg CO₂ Eq.)

Gas/Source	1990	1997	1998	1999	2000	2001	2002	2003
CO₂	5,009.6	5,580.0	5,607.2	5,678.0	5,858.2	5,744.8	5,796.8	5,841.5
Fossil Fuel Combustion	4,711.7	5,263.2	5,278.7	5,345.9	5,545.1	5,448.0	5,501.4	5,551.6
Non-Energy Use of Fuels	108.0	120.3	135.4	141.6	124.7	120.1	118.8	118.0
Iron and Steel Production	85.4	71.9	67.4	64.4	65.7	58.9	55.1	53.8
Cement Manufacture	33.3	38.3	39.2	40.0	41.2	41.4	42.9	43.0
Waste Combustion	10.9	17.8	17.1	17.6	18.0	18.8	18.8	18.8
Ammonia Production and Urea Application	19.3	20.7	21.9	20.6	19.6	16.7	18.6	15.6
Lime Manufacture	11.2	13.7	13.9	13.5	13.3	12.8	12.3	13.0
Natural Gas Flaring	5.8	7.9	6.6	6.9	5.8	6.1	6.2	6.0
Limestone and Dolomite Use	5.5	7.2	7.4	8.1	6.0	5.7	5.9	4.7
Aluminum Production	6.3	5.6	5.8	5.9	5.7	4.1	4.2	4.2
Soda Ash Manufacture and Consumption	4.1	4.4	4.3	4.2	4.2	4.1	4.1	4.1
Petrochemical Production	2.2	2.9	3.0	3.1	3.0	2.8	2.9	2.8
Titanium Dioxide Production	1.3	1.8	1.8	1.9	1.9	1.9	2.0	2.0
Phosphoric Acid Production	1.5	1.5	1.6	1.5	1.4	1.3	1.3	1.4
Ferroalloys	2.0	2.0	2.0	2.0	1.7	1.3	1.2	1.4
Carbon Dioxide Consumption	0.9	0.8	0.9	0.8	1.0	0.8	1.0	1.3
Land-Use Change and Forestry (Sinks)[a]	(1,042.0)	(930.0)	(881.0)	(826.1)	(822.4)	(826.9)	(826.5)	(828.0)
International Bunker Fuels[b]	113.5	109.9	114.6	105.3	101.4	97.9	89.5	84.2
Biomass Combustion[b]	216.7	233.2	217.2	222.3	226.8	200.5	207.2	216.8
CH₄	605.3	579.5	569.1	557.3	554.2	546.8	542.5	545.0
Landfills	172.2	147.4	138.5	134.0	130.7	126.2	126.8	131.2
Natural Gas Systems	128.3	133.6	131.8	127.4	132.1	131.8	130.6	125.9
Enteric Fermentation	117.9	118.3	116.7	116.8	115.6	114.5	114.6	115.0
Coal Mining	81.9	62.6	62.8	58.9	56.2	55.6	52.4	53.8
Manure Management	31.2	36.4	38.8	38.8	38.1	38.9	39.3	39.1
Wastewater Treatment	24.8	31.7	32.6	33.6	34.3	34.7	35.8	36.8
Petroleum Systems	20.0	18.8	18.5	17.8	17.6	17.4	17.1	17.1
Rice Cultivation	7.1	7.5	7.9	8.3	7.5	7.6	6.8	6.9
Stationary Sources	7.8	7.4	6.9	7.1	7.3	6.7	6.4	6.7
Abandoned Coal Mines	6.1	8.1	7.2	7.3	7.7	6.9	6.4	6.4
Mobile Sources	4.8	4.0	3.9	3.6	3.4	3.1	2.9	2.7
Petrochemical Production	1.2	1.6	1.7	1.7	1.7	1.4	1.5	1.5
Iron and Steel Production	1.3	1.3	1.2	1.2	1.2	1.1	1.0	1.0
Agricultural Residue Burning	0.7	0.8	0.8	0.8	0.8	0.8	0.7	0.8
Silicon Carbide Production	(c)	(c)	(c)	(c)	(c)	(c)	(c)	(c)
International Bunker Fuels[b]	0.2	0.1	0.2	0.1	0.1	0.1	0.1	0.1
N₂O	382.0	396.3	407.8	382.1	401.9	385.8	380.5	376.7
Agricultural Soil Management	253.0	252.0	267.7	243.4	263.9	257.1	252.6	253.5
Mobile Sources	43.7	55.2	55.3	54.6	53.2	49.0	45.6	42.1
Manure Management	16.3	17.3	17.4	17.4	17.8	18.0	17.9	17.5
Human Sewage	13.0	14.7	15.0	15.4	15.6	15.6	15.7	15.9
Nitric Acid	17.8	21.2	20.9	20.1	19.6	15.9	17.2	15.8
Stationary Sources	12.3	13.5	13.4	13.5	14.0	13.5	13.5	13.8
Settlements Remaining Settlements	5.5	6.1	6.1	6.2	6.0	5.8	6.0	6.0
Adipic Acid	15.2	10.3	6.0	5.5	6.0	4.9	5.9	6.0
N2O Product Usage	4.3	4.8	4.8	4.8	4.8	4.8	4.8	4.8
Waste Combustion	0.4	0.4	0.3	0.3	0.4	0.4	0.5	0.5
Agricultural Residue Burning	0.4	0.4	0.5	0.4	0.5	0.5	0.4	0.4
Forest Land Remaining Forest Land	0.1	0.3	0.4	0.5	0.4	0.4	0.4	0.4
International Bunker Fuels[b]	1.0	1.0	1.0	0.9	0.9	0.9	0.8	0.8
HFCs, PFCs, and SF₆	91.2	121.7	135.7	134.8	138.9	129.5	138.3	137.0
Substitution of Ozone Depleting Substances	0.4	46.5	56.6	65.8	75.0	83.3	91.5	99.5
Electrical Transmission and Distribution	29.2	21.7	17.1	16.4	15.6	15.4	14.7	14.1
HCFC-22 Production	35.0	30.0	40.1	30.4	29.8	19.8	19.8	12.3
Semiconductor Manufacture	2.9	6.3	7.1	7.2	6.3	4.5	4.4	4.3
Aluminum Production	18.3	11.0	9.1	9.0	9.0	4.0	5.2	3.8
Magnesium Production and Processing	5.4	6.3	5.8	6.0	3.2	2.6	2.6	3.0
Total	6,088.1	6,677.5	6,719.7	6,752.2	6,953.2	6,806.9	6,858.1	6,900.2
Net Emissions (Sources and Sinks)	5,046.1	5,747.5	5,838.8	5,926.1	6,130.8	5,980.1	6,031.6	6,072.2

Notes: Totals may not sum due to independent rounding; Tg CO₂ Eq. - teragrams of carbon dioxide equivalent; (a) Sinks are only included in net emissions total, and are based partially on projected activity data. Parentheses indicate negative values (or sequestration); (b) Emissions from International Bunker Fuels and Biomass Combustion are not included in totals; (c) Does not exceed 0.05 Tg CO₂ Eq.
Source: United States Environmental Protection Agency, Inventory of U.S. Greenhouse Gas Emissions and Sinks: 1990 - 2003

Emissions from energy (Tg CO_2 Eq.)

Gas/Source	1990	1997	1998	1999	2000	2001	2002	2003
CO_2	4,836.4	5,409.1	5,437.7	5,512.1	5,693.5	5,592.9	5,645.3	5,694.3
Fossil Fuel Combustion	4,711.7	5,263.2	5,278.7	5,345.9	5,545.1	5,448.0	5,501.4	5,551.6
Non-Energy Use of Fuels	108.0	120.3	135.4	141.6	124.7	120.1	118.8	118.0
Waste Combustion	10.9	17.8	17.1	17.6	18.0	18.8	18.8	18.8
Natural Gas Flaring	5.8	7.9	6.6	6.9	5.8	6.1	6.2	6.0
Biomass-Wood[a]	212.5	226.3	209.5	214.3	217.6	190.8	195.8	201.0
International Bunker Fuels[a]	113.5	109.9	114.6	105.3	101.4	97.9	89.5	84.2
Biomass-Ethanol[a]	4.2	7.0	7.7	8.0	9.2	9.7	11.5	15.8
CH_4	248.9	234.6	230.9	222.1	224.3	221.6	215.8	212.7
Natural Gas Systems	128.3	133.6	131.8	127.4	132.1	131.8	130.6	125.9
Coal Mining	81.9	62.6	62.8	58.9	56.2	55.6	52.4	53.8
Petroleum Systems	20.0	18.8	18.5	17.8	17.6	17.4	17.1	17.1
Stationary Sources	7.8	7.4	6.9	7.1	7.3	6.7	6.4	6.7
Abandoned Coal Mines	6.1	8.1	7.2	7.3	7.7	6.9	6.4	6.4
Mobile Sources	4.8	4.0	3.9	3.6	3.4	3.1	2.9	2.7
International Bunker Fuels[a]	0.2	0.1	0.2	0.1	0.1	0.1	0.1	0.1
N_2O	56.4	69.1	69.1	68.4	67.5	62.8	59.6	56.4
Mobile Sources	43.7	55.2	55.3	54.6	53.2	49.0	45.6	42.1
Stationary Sources	12.3	13.5	13.4	13.5	14.0	13.5	13.5	13.8
Waste Combustion	0.4	0.4	0.3	0.3	0.4	0.4	0.5	0.5
International Bunker Fuels[a]	1.0	1.0	1.0	0.9	0.9	0.9	0.8	0.8
Total	5,141.7	5,712.8	5,737.7	5,802.6	5,985.3	5,877.3	5,920.7	5,963.4

Notes: *Totals may not sum due to independent rounding; Tg CO_2 Eq. - teragrams of carbon dioxide equivalent; (a) These values are presented for informational purposes only and are not included in totals or are already accounted for in other source categories.*
Source: *United States Environmental Protection Agency, Inventory of U.S. Greenhouse Gas Emissions and Sinks: 1990 - 2003*

Emissions from industrial processes (Tg CO$_2$ Eq.)

Gas/Source	1990	1997	1998	1999	2000	2001	2002	2003
CO$_2$	173.1	170.9	169.4	165.9	164.7	151.8	151.5	147.2
Iron and Steel Production	85.4	71.9	67.4	64.4	65.7	58.9	55.1	53.8
Cement Manufacture	33.3	38.3	39.2	40.0	41.2	41.4	42.9	43.0
Ammonia Manufacture & Urea Application	19.3	20.7	21.9	20.6	19.6	16.7	18.6	15.6
Lime Manufacture	11.2	13.7	13.9	13.5	13.3	12.8	12.3	13.0
Limestone and Dolomite Use	5.5	7.2	7.4	8.1	6.0	5.7	5.9	4.7
Aluminum Production	6.3	5.6	5.8	5.9	5.7	4.1	4.2	4.2
Soda Ash Manufacture and Consumption	4.1	4.4	4.3	4.2	4.2	4.1	4.1	4.1
Petrochemical Production	2.2	2.9	3.0	3.1	3.0	2.8	2.9	2.8
Titanium Dioxide Production	1.3	1.8	1.8	1.9	1.9	1.9	2.0	2.0
Phosphoric Acid Production	1.5	1.5	1.6	1.5	1.4	1.3	1.3	1.4
Ferroalloy Production	2.0	2.0	2.0	2.0	1.7	1.3	1.2	1.4
Carbon Dioxide Consumption	0.9	0.8	0.9	0.8	1.0	0.8	1.0	1.3
CH$_4$	2.5	2.9	2.9	2.9	2.9	2.5	2.5	2.5
Petrochemical Production	1.2	1.6	1.7	1.7	1.7	1.4	1.5	1.5
Iron and Steel Production	1.3	1.3	1.2	1.2	1.2	1.1	1.0	1.0
Silicon Carbide Production	(c)	(c)	(c)	(c)	(c)	(c)	(c)	(c)
N$_2$O	33.0	31.5	26.9	25.6	25.6	20.8	23.1	21.8
Nitric Acid Production	17.8	21.2	20.9	20.1	19.6	15.9	17.2	15.8
Adipic Acid Production	15.2	10.3	6.0	5.5	6.0	4.9	5.9	6.0
HFCs, PFCs, and SF$_6$	91.2	121.7	135.7	134.8	138.9	129.5	138.3	137.0
Substitution of Ozone Depleting Substances	0.4	46.5	56.6	65.8	75.0	83.3	91.5	99.5
Electrical Transmission and Distribution	29.2	21.7	17.1	16.4	15.6	15.4	14.7	14.1
HCFC-22 Production	35.0	30.0	40.1	30.4	29.8	19.8	19.8	12.3
Aluminum Production	18.3	11.0	9.1	9.0	9.0	4.0	5.2	3.8
Semiconductor Manufacture	2.9	6.3	7.1	7.2	6.3	4.5	4.4	4.3
Magnesium Production and Processing	5.4	6.3	5.8	6.0	3.2	2.6	2.6	3.0
Total	299.9	327.1	334.9	329.2	332.1	304.7	315.4	308.6

Notes: Totals may not sum due to independent rounding; Tg CO$_2$ Eq. - teragrams of carbon dioxide equivalent; (c) Does not exceed 0.05 Tg CO$_2$ Eq.
Source: United States Environmental Protection Agency, Inventory of U.S. Greenhouse Gas Emissions and Sinks: 1990 - 2003

Emissions from agriculture (Tg CO$_2$ Eq.)

Gas/Source	1990	1997	1998	1999	2000	2001	2002	2003
CH$_4$	156.9	163.0	164.2	164.6	162.0	161.9	161.5	161.8
Enteric Fermentation	117.9	118.3	116.7	116.8	115.6	114.5	114.6	115.0
Manure Management	31.2	36.4	38.8	38.8	38.1	38.9	39.3	39.1
Rice Cultivation	7.1	7.5	7.9	8.3	7.5	7.6	6.8	6.9
Field Burning of Agricultural Residues	0.7	0.8	0.8	0.8	0.8	0.8	0.7	0.8
N$_2$O	269.6	269.8	285.6	261.3	282.1	275.6	270.9	271.5
Agricultural Soil Management	253.0	252.0	267.7	243.4	263.9	257.1	252.6	253.5
Manure Management	16.3	17.3	17.4	17.4	17.8	18.0	17.9	17.5
Field Burning of Agricultural Residues	0.4	0.4	0.5	0.4	0.5	0.5	0.4	0.4
Total	426.5	432.8	449.8	425.9	444.1	437.5	432.4	433.3

Notes: *Totals may not sum due to independent rounding; Tg CO$_2$ Eq. - teragrams of carbon dioxide equivalent*
Source: *United States Environmental Protection Agency, Inventory of U.S. Greenhouse Gas Emissions and Sinks: 1990 - 2003*

Electricity generation-related greenhouse gas emissions (Tg CO$_2$ Eq.)

Gas/Fuel Type or Source	1990	1997	1998	1999	2000	2001	2002	2003
CO$_2$	1,804.0	2,073.3	2,159.9	2,171.0	2,273.1	2,229.4	2,244.8	2,271.7
CO$_2$ from Fossil Fuel Combustion	1,790.3	2,051.9	2,139.0	2,149.3	2,252.1	2,207.8	2,223.0	2,250.5
Coal	1,513.0	1,758.4	1,786.4	1,792.4	1,880.0	1,817.4	1,839.7	1,876.3
Natural Gas	176.0	218.9	248.0	259.9	280.7	289.1	305.6	277.6
Petroleum	101.0	74.3	104.3	96.7	91.0	100.9	77.4	96.3
Geothermal	0.4	0.4	0.4	0.4	0.4	0.4	0.4	0.3
Waste Combustion	10.9	17.8	17.1	17.6	18.0	18.8	18.8	18.8
Limestone and Dolomite Use	2.8	3.6	3.7	4.0	3.0	2.9	2.9	2.4
CH$_4$	0.6	0.6	0.7	0.7	0.7	0.7	0.7	0.7
Stationary Combustion[a]	0.6	0.6	0.7	0.7	0.7	0.7	0.7	0.7
N$_2$O	8.0	9.0	9.2	9.2	9.6	9.4	9.5	9.8
Stationary Combustion[a]	7.6	8.6	8.9	8.9	9.3	9.0	9.1	9.3
Waste Combustion	0.4	0.4	0.3	0.3	0.4	0.4	0.5	0.5
SF$_6$	29.2	21.7	17.1	16.4	15.6	15.4	14.7	14.1
Electrical Transmission and Distribution	29.2	21.7	17.1	16.4	15.6	15.4	14.7	14.1
Total	1,841.8	2,104.6	2,186.8	2,197.3	2,299.0	2,254.9	2,269.7	2,296.2

Notes: Totals may not sum due to independent rounding; Tg CO$_2$ Eq. - teragrams of carbon dioxide equivalent; (a) Includes only stationary combustion emissions related to the generation of electricity
Source: United States Environmental Protection Agency, Inventory of U.S. Greenhouse Gas Emissions and Sinks: 1990 - 2003

Transportation-related greenhouse gas emissions (Tg CO$_2$ Eq.)

Gas/Vehicle Type	1990	1997	1998	1999	2000	2001	2002	2003
CO$_2$	1,461.7	1,618.0	1,648.7	1,706.2	1,753.1	1,734.2	1,766.4	1,780.7
Passenger Cars	612.5	595.5	613.8	622.4	623.4	625.7	639.5	633.7
Light-Duty Trucks	312.2	421.6	432.1	449.2	452.1	456.2	468.1	478.8
Other Trucks	217.0	279.9	290.4	304.3	320.4	327.5	327.5	341.2
Buses	7.8	9.1	9.3	10.4	10.2	9.6	9.1	8.9
Aircraft[a]	177.2	179.0	181.3	186.7	193.2	183.4	174.9	171.3
Ships and Boats	49.2	38.7	32.4	42.3	63.1	42.7	57.2	57.5
Locomotives	36.3	40.0	40.5	41.7	41.8	42.8	41.0	42.8
Other[b]	49.4	54.2	48.7	49.3	48.9	46.1	49.0	46.6
International Bunker Fuels[c]	93.6	106.1	103.3	102.7	102.2	98.5	89.5	84.2
CH$_4$	4.6	3.8	3.7	3.4	3.2	2.9	2.7	2.4
Passenger Cars	2.6	1.9	1.8	1.7	1.5	1.4	1.2	1.1
Light-Duty Trucks	1.4	1.3	1.3	1.1	1.0	0.9	0.9	0.8
Other Trucks and Buses	0.3	0.3	0.3	0.3	0.2	0.2	0.2	0.2
Aircraft	0.2	0.2	0.1	0.2	0.2	0.1	0.1	0.1
Ships and Boats	0.1	0.1	0.1	0.1	0.1	0.1	0.1	0.1
Locomotives	0.1	0.1	0.1	0.1	0.1	0.1	0.1	0.1
Motorcycles	(e)	(e)	(e)	(e)	(e)	(e)	(e)	(e)
International Bunker Fuels[c]	0.2	0.1	0.2	0.1	0.1	0.1	0.1	0.1
N$_2$O	42.9	54.2	54.4	53.7	52.2	47.9	44.5	40.9
Passenger Cars	25.5	26.7	26.7	25.9	24.7	23.1	21.6	19.9
Light-Duty Trucks	14.1	23.7	23.7	23.6	23.0	20.6	18.6	16.8
Other Trucks and Buses	0.9	1.4	1.6	1.7	1.7	1.7	1.8	1.8
Aircraft	1.7	1.7	1.8	1.8	1.9	1.8	1.7	1.7
Ships and Boats	0.4	0.3	0.3	0.3	0.5	0.3	0.5	0.5
Locomotives	0.3	0.3	0.3	0.3	0.3	0.3	0.3	0.3
Motorcycles	(e)	(e)	(e)	(e)	(e)	(e)	(e)	(e)
International Bunker Fuels[c]	1.0	1.0	1.0	0.9	0.9	0.9	0.8	0.8
HFCs	(e)	19.4	24.4	29.3	33.8	37.4	40.4	42.7
Mobile Air Conditioners[d]	(e)	13.8	17.4	20.8	24.0	26.7	28.8	30.3
Refrigerated Transport	(e)	5.5	7.0	8.5	9.8	10.8	11.5	12.3
Total	1,509.3	1,695.4	1,731.1	1,792.5	1,842.2	1,822.4	1,853.9	1,866.7

Notes: Totals may not sum due to independent rounding; Tg CO$_2$ Eq. - teragrams of carbon dioxide equivalent; (a) Aircraft emissions consist of emissions from all jet fuel (less bunker fuels) and aviation gas consumption; (b) "Other" CO$_2$ emissions include motorcycles, pipelines, and lubricants; (c) Emissions from International Bunker Fuels include emissions from both civilian and military activities, but are not included in totals; (d) Includes primarily HFC-134a; (e) Does not exceed 0.05 Tg CO$_2$ Eq.
Source: United States Environmental Protection Agency, Inventory of U.S. Greenhouse Gas Emissions and Sinks: 1990 - 2003

Emissions of NO_x, CO, NMVOCs, and SO_2 (Gg)

Gas/Activity	1990	1997	1998	1999	2000	2001	2002	2003
NO_x	22,860.0	22,284.0	21,964.0	20,530.0	20,288.0	19,414.0	18,850.0	18,573.0
Stationary Fossil Fuel Combustion	9,884.0	9,578.0	9,419.0	8,344.0	8,002.0	7,667.0	7,523.0	7,222.0
Mobile Fossil Fuel Combustion	12,134.0	11,768.0	11,592.0	11,300.0	11,395.0	10,823.0	10,389.0	10,418.0
Oil and Gas Activities	139.0	130.0	130.0	109.0	111.0	113.0	135.0	124.0
Waste Combustion	82.0	140.0	145.0	143.0	114.0	114.0	134.0	121.0
Industrial Processes	591.0	629.0	637.0	595.0	626.0	656.0	630.0	648.0
Solvent Use	1.0	3.0	3.0	3.0	3.0	3.0	5.0	4.0
Field Burning of Agricultural Residues[a]	28.0	34.0	35.0	34.0	35.0	35.0	33.0	33.0
Waste	0.0	3.0	3.0	3.0	2.0	2.0	2.0	2.0
CO	130,580.0	101,138.0	98,984.0	94,361.0	92,895.0	89,329.0	87,451.0	85,077.0
Stationary Fossil Fuel Combustion	4,999.0	3,927.0	3,927.0	5,024.0	4,340.0	4,377.0	4,020.0	4,454.0
Mobile Fossil Fuel Combustion	119,482.0	90,284.0	87,940.0	83,484.0	83,680.0	79,972.0	78,574.0	75,526.0
Oil and Gas Activities	302.0	333.0	332.0	145.0	146.0	147.0	116.0	125.0
Waste Combustion	978.0	2,668.0	2,826.0	2,725.0	1,670.0	1,672.0	1,672.0	1,674.0
Industrial Processes	4,124.0	3,153.0	3,163.0	2,156.0	2,217.0	2,339.0	2,308.0	2,431.0
Solvent Use	4.0	1.0	1.0	46.0	46.0	45.0	46.0	65.0
Field Burning of Agricultural Residues[a]	689.0	767.0	789.0	767.0	790.0	770.0	707.0	794.0
Waste	1.0	5.0	5.0	13.0	8.0	8.0	8.0	8.0
NMVOCs	20,937.0	16,994.0	16,403.0	15,869.0	15,228.0	15,048.0	14,222.0	13,939.0
Stationary Fossil Fuel Combustion	912.0	1,016.0	1,016.0	1,045.0	1,077.0	1,080.0	926.0	1,007.0
Mobile Fossil Fuel Combustion	10,933.0	7,928.0	7,742.0	7,586.0	7,230.0	6,872.0	6,560.0	6,351.0
Oil and Gas Activities	555.0	442.0	440.0	414.0	389.0	400.0	340.0	345.0
Waste Combustion	222.0	313.0	326.0	302.0	257.0	258.0	281.0	263.0
Industrial Processes	2,426.0	2,038.0	2,047.0	1,813.0	1,773.0	1,769.0	1,725.0	1,711.0
Solvent Use	5,217.0	5,100.0	4,671.0	4,569.0	4,384.0	4,547.0	4,256.0	4,138.0
Field Burning of Agricultural Residues	n/a	n/a	n/a	n/a	n/a	n/a	n/a	n/a
Waste	673.0	157.0	161.0	140.0	119.0	122.0	133.0	125.0
SO_2	20,936.0	17,091.0	17,189.0	15,917.0	14,829.0	14,452.0	13,928.0	14,463.0
Stationary Fossil Fuel Combustion	18,407.0	15,104.0	15,191.0	13,915.0	12,848.0	12,461.0	11,946.0	12,477.0
Mobile Fossil Fuel Combustion	793.0	659.0	665.0	704.0	632.0	624.0	631.0	634.0
Oil and Gas Activities	390.0	312.0	310.0	283.0	286.0	289.0	315.0	293.0
Waste Combustion	39.0	29.0	30.0	30.0	29.0	30.0	24.0	28.0
Industrial Processes	1,306.0	985.0	991.0	984.0	1,031.0	1,047.0	1,009.0	1,029.0
Solvent Use	0.0	1.0	1.0	1.0	1.0	1.0	2.0	2.0
Field Burning of Agricultural Residues	n/a	n/a	n/a	n/a	n/a	n/a	n/a	n/a
Waste	0.0	1.0	1.0	1.0	1.0	1.0	1.0	1.0

Notes: *Totals may not sum due to independent rounding; n/a not available; (a) NOx and CO emission estimates from field burning of agricultural residues were estimated separately, and therefore not taken from EPA (2004).*
Source: *United States Environmental Protection Agency, Inventory of U.S. Greenhouse Gas Emissions and Sinks: 1990 - 2003*

Green Metro Area Rankings by Category

Metro Area	State	Air Quality[1]	Toxic Releases[2]	Superfund Sites[3]	Energy Use[4]	Motor Vehicle Use[5]	Mass Transit Use[6]	Overall Score[7]	Overall Rank[8]
Akron	OH	26	43	42	74	23	72	280	40
Albany	NY	22	36	62	87	72	45	324	70
Albuquerque	NM	22	4	59	46	62	79	272	39
Allentown	PA	31	67	84	65	34	75	356	82
Atlanta	GA	69	60	1	28	87	22	267	36
Austin	TX	16	12	1	27	70	36	162	8
Bakersfield	CA	90	52	44	14	8	52	260	32
Baltimore	MD	76	64	38	51	60	23	312	65
Baton Rouge	LA	73	87	1	15	11	83	270	38
Bergen	NJ	26	5	69	62	2	1	165	10
Birmingham	AL	50	84	18	29	88	86	355	80
Boston	MA	31	26	76	60	33	10	236	30
Buffalo	NY	31	59	49	81	16	54	290	51
Charleston	SC	1	89	67	16	32	81	286	46
Charlotte	NC	38	76	70	33	69	38	324	71
Chicago	IL	46	42	34	83	21	11	237	31
Cincinnati	OH	57	78	20	55	55	46	311	61
Cleveland	OH	74	71	30	73	20	29	297	56
Columbus	OH	50	49	1	64	57	66	287	48
Dallas	TX	79	30	12	38	63	43	265	35
Dayton	OH	54	43	80	69	54	61	361	83
Denver	CO	76	25	54	70	43	26	294	54
Detroit	MI	74	74	31	79	67	63	388	88
El Paso	TX	57	31	1	31	9	48	177	15
Fort Lauderdale	FL	1	10	45	18	49	31	154	6
Fort Worth	TX	82	23	15	36	63	43	262	33
Fresno	CA	88	13	72	19	17	59	268	37
Gary	IN	46	88	77	78	21	11	321	69
Grand Rapids	MI	57	70	88	88	53	70	426	90
Greensboro	NC	26	68	1	39	82	85	301	58
Greenville	SC	16	58	74	30	29	90	297	55
Harrisburg	PA	54	15	47	66	86	78	346	77
Hartford	CT	38	27	52	75	61	62	315	66
Honolulu	HI	12	38	49	26	10	13	148	4
Houston	TX	84	69	57	20	90	34	354	79
Indianapolis	IN	69	72	33	67	85	69	395	89
Jacksonville	FL	1	75	66	10	84	53	289	50
Kansas City	MO	65	46	48	71	76	73	379	86
Knoxville	TN	65	80	25	37	75	89	371	84
Las Vegas	NV	69	83	1	44	81	41	319	68
Little Rock	AR	6	32	51	40	74	82	285	44
Los Angeles	CA	87	21	27	3	41	19	198	19
Louisville	KY	57	73	43	50	68	65	356	81
Memphis	TN	38	65	61	41	66	60	331	73
Miami	FL	6	8	46	22	49	31	162	7
Middlesex	NJ	46	35	89	59	2	1	232	27
Milwaukee	WI	38	51	55	89	56	37	326	72
Minneapolis	MN	6	56	64	90	65	35	316	67
Monmouth	NJ	57	1	86	58	2	1	205	20
Nashville	TN	26	61	1	42	80	76	286	45
Nassau County	NY	63	14	81	57	2	1	218	23
New Orleans	LA	31	86	26	13	1	28	185	17
New York	NY	63	7	11	52	2	1	136	2
Newark	NJ	54	33	82	56	2	1	228	26
Norfolk	VA	16	63	78	34	37	64	292	53
Oakland	CA	38	34	36	4	30	8	150	5
Oklahoma City	OK	16	18	63	47	79	88	311	63
Omaha	NE	12	62	24	85	19	87	289	49
Orange County	CA	81	11	16	2	41	20	171	11
Orlando	FL	6	54	39	11	83	40	233	28

Green Metro Area Rankings by Category - continued

Metro Area	State	Air Quality[1]	Toxic Releases[2]	Superfund Sites[3]	Energy Use[4]	Motor Vehicle Use[5]	Mass Transit Use[6]	Overall Score[7]	Overall Rank[8]
Philadelphia	PA	80	37	85	53	13	16	284	43
Phoenix	AZ	65	20	32	45	45	55	262	34
Pittsburgh	PA	84	85	22	68	27	25	311	62
Portland	OR	1	55	65	24	12	18	175	14
Providence	RI	46	17	83	63	24	49	282	42
Raleigh	NC	31	39	41	35	78	80	304	60
Richmond	VA	26	81	56	43	71	67	344	76
Riverside	CA	89	16	37	8	28	57	235	29
Rochester	NY	14	53	67	82	40	56	312	64
Sacramento	CA	86	9	40	9	15	47	206	21
Saint Louis	MO	65	79	60	61	77	39	381	87
Salt Lake City	UT	69	90	75	72	38	30	374	85
San Antonio	TX	31	29	1	25	46	42	174	13
San Diego	CA	76	19	13	1	48	24	181	16
San Francisco	CA	6	2	14	6	30	8	66	1
San Jose	CA	50	6	87	7	47	27	224	25
Scranton	PA	22	22	71	76	18	77	286	47
Seattle	WA	16	24	53	21	35	14	163	9
Springfield	MA	38	40	28	86	52	58	302	59
Syracuse	NY	21	50	73	84	59	51	338	75
Tacoma	WA	14	41	79	32	35	14	215	22
Tampa	FL	22	48	58	12	73	68	281	41
Toledo	OH	38	82	1	80	58	74	333	74
Tucson	AZ	6	45	19	23	44	50	187	18
Tulsa	OK	38	66	21	48	89	84	346	78
Ventura	CA	83	3	23	5	39	20	173	12
Washington	DC	57	47	35	49	25	7	220	24
West Palm Beach	FL	1	28	17	17	49	31	143	3
Wilmington	DE	50	77	90	54	13	16	300	57
Youngstown	OH	31	57	29	77	26	71	291	52

Note: *The Green Metro Index compares the nation's 90 largest metropolitan areas on measures of environmental quality and performance appropriate to metro areas as a whole. The index is based on federal and private data for six environmental measures including: air quality, toxic releases, superfund sites, energy use, mass transit use and motor vehicle use; The figures above rank how each metro area fared in each category; (1) Based on the number of days the Air Quality Index (AQI) rose above 100 in 2003. Numbers above 100 indicate the air quality was in the unhealthful range; (2) Calculated by adding the total toxic releases for each metro area and dividing by the metro area population. Data is from the Environmental Protection Agency's Toxic Release Inventory for 2003; (3) Based on the per capita number of Superfund Sites located within each metro area. Data is from the Environmental Protection Agency's Superfund National Priorities list (data extracted 7/25/2005); (4) Based on total heating and cooling degree days per year. Data is from Weather America, A Thirty-Year Summary of Statistical Weather Data and Rankings, 2001; (5) Based on the DVMT (daily vehicle-miles of travel) per capita for each urbanized area. Data is from the Department of Transportation's Urbanized Areas: 2003 Selected Characteristics report; (6) Calculated by dividing the total mass transit passenger miles by the population of the urbanized area (defined by the U.S. Census Bureau as a densely settled area containing at least 50,000 people). Data is from the Department of Transportation's 2003 National Transit Summaries and Trends report; (7) The overall score was calculated by combining the rankings of all six environmental indicators, giving equal weight to each indicator; (8) 1=best, 90=worst*
Sources: *U.S Environmental Protection Agency; U.S. Department of Transportation; Grey House Publishing, Weather America, A Thirty-Year Summary of Statistical Weather Data and Rankings, 2001*

Green Metro Area Overall Rankings

Metro Area	State	Overall Rank[1]	Overall Score[2]
San Francisco	CA	1	66
New York	NY	2	136
West Palm Beach	FL	3	143
Honolulu	HI	4	148
Oakland	CA	5	150
Fort Lauderdale	FL	6	154
Miami	FL	7	162
Austin	TX	8	162
Seattle	WA	9	163
Bergen	NJ	10	165
Orange County	CA	11	171
Ventura	CA	12	173
San Antonio	TX	13	174
Portland	OR	14	175
El Paso	TX	15	177
San Diego	CA	16	181
New Orleans	LA	17	185
Tucson	AZ	18	187
Los Angeles	CA	19	198
Monmouth	NJ	20	205
Sacramento	CA	21	206
Tacoma	WA	22	215
Nassau County	NY	23	218
Washington	DC	24	220
San Jose	CA	25	224
Newark	NJ	26	228
Middlesex	NJ	27	232
Orlando	FL	28	233
Riverside	CA	29	235
Boston	MA	30	236
Chicago	IL	31	237
Bakersfield	CA	32	260
Fort Worth	TX	33	262
Phoenix	AZ	34	262
Dallas	TX	35	265
Atlanta	GA	36	267
Fresno	CA	37	268
Baton Rouge	LA	38	270
Albuquerque	NM	39	272
Akron	OH	40	280
Tampa	FL	41	281
Providence	RI	42	282
Philadelphia	PA	43	284
Little Rock	AR	44	285
Nashville	TN	45	286
Charleston	SC	46	286
Scranton	PA	47	286
Columbus	OH	48	287
Omaha	NE	49	289
Jacksonville	FL	50	289
Buffalo	NY	51	290
Youngstown	OH	52	291
Norfolk	VA	53	292
Denver	CO	54	294
Greenville	SC	55	297
Cleveland	OH	56	297
Wilmington	DE	57	300
Greensboro	NC	58	301
Springfield	MA	59	302
Raleigh	NC	60	304

Green Metro Area Overall Rankings - *continued*

Metro Area	State	Overall Rank[1]	Overall Score[2]
Cincinnati	OH	61	311
Pittsburgh	PA	62	311
Oklahoma City	OK	63	311
Rochester	NY	64	312
Baltimore	MD	65	312
Hartford	CT	66	315
Minneapolis	MN	67	316
Las Vegas	NV	68	319
Gary	IN	69	321
Albany	NY	70	324
Charlotte	NC	71	324
Milwaukee	WI	72	326
Memphis	TN	73	331
Toledo	OH	74	333
Syracuse	NY	75	338
Richmond	VA	76	344
Harrisburg	PA	77	346
Tulsa	OK	78	346
Houston	TX	79	354
Birmingham	AL	80	355
Louisville	KY	81	356
Allentown	PA	82	356
Dayton	OH	83	361
Knoxville	TN	84	371
Salt Lake City	UT	85	374
Kansas City	MO	86	379
Saint Louis	MO	87	381
Detroit	MI	88	388
Indianapolis	IN	89	395
Grand Rapids	MI	90	426

Note: *The Green Metro Index compares the nation's 90 largest metropolitan areas on measures of environmental quality and performance appropriate to metro areas as a whole. The index is based on federal data for six environmental measures including: air quality, toxic releases, superfund sites, energy use, mass transit use and motor vehicle use; (1) A lower number indicates better environmental quality or performance; (2) The overall score was calculated by combining the rankings of all six environmental indicators, giving equal weight to each indicator.*

Sources: *U.S Environmental Protection Agency; U.S. Department of Transportation; Grey House Publishing, Weather America, A Thirty-Year Summary of Statistical Weather Data and Rankings, 2001*

Palustrine and estuarine wetlands on nonfederal land and water areas, by land cover/use and farm production region, 2003

Farm Production Region	Cropland, Pastureland, and CRP[1] Land	Forest Land	Rangeland	Other Rural Land	Developed Land	Water Area	Total
Lake States	2,710 (± 11%)	15,480 (± 5%)	0 (n/a)	3,880 (± 8%)	160 (± 20%)	230 (± 11%)	22,460 (± 3%)
Southeast	940 (± 20%)	16,010 (± 2%)	970 (± 18%)	3,460 (± 11%)	420 (± 17%)	560 (± 8%)	22,360 (± 2%)
Delta States	3,240 (± 5%)	11,020 (± 3%)	270 (± 14%)	2,730 (± 7%)	190 (± 20%)	500 (± 7%)	17,950 (± 2%)
Northeast	1,250 (± 8 %)	10,890 (± 5 %)	0 (n/a)	1,550 (± 11%)	240 (± 12%)	220 (± 10%)	14,150 (± 3%)
Northern Plains	3,020 (± 6%)	210 (± 26%)	2,870 (± 9%)	1,090 (± 14%)	80 (± 14%)	370 (± 6%)	7,640 (± 4%)
Appalachian	400 (± 15%)	6,080 (± 3%)	0 (n/a)	570 (± 19%)	110 (± 18%)	300 (± 6%)	7,460 (± 3%)
Southern Plains	970 (± 19%)	2,350 (± 14%)	970 (± 17%)	520 (± 44%)	230 (± 22%)	550 (± 11%)	5,590 (± 6%)
Mountain	1,570 (± 15%)	220 (± 45%)	2,010 (± 17%)	820 (± 34%)	30 (± 62%)	130 (± 15%)	4,780 (± 8%)
Corn Belt	1,330 (± 10%)	2,440 (± 7%)	0 (n/a)	380 (± 23%)	100 (± 25%)	440 (± 6%)	4,690 (± 5%)
Pacific	1,300 (± 21%)	740 (± 21%)	650 (± 25%)	800 (± 27%)	30 (± 53%)	160 (± 11%)	3,680 (± 8%)
Total	16,730 (± 3%)	65,440 (± 2%)	7,740 (± 7%)	15,800 (± 4%)	1,590 (± 6%)	3,460 (± 3%)	110,760 (± 1%)

Notes: All figures are in thousands of acres; percent margin of error appears in parentheses; n/a not applicable; (1) Conservation Reserve Program
Source: United States Department of Agriculture, Natural Resources Conservation Service, 2003 Annual National Resources Inventory

Superfund National Priorities List

St.	City/Area	Site Name	CERCLIS[1] ID	Status	Score[2]
AK	Adak	Adak Naval Air Station	AK4170024323	Final	51.37
AK	Anchorage	Elmendorf Air Force Base	AK8570028649	Final	45.91
AK	Anchorage	Fort Richardson (USARMY)	AK6214522157	Final	50.00
AK	Anchorage	Standard Steel & Metal Salvage Yard (USDOT)	AKD980978787	Deleted	46.25
AK	Fairbanks	Alaska Battery Enterprises	AKD004904215	Deleted	30.98
AK	Fairbanks	Arctic Surplus	AKD980988158	Final	42.24
AK	Fairbanks	Eielson Air Force Base	AK1570028646	Final	48.14
AK	Fort Wainwright	Fort Wainwright	AK6210022426	Final	42.40
AL	Anniston	Anniston Army Depot (Southeast Industrial Area)	AL3210020027	Final	51.91
AL	Axis	Stauffer Chemical Co. (Lemoyne Plant)	ALD008161176	Final	32.34
AL	Bucks	Stauffer Chemical Co. (Cold Creek Plant)	ALD095688875	Final	46.77
AL	Childersburg	Alabama Army Ammunition Plant	AL6210020008	Final	36.83
AL	Greenville	Mowbray Engineering Co.	ALD031618069	Deleted	53.67
AL	Headland	American Brass Inc.	ALD981868466	Final	55.61
AL	Huntsville	USARMY/Nasa Redstone Arsenal	AL7210020742	Final	50.00
AL	Leeds	Interstate Lead Co. (Ilco)	ALD041906173	Final	42.86
AL	Limestone/Morgan	Triana/Tennessee River	ALD983166299	Final	61.42
AL	Mcintosh	Ciba-Geigy Corp. (Mcintosh Plant)	ALD001221902	Final	53.42
AL	Mcintosh	Olin Corp. (Mcintosh Plant)	ALD008188708	Final	39.71
AL	Montgomery	Capitol City Plume	AL0001058056	Proposed	50.00
AL	Montgomery	T.H. Agriculture & Nutrition Co. (Montgomery Plant)	ALD007454085	Final	44.46
AL	Perdido	Perdido Ground Water Contamination	ALD980728703	Final	30.29
AL	Saraland	Redwing Carriers, Inc. (Saraland)	ALD980844385	Final	30.83
AL	Vincent	Alabama Plating Co Inc	ALD004022448	Proposed	30.00
AR	Edmondsen	Gurley Pit	ARD035662469	Deleted	40.13
AR	El Dorado	Popile, Inc.	ARD008052508	Final	50.03
AR	Fort Smith	Industrial Waste Control	ARD980496368	Final	30.31
AR	Jacksonville	Jacksonville Municipal Landfill	ARD980809941	Deleted	29.64
AR	Jacksonville	Rogers Road Municipal Landfill	ARD981055809	Final	29.64
AR	Jacksonville	Vertac, Inc.	ARD000023440	Final	65.46
AR	Mena	Mid-South Wood Products	ARD092916188	Final	45.87
AR	Newport	Cecil Lindsey	ARD980496186	Deleted	35.60
AR	Ola/Birta	Midland Products	ARD980745665	Final	30.77
AR	Omaha	Arkwood, Inc.	ARD084930148	Final	28.95
AR	Paragould	Monroe Auto Equipment Co. (Paragould Pit)	ARD980864110	Final	46.01
AR	Plainview	Mountain Pine Pressure Treating	ARD049658628	Final	41.93
AR	Reader	Ouachita Nevada Wood Treater	ARD042755231	Final	50.00
AR	Walnut Ridge	Frit Industries	ARD059636456	Deleted	39.47
AR	West Memphis	South 8th Street Landfill	ARD980496723	Deleted	50.27
AS	Pago Pago	Taputimu Farm	ASD980637656	Deleted	n/a
AZ	Chandler	Williams Air Force Base	AZ7570028582	Final	37.93
AZ	Glendale	Luke Air Force Base	AZ0570024133	Deleted	37.93
AZ	Globe	Mountain View Mobile Home Estates	AZD980735724	Deleted	n/a
AZ	Goodyear	Phoenix-Goodyear Airport Area	AZD980695902	Final	45.91
AZ	Hassayampa	Hassayampa Landfill	AZD980735666	Final	42.79
AZ	Phoenix	Motorola, Inc. (52nd Street Plant)	AZD009004177	Final	40.83
AZ	Phoenix	Nineteenth Avenue Landfill	AZD980496780	Final	54.27
AZ	Saint David	Apache Powder Co.	AZD008399263	Final	39.09
AZ	Scottsdale	Indian Bend Wash Area	AZD980695969	Final	42.24
AZ	Tucson	Tucson International Airport Area	AZD980737530	Final	57.80
AZ	Yuma	Yuma Marine Corps Air Station	AZ0971590062	Final	32.24
CA	Alameda	Alameda Naval Air Station	CA2170023236	Final	50.00
CA	Alhambra	San Gabriel Valley (Area 3)	CAD980818579	Final	28.90
CA	Alviso	South Bay Asbestos Area	CAD980894885	Final	44.68
CA	Antioch	Gbf, Inc., Dump	CAD980498562	Proposed	51.00
CA	Arvin	Brown & Bryant, Inc. (Arvin Plant)	CAD052384021	Final	53.36
CA	Baldwin Park	San Gabriel Valley (Area 2)	CAD980818512	Final	42.24
CA	Barstow	Barstow Marine Corps Logistics Base	CA8170024261	Final	37.93
CA	Camp Pendleton	Camp Pendleton Marine Corps Base	CA2170023533	Final	33.79
CA	Casmalia	Casmalia Resources	CAD020748125	Final	30.00

Superfund National Priorities List - *continued*

St.	City/Area	Site Name	CERCLIS[1] ID	Status	Score[2]
CA	Clear Lake	Sulphur Bank Mercury Mine	CAD980893275	Final	44.42
CA	Cloverdale	Mgm Brakes	CAD000074120	Final	34.70
CA	Coalinga	Atlas Asbestos Mine	CAD980496863	Final	45.55
CA	Coalinga	Coalinga Asbestos Mine	CAD980817217	Deleted	45.55
CA	Concord	Concord Naval Weapons Station	CA7170024528	Final	50.00
CA	Crescent City	Del Norte Pesticide Storage	CAD000626176	Deleted	35.79
CA	Cupertino	Intersil Inc./Siemens Components	CAD041472341	Final	28.90
CA	Davis	Frontier Fertilizer	CAD071530380	Final	35.04
CA	Davis	Laboratory For Energy-Related Health Research/Old Campus Landfill (USDOE)	CA2890190000	Final	50.00
CA	Edwards Afb	Edwards Air Force Base	CA1570024504	Final	33.62
CA	El Monte	San Gabriel Valley (Area 1)	CAD980677355	Final	42.24
CA	El Toro	El Toro Marine Corps Air Station	CA6170023208	Final	37.43
CA	Fillmore	Pacific Coast Pipe Lines	CAD980636781	Final	46.01
CA	Fresno	Fresno Municipal Sanitary Landfill	CAD980636914	Final	35.57
CA	Fresno	Industrial Waste Processing	CAD980736284	Final	51.13
CA	Fresno	T.H. Agriculture & Nutrition Co.	CAD009106220	Final	42.24
CA	Fullerton	Mccoll	CAD980498695	Final	41.77
CA	Glendale	San Fernando Valley (Area 2)	CAD980894901	Final	42.24
CA	Glendale	San Fernando Valley (Area 3)	CAD980894984	Deleted	42.24
CA	Hoopa	Celtor Chemical Works	CAD980638860	Deleted	30.31
CA	Imperial	Stoker Company	CAD066635442	Proposed	71.00
CA	La Puente	San Gabriel Valley (Area 4)	CAD980817985	Final	28.90
CA	Lathrop	Sharpe Army Depot	CA8210020832	Final	42.24
CA	Livermore	Lawrence Livermore Natl Lab, Main Site (USDOE)	CA2890012584	Final	42.24
CA	Los Angeles	Del Amo	CAD029544731	Final	47.12
CA	Los Angeles	San Fernando Valley (Area 4)	CAD980894976	Final	35.57
CA	Malaga	Purity Oil Sales, Inc.	CAD980736151	Final	43.27
CA	Marina	Fort Ord	CA7210020676	Final	42.24
CA	Markleeville	Leviathan Mine	CAD980673685	Final	50.00
CA	Mather	Mather Air Force Base (AC&W Disposal Site)	CA8570024143	Final	28.90
CA	Maywood	Pemaco Maywood	CAD980737092	Final	45.23
CA	Mcclellan Afb	Mcclellan Air Force Base (Ground Water Contamination)	CA4570024337	Final	57.93
CA	Merced	Castle Air Force Base (6 Areas)	CA3570024551	Final	37.93
CA	Mira Loma	Stringfellow	CAT080012826	Final	n/a
CA	Modesto	Modesto Ground Water Contamination	CAD981997752	Final	28.90
CA	Moffett Field	Moffett Naval Air Station	CA2170090078	Final	29.49
CA	Monterey Park	Operating Industries, Inc., Landfill	CAT080012024	Final	57.22
CA	Mountain View	Cts Printex, Inc.	CAD009212838	Final	33.62
CA	Mountain View	Fairchild Semiconductor Corp. (Mountain View Plant)	CAD095989778	Final	31.94
CA	Mountain View	Intel Corp. (Mountain View Plant)	CAD061620217	Final	29.76
CA	Mountain View	Jasco Chemical Corp.	CAD009103318	Final	35.36
CA	Mountain View	Raytheon Corp.	CAD009205097	Final	29.76
CA	Mountain View	Spectra-Physics, Inc.	CAD009138488	Final	39.92
CA	Mountain View	Teledyne Semiconductor	CAD009111444	Final	35.35
CA	Nevada City	Lava Cap Mine	CAD983618893	Final	33.66
CA	North Hollywood	San Fernando Valley (Area 1)	CAD980894893	Final	42.24
CA	Oakland	Amco Chemical	CA0001576081	Final	50.00
CA	Oroville	Koppers Co., Inc. (Oroville Plant)	CAD009112087	Final	33.73
CA	Oroville	Louisiana-Pacific Corp.	CAD065021594	Deleted	33.73
CA	Oroville	Western Pacific Railroad Co.	CAD980894679	Deleted	39.79
CA	Palo Alto	Hewlett-Packard (620-640 Page Mill Road)	CAD980884209	Final	29.76
CA	Pasadena	Jet Propulsion Laboratory (Nasa)	CA9800013030	Final	50.00
CA	Paso Robles	Klau/Buena Vista Mine	CA1141190578	Proposed	50.00
CA	Petaluma	Sola Optical Usa, Inc.	CAD981171523	Final	33.39
CA	Porterville	Beckman Instruments (Porterville Plant)	CAD048645444	Final	34.21
CA	Rancho Cordova	Aerojet General Corp.	CAD980358832	Final	54.63
CA	Redding	Iron Mountain Mine	CAD980498612	Final	56.16
CA	Richmond	Liquid Gold Oil Corp.	CAT000646208	Deleted	43.32
CA	Richmond	United Heckathorn Co.	CAD981436363	Final	38.49
CA	Riverbank	Riverbank Army Ammunition Plant	CA7210020759	Final	63.94

Superfund National Priorities List - *continued*

St.	City/Area	Site Name	CERCLIS[1] ID	Status	Score[2]
CA	Riverside	Alark Hard Chrome	CAD098229214	Final	50.50
CA	Riverside	March Air Force Base	CA4570024527	Final	31.94
CA	Sacramento	Jibboom Junkyard	CAD980737613	Deleted	28.94
CA	Sacramento	Sacramento Army Depot	CA0210020780	Final	44.46
CA	Salinas	Crazy Horse Sanitary Landfill	CAD980498455	Final	37.93
CA	Salinas	Firestone Tire & Rubber Co. (Salinas Plant)	CAD990793887	Deleted	45.91
CA	San Bernardino	Newmark Ground Water Contamination	CAD981434517	Final	35.57
CA	San Bernardino	Norton Air Force Base (Lndfll #2)	CA4570024345	Final	39.65
CA	San Francisco	Treasure Island Naval Station-Hunters Point Annex	CA1170090087	Final	48.77
CA	San Jose	Fairchild Semiconductor Corp. (South San Jose Plant)	CAD097012298	Final	44.46
CA	San Jose	Lorentz Barrel & Drum Co.	CAD029295706	Final	33.94
CA	Santa Clara	Applied Materials	CAD042728840	Final	31.94
CA	Santa Clara	Intel Corp. (Santa Clara Iii)	CAT000612184	Final	31.94
CA	Santa Clara	Intel Magnetics	CAD092212497	Final	31.94
CA	Santa Clara	National Semiconductor Corp.	CAD041472986	Final	35.57
CA	Santa Clara	Synertek, Inc. (Building 1)	CAD990832735	Final	31.94
CA	Santa Fe Springs	Waste Disposal, Inc.	CAD980884357	Final	34.60
CA	Scotts Valley	Watkins-Johnson Co. (Stewart Division Plant)	CAD980893234	Final	28.90
CA	Selma	Selma Treating Co.	CAD029452141	Final	48.83
CA	South Gate	Cooper Drum Co.	CAD055753370	Final	50.00
CA	Stockton	Mccormick & Baxter Creosoting Co.	CAD009106527	Final	74.86
CA	Sunnyvale	Advanced Micro Devices, Inc.	CAD048634059	Final	37.93
CA	Sunnyvale	Advanced Micro Devices, Inc. (Building 915)	CAT080034234	Final	31.94
CA	Sunnyvale	Monolithic Memories	CAD049236201	Final	35.57
CA	Sunnyvale	Trw Microwave, Inc (Building 825)	CAD009159088	Final	31.94
CA	Sunnyvale	Westinghouse Electric Corp. (Sunnyvale Plant)	CAD001864081	Final	39.93
CA	Torrance	Montrose Chemical Corp.	CAD008242711	Final	32.10
CA	Tracy	Lawrence Livermore Natl Lab (Site 300) (USDOE)	CA2890090002	Final	31.58
CA	Tracy	Tracy Defense Depot (USARMY)	CA4971520834	Final	37.16
CA	Travis Afb	Travis Air Force Base	CA5570024575	Final	29.49
CA	Turlock	Valley Wood Preserving, Inc.	CAD063020143	Final	32.01
CA	Ukiah	Coast Wood Preserving	CAD063015887	Final	44.73
CA	Victorville	George Air Force Base	CA2570024453	Final	33.62
CA	Visalia	Southern California Edison Co. (Visalia Poleyard)	CAD980816466	Final	48.91
CA	Weed	J.H. Baxter & Co.	CAD000625731	Final	34.78
CA	Westminster	Ralph Gray Trucking Co.	CAD981995947	Deleted	35.04
CA	Whittier	Omega Chemical Corporation	CAD042245001	Final	30.94
CO	Adams County	Rocky Mountain Arsenal (USARMY)	CO5210020769	Final	58.15
CO	Aspen	Smuggler Mountain	COD980806277	Deleted	31.31
CO	Aurora	Lowry Landfill	COD980499248	Final	48.36
CO	Boulder	Marshall Landfill	COD980499255	Final	n/a
CO	Canon City	Lincoln Park	COD042167858	Final	31.31
CO	Commerce City	Sand Creek Industrial	COD980717953	Deleted	59.65
CO	Commerce City	Woodbury Chemical Co.	COD980667075	Deleted	44.87
CO	Denver	Asarco, Inc. (Globe Plant)	COD007063530	Proposed	71.00
CO	Denver	Broderick Wood Products	COD000110254	Final	35.13
CO	Denver	Chemical Sales Co.	COD007431620	Final	37.93
CO	Denver	Denver Radium Site	COD980716955	Final	44.11
CO	Denver	Vasquez Boulevard & I-70	CO0002259588	Final	50.00
CO	Golden	Rocky Flats Plant (USDOE)	CO7890010526	Final	64.32
CO	Gunnison National Forest	Standard Mine	CO0002378230	Proposed	50.00
CO	Idaho Springs	Central City, Clear Creek	COD980717557	Final	51.39
CO	Leadville	California Gulch	COD980717938	Final	55.84
CO	Littleton	Air Force Plant Pjks	CO7570090038	Final	42.93
CO	Minturn/Redcliff	Eagle Mine	COD081961518	Final	47.19
CO	Rio Grande County	Summitville Mine	COD983778432	Final	50.00
CO	Salida	Smeltertown Site	COD983769738	Proposed	59.00
CO	Uravan	Uravan Uranium Project (Union Carbide Corp.)	COD007063274	Final	43.53
CO	Ward	Captain Jack Mill	COD981551427	Final	50.56
CT	Barkhamsted	Barkhamsted-New Hartford Landfill	CTD980732333	Final	38.05

Superfund National Priorities List - continued

St.	City/Area	Site Name	CERCLIS[1] ID	Status	Score[2]
CT	Beacon Falls	Beacon Heights Landfill	CTD072122062	Final	46.77
CT	Canterbury	Yaworski Waste Lagoon	CTD009774969	Final	36.72
CT	Cheshire	Cheshire Ground Water Contamination	CTD981067317	Deleted	35.57
CT	Durham	Durham Meadows	CTD001452093	Final	33.94
CT	East Windsor	Broad Brook Mill	CT0002055887	Proposed	54.00
CT	Naugatuck Borough	Laurel Park, Inc.	CTD980521165	Final	n/a
CT	New London	New London Submarine Base	CTD980906515	Final	36.53
CT	Norwalk	Kellogg-Deering Well Field	CTD980670814	Final	39.92
CT	Plainfield	Gallup's Quarry	CTD108960972	Final	46.29
CT	Southington	Old Southington Landfill	CTD980670806	Final	54.35
CT	Southington	Solvents Recovery Service of New England	CTD009717604	Final	44.93
CT	Sterling	Revere Textile Prints Corp.	CTD004532610	Deleted	41.06
CT	Stratford	Raymark Industries, Inc.	CTD001186618	Final	n/a
CT	Vernon	Precision Plating Corp.	CTD051316313	Final	49.10
CT	Waterbury	Scovill Industrial Landfill	CT0002265551	Final	50.00
CT	Wolcott	Nutmeg Valley Road	CTD980669261	Final	42.69
CT	Woodstock	Linemaster Switch Corp.	CTD001153923	Final	33.71
DC	Washington	Washington Navy Yard	DC9170024310	Final	48.57
DE	Cheswold	Coker's Sanitation Service Landfills	DED980704860	Final	52.15
DE	Delaware City	Delaware City Pvc Plant	DED980551667	Final	30.55
DE	Dover	Chem-Solv, Inc.	DED980714141	Final	37.93
DE	Dover	Dover Air Force Base	DE8570024010	Final	35.89
DE	Dover	Dover Gas Light Co.	DED980693550	Final	35.57
DE	Dover	Wildcat Landfill	DED980704951	Deleted	30.61
DE	Kirkwood	Harvey & Knott Drum, Inc.	DED980713093	Final	30.77
DE	Laurel	Sussex County Landfill No. 5	DED980494637	Deleted	28.90
DE	Middletown	Sealand Limited	DED981035520	Deleted	33.10
DE	Millsboro	Ncr Corp. (Millsboro Plant)	DED043958388	Final	38.21
DE	New Castle	Army Creek Landfill	DED980494496	Final	69.92
DE	New Castle	Delaware Sand & Gravel Landfill	DED000605972	Final	46.60
DE	New Castle	Halby Chemical Co.	DED980830954	Final	30.90
DE	New Castle	New Castle Spill	DED058980442	Deleted	38.33
DE	New Castle	New Castle Steel	DED980705255	Deleted	30.40
DE	New Castle	Standard Chlorine of Delaware, Inc.	DED041212473	Final	35.42
DE	New Castle	Tybouts Corner Landfill	DED000606079	Final	n/a
DE	Newport	E.I. Du Pont de Nemours & Co., Inc. (Newport Pigment Plant Landfill)	DED980555122	Final	51.91
DE	Newport	Koppers Co., Inc. (Newport Plant)	DED980552244	Final	33.56
DE	Smyrna	Tyler Refrigeration Pit	DED980705545	Deleted	33.94
FL	Baldwin	Yellow Water Road Dump	FLD980844179	Deleted	30.26
FL	Brandon	Sydney Mine Sludge Ponds	FLD000648055	Final	38.93
FL	Cantonment	Dubose Oil Products Co.	FLD000833368	Deleted	34.18
FL	Clermont	Tower Chemical Co.	FLD004065546	Final	44.03
FL	Cottondale	Sapp Battery Salvage	FLD980602882	Final	47.70
FL	Davie	Davie Landfill	FLD980602288	Final	57.86
FL	Deland	Sherwood Medical Industries	FLD043861392	Final	39.83
FL	Duval County	Hipps Road Landfill	FLD980709802	Final	31.94
FL	Fort Lauderdale	Florida Petroleum Reprocessors	FLD984184127	Final	50.00
FL	Fort Lauderdale	Hollingsworth Solderless Terminal	FLD004119681	Final	44.53
FL	Fort Lauderdale	Wingate Road Municipal Incinerator Dump	FLD981021470	Final	31.72
FL	Gainesville	Cabot/Koppers	FLD980709356	Final	36.69
FL	Galloway	Alpha Chemical Corp.	FLD041495441	Deleted	43.24
FL	Hialeah	B&B Chemical Co., Inc.	FLD004574190	Final	35.35
FL	Hialeah	Northwest 58th Street Landfill	FLD980602643	Deleted	49.43
FL	Hialeah	Standard Auto Bumper Corp.	FLD004126520	Final	42.79
FL	Homestead Air Force Base	Homestead Air Force Base	FL7570024037	Final	42.40
FL	Indiantown	Florida Steel Corp.	FLD050432251	Final	45.92
FL	Jacksonville	Jacksonville Naval Air Station	FL6170024412	Final	32.08
FL	Jacksonville	Pickettville Road Landfill	FLD980556351	Final	42.94
FL	Jacksonville	Usn Air Station Cecil Field	FL5170022474	Final	31.99
FL	Lake Alfred	Callaway & Son Drum Service	FLD094590916	Final	46.22

Superfund National Priorities List - *continued*

St.	City/Area	Site Name	CERCLIS[1] ID	Status	Score[2]
FL	Lake Park	Bmi-Textron	FLD052172954	Deleted	35.34
FL	Lake Park	Trans Circuits, Inc.	FLD091471904	Final	50.00
FL	Lakeland	Landia Chemical Company	FLD042110841	Final	50.00
FL	Live Oak	Brown Wood Preserving	FLD980728935	Deleted	45.51
FL	Madison	Madison County Sanitary Landfill	FLD981019235	Final	37.93
FL	Marianna	United Metals, Inc.	FLD098924038	Final	33.73
FL	Medley	Pepper Steel & Alloys, Inc.	FLD032544587	Final	31.92
FL	Miami	Airco Plating Co.	FLD004145140	Final	42.47
FL	Miami	Anaconda Aluminum Co./Milgo Electronics Corp.	FLD020536538	Deleted	31.03
FL	Miami	Gold Coast Oil Corp.	FLD071307680	Deleted	57.80
FL	Miami	Miami Drum Services	FLD076027820	Final	53.56
FL	Miami	Varsol Spill	FLD980602346	Deleted	44.46
FL	Milton	Whiting Field Naval Air Station	FL2170023244	Final	50.00
FL	Mount Pleasant	Parramore Surplus	FLD041140344	Deleted	37.61
FL	North Miami	Munisport Landfill	FLD084535442	Deleted	32.37
FL	North Miami Beach	Anodyne, Inc.	FLD981014368	Final	31.03
FL	Orlando	Chevron Chemical Co. (Ortho Division)	FLD004064242	Final	50.00
FL	Orlando	City Industries, Inc.	FLD055945653	Final	32.00
FL	Palm Bay	Harris Corp. (Palm Bay Plant)	FLD000602334	Final	35.57
FL	Panama City	Tyndall Air Force Base	FL1570024124	Final	50.00
FL	Pembroke Park	Petroleum Products Corp.	FLD980798698	Final	40.11
FL	Pensacola	Agrico Chemical Co.	FLD980221857	Final	44.98
FL	Pensacola	American Creosote Works, Inc. (Pensacola Plant)	FLD008161994	Final	58.41
FL	Pensacola	Beulah Landfill	FLD980494660	Deleted	38.17
FL	Pensacola	Escambia Wood - Pensacola	FLD008168346	Final	50.00
FL	Pensacola	Pensacola Naval Air Station	FL9170024567	Final	42.40
FL	Plant City	Schuylkill Metals Corp.	FLD062794003	Deleted	59.16
FL	Pompano Beach	Chemform, Inc.	FLD080174402	Deleted	37.93
FL	Pompano Beach	Wilson Concepts of Florida, Inc.	FLD041184383	Deleted	37.93
FL	Port Salerno	Solitron Microwave	FLD045459526	Final	50.00
FL	Princeton	Woodbury Chemical Co. (Princeton Plant)	FLD004146346	Deleted	39.43
FL	Seffner	Taylor Road Landfill	FLD980494959	Final	51.37
FL	Tampa	Alaric Area GW Plume	FLD012978862	Final	41.91
FL	Tampa	Helena Chemical Co. (Tampa Plant)	FLD053502696	Final	30.19
FL	Tampa	Kassauf-Kimerling Battery Disposal	FLD980727820	Deleted	53.42
FL	Tampa	Mri Corp (Tampa)	FLD088787585	Final	37.62
FL	Tampa	Peak Oil Co./Bay Drum Co.	FLD004091807	Final	58.15
FL	Tampa	Reeves Southeastern Galvanizing Corp.	FLD000824896	Final	58.75
FL	Tampa	Sixty-Second Street Dump	FLD980728877	Deleted	49.09
FL	Tampa	Southern Solvents, Inc.	FL0001209840	Final	50.00
FL	Tampa	Stauffer Chemical Co (Tampa)	FLD004092532	Final	59.81
FL	Tampa	Tri-City Oil Conservationist, Inc	FLD070864541	Deleted	39.30
FL	Tarpon Springs	Stauffer Chemical Co. (Tarpon Springs)	FLD010596013	Final	50.00
FL	Temple Terrace	Normandy Park Apartments	FLD984229773	Proposed	50.00
FL	Vero Beach	Piper Aircraft Corp./Vero Beach Water & Sewer Department	FLD004054284	Final	31.13
FL	Warrington	Pioneer Sand Co.	FLD056116965	Deleted	51.97
FL	Whitehouse	Coleman-Evans Wood Preserving Co.	FLD991279894	Final	46.18
FL	Whitehouse	Whitehouse Oil Pits	FLD980602767	Final	52.58
FL	Zellwood	Zellwood Ground Water Contamination	FLD049985302	Final	51.91
FM	Palau	Fed Sta Micronesia, PCB Wastes	FMD980637987	Deleted	n/a
GA	Albany	Firestone Tire & Rubber Co. (Albany Plant)	GAD990855074	Final	30.08
GA	Albany	Marine Corps Logistics Base	GA7170023694	Final	44.65
GA	Albany	T.H. Agriculture & Nutrition Co. (Albany Plant)	GAD042101261	Final	40.93
GA	Athens	Luminous Processes, Inc.	GAD990855819	Deleted	n/a
GA	Augusta	Monsanto Corp. (Augusta Plant)	GAD001700699	Deleted	35.65
GA	Augusta	Peach Orchard Rd Pce Groundwater Plume Site	GAN000407449	Proposed	50.00
GA	Brunswick	Brunswick Wood Preserving	GAD981024466	Final	54.49
GA	Brunswick	Hercules 009 Landfill	GAD980556906	Final	52.58
GA	Brunswick	Lcp Chemicals Georgia	GAD099303182	Final	n/a
GA	Brunswick	Terry Creek Dredge Spoil Areas/Hercules Outfall	GAD982112658	Proposed	50.00

Superfund National Priorities List - *continued*

St.	City/Area	Site Name	CERCLIS[1] ID	Status	Score[2]
GA	Camilla	Camilla Wood Preserving Company	GAD008212409	Final	50.00
GA	Cedartown	Cedartown Industries, Inc.	GAD095840674	Final	42.00
GA	Cedartown	Cedartown Municipal Landfill	GAD980495402	Deleted	33.62
GA	Cedartown	Diamond Shamrock Corp. Landfill	GAD990741092	Final	35.60
GA	Fort Valley	Woolfolk Chemical Works, Inc.	GAD003269578	Final	42.24
GA	Houston County	Robins Air Force Base (Landfill #4/Sludge Lagoon)	GA1570024330	Final	51.66
GA	Kensington	Mathis Brothers Landfill (South Marble Top Road)	GAD980838619	Final	30.78
GA	Peach County	Powersville Site	GAD980496954	Final	35.53
GA	Tifton	Marzone Inc./Chevron Chemical Co.	GAD991275686	Final	30.26
GU	Agana	Ordot Landfill	GUD980637649	Final	n/a
GU	Yigo	Andersen Air Force Base	GU6571999519	Final	50.00
HI	Kunia	Del Monte Corp. (Oahu Plantation)	HID980637631	Final	50.00
HI	Pearl Harbor	Pearl Harbor Naval Complex	HI4170090076	Final	70.82
HI	Schofield	Schofield Barracks (USARMY)	HI7210090026	Deleted	28.90
HI	Wahiawa	Naval Computer & Telecommunications Area Master Station Eastern Pacific	HI0170090054	Final	50.00
IA	Camanche	Lawrence Todtz Farm	IAD000606038	Final	52.11
IA	Cedar Rapids	Electro-Coatings, Inc.	IAD005279039	Final	42.24
IA	Charles City	Labounty Site	IAD980631063	Deleted	70.73
IA	Charles City	Shaw Avenue Dump	IAD980630560	Final	30.01
IA	Charles City	White Farm Equipment Co. Dump	IAD065210734	Deleted	43.40
IA	Council Bluffs	Aidex Corp.	IAD042581256	Deleted	n/a
IA	Des Moines	Des Moines Tce	IAD980687933	Final	42.28
IA	Des Moines	Railroad Avenue Groundwater Contamination Site	IA0001610963	Final	50.00
IA	Dubuque	Peoples Natural Gas Co.	IAD980852578	Final	46.24
IA	Fairfield	Fairfield Coal Gasification Plant	IAD981124167	Final	38.05
IA	Hospers	Farmers' Mutual Cooperative	IAD022193577	Deleted	33.74
IA	Kellogg	Midwest Manufacturing/North Farm	IAD069625655	Final	32.04
IA	Keokuk	Sheller-Globe Corp. Disposal	IAD980630750	Deleted	33.66
IA	Mason City	Mason City Coal Gasification Plant	IAD980969190	Final	69.33
IA	Mason City	Northwestern States Portland Cement Co.	IAD980852461	Deleted	57.80
IA	Middletown	Iowa Army Ammunition Plant	IA7213820445	Final	29.73
IA	Orange City	Vogel Paint & Wax Co.	IAD980630487	Final	31.45
IA	Ottumwa	John Deere (Ottumwa Works Landfills)	IAD005291182	Deleted	42.32
IA	Red Oak	Red Oak City Landfill	IAD980632509	Final	34.13
IA	Sergeant Bluff	Mid-America Tanning Co.	IAD085824688	Deleted	47.91
IA	Waterloo	Waterloo Coal Gasification Plant	IAD984566356	Proposed	50.00
IA	West Point	E.I. Du Pont de Nemours & Co., Inc. (County Road X23)	IAD980685804	Deleted	46.01
ID	Idaho Falls	Idaho National Engineering Laboratory (USDOE)	ID4890008952	Final	51.91
ID	Lemhi County	Blackbird Mine	IDD980725832	Proposed	50.00
ID	Mountain Home	Mountain Home Air Force Base	ID3572124557	Final	57.80
ID	Pocatello	Eastern Michaud Flats Contamination	IDD984666610	Final	57.80
ID	Pocatello	Pacific Hide & Fur Recycling Co.	IDD098812878	Deleted	42.30
ID	Pocatello	Union Pacific Railroad Co.	IDD055030852	Deleted	53.47
ID	Rathdrum	Arrcom (Drexler Enterprises)	IDD000800961	Deleted	29.28
ID	Smelterville	Bunker Hill Mining & Metallurgical Complex	IDD048340921	Final	54.76
ID	Soda Springs	Kerr-Mcgee Chemical Corp. (Soda Springs Plant)	IDD041310707	Final	51.91
ID	Soda Springs	Monsanto Chemical Co. (Soda Springs Plant)	IDD081830994	Final	54.77
ID	St. Maries	St. Maries Creosote	IDSFN1002095	Proposed	50.00
ID	Stibnite	Stibnite/Yellow Pine Mining Area	IDD980665459	Proposed	50.00
IL	Antioch	H.O.D. Landfill	ILD980605836	Final	34.68
IL	Beckemeyer	Circle Smelting Corp.	ILD050231976	Proposed	71.00
IL	Belvidere	Belvidere Municipal Landfill	ILD980497663	Final	28.62
IL	Belvidere	Mig/Dewane Landfill	ILD980497788	Final	49.91
IL	Belvidere	Parsons Casket Hardware Co.	ILD005252432	Final	55.58
IL	Byron	Byron Salvage Yard	ILD010236230	Final	33.93
IL	Carterville	Sangamo Electric Dump/Crab Orchard National Wildlife Refuge (Usdoi)	IL8143609487	Final	43.70
IL	Depue	Depue/New Jersy Zinc/Mobil Chemical Corp.	ILD062340641	Final	70.71
IL	Dupage County	Kerr-Mcgee (Kress Creek/West Branch of Dupage River)	ILD980823991	Final	39.05
IL	East Cape Girardeau	Ilada Energy Co.	ILD980996789	Deleted	34.21
IL	Galesburg	Galesburg/Koppers Co.	ILD990817991	Final	34.78

Superfund National Priorities List - *continued*

St.	City/Area	Site Name	CERCLIS[1] ID	Status	Score[2]
IL	Granite City	Jennison-Wright Corporation	ILD006282479	Final	40.30
IL	Granite City	NI Industries/Taracorp Lead Smelter	ILD096731468	Final	38.11
IL	Greenup	A & F Material Reclaiming, Inc.	ILD980397079	Final	55.49
IL	Hegeler	Hegeler Zinc	ILN000508134	Final	50.00
IL	Joliet	Amoco Chemicals (Joliet Landfill)	ILD002994259	Final	39.44
IL	Joliet	Joliet Army Ammunition Plant (Load-Assembly-Packing Area)	IL0210090049	Final	35.23
IL	Joliet	Joliet Army Ammunition Plant (Manufacturing Area)	IL7213820460	Final	32.08
IL	La Salle	Lasalle Electric Utilities	ILD980794333	Final	42.06
IL	La Salle	Matthiessen & Hegeler Zinc Company	IL0000064782	Final	50.00
IL	Lawrenceville	Indian Refinery-Texaco Lawrenceville	ILD042671248	Final	56.67
IL	Lemont	Lenz Oil Service, Inc.	ILD005451711	Final	42.33
IL	Libertyville	Petersen Sand & Gravel	ILD003817137	Deleted	38.43
IL	Marshall	Velsicol Chemical Corp. (Marshall Plant)	ILD000814673	Final	48.78
IL	Morristown	Acme Solvent Reclaiming, Inc. (Morristown Plant)	ILD053219259	Final	31.98
IL	Ottawa	Ottawa Radiation Areas	ILD980606750	Final	50.00
IL	Pembroke Township	Cross Brothers Pail Recycling (Pembroke)	ILD980792303	Final	42.04
IL	Quincy	Adams County Quincy Landfills 2&3	ILD980607055	Final	34.21
IL	Rantoul	Chanute Air Force Base	IL1570024157	Proposed	48.00
IL	Rockford	Interstate Pollution Control, Inc.	ILT180011975	Final	46.01
IL	Rockford	Pagel's Pit	ILD980606685	Final	45.91
IL	Rockford	Southeast Rockford Ground Water Contamination	ILD981000417	Final	42.24
IL	Rockton	Beloit Corp.	ILD021440375	Final	52.08
IL	Sauget	Sauget & County Landfill (Site Q)	ILD000605790	Proposed	50.00
IL	Sauget	Sauget Area 1	ILD980792006	Proposed	50.00
IL	Savanna	Savanna Army Depot Activity	IL3210020803	Final	42.20
IL	South Elgin	Tri-County Landfill Co./Waste Management of Illinois, Inc.	ILD048306138	Final	42.76
IL	Taylorville	Central Illinois Public Service Co.	ILD981781065	Final	28.95
IL	Warrenville	Dupage County Landfill/Blackwell Forest Preserve	ILD980606305	Final	35.57
IL	Wauconda	Wauconda Sand & Gravel	ILD047019732	Final	53.42
IL	Waukegan	Johns-Manville Corp.	ILD005443544	Final	38.20
IL	Waukegan	Outboard Marine Corp.	ILD000802827	Final	n/a
IL	Waukegan	Yeoman Creek Landfill	ILD980500102	Final	33.23
IL	West Chicago	Kerr-Mcgee (Reed-Keppler Park)	ILD980824007	Final	39.51
IL	West Chicago	Kerr-Mcgee (Residential Areas)	ILD980824015	Final	38.15
IL	West Chicago	Kerr-Mcgee (Sewage Treatment Plant)	ILD980824031	Final	35.20
IL	Winnebago County	Evergreen Manor Ground Water Contamination	ILD984836734	Proposed	50.00
IL	Woodstock	Woodstock Municipal Landfill	ILD980605943	Final	50.10
IN	Bloomington	Bennett Stone Quarry	IND006418651	Final	32.55
IN	Bloomington	Lemon Lane Landfill	IND980794341	Final	29.31
IN	Bloomington	Neal's Landfill (Bloomington)	IND980614556	Final	42.93
IN	Claypool	Lakeland Disposal Service, Inc.	IND064703200	Final	34.10
IN	Columbia City	Wayne Waste Oil	IND048989479	Final	42.33
IN	Columbus	Columbus Old Municipal Landfill #1	IND980607626	Final	45.31
IN	Columbus	Tri-State Plating	IND006038764	Deleted	29.28
IN	East Chicago	U.S. Smelter & Lead Refinery, Inc.	IND047030226	Proposed	71.00
IN	Elkhart	Conrail Rail Yard (Elkhart)	IND000715490	Final	42.24
IN	Elkhart	Himco Dump	IND980500292	Final	42.31
IN	Elkhart	Main Street Well Field	IND980794358	Final	42.49
IN	Evansville	Jacobsville Neighborhood Soil Contamination	INN000508142	Final	35.52
IN	Fort Wayne	Fort Wayne Reduction Dump	IND980679542	Final	42.47
IN	Gary	Lake Sandy Jo (M&M Landfill)	IND980500524	Final	38.21
IN	Gary	Midco I	IND980615421	Final	46.44
IN	Gary	Midco Ii	IND980679559	Final	30.16
IN	Gary	Ninth Avenue Dump	IND980794432	Final	40.32
IN	Griffith	American Chemical Service, Inc.	IND016360265	Final	34.98
IN	Hancock County	Poer Farm	IND980684583	Deleted	37.38
IN	Indianapolis	Carter Lee Lumber Co.	IND016395899	Deleted	35.40
IN	Indianapolis	Reilly Tar & Chemical Corp. (Indianapolis Plant)	IND000807107	Final	34.03
IN	Indianapolis	Southside Sanitary Landfill	IND980607360	Deleted	41.94
IN	Kokomo	Continental Steel Corp.	IND001213503	Final	31.85

Superfund National Priorities List - *continued*

St.	City/Area	Site Name	CERCLIS[1] ID	Status	Score[2]
IN	La Porte	Fisher-Calo	IND074315896	Final	52.05
IN	Lafayette	Tippecanoe Sanitary Landfill, Inc.	IND980997639	Final	42.24
IN	Lebanon	Wedzeb Enterprises, Inc.	IND980794374	Deleted	31.27
IN	Marion	Marion (Bragg) Dump	IND980794366	Final	35.25
IN	Michigan City	Waste, Inc., Landfill	IND980504005	Final	50.63
IN	Mishawaka	Douglass Road/Uniroyal, Inc., Landfill	IND980607881	Final	36.61
IN	Osceola	Galen Myers Dump/Drum Salvage	IND980999635	Final	42.24
IN	Seymour	Seymour Recycling Corp.	IND040313017	Final	n/a
IN	South Bend	Whiteford Sales & Service Inc./Nationalease	IND980999791	Deleted	51.87
IN	Spencer	Neal's Dump (Spencer)	IND980794549	Deleted	36.55
IN	Terre Haute	International Minerals (E. Plant)	INT190010876	Deleted	57.80
IN	Vincennes	Prestolite Battery Division	IND006377048	Final	40.63
IN	Westville	Cam-Or Inc.	IND005480462	Final	58.91
IN	Zionsville	Envirochem Corp.	IND084259951	Final	46.44
IN	Zionsville	Northside Sanitary Landfill, Inc	IND050530872	Final	46.04
KS	Arkansas City	Arkansas City Dump	KSD980500789	Deleted	n/a
KS	Cherokee County	Cherokee County	KSD980741862	Final	58.15
KS	Colby	Ace Services	KSD046746731	Final	50.00
KS	Cowley County	Strother Field Industrial Park	KSD980862726	Final	33.62
KS	Delavan	Tri-County Public Airport	KS0001402320	Proposed	50.00
KS	Desoto	Sunflower Army Ammunition Plant	KS3213820878	Proposed	50.00
KS	El Dorado	Pester Refinery Co.	KSD000829846	Final	30.16
KS	Hutchinson	Obee Road	KSD980631766	Final	33.62
KS	Junction City	Fort Riley	KS6214020756	Final	33.79
KS	Olathe	Chemical Commodities, Inc.	KSD031349624	Final	50.00
KS	Shawnee Mission	Doepke Disposal (Holliday)	KSD980632301	Final	47.46
KS	Topeka	Hydro-Flex Inc.	KSD007135429	Deleted	42.79
KS	Wichita	29th & Mead Ground Water Contamination	KSD007241656	Deleted	35.35
KS	Wichita	Big River Sand Co.	KSD980686174	Deleted	32.56
KS	Wichita	Johns' Sludge Pond	KSD980631980	Deleted	35.94
KS	Wichita Heights	57th & North Broadway Streets Site	KSD981710247	Final	50.00
KS	Wright	Wright Ground Water Contamination	KSD984985929	Final	50.00
KY	Auburn	Caldwell Lace Leather Co., Inc.	KYD045738291	Final	34.21
KY	Brooks	A.L. Taylor (Valley of Drums)	KYD980500961	Deleted	n/a
KY	Brooks	Smith's Farm	KYD097267413	Final	32.69
KY	Calvert City	Airco	KYD041981010	Final	33.29
KY	Calvert City	B.F. Goodrich	KYD006370167	Final	33.01
KY	Dayhoit	National Electric Coil Co./Cooper Industries	KYD985069954	Final	50.00
KY	Hawesville	National Southwire Aluminum Co.	KYD049062375	Final	50.00
KY	Hillsboro	Maxey Flats Nuclear Disposal	KYD980729107	Final	31.71
KY	Howe Valley	Howe Valley Landfill	KYD980501191	Deleted	36.73
KY	Island	Brantley Landfill	KYD980501019	Final	52.73
KY	Jefferson County	Distler Farm	KYD980601975	Final	34.62
KY	Louisville	Lee's Lane Landfill	KYD980557052	Deleted	39.52
KY	Maceo	Green River Disposal, Inc.	KYD980501076	Final	29.12
KY	Mayfield	General Tire & Rubber Co. (Mayfield Landfill)	KYD006371074	Deleted	32.94
KY	Newport	Newport Dump	KYD985066380	Deleted	37.63
KY	Olaton	Fort Hartford Coal Co. Stone Quarry	KYD980844625	Final	43.84
KY	Paducah	Paducah Gaseous Diffusion Plant (USDOE)	KY8890008982	Final	56.95
KY	Peewee Valley	Red Penn Sanitation Co. Landfill	KYD981469794	Deleted	38.10
KY	Shepherdsville	Tri-City Disposal Co.	KYD981028350	Final	33.82
KY	West Point	Distler Brickyard	KYD980602155	Final	44.77
LA	Abbeville	D.L. Mud, Inc.	LAD981058019	Deleted	32.37
LA	Abbeville	Gulf Coast Vacuum Services	LAD980750137	Deleted	42.78
LA	Abbeville	Pab Oil & Chemical Service, Inc.	LAD980749139	Deleted	38.94
LA	Alexandria	Ruston Foundry	LAD985185107	Final	43.17
LA	Ascension Parish	Dutchtown Treatment Plant	LAD980879449	Deleted	36.41
LA	Bayou Sorrel	Bayou Sorrel	LAD980745541	Deleted	34.69
LA	Bossier City	Highway 71/72 Refinery	LAD981054075	Proposed	50.00
LA	Darrow	Old Inger Oil Refinery	LAD980745533	Final	n/a
LA	Denham Springs	Combustion, Inc.	LAD072606627	Final	33.79

Superfund National Priorities List - *continued*

St.	City/Area	Site Name	CERCLIS[1] ID	Status	Score[2]
LA	Doyline	Louisiana Army Ammunition Plant	LA0213820533	Final	30.26
LA	Grand Cheniere	Mallard Bay Landing Bulk Plant	LA0000187518	Final	48.55
LA	Lake Charles	Gulf State Utilities-North Ryan Street	LAD985169317	Proposed	50.00
LA	Madisonville	Madisonville Creosote Works	LAD981522998	Final	48.01
LA	Marion	Marion Pressure Treating	LAD008473142	Final	50.00
LA	New Orleans	Agriculture Street Landfill	LAD981056997	Final	50.00
LA	Ponchatoula	Delatte Metals	LAD052510344	Final	50.00
LA	Scotlandville	Ewell Property-Devil's Swamp	LAD981155872	Proposed	50.00
LA	Scotlandville	Petro-Processors of Louisiana, Inc.	LAD057482713	Final	41.44
LA	Slaughter	Central Wood Preserving Co.	LAD008187940	Final	48.53
LA	Slidell	Bayou Bonfouca	LAD980745632	Final	29.78
LA	Slidell	Southern Shipbuilding	LAD008149015	Deleted	50.00
LA	Sorrento	Cleve Reber	LAD980501456	Deleted	48.80
LA	Winnfield	American Creosote Works, Inc. (Winnfield Plant)	LAD000239814	Final	50.70
MA	Acton	W.R. Grace & Co., Inc. (Acton Plant)	MAD001002252	Final	59.31
MA	Ashland	Nyanza Chemical Waste Dump	MAD990685422	Final	69.22
MA	Bedford	Hanscom Field/Hanscom Air Force Base	MA8570024424	Final	50.00
MA	Bedford	Naval Weapons Industrial Reserve Plant	MA6170023570	Final	50.00
MA	Billerica	Iron Horse Park	MAD051787323	Final	42.93
MA	Bridgewater	Cannon Engineering Corp. (Cec)	MAD079510780	Final	39.89
MA	Concord	Nuclear Metals, Inc.	MAD062166335	Final	58.31
MA	Dartmouth	Re-Solve, Inc.	MAD980520621	Final	47.71
MA	Fairhaven	Atlas Tack Corp.	MAD001026319	Final	42.60
MA	Falmouth	Otis Air National Guard Base/Camp Edwards	MA2570024487	Final	45.92
MA	Fort Devens	Fort Devens	MA7210025154	Final	42.24
MA	Groveland	Groveland Wells	MAD980732317	Final	40.74
MA	Haverhill	Haverhill Municipal Landfill	MAD980523336	Final	30.29
MA	Holbrook	Baird & Mcguire	MAD001041987	Final	66.35
MA	Lanesboro	Rose Disposal Pit	MAD980524169	Final	33.03
MA	Lowell	Silresim Chemical Corp.	MAD000192393	Final	42.72
MA	Mansfield	Hatheway & Patterson	MAD001060805	Final	56.60
MA	Natick	Natick Laboratory Army Research, Development, & Engineering Center	MA1210020631	Final	50.00
MA	New Bedford	New Bedford	MAD980731335	Final	n/a
MA	New Bedford	Sullivan's Ledge	MAD980731343	Final	32.77
MA	Norton/Attleboro	Shpack Landfill	MAD980503973	Final	29.45
MA	Norwood	Norwood PCBs	MAD980670566	Final	29.43
MA	Palmer	Psc Resources	MAD980731483	Final	38.66
MA	Pittsfield	Ge - Housatonic River	MAD002084093	Proposed	71.00
MA	Plymouth	Plymouth Harbor/Cannon Engineering Corp.	MAD980525232	Deleted	54.82
MA	Salem	Salem Acres	MAD980525240	Deleted	34.94
MA	Sudbury	Fort Devens-Sudbury Training Annex	MAD980520670	Deleted	35.57
MA	Tewksbury	Sutton Brook Disposal Area	MAD980520696	Final	57.12
MA	Tyngsborough	Charles-George Reclamation Trust Landfill	MAD003809266	Final	47.20
MA	Walpole	Blackburn & Union Privileges	MAD982191363	Final	50.00
MA	Watertown	Materials Technology Laboratory (USARMY)	MA0213820939	Final	48.57
MA	Westborough	Hocomonco Pond	MAD980732341	Final	44.80
MA	Weymouth	South Weymouth Naval Air Station	MA2170022022	Final	50.00
MA	Woburn	Industri-Plex	MAD076580950	Final	72.42
MA	Woburn	Wells G&H	MAD980732168	Final	42.71
MD	Aberdeen	Aberdeen Proving Ground (Michaelsville Landfill)	MD3210021355	Final	31.09
MD	Abingdon	Bush Valley Landfill	MDD980504195	Final	40.30
MD	Andrews Air Force Base	Andrews Air Force Base	MD0570024000	Final	50.00
MD	Annapolis	Middletown Road Dump	MDD980705099	Deleted	29.36
MD	Baltimore	Chemical Metals Industries, Inc.	MDD980555478	Deleted	n/a
MD	Baltimore	Curtis Bay Coast Guard Yard	MD4690307844	Final	50.00
MD	Baltimore	Kane & Lombard Street Drums	MDD980923783	Final	30.15
MD	Beltsville	Beltsville Agricultural Research Center (Usda)	MD0120508940	Final	50.00
MD	Brandywine	Brandywine Drmo	MD9570024803	Final	50.15
MD	Colora	Woodlawn County Landfill	MDD980504344	Final	48.13

Superfund National Priorities List - *continued*

St.	City/Area	Site Name	CERCLIS[1] ID	Status	Score[2]
MD	Cumberland	Limestone Road	MDD980691588	Final	30.54
MD	Edgewood	Aberdeen Proving Ground (Edgewood Area)	MD2210020036	Final	53.57
MD	Elkton	Sand, Gravel & Stone	MDD980705164	Final	41.08
MD	Elkton	Spectron, Inc.	MDD000218008	Final	51.42
MD	Hagerstown	Central Chemical (Hagerstown)	MDD003061447	Final	50.00
MD	Harmans	Mid-Atlantic Wood Preservers, Inc.	MDD064882889	Deleted	42.31
MD	Hollywood	Southern Maryland Wood Treating	MDD980704852	Deleted	34.21
MD	Indian Head	Indian Head Naval Surface Warfare Center	MD7170024684	Final	50.00
MD	North East	Ordnance Products, Inc.	MDD982364341	Final	32.15
MD	Odenton	Fort George G. Meade	MD9210020567	Final	51.44
MD	Patuxent River	Patuxent River Naval Air Station	MD7170024536	Final	50.00
MD	Rosedale	68th Street Dump/Industrial Enterprises	MDD980918387	Proposed	50.00
ME	Augusta	O'connor Co.	MED980731475	Final	31.86
ME	Brooksville (Cape Rosier)	Callahan Mining Corp	MED980524128	Final	50.00
ME	Brunswick	Brunswick Naval Air Station	ME8170022018	Final	43.38
ME	Corinna	Eastland Woolen Mill	MED980915474	Final	70.71
ME	Gray	Mckin Co.	MED980524078	Final	60.97
ME	Kittery	Portsmouth Naval Shipyard	ME7170022019	Final	50.00
ME	Limestone	Loring Air Force Base	ME9570024522	Final	34.49
ME	Meddybemps	Eastern Surplus	MED981073711	Final	50.00
ME	Plymouth	West Site/Hows Corners	MED985466168	Final	50.00
ME	Saco	Saco Municipal Landfill	MED980504393	Final	29.49
ME	Saco	Saco Tannery Waste Pits	MED980520241	Deleted	43.19
ME	South Hope	Union Chemical Co., Inc.	MED042143883	Final	32.11
ME	Washburn	Pinette's Salvage Yard	MED980732291	Deleted	33.98
ME	Winthrop	Winthrop Landfill	MED980504435	Final	35.62
MI	Adrian	Anderson Development Co.	MID002931228	Deleted	31.02
MI	Albion	Albion-Sheridan Township Landfill	MID980504450	Final	33.79
MI	Albion	Mcgraw Edison Corp.	MID005339676	Final	33.42
MI	Allegan	Rockwell International Corp. (Allegan Plant)	MID006028062	Final	52.15
MI	Battle Creek	Verona Well Field	MID980793806	Final	46.86
MI	Bay City	Bay City Middlegrounds	MID981092935	Proposed	50.00
MI	Belding	H & K Sales	MI0001271535	Deleted	n/a
MI	Benton Harbor	Aircraft Components (D & L Sales)	MI0001119106	Final	n/a
MI	Brighton	Rasmussen's Dump	MID095402210	Final	31.80
MI	Bronson	North Bronson Industrial Area	MID005480900	Final	33.93
MI	Buchanan	Electrovoice	MID005068143	Final	35.36
MI	Cadillac	Kysor Industrial Corp.	MID043681840	Final	33.94
MI	Cadillac	Northernaire Plating	MID020883609	Final	57.93
MI	Charlevoix	Charlevoix Municipal Well	MID980794390	Deleted	37.94
MI	Clare	Clare Water Supply	MID980002273	Final	38.43
MI	Dalton Township	Duell & Gardner Landfill	MID980504716	Final	34.68
MI	Dalton Township	Ott/Story/Cordova Chemical Co.	MID060174240	Final	53.41
MI	Davisburg	Springfield Township Dump	MID980499966	Final	51.97
MI	Detroit	Carter Industrials, Inc.	MID980274179	Deleted	37.79
MI	Filer City	Packaging Corp. of America	MID980794747	Final	51.91
MI	Grand Ledge	Parsons Chemical Works, Inc.	MID980476907	Final	31.32
MI	Grand Rapids	Butterworth #2 Landfill	MID062222997	Final	50.31
MI	Grand Rapids	Folkertsma Refuse	MID980609366	Deleted	33.12
MI	Grand Rapids	H. Brown Co., Inc.	MID017075136	Final	39.88
MI	Grand Rapids	State Disposal Landfill, Inc.	MID980609341	Final	42.24
MI	Grandville	Organic Chemicals, Inc.	MID990858003	Final	32.93
MI	Green Oak Township	Spiegelberg Landfill	MID980794481	Final	53.61
MI	Greilickville	Grand Traverse Overall Supply Co.	MID017418559	Final	35.53
MI	Hartford	Burrows Sanitation	MID980410617	Final	30.59
MI	Highland	Hi-Mill Manufacturing Co.	MID005341714	Final	49.54
MI	Holland	Waste Management of Michigan (Holland Lagoons)	MID060179587	Final	37.20
MI	Houghton County	Torch Lake	MID980901946	Final	46.72
MI	Howard Township	U.S. Aviex	MID980794556	Final	33.66
MI	Howell	Shiawassee River	MID980794473	Final	31.01

Superfund National Priorities List - *continued*

St.	City/Area	Site Name	CERCLIS[1] ID	Status	Score[2]
MI	Ionia	American Anodco, Inc.	MID006029102	Final	57.99
MI	Ionia	Ionia City Landfill	MID980794416	Final	31.31
MI	Kalamazoo	Allied Paper, Inc./Portage Creek/Kalamazoo River	MID006007306	Final	36.41
MI	Kalamazoo	Auto Ion Chemicals, Inc.	MID980794382	Final	32.07
MI	Kalamazoo	Michigan Disposal Service (Cork Street Landfill)	MID000775957	Final	37.93
MI	Kalamazoo	Roto-Finish Co., Inc.	MID005340088	Final	40.70
MI	Kent City	Kent City Mobile Home Park	MID981089915	Deleted	33.62
MI	Kentwood	Kentwood Landfill	MID000260281	Final	35.39
MI	Lake Ann	Metal Working Shop	MID980992952	Deleted	28.82
MI	Lansing	Adam's Plating	MID006522791	Final	29.64
MI	Lansing	Barrels, Inc.	MID017188673	Final	42.24
MI	Lansing Township	Motor Wheel, Inc.	MID980702989	Final	48.91
MI	Macomb Township	South Macomb Disposal Authority (Landfills #9 & #9a)	MID069826170	Final	33.67
MI	Mancelona Township	Tar Lake	MID980794655	Final	48.55
MI	Marquette	Cliff/Dow Dump	MID980608970	Deleted	34.50
MI	Metamora	Metamora Landfill	MID980506562	Final	35.51
MI	Muskegon	Bofors Nobel, Inc.	MID006030373	Final	53.42
MI	Muskegon	Kaydon Corp.	MID006016703	Final	34.21
MI	Muskegon	Peerless Plating Co.	MID006031348	Final	43.94
MI	Muskegon	Thermo-Chem, Inc.	MID044567162	Final	53.36
MI	Muskegon Heights	Sca Independent Landfill	MID000724930	Final	34.75
MI	Oscoda	Hedblum Industries	MID980794408	Final	37.29
MI	Oscoda	Wurtsmith Air Force Base	MI5570024278	Proposed	50.00
MI	Oshtemo Township	K&L Avenue Landfill	MID980506463	Final	38.10
MI	Ossineke	Ossineke Ground Water Contamination	MID980794440	Deleted	33.78
MI	Otisville	Forest Waste Products	MID980410740	Final	38.64
MI	Park Township	Southwest Ottawa County Landfill	MID980608780	Final	39.66
MI	Pere Marquette Twp	Mason County Landfill	MID980794465	Deleted	34.18
MI	Petoskey	Petoskey Municipal Well Field	MID006013049	Final	42.68
MI	Pleasant Plains Twp	Wash King Laundry	MID980701247	Final	40.03
MI	Rochester Hills	J & L Landfill	MID980609440	Final	31.65
MI	Rose Center	Cemetery Dump	MID980794663	Deleted	34.16
MI	Rose Township	Rose Township Dump	MID980499842	Final	50.92
MI	Sault Ste Marie	Cannelton Industries, Inc.	MID980678627	Final	30.16
MI	Sparta Township	Sparta Landfill	MID000268136	Final	32.00
MI	St. Joseph	Bendix Corp./Allied Automotive	MID005107222	Final	37.27
MI	St. Louis	Gratiot County Golf Course	MID980794531	Deleted	40.22
MI	St. Louis	Gratiot County Landfill	MID980506281	Final	n/a
MI	St. Louis	Velsicol Chemical Corp. (Michigan)	MID000722439	Final	52.29
MI	Sturgis	Sturgis Municipal Wells	MID980703011	Final	42.24
MI	Swartz Creek	Berlin & Farro	MID000605717	Deleted	66.74
MI	Temperance	Novaco Industries	MID084566900	Deleted	38.20
MI	Utica	G&H Landfill	MID980410823	Final	49.09
MI	Utica	Liquid Disposal, Inc.	MID067340711	Final	63.28
MI	Whitehall	Muskegon Chemical Co.	MID072569510	Final	34.19
MI	Whitehall	Whitehall Municipal Wells	MID980701254	Deleted	35.45
MI	Wyandotte	Lower Ecorse Creek Dump	MID985574227	Deleted	n/a
MI	Wyoming	Spartan Chemical Co.	MID079300125	Final	41.05
MI	Wyoming Township	Chem Central	MID980477079	Final	38.20
MN	Adrian	Adrian Municipal Well Field	MND980904023	Deleted	33.62
MN	Andover	South Andover Site	MND980609614	Final	35.41
MN	Andover	Waste Disposal Engineering	MND980609119	Deleted	50.92
MN	Baytown Township	Baytown Township Ground Water Plume	MND982425209	Final	35.62
MN	Bemidji	Kummer Sanitary Landfill	MND980904049	Deleted	35.57
MN	Brainerd/Baxter	Burlington Northern (Brainerd/Baxter Plant)	MND000686196	Final	46.77
MN	Brooklyn Center	Joslyn Manufacturing & Supply Co.	MND044799856	Final	44.30
MN	Burnsville	Freeway Sanitary Landfill	MND038384004	Final	45.91
MN	Cannon Falls	Dakhue Sanitary Landfill	MND981191570	Deleted	42.24
MN	Cass Lake	St. Regis Paper Co.	MND057597940	Final	52.88
MN	Dakota County	Pine Bend Sanitary Landfill	MND000245795	Deleted	52.11
MN	East Bethel Township	East Bethel Demolition Landfill	MND981088180	Deleted	28.75

Superfund National Priorities List - *continued*

St.	City/Area	Site Name	CERCLIS[1] ID	Status	Score[2]
MN	Fairview Township	Agate Lake Scrapyard	MND980898068	Deleted	29.68
MN	Faribault	Nutting Truck & Caster Co.	MND006154017	Final	37.87
MN	Fridley	Boise Cascade/Onan Corp./Medtronics, Inc.	MND053417515	Deleted	50.06
MN	Fridley	Fmc Corp. (Fridley Plant)	MND006481543	Final	65.50
MN	Fridley	Fridley Commons Park Well Field	MND985701309	Final	50.00
MN	Fridley	Kurt Manufacturing Co.	MND059680165	Final	31.41
MN	Fridley	Naval Industrial Reserve Ordnance Plant	MN3170022914	Final	30.83
MN	Hermantown	Arrowhead Refinery Co.	MND980823975	Final	43.75
MN	Lagrand Township	Lagrand Sanitary Landfill	MND981090483	Deleted	37.51
MN	Lake Elmo	Washington County Landfill	MND980704738	Deleted	42.24
MN	Lehillier	Lehillier/Mankato	MND980792469	Final	42.49
MN	Long Prairie	Long Prairie Ground Water Contamination	MND980904072	Final	31.94
MN	Minneapolis	General Mills/Henkel Corp.	MND051441731	Final	36.28
MN	Minneapolis	Twin Cities Air Force Reserve Base (Small Arms Range Landfill)	MN8570024275	Deleted	33.62
MN	Minneapolis	Union Scrap Iron & Metal Co.	MND022949192	Deleted	42.63
MN	Minneapolis	Whittaker Corp.	MND006252233	Deleted	40.03
MN	Morris	Morris Arsenic Dump	MND980792287	Deleted	38.27
MN	New Brighton	Macgillis & Gibbs Co./Bell Lumber & Pole Co.	MND006192694	Final	48.33
MN	New Brighton	New Brighton/Arden Hills/Tcaap (USARMY)	MN7213820908	Final	59.16
MN	Oak Grove Township	Oak Grove Sanitary Landfill	MND980904056	Deleted	43.40
MN	Oakdale	Oakdale Dump	MND980609515	Final	55.71
MN	Oronoco	Olmsted County Sanitary Landfill	MND000874354	Deleted	40.70
MN	Perham	Perham Arsenic Site	MND980609572	Final	37.98
MN	Pine Bend	Koch Refining Co./N-Ren Corp.	MND000686071	Deleted	31.14
MN	Rosemount	University of Minnesota (Rosemount Research Center)	MND980613780	Deleted	45.91
MN	Sebeka	Ritari Post & Pole	MND980904064	Final	29.81
MN	St. Augusta Township	St. Augusta Sanitary Landfill/Engen Dump	MND981002256	Deleted	33.85
MN	St. Louis County	St. Louis River Site	MND039045430	Final	32.08
MN	St. Louis Park	NI Industries/Taracorp/Golden Auto	MND097891634	Deleted	39.97
MN	St. Louis Park	Reilly Tar & Chemical Corp. (St. Louis Park Plant)	MND980609804	Final	n/a
MN	St. Paul	Koppers Coke	MND000819359	Final	55.05
MN	Waite Park	Waite Park Wells	MND981002249	Final	31.94
MN	Windom	Windom Dump	MND980034516	Deleted	38.17
MO	Amazonia	Wheeling Disposal Service Co., Inc., Landfill	MOD000830554	Deleted	48.58
MO	Annapolis	Annapolis Lead Mine	MO0000958611	Final	56.67
MO	Bridgeton	Westlake Landfill	MOD079900932	Final	29.85
MO	Cape Girardeau	Kem-Pest Laboratories	MOD980631113	Deleted	33.89
MO	Cape Girardeau	Missouri Electric Works	MOD980965982	Final	31.20
MO	Desloge	Big River Mine Tailings/St. Joe Minerals Corp.	MOD981126899	Final	84.91
MO	Ellisville	Ellisville Site	MOD980633010	Final	n/a
MO	Fredericktown	Madison Mine (Anschutz Mining Corp)	MOD098633415	Final	58.41
MO	Granby	Newton County Mine Tailings	MOD981507585	Final	50.00
MO	Imperial	Minker/Stout/Romaine Creek	MOD980741912	Final	36.78
MO	Independence	Lake City Army Ammunition Plant (Northwest Lagoon)	MO3213890012	Final	33.62
MO	Jasper County	Oronogo-Duenweg Mining Belt	MOD980686281	Final	46.20
MO	Joplin	Newton County Wells	MOD985798339	Final	50.00
MO	Kansas City	Conservation Chemical Co.	MOD000829705	Final	29.85
MO	Liberty	Lee Chemical	MOD980853519	Final	46.81
MO	Malden	Bee Cee Manufacturing Co.	MOD980860522	Final	28.59
MO	Moscow Mills	Shenandoah Stables	MOD980685838	Deleted	30.09
MO	Neosho	Pools Prairie	MO0000958835	Final	50.00
MO	New Haven	Riverfront	MOD981720246	Final	50.00
MO	North Kansas City	Armour Road	MOD046750253	Final	50.00
MO	Oak Grove	Oak Grove Village Well	MOD981717036	Final	50.00
MO	Republic	Solid State Circuits, Inc.	MOD980854111	Final	37.93
MO	Sikeston	Quality Plating	MOD980860555	Final	40.70
MO	Springfield	Fulbright Landfill	MOD980631139	Final	40.60
MO	Springfield	North-U Drive Well Contamination	MOD007163108	Deleted	28.90
MO	St. Charles County	Weldon Spring Former Army Ordnance Works	MO5210021288	Final	58.60
MO	St. Charles County	Weldon Spring Quarry/Plant/Pits (Usdoe/Army)	MO3210090004	Final	30.26
MO	St. Louis	St. Louis Airport/Hazelwood Interim Storage/Futura Coatings Co.	MOD980633176	Final	38.31

Superfund National Priorities List - *continued*

St.	City/Area	Site Name	CERCLIS[1] ID	Status	Score[2]
MO	Times Beach	Times Beach	MOD980685226	Deleted	40.08
MO	Valley Park	Valley Park Tce	MOD980968341	Final	35.57
MO	Verona	Syntex Facility	MOD007452154	Final	43.78
MP	Garapan	PCB Warehouse	MPD980798318	Deleted	n/a
MS	Columbia	Newsom Brothers/Old Reichhold Chemicals, Inc.	MSD980840045	Deleted	45.70
MS	Flowood	Flowood Site	MSD980710941	Deleted	n/a
MS	Greenville	Walcotte Chemical Co. Warehouses	MSD980601736	Deleted	n/a
MS	Gulfport	Chemfax, Inc.	MSD008154486	Proposed	39.00
MS	Hattiesburg	Davis Timber Company	MSD046497012	Final	48.57
MS	Louisville	American Creosote Works Inc	MSD004006995	Final	62.20
MS	Picayune	Picayune Wood Treating Site	MSD065490930	Final	51.03
MS	Wesson	Potter Co.	MSD056029648	Proposed	50.00
MT	Anaconda	Anaconda Co. Smelter	MTD093291656	Final	58.71
MT	Basin	Basin Mining Area	MTD982572562	Final	61.15
MT	Billings	Lockwood Solvent Ground Water Plume	MT0007623052	Final	45.69
MT	Bozeman	Idaho Pole Co.	MTD006232276	Final	38.29
MT	Butte	Montana Pole & Treating	MTD006230635	Final	33.03
MT	Butte	Silver Bow Creek/Butte Area	MTD980502777	Final	63.76
MT	Columbus	Mouat Industries	MTD021997689	Final	31.66
MT	East Helena	East Helena Site	MTD006230346	Final	61.65
MT	Great Falls	Barker Hughesville Mining District	MT6122307485	Final	50.00
MT	Helena	Upper Tenmile Creek Mining Area	MTSFN7578012	Final	50.00
MT	Libby	Libby Asbestos Site	MT0009083840	Final	n/a
MT	Libby	Libby Ground Water Contamination	MTD980502736	Final	37.67
MT	Livingston	Burlington Northern Livingston Shop Complex	MTD986066025	Proposed	50.00
MT	Milltown	Milltown Reservoir Sediments	MTD980717565	Final	43.78
MT	Neihart	Carpenter Snow Creek Mining District	MT0001096353	Final	50.00
NC	210 Miles of Roads	Roadside PCB Spill	NCD980602163	Deleted	n/a
NC	Aberdeen	Aberdeen Pesticide Dumps	NCD980843346	Final	52.70
NC	Aberdeen	Geigy Chemical Corp. (Aberdeen Plant)	NCD981927502	Final	33.02
NC	Arden	Blue Ridge Plating Company	NCD044447589	Proposed	38.67
NC	Belmont	Jadco-Hughes Facility	NCD980729602	Final	42.00
NC	Castle Hayne	Reasor Chemical Company	NCD986187094	Final	32.14
NC	Charlotte	Martin-Marietta, Sodyeco, Inc.	NCD001810365	Final	51.93
NC	Charlotte	Ram Leather Care Site	NCD982096653	Final	40.43
NC	Concord	Bypass 601 Ground Water Contamination	NCD044440303	Final	37.93
NC	Cordova	Charles Macon Lagoon & Drum Storage	NCD980840409	Final	47.10
NC	East Flat Rock	General Electric Co/Shepherd Farm	NCD079044426	Final	70.71
NC	Fayetteville	Cape Fear Wood Preserving	NCD003188828	Final	34.09
NC	Fayetteville	Carolina Transformer Co.	NCD003188844	Final	33.76
NC	Gastonia	Davis Park Road Tce	NCD986175644	Final	33.50
NC	Havelock	Cherry Point Marine Corps Air Station	NC1170027261	Final	70.71
NC	Hazelwood	Benfield Industries, Inc.	NCD981026479	Final	31.67
NC	Jacksonville	Abc One Hour Cleaners	NCD024644494	Final	29.11
NC	Maco	Potter's Septic Tank Service Pits	NCD981023260	Final	29.14
NC	Morrisville	Koppers Co., Inc. (Morrisville Plant)	NCD003200383	Final	41.89
NC	North Belmont	North Belmont Pce	NCD986187128	Final	50.00
NC	Onslow County	Camp Lejeune Military Res. (USNAVY)	NC6170022580	Final	33.13
NC	Oxford	Jfd Electronics/Channel Master	NCD122263825	Final	39.03
NC	Raleigh	North Carolina State University (Lot 86, Farm Unit #1)	NCD980557656	Final	48.36
NC	Raleigh	Ward Transformer	NCD003202603	Final	50.00
NC	Salisbury	National Starch & Chemical Corp.	NCD991278953	Final	46.51
NC	Shelby	Celanese Corp. (Shelby Fiber Operations)	NCD003446721	Final	48.98
NC	Statesville	Fcx, Inc. (Statesville Plant)	NCD095458527	Final	37.93
NC	Statesville	Sigmon's Septic Tank Service	NCD062555792	Final	30.00
NC	Swannanoa	Chemtronics, Inc.	NCD095459392	Final	30.16
NC	Washington	Fcx, Inc. (Washington Plant)	NCD981475932	Final	40.39
NC	Waynesville	Barber Orchard	NCSFN0406989	Final	70.71
NC	Wilmington	New Hanover Cnty Airport Burn Pit	NCD981021157	Final	39.39
ND	Lidgerwood	Wyndmere, RutzArsenic Trioxide Site	NDD980716963	Deleted	n/a
ND	Minot	Minot Landfill	NDD980959548	Deleted	33.58

Superfund National Priorities List - *continued*

St.	City/Area	Site Name	CERCLIS[1] ID	Status	Score[2]
NE	Bruno	Bruno Co-Op Association/Associated Properties	NED981713829	Final	50.00
NE	Columbus	10th Street Site	NED981713837	Final	28.90
NE	Grand Island	Cleburn Street Well	NED981499312	Final	50.00
NE	Grand Island	Parkview Well	NEN000704456	Proposed	50.00
NE	Hall County	Cornhusker Army Ammunition Plant	NE2213820234	Final	51.13
NE	Hastings	Garvey Elevator	NEN000704351	Proposed	50.00
NE	Hastings	Hastings Ground Water Contamination	NED980862668	Final	42.24
NE	Lindsay	Lindsay Manufacturing Co.	NED068645696	Final	47.91
NE	Mead	Nebraska Ordnance Plant (Former)	NE6211890011	Final	31.94
NE	Norfolk	Sherwood Medical Co.	NED084626100	Final	50.00
NE	Ogallala	Ogallala Ground Water Contamination	NED986369247	Final	50.00
NE	Omaha	Omaha Lead	NESFN0703481	Final	50.00
NE	Waverly	Waverly Ground Water Contamination	NED980862718	Final	37.93
NH	Barrington	Tibbetts Road	NHD989090469	Final	41.09
NH	Berlin	Chlor-Alkali Facility (Former)	NHN000103313	Proposed	30.54
NH	Conway	Kearsarge Metallurgical Corp.	NHD062002001	Final	38.45
NH	Dover	Dover Municipal Landfill	NHD980520191	Final	36.98
NH	Epping	Keefe Environmental Services (Kes)	NHD092059112	Final	65.19
NH	Kingston	Ottati & Goss/Kingston Steel Drum	NHD990717647	Final	53.41
NH	Londonderry	Auburn Road Landfill	NHD980524086	Final	36.30
NH	Londonderry	Tinkham Garage	NHD062004569	Final	43.24
NH	Londonderry	Town Garage/Radio Beacon	NHD981063860	Final	31.94
NH	Merrimack	New Hampshire Plating Co.	NHD001091453	Final	50.00
NH	Milford	Fletcher's Paint Works & Storage	NHD001079649	Final	35.39
NH	Milford	Savage Municipal Water Supply	NHD980671002	Final	37.52
NH	Nashua	Mohawk Tannery	NHD981889629	Proposed	52.00
NH	Nashua	Sylvester	NHD099363541	Final	n/a
NH	North Hampton	Coakley Landfill	NHD064424153	Final	29.16
NH	Peterborough	South Municipal Water Supply Well	NHD980671069	Final	35.64
NH	Plaistow	Beede Waste Oil	NHD018958140	Final	70.71
NH	Portsmouth/Newington	Pease Air Force Base	NH7570024847	Final	39.42
NH	Raymond	Mottolo Pig Farm	NHD980503361	Final	40.70
NH	Somersworth	Somersworth Sanitary Landfill	NHD980520225	Final	65.56
NH	Troy	Troy Mills Landfill	NHD980520217	Final	50.00
NJ	Alexandria Township	Crown Vantage Landfill	NJN000204492	Final	50.00
NJ	Asbury Park	M&T Delisa Landfill	NJD085632164	Deleted	32.27
NJ	Atlantic County	Federal Aviation Administration Technical Center (USDOT)	NJ9690510020	Final	39.65
NJ	Bayville	Denzer & Schafer X-Ray Co.	NJD046644407	Deleted	40.36
NJ	Berkeley Township	Beachwood/Berkley Wells	NJD980654123	Deleted	42.24
NJ	Beverly	Cosden Chemical Coatings Corp.	NJD000565531	Final	33.86
NJ	Boonton	Pepe Field	NJD980529598	Deleted	33.83
NJ	Bound Brook	American Cyanamid Co	NJD002173276	Final	50.28
NJ	Bound Brook	Brook Industrial Park	NJD078251675	Final	58.12
NJ	Brick Township	Brick Township Landfill	NJD980505176	Final	58.13
NJ	Bridgeport	Bridgeport Rental & Oil Services	NJD053292652	Final	60.73
NJ	Bridgeport	Chemical Leaman Tank Lines, Inc.	NJD047321443	Final	47.53
NJ	Camden	Martin Aaron, Inc.	NJD014623854	Final	50.00
NJ	Camden & Gloucester Cit	Welsbach & General Gas Mantle (Camden Radiation)	NJD986620995	Final	41.46
NJ	Carlstadt	Scientific Chemical Processing	NJD070565403	Final	55.97
NJ	Chester Township	Combe Fill South Landfill	NJD094966611	Final	45.22
NJ	Cinnaminson Township	Cinnamison Township (Block 702) Ground Water Contamination	NJD980785638	Final	37.93
NJ	Colts Neck	Naval Weapons Station Earle (Site A)	NJ0170022172	Final	29.65
NJ	Dover Township	Dover Municipal Well 4	NJD980654131	Final	28.90
NJ	East Brunswick Township	Fried Industries	NJD041828906	Final	33.61
NJ	East Rutherford	Universal Oil Products (Chemical Division)	NJD002005106	Final	54.63
NJ	Edgewater	Quanta Resources	NJD000606442	Final	50.00
NJ	Edison Township	Chemical Insecticide Corp.	NJD980484653	Final	37.93
NJ	Edison Township	Kin-Buc Landfill	NJD049860836	Final	50.64
NJ	Edison Township	Renora, Inc.	NJD070415005	Deleted	40.44

Superfund National Priorities List - *continued*

St.	City/Area	Site Name	CERCLIS[1] ID	Status	Score[2]
NJ	Egg Harbor Township	Delilah Road	NJD980529002	Final	49.33
NJ	Elizabeth	Chemical Control	NJD000607481	Final	47.13
NJ	Evesham Township	Ellis Property	NJD980529085	Final	34.62
NJ	Fair Lawn	Fair Lawn Well Field	NJD980654107	Final	42.49
NJ	Fairfield	Caldwell Trucking Co.	NJD048798953	Final	58.30
NJ	Florence	Roebling Steel Co.	NJD073732257	Final	41.02
NJ	Florence Township	Florence Land Recontouring, Inc., Landfill	NJD980529143	Deleted	47.39
NJ	Franklin Borough	Metaltec/Aerosystems	NJD002517472	Final	48.95
NJ	Franklin Township	Franklin Burn	NJD986570992	Final	40.67
NJ	Franklin Township	Higgins Farm	NJD981490261	Final	30.47
NJ	Franklin Township	Myers Property	NJD980654198	Final	33.83
NJ	Freehold Township	Lone Pine Landfill	NJD980505424	Final	66.33
NJ	Galloway Township	Emmell's Septic Landfill	NJD980772727	Final	50.00
NJ	Galloway Township	Mannheim Avenue Dump	NJD980654180	Final	36.56
NJ	Galloway Township	Pomona Oaks Residential Wells	NJD980769350	Deleted	31.94
NJ	Gibbsboro	Route 561 Dump	NJ0000453514	Proposed	50.00
NJ	Gibbsboro	United States Avenue Burn	NJ0001120799	Proposed	50.00
NJ	Gibbstown	Hercules, Inc. (Gibbstown Plant)	NJD002349058	Final	40.36
NJ	Glen Ridge	Glen Ridge Radium Site	NJD980785646	Final	49.14
NJ	Gloucester Township	Gems Landfill	NJD980529192	Final	68.53
NJ	Green Village	Rolling Knolls Lf	NJD980505192	Final	58.31
NJ	Hamilton Township	D'imperio Property	NJD980529416	Final	55.79
NJ	Hillsborough Township	Krysowaty Farm	NJD980529838	Deleted	55.14
NJ	Hoboken	Grand Street Mercury	NJ0001327733	Final	n/a
NJ	Howell Township	Bog Creek Farm	NJD063157150	Final	43.23
NJ	Howell Township	Zschiegner Refining	NJD986643153	Final	50.00
NJ	Jackson Township	Jackson Township Landfill	NJD980505283	Deleted	38.11
NJ	Jamesburg/S. Brunswic	Jis Landfill	NJD097400998	Final	45.14
NJ	Jersey City	Pjp Landfill	NJD980505648	Final	28.73
NJ	Kearny	Diamond Head Oil Refinery Div.	NJD092226000	Final	30.00
NJ	Kearny	Standard Chlorine Chem Co Inc	NJD002175057	Proposed	50.00
NJ	Kingston	Higgins Disposal	NJD053102232	Final	30.87
NJ	Kingwood Township	De Rewal Chemical Co.	NJD980761373	Final	35.72
NJ	Lakehurst	Naval Air Engineering Center	NJ7170023744	Final	50.53
NJ	Linden	Lcp Chemicals Inc.	NJD079303020	Final	50.00
NJ	Lodi	Lodi Municipal Well	NJD980769301	Deleted	33.39
NJ	Mantua Township	Helen Kramer Landfill	NJD980505366	Final	72.66
NJ	Manville	Federal Creosote	NJ0001900281	Final	50.00
NJ	Marlboro Township	Burnt Fly Bog	NJD980504997	Final	59.16
NJ	Maywood/Rochelle Park	Maywood Chemical Co.	NJD980529762	Final	51.19
NJ	Middlesex	Middlesex Sampling Plant (USDOE)	NJ0890090012	Final	50.00
NJ	Millington	Asbestos Dump	NJD980654149	Final	39.61
NJ	Millville	Nascolite Corp.	NJD002362705	Final	51.13
NJ	Minotola	Garden State Cleaners Co.	NJD053280160	Final	28.90
NJ	Minotola	South Jersey Clothing Co.	NJD980766828	Final	42.24
NJ	Monroe Township	Monroe Township Landfill	NJD980505671	Deleted	42.37
NJ	Montclair/West Orange	Montclair/West Orange Radium Site	NJD980785653	Final	49.14
NJ	Montgomery Township	Montgomery Township Housing Development	NJD980654164	Final	37.93
NJ	Morganville	Imperial Oil Co., Inc./Champion Chemicals	NJD980654099	Final	33.87
NJ	Mount Holly	Landfill & Development Co.	NJD048044325	Final	33.62
NJ	Mount Olive Township	Combe Fill North Landfill	NJD980530596	Deleted	47.79
NJ	Newark	Diamond Alkali Co.	NJD980528996	Final	35.40
NJ	Newark	White Chemical Corp.	NJD980755623	Final	n/a
NJ	Newfield Borough	Shieldalloy Corp.	NJD002365930	Final	58.75
NJ	Oakland	Witco Chemical Corp. (Oakland Plant)	NJD045653854	Deleted	30.63
NJ	Old Bridge Township	Cps/Madison Industries	NJD002141190	Final	69.73
NJ	Old Bridge Township	Evor Phillips Leasing	NJD980654222	Final	36.64
NJ	Old Bridge Township	Global Sanitary Landfill	NJD063160667	Final	45.92
NJ	Orange	U.S. Radium Corp.	NJD980654172	Final	37.79
NJ	Parsippany	Troy Hills, Sharkey Landfill	NJD980505762	Final	48.85

Superfund National Priorities List - *continued*

St.	City/Area	Site Name	CERCLIS[1] ID	Status	Score[2]
NJ	Pedricktown (Oldmans Town	NI Industries	NJD061843249	Final	52.96
NJ	Pemberton Township	Fort Dix (Landfill Site)	NJ2210020275	Final	37.40
NJ	Pemberton Township	Lang Property	NJD980505382	Final	48.89
NJ	Pennsauken Township	Puchack Well Field	NJD981084767	Final	50.00
NJ	Pennsauken Township	Swope Oil & Chemical Co.	NJD041743220	Final	35.68
NJ	Piscataway	Chemsol, Inc.	NJD980528889	Final	42.69
NJ	Pitman	Lipari Landfill	NJD980505416	Final	75.60
NJ	Pleasant Plains	Reich Farms	NJD980529713	Final	53.48
NJ	Pleasantville	Price Landfill	NJD070281175	Final	n/a
NJ	Plumstead Township	Goose Farm	NJD980530109	Final	47.71
NJ	Plumstead Township	Hopkins Farm	NJD980532840	Deleted	34.09
NJ	Plumstead Township	Pijak Farm	NJD980532808	Deleted	43.48
NJ	Plumstead Township	Spence Farm	NJD980532816	Deleted	45.87
NJ	Plumstead Township	Wilson Farm	NJD980532824	Final	33.93
NJ	Ringwood Borough	Ringwood Mines/Landfill	NJD980529739	Deleted	52.58
NJ	Rockaway Township	Picatinny Arsenal (USARMY)	NJ3210020704	Final	42.92
NJ	Rockaway Township	Radiation Technology, Inc.	NJD047684451	Final	42.56
NJ	Rockaway Township	Rockaway Borough Well Field	NJD980654115	Final	42.34
NJ	Rockaway Township	Rockaway Township Wells	NJD980654214	Final	28.90
NJ	Rocky Hill Borough	Rocky Hill Municipal Well	NJD980654156	Final	37.93
NJ	Saddle Brook Twp	Curcio Scrap Metal, Inc.	NJD011717584	Final	34.37
NJ	Sayreville	Atlantic Resources	NJD981558430	Final	50.00
NJ	Sayreville	Horseshoe Road	NJD980663678	Final	51.37
NJ	Sayreville	Sayreville Landfill	NJD980505754	Final	37.05
NJ	Shamong Township	Ewan Property	NJD980761365	Final	50.19
NJ	South Brunswick	South Brunswick Landfill	NJD980530679	Deleted	53.42
NJ	South Kearny	Syncon Resins	NJD064263817	Final	43.43
NJ	South Plainfield	Cornell Dubilier Electronics Inc.	NJD981557879	Final	50.27
NJ	South Plainfield	Woodbrook Road Dump	NJSFN0204260	Final	50.00
NJ	Sparta Township	A. O. Polymer	NJD030253355	Final	28.91
NJ	Springfield Twp(Jobstown)	Kauffman & Minteer, Inc.	NJD002493054	Final	28.51
NJ	Swainton Middle	Williams Property	NJD980529945	Final	40.45
NJ	Tabernacle Township	Tabernacle Drum Dump	NJD980761357	Final	36.83
NJ	Toms River	Ciba-Geigy Corp.	NJD001502517	Final	50.33
NJ	Upper Deerfield Township	Upper Deerfield Township Sanitary Landfill	NJD980761399	Deleted	33.62
NJ	Upper Freehold Twp	Friedman Property	NJD980532832	Deleted	33.88
NJ	Vineland	Iceland Coin Laundry Area Gw Plume	NJ0001360882	Final	30.30
NJ	Vineland	Vineland Chemical Co., Inc.	NJD002385664	Final	59.16
NJ	Vineland	Vineland State School	NJD980529887	Deleted	40.84
NJ	Voorhees Township	Cooper Road	NJD980761381	Deleted	36.79
NJ	Wall Township	Monitor Devices, Inc./Intercircuits, Inc.	NJD980529408	Final	41.93
NJ	Wall Township	Waldick Aerospace Devices, Inc.	NJD054981337	Final	44.86
NJ	Wall Twp	White Swan Laundry & Cleaner Inc.	NJSFN0204241	Final	41.63
NJ	Wallington Borough	Industrial Latex Corp.	NJD981178411	Deleted	32.38
NJ	Warren County	Pohatcong Valley Ground Water Contamination	NJD981179047	Final	28.90
NJ	Wayne Township	W.R. Grace & Co., Inc./Wayne Interim Storage Site (USDOE)	NJ1891837980	Final	47.14
NJ	Wharton Borough	Dayco Corp./L.E Carpenter Co.	NJD002168748	Final	46.13
NJ	Winslow Township	King of Prussia	NJD980505341	Final	47.19
NJ	Winslow Township	Lightman Drum Company	NJD014743678	Final	42.03
NJ	Wood Ridge Borough	Ventron/Velsicol	NJD980529879	Final	51.38
NJ	Woodland Township	Woodland Route 532 Dump	NJD980505887	Final	34.98
NJ	Woodland Township	Woodland Route 72 Dump	NJD980505879	Final	31.17
NJ	Wrightstown	Mcguire Air Force Base #1	NJ0570024018	Final	47.20
NM	Albuquerque	At&Sf (Albuquerque)	NMD980622864	Final	50.00
NM	Albuquerque	Fruit Avenue Plume	NMD986668911	Final	50.00
NM	Albuquerque	South Valley	NMD980745558	Final	n/a
NM	Carrizozo	Cimarron Mining Corp.	NMD980749378	Final	38.93
NM	Church Rock	United Nuclear Corp.	NMD030443303	Final	30.36

Superfund National Priorities List - continued

St.	City/Area	Site Name	CERCLIS[1] ID	Status	Score[2]
NM	Clovis	At & Sf (Clovis)	NMD043158591	Deleted	33.62
NM	Espanola	North Railroad Avenue Plume	NMD986670156	Final	50.00
NM	Farmington	Lee Acres Landfill (Usdoi)	NMD980750020	Final	39.37
NM	Grants	Grants Chlorinated Solvents	NM0007271768	Final	50.00
NM	Las Cruces	Griggs & Walnut Ground Water Plume	NM0002271286	Final	50.00
NM	Lemitar	Cal West Metals (Ussba)	NMD097960272	Deleted	59.37
NM	Los Lunas	Pagano Salvage	NMD980749980	Deleted	35.57
NM	Milan	Homestake Mining Co.	NMD007860935	Final	34.21
NM	Prewitt	Prewitt Abandoned Refinery	NMD980622773	Final	44.24
NM	Questa	Molycorp, Inc.	NMD002899094	Proposed	50.00
NM	Roswell	Mcgaffey & Main Groundwater Plume	NM0000605386	Final	50.00
NM	Silver City	Cleveland Mill	NMD981155930	Deleted	40.37
NV	Dayton	Carson River Mercury Site	NVD980813646	Final	39.07
NY	Amenia	Sarney Farm	NYD980535165	Final	33.20
NY	Batavia	Batavia Landfill	NYD980507693	Final	50.18
NY	Bohemia	Bioclinical Laboratories, Inc.	NYD980768683	Deleted	32.91
NY	Brant	Wide Beach Development	NYD980652259	Deleted	56.58
NY	Byron Township	Byron Barrel & Drum	NYD980780670	Final	37.27
NY	Caledonia	Jones Chemicals, Inc.	NYD000813428	Final	33.62
NY	Carthage	Crown Cleaners of Watertown Inc.	NYD986965333	Final	49.00
NY	Central Islip	Mackenzie Chemical Works	NYD980753420	Final	50.00
NY	Cheektowaga	Pfohl Brothers Landfill	NYD980507495	Final	50.11
NY	Clayville	Ludlow Sand & Gravel	NYD013468939	Final	36.88
NY	Cold Springs	Marathon Battery Corp.	NYD010959757	Deleted	30.27
NY	Colonie	Mercury Refining, Inc.	NYD048148175	Final	44.58
NY	Conklin	Conklin Dumps	NYD981486947	Deleted	33.93
NY	Copiague	Action Anodizing, Plating, & Polishing Corp.	NYD072366453	Deleted	34.72
NY	Cortland	Rosen Brothers Scrap Yard/Dump	NYD982272734	Final	51.35
NY	Dayton	Peter Cooper Corporation (Markhams)	NYD980592547	Final	30.00
NY	Deer Park	Sms Instruments, Inc.	NYD001533165	Final	37.32
NY	East Farmingdale	Circuitron Corp.	NYD981184229	Final	54.27
NY	East Fishkill	Shenandoah Road Groundwater Contamination	NYSFN0204269	Final	50.00
NY	Ellenville	Ellenville Scrap Iron & Metal	NYSFN0204190	Final	50.27
NY	Elmira	Facet Enterprises, Inc.	NYD073675514	Final	46.67
NY	Farmingdale	Kenmark Textile Corp.	NYD075784165	Deleted	31.72
NY	Farmingdale	Liberty Industrial Finishing	NYD000337295	Final	50.65
NY	Farmingdale	Preferred Plating Corp.	NYD980768774	Final	35.06
NY	Farmingdale	Tronic Plating Co., Inc.	NYD002059517	Deleted	45.14
NY	Franklin Square	Genzale Plating Co.	NYD002050110	Final	33.79
NY	Fulton	Fulton Terminals	NYD980593099	Final	36.50
NY	Garden City	Old Roosevelt Field Contaminated Gw Area	NYSFN0204234	Final	50.00
NY	Glen Cove	Li Tungsten Corp.	NYD986882660	Final	50.00
NY	Glen Cove	Mattiace Petrochemical Co., Inc.	NYD000512459	Final	31.90
NY	Glenwood Landing	Applied Environmental Services	NYD980535652	Final	41.15
NY	Gowanda	Peter Cooper	NYD980530265	Final	50.00
NY	Great Neck	Stanton Cleaners Area Ground Water Contamination	NYD047650197	Final	35.76
NY	Hamilton	C & J Disposal Leasing Co. Dump	NYD981561954	Deleted	35.10
NY	Hauppauge	Computer Circuits	NYD125499673	Final	50.00
NY	Hempstead	Pasley Solvents & Chemicals, Inc.	NYD991292004	Final	39.65
NY	Hewlett	Peninsula Boulevard Groundwater Plume	NYN000204407	Final	50.00
NY	Hicksville	Anchor Chemicals	NYD001485226	Deleted	37.20
NY	Hicksville	Hooker Chemical & Plastics Corp./Ruco Polymer Corp.	NYD002920312	Final	41.60
NY	High Falls	Mohonk Road Industrial Plant	NYD986950012	Final	50.00
NY	Hillburn	Hudson Technologies, Inc.	NY0001392463	Proposed	50.00
NY	Holbrook	Goldisc Recordings, Inc.	NYD980768717	Final	33.39
NY	Holley	Fmc C/O Diaz Chemical C/O Fmc	NYD067532580	Final	50.00
NY	Hopewell Junction	Hopewell Precision	NYD066813064	Final	50.00
NY	Horseheads	Kentucky Avenue Well Field	NYD980650667	Final	39.65
NY	Hudson River	Hudson River Pcbs	NYD980763841	Final	54.66
NY	Hyde Park	Jones Sanitation	NYD980534556	Final	52.52
NY	Islip	Islip Municipal Sanitary Landfill	NYD980506901	Final	33.39

Superfund National Priorities List - *continued*

St.	City/Area	Site Name	CERCLIS[1] ID	Status	Score[2]
NY	Le Roy	Lehigh Valley Railroad	NYD986950251	Final	50.00
NY	Lincklaen	Solvent Savers	NYD980421176	Final	34.78
NY	Lisbon	Sealand Restoration, Inc.	NYD980535181	Final	29.36
NY	Little Valley	Little Valley	NY0001233634	Final	n/a
NY	Malta	Malta Rocket Fuel Area	NYD980535124	Final	33.62
NY	Massena	General Motors (Central Foundry Division)	NYD091972554	Final	40.71
NY	Maybrook	Nepera Chemical Co., Inc.	NYD000511451	Final	39.87
NY	Mineola/North Hempstead	Jackson Steel	NYD001344456	Final	50.00
NY	Moira	York Oil Co.	NYD000511733	Final	47.70
NY	Newburgh	Consolidated Iron & Metal	NY0002455756	Final	50.00
NY	Niagara Falls	Forest Glen Mobile Home Subdivision	NYD981560923	Final	n/a
NY	Niagara Falls	Hooker (102nd Street)	NYD980506810	Deleted	30.48
NY	Niagara Falls	Hooker (Hyde Park)	NYD000831644	Final	34.77
NY	Niagara Falls	Hooker (S Area)	NYD980651087	Final	51.62
NY	Niagara Falls	Love Canal	NYD000606947	Deleted	52.23
NY	North Hempstead	Fulton Avenue	NY0000110247	Final	33.08
NY	North Sea	North Sea Municipal Landfill	NYD980762520	Final	33.74
NY	Noyack/Sag Harbor	Rowe Industries Ground Water Contamination	NYD981486954	Final	31.94
NY	Old Bethpage	Claremont Polychemical	NYD002044584	Final	31.62
NY	Olean	Olean Well Field	NYD980528657	Final	44.46
NY	Oswego	Pollution Abatement Services	NYD000511659	Final	n/a
NY	Oyster Bay	Old Bethpage Landfill	NYD980531727	Final	58.83
NY	Oyster Bay	Syosset Landfill	NYD000511360	Deleted	54.27
NY	Plattekill	Hertel Landfill	NYD980780779	Final	33.62
NY	Plattsburgh	Plattsburgh Air Force Base	NY4571924774	Final	30.34
NY	Port Crane	Tri-Cities Barrel Co., Inc.	NYD980509285	Final	44.06
NY	Port Jefferson Station	Lawrence Aviation Industries, Inc.	NYD002041531	Final	50.00
NY	Port Jervis	Carroll & Dubies Sewage Disposal	NYD010968014	Final	33.74
NY	Port Washington	Port Washington Landfill	NYD980654206	Final	45.46
NY	Putnam County	Brewster Well Field	NYD980652275	Final	37.93
NY	Queens	Radium Chemical Co., Inc.	NYD001667872	Deleted	n/a
NY	Ramapo	Ramapo Landfill	NYD000511493	Final	44.73
NY	Rome	Griffiss Air Force Base (11 Areas)	NY4571924451	Final	34.20
NY	Romulus	Seneca Army Depot	NY0213820830	Final	35.52
NY	Saratoga Springs	Niagara Mohawk Power Corp. (Saratoga Springs Plant)	NYD980664361	Final	35.48
NY	Sidney	Sidney Landfill	NYD980507677	Final	29.36
NY	Sidney Center	Richardson Hill Road Landfill/Pond	NYD980507735	Final	34.86
NY	Smithtown	Smithtown Ground Water Contamination	NY0002318889	Final	50.00
NY	South Cairo	American Thermostat Co.	NYD002066330	Final	33.61
NY	South Glens Falls	Ge Moreau	NYD980528335	Final	58.21
NY	Syracuse	Onondaga Lake	NYD986913580	Final	50.00
NY	Town of Bedford	Katonah Municipal Well	NYD980780795	Deleted	35.35
NY	Town of Colesville	Colesville Municipal Landfill	NYD980768691	Final	30.26
NY	Town of Granby	Clothier Disposal	NYD000511576	Deleted	34.48
NY	Town of Hyde Park	Haviland Complex	NYD980785661	Final	33.62
NY	Town of Johnstown	Johnstown City Landfill	NYD980506927	Final	48.36
NY	Town of Shelby	Fmc Corp. (Dublin Road Landfill)	NYD000511857	Final	32.90
NY	Town of Vestal	Robintech, Inc./National Pipe Co.	NYD002232957	Final	30.75
NY	Town of Volney	Volney Municipal Landfill	NYD980509376	Final	32.89
NY	Union Springs	Cayuga Groundwater Contamination Site	NYN000204289	Final	50.00
NY	Upton	Brookhaven National Laboratory (USDOE)	NY7890008975	Final	39.92
NY	Vestal	Bec Trucking	NYD980768675	Deleted	30.75
NY	Vestal	Vestal Water Supply Well 1-1	NYD980763767	Final	37.93
NY	Vestal	Vestal Water Supply Well 4-2	NYD980652267	Deleted	42.24
NY	Vil of Narrowsburg	Cortese Landfill	NYD980528475	Final	32.11
NY	Village of Endicott	Endicott Village Well Field	NYD980780746	Final	35.57
NY	Village of Sidney	Gcl Tie & Treating Inc.	NYD981566417	Final	48.54
NY	Village of Suffern	Suffern Village Well Field	NYD980780878	Deleted	35.57
NY	Warwick	Warwick Landfill	NYD980506679	Deleted	29.41
NY	Wellsville	Sinclair Refinery	NYD980535215	Final	53.90

Superfund National Priorities List - *continued*

St.	City/Area	Site Name	CERCLIS[1] ID	Status	Score[2]
NY	West Winfield	Hiteman Leather	NYD981560915	Final	50.00
NY	Wheatfield	Niagara County Refuse	NYD000514257	Deleted	39.85
OH	Ashtabula	Fields Brook	OHD980614572	Final	44.95
OH	Beavercreek	Lammers Barrel Factory	OHD981537582	Final	69.33
OH	Circleville	Bowers Landfill	OHD980509616	Deleted	50.49
OH	Cleveland	Chemical & Minerals Reclamation	OHD980614549	Deleted	n/a
OH	Columbus	Air Force Plant 85	OH1170090004	Proposed	50.00
OH	Copley	Copley Square Plaza	OH0000563122	Final	50.00
OH	Darke County	Arcanum Iron & Metal	OHD017506171	Deleted	62.26
OH	Dayton	North Sanitary Landfill	OHD980611875	Final	50.00
OH	Dayton	Powell Road Landfill	OHD000382663	Final	31.62
OH	Dayton	Sanitary Landfill Co. (Industrial Waste Disposal Co., Inc.)	OHD093895787	Final	35.57
OH	Dayton	Wright-Patterson Air Force Base	OH7571724312	Final	57.85
OH	Deerfield Township	Summit National	OHD980609994	Final	52.28
OH	Dover	Dover Chemical Corp.	OHD004210563	Proposed	50.00
OH	Dover	Reilly Tar & Chemical Corp. (Dover Plant)	OHD980610042	Final	31.38
OH	Elyria	Republic Steel Corp. Quarry	OHD980903447	Deleted	29.85
OH	Fernald	Feed Materials Production Center (USDOE)	OH6890008976	Final	57.56
OH	Franklin Township	Coshocton Landfill	OHD980509830	Deleted	39.14
OH	Gnadenhutten	Alsco Anaconda	OHD057243610	Deleted	42.94
OH	Hamilton	Armco Incorporation-Hamilton Plant	OHD074705930	Proposed	69.00
OH	Hamilton	Chem-Dyne	OHD074727793	Final	n/a
OH	Hamilton Township	E.H. Schilling Landfill	OHD980509947	Final	34.56
OH	Hannibal	Ormet Corp.	OHD004379970	Final	46.44
OH	Ironton	Allied Chemical & Ironton Coke	OHD043730217	Final	47.05
OH	Jackson Township	Fultz Landfill	OHD980794630	Final	39.42
OH	Jefferson Township	Laskin/Poplar Oil Co.	OHD061722211	Deleted	35.95
OH	Kings Mills	Peters Cartridge Factory	OHD987051083	Proposed	50.00
OH	Kingsville	Big D Campground	OHD980611735	Final	30.77
OH	Lockbourne	Rickenbacker Air National Guard (Usaf)	OH3571924544	Proposed	50.00
OH	Marietta	Van Dale Junkyard	OHD980794606	Final	33.03
OH	Miamisburg	Mound Plant (USDOE)	OH6890008984	Final	34.61
OH	Minerva	Trw, Inc. (Minerva Plant)	OHD004179339	Final	38.08
OH	Moraine	South Dayton Dump & Landfill	OHD980611388	Proposed	49.00
OH	New Lyme	New Lyme Landfill	OHD980794614	Final	36.70
OH	Painesville	Diamond Shamrock Corp. (Painesville Works)	OHD980611909	Proposed	50.00
OH	Reading	Pristine, Inc.	OHD076773712	Final	35.25
OH	Rock Creek	Old Mill	OHD980510200	Final	35.95
OH	Salem	Nease Chemical	OHD980610018	Final	47.19
OH	South Point	South Point Plant	OHD071650592	Final	46.33
OH	St. Clairsville	Buckeye Reclamation	OHD980509657	Final	35.10
OH	Troy	Miami County Incinerator	OHD980611800	Final	57.84
OH	Troy	United Scrap Lead Co., Inc.	OHD018392928	Final	58.15
OH	Uniontown	Industrial Excess Landfill	OHD000377911	Final	51.13
OH	West Chester	Skinner Landfill	OHD063963714	Final	30.23
OH	Zanesville	Zanesville Well Field	OHD980794598	Final	35.59
OK	Ardmore	Imperial Refining Company	OK0002024099	Final	30.00
OK	Bartlesville	National Zinc Corp.	OKD000829440	Proposed	50.00
OK	Collinsville	Tulsa Fuel & Manufacturing	OKD987096195	Final	50.00
OK	Criner	Hardage/Criner	OKD000400093	Final	51.01
OK	Cushing	Hudson Refinery	OKD082471988	Final	29.34
OK	Cyril	Oklahoma Refining Co.	OKD091598870	Final	46.01
OK	Oklahoma City	Double Eagle Refinery Co.	OKD007188717	Final	30.83
OK	Oklahoma City	Fourth Street Abandoned Refinery	OKD980696470	Final	30.67
OK	Oklahoma City	Mosley Road Sanitary Landfill	OKD980620868	Final	38.06
OK	Oklahoma City	Tenth Street Dump/Junkyard	OKD980620967	Deleted	30.98
OK	Oklahoma City	Tinker Air Force Base (Soldier Creek/Building 3001)	OK1571724391	Final	42.24
OK	Ottawa County	Tar Creek (Ottawa County)	OKD980629844	Final	58.15
OK	Sand Springs	Sand Springs Petrochemical Complex	OKD980748446	Deleted	28.86
OK	Tulsa	Compass Industries (Avery Drive)	OKD980620983	Deleted	36.57
OR	Albany	Teledyne Wah Chang	ORD050955848	Final	54.27

Superfund National Priorities List - *continued*

St.	City/Area	Site Name	CERCLIS[1] ID	Status	Score[2]
OR	Clackamas	Northwest Pipe & Casing/Hall Process Company	ORD980988307	Final	51.09
OR	Corvallis	United Chrome Products, Inc.	ORD009043001	Final	31.07
OR	Hermiston	Umatilla Army Depot (Lagoons)	OR6213820917	Final	31.31
OR	Joseph	Joseph Forest Products	ORD068782820	Deleted	32.60
OR	Lakeview County	Fremont National Forest/White King & Lucky Lass Uranium Mines (Usda)	OR7122307658	Final	50.00
OR	Portland	Allied Plating, Inc.	ORD009051442	Deleted	39.25
OR	Portland	Gould, Inc.	ORD095003687	Deleted	32.12
OR	Portland	Harbor Oil Inc.	ORD071803985	Final	48.00
OR	Portland	Mccormick & Baxter Creosoting Co. (Portland Plant)	ORD009020603	Final	50.00
OR	Portland	Portland Harbor	ORSFN1002155	Final	50.00
OR	Sheridan	Taylor Lumber & Treating	ORD009042532	Final	71.78
OR	The Dalles	Martin-Marietta Aluminum Co.	ORD052221025	Deleted	43.70
OR	The Dalles	Union Pacific Railroad Co. Tie-Treating Plant	ORD009049412	Final	37.93
OR	Troutdale	Reynolds Metals Company	ORD009412677	Final	70.71
PA	Ambler	Ambler Asbestos Piles	PAD000436436	Deleted	34.47
PA	Antis/Logan Twps	Delta Quarries & Disposal, Inc./Stotler Landfill	PAD981038052	Final	41.08
PA	Bally	Bally Ground Water Contamination	PAD061105128	Final	37.93
PA	Bloomsburg	Safety Light Corporation	PAD987295276	Final	71.00
PA	Bridgeton Township	Boarhead Farms	PAD047726161	Final	39.92
PA	Bruin Borough	Bruin Lagoon	PAD980712855	Deleted	73.11
PA	Buffalo Township	Hranica Landfill	PAD980508618	Deleted	51.94
PA	Chambersburg	Letterkenny Army Depot (Se Area)	PA6213820503	Final	34.21
PA	Chester	Wade (Abm)	PAD980539407	Deleted	36.63
PA	Columbia	Ugi Columbia Gas Plant	PAD980539126	Final	50.78
PA	Coraopolis	Breslube-Penn, Inc.	PAD089667695	Final	50.00
PA	Croydon Township	Croydon Tce	PAD981035009	Final	31.60
PA	Darby Twp	Lower Darby Creek Area	PASFN0305521	Final	50.00
PA	Delaware County	Austin Avenue Radiation Site	PAD987341716	Deleted	n/a
PA	Denver	Berkley Products Co. Dump	PAD980538649	Final	30.00
PA	Douglassville	Douglassville Disposal	PAD002384865	Final	55.18
PA	Dublin Borough	Dublin Tce Site	PAD981740004	Final	28.90
PA	Eagleville	Moyers Landfill	PAD980508766	Final	37.62
PA	East Coventry Twp	Recticon/Allied Steel Corp.	PAD002353969	Final	32.06
PA	East Whiteland Township	Foote Mineral Co.	PAD077087989	Final	50.00
PA	Elizabethtown	Elizabethtown Landfill	PAD980539712	Final	28.98
PA	Emmaus Borough	Rodale Manufacturing Co., Inc.	PAD981033285	Final	50.00
PA	Erie	Mill Creek Dump	PAD980231690	Final	49.31
PA	Erie	Presque Isle	PAD980508865	Deleted	40.59
PA	Exton	A.I.W. Frank/Mid-County Mustang	PAD004351003	Final	42.40
PA	Falls Creek	Jackson Ceramix Inc	PAD001222025	Proposed	30.22
PA	Foster Township	C & D Recycling	PAD021449244	Final	43.92
PA	Frackville	Metropolitan Mirror & Glass Co., Inc.	PAD982366957	Final	34.33
PA	Franklin County	Letterkenny Army Depot (Pdo Area)	PA2210090054	Final	37.51
PA	Gettysburg	Westinghouse Elevator Co. Plant	PAD043882281	Final	36.37
PA	Girard Township	Lord-Shope Landfill	PAD980508931	Final	38.89
PA	Glen Rock	Amp, Inc. (Glen Rock Facility)	PAD041421223	Deleted	39.03
PA	Grove City	Osborne Landfill	PAD980712673	Final	54.60
PA	Hamburg	Brown's Battery Breaking	PAD980831812	Final	37.34
PA	Hamburg	Price Battery	PAN000305679	Final	38.00
PA	Harrison Township	Lindane Dump	PAD980712798	Final	51.62
PA	Hatboro	Raymark	PAD039017694	Final	53.42
PA	Hatfield	North Penn - Area 2	PAD002342475	Final	35.57
PA	Haverford	Havertown Pcp	PAD002338010	Final	38.34
PA	Heidelberg Twp.	Ryeland Road Arsenic Site	PAD981033459	Final	60.30
PA	Hellertown	Hellertown Manufacturing Co.	PAD002390748	Final	51.91
PA	Hereford Township	Crossley Farm	PAD981740061	Final	29.66
PA	Hermitage	River Road Landfill (Waste Management, Inc.)	PAD000439083	Deleted	43.12
PA	Hickory Township	Sharon Steel Corp (Farrell Works Disposal Area)	PAD001933175	Final	50.00
PA	Hometown	Eastern Diversified Metals	PAD980830533	Final	31.02

Superfund National Priorities List - *continued*

St.	City/Area	Site Name	CERCLIS[1] ID	Status	Score[2]
PA	Honeybrook Township	Walsh Landfill	PAD980829527	Final	33.64
PA	Hopewell Township	York County Solid Waste & Refuse Authority Landfill	PAD980830715	Deleted	44.26
PA	Jackson Township	Whitmoyer Laboratories	PAD003005014	Final	46.25
PA	Jefferson Borough	Resin Disposal	PAD063766828	Deleted	37.69
PA	Kimberton Borough	Kimberton	PAD980691703	Final	29.44
PA	King of Prussia	Stanley Kessler	PAD014269971	Final	33.89
PA	Lansdale	North Penn - Area 6	PAD980926976	Final	35.57
PA	Lansdowne	Lansdowne Radiation Site	PAD980830921	Deleted	n/a
PA	Lock Haven	Drake Chemical	PAD003058047	Final	38.52
PA	Longswamp Township	Berks Sand Pit	PAD980691794	Final	32.02
PA	Lower Pottsgrove Township	Occidental Chemical Corp./Firestone Tire & Rubber Co.	PAD980229298	Final	45.91
PA	Lower Providence Township	Commodore Semiconductor Group	PAD093730174	Final	42.35
PA	Lower Salford Township	Salford Quarry	PAD980693204	Proposed	50.00
PA	Lower Windsor Twp	Modern Sanitation Landfill	PAD980539068	Final	33.93
PA	Maitland	Jacks Creek/Sitkin Smelting & Refining, Inc.	PAD980829493	Final	40.37
PA	Malvern	Malvern Tce	PAD014353445	Final	46.69
PA	Marcus Hook	East Tenth Street	PAD987323458	Proposed	68.00
PA	Mcadoo Borough	Mcadoo Associates	PAD980712616	Deleted	n/a
PA	Mechanicsburg	Navy Ships Parts Control Center	PA3170022104	Final	50.00
PA	Middletown	Middletown Air Field	PAD980538763	Deleted	35.69
PA	Montgomery Township	North Penn - Area 5	PAD980692693	Final	35.57
PA	Nesquehoning	Tonolli Corp.	PAD073613663	Final	46.58
PA	Neville Island	Ohio River Park	PAD980508816	Final	42.24
PA	Newlin Township	Strasburg Landfill	PAD000441337	Final	30.71
PA	Nockamixon Township	Revere Chemical Co.	PAD051395499	Final	31.31
PA	North Wales	North Penn - Area 7	PAD002498632	Final	35.57
PA	North Whitehall Twp	Heleva Landfill	PAD980537716	Final	50.23
PA	Old Forge	Lackawanna Refuse	PAD980508667	Deleted	36.57
PA	Old Forge	Lehigh Electric & Engineering Co.	PAD980712731	Deleted	30.26
PA	Palmerton	Palmerton Zinc Pile	PAD002395887	Final	42.93
PA	Paoli	Paoli Rail Yard	PAD980692594	Final	32.18
PA	Parker	Craig Farm Drum	PAD980508527	Final	28.72
PA	Philadelphia	Enterprise Avenue	PAD980552913	Deleted	40.80
PA	Philadelphia	Franklin Slag Pile (Mdc)	PASFN0305549	Final	50.20
PA	Philadelphia	Metal Banks	PAD046557096	Final	33.23
PA	Philadephia	Publicker Industries Inc.	PAD981939200	Deleted	59.06
PA	Pittston Township	Butler Mine Tunnel	PAD980508451	Final	49.51
PA	Pocono Summit	Route 940 Drum Dump	PAD981034630	Deleted	44.06
PA	Richland Township	Watson Johnson Landfill	PAD980706824	Final	70.71
PA	Sadsburyville	Old Wilmington Road Gw Contamination	PAD981938939	Final	50.00
PA	Saegertown	Saegertown Industrial Area	PAD980692487	Final	33.62
PA	Scott Township	Aladdin Plating	PAD075993378	Deleted	35.57
PA	Seven Valleys	Old City of York Landfill	PAD980692420	Final	33.93
PA	Sharon	Westinghouse Electric Corp. (Sharon Plant)	PAD005000575	Final	41.33
PA	Souderton	North Penn - Area 1	PAD096834494	Final	35.57
PA	South Montrose	Bendix Flight Systems Division	PAD003047974	Final	33.74
PA	South Whitehall Township	Novak Sanitary Landfill	PAD079160842	Final	42.31
PA	Spring Township	Berks Landfill	PAD000651810	Final	46.10
PA	Springettsbury Township	East Mount Zion	PAD980690549	Final	41.01
PA	State College Borough	Centre County Kepone	PAD000436261	Final	45.09
PA	Straban Township	Hunterstown Road	PAD980830897	Final	48.27
PA	Straban Township	Shriver's Corner	PAD980830889	Final	46.13
PA	Stroudsburg	Brodhead Creek	PAD980691760	Deleted	31.09
PA	Stroudsburg	Butz Landfill	PAD981034705	Final	32.00
PA	Taylor Borough	Taylor Borough Dump	PAD980693907	Deleted	30.94
PA	Terry Township	Bell Landfill	PAD980705107	Final	34.79
PA	Tobyhanna	Tobyhanna Army Depot	PA5213820892	Final	37.93
PA	Union Township	Keystone Sanitation Landfill	PAD054142781	Final	33.76

Superfund National Priorities List - continued

St.	City/Area	Site Name	CERCLIS[1] ID	Status	Score[2]
PA	Upper Macungie Township	Dorney Road Landfill	PAD980508832	Final	46.10
PA	Upper Macungie Twp	Reeser's Landfill	PAD980829261	Deleted	30.35
PA	Upper Merion Township	Crater Resources, Inc./Keystone Coke Co./Alan Wood Steel Co.	PAD980419097	Final	50.00
PA	Upper Merion Township	Henderson Road	PAD009862939	Final	41.69
PA	Upper Merion Twp	Tysons Dump	PAD980692024	Final	63.10
PA	Upper Saucon Twp	Voortman Farm	PAD980692719	Deleted	28.62
PA	Valley Township	Mw Manufacturing	PAD980691372	Final	46.44
PA	Warminster	Fischer & Porter Co.	PAD002345817	Final	29.07
PA	Warminster Township	Naval Air Development Center (8 Waste Areas)	PA6170024545	Final	57.93
PA	Weisenberg Township	Hebelka Auto Salvage Yard	PAD980829329	Deleted	31.94
PA	West Caln Township	Blosenski Landfill	PAD980539985	Final	30.57
PA	West Caln Township	William Dick Lagoons	PAD980537773	Final	36.64
PA	West Hazleton	Valmont Tce Site (Former - Valmont Industrial Park)	PAD982363970	Final	43.16
PA	Westline	Westline	PAD980692537	Deleted	31.71
PA	Williams Township	Industrial Lane	PAD980508493	Final	42.47
PA	Williamsport	Avco Lycoming (Williamsport Division)	PAD003053709	Final	42.24
PA	Willow Grove	Willow Grove Naval Air & Air Reserve Station	PAD987277837	Final	50.00
PA	Worcester	North Penn - Area 12	PAD057152365	Final	28.90
PA	Worman Township	Cryochem, Inc.	PAD002360444	Final	28.58
PR	Almirante Norte Ward	V&M/Albaladejo	PRD987366101	Deleted	50.00
PR	Arecibo	Pesticide Warehouse I	PRD987367349	Proposed	50.00
PR	Barceloneta	Rca Del Caribe	PRD090370537	Deleted	31.14
PR	Barceloneta	Upjohn Facility	PRD980301154	Final	41.92
PR	Candeleria Ward	Scorpio Recycling, Inc.	PRD987376662	Final	50.00
PR	Cidra	Cidra Groundwater Contamination	PRN000204538	Final	50.00
PR	Florida Afuera	Barceloneta Landfill	PRD980509129	Final	41.11
PR	Jobos	Fibers Public Supply Wells	PRD980763783	Final	35.34
PR	Juana Diaz	Ge Wiring Devices	PRD090282757	Deleted	31.24
PR	Juncos	Juncos Landfill	PRD980512362	Final	32.57
PR	Manati	Pesticide Warehouse Iii	PRD987367299	Final	50.00
PR	Rio Abajo	Frontera Creek	PRD980640965	Deleted	43.07
PR	Rio Abajo Ward	Vega Baja Solid Waste Disposal	PRD980512669	Final	50.37
PR	Sabana Seca	Naval Security Group Activity	PR4170027383	Deleted	34.28
PR	Vega Alta	Vega Alta Public Supply Wells	PRD980763775	Final	42.24
PR	Vieques	Atlantic Fleet Weapons Training Area - Vieques	PRN000204694	Final	n/a
RI	Burrillville	Western Sand & Gravel	RID009764929	Final	51.35
RI	Coventry	Picillo Farm	RID980579056	Final	n/a
RI	Glocester	Davis (Gsr) Landfill	RID980731459	Deleted	38.89
RI	Johnston	Central Landfill	RID980520183	Final	46.71
RI	Lincoln/Cumberland	Peterson/Puritan, Inc.	RID055176283	Final	40.10
RI	Newport	Newport Naval Education & Training Center	RI6170085470	Final	32.25
RI	North Kingstown	Davisville Naval Construction Battalion Center	RI6170022036	Final	34.52
RI	North Providence	Centredale Manor Restoration Project	RID981203755	Final	70.71
RI	North Smithfield	Landfill & Resource Recovery, Inc. (L&Rr)	RID093212439	Final	49.58
RI	North Smithfield	Stamina Mills, Inc.	RID980731442	Final	34.07
RI	Smithfield	Davis Liquid Waste	RID980523070	Final	47.25
RI	South Kingstown	Rose Hill Regional Landfill	RID980521025	Final	38.11
RI	South Kingstown	West Kingston Town Dump/Uri Disposal Area	RID981063993	Final	50.00
SC	Aiken	Savannah River Site (USDOE)	SC1890008989	Final	47.70
SC	Barnwell	Shuron Inc.	SCD003357589	Final	68.26
SC	Beaufort	Independent Nail Co.	SCD004773644	Deleted	57.90
SC	Beaufort	Kalama Specialty Chemicals	SCD094995503	Final	57.90
SC	Burton	Wamchem, Inc.	SCD037405362	Final	47.70
SC	Cayce	Lexington County Landfill Area	SCD980558043	Final	37.93
SC	Cayce	Scrdi Dixiana	SCD980711394	Final	40.70
SC	Charleston	Koppers Co., Inc. (Charleston Plant)	SCD980310239	Final	50.00
SC	Columbia	Palmetto Recycling, Inc.	SCD037398120	Deleted	29.46
SC	Columbia	Scrdi Bluff Road	SCD000622787	Final	n/a
SC	Dixiana	Palmetto Wood Preserving	SCD003362217	Final	38.43
SC	Fairfax	Helena Chemical Co. Landfill	SCD058753971	Final	33.89

Superfund National Priorities List - *continued*

St.	City/Area	Site Name	CERCLIS[1] ID	Status	Score[2]
SC	Florence	Koppers Co., Inc. (Florence Plant)	SCD003353026	Final	51.27
SC	Fort Lawn	Carolawn, Inc.	SCD980558316	Final	32.04
SC	Fountain Inn	Beaunit Corp. (Circular Knit & Dyeing Plant)	SCD000447268	Final	32.44
SC	Gaffney	Medley Farm Drum Dump	SCD980558142	Final	31.58
SC	Greer	Aqua-Tech Environmental Inc (Groce Labs)	SCD058754789	Final	50.00
SC	Greer	Elmore Waste Disposal	SCD980839542	Final	31.45
SC	Jefferson	Brewer Gold Mine Dam Failure	SCD987577913	Final	50.00
SC	North Charleston	Macalloy Corporation	SCD003360476	Final	50.00
SC	Parris Island	Parris Island Marine Corps Recruit Depot	SC6170022762	Final	50.00
SC	Pickens	Sangamo Weston, Inc./Twelve-Mile Creek/Lake Hartwell PCB Contamination	SCD003354412	Final	37.63
SC	Pontiac	Townsend Saw Chain Co.	SCD980558050	Final	35.94
SC	Rantoules	Geiger (C & M Oil)	SCD980711279	Final	32.25
SC	Rock Hill	Leonard Chemical Co., Inc.	SCD991279324	Final	47.10
SC	Rock Hill	Rock Hill Chemical Co.	SCD980844005	Final	40.29
SC	Simpsonville	Golden Strip Septic Tank Service	SCD980799456	Deleted	40.30
SC	Simpsonville	Para-Chem Southern, Inc.	SCD002601656	Final	32.94
SC	Travelers Rest	Rochester Property	SCD980840698	Final	36.72
SD	Ellsworth Afb	Ellsworth Air Force Base	SD2571924644	Final	33.62
SD	Lead	Gilt Edge Mine	SDD987673985	Final	50.00
SD	Sioux Falls	Williams Pipe Line Co. Disposal Pit	SDD000823559	Deleted	42.24
SD	Whitewood	Whitewood Creek	SDD980717136	Deleted	n/a
TN	Arlington	Arlington Blending & Packaging	TND980468557	Final	39.03
TN	Chattanooga	Amnicola Dump	TND980729172	Deleted	40.91
TN	Chattanooga	Tennessee Products	TND071516959	Final	n/a
TN	Collierville	Carrier Air Conditioning Co.	TND044062222	Final	48.91
TN	Collierville	Smalley-Piper	TNN000407378	Final	50.00
TN	Gallaway	Gallaway Pits	TND980728992	Deleted	30.77
TN	Jackson	American Creosote Works, Inc. (Jackson Plant)	TND007018799	Final	35.22
TN	Jackson	Icg Iselin Railroad Yard	TND987767795	Deleted	50.00
TN	Lawrenceburg	Murray-Ohio Dump	TND980728836	Final	46.44
TN	Lewisburg	Lewisburg Dump	TND980729115	Deleted	33.45
TN	Memphis	Memphis Defense Depot (Dla)	TN4210020570	Final	58.06
TN	Memphis	North Hollywood Dump	TND980558894	Deleted	n/a
TN	Milan	Milan Army Ammunition Plant	TN0210020582	Final	58.15
TN	Moscow	Chemet Co.	TND987768546	Deleted	50.00
TN	Oak Ridge	Oak Ridge Reservation (USDOE)	TN1890090003	Final	51.13
TN	Rossville	Ross Metals Inc.	TND096070396	Final	37.65
TN	Toone	Velsicol Chemical Corp. (Hardeman County)	TND980559033	Final	47.71
TN	Tullahoma/Manchester	Arnold Engineering Development Center (Usaf)	TN8570024044	Proposed	50.00
TN	Waynesboro	Mallory Capacitor Co.	TND075453688	Final	29.44
TN	Wrigley	Wrigley Charcoal Plant	TND980844781	Final	36.14
TX	Bell County	Rockwool Industries Inc.	TXD066379645	Final	48.00
TX	Bridge City	Bailey Waste Disposal	TXD980864649	Final	53.42
TX	Bridge City	Triangle Chemical Co.	TXD055143705	Deleted	28.75
TX	Conroe	Conroe Creosoting Co.	TXD008091951	Final	48.00
TX	Conroe	United Creosoting Co.	TXD980745574	Final	37.29
TX	Corpus Christi	Brine Service Company	TX0000605264	Final	50.00
TX	Crosby	French, Ltd.	TXD980514814	Final	63.33
TX	Crosby	Sikes Disposal Pits	TXD980513956	Final	61.62
TX	Crystal City	Crystal City Airport	TXD980864763	Deleted	32.26
TX	Dallas	Rsr Corporation	TXD079348397	Final	50.00
TX	Deer Park	Patrick Bayou	TX0000605329	Final	47.83
TX	Fort Worth	Air Force Plant #4 (General Dynamics)	TX7572024605	Final	39.92
TX	Fort Worth	Pesses Chemical Co.	TXD980699656	Deleted	28.86
TX	Freeport	Gulfco Marine Maintenance	TXD055144539	Final	50.00
TX	Friendswood	Brio Refining, Inc.	TXD980625453	Final	50.38
TX	Friendswood	Dixie Oil Processors, Inc.	TXD089793046	Final	34.21
TX	Grand Prairie	Bio-Ecology Systems, Inc.	TXD980340889	Deleted	35.06
TX	Hempstead	Sheridan Disposal Services	TXD062132147	Final	30.16
TX	Highlands	Highlands Acid Pit	TXD980514996	Final	37.77

Superfund National Priorities List - continued

St.	City/Area	Site Name	CERCLIS[1] ID	Status	Score[2]
TX	Houston	Crystal Chemical Co.	TXD990707010	Final	60.90
TX	Houston	Geneva Industries/Fuhrmann Energy	TXD980748453	Final	59.46
TX	Houston	Harris (Farley Street)	TXD980745582	Deleted	33.94
TX	Houston	Jones Road Ground Water Plume	TXN000605460	Final	46.50
TX	Houston	Many Diversified Interests, Inc.	TXD008083404	Final	32.07
TX	Houston	North Cavalcade Street	TXD980873343	Final	37.08
TX	Houston	Sol Lynn/Industrial Transformers	TXD980873327	Final	39.65
TX	Houston	South Cavalcade Street	TXD980810386	Final	38.69
TX	Ingleside	Falcon Refinery	TXD086278058	Proposed	50.00
TX	Jasper	Hart Creosoting Company	TXD050299577	Final	48.00
TX	Jasper	Jasper Creosoting Company Inc.	TXD008096240	Final	50.00
TX	Jefferson County	State Marine of Port Arthur	TXD099801102	Final	48.00
TX	Karnack	Longhorn Army Ammunition Plant	TX6213820529	Final	39.83
TX	La Marque	Motco, Inc.	TXD980629851	Final	n/a
TX	Levelland	State Road 114 Ground Water Plume	TXSFN0605177	Final	42.41
TX	Liberty	Petro-Chemical Systems, Inc. (Turtle Bayou)	TXD980873350	Final	29.94
TX	Longview	Garland Creosoting	TXD007330053	Final	49.10
TX	Odessa	Odessa Chromium #1	TXD980867279	Final	42.24
TX	Odessa	Odessa Chromium #2 (Andrews Highway)	TXD980697114	Deleted	42.24
TX	Odessa	Sprague Road Ground Water Plume	TX0001407444	Final	43.21
TX	Pantex Village	Pantex Plant (USDOE)	TX4890110527	Final	51.22
TX	Pelican Bay	Pelican Bay Public Water System	TXN000605649	Proposed	50.00
TX	Perryton	City of Perryton Well No. 2	TX0001399435	Final	50.00
TX	Point Comfort	Alcoa (Point Comfort)/Lavaca Bay	TXD008123168	Final	50.00
TX	Port Arthur	Palmer Barge Line	TXD068104561	Final	50.00
TX	Port Neches	Star Lake Canal	TX0001414341	Final	50.00
TX	San Antonio	R & H Oil/Tropicana	TXD057577579	Proposed	50.00
TX	Texarkana	Koppers Co., Inc. (Texarkana Plant)	TXD980623904	Final	31.31
TX	Texarkana	Lone Star Army Ammunition Plant	TX7213821831	Final	31.85
TX	Texarkana	Texarkana Wood Preserving Co.	TXD008056152	Final	40.19
TX	Texas City	Malone Service Co - Swan Lake Plant	TXD980864789	Final	50.00
TX	Texas City	Tex-Tin Corp.	TXD062113329	Final	50.00
TX	Waskom	Stewco, Inc.	TXD055337281	Deleted	48.86
UT	Bountiful	Bountiful/Woods Cross 5th S. Pce Plume	UT0001119296	Final	50.00
UT	Bountiful	Intermountain Waste Oil Refinery	UT0001277359	Final	50.00
UT	Copperton	Kennecott (South Zone)	UTD000826404	Proposed	71.00
UT	Eureka	Eureka Mills	UT0002240158	Final	50.00
UT	Hill Afb	Hill Air Force Base	UT0571724350	Final	49.94
UT	Magna	Kennecott (North Zone)	UTD070926811	Proposed	59.00
UT	Midvale	Midvale Slag	UTD081834277	Final	42.47
UT	Midvale	Sharon Steel Corp. (Midvale Tailings)	UTD980951388	Deleted	41.85
UT	Monticello	Monticello Mill Tailings (USDOE)	UT3890090035	Final	35.86
UT	Monticello	Monticello Radioactively Contaminated Properties	UTD980667208	Deleted	35.03
UT	Murray City	Murray Smelter	UTD980951420	Proposed	87.00
UT	Ogden	Ogden Defense Depot (Dla)	UT9210020922	Final	45.10
UT	Park City	Richardson Flat Tailings	UTD980952840	Proposed	50.00
UT	Salt Lake City	Petrochem Recycling Corp./Ekotek Plant	UTD093119196	Deleted	62.81
UT	Salt Lake City	Portland Cement (Kiln Dust 2 & 3)	UTD980718670	Final	54.40
UT	Salt Lake City	Rose Park Sludge Pit	UTD980635452	Deleted	n/a
UT	Salt Lake City	Utah Power & Light/American Barrel Co.	UTD980667240	Final	37.93
UT	Salt Lake City	Wasatch Chemical Co. (Lot 6)	UTD000716399	Final	49.91
UT	Sandy	Davenport & Flagstaff Smelters	UTD988075719	Final	32.50
UT	Stockton	Jacobs Smelter	UT0002391472	Final	50.00
UT	Tooele	International Smelting & Refining	UTD093120921	Final	58.31
UT	Tooele	Tooele Army Depot (North Area)	UT3213820894	Final	53.95
VA	Buckingham	Buckingham County Landfill	VAD089027973	Final	40.70
VA	Chesapeake	St. Juliens Creek Annex (U.S. Navy)	VA5170000181	Final	50.00
VA	Chesterfield County	C & R Battery Co., Inc.	VAD049957913	Final	46.44
VA	Chesterfield County	Defense General Supply Center (Dla)	VA3971520751	Final	33.85
VA	Chuckatuck	Saunders Supply Co.	VAD003117389	Final	36.88
VA	Culpeper	Culpeper Wood Preservers, Inc.	VAD059165282	Final	45.91

609

Superfund National Priorities List - *continued*

St.	City/Area	Site Name	CERCLIS[1] ID	Status	Score[2]
VA	Dahlgren	Naval Surface Warfare Center - Dahlgren	VA7170024684	Final	50.03
VA	Farrington	H & H Inc., Burn Pit	VAD980539878	Final	33.71
VA	Frederick County	Rhinehart Tire Fire Dump	VAD980831796	Final	30.57
VA	Front Royal	Avtex Fibers, Inc.	VAD070358684	Final	35.39
VA	Hampton	Langley Air Force Base/Nasa Langley Research Center	VA2800005033	Final	50.00
VA	Montross	Arrowhead Associates, Inc./Scovill Corp.	VAD042916361	Final	37.15
VA	Newport News	Fort Eustis (Us Army)	VA6210020321	Final	50.00
VA	Newtown	Greenwood Chemical Co.	VAD003125374	Final	53.17
VA	Norfolk	Norfolk Naval Base (Sewells Point Naval Complex)	VA6170061463	Final	50.00
VA	Piney River	U.S. Titanium	VAD980705404	Final	34.78
VA	Pittsylvania County	First Piedmont Corp. Rock Quarry (Route 719)	VAD980554984	Final	30.16
VA	Portsmouth	Abex Corp.	VAD980551683	Final	36.53
VA	Portsmouth	Atlantic Wood Industries, Inc.	VAD990710410	Final	37.14
VA	Portsmouth	Norfolk Naval Shipyard	VA1170024813	Final	50.00
VA	Quantico	Marine Corps Combat Development Command	VA1170024722	Final	50.00
VA	Richmond	Rentokil, Inc. (Virginia Wood Preserving Division)	VAD071040752	Final	30.34
VA	Roanoke	Matthews Electroplating	VAD980712970	Deleted	n/a
VA	Salem	Dixie Caverns County Landfill	VAD980552095	Deleted	35.27
VA	Saltville	Saltville Waste Disposal Ponds	VAD003127578	Final	29.52
VA	Selma	Kim-Stan Landfill	VAD077923449	Final	50.00
VA	Spotsylvania	L.A. Clarke & Son	VAD007972482	Final	34.24
VA	Suffolk	Former Nansemond Ordnance Depot	VAD123933426	Final	70.71
VA	Suffolk	Suffolk City Landfill	VAD980917983	Deleted	35.76
VA	Virginia Beach	Naval Amphibious Base Little Creek	VA5170022482	Final	50.00
VA	Williamsburg	Nws Yorktown - Cheatham Annex	VA3170024605	Final	49.27
VA	York County	Chisman Creek	VAD980712913	Final	47.19
VA	Yorktown	Naval Weapons Station - Yorktown	VA8170024170	Final	50.00
VI	Christiansted	Island Chemical Corp/Virgin Islands Chemical Corp.	VID980651095	Final	50.00
VI	Tutu	Tutu Wellfield	VID982272569	Final	50.00
VT	Bennington	Bennington Municipal Sanitary Landfill	VTD981064223	Final	49.07
VT	Bennington	Tansitor Electronics, Inc.	VTD000509174	Deleted	35.72
VT	Burlington	Pine Street Canal	VTD980523062	Final	n/a
VT	Corinth	Pike Hill Copper Mine	VTD988366720	Final	50.00
VT	Lyndon	Darling Hill Dump	VTD980520118	Deleted	43.92
VT	Lyndon	Parker Sanitary Landfill	VTD981062441	Final	52.29
VT	Pownal	Pownal Tannery	VTD069910354	Final	50.00
VT	Rockingham	Bfi Sanitary Landfill (Rockingham)	VTD980520092	Final	41.92
VT	Springfield	Old Springfield Landfill	VTD000860239	Final	34.79
VT	Strafford	Elizabeth Mine	VTD988366621	Final	50.00
VT	Vershire	Ely Copper Mine	VTD988366571	Final	50.00
VT	Williston	Mitec	VTD098352545	Final	48.00
VT	Woodford	Burgess Brothers Landfill	VTD003965415	Final	52.58
WA	Bainbridge Island	Wyckoff Co./Eagle Harbor	WAD009248295	Final	32.55
WA	Bellingham	Oeser Co.	WAD008957243	Final	69.34
WA	Benton County	Hanford 100-Area (USDOE)	WA3890090076	Final	46.38
WA	Benton County	Hanford 1100-Area (USDOE)	WA4890090075	Deleted	36.34
WA	Benton County	Hanford 200-Area (USDOE)	WA1890090078	Final	69.05
WA	Benton County	Hanford 300-Area (USDOE)	WA2890090077	Final	65.23
WA	Bremerton	Bangor Ordnance Disposal (USNAVY)	WA7170027265	Final	30.42
WA	Bremerton	Puget Sound Naval Shipyard Complex	WA2170023418	Final	50.00
WA	Brush Prairie	Toftdahl Drums	WAD980723506	Deleted	40.22
WA	Centralia	Centralia Municipal Landfill	WAD980836662	Final	36.36
WA	Chehalis	American Crossarm & Conduit Co.	WAD057311094	Final	30.44
WA	Chehalis	Hamilton/Labree Roads Gw Contamination	WASFN1002174	Final	37.65
WA	Everson	Northwest Transformer	WAD980833974	Deleted	33.82
WA	Everson	Northwest Transformer (South Harkness Street)	WAD027315621	Deleted	30.56
WA	Indian Island	Port Hadlock Detachment (USNAVY)	WA4170090001	Deleted	50.00
WA	Kent	Midway Landfill	WAD980638910	Final	54.27
WA	Kent	Seattle Municipal Landfill (Kent Highlands)	WAD980639462	Final	52.19
WA	Kent	Western Processing Co., Inc.	WAD009487513	Final	58.63

Superfund National Priorities List - *continued*

St.	City/Area	Site Name	CERCLIS[1] ID	Status	Score[2]
WA	Keyport	Naval Undersea Warfare Engineering Station (4 Waste Areas)	WA1170023419	Final	33.60
WA	Kitsap County	Jackson Park Housing Complex (USNAVY)	WA3170090044	Final	50.00
WA	Lakewood	Lakewood	WAD050075662	Final	42.49
WA	Loomis	Silver Mountain Mine	WAD980722789	Deleted	29.98
WA	Manchester	Old Navy Dump/Manchester Laboratory (Usepa/Noaa)	WA8680030931	Final	50.00
WA	Maple Valley	Queen City Farms	WAD980511745	Final	34.38
WA	Marysville	Tulalip Landfill	WAD980639256	Deleted	50.00
WA	Mead	Kaiser Aluminum (Mead Works)	WAD000065508	Final	38.07
WA	Mica	Mica Landfill	WAD980511661	Final	34.64
WA	Moses Lake	Moses Lake Wellfield Contamination	WAD988466355	Final	50.00
WA	North Bonneville	Hamilton Island Landfill (Usa/Coe)	WA5210890096	Deleted	51.97
WA	Pasco	Pasco Sanitary Landfill	WAD991281874	Final	44.46
WA	Pierce County	Commencement Bay, Near Shore/Tide Flats	WAD980726368	Final	42.20
WA	Pierce County	Hidden Valley Landfill (Thun Field)	WAD980511539	Final	37.93
WA	Renton	Pacific Car & Foundry Co.	WAD009249210	Final	42.33
WA	Seattle	Harbor Island (Lead)	WAD980722839	Final	34.60
WA	Seattle	Lower Duwamish Waterway	WA0002329803	Final	50.00
WA	Seattle	Pacific Sound Resources	WAD009248287	Final	70.71
WA	Silverdale	Bangor Naval Submarine Base	WA5170027291	Final	55.91
WA	Spokane	Colbert Landfill	WAD980514541	Final	41.59
WA	Spokane	Fairchild Air Force Base (4 Waste Areas)	WA9571924647	Final	31.98
WA	Spokane	General Electric Co. (Spokane Apparatus Service Shop)	WAD001865450	Final	57.80
WA	Spokane	North Market Street	WAD000641548	Final	32.61
WA	Spokane	Northside Landfill	WAD980511778	Final	28.90
WA	Spokane	Old Inland Pit	WAD980982557	Deleted	29.35
WA	Spokane	Spokane Junkyard/Associated Properties	WAD981767296	Deleted	50.00
WA	Spokane County	Greenacres Landfill	WAD980514608	Final	28.90
WA	Tacoma	American Lake Gardens/Mcchord Afb	WAD980833065	Final	28.90
WA	Tacoma	Commencement Bay, South Tacoma Channel	WAD980726301	Final	54.63
WA	Tacoma	Fort Lewis (Landfill No. 5)	WA9214053465	Deleted	33.79
WA	Tacoma	Mcchord Air Force Base (Wash Rack/Treatment Area)	WA8570024200	Deleted	42.24
WA	Tillicum	Fort Lewis Logistics Center	WA7210090067	Final	35.48
WA	Tumwater	Palermo Well Field Ground Water Contamination	WA0000026534	Final	50.00
WA	Vancouver	Alcoa (Vancouver Smelter)	WAD009045279	Deleted	57.80
WA	Vancouver	Bonneville Power Administration Ross Complex (USDOE)	WA1891406349	Deleted	53.67
WA	Vancouver	Boomsnub/Airco	WAD009624453	Final	n/a
WA	Vancouver	Frontier Hard Chrome, Inc.	WAD053614988	Final	57.93
WA	Vancouver	Vancouver Water Station #1 Contamination	WAD988519708	Final	50.00
WA	Vancouver	Vancouver Water Station #4 Contamination	WAD988475158	Final	50.00
WA	Wellpinit	Midnite Mine	WAD980978753	Final	50.00
WA	Whidbey Island	Naval Air Station, Whidbey Island (Ault Field)	WA5170090059	Final	47.58
WA	Whidbey Island	Naval Air Station, Whidbey Island (Seaplane Base)	WA6170090058	Deleted	39.64
WA	Yakima	Fmc Corp. (Yakima Pit)	WAD000643577	Final	38.80
WA	Yakima	Pesticide Lab (Yakima)	WAD120513957	Deleted	29.33
WA	Yakima	Yakima Plating Co.	WAD040187890	Deleted	37.93
WI	Algoma	Algoma Municipal Landfill	WID980610380	Final	39.99
WI	Appleton	N.W. Mauthe Co., Inc.	WID083290981	Final	n/a
WI	Ashippin	Oconomowoc Electroplating Co., Inc.	WID006100275	Final	31.86
WI	Ashland	Ashland/Northern States Power Lakefront	WISFN0507952	Final	50.00
WI	Blooming Grove	Madison Metropolitan Sewerage District Lagoons	WID078934403	Final	32.65
WI	Brookfield	Master Disposal Service Landfill	WID980820070	Final	47.49
WI	Brookfield	Waste Management of Wisconsin, Inc. (Brookfield Sanitary Landfill)	WID980901235	Final	28.90
WI	Caledonia	Hunts Disposal Landfill	WID980511919	Final	31.02
WI	Cleveland Township	Mid-State Disposal, Inc. Landfill	WID980823082	Final	35.23
WI	Daniels	Penta Wood Products	WID006176945	Final	50.00
WI	De Pere	Better Brite Plating Co. Chrome & Zinc Shops	WIT560010118	Final	48.91
WI	Delavan	Delavan Municipal Well #4	WID980820062	Final	28.90
WI	Dunn	City Disposal Corp. Landfill	WID980610646	Final	36.84
WI	Eau Claire	Eau Claire Municipal Well Field	WID980820054	Final	35.57
WI	Eau Claire	National Presto Industries, Inc.	WID006196174	Final	42.39

Superfund National Priorities List - *continued*

St.	City/Area	Site Name	CERCLIS[1] ID	Status	Score[2]
WI	Eau Claire	Waste Research & Reclamation Co.	WID990829475	Deleted	32.13
WI	Excelsior	Sauk County Landfill	WID980610141	Final	34.21
WI	Fond Du Lac County	Ripon City Landfill	WID980610190	Final	39.04
WI	Franklin	Fadrowski Drum Disposal	WID980901227	Final	31.08
WI	Franklin Township	Lemberger Transport & Recycling	WID056247208	Final	34.58
WI	Germantown	Omega Hills North Landfill	WID000808568	Deleted	58.54
WI	Green Bay	Fox River Nrda/PCB Releases	WI0001954841	Proposed	50.00
WI	Harrison	Schmalz Dump	WID980820096	Final	48.92
WI	Janesville	Janesville Ash Beds	WID000712950	Final	57.90
WI	Janesville	Janesville Old Landfill	WID980614044	Final	57.93
WI	Kohler	Kohler Co. Landfill	WID006073225	Final	42.93
WI	La Prairie Township	Wheeler Pit	WID980610620	Deleted	57.80
WI	Medford	Scrap Processing Co., Inc.	WID046536785	Final	34.24
WI	Menomonee Falls	Lauer I Sanitary Landfill	WID058735994	Final	42.69
WI	Middleton	Refuse Hideaway Landfill	WID980610604	Final	34.67
WI	Milwaukee	Moss-American Co., Inc. (Kerr-Mcgee Oil Co.)	WID039052626	Final	32.14
WI	Muskego	Muskego Sanitary Landfill	WID000713180	Final	51.91
WI	Onalaska	Onalaska Municipal Landfill	WID980821656	Final	42.47
WI	Sheboygan	Sheboygan Harbor & River	WID980996367	Final	33.79
WI	Sparta	Northern Engraving Co.	WID006183826	Deleted	38.75
WI	Spencer	Spickler Landfill	WID980902969	Final	44.24
WI	Stoughton	Hagen Farm	WID980610059	Final	32.06
WI	Stoughton	Stoughton City Landfill	WID980901219	Final	35.79
WI	Tomah	Tomah Armory	WID980610299	Final	30.63
WI	Tomah	Tomah Fairgrounds	WID980616841	Deleted	32.87
WI	Tomah	Tomah Municipal Sanitary Landfill	WID980610307	Final	45.91
WI	Wausau	Wausau Ground Water Contamination	WID980993521	Final	28.91
WI	Whitelaw	Lemberger Landfill, Inc.	WID980901243	Final	34.07
WI	Williamstown	Hechimovich Sanitary Landfill	WID052906088	Final	47.91
WV	Fairmont	Big John Salvage - Hoult Road	WVD054827944	Final	48.57
WV	Fairmont	Sharon Steel Corp (Fairmont Coke Works)	WVD000800441	Final	57.08
WV	Follansbee	Follansbee	WVD004336749	Deleted	33.77
WV	Leetown	Leetown Pesticide	WVD980693402	Deleted	36.72
WV	Mineral County	Allegany Ballistics Laboratory (USNAVY)	WV0170023691	Final	50.00
WV	Morgantown	Ordnance Works Disposal Areas	WVD000850404	Final	35.62
WV	Moundsville	Hanlin-Allied-Olin	WVD024185373	Final	53.98
WV	Nitro	Fike Chemical, Inc.	WVD047989207	Final	36.30
WV	Point Pleasant	West Virginia Ordnance (USARMY)	WVD980713036	Final	n/a
WV	Ravenswood	Ravenswood Pce	WVSFN0305428	Final	50.00
WV	Vienna	Vienna Tetrachloroethene	WVD988798401	Final	50.00
WY	Cheyenne	F.E. Warren Air Force Base	WY5571924179	Final	39.23
WY	Evansville	Mystery Bridge Rd/U.S. Highway 20	WYD981546005	Final	32.10
WY	Laramie	Baxter/Union Pacific Tie Treating	WYD061112470	Deleted	37.24

Notes: (1) Comprehensive Environmental Response, Compensation, and Liability Act Information System; (2) Federal Register Hazard Rankings System (HRS) score. The HRS is a model that is used to evaluate the relative threats to human health and the environment posed by actual or potential releases of hazardous substances, pollutants, and contaminants. The HRS criteria take into account the population at risk, the hazard potential of the substances, as well as the potential for contamination of drinking water supplies, direct human contact, destruction of sensitive ecosystems, damage to natural resources affecting the human food chain, contamination of surface water used for recreation or potable water consumption, and contamination of ambient air. The higher the score, the higher the potential threat to human health or the environment.

Source: U.S. Environmental Protection Agency, CERCLIS Hazardous Waste Sites, July 25, 2005.

2003 CERCLA Top 20 list of priority hazardous substances

2003 Rank	Substance Name	Total Points	2001 Rank	CAS Number[1]	Comments
1	Arsenic	1653.61	1	007440-38-2	Exposure to higher than average levels of arsenic happens mostly in the workplace, near hazardous waste sites, or in areas with high natural levels. Arsenic is a powerful poison. At high levels, it can cause death or illness. This chemical has been found in at least 1,014 National Priorities List sites identified by the Environmental Protection Agency.
2	Lead	1528.01	2	007439-92-1	Exposure to lead happens mostly from breathing workplace air or dust, and eating contaminated foods. Children can be exposed from eating lead-based paint chips, or playing in contaminated soil. Lead can damage the nervous system, kidneys, and the immune systems. Lead has been found in at least 1,026 National Priorities List sites identified by the Environmental Protection Agency.
3	Mercury	1503.32	3	007439-97-6	Exposure to mercury occurs from breathing contaminated air, ingesting contaminated water and food, and having dental and medical treatments. Mercury, at high levels, may damage the brain, kidneys, and developing fetus. This chemical has been found in at least 714 National Priorities List sites identified by the Environmental Protection Agency.
4	Vinyl Chloride	1388.65	4	000075-01-4	Exposure to vinyl chloride occurs mainly in the workplace. Breathing high levels of vinyl chloride for short periods of time can cause dizziness, sleepiness, unconsciousness, and at extremely high levels can cause death. Breathing vinyl chloride for long periods of time can result in permanent liver damage, immune reactions, nerve damage, and liver cancer. This substance has been found in at least 496 National Priorities List sites identified by the Environmental Protection Agency
5	Polychlorinated Biphenyls	1364.35	5	001336-36-3	Polychlorinated biphenyls are a mixture of individual chemicals which are no longer produced in the United States, but are still found in the environment. Polychlorinated biphenyls can cause irritation of the nose and throat, and acne and rashes. They have been shown to cause cancer in animal studies. Polychlorinated biphenyls have been found in at least 500 National Priorities List sites identified by the Environmental Protection Agency.
6	Benzene	1356.41	6	000071-43-2	Benzene is a widely used chemical formed from both natural processes and human activities. Breathing benzene can cause drowsiness, dizziness, and unconsciousness; long-term benzene exposure causes effects on the bone marrow and can cause anemia and leukemia. Benzene has been found in at least 813 National Priorities List sites identified by the Environmental Protection Agency.
7	Cadmium	1319.78	7	007440-43-9	Exposure to cadmium happens mostly in the workplace where cadmium products are made. The general population is exposed from breathing cigarette smoke or eating cadmium contaminated foods. Cadmium damages the lungs, can cause kidney disease, and may irritate the digestive tract. Cadmium has been found in at least 776 National Priorities List sites identified by the Environmental Protection Agency.
8	Polycyclic Aromatic Hydrocarbons	1300.73	9	130498-29-2	Exposure to polycyclic aromatic hydrocarbons usually occurs by breathing air contaminated by wild fires or coal tar, or by eating foods that have been grilled. PAHs have been found in at least 600 National Priorities List sites identified by the Environmental Protection Agency.
9	Benzo(a)Pyrene	1303.14	8	000050-32-8	People may be exposed to B[a]P from environmental sources such as air, water, and soil and from cigarette smoke and cooked food. Benzo[a]pyrene is found in the coal tar pitch that industry uses to join electrical parts together. It is also found in creosote, a chemical used to preserve wood. The U.S. Department of Health and Human Services has determined that B[a]P may reasonably be anticipated to be a carcinogen.
10	Benzo(b)Fluoranthene	1271.94	10	000205-99-2	People may be exposed to B[b]F from environmental sources such as air, water, and soil and from cigarette smoke and cooked food. Benzo[a]pyrene is found in the coal tar pitch that industry uses to join electrical parts together. It is also found in creosote, a chemical used to preserve wood. The U.S. Department of Health and Human Services has determined that B[b]F may reasonably be anticipated to be a carcinogen.

613

2001 CERCLA Top 20 list of priority hazardous substances - continued

2003 Rank	Substance Name	Total Points	2001 Rank	CAS Number[1]	Comments
11	Chloroform	1234.42	11	000067-66-3	Exposure to chloroform can occur when breathing contaminated air or when drinking or touching the substance or water containing it. Breathing chloroform can cause dizziness, fatigue, and headaches. Breathing chloroform or ingesting chloroform over long periods of time may damage your liver and kidneys. It can cause sores if large amounts touch your skin. This substance has been found in at least 717 National Priority List sites identified by the Environmental Protection Agency.
12	DDT, p,p'-	1190.24	12	000050-29-3	Exposure to DDT, DDE, and DDD happens mostly from eating contaminated foods, such as root and leafy vegetables, meat, fish, and poultry. At high levels, it can damage the nervous system, causing excitability, tremors, and seizures in people. These chemicals have been found in at least 441 National Priorities List sites identified by the Environmental Protection Agency.
13	Aroclor 1254	1178.24	13	011097-69-1	Polychlorinated biphenyls are a mixture of individual chemicals which are no longer produced in the United States, but are still found in the environment. Polychlorinated biphenyls can cause irritation of the nose and throat, and acne and rashes. They have been shown to cause cancer in animal studies. Polychlorinated biphenyls have been found in at least 383 National Priorities List sites identified by the Environmental Protection Agency.
14	Aroclor 1260	1175.08	14	011096-82-5	Polychlorinated biphenyls are a mixture of individual chemicals which are no longer produced in the United States, but are still found in the environment. Polychlorinated biphenyls can cause irritation of the nose and throat, and acne and rashes. They have been shown to cause cancer in animal studies. Polychlorinated biphenyls have been found in at least 383 National Priorities List sites identified by the Environmental Protection Agency.
15	Dibenzo(a,h)-Anthracene	1159.41	16	000053-70-3	People may be exposed to DB[a,h]A from environmental sources such as air, water, and soil and from cigarette smoke and cooked food. Benzo[a]pyrene is found in the coal tar pitch that industry uses to join electrical parts together. It is also found in creosote, a chemical used to preserve wood. The U.S. Department of Health and Human Services has determined that DB[a,h]A may reasonably be anticipated to be a carcinogen.
16	Trichloroethylene	1160.49	15	000079-01-6	Trichloroethylene is a colorless liquid which is used as a solvent for cleaning metal parts. Drinking or breathing high levels of trichloroethylene may cause nervous system effects, liver and lung damage, abnormal heartbeat, coma, and possibly death. Trichloroethylene has been found in at least 852 National Priorities List sites identified by the Environmental Protection Agency (EPA).
17	Chromium, Hexavalent	1147.80	18	018540-29-9	Exposure to chromium happens mostly from breathing workplace air, or ingesting water or food from soil near waste sites. Chromium can damage the lungs, and cause allergic responses in the skin. Chromium has been found in at least 1,036 National Priorities List sites identified by the Environmental Protection Agency.
18	Dieldrin	1148.51	17	000060-57-1	Exposure to aldrin and dieldrin happens mostly from eating contaminated foods, such as root crops, fish, or seafood. Aldrin and dieldrin build up in the body after years of exposure and can damage the nervous system. Aldrin has been found in at least 207 National Priorities List sites identified by the Environmental Protection Agency. Dieldrin has been found in at least 287 sites.
19	Phosphorus, White	1144.87	24	007723-14-0	White phosphorus is a waxy solid which burns easily and is used in chemical manufacturing and smoke munitions. Exposure to white phosphorus may cause burns and irritation, liver, kidney, heart, lung, or bone damage, and death. White phosphorus has been found in at least 77 of the 1,416 National Priorities List sites identified by the Environmental Protection Agency (EPA).
20	Chlordane	1131.11	19	000057-74-9	Exposure to chlordane occurs mostly from eating contaminated foods, such as root crops, meat, fish, and shellfish, or from touching contaminated soil. High levels of chlordane can cause damage to the nervous system or liver. This chemical has been found in at least 171 National Priorities List sites identified by the Environmental Protection Agency.

Notes: *(1) CAS Number = Chemical Abstracts Service registry number*

The Comprehensive Environmental Response, Compensation, and Liability Act (CERCLA) section 104 (i), as amended by the Superfund Amendments and Reauthorization Act (SARA), requires ATSDR and the EPA to prepare a list, in order of priority, of substances that are most commonly found at facilities on the National Priorities List (NPL) and which are determined to pose the most significant potential threat to human health due to their known or suspected toxicity and potential for human exposure at these NPL sites. CERCLA also requires this list to be revised periodically to reflect additional information on hazardous substances.

This CERCLA priority list is revised and published on a 2-year basis, with a yearly informal review and revision. Each substance on the CERCLA List of Priority Hazardous Substances is a candidate to become the subject of a toxicological profile prepared by ATSDR and subsequently a candidate for the identification of priority data needs. This priority list is based on an algorithm that utilizes the following three components: frequency of occurrence at NPL sites, toxicity, and potential for human exposure to the substances found at NPL sites. This algorithm utilizes data from ATSDR's HazDat database, which contains information from ATSDR's public health assessments and health consultations.

It should be noted that this priority list is not a list of "most toxic" substances, but rather a prioritization of substances based on a combination of their frequency, toxicity, and potential for human exposure at NPL sites. Thus, it is possible for substances with low toxicity but high NPL frequency of occurrence and exposure to be on this priority list. The objective of this priority list is to rank substances across all NPL hazardous waste sites to provide guidance in selecting which substances will be the subject of toxicological profiles prepared by ATSDR.

Source: *Center for Disease Control, Agency for Toxic Substances and Disease Registry, 2003 CERCLA List of Priority Hazardous Substances*

2003 CERCLA list of priority hazardous substances

2003 Rank	Substance Name	Totals Points	2001 Rank	CAS Number[1]
1	Arsenic	1663.11	1	007440-38-2
2	Lead	1531.60	2	007439-92-1
3	Mercury	1506.66	3	007439-97-6
4	Vinyl Chloride	1385.32	4	000075-01-4
5	Polychlorinated Biphenyls	1372.92	5	001336-36-3
6	Benzene	1356.30	6	000071-43-2
7	Cadmium	1319.32	7	007440-43-9
8	Polycyclic Aromatic Hydrocarbons	1317.54	9	130498-29-2
9	Benzo(a)pyrene	1308.71	8	000050-32-8
10	Benzo(b)fluoranthene	1265.26	10	000205-99-2
11	Chloroform	1228.08	11	000067-66-3
12	DDT, p,p'-	1191.57	12	000050-29-3
13	Aroclor 1254	1186.98	13	011097-69-1
14	Aroclor 1260	1176.90	14	011096-82-5
15	Dibenzo(a,h)anthracene	1163.45	16	000053-70-3
16	Trichloroethylene	1161.43	15	000079-01-6
17	Chromium, Hexavalent	1151.98	18	018540-29-9
18	Dieldrin	1148.09	17	000060-57-1
19	Phosphorus, White	1144.87	24	007723-14-0
20	Chlordane	1130.53	19	000057-74-9
21	DDE, p,p'-	1130.20	21	000072-55-9
22	Hexachlorobutadiene	1129.10	20	000087-68-3
23	Coal Tar Creosote	1124.66	22	008001-58-9
24	DDD, p,p'-	1117.38	26	000072-54-8
25	Benzidine	1114.82	25	000092-87-5
26	Aldrin	1111.73	23	000309-00-2
27	Aroclor 1248	1110.73	27	012672-29-6
28	Cyanide	1101.40	28	000057-12-5
29	Aroclor 1242	1095.84	29	053469-21-9
30	Tetrachloroethylene	1086.31	32	000127-18-4
31	Toxaphene	1085.57	31	008001-35-2
32	Hexachlorocyclohexane, Gamma-	1079.78	33	000058-89-9
33	Heptachlor	1067.53	30	000076-44-8
34	1,2-Dibromoethane	1063.22	36	000106-93-4
35	Benzo(a)anthracene	1057.49	34	000056-55-3
36	Disulfoton	1057.16	37	000298-04-4
37	Hexachlorocyclohexane, Beta-	1053.64	35	000319-85-7
38	Beryllium	1044.41	38	007440-41-7
39	Endrin	1040.83	40	000072-20-8
40	Hexachlorocyclohexane, Delta-	1037.91	39	000319-86-8
41	1,2-Dibromo-3-Chloropropane	1034.99	41	000096-12-8
42	Pentachlorophenol	1026.93	43	000087-86-5
43	Carbon Tetrachloride	1024.44	44	000056-23-5
44	Heptachlor Epoxide	1023.70	42	001024-57-3
45	Aroclor 1221	1016.86	46	011104-28-2
46	Aroclor 1016	1014.42	48	012674-11-2
47	DDT, o,p'-	1013.41	50	000789-02-6
48	Di-N-Butyl Phthalate	1011.31	47	000084-74-2
49	Cobalt	1010.49	49	007440-48-4
50	Cis-Chlordane	1008.91	51	005103-71-9
51	Nickel	1006.33	53	007440-02-0
52	Endosulfan Sulfate	1006.02	52	001031-07-8
53	3,3'-Dichlorobenzidine	1004.22	55	000091-94-1
54	Endosulfan	1003.81	54	000115-29-7
55	Trans-Chlordane	1001.83	57	005103-74-2
56	Xylenes, Total	1000.92	56	001330-20-7
57	Endosulfan, Alpha	996.61	45	000959-98-8
58	Dibromochloropropane	993.57	59	067708-83-2
59	Methoxychlor	991.86	58	000072-43-5
60	Aroclor	989.88	61	012767-79-2

2003 CERCLA list of priority hazardous substances - *continued*

2003 Rank	Substance Name	Totals Points	2001 Rank	CAS Number[1]
61	Benzo(k)fluoranthene	984.26	60	000207-08-9
62	Endrin Ketone	979.42	62	053494-70-5
63	Endosulfan, Beta	975.58	63	033213-65-9
64	Chromium(VI) Oxide	967.84	65	001333-82-0
65	Methane	959.39	66	000074-82-8
66	Endrin Aldehyde	954.77	69	007421-93-4
67	Aroclor 1232	954.26	67	011141-16-5
68	Toluene	951.64	68	000108-88-3
69	Benzofluoranthene	949.82	70	056832-73-6
70	2-Hexanone	943.72	71	000591-78-6
71	Acrolein	941.24	72	000107-02-8
72	2,3,7,8-Tetrachlorodibenzo-p-Dioxin	938.30	64	001746-01-6
73	Zinc	929.40	73	007440-66-6
74	Dimethylarsinic Acid	920.42	74	000075-60-5
75	Di(2-Ethylhexyl)Phthalate	919.65	75	000117-81-7
76	Chromium	907.31	76	007440-47-3
77	1,1-Dichloroethene	896.92	79	000075-35-4
78	Naphthalene	896.68	77	000091-20-3
79	Aroclor 1262	893.33	NEW	037324-23-5
80	Methylene Chloride	888.52	78	000075-09-2
81	Aroclor 1240	886.53	80	071328-89-7
82	2,4,6-Trinitrotoluene	877.53	81	000118-96-7
83	2,4,6-Trichlorophenol	875.06	87	000088-06-2
84	Gamma-Chlordene	869.56	175	056641-38-4
85	2,4-Dinitrophenol	868.88	83	000051-28-5
86	Bromodichloroethane	868.28	84	000683-53-4
87	1,2-Dichloroethane	868.24	82	000107-06-2
88	Hydrazine	862.68	86	000302-01-2
89	Bis(2-Chloroethyl) Ether	858.22	85	000111-44-4
90	Thiocyanate	847.54	88	000302-04-5
91	Hexachlorobenzene	844.79	100	000118-74-1
92	Asbestos	843.81	89	001332-21-4
93	Cyclotrimethylenetrinitramine (RDX)	840.67	92	000121-82-4
94	Chlorine	838.86	96	007782-50-5
95	1,1,1-Trichloroethane	838.61	90	000071-55-6
96	2,4-Dinitrotoluene	834.45	101	000121-14-2
97	Uranium	834.01	94	007440-61-1
98	Radium-226	833.83	95	013982-63-3
99	Ethylbenzene	833.62	91	000100-41-4
100	Ethion	832.47	97	000563-12-2
101	Radium	828.32	98	007440-14-4
102	Thorium	825.74	99	007440-29-1
103	4,6-Dinitro-O-Cresol	824.39	93	000534-52-1
104	Radon	818.59	104	010043-92-2
105	Pentachlorobiphenyl	817.89	NEW	025429-29-2
106	1,3,5-Trinitrobenzene	817.85	111	000099-35-4
107	Radium-228	815.49	107	015262-20-1
108	Chlorobenzene	815.45	105	000108-90-7
109	Thorium-230	814.97	108	014269-63-7
110	Barium	814.90	102	007440-39-3
111	n-Nitrosodi-n-Propylamine	814.43	112	000621-64-7
112	Uranium-235	813.99	109	015117-96-1
113	Uranium-234	812.53	113	013966-29-5
114	Diazinon	812.35	114	000333-41-5
115	Fluoranthene	811.22	106	000206-44-0
116	Thorium-228	809.61	115	014274-82-9
117	Radon-222	809.58	116	014859-67-7
118	Hexachlorocyclohexane, Alpha-	807.65	110	000319-84-6
119	Methylmercury	807.18	120	022967-92-6
120	Polonium-210	806.51	121	013981-52-7

2003 CERCLA list of priority hazardous substances - *continued*

2003 Rank	Substance Name	Totals Points	2001 Rank	CAS Number[1]
120	Strontium-90	806.51	118	010098-97-2
122	Plutonium-239	806.47	123	015117-48-3
123	Plutonium-238	806.29	124	013981-16-3
124	Coal Tars	806.25	122	008007-45-2
125	Chrysotile Asbestos	806.06	119	012001-29-5
126	Lead-210	806.02	126	014255-04-0
127	Chlorpyrifos	805.80	125	002921-88-2
128	Plutonium	804.83	131	007440-07-5
129	Thoron (Radon-220)	804.75	128	022481-48-7
130	Americium-241	804.12	127	086954-36-1
131	Manganese	804.07	138	007439-96-5
132	Iodine-131	803.59	130	010043-66-0
132	Tributyltin	803.59	135	000688-73-3
134	Hydrogen Cyanide	803.09	139	000074-90-8
135	Guthion	802.39	134	000086-50-0
136	Neptunium-237	802.17	132	013994-20-2
137	Plutonium-240	801.68	135	014119-33-6
137	Iodine-129	801.68	NEW	015046-84-1
137	Chlordecone	801.68	135	000143-50-0
140	Chrysene	801.36	117	000218-01-9
141	Copper	799.94	129	007440-50-8
142	S,S,S-Tributyl Phosphorotrithioate	796.31	140	000078-48-8
143	Bromine	787.63	141	007726-95-6
144	Polybrominated Biphenyls	787.58	142	067774-32-7
145	Dicofol	786.09	143	000115-32-2
146	Parathion	782.82	145	000056-38-2
147	Selenium	781.84	144	007782-49-2
148	1,1,2,2-Tetrachloroethane	781.51	103	000079-34-5
149	Hexachlorocyclohexane, Technical	773.58	146	000608-73-1
150	n-Nitrosodimethylamine	772.84	237	000062-75-9
151	1,2,3-Trichlorobenzene	771.10	148	000087-61-6
152	Trichlorofluoroethane	769.24	149	027154-33-2
153	Treflan (Trifluralin)	768.72	150	001582-09-8
154	DDD, o,p'-	767.80	152	000053-19-0
155	4,4'-Methylenebis(2-Chloroaniline)	765.16	151	000101-14-4
156	Hexachlorodibenzo-p-Dioxin	759.10	153	034465-46-8
157	Heptachlorodibenzo-p-Dioxin	753.17	154	037871-00-4
158	Pentachlorobenzene	751.95	147	000608-93-5
159	2-Methylnaphthalene	750.45	155	000091-57-6
160	Nitrogen Dioxide	747.82	233	010102-44-0
161	Ammonia	746.01	160	007664-41-7
162	1,1-Dichloroethane	741.30	156	000075-34-3
163	1,1,2-Trichloroethane	732.35	157	000079-00-5
164	1,4-Dichlorobenzene	731.32	161	000106-46-7
165	Acenaphthene	730.35	158	000083-32-9
166	Trichloroethane	724.42	164	025323-89-1
167	1,2,3,4,6,7,8,9-Octachlorodibenzofuran	723.92	159	039001-02-0
168	Hexachlorocyclopentadiene	718.28	165	000077-47-4
169	Heptachlorodibenzofuran	717.32	163	038998-75-3
170	1,2-Diphenylhydrazine	712.62	166	000122-66-7
171	2,3,4,7,8-Pentachlorodibenzofuran	709.06	171	057117-31-4
172	Tetrachlorobiphenyl	707.33	168	026914-33-0
173	Cresol, Para-	707.09	169	000106-44-5
174	Oxychlordane	706.99	170	027304-13-8
175	1,2-Dichloroethene, Trans-	703.86	167	000156-60-5
176	Amosite Asbestos	703.57	133	012172-73-5
177	Carbon Disulfide	703.21	174	000075-15-0
178	Americium	701.63	176	007440-35-9
178	Heptachlorobiphenyl	701.63	NEW	028655-71-2
178	Pentachlorobutadiene	701.63	NEW	055880-77-8

2003 CERCLA list of priority hazardous substances - *continued*

2003 Rank	Substance Name	Totals Points	2001 Rank	CAS Number[1]
181	Tetrachlorophenol	700.97	179	025167-83-3
182	1,2-Dichlorobenzene	700.89	178	000095-50-1
183	Indeno(1,2,3-CD)pyrene	700.86	180	000193-39-5
184	Hexachlorodibenzofuran	699.32	172	055684-94-1
185	Palladium	699.31	173	007440-05-3
186	Phenol	698.68	162	000108-95-2
187	Acetone	694.90	181	000067-64-1
188	Chloroethane	693.46	182	000075-00-3
189	Dibenzofuran	691.84	177	000132-64-9
190	P-Xylene	688.75	183	000106-42-3
191	2,4-Dimethylphenol	688.14	184	000105-67-9
192	Aroclor 1268	684.90	185	011100-14-4
193	Carbon Monoxide	684.55	198	000630-08-0
194	Aluminum	684.18	186	007429-90-5
195	Pentachlorodibenzofuran	672.51	188	030402-15-4
196	Chloromethane	670.70	190	000074-87-3
197	Hydrogen Sulfide	669.09	187	007783-06-4
198	Bis(2-Methoxyethyl) Phthalate	664.55	191	034006-76-3
199	Cresol, Ortho-	659.90	194	000095-48-7
200	Butyl Benzyl Phthalate	654.16	192	000085-68-7
201	2,3,5,6-Tetrachlorophenol	653.47	NEW	000935-95-5
202	Hexachloroethane	652.52	196	000067-72-1
203	Vanadium	649.17	197	007440-62-2
204	1,3-Butadiene	646.15	195	000106-99-0
205	Tetrachloroethane	645.45	189	025322-20-7
206	1,2,4-Trichlorobenzene	645.08	193	000120-82-1
207	Bromoform	644.88	208	000075-25-2
208	Tetrachlorodibenzo-p-Dioxin	635.17	199	041903-57-5
209	1,3-Dichlorobenzene	629.76	200	000541-73-1
210	2,4-Dichlorophenol	627.78	206	000120-83-2
211	n-Nitrosodiphenylamine	625.10	205	000086-30-6
212	Pentachlorodibenzo-p-Dioxin	624.58	201	036088-22-9
213	1,2-Dichloroethylene	624.12	202	000540-59-0
214	2-Butanone	622.63	204	000078-93-3
215	Dibenzothiophene	621.60	NEW	000132-65-0
216	2,3,7,8-Tetrachlorodibenzofuran	621.18	203	051207-31-9
217	Cesium-137	611.86	209	010045-97-3
218	Silver	611.75	207	007440-22-4
219	2-Chlorophenol	611.10	247	000095-57-8
220	Nitrite	611.03	212	014797-65-0
221	Chromium Trioxide	610.93	211	007738-94-5
222	Nitrate	608.36	216	014797-55-8
223	Dinitrotoluene	607.75	213	025321-14-6
224	Potassium-40	607.52	214	013966-00-2
225	Thorium-227	605.50	217	015623-47-9
226	Coal Tar Pitch	605.35	218	065996-93-2
227	Arsenic Acid	604.46	223	007778-39-4
228	2,4,5-Trichlorophenol	604.39	210	000095-95-4
229	Arsenic Trioxide	604.38	220	001327-53-3
230	Antimony	603.25	222	007440-36-0
231	Phorate	603.14	224	000298-02-2
231	Dichloroprop	603.14	NEW	000120-36-5
233	Dimethoate	602.66	225	000060-51-5
234	Strobane	602.58	226	008001-50-1
234	Actinium-227	602.58	226	014952-40-0
236	Pyrethrum	602.52	228	008003-34-7
236	Benzopyrene	602.52	228	073467-76-2
236	4-Aminobiphenyl	602.52	228	000092-67-1
239	Arsine	602.43	231	007784-42-1
240	Naled	602.39	232	000300-76-5

2003 CERCLA list of priority hazardous substances - continued

2003 Rank	Substance Name	Totals Points	2001 Rank	CAS Number[1]
241	Ethoprop	602.17	233	013194-48-4
241	Dibenzofurans, Chlorinated	602.17	233	042934-53-2
243	Alpha Chlordene	601.95	240	056534-02-2
243	Carbophenothion	601.95	236	000786-19-6
245	Dichlorvos	601.68	237	000062-73-7
246	Mercuric Chloride	601.46	240	007487-94-7
246	Uranium-233	601.46	240	013968-55-3
246	Calcium Arsenate	601.46	240	007778-44-1
249	Phenanthrene	601.21	219	000085-01-8
250	Cresols	597.09	244	001319-77-3
251	Formaldehyde	593.74	245	000050-00-0
252	2,4-D Acid	590.62	246	000094-75-7
253	Hydrogen Fluoride	586.62	248	007664-39-3
254	2-Chloroaniline	579.79	NEW	000095-51-2
255	Chlorodibromomethane	578.89	250	000124-48-1
256	1,2,3-Trichloropropane	577.92	NEW	000096-18-4
257	Butylate	576.87	252	002008-41-5
258	Dimethyl Formamide	576.44	255	000068-12-2
259	Pyrene	575.91	249	000129-00-0
260	Dichlorobenzene	575.01	251	025321-22-6
261	Ethyl Ether	571.53	254	000060-29-7
262	Dichloroethane	570.15	253	001300-21-6
263	4-Nitrophenol	568.88	256	000100-02-7
264	1,3-Dichloropropene, Cis-	562.26	257	010061-01-5
265	Phosphine	557.97	260	007803-51-2
266	Trichlorobenzene	555.37	258	012002-48-1
267	2,6-Dinitrotoluene	554.09	263	000606-20-2
268	1,3-Dichloropropene, Trans-	551.34	259	010061-02-6
269	Fluoride	549.11	261	016984-48-8
270	1,2,3,4,6,7,8-Heptachlorodibenzo-p-Dioxin	545.71	271	035822-46-9
271	Methyl Parathion	544.47	262	000298-00-0
272	Carbazole	539.38	266	000086-74-8
273	Bis(2-Ethylhexyl)Adipate	538.73	NEW	000103-23-1
274	Methyl Isobutyl Ketone	533.15	264	000108-10-1
275	Styrene	530.25	265	000100-42-5

Notes: Substances were assigned the same rank when two (or more) substances received equivalent total scores; (1) CAS Number = Chemical Abstracts Service registry number

The Comprehensive Environmental Response, Compensation, and Liability Act (CERCLA) section 104 (i), as amended by the Superfund Amendments and Reauthorization Act (SARA), requires ATSDR and the EPA to prepare a list, in order of priority, of substances that are most commonly found at facilities on the National Priorities List (NPL) and which are determined to pose the most significant potential threat to human health due to their known or suspected toxicity and potential for human exposure at these NPL sites. CERCLA also requires this list to be revised periodically to reflect additional information on hazardous substances.

This CERCLA priority list is revised and published on a 2-year basis, with a yearly informal review and revision. Each substance on the CERCLA List of Priority Hazardous Substances is a candidate to become the subject of a toxicological profile prepared by ATSDR and subsequently a candidate for the identification of priority data needs. This priority list is based on an algorithm that utilizes the following three components: frequency of occurrence at NPL sites, toxicity, and potential for human exposure to the substances found at NPL sites. This algorithm utilizes data from ATSDR's HazDat database, which contains information from ATSDR's public health assessments and health consultations.

It should be noted that this priority list is not a list of "most toxic" substances, but rather a prioritization of substances based on a combination of their frequency, toxicity, and potential for human exposure at NPL sites. Thus, it is possible for substances with low toxicity but high NPL frequency of occurrence and exposure to be on this priority list. The objective of this priority list is to rank substances across all NPL hazardous waste sites to provide guidance in selecting which substances will be the subject of toxicological profiles prepared by ATSDR.

Source: Center for Disease Control, Agency for Toxic Substances and Disease Registry, 2003 CERCLA List of Priority Hazardous Substances

TRI on-site and off-site reported disposed of or otherwise released (in pounds), for facilities in all industries, for all chemicals, by state, U.S., 2003

State	Total On-site Disposal or Other Releases[1]	Total Off-site Disposal or Other Releases[2]	Total On- and Off-site Disposal or Other Releases
Alabama	99,466,982	18,982,161	118,449,143
Alaska	539,444,938	199,327	539,644,265
American Samoa	8,466	0	8,466
Arizona	47,625,926	638,558	48,264,484
Arkansas	34,432,043	6,170,778	40,602,821
California	50,175,151	7,683,197	57,858,347
Colorado	18,044,115	4,474,512	22,518,627
Connecticut	3,775,254	1,651,050	5,426,304
Delaware	9,454,116	4,140,683	13,594,799
District of Columbia	13,482	306	13,788
Florida	123,072,603	3,417,197	126,489,800
Georgia	123,806,928	2,390,117	126,197,045
Guam	236,255	2,048	238,303
Hawaii	2,697,777	419,791	3,117,568
Idaho	60,742,962	584,846	61,327,809
Illinois	100,885,374	31,536,614	132,421,988
Indiana	134,815,824	99,897,387	234,713,210
Iowa	29,949,135	7,484,718	37,433,853
Kansas	25,605,409	3,459,460	29,064,869
Kentucky	82,932,332	7,602,085	90,534,417
Louisiana	121,061,274	5,288,871	126,350,144
Maine	8,509,322	824,734	9,334,055
Maryland	40,760,419	4,739,560	45,499,979
Massachusetts	6,910,270	2,073,014	8,983,284
Michigan	65,289,172	39,045,430	104,334,602
Minnesota	25,815,902	5,622,421	31,438,324
Mississippi	61,587,141	1,489,917	63,077,058
Missouri	94,252,598	8,222,827	102,475,425
Montana	44,529,817	1,120,916	45,650,733
Nebraska	33,749,634	17,722,103	51,471,736
Nevada	402,594,574	758,761	403,353,335
New Hampshire	5,483,433	460,783	5,944,216
New Jersey	16,759,285	6,318,600	23,077,884
New Mexico	17,753,122	146,129	17,899,251
New York	39,572,915	4,507,009	44,079,924
North Carolina	119,553,645	10,292,078	129,845,723
North Dakota	14,425,362	9,213,436	23,638,798
Northern Mariana Isl	6,027	0	6,027
Ohio	204,966,494	45,130,192	250,096,685
Oklahoma	25,317,330	4,621,831	29,939,161
Oregon	40,915,243	1,213,679	42,128,922
Pennsylvania	114,023,038	52,915,509	166,938,547
Puerto Rico	8,065,604	734,652	8,800,256
Rhode Island	634,004	258,276	892,279
South Carolina	61,934,906	24,215,372	86,150,278
South Dakota	10,238,504	94,109	10,332,613
Tennessee	135,432,503	7,179,411	142,611,914
Texas	241,556,148	26,852,811	268,408,959
Utah	238,915,486	3,077,073	241,992,559
Vermont	201,721	144,981	346,702
Virgin Islands	1,323,315	7,916	1,331,231
Virginia	64,812,206	9,454,183	74,266,389
Washington	20,644,737	1,741,529	22,386,266
West Virginia	96,725,971	5,439,311	102,165,281
Wisconsin	31,033,574	19,699,951	50,733,525
Wyoming	18,230,642	1,034,080	19,264,722
Total U.S.	3,920,770,405	522,396,285	4,443,166,690

621

Statistics & Rankings / Hazardous Waste

Notes: *TRI = Toxic Release Inventory; Reporting year 2003 is the most recent TRI data available. (1) On-site disposal or other releases include emissions to the air, discharges to bodies of water, disposal at the facility to land, and disposal in underground injection wells. Disposal or other releases are reported to TRI by media type. On-site disposal or other releases are reported in Section 5 of the TRI Form R. On-site Disposal or Other Releases include Underground Injection to Class I Wells (Section 5.4.1), RCRA Subtitle C Landfills (5.5.1A), Other Landfills (5.5.1B), Fugitive or Non-point Air Emissions (5.1), Stack or Point Air Emissions (5.2), Surface Water Discharges (5.3), Underground Injection to Class II-V Wells (5.4.2), Land Treatment/Application Farming (5.5.2), RCRA Subtitle C Surface Impoundments (5.5.3A), Other Surface Impoundments (5.5.3B), and Other Land Disposal (5.5.4). Off-site Disposal or Other Releases include from Section 6.2 Class I Underground Injection Wells (M81), Class II-V Underground Injection Wells (M82, M71), RCRA Subtitle C Landfills (M65), Other Landfills (M64, M72), Storage Only (M10), Solidification/Stabilization - Metals and Metal Category Compounds only (M41 or M40), Wastewater Treatment (excluding POTWs) - Metals and Metal Category Compounds only (M62 or M61), RCRA Subtitle C Surface Impoundments (M66), Other Surface Impoundments (M67, M63), Land Treatment (M73), Other Land Disposal (M79), Other Off-site Management (M90), Transfers to Waste Broker - Disposal (M94, M91), and Unknown (M99) and, from Section 6.1 Transfers to POTWs (metals and metal category compounds only); (2) An off-site disposal or other release is a discharge of a toxic chemical to the environment that occurs as a result of a facility's transferring a waste containing a TRI chemical off-site for disposal or other release, as reported in Section 6 of the TRI Form R. Certain other types of transfers are also categorized as off-site disposal or other release because, except for location, the outcome of transferring the chemical off-site is the same as disposing of it or releasing it on-site. For each transfer, the amount of the chemical in the waste, type of management activity (chosen from a list of codes referred to as "M" codes) undertaken by the receiving facility, and the address of the receiving site is reported. Off-site disposal or other releases show only net off-site disposal or other releases, that is, off-site disposal or other releases transferred to other TRI facilities reporting such transfers as on-site disposal or other releases are not included to avoid double counting.*

Source: *U.S. Environmental Protection Agency, TRI Explorer, June 8, 2005*

TRI on-site and off-site reported disposed of or otherwise released (in grams), for facilities in all industries, dioxin and dioxin-like compounds, by state, U.S., 2003

State	Total On-site Disposal or Other Releases[1]	Total Off-site Disposal or Other Releases[2]	Total On- and Off-site Disposal or Other Releases
Alabama	260.3	82.9	343.2
Alaska	8.9	0.0	8.9
Arizona	14.8	25.2	40.0
Arkansas	93.3	1.1	94.5
California	110.2	102.9	213.1
Colorado	6.1	0.7	6.8
Connecticut	5.9	1.1	7.0
Delaware	5.9	41,096.8	41,102.7
Florida	87.6	0.4	88.0
Georgia	705.8	0.7	706.5
Guam	0.0	0.0	0.0
Hawaii	4.1	0.9	5.0
Idaho	5.3	24.8	30.1
Illinois	41.6	35.6	77.2
Indiana	104.8	411.7	516.5
Iowa	18.4	0.0	18.4
Kansas	514.3	0.0	514.3
Kentucky	283.6	2,326.5	2,610.1
Louisiana	1,830.4	1,320.4	3,150.8
Maine	8.6	1.4	9.9
Maryland	191.4	0.4	191.7
Massachusetts	10.6	0.4	10.9
Michigan	16,272.7	160.2	16,432.9
Minnesota	798.1	218.4	1,016.5
Mississippi	15,483.3	5.4	15,488.8
Missouri	44.0	0.1	44.1
Montana	5.7	0.1	5.7
Nebraska	2.3	0.0	2.3
Nevada	9.0	0.0	9.0
New Hampshire	176.5	0.0	176.5
New Jersey	18.2	23.3	41.5
New Mexico	2.8	0.0	2.8
New York	40.4	10.6	51.0
North Carolina	887.9	1.6	889.5
North Dakota	12.8	0.0	12.8
Ohio	194.1	160.6	354.7
Oklahoma	25.9	47.8	73.7
Oregon	12.2	0.5	12.7
Pennsylvania	61.6	207.4	269.0
Puerto Rico	15.3	0.2	15.5
South Carolina	56.5	1.9	58.4
South Dakota	4.8	111.9	116.7
Tennessee	1,646.5	198.6	1,845.1
Texas	22,577.5	15,471.7	38,049.1
Utah	4,361.5	0.0	4,361.5
Virgin Islands	2.3	0.0	2.3
Virginia	192.8	2.1	194.8
Washington	30.6	145.1	175.7
West Virginia	60.1	8.6	68.7
Wisconsin	30.4	14.5	44.9
Wyoming	9.3	0.0	9.3
Total	67,346.9	62,224.4	129,571.3

Notes: *TRI = Toxic Release Inventory; Reporting year 2003 is the most recent TRI data available. (1) On-site disposal or other releases include emissions to the air, discharges to bodies of water, disposal at the facility to land, and disposal in underground injection wells. Disposal or other releases are reported to TRI by media type. On-site disposal or other releases are reported in Section 5 of the TRI Form R. On-site Disposal or Other Releases include Underground Injection to Class I Wells (Section 5.4.1), RCRA Subtitle C Landfills (5.5.1A), Other Landfills (5.5.1B), Fugitive or Non-point Air Emissions (5.1), Stack or Point Air Emissions (5.2), Surface Water Discharges (5.3), Underground Injection to Class*

II-V Wells (5.4.2), Land Treatment/Application Farming (5.5.2), RCRA Subtitle C Surface Impoundments (5.5.3A), Other Surface Impoundments (5.5.3B), and Other Land Disposal (5.5.4). Off-site Disposal or Other Releases include from Section 6.2 Class I Underground Injection Wells (M81), Class II-V Underground Injection Wells (M82, M71), RCRA Subtitle C Landfills (M65), Other Landfills (M64, M72), Storage Only (M10), Solidification/Stabilization - Metals and Metal Category Compounds only (M41 or M40), Wastewater Treatment (excluding POTWs) - Metals and Metal Category Compounds only (M62 or M61), RCRA Subtitle C Surface Impoundments (M66), Other Surface Impoundments (M67, M63), Land Treatment (M73), Other Land Disposal (M79), Other Off-site Management (M90), Transfers to Waste Broker - Disposal (M94, M91), and Unknown (M99) and, from Section 6.1 Transfers to POTWs (metals and metal category compounds only); (2) An off-site disposal or other release is a discharge of a toxic chemical to the environment that occurs as a result of a facility's transferring a waste containing a TRI chemical off-site for disposal or other release, as reported in Section 6 of the TRI Form R. Certain other types of transfers are also categorized as off-site disposal or other release because, except for location, the outcome of transferring the chemical off-site is the same as disposing of it or releasing it on-site. For each transfer, the amount of the chemical in the waste, type of management activity (chosen from a list of codes referred to as "M" codes) undertaken by the receiving facility, and the address of the receiving site is reported. Off-site disposal or other releases show only net off-site disposal or other releases, that is, off-site disposal or other releases transferred to other TRI facilities reporting such transfers as on-site disposal or other releases are not included to avoid double counting.

Source: *U.S. Environmental Protection Agency, TRI Explorer, June 8, 2005*

TRI on-site and off-site reported disposed of or otherwise released (in pounds), for facilities in all industries, for all chemicals, top 50 counties, 2003

County	State	Total On-site Disposal or Other Releases[1]	Total Off-site Disposal or Other Releases[2]	Total On- and Off-site Disposal or Other Releases
Northwest Arctic Borough	AK	487,369,936	0	487,369,936
Humboldt	NV	229,066,312	89	229,066,400
Salt Lake	UT	207,587,317	248,673	207,835,990
Elko	NV	99,934,835	2,649	99,937,484
Harris	TX	42,853,740	7,834,469	50,688,209
Juneau Borough	AK	43,855,072	0	43,855,072
Montgomery	IN	201,951	41,799,864	42,001,815
Wayne	MI	9,194,609	31,256,353	40,450,962
Brazoria	TX	39,115,132	169,582	39,284,714
Humphreys	TN	35,328,400	66,533	35,394,933
Peoria	IL	29,854,747	5,398,612	35,253,359
Jefferson	OH	25,300,628	6,274,081	31,574,710
Escambia	FL	31,098,865	167,645	31,266,511
Nye	NV	30,730,096	126,979	30,857,075
Owyhee	ID	30,522,299	127	30,522,426
Berkeley	SC	8,602,420	21,892,677	30,495,098
Beaver	PA	5,120,569	24,719,468	29,840,036
Eureka	NV	29,452,159	5,359	29,457,518
Lake	IN	21,822,204	6,320,797	28,143,001
Jefferson	TX	23,267,161	4,427,203	27,694,364
Gila	AZ	26,024,272	3,055	26,027,327
Reynolds	MO	25,998,172	0	25,998,172
Gilliam	OR	24,562,883	5	24,562,888
Iron	MO	19,280,867	5,265,898	24,546,765
Spencer	IN	22,488,557	640,354	23,128,911
De Kalb	IN	1,444,855	21,438,142	22,882,997
Kings	CA	22,466,838	884	22,467,722
Jefferson	MT	22,197,531	0	22,197,531
Washington	OH	18,113,634	3,963,283	22,076,916
Allen	OH	20,984,963	540,348	21,525,311
Armstrong	PA	20,777,122	3,242	20,780,363
Bartow	GA	19,976,095	41,020	20,017,114
Marshall	WV	19,962,529	11,160	19,973,690
Mobile	AL	7,035,472	12,793,109	19,828,580
Putnam	WV	18,158,380	1,630,560	19,788,940
Ascension	LA	19,049,430	418,577	19,468,007
Baltimore City	MD	18,611,788	201,304	18,813,092
Person	NC	18,598,432	28,863	18,627,296
Mason	WV	18,261,297	37,517	18,298,814
Harrison	MS	18,116,063	28	18,116,091
Marion	IN	3,129,490	14,938,974	18,068,464
Tooele	UT	16,377,275	1,372,871	17,750,146
Adams	OH	17,650,816	11	17,650,827
Hamblen	TN	17,574,634	9,560	17,584,194
Calhoun	TX	17,347,147	12,250	17,359,397
Montgomery	TN	16,477,749	13,773	16,491,522
Cook	IL	5,554,706	10,867,741	16,422,446
St. Charles	LA	16,239,473	76,604	16,316,076
Shoshone	ID	16,179,703	0	16,179,703
Allegheny	PA	8,404,626	7,519,898	15,924,523

Notes: *TRI = Toxic Release Inventory; Reporting year 2003 is the most recent TRI data available. (1) On-site disposal or other releases include emissions to the air, discharges to bodies of water, disposal at the facility to land, and disposal in underground injection wells. Disposal or other releases are reported to TRI by media type. On-site disposal or other releases are reported in Section 5 of the TRI Form R. On-site Disposal or Other Releases include Underground Injection to Class I Wells (Section 5.4.1), RCRA Subtitle C Landfills (5.5.1A), Other Landfills (5.5.1B), Fugitive or Non-point Air Emissions (5.1), Stack or Point Air Emissions (5.2), Surface Water Discharges (5.3), Underground Injection to Class II-V Wells (5.4.2), Land Treatment/Application Farming (5.5.2), RCRA Subtitle C Surface Impoundments (5.5.3A), Other Surface Impoundments (5.5.3B), and Other Land Disposal (5.5.4). Off-site Disposal or Other Releases include from Section 6.2 Class I Underground Injection Wells (M81), Class II-V Underground Injection Wells (M82, M71), RCRA Subtitle C Landfills (M65), Other Landfills (M64, M72), Storage Only (M10), Solidification/Stabilization - Metals and Metal Category Compounds only (M41 or M40), Wastewater Treatment (excluding POTWs) - Metals* and Metal Category Compounds only (M62 or M61), RCRA Subtitle C Surface Impoundments (M66), Other Surface Impoundments (M67, M63), Land Treatment (M73), Other Land Disposal (M79), Other Off-site Management (M90), Transfers to Waste Broker - Disposal (M94, M91), and Unknown (M99) and, from Section 6.1 Transfers to POTWs (metals and metal category compounds only); (2) An off-site disposal or other release is a discharge of a toxic chemical to the environment that occurs as a result of a facility's transferring a waste containing a TRI chemical off-site for disposal or other release, as reported in Section 6 of the TRI Form R. Certain other types of transfers are also categorized as off-site disposal or other release because, except for location, the outcome of transferring the chemical off-site is the same as disposing of it or releasing it on-site. For each transfer, the amount of the chemical in the waste, type of management activity (chosen from a list of codes referred to as "M" codes) undertaken by the receiving facility, and the address of the receiving site is reported. Off-site disposal or other releases show only net off-site disposal or other releases, that is, off-site disposal or other releases transferred to other TRI facilities reporting such transfers as on-site disposal or other releases are not included to avoid double counting.

Source: U.S. Environmental Protection Agency, TRI Explorer, June 8, 2005

TRI on-site and off-site reported disposed of or otherwise released (in grams), for facilities in all industries, dioxin and dioxin-like compounds, top 50 counties, 2003

County	State	Total On-site Disposal or Other Releases[1]	Total Off-site Disposal or Other Releases[2]	Total On- and Off-site Disposal or Other Releases
New Castle	DE	4.8	41,096.8	41,101.6
Brazoria	TX	22,305.5	0.0	22,305.5
Midland	MI	16,255.0	0.0	16,255.0
Harrison	MS	15,045.4	0.0	15,045.5
Harris	TX	218.6	14,787.4	15,006.0
Tooele	UT	4,346.0	0.0	4,346.0
Marshall	KY	1.8	2,326.5	2,328.3
Iberville	LA	1,685.9	278.2	1,964.1
Humphreys	TN	1,604.2	0.0	1,604.2
Sherburne	MN	786.5	0.0	786.5
Montgomery	NC	687.2	0.0	687.2
East Baton Rouge	LA	62.2	532.0	594.2
Chatham	GA	515.0	0.0	515.0
Sedgwick	KS	500.2	0.0	500.2
San Patricio	TX	3.1	478.7	481.8
Rapides	LA	4.5	412.0	416.5
Monroe	MS	298.2	0.0	298.2
Butler	KY	250.6	0.0	250.6
Lancaster	PA	3.7	206.5	210.2
Ramsey	MN	0.1	197.8	197.9
Lake	IN	45.8	141.1	186.9
Wabash	IN	8.2	177.0	185.2
Coos	NH	174.5	0.0	174.5
Calhoun	TX	3.3	161.3	164.6
Hanover	VA	162.5	0.0	162.5
Ashtabula	OH	143.5	14.0	157.5
Crisp	GA	143.0	0.0	143.0
Calcasieu	LA	33.0	102.5	135.5
Branch	MI	1.7	124.9	126.6
Lawrence	SD	0.1	111.9	112.0
Grenada	MS	104.6	3.0	107.6
Baltimore City	MD	106.7	0.0	106.7
Kern	CA	91.4	0.0	91.4
Pierce	WA	2.2	85.7	87.9
Wilkes	NC	86.8	0.0	86.8
Martin	NC	85.6	0.0	85.6
Bibb	AL	75.0	0.0	75.0
San Bernardino	CA	6.2	68.6	74.8
Baltimore	MD	74.6	0.0	74.6
Porter	IN	10.9	63.6	74.5
Tuscarawas	OH	0.8	69.6	70.4
Little River	AR	57.5	0.0	57.5
Loudon	TN	0.7	53.5	54.2
Snohomish	WA	3.2	46.7	49.9
Creek	OK	0.5	47.8	48.3
Dickson	TN	0.6	46.3	47.0
Bedford	TN	0.7	45.1	45.8
Talladega	AL	42.9	0.0	42.9
Tuscaloosa	AL	41.0	0.0	41.0
Butler	OH	2.6	36.2	38.7

Notes: *TRI = Toxic Release Inventory; Reporting year 2003 is the most recent TRI data available. (1) On-site disposal or other releases include emissions to the air, discharges to bodies of water, disposal at the facility to land, and disposal in underground injection wells. Disposal or other releases are reported to TRI by media type. On-site disposal or other releases are reported in Section 5 of the TRI Form R. On-site Disposal or Other Releases include Underground Injection to Class I Wells (Section 5.4.1), RCRA Subtitle C Landfills (5.5.1A), Other Landfills (5.5.1B), Fugitive or Non-point Air Emissions (5.1), Stack or Point Air Emissions (5.2), Surface Water Discharges (5.3), Underground Injection to Class II-V Wells (5.4.2), Land Treatment/Application Farming (5.5.2), RCRA Subtitle C Surface Impoundments (5.5.3A), Other Surface Impoundments (5.5.3B), and Other Land Disposal (5.5.4). Off-site Disposal or Other Releases include from Section 6.2 Class I Underground Injection Wells (M81), Class II-V Underground Injection Wells (M82, M71), RCRA Subtitle C Landfills (M65), Other Landfills (M64, M72), Storage Only (M10), Solidification/Stabilization - Metals and Metal Category Compounds only (M41 or M40), Wastewater Treatment (excluding POTWs) - Metals* *and Metal Category Compounds only (M62 or M61), RCRA Subtitle C Surface Impoundments (M66), Other Surface Impoundments (M67, M63), Land Treatment (M73), Other Land Disposal (M79), Other Off-site Management (M90), Transfers to Waste Broker - Disposal (M94, M91), and Unknown (M99) and, from Section 6.1 Transfers to POTWs (metals and metal category compounds only); (2) An off-site disposal or other release is a discharge of a toxic chemical to the environment that occurs as a result of a facility's transferring a waste containing a TRI chemical off-site for disposal or other release, as reported in Section 6 of the TRI Form R. Certain other types of transfers are also categorized as off-site disposal or other release because, except for location, the outcome of transferring the chemical off-site is the same as disposing of it or releasing it on-site. For each transfer, the amount of the chemical in the waste, type of management activity (chosen from a list of codes referred to as "M" codes) undertaken by the receiving facility, and the address of the receiving site is reported. Off-site disposal or other releases show only net off-site disposal or other releases, that is, off-site disposal or other releases transferred to other TRI facilities reporting such transfers as on-site disposal or other releases are not included to avoid double counting.*

Source: *U.S. Environmental Protection Agency, TRI Explorer, June 8, 2005*

TRI on-site and off-site reported disposed of or otherwise released (in pounds), for facilities in all industries, for all chemicals, all counties, 2003

State	County	Total On-site Disposal or Other Releases[1]	Total Off-site Disposal or Other Releases[2]	Total On- and Off-site Disposal or Other Releases
AK	Aleutians East Borough	69,944.0	0.0	69,944.0
AK	Aleutians West Census Area	8,382.0	0.0	8,382.0
AK	Anchorage Borough	17,198.5	1.0	17,199.5
AK	Bristol Bay Borough	0.1	0.0	0.1
AK	Denali Borough	2,679.3	184,845.0	187,524.3
AK	Fairbanks North Star Borough	5,979,752.4	7,593.3	5,987,345.7
AK	Juneau Borough	43,855,071.8	0.0	43,855,071.8
AK	Kenai Peninsula Borough	2,138,995.2	6,878.4	2,145,873.6
AK	Ketchikan Gateway Borough	0.0	4,180.0	4,180.0
AK	Kodiak Island Borough	368.8	2.6	371.4
AK	Northwest Arctic Borough	487,369,936.0	0.0	487,369,936.0
AK	Southeast Fairbanks Census Area	2,109.6	0.0	2,109.6
AK	Valdez-Cordova Census Area	500.0	0.0	500.0
AL	Autauga	2,935,343.1	0.0	2,935,343.1
AL	Baldwin	21,810.1	9,875.0	31,685.1
AL	Barbour	53,799.0	3,000.0	56,799.0
AL	Bibb	8,699.5	259.4	8,958.8
AL	Blount	23,787.0	0.0	23,787.0
AL	Butler	82,856.0	471.4	83,327.4
AL	Calhoun	546,815.0	341,199.8	888,014.8
AL	Chambers	274,664.5	65,354.8	340,019.4
AL	Cherokee	69.5	1,903.0	1,972.5
AL	Chilton	13,181.1	0.0	13,181.1
AL	Choctaw	1,985,133.2	0.0	1,985,133.2
AL	Clarke	521,456.9	259.7	521,716.6
AL	Clay	149,205.8	273.0	149,478.8
AL	Cleburne	7.0	159.0	166.0
AL	Coffee	983,510.2	20.0	983,530.2
AL	Colbert	7,004,405.5	51,721.2	7,056,126.6
AL	Conecuh	59,013.0	15.2	59,028.2
AL	Coosa	33,997.9	0.0	33,997.9
AL	Covington	47,030.5	256.5	47,287.0
AL	Cullman	1,942,494.0	7,097.5	1,949,591.5
AL	Dale	115,405.5	21,266.0	136,671.5
AL	Dallas	858,242.3	27,449.8	885,692.1
AL	DeKalb	282,528.6	4,241.0	286,769.6
AL	Elmore	71,410.0	3,562.5	74,972.5
AL	Escambia	1,730,383.5	14,871.9	1,745,255.4
AL	Etowah	692,746.5	105,108.4	797,854.8
AL	Fayette	27,274.9	5,669.0	32,943.9
AL	Franklin	339,623.0	4,163.0	343,786.0
AL	Geneva	111,828.0	0.0	111,828.0
AL	Greene	3,489,283.5	0.0	3,489,283.5
AL	Hale	45,869.0	0.0	45,869.0
AL	Henry	12,942.6	0.0	12,942.6
AL	Houston	383,197.1	29,751.8	412,948.8
AL	Jackson	6,527,732.4	13,354.9	6,541,087.3
AL	Jefferson	10,722,598.1	2,207,159.6	12,929,757.7
AL	Lamar	154.2	7.0	161.2
AL	Lauderdale	1,647.0	24,843.0	26,490.0
AL	Lawrence	2,436,470.4	0.0	2,436,470.4
AL	Lee	473,374.4	160,449.1	633,823.5
AL	Limestone	9,484.9	22,755.2	32,240.1
AL	Lowndes	474,319.0	325,500.0	799,819.0
AL	Madison	747,538.8	116,746.1	864,284.9
AL	Marengo	800,512.3	142.6	800,654.8
AL	Marion	692,595.0	10,866.0	703,461.0
AL	Marshall	154,468.6	731.1	155,199.8

**TRI on-site and off-site reported disposed of or otherwise released (in pounds),
for facilities in all industries, for all chemicals, all counties, 2003** - *continued*

State	County	Total On-site Disposal or Other Releases[1]	Total Off-site Disposal or Other Releases[2]	Total On- and Off-site Disposal or Other Releases
AL	Mobile	7,035,471.6	12,793,108.5	19,828,580.1
AL	Monroe	3,813,955.2	716.2	3,814,671.4
AL	Montgomery	51,038.7	49,174.6	100,213.3
AL	Morgan	5,789,420.7	388,462.0	6,177,882.7
AL	Perry	1,391.1	0.0	1,391.1
AL	Pike	2,893,974.2	10,846.0	2,904,820.2
AL	Randolph	9,554.0	3.2	9,557.2
AL	Russell	2,626,898.6	638.0	2,627,536.6
AL	Shelby	9,552,684.5	4,402.0	9,557,086.5
AL	St. Clair	161,598.7	36,478.7	198,077.5
AL	Sumter	7,742,432.1	16,691.0	7,759,123.1
AL	Talladega	1,168,992.5	16,927.9	1,185,920.4
AL	Tallapoosa	33,854.0	0.0	33,854.0
AL	Tuscaloosa	727,532.7	3,667,205.5	4,394,738.2
AL	Walker	5,450,911.1	750.1	5,451,661.2
AL	Washington	2,633,696.6	4,355.0	2,638,051.6
AL	Wilcox	1,761,188.6	0.0	1,761,188.6
AL	Winston	123,479.0	804.6	124,283.6
AR	Arkansas	549,042.5	5.0	549,047.5
AR	Ashley	1,814,612.6	1,950.0	1,816,562.6
AR	Baxter	36,412.0	17,973.7	54,385.7
AR	Benton	810,817.0	127,238.3	938,055.3
AR	Boone	42,643.2	13,252.3	55,895.4
AR	Bradley	52,896.9	369.3	53,266.2
AR	Calhoun	191,201.1	0.0	191,201.1
AR	Carroll	46,555.0	26.0	46,581.0
AR	Clark	168,546.5	3,878.7	172,425.1
AR	Clay	66,300.0	0.0	66,300.0
AR	Cleburne	12,297.2	1,409.7	13,706.9
AR	Columbia	625,780.1	347,638.9	973,418.9
AR	Conway	687,691.9	0.5	687,692.4
AR	Craighead	119,116.7	33,167.6	152,284.3
AR	Crawford	31,994.0	40,554.0	72,547.9
AR	Crittenden	470,298.0	67,936.8	538,234.8
AR	Cross	42,088.8	15,422.2	57,510.9
AR	Dallas	33,799.5	470.8	34,270.3
AR	Desha	650,842.0	0.0	650,842.0
AR	Drew	12,118.0	12.0	12,130.0
AR	Faulkner	258,744.5	71,362.5	330,107.0
AR	Franklin	66,477.0	0.0	66,477.0
AR	Garland	3,452,202.7	54,121.0	3,506,323.7
AR	Grant	106,864.0	42,724.0	149,588.0
AR	Greene	59,374.0	71,369.4	130,743.4
AR	Hempstead	111,797.0	1,750.0	113,547.0
AR	Hot Spring	124,051.5	5,040.3	129,091.8
AR	Howard	98,872.0	16,106.2	114,978.2
AR	Independence	1,485,842.6	65,131.0	1,550,973.6
AR	Jackson	30,496.3	0.0	30,496.3
AR	Jefferson	5,776,139.4	79,489.5	5,855,628.9
AR	Johnson	218,619.0	0.0	218,619.0
AR	Lawrence	2,817.0	0.0	2,817.0
AR	Little River	2,650,547.2	2.7	2,650,549.9
AR	Logan	1,747,909.0	117,997.0	1,865,906.0
AR	Lonoke	45,777.0	51,992.0	97,769.0
AR	Madison	5,417.0	0.0	5,417.0
AR	Marion	290,545.0	0.0	290,545.0
AR	Miller	33,755.0	50,785.0	84,540.0
AR	Mississippi	2,626,991.3	6,285,932.5	8,912,923.8
AR	Nevada	52,890.9	51,934.6	104,825.6

TRI on-site and off-site reported disposed of or otherwise released (in pounds),
for facilities in all industries, for all chemicals, all counties, 2003 - *continued*

State	County	Total On-site Disposal or Other Releases[1]	Total Off-site Disposal or Other Releases[2]	Total On- and Off-site Disposal or Other Releases
AR	Ouachita	104,396.8	136.4	104,533.2
AR	Phillips	69,755.0	105.0	69,860.1
AR	Pike	0.0	0.2	0.2
AR	Poinsett	266.0	5,170.0	5,436.0
AR	Polk	76,099.0	990.0	77,089.0
AR	Pope	77,422.8	2,180.5	79,603.3
AR	Prairie	265.0	510.0	775.0
AR	Pulaski	305,733.7	97,107.7	402,841.4
AR	Randolph	48,161.0	2,039.0	50,200.0
AR	Saline	3,444.2	327,620.4	331,064.5
AR	Scott	10,212.0	638.0	10,850.0
AR	Searcy	12.9	6,490.4	6,503.3
AR	Sebastian	652,477.6	675,087.0	1,327,564.6
AR	Sevier	9,361.0	3.0	9,364.0
AR	St. Francis	18,890.0	6,600.0	25,490.0
AR	Stone	19.0	0.0	19.0
AR	Union	5,438,483.4	1,206,823.2	6,645,306.6
AR	Van Buren	0.6	1,227.0	1,227.6
AR	Washington	496,057.2	77,133.9	573,191.1
AR	White	2,026.7	11,728.1	13,754.9
AR	Woodruff	230,223.6	315.0	230,538.6
AR	Yell	1,177,553.1	260.0	1,177,813.1
AS	Eastern	8,465.8	0.0	8,465.8
AZ	Apache	4,163,626.3	0.0	4,163,626.3
AZ	Cochise	1,043,456.3	0.0	1,043,456.3
AZ	Coconino	1,934,094.2	123.4	1,934,217.6
AZ	Gila	26,024,272.0	3,055.2	26,027,327.2
AZ	Greenlee	4,400,930.2	73,118.6	4,474,048.8
AZ	La Paz	10.0	0.0	10.0
AZ	Maricopa	1,552,356.5	465,290.9	2,017,647.3
AZ	Mohave	152,469.1	126,072.5	278,541.5
AZ	Navajo	1,912,024.7	20.0	1,912,044.8
AZ	Pima	3,363,429.2	143,714.1	3,507,143.3
AZ	Pinal	1,771,762.5	2,322.1	1,774,084.6
AZ	Santa Cruz	596.2	1,575.3	2,171.5
AZ	Yavapai	1,271,913.8	0.0	1,271,913.8
AZ	Yuma	34,985.5	57.5	35,043.0
CA	Alameda	682,031.7	591,244.8	1,273,276.5
CA	Amador	194,365.0	0.0	194,365.0
CA	Butte	810.3	0.3	810.6
CA	Colusa	190,183.0	0.0	190,183.0
CA	Contra Costa	3,588,275.4	337,068.1	3,925,343.5
CA	Del Norte	1,922.0	0.0	1,922.0
CA	Fresno	295,496.2	89,152.0	384,648.2
CA	Glenn	31,425.0	0.0	31,425.0
CA	Humboldt	1,780,911.1	1.0	1,780,912.1
CA	Imperial	331,639.6	4,044.9	335,684.5
CA	Inyo	19,898.3	0.0	19,898.3
CA	Kern	3,771,914.1	14,176.3	3,786,090.4
CA	Kings	22,466,837.8	883.8	22,467,721.6
CA	Lake	1,346.1	0.0	1,346.1
CA	Lassen	47.7	687.0	734.7
CA	Los Angeles	7,584,217.5	3,726,201.9	11,310,419.4
CA	Madera	298,536.3	3,200.0	301,736.3
CA	Mariposa	0.0	0.0	0.0
CA	Mendocino	130.4	251,602.9	251,733.3
CA	Merced	377,654.2	1,126,406.5	1,504,060.6
CA	Mono	624.5	0.0	624.5
CA	Monterey	6,439.0	0.0	6,439.0

TRI on-site and off-site reported disposed of or otherwise released (in pounds), for facilities in all industries, for all chemicals, all counties, 2003 - *continued*

State	County	Total On-site Disposal or Other Releases[1]	Total Off-site Disposal or Other Releases[2]	Total On- and Off-site Disposal or Other Releases
CA	Napa	2.0	104,985.0	104,987.0
CA	Nevada	0.0	1.0	1.0
CA	Orange	1,390,011.6	135,158.0	1,525,169.6
CA	Placer	97,678.7	2,630.5	100,309.2
CA	Plumas	182.0	0.0	182.0
CA	Riverside	353,612.0	5,383.0	358,995.0
CA	Sacramento	630,511.1	91,394.1	721,905.2
CA	San Benito	1,459.3	1,517.3	2,976.6
CA	San Bernardino	1,972,181.5	501,478.1	2,473,659.7
CA	San Diego	1,074,178.1	1,663,197.7	2,737,375.8
CA	San Francisco	33,795.1	7,345.4	41,140.5
CA	San Joaquin	581,610.6	49,522.9	631,133.5
CA	San Luis Obispo	82,880.2	80,264.8	163,144.9
CA	San Mateo	37,356.1	131,295.0	168,651.1
CA	Santa Barbara	12,810.2	4,679.8	17,490.0
CA	Santa Clara	282,136.7	304,052.0	586,188.7
CA	Santa Cruz	443.5	0.0	443.5
CA	Shasta	82,796.6	9,902.0	92,698.7
CA	Siskiyou	113,338.1	1,297.0	114,635.1
CA	Solano	1,244,104.6	33,386.0	1,277,490.5
CA	Sonoma	45,285.2	265.0	45,550.3
CA	Stanislaus	209,491.4	163,962.9	373,454.3
CA	Tehama	585.0	0.0	585.0
CA	Tulare	30,789.6	5,892.0	36,681.6
CA	Ventura	70,025.8	23,144.3	93,170.1
CA	Yolo	161,196.4	638.6	161,835.0
CA	Yuba	41,984.0	0.0	41,984.0
CO	Adams	495,374.5	1,169,473.5	1,664,848.0
CO	Arapahoe	159.6	1,309.5	1,469.1
CO	Boulder	423,841.8	38,750.2	462,592.1
CO	Broomfield	4,761.4	1,669.5	6,430.9
CO	Clear Creek	533,678.0	0.0	533,678.0
CO	Costilla	33,063.0	410.0	33,473.0
CO	Delta	3.5	0.3	3.8
CO	Denver	135,485.6	263,592.4	399,078.0
CO	Douglas	183.2	0.0	183.2
CO	Eagle	0.0	0.0	0.0
CO	El Paso	1,508,617.4	607,506.7	2,116,124.0
CO	Fremont	530,003.8	83,511.0	613,514.8
CO	Garfield	0.0	0.0	0.0
CO	Grand	2,177,403.8	0.0	2,177,403.8
CO	Jackson	174.9	0.0	174.9
CO	Jefferson	567,998.6	234,073.4	802,072.0
CO	Larimer	724,556.3	1,409.9	725,966.3
CO	Logan	0.0	0.0	0.0
CO	Mesa	195,549.8	11,452.0	207,001.7
CO	Moffat	2,369,178.2	2,137,381.2	4,506,559.3
CO	Montrose	39,793.1	27,066.3	66,859.4
CO	Morgan	2,867,153.8	130,833.0	2,997,986.8
CO	Otero	12,359.0	0.0	12,359.0
CO	Prowers	46,206.2	1.8	46,208.0
CO	Pueblo	596,266.2	387,024.5	983,290.7
CO	Rio Blanco	5.6	0.0	5.6
CO	Routt	873,187.9	1.3	873,189.2
CO	Teller	3,184,411.5	0.0	3,184,411.5
CO	Weld	724,698.8	106,746.3	831,445.1
CT	Fairfield	616,829.0	98,572.8	715,401.8
CT	Hartford	528,091.3	953,314.9	1,481,406.2
CT	Litchfield	473,921.8	57,303.8	531,225.6

TRI on-site and off-site reported disposed of or otherwise released (in pounds),
for facilities in all industries, for all chemicals, all counties, 2003 - *continued*

State	County	Total On-site Disposal or Other Releases[1]	Total Off-site Disposal or Other Releases[2]	Total On- and Off-site Disposal or Other Releases
CT	Middlesex	116,816.2	42,442.7	159,258.9
CT	New Haven	1,104,443.6	292,635.8	1,397,079.4
CT	New London	879,064.5	233,494.6	1,112,559.1
CT	Tolland	31,505.6	1,564.8	33,070.4
CT	Windham	24,581.8	66,127.6	90,709.4
DC	District of Columbia	13,482.0	305.9	13,788.0
DE	Kent	96,265.2	10,190.4	106,455.6
DE	New Castle	4,644,824.6	4,129,100.9	8,773,925.6
DE	Sussex	4,713,026.2	1,580.3	4,714,606.5
FL	Alachua	1,767,279.1	138,387.5	1,905,666.6
FL	Baker	11,230.0	11,033.0	22,263.0
FL	Bay	6,982,605.4	99,845.9	7,082,451.3
FL	Bradford	0.0	0.0	0.0
FL	Brevard	901,286.0	50,516.9	951,802.9
FL	Broward	850,834.5	9,152.7	859,987.1
FL	Citrus	13,551,300.5	50.2	13,551,350.7
FL	Clay	66,382.0	9.8	66,391.8
FL	Collier	5,495.0	0.0	5,495.0
FL	Columbia	0.0	0.0	0.0
FL	Dade	368,057.9	39,813.8	407,871.7
FL	DeSoto	55.0	0.0	55.0
FL	Dixie	33,246.5	0.0	33,246.5
FL	Duval	11,718,491.8	610,111.3	12,328,603.1
FL	Escambia	31,098,865.3	167,645.4	31,266,510.7
FL	Flagler	158,446.0	5.0	158,451.0
FL	Gadsden	96.5	0.0	96.5
FL	Glades	0.0	0.0	0.0
FL	Gulf	10,622.0	88.1	10,710.1
FL	Hamilton	401,677.4	0.0	401,677.4
FL	Hardee	96,727.0	250.0	96,977.0
FL	Hendry	1,677,497.9	0.0	1,677,497.9
FL	Hernando	51,316.2	5,648.0	56,964.2
FL	Highlands	140,541.8	438,298.6	578,840.4
FL	Hillsborough	10,083,790.4	188,598.7	10,272,389.1
FL	Indian River	437,247.2	100.0	437,347.2
FL	Jackson	150,495.8	20.1	150,515.9
FL	Lafayette	1,340.0	0.0	1,340.0
FL	Lake	116,251.0	17,381.0	133,632.0
FL	Lee	221,201.4	549.5	221,750.9
FL	Leon	136,736.0	5,837.4	142,573.5
FL	Levy	190,663.0	0.0	190,663.0
FL	Liberty	643.9	750.0	1,393.9
FL	Manatee	2,071,433.4	93,297.0	2,164,730.3
FL	Marion	336,927.4	63.9	336,991.3
FL	Martin	712,747.3	233,812.5	946,559.8
FL	Miami-Dade	765,695.8	180,933.4	946,629.1
FL	Monroe	44.0	0.0	44.0
FL	Nassau	3,517,101.7	0.0	3,517,101.7
FL	Okaloosa	48,341.9	0.1	48,342.0
FL	Okeechobee	0.0	0.0	0.0
FL	Orange	9,849,829.5	15,611.0	9,865,440.5
FL	Osceola	6,674.8	216.0	6,890.8
FL	Palm Beach	1,868,507.7	18,873.6	1,887,381.3
FL	Pasco	878,428.3	17,679.0	896,107.3
FL	Pinellas	1,086,748.8	46,061.0	1,132,809.8
FL	Polk	5,363,761.9	680,665.2	6,044,427.1
FL	Putnam	9,574,341.3	65,857.0	9,640,198.3
FL	Santa Rosa	655,553.0	330,016.0	985,569.0
FL	Sarasota	34,253.5	595.0	34,848.5

**TRI on-site and off-site reported disposed of or otherwise released (in pounds),
for facilities in all industries, for all chemicals, all counties, 2003** - *continued*

State	County	Total On-site Disposal or Other Releases[1]	Total Off-site Disposal or Other Releases[2]	Total On- and Off-site Disposal or Other Releases
FL	Seminole	737.3	10.0	747.3
FL	St. Johns	101,604.6	780.6	102,385.2
FL	St. Lucie	624,228.7	4,435.5	628,664.2
FL	Sumter	84,497.0	260.0	84,757.0
FL	Suwannee	1,656,808.6	2.4	1,656,811.0
FL	Taylor	1,808,715.0	333.4	1,809,048.4
FL	Union	47,538.0	0.0	47,538.0
FL	Volusia	745,009.6	13,340.6	758,350.2
FL	Wakulla	2,651.0	500.0	3,151.0
FL	Walton	0.0	0.0	0.0
GA	Appling	5,368.0	0.0	5,368.0
GA	Atkinson	326,547.7	26,010.8	352,558.6
GA	Bacon	5,611.0	0.0	5,611.0
GA	Baldwin	135,731.0	5.0	135,736.0
GA	Banks	0.0	0.0	0.0
GA	Barrow	44,814.0	3,329.0	48,143.0
GA	Bartow	19,976,094.6	41,019.6	20,017,114.2
GA	Ben Hill	4,876.9	323.0	5,199.9
GA	Berrien	447,167.6	0.0	447,167.6
GA	Bibb	1,809,748.0	119,376.4	1,929,124.4
GA	Brantley	0.0	0.0	0.0
GA	Brooks	138,286.0	219.0	138,505.0
GA	Bryan	932.0	4,800.0	5,732.0
GA	Bulloch	1,523.7	3,840.2	5,363.8
GA	Burke	20,456.0	5,115.0	25,571.0
GA	Butts	0.0	19.1	19.1
GA	Camden	216,011.0	0.0	216,011.0
GA	Carroll	22,982.9	385,011.8	407,994.7
GA	Catoosa	0.0	0.0	0.0
GA	Charlton	32.0	0.0	32.0
GA	Chatham	6,909,559.7	287,660.4	7,197,220.1
GA	Chattahoochee	303,211.0	0.0	303,211.0
GA	Chattooga	236,616.0	594.0	237,210.0
GA	Cherokee	321,126.3	631.4	321,757.7
GA	Clarke	741,988.4	120,264.4	862,252.8
GA	Clayton	330,330.1	18,211.8	348,541.9
GA	Clinch	107,277.0	23,690.0	130,967.0
GA	Cobb	3,078,624.2	108,306.6	3,186,930.8
GA	Coffee	124,896.8	14,660.0	139,556.8
GA	Colquitt	4,094.0	11.0	4,105.0
GA	Columbia	998,015.0	37,116.0	1,035,131.0
GA	Cook	38,780.0	1,127.0	39,907.0
GA	Coweta	4,952,801.3	36,771.1	4,989,572.4
GA	Crawford	0.0	0.0	0.0
GA	Crisp	857,625.7	34,755.0	892,380.7
GA	Dade	2,787.0	7.0	2,794.0
GA	DeKalb	915,455.4	163,767.4	1,079,222.8
GA	Decatur	4,857,660.5	10.8	4,857,671.3
GA	Dooly	499,330.0	227.0	499,557.0
GA	Dougherty	440,754.2	85,679.6	526,433.8
GA	Douglas	12,324.0	0.0	12,324.0
GA	Early	3,477,225.3	112,603.0	3,589,828.3
GA	Effingham	1,449,467.9	10.0	1,449,477.9
GA	Elbert	43,168.4	46,585.0	89,753.4
GA	Evans	6,348.3	3,945.0	10,293.3
GA	Fannin	0.0	0.0	0.0
GA	Fayette	53,403.5	60,698.2	114,101.7
GA	Floyd	5,726,306.3	29,944.5	5,756,250.8
GA	Forsyth	81,497.8	40,332.6	121,830.4

TRI on-site and off-site reported disposed of or otherwise released (in pounds), for facilities in all industries, for all chemicals, all counties, 2003 - continued

State	County	Total On-site Disposal or Other Releases[1]	Total Off-site Disposal or Other Releases[2]	Total On- and Off-site Disposal or Other Releases
GA	Franklin	11,702.0	43.0	11,745.0
GA	Fulton	1,441,214.8	23,132.5	1,464,347.3
GA	Gilmer	3,832.0	0.0	3,832.0
GA	Glynn	2,529,739.7	29,675.0	2,559,414.7
GA	Gordon	53,147.4	9,263.8	62,411.2
GA	Grady	89,876.0	0.0	89,876.0
GA	Greene	322,181.0	0.0	322,181.0
GA	Gwinnett	104,832.3	6,203.2	111,035.4
GA	Habersham	44,498.9	25,283.9	69,782.8
GA	Hall	378,593.6	41,634.4	420,228.0
GA	Haralson	82,178.4	6,486.3	88,664.8
GA	Harris	51,363.0	0.0	51,363.0
GA	Hart	1,071.4	23,974.5	25,045.9
GA	Heard	10,225,204.2	0.0	10,225,204.2
GA	Henry	58,182.6	7,728.7	65,911.4
GA	Houston	400,312.8	27,975.6	428,288.4
GA	Irwin	0.0	0.0	0.0
GA	Jackson	221,065.6	1,820.9	222,886.5
GA	Jasper	222,285.0	0.0	222,285.0
GA	Jefferson	5.0	250.0	255.0
GA	Jenkins	17,449.0	0.0	17,449.0
GA	Lamar	94,403.0	0.0	94,403.0
GA	Laurens	263,564.4	147.9	263,712.3
GA	Lee	19,480.0	0.0	19,480.0
GA	Liberty	696,709.2	1,300.0	698,009.2
GA	Lowndes	2,303,943.7	21,712.0	2,325,655.7
GA	Lumpkin	0.0	21,260.0	21,260.0
GA	Macon	709,402.8	0.0	709,402.8
GA	Marion	66,406.7	0.0	66,406.7
GA	McDuffie	231,767.0	2,253.8	234,020.8
GA	McIntosh	3,955.3	0.0	3,955.3
GA	Meriwether	24,506.5	0.0	24,506.5
GA	Mitchell	11,770.0	40.0	11,810.0
GA	Monroe	15,805,070.5	0.0	15,805,070.5
GA	Morgan	45,387.9	0.0	45,387.9
GA	Murray	0.0	0.0	0.0
GA	Muscogee	112,090.0	128,633.7	240,723.8
GA	Newton	96,530.9	65,580.7	162,111.5
GA	Oconee	12,123.0	35,811.0	47,934.0
GA	Paulding	191.0	8,887.1	9,078.1
GA	Peach	36,402.0	0.0	36,402.0
GA	Pickens	1,195.0	0.0	1,195.0
GA	Pierce	27,665.0	0.0	27,665.0
GA	Polk	403,059.8	3,096.0	406,155.8
GA	Pulaski	49,700.0	13,900.0	63,600.0
GA	Putnam	9,913,909.2	1,200.0	9,915,109.2
GA	Quitman	0.0	0.0	0.0
GA	Rabun	103.6	849.3	952.9
GA	Randolph	72,126.0	0.0	72,126.0
GA	Richmond	10,892,823.9	21,987.8	10,914,811.6
GA	Rockdale	4,258.9	742.3	5,001.1
GA	Schley	20.0	6,615.0	6,635.0
GA	Screven	1,500.0	0.0	1,500.0
GA	Spalding	88,012.4	60,597.7	148,610.1
GA	Stephens	5,760.0	69,201.5	74,961.5
GA	Sumter	925,679.2	110.9	925,790.1
GA	Talbot	0.0	0.0	0.0
GA	Taliaferro	0.0	0.0	0.0
GA	Tattnall	0.0	0.0	0.0

**TRI on-site and off-site reported disposed of or otherwise released (in pounds),
for facilities in all industries, for all chemicals, all counties, 2003** - *continued*

State	County	Total On-site Disposal or Other Releases[1]	Total Off-site Disposal or Other Releases[2]	Total On- and Off-site Disposal or Other Releases
GA	Taylor	0.0	3,451.0	3,451.0
GA	Terrell	92,130.0	0.0	92,130.0
GA	Thomas	5,571.0	250.0	5,821.0
GA	Tift	29,877.3	1,115.0	30,992.3
GA	Toombs	35.0	630.0	665.0
GA	Troup	375,485.2	55,395.1	430,880.3
GA	Upson	204,940.9	2,920.0	207,860.9
GA	Walker	71,813.7	10,368.9	82,182.6
GA	Walton	5,698.6	10,812.8	16,511.3
GA	Ware	271,318.0	750.0	272,068.0
GA	Warren	4.2	0.0	4.2
GA	Washington	1.8	16.5	18.4
GA	Wayne	3,323,048.6	9,800.0	3,332,848.6
GA	White	2.0	3,706.0	3,708.0
GA	Whitfield	519,998.1	27,666.1	547,664.2
GA	Wilkes	1,896.0	0.0	1,896.0
GU	Guam	236,254.6	2,048.0	238,302.6
HI	Hawaii	388,633.3	101.3	388,734.6
HI	Honolulu	2,048,474.8	419,718.3	2,468,193.1
HI	Kauai	16,258.0	0.0	16,258.0
HI	Maui	244,410.7	3.1	244,413.8
IA	Adair	25,049.0	770.0	25,819.0
IA	Adams	11.0	1,735.0	1,746.0
IA	Allamakee	334,846.2	0.0	334,846.2
IA	Appanoose	755.0	90,059.2	90,814.2
IA	Benton	3.0	0.0	3.0
IA	Black Hawk	449,160.7	18,610.5	467,771.2
IA	Boone	0.0	752.0	752.0
IA	Bremer	2,768.0	3.0	2,771.0
IA	Buchanan	68,000.0	0.0	68,000.0
IA	Buena Vista	56,860.0	750.0	57,610.0
IA	Butler	22,664.6	229.0	22,893.6
IA	Calhoun	33,942.0	0.0	33,942.0
IA	Carroll	130,564.0	0.0	130,564.0
IA	Cass	13,016.0	25,548.0	38,564.0
IA	Cedar	143,609.0	0.0	143,609.0
IA	Cerro Gordo	620,489.6	21,887.4	642,377.0
IA	Cherokee	35,268.4	0.0	35,268.4
IA	Chickasaw	9,339.2	1,868.4	11,207.7
IA	Clarke	3,750.0	785.0	4,535.0
IA	Clay	76,590.3	7,035.0	83,625.3
IA	Clayton	28,800.0	0.0	28,800.0
IA	Clinton	3,217,970.5	167,240.2	3,385,210.7
IA	Crawford	58,000.0	63,400.0	121,400.0
IA	Dallas	403,043.6	23,736.0	426,779.6
IA	Davis	255.0	5,157.4	5,412.4
IA	Delaware	4,276.4	8,123.5	12,399.8
IA	Des Moines	106,087.7	18,771.9	124,859.6
IA	Dickinson	19,000.0	0.0	19,000.0
IA	Dubuque	180,686.3	67,150.6	247,836.9
IA	Emmet	53,354.0	0.0	53,354.0
IA	Fayette	65,804.0	13,534.0	79,338.0
IA	Floyd	190,069.0	165,397.2	355,466.2
IA	Franklin	96,840.0	3.0	96,843.0
IA	Fremont	3,480.9	0.0	3,480.9
IA	Greene	0.0	514.2	514.2
IA	Grundy	0.0	0.0	0.0
IA	Guthrie	19,134.0	0.0	19,134.0
IA	Hamilton	430,045.1	171,964.0	602,009.1

TRI on-site and off-site reported disposed of or otherwise released (in pounds), for facilities in all industries, for all chemicals, all counties, 2003 - *continued*

State	County	Total On-site Disposal or Other Releases[1]	Total Off-site Disposal or Other Releases[2]	Total On- and Off-site Disposal or Other Releases
IA	Hancock	211,821.5	685.0	212,506.5
IA	Hardin	164,153.0	0.0	164,153.0
IA	Harrison	1.0	4.0	5.0
IA	Henry	11,390.6	8,995.0	20,385.6
IA	Howard	1,037.0	0.0	1,037.0
IA	Humboldt	7,653.0	6,699.0	14,352.0
IA	Ida	23,326.0	0.0	23,326.0
IA	Iowa	12,147.0	0.0	12,147.0
IA	Jackson	0.0	0.1	0.1
IA	Jasper	404,153.0	24,977.0	429,130.0
IA	Jefferson	26.0	32.0	58.0
IA	Johnson	301,007.6	12,419.4	313,427.0
IA	Jones	82,914.4	6,168.0	89,082.4
IA	Kossuth	88,523.8	2,028.0	90,551.8
IA	Lee	3,133,584.8	192,447.3	3,326,032.1
IA	Linn	1,359,142.9	1,179,915.2	2,539,058.1
IA	Louisa	3,761,953.0	0.0	3,761,953.0
IA	Lyon	0.0	8.7	8.7
IA	Madison	11,616.0	0.0	11,616.0
IA	Mahaska	760.8	60,010.0	60,770.8
IA	Marion	451,881.8	53,296.0	505,177.8
IA	Marshall	18,216.8	60,842.1	79,058.9
IA	Mills	18,640.0	0.0	18,640.0
IA	Monroe	590,180.8	608,190.5	1,198,371.3
IA	Montgomery	120.0	3,723.0	3,843.0
IA	Muscatine	1,286,035.7	9,652,536.2	10,938,571.9
IA	O Brien	133,000.0	0.0	133,000.0
IA	Osceola	0.0	0.0	0.0
IA	Palo Alto	86,000.0	0.0	86,000.0
IA	Plymouth	528.0	0.0	528.0
IA	Pocahontas	0.0	0.0	0.0
IA	Polk	1,191,715.7	58,096.0	1,249,811.8
IA	Pottawattamie	2,311,428.8	858,013.0	3,169,441.8
IA	Poweshiek	32,153.1	0.0	32,153.1
IA	Scott	487,619.6	231,087.4	718,707.0
IA	Shelby	0.0	0.0	0.0
IA	Sioux	108,649.0	1,756.1	110,405.1
IA	Story	499,262.9	3,885.1	503,148.0
IA	Tama	8,410.2	250.0	8,660.2
IA	Taylor	21,708.0	0.0	21,708.0
IA	Union	40,595.2	0.0	40,595.2
IA	Van Buren	500.0	0.0	500.0
IA	Wapello	692,646.9	4,476.0	697,122.9
IA	Washington	3,056.1	1,356.0	4,412.1
IA	Wayne	14.2	0.0	14.3
IA	Webster	417,747.3	2,056.3	419,803.6
IA	Winnebago	74,111.0	746.0	74,857.0
IA	Winneshiek	1,256.8	3,445.5	4,702.3
IA	Woodbury	4,487,398.0	101,593.0	4,588,991.0
IA	Wright	507,515.0	32,565.0	540,080.0
ID	Ada	186,807.1	5,257.3	192,064.4
ID	Bannock	26,436.0	0.0	26,436.0
ID	Benewah	513.0	0.0	513.0
ID	Bingham	27,548.0	129.0	27,677.0
ID	Blaine	0.0	1,205.0	1,205.0
ID	Bonner	43.7	155.0	198.7
ID	Bonneville	1.9	0.0	1.9
ID	Boundary	189.1	0.0	189.1
ID	Butte	27,784.7	76,213.7	103,998.3

637

TRI on-site and off-site reported disposed of or otherwise released (in pounds), for facilities in all industries, for all chemicals, all counties, 2003 - *continued*

State	County	Total On-site Disposal or Other Releases[1]	Total Off-site Disposal or Other Releases[2]	Total On- and Off-site Disposal or Other Releases
ID	Canyon	1,123,255.1	2,331.9	1,125,587.1
ID	Caribou	4,424,448.7	19,556.0	4,444,004.7
ID	Cassia	2,350,332.0	0.0	2,350,332.0
ID	Custer	576,433.9	0.0	576,433.9
ID	Gooding	458,649.0	0.0	458,649.0
ID	Idaho	0.6	0.0	0.6
ID	Jerome	20.0	63,746.0	63,766.0
ID	Kootenai	99,144.1	161,441.5	260,585.6
ID	Latah	13.6	0.0	13.6
ID	Lemhi	39.0	0.0	39.0
ID	Lewis	3.5	0.0	3.5
ID	Lincoln	94,112.0	408.0	94,520.0
ID	Madison	7.1	0.0	7.1
ID	Minidoka	2,676,012.0	0.0	2,676,012.0
ID	Nez Perce	1,229,860.0	388,360.8	1,618,220.8
ID	Owyhee	30,522,299.0	127.0	30,522,426.0
ID	Payette	14,697.0	0.0	14,697.0
ID	Power	251,671.0	33,777.0	285,448.0
ID	Shoshone	16,179,703.3	0.0	16,179,703.3
ID	Twin Falls	472,938.0	240.0	473,178.0
IL	Adams	869,993.3	82,100.2	952,093.5
IL	Alexander	246,111.0	87.0	246,198.0
IL	Bond	621.0	6,600.0	7,221.0
IL	Boone	46,827.4	109,780.1	156,607.5
IL	Bureau	52.4	1,220.0	1,272.4
IL	Carroll	224,465.5	61.0	224,526.5
IL	Cass	1,641,058.0	0.0	1,641,058.0
IL	Champaign	109,862.1	30,996.1	140,858.2
IL	Christian	1,037,355.5	1,268,541.0	2,305,896.5
IL	Clark	24.0	0.0	24.0
IL	Clay	53,236.0	0.0	53,236.0
IL	Clinton	19,293.0	0.0	19,293.0
IL	Coles	603,857.4	98,342.0	702,199.4
IL	Cook	5,554,705.6	10,867,740.6	16,422,446.2
IL	Crawford	635,756.5	320,007.5	955,764.0
IL	DeKalb	71,883.3	1,728.1	73,611.4
IL	Douglas	647,040.4	4,437.9	651,478.3
IL	DuPage	627,111.0	420,842.6	1,047,953.6
IL	Edgar	142,705.0	1,014.0	143,719.0
IL	Effingham	45,866.0	250.0	46,116.0
IL	Fayette	0.0	0.0	0.0
IL	Ford	526,252.0	75.0	526,327.0
IL	Franklin	632,838.0	0.0	632,838.0
IL	Fulton	366,359.7	124.0	366,483.7
IL	Grundy	2,078,115.7	4,403.0	2,082,518.7
IL	Hancock	0.0	0.0	0.0
IL	Hardin	12,388.1	428,348.0	440,736.1
IL	Henry	5,390.0	1,505.0	6,895.0
IL	Iroquois	524,517.0	0.0	524,517.0
IL	Jackson	43,086.0	0.0	43,086.0
IL	Jasper	1,534,134.9	20.0	1,534,154.9
IL	Jefferson	56,972.2	153,550.4	210,522.6
IL	Jo Daviess	1,475,139.3	8,569.0	1,483,708.3
IL	Kane	391,414.6	784,905.3	1,176,319.8
IL	Kankakee	591,431.0	80,309.4	671,740.4
IL	Kendall	13,813.0	3,036.6	16,849.5
IL	Knox	151,062.4	18,926.8	169,989.2
IL	La Salle	1,080,008.4	2,306,752.3	3,386,760.7
IL	Lake	537,818.2	109,970.3	647,788.5

TRI on-site and off-site reported disposed of or otherwise released (in pounds), for facilities in all industries, for all chemicals, all counties, 2003 - *continued*

State	County	Total On-site Disposal or Other Releases[1]	Total Off-site Disposal or Other Releases[2]	Total On- and Off-site Disposal or Other Releases
IL	Lawrence	11,567.0	0.0	11,567.0
IL	Lee	41,971.1	22,500.4	64,471.5
IL	Livingston	127,158.6	42,140.1	169,298.7
IL	Logan	396,678.0	518.0	397,196.0
IL	Macon	4,362,433.6	1,835,818.6	6,198,252.2
IL	Macoupin	0.0	0.0	0.0
IL	Madison	6,489,513.1	5,837,642.3	12,327,155.4
IL	Marion	69,773.0	0.0	69,773.0
IL	Marshall	354,238.0	3,125.0	357,363.0
IL	Mason	1,000,159.9	2,424.7	1,002,584.5
IL	Massac	677,027.9	0.0	677,027.9
IL	McDonough	110,816.1	56,917.2	167,733.3
IL	McHenry	205,341.1	77,230.6	282,571.7
IL	McLean	895,006.6	127,751.1	1,022,757.7
IL	Menard	0.0	0.0	0.0
IL	Monroe	505.0	0.0	505.0
IL	Montgomery	6,195,558.0	195,726.0	6,391,284.0
IL	Morgan	874,860.6	415.0	875,275.6
IL	Moultrie	189,514.1	108.0	189,622.1
IL	Ogle	1,150,598.8	33,829.9	1,184,428.7
IL	Peoria	29,854,746.7	5,398,612.0	35,253,358.7
IL	Perry	15,782.9	202,688.7	218,471.7
IL	Piatt	26,292.0	0.0	26,292.0
IL	Pike	1,431,150.1	0.0	1,431,150.1
IL	Putnam	551,427.0	11.8	551,438.8
IL	Randolph	2,001,321.4	46.5	2,001,367.9
IL	Rock Island	3,314,239.9	65,398.7	3,379,638.6
IL	Saline	2,036.0	0.0	2,036.0
IL	Sangamon	1,142,544.5	53.8	1,142,598.3
IL	Shelby	1,972.0	0.0	1,972.0
IL	St. Clair	3,247,143.3	2,635,855.5	5,882,998.8
IL	Stark	0.0	0.0	0.0
IL	Stephenson	39,856.1	115,445.0	155,301.1
IL	Tazewell	1,911,609.3	1,478,488.7	3,390,098.0
IL	Vermilion	5,716,389.1	14,475.0	5,730,864.1
IL	Wabash	2,383.8	1,242.6	3,626.4
IL	Warren	26,169.0	0.0	26,169.0
IL	Washington	28,479.2	0.0	28,479.2
IL	Wayne	0.0	0.0	0.0
IL	White	7,722.1	1,260.7	8,982.8
IL	Whiteside	14,408.0	4,421,987.0	4,436,395.0
IL	Will	4,561,115.2	337,747.9	4,898,863.1
IL	Williamson	214,636.0	287,797.0	502,433.0
IL	Winnebago	1,022,382.2	1,024,869.7	2,047,251.9
IL	Woodford	283.0	1,282.0	1,565.0
IN	Adams	1,221,694.9	11,057.0	1,232,751.9
IN	Allen	1,987,622.1	491,643.0	2,479,265.0
IN	Bartholomew	290,310.1	56,674.7	346,984.8
IN	Blackford	80,318.9	6,420.0	86,738.9
IN	Boone	18,749.7	18.3	18,768.0
IN	Carroll	59,892.3	48.0	59,940.3
IN	Cass	954,294.7	16,427.3	970,722.0
IN	Clark	430,763.4	0.0	430,763.4
IN	Clay	33,064.0	0.0	33,064.0
IN	Clinton	434,130.0	68.3	434,198.3
IN	Crawford	20.0	0.0	20.0
IN	Daviess	284,898.1	160.8	285,058.9
IN	De Kalb	1,444,854.7	21,438,142.1	22,882,996.8
IN	Dearborn	5,475,926.6	249.2	5,476,175.9

**TRI on-site and off-site reported disposed of or otherwise released (in pounds),
for facilities in all industries, for all chemicals, all counties, 2003** - *continued*

State	County	Total On-site Disposal or Other Releases[1]	Total Off-site Disposal or Other Releases[2]	Total On- and Off-site Disposal or Other Releases
IN	Decatur	15,118.1	110,020.0	125,138.1
IN	Delaware	33,049.6	460,261.0	493,310.6
IN	Dubois	628,021.8	1,659.1	629,680.9
IN	Elkhart	3,595,548.7	82,923.4	3,678,472.1
IN	Fayette	83,640.0	24,672.0	108,312.0
IN	Floyd	3,389,065.5	1,876.0	3,390,941.5
IN	Fountain	25,465.3	25,366.2	50,831.5
IN	Franklin	3.9	0.0	3.9
IN	Fulton	61,392.1	12,788.0	74,180.1
IN	Gibson	12,938,306.3	194,044.3	13,132,350.6
IN	Grant	363,788.3	158,819.8	522,608.1
IN	Greene	85.3	0.0	85.3
IN	Hamilton	101,006.1	22,095.6	123,101.7
IN	Hancock	609,101.0	77,310.6	686,411.6
IN	Harrison	973,066.0	2,533.0	975,599.0
IN	Hendricks	19,824.0	55,555.0	75,379.0
IN	Henry	1,101,390.0	454,567.0	1,555,957.0
IN	Howard	250,789.7	221,222.9	472,012.6
IN	Huntington	352,131.2	17,188.2	369,319.4
IN	Jackson	81,702.5	221,777.5	303,480.0
IN	Jasper	4,430,271.5	68.5	4,430,340.0
IN	Jay	15,079.4	5,519.0	20,598.4
IN	Jefferson	3,946,976.5	16,012.9	3,962,989.4
IN	Jennings	97,410.3	35,881.0	133,291.3
IN	Johnson	427,901.2	289.0	428,190.2
IN	Knox	392,697.9	74,832.1	467,530.0
IN	Kosciusko	1,547,252.5	103,309.3	1,650,561.8
IN	La Porte	1,097,049.4	503,323.2	1,600,372.6
IN	Lagrange	106,249.9	19,435.0	125,684.9
IN	Lake	21,822,204.0	6,320,796.9	28,143,001.0
IN	Lawrence	11,979.0	54,160.0	66,139.0
IN	Madison	219,717.1	25,167.7	244,884.8
IN	Marion	3,129,490.3	14,938,974.1	18,068,464.4
IN	Marshall	1,360,169.0	142,399.1	1,502,568.1
IN	Martin	251,335.9	116,400.4	367,736.3
IN	Miami	38,004.0	0.0	38,004.0
IN	Monroe	470,338.0	38,509.0	508,847.0
IN	Montgomery	201,951.0	41,799,864.0	42,001,815.0
IN	Morgan	993,390.0	0.0	993,390.0
IN	Newton	114,150.0	1,258.0	115,408.0
IN	Noble	225,959.9	18,685.4	244,645.2
IN	Orange	97,994.0	0.0	97,994.0
IN	Owen	124.8	0.0	124.8
IN	Perry	99,144.0	3,229,010.0	3,328,154.0
IN	Pike	3,238,169.4	1,051,804.6	4,289,974.0
IN	Porter	3,023,296.7	375,353.3	3,398,650.0
IN	Posey	2,685,772.9	98,456.1	2,784,229.0
IN	Pulaski	15.0	34,453.0	34,468.0
IN	Putnam	356,661.2	119,000.0	475,661.2
IN	Randolph	189,800.3	10.0	189,810.3
IN	Ripley	331,270.4	29,520.3	360,790.8
IN	Rush	30,559.0	1,176.0	31,735.0
IN	Scott	18,280.7	76,885.0	95,165.7
IN	Shelby	673,373.6	287,835.0	961,208.6
IN	Spencer	22,488,556.8	640,354.0	23,128,910.8
IN	St. Joseph	761,972.4	432,439.8	1,194,412.2
IN	Starke	0.0	2,064.0	2,064.0
IN	Steuben	30,756.0	44,170.0	74,926.0
IN	Sullivan	2,664,051.7	0.0	2,664,051.7

TRI on-site and off-site reported disposed of or otherwise released (in pounds), for facilities in all industries, for all chemicals, all counties, 2003 - *continued*

State	County	Total On-site Disposal or Other Releases[1]	Total Off-site Disposal or Other Releases[2]	Total On- and Off-site Disposal or Other Releases
IN	Switzerland	28,000.0	0.0	28,000.0
IN	Tippecanoe	1,396,818.0	198,940.5	1,595,758.5
IN	Tipton	25,874.7	209.9	26,084.6
IN	Vanderburgh	482,523.9	260,181.8	742,705.7
IN	Vermillion	5,233,254.3	148,105.3	5,381,359.6
IN	Vigo	2,998,915.4	8,705.0	3,007,620.4
IN	Wabash	636,885.7	4,398,718.1	5,035,603.8
IN	Warrick	7,679,445.3	1,126,187.8	8,805,633.1
IN	Washington	100,837.5	8,156.7	108,994.2
IN	Wayne	1,062,191.5	198,664.1	1,260,855.6
IN	Wells	21,669.0	39.0	21,708.0
IN	White	124,729.1	10.0	124,739.1
IN	Whitley	96,268.4	963,999.4	1,060,267.8
KS	Allen	43,655.0	27,510.0	71,165.0
KS	Anderson	52,050.0	0.0	52,050.0
KS	Atchison	466,941.0	179,917.0	646,858.0
KS	Barber	0.8	5.1	5.9
KS	Barton	58,701.0	24,326.0	83,027.0
KS	Bourbon	12,009.0	7,651.0	19,660.0
KS	Brown	12,758.0	0.0	12,758.0
KS	Butler	392,342.6	83,401.8	475,744.4
KS	Cherokee	409,105.1	1,100,522.0	1,509,627.1
KS	Cloud	750.0	0.0	750.0
KS	Cowley	2,525.0	2,060.0	4,585.0
KS	Crawford	102,658.0	27.0	102,685.0
KS	Dickinson	54,025.0	0.0	54,025.0
KS	Doniphan	4,675.1	0.0	4,675.1
KS	Douglas	1,196,815.6	400.0	1,197,215.6
KS	Ellis	59.7	214.4	274.1
KS	Ellsworth	5.0	20.0	25.0
KS	Finney	1,196,819.0	18,215.0	1,215,034.0
KS	Ford	1,392,008.0	1,077.0	1,393,085.0
KS	Franklin	51,833.0	0.0	51,833.0
KS	Geary	117,181.0	0.0	117,181.0
KS	Grant	586,933.3	9.1	586,942.4
KS	Harvey	214,491.1	356.0	214,847.1
KS	Johnson	143,634.7	4,624.0	148,258.8
KS	Kingman	749.1	0.0	749.1
KS	Labette	197,843.0	0.0	197,843.0
KS	Leavenworth	10,559.1	0.0	10,559.1
KS	Linn	2,733,413.8	319.9	2,733,733.7
KS	Lyon	3,802,032.0	15,456.0	3,817,488.0
KS	Marion	5,653.0	0.0	5,653.0
KS	Marshall	225,151.0	0.0	225,151.0
KS	McPherson	482,236.3	22,537.8	504,774.1
KS	Mitchell	0.0	0.0	0.0
KS	Montgomery	2,160,575.2	1,306,972.3	3,467,547.5
KS	Nemaha	57,385.0	0.0	57,385.0
KS	Neosho	344,838.2	253.1	345,091.3
KS	Osage	0.0	0.0	0.0
KS	Osborne	6,578.0	8,249.0	14,827.0
KS	Ottawa	48,765.0	0.0	48,765.0
KS	Phillips	13,666.8	0.0	13,666.8
KS	Pottawatomie	3,008,509.1	10,983.0	3,019,492.1
KS	Reno	203,169.1	582.3	203,751.4
KS	Rice	1,255.0	0.0	1,255.0
KS	Rooks	0.0	0.0	0.0
KS	Rush	92,000.0	0.0	92,000.0
KS	Russell	99,675.2	0.0	99,675.2

TRI on-site and off-site reported disposed of or otherwise released (in pounds), for facilities in all industries, for all chemicals, all counties, 2003 - *continued*

State	County	Total On-site Disposal or Other Releases[1]	Total Off-site Disposal or Other Releases[2]	Total On- and Off-site Disposal or Other Releases
KS	Saline	96,265.3	55,035.4	151,300.8
KS	Scott	0.0	0.0	0.0
KS	Sedgwick	1,894,030.5	160,779.7	2,054,810.2
KS	Seward	31,000.0	0.0	31,000.0
KS	Shawnee	1,240,092.0	223,872.4	1,463,964.5
KS	Sherman	42,933.0	0.0	42,933.0
KS	Smith	37,500.0	0.0	37,500.0
KS	Sumner	0.6	0.0	0.6
KS	Wilson	417,694.4	12,126.5	429,820.9
KS	Wyandotte	1,839,863.0	1,534,372.6	3,374,235.6
KY	Adair	0.0	0.0	0.0
KY	Allen	15,373.0	5,728.0	21,101.0
KY	Anderson	2,174.0	112,011.0	114,185.0
KY	Ballard	1,359,218.4	0.0	1,359,218.4
KY	Barren	39,498.4	1,371,118.6	1,410,617.0
KY	Bath	15.0	750.0	765.0
KY	Bell	393.0	0.0	393.0
KY	Boone	2,438,251.1	13,374.8	2,451,625.8
KY	Bourbon	25,928.0	6,258.0	32,186.0
KY	Boyd	515,689.1	122,542.9	638,232.0
KY	Boyle	25,871.9	67,156.1	93,028.0
KY	Bullitt	255.0	5.0	260.0
KY	Butler	2,359,186.6	0.0	2,359,186.6
KY	Caldwell	88,741.0	7.0	88,748.0
KY	Calloway	29,168.0	8,852.1	38,020.1
KY	Campbell	40,651.0	8.4	40,659.4
KY	Carroll	8,337,573.7	173,163.6	8,510,737.3
KY	Carter	175,000.0	5.0	175,005.0
KY	Christian	835,270.8	25,260.6	860,531.5
KY	Clark	1,704,253.2	472,560.2	2,176,813.3
KY	Clinton	133,466.0	500.0	133,966.0
KY	Daviess	2,182,013.0	336,167.9	2,518,180.9
KY	Fayette	257,309.7	90,866.4	348,176.1
KY	Franklin	112,117.2	48.4	112,165.6
KY	Fulton	6,363.0	2,529.0	8,892.0
KY	Gallatin	8,312.0	18,863.0	27,175.0
KY	Garrard	3,386.0	0.0	3,386.0
KY	Grant	104,210.0	15.0	104,225.0
KY	Graves	226,220.0	1,528.0	227,748.1
KY	Grayson	3.0	225.0	228.0
KY	Green	824.0	0.0	824.0
KY	Greenup	233,088.0	188,691.5	421,779.5
KY	Hancock	3,734,758.6	40,442.5	3,775,201.1
KY	Hardin	368,236.3	192,475.0	560,711.4
KY	Harlan	1.8	0.0	1.8
KY	Harrison	74,730.0	12,170.0	86,900.0
KY	Hart	3,010.0	9,133.0	12,143.0
KY	Henderson	3,523,325.8	55,761.3	3,579,087.1
KY	Henry	7,227.0	36,598.0	43,825.0
KY	Hopkins	962,206.0	429.0	962,635.0
KY	Jefferson	8,784,151.3	664,116.7	9,448,267.9
KY	Jessamine	71,813.0	886,850.0	958,663.0
KY	Johnson	36.0	0.0	36.0
KY	Kenton	9,916.6	40,305.0	50,221.6
KY	Knox	0.0	15.0	15.0
KY	Laurel	3,474.8	19,803.6	23,278.4
KY	Lawrence	6,885,393.1	8,160.3	6,893,553.4
KY	Leslie	1,500.0	0.0	1,500.0
KY	Lewis	145,676.0	0.0	145,676.0

**TRI on-site and off-site reported disposed of or otherwise released (in pounds),
for facilities in all industries, for all chemicals, all counties, 2003** - *continued*

State	County	Total On-site Disposal or Other Releases[1]	Total Off-site Disposal or Other Releases[2]	Total On- and Off-site Disposal or Other Releases
KY	Lincoln	30,805.0	120,675.0	151,480.0
KY	Logan	45,596.2	5,193.0	50,789.2
KY	Lyon	0.0	0.0	0.0
KY	Madison	213,258.9	694,348.1	907,607.0
KY	Marion	40,627.1	3,318.0	43,945.2
KY	Marshall	3,745,928.3	117,151.6	3,863,079.9
KY	Martin	86,016.0	0.0	86,016.0
KY	Mason	6,021,563.3	37,581.4	6,059,144.7
KY	McCracken	3,480,138.0	10,796.0	3,490,934.0
KY	McLean	4,459.8	184.0	4,643.8
KY	Meade	123,830.0	0.0	123,830.0
KY	Mercer	4,020,644.5	824,471.4	4,845,115.9
KY	Metcalfe	41,935.2	983.3	42,918.5
KY	Monroe	0.0	312.8	312.8
KY	Montgomery	172,060.0	26,450.0	198,510.0
KY	Muhlenberg	6,585,802.6	21,635.4	6,607,438.0
KY	Nelson	1.9	2,137.5	2,139.4
KY	Nicholas	0.0	0.0	0.0
KY	Ohio	2,324,984.1	0.0	2,324,984.1
KY	Oldham	68,982.0	0.0	68,982.0
KY	Pendleton	205,616.3	0.0	205,616.3
KY	Perry	2,987.3	223.0	3,210.3
KY	Pike	3,626.9	0.0	3,626.9
KY	Powell	49,640.0	91.0	49,731.0
KY	Pulaski	1,938,276.8	1.0	1,938,277.8
KY	Rowan	124,278.0	121,735.0	246,013.0
KY	Russell	2,718.6	100.7	2,819.3
KY	Scott	830,944.0	95,637.2	926,581.3
KY	Shelby	102,105.3	84,522.9	186,628.2
KY	Simpson	1,034,230.1	33,863.6	1,068,093.7
KY	Taylor	1,557.0	0.0	1,557.0
KY	Todd	3,705.6	1,322.8	5,028.4
KY	Trigg	0.0	48,382.0	48,382.0
KY	Trimble	1,050,624.3	0.0	1,050,624.3
KY	Union	90,919.0	0.0	90,919.0
KY	Warren	4,253,618.5	173,003.6	4,426,622.0
KY	Washington	0.0	8,250.0	8,250.0
KY	Wayne	395.6	2,845.2	3,240.8
KY	Webster	0.0	0.0	0.0
KY	Whitley	92,076.0	68,664.0	160,740.0
KY	Woodford	303,100.5	185,768.5	488,869.1
LA	Acadia	4,631.0	0.0	4,631.0
LA	Allen	15,668.8	0.0	15,668.8
LA	Ascension	19,049,430.0	418,576.6	19,468,006.6
LA	Assumption	235,943.0	176,557.0	412,500.0
LA	Avoyelles	429.0	1,750.0	2,179.0
LA	Beauregard	2,109,556.7	15,928.2	2,125,484.9
LA	Bienville	587.0	270.0	857.0
LA	Bossier	286.4	500.0	786.4
LA	Caddo	1,641,174.1	159,246.5	1,800,420.6
LA	Calcasieu	14,026,516.1	1,022,528.0	15,049,044.2
LA	Cameron	32,792.0	0.0	32,792.0
LA	Concordia	0.0	0.0	0.0
LA	De Soto	11,049,987.5	0.0	11,049,987.5
LA	East Baton Rouge	8,200,258.0	584,741.7	8,784,999.7
LA	East Carroll	0.0	0.0	0.0
LA	Evangeline	124,012.7	277.9	124,290.6
LA	Franklin	3,211.0	0.0	3,211.0
LA	Iberia	167,703.0	80,608.0	248,311.0

**TRI on-site and off-site reported disposed of or otherwise released (in pounds),
for facilities in all industries, for all chemicals, all counties, 2003** - *continued*

State	County	Total On-site Disposal or Other Releases[1]	Total Off-site Disposal or Other Releases[2]	Total On- and Off-site Disposal or Other Releases
LA	Iberville	5,487,024.9	71,511.9	5,558,536.8
LA	Jackson	1,333,331.9	0.0	1,333,331.9
LA	Jefferson	10,639,015.7	6,088.0	10,645,103.7
LA	La Salle	0.0	0.0	0.0
LA	Lafayette	12,507.7	1,291.9	13,799.6
LA	Lafourche	41,326.0	2,312.0	43,638.0
LA	Lincoln	183,755.9	5,089.9	188,845.8
LA	Livingston	64,242.5	10,525.4	74,767.9
LA	Madison	3,720.0	0.0	3,720.0
LA	Morehouse	2,738,560.8	0.0	2,738,560.8
LA	Natchitoches	1,874,384.4	36.0	1,874,420.4
LA	Orleans	171,269.6	155.0	171,424.6
LA	Ouachita	8,499,976.7	121,318.6	8,621,295.3
LA	Plaquemines	631,184.2	468,077.6	1,099,261.8
LA	Pointe Coupee	2,622,913.5	0.0	2,622,913.5
LA	Rapides	1,286,598.6	196,487.2	1,483,085.8
LA	Red River	26.3	0.0	26.3
LA	Richland	12.0	260.1	272.1
LA	Sabine	58,967.9	188.6	59,156.5
LA	St. Bernard	1,804,994.1	7,442.2	1,812,436.3
LA	St. Charles	16,239,472.5	76,603.9	16,316,076.4
LA	St. Helena	4,818.0	0.0	4,818.0
LA	St. James	4,451,790.3	47,761.9	4,499,552.2
LA	St. John the Baptist	1,500,394.7	1,793,215.2	3,293,609.9
LA	St. Landry	143,313.1	1,038.8	144,351.9
LA	St. Martin	32,382.0	0.0	32,382.0
LA	St. Mary	333,969.4	4,345.2	338,314.5
LA	St. Tammany	19,199.0	51,357.0	70,556.0
LA	Tangipahoa	376,746.0	0.0	376,746.0
LA	Terrebonne	47,523.0	0.0	47,523.0
LA	Union	127,388.0	558.0	127,946.0
LA	Vermilion	15,093.6	0.0	15,093.6
LA	Vernon	69,294.0	0.0	69,294.0
LA	Washington	1,843,288.3	10,259.0	1,853,547.3
LA	Webster	80,779.4	83,310.9	164,090.3
LA	West Baton Rouge	697,372.8	127,232.6	824,605.3
LA	West Feliciana	792,451.1	0.0	792,451.1
LA	Winn	169,999.5	332.5	170,332.0
MA	Barnstable	483,117.4	0.0	483,117.4
MA	Berkshire	69,611.2	467.6	70,078.8
MA	Bristol	2,297,446.7	754,592.0	3,052,038.7
MA	Dukes	941.0	0.0	941.0
MA	Essex	983,449.0	178,458.4	1,161,907.4
MA	Franklin	6,905.0	22,590.0	29,495.0
MA	Hampden	1,384,562.9	308,380.1	1,692,942.9
MA	Hampshire	48,650.2	25,054.0	73,704.2
MA	Middlesex	856,561.3	249,335.2	1,105,896.5
MA	Norfolk	129,598.0	281,338.7	410,936.6
MA	Plymouth	81,826.2	44,298.2	126,124.4
MA	Suffolk	129,557.2	27,043.6	156,600.7
MA	Worcester	438,044.3	236,928.3	674,972.6
MD	Allegany	1,943,937.5	2,283,897.0	4,227,834.5
MD	Anne Arundel	217,204.3	388.5	217,592.8
MD	Baltimore	2,804,668.2	161,213.3	2,965,881.5
MD	Baltimore City	18,611,788.2	201,303.8	18,813,092.0
MD	Caroline	30.0	18,150.0	18,180.0
MD	Carroll	447,371.7	327.0	447,698.7
MD	Cecil	85,802.6	1,636.0	87,438.6
MD	Charles	8,013,676.2	853,777.4	8,867,453.6

**TRI on-site and off-site reported disposed of or otherwise released (in pounds),
for facilities in all industries, for all chemicals, all counties, 2003** - *continued*

State	County	Total On-site Disposal or Other Releases[1]	Total Off-site Disposal or Other Releases[2]	Total On- and Off-site Disposal or Other Releases
MD	Dorchester	108,880.1	0.0	108,880.1
MD	Frederick	923,276.7	4,809.0	928,085.7
MD	Garrett	139,138.6	0.0	139,138.6
MD	Harford	170,375.2	29,884.3	200,259.5
MD	Howard	3,206.1	11,505.5	14,711.6
MD	Kent	14,348.0	43.0	14,391.0
MD	Montgomery	1,610,126.9	370,354.6	1,980,481.5
MD	Prince Georges	4,431,105.5	697,574.2	5,128,679.7
MD	Queen Annes	5.0	750.0	755.0
MD	Somerset	0.0	0.0	0.0
MD	St. Marys	6,551.1	39.9	6,591.0
MD	Talbot	18,958.0	0.0	18,958.0
MD	Washington	717,537.3	110,610.1	828,147.4
MD	Wicomico	380,511.0	481.0	380,992.0
MD	Worcester	111,921.0	0.0	111,921.0
ME	Androscoggin	132,018.0	53,908.7	185,926.7
ME	Aroostook	2,642,288.6	106.1	2,642,394.7
ME	Cumberland	271,600.1	133,735.7	405,335.9
ME	Franklin	1,055,129.0	0.0	1,055,129.0
ME	Hancock	425,860.5	610.0	426,470.5
ME	Kennebec	2,936.1	10.0	2,946.1
ME	Knox	31,932.1	0.6	31,932.7
ME	Lincoln	33,214.0	0.0	33,214.0
ME	Oxford	850,595.8	390,709.5	1,241,305.3
ME	Penobscot	527,377.4	101,705.4	629,082.8
ME	Piscataquis	150.1	0.0	150.1
ME	Sagadahoc	47,259.1	0.0	47,259.1
ME	Somerset	1,357,114.5	104,906.8	1,462,021.3
ME	Waldo	10,144.2	28,249.6	38,393.8
ME	Washington	939,239.9	2,326.8	941,566.7
ME	York	182,462.3	86,545.8	269,008.0
MI	Alger	288,535.0	254.0	288,789.0
MI	Allegan	680,580.9	76,545.3	757,126.2
MI	Alpena	825,759.2	35,646.0	861,405.2
MI	Antrim	0.0	5,542.0	5,542.0
MI	Baraga	1,492.8	1,590.7	3,083.5
MI	Barry	71,284.2	35,909.7	107,193.9
MI	Bay	3,800,914.7	32,066.0	3,832,980.7
MI	Benzie	500.0	0.0	500.0
MI	Berrien	103,055.8	49,773.9	152,829.7
MI	Branch	51,119.7	484,385.1	535,504.8
MI	Calhoun	72,886.3	57,091.4	129,977.7
MI	Cass	2,386.0	250.0	2,636.0
MI	Charlevoix	16,159.0	70,207.0	86,366.0
MI	Cheboygan	0.0	5.0	5.0
MI	Clare	794.5	20.0	814.5
MI	Clinton	2,127.6	3,214.6	5,342.2
MI	Crawford	56,906.0	14,076.1	70,982.1
MI	Delta	1,124,575.4	1,300.6	1,125,876.0
MI	Dickinson	542,578.4	211,094.0	753,672.4
MI	Eaton	16,144.1	166.0	16,310.1
MI	Genesee	1,242,751.9	85,380.2	1,328,132.1
MI	Grand Traverse	29,240.0	93,283.5	122,523.5
MI	Gratiot	33,113.1	6.0	33,119.1
MI	Hillsdale	230,814.7	23,093.1	253,907.8
MI	Houghton	496,216.3	28,340.8	524,557.2
MI	Huron	368,999.3	209,034.2	578,033.5
MI	Ingham	3,547,741.6	926,034.9	4,473,776.5
MI	Ionia	108,523.3	28,185.5	136,708.8

TRI on-site and off-site reported disposed of or otherwise released (in pounds), for facilities in all industries, for all chemicals, all counties, 2003 - *continued*

State	County	Total On-site Disposal or Other Releases[1]	Total Off-site Disposal or Other Releases[2]	Total On- and Off-site Disposal or Other Releases
MI	Iosco	2,126.1	2,395.0	4,521.1
MI	Isabella	1,381.5	939.0	2,320.5
MI	Jackson	42,536.0	665,454.8	707,990.9
MI	Kalamazoo	805,445.2	48,096.9	853,542.1
MI	Kalkaska	120,978.0	0.0	120,978.0
MI	Kent	537,029.3	469,914.5	1,006,943.7
MI	Lapeer	118,509.0	59,290.6	177,799.6
MI	Lenawee	94,738.8	481,797.7	576,536.5
MI	Livingston	33,191.5	342,088.0	375,279.5
MI	Luce	70,752.5	10,905.0	81,657.5
MI	Macomb	1,185,935.4	580,797.2	1,766,732.7
MI	Manistee	695,743.5	1,071,031.4	1,766,774.8
MI	Marquette	743,469.7	1,656.5	745,126.2
MI	Mason	37,258.9	64,557.1	101,816.0
MI	Mecosta	107,591.0	9,708.0	117,299.0
MI	Menominee	81,724.8	1,880.3	83,605.1
MI	Midland	676,860.3	8,383.0	685,243.3
MI	Missaukee	40.0	0.0	40.0
MI	Monroe	14,879,720.9	39,681.1	14,919,402.1
MI	Montcalm	236,003.2	10,389.7	246,392.9
MI	Montmorency	7,593.4	0.0	7,593.4
MI	Muskegon	1,495,507.9	342,893.6	1,838,401.5
MI	Newaygo	130,964.0	108.0	131,072.0
MI	Oakland	3,036,485.7	363,826.1	3,400,311.8
MI	Oceana	11,271.0	191.2	11,462.2
MI	Ogemaw	26,316.0	118.0	26,434.0
MI	Ontonagon	425,685.5	72,222.2	497,907.7
MI	Osceola	182,877.0	5.0	182,882.0
MI	Otsego	330,342.3	8,776.5	339,118.8
MI	Ottawa	9,770,156.1	214,062.5	9,984,218.6
MI	Saginaw	684,353.8	144,300.1	828,653.9
MI	Sanilac	118,057.1	635.8	118,692.9
MI	Schoolcraft	0.0	0.0	0.0
MI	Shiawassee	82,788.4	0.0	82,788.4
MI	St. Clair	3,684,722.5	120,623.0	3,805,345.5
MI	St. Joseph	1,005,043.6	126,973.8	1,132,017.4
MI	Tuscola	75,904.0	84,547.0	160,451.0
MI	Van Buren	63,678.0	1,665.0	65,343.0
MI	Washtenaw	149,028.7	69,455.9	218,484.6
MI	Wayne	9,194,608.8	31,256,352.7	40,450,961.5
MI	Wexford	627,552.6	266,739.2	894,291.9
MN	Anoka	127,095.5	67,062.4	194,157.9
MN	Becker	307.0	0.0	307.0
MN	Beltrami	876,689.1	787.6	877,476.7
MN	Benton	71,149.7	14.9	71,164.6
MN	Blue Earth	764,605.9	3,090.4	767,696.3
MN	Brown	8.0	15,580.0	15,588.0
MN	Carlton	703,480.8	37,197.0	740,677.8
MN	Carver	278,065.0	367.9	278,432.9
MN	Cass	7,689.0	0.0	7,689.0
MN	Chippewa	0.0	0.0	0.0
MN	Chisago	8,102.0	0.0	8,102.0
MN	Clay	240,650.2	0.0	240,650.2
MN	Cook	313,106.7	0.0	313,106.7
MN	Cottonwood	9,759.0	0.0	9,759.0
MN	Crow Wing	54,820.8	4,942.0	59,762.8
MN	Dakota	1,164,247.5	604,159.3	1,768,406.8
MN	Dodge	86,796.0	0.0	86,796.0
MN	Douglas	72,965.0	40.0	73,005.0

**TRI on-site and off-site reported disposed of or otherwise released (in pounds),
for facilities in all industries, for all chemicals, all counties, 2003** - *continued*

State	County	Total On-site Disposal or Other Releases[1]	Total Off-site Disposal or Other Releases[2]	Total On- and Off-site Disposal or Other Releases
MN	Faribault	11,473.0	0.0	11,473.0
MN	Fillmore	101,350.6	1,039.8	102,390.5
MN	Freeborn	67,484.0	6,122.9	73,606.9
MN	Goodhue	295,668.2	105,754.0	401,422.2
MN	Hennepin	753,258.7	430,885.8	1,184,144.4
MN	Hubbard	28,227.0	0.0	28,227.0
MN	Isanti	250.0	500.0	750.0
MN	Itasca	2,243,361.3	554.5	2,243,915.8
MN	Jackson	10,450.0	0.0	10,450.0
MN	Kanabec	20,281.0	0.0	20,281.0
MN	Kandiyohi	0.0	0.0	0.0
MN	Koochiching	581,706.4	13,600.0	595,306.4
MN	Lac qui Parle	447,413.1	2,460.0	449,873.1
MN	Lake	34,644.3	1,472.0	36,116.3
MN	Lake of the Woods	2,053.0	0.0	2,053.0
MN	Le Sueur	26,172.0	0.0	26,172.0
MN	Lyon	196,918.5	0.5	196,919.0
MN	Marshall	41,153.0	0.0	41,153.0
MN	Martin	467,296.2	0.0	467,296.2
MN	McLeod	461,829.7	1,160.6	462,990.3
MN	Meeker	15,000.0	1,826.0	16,826.0
MN	Morrison	418,119.0	0.0	418,119.0
MN	Mower	204,263.6	1,186.0	205,449.6
MN	Nicollet	15,905.8	0.2	15,906.0
MN	Nobles	9,656.0	0.0	9,656.0
MN	Olmsted	523,422.3	1,800.0	525,222.3
MN	Otter Tail	298,385.0	4,313.8	302,698.8
MN	Pennington	0.0	0.0	0.0
MN	Pipestone	158,823.0	0.0	158,823.0
MN	Polk	593,374.8	0.0	593,374.8
MN	Ramsey	971,021.4	3,244,059.5	4,215,080.9
MN	Redwood	254,250.0	0.0	254,250.0
MN	Renville	248,772.0	0.0	248,772.0
MN	Rice	165,912.4	10,071.0	175,983.4
MN	Rock	2,307.0	0.0	2,307.0
MN	Roseau	43,000.0	0.0	43,000.0
MN	Scott	62,024.4	401.9	62,426.3
MN	Sherburne	8,126,105.1	3,420.2	8,129,525.3
MN	Sibley	124.0	0.0	124.0
MN	St. Louis	459,309.0	63,541.5	522,850.5
MN	Stearns	683,882.9	31,266.7	715,149.5
MN	Steele	404,598.6	7,786.4	412,385.0
MN	Stevens	24,440.0	0.0	24,440.0
MN	Swift	877.0	20.0	897.0
MN	Todd	268,568.0	516.0	269,084.0
MN	Wabasha	57,090.1	1,676.9	58,767.0
MN	Wadena	16,620.0	0.0	16,620.0
MN	Waseca	9,353.0	299.0	9,652.0
MN	Washington	1,180,190.7	1,192,859.7	2,373,050.4
MN	Watonwan	0.0	28.3	28.3
MN	Winona	21,844.9	70,251.6	92,096.5
MN	Wright	8,135.0	0.0	8,135.0
MO	Audrain	449,296.6	26,396.7	475,693.3
MO	Barry	513,473.8	12,854.0	526,327.8
MO	Barton	0.0	0.0	0.0
MO	Boone	605,308.4	33,787.3	639,095.7
MO	Buchanan	541,469.7	216,490.3	757,960.0
MO	Butler	66,716.3	89,310.3	156,026.6
MO	Callaway	85,887.0	260.0	86,147.0

**TRI on-site and off-site reported disposed of or otherwise released (in pounds),
for facilities in all industries, for all chemicals, all counties, 2003** - *continued*

State	County	Total On-site Disposal or Other Releases[1]	Total Off-site Disposal or Other Releases[2]	Total On- and Off-site Disposal or Other Releases
MO	Camden	10,544.1	750.0	11,294.1
MO	Cape Girardeau	268,269.2	13,586.3	281,855.5
MO	Carroll	25,239.1	0.0	25,239.1
MO	Carter	3,512,016.0	0.0	3,512,016.0
MO	Cass	15,336.1	3,174.0	18,510.1
MO	Cedar	0.0	0.0	0.0
MO	Christian	6,602.4	75.0	6,677.4
MO	Clay	1,708,036.7	105,968.4	1,814,005.2
MO	Clinton	12,968.0	0.0	12,968.0
MO	Cole	10,161.0	3,096.2	13,257.2
MO	Cooper	43,326.0	1,313.0	44,639.0
MO	Crawford	13,414.8	218.0	13,632.8
MO	Daviess	0.0	0.0	0.0
MO	Dent	363,456.0	0.0	363,456.0
MO	Douglas	0.0	76.3	76.3
MO	Dunklin	80,527.0	3,971.0	84,498.0
MO	Franklin	2,595,515.5	109,198.7	2,704,714.2
MO	Greene	558,091.4	74,023.1	632,114.5
MO	Grundy	55.0	4,729.0	4,784.0
MO	Henry	759,342.2	34.0	759,376.2
MO	Holt	83,666.0	0.0	83,666.0
MO	Howard	2,475.0	0.0	2,475.0
MO	Howell	122,321.7	5,951.3	128,273.0
MO	Iron	19,280,867.3	5,265,898.0	24,546,765.3
MO	Jackson	1,685,755.5	55,134.6	1,740,890.1
MO	Jasper	775,913.4	159,824.8	935,738.2
MO	Jefferson	11,173,896.0	429,875.2	11,603,771.2
MO	Johnson	3,601.6	1,154.1	4,755.7
MO	Laclede	151,590.0	5,766.0	157,356.0
MO	Lafayette	9,606.1	0.0	9,606.1
MO	Lawrence	195,082.5	250.0	195,332.5
MO	Lincoln	37,240.0	4,865.0	42,105.0
MO	Livingston	13,990.0	952.8	14,942.8
MO	Macon	505.0	0.0	505.0
MO	Maries	1,306.9	0.0	1,306.9
MO	Marion	272,891.3	1,513.0	274,404.3
MO	McDonald	109,168.0	755.0	109,923.0
MO	Mercer	0.0	0.0	0.0
MO	Miller	1,618.0	546.1	2,164.1
MO	Mississippi	477.0	32,060.0	32,537.0
MO	Moniteau	0.0	0.0	0.0
MO	Monroe	29.4	58.4	87.8
MO	Montgomery	1,706.0	1,000.0	2,706.0
MO	Morgan	0.0	15,312.0	15,312.0
MO	New Madrid	1,553,460.2	116,189.0	1,669,649.2
MO	Newton	78,807.3	6,244.3	85,051.7
MO	Nodaway	728.6	387,584.8	388,313.4
MO	Osage	138,721.0	5.0	138,726.0
MO	Pemiscot	232,322.1	58,883.0	291,205.1
MO	Perry	219,391.0	0.0	219,391.0
MO	Pettis	1,696,251.1	125,761.5	1,822,012.6
MO	Phelps	2,714.3	1,334.4	4,048.7
MO	Pike	2,476,618.3	0.0	2,476,618.3
MO	Platte	629,989.9	2,015.8	632,005.8
MO	Polk	14,216.1	0.0	14,216.1
MO	Pulaski	1,074,277.0	0.0	1,074,277.0
MO	Putnam	0.0	0.0	0.0
MO	Ralls	96,111.6	79,150.7	175,262.3
MO	Randolph	1,217,358.6	30.0	1,217,388.6

TRI on-site and off-site reported disposed of or otherwise released (in pounds), for facilities in all industries, for all chemicals, all counties, 2003 - *continued*

State	County	Total On-site Disposal or Other Releases[1]	Total Off-site Disposal or Other Releases[2]	Total On- and Off-site Disposal or Other Releases
MO	Ray	525.0	18,524.0	19,049.0
MO	Reynolds	25,998,172.3	0.0	25,998,172.3
MO	Saline	6,528.0	1,669.0	8,197.0
MO	Scott	237,996.0	2,986.0	240,982.0
MO	Shannon	2,385,936.0	0.0	2,385,936.0
MO	Shelby	0.0	0.0	0.0
MO	St. Charles	3,234,848.8	19,728.7	3,254,577.5
MO	St. Francois	51,025.0	76.7	51,101.7
MO	St. Louis	1,969,778.1	128,848.7	2,098,626.9
MO	St. Louis City	2,755,688.3	670,884.7	3,426,573.0
MO	Ste. Genevieve	398,102.5	0.0	398,102.5
MO	Stoddard	3,247.0	1,175.6	4,422.6
MO	Sullivan	110,682.0	340.0	111,022.0
MO	Taney	0.3	0.0	0.3
MO	Texas	1,127,376.8	164.9	1,127,541.7
MO	Vernon	217,177.0	24,119.0	241,296.0
MO	Warren	29,942.9	0.0	29,942.9
MO	Washington	819.0	43,415.0	44,234.0
MO	Wayne	2,010.0	500.0	2,510.0
MO	Webster	85,297.0	7,791.0	93,088.0
MO	Wright	37,720.0	750.0	38,470.0
MP	Saipan	6,027.0	0.0	6,027.0
MS	Adams	882,294.6	10,320.3	892,614.8
MS	Alcorn	1,010,940.4	8,453.0	1,019,393.4
MS	Amite	16,356.0	0.0	16,356.0
MS	Attala	4,256.0	56,852.0	61,108.0
MS	Bolivar	88.0	11,060.0	11,148.0
MS	Calhoun	98,046.0	0.0	98,046.0
MS	Chickasaw	10.0	0.0	10.0
MS	Choctaw	1,251,401.7	0.0	1,251,401.7
MS	Claiborne	2.6	131.6	134.2
MS	Clarke	3,195.0	2,706.0	5,901.0
MS	Clay	237,556.0	10.6	237,566.6
MS	Coahoma	544,511.0	10,036.0	554,547.0
MS	Copiah	400,196.3	1,554.0	401,750.3
MS	Covington	1,079,280.0	0.0	1,079,280.0
MS	DeSoto	433,331.3	42,896.6	476,227.9
MS	Forrest	609,702.0	1,043.0	610,745.0
MS	George	13,285.0	0.0	13,285.0
MS	Grenada	510,421.9	20,682.6	531,104.5
MS	Hancock	44,281.6	9,845.6	54,127.1
MS	Harrison	18,116,063.2	27.9	18,116,091.1
MS	Hinds	313,398.5	314,516.7	627,915.3
MS	Holmes	5.0	0.0	5.0
MS	Humphreys	107,718.0	0.0	107,718.0
MS	Itawamba	45,429.2	5,085.7	50,514.9
MS	Jackson	4,843,847.7	226,358.1	5,070,205.8
MS	Jasper	28,245.6	0.0	28,245.6
MS	Jones	1,405,416.7	2,144.0	1,407,560.7
MS	Kemper	5.0	0.0	5.0
MS	Lafayette	0.0	187.1	187.1
MS	Lamar	1,392,426.8	0.0	1,392,426.8
MS	Lauderdale	434,252.6	34,063.9	468,316.5
MS	Lawrence	2,735,440.7	0.0	2,735,440.7
MS	Leake	1,185,969.0	0.0	1,185,969.0
MS	Lee	133,493.8	154,366.7	287,860.5
MS	Leflore	548,312.8	0.0	548,312.8
MS	Lincoln	1,352.5	129,941.5	131,294.0
MS	Lowndes	2,883,596.3	27,372.0	2,910,968.3

**TRI on-site and off-site reported disposed of or otherwise released (in pounds),
for facilities in all industries, for all chemicals, all counties, 2003** - *continued*

State	County	Total On-site Disposal or Other Releases[1]	Total Off-site Disposal or Other Releases[2]	Total On- and Off-site Disposal or Other Releases
MS	Madison	597,771.0	152,805.0	750,576.0
MS	Marion	16,462.0	5.0	16,467.0
MS	Marshall	90,182.0	39.0	90,221.0
MS	Monroe	6,534,312.3	88,499.1	6,622,811.4
MS	Montgomery	1.0	1.0	2.0
MS	Neshoba	26,771.5	10.5	26,782.0
MS	Newton	25,030.6	8,018.0	33,048.6
MS	Noxubee	78,826.1	296.0	79,122.1
MS	Oktibbeha	12,260.0	10,510.7	22,770.7
MS	Panola	287,156.0	226.0	287,382.0
MS	Pearl River	13,070.0	0.0	13,070.0
MS	Perry	1,758,924.2	3.6	1,758,927.8
MS	Pike	723,044.0	0.0	723,044.0
MS	Pontotoc	40,928.0	0.0	40,928.0
MS	Prentiss	14,175.0	77.0	14,252.0
MS	Quitman	387,883.0	0.0	387,883.0
MS	Rankin	85,323.9	53,443.3	138,767.2
MS	Scott	1,944,180.0	1,511.0	1,945,691.0
MS	Simpson	0.0	0.0	0.0
MS	Smith	336,750.8	3,057.4	339,808.2
MS	Stone	723.2	9,229.8	9,953.0
MS	Sunflower	171,028.0	0.0	171,028.0
MS	Tate	315.0	17.0	332.0
MS	Tippah	50.2	17,203.5	17,253.7
MS	Tishomingo	301,261.0	5,067.0	306,328.0
MS	Tunica	186.1	0.0	186.1
MS	Union	128,609.2	19,424.5	148,033.7
MS	Warren	2,146,783.0	1,689.0	2,148,472.0
MS	Washington	902,371.0	949.2	903,320.2
MS	Wayne	1,068.6	33,824.1	34,892.7
MS	Webster	10,532.0	32,895.0	43,427.0
MS	Wilkinson	5.9	252.6	258.5
MS	Winston	223,784.4	13,500.6	237,285.0
MS	Yalobusha	14,945.8	2,008.9	16,954.8
MS	Yazoo	3,398,297.9	0.0	3,398,297.9
MT	Big Horn	5.0	0.0	5.0
MT	Broadwater	70.2	0.0	70.2
MT	Cascade	68,787.9	14.3	68,802.2
MT	Flathead	746,130.0	0.0	746,130.0
MT	Gallatin	11,041.0	0.0	11,041.0
MT	Jefferson	22,197,530.6	0.0	22,197,530.6
MT	Lewis and Clark	14,336.0	10.0	14,346.0
MT	Lincoln	3.4	0.0	3.4
MT	Missoula	1,641,574.0	151,950.4	1,793,524.4
MT	Ravalli	16,588.0	4,534.0	21,122.0
MT	Richland	358,045.6	98,681.8	456,727.4
MT	Roosevelt	433,630.0	0.0	433,630.0
MT	Rosebud	11,298,944.6	208,916.0	11,507,860.6
MT	Sanders	6,721.0	0.0	6,721.0
MT	Silver Bow	6,352,700.0	0.0	6,352,700.0
MT	Stillwater	506,095.7	496,655.0	1,002,750.7
MT	Sweet Grass	180,404.3	0.0	180,404.3
MT	Yellowstone	697,209.5	162,221.3	859,430.8
NC	Alamance	38,405.9	7,363.0	45,768.9
NC	Alexander	0.0	5,940.3	5,940.3
NC	Alleghany	250.0	0.0	250.0
NC	Anson	119,721.1	0.0	119,721.1
NC	Ashe	3,512.0	48,516.0	52,028.0
NC	Beaufort	1,945,147.3	12.0	1,945,159.3

TRI on-site and off-site reported disposed of or otherwise released (in pounds), for facilities in all industries, for all chemicals, all counties, 2003 - *continued*

State	County	Total On-site Disposal or Other Releases[1]	Total Off-site Disposal or Other Releases[2]	Total On- and Off-site Disposal or Other Releases
NC	Bertie	544,566.0	0.0	544,566.0
NC	Bladen	4,918,864.8	0.0	4,918,864.8
NC	Brunswick	2,304,367.6	130,920.0	2,435,287.6
NC	Buncombe	2,984,370.1	37.0	2,984,407.1
NC	Burke	88,071.4	3,562.0	91,633.4
NC	Cabarrus	517,209.0	1,844,789.0	2,361,998.0
NC	Caldwell	1,087,098.3	446.0	1,087,544.3
NC	Carteret	160,627.5	58.7	160,686.2
NC	Caswell	0.0	0.0	0.0
NC	Catawba	15,481,500.0	58,182.3	15,539,682.3
NC	Chatham	2,631,814.7	28,034.4	2,659,849.1
NC	Cherokee	838.3	20,897.0	21,735.3
NC	Chowan	119,028.5	0.0	119,028.5
NC	Clay	141.3	0.0	141.3
NC	Cleveland	616,123.3	86,391.5	702,514.8
NC	Columbus	4,678,929.5	437.9	4,679,367.4
NC	Craven	2,341,051.5	234,474.7	2,575,526.2
NC	Cumberland	1,342,998.4	72,386.0	1,415,384.4
NC	Dare	124.0	0.0	124.0
NC	Davidson	299,806.5	198,841.3	498,647.9
NC	Davie	13,370.0	0.0	13,370.0
NC	Duplin	220,754.6	9,407.0	230,161.5
NC	Durham	23,917.2	118,565.2	142,482.4
NC	Edgecombe	83,504.2	12,794.0	96,298.3
NC	Forsyth	1,225,499.2	73,935.5	1,299,434.7
NC	Franklin	56,099.6	745.0	56,844.6
NC	Gaston	11,631,231.9	331,962.6	11,963,194.5
NC	Graham	136,724.7	0.0	136,724.7
NC	Granville	36,521.6	78,691.6	115,213.1
NC	Greene	0.0	0.0	0.0
NC	Guilford	804,450.0	46,717.8	851,167.9
NC	Halifax	1,169,490.6	185,131.0	1,354,621.6
NC	Harnett	9,972.5	7,321.0	17,293.5
NC	Haywood	2,427,961.6	460,614.0	2,888,575.6
NC	Henderson	285,242.2	1,534.4	286,776.6
NC	Hertford	73,141.3	3,060,419.5	3,133,560.8
NC	Hoke	198.0	672.0	870.0
NC	Iredell	556,707.6	191,179.1	747,886.7
NC	Jackson	66,187.0	99.4	66,286.4
NC	Johnston	88,975.1	16,913.0	105,888.2
NC	Lee	1,270,749.0	80,333.0	1,351,082.0
NC	Lenoir	560,872.5	357.8	561,230.4
NC	Lincoln	244,203.1	32,158.0	276,361.1
NC	Macon	0.0	5.0	5.0
NC	Madison	0.0	0.0	0.0
NC	Martin	1,984,545.6	500.1	1,985,045.7
NC	McDowell	528,211.5	30,505.0	558,716.5
NC	Mecklenburg	389,773.3	616,273.4	1,006,046.7
NC	Mitchell	39,089.0	0.0	39,089.0
NC	Montgomery	47,208.2	10,714.0	57,922.2
NC	Moore	951.8	13.0	964.8
NC	Nash	174,116.5	14,401.1	188,517.6
NC	New Hanover	5,186,428.0	455,462.9	5,641,890.9
NC	Northampton	74,485.6	5,750.0	80,235.6
NC	Onslow	415,533.2	9,917.1	425,450.3
NC	Orange	2,481.0	2,752.0	5,233.0
NC	Pasquotank	0.0	0.0	0.0
NC	Pender	0.0	0.0	0.0
NC	Person	18,598,432.3	28,863.2	18,627,295.5

**TRI on-site and off-site reported disposed of or otherwise released (in pounds),
for facilities in all industries, for all chemicals, all counties, 2003** - *continued*

State	County	Total On-site Disposal or Other Releases[1]	Total Off-site Disposal or Other Releases[2]	Total On- and Off-site Disposal or Other Releases
NC	Pitt	370,200.6	342,309.6	712,510.2
NC	Randolph	82,922.6	351,983.8	434,906.4
NC	Richmond	1,219.0	4,209.0	5,428.0
NC	Robeson	930,447.6	361.0	930,808.6
NC	Rockingham	2,076,598.6	54,891.8	2,131,490.4
NC	Rowan	3,242,838.0	476,437.4	3,719,275.4
NC	Rutherford	5,210,694.4	5,179.6	5,215,874.0
NC	Sampson	57,935.0	4,837.3	62,772.3
NC	Scotland	769,247.0	132,004.4	901,251.4
NC	Stanly	130,104.9	24,287.0	154,391.9
NC	Stokes	10,264,638.0	12.5	10,264,650.5
NC	Surry	973,345.0	103.5	973,448.5
NC	Transylvania	15.2	0.0	15.2
NC	Union	236,920.3	2,108.5	239,028.8
NC	Vance	1,046.5	2,613.0	3,659.5
NC	Wake	335,832.0	72,280.1	408,112.1
NC	Warren	0.0	0.0	0.0
NC	Watauga	5,230.0	34,320.0	39,550.0
NC	Wayne	2,089,671.2	101,880.0	2,191,551.2
NC	Wilkes	1,496,959.9	6,701.6	1,503,661.5
NC	Wilson	626,182.2	346,226.2	972,408.4
NC	Yadkin	0.0	0.0	0.0
ND	Barnes	0.0	32.0	32.0
ND	Benson	286,269.0	0.0	286,269.0
ND	Cass	59,740.7	798.6	60,539.3
ND	Grand Forks	55,373.8	0.0	55,373.8
ND	McHenry	263,849.0	0.0	263,849.0
ND	McLean	4,244,078.9	75,839.8	4,319,918.7
ND	Mercer	4,078,197.9	7,172,179.0	11,250,376.9
ND	Morton	2,266,661.9	25,596.4	2,292,258.3
ND	Oliver	1,711,872.1	1,903,399.0	3,615,271.1
ND	Pembina	368,162.6	24,769.0	392,931.6
ND	Ramsey	43,710.0	0.0	43,710.0
ND	Ransom	394,305.0	0.0	394,305.0
ND	Richland	419,484.3	9,188.1	428,672.4
ND	Rolette	7.2	0.0	7.2
ND	Sargent	30,823.0	0.0	30,823.0
ND	Sheridan	0.0	6.6	6.6
ND	Stark	1,474.2	210.7	1,684.9
ND	Stutsman	29,944.0	1,416.5	31,360.5
ND	Traill	157,082.8	0.0	157,082.8
ND	Williams	14,326.0	0.0	14,326.0
NE	Adams	335,066.7	14,959,165.7	15,294,232.4
NE	Box Butte	255.0	10,115.0	10,370.0
NE	Buffalo	39,590.4	141,741.2	181,331.6
NE	Butler	13,381.5	6,590.0	19,971.5
NE	Cass	231,646.6	0.0	231,646.6
NE	Cedar	250.0	55,506.4	55,756.4
NE	Chase	47,524.8	0.0	47,524.8
NE	Cheyenne	0.0	0.0	0.0
NE	Clay	408,606.0	0.0	408,606.0
NE	Colfax	4,070,505.0	683,092.0	4,753,597.0
NE	Cuming	1,420,020.0	1,466.5	1,421,486.5
NE	Dakota	6,576,054.0	75,480.0	6,651,534.0
NE	Dawson	4,948,671.5	116,612.9	5,065,284.4
NE	Dixon	7,000.0	0.0	7,000.0
NE	Dodge	786,912.6	350,546.1	1,137,458.7
NE	Douglas	1,935,476.2	59,912.7	1,995,388.9
NE	Gage	225,463.0	1,515.0	226,978.0

TRI on-site and off-site reported disposed of or otherwise released (in pounds), for facilities in all industries, for all chemicals, all counties, 2003 - *continued*

State	County	Total On-site Disposal or Other Releases[1]	Total Off-site Disposal or Other Releases[2]	Total On- and Off-site Disposal or Other Releases
NE	Hall	137,508.3	44,763.2	182,271.4
NE	Hamilton	4,050.0	0.0	4,050.0
NE	Holt	2,555.0	0.0	2,555.0
NE	Jefferson	92,143.6	2,585,332.0	2,677,475.6
NE	Kearney	0.0	0.0	0.0
NE	Kimball	344,031.5	52,717.9	396,749.4
NE	Lancaster	1,318,370.0	73,741.6	1,392,111.6
NE	Lincoln	3,858,591.0	0.0	3,858,591.0
NE	Madison	21,673.4	70,529.9	92,203.3
NE	Morrill	0.0	0.0	0.0
NE	Nemaha	100,421.0	10.5	100,431.5
NE	Otoe	4,189,890.0	1.7	4,189,891.7
NE	Phelps	124,443.0	0.0	124,443.0
NE	Pierce	14,237.9	0.0	14,237.9
NE	Platte	224,676.9	57,432.2	282,109.1
NE	Red Willow	1,420.2	62,941.3	64,361.5
NE	Richardson	0.0	0.0	0.0
NE	Saline	1,543,781.2	5,737.0	1,549,518.2
NE	Sarpy	2,075.1	0.0	2,075.1
NE	Scotts Bluff	98,220.5	0.0	98,220.5
NE	Seward	2,755.0	0.0	2,755.0
NE	Stanton	21,591.0	9,686,876.0	9,708,467.0
NE	Thayer	28,541.0	0.0	28,541.0
NE	Thurston	0.0	0.0	0.0
NE	Washington	491,799.7	11,402.4	503,202.1
NE	Wayne	47,285.0	0.0	47,285.0
NE	York	33,150.2	0.8	33,151.0
NH	Belknap	8,376.4	1,511.4	9,887.8
NH	Cheshire	22,874.5	0.0	22,874.5
NH	Coos	517,696.7	33,061.6	550,758.3
NH	Grafton	6,030.4	8,156.0	14,186.4
NH	Hillsborough	242,899.9	121,099.2	363,999.1
NH	Merrimack	3,465,491.1	21,255.6	3,486,746.7
NH	Rockingham	1,183,147.7	257,751.9	1,440,899.6
NH	Strafford	35,959.3	17,450.1	53,409.4
NH	Sullivan	957.0	521.8	1,478.8
NJ	Atlantic	100,294.1	16,023.7	116,317.8
NJ	Bergen	174,147.8	31,526.0	205,673.8
NJ	Burlington	366,392.9	2,717,272.4	3,083,665.3
NJ	Camden	178,397.4	2,917.4	181,314.8
NJ	Cape May	522,126.0	13,367.2	535,493.2
NJ	Cumberland	532,494.5	18,016.4	550,510.9
NJ	Essex	285,578.5	353,541.3	639,119.8
NJ	Gloucester	1,249,930.8	115,607.2	1,365,538.0
NJ	Hudson	3,143,517.4	353,904.8	3,497,422.2
NJ	Hunterdon	10,362.1	69,451.0	79,813.1
NJ	Mercer	2,499,299.2	8,536.0	2,507,835.2
NJ	Middlesex	1,277,015.8	1,611,413.2	2,888,429.0
NJ	Monmouth	23,881.3	3,444.0	27,325.3
NJ	Morris	88,573.6	239,029.9	327,603.5
NJ	Ocean	6,603.2	9,688.5	16,291.7
NJ	Passaic	86,274.7	57,757.9	144,032.6
NJ	Salem	2,968,488.2	449,999.9	3,418,488.2
NJ	Somerset	78,238.6	14,011.4	92,250.1
NJ	Sussex	28,907.0	0.0	28,907.0
NJ	Union	2,725,280.7	202,076.4	2,927,357.1
NJ	Warren	413,480.8	53,249.1	466,729.9
NM	Bernalillo	51,240.5	24,931.3	76,171.8
NM	Chaves	26,176.0	0.0	26,176.0

**TRI on-site and off-site reported disposed of or otherwise released (in pounds),
for facilities in all industries, for all chemicals, all counties, 2003** - *continued*

State	County	Total On-site Disposal or Other Releases[1]	Total Off-site Disposal or Other Releases[2]	Total On- and Off-site Disposal or Other Releases
NM	Colfax	283.5	0.9	284.4
NM	Curry	0.1	0.0	0.1
NM	Dona Ana	123,612.0	14,860.0	138,472.0
NM	Eddy	326,971.7	65,819.1	392,790.7
NM	Grant	9,457,149.7	0.0	9,457,149.7
NM	Lea	277,065.0	22,500.0	299,565.0
NM	Los Alamos	6,142.7	50,853.6	56,996.3
NM	Luna	613.0	0.0	613.0
NM	McKinley	510,571.5	19.8	510,591.3
NM	Otero	110,420.6	0.0	110,420.6
NM	Roosevelt	5,317.0	0.0	5,317.0
NM	San Juan	6,825,840.3	4,948,876.3	11,774,716.6
NM	Sandoval	28,578.7	7,331.7	35,910.4
NM	Santa Fe	0.0	0.0	0.0
NM	Valencia	3,140.0	0.0	3,140.0
NV	Carson City	5,213.0	3,259.0	8,472.0
NV	Churchill	275,956.5	0.0	275,956.5
NV	Clark	2,789,048.3	480,011.5	3,269,059.8
NV	Douglas	4,521.4	0.8	4,522.2
NV	Elko	99,934,835.1	2,649.3	99,937,484.4
NV	Esmeralda	7,651.0	0.0	7,651.0
NV	Eureka	29,452,159.2	5,358.5	29,457,517.7
NV	Humboldt	229,066,311.5	88.6	229,066,400.1
NV	Lander	2,949,098.4	0.0	2,949,098.4
NV	Lyon	147,015.3	143,611.3	290,626.6
NV	Mineral	2.0	0.0	2.0
NV	Nye	30,730,096.1	126,978.9	30,857,075.0
NV	Pershing	6,717,712.8	0.0	6,717,712.8
NV	Storey	1.4	0.0	1.4
NV	Washoe	179,311.3	57,207.3	236,518.6
NV	White Pine	335,640.7	0.0	335,640.7
NY	Albany	731,578.8	138,310.0	869,888.8
NY	Allegany	214,648.0	0.0	214,648.0
NY	Bronx	956.3	1,207.0	2,163.3
NY	Broome	985,056.8	129,635.0	1,114,691.8
NY	Cattaraugus	28,238.2	222,786.6	251,024.8
NY	Cayuga	13,582.7	1,316.4	14,899.1
NY	Chautauqua	2,645,624.3	235,635.9	2,881,260.1
NY	Chemung	227,801.3	250,971.7	478,773.0
NY	Chenango	108,123.9	1,568.3	109,692.2
NY	Clinton	9,409.0	1,000.0	10,409.0
NY	Columbia	6,366.0	907.0	7,273.0
NY	Cortland	57,769.4	160,906.1	218,675.5
NY	Delaware	1,231,870.8	5,088.0	1,236,958.8
NY	Dutchess	1,527,002.2	9,479.5	1,536,481.7
NY	Erie	3,230,235.2	527,184.7	3,757,420.0
NY	Essex	600,837.5	0.0	600,837.5
NY	Franklin	0.0	4,994.0	4,994.0
NY	Fulton	49,377.0	46,765.0	96,142.0
NY	Genesee	0.0	2,085.0	2,085.0
NY	Greene	1,009,236.2	1,542.1	1,010,778.2
NY	Herkimer	2,361.8	5,846.0	8,207.8
NY	Jefferson	327,510.2	2,693.6	330,203.8
NY	Kings	79,460.5	1,932.9	81,393.4
NY	Lewis	43,478.0	416.0	43,894.0
NY	Livingston	23,229.0	185.0	23,414.0
NY	Madison	58,422.3	2,255.2	60,677.5
NY	Monroe	5,257,887.7	727,593.1	5,985,480.7
NY	Montgomery	61,344.0	3,860.0	65,204.0

**TRI on-site and off-site reported disposed of or otherwise released (in pounds),
for facilities in all industries, for all chemicals, all counties, 2003** - *continued*

State	County	Total On-site Disposal or Other Releases[1]	Total Off-site Disposal or Other Releases[2]	Total On- and Off-site Disposal or Other Releases
NY	Nassau	174,094.2	31,058.3	205,152.5
NY	New York	31,596.1	24.4	31,620.5
NY	Niagara	4,207,722.8	378,583.1	4,586,306.0
NY	Oneida	159,000.9	102,378.4	261,379.2
NY	Onondaga	3,345,377.0	176,183.3	3,521,560.3
NY	Ontario	37,097.8	4,566.5	41,664.3
NY	Orange	2,980,778.1	101,421.1	3,082,199.2
NY	Orleans	23,371.0	265.0	23,636.0
NY	Oswego	206,154.3	91,269.4	297,423.6
NY	Otsego	0.0	0.0	0.0
NY	Putnam	6,212.0	54.0	6,266.0
NY	Queens	510,743.4	46,975.9	557,719.3
NY	Rensselaer	120,091.7	112.0	120,203.7
NY	Richmond	23,197.3	19,101.0	42,298.3
NY	Rockland	2,328,183.5	383,130.7	2,711,314.2
NY	Saratoga	865,419.7	76,854.5	942,274.2
NY	Schenectady	347,714.0	24,074.0	371,788.0
NY	Schuyler	57,534.8	0.0	57,534.8
NY	Seneca	167,196.6	9,283.0	176,479.6
NY	St. Lawrence	884,893.4	10,178.8	895,072.2
NY	Steuben	32,451.6	106,000.0	138,451.6
NY	Suffolk	1,288,747.1	574,698.4	1,863,445.5
NY	Tioga	16,792.9	35,721.3	52,514.2
NY	Tompkins	232,618.0	1,055.0	233,673.0
NY	Ulster	22,303.6	27,735.5	50,039.1
NY	Warren	1,215,697.6	20,735.3	1,236,432.9
NY	Washington	3,727.2	54,507.4	58,234.6
NY	Wayne	388,382.2	205,165.7	593,547.9
NY	Westchester	17,853.2	109,432.2	127,285.4
NY	Wyoming	57,433.6	227.9	57,661.5
NY	Yates	1,289,122.0	20,636.0	1,309,758.0
OH	Adams	17,650,816.1	11.0	17,650,827.1
OH	Allen	20,984,963.0	540,348.2	21,525,311.2
OH	Ashland	62,223.0	61,782.0	124,005.0
OH	Ashtabula	11,629,946.8	477,904.5	12,107,851.3
OH	Athens	12,432.0	0.0	12,432.0
OH	Auglaize	519,810.9	447,686.7	967,497.5
OH	Belmont	2,908,952.3	630,039.1	3,538,991.4
OH	Brown	89,657.0	25.6	89,682.6
OH	Butler	1,327,498.0	719,309.5	2,046,807.5
OH	Carroll	3,970.8	52,888.6	56,859.4
OH	Champaign	927.6	6,514.0	7,441.6
OH	Clark	234,355.4	39,290.1	273,645.5
OH	Clermont	7,387,213.8	443,135.9	7,830,349.7
OH	Clinton	5,740.3	21,935.9	27,676.3
OH	Columbiana	89,440.6	583,383.8	672,824.4
OH	Coshocton	12,977,460.5	1,223,139.7	14,200,600.1
OH	Crawford	113,372.6	18,007.2	131,379.8
OH	Cuyahoga	3,941,152.0	2,993,655.2	6,934,807.2
OH	Darke	132,120.2	5,514.0	137,634.2
OH	Defiance	4,493,886.4	13,294.7	4,507,181.1
OH	Delaware	263,172.6	32,037.0	295,209.6
OH	Erie	332,446.4	119,732.5	452,178.9
OH	Fairfield	32,148.0	7,863.9	40,011.9
OH	Fayette	40,250.0	60,686.0	100,936.0
OH	Franklin	1,562,794.3	795,963.7	2,358,758.0
OH	Fulton	105,471.8	51,663.0	157,134.8
OH	Gallia	10,958,678.2	9,253.2	10,967,931.4
OH	Geauga	249,544.2	123,290.3	372,834.4

TRI on-site and off-site reported disposed of or otherwise released (in pounds), for facilities in all industries, for all chemicals, all counties, 2003 - *continued*

State	County	Total On-site Disposal or Other Releases[1]	Total Off-site Disposal or Other Releases[2]	Total On- and Off-site Disposal or Other Releases
OH	Greene	93,395.7	8,641.0	102,036.7
OH	Guernsey	42,767.5	95,151.0	137,918.5
OH	Hamilton	7,866,750.0	623,918.1	8,490,668.0
OH	Hancock	904,938.2	137,337.5	1,042,275.7
OH	Hardin	648,086.0	6,289.0	654,375.0
OH	Harrison	108.2	16.5	124.7
OH	Henry	225,449.0	1,592.0	227,041.0
OH	Highland	155.0	5.0	160.0
OH	Hocking	11,637.0	13,011.1	24,648.1
OH	Holmes	12,778.0	580.2	13,358.2
OH	Huron	615,370.0	177.5	615,547.5
OH	Jackson	257,774.2	156,487.0	414,261.2
OH	Jefferson	25,300,628.3	6,274,081.4	31,574,709.7
OH	Knox	4,980.0	5,177.0	10,157.0
OH	Lake	4,033,988.5	1,083,201.8	5,117,190.3
OH	Lawrence	634,358.0	2.0	634,360.0
OH	Licking	2,480,708.7	39,106.0	2,519,814.6
OH	Logan	213,641.0	71,104.5	284,745.5
OH	Lorain	3,066,746.0	1,324,672.4	4,391,418.3
OH	Lucas	9,842,393.2	675,598.2	10,517,991.4
OH	Madison	123,102.5	1,274,047.0	1,397,149.5
OH	Mahoning	128,959.5	135,080.8	264,040.3
OH	Marion	596,985.4	352,176.3	949,161.7
OH	Medina	259,045.4	58,973.0	318,018.4
OH	Mercer	70.4	6,952.0	7,022.4
OH	Miami	124,415.1	433,913.8	558,328.9
OH	Monroe	824,739.3	174.0	824,913.3
OH	Montgomery	1,594,888.1	858,638.9	2,453,527.1
OH	Morgan	3,952.0	151,979.0	155,931.0
OH	Morrow	4,549.0	0.0	4,549.0
OH	Muskingum	1,473,003.1	2,078,097.2	3,551,100.4
OH	Noble	25,329.0	14,432.0	39,761.0
OH	Ottawa	253,463.5	64,162.5	317,626.0
OH	Paulding	145,445.2	788.6	146,233.8
OH	Perry	110,990.0	822.0	111,812.0
OH	Pickaway	580,497.8	1,041,403.7	1,621,901.5
OH	Pike	187,676.0	68,769.3	256,445.3
OH	Portage	275,965.3	118,193.5	394,158.8
OH	Preble	41,798.8	19,875.0	61,673.8
OH	Putnam	47,875.0	3,604.3	51,479.3
OH	Richland	243,308.1	4,021,486.0	4,264,794.1
OH	Ross	1,064,871.7	201,363.0	1,266,234.7
OH	Sandusky	13,420,222.9	116,992.9	13,537,215.8
OH	Scioto	355,524.2	268,636.0	624,160.2
OH	Seneca	263,944.4	4,606.0	268,550.4
OH	Shelby	434,331.3	57,904.7	492,235.9
OH	Stark	441,824.0	14,421,921.3	14,863,745.3
OH	Summit	1,666,831.6	415,980.1	2,082,811.6
OH	Trumbull	2,836,978.7	74,496.4	2,911,475.1
OH	Tuscarawas	498,288.7	1,277,397.2	1,775,685.8
OH	Union	2,393,766.1	163,490.8	2,557,256.9
OH	Van Wert	381,729.0	131,896.0	513,625.0
OH	Vinton	9,645.0	750.0	10,395.0
OH	Warren	106,336.5	31,934.2	138,270.7
OH	Washington	18,113,633.6	3,963,282.5	22,076,916.1
OH	Wayne	522,112.9	83,554.1	605,667.0
OH	Williams	97,513.2	38,814.1	136,327.3
OH	Wood	269,214.4	591,866.1	861,080.6
OH	Wyandot	676,569.6	64,088.8	740,658.4

TRI on-site and off-site reported disposed of or otherwise released (in pounds), for facilities in all industries, for all chemicals, all counties, 2003 - *continued*

State	County	Total On-site Disposal or Other Releases[1]	Total Off-site Disposal or Other Releases[2]	Total On- and Off-site Disposal or Other Releases
OK	Adair	48,697.7	0.0	48,697.7
OK	Beaver	6,101.0	0.0	6,101.0
OK	Bryan	191,297.0	45.0	191,342.0
OK	Canadian	47,312.0	2,154.0	49,466.0
OK	Carter	358,882.6	720,115.1	1,078,997.6
OK	Cherokee	273.5	7.9	281.4
OK	Choctaw	404,738.2	0.0	404,738.2
OK	Cleveland	4,969.7	8,237.5	13,207.2
OK	Comanche	454,017.9	69,355.7	523,373.6
OK	Craig	0.0	547.0	547.0
OK	Creek	14,606.2	407,796.3	422,402.5
OK	Custer	61,978.0	0.0	61,978.0
OK	Delaware	0.0	0.0	0.0
OK	Dewey	0.0	0.0	0.0
OK	Garfield	3,145,620.0	0.0	3,145,620.0
OK	Garvin	140,375.1	24,052.6	164,427.7
OK	Grady	98,416.9	36,979.0	135,395.9
OK	Hughes	0.0	0.0	0.0
OK	Kay	684,695.8	14,711.5	699,407.2
OK	Kingfisher	10,537.1	0.0	10,537.1
OK	Latimer	0.2	0.0	0.2
OK	Le Flore	185,374.0	575,260.0	760,634.0
OK	Lincoln	6,430.0	0.0	6,430.0
OK	Major	1,692,109.0	1,736.0	1,693,845.0
OK	Marshall	70,502.0	20,014.0	90,516.0
OK	Mayes	3,449,080.0	214.0	3,449,294.0
OK	McCurtain	3,409,152.6	6,720.0	3,415,872.6
OK	McIntosh	24,167.0	0.0	24,167.0
OK	Muskogee	1,034,787.0	213,937.2	1,248,724.2
OK	Noble	318,404.7	358,579.9	676,984.6
OK	Nowata	0.0	0.0	0.0
OK	Oklahoma	831,811.7	51,035.8	882,847.5
OK	Okmulgee	0.0	0.0	0.0
OK	Osage	557,850.0	0.0	557,850.0
OK	Ottawa	107,708.0	0.0	107,708.0
OK	Pawnee	13,954.0	0.0	13,954.0
OK	Payne	272,169.0	165,293.5	437,462.5
OK	Pittsburg	234,915.0	26,700.0	261,615.0
OK	Pontotoc	295,198.2	463.0	295,661.2
OK	Pottawatomie	56,187.2	14,418.0	70,605.2
OK	Pushmataha	0.0	0.0	0.0
OK	Rogers	4,097,686.2	169,102.9	4,266,789.0
OK	Seminole	183,241.1	0.0	183,241.1
OK	Sequoyah	42,253.2	0.0	42,253.2
OK	Stephens	155.0	26,065.0	26,220.0
OK	Texas	15,000.0	0.0	15,000.0
OK	Tillman	6,330.0	11,892.0	18,222.0
OK	Tulsa	1,796,211.2	693,409.9	2,489,621.1
OK	Wagoner	805.0	250.0	1,055.0
OK	Washington	3,029.0	1,128,877.0	1,131,906.0
OK	Woods	14,532.0	0.0	14,532.0
OK	Woodward	925,769.0	669.0	926,438.0
OR	Baker	846.0	0.0	846.0
OR	Benton	1,963.8	637.5	2,601.3
OR	Clackamas	242,605.2	16,726.1	259,331.3
OR	Clatsop	1,061,068.7	206.6	1,061,275.3
OR	Columbia	1,411,880.4	65,273.1	1,477,153.5
OR	Coos	33,340.7	3,137.0	36,477.7
OR	Crook	2,643.0	0.0	2,643.0

TRI on-site and off-site reported disposed of or otherwise released (in pounds), for facilities in all industries, for all chemicals, all counties, 2003 - *continued*

State	County	Total On-site Disposal or Other Releases[1]	Total Off-site Disposal or Other Releases[2]	Total On- and Off-site Disposal or Other Releases
OR	Deschutes	15,727.0	0.0	15,727.0
OR	Douglas	823,671.1	3,319.2	826,990.3
OR	Gilliam	24,562,882.9	5.3	24,562,888.2
OR	Harney	90,880.0	0.0	90,880.0
OR	Hood River	1,010.0	510.0	1,520.0
OR	Jackson	2,031,627.3	1,844.3	2,033,471.6
OR	Jefferson	129,154.0	6,600.0	135,754.0
OR	Josephine	101,547.0	0.0	101,547.0
OR	Klamath	738,477.3	350.0	738,827.3
OR	Lane	2,042,446.4	34,874.4	2,077,320.8
OR	Lincoln	1,182,904.0	0.0	1,182,904.0
OR	Linn	3,021,642.8	105,595.9	3,127,238.7
OR	Malheur	349,227.8	164.0	349,391.8
OR	Marion	168,102.7	3,882.2	171,984.9
OR	Morrow	360,712.6	0.0	360,712.6
OR	Multnomah	1,627,320.8	948,391.1	2,575,712.0
OR	Polk	118,539.4	6,198.4	124,737.8
OR	Tillamook	357.2	710.8	1,068.0
OR	Umatilla	75,313.8	3,647.3	78,961.1
OR	Union	238,635.2	2.0	238,637.2
OR	Wallowa	0.0	0.0	0.0
OR	Wasco	9,810.8	78.2	9,889.0
OR	Washington	192,996.1	7,713.5	200,709.7
OR	Yamhill	277,908.8	6,015,020.5	6,292,929.3
PA	Adams	48,447.8	35,252.0	83,699.8
PA	Allegheny	8,404,625.7	7,519,897.6	15,924,523.3
PA	Armstrong	20,777,121.6	3,241.8	20,780,363.4
PA	Beaver	5,120,568.8	24,719,467.6	29,840,036.4
PA	Bedford	240,473.0	1,763.0	242,236.0
PA	Berks	3,673,908.1	2,644,003.7	6,317,911.9
PA	Blair	141,810.0	41,706.5	183,516.5
PA	Bradford	3,948,304.6	490,116.4	4,438,421.0
PA	Bucks	218,128.0	251,317.6	469,445.6
PA	Butler	5,728,365.7	245,579.7	5,973,945.4
PA	Cambria	340,227.6	1,278,761.5	1,618,989.1
PA	Cameron	1,599.0	7,105.0	8,704.0
PA	Carbon	113,928.1	647,943.3	761,871.4
PA	Centre	137,260.1	395,720.7	532,980.8
PA	Chester	1,398,041.2	87,520.9	1,485,562.1
PA	Clarion	238,923.3	481,189.3	720,112.6
PA	Clearfield	3,599,724.4	20,484.1	3,620,208.5
PA	Clinton	19,021.0	18,262.0	37,283.0
PA	Columbia	26,754.5	82,764.8	109,519.3
PA	Crawford	256,508.8	14,664.1	271,172.9
PA	Cumberland	67,653.3	290,746.3	358,399.6
PA	Dauphin	66,361.7	36,161.6	102,523.3
PA	Delaware	1,585,896.3	79,584.4	1,665,480.7
PA	Elk	691,740.7	83,336.9	775,077.5
PA	Erie	1,169,479.5	366,036.3	1,535,515.8
PA	Fayette	200.0	1,552.0	1,752.0
PA	Franklin	166,252.0	18,253.0	184,505.0
PA	Fulton	19,127.2	1,972.0	21,099.2
PA	Greene	7,939,786.8	3,862.0	7,943,648.8
PA	Huntingdon	54,500.2	20,286.8	74,787.0
PA	Indiana	12,327,056.2	29,042.2	12,356,098.4
PA	Jefferson	91,016.0	64,775.1	155,791.1
PA	Juniata	183,975.0	136.6	184,111.6
PA	Lackawanna	68,620.7	21,290.2	89,910.9
PA	Lancaster	1,193,054.6	3,027,223.9	4,220,278.5

TRI on-site and off-site reported disposed of or otherwise released (in pounds),
for facilities in all industries, for all chemicals, all counties, 2003 - *continued*

State	County	Total On-site Disposal or Other Releases[1]	Total Off-site Disposal or Other Releases[2]	Total On- and Off-site Disposal or Other Releases
PA	Lawrence	2,303,073.4	761,149.9	3,064,223.3
PA	Lebanon	39,895.7	1,477.7	41,373.3
PA	Lehigh	266,216.0	64,246.1	330,462.1
PA	Luzerne	297,108.1	276,847.6	573,955.7
PA	Lycoming	157,097.6	59,656.8	216,754.3
PA	McKean	1,576,531.1	18,975.5	1,595,506.6
PA	Mercer	544,634.5	177,984.5	722,618.9
PA	Mifflin	53,978.6	7,581.6	61,560.2
PA	Monroe	8,115.6	54,483.8	62,599.4
PA	Montgomery	1,006,165.9	199,238.1	1,205,403.9
PA	Montour	5,707,222.0	23,462.0	5,730,684.0
PA	Northampton	4,663,781.2	514,739.7	5,178,520.8
PA	Northumberland	975,793.5	313,287.5	1,289,081.0
PA	Philadelphia	803,970.9	116,448.8	920,419.7
PA	Pike	199.0	0.0	199.0
PA	Potter	441.6	1,154.0	1,595.6
PA	Schuylkill	1,826,464.5	777,731.8	2,604,196.3
PA	Snyder	1,684,455.0	652,144.0	2,336,599.0
PA	Somerset	63,707.3	9,316.4	73,023.7
PA	Susquehanna	0.0	0.0	0.0
PA	Tioga	7,962.0	143,128.1	151,090.1
PA	Union	50,003.5	0.8	50,004.3
PA	Venango	145,751.1	961,705.6	1,107,456.7
PA	Warren	361,213.1	6,464.9	367,678.0
PA	Washington	649,045.8	1,894,995.8	2,544,041.6
PA	Westmoreland	3,570,967.7	1,140,535.5	4,711,503.2
PA	Wyoming	194.9	4.1	198.9
PA	York	7,200,586.9	2,185,560.7	9,386,147.6
PR	Aguada	27.1	0.0	27.1
PR	Aguadilla	225.0	16.0	241.0
PR	Aibonito	250.0	1.0	251.0
PR	Arecibo	79,121.4	8,335.5	87,456.9
PR	Añasco	1,753.0	0.0	1,753.0
PR	Barceloneta	472,710.8	138.0	472,848.8
PR	Bayamón	198,133.4	0.0	198,133.4
PR	Cabo Rojo	0.0	5.0	5.0
PR	Caguas	4,129.6	5.0	4,134.6
PR	Canóvanas	0.0	0.0	0.0
PR	Carolina	28,223.5	8,163.4	36,386.9
PR	Cataño	19.5	0.0	19.5
PR	Cayey	0.0	1,398.0	1,398.0
PR	Ceiba	14,994.7	0.0	14,994.7
PR	Ciales	0.0	2.0	2.0
PR	Cidra	36,954.0	0.0	36,954.0
PR	Coamo	3,058.0	1,848.0	4,906.0
PR	Dorado	78.2	0.0	78.2
PR	Fajardo	5,912.0	0.0	5,912.0
PR	Guayama	550,359.3	635,294.8	1,185,654.1
PR	Guayanilla	2,248,687.5	24,011.9	2,272,699.3
PR	Guaynabo	25,997.0	48.9	26,045.9
PR	Guánica	8,496.0	0.0	8,496.0
PR	Hatillo	0.0	0.0	0.0
PR	Humacao	34,330.3	5,050.0	39,380.3
PR	Isabela	0.2	0.0	0.2
PR	Jayuya	5.0	1,973.0	1,978.0
PR	Juncos	0.0	0.0	0.0
PR	Las Piedras	153,148.0	0.0	153,148.0
PR	Luquillo	0.0	5.0	5.0
PR	Manatí	87,003.0	18,707.4	105,710.4

**TRI on-site and off-site reported disposed of or otherwise released (in pounds),
for facilities in all industries, for all chemicals, all counties, 2003** - *continued*

State	County	Total On-site Disposal or Other Releases[1]	Total Off-site Disposal or Other Releases[2]	Total On- and Off-site Disposal or Other Releases
PR	Maricao	15.7	0.0	15.7
PR	Mayagüez	61,818.7	3,239.0	65,057.7
PR	Naguabo	0.0	0.0	0.0
PR	Patillas	4,637.2	5.0	4,642.2
PR	Peñuelas	344,955.0	0.0	344,955.0
PR	Ponce	15,516.0	244.0	15,760.0
PR	Sabana Grande	1,778.0	0.0	1,778.0
PR	Salinas	2,089,907.2	11,334.4	2,101,241.6
PR	San Germán	23,322.0	0.0	23,322.0
PR	San Juan	336,956.6	9,478.4	346,435.0
PR	San Lorenzo	2,037.0	0.0	2,037.0
PR	Santa Isabel	0.0	1,657.6	1,657.6
PR	Toa Baja	1,011,237.7	7,010.6	1,018,248.3
PR	Vega Alta	1,758.2	547.5	2,305.7
PR	Vega Baja	537.8	41.8	579.5
PR	Vieques	5.0	0.0	5.0
PR	Yabucoa	153,765.5	3,015.0	156,780.5
PR	Yauco	63,740.0	0.0	63,740.0
RI	Bristol	85,499.5	43.0	85,542.5
RI	Kent	123,467.9	6,483.6	129,951.5
RI	Newport	46,176.2	1.8	46,178.0
RI	Providence	194,079.6	209,563.6	403,643.3
RI	Washington	184,780.3	44,607.6	229,387.9
SC	Abbeville	15,903.0	3,448.0	19,351.0
SC	Aiken	1,369,035.4	324,836.5	1,693,871.9
SC	Allendale	147,768.1	0.0	147,768.1
SC	Anderson	2,671,484.0	176,319.5	2,847,803.5
SC	Bamberg	5,581.0	14,484.2	20,065.2
SC	Barnwell	460,871.2	45,092.0	505,963.2
SC	Beaufort	124,543.8	0.0	124,543.8
SC	Berkeley	8,602,420.4	21,892,677.3	30,495,097.7
SC	Calhoun	2,428,177.0	250.0	2,428,427.0
SC	Charleston	2,761,994.5	300,098.8	3,062,093.4
SC	Cherokee	240,406.9	89,427.8	329,834.7
SC	Chester	1,108,550.3	1,018.7	1,109,569.0
SC	Chesterfield	361,766.3	269,827.5	631,593.8
SC	Clarendon	145,463.0	7,764.2	153,227.2
SC	Colleton	2,273,043.5	26,605.0	2,299,648.5
SC	Darlington	2,298,153.4	3,482.0	2,301,635.4
SC	Dillon	59.5	6.7	66.2
SC	Dorchester	212,389.6	21,194.0	233,583.6
SC	Fairfield	15,044.5	20,342.8	35,387.3
SC	Florence	5,039,495.6	124,256.6	5,163,752.2
SC	Georgetown	6,032,961.8	81,490.0	6,114,451.8
SC	Greenville	517,503.7	344,715.6	862,219.4
SC	Greenwood	131,647.0	34,274.3	165,921.3
SC	Hampton	1,798,365.4	0.0	1,798,365.4
SC	Horry	920,563.2	559,717.0	1,480,280.2
SC	Jasper	0.0	0.0	0.0
SC	Kershaw	1,312,751.3	89,204.6	1,401,955.9
SC	Lancaster	367,013.8	47,011.9	414,025.6
SC	Laurens	423,189.3	45,043.6	468,232.9
SC	Lee	295,692.0	102.0	295,794.0
SC	Lexington	1,364,276.9	114,859.1	1,479,135.9
SC	Marion	85,248.9	2,080.0	87,328.9
SC	Marlboro	787,452.9	0.0	787,452.9
SC	McCormick	2.6	0.0	2.6
SC	Newberry	526,713.8	6,116.0	532,829.8
SC	Oconee	133,640.6	146,239.0	279,879.6

**TRI on-site and off-site reported disposed of or otherwise released (in pounds),
for facilities in all industries, for all chemicals, all counties, 2003** - continued

State	County	Total On-site Disposal or Other Releases[1]	Total Off-site Disposal or Other Releases[2]	Total On- and Off-site Disposal or Other Releases
SC	Orangeburg	3,036,226.5	161,687.9	3,197,914.5
SC	Pickens	180,963.1	80,547.4	261,510.4
SC	Richland	8,087,536.0	32,874.0	8,120,409.9
SC	Saluda	0.0	0.0	0.0
SC	Spartanburg	1,793,154.1	2,095,120.6	3,888,274.7
SC	Sumter	515,800.8	144,907.3	660,708.1
SC	Union	18,328.0	136.0	18,464.0
SC	Williamsburg	26,650.0	61.0	26,711.0
SC	York	3,297,073.1	57,734.7	3,354,807.8
SD	Beadle	3,955.0	5.0	3,960.0
SD	Bon Homme	17,355.0	0.0	17,355.0
SD	Brookings	769,052.0	1,160.0	770,212.0
SD	Brown	13,312.0	26.0	13,338.0
SD	Campbell	0.0	0.0	0.0
SD	Codington	23,112.0	19,636.0	42,748.0
SD	Davison	21,119.0	12,540.0	33,658.9
SD	Day	8,255.0	0.0	8,255.0
SD	Deuel	12.0	0.2	12.2
SD	Grant	1,629,431.1	0.0	1,629,431.1
SD	Hamlin	0.0	0.0	0.0
SD	Hutchinson	0.0	0.0	0.0
SD	Lake	46,490.0	1,649.0	48,139.0
SD	Lawrence	4,137,144.3	11,694.9	4,148,839.3
SD	Lincoln	63,271.5	6.0	63,277.4
SD	McCook	14,187.0	0.0	14,187.0
SD	Minnehaha	3,244,177.7	33,685.3	3,277,863.0
SD	Pennington	128,792.3	8,540.2	137,332.6
SD	Perkins	8.0	10.0	18.0
SD	Potter	0.0	0.0	0.0
SD	Sanborn	0.0	0.0	0.0
SD	Turner	5,399.0	0.0	5,399.0
SD	Union	10,738.0	0.0	10,738.0
SD	Yankton	102,693.0	13,509.6	116,202.6
TN	Anderson	5,445,199.6	68,590.1	5,513,789.7
TN	Bedford	75,073.0	198,484.1	273,557.1
TN	Blount	1,801,673.5	147,662.9	1,949,336.5
TN	Bradley	269,771.4	48,059.0	317,830.4
TN	Campbell	374.2	0.0	374.2
TN	Cannon	0.0	0.0	0.0
TN	Carroll	18,923.0	8,864.4	27,787.4
TN	Carter	46,254.2	12,671.6	58,925.9
TN	Cheatham	570,826.0	77,357.0	648,183.0
TN	Chester	12.0	905.0	917.0
TN	Claiborne	41,208.0	0.0	41,208.0
TN	Cocke	90,190.0	2,650.4	92,840.4
TN	Coffee	316,257.5	3,011.0	319,268.5
TN	Crockett	11,500.0	0.0	11,500.0
TN	Cumberland	81,297.0	18.0	81,315.0
TN	Davidson	1,788,508.4	121,706.7	1,910,215.1
TN	DeKalb	39,595.0	505,121.0	544,716.0
TN	Decatur	22,965.1	2.7	22,967.8
TN	Dickson	1,726,411.9	687.0	1,727,098.9
TN	Dyer	755,424.7	116,486.2	871,910.9
TN	Fayette	52,454.0	0.0	52,454.0
TN	Franklin	171,678.0	388.0	172,066.0
TN	Gibson	435,589.0	23,560.0	459,149.0
TN	Giles	13,587.8	334,320.3	347,908.1
TN	Grainger	1,038,672.3	0.0	1,038,672.3
TN	Greene	506,475.1	62,392.2	568,867.3

**TRI on-site and off-site reported disposed of or otherwise released (in pounds),
for facilities in all industries, for all chemicals, all counties, 2003** - *continued*

State	County	Total On-site Disposal or Other Releases[1]	Total Off-site Disposal or Other Releases[2]	Total On- and Off-site Disposal or Other Releases
TN	Hamblen	17,574,633.9	9,560.0	17,584,193.9
TN	Hamilton	556,425.3	176,148.3	732,573.6
TN	Hancock	511.7	0.0	511.7
TN	Hardeman	332,460.0	2,973.0	335,433.0
TN	Hardin	2,467,327.3	0.0	2,467,327.3
TN	Hawkins	2,039,636.0	10,243.5	2,049,879.5
TN	Haywood	9,499.0	31,535.0	41,034.0
TN	Henderson	90,802.7	12,154.1	102,956.8
TN	Henry	144,089.8	37,870.7	181,960.5
TN	Hickman	0.0	0.0	0.0
TN	Humphreys	35,328,400.2	66,532.8	35,394,933.0
TN	Jackson	210,006.0	2,002.0	212,008.0
TN	Jefferson	60,967.0	0.0	60,967.0
TN	Knox	1,045,715.9	14,200.4	1,059,916.3
TN	Lake	0.0	0.0	0.0
TN	Lauderdale	555,665.1	204,058.0	759,723.1
TN	Lawrence	221,829.1	750.0	222,579.1
TN	Lewis	22,681.0	22,177.0	44,858.0
TN	Lincoln	72,932.0	203.0	73,135.0
TN	Loudon	2,849,353.3	276,221.0	3,125,574.3
TN	Macon	24,960.0	0.0	24,960.0
TN	Madison	1,236,547.2	153,184.1	1,389,731.3
TN	Marion	4,430.0	8,100.0	12,530.0
TN	Marshall	373,252.0	1,065.3	374,317.3
TN	Maury	2,268,143.6	130,620.3	2,398,764.0
TN	McMinn	1,244,388.4	1,997,649.3	3,242,037.7
TN	McNairy	2,028,885.8	14,428.9	2,043,314.7
TN	Meigs	10,296.1	659.9	10,956.0
TN	Monroe	897,608.9	32,485.7	930,094.6
TN	Montgomery	16,477,749.3	13,772.6	16,491,521.9
TN	Moore	12.0	0.0	12.0
TN	Morgan	0.0	0.0	0.0
TN	Obion	120,018.7	220,280.1	340,298.9
TN	Overton	118,366.0	71.0	118,437.0
TN	Polk	1,726.1	1,064.0	2,790.1
TN	Putnam	109,211.7	6,971.0	116,182.7
TN	Rhea	99,468.2	72.6	99,540.8
TN	Roane	7,604,023.3	213,372.7	7,817,396.0
TN	Robertson	90,838.3	16,850.0	107,688.3
TN	Rutherford	2,004,335.4	386,780.7	2,391,116.1
TN	Scott	19,818.2	7,805.0	27,623.2
TN	Sequatchie	4,186.0	6.7	4,192.7
TN	Sevier	7,461.0	5,167.0	12,628.0
TN	Shelby	8,777,168.2	970,291.0	9,747,459.2
TN	Smith	1,247,278.5	5,016.5	1,252,295.1
TN	Stewart	2,707,530.0	40,641.8	2,748,171.8
TN	Sullivan	5,675,687.3	242,602.9	5,918,290.2
TN	Sumner	2,882,209.2	90,853.3	2,973,062.5
TN	Tipton	4,413.0	61,066.9	65,479.9
TN	Trousdale	0.0	0.0	0.0
TN	Unicoi	102,421.7	0.0	102,421.7
TN	Union	1,912.0	0.0	1,912.0
TN	Van Buren	0.0	26.0	26.0
TN	Warren	27,474.2	153,229.3	180,703.5
TN	Washington	210,034.0	18,163.1	228,197.1
TN	Wayne	0.1	96.4	96.5
TN	Weakley	69,702.0	83.0	69,785.0
TN	White	4,130.0	5,156.0	9,286.0
TN	Williamson	37,589.0	2,783.0	40,372.0

TRI on-site and off-site reported disposed of or otherwise released (in pounds), for facilities in all industries, for all chemicals, all counties, 2003 - *continued*

State	County	Total On-site Disposal or Other Releases[1]	Total Off-site Disposal or Other Releases[2]	Total On- and Off-site Disposal or Other Releases
TN	Wilson	36,370.8	6,327.4	42,698.3
TX	Andrews	242,154.7	0.0	242,154.7
TX	Angelina	1,205,581.9	38,937.3	1,244,519.3
TX	Aransas	31,836.2	0.0	31,836.2
TX	Atascosa	14,953.8	1,554,352.0	1,569,305.8
TX	Austin	44,748.0	6,112.0	50,860.0
TX	Bastrop	234,352.4	0.0	234,352.4
TX	Bell	1,075,816.4	648.2	1,076,464.7
TX	Bexar	1,002,994.2	240,142.0	1,243,136.2
TX	Bosque	355.4	0.0	355.4
TX	Bowie	234,699.4	118,468.0	353,167.5
TX	Brazoria	39,115,132.3	169,581.6	39,284,713.9
TX	Brazos	1,240,076.8	0.0	1,240,076.8
TX	Brewster	101.0	0.0	101.0
TX	Brown	475,067.9	43,778.8	518,846.7
TX	Burleson	7,985.6	0.0	7,985.6
TX	Burnet	1,017.9	1,650.0	2,667.9
TX	Caldwell	1,045.0	44.0	1,089.0
TX	Calhoun	17,347,147.2	12,250.2	17,359,397.3
TX	Cameron	376,873.5	5,682.0	382,555.5
TX	Camp	787,398.6	2,055.2	789,453.8
TX	Carson	5,784.0	175.0	5,959.0
TX	Cass	4,715,271.0	0.0	4,715,271.0
TX	Castro	241,152.0	68,936.0	310,088.0
TX	Chambers	386,192.7	166,215.0	552,407.7
TX	Cherokee	50,807.2	1,993.0	52,800.2
TX	Coleman	0.0	0.0	0.0
TX	Collin	292,767.5	23,229.1	315,996.6
TX	Colorado	35,673.3	0.0	35,673.3
TX	Comal	59,609.7	0.0	59,609.7
TX	Comanche	0.0	0.0	0.0
TX	Cooke	59,898.0	10,468.0	70,366.0
TX	Dallam	4,760.0	0.0	4,760.0
TX	Dallas	1,475,298.6	1,573,439.3	3,048,737.8
TX	Deaf Smith	1,000.0	0.0	1,000.0
TX	Denton	347,261.4	132,494.3	479,755.7
TX	Dimmit	0.0	0.0	0.0
TX	Eastland	24,442.0	0.0	24,442.0
TX	Ector	2,179,406.4	14,875.5	2,194,281.9
TX	El Paso	487,561.6	802,496.8	1,290,058.5
TX	Ellis	2,355,718.2	208,135.7	2,563,853.9
TX	Erath	52,670.8	3,685.6	56,356.4
TX	Fannin	25.2	908.2	933.4
TX	Fayette	1,228,839.8	5,690.9	1,234,530.7
TX	Fisher	25.8	0.0	25.8
TX	Fort Bend	2,234,522.1	67,175.7	2,301,697.8
TX	Freestone	2,894,285.4	0.0	2,894,285.4
TX	Gaines	206.0	0.0	206.0
TX	Galveston	13,699,267.3	640,599.9	14,339,867.2
TX	Gillespie	548.0	0.0	548.0
TX	Goliad	729,073.0	568.1	729,641.1
TX	Gonzales	10.0	0.0	10.0
TX	Gray	1,139,126.9	23,792.0	1,162,918.9
TX	Grayson	29,411.5	28,963.4	58,374.9
TX	Gregg	423,462.0	182,672.8	606,134.8
TX	Grimes	209,690.0	750.0	210,440.0
TX	Guadalupe	131,336.0	1,435,137.0	1,566,473.0
TX	Hale	772,883.0	0.0	772,883.0
TX	Hardin	27,175.7	23,968.0	51,143.7

**TRI on-site and off-site reported disposed of or otherwise released (in pounds),
for facilities in all industries, for all chemicals, all counties, 2003** - *continued*

State	County	Total On-site Disposal or Other Releases[1]	Total Off-site Disposal or Other Releases[2]	Total On- and Off-site Disposal or Other Releases
TX	Harris	42,853,739.9	7,834,468.7	50,688,208.7
TX	Harrison	6,675,283.5	106,671.5	6,781,955.0
TX	Hays	278,515.8	0.0	278,515.8
TX	Henderson	65,400.0	0.0	65,400.0
TX	Hidalgo	5,230.0	1,780.0	7,010.0
TX	Hockley	0.0	0.0	0.0
TX	Hopkins	0.0	0.0	0.0
TX	Houston	5,417.1	1,192.2	6,609.3
TX	Howard	383,950.3	2,760.2	386,710.4
TX	Hunt	3,434.8	32.9	3,467.7
TX	Hutchinson	4,829,030.0	8,635.3	4,837,665.4
TX	Jack	99,378.0	3,836.0	103,214.0
TX	Jackson	0.0	0.0	0.0
TX	Jasper	1,652,217.9	12,090.3	1,664,308.2
TX	Jefferson	23,267,161.0	4,427,202.9	27,694,363.9
TX	Johnson	557,462.9	62,557.4	620,020.3
TX	Jones	56,861.1	1,973.0	58,834.1
TX	Karnes	46,649.0	0.0	46,649.0
TX	Kaufman	125,544.8	53,400.1	178,944.9
TX	Kerr	3.0	85.0	88.0
TX	Kleberg	662.0	0.0	662.0
TX	Lamar	44,572.5	256.5	44,829.0
TX	Lamb	988,204.1	1,173,697.0	2,161,901.1
TX	Lampasas	5.0	1,501.0	1,506.0
TX	Lavaca	3,512.0	0.0	3,512.0
TX	Lee	217,423.0	0.0	217,423.0
TX	Leon	38,630.4	151,590.0	190,220.4
TX	Liberty	38,413.0	18,010.0	56,423.0
TX	Limestone	4,659,552.5	6.0	4,659,558.5
TX	Live Oak	258,762.4	121,487.2	380,249.6
TX	Lubbock	647,910.0	724.7	648,634.7
TX	Marion	74,790.0	229.0	75,019.0
TX	Matagorda	352,650.8	107.0	352,757.8
TX	Maverick	0.3	0.3	0.6
TX	McCulloch	11,962.0	85,732.0	97,694.0
TX	McLennan	96,997.5	225,501.5	322,499.0
TX	Medina	0.0	0.0	0.0
TX	Midland	43,925.3	15,290.0	59,215.3
TX	Milam	4,535,537.1	3,420.0	4,538,957.1
TX	Montgomery	722,517.4	3,342,060.6	4,064,578.0
TX	Moore	843,087.5	178,428.0	1,021,515.5
TX	Morris	24,091.6	72,944.4	97,036.0
TX	Nacogdoches	692,084.5	2,614.6	694,699.1
TX	Navarro	70,003.0	124,699.6	194,702.5
TX	Nolan	8,458.5	0.0	8,458.5
TX	Nueces	6,474,944.0	471,369.8	6,946,313.9
TX	Orange	5,912,238.3	694,804.3	6,607,042.6
TX	Palo Pinto	720.3	5.0	725.3
TX	Panola	61,391.0	113.0	61,504.0
TX	Parker	80,384.8	6,034.0	86,418.8
TX	Parmer	2,374,891.0	0.0	2,374,891.0
TX	Polk	204,573.7	16,379.9	220,953.6
TX	Potter	1,039,289.1	2,043,609.0	3,082,898.1
TX	Randall	520,312.0	9,652.0	529,964.0
TX	Robertson	489,651.5	1,125.0	490,776.5
TX	Rockwall	1,005.0	0.0	1,005.0
TX	Runnels	0.0	0.0	0.0
TX	Rusk	5,653,807.0	959.3	5,654,766.3
TX	Sabine	60,229.1	0.0	60,229.1

TRI on-site and off-site reported disposed of or otherwise released (in pounds), for facilities in all industries, for all chemicals, all counties, 2003 - continued

State	County	Total On-site Disposal or Other Releases[1]	Total Off-site Disposal or Other Releases[2]	Total On- and Off-site Disposal or Other Releases
TX	San Augustine	0.0	0.0	0.0
TX	San Patricio	958,070.3	1,213.2	959,283.4
TX	Scurry	42.1	0.0	42.1
TX	Shackelford	898.0	0.0	898.0
TX	Shelby	591,739.6	892.5	592,632.1
TX	Smith	622,598.2	109,398.4	731,996.6
TX	Sutton	399.0	0.0	399.0
TX	Tarrant	993,564.6	413,964.8	1,407,529.5
TX	Taylor	110,779.1	15,727.0	126,506.1
TX	Titus	7,923,688.6	37,014.0	7,960,702.6
TX	Tom Green	33,876.0	0.0	33,876.0
TX	Travis	125,364.4	5,966.7	131,331.2
TX	Tyler	132.0	0.0	132.0
TX	Upshur	88,486.0	0.0	88,486.0
TX	Upton	444.0	0.0	444.0
TX	Val Verde	16,967.0	0.0	16,967.0
TX	Victoria	10,168,487.0	2,878.0	10,171,365.0
TX	Walker	14,173.5	1,223.0	15,396.5
TX	Waller	12,130.7	2,878.2	15,008.9
TX	Ward	0.0	0.0	0.0
TX	Washington	357.7	5,700.0	6,057.7
TX	Webb	3,955.0	0.0	3,955.0
TX	Wharton	89,093.3	2,045.4	91,138.6
TX	Wichita	1,260,258.4	36,057.7	1,296,316.1
TX	Wilbarger	307,483.4	639.2	308,122.6
TX	Williamson	83,458.4	545.7	84,004.2
TX	Wilson	6,431.0	364.0	6,795.0
TX	Winkler	554.0	0.0	554.0
TX	Wise	154,587.0	0.0	154,587.0
TX	Wood	93,026.0	3,060.0	96,086.0
TX	Young	8,135.0	0.0	8,135.0
UT	Beaver	0.0	0.0	0.0
UT	Box Elder	1,354,626.2	8,988,032.9	10,342,659.1
UT	Cache	286,707.5	153,102.8	439,810.3
UT	Carbon	593,726.4	4,955.0	598,681.4
UT	Davis	367,053.7	113,450.8	480,504.5
UT	Duchesne	0.0	0.0	0.0
UT	Emery	3,560,853.3	6,356.8	3,567,210.1
UT	Grand	152.0	0.0	152.0
UT	Iron	5,731.7	111,673.3	117,405.0
UT	Juab	48,524.2	0.0	48,524.2
UT	Millard	1,747,284.9	0.0	1,747,284.9
UT	Morgan	53,621.7	0.0	53,621.7
UT	Salt Lake	207,587,316.8	248,673.3	207,835,990.1
UT	San Juan	0.0	0.0	0.0
UT	Sanpete	5,866.0	310.0	6,176.0
UT	Sevier	38.5	0.0	38.5
UT	Tooele	16,377,274.7	1,372,871.0	17,750,145.7
UT	Uintah	1,471,110.3	1,306.1	1,472,416.4
UT	Utah	4,174,557.0	29,657.7	4,204,214.7
UT	Washington	15,248.0	0.0	15,248.0
UT	Weber	1,265,793.1	38,582.3	1,304,375.4
VA	Accomack	2,375,178.7	15.1	2,375,193.8
VA	Albemarle	9,914.2	1,545.0	11,459.2
VA	Alexandria City	2,751,918.7	430,666.5	3,182,585.2
VA	Alleghany	38,011.4	116.6	38,128.0
VA	Amelia	1.7	0.0	1.7
VA	Amherst	0.1	5.7	5.8
VA	Appomattox	7.7	404.0	411.6

TRI on-site and off-site reported disposed of or otherwise released (in pounds), for facilities in all industries, for all chemicals, all counties, 2003 - *continued*

State	County	Total On-site Disposal or Other Releases[1]	Total Off-site Disposal or Other Releases[2]	Total On- and Off-site Disposal or Other Releases
VA	Arlington	1,181.0	0.0	1,181.0
VA	Augusta	479,601.3	94,366.3	573,967.6
VA	Bedford	790,170.9	1,624.0	791,794.9
VA	Bedford City	14,700.0	122,500.0	137,200.0
VA	Bland	44,979.3	0.0	44,979.3
VA	Botetourt	124,859.2	101,928.1	226,787.4
VA	Bristol City	187,480.0	5,648.0	193,128.0
VA	Brunswick	16.0	0.0	16.0
VA	Buchanan	3,578.2	0.0	3,578.2
VA	Buckingham	46,834.8	0.0	46,834.8
VA	Buena Vista City	0.0	1.0	1.0
VA	Campbell	1,269,978.6	160,704.8	1,430,683.5
VA	Caroline	71,965.0	0.0	71,965.0
VA	Carroll	100,354.9	0.0	100,354.9
VA	Charlotte	0.0	0.0	0.0
VA	Chesapeake City	3,294,092.0	331.3	3,294,423.3
VA	Chesterfield	8,061,264.2	33,681.4	8,094,945.6
VA	Clarke	0.0	54.7	54.7
VA	Colonial Heights City	8.0	225.0	233.0
VA	Covington City	4,591,481.1	0.0	4,591,481.1
VA	Culpeper	15.7	1.4	17.2
VA	Cumberland	0.5	0.0	0.5
VA	Danville City	299,270.0	160,398.7	459,668.7
VA	Dinwiddie	18,417.0	16,442.0	34,859.0
VA	Emporia City	44,380.0	61,522.1	105,902.1
VA	Fairfax	34,751.6	18,649.0	53,400.6
VA	Fairfax City	8,264.0	8.0	8,272.0
VA	Fauquier	14,000.0	0.0	14,000.0
VA	Floyd	1,605.0	2,370.0	3,975.0
VA	Fluvanna	1,384,284.0	282.0	1,384,566.0
VA	Franklin	7,818.5	540.0	8,358.5
VA	Frederick	1,074,964.8	576,910.1	1,651,874.9
VA	Fredericksburg City	138,024.4	195,739.0	333,763.3
VA	Galax City	44,326.0	1,754.8	46,080.8
VA	Giles	3,170,049.8	103.5	3,170,153.4
VA	Grayson	840.6	0.0	840.6
VA	Greensville	205,384.5	0.0	205,384.5
VA	Halifax	1,994,762.6	785.0	1,995,547.6
VA	Hampton City	3,026.5	87,507.2	90,533.7
VA	Hanover	163,617.1	306,417.7	470,034.8
VA	Harrisonburg City	5,900.0	0.0	5,900.0
VA	Henrico	1,563,445.4	13,344.2	1,576,789.6
VA	Henry	1,388,691.8	346.1	1,389,037.9
VA	Hopewell City	4,838,484.8	134,850.8	4,973,335.6
VA	Isle of Wight	3,874,729.7	576,410.8	4,451,140.5
VA	James City	612,963.1	14.1	612,977.2
VA	King George	12,779.5	191,747.0	204,526.5
VA	King William	1,836,255.4	327,086.9	2,163,342.3
VA	Lancaster	0.0	15.0	15.0
VA	Loudoun	15,018.0	767.0	15,785.0
VA	Louisa	9,931.7	0.0	9,931.7
VA	Lunenburg	201,835.0	0.0	201,835.0
VA	Lynchburg City	725,917.8	916,561.2	1,642,479.0
VA	Madison	515.0	1,750.0	2,265.0
VA	Manassas City	20,659.0	340.0	20,999.0
VA	Martinsville City	32,058.5	0.0	32,058.5
VA	Mecklenburg	3,262.9	146,124.2	149,387.1
VA	Middlesex	0.0	0.0	0.0
VA	Montgomery	60,963.1	24,451.1	85,414.2

TRI on-site and off-site reported disposed of or otherwise released (in pounds), for facilities in all industries, for all chemicals, all counties, 2003 - *continued*

State	County	Total On-site Disposal or Other Releases[1]	Total Off-site Disposal or Other Releases[2]	Total On- and Off-site Disposal or Other Releases
VA	Nelson	214,219.2	0.0	214,219.2
VA	Newport News City	376,376.2	42,769.4	419,145.6
VA	Norfolk City	951,296.9	49,843.8	1,001,140.7
VA	Northumberland	28,312.2	21.0	28,333.2
VA	Nottoway	0.0	0.0	0.0
VA	Orange	33,367.0	447.0	33,814.0
VA	Page	138.0	16.0	154.0
VA	Patrick	3,254.0	250.0	3,504.0
VA	Petersburg City	475,068.0	18,033.4	493,101.4
VA	Pittsylvania	364,237.2	129,689.9	493,927.2
VA	Portsmouth City	548,075.4	34,007.4	582,082.8
VA	Prince Edward	2,320.0	2.0	2,322.0
VA	Prince George	19,448.9	0.0	19,448.9
VA	Prince William	890,512.5	6.6	890,519.1
VA	Pulaski	63,753.1	108,592.4	172,345.5
VA	Radford City	3,118,421.1	297,171.5	3,415,592.6
VA	Richmond	250.2	674.0	924.2
VA	Richmond City	1,489,091.5	48,676.4	1,537,767.9
VA	Roanoke	154,299.7	13,207.5	167,507.2
VA	Roanoke City	50,770.2	3,806,386.0	3,857,156.2
VA	Rockbridge	9,872.0	740.0	10,612.0
VA	Rockingham	332,696.6	10.9	332,707.5
VA	Russell	3,070,911.3	875.5	3,071,786.8
VA	Salem City	6.1	5.0	11.1
VA	Scott	20.0	0.0	20.0
VA	Shenandoah	477,935.5	35,060.0	512,995.5
VA	Smyth	397,266.1	255.0	397,521.1
VA	South Boston City	0.0	0.0	0.0
VA	Southampton	195,583.0	34,852.0	230,435.0
VA	Spotsylvania	46,232.0	59,898.0	106,130.0
VA	Stafford	0.0	0.0	0.0
VA	Suffolk City	7,501.0	9,837.0	17,338.0
VA	Surry	0.0	0.0	0.0
VA	Sussex	3.6	10.3	13.9
VA	Tazewell	10.0	250.0	260.0
VA	Virginia Beach City	24,004.9	27,340.0	51,344.9
VA	Warren	146,659.2	30,443.5	177,102.7
VA	Washington	100,435.0	11,894.0	112,329.0
VA	Waynesboro City	0.0	0.0	0.0
VA	Westmoreland	0.0	75.1	75.1
VA	Winchester City	252,629.0	688.0	253,317.0
VA	Wythe	11,326.4	16,841.7	28,168.2
VA	York	2,887,112.9	2,707.2	2,889,820.1
VI	St. Croix	1,317,840.0	8,743.0	1,326,583.0
VI	St. John	214.0	0.0	214.0
VI	St. Thomas	5,260.5	0.0	5,260.5
VT	Addison	0.0	78.5	78.5
VT	Bennington	410.0	679.0	1,089.0
VT	Caledonia	8,461.0	1,505.0	9,966.0
VT	Chittenden	152,357.2	13,777.6	166,134.8
VT	Essex	11.6	0.0	11.6
VT	Franklin	19,799.8	5,601.3	25,401.1
VT	Orleans	282.4	0.0	282.4
VT	Rutland	12,837.6	16,237.0	29,074.7
VT	Washington	0.0	57,000.0	57,000.0
VT	Windham	7,561.1	50,727.0	58,288.1
VT	Windsor	0.0	0.0	0.0
WA	Adams	3,190.0	0.0	3,190.0
WA	Asotin	3.5	0.0	3.5

**TRI on-site and off-site reported disposed of or otherwise released (in pounds),
for facilities in all industries, for all chemicals, all counties, 2003** - *continued*

State	County	Total On-site Disposal or Other Releases[1]	Total Off-site Disposal or Other Releases[2]	Total On- and Off-site Disposal or Other Releases
WA	Benton	1,348,045.0	44,282.4	1,392,327.4
WA	Chelan	1,038.0	11,579.0	12,617.0
WA	Clallam	46,065.9	407.1	46,473.0
WA	Clark	961,302.5	39,732.3	1,001,034.8
WA	Cowlitz	3,148,543.4	351,271.5	3,499,814.9
WA	Douglas	0.0	0.0	0.0
WA	Ferry	8,006.6	2.3	8,008.9
WA	Franklin	15,154.0	0.0	15,154.0
WA	Grant	26,223.2	1,227.1	27,450.3
WA	Grays Harbor	232,454.8	35,057.1	267,512.0
WA	Island	0.0	0.0	0.0
WA	Jefferson	637,098.1	0.0	637,098.1
WA	King	804,874.8	591,183.5	1,396,058.3
WA	Kitsap	55,737.9	106,175.0	161,912.9
WA	Klickitat	6,460.0	100.0	6,560.0
WA	Lewis	5,635,874.8	102,286.8	5,738,161.6
WA	Lincoln	0.0	0.0	0.0
WA	Mason	1,158.7	202.0	1,360.7
WA	Pacific	525.0	256.0	781.0
WA	Pend Oreille	132,713.4	6.0	132,719.4
WA	Pierce	1,411,713.8	57,640.5	1,469,354.3
WA	Skagit	1,049,161.8	1,068.9	1,050,230.7
WA	Snohomish	1,514,385.0	172,852.0	1,687,237.0
WA	Spokane	303,110.7	9,658.4	312,769.1
WA	Stevens	44,486.2	1,356.0	45,842.2
WA	Thurston	739,899.0	315.0	740,214.0
WA	Walla Walla	1,309,214.5	230,550.0	1,539,764.5
WA	Whatcom	862,530.2	92,476.1	955,006.3
WA	Whitman	0.0	199.5	199.5
WA	Yakima	345,766.4	1,010.0	346,776.4
WI	Adams	0.0	0.0	0.0
WI	Ashland	43,485.0	287,664.9	331,149.9
WI	Barron	179,433.6	125,306.4	304,739.9
WI	Brown	2,306,735.3	198,244.0	2,504,979.4
WI	Buffalo	920,734.9	291,962.2	1,212,697.1
WI	Burnett	0.0	10.0	10.0
WI	Calumet	160,861.2	36,287.3	197,148.5
WI	Chippewa	66,675.2	4,457.0	71,132.2
WI	Clark	161,318.8	43.6	161,362.4
WI	Columbia	609,225.7	9,454.0	618,679.7
WI	Crawford	4,475.0	1,363.0	5,838.0
WI	Dane	1,045,782.3	123,197.1	1,168,979.4
WI	Dodge	294,010.5	44,998.5	339,009.0
WI	Door	555.4	24.3	579.6
WI	Douglas	23,678.5	4,846.0	28,524.4
WI	Dunn	117,900.0	400.0	118,300.0
WI	Eau Claire	74,440.5	23,624.7	98,065.1
WI	Fond du Lac	1,014,073.3	165,796.6	1,179,869.9
WI	Grant	378,123.1	105,576.4	483,699.5
WI	Green	62,254.3	26,741.1	88,995.4
WI	Green Lake	21,686.0	0.0	21,686.0
WI	Iowa	61.7	0.0	61.7
WI	Jackson	37,093.4	0.0	37,093.4
WI	Jefferson	246,256.3	28,811.0	275,067.3
WI	Juneau	18,831.0	47.1	18,878.1
WI	Kenosha	423,941.3	365,000.1	788,941.4
WI	Kewaunee	93,939.0	461.2	94,400.2
WI	La Crosse	371,763.4	133,979.9	505,743.3
WI	Lafayette	62,190.0	0.0	62,190.0

TRI on-site and off-site reported disposed of or otherwise released (in pounds),
for facilities in all industries, for all chemicals, all counties, 2003 - *continued*

State	County	Total On-site Disposal or Other Releases[1]	Total Off-site Disposal or Other Releases[2]	Total On- and Off-site Disposal or Other Releases
WI	Langlade	4,269.6	423.0	4,692.6
WI	Lincoln	562,732.7	14,093.0	576,825.7
WI	Manitowoc	360,156.9	567,664.0	927,820.9
WI	Marathon	2,263,572.3	282,919.4	2,546,491.7
WI	Marinette	392,832.0	4,359,714.1	4,752,546.1
WI	Marquette	23,684.0	0.0	23,684.0
WI	Milwaukee	1,440,978.4	1,851,866.9	3,292,845.2
WI	Monroe	142,575.3	927.1	143,502.4
WI	Oconto	148,546.0	4,291.0	152,837.0
WI	Oneida	217,313.6	137.6	217,451.3
WI	Outagamie	1,545,791.0	92,583.4	1,638,374.4
WI	Ozaukee	959,983.2	2,986,110.6	3,946,093.8
WI	Pierce	31,138.0	94.3	31,232.3
WI	Polk	122,709.0	0.0	122,709.0
WI	Portage	1,386,984.7	73,772.2	1,460,756.9
WI	Price	318,453.6	20,059.2	338,512.8
WI	Racine	82,848.1	19,459.7	102,307.8
WI	Richland	3,771.5	78,833.5	82,605.0
WI	Rock	491,502.9	193,602.6	685,105.5
WI	Rusk	9,898.3	0.0	9,898.3
WI	Sauk	119,527.0	1,644,478.1	1,764,005.1
WI	Sawyer	252,355.2	5,910.0	258,265.2
WI	Shawano	4,413.0	50,839.5	55,252.5
WI	Sheboygan	513,948.0	1,058,273.7	1,572,221.6
WI	St. Croix	61,580.0	28,820.0	90,400.0
WI	Taylor	50,120.0	235.6	50,355.6
WI	Trempealeau	85,550.9	3.8	85,554.7
WI	Vernon	4,239,350.6	36,487.1	4,275,837.6
WI	Vilas	404.0	48,577.0	48,981.0
WI	Walworth	680,737.3	92,468.2	773,205.5
WI	Washington	139,396.0	494,069.1	633,465.1
WI	Waukesha	478,438.3	227,754.4	706,192.7
WI	Waupaca	199,942.0	4,086,228.0	4,286,170.0
WI	Waushara	3,898.2	0.0	3,898.2
WI	Winnebago	292,932.4	1,173,345.8	1,466,278.2
WI	Wood	4,661,715.1	272,890.4	4,934,605.5
WV	Berkeley	341,762.0	28,973.0	370,735.0
WV	Boone	436,135.3	0.0	436,135.3
WV	Braxton	10,288.1	37,770.6	48,058.7
WV	Brooke	1,277,124.1	698,290.0	1,975,414.1
WV	Cabell	132,013.6	135,804.5	267,818.1
WV	Doddridge	0.0	0.0	0.0
WV	Fayette	694,021.6	0.0	694,021.6
WV	Grant	2,976,382.1	138,117.0	3,114,499.1
WV	Greenbrier	1,000.2	0.1	1,000.3
WV	Hampshire	4,787.7	0.0	4,787.7
WV	Hancock	483,259.2	825,917.9	1,309,177.0
WV	Hardy	1,349,692.9	488.3	1,350,181.3
WV	Harrison	3,091,699.5	1,536.2	3,093,235.7
WV	Jackson	311,539.9	1,075.9	312,615.8
WV	Jefferson	41,575.4	6,570.7	48,146.1
WV	Kanawha	6,704,943.7	1,276,738.1	7,981,681.8
WV	Lewis	19,735.0	0.0	19,735.0
WV	Marion	3,385,146.7	461,189.0	3,846,335.7
WV	Marshall	19,962,529.2	11,160.4	19,973,689.6
WV	Mason	18,261,296.6	37,517.1	18,298,813.7
WV	McDowell	105.4	0.0	105.4
WV	Mercer	19,015.2	15,400.6	34,415.8
WV	Mineral	3,969.6	261.0	4,230.6

**TRI on-site and off-site reported disposed of or otherwise released (in pounds),
for facilities in all industries, for all chemicals, all counties, 2003** - *continued*

State	County	Total On-site Disposal or Other Releases[1]	Total Off-site Disposal or Other Releases[2]	Total On- and Off-site Disposal or Other Releases
WV	Mingo	11,906.6	0.0	11,906.6
WV	Monongalia	7,239,585.7	122,807.2	7,362,392.9
WV	Monroe	20,400.0	0.0	20,400.0
WV	Morgan	10.6	0.0	10.6
WV	Nicholas	199,591.3	373.0	199,964.3
WV	Ohio	7,388.0	5,876.0	13,264.0
WV	Pleasants	6,447,621.8	59,409.0	6,507,030.8
WV	Preston	3,011,896.1	3,767.0	3,015,663.1
WV	Putnam	18,158,380.4	1,630,559.7	19,788,940.1
WV	Raleigh	19,934.2	0.0	19,934.2
WV	Randolph	55,906.6	978.6	56,885.2
WV	Roane	7,560.0	0.0	7,560.0
WV	Taylor	33.0	6,742.0	6,775.0
WV	Tucker	1,718.8	2.0	1,720.8
WV	Tyler	591,805.1	0.0	591,805.1
WV	Upshur	9,833.7	124,027.0	133,860.7
WV	Wayne	87,839.3	23,142.2	110,981.5
WV	Webster	70,174.8	0.0	70,174.8
WV	Wirt	10,599.0	0.0	10,599.0
WV	Wood	1,118,267.7	102,838.3	1,221,106.0
WV	Wyoming	147,494.7	1.0	147,495.7
WY	Albany	115,301.0	0.0	115,301.0
WY	Big Horn	28,556.0	0.0	28,556.0
WY	Campbell	1,563,912.3	857,240.3	2,421,152.6
WY	Carbon	199,478.1	142.6	199,620.7
WY	Converse	1,684,419.2	160,005.3	1,844,424.5
WY	Fremont	4,169.0	0.0	4,169.0
WY	Goshen	55,108.3	0.0	55,108.3
WY	Laramie	8,428,912.0	83.7	8,428,995.7
WY	Lincoln	579,005.1	1,551.9	580,557.0
WY	Natrona	68,567.6	14,589.7	83,157.3
WY	Park	1,050.5	134.8	1,185.3
WY	Platte	2,985,443.5	0.0	2,985,443.5
WY	Sweetwater	2,133,542.8	82.1	2,133,624.9
WY	Teton	333.0	0.0	333.0
WY	Uinta	6,188.6	0.0	6,188.6
WY	Washakie	283,504.0	255.0	283,759.0
WY	Weston	93,151.3	0.0	93,151.3

Notes: *TRI = Toxic Release Inventory; Reporting year 2003 is the most recent TRI data available. (1) On-site disposal or other releases include emissions to the air, discharges to bodies of water, disposal at the facility to land, and disposal in underground injection wells. Disposal or other releases are reported to TRI by media type. On-site disposal or other releases are reported in Section 5 of the TRI Form R. On-site Disposal or Other Releases include Underground Injection to Class I Wells (Section 5.4.1), RCRA Subtitle C Landfills (5.5.1A), Other Landfills (5.5.1B), Fugitive or Non-point Air Emissions (5.1), Stack or Point Air Emissions (5.2), Surface Water Discharges (5.3), Underground Injection to Class II-V Wells (5.4.2), Land Treatment/Application Farming (5.5.2), RCRA Subtitle C Surface Impoundments (5.5.3A), Other Surface Impoundments (5.5.3B), and Other Land Disposal (5.5.4). Off-site Disposal or Other Releases include from Section 6.2 Class I Underground Injection Wells (M81), Class II-V Underground Injection Wells (M82, M71), RCRA Subtitle C Landfills (M65), Other Landfills (M64, M72), Storage Only (M10), Solidification/Stabilization - Metals and Metal Category Compounds only (M41 or M40), Wastewater Treatment (excluding POTWs) - Metals and Metal Category Compounds only (M62 or M61), RCRA Subtitle C Surface Impoundments (M66), Other Surface Impoundments (M67, M63), Land Treatment (M73), Other Land Disposal (M79), Other Off-site Management (M90), Transfers to Waste Broker - Disposal (M94, M91), and Unknown (M99) and, from Section 6.1 Transfers to POTWs (metals and metal category compounds only); (2) An off-site disposal or other release is a discharge of a toxic chemical to the environment that occurs as a result of a facility's transferring a waste containing a TRI chemical off-site for disposal or other release, as reported in Section 6 of the TRI Form R. Certain other types of transfers are also categorized as off-site disposal or other release because, except for location, the outcome of transferring the chemical off-site is the same as disposing of it or releasing it on-site. For each transfer, the amount of the chemical in the waste, type of management activity (chosen from a list of codes referred to as "M" codes) undertaken by the receiving facility, and the address of the receiving site is reported. Off-site disposal or other releases show only net off-site disposal or other releases, that is, off-site disposal or other releases transferred to other TRI facilities reporting such transfers as on-site disposal or other releases are not included to avoid double counting.*
Source: *U.S. Environmental Protection Agency, TRI Explorer, June 8, 2005*

Generation, materials recovery, composting, and discards of municipal solid waste, 1960-2003

	1960	1970	1980	1990	2000	2001	2002	2003
Millions of tons								
Generation	88.1	121.1	151.6	205.2	234.0	231.2	235.5	236.2
Recovery for recycling	5.6	8.0	14.5	29.0	52.4	52.8	53.8	55.4
Recovery for composting[1]	Neg.	Neg.	Neg.	4.2	16.5	16.6	16.7	16.9
Total Materials Recovery	5.6	8.0	14.5	33.2	68.9	69.3	70.5	72.3
Discards after Recovery	82.5	113.0	137.1	172.0	165.1	161.9	165.0	163.9
Pounds per person per day								
Generation	2.68	3.25	3.66	4.50	4.56	4.45	4.48	4.45
Recovery for recycling	0.17	0.22	0.35	0.64	1.02	1.02	1.02	1.04
Recovery for composting[1]	Neg.	Neg.	Neg.	0.09	0.32	0.32	0.32	0.32
Total Materials Recovery	0.17	0.22	0.35	0.73	1.34	1.34	1.34	1.36
Discards after Recovery	2.51	3.03	3.31	3.77	3.22	3.11	3.14	3.09
Population (millions)	179.979	203.984	227.255	249.907	281.422	284.797	287.974	290.810
Percent of total generation								
Generation	100.0	100.0	100.0	100.0	100.0	100.0	100.0	100.0
Recovery for recycling	6.4	6.6	9.6	14.2	22.4	22.8	22.8	23.5
Recovery for composting[1]	Neg.	Neg.	Neg.	2.0	7.0	7.2	7.1	7.1
Total Materials Recovery	6.4	6.6	9.6	16.2	29.4	30.0	29.9	30.6
Discards after Recovery	93.6	93.4	90.4	83.8	70.6	70.0	70.1	69.4

Notes: (1) Composting of yard trimmings and food wastes. Does not include mixed MSW composting or backyard composting; Details may not add to totals due to rounding
Source: United States Environmental Protection Agency, Office of Solid Waste, Municipal Solid Waste in the United States: Facts and Figures for 2003

Generation and recovery of materials in municipal solid waste, 2003
(in millions of tons and percent of generation of each material)

Material	Weight Generated	Weight Recovered	Recovery as a Percent of Generation (%)
Paper and paperboard	83.1	40.0	48.1
Glass	12.5	2.35	18.8
Metals			
Steel	14.0	5.09	36.4
Aluminum	3.23	0.69	21.4
Other nonferrous metals[1]	1.59	1.06	66.7
Total metals	18.8	6.84	36.3
Plastics	26.7	1.39	5.2
Rubber and leather	6.82	1.10	16.1
Textiles	10.6	1.52	14.4
Wood	13.6	1.28	9.4
Other materials	4.32	0.98	22.7
Total Materials in Products	176.4	55.4	31.4
Other wastes			
Food, other[2]	27.6	0.75	2.7
Yard trimmings	28.6	16.1	56.3
Miscellaneous inorganic wastes	3.62	Neg.	Neg.
Total Other Wastes	59.8	16.9	28.2
Total Municipal Solid Waste	236.2	72.3	30.6

Notes: *Includes waste from residential, commercial, and institutional sources; Details may not add to totals due to rounding; Neg.= Less than 50,000 tons or 0.05 percent; (1) Includes lead from lead-acid batteries; (2) Includes recovery of other MSW organics for composting.*
Source: *United States Environmental Protection Agency, Office of Solid Waste, Municipal Solid Waste in the United States: Facts and Figures for 2003*

Generation and recovery of products in municipal solid waste by material, 2003
(in millions of tons and percent of generation of each product)

Product	Weight Generated	Weight Recovered	Recovery as a Percent of Generation (%)
Durable Goods			
Steel	11.2	3.37	30.2
Aluminum	1.06	Neg.	Neg.
Other non-ferrous metals[1]	1.59	1.06	66.7
Total metals	13.8	4.43	32.1
Glass	1.78	Neg.	Neg.
Plastics	8.39	0.33	3.9
Rubber and leather	5.91	1.10	18.6
Wood	5.27	Neg.	Neg.
Textiles	3.03	0.32	10.6
Other materials	1.30	0.98	75.4
Total durable goods	39.5	7.16	18.1
Nondurable Goods			
Paper and paperboard	44.3	18.1	40.8
Plastics	6.35	Neg.	Neg.
Rubber and leather	0.88	Neg.	Neg.
Textiles	7.37	1.20	16.3
Other materials	3.26	Neg.	Neg.
Total nondurable goods	62.1	19.3	31.0
Containers and Packaging			
Steel	2.84	1.72	60.6
Aluminum	1.94	0.69	35.6
Total metals	4.78	2.41	50.4
Glass	10.7	2.35	22.0
Paper and paperboard	39.8	21.9	56.4
Plastics	11.9	1.06	8.9
Wood	8.36	1.28	15.3
Other materials	0.22	Neg.	Neg.
Total containers and packaging	74.8	29.0	38.8
Other wastes			
Food, other[2]	27.6	0.75	2.7
Yard trimmings	28.6	16.1	56.3
Miscellaneous inorganic wastes	3.62	Neg.	Neg.
Total Other Wastes	59.8	16.9	28.2
Total Municipal Solid Waste	236.2	72.3	30.6

Notes: *Includes waste from residential, commercial, and institutional sources; Details may not add to totals due to rounding; Neg.= Less than 50,000 tons or 0.05 percent; (1) Includes lead from lead-acid batteries; (2) Includes recovery of other MSW organics for composting.*
Source: *United States Environmental Protection Agency, Office of Solid Waste, Municipal Solid Waste in the United States: Facts and Figures for 2003*

Reported generation and estimated MSW generated, and rates of MSW recycling, incineration/waste-to-energy, and landfilling[1]

Year	Reported Generation[2] (tons/year)	Estimated MSW Generated[3] (tons/year)	MSW Recycled (%)	MSW Incineration/ Waste-to-Energy[4] (%)	MSW Landfilled (%)
1989	269,000,000	-	8	8	84
1990	293,613,000	-	11.5	11.5	77
1991	280,675,000	-	14	10	76
1992	291,742,000	-	17	11	72
1993	306,866,000	-	19	10	71
1994	322,879,000	-	23	10	67
1995	326,709,000	-	27	10	63
1996	327,460,000	-	28	10	62
1997	340,466,000	-	30	9	61
1998	374,631,000	-	31.5	7.5	61
1999	382,594,000	-	33	7	60
2000	409,029,000	-	32	7	61
2002	482,770,983	369,381,411	26.7	7.7	65.6

Notes: *(1) Alabama, Alaska, and Montana did not report any data for the 2003 "State of Garbage in America" survey. The combined population of these states is 6,039,747 (or two percent of the total U.S. population); (2) Data for 1989-2000 was provided to BioCycle as "MSW generation." Data for 2002 was provided as solid waste generation; (3) MSW generated is computed from reported tonnages of: (landfill + exported landfill + WTE + exported WTE + MSW recycled) - (C&D landfill + industrial landfill + imported landfill + imported WTE); (4) The 2003 "State of Garbage in America" survey only collected data on waste-to-energy combustion. Previous surveys (1990-2000) asked more generally about "incineration."*

Source: *BioCycle, "The State of Garbage in America," January 2004*

Number of municipal solid waste landfills and waste-to-energy plants, average tip fees, and capacity by state for 2002

State	Landfills			Waste-to-Energy (WTE)	
	Number	Average Tip Fee ($/ton)	Remaining Capacity (tons)	Number of Plants	Average Tip Fee ($/ton)
Alabama	n/a	n/a	n/a	n/a	n/a
Alaska	n/a	n/a	n/a	n/a	n/a
Arizona	41	n/a	n/a	0	-
Arkansas	24	28.45	n/a	2	n/a
California	161	13.63	410,501,190	3	n/a
Colorado	65	n/a	n/a	0	-
Connecticut	2	n/a	n/a	6	65.00
Delaware	3	58.50	20,000,000	0	-
District of Columbia	n/a	n/a	n/a	n/a	n/a
Florida		42.47	n/a	13	59.00
Georgia	60	33.50	135,349,274	1	45.00
Hawaii	9	n/a	n/a	1	n/a
Idaho	29	n/a	n/a	0	-
Illinois	51	n/a	212,393,636	0	-
Indiana	35	n/a	52,231,795	1	n/a
Iowa	59	33.25	40,182,628	1	53.00
Kansas	51	28.00	n/a	0	-
Kentucky	25	27.57	36,363,636	1	n/a
Louisiana	24	25.00	n/a	0	-
Maine	8	55.00	3,030,303	4	65.00
Maryland	20	50.00	n/a	3	49.00
Massachusetts	19	72.60	n/a	7	71.00
Michigan	52	n/a	143,939,394	4	76.00
Minnesota	21	50.00	18,700,000	15	50.00
Mississippi	17	26.00	n/a	0	-
Missouri	24	33.54	41,432,836	0	-
Montana	30	32.00	32,727,273	0	-
Nebraska	24	25.00	n/a	0	-
Nevada	23	30.00	60,742,056	0	-
New Hampshire	10	68.00	15,000,000	2	81.00
New Jersey	12	60.00	40,000,000	5	60.00
New Mexico	35	n/a	190,966,142	0	-
New York	26	50.00	90,000,000	10	65.00
North Carolina	41	30.00	100,000,000	1	50.00
North Dakota	14	26.56	n/a	0	-
Ohio	44	32.20	124,079,624	0	-
Oklahoma	40	20.00	n/a	1	n/a
Oregon	30	34.50	n/a	1	68.00
Pennsylvania	49	48.00	298,585,524	6	74.00
Rhode Island	2	41.50	n/a	0	-
South Carolina	19	27.00	109,534,023	4	n/a
South Dakota	15	30.00	16,757,576	0	-
Tennessee	34	28.38	n/a	1	n/a
Texas	175	27.00	970,000,000	2	n/a
Utah	38	n/a	n/a	1	n/a
Vermont	5	80.00	1,453,778	0	-
Virginia	67	n/a	251,810,045	5	n/a
Washington	21	46.48	180,002,767	4	n/a
West Virginia	18	43.00	>5,674,330	0	-
Wisconsin	42	36.43	30,440,024	2	n/a
Wyoming	53	n/a	n/a	0	-
Total	1,767			107	

Notes: (1) Tonnage based on conversion from cubic yards reported (conversion of 3.3 cubic yards/ton); (2) Landfill capacity remaining exceeds ten years; (3) Waste-to-energy plant burns tires for fuel; (4) 2001 data from MSW Management; n/a not available
Source: BioCycle, "The State of Garbage in America," January 2004

Reported solid waste generated, estimated MSW generated, estimated MSW generated per capita and percents of MSW recycled, combusted via WTE, and landfilled for 2002[1]

State	Reported Solid Waste Generated (tons/yr.)	Estimated MSW Generated[2] (tons/yr.)	Estimated MSW Generated per Capita[3] (tons/person)	MSW Recycled (%)	MSW to Waste-to-Energy (%)	MSW Landfilled (%)
Alabama	n/a	n/a	n/a	n/a	n/a	n/a
Alaska	n/a	n/a	n/a	n/a	n/a	n/a
Arizona	4,962,000	6,012,359	1.10	17.5	0.0	82.5
Arkansas	4,061,128	3,838,217	1.42	36.3	1.5	62.3
California[4]	72,000,000	54,429,851	1.55	40.2	1.6	58.1
Colorado	7,673,778	5,051,132	1.12	2.8	0.0	97.2
Connecticut	3,474,981	4,734,132	1.37	18.8	45.0	36.2
Delaware	2,747,205	1,069,042	1.32	20.4	0.0	79.6
District of Columbia	n/a	n/a	n/a	n/a	n/a	n/a
Florida[5]	25,726,175	19,706,584	1.18	24.0	28.2	47.8
Georgia[6]	12,302,534	11,214,006	1.31	8.3	0.5	91.3
Hawaii	1,275,913	1,706,018	1.37	25.2	24.4	50.4
Idaho[7]	1,090,000	1,090,000	0.81	8.4	0.0	91.6
Illinois	15,428,491	15,951,037	1.27	32.5	0.0	67.5
Indiana[8]	16,228,824	9,542,378	1.55	35.0	7.0	58.0
Iowa	3,828,808	3,416,268	1.16	41.7	1.0	57.3
Kansas	7,846,080	4,698,338	1.73	11.5	0.0	88.5
Kentucky	6,529,846	5,465,608	1.34	11.4	0.0	88.6
Louisiana	3,272,331	4,952,900	1.10	8.1	0.0	91.9
Maine	1,844,059	1,327,164	1.03	49.0	33.8	17.2
Maryland	11,172,882	8,904,464	1.63	29.2	16.0	54.8
Massachusetts[9]	12,779,688	8,307,387	1.29	31.1	37.6	31.3
Michigan	19,041,775	16,916,076	1.68	15.1	7.0	77.9
Minnesota	5,881,543	5,043,752	1.00	45.6	25.1	29.3
Mississippi	3,909,508	2,918,407	1.02	0.3	0.0	99.7
Missouri	10,935,989	7,256,744	1.28	38.9	0.3	60.8
Montana	n/a	n/a	n/a	n/a	n/a	n/a
Nebraska[9]	2,395,101	2,395,100	1.39	15.4	0.0	84.6
Nevada	5,313,203	3,365,570	1.55	15.8	0.0	84.2
New Hampshire	1,327,598	1,214,777	0.95	23.7	17.0	59.4
New Jersey[9]	18,865,390	10,606,326	1.23	37.9	9.1	53.1
New Mexico	2,968,729	2,095,052	1.13	6.5	0.0	93.5
New York[10]	24,784,000	24,775,000	1.29	29.8	17.1	53.1
North Carolina	13,500,000	8,981,349	1.08	11.0	1.3	87.6
North Dakota	4,270,000	638,804	1.01	9.4	0.0	90.6
Ohio[9,11]	13,749,996	16,211,198	1.42	23.5	0.0	76.5
Oklahoma	4,489,028	4,489,028	1.28	1.0	0.0	99.0
Oregon	4,772,536	4,074,945	1.16	48.8	4.9	46.3
Pennsylvania	10,881,798	12,675,854	1.03	26.8	16.5	56.7
Rhode Island	1,497,240	1,248,745	1.17	12.8	0.0	87.2
South Carolina	11,464,547	5,973,059	1.45	28.4	3.9	67.7
South Dakota	688,000	518,493	0.68	3.0	0.0	97.0
Tennessee	9,852,194	7,365,920	1.27	26.4	2.0	71.6
Texas[12]	45,300,000	28,531,660	1.31	24.9	0.0	75.1
Utah	3,949,096	2,471,404	1.07	4.8	4.9	90.4
Vermont	700,000	611,617	0.99	29.8	9.2	60.9
Virginia	21,331,253	10,877,723	1.49	29.1	19.8	51.2
Washington[9]	10,470,805	8,666,755	1.43	34.1	5.6	60.2
West Virginia	1,963,791	1,754,523	0.97	6.9	0.0	93.1
Wisconsin	13,542,140	5,592,862	1.03	24.6	3.4	72.0
Wyoming	682,000	693,783	1.39	1.7	0.0	98.3
Total	482,770,983	369,381,411	1.31	26.7	7.7	65.6

Notes: (1) Alabama, Alaska, and Montana did not report any data for the 2003 "State of Garbage in America" survey; (2) Unless otherwise noted, MSW generated is computed from reported tonnages of: (landfill + exported landfill + WTE + exported WTE + MSW recycled) - (C&D landfill + industrial landfill + imported landfill + imported WTE); (3) U.S. per capita generation excludes Alabama, Alaska, and Montana; (4) MSW generation calculated using state population multiplied by 1.55 tons per capita (Nevada's per capita generation rate, chosen because highest rate in neighboring state). State provided tons landfilled and combusted via WTE; (5) 2000 data; (6) MSW generation calculated using state population multiplied by 1.31 tons per capita (national rate). State provided tons landfilled and combusted via WTE; (7) State reported MSW generation and no WTE facilities. 2002 landfill tonnage provided by Chartwell Information (www.wasteinfo.com); (8) MSW generation assumed to be equal to reported tons landfilled +recycled, at same recycling rate as in 2000 (35%); (9) 2001 data; (10) Detailed data for the state provided in New York State Assembly Report, "Where Will the Garbage Go?", 2002; (11) Tons of industrial wastes (10,502,763) were subtracted from reported total tons recycled; (12) MSW generation calculated using state population multiplied by 1.31 tons per capita (national rate). State provided tons landfilled and there are no WTE plants.
Source: BioCycle, "The State of Garbage in America," January 2004

Noise barrier construction by year

Year	Linear Length of Noise Barriers (miles)	Actual Cost ($millions)	Cost in 2001 Dollars ($millions)
1963-1980	224	97	187
1981	38	26	40
1982	25	19	30
1983	41	31	51
1984	54	40	63
1985	45	37	53
1986	65	70	100
1987	54	46	67
1988	107	116	158
1989	102	114	154
1990	64	76	101
1991	99	122	165
1992	142	155	213
1993	88	101	135
1994	89	101	127
1995	129	152	181
1996	54	62	75
1997	88	126	140
1998	129	180	205
1999	51	92	97
2000	76	125	124
2001	69	98	98
Unknown	6	0	0
ALL	1,831	1,986	2,564

Notes: *Six miles of noise barriers can neither be assigned a year of construction nor a cost. Additionally, 28 miles of barriers, while assigned a year of construction, cannot be assigned a cost.*
Source: *U.S. Department of Transportation, Federal Highway Administration, Office of Natural Environment Noise Team, Highway Traffic Noise Barrier Construction Trends, 2003*

Noise barrier construction average unit cost by year

Year	Area in Square Feet (100,000)	Cost in 2001 Dollars ($millions)	Cost per Square Foot ($)
1963-1980	145	187	13
1981	22	40	18
1982	17	30	18
1983	27	51	19
1984	32	63	20
1985	28	53	19
1986	41	100	24
1987	36	67	19
1988	70	158	23
1989	73	154	21
1990	52	101	19
1991	80	165	21
1992	104	213	20
1993	64	135	21
1994	65	127	19
1995	92	181	20
1996	42	75	18
1997	72	140	19
1998	109	205	19
1999	40	97	24
2000	60	124	21
2001	52	98	19
Unknown	4	0	0
ALL	1,324	2,564	19

Source: U.S. Department of Transportation, Federal Highway Administration, Office of Natural Environment Noise Team, Highway Traffic Noise Barrier Construction Trends, 2003

Noise barrier construction material by year

Year	Concrete	Block	Wood	Metal	Berm	Brick	Combination	Absorptive
1963-1990	1,490	1,572	836	177	307	35	878	42
1991	317	217	58	4	1	26	155	20
1992	539	294	55	16	19	5	81	28
1993	327	187	35	1	16	11	30	21
1994	355	133	88	1	12	14	22	15
1995	488	190	21	23	25	10	143	3
1996	244	54	9	0	6	0	56	13
1997	473	196	1	0	4	0	52	0
1998	838	105	14	23	10	0	67	24
1999	298	11	22	6	5	0	52	5
2000	458	29	35	0	11	0	54	7
2001	406	32	34	13	0	12	19	5
Unknown	22	0	0	14	0	0	7	0
ALL	6,256	3,020	1,208	277	417	113	1,616	182

Notes: *There are 1,456,814 square feet of noise barriers constructed with other materials.*
Source: *U.S. Department of Transportation, Federal Highway Administration, Office of Natural Environment Noise Team, Highway Traffic Noise Barrier Construction Trends, 2003*

Noise barrier construction material average unit cost by year

Year	Concrete	Block	Wood	Metal	Berm	Brick	Combination	Absorptive
1963-1990	23	19	16	14	5	25	16	32
1991	23	22	21	24	15	18	14	23
1992	22	19	18	21	8	23	18	24
1993	21	22	21	71	3	23	25	25
1994	20	18	15	16	7	29	22	18
1995	20	19	10	14	6	16	24	15
1996	16	19	15	0	12	0	21	26
1997	23	10	36	0	2	0	24	0
1998	20	14	18	11	4	0	14	28
1999	25	18	20	40	3	0	24	17
2000	21	36	18	0	7	0	12	31
2001	19	17	17	31	0	27	17	16
ALL	21	19	17	16	5	23	17	26

Notes: *Figures are material average unit cost in 2001 dollars per square foot; there are 1,456,814 square feet of noise barriers constructed with other materials costing approximately $29 per square foot.*
Source: *U.S. Department of Transportation, Federal Highway Administration, Office of Natural Environment Noise Team, Highway Traffic Noise Barrier Construction Trends, 2003*

Noise barrier construction material by height

Height	Concrete	Block	Wood	Metal	Berm	Brick	Combination	Absorptive	All Materials
>=30 Feet	84	5	36	0	9	0	98	0	232
27-29 Feet	26	0	0	0	0	0	15	0	41
24-26 Feet	240	5	52	8	3	0	16	5	330
21-23 Feet	505	0	60	0	32	0	44	12	653
18-20 Feet	1,410	46	241	48	26	18	375	75	2,239
15-17 Feet	1,694	422	284	64	58	33	368	46	2,969
12-14 Feet	1,533	1,567	169	81	35	29	412	28	3,855
9-11 Feet	557	816	234	70	187	26	200	15	2,104
6-8 Feet	196	155	131	5	57	9	87	0	641
<6 Feet	11	3	2	2	9	0	1	0	28
ALL	6,256	3,020	1,208	277	417	113	1,616	182	13,090

Notes: Figures are material area in ten thousands of square feet; there are 1,456,814 square feet of noise barriers constructed with other materials

Source: U.S. Department of Transportation, Federal Highway Administration, Office of Natural Environment Noise Team, Highway Traffic Noise Barrier Construction Trends, 2003

Noise barrier construction material average unit cost by height

Height	Concrete	Block	Wood	Metal	Berm	Brick	Combination	Absorptive	All Materials
>=30 Feet	22	6	5	0	1	0	10	0	13
27-29 Feet	20	0	0	0	0	0	14	0	18
24-26 Feet	16	11	19	19	0	0	18	14	16
21-23 Feet	25	0	29	0	3	0	13	21	23
18-20 Feet	24	23	13	12	9	19	16	22	21
15-17 Feet	20	21	18	15	5	25	18	28	20
12-14 Feet	20	17	18	13	5	19	20	29	18
9-11 Feet	22	21	15	22	5	27	17	37	19
6-8 Feet	18	17	18	18	6	27	19	23	17
<6 Feet	21	19	16	29	12	0	41	99	19
ALL	21	19	17	16	5	23	17	26	19

Notes: *Figures are material average unit cost in 2001 dollars per foot squared; there are 1,456,814 square feet of noise barriers constructed with other materials costing approximately $29 per square foot.*
Source: *U.S. Department of Transportation, Federal Highway Administration, Office of Natural Environment Noise Team, Highway Traffic Noise Barrier Construction Trends, 2003*

Noise barrier construction by state, average height, and average unit cost

State	Average Height (feet)	Average Unit Cost (2001 $/foot^2)
Alabama	0	0
Alaska	10	10
Arizona	10	11
Arkansas	13	9
California	11	18
Colorado	10	17
Connecticut	16	12
Delaware	14	5
District of Columbia	0	0
Eastern Dir. Fed.	11	16
Florida	14	20
Georgia	13	14
Hawaii	7	27
Idaho	10	18
Illinois	12	22
Indiana	15	25
Iowa	13	11
Kansas	15	23
Kentucky	20	15
Louisiana	12	15
Maine	14	13
Maryland	18	28
Massachusetts	9	21
Michigan	12	28
Minnesota	16	14
Mississippi	0	0
Missouri	13	17
Montana	0	0
Nebraska	14	20
Nevada	12	16
New Hampshire	15	22
New Jersey	17	27
New Mexico	9	18
New York	14	27
North Carolina	16	12
North Dakota	0	0
Ohio	14	14
Oklahoma	10	14
Oregon	11	12
Pennsylvania	13	30
Puerto Rico	13	32
Rhode Island	0	0
South Carolina	15	12
South Dakota	0	0
Tennessee	15	18
Texas	12	20
Utah	12	11
Vermont	7	21
Virginia	15	19
Washington	11	16
West Virginia	12	14
Wisconsin	17	19
Wyoming	8	15

Source: U.S. Department of Transportation, Federal Highway Administration Office of Natural Environment Noise Team, Highway Traffic Noise Barrier Construction Trends, 2003

Pesticides in streams at agricultural land use sites, 1991-2001

Compound	Max Rep. Limit[1] (ug/L)	No. of Sites	No. of Samp.	Frequency of Detection (%)				Percentiles of Concentration (ug/L)					
				All	>=0.01 ug/L	>=0.1 ug/L	>=1 ug/L	25th	50th	75th	90th	95th	Max
Analyzed by GC/MS													
Acetochlor[a]	0.006	52	1240	31.44	23.41	6.60	1.87	<0.005	<0.005	0.009	0.052	0.165	25.1[c]
Alachlor	0.005	77	1869	37.34	24.11	5.09	0.97	<0.001	0.001	0.009	0.035	0.104	10.9
Atrazine	0.007	76	1852	90.44	80.19	42.97	9.91	0.015	0.071	0.220	0.980	2.86	201[c]
Azinphos-methyl[e]	0.050	75	1800	1.31	>1.20[b]	0.20	0.00	<0.050	<0.050	<0.050	<0.050	<0.050	0.500
Benfluralin	0.010	78	1892	0.77	0.00	0.00	0.00	<0.010	<0.010	<0.010	<0.010	<0.010	0.008
Butylate	0.002	78	1886	3.47	2.03	0.42	0.06	<0.002	<0.002	<0.002	<0.002	<0.002	1.40
Carbaryl[e]	0.041	78	1889	9.17	>5.14[b]	0.77	0.04	<0.041	<0.041	<0.041	<0.041	<0.041	5.20
Carbofuran[e]	0.020	78	1885	9.60	>8.35[b]	3.00	0.47	<0.020	<0.020	<0.020	<0.020	0.043	7.00
Chlorpyrifos	0.005	75	1824	11.44	4.84	0.24	0.00	<0.005	<0.005	<0.005	<0.005	0.009	0.260
Cyanazine	0.018	78	1881	40.38	>35.38[b]	11.20	2.48	<0.001	0.003	0.024	0.116	0.340	160[c]
p,p'-DDE	0.006	78	1885	4.84	0.31	0.00	0.00	<0.006	<0.006	<0.006	<0.006	<0.006	0.062
Dacthal	0.003	78	1890	11.46	2.83	0.91	0.28	<0.003	<0.003	<0.003	<0.003	0.003	40[c]
Deethylatrazine[e]	0.006	76	1830	82.11	61.76	11.92	0.26	0.004	0.018	0.051	0.120	0.186	3.03
Diazinon	0.005	78	1879	13.15	6.67	0.87	0.05	<0.005	<0.005	<0.005	0.005	0.015	2.50
Dieldrin	0.005	78	1887	3.86	1.83	0.00	0.00	<0.005	<0.005	<0.005	<0.005	<0.005	0.064
2,6-Diethylaniline	0.006	78	1891	2.18	0.05	0.00	0.00	<0.004	<0.004	<0.004	<0.004	<0.004	0.018
Disulfoton	0.021	78	1892	0.34	>0.21[b]	0.06	0.00	<0.021	<0.021	<0.021	<0.021	<0.021	0.434
EPTC	0.002	78	1884	14.11	7.80	1.62	0.34	<0.002	<0.002	<0.002	0.006	0.018	7.30
Ethalfluralin	0.009	78	1890	0.98	0.76	0.00	0.00	<0.009	<0.009	<0.009	<0.009	<0.009	0.073
Ethoprop	0.005	78	1883	2.68	1.82	0.29	0.00	<0.005	<0.005	<0.005	<0.005	<0.005	0.455
Fonofos	0.003	78	1889	3.05	1.52	0.22	0.03	<0.003	<0.003	<0.003	<0.003	<0.003	1.20
alpha-HCH	0.005	78	1886	0.24	0.18	0.00	0.00	<0.005	<0.005	<0.005	<0.005	<0.005	0.069
gamma-HCH	0.004	78	1886	0.70	0.55	0.00	0.00	<0.004	<0.004	<0.004	<0.004	<0.004	0.053
Linuron	0.035	78	1890	2.49	>1.96[b]	0.57	0.01	<0.035	<0.035	<0.035	<0.035	<0.035	1.40
Malathion	0.027	78	1892	5.02	>3.55[b]	0.29	0.00	<0.027	<0.027	<0.027	<0.027	<0.027	0.523
Metolachlor	0.013	78	1887	82.74	>68.04[b]	28.38	6.55	0.005	0.029	0.120	0.544	1.38	77.6[c]
Metribuzin	0.006	77	1854	18.39	13.63	3.46	0.58	<0.006	<0.006	<0.006	0.015	0.053	6.61
Molinate	0.004	78	1880	10.67	8.18	4.02	1.82	<0.004	<0.004	<0.004	<0.004	0.042	200[c]
Napropamide	0.007	77	1858	3.25	2.03	0.36	0.00	<0.007	<0.007	<0.007	<0.007	<0.007	0.767
Parathion	0.010	78	1888	0.13	0.11	0.03	0.00	<0.008	<0.008	<0.008	<0.008	<0.008	0.141
Parathion-methyl	0.006	77	1868	1.11	1.00	0.20	0.00	<0.006	<0.006	<0.006	<0.006	<0.006	0.422
Pebulate	0.004	78	1886	0.71	0.18	0.00	0.00	<0.004	<0.004	<0.004	<0.004	<0.004	0.080
Pendimethalin	0.022	78	1884	6.56	>5.57[b]	1.14	0.05	<0.020	<0.020	<0.020	<0.020	<0.020	2.05
cis-Permethrin	0.006	78	1887	0.15	0.12	0.00	0.00	<0.006	<0.006	<0.006	<0.006	<0.006	0.028
Phorate	0.011	77	1879	0.08	>0.08[b]	0.01	0.00	<0.011	<0.011	<0.011	<0.011	<0.011	0.600
Prometon	0.018	77	1874	42.71	>21.42[b]	0.77	0.00	0.005	0.009	0.014	0.021	0.035	0.250
Pronamide	0.004	78	1873	2.15	1.15	0.24	0.00	<0.004	<0.004	<0.004	<0.004	<0.004	0.275
Propachlor	0.010	77	1875	1.52	0.43	0.05	0.00	<0.010	<0.010	<0.010	<0.010	<0.010	0.511
Propanil	0.011	78	1891	1.52	>0.95[b]	0.38	0.03	<0.011	<0.011	<0.011	<0.011	<0.011	2.05
Propargite	0.023	77	1861	3.04	>2.10[b]	0.97	0.11	<0.023	<0.023	<0.023	<0.023	<0.023	2.62
Simazine	0.011	77	1874	56.60	>38.74[b]	7.40	1.06	0.003	0.006	0.022	0.064	0.180	5.76
Tebuthiuron	0.016	74	1803	19.62	>10.30[b]	0.44	0.00	<0.016	<0.016	<0.016	<0.016	0.018	0.950
Terbacil[e]	0.034	77	1858	4.52	>4.40[b]	0.48	0.00	<0.034	<0.034	<0.034	<0.034	<0.034	0.540
Terbufos	0.017	78	1889	0.12	>0.12[b]	0.03	0.00	<0.017	<0.017	<0.017	<0.017	<0.017	0.560
Thiobencarb	0.005	78	1881	3.57	2.37	0.64	0.03	<0.005	<0.005	<0.005	<0.005	<0.005	3.66
Triallate	0.002	77	1864	3.18	1.09	0.06	0.00	<0.001	<0.001	<0.001	<0.001	<0.001	0.650
Trifluralin	0.009	78	1889	13.34	3.72	0.24	0.00	<0.009	<0.009	<0.009	<0.009	<0.009	0.170
Analyzed by HPLC													
Acifluorfen	0.090	48	1233	4.04	(d)	2.06	0.10	<0.040	<0.040	<0.040	<0.040	<0.040	1.10
Aldicarb[e]	0.550	48	1236	0.20	(d)	>0.20[b]	0.00	<0.550	<0.550	<0.550	<0.550	<0.550	0.510
Aldicarb sulfone[e]	0.200	48	1235	0.00	(d)	>0.00[b]	0.00	<0.100	<0.100	<0.100	<0.100	<0.100	<0.100
Aldicarb sulfoxide[e]	0.270	48	1236	0.16	(d)	>0.16[b]	0.06	<0.021	<0.021	<0.021	<0.021	<0.021	1.91[c]
Bentazon	0.050	48	1233	17.41	(d)	7.91	1.21	<0.020	<0.020	<0.020	0.070	0.150	8.60[c]
Bromacil	0.090	48	1236	0.46	(d)	0.18	0.04	<0.040	<0.040	<0.040	<0.040	<0.040	1.90[c]
Bromoxynil	0.070	48	1233	0.49	(d)	0.23	0.00	<0.040	<0.040	<0.040	<0.040	<0.040	0.770

Pesticides in streams at agricultural land use sites, 1991-2001 - *continued*

Compound	Max Rep. Limit[1] (ug/L)	No. of Sites	No. of Samp.	Frequency of Detection (%)				Percentiles of Concentration (ug/L)					
				All	>=0.01 ug/L	>=0.1 ug/L	>=1 ug/L	25th	50th	75th	90th	95th	Max
Chloramben methyl ester	0.420	48	1236	0.00	(d)	>0.00[b]	0.00	<0.420	<0.420	<0.420	<0.420	<0.420	<0.420
Chlorothalonil[e]	0.480	48	1230	0.04	(d)	>0.00[b]	0.00	<0.480	<0.480	<0.480	<0.480	<0.480	0.290
Clopyralid	0.420	48	1233	0.00	(d)	>0.00[b]	0.00	<0.230	<0.230	<0.230	<0.230	<0.230	<0.230
2,4-D	0.160	48	1233	15.35	(d)	>12.84[b]	1.89	<0.150	<0.150	<0.150	0.150	0.350	15.0[c]
Dacthal monoacid	0.070	48	1233	0.18	(d)	0.14	0.00	<0.035	<0.035	<0.035	<0.035	<0.035	0.430
2,4-DB	0.250	48	1233	0.18	(d)	>0.06[b]	0.00	<0.240	<0.240	<0.240	<0.240	<0.240	0.830
Dicamba	0.110	48	1233	1.55	(d)	>0.84[b]	0.04	<0.040	<0.040	<0.040	<0.040	<0.040	1.14
Dichlobenil[e]	1.20	48	1236	0.26	(d)	>0.00[b]	>0.00[b]	<1.20	<1.20	<1.20	<1.20	<1.20	0.010
Dichlorprop	0.120	48	1233	0.94	(d)	>0.69[b]	0.00	<0.032	<0.032	<0.032	<0.032	<0.032	0.260
Dinoseb	0.090	48	1233	0.31	(d)	0.09	0.09	<0.040	<0.040	<0.040	<0.040	<0.040	1.00
Diuron	0.120	48	1235	13.04	(d)	>7.04[b]	1.78	<0.050	<0.050	<0.050	<0.050	0.260	14.0[c]
DNOC[e]	0.420	48	1233	0.12	(d)	>0.12[b]	0.00	<0.420	<0.420	<0.420	<0.420	<0.420	0.190
Fenuron	0.070	48	1236	0.10	(d)	0.00	0.00	<0.050	<0.050	<0.050	<0.050	<0.050	0.020
Fluometuron	0.060	48	1236	7.58	(d)	5.36	1.14	<0.050	<0.050	<0.050	<0.050	0.100	8.60[c]
3-Hydroxycarbofuran	0.110	48	1236	0.00	(d)	>0.00[b]	0.00	<0.014	<0.014	<0.014	<0.014	<0.014	<0.014
MCPA	0.200	48	1233	1.40	(d)	>1.12[b]	0.00	<0.170	<0.170	<0.170	<0.170	<0.170	0.920
MCPB	0.260	48	1232	0.00	(d)	>0.00[b]	0.00	<0.140	<0.140	<0.140	<0.140	<0.140	<0.140
Methiocarb	0.070	48	1235	0.10	(d)	0.10	0.00	<0.030	<0.030	<0.030	<0.030	<0.030	0.100
Methomyl	0.470	48	1234	1.65	(d)	>1.35[b]	0.00	<0.020	<0.020	<0.020	<0.020	<0.020	0.670
Neburon	0.070	48	1236	0.09	(d)	0.00	0.00	<0.015	<0.015	<0.015	<0.015	<0.015	0.030
Norflurazon	0.042	48	1235	3.60	(d)	2.43	0.04	<0.040	<0.040	<0.040	<0.040	<0.040	1.24
Oryzalin	0.310	48	1236	0.49	(d)	>0.34[b]	0.14	<0.310	<0.310	<0.310	<0.310	<0.310	1.80[c]
Oxamyl	0.160	48	1234	0.74	(d)	>0.39[b]	0.00	<0.020	<0.020	<0.020	<0.020	<0.020	0.160
Picloram	0.090	48	1233	0.00	(d)	0.00	0.00	<0.050	<0.050	<0.050	<0.050	<0.050	<0.050
Propham	0.220	48	1236	0.04	(d)	>0.00[b]	0.00	<0.040	<0.040	<0.040	<0.040	<0.040	0.060
Propoxur	0.120	47	1208	0.22	(d)	>0.15[b]	0.00	<0.040	<0.040	<0.040	<0.040	<0.040	0.110
Silvex	0.060	48	1234	0.00	(d)	0.00	0.00	<0.050	<0.050	<0.050	<0.050	<0.050	<0.050
2,4,5-T	0.070	48	1233	0.04	(d)	0.00	0.00	<0.040	<0.040	<0.040	<0.040	<0.040	0.050
Triclopyr	0.250	48	1233	2.27	(d)	>1.74[b]	0.27	<0.250	<0.250	<0.250	<0.250	<0.250	16.0[c]

Notes: *(1) The analytical reporting limit has changed through time. The limit reported in this table is the maximum value routinely used to report nondetections during the period 1991-2001; (a) Acetochlor was not used or analyzed for prior to 1994; (b) The analytical reporting limit for the compound is greater than the threshold concentration used for computing detection frequency. Therefore, the reported percentage may be an underestimate of the actual percentage of time-weighted concentrations that are greater than the threshold concentration; (c) See "Estimated Concentrations" below; (d) The 0.01 ug/L threshold is not applicable for the HPLC method because the reporting limit for most compounds is much higher than the threshold; (e) all concentrations determined for this compound are estimated concentrations.*

Abbreviations: *ug/L, microgram per liter; ">=", greater than or equal to; GC/MS, gas chromatography/mass spectrometry; HPLC, high performance liquid chromatography; NWQL, U.S. Geological Survey National Water Quality Laboratory.*

Frequency of Detection: *Within each statistical summary, four detection frequencies are provided (all detections, >=0.01 ug/L, >=0.1 ug/L, and >=1 ug/L). These detection frequencies provide a measure of "how often" a compound was found. The "all detections" provide the total number of detections for a given compound, but are not comparable among compounds because detection capabilities vary. Because reporting levels varied, detection frequencies were calculated using three common detection thresholds (0.01, 0.1 and 1 ug/L). The use of these detection thresholds facilitates comparisons among compounds by censoring detections to a common reference concentration. Adjustments of this type are essential in order to answer questions like "is compound x found more often than compound y?"*

Percentiles of concentration: *Annual time-weighted concentrations measured for each pesticide are summarized using percentiles. The 25th, 50th, 75th, 90th, and 95th percentiles of concentration by land-use category are provided. Percentiles provide information about the magnitude and duration of concentrations at selected points in the cumulative frequency distribution of the ranked, time-weighted concentrations. For example, concentrations of acetochlor at NAWQA sites draining agricultural land use were less than 0.005 ug/L for 25 percent of the year, less than 0.005 ug/L for 50 percent of the year, less than or equal to 0.009 ug/L for 75 percent of the year, less than or equal to 0.052 ug/L for 90 percent of the year, and less than or equal to 0.165 ug/L for 95 percent of the year. If percentiles are not censored (reported as less than a specified concentration—for example, <0.005 ug/L), the percentiles also may be interpreted as the percentage of the year where concentrations were greater than a given concentration. For example, concentrations of acetochlor at NAWQA sites draining agricultural land use were greater than or equal to 0.009 ug/L for 25 percent of the year, greater than or equal to 0.052 ug/L for 10 percent of the year, and greater than or equal to 0.165 ug/L for 5 percent of the year.*

Maximum Concentrations: *The concentrations in this column are the maximum measured concentrations in the pooled annual data sets for each land-use category. Concentrations greater than the value reported as maximum may have been measured at sites but are not reported because they were not measured during the one-year periods selected to describe the annual distribution of concentrations. In addition, the probability is low that a water sample was collected at the time of the annual maximum concentration for any pesticide. Consequently, the maximum concentrations reported should be interpreted as a lower bound for the true maximum concentrations.*

Estimated concentrations: *All pesticides denoted with "e" have more biased and/or variable analytical performance than the other pesticides in the method. Information on analytical performance is provided in the method reports (Zaugg and others, 1995; Werner and others, 1996) and is summarized for the period 1992-96 in Martin (1999). Individual concentrations that exceed the calibration curve of the analytical method are marked with a "c" to indicate that the uncertainty in these concentrations is greater than for concentrations within the calibration range.*

Source: *U.S. Geological Survey, Pesticides in Streams: Preliminary Results from Cycle I of the National Water Quality Assessment Program (NAWQA), 1992-2001, February 26, 2003*

Pesticides in streams at mixed land use sites, 1991-2001

Compound	Max Rep. Limit[1] (ug/L)	No. of Sites	No. of Samp.	Frequency of Detection (%)				Percentiles of Concentration (ug/L)					
				All	>=0.01 ug/L	>=0.1 ug/L	>=1 ug/L	25th	50th	75th	90th	95th	Max
Analyzed by GC/MS													
Acetochlor[a]	0.006	36	749	9.42	7.14	2.57	0.12	<0.004	<0.004	<0.004	<0.004	0.024	2.00
Alachlor	0.005	47	1019	22.66	12.97	2.55	0.29	<0.005	<0.005	<0.005	0.014	0.041	3.00
Atrazine	0.007	47	971	88.09	70.14	26.61	3.87	0.008	0.026	0.108	0.342	0.804	12.8
Azinphos-methyl[e]	0.050	46	997	0.50	>0.44[b]	0.00	0.00	<0.050	<0.050	<0.050	<0.050	<0.050	0.099
Benfluralin	0.010	47	1022	1.59	0.18	0.10	0.00	<0.010	<0.010	<0.010	<0.010	<0.010	0.205
Butylate	0.002	47	1023	2.36	0.72	0.00	0.00	<0.002	<0.002	<0.002	<0.002	<0.002	0.042
Carbaryl[e]	0.041	47	1017	15.39	>9.73[b]	1.26	0.00	<0.041	<0.041	<0.041	<0.041	<0.041	0.447
Carbofuran[e]	0.020	44	949	3.28	>2.60[b]	0.22	0.00	<0.020	<0.020	<0.020	<0.020	<0.020	0.678
Chlorpyrifos	0.005	47	1015	18.10	8.38	0.23	0.00	<0.005	<0.005	<0.005	0.008	0.015	0.154
Cyanazine	0.018	47	1015	30.26	>23.38[b]	5.59	0.98	<0.018	<0.018	<0.018	0.044	0.130	5.10
p,p'-DDE	0.006	47	1021	6.14	0.00	0.00	0.00	<0.006	<0.006	<0.006	<0.006	<0.006	0.009
Dacthal	0.003	47	1020	15.40	3.48	0.16	0.00	<0.003	<0.003	<0.003	<0.003	0.004	0.179
Deethylatrazine[e]	0.006	47	965	74.89	44.37	2.74	0.00	0.003	0.008	0.024	0.047	0.072	0.828
Diazinon	0.005	46	991	41.97	25.41	2.03	0.00	<0.001	0.003	0.010	0.039	0.066	0.780
Dieldrin	0.005	47	1021	0.74	0.40	0.00	0.00	<0.005	<0.005	<0.005	<0.005	<0.005	0.015
2,6-Diethylaniline	0.006	47	1021	1.10	0.00	0.00	0.00	<0.003	<0.003	<0.003	<0.003	<0.003	0.005
Disulfoton	0.021	47	1022	0.08	>0.08[b]	0.08	0.00	<0.021	<0.021	<0.021	<0.021	<0.021	0.826
EPTC	0.002	47	1000	11.88	4.58	1.99	1.11	<0.002	<0.002	<0.002	0.003	0.009	29.6[c]
Ethalfluralin	0.009	47	1021	0.32	0.08	0.00	0.00	<0.009	<0.009	<0.009	<0.009	<0.009	0.013
Ethoprop	0.005	47	1019	1.81	0.59	0.01	0.00	<0.005	<0.005	<0.005	<0.005	<0.005	0.140
Fonofos	0.003	47	1020	1.20	0.41	0.00	0.00	<0.003	<0.003	<0.003	<0.003	<0.003	0.014
alpha-HCH	0.005	47	1022	0.13	0.00	0.00	0.00	<0.005	<0.005	<0.005	<0.005	<0.005	0.007
gamma-HCH	0.004	45	965	1.39	0.43	0.00	0.00	<0.004	<0.004	<0.004	<0.004	<0.004	0.036
Linuron	0.035	46	1004	1.26	>1.11[b]	0.46	0.00	<0.035	<0.035	<0.035	<0.035	<0.035	0.519
Malathion	0.027	47	1022	7.81	>4.38[b]	0.72	0.00	<0.027	<0.027	<0.027	<0.027	<0.027	0.370
Metolachlor	0.013	47	1023	71.37	>50.73[b]	11.63	1.90	0.002	0.010	0.038	0.112	0.335	9.10
Metribuzin	0.006	45	976	8.98	6.63	0.86	0.00	<0.006	<0.006	<0.006	<0.006	0.015	0.239
Molinate	0.004	47	1019	0.54	0.08	0.00	0.00	<0.004	<0.004	<0.004	<0.004	<0.004	0.047
Napropamide	0.007	46	1004	2.60	1.69	0.09	0.00	<0.007	<0.007	<0.007	<0.007	<0.007	0.332
Parathion	0.010	47	1021	0.14	0.07	0.00	0.00	<0.008	<0.008	<0.008	<0.008	<0.008	0.041
Parathion-methyl	0.006	47	1018	0.35	0.35	0.20	0.00	<0.006	<0.006	<0.006	<0.006	<0.006	0.521
Pebulate	0.004	47	1018	0.20	0.07	0.00	0.00	<0.004	<0.004	<0.004	<0.004	<0.004	0.019
Pendimethalin	0.022	47	1020	6.72	>4.01[b]	0.40	0.00	<0.020	<0.020	<0.020	<0.020	<0.020	0.144
cis-Permethrin	0.006	47	1021	0.52	0.16	0.00	0.00	<0.006	<0.006	<0.006	<0.006	<0.006	0.019
Phorate	0.011	47	1018	0.16	>0.16[b]	0.00	0.00	<0.011	<0.011	<0.011	<0.011	<0.011	0.033
Prometon	0.018	47	1022	60.91	>40.43[b]	3.36	0.04	0.007	0.012	0.022	0.050	0.087	1.35
Pronamide	0.004	46	953	0.77	0.23	0.08	0.00	<0.004	<0.004	<0.004	<0.004	<0.004	0.430
Propachlor	0.010	47	1021	0.65	0.24	0.00	0.00	<0.010	<0.010	<0.010	<0.010	<0.010	0.064
Propanil	0.011	47	1020	0.48	>0.01[b]	0.00	0.00	<0.011	<0.011	<0.011	<0.011	<0.011	0.010
Propargite	0.023	45	968	0.59	>0.40[b]	0.00	0.00	<0.023	<0.023	<0.023	<0.023	<0.023	0.045
Simazine	0.011	47	1024	67.44	>50.90[b]	8.45	0.55	0.004	0.010	0.031	0.086	0.180	2.10
Tebuthiuron	0.016	46	942	32.55	>16.58[b]	0.09	0.00	<0.016	<0.016	<0.016	<0.016	0.017	0.153
Terbacil[e]	0.034	46	996	1.82	>1.48[b]	0.14	0.00	<0.034	<0.034	<0.034	<0.034	<0.034	0.341
Terbufos	0.017	47	1021	0.24	>0.13[b]	0.00	0.00	<0.017	<0.017	<0.017	<0.017	<0.017	0.031
Thiobencarb	0.005	47	1021	0.24	0.00	0.00	0.00	<0.005	<0.005	<0.005	<0.005	<0.005	0.008
Triallate	0.002	46	953	7.17	2.44	0.25	0.00	<0.002	<0.002	<0.002	<0.002	0.003	0.490
Trifluralin	0.009	47	1021	10.62	1.63	0.00	0.00	<0.009	<0.009	<0.009	<0.009	<0.009	0.097
Analyzed by HPLC													
Acifluorfen	0.090	25	561	0.56	(d)	0.00	0.00	<0.035	<0.035	<0.035	<0.035	<0.035	0.040
Aldicarb[e]	0.550	25	562	0.00	(d)	>0.00[b]	0.00	<0.550	<0.550	<0.550	<0.550	<0.550	<0.550
Aldicarb sulfone[e]	0.200	25	560	0.00	(d)	>0.00[b]	0.00	<0.100	<0.100	<0.100	<0.100	<0.100	<0.100
Aldicarb sulfoxide[e]	0.270	25	561	0.00	(d)	>0.00[b]	0.00	<0.021	<0.021	<0.021	<0.021	<0.021	<0.021
Bentazon	0.050	25	561	4.84	(d)	2.11	0.00	<0.025	<0.025	<0.025	<0.025	<0.025	0.690
Bromacil	0.090	25	561	0.22	(d)	0.22	0.05	<0.035	<0.035	<0.035	<0.035	<0.035	4.00[c]
Bromoxynil	0.070	25	561	0.30	(d)	0.15	0.00	<0.035	<0.035	<0.035	<0.035	<0.035	0.600

Pesticides in streams at mixed land use sites, 1991-2001 - *continued*

Compound	Max Rep. Limit[1] (ug/L)	No. of Sites	No. of Samp.	Frequency of Detection (%)				Percentiles of Concentration (ug/L)					
				All	>=0.01 ug/L	>=0.1 ug/L	>=1 ug/L	25th	50th	75th	90th	95th	Max
Chloramben methyl ester	0.420	25	562	0.00	(d)	>0.00[b]	0.00	<0.420	<0.420	<0.420	<0.420	<0.420	<0.420
Chlorothalonil[e]	0.480	25	561	0.00	(d)	>0.00[b]	0.00	<0.480	<0.480	<0.480	<0.480	<0.480	<0.480
Clopyralid	0.420	25	562	0.00	(d)	>0.00[b]	0.00	<0.230	<0.230	<0.230	<0.230	<0.230	<0.230
2,4-D	0.160	25	562	7.79	(d)	>5.52[b]	0.74	<0.150	<0.150	<0.150	<0.150	<0.150	1.40
Dacthal monoacid	0.070	25	561	0.00	(d)	0.00	0.00	<0.017	<0.017	<0.017	<0.017	<0.017	<0.017
2,4-DB	0.250	25	562	0.00	(d)	>0.00[b]	0.00	<0.240	<0.240	<0.240	<0.240	<0.240	<0.240
Dicamba	0.110	25	561	0.80	(d)	>0.80[b]	0.00	<0.035	<0.035	<0.035	<0.035	<0.035	0.390
Dichlobenil[e]	1.20	25	558	0.50	(d)	>0.00[b]	>0.00[b]	<1.20	<1.20	<1.20	<1.20	<1.20	0.080
Dichlorprop	0.120	25	562	0.23	(d)	>0.08[b]	0.00	<0.120	<0.120	<0.120	<0.120	<0.120	0.180
Dinoseb	0.090	25	561	0.03	(d)	0.03	0.00	<0.035	<0.035	<0.035	<0.035	<0.035	0.270
Diuron	0.120	25	557	8.66	(d)	>5.52[b]	0.39	<0.050	<0.050	<0.050	<0.050	0.110	4.15[c]
DNOC[e]	0.420	25	562	0.00	(d)	>0.00[b]	0.00	<0.420	<0.420	<0.420	<0.420	<0.420	<0.420
Fenuron	0.070	25	561	0.11	(d)	0.00	0.00	<0.050	<0.050	<0.050	<0.050	<0.050	0.040
Fluometuron	0.060	25	561	0.00	(d)	0.00	0.00	<0.035	<0.035	<0.035	<0.035	<0.035	<0.035
3-Hydroxycarbofuran	0.110	25	561	0.00	(d)	>0.00[b]	0.00	<0.014	<0.014	<0.014	<0.014	<0.014	<0.014
MCPA	0.200	25	562	2.04	(d)	>1.52[b]	0.31	<0.170	<0.170	<0.170	<0.170	<0.170	18.6[c]
MCPB	0.260	25	562	0.22	(d)	>0.00[b]	0.00	<0.140	<0.140	<0.140	<0.140	<0.140	0.060
Methiocarb	0.070	25	561	0.00	(d)	0.00	0.00	<0.026	<0.026	<0.026	<0.026	<0.026	<0.026
Methomyl	0.470	25	560	0.27	(d)	>0.27[b]	0.00	<0.017	<0.017	<0.017	<0.017	<0.017	0.270
Neburon	0.070	25	561	0.00	(d)	0.00	0.00	<0.015	<0.015	<0.015	<0.015	<0.015	<0.015
Norflurazon	0.042	25	561	0.09	(d)	0.00	0.00	<0.024	<0.024	<0.024	<0.024	<0.024	0.040
Oryzalin	0.310	25	557	0.51	(d)	>0.10[b]	0.00	<0.310	<0.310	<0.310	<0.310	<0.310	0.430
Oxamyl	0.160	25	560	0.00	(d)	>0.00[b]	0.00	<0.018	<0.018	<0.018	<0.018	<0.018	<0.018
Picloram	0.090	25	561	0.12	(d)	0.00	0.00	<0.050	<0.050	<0.050	<0.050	<0.050	0.010
Propham	0.220	25	561	0.00	(d)	>0.00[b]	0.00	<0.035	<0.035	<0.035	<0.035	<0.035	<0.035
Propoxur	0.120	24	544	0.16	(d)	>0.16[b]	0.00	<0.035	<0.035	<0.035	<0.035	<0.035	0.190
Silvex	0.060	25	561	0.00	(d)	0.00	0.00	<0.021	<0.021	<0.021	<0.021	<0.021	<0.021
2,4,5-T	0.070	25	561	0.00	(d)	0.00	0.00	<0.035	<0.035	<0.035	<0.035	<0.035	<0.035
Triclopyr	0.250	25	562	0.19	(d)	>0.19[b]	0.19	<0.250	<0.250	<0.250	<0.250	<0.250	1.04

Notes: *(1) The analytical reporting limit has changed through time. The limit reported in this table is the maximum value routinely used to report nondetections during the period 1991-2001; (a) Acetochlor was not used or analyzed for prior to 1994; (b) The analytical reporting limit for the compound is greater than the threshold concentration used for computing detection frequency. Therefore, the reported percentage may be an underestimate of the actual percentage of time-weighted concentrations that are greater than the threshold concentration; (c) See "Estimated Concentrations" below; (d) The 0.01 ug/L threshold is not applicable for the HPLC method because the reporting limit for most compounds is much higher than the threshold; (e) all concentrations determined for this compound are estimated concentrations.*

Abbreviations: *ug/L, microgram per liter; ">=", greater than or equal to; GC/MS, gas chromatography/mass spectrometry; HPLC, high performance liquid chromatography; NWQL, U.S. Geological Survey National Water Quality Laboratory.*

Frequency of Detection: *Within each statistical summary, four detection frequencies are provided (all detections, >=0.01 ug/L, >=0.1 ug/L, and >=1 ug/L). These detection frequencies provide a measure of "how often" a compound was found. The "all detections" provide the total number of detections for a given compound, but are not comparable among compounds because detection capabilities vary. Because reporting levels varied, detection frequencies were calculated using three common detection thresholds (0.01, 0.1 and 1 ug/L). The use of these detection thresholds facilitates comparisons among compounds by censoring detections to a common reference concentration. Adjustments of this type are essential in order to answer questions like "is compound x found more often than compound y?"*

Percentiles of concentration: *Annual time-weighted concentrations measured for each pesticide are summarized using percentiles. The 25th, 50th, 75th, 90th, and 95th percentiles of concentration by land-use category are provided. Percentiles provide information about the magnitude and duration of concentrations at selected points in the cumulative frequency distribution of the ranked, time-weighted concentrations. For example, concentrations of acetochlor at NAWQA sites draining agricultural land use were less than 0.005 ug/L for 25 percent of the year, less than 0.005 ug/L for 50 percent of the year, less than or equal to 0.009 ug/L for 75 percent of the year, less than or equal to 0.052 ug/L for 90 percent of the year, and less than or equal to 0.165 ug/L for 95 percent of the year. If percentiles are not censored (reported as less than a specified concentration—for example, <0.005 ug/L), the percentiles also may be interpreted as the percentage of the year where concentrations were greater than a given concentration. For example, concentrations of acetochlor at NAWQA sites draining agricultural land use were greater than or equal to 0.009 ug/L for 25 percent of the year, greater than or equal to 0.052 ug/L for 10 percent of the year, and greater than or equal to 0.165 ug/L for 5 percent of the year.*

Maximum Concentrations: *The concentrations in this column are the maximum measured concentrations in the pooled annual data sets for each land-use category. Concentrations greater than the value reported as maximum may have been measured at sites but are not reported because they were not measured during the one-year periods selected to describe the annual distribution of concentrations. In addition, the probability is low that a water sample was collected at the time of the annual maximum concentration for any pesticide. Consequently, the maximum concentrations reported should be interpreted as a lower bound for the true maximum concentrations.*

Estimated concentrations: *All pesticides denoted with "e" have more biased and/or variable analytical performance than the other pesticides in the method. Information on analytical performance is provided in the method reports (Zaugg and others, 1995; Werner and others, 1996) and is summarized for the period 1992-96 in Martin (1999). Individual concentrations that exceed the calibration curve of the analytical method are marked with a "c" to indicate that the uncertainty in these concentrations is greater than for concentrations within the calibration range.*

Source: *U.S. Geological Survey, Pesticides in Streams: Preliminary Results from Cycle I of the National Water Quality Assessment Program (NAWQA), 1992-2001, February 26, 2003*

Pesticides in streams at undeveloped land use sites, 1991-2001

Compound	Max Rep. Limit[1] (ug/L)	No. of Sites	No. of Samp.	Frequency of Detection (%)				Percentiles of Concentration (ug/L)					
				All	>=0.01 ug/L	>=0.1 ug/L	>=1 ug/L	25th	50th	75th	90th	95th	Max
Analyzed by GC/MS													
Acetochlor[a]	0.006	4	60	0.00	0.00	0.00	0.00	<0.004	<0.004	<0.004	<0.004	<0.004	<0.004
Alachlor	0.005	4	60	12.12	2.64	0.00	0.00	<0.002	<0.002	<0.002	<0.002	0.008	0.010
Atrazine	0.007	4	59	60.23	25.85	0.00	0.00	0.001	0.004	0.010	0.030	0.036	0.085
Azinphos-methyl[e]	0.050	3	47	2.65	>2.65[b]	0.00	0.00	<0.050	<0.050	<0.050	<0.050	<0.050	0.014
Benfluralin	0.010	4	60	0.00	0.00	0.00	0.00	<0.010	<0.010	<0.010	<0.010	<0.010	<0.010
Butylate	0.002	4	60	0.00	0.00	0.00	0.00	<0.002	<0.002	<0.002	<0.002	<0.002	<0.002
Carbaryl[e]	0.041	4	60	0.00	>0.00[b]	0.00	0.00	<0.041	<0.041	<0.041	<0.041	<0.041	<0.041
Carbofuran[e]	0.020	4	60	3.97	>3.97[b]	0.00	0.00	<0.020	<0.020	<0.020	<0.020	<0.020	0.034
Chlorpyrifos	0.005	4	60	1.99	0.00	0.00	0.00	<0.005	<0.005	<0.005	<0.005	<0.005	0.008
Cyanazine	0.018	4	60	2.64	>0.00[b]	0.00	0.00	<0.018	<0.018	<0.018	<0.018	<0.018	0.004
p,p'-DDE	0.006	4	60	3.66	0.00	0.00	0.00	<0.006	<0.006	<0.006	<0.006	<0.006	0.002
Dacthal	0.003	4	60	6.34	0.00	0.00	0.00	<0.003	<0.003	<0.003	<0.003	<0.003	0.003
Deethylatrazine[e]	0.006	4	60	44.56	5.75	0.00	0.00	0.001	0.002	0.004	0.005	0.010	0.025
Diazinon	0.005	4	60	4.49	0.00	0.00	0.00	<0.005	<0.005	<0.005	<0.005	<0.005	0.005
Dieldrin	0.005	4	60	0.00	0.00	0.00	0.00	<0.005	<0.005	<0.005	<0.005	<0.005	<0.005
2,6-Diethylaniline	0.006	4	60	0.00	0.00	0.00	0.00	<0.003	<0.003	<0.003	<0.003	<0.003	<0.003
Disulfoton	0.021	4	60	0.00	>0.00[b]	0.00	0.00	<0.021	<0.021	<0.021	<0.021	<0.021	<0.021
EPTC	0.002	4	60	1.64	0.00	0.00	0.00	<0.002	<0.002	<0.002	<0.002	<0.002	0.004
Ethalfluralin	0.009	4	60	0.00	0.00	0.00	0.00	<0.009	<0.009	<0.009	<0.009	<0.009	<0.009
Ethoprop	0.005	4	60	1.40	1.40	0.00	0.00	<0.005	<0.005	<0.005	<0.005	<0.005	0.019
Fonofos	0.003	4	60	0.00	0.00	0.00	0.00	<0.003	<0.003	<0.003	<0.003	<0.003	<0.003
alpha-HCH	0.005	4	60	0.00	0.00	0.00	0.00	<0.005	<0.005	<0.005	<0.005	<0.005	<0.005
gamma-HCH	0.004	4	60	0.00	0.00	0.00	0.00	<0.004	<0.004	<0.004	<0.004	<0.004	<0.004
Linuron	0.035	4	60	0.00	>0.00[b]	0.00	0.00	<0.035	<0.035	<0.035	<0.035	<0.035	<0.035
Malathion	0.027	4	60	6.90	>0.00[b]	0.00	0.00	<0.027	<0.027	<0.027	<0.027	<0.027	0.006
Metolachlor	0.013	4	60	29.11	>5.65[b]	0.00	0.00	<0.013	<0.013	<0.013	<0.013	<0.013	0.027
Metribuzin	0.006	4	60	0.00	0.00	0.00	0.00	<0.006	<0.006	<0.006	<0.006	<0.006	<0.006
Molinate	0.004	4	60	1.21	0.00	0.00	0.00	<0.004	<0.004	<0.004	<0.004	<0.004	0.005
Napropamide	0.007	4	60	1.20	0.00	0.00	0.00	<0.007	<0.007	<0.007	<0.007	<0.007	0.006
Parathion	0.010	4	60	0.00	0.00	0.00	0.00	<0.007	<0.007	<0.007	<0.007	<0.007	<0.007
Parathion-methyl	0.006	4	60	0.00	0.00	0.00	0.00	<0.006	<0.006	<0.006	<0.006	<0.006	<0.006
Pebulate	0.004	4	60	1.40	1.40	0.00	0.00	<0.004	<0.004	<0.004	<0.004	<0.004	0.085
Pendimethalin	0.022	4	60	0.00	>0.00[b]	0.00	0.00	<0.010	<0.010	<0.010	<0.010	<0.010	<0.010
cis-Permethrin	0.006	4	60	0.00	>0.00[b]	0.00	0.00	<0.006	<0.006	<0.006	<0.006	<0.006	<0.006
Phorate	0.011	4	60	0.00	>0.00[b]	0.00	0.00	<0.011	<0.011	<0.011	<0.011	<0.011	<0.011
Prometon	0.018	4	60	8.45	>0.00[b]	0.00	0.00	<0.018	<0.018	<0.018	<0.018	<0.018	0.002
Pronamide	0.004	3	48	0.00	0.00	0.00	0.00	<0.003	<0.003	<0.003	<0.003	<0.003	<0.003
Propachlor	0.010	4	60	0.00	0.00	0.00	0.00	<0.010	<0.010	<0.010	<0.010	<0.010	<0.010
Propanil	0.011	4	60	0.00	>0.00[b]	0.00	0.00	<0.011	<0.011	<0.011	<0.011	<0.011	<0.011
Propargite	0.023	4	60	0.00	>0.00[b]	0.00	0.00	<0.023	<0.023	<0.023	<0.023	<0.023	<0.023
Simazine	0.011	4	60	13.31	>3.11[b]	0.00	0.00	<0.011	<0.011	<0.011	<0.011	<0.011	0.024
Tebuthiuron	0.016	3	44	10.68	>7.71[b]	0.00	0.00	<0.016	<0.016	<0.016	<0.016	<0.016	0.087
Terbacil[e]	0.034	4	60	1.40	>1.40[b]	0.00	0.00	<0.034	<0.034	<0.034	<0.034	<0.034	0.092
Terbufos	0.017	4	60	0.00	>0.00[b]	0.00	0.00	<0.017	<0.017	<0.017	<0.017	<0.017	<0.017
Thiobencarb	0.005	4	60	0.00	0.00	0.00	0.00	<0.005	<0.005	<0.005	<0.005	<0.005	<0.005
Triallate	0.002	3	48	0.00	0.00	0.00	0.00	<0.001	<0.001	<0.001	<0.001	<0.001	<0.001
Trifluralin	0.009	4	60	0.00	0.00	0.00	0.00	<0.009	<0.009	<0.009	<0.009	<0.009	<0.009
Analyzed by HPLC													
Acifluorfen	0.090	1	19	0.00	(d)	0.00	0.00	<0.035	<0.035	<0.035	<0.035	<0.035	<0.035
Aldicarb[e]	0.550	1	19	0.00	(d)	>0.00[b]	0.00	<0.550	<0.550	<0.550	<0.550	<0.550	<0.550
Aldicarb sulfone[e]	0.200	1	19	0.00	(d)	>0.00[b]	0.00	<0.100	<0.100	<0.100	<0.100	<0.100	<0.100
Aldicarb sulfoxide[e]	0.270	1	19	0.00	(d)	>0.00[b]	0.00	<0.021	<0.021	<0.021	<0.021	<0.021	<0.021
Bentazon	0.050	1	19	0.00	(d)	0.00	0.00	<0.014	<0.014	<0.014	<0.014	<0.014	<0.014
Bromacil	0.090	1	19	0.00	(d)	0.00	0.00	<0.035	<0.035	<0.035	<0.035	<0.035	<0.035
Bromoxynil	0.070	1	19	0.00	(d)	0.00	0.00	<0.035	<0.035	<0.035	<0.035	<0.035	<0.035

Pesticides in streams at undeveloped land use sites, 1991-2001 - *continued*

Compound	Max Rep. Limit[1] (ug/L)	No. of Sites	No. of Samp.	Frequency of Detection (%)				Percentiles of Concentration (ug/L)					
				All	>=0.01 ug/L	>=0.1 ug/L	>=1 ug/L	25th	50th	75th	90th	95th	Max
Chloramben methyl ester	0.420	1	19	0.00	(d)	>0.00[b]	0.00	<0.420	<0.420	<0.420	<0.420	<0.420	<0.420
Chlorothalonil[e]	0.480	1	19	0.00	(d)	>0.00[b]	0.00	<0.480	<0.480	<0.480	<0.480	<0.480	<0.480
Clopyralid	0.420	1	19	0.00	(d)	>0.00[b]	0.00	<0.230	<0.230	<0.230	<0.230	<0.230	<0.230
2,4-D	0.160	1	19	0.00	(d)	>0.00[b]	0.00	<0.150	<0.150	<0.150	<0.150	<0.150	<0.150
Dacthal monoacid	0.070	1	19	0.00	(d)	0.00	0.00	<0.017	<0.017	<0.017	<0.017	<0.017	<0.017
2,4-DB	0.250	1	19	0.00	(d)	>0.00[b]	0.00	<0.240	<0.240	<0.240	<0.240	<0.240	<0.240
Dicamba	0.110	1	19	0.00	(d)	>0.00[b]	0.00	<0.035	<0.035	<0.035	<0.035	<0.035	<0.035
Dichlobenil[e]	1.20	1	19	0.00	(d)	>0.00[b]	>0.00[b]	<1.20	<1.20	<1.20	<1.20	<1.20	<1.20
Dichlorprop	0.120	1	19	0.00	(d)	>0.00[b]	0.00	<0.032	<0.032	<0.032	<0.032	<0.032	<0.032
Dinoseb	0.090	1	19	0.00	(d)	0.00	0.00	<0.035	<0.035	<0.035	<0.035	<0.035	<0.035
Diuron	0.120	1	19	9.07	(d)	>0.00[b]	0.00	<0.020	<0.020	<0.020	<0.020	<0.020	0.002
DNOC[e]	0.420	1	19	0.00	(d)	>0.00[b]	0.00	<0.420	<0.420	<0.420	<0.420	<0.420	<0.420
Fenuron	0.070	1	19	0.00	(d)	0.00	0.00	<0.013	<0.013	<0.013	<0.013	<0.013	<0.013
Fluometuron	0.060	1	19	9.07	(d)	0.00	0.00	<0.035	<0.035	<0.035	<0.035	0.080	0.080
3-Hydroxycarbofuran	0.110	1	19	0.00	(d)	>0.00[b]	0.00	<0.014	<0.014	<0.014	<0.014	<0.014	<0.014
MCPA	0.200	1	19	5.62	(d)	>0.00[b]	0.00	<0.170	<0.170	<0.170	<0.170	<0.170	0.020
MCPB	0.260	1	19	0.00	(d)	>0.00[b]	0.00	<0.140	<0.140	<0.140	<0.140	<0.140	<0.140
Methiocarb	0.070	1	19	0.00	(d)	0.00	0.00	<0.026	<0.026	<0.026	<0.026	<0.026	<0.026
Methomyl	0.470	1	19	0.00	(d)	>0.00[b]	0.00	<0.017	<0.017	<0.017	<0.017	<0.017	<0.017
Neburon	0.070	1	19	0.00	(d)	0.00	0.00	<0.015	<0.015	<0.015	<0.015	<0.015	<0.015
Norflurazon	0.042	1	19	9.07	(d)	9.07	0.00	<0.024	<0.024	<0.024	<0.024	0.170	0.170
Oryzalin	0.310	1	19	0.00	(d)	>0.00[b]	0.00	<0.310	<0.310	<0.310	<0.310	<0.310	<0.310
Oxamyl	0.160	1	19	0.00	(d)	>0.00[b]	0.00	<0.018	<0.018	<0.018	<0.018	<0.018	<0.018
Picloram	0.090	1	19	0.00	(d)	0.00	0.00	<0.050	<0.050	<0.050	<0.050	<0.050	<0.050
Propham	0.220	1	19	0.00	(d)	>0.00[b]	0.00	<0.035	<0.035	<0.035	<0.035	<0.035	<0.035
Propoxur	0.120	1	19	0.00	(d)	>0.00[b]	0.00	<0.035	<0.035	<0.035	<0.035	<0.035	<0.035
Silvex	0.060	1	19	0.00	(d)	0.00	0.00	<0.021	<0.021	<0.021	<0.021	<0.021	<0.021
2,4,5-T	0.070	1	19	0.00	(d)	0.00	0.00	<0.035	<0.035	<0.035	<0.035	<0.035	<0.035
Triclopyr	0.250	1	19	0.00	(d)	>0.00[b]	0.00	<0.250	<0.250	<0.250	<0.250	<0.250	<0.250

Notes: *(1) The analytical reporting limit has changed through time. The limit reported in this table is the maximum value routinely used to report nondetections during the period 1991-2001; (a) Acetochlor was not used or analyzed for prior to 1994; (b) The analytical reporting limit for the compound is greater than the threshold concentration used for computing detection frequency. Therefore, the reported percentage may be an underestimate of the actual percentage of time-weighted concentrations that are greater than the threshold concentration; (c) See "Estimated Concentrations" below; (d) The 0.01 ug/L threshold is not applicable for the HPLC method because the reporting limit for most compounds is much higher than the threshold; (e) all concentrations determined for this compound are estimated concentrations.*

Abbreviations: *ug/L, microgram per liter; ">=", greater than or equal to; GC/MS, gas chromatography/mass spectrometry; HPLC, high performance liquid chromatography; NWQL, U.S. Geological Survey National Water Quality Laboratory.*

Frequency of Detection: *Within each statistical summary, four detection frequencies are provided (all detections, >=0.01 ug/L, >=0.1 ug/L, and >=1 ug/L). These detection frequencies provide a measure of "how often" a compound was found. The "all detections" provide the total number of detections for a given compound, but are not comparable among compounds because detection capabilities vary. Because reporting levels varied, detection frequencies were calculated using three common detection thresholds (0.01, 0.1 and 1 ug/L). The use of these detection thresholds facilitates comparisons among compounds by censoring detections to a common reference concentration. Adjustments of this type are essential in order to answer questions like "is compound x found more often than compound y?"*

Percentiles of concentration: *Annual time-weighted concentrations measured for each pesticide are summarized using percentiles. The 25th, 50th, 75th, 90th, and 95th percentiles of concentration by land-use category are provided. Percentiles provide information about the magnitude and duration of concentrations at selected points in the cumulative frequency distribution of the ranked, time-weighted concentrations. For example, concentrations of acetochlor at NAWQA sites draining agricultural land use were less than 0.005 ug/L for 25 percent of the year, less than 0.005 ug/L for 50 percent of the year, less than or equal to 0.009 ug/L for 75 percent of the year, less than or equal to 0.052 ug/L for 90 percent of the year, and less than or equal to 0.165 ug/L for 95 percent of the year. If percentiles are not censored (reported as less than a specified concentration—for example, <0.005 ug/L), the percentiles also may be interpreted as the percentage of the year where concentrations were greater than a given concentration. For example, concentrations of acetochlor at NAWQA sites draining agricultural land use were greater than or equal to 0.009 ug/L for 25 percent of the year, greater than or equal to 0.052 ug/L for 10 percent of the year, and greater than or equal to 0.165 ug/L for 5 percent of the year.*

Maximum Concentrations: *The concentrations in this column are the maximum measured concentrations in the pooled annual data sets for each land-use category. Concentrations greater than the value reported as maximum may have been measured at sites but are not reported because they were not measured during the one-year periods selected to describe the annual distribution of concentrations. In addition, the probability is low that a water sample was collected at the time of the annual maximum concentration for any pesticide. Consequently, the maximum concentrations reported should be interpreted as a lower bound for the true maximum concentrations.*

Estimated concentrations: *All pesticides denoted with "e" have more biased and/or variable analytical performance than the other pesticides in the method. Information on analytical performance is provided in the method reports (Zaugg and others, 1995; Werner and others, 1996) and is summarized for the period 1992-96 in Martin (1999). Individual concentrations that exceed the calibration curve of the analytical method are marked with a "c" to indicate that the uncertainty in these concentrations is greater than for concentrations within the calibration range.*

Source: *U.S. Geological Survey, Pesticides in Streams: Preliminary Results from Cycle I of the National Water Quality Assessment Program (NAWQA), 1992-2001, February 26, 2003*

Pesticides in streams at urban land use sites, 1991-2001

Compound	Max Rep. Limit[1] (ug/L)	No. of Sites	No. of Samp.	Frequency of Detection (%)				Percentiles of Concentration (ug/L)					
				All	>=0.01 ug/L	>=0.1 ug/L	>=1 ug/L	25th	50th	75th	90th	95th	Max
Analyzed by GC/MS													
Acetochlor[a]	0.006	24	640	4.94	3.59	0.65	0.00	<0.004	<0.004	<0.004	<0.004	<0.004	0.311
Alachlor	0.005	33	897	6.90	1.74	0.06	0.00	<0.005	<0.005	<0.005	<0.005	<0.005	0.400
Atrazine	0.007	32	887	74.36	53.45	11.93	1.38	0.003	0.011	0.029	0.119	0.203	3.37
Azinphos-methyl[e]	0.050	32	870	0.35	>0.35[b]	0.23	0.00	<0.050	<0.050	<0.050	<0.050	<0.050	0.171
Benfluralin	0.010	33	904	3.29	0.48	0.00	0.00	<0.010	<0.010	<0.010	<0.010	<0.010	0.022
Butylate	0.002	33	902	0.00	0.00	0.00	0.00	<0.002	<0.002	<0.002	<0.002	<0.002	<0.002
Carbaryl[e]	0.041	33	897	43.79	>35.93[b]	11.83	1.02	<0.001	0.004	0.025	0.134	0.269	5.20
Carbofuran[e]	0.020	33	901	2.12	>1.80[b]	0.00	0.00	<0.020	<0.020	<0.020	<0.020	<0.020	0.062
Chlorpyrifos	0.005	33	882	26.94	13.56	0.59	0.00	<0.005	<0.005	<0.005	0.013	0.026	0.300
Cyanazine	0.018	33	899	3.19	>2.15[b]	0.12	0.00	<0.018	<0.018	<0.018	<0.018	<0.018	0.362
p,p'-DDE	0.006	33	900	1.68	0.00	0.00	0.00	<0.006	<0.006	<0.006	<0.006	<0.006	0.007
Dacthal	0.003	33	902	21.78	3.70	0.00	0.00	<0.003	<0.003	<0.003	0.003	0.007	0.045
Deethylatrazine[e]	0.006	32	871	51.67	20.55	1.44	0.00	0.001	0.003	0.008	0.018	0.043	0.283
Diazinon	0.005	33	870	65.03	54.09	10.26	0.17	0.003	0.013	0.038	0.104	0.190	1.40
Dieldrin	0.005	33	900	3.48	3.15	0.00	0.00	<0.005	<0.005	<0.005	<0.005	<0.005	0.077
2,6-Diethylaniline	0.006	33	902	0.27	0.00	0.00	0.00	<0.003	<0.003	<0.003	<0.003	<0.003	0.006
Disulfoton	0.021	33	904	0.14	>0.03[b]	0.00	0.00	<0.021	<0.021	<0.021	<0.021	<0.021	0.018
EPTC	0.002	33	892	4.81	1.59	0.00	0.00	<0.002	<0.002	<0.002	<0.002	<0.002	0.038
Ethalfluralin	0.009	33	902	0.00	0.00	0.00	0.00	<0.009	<0.009	<0.009	<0.009	<0.009	<0.009
Ethoprop	0.005	33	898	0.80	0.32	0.05	0.00	<0.005	<0.005	<0.005	<0.005	<0.005	0.124
Fonofos	0.003	33	900	0.92	0.51	0.00	0.00	<0.003	<0.003	<0.003	<0.003	<0.003	0.084
alpha-HCH	0.005	32	878	0.11	0.11	0.00	0.00	<0.005	<0.005	<0.005	<0.005	<0.005	0.014
gamma-HCH	0.004	32	875	1.78	0.53	0.00	0.00	<0.004	<0.004	<0.004	<0.004	<0.004	0.048
Linuron	0.035	33	900	0.17	>0.17[b]	0.05	0.00	<0.035	<0.035	<0.035	<0.035	<0.035	0.250
Malathion	0.027	33	903	14.13	>11.46[b]	1.62	0.00	<0.027	<0.027	<0.027	<0.027	0.041	0.634
Metolachlor	0.013	32	885	49.74	>25.13[b]	3.43	0.70	<0.001	0.003	0.010	0.027	0.056	2.42
Metribuzin	0.006	32	846	4.25	3.51	0.79	0.00	<0.006	<0.006	<0.006	<0.006	<0.006	0.358
Molinate	0.004	32	883	0.77	0.66	0.00	0.00	<0.004	<0.004	<0.004	<0.004	<0.004	0.096
Napropamide	0.007	32	876	0.77	0.50	0.00	0.00	<0.007	<0.007	<0.007	<0.007	<0.007	0.040
Parathion	0.010	33	902	0.25	0.25	0.00	0.00	<0.010	<0.010	<0.010	<0.010	<0.010	0.052
Parathion-methyl	0.006	33	901	0.17	0.17	0.00	0.00	<0.006	<0.006	<0.006	<0.006	<0.006	0.061
Pebulate	0.004	33	902	0.18	0.18	0.00	0.00	<0.004	<0.004	<0.004	<0.004	<0.004	0.023
Pendimethalin	0.022	33	897	8.57	>7.55[b]	1.15	0.00	<0.020	<0.020	<0.020	<0.020	<0.020	0.372
cis-Permethrin	0.006	33	903	0.00	0.00	0.00	0.00	<0.006	<0.006	<0.006	<0.006	<0.006	<0.006
Phorate	0.011	33	901	0.00	>0.00[b]	0.00	0.00	<0.011	<0.011	<0.011	<0.011	<0.011	<0.011
Prometon	0.018	32	877	83.99	>64.42[b]	8.88	0.43	0.009	0.018	0.040	0.091	0.148	25.1[c]
Pronamide	0.004	32	865	2.45	2.17	0.09	0.00	<0.003	<0.003	<0.003	<0.003	<0.003	0.131
Propachlor	0.010	33	902	1.00	0.06	0.00	0.00	<0.010	<0.010	<0.010	<0.010	<0.010	0.015
Propanil	0.011	33	903	1.30	>0.66[b]	0.00	0.00	<0.011	<0.011	<0.011	<0.011	<0.011	0.091
Propargite	0.023	32	864	0.18	>0.18[b]	0.00	0.00	<0.023	<0.023	<0.023	<0.023	<0.023	0.042
Simazine	0.011	33	903	63.56	>44.05[b]	11.62	3.24	0.003	0.008	0.029	0.119	0.410	9.03
Tebuthiuron	0.016	30	809	33.12	>26.78[b]	2.89	0.20	<0.016	<0.016	<0.016	0.037	0.072	2.83
Terbacil[e]	0.034	33	896	1.98	>1.72[b]	0.00	0.00	<0.034	<0.034	<0.034	<0.034	<0.034	0.035
Terbufos	0.017	33	902	0.08	>0.08[b]	0.00	0.00	<0.017	<0.017	<0.017	<0.017	<0.017	0.011
Thiobencarb	0.005	33	901	0.36	0.00	0.00	0.00	<0.005	<0.005	<0.005	<0.005	<0.005	0.009
Triallate	0.002	32	869	0.96	0.72	0.00	0.00	<0.001	<0.001	<0.001	<0.001	<0.001	0.036
Trifluralin	0.009	33	902	9.74	0.96	0.00	0.00	<0.009	<0.009	<0.009	<0.009	<0.009	0.037
Analyzed by HPLC													
Acifluorfen	0.090	18	503	0.45	(d)	0.00	0.00	<0.046	<0.046	<0.046	<0.046	<0.046	0.060
Aldicarb[e]	0.550	18	505	0.00	(d)	>0.00[b]	0.00	<0.550	<0.550	<0.550	<0.550	<0.550	<0.550
Aldicarb sulfone[e]	0.200	18	504	0.00	(d)	>0.00[b]	0.00	<0.110	<0.110	<0.110	<0.110	<0.110	<0.110
Aldicarb sulfoxide[e]	0.270	18	504	0.00	(d)	>0.00[b]	0.00	<0.100	<0.100	<0.100	<0.100	<0.100	<0.100
Bentazon	0.050	18	503	0.23	(d)	0.00	0.00	<0.050	<0.050	<0.050	<0.050	<0.050	0.050
Bromacil	0.090	18	504	1.73	(d)	0.96	0.36	<0.046	<0.046	<0.046	<0.046	<0.046	1.55[c]
Bromoxynil	0.070	18	503	0.12	(d)	0.00	0.00	<0.046	<0.046	<0.046	<0.046	<0.046	0.070

Pesticides in streams at urban land use sites, 1991-2001 - *continued*

Compound	Max Rep. Limit[1] (ug/L)	No. of Sites	No. of Samp.	Frequency of Detection (%)				Percentiles of Concentration (ug/L)					
				All	>=0.01 ug/L	>=0.1 ug/L	>=1 ug/L	25th	50th	75th	90th	95th	Max
Chloramben methyl ester	0.420	18	504	0.00	(d)	>0.00[b]	0.00	<0.420	<0.420	<0.420	<0.420	<0.420	<0.420
Chlorothalonil[e]	0.480	18	502	1.26	(d)	>0.00[b]	0.00	<0.480	<0.480	<0.480	<0.480	<0.480	0.080
Clopyralid	0.420	18	504	0.00	(d)	>0.00[b]	0.00	<0.230	<0.230	<0.230	<0.230	<0.230	<0.230
2,4-D	0.160	18	505	15.15	(d)	>12.35[b]	2.19	<0.150	<0.150	<0.150	0.220	0.470	5.53[c]
Dacthal monoacid	0.070	18	503	0.00	(d)	0.00	0.00	<0.050	<0.050	<0.050	<0.050	<0.050	<0.050
2,4-DB	0.250	18	505	0.09	(d)	>0.09[b]	0.00	<0.240	<0.240	<0.240	<0.240	<0.240	0.390
Dicamba	0.110	18	505	0.33	(d)	>0.12[b]	0.00	<0.100	<0.100	<0.100	<0.100	<0.100	0.120
Dichlobenil[e]	1.200	18	505	3.92	(d)	>0.73[b]	>0.11[b]	<1.20	<1.20	<1.20	<1.20	<1.20	1.200
Dichlorprop	0.120	18	505	0.98	(d)	>0.34[b]	0.00	<0.100	<0.100	<0.100	<0.100	<0.100	0.290
Dinoseb	0.090	18	503	0.00	(d)	0.00	0.00	<0.046	<0.046	<0.046	<0.046	<0.046	<0.046
Diuron	0.120	18	505	20.40	(d)	>11.15[b]	1.38	<0.100	<0.100	<0.100	0.120	0.300	11.0[c]
DNOC[e]	0.420	18	504	0.00	(d)	>0.00[b]	0.00	<0.420	<0.420	<0.420	<0.420	<0.420	<0.420
Fenuron	0.070	18	504	0.12	(d)	0.00	0.00	<0.050	<0.050	<0.050	<0.050	<0.050	0.050
Fluometuron	0.060	18	504	0.00	(d)	0.00	0.00	<0.046	<0.046	<0.046	<0.046	<0.046	<0.046
3-Hydroxycarbofuran	0.110	18	505	0.00	(d)	>0.00[b]	0.00	<0.100	<0.100	<0.100	<0.100	<0.100	<0.100
MCPA	0.200	18	505	2.56	(d)	>2.27[b]	0.09	<0.170	<0.170	<0.170	<0.170	<0.170	1.30
MCPB	0.260	18	505	0.00	(d)	>0.00[b]	0.00	<0.140	<0.140	<0.140	<0.140	<0.140	<0.140
Methiocarb	0.070	18	504	0.00	(d)	0.00	0.00	<0.034	<0.034	<0.034	<0.034	<0.034	<0.034
Methomyl	0.470	18	504	0.00	(d)	>0.00[b]	0.00	<0.100	<0.100	<0.100	<0.100	<0.100	<0.100
Neburon	0.070	18	504	0.12	(d)	0.00	0.00	<0.050	<0.050	<0.050	<0.050	<0.050	0.040
Norflurazon	0.042	18	504	0.00	(d)	0.00	0.00	<0.031	<0.031	<0.031	<0.031	<0.031	<0.031
Oryzalin	0.310	18	504	2.84	(d)	>1.72[b]	0.43	<0.310	<0.310	<0.310	<0.310	<0.310	1.90[c]
Oxamyl	0.160	18	504	0.00	(d)	>0.00[b]	0.00	<0.100	<0.100	<0.100	<0.100	<0.100	<0.100
Picloram	0.090	18	502	0.00	(d)	0.00	0.00	<0.066	<0.066	<0.066	<0.066	<0.066	<0.066
Propham	0.220	18	505	0.51	(d)	>0.00[b]	0.00	<0.100	<0.100	<0.100	<0.100	<0.100	0.070
Propoxur	0.120	18	500	0.21	(d)	>0.21[b]	0.00	<0.046	<0.046	<0.046	<0.046	<0.046	0.260
Silvex	0.060	18	503	0.00	(d)	0.00	0.00	<0.028	<0.028	<0.028	<0.028	<0.028	<0.028
2,4,5-T	0.070	18	503	0.00	(d)	0.00	0.00	<0.046	<0.046	<0.046	<0.046	<0.046	<0.046
Triclopyr	0.250	18	505	3.60	(d)	>2.63[b]	0.64	<0.250	<0.250	<0.250	<0.250	<0.250	3.35[c]

Notes: *(1) The analytical reporting limit has changed through time. The limit reported in this table is the maximum value routinely used to report nondetections during the period 1991-2001; (a) Acetochlor was not used or analyzed for prior to 1994; (b) The analytical reporting limit for the compound is greater than the threshold concentration used for computing detection frequency. Therefore, the reported percentage may be an underestimate of the actual percentage of time-weighted concentrations that are greater than the threshold concentration; (c) See "Estimated Concentrations" below; (d) The 0.01 ug/L threshold is not applicable for the HPLC method because the reporting limit for most compounds is much higher than the threshold; (e) all concentrations determined for this compound are estimated concentrations.*

Abbreviations: *ug/L, microgram per liter; ">=", greater than or equal to; GC/MS, gas chromatography/mass spectrometry; HPLC, high performance liquid chromatography; NWQL, U.S. Geological Survey National Water Quality Laboratory.*

Frequency of Detection: *Within each statistical summary, four detection frequencies are provided (all detections, >=0.01 ug/L, >=0.1 ug/L, and >=1 ug/L). These detection frequencies provide a measure of "how often" a compound was found. The "all detections" provide the total number of detections for a given compound, but are not comparable among compounds because detection capabilities vary. Because reporting levels varied, detection frequencies were calculated using three common detection thresholds (0.01, 0.1 and 1 ug/L). The use of these detection thresholds facilitates comparisons among compounds by censoring detections to a common reference concentration. Adjustments of this type are essential in order to answer questions like "is compound x found more often than compound y?"*

Percentiles of concentration: *Annual time-weighted concentrations measured for each pesticide are summarized using percentiles. The 25th, 50th, 75th, 90th, and 95th percentiles of concentration by land-use category are provided. Percentiles provide information about the magnitude and duration of concentrations at selected points in the cumulative frequency distribution of the ranked, time-weighted concentrations. For example, concentrations of acetochlor at NAWQA sites draining agricultural land use were less than 0.005 ug/L for 25 percent of the year, less than 0.005 ug/L for 50 percent of the year, less than or equal to 0.009 ug/L for 75 percent of the year, less than or equal to 0.052 ug/L for 90 percent of the year, and less than or equal to 0.165 ug/L for 95 percent of the year. If percentiles are not censored (reported as less than a specified concentration—for example, <0.005 ug/L), the percentiles also may be interpreted as the percentage of the year where concentrations were greater than a given concentration. For example, concentrations of acetochlor at NAWQA sites draining agricultural land use were greater than or equal to 0.009 ug/L for 25 percent of the year, greater than or equal to 0.052 ug/L for 10 percent of the year, and greater than or equal to 0.165 ug/L for 5 percent of the year.*

Maximum Concentrations: *The concentrations in this column are the maximum measured concentrations in the pooled annual data sets for each land-use category. Concentrations greater than the value reported as maximum may have been measured at sites but are not reported because they were not measured during the one-year periods selected to describe the annual distribution of concentrations. In addition, the probability is low that a water sample was collected at the time of the annual maximum concentration for any pesticide. Consequently, the maximum concentrations reported should be interpreted as a lower bound for the true maximum concentrations.*

Estimated concentrations: *All pesticides denoted with "e" have more biased and/or variable analytical performance than the other pesticides in the method. Information on analytical performance is provided in the method reports (Zaugg and others, 1995; Werner and others, 1996) and is summarized for the period 1992-96 in Martin (1999). Individual concentrations that exceed the calibration curve of the analytical method are marked with a "c" to indicate that the uncertainty in these concentrations is greater than for concentrations within the calibration range.*

Source: *U.S. Geological Survey, Pesticides in Streams: Preliminary Results from Cycle I of the National Water Quality Assessment Program (NAWQA), 1992-2001, February 26, 2003*

Pesticides in ground water from agricultural land use wells, 1991-2001

Compound	Max Rep. Limit[1] (ug/L)	Wells	Frequency of Detection (%)				Percentiles of Concentration (ug/L)					
			All	>=0.01 ug/L	>=0.1 ug/L	>=1 ug/L	25th	50th	75th	90th	95th	Max
Analyzed by GC/MS												
Acetochlor[a]	0.006	900[a]	0.56	0.33	0.0	0.0	<RL	<RL	<RL	<RL	<RL	0.057
Alachlor	0.005	1443	2.70	1.46	0.28	0.0	<RL	<RL	<RL	<RL	<RL	0.946
Atrazine	0.007	1438	41.9	29.0	13.6	1.53	<RL	<RL	0.017	0.169	0.36	4.78
Azinphos-methyl[e]	0.05	1438	0.35	>0.29[b]	0.07	0.0	<RL	<RL	<RL	<RL	<RL	0.18[c]
Benfluralin	0.01	1443	0.14	0.0	0.0	0.0	<RL	<RL	<RL	<RL	<RL	0.004
Butylate	0.002	1443	0.28	0.0	0.0	0.0	<RL	<RL	<RL	<RL	<RL	0.003
Carbaryl[e]	0.041	1443	0.35	>0.14[b]	0.0	0.0	<RL	<RL	<RL	<RL	<RL	0.019[c]
Carbofuran[e]	0.02	1443	1.59	>1.31[b]	0.21	0.07	<RL	<RL	<RL	<RL	<RL	1.3[c]
Chlorpyrifos	0.005	1443	0.83	0.14	0.0	0.0	<RL	<RL	<RL	<RL	<RL	0.021
Cyanazine	0.018	1443	1.25	>0.87[b]	0.14	0.0	<RL	<RL	<RL	<RL	<RL	0.16
p,p'-DDE	0.006	1443	3.26	0.0	0.0	0.0	<RL	<RL	<RL	<RL	<RL	0.008
Dacthal	0.003	1443	1.18	0.14	0.07	0.07	<RL	<RL	<RL	<RL	<RL	10[c]
Deethylatrazine[e]	0.006	1438	42.0	26.7	10.9	0.63	<RL	<RL	0.012	0.12	0.27	2.6[c]
Diazinon	0.005	1443	0.48	0.28	0.0	0.0	<RL	<RL	<RL	<RL	<RL	0.077
Dieldrin	0.005	1438	0.90	0.49	0.0	0.0	<RL	<RL	<RL	<RL	<RL	0.057
2,6-Diethylaniline	0.006	1442	1.25	0.42	0.0	0.0	<RL	<RL	<RL	<RL	<RL	0.049
Disulfoton	0.021	1443	0.0	>0.0[b]	0.0	0.0	<RL	<RL	<RL	<RL	<RL	<RL
EPTC	0.002	1443	1.11	0.49	0.07	0.0	<RL	<RL	<RL	<RL	<RL	0.45
Ethalfluralin	0.009	1443	0.28	0.07	0.0	0.0	<RL	<RL	<RL	<RL	<RL	0.09
Ethoprop	0.005	1443	0.07	0.0	0.0	0.0	<RL	<RL	<RL	<RL	<RL	0.009
Fonofos	0.003	1443	0.07	0.0	0.0	0.0	<RL	<RL	<RL	<RL	<RL	0.009
alpha-HCH	0.005	1443	0.07	0.07	0.0	0.0	<RL	<RL	<RL	<RL	<RL	0.059
gamma-HCH	0.004	1438	0.0	0.0	0.0	0.0	<RL	<RL	<RL	<RL	<RL	<RL
Linuron	0.035	1443	0.21	>0.21[b]	0.0	0.0	<RL	<RL	<RL	<RL	<RL	0.029
Malathion	0.027	1441	0.42	>0.07[b]	0.07	0.0	<RL	<RL	<RL	<RL	<RL	0.1
Metolachlor	0.013	1443	17.0	>8.16[b]	2.91	0.55	<RL	<RL	<RL	<RL	0.022	32.8[c]
Metribuzin	0.006	1443	3.19	2.50	0.35	0.0	<RL	<RL	<RL	<RL	<RL	0.763
Molinate	0.004	1443	0.55	0.14	0.0	0.0	<RL	<RL	<RL	<RL	<RL	0.056
Napropamide	0.007	1443	0.48	0.42	0.0	0.0	<RL	<RL	<RL	<RL	<RL	0.043
Parathion	0.01	1443	0.0	0.0	0.0	0.0	<RL	<RL	<RL	<RL	<RL	<RL
Parathion-methyl	0.006	1443	0.14	0.14	0.0	0.0	<RL	<RL	<RL	<RL	<RL	0.062
Pebulate	0.004	1443	0.21	0.07	0.0	0.0	<RL	<RL	<RL	<RL	<RL	0.052
Pendimethalin	0.022	1443	0.28	>0.21[b]	0.0	0.0	<RL	<RL	<RL	<RL	<RL	0.059
cis-Permethrin	0.006	1443	0.14	0.0	0.0	0.0	<RL	<RL	<RL	<RL	<RL	0.007
Phorate	0.011	1443	0.0	>0.0[b]	0.0	0.0	<RL	<RL	<RL	<RL	<RL	<RL
Prometon	0.018	1443	9.98	>6.44[b]	1.66	0.21	<RL	<RL	<RL	<RL	<RL	40[c]
Pronamide	0.004	1443	0.14	0.14	0.0	0.0	<RL	<RL	<RL	<RL	<RL	0.052
Propachlor	0.01	1443	0.21	0.0	0.0	0.0	<RL	<RL	<RL	<RL	<RL	0.004
Propanil	0.011	1443	0.48	>0.07[b]	0.0	0.0	<RL	<RL	<RL	<RL	<RL	0.015
Propargite	0.023	1443	0.07	>0.0[b]	0.0	0.0	<RL	<RL	<RL	<RL	<RL	0.009
Simazine	0.011	1443	18.4	>10.7[b]	1.52	0.14	<RL	<RL	<RL	<RL	0.027	1.38
Tebuthiuron	0.016	1438	2.02	>1.45[b]	0.21	0.07	<RL	<RL	<RL	<RL	<RL	1.9
Terbacil[e]	0.034	1438	0.76	>0.51[b]	0.28	0.0	<RL	<RL	<RL	<RL	<RL	0.495[c]
Terbufos	0.017	1443	0.07	>0.0[b]	0.0	0.0	<RL	<RL	<RL	<RL	<RL	0.008
Thiobencarb	0.005	1443	0.21	0.14	0.0	0.0	<RL	<RL	<RL	<RL	<RL	0.025
Triallate	0.002	1443	0.21	0.0	0.0	0.0	<RL	<RL	<RL	<RL	<RL	0.002
Trifluralin	0.009	1443	0.69	0.07	0.0	0.0	<RL	<RL	<RL	<RL	<RL	0.014
Analyzed by HPLC												
Acifluorfen	0.09	1223	0.0	(d)	0.0	0.0	<RL	<RL	<RL	<RL	<RL	<RL
Aldicarb[e]	0.55	1226	0.08	(d)	>0.0[b]	0.0	<RL	<RL	<RL	<RL	<RL	0.01[c]
Aldicarb sulfone[e]	0.20	1212	0.08	(d)	>0.08[b]	0.0	<RL	<RL	<RL	<RL	<RL	0.12[c]
Aldicarb sulfoxide[e]	0.27	1212	0.25	(d)	>0.25[b]	0.08	<RL	<RL	<RL	<RL	<RL	1.8[c]
Bentazon	0.05	1217	3.94	(d)	2.63	0.90	<RL	<RL	<RL	<RL	<RL	11.46[c]
Bromacil	0.09	1226	2.53	(d)	2.28	1.14	<RL	<RL	<RL	<RL	<RL	21.9[c]
Bromoxynil	0.07	1219	0.08	(d)	0.0	0.0	<RL	<RL	<RL	<RL	<RL	0.02

Pesticides in ground water from agricultural land use wells, 1991-2001 - *continued*

Compound	Max Rep. Limit[1] (ug/L)	Wells	Frequency of Detection (%)				Percentiles of Concentration (ug/L)					
			All	>=0.01 ug/L	>=0.1 ug/L	>=1 ug/L	25th	50th	75th	90th	95th	Max
Chloramben methyl ester	0.42	1225	0.0	(d)	>0.0[b]	0.0	<RL	<RL	<RL	<RL	<RL	<RL
Chlorothalonil[e]	0.48	1225	0.0	(d)	>0.0[b]	0.0	<RL	<RL	<RL	<RL	<RL	<RL
Clopyralid	0.42	1216	0.0	(d)	>0.0[b]	0.0	<RL	<RL	<RL	<RL	<RL	<RL
2,4-D	0.16	1218	0.58	(d)	>0.25[b]	0.08	<RL	<RL	<RL	<RL	<RL	4.54[c]
Dacthal monoacid	0.07	1217	0.08	(d)	0.08	0.08	<RL	<RL	<RL	<RL	<RL	1.1
2,4-DB	0.25	1222	0.08	(d)	>0.0[b]	0.0	<RL	<RL	<RL	<RL	<RL	0.06
Dicamba	0.11	1218	0.41	(d)	>0.25[b]	0.0	<RL	<RL	<RL	<RL	<RL	0.45[c]
Dichlobenil[e]	1.2	1224	0.24	(d)	>0.16[b]	>0.0[b]	<RL	<RL	<RL	<RL	<RL	0.13[c]
Dichlorprop	0.12	1222	0.24	(d)	>0.16[b]	0.0	<RL	<RL	<RL	<RL	<RL	0.40
Dinoseb	0.09	1223	0.82	(d)	0.41	0.24	<RL	<RL	<RL	<RL	<RL	40[c]
Diuron	0.12	1224	3.84	(d)	>1.22[b]	0.24	<RL	<RL	<RL	<RL	<RL	5.53[c]
DNOC[e]	0.42	1219	0.0	(d)	>0.0[b]	0.0	<RL	<RL	<RL	<RL	<RL	<RL
Fenuron	0.07	1225	0.16	(d)	0.0	0.0	<RL	<RL	<RL	<RL	<RL	0.06
Fluometuron	0.06	1226	0.41	(d)	0.24	0.0	<RL	<RL	<RL	<RL	<RL	0.43
3-Hydroxycarbofuran	0.11	1225	0.0	(d)	>0.0[b]	0.0	<RL	<RL	<RL	<RL	<RL	<RL
MCPA	0.2	1218	0.0	(d)	>0.0[b]	0.0	<RL	<RL	<RL	<RL	<RL	<RL
MCPB	0.26	1222	0.0	(d)	>0.0[b]	0.0	<RL	<RL	<RL	<RL	<RL	<RL
Methiocarb	0.07	1226	0.0	(d)	0.0	0.0	<RL	<RL	<RL	<RL	<RL	<RL
Methomyl	0.47	1212	0.08	(d)	>0.0[b]	0.0	<RL	<RL	<RL	<RL	<RL	0.04
Neburon	0.07	1226	0.0	(d)	0.0	0.0	<RL	<RL	<RL	<RL	<RL	<RL
Norflurazon	0.042	1226	2.12	(d)	1.96	1.06	<RL	<RL	<RL	<RL	<RL	29.2[c]
Oryzalin	0.31	1224	0.0	(d)	>0.0[b]	0.0	<RL	<RL	<RL	<RL	<RL	<RL
Oxamyl	0.16	1211	0.83	(d)	>0.74[b]	0.58	<RL	<RL	<RL	<RL	<RL	2.06[c]
Picloram	0.09	1204	0.17	(d)	0.17	0.08	<RL	<RL	<RL	<RL	<RL	2.2
Propham	0.22	1226	0.0	(d)	>0.0[b]	0.0	<RL	<RL	<RL	<RL	<RL	<RL
Propoxur	0.12	1215	0.08	(d)	>0.0[b]	0.0	<RL	<RL	<RL	<RL	<RL	0.06
Silvex	0.06	1223	0.08	(d)	0.0	0.0	<RL	<RL	<RL	<RL	<RL	0.06
2,4,5-T	0.07	1222	0.08	(d)	0.08	0.0	<RL	<RL	<RL	<RL	<RL	0.61
Triclopyr	0.25	1224	0.0	(d)	>0.0[b]	0.0	<RL	<RL	<RL	<RL	<RL	<RL

Notes: *(1) The analytical reporting limit has changed through time. The limit reported in this table is the maximum value routinely used to report nondetections during the period 1991-2001; (a) Acetochlor was not used or analyzed for prior to 1994; (b) The analytical reporting limit for the compound is greater than the threshold concentration used for computing detection frequency. Therefore, the reported percentage may be an underestimate of the actual percentage of time-weighted concentrations that are greater than the threshold concentration; (c) See "Estimated Concentrations" below; (d) The 0.01 ug/L threshold is not applicable for the HPLC method because the reporting limit for most compounds is much higher than the threshold; (e) all concentrations determined for this compound are estimated concentrations.*

Abbreviations: *ug/L, microgram per liter; ">=", greater than or equal to; <RL, less than the maximum reporting limit; GC/MS, gas chromatography/mass spectrometry; HPLC, high performance liquid chromatography; NWQL, U.S. Geological Survey National Water Quality Laboratory.*

Frequency of Detection: *Within each statistical summary, four detection frequencies are provided (all detections, >=0.01 ug/L, >=0.1 ug/L, and >=1 ug/L). These detection frequencies provide a measure of "how often" a compound was found. The "all detections" provide the total number of detections for a given compound, but are not comparable among compounds because detection capabilities vary. Because reporting levels varied, detection frequencies were calculated using three common detection thresholds (0.01, 0.1 and 1 ug/L). The use of these detection thresholds facilitates comparisons among compounds by censoring detections to a common reference concentration. Adjustments of this type are essential in order to answer questions like "is compound x found more often than compound y?"*

Percentiles of concentration: *Concentrations measured for each pesticide are summarized using percentiles. The 25th, 50th, 75th, 90th, and 95th percentiles of concentration by land-use category are provided. Percentiles provide information about the magnitude of concentrations at selected points in the cumulative frequency distribution of the concentrations. For example, concentrations of atrazine at the 1438 wells in agricultural land-use studies were less than the reporting limit in 50% of the wells, less than or equal to 0.017 ug/L in 75% of the wells, and less than or equal to 0.36 ug/L in 95% of the wells.*

Maximum Concentrations: *Concentrations in this column are the maximum measured concentrations in the condensed (one sample per well) data sets for each land-use category. Concentrations greater than the value reported as maximum may have been measured at sites but are not reported because they were not measured in the sample used for this summary. In addition, the probability is low that these "snapshot" water samples were collected at the time of the maximum concentration for any pesticide. Consequently, the maximum concentrations reported should be interpreted as a lower bound for the true maximum concentrations.*

Estimated concentrations: *All pesticides denoted with "e" have more biased and/or variable analytical performance than the other pesticides in the method. Information on analytical performance is provided in the method reports (Zaugg and others, 1995; Werner and others, 1996) and is summarized for the period 1992-96 in Martin (1999). Individual concentrations that exceed the calibration curve of the analytical method are marked with a "c" to indicate that the uncertainty in these concentrations is greater than for concentrations within the calibration range.*

Source: *U.S. Geological Survey, Pesticides in Streams: Preliminary Results from Cycle I of the National Water Quality Assessment Program (NAWQA), 1992-2001, April 5, 2003*

Pesticides in ground water from mixed land use wells, 1991-2001

Compound	Max Rep. Limit[1] (ug/L)	Wells	Frequency of Detection (%)				Percentiles of Concentration (ug/L)					
			All	>=0.01 ug/L	>=0.1 ug/L	>=1 ug/L	25th	50th	75th	90th	95th	Max
Analyzed by GC/MS												
Acetochlor[a]	0.006	2164[a]	0.14	0.09	0.05	0.0	<RL	<RL	<RL	<RL	<RL	0.114
Alachlor	0.005	2717	0.77	0.48	0.22	0.04	<RL	<RL	<RL	<RL	<RL	1.3
Atrazine	0.007	2712	18.6	9.19	2.58	0.18	<RL	<RL	<RL	0.009	0.035	2.8
Azinphos-methyl[e]	0.05	2717	0.04	>0.0[b]	0.0	0.0	<RL	<RL	<RL	<RL	<RL	0.007
Benfluralin	0.01	2515	0.07	0.0	0.0	0.0	<RL	<RL	<RL	<RL	<RL	0.006
Butylate	0.002	2717	0.18	0.04	0.0	0.0	<RL	<RL	<RL	<RL	<RL	0.045
Carbaryl[e]	0.041	2717	0.77	>0.49[b]	0.07	0.0	<RL	<RL	<RL	<RL	<RL	0.539[c]
Carbofuran[e]	0.02	2716	0.40	>0.33[b]	0.07	0.0	<RL	<RL	<RL	<RL	<RL	0.218[c]
Chlorpyrifos	0.005	2715	0.18	0.0	0.0	0.0	<RL	<RL	<RL	<RL	<RL	0.007
Cyanazine	0.018	2717	0.18	>0.12[b]	0.0	0.0	<RL	<RL	<RL	<RL	<RL	0.066
p,p'-DDE	0.006	2716	2.65	0.0	0.0	0.0	<RL	<RL	<RL	<RL	<RL	0.006
Dacthal	0.003	2717	0.44	0.0	0.0	0.0	<RL	<RL	<RL	<RL	<RL	0.004
Deethylatrazine[e]	0.006	2711	20.6	9.56	2.07	0.11	<RL	<RL	<RL	0.009	0.031	2.0[c]
Diazinon	0.005	2717	1.40	0.74	0.07	0.04	<RL	<RL	<RL	<RL	<RL	19[c]
Dieldrin	0.005	2717	0.85	0.44	0.04	0.0	<RL	<RL	<RL	<RL	<RL	0.809
2,6-Diethylaniline	0.006	2717	0.29	0.0	0.0	0.0	<RL	<RL	<RL	<RL	<RL	0.009
Disulfoton	0.021	2717	0.0	>0.0[b]	0.0	0.0	<RL	<RL	<RL	<RL	<RL	<RL
EPTC	0.002	2717	0.33	0.15	0.04	0.0	<RL	<RL	<RL	<RL	<RL	0.182
Ethalfluralin	0.009	2717	0.0	0.0	0.0	0.0	<RL	<RL	<RL	<RL	<RL	<RL
Ethoprop	0.005	2717	0.0	0.0	0.0	0.0	<RL	<RL	<RL	<RL	<RL	<RL
Fonofos	0.003	2717	0.07	0.0	0.0	0.0	<RL	<RL	<RL	<RL	<RL	0.003
alph-HCH	0.005	2717	0.04	0.04	0.0	0.0	<RL	<RL	<RL	<RL	<RL	0.033
gamma-HCH	0.004	2717	0.07	0.04	0.04	0.0	<RL	<RL	<RL	<RL	<RL	0.152
Linuron	0.035	2717	0.04	>0.04[b]	0.0	0.0	<RL	<RL	<RL	<RL	<RL	0.029
Malathion	0.027	2717	0.40	>0.25[b]	0.0	0.0	<RL	<RL	<RL	<RL	<RL	0.033
Metolachlor	0.013	2717	5.04	>2.26[b]	0.52	0.15	<RL	<RL	<RL	<RL	<RL	2.62
Metribuzin	0.006	2717	0.85	0.52	0.07	0.0	<RL	<RL	<RL	<RL	<RL	0.119
Molinate	0.004	2717	0.40	0.22	0.0	0.0	<RL	<RL	<RL	<RL	<RL	0.053
Napropamide	0.007	2717	0.22	0.11	0.0	0.0	<RL	<RL	<RL	<RL	<RL	0.07
Parathion	0.01	2717	0.0	0.0	0.0	0.0	<RL	<RL	<RL	<RL	<RL	<RL
Parathion-methyl	0.006	2717	0.04	0.0	0.0	0.0	<RL	<RL	<RL	<RL	<RL	0.007
Pebulate	0.004	2717	0.0	0.0	0.0	0.0	<RL	<RL	<RL	<RL	<RL	<RL
Pendimethalin	0.022	2717	0.04	>0.04[b]	0.0	0.0	<RL	<RL	<RL	<RL	<RL	0.098
cis-Permethrin	0.006	2717	0.0	0.0	0.0	0.0	<RL	<RL	<RL	<RL	<RL	<RL
Phorate	0.011	2717	0.0	>0.0[b]	0.0	0.0	<RL	<RL	<RL	<RL	<RL	<RL
Prometon	0.018	2717	5.52	>2.61[b]	0.44	0.04	<RL	<RL	<RL	<RL	<RL	1.3
Pronamide	0.004	2717	0.07	0.0	0.0	0.0	<RL	<RL	<RL	<RL	<RL	0.002
Propachlor	0.01	2717	0.04	0.0	0.0	0.0	<RL	<RL	<RL	<RL	<RL	0.004
Propanil	0.011	2717	0.40	>0.08[b]	0.0	0.0	<RL	<RL	<RL	<RL	<RL	0.015
Propargite	0.023	2707	0.0	>0.0[b]	0.0	0.0	<RL	<RL	<RL	<RL	<RL	<RL
Simazine	0.011	2717	6.18	>2.98[b]	0.26	0.0	<RL	<RL	<RL	<RL	<RL	0.315
Tebuthiuron	0.016	2699	2.56	>2.19[b]	0.59	0.11	<RL	<RL	<RL	<RL	<RL	17.3
Terbacil[e]	0.034	2708	0.26	>0.21[b]	0.11	0.0	<RL	<RL	<RL	<RL	<RL	0.891[c]
Terbufos	0.017	2717	0.04	>0.0[b]	0.0	0.0	<RL	<RL	<RL	<RL	<RL	0.008
Thiobencarb	0.005	2717	0.0	0.0	0.0	0.0	<RL	<RL	<RL	<RL	<RL	<RL
Triallate	0.002	2717	0.15	0.04	0.0	0.0	<RL	<RL	<RL	<RL	<RL	0.021
Trifluralin	0.009	2717	0.33	0.0	0.0	0.0	<RL	<RL	<RL	<RL	<RL	0.009
Analyzed by HPLC												
Acifluorfen	0.09	1476	0.14	(d)	0.14	0.0	<RL	<RL	<RL	<RL	<RL	0.19
Aldicarb[e]	0.55	1480	0.0	(d)	>0.0[b]	0.0	<RL	<RL	<RL	<RL	<RL	<RL
Aldicarb sulfone[e]	0.20	1452	0.14	(d)	>0.07[b]	0.0	<RL	<RL	<RL	<RL	<RL	0.14[c]
Aldicarb sulfoxide[e]	0.27	1454	0.14	(d)	>0.14[b]	0.0	<RL	<RL	<RL	<RL	<RL	0.23[c]
Bentazon	0.05	1476	2.30	(d)	1.02	0.20	<RL	<RL	<RL	<RL	<RL	6.4
Bromacil	0.09	1484	0.61	(d)	0.47	0.07	<RL	<RL	<RL	<RL	<RL	2.24
Bromoxynil	0.07	1475	0.0	(d)	0.0	0.0	<RL	<RL	<RL	<RL	<RL	<RL

Pesticides in ground water from mixed land use wells, 1991-2001 - *continued*

Compound	Max Rep. Limit[1] (ug/L)	Wells	Frequency of Detection (%)				Percentiles of Concentration (ug/L)					
			All	>=0.01 ug/L	>=0.1 ug/L	>=1 ug/L	25th	50th	75th	90th	95th	Max
Chloramben methyl ester	0.42	1485	0.0	(d)	>0.0[b]	0.0	<RL	<RL	<RL	<RL	<RL	<RL
Chlorothalonil[e]	0.48	1478	0.0	(d)	>0.0[b]	0.0	<RL	<RL	<RL	<RL	<RL	<RL
Clopyralid	0.42	1475	0.0	(d)	>0.0[b]	0.0	<RL	<RL	<RL	<RL	<RL	<RL
2,4-D	0.16	1473	0.41	(d)	>0.14[b]	0.0	<RL	<RL	<RL	<RL	<RL	0.15
Dacthal monoacid	0.07	1474	0.0	(d)	0.0	0.0	<RL	<RL	<RL	<RL	<RL	<RL
2,4-DB	0.25	1483	0.0	(d)	>0.0[b]	0.0	<RL	<RL	<RL	<RL	<RL	<RL
Dicamba	0.11	1482	0.07	(d)	>0.0[b]	0.0	<RL	<RL	<RL	<RL	<RL	0.07
Dichlobenil[e]	1.2	1483	0.20	(d)	>0.0[b]	>0.0[b]	<RL	<RL	<RL	<RL	<RL	0.09[c]
Dichlorprop	0.12	1483	0.07	(d)	>0.07[b]	0.0	<RL	<RL	<RL	<RL	<RL	0.1
Dinoseb	0.09	1476	0.34	(d)	0.27	0.20	<RL	<RL	<RL	<RL	<RL	80[c]
Diuron	0.12	1480	2.10	(d)	>0.82[b]	0.14	<RL	<RL	<RL	<RL	<RL	1.92
DNOC[e]	0.42	1476	0.0	(d)	>0.0[b]	0.0	<RL	<RL	<RL	<RL	<RL	<RL
Fenuron	0.07	1477	0.95	(d)	0.07	0.0	<RL	<RL	<RL	<RL	<RL	0.39
Fluometuron	0.06	1484	0.40	(d)	0.20	0.14	<RL	<RL	<RL	<RL	<RL	1.29
3-Hydroxycarbofuran	0.11	1477	0.14	(d)	>0.0[b]	0.0	<RL	<RL	<RL	<RL	<RL	0.07
MCPA	0.2	1483	0.0	(d)	>0.0[b]	0.0	<RL	<RL	<RL	<RL	<RL	<RL
MCPB	0.26	1483	0.0	(d)	>0.0[b]	0.0	<RL	<RL	<RL	<RL	<RL	<RL
Methiocarb	0.07	1480	0.07	(d)	0.0	0.0	<RL	<RL	<RL	<RL	<RL	0.03
Methomyl	0.47	1452	0.07	(d)	>0.07[b]	0.0	<RL	<RL	<RL	<RL	<RL	0.1
Neburon	0.07	1480	0.07	(d)	0.0	0.0	<RL	<RL	<RL	<RL	<RL	0.03
Norflurazon	0.042	1484	0.20	(d)	0.07	0.0	<RL	<RL	<RL	<RL	<RL	0.15
Oryzalin	0.31	1480	0.07	(d)	>0.0[b]	0.0	<RL	<RL	<RL	<RL	<RL	0.03
Oxamyl	0.16	1449	0.08	(d)	>0.0[b]	0.0	<RL	<RL	<RL	<RL	<RL	0.03
Picloram	0.09	1451	0.14	(d)	0.07	0.0	<RL	<RL	<RL	<RL	<RL	0.17
Propham	0.22	1483	0.07	(d)	>0.0[b]	0.0	<RL	<RL	<RL	<RL	<RL	0.04
Propoxur	0.12	1471	0.14	(d)	>0.0[b]	0.0	<RL	<RL	<RL	<RL	<RL	0.06
Silvex	0.06	1476	0.07	(d)	0.0	0.0	<RL	<RL	<RL	<RL	<RL	0.06
2,4,5-T	0.07	1476	0.0	(d)	0.0	0.0	<RL	<RL	<RL	<RL	<RL	<RL
Triclopyr	0.25	1476	0.0	(d)	>0.0[b]	0.0	<RL	<RL	<RL	<RL	<RL	<RL

Notes: *(1) The analytical reporting limit has changed through time. The limit reported in this table is the maximum value routinely used to report nondetections during the period 1991-2001; (a) Acetochlor was not used or analyzed for prior to 1994; (b) The analytical reporting limit for the compound is greater than the threshold concentration used for computing detection frequency. Therefore, the reported percentage may be an underestimate of the actual percentage of time-weighted concentrations that are greater than the threshold concentration; (c) See "Estimated Concentrations" below; (d) The 0.01 ug/L threshold is not applicable for the HPLC method because the reporting limit for most compounds is much higher than the threshold; (e) all concentrations determined for this compound are estimated concentrations.*

Abbreviations: *ug/L, microgram per liter; ">=", greater than or equal to; <RL, less than the maximum reporting limit; GC/MS, gas chromatography/mass spectrometry; HPLC, high performance liquid chromatography; NWQL, U.S. Geological Survey National Water Quality Laboratory.*

Frequency of Detection: *Within each statistical summary, four detection frequencies are provided (all detections, >=0.01 ug/L, >=0.1 ug/L, and >=1 ug/L). These detection frequencies provide a measure of "how often" a compound was found. The "all detections" provide the total number of detections for a given compound, but are not comparable among compounds because detection capabilities vary. Because reporting levels varied, detection frequencies were calculated using three common detection thresholds (0.01, 0.1 and 1 ug/L). The use of these detection thresholds facilitates comparisons among compounds by censoring detections to a common reference concentration. Adjustments of this type are essential in order to answer questions like "is compound x found more often than compound y?"*

Percentiles of concentration: *Concentrations measured for each pesticide are summarized using percentiles. The 25th, 50th, 75th, 90th, and 95th percentiles of concentration by land-use category are provided. Percentiles provide information about the magnitude of concentrations at selected points in the cumulative frequency distribution of the concentrations. For example, concentrations of atrazine at the 1438 wells in agricultural land-use studies were less than the reporting limit in 50% of the wells, less than or equal to 0.017 ug/L in 75% of the wells, and less than or equal to 0.36 ug/L in 95% of the wells.*

Maximum Concentrations: *Concentrations in this column are the maximum measured concentrations in the condensed (one sample per well) data sets for each land-use category. Concentrations greater than the value reported as maximum may have been measured at sites but are not reported because they were not measured in the sample used for this summary. In addition, the probability is low that these "snapshot" water samples were collected at the time of the maximum concentration for any pesticide. Consequently, the maximum concentrations reported should be interpreted as a lower bound for the true maximum concentrations.*

Estimated concentrations: *All pesticides denoted with "e" have more biased and/or variable analytical performance than the other pesticides in the method. Information on analytical performance is provided in the method reports (Zaugg and others, 1995; Werner and others, 1996) and is summarized for the period 1992-96 in Martin (1999). Individual concentrations that exceed the calibration curve of the analytical method are marked with a "c" to indicate that the uncertainty in these concentrations is greater than for concentrations within the calibration range.*

Source: *U.S. Geological Survey, Pesticides in Streams: Preliminary Results from Cycle I of the National Water Quality Assessment Program (NAWQA), 1992-2001, April 5, 2003*

Pesticides in ground water from undeveloped land use wells, 1991-2001

Compound	Max Rep. Limit[1] (ug/L)	Wells	Frequency of Detection (%)				Percentiles of Concentration (ug/L)					
			All	>=0.01 ug/L	>=0.1 ug/L	>=1 ug/L	25th	50th	75th	90th	95th	Max
Analyzed by GC/MS												
Acetochlor[a]	0.006	39[a]	0.0	0.0	0.0	0.0	<RL	<RL	<RL	<RL	<RL	<RL
Alachlor	0.005	67	0.0	0.0	0.0	0.0	<RL	<RL	<RL	<RL	<RL	<RL
Atrazine	0.007	67	13.4	2.98	0.0	0.0	<RL	<RL	<RL	<RL	0.008	0.018
Azinphos-methyl[e]	0.05	67	0.0	>0.0[b]	0.0	0.0	<RL	<RL	<RL	<RL	<RL	<RL
Benfluralin	0.01	67	0.0	0.0	0.0	0.0	<RL	<RL	<RL	<RL	<RL	<RL
Butylate	0.002	67	0.0	0.0	0.0	0.0	<RL	<RL	<RL	<RL	<RL	<RL
Carbaryl[e]	0.041	67	4.48	>4.46[b]	0.0	0.0	<RL	<RL	<RL	<RL	<RL	0.02[c]
Carbofuran[e]	0.02	67	0.0	>0.0[b]	0.0	0.0	<RL	<RL	<RL	<RL	<RL	<RL
Chlorpyrifos	0.005	67	0.0	0.0	0.0	0.0	<RL	<RL	<RL	<RL	<RL	<RL
Cyanazine	0.018	67	0.0	>0.0[b]	0.0	0.0	<RL	<RL	<RL	<RL	<RL	<RL
p,p'-DDE	0.006	67	7.46	0.0	0.0	0.0	<RL	<RL	<RL	<RL	<RL	0.002
Dacthal	0.003	67	0.0	0.0	0.0	0.0	<RL	<RL	<RL	<RL	<RL	<RL
Deethylatrazine[e]	0.006	67	14.9	2.98	0.0	0.0	<RL	<RL	<RL	<RL	0.008	0.034[c]
Diazinon	0.005	67	0.0	0.0	0.0	0.0	<RL	<RL	<RL	<RL	<RL	<RL
Dieldrin	0.005	67	0.0	0.0	0.0	0.0	<RL	<RL	<RL	<RL	<RL	<RL
2,6-Diethylaniline	0.006	67	0.0	0.0	0.0	0.0	<RL	<RL	<RL	<RL	<RL	<RL
Disulfoton	0.021	67	0.0	>0.0[b]	0.0	0.0	<RL	<RL	<RL	<RL	<RL	<RL
EPTC	0.002	67	0.0	0.0	0.0	0.0	<RL	<RL	<RL	<RL	<RL	<RL
Ethalfluralin	0.009	67	0.0	0.0	0.0	0.0	<RL	<RL	<RL	<RL	<RL	<RL
Ethoprop	0.005	67	0.0	0.0	0.0	0.0	<RL	<RL	<RL	<RL	<RL	<RL
Fonofos	0.003	67	0.0	0.0	0.0	0.0	<RL	<RL	<RL	<RL	<RL	<RL
alph-HCH	0.005	67	0.0	0.0	0.0	0.0	<RL	<RL	<RL	<RL	<RL	<RL
gamma-HCH	0.004	67	0.0	0.0	0.0	0.0	<RL	<RL	<RL	<RL	<RL	<RL
Linuron	0.035	67	0.0	>0.0[b]	0.0	0.0	<RL	<RL	<RL	<RL	<RL	<RL
Malathion	0.027	67	0.0	>0.0[b]	0.0	0.0	<RL	<RL	<RL	<RL	<RL	<RL
Metolachlor	0.013	67	1.49	>0.0[b]	0.0	0.0	<RL	<RL	<RL	<RL	<RL	0.005
Metribuzin	0.006	67	0.0	0.0	0.0	0.0	<RL	<RL	<RL	<RL	<RL	<RL
Molinate	0.004	67	1.49	0.0	0.0	0.0	<RL	<RL	<RL	<RL	<RL	0.003
Napropamide	0.007	67	0.0	0.0	0.0	0.0	<RL	<RL	<RL	<RL	<RL	<RL
Parathion	0.01	67	0.0	0.0	0.0	0.0	<RL	<RL	<RL	<RL	<RL	<RL
Parathion-methyl	0.006	67	0.0	0.0	0.0	0.0	<RL	<RL	<RL	<RL	<RL	<RL
Pebulate	0.004	67	0.0	0.0	0.0	0.0	<RL	<RL	<RL	<RL	<RL	<RL
Pendimethalin	0.022	67	0.0	>0.0[b]	0.0	0.0	<RL	<RL	<RL	<RL	<RL	<RL
cis-Permethrins	0.006	67	0.0	0.0	0.0	0.0	<RL	<RL	<RL	<RL	<RL	<RL
Phorate	0.011	67	0.0	>0.0[b]	0.0	0.0	<RL	<RL	<RL	<RL	<RL	<RL
Prometon	0.018	67	1.49	>0.0[b]	0.0	0.0	<RL	<RL	<RL	<RL	<RL	0.008
Pronamide	0.004	67	0.0	0.0	0.0	0.0	<RL	<RL	<RL	<RL	<RL	<RL
Propachlor	0.01	67	0.0	0.0	0.0	0.0	<RL	<RL	<RL	<RL	<RL	<RL
Propanil	0.011	67	0.0	>0.0[b]	0.0	0.0	<RL	<RL	<RL	<RL	<RL	<RL
Propargite	0.023	67	0.0	>0.0[b]	0.0	0.0	<RL	<RL	<RL	<RL	<RL	<RL
Simazine	0.011	67	5.97	>4.62[b]	1.49	0.0	<RL	<RL	<RL	<RL	<RL	0.13
Tebuthiuron	0.016	67	0.0	>0.0[b]	0.0	0.0	<RL	<RL	<RL	<RL	<RL	<RL
Terbacil[e]	0.034	67	0.0	>0.0[b]	0.0	0.0	<RL	<RL	<RL	<RL	<RL	<RL
Terbufos	0.017	67	0.0	>0.0[b]	0.0	0.0	<RL	<RL	<RL	<RL	<RL	<RL
Thiobencarb	0.005	67	0.0	0.0	0.0	0.0	<RL	<RL	<RL	<RL	<RL	<RL
Triallate	0.002	67	0.0	0.0	0.0	0.0	<RL	<RL	<RL	<RL	<RL	<RL
Trifluralin	0.009	67	1.49	0.0	0.0	0.0	<RL	<RL	<RL	<RL	<RL	0.004
Analyzed by HPLC												
Acifluorfen	0.09	47	0.0	(d)	0.0	0.0	<RL	<RL	<RL	<RL	<RL	<RL
Aldicarb[e]	0.55	47	0.0	(d)	>0.0[b]	0.0	<RL	<RL	<RL	<RL	<RL	<RL
Aldicarb sulfone[e]	0.20	47	0.0	(d)	>0.0[b]	0.0	<RL	<RL	<RL	<RL	<RL	<RL
Aldicarb sulfoxide[e]	0.27	47	2.13	(d)	>0.0[b]	0.0	<RL	<RL	<RL	<RL	<RL	0.007[c]
Bentazon	0.05	46	0.0	(d)	0.0	0.0	<RL	<RL	<RL	<RL	<RL	<RL
Bromacil	0.09	47	0.0	(d)	0.0	0.0	<RL	<RL	<RL	<RL	<RL	<RL
Bromoxynil	0.07	46	0.0	(d)	0.0	0.0	<RL	<RL	<RL	<RL	<RL	<RL

Pesticides in ground water from undeveloped land use wells, 1991-2001 - *continued*

Compound	Max Rep. Limit[1] (ug/L)	Wells	Frequency of Detection (%) All	>=0.01 ug/L	>=0.1 ug/L	>=1 ug/L	Percentiles of Concentration (ug/L) 25th	50th	75th	90th	95th	Max
Chloramben methyl ester	0.42	47	0.0	(d)	>0.0[b]	0.0	<RL	<RL	<RL	<RL	<RL	<RL
Chlorothalonil[e]	0.48	46	0.0	(d)	>0.0[b]	0.0	<RL	<RL	<RL	<RL	<RL	<RL
Clopyralid	0.42	46	0.0	(d)	>0.0[b]	0.0	<RL	<RL	<RL	<RL	<RL	<RL
2,4-D	0.16	46	0.0	(d)	>0.0[b]	0.0	<RL	<RL	<RL	<RL	<RL	<RL
Dacthal monoacid	0.07	46	0.0	(d)	0.0	0.0	<RL	<RL	<RL	<RL	<RL	<RL
2,4-DB	0.25	47	0.0	(d)	>0.0[b]	0.0	<RL	<RL	<RL	<RL	<RL	<RL
Dicamba	0.11	46	0.0	(d)	>0.0[b]	0.0	<RL	<RL	<RL	<RL	<RL	<RL
Dichlobenil[e]	1.2	47	0.0	(d)	>0.0[b]	>0.0[b]	<RL	<RL	<RL	<RL	<RL	<RL
Dichlorprop	0.12	47	0.0	(d)	>0.0[b]	0.0	<RL	<RL	<RL	<RL	<RL	<RL
Dinoseb	0.09	47	2.13	(d)	0.0	0.0	<RL	<RL	<RL	<RL	<RL	0.02
Diuron	0.12	47	0.0	(d)	>0.0[b]	0.0	<RL	<RL	<RL	<RL	<RL	<RL
DNOC[e]	0.42	46	0.0	(d)	>0.0[b]	0.0	<RL	<RL	<RL	<RL	<RL	<RL
Fenuron	0.07	47	0.0	(d)	0.0	0.0	<RL	<RL	<RL	<RL	<RL	<RL
Fluometuron	0.06	47	0.0	(d)	0.0	0.0	<RL	<RL	<RL	<RL	<RL	<RL
3-Hydroxycarbofuran	0.11	47	0.0	(d)	>0.0[b]	0.0	<RL	<RL	<RL	<RL	<RL	<RL
MCPA	0.2	46	0.0	(d)	>0.0[b]	0.0	<RL	<RL	<RL	<RL	<RL	<RL
MCPB	0.26	47	0.0	(d)	>0.0[b]	0.0	<RL	<RL	<RL	<RL	<RL	<RL
Methiocarb	0.07	47	0.0	(d)	0.0	0.0	<RL	<RL	<RL	<RL	<RL	<RL
Methomyl	0.47	47	0.0	(d)	>0.0[b]	0.0	<RL	<RL	<RL	<RL	<RL	<RL
Neburon	0.07	47	0.0	(d)	0.0	0.0	<RL	<RL	<RL	<RL	<RL	<RL
Norflurazon	0.042	47	0.0	(d)	0.0	0.0	<RL	<RL	<RL	<RL	<RL	<RL
Oryzalin	0.31	47	0.0	(d)	>0.0[b]	0.0	<RL	<RL	<RL	<RL	<RL	<RL
Oxamyl	0.16	47	0.0	(d)	>0.0[b]	0.0	<RL	<RL	<RL	<RL	<RL	<RL
Picloram	0.09	46	0.0	(d)	0.0	0.0	<RL	<RL	<RL	<RL	<RL	<RL
Propham	0.22	47	0.0	(d)	>0.0[b]	0.0	<RL	<RL	<RL	<RL	<RL	<RL
Propoxur	0.12	47	0.0	(d)	>0.0[b]	0.0	<RL	<RL	<RL	<RL	<RL	<RL
Silvex	0.06	47	0.0	(d)	0.0	0.0	<RL	<RL	<RL	<RL	<RL	<RL
2,4,5-T	0.07	47	0.0	(d)	0.0	0.0	<RL	<RL	<RL	<RL	<RL	<RL
Triclopyr	0.25	47	0.0	(d)	>0.0[b]	0.0	<RL	<RL	<RL	<RL	<RL	<RL

Notes: *(1) The analytical reporting limit has changed through time. The limit reported in this table is the maximum value routinely used to report nondetections during the period 1991-2001; (a) Acetochlor was not used or analyzed for prior to 1994; (b) The analytical reporting limit for the compound is greater than the threshold concentration used for computing detection frequency. Therefore, the reported percentage may be an underestimate of the actual percentage of time-weighted concentrations that are greater than the threshold concentration; (c) See "Estimated Concentrations" below; (d) The 0.01 ug/L threshold is not applicable for the HPLC method because the reporting limit for most compounds is much higher than the threshold; (e) all concentrations determined for this compound are estimated concentrations.*

Abbreviations: *ug/L, microgram per liter; ">=", greater than or equal to; <RL, less than the maximum reporting limit; GC/MS, gas chromatography/mass spectrometry; HPLC, high performance liquid chromatography; NWQL, U.S. Geological Survey National Water Quality Laboratory.*

Frequency of Detection: *Within each statistical summary, four detection frequencies are provided (all detections, >=0.01 ug/L, >=0.1 ug/L, and >=1 ug/L). These detection frequencies provide a measure of "how often" a compound was found. The "all detections" provide the total number of detections for a given compound, but are not comparable among compounds because detection capabilities vary. Because reporting levels varied, detection frequencies were calculated using three common detection thresholds (0.01, 0.1 and 1 ug/L). The use of these detection thresholds facilitates comparisons among compounds by censoring detections to a common reference concentration. Adjustments of this type are essential in order to answer questions like "is compound x found more often than compound y?"*

Percentiles of concentration: *Concentrations measured for each pesticide are summarized using percentiles. The 25th, 50th, 75th, 90th, and 95th percentiles of concentration by land-use category are provided. Percentiles provide information about the magnitude of concentrations at selected points in the cumulative frequency distribution of the concentrations. For example, concentrations of atrazine at the 1438 wells in agricultural land-use studies were less than the reporting limit in 50% of the wells, less than or equal to 0.017 ug/L in 75% of the wells, and less than or equal to 0.36 ug/L in 95% of the wells.*

Maximum Concentrations: *Concentrations in this column are the maximum measured concentrations in the condensed (one sample per well) data sets for each land-use category. Concentrations greater than the value reported as maximum may have been measured at sites but are not reported because they were not measured in the sample used for this summary. In addition, the probability is low that these "snapshot" water samples were collected at the time of the maximum concentration for any pesticide. Consequently, the maximum concentrations reported should be interpreted as a lower bound for the true maximum concentrations.*

Estimated concentrations: *All pesticides denoted with "e" have more biased and/or variable analytical performance than the other pesticides in the method. Information on analytical performance is provided in the method reports (Zaugg and others, 1995; Werner and others, 1996) and is summarized for the period 1992-96 in Martin (1999). Individual concentrations that exceed the calibration curve of the analytical method are marked with a "c" to indicate that the uncertainty in these concentrations is greater than for concentrations within the calibration range.*

Source: *U.S. Geological Survey, Pesticides in Streams: Preliminary Results from Cycle I of the National Water Quality Assessment Program (NAWQA), 1992-2001, April 5, 2003*

Pesticides in ground water from urban land use wells, 1991-2001

Compound	Max Rep. Limit[1] (ug/L)	Wells	Frequency of Detection (%)				Percentiles of Concentration (ug/L)					
			All	>=0.01 ug/L	>=0.1 ug/L	>=1 ug/L	25th	50th	75th	90th	95th	Max
Analyzed by GC/MS												
Acetochlor[a]	0.006	636[a]	0.63	0.47	0.16	0.0	<RL	<RL	<RL	<RL	<RL	0.122
Alachlor	0.005	835	0.48	0.24	0.12	0.0	<RL	<RL	<RL	<RL	<RL	0.146
Atrazine	0.007	833	32.0	17.9	5.28	0.72	<RL	<RL	<RL	0.03	0.105	4.2
Azinphos-methyl[e]	0.05	830	0.0	>0.0[b]	0.0	0.0	<RL	<RL	<RL	<RL	<RL	<RL
Benfluralin	0.01	834	0.24	0.0	0.0	0.0	<RL	<RL	<RL	<RL	<RL	0.006
Butylate	0.002	834	0.24	0.0	0.0	0.0	<RL	<RL	<RL	<RL	<RL	0.007
Carbaryl[e]	0.041	834	1.56	>0.98[b]	0.0	0.0	<RL	<RL	<RL	<RL	<RL	0.031[c]
Carbofuran[e]	0.02	833	0.72	>0.61[b]	0.0	0.0	<RL	<RL	<RL	<RL	<RL	0.093[c]
Chlorpyrifos	0.005	834	0.60	0.0	0.0	0.0	<RL	<RL	<RL	<RL	<RL	0.005
Cyanazine	0.018	833	0.84	>0.62[b]	0.0	0.0	<RL	<RL	<RL	<RL	<RL	0.089
p,p'-DDE	0.006	834	3.96	0.0	0.0	0.0	<RL	<RL	<RL	<RL	<RL	0.005
Dacthal	0.003	834	0.96	0.12	0.0	0.0	<RL	<RL	<RL	<RL	<RL	0.011
Deethylatrazine[e]	0.006	832	31.5	13.7	2.52	0.0	<RL	<RL	<RL	0.018	0.047	0.56[c]
Diazinon	0.005	835	1.92	0.48	0.0	0.0	<RL	<RL	<RL	<RL	<RL	0.023
Dieldrin	0.005	823	5.10	3.04	0.61	0.24	<RL	<RL	<RL	<RL	<RL	5.6
2,6-Diethylaniline	0.006	834	0.12	0.12	0.0	0.0	<RL	<RL	<RL	<RL	<RL	0.031
Disulfoton	0.021	834	0.0	>0.0[b]	0.0	0.0	<RL	<RL	<RL	<RL	<RL	<RL
EPTC	0.002	834	0.72	0.24	0.0	0.0	<RL	<RL	<RL	<RL	<RL	0.02
Ethalfluralin	0.009	834	0.0	0.0	0.0	0.0	<RL	<RL	<RL	<RL	<RL	<RL
Ethoprop	0.005	834	0.0	0.0	0.0	0.0	<RL	<RL	<RL	<RL	<RL	<RL
Fonofos	0.003	835	0.0	0.0	0.0	0.0	<RL	<RL	<RL	<RL	<RL	<RL
alpha-HCH	0.005	835	0.0	0.0	0.0	0.0	<RL	<RL	<RL	<RL	<RL	<RL
gamma-HCH	0.004	823	0.0	0.0	0.0	0.0	<RL	<RL	<RL	<RL	<RL	<RL
Linuron	0.035	834	0.0	>0.0[b]	0.0	0.0	<RL	<RL	<RL	<RL	<RL	<RL
Malathion	0.027	835	0.24	>0.0[b]	0.0	0.0	<RL	<RL	<RL	<RL	<RL	0.01
Metolachlor	0.013	835	8.98	>3.06[b]	0.60	0.12	<RL	<RL	<RL	<RL	<RL	2.09
Metribuzin	0.006	834	1.32	0.48	0.12	0.0	<RL	<RL	<RL	<RL	<RL	0.49
Molinate	0.004	834	0.24	0.0	0.0	0.0	<RL	<RL	<RL	<RL	<RL	0.004
Napropamide	0.007	834	0.0	0.0	0.0	0.0	<RL	<RL	<RL	<RL	<RL	<RL
Parathion	0.01	835	0.0	0.0	0.0	0.0	<RL	<RL	<RL	<RL	<RL	<RL
Parathion-methyl	0.006	834	0.0	0.0	0.0	0.0	<RL	<RL	<RL	<RL	<RL	<RL
Pebulate	0.004	834	0.0	0.0	0.0	0.0	<RL	<RL	<RL	<RL	<RL	<RL
Pendimethalin	0.022	834	0.60	>0.36[b]	0.0	0.0	<RL	<RL	<RL	<RL	<RL	0.084
cis-Permethrin	0.006	834	0.0	0.0	0.0	0.0	<RL	<RL	<RL	<RL	<RL	<RL
Phorate	0.011	834	0.0	>0.0[b]	0.0	0.0	<RL	<RL	<RL	<RL	<RL	<RL
Prometon	0.018	834	22.3	>18.0[b]	5.64	0.60	<RL	<RL	<RL	0.043	0.13	10[c]
Pronamide	0.004	834	0.0	>0.0	0.0	0.0	<RL	<RL	<RL	<RL	<RL	<RL
Propachlor	0.01	834	0.0	0.0	0.0	0.0	<RL	<RL	<RL	<RL	<RL	<RL
Propanil	0.011	834	0.24	>0.0[b]	0.0	0.0	<RL	<RL	<RL	<RL	<RL	0.008
Propargite	0.023	834	0.0	>0.0[b]	0.0	0.0	<RL	<RL	<RL	<RL	<RL	<RL
Simazine	0.011	834	18.8	>11.9[b]	2.40	0.24	<RL	<RL	<RL	0.011	0.043	1.1
Tebuthiuron	0.016	831	6.50	>5.28[b]	1.20	0.36	<RL	<RL	<RL	<RL	<RL	2.12
Terbacil[e]	0.034	830	1.20	>0.49[b]	0.0	0.0	<RL	<RL	<RL	<RL	<RL	0.093[c]
Terbufos	0.017	834	0.12	>0.12[b]	0.0	0.0	<RL	<RL	<RL	<RL	<RL	0.012
Thiobencarb	0.005	834	0.12	0.12	0.0	0.0	<RL	<RL	<RL	<RL	<RL	0.036
Triallate	0.002	834	0.12	0.0	0.0	0.0	<RL	<RL	<RL	<RL	<RL	0.001
Trifluralin	0.009	834	0.60	0.24	0.0	0.0	<RL	<RL	<RL	<RL	<RL	0.014
Analyzed by HPLC												
Acifluorfen	0.09	619	0.0	(d)	0.0	0.0	<RL	<RL	<RL	<RL	<RL	<RL
Aldicarb[e]	0.55	617	0.0	(d)	>0.0[b]	0.0	<RL	<RL	<RL	<RL	<RL	<RL
Aldicarb sulfone[e]	0.20	607	0.0	(d)	>0.0[b]	0.0	<RL	<RL	<RL	<RL	<RL	<RL
Aldicarb sulfoxide[e]	0.27	607	0.0	(d)	>0.0[b]	0.0	<RL	<RL	<RL	<RL	<RL	<RL
Bentazon	0.05	619	2.58	(d)	0.97	0.0	<RL	<RL	<RL	<RL	<RL	0.76
Bromacil	0.09	618	2.75	(d)	1.94	0.48	<RL	<RL	<RL	<RL	<RL	3.0
Bromoxynil	0.07	619	0.0	(d)	0.0	0.0	<RL	<RL	<RL	<RL	<RL	<RL

Pesticides in ground water from urban land use wells, 1991-2001 - *continued*

Compound	Max Rep. Limit[1] (ug/L)	Wells	Frequency of Detection (%)				Percentiles of Concentration (ug/L)					
			All	>=0.01 ug/L	>=0.1 ug/L	>=1 ug/L	25th	50th	75th	90th	95th	Max
Chloramben methyl ester	0.42	619	0.0	(d)	>0.0[b]	0.0	<RL	<RL	<RL	<RL	<RL	<RL
Chlorothalonil[e]	0.48	614	0.16	(d)	>0.16[b]	0.0	<RL	<RL	<RL	<RL	<RL	0.41[c]
Clopyralid	0.42	616	0.0	(d)	>0.0[b]	0.0	<RL	<RL	<RL	<RL	<RL	<RL
2,4-D	0.16	619	0.81	(d)	>0.65[b]	0.16	<RL	<RL	<RL	<RL	<RL	14.8[c]
Dacthal monoacid	0.07	619	0.0	(d)	0.0	0.0	<RL	<RL	<RL	<RL	<RL	<RL
2,4-DB	0.25	619	0.0	(d)	>0.0[b]	0.0	<RL	<RL	<RL	<RL	<RL	<RL
Dicamba	0.11	619	0.48	(d)	>0.32[b]	0.16	<RL	<RL	<RL	<RL	<RL	1.46[c]
Dichlobenil[e]	1.2	619	0.32	(d)	>0.16[b]	>0.0[b]	<RL	<RL	<RL	<RL	<RL	0.21[c]
Dichlorprop	0.12	619	0.16	(d)	>0.16[b]	0.0	<RL	<RL	<RL	<RL	<RL	0.01
Dinoseb	0.09	619	0.0	(d)	0.0	0.0	<RL	<RL	<RL	<RL	<RL	<RL
Diuron	0.12	618	3.24	(d)	>1.78[b]	0.16	<RL	<RL	<RL	<RL	<RL	2.0[c]
DNOC[e]	0.42	619	0.16	(d)	>0.16[b]	0.0	<RL	<RL	<RL	<RL	<RL	0.2[c]
Fenuron	0.07	616	0.65	(d)	0.32	0.0	<RL	<RL	<RL	<RL	<RL	0.57
Fluometuron	0.06	618	0.32	(d)	0.16	0.0	<RL	<RL	<RL	<RL	<RL	0.11
3-Hydroxycarbofuran	0.11	616	0.0	(d)	>0.0[b]	0.0	<RL	<RL	<RL	<RL	<RL	<RL
MCPA	0.2	619	0.0	(d)	>0.0[b]	0.0	<RL	<RL	<RL	<RL	<RL	<RL
MCPB	0.26	619	0.0	(d)	>0.0[b]	0.0	<RL	<RL	<RL	<RL	<RL	<RL
Methiocarb	0.07	619	0.0	(d)	0.0	0.0	<RL	<RL	<RL	<RL	<RL	<RL
Methomyl	0.47	607	0.16	(d)	>0.16[b]	0.0	<RL	<RL	<RL	<RL	<RL	0.38
Neburon	0.07	619	0.0	(d)	0.0	0.0	<RL	<RL	<RL	<RL	<RL	<RL
Norflurazon	0.042	619	0.16	(d)	0.16	0.0	<RL	<RL	<RL	<RL	<RL	0.44
Oryzalin	0.31	619	0.48	(d)	>0.0[b]	0.0	<RL	<RL	<RL	<RL	<RL	0.08
Oxamyl	0.16	607	0.16	(d)	>0.16[b]	0.0	<RL	<RL	<RL	<RL	<RL	0.26[c]
Picloram	0.09	612	0.49	(d)	0.49	0.16	<RL	<RL	<RL	<RL	<RL	3.91[c]
Propham	0.22	618	0.0	(d)	>0.0[b]	0.0	<RL	<RL	<RL	<RL	<RL	<RL
Propoxur	0.12	617	0.16	(d)	>0.16[b]	0.0	<RL	<RL	<RL	<RL	<RL	0.3
Silvex	0.06	619	0.0	(d)	0.0	0.0	<RL	<RL	<RL	<RL	<RL	<RL
2,4,5-T	0.07	619	0.0	(d)	0.0	0.0	<RL	<RL	<RL	<RL	<RL	<RL
Triclopyr	0.25	619	0.0	(d)	>0.0[b]	0.0	<RL	<RL	<RL	<RL	<RL	<RL

Notes: *(1) The analytical reporting limit has changed through time. The limit reported in this table is the maximum value routinely used to report nondetections during the period 1991-2001; (a) Acetochlor was not used or analyzed for prior to 1994; (b) The analytical reporting limit for the compound is greater than the threshold concentration used for computing detection frequency. Therefore, the reported percentage may be an underestimate of the actual percentage of time-weighted concentrations that are greater than the threshold concentration; (c) See "Estimated Concentrations" below; (d) The 0.01 ug/L threshold is not applicable for the HPLC method because the reporting limit for most compounds is much higher than the threshold; (e) all concentrations determined for this compound are estimated concentrations.*

Abbreviations: *ug/L, microgram per liter; ">=", greater than or equal to; <RL, less than the maximum reporting limit; GC/MS, gas chromatography/mass spectrometry; HPLC, high performance liquid chromatography; NWQL, U.S. Geological Survey National Water Quality Laboratory.*

Frequency of Detection: *Within each statistical summary, four detection frequencies are provided (all detections, >=0.01 ug/L, >=0.1 ug/L, and >=1 ug/L). These detection frequencies provide a measure of "how often" a compound was found. The "all detections" provide the total number of detections for a given compound, but are not comparable among compounds because detection capabilities vary. Because reporting levels varied, detection frequencies were calculated using three common detection thresholds (0.01, 0.1 and 1 ug/L). The use of these detection thresholds facilitates comparisons among compounds by censoring detections to a common reference concentration. Adjustments of this type are essential in order to answer questions like "is compound x found more often than compound y?"*

Percentiles of concentration: *Concentrations measured for each pesticide are summarized using percentiles. The 25th, 50th, 75th, 90th, and 95th percentiles of concentration by land-use category are provided. Percentiles provide information about the magnitude of concentrations at selected points in the cumulative frequency distribution of the concentrations. For example, concentrations of atrazine at the 1438 wells in agricultural land-use studies were less than the reporting limit in 50% of the wells, less than or equal to 0.017 ug/L in 75% of the wells, and less than or equal to 0.36 ug/L in 95% of the wells.*

Maximum Concentrations: *Concentrations in this column are the maximum measured concentrations in the condensed (one sample per well) data sets for each land-use category. Concentrations greater than the value reported as maximum may have been measured at sites but are not reported because they were not measured in the sample used for this summary. In addition, the probability is low that these "snapshot" water samples were collected at the time of the maximum concentration for any pesticide. Consequently, the maximum concentrations reported should be interpreted as a lower bound for the true maximum concentrations.*

Estimated concentrations: *All pesticides denoted with "e" have more biased and/or variable analytical performance than the other pesticides in the method. Information on analytical performance is provided in the method reports (Zaugg and others, 1995; Werner and others, 1996) and is summarized for the period 1992-96 in Martin (1999). Individual concentrations that exceed the calibration curve of the analytical method are marked with a "c" to indicate that the uncertainty in these concentrations is greater than for concentrations within the calibration range.*

Source: *U.S. Geological Survey, Pesticides in Streams: Preliminary Results from Cycle I of the National Water Quality Assessment Program (NAWQA), 1992-2001, April 5, 2003*

Residential curbside recycling and yard trimmings composting sites by state for 2002

State	Curbside Programs	Population with Access to Curbside Collection	Yard Trimmings Composting Sites
Alabama	n/a	n/a	n/a
Alaska	n/a	n/a	n/a
Arizona	27	2,570,000	n/a
Arkansas	67	n/a	24
California	396	31,146,000	100
Colorado	22	618,848	5
Connecticut	169	3,460,503	92
Delaware	2	4,000	0
District of Columbia	n/a	n/a	n/a
Florida	333	9,100,000	0
Georgia	184	n/a	63
Hawaii	4	41,000	5
Idaho	12	n/a	n/a
Illinois	n/a	n/a	40
Indiana	79	4,170,000	107
Iowa	627	1,862,314	80
Kansas	113	1,100,000	105
Kentucky	54	1,211,085	30
Louisiana	20	n/a	3
Maine	40	500,000	<25
Maryland	99	4,000,000	37
Massachusetts	160	4,862,806	223
Michigan	347	3,670,072	163
Minnesota	733	3,750,000	n/a
Mississippi	14	325,000	6
Missouri	216	n/a	152
Montana	n/a	n/a	n/a
Nebraska	8	500,000	n/a
Nevada	3	1,963,924	1
New Hampshire	42	>518,000	192
New Jersey	510	7,500,000	170
New Mexico	10	400,000	8
New York	1,500	17,230,000	32
North Carolina	256	3,200,000	120
North Dakota	4	100,000	40
Ohio	459	6,459,072	534
Oklahoma	7	905,790	4
Oregon	133	2,641,136	41
Pennsylvania	945	9,310,252	>300
Rhode Island	26	897,000	15
South Carolina	135	564,552	128
South Dakota	3	60,000	120
Tennessee	58	n/a	n/a
Texas	160	5,000,000	160
Utah	n/a	n/a	20
Vermont	93	545,000	12
Virginia	60	1,144,000	14
Washington	150	4,923,318	41
West Virginia	51	425,134	0
Wisconsin	544	2,695,958	n/a
Wyoming	0	0	15
U.S. Total	8,875	139,374,764	3,227

Notes: n/a not available; (1) 2001 BioCycle, "The State of Garbage in America" data; (2) Based only on datat from 12 cities and/or counties; (3) 2000 data; (4) State reports 140 sites only grinding (i.e., not composting) collected yard trimmings for mulch; (5) 2001 data; (6) 1999 data; (7) May include yard trimmings grinding (only) facilities; (8) 1998 data; (9) Based on conversion of 2.86 people/household; (10) State reports 22 sites only grinding (i.e., not composting) collected yard trimmings for mulch.

Source: BioCycle, "The State of Garbage in America, January 2004

Recycling profiles of the 100 largest cities/towns in the United States

City	State	System Type	Voluntary	Materials Collected	Plastic Types
Akron	OH	curbside	Yes	metal food and beverage cans, aluminum trays and aluminum foil, glass bottles and jars of any color, rigid plastic bottles, empty aerosol cans, newspapers, magazines, cardboard	#1, #2
Albuquerque	NM	curbside, drop-off	Yes	newspapers, magazines, shopping catalogs, junk mail, home office paper, steel/tin (small pieces/containers), aluminum cans, plastic bottles and containers, corrugated cardboard, glass containers (all colors)	#1, #2
Anaheim	CA	curbside, drop-off	Yes	plastic bottles, aluminum cans (CRV & Non-CRV), glass bottles, food and beverage glass bottles, plastic bottles, bi-metal cans, cardboard/kraftpaper, newspapers, computer paper, white/color ledger paper, coated stock, office papers, magazines, mixed paper	#1, #2, #3, #4, #5, #6, #7
Anchorage	AK	curbside, drop-off	Yes	newspapers, corrugated cardboard/paperboard, magazines, catalogs, telephone books, glass, lightbulbs, aluminum cans, other metals, styrofoam, plastic bags, plastic containers, food waste, organic waste, tires/innertubes, batteries, antifreeze, paint	#1, #2
Arlington	TX	curbside, drop-off	Yes	newspapers, magazines, telephone books, junk mail and envelopes, office paper, cardboard, paper bags, jars/bottles (clear, brown and green), plastic jugs and bottles, aluminum/steel/tin cans	#1, #2
Atlanta	GA	curbside	Yes	newspapers, plastic bottles and jars, aluminum/steel/tin cans, glass (brown, clear and green), yard waste, organic waste, Christmas trees, leaves, tree trimmings, brush, grass clippings, weeds	#1, #2
Aurora	CO	curbside (private-no public trash collection system)	Yes	varies	na
Austin	TX	curbside	Yes	newspapers, magazines, catalogs, junk mail and envelopes, home office paper, aluminum/steel/tin cans, glass bottles and jars, plastic bottles, corrugated cardboard	#1, #2
Babylon	NY	drop-off	Yes	cardboard, newspapers, bottles and cans, concrete, tires, metal, and waste oi	na
Bakersfield	CA	curbside, drop-off	Yes	newspaper, cardboard, aluminum beverage containers, tin cans and other aluminum containers, plastic, glass (clear, brown, blue/green)	#1, #2
Baltimore	MD	curbside, drop-off	Yes	glass jars, glass bottles, aluminum/steel/tin cans, plastic, empty aerosol cans, newspapers, magazines, telephone books, ad mail, cardboard/boxes, mixed paper, scrap paper	#1, #2
Baton Rouge	LA	curbside, drop-off	Yes	aluminum beverage cans, foil and pie pans; metal lids, cartons, plastic, steel/tin cans, corrugated cardboard, glass (clear/colored), junk mail, magazines, catalogs, telephone books, newspapers (including inserts), scrap paper, office paper, paperboard	#1, #2
Birmingham	AL	curbside, drop-off	Yes	plastic, aluminum/steel cans, newspaper, glass (clear and green)	#1, #2
Boston	MA	curbside, drop-off	Yes	plastic bottles and jugs, glass (clear, brown and green) bottles and jars, aluminum/steel/tin cans, empty aerosol cans, aluminum foil/foil trays, newspapers, magazines, telephone books, corrugated cardboard, used motor oil, oil-based paints, pesticides	#1, #2, #3, #4, #5, #6, #7

Recycling profiles of the 100 largest cities/towns in the United States - *continued*

City	State	System Type	Voluntary	Materials Collected	Plastic Types
Brookhaven	NY	curbside, drop-off	Yes	glass bottles (clear and colored), aluminum/bimetallic/tin cans, aerosol spray cans, plastic, aluminum foil/containers, newspapers, grocery bags, corrugated cardboard cartons, mixed low grade paper, white page phone books	#1, #2
Buffalo	NY	curbside	Yes	newspapers, magazines, catalogs, telephone books, junk mail, corrugated cardboard, glass bottles(clear, green & brown), metal cans, plastic (food, soap, & beverage containers), milk/juice cartons, drink boxes, containers	na
Charlotte	NC	curbside	Yes	all glass containers, magazines, shopping catalogs, milk jugs, newspapers and inserts, cardboard, plastic soft drink and liquor bottles, spiral paper cans, aluminum/steel/tin cans, telephone books, office paper	#1, #2
Chesapeake	VA	curbside, drop-off	Yes	newspapers, telephone books, aluminum cans, pie plates and foil; glass (clear, brown and green), plastic bottles, steel food cans, household batteries, cardboard, office paper	#1, #2
Chicago	IL	curbside	Yes	newspapers, magazines, junk mail, cardboard, clean food boxes, gift boxes, telephone books, catalogs, brown paper bags, gift wrap, aluminum/steel cans, empty aerosol cans, aluminum foil and pie plates, plastic, glass (clear, green and brown), yard waste	#1, #2
Cincinnati	OH	curbside	Yes	newspapers, brown paper bags, telephone books, magazines, corrugated cardboard, plastic, soda bottles, milk jugs, detergent bottles, shampoo bottles, small mouth drink bottles, glass bottles and jars (clear, brown, green & blue), aluminum/steel/tin cans	#1, #2
Cleveland	OH	curbside	Yes	glass bottles and jars, metal cans, plastic containers, newspapers, cardboard	#1, #2
Colorado Springs	CO	drop-off	Yes	motor oil, paint, hazardous chemicals	na
Columbus	OH	curbside, drop-off	Yes	newspapers with ad slicks, magazines, corrugated cardboard boxes, brown paper bags, telephone books, junk mail, paperboard packaging, glass bottles and jars (all colors and clear), plastic, aluminum/steel cans, aerosol cans, aseptic, gable top cartons	#1, #2
Corpus Christi	TX	curbside, drop-off	Yes	newspapers, magazines, catalogs, corrugated and paperboard boxes, glass jars and bottles (brown, clear and green), plastic bottles, aluminum foil/cans, steel/tin cans	#1, #2
Dallas	TX	curbside, drop-off	Yes	newspapers and inserts, mixed paper, magazines, junk mail, home/office paper, chipboard, glass containers, plastic bottles, aluminum cans, steel/tin food cans, empty aerosol cans	#1, #2, #3
Denver	CO	curbside, drop-off	Yes	newspapers and inserts, mixed paper, magazines, junk mail, home/office paper, glass containers, aluminum cans, steel/tin food cans, empty aerosol cans	na
Des Moines	IA	curbside	Yes	glass bottles & jars (clear, green & brown), newspapers & inserts, brown paper bags, corrugated cardboard, junk mail, telephone books, magazines, catalogs, paperback books, colored paper, folders and window envelopes, all metal cans, plastic containers	#1, #2
Detroit	MI	drop-off	Yes	na	na

Recycling profiles of the 100 largest cities/towns in the United States - *continued*

City	State	System Type	Voluntary	Materials Collected	Plastic Types
El Paso	TX	curbside, drop-off	No	corrugated cardboard, brown paper bags, newspapers and inserts, magazines, junk mail, white/colored bond paper, computer paper, plastic, aluminum/steel/tin cans, aluminum foil and pie plates, copper, brass, iron, aluminum	#1, #2
Fort Wayne	IN	curbside	Yes	newspapers, magazines, catalogues, cardboard, fiberboard, phonebooks, plastic, glass (brown, clear, green), cans (aluminum, bi-metal, tin, steel), aluminum foil & pie pans, and empty steel paint cans	#1, #2
Fort Worth	TX	curbside	Yes	paper, cardboard, catalogs, envelopes, junk mail, magazines, newspapers (all sections), paper bags, telephone books, aluminum cans/baking tins, steel/tin food cans/lids, empty aerosol cans, steel paint cans, glass bottles and jars, plastic bottles/cups	#1, #2, #3, #4, #5, #6, #7
Fremont	CA	curbside, drop-off	Yes	newspapers & inserts, most white/colored paper, magazines, junk mail, brown paper bags, catalogs, window envelopes, paper egg cartons, telephone books, flattened cereal/cracker/shoe boxes, glass bottles & jars, plastic containers, aluminum/steel/tin cans, yard waste, food scraps	#1, #2, #3, #4, #5, #6, #7
Fresno	CA	curbside	No	aluminum/tin cans, cardboard, catalogs, chipboard, glass bottles & jars (all colors), junk mail (including envelopes), magazines, newspapers & inserts, plastic bottles (clear/green plastic soda & water bottles, plastic containers, phone books, yard waste	#1, #2, #3
Garland	TX	curbside, drop-off	Yes	newspapers & inserts, magazines, aluminum/steel/tin cans, empty aerosol cans, glass bottles & jars (clear and colored), plastic, six pack rings, corrugated cardboard, junk mail, brown paper bags, telephone books, white office/computer paper, chipboard	#1, #2
Glendale	AZ	curbside, drop-off	Yes	aluminum cans/foil/foil baking pans, cardboard, cartons, chipboard (without inserts), magazines, junk mail, catalogs, brown paper bags, telephone directories, newspapers and inserts, plastic containers, steel/tin cans	#1, #2
Grand Rapids	MI	curbside	Yes	aluminum/steel cans, glass bottles and jars (all colors), plastic bottles, newspapers and inserts, junk mail, corrugated cardboard, magazines, telephone books, white ledger paper, cereal boxes (with liners removed), colored/mixed paper, yard trimmings	#1, #2
Greensboro	NC	curbside, drop-off	Yes	plastic bottles & jugs, newspapers, magazines, all aluminum beverage cans, office paper, mail, notebook paper, corrugated cardboard, chipboard, brown/gray egg cartons, all steel beverage & food cans, glass (all colors, shapes & sizes), all aerosol cans	#1, #2
Hempstead	NY	curbside, drop-off	Yes	newspapers, cans, plastic, bottles/glass, aluminum foil, corrugated cardboard	#1, #2
Hialeah	FL	curbside	Yes	newspapers, aluminum/steel cans, plastics, glass bottles and jars, plastic coated cardboard, milk and juice containers	#1, #2
Honolulu	HI	drop-off	mandatory for commercial business & single-family homes	aluminum cans, glass bottles and jars, plastic beverage bottles, newspapers, corrugated cardboard, telephone books, colored/white bond paper, rubberbands, magazines, yard waste, appliances, autos, batteries, tires	na
Houston	TX	curbside, drop-off	No	newspapers, magazines, telephone books, aluminum and tin cans, junk mail, corrugated cardboard, plastic soft drink/milk/water containers, used oil	#1, #2

Recycling profiles of the 100 largest cities/towns in the United States - *continued*

City	State	System Type	Voluntary	Materials Collected	Plastic Types
Indianapolis	IN	curbside, drop-off	Yes	steel and aluminum cans, aerosol cans, glass (green, brown and clear), plastics, newspapers	#1, #2
Islip	NY	curbside	Yes	newspapers, cans, plastic, bottles/glass, corrugated cardboard	na
Jacksonville	FL	curbside	Yes	plastic, glass bottles and jars, metal and aluminum cans, newspapers and inserts, magazines, catalogs, phone books, brown paper bags, corrugated cardboard	#1, #2
Jersey City	NJ	curbside, drop-off	No	mixed newspaper, magazines, junk mail, office paper, telephone books, cardboard boxes, corrugated and laundry detergent boxes, glass bottles, aluminum cans, metal cans, milk cartons, drink boxes, water containers, food containers, household cleaner containers, laundry detergent bottles, shampoo bottles, and other related plastic containers	na
Kansas City	MO	drop-off	Yes	aluminum cans, car batteries, household (dry cell) batteries, corrugated cardboard, clothing (dry), foil/pie pans, glass bottles, magazines, newspapers, mixed office paper, paperboard, plastic bottles, tin cans, telephone books, scrap metal, yard waste	#1, #2
Las Vegas	NV	curbside	Yes	newspaper, glass/plastic containers, aluminum/tin cans, corrugated cardboard, phone books	#1, #2
Lexington-Fayette	KY	curbside	No	boxboard, brown paper bags, catalogs, corrugated cardboard, magazines, newspapers/inserts, office paper, telephone books, junk mail, empty aerosol cans, aluminum/steel cans, plastic bottles/jugs only, glass bottles & jars(blue, brown, clear, and green)	na
Lincoln	NE	drop-off	Yes	newspaper, glass containers, aluminum/steel cans, old corrugated cardboard, mixed residential paper, plastic	#1, #2
Long Beach	CA	curbside, drop-off	Yes	glass, aluminum, steel, tin, plastic, aerosol and empty paint cans, mixed paper, bundled newspapers, corrugated cardboard, used motor oil in recycling container	#1, #2, #3, #4
Los Angeles	CA	curbside	Yes	junk mail, telephone books, newspapers, magazines, brown paper bags, catalogs, envelopes (including those with windows), all cardboard boxes and cartons, all metal/aluminum cans, empty paint and aerosol cans, all glass bottles & jars, all plastic bottles	na
Louisville	KY	curbside, drop-off	Yes	junk mail, brown paper bags, telephone books, magazines, catalogs, newspapers and inserts, flattened cardboard, glass bottles and jars (clear, brown, green and blue) aluminum/steel/tin cans, empty aerosol cans, aluminum foil/pie & cake pans, plastic, office paper and envelopes	#1, #2
Lubbock	TX	drop-off	Yes	glass (clear and colored), plastic, aluminum and beverage cans, tin food cans, corrugated cardboard, newspaper, computer paper (green bar or shredded), white ledger paper, yard waste	#1, #2
Madison	WI	curbside	No	newspapers, corrugated cardboard, magazines, catalogs, brown paper bags, telephone books, glass, aluminum, tin/steel cans, plastic containers	#1, #2
Memphis	TN	curbside, drop-off	Yes	aluminum/steel cans, empty aerosol cans, plastic bottles, glass bottles and jars (clear, brown and green), newspapers and inserts, magazines, telephone books, office paper, junk mail, white/colored paper, envelopes, manila folders, stationery	#1, #2

Recycling profiles of the 100 largest cities/towns in the United States - *continued*

City	State	System Type	Voluntary	Materials Collected	Plastic Types
Mesa	AZ	curbside, drop-off	Yes	aluminum, cardboard, chipboard, glass, metal cans, newspapers, magazines, mixed paper, plastic bottles/jugs/jars that have a neck smaller than the base or those with a screw top lid, telephone books	na
Miami	FL	curbside	Yes	glass/plastic bottles, aluminum cans, magazines, office/mixed paper, corrugated cardboard, white goods, yard trash, tires	na
Milwaukee	WI	curbside	Yes	plastic, glass jars and bottles (all colors), food and beverage containers only, aluminum beverage cans, steel food cans, empty aerosol cans	#1, #2
Minneapolis	MN	curbside, drop-off	Yes	dry boxboard, office paper, mail, cans, corrugated cardboard, glass, household batteries, magazines, newspapers, telephone books, plastic bottles	na
Mobile	AL	curbside, drop-off	Yes	plastic beverage containers, aluminum beverage cans, steel cans, corrugated cardboard, newspapers, magazines, junk mail, telephone books, computer paper, cereal boxes, glass jars (brown/amber, green/blue and clear), styrofoam packing peanuts, pine straw	na
Montgomery	AL	curbside, drop-off	Yes	newspaper, white paper, computer paper, chipboard, magazines, brown paper bags, aluminum/steel cans, plastic	#1, #2
Nashville	TN	curbside, drop-off (glass and plastic must be dropped off)	Yes	dry paper, cardboard, paperboard, aluminum/steel cans, glass, plastic, paperboard, brown paper bags, office paper, envelopes, junk mail, catalogs, magazines, telephone books	na
New Orleans	LA	curbside	Yes	newspapers, magazines, telephone books, catalogs, brown paper bags, plastic, aluminum/steel/tin cans, glass (all colors), clean, unsoiled corrugated boxes	#1, #2
New York	NY	curbside, drop-off	No	newspapers, magazines, catalogs, telephone books, paper, mail, envelopes, brown paper bags, soft cover books, smooth cardboard, corrugated cardboard boxes, cans, aluminum foil wraps and trays, household metal objects, bulk metal	#1, #2
Newark	NJ	curbside, drop-off	No	glass bottles and jars, cans, bottles, corrugated cardboard, newspapers, magazines, mixed high-grade white paper, aluminum and bimetal cans, used motor oil, leaves	na
Norfolk	VA	curbside, drop-off	Yes	aluminum cans, glass bottles and jars (clear/brown), corrugated cardboard, household batteries, newspapers, plastic soda/water bottles, plastic milk/water/detergent jugs, steel cans, mixed office paper	#1, #2, #3, #4, #5, #6, #7
North Hempstead	NY	curbside, drop-off	Yes	newspapers/inserts, magazines, direct mail, catalogs, construction/wrapping paper, index/greeting cards, paperback books, plastic, food/beverage cans, aluminum foil/pie tins, glass (clear and colored)	#1, #2
Oakland	CA	curbside, drop-off	Yes	glass bottles and jars, narrow neck plastic bottles and jugs (where the neck is smaller than the base), tin and aluminum cans/foil/pie plates, milk cartons, empty spray cans, empty and dry metal latex paint containers, newspapers, catalogs, magazines, co	na
Oklahoma City	OK	curbside	Yes	plastic milk jugs and beverage bottles, aluminum and steel food and beverage cans, glass food and beverage jars and bottles, newspapers and inserts, magazines	na

Recycling profiles of the 100 largest cities/towns in the United States - *continued*

City	State	System Type	Voluntary	Materials Collected	Plastic Types
Omaha	NE	curbside, drop-off	Yes	newspapers & inserts, magazines, catalogs, telephone books, junk mail, detergent boxes, wrapping paper, paperback books, office/school paper, plastic, glass bottles & jars (clear, green, brown & blue), aluminum/steel/tin cans, corrugated cardboard	#1, #2
Oyster Bay	NY	curbside, drop-off	Yes	newspapers, cans, plastics, bottles/glass	na
Philadelphia	PA	curbside, drop-off (plastic must be dropped off)	No	newspapers and inserts, junk mail, envelopes w/ or w/out windows, telephone books, magazines, catalogs, aluminum/metal cans, empty aerosol cans, empty paint cans air dried, paint can lids separated from the paint cans, glass bottles and jars	#1, #2
Phoenix	AZ	curbside, drop-off	Yes	telephone books, plastic, food/glass bottles/jars, office paper, newspapers, magazines, cardboard, chipboard, milk/juice cartons, juice boxes, junk mail, aluminum cans/pie plates/foil, steel cans, metal hangers, scrap metal, aerosol cans	#1, #2, #6
Pittsburgh	PA	curbside, drop-off	Yes	plastic/metal containers, newspapers & inserts, corrugated cardboard, magazines, catalogs, glossy paper, white office paper, glass (clear & colored), all metal cans, empty aerosol & paint cans, aluminum foil/containers, plastic bottles, jugs, and jars	#1, #2, #3, #4, #5
Plano	TX	curbside, drop-off	Yes	newspapers, magazines, catalogs, junk mail, telephone books, brown paper bags, chipboard/boxboard, corrugated cardboard boxes, aluminum/steel/tin cans, plastic, aerosol cans, glass jars/containers/dishes/drinking glasses/vases (any color), office paper	#1, #2
Portland	OR	curbside	No	newspapers, scrap paper, glass bottles and jars, magazines, corrugated cardboard, Kraft paper, plastic bottles (including milk jugs), steel/tin cans	na
Raleigh	NC	curbside, drop-off	Yes	newspapers and inserts, magazines, catalogs, junk mail, food and beverage cans, plastic drink bottles, glass food and beverage containers	#1, #2
Richmond	VA	curbside, drop-off	Yes	newspapers, mixed paper, aluminum cans/foil, steel cans, glass bottles and jars, milk cartons, juice boxes, plastic	#1, #2
Riverside	CA	curbside, drop-off	Yes	aluminum/tin cans, ferrous, non-ferrous, car batteries, appliances, electronics, CRV and non-CRV glass, polystyrene, newspapers, white office paper, computer paper, cardboard, magazines, junk mail, gas run equipment, air conditioners, refrigerators	#1, #2
Rochester	NY	curbside	No	newspapers and inserts, magazines, glossy catalogs, corrugated cardboard, glass (clear, green and brown), plastic food and beverage containers, milk/juice cartons, empty aerosol cans, aluminum/tin/bi-metal cans, hi-grade paper	#1, #2
Sacramento	CA	curbside	Yes	glass bottles and jars (all colors), newspapers, junk mail, envelopes, telephone books, magazines, catalogs, brown paper bags, paper egg cartons, shoe boxes, computer/colored paper, paper 6-pk containers and boxes, cardboard, aluminum/tin cans, plastic	#1, #2
Saint Louis	MO	curbside, drop-off	Yes	aluminum cans, glass bottles and jars (clear, green and brown), newspapers and inserts, magazines, thin catalogs, steel food cans, plastic bottles and jugs, paperboard, office paper, corrugated cardboard	#1, #2

Recycling profiles of the 100 largest cities/towns in the United States - *continued*

City	State	System Type	Voluntary	Materials Collected	Plastic Types
Saint Paul	MN	curbside, drop-off	Yes	newspapers/inserts, mail, envelopes, magazines, catalogs, office/school paper, boxboard, corrugated cardboard boxes, glass bottles and jars (clear/colors), metal cans/lids, jar lids, bottle caps, aluminum foil/trays, good clothes/linens	#1, #2
Saint Petersburg	FL	drop-off	Yes	newspapers, glass/plastic bottles, aluminum cans, corrugated cardboard boxes	na
San Antonio	TX	curbside	Yes	newspapers and the supplements, glass jars and bottles (clear, green and brown), aluminum cans only, plastic household jars and bottles, aerosol cans, steel and tin household containers	#1, #2
San Diego	CA	curbside	Yes	all paper, cardboard, bottles and cans	#1, #2
San Francisco	CA	curbside, drop-off	Yes	glass bottles/containers/jars, aluminum foil/cans/pie tins, tin/steel cans, all plastic bottles, empty paint/aerosol cans, newspapers, junk mail, magazines, catalogs, flattened cardboard, telephone books, paper bags/packaging, organic material, paper	#2, #4, #5,
San Jose	CA	curbside	Yes	metal cans, milk and juice cartons, glass bottles and jars (brown, clear and green), plastic, carbonless paper, cardboard, catalogs, envelopes, junk mail, magazines, newspapers and inserts, paper bags, polystyrene, scrap metals, textiles	#1, #2, #3, #4, #5, #6, #7
Santa Ana	CA	curbside, drop-off	Yes	newspaper, mixed/white paper, cardboard, junk mail, magazines, telephone books, cereal/tissue boxes, glass/plastic bottles and jars, aluminum/steel/tin cans, empty aerosol cans, pie tins, plastic milk containers	na
Scottsdale	AZ	curbside, drop-off	Yes	aseptic boxes, corrugated cardboard/chipboard, glass (clear, green, amber), magazines, telephone books, aluminum/steel/tin cans, empty aerosol cans, newspapers/inserts, junk mail, brown paper bags, plastic jugs/bottles (regardless of recycling #)	na
Seattle	WA	curbside, drop-off	Yes	cardboard, magazines, junk mail, envelopes, aseptic packaging, telephone books, glass bottles and jars (all colors), steel/tin cans, plastic bottles/jugs/jars, milk and juice cartons, shredded paper and aluminum	na
Shreveport	LA	drop-off	Yes	paper, plastic, metal, glass	na
Stockton	CA	curbside, drop-off	Yes	aluminum beverage containers, steel/tin cans, glass bottles/jars, plastic, newspapers, telephone books	#1, #2
Tampa	FL	curbside	Yes	glass bottles and jars (brown, green and clear), newspapers, aluminum cans, milk and juice cartons made from paper, plastic bottles	na
Toledo	OH	curbside	Yes	junk mail, paper, boxboard, corrugated cardboard, newspapers, magazines, glass bottles, tin and aluminum cans	na
Tucson	AZ	curbside, drop-off	Yes	white/colored paper, envelopes, junk mail, magazines, catalogs, paperboard/chipboard, phone books, fiberboard, cartons, newspapers, brown paper bags, corrugated cardboard, plastic, steel/tin cans, aluminum cans/foil/baking pans, glass food/beverage containers	#1, #2
Tulsa	OK	curbside, drop-off	Yes	glass jars and bottles (green, brown and clear), plastic bottles, aluminum/steel cans, newspapers and inserts, magazines, office paper, household and auto batteries, motor oil and antifreeze	#1, #2
Virginia Beach	VA	curbside, drop-off	Yes	newspapers, cardboard, chipboard, junk mail, catalogs, magazines, telephone books, glass bottles and jars (clear, green and brown), plastic bottles with a spout only, aluminum/steel/tin cans	na

Recycling profiles of the 100 largest cities/towns in the United States - *continued*

City	State	System Type	Voluntary	Materials Collected	Plastic Types
Washington	DC	curbside	Yes	newspapers, corrugated cardboard, computer and office paper, metal food and beverage cans, magazines, catalogs, telephone books, plastic bottles and jugs, glass jars and bottles	#1, #2
Wichita	KS	curbside (private-no public trash collection system)	Yes	varies	na

Source: *Independent research by the editors, July 2005*

Occupational and environmental exposures capable of causing illness

AOEC Code[1]	Primary Name	Synonym	Use[2]	Asthma Inducer?	CAS Number[3]
190.12	1,1,2-Trichloroethane	Vinyl Trichloride	S		79-00-5
190.12	1,1,2-Trichloroethane	1,1,2-Trichloroethane	S		79-00-5
190.12	1,1,2-Trichloroethane	Ethane Trichloride	S		79-00-5
180.06	1,2,3-Trihydroxybenzene	1,2,3-Trihydroxybenzene			87-66-1
180.06	1,2,3-Trihydroxybenzene	Pyrogallol			87-66-1
180.06	1,2,3-Trihydroxybenzene	Pyrogallic Acid			87-66-1
201.02	1,2,4-Trichlorobenzene	1,2,4-Trichlorobenzene			120-82-1
190.06	1,2-Dichloroethylene	1,2-Dichloroethylene	S		540-59-0
190.06	1,2-Dichloroethylene	Acetylene Dichloride	S		540-59-0
080.04	1,3-Butylene Glycol	1,3-Butylene Glycol			107-88-0
190.17	1,3-Dichloro-2-Propanol	1,3-Dichloro-2-Propanol			96-23-1
191.04	1-Bromo-3-chloro-5,5-dimethylhydantoin	BCDMH			16079-88-2
191.04	1-Bromo-3-chloro-5,5-dimethylhydantoin	di-Halo			16079-88-2
200.10	2,6-Dichlorobenzonitrile	Dichlobenil	P		1194-65-6
200.10	2,6-Dichlorobenzonitrile	2,6-Dichlorobenzonitrile	P		1194-65-6
141.05	2-Ethoxyethyl Acetate	Glycol Monoethyl Ether Acetate			111-15-9
141.05	2-Ethoxyethyl Acetate	2-Ethoxyethyl Acetate			111-15-9
142.09	2-Hydroxypropyl Acrylate	Acrylic Acid, 2-Hydroxypropylester			999-61-1
142.09	2-Hydroxypropyl Acrylate	2-Hydroxypropyl Acrylate			999-61-1
250.13	3,3'-Dichlorobenzidene & Salts	3,3'-Dichlorobenzidene & Salts			91-94-1
232.05	3-DMAPA	3-DMAPA		Yes	
232.05	3-DMAPA	3-Dimethylamino Propylamine		Yes	
310.13	3-mercaptopropionate	propionic acid			7575-23-7
310.13	3-mercaptopropionate	3-mercapto-			7575-23-7
310.13	3-mercaptopropionate	Pentaerythritol tetrakis			7575-23-7
310.12	3-mercaptopropionic acid	propionic acid			107-96-0
310.12	3-mercaptopropionic acid	mercaptopropionic acid			107-96-0
310.12	3-mercaptopropionic acid	3-mercaptopropionic acid			107-96-0
310.12	3-mercaptopropionic acid	beta-thiopropionic acid			107-96-0
310.12	3-mercaptopropionic acid	3-thiopropionic acid			107-96-0
310.12	3-mercaptopropionic acid	beta-mercaptopropionic acid			107-96-0
250.03	4,4'-Methylenebis(2-Chloroaniline)	ME-MDA			101-14-4
250.03	4,4'-Methylenebis(2-Chloroaniline)	4,4'-Methylenebis(2-Chloroaniline)			101-14-4
250.03	4,4'-Methylenebis(2-Chloroaniline)	MOCA			101-14-4
060.11	4-PC	4-Phenylcyclohexene			
060.11	4-PC	4-PC			
060.11	4-PC	New Carpet Odor			
250.02	8-Hydroxyquinoline	8-Quinolinol			
250.02	8-Hydroxyquinoline	8-Hydroxyquinoline			148-24-3
373.08	Abiruana	Abiruana		Yes	
373.08	Abiruana	Pouteria		Yes	
010.01	Abrasives, NOS	Grinding Dust			
010.01	Abrasives, NOS	Abrasives, NOS			
382.11	Acarian	Acarian		Yes	
291.11	Acephate	Acephate	P	Yes	30560-19-1
291.11	Acephate	Orthene	P	Yes	30560-19-1
120.01	Acetaldehyde	Acetaldehyde			75-07-0
260.01	Acetamide	Acetamide			60-35-5
141.00	Acetates, NOS	Acetates, NOS			
050.34	Acetic Acid	Glacial Acetic Acid		Yes	64-19-7
050.34	Acetic Acid	Acetic Acid			64-19-7
151.09	Acetic Anhydride	Acetic Anhydride			108-24-7
151.09	Acetic Anhydride	Acetic Oxide			108-24-7
130.01	Acetone	Dimethyl Ketone	S		67-64-1
130.01	Acetone	2-Propanone	S		67-64-1
130.01	Acetone	Acetone	S		67-64-1
210.01	Acetonitrile	Acetonitrile			75-05-8
060.01	Acetylene	Ethine			74-86-2
060.01	Acetylene	Acetylene			74-86-2

Occupational and environmental exposures capable of causing illness - *continued*

AOEC Code[1]	Primary Name	Synonym	Use[2]	Asthma Inducer?	CAS Number[3]
060.01	Acetylene	Ethyne			74-86-2
050.33	Acid & Base Mixture	Acid & Base Mixture			
050.26	Acid Solder	Acid Solder			
050.25	Acid Stripper	Acid Stripper			
050.00	Acids, Bases, Oxidizers, NOS	Acids, Bases, Oxidizers, NOS			
050.00	Acids, Bases, Oxidizers, NOS	Inorganic Acids, NOS			
250.15	Acridine	Acridine			260-94-6
120.02	Acrolein	2-Propenal			107-02-8
120.02	Acrolein	Acrylic Aldehyde			107-02-8
120.02	Acrolein	Acrolein			107-02-8
260.02	Acrylamide	Propenamide			79-06-1
260.02	Acrylamide	Acrylic Amide			79-06-1
260.02	Acrylamide	Acrylamide			79-06-1
142.00	Acrylates, NOS	Acrylates, NOS			
150.08	Acrylic Acid	Acrylic Acid			79-10-7
142.01	Acrylic Monomer	Acrylate			
142.01	Acrylic Monomer	Acrylic Monomer			
270.01	Acrylics	Acrylic Resins			
270.01	Acrylics	Acrylic Acid Polymer			
270.01	Acrylics	Acrylics			
270.01	Acrylics	Paint, Acrylic			
210.02	Acrylonitrile	Vinyl Cyanide			107-13-1
210.02	Acrylonitrile	Acrylonitrile			107-13-1
210.02	Acrylonitrile	Propenenitrile			107-13-1
270.29	Acrylonitrile-Butadiene-Styrene Copolymer	Acrylonitrile-Butadiene-Styrene Copolymer			9003-56-9
270.29	Acrylonitrile-Butadiene-Styrene Copolymer	ABS Copolymer			9003-56-9
270.30	Acrylonitrile-Butadiene-Styrene-Polyvinyl Chloride	Acrylonitrile-Butadiene-Styrene-Polyvinyl Chloride			
110.06	Adhesive, Epoxy	Adhesive, Epoxy		Yes	
373.09	African Maple	Triplochiton Scleroxylon		Yes	
373.09	African Maple	African Maple		Yes	
373.13	African Zebrawood	Microberlinia		Yes	
373.13	African Zebrawood	African Zebrawood		Yes	
200.01	Agent Orange	2,4-D	P		94-75-7
200.01	Agent Orange	2,4,5-T	P		93-76-5
200.01	Agent Orange	Agent Orange	P		39277-47-9
320.01	Air Pollutants, Indoor	Air Pollutants, Indoor			
320.01	Air Pollutants, Indoor	Sick Building			
320.01	Air Pollutants, Indoor	Ventilation, Inadequate			
320.02	Air Pollutants, Outdoor	Air Pollutants, Outdoor			
350.07	Air Pressure, Changes	Air Pressure, Changes			
350.07	Air Pressure, Changes	Barometric Pressure, Changes			
070.00	Alcohols, NOS	Alcohols, NOS	S		
280.01	Aldrin	Aldrin	P		309-00-2
060.00	Aliphatic Hydrocarbons, NOS	Alicyclic Hydrocarbons, NOS			
060.00	Aliphatic Hydrocarbons, NOS	Aliphatic Hydrocarbons, NOS			
270.21	Alkyd Resins	Alkyd Resins			63148-69-6
070.10	Alkyl Aryl Polyether Alcohol/Polypropylene Glycol	Alkyl Aryl Polyether Alcohol/Polypropylene Glycol Mixt.		Yes	
070.01	Allyl Alcohol	Vinyl Carbinol			107-18-6
070.01	Allyl Alcohol	2-Propen-1-ol			107-18-6
070.01	Allyl Alcohol	Allyl Alcohol			107-18-6
251.04	Alpha-Naphthylamine	Alpha-Naphthylamine			134-32-7
391.12	Alternaria	Alternaria alternata		Yes	
391.12	Alternaria	Alternaria spp		Yes	
391.12	Alternaria	Alternaria alternata toxin		Yes	
020.01	Aluminum	Aluminum		Yes	7429-90-5
020.01	Aluminum	Aluminum Compounds		Yes	
050.30	Aluminum Hydroxide	Aluminum Hydroxide			21645-51-2
020.02	Aluminum Oxide	Corundum			1302-74-5

Occupational and environmental exposures capable of causing illness - *continued*

AOEC Code[1]	Primary Name	Synonym	Use[2]	Asthma Inducer?	CAS Number[3]
020.02	Aluminum Oxide	Alumina			1344-28-1
020.02	Aluminum Oxide	Aluminum Oxide			1344-28-1
040.31	Aluminum Phosphide	Aluminum Phosphide	P		20859-73-8
040.31	Aluminum Phosphide	Phostoxin	P		20859-73-8
252.01	Amaranth	Amaranth			915-67-3
252.01	Amaranth	FD&C Red No. 2			915-67-3
250.19	Amino Acids, NOS	Amino Acids, NOS			
270.22	Amino Resins	Amino Resins			
231.03	Aminoethyl Ethanolamine	Aminoethyl Ethanolamine		Yes	
231.03	Aminoethyl Ethanolamine	Ethanol Ethylene Diamine		Yes	
052.01	Ammonia Gas	Ammonia Gas			7664-41-7
052.01	Ammonia Gas	Ammonia, Anhydrous			7664-41-7
322.09	Ammonia Solution (10%)	Ammonia Solution (10%)			1336-21-6
322.09	Ammonia Solution (10%)	Ammonia, Household			1336-21-6
322.08	Ammonia Solution (29%)	Ammonia Solution (29%)			1336-21-6
322.07	Ammonia Solution, NOS	Ammonia Solution, NOS			1336-21-6
322.07	Ammonia Solution, NOS	Ammonium Hydroxide, NOS			1336-21-6
022.02	Ammonium Bichromate	Ammonium Bichromate		Yes	7789-09-5
024.01	Ammonium Hexachloroplatinate (IV)	Ammonium Hexachloroplatinate (IV)		Yes	16919-58-7
052.04	Ammonium Persulfate	Ammonium Persulfate			7727-54-0
040.32	Ammonium Phosphate	Ammonium Phosphate	P		7722-76-1
040.32	Ammonium Phosphate	Monoammonium Phosphate	P		7722-76-1
040.32	Ammonium Phosphate	MAP	P		7722-76-1
052.03	Ammonium Salts	Ammonium Salts			
052.03	Ammonium Salts	Ammonium Chloride			12125-02-9
321.11	Amprolium	Amprolium		Yes	
141.01	Amyl Acetate	N-Pentyl Acetate			628-63-7
141.01	Amyl Acetate	Amyl Acetate			628-63-7
141.01	Amyl Acetate	N-Amyl Acetate			628-63-7
070.02	Amyl Alcohol	Amyl Alcohol			71-41-0
070.02	Amyl Alcohol	N-Amyl Alcohol			71-41-0
070.02	Amyl Alcohol	1-Pentanol			71-41-0
260.27	Amyl Nitrite	Snappers			110-46-3
260.27	Amyl Nitrite	Amyl Nitrite			110-46-3
260.27	Amyl Nitrite	Isoamyl Nitrite			110-46-3
260.27	Amyl Nitrite	Poppers			110-46-3
190.15	Anesthetic Gases, Halogenated	Enflurane		Yes	13838-16-9
190.15	Anesthetic Gases, Halogenated	Halothane			151-67-7
190.15	Anesthetic Gases, Halogenated	Methoxyflurane			76-38-0
190.15	Anesthetic Gases, Halogenated	Anesthetic Ethers, NOS			
190.15	Anesthetic Gases, Halogenated	Anesthetic Gases, Halogenated			
320.03	Anesthetic Gases, NOS	Anesthetic Gases, NOS			
250.01	Aniline	Aminophen			62-53-3
250.01	Aniline	Aminobenzene			62-53-3
250.01	Aniline	Phenylamine			62-53-3
250.01	Aniline	Aniline			62-53-3
251.00	Aniline Dyes, NOS	Aniline Dyes, NOS			
380.00	Animal Material, NOS	Laboratory Animals, NOS (see specific animals)			
380.00	Animal Material, NOS	Animal Material, NOS (see specific animals)			
161.01	Anthracene	Anthracene			120-12-7
380.01	Antigens, Animal	Antigens, Animal (see specific animals)			
020.03	Antimony	Antimony			7440-36-0
020.03	Antimony	Antimony Compounds			
040.01	Argon	Argon			7440-37-1
160.00	Aromatic Hydrocarbons, NOS	Aromatic Hydrocarbons, NOS	S		
160.00	Aromatic Hydrocarbons, NOS	Aromatic Solvents, NOS	S		
020.05	Arsenic	Arsenic Compounds			
020.05	Arsenic	Arsenic			7440-38-2

Occupational and environmental exposures capable of causing illness - *continued*

AOEC Code[1]	Primary Name	Synonym	Use[2]	Asthma Inducer?	CAS Number[3]
020.06	Arsine	Arsine			7784-42-1
010.02	Asbestos	Asbestos			1332-21-4
010.02	Asbestos	Amosite			1332-21-4
010.02	Asbestos	Tremolite			77536-68-6
010.02	Asbestos	Crocidolite			12001-28-4
010.02	Asbestos	Chrysotile			12001-29-5
010.16	Ash, NOS	Ash, NOS			
010.16	Ash, NOS	Wood Ash			
373.17	Ashwood	Fraxinus Americana		Yes	
373.17	Ashwood	Ashwood		Yes	
391.02	Aspergillus	Aspergillus			
391.02	Aspergillus	Aspergillus flavus			
391.02	Aspergillus	Aspergillus fumigatus			
391.02	Aspergillus	Aspergillus glaucus			
391.02	Aspergillus	Aspergillus versicolor			
391.02	Aspergillus	Aspergillus niger			
061.07	Asphalt	Road Tar			
061.07	Asphalt	Asphalt			8052-42-4
061.07	Asphalt	Petroleum			
061.07	Asphalt	Road Pitch			
040.25	Asphyxiant Gases, NOS	Asphyxiant Gases, NOS			
353.12	Assault, Physical	Violence, Physical Assault			
353.12	Assault, Physical	Assault, Physical			
251.02	Auramine	Auramine			492-80-8
380.16	Avian Material, NOS	Avian Material, NOS			
252.00	Azo Compounds, NOS	Azo Compounds, NOS			
252.00	Azo Compounds, NOS	Azo Dyes, NOS			
260.18	Azodicarbamide	1,1-Azobisformamide		Yes	123-77-3
260.18	Azodicarbamide	Azobisformamide		Yes	123-77-3
260.18	Azodicarbamide	Azodicarbamide		Yes	123-77-3
370.17	Baby's Breath	Baby's Breath		Yes	
370.17	Baby's Breath	Gypsophilia Paniculata		Yes	
324.01	Bacillus Subtilis	Bacillus Subtilis		Yes	
390.14	Bacillus Thurigiensis	Bacillus Thurigiensis	P		23526-02-5
390.14	Bacillus Thurigiensis	Bt	P		23526-02-5
020.07	Barium	Barium Compounds			
020.07	Barium	Barium			7440-39-3
382.15	Barn Mite	Barn Mite		Yes	
380.12	Bat Guano	Bat Guano		Yes	
292.04	Baygon	Baygon	P		114-26-1
292.04	Baygon	Propoxur	P		114-26-1
382.05	Bee Moth	Bee Moth		Yes	
260.03	Benlate	Benlate Fungicide	P		17804-35-2
260.03	Benlate	Benlate	P		17804-35-2
160.01	Benzene	Coal Naphtha	S		71-43-2
160.01	Benzene	Mineral Naphtha	S		71-43-2
160.01	Benzene	Benzene	S		71-43-2
160.01	Benzene	Benzol	S		71-43-2
251.03	Benzidine	Benzidine			92-87-5
251.03	Benzidine	4,4-Aminodiphenyl			92-87-5
150.02	Benzoic Acid	Benzoic Acid			65-85-0
200.09	Benzoyl Chloride	Benzoyl Chloride			98-88-4
050.27	Benzoyl Peroxide	Benzoyl Peroxide			94-36-0
200.02	Benzyl Chloride	Alpha-Chlorotoluene			100-44-7
200.02	Benzyl Chloride	Benzyl Chloride			100-44-7
020.08	Beryllium	Beryllium			7440-41-7
020.08	Beryllium	Beryllium Compounds			
070.03	Beta-Chloroethyl Alcohol	Ethylene Chlorohydrin			107-07-3
070.03	Beta-Chloroethyl Alcohol	Beta-Chloroethyl Alcohol			107-07-3
070.03	Beta-Chloroethyl Alcohol	2-Chloroethanol			107-07-3

Occupational and environmental exposures capable of causing illness - *continued*

AOEC Code[1]	Primary Name	Synonym	Use[2]	Asthma Inducer?	CAS Number[3]
251.05	Beta-Naphthylamine	Beta-Naphthylamine			
251.05	Beta-Naphthylamine	2-Naphthylamine			
320.04	Beta-Propiolactone	2-Oxetanone			57-57-8
320.04	Beta-Propiolactone	Beta-Propiolactone			57-57-8
320.04	Beta-Propiolactone	BPL			57-57-8
100.01	Bis Chloromethyl Ether	Dichloromethyl Ether			542-88-1
100.01	Bis Chloromethyl Ether	Chloromethyl Ether			542-88-1
100.01	Bis Chloromethyl Ether	Bis Chloromethyl Ether			542-88-1
100.01	Bis Chloromethyl Ether	BCME			542-88-1
020.09	Bismuth	Bismuth Compounds			
020.09	Bismuth	Bismuth			7440-69-9
322.10	Bleach	Bleach			7681-52-9
322.10	Bleach	Sodium Hypochlorite			7681-52-9
322.11	Bleach plus Acid (mixture)	Bleach plus Acid (mixture)			
322.12	Bleach plus Ammonia (mixture)	Bleach plus Ammonia (mixture)			
382.18	Bloodworm	Bloodworm			
360.07	Bodily Reaction	Bodily Reaction			
390.16	Body Fluid Exposure (Unknown Infection Status)	Body Fluid Exposure (Unknown Exposure Status)			
390.16	Body Fluid Exposure (Unknown Infection Status)	Blood Exposure (Unknown Infection Status)			
020.10	Boron	Boron			7440-42-8
020.10	Boron	Boric Acid			1043-35-3
020.10	Boron	Boron Compounds			
390.15	Borrelium Virus	Borrelium Virus			
380.13	Bovine Serum	Bovine Serum			
020.43	Brass	Brass			
324.07	Bromelin	Bromelain		Yes	
324.07	Bromelin	Bromelin		Yes	
192.03	Brominated Fluorocarbon	Brominated Fluorocarbon			
191.00	Brominated Pesticides, NOS	Brominated Pesticides, NOS	P		
030.01	Bromine	Bromine			7726-95-6
371.01	Buckwheat	Buckwheat		Yes	
271.01	Butadiene & Styrene	SBR			
271.01	Butadiene & Styrene	Styrene-Butadiene Copolymer			
271.01	Butadiene & Styrene	Butadiene & Styrene			
141.02	Butyl Acetate	Butyl Acetate	S		123-86-4
141.02	Butyl Acetate	N-Butyl Acetate	S		123-86-4
142.02	Butyl Acrylate	N-Butyl Acrylate			141-32-2
142.02	Butyl Acrylate	Acrylic Acid Butyl Ester			141-32-2
142.02	Butyl Acrylate	Butyl Acrylate			141-32-2
130.06	Butyl Ketone	Butyl Ketone	S		502-56-7
130.06	Butyl Ketone	5-Nonanone	S		502-56-7
373.23	Cabreuva	Cabreuva		Yes	
373.23	Cabreuva	Myrocarpus Fastigiatus Fr. All.		Yes	
370.32	Cacoon Seed	Cacoon Seed		Yes	
020.12	Cadmium	Cadmium			7440-43-9
020.12	Cadmium	Cadmium Salts			
020.12	Cadmium	Cadmium Compounds			
050.03	Calcium Bisulfite	Calcium Bisulfite			13780-03-5
050.36	Calcium Carbide	Calcium Carbide			75-20-7
050.35	Calcium Carbonate	Calcium Carbonate			1317-65-3
041.01	Calcium Chloride	Calcium Chloride			10043-52-4
260.04	Calcium Cyanamide	Calcium Cyanamide			156-62-7
322.13	Calcium Hypochloride	Calcium Hypochloride			7778-54-3
322.13	Calcium Hypochloride	Bleaching Powder			7778-54-3
322.13	Calcium Hypochloride	Lime Chloride			7778-54-3
050.05	Calcium Oxide	Calcium Oxide			1305-78-8
050.05	Calcium Oxide	Lime			1305-78-8
050.05	Calcium Oxide	Burnt Lime			1305-78-8

Occupational and environmental exposures capable of causing illness - continued

AOEC Code[1]	Primary Name	Synonym	Use[2]	Asthma Inducer?	CAS Number[3]
050.05	Calcium Oxide	Quicklime			1305-78-8
040.02	Calcium Salts, NOS	Calcium Salts, NOS			
373.02	California Redwood	Sequoia Sempervirens		Yes	
373.02	California Redwood	California Redwood		Yes	
370.35	Capsicum	Capsicum			
370.35	Capsicum	Pepper Spray			
260.05	Captan	Captan	P		133-06-2
292.00	Carbamate Pesticides, NOS	Carbamate Pesticides, NOS	P		
010.04	Carbon Black	Carbon Black			1333-86-4
010.04	Carbon Black	Actibon			
010.04	Carbon Black	Toner, Copier			
010.04	Carbon Black	Activated Carbon			7440-44-0
040.03	Carbon Dioxide	Carbon Dioxide			124-38-9
310.01	Carbon Disulfide	Carbon Disulfide			75-15-0
040.04	Carbon Monoxide	Carbon Monoxide			630-08-0
040.04	Carbon Monoxide	CO			630-08-0
190.01	Carbon Tetrachloride	Carbon Tetrachloride	S		56-23-5
190.01	Carbon Tetrachloride	Tetrachloromethane	S		56-23-5
320.05	Carbonless Paper	NCR Paper			
320.05	Carbonless Paper	Carbonless Paper			
270.36	Carbopol, NOS	Carbopol			
270.36	Carbopol, NOS	Carbopol, NOS			
320.24	Carmine	Carmine			1260-17-9
320.24	Carmine	Carminic Acid			1260-17-9
320.35	Carpet Dust	Carpet Fibers			
320.35	Carpet Dust	Carpet Dust			
380.11	Casein	Casein		Yes	
370.13	Castor Bean	Castor Bean		Yes	
380.22	Cat	Fel d			
353.10	Caught In or Between Objects	Caught In or Between Objects			
321.26	Ceclor	Ceclor			53994-73-3
321.26	Ceclor	Cefaclor			53994-73-3
373.03	Cedar of Lebanon	Cedar of Lebanon		Yes	
373.03	Cedar of Lebanon	Cedrus Libani		Yes	
010.03	Cement Dust	Cement Dust			65997-15-1
010.03	Cement Dust	Portland Cement			65997-15-1
373.11	Central American Walnut	Juglans Olanchana		Yes	
373.11	Central American Walnut	Central American Walnut		Yes	
321.06	Cephalosporins	Cephalosporins		Yes	
020.13	Cerium	Cerium Compounds			
020.13	Cerium	Cerium			
320.06	Chemicals, NOS	Chemicals, NOS			
320.06	Chemicals, NOS	Multiple Chemicals			
320.06	Chemicals, NOS	Chemicals, Unknown			
320.06	Chemicals, NOS	Chemical Dust, NOS			
321.01	Chemotherapeutic Drugs	Chemotherapeutic Drugs			
380.07	Chicken	Chicken		Yes	
370.18	Chicory	Chicory		Yes	
320.31	Chil-Perm CP-30	Chil-Perm CP-30	S		
050.06	Chloramine	Chloramine			55-86-7
260.21	Chloramine T	Chloramine T		Yes	127-65-1
280.02	Chlordane	Chlordane	P		57-74-9
200.08	Chlorhexidine	Chlorhexidine		Yes	
201.00	Chlorinated Benzenes, NOS	Chlorinated Benzenes, NOS			
200.04	Chlorinated Dibenzodioxins	2,3,7,8-TCDD	P		1746-01-6
200.04	Chlorinated Dibenzodioxins	Chlorinated Dibenzodioxins	P		
200.04	Chlorinated Dibenzodioxins	Dioxin	P		
190.00	Chlorinated Hydrocarbons, NOS	Chlorinated Hydrocarbons, NOS	S		
190.00	Chlorinated Hydrocarbons, NOS	Chlorinated Solvents, NOS	S		

Occupational and environmental exposures capable of causing illness - *continued*

AOEC Code[1]	Primary Name	Synonym	Use[2]	Asthma Inducer?	CAS Number[3]
200.03	Chlorinated Naphthalene	Halowax			
200.03	Chlorinated Naphthalene	Chlorinated Naphthalene			
181.00	Chlorinated Phenols, NOS	Chlorinated Phenols, NOS			
030.02	Chlorine	Chlorine		Yes	7782-50-5
030.06	Chlorine Dioxide	Chlorine Dioxide			10049-04-4
030.05	Chlorine Trifluoride	Chlorine Trifluoride			7790-91-2
201.04	Chlorobenzene	Phenyl Chloride	S		108-90-7
201.04	Chlorobenzene	Chlorobenzene	S		108-90-7
201.04	Chlorobenzene	Chlorbenzol	S		108-90-7
190.02	Chloroethane	Chloroethane	S		75-00-3
190.02	Chloroethane	Monochloroethane	S		75-00-3
190.02	Chloroethane	Ethyl Chloride	S		75-00-3
192.02	Chlorofluorocarbon, NOS	CFC			
192.02	Chlorofluorocarbon, NOS	Chlorofluorocarbon, NOS			
190.03	Chloroform	Chloroform			67-66-3
190.03	Chloroform	Trichloromethane			67-66-3
100.02	Chloromethyl Methyl Ether	CMME			107-30-2
100.02	Chloromethyl Methyl Ether	Chloromethyl Methyl Ether			107-30-2
200.11	Chlorophenoxy Herbicides, NOS	Chlorophenoxy Herbicides, NOS	P		
260.06	Chloropicrin	Trichloronitromethane	P		76-06-2
260.06	Chloropicrin	Chloropicrin	P		76-06-2
190.04	Chloroprene	Chlorobutadiene			126-99-8
190.04	Chloroprene	Chloroprene			126-99-8
190.05	Chloropropane	1-Propyl Chloride	S		540-54-5
190.05	Chloropropane	Chloropropane	S		540-54-5
190.05	Chloropropane	N-Chloropropane	S		540-54-5
210.03	Chlorothalonil	Tetrachloro-Isophthalonitrile	P	Yes	1897-45-6
210.03	Chlorothalonil	Chlorothalonil	P	Yes	1897-45-6
291.05	Chlorpyrifos	Dursban	P		2921-88-2
291.05	Chlorpyrifos	Chlorpyrifos	P		2921-88-2
370.36	Chorella Algae	Chorella Algae		Yes	
022.04	Chromic Acid	Chromic Acid		Yes	1333-82-0
022.00	Chromium, Hexavalent, NOS	Chromium, Hexavalent, NOS		Yes	
020.14	Chromium, Not Hexavalent	Chromium Metal		Yes	7440-47-3
020.14	Chromium, Not Hexavalent	Chromium, Not Hexavalent		Yes	7440-47-3
382.22	Chrysoperla carnea	Green Lacewing		Yes	
382.22	Chrysoperla carnea	Chrysoperla carnea		Yes	
382.22	Chrysoperla carnea	C. carnea		Yes	
325.03	Cibachrome Brilliant Scarlet 32	Cibachrome Brilliant Scarlet 32		Yes	
330.01	Cigarette Smoke	Cigarette Smoke			
330.01	Cigarette Smoke	Environmental Tobacco Smoke			
330.01	Cigarette Smoke	ETS			
321.17	Cimetidine	Cimetidine		Yes	51481-61-9
120.08	Cinnamic Aldehyde	Cinnamic Aldehyde			104-55-2
120.08	Cinnamic Aldehyde	Cinnamon Oil			8007-80-5
373.22	Cinnamon	Cinnamon		Yes	
373.22	Cinnamon	Cinnamomum Zeylanicum		Yes	
050.07	Citric Acid	Citric Acid			77-92-9
381.10	Clam	Clam		Yes	
010.05	Clay	Terra Cotta			
010.05	Clay	Clay			1332-58-7
322.14	Cleaners, Abrasive	Cleaners, Abrasive			
322.15	Cleaners, Acid	Cleaners, Acid			
322.16	Cleaners, Carpet	Cleaners, Carpet			
322.17	Cleaners, Caustic (excluding Lye)	Cleaners, Caustic (excluding Lye)			
322.18	Cleaners, Detergent, NOS	Cleaners, Detergent, NOS			
322.19	Cleaners, Disinfectant, NOS	Cleaners, Disinfectant, NOS			
322.20	Cleaners, Drain	Cleaners, Drain			
322.21	Cleaners, Floor Stripping	Cleaners, Floor Stripping			

Occupational and environmental exposures capable of causing illness - continued

AOEC Code[1]	Primary Name	Synonym	Use[2]	Asthma Inducer?	CAS Number[3]
322.22	Cleaners, Graffiti Removing	Cleaners, Graffiti Removing	S		
322.04	Cleaners, Household, General Purpose	Cleaners, Household, General Purpose			
322.23	Cleaners, Laundry Soap/Detergent	Cleaners, Laundry Soap/Detergent			
322.24	Cleaners, Lye	Cleaners, Lye			
322.25	Cleaners, Oven	Cleaners, Oven			
322.26	Cleaners, Pine Oil	Cleaners, Pine Oil			
322.06	Cleaners, Solvent-Based	Cleaners, Solvent-Based	S		
322.06	Cleaners, Solvent-Based	ILA Soap	S		
322.27	Cleaners, Tile	Cleaners, Tile			
322.28	Cleaners, Toilet Bowl	Cleaners, Toilet Bowl			
322.29	Cleaners, Wallpaper	Cleaners, Wallpaper			
322.03	Cleaning Fluids, Photocopier	Cleaning Fluids, Photocopier			
322.30	Cleaning Fluids/Spot Removers	Cleaning Fluids/Spot Removers	S		
322.00	Cleaning Materials, NOS	Cleaning Materials, NOS			
322.31	Cleaning Mixtures (excluding Bleach plus Acid or Ammonia)	Cleaning Mixtures (excluding Bleach plus Acid or Ammonia)			
010.06	Coal	Coal			
020.15	Cobalt	Cobalt		Yes	7440-48-4
020.15	Cobalt	Cobalt Compounds			
373.04	Cocabolla	Cocobolo		Yes	
373.04	Cocabolla	Dalbergia Retusa		Yes	
373.04	Cocabolla	Cocabolla		Yes	
370.47	Cocoa Bean	Cocoa			
370.47	Cocoa Bean	Cocoa Bean			
370.12	Coffee Bean	Coffee Bean		Yes	
370.12	Coffee Bean	Green Coffee Bean		Yes	
350.02	Cold	Low Temperature			
350.02	Cold	Cold			
270.39	Collodion	Collodium			
023.05	Colophony	Colophony		Yes	
360.01	Contact Pressure	Contact Pressure			
360.01	Contact Pressure	Mechanical Pressure			
360.01	Contact Pressure	Skin Contact			
020.16	Copper	Copper Compounds			
020.16	Copper	Copper			7440-50-8
020.49	Copper Phthalocyanine	Copper Phthalocyanine			147-14-8
020.49	Copper Phthalocyanine	Cuprilinic Blue			147-14-8
020.51	Copper Sulfate	Copper Sulfate			7758-98-7
020.51	Copper Sulfate	Copper Sulfate (Anhydrous)			7758-98-7
370.51	Corn Dust	Corn Dust			
371.06	Corn Starch	Corn Starch			9005-25-8
320.28	Cosmetics, NOS	Cosmetology Chemicals, NOS		Yes	
320.28	Cosmetics, NOS	Cosmetics, NOS			
370.02	Cotton Dust	Cotton Dust			
321.03	Coumarin	Coumarin			
381.02	Crab	Crab		Yes	
180.01	Creosote	Brick Oil			8001-58-9
180.01	Creosote	Coal Tar Oil			8001-58-9
180.01	Creosote	Creosote			8001-58-9
180.02	Cresol	Cresol			1319-77-3
180.02	Cresol	Methylphenol			1319-77-3
180.02	Cresol	Cresylic Acid			1319-77-3
382.04	Cricket	Cricket		Yes	
120.09	Crotonaldehyde	Crotonaldehyde			4170-30-3
170.01	Cutting Oils	Oil Mist		Yes	
170.01	Cutting Oils	Metal Working Fluids		Yes	
170.01	Cutting Oils	Cutting Oils		Yes	
353.11	Cutting or Piercing Object, Except Blood-Contam. Sharps	Cutting or Piercing Object, Except Blood-Contam. Sharps			
381.06	Cuttlefish	Cuttlefish		Yes	

Occupational and environmental exposures capable of causing illness - *continued*

AOEC Code[1]	Primary Name	Synonym	Use[2]	Asthma Inducer?	CAS Number[3]
211.00	Cyanides, NOS	Cyanides, NOS			
142.07	Cyanoacrylates, NOS	Cyanoacrylates, NOS		Yes	
142.07	Cyanoacrylates, NOS	Alkylcyanoacrylates, NOS		Yes	
060.14	Cyclohexane	Cyclohexane	S		110-82-7
130.07	Cyclohexanone	Cyclohexanone	S		108-94-1
060.13	Cyclopentadiene	Cyclopentadiene			542-92-7
321.29	Cyclophosphamide	Neosar			50-18-0
321.29	Cyclophosphamide	Cyclophosphamide			50-18-0
321.29	Cyclophosphamide	Cytoxan			50-18-0
060.18	D-Limonene	D-Limonene	S		5989-27-5
380.04	Dander, Animal	Cow Dander		Yes	
380.04	Dander, Animal	Dander, Animal			
382.10	Daphnia	Daphnia		Yes	
280.03	DDT	Dichlorodiphenyltrichloroethane	P		50-29-3
280.03	DDT	DDT	P		50-29-3
171.04	Degreaser, NOS	Degreaser, NOS			
320.37	Denatonium benzoate	Bitrex			3734-33-6
320.06	Deodorant, Aerosol	Deodorant, Aerosol			
291.03	Diazinon	Diazinon	P	Yes	333-41-5
260.23	Diazomethane	Diazomethane			334-88-3
252.03	Diazonium Salt	Diazonium Salt		Yes	
020.11	Diborane	Diborane			19287-45-7
020.11	Diborane	Boron Hydride			19287-45-7
201.01	Dichlorobenzene	Mixed DCB	P		
201.01	Dichlorobenzene	P-Dichlorobenzene	P		106-46-7
201.01	Dichlorobenzene	Dichlorobenzene	P		
201.01	Dichlorobenzene	DCB	P		
100.03	Dichloroethyl Ether	Dichloroethyl Ether			111-44-4
100.03	Dichloroethyl Ether	Chlorex			111-44-4
100.03	Dichloroethyl Ether	Bis 2-Chloroethyl Ether			111-44-4
040.27	Dichlorosilane	Dichlorosilane			
291.10	Dichlorvos	Dichlorvos	P	Yes	62-73-7
280.04	Dieldrin	Dieldrin	P		60-57-1
331.01	Diesel Exhaust	Diesel Exhaust			
061.06	Diesel Fuel	No. 2 Fuel Oil	S		
061.06	Diesel Fuel	Diesel Oil	S		68334-30-5
061.06	Diesel Fuel	Diesel Fuel	S		
231.04	Diethanolamine	DEA		Yes	111-42-2
260.24	Diethyl Formamide	Diethyl Formamide	S		617-84-5
140.02	Diethyl Silicate	Diethyl Silicate			
310.02	Diethyl Sulfate	Ethyl Sulfate			64-67-5
310.02	Diethyl Sulfate	Sulfuric Acid Diethyl Ester			64-67-5
310.02	Diethyl Sulfate	Diethyl Sulfate			64-67-5
230.02	Diethylamine	Diethylamine			109-89-7
142.06	Diethylaminoethyl Acrylate	Acrylic Acid, 2(Diethylamino)ethyl Ester			2426-54-2
142.06	Diethylaminoethyl Acrylate	Diethylaminoethyl Acrylate			2426-54-2
250.18	Diethylaniline	DEA			91-66-7
250.18	Diethylaniline	Diethylaniline			91-66-7
080.02	Diethylene Glycol	Diethylene Glycol			111-46-6
232.07	Diethylenetriamine	Diethylenetriamine			111-40-0
192.05	Difluoroethane	Difluoroethane			25497-28-3
130.08	Dihydroxyacetone	DHA			96-26-4
130.08	Dihydroxyacetone	1,3-Dihydroxy-2-propanone			96-26-4
291.07	Dimethoate	Dimethoate	P	Yes	60-51-5
310.03	Dimethyl Sulfate	DMS			77-78-1
310.03	Dimethyl Sulfate	Dimethyl Sulfate			77-78-1
310.03	Dimethyl Sulfate	Sulfuric Acid Methyl Ester			
230.04	Dimethylamine	Dimethylamine			124-40-3
231.04	Dimethylethanolamine	Dimethylaminoethanol		Yes	108-01-0

717

Occupational and environmental exposures capable of causing illness - *continued*

AOEC Code[1]	Primary Name	Synonym	Use[2]	Asthma Inducer?	CAS Number[3]
231.04	Dimethylethanolamine	Dimethylethanolamine		Yes	108-01-0
260.09	Dimethylhydrazine	Dimethylhydrazine			57-14-7
250.06	Dinitro-o-Cresol	Dinitrol			
250.06	Dinitro-o-Cresol	DNOC			
250.06	Dinitro-o-Cresol	Dinitro-o-Cresol			1335-85-9
250.05	Dinitrobenzene	Dinitrobenzene			
250.05	Dinitrobenzene	Dinitrobenzol			
250.12	Dinitrophenol	Dinitrophenol			25550-58-7
250.07	Dinitrotoluene	DNT			
250.07	Dinitrotoluene	Dinitrotoluol			
250.07	Dinitrotoluene	Dinitrotoluene			
140.03	Dioctylphthalate	Dioctylphthalate			
370.39	Dioscorea batatas	Dioscorea batatas		Yes	
100.04	Dioxane	1,4-Dioxane	S		123-91-1
100.04	Dioxane	Dioxane	S		123-91-1
310.14	DMSO	Dimethyl sulfoxide			67-68-5
310.14	DMSO	Methyl Sulfoxide			67-68-5
310.14	DMSO	Dermasorb			67-68-5
180.08	Dodecylphenol	Dodecylphenol			7193-86-8
325.04	Drimaren Brilliant Blue K-BL	Drimaren Brilliant Blue K-BL		Yes	
325.02	Drimaren Brilliant Yellow K-3GL	Drimaren Brilliant Yellow-K-3GL		Yes	
322.34	Dry Cleaning Fluid, NOS	Dry Cleaning Fluid, NOS	S		
010.00	Dust, NOS	Dust, NOS			
010.00	Dust, NOS	Inorganic Dust, NOS			
250.17	Dyes, NOS	Dyes, NOS			
250.17	Dyes, NOS	Dye Intermediates, NOS			
373.19	Eastern White Cedar	Thuja Occidentalis		Yes	
373.19	Eastern White Cedar	Eastern White Cedar		Yes	
373.20	Ebony	Ebony		Yes	
373.20	Ebony	Diospyros Crassiflora		Yes	
191.01	EDB	EDB	P		106-93-4
191.01	EDB	1,2-Dibromoethane	P		106-93-4
191.01	EDB	Ethylene Dibromide	P		106-93-4
091.03	EGBE	2-Butoxyethanol	S		111-76-2
091.03	EGBE	Butyl-Cellosolve	S		111-76-2
091.03	EGBE	EGBE	S		
091.03	EGBE	Ethylene Glycol Monobutyl Ether	S		111-76-2
091.02	EGEE	Ethylene Glycol Monoethyl Ether	S		110-80-5
091.02	EGEE	EGEE	S		109-86-4
091.02	EGEE	Cellosolve	S		110-80-5
091.02	EGEE	2-Ethoxyethanol	S		110-80-5
324.08	Egg Lysozyme	Egg Lysozyme		Yes	
380.06	Egg Protein	Egg Protein		Yes	
091.01	EGME	Methoxyethanol	S		109-86-4
091.01	EGME	EGME	S		109-86-4
091.01	EGME	Ethylene Glycol Monomethyl Ether	S		109-86-4
091.01	EGME	Methyl Cellosolve	S		109-86-4
353.01	Electrical Shock	Electrical Shock			
353.01	Electrical Shock	Electricity			
320.07	Electroplating Chemicals, NOS	Electroplating Chemicals, NOS			
280.05	Endrin	Endrin	P		72-20-8
331.02	Engine Exhaust	Engine Exhaust			
331.02	Engine Exhaust	Gasoline Exhaust			
324.00	Enzymes, NOS	Proteolytic Enzymes, NOS			
324.00	Enzymes, NOS	Enzymes, NOS			
324.00	Enzymes, NOS	Catalysts, NOS			
382.25	Ephestia kuehniella	E. kuehniella		Yes	
382.25	Ephestia kuehniella	Ephestia kuehniella		Yes	
110.01	Epichlorohydrin	1-Chloro-2,3-Epoxypropane			106-89-8

Occupational and environmental exposures capable of causing illness - *continued*

AOEC Code[1]	Primary Name	Synonym	Use[2]	Asthma Inducer?	CAS Number[3]
110.01	Epichlorohydrin	Epichlorohydrin			106-89-8
232.06	EPO 60	Polyamine EPO 60		Yes	
232.06	EPO 60	EPO 60		Yes	
110.02	Epoxy Resins	Epoxies		Yes	
110.02	Epoxy Resins	Epoxy Resins		Yes	
110.02	Epoxy Resins	Epoxy Resin Hardeners		Yes	
360.00	Ergonomic Factors, NOS	Ergonomic Factors, NOS			
324.10	Esperase	Esperase		Yes	
321.02	Estrogens	Estrogens			
070.05	Ethanol	Alcohol	S		
070.05	Ethanol	Methyl Carbinol	S		64-17-5
070.05	Ethanol	Grain Alcohol	S		64-17-5
070.05	Ethanol	Ethyl Alcohol	S		64-17-5
070.05	Ethanol	Ethanol	S		64-17-5
231.00	Ethanolamines, NOS	Ethanolamines, NOS			
291.12	Ethephon	2-Chloroethylphosphonic Acid	P		16672-87-0
291.12	Ethephon	Ethephon	P		16672-87-0
141.03	Ethyl Acetate	Ethyl Acetate	S		141-78-6
141.03	Ethyl Acetate	Acetic Acid Ethyl Ester	S		141-78-6
141.03	Ethyl Acetate	Acetic Ether	S		141-78-6
142.05	Ethyl Acrylate	Ethyl Acrylate			140-88-5
160.07	Ethyl Benzene	Ethyl Benzene	S		100-41-4
100.05	Ethyl Ether	Ethyl Ether			60-29-7
100.05	Ethyl Ether	Ethyl Oxide			
100.05	Ethyl Ether	Diethyl Ether			60-29-7
100.05	Ethyl Ether	Ethoxyethane			60-29-7
100.05	Ethyl Ether	Ether			60-29-7
142.03	Ethyl Methacrylate	Ethyl Methacrylate			97-63-2
100.06	Ethyl Methyl Ether	Ethyl Methyl Ether			540-67-0
100.06	Ethyl Methyl Ether	Ethoxymethane			
190.07	Ethylene Dichloride	1,2-Dichloroethane	S		107-06-2
190.07	Ethylene Dichloride	Ethylene Dichloride	S		107-06-2
080.01	Ethylene Glycol	Ethylene Glycol			107-21-1
080.01	Ethylene Glycol	1,2-Ethanediol			107-21-1
080.01	Ethylene Glycol	Antifreeze			
091.00	Ethylene Glycol Ethers, NOS	Ethylene Glycol Ethers, NOS	S		
110.03	Ethylene Oxide	ETO		Yes	
110.03	Ethylene Oxide	Ethylene Oxide		Yes	75-21-8
232.01	Ethylenediamine	Ethylenediamine		Yes	107-15-3
260.10	Ethylenimine	Aziridine			151-56-4
260.10	Ethylenimine	Ethylenimine			151-56-4
320.36	Eucalyptus Oil	Eucalyptus Scent			8000-48-4
320.36	Eucalyptus Oil	Eucalyptus Scent			8000-48-4
100.07	Eugenol	Allyl Guaiacol			97-53-0
100.07	Eugenol	Eugenol			97-53-0
100.07	Eugenol	1-Allyl-3-Methoxy-4-Hydroxybenzene			97-53-0
360.06	Exercise	Exercise			
331.00	Exhaust, NOS	Exhaust, NOS			
353.02	Explosion	Explosion			
EX_CODE	EX_PRIMARY	EX_SYNONYM	E	E	EX_CAS
353.03	Fall, NOS	Fall, NOS			
353.03	Fall, NOS	Slip, Trip, or Fall on Same Level			
353.03	Fall, NOS	Struck By/Against Object as Result of Fall			
353.03	Fall, NOS	Fall From Height			
370.42	Fenugreek	Fenugreek		Yes	68990-15-8
373.16	Fernambouc	Caesalpinia Echinata		Yes	
373.16	Fernambouc	Fernambouc		Yes	
320.09	Fertilizers, NOS	Fertilizers, NOS			

Occupational and environmental exposures capable of causing illness - continued

AOEC Code[1]	Primary Name	Synonym	Use[2]	Asthma Inducer?	CAS Number[3]
292.03	Ficam	Ficam	P		22781-23-3
320.34	Fingerprint Powder	Fingerprint Powder			
320.34	Fingerprint Powder	Lightning Powder			
320.10	Fire Extinguisher Discharge	Fire Extinguisher Discharge			
381.09	Fish Feed	Echinodorus Larva		Yes	
381.09	Fish Feed	Fish Feed			
381.09	Fish Feed	Fish Feed, Mosquito Larva		Yes	
324.06	Flaviastase	Flaviastase		Yes	
322.35	Floor wax	Floor finish			
371.00	Flour, NOS	Flour, NOS		Yes	
030.03	Fluorine	Fluorine		Yes	7782-41-5
192.00	Fluorocarbons, NOS	Fluorocarbons, NOS			
040.05	Fluxes, NOS	Fluxes, NOS			
010.07	Fly Ash	Fly Ash			
361.01	Forceful Movements, NOS	Forceful Movements, NOS			
361.01	Forceful Movements, NOS	High Force			
361.01	Forceful Movements, NOS	Pulling			
361.01	Forceful Movements, NOS	Pushing			
120.03	Formaldehyde	Formalin		Yes	50-00-0
120.03	Formaldehyde	Formaldehyde		Yes	50-00-0
140.01	Formates, NOS	Formates, NOS			
150.01	Formic Acid	Formic Acid			64-18-6
382.14	Fowl Mite	Fowl Mite		Yes	
370.30	Freesia	Freesia		Yes	
192.06	Freon, Heated	Freon, Heated		Yes	
192.06	Freon, Heated	Welding Freon		Yes	
192.01	Freon, NOS	Freon, NOS			75-45-6
192.07	Freon, Unheated	Freon, Unheated			75-45-6
380.09	Frog	Frog		Yes	
382.07	Fruit Fly	Fruit Fly		Yes	
370.05	Fruit Juices	Fruit Juices			
370.05	Fruit Juices	Raspberries			
370.05	Fruit Juices	Vegetable Juices			
324.09	Fungal Amylase	Fungal Amylase		Yes	
324.11	Fungal Amyloglucosidade	Fungal Amyloglucosidade		Yes	
324.12	Fungal Hemicellulase	Fungal Hemicellulase		Yes	
320.08	Fungicide, NOS	Fungicide, NOS	P		
120.04	Furfural	Furfural			98-01-1
070.09	Furfuryl Alcohol	2-Furanmethanol		Yes	98-00-0
070.09	Furfuryl Alcohol	Furfuryl Alcohol		Yes	98-00-0
370.21	Garlic Dust	Garlic Dust		Yes	
260.25	Garlon 4	Garlon 4	P		64700-56-7
023.07	Gas Metal Arc Welding on Uncoated Mild Steel	Gas Metal Arc Welding on Uncoated Mild Steel		Yes	
061.04	Gasoline	Gasoline	S		
061.04	Gasoline	Petrol	S		
020.17	Germanium	Germanium Compounds			
020.17	Germanium	Germanium			7440-56-4
320.11	Glues, NOS	Glues, NOS			
320.11	Glues, NOS	Adhesive, NOS			
120.05	Glutaraldehyde	Glutaraldehyde		Yes	111-30-8
371.02	Gluten	Gluten		Yes	
090.00	Glycol Ethers, NOS	Glycol Ethers, NOS	S		
020.18	Gold	Gold Compounds			
020.18	Gold	Gold			7440-57-5
370.03	Grain Dust	Grain Dust		Yes	
382.16	Grain Mite	Grain Mite		Yes	
382.16	Grain Mite	Grain Parasite		Yes	
011.01	Granite	Granite			
010.08	Graphite	Graphite			7440-44-0

Occupational and environmental exposures capable of causing illness - *continued*

AOEC Code[1]	Primary Name	Synonym	Use[2]	Asthma Inducer?	CAS Number[3]
370.07	Grass Cuttings	Plant Waste			
370.07	Grass Cuttings	Grass Cuttings			
370.37	Green Beans	Green Beans		Yes	
361.03	Gripping, Forceful	Gripping, Forceful			
361.03	Gripping, Forceful	Handwriting			
361.03	Gripping, Forceful	Pinching			
321.27	Griseofulvin	Grisactin-Ultra			126-07-8
321.27	Griseofulvin	Fulvin P/G			126-07-8
321.27	Griseofulvin	Gris-PEG			126-07-8
321.27	Griseofulvin	Griseoufulvin			126-07-8
100.09	Guaiacol	Methylcatechol			90-05-1
100.09	Guaiacol	o-methoxyphenol			90-05-1
100.09	Guaiacol	1-Hydroxy-2-methoxybenzene			90-05-1
100.09	Guaiacol	Guaiacol			90-05-1
372.04	Guar	Guar		Yes	9000-30-0
380.17	Guinea Pig Antigens	Guinea Pig Antigens		Yes	
372.01	Gum Acacia	Gum Acacia		Yes	
372.01	Gum Acacia	Acacia		Yes	
372.05	Gum Arabic	Gum Arabic			9000-01-5
372.06	Gutta-percha	Gutta-percha		Yes	
010.17	Gypsum	Gypsum			13397-24-5
010.17	Gypsum	Plaster of Paris			26499-65-0
320.12	Hair Products	Hair Spray			
320.12	Hair Products	Hair Products			
320.12	Hair Products	Hair Solutions			
020.45	Hard Metal	Hard Metal			
020.45	Hard Metal	Cobalt/Tungsten Carbide			
373.24	Hardwood, Tropical, NOS	Hardwood, Tropical, NOS			
370.08	Hay	Hay			
000.00	Hazard Not On File	Hazard Not On File			
221.08	HDI Prepolymers	HDI Prepolymers		Yes	
350.03	Heat	High Temperature			
350.03	Heat	Thermal Energy			
350.03	Heat	Heat			
350.03	Heat	Steam			
350.03	Heat	Hot Liquid			
326.02	Heat Shrink Wrapping	Heat Shrink Wrapping			
020.46	Heavy Metals, NOS	Heavy Metals, NOS			
370.27	Henna	Henna		Yes	83-72-7
390.09	Hepatitis B	Hepatitis B			
390.12	Hepatitis C	Hepatitis C			
280.08	Heptachlor	Haptachlorine	P		76-44-8
280.08	Heptachlor	Heptachlor	P		76-44-8
060.03	Heptane	N-Heptane	S		142-82-5
060.03	Heptane	Heptane	S		142-82-5
320.13	Herbicides, NOS	Herbicides, NOS	P		
201.03	Hexachlorobenzene	Hexachlorobenzene	P		118-74-1
201.03	Hexachlorobenzene	Perchlorobenzene Fungicide	P		
181.02	Hexachlorophene	Hexachlorophene		Yes	70-30-4
151.08	Hexahydrophthalic Anhydride	Hexahydrophthalic Anhydride		Yes	
221.04	Hexamethylene Diisocyanate	1,6-Diisocyanato-Hexane		Yes	822-06-0
221.04	Hexamethylene Diisocyanate	Hexamethylene Diisocyanate		Yes	822-06-0
221.04	Hexamethylene Diisocyanate	HDI		Yes	822-06-0
232.02	Hexamethylenetetramine	Hexamethylenetetramine		Yes	100-97-0
060.02	Hexane	Hexane	S		110-54-3
060.02	Hexane	N-Hexane	S		110-54-3
080.03	Hexylene Glycol	Hexylene Glycol			107-41-5
151.07	Himic Anhydride	Himic Anhydride		Yes	
390.03	Histoplasma Capsulatum	Histoplasma Capsulatum			

Occupational and environmental exposures capable of causing illness - continued

AOEC Code[1]	Primary Name	Synonym	Use[2]	Asthma Inducer?	CAS Number[3]
390.08	HIV Exposure	HIV Exposure			
390.08	HIV Exposure	AIDS Exposure			
390.08	HIV Exposure	Sharps			
382.08	Honeybee	Honeybee		Yes	
370.16	Hops	Hops		Yes	
381.04	Hoya	Hoya		Yes	
350.05	Humidity, High	Humidity, High			
350.05	Humidity, High	Moisture			
350.05	Humidity, High	Wet Weather			
350.04	Humidity, Low	Dry Air			
350.04	Humidity, Low	Humidity, Low			
321.14	Hydralazine	Hydralazine		Yes	86-54-4
260.11	Hydrazine	Hydrazine			302-01-2
260.11	Hydrazine	Hydrazine Derivatives			
050.08	Hydrazoic Acid	Hydrazoic Acid			7782-79-8
050.09	Hydrobromic Acid	Hydrogen Bromide			10035-10-6
050.09	Hydrobromic Acid	Hydrobromic Acid			10035-10-6
050.09	Hydrobromic Acid	HBR			10035-10-6
170.00	Hydrocarbons, NOS	Organic Chemicals, NOS			
170.00	Hydrocarbons, NOS	Hydrocarbons, NOS			
170.00	Hydrocarbons, NOS	VOC, NOS			
170.00	Hydrocarbons, NOS	Petrochemicals, NOS			
050.10	Hydrochloric Acid	HCL			7647-01-0
050.10	Hydrochloric Acid	Hydrochloric Acid			7647-01-0
050.10	Hydrochloric Acid	Muriatic Acid			7647-01-0
050.10	Hydrochloric Acid	Hydrogen Chloride			7647-01-0
050.11	Hydrofluoric Acid	Hydrogen Fluoride			7664-39-3
050.11	Hydrofluoric Acid	Hydrofluoric Acid			7664-39-3
050.11	Hydrofluoric Acid	HF			7664-39-3
050.44	Hydrofluosilicic Acid	Fluosilicic Acid			16961-83-4
050.44	Hydrofluosilicic Acid	Hydrofluosilicic Acid			16961-83-4
211.01	Hydrogen Cyanide	Hydrogen Cyanide			74-90-8
211.01	Hydrogen Cyanide	HCN			74-90-8
050.12	Hydrogen Peroxide	Hydrogen Peroxide			
040.06	Hydrogen Sulfide	Hydrogen Sulfide			7783-06-4
040.06	Hydrogen Sulfide	H2S			7783-06-4
040.06	Hydrogen Sulfide	Sewer Gas			
180.03	Hydroquinone	P-Hydroxyphenol			123-31-9
180.03	Hydroquinone	Photo Developer, Black & White			
180.03	Hydroquinone	Hydroquinone			123-31-9
180.03	Hydroquinone	X-Ray Developer			
180.03	Hydroquinone	P-Dihydroxybenzene			123-31-9
353.04	Hypoxia	Suffocation			
353.04	Hypoxia	Hypoxia			
353.04	Hypoxia	Oxygen Deficiency			
321.20	Imipenem	Imipenem			64221-86-9
330.07	Incense Smoke	Incense Smoke			
330.04	Incinerator Fume	Incinerator Fume, NOS			
020.19	Indium	Indium Solder			
020.19	Indium	Indium			7440-74-6
320.33	Indoor Air Pollutants from Building Renovation	Indoor Air Pollutants from Building Renovation			
320.33	Indoor Air Pollutants from Building Renovation	Building Renovation			
320.33	Indoor Air Pollutants from Building Renovation	Air Pollutants, Indoor, from Building Renovation			
390.07	Infectious Agents, NOS	Infectious Agents, NOS			
390.07	Infectious Agents, NOS	Bacteria			
390.07	Infectious Agents, NOS	Viruses			
352.05	Infrared Light	IR			

Occupational and environmental exposures capable of causing illness - *continued*

AOEC Code[1]	Primary Name	Synonym	Use[2]	Asthma Inducer?	CAS Number[3]
352.05	Infrared Light	IR Light			
352.05	Infrared Light	Infrared Light			
170.02	Inks	Inks			
040.00	Inorganic Compounds, NOS	Inorganic Compounds, NOS			
382.21	Insect Bite, NOS	Insect Bite, NOS			
382.21	Insect Bite, NOS	Spider Bite			
382.21	Insect Bite, NOS	Mosquito Bite			
382.00	Insect, NOS	Arthropod, NOS			
382.00	Insect, NOS	Insect, NOS			
030.04	Iodine	Iodine			7553-56-2
322.05	Iodophors	Iodophors			8037-86-3
321.16	Ipecacuanha	Ipecac		Yes	
321.16	Ipecacuanha	Ipecacuanha		Yes	
373.05	Iroko	Iroko		Yes	
373.05	Iroko	Chlorophora Excelsa		Yes	
020.20	Iron	Iron Compounds			
020.20	Iron	Ferric Chloride			7705-08-0
020.20	Iron	Iron			7439-89-6
040.24	Irritant Gases, NOS	Irritant Gases, NOS			
060.04	Isobutane	Isobutane			75-28-5
060.04	Isobutane	2-Methylpropane			75-28-5
221.00	Isocyanates, NOS	Diisocyanates, NOS		Yes	
221.00	Isocyanates, NOS	Isocyanates, NOS		Yes	
321.13	Isonicotinic Acid Hydrazide	Isoniazid		Yes	
321.13	Isonicotinic Acid Hydrazide	Isonicotinic Acid Hydrazide		Yes	
321.13	Isonicotinic Acid Hydrazide	INH		Yes	
060.12	Isooctane	Isooctane			540-84-1
221.05	Isophorone Diisocyanate	IPDI		Yes	4098-71-9
221.05	Isophorone Diisocyanate	Isophorone Diisocyanate		Yes	4098-71-9
200.05	Isophthaloyl Chloride	Phthaloyl Chloride			99-63-8
200.05	Isophthaloyl Chloride	Isophthaloyl Chloride			99-63-8
200.05	Isophthaloyl Chloride	Isophthalic Acid Chloride			99-63-8
070.06	Isopropyl Alcohol	Sec-Propyl Alcohol	S		67-63-0
070.06	Isopropyl Alcohol	2-Propanol	S		67-63-0
070.06	Isopropyl Alcohol	Isopropyl Alcohol	S		67-63-0
070.06	Isopropyl Alcohol	Propanol	S		67-63-0
070.06	Isopropyl Alcohol	Isopropanol	S		67-63-0
230.03	Isopropylamine	Isopropylamine			
230.03	Isopropylamine	2-Aminopropane			
370.46	Jalapeno Pepper	Jalapeno Pepper			
331.03	Jet Exhaust	Jet Exhaust			
061.05	Jet Fuel	Jet Fuel	S		
370.43	Kapok	Kapok		Yes	
370.43	Kapok	Ceiba pentandra Gaertner		Yes	
372.03	Karaya	Karaya		Yes	
373.12	Kejaat	Pterocarpus Angolensis		Yes	
373.12	Kejaat	Kejaat		Yes	
061.03	Kerosene	Astral Oil	S		
061.03	Kerosene	Kerosene	S		8008-20-6
061.03	Kerosene	No. 1 Fuel Oil	S		
061.03	Kerosene	Kerosine	S		8008-20-6
130.00	Ketones, NOS	Ketones, NOS	S		
270.35	Kevlar	Poly (P-Phenylenediamine)			26125-61-1
270.35	Kevlar	Kevlar			26125-61-1
360.02	Keyboard Use	Typing			
360.02	Keyboard Use	Computer Mouse Use			
360.02	Keyboard Use	Typewriter			
360.02	Keyboard Use	Keyboard Use			
360.02	Keyboard Use	Word Processing			
360.02	Keyboard Use	Adding Machine			

Occupational and environmental exposures capable of causing illness - *continued*

AOEC Code[1]	Primary Name	Synonym	Use[2]	Asthma Inducer?	CAS Number[3]
360.02	Keyboard Use	Computer Keyboard			
360.02	Keyboard Use	Calculator			
360.02	Keyboard Use	VDT Keyboard			
360.02	Keyboard Use	VDT Typing			
360.02	Keyboard Use	Key Punching			
360.02	Keyboard Use	Keyboard and Mouse Use			
373.21	Kotibe	Kotibe		Yes	
373.21	Kotibe	Nesorgordonia Papaverifera		Yes	
382.19	L. Caesar Larva	L. Caesar Larva		Yes	
171.07	Lacquer	Varnish			
171.07	Lacquer	Lacquer			
324.14	Lactase	B-Galactosidase			9031-11-2
324.14	Lactase	Lactase			9031-11-2
324.14	Lactase	Beta-Galactosidase			9031-11-2
324.14	Lactase	Beta-Lactosidase			9031-11-2
380.10	Lactoserum	Lactoserum		Yes	
325.09	Lanasol Yellow 4G	Lanasol Yellow 4G		Yes	
352.02	Lasers	Lasers			
270.02	Latex, Natural Rubber	Latex, Natural Rubber		Yes	
270.02	Latex, Natural Rubber	Latex Gloves, NOS		Yes	
270.03	Latex, Synthetic	Polydimethyl Siloxane			9016-00-6
270.03	Latex, Synthetic	Latex, Synthetic			
370.29	Lathyrus Sativus	Lathyrus Sativus		Yes	
323.02	Leachate	Leachate			
020.21	Lead, Inorganic	Lead, Inorganic			7439-92-1
020.21	Lead, Inorganic	Lead-based Paint			
020.21	Lead, Inorganic	Inorganic Lead Compounds			
020.22	Lead, Organic	Lead, Organic			
020.22	Lead, Organic	Organic Lead Compounds			
020.22	Lead, Organic	Tetraethyl Lead			78-00-2
380.03	Leather Dust	Leather Dust			
382.23	Leptinotarsa decemlineata	Colorado Potato Beetle		Yes	73468-57-2
382.23	Leptinotarsa decemlineata	L. decemlineata		Yes	73468-57-2
382.23	Leptinotarsa decemlineata	Leptinotarsa decemlineata		Yes	73468-57-2
382.09	Lesser Mealworm	Lesser Mealworm		Yes	
325.01	Levafix Brilliant Yellow E36	Levafix Brilliant Yellow E36		Yes	
361.02	Lifting	Lifting			
361.02	Lifting	Heavy Lifting			
361.02	Lifting	Carrying			
361.02	Lifting	Repetitive Lifting			
060.05	Limonene	Dipentene	S		138-86-3
060.05	Limonene	Limonene	S		138-86-3
060.05	Limonene	Methyl-4-Isopropenyl Cyclohexene-1	S		
060.05	Limonene	P-Mentha-1,8-Diene	S		
370.38	Limonium tataricum	Limonium tataricum		Yes	
280.06	Lindane	Lindane	P		58-89-9
060.15	Linseed Oil	Linseed Oil			8001-26-1
382.02	Locust	Locust		Yes	
320.14	Lubricants, NOS	Coolants			
320.14	Lubricants, NOS	Hydraulics			
320.14	Lubricants, NOS	Lubricants, NOS			
320.14	Lubricants, NOS	Transmission Fluid			
370.25	Lycopodium	Lycopodium		Yes	
320.27	Mace	Mace			532-27-4
320.27	Mace	Tear Gas			532-27-4
320.27	Mace	Acetophenone			532-27-4
320.27	Mace	2-Chloroacetophenone			532-27-4
251.07	Magenta	Magenta			632-99-5
020.23	Magnesium	Magnesium Compounds			

Occupational and environmental exposures capable of causing illness - continued

AOEC Code[1]	Primary Name	Synonym	Use[2]	Asthma Inducer?	CAS Number[3]
020.23	Magnesium	Magnesium Aluminum Silicate			1327-43-1
020.23	Magnesium	Magnesium			7439-95-4
373.07	Mahogany	Mahogany		Yes	
291.01	Malathion	Malathion	P	Yes	121-75-5
151.02	Maleic Anhydride	Maleic Anhydride		Yes	108-31-6
010.09	Man-Made Mineral Fibers	Fibrous Glass			
010.09	Man-Made Mineral Fibers	Fiberglass			
010.09	Man-Made Mineral Fibers	Rock Wool			
010.09	Man-Made Mineral Fibers	Mineral Wool			
010.09	Man-Made Mineral Fibers	MMMF			
010.09	Man-Made Mineral Fibers	Refractory Ceramic Fiber			
010.09	Man-Made Mineral Fibers	Man-Made Mineral Fibers			
020.24	Manganese	Manganese Compounds			
020.24	Manganese	Manganese			7439-96-5
380.02	Manure	Manure			
380.02	Manure	Barn Dust			
330.05	Marijuana Smoke	Marijuana Smoke			
250.16	MDA	MDA			101-77-9
250.16	MDA	Methylenedianiline			101-77-9
250.16	MDA	4,4-Diaminodiphenylmethane			101-77-9
130.02	MEK Peroxide	MEK Peroxide			1338-23-4
130.02	MEK Peroxide	Methyl Ethyl Ketone Peroxide			1338-23-4
130.02	MEK Peroxide	Peroxide 2-Butanone			1338-23-4
260.28	Melamine	Cyanurotramide			108-78-1
311.00	Mercaptans, NOS	Mercaptans, NOS			
310.04	Mercaptoethanol	Thioglycol			60-24-2
310.04	Mercaptoethanol	Mercaptoethanol			60-24-2
310.04	Mercaptoethanol	B-Mercaptoethanol			60-24-2
310.04	Mercaptoethanol	2-Hydroxyethanethiol			60-24-2
020.25	Mercury, Inorganic	Mercuric Salts			
020.25	Mercury, Inorganic	Mercuric Chloride			7487-94-7
020.25	Mercury, Inorganic	Mercury, Inorganic			
020.25	Mercury, Inorganic	Inorganic Mercury Compounds			
020.26	Mercury, Organic	Organic Mercury Compounds			
020.26	Mercury, Organic	Mercury, Organic			
020.44	Metal Carbonyls	Metal Carbonyls			
021.00	Metal Dust, NOS	Metal Dust, NOS			
020.00	Metal Fumes, NOS	Metallic Oxides, NOS			
020.00	Metal Fumes, NOS	Metal Fumes, NOS			
322.02	Metal Polish, Tarnish Remover, or Preventative	Metal Polish, Tarnish Remover, or Preventative			
020.47	Metals, NOS	Metals, NOS			
292.01	Metam Sodium	Metam Sodium	P		137-42-8
292.01	Metam Sodium	Metham Sodium	P		137-42-8
060.06	Methane	Methane			74-82-8
070.07	Methanol	Wood Alcohol	S		67-56-1
070.07	Methanol	Carbinol	S		67-56-1
070.07	Methanol	Wood Spirits	S		67-56-1
070.07	Methanol	Methyl Alcohol	S		67-56-1
070.07	Methanol	Methanol	S		67-56-1
280.07	Methoxychlor	Methoxychlor	P		72-43-5
325.10	Methyl Blue	Methyl Blue		Yes	28983-56-4
191.03	Methyl Bromide	Monobromomethane	P		74-83-9
191.03	Methyl Bromide	Methyl Bromide	P		74-83-9
191.03	Methyl Bromide	Bromomethane	P		74-83-9
190.08	Methyl Chloroform	1,1,1-Trichloroethane	S		71-55-6
190.08	Methyl Chloroform	Methyl Chloroform	S		71-55-6
130.03	Methyl Ethyl Ketone	2-Butanone	S		78-93-3
130.03	Methyl Ethyl Ketone	Methyl Ethyl Ketone	S		78-93-3

Occupational and environmental exposures capable of causing illness - continued

AOEC Code[1]	Primary Name	Synonym	Use[2]	Asthma Inducer?	CAS Number[3]
130.03	Methyl Ethyl Ketone	MEK	S		78-93-3
130.04	Methyl Isobutyl Ketone	MIBK	S		108-10-1
130.04	Methyl Isobutyl Ketone	Hexone	S		108-10-1
130.04	Methyl Isobutyl Ketone	4-Methyl-2-Pentanone	S		108-10-1
130.04	Methyl Isobutyl Ketone	Methyl Isobutyl Ketone	S		108-10-1
220.01	Methyl Isocyanate	Methyl Isocyanate			624-83-9
220.01	Methyl Isocyanate	Methyl Ester Isocyanic Acid			
220.01	Methyl Isocyanate	MIC			624-83-9
220.02	Methyl Isothiocyanate	Methyl Isothiocyanate	P		556-61-6
142.04	Methyl Methacrylate	Methyl Methacrylate		Yes	80-62-6
130.05	Methyl N-Butyl Ketone	Methyl N-Butyl Ketone	S		591-78-6
130.05	Methyl N-Butyl Ketone	MNBK	S		
260.12	Methyl Pyrrolidone	1-Methyl-2-Pyrrolidinone			872-50-4
260.12	Methyl Pyrrolidone	Methyl Pyrrolidone			51013-18-4
140.05	Methyl Salicylate	Methyl Salicylate			119-36-8
140.05	Methyl Salicylate	Oil of Wintergreen			119-36-8
140.05	Methyl Salicylate	Methyl 2-Hydroxybenzoate			119-36-8
100.08	Methyl Tertiary Butyl Ether	MTBE			1634-04-4
100.08	Methyl Tertiary Butyl Ether	Methyl Tertiary Butyl Ether			1634-04-4
151.05	Methyl Tetrahydrophthalic Anhydride	Methyl Tetrahydrophthalic Anhydride		Yes	
321.09	Methyldopa	Methyldopa		Yes	555-30-6
190.09	Methylene Chloride	Dichloromethane	S		75-09-2
190.09	Methylene Chloride	Methylene Chloride	S		75-09-2
190.09	Methylene Chloride	Methylene Dichloride	S		75-09-2
221.02	Methylene Diisocyanate	MDI		Yes	101-68-8
221.02	Methylene Diisocyanate	Methylene Bisphenyl Isocyanate		Yes	101-68-8
221.02	Methylene Diisocyanate	Methylene Diisocyanate		Yes	101-68-8
221.02	Methylene Diisocyanate	Diphenylmethane Diisocyanate		Yes	101-68-8
382.06	Mexican Bean Weevil	Mexican Bean Weevil		Yes	
380.14	Mice	Mice		Yes	
321.30	Micotil	Micotil			137330-133
321.30	Micotil	Tilmicosin Phosphate			137330-133
390.00	Microorganisms, NOS	Bioaerosols			
390.00	Microorganisms, NOS	Microorganisms, NOS			
060.07	Mineral Oil	Petrolatum Liquid			
060.07	Mineral Oil	Paraffin Oil			8012-95-1
060.07	Mineral Oil	Mineral Oil			8012-95-1
060.07	Mineral Oil	White Oil			
060.07	Mineral Oil	Mineral Oil Mist			8012-95-1
382.13	Mites, NOS	Mites, NOS		Yes	
391.01	Mold, NOS	Fungi, NOS			
391.01	Mold, NOS	Mold, NOS			
390.01	Mold, NOS (new code 391.01)	Mold, NOS			
020.27	Molybdenum	Molybdenum			7439-98-7
020.27	Molybdenum	Molybdenum Compounds			
231.01	Monoethanolamine	Monoethanolamine		Yes	141-43-5
231.01	Monoethanolamine	2-Aminoethanol		Yes	141-43-5
353.05	Motor Vehicle Accident	Auto Accident			
353.05	Motor Vehicle Accident	Motor Vehicle Accident			
353.05	Motor Vehicle Accident	Car Accident			
353.05	Motor Vehicle Accident	Car Crash			
391.10	Mushrooms, NOS	Mushrooms, NOS		Yes	
190.16	Mustard Gas	Bis-2-Chloroethyl Sulfide			505-60-2
190.16	Mustard Gas	Mustard Gas			505-60-2
321.28	Mustargen	Nitrogen Mustard Hydrochloride			55-86-7
321.28	Mustargen HCL	Mustargen HCL			55-86-7
391.09	Mycotoxins	Mycotoxins			
391.09	Mycotoxins	Aflatoxin			1402-68-2
260.26	N,N-diethyl-m-touamide	Benzamide	P		134-62-3

Occupational and environmental exposures capable of causing illness - *continued*

AOEC Code[1]	Primary Name	Synonym	Use[2]	Asthma Inducer?	CAS Number[3]
260.26	N,N-diethyl-m-touamide	DEET	P		134-62-3
260.26	N,N-diethyl-m-touamide	m-Delphene	P		134-62-3
260.26	N,N-diethyl-m-touamide	N,N-diethyl-3-methyl-	P		134-62-3
260.26	N,N-diethyl-m-touamide	Diethyltoluamide	P		134-62-3
260.07	N,N-Dimethylacetamide	Dimethylacetamide			127-19-5
260.07	N,N-Dimethylacetamide	dmac			127-19-5
260.07	N,N-Dimethylacetamide	N,N-Dimethylacetamide			127-19-5
260.08	N,N-Dimethylformamide	DMF	S		68-12-2
260.08	N,N-Dimethylformamide	N,N-Dimethylformamide	S		68-12-2
070.04	N-Butyl Alcohol	1-Butanol	S		71-36-3
070.04	N-Butyl Alcohol	N-Butyl Alcohol	S		71-36-3
311.01	N-Butyl Mercaptan	N-Butyl Mercaptan			109-79-5
230.01	N-Butylamine	N-Butylamine			109-73-9
260.20	N-Methylmorpholine	N-Methylmorpholine		Yes	109-02-4
240.01	N-Nitrosodimethylamine	DMN			62-75-9
240.01	N-Nitrosodimethylamine	N,N-Dimethylnitrosamine			62-75-9
240.01	N-Nitrosodimethylamine	DMNA			62-75-9
240.01	N-Nitrosodimethylamine	N-Nitrosodimethylamine			62-75-9
320.30	N-Octyl Bicycloheptene Dicarboximide	MGK 264	P		113-48-4
320.30	N-Octyl Bicycloheptene Dicarboximide	N-Octyl Bicycloheptene Dicarboximide	P		113-48-4
320.30	N-Octyl Bicycloheptene Dicarboximide	Octacide 264	P		113-48-4
370.22	Nacre Dust	Nacre Dust		Yes	
061.02	Naphtha	Petroleum Naphtha	S		8030-30-6
061.02	Naphtha	Naphtha	S		8030-30-6
160.05	Naphthalene	Tar Camphor			91-20-3
160.05	Naphthalene	Naphthalin			91-20-3
160.05	Naphthalene	White Tar			91-20-3
160.05	Naphthalene	Naphthalene			91-20-3
221.03	Naphthalene Diisocyanate	Naphthalene Diisocyanate		Yes	
221.03	Naphthalene Diisocyanate	1,5-Naphthylene Ester Isocyanic Acid		Yes	
221.03	Naphthalene Diisocyanate	NDI		Yes	
060.08	Natural Gas	Natural Gas			74-82-8
291.06	Nemacur	Nemacur	P	Yes	22224-92-6
271.02	Neoprene	Neoprene			9010-98-4
271.02	Neoprene	Polychlorobutadiene			
271.02	Neoprene	Polychloroprene			
291.04	Nerve Gas	Nerve Gas			
390.11	Neurospora	Neurospora		Yes	
391.04	Neurospora				
382.20	New Mexico Range Moth Caterpillar	New Mexico Range Moth Caterpillar		Yes	
020.28	Nickel	Nickel Compounds		Yes	
020.28	Nickel	Nickel		Yes	7440-02-0
260.13	Nicotine Sulfate	Nicotine Sulfate	P		65-30-5
320.25	Ninhydrin	Ninhydrin			485-47-2
050.13	Nitric Acid	Nitric Acid			7697-37-2
250.04	Nitrobenzene	Oil of Mirbane			98-95-3
250.04	Nitrobenzene	Nitrobenzol			98-95-3
250.04	Nitrobenzene	Nitrobenzene			98-95-3
040.07	Nitrogen	Nitrogen			7727-37-9
040.08	Nitrogen Oxides	NO			10102-43-9
040.08	Nitrogen Oxides	Nitrogen Dioxide			10102-44-0
040.08	Nitrogen Oxides	Nitric Oxide			10102-43-9
040.08	Nitrogen Oxides	NO2			10102-44-0
040.08	Nitrogen Oxides	Nitrogen Oxides			
261.01	Nitroglycerin	Glyceryl Trinitrate			55-63-0
261.01	Nitroglycerin	Trinitroglycerin			55-63-0
261.01	Nitroglycerin	Nitroglycerin			55-63-0
260.14	Nitroparaffins	Nitromethane			75-52-5
260.14	Nitroparaffins	Nitroethane			79-24-3

Occupational and environmental exposures capable of causing illness - *continued*

AOEC Code[1]	Primary Name	Synonym	Use[2]	Asthma Inducer?	CAS Number[3]
260.14	Nitroparaffins	2-Nitropropane			79-46-9
260.14	Nitroparaffins	Nitroparaffins			
250.11	Nitrophenol	Nitrophenol			
040.09	Nitrous Oxide	N2O			10024-97-2
040.09	Nitrous Oxide	Nitrous Oxide			10024-97-2
350.01	Noise	Noise			
351.02	Nuclear Reactor Release	Nuclear Reactor Release			
270.17	Nylon	Nylon			63428-83-1
010.22	Nylon Flock	Polyamide fibers			
373.06	Oak	Quercus Rubra		Yes	
373.06	Oak	Oak		Yes	
320.15	Odors	Odors			
160.06	Oil of Clove	Clove Oil			
160.06	Oil of Clove	Oil of Clove			
252.02	Oil Orange SS	Oil Orange SS			2646-17-5
170.03	Oils, NOS	Fats			
170.03	Oils, NOS	Greases			
170.03	Oils, NOS	Oils, NOS			
370.34	Oils, Vegetable	Oils, Vegetable			
310.06	Omite	Omite	P		2312-35-8
310.06	Omite	Omite Cr	P		
321.18	Opiate Compounds	Morphine			57-27-2
321.18	Opiate Compounds	Opiate Compounds		Yes	57-27-2
300.00	Organic Phosphates, Nonpesticide	Organic Phosphates, Nonpesticide			
280.00	Organochlorine Pesticides, NOS	Organochlorine Insecticides, NOS	P		
280.00	Organochlorine Pesticides, NOS	Organochlorine Pesticides, NOS	P		
291.00	Organophosphate Pesticides, NOS	Organophosphate Pesticides, NOS	P	Yes	
180.07	Orthophenylphenol	Orthophenylphenol	P		90-43-7
020.29	Osmium	Osmium Compounds			
020.29	Osmium	Osmium			7440-04-2
382.24	Ostrinia nubilalis	Ostrinia nubilalis		Yes	
382.24	Ostrinia nubilalis	European Corn Borer		Yes	
382.24	Ostrinia nubilalis	O. nubilalis		Yes	
150.03	Oxalic Acid	Ethanedioic Acid			144-62-7
150.03	Oxalic Acid	Oxalic Acid			144-62-7
040.10	Oxygen, Liquid	Oxygen, Liquid			7782-44-7
040.11	Ozone	Ozone			10028-15-6
171.01	Paint	Paint	S		
110.05	Paint, Epoxy	Paint, Epoxy			
110.05	Paint, Epoxy	Coating, Epoxy			
171.05	Paint, Latex	Paint, Latex			
171.06	Paint, Oil-Based	Paint, Oil-Based			
324.05	Pancreatin	Pancreatin		Yes	8049-47-6
324.03	Papain	Papain		Yes	9001-73-4
370.01	Paper Dust	Paper Dust			
370.31	Paprika	Paprika		Yes	
251.01	Para-Aminophenol	Activol			123-30-8
251.01	Para-Aminophenol	PAP			123-30-8
251.01	Para-Aminophenol	Para-Aminophenol			123-30-8
251.01	Para-Aminophenol	4-Hydroxyaniline			123-30-8
060.09	Paraffin	Paraffin Wax			
060.09	Paraffin	Paraffin			8002-74-2
120.06	Paraformaldehyde	Paraformaldehyde			30525-89-4
120.07	Paraldehyde	Paraldehyde			
260.15	Paraquat	Paraquat	P		4685-14-7
382.17	Parasites, NOS	Parasites, NOS			
291.02	Parathion	Parathion	P	Yes	56-38-2
373.18	Pau Marfim	Pau Marfim		Yes	
373.18	Pau Marfim	Balfourodendron Riedelianum		Yes	
200.07	PCBs	PCBs			1336-36-3

Occupational and environmental exposures capable of causing illness - *continued*

AOEC Code[1]	Primary Name	Synonym	Use[2]	Asthma Inducer?	CAS Number[3]
200.07	PCBs	Kanechlor			1336-36-3
200.07	PCBs	Chlorodiphenyl			1336-36-3
200.07	PCBs	Polychlorinated Biphenyls			1336-36-3
200.07	PCBs	Arochlor			1336-36-3
370.23	Pectin	Pectin		Yes	9000-69-5
321.05	Penicillamine	Penicillamine		Yes	
321.04	Penicillins	Ampicillin		Yes	69-53-4
321.04	Penicillins	Penicillins		Yes	69-53-4
391.03	Penicillium	Penicillium		Yes	
200.06	Pentachloronitrobenzene	Pentachloronitrobenzene	P		82-68-8
181.01	Pentachlorophenol	Pentachlorophenol			87-86-5
181.01	Pentachlorophenol	PCP			87-86-5
261.02	Pentaerythritol Tetranitrate	PETN			78-11-5
261.02	Pentaerythritol Tetranitrate	Pentaerythritol Tetranitrate			78-11-5
321.23	Pentamidine	Aerosolized Pentamidine			100-33-4
321.23	Pentamidine	Pentamidine			100-33-4
060.19	Pentane	Pentane	S		109-66-0
060.19	Pentane	N-Pentane	S		109-66-0
324.04	Pepsin	Pepsin		Yes	9001-75-6
190.10	Perchlorethylene	Perchlorethylene	S		127-18-4
190.10	Perchlorethylene	Perc	S		127-18-4
190.10	Perchlorethylene	Perchloroethylene	S		127-18-4
190.10	Perchlorethylene	Tetrachloroethylene	S		127-18-4
320.23	Perfume, NOS	Perfume, NOS			
010.15	Perlite	Perlite			93763-70-3
324.13	Peroxidase Catalyst	Peroxidase Catalyst			
050.14	Peroxides	Peroxides			
050.42	Peroxyacetic Acid	Peroxyacetic Acid		Yes	79-21-0
040.26	Persulfate Salts	Persulfate Salts		Yes	
320.16	Pesticides, NOS	Pesticides, NOS	P		
061.00	Petroleum Fractions, NOS	Petroleum Fractions, NOS	S		
061.00	Petroleum Fractions, NOS	Petroleum Distillates, NOS			
061.01	Petroleum Spirits	Solvent Naptha	S		
061.01	Petroleum Spirits	Refined Petroleum Solvent	S		
061.01	Petroleum Spirits	Stoddard Solvent	S		8052-41-3
061.01	Petroleum Spirits	Varnish Makers' & Painters' Naptha	S		
061.01	Petroleum Spirits	Mineral Thinner	S		
061.01	Petroleum Spirits	Mineral Turpentine	S		
061.01	Petroleum Spirits	White Spirits	S		
061.01	Petroleum Spirits	Petroleum Spirits	S		64475-85-0
061.01	Petroleum Spirits	Vm & P Naptha	S		
061.01	Petroleum Spirits	Varsol	S		
061.01	Petroleum Spirits	Mineral Spirits	S		64475-85-0
321.00	Pharmaceuticals, NOS	Pharmaceuticals, NOS			
180.04	Phenol	Phenol			108-95-2
180.04	Phenol	Carbolic Acid			108-95-2
180.04	Phenol	Hydroxybenzene			108-95-2
270.04	Phenolics	Phenolic Resins			
270.04	Phenolics	Phenolics			
180.00	Phenols, NOS	Phenols, NOS			
251.06	Phenylene Diamine	Phenylene Diamine		Yes	25265-76-3
251.06	Phenylene Diamine	Paraphenylene Diamine		Yes	25265-76-3
251.06	Phenylene Diamine	Diaminobenzene		Yes	25265-76-3
321.07	Phenylglycine Acid Chloride	Phenylglycine Acid Chloride		Yes	
040.12	Phosgene	Phosgene			75-44-5
140.06	Phosphate Ester, NOS	Phosphate Ester, NOS			
040.13	Phosphine	Phosphine			7803-51-2
050.15	Phosphoric Acid	Phosphoric Acid			7664-38-2
050.32	Phosphorus Bromide	Phosphorus Bromide			

Occupational and environmental exposures capable of causing illness - continued

AOEC Code[1]	Primary Name	Synonym	Use[2]	Asthma Inducer?	CAS Number[3]
050.32	Phosphorus Bromide	Phosphorus Tribromide			
040.14	Phosphorus Pentasulfide	Phosphorus Pentasulfide			1314-80-3
040.14	Phosphorus Pentasulfide	Phosphorus Sulfide			1314-80-3
040.14	Phosphorus Pentasulfide	Thiophosphoric Anhydride			
050.16	Phosphorus Trichloride	Phosphorus Trichloride			7719-12-2
320.17	Photo Developing Chemicals, NOS	Fixer			
320.17	Photo Developing Chemicals, NOS	Photo Developing Chemicals, NOS			
320.17	Photo Developing Chemicals, NOS	Photo Developer, Color			
140.04	Phthalate Ester	Phthalate Ester			
151.01	Phthalic Anhydride	Phthalic Anhydride		Yes	85-44-9
151.01	Phthalic Anhydride	Phthalic Acid Anhydride		Yes	85-44-9
350.00	Physical Factors, NOS	Physical Factors, NOS			
326.01	Pickle Processing (Unknown Causal Agent)	Pickle Processing (Unknown Causal Agent)			
250.09	Picric Acid	Picric Acid			88-89-1
250.09	Picric Acid	Carbazotic Acid			88-89-1
250.09	Picric Acid	2,4,6-Trinitrophenol			88-89-1
380.08	Pig	Pig		Yes	
380.21	Pigeon Droppings	Pigeon Droppings			
370.40	Pinellia ternata	Pinellia ternata		Yes	
260.19	Piperazine Hydrochloride	Piperazine Hydrochloride		Yes	142-64-3
290.01	Piperonyl Butoxide	Piperonyl Butoxide	P		51-03-6
370.00	Plant Material, NOS	Plant Material, NOS			
370.00	Plant Material, NOS	Organic Dusts, NOS			
370.00	Plant Material, NOS	Granola, NOS			
370.00	Plant Material, NOS	Herbal Tea, NOS		Yes	
391.05	Plasmopara	Plasmopara viticola			
010.10	Plaster	Plaster			
330.02	Plastic Smoke	Vinyl Fumes			
330.02	Plastic Smoke	Plastic Smoke			
330.02	Plastic Smoke	Polymer Fume			
270.13	Plasticizers	Plasticizers			
020.30	Platinum	Platinum Compounds			
020.30	Platinum	Platinum			7440-06-4
270.18	Plexiglass Dust	Plexiglass Dust		Yes	
351.01	Plutonium	Plutonium			
370.06	Poisonous Plants	Poison Sumac			
370.06	Poisonous Plants	Poisonous Plants			
370.06	Poisonous Plants	Poison Oak			
370.06	Poisonous Plants	Poison Ivy			
370.10	Pollen	Pollen			
232.00	Polyamines, NOS	Polyamines, NOS			
161.00	Polycyclic Aromatic Hydrocarbons, NOS	Producer Gas Residue			
161.00	Polycyclic Aromatic Hydrocarbons, NOS	PAH			
161.00	Polycyclic Aromatic Hydrocarbons, NOS	Polynuclear Aromatics			
161.00	Polycyclic Aromatic Hydrocarbons, NOS	Polycyclic Aromatic Hydrocarbons, NOS			
161.00	Polycyclic Aromatic Hydrocarbons, NOS	Coal Tar			
161.00	Polycyclic Aromatic Hydrocarbons, NOS	PNA			
161.00	Polycyclic Aromatic Hydrocarbons, NOS	Coke Oven Emissions			
270.05	Polyester Resin	Polyester Resin			
270.26	Polyethylene Terephthalate	Polyethylene Terephthalate			25038-59-9
270.26	Polyethylene Terephthalate	Mylar			25038-59-9
270.25	Polyethylene Terephthalate/Polybutylene Terephthal	Heated Electrostatic Polyester Paint		Yes	
270.25	Polyethylene Terephthalate/Polybutylene Terephthal	Polyethylene Terephthalate/Polybutylene Terephthalate		Yes	
270.31	Polyethylene, Heated	Polyethylene, Heated		Yes	
270.06	Polyethylene, NOS	Polyethylene, NOS			
270.06	Polyethylene, NOS	PE, NOS			
270.32	Polyethylene, Unheated	Polyethylene, Unheated			9002-88-4

Occupational and environmental exposures capable of causing illness - *continued*

AOEC Code[1]	Primary Name	Synonym	Use[2]	Asthma Inducer?	CAS Number[3]
270.19	Polyimides	Polyimides			
270.00	Polymers, NOS	Polymers, NOS			
270.00	Polymers, NOS	Plastics, NOS			
270.00	Polymers, NOS	Plastics, Pre-Polymer			
221.06	Polymethylene Polyphenylisocyanate	Polymethylene Polyphenylisocyanate		Yes	9016-87-9
221.06	Polymethylene Polyphenylisocyanate	PPI		Yes	9016-87-9
270.33	Polypropylene, Heated	Polypropylene, Heated		Yes	
270.24	Polypropylene, NOS	Polypropylene, NOS			9003-07-0
270.34	Polypropylene, Unheated	Polypropylene, Unheated			9003-07-0
270.20	Polystyrene	Polystyrene			9003-53-6
270.37	Polytetrafluoroethylene, Thermal Decomposition Products	Polytetrafluoroethylene, Thermal Decomposition Products			
270.37	Polytetrafluoroethylene, Thermal Decomposition Products	Teflon, Thermal Decomposition Products			
270.07	Polyurethane	Varethane Paint			
270.07	Polyurethane	Urethane			51-79-6
270.07	Polyurethane	Polyurethane			
270.07	Polyurethane	Polycarbamate			64440-88-6
270.07	Polyurethane	Urethane Enamel Paint			
270.08	Polyvinyl Alcohol	PVA			9002-89-5
270.08	Polyvinyl Alcohol	Polyvinyl Alcohol			9002-89-5
270.09	Polyvinyl Chloride	PVC		Yes	9002-86-2
270.09	Polyvinyl Chloride	Polyvinyl Chloride		Yes	9002-86-2
270.38	Polyvinyl Chloride, Thermal Decomposition Products	Polyvinyl Chloride, Thermal Decomposition Products	A		
270.14	Polyvinyl Pyrrolidone	Polyvinyl Pyrrolidone			
270.28	Polyvinylbutyral	Polyvinylbutyral			63148-65-2
270.28	Polyvinylbutyral	Polyvinyl Butyral Resins			63148-65-2
010.11	Porcelain	Porcelain			
362.03	Posture, Body - Dynamic	Posture, Body - Dynamic			
362.03	Posture, Body - Dynamic	Twisting			
362.03	Posture, Body - Dynamic	Bending			
362.03	Posture, Body - Dynamic	Stooping			
362.02	Posture, Body - Static	Sitting			
362.02	Posture, Body - Static	Standing			
362.02	Posture, Body - Static	Kneeling			
362.02	Posture, Body - Static	Prolonged Position			
362.02	Posture, Body - Static	Posture, Body - Static			
362.00	Posture, NOS	Posture, NOS			
362.01	Posture, Upper Extremity	Hand-Arm Posture			
362.01	Posture, Upper Extremity	Posture, Upper Extremity			
362.01	Posture, Upper Extremity	Upper Extremity Awkward Positions			
050.19	Potassium Bicarbonate	Potassium Bicarbonate			
050.41	Potassium Carbonate	Potassium Carbonate			584-08-7
040.16	Potassium Chlorate	Potassium Chlorate			3811-04-9
211.02	Potassium Cyanide	Potassium Cyanide			151-50-8
022.01	Potassium Dichromate	Potassium Dichromate			7778-50-9
050.17	Potassium Hydroxide	KOH			1310-58-3
050.17	Potassium Hydroxide	Potassium Hydroxide			1310-58-3
042.01	Potassium Nitrate	Potassium Nitrate			7757-79-1
050.39	Potassium Permanganate	Potassium Permanganate			7722-64-7
040.15	Potassium Salts, NOS	Potassium Salts, NOS			
381.03	Prawn	Prawn		Yes	
320.29	Printing Chemicals, NOS	Printing Chemicals, NOS			
060.20	Propane	LPG			74-98-6
060.20	Propane	Dimethyl methane			74-98-6
060.20	Propane	Propyl hydride			74-98-6
070.08	Propyl Alcohol	Propyl Alcohol	S		71-23-8
070.08	Propyl Alcohol	1-Propanol	S		71-23-8
190.11	Propylene Dichloride	1,2-Dichloropropane	S		78-87-5

Occupational and environmental exposures capable of causing illness - *continued*

AOEC Code[1]	Primary Name	Synonym	Use[2]	Asthma Inducer?	CAS Number[3]
190.11	Propylene Dichloride	Propylene Dichloride	S		78-87-5
090.01	Propylene Glycol Ethers	Dipropylene Glycol Methyl Ether	S		34590-94-8
090.01	Propylene Glycol Ethers	Propylene Glycol Ethers	S		
110.07	Propylene Oxide	Propylene Oxide			75-56-9
390.13	Protozoa Giardia	Protozoa Giardia			
321.31	Proventil	Albuterol Aerosol			18559-94-9
321.31	Proventil	Proventil			18559-94-9
321.08	Psyllium	Psyllium		Yes	
320.18	Pyrethrins	Pyrethroids	P		
320.18	Pyrethrins	Pyrethrins	P		
291.09	Pyrfon	Pyrfon	P	Yes	25311-71-1
250.14	Pyridine	Azabenzene			110-86-1
250.14	Pyridine	Azine			110-86-1
250.14	Pyridine	Pyridine			110-86-1
151.04	Pyromellitic Dianhydride	Pyromellitic Dianhydride		Yes	89-32-7
151.04	Pyromellitic Dianhydride	Pyromellitic Acid Dianhydride		Yes	89-32-7
322.32	Quaternary Ammonium Compounds, NOS	Quaternary Ammonium Compounds, NOS			
322.32	Quaternary Ammonium Compounds, NOS	Benzalkonium Chloride		Yes	8001-54-5
322.32	Quaternary Ammonium Compounds, NOS	Lauryl Dimethyl Benzyl Ammonium Chloride		Yes	139-07-1
322.32	Quaternary Ammonium Compounds, NOS	Dodecyl-dimethyl-benzylammonium Chloride		Yes	139-07-1
373.15	Quillaja Bark	Quillaja Bark		Yes	
373.15	Quillaja Bark	Soapbark		Yes	
180.05	Quinone	Quinone			106-51-4
380.15	Rabbit Antigens	Rabbit Antigens		Yes	
352.01	Radiation, Electromagnetic	ELF			
352.01	Radiation, Electromagnetic	CRT Radiation			
352.01	Radiation, Electromagnetic	Electromagnetic Fields			
352.01	Radiation, Electromagnetic	VDT Radiation			
352.01	Radiation, Electromagnetic	EMF			
352.01	Radiation, Electromagnetic	Radiation, Electromagnetic			
352.01	Radiation, Electromagnetic	Extremely Low Frequency Elctromagnetic Radiation			
351.00	Radiation, Ionizing, NOS	Radiation, Ionizing, NOS			
352.03	Radiation, Microwave	Radio Frequency Radiation			
352.03	Radiation, Microwave	Radiation, Microwave			
352.00	Radiation, Nonionizing, NOS	Radiation, Nonionizing, NOS			
352.04	Radiation, Ultraviolet	Radiation, Ultraviolet			
352.04	Radiation, Ultraviolet	UV Light			
352.04	Radiation, Ultraviolet	UV Radiation			
320.32	Radiographic Fixative	Radiographic Fixative		Yes	
351.03	Radon	Radon			10043-92-2
373.14	Ramin	Gonystylus Bancanus		Yes	
373.14	Ramin	Ramin		Yes	
380.18	Rat Antigens	Rat Antigens		Yes	
380.20	Rat Feces	Rat Feces			
325.00	Reactive Dyes, NOS	Reactive Dyes, NOS		Yes	
381.05	Red Soft Coral	Red Soft Coral		Yes	
360.03	Repetitive Motion	Repetitive Trauma			
360.03	Repetitive Motion	Repetitive Motion			
270.15	Resin Systems, NOS	Resin Systems, NOS			
321.32	Ribaviran	Virazole			36791-04-5
321.32	Ribaviran	1,2,4-Triazole-3-carboxamide			36791-04-5
321.32	Ribaviran	1-beta-D-ribofuranosyl-			36791-04-5
370.48	Rice Dust	Rice Dust		Yes	
371.07	Rice Flour	Rice Flour			
325.05	Rifacion Orange HE 2G	Rifacion Orange HE 2G		Yes	
325.05	Rifacion Orange HE 2G	Color Index No. 0-20		Yes	
325.06	Rifafix Yellow 3 RN	Rifafix Yellow 3 RN		Yes	

Occupational and environmental exposures capable of causing illness - *continued*

AOEC Code[1]	Primary Name	Synonym	Use[2]	Asthma Inducer?	CAS Number[3]
325.08	Rifazol Black GR	Color Index No. BK-5		Yes	
325.08	Rifazol Black GR	Rifazol Black GR		Yes	
325.07	Rifazol Brilliant Orange 3R	Color Index No. 0-16		Yes	
325.07	Rifazol Brilliant Orange 3R	Rifazol Brilliant Orange 3R		Yes	
011.00	Rock, NOS	Rock, NOS			
370.19	Rose Hips	Rose Hips		Yes	
110.04	Rosin	Rosin		Yes	
110.04	Rosin	Tall Oil		Yes	
271.00	Rubber, NOS	Rubber, NOS			
370.49	Rye Dust	Rye Dust			
371.03	Rye Flour	Rye Flour		Yes	
320.38	Saccharin	1,2-Benzisothiazol-3(2H)one			81-07-2
291.08	Safrotin	Propetamphos	P	Yes	31218-83-4
291.08	Safrotin	Safrotin	P	Yes	31218-83-4
321.21	Salbutamol Intermediate	Salbutamol Intermediate		Yes	
321.25	Salicylic Acid	Salicylic Acid			63-36-5
321.25	Salicylic Acid	Salicylate			63-36-5
382.03	Screw Worm Fly	Screw Worm Fly		Yes	
020.31	Selenium	Selenium			7782-49-2
020.31	Selenium	Selenium Compounds			
321.19	Senna	Senna			8013-11-4
370.26	Sericin	Sericin		Yes	
370.45	Sesame Seed Dust	Sesame Seed Dust			
292.02	Sevin	Sevin	P		63-25-2
292.02	Sevin	Carbaryl	P		63-25-2
323.03	Sewer Water	Sewer Water			
323.03	Sewer Water	Sewage			
061.08	Shale	Shale			
381.11	Shark Cartilage	Shark Cartilage			
382.12	Sheep Blowfly	Sheep Blowfly		Yes	
381.01	Shellfish	Shellfish			
381.07	Shrimp Meal	Shrimp Meal		Yes	
381.07	Shrimp Meal	Shrimp		Yes	
391.11	Silage Microorganisms, NOS	Silage Microorganisms, NOS			
010.12	Silica, Amorphous	Diatomaceous Earth			61790-53-2
010.12	Silica, Amorphous	Silica, Amorphous			61790-53-2
010.13	Silica, Crystalline	Quartz Dust			
010.13	Silica, Crystalline	Silica, Crystalline			
010.13	Silica, Crystalline	Silica Flour			
010.13	Silica, Crystalline	Silica Sand			
010.13	Silica, Crystalline	Feldspar Dust			60676-86-0
010.13	Silica, Crystalline	Silicon Dioxide			409-21-2
010.18	Silicon Carbide	Silicon Carbide			409-21-2
010.18	Silicon Carbide	Carborundum			
270.10	Silicone	Siloxanes			
270.10	Silicone	Silicone			
270.10	Silicone	Silicone Rubber			
270.10	Silicone	Silicone Fluid			
382.01	Silkworm	Silkworm Larva		Yes	
382.01	Silkworm	Silkworm		Yes	
020.32	Silver	Silver Compounds			
020.32	Silver	Silver			7440-22-4
020.32	Silver	Silver Sulfate			
391.06	Slime Mold	Slime Mold Dictyostelium discoideum			
330.06	Smoke, Lead-Containing	Smoke, Lead-Containing			
330.03	Smoke, NOS	Fumes, NOS			
330.03	Smoke, NOS	Smoke Inhalation			
330.03	Smoke, NOS	Smoke, NOS			
330.03	Smoke, NOS	Combustion Products, NOS			
322.01	Soap, excluding Laundry Soap/Detergent	Soap (excluding Laundry Soap/Detergent)			

Occupational and environmental exposures capable of causing illness - *continued*

AOEC Code[1]	Primary Name	Synonym	Use[2]	Asthma Inducer?	CAS Number[3]
211.04	Sodium Azide	Sodium Azide			26628-22-8
050.31	Sodium Benzoate	Sodium Benzoate			
050.40	Sodium Bisulfate	Sodium Bisulfate			7681-38-1
040.18	Sodium Bisulfite	Sodium Sulfite			7631-90-5
040.18	Sodium Bisulfite	Sodium Bisulfite			7631-90-5
040.29	Sodium Borohydride	Sodium Borohydride			
050.21	Sodium Carbonate	Soda Ash			497-19-8
050.21	Sodium Carbonate	Sodium Carbonate			
270.11	Sodium Carboxymethyl Cellulose	Sodium Carboxymethyl Cellulose			9004-32-4
041.02	Sodium Chloride	Sodium Chloride			7647-14-5
041.03	Sodium Chlorite	Sodium Chlorite			7758-19-2
211.03	Sodium Cyanide	Sodium Cyanide			143-33-9
022.03	Sodium Dichromate	Sodium Dichromate			10588-01-9
050.18	Sodium Hydroxide	Caustic Soda			1310-73-2
050.18	Sodium Hydroxide	Lye			1310-73-2
050.18	Sodium Hydroxide	NaOH			1310-73-2
050.18	Sodium Hydroxide	Sodium Hydroxide			1310-73-2
040.28	Sodium Metabisulfite	Sodium Metabisulfite		Yes	7681-57-4
050.22	Sodium Metasilicate	Sodium Metasilicate			6834-92-0
042.02	Sodium Nitrate	Sodium Nitrate			
040.17	Sodium Salts, NOS	Sodium Salts, NOS			
050.23	Sodium Silicate	Sodium Silicate			6834-92-0
040.19	Sodium Sulfide	Sodium Sulfide			16721-80-5
050.38	Sodium Tripolyphosphate	Sodium Tripolyphosphate			13573-18-7
023.04	Soldering Flux, NOS	Soldering Flux, NOS			
023.08	Soldering Flux, Zinc Chloride/Ammonium Chloride	Soldering Flux, Zinc Chloride/Ammonium Chloride		Yes	
024.00	Soluble Halogenated Platinum Compounds, NOS	Soluble Halogenated Platinum Compounds, NOS		Yes	
171.00	Solvents, NOS	Unspecified Solvents	S		
171.00	Solvents, NOS	Solvents, NOS	S		
171.00	Solvents, NOS	Multiple Solvents	S		
371.05	Soya Flour	Soya Flour		Yes	
370.41	Soybean Lecithin	Soybean Lecithin		Yes	90320-57-3
370.41	Soybean Lecithin	Soybean Lectin		Yes	90320-57-3
321.10	Spiramycin	Spiramycin		Yes	8025-81-8
391.07	Stachybotrys	Stachybotrys			
150.05	Stearic Acid	Stearic Acid			57-11-4
020.04	Stibine	Stibine			7803-52-3
020.04	Stibine	Antimony Hydride			7803-52-3
360.04	Stress	Psychological Factors			
360.04	Stress	Stress			
360.04	Stress	Job Demand			
360.04	Stress	Rotating Shifts			
360.04	Stress	Job Control			
360.04	Stress	Mental Factors			
171.03	Stripper	Stripper	S		
020.50	Strontium	Strontium			
020.50	Strontium	Strontium Compounds			
353.09	Struck Against/Struck By Objects or Persons	Struck Against/Struck By Objects or Persons			
353.08	Struck by Falling Object	Struck by Falling Object			
353.06	Struck by Motor Vehicle (Road)	Struck by Motor Vehicle (Road)			
353.07	Struck by Vehicle or Equipment (Non-road)	Struck by Vehicle or Equipment (Non-road)			
260.16	Strychnine	Strychnine	P		57-24-9
160.04	Styrene	Styrene Monomer	S	Yes	100-42-5
160.04	Styrene	Vinylbenzene	S	Yes	100-42-5
160.04	Styrene	Styrene	S	Yes	100-42-5
160.04	Styrene	Cinnamene	S	Yes	100-42-5
270.23	Styrene-Maleic Anhydride Resin	Styrene-Maleic Anhydride Polymer			9011-13-6

Occupational and environmental exposures capable of causing illness - *continued*

AOEC Code[1]	Primary Name	Synonym	Use[2]	Asthma Inducer?	CAS Number[3]
270.23	Styrene-Maleic Anhydride Resin	Styrene-Maleic Anhydride Resin			9011-13-6
310.10	Sulfites, NOS	Sulfites, NOS			
322.33	Sulfonates, NOS	Isononanyl Oxybenzene Sulfonate		Yes	
322.33	Sulfonates, NOS	Sulfonates, NOS			
040.22	Sulfur Chloride	Sulfur Monochloride			10025-67-9
040.22	Sulfur Chloride	Sulfur Chloride			10025-67-9
040.21	Sulfur Gas	Sulfur Gas			
040.20	Sulfur Oxides	Sulfur Oxides			
040.20	Sulfur Oxides	SO2			7446-09-5
040.20	Sulfur Oxides	Sulfur Dioxide			7446-09-5
010.21	Sulfur, Elemental	Sulfur			7704-34-9
050.24	Sulfuric Acid	Sulfuric Acid		Yes	7664-93-9
050.24	Sulfuric Acid	Hydrogen Sulfate		Yes	7664-93-9
370.20	Sunflower	Sunflower		Yes	
320.19	Surfactants, NOS	Nonoxynol			
320.19	Surfactants, NOS	Surfactants, NOS			
320.19	Surfactants, NOS	Polyethylene Glycol Stearates			
012.00	Talc	Talc			
012.01	Talc, Fibrous	Talc, Fibrous			
012.02	Talc, Nonasbestiform	Talc, Nonasbestiform			14807-96-6
373.10	Tanganyika Aningre	Tanganyika Aningre		Yes	
150.06	Tannic Acid	Tannic Acid			1401-55-4
150.07	Tartaric Acid	Tartaric Acid			87-69-4
221.07	TDI Prepolymers	TDI Prepolymers		Yes	
370.14	Tea	Tea		Yes	102-71-6
270.12	Teflon	Polytetrafluoroethylene			9002-84-0
270.12	Teflon	PTFE			9002-84-0
270.12	Teflon	Teflon			9002-84-0
020.33	Tellurium	Tellurium			13494-80-9
020.33	Tellurium	Tellurium Compounds			
060.17	Terpene	Terpene	S	Yes	9003-74-1
151.06	Tetrachlorophthalic Anhydride	Tetrachlorophthalic Anhydride		Yes	117-08-8
321.12	Tetracycline	Tetracycline		Yes	60-54-8
060.16	Tetrahydrofuran	Tetrahydrofuran			109-99-9
260.22	Tetrazene	Tetrazene		Yes	70816-59-0
250.10	Tetryl	Tetryl			479-45-8
250.10	Tetryl	Nitramine			479-45-8
320.22	Textile Dust, NOS	Textile Dust, NOS			
040.23	Thallium Salts	Thallium Compounds	P		
040.23	Thallium Salts	Thallium Salts	P		
370.33	Thapsigargin	Thapsigargin			67526-95-8
370.33	Thapsigargin	Thapsia Garganica L			67526-95-8
320.21	Theatrical Fog, Glycol-Based	Theatrical Fog, Glycol-Based			
320.20	Theatrical Fog, NOS	Theatrical Fog, NOS			
321.24	Theophylline	Theophylline			58-55-9
351.04	Therapeutic Radiation	Xrays			
351.04	Therapeutic Radiation	Therapeutic Radiation			
351.04	Therapeutic Radiation	Radioisotopes			
390.05	Thermophilic Actinomyces	Thermophilic Actinomyces			
171.02	Thinner	Lacquer Thinner	S		
171.02	Thinner	Enamel Thinner	S		
171.02	Thinner	Paint Thinner	S		
171.02	Thinner	Thinner	S		
310.07	Thiourea	THU			62-56-6
310.07	Thiourea	Thiourea			62-56-6
310.07	Thiourea	2-Thiocarbamide			
310.08	Thiuram	Thiram			137-26-8
310.08	Thiuram	Thiuram			137-26-8
020.34	Thorium	Thorium			7440-29-1

Occupational and environmental exposures capable of causing illness - *continued*

AOEC Code[1]	Primary Name	Synonym	Use[2]	Asthma Inducer?	CAS Number[3]
020.34	Thorium	Thorium Compounds			
020.35	Tin, Inorganic	Inorganic Tin Compounds			
020.35	Tin, Inorganic	Tin, Inorganic			
020.36	Tin, Organic	Tributyl Tin Oxide	P	Yes	56-35-9
020.36	Tin, Organic	Organic Tin Compounds			
020.36	Tin, Organic	Tin, Organic			
020.48	Titanium	Titanium			7440-32-6
020.48	Titanium	Titanium Compounds			
370.15	Tobacco Leaf	Tobacco Leaf		Yes	
160.02	Toluene	Methylbenzene	S		108-88-3
160.02	Toluene	Phenylmethane	S		108-88-3
160.02	Toluene	Toluene	S		108-88-3
160.02	Toluene	Toluol	S		108-88-3
221.01	Toluene Diisocyanate	TDI		Yes	584-84-9
221.01	Toluene Diisocyanate	Toluene-2,4-Diisocyanate		Yes	584-84-9
221.01	Toluene Diisocyanate	Toluene Diisocyanate		Yes	584-84-9
010.19	Tooth Enamel Dust	Tooth Enamel Dust			
372.02	Tragacanth	Tragacanth		Yes	9000-65-1
353.00	Trauma, Acute, NOS	Trauma, Acute, NOS			
321.22	Trental	Trental			6493-05-6
260.17	Triazine	Triazine	P		101-05-3
260.17	Triazine	Triazine Herbicides	P		
050.43	Trichloroacetic Acid	Trichloroacetic Acid			76-03-9
190.13	Trichloroethylene	Trichloroethylene	S		79-01-6
190.13	Trichloroethylene	Acetylene Trichloride	S		79-01-6
192.04	Trichlorotrifluoroethane	Trifluorotrichloroethane			76-13-1
192.04	Trichlorotrifluoroethane	Trichlorotrifluoroethane			76-13-1
192.04	Trichlorotrifluoroethane	FC 113			76-13-1
391.08	Trichoderma	Trichoderma koningii			
181.03	Triclosan	Triclosan			3380-34-5
181.03	Triclosan	Irgasan			3380-34-5
181.03	Triclosan	5-chloro-2-(2,4-dichlorohydroxy) phenol			3380-34-5
231.02	Triethanolamine	Triethanolamine		Yes	102-71-6
232.03	Triethylene Tetramine	Triethylene Tetramine		Yes	
050.37	Trifluoroacetic Acid	Trifluoroacetic Acid			76-05-1
151.03	Trimellitic Anhydride	TMA		Yes	552-30-7
151.03	Trimellitic Anhydride	1,2,4-Benzenetricarboxylic Acid 1,2-Anhydride		Yes	552-30-7
151.03	Trimellitic Anhydride	Trimellitic Anhydride		Yes	552-30-7
232.04	Trimethylhexanediamine/Isophorondiamine Mixture	Trimethylhexanediamine/Isophorondiamine Mixture		Yes	
142.08	Trimethylolpropane Triacrylate	Acrylic Acid, 1,1,1-(Trihydroxymethyl) Propane Triester			15625-89-5
142.08	Trimethylolpropane Triacrylate	Trimethylolpropane Triacrylate			15625-89-5
142.08	Trimethylolpropane Triacrylate	TMPTA			15625-89-5
142.10	Trimethylolpropane Triacrylate/2-Hydroxypropyl Acrylate	Trimethylolpropane Triacrylate/2-Hydroxypropyl Acrylate		Yes	
250.08	Trinitrotoluene	Trinitrotoluene			118-96-7
250.08	Trinitrotoluene	Trinitrotoluol			
250.08	Trinitrotoluene	TNT			
300.01	Tris	Tris			126-72-7
300.01	Tris	Tris 2,3-Dibromopropyl Phosphate			126-72-7
050.20	Trisodium Phosphate	Trisodium Phosphate			7601-54-9
381.08	Trout	Trout		Yes	
324.02	Trypsin	Trypsin		Yes	9002-07-7
390.10	Tuberculosis	Tuberculosis			
020.52	Tungsten	Tungsten compounds			
020.37	Tungsten Carbide	Tungsten Carbide		Yes	12070-12-1
060.10	Turpentine	Turpentine	S		8006-64-2
060.10	Turpentine	Spirits of Turpentine	S		8006-64-2

Occupational and environmental exposures capable of causing illness - *continued*

AOEC Code[1]	Primary Name	Synonym	Use[2]	Asthma Inducer?	CAS Number[3]
321.15	Tylosin Tartrate	Tylosin Tartrate		Yes	1405-54-5
350.06	Ultrasound	Ultrasound			
351.05	Uranium	Uranium			7440-61-1
270.16	Urea Formaldehyde	Urea Formaldehyde Resin		Yes	
270.16	Urea Formaldehyde	Urea Formaldehyde		Yes	
020.38	Vanadium	Vanadium			7440-62-2
020.38	Vanadium	Vanadium Compounds			
040.30	Vanadium Hydroxide Oxide Phosphate	Vanadium Hydroxide Oxide Phosphate			
360.05	VDT Screen/Visual	VDT Screen/Visual			
370.04	Vegetable Dust	Vegetable Dust			
380.05	Venom	Snake Bite			
380.05	Venom	Bee Sting			
380.05	Venom	Venom			
010.20	Vermiculite	Vermiculite			1318-00-9
354.00	Vibration, NOS	Vibration, NOS			
354.01	Vibration, Regional	Vibration, Regional			
354.01	Vibration, Regional	Local Vibration			
354.01	Vibration, Regional	Hand-Arm Vibration			
354.02	Vibration, Whole Body	Vibration, Whole Body			
370.11	Vicia Sativa	Vetch		Yes	
370.11	Vicia Sativa	Vicia Sativa		Yes	
141.04	Vinyl Acetate	Vinyl Acetate			108-05-4
190.14	Vinyl Chloride Monomer	Vinyl Chloride Monomer			75-01-4
190.14	Vinyl Chloride Monomer	Chloroethene			75-01-4
190.14	Vinyl Chloride Monomer	Vinyl Monomer			75-01-4
190.14	Vinyl Chloride Monomer	Monochloroethylene			75-01-4
010.14	Vinyl Dust	Plastic Dust			
010.14	Vinyl Dust	Rubber Dust			
010.14	Vinyl Dust	PVC Dust			
010.14	Vinyl Dust	Vinyl Dust			9002-86-2
270.27	Vinyl Plastic Wrap	Vinyl Plastic Wrap			25013-15-4
160-08	Vinyl Toluene	Methylstyrene	S		25013-15-4
160.08	Vinyl Toluene	Methylvinylbenzene	S		25013-15-4
160.08	Vinyl Toluene	Vinyl Toluene	S		25013-15-4
353.13	Violence, Other than Physical Assault	Violence, Other than Physical Assault			
360.08	Walking	Walking			
323.01	Waste, Hazardous	Waste, Hazardous Acid			
323.01	Waste, Hazardous	Waste, Hazardous			
323.00	Waste, NOS	Waste, NOS			
323.04	Waste, Treated Human Sludge	Waste, Treated Human Sludge			
327.01	Water Chlorination Byproducts	Water Chlorination Byproducts			
327.02	Water Contamination, Inorganic	Water Contamination, Inorganic			
327.00	Water Contamination, NOS	Water Contamination, NOS			
327.03	Water Contamination, Organic	Water Contamination, Organic			
170.06	Waxes, NOS	Waxes, NOS			
370.24	Weeping Fig	Weeping Fig		Yes	
023.03	Welding Fume, Copper/Nickel	Welding Fume, Copper/Nickel			
023.06	Welding Fume, Galvanized Metal	Welding Fume, Galvanized Metal			
023.02	Welding Fume, Iron or Steel	Welding Fume, Iron or Steel			
023.01	Welding Fume, Stainless Steel	Welding Fume, Stainless Steel		Yes	
023.00	Welding, NOS	Welding Fume, NOS			
023.00	Welding, NOS	Brazing, NOS			
023.00	Welding, NOS	Burning Fume			
023.00	Welding, NOS	Welding, NOS			
023.00	Welding, NOS	Soldering, NOS			
373.01	Western Red Cedar	Thuja Plicata		Yes	
373.01	Western Red Cedar	Western Red Cedar		Yes	
370.50	Wheat Dust	Wheat Dust			
371.04	Wheat Flour	Wheat Flour		Yes	

Occupational and environmental exposures capable of causing illness - continued

AOEC Code[1]	Primary Name	Synonym	Use[2]	Asthma Inducer?	CAS Number[3]
373.00	Wood Dust, NOS	Wood Bark, NOS			
373.00	Wood Dust, NOS	Wood Dust, NOS			
370.44	Wool Dust	Wool Dust			
320.39	World Trade Center Pollution	World Trade Center Dust			
320.39	World Trade Center Pollution	Pollution from Acts of Terrorism/War			
320.39	World Trade Center Pollution	World Trade Center Pollution			
160.03	Xylene	Xylol	S		1330-20-7
160.03	Xylene	Xylene	S		1330-20-7
160.03	Xylene	Dimethylbenzene	S		1330-20-7
370.28	Yeast	Yeast			
324.15	Zeolite	Zeolite			1318-02-1
020.39	Zinc	Zinc		Yes	7440-66-6
020.39	Zinc	Zinc Compounds		Yes	
020.39	Zinc	Zinc Oxide			1314-13-2
020.40	Zinc Chloride	Zinc Chloride			7646-85-7
020.41	Zirconium	Zirconium Compounds			
020.41	Zirconium	Zirconium			7440-67-7

Notes: The items above cover the entire range of occupational and environmental exposures capable of causing illness;
(1) Association of Occupational & Environmental Clinics code; (2) S=Solvent; P=Pesticide; (3) Chemical Abstracts Service registry number
Source: Association of Occupational & Environmental Clinics, May 20, 2005

Selected percentiles and geometric means of blood and urine levels of environmental chemicals (or metabolites)

Chemical Name (Age/Gender/Race Tested)	Units	Geometric Mean (95% CI[1])	Selected Percentiles (95% Confidence Interval)						Sample Size
			10th	25th	50th	75th	90th	95th	
Herbicides[2]									
2,4-Dichlorophenol (Age 6-59)	ug/L	1.11 (.883-1.40)	<LOD	<LOD	.750 (.600-1.00)	2.90 (1.80-4.70)	11.0 (6.40-17.0)	22.0 (17.0-31.0)	1990
Atrazine mercapturate[5] (Age 6-59)	ug/L	NC	<LOD	<LOD	<LOD	<LOD	<LOD	<LOD	1878
Alachlor mercapturate[6] (Age 6-59)	ug/L	NC	<LOD	<LOD	<LOD	<LOD	<LOD	<LOD	1942
Metals[3]									
Cadmium (Age 1 and older)	ug/L	.412 (.386-.439)	<LOD	<LOD	.300 (.300-.400)	.600 (.500-.600)	1.00 (.900-1.00)	1.30 (1.20-1.40)	7970
Lead (Age 1 and older)	ug/dL	1.66 (1.58-1.73)	.800 (.700-.800)	1.00 (1.00-1.10)	1.60 (1.50-1.60)	2.40 (2.30-2.60)	3.80 (3.50-4.00)	4.90 (4.50-5.50)	7970
Mercury (Children 1-5 years)	ug/L	.343 (.299-.393)	<LOD	<LOD	.300 (.200-.300)	.500 (.500-.600)	1.40 (1.10-2.00)	2.30 (1.40-3.20)	705
Mercury (Females 16-49 years)	ug/L	1.02 (.860-1.22)	.200 (<LOD-.200)	.400 (.400-.600)	.900 (.800-1.20)	2.00 (1.60-2.70)	4.90 (4.00-6.10)	7.10 (5.60-9.90)	1709
Metals[2]									
Antimony (Age 6 and older)	ug/L	.128 (.116-.140)	.050 (<LOD-.060)	.070 (.070-.090)	.130 (.120-.140)	.210 (.200-.230)	.330 (.290-.350)	.420 (.390-.470)	2276
Barium (Age 6 and older)	ug/L	1.48 (1.36-1.61)	.200 (.200-.300)	.800 (.700-.900)	1.50 (1.40-1.70)	3.00 (2.80-3.30)	5.40 (4.70-6.00)	6.80 (6.20-8.40)	2180
Beryllium (Age 6 and older)	ug/L	NC	<LOD	<LOD	<LOD	<LOD	<LOD	<LOD	2465
Cadmium (Age 6 and older)	ug/L	.326 (.306-.347)	.110 (.090-.110)	.190 (.170-.200)	.330 (.310-.350)	.590 (.550-.640)	1.01 (.910-1.11)	1.36 (1.21-1.53)	2465
Cesium (Age 6 and older)	ug/L	4.35 (4.08-4.64)	1.60 (1.30-1.80)	2.90 (2.70-3.30)	4.80 (4.50-5.30)	7.10 (6.80-7.40)	9.60 (8.90-10.2)	11.4 (10.3-12.5)	2464
Cobalt (Age 6 and older)	ug/L	.372 (.347-.399)	.130 (.110-.140)	.220 (.200-.250)	.400 (.370-.420)	.630 (.580-.660)	.940 (.880-1.06)	1.32 (1.16-1.45)	2465
Lead (Age 6 and older)	ug/L	.758 (.711-.808)	.200 (.100-.200)	.500 (.400-.500)	.800 (.700-.800)	1.30 (1.30-1.40)	2.10 (1.90-2.30)	2.90 (2.50-3.20)	2465
Mercury (Females 16-49 years)	ug/L	.720 (.642-.808)	<LOD	.310 (.260-.370)	.770 (.650-.880)	1.62 (1.46-1.84)	3.15 (2.68-3.58)	5.00 (3.86-5.55)	1748
Molybdenum (Age 6 and older)	ug/L	45.9 (42.0-50.1)	12.6 (11.3-15.0)	26.7 (22.4-30.8)	50.7 (46.5-56.7)	84.9 (79.8-91.2)	134 (126-145)	178 (157-216)	2257
Platinum (Age 6 and older)	ug/L	NC	<LOD	<LOD	<LOD	<LOD	<LOD	<LOD	2465
Thallium (Age 6 and older)	ug/L	.176 (.167-.186)	.060 (.060-.070)	.110 (.110-.130)	.200 (.180-.200)	.280 (.270-.300)	.400 (.380-.420)	.450 (.420-.470)	2413
Tungsten (Age 6 and older)	ug/L	.085 (.077-.093)	<LOD	<LOD	.090 (.080-.090)	.180 (.150-.200)	.320 (.270-.370)	.500 (.410-.550)	2338

Selected percentiles and geometric means of blood and urine levels of environmental chemicals (or metabolites) - *continued*

Chemical Name (Age/Gender/Race Tested)	Units	Geometric Mean (95% CI[1])	Selected Percentiles (95% Confidence Interval)						Sample Size
			10th	25th	50th	75th	90th	95th	
Uranium (Age 6 and older)	ug/L	.007 (.006-.008)	<LOD	<LOD	.007 (.006-.007)	.013 (.011-.015)	.026 (.022-.036)	.046 (.036-.054)	2464
Pest Repellents and Disinfectants[2]									
2-Naphthol[7] (Age 6-59)	ug/L	.471 (.327-.680)	<LOD	<LOD	.370 (<LOD-.740)	2.00 (1.10-3.30)	7.90 (4.00-12.0)	15.0 (9.90-19.3)	1993
2,5-Dichlorophenol[8] (Age 6-59)	ug/L	6.01 (4.22-8.57)	<LOD	1.40 (.710-2.10)	6.50 (4.60-9.90)	37.8 (23.0-52.0)	144 (88.0-240)	440 (240-700)	1989
N,N-diethyl-3-methylbenzamide (DEET) (Age 6-59)	ug/L	NC	<LOD	<LOD	<LOD	<LOD	<LOD	<LOD	1977
ortho-Phenylphenol (Age 6-59)	ug/L	.494 (.412-.593)	<LOD	<LOD	.490 (.300-.590)	.850 (.650-1.10)	1.46 (1.10-1.80)	2.00 (1.60-2.50)	1991
Pesticides, Carbamate[2]									
1-Naphthol[9] (Age 6-59)	ug/L	1.70 (1.38-2.09)	<LOD	<LOD	1.22 (1.00-1.60)	2.72 (1.90-3.76)	6.20 (4.10-9.60)	12.0 (7.20-19.0)	1998
2-Isopropoxyphenol[10] (Age 6-59)	ug/L	NC	<LOD	<LOD	<LOD	<LOD	<LOD	<LOD	1917
Carbofuranphenol[11] (Age 6-59)	ug/L	NC	<LOD	<LOD	<LOD	<LOD	<LOD	.740 (<LOD-1.30)	1994
Pesticides, Organochlorine[4]									
Hexachlorobenzene* (Age 12 and older)	ng/g	NC	<LOD	<LOD	<LOD	<LOD	<LOD	<LOD	1702
Beta-hexachlorocyclohexane* (Age 12 and older)	ng/g	9.68 (<LOD-10.4)	<LOD	<LOD	<LOD	19.0 (17.0-20.7)	42.0 (35.9-47.1)	68.9 (52.7-80.5)	1893
p,p'-DDT* (Age 12 and older)	ng/g	NC	<LOD	<LOD	<LOD	<LOD	<LOD	27.0 (<LOD-34.0)	1679
p,p'-DDE* (Age 12 and older)	ng/g	260 (234-289)	74.2 (66.1-84.2)	114 (99.8-129)	226 (191-267)	538 (485-609)	1120 (991-1290)	1780 (1520-2230)	1964
Oxychlordane* (Age 12 and older)	ng/g	NC	<LOD	<LOD	<LOD	21.4 (18.6-23.5)	35.7 (30.5-41.3)	44.8 (41.4-49.6)	1661
trans-Nonachlor* (Age 12 and older)	ng/g	18.3 (16.9-19.7)	<LOD	<LOD	18.0 (16.4-20.4)	32.7 (29.5-36.0)	54.6 (47.4-64.5)	77.1 (65.9-84.6)	1933
Pesticides, Organophosphate, Dialkyl Phosphate Metabolites[2]									
Dimethylphosphate (Age 6-59)	ug/L	NC	<LOD	<LOD	.740 (<LOD-1.30)	2.80 (2.10-3.90)	7.90 (5.90-9.50)	13.0 (9.50-21.0)	1949
Dimethylthiophosphate (Age 6-59)	ug/L	1.82 (1.43-2.32)	<LOD	<LOD	2.70 (1.50-3.80)	10.0 (8.00-16.0)	38.0 (21.0-38.0)	46.0 (38.0-60.0)	1948
Dimethyldithiophosphate (Age 6-59)	ug/L	NC	<LOD	<LOD	<LOD	2.30 (1.40-3.60)	12.0 (5.40-17.0)	19.0 (17.0-37.0)	1949

Selected percentiles and geometric means of blood and urine levels of environmental chemicals (or metabolites) - *continued*

Chemical Name (Age/Gender/Race Tested)	Units	Geometric Mean (95% CI[1])	Selected Percentiles (95% Confidence Interval)						Sample Size
			10th	25th	50th	75th	90th	95th	
Diethylphosphate (Age 6-59)	ug/L	1.03 (.757-1.40)	<LOD	<LOD	1.20 (.800-1.50)	3.10 (2.40-4.60)	7.50 (5.20-11.0)	13.0 (8.00-21.0)	1949
Diethylthiophosphate (Age 6-59)	ug/L	NC	<LOD	<LOD	.490 (<LOD-.620)	.760 (.660-.910)	1.30 (1.20-1.60)	2.20 (1.70-2.80)	1949
Diethyldithiophosphate (Age 6-59)	ug/L	NC	<LOD	<LOD	.080 (<LOD-.110)	.200 (.150-.290)	.470 (.390-.630)	.870 (.650-1.00)	1949
Pesticides, Organophosphate, Specific Metabolites[2]									
Malathion dicarboxylic acid[12] (Age 6-59)	ug/L	NC	<LOD	<LOD	<LOD	<LOD	<LOD	<LOD	1920
para-Nitrophenol[13] (Age 6-59)	ug/L	NC	<LOD	<LOD	<LOD	<LOD	2.40 (1.70-3.80)	5.00 (3.30-9.00)	1989
3,5,6-Trichloro-2-pyridinol[14] (Age 6-59)	ug/L	1.77 (1.56-2.01)	<LOD	.870 (.770-.990)	1.70 (1.50-2.00)	3.50 (2.70-4.50)	7.30 (5.40-9.40)	9.90 (7.60-14.0)	1994
2-Isopropyl-4-methyl-6-hydroxypyrimidine[15] (Age 6-59)	ug/L	NC	<LOD	<LOD	<LOD	<LOD	<LOD	<LOD	1842
Phthalate Metabolites[2]									
Mono-2-ethylhexyl phthalate (Age 6 and older)	ug/L	3.43 (3.19-3.69)	<LOD	1.20 (<LOD-1.40)	3.20 (2.90-3.50)	7.60 (6.80-8.20)	14.8 (13.6-17.3)	23.8 (19.2-28.6)	2541
Mono-benzyl phthalate (Age 6 and older)	ug/L	15.3 (14.0-16.8)	2.80 (2.30-3.50)	6.90 (6.00-8.20)	17.0 (15.3-18.6)	35.3 (32.8-38.9)	67.1 (56.8-80.7)	103 (90.3-123)	2541
Mono-butyl phthalates (Age 6 and older)	ug/L	24.6 (22.7-26.6)	5.70 (4.80-6.40)	12.6 (11.0-14.3)	26.0 (24.2-28.5)	51.6 (46.9-56.1)	98.6 (90.7-114)	149 (126-167)	
Mono-cyclohexyl phthalate (Age 6 and older)	ug/L	NC	<LOD	<LOD	<LOD	<LOD	<LOD	1.00 (<LOD-1.50)	2541
Mono-ethyl phthalate (Age 6 and older)	ug/L	179 (159-201)	28.9 (23.6-34.2)	61.4 (54.0-69.7)	164 (142-192)	450 (378-523)	1260 (988-1490)	2840 (2020-4070)	2536
Mono-isononyl phthalate (Age 6 and older)	ug/L	NC	<LOD	<LOD	<LOD	<LOD	<LOD	3.50 (<LOD-11.9)	2541
Mono-n-octyl phthalate (Age 6 and older)	ug/L	NC	<LOD	<LOD	<LOD	<LOD	1.60 (1.20-2.10)	2.90 (2.20-3.40)	2541
Polychlorinated Biphenyls, Non-coplanar[4]									
2,2',4,4',5,5'-Hexachloro-biphenyl (PCB 153)* (Age 12 and older)	ng/g	NC	<LOD	<LOD	<LOD	<LOD	77.8 (67.9-88.8)	112 (93.0-128)	1926
Polychlorinated Dibenzo-p-dioxins[4]									
2,3,7,8-Tetrachlorodibenzo-p-dioxin (TCDD)* (Age 12 and older)	pg/g	NC	<LOD	<LOD	<LOD	<LOD	<LOD	<LOD	1898

Selected percentiles and geometric means of blood and urine levels of environmental chemicals (or metabolites) - *continued*

Chemical Name (Age/Gender/Race Tested)	Units	Geometric Mean (95% CI[1])	Selected Percentiles (95% Confidence Interval)						Sample Size
			10th	25th	50th	75th	90th	95th	
Polycyclic Aromatic Hydrocarbons (PAHs)[2]									
1-Hydroxybenz[a]anthracene[16] (Age 6 and older)	ng/L	NC	<LOD	<LOD	<LOD	<LOD	<LOD	<LOD	2084
3-Hydroxybenz[a]anthracene[16] (Age 6 and older)	ng/L	NC	<LOD	<LOD	<LOD	<LOD	7.50 (6.00-9.90)	11.6 (9.10-14.5)	2152
Tobacco Smoke (Cotinine)[4]									
Age group									
Age 3 and older	ng/mL	NC	<LOD	<LOD	.059 (<LOD-.070)	.236 (.180-.310)	1.02 (.740-1.27)	1.96 (1.64-2.56)	5999
3-11 years	ng/mL	NC	<LOD	<LOD	.109 (.064-.180)	.500 (.290-1.02)	1.88 (1.19-3.09)	3.37 (1.79-4.23)	1174
12-19 years	ng/mL	NC	<LOD	<LOD	.107 (.080-.163)	.540 (.371-.762)	1.65 (1.25-2.11)	2.56 (2.35-3.23)	1773
20 years and older	ng/mL	NC	<LOD	<LOD	<LOD	.167 (.137-.200)	.630 (.520-.863)	1.48 (1.23-1.77)	3052
Gender									
Males	ng/mL	NC	<LOD	<LOD	.080 (.060-.100)	.302 (.220-.390)	1.20 (.890-1.56)	2.39 (1.78-3.06)	2789
Females	ng/mL	NC	<LOD	<LOD	<LOD	.179 (.135-.250)	.850 (.590-1.14)	1.85 (1.41-2.37)	3210
Race/Ethnicity									
Mexican Americans	ng/mL	NC	<LOD	<LOD	<LOD	.139 (.107-.182)	.506 (.340-.813)	1.21 (.813-1.84)	2242
Non-Hispanic blacks	ng/mL	NC	<LOD	<LOD	.131 (.110-.150)	.505 (.400-.625)	1.43 (1.22-1.66)	2.34 (1.89-2.97)	1333
Non-Hispanic whites	ng/mL	NC	<LOD	<LOD	.050 (<LOD-.070)	.210 (.150-.313)	.950 (.621-1.40)	1.92 (1.54-2.74)	1949

Notes: *<LOD means below the limit of detection of the analytical method; NC indicates value was not calculated. Proportion of results below the limit of detection was too high to provide a valid result; * Lipid adjusted; (1) Confidence interval; (2) Urine concentrations; (3) Blood concentrations; (4) Serum concentrations; (5) Metabolite of atrazine; (6) Metabolite of alachlor; (7) Metabolite of naphthalene; (8) Metabolite of p-dichlorobenzene; (9) Metabolite of carbaryl and other chemicals; (10) Metabolite of propoxur; (11) Metabolite of benfuracarb, carbofuran and other chemicals; (12) Metabolite of malathion; (13) Metabolite of parathion and other pesticides; (14) Metabolite of chlorpyrifos; (15) Metabolite of diazinon; (16) Metabolite of benz[a]anthracene*

Source: *Centers for Disease Control and Prevention, Second National Report on Human Exposure to Environmental Chemicals, January 2003*

Lead in blood: Geometric mean and selected percentiles of blood concentrations (in µg/dL) for the U.S. population aged 1 year and older

	Survey years	Geometric mean (95% conf. interval)	Selected percentiles (95% confidence interval)				Sample size
			50th	75th	90th	95th	
Total, age 1 and older	99-00	**1.66** (1.60-1.72)	**1.60** (1.50-1.60)	**2.40** (2.30-2.60)	**3.80** (3.60-3.90)	**4.90** (4.60-5.30)	7970
	01-02	**1.45** (1.39-1.51)	**1.40** (1.30-1.40)	**2.20** (2.10-2.20)	**3.40** (3.10-3.50)	**4.40** (4.20-4.70)	8945
Age group							
1-5 years	99-00	**2.23** (1.96-2.53)	**2.20** (1.90-2.50)	**3.30** (2.80-3.80)	**4.80** (4.00-6.60)	**7.00** (6.10-8.30)	723
	01-02	**1.70** (1.55-1.87)	**1.50** (1.40-1.70)	**2.50** (2.20-2.80)	**4.10** (3.40-5.00)	**5.80** (4.70-6.90)	898
6-11 years	99-00	**1.51** (1.36-1.66)	**1.30** (1.20-1.50)	**2.00** (1.70-2.40)	**3.30** (2.70-3.60)	**4.50** (3.40-6.20)	905
	01-02	**1.25** (1.14-1.36)	**1.10** (1.00-1.30)	**1.60** (1.50-1.80)	**2.70** (2.40-3.00)	**3.70** (3.00-4.70)	1044
12-19 years	99-00	**1.10** (1.04-1.17)	**1.00** (.900-1.10)	**1.40** (1.30-1.60)	**2.30** (2.10-2.30)	**2.80** (2.60-3.00)	2135
	01-02	**.942** (.899-.986)	**.800** (.800-.900)	**1.20** (1.20-1.30)	**1.90** (1.80-2.00)	**2.70** (2.30-2.90)	2231
20 years and older	99-00	**1.75** (1.68-1.81)	**1.70** (1.60-1.70)	**2.50** (2.50-2.60)	**3.90** (3.70-4.00)	**5.20** (4.80-5.50)	4207
	01-02	**1.56** (1.49-1.62)	**1.60** (1.50-1.60)	**2.20** (2.20-2.30)	**3.60** (3.30-3.70)	**4.60** (4.20-4.90)	4772
Gender							
Males	99-00	**2.01** (1.93-2.09)	**1.80** (1.80-1.90)	**2.90** (2.80-3.00)	**4.40** (4.10-4.80)	**6.00** (5.40-6.40)	3913
	01-02	**1.78** (1.71-1.86)	**1.70** (1.70-1.80)	**2.70** (2.50-2.80)	**3.90** (3.70-4.10)	**5.30** (5.00-5.50)	4339
Females	99-00	**1.37** (1.32-1.43)	**1.30** (1.20-1.30)	**1.90** (1.90-2.10)	**3.00** (2.90-3.20)	**4.00** (3.70-4.20)	4057
	01-02	**1.19** (1.14-1.25)	**1.10** (1.10-1.20)	**1.80** (1.70-1.80)	**2.60** (2.40-2.70)	**3.60** (3.00-3.80)	4606
Race/ethnicity							
Mexican Americans	99-00	**1.83** (1.75-1.91)	**1.80** (1.60-1.80)	**2.70** (2.60-2.90)	**4.20** (3.90-4.50)	**5.80** (5.10-6.60)	2742
	01-02	**1.46** (1.34-1.60)	**1.50** (1.30-1.60)	**2.20** (2.00-2.60)	**3.60** (3.30-4.00)	**5.40** (4.40-6.60)	2268
Non-Hispanic blacks	99-00	**1.87** (1.75-2.00)	**1.70** (1.60-1.90)	**2.80** (2.50-2.90)	**4.20** (4.00-4.60)	**5.70** (5.20-6.10)	1842
	01-02	**1.65** (1.52-1.80)	**1.60** (1.40-1.70)	**2.50** (2.30-2.80)	**4.20** (3.80-4.60)	**5.70** (5.30-6.50)	2219
Non-Hispanic whites	99-00	**1.62** (1.55-1.69)	**1.60** (1.50-1.60)	**2.40** (2.30-2.40)	**3.60** (3.40-3.70)	**5.00** (4.40-5.70)	2716
	01-02	**1.43** (1.37-1.48)	**1.40** (1.30-1.40)	**2.10** (2.10-2.20)	**3.10** (3.00-3.40)	**4.10** (3.90-4.50)	3806

Source: *Centers for Disease Control and Prevention, Third National Report on Human Exposure to Environmental Chemicals, July 2005*

Statistics & Rankings / Toxic Environmental Exposures

Lead in urine (creatinine corrected): Geometric mean and selected percentiles of urine concentrations (in µg/g of creatinine) for the U.S. population aged 6 years and older

	Survey years	Geometric mean (95% conf. interval)	Selected percentiles (95% confidence interval)				Sample size
			50th	75th	90th	95th	
Total, age 6 and older	99-00	.721 (.700-.742)	.700 (.677-.725)	1.11 (1.06-1.15)	1.70 (1.62-1.85)	2.37 (2.21-2.76)	2465
	01-02	.639 (.603-.677)	.634 (.586-.676)	1.03 (.962-1.08)	1.52 (1.42-1.60)	2.03 (1.89-2.22)	2689
Age group							
6-11 years	99-00	1.17 (.975-1.41)	1.06 (.918-1.22)	1.55 (1.22-1.97)	2.71 (1.67-4.66)	4.66 (1.97-18.0)	340
	01-02	.918 (.841-1.00)	.870 (.798-.933)	1.26 (1.12-1.43)	2.33 (1.59-3.64)	3.64 (1.83-5.56)	368
12-19 years	99-00	.496 (.460-.535)	.469 (.408-.508)	.702 (.655-.828)	1.10 (.981-1.28)	1.65 (1.15-2.78)	719
	01-02	.404 (.380-.428)	.373 (.342-.400)	.602 (.541-.702)	.990 (.882-1.18)	1.41 (1.07-1.63)	762
20 years and older	99-00	.720 (.683-.758)	.712 (.667-.739)	1.10 (1.02-1.18)	1.69 (1.53-1.87)	2.31 (2.11-2.62)	1406
	01-02	.658 (.617-.703)	.649 (.608-.702)	1.04 (.992-1.11)	1.51 (1.40-1.61)	2.00 (1.85-2.19)	1559
Gender							
Males	99-00	.720 (.679-.763)	.693 (.645-.734)	1.10 (.991-1.22)	1.68 (1.50-2.09)	2.43 (2.15-3.03)	1227
	01-02	.639 (.607-.673)	.638 (.586-.684)	1.01 (.957-1.08)	1.55 (1.41-1.61)	2.06 (1.88-2.43)	1334
Females	99-00	.722 (.681-.765)	.706 (.667-.746)	1.11 (1.05-1.16)	1.74 (1.50-2.02)	2.38 (2.03-2.88)	1238
	01-02	.639 (.594-.688)	.625 (.571-.670)	1.03 (.938-1.11)	1.50 (1.39-1.61)	1.98 (1.85-2.15)	1355
Race/ethnicity							
Mexican Americans	99-00	.940 (.876-1.01)	.882 (.796-1.02)	1.43 (1.36-1.56)	2.38 (2.08-2.77)	3.31 (2.78-4.18)	884
	01-02	.810 (.731-.898)	.769 (.702-.893)	1.28 (1.09-1.44)	2.05 (1.75-2.50)	2.78 (2.55-3.33)	682
Non-Hispanic blacks	99-00	.722 (.659-.790)	.667 (.583-.753)	1.11 (.988-1.20)	1.98 (1.56-2.51)	2.83 (2.20-3.88)	568
	01-02	.644 (.559-.742)	.606 (.510-.710)	.962 (.853-1.19)	1.79 (1.36-2.33)	2.75 (2.04-3.98)	667
Non-Hispanic whites	99-00	.696 (.668-.725)	.677 (.645-.718)	1.07 (.997-1.14)	1.66 (1.50-1.83)	2.31 (1.94-2.82)	822
	01-02	.615 (.579-.654)	.621 (.571-.667)	1.00 (.930-1.06)	1.46 (1.37-1.52)	1.88 (1.61-2.00)	1132

Source: Centers for Disease Control and Prevention, Third National Report on Human Exposure to Environmental Chemicals, July 2005

Mercury in blood: Geometric mean and selected percentiles of blood concentrations (in µg/L) for males and females aged 1 to 5 years and females aged 16 to 49 years in the U.S. population

	Survey years	Geometric mean (95% conf. interval)	Selected percentiles (95% confidence interval)				Sample size
			50th	75th	90th	95th	
Age Group							
1-5 years (females and males)	99-00	**.343** (.297-.395)	**.300** (.200-.300)	**.500** (.500-.600)	**1.40** (1.00-2.30)	**2.30** (1.20-3.50)	705
	01-02	**.318** (.268-.377)	**.300** (.200-.300)	**.700** (.500-.800)	**1.20** (.900-1.60)	**1.90** (1.40-2.90)	872
Females	99-00	**.377** (.299-.475)	**.200** (.200-.300)	**.800** (.500-1.10)	**1.60** (1.00-2.80)	**2.70** (1.30-5.50)	318
	01-02	**.329** (.265-.407)	**.300** (.200-.300)	**.700** (.500-.800)	**1.30** (1.00-2.10)	**2.60** (1.30-4.90)	432
Males	99-00	**.317** (.269-.374)	**.200** (.200-.300)	**.500** (.500-.600)	**1.10** (.800-1.60)	**2.10** (1.10-3.50)	387
	01-02	**.307** (.256-.369)	**.300** (.200-.300)	**.600** (.400-.700)	**1.30** (.900-1.70)	**1.70** (1.40-2.00)	440
16-49 years (females only)	99-00	**1.02** (.825-1.27)	**.900** (.800-1.20)	**2.00** (1.50-3.00)	**4.90** (3.70-6.30)	**7.10** (5.30-11.3)	1709
	01-02	**.833** (.738-.940)	**.700** (.700-.800)	**1.70** (1.40-1.90)	**3.00** (2.70-3.50)	**4.60** (3.70-5.90)	1928
Race/ethnicity (females, 16-49 years)							
Mexican Americans	99-00	**.820** (.664-1.01)	**.900** (.700-1.00)	**1.40** (1.20-2.00)	**2.60** (2.00-3.60)	**4.00** (2.70-5.50)	579
	01-02	**.667** (.541-.824)	**.700** (.500-.800)	**1.10** (1.00-1.40)	**2.10** (1.70-3.00)	**3.50** (2.30-4.40)	527
Non-hispanic blacks	99-00	**1.35** (1.06-1.73)	**1.30** (1.10-1.70)	**2.60** (1.80-3.40)	**4.80** (3.30-6.60)	**5.90** (4.20-11.7)	370
	01-02	**1.06** (.871-1.29)	**1.10** (.800-1.20)	**1.80** (1.50-2.20)	**3.20** (2.20-3.90)	**4.10** (3.30-6.00)	436
Non-hispanic whites	99-00	**.944** (.726-1.23)	**.900** (.700-1.10)	**1.90** (1.30-3.30)	**5.00** (3.00-6.90)	**6.90** (4.50-12.0)	588
	01-02	**.800** (.697-.919)	**.800** (.700-.800)	**1.50** (1.30-2.00)	**3.00** (2.20-3.70)	**4.60** (3.30-6.80)	806

Source: *Centers for Disease Control and Prevention, Third National Report on Human Exposure to Environmental Chemicals, July 2005*

Mercury in urine (creatinine corrected): Geometric mean and selected percentiles of urine concentrations (in µg/g of creatinine) for females aged 16 to 49 years in the U.S. population

	Survey years	Geometric mean (95% conf. interval)	Selected percentiles (95% confidence interval)				Sample size
			50th	75th	90th	95th	
Age group (females)							
16-49 years	99-00	**.710** (.624-.806)	**.723** (.636-.833)	**1.41** (1.24-1.65)	**2.48** (2.10-2.97)	**3.27** (2.85-3.92)	1748
	01-02	**.620** (.579-.664)	**.650** (.582-.709)	**1.27** (1.15-1.42)	**2.30** (2.07-2.45)	**3.00** (2.68-3.39)	1960
Race/ethnicity (females, 16-49 years)							
Mexican Americans	99-00	**.685** (.580-.809)	**.639** (.508-.790)	**1.45** (1.27-1.61)	**2.89** (2.21-3.42)	**4.51** (3.07-5.68)	595
	01-02	**.600** (.526-.686)	**.596** (.426-.709)	**1.32** (1.04-1.47)	**2.41** (2.14-2.77)	**3.21** (2.65-4.46)	531
Non-Hispanic blacks	99-00	**.658** (.520-.831)	**.615** (.475-.892)	**1.22** (.909-1.79)	**2.56** (1.69-3.99)	**3.99** (2.76-5.14)	381
	01-02	**.522** (.410-.665)	**.516** (.387-.664)	**1.03** (.742-1.47)	**1.97** (1.42-3.25)	**3.21** (1.87-4.44)	442
Non-Hispanic whites	99-00	**.706** (.605-.824)	**.721** (.631-.846)	**1.41** (1.23-1.72)	**2.46** (1.99-2.97)	**3.05** (2.46-4.00)	594
	01-02	**.632** (.578-.691)	**.655** (.569-.744)	**1.28** (1.14-1.45)	**2.30** (2.03-2.56)	**2.95** (2.45-3.53)	826

Source: *Centers for Disease Control and Prevention, Third National Report on Human Exposure to Environmental Chemicals, July 2005*

Cotinine (Tobacco Smoke): Geometric mean and selected percentiles of serum concentrations (in ng/mL) for the non-smoking U.S. population aged 3 years and older

	Survey years	Geometric mean (95% conf. interval)	Selected percentiles (95% confidence interval)				Sample size
			50th	75th	90th	95th	
Total, age 3 and older	99-00	*	.059 (<LOD-.070)	.236 (.190-.300)	1.02 (.750-1.25)	1.96 (1.60-2.62)	5999
	01-02	.062 (.050-.077)	†	.163 (.123-.224)	.932 (.737-1.17)	2.19 (1.83-2.44)	6813
Age group							
3-11 years	99-00	*	.109 (.063-.180)	.500 (.259-1.09)	1.88 (.997-3.44)	3.37 (1.42-4.79)	1174
	01-02	.110 (.076-.160)	.071 (<LOD-.124)	.570 (.306-1.01)	2.23 (1.60-2.78)	3.21 (2.53-4.01)	1414
12-19 years	99-00	*	.107 (.080-.160)	.540 (.428-.660)	1.65 (1.48-1.92)	2.56 (2.09-3.39)	1773
	01-02	.086 (.059-.126)	.051 (<LOD-.109)	.352 (.189-.580)	1.53 (1.09-2.12)	3.12 (2.47-3.99)	1902
20 years and older	99-00	*	< LOD	.167 (.140-.193)	.630 (.530-.810)	1.48 (1.28-1.66)	3052
	01-02	.052 (<LOD-.063)	†	.113 (.090-.150)	.623 (.465-.770)	1.38 (1.11-1.84)	3497
Gender							
Males	99-00	*	.080 (.059-.109)	.302 (.220-.394)	1.20 (.950-1.49)	2.39 (1.66-3.22)	2789
	01-02	.075 (.059-.095)	†	.230 (.165-.316)	1.17 (.932-1.42)	2.44 (2.23-2.97)	3149
Females	99-00	*	< LOD	.179 (.148-.220)	.850 (.597-1.14)	1.85 (1.33-2.45)	3210
	01-02	.053 (<LOD-.066)	†	.123 (.092-.180)	.711 (.537-.990)	1.76 (1.32-2.16)	3664
Race/ethnicity							
Mexican Americans	99-00	*	< LOD	.138 (.110-.176)	.506 (.370-.726)	1.21 (.900-1.70)	2241
	01-02	.060 (<LOD-.084)	†	.157 (.080-.308)	.727 (.452-1.19)	2.11 (1.14-2.98)	1877
Non-Hispanic blacks	99-00	*	.131 (.111-.150)	.505 (.400-.625)	1.43 (1.18-1.75)	2.34 (1.84-3.50)	1333
	01-02	.164 (.136-.197)	.132 (.106-.161)	.570 (.436-.760)	1.77 (1.54-2.01)	3.12 (2.47-4.25)	1599
Non-Hispanic whites	99-00	*	.050 (<LOD-.070)	.210 (.150-.310)	.950 (.621-1.40)	1.92 (1.48-3.02)	1950
	01-02	.052 (<LOD-.068)	†	.119 (.087-.180)	.800 (.571-1.11)	1.88 (1.49-2.30)	2845

< LOD means less than the limit of detection, which may vary for some chemicals by year and by individual sample. See Appendix A for LODs.

* Not calculated. Proportion of results below limit of detection was too high to provide a valid result.

† 83% of measurements had a LOD of 0.015 ng/mL and 17% of measurements had a LOD of 0.050 ng/mL. See note in text.

Source: Centers for Disease Control and Prevention, Third National Report on Human Exposure to Environmental Chemicals, July 2005

UV Index for 57 U.S. Cities

City	State	Clear Sky UV Index					UV Index Forecast				
		Extreme	Very High	High	Moderate	Low	Extreme	Very High	High	Moderate	Low
Albuquerque	NM	17	139	49	109	48	1	130	56	91	84
Anchorage	AK	0	0	0	121	241	0	0	0	69	293
Atlanta	GA	2	147	55	104	54	0	15	120	117	110
Atlantic City	NJ	0	76	87	87	112	0	3	58	135	166
Baltimore	MD	0	81	82	90	109	0	4	61	133	164
Billings	MT	0	56	74	90	142	0	9	74	99	180
Bismarck	ND	0	21	94	95	152	0	0	54	105	203
Boise	ID	0	80	55	97	130	0	41	66	91	164
Boston	MA	0	48	92	81	141	0	1	42	132	187
Buffalo	NY	0	41	97	79	145	0	2	39	119	202
Burlington	VT	0	21	107	81	153	0	1	34	118	209
Charleston	SC	0	148	58	113	43	0	23	106	132	101
Charleston	WV	0	96	73	92	101	0	5	65	128	164
Cheyenne	WY	1	108	59	74	120	0	66	48	107	141
Chicago	IL	0	43	104	81	134	0	2	58	118	184
Cleveland	OH	0	52	99	78	133	0	3	46	125	188
Concord	NH	0	41	94	79	148	0	0	43	118	201
Dallas	TX	1	151	58	116	36	0	68	94	96	104
Denver	CO	0	114	60	84	104	0	68	76	87	131
Des Moines	IA	0	61	89	81	131	0	1	76	110	175
Detroit	MI	0	44	100	77	141	0	3	46	119	194
Dover	DE	0	77	89	88	108	0	3	58	135	166
Hartford	CT	0	57	89	77	139	0	1	50	120	191
Honolulu[1,2]	HI	123	128	78	32	0	123	128	78	32	0
Houston	TX	9	164	56	133	0	0	50	132	111	69
Indianapolis	IN	0	72	86	88	116	0	3	57	133	169
Jacksonville	FL	5	162	63	128	4	0	26	130	155	51
Jacksonville	MS	2	153	59	121	27	0	37	126	107	92
Las Vegas	NV	0	124	68	95	75	0	97	72	81	112
Little Rock	AR	1	134	55	101	71	0	17	113	111	121
Los Angeles	CA	0	131	69	112	50	0	81	89	109	83
Louisville	KY	0	97	69	95	101	0	5	82	124	151
Memphis	TN	0	131	55	103	73	0	19	91	127	125
Miami	FL	41	177	54	90	0	0	132	84	144	2
Milwaukee	WI	0	33	112	74	143	0	1	51	116	194
Minneapolis	MN	0	26	98	89	149	0	0	47	111	204
Mobile	AL	4	163	58	128	9	0	34	136	120	72
New Orleans	LA	7	164	60	129	2	0	35	145	124	58
New York	NY	0	64	93	76	129	0	2	52	126	182
Norfolk	VA	0	109	68	92	93	0	10	68	139	145
Oklahoma City	OK	0	137	52	103	70	0	55	94	95	118
Omaha	NE	0	73	77	84	128	0	7	78	111	166
Philadelphia	PA	0	74	87	84	117	0	3	59	129	171
Phoenix	AZ	1	149	60	117	35	0	122	72	111	57
Pittsburgh	PA	0	71	85	81	125	0	3	41	135	183
Portland	ME	0	37	94	82	149	0	0	38	124	200
Portland	OR	0	30	80	99	153	0	2	50	105	205
Providence	RI	0	58	92	75	137	0	1	48	127	186
Raleigh	NC	0	123	62	97	80	0	13	79	138	132
St. Louis	MO	0	92	71	91	108	0	9	85	114	154
Salt Lake City	UT	0	105	63	71	123	0	69	56	97	140
San Francisco	CA	0	104	72	85	101	0	43	96	102	121
Seattle	WA	0	11	87	101	163	0	0	39	107	216
Sioux Falls	SD	0	50	96	77	139	0	1	75	100	186
Tampa	FL	23	174	52	113	0	0	101	97	157	7
Washington	DC	0	83	82	92	105	0	5	61	133	163
Wichita	KS	0	117	58	91	96	0	35	93	102	132

Notes: *Figures are the number of days in each exposure category; The days may not add up to 365 due to missing data; (1) Honolulu does not have cloud information provided so as to compute the cloud attenuation, hence the UV Index forecast is the same as the "clear sky" UV Index; (2) 2001 data*
Source: *NOAA, Climate Prediction Center, UV Index: Annual Time Series, 2003*

Glossary of Terms

A

Abandoned Well: A well whose use has been permanently discontinued or which is in a state of such disrepair that it cannot be used for its intended purpose.

Abatement: Reducing the degree or intensity of, or eliminating, pollution.

Abatement Debris: Waste from remediation activities.

Absorbed Dose: In exposure assessment, the amount of a substance that penetrates an exposed organism's absorption barriers (e.g. skin, lung tissue, gastrointestinal tract) through physical or biological processes. The term is synonymous with internal dose.

Absorption: The uptake of water, other fluids, or dissolved chemicals by a cell or an organism (as tree roots absorb dissolved nutrients in soil.)

Absorption Barrier: Any of the exchange sites of the body that permit uptake of various substances at different rates (e.g. skin, lung tissue, and gastrointestinal-tract wall)

Accident Site: The location of an unexpected occurrence, failure or loss, either at a plant or along a transportation route, resulting in a release of hazardous materials.

Acclimatization: The physiological and behavioral adjustments of an organism to changes in its environment.

Acid Aerosol: Acidic liquid or solid particles small enough to become airborne. High concentrations can irritate the lungs and have been associated with respiratory diseases like asthma.

Acid Deposition: A complex chemical and atmospheric phenomenon that occurs when emissions of sulfur and nitrogen compounds and other substances are transformed by chemical processes in the atmosphere, often far from the original sources, and then deposited on earth in either wet or dry form. The wet forms, popularly called "acid rain," can fall to earth as rain, snow, or fog. The dry forms are acidic gases or particulates.

Acid Mine Drainage: Drainage of water from areas that have been mined for coal or other mineral ores. The water has a low pH because of its contact with sulfur-bearing material and is harmful to aquatic organisms.

Acid Neutralizing Capacity: Measure of ability of a base (e.g. water or soil) to resist changes in pH.

Acid Rain: (See: acid deposition)

Acidic: The condition of water or soil that contains a sufficient amount of acid substances to lower the pH below 7.0.

Action Levels: 1. Regulatory levels recommended by EPA for enforcement by FDA and USDA when pesticide residues occur in food or feed commodities for reasons other than the direct application of the pesticide. As opposed to "tolerances" which are established for residues occurring as a direct result of proper usage, action levels are set for inadvertent residues resulting from previous legal use or accidental contamination. 2. In the Superfund program, the existence of a contaminant concentration in the environment high enough to warrant action or trigger a response under SARA and the National Oil and Hazardous Substances Contingency Plan. The term is also used in other regulatory programs. (See: tolerances)

Activated Carbon: A highly adsorbent form of carbon used to remove odors and toxic substances from liquid or gaseous emissions. In waste treatment, it is used to remove dissolved organic matter from waste drinking water. It is also used in motor vehicle evaporative control systems.

Activated Sludge: Product that results when primary effluent is mixed with bacteria-laden sludge and then agitated and aerated to promote biological treatment, speeding the breakdown of organic matter in raw sewage undergoing secondary waste treatment.

Activator: A chemical added to a pesticide to increase its activity.

Active Ingredient: In any pesticide product, the component that kills, or otherwise controls, target pests. Pesticides are regulated primarily on the basis of active ingredients.

Activity Plans: Written procedures in a school's asbestos-management plan that detail the steps a Local Education Agency (LEA) will follow in performing the initial and additional cleaning, operation and maintenance-program tasks; periodic surveillance; and reinspection required by the Asbestos Hazard Emergency Response Act (AHERA).

Acute Exposure: A single exposure to a toxic substance which may result in severe biological harm or death. Acute exposures are usually characterized as lasting no longer than a day, as compared to longer, continuing exposure over a period of time.

Acute Toxicity: The ability of a substance to cause severe biological harm or death soon after a single exposure or dose. Also, any poisonous effect resulting from a single short-term exposure to a toxic substance. (See: chronic toxicity, toxicity)

Adaptation: Changes in an organism's physiological structure or function or habits that allow it to survive in new surroundings.

Add-on Control Device: An air pollution control device such as carbon absorber or incinerator that reduces the pollution in an exhaust gas. The control device usually does not affect the process being controlled and thus is "add-on" technology, as opposed to a scheme to control pollution through altering the basic process itself.

Adequately Wet: Asbestos containing material that is sufficiently mixed or penetrated with liquid to prevent the release of particulates.

Administered Dose: In exposure assessment, the amount of a substance given to a test subject (human or animal) to determine dose-response relationships. Since exposure to chemicals is usually inadvertent, this quantity is often called potential dose.

Administrative Order: A legal document signed by EPA directing an individual, business, or other entity to take corrective action or refrain from an activity. It describes the violations and actions to be taken, and can be enforced in court. Such orders may be issued, for example, as a result of an administrative complaint whereby the respondent is ordered to pay a penalty for violations of a statute.

Administrative Order On Consent: A legal agreement signed by EPA and an individual, business, or other entity through which the violator agrees to pay for correction of violations, take the required corrective or cleanup actions, or refrain from an activity. It describes the actions to be taken, may be subject to a comment period, applies to civil actions, and can be enforced in court.

Administrative Procedures Act: A law that spells out procedures and requirements related to the promulgation of regulations.

Administrative Record: All documents which EPA considered or relied on in selecting the response action at a Superfund site, culminating in the record of decision for remedial action or, an action memorandum for removal actions.

Adsorption: Removal of a pollutant from air or water by collecting the pollutant on the surface of a solid material; e.g., an advanced method of treating waste in which activated carbon removes organic matter from waste-water.

Adulterants: Chemical impurities or substances that by law do not belong in a food, or pesticide.

Adulterated: 1. Any pesticide whose strength or purity falls below the quality stated on its label. 2. A food, feed, or product that contains illegal pesticide residues.

Advanced Treatment: A level of wastewater treatment more stringent than secondary treatment; requires an 85-percent reduction in conventional pollutant concentration or a significant reduction in non-conventional pollutants. Sometimes called tertiary treatment

Advanced Wastewater Treatment: Any treatment of sewage that goes beyond the secondary or biological water treatment stage and includes the removal of nutrients such as phosphorus and nitrogen and a high percentage of suspended solids. (See primary, secondary treatment.)

Adverse Effects Data: FIFRA requires a pesticide registrant to submit data to EPA on any studies or other information regarding unreasonable adverse effects of a pesticide at any time after its registration.

Advisory: A non-regulatory document that communicates risk information to those who may have to make risk management decisions.

Aerated Lagoon: A holding and/or treatment pond that speeds up the natural process of biological decomposition of organic waste by stimulating the growth and activity of bacteria that degrade organic waste.

Aeration: A process which promotes biological degradation of organic matter in water. The process may be passive (as when waste is exposed to air), or active (as when a mixing or bubbling device introduces the air).

Aeration Tank: A chamber used to inject air into water.

Glossary of Terms

Aerobic: Life or processes that require, or are not destroyed by, the presence of oxygen. (See: anaerobic.)

Aerobic Treatment: Process by which microbes decompose complex organic compounds in the presence of oxygen and use the liberated energy for reproduction and growth. (Such processes include extended aeration, trickling filtration, and rotating biological contactors.)

Aerosol: 1. Small droplets or particles suspended in the atmosphere, typically containing sulfur. They are usually emitted naturally (e.g. in volcanic eruptions) and as the result of anthropogenic (human) activities such as burning fossil fuels. 2. The pressurized gas used to propel substances out of a container.

Aerosol: A finely divided material suspended in air or other gaseous environment.

Affected Landfill: Under the Clean Air Act, landfills that meet criteria for capacity, age, and emissions rates set by the EPA. They are required to collect and combust their gas emissions.

Affected Public: 1.The people who live and/or work near a hazardous waste site. 2. The human population adversely impacted following exposure to a toxic pollutant in food, water, air, or soil.

Afterburner: In incinerator technology, a burner located so that the combustion gases are made to pass through its flame in order to remove smoke and odors. It may be attached to or be separated from the incinerator proper.

Age Tank: A tank used to store a chemical solution of known concentration for feed to a chemical feeder. Also called a day tank.

Agent: Any physical, chemical, or biological entity that can be harmful to an organism (synonymous with stressors.)

Agent Orange: A toxic herbicide and defoliant used in the Vietnam conflict, containing 2,4,5-trichlorophen-oxyacetic acid (2,4,5-T) and 2-4 dichlorophenoxyacetic acid (2,4-D) with trace amounts of dioxin.

Agricultural Pollution: Farming wastes, including runoff and leaching of pesticides and fertilizers; erosion and dust from plowing; improper disposal of animal manure and carcasses; crop residues, and debris.

Agroecosystem: Land used for crops, pasture, and livestock; the adjacent uncultivated land that supports other vegetation and wildlife; and the associated atmosphere, the underlying soils, groundwater, and drainage networks.

AHERA Designated Person (ADP): A person designated by a Local Education Agency to ensure that the AHERA requirements for asbestos management and abatement are properly implemented.

Air Binding: Situation where air enters the filter media and harms both the filtration and backwash processes.

Air Changes Per Hour (ACH): The movement of a volume of air in a given period of time; if a house has one air change per hour, it means that the air in the house will be replaced in a one-hour period.

Air Cleaning: Indoor-air quality-control strategy to remove various airborne particulates and/or gases from the air. Most common methods are particulate filtration, electrostatic precipitation, and gas sorption.

Air Contaminant: Any particulate matter, gas, or combination thereof, other than water vapor. (See: air pollutant.)

Air Curtain: A method of containing oil spills. Air bubbling through a perforated pipe causes an upward water flow that slows the spread of oil. It can also be used to stop fish from entering polluted water.

Air Exchange Rate: The rate at which outside air replaces indoor air in a given space.

Air Gap: Open vertical gap or empty space that separates drinking water supply to be protected from another water system in a treatment plant or other location. The open gap protects the drinking water from contamination by backflow or back siphonage.

Air Handling Unit: Equipment that includes a fan or blower, heating and/or cooling coils, regulator controls, condensate drain pans, and air filters.

Air Mass: A large volume of air with certain meteorological or polluted characteristics—e.g., a heat inversion or smogginess—while in one location. The characteristics can change as the air mass moves away.

Air Monitoring: (See: monitoring.)

Air/Oil Table: The surface between the vadose zone and ambient oil; the pressure of oil in the porous medium is equal to atmospheric pressure.

Air Padding: Pumping dry air into a container to assist with the withdrawal of liquid or to force a liquefied gas such as chlorine out of the container.

Air Permeability: Permeability of soil with respect to air. Important to the design of soil-gas surveys. Measured in darcys or centimeters-per-second.

Air Plenum: Any space used to convey air in a building, furnace, or structure. The space above a suspended ceiling is often used as an air plenum.

Air Pollutant: Any substance in air that could, in high enough concentration, harm man, other animals, vegetation, or material. Pollutants may include almost any natural or artificial composition of airborne matter capable of being airborne. They may be in the form of solid particles, liquid droplets, gases, or in combination thereof. Generally, they fall into two main groups: (1) those emitted directly from identifiable sources and (2) those produced in the air by interaction between two or more primary pollutants, or by reaction with normal atmospheric constituents, with or without photoactivation. Exclusive of pollen, fog, and dust, which are of natural origin, about 100 contaminants have been identified. Air pollutants are often grouped in categories for ease in classification; some of he categories are: solids, sulfur compounds, volatile organic chemicals, particulate matter, nitrogen compounds, oxygen compounds, halogen compounds, radioactive compound, and odors.

Air Pollution: The presence of contaminants or pollutant substances in the air that interfere with human health or welfare, or produce other harmful environmental effects.

Air Pollution Control Device: Mechanism or equipment that cleans emissions generated by a source (e.g. an incinerator, industrial smokestack, or an automobile exhaust system) by removing pollutants that would otherwise be released to the atmosphere.

Air Pollution Episode: A period of abnormally high concentration of air pollutants, often due to low winds and temperature inversion, that can cause illness and death. (See: episode, pollution.)

Air Quality Control Region:

Air Quality Criteria: The levels of pollution and lengths of exposure above which adverse health and welfare effects may occur.

Air Quality Standards: The level of pollutants prescribed by regulations that are not be exceeded during a given time in a defined area.

Air Sparging: Injecting air or oxygen into an aquifer to strip or flush volatile contaminants as air bubbles up through The ground water and is captured by a vapor extraction system.

Air Stripping: A treatment system that removes volatile organic compounds (VOCs) from contaminated ground water or surface water by forcing an airstream through the water and causing the compounds to evaporate.

Air Toxics: Any air pollutant for which a national ambient air quality standard (NAAQS) does not exist (i.e. excluding ozone, carbon monoxide, PM-10, sulfur dioxide, nitrogen oxide) that may reasonably be anticipated to cause cancer; respiratory, cardiovascular, or developmental effects; reproductive dysfunctions, neurological disorders, heritable gene mutations, or other serious or irreversible chronic or acute health effects in humans.

Airborne Particulates: Total suspended particulate matter found in the atmosphere as solid particles or liquid droplets. Chemical composition of particulates varies widely, depending on location and time of year. Sources of airborne particulates include: dust, emissions from industrial processes, combustion products from the burning of wood and coal, combustion products associated with motor vehicle or non-road engine exhausts, and reactions to gases in the atmosphere.

Airborne Release: Release of any pollutant into the air.

Alachlor: A herbicide, marketed under the trade name Lasso, used mainly to control weeds in corn and soybean fields.

Alar: Trade name for daminozide, a pesticide that makes apples redder, firmer, and less likely to drop off trees before growers are ready to pick them. It is also used to a lesser extent on peanuts, tart cherries, concord grapes, and other fruits.

Aldicarb: An insecticide sold under the trade name Temik. It is made from ethyl isocyanate.

Algae: Simple rootless plants that grow in sunlit waters in proportion to the amount of available nutrients. They can affect water quality adversely by lowering the dissolved oxygen in the water. They are food for fish and small aquatic animals.

Algal Blooms: Sudden spurts of algal growth, which can affect water quality adversely and indicate potentially hazardous changes in local water chemistry.

Algicide: Substance or chemical used specifically to kill or control algae.

Aliquot: A measured portion of a sample taken for analysis. One or more aliquots make up a sample. (See: duplicate.)

Alkaline: The condition of water or soil which contains a sufficient amount of alkali substance to raise the pH above 7.0.

Alkalinity: The capacity of bases to neutralize acids. An example is lime added to lakes to decrease acidity.

Allergen: A substance that causes an allergic reaction in individuals sensitive to it.

Alluvial: Relating to and/or sand deposited by flowing water.

Alternate Method: Any method of sampling and analyzing for an air or water pollutant that is not a reference or equivalent method but that has been demonstrated in specific cases-to EPA's satisfaction-to produce results adequate for compliance monitoring.

Alternative Compliance: A policy that allows facilities to choose among methods for achieving emission-reduction or risk-reduction instead of command-and control regulations that specify standards and how to meet them. Use of a theoretical emissions bubble over a facility to cap the amount of pollution emitted while allowing the company to choose where and how (within the facility) it complies.(See: bubble, emissions trading.)

Alternative Fuels: Substitutes for traditional liquid, oil-derived motor vehicle fuels like gasoline and diesel. Includes mixtures of alcohol-based fuels with gasoline, methanol, ethanol, compressed natural gas, and others.

Alternative Remedial Contract Strategy Contractors: Government contractors who provide project management and technical services to support remedial response activities at National Priorities List sites.

Ambient Air: Any unconfined portion of the atmosphere: open air, surrounding air.

Ambient Air Quality Standards: (See: Criteria Pollutants and National Ambient Air Quality Standards.)

Ambient Measurement: A measurement of the concentration of a substance or pollutant within the immediate environs of an organism; taken to relate it to the amount of possible exposure.

Ambient Medium: Material surrounding or contacting an organism (e.g. outdoor air, indoor air, water, or soil, through which chemicals or pollutants can reach the organism. (See: biological medium, environmental medium.)

Ambient Temperature: Temperature of the surrounding air or other medium.

Amprometric Titration: A way of measuring concentrations of certain substances in water using an electric current that flows during a chemical reaction.

Anaerobic: A life or process that occurs in, or is not destroyed by, the absence of oxygen.

Anaerobic Decomposition: Reduction of the net energy level and change in chemical composition of organic matter caused by microorganisms in an oxygen-free environment.

Animal Dander: Tiny scales of animal skin, a common indoor air pollutant.

Animal Studies: Investigations using animals as surrogates for humans with the expectation that the results are pertinent to humans.

Anisotropy: In hydrology, the conditions under which one or more hydraulic properties of an aquifer vary from a reference point.

Annular Space, Annulus: The space between two concentric tubes or casings, or between the casing and the borehole wall.

Antagonism: Interference or inhibition of the effect of one chemical by the action of another.

Antarctic "Ozone Hole": Refers to the seasonal depletion of ozone in the upper atmosphere above a large area of Antarctica. (See: Ozone Hole.)

Anti-Degradation Clause: Part of federal air quality and water quality requirements prohibiting deterioration where pollution levels are above the legal limit.

Anti-Microbial: An agent that kills microbes.

Applicable or Relevant and Appropriate Requirements (ARARs): Any state or federal statute that pertains to protection of human life and the environment in addressing specific conditions or use of a particular cleanup technology at a Superfund site,

Applied Dose: In exposure assessment, the amount of a substance in contact with the primary absorption boundaries of an organism (e.g. skin, lung tissue, gastrointestinal track) and available for absorption.

Aqueous: Something made up of water.

Aqueous Solubility: The maximum concentration of a chemical that will dissolve in pure water at a reference temperature.

Aquifer: An underground geological formation, or group of formations, containing water. Are sources of groundwater for wells and springs.

Aquifer Test: A test to determine hydraulic properties of an aquifer.

Aquitard: Geological formation that may contain groundwater but is not capable of transmitting significant quantities of it under normal hydraulic gradients. May function as confining bed.

Architectural Coatings: Coverings such as paint and roof tar that are used on exteriors of buildings.

Area of Review: In the UIC program, the area surrounding an injection well that is reviewed during the permitting process to determine if flow between aquifers will be induced by the injection operation.

Area Source: Any source of air pollution that is released over a relatively small area but which cannot be classified as a point source. Such sources may include vehicles and other small engines, small businesses and household activities, or biogenic sources such as a forest that releases hydrocarbons.

Aromatics: A type of hydrocarbon, such as benzene or toluene, with a specific type of ring structure. Aromatics are sometimes added to gasoline in order to increase octane. Some aromatics are toxic.

Arsenicals: Pesticides containing arsenic.

Artesian (Aquifer or Well): Water held under pressure in porous rock or soil confined by impermeable geological formations.

Asbestos: A mineral fiber that can pollute air or water and cause cancer or asbestosis when inhaled. EPA has banned or severely restricted its use in manufacturing and construction.

Asbestos Abatement: Procedures to control fiber release from asbestos-containing materials in a building or to remove them entirely, including removal, encapsulation, repair, enclosure, encasement, and operations and maintenance programs.

Asbestos Assessment: In the asbestos-in-schools program, the evaluation of the physical condition and potential for damage of all friable asbestos containing materials and thermal insulation systems.

Asbestos Program Manager: A building owner or designated representative who supervises all aspects of the facility asbestos management and control program.

Asbestos-Containing Waste Materials (ACWM): Mill tailings or any waste that contains commercial asbestos and is generated by a source covered by the Clean Air Act Asbestos NESHAPS.

Asbestosis: A disease associated with inhalation of asbestos fibers. The disease makes breathing progressively more difficult and can be fatal.

Ash: The mineral content of a product remaining after complete combustion.

Assay: A test for a specific chemical, microbe, or effect.

Assessment Endpoint: In ecological risk assessment, an explicit expression of the environmental value to be protected; includes both an ecological entity and specific attributed thereof. entity (e.g. salmon are a valued ecological entity; reproduction and population maintenance—the attribute—form an assessment endpoint.)

Assimilation: The ability of a body of water to purify itself of pollutants.

Assimilative Capacity: The capacity of a natural body of water to receive wastewaters or toxic materials without deleterious effects and without damage to aquatic life or humans who consume the water.

Association of Boards of Certification: An international organization representing boards which certify the operators of waterworks and wastewater facilities.

Attainment Area: An area considered to have air quality as good as or better than the national ambient air quality standards as defined in the Clean Air Act. An area may be an attainment area for one pollutant and a non-attainment area for others.

Attenuation: The process by which a compound is reduced in concentration over time, through absorption, adsorption, degradation, dilution, and/or transformation. an also be the decrease with distance of sight caused by attenuation of light by particulate pollution.

Attractant: A chemical or agent that lures insects or other pests by stimulating their sense of smell.

Attrition: Wearing or grinding down of a substance by friction. Dust from such processes contributes to air pollution.

Availability Session: Informal meeting at a public location where interested citizens can talk with EPA and state officials on a one-to-one basis.

Glossary of Terms

Available Chlorine: A measure of the amount of chlorine available in chlorinated lime, hypochlorite compounds, and other materials used as a source of chlorine when compared with that of liquid or gaseous chlorines.

Avoided Cost: The cost a utility would incur to generate the next increment of electric capacity using its own resources; many landfill gas projects' buy back rates are based on avoided costs.

A-Scale Sound Level: A measurement of sound approximating the sensitivity of the human ear, used to note the intensity or annoyance level of sounds.

B

Back Pressure: A pressure that can cause water to backflow into the water supply when a user's waste water system is at a higher pressure than the public system.

Backflow/Back Siphonage: A reverse flow condition created by a difference in water pressures that causes water to flow back into the distribution pipes of a drinking water supply from any source other than the intended one.

Background Level: 1. The concentration of a substance in an environmental media (air, water, or soil) that occurs naturally or is not the result of human activities. 2. In exposure assessment the concentration of a substance in a defined control area, during a fixed period of time before, during, or after a data-gathering operation..

Backwashing: Reversing the flow of water back through the filter media to remove entrapped solids.

Backyard Composting: Diversion of organic food waste and yard trimmings from the municipal waste stream by composting hem in one's yard through controlled decomposition of organic matter by bacteria and fungi into a humus-like product. It is considered source reduction, not recycling, because the composted materials never enter the municipal waste stream.

Barrel Sampler: Open-ended steel tube used to collect soil samples.

BACT - Best Available Control Technology: An emission limitation based on the maximum degree of emission reduction (considering energy, environmental, and economic impacts) achievable through application of production processes and available methods, systems, and techniques. BACT does not permit emissions in excess of those allowed under any applicable Clean Air Act provisions. Use of the BACT concept is allowable on a case by case basis for major new or modified emissions sources in attainment areas and applies to each regulated pollutant.

Bacteria: (Singular: bacterium) Microscopic living organisms that can aid in pollution control by metabolizing organic matter in sewage, oil spills or other pollutants. However, bacteria in soil, water or air can also cause human, animal and plant health problems.

Baffle: A flat board or plate, deflector, guide, or similar device constructed or placed in flowing water or slurry systems to cause more uniform flow velocities to absorb energy and to divert, guide, or agitate liquids.

Baffle Chamber: In incinerator design, a chamber designed to promote the settling of fly ash and coarse particulate matter by changing the direction and/or reducing the velocity of the gases produced by the combustion of the refuse or sludge.

Baghouse Filter: Large fabric bag, usually made of glass fibers, used to eliminate intermediate and large (greater than 20 PM in diameter) particles. This device operates like the bag of an electric vacuum cleaner, passing the air and smaller particles while entrapping the larger ones.

Bailer: A pipe with a valve at the lower end, used to remove slurry from the bottom or side of a well as it is being drilled, or to collect groundwater samples from wells or open boreholes. 2. A tube of varying length.

Baling: Compacting solid waste into blocks to reduce volume and simplify handling.

Ballistic Separator: A machine that sorts organic from inorganic matter for composting.

Band Application: The spreading of chemicals over, or next to, each row of plants in a field.

Banking: A system for recording qualified air emission reductions for later use in bubble, offset, or netting transactions. (See: emissions trading.)

Bar Screen: In wastewater treatment, a device used to remove large solids.

Barrier Coating(s): A layer of a material that obstructs or prevents passage of something through a surface that is to be protected; e.g., grout, caulk, or various sealing compounds; sometimes used with polyurethane membranes to prevent corrosion or oxidation of metal surfaces, chemical impacts on various materials, or, for example, to prevent radon infiltration through walls, cracks, or joints in a house.

Basal Application: In pesticides, the application of a chemical on plant stems or tree trunks just above the soil line.

Basalt: Consistent year-round energy use of a facility; also refers to the minimum amount of electricity supplied continually to a facility.

Bean Sheet: Common term for a pesticide data package record.

Bed Load: Sediment particles resting on or near the channel bottom that are pushed or rolled along by the flow of water.

BEN: EPA's computer model for analyzing a violator's economic gain from not complying with the law.

Bench-scale Tests: Laboratory testing of potential cleanup technologies (See: treatability studies)

Benefit-Cost Analysis: An economic method for assessing the benefits and costs of achieving alternative health-based standards at given levels of health protection.

Benthic/Benthos: An organism that feeds on the sediment at the bottom of a water body such as an ocean, lake, or river.

Bentonite: A colloidal clay, expansible when moist, commonly used to provide a tight seal around a well casing.

Beryllium: A metal hazardous to human health when inhaled as an airborne pollutant. It is discharged by machine shops, ceramic and propellant plants, and foundries.

Best Available Control Measures (BACM): A term used to refer to the most effective measures (according to EPA guidance) for controlling small or dispersed particulates and other emissions from

sources such as roadway dust, soot and ash from woodstoves and open burning of rush, timber, grasslands, or trash.

Best Available Control Technology (BACT): For any specific source, the currently available technology producing the greatest reduction of air pollutant emissions,taking into account energy, environmental, economic, and other costs.

Best Available Control Technology (BACT): The most stringent technology available for controlling emissions; major sources are required to use BACT, unless it can be demonstrated that it is not feasible for energy, environmental, or economic reasons.

Best Demonstrated Available Technology (BDAT): As identified by EPA, the most effective commercially available means of treating specific types of hazardous waste. The BDATs may change with advances in treatment technologies.

Best Management Practice (BMP): Methods that have been determined to be the most effective, practical means of preventing or reducing pollution from non-point sources.

Bimetal: Beverage containers with steel bodies and aluminum tops; handled differently from pure aluminum in recycling.

Bioaccumulants: Substances that increase in concentration in living organisms as they take in contaminated air, water, or food because the substances are very slowly metabolized or excreted. (See: biological magnification.)

Bioassay: A test to determine te relative strength of a substance by comparing its effect on a test organism with that of a standard preparation.

Bioavailabiliity: Degree of ability to be absorbed and ready to interact in organism metabolism.

Biochemical Oxygen Demand (BOD): A measure of the amount of oxygen consumed in the biological processes that break down organic matter in water. The greater the BOD, the greater the degree of pollution.

Bioconcentration: The accumulation of a chemical in tissues of a fish or other organism to levels greater than in the surrounding medium.

Biodegradable: Capable of decomposing under natural conditions.

Biodiversity: Refers to the variety and variability among living organisms and the ecological complexes in which they occur. Diversity can be defined as the number of different items and their relative frequencies. For biological diversity, these items are organized at many levels, ranging from complete ecosystems to the biochemical structures that are the molecular basis of heredity. Thus, the term encompasses different ecosystems, species, and genes.

Biological Contaminants: Living organisms or derivates (e.g. viruses, bacteria, fungi, and mammal and bird antigens) that can cause harmful health effects when inhaled, swallowed, or otherwise taken into the body.

Biological Control: In pest control, the use of animals and organisms that eat or otherwise kill or out-compete pests.

Biological Integrity: The ability to support and maintain balanced, integrated, functionality in the natural habitat of a given region. Concept is applied primarily in drinking water management.

Biological Magnification: Refers to the process whereby certain substances such as pesticides or heavy metals move up the food chain, work their way into rivers or lakes, and are eaten by aquatic organisms such as fish, which in turn are eaten by large birds, animals or humans. The substances become concentrated in tissues or internal organs as they move up the chain. (See: bioaccumulants.)

Biological Measurement: A measurement taken in a biological medium. For exposure assessment, it is related to the measurement is taken to related it to the established internal dose of a compound.

Biological Medium: One of the major component of an organism; e.g. blood, fatty tissue, lymph nodes or breath, in which chemicals can be stored or transformed. (See: ambient medium, environmental medium.)

Biological Oxidation: Decomposition of complex organic materials by microorganisms. Occurs in self-purification of water bodies and in activated sludge wastewater treatment.

Biological Oxygen Demand (BOD): An indirect measure of the concentration of biologically degradable material present in organic wastes. It usually reflects the amount of oxygen consumed in five days by biological processes breaking down organic waste.

Biological Stressors: Organisms accidentally or intentionally dropped into habitats in which they do not evolve naturally; e.g. gypsy moths, Dutch elm disease, certain types of algae, and bacteria.

Biological Treatment: A treatment technology that uses bacteria to consume organic waste.

Biologically Effective Dose: The amount of a deposited or absorbed compound reaching the cells or target sites where adverse effect occur, or where the chemical interacts with a membrane.

Biologicals: Vaccines, cultures and other preparations made from living organisms and their products, intended for use in diagnosing, immunizing, or treating humans or animals, or in related research.

Biomass: All of the living material in a given area; often refers to vegetation.

Biome: Entire community of living organisms in a single major ecological area. (See: biotic community.)

Biomonitoring: 1. The use of living organisms to test the suitability of effluents for discharge into receiving waters and to test the quality of such waters downstream from the discharge. 2. Analysis of blood, urine, tissues, etc. to measure chemical exposure in humans.

Bioremediation: Use of living organisms to clean up oil spills or remove other pollutants from soil, water, or wastewater; use of organisms such as non-harmful insects to remove agricultural pests or counteract diseases of trees, plants, and garden soil.

Biosensor: Analytical device comprising a biological recognition element (e.g. enzyme, receptor, DNA, antibody, or microorganism) in intimate contact with an electrochemical, optical, thermal, or acoustic signal transducer that together permit analyses of chemical properties or quantities. Shows potential development in some areas, including environmental monitoring.

Biosphere: The portion of Earth and its atmosphere that can support life.

Biostabilizer: A machine that converts solid waste into compost by grinding and aeration.

Biota: The animal and plant life of a given region.

Biotechnology: Techniques that use living organisms or parts of organisms to produce a variety of products (from medicines to industrial enzymes) to improve plants or animals or to develop microorganisms to remove toxics from bodies of water, or act as pesticides.

Biotic Community: A naturally occurring assemblage of plants and animals that live in the same environment and are mutually sustaining and interdependent. (See: biome.)

Biotransformation: Conversion of a substance into other compounds by organisms; includes biodegredation.

Blackwater: Water that contains animal, human, or food waste.

Blood Products: Any product derived from human blood, including but not limited to blood plasma, platelets, red or white corpuscles, and derived licensed products such as interferon.

Bloom: A proliferation of algae and/or higher aquatic plants in a body of water; often related to pollution, especially when pollutants accelerate growth.

BOD5: The amount of dissolved oxygen consumed in five days by biological processes breaking down organic matter.

Body Burden: The amount of a chemical stored in the body at a given time, especially a potential toxin in the body as the result of exposure.

Bog: A type of wetland that accumulates appreciable peat deposits. Bogs depend primarily on precipitation for their water source, and are usually acidic and rich in plant residue with a conspicuous mat of living green moss.

Boiler: A vessel designed to transfer heat produced by combustion or electric resistance to water. Boilers may provide hot water or steam.

Boom: 1. A floating device used to contain oil on a body of water. 2. A piece of equipment used to apply pesticides from a tractor or truck.

Borehole: Hole made with drilling equipment.

Botanical Pesticide: A pesticide whose active ingredient is a plant-produced chemical such as nicotine or strychnine. Also called a plant-derived pesticide.

Bottle Bill: Proposed or enacted legislation which requires a returnable deposit on beer or soda containers and provides for retail store or other redemption. Such legislation is designed to discourage use of throw-away containers.

Bottom Ash: The non-airborne combustion residue from burning pulverized coal in a boiler; the material which falls to the bottom of the boiler and is removed mechanically; a concentration of non-combustible materials, which may include toxics.

Bottom Land Hardwoods: Forested freshwater wetlands adjacent to rivers in the southeastern United States, especially valuable for wildlife breeding, nesting and habitat.

Bounding Estimate: An estimate of exposure, dose, or risk that is higher than that incurred by the person in the population with the currently highest exposure, dose, or risk. Bounding estimates are useful in developing statements that exposures, doses, or risks are not greater than an estimated value.

Brackish: Mixed fresh and salt water.

Breakpoint Chlorination: Addition of chlorine to water until the chlorine demand has been satisfied.

Breakthrough: A crack or break in a filter bed that allows the passage of floc or particulate matter through a filter; will cause an increase in filter effluent turbidity.

Breathing Zone: Area of air in which an organism inhales.

Brine Mud: Waste material, often associated with well-drilling or mining, composed of mineral salts or other inorganic compounds.

British Thermal Unit: Unit of heat energy equal to the amount of heat required to raise the temperature of one pound of water by one degree Fahrenheit at sea level.

Broadcast Application: The spreading of pesticides over an entire area.

Brownfields: Abandoned, idled, or under used industrial and commercial facilities/sites where expansion or redevelopment is complicated by real or perceived environmental contamination. They can be in urban, suburban, or rural areas. EPA's Brownfields initiative helps communities mitigate potential health risks and restore the economic viability of such areas or properties.

Bubble: A system under which existing emissions sources can propose alternate means to comply with a set of emissions limitations; under the bubble concept, sources can control more than required at one emission point where control costs are relatively low in return for a comparable relaxation of controls at a second emission point where costs are higher.

Bubble Policy: (See: emissions trading.)

Buffer: A solution or liquid whose chemical makeup is such that it minimizes changes in pH when acids or bases are added to it.

Buffer Strips: Strips of grass or other erosion-resisting vegetation between or below cultivated strips or fields.

Building Cooling Load: The hourly amount of heat that must be removed from a building to maintain indoor comfort (measured in British thermal units (Btus).

Building Envelope: The exterior surface of a building's construction—the walls, windows, floors, roof, and floor. Also called building shell.

Building Related Illness: Diagnosable illness whose cause and symptoms can be directly attributed to a specific pollutant source within a building (e.g. Legionnaire's disease, hypersensitivity, pneumonitis.) (See: sick building syndrome.)

Bulk Sample: A small portion (usually thumbnail size) of a suspect asbestos-containing building material collected by an asbestos inspector for laboratory analysis to determine asbestos content.

Bulky Waste: Large items of waste materials, such as appliances, furniture, large auto parts, trees, stumps.

Burial Ground (Graveyard): A disposal site for radioactive waste materials that uses earth or water as a shield.

Glossary of Terms

Buy-Back Center: Facility where individuals or groups bring recyclables in return for payment.

By-product: Material, other than the principal product, generated as a consequence of an industrial process or as a breakdown product in a living system.

C

Cadmium (Cd): A heavy metal that accumulates in the environment.

Cancellation: Refers to Section 6 (b) of the Federal Insecticide, Fungicide and Rodenticide Act (FIFRA) which authorizes cancellation of a pesticide registration if unreasonable adverse effects to the environment and public health develop when a product is used according to widespread and commonly recognized practice, or if its labeling or other material required to be submitted does not comply with FIFRA provisions.

Cap: A layer of clay, or other impermeable material installed over the top of a closed landfill to prevent entry of rainwater and minimize leachate.

Capacity Assurance Plan: A statewide plan which supports a state's ability to manage the hazardous waste generated within its boundaries over a twenty year period.

Capillary Action: Movement of water through very small spaces due to molecular forces called capillary forces.

Capillary Fringe: The porous material just above the water table which may hold water by capillarity (a property of surface tension that draws water upwards) in the smaller void spaces.

Capillary Fringe: The zone above he water table within which the porous medium is saturated by water under less than atmospheric pressure.

Capture Efficiency: The fraction of organic vapors generated by a process that are directed to an abatement or recovery device.

Carbon Absorber: An add-on control device that uses activated carbon to absorb volatile organic compounds from a gas stream. (The VOCs are later recovered from the carbon.)

Carbon Adsorption: A treatment system that removes contaminants from ground water or surface water by forcing it through tanks containing activated carbon treated to attract the contaminants.

Carbon Monoxide (CO): A colorless, odorless, poisonous gas produced by incomplete fossil fuel combustion.

Carbon Tetrachloride (CC14): Compound consisting of one carbon atom ad four chlorine atoms, once widely used as a industrial raw material, as a solvent, and in the production of CFCs. Use as a solvent ended when it was discovered to be carcinogenic.

Carboxyhemoglobin: Hemoglobin in which the iron is bound to carbon monoxide(CO) instead of oxygen.

Carcinogen: Any substance that can cause or aggravate cancer.

Carrier: 1.The inert liquid or solid material in a pesticide product that serves as a delivery vehicle for the active ingredient. Carriers do not have toxic properties of their own. 2. Any material or system that can facilitate the movement of a pollutant into the body or cells.

Carrying Capacity: 1. In recreation management, the amount of use a recreation area can sustain without loss of quality. 2. In wildlife management, the maximum number of animals an area can support during a given period.

CAS Registration Number: A number assigned by the Chemical Abstract Service to identify a chemical.

Case Study: A brief fact sheet providing risk, cost, and performance information on alternative methods and other pollution prevention ideas, compliance initiatives, voluntary efforts, etc.

Cask: A thick-walled container (usually lead) used to transport radioactive material. Also called a coffin.

Catalyst: A substance that changes the speed or yield of a chemical reaction without being consumed or chemically changed by the chemical reaction.

Catalytic Converter: An air pollution abatement device that removes pollutants from motor vehicle exhaust, either by oxidizing them into carbon dioxide and water or reducing them to nitrogen.

Catalytic Incinerator: A control device that oxidizes volatile organic compounds (VOCs) by using a catalyst to promote the combustion process. Catalytic incinerators require lower temperatures than conventional thermal incinerators, thus saving fuel and other costs.

Categorical Exclusion: A class of actions which either individually or cumulatively would not have a significant effect on the human environment and therefore would not require preparation of an environmental assessment or environmental impact statement under the National Environmental Policy Act (NEPA).

Categorical Pretreatment Standard: A technology-based effluent limitation for an industrial facility discharging into a municipal sewer system. Analogous in stringency to Best Availability Technology (BAT) for direct dischargers.

Cathodic Protection: A technique to prevent corrosion of a metal surface by making it the cathode of an electrochemical cell.

Cavitation: The formation and collapse of gas pockets or bubbles on the blade of an impeller or the gate of a valve; collapse of these pockets or bubbles drives water with such force that it can cause pitting of the gate or valve surface.

Cells: 1. In solid waste disposal, holes where waste is dumped, compacted, and covered with layers of dirt on a daily basis. 2. The smallest structural part of living matter capable of functioning as an independent unit.

Cementitious: Densely packed and nonfibrous friable materials.

Central Collection Point: Location were a generator of regulated medical waste consolidates wastes originally generated at various locations in his facility. The wastes are gathered together for treatment on-site or for transportation elsewhere for treatment and/or disposal. This term could also apply to community hazardous waste collections, industrial and other waste management systems.

Centrifugal Collector: A mechanical system using centrifugal force to remove aerosols from a gas stream or to remove water from sludge.

Channelization: Straightening and deepening streams so water will move faster, a marsh-drainage tactic that can interfere with waste assimilation capacity, disturb fish and wildlife habitats, and aggravate flooding.

Characteristic: Any one of the four categories used in defining hazardous waste: ignitability, corrosivity, reactivity, and toxicity.

Characterization of Ecological Effects: Part of ecological risk assessment that evaluates ability of a stressor to cause adverse effects under given circumstances.

Characterization of Exposure: Portion of an ecological risk assessment that evaluates interaction of a stressor with one or more ecological entities.

Check-Valve Tubing Pump: Water sampling tool also referred to as a water Pump.

Chemical Case: For purposes of review and regulation, the grouping of chemically similar pesticide active ingredients (e.g. salts and esters of the same chemical) into chemical cases.

Chemical Compound: A distinct and pure substance formed by the union or two or more elements in definite proportion by weight.

Chemical Element: A fundamental substance comprising one kind of atom; the simplest form of matter.

Chemical Oxygen Demand (COD): A measure of the oxygen required to oxidize all compounds, both organic and inorganic, in water.

Chemical Stressors: Chemicals released to the environment through industrial waste, auto emissions, pesticides, and other human activity that can cause illnesses and even death in plants and animals.

Chemical Treatment: Any one of a variety of technologies that use chemicals or a variety of chemical processes to treat waste.

Chemnet: Mutual aid network of chemical shippers and contractors that assigns a contracted emergency response company to provide technical support if a representative of the firm whose chemicals are involved in an incident is not readily available.

Chemosterilant: A chemical that controls pests by preventing reproduction.

Chemtrec: The industry-sponsored Chemical Transportation Emergency Center; provides information and/or emergency assistance to emergency responders.

Child Resistant Packaging (CRP): Packaging that protects children or adults from injury or illness resulting from accidental contact with or ingestion of residential pesticides that meet or exceed specific toxicity levels. Required by FIFRA regulations. Term is also used for protective packaging of medicines.

Chiller: A device that generates a cold liquid that is circulated through an air-handling unit's cooling coil to cool the air supplied to the building.

Chilling Effect: The lowering of the Earth's temperature because of increased particles in the air blocking the sun's rays. (See: greenhouse effect.)

Chisel Plowing: Preparing croplands by using a special implement that avoids complete inversion of the soil as in conventional plowing. Chisel plowing

Glossary of Terms

can leave a protective cover or crops residues on the soil surface to help prevent erosion and improve filtration.

Chlorinated Hydrocarbons: 1. Chemicals containing only chlorine, carbon, and hydrogen. These include a class of persistent, broad-spectrum insecticides that linger in the environment and accumulate in the food chain. Among them are DDT, aldrin, dieldrin, heptachlor, chlordane, lindane, endrin, Mirex, hexachloride, and toxaphene. Other examples include TCE, used as an industrial solvent. 2. Any chlorinated organic compounds including chlorinated solvents such as dichloromethane, trichloromethylene, chloroform.

Chlorinated Solvent: An organic solvent containing chlorine atoms(e.g. methylene chloride and 1,1,1-trichloromethane). Uses of chlorinated solvents are include aerosol spray containers, in highway paint, and dry cleaning fluids.

Chlorination: The application of chlorine to drinking water, sewage, or industrial waste to disinfect or to oxidize undesirable compounds.

Chlorinator: A device that adds chlorine, in gas or liquid form, to water or sewage to kill infectious bacteria.

Chlorine-Contact Chamber: That part of a water treatment plant where effluent is disinfected by chlorine.

Chlorofluorocarbons (CFCs): A family of inert, nontoxic, and easily liquefied chemicals used in refrigeration, air conditioning, packaging, insulation, or as solvents and aerosol propellants. Because CFCs are not destroyed in the lower atmosphere they drift into the upper atmosphere where their chlorine components destroy ozone. (See: fluorocarbons.)

Chlorophenoxy: A class of herbicides that may be found in domestic water supplies and cause adverse health effects.

Chlorosis: Discoloration of normally green plant parts caused by disease, lack of nutrients, or various air pollutants.

Cholinesterase: An enzyme found in animals that regulates nerve impulses by the inhibition of acetylcholine. Cholinesterase inhibition is associated with a variety of acute symptoms such as nausea, vomiting, blurred vision, stomach cramps, and rapid heart rate.

Chromium: (See: heavy metals.)

Chronic Effect: An adverse effect on a human or animal in which symptoms recur frequently or develop slowly over a long period of time.

Chronic Exposure: Multiple exposures occurring over an extended period of time or over a significant fraction of an animal's or human's lifetime (Usually seven years to a lifetime.)

Chronic Toxicity: The capacity of a substance to cause long-term poisonous health effects in humans, animals, fish, and other organisms. (See: acute toxicity.)

Circle of Influence: The circular outer edge of a depression produced in the water table by the pumping of water from a well. (See: cone of depression.)

Cistern: Small tank or storage facility used to store water for a home or farm; often used to store rain water.

Clarification: Clearing action that occurs during wastewater treatment when solids settle out. This is often aided by centrifugal action and chemically induced coagulation in wastewater.

Clarifier: A tank in which solids settle to the bottom and are subsequently removed as sludge.

Class I Area: Under the Clean Air Act. a Class I area is one in which visibility is protected more stringently than under the national ambient air quality standards; includes national parks, wilderness areas, monuments, and other areas of special national and cultural significance.

Class I Substance: One of several groups of chemicals with an ozone depletion potential of 0.2 or higher, including CFCS, Halons, Carbon Tetrachloride, and Methyl Chloroform (listed in the Clean Air Act), and HBFCs and Ethyl Bromide (added by EPA regulations). (See: Global warming potential.)

Class II Substance: A substance with an ozone depletion potential of less than 0.2. All HCFCs are currently included in this classification. (See: Global warming potential.)

Clay Soil: Soil material containing more than 40 percent clay, less than 45 percent sand, and less than 40 percent silt.

Clean Coal Technology: Any technology not in widespread use prior to the Clean Air Act Amendments of 1990. This Act will achieve significant reductions in pollutants associated with the burning of coal.

Clean Fuels: Blends or substitutes for gasoline fuels, including compressed natural gas, methanol, ethanol, and liquified petroleum gas.

Cleaner Technologies Substitutes Assessment: A document that systematically evaluates the relative risk, performance, and cost trade-offs of technological alternatives; serves as a repository for all the technical data (including methodology and results) developed by a DfE or other pollution prevention or education project.

Cleanup: Actions taken to deal with a release or threat of release of a hazardous substance that could affect humans and/or the environment. The term "cleanup" is sometimes used interchangeably with the terms remedial action, removal action, response action, or corrective action.

Clear Cut: Harvesting all the trees in one area at one time, a practice that can encourage fast rainfall or snowmelt runoff, erosion, sedimentation of streams and lakes, and flooding, and destroys vital habitat.

Clear Well: A reservoir for storing filtered water of sufficient quantity to prevent the need to vary the filtration rate with variations in demand. Also used to provide chlorine contact time for disinfection.

Climate Change (also referred to as 'global climate change'): The term 'climate change' is sometimes used to refer to all forms of climatic inconsistency, but because the Earth's climate is never static, the term is more properly used to imply a significant change from one climatic condition to another. In some cases, 'climate change' has been used synonymously with the term, 'global warming'; scientists however, tend to use the term in the wider sense to also include natural changes in climate. (See: global warming.)

Cloning: In biotechnology, obtaining a group of genetically identical cells from a single cell; making identical copies of a gene.

Closed-Loop Recycling: Reclaiming or reusing wastewater for non-potable purposes in an enclosed process.

Closure: The procedure a landfill operator must follow when a landfill reaches its legal capacity for solid ceasing acceptance of solid waste and placing a cap on the landfill site.

Co-fire: Burning of two fuels in the same combustion unit; e.g., coal and natural gas, or oil and coal.

Coagulation: Clumping of particles in wastewater to settle out impurities, often induced by chemicals such as lime, alum, and iron salts.

Coal Cleaning Technology: A precombustion process by which coal is physically or chemically treated to remove some of its sulfur so as to reduce sulfur dioxide emissions.

Coal Gasification: Conversion of coal to a gaseous product by one of several available technologies.

Coastal Zone: Lands and waters adjacent to the coast that exert an influence on the uses of the sea and its ecology, or whose uses and ecology are affected by the sea.

Code of Federal Regulations (CFR): Document that codifies all rules of the executive departments and agencies of the federal government. It is divided into fifty volumes, known as titles. Title 40 of the CFR (referenced as 40 CFR) lists all environmental regulations.

Coefficient of Haze (COH): A measurement of visibility interference in the atmosphere.

Cogeneration: The consecutive generation of useful thermal and electric energy from the same fuel source.

Coke Oven: An industrial process which converts coal into coke, one of the basic materials used in blast furnaces for the conversion of iron ore into iron.

Cold Temperature CO: A standard for automobile emissions of carbon monoxide (CO) emissions to be met at a low temperature (i.e. 20 degrees Fahrenheit). Conventional automobile catalytic converters are not efficient in cold weather until they warm up.

Coliform Index: A rating of the purity of water based on a count of fecal bacteria.

Coliform Organism: Microorganisms found in the intestinal tract of humans and animals. Their presence in water indicates fecal pollution and potentially adverse contamination by pathogens.

Collector: Public or private hauler that collects nonhazardous waste and recyclable materials from residential, commercial, institutional and industrial sources. (See: hauler.)

Collector Sewers: Pipes used to collect and carry wastewater from individual sources to an interceptor sewer that will carry it to a treatment facility.

Colloids: Very small, finely divided solids (that do not dissolve) that remain dispersed in a liquid for a long time due to their small size and electrical charge.

Combined Sewer Overflows: Discharge of a mixture of storm water and domestic waste when the flow capacity of a sewer system is exceeded during rainstorms.

755

Glossary of Terms

Combined Sewers: A sewer system that carries both sewage and storm-water runoff. Normally, its entire flow goes to a waste treatment plant, but during a heavy storm, the volume of water may be so great as to cause overflows of untreated mixtures of storm water and sewage into receiving waters. Storm-water runoff may also carry toxic chemicals from industrial areas or streets into the sewer system.

Combustion: 1. Burning, or rapid oxidation, accompanied by release of energy in the form of heat and light. 2. Refers to controlled burning of waste, in which heat chemically alters organic compounds, converting into stable inorganics such as carbon dioxide and water.

Combustion Chamber: The actual compartment where waste is burned in an incinerator.

Combustion Product: Substance produced during the burning or oxidation of a material.

Command Post: Facility located at a safe distance upwind from an accident site, where the on-scene coordinator, responders, and technical representatives make response decisions, deploy manpower and equipment, maintain liaison with news media, and handle communications.

Command-and-Control Regulations: Specific requirements prescribing how to comply with specific standards defining acceptable levels of pollution.

Comment Period: Time provided for the public to review and comment on a proposed EPA action or rulemaking after publication in the Federal Register.

Commercial Waste: All solid waste emanating from business establishments such as stores, markets, office buildings, restaurants, shopping centers, and theaters.

Commercial Waste Management Facility: A treatment, storage, disposal, or transfer facility which accepts waste from a variety of sources, as compared to a private facility which normally manages a limited waste stream generated by its own operations.

Commingled Recyclables: Mixed recyclables that are collected together.

Comminuter: A machine that shreds or pulverizes solids to make waste treatment easier.

Comminution: Mechanical shredding or pulverizing of waste. Used in both solid waste management and wastewater treatment.

Common Sense Initiative: Voluntary program to simplify environmental regulation to achieve cleaner, cheaper, smarter results, starting with six major industry sectors.

Community: In ecology, an assemblage of populations of different species within a specified location in space and time. Sometimes, a particular subgrouping may be specified, such as the fish community in a lake or the soil arthropod community in a forest.

Community Relations: The EPA effort to establish two-way communication with the public to create understanding of EPA programs and related actions, to ensure public input into decision-making processes related to affected communities, and to make certain that the Agency is aware of and responsive to public concerns. Specific community relations activities are required in relation to Superfund remedial actions.

Community Water System: A public water system which serves at least 15 service connections used by year-round residents or regularly serves at least 25 year-round residents.

Compact Fluorescent Lamp (CFL): Small fluorescent lamps used as more efficient alternatives to incandescent lighting. Also called PL, CFL, Twin-Tube, or BIAX lamps.

Compaction: Reduction of the bulk of solid waste by rolling and tamping.

Comparative Risk Assessment: Process that generally uses the judgement of experts to predict effects and set priorities among a wide range of environmental problems.

Complete Treatment: A method of treating water that consists of the addition of coagulant chemicals, flash mixing, coagulation-flocculation, sedimentation, and filtration. Also called conventional filtration.

Compliance Coal: Any coal that emits less than 1.2 pounds of sulfur dioxide per million Btu when burned. Also known as low sulfur coal.

Compliance Coating: A coating whose volatile organic compound content does not exceed that allowed by regulation.

Compliance Cycle: The 9-year calendar year cycle, beginning January 1, 1993, during which public water systems must monitor. Each cycle consists of three 3-year compliance periods.

Compliance Monitoring: Collection and evaluation of data, including self-monitoring reports, and verification to show whether pollutant concentrations and loads contained in permitted discharges are in compliance with the limits and conditions specified in the permit.

Compliance Schedule: A negotiated agreement between a pollution source and a government agency that specifies dates and procedures by which a source will reduce emissions and, thereby, comply with a regulation.

Composite Sample: A series of water samples taken over a given period of time and weighted by flow rate.

Compost: The relatively stable humus material that is produced from a composting process in which bacteria in soil mixed with garbage and degradable trash break down the mixture into organic fertilizer.

Composting: The controlled biological decomposition of organic material in the presence of air to form a humus-like material. Controlled methods of composting include mechanical mixing and aerating, ventilating the materials by dropping them through a vertical series of aerated chambers, or placing the compost in piles out in the open air and mixing it or turning it periodically.

Composting Facilities: 1. An offsite facility where the organic component of municipal solid waste is decomposed under controlled conditions; 2.an aerobic process in which organic materials are ground or shredded and then decomposed to humus in windrow piles or in mechanical digesters, drums, or similar enclosures.

Compressed Natural Gas (CNG): An alternative fuel for motor vehicles; considered one of the cleanest because of low hydrocarbon emissions and its vapors are relatively non-ozone producing. However, vehicles fueled with CNG do emit a significant quantity of nitrogen oxides.

Concentration: The relative amount of a substance mixed with another substance. An example is five ppm of carbon monoxide in air or 1 mg/l of iron in water.

Condensate: 1.Liquid formed when warm landfill gas cools as it travels through a collection system. 2. Water created by cooling steam or water vapor.

Condensate Return System: System that returns the heated water condensing within steam piping to the boiler and thus saves energy.

Conditional Registration: Under special circumstances, the Federal Insecticide, Fungicide, and Rodenticide Act (FIFRA) permits registration of pesticide products that is "conditional" upon the submission of additional data. These special circumstances include a finding by the EPA Administrator that a new product or use of an existing pesticide will not significantly increase the risk of unreasonable adverse effects. A product containing a new (previously unregistered) active ingredient may be conditionally registered only if the Administrator finds that such conditional registration is in the public interest, that a reasonable time for conducting the additional studies has not elapsed, and the use of the pesticide for the period of conditional registration will not present an unreasonable risk.

Conditionally Exempt Generators (CE): Persons or enterprises which produce less than 220 pounds of hazardous waste per month. Exempt from most regulation, they are required merely to determine whether their waste is hazardous, notify appropriate state or local agencies, and ship it by an authorized transporter to a permitted facility for proper disposal. (See : small quantity generator.)

Conductance: A rapid method of estimating the dissolved solids content of water supply by determining the capacity of a water sample to carry an electrical current. Conductivity is a measure of the ability of a solution to carry and electrical current.

Conductivity: A measure of the ability of a solution to carry an electrical current.

Cone of Depression: A depression in the water table that develops around a pumped well.

Cone of Influence: The depression, roughly conical in shape, produced in a water table by the pumping of water from a well.

Cone Penterometer Testing (CPT): A direct push system used to measure lithology based on soil penetration resistance. Sensors in the tip of the cone of the DP rod measure tip resistance and side-wall friction, transmitting electrical signals to digital processing equipment on the ground surface. (See: direct push.)

Confidential Business Information (CBI): Material that contains trade secrets or commercial or financial information that has been claimed as confidential by its source (e.g. a pesticide or new chemical formulation registrant). EPA has special procedures for handling such information.

Confidential Statement of Formula (CSF): A list of the ingredients in a new pesticide or chemical formulation. The list is submitted at the time for application for registration or change in formulation.

Confined Aquifer: An aquifer in which ground water is confined under pressure which is significantly greater than atmospheric pressure.

Confluent Growth: A continuous bacterial growth covering all or part of the filtration area of a membrane filter in which the bacteria colonies are not discrete.

Consent Decree: A legal document, approved by a judge, that formalizes an agreement reached between EPA and potentially responsible parties (PRPs) through which PRPs will conduct all or part of a cleanup action at a Superfund site; cease or correct actions or processes that are polluting the environment; or otherwise comply with EPA initiated regulatory enforcement actions to resolve the contamination at the Superfund site involved. The consent decree describes the actions PRPs will take and may be subject to a public comment period.

Conservation: Preserving and renewing, when possible, human and natural resources. The use, protection, and improvement of natural resources according to principles that will ensure their highest economic or social benefits.

Conservation Easement: Easement restricting a landowner to land uses that that are compatible with long-term conservation and environmental values.

Constituent(s) of Concern: Specific chemicals that are identified for evaluation in the site assessment process

Construction and Demolition Waste: Waste building materials, dredging materials, tree stumps, and rubble resulting from construction, remodeling, repair, and demolition of homes, commercial buildings and other structures and pavements. May contain lead, asbestos, or other hazardous substances.

Construction Ban: If, under the Clean Air Act, EPA disapproves an area's planning requirements for correcting nonattainment, EPA can ban the construction or modification of any major stationary source of the pollutant for which the area is in nonattainment.

Consumptive Water Use: Water removed from available supplies without return to a water resources system, e.g. water used in manufacturing, agriculture, and food preparation.

Contact Pesticide: A chemical that kills pests when it touches them, instead of by ingestion. Also, soil that contains the minute skeletons of certain algae that scratch and dehydrate waxy-coated insects.

Contaminant: Any physical, chemical, biological, or radiological substance or matter that has an adverse effect on air, water, or soil.

Contamination: Introduction into water, air, and soil of microorganisms, chemicals, toxic substances, wastes, or wastewater in a concentration that makes the medium unfit for its next intended use. Also applies to surfaces of objects, buildings, and various household and agricultural use products.

Contamination Source Inventory: An inventory of contaminant sources within delineated State Water-Protection Areas. Targets likely sources for further investigation.

Contingency Plan: A document setting out an organized, planned, and coordinated course of action to be followed in case of a fire, explosion, or other accident that releases toxic chemicals, hazardous waste, or radioactive materials that threaten human health or the environment. (See: National Oil and Hazardous Substances Contingency Plan.)

Continuous Discharge: A routine release to the environment that occurs without interruption, except for infrequent shutdowns for maintenance, process changes, etc.

Continuous Sample: A flow of water, waste or other material from a particular place in a plant to the location where samples are collected for testing. May be used to obtain grab or composite samples.

Contour Plowing: Soil tilling method that follows the shape of the land to discourage erosion.

Contour Strip Farming: A kind of contour farming in which row crops are planted in strips, between alternating strips of close-growing, erosion-resistant forage crops.

Contract Labs: Laboratories under contract to EPA, which analyze samples taken from waste, soil, air, and water or carry out research projects.

Control Technique Guidelines (CTG): EPA documents designed to assist state and local pollution authorities to achieve and maintain air quality standards for certain sources (e.g. organic emissions from solvent metal cleaning known as degreasing) through reasonably available control technologies (RACT).

Controlled Reaction: A chemical reaction under temperature and pressure conditions maintained within safe limits to produce a desired product or process.

Conventional Filtration: (See: complete treatment.)

Conventional Pollutants: Statutorily listed pollutants understood well by scientists. These may be in the form of organic waste, sediment, acid, bacteria, viruses, nutrients, oil and grease, or heat.

Conventional Site Assessment: Assessment in which most of the sample analysis and interpretation of data is completed off-site; process usually requires repeated mobilization of equipment and staff in order to fully determine the extent of contamination.

Conventional Systems: Systems that have been traditionally used to collect municipal wastewater in gravity sewers and convey it to a central primary or secondary treatment plant prior to discharge to surface waters.

Conventional Tilling: Tillage operations considered standard for a specific location and crop and that tend to bury the crop residues; usually considered as a base for determining the cost effectiveness of control practices.

Conveyance Loss: Water loss in pipes, channels, conduits, ditches by leakage or evaporation.

Cooling Electricity Use: Amount of electricity used to meet the building cooling load. (See: building cooling load.)

Cooling Tower: A structure that helps remove heat from water used as a coolant; e.g., in electric power generating plants.

Cooling Tower: Device which dissipates the heat from water-cooled systems by spraying the water through streams of rapidly moving air.

Cooperative Agreement: An assistance agreement whereby EPA transfers money, property, services or anything of value to a state, university, non-profit, or not-for-profit organization for the accomplishment of authorized activities or tasks.

Core: The uranium-containing heart of a nuclear reactor, where energy is released.

Core Program Cooperative Agreement: An assistance agreement whereby EPA supports states or tribal governments with funds to help defray the cost of non-item-specific administrative and training activities.

Corrective Action: EPA can require treatment, storage and disposal (TSDF) facilities handling hazardous waste to undertake corrective actions to clean up spills resulting from failure to follow hazardous waste management procedures or other mistakes. The process includes cleanup procedures designed to guide TSDFs toward in spills.

Corrosion: The dissolution and wearing away of metal caused by a chemical reaction such as between water and the pipes, chemicals touching a metal surface, or contact between two metals.

Corrosive: A chemical agent that reacts with the surface of a material causing it to deteriorate or wear away.

Cost/Benefit Analysis: A quantitative evaluation of the costs which would have incurred by implementing an environmental regulation versus the overall benefits to society of the proposed action.

Cost Recovery: A legal process by which potentially responsible parties who contributed to contamination at a Superfund site can be required to reimburse the Trust Fund for money spent during any cleanup actions by the federal government.

Cost Sharing: A publicly financed program through which society, as a beneficiary of environmental protection, shares part of the cost of pollution control with those who must actually install the controls. In Superfund, for example, the government may pay part of the cost of a cleanup action with those responsible for the pollution paying the major share.

Cost-Effective Alternative: An alternative control or corrective method identified after analysis as being the best available in terms of reliability, performance, and cost. Although costs are one important consideration, regulatory and compliance analysis does not require EPA to choose the least expensive alternative. For example, when selecting or approving a method for cleaning up a Superfund site, the Agency balances costs with the long-term effectiveness of the methods proposed and the potential danger posed by the site.

Cover Crop: A crop that provides temporary protection for delicate seedlings and/or provides a cover canopy for seasonal soil protection and improvement between normal crop production periods.

Cover Material: Soil used to cover compacted solid waste in a sanitary landfill.

Cradle-to-Grave or Manifest System: A procedure in which hazardous materials are identified and followed as they are produced, treated, transported, and disposed of by a series of permanent, linkable, descriptive documents (e.g. manifests). Commonly referred to as the cradle-to-grave system.

Criteria: Descriptive factors taken into account by EPA in setting standards for various pollutants. These factors are used to determine limits on allowable concentration levels, and to limit the number of violations per year. When issued by EPA, the criteria provide guidance to the states on how to establish their standards.

Glossary of Terms

Criteria Pollutants: The 1970 amendments to the Clean Air Act required EPA to set National Ambient Air Quality Standards for certain pollutants known to be hazardous to human health. EPA has identified and set standards to protect human health and welfare for six pollutants: ozone, carbon monoxide, total suspended particulates, sulfur dioxide, lead, and nitrogen oxide. The term, "criteria pollutants" derives from the requirement that EPA must describe the characteristics and potential health and welfare effects of these pollutants. It is on the basis of these criteria that standards are set or revised.

Critical Effect: The first adverse effect, or its known precursor, that occurs as a dose rate increases. Designation is based on evaluation of overall database.

Crop Consumptive Use: The amount of water transpired during plant growth plus what evaporated from the soil surface and foliage in the crop area.

Crop Rotation: Planting a succession of different crops on the same land rea as opposed to planting the same crop time after time.

Cross Contamination: The movement of underground contaminants from one level or area to another due to invasive subsurface activities.

Cross-Connection: Any actual or potential connection between a drinking water system and an unapproved water supply or other source of contamination.

Crumb Rubber: Ground rubber fragments the size of sand or silt used in rubber or plastic products, or processed further into reclaimed rubber or asphalt products.

Cryptosporidium: A protozoan microbe associated with the disease cryptosporidiosis in man. The disease can be transmitted through ingestion of drinking water, person-to-person contact, or other pathways, and can cause acute diarrhea, abdominal pain, vomiting, fever, and can be fatal as it was in the Milwaukee episode.

Cubic Feet Per Minute (CFM): A measure of the volume of a substance flowing through air within a fixed period of time. With regard to indoor air, refers to the amount of air, in cubic feet, that is exchanged with outdoor air in a minute's time; i.e. the air exchange rate.

Cullet: Crushed glass.

Cultural Eutrophication: Increasing rate at which water bodies "die" by pollution from human activities.

Cultures and Stocks: Infectious agents and associated biologicals including cultures from medical and pathological laboratories; cultures and stocks of infectious agents from research and industrial laboratories; waste from the production of biologicals; discarded live and attenuated vaccines; and culture dishes and devices used to transfer, inoculate, and mix cultures. (See: regulated medical waste.)

Cumulative Ecological Risk Assessment: Consideration of the total ecological risk from multiple stressors to a given eco-zone.

Cumulative Exposure: The sum of exposures of an organism to a pollutant over a period of time.

Cumulative Working Level Months (CWLM): The sum of lifetime exposure to radon working levels expressed in total working level months.

Curb Stop: A water service shutoff valve located in a water service pipe near the curb and between the water main and the building.

Curbside Collection: Method of collecting recyclable materials at homes, community districts or businesses.

Cutie-Pie: An instrument used to measure radiation levels.

Cuttings: Spoils left by conventional drilling with hollow stem auger or rotary drilling equipment.

Cyclone Collector: A device that uses centrifugal force to remove large particles from polluted air.

D

Data Call-In: A part of the Office of Pesticide Programs (OPP) process of developing key required test data, especially on the long-term, chronic effects of existing pesticides, in advance of scheduled Registration Standard reviews. Data Call-In from manufacturers is an adjunct of the Registration Standards program intended to expedite re-registration.

Data Quality Objectives (DQOs): Qualitative and quantitative statements of the overall level of uncertainty that a decision-maker will accept in results or decisions based on environmental data. They provide the statistical framework for planning and managing environmental data operations consistent with user's needs.

Day Tank: Another name for deaerating tank. (See: age tank.)

DDT: The first chlorinated hydrocarbon insecticide chemical name: Dichloro-Diphenyl-Trichloroethane. It has a half-life of 15 years and can collect in fatty tissues of certain animals. EPA banned registration and interstate sale of DDT for virtually all but emergency uses in the United States in 1972 because of its persistence in the environment and accumulation in the food chain.

Dead End: The end of a water main which is not connected to other parts of the distribution system.

Deadmen: Anchors drilled or cemented into the ground to provide additional reactive mass for DP sampling rigs.

Decant: To draw off the upper layer of liquid after the heaviest material (a solid or another liquid) has settled.

Decay Products: Degraded radioactive materials, often referred to as "daughters" or "progeny"; radon decay products of most concern from a public health standpoint are polonium-214 and polonium-218.

Dechlorination: Removal of chlorine from a substance.

Decomposition: The breakdown of matter by bacteria and fungi, changing the chemical makeup and physical appearance of materials.

Decontamination: Removal of harmful substances such as noxious chemicals, harmful bacteria or other organisms, or radioactive material from exposed individuals, rooms and furnishings in buildings, or the exterior environment.

Deep-Well Injection: Deposition of raw or treated, filtered hazardous waste by pumping it into deep wells, where it is contained in the pores of permeable subsurface rock.

Deflocculating Agent: A material added to a suspension to prevent settling.

Defluoridation: The removal of excess flouride in drinking water to prevent the staining of teeth.

Defoliant: An herbicide that removes leaves from trees and growing plants.

Degasification: A water treatment that removes dissolved gases from the water.

Degree-Day: A rough measure used to estimate the amount of heating required in a given area; is defined as the difference between the mean daily temperature and 65 degrees Fahrenheit. Degree-days are also calculated to estimate cooling requirements.

Delegated State: A state (or other governmental entity such as a tribal government) that has received authority to administer an environmental regulatory program in lieu of a federal counterpart. As used in connection with NPDES, UIC, and PWS programs, the term does not connote any transfer of federal authority to a state.

Delist: Use of the petition process to have a facility's toxic designation rescinded.

Demand-side Waste Management: Prices whereby consumers use purchasing decisions to communicate to product manufacturers that they prefer environmentally sound products packaged with the least amount of waste, made from recycled or recyclable materials, and containing no hazardous substances.

Demineralization: A treatment process that removes dissolved minerals from water.

Denitrification: The biological reduction of nitrate to nitrogen gas by denitrifying bacteria in soil.

Dense Non-Aqueous Phase Liquid (DNAPL): Non-aqueous phase liquids such as chlorinated hydrocarbon solvents or petroleum fractions with a specific gravity greater than 1.0 that sink through the water column until they reach a confining layer. Because they are at the bottom of aquifers instead of floating on the water table, typical monitoring wells do not indicate their presence.

Density: A measure of how heavy a specific volume of a solid, liquid, or gas is in comparison to water. depending on the chemical.

Depletion Curve: In hydraulics, a graphical representation of water depletion from storage-stream channels, surface soil, and groundwater. A depletion curve can be drawn for base flow, direct runoff, or total flow.

Depressurization: A condition that occurs when the air pressure inside a structure is lower that the air pressure outdoors. Depressurization can occur when household appliances such as fireplaces or furnaces, that consume or exhaust house air, are not supplied with enough makeup air. Radon may be drawn into a house more rapidly under depressurized conditions.

Dermal Absorption/Penetration: Process by which a chemical penetrates the skin and enters the body as an internal dose.

Dermal Exposure: Contact between a chemical and the skin.

Dermal Toxicity: The ability of a pesticide or toxic chemical to poison people or animals by contact with the skin. (See: contact pesticide.)

DES: A synthetic estrogen, diethylstilbestrol is used as a growth stimulant in food animals. Residues in meat are thought to be carcinogenic.

Desalination: [Desalinization] (1) Removing salts from ocean or brackish water by using various technologies. (2) Removal of salts from soil by artificial means, usually leaching.

Desiccant: A chemical agent that absorbs moisture; some desiccants are capable of drying out plants or insects, causing death.

Design Capacity: The average daily flow that a treatment plant or other facility is designed to accommodate.

Design Value: The monitored reading used by EPA to determine an area's air quality status; e.g., for ozone, the fourth highest reading measured over the most recent three years is the design value.

Designated Pollutant: An air pollutant which is neither a criteria nor hazardous pollutant, as described in the Clean Air Act, but for which new source performance standards exist. The Clean Air Act does require states to control these pollutants, which include acid mist, total reduced sulfur (TRS), and fluorides.

Designated Uses: Those water uses identified in state water quality standards that must be achieved and maintained as required under the Clean Water Act. Uses can include cold water fisheries, public water supply, and irrigation.

Designer Bugs: Popular term for microbes developed through biotechnology that can degrade specific toxic chemicals at their source in toxic waste dumps or in ground water.

Destination Facility: The facility to which regulated medical waste is shipped for treatment and destruction, incineration, and/or disposal.

Destratification: Vertical mixing within a lake or reservoir to totally or partially eliminate separate layers of temperature, plant, or animal life.

Destroyed Medical Waste: Regulated medical waste that has been ruined, torn apart, or mutilated through thermal treatment, melting, shredding, grinding, tearing, or breaking, so that it is no longer generally recognized as medical waste, but has not yet been treated (excludes compacted regulated medical waste).

Destruction and Removal Efficiency (DRE): A percentage that represents the number of molecules of a compound removed or destroyed in an incinerator relative to the number of molecules entering the system (e.g. a DRE of 99.99 percent means that 9,999 molecules are destroyed for every 10,000 that enter; 99.99 percent is known as "four nines." For some pollutants, the RCRA removal requirement may be as stringent as "six nines").

Destruction Facility: A facility that destroys regulated medical waste.

Desulfurization: Removal of sulfur from fossil fuels to reduce pollution.

Detectable Leak Rate: The smallest leak (from a storage tank), expressed in terms of gallons- or liters-per-hour, that a test can reliably discern with a certain probability of detection or false alarm.

Detection Criterion: A predetermined rule to ascertain whether a tank is leaking or not. Most volumetric tests use a threshold value as the detection criterion. (See: volumetric tank tests.)

Detection Limit: The lowest concentration of a chemical that can reliably be distinguished from a zero concentration.

Detention Time: 1. The theoretical calculated time required for a small amount of water to pass through a tank at a given rate of flow. 2. The actual time that a small amount of water is in a settling basin, flocculating basin, or rapid-mix chamber. 3. In storage reservoirs, the length of time water will be held before being used.

Detergent: Synthetic washing agent that helps to remove dirt and oil. Some contain compounds which kill useful bacteria and encourage algae growth when they are in wastewater that reaches receiving waters.

Development Effects: Adverse effects such as altered growth, structural abnormality, functional deficiency, or death observed in a developing organism.

Dewater: 1. Remove or separate a portion of the water in a sludge or slurry to dry the sludge so it can be handled and disposed of. 2. Remove or drain the water from a tank or trench.

Diatomaceous Earth (Diatomite): A chalk-like material (fossilized diatoms) used to filter out solid waste in wastewater treatment plants; also used as an active ingredient in some powdered pesticides.

Diazinon: An insecticide. In 1986, EPA banned its use on open areas such as sod farms and golf courses because it posed a danger to migratory birds. The ban did not apply to agricultural, home lawn or commercial establishment uses.

Dibenzofurans: A group of organic compounds, some of which are toxic.

Dicofol: A pesticide used on citrus fruits.

Diffused Air: A type of aeration that forces oxygen into sewage by pumping air through perforated pipes inside a holding tank.

Diffusion: The movement of suspended or dissolved particles (or molecules) from a more concentrated to a less concentrated area. The process tends to distribute the particles or molecules more uniformly.

Digester: In wastewater treatment, a closed tank; in solid-waste conversion, a unit in which bacterial action is induced and accelerated in order to break down organic matter and establish the proper carbon to nitrogen ratio.

Digestion: The biochemical decomposition of organic matter, resulting in partial gasification, liquefaction, and mineralization of pollutants.

Dike: A low wall that can act as a barrier to prevent a spill from spreading.

Diluent: Any liquid or solid material used to dilute or carry an active ingredient.

Dilution Ratio: The relationship between the volume of water in a stream and the volume of incoming water. It affects the ability of the stream to assimilate waste.

Dimictic: Lakes and reservoirs that freeze over and normally go through two stratifications and two mixing cycles a year.

Dinocap: A fungicide used primarily by apple growers to control summer diseases. EPA proposed restrictions on its use in 1986 when laboratory tests found it caused birth defects in rabbits.

Dinoseb: A herbicide that is also used as a fungicide and insecticide. It was banned by EPA in 1986 because it posed the risk of birth defects and sterility.

Dioxin: Any of a family of compounds known chemically as dibenzo-p-dioxins. Concern about them arises from their potential toxicity as contaminants in commercial products. Tests on laboratory animals indicate that it is one of the more toxic anthropogenic (man-made) compounds.

Direct Discharger: A municipal or industrial facility which introduces pollution through a defined conveyance or system such as outlet pipes; a point source.

Direct Filtration: A method of treating water which consists of the addition of coagulent chemicals, flash mixing, coagulation, minimal flocculation, and filtration. Sedimentation is not uses.

Direct Push: Technology used for performing subsurface investigations by driving, pushing, and/or vibrating small-diameter hollow steel rods into the ground/ Also known as direct drive, drive point, or push technology.

Direct Runoff: Water that flows over the ground surface or through the ground directly into streams, rivers, and lakes.

Discharge: Flow of surface water in a stream or canal or the outflow of ground water from a flowing artesian well, ditch, or spring. Can also apply tp discharge of liquid effluent from a facility or to chemical emissions into the air through designated venting mechanisms.

Disinfectant: A chemical or physical process that kills pathogenic organisms in water, air, or on surfaces. Chlorine is often used to disinfect sewage treatment effluent, water supplies, wells, and swimming pools.

Disinfectant By-Product: A compound formed by the reaction of a disinfectant such as chlorine with organic material in the water supply; a chemical byproduct of the disinfection process..

Disinfectant Time: The time it takes water to move from the point of disinfectant application (or the previous point of residual disinfectant measurement) to a point before or at the point where the residual disinfectant is measured. In pipelines, the time is calculated by dividing the internal volume of the pipe by he maximum hourly flow rate; within mixing basins and storage reservoirs it is determined by tracer studies of an equivalent demonstration.

Dispersant: A chemical agent used to break up concentrations of organic material such as spilled oil.

Displacement Savings: Saving realized by displacing purchases of natural gas or electricity from a local utility by using landfill gas for power and heat.

Disposables: Consumer products, other items, and packaging used once or a few times and discarded.

Disposal: Final placement or destruction of toxic, radioactive, or other wastes; surplus or banned pesticides or other chemicals; polluted soils; and drums containing hazardous materials from removal actions or accidental releases. Disposal may be accomplished through use of approved secure landfills, surface impoundments, land farming, deep-well injection, ocean dumping, or incineration.

Glossary of Terms

Disposal Facilities: Repositories for solid waste, including landfills and combustors intended for permanent containment or destruction of waste materials. Excludes transfer stations and composting facilities.

Dissolved Oxygen (DO): The oxygen freely available in water, vital to fish and other aquatic life and for the prevention of odors. DO levels are considered a most important indicator of a water body's ability to support desirable aquatic life. Secondary and advanced waste treatment are generally designed to ensure adequate DO in waste-receiving waters.

Dissolved Solids: Disintegrated organic and inorganic material in water. Excessive amounts make water unfit to drink or use in industrial processes.

Distillation: The act of purifying liquids through boiling, so that the steam or gaseous vapors condense to a pure liquid. Pollutants and contaminants may remain in a concentrated residue.

Disturbance: Any event or series of events that disrupt ecosystem, community, or population structure and alters the physical environment.

Diversion: 1. Use of part of a stream flow as water supply. 2. A channel with a supporting ridge on the lower side constructed across a slope to divert water at a non-erosive velocity to sites where it can be used and disposed of.

Diversion Rate: The percentage of waste materials diverted from traditional disposal such as landfilling or incineration to be recycled, composted, or re-used.

DNA Hybridization: Use of a segment of DNA, called a DNA probe, to identify its complementary DNA; used to detect specific genes.

Dobson Unit (DU): Units of ozone level measurement. measurement of ozone levels. If, for example, 100 DU of ozone were brought to the earth's surface they would form a layer one millimeter thick. Ozone levels vary geographically, even in the absence of ozone depletion.

Domestic Application: Pesticide application in and around houses, office buildings, motels, and other living or working areas.(See: residential use.)

Dosage/Dose: 1. The actual quantity of a chemical administered to an organism or to which it is exposed. 2. The amount of a substance that reaches a specific tissue (e.g. the liver). 3. The amount of a substance available for interaction with metabolic processes after crossing the outer boundary of an organism. (See: absorbed dose, administered dose, applied dose, potential dose.)

Dose Equivalent: The product of the absorbed dose from ionizing radiation and such factors as account for biological differences due to the type of radiation and its distribution in the body in the body.

Dose Rate: In exposure assessment, dose per time unit (e.g. mg/day), sometimes also called dosage.

Dose Response: Shifts in toxicological responses of an individual (such as alterations in severity) or populations (such as alterations in incidence) that are related to changes in the dose of any given substance.

Dose Response Curve: Graphical representation of the relationship between the dose of a stressor and the biological response thereto.

Dose-Response Assessment: 1. Estimating the potency of a chemical. 2. In exposure assessment, the process of determining the relationship between the dose of a stressor and a specific biological response. 3. Evaluating the quantitative relationship between dose and toxicological responses.

Dose-Response Relationship: The quantitative relationship between the amount of exposure to a substance and the extent of toxic injury or disease produced.

Dosimeter: An instrument to measure dosage; many so-called dosimeters actually measure exposure rather than dosage. Dosimetry is the process or technology of measuring and/or estimating dosage.

DOT Reportable Quantity: The quantity of a substance specified in a U.S. Department of Transportation regulation that triggers labeling, packaging and other requirements related to shipping such substances.

Downgradient: The direction that groundwater flows; similar to "downstream" for surface water.

Downstream Processors: Industries dependent on crop production (e.g. canneries and food processors).

DP Hole: Hole in the ground made with DP equipment. (See: direct push.)

Draft: 1. The act of drawing or removing water from a tank or reservoir. 2. The water which is drawn or removed.

Draft Permit: A preliminary permit drafted and published by EPA; subject to public review and comment before final action on the application.

Drainage: Improving the productivity of agricultural land by removing excess water from the soil by such means as ditches or subsurface drainage tiles.

Drainage Basin: The area of land that drains water, sediment, and dissolved materials to a common outlet at some point along a stream channel.

Drainage Well: A well drilled to carry excess water off agricultural fields. Because they act as a funnel from the surface to the groundwater below. Drainage wells can contribute to groundwater pollution.

Drawdown: 1. The drop in the water table or level of water in the ground when water is being pumped from a well. 2. The amount of water used from a tank or reservoir. 3. The drop in the water level of a tank or reservoir.

Dredging: Removal of mud from the bottom of water bodies. This can disturb the ecosystem and causes silting that kills aquatic life. Dredging of contaminated muds can expose biota to heavy metals and other toxics. Dredging activities may be subject to regulation under Section 404 of the Clean Water Act.

Drilling Fluid: Fluid used to lubricate the bit and convey drill cuttings to the surface with rotary drilling equipment. Usually composed of bentonite slurry or muddy water. Can become contaminated, leading to cross contamination, and may require special disposal. Not used with DP methods

Drinking Water Equivalent Level: Protective level of exposure related to potentially non-carcinogenic effects of chemicals that are also known to cause cancer.

Drinking Water State Revolving Fund: The Fund provides capitalization grants to states to develop drinking water revolving loan funds to help finance system infrastructure improvements, assure source-water protection, enhance operation and management of drinking-water systems, and otherwise promote local water-system compliance and protection of public health.

Drive Casing: Heavy duty steel casing driven along with the sampling tool in cased DP systems. Keeps the hole open between sampling runs and is not removed until last sample has been collected.

Drive Point Profiler: An exposed groundwater DP system used to collect multiple depth-discrete groundwater samples. Ports in the tip of the probe connect to an internal stainless steel or teflon tube that extends to the surface. Samples are collected via suction or airlift methods. Deionized water is pumped down through the ports to prevent plugging while driving the tool to the next sampling depth.

Drop-off: Recyclable materials collection method in which individuals bring them to a designated collection site.

Dual-Phase Extraction: Active withdrawal of both liquid and gas phases from a well usually involving the use of a vacuum pump.

Dump: A site used to dispose of solid waste without environmental controls.

Duplicate: A second aliquot or sample that is treated the same as the original sample in order to determine the precision of the analytical method. (See: aliquot.)

Dustfall Jar: An open container used to collect large particles from the air for measurement and analysis.

Dynamometer. A device used to place a load on an engine and measure its performance.

Dystrophic Lakes: Acidic, shallow bodies of water that contain much humus and/or other organic matter; contain many plants but few fish.

E

Ecological Entity: In ecological risk assessment, a general term referring to a species, a group of species, an ecosystem function or characteristic, or a specific habitat or biome.

Ecological/Environmental Sustainability: Maintenance of ecosystem components and functions for future generations.

Ecological Exposure: Exposure of a non-human organism to a stressor.

Ecological Impact: The effect that a man-caused or natural activity has on living organisms and their non-living (abiotic) environment.

Ecological Indicator: A characteristic of an ecosystem that is related to, or derived from, a measure of biotic or abiotic variable, that can provide quantitative information on ecological structure and function. An indicator can contribute to a measure of integrity and sustainability.

Ecological Integrity: A living system exhibits integrity if, when subjected to disturbance, it sustains and organizes self-correcting ability to recover toward a biomass end-state that is normal for that system. End-states other than the pristine or naturally whole may be accepted as normal and good.

Ecological Risk Assessment: The application of a formal framework, analytical process, or model to estimate the effects of human actions(s) on a natural resource and to interpret the significance of those effects in light of the uncertainties identified in each component of the assessment process. Such analysis includes initial hazard identification, exposure and dose-response assessments, and risk characterization.

Ecology: The relationship of living things to one another and their environment, or the study of such relationships.

Economic Poisons: Chemicals used to control pests and to defoliate cash crops such as cotton.

Ecosphere: The "bio-bubble" that contains life on earth, in surface waters, and in the air. (See: biosphere.)

Ecosystem: The interacting system of a biological community and its non-living environmental surroundings.

Ecosystem Structure: Attributes related to the instantaneous physical state of an ecosystem; examples include species population density, species richness or evenness, and standing crop biomass.

Ecotone: A habitat created by the juxtaposition of distinctly different habitats; an edge habitat; or an ecological zone or boundary where two or more ecosystems meet.

Effluent: Wastewater—treated or untreated—that flows out of a treatment plant, sewer, or industrial outfall. Generally refers to wastes discharged into surface waters.

Effluent Guidelines: Technical EPA documents which set effluent limitations for given industries and pollutants.

Effluent Limitation: Restrictions established by a state or EPA on quantities, rates, and concentrations in wastewater discharges.

Effluent Standard: (See: effluent limitation.)

Ejector: A device used to disperse a chemical solution into water being treated.

Electrodialysis: A process that uses electrical current applied to permeable membranes to remove minerals from water. Often used to desalinize salty or brackish water.

Electromagnetic Geophysical Methods: Ways to measure subsurface conductivity via low-frequency electromagnetic induction.

Electrostatic Precipitator (ESP): A device that removes particles from a gas stream (smoke) after combustion occurs. The ESP imparts an electrical charge to the particles, causing them to adhere to metal plates inside the precipitator. Rapping on the plates causes the particles to fall into a hopper for disposal.

Eligible Costs: The construction costs for wastewater treatment works upon which EPA grants are based.

EMAP Data: Environmental monitoring data collected under the auspices of the Environmental Monitoring and Assessment Program. All EMAP data share the common attribute of being of known quality, having been collected in the context of explicit data quality objectives (DQOs) and a consistent quality assurance program.

Emergency (Chemical): A situation created by an accidental release or spill of hazardous chemicals that poses a threat to the safety of workers, residents, the environment, or property.

Emergency Episode: (See: air pollution episode.)

Emergency Exemption: Provision in FIFRA under which EPA can grant temporary exemption to a state or another federal agency to allow the use of a pesticide product not registered for that particular use. Such actions involve unanticipated and/or severe pest problems where there is not time or interest by a manufacturer to register the product for that use. (Registrants cannot apply for such exemptions.)

Emergency Removal Action: 1. Steps take to remove contaminated materials that pose imminent threats to local residents (e.g. removal of leaking drums or the excavation of explosive waste.) 2. The state record of such removals.

Emergency Response Values: Concentrations of chemicals, published by various groups, defining acceptable levels for short-term exposures in emergencies.

Emergency Suspension: Suspension of a pesticide product registration due to an imminent hazard. The action immediately halts distribution, sale, and sometimes actual use of the pesticide involved.

Emission: Pollution discharged into the atmosphere from smokestacks, other vents, and surface areas of commercial or industrial facilities; from residential chimneys; and from motor vehicle, locomotive, or aircraft exhausts.

Emission Cap: A limit designed to prevent projected growth in emissions from existing and future stationary sources from eroding any mandated reductions. Generally, such provisions require that any emission growth from facilities under the restrictions be offset by equivalent reductions at other facilities under the same cap. (See: emissions trading.)

Emission Factor: The relationship between the amount of pollution produced and the amount of raw material processed. For example, an emission factor for a blast furnace making iron would be the number of pounds of particulates per ton of raw materials.

Emission Inventory: A listing, by source, of the amount of air pollutants discharged into the atmosphere of a community; used to establish emission standards.

Emission Standard: The maximum amount of air polluting discharge legally allowed from a single source, mobile or stationary.

Emissions Trading: The creation of surplus emission reductions at certain stacks, vents or similar emissions sources and the use of this surplus to meet or redefine pollution requirements applicable to other emissions sources. This allows one source to increase emissions when another source reduces them, maintaining an overall constant emission level. Facilities that reduce emissions substantially may "bank" their "credits" or sell them to other facilities or industries.

Emulsifier: A chemical that aids in suspending one liquid in another. Usually an organic chemical in an aqueous solution.

Encapsulation: The treatment of asbestos-containing material with a liquid that covers the surface with a protective coating or embeds fibers in an adhesive matrix to prevent their release into the air.

Enclosure: Putting an airtight, impermeable, permanent barrier around asbestos-containing materials to prevent the release of asbestos fibers into the air.

End User: Consumer of products for the purpose of recycling. Excludes products for re-use or combustion for energy recovery.

End-of-the-pipe: Technologies such as scrubbers on smokestacks and catalytic convertors on automobile tailpipes that reduce emissions of pollutants after they have formed.

End-use Product: A pesticide formulation for field or other end use. The label has instructions for use or application to control pests or regulate plant growth. The term excludes products used to formulate other pesticide products.

Endangered Species: Animals, birds, fish, plants, or other living organisms threatened with extinction by anthropogenic (man-caused) or other natural changes in their environment. Requirements for declaring a species endangered are contained in the Endangered Species Act.

Endangerment Assessment: A study to determine the nature and extent of contamination at a site on the National Priorities List and the risks posed to public health or the environment. EPA or the state conducts the study when a legal action is to be taken to direct potentially responsible parties to clean up a site or pay for it. An endangerment assessment supplements a remedial investigation.

Endrin: A pesticide toxic to freshwater and marine aquatic life that produces adverse health effects in domestic water supplies.

Energy Management System: A control system capable of monitoring environmental and system loads and adjusting HVAC operations accordingly in order to conserve energy while maintaining comfort.

Energy Recovery: Obtaining energy from waste through a variety of processes (e.g. combustion).

Enforceable Requirements: Conditions or limitations in permits issued under the Clean Water Act Section 402 or 404 that, if violated, could result in the issuance of a compliance order or initiation of a civil or criminal action under federal or applicable state laws. If a permit has not been issued, the term includes any requirement which, in the Regional Administrator's judgement, would be included in the permit when issued. Where no permit applies, the term includes any requirement which the RA determines is necessary for the best practical waste treatment technology to meet applicable criteria.

Enforcement: EPA, state, or local legal actions to obtain compliance with environmental laws, rules, regulations, or agreements and/or obtain penalties or criminal sanctions for violations. Enforcement procedures may vary, depending on the requirements of different environmental laws and related implementing regulations. Under CERCLA, for example, EPA will seek to require potentially responsible parties to clean up a Superfund site, or pay for the cleanup, whereas under the Clean Air Act the Agency may invoke sanctions against cities failing to meet ambient air quality standards that could prevent certain types of construction or federal funding. In other situations, if investigations by EPA and state agencies uncover willful violations, criminal trials and penalties are sought.

Glossary of Terms

Enforcement Decision Document (EDD): A document that provides an explanation to the public of EPA's selection of the cleanup alternative at enforcement sites on the National Priorities List. Similar to a Record of Decision.

Engineered Controls: Method of managing environmental and health risks by placing a barrier between the contamination and the rest of the site, thus limiting exposure pathways.

Enhanced Inspection and Maintenance (I&M): An improved automobile inspection and maintenance program—aimed at reducing automobile emissions—that contains, at a minimum, more vehicle types and model years, tighter inspection, and better management practices. It may also include annual computerized or centralized inspections, under-the-hood inspection—for signs of tampering with pollution control equipment—and increased repair waiver cost.

Enrichment: The addition of nutrients (e.g. nitrogen, phosphorus, carbon compounds) from sewage effluent or agricultural runoff to surface water, greatly increases the growth potential for algae and other aquatic plants.

Entrain: To trap bubbles in water either mechanically through turbulence or chemically through a reaction.

Environment: The sum of all external conditions affecting the life, development and survival of an organism.

Environmental Assessment: An environmental analysis prepared pursuant to the National Environmental Policy Act to determine whether a federal action would significantly affect the environment and thus require a more detailed environmental impact statement.

Environmental Audit: An independent assessment of the current status of a party's compliance with applicable environmental requirements or of a party's environmental compliance policies, practices, and controls.

Environmental/Ecological Risk: The potential for adverse effects on living organisms associated with pollution of the environment by effluents, emissions, wastes, or accidental chemical releases; energy use; or the depletion of natural resources.

Environmental Equity/Justice: Equal protection from environmental hazards for individuals, groups, or communities regardless of race, ethnicity, or economic status. This applies to the development, implementation, and enforcement of environmental laws, regulations, and policies, and implies that no population of people should be forced to shoulder a disproportionate share of negative environmental impacts of pollution or environmental hazard due to a lack of political or economic strength levels.

Environmental Exposure: Human exposure to pollutants originating from facility emissions. Threshold levels are not necessarily surpassed, but low-level chronic pollutant exposure is one of the most common forms of environmental exposure (See: threshold level).

Environmental Fate: The destiny of a chemical or biological pollutant after release into the environment.

Environmental Fate Data: Data that characterize a pesticide's fate in the ecosystem, considering factors that foster its degradation (light, water, microbes), pathways and resultant products.

Environmental Impact Statement: A document required of federal agencies by the National Environmental Policy Act for major projects or legislative proposals significantly affecting the environment. A tool for decision making, it describes the positive and negative effects of the undertaking and cites alternative actions.

Environmental Indicator: A measurement, statistic or value that provides a proximate gauge or evidence of the effects of environmental management programs or of the state or condition of the environment.

Environmental Lien: A charge, security, or encumbrance on a property's title to secure payment of cost or debt arising from response actions, cleanup, or other remediation of hazardous substances or petroleum products.

Environmental Medium: A major environmental category that surrounds or contacts humans, animals, plants, and other organisms (e.g. surface water, ground water, soil or air) and through which chemicals or pollutants move. (See: ambient medium, biological medium.)

Environmental Monitoring for Public Access and Community Tracking: Joint EPA, NOAA, and USGS program to provide timely and effective communication of environmental data and information through improved and updated technology solutions that support timely environmental monitoring reporting, interpreting, and use of the information for the benefit of the public. (See: real-time monitoring.)

Environmental Response Team: EPA experts located in Edison, N.J., and Cincinnati, OH, who can provide around-the-clock technical assistance to EPA regional offices and states during all types of hazardous waste site emergencies and spills of hazardous substances.

Environmental Site Assessment: The process of determining whether contamination is present on a parcel of real property.

Environmental Sustainability: Long-term maintenance of ecosystem components and functions for future generations.

Environmental Tobacco Smoke: Mixture of smoke from the burning end of a cigarette, pipe, or cigar and smoke exhaled by the smoker. (See: passive smoking/secondhand smoke.)

Epidemiology: Study of the distribution of disease, or other health-related states and events in human populations, as related to age, sex, occupation, ethnicity, and economic status in order to identify and alleviate health problems and promote better health.

Epilimnion: Upper waters of a thermally stratified lake subject to wind action.

Episode (Pollution): An air pollution incident in a given area caused by a concentration of atmospheric pollutants under meteorological conditions that may result in a significant increase in illnesses or deaths. May also describe water pollution events or hazardous material spills.

Equilibrium: In relation to radiation, the state at which the radioactivity of consecutive elements within a radioactive series is neither increasing nor decreasing.

Equivalent Method: Any method of sampling and analyzing for air pollution which has been demonstrated to the EPA Administrator's satisfaction to be, under specific conditions, an acceptable alternative to normally used reference methods.

Erosion: The wearing away of land surface by wind or water, intensified by land-clearing practices related to farming, residential or industrial development, road building, or logging.

Established Treatment Technologies: Technologies for which cost and performance data are readily available. (See: Innovative treatment technologies.)

Estimated Environmental Concentration: The estimated pesticide concentration in an ecosystem.

Estuary: Region of interaction between rivers and near-shore ocean waters, where tidal action and river flow mix fresh and salt water. Such areas include bays, mouths of rivers, salt marshes, and lagoons. These brackish water ecosystems shelter and feed marine life, birds, and wildlife. (See: wetlands.)

Ethanol: An alternative automotive fuel derived from grain and corn; usually blended with gasoline to form gasohol.

Ethylene Dibromide (EDB): A chemical used as an agricultural fumigant and in certain industrial processes. Extremely toxic and found to be a carcinogen in laboratory animals, EDB has been banned for most agricultural uses in the United States.

Eutrophic Lakes: Shallow, murky bodies of water with concentrations of plant nutrients causing excessive production of algae. (See: dystrophic lakes.)

Eutrophication: The slow aging process during which a lake, estuary, or bay evolves into a bog or marsh and eventually disappears. During the later stages of eutrophication the water body is choked by abundant plant life due to higher levels of nutritive compounds such as nitrogen and phosphorus. Human activities can accelerate the process.

Evaporation Ponds: Areas where sewage sludge is dumped and dried.

Evapotranspiration: The loss of water from the soil both by evaporation and by transpiration from the plants growing in the soil.

Exceedance: Violation of the pollutant levels permitted by environmental protection standards.

Exclusion: In the asbestos program, one of several situations that permit a Local Education Agency (LEA) to delete one or more of the items required by the Asbestos Hazard Emergency Response Act (AHERA); e.g. records of previous asbestos sample collection and analysis may be used by the accredited inspector in lieu of AHERA bulk sampling.

Exclusionary Ordinance: Zoning that excludes classes of persons or businesses from a particular neighborhood or area.

Exempt Solvent: Specific organic compounds not subject to requirements of regulation because they are deemed by EPA to be of negligible photochemical reactivity.

Exempted Aquifer: Underground bodies of water defined in the Underground Injection Control program as aquifers that are potential sources of

762

drinking water though not being used as such, and thus exempted from regulations barring underground injection activities.

Exemption: A state (with primacy) may exempt a public water system from a requirement involving a Maximum Contaminant Level (MCL), treatment technique, or both, if the system cannot comply due to compelling economic or other factors, or because the system was in operation before the requirement or MCL was instituted; and the exemption will not create a public health risk. (See: variance.)

Exotic Species: A species that is not indigenous to a region.

Experimental Use Permit: Obtained by manufacturers for testing new pesticides or uses thereof whenever they conduct experimental field studies to support registration on 10 acres or more of land or one acre or more of water.

Experimental Use Permit: A permit granted by EPA that allows a producer to conduct tests of a new pesticide, product and/or use outside the laboratory. The testing is usually done on ten or more acres of land or water surface.

Explosive Limits: The amounts of vapor in the air that form explosive mixtures; limits are expressed as lower and upper limits and give the range of vapor concentrations in air that will explode if an ignition source is present.

Exports : In solid waste program, municipal solid waste and recyclables transported outside the state or locality where they originated.

Exposure: The amount of radiation or pollutant present in a given environment that represents a potential health threat to living organisms.

Exposure Assessment: Identifying the pathways by which toxicants may reach individuals, estimating how much of a chemical an individual is likely to be exposed to, and estimating the number likely to be exposed.

Exposure Concentration: The concentration of a chemical or other pollutant representing a health threat in a given environment.

Exposure Indicator: A characteristic of the environment measured to provide evidence of the occurrence or magnitude of a response indicator's exposure to a chemical or biological stress.

Exposure Level: The amount (concentration) of a chemical at the absorptive surfaces of an organism.

Exposure Pathway: The path from sources of pollutants via, soil, water, or food to man and other species or settings.

Exposure Route: The way a chemical or pollutant enters an organism after contact; i.e. by ingestion, inhalation, or dermal absorption.

Exposure-Response Relationship: The relationship between exposure level and the incidence of adverse effects.

Extraction Procedure (EP Toxic): Determining toxicity by a procedure which simulates leaching; if a certain concentration of a toxic substance can be leached from a waste, that waste is considered hazardous, i.e."EP Toxic."

Extraction Well: A discharge well used to remove groundwater or air.

Extremely Hazardous Substances: Any of 406 chemicals identified by EPA as toxic, and listed under SARA Title III. The list is subject to periodic revision.

F

Fabric Filter: A cloth device that catches dust particles from industrial emissions.

Facilities Plans: Plans and studies related to the construction of treatment works necessary to comply with the Clean Water Act or RCRA. A facilities plan investigates needs and provides information on the cost-effectiveness of alternatives, a recommended plan, an environmental assessment of the recommendations, and descriptions of the treatment works, costs, and a completion schedule.

Facility Emergency Coordinator: Representative of a facility covered by environmental law (e.g, a chemical plant) who participates in the emergency reporting process with the Local Emergency Planning Committee (LEPC).

Facultative Bacteria: Bacteria that can live under aerobic or anaerobic conditions.

Feasibility Study: 1. Analysis of the practicability of a proposal; e.g., a description and analysis of potential cleanup alternatives for a site such as one on the National Priorities List. The feasibility study usually recommends selection of a cost-effective alternative. It usually starts as soon as the remedial investigation is underway; together, they are commonly referred to as the "RI/FS". 2. A small-scale investigation of a problem to ascertain whether a proposed research approach is likely to provide useful data.

Fecal Coliform Bacteria: Bacteria found in the intestinal tracts of mammals. Their presence in water or sludge is an indicator of pollution and possible contamination by pathogens.

Federal Implementation Plan: Under current law, a federally implemented plan to achieve attainment of air quality standards, used when a state is unable to develop an adequate plan.

Federal Motor Vehicle Control Program: All federal actions aimed at controlling pollution from motor vehicles by such efforts as establishing and enforcing tailpipe and evaporative emission standards for new vehicles, testing methods development, and guidance to states operating inspection and maintenance programs. Federally designated area that is required to meet and maintain federal ambient air quality standards. May include nearby locations in the same state or nearby states that share common air pollution problems.

Feedlot: A confined area for the controlled feeding of animals. Tends to concentrate large amounts of animal waste that cannot be absorbed by the soil and, hence, may be carried to nearby streams or lakes by rainfall runoff.

Fen: A type of wetland that accumulates peat deposits. Fens are less acidic than bogs, deriving most of their water from groundwater rich in calcium and magnesium. (See: wetlands.)

Ferrous Metals: Magnetic metals derived from iron or steel; products made from ferrous metals include appliances, furniture, containers, and packaging like steel drums and barrels. Recycled products include processing tin/steel cans, strapping, and metals from appliances into new products.

FIFRA Pesticide Ingredient: An ingredient of a pesticide that must be registered with EPA under the Federal Insecticide, Fungicide, and Rodenticide Act. Products making pesticide claims must register under FIFRA and may be subject to labeling and use requirements.

Fill: Man-made deposits of natural soils or rock products and waste materials.

Filling: Depositing dirt, mud or other materials into aquatic areas to create more dry land, usually for agricultural or commercial development purposes, often with ruinous ecological consequences.

Filter Strip: Strip or area of vegetation used for removing sediment, organic matter, and other pollutants from runoff and wastewater.

Filtration: A treatment process, under the control of qualified operators, for removing solid (particulate) matter from water by means of porous media such as sand or a man-made filter; often used to remove particles that contain pathogens.

Financial Assurance for Closure: Documentation or proof that an owner or operator of a facility such as a landfill or other waste repository is capable of paying the projected costs of closing the facility and monitoring it afterwards as provided in RCRA regulations.

Finding of No Significant Impact: A document prepared by a federal agency showing why a proposed action would not have a significant impact on the environment and thus would not require preparation of an Environmental Impact Statement. An FNSI is based on the results of an environmental assessment.

Finished Water: Water is "finished" when it has passed through all the processes in a water treatment plant and is ready to be delivered to consumers.

First Draw: The water that comes out when a tap is first opened, likely to have the highest level of lead contamination from plumbing materials.

Fix a Sample: A sample is "fixed" in the field by adding chemicals that prevent water quality indicators of interest in the sample from changing before laboratory measurements are made.

Fixed-Location Monitoring: Sampling of an environmental or ambient medium for pollutant concentration at one location continuously or repeatedly.

Flammable: Any material that ignites easily and will burn rapidly.

Flare: A control device that burns hazardous materials to prevent their release into the environment; may operate continuously or intermittently, usually on top of a stack.

Flash Point: The lowest temperature at which evaporation of a substance produces sufficient vapor to form an ignitable mixture with air.

Floc: A clump of solids formed in sewage by biological or chemical action.

Flocculation: Process by which clumps of solids in water or sewage aggregate through biological or chemical action so they can be separated from water or sewage.

Floodplain: The flat or nearly flat land along a river or stream or in a tidal area that is covered by water during a flood.

Glossary of Terms

Floor Sweep: Capture of heavier-than-air gases that collect at floor level.

Flow Rate: The rate, expressed in gallons -or liters-per-hour, at which a fluid escapes from a hole or fissure in a tank. Such measurements are also made of liquid waste, effluent, and surface water movement.

Flowable: Pesticide and other formulations in which the active ingredients are finely ground insoluble solids suspended in a liquid. They are mixed with water for application.

Flowmeter: A gauge indicating the velocity of wastewater moving through a treatment plant or of any liquid moving through various industrial processes.

Flue Gas: The air coming out of a chimney after combustion in the burner it is venting. It can include nitrogen oxides, carbon oxides, water vapor, sulfur oxides, particles and many chemical pollutants.

Flue Gas Desulfurization: A technology that employs a sorbent, usually lime or limestone, to remove sulfur dioxide from the gases produced by burning fossil fuels. Flue gas desulfurization is current state-of-the art technology for major SO2 emitters, like power plants.

Fluidized: A mass of solid particles that is made to flow like a liquid by injection of water or gas is said to have been fluidized. In water treatment, a bed of filter media is fluidized by backwashing water through the filter.

Fluidized Bed Incinerator: An incinerator that uses a bed of hot sand or other granular material to transfer heat directly to waste. Used mainly for destroying municipal sludge.

Flume: A natural or man-made channel that diverts water.

Fluoridation: The addition of a chemical to increase the concentration of fluoride ions in drinking water to reduce the incidence of tooth decay.

Fluorides: Gaseous, solid, or dissolved compounds containing fluorine that result from industrial processes. Excessive amounts in food can lead to fluorosis.

Fluorocarbons (FCs): Any of a number of organic compounds analogous to hydrocarbons in which one or more hydrogen atoms are replaced by fluorine. Once used in the United States as a propellant for domestic aerosols, they are now found mainly in coolants and some industrial processes. FCs containing chlorine are called chlorofluorocarbons (CFCs). They are believed to be modifying the ozone layer in the stratosphere, thereby allowing more harmful solar radiation to reach the Earth's surface.

Flush: 1. To open a cold-water tap to clear out all the water which may have been sitting for a long time in the pipes. In new homes, to flush a system means to send large volumes of water gushing through the unused pipes to remove loose particles of solder and flux. 2. To force large amounts of water through a system to clean out piping or tubing, and storage or process tanks.

Flux: 1. A flowing or flow. 2. A substance used to help metals fuse together.

Fly Ash: Non-combustible residual particles expelled by flue gas.

Fogging: Applying a pesticide by rapidly heating the liquid chemical so that it forms very fine droplets that resemble smoke or fog. Used to destroy mosquitoes, black flies, and similar pests.

Food Chain: A sequence of organisms, each of which uses the next, lower member of the sequence as a food source.

Food Processing Waste: Food residues produced during agricultural and industrial operations.

Food Waste: Uneaten food and food preparation wastes from residences and commercial establishments such as grocery stores, restaurants, and produce stands, institutional cafeterias and kitchens, and industrial sources like employee lunchrooms.

Food Web: The feeding relationships by which energy and nutrients are transferred from one species to another.

Formaldehyde: A colorless, pungent, and irritating gas, CH20, used chiefly as a disinfectant and preservative and in synthesizing other compounds like resins.

Formulation: The substances comprising all active and inert ingredients in a pesticide.

Fossil Fuel: Fuel derived from ancient organic remains; e.g. peat, coal, crude oil, and natural gas.

Fracture: A break in a rock formation due to structural stresses; e.g. faults, shears, joints, and planes of fracture cleavage.

Free Product: A petroleum hydrocarbon in the liquid free or non aqueous phase. (See: non-aqueous phase liquid.)

Freeboard: 1. Vertical distance from the normal water surface to the top of a confining wall. 2. Vertical distance from the sand surface to the underside of a trough in a sand filter.

Fresh Water: Water that generally contains less than 1,000 milligrams-per-liter of dissolved solids.

Friable: Capable of being crumbled, pulverized, or reduced to powder by hand pressure.

Friable Asbestos: Any material containing more than one-percent asbestos, and that can be crumbled or reduced to powder by hand pressure. (May include previously non-friable material which becomes broken or damaged by mechanical force.)

Fuel Economy Standard: The Corporate Average Fuel Economy Standard (CAFE) effective in 1978. It enhanced the national fuel conservation effort imposing a miles-per-gallon floor for motor vehicles.

Fuel Efficiency: The proportion of energy released by fuel combustion that is converted into useful energy.

Fuel Switching: 1. A precombustion process whereby a low-sulfur coal is used in place of a higher sulfur coal in a power plant to reduce sulfur dioxide emissions. 2. Illegally using leaded gasoline in a motor vehicle designed to use only unleaded.

Fugitive Emissions: Emissions not caught by a capture system.

Fume: Tiny particles trapped in vapor in a gas stream.

Fumigant: A pesticide vaporized to kill pests. Used in buildings and greenhouses.

Functional Equivalent: Term used to describe EPA's decision-making process and its relationship to the environmental review conducted under the National Environmental Policy Act (NEPA). A review is considered functionally equivalent when it addresses the substantive components of a NEPA review.

Fungicide: Pesticides which are used to control, deter, or destroy fungi.

Fungistat: A chemical that keeps fungi from growing.

Fungus (Fungi): Molds, mildews, yeasts, mushrooms, and puffballs, a group of organisms lacking in chlorophyll (i.e. are not photosynthetic) and which are usually non-mobile, filamentous, and multicellular. Some grow in soil, others attach themselves to decaying trees and other plants whence they obtain nutrients. Some are pathogens, others stabilize sewage and digest composted waste.

Furrow Irrigation: Irrigation method in which water travels through the field by means of small channels between each groups of rows.

Future Liability: Refers to potentially responsible parties' obligations to pay for additional response activities beyond those specified in the Record of Decision or Consent Decree.

G

Game Fish: Species like trout, salmon, or bass, caught for sport. Many of them show more sensitivity to environmental change than "rough" fish.

Garbage: Animal and vegetable waste resulting from the handling, storage, sale, preparation, cooking, and serving of foods.

Gas Chromatograph/Mass Spectrometer: Instrument that identifies the molecular composition and concentrations of various chemicals in water and soil samples.

Gasahol: Mixture of gasoline and ethanol derived from fermented agricultural products containing at least nine percent ethanol. Gasohol emissions contain less carbon monoxide than those from gasoline.

Gasification: Conversion of solid material such as coal into a gas for use as a fuel.

Gasoline Volatility: The property of gasoline whereby it evaporates into a vapor. Gasoline vapor is a mixture of volatile organic compounds.

General Permit: A permit applicable to a class or category of dischargers.

General Reporting Facility: A facility having one or more hazardous chemicals above the 10,000 pound threshold for planning quantities. Such facilities must file MSDS and emergency inventory information with the SERC, LEPC, and local fire departments.

Generally Recognized as Safe (GRAS): Designation by the FDA that a chemical or substance (including certain pesticides) added to food is considered safe by experts, and so is exempted from the usual FFDCA food additive tolerance requirements.

Generator: 1. A facility or mobile source that emits pollutants into the air or releases hazardous waste into water or soil. 2. Any person, by site, whose act or process produces regulated medical waste or whose act first causes such waste to become subject to regulation. Where more than one person (e.g. doctors with separate medical practices) are located in the same building, each business entity is a separate generator.

Genetic Engineering: A process of inserting new genetic information into existing cells in order to modify a specific organism for the purpose of changing one of its characteristics.

Genotoxic: Damaging to DNA; pertaining to agents known to damage DNA.

Geographic Information System (GIS): A computer system designed for storing, manipulating, analyzing, and displaying data in a geographic context.

Geological Log: A detailed description of all underground features (depth, thickness, type of formation) discovered during the drilling of a well.

Geophysical Log: A record of the structure and composition of the earth encountered when drilling a well or similar type of test hold or boring.

Geothermal/Ground Source Heat Pump: These heat pumps are underground coils to transfer heat from the ground to the inside of a building. (See: heat pump; water source heat pump)

Germicide: Any compound that kills disease-causing microorganisms.

Giardia Lamblia: Protozoan in the feces of humans and animals that can cause severe gastrointestinal ailments. It is a common contaminant of surface waters.

Glass Containers: For recycling purposes, containers like bottles and jars for drinks, food, cosmetics and other products. When being recycled, container glass is generally separated into color categories for conversion into new containers, construction materials or fiberglass insulation.

Global Warming: An increase in the near surface temperature of the Earth. Global warming has occurred in the distant past as the result of natural influences, but the term is most often used to refer to the warming predicted to occur as a result of increased emissions of greenhouse gases. Scientists generally agree that the Earth's surface has warmed by about 1 degree Fahrenheit in the past 140 years. The Intergovernmental Panel on Climate Change (IPCC) recently concluded that increased concentrations of greenhouse gases are causing an increase in the Earth's surface temperature and that increased concentrations of sulfate aerosols have led to relative cooling in some regions, generally over and downwind of heavily industrialized areas. (See: climate change)

Global Warming Potential: The ratio of the warming caused by a substance to the warming caused by a similar mass of carbon dioxide. CFC-12, for example, has a GWP of 8,500, while water has a GWP of zero. (See: Class I Substance and Class II Substance.)

Glovebag: A polyethylene or polyvinyl chloride bag-like enclosure affixed around an asbestos-containing source (most often thermal system insulation) permitting the material to be removed while minimizing release of airborne fibers to the surrounding atmosphere.

Gooseneck: A portion of a water service connection between the distribution system water main and a meter. Sometimes called a pigtail.

Grab Sample: A single sample collected at a particular time and place that represents the composition of the water, air, or soil only at that time and place.

Grain Loading: The rate at which particles are emitted from a pollution source. Measurement is made by the number of grains per cubic foot of gas emitted.

Granular Activated Carbon Treatment: A filtering system often used in small water systems and individual homes to remove organics. Also used by municipal water treatment plantsd. GAC can be highly effective in lowering elevated levels of radon in water.

Grasscycling: Source reduction activities in which grass clippings are left on the lawn after mowing.

Grassed Waterway: Natural or constructed watercourse or outlet that is shaped or graded and established in suitable vegetation for the disposal of runoff water without erosion.

Gray Water: Domestic wastewater composed of wash water from kitchen, bathroom, and laundry sinks, tubs, and washers.

Greenhouse Effect: The warming of the Earth's atmosphere attributed to a buildup of carbon dioxide or other gases; some scientists think that this build-up allows the sun's rays to heat the Earth, while making the infra-red radiation atmosphere opaque to infra-red radiation, thereby preventing a counterbalancing loss of heat.

Greenhouse Gas: A gas, such as carbon dioxide or methane, which contributes to potential climate change.

Grinder Pump: A mechanical device that shreds solids and raises sewage to a higher elevation through pressure sewers.

Gross Alpha/Beta Particle Activity: The total radioactivity due to alpha or beta particle emissions as inferred from measurements on a dry sample.

Gross Power-Generation Potential: The installed power generation capacity that landfill gas can support.

Ground Cover: Plants grown to keep soil from eroding.

Ground Water: The supply of fresh water found beneath the Earth's surface, usually in aquifers, which supply wells and springs. Because ground water is a major source of drinking water, there is growing concern over contamination from leaching agricultural or industrial pollutants or leaking underground storage tanks.

Ground Water Under the Direct Influence (UDI) of Surface Water: Any water beneath the surface of the ground with: 1. significant occurence of insects or other microorganisms, algae, or large-diameter pathogens; 2. significant and relatively rapid shifts in water characteristics such as turbidity, temperature, conductivity, or pH which closely correlate to climatological or surface water conditions. Direct influence is determined for individual sources in accordance with criteria established by a state.

Ground-Penetrating Radar: A geophysical method that uses high frequency electromagnetic waves to obtain subsurface information.

Ground-Water Discharge: Ground water entering near coastal waters which has been contaminated by landfill leachate, deep well injection of hazardous wastes, septic tanks, etc.

Ground-Water Disinfection Rule: A 1996 amendment of the Safe Drinking Water Act requiring EPA to promulgate national primary drinking water regulations requiring disinfection as for all public water systems, including surface waters and ground water systems.

Gully Erosion: Severe erosion in which trenches are cut to a depth greater than 30 centimeters (a foot). Generally, ditches deep enough to cross with farm equipment are considered gullies.

H

Habitat: The place where a population (e.g. human, animal, plant, microorganism) lives and its surroundings, both living and non-living.

Habitat Indicator: A physical attribute of the environment measured to characterize conditions necessary to support an organism, population, or community in the absence of pollutants; e.g. salinity of estuarine waters or substrate type in streams or lakes.

Half-Life: 1. The time required for a pollutant to lose one-half of its original coconcentrationor example, the biochemical half-life of DDT in the environment is 15 years. 2. The time required for half of the atoms of a radioactive element to undergo self-transmutation or decay (half-life of radium is 1620 years). 3. The time required for the elimination of half a total dose from the body.

Halogen: A type of incandescent lamp with higher energy-efficiency that standard ones.

Halon: Bromine-containing compounds with long atmospheric lifetimes whose breakdown in the stratosphere causes depletion of ozone. Halons are used in firefighting.

Hammer Mill: A high-speed machine that uses hammers and cutters to crush, grind, chip, or shred solid waste.

Hard Water: Alkaline water containing dissolved salts that interfere with some industrial processes and prevent soap from sudsing.

Hauler: Garbage collection company that offers complete refuse removal service; many will also collect recyclables.

Hazard: 1. Potential for radiation, a chemical or other pollutant to cause human illness or injury. 2. In the pesticide program, the inherent toxicity of a compound. Hazard identification of a given substances is an informed judgment based on verifiable toxicity data from animal models or human studies.

Hazard Assessment: Evaluating the effects of a stressor or determining a margin of safety for an organism by comparing the concentration which causes toxic effects with an estimate of exposure to the organism.

Hazard Communication Standard: An OSHA regulation that requires chemical manufacturers, suppliers, and importers to assess the hazards of the chemicals that they make, supply, or import, and to inform employers, customers, and workers of these hazards through MSDS information.

Hazard Evaluation: A component of risk evaluation that involves gathering and evaluating data on the types of health injuries or diseases that may be produced by a chemical and on the conditions of exposure under which such health effects are produced.

Glossary of Terms

Hazard Identification: Determining if a chemical or a microbe can cause adverse health effects in humans and what those effects might be.

Hazard Quotient: The ratio of estimated site-specific exposure to a single chemical from a site over a specified period to the estimated daily exposure level, at which no adverse health effects are likely to occur.

Hazard Ratio: A term used to compare an animal's daily dietary intake of a pesticide to its LD 50 value. A ratio greater than 1.0 indicates that the animal is likely to consume an a dose amount which would kill 50 percent of animals of the same species. (See: LD 50 /Lethal Dose.)

Hazardous Air Pollutants: Air pollutants which are not covered by ambient air quality standards but which, as defined in the Clean Air Act, may present a threat of adverse human health effects or adverse environmental effects.Such pollutants include asbestos, beryllium, mercury, benzene, coke oven emissions, radionuclides, and vinyl chloride.

Hazardous Chemical: An EPA designation for any hazardous material requiring an MSDS under OSHA's Hazard Communication Standard. Such substances are capable of producing fires and explosions or adverse health effects like cancer and dermatitis. Hazardous chemicals are distinct from hazardous waste.(See: Hazardous Waste.)

Hazardous Ranking System: The principal screening tool used by EPA to evaluate risks to public health and the environment associated with abandoned or uncontrolled hazardous waste sites. The HRS calculates a score based on the potential of hazardous substances spreading from the site through the air, surface water, or ground water, and on other factors such as density and proximity of human population. This score is the primary factor in deciding if the site should be on the National Priorities List and, if so, what ranking it should have compared to other sites on the list.

Hazardous Substance: 1. Any material that poses a threat to human health and/or the environment. Typical hazardous substances are toxic, corrosive, ignitable, explosive, or chemically reactive. 2. Any substance designated by EPA to be reported if a designated quantity of the substance is spilled in the waters of the United States or is otherwise released into the environment.

Hazardous Waste: By-products of society that can pose a substantial or potential hazard to human health or the environment when improperly managed. Possesses at least one of four characteristics (ignitability, corrosivity, reactivity, or toxicity), or appears on special EPA lists.

Hazardous Waste Landfill: An excavated or engineered site where hazardous waste is deposited and covered.

Hazardous Waste Minimization: Reducing the amount of toxicity or waste produced by a facility via source reduction or environmentally sound recycling.

Hazards Analysis: Procedures used to (1) identify potential sources of release of hazardous materials from fixed facilities or transportation accidents; (2) determine the vulnerability of a geographical area to a release of hazardous materials; and (3) compare hazards to determine which present greater or lesser risks to a community.

Hazards Identification: Providing information on which facilities have extremely hazardous substances, what those chemicals are, how much there is at each facility, how the chemicals are stored, and whether they are used at high temperatures.

Headspace: The vapor mixture trapped above a solid or liquid in a sealed vessel.

Health Advisory Level: A non-regulatory health-based reference level of chemical traces (usually in ppm) in drinking water at which there are no adverse health risks when ingested over various periods of time. Such levels are established for one day, 10 days, long-term and life-time exposure periods. They contain a wide margin of safety.

Health Assessment: An evaluation of available data on existing or potential risks to human health posed by a Superfund site. The Agency for Toxic Substances and Disease Registry (ATSDR) of the Department of Health and Human Services (DHHS) is required to perform such an assessment at every site on the National Priorities List.

Heat Island Effect: A "dome" of elevated temperatures over an urban area caused by structural and pavement heat fluxes, and pollutant emissions.

Heat Pump: An electric device with both heating and cooling capabilities. It extracts heat from one medium at a lower (the heat source) temperature and transfers it to another at a higher temperature (the heat sink), thereby cooling the first and warming the second. (See: geothermal, water source heat pump.)

Heavy Metals: Metallic elements with high atomic weights; (e.g. mercury, chromium, cadmium, arsenic, and lead); can damage living things at low concentrations and tend to accumulate in the food chain.

Heptachlor: An insecticide that was banned on some food products in 1975 and in all of them 1978. It was allowed for use in seed treatment until 1983. More recently it was found in milk and other dairy products in Arkansas and Missouri where dairy cattle were illegally fed treated seed.

Herbicide: A chemical pesticide designed to control or destroy plants, weeds, or grasses.

Herbivore: An animal that feeds on plants.

Heterotrophic Organisms: Species that are dependent on organic matter for food.

High End Exposure (dose) Estimate: An estimate of exposure, or dose level received anyone in a defined population that is greater than the 90th percentile of all individuals in that population, but less than the exposure at the highest percentile in that population. A high end risk descriptor is an estimate of the risk level for such individuals. Note that risk is based on a combination of exposure and susceptibility to the stressor.

High Intensity Discharge: A generic term for mercury vapor, metal halide, and high pressure sodium lamps and fixtures.

High-Density Polyethylene: A material used to make plastic bottles and other products that produces toxic fumes when burned.

High-Level Nuclear Waste Facility: Plant designed to handle disposal of used nuclear fuel, high-level radioactive waste, and plutonium waste.

High-Level Radioactive Waste (HLRW): Waste generated in core fuel of a nuclear reactor, found at nuclear reactors or by nuclear fuel reprocessing; is a serious threat to anyone who comes near the waste without shielding. (See: low-level radioactive waste.)

High-Line Jumpers: Pipes or hoses connected to fire hydrants and laid on top of the ground to provide emergency water service for an isolated portion of a distribution system.

High-Risk Community: A community located within the vicinity of numerous sites of facilities or other potential sources of envienvironmental exposure/health hazards which may result in high levels of exposure to contaminants or pollutants.

High-to-Low-Dose Extrapolation: The process of prediction of low exposure risk to humans and animals from the measured high-exposure-high-risk data involving laboratory animals.

Highest Dose Tested: The highest dose of a chemical or substance tested in a study.

Holding Pond: A pond or reservoir, usually made of earth, built to store polluted runoff.

Holding Time: The maximum amount of time a sample may be stored before analysis.

Hollow Stem Auger Drilling: Conventional drilling method that uses augurs to penetrate the soil. As the augers are rotated, soil cuttings are conveyed to the ground surface via augur spirals. DP tools can be used inside the hollow augers.

Homeowner Water System: Any water system which supplies piped water to a single residence.

Homogeneous Area: In accordance with Asbestos Hazard and Emergency Response Act (AHERA) definitions, an area of surfacing materials, thermal surface insulation, or miscellaneous material that is uniform in color and texture.

Hood Capture Efficiency: Ratio of the emissions captured by a hood and directed into a control or disposal device, expressed as a percent of all emissions.

Host: 1. In genetics, the organism, typically a bacterium, into which a gene from another organism is transplanted. 2. In medicine, an animal infected or parasitized by another organism.

Household Hazardous Waste: Hazardous products used and disposed of by residential as opposed to industrial consumers. Includes paints, stains, varnishes, solvents, pesticides, and other materials or products containing volatile chemicals that can catch fire, react or explode, or that are corrosive or toxic.

Household Waste (Domestic Waste): Solid waste, composed of garbage and rubbish, which normally originates in a private home or apartment house. Domestic waste may contain a significant amount of toxic or hazardous waste.

Human Equivalent Dose: A dose which, when administered to humans, produces an effect equal to that produced by a dose in animals.

Human Exposure Evaluation: Describing the nature and size of the population exposed to a substance and the magnitude and duration of their exposure.

Human Health Risk: The likelihood that a given exposure or series of exposures may have damaged or will damage the health of individuals.

Hydraulic Conductivity: The rate at which water can move through a permeable medium. (i.e. the coefficient of permeability.)

Hydraulic Gradient: In general, the direction of groundwater flow due to changes in the depth of the water table.

Hydrocarbons (HC): Chemical compounds that consist entirely of carbon and hydrogen.

Hydrogen Sulfide (H2S): Gas emitted during organic decomposition. Also a by-product of oil refining and burning. Smells like rotten eggs and, in heavy concentration, can kill or cause illness.

Hydrogeological Cycle: The natural process recycling water from the atmosphere down to (and through) the earth and back to the atmosphere again.

Hydrogeology: The geology of ground water, with particular emphasis on the chemistry and movement of water.

Hydrologic Cycle: Movement or exchange of water between the atmosphere and earth.

Hydrology: The science dealing with the properties, distribution, and circulation of water.

Hydrolysis: The decomposition of organic compounds by interaction with water.

Hydronic: A ventilation system using heated or cooled water pumped through a building.

Hydrophilic: Having a strong affinity for water.

Hydrophobic: Having a strong aversion for water.

Hydropneumatic: A water system, usually small, in which a water pump is automatically controlled by the pressure in a compressed air tank.

Hypersensitivity Diseases: Diseases characterized by allergic responses to pollutants; diseases most clearly associated with indoor air quality are asthma, rhinitis, and pneumonic hypersensitivity.

Hypolimnion: Bottom waters of a thermally stratified lake. The hypolimnion of a eutrophic lake is usually low or lacking in oxygen.

Hypoxia/Hypoxic Waters: Waters with dissolved oxygen concentrations of less than 2 ppm, the level generally accepted as the minimum required for most marine life to survive and reproduce.

I

Identification Code or EPA I.D. Number: The unique code assigned to each generator, transporter, and treatment, storage, or disposal facility by regulating agencies to facilitate identification and tracking of chemicals or hazardous waste.

Ignitable: Capable of burning or causing a fire.

IM240: A high-tech, transient dynamometer automobile emissions test that takes up to 240 seconds.

Imhoff Cone: A clear, cone-shaped container used to measure the volume of settleable solids in a specific volume of water.

Immediately Dangerous to Life and Health (IDLH): The maximum level to which a healthy individual can be exposed to a chemical for 30 minutes and escape without suffering irreversible health effects or impairing symptoms. Used as a "level of concern." (See: level of concern.)

Imminent Hazard: One that would likely result in unreasonable adverse effects on humans or the environment or risk unreasonable hazard to an endangered species during the time required for a pesticide registration cancellation proceeding.

Imminent Threat: A high probability that exposure is occurring.

Immiscibility: The inability of two or more substances or liquids to readily dissolve into one another, such as soil and water. Immiscibility The inability of two or more substances or liquids to readily dissolve into one another, such as soil and water.

Impermeable: Not easily penetrated. The property of a material or soil that does not allow, or allows only with great difficulty, the movement or passage of water.

Imports: Municipal solid waste and recyclables that have been transported to a state or locality for processing or final disposition (but that did not originate in that state or locality).

Impoundment: A body of water or sludge confined by a dam, dike, floodgate, or other barrier.

In Situ: In its original place; unmoved unexcavated; remaining at the site or in the subsurface.

In-Line Filtration: Pre-treattment method in which chemicals are mixed by the flowing water; commonly used in pressure filtration installations. Eliminates need for flocculation and sedimentation.

In-Situ Flushing: Introduction of large volumes of water, at times supplemented with cleaning compounds, into soil, waste, or ground water to flush hazardous contaminants from a site.

In-Situ Oxidation: Technology that oxidizes contaminants dissolved in ground water, converting them into insoluble compounds.

In-Situ Stripping: Treatment system that removes or "strips" volatile organic compounds from contaminated ground or surface water by forcing an airstream through the water and causing the compounds to evaporate.

In-Situ Vitrification: Technology that treats contaminated soil in place at extremely high temperatures, at or more than 3000 degrees Fahrenheit.

In Vitro: Testing or action outside an organism (e.g. inside a test tube or culture dish.)

In Vivo: Testing or action inside an organism.

Incident Command Post: A facility located at a safe distance from an emergency site, where the incident commander, key staff, and technical representatives can make decisions and deploy emergency manpower and equipment.

Incident Command System (ICS): The organizational arrangement wherein one person, normally the Fire Chief of the impacted district, is in charge of an integrated, comprehensive emergency response organization and the emergency incident site, backed by an Emergency Operations Center staff with resources, information, and advice.

Incineration: A treatment technology involving destruction of waste by controlled burning at high temperatures; e.g., burning sludge to remove the water and reduce the remaining residues to a safe, non-burnable ash that can be disposed of safely on land, in some waters, or in underground locations.

Incineration at Sea: Disposal of waste by burning at sea on specially-designed incinerator ships.

Incinerator: A furnace for burning waste under controlled conditions.

Incompatible Waste: A waste unsuitable for mixing with another waste or material because it may react to form a hazard.

Indemnification: In the pesticide program, legal requirement that EPA pay certain end-users, dealers, and distributors for the cost of stock on hand at the time a pesticide registration is suspended.

Indicator: In biology, any biological entity or processes, or community whose characteristics show the presence of specific environmental conditions. 2. In chemistry, a substance that shows a visible change, usually of color, at a desired point in a chemical reaction. 3.A device that indicates the result of a measurement; e.g. a pressure gauge or a moveable scale.

Indirect Discharge: Introduction of pollutants from a non-domestic source into a publicly owned waste-treatment system. Indirect dischargers can be commercial or industrial facilities whose wastes enter local sewers.

Indirect Source: Any facility or building, property, road or parking area that attracts motor vehicle traffic and, indirectly, causes pollution.

Indoor Air: The breathable air inside a habitable structure or conveyance.

Indoor Air Pollution: Chemical, physical, or biological contaminants in indoor air.

Indoor Climate: Temperature, humidity, lighting, air flow and noise levels in a habitable structure or conveyance. Indoor climate can affect indoor air pollution.

Industrial Pollution Prevention: Combination of industrial source reduction and toxic chemical use substitution.

Industrial Process Waste: Residues produced during manufacturing operations.

Industrial Sludge: Semi-liquid residue or slurry remaining from treatment of industrial water and wastewater.

Industrial Source Reduction: Practices that reduce the amount of any hazardous substance, pollutant, or contaminant entering any waste stream or otherwise released into the environment. Also reduces the threat to public health and the environment associated with such releases. Term includes equipment or technology modifications, substitution of raw materials, and improvements in housekeeping, maintenance, training or inventory control.

Industrial Waste: Unwanted materials from an industrial operation; may be liquid, sludge, solid, or hazardous waste.

Inert Ingredient: Pesticide components such as solvents, carriers, dispersants, and surfactants that are not active against target pests. Not all inert ingredients are innocuous.

Inertial Separator: A device that uses centrifugal force to separate waste particles.

Glossary of Terms

Infectious Agent: Any organism, such as a pathogenic virus, parasite, or or bacterium, that is capable of invading body tissues, multiplying, and causing disease.

Infectious Waste: Hazardous waste capable of causing infections in humans, including: contaminated animal waste; human blood and blood products; isolation waste, pathological waste; and discarded sharps (needles, scalpels or broken medical instruments).

Infiltration: 1. The penetration of water through the ground surface into sub-surface soil or the penetration of water from the soil into sewer or other pipes through defective joints, connections, or manhole walls. 2. The technique of applying large volumes of waste water to land to penetrate the surface and percolate through the underlying soil. (See: percolation.)

Infiltration Gallery: A sub-surface groundwater collection system, typically shallow in depth, constructed with open-jointed or perforated pipes that discharge collected water into a watertight chamber from which the water is pumped to treatment facilities and into the distribution system. Usually located close to streams or ponds.

Infiltration Rate: The quantity of water that can enter the soil in a specified time interval.

Inflow: Entry of extraneous rain water into a sewer system from sources other than infiltration, such as basement drains, manholes, storm drains, and street washing.

Influent: Water, wastewater, or other liquid flowing into a reservoir, basin, or treatment plant.

Information Collection Request (ICR): A description of information to be gathered in connection with rules, proposed rules, surveys, and guidance documents that contain information-gathering requirements. The ICR describes what information is needed, why it is needed, how it will be collected, and how much collecting it will cost. The ICR is submitted by the EPA to the Office of Management and Budget (OMB) for approval.

Information File: In the Superfund program, a file that contains accurate, up-to-date documents on a Superfund site. The file is usually located in a public building (school, library, or city hall) convenient for local residents.

Inhalable Particles: All dust capable of entering the human respiratory tract.

Initial Compliance Period (Water): The first full three-year compliance period which begins at least 18 months after promulgation.

Injection Well: A well into which fluids are injected for purposes such as waste disposal, improving the recovery of crude oil, or solution mining.

Injection Zone: A geological formation receiving fluids through a well.

Innovative Technologies: New or inventive methods to treat effectively hazardous waste and reduce risks to human health and the environment.

Innovative Treatment Technologies: Technologies whose routine use is inhibited by lack of data on performance and cost. (See: Established treatment technologies.)

Inoculum: 1. Bacteria or fungi injected into compost to start biological action. 2. A medium containing organisms, usually bacteria or a virus, that is introduced into cultures or living organisms.

Inorganic Chemicals: Chemical substances of mineral origin, not of basically carbon structure.

Insecticide: A pesticide compound specifically used to kill or prevent the growth of insects.

Inspection and Maintenance (I/M): 1. Activities to ensure that vehicles' emission controls work properly. 2. Also applies to wastewater treatment plants and other anti-pollution facilities and processes.

Institutional Waste: Waste generated at institutions such as schools, libraries, hospitals, prisons, etc.

Instream Use: Water use taking place within a stream channel; e.g., hydro-electric power generation, navigation, water quality improvement, fish propagation, recreation.

Integrated Exposure Assessment: Cumulative summation (over time) of the magnitude of exposure to a toxic chemical in all media.

Integrated Pest Management (IPM): A mixture of chemical and other, non-pesticide, methods to control pests.

Integrated Waste Management: Using a variety of practices to handle municipal solid waste; can include source reduction, recycling, incineration, and landfilling.

Interceptor Sewers: Large sewer lines that, in a combined system, control the flow of sewage to the treatment plant. In a storm, they allow some of the sewage to flow directly into a receiving stream, thus keeping it from overflowing onto the streets. Also used in separate systems to collect the flows from main and trunk sewers and carry them to treatment points.

Interface: The common boundary between two substances such as a water and a solid, water and a gas, or two liquids such as water and oil.

Interfacial Tension: The strength of the film separating two immiscible fluids (e.g. oil and water) measured in dynes per, or millidynes per centimeter.

Interim (Permit) Status: Period during which treatment, storage and disposal facilities coming under RCRA in 1980 are temporarily permitted to operate while awaiting a permanent permit. Permits issued under these circumstances are usually called "Part A" or "Part B" permits.

Internal Dose: In exposure assessment, the amount of a substance penetrating the absorption barriers (e.g. skin, lung tissue, gastrointestinal tract) of an organism through either physical or biological processes. (See: absorbed dose)

Interstate Carrier Water Supply: A source of water for drinking and sanitary use on planes, buses, trains, and ships operating in more than one state. These sources are federally regulated.

Interstate Commerce Clause: A clause of the U.S. Constitution which reserves to the federal government the right to regulate the conduct of business across state lines. Under this clause, for example, the U.S. Supreme Court has ruled that states may not inequitably restrict the disposal of out-of-state wastes in their jurisdictions.

Interstate Waters: Waters that flow across or form part of state or international boundaries; e.g. the Great Lakes, the Mississippi River, or coastal waters.

Interstitial Monitoring: The continuous surveillance of the space between the walls of an underground storage tank.

Intrastate Product: Pesticide products once registered by states for sale and use only in the state. All intrastate products have been converted to full federal registration or canceled.

Inventory (TSCA): Inventory of chemicals produced pursuant to Section 8 (b) of the Toxic Substances Control Act.

Inversion: A layer of warm air that prevents the rise of cooling air and traps pollutants beneath it; can cause an air pollution episode.

Ion: An electrically charged atom or group of atoms.

Ion Exchange Treatment: A common water-softening method often found on a large scale at water purification plants that remove some organics and radium by adding calcium oxide or calcium hydroxide to increase the pH to a level where the metals will precipitate out.

Ionization Chamber: A device that measures the intensity of ionizing radiation.

Ionizing Radiation: Radiation that can strip electrons from atoms; e.g. alpha, beta, and gamma radiation.

IRIS: EPA's Integrated Risk Information System, an electronic data base containing the Agency's latest descriptive and quantitative regulatory information on chemical constituents.

Irradiated Food: Food subject to brief radioactivity, usually gamma rays, to kill insects, bacteria, and mold, and to permit storage without refrigeration.

Irradiation: Exposure to radiation of wavelengths shorter than those of visible light (gamma, x-ray, or ultra-violet), for medical purposes, to sterilize milk or other foodstuffs, or to induce polymerization of monomers or vulcanization of rubber.

Irreversible Effect: Effect characterized by the inability of the body to partially or fully repair injury caused by a toxic agent.

Irrigation: Applying water or wastewater to land areas to supply the water and nutrient needs of plants.

Irrigation Efficiency: The amount of water stored in the crop root zone compared to the amount of irrigation water applied.

Irrigation Return Flow: Surface and subsurface water which leaves the field following application of irrigation water.

Irritant: A substance that can cause irritation of the skin, eyes, or respiratory system. Effects may be acute from a single high level exposure, or chronic from repeated low-level exposures to such compounds as chlorine, nitrogen dioxide, and nitric acid.

Isoconcentration: More than one sample point exhibiting the same isolate concentration.

Isopleth: The line or area represented by an isoconcentration.

Isotope: A variation of an element that has the same atomic number of protons but a different weight because of the number of neutrons. Various isotopes of the same element may have different radioactive behaviors, some are highly unstable..

Isotropy: The condition in which the hydraulic or other properties of an aquifer are the same in all directions.

J

Jar Test: A laboratory procedure that simulates a water treatment plant's coagulation/flocculation units with differing chemical doses, mix speeds, and settling times to estimate the minimum or ideal coagulant dose required to achieve certain water quality goals.

Joint and Several Liability: Under CERCLA, this legal concept relates to the liability for Superfund site cleanup and other costs on the part of more than one potentially responsible party (i.e. if there were several owners or users of a site that became contaminated over the years, they could all be considered potentially liable for cleaning up the site.)

K

Karst: A geologic formation of irregular limestone deposits with sinks, underground streams, and caverns.

Kinetic Energy: Energy possessed by a moving object or water body.

Kinetic Rate Coefficient: A number that describes the rate at which a water constituent such as a biochemical oxygen demand or dissolved oxygen rises or falls, or at which an air pollutant reacts.

L

Laboratory Animal Studies: Investigations using animals as surrogates for humans.

Lagoon: 1. A shallow pond where sunlight, bacterial action, and oxygen work to purify wastewater; also used for storage of wastewater or spent nuclear fuel rods. 2. Shallow body of water, often separated from the sea by coral reefs or sandbars.

Land Application: Discharge of wastewater onto the ground for treatment or reuse. (See: irrigation.)

Land Ban: Phasing out of land disposal of most untreated hazardous wastes, as mandated by the 1984 RCRA amendments.

Land Disposal Restrictions: Rules that require hazardous wastes to be treated before disposal on land to destroy or immobilize hazardous constituents that might migrate into soil and ground water.

Land Farming (of Waste): A disposal process in which hazardous waste deposited on or in the soil is degraded naturally by microbes.

Landfills: 1. Sanitary landfills are disposal sites for non-hazardous solid wastes spread in layers, compacted to the smallest practical volume, and covered by material applied at the end of each operating day. 2. Secure chemical landfills are disposal sites for hazardous waste, selected and designed to minimize the chance of release of hazardous substances into the environment.

Landscape: The traits, patterns, and structure of a specific geographic area, including its biological composition, its physical environment, and its anthropogenic or social patterns. An area where interacting ecosystems are grouped and repeated in similar form.

Landscape Characterization: Documentation of the traits and patterns of the essential elements of the landscape.

Landscape Ecology: The study of the distribution patterns of communities and ecosystems, the ecological processes that affect those patterns, and changes in pattern and process over time.

Landscape Indicator: A measurement of the landscape, calculated from mapped or remotely sensed data, used to describe spatial patterns of land use and land cover across a geographic area. Landscape indicators may be useful as measures of certain kinds of environmental degradation such as forest fragmentation.

Langelier Index (LI): An index reflecting the equilibrium pH of a water with respect to calcium and alkalinity; used in stabilizing water to control both corrosion and scale deposition.

Large Quantity Generator: Person or facility generating more than 2200 pounds of hazardous waste per month. Such generators produce about 90 percent of the nation's hazardous waste, and are subject to all RCRA requirements.

Large Water System: A water system that services more than 50,000 customers.

Laser Induced Fluorescence: A method for measuring the relative amount of soil and/or groundwater with an in-situ sensor.

Latency: Time from the first exposure of a chemical until the appearance of a toxic effect.

Lateral Sewers: Pipes that run under city streets and receive the sewage from homes and businesses, as opposed to domestic feeders and main trunk lines.

Laundering Weir: Sedimention basin overflow weir.

LC 50/Lethal Concentration: Median level concentration, a standard measure of toxicity. It tells how much of a substance is needed to kill half of a group of experimental organisms in a given time. (See: LD 50.)

LD 50/ Lethal Dose: The dose of a toxicant or microbe that will kill 50 percent of the test organisms within a designated period. The lower the LD 50, the more toxic the compound.

Ldlo: Lethal dose low; the lowest dose in an animal study at which lethality occurs.

Leachate: Water that collects contaminants as it trickles through wastes, pesticides or fertilizers. Leaching may occur in farming areas, feedlots, and landfills, and may result in hazardous substances entering surface water, ground water, or soil.

Leachate Collection System: A system that gathers leachate and pumps it to the surface for treatment.

Leaching: The process by which soluble constituents are dissolved and filtered through the soil by a percolating fluid. (See: leachate.)

Lead (Pb): A heavy metal that is hazardous to health if breathed or swallowed. Its use in gasoline, paints, and plumbing compounds has been sharply restricted or eliminated by federal laws and regulations. (See: heavy metals.)

Lead Service Line: A service line made of lead which connects the water to the building inlet and any lead fitting connected to it.

Legionella: A genus of bacteria, some species of which have caused a type of pneumonia called Legionaires Disease.

Level of Concern (LOC): The concentration in air of an extremely hazardous substance above which there may be serious immediate health effects to anyone exposed to it for short periods

Life Cycle of a Product: All stages of a product's development, from extraction of fuel for power to production, marketing, use, and disposal.

Lifetime Average Daily Dose: Figure for estimating excess lifetime cancer risk.

Lifetime Exposure: Total amount of exposure to a substance that a human would receive in a lifetime (usually assumed to be 70 years).

Lift: In a sanitary landfill, a compacted layer of solid waste and the top layer of cover material.

Lifting Station: (See: pumping station.)

Light Non-Aqueous Phase Liquid (LNAPL): A non-aqueous phase liquid with a specific gravity less than 1.0. Because the specific gravity of water is 1.0, most LNAPLs float on top of the water table. Most common petroleum hydrocarbon fuels and lubricating oils are LNAPLs.

Light-Emitting Diode: A long-lasting illumination technology used for exit signs which requires very little power

Limestone Scrubbing: Use of a limestone and water solution to remove gaseous stack-pipe sulfur before it reaches the atmosphere.

Limit of Detection (LOD): The minimum concentration of a substance being analyzed test that has a 99 percent probability of being identified.

Limited Degradation: An environmental policy permitting some degradation of natural systems but terminating at a level well beneath an established health standard.

Limiting Factor: A condition whose absence or excessive concentration, is incompatible with the needs or tolerance of a species or population and which may have a negative influence on their ability to thrive.

Limnology: The study of the physical, chemical, hydrological, and biological aspects of fresh water bodies.

Lindane: A pesticide that causes adverse health effects in domestic water supplies and is toxic to freshwater fish and aquatic life.

Liner: 1. A relatively impermeable barrier designed to keep leachate inside a landfill. Liner materials include plastic and dense clay. 2. An insert or sleeve for sewer pipes to prevent leakage or infiltration.

Lipid Solubility: The maximum concentration of a chemical that will dissolve in fatty substances. Lipid soluble substances are insoluble in water. They will very selectively disperse through the environment via uptake in living tissue.

Glossary of Terms

Liquefaction: Changing a solid into a liquid.

Liquid Injection Incinerator: Commonly used system that relies on high pressure to prepare liquid wastes for incineration by breaking them up into tiny droplets to allow easier combustion.

List: Shorthand term for EPA list of violating facilities or firms debarred from obtaining government contracts because they violated certain sections of the Clean Air or Clean Water Acts. The list is maintained by The Office of Enforcement and Compliance Monitoring.

Listed Waste: Wastes listed as hazardous under RCRA but which have not been subjected to the Toxic Characteristics Listing Process because the dangers they present are considered self-evident.

Lithology: Mineralogy, grain size, texture, and other physical properties of granular soil, sediment, or rock.

Litter: 1. The highly visible portion of solid waste carelessly discarded outside the regular garbage and trash collection and disposal system. 2. leaves and twigs fallen from forest trees.

Littoral Zone: 1. That portion of a body of fresh water extending from the shoreline lakeward to the limit of occupancy of rooted plants. 2. A strip of land along the shoreline between the high and low water levels.

Local Education Agency (LEA): In the asbestos program, an educational agency at the local level that exists primarily to operate schools or to contract for educational services, including primary and secondary public and private schools. A single, unaffiliated school can be considered an LEA for AHERA purposes.

Local Emergency Planning Committee (LEPC): A committee appointed by the state emergency response commission, as required by SARA Title III, to formulate a comprehensive emergency plan for its jurisdiction.

Low Density Polyethylene (LOPE): Plastic material used for both rigid containers and plastic film applications.

Low Emissivity (low-E) Windows: New window technology that lowers the amount of energy loss through windows by inhibiting the transmission of radiant heat while still allowing sufficient light to pass through.

Low NOx Burners: One of several combustion technologies used to reduce emissions of Nitrogen Oxides (NOx.)

Low-Level Radioactive Waste (LLRW): Wastes less hazardous than most of those associated with a nuclear reactor; generated by hospitals, research laboratories, and certain industries. The Department of Energy, Nuclear Regulatory Commission, and EPA share responsibilities for managing them. (See: high-level radioactive wastes.)

Lower Detection Limit: The smallest signal above background noise an instrument can reliably detect.

Lower Explosive Limit (LEL): The concentration of a compound in air below which the mixture will not catch on fire.

Lowest Acceptable Daily Dose: The largest quantity of a chemical that will not cause a toxic effect, as determined by animal studies.

Lowest Achievable Emission Rate: Under the Clean Air Act, the rate of emissions that reflects (1) the most stringent emission limitation in the implementation plan of any state for such source unless the owner or operator demonstrates such limitations are not achievable; or (2) the most stringent emissions limitation achieved in practice, whichever is more stringent. A proposed new or modified source may not emit pollutants in excess of existing new source standards.

Lowest Observed Adverse Effect Level (LOAEL): The lowest level of a stressor that causes statistically and biologically significant differences in test samples as compared to other samples subjected to no stressor.

M

Macropores: Secondary soil features such as root holes or desiccation cracks that can create significant conduits for movement of NAPL and dissolved contaminants, or vapor-phase contaminants.

Magnetic Separation: Use of magnets to separate ferrous materials from mixed municipal waste stream.

Major Modification: This term is used to define modifications of major stationary sources of emissions with respect to Prevention of Significant Deterioration and New Source Review under the Clean Air Act.

Major Stationary Sources: Term used to determine the applicability of Prevention of Significant Deterioration and new source regulations. In a nonattainment area, any stationary pollutant source with potential to emit more than 100 tons per year is considered a major stationary source. In PSD areas the cutoff level may be either 100 or 250 tons, depending upon the source.

Majors: Larger publicly owned treatment works (POTWs) with flows equal to at least one million gallons per day (mgd) or servicing a population equivalent to 10,000 persons; certain other POTWs having significant water quality impacts. (See: minors.)

Man-Made (Anthropogenic) Beta Particle and Photon Emitters: All radionuclides emitting beta particles and/or photons listed in Maximum Permissible Body Burdens and Maximum Permissible Concentrations of Radonuclides in Air and Water for Occupational Exposure.

Management Plan: Under the Asbestos Hazard Emergency Response Act (AHERA), a document that each Local Education Agency is required to prepare, describing all activities planned and undertaken by a school to comply with AHERA regulations, including building inspections to identify asbestos-containing materials, response actions, and operations and maintenance programs to minimize the risk of exposure.

Managerial Controls: Methods of nonpoint source pollution control based on decisions about managing agricultural wastes or application times or rates for agrochemicals.

Mandatory Recycling: Programs which by law require consumers to separate trash so that some or all recyclable materials are recovered for recycling rather than going to landfills.

Manifest: A one-page form used by haulers transporting waste that lists EPA identification numbers, type and quantity of waste, the generator it originated from, the transporter that shipped it, and the storage or disposal facility to which it is being shipped. It includes copies for all participants in the shipping process.

Manifest System: Tracking of hazardous waste from "cradle-to-grave" (generation through disposal) with accompanying documents known as manifests.(See: cradle to grave.)

Manual Separation: Hand sorting of recyclable or compostable materials in waste.

Manufacturer's Formulation: A list of substances or component parts as described by the maker of a coating, pesticide, or other product containing chemicals or other substances.

Manufacturing Use Product: Any product intended (labeled) for formulation or repackaging into other pesticide products.

Margin of Safety: Maximum amount of exposure producing no measurable effect in animals (or studied humans) divided by the actual amount of human exposure in a population.

Margin of Exposure (MOE): The ratio of the no-observed adverse-effect-level to the estimated exposure dose.

Marine Sanitation Device: Any equipment or process installed on board a vessel to receive, retain, treat, or discharge sewage.

Marsh: A type of wetland that does not accumulate appreciable peat deposits and is dominated by herbaceous vegetation. Marshes may be either fresh or saltwater, tidal or non-tidal. (See: wetlands.)

Material Category: In the asbestos program, broad classification of materials into thermal surfacing insulation, surfacing material, and miscellaneous material.

Material Safety Data Sheet (MSDS): A compilation of information required under the OSHA Communication Standard on the identity of hazardous chemicals, health, and physical hazards, exposure limits, and precautions. Section 311 of SARA requires facilities to submit MSDSs under certain circumstances.

Material Type: Classification of suspect material by its specific use or application; e.g., pipe insulation, fireproofing, and floor tile.

Materials Recovery Facility (MRF): A facility that processes residentially collected mixed recyclables into new products available for market.

Maximally (or Most) Exposed Individual: The person with the highest exposure in a given population.

Maximum Acceptable Toxic Concentration: For a given ecological effects test, the range (or geometric mean) between the No Observable Adverse Effect Level and the Lowest Observable Adverse Effects Level.

Maximum Available Control Technology (MACT): The emission standard for sources of air pollution requiring the maximum reduction of hazardous emissions, taking cost and feasibility into account. Under the Clean Air Act Amendments of 1990, the MACT must not be less than the average emission level achieved by controls on the best performing 12 percent of existing sources, by category of industrial and utility sources.

Maximum Contaminant Level: The maximum permissible level of a contaminant in water delivered to any user of a public system. MCLs are enforceable standards.

Maximum Contaminant Level Goal (MCLG): Under the Safe Drinking Water Act, a non-enforceable concentration of a drinking water contaminant, set at the level at which no known or anticipated adverse effects on human health occur and which allows an adequate safety margin. The MCLG is usually the starting point for determining the regulated Maximum Contaminant Level. (See: maximum contaminant level.)

Maximum Exposure Range: Estimate of exposure or dose level received by an individual in a defined population that is greater than the 98th percentile dose for all individuals in that population, but less than the exposure level received by the person receiving the highest exposure level.

Maximum Residue Level: Comparable to a U.S. tolerance level, the Maximum Residue Level the enforceable limit on food pesticide levels in some countries. Levels are set by the Codex Alimentarius Commission, a United Nations agency managed and funded jointly by the World Health Organization and the Food and Agriculture Organization.

Maximum Tolerated Dose: The maximum dose that an animal species can tolerate for a major portion of its lifetime without significant impairment or toxic effect other than carcinogenicity.

Measure of Effect/ Measurement Endpoint: A measurable characteristic of ecological entity that can be related to an assessment endpoint; e.g. a laboratory test for eight species meeting certain requirements may serve as a measure of effect for an assessment endpoint, such as survival of fish, aquatic, invertebrate or algal species under acute exposure.

Measure of Exposure: A measurable characteristic of a stressor (such as the specific amount of mercury in a body of water) used to help quantify the exposure of an ecological entity or individual organism.

Mechanical Aeration: Use of mechanical energy to inject air into water to cause a waste stream to absorb oxygen.

Mechanical Separation: Using mechanical means to separate waste into various components.

Mechanical Turbulence: Random irregularities of fluid motion in air caused by buildings or other nonthermal, processes.

Media: Specific environments—air, water, soil—which are the subject of regulatory concern and activities.

Medical Surveillance: A periodic comprehensive review of a worker's health status; acceptable elements of such surveillance program are listed in the Occupational Safety and Health Administration standards for asbestos.

Medical Waste: Any solid waste generated in the diagnosis, treatment, or immunization of human beings or animals, in research pertaining thereto, or in the production or testing of biologicals, excluding hazardous waste identified or listed under 40 CFR Part 261 or any household waste as defined in 40 CFR Sub-section 261.4 (b)(1).

Medium-size Water System: A water system that serves 3,300 to 50,000 customers.

Meniscus: The curved top of a column of liquid in a small tube.

Mercury (Hg): Heavy metal that can accumulate in the environment and is highly toxic if breathed or swallowed. (See:heavy metals.)

Mesotrophic: Reservoirs and lakes which contain moderate quantities of nutrients and are moderately productive in terms of aquatic animal and plant life.

Metabolites: Any substances produced by biological processes, such as those from pesticides.

Metalimnion: The middle layer of a thermally stratified lake or reservoir. In this layer there is a rapid decrease in temperature with depth. Also called thermocline.

Methane: A colorless, nonpoisonous, flammable gas created by anaerobic decomposition of organic compounds. A major component of natural gas used in the home.

Methanol: An alcohol that can be used as an alternative fuel or as a gasoline additive. It is less volatile than gasoline; when blended with gasoline it lowers the carbon monoxide emissions but increases hydrocarbon emissions. Used as pure fuel, its emissions are less ozone-forming than those from gasoline. Poisonous to humans and animals if ingested.

Method 18: An EPA test method which uses gas chromatographic techniques to measure the concentration of volatile organic compounds in a gas stream.

Method 24: An EPA reference method to determine density, water content and total volatile content (water and VOC) of coatings.

Method 25: An EPA reference method to determine the VOC concentration in a gas stream.

Method Detection Limit (MDL): See limit of detection.

Methoxychlor: Pesticide that causes adverse health effects in domestic water supplies and is toxic to freshwater and marine aquatic life.

Methyl Orange Alkalinity: A measure of the total alkalinity in a water sample in which the color of methyl orange reflects the change in level.

Microbial Growth: The amplification or multiplication of microorganisms such as bacteria, algae, diatoms, plankton, and fungi.

Microbial Pesticide: A microorganism that is used to kill a pest, but is of minimum toxicity to humans.

Microclimate: 1. Localized climate conditions within an urban area or neighborhood. 2. The climate around a tree or shrub or a stand of trees.

Microenvironmental Method: A method for sequentially assessing exposure for a series of microenvironments that can be approximated by constant concentrations of a stressor.

Microenvironments: Well-defined surroundings such as the home, office, or kitchen that can be treated as uniform in terms of stressor concentration.

Million-Gallons Per Day (MGD): A measure of water flow.

Minimization: A comprehensive program to minimize or eliminate wastes, usually applied to wastes at their point of origin. (See: waste minimization.)

Mining of an Aquifer: Withdrawal over a period of time of ground water that exceeds the rate of recharge of the aquifer.

Mining Waste: Residues resulting from the extraction of raw materials from the earth.

Minor Source: New emissions sources or modifications to existing emissions sources that do not exceed NAAQS emission levels.

Minors: Publicly owned treatment works with flows less than 1 million gallons per day. (See: majors.)

Miscellaneous ACM: Interior asbestos-containing building material or structural components, members or fixtures, such as floor and ceiling tiles; does not include surfacing materials or thermal system insulation.

Miscellaneous Materials: Interior building materials on structural components, such as floor or ceiling tiles.

Miscible Liquids: Two or more liquids that can be mixed and will remain mixed under normal conditions.

Missed Detection: The situation that occurs when a test indicates that a tank is "tight" when in fact it is leaking.

Mist: Liquid particles measuring 40 to 500 micrometers (pm), are formed by condensation of vapor. By comparison, fog particles are smaller than 40 micrometers (pm).

Mitigation: Measures taken to reduce adverse impacts on the environment.

Mixed Funding: Settlements in which potentially responsible parties and EPA share the cost of a response action.

Mixed Glass: Recovered container glass not sorted into categories (e.g. color, grade).

Mixed Liquor: A mixture of activated sludge and water containing organic matter undergoing activated sludge treatment in an aeration tank.

Mixed Metals: Recovered metals not sorted into categories such as aluminum, tin, or steel cans or ferrous or non-ferrous metals.

Mixed Municipal Waste: Solid waste that has not been sorted into specific categories (such as plastic, glass, yard trimmings, etc.)

Mixed Paper: Recovered paper not sorted into categories such as old magazines, old newspapers, old corrugated boxes, etc.

Mixed Plastic: Recovered plastic unsorted by category.

Mobile Incinerator Systems: Hazardous waste incinerators that can be transported from one site to another.

Mobile Source: Any non-stationary source of air pollution such as cars, trucks, motorcycles, buses, airplanes, and locomotives.

Model Plant: A hypothetical plant design used for developing economic, environmental, and energy impact analyses as support for regulations or regulatory guidelines; first step in exploring the economic impact of a potential NSPS.

Modified Bin Method: Way of calculating the required heating or cooling for a building based on determining how much energy the system would use if outdoor temperatures were within a certain temperature interval and then multiplying the energy use by the time the temperature interval typically occurs.

Glossary of Terms

Modified Source: The enlargement of a major stationary pollutant sources is often referred to as modification, implying that more emissions will occur.

Moisture Content: 1.The amount of water lost from soil upon drying to a constant weight, expressed as the weight per unit of dry soil or as the volume of water per unit bulk volume of the soil. For a fully saturated medium, moisture content indicates the porosity. 2. Water equivalent of snow on the ground; an indicator of snowmelt flood potential.

Molecule: The smallest division of a compound that still retains or exhibits all the properties of the substance.

Molten Salt Reactor: A thermal treatment unit that rapidly heats waste in a heat-conducting fluid bath of carbonate salt.

Monitoring: Periodic or continuous surveillance or testing to determine the level of compliance with statutory requirements and/or pollutant levels in various media or in humans, plants, and animals.

Monitoring Well: 1. A well used to obtain water quality samples or measure groundwater levels. 2. A well drilled at a hazardous waste management facility or Superfund site to collect ground-water samples for the purpose of physical, chemical, or biological analysis to determine the amounts, types, and distribution of contaminants in the groundwater beneath the site.

Monoclonal Antibodies (Also called MABs and MCAs): 1. Man-made (anthropogenic) clones of a molecule, produced in quantity for medical or research purposes. 2. Molecules of living organisms that selectively find and attach to other molecules to which their structure conforms exactly. This could also apply to equivalent activity by chemical molecules.

Monomictic: Lakes and reservoirs which are relatively deep, do not freeze over during winter, and undergo a single stratification and mixing cycle during the year (usually in the fall).

Montreal Protocol: Treaty, signed in 1987, governs stratospheric ozone protection and research, and the production and use of ozone-depleting substances. It provides for the end of production of ozone-depleting substances such as CFCS. Under the Protocol, various research groups continue to assess the ozone layer. The Multilateral Fund provides resources to developing nations to promote the transition to ozone-safe technologies.

Moratorium: During the negotiation process, a period of 60 to 90 days during which EPA and potentially responsible parties may reach settlement but no site response activities can be conducted.

Morbidity: Rate of disease incidence.

Mortality: Death rate.

Most Probable Number: An estimate of microbial density per unit volume of water sample, based on probability theory.

Muck Soils: Earth made from decaying plant materials.

Mudballs: Round material that forms in filters and gradually increases in size when not removed by backwashing.

Mulch: A layer of material (wood chips, straw, leaves, etc.) placed around plants to hold moisture, prevent weed growth, and enrich or sterilize the soil.

Multi-Media Approach: Joint approach to several environmental media, such as air, water, and land.

Multiple Chemical Sensitivity: A diagnostic label for people who suffer multi-system illnesses as a result of contact with, or proximity to, a variety of airborne agents and other substances.

Multiple Use: Use of land for more than one purpose; e.g., grazing of livestock, watershed and wildlife protection, recreation, and timber production. Also applies to use of bodies of water for recreational purposes, fishing, and water supply.

Multistage Remote Sensing: A strategy for landscape characterization that involves gathering and analyzing information at several geographic scales, ranging from generalized levels of detail at the national level through high levels of detail at the local scale.

Municipal Discharge: Discharge of effluent from waste water treatment plants which receive waste water from households, commercial establishments, and industries in the coastal drainage basin. Combined sewer/separate storm overflows are included in this category.

Municipal Sewage: Wastes (mostly liquid) orginating from a community; may be composed of domestic wastewaters and/or industrial discharges.

Municipal Sludge: Semi-liquid residue remaining from the treatment of municipal water and wastewater.

Municipal Solid Waste: Common garbage or trash generated by industries, businesses, institutions, and homes.

Mutagen/Mutagenicity: An agent that causes a permanent genetic change in a cell other than that which occurs during normal growth. Mutagenicity is the capacity of a chemical or physical agent to cause such permanent changes.

N

National Ambient Air Quality Standards (NAAQS): Standards established by EPA that apply for outdoor air throughout the country. (See: criteria pollutants, state implementation plans, emissions trading.)

National Emissions Standards for Hazardous Air Pollutants (NESHAPS): Emissions standards set by EPA for an air pollutant not covered by NAAQS that may cause an increase in fatalities or in serious, irreversible, or incapacitating illness. Primary standards are designed to protect human health, secondary standards to protect public welfare (e.g. building facades, visibility, crops, and domestic animals).

National Environmental Performance Partnership Agreements: System that allows states to assume greater responsibility for environmental programs based on their relative ability to execute them.

National Estuary Program: A program established under the Clean Water Act Amendments of 1987 to develop and implement conservation and management plans for protecting estuaries and restoring and maintaining their chemical, physical, and biological integrity, as well as controlling point and nonpoint pollution sources.

National Municipal Plan: A policy created in 1984 by EPA and the states in 1984 to bring all publicly owned treatment works (POTWs) into compliance with Clean Water Act requirements.

National Oil and Hazardous Substances Contingency Plan (NOHSCP/NCP): The federal regulation that guides determination of the sites to be corrected under both the Superfund program and the program to prevent or control spills into surface waters or elsewhere.

National Pollutant Discharge Elimination System (NPDES): A provision of the Clean Water Act which prohibits discharge of pollutants into waters of the United States unless a special permit is issued by EPA, a state, or, where delegated, a tribal government on an Indian reservation.

National Priorities List (NPL): EPA's list of the most serious uncontrolled or abandoned hazardous waste sites identified for possible long-term remedial action under Superfund. The list is based primarily on the score a site receives from the Hazard Ranking System. EPA is required to update the NPL at least once a year. A site must be on the NPL to receive money from the Trust Fund for remedial action.

National Response Center: The federal operations center that receives notifications of all releases of oil and hazardous substances into the environment; open 24 hours a day, is operated by the U.S. Coast Guard, which evaluates all reports and notifies the appropriate agency.

National Response Team (NRT): Representatives of 13 federal agencies that, as a team, coordinate federal responses to nationally significant incidents of pollution—an oil spill, a major chemical release, or a - superfund response action—and provide advice and technical assistance to the responding agency(ies) before and during a response action.

National Secondary Drinking Water Regulations: Commonly referred to as NSDWRs.

Navigable Waters: Traditionally, waters sufficiently deep and wide for navigation by all, or specified vessels; such waters in the United States come under federal jurisdiction and are protected by certain provisions of the Clean Water Act.

Necrosis: Death of plant or animal cells or tissues. In plants, necrosis can discolor stems or leaves or kill a plant entirely.

Negotiations (Under Superfund): After potentially responsible parties are identified for a site, EPA coordinates with them to reach a settlement that will result in the PRP paying for or conducting the cleanup under EPA supervision. If negotiations fail, EPA can order the PRP to conduct the cleanup or EPA can pay for the cleanup using Superfund monies and then sue to recover the costs.

Nematocide: A chemical agent which is destructive to nematodes.

Nephelometric: Method of of measuring turbidity in a water sample by passing light through the sample and measuring the amount of the light that is deflected.

Netting: A concept in which all emissions sources in the same area that owned or controlled by a single company are treated as one large source, thereby allowing flexibility in controlling individual sources in order to meet a single emissions standard. (See: bubble.)

Neutralization: Decreasing the acidity or alkalinity of a substance by adding alkaline or acidic materials, respectively.

New Source: Any stationary source built or modified after publication of final or proposed regulations that prescribe a given standard of performance.

New Source Performance Standards (NSPS): Uniform national EPA air emission and water effluent standards which limit the amount of pollution allowed from new sources or from modified existing sources.

New Source Review (NSR): A Clean Air Act requirement that State Implementation Plans must include a permit review that applies to the construction and operation of new and modified stationary sources in nonattainment areas to ensure attainment of national ambient air quality standards.

Nitrate: A compound containing nitrogen that can exist in the atmosphere or as a dissolved gas in water and which can have harmful effects on humans and animals. Nitrates in water can cause severe illness in infants and domestic animals. A plant nutrient and inorganic fertilizer, nitrate is found in septic systems, animal feed lots, agricultural fertilizers, manure, industrial waste waters, sanitary landfills, and garbage dumps.

Nitric Oxide (NO): A gas formed by combustion under high temperature and high pressure in an internal combustion engine; it is converted by sunlight and photochemical processes in ambient air to nitrogen oxide. NO is a precursor of ground-level ozone pollution, or smog..

Nitrification: The process whereby ammonia in wastewater is oxidized to nitrite and then to nitrate by bacterial or chemical reactions.

Nitrilotriacetic Acid (NTA): A compound now replacing phosphates in detergents.

Nitrite: 1. An intermediate in the process of nitrification. 2. Nitrous oxide salts used in food preservation.

Nitrogen Dioxide (NO2): The result of nitric oxide combining with oxygen in the atmosphere; major component of photochemical smog.

Nitrogen Oxide (NOx): The result of photochemical reactions of nitric oxide in ambient air; major component of photochemical smog. Product of combustion from transportation and stationary sources and a major contributor to the formation of ozone in the troposphere and to acid deposition.

Nitrogenous Wastes: Animal or vegetable residues that contain significant amounts of nitrogen.

Nitrophenols: Synthetic organopesticides containing carbon, hydrogen, nitrogen, and oxygen.

No Further Remedial Action Planned: Determination made by EPA following a preliminary assessment that a site does not pose a significant risk and so requires no further activity under CERCLA.

No Observable Adverse Effect Level (NOAEL): An exposure level at which there are no statistically or biologically significant increases in the frequency or severity of adverse effects between the exposed population and its appropriate control; some effects may be produced at this level, but they are not considered as adverse, or as precurors to adverse effects. In an experiment with several NOAELs, the regulatory focus is primarily on the highest one, leading to the common usage of the term NOAEL as the highest exposure without adverse effects.

No Till: Planting crops without prior seedbed preparation, into an existing cover crop, sod, or crop residues, and eliminating subsequent tillage operations.

No-Observed-Effect-Level (NOEL): Exposure level at which there are no statistically or biological significant differences in the frequency or severity of any effect in the exposed or control populations.

Noble Metal: Chemically inactive metal such as gold; does not corrode easily.

Noise: Product-level or product-volume changes occurring during a test that are not related to a leak but may be mistaken for one.

Non-Aqueous Phase Liquid (NAPL): Contaminants that remain undiluted as the original bulk liquid in the subsurface, e.g. spilled oil. (See: fee product.)

Non-Attainment Area: Area that does not meet one or more of the National Ambient Air Quality Standards for the criteria pollutants designated in the Clean Air Act.

Non-Binding Allocations of Responsibility (NBAR): A process for EPA to propose a way for potentially responsible parties to allocate costs among themselves.

Non-Community Water System: A public water system that is not a community water system; e.g. the water supply at a camp site or national park.

Non-Compliance Coal: Any coal that emits greater than 3.0 pounds of sulfur dioxide per million BTU when burned. Also known as high-sulfur coal.

Non-Contact Cooling Water: Water used for cooling which does not come into direct contact with any raw material, product, byproduct, or waste.

Non-Conventional Pollutant: Any pollutant not statutorily listed or which is poorly understood by the scientific community.

Non-Degradation: An environmental policy which disallows any lowering of naturally occurring quality regardless of preestablished health standards.

Non-Ferrous Metals: Nonmagnetic metals such as aluminum, lead, and copper. Products made all or in part from such metals include containers, packaging, appliances, furniture, electronic equipment and aluminum foil.

Non-ionizing Electromagnetic Radiation: 1. Radiation that does not change the structure of atoms but does heat tissue and may cause harmful biological effects. 2. Microwaves, radio waves, and low-frequency electromagnetic fields from high-voltage transmission lines.

Non-Methane Hydrocarbon (NMHC): The sum of all hydrocarbon air pollutants except methane; significant precursors to ozone formation.

Non-Methane Organic Gases (NMOG): The sum of all organic air pollutants. Excluding methane; they account for aldehydes, ketones, alcohols, and other pollutants that are not hydrocarbons but are precursors of ozone.

Non-Point Sources: Diffuse pollution sources (i.e. without a single point of origin or not introduced into a receiving stream from a specific outlet). The pollutants are generally carried off the land by storm water. Common non-point sources are agriculture, forestry, urban, mining, construction, dams, channels, land disposal, saltwater intrusion, and city streets.

Non-potable: Water that is unsafe or unpalatable to drink because it contains pollutants, contaminants, minerals, or infective agents.

Non-Road Emissions: Pollutants emitted by combustion engines on farm and construction equipment, gasoline-powered lawn and garden equipment, and power boats and outboard motors.

Non-Transient Non-Community Water System: A public water system that regularly serves at least 25 of the same non-resident persons per day for more than six months per year.

Nondischarging Treatment Plant: A treatment plant that does not discharge treated wastewater into any stream or river. Most are pond systems that dispose of the total flow they receive by means of evaporation or percolation to groundwater, or facilities that dispose of their effluent by recycling or reuse (e.g. spray irrigation or groundwater discharge).

Nonfriable Asbestos-Containing Materials: Any material containing more than one percent asbestos (as determined by Polarized Light Microscopy) that, when dry, cannot be crumbled, pulverized, or reduced to powder by hand pressure.

Nonhazardous Industrial Waste: Industrial process waste in wastewater not considered municipal solid waste or hazardous waste under RARA.

Notice of Deficiency: An EPA request to a facility owner or operator requesting additional information before a preliminary decision on a permit application can be made.

Notice of Intent to Cancel: Notification sent to registrants when EPA decides to cancel registration of a product containing a pesticide.

Notice of Intent to Deny: Notification by EPA of its preliminary intent to deny a permit application.

Notice of Intent to Suspend: Notification sent to a pesticide registrant when EPA decides to suspend product sale and distribution because of failure to submit requested data in a timely and/or acceptable manner, or because of imminent hazard. (See: emergency suspension.)

Nuclear Reactors and Support Facilities: Uranium mills, commercial power reactors, fuel reprocessing plants, and uranium enrichment facilities.

Nuclear Winter: Prediction by some scientists that smoke and debris rising from massive fires of a nuclear war could block sunlight for weeks or months, cooling the earth's surface and producing climate changes that could, for example, negatively affect world agricultural and weather patterns.

Nuclide: An atom characterized by the number of protons, neturons, and energy in the nucleus.

Nutrient: Any substance assimilated by living things that promotes growth. The term is generally applied to nitrogen and phosphorus in wastewater, but is also applied to other essential and trace elements.

Nutrient Pollution: Contamination of water resources by excessive inputs of nutrients. In surface waters, excess algal production is a major concern.

O

Ocean Discharge Waiver: A variance from Clean Water Act requirements for discharges into marine waters.

Glossary of Terms

Odor Threshold: The minimum odor of a water or air sample that can just be detected after successive dilutions with odorless water. Also called threshold odor.

OECD Guidelines: Testing guidelines prepared by the Organization of Economic and Cooperative Development of the United Nations. They assist in preparation of protocols for studies of toxicology, environmental fate, etc.

Off-Site Facility: A hazardous waste treatment, storage or disposal area that is located away from the generating site.

Office Paper: High grade papers such as copier paper, computer printout, and stationary almost entirely made of uncoated chemical pulp, although some ground wood is used. Such waste is also generated in homes, schools, and elsewhere.

Offsets: A concept whereby emissions from proposed new or modified stationary sources are balanced by reductions from existing sources to stabilize total emissions. (See: bubble, emissions trading, netting)

Offstream Use: Water withdrawn from surface or groundwater sources for use at another place.

Oil and Gas Waste: Gas and oil drilling muds, oil production brines, and other waste associated with exploration for, development and production of crude oil or natural gas.

Oil Desulfurization: Widely used precombustion method for reducing sulfur dioxide emissions from oil-burning power plants. The oil is treated with hydrogen, which removes some of the sulfur by forming hydrogen sulfide gas.

Oil Fingerprinting: A method that identifies sources of oil and allows spills to be traced to their source.

Oil Spill: An accidental or intentional discharge of oil which reaches bodies of water. Can be controlled by chemical dispersion, combustion, mechanical containment, and/or adsorption. Spills from tanks and pipelines can also occur away from water bodies, contaminating the soil, getting into sewer systems and threatening underground water sources.

Oligotrophic Lakes: Deep clear lakes with few nutrients, little organic matter and a high dissolved-oxygen level.

On-Scene Coordinator (OSC): The predesignated EPA, Coast Guard, or Department of Defense official who coordinates and directs Superfund removal actions or Clean Water Act oil- or hazardous-spill response actions.

On-Site Facility: A hazardous waste treatment, storage or disposal area that is located on the generating site.

Onboard Controls: Devices placed on vehicles to capture gasoline vapor during refueling and route it to the engines when the vehicle is starting so that it can be efficiently burned.

Onconogenicity: The capacity to induce cancer.

One-hit Model: A mathematical model based on the biological theory that a single "hit" of some minimum critical amount of a carcinogen at a cellular target such as DNA can start an irreversible series events leading to a tumor.

Opacity: The amount of light obscured by particulate pollution in the air; clear window glass has zero opacity, a brick wall is 100 percent opaque. Opacity is an indicator of changes in performance of particulate control systems.

Open Burning: Uncontrolled fires in an open dump.

Open Dump: An uncovered site used for disposal of waste without environmental controls. (See: dump.)

Operable Unit: Term for each of a number of separate activities undertaken as part of a Superfund site cleanup. A typical operable unit would be removal of drums and tanks from the surface of a site.

Operating Conditions: Conditions specified in a RCRA permit that dictate how an incinerator must operate as it burns different waste types. A trial burn is used to identify operating conditions needed to meet specified performance standards.

Operation and Maintenance: 1. Activities conducted after a Superfund site action is completed to ensure that the action is effective. 2. Actions taken after construction to ensure that facilities constructed to treat waste water will be properly operated and maintained to achieve normative efficiency levels and prescribed effluent limitations in an optimum manner. 3. On-going asbestos management plan in a school or other public building, including regular inspections, various methods of maintaining asbestos in place, and removal when necessary.

Operator Certification: Certification of operators of community and nontransient noncommunity water systems, asbestos specialists, pesticide applicators, hazardous waste transporter, and other such specialists as required by the EPA or a state agency implementing an EPA-approved environmental regulatory program.

Optimal Corrosion Control Treatment: An erosion control treatment that minimizes the lead and copper concentrations at users' taps while also ensuring that the treatment does not cause the water system to violate any national primary drinking water regulations.

Oral Toxicity: Ability of a pesticide to cause injury when ingested.

Organic: 1. Referring to or derived from living organisms. 2. In chemistry, any compound containing carbon.

Organic Chemicals/Compounds: Naturally occuring (animal or plant-produced or synthetic) substances containing mainly carbon, hydrogen, nitrogen, and oxygen.

Organic Matter: Carbonaceous waste contained in plant or animal matter and originating from domestic or industrial sources.

Organism: Any form of animal or plant life.

Organophosphates: Pesticides that contain phosphorus; short-lived, but some can be toxic when first applied.

Organophyllic: A substance that easily combines with organic compounds.

Organotins: Chemical compounds used in anti-foulant paints to protect the hulls of boats and ships, buoys, and pilings from marine organisms such as barnacles.

Original AHERA Inspection/Original Inspection/Inspection: Examination of school buildings arranged by Local Education Agencies to identify asbestos-containing-materials, evaluate their condition, and take samples of materials suspected to contain asbestos; performed by EPA-accredited inspectors.

Original Generation Point: Where regulated medical or other material first becomes waste.

Osmosis: The passage of a liquid from a weak solution to a more concentrated solution across a semipermeable membrane that allows passage of the solvent (water) but not the dissolved solids.

Other Ferrous Metals: Recyclable metals from strapping, furniture, and metal found in tires and consumer electronics but does not include metals found in construction materials or cars, locomotives, and ships. (See: ferrous metals.)

Other Glass: Recyclable glass from furniture, appliances, and consumer electronics. Does not include glass from transportation products (cars trucks or shipping containers) and construction or demolition debris. (See: glass.)

Other Nonferrous Metals: Recyclable nonferrous metals such as lead, copper, and zinc from appliances, consumer electronics, and nonpackaging aluminum products. Does not include nonferrous metals from industrial applications and construction and demolition debris. (See: nonferrous metals.)

Other Paper: For Recyclable paper from books, third-class mail, commercial printing, paper towels, plates and cups; and other nonpackaging paper such as posters, photographic papers, cards and games, milk cartons, folding boxes, bags, wrapping paper, and paperboard. Does not include wrapping paper or shipping cartons.

Other Plastics: Recyclable plastic from appliances, eating utensils, plates, containers, toys, and various kinds of equipment. Does not include heavy-duty plastics such as yielding materials.

Other Solid Waste: Recyclable nonhazardous solid wastes, other than municipal solid waste, covered under Subtitle D of RARA. (See: solid waste.)

Other Wood: Recyclable wood from furniture, consumer electronics cabinets, and other nonpackaging wood products. Does not include lumber and tree stumps recovered from construction and demolition activities, and industrial process waste such as shavings and sawdust.

Outdoor Air Supply: Air brought into a building from outside.

Outfall: The place where effluent is discharged into receiving waters.

Overburden: Rock and soil cleared away before mining.

Overdraft: The pumping of water from a groundwater basin or aquifer in excess of the supply flowing into the basin; results in a depletion or "mining" of the groundwater in the basin. (See: groundwater mining)

Overfire Air: Air forced into the top of an incinerator or boiler to fan the flames.

Overflow Rate: One of the guidelines for design of the settling tanks and clarifers in a treatment plant; used by plant operators to determine if tanks and clarifers are over or under-used.

Overland Flow: A land application technique that cleanses waste water by allowing it to flow over a sloped surface. As the water flows over the surface, contaminants are absorbed and the water is collected at the bottom of the slope for reuse.

Oversized Regulated Medical Waste: Medical waste that is too large for plastic bags or standard containers.

Overturn: One complete cycle of top to bottom mixing of previously stratified water masses. This phenomenon may occur in spring or fall, or after storms, and results in uniformity of chemical and physical properties of water at all depths.

Oxidant: A collective term for some of the primary constituents of photochemical smog.

Oxidation Pond: A man-made (anthropogenic) body of water in which waste is consumed by bacteria, used most frequently with other waste-treatment processes; a sewage lagoon.

Oxidation: The chemical addition of oxygen to break down pollutants or organizac waste; e.g., destruction of chemicals such as cyanides, phenols, and organic sulfur compounds in sewage by bacterial and chemical means.

Oxidation-Reduction Potential: The electric potential required to transfer electrons from one compound or element (the oxidant) to another compound (the reductant); used as a qualitative measure of the state of oxidation in water treatment systems.

Oxygenated Fuels: Gasoline which has been blended with alcohols or ethers that contain oxygen in order to reduce carbon monoxide and other emissions.

Oxygenated Solvent: An organic solvent containing oxygen as part of the molecular structure. Alcohols and ketones are oxygenated compounds often used as paint solvents.

Ozonation/Ozonator: Application of ozone to water for disinfection or for taste and odor control. The ozonator is the device that does this.

Ozone (O3): Found in two layers of the atmosphere, the stratosphere and the troposphere. In the stratosphere (the atmospheric layer 7 to 10 miles or more above the earth's surface) ozone is a natural form of oxygen that provides a protective layer shielding the earth from ultraviolet radiation.In the troposphere (the layer extending up 7 to 10 miles from the earth's surface), ozone is a chemical oxidant and major component of photochemical smog. It can seriously impair the respiratory system and is one of the most wide- spread of all the criteria pollutants for which the Clean Air Act required EPA to set standards. Ozone in the troposphere is produced through complex chemical reactions of nitrogen oxides, which are among the primary pollutants emitted by combustion sources; hydrocarbons, released into the atmosphere through the combustion, handling and processing of petroleum products; and sunlight.

Ozone Depletion: Destruction of the stratospheric ozone layer which shields the earth from ultraviolet radiation harmful to life. This destruction of ozone is caused by the breakdown of certain chlorine and/or bromine containing compounds (chlorofluorocarbons or halons), which break down when they reach the stratosphere and then catalytically destroy ozone molecules.

Ozone Hole: A thinning break in the stratospheric ozone layer. Designation of amount of such depletion as an "ozone hole" is made when the detected amount of depletion exceeds fifty percent.

Seasonal ozone holes have been observed over both the Antarctic and Arctic regions, part of Canada, and the extreme northeastern United States.

Ozone Layer: The protective layer in the atmosphere, about 15 miles above the ground, that absorbs some of the sun's ultraviolet rays, thereby reducing the amount of potentially harmful radiation that reaches the earth's surface.

P

Packaging: The assembly of one or more containers and any other components necessary to ensure minimum compliance with a program's storage and shipment packaging requirements. Also, the containers, etc. involved.

Packed Bed Scrubber: An air pollution control device in which emissions pass through alkaline water to neutralize hydrogen chloride gas.

Packed Tower: A pollution control device that forces dirty air through a tower packed with crushed rock or wood chips while liquid is sprayed over the packing material. The pollutants in the air stream either dissolve or chemically react with the liquid.

Packer: An inflatable gland, or balloon, used to create a temporary seal in a borehole, probe hole, well, or drive casing. It is made of rubber or non-reactive materials.

Palatable Water: Water, at a desirable temperature, that is free from objectionable tastes, odors, colors, and turbidity.

Pandemic: A widespread epidemic throughout an area, nation or the world.

Paper: In the recycling business, refers to products and materials, including newspapers, magazines, office papers, corrugated containers, bags and some paperboard packaging that can be recycled into new paper products.

Paper Processor/Plastics Processor: Intermediate facility where recovered paper or plastic products and materials are sorted, decontaminated, and prepared for final recycling.

Parameter: A variable, measurable property whose value is a determinant of the characteristics of a system; e.g. temperature, pressure, and density are parameters of the atmosphere.

Paraquat: A standard herbicide used to kill various types of crops, including marijuana. Causes lung damage if smoke from the crop is inhaled..

Parshall Flume: Device used to measure the flow of water in an open channel.

Part A Permit, Part B Permit: (See: Interim Permit Status.)

Participation Rate: Portion of population participating in a recycling program.

Particle Count: Results of a microscopic examination of treated water with a special "particle counter" that classifies suspended particles by number and size.

Particulate Loading: The mass of particulates per unit volume of air or water.

Particulates: 1. Fine liquid or solid particles such as dust, smoke, mist, fumes, or smog, found in air or emissions. 2. Very small solids suspended in water;

they can vary in size, shape, density and electrical charge and can be gathered together by coagulation and flocculation.

Partition Coefficient: Measure of the sorption phenomenon, whereby a pesticide is divided between the soil and water phase; also referred to as adsorption partition coefficient.

Parts Per Billion (ppb)/Parts Per Million (ppm): Units commonly used to express contamination ratios, as in establishing the maximum permissible amount of a contaminant in water, land, or air.

Passive Smoking/Secondhand Smoke: Inhalation of others' tobacco smoke.

Passive Treatment Walls: Technology in which a chemical reaction takes place when contaminated ground water comes in contact with a barrier such as limestone or a wall containing iron filings.

Pathogens: Microorganisms (e.g., bacteria, viruses, or parasites) that can cause disease in humans, animals and plants.

Pathway: The physical course a chemical or pollutant takes from its source to the exposed organism.

Pay-As-You-Throw/Unit-Based Pricing: Systems under which residents pay for municipal waste management and disposal services by weight or volume collected, not a fixed fee.

Peak Electricity Demand: The maximum electricity used to meet the cooling load of a building or buildings in a given area.

Peak Levels: Levels of airborne pollutant contaminants much higher than average or occurring for short periods of time in response to sudden releases.

Percent Saturatiuon: The amount of a substance that is dissolved in a solution compared to the amount that could be dissolved in it.

Perched Water: Zone of unpressurized water held above the water table by impermeable rock or sediment.

Percolating Water: Water that passes through rocks or soil under the force of gravity.

Percolation: 1. The movement of water downward and radially through subsurface soil layers, usually continuing downward to ground water. Can also involve upward movement of water. 2. Slow seepage of water through a filter.

Performance Bond: Cash or securities deposited before a landfill operating permit is issued, which are held to ensure that all requirements for operating ad subsequently closing the landfill are faithful performed. The money is returned to the owner after proper closure of the landfill is completed. If contamination or other problems appear at any time during operation, or upon closure, and are not addressed, the owner must forfeit all or part of the bond which is then used to cover clean-up costs.

Performance Data (For Incinerators): Information collected, during a trial burn, on concentrations of designated organic compounds and pollutants found in incinerator emissions. Data analysis must show that the incinerator meets performance standards under operating conditions specified in the RCRA permit. (See: trial burn; performance standards.)

Performance Standards: 1. Regulatory requirements limiting the concentrations of designated organic compounds, particulate matter,

Glossary of Terms

and hydrogen chloride in emissions from incinerators. 2. Operating standards established by EPA for various permitted pollution control systems, asbestos inspections, and various program operations and maintenance requirements.

Periphyton: Microscopic underwater plants and animals that are firmly attached to solid surfaces such as rocks, logs, and pilings.

Permeability: The rate at which liquids pass through soil or other materials in a specified direction.

Permissible Dose: The dose of a chemical that may be received by an individual without the expectation of a significantly harmful result.

Permit: An authorization, license, or equivalent control document issued by EPA or an approved state agency to implement the requirements of an environmental regulation; e.g. a permit to operate a wastewater treatment plant or to operate a facility that may generate harmful emissions.

Persistence: Refers to the length of time a compound stays in the environment, once introduced. A compound may persist for less than a second or indefinitely.

Persistent Pesticides: Pesticides that do not break down chemically or break down very slowly and remain in the environment after a growing season.

Personal Air Samples: Air samples taken with a pump that is directly attached to the worker with the collecting filter and cassette placed in the worker's breathing zone (required under OSHA asbestos standards and EPA worker protection rule).

Personal Measurement: A measurement collected from an individual's immediate environment.

Personal Protective Equipment: Clothing and equipment worn by pesticide mixers, loaders and applicators and re-entry workers, hazmat emergency responders, workers cleaning up Superfund sites, et. al., which is worn to reduce their exposure to potentially hazardous chemicals and other pollutants.

Pest: An insect, rodent, nematode, fungus, weed or other form of terrestrial or aquatic plant or animal life that is injurious to health or the environment.

Pest Control Operator: Person or company that applies pesticides as a business (e.g. exterminator); usually describes household services, not agricultural applications.

Pesticide: Substances or mixture there of intended for preventing, destroying, repelling, or mitigating any pest. Also, any substance or mixture intended for use as a plant regulator, defoliant, or desiccant.

Pesticide Regulation Notice: Formal notice to pesticide registrants about important changes in regulatory policy, procedures, regulations.

Pesticide Tolerance: The amount of pesticide residue allowed by law to remain in or on a harvested crop. EPA sets these levels well below the point where the compounds might be harmful to consumers.

PETE (Polyethylene Terepthalate): Thermoplastic material used in plastic soft drink and rigid containers.

Petroleum: Crude oil or any fraction thereof that is liquid under normal conditions of temperature and pressure. The term includes petroleum-based substances comprising a complex blend of hydrocarbons derived from crude oil through the process of separation, conversion, upgrading, and finishing, such as motor fuel, jet oil, lubricants, petroleum solvents, and used oil.

Petroleum Derivatives: Chemicals formed when gasoline breaks down in contact with ground water.

pH: An expression of the intensity of the basic or acid condition of a liquid; may range from 0 to 14, where 0 is the most acid and 7 is neutral. Natural waters usually have a pH between 6.5 and 8.5.

Pharmacokinetics: The study of the way that drugs move through the body after they are swallowed or injected.

Phenolphthalein Alkalinity: The alkalinity in a water sample measured by the amount of standard acid needed to lower the pH to a level of 8.3 as indicated by the change of color of the phenolphthalein from pink to clear.

Phenols: Organic compounds that are byproducts of petroleum refining, tanning, and textile, dye, and resin manufacturing. Low concentrations cause taste and odor problems in water; higher concentrations can kill aquatic life and humans.

Phosphates: Certain chemical compounds containing phosphorus.

Phosphogypsum Piles (Stacks): Principal byproduct generated in production of phosphoric acid from phosphate rock. These piles may generate radioactive radon gas.

Phosphorus: An essential chemical food element that can contribute to the eutrophication of lakes and other water bodies. Increased phosphorus levels result from discharge of phosphorus-containing materials into surface waters.

Phosphorus Plants: Facilities using electric furnaces to produce elemental phosphorous for commercial use, such as high grade phosphoric acid, phosphate-based detergent, and organic chemicals use.

Photochemical Oxidants: Air pollutants formed by the action of sunlight on oxides of nitrogen and hydrocarbons.

Photochemical Smog: Air pollution caused by chemical reactions of various pollutants emitted from different sources. (See: photochemical oxidants.)

Photosynthesis: The manufacture by plants of carbohydrates and oxygen from carbon dioxide mediated by chlorophyll in the presence of sunlight.

Physical and Chemical Treatment: Processes generally used in large-scale wastewater treatment facilities. Physical processes may include air-stripping or filtration. Chemical treatment includes coagulation, chlorination, or ozonation. The term can also refer to treatment of toxic materials in surface and ground waters, oil spills, and some methods of dealing with hazardous materials on or in the ground.

Phytoplankton: That portion of the plankton community comprised of tiny plants; e.g. algae, diatoms.

Phytoremediation: Low-cost remediation option for sites with widely dispersed contamination at low concentrations.

Phytotoxic: Harmful to plants.

Phytotreatment: The cultivation of specialized plants that absorb specific contaminants from the soil through their roots or foliage. This reduces the concentration of contaminants in the soil, but incorporates them into biomasses that may be released back into the environment when the plant dies or is harvested.

Picocuries Per Liter pCi/L): A unit of measure for levels of radon gas; becquerels per cubic meter is metric equivalent.

Piezometer: A nonpumping well, generally of small diameter, for measuring the elevation of a water table.

Pilot Tests: Testing a cleanup technology under actual site conditions to identify potential problems prior to full-scale implementation.

Plankton: Tiny plants and animals that live in water.

Plasma-Arc Reactor: An incinerator that operates at extremely high temperatures; treats highly toxic wastes that do not burn easily.

Plasmid: A circular piece of DNA that exists apart from the chromosome and replicates independently of it. Bacterial plasmids carry information that renders the bacteria resistant to antibiotics. Plasmids are often used in genetic engineering to carry desired genes into organisms.

Plastics: Non-metallic chemoreactive compounds molded into rigid or pliable construction materials, fabrics, etc.

Plate Tower Scrubber: An air pollution control device that neutralizes hydrogen chloride gas by bubbling alkaline water through holes in a series of metal plates.

Plug Flow: Type of flow the occurs in tanks, basins, or reactors when a slug of water moves through without ever dispersing or mixing with the rest of the water flowing through.

Plugging: Act or process of stopping the flow of water, oil, or gas into or out of a formation through a borehole or well penetrating that formation.

Plume: 1. A visible or measurable discharge of a contaminant from a given point of origin. Can be visible or thermal in water, or visible in the air as, for example, a plume of smoke. 2 The area of radiation leaking from a damaged reactor. 3. Area downwind within which a release could be dangerous for those exposed to leaking fumes.

Plutonium: A radioactive metallic element chemically similar to uranium.

PM-10/PM-2.5: PM-10 is measure of particles in the atmosphere with a diameter of less than ten or equal to a nominal 10 micrometers. PM-2.5 is a measure of smaller particles in the air. PM-10 has been the pollutant particulate level standard against which EPA has been measuring Clean Air Act compliance. On the basis of newer scientific findings, the Agency is considering regulations that will make PM-2.5 the new "standard".

Pneumoconiosis: Health conditions characterized by permanent deposition of substantial amounts of particulate matter in the lungs and by the tissue reaction to its presence; can range from relatively harmless forms of sclerosis to the destructive fibrotic effect of silicosis.

Point Source: A stationary location or fixed facility from which pollutants are discharged; any single identifiable source of pollution; e.g. a pipe, ditch, ship, ore pit, factory smokestack.

776

Point-of-Contact Measurement of Exposure: Estimating exposure by measuring concentrations over time (while the exposure is taking place) at or near the place where it is occurring.

Point-of-Disinfectant Application: The point where disinfectant is applied and water downstream of that point is not subject to recontamination by surface water runoff.

Point-of-Use Treatment Device: Treatment device applied to a single tap to reduce contaminants in the drinking water at the one faucet.

Pollen: The fertilizing element of flowering plants; background air pollutant.

Pollutant: Generally, any substance introduced into the environment that adversely affects the usefulness of a resource or the health of humans, animals, or ecosystems..

Pollutant Pathways: Avenues for distribution of pollutants. In most buildings, for example, HVAC systems are the primary pathways although all building components can interact to affect how air movement distributes pollutants.

Pollutant Standard Index (PSI): Indicator of one or more pollutants that may be used to inform the public about the potential for adverse health effects from air pollution in major cities.

Pollution: Generally, the presence of a substance in the environment that because of its chemical composition or quantity prevents the functioning of natural processes and produces undesirable environmental and health effects.Under the Clean Water Act, for example, the term has been defined as the man-made or man-induced alteration of the physical, biological, chemical, and radiological integrity of water and other media.

Pollution Prevention: 1. Identifying areas, processes, and activities which create excessive waste products or pollutants in order to reduce or prevent them through, alteration, or eliminating a process. Such activities, consistent with the Pollution Prevention Act of 1990, are conducted across all EPA programs and can involve cooperative efforts with such agencies as the Departments of Agriculture and Energy. 2. EPA has initiated a number of voluntary programs in which industrial, or commercial or "partners" join with EPA in promoting activities that conserve energy, conserve and protect water supply, reduce emissions or find ways of utilizing them as energy resources, and reduce the waste stream. Among these are: Agstar, to reduce methane emissions through manure management. Climate Wise, to lower industrial greenhouse-gas emissions and energy costs. Coalbed Methane Outreach, to boost methane recovery at coal mines. Design for the Environment, to foster including environmental considerations in product design and processes. Energy Star programs, to promote energy efficiency in commercial and residential buildings, office equipment, transformers, computers, office equipment, and home appliances. Environmental Accounting, to help businesses identify environmental costs and factor them into management decision making. Green Chemistry, to promote and recognize cost-effective breakthroughs in chemistry that prevent pollution. Green Lights, to spread the use of energy-efficient lighting technologies. Indoor Environments, to reduce risks from indoor-air pollution. Landfill Methane Outreach, to develop landfill gas-to-energy projects. Natural Gas Star, to reduce methane emissions from the natural gas industry. Ruminant Livestock Methane, to reduce methane emissions from ruminant livestock. Transportation Partners, to reduce carbon dioxide emissions from the transportation sector. Voluntary Aluminum Industrial Partnership, to reduce perfluorocarbon emissions from the primary aluminum industry. WAVE, to promote efficient water use in the lodging industry. Wastewi$e, to reduce business-generated solid waste through prevention, reuse, and recycling. (See: Common Sense Initiative and Project XL.)

Portal-of-Entry Effect: A local effect produced in the tissue or organ of first contact between a toxicant and the biological system.

Polonium: A radioactive element that occurs in pitchblende and other uranium-containing ores.

Polyelectrolytes: Synthetic chemicals that help solids to clump during sewage treatment.

Polymer: A natural or synthetic chemical structure where two or more like molecules are joined to form a more complex molecular structure (e.g. polyethylene in plastic).

Polyvinyl Chloride (PVC): A tough, environmentally indestructible plastic that releases hydrochloric acid when burned.

Population: A group of interbreeding organisms occupying a particular space; the number of humans or other living creatures in a designated area.

Population at Risk: A population subgroup that is more likely to be exposed to a chemical, or is more sensitive to the chemical, than is the general population.

Porosity: Degree to which soil, gravel, sediment, or rock is permeated with pores or cavities through which water or air can move.

Post-Chlorination: Addition of chlorine to plant effluent for disinfectant purposes after the effluent has been treated.

Post-Closure: The time period following the shutdown of a waste management or manufacturing facility; for monitoring purposes, often considered to be 30 years.

Post-Consumer Materials/Waste: Recovered materials that are diverted from municipal solid waste for the purpose of collection, recycling, and disposition.

Post-Consumer Recycling: Use of materials generated from residential and consumer waste for new or similar purposes; e.g. converting wastepaper from offices into corrugated boxes or newsprint.

Potable Water: Water that is safe for drinking and cooking.

Potential Dose: The amount of a compound contained in material swallowed, breathed, or applied to the skin.

Potentially Responsible Party (PRP): Any individual or company—including owners, operators, transporters or generators—potentially responsible for, or contributing to a spill or other contamination at a Superfund site. Whenever possible, through administrative and legal actions, EPA requires PRPs to clean up hazardous sites they have contaminated.

Potentiation: The ability of one chemical to increase the effect of another chemical.

Potentiometric Surface: The surface to which water in an aquifer can rise by hydrostatic pressure.

Precautionary Principle: When information about potential risks is incomplete, basing decisions about the best ways to manage or reduce risks on a preference for avoiding unnecessary health risks instead of on unnecessary economic expenditures.

Pre-Consumer Materials/Waste: Materials generated in manufacturing and converting processes such as manufacturing scrap and trimmings and cuttings. Includes print overruns, overissue publications, and obsolete inventories.

Pre-Harvest Interval: The time between the last pesticide application and harvest of the treated crops.

Prechlorination: The addition of chlorine at the headworks of a treatment plant prior to other treatment processes. Done mainly for disinfection and control of tastes, odors, and aquatic growths, and to aid in coagulation and settling,

Precipitate: A substance separated from a solution or suspension by chemical or physical change.

Precipitation: Removal of hazardous solids from liquid waste to permit safe disposal; removal of particles from airborne emissions as in rain (e.g. acid precipitation).

Precipitator: Pollution control device that collects particles from an air stream.

Precursor: In photochemistry, a compound antecedent to a pollutant. For example, volatile organic compounds (VOCs) and nitric oxides of nitrogen react in sunlight to form ozone or other photochemical oxidants. As such, VOCs and oxides of nitrogen are precursors.

Preliminary Assessment: The process of collecting and reviewing available information about a known or suspected waste site or release.

Prescriptive: Water rights which are acquired by diverting water and putting it to use in accordance with specified procedures; e.g. filing a request with a state agency to use unused water in a stream, river, or lake.

Pressed Wood Products: Materials used in building and furniture construction that are made from wood veneers, particles, or fibers bonded together with an adhesive under heat and pressure.

Pressure Sewers: A system of pipes in which water, wastewater, or other liquid is pumped to a higher elevation.

Pressure, Static: In flowing air, the total pressure minus velocity pressure, pushing equally in all directions.

Pressure, Total: In flowing air, the sum of the static and velocity pressures.

Pressure, Velocity: In flowing air, the pressure due to velocity and density of air.

Pretreatment: Processes used to reduce, eliminate, or alter the nature of wastewater pollutants from non-domestic sources before they are discharged into publicly owned treatment works (POTWs).

Prevalent Level Samples: Air samples taken under normal conditions (also known as ambient background samples).

Prevalent Levels: Levels of airborne contaminant occurring under normal conditions.

Prevention of Significant Deterioration (PSD): EPA program in which state and/or federal permits are required in order to restrict emissions from new or modified sources in places where air quality already meets or exceeds primary and secondary ambient air quality standards.

Glossary of Terms

Primacy: Having the primary responsibility for administering and enforcing regulations.

Primary Drinking Water Regulation: Applies to public water systems and specifies a contaminant level, which, in the judgment of the EPA Administrator, will not adversely affect human health.

Primary Effect: An effect where the stressor acts directly on the ecological component of interest, not on other parts of the ecosystem. (See: secondary effect.)

Primary Standards: National ambient air quality standards designed to protect human health with an adequate margin for safety. (See: National Ambient Air Quality Standards, secondary standards.)

Primary Waste Treatment: First steps in wastewater treatment; screens and sedimentation tanks are used to remove most materials that float or will settle. Primary treatment removes about 30 percent of carbonaceous biochemical oxygen demand from domestic sewage.

Principal Organic Hazardous Constituents (POHCs): Hazardous compounds monitored during an incinerator's trial burn, selected for high concentration in the waste feed and difficulty of combustion.

Prions: Microscopic particles made of protein that can cause disease.

Prior Appropriation: A doctrine of water law that allocates the rights to use water on a first-come, first-served basis.

Probability of Detection : The likelihood, expressed as a percentage, that a test method will correctly identify a leaking tank.

Process Variable: A physical or chemical quantity which is usually measured and controlled in the operation of a water treatment plant or industrial plant.

Process Verification: Verifying that process raw materials, water usage, waste treatment processes, production rate and other facts relative to quantity and quality of pollutants contained in discharges are substantially described in the permit application and the issued permit.

Process Wastewater: Any water that comes into contact with any raw material, product, byproduct, or waste.

Process Weight: Total weight of all materials, including fuel, used in a manufacturing process; used to calculate the allowable particulate emission rate.

Producers: Plants that perform photosynthesis and provide food to consumers.

Product Level: The level of a product in a storage tank.

Product Water: Water that has passed through a water treatment plant and is ready to be delivered to consumers.

Products of Incomplete Combustion (PICs): Organic compounds formed by combustion. Usually generated in small amounts and sometimes toxic, PICs are heat-altered versions of the original material fed into the incinerator (e.g. charcoal is a P.I.C. from burning wood).

Project XL: An EPA initiative to give states and the regulated community the flexibility to develop comprehensive strategies as alternatives to multiple current regulatory requirements in order to exceed compliance and increase overall environmental benefits.

Propellant: Liquid in a self-pressurized pesticide product that expels the active ingredient from its container.

Proportionate Mortality Ratio (PMR): The number of deaths from a specific cause in a specific period of time per 100 deaths from all causes in the same time period.

Proposed Plan: A plan for a site cleanup that is available to the public for comment.

Proteins: Complex nitrogenous organic compounds of high molecular weight made of amino acids; essential for growth and repair of animal tissue. Many, but not all, proteins are enzymes.

Protocol: A series of formal steps for conducting a test.

Protoplast: A membrane-bound cell from which the outer wall has been partially or completely removed. The term often is applied to plant cells.

Protozoa: One-celled animals that are larger and more complex than bacteria. May cause disease.

Public Comment Period: The time allowed for the public to express its views and concerns regarding an action by EPA (e.g. a Federal Register Notice of proposed rule-making, a public notice of a draft permit, or a Notice of Intent to Deny).

Public Health Approach: Regulatory and voluntary focus on effective and feasible risk management actions at the national and community level to reduce human exposures and risks, with priority given to reducing exposures with the biggest impacts in terms of the number affected and severity of effect.

Public Health Context: The incidence, prevalence, and severity of diseases in communities or populations and the factors that account for them, including infections, exposure to pollutants, and other exposures or activities.

Public Hearing: A formal meeting wherein EPA officials hear the public's views and concerns about an EPA action or proposal. EPA is required to consider such comments when evaluating its actions. Public hearings must be held upon request during the public comment period.

Public Notice: 1. Notification by EPA informing the public of Agency actions such as the issuance of a draft permit or scheduling of a hearing. EPA is required to ensure proper public notice, including publication in newspapers and broadcast over radio and television stations. 2. In the safe drinking water program, water suppliers are required to publish and broadcast notices when pollution problems are discovered.

Public Water System: A system that provides piped water for human consumption to at least 15 service connections or regularly serves 25 individuals.

Publicly Owned Treatment Works (POTWs): A waste-treatment works owned by a state, unit of local government, or Indian tribe, usually designed to treat domestic wastewaters.

Pumping Station: Mechanical device installed in sewer or water system or other liquid-carrying pipelines to move the liquids to a higher level.

Pumping Test: A test conducted to determine aquifer or well characteristics.

Purging: Removing stagnant air or water from sampling zone or equipment prior to sample collection.

Putrefaction: Biological decomposition of organic matter; associated with anaerobic conditions.

Putrescible: Able to rot quickly enough to cause odors and attract flies.

Pyrolysis: Decomposition of a chemical by extreme heat.

Q

Qualitative Use Assessment: Report summarizing the major uses of a pesticide including percentage of crop treated, and amount of pesticide used on a site.

Quality Assurance/Quality Control: A system of procedures, checks, audits, and corrective actions to ensure that all EPA research design and performance, environmental monitoring and sampling, and other technical and reporting activities are of the highest achievable quality.

Quench Tank: A water-filled tank used to cool incinerator residues or hot materials during industrial processes.

R

Radiation: Transmission of energy though space or any medium. Also known as radiant energy.

Radiation Standards: Regulations that set maximum exposure limits for protection of the public from radioactive materials.

Radio Frequency Radiation: (See non-ionizing electromagnetic radiation.)

Radioactive Decay: Spontaneous change in an atom by emission of of charged particles and/or gamma rays; also known as radioactive disintegration and radioactivity.

Radioactive Substances: Substances that emit ionizing radiation.

Radioisotopes: Chemical variants of radioactive elements with potentially oncogenic, teratogenic, and mutagenic effects on the human body.

Radionuclide: Radioactive particle, man-made (anthropogenic) or natural, with a distinct atomic weight number. Can have a long life as soil or water pollutant.

Radius of Vulnerability Zone: The maximum distance from the point of release of a hazardous substance in which the airborne concentration could reach the level of concern under specified weather conditions.

Radius of Influence: 1. The radial distance from the center of a wellbore to the point where there is no lowering of the water table or potentiometric surface (the edge of the cone of depression); 2. the radial distance from an extraction well that has adequate air flow for effective removal of contaminants when a vacuum is applied to the extraction well.

Radon: A colorless naturally occurring, radioactive, inert gas formed by radioactive decay of radium atoms in soil or rocks.

Radon Daughters/Radon Progeny: Short-lived radioactive decay products of radon that decay into longer-lived lead isotopes that can attach themselves to airborne dust and other particles and, if inhaled, damage the linings of the lungs.

Radon Decay Products: A term used to refer collectively to the immediate products of the radon decay chain. These include Po-218, Pb-214, Bi-214, and Po-214, which have an average combined half-life of about 30 minutes.

Rainbow Report: Comprehensive document giving the status of all pesticides now or ever in registration or special reviews. Known as the "rainbow report" because chapters are printed on different colors of paper.

Rasp: A machine that grinds waste into a manageable material and helps prevent odor.

Raw Agricultural Commodity: An unprocessed human food or animal feed crop (e.g., raw carrots, apples, corn, or eggs.)

Raw Sewage: Untreated wastewater and its contents.

Raw Water: Intake water prior to any treatment or use.

Re-entry: (In indoor air program) Refers to air exhausted from a building that is immediately brought back into the system through the air intake and other openings.

Reaeration: Introduction of air into the lower layers of a reservoir. As the air bubbles form and rise through the water, the oxygen dissolves into the water and replenishes the dissolved oxygen. The rising bubbles also cause the lower waters to rise to the surface where they take on oxygen from the atmosphere.

Real-Time Monitoring: Monitoring and measuring environmental developments with technology and communications systems that provide time-relevant information to the public in an easily understood format people can use in day-to-day decision-making about their health and the environment.

Reasonable Further Progress: Annual incremental reductions in air pollutant emissions as reflected in a State Implementation Plan that EPA deems sufficient to provide for the attainment of the applicable national ambient air quality standards by the statutory deadline.

Reasonable Maximum Exposure: The maximum exposure reasonably expected to occur in a population.

Reasonable Worst Case: An estimate of the individual dose, exposure, or risk level received by an individual in a defined population that is greater than the 90th percentile but less than that received by anyone in the 98th percentile in the same population.

Reasonably Available Control Measures (RACM): A broadly defined term referring to technological and other measures for pollution control.

Reasonably Available Control Technology (RACT): Control technology that is reasonably available, and both technologically and economically feasible. Usually applied to existing sources in nonattainment areas; in most cases is less stringent than new source performance standards.

Recarbonization: Process in which carbon dioxide is bubbled into water being treated to lower the pH.

Receiving Waters: A river, lake, ocean, stream or other watercourse into which wastewater or treated effluent is discharged.

Receptor: Ecological entity exposed to a stressor.

Recharge: The process by which water is added to a zone of saturation, usually by percolation from the soil surface; e.g., the recharge of an aquifer.

Recharge Area: A land area in which water reaches the zone of saturation from surface infiltration, e.g., where rainwater soaks through the earth to reach an aquifer.

Recharge Rate: The quantity of water per unit of time that replenishes or refills an aquifer.

Reclamation: (In recycling) Restoration of materials found in the waste stream to a beneficial use which may be for purposes other than the original use.

Recombinant Bacteria: A microorganism whose genetic makeup has been altered by deliberate introduction of new genetic elements. The offspring of these altered bacteria also contain these new genetic elements; i.e. they "breed true."

Recombinant DNA: The new DNA that is formed by combining pieces of DNA from different organisms or cells.

Recommended Maximum Contaminant Level (RMCL): The maximum level of a contaminant in drinking water at which no known or anticipated adverse effect on human health would occur, and that includes an adequate margin of safety. Recommended levels are nonenforceable health goals. (See: maximum contaminant level.)

Reconstructed Source: Facility in which components are replaced to such an extent that the fixed capital cost of the new components exceeds 50 percent of the capital cost of constructing a comparable brand-new facility. New-source performance standards may be applied to sources reconstructed after the proposal of the standard if it is technologically and economically feasible to meet the standards.

Reconstruction of Dose: Estimating exposure after it has occurred by using evidence within an organism such as chemical levels in tissue or fluids.

Record of Decision (ROD): A public document that explains which cleanup alternative(s) will be used at National Priorities List sites where, under CERCLA, Trust Funds pay for the cleanup.

Recovery Rate: Percentage of usable recycled materials that have been removed from the total amount of municipal solid waste generated in a specific area or by a specific business.

Recycle/Reuse: Minimizing waste generation by recovering and reprocessing usable products that might otherwise become waste (.i.e. recycling of aluminum cans, paper, and bottles, etc.).

Recycling and Reuse Business Assistance Centers: Located in state solid-waste or economic-development agencies, these centers provide recycling businesses with customized and targeted assistance.

Recycling Economic Development Advocates: Individuals hired by state or tribal economic development offices to focus financial, marketing, and permitting resources on creating recycling businesses.

Recycling Mill: Facility where recovered materials are remanufactured into new products.

Recycling Technical Assistance Partnership National Network: A national information-sharing resource designed to help businesses and manufacturers increase their use of recovered materials.

Red Bag Waste: (See: infectious waste.)

Red Border: An EPA document undergoing review before being submitted for final management decision-making.

Red Tide: A proliferation of a marine plankton toxic and often fatal to fish, perhaps stimulated by the addition of nutrients. A tide can be red, green, or brown, depending on the coloration of the plankton.

Redemption Program: Program in which consumers are monetarily compensated for the collection of recyclable materials, generally through prepaid deposits or taxes on beverage containers. In some states or localities legislation has enacted redemption programs to help prevent roadside litter. (See: bottle bill.)

Reduction: The addition of hydrogen, removal of oxygen, or addition of electrons to an element or compound.

Reentry Interval: The period of time immediately following the application of a pesticide during which unprotected workers should not enter a field.

Reference Dose (RfD): The RfD is a numerical estimate of a daily oral exposure to the human population, including sensitive subgroups such as children, that is not likely to cause harmful effects during a lifetime. RfDs are generally used for health effects that are thought to have a threshold or low dose limit for producing effects.

Reformulated Gasoline: Gasoline with a different composition from conventional gasoline (e.g., lower aromatics content) that cuts air pollutants.

Refueling Emissions: Emissions released during vehicle re-fueling.

Refuse: (See: solid waste.)

Refuse Reclamation: Conversion of solid waste into useful products; e.g., composting organic wastes to make soil conditioners or separating aluminum and other metals for recycling.

Regeneration: Manipulation of cells to cause them to develop into whole plants.

Regional Response Team (RRT): Representatives of federal, local, and state agencies who may assist in coordination of activities at the request of the On-Scene Coordinator before and during a significant pollution incident such as an oil spill, major chemical release, or Superfund response.

Registrant: Any manufacturer or formulator who obtains registration for a pesticide active ingredient or product.

Registration: Formal listing with EPA of a new pesticide before it can be sold or distributed. Under the Federal Insecticide, Fungicide, and Rodenticide Act, EPA is responsible for registration (pre-market licensing) of pesticides on the basis of data demonstrating no unreasonable adverse effects on human health or the environment when applied according to approved label directions.

Registration Standards: Published documents which include summary reviews of the data available on a pesticide's active ingredient, data gaps, and the Agency's existing regulatory position on the pesticide.

Glossary of Terms

Regulated Asbestos-Containing Material (RACM): Friable asbestos material or nonfriable ACM that will be or has been subjected to sanding, grinding, cutting, or abrading or has crumbled, or been pulverized or reduced to powder in the course of demolition or renovation operations.

Regulated Medical Waste: Under the Medical Waste Tracking Act of 1988, any solid waste generated in the diagnosis, treatment, or immunization of human beings or animals, in research pertaining thereto, or in the production or testing of biologicals. Included are cultures and stocks of infectious agents; human blood and blood products; human pathological body wastes from surgery and autopsy; contaminated animal carcasses from medical research; waste from patients with communicable diseases; and all used sharp implements, such as needles and scalpels, and certain unused sharps. (See: treated medical waste; untreated medical waste; destroyed medical waste.)

Relative Ecological Sustainability: Ability of an ecosystem to maintain relative ecological integrity indefinitely.

Relative Permeability: The permeability of a rock to gas, NAIL, or water, when any two or more are present.

Relative Risk Assessment: Estimating the risks associated with different stressors or management actions.

Release: Any spilling, leaking, pumping, pouring, emitting, emptying, discharging, injecting, escaping, leaching, dumping, or disposing into the environment of a hazardous or toxic chemical or extremely hazardous substance.

Remedial Action (RA): The actual construction or implementation phase of a Superfund site cleanup that follows remedial design.

Remedial Design: A phase of remedial action that follows the remedial investigation/feasibility study and includes development of engineering drawings and specifications for a site cleanup.

Remedial Investigation: An in-depth study designed to gather data needed to determine the nature and extent of contamination at a Superfund site; establish site cleanup criteria; identify preliminary alternatives for remedial action; and support technical and cost analyses of alternatives. The remedial investigation is usually done with the feasibility study. Together they are usually referred to as the "RI/FS".

Remedial Project Manager (RPM): The EPA or state official responsible for overseeing on-site remedial action.

Remedial Response: Long-term action that stops or substantially reduces a release or threat of a release of hazardous substances that is serious but not an immediate threat to public health.

Remediation: 1. Cleanup or other methods used to remove or contain a toxic spill or hazardous materials from a Superfund site; 2. for the Asbestos Hazard Emergency Response program, abatement methods including evaluation, repair, enclosure, encapsulation, or removal of greater than 3 linear feet or square feet of asbestos-containing materials from a building.

Remote Sensing: The collection and interpretation of information about an object without physical contact with the object; e.g., satellite imaging, aerial photography, and open path measurements.

Removal Action: Short-term immediate actions taken to address releases of hazardous substances that require expedited response. (See: cleanup)

Renewable Energy Production Incentive (REPI): Incentive established by the Energy Policy Act available to renewable energy power projects owned by a state or local government or nonprofit electric cooperative.

Repeat Compliance Period: Any subsequent compliance period after the initial one.

Reportable Quantity (RQ): Quantity of a hazardous substance that triggers reports under CERCLA. If a substance exceeds its RQ, the release must be reported to the National Response Center, the SERC, and community emergency coordinators for areas likely to be affected.

Repowering: Rebuilding and replacing major components of a power plant instead of building a new one.

Representative Sample: A portion of material or water that is as nearly identical in content and consistency as possible to that in the larger body of material or water being sampled.

Reregistration: The reevaluation and relicensing of existing pesticides originally registered prior to current scientific and regulatory standards. EPA reregisters pesticides through its Registration Standards Program.

Reserve Capacity: Extra treatment capacity built into solid waste and wastewater treatment plants and interceptor sewers to accommodate flow increases due to future population growth.

Reservoir: Any natural or artificial holding area used to store, regulate, or control water.

Residential Use: Pesticide application in and around houses, office buildings, apartment buildings, motels, and other living or working areas.

Residential Waste: Waste generated in single and multi-family homes, including newspapers, clothing, disposable tableware, food packaging, cans, bottles, food scraps, and yard trimmings other than those that are diverted to backyard composting. (See: Household hazardous waste.)

Residual: Amount of a pollutant remaining in the environment after a natural or technological process has taken place; e.g., the sludge remaining after initial wastewater treatment, or particulates remaining in air after it passes through a scrubbing or other process.

Residual Risk: The extent of health risk from air pollutants remaining after application of the Maximum Achievable Control Technology (MACT).

Residual Saturation: Saturation level below which fluid drainage will not occur.

Residue: The dry solids remaining after the evaporation of a sample of water or sludge.

Resistance: For plants and animals, the ability to withstand poor environmental conditions or attacks by chemicals or disease. May be inborn or acquired.

Resource Recovery: The process of obtaining matter or energy from materials formerly discarded.

Response Action: 1. Generic term for actions taken in response to actual or potential health-threatening environmental events such as spills, sudden releases, and asbestos abatement/management problems. 2. A CERCLA-authorized action involving either a short-term removal action or a long-term removal response. This may include but is not limited to: removing hazardous materials from a site to an EPA-approved hazardous waste facility for treatment, containment or treating the waste on-site, identifying and removing the sources of ground-water contamination and halting further migration of contaminants. 3. Any of the following actions taken in school buildings in response to AHERA to reduce the risk of exposure to asbestos: removal, encapsulation, enclosure, repair, and operations and maintenance. (See: cleanup.)

Responsiveness Summary: A summary of oral and/or written public comments received by EPA during a comment period on key EPA documents, and EPA's response to those comments.

Restoration: Measures taken to return a site to pre-violation conditions.

Restricted Entry Interval: The time after a pesticide application during which entry into the treated area is restricted.

Restricted Use: A pesticide may be classified (under FIFRA regulations) for restricted use if it requires special handling because of its toxicity, and, if so, it may be applied only by trained, certified applicators or those under their direct supervision.

Restriction Enzymes: Enzymes that recognize specific regions of a long DNA molecule and cut it at those points.

Retrofit: Addition of a pollution control device on an existing facility without making major changes to the generating plant. Also called backfit.

Reuse: Using a product or component of municipal solid waste in its original form more than once; e.g., refilling a glass bottle that has been returned or using a coffee can to hold nuts and bolts.

Reverse Osmosis: A treatment process used in water systems by adding pressure to force water through a semi-permeable membrane. Reverse osmosis removes most drinking water contaminants. Also used in wastewater treatment. Large-scale reverse osmosis plants are being developed.

Reversible Effect: An effect which is not permanent; especially adverse effects which diminish when exposure to a toxic chemical stops.

Ribonucleic Acid (RNA): A molecule that carries the genetic message from DNA to a cellular protein-producing mechanism.

Rill: A small channel eroded into the soil by surface runoff; can be easily smoothed out or obliterated by normal tillage.

Ringlemann Chart: A series of shaded illustrations used to measure the opacity of air pollution emissions, ranging from light grey through black; used to set and enforce emissions standards.

Riparian Habitat: Areas adjacent to rivers and streams with a differing density, diversity, and productivity of plant and animal species relative to nearby uplands.

Riparian Rights: Entitlement of a land owner to certain uses of water on or bordering the property, including the right to prevent diversion or misuse of upstream waters. Generally a matter of state law.

Risk: A measure of the probability that damage to life, health, property, and/or the environment will occur as a result of a given hazard.

Risk (Adverse) for Endangered Species: Risk to aquatic species if anticipated pesticide residue levels equal one-fifth of LD10 or one-tenth of LC50; risk to terrestrial species if anticipated pesticide residue levels equal one-fifth of LC10 or one-tenth of LC50.

Risk Assessment: Qualitative and quantitative evaluation of the risk posed to human health and/or the environment by the actual or potential presence and/or use of specific pollutants.

Risk Characterization: The last phase of the risk assessment process that estimates the potential for adverse health or ecological effects to occur from exposure to a stressor and evaluates the uncertainty involved.

Risk Communication: The exchange of information about health or environmental risks among risk assessors and managers, the general public, news media, interest groups, etc.

Risk Estimate: A description of the probability that organisms exposed to a specific dose of a chemical or other pollutant will develop an adverse response, e.g., cancer.

Risk Factor: Characteristics (e.g., race, sex, age, obesity) or variables (e.g., smoking, occupational exposure level) associated with increased probability of a toxic effect.

Risk for Non-Endangered Species: Risk to species if anticipated pesticide residue levels are equal to or greater than LC50.

Risk Management: The process of evaluating and selecting alternative regulatory and non-regulatory responses to risk. The selection process necessarily requires the consideration of legal, economic, and behavioral factors.

Risk-based Targeting: The direction of resources to those areas that have been identified as having the highest potential or actual adverse effect on human health and/or the environment.

Risk-Specific Dose: The dose associated with a specified risk level.

River Basin: The land area drained by a river and its tributaries.

Rodenticide: A chemical or agent used to destroy rats or other rodent pests, or to prevent them from damaging food, crops, etc.

Rotary Kiln Incinerator: An incinerator with a rotating combustion chamber that keeps waste moving, thereby allowing it to vaporize for easier burning.

Rough Fish: Fish not prized for sport or eating, such as gar and suckers. Most are more tolerant of changing environmental conditions than are game or food species.

Route of Exposure: The avenue by which a chemical comes into contact with an organism, e.g., inhalation, ingestion, dermal contact, injection.

Rubbish: Solid waste, excluding food waste and ashes, from homes, institutions, and workplaces.

Run-Off: That part of precipitation, snow melt, or irrigation water that runs off the land into streams or other surface-water. It can carry pollutants from the air and land into receiving waters.

Running Losses: Evaporation of motor vehicle fuel from the fuel tank while the vehicle is in use.

S

Sacrifical Anode: An easily corroded material deliberately installed in a pipe or intake to give it up (sacrifice it) to corrosion while the rest of the water supply facility remains relatively corrosion-free.

Safe: Condition of exposure under which there is a practical certainty that no harm will result to exposed individuals.

Safe Water: Water that does not contain harmful bacteria, toxic materials, or chemicals, and is considered safe for drinking even if it may have taste, odor, color, and certain mineral problems.

Safe Yield: The annual amount of water that can be taken from a source of supply over a period of years without depleting that source beyond its ability to be replenished naturally in "wet years."

Safener: A chemical added to a pesticide to keep it from injuring plants.

Salinity: The percentage of salt in water.

Salt Water Intrusion: The invasion of fresh surface or ground water by salt water. If it comes from the ocean it may be called sea water intrusion.

Salts: Minerals that water picks up as it passes through the air, over and under the ground, or from households and industry.

Salvage: The utilization of waste materials.

Sampling Frequency: The interval between the collection of successive samples.

Sanctions: Actions taken by the federal government for failure to provide or implement a State Implementation Plan (SIP). Such action may include withholding of highway funds and a ban on construction of new sources of potential pollution.

Sand Filters: Devices that remove some suspended solids from sewage. Air and bacteria decompose additional wastes filtering through the sand so that cleaner water drains from the bed.

Sanitary Landfill: (See: landfills.)

Sanitary Sewers: Underground pipes that carry off only domestic or industrial waste, not storm water.

Sanitary Survey: An on-site review of the water sources, facilities, equipment, operation and maintenance of a public water system to evaluate the adequacy of those elements for producing and distributing safe drinking water.

Sanitary Water (Also known as gray water): Water discharged from sinks, showers, kitchens, or other non-industrial operations, but not from commodes.

Sanitation: Control of physical factors in the human environment that could harm development, health, or survival.

Saprolite: A soft, clay-rich, thoroughly decomposed rock formed in place by chemical weathering of igneous or metamorphic rock. Forms in humid, tropical, or subtropical climates.

Saprophytes: Organisms living on dead or decaying organic matter that help natural decomposition of organic matter in water.

Saturated Zone: The area below the water table where all open spaces are filled with water under pressure equal to or greater than that of the atmosphere.

Saturation: The condition of a liquid when it has taken into solution the maximum possible quantity of a given substance at a given temperature and pressure.

Science Advisory Board (SAB): A group of external scientists who advise EPA on science and policy.

Scrap: Materials discarded from manufacturing operations that may be suitable for reprocessing.

Scrap Metal Processor: Intermediate operating facility where recovered metal is sorted, cleaned of contaminants, and prepared for recycling.

Screening: Use of screens to remove coarse floating and suspended solids from sewage.

Screening Risk Assessment: A risk assessment performed with few data and many assumptions to identify exposures that should be evaluated more carefully for potential risk.

Scrubber: An air pollution device that uses a spray of water or reactant or a dry process to trap pollutants in emissions.

Secondary Drinking Water Regulations: Non-enforceable regulations applying to public water systems and specifying the maximum contamination levels that, in the judgment of EPA, are required to protect the public welfare. These regulations apply to any contaminants that may adversely affect the odor or appearance of such water and consequently may cause people served by the system to discontinue its use.

Secondary Effect: Action of a stressor on supporting components of the ecosystem, which in turn impact the ecological component of concern. (See: primary effect.)

Secondary Materials: Materials that have been manufactured and used at least once and are to be used again.

Secondary Standards: National ambient air quality standards designed to protect welfare, including effects on soils, water, crops, vegetation, man-made (anthropogenic) materials, animals, wildlife, weather, visibility, and climate; damage to property; transportation hazards; economic values, and personal comfort and well-being.

Secondary Treatment: The second step in most publicly owned waste treatment systems in which bacteria consume the organic parts of the waste. It is accomplished by bringing together waste, bacteria, and oxygen in trickling filters or in the activated sludge process. This treatment removes floating and settleable solids and about 90 percent of the oxygen-demanding substances and suspended solids. Disinfection is the final stage of secondary treatment. (See: primary, tertiary treatment.)

Secure Chemical Landfill: (See:landfills.)

Secure Maximum Contaminant Level: Maximum permissible level of a contaminant in water delivered to the free flowing outlet of the ultimate user, or of contamination resulting from corrosion of piping and plumbing caused by water quality.

Sediment Yield: The quantity of sediment arriving at a specific location.

Glossary of Terms

Sedimentation: Letting solids settle out of wastewater by gravity during treatment.

Sedimentation Tanks: Wastewater tanks in which floating wastes are skimmed off and settled solids are removed for disposal.

Sediments: Soil, sand, and minerals washed from land into water, usually after rain. They pile up in reservoirs, rivers and harbors, destroying fish and wildlife habitat, and clouding the water so that sunlight cannot reach aquatic plants. Careless farming, mining, and building activities will expose sediment materials, allowing them to wash off the land after rainfall.

Seed Protectant: A chemical applied before planting to protect seeds and seedlings from disease or insects.

Seepage: Percolation of water through the soil from unlined canals, ditches, laterals, watercourses, or water storage facilities.

Selective Pesticide: A chemical designed to affect only certain types of pests, leaving other plants and animals unharmed.

Semi-Confined Aquifer: An aquifer partially confined by soil layers of low permeability through which recharge and discharge can still occur.

Semivolatile Organic Compounds: Organic compounds that volatilize slowly at standard temperature (20 degrees C and 1 atm pressure).

Senescence: The aging process. Sometimes used to describe lakes or other bodies of water in advanced stages of eutrophication. Also used to describe plants and animals.

Septic System: An on-site system designed to treat and dispose of domestic sewage. A typical septic system consists of tank that receives waste from a residence or business and a system of tile lines or a pit for disposal of the liquid effluent (sludge) that remains after decomposition of the solids by bacteria in the tank and must be pumped out periodically.

Septic Tank: An underground storage tank for wastes from homes not connected to a sewer line. Waste goes directly from the home to the tank. (See: septic system.)

Service Connector: The pipe that carries tap water from a public water main to a building.

Service Line Sample: A one-liter sample of water that has been standing for at least 6 hours in a service pipeline and is collected according to federal regulations.

Service Pipe: The pipeline extending from the water main to the building served or to the consumer's system.

Set-Back: Setting a thermometer to a lower temperature when the building is unoccupied to reduce consumption of heating energy. Also refers to setting the thermometer to a higher temperature during unoccupied periods in the cooling season.

Settleable Solids: Material heavy enough to sink to the bottom of a wastewater treatment tank.

Settling Chamber: A series of screens placed in the way of flue gases to slow the stream of air, thus helping gravity to pull particles into a collection device.

Settling Tank: A holding area for wastewater, where heavier particles sink to the bottom for removal and disposal.

7Q10: Seven-day, consecutive low flow with a ten year return frequency; the lowest stream flow for seven consecutive days that would be expected to occur once in ten years.

Sewage: The waste and wastewater produced by residential and commercial sources and discharged into sewers.

Sewage Lagoon: (See: lagoon.)

Sewage Sludge: Sludge produced at a Publicly Owned Treatment Works, the disposal of which is regulated under the Clean Water Act.

Sewer: A channel or conduit that carries wastewater and storm-water runoff from the source to a treatment plant or receiving stream. "Sanitary" sewers carry household, industrial, and commercial waste. "Storm" sewers carry runoff from rain or snow. "Combined" sewers handle both.

Sewerage: The entire system of sewage collection, treatment, and disposal.

Shading Coefficient: The amount of the sun's heat transmitted through a given window compared with that of a standard 1/8- inch-thick single pane of glass under the same conditions.

Sharps: Hypodermic needles, syringes (with or without the attached needle), Pasteur pipettes, scalpel blades, blood vials, needles with attached tubing, and culture dishes used in animal or human patient care or treatment, or in medical, research or industrial laboratories. Also included are other types of broken or unbroken glassware that were in contact with infectious agents, such as used slides and cover slips, and unused hypodermic and suture needles, syringes, and scalpel blades.

Shock Load: The arrival at a water treatment plant of raw water containing unusual amounts of algae, colloidal matter. color, suspended solids, turbidity, or other pollutants.

Short-Circuiting: When some of the water in tanks or basins flows faster than the rest; may result in shorter contact, reaction, or settling times than calculated or presumed.

Sick Building Syndrome: Building whose occupants experience acute health and/or comfort effects that appear to be linked to time spent therein, but where no specific illness or cause can be identified. Complaints may be localized in a particular room or zone, or may spread throughout the building. (See: building-related illness.)

Signal: The volume or product-level change produced by a leak in a tank.

Signal Words: The words used on a pesticide label—Danger, Warning, Caution—to indicate level of toxicity.

Significant Deterioration: Pollution resulting from a new source in previously "clean" areas. (See: prevention of significant deterioration.)

Significant Municipal Facilities: Those publicly owned sewage treatment plants that discharge a million gallons per day or more and are therefore considered by states to have the potential to substantially affect the quality of receiving waters.

Significant Non-Compliance: (See significant violations.)

Significant Potential Source of Contamination: A facility or activity that stores, uses, or produces compounds with potential for significant contaminating impact if released into the source water of a public water supply.

Significant Violations: Violations by point source dischargers of sufficient magnitude or duration to be a regulatory priority.

Silt: Sedimentary materials composed of fine or intermediate-sized mineral particles.

Silviculture: Management of forest land for timber.

Single-Breath Canister: Small one-liter canister designed to capture a single breath. Used in air pollutant ingestion research.

Sink: Place in the environment where a compound or material collects.

Sinking: Controlling oil spills by using an agent to trap the oil and sink it to the bottom of the body of water where the agent and the oil are biodegraded.

SIP Call: EPA action requiring a state to resubmit all or part of its State Implementation Plan to demonstrate attainment of the require national ambient air quality standards within the statutory deadline. A SIP Revision is a revision of a SIP altered at the request of EPA or on a state's initiative. (See: State Implementation Plan.)

Site: An area or place within the jurisdiction of the EPA and/or a state.

Site Assessment Program: A means of evaluating hazardous waste sites through preliminary assessments and site inspections to develop a Hazard Ranking System score.

Site Inspection: The collection of information from a Superfund site to determine the extent and severity of hazards posed by the site. It follows and is more extensive than a preliminary assessment. The purpose is to gather information necessary to score the site, using the Hazard Ranking System, and to determine if it presents an immediate threat requiring prompt removal.

Site Safety Plan: A crucial element in all removal actions, it includes information on equipment being used, precautions to be taken, and steps to take in the event of an on-site emergency.

Siting: The process of choosing a location for a facility.

Skimming: Using a machine to remove oil or scum from the surface of the water.

Slow Sand Filtration: Passage of raw water through a bed of sand at low velocity, resulting in substantial removal of chemical and biological contaminants.

Sludge: A semi-solid residue from any of a number of air or water treatment processes; can be a hazardous waste.

Sludge Digester: Tank in which complex organic substances like sewage sludges are biologically dredged. During these reactions, energy is released and much of the sewage is converted to methane, carbon dioxide, and water.

Slurry: A watery mixture of insoluble matter resulting from some pollution control techniques.

Small Quantity Generator (SQG-sometimes referred to as "Squeegee"): Persons or enterprises that produce 220-2200 pounds per month of hazardous waste; they are required to keep more records than conditionally exempt generators. The largest category of hazardous waste generators, SQGs, include automotive shops, dry cleaners, photographic developers, and many other small businesses. (See: conditionally exempt generators.)

Smelter: A facility that melts or fuses ore, often with an accompanying chemical change, to separate its metal content. Emissions cause pollution. "Smelting" is the process involved.

Smog: Air pollution typically associated with oxidants. (See: photochemical smog.)

Smoke: Particles suspended in air after incomplete combustion.

Soft Detergents: Cleaning agents that break down in nature.

Soft Water: Any water that does not contain a significant amount of dissolved minerals such as salts of calcium or magnesium.

Soil Adsorption Field: A sub-surface area containing a trench or bed with clean stones and a system of piping through which treated sewage may seep into the surrounding soil for further treatment and disposal.

Soil and Water Conservation Practices: Control measures consisting of managerial, vegetative, and structural practices to reduce the loss of soil and water.

Soil Conditioner: An organic material like humus or compost that helps soil absorb water, build a bacterial community, and take up mineral nutrients.

Soil Erodibility: An indicator of a soil's susceptibility to raindrop impact, runoff, and other erosive processes.

Soil Gas: Gaseous elements and compounds in the small spaces between particles of the earth and soil. Such gases can be moved or driven out under pressure.

Soil Moisture: The water contained in the pore space of the unsaturated zone.

Soil Sterilant: A chemical that temporarily or permanently prevents the growth of all plants and animals,

Solder: Metallic compound used to seal joints between pipes. Until recently, most solder contained 50 percent lead. Use of solder containing more than 0.2 percent lead in pipes carrying drinking water is now prohibited.

Sole-Source Aquifer: An aquifer that supplies 50-percent or more of the drinking water of an area.

Solid Waste: Non-liquid, non-soluble materials ranging from municipal garbage to industrial wastes that contain complex and sometimes hazardous substances. Solid wastes also include sewage sludge, agricultural refuse, demolition wastes, and mining residues. Technically, solid waste also refers to liquids and gases in containers.

Solid Waste Disposal: The final placement of refuse that is not salvaged or recycled.

Solid Waste Management: Supervised handling of waste materials from their source through recovery processes to disposal.

Solidification and Stabilization: Removal of wastewater from a waste or changing it chemically to make it less permeable and susceptible to transport by water.

Solubility: The amount of mass of a compound that will dissolve in a unit volume of solution. Aqueous Solubility is the maximum concentration of a chemical that will dissolve in pure water at a reference temperature.

Soot: Carbon dust formed by incomplete combustion.

Sorption: The action of soaking up or attracting substances; process used in many pollution control systems.

Source Area: The location of liquid hydrocarbons or the zone of highest soil or groundwater concentrations, or both, of the chemical of concern.

Source Characterization Measurements: Measurements made to estimate the rate of release of pollutants into the environment from a source such as an incinerator, landfill, etc.

Source Reduction: Reducing the amount of materials entering the waste stream from a specific source by redesigning products or patterns of production or consumption (e.g., using returnable beverage containers). Synonymous with waste reduction.

Source Separation: Segregating various wastes at the point of generation (e.g., separation of paper, metal and glass from other wastes to make recycling simpler and more efficient).

Source-Water Protection Area: The area delineated by a state for a Public Water Supply or including numerous such suppliers, whether the source is ground water or surface water or both.

Sparge or Sparging: Injection of air below the water table to strip dissolved volatile organic compounds and/or oxygenate ground water to facilitate aerobic biodegradation of organic compounds.

Special Local-Needs Registration: Registration of a pesticide product by a state agency for a specific use that is not federally registered. However, the active ingredient must be federally registered for other uses. The special use is specific to that state and is often minor, thus may not warrant the additional cost of a full federal registration process. SLN registration cannot be issued for new active ingredients, food-use active ingredients without tolerances, or for a canceled registration. The products cannot be shipped across state lines.

Special Review: Formerly known as Rebuttable Presumption Against Registration (RPAR), this is the regulatory process through which existing pesticides suspected of posing unreasonable risks to human health, non-target organisms, or the environment are referred for review by EPA. Such review requires an intensive risk/benefit analysis with opportunity for public comment. If risk is found to outweigh social and economic benefits, regulatory actions can be initiated, ranging from label revisions and use-restriction to cancellation or suspended registration.

Special Waste: Items such as household hazardous waste, bulky wastes (refrigerators, pieces of furniture, etc.) tires, and used oil.

Species: 1. A reproductively isolated aggregate of interbreeding organisms having common attributes and usually designated by a common name.2. An organism belonging to belonging to such a category.

Specific Conductance: Rapid method of estimating the dissolved solid content of a water supply by testing its capacity to carry an electrical current.

Specific Yield: The amount of water a unit volume of saturated permeable rock will yield when drained by gravity.

Spill Prevention, Containment, and Countermeasures Plan (SPCP): Plan covering the release of hazardous substances as defined in the Clean Water Act.

Spoil: Dirt or rock removed from its original location—destroying the composition of the soil in the process—as in strip-mining, dredging, or construction.

Sprawl: Unplanned development of open land.

Spray Tower Scrubber: A device that sprays alkaline water into a chamber where acid gases are present to aid in neutralizing the gas.

Spring: Ground water seeping out of the earth where the water table intersects the ground surface.

Spring Melt/Thaw: The process whereby warm temperatures melt winter snow and ice. Because various forms of acid deposition may have been stored in the frozen water, the melt can result in abnormally large amounts of acidity entering streams and rivers, sometimes causing fish kills.

Stabilization: Conversion of the active organic matter in sludge into inert, harmless material.

Stabilization Ponds: (See: lagoon.)

Stable Air: A motionless mass of air that holds, instead of dispersing, pollutants.

Stack: A chimney, smokestack, or vertical pipe that discharges used air.

Stack Effect: Air, as in a chimney, that moves upward because it is warmer than the ambient atmosphere.

Stack Effect: Flow of air resulting from warm air rising, creating a positive pressure area at the top of a building and negative pressure area at the bottom. This effect can overpower the mechanical system and disrupt building ventilation and air circulation.

Stack Gas: (See: flue gas.)

Stage II Controls: Systems placed on service station gasoline pumps to control and capture gasoline vapors during refuelling.

Stagnation: Lack of motion in a mass of air or water that holds pollutants in place.

Stakeholder: Any organization, governmental entity, or individual that has a stake in or may be impacted by a given approach to environmental regulation, pollution prevention, energy conservation, etc.

Standard Sample: The part of finished drinking water that is examined for the presence of coliform bacteria.

Standards: Norms that impose limits on the amount of pollutants or emissions produced. EPA establishes minimum standards, but states are allowed to be stricter.

Glossary of Terms

Start of a Response Action: The point in time when there is a guarantee or set-aside of funding by EPA, other federal agencies, states or Principal Responsible Parties in order to begin response actions at a Superfund site.

State Emergency Response Commission (SERC): Commission appointed by each state governor according to the requirements of SARA Title III. The SERCs designate emergency planning districts, appoint local emergency planning committees, and supervise and coordinate their activities.

State Environmental Goals and Indication Project: Program to assist state environmental agencies by providing technical and financial assistance in the development of environmental goals and indicators.

State Implementation Plans (SIP): EPA approved state plans for the establishment, regulation, and enforcement of air pollution standards.

State Management Plan: Under FIFRA, a state management plan required by EPA to allow states, tribes, and U.S. territories the flexibility to design and implement ways to protect ground water from the use of certain pesticides.

Static Water Depth: The vertical distance from the centerline of the pump discharge down to the surface level of the free pool while no water is being drawn from the pool or water table.

Static Water Level: 1. Elevation or level of the water table in a well when the pump is not operating. 2. The level or elevation to which water would rise in a tube connected to an artesian aquifer or basin in a conduit under pressure.

Stationary Source: A fixed-site producer of pollution, mainly power plants and other facilities using industrial combustion processes. (See: point source.)

Sterilization: The removal or destruction of all microorganisms, including pathogenic and other bacteria, vegetative forms, and spores.

Sterilizer: One of three groups of anti-microbials registered by EPA for public health uses. EPA considers an antimicrobial to be a sterilizer when it destroys or eliminates all forms of bacteria, viruses, and fungi and their spores. Because spores are considered the most difficult form of microorganism to destroy, EPA considers the term sporicide to be synonymous with sterilizer.

Storage: Temporary holding of waste pending treatment or disposal, as in containers, tanks, waste piles, and surface impoundments.

Storm Sewer: A system of pipes (separate from sanitary sewers) that carries water runoff from buildings and land surfaces.

Stratification: Separating into layers.

Stratigraphy: Study of the formation, composition, and sequence of sediments, whether consolidated or not.

Stratosphere: The portion of the atmosphere 10-to-25 miles above the earth's surface.

Stressors: Physical, chemical, or biological entities that can induce adverse effects on ecosystems or human health.

Strip-Cropping: Growing crops in a systematic arrangement of strips or bands that serve as barriers to wind and water erosion.

Strip-Mining: A process that uses machines to scrape soil or rock away from mineral deposits just under the earth's surface.

Structural Deformation: Distortion in walls of a tank after liquid has been added or removed.

Subchronic: Of intermediate duration, usually used to describe studies or periods of exposure lasting between 5 and 90 days.

Subchronic Exposure: Multiple or continuous exposures lasting for approximately ten percent of an experimental species lifetime, usually over a three-month period.

Submerged Aquatic Vegetation: Vegetation that lives at or below the water surface; an important habitat for young fish and other aquatic organisms.

Subwatershed: Topographic perimeter of the catchment area of a stream tributary.

Sulfur Dioxide (SO2): A pungent, colorless, gasformed primarily by the combustion of fossil fuels; becomes a pollutant when present in large amounts.

Sump: A pit or tank that catches liquid runoff for drainage or disposal.

Superchlorination: Chlorination with doses that are deliberately selected to produce water free of combined residuals so large as to require dechlorination.

Supercritical Water: A type of thermal treatment using moderate temperatures and high pressures to enhance the ability of water to break down large organic molecules into smaller, less toxic ones. Oxygen injected during this process combines with simple organic compounds to form carbon dioxide and water.

Superfund: The program operated under the legislative authority of CERCLA and SARA that funds and carries out EPA solid waste emergency and long-term removal and remedial activities. These activities include establishing the National Priorities List, investigating sites for inclusion on the list, determining their priority, and conducting and/or supervising cleanup and other remedial actions.

Superfund Innovative Technology Evaluation (SITE) Program: EPA program to promote development and use of innovative treatment and site characterization technologies in Superfund site cleanups.

Supplemental Registration: An arrangement whereby a registrant licenses another company to market its pesticide product under the second company's registration.

Supplier of Water: Any person who owns or operates a public water supply.

Surface Impoundment: Treatment, storage, or disposal of liquid hazardous wastes in ponds.

Surface Runoff: Precipitation, snow melt, or irrigation water in excess of what can infiltrate the soil surface and be stored in small surface depressions; a major transporter of non-point source pollutants in rivers, streams, and lakes..

Surface Uranium Mines: Strip mining operations for removal of uranium-bearing ore.

Surface Water: All water naturally open to the atmosphere (rivers, lakes, reservoirs, ponds, streams, impoundments, seas, estuaries, etc.)

Surface-Water Treatment Rule: Rule that specifies maximum contaminant level goals for Giardia lamblia, viruses, and Legionella and promulgates filtration and disinfection requirements for public water systems using surface-water or ground-water sources under the direct influence of surface water. The regulations also specify water quality, treatment, and watershed protection criteria under which filtration may be avoided.

Surfacing ACM: Asbestos-containing material that is sprayed or troweled on or otherwise applied to surfaces, such as acoustical plaster on ceilings and fireproofing materials on structural members.

Surfacing Material: Material sprayed or troweled onto structural members (beams, columns, or decking) for fire protection; or on ceilings or walls for fireproofing, acoustical or decorative purposes. Includes textured plaster, and other textured wall and ceiling surfaces.

Surfactant: A detergent compound that promotes lathering.

Surrogate Data: Data from studies of test organisms or a test substance that are used to estimate the characteristics or effects on another organism or substance.

Surveillance System: A series of monitoring devices designed to check on environmental conditions.

Susceptibility Analysis: An analysis to determine whether a Public Water Supply is subject to significant pollution from known potential sources.

Suspect Material: Building material suspected of containing asbestos; e.g., surfacing material, floor tile, ceiling tile, thermal system insulation.

Suspended Loads: Specific sediment particles maintained in the water column by turbulence and carried with the flow of water.

Suspended Solids: Small particles of solid pollutants that float on the surface of, or are suspended in, sewage or other liquids. They resist removal by conventional means.

Suspension: Suspending the use of a pesticide when EPA deems it necessary to prevent an imminent hazard resulting from its continued use. An emergency suspension takes effect immediately; under an ordinary suspension a registrant can request a hearing before the suspension goes into effect. Such a hearing process might take six months.

Suspension Culture: Cells growing in a liquid nutrient medium.

Swamp: A type of wetland dominated by woody vegetation but without appreciable peat deposits. Swamps may be fresh or salt water and tidal or non-tidal. (See: wetlands.)

Synergism: An interaction of two or more chemicals that results in an effect greater than the sum of their separate effects.

Synthetic Organic Chemicals (SOCs): Man-made (anthropogenic) organic chemicals. Some SOCs are volatile; others tend to stay dissolved in water instead of evaporating.

System With a Single Service Connection: A system that supplies drinking water to consumers via a single service line.

Systemic Pesticide: A chemical absorbed by an organism that interacts with the organism and makes the organism toxic to pests.

T

Tail Water: The runoff of irrigation water from the lower end of an irrigated field.

Tailings: Residue of raw material or waste separated out during the processing of crops or mineral ores.

Tailpipe Standards: Emissions limitations applicable to mobile source engine exhausts.

Tampering: Adjusting, negating, or removing pollution control equipment on a motor vehicle.

Technical Assistance Grant (TAG): As part of the Superfund program, Technical Assistance Grants of up to $50,000 are provided to citizens' groups to obtain assistance in interpreting information related to clean-ups at Superfund sites or those proposed for the National Priorities List. Grants are used by such groups to hire technical advisors to help them understand the site-related technical information for the duration of response activities.

Technical-Grade Active Ingredient (TGA): A pesticide chemical in pure form as it is manufactured prior to being formulated into an end-use product (e.g. wettable powders, granules, emulsifiable concentrates). Registered manufactured products composed of such chemicals are known as Technical Grade Products.

Technology-Based Limitations: Industry-specific effluent limitations based on best available preventive technology applied to a discharge when it will not cause a violation of water quality standards at low stream flows. Usually applied to discharges into large rivers.

Technology-Based Standards: Industry-specific effluent limitations applicable to direct and indirect sources which are developed on a category-by-category basis using statutory factors, not including water-quality effects.

Teratogenesis: The introduction of nonhereditary birth defects in a developing fetus by exogenous factors such as physical or chemical agents acting in the womb to interfere with normal embryonic development.

Terracing: Dikes built along the contour of sloping farm land that hold runoff and sediment to reduce erosion.

Tertiary Treatment: Advanced cleaning of wastewater that goes beyond the secondary or biological stage, removing nutrients such as phosphorus, nitrogen, and most BOD and suspended solids.

Theoretical Maximum Residue Contribution: The theoretical maximum amount of a pesticide in the daily diet of an average person. It assumes that the diet is composed of all food items for which there are tolerance-level residues of the pesticide. The TMRC is expressed as milligrams of pesticide/kilograms of body weight/day.

Therapeutic Index: The ratio of the dose required to produce toxic or lethal effects to the dose required to produce nonadverse or therapeutic response.

Thermal Pollution: Discharge of heated water from industrial processes that can kill or injure aquatic organisms.

Thermal Stratification: The formation of layers of different temperatures in a lake or reservoir.

Thermal System Insulation (TSI): Asbestos-containing material applied to pipes, fittings, boilers, breeching, tanks, ducts, or other interior structural components to prevent heat loss or gain or water condensation.

Thermal Treatment: Use of elevated temperatures to treat hazardous wastes. (See: incineration; pyrolysis.)

Thermocline: The middle layer of a thermally stratified lake or reservoir. In this layer, there is a rapid decrease in temperatures in a lake or reservoir.

Threshold: The lowest dose of a chemical at which a specified measurable effect is observed and below which it is not observed.

Threshold: The dose or exposure level below which a significant adverse effect is not expected.

Threshold Level: Time-weighted average pollutant concentration values, exposure beyond which is likely to adversely affect human health. (See: environmental exposure)

Threshold Limit Value (TLV): The concentration of an airborne substance to which an average person can be repeatedly exposed without adverse effects. TLVs may be expressed in three ways: (1) TLV-TWA—Time weighted average, based on an allowable exposure averaged over a normal 8-hour workday or 40-hour work- week; (2) TLV-STEL—Short-term exposure limit or maximum concentration for a brief specified period of time, depending on a specific chemical (TWA must still be met); and (3) TLV-C—Ceiling Exposure Limit or maximum exposure concentration not to be exceeded under any circumstances. (TWA must still be met.)

Threshold Odor: (See: Odor threshold)

Threshold Planning Quantity: A quantity designated for each chemical on the list of extremely hazardous substances that triggers notification by facilities to the State Emergency Response Commission that such facilities are subject to emergency planning requirements under SARA Title III.

Thropic Levels: A functional classification of species that is based on feeding relationships (e.g. generally aquatic and terrestrial green plants comprise the first thropic level, and herbivores comprise the second.)

Tidal Marsh: Low, flat marshlands traversed by channels and tidal hollows, subject to tidal inundation; normally, the only vegetation present is salt-tolerant bushes and grasses. (See: wetlands.)

Tillage: Plowing, seedbed preparation, and cultivation practices.

Time-weighted Average (TWA): In air sampling, the average air concentration of contaminants during a given period.

Tire Processor: Intermediate operating facility where recovered tires are processed in preparation for recycling.

Tires: As used in recycling, passenger car and truck tires (excludes airplane, bus, motorcycle and special service military, agricultural, off-the-road and slow speed industrial tires). Car and truck tires are recycled into rubber products such as trash cans, storage containers, rubberized asphalt or used whole for playground and reef construction.

Tolerance Petition: A formal request to establish a new tolerance or modify an existing one.

Tolerances: Permissible residue levels for pesticides in raw agricultural produce and processed foods. Whenever a pesticide is registered for use on a food or a feed crop, a tolerance (or exemption from the tolerance requirement) must be established. EPA establishes the tolerance levels, which are enforced by the Food and Drug Administration and the Department of Agriculture.

Tonnage: The amount of waste that a landfill accepts, usually expressed in tons per month. The rate at which a landfill accepts waste is limited by the landfill's permit.

Topography: The physical features of a surface area including relative elevations and the position of natural and man-made (anthropogenic) features.

Total Dissolved Phosphorous: The total phosphorous content of all material that will pass through a filter, which is determined as orthophosphate without prior digestion or hydrolysis. Also called soluble P. or ortho P.

Total Dissolved Solids (TDS): All material that passes the standard glass river filter; now called total filtrable residue. Term is used to reflect salinity.

Total Petroleum Hydrocarbons (TPH): Measure of the concentration or mass of petroleum hydrocarbon constituents present in a given amount of soil or water. The word "total" is a misnomer—few, if any, of the procedures for quantifying hydrocarbons can measure all of them in a given sample. Volatile ones are usually lost in the process and not quantified and non-petroleum hydrocarbons sometimes appear in the analysis.

Total Recovered Petroleum Hydrocarbon: A method for measuring petroleum hydrocarbons in samples of soil or water.

Total Suspended Particles (TSP): A method of monitoring airborne particulate matter by total weight.

Total Suspended Solids (TSS): A measure of the suspended solids in wastewater, effluent, or water bodies, determined by tests for "total suspended non-filterable solids." (See: suspended solids.)

Toxaphene: Chemical that causes adverse health effects in domestic water supplies and is toxic to fresh water and marine aquatic life.

Toxic Chemical: Any chemical listed in EPA rules as "Toxic Chemicals Subject to Section 313 of the Emergency Planning and Community Right-to-Know Act of 1986."

Toxic Chemical Release Form: Information form required of facilities that manufacture, process, or use (in quantities above a specific amount) chemicals listed under SARA Title III.

Toxic Chemical Use Substitution: Replacing toxic chemicals with less harmful chemicals in industrial processes.

Toxic Cloud: Airborne plume of gases, vapors, fumes, or aerosols containing toxic materials.

Toxic Concentration: The concentration at which a substance produces a toxic effect.

Toxic Dose: The dose level at which a substance produces a toxic effect.

Glossary of Terms

Toxic Pollutants: Materials that cause death, disease, or birth defects in organisms that ingest or absorb them. The quantities and exposures necessary to cause these effects can vary widely.

Toxic Release Inventory: Database of toxic releases in the United States compiled from SARA Title III Section 313 reports.

Toxic Substance: A chemical or mixture that may present an unreasonable risk of injury to health or the environment.

Toxic Waste: A waste that can produce injury if inhaled, swallowed, or absorbed through the skin.

Toxicant: A harmful substance or agent that may injure an exposed organism.

Toxicity: The degree to which a substance or mixture of substances can harm humans or animals. Acute toxicity involves harmful effects in an organism through a single or short-term exposure. Chronic toxicity is the ability of a substance or mixture of substances to cause harmful effects over an extended period, usually upon repeated or continuous exposure sometimes lasting for the entire life of the exposed organism. Subchronic toxicity is the ability of the substance to cause effects for more than one year but less than the lifetime of the exposed organism.

Toxicity Assessment: Characterization of the toxicological properties and effects of a chemical, with special emphasis on establishment of dose-response characteristics.

Toxicity Testing: Biological testing (usually with an invertebrate, fish, or small mammal) to determine the adverse effects of a compound or effluent.

Toxicological Profile: An examination, summary, and interpretation of a hazardous substance to determine levels of exposure and associated health effects.

Transboundary Pollutants: Air pollution that travels from one jurisdiction to another, often crossing state or international boundaries. Also applies to water pollution.

Transfer Station: Facility where solid waste is transferred from collection vehicles to larger trucks or rail cars for longer distance transport.

Transient Water System: A non-community water system that does not serve 25 of the same nonresidents per day for more than six months per year.

Transmission Lines: Pipelines that transport raw water from its source to a water treatment plant, then to the distribution grid system.

Transmissivity: The ability of an aquifer to transmit water.

Transpiration: The process by which water vapor is lost to the atmosphere from living plants. The term can also be applied to the quantity of water thus dissipated.

Transportation Control Measures (TCMs): Steps taken by a locality to reduce vehicular emission and improve air quality by reducing or changing the flow of traffic; e.g. bus and HOV lanes, carpooling and other forms of ride-sharing, public transit, bicycle lanes.

Transporter: Hauling firm that picks up properly packaged and labeled hazardous waste from generators and transports it to designated facilities

for treatment, storage, or disposal. Transporters are subject to EPA and DOT hazardous waste regulations.

Trash: Material considered worthless or offensive that is thrown away. Generally defined as dry waste material, but in common usage it is a synonym for garbage, rubbish, or refuse.

Trash-to-Energy Plan: Burning trash to produce energy.

Treatability Studies: Tests of potential cleanup technologies conducted in a laboratory (See: bench-scale tests.)

Treated Regulated Medical Waste: Medical waste treated to substantially reduce or eliminate its pathogenicity, but that has not yet been destroyed.

Treated Wastewater: Wastewater that has been subjected to one or more physical, chemical, and biological processes to reduce its potential of being health hazard.

Treatment: (1) Any method, technique, or process designed to remove solids and/or pollutants from solid waste, waste-streams, effluents, and air emissions. (2) Methods used to change the biological character or composition of any regulated medical waste so as to substantially reduce or eliminate its potential for causing disease.

Treatment Plant: A structure built to treat wastewater before discharging it into the environment.

Treatment, Storage, and Disposal Facility: Site where a hazardous substance is treated, stored, or disposed of. TSD facilities are regulated by EPA and states under RCRA.

Tremie: Device used to place concrete or grout under water.

Trial Burn: An incinerator test in which emissions are monitored for the presence of specific organic compounds, particulates, and hydrogen chloride.

Trichloroethylene (TCE): A stable, low boiling-point colorless liquid, toxic if inhaled. Used as a solvent or metal degreasing agent, and in other industrial applications.

Trickle Irrigation: Method in which water drips to the soil from perforated tubes or emitters.

Trickling Filter: A coarse treatment system in which wastewater is trickled over a bed of stones or other material covered with bacteria that break down the organic waste and produce clean water.

Trihalomethane (THM): One of a family of organic compounds named as derivative of methane. THMs are generally by-products of chlorination of drinking water that contains organic material.

Troposphere: The layer of the atmosphere closest to the earth's surface.

Trust Fund (CERCLA): A fund set up under the Comprehensive Environmental Response, Compensation and Liability Act (CERCLA) to help pay for cleanup of hazardous waste sites and for legal action to force those responsible for the sites to clean them up.

Tube Settler: Device using bundles of tubes to let solids in water settle to the bottom for removal by conventional sludge collection means; sometimes used in sedimentation basins and clarifiers to improve particle removal.

Tuberculation: Development or formation of small mounds of corrosion products on the inside of iron pipe. These tubercules roughen the inside of the pipe, increasing its resistance to water flow.

Tundra: A type of treeless ecosystem dominated by lichens, mosses, grasses, and woody plants. Tundra is found at high latitudes (arctic tundra) and high altitudes (alpine tundra). Arctic tundra is underlain by permafrost and is usually water saturated. (See: wetlands.)

Turbidimeter: A device that measures the cloudiness of suspended solids in a liquid; a measure of the quantity of suspended solids.

Turbidity: 1. Haziness in air caused by the presence of particles and pollutants. 2. A cloudy condition in water due to suspended silt or organic matter.

U

Ultra Clean Coal (UCC): Coal that is washed, ground into fine particles, then chemically treated to remove sulfur, ash, silicone, and other substances; usually briquetted and coated with a sealant made from coal.

Ultraviolet Rays: Radiation from the sun that can be useful or potentially harmful. UV rays from one part of the spectrum (UV-A) enhance plant life. UV rays from other parts of the spectrum (UV-B) can cause skin cancer or other tissue damage. The ozone layer in the atmosphere partly shields us from ultraviolet rays reaching the earth's surface.

Uncertainty Factor: One of several factors used in calculating the reference dose from experimental data. UFs are intended to account for (1) the variation in sensitivity among humans; (2) the uncertainty in extrapolating animal data to humans; (3) the uncertainty in extrapolating data obtained in a study that covers less than the full life of the exposed animal or human; and (4) the uncertainty in using LOAEL data rather than NOAEL data.

Unconfined Aquifer: An aquifer containing water that is not under pressure; the water level in a well is the same as the water table outside the well.

Underground Injection Control (UIC): The program under the Safe Drinking Water Act that regulates the use of wells to pump fluids into the ground.

Underground Injection Wells: Steel- and concrete-encased shafts into which hazardous waste is deposited by force and under pressure.

Underground Sources of Drinking Water: Aquifers currently being used as a source of drinking water or those capable of supplying a public water system. They have a total dissolved solids content of 10,000 milligrams per liter or less, and are not "exempted aquifers." (See: exempted aquifer.)

Underground Storage Tank (UST): A tank located at least partially underground and designed to hold gasoline or other petroleum products or chemicals.

Unreasonable Risk: Under the Federal Insecticide, Fungicide, and Rodenticide Act (FIFRA), "unreasonable adverse effects" means any unreasonable risk to man or the environment, taking into account the medical, economic, social, and environmental costs and benefits of any pesticide.

Unsaturated Zone: The area above the water table where soil pores are not fully saturated, although some water may be present.

Upper Detection Limit: The largest concentration that an instrument can reliably detect.

Uranium Mill Tailings Piles: Former uranium ore processing sites that contain leftover radioactive materials (wastes), including radium and unrecovered uranium.

Uranium Mill-Tailings Waste Piles: Licensed active mills with tailings piles and evaporation ponds created by acid or alkaline leaching processes.

Urban Runoff: Storm water from city streets and adjacent domestic or commercial properties that carries pollutants of various kinds into the sewer systems and receiving waters.

Urea-Formaldehyde Foam Insulation: A material once used to conserve energy by sealing crawl spaces, attics, etc.; no longer used because emissions were found to be a health hazard.

Use Cluster: A set of competing chemicals, processes, and/or technologies that can substitute for one another in performing a particular function.

Used Oil: Spent motor oil from passenger cars and trucks collected at specified locations for recycling (not included in the category of municipal solid waste).

User Fee: Fee collected from only those persons who use a particular service, as compared to one collected from the public in general.

Utility Load: The total electricity demand for a utility district.

V

Vadose Zone: The zone between land surface and the water table within which the moisture content is less than saturation (except in the capillary fringe) and pressure is less than atmospheric. Soil pore space also typically contains air or other gases. The capillary fringe is included in the vadose zone. (See: Unsaturated Zone.)

Valued Environmental Attributes/Components: Those aspects(components/processes/functions) of ecosystems, human health, and environmental welfare considered to be important and potentially at risk from human activity or natural hazards. Similar to the term "valued environmental components" used in environmental impact assessment.

Vapor Capture System: Any combination of hoods and ventilation system that captures or contains organic vapors so they may be directed to an abatement or recovery device.

Vapor Dispersion: The movement of vapor clouds in air due to wind, thermal action, gravity spreading, and mixing.

Vapor Plumes: Flue gases visible because they contain water droplets.

Vapor Pressure: A measure of a substance's propensity to evaporate, vapor pressure is the force per unit area exerted by vapor in an equilibrium state with surroundings at a given pressure. It increases exponentially with an increase in temperature. A relative measure of chemical volatility, vapor pressure is used to calculate water partition coefficients and volatilization rate constants.

Variance: Government permission for a delay or exception in the application of a given law, ordinance, or regulation.

Vector: 1. An organism, often an insect or rodent, that carries disease. 2. Plasmids, viruses, or bacteria used to transport genes into a host cell. A gene is placed in the vector; the vector then "infects" the bacterium.

Vegetative Controls: Non-point source pollution control practices that involve vegetative cover to reduce erosion and minimize loss of pollutants.

Vehicle Miles Travelled (VMT): A measure of the extent of motor vehicle operation; the total number of vehicle miles travelled within a specific geographic area over a given period of time.

Ventilation Rate: The rate at which indoor air enters and leaves a building. Expressed as the number of changes of outdoor air per unit of time (air changes per hour (ACH), or the rate at which a volume of outdoor air enters in cubic feet per minute (CFM).

Ventilation/Suction: The act of admitting fresh air into a space in order to replace stale or contaminated air; achieved by blowing air into the space. Similarly, suction represents the admission of fresh air into an interior space by lowering the pressure outside of the space, thereby drawing the contaminated air outward.

Venturi Scrubbers: Air pollution control devices that use water to remove particulate matter from emissions.

Vinyl Chloride: A chemical compound, used in producing some plastics, that is believed to be oncogenic.

Virgin Materials: Resources extracted from nature in their raw form, such as timber or metal ore.

Viscosity: The molecular friction within a fluid that produces flow resistance.

Volatile: Any substance that evaporates readily.

Volatile Liquids: Liquids which easily vaporize or evaporate at room temperature.

Volatile Organic Compound (VOC): Any organic compound that participates in atmospheric photochemical reactions except those designated by EPA as having negligible photochemical reactivity.

Volatile Solids: Those solids in water or other liquids that are lost on ignition of the dry solids at 550ø centigrade.

Volatile Synthetic Organic Chemicals: Chemicals that tend to volatilize or evaporate.

Volume Reduction: Processing waste materials to decrease the amount of space they occupy, usually by compacting, shredding, incineration, or composting.

Volumetric Tank Test: One of several tests to determine the physical integrity of a storage tank; the volume of fluid in the tank is measured directly or calculated from product-level changes. A marked drop in volume indicates a leak.

Vulnerability Analysis: Assessment of elements in the community that are susceptible to damage if hazardous materials are released.

Vulnerable Zone: An area over which the airborne concentration of a chemical accidentally released could reach the level of concern.

W

Waste: 1. Unwanted materials left over from a manufacturing process. 2. Refuse from places of human or animal habitation.

Waste Characterization: Identification of chemical and microbiological constituents of a waste material.

Waste Exchange: Arrangement in which companies exchange their wastes for the benefit of both parties.

Waste Feed: The continuous or intermittent flow of wastes into an incinerator.

Waste Generation: The weight or volume of materials and products that enter the waste stream before recycling, composting, landfilling, or combustion takes place. Also can represent the amount of waste generated by a given source or category of sources.

Waste Load Allocation: 1. The maximum load of pollutants each discharger of waste is allowed to release into a particular waterway. Discharge limits are usually required for each specific water quality criterion being, or expected to be, violated. 2. The portion of a stream's total assimilative capacity assigned to an individual discharge.

Waste Minimization: Measures or techniques that reduce the amount of wastes generated during industrial production processes; term is also applied to recycling and other efforts to reduce the amount of waste going into the waste stream.

Waste Piles: Non-containerized, lined or unlined accumulations of solid, nonflowing waste.

Waste Reduction: Using source reduction, recycling, or composting to prevent or reduce waste generation.

Waste Stream: The total flow of solid waste from homes, businesses, institutions, and manufacturing plants that is recycled, burned, or disposed of in landfills, or segments thereof such as the "residential waste stream" or the "recyclable waste stream."

Waste Treatment Lagoon: Impoundment made by excavation or earth fill for biological treatment of wastewater.

Waste Treatment Plant: A facility containing a series of tanks, screens, filters and other processes by which pollutants are removed from water.

Waste Treatment Stream: The continuous movement of waste from generator to treater and disposer.

Waste-Heat Recovery: Recovering heat discharged as a byproduct of one process to provide heat needed by a second process.

Waste-to-Energy Facility/Municipal-Waste Combustor: Facility where recovered municipal solid waste is converted into a usable form of energy, usually via combustion.

Wastewater: The spent or used water from a home, community, farm, or industry that contains dissolved or suspended matter.Water Pollution: The presence in water of enough harmful or objectionable material to damage the water's quality.

787

Glossary of Terms

Wastewater Infrastructure: The plan or network for the collection, treatment, and disposal of sewage in a community. The level of treatment will depend on the size of the community, the type of discharge, and/or the designated use of the receiving water.

Wastewater Operations and Maintenance: Actions taken after construction to ensure that facilities constructed to treat wastewater will be operated, maintained, and managed to reach prescribed effluent levels in an optimum manner.

Water Purveyor: A public utility, mutual water company, county water district, or municipality that delivers drinking water to customers.

Water Quality Criteria: Levels of water quality expected to render a body of water suitable for its designated use. Criteria are based on specific levels of pollutants that would make the water harmful if used for drinking, swimming, farming, fish production, or industrial processes.

Water Quality Standards: State-adopted and EPA-approved ambient standards for water bodies. The standards prescribe the use of the water body and establish the water quality criteria that must be met to protect designated uses.

Water Quality-Based Limitations: Effluent limitations applied to dischargers when mere technology-based limitations would cause violations of water quality standards. Usually applied to discharges into small streams.

Water Quality-Based Permit: A permit with an effluent limit more stringent than one based on technology performance. Such limits may be necessary to protect the designated use of receiving waters (e.g. recreation, irrigation, industry or water supply).

Water Solubility: The maximum possible concentration of a chemical compound dissolved in water. If a substance is water soluble it can very readily disperse through the environment.

Water Storage Pond: An impound for liquid wastes designed to accomplish some degree of biochemical treatment.

Water Supplier: One who owns or operates a public water system.

Water Supply System: The collection, treatment, storage, and distribution of potable water from source to consumer.

Water Table: The level of groundwater.

Water Treatment Lagoon: An impound for liquid wastes designed to accomplish some degree of biochemical treatment.

Water Well: An excavation where the intended use is for location, acquisition, development, or artificial recharge of ground water.

Water-Soluble Packaging: Packaging that dissolves in water; used to reduce exposure risks to pesticide mixers and loaders.

Water-Source Heat Pump: Heat pump that uses wells or heat exchangers to transfer heat from water to the inside of a building. Most such units use ground water. (See: groundsource heat pump; heat pump.)

Waterborne Disease Outbreak: The significant occurence of acute illness associated with drinking water from a public water system that is deficient in treatment, as determined by appropriate local or state agencies.

Watershed: The land area that drains into a stream; the watershed for a major river may encompass a number of smaller watersheds that ultimately combine at a common point.

Watershed Approach: A coordinated framework for environmental management that focuses public and private efforts on the highest priority problems within hydrologically-defined geographic areas taking into consideration both ground and surface water flow.

Watershed Area: A topographic area within a line drawn connecting the highest points uphill of a drinking waterintake into which overland flow drains.

Weight of Scientific Evidence: Considerations in assessing the interpretation of published information about toxicity—quality of testing methods, size and power of study design, consistency of results across studies, and biological plausibility of exposure-response relationships and statistical associations.

Weir: 1. A wall or plate placed in an open channel to measure the flow of water. 2. A wall or obstruction used to control flow from settling tanks and clarifiers to ensure a uniform flow rate and avoid short-circuiting. (See: short-circuiting.)

Well: A bored, drilled, or driven shaft, or a dug hole whose depth is greater than the largest surface dimension and whose purpose is to reach underground water supplies or oil, or to store or bury fluids below ground.

Well Field: Area containing one or more wells that produce usable amounts of water or oil.

Well Injection: The subsurface emplacement of fluids into a well.

Well Monitoring: Measurement by on-site instruments or laboratory methods of well water quality.

Well Plug: A watertight, gastight seal installed in a bore hole or well to prevent movement of fluids.

Well Point: A hollow vertical tube, rod, or pipe terminating in a perforated pointed shoe and fitted with a fine-mesh screen.

Wellhead Protection Area: A protected surface and subsurface zone surrounding a well or well field supplying a public water system to keep contaminants from reaching the well water.

Wetlands: An area that is saturated by surface or ground water with vegetation adapted for life under those soil conditions, as swamps, bogs, fens, marshes, and estuaries.

Wettability: The relative degree to which a fluid will spread into or coat a solid surface in the presence of other immiscible fluids.

Wettable Powder: Dry formulation that must be mixed with water or other liquid before it is applied.

Wheeling: The transmission of electricity owned by one entity through the facilities owned by another (usually a utility).

Whole-Effluent-Toxicity Tests: Tests to determine the toxicity levels of the total effluent from a single source as opposed to a series of tests for individual contaminants.

Wildlife Refuge: An area designated for the protection of wild animals, within which hunting and fishing are either prohibited or strictly controlled.

Wire-to-Wire Efficiency: The efficiency of a pump and motor together.

Wood Packaging: Wood products such as pallets, crates, and barrels.

Wood Treatment Facility: An industrial facility that treats lumber and other wood products for outdoor use. The process employs chromated copper arsenate, which is regulated as a hazardous material.

Wood-Burning-Stove Pollution: Air pollution caused by emissions of particulate matter, carbon monoxide, total suspended particulates, and polycyclic organic matter from wood-burning stoves.

Working Level (WL): A unit of measure for documenting exposure to radon decay products, the so-called "daughters." One working level is equal to approximately 200 picocuries per liter.

Working Level Month (WLM): A unit of measure used to determine cumulative exposure to radon.

X

Xenobiota: Any biotum displaced from its normal habitat; a chemical foreign to a biological system.

Y

Yard Waste: The part of solid waste composed of grass clippings, leaves, twigs, branches, and other garden refuse.

Yellow-Boy: Iron oxide flocculant (clumps of solids in waste or water); usually observed as orange-yellow deposits in surface streams with excess iron content. (See: floc, flocculation.)

Yield: The quantity of water (expressed as a rate of flow or total quantity per year) that can be collected for a given use from surface or groundwater sources.

Z

Zero Air: Atmospheric air purified to contain less than 0.1 ppm total hydrocarbons.

Zooplankton: Small (often microscopic) free-floating aquatic plants or animals.

Source: U.S. Environmental Protection Agency, "Terms of Environment," Revised June 20, 2005 (http://www.epa.gov/OCEPAterms/intro.htm)

Abbreviations & Acronyms

A

A&I: Alternative and Innovative (Wastewater Treatment System)

AA: Accountable Area; Adverse Action; Advices of Allowance; Assistant Administrator; Associate Administrator; Atomic Absorption

AAEE: American Academy of Environmental Engineers

AANWR: Alaskan Arctic National Wildlife Refuge

AAP: Asbestos Action Program

AAPCO: American Association of Pesticide Control Officials

AARC: Alliance for Acid Rain Control

ABEL: EPA's computer model for analyzing a violator's ability to pay a civil penalty.

ABES: Alliance for Balanced Environmental Solutions

AC: Actual Commitment. Advisory Circular

A&C: Abatement and Control

ACA: American Conservation Association

ACBM: Asbestos-Containing Building Material

ACE: Alliance for Clean Energy

ACE: Any Credible Evidence

ACEEE: American Council for an Energy Efficient Economy

ACFM: Actual Cubic Feet Per Minute

ACL: Alternate Concentration Limit. Analytical Chemistry Laboratory

ACM: Asbestos-Containing Material

ACP: Agriculture Control Program (Water Quality Management); ACP: Air Carcinogen Policy

ACQUIRE: Aquatic Information Retrieval

ACQR: Air Quality Control Region

ACS: American Chemical Society

ACT: Action

ACTS: Asbestos Contractor Tracking System

ACWA: American Clean Water Association

ACWM: Asbestos-Containing Waste Material

ADABA: Acceptable Data Base

ADB: Applications Data Base

ADI: Acceptable Daily Intake

ADP: AHERA Designated Person; Automated Data Processing

ADQ: Audits of Data Quality

ADR: Alternate Dispute Resolution

ADSS: Air Data Screening System

ADT: Average Daily Traffic

AEA: Atomic Energy Act

AEC: Associate Enforcement Counsels

AEE: Alliance for Environmental Education

AEERL: Air and Energy Engineering Research Laboratory

AEM: Acoustic Emission Monitoring

AERE: Association of Environmental and Resource Economists

AES: Auger Electron Spectrometry

AFA: American Forestry Association

AFCA: Area Fuel Consumption Allocation

AFCEE:Air Force Center for Environmental Excellence

AFS: AIRS Facility Subsystem

AFUG: AIRS Facility Users Group

AH: Allowance Holders

AHERA: Asbestos Hazard Emergency Response Act

AHU: Air Handling Unit

AI: Active Ingredient

AIC: Active to Inert Conversion

AICUZ: Air Installation Compatible Use Zones

AID: Agency for International Development

AIHC: American Industrial Health Council

AIP: Auto Ignition Point

AIRMON: Atmospheric Integrated Research Monitoring Network

AIRS: Aerometric Information Retrieval System

AL: Acceptable Level

ALA: Delta-Aminolevulinic Acid

ALA-O: Delta-Aminolevulinic Acid Dehydrates

ALAPO: Association of Local Air Pollution Control Officers

ALARA: As Low As Reasonably Achievable

ALC: Application Limiting Constituent

ALJ: Administrative Law Judge

ALMS: Atomic Line Molecular Spectroscopy

ALR: Action Leakage Rate

AMBIENS: Atmospheric Mass Balance of Industrially Emitted and Natural Sulfur

AMOS: Air Management Oversight System

AMPS: Automatic Mapping and Planning System

AMSA: Association of Metropolitan Sewer Agencies

ANC: Acid Neutralizing Capacity

ANPR: Advance Notice of Proposed Rulemaking

ANRHRD: Air, Noise, & Radiation Health Research Division/ORD

ANSS: American Nature Study Society

AOAC: Association of Official Analytical Chemists

AOC: Abnormal Operating Conditions

AOD: Argon-Oxygen Decarbonization

AOML: Atlantic Oceanographic and Meteorological Laboratory

AP: Accounting Point

APA: Administrative Procedures Act

APCA: Air Pollution Control Association

APCD: Air Pollution Control District

APDS: Automated Procurement Documentation System

APHA: American Public Health Association

APRAC: Urban Diffusion Model for Carbon Monoxide from Motor Vehicle Traffic

APTI: Air Pollution Training Institute

APWA: American Public Works Association

AQ-7: Non-reactive Pollutant Modelling

AQCCT: Air-Quality Criteria and Control Techniques

AQCP: Air Quality Control Program

AQCR: Air-Quality Control Region

AQD: Air-Quality Digest

AQDHS: Air-Quality Data Handling System

AQDM: Air-Quality Display Model

AQMA: Air-Quality Maintenance Area

AQMD: Air Quality Management District

AQMP: Air-Quality Maintenance Plan; Air-Quality Management Plan

AQSM: Air-Quality Simulation Model

AQTAD: Air-Quality Technical Assistance Demonstration

Abbreviations & Acronyms

AR: Administrative Record

A&R: Air and Radiation

ARA: Assistant Regional Administrator; Associate Regional Administrator

ARAC: Acid Rain Advisory Committee

ARAR: Applicable or Relevant and Appropriate Standards, Limitations, Criteria, and Requirements

ARB: Air Resources Board

ARC: Agency Ranking Committee

ARCC: American Rivers Conservation Council

ARCS: Alternative Remedial Contract Strategy

ARG: American Resources Group

ARIP: Accidental Release Information Program

ARL: Air Resources Laboratory

ARM: Air Resources Management

ARNEWS: Acid Rain National Early Warning Systems

ARO: Alternate Regulatory Option

ARRP: Acid Rain Research Program

ARRPA: Air Resources Regional Pollution Assessment Model

ARS: Agricultural Research Service

ARZ: Auto Restricted Zone

AS: Area Source

ASC: Area Source Category

ASDWA: Association of State Drinking Water Administrators

ASHAA: Asbestos in Schools Hazard Abatement Act

ASHRAE: American Society of Heating, Refrigerating, and Air-Conditioning Engineers

ASIWCPA: Association of State and Interstate Water Pollution Control Administrators

ASMDHS: Airshed Model Data Handling System

ASRL: Atmospheric Sciences Research Laboratory

AST: Advanced Secondary (Wastewater) Treatment

ASTHO: Association of State and Territorial Health Officers

ASTM: American Society for Testing and Materials

ASTSWMO: Association of State and Territorial Solid Waste Management Officials

AT: Advanced Treatment. Alpha Track Detection

ATERIS: Air Toxics Exposure and Risk Information System

ATS: Action Tracking System; Allowance Tracking System

ATSDR: Agency for Toxic Substances and Disease Registry

ATTF: Air Toxics Task Force

AUSM: Advanced Utility Simulation Model

A/WPR: Air/Water Pollution Report

AWRA: American Water Resources Association

AWT: Advanced Wastewater Treatment

AWWA: American Water Works Association

AWWARF: American Water Works Association Research Foundation.

B

BAA: Board of Assistance Appeals

BAC: Bioremediation Action Committee; Biotechnology Advisory Committee

BACM: Best Available Control Measures

BACT: Best Available Control Technology

BADT: Best Available Demonstrated Technology

BAF: Bioaccumulation Factor

BaP: Benzo(a)Pyrene

BAP: Benefits Analysis Program

BART: Best Available Retrofit Technology

BASIS: Battelle's Automated Search Information System

BAT: Best Available Technology

BATEA: Best Available Treatment Economically Achievable

BCT: Best Control Technology

BCPCT: Best Conventional Pollutant Control Technology

BDAT: Best Demonstrated Achievable Technology

BDCT: Best Demonstrated Control Technology

BDT: Best Demonstrated Technology

BEJ: Best Engineering Judgement. Best Expert Judgment

BF: Bonafide Notice of Intent to Manufacture or Import (IMD/OTS)

BID: Background Information Document. Buoyancy Induced Dispersion

BIOPLUME: Model to Predict the Maximum Extent of Existing Plumes

BMP: Best Management Practice(s)

BMR: Baseline Monitoring Report

BO: Budget Obligations

BOA: Basic Ordering Agreement (Contracts)

BOD: Biochemical Oxygen Demand. Biological Oxygen Demand

BOF: Basic Oxygen Furnace

BOP: Basic Oxygen Process

BOPF: Basic Oxygen Process Furnace

BOYSNC: Beginning of Year Significant Non-Compliers

BP: Boiling Point

BPJ: Best Professional Judgment

BPT: Best Practicable Technology. Pest Practicable Treatment

BPWTT: Best Practical Wastewater Treatment Technology

BRI: Building-Related Illness

BRS: Bibliographic Retrieval Service

BSI: British Standards Institute

BSO: Benzene Soluble Organics

BTZ: Below the Treatment Zone

BUN: Blood Urea Nitrogen

C

CA: Citizen Act. Competition Advocate. Cooperative Agreements. Corrective Action

CAA: Clean Air Act; Compliance Assurance Agreement

CAAA: Clean Air Act Amendments

CAER: Community Awareness and Emergency Response

CAFE: Corporate Average Fuel Economy

CAFO: Concentrated Animal Feedlot; Consent Agreement/Final Order

CAG: Carcinogenic Assessment Group

CAIR: Clean Air Interstate Rule: Comprehensive Assessment of Information Rule

CALINE: California Line Source Model

CAM: Compliance Assurance Monitoring rule; Compliance Assurance Monitoring

CAMP: Continuous Air Monitoring Program

CAN: Common Account Number

CAO: Corrective Action Order

CAP: Corrective Action Plan. Cost Allocation Procedure. Criteria Air Pollutant

CAPMoN: Canadian Air and Precipitation Monitoring Network

CAR: Corrective Action Report

CAS: Center for Automotive Safety; Chemical Abstract Service

CASAC: Clean Air Scientific Advisory Committee

CASLP: Conference on Alternative State and Local Practices

CASTNet: Clean Air Status and Trends Network

CATS: Corrective Action Tracking System

CAU: Carbon Adsorption Unit; Command Arithmetic Unit

CB: Continuous Bubbler

CBA: Chesapeake Bay Agreement. Cost Benefit Analysis

CBD: Central Business District

CBEP: Community Based Environmental Project

CBI: Compliance Biomonitoring Inspection; Confidential Business Information

CBOD: Carbonaceous Biochemical Oxygen Demand

CBP: Chesapeake Bay Program; County Business Patterns

CCA: Competition in Contracting Act

CCAA: Canadian Clean Air Act

CCAP: Center for Clean Air Policy; Climate Change Action Plan

CCEA: Conventional Combustion Environmental Assessment

CCHW: Citizens Clearinghouse for Hazardous Wastes

CCID: Confidential Chemicals Identification System

CCMS/NATO: Committee on Challenges of a Modern Society/North Atlantic Treaty Organization

CCP: Composite Correction Plan

CC/RTS:Chemical Collection/ Request Tracking System

CCTP: Clean Coal Technology Program

CD: Climatological Data

CDB: Consolidated Data Base

CDBA: Central Data Base Administrator

CDBG: Community Development Block Grant

CDD: Chlorinated dibenzo-p-dioxin

CDF: Chlorinated dibenzofuran

CDHS: Comprehensive Data Handling System

CDI: Case Development Inspection

CDM: Climatological Dispersion Model; Comprehensive Data Management

CDMQC: Climatological Dispersion Model with Calibration and Source Contribution

CDNS: Climatological Data National Summary

CDP: Census Designated Places

CDS: Compliance Data System

CE: Categorical Exclusion. Conditionally Exempt Generator

CEA: Cooperative Enforcement Agreement; Cost and Economic Assessment

CEAT: Contractor Evidence Audit Team

CEARC: Canadian Environmental Assessment Research Council

CEB: Chemical Element Balance

CEC: Commission for Environmental Cooperation

CECATS: CSB Existing Chemicals Assessment Tracking System

CEE: Center for Environmental Education

CEEM: Center for Energy and Environmental Management

CEI: Compliance Evaluation Inspection

CELRF: Canadian Environmental Law Research Foundation

CEM: Continuous Emission Monitoring

CEMS: Continuous Emission Monitoring System

CEPA: Canadian Environmental Protection Act

CEPP: Chemical Emergency Preparedness Plan

CEQ: Council on Environmental Quality

CERCLA: Comprehensive Environmental Response, Compensation, and Liability Act (1980)

CERCLIS: Comprehensive Environmental Response, Compensation, and Liability Information System

CERT: Certificate of Eligibility

CESQG: Conditionally Exempt Small Quantity Generator

CEST: Community Environmental Service Teams

CF: Conservation Foundation

CFC: Chlorofluorocarbons

CFM: Chlorofluoromethanes

CFR: Code of Federal Regulations

CHABA: Committee on Hearing and Bio-Acoustics

CHAMP: Community Health Air Monitoring Program

CHEMNET: Chemical Industry Emergency Mutual Aid Network

CHESS: Community Health and Environmental Surveillance System

CHIP: Chemical Hazard Information Profiles

CI: Compression Ignition. Confidence Interval

CIAQ: Council on Indoor Air Quality

CIBL: Convective Internal Boundary Layer

CICA: Competition in Contracting Act

CICIS: Chemicals in Commerce Information System

CIDRS: Cascade Impactor Data Reduction System

CIMI: Committee on Integrity and Management Improvement

CIS: Chemical Information System. Contracts Information System

CKD: Cement Kiln Dust

CKRC: Cement Kiln Recycling Coalition

CLC: Capacity Limiting Constituents

CLEANS: Clinical Laboratory for Evaluation and Assessment of Toxic Substances

CLEVER: Clinical Laboratory for Evaluation and Validation of Epidemiologic Research

CLF: Conservation Law Foundation

CLI: Consumer Labelling Initiative

CLIPS: Chemical List Index and Processing System

CLP: Contract Laboratory Program

CM: Corrective Measure

CMA: Chemical Manufacturers Association

CMB: Chemical Mass Balance

CME: Comprehensive Monitoring Evaluation

CMEL: Comprehensive Monitoring Evaluation Log

CMEP: Critical Mass Energy Project

CNG:Compressedd Natural Gas

COCO: Contractor-Owned/ Contractor-Operated

COD: Chemical Oxygen Demand

COH: Coefficient Of Haze

CPDA: Chemical Producers and Distributor Association

CPF: Carcinogenic Potency Factor

CPO: Certified Project Officer

CQA: Construction Quality Assurance

CR: Continuous Radon Monitoring

CROP: Consolidated Rules of Practice

CRP: Child-Resistant Packaging; Conservation Reserve Program

CRR: Center for Renewable Resources

CRSTER: Single Source Dispersion Model

CSCT: Committee for Site Characterization

CSGWPP: Comprehensive State Ground Water Protection Program

CSI: Common Sense Initiative; Compliance Sampling Inspection

CSIN: Chemical Substances Information Network

CSMA: Chemical Specialties Manufacturers Association

CSO: Combined Sewer Overflow

CSPA: Council of State Planning Agencies

CSRL: Center for the Study of Responsive Law

Abbreviations & Acronyms

CTARC: Chemical Testing and Assessment Research Commission

CTG: Control Techniques Guidelines

CTSA: Cleaner TechnologiesSubstitutess Assessment

CV: Chemical Vocabulary

CVS: Constant Volume Sampler

CW: Continuous working-level monitoring

CWA: Clean Water Act (aka FWPCA)

CWAP: Clean Water Action Project

CWTC: Chemical Waste Transportation Council

CZMA: Coastal Zone Management Act

CZARA: Coastal Zone Management Act Reauthorization Amendments

D

DAPSS: Document and Personnel Security System (IMD)

DBP: Disinfection By-Product

DCI: Data Call-In

DCO: Delayed Compliance Order

DCO: Document Control Officer

DDT: DichloroDiphenylTrichloroethane

DERs: Data Evaluation Records

DES: Diethylstilbesterol

DfE: Design for the Environment

DI: Diagnostic Inspection

DMR: Discharge Monitoring Report

DNA: Deoxyribonucleic acid

DNAPL: Dense Non-Aqueous Phase Liquid

DO: Dissolved Oxygen

DOW: Defenders Of Wildlife

DPA: Deepwater Ports Act

DPD: Method of Measuring Chlorine Residual in Water

DQO: Data Quality Objective

DRE: Destruction and Removal Efficiency

DRES: Dietary Risk Evaluation System

DRMS: Defense Reutilization and Marketing Service

DRR: Data Review Record

DS: Dichotomous Sampler

DSAP: Data Self Auditing Program

DSCF: Dry Standard Cubic Feet

DSCM: Dry Standard Cubic Meter

DSS: Decision Support System; Domestic Sewage Study

DT: Detectors (radon) damaged or lost; Detention Time

DU: Decision Unit. Ducks Unlimited; Dobson Unit

DUC: Decision Unit Coordinator

DWEL: Drinking Water Equivalent Level

DWS: Drinking Water Standard

DWSRF: Drinking Water State Revolving Fund

E

EA: Endangerment Assessment; Enforcement Agreement; Environmental Action; Environmental Assessment;. Environmental Audit

EAF: Electric Arc Furnaces

EAG: Exposure Assessment Group

EAP: Environmental Action Plan

EAR: Environmental Auditing Roundtable

EASI: Environmental Alliance for Senior Involvement

EB: Emissions Balancing

EC: Emulsifiable Concentrate; Environment Canada; Effective Concentration

ECA: Economic Community for Africa

ECAP: Employee Counselling and Assistance Program

ECD: Electron Capture Detector

ECHH: Electro-Catalytic Hyper-Heaters

ECL: Environmental Chemical Laboratory

ECOS: Environmental Council of the States

ECR: Enforcement Case Review

ECRA: Economic Cleanup Responsibility Act

ED: Effective Dose

EDA: Emergency Declaration Area

EDB: Ethylene Dibromide

EDC: Ethylene Dichloride

EDD: Enforcement Decision Document

EDF: Environmental Defense Fund

EDRS: Enforcement Document Retrieval System

EDS: Electronic Data System; Energy Data System

EDTA: Ethylene Diamine Triacetic Acid

EDX: Electronic Data Exchange

EDZ: Emission Density Zoning

EEA: Energy and Environmental Analysis

EECs: Estimated Environmental Concentrations

EER: Excess Emission Report

EERL: Eastern Environmental Radiation Laboratory

EERU: Environmental Emergency Response Unit

EESI: Environment and Energy Study Institute

EESL: Environmental Ecological and Support Laboratory

EETFC: Environmental Effects, Transport, and Fate Committee

EF: Emission Factor

EFO: Equivalent Field Office

EFTC: European Fluorocarbon Technical Committee

EGR: Exhaust Gas Recirculation

EH: Redox Potential

EHC: Environmental Health Committee

EHS: Extremely Hazardous Substance

EI: Emissions Inventory

EIA: Environmental Impact Assessment. Economic Impact Assessment

EIL: Environmental Impairment Liability

EIR: Endangerment Information Report; Environmental Impact Report

EIS: Environmental Impact Statement; Environmental Inventory System

EIS/AS: Emissions Inventory System/Area Source

EIS/PS: Emissions Inventory System/Point Source

EKMA: Empirical Kinetic Modeling Approach

EL: Exposure Level

ELI: Environmental Law Institute

ELR: Environmental Law Reporter

EM: Electromagnetic Conductivity

EMAP: Environmental Mapping and Assessment Program

EMAS: Enforcement Management and Accountability System

EMR: Environmental Management Report

EMS: Enforcement Management System

EMSL: Environmental Monitoring Support Systems Laboratory

EMTS: Environmental Monitoring Testing Site; Exposure Monitoring Test Site

EnPA: Environmental Performance Agreement

EO: Ethylene Oxide

EOC: Emergency Operating Center

EOF: Emergency Operations Facility (RTP)

EOP: End Of Pipe

EOT: Emergency Operations Team

EP: Earth Protectors; Environmental Profile; End-use Product; Experimental Product; Extraction Procedure

EPAA: Environmental Programs Assistance Act

EPAAR: EPA Acquisition Regulations

EPCA: Energy Policy and Conservation Act

EPACT: Environmental Policy Act

EPACASR: EPA Chemical Activities Status Report

EPCRA: Emergency Planning and Community Right to Know Act

EPD: Emergency Planning District

EPI: Environmental Policy Institute

EPIC: Environmental Photographic Interpretation Center

EPNL: Effective Perceived Noise Level

EPRI: Electric Power Research Institute

EPTC: Extraction Procedure Toxicity Characteristic

EQIP: Environmental Quality Incentives Program

ER: Ecosystem Restoration; Electrical Resistivity

ERA: Economic Regulatory Agency

ERAMS: Environmental Radiation Ambient Monitoring System

ERC: Emergency Response Commission. Emissions Reduction Credit, Environmental Research Center

ERCS: Emergency Response Cleanup Services

ERDA: Energy Research and Development Administration

ERD&DAA: Environmental Research, Development and Demonstration Authorization Act

ERL: Environmental Research Laboratory

ERNS: Emergency Response Notification System

ERP: Enforcement Response Policy

ERT: Emergency Response Team

ERTAQ: ERT Air Quality Model

ES: Enforcement Strategy

ESA: Endangered Species Act. Environmentally Sensitive Area

ESC: Endangered Species Committee

ESCA: Electron Spectroscopy for Chemical Analysis

ESCAP: Economic and Social Commission for Asia and the Pacific

ESECA: Energy Supply and Environmental Coordination Act

ESH: Environmental Safety and Health

ESP: Electrostatic Precipitators

ET: Emissions Trading

ETI: Environmental Technology Initiative

ETP: Emissions Trading Policy

ETS: Emissions Tracking System; Environmental Tobacco Smoke

ETV: Environmental Technology Verification Program

EUP: End-Use Product; Experimental Use Permit

EWCC: Environmental Workforce Coordinating Committee

EXAMS: Exposure Analysis Modeling System

ExEx: Expected Exceedance

F

FACA: Federal Advisory Committee Act

FAN: Fixed Account Number

FATES: FIFRA and TSCA Enforcement System

FBC: Fluidized Bed Combustion

FCC: Fluid Catalytic Converter

FCCC: Framework Convention on Climate Change

FCCU: Fluid Catalytic Cracking Unit

FCO: Federal Coordinating Officer (in disaster areas); Forms Control Officer

FDF: Fundamentally Different Factors

FDL: Final Determination Letter

FDO: Fee Determination Official

FE: Fugitive Emissions

FEDS: Federal Energy Data System

FEFx: Forced Expiratory Flow

FEIS: Fugitive Emissions Information System

FEL: Frank Effect Level

FEPCA: Federal Environmental Pesticide Control Act; enacted as amendments to FIFRA.

FERC: Federal Energy Regulatory Commission

FES: Factor Evaluation System

FEV: Forced Expiratory Volume

FEV1: Forced Expiratory Volume—one second; Front End Volatility Index

FF: Federal Facilities

FFAR: Fuel and Fuel Additive Registration

FFDCA: Federal Food, Drug, and Cosmetic Act

FFF: Firm Financial Facility

FFFSG: Fossil-Fuel-Fired Steam Generator

FFIS: Federal Facilities Information System

FFP: Firm Fixed Price

FGD: Flue-Gas Desulfurization

FID: Flame Ionization Detector

FIFRA: Federal Insecticide, Fungicide, and Rodenticide Act

FIM: Friable Insulation Material

FINDS: Facility Index System

FIP: Final Implementation Plan

FIPS: Federal Information Procedures System

FIT: Field Investigation Team

FLETC: Federal Law Enforcement Training Center

FLM: Federal Land Manager

FLP: Flash Point

FLPMA: Federal Land Policy and Management Act

FMAP: Financial Management Assistance Project

F/M: Food to Microorganism Ratio

FML: Flexible Membrane Liner

FMP: Facility Management Plan

FMP: Financial Management Plan

FMS: Financial Management System

FMVCP: Federal Motor Vehicle Control Program

FOE: Friends Of the Earth

FOIA: Freedom Of Information Act

FOISD: Fiber Optic Isolated Spherical Dipole Antenna

FONSI: Finding Of No Significant Impact

FORAST: Forest Response to Anthropogenic Stress

FP: Fine Particulate

FPA: Federal Pesticide Act

FPAS: Foreign Purchase Acknowledgement Statements

FPD: Flame Photometric Detector

FPEIS: Fine Particulate Emissions Information System

FPM: Federal Personnel Manual

FPPA: Federal Pollution Prevention Act

FPR: Federal Procurement Regulation

FPRS: Federal Program Resources Statement; Formal Planning and Supporting System

FQPA: Food Quality Protection Act

FR: Federal Register. Final Rulemaking

FRA: Federal Register Act

Abbreviations & Acronyms

FREDS: Flexible Regional Emissions Data System

FRES: Forest Range Environmental Study

FRM: Federal Reference Methods

FRN: Federal Register Notice. Final Rulemaking Notice

FRS: Formal Reporting System

FS: Feasibility Study

FSA: Food Security Act

FSS: Facility Status Sheet; Federal Supply Schedule

FTP: Federal Test Procedure (for motor vehicles)

FTS: File Transfer Service

FTTS: FIFRA/TSCA Tracking System

FUA: Fuel Use Act

FURS: Federal Underground Injection Control Reporting System

FVMP: Federal Visibility Monitoring Program

FWCA: Fish and Wildlife Coordination Act

FWPCA: Federal Water Pollution and Control Act (aka CWA). Federal Water Pollution and Control Administration

G

GAAP: Generally Accepted Accounting Principles

GAC: Granular Activated Carbon

GACT: Granular Activated Carbon Treatment

GAW: Global Atmospheric Watch

GCC: Global Climate Convention

GC/MS: Gas Chromatograph/ Mass Spectograph

GCVTC: Grand Canyon Visibility Transport Commission

GCWR: Gross Combination Weight Rating

GDE: Generic Data Exemption

GEI: Geographic Enforcement Initiative

GEMI: Global Environmental Management Initiative

GEMS: Global Environmental Monitoring System; Graphical Exposure Modeling System

GEP: Good Engineering Practice

GFF: Glass Fiber Filter

GFO: Grant Funding Order

GFP: Government-Furnished Property

GICS: Grant Information and Control System

GIS: Geographic Information Systems; Global Indexing System

GLC: Gas Liquid Chromatography

GLERL: Great Lakes Environmental Research Laboratory

GLNPO: Great Lakes National Program Office

GLP: Good Laboratory Practices

GLWQA: Great Lakes Water Quality Agreement

GMCC: Global Monitoring for Climatic Change

G/MI: Grams per mile

GOCO: Government-Owned/ Contractor-Operated

GOGO: Government-Owned/ Government-Operated

GOP: General Operating Procedures

GOPO: Government-Owned/ Privately-Operated

GPAD: Gallons-per-acre per-day

GPG: Grams-per-Gallon

GPR: Ground-Penetrating Radar

GPS: Groundwater Protection Strategy

GR: Grab Radon Sampling

GRAS: Generally Recognized as Safe

GRCDA: Government Refuse Collection and Disposal Association

GRGL: Groundwater Residue Guidance Level

GT: Gas Turbine

GTN: Global Trend Network

GTR: Government Transportation Request

GVP: Gasoline Vapor Pressure

GVW: Gross Vehicle Weight

GVWR: Gross Vehicle Weight Rating

GW: Grab Working-Level Sampling. Groundwater

GWDR: Ground Water Disinfection Rule

GWM: Groundwater Monitoring

GWP: Global Warming Potential

GWPC: Ground Water Protection Council

GWPS: Groundwater Protection Standard; Groundwater Protection Strategy

H

HA: Health Advisory

HAD: Health Assessment Document

HAP: Hazardous Air Pollutant

HAPEMS: Hazardous Air Pollutant Enforcement Management System

HAPPS: Hazardous Air Pollutant Prioritization System

HATREMS: Hazardous and Trace Emissions System

HAZMAT: Hazardous Materials

HAZOP: Hazard and Operability Study

HBFC: Hydrobromofluorocarbon

HC: Hazardous Constituents; Hydrocarbon

HCCPD: Hexachlorocyclo-pentadiene

HCFC: Hydrochlorofluorocarbon

HCP: Hypothermal Coal Process

HDD: Heavy-Duty Diesel

HDDT: Heavy-duty Diesel Truck

HDDV: Heavy-Duty Diesel Vehicle

HDE: Heavy-Duty Engine

HDG: Heavy-Duty Gasoline-Powered Vehicle

HDGT: Heavy-Duty Gasoline Truck

HDGV: Heavy-Duty Gasoline Vehicle

HDPE: High Density Polyethylene

HDT: Highest Dose Tested in a study. Heavy-Duty Truck

HDV: Heavy-Duty Vehicle

HEAL: Human Exposure Assessment Location

HECC: House Energy and Commerce Committee

HEI: Health Effects Institute

HEM: Human Exposure Modeling

HEPA: High-Efficiency Particulate Air

HEPA: Highly Efficient Particulate Air Filter

HERS: Hyperion Energy Recovery System

HFC: Hydrofluorocarbon

HHDDV: Heavy Heavy-Duty Diesel Vehicle

HHE: Human Health and the Environment

HHV: Higher Heating Value

HI: Hazard Index

HI-VOL: High-Volume Sampler

HIWAY: A Line Source Model for Gaseous Pollutants

HLRW: High Level Radioactive Waste

HMIS: Hazardous Materials Information System

HMS: Highway Mobile Source

HMTA: Hazardous Materials Transportation Act

HMTR: Hazardous Materials Transportation Regulations

HOC: Halogenated Organic Carbons

HON: Hazardous Organic NESHAP

HOV: High-Occupancy Vehicle

Abbreviations & Acronyms

HP: Horse Power

HPLC: High-Performance Liquid Chromatography

HPMS: Highway Performance Monitoring System

HPV: High Priority Violator

HQCDO: Headquarters Case Development Officer

HRS: Hazardous Ranking System

HRUP: High-Risk Urban Problem

HSDB: Hazardous Substance Data Base

HSL: Hazardous Substance List

HSWA: Hazardous and Solid Waste Amendments

HT: Hypothermally Treated

HTP: High Temperature and Pressure

HVAC: Heating, Ventilation, and Air-Conditioning system

HVIO: High Volume Industrial Organics

HW: Hazardous Waste

HWDMS: Hazardous Waste Data Management System

HWGTF: Hazardous Waste Groundwater Task Force; Hazardous Waste Groundwater Test Facility

HWIR: Hazardous Waste Identification Rule

HWLT: Hazardous Waste Land Treatment

HWM: Hazardous Waste Management

HWRTF: Hazardous Waste Restrictions Task Force

HWTC: Hazardous Waste Treatment Council

I

I/A: Innovative/Alternative

IA: Interagency Agreement

IAAC: Interagency Assessment Advisory Committee

IADN: Integrated Atmospheric Deposition Network

IAG: Interagency Agreement

IAP: Incentive Awards Program. Indoor Air Pollution

IAQ: Indoor Air Quality

IARC: International Agency for Research on Cancer

IATDB: Interim Air Toxics Data Base

IBSIN: Innovations in Building Sustainable Industries

IBT: Industrial Biotest Laboratory

IC: Internal Combustion

ICAIR: Interdisciplinary Planning and Information Research

ICAP: Inductively Coupled Argon Plasma

ICB: Information Collection Budget

ICBN: International Commission on the Biological Effects of Noise

ICCP: International Climate Change Partnership

ICE: Industrial Combustion Emissions Model. Internal Combustion Engine

ICP: Inductively Coupled Plasma

ICR: Information Collection Request

ICRE: Ignitability, Corrosivity, Reactivity, Extraction

ICRP: International Commission on Radiological Protection

ICRU: International Commission of Radiological Units and Measurements

ICS: Incident Command System. Institute for Chemical Studies; Intermittent Control Strategies.; Intermittent Control System

ICWM: Institute for Chemical Waste Management

IDLH: Immediately Dangerous to Life and Health

IEB: International Environment Bureau

IEMP: Integrated Environmental Management Project

IES: Institute for Environmental Studies

IFB: Invitation for Bid

IFCAM: Industrial Fuel Choice Analysis Model

IFCS: International Forum on Chemical Safety

IFIS: Industry File Information System

IFMS: Integrated Financial Management System

IFPP: Industrial Fugitive Process Particulate

IGCC: Integrated Gasification Combined Cycle

IGCI: Industrial Gas Cleaning Institute

IIS: Inflationary Impact Statement

IINERT: In-Place Inactivation and Natural Restoration Technologies

IJC: International Joint Commission (on Great Lakes)

I/M: Inspection/Maintenance

IMM: Intersection Midblock Model

IMPACT: Integrated Model of Plumes and Atmosphere in Complex Terrain

IMPROVE: Interagency Monitoring of Protected Visual Environment

INPUFF: Gaussian Puff Dispersion Model

INT: Intermittent

IOB: Iron Ore Beneficiation

IOU: Input/Output Unit

IPCS: International Program on Chemical Safety

IP: Inhalable Particles

IPM: Inhalable Particulate Matter. Integrated Pest Management

IPP: Implementation Planning Program. Integrated Plotting Package; Inter-media Priority Pollutant (document); Independent Power Producer

IPCC: Intergovernmental Panel on Climate Change

IPM: Integrated Pest Management

IRG: Interagency Review Group

IRLG: Interagency Regulatory Liaison Group (Composed of EPA, CPSC, FDA, and OSHA)

IRIS: Instructional Resources Information System. Integrated Risk Information System

IRM: Intermediate Remedial Measures

IRMC: Inter-Regulatory Risk Management Council

IRP: Installation Restoration Program

IRPTC: International Register of Potentially Toxic Chemicals

IRR: Institute of Resource Recovery

IRS: International Referral Systems

IS: Interim Status

ISAM: Indexed Sequential File Access Method

ISC: Industrial Source Complex

ISCL: Interim Status Compliance Letter

ISCLT: Industrial Source Complex Long Term Model

ISCST: Industrial Source Complex Short Term Model

ISD: Interim Status Document

ISE: Ion-specific electrode

ISMAP: Indirect Source Model for Air Pollution

ISO: International Organization for Standardization

ISPF: (IBM) Interactive System Productivity Facility

ISS: Interim Status Standards

ITC:Innovative Technology Council

ITC: Interagency Testing Committee

ITRC: Interstate Technology Regulatory Coordination

ITRD: Innovative Treatment Remediation Demonstration

IUP: Intended Use Plan

IUR: Inventory Update Rule

IWC: In-Stream Waste Concentration

IWS: Ionizing Wet Scrubber

J

JAPCA: Journal of Air Pollution Control Association

JCL: Job Control Language

Abbreviations & Acronyms

JEC: Joint Economic Committee

JECFA: Joint Expert Committee of Food Additives

JEIOG: Joint Emissions Inventory Oversight Group

JLC: Justification for Limited Competition

JMPR: Joint Meeting on Pesticide Residues

JNCP: Justification for Non-Competitive Procurement

JOFOC: Justification for Other Than Full and Open Competition

JPA: Joint Permitting Agreement

JSD: Jackson Structured Design

JSP: Jackson Structured Programming

JTU: Jackson Turbidity Unit

L

LAA: Lead Agency Attorney

LADD: Lifetime Average Daily Dose; Lowest Acceptable Daily Dose

LAER: Lowest Achievable Emission Rate

LAI: Laboratory Audit Inspection

LAMP: Lake Acidification Mitigation Project

LC: Lethal Concentration. Liquid Chromatography

LCA: Life Cycle Assessment

LCD: Local Climatological Data

LCL: Lower Control Limit

LCM: Life Cycle Management

LCRS: Leachate Collection and Removal System

LD: Land Disposal. Light Duty

LD L0: The lowest dosage of a toxic substance that kills test organisms.

LDC: London Dumping Convention

LDCRS: Leachate Detection, Collection, and Removal System

LDD: Light-Duty Diesel

LDDT: Light-Duty Diesel Truck

LDDV: Light-Duty Diesel Vehicle

LDGT: Light-Duty Gasoline Truck

LDIP: Laboratory Data Integrity Program

LDR: Land Disposal Restrictions

LDRTF: Land Disposal Restrictions Task Force

LDS: Leak Detection System

LDT: Lowest Dose Tested. Light-Duty Truck

LDV: Light-Duty Vehicle

LEL: Lowest Effect Level. Lower Explosive Limit

LEP: Laboratory Evaluation Program

LEPC: Local Emergency Planning Committee

LERC: Local Emergency Response Committee

LEV: Low Emissions Vehicle

LFG: Landfill Gas

LFL: Lower Flammability Limit

LGR: Local Governments Reimbursement Program

LHDDV: Light Heavy-Duty Diesel Vehicle

LI: Langelier Index

LIDAR: Light Detection and Ranging

LIMB: Limestone-Injection Multi-Stage Burner

LLRW: Low Level Radioactive Waste

LMFBR: Liquid Metal Fast Breeder Reactor

LMOP: Landfill Methane Outreach Program

LNAPL: Light Non-Aqueous Phase Liquid

LOAEL: Lowest-Observed-Adverse-Effect-Level

LOD: Limit of Detection

LQER: Lesser Quantity Emission Rates

LQG: Large Quantity Generator

LRTAP: Long Range Transboundary Air Pollution

LUIS: Label Use Information System

M

MAC: Mobile Air Conditioner

MAPSIM: Mesoscale Air Pollution Simulation Model

MATC: Maximum Acceptable Toxic Concentration

MBAS: Methylene-Blue-Active Substances

MCL: Maximum Contaminant Level

MCLG: Maximum Contaminant Level Goal

MCS: Multiple Chemical Sensitivity

MDL: Method Detection Limit

MEC: Model Energy Code

MEI: Maximally (or most) Exposed Individual

MEP: Multiple Extraction Procedure

MHDDV: Medium Heavy-Duty Diesel Vehicle

MOBILE5A: Mobile Source Emission Factor Model

MOE: Margin Of Exposure

MOS: Margin of Safety

MP: Manufacturing-use Product; Melting Point

MPCA: Microbial Pest Control Agent

MPI: Maximum Permitted Intake

MPN: Maximum Possible Number

MPWC: Multiprocess Wet Cleaning

MRF: Materials Recovery Facility

MRID: Master Record Identification number

MRL: Maximum-Residue Limit (Pesticide Tolerance)

MSW: Municipal Solid Waste

MTD: Maximum Tolerated Dose

MUP: Manufacturing-Use Product

MUTA: Mutagenicity

MWC: Machine Wet Cleaning

N

NAA: Nonattainment Area

NAAEC: North American Agreement on Environmental Cooperation

NAAQS: National Ambient Air Quality Standards

NACA: National Agricultural Chemicals Association

NACEPT: National Advisory Council for Environmental Policy and Technology

NADP/NTN: National Atmospheric Deposition Program/National Trends Network

NAMS: National Air Monitoring Stations

NAPAP: National Acid Precipitation Assessment Program

NAPL: Non-Aqueous Phase Liquid

NAPS: National Air Pollution Surveillance

NARA: National Agrichemical Retailers Association

NARSTO: North American Research Strategy for Tropospheric Ozone

NAS: National Academy of Sciences

NASDA: National Association of State Departments of Agriculture

NCAMP: National Coalition Against the Misuse of Pesticides

NCEPI: National Center for Environmental Publications and Information

NCWS: Non-Community Water System

NEDS: National Emissions Data System

NEPI: National Environmental Policy Institute

NEPPS: National Environmental Performance Partnership System

NESHAP: National Emission Standard for Hazardous Air Pollutants

NIEHS: National Institute for Environmental Health Sciences

NETA: National Environmental Training Association

NFRAP: No Further Remedial Action Planned

NICT: National Incident Coordination Team

NIOSH: National Institute of Occupational Safety and Health

NIPDWR: National Interim Primary Drinking Water Regulations

NISAC: National Industrial Security Advisory Committee

NMHC: Nonmethane Hydrocarbons

NMOC: Non-Methane Organic Component

NMVOC: Non-methane Volatile Organic Chemicals

NO: Nitric Oxide

NOA: Notice of Arrival

NOAA: National Oceanographic and Atmospheric Agency

NOAC: Nature of Action Code

NOAEL: No Observable Adverse Effect Level

NOEL: No Observable Effect Level

NOIC: Notice of Intent to Cancel

NOIS: Notice of Intent to Suspend

N2O: Nitrous Oxide

NOx: Nitrogen Oxides

NORM: Naturally Occurring Radioactive Material

NPCA: National Pest Control Association

NPDES: National Pollutant Discharge Elimination System

NPHAP: National Pesticide Hazard Assessment Program

NPIRS: National Pesticide Information Retrieval System

NPTN: National Pesticide Telecommunications Network

NRD: Natural Resource Damage

NRDC: Natural Resources Defense Council

NSDWR: National Secondary Drinking Water Regulations

NSEC: National System for Emergency Coordination

NSEP: National System for Emergency Preparedness

NSPS: New Source Performance Standards

NSR: New Source Review

NTI: National Toxics Inventory

NTIS: National Technical Information Service

NTNCWS: Non-Transient Non-Community Water System

NTP: National Toxicology Program

NTU: Nephlometric Turbidity Unit

O

O3: Ozone

OCD: Offshore and Coastal Dispersion

ODP: Ozone-Depleting Potential

ODS: Ozone-Depleting Substances

OECD: Organization for Economic Cooperation and Development

OF: Optional Form

OLTS: On Line Tracking System

O&M: Operations and Maintenance

ORM: Other Regulated Material

ORP: Oxidation-Reduction Potential

OTAG: Ozone Transport Assessment Group

OTC: Ozone Transport Commission

OTR: Ozone Transport Region

P

P2: Pollution Prevention

PAG: Pesticide Assignment Guidelines

PAH: Polynuclear Aromatic Hydrocarbons

PAI: Performance Audit Inspection (CWA); Pure Active Ingredient compound

PAM: Pesticide Analytical Manual

PAMS: Photochemical Assessment Monitoring Stations

PAT: Permit Assistance Team (RCRA)

PATS: Pesticide Action Tracking System; Pesticides Analytical Transport Solution

Pb: Lead

PBA: Preliminary Benefit Analysis (BEAD)

PCA: Principle Component Analysis

PCB: Polychlorinated Biphenyl

PCE: Perchloroethylene

PCM: Phase Contrast Microscopy

PCN: Policy Criteria Notice

PCO: Pest Control Operator

PCSD: President's Council on Sustainable Development

PDCI: Product Data Call-In

PFC: Perfluorated Carbon

PFCRA: Program Fraud Civil Remedies Act

PHC: Principal Hazardous Constituent

PHI: Pre-Harvest Interval

PHSA: Public Health Service Act

PI: Preliminary Injunction. Program Information

PIC: Products of Incomplete Combustion

PIGS: Pesticides in Groundwater Strategy

PIMS: Pesticide Incident Monitoring System

PIN: Pesticide Information Network

PIN: Procurement Information Notice

PIP: Public Involvement Program

PIPQUIC: Program Integration Project Queries Used in Interactive Command

PIRG: Public Interest Research Group

PIRT: Pretreatment Implementation Review Task Force

PIT: Permit Improvement Team

PITS: Project Information Tracking System

PLIRRA: Pollution Liability Insurance and Risk Retention Act

PLM: Polarized Light Microscopy

PLUVUE: Plume Visibility Model

PM: Particulate Matter

PMAS: Photochemical Assessment Monitoring Stations

PM2.5: Particulate Matter Smaller than 2.5 Micrometers in Diameter

PM10: Particulate Matter (nominally 10m and less)

PM15: Particulate Matter (nominally 15m and less)

PMEL: Pacific Marine Environmental Laboratory

PMN: Premanufacture Notification

PMNF: Premanufacture Notification Form

PMR: Pollutant Mass Rate

PMR: Proportionate Mortality Ratio

PMRS: Performance Management and Recognition System

PMS: Program Management System

PNA: Polynuclear Aromatic Hydrocarbons

PO: Project Officer

POC: Point Of Compliance

POE: Point Of Exposure

POGO: Privately-Owned/ Government-Operated

POHC: Principal Organic Hazardous Constituent

POI: Point Of Interception

POLREP: Pollution Report

Abbreviations & Acronyms

POM: Particulate Organic Matter. Polycyclic Organic Matter

POP: Persistent Organic Pollutant

POR: Program of Requirements

POTW: Publicly Owned Treatment Works

POV: Privately Owned Vehicle

PP: Program Planning

PPA: Planned Program Accomplishment

PPB: Parts Per Billion

PPE: Personal Protective Equipment

PPG: Performance Partnership Grant

PPIC: Pesticide Programs Information Center

PPIS: Pesticide Product Information System; Pollution Prevention Incentives for States

PPMAP: Power Planning Modeling Application Procedure

PPM/PPB: Parts per million/ parts per billion

PPSP: Power Plant Siting Program

PPT: Parts Per Trillion

PPTH: Parts Per Thousand

PQUA: Preliminary Quantitative Usage Analysis

PR: Pesticide Regulation Notice; Preliminary Review

PRA: Paperwork Reduction Act; Planned Regulatory Action

PRATS: Pesticides Regulatory Action Tracking System

PRC: Planning Research Corporation

PRI: Periodic Reinvestigation

PRM: Prevention Reference Manuals

PRN: Pesticide Registration Notice

PRP: Potentially Responsible Party

PRZM: Pesticide Root Zone Model

PS: Point Source

PSAM: Point Source Ambient Monitoring

PSC: Program Site Coordinator

PSD: Prevention of Significant Deterioration

PSES: Pretreatment Standards for Existing Sources

PSI: Pollutant Standards Index; Pounds Per Square Inch; Pressure Per Square Inch

PSIG: Pressure Per Square Inch Gauge

PSM: Point Source Monitoring

PSNS: Pretreatment Standards for New Sources

PSU: Primary Sampling Unit

PTDIS: Single Stack Meteorological Model in EPA UNAMAP Series

PTE: Potential to Emit

PTFE: Polytetrafluoroethylene (Teflon)

PTMAX: Single Stack Meteorological Model in EPA UNAMAP series

PTPLU: Point Source Gaussian Diffusion Model

PUC: Public Utility Commission

PV: Project Verification

PVC: Polyvinyl Chloride

PWB: Printed Wiring Board

PWS: Public Water Supply/ System

PWSS: Public Water Supply System

Q

QAC: Quality Assurance Coordinator

QA/QC: Quality Assistance/ Quality Control

QAMIS: Quality Assurance Management and Information System

QAO: Quality Assurance Officer

QAPP: Quality Assurance Program (or Project) Plan

QAT: Quality Action Team

QBTU: Quadrillion British Thermal Units

QC: Quality Control

QCA: Quiet Communities Act

QCI: Quality Control Index

QCP: Quiet Community Program

QL: Quantification Limit

QNCR: Quarterly Noncompliance Report

QUA: Qualitative Use Assessment

QUIPE: Quarterly Update for Inspector in Pesticide Enforcement

R

RA: Reasonable Alternative; Regulatory Alternatives; Regulatory Analysis; Remedial Action; Resource Allocation; Risk Analysis; Risk Assessment

RAATS: RCRA Administrate Action Tracking System

RAC: Radiation Advisory Committee. Raw Agricultural Commodity; Regional Asbestos Coordinator. Response Action Coordinator

RACM: Reasonably Available Control Measures

RACT: Reasonably Available Control Technology

RAD: Radiation Adsorbed Dose (unit of measurement of radiation absorbed by humans)

RADM: Random Walk Advection and Dispersion Model; Regional Acid Deposition Model

RAM: Urban Air Quality Model for Point and Area Source in EPA UNAMAP Series

RAMP: Rural Abandoned Mine Program

RAMS: Regional Air Monitoring System

RAP: Radon Action Program; Registration Assessment Panel; Remedial Accomplishment Plan; Response Action Plan

RAPS: Regional Air Pollution Study

RARG: Regulatory Analysis Review Group

RAS: Routine Analytical Service

RAT: Relative Accuracy Test

RB: Request for Bid

RBAC: Re-use Business Assistance Center

RBC: Red Blood Cell

RC: Responsibility Center

RCC: Radiation Coordinating Council

RCDO: Regional Case Development Officer

RCO: Regional Compliance Officer

RCP: Research Centers Program

RCRA: Resource Conservation and Recovery Act

RCRIS: Resource Conservation and Recovery Information System

RD/RA: Remedial Design/ Remedial Action

R&D: Research and Development

RD&D: Research, Development and Demonstration

RDF: Refuse-Derived Fuel

RDNA: Recombinant DNA

RDU: Regional Decision Units

RDV: Reference Dose Values

RE: Reasonable Efforts; Reportable Event

REAP: Regional Enforcement Activities Plan

REE: Rare Earth Elements

REEP: Review of Environmental Effects of Pollutants

RECLAIM: Regional Clean Air Initiatives Marker

RED: Reregistration Eligibility Decision Document

REDA: Recycling Economic Development Advocate

ReFIT: Reinvention for Innovative Technologies

REI: Restricted Entry Interval

REM: (Roentgen Equivalent Man)

REM/FIT: Remedial/Field Investigation Team

REMS: RCRA Enforcement Management System

REP: Reasonable Efforts Program

REPS: Regional Emissions Projection System

RESOLVE: Center for Environmental Conflict Resolution

RF: Response Factor

RFA: Regulatory Flexibility Act

RFB: Request for Bid

RfC: Reference Concentration

RFD: Reference Dose Values

RFI: Remedial Field Investigation

RFP: Reasonable Further Programs. Request for Proposal

RHRS: Revised Hazard Ranking System

RI: Reconnaissance Inspection

RI: Remedial Investigation

RIA: Regulatory Impact Analysis; Regulatory Impact Assessment

RIC: Radon Information Center

RICC: Retirement Information and Counseling Center

RICO: Racketeer Influenced and Corrupt Organizations Act

RI/FS: Remedial Investigation/ Feasibility Study

RIM: Regulatory Interpretation Memorandum

RIN: Regulatory Identifier Number

RIP: RCRA Implementation Plan

RISC: Regulatory Information Service Center

RJE: Remote Job Entry

RLL: Rapid and Large Leakage (Rate)

RMCL: Recommended Maximum Contaminant Level (this phrase being discontinued in favor of MCLG)

RMDHS: Regional Model Data Handling System

RMIS: Resources Management Information System

RNA: Ribonucleic Acid

ROADCHEM: Roadway Version that Includes Chemical Reactions of BI, NO2, and O3

ROADWAY: A Model to Predict Pollutant Concentrations Near a Roadway

ROC: Record Of Communication

RODS: Records Of Decision System

ROG: Reactive Organic Gases

ROLLBACK: A Proportional Reduction Model

ROM: Regional Oxidant Model

ROMCOE: Rocky Mountain Center on Environment

ROP: Rate of Progress; Regional Oversight Policy

ROPA: Record Of Procurement Action

ROSA: Regional Ozone Study Area

RP: Radon Progeny Integrated Sampling. Respirable Particulates. Responsible Party

RPAR: Rebuttable Presumption Against Registration

RPM: Reactive Plume Model. Remedial Project Manager

RQ: Reportable Quantities

RRC: Regional Response Center

RRT: Regional Response Team; Requisite Remedial Technology

RS: Registration Standard

RSCC: Regional Sample Control Center

RSD: Risk-Specific Dose

RSE: Removal Site Evaluation

RTCM: Reasonable Transportation Control Measure

RTDF: Remediation Technologies Development Forum

RTDM: Rough Terrain Diffusion Model

RTECS: Registry of Toxic Effects of Chemical Substances

RTM: Regional Transport Model

RTP: Research Triangle Park

RUP: Restricted Use Pesticide

RVP: Reid Vapor Pressure

RWC: Residential Wood Combustion

S

S&A: Sampling and Analysis. Surveillance and Analysis

SAB: Science Advisory Board

SAC: Suspended and Cancelled Pesticides

SAEWG: Standing Air Emissions Work Group

SAIC: Special-Agents-In-Charge

SAIP: Systems Acquisition and Implementation Program

SAMI: Southern Appalachian Mountains Initiative

SAMWG: Standing Air Monitoring Work Group

SANE: Sulfur and Nitrogen Emissions

SANSS: Structure and Nomenclature Search System

SAP: Scientific Advisory Panel

SAR: Start Action Request. Structural Activity Relationship (of a qualitative assessment)

SARA: Superfund Amendments and Reauthorization Act of 1986

SAROAD: Storage and Retrieval Of Aerometric Data

SAS: Special Analytical Service. Statistical Analysis System

SASS: Source Assessment Sampling System

SAV: Submerged Aquatic Vegetation

SBC: Single Breath Cannister

SC: Sierra Club

SCAP: Superfund Consolidated Accomplishments Plan

SCBA: Self-Contained Breathing Apparatus

SCC: Source Classification Code

SCD/SWDC: Soil or Soil and Water Conservation District

SCFM: Standard Cubic Feet Per Minute

SCLDF: Sierra Club Legal Defense Fund

SCR: Selective Catalytic Reduction

SCRAM: State Consolidated RCRA Authorization Manual

SCRC: Superfund Community Relations Coordinator

SCS: Supplementary Control Strategy/System

SCSA: Soil Conservation Society of America

SCSP: Storm and Combined Sewer Program

SCW: Supercritical Water Oxidation

SDC: Systems Decision Plan

SDWA: Safe Drinking Water Act

SDWIS: Safe Drinking Water Information System

SBS: Sick Building Syndrome

SEA: State Enforcement Agreement

SEA: State/EPA Agreement

SEAM: Surface, Environment, and Mining

SEAS: Strategic Environmental Assessment System

SEDS: State Energy Data System

SEGIP: State Environmental Goals and Improvement Project

SEIA: Socioeconomic Impact Analysis

SEM: Standard Error of the Means

SEP: Standard Evaluation Procedures

SEP: Supplementary Environmental Project

SEPWC: Senate Environment and Public Works Committee

SERC: State Emergency Planning Commission

SES: Secondary Emissions Standard

Abbreviations & Acronyms

SETAC: Society for Environmental Toxicology and Chemistry

SETS: Site Enforcement Tracking System

SF: Standard Form. Superfund

SFA: Spectral Flame Analyzers

SFDS: Sanitary Facility Data System

SFFAS: Superfund Financial Assessment System

SFIREG: State FIFRA Issues Research and Evaluation Group

SFS: State Funding Study

SHORTZ: Short Term Terrain Model

SHWL: Seasonal High Water Level

SI: International System of Units. Site Inspection. Surveillance Index. Spark Ignition

SIC: Standard Industrial Classification

SICEA: Steel Industry Compliance Extension Act

SIMS: Secondary Ion-Mass Spectrometry

SIP: State Implementation Plan

SITE: Superfund Innovative Technology Evaluation

SLAMS: State/Local Air Monitoring Station

SLN: Special Local Need

SLSM: Simple Line Source Model

SMART: Simple Maintenance of ARTS

SMCL: Secondary Maximum Contaminant Level

SMCRA: Surface Mining Control and Reclamation Act

SME: Subject Matter Expert

SMO: Sample Management Office

SMOA: Superfund Memorandum of Agreement

SMP: State Management Plan

SMR: Standardized Mortality Ratio

SMSA: Standard Metropolitan Statistical Area

SNA: System Network Architecture

SNAAQS: Secondary National Ambient Air Quality Standards

SNAP: Significant New Alternatives Project; Significant Noncompliance Action Program

SNARL: Suggested No Adverse Response Level

SNC: Significant Noncompliers

SNUR: Significant New Use Rule

SO: Sulfur Dioxide

SOC: Synthetic Organic Chemicals

SOCMI: Synthetic Organic Chemicals Manufacturing Industry

SOFC: Solid Oxide Fuel Cell

SOTDAT: Source Test Data

SOW: Scope Of Work

SPAR: Status of Permit Application Report

SPCC: Spill Prevention, Containment, and Countermeasure

SPE: Secondary Particulate Emissions

SPF: Structured Programming Facility

SPI: Strategic Planning Initiative

SPLMD: Soil-pore Liquid Monitoring Device

SPMS: Strategic Planning and Management System; Special Purpose Monitoring Stations

SPOC: Single Point Of Contact

SPS: State Permit System

SPSS: Statistical Package for the Social Sciences

SPUR: Software Package for Unique Reports

SQBE: Small Quantity Burner Exemption

SQG: Small Quantity Generator

SR: Special Review

SRAP: Superfund Remedial Accomplishment Plan

SRC: Solvent-Refined Coal

SRF: State Revolving Fund

SRM: Standard Reference Method

SRP: Special Review Procedure

SRR: Second Round Review. Submission Review Record

SRTS: Service Request Tracking System

SS: Settleable Solids. Superfund Surcharge. Suspended Solids

SSA: Sole Source Aquifer

SSAC: Soil Site Assimilated Capacity

SSC: State Superfund Contracts

SSD: Standards Support Document

SSEIS: Standard Support and Environmental Impact Statement;. Stationary Source Emissions and Inventory System.

SSI: Size Selective Inlet

SSMS: Spark Source Mass Spectrometry

SSO: Sanitary Sewer Overflow; Source Selection Official

SSRP: Source Reduction Review Project

SSTS: Section Seven Tracking System

SSURO: Stop Sale, Use and Removal Order

STALAPCO: State and Local Air-Pollution Control Officials

STAPPA: State and Territorial Air Pollution

STAR: Stability Wind Rose. State Acid Rain Projects

STARS: Strategic Targeted Activities for Results System

STEL: Short Term Exposure Limit

STEM: Scanning Transmission-Electron Microscope

STN: Scientific and Technical Information Network

STORET: Storage and Retrieval of Water-Related Data

STP: Sewage Treatment Plant. Standard Temperature and Pressure

STTF: Small Town Task Force (EPA)

SUP: Standard Unit of Processing

SURE: Sulfate Regional Experiment Program

SV: Sampling Visit; Significant Violater

SW: Slow Wave

SWAP: Source Water Assessment Program

SWARF: Waste from Metal Grinding Process

SWC: Settlement With Conditions

SWDA: Solid Waste Disposal Act

SWIE: Southern Waste Information Exchange

SWMU: Solid Waste Management Unit

SWPA: Source Water Protection Area

SWQPPP: Source Water Quality Protection Partnership Petitions

SWTR: Surface Water Treatment Rule

SYSOP: Systems Operator

T

TAD: Technical Assistance Document

TAG: Technical Assistance Grant

TALMS: Tunable Atomic Line Molecular Spectroscopy

TAMS: Toxic Air Monitoring System

TAMTAC: Toxic Air Monitoring System Advisory Committee

TAP: Technical Assistance Program

TAPDS: Toxic Air Pollutant Data System

TAS: Tolerance Assessment System

TBT: Tributyltin

TC: Target Concentration. Technical Center. Toxicity Characteristics. Toxic Concentration:

TCDD: Dioxin (Tetrachlorodibenzo-p-dioxin)

TCDF: Tetrachlorodi-benzofurans

TCE: Trichloroethylene

TCF: Total Chlorine Free

TCLP: Total Concentrate Leachate Procedure. Toxicity Characteristic Leachate Procedure

TCM: Transportation Control Measure

TCP: Transportation Control Plan; Trichloropropane;

TCRI: Toxic Chemical Release Inventory

TD: Toxic Dose

TDS: Total Dissolved Solids

TEAM: Total Exposure Assessment Model

TEC: Technical Evaluation Committee

TED: Turtle Excluder Devices

TEG: Tetraethylene Glycol

TEGD: Technical Enforcement Guidance Document

TEL: Tetraethyl Lead

TEM: Texas Episodic Model

TEP: Typical End-use Product. Technical Evaluation Panel

TERA: TSCA Environmental Release Application

TES: Technical Enforcement Support

TEXIN: Texas Intersection Air Quality Model

TGO: Total Gross Output

TGAI: Technical Grade of the Active Ingredient

TGP: Technical Grade Product

THC: Total Hydrocarbons

THM: Trihalomethane

TI: Temporary Intermittent

TI: Therapeutic Index

TIBL: Thermal Internal Boundary Layer

TIC: Technical Information Coordinator. Tentatively Identified Compounds

TIM: Technical Information Manager

TIP: Technical Information Package

TIP: Transportation Improvement Program

TIS: Tolerance Index System

TISE: Take It Somewhere Else

TITC: Toxic Substance Control Act Interagency Testing Committee

TLV: Threshold Limit Value

TLV-C: TLV-Ceiling

TLV-STEL: TLV-Short Term Exposure Limit

TLV-TWA: TLV-Time Weighted Average

TMDL: Total Maximum Daily Limit; Total Maximum Daily Load

TMRC: Theoretical Maximum Residue Contribution

TNCWS: Transient Non-Community Water System

TNT: Trinitrotoluene

TO: Task Order

TOA: Trace Organic Analysis

TOC: Total Organic Carbon/ Compound

TOX: Tetradichloroxylene

TP: Technical Product; Total Particulates

TPC: Testing Priorities Committee

TPI: Technical Proposal Instructions

TPQ: Threshold Planning Quantity

TPSIS: Transportation Planning Support Information System

TPTH: Triphenyltinhydroxide

TPY: Tons Per Year

TQM: Total Quality Management

T-R: Transformer-Rectifier

TRC: Technical Review Committee

TRD: Technical Review Document

TRI: Toxic Release Inventory

TRIP: Toxic Release Inventory Program

TRIS: Toxic Chemical Release Inventory System

TRLN: Triangle Research Library Network

TRO: Temporary Restraining Order

TSA: Technical Systems Audit

TSCA: Toxic Substances Control Act

TSCATS: TSCA Test Submissions Database

TSCC: Toxic Substances Coordinating Committee

TSD: Technical Support Document

TSDF: Treatment, Storage, and Disposal Facility

TSDG: Toxic Substances Dialogue Group

TSI: Thermal System Insulation

TSM: Transportation System Management

TSO: Time Sharing Option

TSP: Total Suspended Particulates

TSS: Total Suspended (non-filterable) Solids

TTFA: Target Transformation Factor Analysis

TTHM: Total Trihalomethane

TTN: Technology Transfer Network

TTO: Total Toxic Organics

TTY: Teletypewriter

TVA: Tennessee Valley Authority

TVOC: Total Volatile Organic Compounds

TWA: Time Weighted Average

TWS: Transient Water System

TZ: Treatment Zone

U

UAC: User Advisory Committee

UAM: Urban Airshed Model

UAO: Unilateral Administrative Order

UAPSP: Utility Acid Precipitation Study Program

UAQI: Uniform Air Quality Index

UARG: Utility Air Regulatory Group

UCC: Ultra Clean Coal

UCCI: Urea-Formaldehyde Foam Insulation

UCL: Upper Control Limit

UDMH: Unsymmetrical Dimethyl Hydrazine

UEL: Upper Explosive Limit

UF: Uncertainty Factor

UFL: Upper Flammability Limit

ug/m3: Micrograms Per Cubic Meter

UIC: Underground Injection Control

ULEV: Ultra Low Emission Vehicles

UMTRCA: Uranium Mill Tailings Radiation Control Act

UNAMAP: Users' Network for Applied Modeling of Air Pollution

UNECE: United Nations Economic Commission for Europe

UNEP: United Nations Environment Program

USC: Unified Soil Classification

USDA: United States Department of Agriculture

USDW: Underground Sources of Drinking Water

USFS: United States Forest Service

UST: Underground Storage Tank

UTM: Universal Transverse Mercator

UTP: Urban Transportation Planning

UV: Ultraviolet

UVA, UVB, UVC: Ultraviolet Radiation Bands

UZM: Unsaturated Zone Monitoring

Abbreviations & Acronyms

V

VALLEY: Meteorological Model to Calculate Concentrations on Elevated Terrain

VCM: Vinyl Chloride Monomer

VCP: Voluntary Cleanup Program

VE: Visual Emissions

VEO: Visible Emission Observation

VHS: Vertical and Horizontal Spread Model

VHT: Vehicle-Hours of Travel

VISTTA: Visibility Impairment from Sulfur Transformation and Transport in the Atmosphere

VKT: Vehicle Kilometers Traveled

VMT: Vehicle Miles Traveled

VOC: Volatile Organic Compounds

VOS: Vehicle Operating Survey

VOST: Volatile Organic Sampling Train

VP: Vapor Pressure

VSD: Virtually Safe Dose

VSI: Visual Site Inspection

VSS: Volatile Suspended Solids

W

WA: Work Assignment

WADTF: Western Atmospheric Deposition Task Force

WAP: Waste Analysis Plan

WAVE: Water Alliances for Environmental Efficiency

WB: Wet Bulb

WCED: World Commission on Environment and Development

WDROP: Distribution Register of Organic Pollutants in Water

WENDB: Water Enforcement National Data Base

WERL: Water Engineering Research Laboratory

WET: Whole Effluent Toxicity test

WHO: World Health Organization

WHP: Wellhead Protection Program

WHPA: Wellhead Protection Area

WHWT: Water and Hazardous Waste Team

WICEM: World Industry Conference on Environmental Management

WL: Warning Letter; Working Level (radon measurement)

WLA/TMDL: Wasteload Allocation/Total Maximum Daily Load

WLM: Working Level Months

WMO: World Meteorological Organization

WP: Wettable Powder

WPCF: Water Pollution Control Federation

WQS: Water Quality Standard

WRC: Water Resources Council

WRDA: Water Resources Development Act

WRI: World Resources Institute

WS: Work Status

WSF: Water Soluble Fraction

WSRA: Wild and Scenic Rivers Act

WSTB: Water Sciences and Technology Board

WSTP: Wastewater Sewage Treatment Plant

WWEMA: Waste and Wastewater Equipment Manufacturers Association

WWF: World Wildlife Fund

WWTP: Wastewater Treatment Plant

WWTU: Wastewater Treatment Unit

Z

ZEV: Zero Emissions Vehicle

ZHE: Zero Headspace Extractor

ZOI: Zone Of Incorporation

ZRL: Zero Risk Level

Note: Some acronyms have more than one meaning. Multiple meanings are listed, separated by semi-colons.

Source: U.S. Environmental Protection Agency, "Terms of Environment," Revised June 20, 2005 (http://www.epa.gov/OCEPAterms/intro.htm)

A

F

G

H

I

J

Outdoor Writers Association of America, 1589
Outdoors Unlimited, 628, 4297
Outer Continental Shelf Advisory Board Minerals Management Service, 2726
Outside Chicagoland: Iowa-Illinois Safety Council, 986
Overbrook Foundation, 2388
Overflow National Wildlife Refuge, 3513
Oversight of Implementation of the Clean Air Act Amendments of 1990, 2274
Owen Engineering and Management Consultants, 1994
Ozark Environmental Laboratories, 5349
Ozark National Scenic Riverways, 3714, 4458
Ozark Society, 453, 1590
Ozark-St. Francis National Forest, 3514
Ozarks Resource Center, 454
Ozone Diplomacy: New Directions in Safeguarding the Planet, 4551

P

P&P Laboratories, 5350
P.I.N.E.S., 5990
PACE, 5351
PACE Analytical Services, 1995, 5352
PACE Environmental Products, 5353
PACE Resources, Incorporated, 5354
PAR Environmental, 1996
PARS Environmental, 5355
PBR HAWAII, 1997
PBS Environmental Building Consultants, 1998
PCCI, 5356
PDC Laboratories, 5357
PE LaMoreaux & Associates, 5358
PE LaMoreaux and Associates, 1999
PEER Consultants, 2000
PEI Associates, 5359
PELA, 5360
POWS Wildlife Rehabilitation Center, 289
PRC Environmental Management, 5361
PRD Tech, 5362
PSC Environmental Services, 5363
PSI, 5364
Pablo National Wildlife Refuge, 3736
Pace, 5365
Pace Laboratory, 5366
Pace New Jersey, 5367
Pacific Fishery Management Council Conferences, 1668
Pacific Fishery Management Council Newsletter, 4298
Pacific Gamefish Research Foundation, 5772
Pacific Institute for Studies in Development, Environment and Security, 596
Pacific Materials Laboratory, 5368
Pacific Northwest National Lab, 5369
Pacific Northwest Research Institute, 5773
Pacific Nuclear, 5370
Pacific Rivers Council, 1333
Pacific Rivers Council: Freeflow, 4459
Pacific Soils Engineering, 2001
Pacific States Marine Fisheries Commission Newsletter, 1334, 1669, 3294, 4299
Pacific Whale Foundation, 290, 2389
Package Research Laboratory, 5371
Padre Island National Seashore, 3871
Pahranagat National Wildlife Refuge, 3752

Palmer Office: Alaska Department of Environmental Conservation, 2817
Pan American Laboratories, 5372
Pan Earth Designs, 5373
Panos Institute, 597
Panther Swamp National Wildlife Refuge, 3711
Paper Stock Institute, 541
Par Enterprises, 5374
Par Environmental Services, 5375
Parish, Weiner and Maffia, 2002
Parks & Wildlife: Conservation Communication, 3351
Parks & Wildlife: Public Lands Division, 3352
Parks & Wildlife: Resource Protection Division, 3353
Parks and Recreation Magazine, 4428
Parks and Trails Council of Minnesota Annual Meeting, 1138, 1670
Parks and Trails Council of Minnesota: Newsletter, 4300
Parsons Engineering Science, 5376
Partners in Parks, 455
Passaic River Coalition, 1219
Patrick and Anna Cudahy Fund, 2390
Patuxent Research Refuge, 3680
Pavia-Byrne Engineering Corporation, 2003
Payette National Forest, 3651
Pea Island River National Wildlife Refuge, 3788
Peace Corps, 2727
Pearl Harbor National Wildlife Refuge, 3638
Pecos River Commission, 3354
Pediatric Annals: A Journal of Continuing Pediatric Education, 2214
Pee Dee National Wildlife Refuge, 3789
Pelican Island National Wildlife Refuge, 3616
Pelican Man s Bird Sanctuary, 900
Peninsula Conservation Foundation Library of the Environment, 4678
Penn State Institutes of the Environment, 1346
Penniman & Browne, 5377
Pennsylvania Association of Accredited Environmental Laboratories, 1347
Pennsylvania Association of Conservation Districts, 1348
Pennsylvania Association of Environmental Professionals, 1591
Pennsylvania BASS Chapter Federation, 1349
Pennsylvania Cooperative Fish & Wildlife Research Unit, 5774
Pennsylvania Department of Conservation and Natural Resources, 3308
Pennsylvania Environmental Council, 1350
Pennsylvania Fish and Boat Commission: Northeast Region, 3309
Pennsylvania Forest Stewardship Program, 3310
Pennsylvania Forestry Association, 1351
Pennsylvania Game Commission, 3311
Pennsylvania Resources Council, 1352
People for Puget Sound, 1478
People for the Ethical Treatment of Animals, 291, 2275
People, Food and Land Foundation Library, 4679
Peoria Disposal Company, 5378
Peregrine Fund, 456
Perkiomen Watershed Conservancy, 6134

Permaculture Gap Mountain, 5775
Perry-Carrington Engineering Corporation, 2004
Pesticide Action Network North America, 809, 2166, 6634
Pesticide Directory: A Guide to Producers and Products, Regulators, and Researchers, 3965, 3989, 4040
Pesticide Education Center, 810, 2167
Pesticide Regulation, Environmental Monitoring and Pesticide Management, 2872
Pesticide Research Center, 5776
Pesticide Research Laboratory & Graduate Study Center, 5777
Pesticides and the Immune System: Public Health Risks, 2215
Pesticides in Children s Food, 2216
Petra Environmental, 2005
Petrified Forest National Park, 3497
Pew Charitable Trusts, 2391
Pharmaco LSR, 5379
Phase One, 2006
Pheasants Forever, 292
Philip Environmental Services, 5380
Phoenix District Advisory Council Bureau of Land Management, 2842
Physical Sciences, 5381
Physicians for Social Responsibility, 232, 2168
Pictured Rocks National Lakeshore, 3688
Piedmont National Wildlife Refuge, 3623
Pike National Forest, 3592
Pinchot Institute for Conservation, 598
Pinchot Letter, 4406
Pinckney Island National Wild Refuge, 3624
Pine Bluff Cooperative Fishery Research Project, 2849
Pinnacles National Monument: National Park Service, Department of the Interior, 3555
Pipeline, 4460
Pittsburgh Mineral & Environmental Technology, 5382
Pixley National Wildlife Refuge, 3556
Planning Concepts, 5383
Planning Design & Research Engineers, 5384
Planning Institute, 5778
Planning Resources, 2007, 5385
Plant Research Technologies, 5386
Plant and Facilities Expo, 1671
Plasma Science & Fusion Center, 5387
Plastic: America s Packaging Dilemma, 4578
Plastics Recycling Foundation, 542
Plastics Recycling Update, 4368
Plumas National Forest, 3557
Pocono Environmental Education Center, 1353
Pocosin Lakes National Wildlife Refuge, 3790
Polaroid Corporation, 5388
Policy Management Office of Environmental Affairs: Pacific Northwest, 3295
Policy Review Board of the Gulf of Mexico: Gulf of Mexico Program, 3167
Pollution Abstracts, 3980, 4041
Pollution Engineering, 4179
Pollution Equipment News, 4369
Pollution Knows No Frontiers, 4579
Pollution Prevention News, 4370
Pollution Prevention Trust Fund, 6135
Polyengineering, 5389
Polytechnic, 5390
Pond Creek National Wildlife Refuge, 3515

833

Q

R

S

Trinity River Basin Fish & Wildlife Task Force Mid-Pacific Region, 2880
Triumvirate Environmental, 6156
Tropical Forest Foundation, 6014
Tropical Rainforest, 6725
Tropical Rainforests Under Fire, 6726
Troppe Environmental Consulting, 2063
Trout Unlimited, 301, 819, 841, 854, 959, 1098, 1123, 1165, 1182, 1205, 1267, 1284, 1317, 1392, 1427, 1482, 1507, 1519, 1600, 2418
Trout Unlimited Chapter and Council Handbook, 4018
True North Foundation, 2419
Trumpeter Swan Society, 302
Trust for Public Land, 479
Trust for the Future, 480
Trustees for Alaska, 732
Tug Hill Tomorrow Land Trust, 1268
Tulane Environment Law Clinc, 1047
Tulane University, 6015
Tule Lake National Wildlife Refuge, 3579
Turkey Call, 4217
Turner Foundation, 2420
Turner Laboratories, 5541
Turner Technologies, 2064
Turner, Collie and Braden Library, 4695
Turning the Tide: Saving the Chesapeake Bay, 4602
Turtle Help Network, 4019
20/20 Vision, 5847

U

U.S. Global Change Research Information Office, 91
UEC Industrial Hygiene and Environmental Health Laboratories, 5542
UIS Department of Environmental Studies, 6016
URI Cooperative Extension Education Center, 3320
URS, 5543
US Agency for International Development Information Center, 2749
US Army Corps of Engineers, 1601
US Bureau of Land Management, 4696
US Bureau of Land Management Library, 4697
US Consumer Product Safety Commission, 2180, 2750
US Customs Service, 2751
US Department of Agriculture, 2752
US Department of Agriculture: National Agricultural Library, Water Quality Info Center, 4698
US Department of Education, 2753
US Department of Energy, 240, 1602
US Department of Housing and Urban Development, 2754
US Department of Housing and Urban Development, 2755
US Department of Labor, 2756
US Department of Treasury, 2757
US Department of the Air Force Major Air Commands, 2758
US Department of the Army: Office of Public Affairs, 2759
US Department of the Interior, 1603
US Department of the Interior: Bureau of Indian Affairs, 2760

US Department of the Navy: US Marine Corps, 2761
US Dye Manufacturers Operating Committee, 92
US Environmental Protection Agency, 1604, 2205, 2762
US Environmental Protection Agency Office of Children s Health Protection, 2763
US Environmental Protection Agency: Clean Air Markets Division, 2765
US Environmental Protection Agency: Environmental Monitoring and Assessment Program, 2764, 6565
US Environmental Protection Agency: Great Lakes National Program Office, 6157
US Environmental Protection Agency: Office of Air and Radiation, 2766
US Environmental Protection Agency: Office of Environmental Justice, 2767
US Filter/Control Systems, 5544
US Fish & Wildlife Service Tishomingo National Wildlife Refuge, 3480, 3806
US Forest Service, 2768
US Forest Service: Wildlife Habitat & Silviculture Laboratory, 5819
US Geological Survey: Great Lakes Science Center, 4699
US Geological Survey: National Wetlands Research Center, 4700
US Geological Survey: Upper Midwest Environmental Sciences Center Library, 4701
US Geological Survey: Water Resources Division Library, 4702
US Global Change Data and Information System, 6650
US Global Research Information Office, 6651
US National Committee for Man and the Biosphere, 2769
US Nuclear Regulatory Commission, 2181, 2770
US Public Interest Research Group, 2206, 2254, 5545
US Sportsman s Alliance, 303
US Sportsmen s Alliance, 304
US-Japan Foundation, 2421
US-Mexico Border Health Association, 2207
USDA Forest Service: North Central Research Station, 5820
USDA Forest Service: Pacific Southwest Researc Station, 5546
USDA Forest Service: Rocky Mountain Research Station, 5821
USDA Forest Service: Watershed and Air Management, 2771
USF and G Foundation, 2422
UVA Institute for Environmental Negotiation, 5822
Uinta National Forest, 3887
Umatilla National Forest, 3831
Umatilla National Wildlife Refuge, 3832
Umpqua National Forest, 3833
Umpqua Research Company, 5547
Underground Injection Practices Council, 1318
Underground Tank Testing & Service, 5548
Understanding Environmental Administration and Law, 2281
Underwater Society of America, 1605
Unexpected Wildlife Refuge Library, 4703
Union of Concerned Scientists, 2423
Unitarian Universalist Veatch Program at Shelter Rock, 2424

United Citizens Coastal Protection League, 703
United Environment Systems, 5549
United Nations Environment Programme New York Office, 93
United States Army Engineer District: Memphis, 3341
United States Department of Agriculture United States Forest Service, 2881, 2998, 3153, 3440
United States Department of Treasury US Customs Service, 2931
United States Department of the Air Force Major Air Commands, 2781, 2905, 2958, 2984, 3276
United States Department of the Army, 2823, 2882, 2946, 2999, 3115, 3154, 3168, 3174, 3198, 3255, 3262, 3314, 3328, 3397
United States Department of the Army: US Army Corps of Engineers, 3061, 3277, 3299, 3364
United States Department of the Interior Bureau of Land Management, 2883, 2906, 2918, 2985
United States Department of the Interior: United States Fish and Wildlife Service, 3239
United States Fish and Wildlife Service, 2947
United States Forest Service: Southwestern R-3, 3240
United States Forest Service: United States Department of Agriculture, 2907
United States Geological Survey: National Earthquake Information Centre, 6566
United States Land, Air, and Water Environment Industries, 5550
Unity College in Maine, 6017
Universal Environmental Technologies, 5551
University Forest, 5823
University of California, 4704
University of California at Los Angeles, 6018
University of California at Riverside, 6019
University of California: Santa Cruz, 6020
University of Colorado, 6021
University of Colorado: Boulder, 2500
University of Florida Coastal Engineering Archives, 4705
University of Georgia, 6022
University of Hawaii at Manoa Water Resources Center, 4706
University of Idaho, 6023
University of Illinois at Chicago, 4707
University of Illinois at Urbana, 6024
University of Kansas Field Stations, 6025
University of Maine Cooperative Extension Forestry & Wildlife Office, 3098
University of Maine at Fort Kent, 6026
University of Maryland at Baltimore, 6027
University of Maryland at College Park, 6028
University of Maryland: Center for Environmental Science Chesapeake Biological Lab, 4708
University of Minnesota at St. Paul, 6029
University of Montana, 6030
University of Montana Wilderness Institute Library, 4709
University of Nebraska, 6031
University of Nevada at Las Vegas, 6032
University of New Hampshire Cooperative Extension, 3217
University of New Haven, 6033
University of North Carolina at Chapel Hill, 6034

X

Y

Z

Arizona BASS Chapter Federation, 740
Arizona Chapter, National Safety Council, 741
Arizona Department of Agriculture: Animal Services Division, 2825
Arizona Department of Environmental Protection, 2826
Arizona Department of Environmental Quality, 2827
Arizona Department of Health Services: Office of Environmental Health, 2828
Arizona Environmental Quality Department, 2829
Arizona Game & Fish Department, 2830
Arizona Game & Fish Department: Region I, 2831
Arizona Game & Fish Department: Region II, 2832
Arizona Game & Fish Department: Region III, 2833
Arizona Game & Fish Department: Region IV, 2834
Arizona Game & Fish Department: Region V, 2835
Arizona Game & Fish Department: Region VI, 2836
Arizona Geological Survey, 2837
Arizona Solar Energy Industries Association, 742
Arizona Solar Update, 743
Arizona State Parks, 2838
Arizona State Parks Board, 2839
Arizona State University, 5859
Arizona Water Well Association, 744
Arizona-Sonora Desert Museum, 745
Atlas Weathering Services Group, 4835
Bill Williams National Wildlife Refuge, 3485
Buenos Aires National Wildlife Refuge, 3486
CO2 Science, 5871
Cabeze Prieta National Wildlife Refuge, 3487
Chiricahua National Monument, 3488
Cibola National Wildlife Refuge, 3525
Coconino National Forest, 3489
Copper State Analytical Lab, 4934
Coronado National Forest, 3490
Earth First!, 37
Electron Microprobe Laboratory, 4999
Environmental and Analytical Chemistry Laboratory, 2840
Foresta Institute for Ocean and Mountain Studies, 666
Gila Box Riparian National Conservation Area BLM Safford District Office, 2841
Glen Canyon National Recreation Area, 3491
Grand Canyon National Park, 3492
Hasbrouck Geophysics, 1937
Heinrichs Geoexploration Company, 5155
IHI Environmental, 5178
Imperial National Wildlife Refuge, 3493
Inter Ag Services IAS Laboratories, 5191
International Society for the Protection of Mustangs and Burros, 401
JABA, 5201
Kaibab National Forest, 3494
Kofa National Wildlife Refuge, 3495
National Environmental, Safety and Health Trai ning Association Show, 1650
Native Seeds/SEARCH, 591
North American Bear Society, 286
Northern Arizona University, 5985
Organ Pipe Cactus National Monument, 3496
Petrified Forest National Park, 3497

Phoenix District Advisory Council Bureau of Land Management, 2842
Prescott College, 5992
Prescott National Forest, 3498
SHB AGRA, 5442
Safari Club International, 631
Safari Club International Conservation Fund, 2403
Saguaro National Park, 3499
San Bernardino/Leslie Canyon National Refuge, 3500
Society for Ecological Restoration, 610, 5802
Sonoran/Rincon Institutes, 472
Southwest Environmental Health Sciences Center: University of Arizona, 6003
Southwest Environmental Health Sciences: Community Outreach and Education Program, 6153
Sunset Crater Volcano National Monument, 3501
Threshold, 475
Tonto National Forest, 3502
Turner Laboratories, 5541
Walnut Canyon National Monument, 3503
Water Resources Associates, 5568
Wild Horses of America Registry, 486

Arkansas

American Lung Association of Arkansas, 748
Anderson Engineering Consultants, 4806
Arkansas Association of Conservation Districts Annual Conference, 749, 1620
Arkansas Department of Parks and Tourism, 2843
Arkansas Environmental Education Association, 750
Arkansas Environmental Federation, 751
Arkansas Fish and Game Commission, 2844
Arkansas Natural Heritage Commission, 2845
Arkansas State Plant Board, 2846
Arkansas Tech University, 5860
Bald Knob National Wildlife Refuge, 3504
Big Lake National Wildlife Refuge Northeast Arkansas Refuges, 3505
Bioengineering Resources, 4864
Buffalo National River, 3506
Cache River National Wildlife Refuge Northeast Arkansas Refuges, 3507
Challenge Environmental Laboratories, 1845
Data Testing, 4951
Department of Environment Quality, 2847
Engineering Research, 5010
Felsenthal National Wildlife Refuge, 3508
Game and Fish Commission Wildlife Management Division, 2848
Greers Ferry National Fish Hatchery, 3509
Holla Bend National Wildlife Refuge, 3510
Hot Springs National Park, 3511
KE Sorrells Research Associates, 5221
Ouachita National Forest, 3512
Overflow National Wildlife Refuge, 3513
Ozark Society, 453
Ozark-St. Francis National Forest, 3514
Pine Bluff Cooperative Fishery Research Project, 2849
Pond Creek National Wildlife Refuge, 3515
Spears Professional Environmental & Archeological Research Service, 5484
Thermoenergy Corporation, 5524
Wapanocca National Wildlife Refuge, 3516

Water Center, 704
White River National Wildlife Refuge, 3517

California

AAA Lead Consultants and Inspections, 1698
ACC Environmental Consultants, 1701
APC Lab, 4755
ARCO Foundation, 2283
ASLA: California-Sierra Chapter, 754
ASLA: Northern California Chapter, 755
ASLA: San Diego Chapter, 756
ASLA: Southern California Chapter, 757
Abelard Foundation, 2284
Acorn Foundation, 2286
Acumen Industrial Hygiene, 1717
AeroVironment, 4776
Allied Engineers, 1729
Allwest Environmental, 1731
Alton Geoscience, 4788
American Bass Association, 242
American Cetacean Society, 2850
American Fisheries Society-Fish Health Section, 758
American Land Conservancy, 319
American Lung Association of California, 759
American Lung Association of California-Redwood Empire Branch, 760
American Lung Association of California-East Bay Branch, 761
American Lung Association of California Superior Branch, 762
American Lung Association of Central California, 763
American Lung Association of Los Angeles County, 764
American Lung Association of Orange County - Santa Ana, 765
American Lung Association of Sacramento-Emigrant Trails, 766
American Lung Association of San Diego & Imperial Counties, 767
American Lung Association of San Francisco & San Mateo Counties, 768
American Lung Association of Santa Barbara and Ventura Counties, 769
American Lung Association of Santa Clara - San Benito Counties, 770
American Lung Association of the Central Coast, 771
American Lung Association of the Inland Counties, 772
American Research & Testing, 4793
Anametrix, 4804
Ancient Forest International, 555, 642
Angeles National Forest, 3518
Animal Protection Institute, 253
Anlab, 4809
Antioch Dunes National Wildlife Refuge, 3519
Apple Computer Earth Grants: Community Affairs Department, 2457
Applied P&C Laboratory, 4815
Ardea Consulting, 1753
Arete Associates, 4823
Asian Pacific Environmental Network, 773, 2095
Atkins Environmental HELP, 1761
Atlas Environmental Engineering, 1763
BC Analytical, 4839

BC Laboratories, 4840
Bakersfield Field Office, 2851
Beak Consultants, 1779
Bear Valley National Wildlife Refuge, 3520
Bendix Environmental Research, 1785, 4855
Benson C Stone & Associates, 4856
Beyaz and Patel, 1788
Bio Integral Resource Center, 774, 2101
Bio-Integral Resource Center, 5607
Bio-Science Research Institute, 4860
Biological Frontiers Institute, 1792
Bishop Resource Area, 6075
Bitter Creek National Wildlife Refuge, 3521
Block Environmental Services, 1802
Blue Ridge National Wildlife Refuge, 3522
Brown, Vence and Associates, 1810
Business for Social Responsibility, 29
C L Technology, 4888
CS Fund, 2460
CTL Environmental Services, 1826
California Academy of Sciences Library, 775
California Air Resources Board, 776
California Association of Environmental
Health, 777
California Association of Resource
Conservation Districts, 778
California BASS Chapter Federation, 779
California Birth Defects Monitoring
Program, 780
California Certified Organic Farmers, 781
California Council for Environmental and
Economic Balance, 783
California Department of Education Office
of Environmental Education, 2852
California Department of Fish and Game,
2853
California Department of Water Resources,
2854
California Desert District Advisory Council
Bureau of Land Management, 2855
California Environmental, 1829
California Environmental Protection Agency,
2856
California Geo-Systems, 1830
California Institute of Public Affairs, 2857
California Pollution Control Financing
Authority, 2858
California Polytechnic State University City
and Regional Planning Department, 5872
California Renewable Fuels Council, 784
California Sea Grant College Program, 5610
California Solar Energy Industries
Association, 785
California State University at Fullerton, 5873
California State University: Monterey Bay
Capstone Project, 5611, 5874
California Trappers Association (CTA), 786
California Waterfowl Association, 787
California Wildlife Federation, 788
Californians for Population Stabilization
(CAPS), 789
Castle Rock National Wildlife Refuge, 3523
Center for Community Action and
Environmental Justice, 2221
Center for Ecoliteracy, 5876
Center for Environmental Science and
Policy, 5880
Center for Investigative Reporting, 2224
Center for Population Biology, 5635
Channel Islands National Park, 3524
Chemical Data Management Systems, 1848
Chevron Corporation, 2310
Children of the Green Earth, 182

Children s Health Environmental Coalition,
2195
Clear Lake National Wildlife Refuge, 3526
Cleveland National Forest, 3527
Coachella Valley National Wildlife Refuge,
3528
Coalition Against Pipeline Pollution, 520
Coalition for Clean Air, 2196
Coastal Lawyer, 1856
Cohen Group, 1858
Cohrssen Environmental, 1859
Colorado River Board of California, 790
Committee for the Preservation of the Tule
Elk, 348
Commonweal, 186
Communities for a Better Environment, 791,
2110, 2225, 2225
Community Environmental Council, 2226
Compton Foundation, 2313
Concerned Citizens of South Central Los
Angeles, 792
Conservtech, 1870
Consumer Pesticide Project, 187
Continental Systems, 4930
Coral Reef Alliance, 662
Council for Planning and Conservation, 793
Crosby & Overton, 4940
Curtis & Tompkins, 4941
Custom Environmental Services, 1880
Death Valley National Park, 3529
Delevan National Wildlife Refuge, 3530
Dellavalle Laboratory, 4954
Department of Agriculture: Forest Service,
Pacific Southwest Region, 2859
Department of Justice Environment and
Natural Resources, 2860
Department of the Interior: Water Resources
Western, 2862
Desert Fishes Council, 356
Desert Protective Council, 357
Desert Tortoise Council, 358
Desert Tortoise Preserve Committee, 794
Devil s Postpile National Monument, 3531
Donald Bren School of Environmental
Science, 5906
EMS Laboratories, 4969
Eagle Lake Biological Field Station, 5648
Earth Ecology Foundation, 362
Earth Island Institute, 363, 559, 663, 663,
2466
Earth Regeneration Society, 191, 4982, 5910,
5910
Earth Science Associates, 1894
Earth Technology Corporation, 4984
Earthjustice, 2229
EcoLogic Systems, 1896
Economic Development/Marketing
California Environmental Business
Council, 6094
Eldorado National Forest, 3532
Ellicott Slough National Wildlife Refuge,
3533
Energy Foundation, 2330
Energy Resources Conservation &
Development Commission, 2863
Energy Technology Consultants, 1901
Environment Associates, 5025
Environmental Defense Center, 796
Environmental Exposure Laboratory, 5656
Environmental Forum of Marin, 6098
Environmental Health Coalition, 797, 2114
Environmental Health Network, 194, 798,
2116, 2116

Environmental Management, 2624, 5923
Environmental Measurements, 5047
Environmental Media Association, 5926
Environmental Policy Center Global Cities
Project, 561
Environmental Protection Office: Hazard
Identification, 2865
Environmental Protection Office: Toxic
Substance Control Department, 2866
Environmental Science Associates, 1910
Environmental Studies Institute, 5666, 5937
Environmental Studies and Planning
Department, 5940
Enviropro, 5064
Eureka Laboratories, 5070
Farallon National Wildlife Refuge, 3534
Fossil Fuels Policy Action Institute, 127
Friends of the River, 799, 2123
Friends of the Sea Lion Marine Mammal
Center, 372
Friends of the Sea Otter, 373
GLOBE, 6107
Geo-Marine Technology, 1922
Geo-Research/Geo-Test, 5103
Get Oil Out, 526
Giblin Associates, 5116
Global Action Network, 564
Global Action and Information Network,
5948
Global Geochemistry Corporation, 5118
Golden Gate National Recreation Area, 2867
Green TV, 5955
Gruen, Gruen & Associates, 5131
Guanterra Environmental Services, 5132
Havasu National Wildlife Refuge, 3535
Hazardous Waste Reduction Loan Program,
2480
Heller Charitable and Educational Fund, 2481
Hidden Villa, 6112
Hopper Mountain National Wildlife Refuge,
3536
Humboldt Bay National Wildlife Refuge San
Francisco Bay National Wildlife Refuge,
3537
Huntingdon Engineering & Environmental,
5171
Hydrocomp, 5174
ICS Radiation Technologies, 5177
Ike Yen Associates, 5180
Institute for Applied Research, 5698
Institute for Biopsychological Studies of
Color, Light, Radiation, Health, 5699
Institute for Crustal Studies, 5700
Institute for the Human Environment, 208,
800
Institute of Ecology, 5710
Inter-American Tropical Tuna Commission,
2868
International Association for Bear Research
and Management, 265, 390
International Bird Rescue Research Center,
392
International Center for Earth Concerns,
5965, 6115
International Maritime, Inc, 5194
International Mountain Society, 570
International Rivers Network, 675
International Society for the Preservation of
the Tropical Rainforest, 400
International Society of Chemical Ecology,
5196
International Wildlife Rehabilitation
Council, 404

Colorado

San Juan National Forest, 3596
Simons and Associates, 5466
Slosky & Company, 2049
Solar Energy Research Institute, 5473
Solar Power Engineering Company, 5474
State Forest Service, 2904
Thorne Ecological Institute, 6010
Trace Minerals International, 5536
True North Foundation, 2419
USDA Forest Service: Rocky Mountain
    Research Station, 5821
United States Forest Service: United States
    Department of Agriculture, 2907
University of Colorado, 6021
University of Colorado: Boulder, 2500
Whale Scientific, 5580
White River National Forest, 3597
Wilderness Education Association, 632,
    6059, 6159, 6159
Windstar Foundation, 497
Yellowstone Grizzly Foundation, 843

Connecticut

APS Technology, 4756
ASLA: Connecticut Chapter, 844
ASW Environmental Consultants, 4760
Aaron Environmental, 4766
Abacus Environmental, 1711
American Association in Support of
    Ecological Initiatives, 845, 2078
American Lung Association of Connecticut,
    846
Aqualogic, 1748
Axiom Laboratories, 4837
Baron Consulting Company, 1774
Bollyky Associates, 1804, 4873
Bolt Technology Corporation, 4874
Brooks Companies, 4881
Brooks Laboratories, 1809, 4882
Business & Legals Reports, 2220
Cetacean Society International, 255, 6082
Connecticut Audubon Society, 847
Connecticut Botanical Society, 848
Connecticut Department of Agriculture, 2908
Connecticut Department of Environmental
    Protection, 2909
Connecticut Department of Public Health,
    2910
Connecticut Forest and Park Association
    Annual Meeting, 849, 1625
Connecticut Fund for the Environment, 850
DRB Communications, 2463
EMCO Testing & Engineering, 4966
EnviroAnalytical, 5018
Environmental Consulting Laboratories,
    5027
Environmental Compliance in Your State,
    2230
Environmental Consulting Laboratories, 5036
Environmental Data Resources, 5040, 6095
Environmental Laboratories, 5044
Environmental Monitoring Laboratory, 5048
Environmental Risk Limited, 1908, 5052
Evans Cooling Systems, 5072
Financial Support for Graduate Work, 2472
Friends of Animals, 851
FuelCell Energy, 5083
Game Conservancy USA, 261
Geoteknika, 5113

Goodwin-Niering Ctr for Conservation
    Biology and Environmental Studies:
    Connecticut College, 5952
Henry Souther Laboratories, 5156
Institute of Analytical and Environmental
    Chemistry, 5708
Keep America Beautiful, 2359
Laticrete International, 5249
Litchfield Environmental Council: Berkshire,
    852
Managing Planet Earth: Perspectives on
    Population, Ecology and the Law, 2271
Mystic Air Quality Consultants, 5313
National Association for Humane and
    Environmental Education, 59
National Shooting Sports Foundation, 280
Project Oceanology, 6137
RPM Systems, 5408
Redniss and Mead, 2021
SPECTROGRAM Corporation, 5446
Save the Sound, 853
Southern Connecticut State University, 6002
Stanford Technology Corporation, 5491
Steven Winter Associates, 5493
Stewart B McKinney National Wildlife
    Refuge: Ninigret National Wildlife
    Refuge, 3598
Sunsearch, 5498
TRC Environmental Corporation, 5507
University of New Haven, 6033
WasteExpo, 1686
Women s Seamen s Friend Society of
    Connecticut, 2506
Yale Institute for Biospheric Studies (YIBS),
    2507, 5846
Yale University, 6063

Delaware

ASTB Analytical Services, 4759
American Lung Association of Delaware, 860
Artesian Laboratories, 4826
Ashland Nature Center, 6072
Atlantic Waterfowl Council, 861
Batta Environmental Associates, 1776
Bombay Hook National Wildlife Refuge,
    3599
Cabe Associates, 1828
Clean Technologies, 1854
Corrosion Testing Laboratories, 4937
Delaware Association of Conservation
    Districts, 2912
Delaware Association of Conservation
    Districts, 862
Delaware BASS Chapter Federation, 863
Delaware Cooperative Extension, 2913
Delaware Department of Agriculture, 2914
Delaware Department of Natural Resources
    and Environmental Control, 2915
Delaware Greenways, 864
Delaware Sea Grant Program, 2916
Du Pont de Nemours and Company, 2465
Hummingbird Society, 387
Mid-Atlantic Fishery Management Council,
    2917
Prime Hook National Wildlife Refuge, 3600
Save Wetlands and Bays, 867
Tetra Tech, 2057, 5520
Wik Associates, 5582

District of Columbia

A Guide to Environmental Law in
    Washington DC, 2256
ASLA: Arkansas Chapter, 747
ASLA: Lousiana Chapter, 1040
ASLA: Potomac Chapter, 855
Action on Smoking and Health, 157, 2072
Advisory Committee on Nuclear Waste, 2508
Advisory Council on Historic Preservation,
    2509
Advisory Panel Ecosystem Studies, 2510
African Wildlife Foundation, 2287
Air and Radiation Research Committee, 2513
Alliance for Acid Rain Control and Energy
    Policy, 160, 510, 2075, 2075
Alliance for a Clean Rural Environment, 634
Alliance to End Childhood Lead Poisoning,
    2186
Alliance to Save Energy, 112
Aluminum Recycling Association, 511
America the Beautiful Fund, 313
American Association for the Advancement
    of Science, 2450
American Association of Poison Control
    Centers, 162, 2079
American Chemical Society, 5852
American Conservation Association, 316
American Council for an Energy-Efficient
    Economy, 115
American Council on the Environment, 317
American Crop Protection Association, 549
American Farmland Trust, 2514
American Federation of Teachers, 20, 2086
American Forest & Paper Association, 21
American Forest Foundation, 550
American Forests, 551
American Gas Association, 116
American Geophysical Union Member
    Programs Division, 2452
American Indian Environmental Office,
    2087, 2515
American Industrial Health Council, 168
American Insitute of Biological Sciences,
    5854
American Lands, 320
American Lung Association, 170, 2089
American Lung Association of the District of
    Columbia, 856
American Petroleum Institute, 117, 5598
American Public Health Association, 171,
    2091
American Public Power Association, 118
American Public Works Association, 23
American Recreation Coalition, 618
American Rivers, 638, 2291
American Society for Microbiology, 172,
    2092
American Society of Landscape Architects,
    95
Americans for the Environment, 512
Animal and Plant Health Inspection Service
    Protection Quarantine, 2516
Antarctica Project, 323, 2517
Army Corps of Engineers, 2518
Aspen Institute, 2519
Association of American Geographers, 5861
Association of Ecosystem Research Centers,
    4830
Association of Environmental and Resource
    Economists, 5862

Association of Local Air Pollution Control
Officials, 4

Association of Maternal and Child Health
Program, 2189

Association of Metropolitan Sewerage
Agencies, 644

Association of Metropolitan Water Agencies,
645

Association of State and Interstate Water
Pollution Control Administrators, 648

Association of State and Territorial Health
Officials, 176

Association of State and Territorial Solid
Waste Management Officials, 517

Asthma and Allergy Foundation of America,
178, 2099, 2294, 2294

Atlantic States Marine Fisheries
Commission, 1621, 2520

BLM Environmental Education Program,
5865

Bank Information Center, 27

Battleground National Cemetery, 3601

Bauman Foundation, 2299

Benton Foundation: Kids Campaigns, 2190

Beyond Pesticides/National Coalition
Against the Misuse of Pesticides, 179,
2100

Biodiversity Support Program, 2521

Bioengineering & Environmental Systems
Section, 2522

Biotechnology Industry Organization, 28

Blue Mountain Natural Resource Institute
Advisory Board, 2523

Bureau of Economic Analysis, 2525

Bureau of Land Management, Land &
Renewable Resources, 2526

Bureau of Oceans International
Environmental & Scientific Affairs, 2527

Buying Green: Federal Purchasing Practices
and the Environment, 2257

CO-OP America, 31

CONCERN, 557

Carrying Capacity Network, 343

Center for Clean Air Policy, 5

Center for Environmental Strategies:
Academy for Educational Development,
5881

Center for International Environmental Law,
2223

Center for Marine Conservation, 651

Center for Policy Alternatives, 32

Center for Science in the Public Interest, 181,
2105

Chemical Emergency Preparedness Program,
101

Chief of Engineers Environmental Advisory
Board, 2532

Children s Defense Fund, 2193, 2210

Children s Environmental Health Network,
2194

Children s Environmental Health: Research,
Practice, Prevention and Policy, 1622

Chlorine Institute, 34, 183

Civil Division: Consumer Litigation Office,
2533

Clean Air Scientific Advisory Committee,
2534

Clean Water Act Twenty Years Later, 2258

Clean Water Action, 655, 2311

Clean Water Fund, 656

Clean Water Network, 657

Climate Institute, 7

Coast Alliance, 658

Coast Guard, 2535

Coastal States Organization, 2536

Committee for Conservation and Care of
Chimpanzees, 347

Committee for Environmentally Effective
Packaging, 1862

Committee of State Foresters, 2537

Committee on Agriculture, 2538

Committee on Agriculture, Nutrition, and
Forestry, 2539

Committee on Appropriations, 2540

Committee on Commerce, 2541

Committee on Commerce, Science, and
Transportation, 2542

Committee on Education and the Workforce,
2543

Committee on Energy and Natural
Resources, 2544

Committee on Environment and Public
Works Republicans, 2545

Committee on Glen Canyon Environmental
Studies Review, 2546

Committee on Government Reform and
Oversight, 2547

Committee on Labor and Human Resources,
2548

Committee on Resources, 2549

Committee on Science, 2550

Committee on Small Business: House of
Representatives, 2551

Committee on Small Business: US Senate,
2552

Committee on Transportation and
Infrastructure, 2553

Community Planning and Development
Planning and Community Viability, 2555

Community Rights Counsel, 2227

Conservation Foundation, 4925

Conservation International, 350, 2314

Consumer Energy Council of America
Research Foundation, 124

Consumer Specialty Products Association, 35

Cooperative Forestry Research Advisory
Council, 2556

Council on Environmental Quality, 2557

Counterpart International, 353

Danaher Corporation, 4950

Dangerous Goods Advisory Council, 189,
2112, 2558, 2558

Defenders of Wildlife, 354, 2322

Department of Agriculture, 2559

Department of Agriculture: Research
Department, Forest Environment
Research, 2560

Department of Agriculture: Forest Inventory,
Economics, 2561

Department of Agriculture: Forest Service
Public Affairs, 2562

Department of Agriculture: National Forest
Watershed and Air Management, 2563

Department of Agriculture: National Forest
Watershed and Soil Resource, 2564

Department of Agriculture: Research
Department, 2566

Department of Agriculture: Research
Department Fire Sciences Program, 2567

Department of Agriculture: Soil
Conservation, 2568

Department of Agriculture: State & Private
Forestry, 2569

Department of Energy: Office of Alcohol
Fuels, 2574

Department of Energy: Office of NEPA
Policy and Compliance, 2575

Department of Energy: Transportation and
Alternative Fuels, 2576

Department of Energy: Utility Technologies
Department, 2577

Department of Justice: Environment and
Natural Resources Division, 2578

Department of Justice: Environment and
Resources, Environmental Defense, 2579

Department of State: Bureau of Economic
and Business, 2580

Department of State: Bureau of Oceans and
International Environmental and Scientific
Affairs, 2581

Department of State: Energy Resources and
Food Policy Department, 2582

Department of State: National Office of
Environmental Protection, 2583

Department of State: Ocean and Fisheries
Affairs, 2584

Department of State: Office of Ecology,
Health, and Conservation State
Department, 2585

Department of State: Office of Global
Change, 2586

Department of Transportation: Administrator
for Pipeine Safety, Alaska, 2587

Department of Transportation: Associate
Administrator for Pipeline Safety, 2588

Department of Transportation: Office of
Marine Safety and Environmental, 2589

Department of Transportation: Office of
Marine Safety, Security & Environmental,
2590

Department of the Interior, 2591

Department of the Interior, U.S. Fish &
Wildlife Service, 2592

Department of the Interior: National Parks
Service, 2593

Department of the Interior: Bureau of Land
Management, 2594

Department of the Interior: Division of Parks
and Wildlife, 2595

Department of the Interior: Land &
Renewable Wildlife & Fisheries Division,
2596

Department of the Interior: Land and
Renewable Forestry Division, 2597

Department of the Interior: National Park
Serv ices, 2598

Department of the Interior: National
Resources Department, 2599

Department of the Interior: Office of the
Solicitor, 2600

Department of the Interior: Policy,
Management Office of Environmental
Affairs, 2601

Department of the Interior: Soil, Water &
Air, 2602

Department of the Interior: US Fish &
Wildlife, Fisheries Department, 2603

Department of the Interior: US Fish &
Wildlife Services, 2604

Department of the Interior: Water and
Science Bureau of Reclamation, 2607

Department of the Interior: Wild Horses and
Burros, 2608

Department of the Interior: Wildlife and
Vegetation, 2609

Dept. of Agriculture: National Forest
Watershed and Hydrology, 2610

Dialogue Committee on Phosphoric Acid
Product Consensus and Dispute
Resolution, 2611

District of Columbia State Extension
Services, 2911
EPA: Border Environmental Plan, 2612
EPA: Department of Solid Chemical
Emergency Preparedness & Prevention,
2613
EPA: Office of Solid Waste, Municipal & &
Industrial Solid Waste, 2614
Earth Day Network, 6091
Earth Share, 38, 2325
Earth System Science Community
Curriculum, 5911
Ecological Society of America, 39, 664,
5915, 5915
Economists, 4994
Endangered Species Coalition, 366
Endangered Species Committee, 2616
Energy Research Office, 2617
Environment Safety and Health Office, 2618
Environment and Natural Resources:
Environmental Crimes Section, 2619
Environment, Safety and Health, 2620
Environmental Change and Security
Program: Woodrow Wilson International
Center for Scholars, 2621
Environmental Council of the States, 41
Environmental Enforcement, 2622
Environmental Financial Advisory Board
(EFAB), 3245
Environmental Industry Association, 42, 524
Environmental Law Institute, 2233, 2331,
5661, 5661
Environmental Management Advisory
Board, 2625
Environmental Monitoring Management
Council, 2626
Environmental Politics and Policy, 2265
Environmental Protection Agency Global
Change Division, 2629
Environmental Protection Agency Ground
Water and Drinking Water, 2630
Environmental Protection Agency Resource
Conservation and Recovery Act, 2631
Environmental Protection Agency: Health &
Environmental Review Division, 2632
Environmental Protection Agency: Air &
Radiation, 2633
Environmental Protection Agency: Grants
Administration Division, 2469
Environmental Protection Agency: Health
Effects Division, 2634
Environmental Protection Agency: Indoor
Air Division, 2635
Environmental Protection Agency: Office of
Pollution Prevention, 2636
Environmental Protection Agency: Office of
Health & Environmental Assessment, 2637
Environmental Security, 2640
Environmental Support Center, 2235
Environmental Technologies Exports, 6101
Environmental Technology Council, 525
Environmental Working Group, 857, 2119,
5062, 5062, 5941
Environmental and Energy Study Institute,
562, 5942
Environmental and Engineering Fellowship,
2471
Federal Aviation Administration, 2641
Federal Energy Regulatory Commission,
2642
Federal Facilities Environmental Restoration,
2643
Federal Highway Administration, 2644

Federal Highway Administration
Recreational Trails Programs, 2645
Federal Highway Administration
Transportation Enhancement Activities,
2646
Federal Railroad Administration, 2647
Federal Task Force on Environmental
Education, 2648
Federal Transit Administration, 2649
Formaldehyde Institute, 5676
Friends of the Earth, 259, 563
Friends of the Earth Foundation, 2338
General Services Administration, 2652
George Washington University, 5946
Georgetown University, 5947
German Marshall Fund of the United States,
2477
Global Climate Coalition, 8, 47
Global Environmental Management
Initiative, 48
Global Tomorrow Coalition, 566
Green Media Toolshed, 5954
Green Seal, 49
Greenhouse Crisis Foundation, 9
Greenpeace, 2239
Greenpeace USA, 50
H John Heinz III Center for Science
Economics and the Environment, 5136
Hazardous Waste Resource Center
Environmental Technology Council, 200,
528, 2125, 2125
Heinz Center, 52
Historic Trust for Historic Preservation, 384
House Committee on Agriculture
Operations, Oversight, Nutrition, 2654
House Committee on Commerce
Telecommunications, Trade and
Consumer Protection, 2655
House Committee on International Relations,
2656
House Committee on Transportation and
Infrastructure, 2657
Household Products Disposal Council, 204
Human Environment Center, 206, 858
Humane Society of the United States, 264,
386
Indian Affairs Office of Management
Support Services, 2658
Installation Management, 2659
Installations and Environment, 2660
Institute for Conservation Leadership, 389
Institute of Clean Air Companies, 11, 54,
529, 529
Institute of Medicine: Board of Children,
Youth and Families, 2199
Institute of Scrap Recycling Industries
Convention, 530, 1635
Institute of Scrap Recycling Industries:
Seaboard Chapter, 1444
International Association of Fish and
Wildlife Agencies, 569
International Center for Development Policy,
2662
International Council for Bird Preservation,
US Section, 393
International Joint Commission, 2663
International Trade Administration Trade
Development, 2664
International Union for Conservation of
Nature and Natural Resources, 402
International Union for the Conservation of
Nature s Primate Specialist Group, 403
Interstate Council on Water Policy, 677
Island Resources Foundation, 574

Joint Oceanographic Institutions, 2485
Kangaroo Protection Foundation, 2358
Land & Renewable Recreation & Wilderness
Resources: Wilderness Branch, 2665
Land Trust Alliance, 578
Land and Minerals Management, 2666
Land and Minerals Office of Surface Mining
Reclamation & Enforcement, 2667
League of Conservation Voters, 2243
League of Women Voters of the United
States, 2244
Management Advisory Group to the
Assistant Administrator, 2668
Management and Budget Office: Natural
Resources, Energy and Science, 2669
Manpower, Reserve Affairs, Installations and
Environment, 2670
Manufacturers of Emission Controls
Association, 532, 580
Maritime Administration, 2672
Marketing and Regulatory Programs and
Support Services Agricultural Marketing,
2673
McIntosh Foundation, 2374
Mineral Policy Center, 411
Miranda Associates, 5308
Monitor Consortium of Conservation
Groups, 412
Municipal Waste Management Association,
533
National Academy of Sciences, 2490
National Aeronautics and Space
Administration, 2679
National Alliance for Hispanic Health, 214,
2137
National Animal Damage Control Advisory
Committee, 2680
National Association for Environmental
Management, 58
National Association of Chemical Recyclers,
535
National Association of City and County
Health Officials, 215
National Association of City and County
Health Officials, 2138
National Association of Flood and Storm
Water Management Agencies, 105, 680
National Association of Service and
Conservation Corps, 63
National Association of State Departments of
Agriculture, 582
National BioEnergy Industries Association,
134
National Conservation Foundation, 417
National Council for Science and the
Environment, 65
National Council of State Tourism Directors,
627
National Education Association Health
Information Network, 222, 2147
National Endangered Species Act Reform
Coalition, 275
National Environmental Development
Association, 67, 585
National Environmental Education, 68
National Environmental Justice Advisory
Council, 2685
National Environmental Trust, 69, 2150
National Fish and Wildlife Foundation, 419,
2377
National Forest Foundation, 420
National Geographic Society, 71
National Geographic Society Education
Foundation, 2378

World Resources Institute, 2443, 6062
World Wildlife Fund, 505, 2445
Worldwatch Institute, 615
Worldwide Network, 506

Florida

AB2MT Consultants, 1699
ABC Research Corporation, 4745
ASLA: Florida Chapter, 869
Air Consulting and Engineering, 1723
Alpha-Omega Environmental Services, 1732
American Fisheries Society: Agriculture Economics Section, 870
American Lung Association of Florida, 871
American Lung Association: Central Area Office, 872
American Lung Association: Gulfcoast Area, 873
American Lung Association: Gulfcoast Area Sou thwest Office, 874
American Lung Association: North Area-Northwest Office, 875
American Lung Association: North Area-Big Bend Office, 876
American Lung Association: North Area-Daytona Office, 877
American Lung Association: South Area Office, 878
American Lung Association: Southeast Area Office, 879
American Lung Association: Southeast Area-Belle Glade Office, 880
American Lung Association:Gulfcoast Area-Nature Coast Office, 881
American Lung Association:Gulfcoast Area-East Bay Office, 882
American Lung Association:Gulfcoast Area Sou th Bay Office, 883
American Society of Ichthyologists and Herpetologists, 251
Apalachicola National Forest, 3602
Applied Marine Ecology, 1744
Ardaman & Associates, 4822
Arthur R Marshall Loxahatchee National Wildlife Refuge, 3603
Association of Battery Recyclers, 514, 884, 2098, 2098
Astro-Pure Water Purifiers, 4832
Audubon of Florida, 885
BCI Engineers and Scientists, 1766
Big Cypress National Preserve, 3604
Biological Research Associates, 1794, 4865
Biscayne National Park, 3605
Bromwell & Carrier, 4880
CRB Geological and Environmental Services, 1823
CZR, 1827
Canaveral National Seashore, 3606
Canin Associates, 1836
Caribbean Conservation Corporation, 342, 2305
Center for Aquatic Research and Resource Management, 5615
Center for Mathematical Services, 6081
Center for Solid & Hazardous Waste Management, 5638
Center for Tropical Agriculture, 5640
Central and North Florida Chapter, National Safety Council, 886

Chassahowitzka National Wildlife Refuge Complex, 3607
Citizens for a Scenic Florida, 621
Clean Energy Research Institute, 5643
Coastal Planning and Engineering, 1857
Consumer Testing Laboratories, 4926
Continental Shelf Associates, 4929
Dennis Breedlove and Associates, 1887
Department of Commerce National Oceanic & Atlantic Oceanographic & Meteorological Laboratory, 2919
Dry Torgus National Park/Everglades National Park, 3608
Dunspaugh-Dalton Foundation, 2324
Egmont National Wildlife Refuge, 3609
Energy Research Institute, 125
EnviroTech Southeast, 5020
Enviroclean Technology, 5022
Environmental Education K-12, 6097
Environmental Human Toxicology, 5659
Environmental Science, 5930
Environmental Systems Engineering Institute, 5668
Everglades Laboratories, 5073
Everglades National Park, 3610
First Coast Environmental Laboratory, 5075
Florida Center for Environmental Studies, 5944
Florida Cooperative Extension Service, 2920
Florida Cooperative Fish and Wildlife Research Unit, 5673
Florida Defenders of the Environment, 887
Florida Department of Agriculture & Consumer Service, 2921
Florida Department of Environmental Protection, 2922
Florida Department of Health and Rehabilitative Services, 2923
Florida Environmental Health Association, 888
Florida Forestry Association, 889
Florida Keys Wild Bird Rehabilitation Center, 890
Florida Museum of Natural History, 5674
Florida Ornithological Society, 891
Florida Panther National Wildlife Refuge, 3611
Florida Public Interest Research Group, 892
Florida Solar Energy Industries Association, 893
Florida State Department of Health, 2924
Florida State University, 5945
Florida Trail Association, 894
Flowers Chemical Laboratories, 5077
Game & Fresh Water Fish Commission Informational Services, 2925
Gannett Energy Laboratory, 5678
Geosyntec Consultants, 5111
Global Environmental Institute, 5950
Gopher Tortoise Council, 378
Gradient Corporation, 1926, 5123
Gulf of Mexico Fishery Management Council, 2926
Hobe Sound National Wildlife Refuge, 3612
Inprimis, 5189
Interdisciplinary Center for Aeronomy & Other Atmospheric Sciences, 5712
International Association for Hydrogen Energy, 895
International Game Fish Association, 896
International Oceanographic Foundation, 674
International Osprey Foundation, 398
Island Conservation Effort, 405

JN Darling National Wildlife Refuge, 3613
Jessie Ball duPont Religious, Charitable and Educational Fund, 2483
John D MacArthur Agro Ecology Research Center, 5717
Joyce Environmental Consultants, 5218
Keep Florida Beautiful, 897
Kimre, 1953
Lake Woodruff National Wildlife Refuge, 3614
Land Research Management, 5247
Lee County Parks & Recreation, 2927
Legal Environmental Assistance Foundation (LEAF), 898, 2246
Marine and Freshwater Biomedical Sciences Center, 5737
Miami Marine Research & Testing Station, 5295
Millar Wilson Laboratory for Chemical Research, 5742
Mirage Systems, 5307
Mote Environmental Services, 6125
National Audubon Society: Everglades Campaign, 584
National Key Deer Refuge, 3615
National Wildlife Federation: Everglades Project, 899
Natural Resources Department: Beaches & Division, 2928
Natural Resources Department: Recreation & Parks Division, 2929
Occupational Health Conservation, 5335
Occupational Safety and Health Consultants, 1989
Orlando Laboratories, 5346
Pelican Island National Wildlife Refuge, 3616
Pelican Man s Bird Sanctuary, 900
Pollution Prevention Trust Fund, 6135
Post, Buckley, Schuh and Jernigan, 2008
Professional Service Industries/Jammal & Associates Division, 5398
Q-Lab, 5401
Reef Relief, 901
Rosenstiel School of Marine and Atmospheric Science, 5996
Sanibel-Captiva Conservation Foundation, 902
Save the Manatee Club, 464
School of Forest Resources and Conservation, 5999
Seacamp Association, Inc, 6148
Society of Environmental Toxicology and Chemistry, 904, 2177, 6152, 6152
Solar Energy and Energy Conversion Laboratory, 5806
South Florida Chapter, National Safety Council, 905
Southeastern Association of Fish and Wildlife Agencies Annual Meeting, 906, 1039, 1676, 1676
Southeastern Engineering & Testing Laboratories, 5479
Southwest Florida Water Management District, 2930
Spectrum Sciences and Software, 5487
St. Marks National Wildlife Refuge, 3617
St. Vincent National Wildlife Refuge, 3618
Suncoast Seabird Sanctuary, 907
Tall Timbers Research Station, 473
Tallahassee Museum of History and Natural Science, 908
Taylor Engineering, 5511
Technos, 2055

Thornton Laboratories, 5527
Timucuan Ecological & Historic Preserve, 3619
United States Department of Treasury US Customs Service, 2931
University of West Florida, 6042
Vara International: Division of Calgon Corporation, 5554
Villanueva Associates, 5556
Water and Air Research, 2066, 5569
Wildlife Foundation of Florida, 909

Georgia

ASLA: Georgia Chapter, 910
Advanced Chemistry Labs, 1718
Agency for Toxic Substances and Disease Registry, 911, 2073, 2512, 2512
American Academy of Sanitarians, 912
American Cancer Society, 165, 913, 2082, 2082
American Lung Association of Georgia, 914
Analytical Services, 4803
Andersen 2000 Inc/Crown Andersen, 1740
Applied Technical Services, 4817
Association of Energy Engineers, 149, 515
Blackbeard Island National Wildlife Refuge, 2932
Board of Scientific Counselors: Agency for Toxic Substance and Disease Registry, 2933
Cape Environmental Management, 1837
Case Studies in Environmental Medicine, 2208
Center for Disease Control and Prevention, 2528
Center for a Sustainable Coast, 915
Centers for Disease Control & Prevention, 2107
Centers for Disease Control and Prevention, 916, 2108, 2530, 2530
Chattahoochee River National Recreation Area: Advisory Commission, 2934
Chattahoochee-Ocnee National Forest, 3620
Coastal Conservation Association of Georgia, 917
Compass Environmental, 1866
Coosa River Basin Initiative, 918
Council of State and Territorial Epidemiologists, 919, 2111
Cumberland Island National Seashore, 3621
Earth Share of Georgia, 920
Environmental Justice Resource Center, 921, 2117
Environmental Technology Expo, 1628
Georgia Association of Conservation District Supervisors, 922
Georgia Chapter, National Safety Council, 923
Georgia Conservancy, 924
Georgia Department of Agriculture, 2935
Georgia Department of Education, 2936
Georgia Department of Natural Resources: Historic Preservation Division, 2937
Georgia Department of Natural Resources: Pollution Prevention Assistance Division, 2938
Georgia Environmental Health Association, 925
Georgia Environmental Organization, 926
Georgia Federation of Forest Owners, 927

Georgia Pacific Foundation, 2344
Georgia Sea Grant College Program, 2939
Georgia Trappers Association, 928
Georgia Water and Pollution Control Association, 929
Georgia Wildlife Federation, 930
GlobalCon, 1632
Greenprints: Sustainable Communities by Design, 1633
Human Ecology Action League (HEAL), 205, 931, 2127, 2127
Insider s Guide to Environmental Negotiation, 2268
Law & Company, 5250
Law Environmental, 1957
McWhorter and Associates, 5288
Mountain Conservation Trust of Georgia, 932
National Center for Environmental Health, 219, 2143, 2940, 2940
National Coalition for Marine Conservation, 683
National Park Service Cooperative Unit: Athens, 5756
Natural Resources Department: Air Protection, 2942
Natural Resources Department: Coastal Division, 2943
Natural Resources Department: Environmental Protection Division, 2944
Natural Resources Department: Land Protection, 2945
Nature Conservancy: Georgia Chapter, 934
North American Science Association, 5326
Okefenokee National Wildlife Refuge, 3622
Piedmont National Wildlife Refuge, 3623
Pinckney Island National Wild Refuge, 3624
Plant and Facilities Expo, 1671
QORE, 2014
Savannah Laboratories, 5452
Savannah National Wildlife Refuge: Savannah Coastal Refuges, 3625
Sierra Club: Georgia Chapter, 935
Southface Energy Institute, 611
TRC Garrow Associates, 5508
Timber Products Inspection and Testing, 5529
Trees Atlanta, 936
Tribble & Richardson, 5540
Turner Foundation, 2420
United States Fish and Wildlife Service, 2947
Upper Chattahoochee Riverkeeper, 937
Wassaw National Wildlife Refuge, 2948
Westinghouse Remediation Services, 2069, 5578
World Energy Engineering Congress, 1696

Hawaii

AECOS, 1704
ASLA: Hawaii Chapter, 939
American Lung Association of Hawaii, 940
American Lung Association of Hawaii- Maui Office, 941
American Lung Association of Hawaii: East Hawaii Office, 942
American Lung Association of Hawaii: Kauai Office, 943
American Lung Association of Hawaii: West Hawaii Office, 944
Atherton Family Foundation, 2295

Baker, Howland, & Jarvis Islands & Johnston Atoll National Wildlife Refuge, 3626
Big Island Rain Forest Action Group, 945
College of Tropical Agriculture and Human Resources, 2950
Department of Land and Natural Resources Division of Water Resource Management, 2951
Department of Oceanography, 5905
EarthTrust, 946
Environmental Center: University of Hawaii, 2952, 5654
First Hawaiian Foundation, 2333
Flipper Foundation, 947
Greenpeace Foundation, 948
Hakalau Forest National Wildlife Refuge Hawaiian & Pacfic Islands Complex, 3627
Haleakala National Park, 3628
Hanalei National Wildlife Refuge: Kauai, 3629
Hawaii Association of Conservation Districts, 949
Hawaii Cooperative Fishery Research Unit, 5689
Hawaii Department of Agriculture, 2953
Hawaii Institute of Marine Biology University of Hawaii, 2954
Hawaii Nature Center, 950
Hawaii Undersea Research Laboratory, 5690
Hawaii Volcanoes National Park, 3630
Hawaiian Botanical Society, 951
Hawaiian Island & Midway Atoll National Wildlife Refuge, 3631
Health Department: Environmental Quality Control, 2955
Health Department: Noise & Radiation Branch, 2956
Huleia National Wildlife Refuge: Kauai, 3632
James Campbell National Wildlife Refuge, 3633
Kakahaua National Wildlife Refuge, 3634
Kealia Pond National Wildlife Refuge, 3635
Kilauea Point National Wildlife Refuge, 3636
Life of the Land, 408
National Undersea Research Center, 5758
Natural Energy Laboratory of Hawaii Authority, 5759
Nature Conservancy: Hawaii Chapter, 952
Oahu National Wildlife Refuge, 3637
Oceanic Institute, 5768
PBR HAWAII, 1997
Pacific Gamefish Research Foundation, 5772
Pacific Whale Foundation, 290, 2389
Pearl Harbor National Wildlife Refuge, 3638
Sierra Club: Hawaii Chapter, 953
State of Hawaii: Department of Land and Natural Resources, 2957
Water Resources Research Center University of Hawaii, 2959

Idaho

American Lung Association of Idaho/Nevada: Boise Office, 954
Bear Lake National Wildlife Refuge Southeast Idaho Complex, 3639
Boise Cascade Corporation, 2302
Boise National Forest, 3640
Camas National Wildlife Refuge Southeast Idaho Refuge Complex, 3641

Caribou-Targhee National Forest, 3642
Clearwater National Forest, 3643
Coeur d Alene District Advisory Council, 2960
Craters of the Moon National Monument, 3644
Deer Flat National Wildlife Refuge, 3645
Department of Lands, 2961
Eco-Analysts, 4990
Environmental Research Institute, 5663
Environmental Resource Center, 5664, 5929
Grays Lake National Wildlife Refuge Southeast Idaho Refuge Complex, 3646
Health & Welfare Department: Legal Services, Environmental Quality, 2962
Idaho Association of Soil Conservation Districts, 955, 2963
Idaho Conservation League, 956
Idaho Cooperative Extension, 2964
Idaho Cooperative Fish & Wildlife Research Unit, 5696
Idaho Department of Environemtal Quality: Idaho Falls Regional Office, 2965
Idaho Department of Environmental Quality: Pocatello Regional Office, 2966
Idaho Department of Environmental Quality: State Office, 2967
Idaho Department of Fish & Game: Clearwater Region, 2968
Idaho Department of Fish & Game: Headquarters, 2969
Idaho Department of Fish & Game: Magic Valley Region, 2970
Idaho Department of Fish & Game: McCall, 2971
Idaho Department of Fish & Game: Panhandle Region, 2972
Idaho Department of Fish & Game: Salmon Region, 2973
Idaho Department of Fish & Game: Southeast Region, 2974
Idaho Department of Fish & Game: Southwest Region, 2975
Idaho Department of Fish & Game: Upper Snake Region, 2976
Idaho Department of State Parks and Recreation, 2977
Idaho Department of Water Resources, 2978
Idaho Falls District Advisory Council Bureau of Land Management, 2979
Idaho Forest Owners Association, 957
Idaho Geological Survey, 2980
Idaho Panhandle National Forest: Kaniksu, 3647
Idaho State Department of Agriculture, 2981
Idaho State University, 5958
Kootenai National Wildlife Refuge, 3648
Lands Department: Soil Conservation Commission, 2982
Minidoka National Wildlife Refuge Southeast Idaho Refuge Complex, 3649
Nez Perce National Forest, 3650
Payette National Forest, 3651
Peregrine Fund, 456
Salmon District Advisory Council Bureau of Land Management, 2983
Salmon-Challis National Forest: Forest Service Building, 3652
Sawtooth National Forest, 3653
Southeast Idaho National Wildlife Refuge, 3654
Targhee National Forest, 3655

United States Department of the Interior Bureau of Land Management, 2883, 2906, 2918, 2918, 2985
University of Idaho, 6023
Wilderness Research Center, 5842
Wolf Education and Research Center, 498

Illinois

ARDL, 4758
ASLA: Illinois Chapter, 961
Abandoned Mined Lands Reclamation Council, 507
Accurate Engineering Laboratories, 4767
Aires Consulting Group, 1725
Alar Engineering Corporation, 4782
Allied Laboratories, 4784
American Academy of Pediatrics: Committee on Environment Health, 2188
American College of Occupational and Environmental Medicine, 962, 2083
American Fishing Tackle Manufacturers Association, 2451
American Laboratory for Environmental Excellence, 4792
American Lung Association of Illinois-Iowa, 963
American Lung Association of Metropolitan Chicago, 964
American Lung Association: Chicagoland Collar Counties, 965
American Lung Association: Northern Illinois, 966
American Lung Association: Southwestern Illinois, 967
American Medical Association, 968, 2090
American Nuclear Society, 2455
American Planning Association, 552
American Society of Safety Engineers, 145, 2093
American Waste Processing, 4796
Amoco Foundation, 2292
Argonne National Laboratory, 5603, 6071
Arro, 4825
Arro Laboratory, 1756
Arthur L Conn & Associates, 4827
Association for Conservation Information Illinois Department of Conservation, 325, 2986
Association of Illinois Soil and Water Conservation Districts, 2987
Audubon Council of Illinois, 969
Baxter and Woodman, 1777, 4851
Boelter and Yates, 1803
Bottom Line Consulting, 1805
Bradley University, 5868
C&D Debris Recycling, 518
CTE Engineers, 1824
Camiros Limited, 1833
Caterpillar Foundation, 2307
Center for the Great Lakes, 653
Central Midwest Interstate Low-Level Radioactive Waste Commission, 2988
Central States Environmental Services, 1844
Chelsea Group, 1847
Chicago Botanic Garden, 6083
Chicago Chapter, National Safety Council, 970
Chicago Chem Consultants Corporation, 1849
Chicago Zoological Society, 971

Clean Air Engineering, 1852, 4912
Clean World Engineering, 1855
Conservation Deartment: Resource Marketing and Education, 2989
Conservation Department, 2990
Consoer Townsend Envirodyne Engineers, 1871
Construction Engineering Research Laboratory, 2991
Curtis and Edith Munson Foundation, 2320
Daily Analytical Laboratories, 4948
ESTECH, 1626
Eagle Nature Foundation, 361, 523, 972, 972
Eastern Illinois University, 5913
Eichleay Corporation of Illinois, 4997
Elsa Wild Animal Appeal USA, 365
Energy Resources Center, 5651
Envirodyne Engineers, 5023
Environmental Law and Policy Center of the Midwest, 2234
Environmental Protection Agency: Region 5, 2992
Environmental Protection Agency: Water Pollution Control, 2639, 2993
Environmental Science & Engineering, 5054
GL Applied Research, 5090
Gabriel Laboratories, 5092
Gas Technology Institute, 5096
Gaynes Labs, 5097
Global Warming International Conference and Expo, 1631
Globetrotters Engineering Corporation, 5119
Great Lakes Protection Fund, 2478
Great Lakes Sport Fishing Council, 974
Gulf Coast Laboratories, 5134
Handbook of Pediatric Environmental Health, 2211
Hermann Associates, 1940
Huff and Huff, 1941
Illinois Association of Conservation Districts, 975
Illinois Association of Environmental Professionals, 976
Illinois Audubon Society, 977
Illinois Department of Agriculture Bureau of Land and Water Resources, 2994
Illinois Department of Transportation, 2995
Illinois Dept. Commerce & Econimic Opportunity Division of Recycling & Waste Reduction, 2996
Illinois Environmental Council, 978
Illinois Nature Preserves Commission, 2997
Illinois Prairie Path, 979
Illinois Recycling Association, 980
Illinois Solar Energy Association, 981
Institute for Regional and Community Studies, 5706
Institute of Environmental Sciences and Technology, 5963
International Association for the Advancement of Earth and Environmental Sciences, 5964
International Certification Accreditation Board, 1946
International Society for Environmental Toxicology and Cancer, 210
International Society of Arboriculture, 571
International Water Resources Association, 676
Invensys Climate Controls, 5198
John D and Catherine T MacArthur Foundation, 2356

Indiana

Iowa

Western Research Farm, 5839
World Association of Soil and Water
Conservation, 711

Kansas

American Academy of Environmental
Medicine, 161, 1616, 2077, 2077
American Association of Zoo Keepers, 18
American Lung Association of Kansas, 1019
American Society of Mammalogists, 252
Audubon of Kansas, 1020
Black and Veatch Engineers: Architects, 1799
Center for Hazardous Substance Research,
5629
Cook Flatt and Strobel Engineers, 1874
DPRA, 1883, 4946
ENVIRO Tech Services, 4973
Emporia Research and Survey Office Kansas
Department of Wildlife & Parks, 3012
Emporia State University, 5917
Environmental Affairs Office, 3013
Environmental Protection Agency: Region 7,
Air & Toxics Division, 3014
Fitch Natural History Reservation, 5672
Gordon Piatt Energy Group, 5122
Grassland Heritage Foundation, 380
Great Plains: Rocky Mountain Hazardous
Substance Research Center, 5685
Health & Environment Department, 3015
Health & Environment Department: Air &
Radiation, 3016
Health & Environment Department:
Environment Division, 3017
Health & Environment Department: Waste
Management, 3018
Hydro-logic, 5173
Information Systems Division Public
Information Services, 3019
Kansas Academy of Science, 1022
Kansas Association for Conservation and
Environmental Education, 1023
Kansas Association of Soil Conservation
Districts, 3020
Kansas City Testing Laboratory, 5224
Kansas Cooperative Fish & Wildlife
Research, 3021
Kansas Corporation Commission
Conservation Division, 3022
Kansas Department of Wildlife & Parks,
3023
Kansas Department of Wildlife & Parks
Region 5, 3024, 3025, 3026, 3026, 3027,
3028
Kansas Geological Survey, 3029
Kansas Health & Environmental
Laboratories, 3030
Kansas Natural Resources Council, 1024
Kansas Rural Center, 5722
Kansas State University, 5970
Kansas Water Office, 3031
Kansas Wildflower Society, 1025
Kansas Wildscape Foundation, 1026
Kirwin National Wildlife Refuge, 3663
Land Institute, 577
Mayhew Environmental Training Associates
(META), 5282
North American Falconers Association, 446
North Dakota Natural Science Society, 1027
Pratt Operations Office Kansas Department
of Wildlife & Parks, 3032

Professional Service Industries, 5396
Purewater Corporation, 5399
Systematics Museums, 5813
Trees for Life, 478
University of Kansas Field Stations, 6025
Water Resources Center, 5831
Wildlife Disease Association, 491
Wildlife and Parks Department
Environmental Affairs Office, 3033

Kentucky

Agriculture Department Consumer Safety:
Office Pesticides Division, 3034
American Cave Conservation Association,
315
American Lung Association of Kentucky,
1030
Attorney General s Office Civil and
Environmental Law Division, 3035
Biosystems and Agricultural Engineering,
6074
Breaks Interstate Park Commission, 3036
Business Health Environmental Lab, 4887
Center for Applied Energy Research, 5613
Center for Cave and Karst Studies Western
Kentucky University: Geography &
Geology, 5616
Central/Southern Indiana: National Safety
Council, Kentucky Office, 997
Commonwealth Technology, 4922
Conjun Laboratories, 4924
Council of State Governments, 36
Daniel Boone National Forest, 3664
Department for Environmental Protection,
3037
Economic Development Cabinet:
Community Development Department
Brokerage Division, 3038
Ed Caicedo Engineers/Consultants, 1898
Energy & Environmental Technology, 5003
Environmental Education Department, 3039
Environmental Protection Department:
Management Services Branch, 3040
Environmental Protection Cabinet: Law
Department, 3041
Environmental Protection Department:
Environmental Services Division, 3042
Environmental Protection Department:
Waste Management Division, 3043
Environmental Protection Department: Water
Division, 3044
Fish and Wildlife Resources Department:
Fisheries Division, 3045
Fish and Wildlife Resources Department:
Conservation Education Division, 3046
Geological Sciences & Laboratories, 5106
Heritage Remediation Engineering, 5158
Kentucky Association for Environmental
Education, 1031
Kentucky Audubon Council, 1032
Kentucky Department for Public Health, 3047
Kentucky Environmental and Public
Protection C abinet, 3048
Kentucky Natural Resources and
Environmental Protection Cabinet, 3049
Kentucky Resource Laboratory, 5229
Kentucky Resources Council, 1033
Kentucky State Cooperative Extension
Services, 3050

Kentucky State Nature Preserves
Commission, 3051
Kenvirons, 5230
Land Between the Lakes Association, 1034
Louisville Testing Laboratory, 5260
Mammoth Cave National Park, 3665
McCoy & McCoy Laboratories, 5283
Metro Services Laboratories, 5294
National Council for Environmental Balance,
418
National Safety Council, Kentucky Office:
Central/Southern Indiana & Cincinnati,
1035
Natural Resources & Environmental Project
Cabinet: Department for Natural
Resources, 3052
Natural Resources & Environmental
Protection Cabinet, 3053
Natural Resources Conservation and
Management, 6130
Natural Resources Department: Conservation
Division, 3054
Natural Resources Department: Division of
Fore stry, 3055
Natural Resources Department: Energy
Division, 3056
Natural Resources and Environment
Protection Cabinet: Environmental Quality
Commission, 3057
Natural Resources and Environmental
Protection Cabinet, 3058
Nature Conservancy: Kentucky Chapter,
1036
PRD Tech, 5362
Presnell Associates, 2009
Reclamation Services Unlimited, 2020
Scenic Kentucky, 1037
Sportsmans Network, 300
Surface Mining Reclamation and
Enforcement Department, 3059
Theil Consulting, 2058
Tourism Cabinet: Parks Department, 3060
Water Testing Laboratory, 5835

Louisiana

AF Meyer and Associates, 4751
Agriculture & Forestry: Soil & Water
Conservation, 3062
American Lung Association of Louisiana,
1041
Atchafalaya National Wildlife Refuge
Sherburne Wildlife Management Area,
3666
Barataria Preserve, 3667
Best Environmental, Safety and Industrial
Products, 4857
Burk-Kleinpeter, 1812
Calcasieu Parish Animal Control and
Protection Department, 1042
Cameron Prairie National Wildlife Refuge,
3668
Catahoula National Wildlife Refuge, 3669
Consultox, 1872
Culture, Recreation and Tourism Department
State Parks Office, 3064
D Arbonne National Wildlife Refuge, 3670
Delta Region Preservation Commission, 3065
Department of Natural Resources: Office of
Mineral Resources, 3066
EarthNet Laboratories, 4985

EnviroMed Laboratories, 5019
Environmental Health Sciences Research Laboratory, 5658
Eustis Engineering Company, 5071
Feed and Fertilizer Laboratory, 5671
Geospec, 1925
Inchcape Testing Services, 5184
Institute for Ecological Infrastructure Engineering, 5701
Kisatchie National Forest, 3671
Lacassine National Wildlife Refuge, 3672
Louisana Department of Natural Resources, 3067
Louisiana Association of Conservation Districts, 1043
Louisiana BASS Chapter Federation, 1044
Louisiana Department of Health and Hospitals: Office of Public Health, 3068
Louisiana Department of Natural Resources Office of Coastal Restoration and Management, 3069
Louisiana Land and Exploration Company, 2365
Louisiana State Extension Services, 3070
Louisiana Tech University, 5974
Louisiana Wildlife Federation, 1045
National Institute for Global Environmental Change: South Central Regional Center, 73, 805, 2151, 2151, 5751
National Wetlands Technical Council Research Center, 692
Natural Resources: Coastal Management Division, 3071
Natural Resources: Conservation Office, 3072
Natural Resources: Geological Oil & Gas Division, 3073
Natural Resources: Injection & Mining Division, 3074
Nature Conservancy: Louisiana Chapter, 1046
Omega Waste Management, 1993
Pan American Laboratories, 5372
Pavia-Byrne Engineering Corporation, 2003
Sabine National Wildlife Refuge, 3673
Sherry Laboratories, 5461
Soil and Water Research, 5804
Southern Arkansas: National Safety Council, Ark-La-Tex Chapter, 753
Tulane Environment Law Clinc, 1047
Tulane University, 6015
Wetland Biogeochemistry Institute, 5840
Whooping Crane Conservation Association, 483
World Aquaculture Society, 710

Maine

ABB Environmental Services, 1700
Acadia National Park Advisory Commission, 3674
Acheron Engineering Services, 1716
American Lung Association of Maine, 1049
Aroostook Testing & Consulting Laboratory, 4824
Association of Field Ornithologists, 329
Atlantic Salmon Federation, 1050
Conservation Department: Waste Reduction and Recycling, 3075
Controlled Environment Corporation, 4931
Cross Island National Wildlife Refuge, 3675

Ecology and Environmental Sciences, 6093
Elmina Sewall Foundation, 2329
Friends of Acadia, 371
ImmuCell Corporation, 5182
James W Sewall Company, 1950
Lowry Engineering, 5261
Lowry Systems, 5262
Maine Association of Conservation Commissions, 1051
Maine Association of Conservation Districts, 1052
Maine Audubon, 1053
Maine Coast Heritage Trust, 1054
Maine Cooperative Fish & Wildlife Research Unit, 3076
Maine Department of Environmental Protection: Augusta, 3077
Maine Department of Conservation, 3078
Maine Department of Conservation: Ashland Regional Office, 3079
Maine Department of Conservation: Bangor Regional Office, 3080
Maine Department of Conservation: Bolton Hill Regional Office, 3081
Maine Department of Conservation: Bureau of Parks & Lands, 3082
Maine Department of Conservation: Entomology Laboratory, 3083
Maine Department of Conservation: Farmington Regional Office, 3084
Maine Department of Conservation: Greenville Regional Office, 3085
Maine Department of Conservation: Hallowell Regional Office, 3086
Maine Department of Conservation: Jonesboro Regional Office, 3087
Maine Department of Conservation: Land Use Regulation Commission, 3088
Maine Department of Conservation: Millinocket Regional Office, 3089
Maine Department of Conservation: Old Town Regional Office, 3090
Maine Department of Conservation: Rangeley Regional Office, 3091
Maine Department of Environmental Protection: Presque Isle, 3092
Maine Department of Environmental Protection: Portland, 3093
Maine Inland Fisheries & Wildlife Department, 3094
Maine Natural Areas Program, 3095
Maine Sea Grant College Program, 3096
National Council of Forestry Association Executives, 66
Northeast Test Consultants, 5327
Northeastern Forest Fire Protection Commission, 3215
Oak Creek, 1987
Pond Island National Wildlife Refuge, 3676
Small Woodland Owners Association of Maine, 1056
St Croix International Waterway Commission, 3097
Sunkhaze Meadows National Wildlife Refuge, 3677
Unity College in Maine, 6017
University of Maine Cooperative Extension Forestry & Wildlife Office, 3098
University of Maine at Fort Kent, 6026
Woods End Research Laboratory, 5585

Maryland

20/20 Vision, 5847
AMA Analytical Services, 4753
ASLA: Maryland Chapter, 1076
Aarcher, 1710
Aerosol Monitoring & Analysis, 4777
Aerosol Monitoring and Analysis, 1721
Alliance for the Chesapeake Bay: Baltimore Office, 1077
American Academy of Environmental Engineers, 139
American Bass Association of Maryland, 1078
American Clean Water Association, 636
American Environmental Network, 4791
American Federation of Mineralogical Societies, 140
American Fisheries Society, 244
American Fisheries Society: Equal Opportunities, 994
American Hiking Society, 617
American Institute of Fishery Research Biologists, 318
American Lung Association of Maryland, 1079
American Zoo and Aquarium Association, 25, 620
Analyte Laboratories, 4799
Antietam National Battlefield, 3678
Appalachian States Low-Level Radioactive Waste Commission, 173
Assateague Island National Seashore, 3679
Association of Partners for Public Lands, 331
Athena Environmental Sciences, 4833
Audubon Naturalist Society of the Central Atlantic States, 1080, 2296
Baltimore Gas & Electric Foundation, 2298
Barco Enterprises, 1772
Biospherics, 4869
Bregman and Company, 1807, 4877
Brotherhood of the Jungle Cock, 338
Center for Chesapeake Communities, 1081
Center for Watershed Protection, 652
Chesapeake Bay Critical Areas Commission, 2531
Chesapeake Bay Executive Council, 3117
Chesapeake Bay Foundation, 1082, 1440, 2309, 2309
Chesapeake Wildlife Heritage, 1083
Clean Fuels Development Coalition, 121
Coastal Resources, 4916
Colonial Waterbird Society: United States Fish & Wildlife Service, 1084
Conservation Treaty Support Fund, 352, 2315
Conservation and Renewable Energy Inquiry and Referral Service, 123
Datanet Engineering, 1884
Delaware: Chesapeake Region Safety Council, 865
Department of Commerce: National Oceanic & Atmospheric Administration, 2570
Department of Commerce: National Marine Office of Habitat Protection Chesapeake Bay, 2571, 3118
Department of Commerce: National Ocean Service, 2572
Department of Commerce: Ocean Observation Division, 2573
Department of Geography and Environmental Engineering, 5903
Dynamac Corporation, 4959

Massachusetts

Michigan

Minnesota

Raptor Center, 1139
Rice Lake National Wildlife Refuge, 3701
Rydell National Wildlife Refuge, 3702
Schoell and Madson, 2042
Sherburne National Wildlife Refuge, 3703
Soil Engineering Testing, 5472
Superior National Forest, 3704
Tamarac National Wildlife Refuge, 3705
Trumpeter Swan Society, 302
US Filter/Control Systems, 5544
University of Minnesota at St. Paul, 6029
Voyageurs National Park, 3706
Water Resource Center, 5830
Wenck Associates, 2068
Wildlife Forever, 492
Winona District National Wildlife Refuge
    Upper Mississippi River National Wildlife
    and Fish, 3707

Mississippi

ASLA: Mississippi Chapter, 1141
American Lung Association of Mississippi,
    1142
Army Coastal Engineering Research Board,
    3155
Bienville National Forest, 3708
Crosby Arboretum, 1143
Davis Research, 4953
Delta Wildlife, 355
Environmental Quality Protection Systems
    Company, 5049
Gulf Coast Research Laboratory, 3156
Gulf of Mexico Fisheries Commission, 3157
Karl A Riggs PhD, 5226
Land & Water Resources Bureau, 3158
Mississippi Alabama Sea Grant Consortium,
    3159
Mississippi BASS Chapter Federation, 1144
Mississippi Cooperative Fish & Wildlife
    Research Unit, 5744
Mississippi Department Agriculture &
    Commerce, 3160
Mississippi Department of Wildlife,
    Fisheries and Parks, 3161
Mississippi Forestry Commission, 3162
Mississippi Native Plant Society, 1145
Mississippi Sandhill Crane National Wildlife
    Refuge: US Fish & Wildlife Service, 3709
Mississippi Sea Grant Program, 3163
Mississippi State Chemical Laboratory, 5745
Mississippi State Department of Health
    Bureau of Child/Adolescent Health, 3164
Mississippi State Health Department
    Environmental Health Bureau, 3165
Mississippi Wildlife Federation, 1146
Mississippi: Environmental Quality Pollution
    Control Department, 3166
Natural Resources and Environmental
    Management, 5982
Noxubee National Wildlife Refuge, 3710
Panther Swamp National Wildlife Refuge,
    3711
Policy Review Board of the Gulf of Mexico:
    Gulf of Mexico Program, 3167
Program in Freshwater Biology, 5781
University of Southern Mississippi, 6040

Missouri

ASLA: Saint Louis Chapter, 1149
AZTEC Laboratories, 4765
American Fisheries Society: North Central
    Division, 1150
American Lung Association of Missouri,
    1151
American Lung Association of Missouri-
    Southeast Missouri Office, 1152
American Lung Association of Missouri
    Southwest Missouri Office, 1153
American Lung Association of
    Missouri K ansas City Office, 1154
Association for Natural Resources
    Enforcement Training, 327
Association of Conservation Engineers, 148
Baird Scientific, 4846
Brotcke Engineering Company, 4883
Burns and McDonnell, 1813
Camp Fire Conservation Fund, 339
Camp Fire USA, 6077
Center for Plant Conservation, 344
Chemir Analytical Services, 4905
Conservation Education Association, 5894
DW Ryckman and Associates, 4947
Deer Creek Foundation, 2321
Environmental Analysis, 5030
Environmental Sciences, 5933, 6100
Environmetrics, 5063
Greenley Memorial Research Center, 5686
Horner & Shifrin, 5164
Household Hazardous Waste Project, 203
HydroVision, 1634
International Center for Tropical Ecology,
    5966
Kansas BASS Chapter Federation, 1155
Mark Twain National Forest, 3712
May Stores Foundation, 2373
Midwest Research Institute, 5303
Mingo National Wildlife Refuge, 3713
Missouri Audubon Council, 1156
Missouri Conservation Department, 3169
Missouri Cooperative Fish & Wildlife
    Research Unit, 5746
Missouri Department of Natural Resources,
    3170
Missouri Forest Products Association, 1157
Missouri Prairie Foundation, 1158
Missouri Public Interest Research Group,
    1159
Missouri Stream Team: Missouri
    Department of Conservation, 1160
National Association Civilian Conservation
    Corps Alumni, 1647
National Garden Clubs, 421
Natural Resources Department: Pollution
    Control, 3171
Natural Resources: Energy Division, 3172
Natural Resources: Environmental
    Improvement and Energy Resources, 3173
OCCU-TECH, 1986
Ozark Environmental Laboratories, 5349
Ozark National Scenic Riverways, 3714
Ozarks Resource Center, 454
Prarie Gateway Chapter, 1161
Professional Service Industries Laboratory,
    5397
RMC Corporation Laboratories, 5406
React Environmental Engineers, 2019
Safety & Health Council of Western
    Missouri & Kansas, 1028

Scenic Missouri, 1163
Shell Engineering and Associates, 2046, 5460
Society for Environmental Geochemistry and
    Health, 239, 1164
St. Louis Metropolitan Area: Safety Council
    of Greater St. Louis, MO, 990
St. Louis Testing Laboratories, 5489
Swan Lake National Wildlife Refuge, 3715
University Forest, 5823
Wild Canid Survival and Research Center,
    484
World Bird Sanctuary, 502

Montana

Alternative Energy Resources Organization,
    113
American Lung Association of the Northern
    Rockies, 1166
American Rivers Montana Field Office, 1167
Beaverhead National Forest, 3716
Benton Lake National Wildlife Refuge, 3717
Bighorn Canyon National Recreation Area,
    3718
Bison Engineering, 1797, 4870
Bitterroot National Forest, 3719
Boone and Crockett Club, 337
Bowdoin National Wildlife Refuge: Refuge
    Manager, 3720
Butte District Advisory Council, 3175
Center for Resourceful Building Technology
    NCAT, 97
Charles M Russell National Wildlife Refuge:
    San Creek Wildlife Station, 3721
Charles M Russell National Wildlife Refuge:
    Jordan Wildlife Station, 3722
Charles M Russell National Wildlife Refuge,
    3723
Charles M Russell National Wildlife Refuge:
    Fort Peck Wildlife Station, 3724
Chemical Injury Information Network, 1168,
    2109
ChromatoChem, 4909
Craighead Environmental Research Institute,
    1169
Craighead Wildlife: Wetlands Institute, 1170
Crown of the Continent Research Learning
    Center - Glacier National Park, 3176
Custer National Forest, 3725
Deerlodge National Forest, 3726
Environmental Engineering Department,
    5921
Environmental Quality Council, 3177
Flathead National Forest, 3727
Foundation for Research on Economics and
    the Environment (FREE), 5677
Gallatin National Forest, 3728
Glacier Institute, 6108
Great Bear Foundation, 381
Greater Yellowstone Coalition, 1171
Helena National Forest, 3729
Humphrey Energy Enterprises, 5168
Interagency Grizzly Bear Committee
    USFWS NS 312, 3178
Kootenai National Forest, 3730
Lee Metcalf National Wildlife Refuge, 3731
Lewis & Clark National Forest, 3732
Lewiston District Advisory Council Bureau
    of Land Management, 3179
Lighthawk: Northern Rocky Mountain Field
    Office, 1172

Lolo National Forest, 3733
MSE, 5271
Medicine Lake National Wildlife Refuge
Refuge Manager, 3734
Montana Association of Conservation
Districts Annual Meeting, 1173, 1644
Montana Audubon, 1174
Montana Cooperative Fishery Research Unit,
5747
Montana Department of Agriculture, 3180
Montana Environmental Information Center,
1175
Montana Environmental Training Center,
5980
Montana Land Reliance, 1176
Montana Natural Heritage Program, 3181
Montana Water Environment Association
Annual Meeting, 1177, 1645
Montana Wildlife Federation, 1178
Mycotech, 5311
National Bison Range National Wildlife
Refuge, 3735
Natural Resources & Conservation
Department, 3182
Natural Resources & Conservation
Department: Conservation & Resource
Development Division, 3183
Natural Resources & Conservation: Oil &
Gas, 3184
Natural Resources & Conservation: Water
Resource, 3185
Northern Lights Institute, 5763
OEA Research, 5333
Pablo National Wildlife Refuge, 3736
Red Rocks Lakes National Wildlife Refuge,
3737
Rocky Mountain Elk Foundation, 958, 1162,
1180, 1180
SEEK, 6142
Swan River National Wildlife Refuge, 3738
Theodore Roosevelt Conservation Alliance,
474
University of Montana, 6030
Western Montana College Environmental
Science, 6058
Wilderness Institute: University of Montana,
5841
Wilderness Watch, 488
Yellowstone Environmental Science, 5587
Yes Technologies, 5588

Nebraska

ATC Associates: Omaha, 1708
Agate Fossil Beds National
Monument/Scotts Bluff National
Monument, 3739
Agricultural Research and Development
Center, 5593
American Lung Association of Nebraska,
1184
Central Interstate Low-Level Radioactive
Waste Commission, 3063
Crescent Lake National Wildlife Refuge,
3740
Department of Agriculture: Natural
Resources Conservation Service, 2565,
3186
Department of Natural Resources, 3187
Food Safety and Inspection Service, 2121,
2650

Fort Niobrara National Wildlife Refuge:
Valentine, 3741
Games & Parks Commission: Nebraska, 3188
Groundwater Foundation, 671, 6110
HWS Consulting Group, 5140
Harris Laboratories, 5147
Health Department: Drinking Water and
Environmental Sanitation Division, 3189
Inland Bird Banding Association, 388
Iowa Prairie Network, 1185
Midwest Laboratories, 5302
Mississippi River Corridor Study
Commission Planning & Resources
Preservation Natl. Park Svc., 3190
Missouri National Recreation River/Niobrara
Missouri National Scenic Riverways, 3742
Nature Conservancy: Nebraska Chapter, 1186
Nebraska Association of Resource Districts,
1187
Nebraska Association of Resources Districts
Annual Meeting, 1654
Nebraska BASS Chapter Federation, 1188
Nebraska Department of Agriculture, 3191
Nebraska Department of Environmental
Quality, 3192
Nebraska Department of Natural Resources,
3193
Nebraska Ethanol Board, 3194
Nebraska Game & Parks Commission:
Fisheries Division, 3195
Nebraska Games & Parks Commission, 3196
Nebraska Games & Parks: Wildlife Division,
3197
Nebraska National Forest, 3743
Nebraska Testing Corporation, 5317
Nebraska Wildlife Federation, 1189
North Platte National Wildlife Refuge
Crescent Lake National Wildlife Refuge,
3744
RDG Geoscience and Engineering, 2015
The National Arbor Day Foundation, 2412
University of Nebraska, 6031

Nevada

American Lung Association of
Idaho/Nevada, 1191
American Lung Association of
Idaho/Nevada: Las Vegas Office, 1192
Anaho Islands National Wildlife Refuge,
3745
Ash Meadows National Wildlife Refuge,
3746
Bureau of Land Management, 2884, 3199
Carson City Field Office Advisory Council,
3200
Conservation and Natural Resources
Department, 2779, 3202
Conservation and Natural Resources: Water
Resources Division, 3203
Department of the Interior: Bureau of
Reclamation, 3204
Desert National Wildlife Range, 3747
Elko District Advisory Council: Bureau of
Land Management, 3205
Environmental Engineering University of
Nevada, Reno, 5920
First Interstate Bank of Nevada Foundation,
2334
Forensic Engineering, 5079
Great Basin National Park, 3748

Harry Reid Center for Environmental
Studies, 5688
Humboldt National Forest, 3749
Immunotox, 5183
Lake Mead National Recreation Area
(NRA), 3750
Las Vegas Bureau of Land Management,
3206
Moapa Valley National Wildlife Refuge,
3751
Nevada Bureau of Mines & Geology, 3207
Nevada Department of Conservation and
Natural Resources, 3208
Nevada Department of Conservation:
Wildlife Division, 3209
Nevada Natural Heritage Program, 3210
Nevada Wildlife Federation, 1193
Pahranagat National Wildlife Refuge, 3752
Resource Concepts, 2026
Ruby Lake National Wildlife Refuge, 3753
Tahoe Regional Planning Agency, 1195
Tahoe Regional Planning Agency (TRPA)
Advisory Planning Commission, 3211
Toiyabe National Forest, 3754
University of Nevada at Las Vegas, 6032
Wild Horse Organized Assistance, 485

New Hampshire

Advance Pump and Filter Company, 4773
American Bass Association of New
Hampshire, 1196
American Ground Water Trust, 637
American Lung Association of New
Hampshire, 1197
Antioch New England Graduate School, 5857
Audubon Society of New Hampshire, 1198
Center for Environmental Education, 5877
Center for Marine Biology, 5634
Comprehensive Environmental, 1867
Dartmouth College, 5901
Elm Research Institute, 5000
Enviro Systems, 5014
Environmental Hazards Management
Institute, 193
Great Bay National Wildlife Refuge, 3755
Institute for Community and Environment,
5959
John Hay National Wildlife Refuge, 3756
KAI Technologies, 5219
Keene State College, 5971
Lake Umbagog National Wildlife Refuge,
3757
Les A Cartier and Associates, 1960
Nature Conservancy: New Hampshire
Chapter, 1199
New England Coalition for Sustainable
Population, 595
New Hampshire Agriculture Markets &
Foods: Department Conservation
Committee, 3212
New Hampshire Association of Conservation
Districts, 1200
New Hampshire Association of Conservation
Commissions Annual Meeting, 1201, 1657
New Hampshire Department of
Environmental Services, 3213
New Hampshire Fish and Game Department,
3214
New Hampshire Lakes Association (NHLA),
1202

New Hampshire Sea Grant Program, 5760
New Hampshire Wildlife Federation, 1203
Nobis Engineering, 5323
North American Loon Fund, 447
North American Loon Fund Grants, 2495
Northeast Resource Recovery Association, 1204
Northeast Resource Recovery Association Annual Conference, 1666
Ocean Engineering Center, 5767
Permaculture Gap Mountain, 5775
Resources & Development Council: State Planning, 3216
Seacoast Anti-Pollution League, 701
Student Conservation Association, 1451, 6155
Switzer Foundation New Hampshire Charitable Foundation, 2409
Universal Environmental Technologies, 5551
University of New Hampshire Cooperative Extension, 3217
White Mountain National Forest, 3758
Wildcat River Advisory Commission White Mountain National Forest, 3218

New Jersey

ASLA: New Jersey Chapter, 1206
AccuTech Environmental Services, 1714
Accutest Laboratories, 1715, 4768
Allstate Power Vac, 1730
American Bass Association of Eastern Pennsylvania/New Jersey, 1207
American Littoral Society, 321
American Lung Association of New Jersey M ain Office, 1208
Analab, 4798
Andrea Aromatics, 4807
Anoroc Scientific, 4810
Aqua Survey, 1747
Association of New Jersey Environmental Commissions, 1209
Association of University Environmental Health/Sciences Centers, 177
Beaver Defenders Unexpected Wildlife Refuge, 334
Biomass Users Network, 180
Biomass Users Network: Central America, 1210
Brinkerhoff Environmental Services, 1808
Buck, Seifert and Jost, 1811
C&H Environmental, 1814
Cape Branch Foundation, 2303
Cape May National Wildlife Refuge, 3759
Center for Environmental Communications (CEC), 5619
Center for Sustainable Development in the Americas, 2529
Chyun Associates, 4910
Clean Harbors Cooperative, 654
Clean Ocean Action, 6084
Converse Consultants, 1873, 4933
Dan Raviv Associates, 4949
Detail Associates: Environmental Engineering Consultants, 1888
Ecological & Environmental Learning Services, 6092
Edison Facilities, 1211
Emergency Committee to Save America s Marine Resources, 665
Enviro-Sciences, 5017

Environmental Protection and Energy: Fish and Wildlife, 3219
Environmental Risk: Clifton Division, 5053
Environmental Testing Laboratories, 5059
Environmental and Occupational Health Science Institute, 1212, 2120, 5670, 5670
Enviroplan, 1914
Excel Environmental Resources, 5074
Geraldine R. Dodge Foundation, 2345
Great Swamp National Wildlife Refuge, 3760
Handex Environmental Recovery, 5146
Hatch Mott MacDonald, 5149
Hillmann Environmental Company, 5161
Hoffman-La Roche Foundation, 2350
Hydro Science Laboratories, 5172
International Asbestos Testing Laboratories, 5193
J Dallon and Associates, 5199
JR Henderson Labs, 5208
Laboratory Resources, 5241
Ledoux and Company, 5254
Louis Berger Group, 1961
Marine Mammal Stranding Center, 104
Merck & Company, 5291
Miceli Kulik Williams and Associates, 1967
Mikropul Environmental Systems, 5304
National Association of Noise Control Officials, 216, 2139
New Jersey Association of Conservation Districts, 1213
New Jersey BASS Chapter Federation, 1214
New Jersey Department of Agriculture State Soil & Conservation Committee, 3220
New Jersey Department of Environmental Protection, 3221
New Jersey Department of Health and Senior Services, 1215
New Jersey Dept. Environmental Protection: Division of Publicly Funded Site Remediation, 3222
New Jersey Environmental Lobby, 1216
New Jersey Geological Survey Department Environmental Protection, 3223
New Jersey Pinelands Commission, 3224
New Jersey Public Interest Research Group, 1217
New Jersey Society for Environmental Economic Development Annual Conference, 1218, 1658
New Jersey Water Environment Association Conference, 1659
New Jersey: Department Environmental Protection Law Enforcement, 3225
New York Turtle and Tortoise Society, 1257
Northeastern Analytical Corporation, 5328
Ocean City Research, 1992
Omega Thermal Technologies, 5342
Ostergaard Acoustical Associates, 5348
P&P Laboratories, 5350
P.I.N.E.S., 5990
PARS Environmental, 5355
Pace New Jersey, 5367
Package Research Laboratory, 5371
Passaic River Coalition, 1219
Pharmaco LSR, 5379
Population Resource Center, 604
Recon Environmental Corporation, 5416
Recon Systems, 5417
Shaw Environmental, 2045
Sierra Club: NJ Chapter, 1220
Spectrochem Laboratories, 5485
Supawna Meadows National Wildlife Refuge, 3761

THP, 2054
Teledyne Isotopes, 5515
Terryn Barill, 2056
Townley Laboratories, 5533
United Environment Systems, 5549
United States Land, Air, and Water Environment Industries, 5550
Victoria Foundation, 2425
Whibco, 5581

New Mexico

Albuquerque Bureau of Land Management, 3226
American Indian Science and Engineering Society, 2453
Applied Physics, 4816
Attorney General, 3227
Bitter Lake National Wildlife Refuge, 3762
Bosque del Apache National Wildlife Refuge, 3763
Canadian River Commission, 3228
Capulin Volcano National Monument National Park Service, 3764
Carlsbad Caverns National Park, 3765
Carson National Park, 3766
Center for Holistic Resource Management, 1221, 2104
Cibola National Forest, 3767
Controls for Environmental Pollution, 4932
Department of Geological Sciences, 5904
Eberline Services, 4989
El Malpais National Monument, 3768
Energy, Minerals & Natural Resources: Energy Conservation Management, 3229
Energy, Minerals and Natural Resources Department, 3230
Environmental Control, 5037
Environmental Technology & Education Center, 5058
Forest Trust, 370
Frost Foundation, 2339
Kramer & Associates, 5236
Las Cruces District Advisory Bureau of Land Management, 3231
Las Vegas National Wildlife Refuge, 3769
Lee Wilson and Associates, 5255
Lincoln National Forest, 3770
Los Alamos Technical Associates, 5259
Maxwell National Wildlife Refuge, 3771
Nature Conservancy: New Mexico Chapter, 1223
New Mexico Association of Conservation Districts, 1224
New Mexico Association of Soil and Water Conservation Annual Conference, 1225, 1660
New Mexico Bureau of Geology & Mineral Resources, 3232
New Mexico Center for Wildlife Law, 1226
New Mexico Cooperative Fish & Wildlife Research Unit, 3233
New Mexico Department of Game & Fish, 3234
New Mexico Department of Game and Fish, 3235
New Mexico Environment Department, 3236
New Mexico Environmental Law Center, 2250
New Mexico Environmental Law Center: Green Fire Report, 2273

New Mexico Rural Water Association, 1227
New Mexico Soil & Water Conservation
  Commission, 3237
Roswell District Advisory Council: Bureau
  of Land Management, 3238
San Andres National Wildlife Refuge, 3772
Sevilleta National Wildlife Refuge, 3773
Solar Resources, 5475
Southwest Consortium on Plant Genetics &
  Water Resources, 5809
TECH, 5504
Tech Reps, 5512
United States Department of the Interior:
  United States Fish and Wildlife Service,
  3239
United States Forest Service: Southwestern
  R-3, 3240
WERC Undergraduate Fellowships, 2502
WERC: Consortium for Environmental
  Education & Technology Development,
  5560
Waste Management Education & Research
  Consortium, 5827
White Sands National Monument, 3774

New York

A Closer Look at Plant Life, 6064
A Closer Look at Pondlife - CD-ROM, 6065
ASLA: New York Chapter, 1230
ASLA: New York Upstate Chapter, 1231
Acoustical Society of America, 508
Acts Testing Labs, 4769
Adelaide Associates, 4770
Adelaide Environmental Health Associates,
  4771
Adirondack Council, 1232
Adirondack Ecological Center: SUNY
  College of Environmental Science &
  Forestry, 5591
Adirondack Environmental Services, 4772
Adirondack Lakes Survey Corporation, 5592
Adirondack Land Trust, 1233
Adirondack Park Agency, 3241
Airtek Environmental Corporation, 1726
Allee, King, Rosen and Fleming, 1728
Amax Foundation, 2288
American Council on Science and Health,
  167, 1234, 2085, 2085
American Insitute of Chemical Engineers,
  141
American Lung Association of Mid-New
  York, 1235
American Lung Association of New York
  State, 1236
American Lung Association of New York
  State- Northeastern Region, 1237
American Museum of Natural History, 2454,
  6067
American National Standards Institute, 22
American Society of Naturalists, 2456
Andco Environmental Processes, 1739
Andrew D Sauter Consulting, 4808
Andrew W. Mellon Foundation, 2293
Animal Tracks and Signs, 6068
Annotated Invertebrate Clipart CD-ROM,
  6069
Annotated Vertebrate Clipart CD-ROM, 6070
Applied Biomathematics, 4813
Association of State Wetland Managers, 26,
  332, 647, 647

Atlantic States Legal Foundation, 2219
Atlantic Testing Laboratories, 1762, 4834
Baltec Associates, 1771
Barer Engineering, 1773
Barton & Loguidice, 4850
Bay Foundation, 2300
Beldon Fund, 2458
Beltran Associates, 4853
Blasland, Bouck and Lee, 1801
Bog Ecology, 6076
Brooklyn College Environmental Studies
  Program, 5869
Buffalo Testing Laboratories, 4885
CA Rich, 1815
CEDAM International, 649
CPAC, 4893
Camo Pollution Control, 1834
Carpenter Environmental Associates, 1840
Catskill Forest Association, 1238
Center for Earth & Environmental Science,
  5618
Center for Environmental Information, 5878
Center for Environmental Research
  Education, 5622
Center for Environmental Research and
  Conservation, 6079
Center for Great Lakes Environmental
  Education, 5884
Charles Engelhard Foundation, 2308
Chemical Waste Disposal Corporation, 4904
Chesner Engineering, 4906
Child Health and the Environment, 2209
Chopra-Lee, 4908
City University of New York, 5888
Coalition for Education in the Outdoors, 5890
Comet Halley: Once in a Lifetime!, 6087
Comparative Environmental Law and
  Regulation, 2259
Conestoga-Rovers and Associates, 1869
Container Testing Laboratory, 4927
Cornell Center for the Environment, 5899
Cornell Lab of Ornithology, 1239
Cornell University, 5900
Cornell Waste Management Institute, 5647
Corning, 4935
Cyberchron Corporation, 4943
Donald Friedlander, 1889
Dragonfly Society of the Americas, 359
Dunn Corporation, 1890
Earth Dimensions, 4981
Earth Society Foundation, 2326
EarthSave Foundation, 192
Echoing Green, 2327
EcoTest Laboratories, 4991
Ecology and Environment, 1897, 4993
Eder Associates, 4996
Edward John Noble Foundation, 2328
Entek Environmental & Technical Services,
  5011
Enviro Equipment Sales, 1902
Enviro-Lab, 5016
EnviroTest Laboratories, 1903
Environment/One Corporation, 5026
Environmental Action Coalition, 1240
Environmental Conservation, 3242
Environmental Conservation Department:
  Air Resources, 3243
Environmental Conservation: Marine
  Resources, 3244
Environmental Defense, 2261, 2467
Environmental Defense Fund, 2113, 2231
Environmental Defense Newsletter, 2262

Environmental Grantmakers Association,
  2468
Environmental Health Sciences Center:
  University of Rochester Medical Center,
  5657
Environmental Law Lexicon, 2263
Environmental Law and Compliance
  Methods, 2264
Environmental Management and Policy
  Program Rensselaer Polytechnic Institute,
  5925
Environmental Science & Engineering
  Program, 5665
Environmental Sciences Program, 5934
Environmental Studies Department, 5936
Environmental Technology Seminar, 1241
Exploring Animal Life - CD-ROM, 6102
Exploring Environmental Science Topics
  CD-ROM, 6103
Exploring Freshwater Communities -
  CD-ROM, 6104
Federation of New York State Bird Clubs,
  367, 1242
Fire Island National Seashore, 3775
Ford Foundation, 368
Frank Weeden Foundation, 2337
Friends of the Australian Koala Foundation,
  258
Fund for Animals, 260, 2237, 2340, 2340
Gaia Institute, 669
Galson Corporation, 1921
Galson Laboratories, 5093
Garden Club of America, 375, 2475
Gateway National Recreation Area, 3776
George W Perkins Memorial Foundation,
  2343
Geraghty & Miller, 5114
Global Committee of Parliamentarians on
  Population and Development, 565
Global Coral Reef Alliance, 377
Global and Environmental Education
  Resources, 5951
Great Lakes United, 670, 1243
Greensward Foundation, 2346
H2M Group: Holzmacher, McLendon &
  Murrell, 5137
HKH Foundation, 2347
Hamilton Research, Ltd., 5144
Handle with Care: Children and
  Environmental Carcinogens, 2212
Harold I Zeliger PhD, 1935
HazMat Environmental Group, 1938, 5154
Hazardous Chemicals: Handle With Care,
  6111
Healthy Schools Network, 2198
Holy Land Conservation Fund, 385
How Wet is a Wetland?: The Impacts of the
  Proposed Revisions to the Federal
  Wetlands Manual, 2267
Hudsonia, 5693
Hudsonia Limited, 1244
Huntingdon Analytical Services, 5169
IC Laboratories, 5176
INFORM, 207, 1245, 2128, 2128, 2351,
  5179, 5695
Ice Age Relicts: Living Glaciers and Signs of
  Ancient Ice Sheets, 6113
In-Flight Radiation Protection Services, 1943
Innovative Biotechnologies International,
  5188
Institute for Environmental Policy &
  Planning, 5703
Institute for Urban Ports and Harbors, 5707

North Carolina

Oklahoma

Oregon

University of Oregon Environmental Studies Center, 1335
Vale District Advisory Council: Bureau of Land Management, 3300
Wallowa-Whitman National Forest, 3834
Western Environmental Law Center, 2255
Western Forestry and Conservation Association Conference, 1689
Western Region Hazardous Substance Research Center, 5838
William L Finley National Wildlife Refuge, 3835
Winema National Forest, 3836
World Forestry Center, 503
Xerces Society, 309, 2446

Pennsylvania

AMETEK Foundation, 2282
ASLA: Pennsylvania/Delaware Chapter, 1337
ATS-Chester Engineers, 1709
Agronomy and Environmental Science, 5848
Air and Waste Management Association, 1, 158, 509, 509, 1338, 1615, 2074, 6066
All 4 Inc., 159
Allegheny College, 5849
Allegheny National Forest, 3837
Alliance for the Chesapeake Bay Harrisburg Office, 1339
American Association of Botanical Gardens and Arboreta, 314
American Canal Society, 635
American Institute of Chemists, 142
American Medical Fly Fishing Association, 249
American Nature Study Society, 5855
American Society for Testing and Materials, 24
American Society of Landscape Architects , 96
Anderson Consulting Group, 1741
Applied Geoscience and Engineering, 1743
Arro Consulting, 1755
Astorino Branch Environmental, 1760
Audubon Society of Western Pennsylvania Beechwood Farms Nature Reserve, 1340
BCM Engineers, 4842
BCR National Laboratory, 4843
Baker Environmental, 4847
Beaumont Environmental Systems, 1782
Benchmark Analytics, 4854
BioTest, 4862
Bioscience, 4868
Brandywine Conservancy, 1341
Buchart-Horn, 4884
Burt Hill Kosar Rittelmann Associates, 4886
CBA Environmental Services, 1816
CDS Laboratories, 1817, 4889
CIH Environmental, 1819
CONSAD Research Corporation, 4892
California University of Pennsylvania, 5875
Camtech, 1835
Cedar Grove Environmental Laboratories, 4895
Center for Hazardous Materials Research, 519
Center for Statistical Ecology & Environmental Statistics, 5886
Centre Analytical Laboratories, 4899
Childhood Lead Poisoning Prevention Program, 2191, 3268, 3301, 3301

Citizens Advisory Council, 3302
Clean Air Council, 6
Cobbs Creek Community Environment Educational Center (CCCEEC), 5644
Combustion Unlimited, 1861
Commonwealth Engineering and Technology CET Engineering Services, 1863
Conservation Leadership School, 5896
Conti Testing Laboratories, 4928
Crane Environmental, 4939
Crouse & Company, 1877
Cyrus Rice Consulting Group, 4944
D Appolonia, 1881
Delaware National Scenic River/Delaware Water Gap National Recreation Area, 3838
Delaware Water Gap National Recreation Area Citizen Advisory Commission, 3303
Department of Energy and Geo-Environmental Engineering, 6090
Department of the Interior: National Parks, 2861, 3304
Duquesne University, 5907
EADS Group, 4962
Eastern Laboratory Service Associates, 4987
Enviro-Bio-Tech, 5015
Environmental Coalition on Nuclear Power, 126
Environmental Institute, 5922
Environmental Protection Agency: Region 3 Hazardous Waste Management, 3305
Environmental Protection Agency: Region III, 3306
Environmental Research Associates, 5050
Enviroscan Inc, 5065
Erie National Wildlife Refuge, 3839
Five Winds International, 6105
Forestry Conservation Communications Association Annual Meeting, 45, 1630
Free-Col Laboratories: a Division of Modern Industries, 5081
GFG Environmental, 5088
Gas Desulfurization Corporation, 5094
GemNet Global Education Motivators, 1342
Geo-Con, 5102
Gerhart Laboratories, 5115
Gettysburg National Military Park, 3840
Gilbert/Commonwealth, 5117
Granville Composite Products Corporation, 1927
Greeley-Polhemus Group, 1929, 5126
Hawk Migration Association of North America, 262
Hawk Mountain Sanctuary Association, 1343, 2479
Helen Clay Frick Foundation, 2348
Informatics Division of Bio-Rad, 5187
International Conference on Solid Waste, 1637
JM Best, 5207
JWS Delavau Company, 5209
John Heinz National Wildlife Refuge at Tinicum, 5720
Lacawac Sanctuary Foundation, 5725
Lancaster Laboratories, 5243
Lancy Environmental, 5244
Land Management Decisions, 5245
Lawrence G Spielvogel, 5253
Limnological Research Unit, 5729
Mack Laboratories, 5274
Mateson Chemical Corporation, 5281
Michael Baker Corporation, 1968
Michael Baker Jr: Civil and Water Division, 5296

Michael Baker Jr: Environmental Division, 5297
Microseeps, Inc, 5299
National Mine Land Reclamation Center: Eastern Region, 5753
Nature Conservancy: Pennsylvania Chapter, 1344
North American Native Fishes Association, 287
Northeast Conservation Law Enforcement Chiefs Association, 1345
Northern Allegheny National Wild & Scenic River/Allegheny National Forest, 3307
PACE, 5351
PACE Environmental Products, 5353
PACE Resources, Incorporated, 5354
PSC Environmental Services, 5363
Partners in Parks, 455
Penn State Institutes of the Environment, 1346
Pennsylvania Association of Accredited Environmental Laboratories, 1347
Pennsylvania Association of Conservation Districts, 1348
Pennsylvania BASS Chapter Federation, 1349
Pennsylvania Cooperative Fish & Wildlife Research Unit, 5774
Pennsylvania Department of Conservation and Natural Resources, 3308
Pennsylvania Environmental Council, 1350
Pennsylvania Fish and Boat Commission: Northeast Region, 3309
Pennsylvania Forest Stewardship Program, 3310
Pennsylvania Forestry Association, 1351
Pennsylvania Game Commission, 3311
Pennsylvania Resources Council, 1352
Perkiomen Watershed Conservancy, 6134
Pesticide Research Laboratory & Graduate Study Center, 5777
Pew Charitable Trusts, 2391
Pittsburgh Mineral & Environmental Technology, 5382
Plastics Recycling Foundation, 542
Pocono Environmental Education Center, 1353
Porter Consultants, 5391
Professional Analytical and Consulting Services (PACS), 2012
Puricons, 5400
Purple Martin Conservation Association, 294
QC, 5402
RARE Center for Tropical Bird Conservation, 2395
Regional Center for Environmental Information US EPA Region 3, 3312
Resource Technologies Corporation, 5424
Richard King Mellon Foundation, 2399
Rodale Institute, 236, 1354, 2172, 2172
Ruffed Grouse Society, 462
SJS Archeological Services, 5443
SMC Martin, 5444
Sierra Club: Pennsylvania Chapter, 1355
Slippery Rock University, 6000
Spotts, Stevens and McCoy, 5488
Steel Recycling Institute, 547
Stroud Water Research Center, 5811
Student Environmental Action Coalition, 2253
Susquehanna River Basin Commission, 3313
UEC Industrial Hygiene and Environmental Health Laboratories, 5542

University of Pennsylvania, 6035
University of Pittsburgh Department of Geology and Planetary Science, 6036
Upper Delaware Scenic & Recreational River, 3841
Weavertown Group Optimal Technologies, 2067
West More Mechanical Testing and Research, 5574
Western Pennsylvania Conservancy, 1356
Westinghouse Electric Company, 5577
Weston Institute, 2504
Weston Solutions, Inc, 2070, 5579
Wildlands Conservancy, 709
Wildlife Information Center, 494
Wildlife Preservation Trust International, 2434
William Penn Foundation, 2437

Rhode Island

American Lung Association of Rhode Island, 1357
Applied Science Associates, 1745
Audubon Society of Rhode Island, 1358
Brown University Center for Environmental Studies, 5870
Ceimic Corporation, 4896
Coastal Resources Center, 6086
Environmental Management: Division of Fish and Wildlife, 3315
Environmental Management: Parks and Recreation, 3316
Environmental Science Services, 1911
Environmental Studies and Environmental Science Programs, 5939
Eppley Laboratory, 5067
Geo Environmental Technologies, 5101
National Network of Forest Practitioners, 75
Nature Conservancy: Northeast Division Office, 1359
Nature Conservancy: Rhode Island Chapter, 1360
New England Testing Laboratory, 5320
Providence Journal Charitable Foundation, 2392
Rhode Island BASS Chapter Federation, 1361
Rhode Island Department of Environmental Management, 3317
Rhode Island Department of Evironmental Management: Forest Environment, 3318
Rhode Island National Wildlife Refuge Complex, 3842
Rhode Island State Association of Conservation, 1362
Rhode Island State Water Resources Board, 3319
Roger Williams University, 5995
Science Applications International Corporation, 5453
URI Cooperative Extension Education Center, 3320

South Carolina

ASLA: South Carolina Chapter, 1363
Ace Basin National Wildlife Refuge, 3843
Agromedicine Program Medical University of South Carolina, 5594

Alpha Manufacturing Company, 4786
American Lung Association of South Carolina, 1364
American Lung Association of South Carolina- Coastal Region, 1365
American Lung Association of South Carolina Upstate Region, 1366
Belle W Baruch Institute for Marine Biology and Coastal Research, 5606
Cape Romain National Wildlife Refuge, 3844
Carolina Sandhills National Wildlife Refuge, 3845
Clemson University, 5889
Department of Interior: South Carolina Cooperative Fish and Wildlife, 3321
Department of Parks: Recreation and Tourism, 3322
Francis Marion-Sumter National Forest, 3846
Friends of the Reedy River, 1367
General Engineering Labs, 5098
International Primate Protection League, 2352
JL Rogers & Callcott Engineers, 5206
National Association of Recreation Resource Planners, 625
Nature Conservancy: South Carolina Chapter, 1368
Normandeau Associates, 1983, 5324
North American Benthological Society, 136
Office of Environmental Laboratory Certification, 3323
Priester and Associates, 2010
Quail Unlimited, 458
RMT Inc., 233
Research Planning, 5423
Resource Management, 2029
Santee National Wildlife Refuge, 3847
South Atlantic Fishery Management Council, 1369, 3324
South Carolina BASS Chapter Federation, 1370
South Carolina Department of Health and Environmental Control, 3325
South Carolina Department of Natural Resources, 3326
South Carolina Forestry Commission, 3327
South Carolina Native Plant Society, 1371
South Carolina Sea Grant Consortium, 5807
Southern Appalachian Botanical Society, 1372
Strom Thurmond Institute of Government & Public Affairs, Regional Development Group, 5810
University of Georgia, 6022
University of South Carolina Baruch Marine Field Lab, 6038
Waterfowl USA, 305
Wildlife Action, 489

South Dakota

ASLA: Great Plains Chapter, 1373
ATC Environmental, 4762
American Lung Association of South Dakota, 1374
Attorney General s Office Natural Resources Division, 3329
Badlands National Park, 3848
Black Hills National Forest, 3849
Department of Environment & Natural Resources, 3330

Department of Wildlife and Fishery Sciences, 3331
Great Plains Native Plant Society, 1375
Huron Wetland Management District, 3850
International Association of Wildland Fires, 103
Jewel Cave National Monument, 3851
Nature Conservancy: South Dakota Chapter, 1376
RE/SPEC, 5405
Respec Engineering, 2031
Sand Lake National Wildlife Refuge, 3852
South Dakota Association of Conservation Districts Conference, 1377, 1674
South Dakota Department of Game, Fish & Parks, 3332
South Dakota Environmental Health Association, 1675, 3333
South Dakota Ornithologists Union, 1378
South Dakota State Extension Services, 3334
South Dakota Wildlife Federation, 1379
Wind Cave National Park, 3853

Tennessee

Advanced Waste Management Systems, 1720
Alexander Hollaender Distinguished Postdoctoral Fellowships, 2449
American Lung Association of Tennessee, 1381
American Lung Association of Tennessee- Southeast Office, 1382
Bhate Environmental Associates, 4858
Big South Fork National River Recreation Area, 3854
Carbon Dioxide Information Analysis Center, 3335
Center for Energy and Environmental Analysis Oak Ridge Laboratory, 1843
Center for Field Biology, 5626
Center for Geography and Environmental Education, 6080
Center for the Management, Utilization and Protection of Water Resources, 5641
Cherokee National Forest, 3855
Chickasaw National Wildlife Refuge, 3856
Cross Creeks National Wildlife Refuge, 3857
Ducks Unlimited, 360, 522
Earth Science Associates/ESA Consultants, 1895
En Safe, 1900
Energy, Environment & Resource Center, 5652
Environmental Systems Corporation, 5056
Environmental Testing and Consulting, 1913, 5061
Forestry Experiment Stations and Arboretum, 5675
Great Smokey Mountains National Park, 3858
Hatchie National Wildlife Refuge, 3859
Hess Environmental Services, 5159
Integrated Environmental Management, 1945
Kentucky-Tennessee Society of American Foresters, 1383
Kids for a Clean Environment, 211, 1384, 2203, 2203
Lower Hatchie National Wildlife Refuge, 3860
Narrow Ridge Earth Literacy Center, 6126

Texas

Utah

Vermont

Virginia

American Pheasant and Waterfowl Society, 250

American Society of Agronomy, 554

Analytical Process Laboratories, 4801

Apostle Islands National Lakeshore, 3913

Applied Ecological Services, 1742

Association of State Floodplain Managers, 646

Ayres Associates, 1764

Badger Laboratories & Engineering Company, 4845

Badger Laboratories and Engineering Company, 1770

Becher-Hoppe Associates, 1783

Botanical Club of Wisconsin, 1499

Cardinal Environmental, 1839

Center for Alternative Mining Development Policy, 1500

Center for Resource Policy Studies, 5637

Central Wisconsin Environmental Station (CWES), 1501

Chequamegon National Forest, 3914

Citizens for Animals: Resources and Environment, 1502

Cleaner and Greener Environment, 6085

Community Conservation Consultants Howlers Forever, 1865

Earth Tech, 4983

Environmental Audits, 5032

Environmental Chemistry and Technology Program, 5655

Environmental Compliance Consulting, 1905

Environmental Innovations, 5043

Environmental Remote Sensing Center, 5662

Environmental Resources, 6099

Environmental Toxicology Center, 5669

Federation of Environmental Technologists, 44, 1629, 2236, 2236

Great Lakes Indian Fish and Wildlife Commission, 3438

Ice Age National Scientific Reserve, 3915

Institute for Environmental Studies, 5961, 6114

International Crane Foundation, 394

JJ Keller and Associates, 5204

Johnson Controls, 5215

Johnson s Wax Fund, 2484

Kag Laboratories International, 5223

MacKenzie Environmental Education Center, 6122

Marshall and Ilsley Foundation, 2369

Micro Controls Systems, 5298

Miller Engineers, 5305

National Association of State Outdoor Recreation Liason Officers, 626

Natural Resources Department, 3439

Nicolet National Forest, 3916

North American Lake Management Society International Symposium, 694, 1662

Northland College, 5987

Patrick and Anna Cudahy Fund, 2390

Perry-Carrington Engineering Corporation, 2004

Petra Environmental, 2005

RMT, 2017, 5407

River Alliance of Wisconsin, 1503

River Studies Center, 5791

S-F Analytical Laboratories, 5438

Schlitz Audubon Nature Center, 6145

Sierra Club-John Muir Chapter (Wisconsin), 1504

Sixteenth Street Community Health Center, 1505

Society of Tympanuchus Cupido Pinnatus, 299

Solsorce, Inc, 5478

St Croix National Scenic Riverway, 3917

Trees for Tomorrow, 6013

Trees for Tomorrow Natural Resources Educational Center, 1506

United States Department of Agriculture United States Forest Service, 2881, 2998, 3153, 3153, 3440

University of Wisconsin at Green Bay, 6043

University of Wisconsin at Madison, 6044

University of Wisconsin at Stevens Point, 6045

Water Resources Institute, 5832

Water Resources Management, 6158

Whitetails Unlimited, 306

Wildlife Society, 495, 734, 746, 746, 842, 938, 960, 1004, 1029, 1048, 1101, 1148, 1183, 1190, 1229, 1285, 1309

William T Lorenz & Company, 5583

Wisconsin Applied Water Pollution Research Consortium: University of Wisconsin-Madison, 5843

Wisconsin Association for Environmental Education Annual Conference, 1509, 1693

Wisconsin Association of Lakes, 1510

Wisconsin Cooperative Fishery Research Unit, 3441

Wisconsin Department of Agriculture Trade and Consumer Protection: Land & Water Resources Bureau, 3442

Wisconsin Energy Corporation Foundation, 2440

Wisconsin Geological & Natural History Survey, 3443

Wisconsin Land and Water Conservation Association Annual Conference, 1511, 1694

Wisconsin Rural Development Center, 5844

Wisconsin Sea Grant Institute, 5845

Wisconsin Society for Ornithology, 1512

Wisconsin State Extension Services Community Natural Resources & Economic Development, 3444

Wisconsin Wildlife Federation, 1513

Wisconsin Woodland Owners Association, 1514

Wisconsin Woodland Owners Association Annual Conference, 1695

Zimpro Environmental, 5589

Wyoming

Bighorn National Forest, 3918

Bridger-Teton National Forest, 3919

Casper District Advisory Council, 3445

Devils Tower National Monument, 3920

Energy Laboratories, 5005

Environmental Quality Department, 3446

Fossil Butte National Monument, 3921

Foundation for North American Wild Sheep Headquarters, 257

George B Storer Foundation, 2342

Grand Teton National Park, 3922

Inter-Mountain Laboratories, 5192

Jackson Hole Conservation Alliance, 1515

Lighthawk, 409, 624

Lighthawk, Southern Rocky Mountain Field Office, 837

Medicine Bow National Forest, 3923

National Elk Refuge, 3924

Nature Conservancy: Wyoming Chapter, 1516

Powder River Basin Resource Council, 1517

Resource Technology Corporation, 2030

Rock Springs Field Office Bureau of Land Management, 3447

Seedskadee National Wildlife Refuge, 3925

Shoshone National Forest, 3926

State Parks & Cultural Resources Division of State Parks & Historical Sites, 3448

Water Quality Laboratory, 5829

Western Environmental Services, 5575

Wolf Fund, 499

Wyoming Association of Conservation Districts, 1521

Wyoming Board of Land Commissioners, 3449

Wyoming Cooperative Fishery and Wildlife Research Unit, 3450

Wyoming Native Plant Society, 1522

Wyoming State Forestry Division, 3451

Wyoming State Geological Survey, 3452

Wyoming Wildlife Federation, 1523

Yellowstone National Park, 3927

Canada

BC Research, 4841

CENSOL, 4890

California Chapter, National Safety Council, 782

Northern Arkansas: Safety Council of the Ozarks, 752

Western Association of Fish and Wildlife Agencies Annual Meeting, 1520, 1688

Antarctica

Arctic Network, 324

Aquaculture

Aqua Tech Environmental Laboratory, 4819
Aquatec Chemical International, 4820
Desert Fishes Council, 356
Edwards Aquifer Research and Data Center, 6296
Proceedings of the Desert Fishes Council, 4122
World Aquaculture Society, 710

Aquatic ecology (See also: Marine ecology)

American Zoo and Aquarium Association, 25, 620
Aquatic Nuisance Species Task Force, 513
Aquatic Research Laboratory: Lake Superior State University, 5601
Center for Aquatic Research and Resource Management, 5615
Educational Images, 6686
Freshwater and Saltwater Marshes, 6704
Friends of the Sea Otter, 373
HT Peters Aquatic Biology Laboratory, 5687
Salt Marshes-A Special Resource, 6720
Sea Turtle Restoration Project, 297
World Fish Center, 6570

Architecture and energy conservation

Abonmarche Environmental, 1713
Architectural Energy Corporation, 1751, 4821
Architecture Research Laboratory, 5602

Arctic ecology

Great Bear Foundation, 381

Arid regions

Arizona-Sonora Desert Museum, 745
Chihuahuan Desert Research Institute, 3348, 4907
Desert Protective Council, 357
Desert Tortoise Council, 358
I Walk in the Desert, 6708
Sand Dune Ecology and Formation, 6721

Asbestos

Abacus Environmental, 1711
Argus/King Environmental Limited, 1754
Asbestos & Lead Abatement Report, 4151, 4187
Asbestos Information Association of North America, 174, 2094
EMS Laboratories, 4969
Environmental Analytical Laboratory, 5031
Environmental Management: Waltham, 5046
Environmental Testing Services, 5060
Hub Testing Laboratory, 5167
International Asbestos Testing Laboratories, 5193
Northeast Test Consultants, 5327

Atmospheric sciences

American Meteorological Society, 2
British Atmospheric Data Centre, 6213
Greenhouse Crisis Foundation, 9
Institute of Clean Air Companies, 11, 54, 529
Interdisciplinary Center for Aeronomy & Other Atmospheric Sciences, 5712
Kola Ecogeochemistry, 6430
Michigan Atmospheric Deposition Laboratory, 5741
National Center for Appropriate Technology, 6492
National Center for Atmospheric Research, 6493
National Climatic Data Center, 2683
National Environmental Satellite Data & Information Service, 2686
National Oceanic & Atmospheric Administration, 2155, 2694
Oceanic and Atmospheric Research Office, 2710
Rosenstiel School of Marine and Atmospheric Science, 5996

Biology-Ecology

AECOS, 1704
Advisory Panel for Ecology, 2511
American Inistute of Biological Sciences, 5854
American Society of Agricultural Engineers, 143
Applied Biomathematics, 4813
Bio/West, 4861
BioTest, 4862
BioTrol, 4863
Bioengineering Resources, 4864
Biological Research Associates, 4865
Biology: Careers and Jobs, 6211
Bionetics Corporation Analytical Laboratories, 4867
Biospherics, 4869
Bog Ecology, 6662
Center for Field Biology, 5626
Center for Population Biology, 5635
Coweeta LTER Site, 6271
Department of State: Office of Ecology, Health, and Conservation State Department, 2585
Ecological Society of America, 39
Florida Center for Environmental Studies, 5944, 6341
LIFE, 6431
LTER (US Long-Term Ecological Research), 6432
National Association of Biology Teachers, 60
Program in Freshwater Biology, 5781
SGS Environmental Services Inc, 5441
Tennessee Technological University, 6007
Teratology Society, 1452, 2178
Terrain Magazine, 4313
Western Society of Naturalists, 482
Western Society of Naturalists Annual Meeting, 1690

Biomass energy

Biomarine, 4866
Biomass Users Network, 180
Biomass Users Network: Central America, 1210

Biotechnology

Baron Consulting Company, 1774
Biocatalysis/Biodegradation Database, 6209
Biotechnology Industry Organization, 28
Center for Environmental Biotechnology, 6233
ChromatoChem, 4909
Environmental Research Institute, 5663
National Center for Biotechnology Information, 6620

Biotic communities (See also: Plant communities)

Association of Ecosystem Research Centers, 4830
Natural Resource Management, 5981

Botanical ecology

Botanical Society of America, 1545
Chicago Botanic Garden, 6083
Connecticut Botanical Society, 848

National Endangered Species Act Reform Coalition, 275
North American Bear Society, 286
North American Wolf Society, 449
North Dakota State University, 5984
Save the Manatee Club, 464
Scientists Center for Animal Welfare, 295
Society for the Conservation of Bighorn Sheep, 470
Trumpeter Swan Society, 302
Whooping Crane Conservation Association, 483
Wild Canid Survival and Research Center, 484
Wolf Education and Research Center, 498
Wolf Fund, 499
Wolf Haven International, 500

Endangered species, Plants (See also: charts on pages 475-530)

American Association of Botanical Gardens and Arboreta, 314
Biological Frontiers Institute, 1792
Rare and/or Endangered Species Research Center, 5782

Energy conservation

Alternative Energy Network Online, 3968
American Coal Ash Association, 114
Cut Your Home Heating Cost, 6675
Ecology Action Centre, 6294
Energy & Environmental Research Center (EERC), 6300
Energy Engineering, 4136
Energy Foundation, 2330
Environmental and Energy Study Institute, 562, 5942
Institute of Scrap Recycling Industries Convention, 1635
International Bicycle Fund, 1470
Legacy International, 6119
Maryland Recyclers Coalition Annual Conference, 1638
National Recycling Congress Show, 1652
Northeast Recycling Council Conference, 1665
Office of Energy Efficiency, 6548
Powder River Basin Resource Council Energy Conservation Education, 6635
Take it Back, 1677
Women s Council on Energy and the Environment Membership Directory, 3975
World Energy Engineering Congress, 1696
Zurn Industries, 5590

Energy consumption (See charts on pages 531-556)

Energy conversion

Advanced Resources International, 1719
AeroVironment, 4776
Alternative Energy Resources Organization, 113
Alternative Energy Resources Organization Newsletter, 4132, 4389
Association of Conservation Engineers: Membership Directory, 3976
CII Engineered Systems, 1821
Compressed Gas Association, 122
Energy Conversion Devices, 5004
Fossil Fuels Policy Action Institute, 127
FuelCell Energy, 5083
James W Bunger and Associates, 5212
Mikropul Environmental Systems, 5304
Natural Energy Laboratory of Hawaii, 6517
RPM Systems, 5408

Energy economics

Apollo Energy Systems, 4812
Commerce and Trade: Energy Division, 3385
Commerce and Trade: Gas and Oil Division, 3386
Committee on Commerce, 2541
Committee on Commerce, Science, and Transportation, 2542

Committee on Transportation and Infrastructure, 2553
Environmental Systems Corporation, 5056
Gas Technology Institute, 5096
Gordon Piatt Energy Group, 5122
International Association for Energy Economics, 129, 6391
International Association for Energy Economics Conference, 1636
International Association for Hydrogen Energy, 895
International Journal of Hydrogen Energy, 4139
Management and Budget Office: Natural Resources, Energy and Science, 2669
Research, Education, and Economics Economic Research Service Energy Office, 2733

Energy management

Alliance for Acid Rain Control and Energy Policy, 160, 510, 2075
Clean Energy Research Institute, 5643
Ecotope, 4995
Michael Baker Corporation, 1968
Rural Utilities Service, 2736
Safe Energy Communication Council, 138

Energy policy

American Gas Association, 116
CARIERS US Department of Energy, 6578
California Energy Commission, 6221
Civil Engineering Research Foundation, 120
Clean Fuels Development Coalition, 121
Department of Energy Annual Procurement and Financial Assistance Report, 3970
Energy Engineering: Directory of Software for Energy Managers and Engineers, 3977
Federal Energy Regulatory Commission, 2642
Office of the Secretary of Energy, 2721
State Energy Advisory Board: Office of the Secretary Department of Energy, 2745

Energy resources (See also: Biomass energy)

Alliance to Save Energy, 112
Alternative Energy Institute, 5597
American Solar Energy Society, 119, 6190
Aspen Institute, 2519
Associates in Rural Development, 1758
Audubon of Florida, 885
Center for Applied Energy Research, 5613
Consumer Energy Council of America Research Foundation, 124, 6591
Department of State: Energy Resources and Food Policy Department, 2582
Edison Electric Institute, 6295
Energy Ideas Clearinghouse, 6301
Energy Now, 6693
Energy Research Office, 2617
Energy to Go Around, 6694
Forms of Energy, 6703
Institute for Resource Management, 53
International Centre for Gas Technology Information, 6395
Louisiana Energy & Environmental Resources & Information Center, 6441
Minority Energy Information Clearinghouse: Office of Minority Economic Inpact, US Dept Energy, 6616
National Energy Foundation, 135, 6496
Natural Resources: Environmental Improvement and Energy Resources, 3173
Tennessee Valley Authority, 3340
Virginia Center for Coal & Energy Research, 5825
Wisconsin Energy Information Clearinghouse Wisconsin Division of Energy, 6656

Energy technology

American Council for an Energy-Efficient Economy, 115, 6184
American Petroleum Institute, 117
Argonne National Laboratory, 6071, 6194
Brookhaven National Laboratory, 6214
Centre for Alternative Transportation Fuels, 6243
Centre for the Analysis and Dissemination of Demonstrated Energy Technologies, 6244
Cleaner and Greener Environment, 6085
Current Alternative Energy Research and Development in Illinois, 3969
Defense Technical Information Center (DTIC), 6275
Department of Energy, 6279
Department of Energy: Utility Technologies Department, 2577
Electric Power Research Institute (EPRI), 6297
Energy Research Institute, 125
Energy Science and Technology, 3972
Energy Statistics Spreadsheets, 3973
Energy Technology Consultants, 1901
Energy Technology Data Exchange, 6302
Energy and Environmental Analysis, 5006
Energy and Environmental Engineering, 5007
Engineering & Environmental Management Group, 5008
Federal Energy Technology Center, 3431
Greenhouse Gas Technology Information Exchange GREENTIE, 6366
Household Products Disposal Council, 204
National BioEnergy Industries Association, 134, 6491
Northeast Advanced Vehicle Consortium, 6542
Northeast Sustainable Energy Association Conferences, 1667
Rocky Mountain Institute, 85

Environmental chemistry (See also: Air pollution)

Chopra-Lee, 4908
Eastern Technical Associates, 4988
Eastern Technical Associates Library, 4634
University of Maryland at College Park, 6028

Environmental compliance

ABS Consulting Training Services, 6162
Aerosol Monitoring & Analysis, 4777
Applied Technical Services, 4817
Association of Local Air Pollution Control Officials, 4
Business & Legal Reports, 6219
Center for Environmental and Regulatory Information Systems: Purdue University, 6582
Danaher Corporation, 4950
Economists, 4994
Environmental Compliance Assistance Center, 6314
Environmental Control Technology, 5039
Environmental Technology Council, 525
Environmental and Engineering Geophysical Society, 152
Enviropro, 5064
Minnesota Pollution Control Agency, 6471
Open Space Institute, 450
Openlands Project, 451
TSCA Assistance Information Service Hotline Environmental Assistance Division, 6647
Touch Vision Interactive Kiosks, 5531

Environmental design

AM Kinney, 4752
Advanced Technologies for Commercial Buildings, 6169
Air Force Center for Environmental Excellence, 6174
Alloway Testing, 4785
America the Beautiful Fund, 313
AndCare, 4805
Andrew D Sauter Consulting, 4808

Applied Physics, 4816
CA Rich, 1815
Center for Environmental Design Research College of Environmental Design, 6235
Columbus Instruments International, 4919
Controlled Environment Corporation, 4931
Cyberchron Corporation, 4943
Designing the Environment, 6676
El Dorado Engineering, 4998
Environ Laboratories, 5024
Environment/One Corporation, 5026
Environmental Abatement, 5028
Environmental Design Research Association, 99
Evans Cooling Systems, 5072
Global Environmental Options (GEO), 6356
Greenprints: Sustainable Communities by Design, 1633
Marin Conservation League, 410
Southface Energy Institute, 611
Southface Journal of Sustainable Building, 4415

Environmental economics (See also: Energy economics)

American Chemistry Council, 19
Anderson Engineering Consultants, 4806
Axiom Laboratories, 4837
B&P Laboratories, 4838
Bureau of Economic Analysis, 2525
CERES (Coalition for Environmentally Responsible Economies), 30
CPAC, 4893
California Council for Environmental and Economic Balance, 783
Center for Clean Air Policy, 5
Clean Sites, 346
Cornell Center for the Environment, 5899
Cornell University Center for the Environment, 6268
Department of State: Bureau of Economic and Business, 2580
Get America Working!, 46
H John Heinz III Center for Science Economics and the Environment, 5136
International Association for Impact Assessment, 1287
Management Advisory Group to the Assistant Administrator, 2668
National Garden Clubs, 421
National Gardener Magazine, 4288
Resource Decisions, 2028
Women s Environment and Development Organization, 1270

Environmental education

AIMS Multimedia, 6657
APC Lab, 4755
ARCADIS, 4757
ASTB Analytical Services, 4759
ASW Environmental Consultants, 4760
ATC Environmental, 4762
AZTEC Laboratories, 4765
Aaron Environmental, 4766
Adirondack Environmental Services, 4772
Agronomy and Environmental Science, 5848
Alliance for Environmental Education, 5850
Alpha Manufacturing Company, 4786
Alternatives for Community and Environment Roxbury Environmental Empowerment Project, 2076
Aluminum Recycling Association, 511
Amazon Center for Environmental Education Research, 5851
American Analytical Laboratories, 4790
American Council on the Environment, 317
American Environmental Network, 4791
American Geophysical Union, 6187
American Laboratory for Environmental Excellence, 4792
American Museum of Natural History, 6067
American Nature Study Society, 5855
Americans for the Environment, 512
Ana-Lab Corporation, 1738, 4797

Environmental engineering (See also: Environmental design)

University of Nevada at Las Vegas, 6032
University of Southern California, 6039
Upstate Laboratories, 5552
Vapex Environmental Technologies, 5553
Vara International: Division of Calgon Corporation, 5554
Versar, 5555
Villanueva Associates, 5556
Virginia Polytechnic Institute Center, 6051
Vista Research, 5557
Volumetric Techniques, 5558
WCH Industries, 5559
WQS Environmental Laboratory, 5561
Wadsworth/Alert Laboratories, 5562
Waid & Associates, 5563
Wallgren Environmental Services, 5564
West Michigan Testing, 5573
Weston Solutions, Inc, 2070, 5579
Wik Associates, 5582

Environmental ethics

American Society for Testing and Materials, 24
CO-OP America, 31
Canyonlands Field Institute, 341
Federation of Environmental Technologists, 44, 1629, 4161
Global Response, 102
US Dye Manufacturers Operating Committee, 92
US Public Interest Research Group, 2206, 5545
3D/International, 1697

Environmental finances (See charts on pages 557-566)

Environmental health (See also: Air pollution)

ACC Environmental Consultants, 1701
AccuTech Environmental Services, 1714
Action on Smoking and Health, 157, 2072
Adelaide Associates, 4770
Adelaide Environmental Health Associates, 4771
Alliance to End Childhood Lead Poisoning, 2186
Alternatives for Community and Environment, 1058
Ambulatory Pediatric Association, 2187
American Academy of Pediatrics: Committee on Environment Health, 2188
American Association of Poison Control Centers, 162, 2079
American Board of Environmental Medicine, 163, 2080
American Cancer Society, 165, 913, 2082
American Council on Science and Health, 167, 1234, 2085
American Lung Association, 170, 2089
American Medical Association, 968, 2090
American Public Health Association, 171, 2091
Arizona Department of Health Services: Office of Environmental Health, 2828
Association for Environmental Health of Soils, 147, 326, 1059, 2096
Association for Environmental Health of Soils (AEHS): Magazine, Newsletter, 4152
Association of Environmental Engineering Professors, 150
Association of Maternal and Child Health Program, 2189
Asthma and Allergy Foundation of America, 178, 2099
Atkins Environmental HELP, 1761
Board of Scientific Counselors: National Institute of Environmental Health, 2524
Business Health Environmental Lab, 4887
CQS Health and Environmental, 6579
California Association of Environmental Health, 777
California Birth Defects Monitoring Program, 780
California Environmental Protection Agency, 2856, 6222
Cambridge Environmental, 1832
Center for Disease Control, 6232
Center for Holistic Resource Management, 1221, 2104
Centers for Disease Control and Prevention, 916, 2108, 2530

Chemical Injury Information Network, 1168, 2109
Childhood Lead Poisoning Prevention Program, 2191, 3268, 3301
Childhood Lead Poisoning Program, 2192, 3278
Children s Defense Fund, 2193, 2210
Children s Environmental Health Network, 2194
Children s Health Environmental Coalition, 2195
Chlorine Chemistry Council, 6252
Connecticut Department of Public Health, 2910
Consortium for Environmental Education in Medicine, 5898
Consumer Product Safety Commission Hotline, 6592
Cornerstone Environmental, Health and Safety, 1875
Department of Energy: Office of NEPA Policy and Compliance, 2575
Department of State: Bureau of Oceans and International Environmental and Scientific Affairs, 2581
EcoLogic Systems, 1896
Enviro Dynamics, 5013
Environment Safety and Health Office, 2618
Environment, Safety and Health, 2620
Environmental Defense Fund, 2113, 2231
Environmental Health Information Service, 4196
Environmental Health Letter, 4197
Environmental Health Network, 194, 798, 2116
Environmental Health Sciences Center: University of Rochester Medical Center, 5657
Environmental Health Sciences Research Laboratory, 5658
Environmental Health Sciences Review Committee, 2623
Environmental Protection Agency: Office of Health & Environmental Assessment, 2637
Environmental and Occupational Health Science Institute, 1212, 2120, 5670
Ernaco, 5069
Florida Journal of Environmental Health, 4198
Food Safety and Inspection Service, 2650
Food Safety and Inspection Service US Department of Agriculture, 2121
Food and Drug Administration, 2122, 2651
Global Environmental Management Initiative, 48
Great Lakes/Mid-Atlantic Hazardous Substance Research Center, 5684
Harvard Medical School Center for Health and the Global Environment, 5956
Healthy Mothers, Healthy Babies, 201, 2126
Healthy Schools Network, 2198
House Committee on Commerce Telecommunications, Trade and Consumer Protection, 2655
Human Ecology Action League (HEAL), 205, 931, 2127
Immunotox, 5183
Institute for Food and Development Policy, 4726
Institute of Medicine: Board of Children, Youth and Families, 2199
Journal of Environmental Health, 4203
Keep America Beautiful, 1567, 2359
Kids Count Data Book: State Profiles of Child Well-Being, 2213
Kresge Center for Environmental Health, 5724
Land Improvement Contractors of America, 57, 100
Lead Safe California, 802
Liz Claiborne and Art Ortenberg Foundation, 2363
Mabbett & Associates: Environmental Consultants and Engineers, 5273
March of Dimes Birth Defects Foundation, 2204
Marine/Freshwater Biomedical Center, 5976
Michigan Department of Environment Quality, 6460
Mine Safety and Health Administration, 2676
NEPA Lessons Learned, 3978
National Alliance for Hispanic Health, 214, 2137
National Association of City and County Health Officials, 215, 2138
National Association of Environmental Risk Auditors, 62
National Association of Physicians for the Environment, 217, 2140
National Association of School Nurses, 1055, 2141
National Cancer Institute, 218, 2142
National Cancer Institute: Cancer Epidemiology and Genetics Division, 2681

Environmental impact analysis

Environmental monitoring

Environmental policy

Environmental protection

Environmental quality

Environmental sciences

Estuaries

Fertilizers

Fire ecology

Fish feeding and feeds

Fish habitat

Fish reproduction and growth

Peregrine Fund, 456
Salt Plains National Wildlife Refuge, 3281
San Francisco Bay National Wildlife Refuge, 3563
Sterling College, 6006
Three Circles Center for Multicultural Environmental Education, 6011

Harbors

Clean Harbors Cooperative, 654

Hazardous substances

Advanced Technology Environmental Education Center, 6170
Aptus, 4818
Asbestos Ombudsman Clearinghouse/Hotline, 6576
Association of American Pesticide Control Officials, 175, 2097
Association of Responsible Recyclers, 516
Beyond Pesticides/National Coalition Against the Misuse of Pesticides, 179, 2100
Board of Scientific Counselors: Agency for Toxic Substance and Disease Registry, 2933
Bureau of Explosives Hotline, 6577
Case Studies in Environmental Medicine, 2208
Conservtech, 1870
Dangerous Goods Advisory Council, 189, 2112, 2558
Environmental Protection Agency, 2627
Hazardous Substance Research Centers, 6375
MCS Referral and Resources, 2132
Montefiore Medical Center Lead Poisoning Prevention Program, 1246, 2134
Multiple Chemical Sensitivity Referral and Resources, 1094
National Center for Lead-Safe Housing, 220, 2144
National Environmental, 1976
National Lead Information Center, 2692
Office of Research & Engineering Hazardous Materials, 2713
RMC Medical Inc, 6717
US Department of Housing and Urban Development, 2755

Hazardous waste (See also: charts on pages 586-629)

Aguirre Engineers, 1722
Ames Laboratory: Environment Technology Department, 6191
Barco Enterprises, 1772
Captain Planet & the Planeteers: Toxic Terror, 6663
Center for Hazardous Substance Research, 5629
Columbia Analytical Services, 1860
Copper State Analytical Lab, 4934
Environmental Industry Association, 42, 524
Environmental Protection Agency Resource Conservation and Recovery Act, 2631
Hazard Technology, 4129
Hazardous Materials Intelligence Report, 4163
Hazardous Materials Newsletter, 4130
Hazardous Materials Regulations Guide, 3963, 4033
Hazardous Waste Ombudsman Program: US EPA, 6607
Hazardous Waste Reduction Loan Program, 2480
Hazmat Transport News, 4164
Idaho State University, 5958
Industrial Health and Hazards Update, 4202
Louis Defilippi, 1962
Mateson Chemical Corporation, 5281
McCoy s Hazardous Waste Regulatory Update Service, 4175
Medical Waste News, 4177
NIREX, 6480
Natural Hazards Research and Applications Information Center Newsletter, 109, 4131
Neilson Research Corporation, 5318
Ninyo and Moore, 1981
Office of Environmental Laboratory Certification, 3323
Peoria Disposal Company, 5378
Shaw Environmental, 2045

Solid Hazardous Waste Bureau, 3370
Spill Control Association of America, 1122
Suburban Laboratories, 5497
Townley Laboratories, 5533
Tox Scan, 5534
Toxic Substances: Hazardous Waste Management Division, 2878
Tracking Toxic Wastes in CA: A Guide to Federal and State Government Information Sources, 3967, 4047
Triumvirate Environmental, 6156
Western Region Hazardous Substance Research Center, 5838

Hazardous waste management

Asbestos Management, 4828
Best Environmental, Safety and Industrial Products, 4857
Brinkerhoff Environmental Services, 1808
Center for Hazardous Materials Research, 519
Citizens Clearinghouse for Hazardous Waste, 1441
Communities for a Better Environment, 791, 2110, 2225
Comprehensive Environmental, 1867
Department of Defense Environmental Cleanup Home Page, 6278
Environmental Assessment Association, 6311
Environmental Protection Agency Emergency Hazardous Site Control, 2628
Hazardous Waste Clean-Up Information (CLU-IN), 6376
Hazardous Waste Resource Center Environmental Technology Council, 200, 528, 2125
Institute of Hazardous Materials Management, 55, 209, 1088, 2130
Les A Cartier and Associates, 1960
National Association of Chemical Recyclers, 535
Resource Management, 2029
Titan Corp. Ship and Aviation Engineering Group, 2059

Human ecology (See also: Conservation of natural resources)

Association of State and Territorial Health Officials, 176
Center for Disease Control and Prevention, 2528
Centers for Disease Control & Prevention, 2107
Commonweal, 186
Commonwealth Engineering and Technology CET Engineering Services, 1863
Commonwealth Laboratory, 1864, 4921
Fossil Rim Wildlife Center, 6106
Genesis Fund/National Birth Defects Center: Pregnancy Environmental Hotline, 1063, 2124
Institute for the Human Environment, 208, 800
National Governors Association, 72
Negative Population Growth, 594
Population Communications International, 599
Population Connection, 600
Population Crisis Committee, 601
Population Institute, 602
Population Institute Newsletter, 4407
Population Reference Bureau, 603
Population Reference Bureau: World Population Data Sheet, 4408
Population Resource Center, 604
Population: Environment Balance, 605
Population: Environmental Council, 606
Rene Dubos Center for Human Environments, 235, 1261, 2171
Reporter, 4409
World Population Society, 614
Worldwatch Institute, 615
Worldwatch Institute: State of the World, 4422
Worldwatch Institute: Vital Signs, 4423
Worldwatch Institute: World Watch, 4424
Worldwatch Institute: Worldwatch Papers, 4425

Hydrocarbons

Membrane Technology & Research Corporate Headquarters, 5290

Hydrogen as fuel

Chlorine Institute, 34, 183

Hydrology

Department of Environmental Sciences, 5902
Dept. of Agriculture: National Forest Watershed and Hydrology, 2610
Environmental Simulations, 6326
International Association for Environmental Hydrology, 672

Industrial hygiene

Acumen Industrial Hygiene, 1717
Aerosol Monitoring and Analysis, 1721
Albrook Hydraulics Laboratory, 5596
American Industrial Hygiene Association Conference and Exposition, 1617
Astorino Branch Environmental, 1760
CIH Environmental, 1819
Challenge Environmental Laboratories, 1845
Chemical Data Management Systems, 1848
Cigna Loss Control Services, 1850
Cohen Group, 1858
Cohrssen Environmental, 1859
Compass Environmental, 1866
Comprehensive Environmental Strategies, 1868
Network Environmental Systems, 1980
Norton Associates, 1984
Occupational Health and Safety Management, 1988
Occupational Safety and Health Consultants, 1989
Professional Conference on Industrial Hygiene, 1672
RGA Environmental, 2016
Raterman Group, 2018
UEC Industrial Hygiene and Environmental Health Laboratories, 5542

Industrial safety

American Board of Industrial Hygiene, 164, 2081
California Environmental, 1829
Manpower, Reserve Affairs, Installations and Environment, 2670
Radian Corporation, 5411
University of Southern Mississippi, 6040

Inhalation toxicology

National Air Toxics Information Clearinghouse: US EPA, 6618

Lake renewal

Lake Pontchartrain Basin Foundation, 6433
League to Save Lake Tahoe, 803

Lakes

Center for Great Lakes Environmental Education, 5884
Great Lakes Information Network (GLIN), 6361
Great Lakes United, 670
International Joint Commission, 2663
North American Lake Management Society, 694, 6538
Wisconsin Sea Grant Program, 6568

Land use (See also: Agricultural conservation)

Adirondack Council, 6166
Advanced Forest Technologies Program, 6167
Bishop Resource Area, 6075
Bureau of Land Management, 3199
Bureau of Land Management, Land & Renewable Resources, 2526

Dobbin Milus International, 4955
Elko District Advisory Council: Bureau of Land Management, 3205
Germinal Project, 6352
Land Conservancy of San Luis Obispo County, 6434

Law, Environmental (See also: Acid rain; Air pollution)

A Guide to Environmental Law in Washington DC, 2256
Analysis of the Stockholm Convention on Persistent Organic Pollutants, 4023
Center for Community Action and Environmental Justice, 2221
Center for Health, Environment, and Justice, 2222
Community Rights Counsel, 2227
Comparative Environmental Law and Regulation, 2259
Environmental Compliance Update, 4108
Environmental Compliance in Your State, 2230
Environmental Law Institute, 2233, 2331, 5661
Environmental Law Lexicon, 2263
Environmental Law and Compliance Methods, 2264
Environmental Policy Alert, 4081
Harvard Environmental Law Society, 2240
International Protection of the Environment, 4540
LEXIS Environmental Law Library, 4009
Trade and the Environment: Law, Economics and Policy, 2280
US Environmental Protection Agency: Office of Environmental Justice, 2767
Understanding Environmental Administration and Law, 2281

Limnology (See also: Aquatic ecology)

Limnological Research Unit, 5729

Litter (Trash)

American Littoral Society, 321

Livestock

ALBC News, 4093
American Livestock Breeds Conservancy, 248

Marine biology (See also: Marine ecology)

Center for Marine Biology, 5634
Living Marine Resources Institute, 5730
Marine Biological Laboratory, 6123
Roger Williams University, 5995

Marine ecology (See also: Coral reef ecology)

Antarctic and Southern Ocean Coalition, 6192
Applied Marine Ecology, 1744
Crosby & Overton, 4940
Manatees: A Living Resource, 6713
Seacamp Association, Inc, 6148

Marine engineering

Belle W Baruch Institute for Marine Biology and Coastal Research, 5606
PCCI, 5356

Marine mammals (See also: Dolphins; Whales)

American Cetacean Society, 2850
Cascadia Research, 4894
Cetacean Society International, 255, 6082, 6245
Cousteau Odyssey, Vol. 10: Warm-Blooded Sea, Mammals of the Deep, 6673
Marine Mammal Commission, 2671

Natural resources development

Noise pollution (See also: charts on pages 677-683)

Nuclear energy

Nuclear engineering

Nuclear safety

Occupational diseases

Ocean

Coral Reef Alliance, 662, 6267
Department of Commerce: National Ocean Service, 2572
Department of Commerce: National Oceanic & Atmospheric Administration, 2570
Department of Commerce: Ocean Observation Division, 2573
Harbor Branch Oceanographic Institution, 6368
International Council for the Exploration of the Sea(ICES), 6396
International Year of the Ocean -1998, 6417
Marine Biological Association, 6449
Marine Technology Society, 678, 6453
National Audubon Society: Living Oceans Program, 681
National Response Center: US Coast Guard Headquarters, 6626
National Science Foundation Directorate for Geosciences, 2700
Ocean Conservancy, 696
Oceana, 697
Oceanic Society, 698
Project Oceanology, 6137
Scientific Committee on Oceanic Research: Department of Earth and Planetary Science, 700

Ocean engineering

Bioengineering & Environmental Systems Section, 2522
Juneau Center School of Fisheries & Ocean Sciences, 5721
Ocean & Coastal Policy Center, 5766
Ocean Engineering Center, 5767

Ocean-Atmosphere interaction

Carbon Dioxide Information Analysis Center, 3335

Oceanography

Australian Oceanographic Data Centre, 6203
International Oceanographic Foundation, 674
MBC Applied Environmental Sciences, 5269
Mediterranean Oceanic Data Base, 6457
NEMO: Oceanographic Data Server, 6479

Organic gardening

California Certified Organic Farmers: Membership Directory, 3933

Ornithology

Alaska Chilkat Bald Eagle Preserve, 6177
Atlantic Waterfowl Council, 861
Birding on the Web, 6212
California Waterfowl Association, 787
Guardians of the Cliff: The Peregrine Falcon Story, 6706
International Crane Foundation, 394, 6397
Learning about Backyard Birds, 6613
Migratory Bird Conservation Commission, 2674
Migratory Bird Regulations Committee Office of Migratory Bird Management, 2675
Milton Keynes Wildlife Hospital, 6468
Western Hemisphere Shorebird Reserve Network, 481
Windrifters: The Bald Eagle Story, 6732

Ozone

Earth Technologies Forum: Conference on Climate Change and Ozone Protection, 1627
Stratospheric Ozone Information Hotline/The Bruce Company, 6645

Parks (See also: National parks and reserves)

California Resources Agency, 6225

Connecticut Forest and Park Association, 849
Conservation Fund, 558, 6261
Department of the Interior, 2591, 6280
Department of the Interior: National Park Services, 2598
George Wright Society, 376
National Park Service: Fish, Wildlife and Parks, 2696
National Park System Advisory Board, 2697
National Park and Conservation Association, 6505
World Parks Endowment, 2441

Pest control, Integrated

Bio Integral Resource Center, 774, 2101
Pesticide Action Network North America, 809, 2166
Safer Pest Control Project, 2173

Pesticides (See also: charts on pages 684-698)

Citizens for Alternatives to Chemicals, 184
Handle with Care: Children and Environmental Carcinogens, 2212
Idaho State Department of Agriculture, 2981
Journal of Pesticide Reform, 4205
National Pesticide Information Center, 227, 2157
National Pesticide Information Retrieval System, 6622
National Pesticide Telecommunications Network Dept. of Preventive Medicine & Community Health, 6623
Northwest Coalition for Alternatives to Pesticides, 1325, 2164
Pesticide Directory: A Guide to Producers and Products, Regulators, and Researchers, 3965, 3989, 4040
Pesticide Research Center, 5776
Pesticide Research Laboratory & Graduate Study Center, 5777

Pesticides and the environment

Chemical Producers and Distributors Association, 33
Consumer Pesticide Project, 187
Food and Water, 197
Pesticide Regulation, Environmental Monitoring and Pesticide Management, 2872
Rachel Carson Council, 234, 1096, 2170, 6556, 6639

Petroleum

American Association of Petroleum Geologists Foundation, 2289
Cousteau Collection, Vol. 1: Alaska, Outrage at Valdez, 6672
Department of Transportation: Administrator for Pipeine Safety, Alaska, 2587
Department of Transportation: Associate Administrator for Pipeline Safety, 2588
Legacy of an Oil Spill: Ten Years After Exxon Valdez, 6711
National Petroleum Council, 2698
Natural Resources: Geological Oil & Gas Division, 3073

Plant communities

American Crop Protection Association, 549

Plant conservation

Abundant Life Seed Foundation, 310
Boone and Crockett Club, 337
Center for Plant Conservation, 344, 6239
Great Plains Native Plant Society, 1375
New England Wild Flower Society, 1587, 6526

Plants, Protection of (See also: Plant conservation)

Hawaii Biological Survey, 6370
National Wildflower Research Center, 434

Plastics

Magma-Seal, 5275
Scrap Plastics Markets Directory, 4046

Polar regions (See also: Arctic ecology)

Antarctica Project, 323, 2517
National Science Foundation Office of Polar Programs, 2701
Natural Environmental Research Council, 6518

Pollution (See also: Air pollution)

Allied Engineers, 1729
American Services Associates, 1737
Andersen 2000 Inc/Crown Andersen, 1740
Criteria Pollutant Point Source Directory, 4024
Environmental Protection Agency: Office of Pollution Prevention, 2636
Frank A Chambers Award, 1554
International Handbook of Pollution Control, 4036
Kansas State University, 5970
Keep North Carolina Beautiful, 1568
Lyman A Ripperton Award, 1570
Mercury Technology Services, 1965
Midwest Center for Environmental Science and Public Policy, 212, 984, 2133, 6466
Pollution Abstracts, 3980, 4041
Pollution Engineering, 4179
Rodriguez, Villacorta and Weiss, 2037

Pollution control

Agency for Toxic Substances and Disease Registry, 911, 2073, 2512, 6172
Barer Engineering, 1773
CZR, 1827
California Pollution Control Financing Authority, 2858
Canadian Council of Ministers of the Environment (CCME), 6227
Center for Sustainable Systems University of Michigan, 6584
Chapman Environmental Control, 1846
Clean-Up Information Bulletin Board System: US EPA Technology Innovation Office, 6588
Coalition Against Pipeline Pollution, 520
Community Greens, 2554
Consortium on Green Design and Manufacturing (CGDM), 6264
Consumer Specialty Products Association, 35
Control Technology Center Emission Standards Division, 6593
Curt B Beck Consulting Engineer, 1879
EPA: Department of Solid Chemical Emergency Preparedness & Prevention, 2613
Earth Options, 6740
Earth Science, 6741
Eco-Store, 6742
Ecology Store, 6743
Emission Factor Clearinghouse: US EPA, 6601
Energy Efficient Environments, 6744
Enviroene, 6303
Epcon Industrial Systems NV, Ltd, 1915
Erlander s Natural Products, 6746
Federal Facilities Environmental Restoration, 2643
GAIA Clean Earth Products, 6747
Green Hotels Association, 623
Green Seal, 49, 6363
Greenpeace, 2239, 6748
Installations and Environment, 2660
Inter-American Foundation, 2661
Jason Natural Cosmetics, 6749
Legal Environmental Assistance Foundation (LEAF), 898, 2246
Living on Earth, 6438
Mangrove Replenishment Initiative, 6447
Michigan Pulp and Paper Pollution Preservation Program, 6464

Mineral Policy Center, 411, 6469
National Association of Noise Control Officials, 216, 2139
National Pollutant Inventory, 6506
National Pollution Prevention Center for Higher Education (NPPC), 6507
Pollution Prevention Trust Fund, 6135
Rich Tech, 2033
Slippery Rock University, 6000
Sparky Boy Enterprises, 6753
Sunrise Lane Products, 6754
Toxic Substances: Pollution Prevention and Environmental Technology, 2879
Williams Distributors, 6755

Pollution, Environmental effects of

CTE Engineers, 1824
Get Oil Out, 526
HazDat-Hazardous Substance Release: Health Effects Database, 6374

Population

Californians for Population Stabilization (CAPS), 789
Council of State and Territorial Epidemiologists, 919, 2111
Environmental Change and Security Program: Woodrow Wilson International Center for Scholars, 2621
Global Committee of Parliamentarians on Population and Development, 565
Population Reference Bureau: Household Transportation Use and Urban Pollution, 4085
Population Reference Bureau: Population & Environment Dynamics, 4086
Population Reference Bureau: Water, 4087
Williams College, 6061

Prairie ecology

Grassland Heritage Foundation, 380

Public awareness/information programs

APEC-AM Environmental Consultants, 1706
Aarcher, 1710
Abco Engineering Corporation, 1712
Acheron Engineering Services, 1716
Advisory Council on Historic Preservation, 2509
American College of Occupational and Environmental Medicine, 962, 2083
American Conference of Governmental Industrial Hygienists, 166, 1530, 2084
American Environmental Health Foundation, 736
American Federation of Teachers, 20, 2086
American Geological Institute, 5853
American Humane Association, 247
American National Standards Institute, 22
American Public Works Association, 23
American Society of Civil Engineers, 144
American Society of Safety Engineers, 145, 2093
Animal Protection Institute, 253
Arro Consulting, 1755
Benton Foundation, 6207
Benton Foundation: Kids Campaigns, 2190
Bureau of Transportation Statistics, 6218
Center for Ecoliteracy, 5876
Center for Environmental Citizenship, 6234
Center for Environmental Information, 5878
Center for Environmental Strategies: Academy for Educational Development, 5881
Center for Environmental Study, 5882
Center for Health, Environment and Justice, 1438, 2103, 6583
Central European Environmental Data Request Facility, 6242

Pulp and paper technology

Radiation effects

Radioactive pollution (See also: Radioactive wastes)

Radioactive wastes (See also: Hazardous waste management)

Radon

Rain forest

Range management

Recreation areas

Trust for Public Land, 479

Recycling (See also: Energy conservation, and charts on pages 700-701)

Advanced Recovery, 6168
American Recycling Market Directory: Reference Manual, 4022
Arizona Automotive Recyclers Association, 739
Bio-Sun Systems, 6738
Bottom Line Consulting, 1805
Bureau of International Recycling, 6216
C&D Debris Recycling, 518
Container Recycling Institute, 521
Directory of Key Recycling Contacts, 4025
Environment Friendly Papers, 6745
Fibre Market News: Paper Recycling Markets Directory, 3944, 4032
Gaia Institute, 669
Governmental Refuse Collection and Disposal Association, 527
Granville Composite Products Corporation, 1927
Illinois Recycling Association, 6386
Institute of Scrap Recycling Industries, 530
National Association for Pet Container Resources, 6486
National Association for Plastic Container Recovery (NAPCOR), 534
National Office Paper Recycling Project, 537
National Recycling Coalition, 538, 1580
Nebraska Recycling Resource Directory, 4039
New York Association for Reduction, Reuse and Recycling, 6535
Paper Stock Institute, 541
Plastics Recycling Foundation, 542
Real Goods Trading Company, 6751
Recycling Related Newsletters, Publications and Periodicals, 4043
Recycling Today: Recycling Products & Services Buyers Guide, 4044
Resource Conversion Systems, 2027
Secondary Materials and Recycled Textiles Association, 544
Steel Recycling Institute, 547, 6561

Remediation

Together Foundation, 90

Renewable energy sources (See also: Solar energy, Wind energy)

Association of Battery Recyclers, 514, 884, 2098
California Renewable Fuels Council, 784
Colorado River Board of California, 790
Energy: The Alternatives, 6695
IISDnet: International Institute for Sustainable Development, 6385
Midwest Renewable Energy Association (MREA), 6467
Minnesotans for An Energy-Efficient Economy (ME3), 6472
National Renewable Energy Laboratory (NREL), 6508
National Renewable Energy Laboratory Technical Inquiry Service, 6625
New Forest Project, 6528
Northeast Sustainable Energy Association, 6543
Renewable Fuels Association, 84, 137, 6557

Renewable natural resources

Acorn Designs, 6733
Association for Natural Resources Enforcement Training, 327
President s Council on Sustainable Development, 2728
Renewable Natural Resources Foundation, 6558

Reservoirs (See also: Water resources)

North American Lake Management Society International Symposium, 1662

Risk

American Academy of Environmental Medicine, 161, 2077
Center for the Evaluation of Risks to Human Reproduction, 1439, 2106
Environmental Data Resources, 5040, 6095
National Safety Council, 228, 2159
Pesticides and the Immune System: Public Health Risks, 2215
Risk Management Internet Services, 6140

Rivers (See also: Estuaries; Water pollution; Watersheds)

Allegheny National Forest, 3837
American Rivers, 638, 2291, 6189
American Rivers Montana Field Office, 1167
American Rivers Northwest Regional Office, 1465
American Rivers, Southwest Regional Office, 738
Arkansas River Compact Administration, 825
Earth Observing System Amazon Project, 6286
Flooding River, 6700
Friends of the River, 799
National Association for State and Local River Conservation Programs, 679
Peace Corps, 2727
Pecos River Commission, 3354
Rio Grande Compact Commission, 3355
River Network, 699
Susquehanna River Basin Commission, 3313

Salmon

Connecticut River Atlantic Salmon Commission, 3099
Trout Unlimited, 301, 1600, 2418

Sand dune ecology

American Shore and Beach Preservation Association Directory, 322, 639, 4053
Save the Dunes Council, 463
Shore and Beach, 4308

Sanitary engineering (See also: Pollution; Water resources)

Inter-American Association of Sanitary Engineering and Environmental Sciences, 155
Pavia-Byrne Engineering Corporation, 2003

Sewage sludge

Environmental Technical Services, 5057

Soil conservation

Colorado Association of Conservation Districts, 828
Department of Agriculture: National Forest Watershed and Soil Resource, 2563, 2564
Department of Agriculture: Soil Conservation, 2568
International Erosion Control Association, 396
Kar Laboratories, 5225
Metro Services Laboratories, 5294
North Dakota Association of Soil Conservation Districts Annual Conference, 1664
PACE Analytical Services, 5352
RDG Geoscience and Engineering, 2015
Soil Engineering Testing, 5472
Woods End Research Laboratory, 5585
World Association of Soil and Water Conservation, 711

Soil pollution

Association for Environmental Health and Sciences, 6073

Continental Systems, 4930
Geo-Con, 5102

Solar energy

American Solar Energy Society Conference, 1618
Arizona Solar Energy Industries Association, 742
Arizona Solar Update, 743
California Solar Energy Industries Association, 785
Colorado Solar Energy Industries Association, 832
Directory of Solar-Terrestrial Physics Monitoring Stations, 3971
Double Envelope Solar House: Living with Tomorrow, 6679
International Energy Agency Solar Heating and Cooling Programme, 6398
International Solar Energy Society, 6413
Joint Center for Energy Management (JCEM), 6425
Solar Energy, 4146
Solar Energy Group, 5805
Solar Energy Industries Association, 1599
Solar Energy Report, 4147
Solar Energy Research Institute, 5473
Solar Energy and Energy Conversion Laboratory, 5806
Solar Power Engineering Company, 5474
Solar Resources, 5475
Solar Testing Laboratories, 5476
Solar Today, 4148
TXSES Newsletter, 4149
Texas Solar Energy Society Annual Conference, 1679

Solid waste

Alternative Resources, 1733
Center for Solid & Hazardous Waste Management, 5638
Directory of Municipal Solid Waste Management Facilities, 4026
EPA: Office of Solid Waste, Municipal & & Industrial Solid Waste, 2614
Office of Solid Waste Management & Emergency Response, 2714
Pittsburgh Mineral & Environmental Technology, 5382
Solid Waste Information Clearinghouse and Hotline, 6644
Solid Waste Report, 4183
Widener University: International Conference on Solid Waste Magazine, 4185

Solid waste management

American Waste Processing, 4796
Association of Metropolitan Sewerage Agencies, 644
Association of State and Territorial Solid Waste Management Officials, 517
Better Management Corporation of Ohio, 1787
Community Environmental Council, 2226
Floyd Browne Associates, 1916
International Conference on Solid Waste, 1637
Regional Services Corporation, 2023
Solid Waste Association of North America, 545

Speleology

Journal of Caves & Karst Studies, 4277
NSS News, 4120
National Speleological Society, 430

Sustainable agriculture

California Certified Organic Farmers, 781
Center for Sustainable Development in the Americas, 2529
Columbia Earth Institute: Columbia University, 6259
Consultative Group on International Agricultural Research (CGIAR), 6265
CropLife Canada, 6272
EarthSave Foundation, 192
Institute for Agriculture and Trade Policy, 567, 1129, 2129

Napa County Resource Conservation, 6482

Technology and the environment

AFE Journal, 4092
Association for Facilities Engineering, 1294
Enviro-Access, 6304
Global Network of Environment & Technology, 6358
Technology Administration: National Institute of Standards & Technology, 2748

Thermal energy

Thermo Electron Corporation, 5522
ThermoRetec Corporation, 5523
Thermotron Industries, 5525

Tourism and recreation

Culture, Recreation and Tourism Department State Parks Office, 3064

Toxicology (See also: charts on pages 709-747)

American College of Toxicology, 4186
Aroostook Testing & Consulting Laboratory, 4824
Bio-Integral Resource Center, 5607
Block Environmental Services, 1802
Center for Environmental Toxicology and Technology, 5624
Center for Health Effects of Environmental Contamination, 6236
Consultox, 1872
Environmental Contaminants Encyclopedia, 6315
Environmental Human Toxicology, 5659
Environmental Mutagen Society, 5927
Environmental Toxicology Center, 5669
Extension Toxicology Network ETOXNET, 6335
Franklin D Aldrich MD, PhD, 1918
Institute of Chemical Toxicology, 5709
Kinnetic Laboratories, 5232
National Capital Poison Center Georgetown University Hospital, 6619
Oneil M Banks, 5343
P&P Laboratories, 5350
Resource Technologies Group, 5425
Scientific & Technical Resources, 5454
Society of Environmental Toxicology and Chemistry, 904, 2177, 6152
Toxic Chemicals Laboratory, 5818
Toxicology Information Response Center, 1391, 2179
Toxicology Information Response Center Oak Ridge National Laboratory, 6648

Tropical ecology (See also: Ecology, Tropical)

Island Resources Foundation, 574
North American Loon Fund, 447
Smithsonian Tropical Research Institute, 2744

Tropical forestry

International Society of Tropical Foresters, 572
International Society of Tropical Foresters: Membership Directory, 4007
RARE Center for Tropical Conservation, 459
Tropical Rainforest, 6725
Tropical Rainforests Under Fire, 6726

UV index (See chart on page 748)

Urban ecology (Biology)

Urban Habitat Program, 820
Urban Initiatives, 612

Waste disposal

Andco Environmental Processes, 1739
Beaumont Environmental Systems, 1782
Environmental Enterprises, 1297, 5042

Waste management

Advanced Waste Management Systems, 1720
Air and Waste Management Association, 1, 158, 509, 1338, 1527, 2074, 6066, 6175
Air and Waste Management Association s Magazine for Environmental Managers, 4150
Allstate Power Vac, 1730
Brown, Vence and Associates, 1810
Cornell Waste Management Institute, 5647
Custom Environmental Services, 1880
EDIE: Environmental Data Interactive Exchange, 6282
Eberline Services, 4989
El Environmental Services Directory, 4027
El Environmental Services Directory Online, 3960
Environmental Control Laboratories, 5038
Heritage Environmental Services, 1939
Institute of Chemical Waste Management Directory of Hazardous Waste Treatment and Dispos., 3964, 4035
Integrated Waste Management, 4165
Iowa Waste Reduction Center, 5715
Journal of Air and Waste Management Association, 4167
Journal of the Air and Waste Management Association, 4084
Kentucky Partners: State Waste Reduction Center, 6612
MBA Labs, 5267
Municipal Waste Management Association, 533
Natural Resources: Waste Management Division, 3142
Northeast Waste Management Officials Association: NE Multi-Media Pollution Prevention, 6630
Office of Environmental Restoration & Waste Management, 2712
Ozark Environmental Laboratories, 5349
PDC Laboratories, 5357
Parsons Engineering Science, 5376
Petra Environmental, 2005
RMC Corporation Laboratories, 5406
STS Consultants, 5448
Systech Environmental Corporation, 5502
Tri-Tech Laboratories, 5539
Waste Age: Resource Recovery Acitivities Update Issue, 4048
Waste Age: Waste Industry Buyer Guide, 4049
Waste Compliance Services, 5565
Waste Management, 2430
Waste Management Education & Research Consortium, 5827
Waste Management Research & Education Institute, 5828
Waste Manifest Software Report, 4050
Waste News, 4184
Waste Water Engineers, 5566
Wastes to Resources: Appropriate Technologies for Sewage Treatment and Conversion, 4051

Waste products (See also: Recycling [Waste, etc.])

Clean Air Council, 6
Northeast Industrial Waste Exchange, 6629
Waste Exchange Clearinghouse Waste Systems Institute of Michigan, 6652
Waste Expo, 1685
WasteExpo, 1686

Waste treatment

Chemical Waste Disposal Corporation, 4904
Owen Engineering and Management Consultants, 1994

Wastewater treatment

ADS Corporation, 1703
Aqua Sierra, 1746
Arro Laboratory, 1756
BBS Corporation, 1765
Bioscience, 4868
Bjaam Environmental, 1798
Buck, Seifert and Jost, 1811
Earth Tech, 4983
EnviroTech Southeast, 5020
Lenox Institute of Water Technology, 1959
Montana Environmental Training Center, 5980
National Small Flows Clearinghouse- West Virginia University, 6627
New England Water Environment Association Annual Meeting, 1656
New Jersey Water Environment Association Conference, 1659
New York Water Environment Association Semi- Annual Conferences, 1661
Resource Technology Corporation, 2030

Water conservation (See also: Water quality control)

Artesian Laboratories, 4826
Blasland, Bouck and Lee, 1801
Catlin Engineers and Scientists, 1842
Clean Water Action, 655, 2311
Environmental Chemistry and Technology Program, 5655
Great Lakes Commission, 1559
Great Plains: Rocky Mountain Hazardous Substance Research Center, 5685
ImmuCell Corporation, 5182
Izaak Walton League, 6421
MWH Laboratories, 5272
Maine Department of Conservation, 6444
Minnesota Association of Soil and Water Conservation Districts Annual Meeting, 1643
National Water Resources Association, 688, 1582
Nature Conservancy, 839, 866, 1446, 1586, 2381, 6520
New Mexico Association of Soil and Water Conservation Annual Conference, 1660
Nordlund and Associates, 1982
North Carolina Association of Soil and Water Conservation Districts Annual Conference, 1663
Simons and Associates, 5466
Soil and Water Conservation Society, 1017, 1598
Texas Water Conservation Association Annual Conference, 1680
Thompson Engineering Testing, 5526
Tri-State Laboratories, 5538
Turner Laboratories, 5541
Virginia Association of Soil and Water Conservation Districts Annual Conference, 1682
Water Environment & Technology Buyer s Guide and Yearbook, 3955, 4076
Water Quality Services, 5567
Water Resources Associates, 5568
Water Testing Laboratory, 5835
William T Lorenz & Company, 5583
Wisconsin Land and Water Conservation Association Annual Conference, 1694

Water pollution (See also: Acid rain)

Association of State and Interstate Water Pollution Control Administrators, 648
Environmental Technology Expo, 1628

Kimre, 1953
List of Water Pollution Control Administrators, 4038, 4074
Massachusetts Water Pollution Control Association Annual Conference, 1640
Perry-Carrington Engineering Corporation, 2004
Wisconsin Applied Water Pollution Research Consortium: University of Wisconsin-Madison, 5843

Water quality (See also: Water pollution; Water quality control)

Arizona Water Well Association, 744
Astro-Chem Services, 4831
Astro-Pure Water Purifiers, 4832
Bioengineering Group, 1790
Canadian Chlorine Chemistry Council, 6226
Citizen s Directory for Water Quality Abuses, 4069
Environmental Protection Agency: Water, 2639
Fresh-Water Foundation, 667
National Water Supply Improvement Association, 689
Office of Civil Water Enforcement Division, 2711
Ohio River Valley Water Sanitation Commission, 3274
Safe Drinking Water Hotline, 6642
Southern Environmental Law Center, 2251
WEFTEC Show, 1684
Water Environment Federation, 705, 1459, 1607, 2182
Water Quality Laboratory, 5829

Water quality control (See also: Water reclamation)

American Society of Sanitary Engineering, 146
American Water Works Association, 641, 4608
Ceimic Corporation, 4896
Century Testing Laboratories, 4900
Clean Harbors, 4913
Clean Water Systems, 4914
Columbus Water and Chemical Testing Laboratory, 4920
New England Interstate Water Pollution Control Commission, 3114
Professional Analytical and Consulting Services (PACS), 2012
University of Wisconsin at Stevens Point, 6045

Water reclamation (See also: Water use)

Department of the Interior: Bureau of Reclamation, 3204
Department of the Interior: Water and Science Bureau of Reclamation, 2607
International Desalination Association, 673
National Xeriscape Council, 693

Water resources (See also: Groundwater; Reservoirs)

American Water Resources Association, 1538
American Water Resources Association Conference, 1619
Arizona Geological Survey, 6195
Arkansas Environmental Federation, 751
Association of Metropolitan Water Agencies, 645
Baystate Environmental Consultants, Inc, 1778
Beals and Thomas, 1780
Black Waters, 6661
Bureau of Reclamation, 6217
Center for Streamside Studies, 5639
Center for the Management, Utilization and Protection of Water Resources, 5641
Chesapeake Bay Critical Areas Commission, 2531
Colorado Water Congress, 834
Colorado Water Congress Annual Meeting, 1624
Coordination Directory of State and Federal Agency Water Resources Officials: Missouri Basin, 4070
Department of the Interior: Office of the Solicitor, 2600
Department of the Interior: Soil, Water & Air, 2602
Department of the Interior: Water Resources Western, 2862

Department of the Interior: Water and Science, Water Resources Division, 2606
Directory of Water Resources Expertise, 4071
Environmental Defense, 2261, 6316
Environmental Protection Agency Ground Water and Drinking Water, 2630
Enviroscan Inc, 5065
Geological Survey, 3395
International Water Resources Association, 676
Kentucky Water Resources Research Institute, 6429
Masschusetts Water Resources Research Center, 5740
National Institutes for Water Resources, 686
National Water Resources Association Annual Conference, 1653
National Waterways Conference, 691
Natural Resources Consulting Engineers, 1979
Natural Resources and Environmental Management, 5982
New York State Water Resources Institute, 5762
Ohio Water Development Authority, 3275
Sacramento River Discovery Center, 6143
Saint Lawrence Seaway Development Corporation, 2737
Soil and Water Research, 5804
Southwest Consortium on Plant Genetics & Water Resources, 5809
St Croix International Waterway Commission, 3097
Stroud Water Research Center, 5811
Texas Water Resources Institute, 5816
Upper Colorado River Commission, 3371
Water Center, 704
Water Quality Association, 706
Water Resource Center, 5830
Water Resources Congress, 707
Water Resources Institute, 5832
Water Resources Management, 6158
Water Resources Research Institute at Kent University, 5833
Water Resources Research of the University of North Carolina, 5834
Water Resources Videos, 6728
West Virginia Water Research Institute, 5837

Water treatment (See also: Water reclamation)

American Clean Water Association, 636
Clean Water Network, 657
Clean Water Network: CWN Status Water Report, 4438
Crane Environmental, 4939
Ground-Water Remediation Technologies Analysis Center, 6367
Wastewater Treatment Information Exhange National Small Flows Clearinghouse, 6653

Water use (See also: Water reclamation)

American Whitewater, 619

Waterfowl

National Waterfowl Council, 433
Wildfowl Trust of North America, 1100

Watersheds

Adopt-A-Stream Foundation, 633
Center for Earth & Environmental Science, 5618
Center for Watershed Protection, 652
Connecticut River Watershed Council, 1547
Cook Inlet Keeper, 661
Maryland Department of Natural Resources Chesapeake Bay & Watershed, 3125
Medomak Valley Land Trust, 6458
Save the Sound, 853
USDA Forest Service: Watershed and Air Management, 2771
Watershed: Canada s Threatened Rainforest, 6729

Wetlands (See also: Estuaries)

Whales

Wilderness area

Wildlife (See also: Animal ecology; Wildlife management)

Seals, 6722
Sierra Club Guide to the Natural Areas of California, 4065
Society for the Preservation of Birds of Prey, 471
Southeastern Association of Fish and Wildlife Agencies Annual Meeting, 1676
Southwestern Herpetologists Society, 818
Trinity River Basin Fish & Wildlife Task Force Mid-Pacific Region, 2880
Turkey Call, 4217
US Forest Service: Wildlife Habitat & Silviculture Laboratory, 5819
United States Department of the Interior United States Fish and Wildlife Service, 2883, 2906, 2918
United States Department of the Interior: United States Fish and Wildlife Service, 3239
United States Fish and Wildlife Service, 2947
Western Association of Fish and Wildlife Agencies Annual Meeting, 1688
Wheelin Sportsmen, 4218
Wilderness Society, 487, 1610, 2431
Wildlife Action, 489
Wildlife Conservation Fund of America, 2433
Wildlife Conservation Society, 490, 2505
Wildlife Disease Association, 491
Wildlife Federation of Alaska, 733
Wildlife Forever, 492
Wildlife Habitat Enhancement Council, 493
Wildlife Information Center, 494
Wildlife Law News Quarterly, 4328
Wildlife Society Annual Conference, 1692
Wildlife Trust, 308
Willowbrook Wildlife Haven Preservation, 1612
Women in the Outdoors, 4219
World Society for the Protection of Animals, 2444
World Wildlife Fund, 505, 1614, 2445
Yellowstone Grizzly Foundation, 843
Young Entomologists Society, 6161

Wildlife management

Ardea Consulting, 1753
Birds of Prey Rehabilitation Association, 254, 336
Connecticut Forest and Park Association Annual Meeting, 1625
Federal Wildlife Association, 43
Fish and Wildlife Information Exchange Homepage, 6340
Friends of the Sea Lion Marine Mammal Center, 372
National Hunters Association, 277
National Rifle Association of America, 279
National Shooting Sports Foundation, 280
Native American Fish and Wildlife Society, 285
Pope and Young Club, 293
Quail Unlimited, 458
Ruffed Grouse Society, 462
Sportsmans Network, 300
US Sportsman s Alliance, 303
US Sportsmen s Alliance, 304
Whitetails Unlimited, 306
Wildlife Management Institute, 6060

Wind energy

American Hydrogen Association, 6188
Association of Energy Engineers, 149, 515, 6198
Center for Renewable Energy and Sustainable Technology, 6240

The Environmental Resource Handbook

Available Formats

Online Database

The Environmental Resource Handbook – Online Database is the most up-to-date and comprehensive source for Environmental Resources and Statistics. With your subscription to **The Environmental Resource Handbook – Online Database** you'll have immediate access to thousands of Environmental Resources – with detailed contact information, including hot links to web sites and e-mail addresses. Plus – you'll also have access to over 100 statistical and ranking tables available in pdf format. These informative tables provide at-a-glance access to data on critical environmental issues. **The Environmental Resource Handbook – Online Database** is a necessary tool for the reference department of any Public Library, Academic Library, as well as a perfect research tool for any organization with a primary focus on the environment. Provide your patrons or your research department with the details needed to quickly and easily find Environmental Resources and Statistics and subscribe today! Visit www.greyhouse.com or call (800) 562-2139 for more information.

Mailing List Information

This directory is available in mailing list form on mailing labels or diskettes. Call (800) 562-2139 to place an order or inquire about counts. There are a number of ways we can segment the database to meet your mailing list requirements.

Licensable Database on Disk

The database of this directory is available on diskette in an ASCII text file, delimited or fixed fielded. Call (800) 562-2139 for more details.

Call (800) 562-2139 for more information

Grey House Publishing
Business Directories

The Directory of Business Information Resources, 2006

With 100% verification, over 1,000 new listings and more than 12,000 updates, this 2006 edition of *The Directory of Business Information Resources* is the most up-to-date source for contacts in over 98 business areas – from advertising and agriculture to utilities and wholesalers. This carefully researched volume details: the Associations representing each industry; the Newsletters that keep members current; the Magazines and Journals - with their "Special Issues" - that are important to the trade, the Conventions that are "must attends," Databases, Directories and Industry Web Sites that provide access to must-have marketing resources. Includes contact names, phone & fax numbers, web sites and e-mail addresses. This one-volume resource is a gold mine of information and would be a welcome addition to any reference collection.

"This is a most useful and easy-to-use addition to any researcher's library." –The Information Professionals Institute

2,500 pages; Softcover ISBN 1-59237-078-0, $195.00 ♦ Online Database $495.00

Nations of the World, 2006 A Political, Economic and Business Handbook

This completely revised edition covers all the nations of the world in an easy-to-use, single volume. Each nation is profiled in a single chapter that includes Key Facts, Political & Economic Issues, a Country Profile and Business Information. In this fast-changing world, it is extremely important to make sure that the most up-to-date information is included in your reference collection. This edition is just the answer. Each of the 200+ country chapters have been carefully reviewed by a political expert to make sure that the text reflects the most current information on Politics, Travel Advisories, Economics and more. You'll find such vital information as a Country Map, Population Characteristics, Inflation, Agricultural Production, Foreign Debt, Political History, Foreign Policy, Regional Insecurity, Economics, Trade & Tourism, Historical Profile, Political Systems, Ethnicity, Languages, Media, Climate, Hotels, Chambers of Commerce, Banking, Travel Information and more. Five Regional Chapters follow the main text and include a Regional Map, an Introductory Article, Key Indicators and Currencies for the Region. An all-inclusive CD-ROM is available as a companion to the printed text. Noted for its sophisticated, up-to-date and reliable compilation of political, economic and business information, this brand new edition will be an important acquisition to any public, academic or special library reference collection.

"A useful addition to both general reference collections and business collections." –RUSQ

1,700 pages; Softcover ISBN 1-59237-079-9, $145.00

New York State Directory, 2005/06

The New York State Directory, published annually since 1983, is a comprehensive and easy-to-use guide to accessing public officials and private sector organizations and individuals who influence public policy in the state of New York. *The New York State Directory* includes important information on all New York state legislators and congressional representatives, including biographies and key committee assignments. It also includes staff rosters for all branches of New York state government and for federal agencies and departments that impact the state policy process. Following the state government section are 25 chapters covering policy areas from agriculture through veterans' affairs. Each chapter identifies the state, local and federal agencies and officials that formulate or implement policy. In addition, each chapter contains a roster of private sector experts and advocates who influence the policy process. The directory also offers appendices that include statewide party officials; chambers of commerce; lobbying organizations; public and private universities and colleges; television, radio and print media; and local government agencies and officials.

New York State Directory - 800 pages; Softcover ISBN 1-59237-093-4; $129.00

Profiles of New York, 2005/06 ♦ Profiles of Florida, 2005/06 ♦ Profiles of Texas, 2005/06

Packed with over 50 pieces of data that make up a complete, user-friendly profile of each state, these directories go even further by then pulling selected data and providing it in ranking list form for even easier comparisons between the 100 largest towns and cities! The careful layout gives the user an easy-to-read snapshot of every single place and county in the state, from the biggest metropolis to the smallest unincorporated hamlet. Here is a look at just a few of the data sets you'll find in each profile: History, Geography, Climate, Population, Vital Statistics, Economy, Income, Taxes, Education, Housing, Health & Environment, Public Safety, Newspapers, Transportation, Presidential Election Results, Information Contacts and Chambers of Commerce. As an added bonus, there is a section on Selected Statistics, where data from the 100 largest towns and cities is arranged into easy-to-use charts. Each of 22 different data points has its own two-page spread with the cities listed in alpha order so researchers can easily compare and rank cities. A remarkable compilation that offers overviews and insights into each corner of the state, *Profiles of New York*, *Profiles of Florida* and *Profiles of Texas* go beyond Census statistics, beyond metro area coverage, beyond the 100 best places to live. Drawn from official census information, other government statistics and original research, you will have at your fingertips data that's available nowhere else in one single source. Data will be published on additional states in 2006 and 2007.

Profiles of New York, 2005/06: 800 pages; Softcover ISBN 1-59237-108-6; $149.00 ♦ Profiles of Florida, 2005/06: 800 pages; Softcover ISBN 1-59237-110-8; $129.00 ♦ Profies of Texas, 2005/06: 800 pages; Softcover ISBN 1-59237-111-6; $149.00

To preview any of our Directories Risk-Free for 30 days, call (800) 562-2139 or fax to (518) 789-0556

The Grey House Performing Arts Directory, 2005

The Grey House Performing Arts Directory is the most comprehensive resource covering the Performing Arts. This important directory provides current information on over 8,500 Dance Companies, Instrumental Music Programs, Opera Companies, Choral Groups, Theater Companies, Performing Arts Series and Performing Arts Facilities. Plus, this edition now contains a brand new section on Artist Management Groups. In addition to mailing address, phone & fax numbers, e-mail addresses and web sites, dozens of other fields of available information include mission statement, key contacts, facilities, seating capacity, season, attendance and more. This directory also provides an important Information Resources section that covers hundreds of Performing Arts Associations, Magazines, Newsletters, Trade Shows, Directories, Databases and Industry Web Sites. Five indexes provide immediate access to this wealth of information: Entry Name, Executive Name, Performance Facilities, Geographic and Information Resources. *The Grey House Performing Arts Directory* pulls together thousands of Performing Arts Organizations, Facilities and Information Resources into an easy-to-use source – this kind of comprehensiveness and extensive detail is not available in any resource on the market place today.

"Immensely useful and user-friendly ... recommended for public, academic and certain special library reference collections." –Booklist

1,500 pages; Softcover ISBN 1-59237-023-3, $185.00 ♦ Online Database $335.00

The Directory of Venture Capital & Private Equity Firms, 2005

This edition has been extensively updated and broadly expanded to offer direct access to over 2,800 Domestic and International Venture Capital Firms, including address, phone & fax numbers, e-mail addresses and web sites for both primary and branch locations. Entries include details on the firm's Mission Statement, Industry Group Preferences, Geographic Preferences, Average and Minimum Investments and Investment Criteria. You'll also find details that are available nowhere else, including the Firm's Portfolio Companies and extensive information on each of the firm's Managing Partners, such as Education, Professional Background and Directorships held, along with the Partner's E-mail Address. *The Directory of Venture Capital & Private Equity Firms* offers five important indexes: Geographic Index, Executive Name Index, Portfolio Company Index, Industry Preference Index and College & University Index. With its comprehensive coverage and detailed, extensive information on each company, *The Directory of Venture Capital & Private Equity Firms* is an important addition to any finance collection.

"The sheer number of listings, the descriptive information provided and the outstanding indexing make this directory a better value than its principal competitor, Pratt's Guide to Venture Capital Sources. Recommended for business collections in large public, academic and business libraries." –Choice

1,300 pages; Softcover ISBN 1-59237-062-4, $450.00 ♦ Online Database (includes a free copy of the directory) $889.00

The Directory of Mail Order Catalogs, 2005

Published since 1981, this 2005 edition features 100% verification of data and is the premier source of information on the mail order catalog industry. Details over 12,000 consumer catalog companies with 44 different product chapters from Animals to Toys & Games. Contains detailed contact information including e-mail addresses and web sites along with important business details such as employee size, years in business, sales volume, catalog size, number of catalogs mailed and more. Four indexes provide quick access to information: Catalog & Company Name Index, Geographic Index, Product Index and Web Sites Index.

"This is a godsend for those looking for information." –Reference Book Review

1,700 pages; Softcover ISBN 1-59237-066-7 $250.00 ♦ Online Database (includes a free copy of the directory) $495.00

The Directory of Business to Business Catalogs, 2005

The completely updated 2005 *Directory of Business to Business Catalogs,* provides details on over 6,000 suppliers of everything from computers to laboratory supplies... office products to office design... marketing resources to safety equipment... landscaping to maintenance suppliers... building construction and much more. Detailed entries offer mailing address, phone & fax numbers, e-mail addresses, web sites, key contacts, sales volume, employee size, catalog printing information and more. Jut about every kind of product a business needs in its day-to-day operations is covered in this carefully-researched volume. Three indexes are provided for at-a-glance access to information: Catalog & Company Name Index, Geographic Index and Web Sites Index.

"An excellent choice for libraries... wishing to supplement their business supplier resources." –Booklist

800 pages; Softcover ISBN 1-59237-064-0, $165.00 ♦ Online Database (includes a free copy of the directory) $325.00

To preview any of our Directories Risk-Free for 30 days, call (800) 562-2139 or fax to (518) 789-0556

Thomas Food and Beverage Market Place, 2005

Thomas Food and Beverage Market Place is bigger and better than ever with thousands of new companies, thousands of updates to existing companies and two revised and enhanced product category indexes. This comprehensive directory profiles over 18,000 Food & Beverage Manufacturers, 12,000 Equipment & Supply Companies, 2,200 Transportation & Warehouse Companies, 2,000 Brokers & Wholesalers, 8,000 Importers & Exporters, 900 Industry Resources and hundreds of Mail Order Catalogs. Listings include detailed Contact Information, Sales Volumes, Key Contacts, Brand & Product Information, Packaging Details and much more. *Thomas Food and Beverage Market Place* is available as a three-volume printed set, a subscription-based Online Database via the Internet, on CD-ROM, as well as mailing lists and a licensable database.

"An essential purchase for those in the food industry but will also be useful in public libraries where needed. Much of the information will be difficult and time consuming to locate without this handy three-volume ready-reference source." –ARBA

8,500 pages, 3 Volume Set; Softcover ISBN 1-59237-058-6, $495.00 ◆ CD-ROM $695.00 ◆ CD-ROM & 3 Volume Set Combo $895.00 ◆ Online Database $695.00 ◆ Online Database & 3 Volume Set Combo, $895.00

Sports Market Place Directory, 2005

For over 20 years, this comprehensive, up-to-date directory has offered direct access to the Who, What, When & Where of the Sports Industry. With over 20,000 updates and enhancements, the *Sports Market Place Directory* is the most detailed, comprehensive and current sports business reference source available. In 1,800 information-packed pages, *Sports Market Place Directory* profiles contact information and key executives for: Single Sport Organizations, Professional Leagues, Multi-Sport Organizations, Disabled Sports, High School & Youth Sports, Military Sports, Olympic Organizations, Media, Sponsors, Sponsorship & Marketing Event Agencies, Event & Meeting Calendars, Professional Services, College Sports, Manufacturers & Retailers, Facilities and much more. *The Sports Market Place Directory* provides organization's contact information with detailed descriptions including: Key Contacts, physical, mailing, email and web addresses plus phone and fax numbers. Plus, nine important indexes make sure that you can find the information you're looking for quickly and easily: Entry Index, Single Sport Index, Media Index, Sponsor Index, Agency Index, Manufacturers Index, Brand Name Index, Facilities Index and Executive/Geographic Index. For over twenty years, *The Sports Market Place Directory* has assisted thousands of individuals in their pursuit of a career in the sports industry. Why not use "THE SOURCE" that top recruiters, headhunters and career placement centers use to find information on or about sports organizations and key hiring contacts.

1,800 pages; Softcover ISBN 1-59237-077-2, $225.00 ◆ CD-ROM $479.00

Research Services Directory: Commercial & Corporate Research Centers

This Ninth Edition provides access to well over 8,000 independent Commercial Research Firms, Corporate Research Centers and Laboratories offering contract services for hands-on, basic or applied research. *Research Services Directory* covers the thousands of types of research companies, including Biotechnology & Pharmaceutical Developers, Consumer Product Research, Defense Contractors, Electronics & Software Engineers, Think Tanks, Forensic Investigators, Independent Commercial Laboratories, Information Brokers, Market & Survey Research Companies, Medical Diagnostic Facilities, Product Research & Development Firms and more. Each entry provides the company's name, mailing address, phone & fax numbers, key contacts, web site, e-mail address, as well as a company description and research and technical fields served. Four indexes provide immediate access to this wealth of information: Research Firms Index, Geographic Index, Personnel Name Index and Subject Index.

"An important source for organizations in need of information about laboratories, individuals and other facilities." –ARBA

1,400 pages; Softcover ISBN 1-59237-003-9, $395.00 ◆ Online Database (includes a free copy of the directory) $850.00

International Business and Trade Directories

Completely updated, the Third Edition of *International Business and Trade Directories* now contains more than 10,000 entries, over 2,000 more than the last edition, making this directory the most comprehensive resource of the worlds business and trade directories. Entries include content descriptions, price, publisher's name and address, web site and e-mail addresses, phone and fax numbers and editorial staff. Organized by industry group, and then by region, this resource puts over 10,000 industry-specific business and trade directories at the reader's fingertips. Three indexes are included for quick access to information: Geographic Index, Publisher Index and Title Index. Public, college and corporate libraries, as well as individuals and corporations seeking critical market information will want to add this directory to their marketing collection.

"Reasonably priced for a work of this type, this directory should appeal to larger academic, public and corporate libraries with an international focus." –Library Journal

1,800 pages; Softcover ISBN 1-930956-63-0, $225.00 ◆ Online Database (includes a free copy of the directory) $450.00

To preview any of our Directories Risk-Free for 30 days, call (800) 562-2139 or fax to (518) 789-0556

The Grey House Safety & Security Directory, 2005

The Grey House Safety & Security Directory is the most comprehensive reference tool and buyer's guide for the safety and security industry. Arranged by safety topic, each chapter begins with OSHA regulations for the topic, followed by Training Articles written by top professionals in the field and Self-Inspection Checklists. Next, each topic contains Buyer's Guide sections that feature related products and services. Topics include Administration, Insurance, Loss Control & Consulting, Protective Equipment & Apparel, Noise & Vibration, Facilities Monitoring & Maintenance, Employee Health Maintenance & Ergonomics, Retail Food Services, Machine Guards, Process Guidelines & Tool Handling, Ordinary Materials Handling, Hazardous Materials Handling, Workplace Preparation & Maintenance, Electrical Lighting & Safety, Fire & Rescue and Security. The Buyer's Guide sections are carefully indexed within each topic area to ensure that you can find the supplies needed to meet OSHA's regulations. Six important indexes make finding information and product manufacturers quick and easy: Geographical Index of Manufacturers and Distributors, Company Profile Index, Brand Name Index, Product Index, Index of Web Sites and Index of Advertisers. This comprehensive, up-to-date reference will provide every tool necessary to make sure a business is in compliance with OSHA regulations and locate the products and services needed to meet those regulations.

"Presents industrial safety information for engineers, plant managers, risk managers, and construction site supervisors..." –Choice

1,500 pages, 2 Volume Set; Softcover ISBN 1-59237-067-5, $225.00

The Grey House Homeland Security Directory, 2005

This updated edition features the latest contact information for government and private organizations involved with Homeland Security along with the latest product information and provides detailed profiles of nearly 1,000 Federal & State Organizations & Agencies and over 3,000 Officials and Key Executives involved with Homeland Security. These listings are incredibly detailed and include Mailing Address, Phone & Fax Numbers, Email Addresses & Web Sites, a complete Description of the Agency and a complete list of the Officials and Key Executives associated with the Agency. Next, *The Grey House Homeland Security Directory* provides the go-to source for Homeland Security Products & Services. This section features over 2,000 Companies that provide Consulting, Products or Services. With this Buyer's Guide at their fingertips, users can locate suppliers of everything from Training Materials to Access Controls, from Perimeter Security to BioTerrorism Countermeasures and everything in between – complete with contact information and product descriptions. A handy Product Locator Index is provided to quickly and easily locate suppliers of a particular product. Lastly, an Information Resources Section provides immediate access to contact information for hundreds of Associations, Newsletters, Magazines, Trade Shows, Databases and Directories that focus on Homeland Security. This comprehensive, information-packed resource will be a welcome tool for any company or agency that is in need of Homeland Security information and will be a necessary acquisition for the reference collection of all public libraries and large school districts.

"Compiles this information in one place and is discerning in content. A useful purchase for public and academic libraries." –Booklist

800 pages; Softcover ISBN 1-59237-057-8, $195.00 ♦ Online Database (includes a free copy of the directory) $385.00

The Grey House Transportation Security Directory & Handbook, 2005

This brand new title is the only reference of its kind that brings together current data on Transportation Security. With information on everything from Regulatory Authorities to Security Equipment, this top-flight database brings together the relevant information necessary for creating and maintaining a security plan for a wide range of transportation facilities. With this current, comprehensive directory at the ready you'll have immediate access to: Regulatory Authorities & Legislation; Information Resources; Sample Security Plans & Checklists; Contact Data for Major Airports, Seaports, Railroads, Trucking Companies and Oil Pipelines; Security Service Providers; Recommended Equipment & Product Information and more. Using the *Grey House Transportation Security Directory & Handbook*, managers will be able to quickly and easily assess their current security plans; develop contacts to create and maintain new security procedures; and source the products and services necessary to adequately maintain a secure environment. This valuable resource is a must for all Security Managers at Airports, Seaports, Railroads, Trucking Companies and Oil Pipelines.

800 pages; Softcover ISBN 1-59237-075-6, $195

To preview any of our Directories Risk-Free for 30 days, call (800) 562-2139 or fax to (518) 789-0556

Universal Reference Publications
Statistical & Demographic Reference Books

The Asian Databook: Statistics for all US Counties & Cities with Over 10,000 Population

This is the first-ever resource that compiles statistics and rankings on the US Asian population. *The Asian Databook* presents over 20 statistical data points for each city and county, arranged alphabetically by state, then alphabetically by place name. Data reported for each place includes Population, Languages Spoken at Home, Foreign-Born, Educational Attainment, Income Figures, Poverty Status, Homeownership, Home Values & Rent, and more. Next, in the Rankings Section, the top 75 places are listed for each data element. These easy-to-access ranking tables allow the user to quickly determine trends and population characteristics. This kind of comparative data can not be found elsewhere, in print or on the web, in a format that's as easy-to-use or more concise. A useful resource for those searching for demographics data, career search and relocation information and also for market research. With data ranging from Ancestry to Education, *The Asian Databook* presents a useful compilation of information that will be a much-needed resource in the reference collection of any public or academic library along with the marketing collection of any company whose primary focus in on the Asian population.

1,000 pages; Softcover ISBN 1-59237-044-6 $150.00

The Hispanic Databook: Statistics for all US Counties & Cities with Over 10,000 Population

Previously published by Toucan Valley Publications, this second edition has been completely updated with figures from the latest census and has been broadly expanded to include dozens of new data elements and a brand new Rankings section. The Hispanic population in the United States has increased over 42% in the last 10 years and accounts for 12.5% of the total US population. For ease-of-use, *The Hispanic Databook* presents over 20 statistical data points for each city and county, arranged alphabetically by state, then alphabetically by place name. Data reported for each place includes Population, Languages Spoken at Home, Foreign-Born, Educational Attainment, Income Figures, Poverty Status, Homeownership, Home Values & Rent, and more. Next, in the Rankings Section, the top 75 places are listed for each data element. These easy-to-access ranking tables allow the user to quickly determine trends and population characteristics. This kind of comparative data can not be found elsewhere, in print or on the web, in a format that's as easy-to-use or more concise. A useful resource for those searching for demographics data, career search and relocation information and also for market research. With data ranging from Ancestry to Education, *The Hispanic Databook* presents a useful compilation of information that will be a much-needed resource in the reference collection of any public or academic library along with the marketing collection of any company whose primary focus in on the Hispanic population.

"This accurate, clearly presented volume of selected Hispanic demographics is recommended for large public libraries and research collections."-Library Journal

1,000 pages; Softcover ISBN 1-59237-008-X, $150.00

Ancestry in America: A Comparative Guide to Over 200 Ethnic Backgrounds

This brand new reference work pulls together thousands of comparative statistics on the Ethnic Backgrounds of all populated places in the United States with populations over 10,000. Never before has this kind of information been reported in a single volume. Section One, Statistics by Place, is made up of a list of over 200 ancestry and race categories arranged alphabetically by each of the 5,000 different places with populations over 10,000. The population number of the ancestry group in that city or town is provided along with the percent that group represents of the total population. This informative city-by-city section allows the user to quickly and easily explore the ethnic makeup of all major population bases in the United States. Section Two, Comparative Rankings, contains three tables for each ethnicity and race. In the first table, the top 150 populated places are ranked by population number for that particular ancestry group, regardless of population. In the second table, the top 150 populated places are ranked by the percent of the total population for that ancestry group. In the third table, those top 150 populated places with 10,000 population are ranked by population number for each ancestry group. These easy-to-navigate tables allow users to see ancestry population patterns and make city-by-city comparisons as well. Plus, as an added bonus with the purchase of *Ancestry in America*, a free companion CD-ROM is available that lists statistics and rankings for all of the 35,000 populated places in the United States. This brand new, information-packed resource will serve a wide-range or research requests for demographics, population characteristics, relocation information and much more. *Ancestry in America: A Comparative Guide to Over 200 Ethnic Backgrounds* will be an important acquisition to all reference collections.

"This compilation will serve a wide range of research requests for population characteristics … it offers much more detail than other sources." –Booklist

1,500 pages; Softcover ISBN 1-59237-029-2, $225.00

To preview any of our Directories Risk-Free for 30 days, call (800) 562-2139 or fax to (518) 789-0556

The Value of a Dollar 1860-2004, Third Edition

A guide to practical economy, *The Value of a Dollar* records the actual prices of thousands of items that consumers purchased from the Civil War to the present, along with facts about investment options and income opportunities. This brand new Third Edition boasts a brand new addition to each five-year chapter, a section on Trends. This informative section charts the change in price over time and provides added detail on the reasons prices changed within the time period, including industry developments, changes in consumer attitudes and important historical facts. Plus, a brand new chapter for 2000-2004 has been added. Each 5-year chapter includes a Historical Snapshot, Consumer Expenditures, Investments, Selected Income, Income/Standard Jobs, Food Basket, Standard Prices and Miscellany. This interesting and useful publication will be widely used in any reference collection.

"Recommended for high school, college and public libraries." –ARBA

600 pages; Hardcover ISBN 1-59237-074-8, $135.00

The Value of a Dollar 1600-1859, The Colonial Era to The Civil War

Following the format of the widely acclaimed, T*he Value of a Dollar, 1860-2004*, *The Value of a Dollar 1600-1859, The Colonial Era to The Civil War* records the actual prices of thousands of items that consumers purchased from the Colonial Era to the Civil War. Our editorial department had been flooded with requests from users of our Value of a Dollar for the same type of information, just from an earlier time period. This new volume is just the answer – with pricing data from 1600 to 1859. Arranged into five-year chapters, each 5-year chapter includes a Historical Snapshot, Consumer Expenditures, Investments, Selected Income, Income/Standard Jobs, Food Basket, Standard Prices and Miscellany. There is also a section on Trends. This informative section charts the change in price over time and provides added detail on the reasons prices changed within the time period, including industry developments, changes in consumer attitudes and important historical facts. This fascinating survey will serve a wide range of research needs and will be useful in all high school, public and academic library reference collections.

600 pages; Hardcover ISBN 1-59237-094-2, $135.00

Working Americans 1880-1999
Volume I: The Working Class, Volume II: The Middle Class, Volume III: The Upper Class

Each of the volumes in the *Working Americans 1880-1999* series focuses on a particular class of Americans, The Working Class, The Middle Class and The Upper Class over the last 120 years. Chapters in each volume focus on one decade and profile three to five families. Family Profiles include real data on Income & Job Descriptions, Selected Prices of the Times, Annual Income, Annual Budgets, Family Finances, Life at Work, Life at Home, Life in the Community, Working Conditions, Cost of Living, Amusements and much more. Each chapter also contains an Economic Profile with Average Wages of other Professions, a selection of Typical Pricing, Key Events & Inventions, News Profiles, Articles from Local Media and Illustrations. The *Working Americans* series captures the lifestyles of each of the classes from the last twelve decades, covers a vast array of occupations and ethnic backgrounds and travels the entire nation. These interesting and useful compilations of portraits of the American Working, Middle and Upper Classes during the last 120 years will be an important addition to any high school, public or academic library reference collection.

"These interesting, unique compilations of economic and social facts, figures and graphs will support multiple research needs. They will engage and enlighten patrons in high school, public and academic library collections." –Booklist

Volume I: The Working Class ◆ 558 pages; Hardcover ISBN 1-891482-81-5, $145.00
Volume II: The Middle Class ◆ 591 pages; Hardcover ISBN 1-891482-72-6; $145.00
Volume III: The Upper Class ◆ 567 pages; Hardcover ISBN 1-930956-38-X, $145.00

Working Americans 1880-1999 Volume IV: Their Children

This Fourth Volume in the highly successful *Working Americans 1880-1999* series focuses on American children, decade by decade from 1880 to 1999. This interesting and useful volume introduces the reader to three children in each decade, one from each of the Working, Middle and Upper classes. Like the first three volumes in the series, the individual profiles are created from interviews, diaries, statistical studies, biographies and news reports. Profiles cover a broad range of ethnic backgrounds, geographic area and lifestyles – everything from an orphan in Memphis in 1882, following the Yellow Fever epidemic of 1878 to an eleven-year-old nephew of a beer baron and owner of the New York Yankees in New York City in 1921. Chapters also contain important supplementary materials including News Features as well as information on everything from Schools to Parks, Infectious Diseases to Childhood Fears along with Entertainment, Family Life and much more to provide an informative overview of the lifestyles of children from each decade. This interesting account of what life was like for Children in the Working, Middle and Upper Classes will be a welcome addition to the reference collection of any high school, public or academic library.

600 pages; Hardcover ISBN 1-930956-35-5, $145.00

Working Americans 1880-2003 Volume V: Americans At War

Working Americans 1880-2003 Volume V: Americans At War is divided into 11 chapters, each covering a decade from 1880-2003 and examines the lives of Americans during the time of war, including declared conflicts, one-time military actions, protests, and preparations for war. Each decade includes several personal profiles, whether on the battlefield or on the homefront, that tell the stories of civilians, soldiers, and officers during the decade. The profiles examine: Life at Home; Life at Work; and Life in the Community. Each decade also includes an Economic Profile with statistical comparisons, a Historical Snapshot, News Profiles, local News Articles, and Illustrations that provide a solid historical background to the decade being examined. Profiles range widely not only geographically, but also emotionally, from that of a girl whose leg was torn off in a blast during WWI, to the boredom of being stationed in the Dakotas as the Indian Wars were drawing to a close. As in previous volumes of the *Working Americans* series, information is presented in narrative form, but hard facts and real-life situations back up each story. The basis of the profiles come from diaries, private print books, personal interviews, family histories, estate documents and magazine articles. For easy reference, *Working Americans 1880-2003 Volume V: Americans At War* includes an in-depth Subject Index. The *Working Americans* series has become an important reference for public libraries, academic libraries and high school libraries. This fifth volume will be a welcome addition to all of these types of reference collections.

600 pages; Hardcover ISBN 1-59237-024-1; $145.00
Five Volume Set (Volumes I-V), Hardcover ISBN 1-59237-034-9, $675.00

Working Americans 1880-2005 Volume VI: Women at Work

Unlike any other volume in the *Working Americans* series, this Sixth Volume, is the first to focus on a particular gender of Americans. *Volume VI: Women at Work*, traces what life was like for working women from the 1860's to the present time. Beginning with the life of a maid in 1890 and a store clerk in 1900 and ending with the life and times of the modern working women, this text captures the struggle, strengths and changing perception of the American woman at work. Each chapter focuses on one decade and profiles three to five women with real data on Income & Job Descriptions, Selected Prices of the Times, Annual Income, Annual Budgets, Family Finances, Life at Work, Life at Home, Life in the Community, Working Conditions, Cost of Living, Amusements and much more. For even broader access to the events, economics and attitude towards women throughout the past 130 years, each chapter is supplemented with News Profiles, Articles from Local Media, Illustrations, Economic Profiles, Typical Pricing, Key Events, Inventions and more. This important volume illustrates what life was like for working women over time and allows the reader to develop an understanding of the changing role of women at work. These interesting and useful compilations of portraits of women at work will be an important addition to any high school, public or academic library reference collection.

600 pages; Hardcover ISBN 1-59237-063-2; $145.00
Six Volume Set (Volumes I-VI), Hardcover ISBN 1-59237-063-2, $810.00

America's Top-Rated Cities, 2005

America's Top-Rated Cities provides current, comprehensive statistical information and other essential data in one easy-to-use source on the 100 "top" cities that have been cited as the best for business and living in the U.S. This handbook allows readers to see, at a glance, a concise social, business, economic, demographic and environmental profile of each city, including brief evaluative comments. In addition to detailed data on Cost of Living, Finances, Real Estate, Education, Major Employers, Media, Crime and Climate, city reports now include Housing Vacancies, Tax Audits, Bankruptcy, Presidential Election Results and more. This outstanding source of information will be widely used in any reference collection.

"The only source of its kind that brings together all of this information into one easy-to-use source. It will be beneficial to many business and public libraries." –ARBA

2,500 pages, 4 Volume Set; Softcover ISBN 1-59237-076-4, $195.00

America's Top-Rated Smaller Cities, 2004/05

A perfect companion to *America's Top-Rated Cities*, *America's Top-Rated Smaller Cities* provides current, comprehensive business and living profiles of smaller cities (population 25,000-99,999) that have been cited as the best for business and living in the United States. Sixty cities make up this 2004 edition of *America's Top-Rated Smaller Cities*, all are top-ranked by Population Growth, Median Income, Unemployment Rate and Crime Rate. City reports reflect the most current data available on a wide-range of statistics, including Employment & Earnings, Household Income, Unemployment Rate, Population Characteristics, Taxes, Cost of Living, Education, Health Care, Public Safety, Recreation, Media, Air & Water Quality and much more. Plus, each city report contains a Background of the City, and an Overview of the State Finances. *America's Top-Rated Smaller Cities* offers a reliable, one-stop source for statistical data that, before now, could only be found scattered in hundreds of sources. This volume is designed for a wide range of readers: individuals considering relocating a residence or business; professionals considering expanding their business or changing careers; general and market researchers; real estate consultants; human resource personnel; urban planners and investors.

"Provides current, comprehensive statistical information in one easy-to-use source... Recommended for public and academic libraries and specialized collections." –Library Journal

1,100 pages; Softcover ISBN 1-59237-043-8, $160.00

To preview any of our Directories Risk-Free for 30 days, call (800) 562-2139 or fax to (518) 789-0556

Crime in America's Top-Rated Cities, 2000

This volume includes over 20 years of crime statistics in all major crime categories: violent crimes, property crimes and total crime. *Crime in America's Top-Rated Cities* is conveniently arranged by city and covers 76 top-rated cities. *Crime in America's Top-Rated Cities* offers details that compare the number of crimes and crime rates for the city, suburbs and metro area along with national crime trends for violent, property and total crimes. Also, this handbook contains important information and statistics on Anti-Crime Programs, Crime Risk, Hate Crimes, Illegal Drugs, Law Enforcement, Correctional Facilities, Death Penalty Laws and much more. A much-needed resource for people who are relocating, business professionals, general researchers, the press, law enforcement officials and students of criminal justice.

"Data is easy to access and will save hours of searching." –Global Enforcement Review

832 pages; Softcover ISBN 1-891482-84-X, $155.00

Profiles of America: Facts, Figures & Statistics for Every Populated Place in the United States

Profiles of America is the only source that pulls together, in one place, statistical, historical and descriptive information about every place in the United States in an easy-to-use format. This award winning reference set, now in its second edition, compiles statistics and data from over 20 different sources – the latest census information has been included along with more than nine brand new statistical topics. This Four-Volume Set details over 40,000 places, from the biggest metropolis to the smallest unincorporated hamlet, and provides statistical details and information on over 50 different topics including Geography, Climate, Population, Vital Statistics, Economy, Income, Taxes, Education, Housing, Health & Environment, Public Safety, Newspapers, Transportation, Presidential Election Results and Information Contacts or Chambers of Commerce. Profiles are arranged, for ease-of-use, by state and then by county. Each county begins with a County-Wide Overview and is followed by information for each Community in that particular county. The Community Profiles within the county are arranged alphabetically. *Profiles of America* is a virtual snapshot of America at your fingertips and a unique compilation of information that will be widely used in any reference collection.

A Library Journal Best Reference Book *"An outstanding compilation." –Library Journal*

10,000 pages; Four Volume Set; Softcover ISBN 1-891482-80-7, $595.00

The Comparative Guide to American Suburbs, 2005

The Comparative Guide to American Suburbs is a one-stop source for Statistics on the 2,000+ suburban communities surrounding the 50 largest metropolitan areas – their population characteristics, income levels, economy, school system and important data on how they compare to one another. Organized into 50 Metropolitan Area chapters, each chapter contains an overview of the Metropolitan Area, a detailed Map followed by a comprehensive Statistical Profile of each Suburban Community, including Contact Information, Physical Characteristics, Population Characteristics, Income, Economy, Unemployment Rate, Cost of Living, Education, Chambers of Commerce and more. Next, statistical data is sorted into Ranking Tables that rank the suburbs by twenty different criteria, including Population, Per Capita Income, Unemployment Rate, Crime Rate, Cost of Living and more. *The Comparative Guide to American Suburbs* is the best source for locating data on suburbs. Those looking to relocate, as well as those doing preliminary market research, will find this an invaluable timesaving resource.

"Public and academic libraries will find this compilation useful... The work draws together figures from many sources and will be especially helpful for job relocation decisions." – Booklist

1,700 pages; Softcover ISBN 1-59237-004-7, $130.00

The American Tally: Statistics & Comparative Rankings for U.S. Cities with Populations over 10,000

This important statistical handbook compiles, all in one place, comparative statistics on all U.S. cities and towns with a 10,000+ population. *The American Tally* provides statistical details on over 4,000 cities and towns and profiles how they compare with one another in Population Characteristics, Education, Language & Immigration, Income & Employment and Housing. Each section begins with an alphabetical listing of cities by state, allowing for quick access to both the statistics and relative rankings of any city. Next, the highest and lowest cities are listed in each statistic. These important, informative lists provide quick reference to which cities are at both extremes of the spectrum for each statistic. Unlike any other reference, *The American Tally* provides quick, easy access to comparative statistics – a must-have for any reference collection.

"A solid library reference." -Bookwatch

500 pages; Softcover ISBN 1-930956-29-0, $125.00

To preview any of our Directories Risk-Free for 30 days, call (800) 562-2139 or fax to (518) 789-0556

The Comparative Guide to American Elementary & Secondary Schools, 2004/05

The only guide of its kind, this award winning compilation offers a snapshot profile of every public school district in the United States serving 1,500 or more students – more than 5,900 districts are covered. Organized alphabetically by district within state, each chapter begins with a Statistical Overview of the state. Each district listing includes contact information (name, address, phone number and web site) plus Grades Served, the Numbers of Students and Teachers and the Number of Regular, Special Education, Alternative and Vocational Schools in the district along with statistics on Student/Classroom Teacher Ratios, Drop Out Rates, Ethnicity, the Numbers of Librarians and Guidance Counselors and District Expenditures per student. As an added bonus, *The Comparative Guide to American Elementary and Secondary Schools* provides important ranking tables, both by state and nationally, for each data element. For easy navigation through this wealth of information, this handbook contains a useful City Index that lists all districts that operate schools within a city. These important comparative statistics are necessary for anyone considering relocation or doing comparative research on their own district and would be a perfect acquisition for any public library or school district library.

"This straightforward guide is an easy way to find general information. Valuable for academic and large public library collections." –ARBA

2,400 pages; Softcover ISBN 1-59237-047-0, $125.00

Weather America, A Thirty-Year Summary of Statistical Weather Data and Rankings

This valuable resource provides extensive climatological data for over 4,000 National and Cooperative Weather Stations throughout the United States. *Weather America* begins with a new Major Storms section that details major storm events of the nation and a National Rankings section that details rankings for several data elements, such as Maximum Temperature and Precipitation. The main body of *Weather America* is organized into 50 state sections. Each section provides a Data Table on each Weather Station, organized alphabetically, that provides statistics on Maximum and Minimum Temperatures, Precipitation, Snowfall, Extreme Temperatures, Foggy Days, Humidity and more. State sections contain two brand new features in this edition – a City Index and a narrative Description of the climatic conditions of the state. Each section also includes a revised Map of the State that includes not only weather stations, but cities and towns.

"Best Reference Book of the Year." –Library Journal

2,013 pages; Softcover ISBN 1-891482-29-7, $175.00

To preview any of our Directories Risk-Free for 30 days, call (800) 562-2139 or fax to (518) 789-0556

Sedgwick Press
Health Directories

The Complete Directory for People with Disabilities, 2005

A wealth of information, now in one comprehensive sourcebook. Completely updated for 2005, this edition contains more information than ever before, including thousands of new entries and enhancements to existing entries and thousands of additional web sites and e-mail addresses. This up-to-date directory is the most comprehensive resource available for people with disabilities, detailing Independent Living Centers, Rehabilitation Facilities, State & Federal Agencies, Associations, Support Groups, Periodicals & Books, Assistive Devices, Employment & Education Programs, Camps and Travel Groups. Each year, more libraries, schools, colleges, hospitals, rehabilitation centers and individuals add *The Complete Directory for People with Disabilities* to their collections, making sure that this information is readily available to the families, individuals and professionals who can benefit most from the amazing wealth of resources cataloged here.

"No other reference tool exists to meet the special needs of the disabled in one convenient resource for information." –Library Journal

1,200 pages; Softcover ISBN 1-59237-054-3, $165.00 ♦ Online Database $215.00 ♦ Online Database & Directory Combo $300.00

The Complete Directory for People with Chronic Illness, 2005/06

Thousands of hours of research have gone into this completely updated 2005/06 edition – several new chapters have been added along with thousands of new entries and enhancements to existing entries. Plus, each chronic illness chapter has been reviewed by an medical expert in the field. This widely-hailed directory is structured around the 90 most prevalent chronic illnesses – from Asthma to Cancer to Wilson's Disease – and provides a comprehensive overview of the support services and information resources available for people diagnosed with a chronic illness. Each chronic illness has its own chapter and contains a brief description in layman's language, followed by important resources for National & Local Organizations, State Agencies, Newsletters, Books & Periodicals, Libraries & Research Centers, Support Groups & Hotlines, Web Sites and much more. This directory is an important resource for health care professionals, the collections of hospital and health care libraries, as well as an invaluable tool for people with a chronic illness and their support network.

"A must purchase for all hospital and health care libraries and is strongly recommended for all public library reference departments." –ARBA

1,200 pages; Softcover ISBN 1-59237-081-0, $165.00 ♦ Online Database $215.00 ♦ Online Database & Directory Combo $300.00

The Complete Learning Disabilities Directory, 2005

The Complete Learning Disabilities Directory is the most comprehensive database of Programs, Services, Curriculum Materials, Professional Meetings & Resources, Camps, Newsletters and Support Groups for teachers, students and families concerned with learning disabilities. This information-packed directory includes information about Associations & Organizations, Schools, Colleges & Testing Materials, Government Agencies, Legal Resources and much more. For quick, easy access to information, this directory contains four indexes: Entry Name Index, Subject Index and Geographic Index. With every passing year, the field of learning disabilities attracts more attention and the network of caring, committed and knowledgeable professionals grows every day. This directory is an invaluable research tool for these parents, students and professionals.

"Due to its wealth and depth of coverage, parents, teachers and others... should find this an invaluable resource." –Booklist

900 pages; Softcover ISBN 1-59237-092-6, $145.00 ♦ Online Database $195.00 ♦ Online Database & Directory Combo $280.00

The Complete Mental Health Directory, 2004/05

This is the most comprehensive resource covering the field of behavioral health, with critical information for both the layman and the mental health professional. For the layman, this directory offers understandable descriptions of 25 Mental Health Disorders as well as detailed information on Associations, Media, Support Groups and Mental Health Facilities. For the professional, *The Complete Mental Health Directory* offers critical and comprehensive information on Managed Care Organizations, Information Systems, Government Agencies and Provider Organizations. This comprehensive volume of needed information will be widely used in any reference collection.

"... the strength of this directory is that it consolidates widely dispersed information into a single volume." –Booklist

800 pages; Softcover ISBN 1-59237-046-2, $165.00 ♦ Online Database $215.00 ♦ Online & Directory Combo $300.00

To preview any of our Directories Risk-Free for 30 days, call (800) 562-2139 or fax to (518) 789-0556

Older Americans Information Directory, 2004/05

Completely updated for 2004/05, this Fifth Edition has been completely revised and now contains 1,000 new listings, over 8,000 updates to existing listings and over 3,000 brand new e-mail addresses and web sites. You'll find important resources for Older Americans including National, Regional, State & Local Organizations, Government Agencies, Research Centers, Libraries & Information Centers, Legal Resources, Discount Travel Information, Continuing Education Programs, Disability Aids & Assistive Devices, Health, Print Media and Electronic Media. Three indexes: Entry Index, Subject Index and Geographic Index make it easy to find just the right source of information. This comprehensive guide to resources for Older Americans will be a welcome addition to any reference collection.

"Highly recommended for academic, public, health science and consumer libraries..." –Choice

1,200 pages; Softcover ISBN 1-59237-037-3, $165.00 ◆ Online Database $215.00 ◆ Online Database & Directory Combo $300.00

The Complete Directory for Pediatric Disorders, 2004/05

This important directory provides parents and caregivers with information about Pediatric Conditions, Disorders, Diseases and Disabilities, including Blood Disorders, Bone & Spinal Disorders, Brain Defects & Abnormalities, Chromosomal Disorders, Congenital Heart Defects, Movement Disorders, Neuromuscular Disorders and Pediatric Tumors & Cancers. This carefully written directory offers: understandable Descriptions of 15 major bodily systems; Descriptions of more than 200 Disorders and a Resources Section, detailing National Agencies & Associations, State Associations, Online Services, Libraries & Resource Centers, Research Centers, Support Groups & Hotlines, Camps, Books and Periodicals. This resource will provide immediate access to information crucial to families and caregivers when coping with children's illnesses.

"Recommended for public and consumer health libraries." –Library Journal

1,200 pages; Softcover ISBN 1-59237-045-4, $165.00 ◆ Online Database $215.00 ◆ Online Database & Directory Combo $300.00

The Complete Directory for People with Rare Disorders

This outstanding reference is produced in conjunction with the National Organization for Rare Disorders to provide comprehensive and needed access to important information on over 1,000 rare disorders, including Cancers and Muscular, Genetic and Blood Disorders. An informative Disorder Description is provided for each of the 1,100 disorders (rare Cancers and Muscular, Genetic and Blood Disorders) followed by information on National and State Organizations dealing with a particular disorder, Umbrella Organizations that cover a wide range of disorders, the Publications that can be useful when researching a disorder and the Government Agencies to contact. Detailed and up-to-date listings contain mailing address, phone and fax numbers, web sites and e-mail addresses along with a description. For quick, easy access to information, this directory contains two indexes: Entry Name Index and Acronym/Keyword Index along with an informative Guide for Rare Disorder Advocates. The Complete Directory for People with Rare Disorders will be an invaluable tool for the thousands of families that have been struck with a rare or "orphan" disease, who feel that they have no place to turn and will be a much-used addition to the reference collection of any public or academic library.

"Quick access to information... public libraries and hospital patient libraries will find this a useful resource in directing users to support groups or agencies dealing with a rare disorder." –Booklist

726 pages; Softcover ISBN 1-891482-18-1, $165.00

The Directory of Drug & Alcohol Residential Rehabilitation Facilities

This brand new directory is the first-ever resource to bring together, all in one place, data on the thousands of drug and alcohol residential rehabilitation facilities in the United States. *The Directory of Drug & Alcohol Residential Rehabilitation Facilities* covers over 1,000 facilities, with detailed contact information for each one, including mailing address, phone and fax numbers, email addresses and web sites, mission statement, type of treatment programs, cost, average length of stay, numbers of residents and counselors, accreditation, insurance plans accepted, type of environment, religious affiliation, education components and much more. It also contains a helpful chapter on General Resources that provides contact information for Associations, Print & Electronic Media, Support Groups and Conferences. Multiple indexes allow the user to pinpoint the facilities that meet very specific criteria. This time-saving tool is what so many counselors, parents and medical professionals have been asking for. *The Directory of Drug & Alcohol Residential Rehabilitation Facilities* will be a helpful tool in locating the right source for treatment for a wide range of individuals. This comprehensive directory will be an important acquisition for all reference collections: public and academic libraries, case managers, social workers, state agencies and many more.

"This is an excellent, much needed directory that fills an important gap..." –Booklist

300 pages; Softcover ISBN 1-59237-031-4, $135.00

To preview any of our Directories Risk-Free for 30 days, call (800) 562-2139 or fax to (518) 789-0556

Sedgwick Press
Hospital & Health Plan Directories

The Comparative Guide to American Hospitals

This brand new title is the first ever resource to compare all of the nation's hospitals by 17 measures of quality in the treatment of heart attack, heart failure and pneumonia. This data is based on the recently announced Hospital Compare, produced by Medicare, and is available in print and in a unique and user-friendly format from Grey House Publishing, along with extra contact information from Grey House's *Directory of Hospital Personnel*. *The Comparative Guide to American Hospitals* provides a snapshot profile of each of the nations 6,000 hospitals. These informative profiles illustrate how the hospital rates in 17 important areas: Heart Attack Care (% who receive Aspirin at Arrival, Aspirin at Discharge, ACE Inhibitor for LVSD, Beta Blocker at Arrival, Beta Blocker at Discharge, Thrombolytic Agent Received, PTCA Received and Adult Smoking Cessation Advice); Heart Failure (% who receive LVF Assessment, ACE Inhibitor for LVSD, Discharge Instructions, Adult Smoking Cessation Advice); and Pneumonia (% who receive Initial Antibiotic Timing, Pneumococcal Vaccination, Oxygenation Assessment, Blood Culture Performed and Adult Smoking Cessation Advice). Each profile includes the raw percentage for that hospital, the state average, the US average and data on the top hospital. For easy access to contact information, each profile includes the hospitals address, phone and fax numbers, email and web addresses, type and accreditation along with 5 top key administrations. These profiles will allow the user to quickly identify the quality of the hospital and have the necessary information at their fingertips to make contact with that hospital. Most importantly, *The Comparative Guide to American Hospitals* provides an easy-to-use Ranking Table for each of the data elements to allow the user to quickly locate the hospitals with the best level of service. This brand new title will be a must for the reference collection at all public, medical and academic libraries.

2,500 pages; Softcover ISBN 1-59237-109-4 $175.00

The Directory of Hospital Personnel, 2005

The Directory of Hospital Personnel is the best resource you can have at your fingertips when researching or marketing a product or service to the hospital market. A "Who's Who" of the hospital universe, this directory puts you in touch with over 150,000 key decision-makers. With 100% verification of data you can rest assured that you will reach the right person with just one call. Every hospital in the U.S. is profiled, listed alphabetically by city within state. Plus, three easy-to-use, cross-referenced indexes put the facts at your fingertips faster and more easily than any other directory: Hospital Name Index, Bed Size Index and Personnel Index. *The Directory of Hospital Personnel* is the only complete source for key hospital decision-makers by name. Whether you want to define or restructure sales territories... locate hospitals with the purchasing power to accept your proposals... keep track of important contacts or colleagues... or find information on which insurance plans are accepted, *The Directory of Hospital Personnel* gives you the information you need – easily, efficiently, effectively and accurately.

"Recommended for college, university and medical libraries." –ARBA

2,500 pages; Softcover ISBN 1-59237-065-9 $275.00 ♦ Online Database $545.00 ♦ Online Database & Directory Combo, $650.00

The Directory of Health Care Group Purchasing Organizations

This comprehensive directory provides the important data you need to get in touch with over 800 Group Purchasing Organizations. By providing in-depth information on this growing market and its members, *The Directory of Health Care Group Purchasing Organizations* fills a major need for the most accurate and comprehensive information on over 800 GPOs – Mailing Address, Phone & Fax Numbers, E-mail Addresses, Key Contacts, Purchasing Agents, Group Descriptions, Membership Categorization, Standard Vendor Proposal Requirements, Membership Fees & Terms, Expanded Services, Total Member Beds & Outpatient Visits represented and more. Five Indexes provide a number of ways to locate the right GPO: Alphabetical Index, Expanded Services Index, Organization Type Index, Geographic Index and Member Institution Index. With its comprehensive and detailed information on each purchasing organization, *The Directory of Health Care Group Purchasing Organizations* is the go-to source for anyone looking to target this market.

"The information is clearly arranged and easy to access...recommended for those needing this very specialized information." –ARBA

1,000 pages; Softcover ISBN 1-59237-036-5, $325.00 ♦ Online Database, $650.00 ♦ Online Database & Directory Combo, $750.00

To preview any of our Directories Risk-Free for 30 days, call (800) 562-2139 or fax to (518) 789-0556

The HMO/PPO Directory, 2005

The HMO/PPO Directory is a comprehensive source that provides detailed information about Health Maintenance Organizations and Preferred Provider Organizations nationwide. This comprehensive directory details more information about more managed health care organizations than ever before. Over 1,100 HMOs, PPOs and affiliated companies are listed, arranged alphabetically by state. Detailed listings include Key Contact Information, Prescription Drug Benefits, Enrollment, Geographical Areas served, Affiliated Physicians & Hospitals, Federal Qualifications, Status, Year Founded, Managed Care Partners, Employer References, Fees & Payment Information and more. Plus, five years of historical information is included related to Revenues, Net Income, Medical Loss Ratios, Membership Enrollment and Number of Patient Complaints. Five easy-to-use, cross-referenced indexes will put this vast array of information at your fingertips immediately: HMO Index, PPO Index, Other Providers Index, Personnel Index and Enrollment Index. *The HMO/PPO Directory* provides the most comprehensive information on the most companies available on the market place today.

"Helpful to individuals requesting certain HMO/PPO issues such as co-payment costs, subscription costs and patient complaints. Individuals concerned (or those with questions) about their insurance may find this text to be of use to them." –ARBA

600 pages; Softcover ISBN 1-59237-057-8, $275.00 ◆ Online Database, $495.00 ◆ Online Database & Directory Combo, $600.00

The Directory of Independent Ambulatory Care Centers

This first edition of *The Directory of Independent Ambulatory Care Centers* provides access to detailed information that, before now, could only be found scattered in hundreds of different sources. This comprehensive and up-to-date directory pulls together a vast array of contact information for over 7,200 Ambulatory Surgery Centers, Ambulatory General and Urgent Care Clinics, and Diagnostic Imaging Centers that are not affiliated with a hospital or major medical center. Detailed listings include Mailing Address, Phone & Fax Numbers, E-mail and Web Site addresses, Contact Name and Phone Numbers of the Medical Director and other Key Executives and Purchasing Agents, Specialties & Services Offered, Year Founded, Numbers of Employees and Surgeons, Number of Operating Rooms, Number of Cases seen per year, Overnight Options, Contracted Services and much more. Listings are arranged by State, by Center Category and then alphabetically by Organization Name. Two indexes provide quick and easy access to this wealth of information: Entry Name Index and Specialty/Service Index. *The Directory of Independent Ambulatory Care Centers* is a must-have resource for anyone marketing a product or service to this important industry and will be an invaluable tool for those searching for a local care center that will meet their specific needs.

"Among the numerous hospital directories, no other provides information on independent ambulatory centers. A handy, well-organized resource that would be useful in medical center libraries and public libraries." –Choice

986 pages; Softcover ISBN 1-930956-90-8, $185.00 ◆ Online Database, $365.00 ◆ Online Database & Directory Combo, $450.00

Sedgwick Press
Education Directories

Educators Resource Directory, 2005/06

Educators Resource Directory is a comprehensive resource that provides the educational professional with thousands of resources and statistical data for professional development. This directory saves hours of research time by providing immediate access to Associations & Organizations, Conferences & Trade Shows, Educational Research Centers, Employment Opportunities & Teaching Abroad, School Library Services, Scholarships, Financial Resources, Professional Consultants, Computer Software & Testing Resources and much more. Plus, this comprehensive directory also includes a section on Statistics and Rankings with over 100 tables, including statistics on Average Teacher Salaries, SAT/ACT scores, Revenues & Expenditures and more. These important statistics will allow the user to see how their school rates among others, make relocation decisions and so much more. For quick access to information, this directory contains four indexes: Entry & Publisher Index, Geographic Index, a Subject & Grade Index and Web Sites Index. *Educators Resource Directory* will be a well-used addition to the reference collection of any school district, education department or public library.

"Recommended for all collections that serve elementary and secondary school professionals." –Choice

1,000 pages; Softcover ISBN 1-59237-080-2, $145.00 ◆ Online Database $195.00 ◆ Online Database & Directory Combo $280.00

To preview any of our Directories Risk-Free for 30 days, call (800) 562-2139 or fax to (518) 789-0556